The Occupational Environment – *Its Evaluation and Control*

The Occupational Environment – *Its Evaluation and Control*

Salvatore R. DiNardi, Editor

A Publication of the
American Industrial Hygiene Association
Fairfax, Virginia

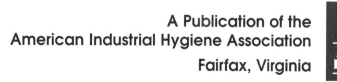

The information presented in this book was developed by occupational hygiene professionals with backgrounds, training, and experience in occupational and environmental health and safety, working with information and conditions existing at the time of publication. The American Industrial Hygiene Association (AIHA), as publisher, and the authors have been diligent in ensuring that the materials and methods addressed in this book reflect prevailing occupational health and safety and industrial hygiene practices. It is possible, however, that certain procedures discussed will require modification because of changing federal, state, and local regulations, or heretofore unknown developments in research. As the body of knowledge is expanded, improved solutions to workplace hazards will become available. Readers should consult a broad range of sources of information before developing workplace health and safety programs.

AIHA and the authors disclaim any liability, loss, or risk resulting directly or indirectly from the use of the practices and/or theories discussed in this book. Moreover, it is the reader's responsibility to stay informed of any changing federal, state, or local regulations that might affect the material contained herein, and the policies adopted specifically in the reader's workplace.

Specific mention of manufacturers and products in this book does not represent an endorsement by AIHA.

ISBN 0-932627-82-X

AIHA Press
American Industrial Hygiene Association
2700 Prosperity Avenue, Suite 250
Fairfax, Virginia 22031

Tel.: (703) 849-8888
Fax: (703) 207-3561
http://www.aiha.org
e-mail: infonet@aiha.org

AIHA Stock No. 252-BP-97

A special thank you to
Jeffrey S. Lee, Ph.D.,CIH, and
Vernon E. Rose, Dr PH, CIH, whose vision
and efforts resulted in this publication.

— the AIHA Board of Directors

Foreword

For the past 25 years, the 1973 edition of *The Industrial Environment — Its Evaluation & Control* served as the cornerstone of occupational safety and health information. Numerous students and practitioners used this book as both a text and a reference. In addition, many workers benefited from the prevention and control measures they learned from this document, which was one of NIOSH's most widely requested publications. Some of the most knowledgeable practitioners and researchers in the field contributed to the 1973 edition. Their contributions were clear, thorough, and informative.

In the past 25 years, the workplace has changed in many ways. To meet the needs of today's workplace, the American Industrial Hygiene Association (AIHA) is publishing a new edition titled *The Occupational Environment — Its Evaluation and Control.* This new edition addresses hazards faced by contemporary workers and new approaches to prevention and control; it is intended as a source book for much of the knowledge needed to protect current and future workers from occupational illnesses and injuries. New chapters have therefore been added to cover contemporary workplace issues such as computers in occupational hygiene, risk communication and assessment, hazardous waste management, indoor air, nonionizing radiation, ergonomics, and worker education and training.

NIOSH appreciates the efforts of the many experts who have contributed to this new edition. Their contributions will influence many scientists and workers who will shape the future of occupational hygiene. We also acknowledge the American Industrial Hygiene Association for developing this new edition, which will benefit working men and women and their employers.

Linda Rosenstock, M.D., MPH
Director, National Institute for
Occupational Safety and Health
Centers for Disease Control and Prevention

September 1997

Preface

It is an honor and privilege to be the editor for the flagship AIHA book, *The Occupational Environment — Its Evaluation and Control*. Being asked by AIHA to summarize the profession in one reference book is a challenge. This book continues a tradition started in 1914 by the Occupational Health Activity of the U.S. Public Health Service that continued to grow and develop, with NIOSH leadership, culminating in *The Industrial Environment — Its Evaluation & Control*, 3rd Edition. This is the so-called "White Book" that every practicing occupational hygienist cherishes. A couple of years ago, two members of the AIHA Board of Directors, Jeff Lee and Vern Rose, approached NIOSH about having AIHA take on the fourth edition. NIOSH graciously passed the gauntlet to AIHA to continue this 83-year tradition.

My goal for this project was to make this book the primary reference for anyone interested in occupational hygiene. The book provides coverage of disciplines germane to our field. The book contains six major sections representing the practice model familiar to all occupational hygienists and reviews many special topics.

My guiding philosophy as editor was to design the book to serve the needs of practicing professionals and students at all levels.

In offering readers a new edition of this important work, I feel an obligation to explain why this book does not contain chapters on basic science and mathematics and statistics.

I believe that anyone lacking these skills must acquire them through regular academic routes. Although these topics are important to occupational hygiene practice, there are abundant reference materials that do more than just review the subject in a few dozen pages. Authors include references to these topics and other related topics by citation and examples in their chapters. Integrating basic sciences into each chapter highlights the importance of these disciplines.

As we enter the next millennium, the occupational hygiene profession continues to grow and change. The most obvious change is referring to our profession as occupational hygiene throughout the book. The book's title contains the key term *Occupational Environment*, reflecting a major shift in our professional practice. Today, the professional occupational hygienist practices in comprehensive arenas such as traditional industrial hygiene, environmental compliance, and safety management. This growing change prompted the chapters on *Epidemiological Surveillance, Modeling Inhalation Exposure, Biohazards in the Workplace, Indoor Air Quality, and Risk Assessment in the Workplace* as well as the section called *The Human Environment at Work*.

Not surprisingly, spirited discussions ensued among many AIHA members and various committees and advisory groups to eliminate certain chapters and drastically modify others so that it looked more like "traditional industrial hygiene." To do this, I believe, would be the kiss of death for a book that will survive well into the next century.

My strategy in designing this book is to emphasize a competency-based approach for each chapter. The outcome competencies, written in behavioral terms, can be measured via professional practice or by academic examination.

These competencies were prepared using Bloom's Taxonomy and written in behavioral terms that can be evaluated by examination or practice. The competencies use specific verbs to describe the desired competency. Each set of verbs addresses a specific competency.

It is with great modesty that I can say this is AIHA's flagship publication — one the entire membership can feel a part of and use with pride. The authors are experts in their respective fields and bring new ideas to this effort because they integrate the practice perspective into their work.

In closing I must say it has been a privilege to work with dedicated and committed authors, technical committees, AIHA headquarters staff, and the membership on this important and visible project. Editing the seminal work in occupational hygiene for AIHA is a daunting task. The time line for this first edition was less than two years. AIHA technical committees and peer reviewers were an invaluable part of the process. I could easily say the errors of omissions and commission are due to the time constraints placed on this project or those involved in the project. The truth is the authors, reviewers, and AIHA staff are responsible for the good parts, and I, as editor, am solely responsible for the bad and ugly parts.

Finally, as you read this book and find something that you just plain disagree with, let me know; do not damn the whole book. You should view this as a dynamic document that will only evolve into a "perfect" document over time and with comments from the membership and readers.

Salvatore R. DiNardi, Ph.D., CIH
Professor and Chair
Environmental Health Sciences Department
School of Public Health and Health Sciences
University of Massachusetts
October 1997

Acknowledgments

AIHA Executive Director **O. Gordon Banks, CAE**, took a risk and placed the staff and financial resources of the American Industrial Hygiene Association at my disposal. His coaching management style allowed us all to grow and do our personal best. He was there when I needed moral support and encouraged me when I was down. His subtle leadership style kept this project on a focused time line. When the time line slipped a few months, his confidence in my ability to continue was undaunted.

Many of us believe that Alice Hamilton is the mother of the industrial hygiene profession. I know AIHA Director of Communications **Anne Dees** is the mother and the ongoing force behind this book. She was unrelenting in her search for excellence in the quality of the Communications Division staff that she assigned to this project. Anne worked tirelessly behind the scenes to resolve some very complicated issues with authors, reviewers, committee chairs, and board members. Anne directed the massive communication effort with authors, reviewers, technical committees, and members. She smoothed the inevitable ruffled feathers a project of this magnitude is bound to generate. Without Anne's formidable talents and velvet glove, my job as editor would be impossible and this book would be a long time in gestation. Anne did not just do her job as an AIHA staffer, she took on this project with a deep personal commitment to see it through to the final publication.

I want to acknowledge and thank the following members of the AIHA Communications Division staff for their meticulous editing, proofreading, and constant inquiries: **Faythe Benson**, Manager of Publishing; **Elizabeth Simon**, Editor; **Christopher Howland**, Editor/Books; **James Myers**, Manager of Design and Production; **Joni Lucas**, Senior Editor; **Deborah Williams**, Production Editor; **Sheila Brown**, Editorial Assistant; **Leslie Miller**, Assistant Editor; **Natalie Komitsky**, Production Assistant; **Rebecca Howland**; Editorial Assistant; and **Denell Deavers**, Publications Assistant.

Denell Deavers is the glue that kept this project together. Her soft-spoken and pleasant telephone skills were essential to tease the last shred of paper from authors and reviewers. Her unwavering tenacity and personal endurance made her work, which was onerous at times, flow continuously and smoothly. Denell made hundreds of telephone calls, sent voluminous e-mail messages, and managed all the paper for this project — and never lost her sense of humor, a piece of paper, or a message. Her willingness to work nights and weekends kept me on target. Managing this project is similar to herding cats and Denell is the shepherd that made it happen. She herded the book across the finish line.

I also want to acknowledge and thank the following members of the AIHA Scientific Affairs Division: **Manuel Gomez,** Director, Scientific Affairs; **John Meagher**, CIH, Manager, Scientific Affairs; and **Fred Grunder**, CIH, Manager, Laboratory Programs.

I want to acknowledge and thank the **AIHA Board of Directors** for their confidence in me and my ability to bring this project to completion. The board fully supported my competency-based approach to organizing the book. Without its support, this project could not have gotten off the ground.

I want to acknowledge and thank all the **AIHA technical committees** and the **Publications Committee** for providing timely input and supplying the never-ending list of peer reviewers.

I want to acknowledge and thank the **peer reviewers**. Without their commitment to technical excellence and constant questioning to clarify the written material, this project could not succeed.

When I think about the **chapter authors** and their commitment to this project, I am reminded of Sisyphus. Sisyphus was condemned forever to roll a huge stone up a hill only to have it roll down the hill again when he neared the top. Just as the authors completed their writing and satisfying the editor's imposed deadline, the peer reviewers' comments came back. This was followed by author queries from the technical editors. And as the authors thought they were done, the Communications Division staff proofreaders and editors had yet more questions that needed answers.

Stephen H. Gehlbach, the Dean of the University of Massachusetts, School of Public Health and Health Sciences in Amherst, Massachusetts, is a constant source of inspiration for me. His patience with my devotion to this project is second to none. He allowed me the time and the space needed to concentrate on this work. My department could not have survived this difficult project without his leadership and direction.

The graduate students in the Amherst Environmental Health Sciences Department, and the occupational hygiene program, are a special group of aspiring professionals that tolerated my distraction from time to time as the project began to consume me. Thank you all for your patients and understanding.

This project overlapped my Intergovernmental Personnel Act assignment to the **U.S. Naval Submarine Medical Research Laboratory** in Groton, Connecticut.

I want to acknowledge and thank the staff of the laboratory for carrying my share of the workload when the never-ending trips to AIHA headquarters demanded that I leave them: **Commanding Officers Captains Walter and Wooster; Technical Director Captain Curley; Bio-Medical Science Department Chair Commander Carlson;** and **the Submarine Atmosphere Health Assessment Program staff — Captain Raymond Woolrich (U.S. Navy retired), Dale Greenwell, and Petty Officer Warnock** — and the **NSMRL administrative staff.**

I cannot forget my children, **Peter and Christopher**, and their willingness to pick up the pieces when I was distracted by the demands of this project.

Joan M. Culley, my life support and partner, taught me the virtues of an organized approach to teaching and learning. I am forever grateful to her.

Sal DiNardi
October 1997

About the Editor

Salvatore Robert DiNardi, Ph.D., CIH
Professor and Chair
Environmental Health Sciences Department
School of Public Health and Health Sciences
University of Massachusetts

In 1970, Dr. DiNardi started his academic career in public health at the University of Massachusetts in Amherst. He is the chair of the Environmental Health Sciences Department and director of the Industrial Hygiene Program. He teaches Principles of Industrial Ventilation, Principles of Occupational Health, Quantitative Methods in Environmental Health, Industrial Hygiene Laboratory, and Aerosol Science.

As the scientific director for PHRA, Inc., Dr. DiNardi designs and implements indoor air quality surveys to recognize, evaluate, and control the sources of building-related illness in the nonindustrial workplace. He is also active in professional education and offers corporations on-site continuing education programs based on his university courses.

Dr. DiNardi is a ventilation design and indoor air quality training consultant for AT&T, Eastman Kodak, Northeast Utilities, W.R. Grace, Digital Equipment Corporation, DuPont, IBM, ITT, and American Cyanamid. He is also a continuing education consultant in ventilation design, industrial hygiene, and indoor air quality for the University of Massachusetts, the University of North Carolina, George Washington University, and the University of California-Irvine.

Dr. DiNardi has hands-on experience with indoor air quality problems, especially epidemics of building-related illness. He provides solutions to these epidemics and training and advice on the subject for workers, supervisors, and managers in many Fortune 500 companies and smaller companies and health care institutions. He has completed indoor air quality surveys in more than 10 million square feet of industrial/commercial spaces throughout the northeast quadrant of the United States.

Dr. DiNardi is a member of the American Industrial Hygiene Association (AIHA) and the American Society of Heating, Refrigerating and Air-Conditioning Engineers (ASHRAE). For five years he was on the board of directors of the New England Section of the Air Pollution Control Association (now the Air and Waste Management Association). He is certified in the comprehensive practice of industrial hygiene and a Diplomate of the American Board of Industrial Hygiene (ABIH).

Dr. DiNardi has published more than 40 articles in journals of environmental health, industrial hygiene, and chemistry. Van Nostrand Reinhold published his book *Calculation Methods for Industrial Hygiene* in March 1995. He holds a Ph.D. in physical chemistry from the University of Massachusetts.

One of Dr. DiNardi's goals is to find effective ways of protecting and enhancing the public health through the education of both practicing professionals and concerned citizens. In this regard, he has worked closely with continuing education programs nationally to provide practical, current, and lively courses. He has developed innovative ways of meeting the educational needs of people whose activities have a direct impact on the health of individuals, industries, communities, governments, and society. Alpha Sigma Lambda, the national continuing education honor society, acknowledged Dr. DiNardi's service and commitment to continuing education by awarding him honorary membership.

Contents

Part VI. Program Management

Section 1

Introduction and Background

Chapter 1

Outcome Competencies

After completing this chapter, the user should be able to:

1. Define underlined terms used in this chapter.
2. Describe the contribution of key individuals in the development of the occupational hygiene profession.
3. Explain the role that the history of civilization plays in the development of the occupational hygiene profession.
4. Discuss the role of governmental entities in the history of occupational hygiene.
5. Describe how the public health profession relates to occupational hygiene.
6. Implement the code of ethics in the practice of occupational hygiene.
7. Explain the occupational hygiene paradigm of anticipate, recognize, evaluate, and control.
8. Recognize the conflicts around, "whom do we serve?"

Key Terms

Agricola • American Academy of Industrial Hygiene • American Board of Industrial Hygiene • American Conference of Governmental Industrial Hygienists • American Industrial Hygiene Association • certified industrial hygienist • Code of Professional Ethics for the Practice of Industrial Hygiene • Ulrich Ellenbog • Dr. Alice Hamilton • Benjamin W. McCready • Mine Safety and Health Administration • National Institute for Occupational Safety and Health • Occupational Safety and Health Act • Occupational Safety and Health Administration • permissible exposure limits • Pliny the Elder • Sir Percival Pott • Bernardino Ramazzini • Charles T. Thackrah • threshold limit values

Prerequisite Knowledge

None

Key Topics

I. Origins of Industrial Hygiene

II. The U.S. Experience
   A. Recognition
   B. Evaluation
   C. Control
   D. Education and Professional Organizations

III. Public Health Roots

IV. Professional Recognition

V. Professional Code of Ethics

VI. Whom Do We Serve?

1

History and Philosophy of Industrial Hygiene

Vernon E. Rose, Dr PH, CIH, CSP, PE

Origins of Industrial Hygiene

As with most professions, identifying the origin of the practice of industrial hygiene is difficult, if not impossible. Might we designate as founders of the profession the early chroniclers of occupational hazards and control measures such as Agricola, who, in 1556, described the prevalent diseases and accidents in mining, smelting, and refining and prevention measures including ventilation? In that case, the contributions of Plinius Secundus (Pliny the Elder) should also be considered, who, in the first century AD, wrote "minimum refiners . . . envelop their faces with loose bladders, which enable them to see without inhaling the fatal dust."[(1)] If their works were read, thus influencing others to control work hazards, they deserve the title, at least posthumously, of industrial hygienists.

But what of those who simply identified problems? It has been suggested that a special honor in the field of occupational medicine is owed to Hippocrates (c. 460–370 BC). His writings include the first recorded mention of occupational diseases (e.g., lead poisoning in miners and metallurgists) and provide more frequent allusions to this class of ailments than those of any other author prior to Ramazzini.[(2)]

In 1713 Bernardino Ramazzini published the first book that could be considered a complete treatise on occupational diseases, *De Morbis Artificum Diatriba*.[(3)] From his own observations he accurately described scores of occupations, their hazards, and resulting diseases. Although he recommended some specific as well as general preventive measures (workers should cover their faces to avoid breathing dust), most of his control recommendations were therapeutic and curative. While he had a vast knowledge of the literature of his time, it has been suggested that many of the works he cited were of questionable scientific validity, and some were more myth than science and should have been recognized as such even in Ramazzini's time. Because of his prestige these "fanciful notions must have received wide acceptance . . . and because his book was so admired, Ramazzini's influence may have stifled progress in his field during a period when great advances were being made in other branches of medicine."[(2)]

Nevertheless, his cautions to protect workers and his admonition that any doctor called on to treat patients of the working class ask "What occupation does he follow?" earned him the appellation "Father of Industrial Medicine."

For more than 100 years following Ramazzini's work, no significant additions to the literature on occupational medicine were published. In the nineteenth century two physicians, Charles T. Thackrah in England and Benjamin W. McCready in America, began the modern literature on the recognition of occupational diseases. McCready's book, *On the Influence of Trades, Professions, and Occupations in the United States, in the Production of Disease*, is generally recognized as the first work on occupational medicine published in the United States.[(4)]

The recognition of a causal link between workplace hazards and disease was a key step in the development of the practice of industrial hygiene. The observations by physicians, from Hippocrates to Ramazzini and extending into the twentieth century, of the relationship between work and disease

Figure 1.1 — A portrait of Bernardino Ramazzini — the father of occupational medicine.

are the foundation of the profession. But recognition of hazards without intervention and control, i.e., without prevention of disease, should not qualify one as an industrial hygienist.

The crystallization of the practice of the profession can be traced to simultaneous developments in Great Britain and the United States in the late nineteenth and early twentieth centuries. While legislation controlling working conditions was enacted in England beginning in 1802, the early laws were considered totally ineffective, as "no proper system of inspection or enforcement was provided."[5] The British Factories Act of 1864, however, required the use of dilution ventilation to reduce air contaminants, while the 1878 version specified the use of exhaust ventilation by fans. The real watershed in industrial medicine and hygiene, however, came in the British Factories Act of 1901, which provided for the creation of regulations to control dangerous trades. The development of regulations created the impetus for investigation of workplace hazards and enforcement of control measures. In the United States in 1905, the Massachusetts Health Department appointed health inspectors to evaluate dangers of occupations, thus establishing government's role in the nascent field of occupational health.

It has been suggested that industrial hygiene did not "emerge as a unique field of endeavor until quantitative measurements of the environment became available."[6] But in 1910 when Dr. Alice Hamilton went, in her own words, "as a pioneer into a new, unexplored field of American medicine, the field of industrial disease,"[7] worker exposures to many hazards (e.g., lead and silica) were so excessive and resulting diseases so acute and obvious, the "evaluation" step of industrial hygiene practice required only the sense of sight and an understanding of the concept of cause and effect. This "champion of social responsibility" for worker health and welfare not only presented substantial evidence of a relationship between exposure to toxins and ill health, but also proposed concrete solutions to the problems she encountered.[8] On an individual basis, Dr. Hamilton's work, which comprised not only the recognition of occupational disease, but the evaluation and control of the causative agents, should be considered as the initial practice of industrial hygiene, at least in the United States.

Figure 1.2 — Dr. Alice Hamilton.

It should be appreciated that many of the early practitioners of industrial hygiene were physicians who, like Alice Hamilton, were interested not only in the diagnosis and treatment of illnesses in industrial workers, but also in hazard control to prevent further cases. These physicians working with engineers and other scientists interested in public health and environmental hazards took the knowledge and insights developed over several millennia from Hippocrates to Ramazzini, Thackrah and McCready, and began the process of deliberately changing the work environment with the goal of preventing occupational diseases.

What or who then can be designated as representing the origin of the profession? Is there any one person who deserves the title "Founder of Industrial Hygiene"? Certainly, if the name of one individual is sought, that of Alice Hamilton shines like a beacon. But think back to more than 10,000 years ago at the end of the Stone Age, when occupations began to form with the grinding of stone, horn, bone, and ivory tools with sandstone, and with pottery making and linen weaving. Envision a thoughtful worker who suffered from the musculoskeletal problems associated with grinding, made adjustments to his working conditions, and passed the ideas on to co-workers. Recognizing ergonomic problems and solving them would qualify him as an early industrial hygiene practitioner. If that scenario can be imagined, perhaps it is also conceivable that tens of thousands of years ago there was a huntress who recognized the signs and symptoms of anthrax in the bison her group had killed and who made the connection between earlier kills of diseased animals and sickness in members of her tribe. If she then warned her companions of the hazard involved and sought to avoid diseased animals, would she not qualify as one of the founders of the industrial hygiene profession?

If the basic philosophy of the profession is understood—protection of the health and well being of workers and the public through anticipation, recognition, evaluation, and control of hazards arising in or from the workplace—then the rich tapestry that chronicles the history of industrial hygiene can be imagined. It began when one person recognized a work hazard and took steps not only for self protection, but also for protection of fellow workers. This is the origin and essence of the profession of industrial hygiene.

The U.S. Experience

The events presented in Table 1.1 illustrate the history of industrial hygiene. While the concepts that formed the art and science of industrial hygiene flourished in many countries, the United States provided the fertile ground for the development of the profession as it exists today.

As the industrial revolution, propelled by the Civil War, progressed in the nineteenth century, individuals began to observe serious health and safety problems (recognition). They also considered the effects on workers (evaluation) and made changes in the work environment (control) to lessen the effects observed. Although these efforts may have resulted in improved worker health and safety, their application was not recognized as the practice of industrial hygiene until the early 1900s. In addition to a chronological listing, these activities also illustrate the concepts of the profession—i.e., recognition, evaluation, and control—which may help to better describe today's practice of industrial hygiene.

Recognition

Recognition, as well as anticipation, of the potential for occupational health problems is a prerequisite for the implementation of occupational hygiene activities. Therefore, early attempts at defining the scope and magnitude of occupational health problems were very important to the subsequent efforts of evaluation and control. The Illinois Occupational Disease Commission's survey of the extent of occupational health problems in the state of Illinois in 1910 was the first such survey undertaken in the United States.[9] Dr. Alice Hamilton was a member of that commission and served as its chief investigator. Two years later she presented her survey of lead hazards in American industry. Although other states formed commissions to identify problems, it was many years before there was an organized effort to develop information as to the scope of the industrial health problems in the working population in America.

The Division of Industrial Hygiene in the United States Public Health Service (USPHS), later to become the National Institute for Occupational Safety and Health (NIOSH), revived interest in assessing the extent of health hazards. Beginning in the 1960s the USPHS conducted surveys in a number of states and metropolitan areas for the purpose of identifying the extent of worker exposure to occupational health hazards. The results of these studies were used to determine the need for occupational health specialists in the governmental agencies and for setting priorities for government inspection and consultation

Table 1.1 —
Historical Events in Occupational Health and Safety

1,000,000 BC	Australopithecus used stones as tools and weapons. Flint knappers suffered cuts and eye injuries; bison hunters contracted anthrax.
10,000 BC	Neolithic man began food-producing economy and the urban revolution in Mesopotamia. At end of Stone Age, grinding of stone, horn, bone, and ivory tools with sandstone; pottery making, linen weaving. Beginning of the history of occupations.
5000 BC	Copper and Bronze Age—metal workers released from food production. Metallurgy—the first specialized craft.
370 BC	Hippocrates dealt with the health of citizens, not workers, but did identify lead poisoning in miners and metallurgists.
50 AD	Plinius Secundus (Pliny the Elder) identified use of animal bladders intended to prevent inhalation of dust and lead fume.
200 AD	Galen visited a copper mine, but his discussions on public health did not include workers' disease.
Middle Ages	No documented contributions to the study of occupational diseases.
1473	Ellenbog recognized that the vapors of some metals were dangerous and described the symptoms of industrial poisoning from lead and mercury with suggested preventive measures.
1500	In *De Re Metallica* (1556), Georgius Agricola described every facet of mining, smelting, and refining, noting prevalent diseases and accidents, and means of prevention including the need for ventilation. Paracelsus (1567) described respiratory diseases among miners with an excellent description of mercury poisoning. Remembered as the father of toxicology. "All substances are poisons . . . the right dose differentiates a poison and a remedy."
1665	Workday for mercury miners at Idria shortened.
1700	Bernardino Ramazzini, "father of occupational medicine," published *De Morbis Artificum Diatriba*, (Diseases of Workers) and examined occupational diseases and "cautions." He introduced the question, "Of what trade are you?"
1775	Percival Pott described occupational cancer among English chimney sweeps, identifying soot and the lack of hygiene measures as a cause of scrotal cancer. The result was the Chimney-Sweeps Act of 1788.
1830	Charles Thackrah authored the first book on occupational diseases to be published in England. His views on disease and prevention helped stimulate factory and health legislation. Medical inspection and compensation were established in 1897.
1900s	Dr. Alice Hamilton investigated many dangerous occupations and had tremendous influence on early regulation of occupational hazards in the United States. In 1919 she became the first woman faculty member at Harvard University and wrote *Exploring the Dangerous Trades.*

Figure 1.3 — A portrait of Paracelsus (1567) who is remembered as the father of toxicology.

Table 1.1 (continued) — Historical Events in Occupational Health and Safety

1902–1911	Federal and then state (Washington) legislation covering workers' compensation. By 1948 all states covered occupational diseases. First survey in the United States of the extent of occupational disease conducted by the Illinois Occupational Disease Commission. Massachusetts appointed health inspectors to evaluate dangers of occupations.
1910	First national conference on industrial diseases in the United States.
1912	U.S. Congress levied prohibitive tax on the use of white phosphorus in making matches.
1913	National Safety Council organized. New York and Ohio established first state industrial hygiene agencies.
1914	USPHS organized a Division of Industrial Hygiene and Sanitation. American Public Health Association organized section on industrial hygiene.
1916	American Association of Industrial Physicians and Surgeons formed. American Medical Association held first symposium on industrial hygiene and medicine.
1922	Harvard established industrial hygiene degree program.
1928–1932	Bureau of Mines conducted toxicological research on solvents, vapors, and gases.
1936	Walsh-Healy Act required companies supplying goods to government to maintain safe and healthful workplaces.
1938	National (later American) Conference of Governmental Industrial Hygienists formed.
1939	American Industrial Hygiene Association organized. American Standards Association and ACGIH prepared first list (maximum allowable concentrations) of standards for chemical exposures in industry.
1941–1945	Expanded industrial hygiene programs in states.
1941	Bureau of Mines authorized to inspect mines.
1960	American Board of Industrial Hygiene organized by AIHA and ACGIH.
1966	Metal and Nonmetallic Mine Safety Act.
1968	Professional Code of Ethics drafted by AAIH. Code adopted by all four industrial hygiene associations by 1981.
1969	Coal Mine Health and Safety Act.
1970	Occupational Safety and Health Act.
1977	Federal Mine Safety and Health Act.
1992–present	Efforts to significantly amend OSHAct.
1995	Revised Professional Code of Ethics adopted by all four industrial hygiene associations.

programs.[10] These efforts culminated in NIOSH's National Occupational Hazard Survey conducted in the early 1970s. This project involved walk-through surveys in more than 5000 randomly selected workplaces in the United States and for the first time provided a comprehensive assessment of workplace hazards and the extent of their control for a large segment of the working population. A subsequent NIOSH study, the National Occupational Exposure Survey, provided additional insight on the magnitude of worker exposure to hazards.[11] These studies were used by NIOSH to set priorities for research and for developing recommended standards.

Concurrent with the development of occupational hazard information was the recognition that information on the incidence of occupational illnesses in the U.S. work force was not available. Prior to the passage of the Occupational Safety and Health Act (OSH Act) in 1970, the sources of occupational disease incidence data were limited primarily to the information developed in several state health and labor departments such as California, New York, and Michigan. The California reports of occupational illnesses were thought to be the most comprehensive and reliable; consequently, they were often used for projections of the national problem. The passage of the OSH Act gave the U.S. Department of Labor the responsibility for developing a national occupational injury and illness reporting system. The system includes a requirement that employers record occupationally related injuries and illnesses with separate categories for dermatitis, lung disorders due to dusts, lung disorders due to chemicals, systemic effects, physical agent disorders, repeated trauma disorders, and "all other" illnesses. A sample of employers report their experience, which is used to develop state and national estimates. For many years skin disorders were the most frequently reported illnesses, but in the late 1980s repeated trauma, which includes musculoskeletal problems such as carpal tunnel syndrome and back disorders, became the leading category of illness.

It is widely recognized that occupational diseases are underreported in this system for a variety of reasons beyond a conscious decision not to record. Many occupationally related diseases can also be caused by nonoccupational exposures (e.g., lung cancer and tobacco). Also, when a disease has a long latency period before it can be diagnosed, its relationship to early work exposures may be obscure. In some cases work factors may worsen a preexisting condition. And, as a relatively new discipline, the recognition of workplace illnesses can be hampered by a limited knowledge base. It is unlikely that a worker's illness will be recognized as work related if epidemiologic or toxicologic studies have not documented cause and effect, i.e., exposure to a toxin and illness.

Concurrent with studies to estimate hazard distribution and incidence of occupational illness were specific epidemiologic studies designed to link cause and effect. Two key studies firmly established the specialty of occupational epidemiology first practiced by Sir Percival Pott in the late eighteenth century. One study could be considered an extension of his work, which linked a byproduct of coal combustion (soot) with scrotal cancer in chimney sweeps. In the twentieth-century study, the epidemiologists who looked at the mortality distribution of U.S. steelworkers found a population whose health was better than expected when compared with the overall U.S. population (the healthy worker effect).[(12)] However, when subpopulations, specifically coke oven workers, were considered, excesses of respiratory and kidney cancer deaths were uncovered. The complexity of environmental exposures surrounding these workers precluded the identification of any one specific causative agent and led to the designation of coal tar pitch volatiles (CTPV) as the surrogate hazard to be controlled. Documentation of excess mortality in coke oven workers led to promulgation of the Occupational Safety and Health Administration (OSHA) regulation on coke oven emissions.[(13)]

The second major occupational epidemiologic study focused on asbestos. Reports of cases of lung disease due to asbestos exposure began to accumulate beginning in 1906. By 1938 the USPHS had studied workers in asbestos textile plants and had recommended a tentative limit for asbestos dust in the textile industry of 5 million particles per cubic foot (mp/ft^3), determined by the impinger technique.[(14)] In the late 1940s workers manufacturing asbestos products in England were observed to have a frequency of bronchogenic cancer greater than that expected as compared with the general male population. These findings led to the study of U.S. insulation workers exposed to airborne asbestos fibers and the documentation of an excess of bronchogenic cancers in this population.[(15)] These and other studies led to the promulgation of OSHA's first emergency temporary standard in 1971 and first complete health standard[(16)] in 1972. The hazards of asbestos also firmly established toxic tort (product liability) lawsuits as a force to control workplace hazards.

In occupational epidemiologic studies, one test for concluding the existence of cause and effect is the presence of an exposure-response relationship. When industrial hygiene exposure data are not available, substitute measures such as "high, medium, and low" are relied on. However, where historical exposure data are available or can be estimated, risk-assessment evaluation can be more objective and lead to occupational exposure limits that define the expected reduction in the incidence of illness. Use of risk-assessment tools to estimate illness reduction from lowering workplace exposure limits has been applied by OSHA since the benzene standard[(17)] was promulgated in 1987.

At the present time, occupational hygiene efforts in the United States are guided by hazard rather than disease considerations. At the national and state levels, information on worker exposure to hazards by various industry categories is available and is used to set priorities for governmental investigations and research. Within industry, the concept of hazard recognition vis-a-vis illness is seen as important in developing programs that focus on prevention. Consequently, the emphasis on anticipation and recognition of potential occupational health problems primarily involves the industrial hygiene practice of hazard determination, where hazard combines the inherent toxicity of a substance or agent and the likelihood for exposure.

Evaluation

Although the use of the senses, including sight, smell, and sometimes taste, were important in the early years of the practice of industrial hygiene, the transition to a science required the development of more sophisticated sampling methods to aid in the evaluation of problems. One of the first such sampling methods, developed by researchers at Harvard University in 1917, was the detector tube (color-indicating device) for measuring airborne levels of carbon monoxide.[(18)] Dust exposure in mining and other industries was an early industrial hygiene concern and generated the need to measure airborne concentrations of particulates. In 1922 Greenberg and Smith developed their impinger, and in 1938 Littlefield and Schrenk modified the design and developed the midget impinger.[(6)] The subsequent development of the hand-operated pump for dust sampling with midget impingers gave the industrial hygienist flexibility in collecting breathing zone samples to better characterize occupational exposures. The associated analytical method of counting and sizing particles with a microscope, thus yielding concentrations of million particles per cubic foot, was the standard method for characterizing particulate exposures until the application of the membrane filter in 1953 allowed exposures to be evaluated on a mass per volume basis.[(19)]

Figure 1.4 — A London chimney sweep (mid-1700s).

Figure 1.5 — A German chimney sweep (mid-1700s). Note the tight-fitting, personal protective clothing. There were no reported problems of scrotal cancer among the German workers.

Figure 1.6 — A woodcut of Georgius Agricola, author of *De Re Metallica.*

Figure 1.7 — Woodcuts depicting foundry workers at a smeltery.

The early application of impingers, using water as the collection medium, was for dust sampling. Gas and vapor monitoring required the development of a variety of sampling media for use in midget impingers and later in the more efficient fritted-glass bubblers. In 1970 a major breakthrough in sampling methodology occurred when NIOSH developed the charcoal sampling tube and provided support for development of the battery-operated pump.[20] Concurrent with the development of these active sampling devices was Palmes' work in 1973 on a passive dosimeter for nitrogen dioxide. Subsequent commercialization of the passive dosimeter concept led to a modest revolution in the scope of industrial hygiene sampling. These technological developments greatly enhanced the art of personal sampling and allowed the industrial hygienist even greater flexibility in characterizing worker exposure to hazardous conditions.[21]

At the same time sampling methods were being developed, the application of new analytical technology to industrial hygiene assessments was taking place. In the early 1930s technical articles described the use of gas chromatography for evaluating samples of airborne organic vapors. In later years other forms of technology were employed at a rapid pace. Today hygienists use atomic absorption, high pressure liquid chromatography, mass spectroscopy, and other sophisticated instrumentation and techniques.

As industrial hygienists learned more about the environment and further refined their techniques for measuring hazardous exposures in the workplace, the need to compare measurements with unacceptable exposure levels became apparent. In 1929 several industrial hygienists in the USPHS recommended upper limits for exposure to quartz-bearing industrial dusts based on studies in the Vermont granite industry.[22] The publication of workplace exposure limits was greatly enhanced by the formation of the American Conference of Governmental Industrial Hygienists (ACGIH) in 1938. In 1939 ACGIH, in cooperation with the American Standards Association, developed the first list of maximum allowable concentrations (MACs) to limit worker exposure to airborne contaminants. In 1943 Dr. James Sterner published the MAC list in *Industrial Medicine*, followed by Warren Cook's publication (also in *Industrial Medicine*) in 1945 of a MAC list for 140 substances with sources and bases for the recommendations. In 1947 ACGIH began publication of its MAC list, and converted to the term threshold limit values (TLVs®) in 1948.[23] Today, even with major responsibility for standards-setting vested with the federal government, the role of ACGIH in developing exposure guidelines is a valuable tool in protecting the health of workers.

Control

Control of occupational health problems can take several forms. Industrial hygienists most often employ the technological approach, i.e., engineering measures such as substitution with less hazardous substances or local exhaust ventilation. Where these techniques cannot eliminate or reduce the hazard sufficiently, administrative measures and personal protective equipment are also relied on. These concepts, which can also be reshuffled to the categories of control "at the source, in the environment, and at the worker," were first introduced, in a comprehensive form, in 1473 by Ulrich Ellenbog.[6] He suggested three methods of control still applied today: use dry coal instead of wet coal to avoid production of toxic "fumes," work with windows open, and cover the mouth to prevent inhalation of noxious "fumes."[6] The history of two specific control measures, industrial ventilation and respirators, should be of special interest to practicing industrial hygienists.

Agricola, in his 1561 publication *De Re Metallica*, emphasized the need for ventilation of mines and included many illustrations of devices to force air below ground.[24] The first recorded design of a local exhaust ventilation system, however, was by the Frenchman D'Arcet in the early 1800s. To control noxious fumes he led an exhaust duct from a hood at a furnace into a chimney that had a strong draft. The induced airflow carried the fumes away from the source.

The British window tax of 1696, which was not repealed until 1851, resulted in dark and underventilated factories. The first legislation regulating conditions in factories was the British Factory and Workshops Act of 1802, which required ventilation in workplaces. The Act of 1864 required sufficient ventilation to render gases and dusts as harmless as possible, but it was not until 1867 that an inspector was given authority to require fans or other mechanical devices to control dust. Although the British Factories Act of 1897 required the use of ventilation for certain operations, little was published on techniques until the late 1930s. It has been suggested that the main reasons for the lack of published information probably were "attempts on the part of industry to keep their trade secrets, and the lack of interest on the part of universities and colleges."[6] In 1951 ACGIH published the first edition of *Industrial Ventilation: A Manual of Recommended Practice*. The manual, now in its 22nd edition, is a compilation on design, maintenance, and evaluation of industrial exhaust ventilation systems. The manual has found wide acceptance as a guide for official agencies, as a standard for industrial ventilation designers, and as a textbook for industrial hygiene courses.

As noted earlier, the concept of respiratory protective devices (e.g., animal bladders) to reduce exposure to airborne contaminants dates back to at least 50 AD. There is no record of worker acceptance of these early respirators nor of their effectiveness as personal protective devices, but in all likelihood they did not score high in either category. The same can probably be said for other devices that have fallen by the wayside over the centuries: scarves, shawls, handkerchiefs, magnetic mouthpieces, magnetized screens, wet sponges, and breathing tubes. It should come as no surprise that Leonardo da Vinci (1452-1519) considered the problems of respiratory protection when he recommended the use of a wet cloth to protect against chemical warfare agents. He also devised two underwater breathing devices, one being a snorkel consisting of a breathing tube with an attached float.[25] Ramazzini wrote a critical review of the inadequate respiratory protection available in his time (c. 1700). Shortly thereafter, the first descriptions appeared of the ancestors of today's atmosphere-supplying devices, such as open- and closed-circuit self-contained breathing apparatus and hose masks.

In the 1800s the realization of the separate natures of particles and gases or vapors led to great advances in respirators. In 1814 a particulate-removing filter encased in a rigid container was developed and was the predecessor of modern filters for air-purifying respirators. The ability of activated charcoal to remove organic vapors from air was discovered in 1854 and was almost immediately put to use in respirators. The most rapid advances in respiratory protection grew out of the use of chemical warfare agents in World War I. Research on gas sorbents for use in military masks and high efficiency particulate filters was accelerated by the introduction of different gases and highly toxic particulate matter on the battlefield. Since the 1920s the major advances in respirator design include resin-impregnated dust filters, which use electrostatic force fields to remove dust particles from air, and the ultra high-efficiency filter from paper containing very fine glass fibers.[25] Other advances include the use of more flexible and durable (plastic) materials for facepieces, and the combination of a battery-operated air mover with a respirator for use as a lightweight air-supplied respirator (i.e., the powered air purifying respirator).

Figure 1.8 — Woodcuts showing ventilation systems for mines.

A different approach to control involves the application of governmental powers to assist, motivate, and require employers to maintain safe and healthful work environments. In the United States, legal protection of industrial workers was, until 1971, the responsibility of state and local governments. The development of state governmental responsibilities in occupational health took place as early as 1905, when inspectors in the Massachusetts Health Department investigated workplace dangers. In 1913 the first formal governmental program, the New York Department of Labor's Division of Industrial Hygiene, was established. In 1914 the Office of Industrial Hygiene and Sanitation was formed in the USPHS and subsequently underwent many reorganizations before becoming NIOSH in 1971.[26]

During the 1920s and early 1930s industrial hygiene activities were initiated in five state health departments (Connecticut, Maryland, Mississippi, Ohio, and Rhode Island). The Social Security Act of 1935 made federal resources (money and industrial hygienists) available to states to aid in the development of industrial hygiene programs.[27] By 1936, 12 more state health department programs were initiated, and the USPHS recommended that a "large industrial state" should have at least one chief industrial hygienist with a salary of $6,000. The minimum qualifications of this specialist called for a chemical engineering degree, two years' graduate work in industrial hygiene, three years of experience, and in addition to a wide range of scientific and technical knowledge, "the ability to establish contacts with plant executives; ability to enlist cooperation of plant executives, foremen and laborers; initiative; tact; good judgement; and good address."[28]

World War II provided significant impetus for the development of state and local government industrial hygiene programs. By 1946, 52 programs were operational in 41 states; however, the withdrawal of federal resources after the war led to a steady decline in both number and activity of these programs. By the late 1960s, while there were a number of states with programs, most involved only one or two professionals. The exceptions to this situation were found in the large industrial states such as Massachusetts, New York, Pennsylvania, Michigan, and California.

As in most countries, national legislation to control hazards in the workplace first focused on mining. The U.S. Bureau of Mines was established shortly after the turn of the century, but it was not until 1941 that the Bureau was authorized to conduct inspections in mines. A number of mining tragedies in the mid-1960s led to the passage of the Metal and Nonmetallic Mine Safety Act and the Coal Mine Safety and Health Act in 1966 and 1969, respectively. These mining laws were superseded by the comprehensive Mine Safety and Health Act of 1977. This law created the Mine Safety and Health Administration (MSHA) within the U.S. Department of Labor. MSHA regulates and conducts inspections in mines and related industries.

Federal safety and health activities in general industry were not initiated until passage of the Walsh-Healey Public Contracts Act in 1936. This legislation authorized federal occupational safety and health standards for government contractors. The Department of Labor adopted existing American Standards Association (now the American National Standards Institute, or ANSI) safety and health standards and the ACGIH TLVs as Walsh-Healey standards.

In the late 1960s the U.S. Congress became concerned with the comprehensive problem of occupational safety and health in the workplace. A number of congressional hearings resulted in the documentation of the seriousness of workplace deaths, injuries, and illnesses, and the lack of consistent and comprehensive safety and health programs at the state and local level to prevent such problems. Consequently, the Occupational Safety and Health Act[29] was passed in 1970 and created OSHA within the U.S. Department of Labor and NIOSH in the Department of Health and Human Services. OSHA was given the responsibility to promulgate standards and conduct inspections in most workplaces, while NIOSH was to conduct research and recommend health and safety standards to OSHA.

The initial health standards adopted by OSHA were the existing Walsh-Healey standards. These included the 1968 ACGIH TLVs[30] minus the 21 chemicals for which there were ANSI standards.[31] The exposure limits for these substances are known as permissible exposure limits (PELs). Subsequent to these initial standards, OSHA began promulgating comprehensive health standards for a variety of chemicals including asbestos, benzene, coke oven emissions, and lead. In addition to a PEL, these comprehensive standards include requirements such as medical monitoring, administrative control measures, respirator selection, training, and record keeping.

Challenges to the regulation of two hazards ultimately led to U.S. Supreme Court decisions on the need for OSHA to quantify risk reduction and the cost-benefit of its regulatory actions. OSHA's 1978 regulation to reduce the PEL for benzene from 10 to 1 ppm was remanded by the Court because of its belief that OSHA did not justify that the PEL reduction would substantially reduce the risk to the health of exposed workers. The Court rejected the notion that lower exposure is better and required OSHA to quantify the risk at current conditions of exposure and then document that the proposed standard would substantially reduce that risk to an acceptable level. In describing an acceptable level of risk, the Court suggested that the risk of job-related death caused by injury or illness over a working lifetime would be an appropriate level of acceptability. This risk was estimated to be 1 in 1000, and this value has become OSHA's target in determining acceptable PELs for chemical substances. In reviewing the 1978 OSHA standard for cotton dust, the U.S. Supreme Court decided that the OSH Act required OSHA to consider the technical and economic feasibility of new standards, but that a cost-benefit analysis was not required.

In January 1989 OSHA promulgated a new regulation amending its existing Air Contaminants standard.[32] This reduced 212 PELs and established 164 new ones. OSHA stated that it had reviewed health, risk, and technical/economic feasibility for the 428 substances considered in this rulemaking and found that the new PELs substantially reduced a significant risk of material health impairment among American workers and were technologically and economically feasible. OSHA estimated that benefits would accrue to 4.5 million workers and would result in the reduction of over 55,000 occupational illness cases annually. If not prevented, it was projected that these illness cases would eventually result in approximately 700 fatalities each year.

Unfortunately, several industrial organizations and the AFL/CIO had a different perspective and successfully convinced the U.S. Court of Appeals that the regulations should be vacated and remanded to OSHA. The court ruled that

> *(1) OSHA failed to establish that existing exposure limits in the workplace represent significant risk of material health impairment or that new standards eliminated or substantially lessened the risk; (2) OSHA did not meet its burden of establishing that the new PELs were either economically or technologically feasible; and (3) there was insufficient explanation in the record to support across-the-board, four-year delay in implementation of rule.*[33]

The court stated that "the inadequate explanation made it virtually impossible for a reviewing Court to determine if sufficient evidence supports the agency's conclusion."[33] The court indicated that it could easily believe OSHA's claim that going through detailed analysis for each of the substances was not possible given the time constraints set by the agency for this rulemaking. Unfortunately, OSHA's approach to this rulemaking was not consistent with the requirements of the OSH Act.

The end result was a major setback to efforts to formally update the PELs. Benefits from the rulemaking effort that remain after the court decision include (1) adoption and continued use of the revised PELs by several state health protection programs; (2) informal adoption by some industrial organizations; (3) sensitization by many parties to the inadequacies and obsolescence of the existing PELs; and (4) plans by OSHA to initiate another effort to revise PELs for a group of substances (fewer than 428) where some commonality can be identified, and where the objections identified by the court can be avoided.

Reestablishing state authority for occupational safety and health was also an important goal of the OSH Act. States were encouraged to enact their own legislation and to develop programs at least as stringent as OSHA's. State programs approved by OSHA would receive 50% of their operational costs from the federal government. However, 25 years later fewer than half of the states have reasserted their control over workplace hazards.

In addition to standards setting and enforcement, the OSH Act also included a provision to provide free consultation services to small businesses. These services are found in all 50 states and offer management confidential advice on safety and health problems.

While some may argue with the details of the content or the implementation of the OSH Act, most would agree with an early OSHA administrator that "its impact on working life in the United States cannot be overestimated."[34] One only has to look at the growth in the membership of the American Industrial Hygiene Association (AIHA) to see the impact on the number of practitioners in the profession. In 1970 AIHA membership numbered 1649 and within 10 years had tripled. It cannot be scientifically proven that these additional practitioners (as well as those who entered the occupational safety profession) have improved the health and well-being of the American worker. But we can assume that the attention paid to workplace hazards by this greatly expanded number of professionals is one factor contributing to a safer U.S. workplace.

Education and Professional Organizations

Discussion of the evolution of the practice of industrial hygiene in the United States would be incomplete without a brief review of the growth of educational programs and professional organizations associated with the field. Although the first course in industrial hygiene was taught at the Massachusetts Institute of Technology, Harvard University is recognized as having developed, in 1922, the first educational and research program leading to an advanced degree in industrial hygiene. While other universities (e.g., Johns Hopkins, the University of Michigan, and the University of North Carolina) implemented educational programs for industrial hygiene in subsequent years, the great leap forward came in 1977 with the NIOSH Educational Resource Centers program. These centers stress interdisciplinary education in the occupational aspects of hygiene, medicine, nursing, and safety. As a result of federal support for the development of an adequate supply of professionals to control occupational health and safety hazards, there are now 15 industrial hygiene programs at NIOSH Educational Resource Centers, with at least 14 more individual industrial hygiene education programs being supported by NIOSH. Industrial hygiene educational programs not supported by NIOSH can be found at both the graduate and undergraduate levels at other universities. A list of industrial hygiene academic programs is available from AIHA.

It has been suggested "one can date the emergence of the profession of industrial hygiene by the formation of our professional societies."[35] In 1938, 76 industrial hygienists representing 24 states, 3 cities, 1 university, the USPHS, the U.S. Bureau of Mines, and the Tennessee Valley Authority convened in Washington, D.C., to formally establish the National (later to become the American) Conference of Governmental Industrial Hygienists. ACGIH was established to "coordinate activities in federal, state, local, and territorial organizations and agencies; to help the public health service carry out its mission; and to develop state industrial hygiene units in a rational manner."[36] By 1996 ACGIH membership had reached 5400, and the association had a staff of 25 with an operating budget of almost $4 million.

In 1939 AIHA was formed. At the first meeting of the AIHA Board of Directors, four major goals of the association were enunciated.

1. The advancement and application of industrial hygiene and sanitation through the interchange and dissemination of technical knowledge on these subjects.
2. The furthering of the study and control of industrial health hazards through determination and elimination of excessive exposures.
3. The correlation of such activities as conducted by diverse individuals and agencies throughout industry, educational, and governmental groups.
4. The uniting of persons with these interests.[37]

AIHA membership has grown from 160 members in 1940 to more than 13,000 members in 1996. AIHA also has 93 local sections throughout the United States and in 3 other countries, which foster the interaction of industrial hygienists. The annual American Industrial Hygiene Conference and Exposition, sponsored by AIHA and ACGIH, provides the opportunity for members and others interested in the field to come together and exchange experiences and information. Both professional organizations also sponsor technical committees, develop technical publications, and publish journals. The *American Industrial Hygiene Association Journal* first appeared in 1946, while ACGIH's *Applied Occupational and Environmental Hygiene* (originally appearing as *Applied Industrial Hygiene*) debuted in 1986. AIHA and ACGIH were also founding members of the International Occupational Hygiene Association (IOHA). Membership in IOHA is limited to occupational hygiene professional associations, which by 1996 included associations located in Australia, Belgium, Brazil, Canada, Denmark, Finland, France, Germany, Hong Kong, Italy, Japan, The Netherlands, Norway, Spain, Sweden, Switzerland, the United Kingdom, and the United States (AIHA and ACGIH).

In 1960 AIHA and ACGIH created the American Board of Industrial Hygiene (ABIH) to develop voluntary professional certification standards for industrial hygienists and to implement a certification program.[38] The initial group of certified industrial hygienists (CIHs) has grown from 18 in 1960 to more than 7000 in 1996. In 1966 the diplomates of the ABIH certification program activated the American Academy of Industrial Hygiene (AAIH) as a professional organization. As stated in their bylaws, the purpose of AAIH is to

> *provide leadership in advancing the professional field of industrial hygiene, by raising the level of competence of industrial hygienists and by securing wide recognition of the need for high quality industrial hygiene practice to ensure healthful work conditions in the various occupations and industries.*[39]

The major activities of AAIH include promotion of the recognition of industrial hygiene as a profession by individuals, employers, and regulatory agencies and the accreditation of academic programs in industrial hygiene in cooperation with the Accreditation Board of Engineering and Technology. AAIH sponsors the annual Professional Conference on Industrial Hygiene to provide a forum for exploring professional issues, especially those encountered by more experienced industrial hygienists.

Summary

This brief overview of the development of industrial hygiene in the United States has been presented from a chronological as well as from a practice point of view. Growth has been rapid, with at least three major phases involving the early years prior to 1930, the dramatic increase between 1935 and 1945 due to social security legislation and World War II, and the more recent phase of growth resulting from the passage of the federal OSH Act in 1970.

Over the past few years it would appear that, at least in the United States, occupational hygienists are in the midst of a fourth period of change. There are at least three major driving forces involved with this phase. The first involves the "reinvention" of much of the government, including OSHA, to focus less on a regulatory approach to problem solving and to enhance cooperation and partnerships with those considered to be stakeholders. Although OSHA remains a regulatory agency, the pace of regulation and inspection has slowed. The second force is the downsizing of corporate America and increased outsourcing of services, including occupational hygiene. In 1986 approximately 15% of AIHA's 6500 members identified themselves as consultants. By 1996, of the 13,000 members, 30% were listed as consultants. The other major factor affecting the profession is the shift of the American economy from a manufacturing to a service base. Although not necessarily less hazardous, service industries often have problems much different from those of a traditional manufacturing site. While service industries have many of the same chemical, physical, and biological hazards found in manufacturing, the awareness of the need for industrial hygiene services is usually low.

Recognition of the need to address occupational health problems in nontraditional workplaces is not new. Jack Bloomfield, one of the first industrial hygienists in the USPHS, noted in 1938 that

> *industrial hygienists have concentrated their efforts on the so-called* industrial population—*that group of workers engaged in the manufacture, mechanical and mining industries. Although it is probably true that the bulk of the occupational diseases occurs in these industries, nevertheless, the 10 million workers in agriculture, the 4 million persons employed in transportation and communication, and the large number of workers in domestic and personal services, all have health problems deserving attention.*[27]

To not only remain viable but also to achieve increased recognition and acceptance of their work, it is apparent that occupational hygienists must achieve two goals. First, hygienists must continue to expand the scope of their practice to include environmental health considerations, especially those arising from the workplace. The skills and knowledge involved in the recognition, evaluation, and control of air, and even water, pollution problems as well as those of hazardous wastes as they affect the health and well-being of those in the community are not dissimilar to the skills and knowledge applied to traditional industrial hygiene problems. Indeed, many of the early environmental health practitioners, especially addressing air pollution, radiologic health and hazardous waste problems, were industrial hygienists.[40] The first air pollution disaster in the United States, the Donora, Penn., smog of October 1948, resulted in the death of 20 people and to some degree affected almost 6000 persons.[41] The investigation of the incident involved a multidisciplinary team of physicians, nurses, engineers, chemists, meteorologists, housing experts, veterinarians, and dentists under the direction of a USPHS industrial hygienist—George D. Clayton. Occupational hygienists must become more effective in providing comprehensive rather than one-issue services.

Secondly, as a profession and as individuals, occupational hygienists must demonstrate that their services, in addition to preventing occupational and environmental health problems, provide a positive contribution to the individual worker, the community, and the employer.

Public Health Roots

It has been postulated that the "[f]irst Public Health Revolution unfolded in nineteenth-century Europe as that society sought to address the adverse health effects of squalid living conditions: poor sanitation, poor housing, dangerous work environments and air pollution."[42]

The Institute of Medicine has defined public health as "what we, as a society, do collectively to assure the conditions in which people can be healthy."[43] Thus, when industrial hygienists implement a control measure to reduce a worker's exposure to a toxin, they are practicing the art and science of public health. Those who practice in the field of public health are well aware that throughout the history of public and especially occupational health, two major factors have shaped our solutions: the availability of scientific and technical knowledge and the content of public values and opinions. The Institute of Medicine has recognized the lack of agreement about the public health mission as reflected in the diversion in some states of traditional public health functions "such as water and air pollution control, to separate departments of environmental services, where the health effects of pollutants often receive less notice."[43] Although industrial hygiene functions are now most often found combined with safety programs in labor rather than public health departments, the concepts of injury prevention, whether on or off the job, have strong public health underpinnings. Thus, safety and hygiene are intertwined.

The first professional journal to address industrial hygiene concerns was the *Journal of the American Public Health Association*. In 1914 the journal announced a new department on industrial hygiene and sanitation to put the readers "in touch with the latest information in this very recently new field of Public Health work," and a review article on industrial hygiene and sanitation was published.[44] In the same year, the association created an industrial hygiene section, and at the annual meeting a special symposium on industrial hygiene was held. Alice Hamilton served as vice chairman of the new section and succeeded to the chairmanship in 1916.[45]

As noted earlier, in 1914 the USPHS established an office of industrial hygiene and sanitation, which 57 years later became NIOSH. NIOSH is part of the federal Centers for Disease Control and Prevention, which encompasses most of the public health functions of the federal government. Prior to the OSH Act of 1970, most state and local government industrial hygiene units were found in public health agencies.

Industrial hygiene graduate level academic training in the United States has since its inception been associated with schools of public health. In 1918 the Harvard Medical School established a Department of Applied Physiology, which in 1922 became the Department of Physiology and the Department of Industrial Hygiene in the School of Public Health. In the late 1930s state public health agencies used Social Security funds to train industrial hygienists in schools of public health at Harvard, the University of California, Columbia University, The Johns Hopkins University, the University of Michigan, and Yale University. Today industrial hygiene academic programs are found in at least 18 of the 27 accredited schools of public health.

There is ample evidence that the roots of industrial hygiene are in the field of public health. The public health philosophy of protection and enhancement of the health and well being of groups of people through preventive rather than curative measures applies as well to industrial hygiene.

Professional Recognition

As the profession of industrial hygiene and occupational health and safety changes, so do the perceptions of what constitutes an individual who is qualified to practice industrial hygiene. It is an issue that has been debated within the profession for many years.

In 1994 the AIHA Board of Directors determined the time had come to become more proactive on this issue. This increased interest was a result of several events, most notably an increasing requirement by federal and state regulators mandating that all individuals have additional training and certification for many single-substance issues (e.g., asbestos and lead).

In addition, over the course of several years, the profession witnessed the creation of so-called credentialing organizations that seized on the opportunity to bestow new credentials to individuals in almost every imaginable field of occupational health and safety. Many of these new credentials can be readily obtained either by submitting a small fee for the honor or by taking a mail-order examination. Some credentials are even granted for life, without requiring any demonstration of maintenance of competency.

These attempts to confuse policy makers and the public forced AIHA to become more involved in protecting industrial hygiene titles. Starting with the adoption of an official position statement on title protection in 1994, AIHA began working to assist local sections and state government affairs organizations with efforts to enact title protection legislation in the individual states. AIHA's model bill includes definitions for CIH and IHIT. While title protection will not eliminate new credentialing organizations, it provides the industrial hygiene professional with legal recognition and protection. It will also be of assistance in future regulatory efforts on both the federal and state levels as AIHA seeks to have this profession recognized as a leader in the field of protecting workers.

The first states to actively seek some form of title protection were California and Tennessee. Both enacted legislation in 1993 recognizing the profession and were the impetus for AIHA's subsequent proactive efforts. In addition to California and Tennessee, the state of Illinois enacted a form of voluntary licensing for CIHs in 1993 that is still the most far-reaching and somewhat restrictive professional recognition legislation.

In mid-1995 AIHA adopted a joint position statement on the issue with the American Society of Safety Engineers. This joint statement promotes cooperation between the two professions and suggests a cooperative venture, when possible, on enacting legislation that not only protects the titles of industrial hygiene, but also the titles of the safety profession.

State legislative efforts in 1995 resulted in legislation enacted in Alaska and Nevada. In 1996 the states of Connecticut and Indiana were added to the list of successful efforts and a half dozen additional states are considering title protection legislation. In 1997 AIHA expects more than 10 states will consider bills supported by AIHA. It is possible that by the end of the century, as many as 15 states will have enacted title protection legislation.

On the federal level, AIHA works to see that federal regulations and legislative efforts do not restrict the profession. All regulatory and congressional activities are monitored to assure that the profession is protected. When possible, AIHA seeks to add language that recognizes the certified industrial hygienist as a competent, qualified, and knowledgeable occupational health and safety professional.

Professional Code of Ethics

In 1968 the Code of Ethics for Professional Practice was developed by the AAIH Ethics Committee. The officers and councillors at that time accepted the code and committee report without taking further action. In 1973–1974 there was renewed interest in a professional code of ethics by not only AAIH but also by ABIH. In 1975 AIHA also began work on a code of ethics; the work of AIHA's Law Committee was completed in 1977.

In 1978 the AAIH draft code was mailed to the membership for comment. Approximately 100 responses were received and were used to prepare a redraft of the code, which was submitted for approval by the membership. Of the 743 respondents (67.5% of all members), 712 (95.8%) voted to accept and 31 (4.2%) voted to reject. The code provided standards of ethical conduct to be followed by industrial hygienists as they practice their profession and serve employees, employers and clients, and the general public. By 1981 AIHA and ACGIH also adopted the code of ethics, thus extending coverage to most industrial hygiene professionals.

In 1991 the four industrial hygiene organizations chartered the Code of Ethics Task Force, whose charge was to (1) review and revise the industrial hygiene code of ethics and to supplement it with supporting interpretive guidelines, and (2) recommend methods to educate members about ethical conduct and to recommend disciplinary procedures and mechanisms for enforcement.[46]

The task force determined that the original code of ethics, while presenting general principles of ethical conduct, lacked interpretive guidelines. Of special concern was the change in the scope of industrial hygiene practice since 1978, especially the increase in the number of consultants. The most difficult issues considered by the task force were disciplinary procedures and mechanisms for enforcement. The task force surveyed 12 similar professional organizations and determined that while the majority had developed a code of ethics, only two had charters to enforce their codes. It was, however, recognized that 6 of the 12 organizations involved professions (physicians,

Code of Ethics
for the Professional Practice of Industrial Hygiene

At its meeting, Dec. 14, 1994, the American Industrial Hygiene Association Board of Directors adopted, and commends to every member of AIHA, the following Code of Ethics. This Code of Ethics was developed jointly with the American Conference of Governmental Industrial Hygienists, the American Board of Industrial Hygiene, and the American Academy of Industrial Hygiene and has also been approved by the other IH groups.

Objective
This canon provides standards of ethical conduct for Industrial Hygienists as they practice their profession and exercise their primary mission to protect the health and well-being of working people and the public from chemical, microbiological, and physical health hazards present at, or emanating from, the workplace.

Canons of Ethical Conduct
Industrial Hygienists shall:

- Practice their profession following recognized scientific principles with the realization that the lives, health, and well-being of people may depend upon their professional judgment and that they are obligated to protect the health and well-being of people.
- Counsel affected parties factually regarding potential health risks and precautions necessary to avoid adverse health effects.
- Keep confidential personal and business information obtained during the exercise of industrial hygiene activities, except when required by law or overriding health and safety considerations.
- Avoid circumstances where a compromise of professional judgment or conflict of interest may arise.
- Perform services only in the areas of their competence.
- Act responsibly to uphold the integrity of the profession.

nurses, engineers) whose practice was regulated through state licensure.

The task force reviewed the pros and cons of six enforcement options: no enforcement, education, mediation, arbitration, title licensing, and a formal enforcement procedure. The task force recommended that regardless of the choice made by the professional organizations, education in and communication of ethical concepts were needed, and that the four organizations should consider establishing one joint ethics committee.

The report of the task force was presented to the four boards in October 1993, a final draft of a revised code of ethics was presented at the 1994 American Industrial Hygiene Conference & Exposition, and by January 1995 all four organizations had approved the new code. In 1995 the industrial hygiene organizations approved the formation of the Joint Industrial Hygiene Ethics Education Committee, the purpose of which is to conduct educational activities for industrial hygienists and interested parties that will assist in promoting the code of ethics. Enforcement of the code of ethics is a question yet to be answered by the professional organizations.

The new Code of Professional Ethics for the Practice of Industrial Hygiene (Figure 1) comprises six canons with interpretive guidelines. The canons of ethical conduct require industrial hygienists to (1) practice their profession by applying scientific principles, (2) counsel affected parties regarding risks and protective measures, (3) keep information obtained confidential except under special circumstances, (4) avoid compromise of professional judgment and conflict of interest situations, (5) practice only in their areas of competence, and (6) uphold the integrity of the profession.

In developing the interpretive guidelines, the task force emphasized that they were not rules of practice, did not necessarily define right from wrong, were not intended to carry the weight of law, and should not be considered to be all inclusive. They are designed to give professionals guidance in deciding what constitutes ethical practices.

Whom Do We Serve?

One of the questions often facing occupational health professionals is, "Who are our clients?" Simply defined, do occupational hygienists owe allegiance to those who pay them or to those whose health and well-being they are entrusted to protect? Ideally, the answer to that question is "both." To those who pay for their services, as either employer or client, industrial hygienists owe their best professional effort to anticipate, recognize, evaluate, and control workplace hazards. If they meet that goal, they also discharge their responsibilities to the workers and those in the community who may be adversely affected by the hazards they address.

But what of those situations where occupational hygienists are asked to compromise or at least soften or shade their findings? While they may not be asked by a manager to change their recommendations, what are the responsibilities of occupational hygienists if they believe needed change will not be made? Do they say to themselves, "I have discharged my responsibilities by recognizing, evaluating, and recommending controls—it is someone else's responsibility to see that they are implemented"? Or should hygienists believe that their ultimate responsibility is to see that worker and community health is protected? If those who create a hazard do not redress it, is it the responsibility of occupational hygienists to inform those whose health is likely to be adversely affected, so that they can take steps to protect themselves?

What of those situations where management has implemented controls, but hygienists see workers not following good work practices? Do they say to themselves, "I have discharged my responsibilities by recognizing, evaluating, and recommending controls. It is someone else's responsibility to see that they are implemented"? Or should industrial hygienists believe that they play a part in the implementation and owe it to their clients and to the workers to see that good work practices are followed or workers disciplined when they are not?

To answer these and similar questions, each occupational hygienist must define the principles that will guide the practice of the profession. The code of ethics does not come alive until it is used to set the compass to guide professional lives.

But when occupational hygienists seek to define the raison d'être of their profession, the final words of the first canon of the ethics code could serve them well, in all circumstances: "(We) are obligated to protect the health and well-being of people."

Summary

This chapter reviews the history of industrial hygiene with an emphasis on the U.S. experience. In addition to a chronological listing of developments, the profession's history is also viewed in the context of its tenants of practice: recognition, evaluation, and control. The considerable impact of the OSH Act on the growth of the profession and its practice is described. A historical review of the professional organizations—AIHA, ACGIH, AAIH, and ABIH—is presented. In considering the philosophy of the practice of the profession, its public health roots are apparent. The public health philosophy of protection and enhancement of the health and well-being of groups of people through preventive rather than curative measures applies as well to the practice of industrial hygiene. As the importance of the profession grew, AIHA recognized the need to ensure that the practice of industrial hygiene was limited to qualified professionals. AIHA's effort in title enhancement and protection has resulted in increased recognition of the profession by federal and state legislators and regulators. The question of who is served by industrial hygienists—employers or the people whose health and well-being hygienists protect—can be perplexing. To help professionals answer this and similar questions, the four industrial hygiene organizations developed the Code of Professional Ethics for the Practice of Industrial Hygiene. The code consists of six canons with interpretive guidelines that can be used by industrial hygienists to guide the practice of their profession.

References

1. **Patty, F.A.:** Industrial hygiene: retrospect and prospect. In *Patty's Industrial Hygiene and Toxicology*, vol. 1, 3rd ed. New York: John Wiley & Sons, 1978. pp. 1–21.
2. **Goldwater, L.J.:** *Historical Highlights in Occupational Medicine.* (Readings and Perspectives in Medicine, Booklet 9). Durham, NC: Duke University Medical Center, 1985. pp. 1–14.
3. **Ramazzini, B.:** *Diseases of Workers.* New York: Hafner Publishing Co., 1964.
4. **McReady, B.W.:** *On the Influence of Trades, Professions and Occupations in the United State, in the Production of Disease* (transactions of the Medical Society of the State of New York, vol. III). Albany, NY: Medical Society of the State of New York, 1837.
5. **Luxon, S.G.:** A history of industrial hygiene. *Am. Ind. Hyg. Assoc. J. 45:*731–739 (1984).
6. **Brown, H.V.:** This history of industrial hygiene: a review with special reference to silicosis. *Am. Ind. Hyg. Assoc. J. 26:*212–226 (1965).
7. **Hamilton, A.:** *Exploring the Dangerous Trades.* Fairfax, VA: American Industrial Hygiene Association, 1995. p. 1.
8. **Clayton, G.D.:** Industrial hygiene: Retrospect and prospect. In *Patty's Industrial Hygiene and Toxicology*, vol. 1, part A, 4th ed. New York: John Wiley & Sons, 1991. p. 1–13.
9. **Corn, J.K.:** Historical aspects of industrial hygiene—I: Changing attitudes toward occupational health. *Am. Ind. Hyg. Assoc. J. 39:*695–699 (1978).
10. **Rose, V.E.:** The development of occupational hygiene in the United States—a history. *Ann. Am. Conf. Gov. Ind. Hyg. 15:*5–8 (1988).
11. **Greife, A., R. Young, M. Carrole, W.K. Sieber, et al.:** National Institute for Occupational Safety and Health—general industry occupational exposure databases: their structure, capabilities, and limitations. *Appl. Occup. Environ. Hyg. 10:*264–269 (1995).
12. **Redmond, S.K., A. Ciocco, J.W. Lloyd, and H.W. Rush:** Long-term mortality of steel workers. VI. Mortality from malignant neoplasms among coke oven workers. *J. Occup. Med. 13:*53–68 (1971).
13. "Coke Oven Emissions," *Code of Federal Regulations*, Title 29, Section 1910.1029. 1977.
14. **National Institute for Occupational Safety and Health (NIOSH):** *Criteria for a Recommended Standard . . . Occupational Exposure to Asbestos* (DHEW/NIOSH pub. HSM72-10267). Washington, DC: Government Printing Office, 1972. p. V-4.
15. **Selikoff, I.J., J. Chung, and E.C. Hammond:** Asbestos exposure and neoplasia. *J. Am. Med. Assoc. 188:*22 (1964).
16. "Asbestos," *Code of Federal Regulations*, Title 29, Section 1910.1001. 1972.
17. "Benzene," *Code of Federal Regulations*, Title 29, Section 1910.1028. 1987.
18. **Lamb, A.B. and C.R. Hoover:** U.S. Patent 1321 062 (1919).
19. **Goetz, A.:** Application of molecular filter membranes to the analysis of aerosols. *Am. J. Pub. Health 43:*150–159 (1953).

20. **White, L.D., D.G. Taylor, P.A. Mauer, and R.E. Kupel:** A convenient optimized method for the analysis of selected solvent vapors in the industrial atmosphere. *Am. Ind. Hyg. Assoc. J. 31:*225–232 (1970).
21. **Rose, V.E. and J.L. Perkins:** Passive dosimetry—a state of the art review. *Am. Ind. Hyg. Assoc. J. 43:*605–621 (1982).
22. **National Institute for Occupational Safety and Health (NIOSH):** *Criteria for a Recommended Standard. . . Occupational Exposure to Crystalline Silica* (DHEW/NIOSH pub. 75-120). Washington, DC: Government Printing Office, 1975. p. 62.
23. **Baetzer, A.M.:** The early days of industrial hygiene—their contribution to current problems. *Am. Ind. Hyg. Assoc. J. 41:*773–777 (1980).
24. **Felton, J.S.:** History. In *Occupational Health & Safety*, 2nd ed., J. LaDou, ed. Itasca, IL: National Safety Council. 1994. pp. 17–31.
25. **National Institute for Occupational Safety and Health (NIOSH):** *A Guide to Industrial Respiratory Protection*, by J.A. Pritchard. (DHEW/NIOSH pub. 76-189). Washington, DC: Government Printing Office, 1976. p. 5–8.
26. **Cralley, L.J.:** Industrial hygiene in the U.S. Public Health Service (1914–1968). *Appl. Occup. Environ. Hyg. 11:*147–155 (1996).
27. **Bloomfield, J.J.:** Development of industrial hygiene in the United States. *Am. J. Pub. Health 28:*1388–1397 (1938).
28. **Sayers, R.R. and J.J. Bloomfield:** Industrial hygiene activities in the United States. *Am. J. Pub. Health 26:*1087–1096 (1936).
29. "Occupational Safety and Health Act," Pub. Law 91-596, Section 2193. 91st Congress, Dec. 29, 1970; as amended, Pub. Law 101-552, Section 3101, Nov. 5, 1990.
30. "Limits for Air Contaminants," *Code of Federal Regulations*, Title 29, Section 1910.1000. 1971. Tables Z-1 (chemicals) and Z-3 (mineral dusts)
31. "Limits for Air Contaminants," *Code of Federal Regulations*, Title 29, Section 1910.1000. 1971. Table Z-2.
32. "Limits for Air Contaminants; Final Rule." *Federal Register*, January 19, 1989.
33. *AFL/CIO v. OSHA*, U.S. Court of Appeals, 11th Cir., July 7, 1992.
34. **Corn, M.:** The progression of industrial hygiene. *Appl. Ind. Hyg. 4:*153–157 (1989).
35. **Corn, J.K.:** Historical review of industrial hygiene. *Ann. Am. Conf. Gov. Ind. Hyg. 5:*13–17 (1983).
36. **Corn, J.K.:** *Protecting the Health of Workers: The American Conference of Governmental Industrial Hygienists 1938-1988*. Cincinnati, OH: American Conference of Governmental Industrial Hygienists, 1989. p. 8.
37. **American Industrial Hygiene Association (AIHA):** *The American Industrial Hygiene Association. Its History and Personalities 1939–1990.* Fairfax, VA: AIHA, 1994. p. 3.
38. **American Board of Industrial Hygiene (ABIH):** *Bulletin of the American Board of Industrial Hygiene.* Lansing, MI: ABIH, 1996.
39. **American Board of Industrial Hygiene (ABIH):** *Roster of the American Board of Industrial Hygiene.* Lansing, MI: ABIH, 1996.
40. **Rose, V.E.:** Industrial hygiene—coping with change. *Am. Ind. Hyg. Assoc. J. 56:*853 (1995).
41. **Clayton, G.D.:** Air pollution; U.S. Public Health Service pioneering studies; genesis of the Environmental Protection Agency. *Appl. Occup. Environ. Hyg. 12:*7–10 (1997).
42. **McMichael, A.J.:** *Planetary Overload.* Cambridge, UK: Cambridge University Press, 1993. p. 63.
43. **Institute of Medicine:** *The Future of Public Health.* Washington, DC: National Academy Press, 1988. p. 1.
44. Industrial Hygiene and Sanitation. [Announcement] *Am. J. Pub. Health,* vol. 4 (1914).
45. **Harris, R.L.:** "The Public Health Roots of Industrial Hygiene." Paper presented at the Professional Conference in Industrial Hygiene, Nashville, TN, October 14, 1996.
46. **Farrar, A.C.:** Industrial hygiene ethics in the '90s: a professional challenge. *Am. Ind. Hyg. Assoc. J. 34:*403–407 (1993).

From the Collections of Henry Ford Museum & Greenfield Village.

Outcome Competencies

After completing this chapter, the user should be able to:

1. Define underlined terms used in this chapter that are germane to understanding occupational exposure limits (OELs).
2. Select resources for obtaining OEL values.
3. Describe the differences between the various OELs.
4. Summarize the factors evaluated in setting an OEL.
5. Quote limitations of OELs.
6. Calculate workplace exposures.

Key Terms

action level • airborne particulate matter • biological exposure indeces (BEIs) • carcinogen classification system • ceiling • maximum allowable concentrations (MAKs) • new chemical exposure limits (NCELs) • notice of intended change (NIC) • permissible exposure limit (PEL) • recommended exposure limit (REL) • short-term exposure limit (STEL) • skin notation • threshold limit value (TLV) • time-weighted average (TWA) • workplace environmental exposure limit (WEEL)

Prerequisite Knowledge

Prior to beginning this chapter, the user should review the following references for general content:

Academic Discipline	Reference Citation
Occupational hygiene	*Threshold Limit Values for Chemical Substances and Physical Agents—Biological Exposure Indices*, ACGIH
Toxicology	*Toxicology Primer*, Kamrin
Toxicology	*Casarett & Doull's Toxicology—the Basic Science of Poisons*, Klaassen

Prior to beginning this chapter, the reader should review the following chapters.

Chapter Number	Chapter Topic
1	History and the Philosophy of Industrial Hygiene
4	Environmental and Occupational Toxicology
5	Epidemiological Surveillance
15	Comprehensive Exposure Assessment
16	Modeling Inhalation Exposure
17	Risk Assessment in the workplace

Key Topics

I. Introduction
   A. The Question from the Safety Team
   B. Objectives and Terminology
   C. Physiology of the Respiratory Tract and Skin

II. Goals and Limitations of OELs

III. Terminology Used in OELs

IV. Groups That Recommend OELs
   A. Inhalation Exposure Limits
   B. Biological Exposure Limits
   C. Physical Agent Exposure Limits

V. Important Considerations in the Development of OELs
   A. Identification of the Hazard
   B. Routes of Exposure
   C. Review and Evaluation of the Chemical-Specific Toxicology Data
   D. Carcinogen Classification System and Risk Models

VI. Other Issues Impacting OELS

2

Occupational Exposure Limits

Gordon C. Miller, CIH, CHCM

Dennis R. Klonne, Ph.D., DABT

Introduction

The Question from the Safety Team

The toughest question faced by the occupational safety, health, and environmental affairs professional is, "We know we are exposed to this chemical (or physical or biological) agent, but is it safe for us at this level?"

How many times has this question been raised at safety meetings, training sessions, new process reviews, high-level management meetings, and other gatherings both inside and outside the workplace? Too many times to be counted is the best answer, because it is so often asked. Although this question may be difficult to answer, it deserves a response. However, the answer must always be given with the caveat that it is the best answer that can be given based on current data and the application of best professional judgment. There is no black and white answer to this question.

Objectives and Terminology

The objective of this chapter is to make the reader aware of resources for occupational exposure limits (OELs), to understand the basic terminology used in OELs, and to develop an appreciation of the factors that are evaluated in setting an OEL. The reader should not expect to be able to develop an OEL after reading this chapter, since that is best left to teams of occupational hygienists, toxicologists, occupational physicians, etc., who have many years of experience and have developed expertise in this area.

Following is some terminology that is defined for its specific use in this chapter.

OEL. Here this term is used in a generic sense, as opposed to limits with specific acronyms that are set by specific groups, such as the TLV® (threshold limit value) set by ACGIH (the American Conference of Governmental Industrial Hygienists) or the WEEL (workplace environmental exposure level) set by AIHA (the American Industrial Hygiene Association).

Agent. An agent may be physical (e.g., radiation, noise), chemical, or biological (e.g., virus, bacteria) in nature. While in this chapter chemical agents principally are discussed since they are the most common agent in the industrial setting, the same principles generally apply to physical or biological agents.

Workplace Health Professional. As used here, this is a generic term encompassing many different disciplines, including but not limited to occupational hygiene, safety, toxicology, industrial medicine, environmental and process engineering, etc.

With regard to resources for obtaining information on specific chemicals, there are myriad books available to obtain such information, many of which will be cited in this chapter. However, the use of computerized online databases should be included in any serious search since the most recent published data will likely be cited in the journals included in those databases. The National Library of Medicine's (NLM's) MEDLARS® system has over 40 online databases containing over 18 million references. In the Additional Resources section of this chapter there are further descriptions of some pertinent databases as well as phone numbers and World Wide Web addresses to obtain additional information. However, to obtain data from the NLM databases you must have an account number. Therefore, if your site does not have a library with online capabilities, data can be obtained through most university libraries or NLM, usually for nominal charges. Additionally, there are many consulting companies that will perform this service for a fee.

Physiology of the Respiratory Tract and Skin

A review of the basic physiology of the respiratory tract and skin can be found elsewhere in this book as well as in several references.[(1–11)] Therefore, basic knowledge in these areas is assumed, and only some specific ideas, directly applicable to topics in this chapter, are discussed here.

Physiology of the Respiratory Tract

The upper respiratory tract consists of the nasopharynx and is the portal of entry for many toxicants. Its geometry provides for turbulent airflow, which helps to trap particles greater than about 5 microns by inertial impaction. Because its surface area is covered by a mucus layer and because the incoming air is humidified to approximately 100% relative humidity in this region, it is also a site where soluble gases and vapors may be trapped. The olfactory epithelium also provides an early warning system with its ability to detect odors. However, individual ability to detect odor varies greatly, and a phenomenon known as olfactory fatigue decreases the ability to detect odors over time. That is, the threshold for detecting the odor is raised over time. It should further be noted that odor does not correlate with toxicity. That is, compounds with a very low odor threshold and very disagreeable odor, such as mercaptans or thiols, may be detected at levels very much lower than that producing toxicity, while compounds such as carbon monoxide are never detected by odor even up to concentrations that cause death. For further information on odor thresholds, see references 12 and 13. Another important component of the nasopharyngeal region (and in the conducting airways of the lung) is the cilium. The hair-like cilia keep mucus flowing upward to clear the upper respiratory tract. However, some chemicals (e.g., cigarette smoke) may paralyze the cilia, resulting in increased accumulation of mucus and decreased clearance of toxicants.

The conducting airways, consisting of the trachea and bronchi, have many bifurcations (splits) that are potential sites of impact and deposition for toxicants. The increased branching of these airways into the lung results in a decreased airflow due to increasing total cross-sectional area with resultant differences in particle deposition throughout the lower respiratory tract. Particles in the 1–5 micron range are deposited in this region of the lung primarily by sedimentation. Again, mucociliary transport is an important route of toxicant elimination that works by trapping the vapor or particle in the mucus layer then moving it upward to be expectorated or swallowed. If swallowed, the route of exposure then becomes oral, instead of the initial inhalation route of exposure.

The lungs are designed for maximum transport and absorption. The alveolar/blood barrier is only a few cells thick, and the lung is highly vascularized and perfused. Furthermore, the lung has a very large surface area (approximately 300 to 1000 ft^2 [27.87 to 92.9 m^2]), especially in comparison with that of the skin (approximately 20 ft^2 [1.86 m^2]). Submicron particles and vapors are deposited in this alveolar region of the lung by simple diffusion. For most calculations of respiratory volume during a normal workday, a value of 10 m^3 can be used.

Physiology of the Skin

The outer skin (epidermis) is the principal barrier (approximately 0.2 mm thick) blocking entry of foreign chemicals into the body. The layer of dead cells (stratum corneum) is the principal component in this role and is replaced about every 2–4 weeks. Passage of chemicals through this layer to the dermis (which is approximately 2 mm thick) can result in the systemic uptake of the chemical and its distribution through the body. Chemicals that have both lipid and water solubility pass through the skin the easiest; those that are mostly lipid-soluble pass easier than those that are mostly water-soluble. Many chemicals are known to pass through the skin readily and are used as carriers for drugs (e.g., dimethyl sulfoxide). However, chemicals that pass through the dermal barrier can carry other toxic chemicals with them, producing illness in workers despite the fact that the workplace concentrations are well below the OEL. Similarly, chemicals that de-fat or dry out the skin can also compromise the dermal barrier and thereby allow other chemicals to pass into the body in toxic amounts. The most common skin-related complaints in the workplace are dermatitis, which is a localized inflammation and is reversible; corrosion, which is a destruction of the skin and results in irreversible scarring; and sensitization, which is a reversible allergic reaction but which produces potentially irreversible changes in immune cells thereby causing future reactions to the chemical at extremely low exposures. Phototoxicity and photosensitization occur less frequently and result from the interaction of chemicals with sunlight (ultraviolet radiation) and the cells in the skin.

Goals and Limitations of Occupational Exposure Limits

The question asked at the opening of this chapter, whether a chemical is safe at a particular level, is the reason for OELs. While some people think that only a zero exposure level is safe or question the need for OELs, the real world dictates that we answer the question as best we can. The OELs are quite simply there to protect worker health. While zero exposure should be a goal to strive for, it is generally not a reality of the industrial environment. In cases where no safe level is thought to exist (e.g., very potent carcinogens, lethal viruses), then extreme measures such as engineering controls and use of personal protective equipment must be taken to ensure that there is indeed no exposure.

The Occupational Safety and Health Act of 1970 (OSH Act)[10,11,14,15] was enacted

> *to assure so far as possible every working man and woman in the nation safe and healthful working conditions by authorizing the Secretary of Labor to set mandatory occupational safety and health standards . . . (and) by providing medical criteria which will assure insofar as practicable that no employee will suffer diminished health, functional capacity, or life expectancy as a result of his work experience.*

While this is indeed a desirable goal, ACGIH[16] recognizes that OELs may not protect all people in all cases.

> *Threshold Limit Values (TLVs®) refer to airborne concentrations of substances and represent conditions under which it is believed that nearly all workers may be repeatedly exposed day after day without adverse health effects. Because of wide variation in individual susceptibility, however, a small percentage of workers may experience discomfort from some substances at concentrations at or below the threshold limit; a smaller percentage may be affected more seriously by aggravation of a pre-existing condition or by development of an occupational illness. ... Individuals may also be hypersusceptible or otherwise unusually responsive to some industrial chemicals because of genetic factors, age, personal habits (smoking, alcohol, or other drugs), medication or previous exposures. Such workers may not be adequately protected from adverse health effects from certain chemicals at concentrations at or below the threshold limits.*

Thus, it is already stated that not all people are protected all the time. There are even many instances where a single individual is protected until a change in physiological state suddenly occurs (e.g., pregnancy, acute liver disease), and then the OEL is no longer protective. There are also examples where hobbies can render an OEL unprotective, such as a foundry worker who likes to target shoot in the evening, or a person who works in a solvent production plant by day and strips and refinishes furniture at night.

Being cognizant of the limitations of OELs requires the practicing workplace health professional to be ever vigilant for the individual(s) who may suffer adverse effects, even if the OEL is always maintained. It should always be remembered that OELs such as the TLV or WEEL do not have the force of law and are recommendations or guidelines, not boundaries between safe and unsafe conditions for all workers. Their proper implementation requires people with appropriate training to continually observe and monitor both the employees and the work environment.

The goal of OELs is to protect workers over their entire working lifetime, which is approximately 40 years. This has traditionally been applied to the 8 hours/day, 5 days/week work schedule. With the concept of the flexible workweek (e.g., four 10-hour workdays), the traditional workweek may well apply to a smaller percentage of the work force. When this is the case, people engaged in these unusual work shifts should be monitored carefully, and in some instances the OEL may have to be adjusted to a lower level to account for the increased time of continuous exposure and the decreased time for metabolism and elimination between exposures. The reader is referred to the model by Brief and Scala[17] for further discussion of this topic.

It is also very important to remember that the basis for the OEL varies from chemical to chemical. The goal for one chemical may be simply to avoid unpleasant odor, while for another it may be to prevent irreversible reproductive effects or cancer. This should always be considered by the workplace health professional since the relative importance of overexposures is very much affected by the toxicity endpoint. Thus, it is of utmost importance for the practicing professional to read the documentation for the OEL so that an understanding is gained with regard to the basis of the level and the strength of the supporting data. It is strongly recommended that every practicing occupational hygienist read the introduction section of the TLV booklet[16] since it eloquently describes both the goals and limitations of OELs.

One final point with regard to OELs. OELs are typically meant to apply to a work force that, on average, is healthier than the general public. The need to have people consistently perform activities that require some degree of physical exertion generally requires a subpopulation that is healthier than the U.S. population in general. Furthermore, in industrial settings it has been the practice to exclude the very young and very old, as well as those with infirmities and physical impairment, because of the nature of the work. This phenomenon is typically referred to as the healthy worker effect. Therefore, OELs are not meant to apply to the general population and are not to be used as community-based standards. Furthermore, the application of some set safety factor (e.g., divide the OEL by 100) to produce a community-based standard from an OEL is equally inappropriate.

Terminology Used in OELs

A fairly standard terminology has come to be used with regard to OELs. All the terms below (except action level) are derived from the definitions given in the TLV booklet[16] but are provided here for the sake of completeness and as points of discussion. Additionally, Appendix A contains definitions of terminology commonly found in the explanatory documents for OELs.[18] The reader is also referred to glossaries in Plog[11] and in Accrocco.[19]

Time-Weighted Average (TWA). This is the fundamental concept of most OELs. It is usually presented as the average concentration over an 8-hour workday for a 40-hour workweek. However, this implies that concentrations will be both above and below the average value. The ACGIH TLV committee has recommended excursion limits to prevent concentrations from severely exceeding the average value. The proposed excursion limits are that exposures should typically not exceed the TWA by more than threefold and for a period not exceeding 30 minutes during the workday. Even if the TWA is not exceeded for the work shift, in no case should the excursion be more than fivefold the TWA value. Examples of calculations germane to the concept of the TWA appear in Appendix A. See also Chapters 7 and 8.

Short-Term Exposure Limit (STEL). STELs are recommended when exposures of even short duration to high concentrations of a chemical are known to produce acute toxicity. It is the concentration to which workers can be exposed continuously for a short period of time without suffering from (1) irritation, (2) chronic or irreversible tissue damage, or (3) narcosis of sufficient degree to increase the likelihood of accidental injury, impaired self-rescue, or reduced work efficiency. A STEL is defined as a 15-minute TWA exposure that should not be exceeded at any time during a workday, even if the overall 8-hour TWA is within limits, and it should not occur more than four times per day. There should be at least 60 minutes between successive exposures in this range. If warranted, an averaging period other than 15 minutes can also be used.

Ceiling (C). The concentration that should not be exceeded during any part of the working exposure. In conventional occupational hygiene practice, if instantaneous monitoring is not feasible, then the ceiling can be assessed by sampling over a 15-minute period, except for chemicals that may cause immediate irritation, even with exposures of extremely short duration.

Action Level. This is the concentration or level of an agent at which it is deemed that some specific action should be taken. The action can range from more closely monitoring the exposure atmosphere to making engineering adjustments. In general practice the action level is usually set at one-half of the TLV.

Skin Notation (skin). The skin notation denotes the possibility that dermal absorption may be a significant contribution to the overall body burden of the chemical. That is, the airborne OEL may not be adequate to protect the worker because the compound also readily penetrates the skin. Other toxicity endpoints on skin such as irritation, dermatitis, and sensitization are not sufficient to warrant the skin notation. In practice, the skin notation is given to compounds with a dermal LD_{50} less than 1000 mg/kg, or if there are other data indicating that repeated dermal exposure results in systemic toxicity.

Airborne Particulate Matter. The ACGIH TLV committee has divided this general category into three classes based on the likely deposition within the respiratory tract. While past practice was to provide TLVs in terms of total particulate mass, the recent approach is to take into account the aerodynamic diameter of the particle and its site of action. Inhalable particulate mass (IPM) TLVs are designated for compounds that are toxic if deposited at any site within the respiratory tract. The typical size for these particles can range from submicron size to approximately 100 microns. Thoracic particulate mass (TPM) TLVs are designated for compounds that are toxic if deposited either within the airways of the lung or the gas-exchange region. The typical size for these particles can range from approximately 5 to 15 microns. Respirable particulate mass (RPM) TLVs are designated for those compounds that are toxic if deposited within the gas-exchange region of the lung. The typical size for these particles is approximately 5 microns or less. It should also be noted that the term "nuisance dust" is no longer used since all dusts have biological effects at some dose. The term particulates not otherwise classified is now being used in place of nuisance dusts. However, the TWA of 10 mg/m^3 for IPM is still used while a value of 3 mg/m^3 for RPM is now recommended.

Notice of Intended Change (NIC). This term is unique to ACGIH. Chemicals appearing on the NIC list for at least 1 year serve as notice that a chemical has a TLV proposed for the first time or that a current TLV is being changed. This procedure allows ample time for those with data or comments to come forth.

Again, the reader is strongly advised to consult the TLV booklet[16] for a more complete discussion of all the above terms and several other terms that are not discussed here.

Groups That Recommend OELs

Many sources of OELs for chemicals are available to the practicing workplace health professional. It is a good idea to gather as many of the lists as possible to keep a compendium of OELs for quick reference. This section will briefly describe some of the groups that recommend OELs and provide an indication of whether they have the force of law in the United States. Of course, OELs recommended by groups outside the United States do not have the force of law here but provide guidance of a similar quality to that of groups based within the United States.

One other point should be noted about the OELs that do not have the force of law behind them. While they are merely recommendations and are not legally binding, one would be well advised to have a very strong case regarding why those recommendations were not followed if litigation arose because the limits were routinely exceeded. It should be remembered that these groups are considered to be learned bodies consisting of individuals from many different health-related professions, that come together to reach some level of consensus and recommend the OEL. They therefore carry much more weight in a litigation situation than one or two health professionals within a single company or agency.

Inhalation Exposure Limits

Table 2.1 summarizes several of the groups that recommend OELs. Additionally, several of the groups that recommend community-based limits that are not necessarily applicable to the occupational setting are discussed to alleviate confusion.

The original list of approximately 400 permissible exposure limits (PELs) adopted in 1970 under the OSH Act came from the 1968

Table 2.1 — Summary of Various Inhalation Exposure Limits

Type of Limit	*Recommending Body*	*Legally Binding?*
Permissible exposure limit	Occupational Safety and Health Administration	Yes
Recommended exposure limit	National Institute for Occupational Safety and Health	No
Threshold limit value	American Conference of Governmental Industrial Hygienists	No
Workplace Environmental Exposure Level	American Industrial Hygiene Association (AIHA)	No
New chemical exposure limit	Environmental Protection Agency (EPA)	Yes
Maximum allowable concentration (translated)	Deutsche Forschungsgemeinschaft (Commission for the Investigation of Health Hazards of Chemical Compounds in the Work Area) (Germany)	No
Occupational exposure limit	Health and Safety Commission & Health and Safety Executive (Britain)	No
Emergency Response Planning Guide	AIHA (community-based standard, not an OEL)	No
Reference concentration	EPA (community-based standard, not an OEL)	Yes

list of TLVs and the standards of the American National Standards Institute (ANSI).[9,11,20,21] However, since that time only about two dozen limits have been adopted because the political process of having these values made into law is very difficult. An attempt was made in 1989 to adopt 428 chemicals from the 1989 TLV list as legally binding PELs, but legal proceedings by various groups ultimately resulted in the overturning of adopted values in 1992.

The recommended exposure limits (RELs) are published as criteria documents that could serve as the scientific basis for compounds that could then be evaluated by the Occupational Safety and Health Administration (OSHA) for setting PELs. However, because of the breakdown in the PEL process, these documents remain as recommended limits without the force of law. Nevertheless, they are comprehensive documents that provide a wide spectrum of use and toxicology information. The *Pocket Guide to Chemical Hazards* published by the National Institute for Occupational Safety and Health (NIOSH) is indispensable to anyone dealing with OELs or chemical exposures and provides a concise summary of such information as physico-chemical properties, respirator selection and personal protective equipment, health hazards, etc., on about 400 chemicals.[22] It should also be noted that NIOSH states that its RELs given as TWAs are appropriate for a 10-hour workday during a 40-hour workweek.

ACGIH is undoubtedly the first and foremost authority on OELs. Their publications include the TLV booklet,[16] the *Documentation of the TLVs*,[23] and a compendium of exposure values from ACGIH, OSHA, NIOSH, and the German maximum allowable concentrations (MAKs).[24] The TLV committee started in 1942 and has developed TLVs for approximately 700 chemicals. Again, these ACGIH publications are a must for anyone dealing with exposures to chemical or physical agents.

The AIHA WEEL committee started in 1980 with the goal of providing OELs, on request of AIHA members, for chemicals for which the ACGIH TLV committee was not considering setting TLVs. There are currently approximately 80 WEELs, and AIHA provides a summary list of the limits and a volume of documentation for them.[25]

Environmental Protection Agency (EPA) new chemical exposure limits (NCELs) are promulgated under Section 5 of the Toxic Substances Control Act. They are set by an individual company (or group of companies) entering into an agreement with EPA for chemicals to be produced for commerce under the Significant New Use Rules or Pre-Manufacturing Notification.[26] The authors of this chapter are not aware of any published list of these values.

The German Commission for the Investigation of Health Hazards of Chemical Compounds in the Work Area, which develops the MAKs, is probably the best known foreign group recommending OELs. Their publication and thought process is similar to that of ACGIH, but some significant differences do exist. Some examples of differences are the designation of compounds causing chemical sensitization, presenting potential for embryo/fetal toxicity, or which are considered to be mutagens. There are currently about 800 chemicals reviewed in the handbook.[27] Again, this publication is a must for anyone dealing with chemical exposures both because of the number of compounds reviewed and the additional designations just described.

Great Britain has several hundred OELs, some approved by the Health and Safety Commission having the force of law, while others serve as recommendations made by Britain's Health and Safety Executive.[28] These OELs also carry designations such as the skin notation and sensitization potential.

AIHA's Emergency Response Planning Guidelines (ERPGs) are community-based values that are recommended in the event of a large chemical spill that may result in exposure of the general public.[29] While they provide an excellent summary of the expected inhalation toxicity of compounds at various concentrations, they are not workplace recommendations and will not be discussed further here.

EPA Reference Concentrations (RfCs) are also community-based values that take into account the entire spectrum of the general population assumed to be exposed to the compound for 24 hours/day for a lifetime. The values can be found in the EPA's Integrated Risk Information System (IRIS) database, which can be obtained from EPA[30] or accessed online through NLM. Data for 189 chemicals listed on the priority air pollutant list will ultimately be reviewed. Since the RfCs are not workplace limits, they will not be discussed further here.

One additional source of OELs is the manufacturers/formulators of chemicals. Larger chemical companies may have an internal group that sets OELs for the compounds manufactured or formulated in their plants. These internal OELs should appear on the company material safety data sheet. If a company purchases a chemical for formulation, etc., the workplace health professional responsible for exposures to the chemical within your plant should contact the manufacturer and request information on the documentation of its internal OEL.

Biological Exposure Limits

ACGIH has recommended biological exposure indices (BEIs) for nearly 40 chemicals or classes of chemicals.[16,23] A BEI has been defined by ACGIH as a level of a determinant (which is either the actual chemical, a metabolite, or a biochemical change produced by the chemical) that is likely to be observed in a specimen (such as blood, urine, or air) collected from a worker who was exposed to a chemical and who has similar levels of the determinant as if he or she had been exposed to the chemical at the TLV. That is, because the BEI takes into account exposures from all sources (inhalation, dermal, and oral), it is indexed to reflect the level of the determinant if a person were exposed only via inhalation. As with airborne OELs, the values are not exact numbers that dictate healthy versus unhealthy exposures or exposure levels. As with the TLVs, the BEIs are recommended for 8-hour days and 40-hour workweeks.

BEIs are the best technique available for determining the actual body burden of a chemical by all routes of exposure. This is particularly useful where skin exposures to a chemical are significant and where merely monitoring the airborne concentration of a chemical would not provide a reliable indicator of potential exposure. However, BEIs require much metabolism and pharmacokinetic data from humans to be established.

Other issues also impact the BEIs. The collection of the specimen has much better acceptance if it is noninvasive (urine or breath) than if it is invasive (blood collection). The sources and routes of the exposure are also not known, since an endpoint of total exposure is being evaluated. For instance, inhibition of cholinesterase from an organophosphate could occur via air and skin during production at the workplace, via air and skin at home during application to lawns or gardens, and orally from water or food that have trace residues in them. The workplace health professional must then conduct an investigation to determine what the various sources of the exposure are and then make recommendations on how to limit the total exposure.

Physical Agent Exposure Limits

ACGIH has recommendations for several physical agents listed in their TLV booklet.[16] Additional information (and references) on several physical agents such as ionizing and nonionizing radiation, biological agents, heat stress, noise, electromagnetic fields, etc. is also available.[7,9,11,31–33] Unlike many of the TLVs for chemicals, the recommendations for physical agents are primarily derived from human data. Extensive epidemiological data for agents such as ionizing radiation have been gathered from large populations for some 40 years (due to the sad chapter in human history when World War II raged and the Japanese cities of Nagasaki and Hiroshima were bombed with atomic weapons), while other data have been gathered due to more recent health concerns (e.g., microwave radiation).

Important Considerations in the Development of OELs

The purpose of this section is to help the reader gain an appreciation for the types of data evaluated in setting an OEL and to be aware of the issues surrounding the use of these data. In gaining this awareness the reader should then be more able to appraise the strengths and weaknesses of the data used to set an OEL. Again, it is not the intent of this section or this chapter to provide the reader the wherewithal to begin setting OELs. Additional references are also available that provide historical perspectives as well as information on the level setting process.[1,11,20,34–37]

Identification of the Hazard

Although hazard identification may at first seem to be apparent to the point of being overly simplistic, in practice it can be the source of significant consternation. It should be recognized that the type of toxicity hazard changes with the physical form of the chemical, which is in turn a function of the chemical process that it is involved in or its end use. For example, an isocyanate monomer in paint may produce a vapor hazard during production or high temperature curing; an aerosol inhalation hazard (spray process) or dermal hazard (brush-on) at room temperature; and during these various processes one must be cognizant that the toxicity of the prepolymer or polymer may be significantly different from that of the monomer. Another factor to consider is that the primary chemical of interest may present very different hazards when in different formulations. Every chemical mixture has its unique physical and chemical hazards due to differences in solvents, stabilizers, surfactants, and so forth that result in differences in vapor pressures, impurities, reactivity, etc. Exacerbation (potentiation) of toxicity, masking of toxicity, or different manifestations of toxicity can occur for the same chemical of interest when put in different formulations. Again, constant monitoring and observation of exposed workers by the workplace health professional can help prevent problems of a serious nature.

Related to this is the situation where two or more chemicals, each with proposed OELs, are present together in the exposure atmosphere. Again, the combination of the chemicals may have no untoward effect at all, the effect may be additive, there may be potentiation of the effect (an effect that is greater than additive), or in rare instances the presence of one chemical may actually decrease the toxic effects of another. However, there is a standard practice of adjusting the total atmosphere based on whether the multiple chemicals are expected to have additive or independent effects. ACGIH has provided detailed examples in the TLV booklet,[16] and examples of these calculations are provided in Appendix A.

Routes of Exposure

The primary route of exposure in an occupational setting is generally via inhalation, but this obviously depends on such factors as the production method, uses, physicochemical properties, and so on. Some chemicals primarily have their effect at the portal of entry (e.g., upper respiratory tract irritants) and may produce a spectrum of effects ranging from irritation to tumors. Other chemicals exert their toxic effects at a remote site from the portal of entry, that is, systemically. Examples of this include chemicals that produce neurotoxic effects, effects on the fetus, liver tumors, and so forth when the route of exposure is either via inhalation or the skin. Even when the apparent route is via inhalation, the skin may actually serve as the principal route of exposure. Compounds with the skin notation in the TLV booklet[16] should be monitored carefully in the workplace as they have demonstrated the potential to be absorbed dermally in toxicologically significant amounts. Effects at the portal of entry include irritation and necrosis, while allergic responses are actually systemically mediated.

Review and Evaluation of the Chemical-Specific Toxicology Data

Data that are used in setting OELs for human workers are generally derived from laboratory animals. A discussion of toxicological principles and the use of animals for extrapolation to humans can be found later in this book as well as in other sources, ranging from basic treatment of the topics[1,4,20,38,39] to very detailed discussions.[2,7,8,40–43] Therefore, no lengthy discussion on this topic will ensue here, but the reader is reminded that not all effects in animals can be extrapolated to man and that experience and professional judgment are necessary to weigh the relative importance of these effects. The detailed methodology or protocol typically used for conducting these studies also will not be discussed here. However, the reader is referred to information found elsewhere in this book and also to several other sources for more complete details.[8,44–46]

Physico-Chemical Properties

Some important information can be gleaned just from the physico-chemical data on the chemical. For instance, vapor pressure data provide information on the expected saturated vapor concentration, and an assessment of hazard via this route can be gained by comparing the value to inhalation data. Octanol:water partition coefficients provide some insight as to the potential for the compound to accumulate in biological systems. Solubility data provide an indication of the potential to cross the dermal barrier. Odor information provides some indication of whether the chemical has potential warning properties.

Acute Toxicity and Irritation Data

Acute toxicity data primarily provide a relatively crude estimate of toxicity, with death as the principal endpoint, usually being assessed for 14 days following a single dose of the chemical. Data are typically available for oral and dermal routes of exposure, with data via the inhalation route available somewhat less frequently. Acute toxicity data are often performed to meet regulatory requirements for classification of a chemical for labeling, transport, and commerce. The data serve as a rough comparison of the relative toxicity of chemicals to one another, and this may often be the only information available for low-production or new chemicals or chemical intermediates. Comparison of acute toxicity values for a single chemical by various routes also gives an indication of the relative absorption by those routes. Acute toxicity information, particularly the inhalation data, may also be useful in the assessment of whether a chemical should be considered for a ceiling or STEL. Listings of different toxicity classification schemes are presented in Appendix B.

Irritation studies are performed to determine the potential for a chemical to produce damage to the eye or skin from direct contact. They are typically performed in rabbits and involve material applied directly to the eye or skin, with measures employed to maintain contact, then a follow-up period of up to 21 days. These data are often used for hazard classification for handling and transport. They can also impact the decision on whether to set a ceiling or STEL for a chemical.

Sensitization Studies

These studies are used to determine whether a chemical has the potential to cause an allergic reaction, most often in the skin. There are several different protocols for doing these studies in guinea pigs, ranging from painting the chemical on the skin to mimic the workplace, to using subcutaneous injections as a stringent evaluation for any sensitizing potential. This information can be very important because sensitized individuals may not be protected by the OEL, and they may react at much lower concentrations of the chemical or biological agent. The reaction is not that considered consistent with the expected toxicity of the agent, but is instead consistent with a wide spectrum of agents that produce allergic reactions. The reaction can range from mild dermatitis to anaphylactic shock and death. Thus, the workplace health professional should be vigilant for allergic-type reactions in workers and immediately remove those individuals from exposure. If this is more than a very rare event, then the OEL and occupational hygiene practices must be closely scrutinized.

Metabolism and Pharmacokinetics

Metabolism and pharmacokinetic data provide information on the uptake, distribution, excretion, and biotransformation of a chemical. Information can be obtained on whether a chemical is rapidly metabolized and eliminated, or whether it is bioaccumulated (i.e., is 16 hours without exposure at work enough time to eliminate the chemical from the body); whether the chemical produces toxic metabolites and their possible relevance for humans; what the critical doses at target organs are; and what doses exceed the normal metabolic capability and thereby produce nonphysiological responses. These studies are critical to understanding and interpreting many of the toxicology studies performed on a chemical. Unfortunately, while they are often performed to some degree for agricultural and pharmaceutical chemicals, these studies are only rarely performed for commodity or industrial chemicals. This is related to the technical difficulty and high cost of performing them.

Genotoxicity

These data are derived from a wide variety of studies using models as diverse as bacteria, mammalian cells, insects, and whole mammals. Many of the test systems have been modified to be very sensitive in order to minimize false negatives. However, the disadvantage to this approach is that it may lead to an increased number of false positives. Thus, evaluation of genotoxicity data requires experience and expertise to know which assays are designed to demonstrate any genotoxic potential by amplifying effects and circumventing normal uptake and metabolic processes. Often, a weight-of-the-evidence approach is used to evaluate the genotoxicity data as a whole, since there are bound to be a few equivocal or even positive findings when several of these studies have been performed. However, these data can have a substantial impact on the determination of the relevance of tumorigenic effects. These data may determine whether a threshold (or nongenotoxic) mechanism is assumed for a carcinogenic effect with the resultant use of a safety factor approach, or if a nonthreshold (genotoxic) mechanism is assumed and a more severe model is used for the risk assessment (see later section on Carcinogen Classification System and Risk Models for further explanation of these terms and this approach).

Reproductive/Developmental Toxicity

The severe and debilitating birth defects resulting from the administration of thalidomide to pregnant women served as the impetus for the routine evaluation of the teratogenic potential of chemicals. This was an example of a drug, taken voluntarily, that was therapeutic to the expectant mother but uniquely toxic to the fetus. With females of reproductive age now being commonly employed in the workplace, the impact of these studies is obvious. Developmental and reproductive toxicity studies are performed to evaluate the potential of agents to produce structural or functional deficits in the offspring during pregnancy and the postnatal period, until development is complete, as well as to evaluate the behavioral and functional aspects of parental animals and their offspring to successfully mate and reproduce.

Teratology studies are most commonly conducted in rats and rabbits and employ a forced oral administration of the chemical to the dam during the time when embryo/fetal toxicity is likely to produce a teratogenic effect (during the period of organogenesis). Endpoints for the dam include general growth and appearance and an assessment of successful maintenance of the pregnancy (abortions, early deliveries, and dead or resorbed fetuses). The fetuses are collected just before actual birth and evaluated for such endpoints as growth and the development and appearance of internal organs and bones. One aspect of these studies is the requirement that the highest dose administered be maternally toxic. This may produce some degree of toxicity to the fetus, often exhibited as decreased maturity of the skeleton (delayed ossification), extra or fused ribs, etc. This effect should be factored into the evaluation of these studies and its impact on the OEL. However, many agents known or suspected to be human teratogens or developmental toxicants, producing either physical defects, growth retardation, or learning or behavioral deficits are readily encountered outside the workplace (radiation, therapeutic drugs, alcohol, cigarettes, viruses such as rubella and herpes, excesses of vitamin A or D, lead, etc.). Additionally, nearly 10% of all pregnancies are expected to result in some sort of birth defect. This makes the evaluation of workplace exposures and their contribution to normal pregnancy outcome an extremely difficult task.

Reproductive effects of chemicals are often evaluated in rats, with the exposure customarily being via the diet or water for two consecutive generations. The endpoints evaluated include the reproductive success of parental males and females, the care of the neonates, the viability and development of the offspring, their reproductive success, and the development of their offspring through weaning. Other endpoints include observations of growth and appearance and histological evaluations. Often concern is highest for the female for reproductive effects of agents, but it should be remembered that effects on the male are also well documented in the workplace setting (e.g., dibromochloropropane).

Neurotoxicity

Although behavioral evaluations for neurotoxicity have been informally performed for some time,[47] formalized guidelines have only recently been promulgated by EPA.[48] This study battery consists of evaluations for effects from acute (single dose) and subchronic (90-day) dosing as well as evaluations for developmental effects, with dams being dosed from gestation to 10 days after birth and pups evaluated through about 10 weeks of age. The evaluated endpoints include effects on growth, behavior, development, activity levels, and nervous system morphology. Considerations in the review of these studies are that the top dose must result in toxicity in all the studies, so some limited effects may be expected in the animals or offspring at the top dose. While many chemicals have neurotoxic effects (e.g., narcosis due to solvents), these effects are ordinarily readily reversible. The objective of these studies is really to identify chemicals that produce irreversible effects on adults or offspring. True neurotoxic effects have previously been demonstrated for such chemicals as methanol and methyl isobutyl ketone. Lead serves as an example of a chemical that is still being feverishly evaluated for its potential to affect the learning and behavior of children following fetal exposure, and some researchers propose that there may be many chemicals, as yet unscreened, that also have this potential.

Subacute/Subchronic Toxicity

Subacute studies are defined here as studies with exposure lasting up to 2 weeks; subchronic studies as studies ordinarily lasting from about 15 days up to 6 months. These studies are generally 14, 28, or 90 days in duration and are conducted in mice, rats, or dogs. The endpoints of toxicity generally include survival, general health parameters (e.g., growth and clinical signs), extensive blood evaluations (hematological and biochemical), urinary evaluations, and the general appearance of internals organs coupled with fairly extensive microscopic evaluations. The objective of these studies is to provide information on target organs, reversibility of effects, potential for bioaccumulation, and so on. These studies very often serve as the longest term studies available for a chemical, and as such they are often the critical study on which the

OEL is based. Ninety days has been a common duration for subchronic studies for several decades, but because the state of the science has evolved greatly over this period there is a wide spectrum in the quality of these studies.

Chronic Toxicity and Oncogenicity

As a rule, the longest duration animal toxicology studies that are typically performed on chemicals are the chronic studies. These studies are almost always performed for agricultural chemicals, quite often for pharmaceuticals, and also for many of the high-volume industrial or commodity chemicals. They are much less frequently performed on the tens of thousands of industrial chemicals that have more limited distribution or lesser economic value. These studies may include a 1- or 2-year toxicity study with dogs, an 18- or 24-month oncogenicity study with mice, and/or a 24-month chronic toxicity and oncogenicity study with rats. The endpoints generally include survival, general health parameters (e.g., growth and clinical signs), extensive blood evaluations (hematological and biochemical), urinary evaluations, and extensive microscopic evaluations of internal organs. The objective of these studies is to determine if repeated exposures over long durations, approaching the complete lifetime for the rodents, produce cancer or other types of toxicity.

While these studies are the best available information for assessing the toxic potential of lifetime exposures to chemicals, they are not without their limitations.[7,43,49,50] The reader should be aware of the various limitations and realize that they may have a very significant impact on the strength of the data set for the OEL. Some of the issues that affect these studies include the use of excessively high doses that may alter the normal metabolism, distribution, and excretion of the chemical; molecular mechanisms in the animal that do not directly extrapolate to humans, such as kidney tumors in male rodents from alpha-2u-globulin; liver tumors from peroxisomal proliferators; and differences in metabolism between animals and humans. To those without an appreciation or understanding of these issues, an OEL can have the appearance of applying too small a safety factor.

Human Use and Experience

While data from human use would be the most useful for setting an OEL, there is often a paucity of information and/or the quality of the information is not sufficient to be of practical use. Except for rare cases where adequate human data exist to factor heavily in setting an OEL for a chemical with systemic toxicity, data from humans are probably most often used for OELs based on avoidance of irritation. That is, in the course of normal use and production there is enough information to determine an irritation threshold, incorporate a safety factor, then determine whether more complaints arise.

Epidemiology data are another source of human toxicity information. However, the number of well-controlled epidemiology studies useful for setting an OEL is surprisingly small. Many of the older studies, or even more recent reports for that matter, have poor control group data and too many confounding variables, since people in the workplace are rarely exposed to a single chemical and rarely have homogeneous personal habits. Additionally, there are often poor exposure concentration data to correlate with the human findings, and the myriad anecdotal reports shed little light on the topic. Somewhat related to this are the situations where accidental exposures have occurred in the workplace or in the general public. Again, there are generally little or no reliable exposure concentration data, too many anecdotal reports, and when the exposure concentrations are available, they are typically so high that they are not useful for setting OELs.

References

References are often an overlooked factor in the evaluation of the strength of the data that support an OEL. The references should be judged on such considerations as whether they appeared in refereed journals, the studies reported were performed at reputable labs, the work was amply described or performed to specific guidelines or regulations, and so forth.

Rationale

The rationale is ultimately the most important section of the OEL document because it brings together all the data; it is where the data are discussed and weighted for their relative importance, and the OEL is derived. Unfortunately, it is too often the only section of the document that gets read, if anything other than the actual value is scrutinized at all. This section should summarize the pertinent physico-chemical properties of the compound (e.g., warning properties from odor, vapor pressure, potential for bioaccumulation); note the location of the effect (portal or systemic); show important toxicity endpoints or types of effects (e.g., odor or irritation versus teratogenic effect); determine the relevance of the routes of exposure, doses, and so forth used in the animal studies (e.g., whether all studies are via gavage doses, or at or above the maximum tolerated dose); determine the relevance of the effects observed in animals for humans (e.g., certain kidney tumors in male rats, whether genotoxic or nongenotoxic); differences in the susceptibility between animals and humans, and so on. Additionally, such factors as uses of the chemical and structural similarities to other chemicals may have a significant impact in the evaluation. Finally, experience and professional judgment are employed to weight all the information, apply an appropriate safety factor based on the strength of the available data, and recommend the OEL.

Carcinogen Classification Systems and Risk Models

Several different carcinogen classifications have been developed by organizations such as EPA, the International Agency for Research on Cancer, ACGIH, etc.[49,51] However, for the discussion here the system used by ACGIH for the TLVs[16] is selected because it has an expanded classification system that uses all the available data and also considers the lack of data, in an attempt to account better for the weight of the evidence. The system used by ACGIH can be summarized as follows.

A1: Confirmed Human Carcinogen. To have this designation a chemical must have human data to support its classification. Examples include nickel subsulfide, bis(chloromethyl) ether, and chromium VI compounds.

A2: Suspected Human Carcinogen. This designation requires relevant animal data in the face of conflicting or insufficient human data. Examples include diazomethane, chloromethyl methyl ether, and carbon tetrachloride.

A3: Animal Carcinogen. To have this designation a chemical must have caused cancer in animal studies by nonrelevant routes of exposure or mechanisms (e.g., kidney tumors produced in male rats from many hydrocarbons, para-dichlorobenzene, d-limonene, etc.; hormone-mediated thyroid tumors from compounds such as the ethylene bisdithiocarbamates) or at excessive doses, etc. Furthermore, there are human data available that are in contradiction to the results in animals. Examples include nitrobenzene, crotonaldehyde, and gasoline.

A4: Not Classifiable as a Human Carcinogen. This classification is given to chemicals for which there are inadequate data to say whether it is a potential human carcinogen. It is typically applied to chemicals for which an issue of carcinogenicity has been raised but for which there are insufficient data to answer the question. Examples include pentachloronitrobenzene, phthalic anhydride, and acetone.

A5: Not Suspected as a Human Carcinogen. This classification is given to chemicals that have strong supporting data in humans to show that they are not carcinogens. Data from animals can also be used to support this classification. To date, the only example of a chemical receiving this designation is trichloroethylene.

There are even more risk models from which to choose than there are carcinogen classification systems. However, a few methods are more common than the others.[2,8,40,41,49,50,52–54] The most common approach for setting OELs has been the use of a safety factor (SF) of some multiple applied to the no-observable-effect level (NOEL) of interest. This approach involves determining the lowest NOEL or NOAEL (no-observable-adverse-effect level) in the most sensitive species (or if the data allow, in the most appropriate species), strain, and sex from a critical study or set of studies. From this NOEL a safety factor (often ranging from 10 to 1000) is divided into the NOEL to determine the OEL for workers. This additional SF is applied to obtain greater confidence and further assurance that the health of the worker will be protected. The actual SF depends on many considerations, but the type of toxic effect and the duration of the longest study often have the most impact. This SF approach is also used for air, food, and water contaminants.

Another method currently being assessed for use by several regulatory agencies, most notably EPA, is the so-called benchmark dose approach. This approach fits a curve to all the available data from a given study type (e.g., subchronic studies in rats). A benchmark dose is the dose that is calculated from the curve to affect some predetermined percentage of animals, ordinarily from 1 to 10% of the animals (effective dose ED_1 or ED_{10}). The lower-bound confidence value for that benchmark dose will then have the safety factors applied to it to estimate the acceptable exposure level. This method has the advantage of using all the data and fitting a curve to it instead of using just a single NOEL value to select the acceptable exposure level. While this method may become popular in succeeding years, it is not known to these authors that it has been used in the calculation of any published OELs.

An approach used for some carcinogens that are assumed not to have a threshold (i.e., a genotoxic chemical for which only zero exposure results in zero risk) a model called the linearized multistage is employed. Since zero exposure is often not achievable, a level of acceptable or negligible risk must be decided on. For the general public, a risk of 1×10^{-6}, or 1 in 1,000,000, is considered negligible by EPA. For workers, OSHA has used a range of 10^{-3} to 10^{-4}. When evaluating these risk calculations from the linearized multistage approach, it should be remembered that, assuming that it is the appropriate calculation to use (i.e., the chemical is a potent genotoxin that is relevant for humans), there are multiple conservative assumptions built into the model. It has been estimated that the actual risk number is nearer to one hundred- to one thousand-fold lower than the calculated value. That is, the linearized multistage model is thought to overstate the actual risk by up to several orders of magnitude.

Other Issues Impacting OELs

Setting workplace OELs is not an exact science. It requires knowledge, experience, and professional judgment. There is plenty of room for honest disagreement. However, knowing that workplace exposures are not likely to approach zero in the foreseeable future, it represents the best mechanism currently available to protect worker health and safety.

OELs are not values that are etched in stone. Always staying below the value does not guarantee good health for all workers, and going above the limit does not mean that workers will necessarily suffer toxic effects. There is sufficient biological variability in the workforce, sufficient analytical variability in the instrumentation used for the collection and analysis of samples, sufficient lack of reliable data, and sufficient uncertainty on the part of those making the recommendations to state with confidence that the values will not always be 100% correct. Thus, it is always incumbent on the practicing workplace health professional to carefully observe the work force and not just blindly rely on values that have been proposed by various groups of professionals.

It quite often happens that new toxicology data become available for chemicals over the course of many years after the OEL is recommended. It may be that none of the data are that significant alone and therefore may not warrant immediate attention. However, the total of available new data may have an impact on an OEL and require adjustment of the value. While all OEL-setting groups will re-evaluate an agent if significant new data arise, it is sometimes difficult to tell exactly when that point does occur. It is obvious when some data grab headlines, encourage sudden extensive publications, or prompt symposia. It is less obvious in other situations. Therefore, it would be desirable for all the OELs to be updated on some routine basis, with complete new literature searches and documentation prepared. While this is the procedure followed by the AIHA WEEL committee (review required every 10 years) with only about 100 OELs, it is obvious that it would be a daunting task for ACGIH or the MAK commission with more than 700 chemicals. Thus, the important point is that some OELs may be quite old, and only by reviewing the documentation for the OEL can one gain an appreciation for its age, the strength of the data supporting it, and the approach used to set a safety factor and ultimately the OEL. All of these factors may change over time and could have a significant impact on an OEL that would be recommended according to present-day practice.

There were 78,302 chemicals on the Toxic Substances Control Act inventory as of December 1996,[(55)] and the number grows every year. There are only a few thousand OELs at most to help guide the practicing workplace health professional. Thus, the question arises as to how to approach some sort of level setting for new chemicals or existing chemicals without much available toxicity data. Some companies or groups have proposed using systems where the amount and quality of the toxicity data can have very drastic effects on the OEL. Although there can be many variations, the system works roughly as follows. A chemical that is being prepared for commercial production must have a base set of data such as the acute toxicity and irritation battery and some genotoxicity tests. The safety factor used in setting the provisional OEL would be extremely high, resulting in an extremely low OEL. When the chemical begins commercial production, certain types of toxicology studies are required based on the production volume of the chemical. As production volume increases and more revenue is generated and more people become involved in production, more toxicology data must be generated. Additionally, as more toxicity data become available and confidence increases in the provisional OEL (and the safety factor should typically decrease with the result that the OEL becomes higher), then the engineering controls and personal protective equipment requirements should become less onerous and cumbersome. Thus, in this system there is linkage between the production volume, the amount of available toxicology data, and the safety factors applied in setting the OEL. This is similar to the approach now being used in Europe whereby certain toxicology data sets are required based on the production volume of the chemical. Another system similar to this that is being used by several pharmaceutical companies is the performance-based exposure control limit. Data on toxicity, structure-activity relationships, and so forth are used to assign chemicals to various health hazard categories that have defined engineering control strategies. A seminal paper on this approach has been published recently by Nauman et al.[56]

Summary

The makeup of the workplace has changed substantially over the last several decades. Women proved during the war years that they were every bit as capable as their male counterparts in the workplace environment of heavy industry. Their numbers have increased over the years in all workplace settings. Furthermore, approaching the workplace as simply consisting of production plants is also no longer possible. Office environments contained within sealed buildings have added a new dimension to the role of the workplace health professional. Additionally, laws mandating equal access for the physically infirm or disabled have substantially changed the makeup of the work force. Thus, the idea that the work force is made up entirely of healthy adult males is no longer valid in many scenarios. These changes have raised many questions regarding the appropriateness of current OELs in protecting all workers. It is obvious that issues regarding reproduction and effects on the fetus and the newborn have become a primary concern in the workplace. Moreover, issues regarding the indoor environment of office buildings and the broader spectrum of the health status of this work force have spawned numerous groups concerned with indoor air quality. Now and in the future, all these issues will have to be dealt with by groups recommending OELs, as well as the workplace health professional who is responsible for the health of the work force in these environments.

Additional Sources

Online Databases

Medlars® (Medical Literature Analysis and Retrieval System) is the computerized system of databases and data banks offered by the NLM. It is comprised of two computer subsystems: ELHILL® and TOXNET®. Some databases of interest include Cancerlit® (Cancer Literature), ChemID® (Chemical Identification), CHEMLINE® (Chemical Dictionary Online), TOXLINE® (Toxicology Information online), TOXLIT® (Toxicology Literature from special sources), CCRIS (Chemical Carcinogenesis Research Information System), DART® (Development And Reproductive Toxicology), EMIC and EMICBACK (Environmental Mutagen Information Center [and back file]), ETICBACK (Environmental Teratology Information Center Back File), GENE-TOX (Genetic Toxicology), HSDB® (Hazardous Substances Data Bank), IRIS (Integrated Risk Information System), RTECS® (Registry of Toxic Effects of Chemical Substances), and TRI (Toxic Chemical Release Inventory) series. For more information contact Medlars Management Section, National Library of Medicine, 8600 Rockville Pike, Bethesda, MD 20894; (800) 638-8480, or for NLM (888) 346-3656.

National Library of Medicine (NLM), National Institutes of Health–NLM online databases and data banks are located on the World Wide Web at (http://www.nlm.nih.gov/databases/databases_txt.html).

References

1. **Schaper, M.M.:** "General Toxicology." Paper presented at the 11th Annual AIHA Toxicology Symposium, San Antonio, TX, 1996.
2. **Klaassen, C.D.:** "Mid-America Toxicology Course." University of Kansas Medical Center, Kansas City, KS, 1997.

3. **West, J.B.:** *Respiratory Physiology—The Essentials*, 2nd ed. Baltimore, MD: Williams & Wilkins Co., 1979.
4. **Loomis, T.A.:** *Essentials of Toxicology*, 3rd ed. Philadelphia, PA: Lea & Febiger, 1978.
5. **Casarett, L.J. and J. Doull:** *Toxicology—The Basic Science of Poisons*. New York: Macmillan Publishing Co., 1975.
6. **Hayes, A.W.:** *Principles and Methods of Toxicology*. New York: Raven Press, 1982.
7. **Klaassen, C.D.:** Casarett & Doull's *Toxicology—The Basic Science of Poisons*, 5th ed. New York: McGraw-Hill, 1996.
8. **Hayes, A.W.:** *Principles and Methods of Toxicology*, 3rd ed. New York: Raven Press, 1994.
9. **Clayton, G.D. and F.E. Clayton:** *Patty's Industrial Hygiene and Toxicology*, vol. 1, 4th ed. New York: John Wiley & Sons, 1991.
10. **National Institute for Occupational Safety and Health:** *The Industrial Environment—Its Evaluation and Control*. Washington, D.C.: U.S. Government Printing Office, 1973.
11. **Plog, B.A.:** *Fundamentals of Industrial Hygiene*, 3rd ed. Chicago, IL: National Safety Council, 1988.
12. **American Industrial Hygiene Association (AIHA):** *Odor Thresholds for Chemicals with Established Occupational Health Standards*. Fairfax, VA: AIHA, 1993.
13. **Amoore, J.E. and E. Hautala:** Odor as an aid to chemical safety: odor thresholds compared with threshold limit values and volatilities for 214 industrial chemicals in air and water dilution. *J. Appl. Toxicol. 3*:272–290 (1983).
14. **Miller, J.:** "Workshop on Setting Exposure Limits." Course presented at the 9th Annual AIHA Toxicology Symposium, Baltimore, MD, 1994.
15. Occupational Safety and Health Act of 1970, Pub. L. 91-596, 91st Congress December 29, 1970.
16. **American Conference of Governmental Industrial Hygienists (ACGIH):** *1996 TLVs® and BEIs® - Threshold Limit Values for Chemical Substances and Physical Agents - Biological Exposure Indices*. Cincinnati, Ohio: ACGIH, 1996.
17. **Cralley, L.J. and L.V. Cralley:** *Patty's Industrial Hygiene and Toxicology*, vol. 3A, 2nd ed. New York: John Wiley & Sons, 1985. pp. 150–153, 220.
18. **American Industrial Hygiene Association (AIHA):** *Emergency Response Planning Guidelines and Workplace Environmental Exposure Level Guides Handbook*. Fairfax, VA: AIHA, 1996.
19. **Accrocco, J.O.:** *The MSDS Pocket Dictionary*. Schenectady, NY: Genium Publishing Corp., 1988.
20. **Hathaway, G.J., N.H. Proctor, J.P. Hughes, and M.L. Fischman:** *Proctor and Hughes' Chemical Hazards of the Workplace*, 3rd ed. New York: Van Nostrand Reinhold, 1991.
21. **Mackison, F.W., R.S. Stricoff, and L.J. Partridge Jr.:** *Occupational Health Guidelines for Chemical Hazards* (DHHS [NIOSH] pub. 81-123). Cincinnati, OH: National Institute for Occupational Safety and Health, 1981.
22. **National Institute for Occupational Safety and Health (NIOSH):** *NIOSH Pocket Guide to Chemical Hazards*. Cincinnati, Ohio: NIOSH, 1990.
23. **American Conference of Governmental Industrial Hygienists (ACGIH):** *Documentation of the Threshold Limit Values Biological Exposure Indices*, 6th ed. Cincinnati, OH: ACGIH, 1991.
24. **American Conference of Governmental Industrial Hygienists (ACGIH):** *Guide to Occupational Exposure Values—1993*. Cincinnati, Ohio: ACGIH, 1993.
25. **Workplace Environmental Exposure Level Committee:** *Workplace Environmental Exposure Level Guides*. Fairfax, VA: American Industrial Hygiene Association, 1996.
26. **Abel, M.T., S. Ahir, M.C. Fehrenbacher, R.S. Holmes, J.B. Moran, and M.A. Puskar:** "EPA's New Chemical Exposure Limits Program." Presented at the American Industrial Hygiene Conference and Exposition, Kansas City, MO, 1995. [Roundtable discussion]
27. **Deutsche Forschungsgemeinschaft:** *MAK and BAT Values 1992* (Report 28). Weinheim, Germany: Commission for the Investigation of Health Hazards of Chemical Compounds in the Work Area, 1992. Available in the United States from VCH, 220 E. 23rd St., New York, NY 10010-4606.
28. **Health & Safety Executive:** *Occupational Exposure Limits 1997* (EH40/97). Sudbury, Suffolk, UK: Health and Safety Executive Books, 1997.

29. **Emergency Response Planning Committee:** *Emergency Response Planning Guidelines.* Fairfax, VA: American Inudstrial Hygiene Association, 1996.
30. Risk Information Hotline; phone (513) 569-7254; fax (513) 569-7159; e-mail RIH.IRIS@EPAMAIL.EPA.GOV
31. **Clayton, G.D. and F.E. Clayton:** *Patty's Industrial Hygiene and Toxicology,* vol. 1, part B, 4th ed. New York: John Wiley & Sons, 1991.
32. **Goldberg, R.B.:** "Electromagnetic Fields (EMFs): Bioeffects and Potential Health Concerns." Paper presented at the 11th Annual AIHA Toxicology Symposium, San Antonio, TX, 1996.
33. **International Agency for Research on Cancer (IARC):** *Solar and Ultraviolet Radiation.* (IARC monographs on the evaluation of carcinogenic risks to humans, vol. 55). Lyon, France: IARC, 1992.
34. **Klonne, D.R.:** "Occupational Exposure Limits—the Practical Implementation of Toxicology in the Workplace." Paper presented at the 8th Annual AIHA Toxicology Symposium, Asheville, NC, 1993.
35. **Klonne, D.R.:** "Inhalation Exposure Assessment and the Role of Occupational Exposure Limits." Paper presented at the 10th Annual AIHA Toxicology Symposium, Boston, MA, 1995.
36. **Doull, D., G.L. Kennedy Jr., L.K. Loury, R.S. Ratney, D.H. Sliney, and W.D. Wagner:** "Threshold Limit Values (TLVs): How, Why, and Current Issues." Course presented at the American Industrial Hygiene Conference and Exposition, New Orleans, LA, 1993. [Professional development course]
37. **Workplace Environmental Exposure Level Committee:** "Establishing, Interpreting and Applying Occupational Exposure Limits: Current Practices and Future Directions." Course presented at the American Industrial Hygiene Conference and Exposition, Dallas, TX, 1997. [Professional development course]
38. **Kamrin, M.K.:** *Toxicology—A Primer on Toxicology Principles and Applications.* Chelsea, MI: Lewis Publishers, 1988.
39. **Ottoboni, M.A.:** *The Dose Makes the Poison—A Plain Language Guide to Toxicology.* Berkeley, CA: Vincente Books, 1984.
40. **Cralley, L.J., L.V. Cralley, and J.S. Bus:** *Patty's Industrial Hygiene and Toxicology,* vol. 3, part B, 3rd ed. New York: John Wiley & Sons, 1995.
41. **Calabrese, E.J.:** *Principles of Animal Extrapolation.* New York: John Wiley & Sons, 1983.
42. **Hardman, J.G., L.E. Limbird, P.B. Molinoff, R.W. Ruddon, et al.:** *Goodman & Gilman's—The Pharmacological Basis of Therapeutics,* 9th ed. New York: McGraw-Hill, 1996.
43. **Homburger, F.:** *Safety Evaluation and Regulation of Chemicals 2—Impact of Regulations—Improvment of Methods.* Basel, Switzerland: Karger, 1985.
44. **Environmental Protection Agency (EPA):** *Pesticide Assessment Guidelines,* rev. ed., by B. Jaeger (TS-769C). Washington, DC: EPA, 1984.
45. **Environmental Protection Agency (EPA):** *New and Revised Health Effects Test Guidelines* (TS-792). Washington, DC: EPA, 1983.
46. **Organization for Economic Cooperation and Development (OECD):** *OECD Guidelines for Testing of Chemicals.* Paris, France: OECD, 1981.
47. **Irwin, S.:** Comprehensive observational assessment: Ia. A systematic quantitative procedure for assessing the behavioral and physiologic state of the mouse. *Psychopharmacologia 13*:222–257 (1968).
48. **Environmental Protection Agency (EPA):** *Pesticide Assessment Guidelines* (Neurotoxicity series 81, 82, and 83). Washington, DC: EPA, 1991.
49. **Clayton, G.D. and F.E. Clayton:** *Patty's Industrial Hygiene and Toxicology,* vol. 2, part A, 4th ed. New York: John Wiley & Sons, 1993.
50. **National Research Council:** *Issues in Risk Assessment.* Washington, DC: National Academy Press, 1993.
51. **International Agency for Research on Cancer (IARC):** *IARC Monographs on the Evaluation of Carcinogenic Risks to Humans. Supplement 7—Overall Evaluations of Carcinogenicity: an Updating of IARC Monographs, Volumes 1 to 42.* Lyon, France: IARC, 1987.
52. **Baker, S.:** "Fundamentals of Risk Assessment." Paper presented at the 9th Annual AIHA Toxicology Symposium, Baltimore, MD, 1994.
53. **Cohen, S.M.:** "Chemical Carcinogenesis: Implications for Dose and Species Extrapolation." Paper

presented at the 11th Annual AIHA Toxicology Symposium, San Antonio, TX, 1996.

54. **Keller, J.G.:** "Risk Assessment for Industrial Hygienists." Paper presented at the 11th Annual AIHA Toxicology Symposium, San Antonio, TX, 1996.
55. **Chemical Abstracts Service (CAS):** *CHEMLIST*. Columbus, OH: CAS, 1997. [Online STN database]
56. **Naumann, B.D., E.V. Sargent, B.S. Starkman, W.J. Fraser, et al.:** Performance-based exposure control limits for pharmaceutical active ingredients. *Am. Ind. Hyg. Assoc. J. 57*:33–42, 1996.
57. **DiNardi, S.R.:** *Calculation Methods for Industrial Hygiene*. New York: Van Nostrand Reinhold, 1995.

Appendix A

Calculations for OELs

This appendix contains calculations often required for assessing contaminant levels in workplace atmospheres. The first equation converts ppm to mg/m^3 for vapors.(16,57)

Changing Parts Per Million to Milligrams per Cubic Meter of Air

$$\text{OEL}\ \frac{(\text{mg})}{(\text{m}^3)} = \frac{\text{OEL (ppm) molecular weight (grams)}}{24.45}$$

or

Changing Milligrams Per Cubic Meter to Parts Per Million

$$\text{OEL (ppm)} = \frac{\text{OEL (mg/m}^3\text{) } 24.45}{\text{molecular weight (grams)}}$$

Example

If the measurement of the vapor phase of toluene is 25 ppm, what is the measurement in mg/m^3?

Assume 25 ppm of toluene and that 92 g/mole is the gram molecular weight of toluene.

$$\begin{aligned} \text{mg/m}^3 &= \text{(ppm)(gram molecular weight of substance)}/24.45 \\ &= (25\ \text{ppm})(92)/24.45 \\ &= 94\ \text{mg/m}^3 \end{aligned}$$

Approaches for Multiple Agents in the Workplace

As is often the situation in the work environment, systems are very dynamic. Often, exposure to more than one agent is present. Different approaches are used to evaluate the following situations: additive effects—for agents with similar toxicological effects; independent effects—for agents with different toxicological effects; and a mixed atmosphere assumed to be similar to a liquid mixture.

Additive Effects

If it is reasonable to conclude that the chemicals present in the workplace could add, one on the other, to the total effect, then it is also reasonable to consider adding the exposure assessments to derive a total exposure assessment. An example would be the presence of three chemicals, X, Y, and Z, each having a similar toxicological effect on the same target organ, the liver.

Example

$$\text{total value} = \text{concentration}_1/\text{TLV}_1 + \text{concentration}_2/\text{TLV}_2 \ldots \text{concentration}_n/\text{TLV}_n$$

or sometimes shortened to

$$TV = C_1/TLV_1 + C_2/TLV_2 \ldots C_n/TLV_n$$

Assume 25 ppm of toluene with a TLV of 50 ppm; 25 ppm of mixed xylenes with a TLV of 100 ppm; and 75 ppm, of ethylbenzene with a TLV of 100 ppm. Substituting into the equation, it becomes

$$\begin{aligned} TV &= 25\ \text{ppm}_{(\text{toluene})}/50\ \text{ppm} + 25\ \text{ppm}_{(\text{xylene})}/100\ \text{ppm} + 75\ \text{ppm}_{(\text{ethylbenzene})}/100\ \text{ppm} \\ &= 0.50 + 0.25 + 0.75 \\ &= 1.50 \end{aligned}$$

The generally used standard for the total value is 1.00. It could be judged from this evaluation that the exposure may have an additive impact above acceptable limits, and the exposure should be reduced.

Independent Effects

Workplace atmospheric mixtures may also be comprised of agents with different toxicological effects and target organs. That is, the effects are not considered to be additive. In this situation the exposure assessment is simpler but still uses the acceptable total value of 1.0.

$$\text{total value}_1 = \text{concentration}_1/\text{TLV}_1$$

$$\text{total value}_2 = \text{concentration}_2/\text{TLV}_2$$

$$\text{total value}_n = \text{concentration}_n/\text{TLV}_n$$

or sometimes shortened to

$$TV_1 = C_1/TLV_1$$

$$TV_2 = C_2/TLV_2$$

$$TV_n = C_n/TLV_n$$

Assume 25 ppm of toluene with a TLV of 50 ppm and 0.1 mg/m^3 of cotton dust with a TLV of 0.2 mg/m^3. Substituting into the equation, it becomes

$$TV_1 = 25\ \text{ppm}_{(\text{toluene})}/100\ \text{ppm}$$

$$TV_1 = 0.25$$

$$TV_2 = 0.1 \text{ mg/m}^3_{\text{(cotton dust)}}/0.2 \text{ mg/m}^3$$

$$TV_2 = 0.50$$

It could be judged from this evaluation that the exposures may not have exceeded acceptable limits.

Synthetic Limit for Mixtures

Another approach to the issue of mixtures is to create a synthetic OEL. This is done for mixtures of liquids where the relative percentage of the components is known and where it can be assumed that all the components evaporate so that the atmospheric vapor concentration is similar in relative composition to the liquid.

Example

$OEL = 1/(\text{decimal fraction}_a/TLV_a) + (\text{decimal fraction}_b/TLV_b) + \ldots (\text{decimal fraction}_n/TLV_n)$

or sometimes shortened to

$OEL = 1/(f_a/TLV_a) + (f_b/TLV_b) + \ldots (f_n/TLV_n)$

Assume 25% component by weight of toluene with TLV of 188 mg/m³, 25% component by weight of mixed xylenes with a TLV of 434 mg/m³, and 50% component by weight of ethylbenzene with a TLV of 434 mg/m³. Substituting into the equation, it becomes

$$\begin{aligned} OEL &= 1/(0.25_{\text{(toluene)}}/180 \text{ mg/m}^3) + (0.25_{\text{(xylenes)}}/434 \text{ mg/m}^3) + (0.50_{\text{(ethylebenzene)}}/434 \text{ mg/m}^3) \\ &= 1/(0.001389 + 0.00058 + 0.00115) \\ &= 1/(0.00312) \\ &= 320 \text{ mg/m}^3 \end{aligned}$$

Therefore, each component contributes the following amount to the overall occupational exposure limit of 320 mg/m³.

toluene ➠ $0.25 \times 320 \text{ mg/m}^3 = 80 \text{ mg/m}^3 = 22.2$ ppm
xylenes ➠ $0.25 \times 320 \text{ mg/m}^3 = 80 \text{ mg/m}^3 = 18.4$ ppm
ethylbenzene ➠ $0.50 \times 320 \text{ mg/m}^3 = 160 \text{ mg/m}^3 = 36.9$ ppm

or an OEL of approximately 77.5 ppm.

Appendix B

Toxicity Classification Schemes

There are many different toxicity classification schemes available throughout the world. Even within the United States there are several different schemes for the various regulatory agencies that deal with shipping and handling of various classes of chemicals. Toxicity classification schemes are inexact, at best, at categorizing the acute toxicity and primary irritation of chemicals because of the different protocols used, the number of dose groups used, animal variability, vehicles for dosing, and so forth. However, they do provide at least some indication of comparative acute toxicity and/or irritation and provide some useful information to those unfamiliar with the toxicity of the chemicals that they handle as part of their job.

Provided herein are four classification schemes, some of which are used by regulatory agencies. They at least provide some indication of the relative differences in the approaches used, and they are schemes that will likely be encountered by workplace health professionals.

Table 2.B1 — Combined Tabulation of Toxicity Classes

Toxicity Rating	*Common Term*	*LD_{50} Single Oral Dose (Rats)*	*Vapor Exposure Mortality*[A]	*LD_{50} Skin (Rabbits)*	*Probable Lethal Dose for Humans*
1	extremely toxic	1 mg or less/kg	<10 ppm	5 mg or less/kg	a taste, 1 grain
2	highly toxic	1–50 mg	10–100	5–43 mg/kg	1 teaspoon, 4 cc
3	moderately toxic	50–500 mg	100–1000	44–340 mg/kg	1 ounce, 30 g
4	slightly toxic	0.5–5 g	1000–10,000	0.35–2.81 g/kg	1 pint, 250 g
5	practically nontoxic	5–15 g	10,000–100,000	2.82–22.59 g/kg	1 quart
6	relatively harmless	15 g and more	>100,000	22.6 or more g/kg	>1 quart

Source: **Hodge, H.C. and J.H. Sterner:** Tabulation of toxicity classes. *Am. Ind. Hyg. Assoc. Q. 10*:93 (1949).
[A]Inhalation 4 hours, 2/6–4/6 rats

Table 2.B2 — Warnings and Precautionary Statements — EPA (FIFRA)

Hazard Indicators	*Toxicity Categories* I	*II*	*III*	*IV*
Oral LD_{50}	up to and including 50 mg/kg	from 50–500 mg/kg	from 500–5000 mg/kg	>5000 mg/kg
Inhalation LC_{50}	up to and including 0.2 mg/L	from 0.2–2 mg/L	from 2–20 mg/L	>20 mg/L
Dermal LD_{50}	up to and including 200 mg/kg	from 200–2000 mg/kg	from 2000–20,000 mg/kg	>20,000 mg/kg
Eye effects	corrosive; corneal opacity not reversible within 7 days	corneal opacity reversible within 7 days; irrigation persisting for 7 days	no corneal opacity; irritation reversible within 7 days	no irritation
Skin effects	corrosive	severe irritation at 72 hrs	moderate irritation at 72 hrs	mild or slight irritation at 72 hrs

Source: *Code of Federal Regulations* Title 40, Part 162.10 (h)

Table 2.B3 — OSHA Health Hazard Definitions

Acute oral LD_{50}	highly toxic; $LD_{50} \leq 50$ mg/kg	toxic; 50 mg/kg < $LD_{50} \leq 500$ mg/kg
Acute dermal LD_{50}	highly toxic; $LD_{50} \leq 200$ mg/kg; 24 hrs	toxic; 200 mg/kg < $LD_{50} \leq 1000$ mg/kg
LC_{50} inhalation	highly toxic; $LC_{50} \leq 200$ ppm $LC_{50} \leq 2$ mg/L; 1 hr	toxic; 200 ppm; < $LC_{50} \leq 2000$ ppm 2 mg/L ≤ LC_{50} < 20 mg/L
Carcinogen	if IARC "carcinogen" or "potential carcinogen"; or if National Toxicology Program "carcinogen" or "potential carcinogen"; or OSHA regulated carcinogen	
Corrosive	visible destruction of, or irreversible alterations in living tissue by contact; 4 hrs	
Irritant	not corrosive; reversible inflammatory effect on living tissue by contact; 4 hrs; skin score ≥ 5	
Sensitizer	substantial portion of exposed people or animals develop allergic reaction	

Source: *Code of Federal Regulations* Title 29, 1910.1200. Appendix A

Table 2.B4 — Acute Toxicity Rating Criteria

Acute Toxicity Rating	*Oral[A] Liquids, Solids*	*Dermal Liquids, Solids*	*Inhalation Gases, Vapors (ppm) Dusts, Fumes, Mists (mg/L)*	*Skin Irritation Liquids, Solids*	*Eye Irritation Liquids, Solids*
	LD_{50} Rat (mg/kg)	LD_{50} Rabbit (mg/kg)	LC_{50}Rat 1-hr Exposure	4-hr Exposure[B]	
4	0–1	0–20	0–0.2 mg/L	not applicable	not applicable
3	>1–50	>20–200	>0.2–2 mg/L >20–200 ppm >20–200 ppm	severely irritating and/or corrosive	corrosive; irreversible corneal opacity
2	>50–500	>200–1000	>2–20 mg/L >200–2,000 ppm	primary irritant sensitizer	irritating or moderately persisting > 7 days with reversible corneal opacity
1	>500–5000	>1000–5000	>20–200 mg/L >2000–10,000 ppm	slightly irritating	slightly irritating but reversible within 7 days
0	>5000	>5000	>200 mg/L >10,000 ppm	essentially nonirritating	essentially nonirritating

[A]The oral route of exposure is highly unlikely in a workplace setting. If situations are encountered through where the oral LD_{50} value would indicate a significantly different rating, toxicity values for the other routes of entry may be considered more appropriate when assigning the rating.
[B]Note animal species and duration of exposure if different from that recommended.

Source: National Paint & Coatings Association (Hazardous Materials Identification System® HMIS®; Label Master); Washington, DC

Chapter 3

Outcome Competencies

After completing this chapter, the user should be able to:

1. Define underlined terms used in this chapter.
2. Identify conditions that caused the development of the OSH Act.
3. Recognize legal aspects of industrial hygiene practice.
5. Relate the role of occupational hygiene in providing professional services to employers.
6. Recognize the role of the occupational hygienist as an expert witness.
7. Recall the general duty clause.
8. Explain the phrase ". . . free from recognized hazards."

Prerequisite Knowledge

None

Key Terms

general duty clause • good samaritan doctrine • health inspections • occupational safety and health (OSH) standards • Occupational Safety and Health Act • Occupational Safety and Health Review Commission • performance standards • safety inspections • specification standards • vertical standards

Key Topics

I. Employer Duties
   A. OSH Standards
   B. The General Duty Requirement

II. Workplace Inspections
   A. OSHA Inspection Priorities
   B. The Inspection Process

III. The Citation Process

IV. The Contest
   A. The Safety and Health Professional as a Counselor
   B. The Safety and Health Professional as Witness

3

Legal Aspects of the Occupational Environment

W. Scott Railton, JD

Introduction

The practice of occupational hygiene requires the professional to have some basic knowledge of the laws that he or she may encounter. The purpose of this chapter is to provide some background knowledge concerning the federal Occupational Safety and Health Act (OSH Act) of 1990. A familiarity with the requirements imposed by this statute on employers and employees is necessary since the professional occupational hygienist will often be called on to render compliance advice and, indeed, may be the primary resource for developing Occupational Safety and Health Administration (OSHA) compliance plans.

The Act is used as the basis for this chapter because most safety and health professionals will spend a good portion of their careers rendering advice with regard to activities under this statute. However, the chapter also addresses the role of the safety and health professional with respect to other activities in which they engage and which arise under many different laws, such as workers' compensation statutes and tort laws.

In principal part, the rendering of compliance advice means that the professional must have knowledge of the compliance obligations that the law imposes on his or her client or employer. Thus, the professional occupational hygienist must be familiar with occupational safety and health standards and how they are developed and interpreted, because the primary obligation of employers and employees under the Act and state clones of the Act is to comply with these standards. Such standards should be considered a regulatory floor and not necessarily the mark of a good safety and health program.

The rendering of compliance advice also entails an obligation to advise employers concerning compliance problems they may have under a general provision of the law that is known as a "general duty." This duty requires employers to rid jobs and their workplaces of serious recognized hazards.

The rendering of this kind of advice also will require the occupational hygienist to have some familiarity with the manner in which the law is enforced and the role that the safety and health professional plays in such enforcement and other areas.

The Act was passed principally because state workers' compensation statutes were not preventing fatalities and traumatic injuries in the workplace. Virtually nothing was known about occupational illnesses that might result from the thousands of new chemicals introduced into the marketplace each year.[1] While some occupational illnesses had been identified (e.g., silicosis), most state workers' compensation systems did not recognize or compensate for such illnesses, and record keeping was practically nonexistent.

The Act created three new agencies, including OSHA within the U.S. Department of Labor, the National Institute for Occupational Safety and Health (NIOSH) in the U.S. Department of Health and Human Services, and the U.S. Occupational Safety and Health Review Commission (OSHRC).

OSHA was given two primary responsibilities. It was (1) empowered to promulgate occupational safety and health standards,[2] and (2) authorized to inspect workplaces and issue citations and civil penalties[3] to employers it determines are not in compliance with the Act. OSHA and the Department of Labor also are authorized to prosecute employers who contest citations and penalties issued by the agency.[4]

For its part, OSHRC was set up as an independent agency having the responsibility to try OSHA cases and to hear appeals from decisions of administrative law judges (ALJs).[5] OSHRC cases are tried before ALJs pursuant to procedural and evidentiary rules of the kind used in federal civil courts.[6] Appeals from ALJ decisions are taken to the members of the OSHRC, who are U.S. commissioners appointed to their positions by the president with the advice and consent of the senate. Decisions of the commissioners may be appealed to the U.S. courts of appeal.

NIOSH is the worker safety and health research branch of government. This agency, like OSHA, is authorized to investigate workplaces to determine the existence of health hazards.[7] NIOSH also has the responsibility to recommend new standards to OSHA and to undertake basic occupational safety and health research projects. NIOSH also supplies expert testimony in OSHA rulemaking proceedings.

The following states and territories operate their own safety and health plans covering most private employers within their boundaries.

Alaska
Arizona
California
Hawaii
Indiana
Iowa
Kentucky
Maryland
Michigan
Minnesota
Nevada
New Mexico
North Carolina
Puerto Rico
Oregon
South Carolina
Tennessee
Utah
Vermont
Virgin Islands
Virginia
Washington
Wyoming

Some of these states exclude some private employers. For example, the plan for Alaska does not include private sector maritime employments. Those employments are covered by federal OSHA and other federal agencies. Some states exclude private employments on federal property or in federal enclaves.

The enactment of the Act carried with it a provision that pre-empted state occupational safety and health laws.[8] The states, however, were authorized to regain primary authority by developing state plans at least as effective as the federal program.[9] Approximately 23 states and territories have elected to pursue their own safety and health programs either in whole or in part. The remaining states and territories are under the federal program.

When Congress enacted the statute, it also determined not to duplicate or pre-empt the work of other federal agencies in this field. Thus, OSHA does not cover miners because the safety and health of these workers is entrusted to the Mine Safety and Health Administration. Similarly, the Coast Guard regulates the safety of seamen, the Federal Railway Administration regulates safety of railroad employees, and the Federal Aviation Administration regulates the safety of flight crews and mechanics. Other agencies regulate specific industries.[10]

Employer Duties

Virtually all employers in the nation are covered either by the Act or state clones of that act. The Act does exclude governments at all levels from coverage. Federal agencies are required to adopt effective and comprehensive safety and health programs pursuant to the provisions of Section 19 of the Act (implemented by Executive Order 12196). In addition, the states and territories operating under state plans usually provide for coverage of otherwise exempt state and local governments.

Every other employer engaged in a business affecting commerce is covered and obliged to comply with the duty requirements imposed by the Act. OSHA regulations are set out in various parts published in Title 29 of the *Code of Federal Regulations* (CFR). Part 1975 of Title 29 describes the extent to which OSHA believes Congress used its power under the commerce clause of Article 1 of the Constitution to require employer compliance. Thus, OSHA believes that the secular activities of churches are covered, as are doctors, lawyers, and other professional and charitable organizations, private educational institutions, and even activities of Indians whether on or off the reservation.[11]

Employers are required to comply (1) with occupational safety and health (OSH) standards;[12] (2) a general statutory duty to render workplaces and employments free of serious recognized hazards, as in Section 5(a)(1) of the Act; and (3) regulations prescribed by the secretaries of labor and/or health and human services.[13]

OSH Standards

OSHA prefers to adopt standards with broad application across industry lines. Thus, in 29 CFR 1910, OSHA has published OSH standards said to be general industry standards. It is in this part that the occupational hygienist will find health standards, such as the standard for inorganic lead (29 CFR 1910.1025) and safety standards covering diverse subjects such as falling hazards (e.g., 29 CFR 1910.23), hazards associated with cranes (e.g., 29 CFR 1910.179), and electrical hazards (29 CFR 1910, Subpart S). According to the rules of interpretation, these standards apply unless they are pre-empted by standards specific to individual industries.

Industry-specific standards are characterized as "vertical" standards. The standards specified for the construction industry (29 CFR 1926) are principal examples.[14] Maritime industries such as longshoring are also regulated by vertical standards.[15] It is important to understand, however, that an industry regulated under a vertical standard is also subject to general industry standards to the extent that the vertical standard fails to cover specific hazards.[16]

OSH standards for all industries fall into two broad categories, characterized as either performance or specification standards. Performance standards state the object to be obtained or the hazard to be abated. They do not specify the method of abatement or control. OSHA has, since the early 1980s, preferred the adoption of performance-oriented standards over specification standards.

Specification standards, however, describe the specific means of hazard abatement. The guardrail standards of 29 CFR 1910.23 provide an example. Thus, guardrails are to be erected at a height of 42 inches, and they must be capable of withstanding a force of 200 lbs applied in any direction.[17] In contrast, the bloodborne pathogen standard at 29 CFR 1910.1030 is a performance standard requiring employers to develop written abatement plans to control exposures to bloodborne infectious materials such as the HIV virus. While the plan must cover specific subjects such as universal precautions, the standard does not say how the plan must be written. Similarly, standards for airborne toxic substances specify permissible exposure limits and require that feasible engineering controls be used to achieve those limits; they leave it to the employer to determine the controls used to achieve compliance.

Most specification standards were adopted in the early to mid-1970s and were derived from standards promulgated by the American National Standards Institute (ANSI). Congress authorized OSHA from 1971–1973 to adopt national consensus standards (e.g., ANSI standards[18]) and "established federal"[19] standards as OSHA standards by publication.[20] That is, OSHA did not have to submit these standards for public rulemaking because established federal standards had already been the subject of federal rulemaking proceedings. For example, the general machine guarding standard at 29 CFR 1910.212(a) was originally adopted pursuant to notice and comment rulemaking under the Walsh-Healey Public Contracts Act. ANSI standards (and standards of the National Fire Protection Association) were considered permissible for adoption by publication because they were adopted on a consensus basis by affected industries.

ANSI standards and the activities of the American Conference of Governmental Industrial Hygienists (ACGIH) have been, and continue to be, important sources of advice to OSHA for its ongoing efforts to adopt and update OSH standards. For example, while in 1995–1996 OSHA was barred from considering adoption of an OSH standard to regulate ergonomic hazards,[21] the ANSI Z-365 committee was hard at work on the development of a consensus standard to protect workers against work-related cumulative trauma disorders. ANSI standards are drafted and approved by committees drawn from various interested and affected parties, including business, academia, insurance companies, and government.

ANSI standards have been used as the starting point for the development of new OSHA standards, such as ANSI Z-244.1 Lockout/Tagout of Energy Sources.[22] OSHA also uses NIOSH recommendations as the basis for promulgating OSH standards. The bloodborne pathogens standard was in large part based on NIOSH/Centers for Disease Control and Prevention (CDC) guidelines.

New OSH standards are adopted by OSHA using what are essentially informal federal rulemaking procedures as described below. It is important to note that professionals having client responsibility for safety and health are well advised to monitor OSHA's rulemaking efforts. There are many sources of information available. OSHA publishes its notices in the *Federal Register*, usually on Tuesdays and Fridays. The Bureau of National Affairs (BNA), Commerce Clearinghouse (CCH), and *Inside Washington* all publish news reports concerning the agency's rulemaking activities.

There are at least two fundamental reasons for monitoring OSHA rulemaking activities. First, the law presumes that publication in the *Federal Register* is sufficient to put affected persons including businesses, labor organizations, and others on notice concerning OSHA's intent to adopt new or amended OSH standards. That means the public as a whole is provided with the opportunity to comment on the proposed OSH standard. Second, for enforcement purposes OSHA will assume that employers know their compliance obligations under new OSH standards because these standards have been published in the *Federal Register*. There are other ramifications as discussed below.

Presumably, the professional will have knowledge of the potential impact a proposed OSH standard will have on his clients or employer and can assist in deciding whether the clients or employers should participate in the rulemaking proceeding. Similarly, professionals usually have responsibility for rendering compliance advice to their clients or employers after new OSH standards are adopted by OSHA.

The Act specifies the procedural steps OSHA must take to adopt or promulgate a new or amended OSH standard.[23] Specifically, the agency is required to publish a notice of proposed rulemaking in the *Federal Register*.[24] The notice must specify the requirements of the proposed standard, and the public must be given at least 30 days to file written comments.[25] The agency is also required to hold a public hearing whenever any person affected by the proposed standard requests it.[26] OSHA has by rule added other procedural steps.[27]

Under the rules, an ALJ is appointed to preside over the hearing, accept evidence, regulate the course of the hearing, and permit cross-examination of witnesses. The judge also has the power to order and accept posthearing comments and arguments.[28] Thereafter, the record is closed, and recommendations are made and submitted to the OSHA assistant secretary for a decision regarding whether and what should be promulgated as a final OSH standard.

It is worth noting at this juncture that new OSH standards must be based on the record developed during the rulemaking[29] proceeding. OSHA is permitted to vary the terms of the final rule from those contained in the proposed rule. For example, the agency is not required to adopt the permissible exposure limit (PEL) it proposed in the notice of proposed rulemaking, but may adopt whatever level seems to be dictated by the evidence in the rulemaking record, subject to other considerations as discussed below.[30] Moreover, OSHA is not bound by existing science, but instead may adopt, for example, PELs that are suggested by the evidence of record.

In general, OSHA rulemaking and OSH standards are constrained by law, however. The agency must demonstrate the need for a new standard based on risk factors.[31] The lead standard provides an example. OSHA was able to demonstrate that the 200 $\mu g/m^3$ PEL that predated the 1978 standard was too high. In essence, the agency concluded that the existing PEL did not account for the precursor effects of lead that, while not rising to the level of disease, did indicate that significant changes were occurring in exposed workers.

Accordingly, as the courts later concluded, OSHA was authorized to adopt a more protective standard than it had proposed. OSHA chose to adopt the 50 $\mu g/m^3$ PEL, even though ACGIH and a good number of foreign countries did not agree. However, according to OSHA this standard was not the most protective standard that could be dictated by adverse health effects of lead exposure in adults. OSHA standards are not the most protective because the agency is limited by feasibility concerns under Section 6(b)(5) of the Act.

In general, a standard is "feasible" if its requirements are capable of being achieved over a period of time. In other words, the evidence in the rulemaking record must demonstrate that the technology exists or may be developed to achieve compliance, basically, through the use of engineering controls.[32]

In addition, the courts have held that new OSHA standards must be economically feasible. It is not at all clear what the test of economic feasibility really means, but on one hand it may mean that OSHA cannot adopt a standard so costly that it will bankrupt an entire industry. On the other hand, the costs of compliance with a new OSHA standard may be so expensive that the so-called laggards in an industry may be driven out of business.[33]

The courts also have created escape valves by recognizing that OSHA's conclusions made during rulemaking proceedings may be proven wrong. That is, the courts have recognized that abatement technology may not in practice be feasible or may in fact be too expensive. Accordingly, employers may challenge the feasibility of standards during enforcement proceedings conducted before the OSHRC.[34]

Safety and health professionals can and should play vital roles in both rulemaking and enforcement proceedings. The professional's role is at least twofold in rulemaking: Occupational hygienists can provide input on whether a new standard is necessary from a health perspective, and can also play a vital role concerning the feasibility of engineering control technology for proposed PELs.

The role of the professional is also important in enforcement proceedings. Employers can defend against OSHA citations on the basis that the standard cited against them is infeasible as applied to their operations. Employers who raise this defense of invalidity must plead it, and they have the burden of proof.

Enforcement cases focus on the cited employer and in particular on the cited facility.[35] There have been very few cases in which the infeasibility defense has been raised, so there is virtually no decisional law in this area. The employer has the heavy burden of establishing the defense, and in toxic substance cases, that burden will be carried by the expert technical witness or witnesses.

OSHA standards are presumed to be valid in prosecutions conducted to enforce their terms. That means the OSHRC will presume the standard cited against an employer is valid, and that, for example, engineering technology exists and can be used to achieve the PEL set by the cited standard. It is up to the employer who contests the citation to overcome that presumption. The employer using expert witnesses, for example, may show that engineering technology does not exist or cannot be imported from another industry and thus is not available to reduce airborne values to the PEL in the standard the employer was cited as having violated.

Although such proof will not render the standard per se invalid, it can establish that engineering technology will reduce airborne values only to some value that exceeds the PEL. That by itself can be a good result for the employer. The alternative is that the employer will be under an indefinite and continuing duty to experiment with technology by which it might eventually achieve the PEL.

An expert technical witness may assist in demonstrating that the costs of implementing existing technology are prohibitively expensive. As mentioned above, there is much uncertainty regarding what the courts mean by economic feasibility. In theory, however, it is conceivable that as the typical plant in an industry attempts to comply it may determine that the costs of compliance cannot be borne. The expert technical witness for the employer will be expected to provide testimony supporting the proposition.

The expert witness must realize that OSHA and possibly the collective bargaining agent[36] will oppose his or her views. That opposition often takes the form of testimony presented by other expert witnesses.

While the foregoing possibilities are presented in the context of a case being litigated before OSHRC, it is also at least theoretically possible to present them in the context of a variance application. The Act, in Section 6(d), authorizes OSHA to grant permanent variances from the terms of OSHA standards. A variance is just what its name suggests: a variation of the terms of the root OSH standard. The essential requirement for a variance is that it provide for "employment and places of employment ... which are as safe and healthful as those which would prevail" if the employer complied with the terms of the standard from which the variance is sought.

An obvious problem with the statutory language arises when OSH standards are stated in terms of the performance they require. OSHA can reply to the application for a variance by arguing that the employer can comply simply by choosing any means to effect safe and healthful conditions.

Variance applications are processed using formal adjudicative proceedings under the Administrative Procedures Act (see 29 CFR 1905) as they are in contested cases before OSHRC. Expert witnesses are used in essentially the manner they are used in contested penalty cases.

The General Duty Requirement

Employers must also comply with a statutory catch-all provision commonly referred to as the general duty clause. Specifically, Section 5(a)(1) of the Act imposes the following general duty.

> *Each employer—(1) shall furnish to each of his employees employment and a place of employment which are free from recognized hazards that are causing or are likely to cause death or serious physical harm to his employees.*

This general duty is unlike the employer's specific duty stated in Section 5(a)(2) to comply with occupational safety and health standards. The general duty obligation is limited in three respects. First, the obligation runs from employer to employees. In contrast, the special duty to comply with OSH standards has been construed to run not only to an employer's own employees but also to employees of other employers.[37] Second, the general duty obligation applies only to serious hazards, where death or serious physical harm may be incurred if a worker is injured or becomes ill because of the hazard.[38] Third, the general duty is limited to recognized hazards.

Hazards that are recognized include obvious hazards (bricks falling as the result of a demolition project would be an example),[39] those personally recognized by the cited employer,[40] and those generally recognized by the employer's industry.

What constitutes industry recognition of a hazard is an issue that is frequently litigated. OSHA has the burden of proof on this issue in contested cases. That means OSHA must prove it is more likely than not that the standard of knowledge in the affected industry recognizes the existence of the hazard. One method for meeting this burden is to use the testimony of an expert witness who is familiar with the industry and its hazards.[41]

OSHA uses other means to prove that the affected industry recognizes the existence of hazards. Thus, public health warnings such as those issued by the CDC have been used in OSHA's attempt to prove this element of the alleged violation against health industry employers. CDC guidelines for bloodborne pathogens were used in this manner prior to the adoption by OSHA of the bloodborne pathogen standard.[42] Similarly, CDC guidelines for tuberculosis have been used to prosecute OSHA citations in the mid-1990s.

ANSI standards are frequently used to prove recognition by an industry of a cited hazard. Industry safety manuals are also used to supply this element of proof.[43]

In addition to proving that a hazard is recognized, OSHA also has the burden of proving that a feasible means is available to abate the hazard. As with proving hazard recognition, expert testimony can be used to prove that abatement is or is not feasible. What is feasible abatement will, of course, be dictated largely by the nature of the recognized hazard.

In one case, for example, Freon™ was used to clean the interiors of M-1 tanks. The employer recognized the hazard that Freon would displace oxygen in the confined space, but the OSHRC found that the problem was not solved with increased ventilation and decreased use of Freon. The OSHRC also found that the hazard could be abated by instituting what essentially were confined space procedures and training.[44]

Workplace Inspections

The secretary of labor is authorized by Section 8(a) of the Act to inspect plants, workplaces, construction sites, and the accompanying machinery, apparatus, equipment, materials and so forth at reasonable times and within reasonable limits. The secretary is authorized by § 8(b) to require the attendance and testimony of witnesses and the production of documents, by administrative subpoena, if necessary.

Similarly, to carry out his or her functions under the Act, the secretary of health and human services also has the right to inspect workplaces, as provided for by Section 20(b) of the Act. The secretary of labor makes compliance inspections whereas the secretary of health and human services is required to inspect workplaces in response to requests by employees and employers to determine the presence of health hazards.

Employers do have some limited privacy rights and can refuse an inspection if the inspector(s), whether from OSHA or the Department of Health and Human Services (HHS), do not have a warrant. OSHA and HHS must have a probable cause basis for obtaining inspection warrants.[45] In general, employee complaints provide limited probable cause, as do accidents, including those involving fatalities.[46] OSHA also generates administrative inspection plans periodically, and these plans have been found to support inspection warrants.

In general, however, employers recognize that OSHA and HHS usually have a basis for making inspections; thus, permission to make the inspection is the normal practice. This is not to say that special circumstances may not dictate an employer's decision to require a warrant. For example, labor-relations problems may play a role in the decision. Thus, some companies have gone so far as to refuse entry to inspectors when they believe a union is using the agency to further its own agenda.[47] These are unusual cases.

OSHA Inspection Priorities

OSHA prioritizes workplace inspections into four categories: imminent dangers, fatality/catastrophe investigations, complainants/referral investigations, and programmed inspections.[48] An imminent danger is defined in Section 13(d) of the Act as an existing danger

> *which could reasonably be expected to cause death or serious physical harm immediately or before the imminence of such danger can be eliminated through the enforcement procedures otherwise provided by the Act.*

A fatality will result in an inspection. Similarly, a catastrophe is defined as a workplace event that results in the hospitalization of three or more employees, and the reporting of a catastrophe will produce an inspection.[49] Complaint-based inspections are required when employees or unions representing employees file formal complaints, defined by Section 8(f)(1) of the Act as written complaints signed by an employee or a representative of employees such as a collective bargaining agent. OSHA may also conduct inspections based on any other kinds of complaints, oral or written, filed by anybody. Referrals are made by any other sources, such as hospitals, police, emergency services and the like.[50] All other inspections are programmed.

Programmed inspections may be derived from schedules produced by the national office or they may result from special emphasis programs either from the national office or local offices. The national office has imposed a number of special emphasis programs, such as those targeting nursing homes and ergonomic injuries,[51] and local OSHA area offices can obtain approval for local special emphasis programs.

The Inspection Process

In general, OSHA inspections are made to determine whether employers are in compliance with OSH standards, OSH regulations, and the general duty. OSH inspectors are expected to recommend that citations be issued for any apparent violation, and they are expected to recommend penalties as the *Field Inspection Reference Manual* instructs. Apparent violations may be either of a willful, repeat, serious, other-than-serious, or failure-to-abate nature as defined in § 17 of the Act. Penalties may be as much as $70,000 for willful and repeat violations and up to $7,000 for each serious or other-than-serious violation. Failure to abate penalties may be as much as $7,000 per day.

OSHA's procedures for making inspections, issuing citations, and calculating monetary penalties are set out in Chapters II–IV of the *Field Inspection Reference Manual*. All types of OSHA inspections have three phases: the opening conference, the walkaround/investigation, and the closing conference. Citations and penalties issue at some date after the closing conference is held. In general, the latest date for issuing citations and penalties is no later than 6 months following the opening conference.[52]

The Opening Conference

The secretary's regulations require that OSHA compliance personnel hold an opening conference with employers and representatives of employees before the walkaround inspection of the workplace or job site is performed.[53] The compliance safety and health officer will present his or her credentials (persons have been known to falsely represent themselves as OSHA compliance personnel) and explain the purpose of the inspection. It is important that all persons having an interest in the inspection be present at the opening conference. All employers should be represented, and it is preferable that all parties be represented by safety and health professionals. The inspector will ask if the employees are represented by a union, and if so, that a representative of the union attend the opening conference.[54]

The inspection will be broadly focused if it is based on OSHA's general safety or health schedules. OSHA maintains separate general inspection schedules for safety and for health. In general, the schedules are generated from industry data from the Bureau of Labor Statistics and from histories of industry-based citations and focus on high hazard industries. Inspections of the construction industry, considered a high hazard industry, are made up from lists developed by a contractor.

Safety inspections are usually made by safety specialists, and health inspections are conducted by inspectors who are classified as occupational hygienists. Occupational hygienists who work for OSHA usually hold degrees but not always in occupational hygiene; few are certified industrial hygienists. Biologists and chemists are hired into these jobs and receive occupational hygiene training from OSHA.

OSHA usually conducts a review of documents during all of its inspections and particularly during all general schedule inspections. The agency reviews appropriate written plan materials. For example, OSHA inspectors always ask to see the employer's written hazard communications plan required by the hazard communications standard.[55] Also reviewed will be OSHA record- keeping forms such as the OSHA 200, which is used to record illnesses and workplace injuries and fatalities. Under OSHA's record-keeping requirements, employers are required to record lost workday injuries, injuries that result in restricted activity, and injuries that require medical treatment (but not those requiring only first aid). Similarly, illnesses must be recorded.[56]

Other written programs often called for by OSHA inspectors include lockout/tagout programs,[57] bloodborne pathogen programs,[58] emergency evacuation plans,[59] and the company's safety and health program.

There currently is no requirement that a company have a written safety and health program. However, in the mid-1990s OSHA developed a number of programs designed to encourage employers to develop such programs. For example, OSHA tested a program called PEP (Program Evaluation Profile). PEP requires compliance officers to grade the company's overall safety and health program, and one of the elements evaluated includes the employer's workplace hazard analysis and the program devised for hazard abatement. Good programs will result in lower penalties.

In some investigations, OSHA will inquire as to the existence of other company written materials including company audits and other records that may demonstrate employer knowledge of the existence of hazards. These kinds of records are called for in inspections under OSHA's instance-by-instance penalty policy and may be called for in fatality/catastrophe investigations.[60]

During the opening conference the compliance safety and health officer will explain the procedures he or she plans to use during the walkaround. OSHA inspectors photograph and videotape apparent violations. They normally seek interviews with a representative number of employees, usually in confidence. Inspectors conducting health inspections may monitor the workplace and jobs for toxic substance and noise exposures as may be appropriate. OSHA occupational hygienists often perform an initial walkdown to determine the monitoring strategy they will use during the inspection. They also use this tour to gather evidence of apparent safety violations.

The employer should use the opening conference to inform the compliance officer of areas that are restricted because of trade secrets or national security matters. OSHA will use officers who have security clearances to inspect areas that are subject to national security and also will maintain confidentiality for trade secrets.

The Walkthrough Inspection

The walkthrough inspection is commonly referred to as an evidence gathering session. The pictures and videotapes taken by compliance officers are intended as records of their observations of apparent violations. Still photographs are processed and mounted on forms on which the compliance officer usually makes notations concerning what was being photographed. Some area offices do not process the film unless a notice of contest is filed; still pictures are always mounted in contested cases. Compliance officers record their impressions on the audio portion of the videotape. Copies of these materials and the inspection file can be obtained if the citations OSHA issues following the inspection are contested. The inspectors also usually note their observations either on paper, in their laptop computers, or both.

Employers are well advised to do their own taping and photographing during the walkthrough. Employers and a representative of affected employees have a statutory right to accompany the inspector, and the rule of thumb used by most informed employers is to have a representative accompany each OSHA compliance officer everywhere on the employer's premises. It is desirable that each inspector be accompanied by a safety and health professional.

The purpose of having representatives on the inspection is to obtain notice of those apparent violations that may later be alleged as violations in a citation. The safety and health professional working for the employer might well agree with the OSHA inspector that an observed condition is hazardous, and he or she may want to take steps to abate the hazard during the inspection. (The term "hazard" is used instead of the term "violation" since the employer's representative may disagree with the inspector on characterizing the condition as a violation and should make this clear to the inspector. Failure to do so might be considered an agreement with the inspector's viewpoint.) Under programs being instituted by OSHA during the mid-1990s, prompt, voluntary abatement of hazardous conditions may be taken into consideration by OSHA in its penalty determinations.

The employer's representatives should also bear in mind that anything they say to the inspectors that may be understood as a concession can be used against the employer in a trial. Similarly, statements made by employees who are interviewed by the inspector can also be used by OSHA against the employer, though OSHA will only reveal the identity of its informant if his or her statement is used at a trial. All statements made by an employee are usable as "admissions against interest" so long as they are made concerning the work the employee performs.[61] That means the OSHA inspector can relate employee statements at a trial, therefore OSHA does not have to bring the employees to the trial to testify. The employer, however, may want to use the employees to rebut the testimony of the compliance officer.

It should be noted that OSHA compliance officers usually only make notes of what they are told by employees. However, in fatality cases and in egregious cases the inspector may write down the entire statement given by an employee, have the employee sign it, and have it witnessed. The purpose of having the employee sign his or her statement is to create a piece of evidence that is not easily retracted. Obviously, if the inspector only makes notes of a conversation, it may later be difficult to recall the precise nature and content of the statement.

As for health inspections, the employer should consider taking its own noise dosimetry and airborne monitoring samples, which are useful to compare with the results OSHA may use in support of its citations. Otherwise, the employer may not have a record of air quality and noise levels. OSHA uses its own laboratory to process samples from compliance officers. Because samples are often batched for processing with samples taken by other officers, errors may be introduced that appear in the final values OSHA uses for citations. Of course, the employer who has safety and health professionals on staff should know of the presence of any airborne toxic substances in the workplace or if employees are working in high noise areas.

Employers are cautioned, however, that the samples they take during the inspection are discoverable by OSHA if the citation is contested.

Walkthrough May Result in Private Lawsuits

OSHA inspections and other inspections including those conducted by consultants, insurance companies, and NIOSH, may result in private lawsuits brought against the government, consultants, and the insurance companies.

In general, the courts have uniformly held that the Act does not give individuals the right to bring a personal injury action.[62] Therefore, persons who want to bring such actions against professionals who perform safety and health inspections, or the employers of such persons, must concoct a legal theory that will survive motions to have the case thrown out of court. Typically, the plaintiff (injured person or estate) must show that the defendant (consultant, insurance company, government, or fellow employee) owed a duty to the plaintiff, that they negligently failed to discharge that duty, and the plaintiff's injury was caused by such negligence.

The most common theory used in such cases is known as the good samaritan doctrine. Generally, this doctrine applies in situations where the defendant normally does not owe a duty of care to the plaintiff. A frequently cited decision involved a situation in which a parent company engaged the services of an engineering firm to perform safety inspections of a subsidiary, among other things.[63] The plaintiff, a plant manager for the subsidiary, brought suit against the engineering firm, among others. The court concluded that under state statutes the engineering firm was a good samaritan and because of its undertaking for the parent corporation, it owed a duty of care to employees of the subsidiary.

Similarly, supervising engineers who by contract assume the obligation to enforce safety and health regulations on a construction site have been characterized as good samaritans to all workers on the site.[64]

Most recently, the U.S. court of appeals in Boston and U.S. district court in New Hampshire have concluded that the government may be held liable when OSHA inspectors failed to discover an unguarded machine during a general schedule inspection.[65] A worker was injured by a machine and brought suit against the federal government under the Federal Tort Claims Act to recover for her injuries. In that case, OSHA performed a comprehensive or wall-to-wall safety inspection of a manufacturing facility. Normally, the government cannot be sued for its negligence as to discretionary tasks performed by its inspectors. In this case, however, the court concluded that the inspectors had no discretion; OSHA requires its inspectors to find and cite violations. The court concluded, therefore, that the government was not exempt, and the plaintiff could bring suit under the Federal Tort Claims Act.

It also should be noticed that normally the employer of the injured person is protected against personal injury lawsuits by workers' compensation statutes. Those statutes provide what is supposed to be an exclusive remedy to the injured employee for the employer's negligent acts that result in injury. In most but not all states the protection afforded by these statutes extends to certain other persons. Fellow employees usually enjoy this protection. That means professionals on the employer's staff are usually not at risk of being found personally liable for ordinary negligence (although some states do not extend the shield of exclusivity provided by workers' compensation statutes when the employer or fellow employee's negligence is considered willful).

The protection afforded by workers' compensation statutes usually does not extend to intentional torts (wrongs or acts done deliberately to injure another). If an injured worker can convince a court that the injury resulted from an intentional tort, then the worker may sue the employer, co-workers, and others. In general, an employer's willful noncompliance with OSHA regulations may in some states be enough to convince courts that the injury was the result of intentional misconduct.[66]

The Closing Conference

After the physical inspection of the workplace is completed, the inspector(s) will hold a closing conference with the employer. The conference may not occur for several weeks. There may be a number of reasons for the delay. For example, it usually takes OSHA's laboratory 6 to 7 weeks to process air contaminant samples. A long delay could also mean that the inspector or area office has consulted either with the regional or national office of the agency or both. These consultations usually take place if the agency intends to issue willful citations or citations under the general duty clause.

The closing conference is used by OSHA to explain or identify conditions that likely will be cited, about which the agency has generally made up its mind by the time of the conference. The compliance officer may also use the closing conference as a means of obtaining employer and union input on abatement. In general, the employer who agrees voluntarily to abate violations mentioned during the conference may receive favorable treatment on penalties. The inspector may also invite input on the question of how long it will take to abate violations. The longest abatement period usually given by OSHA, provided for complex health abatement issues, is 1 year. The statute provides that an employer may petition for an extension of the abatement period.

Employers should exercise good judgment concerning what they agree to during the closing conference. Generally speaking, if complex or expensive abatement may be called for to abate violations, the employer may be well advised to tell the compliance officer that it will study the abatement question. This is particularly true if, for example, ergonomic abatement or abatement of airborne toxic substances will be called for by the citation.

Ergonomic cases can be particularly complex. OSHA citations often call for some experimentation in arriving at solutions.(67) These cases are also difficult to document. Usually, OSHA will videotape workers performing their jobs during the inspection and cross-reference the job functions of these employees to entries on the OSHA 200.

Similarly, abatement of airborne toxic substances can be difficult. Engineering controls are the preferred means of abatement. OSHA usually takes the position that compliance with PELs or threshold limit values can be accomplished using feasible engineering controls. OSHA defines the word "feasible" to mean capable of being done.(68) The OSHRC disagreed at one time but reversed position and aligned itself with OSHA at least to the extent that both agencies define "economic feasibility" as meaning achievable.(69)

The Citation Process

The statute requires that each citation issue "with reasonable promptness, be in writing, and describe with particularity the nature of the violation, including a reference to the provision of the Act, standard, rule, regulation, or order alleged to have been violated.(70) A citation becomes final as a matter of law if the employer fails to contest it within 15 working days of receipt of the penalty notice.(71) The 15 working-day period technically runs from the date the employer receives notice of the penalty. OSHA issues citations on a form that states the charge, the proposed penalty, and the date by which the violation must be abated. The citation is mailed with a return receipt post card. OSHA knows, therefore, the day the employer received the citation. Weekends and federal holidays are not counted as working days.

The OSHRC has only on the rarest of occasions forgiven employers who fail to file notices of contest on time.(72)

OSHA and OSHRC cannot change the citation or penalty if the contest period passes and no contest is filed. The abatement period can be changed, but only if a petition to modify the period is filed before the abatement period set in the citation expires. If the citation calls for immediate abatement of an alleged violation, however, and the citation is not contested, any failure to effect immediate abatement is technically subject to daily penalties.(73)

The citation package includes useful information. The employer is advised of its rights under the Act. The employer and affected employees, or their collective bargaining agent, are invited to request an informal conference during the 15 working-day period. The employers also are usually provided with forms that may be used to report their abatement efforts to the OSHA area office.

The informal conference can be used to gain insight into OSHA's concerns as described in the citation. Employers can determine what it is that the area office requires in regard to abatement. The conference is also used to discuss the employer's concerns regarding what was cited and the proposed penalties. Sometimes OSHA can be convinced that the citation was or some items in it were issued in error. Usually, the conferences are conducted either by the area director or an assistant area director, and often OSHA will offer to reduce the proposed penalties. OSHA may also recharacterize the alleged violations or some of them.[74]

The informal conference obviously is a setting in which a safety and health professional may make his or her case as spokesperson for the employer or union. The employer should not, however, have great expectations that the informal conference will result in an OSHA decision to withdraw citations. The conference should result in a better understanding of the expectations and positions of each side participating in the conference. The settlement process started at an informal conference can be continued if a notice of contest is filed. It cannot be continued if no notice is filed.

The Contest

The employer who files a notice of contest may do it either because a longer period to negotiate a settlement is needed or because there is a good faith reason to believe the employer must have the citation(s) vacated. The Act also authorizes employee contests; however, such contests are limited "to the [question whether the] period [of] time fixed in the citation for abatement of the violation is unreasonable."[75] Contests that proceed through trial are not to be taken lightly. What is involved in such contests are formal proceedings conducted much like trials in the federal courts.[76] Most contests, however, are resolved by settlements. The safety and health professional has an important role to play in all contests.

Contests are marked by legal procedures. The parties, including affected employees in cases where they intervene, are required to file formal pleadings. They may also engage in discovery by requesting documents, exchanging interrogatories, requesting admissions, and deposing witnesses.[77] The trial is conducted before an ALJ, and the Federal Rules of Evidence apply in OSHRC proceedings. Safety and health professionals have at least two essential roles in these proceedings.

The Safety and Health Professional as Counselor

The professional—whether an OSHA inspector, consultant, employee of the cited employer, or employee or consultant to a union—is the principal adviser to his client or employer and the client's attorney. Clients and attorneys look to the professional for expert advice.

That advice in a contest setting has to do with questions concerning the merits of the position taken by each party and the abatement required by the citation. The company must first know if it violated the Act as alleged in the citation. It must also know what abatement is required by the citation, what it will cost, whether it is feasible, whether it should be done, and whether it is required by the cited regulation. Ideally, this information should be compiled and evaluated by the safety and health professional before the decision to contest is made. However, time often is too short to permit a calculated analysis, particularly during the 15-day contest period.

Other issues also arise with regard to evaluating abatement as required by the citation. For example, the professional needs to know whether violations of a similar kind will or may occur in the future. The question should be asked, is this a recurring problem for our company? The professional needs to know whether the abatement called for by the citation is feasible. The OSHA inspector may anticipate questions on this issue at settlement conferences and at trial, and the company's professional has a need to know the answer as well. The answers to these questions may well dictate whether a settlement can be achieved or whether the case should go to trial.

The answers also have great value in injury and fatality cases. Those cases are concerned with questions of causation, preventability, and relative fault. In sum, they involve negligence, causation, and damages. Negligence may occur if the defendant owed a duty to the plaintiff and if the defendant violated that duty as, for example, by violating an OSH standard. The fact that an OSH standard may have been violated is not by itself enough to make the employer liable. The negligent acts or omissions of the employer must also have caused the injury or illness suffered by the plaintiff. Obviously, the safety and health professional has an important role to render in providing advice in these kinds of cases.

Examples of persons owing duties to other persons have been given. An employer, for example, owes a duty to its employees to provide a safe and healthful workplace. This duty was owed at common law. Prior to the enactment of workers' compensation statutes, employees routinely sued their employers when they were injured on the job. The theory of those suits was that the employers failed to discharge the duty of care to provide a safe and healthful workplace. the various workers' compensation statutes abrogated employee rights to bring simple negligence suits, in general. A good samaritan is a volunteer who by his or her actions assumes a duty of care to the beneficiaries of his or her good acts.

The lawyers will listen to the safety and health professionals on these issues. They will make their determination and render their advice to the employer/defendant in reliance on what the professional has to say. The lawyers will assess their cases based on their understanding of the facts and the opinions of safety and health professionals. They will advise their client as to the strategy it should purse, and they will make assessments concerning the merits of the client's case.

It is important that the safety and health professional educate both the employer and the employer's lawyer in these situations. The safety and health professional must communicate and do so in language that is understandable to persons who are not trained in safety and health.

The Safety and Health Professional as Witness

The other principal role played by the safety and health professional is that of witness. The professional may be called on to testify in an OSHA hearing, a workers' compensation hearing, or a lawsuit brought by an injured or ill worker. Though the legal issues will differ, the role of the safety and health professional in all such proceedings will not be materially different.

In general, witnesses either are fact witnesses, expert witnesses, or both. Fact witnesses are persons who testify as to things they did or did not do and things they observed. In OSHA cases a safety and health professional may testify concerning sampling techniques used to determine the existence of airborne toxic substances and worker exposure levels. Typically, the witness will testify as to the equipment used, pre- and postsampling calibrations, the conditions under which the samples were taken, the workers sampled, periodic checks that were made during the sampling period, observations of the workers, and what the occupational hygienist did to ensure the integrity of the samples for shipment to the laboratory. This kind of testimony is about facts.

Similarly, the professional may be called on to describe point-of-operation guards on machines, how the guards should be used, and procedures to lockout or tagout the machine for servicing and maintenance.

As in any trial, fact witnesses may be and usually are cross-examined by counsel for the opposing party. Judges usually allow the attorney who is doing the cross-examination wide latitude. That means the witness often will be examined on issues not raised during direct examination. It also means that the attorney performing the cross-examination will try to get the witness to agree to the attorney's version of the facts and to get the witness to recant the version of the facts related on direct examination. The attorney will also seek to show that the testimony of the witness is not worthy of belief (called impeachment).

To illustrate using a defense that is often raised in OSHA contests, we will assume the witness is testifying for the employer. We will assume further that the employer was cited for a violation of the machine guarding standard at 29 CFR 1910.212(a)(1). The compliance officer testified during the secretary of labor's case that he did not see a guard being used at the point of operation of a hydraulic press, and further testified that the operator of the press told him that the guard was never used.

The compliance officer is allowed to testify about his conversation with the press operator only to the extent that the conversation concerned the normal work activity of the operator. Courts normally will not allow hearsay evidence. For example, if the compliance officer testified that he was told by the press operator that operators of band saws in another department did not use blade guards, such evidence might be excluded as hearsay.

Assuming this testimony, the witness for the employer may present facts that concern the employee misconduct defense, essentially arguing that the employee violated the employer's workplace rules. The witness for the employer is presented to prove (1) the existence of the rule, (2) that the rule was made known to the employee to prove the employee violated the rule, and (3) that the employer enforced the rule in the workplace.

In the best case scenario, the witness may be the safety professional or part of the employer's safety and health operation who was responsible for drafting the rule, communicating the rule to employees, policing the workplace to see that the rules were followed, and recommend discipline for employees who broke the rules.[78] All of these elements of the defense must be established to prove the employer's case.[79]

The employer's witness will testify as to the existence of the rule (i.e., point-of-operation guards must be used when hydraulic presses are operated). Usually the rule will be in writing, and the attorney for the employer will have the witness identify the document and place it in evidence. The witness will also have to testify as to the manner in which the rule was made known to operators. Assuming the witness is a safety and health professional, he or she will also testify as to inspection tours of the workplace. Ideally, the professional will testify at least to the fact that the workplace is inspected on a regular schedule to determine whether employees are obeying the rules.

Employers who have effective safety and health programs usually have a reporting system of some kind by which the professional reports infractions to management. The system ensures that discipline is imposed. In addition, company safety and health inspectors are, in a good system, empowered to correct serious violations on the spot. The employer's witness will testify to all of these things. When that examination is completed, the attorneys for OSHA and the union, if it is participating, may cross-examine the witness.

The attorney for OSHA will attempt to demonstrate that at least one of the elements of the employee misconduct defense does not exist or is suspect for other reasons. In some cases, the employee misconduct defense has not been allowed because the employer's rule was too imprecise. Thus, it is not enough for the employer's rule to say "employees shall use fall protection"; the employer's rule must specify the kind of protection to be used at specific elevations.[80]

In other cases the defense has failed because the employer's rule, although sufficiently precise, was not communicated to employees. Generally, however, the defense fails because the attorney for OSHA demonstrates either that the rule was not followed[81] or discipline for an infraction was not imposed. The attorneys for OSHA and the union, if present, will attempt to get the employer's witness to agree that the employer's safety rules were not effective in any of these areas.

The safety and health professional may also be called on to present expert testimony. Expert testimony usually means the witness is presented to render expert opinion evidence.

Expert opinion evidence should not be confused with opinion evidence, which may be offered by a lay witness. Opinion evidence may be rendered by a nonexpert witness under Rule 701 of the Federal Rules of Evidence. Such opinions must be rationally related to the perception of the witness. Thus, a machine operator may have acceptable opinions concerning the adequacy of guards used on the machine or machines he or she operates.

The rules of evidence give the court or OSHRC judge discretion to refuse to hear expert testimony. Rule 702 states that such testimony may be allowed if "specialized knowledge will assist the trier of fact to understand the evidence or to determine a fact in issue." The attorney who wants to present expert testimony must (1) convince the court that the evidence will assist the court, and (2) specifically define the area of expertise that will be useful to the court.

Usually, the attorney will define the expertise required to prove his case and will look for an expert who is credentialed in that field. Obviously, there are varying levels of expertise, and the attorney will choose his expert accordingly. In the context of OSHA, most compliance officers, including OSHA occupational hygienists, are generalists. The nature of their jobs requires them to have a broad base of safety and health knowledge, but that generally means that they do not develop special areas of expertise. OSHA does not employ many certified industrial hygienists.

As an example, assume that an essential issue in a case concerns airborne toxic substance levels that OSHA cited when the agency issued a citation to an employer for alleged violations of the lead standard. Assume further that the employees OSHA named as being exposed at levels in excess of the 50 $\mu g/m^3$ PEL are burners/welders. An issue that has arisen in several cases of this kind[82] concerns the location of the sampling filter cassette. Typically, compliance officers have fastened the cassette to the collars of employees outside the burner/welder's helmet. OSHA's *Industrial Hygiene Technical Manual* and NIOSH instructions both state that the cassette must be placed inside the helmet to obtain an accurate measurement for a breathing zone exposure value. The question for the judge in a contested OSHA case is whether the lead exposure value obtained as a result of the compliance officer's measurement is enough to prove that the employee was overexposed to airborne lead.

The cited employer must make an issue of the location of the filter cassette if the employer is to mount a potentially successful defense. The most appropriate means of raising the defense is through the use of testimony from a professional occupational hygienist who by education and experience is knowledgeable about airborne sampling methodology.

To further illustrate the differing levels of expertise that may be available, one might contrast the expertise of an occupational hygienist who performs field surveys with an occupational hygienist whose strengths are in laboratory work. the latter may not necessarily be qualified to give an opinion concerning the placement of the filter cassette.

Typically, this professional will have a curriculum vitae spelling out his or her education and work experience, listing honors earned, memberships in professional associations, published works authored, and representative speeches delivered (in particular, at professional meetings and in academic settings). The attorney will present the expert witness, introduce the curriculum vitae in evidence, and have the witness testify concerning its highlights. The testimony rendered by the expert may be wholly of an opinion nature, or it may consist of a mixture of fact and opinion testimony.

The witness may present fact testimony based on observations or measurements or both. Thus, in the example posed above, the expert might actually measure the air inside the welding helmet to obtain airborne levels of lead under circumstances like those that existed when the compliance officer took his samples. There is, of course, an element of risk in defending on this basis. The expert may well prove some elements of OSHA's case. While defense lawyers may reject this approach in an OSHA case, it should not be overlooked by the professional when assessing the employer's compliance obligations. More likely, the witness will be asked to opine that the sampling technique used by the compliance officer was improper such that his measurements are not representative of employee exposure levels.

The attorney for the employer might also use the expert witness to identify documentary evidence that supports the proposition that the filter cassettes were not placed properly. The expert witness may be asked to testify or opine as to the reasons why the filter's placement outside of the welding helmet is improper. Finally, the expert may be asked to opine concerning the value, if any, of the measurements obtained by the compliance officer with respect to the ultimate issue. That is, do the compliance officer's exposure values prove that the lead standard was violated?[83]

The role of professionals as testimonial witnesses has been presented from the viewpoint of rendering testimony in an OSHA case. These roles are not remarkably different when they are used in workers' compensation and in tort liability cases. What changes are the legal standards that apply in such cases. Thus, where the issue in an OSHA case may concern the alleged violation of an OSH standard, the issues in a tort case are whether that violation was the proximate cause of injury to the plaintiff. There are, of course, many other issues in each kind of case, but usually the testimony of safety and health professionals is requested for three basic issues: (1) was there a violation, (2) was that violation a proximate cause of injury, and (3) did it arise in the workplace? To this one might add workers' compensation issues, such as whether the injury was work-related.

Summary

This chapter generally describes the role of an occupational hygienist in providing professional services either as an employee or a consultant to employers who are obliged to comply with the provisions of the Occupational Safety and Health Act. The occupational hygienist will provide pre-OSHA inspection compliance advice concerning an employer's obligations to comply with relevant OSHA standards and with the duty to render workplaces and jobs free of recognized hazards. The chapter describes the responsibilities of the occupational hygienist during OSHA inspections and in postinspection proceedings, including contests of citation. The role of the occupational hygienist as an advisor and as a witness is also discussed. The chapter, while using the Occupational Safety and Health Act as the principal vehicle for discussion, also touches on the responsibilities occupational hygienists may have under workers' compensation and negligence laws.

References

1. See generally, Staff of Senate Subcommittee on Labor, 92nd Cong., 1st Sess. Legislative History of the Occupational Safety and Health Act of 1970. (Comm. Print pp. 142–145)
2. 29 USCA § 655.
3. 29 USCA §§ 658(a), 666.
4. 29 USCA § 659.
5. 29 USCA §§ 659, 661.
6. See 29 CFR § 2200.
7. 29 USCA § 669.
8. 29 USCA § 667(a).
9. 29 USCA §§ 667(b) & (c).
10. See 29 USCA § 653(a)(1).

11. 29 CFR § 1975.4. See, e.g., *Secretary v. Mashantucket Sand & Gravel*, 17 BNA OSHC 1747 (2d Cir 1996). The Bureau of National Affairs (BNA) publishes the *Occupational Safety and Health Reporter* including decisions of the OSHRC and the courts. Commerce Clearing House (CCH) publishes the *Employment Safety and Health Guide*, including decisions of OSHRC and the courts. For ease of reference and to the extent possible, citations to decisions of the courts and of OSHRC will be made to the BNA *Occupational Safety and Health Case* (OSHC) reporter. Readers may locate the same decisions reported in the CCH *Employment Safety and Health Guide* (ESHG) by use of the case name and date of the decision.
12. Occupational Safety and Health standards are defined at § 3(8) of the Act to mean standards that require the adoption of "practices, means, methods, operations, or processes, reasonably necessary or appropriate to provide safe or healthful employment and places of employment." The employer's duty to comply with such standards is specified by § 5(a)(2) of the Act.
13. Section 8 of the Act authorizes the secretaries to promulgate various types of regulations. A § 8(a) regulation differs from an OSH standard in the sense that the rules used to adopt regulations are more relaxed than the rules used to adopt OSH standards. OSH standards usually are the subject of public hearings. There is no public hearing mandate for § 8(c) regulations, which include, for example, the OSHA record-keeping requirements of 29 CFR § 1904.
14. Some vertical standards are published in § 1910. Thus, the telecommunications industry is specifically regulated by 29 CFR § 1910.268 and the electrical power generation industry by 29 CFR § 1910.269.
15. 29 CFR §§ 1910.1015–1019.
16. See 29 CFR § 1910.05.
17. 29 CFR § 1910.23(e).
18. The term national consensus standard is defined by § 3(9) of the Act. Practically speaking, only ANSI standards and standards promulgated by the National Fire Protection Association meet the definition.
19. Established federal standards were governmental standards in effect on the date of enactment of the Act; 29 U.S.C. § 652(10). The term established federal standard is defined by § 3(10) of the Act.
20. 29 U.S.C. § 655(a). The authority to adopt standards under this section expired in 1973.
21. NIOSH, however, is proceeding with a study of the science of ergonomics. It has been reported that the agency believes that substantial, reliable science supports the conclusion that a causal relationship exists between some work activities and musculoskeletal disorders of the back, neck, and upper extremities. See *Inside OSHA 17*:1 (August 12, 1996).
22. See 53 Fed. Reg. 15497 (1988).
23. The procedural provisions are contained in Section 6(b) of the Act.
24. In special circumstances, the notice of proposed rulemaking can be initiated by the adoption and publication of a temporary emergency standard. OSHA is authorized to issue temporary emergency standards under the provisions of § 6(c) of the Act and is required to adopt a permanent standard within 6 months of publication of the temporary emergency standard. An emergency standard is unusual, and OSHA has rarely used the authority granted by § 6(c). Perhaps the reason is that the courts have generally refused to enforce this kind of standard. See *Florida Peach Growers Ass'n, Inc. v. Department of Labor*, 1 BNA OSHC 1472, 489 F2d 120 (5th Cir 1974).
25. OSHA usually provides for a 60- or 90-day period to file written comments. Frequently, the agency will extend the comment period if a timely request supported by reasons is made. See, e.g., 61 Fed. Reg. 27850 (1996).
26. OSHA virtually always schedules public hearings at the time it proposes a new rule. See, e.g., 61 Fed. Reg. 4030 (1996) in which OSHA proposed revisions to the record-keeping requirements of 29 CFR § 1904 and scheduled public meetings.
27. See 29 CFR § 1911.
28. See 29 CFR §§ 1911.15–17.
29. See, e.g., *Synthetic Organic Chemical Manufacturers v. Brennan*, 2 BNA OSHC 1159 (3d Cir 1974); *Society of the Plastics Industry, Inc. v. OSHA*, 2 BNA OSHC 1496 (2d Cir 1975). These two courts had to wrestle with what § 6(f) of the Act means when it states that OSHA rules must be based on "substantial evidence in the record

considered as a whole." The test for determining whether OSHA's rulemaking action passes judicial muster cannot be taken at face value as Mr. Justice Clark (retired) discussed in the Plastics Industry decision. Ultimately, the courts have settled on a rule that involves a searching review of the record to determine whether the new OSH standard resulted "from a process of reasoned decision making." See Bokat and Thompson (eds.): *Occupational Safety and Health Law*. Washington, DC: BNA, 1988. pp. 582–583.

30. One of the more notable examples of a variance in this regard occurred in 1978 when OSHA adopted the OSH standard for inorganic lead. The proposed standard called for reductions in the 8-hour PEL from 200 μg/m³ to 100 μg/m³. The new standard adopted a PEL of 50 μg/m³. See 29 CFR § 1910.1025(c)(1) and *United Steelworkers of America v. Marshall*, 647 F2d 1189 (DC Cir 1980).
31. The question whether OSHA had a mandate to eliminate all risk in the workplace was answered in the negative by the Supreme Court in 1980. *Industrial Union Dept., AFL-CIO American Petroleum Institute*, 8 BNA OSHC (1980). This decision, commonly referred to as the Benzene decision, established the rule that, as a threshold, OSHA must determine that a significant risk at the current level of regulation exists to justify its rulemaking efforts.
32. The lead standard is the prime example of the extent to which OSHA has authority to adopt regulations. It was clear that major impacted industries such as the primary and secondry lead smelting industries could not comply with the 50 μg/m³ PEL given the existing state of engineering abatement technology. The record did indicate, however, that new technology might be developed to replace existing smelting processes. That was enough for the reviewing court to conclude that a presumption of feasibility existed in OSHA's favor. See, e.g., the decision on *United Steelworkers v. Marshall*, 8 BNA OSHC 1810, 1874 (DC Cir 1980).
33. See *United Steelworkers*, 8 BNA OSHC at 1868 n.119.
34. *United Steelworkers*, 8 BNA OSHC at 1868; *Atlantic & Gulf Stevedores, Inc. v. OSHRC*, 4 BNA OSHC 1061 (3d Cir 1976).
35. OSHA inspections are conducted at particular facilities, and penalties which issue as a result of the inspections are directed to the employer or employers who were conducting operations at the inspected facility. The citations and penalties are limited to such violations as OSHA believes exist at the facility. In other words, OSHA citations do not apply to all facilities operated by the cited employer. However, OSHA maintains a database of all citations and penalties it issues. Data entries are made by the name of the cited employers under their SIC codes. Citations affirmed for any reason at any facility operated by the employer are used as a history for the employer. They can be used to cite the employer for a similar repeat violation at any other facility it operates wherever that facility is located. Repeat citations carry enhanced penalties. OSHA operates on the principle that a violation at one facility provides notice to the cited employer to eliminate that violation at all its other facilities. See *George Hyman Constr. Co. v. OSHRC*, 7 BNA OSHC 204 (4th Cir 1979).
36. OSHRC rules provide that the representative of affected employees, which may be a union, can appear as a party in OSHRC proceedings. 29 CFR § 2200.22. All parties to a contested case before OSHRC have the right to present their own witnesses and to cross-examine witnesses of the other parties.
37. In general, an employer who creates a hazard by violating the terms of an OSH standard is responsible regardless of whose employees are exposed or have access to the hazard. Moreover, employers who are innocent of having created the hazard are nevertheless responsible (and citable) if their employees are exposed or have access to the hazard. See, e.g., *Anning-Johnson Co.*, 4 BNA OSHC 1193 (Rev. Comm'n 1976).
38. See, e.g., *R.L. Sanders Roofing Co.*, 7 BNA OSHC 1566 (Rev. Comm'n. 1979).
39. See *Usery v. Marquette Cement Mfg., Co.*, 568 F2d 902 (2d Cir 1975). Also see *Donovan v. Farmers Ass'n*, 674 F2d 690 (8th Cir 1982), in which an enclosed pit not having a means of exit was thought to be an obvious hazard.

40. In *Brennan v. OSHRC (Vy/Lactos)*, 494 F2d 460, 464 (8th Cir 1974), the employer was personally aware that decomposing fish produces hydrogen sulfide gas, which can be hazardous in a confined space.
41. The hazard may be recognized even if the cited employer is ignorant of its existence. *National Realty & Constr. Co. v. OSHRC*, 1 BNA OSHC 1422 (DC Cir 1973). Moreover, the hazard may be hidden in the sense that it can be detected only by instrumentation. *American Smelting & Ref. Co. v. OSHRC*, 2 BNA OSHC 1041 (5th Cir 1974). Employee exposures to toxic substances frequently fall into these categories.
42. 29 CFR § 1910.1030; see *Waldon Healthcare Center*, 16 BNA OSHC 1502 (Rev. Comm'n. 1993).
43. See, e.g., *St. Joe Mineral Corp. v. OSHRC*, 647 F2d 840, 845 n. 8 (8th Cir 1981); *Betten Processing Corp.*, 2 BNA OSHC 1724 (Rev. Comm'n 1975); *Puffer's Hardware v. Secretary of Labor*, 11 BNA OSHC 2197 (1st Cir 1984).
44. See *General Dynamics Land Systems Div., Inc.*, 15 BNA OSHC 1275 (Rev. Comm'n 1991). An earlier version of this case provides interesting reading concerning the interplay of the general duty with the obligation to comply with OSH standards. See *Auto Workers v. General Dynamics Land Systems Div.*, 13 BNA OSHC 1201 (DC Cir 1987). In that case, the court overturned an OSHRC decision. OSHRC had, pursuant to its usual rule, determined that the general duty obligation was displaced by the 8-hour TWA standard published at 29 CFR § 1910.1000 (Z tables); 12 BNA OSHC 1514. General Dynamics, however, had recognized that the hazard was one associated with a confined space. That was enough for the court.
45. See *Marshall v. Barlows, Inc.*, 4 BNA OSHC 1991 (US 1977).
46. Employers are required to report workplace fatalities to OSHA within 8 hours of occurrence. 29 CFR § 1904.8.
47. For insight into how labor relations may effect the employer's decision, see, e.g., *Secretary of HHS v. Caterpillar, Inc.*, 17 BNA OSHC 1610 (MD Pa. 1996). In this case a union requested a health hazard evaluation, and the company refused the inspection. The company was found in contempt of court and purged the contempt by permitting the inspection after being threatened with daily penalties of $10,000 per day. See BNA, 25 *OSH Reporter* 1657 (1996).
48. **Occupational Safety and Health Administration (OSHA):** *Field Inspection Reference Manual.* Washington, DC: OSHA. pp. 2–4, OSHA Instruction CPL 2.103.
49. The reporting requirements are published at 29 CFR § 1904.8.
50. **Occupational Safety and Health Administration (OSHA):** *Field Inspection Reference Manual.* Washington, DC: OSHA. pp. 1–6.
51. For example, on August 7, 1996, OSHA announced a special emphasis program focused on nursing homes located in the states of Pennsylvania, New York, Missouri, Ohio, Illinois, Florida, and Massachusetts. The program reportedly has a focus on ergonomically related injuries and on bloodborne pathogens. 26 *BNA OSH Reporter* p. 251 (August 7, 1996). Similarly, OSHA instituted a special emphasis program for silicosis on May 2, 1996. See 26 *BNA OSH Reporter* pp. 1749–1763 (May 15, 1996). A number of other special emphasis programs are in effect.
52. Section 9(c) of the Act.
53. 29 CFR § 1903.7(a). The regulation is addressed to OSHA compliance personnel and creates no inherent rights for employers. That is, the OSHRC is not likely to vacate a citation when a compliance officer fails to hold an opening conference.
54. The union will be invited by OSHA to participate in all phases of the inspection and postcitation meetings or proceedings. The OSH Act specifically authorizes union participation in the inspection and a union's right to make complaints to OSHA. 29 USC § 657(e) and 657(f).
55. 29 CFR § 1910.1200. This standard requires, among other things, that employers develop written plans identifying hazardous chemicals used in their workplaces. The plan must describe how employees are to be trained, material safety data sheets, labels used on containers, and more. The written plan was the most frequently cited OSHA standard in fiscal years 1990-1995.
56. Under the state of the law as it existed in 1996, recordings were to be made as called for by *Recordkeeping Guidelines for Occupational Injuries and*

Illnesses published in 1986 by the Bureau of Labor Statistics.

57. 29 CFR § 1910.147(c)(1), (c)(4). The lockout/tagout standard also requires that employers maintain records of audits and training performed in accordance with requirements imposed by the standard.
58. 29 CFR § 1910.1030(c).
59. 29 CFR § 1910.38.
60. OSHA's egregious policy or instance-by-instance policy is applied to employers who have a history of prior violations and who are candidates for willful citations. See OSHA CPL 2.80. In general, willful violators are those employers who deliberately violate OSHA standards or those employers who act with plain indifference to the requirements of the statute. See Bokat & Thompson, *Occupational Safety and Health Law*, Chap. 9 at pp. 270–271 (BNA, 1988).
61. See Fed. R. Evid. 801(d)(2). These rules are used in OSHRC proceedings. 29 CFR § 2200.71.
62. See e.g., *Byrd v. Fieldcrest Mills, Inc.*, 1 BNA OSHC 13232 (4th Cir 1974); *National Marine Serv. v. Gulf Oil Co.*, 433 FSupp 913 (ED La 1977), aff'd. 608 F2d 522 (5th Cir 1979).
63. *Santillo v. Chambersburg Eng'rg Co.*, 603 FSupp 211 (ED Pa 1985), aff'd, 802 F2d 448 (3d Cir 1986).
64. See, *Riggins v. Bechtel Power Corp.*, 722 P2d 819 (Wn App 1986); *Duncan v. Pennington Cty Hous. Auth.*, 283 NW2d 546 (SD 1979).
65. See *Irving v. United States*, 17 BNA OSHC 1705 (DNH 1996) on remand from 14 BNA OSHC 1705 (1st Cir 1990).
66. See, e.g., *Jones v. VIP Dev. Co.*, 472 NE2d 1046 (Ohio 1984); *Adams v. Aluchen*, 604 NE2d 254 (Ohio App 1992). Also See *Mandolidis v. Elkins Indus.*, 261 SE2d 907 (WVa 1978).
67. Ergonomic citations in the mid-1990s issue as general duty clause violations. Two cases have resulted in ALJ decisions. See *Secretary v. Beverly Enterprises, Inc.*, 1995 ESHG ¶ 30,929 (Synopsis) (ALJ 1995); *Secretary v. Pepperidge Farm, Inc.*, 1993 ESHG ¶ 30,205 (Synopsis) (ALJ 1993). In *Beverly*, the judge concluded that the secretary failed to establish a recognized hazard. In *Pepperidge Farm* the judge concluded that the secretary, in calling for experimentation, failed to prove that abatement was feasible. Both cases are on review before the Commissioners.
68. This definition was one of several definitions for the word "feasible" that was presented to the Supreme Court. See *American Textile Manufacturers Institute, Inc. v. Donovan*, 9 BNA OSHC 1913 (1981). The court adopted the definition given in the text.
69. OSHA has been careful when promulgating substance-specific standards to make clear that the word "feasible" means capable of being done. That was the sense of the word as used in the cotton dust and lead standards. See *American Textile Manufacturers Institute v. Donovan*, 452 US 490 (1981); *United Steelworkers of America v. Marshall*, *supra*. Also, see *Sun Ship Inc.*, 11 BNA OSHC 1028 (Rev. Comm'n 1982). There is at least one court decision that takes a different view. See *Donovan v. Castle & Cooke Foods*, 10 BNA OSHC 2169 (9th Cir 1982).
70. The requirements that the citation issue with reasonable promptness and state the violation with "particularity" are technical requirements. The OSHRC normally does not vacate citations on technical grounds; an employer must be prejudiced to have a citation vacated on these grounds. If, for example, the description of the violation is so vague that it cannot be understood, then an employer might be prejudiced if it cannot defend against the citation. This situation is very rare. OSHA uses standardized descriptions of alleged violations called "SAVE" wherever possible.
71. 29 USC §659(a).
72. See e.g., *Secretary v. Louisiana Pacific Corp.*, 13 BNA OSHC 2020 (Rev. Comm'n. 1989).
73. OSHA is expected to finalize a regulation in 1997 that, as proposed, would require employers to report the date abatement is actually accomplished. Failure to comply with the requirement would be a citable offense. What this means, of course, is that OSHA will require employers to report untimely abatement.
74. The employer should know that recharacterization of citation items from serious to other-than-serious does not change the value of the citation for repeat citation purposes. A recharacterization of this kind will,

however, result in a lower repeat penalty. See *Field Inspection Reference Manual* at IV-16,17. More importantly, a willful citation may be recharacterized as a § 17 violation. This recharacterization may have value. Some states void the protection against private lawsuits provided by workers' compensation laws in injury cases where OSHA issues a willful citation. That is, in these states an employer may be sued in a tort case if the injury suffered by its employee resulted from a willful violation of an OSHA standard. In addition, some governments and some construction industry firms require that employers supply them with their history of OSHA violations during contract negotiations or bid processes. The presumption is that a history of willful violations cited by OSHA may result in the denial of a contract or bid.

75. The courts have narrowly construed this right, which is specified in §10(c) of the Act. See, e.g., *Automobile Workers Local 588 v. OSHRC*, 5 BNA OSHC 1525 (7th Cir 1977).
76. The OSHRC has a trial procedure called E-Z Trial, which is being tested in 1996–1997. The procedure allows simple cases to be tried quickly and informally before an administrative law judge. There are no formal pleadings and no discovery and no briefs unless the judge permits briefs. These procedures are for simple cases. No party should agree to E-Z Trial if the case is important because the constitutional right to due process may not be preserved by these procedures.
77. 29 CFR § 2200, Subpart D.
78. The case books are full of decisions in which employers have raised the employee misconduct defense. See, e.g., *Secretary v. Dover Elevator Co.*, 16 BNA OSHC 1281 (Rev. Comm'n 1993); *New York State Gas & Electric Corp. v. Secretary of Labor*, 17 BNA OSHC 1650 (2d Cir 1996).
79. The element of "recommend discipline" is understated. The safety and health professional may lack authority to impose disciplinary measures on employees. Line supervisors or the human resources department usually have this authority. However, disciplinary steps must be taken if the defense is to succeed. Moreover, the employer must discipline the errant employee as soon as possible. It is not enough for the employer to wait and see whether OSHA will cite a violation. *CF&T Available Concrete Pumping, Inc. v. Secretary*, 15 BNA OSHC 2195 (Rev. Comm'n 1993). The point is that the employee has broken a safety rule imposed by the employer, and the punishment is for this infraction.
80. See e.g., *Secretary v. Regina Constr. Co.*, 15 BNA OSHC 1044 (Rev. Comm'n 1991); *Secretary v. Superior Electric Co.*, 17 BNA OSHC 1635 (Rev. Comm'n. 1996).
81. The attorney for OSHA often demonstrates, for example, that press operators routinely fail to use point-of-operation guards and that failure was observed by or otherwise known by their foreman. See *Superior Electric Co., supra.*
82. See, e.g., *Secretary v. Omaha Steel Casting Co.*, 12 BNA OSHC 1804 (Rev. Comm'n 1986).
83. The Federal Rules of Evidence permit expert testimony on the ultimate issue to be decided by the court. See Rule 704(a).

Outcome Competencies

After completing this chapter, the user should be able to:

1. Define underlined terms used in this chapter.
2. Apply the fundamentals of chemistry and biology to the specialized discipline of toxicology as related to environmental and occupational health.
3. Classify toxic materials based on physical states and potential adverse interactions or effects (toxic responses).
4. Describe the concepts of dose-response and time-response and the factors that influence the potential for toxic responses to occur.
5. Summarize the fate of toxic foreign substances (xenobiotics) that contact and/or enter the body through environmental and occupational exposures (toxicokinetics).
6. Describe the potential mechanisms of adverse effects of toxic foreign substances (xenobiotics) that contact and/or enter the body through environmental exposures (toxicodynamics).
7. Discuss the application of toxicological data to establish environmental and occupational exposure limits.

Key Terms

absorption • action potentials • airway obstruction • antibodies • antigens • asphyxia • autoimmunity • autonomic nervous system • biological monitoring • biotransformation • bronchoconstriction • carcinogenesis • central nervous system • cholestasis • cirrhosis • demyelination • disease cluster • distribution • dose-response relationship • elimination • embryogenesis • encephalopathy • excretion • genotoxic chemicals • haptens • hepatitis • hypersensitivity • hypoxia • immunosuppression • lung restriction • metabolism • methemoglobinemia • necrosis • neurotransmitter • no-effect levels • nongenotoxic chemicals • peripheral nervous system • peripheral neuropathy • portals of entry • pulmonary irritation • risk assessment • routes of entry • sensory irritation • SKIN designation • somatic nervous system • steatosis • target systems • teratogenesis • threshold • time-response relationship • uptake

Prerequisite Knowledge

Prior to beginning this chapter, the user should review the following chapters:

Chapter Number	Chapter Topic
2	Occupational Exposure Limits
5	Epidemiological Surveillance
13	Biological Monitoring
14	Dermal Exposure and Occupational Dermatoses
16	Modeling Inhalation Exposure
17	Risk Assessment
46	Collaborating with the Occupational Physician

Key Topics

I. Fundamental Concepts in Toxicology
   A. Dose-Response Relationship
   B. Role of Time in Toxicology
   C. Nature of Experimental Toxicology Studies
   D. Concerns with Toxicological Data
   E. Occupational Diseases

II. Disposition of Chemicals in the Body
   A. Step 1: Absorption of Chemicals
   B. Step 2: Transport of Chemicals
   C. Step 3: Biotransformation of Chemicals
   D. Step 4: Elimination of Chemicals

III. Use of Biological Monitoring and Biomarkers

IV. Target Systems
   A. Respiratory System
   B. Circulatory System
   C. Immune System
   D. Hepatic System
   E. Renal System
   F. Nervous System
   G. Reproductive System

V. Disease Clusters

VI. Goals in Chemical Selection

VII. Risk Assessment: What Should Be Done with All the Data?

4

Environmental and Occupational Toxicology

Michelle M. Schaper, Ph.D.

Michael S. Bisesi, Ph.D., CIH

Introduction

Toxicology has been described as the "science of poisons." In past centuries, intentional poisonings were carried out using arsenic or cyanide, and one often reads of these tales in English, French, Greek, and Italian literature. Kings and emperors, for example, had lower members of their courts serve as tasters before indulging themselves with food or drink. In many cultures family members have arranged premature deaths of relatives by lacing their food with such poisons in an effort to inherit money, real estate, and/or property.

Today it may be more appropriate to describe toxicology as the study of adverse effects of chemicals on living organisms. Toxicologists are concerned with the action of chemicals on vegetation, insects, fish, animals and, of course, humans. Since the 1970s (following the passage of the Occupational Safety and Health Act), there has been a great deal of attention directed toward workers and the chemicals and processes in the workplace. In occupational environments the mission of health professionals is to prevent, anticipate, recognize, evaluate, and control exposures. Toxicological data are essential to perform these steps thoroughly.

Over the past 10 years there has also been a growing concern with nonoccupational environments; indoor air quality has become a hotly debated topic. Nonoccupational settings may include residential dwellings, hospitals, nursing homes, airplane cabins, schools, and offices. Exposed individuals may include infants, toddlers, immunocompromised patients, and elderly family members. Past exposure guidelines and practices, established for workers, may not be applicable for such a broad-based population. Again, toxicological data are needed to address this issue.

Current activities of toxicologists are diverse, ranging from basic research to the proposition and implementation of regulations. Within the Society of Toxicology (headquartered in Reston, Va.), there are 13 specialty sections (carcinogenesis, food safety, immunotoxicology, inhalation, in vitro, mechanisms, metals, molecular biology, neurotoxicology, regulatory and safety evaluation, reproductive and developmental, risk assessment, veterinary), reflecting the many areas in which toxicologists are involved. Molecular and mechanistic toxicology have become active areas of research, particularly over the past five years.

Fundamental Concepts in Toxicology

Dose-Response Relationship

The quote of Paracelsus (1493-1541), "sola dosis facit venenum," has been cited in many toxicology textbooks. It has been translated as "only the dose makes the poison." The point is that any chemical, when administered in sufficient quantities, is capable of evoking adverse effects. Thus, even chemicals thought to be innocuous, such as water or oxygen, may produce unacceptable human health effects at specified doses (or concentrations). For example, exposure to low levels of oxygen (i.e., <10% v/v) may result in severe hypoxia and exposure to high levels (i.e., 100% v/v) may result in pulmonary edema (i.e., fluid in the lung).[1] The amount of a chemical, or dose, ultimately delivered to an organ (tissue) is critical in defining its toxicity. Typically, dose is expressed in milligrams per kilogram (mg/kg), although for some dermal toxicants, it may be expressed as milligrams per square centimeter (mg/cm^2). For inhaled chemicals we often speak of exposure concentrations that are expressed in parts per million (ppm) or in milligrams per cubic meter (mg/m^3). These values do not represent inhaled doses and should not be confused as such. However, the dose of an inhaled chemical may be estimated using the exposure concentration, length of exposure, respiratory frequency, tidal volume, and body weight of the exposed subject. When working with vapors, their solubility and reactivity must also be taken into account, while with aerosols, the fractional deposition, which depends on particle size, must be considered.

With increasing dose, greater biological effects (i.e., responses) will be elicited; that is to say a dose-response relationship can be demonstrated. For example, with exposure to formaldehyde, workers may experience burning of the eyes, nose, and throat (i.e., sensory irritation), depending on the airborne levels of this vapor. The level of sensory irritation increases as exposure concentration is raised. As the formaldehyde concentration approaches 3 ppm, the work environment would become intolerable.[2] Irritation is only one endpoint that has been reported with chemical exposures, although it is among the most common complaints of exposed workers. It is also a recurrent problem with indoor environments.

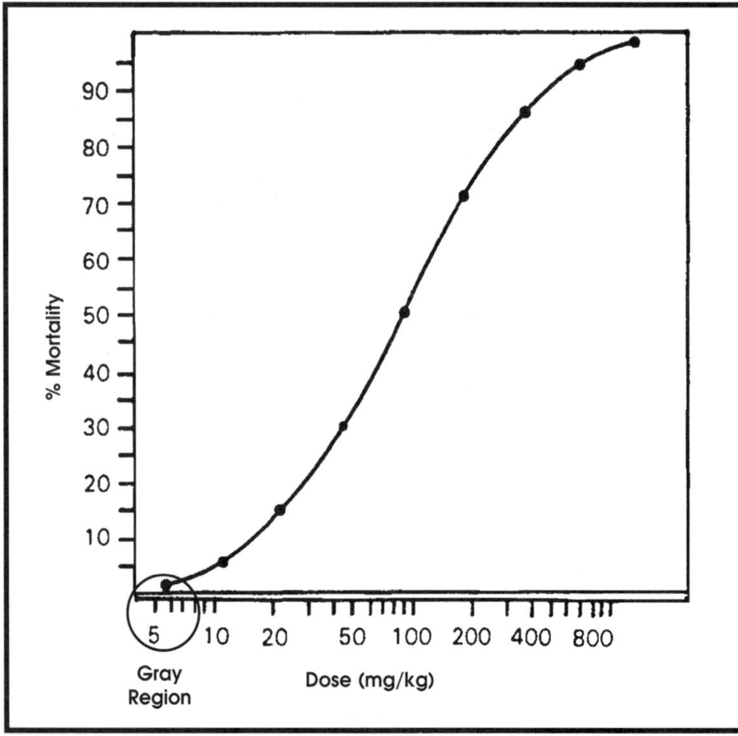

Figure 4.1 — The dose-response relationship (from C.D. Klaassen, *Casarett and Doull's Toxicology*, 5th ed., New York: McGraw-Hill, 1996; reproduced with permission).

As discussed in greater detail later in this chapter, chemicals may produce toxicity in many target systems, including the immune system (immunotoxicity), the kidney (nephrotoxicity), the liver (hepatotoxicity), the lung (pneumotoxicity), the nervous system (neurotoxicity), and the reproductive system (reproductive/developmental toxicity). Also, the types of responses produced by a chemical may be extensive. For example, mortality (lethality), mutations, or cancer may be assessed following exposures. These are common endpoints that are evaluated by toxicologists. Responses to exposures may be expressed in absolute units (e.g., number of tumors) or in relative terms (e.g., percentage tumor incidence).

In toxicological studies, the logarithm of dose (or concentration) is plotted against the level of response. As seen in Figure 4.1, where there are numerous data points the dose-response curve may be described as sigmoid or S-shaped. There is a relatively flat area at the bottom of the curve (i.e., low slope) where little response is observed as dose is increased. The midsection of the curve has a steeper slope, and a straight line may be fitted to the data points in this area. At the top of the curve a plateau is seen (i.e., low slope), and again little additional response is observed as dose is increased. It has been postulated that the lowest portion of the curve represents hypersusceptible members of the tested population, while the uppermost portion of the curve represents those who are hyposusceptible. In practice, it is unusual to find such a completely defined dose-response curve for a chemical as shown in Figure 4.1. Rather, there may be as few as three tested doses (low, medium, high), and we may only have the midportion of the curve.

The dose-response relationship may provide other important information on the chemical of interest. First, it may be possible to define a threshold for the chemical. The threshold is the smallest dose capable of eliciting a response. Below the threshold, no response occurs. In the case of airway irritation or reproductive/developmental effects, the concept of a threshold is plausible. However, for cancer there are various schools of thought on this issue. Some scientists have argued that there are no thresholds for carcinogens; a "single hit" may cause cancer. When there are few data points in the lowest region of the dose-response curve, downward extrapolation is needed to arrive at threshold levels and no-effect levels. Thus, it is understandable that controversy may arise when working in this gray region of the dose-response relationship.

Secondly, the dose-response curve may be used to define the dose (or concentration) that produces a 50% response. This response is known as the LD50 (or LC50), and it is a statistically estimated value. The LD50 represents the midpoint of the dose-response curve, and it is also the point at which the 95% confidence intervals are the tightest for the entire curve. LD50 and LC50 values are commonly reported in the literature and comparisons of relative potency may be made between chemicals on this basis. For example, the magnitude of LD50 values may range from 10^{-3} mg/kg for dioxin to 10^{4} mg/kg for ethanol. Thus, one can conclude that dioxin is much more acutely toxic than ethanol.

Other statistically estimated doses may be derived from dose-response curves. They include the no observable effect level (NOEL), the no observable adverse effect level (NOAEL), the lowest observable effect level (LOEL), the lowest observable adverse effect level (LOAEL), and the benchmark dose (BMD). These values are often used when conducting risk assessments.[3]

Role of Time in Toxicology

Although the amount of absorbed chemical determines its toxicity, researchers often report the time needed to produce a given effect. Thus, time is a secondary concern in toxicology. For example, death may occur on exposure to small doses (or concentrations) of the pesticide paraquat. However, these deaths occur several days to perhaps a week following a single exposure. This information may be useful in situations where workplace accidents occur and antidotes/treatments exist. Medical attention can be given promptly. In contrast, with a toxicant such as cyanide, its effects are produced rapidly (within minutes), resulting in immediate death. There is little or no time for medical intervention.

The duration of an exposure is another commonly reported variable in toxicology. Acute exposures are conducted over a period of hours or days, while subacute exposures may approach a period of one month. The length of subchronic exposures is generally one to three months and chronic studies will last beyond three months and up to two years (i.e., life span of a rodent). Single or repeated exposures may be conducted in these same time frames.

When a particular toxic effect is closely monitored as a function of time, it is possible to develop a time-response relationship. This curve, like the dose-response curve, contains valuable information and can reveal the onset of response, the time of maximum response, the duration of the response (i.e., sustained or transient), and possible recovery from the exposure. If, for example, a chemical such as calcium hydroxide induces delayed ocular toxicity (e.g., corneal opacity), we would want to know when this effect will be the greatest and whether we can expect permanent eye damage. Thus, without recovery, vision impairment would be imminent.

Nature of Experimental Toxicology Studies

As noted above, single or repeated exposures to toxicants may be conducted on an acute or chronic basis. Rats are typically used. Because of cost, one commonly finds data in the literature on single acute exposures to chemicals. Also, most of the data will be on single chemicals, tested at levels where a statistically significant level of response is elicited (i.e., higher doses or concentrations).

These points are emphasized because of the profound differences with respect to many, if not most, environmental and occupational exposures of humans. In contrast to the animal experiments, humans are routinely exposed to multiple chemicals, at low concentrations, and for prolonged periods of time. Reports are rare, if not impossible to find, in which animal studies have been conducted utilizing this type of protocol. Even with chemical mixtures, the data are limited.[4,5] Health and safety professionals need to keep such issues in mind when assessing exposures and their potential biological consequences.

When animals or humans are exposed to more than one chemical at a time, interactions may occur. The effects of multiple chemicals may be simply additive, and indeed, this concept has been a working hypothesis for some investigators studying mixtures. Alternatively, there may be synergistic effects of two or more chemicals administered together. That is, observed effects of the mixture exceed the sum of the individual effects of the components. This phenomenon has been reported with ethanol and chloroform exposure. Another possibility also exists: there may be antagonistic effects of two or more chemicals administered together such that observed effects are less than those predicted on the basis of the components. This phenomenon has been reported with ethanol and methanol exposure.

Concerns With Toxicological Data

As noted above, toxicological data are generally obtained in animal studies. Other data have come from accidents (e.g., methyl isocyanate release in India), suicides/case reports (e.g., solvent "huffing" among teenagers), epidemiology studies (e.g., black lung in coal miners), and clinical studies (e.g., reproductive hazards of dibromochloropropane exposure). In evaluating the results of any study, there are several issues to keep in mind.

Sample Size. Many studies have a small sample size. For example, there may be only three or four animals in an exposure group. Thus, if an effect has been reported in 0.01% of the human population, and researchers are trying to reproduce it in animals, it may be undetectable simply because there are too few animals in the study. Also, there are many single case reports in humans. Thus, interpretation of these results may also be difficult.

Control Groups. In a good study there are carefully matched controls. That is, a group of subjects (animals or humans) is not exposed to the agent of interest. Controls permit a proper interpretation to be made when a response is observed in the exposed group. Human studies often have confounding variables (e.g., age, smoking) that make it difficult to assemble a good set of controls. Unfortunately, some investigators who conduct animal studies fail to include control groups. This practice is a known pitfall in toxicology.

Characteristics of the Exposure. In general, animal studies are closely controlled. The exact chemical, the length of exposure, and the dose (or concentration) are reported. This information is essential, particularly if there are serious effects of the exposure and there is an interest in reproducing the study. With human studies (e.g., accidents, suicides/case reports, epidemiology studies, clinical studies), one may not be so lucky. Indeed, exposure assessment data are frequently lacking in such studies, and it may be hard to establish a causal relationship between a measured effect and an exposure.

Occupational Diseases

Toxicological data obtained from animal bioassays are often used to predict the effects of chemicals in workers. However, responses occurring in animals may not always match those of humans (i.e., false positives, false negatives). Thus, errors may be made in the extrapolation process. Moreover, occupational exposures may occur over many decades, whereas animal exposures are generally conducted on the order of hours or days. One must also take into account that workers are exposed to mixtures of chemicals, and that they may be exposed to additional chemicals in the environment and at home. Thus, the determination as to whether an individual worker or a group of workers has developed a chemically related disease may be difficult. Many chemically related conditions and illnesses are characterized by signs and symptoms that parallel those in the normal diseases of life. The problem lies in recognizing and ranking the contribution of each factor, from biological to lifestyle.

While there are overlapping situations that may cause disease-like symptoms, many chemically associated diseases form a pattern or produce clues as to causation. Some of these patterns are shown below.

- The appearance of a disease in a previously healthy individual that is not explained by traditional diagnostic criteria
- A population with common chemical exposures, showing similar symptoms characteristic of a disease
- A particular disease or cause of death that appears to be clustering within a small population with similar chemical exposures
- Any unusual or uncommon disease patterns in a specific group
- Symptoms or diseases occurring in a worker population that are similar to animal results from toxicology tests
- The appearance or disappearance of a disease or set of symptoms temporarily associated with a change of occupation or environmental setting

Whenever one or more of these patterns occurs in an individual worker or in a group of workers, the possibility of finding a chemically associated disease increases. Because of the difficulties in diagnosing occupational diseases, there must be a coordinated effort by occupational health and safety professionals (e.g., occupational hygienists, toxicologists, occupational physicians) to reach a conclusion.

Some examples of occupational diseases are given below.

- A group of agricultural workers is experiencing nausea, vomiting, diarrhea, and lacrimation. These workers may have been overexposed to an organophosphate pesticide such as parathion or dichlorvos.
- A painter has complained of weakness, headache, and abdominal pain that appear to have developed over a long period of time. This worker may have been using leaded paints. In addition, volatile solvents in the paint and paint thinners may exacerbate or mask the symptoms of lead toxicity.
- An employee at a plastics manufacturing plant suddenly has difficulty in breathing. This worker may have become sensitized to an isocyanate such as toluene or diphenylmethane diisocyanate.
- An engineer mechanic has headaches and flushing of the face. This worker may have been overexposed to carbon monoxide or trichloroethylene.

Disposition of Chemicals in the Body

Exposure to a chemical need not result in toxicity. First, a chemical must enter the body before any actions may be exerted. This passage through biological membranes is called absorption (or uptake). Once a chemical has entered the body, it may need to be transported to other sites to induce toxic effects. This process is distribution. Normal bodily functions then attempt to remove foreign materials even when they reach various organ systems. Biotransformation (or metabolism) occurs, which is typically followed by elimination (or excretion). Figure 4.2 summarizes these steps.

Step 1: Absorption of Chemicals

In environmental and occupational settings, the two principal modes of exposure are inhalation (IH) and dermal (D). To a lesser extent, ingestion (IG) of chemicals may occur. Thus, the lung (respiratory tract), skin, and/or gut may serve as routes of entry or portals of entry. When experiments are conducted with laboratory animals, other modes of exposure may also be used, including infusion (IV), intraperitoneal injection (IP), subcutaneous injection

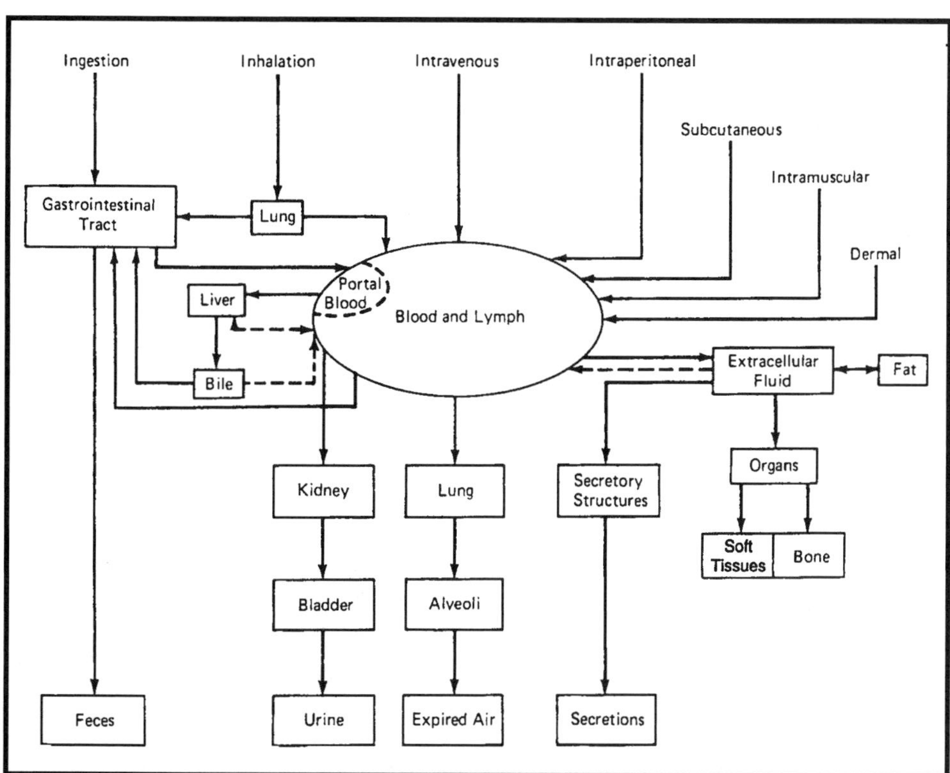

Figure 4.2 — The disposition of chemicals entering the body (from C.D. Klaassen, *Casarett and Doull's Toxicology*, 5th ed., New York: McGraw-Hill, 1996; reproduced with permission).

(SC), intramuscular injection (IM), and intradermal injection (ID). The relative rate of absorption has been estimated as follows.

IV > IH > IP > SC > IM > ID > IG >D

Inhalation exposures are ubiquitous in environmental and occupational settings. Gases, vapors, and/or aerosols may be involved. Examples of gases include carbon monoxide (CO), hydrogen cyanide (HCN), and phosgene. In the workplace, solvents are used in a variety of operations. Depending on their vapor pressure and the temperature, volatilization may occur, thus creating a potential for inhalation exposures of workers. Acetonitrile, methylene chloride, and trichloroethylene are only a few of the thousands of available solvents whose vapors may pose health hazards in the workplace. Examples of aerosols include cotton dust, sulfuric acid mist, welding fumes, and oil mists. These particles or droplets, which are suspended in the air, may be inhaled.

Regardless of the nature of an inhalation exposure, chemicals enter the respiratory tract through the nose and/or mouth and may pass through the branching network of airways, ultimately reaching the alveolar region (or deep lung).[1] As seen in Figure 4.3, the major divisions of the respiratory tract are the naso-pharyngeal region, the tracheal-bronchial region, and the pulmonary compartment. Once a chemical reaches the P compartment, it is approximately 0.5 micron (µm) from the pulmonary circulation. Once diffusion occurs through this blood-gas barrier, the chemical may be transported throughout the body.

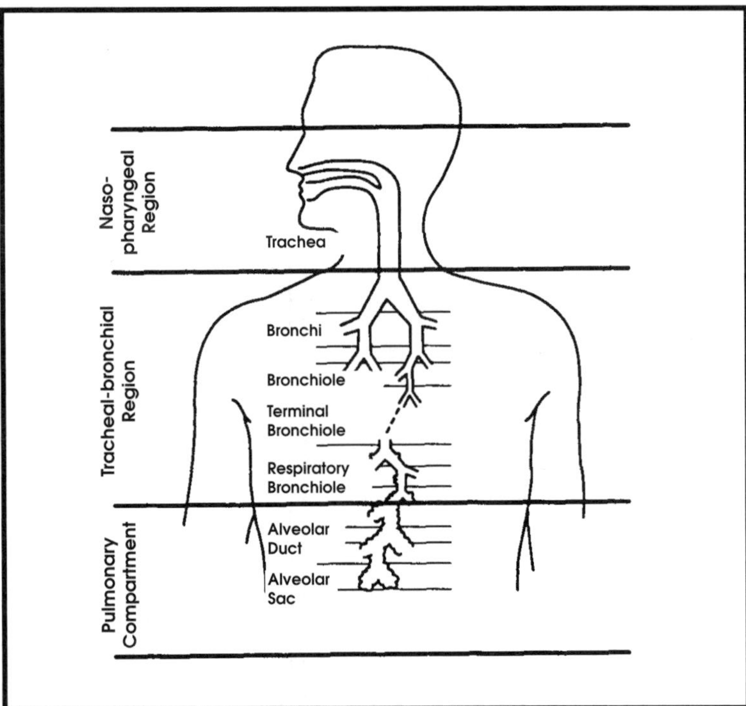

Figure 4.3 — The major divisions of the respiratory tract (adapted from J.B. West, *Respiratory Physiology*, 4th ed. Baltimore, MD: Williams and Wilkins, 1990).

Dermal (or percutaneous) absorption may involve solids, liquids, or gases. Examples of solids that may pass through the skin include ointments (e.g., zinc oxide) and metals (e.g., mercury, tetraethyl lead). Liquids such as malathion, triorthocresol phosphate, and salt solutions of cyanide (e.g., calcium cyanide, potassium cyanide, sodium cyanide) may also penetrate the skin. As seen with inhalation exposures, numerous solvents (in the liquid state) and their vapors (e.g., methyl n-butyl ketone, methanol, toluene) may be absorbed percutaneously. The American Conference of Governmental Industrial Hygienists (ACGIH) has recognized the importance of dermal absorption in establishing Threshold Limit Values (TLVs®) and has given the SKIN designation to many chemicals listed in the TLV® booklet.[6] This special designation emphasizes the role of the skin in contributing to overall exposures to chemicals and does not denote toxic effects on the skin. It may be noted that acids or bases that may produce burns do not have the SKIN designation. Therefore, this designation has often been a confusing point for students.

As a general rule, when substances are attracted to lipids (fats) rather than to water (i.e., lipophilic substances), the more readily they are absorbed through the skin. This rule makes sense considering that the skin is itself a phospholipid bilayer and, thus, "like dissolves like." One can easily understand why so many solvents and their lipophilic vapors have the SKIN designation. After passing through the epidermis (outermost layer of the skin), such chemicals pass through the dermis to reach blood vessels, where they are transported to other parts of the body. Because of the multiple skin layers through which a chemical must pass to gain access to the circulation, dermal absorption is slower than any other form of absorption.

A variety of factors, including molecular size, molecular lipophilicity, and the age/health status of the exposed subject may influence respiratory or dermal uptake of a chemical. For example, a baby or young child may absorb and respond to a toxicant that would not affect others in the general population. Age-related differences in biotransformation could exist and thus account for such a finding. Also, when disease is present in an individual, the effects of a toxicant may be magnified. Generally speaking, smaller and lipophilic chemicals enter the body more readily than larger ones. The advantages of smaller and lipophilic molecules are that they can diffuse through cellular membranes and tend to fit into receptors within the body. For example, some pesticides that are cutaneously absorbed may act on receptors at the neuromuscular junction, mimicking endogenous neurotransmitters (e.g., acetylcholine).

Step 2: Transport of Chemicals

Once chemicals have been absorbed into the blood stream, they may bind to plasma proteins (e.g., albumin, ceruloplasmin, transferrin). Whether they are bound or unbound to proteins, chemicals may then be moved from the portal of entry and distributed to other, perhaps distal, organ systems. With many chemicals, the distribution process is essential if there is to be toxicity. Also, the portal of entry need not be the ultimate target system. For example, inhaled hexane vapors may reach the lungs, but must be distributed to the brain and nervous system before neuropathy will occur. Likewise, vinyl chloride vapors must be transported from the lungs to initiate the production of tumors in the liver.

Two of the factors that greatly affect the distribution of a chemical are its affinity in an organ system and the perfusion through that organ system. Tissues that are well-perfused, such as those within the brain, kidney, and liver, are more likely to be exposed to circulating chemicals than those that are not well-perfused. The liver, like the skin, is also highly lipophilic and thus has an affinity for many organic chemicals. For these reasons, not surprisingly, the liver is involved in many toxic reactions, irrespective of the route of exposure.

In some cases, the tissues in which a chemical has been distributed may also serve as storage for them. Metals, for example, may be stored in the liver or kidney (e.g., cadmium, zinc) or in bones (e.g., fluoride, lead, strontium). With organochlorine compounds (e.g., DDT, polychlorinated biphenyls or PCBs), there may be long-term storage in the fat depots of the body. The release of such metals or pesticides back into the bloodstream occurs slowly, and thus, complete elimination is also slow.

There are several special barriers to chemical distribution that merit discussion here. Two of these are the blood-brain barrier and the placenta (mother-fetus barrier). The blood-brain barrier has tight junctions between capillary epithelial cells and a low protein content; it is less permeable than other cellular membranes. For physiological purposes, oxygen and carbon dioxide must be able to diffuse into and out of the brain. In addition to these vital gases, some of the more noteworthy toxicants that are also known to cross the blood-brain barrier are ethanol, methyl mercury, and opiates; once across, they have gained access to the nervous system. Thus, adverse effects may occur within the brain itself, in the central nervous system, or in the peripheral nervous system. In contrast to the blood-brain barrier, the placental barrier allows many more chemicals to pass from the mother to the developing embryo or fetus. However, at one time, the placenta was thought to be effective in limiting the access of chemicals to in utero babies. However, developmental and reproductive toxicologists have widely demonstrated that both biological as well as chemical agents may pass through the placenta and may adversely affect the developing embryo or fetus. Several examples of chemicals that are known to pass from the mother to the developing embryo or fetus are ethanol, opiates, and thalidomide.

Step 3: Biotransformation of Chemicals

Biotransformation (or metabolism) is the process through which toxicants are chemically converted, generally reducing their lipophilicity and increasing their hydrophilicity. By creating less toxic and more water-soluble metabolites, this process typically enhances elimination of toxicants from the body. However, there are some chemicals that, when biotransformed, yield more toxic intermediates or end-products than their parent compounds. Benzene, for example, is ultimately metabolized to phenol, but numerous intermediates are formed during the biotransformation process. Indeed, researchers believe that one or more of such products is responsible for the hemopoietic effects (i.e., leukemia) associated with benzene exposure.

Although most, if not all, organ systems have some ability to metabolize toxicants, the liver is the most important in this regard. Enzymes (e.g., cytochrome P450s) that are found in the liver are used to catalyze conversion reactions. Biotransformation is generally a two-step process, occurring via Phase I and Phase II reactions. In Phase I reactions (Figure 4.4), there is oxidation, reduction, and/or hydrolysis of toxicants that serves to add or expose their functional groups. In Phase II (Figure 4.5), the products of Phase I reactions are conjugated to an endogenous molecule to facilitate elimination. Common conjugates are glutathione *S*- transferases and glucuronic acid.

As in absorption and distribution, there are many factors that will affect the metabolism of chemicals. First, properties of the chemical itself (e.g., affinity for the enzyme system, binding to proteins, lipophilicity, storage) will influence the degree to which it may be metabolized. As mentioned above, chemicals that are bound to plasma proteins or stored in particular body sites will not be readily available for Phase I or Phase II reactions. They are slowly released into the blood (i.e., unbound) before such reactions can ensue. Secondly, characteristics of the exposed subject (human or animal) cannot be underestimated. Age, gender, health status, and other genetic-based information should be considered. Researchers have reported that some individuals lack specific enzymes or fail to produce sufficient quantities of them. If, for example, the enzyme alcohol dehydrogenase is lacking, then an individual cannot metabolize ethanol properly. Likewise, if there is a lactase deficiency, the individual will be unable to digest dairy products. Finally, the importance of enzyme systems in metabolic processes has been emphasized. Specificity, competition, saturation, and inductibility of enzymes will all influence differences that can be found between individuals. Research has suggested that toxicity may result once an enzyme has become saturated and can no longer act on a chemical to detoxify it. If, for example, a chemical has been distributed to the liver but can no longer be metabolized as a result of enzyme saturation, one can easily envision how hepatic damage might follow.

Enzyme (E) + Nonpolar Toxicant (RH) →

Enzyme-Toxicant Complex (E-RH) →

Enzyme (E) + Polar Metabolite (ROH)

Figure 4.4 — Generic model of enzymatic catalyzed biotransformation.

Generic Phase I Reactions:

Oxidation: E + RH → E-RH → E + ROH
Reduction: E + ROH → E-ROH → E + RH
Hydrolysis: E + RO(CO)R → E-RO(CO)R → E + ROH + RCOOH

Generic Phase II Reactions:

Phase I Metabolite + Conjugating Agent + Enzyme →
Conjugated Phase II Metabolite

Figure 4.5 — Generic models of Phase I and Phase II biotransformation.

Although most metabolites are reflective of detoxification processes, some biotransformation reactions form active metabolites that are more toxic than the original parent compound absorbed into the body. Protoxic chemicals refer to original or parent compounds that require biotransformation to an active metabolite following absorption and internal exposure. Many or most cancer-causing chemicals or carcinogens must be biotransformed to an active or ultimate carcinogen, in which case the parent compounds are called procarcinogens.

Step 4: Elimination of Chemicals

Elimination (or excretion) is the process through which chemicals and/or their metabolites are removed from the body. The major routes of elimination are biliary (e.g., liver) and renal (e.g., kidney). These pathways are not surprising given that so many chemicals are biotransformed in the liver and may be excreted with bile and ultimately eliminated from the body in feces. Also, because of the decrease in lipophilicity and increase in hydrophilicity, not surprisingly the urinary excretion of metabolites of toxicants is very common. As discussed below, the uptake and fate of a chemical may be monitored using urine samples.

Some chemicals are able to pass through the blood-gas barrier (e.g., ethanol, methylene chloride, toluene) and may simply be exhaled. These latter chemicals need not be biotransformed for this to occur. Ethanol, for example, may be detected on the breath of someone who has been drinking. Because of the partitioning of alcohol between blood and air, exhaled air samples have been used to determine blood alcohol levels. Likewise, vapors from solvents may move from blood into the lung and may be eliminated on exhalation.

Other forms of elimination include milk, sweat, saliva, hair, and nails. Milk is an important concern for several reasons. First, a lactating mother may pass toxicants and/or their metabolites to her offspring. Secondly, people may be exposed to chemicals and/or their metabolites in dairy products. Cattle, for example, may ingest grain that has pesticide residues, and these residues may end up in their milk.

Some of the factors that affect elimination have been discussed above in the absorption, distribution, and biotransformation steps. Age, health status, molecular size, and protein binding are just a few examples. Chemicals with smaller molecular size (< 60,000 molecular weight) may be filtered through the glomerular apparatus of the kidney and are excreted in the urine. In contrast, larger chemicals are unable to pass through the kidney and may be excreted in bile.

Use of Biological Monitoring and Biomarkers

Occupational hygienists typically measure airborne concentrations of chemicals and use the data to determine whether there is a potential for adverse human health effects to occur. Occupational hygienists may also assess exposures to chemicals by monitoring workers themselves, as internal dosimeters. As stated by ACGIH, biological monitoring should be complementary to air monitoring.[6]

Biological monitoring can be performed using specimens such as air, blood, or urine (i.e., determinants) and the biomarkers of interest (chemicals and/or their metabolites) are quantitated.[7–9] End-exhaled air is often a determinant, since it is collected from the deep lung and thus represents blood levels. Whole-blood, serum, plasma, or erythrocytes may be used when analyzing blood samples. Knowledge of the absorption, distribution, biotransformation, and elimination of chemicals is essential to determine appropriate biomarkers and to properly conduct such monitoring in workers.

ACGIH has developed Biological Exposure Indices (BEIs®) for approximately 40 industrial chemicals, and there are approximately 10 additional chemicals under study by ACGIH for which BEIs may be developed.[6] BEIs are defined as reference values that are intended as guidelines to evaluate potential health hazards. When such values are exceeded, there may be increased health risks for exposed workers. Also, additional workplace evaluation and perhaps controls (e.g., ventilation, personal protective equipment) may be warranted.

For example, styrene exposure may be assessed via measurements of its concentration in the workplace. The current TLV for styrene is 50 ppm, and it has the ACGIH SKIN designation[4] as described earlier in this chapter. To complement the air monitoring (1) styrene in venous blood, (2) mandelic acid in urine (its principal metabolite), and/or (3) phenylglyoxylic acid in urine can also be monitored in workers. The BEIs are expressed in milligrams per gram of creatinine (urine samples) or in milligrams per liter (blood samples). Creatinine is a major end-product of normal metabolism within the body, and it is excreted in the urine. Because there may be variation in urine volume (output) when conducting biological monitoring, the concentrations of some determinants (e.g., mandelic acid, phenylglyoxylic acid) are given in relation to that of creatinine. In essence, creatinine serves as an internal standard. For each determinant the appropriate sampling times are also specified (e.g., end of shift, prior to next shift). An example is given in Table 4.1.[6]

As noted in the far right column of Table 4.1, ACGIH also provides several notations with BEIs. They are Sc (susceptibility), B (background), Ns (nonspecific), and Sq (semiquantitative).[4] In the above example, mandelic acid and phenylglyoxylic acid measurements in urine are considered Ns for styrene exposure. These metabolites may also be found in workers consuming ethanol or in those exposed to other solvents such as acetophenone, ethylbenzene, or styrene glycol, and thus, they are not specific for styrene exposure.[10] Styrene in mixed venous blood may be a good indicator of exposure to styrene, but it may not reflect the original air concentration to which the worker was exposed. As emphasized in the previous section on chemical disposition, metabolism will occur, and this will greatly affect the amount of styrene that remains in the blood. Also,

Table 4.1 —
Monitoring for Styrene in the Workplace 1996-1997 TLV 50 ppm (SKIN)

Determinant	*Sampling Time*	*BEI*	*Notation*
Mandelic acid (urine)	end of shift	800 mg/g creatinine	Ns
	prior to next shift	300 mg/g creatinine	Ns
Phenylglyoxylic acid (urine)	end of shift	240 mg/g creatinine	Ns
	prior to next shift	100 mg/g creatinine	Ns
Styrene (mixed venous blood)	end of shift	0.55 mg/L	Sq
	prior to next shift	0.02 mg/L	Sq

blood levels of styrene will vary from individual to individual and will vary with time. Thus, as stated by ACGIH, the quantitative interpretation of this measurement should be considered ambiguous. Further discussion of these notations may be found in the ACGIH TLV booklet.[6]

Target Systems

Assuming that a sufficient amount of a chemical has reached a particular target site and has not been detoxified via biotransformation, toxic substances can ultimately alter a system and, in turn, the overall health of an exposed individual. As discussed earlier, adverse or toxic effects can involve a variety of possible systems, including the lungs (pneumotoxicity), blood (hematotoxicity), immune system (immunotoxicity), liver (hepatotoxicity), kidneys (nephrotoxicity), nervous system (neurotoxicity), and embryos or fetuses (teratogenicity).

A descending list of the components of a system are organs, tissues, cells, and biochemicals. Indeed, the actual toxic interaction with a system begins at the molecular and cellular levels. Alteration of the structure, function, and/or synthesis of the major endogenous biochemicals classified as proteins, lipids, carbohydrates, and nucleic acids can affect cellular structure, function, and synthesis. Alteration at these molecular and cellular levels, if not corrected, reversed, or compensated for, can consequently alter the structure of tissues and function of organs and entire systems.

Nongenotoxic chemicals interact with endogenous biochemicals other than the nucleic acid component of genes, deoxyribonucleic acid (DNA). The nongenotoxic chemicals can interact with macromolecules such as proteins and lipids causing biochemical alterations leading to adverse physiological responses. For example, disruption of lipid metabolism and synthesis can disrupt important cellular constituents involved in the structure of cell membranes and many cellular and metabolic processes. Interference with lipid metabolism can cause lipid accumulation in the liver and disruption of the synthesis of cell membranes. Disruption of cell membranes can alter transport systems that control transfer of substances in and out of cells. Disrupted synthesis of proteins can adversely affect levels of proteinaceous antibodies and enzymes. Alternatively, some nongenotoxic substances interact by increasing the magnitude of toxic interaction when combined with another substance. In all of these examples of nongenotoxicity, direct alteration or disruption of nucleic acids such as DNA does not occur. Toxic substances that interact directly with DNA are called genotoxic.

Genotoxic chemicals are electrophilic (i.e., electron-deficient) compounds that have an affinity for genetic information, specifically the electron-dense (i.e., nucleophilic) DNA. Damage to the DNA, specifically a nitrogenous base on a nucleotide, can subsequently result in a mutation. DNA is altered, subsequently leading to mutation, via several possible mechanisms following exposure to a mutagenic chemical. Formation of DNA adducts involves covalent bonding of an electrophilic compound with nucleophilic DNA, forming an adduct. Adduct formation increases the probability of (1) mispairing of nitrogenous bases, (2) inhibiting a nitrogenous base to pair at all, or (3) causing breakage of a DNA strand. If the damaged or altered DNA is repaired, a mutation can be avoided. Mutated DNA can be expressed in the progeny cells following cell division.

The manifestations of the mutagenic events are somatic and genetic. Somatic mutations are characterized by manifestation of the damage in the exposed individual due to alteration of somatic or body cells. Genetic mutations, however, involve expression of the aberration in the offspring of the exposed individual due to alteration of germ or sex cells (male spermatozoa and/or female ova).

Although cancer can be caused by either nongenotoxic or genotoxic mechanisms, most cancer-causing or carcinogenic events are a result of genotoxic interaction of a carcinogen with DNA. Carcinogenesis refers to development of malignant tumors or neoplasms composed of abnormal cells exhibiting uncontrolled growth, invasiveness, and metastasis. The genotoxic mechanism involves exposure of cellular DNA to an electrophilic carcinogen. The formation of the DNA-carcinogen adduct is the first step of carcinogenesis called initiation. Subsequently, the mutated precancerous cell can divide and proliferate, via a step called promotion, and is characterized by expression of the mutation in the progeny cells. Finally, progression of the mutation and cell proliferation results in the formation of a malignant or cancerous neoplasm or tumor.

There are a variety of nongenotoxic and genotoxic effects that may occur. Some of the major target systems in which toxicity is seen are reviewed below and in books listed in the Additional Sources section.

Respiratory System

The major function of the lungs is gas exchange. Additionally, this organ system is involved with biotransformation of some chemicals. The lungs receive the entire cardiac output from the right heart via the pulmonary artery. Oxygenated blood then leaves the lungs and returns to the left heart via the pulmonary vein. The lungs serve as a filter for blood, thereby removing clots that could be life-threatening if they were to reach other vital organs such as the heart or brain.

As shown in Figure 4.3, the respiratory tract is divided into the naso-pharyngeal region, the tracheal-bronchial region, and the pulmonary compartment. Inspired materials that enter the naso-pharyngeal region may pass through the tracheal-bronchial region (i.e., bronchi, bronchioles). Excluding the respiratory bronchioles, these upper-mid airways represent the conducting zone of the lung, and no gas exchange occurs here. The respiratory bronchioles and the pulmonary compartment (i.e., alveolar ducts and alveolar sacs) comprise the gas exchange zone at which diffusion occurs. Going from the naso-pharyngeal region into the pulmonary compartment of the lung, the diameters of airways become smaller, air velocity diminishes, and the surface area increases dramatically. These morphometric features of the lung make it ideal for gas diffusion at the respiratory bronchioles and pulmonary compartment.[11]

For particulates entering the respiratory tract, some of factors that affect deposition include particle size, shape, and density.[12,13] For example, particles that are 10 µm or above will be deposited in the naso-pharyngeal region of the respiratory tract. Particles that are 1–10 µm are deposited throughout the respiratory tract, but as the size decreases toward 1 µm, more of these particles can reach the tracheal-bronchial region and pulmonary compartment. For 0.5 µm particles there is minimal probability of deposition in any region of the respiratory tract. However, as the size diminishes below 0.5 µm, the probability of deposition again increases, particularly in the pulmonary compartment and to a lesser extent in the tracheal-bronchial region. The three principal mechanisms for particle deposition are sedimentation, impaction, and diffusion.[12,13] Of lesser importance are electrostatic attraction and interception. Sedimentation is due to the force of gravity, which causes particles to settle on lower surfaces of the airways. All particles with a density greater than air experience a downward force because of gravity. Particles that are above 1 µm are likely to sediment in the respiratory tract. With impaction (or inertia), particles that are traveling along a straight line will be deposited on any surfaces that serve as obstacles. Impaction is influenced by air velocity, density of the particle, and most importantly, by the square root of the particle diameter. Particles that are 3 µm or above may impact in the naso-pharyngeal region and in the tracheal-bronchial region of the respiratory tract. With diffusion (or Brownian movement), there is random collision of particles with each other or with any surface. Particles under 0.5 µm are affected in this manner. Electrostatic forces may be important if the particles have a charge, and interception pertains to fibers coming in close contact with the wall of the respiratory tract.

Within the conducting zone of the lung, there is cartilage in the trachea and then smooth muscle in the bronchi. Cilia and goblet cells are also found in the bronchial region of the lung. Mucus is secreted from the goblet cells, and the cilia beat rhythmically to move foreign materials upward (toward the throat) from this region of the lung. The mucus may then be swallowed or coughed.[14] In many texts this has been described as the mucociliary escalator.

Within the gas exchange region, there are several important types of cells. Type I cells are epithelial cells along the surface of the alveoli. They may be destroyed by a variety of pneumotoxicants. Type II cells are responsible for the production of surfactant, the lipoprotein that lines the airways and mechanically stabilizes them, thus preventing their collapse. There are also pulmonary macrophages in this region of the lung, which are large wandering cells that engulf foreign materials and remove them via the blood or lymphatics.

Some of the important types of receptors/nerve endings that are involved in mediating responses to inhaled materials are summarized in Table 4.2.[15]

When chemicals act on trigeminal nerve endings in the eyes and upper airways (e.g., nose, throat), sensory irritation occurs. This is one of the most common effects produced by industrial chemicals. Humans exposed to sensory irritants experience a painful burning sensation, which if severe will make it intolerable to work or to remain in a given environment. Good examples of sensory irritants are acrolein, formaldehyde, and 2,4-toluene diisocyanate.[2,15] There are many

Table 4.2 —
Receptors Involved in Mediating Responses to Airborne Chemicals

Receptor/Nerve Endings	Location	Principal Response
Trigeminal	eyes, nose, throat	burning sensation
Laryngeal	throat	cough
Stretch, irritant	bronchi	airways constriction
Type J, C-fiber	alveoli	rapid, shallow breathing apnea

chemicals whose TLVs have been set to protect workers from this effect.[10] In general, most chemicals are capable of producing sensory irritation, but as noted above, the level of sensory irritation will depend on the exposure concentration and on the duration of exposure. With exposure to only small concentrations (e.g., 1–5 ppm), humans will respond rapidly to the sensory irritating effects of acrolein or formaldehyde. In contrast, similar concentrations of 2,4-toluene diisocyanate will also evoke this type of respiratory response, but the response occurs more slowly than that seen with acrolein or formaldehyde. There have also been a large number of studies conducted in mice that have attempted to identify sensory irritants, to estimate their potency, and to suggest occupational exposure limits for such chemicals. Indeed, a good correlation has been demonstrated between the responses in mice (i.e., RD50s) and the ACGIH TLVs for some 89 chemicals.[2]

Other chemicals may produce stimulation of laryngeal nerve endings, resulting in coughing. Two examples of chemicals evoking this type of reaction are sodium lauryl sulfate and citric acid. Many sensory irritants are also capable of producing coughing. Studies have been conducted using guinea pigs to study the cough reflex.[16]

In the midairways, bronchoconstriction may occur on exposure to chemicals. This reaction may occur directly (e.g., aerosols of carbamylcholine, histamine, or sulfuric acid) or may occur following sensitization (see Immune System, below) to industrial chemicals (e.g., toluene diisocyanate, trimellitic anhydride). In either case, the individual experiences a narrowing of the airways (i.e., increased pulmonary resistance, decreased lung compliance) that makes it difficult to breathe. The constriction may become so intense that the individual collapses and medical intervention is needed. More importantly, with workers who have become sensitized to such chemicals and experience this reaction, it may be necessary to transfer them from a particular job. A large number of studies have been conducted in which guinea pigs have been exposed to bronchoconstrictors.[17] Additionally, investigators have attempted to characterize the immunological responses to known sensitizers.[18–20]

In the lower respiratory tract, pulmonary irritation may be evoked.[6] Examples of chemicals that are pure pulmonary irritants are ozone and phosgene. Humans exposed to these chemicals do not experience burning of the eyes, nose, and throat. In fact, there may be no awareness that an exposure is occurring or has occurred. This type of irritant is capable of passing into the deep lung, where it may cause congestion, hemorrhage, and edema. The Type I cells may be destroyed on exposure to pulmonary irritants, and there may be a proliferation of Type II cells. Rapid, shallow breathing and apnea (i.e., pausing between breaths) may be seen in humans (or animals) that have been exposed to pulmonary irritants. However, if fluid enters the airspace (i.e., alveolar edema), gas exchange is impaired and death may occur. Thus, from an environmental and occupational standpoint, such agents are of much greater concern than are sensory irritants or bronchoconstrictors. Also in recent times, ozone has received a great deal of attention for environmental reasons. In large cities such as Los Angeles, Calif., ozone concentrations are routinely monitored, and ozone alerts are broadcast on the radio and television. Additionally, many animal studies have focused on exposures to ozone.[20–22]

The above respiratory reactions are acute, and there is a possible recovery from each reaction, if it is not severe. Some of the chronic respiratory responses to chemicals include airway obstruction, lung restriction, and lung cancer.[1] These conditions can be rapidly debilitating and may lead to premature death. With chronic obstructive pulmonary disease, the airways are occluded but the effect is permanent. Such occlusion may result from airway secretions, narrowing of the airways, and/or a loss of airway recoil. In contrast, with restrictive lung disease, the lungs are stiffer than normal. In some reports this effect is called fibrosis. Pure obstructive or restrictive lung disease can be differentiated through pulmonary function tests. However, in some diseased individuals, there are features of both obstruction and restriction. Irrespective of the type of disease, the individual will have difficulty in breathing, and this will be magnified by work or exercise.[23] Thus, some workers may be unable to perform their jobs adequately under such conditions.

Pneumoconiosis is a term that has been used to describe the accumulation of dust in the lungs and the tissue reaction to its presence. This condition is usually caused by inert dust or similar chemicals of low toxicity. Nuisance particles such as carbon black or iron oxide cause minor localized reactions that do not cause respiratory symptoms and are usually cleared from the lungs over time. This condition is referred to as benign pneumoconiosis and has been seen in foundry workers, stone processors, sandblasters, miners, and glass/ceramic workers.

Inert dust that has adsorbed toxic chemicals or other inorganic dusts stimulate a more widespread response in the lungs and can lead to irreversible changes including obstruction, restriction, or even cancer. The most important chemicals in this group include asbestos, coal dust, and silica. Asbestosis, a restrictive lung disease, is produced by inhaling large quantities of asbestos fibers of a particular size and shape. Evidence of this disease does not appear until 20 or more years after exposure. Coal worker's pneumoconiosis (miner's black lung) is an obstructive lung disease. It has been seen in carbon black, graphite, and carbon electrode workers. Again, there is a long latency period (i.e., 20–30 years) until this disease is seen. While the lung has a great capacity for particle removal, moderate overexposure to silica for a period of 20–45 years can progress to the development of silicosis. This is also a restrictive lung disease.

For occupational hygienists, it is important to control dust levels and thus minimize worker exposures. With this approach, the long-term respiratory effects described above can be avoided. Also, medical surveillance is of importance for workers exposed to dusts. This allows the early stages of pneumoconiosis to be detected and treated.

Lung cancer is among the leading causes of death in the United States. Although many such deaths have been associated with tobacco smoking, there are other environmental and occupational agents that are known to produce lung cancer. Additionally, there may be chemical interactions between agents used in the workplace and those found in cigarette smoke.[24] Like obstructive and restrictive lung disease, cancer is typically seen after many years of exposure and after a long latency period (e.g., 20–30 years). ACGIH has provided special cancer designations (A1, A2, A3, A4, A5) for chemicals listed in the TLV booklet.[6] The numbering system (1–5) is related to the weight of evidence for cancer causation. Chemicals having the A1 designation are considered "Confirmed Human Carcinogens," and examples include asbestos, bis(chloromethyl) ether, and chromium VI compounds. Chemicals having the A2 designation are considered "Suspected Human Carcinogens," and examples include acrylonitrile, ethylene oxide, and lead chromates. "Animal Carcinogens" are denoted by the A3 designation. For A3 chemicals, available epidemiologic data do not confirm an increased risk of cancer in exposed workers. Aniline, methylene chloride, and pentachlorophenol are in this category of carcinogens. When there are inadequate human and/or animal data on the carcinogenicity of a chemical, it is given the A4 designation. Acetonitrile, caprolactam, and methyl methacrylate carry the A4 designation. Finally, when there are no data from properly conducted epidemiology studies to suggest that a chemical is carcinogenic, the A5 designation is applied. Trichloroethylene, for example, is "Not Suspected as a Human Carcinogen."[6]

Circulatory System

The major functions of the blood are to provide nutrients to the tissues and to remove metabolic waste products from the tissues. Environmental and occupational toxicants (and their possible metabolites) are also distributed to and from tissues by the circulation. At the cellular level, the blood assists in the regulation of electrolytic (e.g, potassium ions, sodium ions) and osmotic balance. Additionally, clotting factors and cells (e.g., leukocytes) that are integral in the defense mechanisms of the body are found in the circulation.

Blood vessels, like airways, form a branching network characterized by changes in diameter of the vessels throughout the body. The largest vessel that carries oxygenated blood is the aorta, which exits from the left heart. The aorta then divides into arteries, arterioles, and capillaries, which travel away from the heart. Through these vessels, oxygen and other nutrients can be delivered to the peripheral tissues. Venules and veins return deoxygenated blood to the vena cava, which leads back to the right heart. The circulatory system is affected by a number of neurotransmitters, including norepinephrine and acetylcholine. When released, these agents may act on receptors/nerve endings in the blood vessels, causing vasoconstriction or vasodilation. Thus, changes in blood flow may occur at peripheral tissues.

Adults have approximately 5 L of blood, which constitutes 7% of body weight. Of this volume, 55% is water (i.e., plasma), and 45% is cellular. Cells that are suspended within the plasma include erythrocytes (i.e., red blood cells), platelets, and leukocytes (i.e., white blood cells). Leukocytes may be subclassified as phagocytes (e.g., neutrophils, eosinophils, basophils, monocytes) and immunocytes (e.g., T cells, B cells). Further discussion of the importance of each of these cell types is given in the next section (see Immune System, below). Stem cells, which are precursors of these cellular components of the blood, are formed in bone marrow. The stem cells undergo maturation and differentiation, eventually producing the erythrocytes, platelets, and leukocytes.

As noted above, one of the cardinal functions of the circulation is to provide nutrients to the tissues. Oxygen, which is needed to sustain human and animal life, is carried in the blood. Although a small amount is physically dissolved in the plasma (e.g., 0.3 mL per 100 mL of blood when the partial pressure of oxygen (pO_2) is 100 mmHg), the majority of oxygen is transported with the oligomeric protein hemoglobin (Hb). Normal adult hemoglobin is composed of a heme group containing ferrous iron (i.e., Fe^{++}) at its center and the protein, globin, which has four polypeptide chains (2α, 2β). Oxygen, which diffuses across the blood- gas barrier, passes into red blood cells where it forms a bond with the heme groups in hemoglobin. Oxygen is then released at the peripheral tissues. The normal oxygen level in inspired air is 20 to 21% (v/v) or 150 mm Hg (pO_2). It is slightly diluted on passing into the alveolar region, and the pO_2 drops to approximately 100 mm Hg. In the pulmonary artery, which carries mixed-venous blood (i.e., deoxygenated) into the lung from the peripheral tissues, the pO_2 is approximately 40 mm Hg. Because of this pressure gradient (i.e., 60 mm Hg), diffusion of oxygen from the alveoli to the red blood cells is highly favored. Similarly, there is a pressure gradient between peripheral blood vessels and tissues, enhancing the movement of oxygen into the cells.

When there is an insufficient amount of oxygen delivered to the tissues, this is called hypoxia. An insufficient amount of oxygen in arterial blood is called hypoxemia. When this condition becomes life-threatening, it may be called asphyxia. In environmental and occupational settings, hypoxia/asphyxia may be produced acutely by a number of agents. For example, simple asphyxiants such as acetylene, argon, and methane displace oxygen from the inspired air resulting in an inadequate level of oxygen diffusing into the blood (i.e., arterial hypoxemia). In general, these gases produce no other significant physiological effects in humans. Within confined spaces, oxygen deprivation may be rapidly fatal, and for this reason special sensors may be used for continuously monitoring oxygen levels. In the TLV booklet, ACGIH has provided a special footnote ("c") when listing the simple asphyxiants.[6] In addition to the health-related concerns, ACGIH emphasizes that some simple asphyxiants also introduce safety hazards. For example, there may be fires or explosions at increasing concentrations of acetylene or methane. For the simple asphyxiants, ACGIH has recommended that the minimum oxygen level should be 18% (v/v) or 135 mm Hg (pO_2).

A number of chemicals that are encountered environmentally and occupationally may also react with hemoglobin in the red blood cells, thereby producing hypoxia at the tissues. Carbon monoxide (CO) is a colorless and odorless gas that is generated as a result of incomplete combustion. It is a common air pollutant that is emitted from automobiles, heaters, cigarettes, and fires.[25] This gas competes with oxygen for the hemoglobin in red blood cells and carboxyhemoglobin (COHb) is formed instead of oxyhemoglobin. The affinity of hemoglobin for CO is nearly 210 times that for oxygen. Thus, CO avidly binds with hemoglobin, rendering less available hemoglobin for binding with oxygen and creating a lack of oxygen at the tissues. This condition has been described as anemic hypoxia. In essence, the victim of CO poisoning has the same problem as the anemic patient. There is an inability to transport the proper amount of oxygen, even if the inspired air has 20 to 21% oxygen. ACGIH has established a TLV (25 ppm) and BEIs (COHb, CO in end-exhaled air) for this asphyxiant.[6] At a concentration of 25 ppm, the COHb level would not exceed 4%, and no significant adverse health effects would be expected in normal healthy adult workers.[10]

Other chemicals may also produce COHb, most notably methylene chloride. While this chlorinated hydrocarbon is similar to others in its class such as tri- and tetrachloroethylene, its metabolism is unique in that CO is generated, and thus COHb may be formed. Metabolic differences also appear to be important with respect to the carcinogenicity of this solvent.

Methylene chloride has been shown to induce cancer in mice (and in other species), yet not in humans as found in various epidemiologic studies. Because mice metabolize methylene chloride differently than do humans, this finding appears to account for the lack of a carcinogenic effect in humans.

Another acute reaction that may occur within the red blood cells involves the oxidation of heme (from Fe^{++} to Fe^{+++}). This is known as methemoglobinemia; examples of agents that may produce this effect include aniline, nitrobenzene, and nitrites.[6,10] Oxygen cannot reversibly bind to methemoglobin and again, humans will experience symptoms of hypoxia. This condition has also been compared to anemia.

The term sulfhemoglobin has been used to describe partially denatured hemoglobins that are often part of a condition involving Heinz bodies and hemolysis. Some chemicals that are associated with this condition are arsine, chlorates, methylene blue, naphthalene, and phenylhydrazine. These effects are most noted in glucose-6-phosphate dehydrogenase deficient cells.

At the tissues, it is possible for toxicants to interfere with oxygen utilization. This condition has been termed cytotoxic hypoxia or histotoxic hypoxia. Examples of chemicals that are recognized for this acute effect are acetonitrile (CH_3CN), cyanide salts (calcium cyanide, potassium cyanide, sodium cyanide), hydrogen cyanide (HCN), and hydrogen sulfide (H_2S). The most noteworthy compounds are hydrogen cyanide and the cyanide salts, which may be found in electroplating, dying, and tanning operations. Additionally, hydrogen cyanide may be generated as a result of incomplete combustion of nitrogen-containing polymers.[25] With histotoxic hypoxia, the blood may be carrying plenty of oxygen, but there is an inability of the cells to use the available oxygen. Death may occur rapidly in these individuals. ACGIH has recognized the acute toxicity and fast-action of hydrogen cyanide and its salts by assigning ceiling limits (C) with their TLVs. This means that the exposure levels may fluctuate but may never exceed the designated value (5 mg/m^3).[6,10]

Numerous industrial chemicals and pharmaceutical products affect the blood cells, thereby producing blood dyscrasias. Such chemicals and products have been well-summarized along with the types of disorders that they induce, including thrombocytopenia, agranulocytosis, aplastic anemia, and pancytopenia.[26]

Up to this point, this section on the circulatory system has focused on chemicals that evoke acute reactions. One chemical known for its chronic effects on the blood is benzene. This solvent is used as a building block in the synthesis of many organic compounds. Benzene was widely used by ink and rubber manufacturers. In environmental and occupational settings the vapors from benzene may be inhaled or absorbed percutaneously. As a result of exposure to benzene, bone marrow damage has been observed in animals and humans. Because stem cells are formed in the bone marrow, it is understandable that damage to these precursors of the cellular components of blood may lead to hemopoietic disorders. Anemia, leukopenia, and thrombocytopenia have been reported. In particular, a great deal of attention has been devoted to the carcinogenic potential (i.e., leukemia) of benzene. The TLV and BEIs for benzene have been set to guard against these hemopoietic effects.[10] As noted earlier in this chapter, researchers believe that one or more intermediate metabolites of benzene is responsible for its toxicity.

Immune System

The primary function of the immune system is to protect the body from foreign biological agents, such as potentially infectious bacteria, fungi, and viruses. The system also acts to destroy mutated and otherwise altered and abnormal cells, such as those associated with development of malignant neoplasms (tumors), that originate within the body. The same immune responses that react to foreign infectious agents and abnormal cells, however, can be activated following absorption of exogenous chemical agents.

The immune system is composed of specialized blood cells that commonly work in concert with endogenous biochemicals to recognize and interact with foreign substances and abnormal cells. Major cells directly and indirectly involved with normal immune response are lymphocytes as T-cells and B-cells, phagocytic cells as macrophages and neutrophils, and cells called natural killer cells. Production of the cells is initiated in the bone marrow which, along with the thymus gland, lymph nodes, and spleen are major storage sites. The cells are continuously discharged into the circulatory system for distribution to body tissues when needed. Noncellular biochemicals called antibodies, cytokines, and complement also play major roles in immune function.

Natural killer cells and phagocytes are associated with innate immunity that is present at the time of birth. The innate immune response is characterized as non-specific and the magnitude unrelated to prior exposures. Innate immunity can be considered as the fundamental or primary immune response to potentially infectious microbes that enter the body and to abnormal cells (e.g., neoplastic cells) that originate in the body. The natural killer cells recognize and bind to foreign substances and abnormal cells ultimately causing destruction. Phagocytic cells can engulf and enzymatically destroy or denature other substances via a process known as phagocytosis. Biochemicals called complement are specialized proteins in the blood that enhance the innate immune response. When an antibody binds to an antigen projecting from an invading microorganism or an indigenous abnormal or foreign cell, it exposes sites on the cell on which the complement can bind. The proteinaceous complement acts enzymatically to catalyze a cascade of reactions ultimately leading to destruction of the cell.

Acquired immunity differs from innate in that it develops with time as a result of prior exposure. Accordingly, the acquired immune response is characterized as specific, and the magnitude is related to memory of prior exposure. Specificity and memory are related to unique individual characteristics of foreign substances referred to as antigens and synthesis of a corresponding antigen-specific antibody. Antigens are the exogenous materials that enter the body and are recognized for binding by antibodies. Antigens are typically organic in nature and, more specifically, are biochemically composed of a protein, carbohydrate, lipid, or nucleic acid. Some inorganic substances such as metals can interact with the immune system. They are classified as haptens, and following absorption into the body they first bond with proteins in the blood to form hapten-protein or antigen conjugates.

Acquired immunity is categorized as humoral immunity and cell-mediated immunity. The humoral mechanism of acquired immunity involves synthesis and response of antibodies. B-cells are noted mainly for producing antibodies. Antibodies are large Y-shaped protein structures called immunoglobulins that selectively combine with exogenous substances, antigens, or antigen (hapten-protein) conjugates that enter the body. Immunoglobulins are identified with a letter. Three examples are immunoglobulin G (IgG), which is the major antibody in the blood; M (IgM), which is the major antibody synthesized following initial exposure to an antigen; and E (IgE), which is associated with allergic immune responses. When an antibody binds to an antigen, it can result in reduced toxicity of chemicals and increased potential for phagocytosis of invading microbes such as bacteria.

The cell-mediated mechanism of acquired immunity involves T-cells. T-cells synthesize and release a biochemical cytokine specifically known as lymphokine. When an antigen binds with a T-cell, it causes the cell to interact directly with the antigen or undergo transformations and reactions leading to release of lymphokines into surrounding areas. Depending on the type of lymphokine (e.g., Interleukin-1), the chemicals either attract macrophages to a specific site for necessary phagocytic activity, inhibit migration of macrophages, or activate macrophages to conduct phagocytosis.

Exposure to infectious biological and toxic chemical agents can cause numerous immune responses, including hypersensitivity, immunosuppression, and autoimmunity.

Toxic interactions with the immune system resulting in exaggerated responses are known as hypersensitivity or allergic reactions. Initial exposure to an antigen or to an agent (hapten) that forms a hapten-protein conjugate results in allergic or exaggerated reactions during subsequent exposures. The hypersensitivity reactions are divided into four types and are associated with exposures to either biological or chemical agents. Type I, II, and III reactions involve synthesis of specific immunoglobulins (i.e., antibodies), and Type IV involves synthesis of memory T-cells.

Type I atopic or anaphylactic reactions involve IgE and are characterized by recognition of an antigen by T-cells which, when activated, release a variety of chemical mediators contained within the cell. Simultaneously, B-cells are stimulated to synthesize and release IgE antibodies. Local respiratory responses are associated with excessive histamine release and are characterized by rhinitis (i.e., runny eyes and nose), conjunctivitis, and asthma-related bronchoconstriction. Extreme systemic reactions can cause potentially lethal respiratory and heart failure due to anaphylactic shock. Type I reactions also may involve the skin, resulting in eruption of dermal hives or urticaria. Dermal contact with natural plant toxins and synthetic materials such as latex has caused urticaria in exposed individuals.

Type II cytotoxic reactions involve IgG or IgM plus the proteinaceous complement and are characterized by response against cell-associated antigens. The antibodies bind directly to specific antigens exposed on the extracellular surface of a foreign or abnormal cell. Cellular lysis occurs as a result of interaction with enzymatic complement or phagocytosis. Type II reactions causing lysis of blood cells can cause hemolytic anemia.

Type III immune complex reactions involve IgA and IgG antibodies, complement, and phagocytes and are associated with blood plasma-soluble antigens. The antigen-antibody complexes can be transported via blood to tissues, such as the kidneys or joints, causing inflammation due to activation of complement.

Type IV delayed hypersensitivity reactions are cell-mediated responses involving T-cells and phagocytes. Responses may be delayed for 24 to 48 hours following contact and entry of an antigen or hapten. Effects include fibrotic changes to the lung or the more commonly observed dermal eruptions associated with allergic contact dermatitis.

Toxic interactions can cause suppression of the immune response. Immunosuppression decreases an individual's resistance and increases vulnerability to infection and proliferation of neoplastic or other mutated cells.

An autoimmune response also can occur following exposure to a toxic chemical that acts as a hapten and binds to an indigenous protein forming a hapten-protein conjugate. As discussed earlier, this reaction is the initial hapten-protein interaction and is a natural immune response to initiate immunity against the absorbed toxic chemical. Unfortunately, immune responses also may recognize the entire hapten-protein conjugate as foreign. If the proteins are associated with endogenous cells, destruction of otherwise normal cells can occur.

A variety of chemicals that are present in environmental and occupational settings are associated with immunotoxicity.[27,28] Hypersensitivity reactions of the lung or skin are associated with isocyanates (e.g., toluene diisocyanate), acid anhydrides (e.g., trimellitic acid anhydride), metals (e.g., beryllium, chromium, cobalt, nickel), and aldehydes (e.g., formaldehyde). Immunosuppression has resulted from exposures to halogenated aromatic chemicals (e.g., polychlorinated biphenyls), polycyclic aromatic chemicals (e.g., benzo(a)pyrene), metals (e.g., arsenic, beryllium, cadmium, lead, mercury), pesticides (e.g., organophosphates), and organic solvents (e.g., benzene), among others. Autoimmune responses have been initiated by chemicals, including crystalline silica and vinyl chloride.

Hepatic System

The liver is the largest organ in the body and has metabolic, vascular, and secretory functions. In toxicology there is a great emphasis on its metabolic function. As noted above, many toxicants are biotransformed in the liver, producing more water-soluble compounds, which may then be excreted. One must remember that metabolic reactions involving ingested nutrients, vitamins, hormones, and essential metals also occur within the liver. For example, there is deamination of amino acids, conversion of fructose and galactose to glucose, and degradation of estrogen and insulin. Additionally, there is synthesis of urea, cholesterol, and phospholipids. With respect to vascular functions, the liver may serve as a reservoir for blood, and like the lung, serves as a filter for blood to remove clots. Another vascular function of the liver involves hemoglobin. When red blood cells die, hemoglobin is broken into its components, heme and globin. Heme is then metabolized in the liver to produce bilirubin. The clearance of this pigment may be used to assess liver function. Finally, bile is produced and secreted by the liver. Bile consists primarily of water (>97%), with smaller quantities of bilirubin, bile salts, cholesterol, fatty acids, lecithin, and electrolytes (e.g., chloride ions, sodium ions). This complex mixture normally moves into the duodenum where it assists in digestion and in the absorption of nutrients. While moving into the duodenum, bile may be stored in the gall bladder. As noted above, one of the major ways in which toxicants are removed from the body is via biliary excretion.

The liver, with its two functional lobes, has a marked capacity for regeneration in comparison with other organs. Researchers have estimated that nearly 90% of the liver may be surgically removed and yet the remaining tissue will redivide to produce the original sized organ. The liver is also well-perfused in comparison to other organs. As mentioned earlier, many toxicants may pass into the liver for this reason. The two principal blood vessels carrying blood to the liver are the hepatic artery (HA) and the portal vein (PV). Blood flows through sinusoids (i.e., channels), along the hepatocytes (i.e., parenchymal cells of the liver). Within the sinusoids, there are several important types of cells. Endothelial cells line the sinusoids, while Kupffer cells are wandering cells that engulf foreign material (i.e., liver macrophages). Ito cells lie below the endothelial cells and serve to store fat. Once blood passes through the sinusoids, it is collected in the central vein (CV) (also known as the terminal hepatic artery, THA). From this point, the blood passes back into the heart. Two other important structures within the liver are the bile ducts (BD) and the lymphatic vessels (LV). BD allow bile, produced in the internal canaliculi, to move out of the liver, while LV allow lymph to drain from the liver.

Within the liver, there are subdivisions or units called lobules or acini. In the older scientific literature, a hexagonal model of the liver lobule was proposed. At each corner of the hexagonal unit, there were four structures (HA, PV, BD, LV), referred to as the portal space (PS). Blood was thought to move from these six corners, through the liver parenchyma, and eventually to the CV. This model is not favored today, since it has not been possible to demonstrate such hexagonal structures microscopically. Instead, an acinar model is preferred, as shown in Figure 4.6. As in the hexagonal model, blood flows from the PS to the CV. However, the acinar model defines three zones in which the hepatocytes are located and through which the blood must pass to reach the CV. The older scientific literature has described the zones as periportal (PZ), mid (MZ), and centrilobular (CZ). Blood leaving the PS first enters the PZ, and then continues through the MZ and CZ before draining into the CV. In the newer scientific literature, the zones are numbered 1 (PZ), 2 (MZ), or 3 (CZ). As discussed below, hepatotoxicants may be classified on the basis of the zone(s) that they affect.

In humans and in animals exposed to environmental and occupational agents, several types of acute reactions may occur within the liver.[29–31] First, lipid droplets may accumulate and produce what is known as fatty liver or steatosis. Examples of chemicals that may produce steatosis include antimony, chloroform, and carbon tetrachloride. A second effect commonly seen following exposure to many chemicals is necrosis (i.e., cell death) of the hepatocytes. Such cellular necrosis may or may not be zone-specific. Likewise, these effects may or may not be organelle-specific. For example, zone-specific hepatoxicity has been reported with allyl formate (Zone 1),

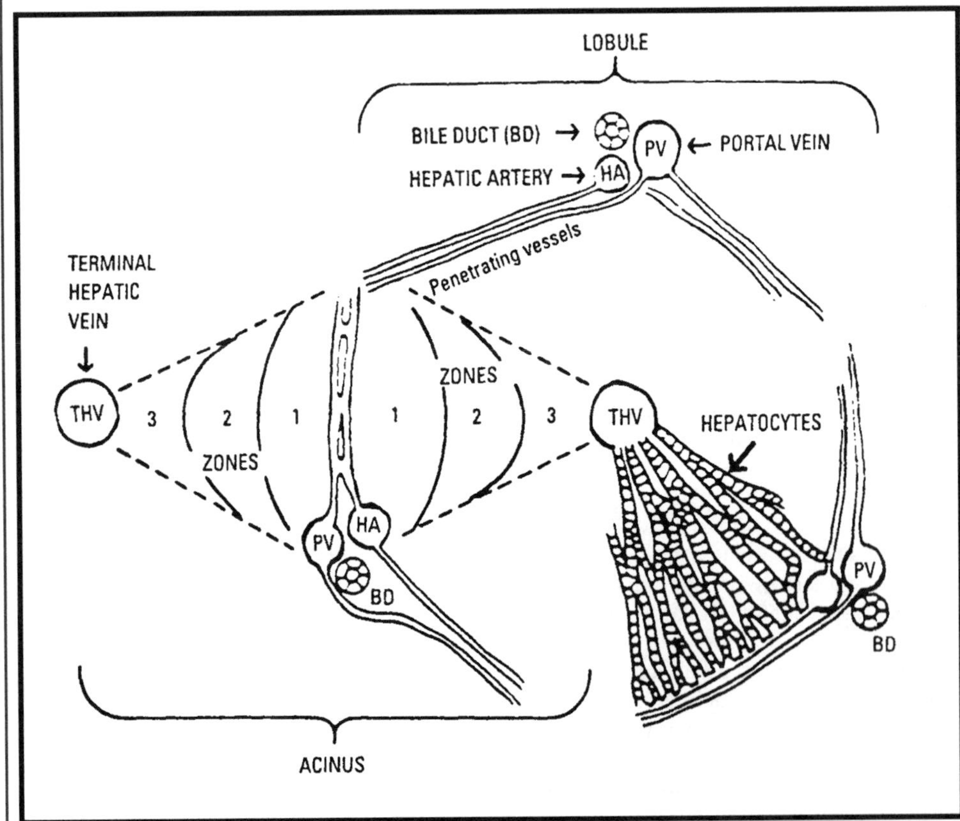

Figure 4.6 — The hepatic acinus (from C.D. Klaassen, *Casarett and Doull's Toxicology*, 5th ed., New York: McGraw-Hill, 1996; reproduced with permission).

beryllium (Zone 2), and chloroprene (Zone 3). Furthermore, researchers have reported that beryllium toxicity is not only zone-specific (Zone 2), but is also organelle-specific (e.g., nucleus, lysosomes). In contrast, cadmium and selenium are capable of producing massive damage throughout all three liver zones. A third type of effect that is induced by some chemicals is cholestasis. In this condition, the production and/or secretion of bile is impaired. Manganese compounds and toluene diamine may induce cholestasis. Hepatitis, which is an inflammation of the liver, is a fourth type of acute hepatic reaction. Since hepatitis does not often occur following exposure to environmental or occupational agents, its etiology is not well-understood. Most of the cited cases of hepatitis have involved drugs (e.g., halothane, indomethacin) rather than industrial chemicals. Finally, chemicals may be involved in the production of more than one of the above reactions. For example, chloroform and carbon tetrachloride may produce steatosis; they may also produce necrosis of the hepatocytes, particularly in Zone 3.

As described with other target systems, chronic injury may also occur within the liver. The most noteworthy types of chronic hepatotoxicity are cirrhosis and cancer. Cirrhosis is a condition in which the liver has become hardened and its physiological functions are highly impaired. Over a long period of time (i.e., years) there is an excessive deposition of fibrous tissue in the liver and the hepatic cells are destroyed. Ethanol is well-recognized for its ability to cause cirrhosis. Like cirrhosis, there is a lengthy latency period before liver cancer is seen. Vinyl chloride and aflatoxins are examples of hepatocarcinogens. Interestingly, during the development of ethanol-induced cirrhosis and aflatoxin-induced liver cancer, there may be some synergistic effects of malnutrition and the causative agent.

Renal System

The function of the renal system is to filter, reabsorb, and secrete water and solutes from the blood and form urine, which is readily excreted from the body. Two kidneys serve as the filtration, reabsorption, and secretion components, and each is connected to a ureter that leads into the bladder.

Each kidney is actually a heterogenous complex of components, including the gross anatomical areas called the renal capsule, cortex, and medulla. The majority (i.e., 80 to 90%) of blood flow to the kidneys is to the cortex, which extends from the exterior capsule to the renal interior where it surrounds numerous lobes. The medulla is the center region of each lobe and within each are renal pyramids. Nephrons, the functional components of the cortex and medulla, are microscopic units that account for the bulk of the kidneys. Nephrons are composed of renal corpuscles and tubules. The renal corpuscles of each nephron consist of a network of capillaries called the glomerulus positioned within a structure called Bowman's capsule. As seen in Figure 4.7, the renal tubules consist of the proximal convoluted tubule, the Loop of Henle (ascending and descending), the distal convoluted tubule, and the collecting tubule.

Renal excretion is the major mechanism for eliminating excess water, solute as metabolic waste products, and toxic parent compounds and metabolites from the body via suspension and dissolution in urine. There are more than one million nephrons per kidney, and they are the sites for the three physiological processes of filtration, reabsorption, and secretion that are required for urine production.

Blood from the body perfuses the kidneys. As blood flows through the glomerular capillaries of individual nephrons, water and solutes (i.e., salts, nutrients, metabolic wastes) are filtered out via passive diffusion

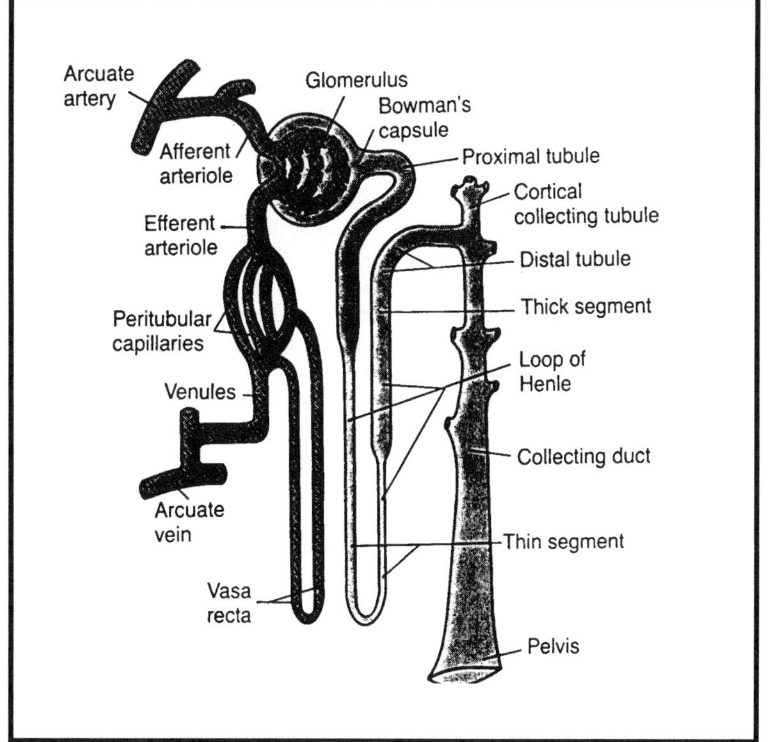

Figure 4.7 — The structure of the nephron (from A.C. Guyton, *Textbook of Medical Physiology*, Philadelphia, W.B. Saunders Co.; 1991; reproduced with permission.)

and passed into the Bowman's capsule of each nephron. Filtration and movement of water and substances through the glomerular capillaries is facilitated due to the membrane pores and relatively high hydrostatic pressure associated with blood pressure. As metabolic wastes are filtered through the glomerular capillaries and removed from the blood, water and solutes (i.e., salts, nutrients) considered essential for maintaining normal body electrolyte balance and function are passively diffused and actively transported or reabsorbed back into the blood via the renal tubules (proximal convoluted tubule, Loop of Henle, distal convoluted tubule, and collecting tubule) at peritubular capillaries. The major site of reabsorption activity is the proximal convoluted tubule of each nephron. Whereas polar or hydrophilic parent compounds and biotransformed metabolites are readily excreted from the body via urine, the nonpolar or lipophilic compounds are more likely to be reabsorbed from the tubules back into the blood for continued distribution to potential target and storage sites in the body. Tubules are also involved in passive diffusion and active transport or secretion of excess water and solutes from the blood to the kidneys to subsequently form urine. The ureters drain urine from the kidneys for temporary storage in the bladder. Ultimately, the urethra drains urine from the bladder for excretion from the body.

The renal system processes a very high volume of blood and, consequently, is exposed and vulnerable to an array of toxic parent compounds and metabolites that may be present.[32] The nephrons are potentially most vulnerable due to a high surface area of exposed cells and since they are the major functional unit of the kidneys. Mechanistically, a relatively nontoxic or low concentration of a toxicant in blood plasma can be concentrated in the kidneys due to the normal physiologic processes of filtration, reabsorption, and secretion during urine production. The presence of a concentrated amount of toxic chemical can create and increase a concentration gradient across nephronal cell membranes. Consequently, there would be increased passive diffusion of toxicants into nephronal cells. Also, active transport processes necessary for regulating cross-membrane transport of nutrients and wastes can be altered due to toxic interference(s) with membrane enzymes.

Accumulation of toxic substances in the kidneys can cause cell injury or cell death (i.e., necrosis). Toxic substances such as metals can accumulate by first bonding with endogenous proteins, which may or may not reduce the risk of toxic interactions with cells. For example, the protein, metallothionein, has sulfhydryl groups to which metals such as arsenic, cadmium, and mercury may readily bond. Thus, a metal-protein complex may serve as a storage reservoir and protective mechanism, or damage may possibly result from storage and transport of the complexes. Disruption of filtration is a possible effect resulting from decreased nephron function due to glomerular and/or tubule damage. Indeed, since the proximal convoluted tubule is a major site of diffusion and active transport for reabsorption, researchers have shown it to be a major target site for nephrotoxicity.

Metals, including arsenic, cadmium, chromium, lead, mercury, and uranium, and nonmetals such as carbon tetrachloride, chloroform, and ethylene glycol, can cause renal disease.[33,34] Acute renal disease can involve glomerular and tubule damage, proximal tubule damage, tubular cell necrosis, and acute renal failure. Acute disease can be initiated by exposure, accumulation, and concentration of a toxicant. The effects range from minor damage and functional effects, such as altered cellular transport activity, to necrosis and acute renal failure. If exposure discontinues and damage is not too extensive, recovery is possible. Prolonged exposure and excessive accumulation of toxicants in the kidneys, however, can lead to chronic disease. Chronic renal disease involves tubule damage and glomerular damage with the ultimate potential for chronic renal failure and premature death.

Neoplastic changes also are possible. Arsenic and lead, for example, may be associated with renal cancer. The bladder is particularly vulnerable to stored aromatic amines, such as naphthylamine and benzidine, which can accumulate and cause bladder cancer.

Manifestations of renal damage include decreased glomerular filtration rate causing an increase in the biomarkers, blood urea nitrogen and serum creatinine.[32] Increased excretion of normally reabsorbed biochemical metabolites including glucose and phosphates is associated with damage to the proximal tubular epithelium. Also, excessive concentrations of high and low molecular weight proteins in the urine are an indicator of renal damage. High molecular weight proteins may be associated with chronic interstitial nephropathy. Low molecular weight proteins may indicate tubular damage. Enzymes in urine are another biomarker of renal damage and are specifically associated with nephronal cell damage.

Nervous System

The nervous system is the major communication network in the body. This system is divided into the central and peripheral nervous systems. The central nervous system (CNS) consists of the brain and spinal cord and functions as the major signal receptor and integration site for the nervous system. Nerves connected to and extending from the CNS form the peripheral nervous system (PNS). The peripheral nerves extend and connect to various receptor structures, such as muscles and glands. The sensory function of the PNS is detection of stimuli and conduction and transmission of signals to the CNS where the signal is received and integrated. This process is referred to as afferent conduction. Subsequently, the CNS responds via the efferent peripheral nerve pathway to conduct and transmit a response back to a muscle or gland. This process is referred to as efferent conduction.

A subdivision of the CNS is referred to the somatic nervous system, which transmits signals from the CNS to the skeletal muscles for movement, and the autonomic nervous system, which transmits signals to the smooth and cardiac muscles associated with the viscera or organ systems. Stimulation of organs via the autonomic nervous system is related to the sympathetic pathway, whereas decreased function is associated with the parasympathetic pathway.

As seen in Figure 4.8, the nerve cells or neurons consist of a cell body from which project shorter extensions called dendrites and a longer extension called the axon. Neurons range in length from millimeters to over a meter and serve as cellular structures for receiving, conducting, and transmitting signals associated with sensory stimuli and motor responses. Typically, the distal receptors of the dendrites receive sensory input and conduct a signal toward the cell body. From there, the signal is further conducted away from the cell body via the axon.

The signals or impulses are created in a neuron initially due to a stimulus and are subsequently conducted due to action potentials (AP) across and down the neuronal membranes. Conduction of an impulse results from a change in the electrochemical charge across the neuronal membrane due to an imbalance of positive charges associated with sodium (Na^+) and potassium (K^+) cations and negative charges from chloride (Cl^-) anions. During a resting state, a neuron has a net negative charge across the neuronal membrane and is polarized. A stimulus (e.g., auditory, chemical, electrical, mechanical) causes the membrane to depolarize due to a change in membrane permeability and an associated influx of Na^+ cations. The polarity changes from negative and reaches a depolarized state. As Na^+ cations continue to enter the neuron, it repolarizes because of a net positive charge inside the neuronal membrane relative to a negative exterior because of Cl^- anions. A local current is established and acts as a stimulus for continued progressive conduction of the signal down the length of the axon. Indeed, the action potential is self-propagating such that

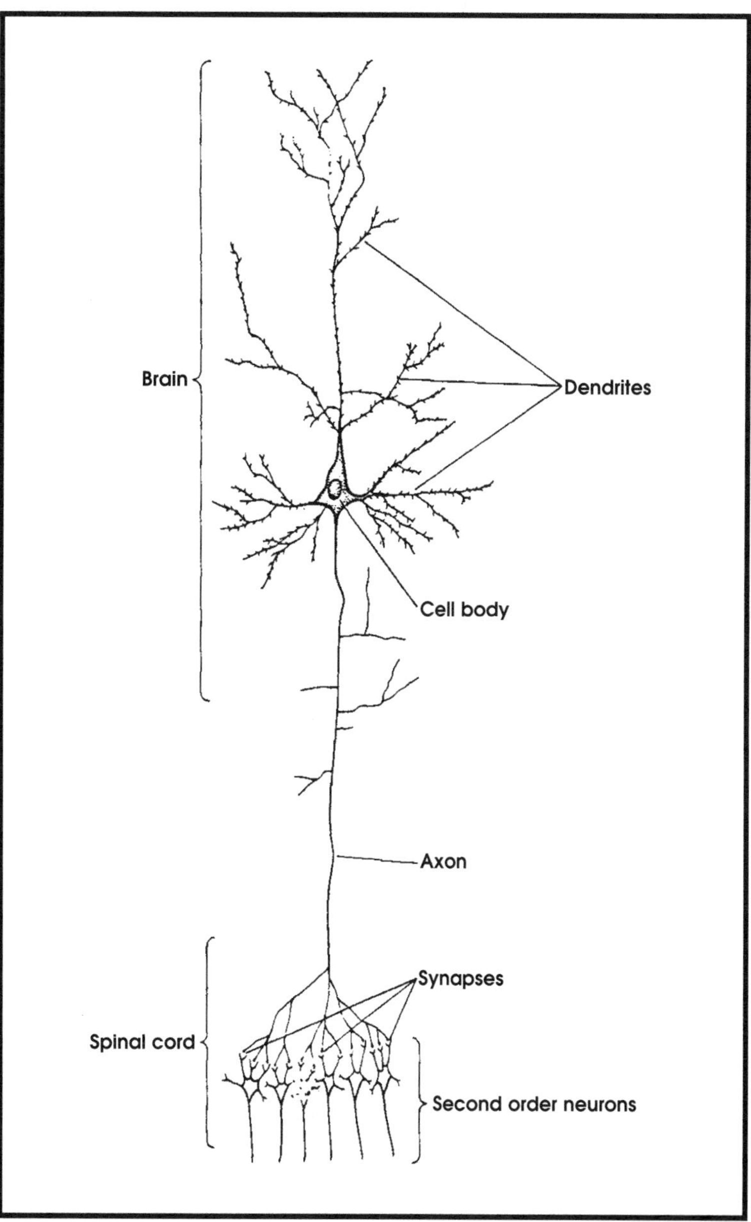

Figure 4.8 — The structure of the neuron and synapse (from A.C. Guyton, *Textbook of Medical Physiology*, Philadelphia, W.B. Saunders Co.; 1991; reproduced with permission.)

cycles of polarization and depolarization continue until the signal is conducted to the terminal point of the axon known as the presynaptic region.

The signal conducted along the axon is enhanced due to the presence of an insulative myelin sheath surrounding neurons. Myelin is composed of predominantly phospholipid molecules and is synthesized from the neuroglial structures that are closely associated with the surrounding neuron areas. Astrocytes are neuroglia cells located in the CNS and connect blood vessels with neurons. Oligodendroglia are glial cells that produce the myelin sheath in CNS neurons, and Schwann cells produce the myelin sheath in the PNS.

The signal is transmitted from neuron to neuron or from neuron to receptor (e.g., muscle, gland) by transfer of a chemical neurotransmitter across an interstructure space called the synapse. Chemical synapses are channels between the terminal axon presynaptic end and the postsynaptic receptors of neuronal dendrites (neural-neural), muscles (neural-muscular), and glands (neural-glandular).

A neurotransmitter, such as acetylcholine, is released from terminal vesicles located in the presynaptic region of an axon and diffuses across the synapse, where it binds to postsynaptic receptor molecules. Accordingly, the signal is transmitted to the receiving structure, such as a muscle, to stimulate motor activity. Following receipt of the neurotransmitter and stimulation of the receptor structure, the enzyme acetylcholine esterase (AChE) is released to deactivate the acetylcholine via hydrolysis. The resulting acetate ion and choline molecules are transferred back to the presynaptic axonal region where acetylcholine is resynthesized and stored in vesicles. The release and deactivation of the neurotransmitter discontinues stimulation at the postsynaptic receptors.

As discussed previously under Transport of Chemicals, the blood-brain barrier is a defense mechanism for the CNS. The blood-brain barrier is actually an arrangement of densely compacted capillary endothelial cells and astrocytic glial cells. It permits selective entry of substances and is somewhat impervious to relatively larger water-soluble toxicants and to ions. Lipid soluble molecules such as organic solvents, however, can more readily pass across the barrier via passive diffusion. A similar anatomical arrangement of cells called the choroid plexus limits movement of toxicants from blood to cerebrospinal fluid.

The peripheral nerves are protected to some degree by a blood-nerve barrier. This barrier is an anatomical arrangement of densely compacted capillary endothelial cells. The myelin sheath exterior associated with the surrounding Schwann cells also acts to protect neurons. Nonetheless, the defenses for both the CNS and PNS are vulnerable to lipid-soluble molecules and their breach can result in adverse neurotoxic responses.

Most organic solvents and numerous metals and pesticides have been shown to cause neurotoxicity in humans and in animals.[35–37] Damage to the CNS includes mental impairment such as behavior changes (e.g., mood, emotions), attention deficit, loss of short-term memory, and interference with calculation.[36] In addition, damage to cranial nerves can cause hearing loss, facial numbness, and altered motor function resulting in decreased muscle strength, poor coordination, and impaired gait. Acute toxic encephalopathy is characterized by headaches, irritability, poor coordination, seizures, coma, and death. Causative agents include carbon monoxide, organic solvents such as carbon disulfide, and metals, including lead and manganese. Chronic toxic encephalopathy is characterized by a gradual loss of memory and psychomotor control, dementia, and motor disorder. Associated toxicants include arsenic, lead, manganese, and mercury.

The majority of toxic interactions with the PNS involves the neuronal axon, while less frequently the Schwann cells are damaged.[37] Most commonly, axonal degeneration occurs concomitantly with an observed degeneration of myelin starting at the distal end of the axon. Prolonged or repeated exposure can cause the axon to continue to degenerate from the distal to the proximal end, a condition referred to as dying back. If exposure is discontinued, there can be some degree of axonal regeneration. Otherwise, the entire cell body can be damaged, and the neuron can be adversely altered.

Demyelination is another major impact that toxic substances can cause. This toxic effect is characterized by the axon remaining intact while the myelin degenerates. This effect is caused either by damaging the myelin or the Schwann cell. Axonal and myelin degeneration can occur through exposure to carbon disulfide and the biotransformed product, 2,5 hexanedione, the hexacarbon compounds n-hexane and 2-hexanone (methyl-n-butyl ketone). Indeed, hexane is an example of a parent compound that can follow either detoxifying or intoxicating pathways. One oxidation pathway ultimately transforms the nonpolar parent n-hexane to the more polar and more readily excreted carboxylic acid metabolites. Another oxidation pathway, however, biotransforms n-hexane to the ultimate neurotoxic 2,5-hexanedione. The resulting peripheral neuropathy is characterized by muscle weakness and numbness of hands and feet. Lead and various organic solvents can cause separation of myelin from the axon and demyelination.

Although various forms of mercury are neurotoxic, it is most neurotoxic in the elemental (Hg^0) and alkyl (R-Hg) forms. Elemental mercury poses a major inhalation hazard, more so than ingestion, as a result of its high volatility and efficient absorption across alveolar-capillary membranes. Elemental mercury is readily transported across the blood-brain barrier into the CNS, where early symptoms include fatigue, weakness, and memory loss. Chronic exposure can cause muscular tremors and psychological disturbances. Methyl and ethyl mercury are the most toxic forms of organic mercury. Following slow absorption from blood to brain and subsequent accumulation, the alkyl mercuries can cause disturbances of vision, speech, hearing, and gait plus signs of tremor and muscular rigidity.

Lead is a major neurotoxic metal. This heavy metal damages peripheral nerves by causing axonal degeneration and segmental demyelination. The resulting peripheral neuropathy is associated with impaired motor function. The CNS is affected mostly by behavioral impairment, including alteration of cognitive processes. Other effects include changes in sensory functions, including impaired hearing and vision.

Organophosphate and carbamate compounds cause neurotoxicity by interfering with transmission of a signal via the neurotransmitter, acetylcholine, across a synapse. Representative compounds exhibit anticholinesterase activity. Some organophosphates and carbamates have a high affinity for the AChE enzyme and bond to it, resulting in catalytic inhibition. As a consequence, the acetylcholine that transmitted the signal across the synapse remains bonded to the postsynaptic receptors, causing continued stimulation. Prolonged stimulation of receptor muscle cells can cause tremors, muscular fatigue, paralysis, and if not reversed, ultimately death.

Reproductive System

The ultimate function of the female and male reproductive systems is to conceive and propagate future generations of offspring. Each system contains gonads, which serve the purposes of production of sex cells or germ cells, and synthesis of estrogenic (female) and androgenic (male) steroid hormones.

The female reproductive system consists of the vagina, cervix, uterus, oviducts, and ovaries. A primary function of the ovaries, the female gonads, is to synthesize germ cells. Oogenesis is the process of producing female germ cells, which are known as oocytes, ova, or simply eggs. The two ovaries contain follicles, each consisting of an egg surrounded by a layer of granulosa and theca cells. This process of ova or egg production begins perinatally; that is, during embryonic and fetal development. Accordingly, prior to birth, females synthesize and contain ova necessary for reproduction during their lifetime. Follicles in various stages of development do not mature and release ova until the female reaches puberty and the cyclical physiological process of ovulation begins.

Ovaries are endocrine glands that secrete hormones. They are components of a nervous system-endocrine system loop (or neuroendocrine loop). There is hormonal synthesis, secretion, transport, and communication between the hypothalamus and pituitary glands in the brain and subsequently between the pituitary gland and the peripheral ovaries. The hypothalamus gland secretes a gonadotropin-releasing hormone for venous transport to the pituitary gland. Stimulated receptors called gonadotrophs in the pituitary gland release luteinizing hormone (LH) and follicle stimulating hormone (FSH) into the circulation for transport to the ovaries. The FSH binds to the estrogen-secreting granulosa cells of follicles in the ovaries, which in turn stimulates secretion of estrogen by the theca cells. The LH also binds to follicles and stimulates ovulation, which is the release of an ovum from a follicle into a fallopian tube. The remaining components of the follicle are converted to form progesterone-secreting luteal cells, which eventually form the corpus luteum. The corpus luteum secretes estrogen plus the hormone, progesterone, which is necessary following conception for implantation of a fertilized egg into the uterine wall and subsequent development of the placenta and fetus.

The hypothalamus, pituitary, and ovarian glands are major organs involved in normal female reproductive physiology and are potential target sites for absorbed toxicants.[38,39] For example, cadmium as cadmium chloride salt has been shown to alter the pituitary and ovarian glands in rats. Chemicals such as some chlorinated hydrocarbon pesticides (e.g., DDT, methoxychlor, chlordecone, dibromochloropropane), and some polyhalogenated biphenyls (e.g., PCBs, PBBs) have been shown to mimic estrogen and exhibit estrogenic properties, while other chemicals such as lindane exhibit antiestrogenic properties. Chemicals exhibiting estrogenic or antiestrogenic properties can cause adverse reproductive effects through interference with normal neuroendocrine physiology. At the cellular level within the ovaries, potential target sites include the ova, and granulosa, theca, and luteal cells. Based on animal studies, several metals including cadmium, lead, and mercury can cause ovarian toxicity.

Since females do not synthesize new ova, damage from chemical or radiation exposure can permanently render a female infertile. In addition, damage to the DNA or chromosomes of ova can result in genetic mutations, and, following fertilization, adverse effects can be manifested in the progeny.

The main components of the male reproductive system are the penis and two testicles. Accessory organs include the epididymis, the seminal vesicles, the Cowpers gland, and the prostate. The testicles are the male gonads and consist of seminiferous tubules lined with Sertoli cells and germ cells. Adjacent to the tubules are interstitial Leydig cells. The gonads are encased in a fibrous capsule called the tunica albuginea. The primary function of the testicles is to synthesize male germ cells called spermatozoa and secrete applicable hormones. Unlike females, development of male germ cells, a process called spermatogenesis, does not begin until puberty and continues throughout a man's lifetime.

Similar to female gonads or ovaries, the male gonads or testicles are endocrine glands that secrete hormones and are a component of the neuroendocrine system loop. There is hormonal transport and communication between the hypothalamus and pituitary gland and subsequently, the pituitary gland and testicles. The hypothalamus secretes gonadotropin-releasing hormone for venous transport to the pituitary gland. Stimulated receptors called gonadotrophs in the pituitary gland release LH and FSH into the circulation for transport to the testicles. The FSH binds to the Sertoli cells, and LH binds to Leydig cells in the seminiferous tubules of the testicles. This binding activates spermatogenesis by the Sertoli cells and stimulates synthesis and secretion of testosterone by Leydig cells.

The hypothalamus, pituitary, and testicular glands are major organs involved in normal male reproductive physiology and are potential target sites for absorbed toxicants.[38,39] Cellular target sites of the male reproductive system include the spermatozoa, Sertoli cells, and Leydig cells. Phthalate esters have been shown in animal studies to damage Leydig cells, interfere with synthesis of testosterone, and cause testicular atrophy. Other toxicants, including triorthocresolphosphate and the metals cadmium, cobalt, and lead, may interact with Leydig and Sertoli cells, disrupting spermatogenesis and causing infertility. Interactions with sperm have been shown for the chemicals 7,12-dimethylbenzanthracene, benzo(a)pyrene, lead, and carbon disulfide. The fumigant dibromochloropropane is a potent, selective male reproductive toxicant that adversely affects the seminiferous tubes and Sertoli cells and was determined to be the cause of sterility in exposed workers.[40] Damage to accessory organs, such as the prostate, also can occur and has been associated with cadmium.

Tightly anatomically positioned endothelial cells form a blood-testicular barrier. This barrier controls access of some water-soluble exogenous chemicals to the testicles, but is not totally impervious, especially to nonpolar lipophilic agents.

Thus, major somatic impacts to male and female reproduction include alteration of normal hormonal and/or cellular function leading to infertility. Genetic effects, however, include altered development of offspring. Genetic effects occur through damage to DNA or chromosomes resulting in mutation of the male or female germ cells, sperm and ova, respectively. Conception of a mutated sperm or mutated ovum can result in abnormal progeny.

Reproductive toxicity can also directly affect a developing embryo or fetus. The normal human karyotype consists of 23 pairs of chromosomes, including 1 pair of sex chromosomes, XX for females or XY for males, equivalent to a total of 46 chromosomes per diploid cell. Conception occurs when an egg is fertilized by sperm and results in the combining of 23 chromosomes from the mother and 23 chromosomes from the father to form a 46-chromosome (23-pair) diploid cell. During early human development, several cellular phases occur during a process known as embryogenesis.

The phases include cell proliferation, cell differentiation, cell migration, and organogenesis. The stages of development that occur during the normal human gestation period of approximately 9 months include formation of a zygote (first 17 days), an embryo (next 38 days), and a fetus (final days until delivery).

During phases and stages of development, the perinate is relatively vulnerable to exposure to some toxic chemicals.[41,42] Teratogenesis refers to the process by which toxic compounds called teratogens can induce perinatal aberrations in an exposed embryo or fetus, yet because of relatively low concentrations, may pose no significant hazard of toxicity to the mother. Possible teratogens include ionizing radiation, 2,3,7,8-tetrachloro-*p*-dibenzodioxin, ethanol, and lead. The embryonic and fetal stages of development are the most vulnerable to intoxication. However, damage at the zygote stage could result in termination of pregnancy and spontaneous abortion. Embryonic exposures are associated with morphological aberrations, and fetal exposures appear more susceptible to morphological and physiological alterations.

As mentioned earlier in the chapter under Transport of Chemicals, the placenta serves as a semipermeable membrane between maternal and fetal blood streams. There is no direct mixing of maternal and fetal blood, but the membrane permits transfer of respiratory gases, nutrients, and wastes. Unfortunately, some toxic compounds, including carcinogens, can also cross the placenta. In addition, chemicals including cadmium and mercury may directly alter placental membrane transport of nutrients from the maternal blood, and consequently may adversely affect the developing embryo or fetus. Furthermore, it is suggested that placental tissue contains enzymes that can possibly catalyze the transformation of some chemicals into active metabolites.

Disease Clusters

In occupational settings, occurrences of disease may be sporadic and may not obviously be related to chemical exposure. In some cases, there are known disease/chemical relationships that occur and can be corrected to eliminate the symptoms; these are usually acute exposures. However, there are occurrences within small populations of questionable diseases/deaths that may be related to chemical exposure. These cases tend to be difficult to analyze and are referred to as clusters.

The definition of a disease cluster is an apparent increase in specific outcomes (e.g., disease) among individuals linked in time or space (i.e., cluster) or by exposure characteristics. When examining a population, a rate of disease occurrence is usually determined by measuring the frequency or the ratio of disease events in a population. All rates or ratios are calculated by dividing the number of events (e.g., diseases or deaths) by the population or the denominator.

In cluster analysis, the numerator, or number of unusual events, is noticed first. In epidemiology, determining the numerator first is a backward procedure. Generally, epidemiologists determine the total population (i.e., denominator) first, and then look at the disease (or mortality) second. Therefore, analysis of a cluster event can be difficult in part because one may be linking to a denominator that may or may not be appropriate for the diseased population.

Moreover, the disease cluster is usually noticed by an individual who is closely working with the data and who then formulates a potential correlation. Performing an analysis generally involves a set of workers who are very concerned about the correlations that may be generated.

The investigator of clusters must be aware of the problems mentioned below.

- When the number of workers is small, statistical analysis may be meaningless.
- If statistical analysis is performed, then one must use methods that are specific for clusters. (For example, Cluster Software, Agency for Toxic Substances and Disease Registry, Atlanta, GA 30333, telephone 404-639-6200).
- Because of the lack of a denominator, a control population may be difficult to generate, and thus, comparisons may be very difficult.
- One typically tests the hypothesis that there is no statistical difference between the response of a control group versus that of an exposed group. However, clusters are investigated because they are unusual, and thus, there is an underlying assumption that these individuals are different from others. This bias may interfere with proper statistical testing.

While clusters of disease or deaths may be of interest and may appear to have a cause, statistically most clusters have no single cause. Paradoxically, often what is known about etiological agents has come from clusters. Some important examples of clusters showing a new effect include aplastic anemia from benzene, bladder cancer from dyes, mesothelioma from asbestos, angiosarcoma from vinyl chloride, and depressed male reproductive ability from dibromochloropropane. Since 1970 an average of two clusters per year has yielded important new understandings about the relationships between disease episodes and specific exposure conditions, the most noteworthy being hypersensitivity and asthma responses.[43–45]

Goals in Chemical Selection

When evaluating processes that involve chemicals, occupational hygienists and other health and safety professionals should attempt to meet the following goals.

Function. The function of each chemical used in a process should be considered. It may be possible to eliminate a chemical from a process. If this is not possible, then physical agents might be used to replace chemical agents. For example, from a toxicity standpoint, it may be advantageous to use bead-blasting in place of an organic solvent.

Use of Water. The use of water as a replacement or as a diluent should be considered. Given the minimal toxicity of water and solubility of many compounds in water, it could serve as a good replacement for another solvent.

Chemical Replacements. The use of other chemical replacements should be considered. It may be possible to identify less toxic and physically safe (e.g., nonflammable) chemicals for use in a process. Also, chemicals may be selected that produce less hazardous or nonhazardous waste.

Personal Protective Equipment. The use of personal protective equipment should be viewed as a last resort. Engineering and administrative controls should be evaluated and employed before considering any personal protective equipment.

Risk Assessment: What Should Be Done with All the Data?

Risk assessment is the systematic process for describing and quantifying the risks associated with hazardous substances, processes, actions, or events.[3] There are four basic steps when conducting a risk assessment: hazard evaluation, dose-response assessment, exposure assessment, and risk characterization.

In hazard evaluation one must determine whether substances, processes, actions, or events have the potential to pose a risk. In some instances the risk may not be related to the toxicity of the chemical, but rather to its physical or chemical properties (e.g., flammability, reactivity). In other cases there is so little known about a chemical's toxicological properties that one must rely on structure-activity relationships, molecular configuration, or potential reactive sites. Predictions are then made as to the response(s) if the chemical is absorbed by the body. When possible, a toxicologist should be consulted to evaluate the chemical in such cases.

The second step in a risk assessment is dose-response assessment. As noted earlier in this chapter, dose is the amount of a chemical absorbed by the body, and the effects from chemicals are dose-related. Thus, it is critical to determine or estimate how much of a chemical is needed to elicit a specific adverse effect. This estimation may be made for both acute and chronic exposures.

Most occupational hygienists are involved in the exposure assessment phase of risk assessments. Ideally, the concentrations of chemicals in appropriate environmental and occupational samples (e.g., air, soil, water) are determined analytically. When such measurements cannot be made, it may be necessary to use mathematical models to predict the concentrations of chemicals in air, water, and soil as well as to predict the fate and transport of these chemicals. Also, when conducting exposure assessments, the frequency and duration of exposures must be estimated.

Finally, occupational hygienists in conjunction with toxicologists and other health and safety professionals work to tie all of the above information together to arrive at an estimate of risk. In some instances the toxicological endpoint of interest may be cancer, while in other instances, it may be a noncancer hazard (e.g., reproduction). For cancer, one is typically concerned when the risk is greater than 10^6 (i.e., greater than one in a million), while for other toxicological endpoints, there is concern when the hazard index exceeds 1.0. Guidelines (including equations, reference values, etc.) have been published that may assist those individuals involved in risk characterization.

While a risk assessment is being conducted and after it is completed, there is a continuing need for risk management and risk communication. Occupational hygienists, toxicologists, and other health and safety professionals often participate in these activities. Risk management is the process by which policy actions are taken to deal with the hazardous substances, processes, actions, or events identified in the risk assessment. For example, an occupational hygienist may need to establish an occupational exposure limit for a hazardous substance or may need to automate an operation to reduce the risks associated with it. Risk assessment and risk management must be understandable to other parties (e.g., workers, general public). Occupational hygienists may be asked to communicate information verbally (e.g., meetings, training sessions) or through other types of media (e.g., material safety data sheets, computer internet). The use of appropriate terminology for the audience is crucial and will affect the outcome of these interactions. Risk communication has become an important issue in itself and has been addressed at national as well as international meetings.

Summary

Humans are exposed to individual and, more commonly, combinations of chemicals in environmental and occupational settings. Such chemicals enter the body via the respiratory tract, skin, and gastrointestinal tract and may be distributed throughout the body. Subsequently, adverse effects may be produced in a variety of target systems. These effects may be seen after only brief exposures or after many years of exposure. Health and safety professionals have the responsibilities of preventing, anticipating, recognizing, evaluating, and controlling such exposures, minimizing these effects in humans. In this chapter the principles and applications of toxicology have been reviewed to assist health and safety professionals in understanding some concepts related to these responsibilities.

Additional Sources

Amdur, M.O., J. Doull, and C.D. Klaassen (eds.): *Casarett and Doull's Toxicology—The Basic Science of Poisons.* 4th ed. New York: Pergamon Press, 1991.

Clayton, G.D. and F.E. Clayton (eds.): *Patty's Industrial Hygiene and Toxicology.* New York: John Wiley & Sons, 1991.

Craighead, J.E. (ed.): *Pathology of Environmental and Occupational Disease.* St. Louis, MO: Mosby, 1995.

Guyton, A.C.: *Textbook of Medical Physiology.* 8th ed. Philadelphia: W.B. Saunders Co., 1991.

Hall, S.K., J. Chakraborty and R.J. Ruch (eds.): *Chemical Exposure and Toxic Responses.* Boca Raton, FL: CRC/Lewis Publishers, 1996. pp. 159–190.

Hodgson, E. and P.E. Levi: *Introduction to Biochemical Toxicology.* Norwalk, CT: Appleton and Lange, 1994.

Klaassen, C.D. (ed.): *Casarett and Doull's Toxicology—The Basic Science of Poisons.* 5th ed. New York: McGraw-Hill Health Professions Division, 1996.

Lu, F.C.: *Basic Toxicology.* 3rd ed. Washington, DC: Taylor and Francis Publishers, 1996.

Paul, M. (ed.): *Occupational and Environmental Reproductive Hazards: A Guide for Clinicians.* Baltimore, MD: Williams & Wilkins, 1993.

Rosenstock, L. and M.R. Cullen (eds.): *Clinical Occupational Medicine.* Philadelphia: W.B. Saunders Co., 1986.

West, J.B.: *Pulmonary Pathophysiology.* 4th ed. Baltimore, MD: Williams and Wilkins, 1992.

References

1. **West, J.B.:** *Respiratory Physiology,* 4th ed. Baltimore, MD: Williams and Wilkins, 1990.
2. **Schaper, M.:** Development of a database for sensory irritants and its use in establishing occupational exposure limits. *Am. Ind. Hyg. Assoc. J. 54*:488–545 (1993).
3. **Covello, V.T. and M.W. Merkhofer:** *Risk Assessment Methods: Approaches for Assessing Health and Environmental Risks.* New York: Plenum Press, 1993.
4. **Alarie, Y., M. Schaper, G.D. Nielsen, and M.H. Abraham:** Estimating the sensory irritating potency of airborne nonreactive volatile organic chemicals and their mixtures. *Structure Activity Relationships and Quantitative Structure Activity Relationships in Environ. Res. 5*:151–165 (1996).
5. **Oehme, F.W., R.W. Coppock, M.S. Mostrom and A.A. Khan:** A review of air pollutants: toxicology of chemical mixtures. *Vet. Hum. Toxicol. 38*:371–377 (1996).
6. **American Conference of Governmental Industrial Hygienists (ACGIH):** *1996–1997 TLVs® and*

BEIs®. Threshold Limit Values for Chemical Substances and Physical Agents, Biological Exposure Indices. Cincinnati, OH: ACGIH, 1996.

7. **Derosa, C.T., Y.W. Stevens, J.D. Wilson, A.A. Ademoyero, et al.:** The Agency for Toxic Substances and Disease Registry's role in development and application of biomarkers in public health practice. *Toxicol. Ind. Health 9*:979–994 (1993).
8. **Rappaport, S.M.:** Biological monitoring and standard setting in the USA: a critical appraisal. *Toxicol. Lett. 77*:171–182 (1995).
9. **Talaska, G., J. Roh, and Q. Zhou:** Molecular biomarkers of occupational lung cancer. *Yonsei Med. J. 37*:1–18 (1996).
10. **American Conference of Governmental Industrial Hygienists (ACGIH):** *Documentation of TLVs® and BEIs®*, 6th ed. Cincinnati, OH: ACGIH, 1991.
11. **Hyde, D.M., R.P. Bolender, J.R. Harkema, and C.G. Plopper:** Morphometric approaches for evaluating pulmonary toxicity in mammals: implications for risk assessment. *Risk Anal. 14*:293–302 (1994).
12. **Silverman, L., C.E. Billings, and M.W. First:** *Particle Size Analysis in Industrial Hygiene.* New York: Academic Press, 1971.
13. **Mercer, T.T.:** *Aerosol Technology in Hazard Evaluation.* New York: Academic Press, 1973.
14. **Samet, J.M. and P.W. Cheng:** The role of airway mucus in pulmonary toxicology. *Environ. Health Perspect. 102 (Suppl. 2)*:89–103 (1994).
15. **Alarie, Y.:** Sensory irritation by airborne chemicals. *CRC Crit. Rev. Toxicol. 2*:299–363 (1973).
16. **Zelenak, J.P., Y. Alarie, and D.A. Weyel:** Assessment of the cough reflex caused by inhalation of sodium lauryl sulfate and citric acid aerosols. *Fundam. Appl. Toxicol. 2*:177–180.
17. **Amdur, M.O. and J. Mead:** A method for studying the mechanical properties of the lungs of unanesthetized animals: application to the study of respiratory irritants. In *Proceedings of the Inhalation Toxicology Technical Symposium*, Ann Arbor, MI: Ann Arbor Science Publishers, 1955. pp. 207–231.
18. **Karol, M.:** Comparison of clinical and experimental data from an animal model of pulmonary immunologic sensitivity. *Ann. Allergy 66*:485–489 (1991).
19. **Karol, M.:** Animal models of occupational asthma. *Eur. Resp. J. 7*:555–568 (1994).
20. **Witschi, H.:** Ozone, nitrogen dioxide and lung cancer: a review of some recent issues and problems. *Toxicology 48*:1–20 (1988).
21. **Lippmann, M.:** Health effects of ozone: a critical review. *J. Air Pollut. Control Assoc. 39*:671–695 (1989).
22. **Folinsbee, L.J.:** Human health effects of air pollution. *Environ. Health Perspect. 100*:45–56 (1993).
23. **Alarie, Y. and M. Schaper:** Pulmonary performance in laboratory animals exposed to toxic agents and correlations with lung diseases in humans. In *Lung Biology in Health and Disease: Pathophysiology and Treatment of Inhalation Injuries.* J. Loke (ed.). New York: Marcel Dekker Inc., 1988. pp. 67–122.
24. **Burns, D.M., J.R. Froines, and M.E. Jarvik:** Biologic interactions between smoking and occupational exposures. *Am. J. Ind. Med. 13*:169–179 (1988).
25. **Breen, P.H., S.A. Isserles, J. Westley, M.F. Roizen, et al.:** Combined carbon monoxide and cyanide poisoning: a place for treatment. *Anesth. Analg. 80*:671–677 (1995).
26. **Williams, P.L. and J.L. Burson (eds.):** *Industrial Toxicology. Safety and Health Applications in the Workplace.* New York: Van Nostrand Reinhold, 1985.
27. **Burns, L.A., B.J. Meade, and A.E. Munson:** Toxic responses of the immune system. In *Casarett and Doull's Toxicology: The Basic Science of Poisons*, 5th ed., C.D. Klaassen (ed.). New York: McGraw-Hill, 1996. pp. 355–402.
28. **Huber, S.A.:** Immunopathology. In *Pathology of Environmental and Occupational Disease*, J.E. Craighead (ed.). St. Louis, MO: Mosby, 1995. pp. 397–409.
29. **Plaa, G.L. and W.R. Hewitt (eds.):** *Toxicology of the Liver* (Target Organ Series). New York: Raven Press, 1982.
30. **Lee, W.M.:** Review article: drug-induced hepatotoxicity. *Aliment. Pharmacol. Ther. 7*:477–485 (1993).
31. **Mehendale, H.M., R.A. Roth, A.J. Gandolfi, J.E. Klaunig, et al.:** Novel mechanisms in chemically induced hepatotoxicity. *Fed. Am. Soc. Exp. Biol. J. 8*:1285–1295 (1994).
32. **Tarloff, J.B. and R.S. Goldstein:** Biochemical mechanisms of renal toxicity. In *Introduction to Biochemical Toxicology*, E. Hodgson and P.E. Levi

(eds.). Norwalk, CT: Appleton and Lange, 1994. pp. 519–546.

33. **Goyer, R.A., C.D. Klaassen, and M.P. Waalkes (eds.):** *Metal Toxicology.* San Diego, CA: Academic Press, 1995.
34. **Wedeen, R.P.:** Renal diseases of occupational origin. *Occup. Med. 7(3)*:449–463 (1992).
35. **Chang, L.W. and R.S. Dyer (eds.):** *Handbook of Neurotoxicology.* New York: Marcel Dekker, Inc., 1995.
36. **Mailman, R., C.P. Lawler, and P. Martin:** Biochemical toxicology of the central nervous system. In *Introduction to Biochemical Toxicology,* E. Hodgson and P.E. Levi (eds.). Norwalk, CT: Appleton and Lange, 1994. pp. 431–457.
37. **Morell, P., J.F. Goodrum, and T.W. Bouldin:** Biochemical toxicology of the peripheral nervous system. In *Introduction to Biochemical Toxicology,* E. Hodgson and P.E. Levi (eds.). Norwalk, CT: Appleton and Lange, 1994. pp. 415–429.
38. **Richardson, M.:** *Reproductive Toxicology.* New York: VCH Publishers, 1993.
39. **Witorsch, R.J. (ed.):** *Target Organ Toxicology Series: Reproductive Toxicology,* 2nd ed. New York: Raven Press, 1995.
40. **Whorton, D., R.M. Krauss, S. Marshall, and T.H. Milby:** Infertility in male pesticide workers. *Lancet* 2:1259–1260 (1977).
41. **Sever, L.E.:** Congenital malformations related to occupational reproductive hazards. *Occup. Med. 9(3)*:471–494 (1994).
42. **Needleman, H.L. and D. Bellinger (eds.):** *Prenatal Exposure to Toxicants.* Baltimore, MD: The John Hopkins University Press, 1994.
43. **Centers for Disease Control:** Guidelines for investigating clusters of health events. *Morb. Mort. Wkly. Rep. 39 (RR-11)*:1–23 (1990).
44. **Fleming, L.E., A.M. Ducatman, and S.L. Shalat:** Disease clusters in occupational medicine: a protocol for their investigation in the workplace. *Am. J. Ind. Med. 22*:33–47 (1992).
45. **Hernberg, S.:** *Introduction to Occupational Epidemiology.* Chelsea, MI: Lewis Publishers, 1992.

Outcome Competencies

After completing this chapter, the user should be able to:

1. Define underlined terms used in this chapter.
2. Recognize the importance of epidemiological surveillance of the workplace.
3. Identify the significance of epidemiological surveillance in the identification of potential work-related health problems.
4. Learn how to use routinely collected data for surveillance.
5. Describe several strategies for epidemiological surveillance.

Key Terms

epidemiological surveillance • exposure surveillance • medical surveillance • medical screening • routinely collected data • validity

Prerequisite Knowledge

Prior to beginning this chapter, the user should review the following references for general content:

Academic Discipline	Reference Citation
Epidemiological Methods	**Friis, R.H., and T.A. Sellers:** *Epidemiology for Public Health Practice.* Gaithersburg, MD: Aspen, 1996.
Occupational Epidemiology	**Hernberg, S.:** *Introduction to Occupational Epidemiology.* Chelsea, MI: Lewis Publishers, 1992.
Epidemiological Terminology	**Last, J.M.:** *A Dictionary of Epidemiology.* New York: Oxford University Press, 1988.

Prior to beginning this chapter, the user should review the following chapters:

Chapter Number	Chapter Topic
2	Occupational Exposure Limits
6	Principles of Evaluating Worker Exposure
37	Program Management
41	Risk Communication in the Workplace
43	Use of Computers in Occupational Hygiene
46	Collaborating with the Occupational Physician

Key Topics

I. Historical Aspects

II. Objectives of Epidemiological Surveillance

III. Choice of Terms and Meanings

IV. Definitions

V. Strategies for Epidemiological Surveillance

VI. Core and Dynamic Elements of Epidemiological Surveillance
   A. Core Dimension I: Demographic Information
   B. Core Dimension II: Employment/Exposure Information
   C. Core Dimension III: Information on Potential Confounders
   D. Core Dimension IV: Health Outcome Information
   E. Dynamic Dimension I: Formulation of Group Data
   F. Dynamic Dimension II: Time
   G. Purposes for which Health Data are Collected

VII. Questioning the Validity of Data: A Requirement of Surveillance
   A. Using Routinely Collected Data for Surveillance
   B. Individuals and Groups

VIII. Sample of Existing Strategies

5

Epidemiological Surveillance

Harris Pastides, Ph.D.

Kenneth A. Mundt, Ph.D.

Introduction

One of the key figures responsible for the eradication of smallpox, D.A. Henderson, once said that "the most powerful, effective and under-rated tool in communicable disease control is the technique of surveillance. In essence, it represents organically the brain and the nervous system in a management process."[1] This clever paradigm is readily transferable to the modern practice of occupational hygiene and occupational medicine, which have evolved dramatically during the past few decades. More than ever it is recognized that the primary, secondary, and tertiary prevention of occupational illness and injury demands a broad perspective and a working knowledge of a wide array of subjects. While technological advances in diagnostic medicine and in exposure assessment now facilitate closer examination of potential associations between work and health, the cornerstone of a successful occupational health program continues to be the careful observation, enumeration, and interpretation of data. These traits comprise the required functional elements of an occupational health surveillance program.

The purpose of this chapter is to describe the rationale for establishing an occupational health surveillance program and the methods for achieving a successful program. It is meant to be a reference document that can offer practical assistance for occupational hygienists and other health professionals in industry who are committed to establishing or improving their own health surveillance programs. By offering numerous examples from the published literature, the authors hope to illustrate the important set of principles being put forth. Indeed, this chapter overlaps only slightly with the wealth of published literature in occupational health because it has a specific focus on occupational epidemiological surveillance. This chapter is not intended to serve as an introduction to epidemiological methods, which can be found elsewhere. The science and practice of epidemiology, like occupational hygiene, is complex, and therefore cannot be addressed in a single chapter. However, key epidemiological methods, as applied to the workplace environment in the form of surveillance, can serve as a practical introduction to epidemiology. Furthermore, some of the ideas presented are new and deserve to be discussed among occupational health experts.

Representative examples are presented in sidebars on the pages with the principles or statements they illustrate. This enables a reader to skip over the examples on the first reading rather than interfere with the flow of text, or if the applied meaning of statements is not readily apparent, the illustration may be read to clarify the principle or activity being described.

Historical Aspects

The obvious advantages of examining the illness experience of individuals relative to their environmental and occupational circumstances was stated in the early days of recorded history by Hippocrates and Herodotus. Although these thinkers are often credited with this seminal realization, it is sometimes overlooked that Hippocrates also strongly promoted the idea of observing, recording, and interpreting facts to direct actions at disease prevention. This may be considered the conceptual origin of health surveillance.

As has been reviewed by Eylenbosch and Noah,[2] health surveys and population censuses are known to have been conducted during the rule of Europe by the Romans and again during the Renaissance. However, it was not until the publication of the Bills of Mortality in London and John Graunt's analysis of them (1662) that a quantitative approach to surveillance was demonstrated. Elsewhere, Bernadino Ramazzini (1633–1717) of Modena inspired the systematic observation and analysis of data related to work-related illness and mortality in several countries of central Europe.[2] Two centuries later, William Farr (1807–1883), Compiler of Statistical Abstracts in the General Register Office in London, linked these mortality statistics with population census data for the purpose of calculating mortality rates. In an 1864 report, Farr constructed age-specific mortality rates for metal miners. By comparing these rates with those of nonminers and of coal miners, he showed that the metal miners were at greatly increased risk of death from pulmonary disease. Farr, therefore, demonstrated that an important medical discovery could be made through tabulation, data linkage, and quantitative analysis, and without the involvement of laboratory scientists.[2]

In the United States, the pioneering work of Lemuel Shattuck of Boston (1793–1859) established a health agenda leading to the decennial census, standardization of disease nomenclature, and the recording of occupational code on certain disease records. The 20th century witnessed the establishment of the National Health Survey (1935), numerous (about 40) state cancer registries (the first in 1937), the Framingham Heart Study (1950), and periodic National Institute for Occupational Safety and Health (NIOSH) exposure surveys, all of which served to further the development of occupational epidemiological surveillance.[2]

Objectives of Epidemiological Surveillance

Limiting activities to ad hoc epidemiological studies when there is some hint of an adverse effect among workers is a shortsighted view of the potential contributions of epidemiology to an occupational health program. Since various types of exposure and health data are collected routinely in the industrial setting, it makes sense to integrate and analyze them periodically. This activity is usually most productive when organized around some central objectives or questions. Objectives that can be met fully or partially by implementing an epidemiological surveillance program include the following.

Evaluating the Health Experience of Employees Exposed to a Known Hazard. In the event that exposure to low levels of potential chemical hazards is sustained, a surveillance program that examines changes in illness rates (or markers of exposure) can help a company evaluate the adequacy of control measures.

Estimation of Baseline Rates of Illness and Mortality. This is most useful for determining a baseline against which changes in health or disease markers may be evaluated. It is also helpful for comparing results of special studies that may be done to address specific health concerns. These may be used in place of standard rates derived from general population groups. In this way the potential for the healthy worker effect may be mitigated.

Screening Mechanism for Identifying Excess Risk of Illness. Through the periodic calculation of morbidity and mortality rates, a company may be able to identify, at an earlier time than otherwise possible, an excess rate of disease occurrence. Furthermore, by comparing rates of disease by process areas or by levels of monitored exposure, these excesses may be more readily attributable to specific workplace exposures. From this information, estimates of relative risk and excess risk may be derived and appropriate interventions considered.

Providing Assistance in the Design and Interpretation of Special Studies. Evaluation of routine morbidity and mortality statistics can serve to highlight the need for initiating an analytic study. This may be thought of as a hypothesis generating function. Also, these data may be useful in helping to estimate sample size or other quantitative parameters needed in planning a large scale study. Analogously, data from routine surveillance that indicate no excess risk may be used as evidence against the need for a large scale study. This would allow the resources to be directed to more productive activities.

Prompt Response to Health-Related Inquiries. By referring to a functional, comprehensive epidemiological database, queries about possible health effects or rates of specific illnesses from employees, community physicians, reporters, and other groups may be addressed in a more prompt and efficient manner than if such a resource did not exist.

Application to a Wide Range of Interests. A comprehensive health database, and the availability of staff who are experienced in manipulating the database for addressing relevant questions, should be an asset to sectors of a company responsible for medical insurance cost-containment, health promotion programs, environmental health and safety, and others.

Surveillance System a Success at Shell

Since 1979, a health surveillance system has been used at Shell Oil Company for the collection of information on employee absences in excess of five days. For the period 1981 to 1988, illness/absence information was extracted from the database in order to calculate annual morbidity rates among 14,710 Shell refinery and petrochemical employees. Among male workers, five conditions accounted for the majority of morbidity and subsequent work days lost. Injuries were identified as the leading cause of illness/absences (25%), followed by respiratory illness (17%); disorders of the musculoskeletal system (14%); digestive illnesses (9%); and heart disease (7%). Similar rates were noted among women. In general, morbidity patterns were highest in production workers, women, and older age groups.

Source: **Tsai, C.P., C.M. Dowd, S.R. Cowless, and C.E. Ross:** Prospective Morbidity Surveillance of Shell Refinery and Petrochemical Workers. *Br. J. Ind. Med. 48*:155-163 (1991).

Choice of Terms and Meanings

There are numerous definitions for epidemiological surveillance and related activities, probably because of the diversity of backgrounds, purposes, and perspectives of those generating and using the data. The lack of a standard set of definitions presents some problems, especially since epidemiological surveillance generally requires professionals from various fields to communicate with each other. Several attempts to standardize terminology have been made, and there appears to be some consistency emerging.

Since surveillance, medical surveillance, and epidemiological surveillance may mean different things to different people — even those in the same profession — it will be important to describe precisely what is intended when each of these terms is used. Because surveillance has meanings and connotations outside of the field of public health (most notably, espionage), a modifier such as epidemiological, occupational, or health will be used.

Because the primary goal of epidemiological surveillance is not necessarily to identify individuals for treatment, the term medical as a general modifier of surveillance will be avoided unless medical surveillance is specifically intended. In contrast to medical surveillance, epidemiological surveillance refers to a process by which the health and disease experience of groups of individuals is pooled and compared by various categories of risk factors (or exposures, or exposure levels), often taking other possible explanatory factors (potential confounders) into account.

When epidemiological surveillance is conducted in the workplace, or when one's job determines the comparison group in which one's health and disease experience are counted, the activity might be considered occupational epidemiological surveillance. In this chapter, however, the term "epidemiological surveillance" is used preferentially for two reasons: it is less cumbersome than occupational epidemiological surveillance, and the occupational context of this entire chapter should be otherwise apparent.

Definitions

Now that some principles of epidemiological surveillance have been presented, more formal definitions of common terms should be useful. The following definitions have been used fairly consistently and reflect the principles described above.

Exposure surveillance (or hazard surveillance) may be defined as a systematic and ongoing characterization of chemical or physical agents in the occupational setting, often to determine how many employees have been exposed and to what extent over time. This approach may be especially useful in those instances where there is an absence of health outcomes known to be associated with potential hazardous exposures in the workplace. On the basis of their exposures to known hazards, groups of employees may participate in medical surveillance.

Epidemiological surveillance (or rate-based surveillance) within an occupational setting may be defined as an ongoing systematic analysis and interpretation of the distribution and trends of illness, injury, and/or mortality in a defined population, relative to one or more indicators of workplace hazards or risks.

Surveillance may be thought of somewhat differently from traditional epidemiological studies, in which studies are specifically designed to address narrow hypotheses pertaining to the etiology of diseases or causes of death, or evaluation of clinical services or environmental or other interventions. Surveillance, on the other hand,

Males More at Risk

In a nurse-administered screening program, employees with increased coronary risk factors were identified through the assessment of blood cholesterol, body mass index, blood pressure, and smoking history. The results indicated that male employees had the higher prevalence of risk factors including hypercholesterolemia, obesity, and smoking. Since both blood pressure and cholesterol increased with increasing body weight, health promotion activities targeting weight loss along with smoking cessation could provide effective primary prevention of coronary disease.

Source: **Deacon, S.P.:** Screening for Coronary Risk Factors in Occupational Health Practice. *J. Soc. Occup. Med. 41*:126–128 (1991).

casts a broad net to discover, as early as possible, signals suggesting changes in health indicators. Still, it is necessary to maintain an ability to design and conduct, or to manage, ad hoc studies of the relationship between workplace exposures and potential adverse health outcomes, particularly in response to the clues generated by surveillance. In fact, the same meticulous attention to issues of validity, bias, data analysis, and interpretation that is required in analytic studies must be given to all surveillance efforts.

Medical surveillance is a system for the identification and management of individual cases of illness, as well as for the early recognition of latent disease (screening) and for the provision of health promotion activities. Medical surveillance is usually structured and administered in such a way as to be useful to patients and to plant physicians and nurses who are responsible for providing patient care. Medical surveillance programs are not necessarily designed with the integration of exposure and health information in mind.

Medical screening is the performance of tests or procedures aimed at the early identification of subclinical or clinical disease. Usually, medical screening consists of the application of tests or procedures that provide an indication of the presence or absence of disease but are not as sensitive or specific as diagnostic tests.

Clinics a Useful Source of Health and Hazard Data

Occupational medicine clinics are a source of health and hazard data useful for the surveillance of work-related injuries. Despite the prevalence of injuries in the workplace, data on their rates and circumstances are often lacking. For the calendar year 1988, Brewer et al. obtained 14,156 admission reports (for minor work-related trauma) from five occupational medicine clinics in the Chicago area. Analysis of the data indicated that the majority (78%) of these injuries were lacerations, sprains or strains, or contusions caused by metal items, boxes, machines, or working surfaces. Also, the reports allowed for the identification of the distribution of injury by industry, occupation, injury type, and body part affected.

Source: **Brewer, R.D., D.M. Oleske, J. Hahn, and M. Leibold:** A Model for Occupational Injury Surveillance by Occupational Health Centers. *J. Occup. Med. 32*:698–702 (1990).

Strategies for Epidemiological Surveillance

There are three basic approaches to the organization of an epidemiological surveillance system.

Disease Registry (Sentinel Events). An enumeration system for illnesses of particular concern to a company is established. The goal is to identify all new cases of this disease (and deaths from it) and to have valid denominator data as well; that is, the number of workers at risk for developing this disease. Denominator data about the numbers of individuals exposed to different chemical or physical agents should be available from a company's personnel and/or occupational hygiene database. Rates of illness, including a wide range of potentially relevant diagnoses or injuries, and mortality can then be calculated at some regular interval to provide an indication of the risks in the workplace. A related approach to this type of surveillance involves the systematic monitoring of targeted disease outcomes or sentinel events. This strategy is sometimes referred to as case-based surveillance.

Exposure Surveillance. A database is established that characterizes, based on personal and area monitoring, the potential exposure levels sustained in defined production or work areas or by representative individuals from discrete exposure zones. Priority substances selected as targets for surveillance are usually those with a relatively high hazard index, or they may be commercially important to a company. Exposures should be monitored over time and during normal operations as well as down times and under upset conditions. Persons determined to have been exposed at or above some action level may then be followed over time to investigate the possible occurrence of illness; however, if health effects are not known or are not known to occur below certain thresholds, such efforts are not likely to be productive. For exposures with known health effects (such as vinyl chloride) hazard-based medical surveillance may be required or simply prudent.

Disease and Exposure Integrated Surveillance. A comprehensive model that contains both exposure data and illness reports and has the ability to link these two components for individual workers, as well as for categories of workers (e.g., by job title or work area).

Although both the disease registry and exposure surveillance approaches have much to offer an occupational health program, either alone should be thought of as merely a starting point for arriving at an integrated model. Furthermore, in any of these approaches accurate record keeping is a core requirement, so that information collected at the present time will be comprehensible to a future generation of occupational hygienists and allied health professionals.

Detecting Disease Among the Asymptomatic

A cohort of 200 asymptomatic employees with known exposure to 4,4'-methylene-bis (2-chloroaniline) were examined by cystoscopy for the presence of bladder tumor. In this high-risk cohort two of the three workers found to have low-grade papillary tumors had completely normal results in a previous urine screening examination. Thus, for workers exposed to known or potential bladder carcinogens, cystoscopy screening allowed for the detection of disease in asymptomatic employees.

Source: **Ward, E., W. Halperin, M. Thun, H.B. Grossman, et al.:** Screening Workers Exposed to 4,4'-methylenebis (2-chloroaniline) for Bladder Cancer by Cystoscopy. *J. Occup. Med. 32*:865–868 (1990).

Core and Dynamic Elements of Epidemiological Surveillance

To understand better the rationale behind some of the more common definitions, it may be useful to identify the main elements, or dimensions, of epidemiological surveillance. It is proposed that there are four distinct core dimensions to an epidemiological surveillance model, coupled with two dynamic dimensions. Core dimensions refer to information groups that are usually collected at distinct points in time (cross sectionally). These points are sometimes scheduled on a periodic basis. Dynamic dimensions refer to the ways in which a surveillance database is organized. Specifically, they comprise the rational grouping of data collected about individuals (or exposures) and the systematic and periodic arrangement of data over defined time periods.

The four core dimensions represent the categories of information commonly required when conducting occupational epidemiological studies. These are (1) demographic information; (2) characterization of workplace exposures; (3) potential alternative explanations (confounders); and (4) defined health outcomes.

The two dynamic dimensions are (1) the arrangement of information about individuals into a group structure and (2) the arrangement of cross-sectional information into a time-dependent format.

Note that all strategies of epidemiological surveillance may not necessarily address each of these elements to the same extent; however, each usually is represented or implied in some manner. For example, even the simplest model of comparing the mortality experience of an occupational group with that of the general population touches on each dimension: individual vital status and cause of death must be known; mortality rates (all cause or cause-specific) are estimated for the entire occupational group or for subgroups; exposures or risk factors associated with the occupational group or subgroup are implied, in that the occupational group's mortality is evaluated to see if it differs from that of the general (e.g., U.S. or state) population; confounders should be taken into account to the degree information is available; and for each group being compared, a period during which all deaths are considered must be specified (which also implies a time period of occupational risk; i.e., some past exposure period). A brief consideration of each of the basic dimensions follows.

Core Dimension I: Demographic Information

It should be clear that the interpretation of an association between an exposure and an illness must always be considered in light of the personal attributes of an individual. For example, a physician performing a differential diagnosis on a patient during a clinical encounter must consider the age, gender, and personal background of the patient. Similarly, a surveillance system will need to capture fundamental data on the individual employee; this may require cooperation from a company's personnel or human resources department.

Core Dimension II: Employment/Exposure Information

The basic analytical approach to surveillance is to compare the disease or mortality experience of groups of employees with different occupational exposures. In the chemical industrial setting the focus would, of course, generally be chemicals. Even if an occupational group is compared with a group of individuals from the general population (with the same age and gender distribution) it is implied that the occupational group has a different risk factor profile by virtue of employment. The exposure dimension may be considered by some to be the heart of the surveillance system because it defines the need for the system. For this reason, close collaboration with professionals (technicians and managers) from a company's occupational hygiene program is usually required.

Finding a Foundation for Hazard Surveillance

To identify work-related hazards, a strategy has been developed for hazard surveillance using information in the management information system (MIS) from the Occupational Safety and Health Administration (OSHA). Based on reports from OSHA inspections, the MIS database contains actual exposure measurements as well as exposure levels by job title. Through the use of these data, hazard profiles and exposure rankings were developed for a variety of industries defined by their four-digit Standard Industrial Classification. Thus, the MIS data files provide a foundation for hazard surveillance in the absence of comprehensive exposure data.

Source: **Froines, J.R., C.A. Dellenbaugh, and D.H. Wegman:** Occupational Health Surveillance: A Means to Identify Work-Related Risks. *Am. J. Public Health 76*:1089–1096 (1986).

It should also be affirmed that many successful occupational health research programs can begin without a fully integrated, detailed exposure database (of the kind usually kept in the occupational hygiene department).

Core Dimension III: Information on Potential Confounders

A substantial contribution of epidemiology to the advancement of medical research over the past few decades is its application in the area of data analysis, and specifically in the statistical control of confounding. Confounding is a type of bias that adversely affects the understanding of an exposure/disease relationship. Confounding may occur when an extraneous variable either (1) obscures a true relationship between an exposure and a disease or (2) creates an artifactual (spurious) association. For example, it has commonly been observed that chronic gamblers have higher death rates than the general population. This association would quite likely disappear, however, if smoking and alcohol consumption were in some manner taken into account. On the other hand, a true underlying relationship between smoking and the risk of mortality might be missed if there were different age distributions in the groups of smokers and nonsmokers.

OCISS Provides Answers

Using pooled data from the Occupational Cancer Incidence Surveillance Study (OCISS), newly diagnosed cases of lung cancer were identified in the Detroit metropolitan area. Through telephone interviews, information on occupation and smoking history was obtained for 5935 incident lung cancer cases. Analysis of the data according to 43 usual occupational groups and 48 usual industry groups identified numerous occupations and industries in which the risk of lung cancer was elevated significantly. Often, probable exposure to diesel exhaust occurred in the occupations with excess lung cancer.

Source: **Burns, P.B., and G.M. Swanson:** The Occupational Cancer Incidence Surveillance Study (OCISS): Risk of Lung Cancer by Usual Occupational and Industry in the Detroit Metropolitan Area. *Am. J. Ind. Med. 19*:655–671 (1991).

The control of confounding may be accomplished in a variety of ways; most common is the use of computer-based statistical techniques. More important than the method of control, however, is the commitment to anticipate the range of potential confounding variables that would be of interest in the analysis of surveillance data. In turn, this should lead to inclusion of the most important of these variables in a data collection effort that would eventually become part of the surveillance database.

Core Dimension IV: Health Outcome Information

Many of the landmark efforts of epidemiological surveillance in the chemical industry were based on examining mortality and/or cancer trends of employees. Similarly, safety departments have a long tradition of monitoring injury trends. Recently, however, there has been growing appreciation for a broadening of perspective that would include the examination of a variety of less traditional health endpoints such as cardiovascular disease, neurological signs and symptoms, hematologic markers of disease, conditions related to repetitive motions, psychosocial parameters, and others. In any surveillance model a balance will have to be arrived at by considering (1) the available scientific knowledge about known exposure/disease relationships; (2) conditions that may have little direct evidence for associations with occupational exposures but may be plausible; and (3) the complexity and inefficiency of attempting to conduct surveillance on too many, or on impractical, endpoints.

Dynamic Dimension I: Formulation of Group Data

Individual employees represent the fundamental unit of measure in epidemiological surveillance. Some basic facts or information must be known about each individual, such as employment and vital status, but also possibly including more detailed information such as health history, occupational history, exposure measures, tobacco use, possible exposures outside the workplace, etc.

Also, individual employees may be screened for disease or for body burdens of chemical or other exposures, monitored (over time) for indications of disease, or managed as patients under physicians' care. In turn, the information derived from these interventions can be integrated into the surveillance database.

Individual employees can also serve as sentinels of events or disease processes that have occurred or that will occur in the future, thus possibly allowing some further investigation or intervention.

Surveillance, however, cannot be accomplished at the individual level, but becomes possible only when data from groups of employees are pooled and evaluated. This group perspective is the fundamental attribute that launched epidemiology into the forefront of communicable disease research, and it continues to be a cornerstone of the profession today. Typically, the rate or prevalence of disease, or prevalence of some indicator of exposure or risk, are compared among groups of employees or with some other referent group (such as the general population).

Dynamic Dimension II: Time

Time plays several roles in epidemiological surveillance: rates of disease or death are compared among groups over specified periods of time (such as annual cancer incidence rates); employees may be monitored for changes in health occurring over time; and for each case of disease detected, an appropriate window of time may be identified to see whether the employee might have sustained some particular exposure during it. For example, many cancers remain undiscovered for several years after tumor initiation, and therefore, the exposures sustained during the years immediately preceding diagnosis would be irrelevant to the formation of that specific tumor. For other conditions, the most recent exposures will be most relevant (e.g., skin rash). Each disease has its own distinct latency period (time between inception and clinical detection); there is even some interindividual difference in the rate of disease progression.

Purposes for which Health Data are Collected

Health-related data are gathered for diverse reasons and may be put to a number of uses, including uses not necessarily intended or thought of at the time they were first collected. Whether or not any of the uses is valid is a critical concern. This concern includes the purpose for which the data were originally intended.

For example, the context in which an employee's blood pressure is measured may be any one or more of the following:

- Annual plant-wide high blood pressure identification program;
- Routine physical exam ("well-visit");
- Sick employee seeking medical advice ("sick-visit");
- Tracking effectiveness of a high blood pressure control program; and
- Ad hoc program to detect cardiovascular effects, including elevated blood pressure, among employees exposed to a specific chemical.

Although the blood pressure measurement may be performed exactly the same way regardless of the intended purpose, the underlying reason it is measured varies with the context. Consider some of the scenarios above.

In a plant-wide high blood pressure identification program, the purpose of the blood pressure measurement is to sort individuals at one point in time into two groups: those with acceptably low blood pressure who require no further action, and those with apparently elevated blood pressure who require some follow-up or referral to a physician.

In contrast, the blood pressure measurement from the high blood pressure treatment tracking program represents one of a series of measurements. Soon after treatment begins, blood pressure readings are expected to drop, but still may be considered elevated.

In the program designed to detect changes in blood pressure related to chemical exposures, individual changes over time, although possibly quite small, may be evaluated at a group level and related to changes in some measure of exposure (such as occupational hygiene measurements).

Questioning the Validity of Data: A Requirement of Surveillance

For each of these purposes, the validity (accuracy) of the blood pressure reading is important, and lack of validity has some implications. If in the high blood pressure detection program the blood pressure measurement device always overestimates the true blood pressure, some proportion of participants will be unnecessarily referred for follow-up (these are often called false-positives). Conversely, if the device underestimates true blood pressure, some with high blood pressure will be missed (false-negatives). If the error is random, sometimes overestimating and sometimes underestimating blood pressure, then there will be both false-positives and false-negatives.

In the program looking for changes in blood pressure over time related to exposures, the actual value of the blood pressure is not as important, but rather the reliability is more critical (ability to give the same answer for a given blood pressure). For example, if the readings always overestimate blood pressure by 7%, an accurate picture of changes across time is still possible. If, however, the error is moderate in size but random, changes over time might simply result from chance, and true changes might be masked.

Similarly, turning to surveillance, data may be collected for one purpose but have additional uses. Data may be collected primarily for surveillance purposes, or may be routinely collected for another purpose but secondarily used for surveillance. Regardless of the underlying purpose, the central concern, as in all of epidemiology, is whether the data are valid for the intended use, primary or secondary. Further, lack of validity, either because of lack of overall accuracy or lack of precision (reliability or repeatability of individual measurements), may have various implications depending on how the data are subsequently used.

Using Routinely Collected Data for Surveillance

Consider the scenarios above once again, and the various reasons that a blood pressure might be measured. Also consider how that measurement could be used for epidemiological surveillance purposes. The routine physical (well-visit) and the sick employee visit (sick-visit) represent medical services provided directly to an employee to address his or her individual health needs, including prevention, early detection, and treatment of disease. If blood pressure measurements are taken during this process and recorded in the medical record system, these data may serve a variety of purposes, including early detection of the development of hypertension, insights into the diagnosis of the current illness, etc. To use the same data for surveillance, a number of key issues must be addressed.

- Are these measurements valid estimates of the true blood pressure at the time of measurement?
- Do all or a substantial proportion of employees have blood pressure measurements taken regularly? Are these measurements made using similar equipment and procedures?
- Can these data (recorded in a medical record) be accessed for any other purpose without violating confidentiality agreements?
- If accessible, are these data retained in a standard format on an easily retrievable record/medium?

These points suggest that information obtained during medical visits may not directly be usable for surveillance purposes.

On the other hand, data obtained in some of the other scenarios above may be more easily used for epidemiological surveillance. The annual plant-wide high blood pressure identification program, for example, may provide a better opportunity for surveillance for the following reasons.

First, if the screening program is indeed plant-wide, and all or a large proportion of employees participate, chances are improved that the participants are representative of the group of all employees, reducing the chances of obtaining a biased (or erroneous) impression based on data collected on a select subgroup of volunteers.

Second, depending on how the program is run, it is likely that the blood pressure measurement is made using a common protocol — including the same equipment (or same model/manufacturer) and procedure for all participants. This avoids problems of obtaining different results depending on the equipment or technique used.

Third, the date of screening, results, and other relevant information about the participants, such as age, sex, weight, etc., may be recorded on a standard data collection instrument for all participants.

Fourth, the annual nature of the program allows for efficient tracking of individuals over time. This is especially important if meaningful increases in blood pressure, rather than attainment of some arbitrary cut point, are used as the indicator of change in health status.

In this example the missing element of surveillance, as described above, is some indicator of exposure or other risk factor on which individuals may be combined into groups and compared. For example, employees involved in manufacturing Chemical A may show a pattern of increasing blood pressure different from those involved in manufacturing Chemical B, even if the two groups are otherwise similar (with respect to other risk factors for elevations in blood pressure).

Without some indicator of risk, the activity remains a high blood pressure screening program, intended to identify individuals requiring medical attention, regardless of the reason(s) for the condition. Epidemiological surveillance provides an additional dimension in which potential workplace (and possibly lifestyle) factors are examined to determine their possible role in causing or increasing the chances of that disease occurring. Identification of these factors makes it possible to intervene to prevent subsequent disease from occurring or to further study the apparent association in a more efficient and definitive manner.

Individuals and Groups

Tracking the effectiveness of a high blood pressure control program may be quite analogous to some forms of epidemiological surveillance. Presumably, a group of employees with high blood pressure is identified, and one or more interventions are assigned (such as weight control, exercise, dietary modifications, medications, etc.). Individual blood pressures are measured periodically to see whether the regimen is effective for each individual. The changes in blood pressure are considered for the entire group, and, if the effectiveness of two or more different programs is being studied, these group-level measurements would be compared to see which treatment was more effective.

In this example, however, there is only one group (the "exposed" or treated group), and it is expected that each employee will experience some decrease in blood pressure. However, the overall change and rate of change may differ from individual to individual. Also, for various reasons such as lack of compliance, effectiveness, etc., some individuals may not demonstrate any improvement in blood pressure. Nevertheless, the blood pressure control program may still be considered a clear success if a large majority of participants derived some benefit (as reflected in some summary measurement of the group's experience, such as a decrease in the average blood pressure — which is the same as the average of all individual changes in blood pressure). This underscores the importance of looking at health events and indicators at a group level, as in epidemiological surveillance.

The final scenario from above, the special program for the detection of cardiovascular effects among employees exposed to a specific chemical, may be seen as a specific type of epidemiological surveillance. The expected results, or findings of the program, are that the exposed employees experience cardiovascular effects such as increased blood pressure at a rate no higher than among comparable unexposed individuals. If an excessive

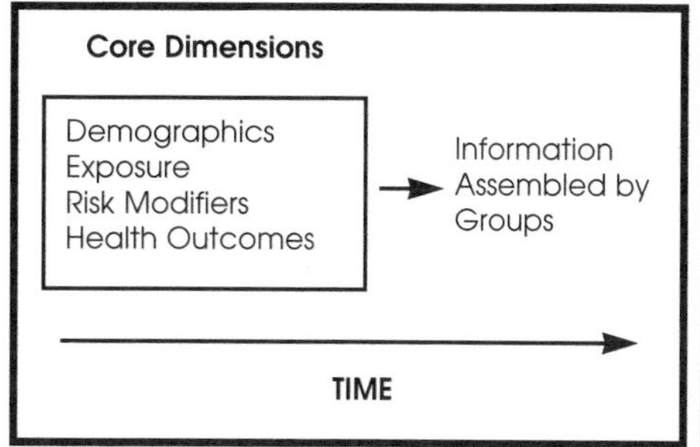

Figure 5.1 — Schematic of epidemiological surveillance.

rate of increase in blood pressure or other signs or symptoms of a disease process are detected, further investigation with possible intervention may be necessary.

The simple schematic in Figure 5.1 represents the conceptual requirements of epidemiological surveillance that were introduced in this chapter. Fundamentally, core dimensions are data that are typically collected, more or less, on some routine basis in the occupational setting. The dynamic dimensions refer to (1) grouping of data collected on individuals into relevant categories of interest with respect to exposure, job type, demographic characteristics, etc., and (2) following individuals and groups systematically and periodically over time.

Sample of Existing Strategies

In 1982 an issue of the *Journal of Occupational Medicine* was devoted to the topic of medical information systems.(3) Although some of the material presented is now obviously dated, an interesting historical perspective about surveillance implementation in chemical and other companies may be gained by a review of that issue. Also, some insights about the more mechanical steps of surveillance implementation are presented in that issue.

The issue included presentations from Diamond Shamrock (COHESS); Monsanto (MEHI System); Standard Oil (Amoco System); Standard Oil-California (OHIS); Standard Oil-Ohio; General Foods (MEHS System); SmithKline; Phillips Petroleum (CHESS); Upjohn (OHSS); Shell Oil (HSS); New York Telephone (MIS); Ford (EHSS); Control Data (MSM); Digital Equipment (IHMS); Exxon (HIS); DuPont (PERS/MERS); and IBM (ECHOES).

Certainly, all of the systems described have evolved beyond the descriptions given; some have even been abandoned or replaced. Nevertheless, in perusing this

Flexibility	Economy
Interaction	Innovation
User Friendliness	Key Staff
Modular Design	Commitment
Valid Database	Phased Approach

Figure 5.2 — Common factors of epidemiological surveillance systems. (*Source:* **Joiner, R.L.:** Occupational Health and Environmental Information Systems: Basic Considerations. *J. Occup. Med. 24*:863-867 (1982).)

issue, one can appreciate the similarity in goals and efforts exhibited by diverse companies. Furthermore, one gains a clear respect for the painstaking effort invested in setting goals, developing and testing a new system, and implementing and evaluating it. The common factors of epidemiological surveillance systems expressed in a roundtable organized by the Committee on Medical Information Systems of the American Occupational Medical Association (which led to this issue of the *Journal of Occupational Medicine*) are found in Figure 5.2.

Summary

Occupational epidemiology relies on the integration of environmental exposure data with health indicator or outcome data to assess their interrelationship. Although traditional epidemiological studies are conducted frequently, a method of growing usefulness in the modern workplace is epidemiological surveillance — the systematic and ongoing analysis of illness or injury data relative to various indicators of workplace hazards.

The occupational hygienist today is increasingly likely to participate in some form of epidemiological surveillance and needs to be familiar with the basic concepts, terminology, and practices. This chapter provides an overview of the topic, practical strategies for establishing epidemiological surveillance, and examples of several approaches to this powerful occupational health tool.

References

1. **Henderson, D.A.:** Surveillance of Smallpox. *Int. J. Epidemiol. 5*:19-28 (1976).
2. **Eylenbosch, W.J., and N.D. Noah:** Historical Aspects. In *Surveillance in Health and Disease*, edited by W.J. Eylenbosch and N.D. Noah. Oxford: Oxford University Press, 1988.
3. **Anonymous:** Medical Information Systems. *J. Occup. Med. 24(10)* (1982). [Special issue.]

Section 2

Hazard Recognition and Evaluation

Outcome Competencies

After completing this chapter, the user should be able to:

1. Define underlined terms used in this chapter.
2. Employ the general principles of evaluation of worker exposure.
3. Perform the four primary responsibilities of the hygienist: anticipation, recognition, evaluation, and control.
4. Analyze and evaluate the purposes of the workplace survey.
5. Plan a thorough survey of the workplace prior to beginning any testing.
6. Describe the basic tenets of sampling, including the who, what, when, where, and how many.
7. Interpret sampling results, based on standard comparison criteria.
8. Explain other workplace evaluations, such as ambient and biological monitoring.
9. Apply good judgment to the application of scientific inquiry.

Key Terms

administrative controls • agents • ambient monitoring • anticipation • biological effect monitoring • biological monitoring • chain of custody • compliance • control • control measures • engineering controls • evaluation • "Hawthorne Effect" • hazard • health surveillance • noncompliance • odor threshold concentration • olfactory fatigue • personal protective equipment • phased sampling schemes • program audits • prospective epidemiology • recognition • regulatory standards • risk assessment • sampling strategy • short-term exposure limit (STEL) • significance • stresses • survey • technical audits • time-weighted average (TWA) • toxicity • voluntary guidelines • work practices

Prerequisite Knowledge

Prior to beginning this chapter, the user should review the following chapters:.

Chapter Number	Chapter Topic
1	History and Philosophy of Industrial Hygiene
2	Occupational Exposure Limits
3	Legal Aspects of the Occupational Environment
4	Environmental and Occupational Toxicology
29	Psychology and Occupational Health
30	Worker Education and Training

Key Topics

I. General Principles of Occupational Hygiene
   A. Anticipation
   B. Recognition
   C. Evaluation
   D. Control

II. Evaluation Process: The Occupational Hygiene Survey
   A. Determining the Purpose and Scope of the Survey
   B. Familiarization with Process Operations
   C. Preliminary Survey (Qualitative)
   D. Occupational Hygiene Field Survey (Quantitative)

III. Sampling/Analytical Procedures
   A. Proximity to Breathing Zone
   B. Accuracy

IV. Interpretation of Sampling Results
   A. Time-Weighted Average Exposures
   B. Analysis of Short-Term Exposures
   C. Comparing Sampling Results with Appropriate Standards
   D. Comparing Results with Previous Data

V. Other Evaluations Involving the Occupational Hygienist

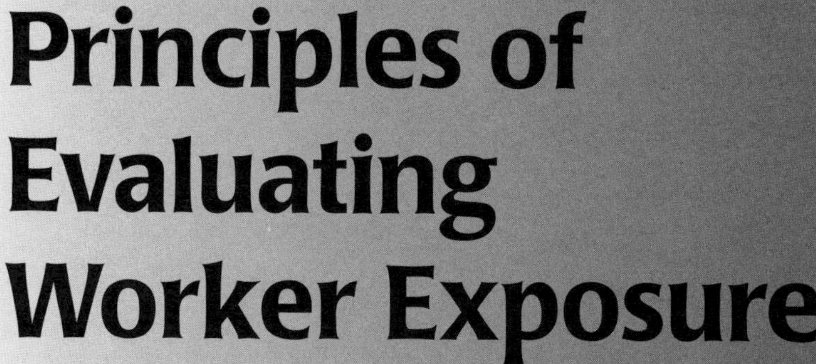

6 Principles of Evaluating Worker Exposure

Ronald G. Conrad, CIH

Robert D. Soule, Ed.D., CIH, CSP, PE

Introduction

The purpose of this chapter is to provide the occupational hygienist with an understanding of the general principles of evaluating worker exposures in the industrial environment. Evaluation is one of the four primary responsibilities of the occupational hygienist: anticipation, recognition, evaluation and control of environmental stresses arising from the workplace. When a potential hazard is anticipated and/or has been recognized, it is incumbent upon the occupational hygienist to determine the magnitude of the stress by appropriate evaluation of the environment. The ultimate responsibility (i.e., effective control of the work environment) can be accomplished most directly only when based on adequate evaluation.

The general principles discussed in this chapter are drawn from traditional evaluation methodologies and, it is believed, properly reflect the prudent practice of occupational hygiene today and in the near future. This chapter does not address in any detail new technologies that represent "cutting-edge" occupational hygiene evaluation methods; these are discussed in other chapters in this book. This chapter does not provide checklists and how-to methods; again, this information is provided elsewhere in the book. Instead, this chapter focuses on the evaluation process, allowing for interjection of technologies of all types.

Traditionally, the most effective evaluations of work environments have used a multidisciplinary, knowledge-based approach. Common examples include occupational hygiene, chemistry, engineering, health physics, medicine, epidemiology, toxicology, and nursing, as well as management and manufacturing expertise. A most successful approach to evaluating the work environment would integrate the knowledge of these various disciplines for evaluation and control of potential risks.

Obviously, it is not always possible, or even practical, to assemble such a multidisciplinary group of highly skilled professionals in most occupational settings. However, it is essential that the occupational hygienist be knowledgeable of these professional disciplines and the potential contributions to the solution of specific problems that each of them represents. In evaluation of the occupational environment, the occupational hygienist often interacts closely with other professionals.

General Principles of Occupational Hygiene

Occupational hygiene is the science and art "dedicated to the anticipation, recognition, evaluation, and control of environmental factors arising in or from the workplace that may result in injury, illness, impairment, or affect the well-being of workers and members of the community."[1] The process for systematic elimination or control of health hazards in the work environment has been well-established in the occupational hygiene profession. The sequence of responsibilities in the above definition also suggests a sequence in the flow of information among the four responsibilities, or phases, of occupational hygiene decision making. Assuming a hazard is anticipated effectively, it becomes recognized and its magnitude is determined. The hazard is then evaluated and, when the level of hazard is unacceptable, necessary control of the environment is implemented. In the ideal case, this flow of information is mainline, direct, and sequential. However, the overall process allows for, even demands, that multiple channels of feedback be maintained. For example, in the recognition phase, hazards may be found that were not anticipated or those anticipated may not be found. Exposure estimates made for purposes of prioritizing activities will be refined by actual exposure measurements. New control measures should be evaluated and the results used to judge control effectiveness and, perhaps, reset priorities. In an effective, continuing occupational hygiene program, information collection, analysis, and decision making go on continuously and simultaneously.

Anticipation

The difficulty and complexity of health hazard control have come to be so great that it is necessary to anticipate what problems may occur before a plant, process, or product is introduced. Problems discovered later may cost too much to fix, both technologically and economically, to continue the process. Had these problems been anticipated, the project might have been abandoned at an earlier stage of development. Therefore, it is necessary to anticipate potential problems and possible solutions early in the development stage before major commitments have been made. This particular addition to the occupational hygienist's responsibilities is most difficult for the entry-level, inexperienced occupational hygienist.

Anticipation depends on and extends the ability to recognize, coupled with a broad and current awareness of developments in the organization and its business, in scientific developments and new technologies, in regulatory areas bearing on the organization's activities, and in other activities that have an impact on the health of workers. Particularly among those businesses that invest heavily in time and other resources for research and development of new chemicals or processes, the direct involvement of the occupational hygienist is moving more and more into the laboratory phase of development. This is being done, in part, to allow the occupational hygienist to be able to anticipate more fully what the health impact of the new product or process constitute.

Recognition

The line separating "anticipation" and "recognition" is not always a clear one. Some have distinguished them on the basis of whether the situation being examined actually exists. If it is still in a conceptual phase, the process being applied is considered to be "anticipation." Then it is presumed that in the recognition phase the facility exists. This is a somewhat arbitrary distinction; anticipation of hazards can and does occur with existing facilities, and recognition of hazards can take place when the facility is in a planning stage.

Whenever it is done, recognition of health hazards typically requires use of available information and application of occupational hygiene fundamentals such as exposure elements, toxic response characteristics, etc. The recognition of potential or existing hazards evolves as one becomes familiar with the processes, creating and maintaining an inventory of physical, chemical, and biological agents encountered, reviewing the different job activities of a work area, and studying the existing control measures. These activities are discussed in detail later in this chapter.

From this information, characterization of the workplace is derived. It may be based on workplace (the physical environment), work force (workers) or agents (chemicals, physical or biological agents). When examining a large population, characterization often results in defining "exposure groups" (i.e., workers having similar exposure profiles).

Whether it occurs in anticipation or recognition of hazards, the process commonly referred to as risk assessment is appropriate at this point because of the potential impact of health considerations on project, process, and product viability. Risk assessment is covered in detail in Chapter 17. Risk assessment, in this context, typically results in estimates of chemical toxicity and/or future exposures. If a new or uncommon chemical is being considered, chances are there is little toxicological information available. It may be necessary to conduct preliminary testing or to estimate toxicity by chemical constituent activity or by analogy to other substances. Although most of this effort applies for new chemicals, it is equally important to assess the effects of new or modified nonchemical processes, which may use and release chemicals or generate physical or biological hazards.

The exposure element of the risk assessment should be estimated for both production and downstream users. Techniques used in this assessment include prediction from first principles, such as heat and mass transfer, modeling, and analogy. Generally, occupational hygienists use all three techniques. For either toxicity or exposure estimation, it is important to reduce uncertainty in the data since the greater the uncertainty the more conservative the conclusion must be.

For both chemical and process risk assessments, there are four possible outcomes of the anticipation/recognition phase (the probable resulting decision is provided parenthetically):

1. It is likely that no problems or only small problems that can be controlled without having an impact on the project economically will develop [Proceed with the project.]
2. Project costs, including those for control of hazards, are uncertain but are potentially large enough to affect the project economics significantly [Obtain more data.]
3. Control costs are known reliably and are high enough to have an effect on the project viability [Factor control costs into the economic analysis.]
4. The risks/uncertainties are so great that the project is not acceptable [Modify the project to avoid/minimize risk or cancel the project.]

Evaluation

Evaluation means the examination and judgment of the amount, degree, significance, worth, or condition of something and perhaps uses more "art" in its implementation and use than any of the other occupational hygiene responsibilities. Key skills in evaluation are observation and judgment, both of which are developed and refined as a result of years of training and experience. This subjective application of one's experience is used together with the objective measurements to determine the magnitude of a particular stress. The whole body of information then is analyzed, synthesized, and tested for significance. Judgment, born of experience again, is applied to determine the significance of findings in order to arrive at appropriate conclusions and formulate meaningful recommendations. This overall process comprises what is referred to in this chapter as "evaluation."

In the early years of the occupational hygiene profession, evaluation was directed at the "dangerous trades" where frank occupational agent stress factors and diseases were rampant. Today, occupational hygiene embraces all workplaces, including such subtle issues as indoor air quality in office environments, repetitive motion problems with keyboarding activities, and management of hazardous waste operations. The primary objective of evaluation is to determine the magnitude and significance of health hazards, defined broadly as absence of well-being. The complex, diverse, and variable characteristics of the modern occupational environment add to the difficulties of the evaluation process. There may be health hazards in the workplace as a result of exposure to gases, vapors or aerosols, biological agents, noise, ionizing or nonionizing radiation, temperature extremes, or the physical and psychological stress imposed in the form of ergonomic factors at the human-machine interface. The remainder of this chapter presents the guidelines to be considered by an occupational hygienist in developing a process for evaluation of the occupational environment.

Control

As indicated in the preceding section, this chapter focuses on the evaluation aspect of the occupational hygienist's responsibilities. Control is mentioned here for purposes of completeness of the discussion of the overall occupational hygiene effort. "Control" is the culmination of the effort in addressing the primary objective of the occupational hygienist: providing a healthful work environment. Current occupational hygiene practice recognizes a hierarchy of controls; in priority order, these are engineering controls, work practices, administrative controls, and, as a last resort, use of personal protective equipment. These are addressed in detail in Chapters 31, 32, 35, and 36 of this book and the reader is referred to them for further understanding of occupational hygiene controls.

Evaluation Process: The Occupational Hygiene Survey

With few exceptions, the process associated with occupational hygiene evaluations has taken the form of the "survey" (i.e., an episodic investigation of a particular situation and, usually, for a specific purpose). Although all aspects of occupational hygiene (i.e., anticipation, recognition, evaluation, and control) may be involved, it has become common to think of the survey itself as more narrowly limited to, or focused on, evaluation.

Within the context of this chapter, then, the occupational hygiene survey is defined as consisting of the following steps: 1) determination of the purpose and scope; 2) familiarization with process operations; 3) a preliminary, walk-through (qualitative) survey; and 4) the actual field (quantitative) survey. Surveys are, in a sense, technical audits that examine present status of health hazards; these are distinguished from program audits that look at the management system in place to ensure that health hazards remain under continuing control. For some applications, surveys are repetitive and can be used to build a comprehensive exposure data base. Interpretation of the results of surveys, including formulation of any strategy for corrective action dictated by the results, are discussed here as activities that occur following the survey.

Determining the Purpose and Scope of the Survey

The reasons for performing an occupational hygiene survey are many and varied. The purpose for which the survey is desired, and the context in which it is performed, are major factors in determining the depth and breadth of the evaluation. Reasons for conducting occupational hygiene evaluations include: 1) identification and quantification of specific contaminants that may be present in the environment (this can be done as part of implementation of a comprehensive occupational hygiene program or in response to specific, limited concerns); 2) assessment of compliance status with respect to various occupational health standards; 3) determinations of exposure in response to complaints; 4) generation of environmental data in conjunction with an interface with medical and/or epidemiological efforts; and 5) evaluation of the effectiveness of engineering controls.

Comprehensive Occupational Hygiene Survey

A large portion of the total occupational hygiene evaluation effort involves the ongoing monitoring activities that are critical components of the administration of good occupational hygiene management programs. In these programs, there are regular, periodic exposure measurements of representative workers to selected agents. This monitoring usually documents the average exposure condition of a population of workers being tracked over extended periods of time, up to their working life times. These monitoring programs can be conducted in a manner that makes it possible for the evaluations to serve multiple purposes. For example, most routine monitoring of workers' exposures to lead is conducted in a manner that complies with OSHA requirements for periodic monitoring, even though the original and primary purpose for the monitoring was to track worker exposure as part of a good occupational hygiene program. The primary objective, then, of comprehensive, ongoing occupational hygiene evaluations is to create and maintain a database describing the exposure of individual workers over time.

Specific, Limited Survey

The occupational hygiene evaluation described above usually is done so as to be representative of the average condition of a work population or other similar exposure group. As a consequence, unusual exposures may not be evaluated. These exposures include those that result from infrequent activities; from cyclic or periodic operations or those associated with batch operations with extended cycle time; from accidental or inadvertent releases of substances; from experimental, pilot-plant or research operations; and a variety of other nonroutine activities. When investigating accidental exposures, claims of past exposures, or other situations for which the current situation may not represent the actual exposure of interest, it might be desirable to attempt to simulate the conditions present at the time of the incident. In all of these scenarios, there is a need for a specific, fairly limited occupational hygiene evaluation, designed with specific, narrowly defined objectives.

Frequently, the limited, specific surveys result from the findings of the comprehensive, ongoing monitoring program. Any result of the routine evaluation effort that appears to deviate from the historical pattern warrants appropriate follow-up with a more focused evaluation of the specific situation. In other words, the comprehensive and the specific surveys should not be separate efforts but instead complimentary aspects of good occupational hygiene management.

Compliance Survey

Many of the occupational hygiene evaluations performed in the United States are done to satisfy some aspect of a legal obligation. Most common, of course, are the efforts mandated by regulations from agencies such as the Mine Safety and Health Administration (MSHA) and Occupational Safety and Health Administration (OSHA). Interestingly, the fundamental objective of the evaluator depends on his/her role in the evaluation process. The occupational hygienist working on behalf of the organization will be interested in documenting compliance; the compliance officer takes on the evaluation looking for evidence of noncompliance. When standards are interpreted to mean that an exposure concentration over the appropriate averaging time is never to exceed the limit, then demonstration of noncompliance requires a sample set that is over the limit with an acceptable degree of certainty. While there are certain practical reasons why inspection programs rely on this "traffic cop" approach, it is clear that the result does not necessarily contribute to a process of judging health risks fully. Similarly, the employer's objective is to demonstrate confidence that overexposures are not occurring. This means that results of evaluations need to be far enough below the exposure criteria that the employer can be reasonably sure that all exposures, whether evaluated or not, are below the acceptable limit.

The measurements made by a compliance officer, in order to detect noncompliance as defined by regulations, usually are made in situations where such noncompliance is most likely to be found. It is not intended, nor necessary, to conclude that the results of the evaluation are descriptive of typical conditions. A significant overexposure identified by a compliance officer is, of course, a reason for concern because of the possible citation and penalty that could result. More important, however, the noncompliance suggests a possible weakness in the surveillance, evaluation and control activities that make up the occupational hygiene program. Similarly, it could be that the apparent overexposure documented by the compliance officer is a rare, highly atypical event. In either case, the finding could and should have impact on the administration of both of the purposes for conducting occupational hygiene surveys: comprehensive ongoing evaluations and specific focused surveys. Current occupational hygiene standards allow re-evaluation of a finding of noncompliance when the employer can document that ongoing evaluations of exposure indicate that health hazards are under control as defined on a statistically rigorous basis. Obviously, this process would be an added incentive to organizations to administer ongoing comprehensive evaluations as a vital component of their occupational hygiene programs.

Response to Specific Complaints

Even where there is comprehensive, ongoing monitoring of worker exposure, and specific surveys and compliance monitoring is done faithfully, there are likely to be occasional complaints of specific conditions that may constitute an overexposure. In all probability, a continuing responsibility of the occupational hygienist will be the need to respond to these "brush fires." Indeed, employee complaints should be evaluated—not only because they may be real and warrant corrective action—but because they are believed to be real by the complainant and therefore cause anxiety. Increasingly, an occupational hygienist must be concerned not only with the physical facts but also with the social and psychological aspects of occupational health. When evaluating a health hazard based on a complaint, it is best to talk directly to the complainant if at all possible. The purpose of this is to demonstrate that the complaint is being taken seriously and to obtain the facts directly from the person experiencing the effects. In evaluating a complaint, it is important to recognize that the complaint may be valid even when the worker perceives the basis for the complaint incorrectly. For example, an individual may claim to be affected by one substance when another substance is actually the problem. It is not uncommon for complaints to be registered in situations where the actual exposure is entirely within acceptable limits; this is particularly true where the substance being handled has an unusual or unpleasant odor or a low odor threshold. Whatever the outcome of the evaluation, the complainant should be informed so that it is demonstrated that all possible health hazards are taken seriously.

Medical/Epidemiological Interface

As the occupational hygienist moves toward a closer working relationship with the medical health professional, the contribution of occupational hygiene evaluations to the overall worker health management process will increase. Epidemiologists have always needed to know what workers were exposed to in the past in order to determine if there is an association between the past exposure and the current health condition. Since evaluations may not have been made or the data were incomplete, it sometimes is necessary to develop estimates of past exposures by some appropriate modeling scheme. Historically, this process has often been very qualitative, even subjective, in nature. For example, estimates of the concentrations of substances to which workers were exposed in the past sometimes were based on the workers' descriptions of symptoms or by reference to job titles. In most cases, the quantification of past exposures has been, and remains, a difficult process.

This epidemiological application of occupational hygiene data is by definition a retrospective process, one in which there are attempts to explain the health status seen in individuals today by constructing exposure profiles from the past. An encouraging development in overall worker protection programs during the past several decades is a process that might be thought of as "prospective epidemiology." If epidemiology involves the correlation of cause-and-effect, or dose-response, data, then those organizations that are maintaining comprehensive occupational hygiene and medical surveillance programs are in a position to perform epidemiological studies of their workers at any point in the future. The occupational hygiene evaluations represent the "cause" or the "dose" and the health status of workers documented by medical surveillance represents the "effect" or the "response," respectively. The contribution of ongoing comprehensive occupational hygiene evaluations to this effort is obvious.

Evaluation of Effectiveness of Engineering Controls

Although there may be many other reasons for undertaking occupational hygiene evaluations, the last discussed in this chapter is evaluating the apparent effectiveness of engineering controls. In the simplest of cases, this might be accomplished satisfactorily simply by having a limited number of "before and after" air samples. For instance, a reasonable estimate of the effectiveness of a newly installed local exhaust ventilation system at a work station might be made by examining pairs of samples taken before and after installation, or with and without the ventilation system operating. In other cases or for other purposes, documentation of engineering control effectiveness might require a more elaborate array of sampling sites. To determine the effectiveness of a general ventilation system serving an area, for instance, it might be desirable to position sampling units not only in the occupied spaces in the facility but within the ventilation system itself (e.g., before and after filters, in the return air plenum, in the exhaust duct and in the make-up air inlet from outside the building). Although occupational hygiene evaluations for purposes of documenting effectiveness of engineering controls are relatively uncommon, the design of engineering controls for purposes of improving environmental conditions in the workplace should always be preceded by, and be based on, the results of occupational hygiene evaluations.

Familiarization with Process Operations

Once the reason for conducting the occupational hygiene study is understood fully, the next step in the evaluation process is to become as familiar as possible with the process that is to be the focus of the survey. Unfortunately, typical occupational hygienists, particularly those working in entry-level positions, do not devote as much time as they should to this phase of the evaluation. Reasons for this are many: there may be a sense of urgency to the evaluation and spending resources on this preliminary activity is looked at as a waste of time; this effort may be discouraged when work is being performed by a consultant or on a charge-back basis where "time is money"; or, in some cases, the importance of this step to the success of the overall evaluation effort simply is not realized.

Becoming familiar with the process requires the accumulation and comprehension of information from and about various aspects of the process's exposure conditions. These include a reasonable understanding of: 1) the fundamental process being operated by workers whose exposures are being evaluated; 2) the physical facilities in which the process is housed or operated; 3) all chemical substances associated with the process, including raw materials, intermediates, by-products, and products; 4) the health hazards associated with all of the chemicals; 5) the nature of the jobs and job duties required of the workers in the plant; 6) controls in place to protect workers; 7) the health status of workers associated with the process; 8) results of past evaluations; and 9) any other hazards associated with the operation. Each of these inputs to the familiarization process is described in the following sections.

Physical Plant Layout

An initial orientation to an industrial process should begin with an understanding of the general physical layout of the facilities. The location of the particular building or buildings of interest in relation to other structures in the area, the general terrain surrounding the facilities, and the physical arrangement of the plant itself all allow the occupational hygienist to put the general environmental condition of the plant in perspective.

Closer examination of the buildings is then warranted. This can be facilitated by the use of blueprints, engineering drawings, piping and electrical diagrams, and other documents. It is of particular value at this time to become familiar with the general scheme for ventilation of the structure. The desired airflow patterns throughout the building can be determined, based on examination of the mechanical supply and exhaust of air from the various rooms or other subdivisions of the building. Meaningful information can be obtained simply by observing the physical arrangement of ventilation equipment, particularly on the outside (usually the roof) of the building. It is not uncommon, even in modern buildings, to discover arrangements of air supply systems in the vicinity of the exhaust streams from separate locations within the building, making recirculation of contaminated air a potential problem.

Although in most plants the information referred to above is readily available through the organization's engineering function, it may be necessary for the occupational hygienist to create his or her own sketches and notes, based on personal observation and interviews with knowledgeable people.

Understanding the Process

The occupational hygienist must become familiar with all processes used in a particular plant or establishment. Again, engineering drawings and specifications may be available; however, in general these are much more detailed than is necessary for the occupational hygienist's purpose. Simple block flow diagrams with sufficient description of the individual activities making up the process are usually sufficient. Basically, the occupational hygienist must know what materials are being used and produced and the fundamental processes by which they are used. This information may be obtained by asking questions during interviews, by personal visual observation, by reviewing purchase, inventory and other records, and from technical materials that describe the process, operations, and materials.

Once a general understanding of the process is obtained, a walk-through tour of the facilities is important so that a general understanding of the process can be applied to the facility. This provides an opportunity for the occupational hygienist to verify and, if necessary, modify his or her understanding. Inspecting the facilities also provides the occupational hygienist with an opportunity to look for potential sources of contaminants. Many potentially hazardous operations and sources of contaminants can be detected by visual observation. Dusty operations can be spotted easily, although this does not necessarily mean they are hazardous: Dust particles that cannot be seen by the unaided eye are more hazardous than easily visible ones since they are more likely of respirable size (see Chapters 9 and 12). Furthermore, dust concentrations must reach relatively high levels before they are readily visible. Thus, the absence of a visible dust cloud does not mean necessarily that a "dust-free" atmosphere exists. However, those operations that generate fumes, such as welding, can be spotted visually.

Many of the air contaminants found in occupational environments were generated there as a result of basic industrial processes. The following are some examples:

- *Welding fumes.* In addition to the metal fumes and oxides generated by the welding action, contaminants of concern include welding rod and rod coating decomposition products, oxides of nitrogen, ozone, combustion products
- *Combustion products.* A wide range of substances make up this category: carbon monoxide, oxides of nitrogen, diesel exhaust particulates, ash, smoke, acrolein, acid gases, polynuclear aromatic hydrocarbons.
- *Foundry emissions.* Common among these are siliceous dust, oil mist, core and core rosin decomposition products, metal fume, carbon monoxide, any polynuclear aromatic hydrocarbons.
- *Smelting.* Similar to foundry operations, these include combustion products, metal fumes, sulfur dioxide, and specific metals (e.g., arsenic).

In addition to the general categories of chemical substances alluded to above, physical agents should be included in this preliminary review. Some of the more common physical agents and their respective sources are as follows:

- *Noise.* A variety of industrial machinery, engines, furnaces, pneumatic systems and other compressed air tools, and releases make up sources creating excessive noise (see Chapter 20).

- *Ionizing radiation.* This physical stress could result from use of radioisotopes, high voltage machinery, X-ray equipment, process control equipment (e.g., level controls and thickness gauges) as well as natural radiation (see Chapter 22).
- *Nonionizing radiation.* Sources contributing this physical stress include welding and other sources of ultraviolet energy, thermal sources of infrared radiation, radar equipment, microwave ovens and other applications, and various high frequency sources (see Chapter 21).
- *Heat.* Sources of thermal stresses include furnaces, smelting and casting operations, drying ovens, glassmaking, and other activities (see Chapters 24 and 25).
- *Barometric hazards* (see Chapter 23).

Inventory of Raw Materials, Intermediates, Byproducts, and Products

It is of utmost importance that a complete list of all chemicals, raw materials, intermediates, byproducts, and products be obtained for appropriate reference during the evaluation. The number and variety of substances capable of causing occupational hazard and disease potential increase steadily.

New products are being introduced constantly; these require use of new raw materials or, at least, new combinations and applications of older substances, and new processes. Some estimates suggest that new substances are introduced into industry at the rate of one every 20 minutes and that nearly 70,000 such materials are in use today.

New uses for physical agents in industrial processes are increasing at a rapid rate, as well. Examples include the expanding use of lasers, microwaves and other forms of nonionizing radiation. These, too, are potentially hazardous unless management institutes proper control measures and, therefore, should be included in any inventory of potential stresses.

The input to the inventory of stresses should come from the various sources mentioned above as they are encountered in the preliminary familiarization process. Inventories of chemicals on site are relatively straightforward to obtain and/or compile these days. This is due, in part, to the requirement to maintain such information under the various workplace safety regulations (e.g., the hazard communication standard at the federal level and the various "right-to-know" laws promulgated in many states). A fairly complete set of material safety data sheets (MSDSs) is probably available at most operating locations; that would serve as a good starting point for the inventory of stresses. A word of caution and advice, however, would suggest that these inventories should be verified by direct observation of the processes that use the chemicals. Ideally, all chemicals seen on the production floor should have a MSDSs on file and all MSDSs on file should be represented by chemicals in use somewhere in the plant.

In developing an inventory there is a tendency to begin by focusing on the chemicals that are present in greatest quantity and pay less attention to those present in only small amounts. Strict adherence to the occupational hygiene process will help resist this tendency; experience has shown that high volume materials present only in closed systems may be much less of a problem than small quantities used in the open. For example, the large number and variety of materials used in support of maintenance functions should be reviewed carefully. This must be a continuing effort since new coatings, cleansers, and other common chemicals are tried continually and the formulations of materials already in use may change, often without adequate notice to users.

A complete inventory may also include trace materials in process streams; these materials may not even appear on the description of the process stream if they are not of significance to the process. When a product is purified, however, these trace contaminants may be concentrated and the small quantities then involved may be handled in semi-open systems as waste or byproduct. Similarly, it is possible for some trace contaminants to accumulate inside equipment and result in inadvertent exposure when the equipment is opened for maintenance or other purpose.

Given all of these considerations, it is evident that while some industries that only use chemicals—and do not intentionally produce them—can have as large an assortment of chemicals present as a moderately-complex chemical plant. Inventorying these chemicals often is difficult since the materials themselves are dispersed throughout the plant and purchase records, and other documentation of their existence may be decentralized and unclear. Suffice to say, developing the inventory of stresses is not necessarily a simple task and should be taken very seriously.

Review Toxicological Information. After the complete inventory of chemicals used and/or produced is obtained, it is necessary for the occupational hygienist to determine the relative order of toxicities represented by substances on the list. Other factors being equal, higher priority would then be given to those operations using the substances of higher toxicity.

For the record, it is important for the occupational hygienist to remember that the toxicity of a substance is not the sole criterion, or usually even the most important one, in judging the nature of the health hazard associated with a particular operation. Put very basically, "toxicity" and "hazard" are not synonymous. Toxicity is the ability of a substance to cause damage to living tissue by other-than-mechanical means. It is a "property," like the substance's boiling point, vapor pressure, or viscosity. All substances are toxic. Hazard, on the other hand, is the likelihood of injury associated with use of the particular substance and, as such, is much more influenced by conditions of use than by toxicity. Highly toxic substances can be handled quite safely with little hazard; low-toxicity materials can be used in poorly controlled ways that result in a moderate hazard.

The nature of the process in which the substance is used or generated, the possibility of reaction with other agents (chemical and physical), the individual worker's work practices, the degree of effective ventilation control or the extent of enclosure of the process materials all relate to the potential hazard associated with each use of a given chemical agent. Such an assessment must be made along with due consideration of the type and degree of toxic responses the agent may elicit in both the average and, possibly, susceptible or hypersusceptible worker.

Hazard data should be obtained for each entry on the inventory of potential hazards. For physical and biological agents and for identified pure chemicals, this information is readily available in a number of standard textbooks and secondary source documents. For example, such information can be found in most of the current reference books and in scientific journals; the *NIOSH Pocket Guide to Chemical Hazards;* the threshold limit values (TLVs®) published by the American Conference of Governmental Industrial Hygienists (ACGIH) and the Workplace Environmental Exposure Level guides published by the American Industrial Hygiene Association (AIHA); the Z–37 standards published by the American National Standards Institute (ANSI); and by direct interaction with toxicologists, technical information centers, and manufacturers. The Toxic Substances List, published by the National Institute for Occupational Safety and Health (NIOSH), contains more than 8000 substances along with abbreviated summaries of toxicological information. See Chapter 4 for a more comprehensive survey on toxicology.

There are extensive computer-aided libraries of chemical hazard data. Increasingly, the Internet is providing access to a wide range of sites that contribute and maintain information on hazardous properties of substances. The Internet has become a useful process for locating the information for mixtures and other substances or formulations without specific chemical identifiers or names.

In the absence of any specific information on mixtures or formulations, it may be necessary to rely on information provided by the supplier or manufacturer in the form of the material safety data sheet. If completed correctly, these sheets should give the information on effects of overexposure to the extent that they are known.

Job Classifications

Since the occupational hygienist's efforts will be devoted to evaluating the potential exposure of specific workers to specific stresses while engaged in specific tasks, it is to become familiar with the various categories of jobs, the activities required of them, and the potential stresses associated with them.

Job Activities. Ultimately, exposure is related directly to the job activities performed by the worker. As part of the familiarization process, one must become knowledgeable about the workers' job activities and the consequent potential for exposure. A logical starting point for this effort is the formal job description; however, in many situations a formal job description does not exist and even it does it lacks the needed detail. Even with a formal, well-defined job description in hand, it is wise to talk to the employee and the employee's supervisor.

Although these sources would be presumed to be reliably able to describe work activities, the supervisor might not be aware of individual worker routines, and the worker may have difficulty knowing what the occupational hygienist really needs. Actually, it is fairly common to end up with three different versions of a job based on inputs from the official job description, the worker, and the worker's supervisor.

It might be necessary for the occupational hygienist to help develop the job description needed for purposes of the evaluation. Adapting the job safety and health analysis can be useful for this. This approach, perhaps more common among safety professionals than occupational hygienists, uses a standard procedure for reviewing job methods to detect hazards. Although being discussed as a part of evaluation, it has excellent applicability in both the anticipation and recognition aspects of occupational hygiene as well.

Briefly stated, the procedure for conducting a job analysis with this approach consists of the following steps:

1. *Select the job to be studied.* Criteria that can be used to select a job will depend on the purpose of the study. These include the number, frequency, and severity of accidents, injuries or illnesses, whether the jobs are new or recently modified (e.g., personnel, equipment or procedures, and what is known historically about the exposures).
2. *Select an employee who performs the job.* For purposes of evaluating the job, it is desirable to use a worker who is familiar with the job, rather than a new employee.
3. *Inform the employee of the purpose of the analysis* (i.e., to develop a description and time line of activities that will assist in determining the nature, frequency, and duration of potential exposures that might occur as a consequence of the job). If the occupational hygienist is relying on the employee to faithfully perform the job, it is important that the employee fully understand this purpose.

   Employees being observed have a tendency to do the job in a nonroutine manner (i.e., the "Hawthorne Effect"). For proper use by the occupational hygienist, it is important that the job be done as close to the worker's normal way of performing it as possible.
4. *Observe the job.* The primary input to this process is the observation of the occupational hygienist. Maintain a position that permits visual access to all of the key job actions, paying attention to the details from a safety/health view.
5. *Document the steps and timetable for the job.* It is particularly important to identify and differentiate among the various activities that make up the overall job so as to be able to define exposure potential. Although the occupational hygienist will focus on those aspects of the job that contribute to potential for health hazards, assessment of safety hazard potential can be accomplished at the same time with little additional effort.
6. *Review the steps and timetable with the worker(s) and supervisor(s).* The review is important in order to ensure that something has not been missed. The worker and supervisor are most knowledgeable about the job details and may be able to correct, modify, or refine the job analysis created by the occupational hygienist.
7. *Summarize the job descriptions and analyses.* This information generated during this activity should be put in a format that allows easy correlation with the other information being obtained during the familiarization process. The bottom-line purpose of this activity is to be able to anticipate, recognize, and evaluate the various health hazards associated with the job analyzed.

Inventory of Stresses and OELs. On the basis of the job descriptions and the prior understanding of the process and facilities, the various stresses that represent potential adverse impacts on the health of the workers can be defined. The evaluation of health hazard risk eventually requires use of one or more of several occupational exposure limits. These limits exist in two basic forms: regulatory standards and voluntary guidelines.

Regulatory standards are issued by a governmental body that is authorized by law to issue, and usually enforce compliance with, such standards; there is no voluntary or discretionary option for the user to determine whether or not to comply. Guidelines are published by consensus organizations and are published for general voluntary use; the decision to use the guidelines is totally a discretionary one considered by the user.

Occupational exposure limits for potentially hazardous substances, either as guidelines or regulatory standards, can be identified where they exist, but in many cases the chemicals present are not among those for which such guidelines have been established. Where anticipated exposure is significant and no exposure limits are available, it might be necessary to create "local" limits for use in exposure evaluation. When there is no hazard data or when information is unreliable, review of the toxicological literature—or even commitment to new testing and studies—may be the only alternative for proper risk assessment.

The relative uncertainty about the toxicity of some entries on the list of substances creates a common difficulty in relating the chemical inventory to worker activities. Although pure substances, purchased materials, or products are easy to identify, this is not often the case with intermediates, byproducts and other wastes, and other impurities in a process. Sometimes, even the purchased materials are not identifiable easily; many are sold under a brand name with trade-secret protection of the composition. However, if the toxicity of the mixture overall is known, it is still possible to complete the risk assessment, even though the individual ingredients are not specified. If the risk is high and exposure measurements are needed in the evaluation, some information about composition is essential. Provisions within the federal hazard communication standard and other similar regulations create a mechanism and process that allow health professionals to obtain information.

Review Worker Health and Medical Records

A source of information often overlooked as part of the familiarization with a particular work environment is what is known about the workers from a general perspective of overall health status, including the records of first aid treatments and other related information. Even in organizations where there is faithful, periodic evaluation of workers' health and the results are shared with individual workers, there may not be a mechanism for reviewing the overall data for evidence of findings in common, trends, or other indications of a pattern to the relationships between work environment and worker health. To the extent that this information is or can be made available to the occupational hygienist, it is a potentially useful database from which any picture of an overall "cause-and-effect" relationship can be formed.

It is not the intent at this stage to examine and compare the environmental exposures and health status for individual workers. Rather, it is desirable to have an impression of the "bigger picture" (i.e., what is going on in the plant overall or within identifiable departments or other subdivisions of the plant). The process may lead to an interest in examining specific workers' situations and, at the end of the evaluation, it may be important to return to examinations of individual cases. However, review of the medical "condition" of the plant population as a whole would be valuable input to the planning of the occupational hygiene evaluation.

Review Controls in Place

The familiarization process described here would not be complete unless the various control measures in use, and their apparent effectiveness, were noted. Control measures here means the overall strategy for controlling the environment as well as the specific components that make up that strategy. These include local exhaust and general ventilation, process isolation or enclosure, shielding from heat, ionizing radiation, ultraviolet light, or any other forms of radiant energy, protective clothing, and respiratory protective devices, and other controls.

During this familiarization phase, it might be desirable to conduct simple checks on the ventilation systems (e.g., airflow measurements to determine whether sufficient quantities of air are being moved), although full evaluation of any engineering control system should be an integral part of the occupational hygiene evaluation, depending on the purpose for the survey. There are many useful guides available in the literature that for describe test procedures and examples of recommended practice for local exhaust ventilation of various specific process equipment (see Chapter 33 on ventilation system management).

At the very least, the familiarization phase should include a walk-through observation of the process area, during which evidence of apparent control effectiveness is sought. Examples of this evidence would be the presence or absence of dust on floors, ledges or other horizontal surfaces, holes in ductwork or other signs of deterioration, fans not operating or rotating in the wrong direction, or the manner in which personal protective equipment is used by the workers.

Review Past Reports

As a final resource in the familiarization process, any past occupational hygiene or related studies should be reviewed. This is particularly critical to the process if since the most recent study: 1) a significant period of time elapsed; 2) there have been substantial changes in process, equipment or workforce; 3) significant problems, potential or real, had been identified; or 4) indicators of problems (e.g., worker or neighborhood complaints, increased absenteeism, etc.) have increased. Ideally, information used in this effort would be in a single, centralized source. However, that normally is not the case: occupational hygiene reports may be maintained by the Environment, Safety and Health office, health records may be kept in the medical department, employee complaints may be filed by human resources, if at all, and complaints from outside the work force (such as residents in the neighborhood) may be kept by public relations or other unit. The complete and effective accumulation of information that would contribute to an understanding of the recent history of a relationship between environment exposures within the plant and the health of workers is not a trivial exercise.

Determine Potential Hazards Associated with Jobs

The occupational hygienist can now apply the fundamentals of the profession to define potential hazards associated with the jobs about to be evaluated. In general, these have included facilities, chemicals, job activities, work procedures, controls, and personal protective equipment.

As part of the summation of this process, the occupational hygienist should interview selected employees and supervisors to solicit their input as well as to confirm his or her understanding of these items. In addition to assisting in the design of the occupational hygiene survey, these discussions continue and extend the employee involvement in such activities, a process of increasing value in terms of overall labor relations. Getting employees involved at the familiarization stage of a survey will facilitate ultimate acceptance of findings and implementation of any recommendations warranted by evaluation results.

Preliminary Survey (Qualitative)

An experienced occupational hygienist often can evaluate—quite accurately and in some detail—the magnitude of certain chemical and physical stresses associated with an operation without benefit of any instrumentation. In fact, the professional uses this qualitative evaluation every time a survey is made, whether it is intended to be the total effort of the work or a preliminary inspection prior to actual sampling and analysis of potential stresses. Qualitative evaluation can be applied by anyone familiar with an operation, from the worker to the professional investigator, to ascertain some of the potential problems associated with work activities.

During a qualitative walk-through evaluation, many potentially hazardous operations can be detected visually. Operations that produce large amounts of dusts and fumes can be spotted. As discussed earlier, however, "visible" does not necessarily mean "hazardous" levels of a dust exist. In fact, airborne dust particles that *cannot* be seen by the unaided eye normally are more hazardous, since they are more likely to be capable of being inhaled into the lungs. Generally, activities that generate dust that can be spotted visually are likely to warrant implementation of additional controls.

In addition to sight, the sense of smell can be used to detect the presence of many vapors and gases. Trained observers are able to estimate rather accurately the concentration of various gases and solvent vapors present in the workroom air. Unfortunately, the odor threshold concentration—that is, the lowest concentration that can be detected by smell—is for many substances greater than the permissible exposure limit. When such substances can be detected by their odors, that indicates excessive levels. Some substances, notably hydrogen sulfide, can cause olfactory fatigue (i.e., numbing of the olfactory nerve endings) to the extent that even dangerously high concentrations cannot be detected by odor. Suffice to say, odor cannot be relied on as a "warning property" for most materials.

Although it is usually possible to determine the presence or absence of potentially hazardous physical agents at the time of the qualitative evaluation, rarely can the potential hazard be evaluated without the aid of special instruments. As a minimum, however, the sources of physical agents such as radiant heat, abnormal temperatures and humidities, excessive noise, improper or inadequate illumination, ultraviolet radiation, microwaves, and various other forms of radiation, can be noted.

An important aspect of the qualitative evaluation is an inspection of the control measures in use at a particular operation. In general, the control measures include such features as shielding from radiant or ultraviolet energy, local exhaust and general ventilation provision, respiratory protection devices, and other personal protective measures. General measures of the relative effectiveness of these controls are the presence or absence of accumulated dust on floors, ledges, and other work surfaces, the condition of ventilation ductwork (are holes or badly damaged sections of ductwork?), whether the system appears to provide adequate control of contaminants, and the manner in which personal protective measures are accepted and used by workers.

Occupational Hygiene Field Survey (Quantitative)

Although the information obtained during a qualitative evaluation or walk-through inspection of a facility is important and always useful, only by measurement can the occupational hygienist document actual levels of chemical or physical agents. Of course, the strategy used for any given air sampling program depends to a great extent on the purpose of the study, as discussed earlier. Briefly, these survey objectives may include one or more of the following:

- Identification of the source(s) of contaminant release;
- Assistance in design and/or evaluation of control systems;
- Development of an environmental exposure conditions record;
- Correlation of disease or injury with exposures to specific stresses; and
- Documentation of health and safety regulation compliance.

These objectives can be condensed into two major categories: sampling for engineering testing, surveillance, or control; and sampling for compliance, health research, or epidemiological purposes. A sampling program for engineering purposes should be designed to yield the specific information desired. For example, one might need only a single sample before and after a change in ventilation to determine whether the change has had the desired effect. On the other hand, occupational hygiene primarily is concerned with health effects of exposure, comparing sampling results with exposure guides, determining compliance with health codes or regulations, or defining as precisely as possible environmental factors for comparison with observed health effects.

The purpose of evaluation—the occupational hygiene field survey—is to develop information that will lead to decisions that will reduce or eliminate health risks to the workers. Sampling only produces data; data must be converted into information if it is to be of value in the decision-making process. This conversion starts with setting the model to which the data are to be applied, which in turn relates directly to the question to be answered. The most important question to be dealt with by the occupational hygienist at this point is, "what is the question to be answered by this field survey, this sampling exercise?" This points out the critical nature of the first step in the process, determining the purpose of the survey. Once this basic question is answered satisfactorily, the details of the sampling strategy, discussed later in this chapter, can be addressed. Although it might seem that an inordinate amount of time is devoted to discussion of the planning activities here, *the planning of an occupational hygiene survey is, for many reasons, the most important step in the process.*

The fundamental question to be answered is defined by a series of circular exercises that finally set the stage from which all subsequent sampling issues can be decided. Too often, this initial question is phrased simply (such as, "what is the welder's exposure?") This is not detailed enough to design a strategy or sampling protocol. A more meaningful question would be, "what is the exposure distribution of the welders in Department A on the first shift doing MIG-welding on stainless steel A10 materials?" Here there is differentiation between exposure distribution and compliance determination, definition of similar exposure groups, location descriptions in time and space, and definitions of the task and relevant processing conditions and materials. Chapters 7–19 discuss specific evaluation methods to be used during an occupational hygiene survey.

Select Stresses to be Evaluated

The familiarization process should provide all of the information necessary to identify the potential stresses to which the workers might be exposed and to prioritize them in terms of significance of exposure. Prioritization of the stresses in terms of need for evaluation usually is based on one or both of two criteria, the probability of overexposure or the consequence of overexposure should it occur. Stress may include chemical and physical agents as well as ergonomic and psycological stress (see Chapters 21–29).

Estimate Range of Contaminant Concentrations

It is helpful to both the field investigator and the laboratory analyst if a reasonable estimate of the concentrations of the contaminants to be measured can be made. The evaluator's walk-through survey, discussions with plant personnel, degree of familiarity with similar operations, and examination of past data from the facility, all serve as input to the estimate of contaminant concentrations. By knowing the probable range of contaminants, the occupational hygienist can select sampling equipment, size and/or type of collection media, and plan sampling times in accordance with the amount of material likely to be collected. Knowing the likely levels of contaminants also may facilitate the occupational hygienist's selection of specific monitoring equipment or the choice of equipment for collection of samples for subsequent analysis.

Review Sampling and Analytical Methods Available

Except for those generated by direct-reading instrumentation, the typical occupational hygienist does not actually analyze most of the environmental samples collected in the field. And yet, the use of accurate, sensitive, specific, and reproducible analytical methods is equally as important as the proper calibration of sampling equipment. Thus, it is important for the "field person" and the "analyst" to discuss the details of the evaluation so that each will understand the other's expectations. For example, the field person should be aware of the analytical limitations regarding the minimum amount of material needed to be effectively analyzed; control of volumetric sampling rates and duration of sampling can be used to accommodate the analyst. Similarly, if the field person is aware of the presence of materials in the environment that may interfere with the analytical procedure, the analyst may be able to eliminate them prior to analysis. In any event, it is important that field person and analyst begin communications long before the field portion of the evaluation is completed. In evaluating the occupational environment, the concentration of contaminant in the air generally is quite small. In fact, the direct-reading instruments and other devices used to collect samples for subsequent analysis are required to detect quantities of substances often in the nanogram or parts-per-billion (ppb) range. Thus, a sufficient quantity of sample may have to be collected to enable the analyst to determine accurately this small amount of substance. It is beyond the scope of this chapter to describe specific analytical principles or procedures; those are presented elsewhere in this book. This chapter refers to specific analytical procedures as examples only.

Principles. When available, validated methods of analysis should always be used. Fortunately, a number of such methods have been tested under various situations and conditions by various recognized agencies. NIOSH has developed and validated a series of methods for most of the substances for which acceptable exposure limits have been established; these have published in their "Manual of Analytical Methods." Similarly, the OSHA laboratory in Salt Lake City has compiled a set of sampling and analytical methods optimized for their compliance officers to use with the OSHA enforcement program. These are good methods, particularly for purposes of determining compliance with the OSHA standards.

Many of the data sheets and guides from various organizations contain recommended methods of analysis from the perspective of the issuing organization. The American Public Health Association (APHA) has published a manual on methods of air sampling and analysis; ACGIH publishes references on air sampling instrumentation; and the American Society for Testing and Materials (ASTM) has undertaken to validate testing methods in evaluation of ambient air quality.

All of these methods, even those focused on air pollution applications, can be used with minor modification for evaluation of contaminant levels in the workplace. The user is encouraged to refer to the latest publications of those organizations mentioned above and others involved in development and use of methods of air sampling and analysis.

This does not mean that methods other than the standard and proven methods cannot be used. In fact, it often is desirable and productive to explore development of specialized methods that suit the particular needs of the investigator; it is from such efforts that new methodologies emerge. The problem of using other-than-standard methods is that they may not have been validated under conditions that are representative of the conditions in the occupational environment. Thus, other methods must also be tested, standardized, calibrated against known concentrations, and validated in order to have a reliable means of documenting the amounts of specific contaminants in the environment. Many analytical methods have been published in the various discipline-specific literature, and such methods should be evaluated properly prior to use.

Specificity, Selectivity, and Other Considerations. In evaluating a worker's exposure, or the environment in which he or she works, it is necessary to have an instrument that will provide the necessary sensitivity, selectivity, accuracy, and reproducibility, preferably with instantaneous, or at least rapid, results. Detailed discussions of instruments used for sampling for particulates, for gases and vapors, and as direct-reading instruments for measuring airborne chemicals, as well as for noise and other physical agents, appear in separate chapters elsewhere in this book. ACGIH also publishes a reference book on air sampling instruments.

The use of continuous monitoring devices to evaluate the work environment has increased tremendously in recent years. While these devices normally are not designed for field use, many are available in sizes that may be convenient for this purpose. In general, however, many industries install these instruments in areas where exposures to certain gases or vapors may vary considerably or where the presence of the contaminant would be of concern. In the latter application, the instrument serves more as a "leak detector" or alarm than as an exposure monitor. Examples include the use of continuous monitors for carbon monoxide in tunnels or plant areas where this gas is produced, monitors for chlorinated hydrocarbons such as in the production of trichloroethylene, and monitors for certain, specific alcohols. Many of these continuous detecting and recording instruments can be equipped to sample at several remote locations in a plant and record the general air concentrations to which workers may be exposed during a work shift. Many large plants have added computerized equipment to the recorders so that the data can be made available readily and summarized for instant review. Of course, as is the case with all other instruments, continuous monitors must be calibrated periodically and any interferences known (see Chapters 9 and 10).

Limitations. Apart from deliberate tampering with a sample—which usually is easy to detect by an experienced occupational hygienist—other factors can influence or otherwise affect the representativeness of samples and, therefore, their usefulness in making decisions based on them. Compromises of any of the criteria discussed in the preceding section may limit the interpretation and use of a sampling/analytical result. For example, if it has been necessary to use an analytical method that produces a result that is specific for a particular constituent on a chemical (e.g., the carboxyl group) rather than a specific chemical (e.g., methyl ethyl ketone) the occupational hygienist must take into consideration the likely presence of chemicals other than the one of specific interest, but having the same constituent.

The basic question "What was measured?" must be kept in mind to know whether the result can be applied as intended. If the measurement represents an unusual event, it is probably not useful for characterizing the long-term average exposure of workers. However, it may be useful in deciding if engineering controls are needed to prevent an infrequent but excessive exposure. If an evaluation is performed during summer months in warm or hot weather with all windows opened, it should be remembered that the natural ventilation probably will be less, and concentrations higher, in the winter months when the building is closed up. Simply stated, it is necessary to combine observation—an understanding of the observation—with the measurement in order to be able to interpret the results properly.

Selection of Equipment

The choice of a particular sampling instrument depends on a number of factors: 1) type of analysis or information required; 2) efficiency of the equipment or device; 3) reliability of the equipment under various conditions of field use; 4) portability and ease of use; and 5) personal choice based on past experience and other factors. No single, universal sampling instrument is available, and it is doubtful that one will ever be developed. The present trend, in fact, is to develop greater numbers of specialized instruments such as direct-reading gas and vapor detectors or general analysis equipment based on chromatographic or other separation techniques with detection of the separated components by specialized detectors. Many of the latter now include microprocessors that can identify and quantify individual components. Needless to say, these sophisticated instruments are still quite expensive and, even with their multicomponent capability, they are not universal samplers (see Chapters 7–13).

Calibration. The instruments and detailed techniques used in calibration of sampling equipment are discussed elsewhere in this book. This brief discussion is included to stress the necessity for following recommended procedures in order that the data resulting from the analysis of field samples, by whatever method, will truly represent concentrations in the environment, particularly concentrations to which the worker was exposed (see Chapters 7 and 8).

Since the amount of sample, whether gathered by means of a collection device or medium or indicated by a direct-reading instrument, depends on the volume of air sampled and in some cases the duration of sampling, it is essential that the device operate at a known rate of airflow. The equipment must be calibrated against a standard airflow measuring device both before and after use in the field. The exact rate of airflow occurring in the sampling train must be recorded so that when it is multiplied by the sampling time, the correct total volume of air sampled or collected will be known. This volume of air is used to estimate the concentration of contaminant to which the worker was exposed. Temperature and pressure variations between calibration conditions and sampling conditions must be taken into account also. Final airborne concentration results must be corrected to standard conditions for comparison to the exposure limits being used.

Direct-reading instruments and colorimetric indicator (i.e., detector) tubes must be calibrated against a known concentration of the substance for which they are used. Results obtained during a survey or study are no more accurate than the instruments used to obtain the data. In some situations (e.g., periodic checks of the flowrates of sampling pumps) the individual doing the sampling must do his own calibrating—more often, and for more sophisticated direct-reading instruments, this may be done by others at a central laboratory (see Chapter 39).

Selection of Personal Protective Equipment

The final category of equipment to be selected by the occupational hygienist is personal protective equipment. Too often, field investigators arrive ready to undertake extensive air sampling without any provision made for personal protection. Absence of such equipment can be rationalized by saying that the investigator, if exposed to excessive concentrations of a contaminant, is in the area for "only a short time" or will not be put in a position of overexposure. However, it is somewhat hypocritical for the occupational hygienist to work in an area where personal protection is required, or is being recommended by the occupational hygienist, and not wear this equipment himself. Perhaps more important, the message given to workers in the area is mixed: signs or policy state clearly that protection is required and yet a health professional is working in the area without it. The occupational hygienist should therefore include appropriate personal protective equipment as part of the preparation for the field investigation (see Chapters 35 and 36).

Preparation of a Tentative Sampling Strategy

"Sampling strategy" refers to an overall plan or framework prepared so that ultimate decision making will be based on an occupational hygiene survey than is correct and complete. Individual monitoring efforts should collect exactly those samples that will enable the decision making about controls to be made at the required level of confidence and with minimum cost and effort. Data too often are collected with insufficient thought given to how they will be used. Basic decisions for strategy development are the "who, what, why, when and where," discussed later in this chapter. Detailed discussion of the statistical basis for sampling strategy and design of sampling programs are covered in Chapter 15 of this book.

The tremendous variety and variability of occupational exposures provides the basis for the complexity of sampling strategy development and refinement. For example, the concentration of an air contaminant in the environment of a workplace varies with time, over both short and long periods. Workers move in varying patterns through an environment where the contaminant concentration varies with location. The activities of the workers themselves may cause the concentrations to vary. All of these contributors to the variability lead to an exposure profile that usually is described statistically by a lognormal distribution, typically with geometric standard deviations between 2 and 5, or even greater. Simply stated, this means that those results in the upper (i.e., 95th percentile) may be as much as two to five times different—that is, beyond—the mean. This variability is compounded by the problem of estimating the exposure of a group of workers with differing exposure levels to find the most-exposed worker or group of workers. Compared to this environmental variability, in most situations the variability introduced by the sampling and analytical error is small, even for those methods such as asbestos counting that are relatively imprecise.

The purpose of this discussion is to describe the factors that dictate decision-making for individual monitoring situations. Within the context of the established sampling strategy, each sampling situation should be supported by the design of a sampling protocol that takes into account the specific characteristics associated with the work environment to be evaluated. The protocol is designed by answering a set of questions beginning with, "What is the question to be answered?" (i.e., "What is the purpose of the evaluation?").

Conducting the Field Survey

The planning of an evaluation sampling exercise should be complete before any actual measurements are made; conducting the field survey is, in a real sense, less demanding than proper planning. The plan describes how to get the required information efficiently in the required time. As such, it should include the elements of "who," "what," "when," "where," and "how many"; presumably, the "why" was predetermined. Also, the plan should include choice of sampling and analytical methods and how the data will be analyzed and tested to arrive at the point for decision making. Weakly supported data will not support a major decision, while some decisions require no data at all.

Regardless of the objectives of the sampling program, the investigating occupational hygienist must answer the following questions to be able to implement the correct strategy:

- Where should sampling be done?
- For how long and over what period should the samples be taken?
- Who should be sampled?
- How many samples are needed?
- How should the sample be obtained?

Where to Sample

Three general locations are used for collection of air samples: at a specific operation, in the general workroom air, and in a worker's breathing zone. The choice of location is dictated by the type of information desired; in some cases, a combination of the three types of sampling may be necessary. Probably the most frequent occupational hygiene sampling determines the levels of exposure of workers to a given contaminant throughout a work day. To obtain this type of information, it is necessary to collect samples at the worker's breathing zone as well as in the areas adjacent to his particular activities. On the other hand, when the purpose of the survey is to determine sources of contamination or to evaluate engineering controls, a strategic network of area sampling would be more appropriate.

When to Sample

Again, the type of information desired and the particular operations under study will determine when sampling should be done. If the operation continues for more than one shift, it usually is desirable to collect air samples during each shift, since the airborne concentrations may differ for each shift. Similarly, it is desirable to obtain samples during both summer and winter months, particularly in plants located in areas where large temperature variations occur.

How Long to Sample

In most cases, minimum sampling time that which is necessary to obtain a sufficient amount of the material for accurate analysis. The duration of the sampling period, therefore, is based on the sensitivity of the analytical procedure and the acceptable concentration of the particular contaminant in air. Preferably, the sampling period should represent some identifiable period of time of the worker's exposure, usually a minimum of one complete cycle of activity. This is particularly important in studying nonroutine or batch activities, which are characteristic of many industrial operations. Exceptions include operations that are automated and enclosed operations where the processing is done automatically and the operator's exposure is relatively uniform throughout the workday. In many cases it is desirable to sample the worker's breathing zone for the duration of the full shift, particularly if sampling is being done to determine compliance with occupational health standards.

Whom to Sample

Logically, samples should be collected in the vicinity of workers directly exposed to contaminants generated by their own activities. In addition, however, samples should be taken in the breathing zones of workers in nearby work areas not directly involved in these activities, and in those of any workers remote from the activities who either have complained or have reason to suspect that contaminants have been drawn into their work areas.

How Many Samples to Collect

The number of samples needed depends to a great extent on the purpose of sampling. Two samples may be sufficient to estimate the relative efficiency of control methods, one sample being taken while the control method is in operation and the other while it is off. On the other hand, several dozen samples may be necessary to define accurately the average daily exposure of a worker who performs a variety of tasks. The number of samples also depends to some extent on the concentrations encountered. If the concentration is quite high, a single sample may be sufficient to warrant further action. If the concentration is near the acceptable level, a minimum of three to five samples usually is desirable for each operation being studied. There is no set rule regarding the duration of sampling or the number of samples needed. These decisions usually can be reached quickly and reliably only after much experience.

How Should the Sample be Obtained

In general, the choice of instrumentation to sample for a particular substance depends on a number of factors including the portability and ease of use, efficiency of the instrument or method, reliability of the equipment under various conditions of use, type of analysis or information desired, and other factors. As discussed previously, the decision on how to collect the samples should be made in close collaboration with the analyst. Choice of methods may ultimately be dictated by the capabilities and limitations of the analytical laboratory that will be responsible for processing the samples.

The Field Survey

Once the evaluation plan is set, it should be followed to the extent possible, since divergence may bias the result and even compromise the integrity of the conclusions. However, if planning assumptions turn out to be incorrect, it may be necessary to revise the plan. For example, when it is obvious that the sampling and analysis are not working because of interference, it would be useless to continue; a new method must be found and the strategy altered. Also, some plans have built-in decision points such as phased sampling schemes in which a decision to conduct a second sequential set of sampling is based on results of a first set.

All of the sampling procedures must be understood by the field investigator. All necessary equipment and supplies must be available at the sampling site. An occupational hygienist stumbling around trying to figure out what to do or how to do it presents an image of incompetence, at best, and compromises the future interactions with affected workers and supervisors, at worst.

During the field work, the occupational hygienist should be aware of the impact the sampling effort has on the work environment. Sampling should always be coordinated in advance with the persons responsible for the work being performed, both the worker and the supervisor. The purpose and activities should be explained and assistance of the worker should be solicited to make sure that the end result is truly representative of the work conditions. The occupational hygienist should be conscious of what impact the sampling equipment may have on the worker's routine and mount any equipment to minimize the impact. For instance, the worker should be advised that the pump is set for the desired characteristics and deviation from them may nullify the results. It might be desirable to make the workers part of the activity by asking them to check the sampling equipment routinely to make sure it is operating and to inform the occupational hygienist immediately when any pump stoppage or other deviation occurs.

During the sampling, observe the worker and record what is happening. Without this information, complete interpretation of survey results is difficult, particularly if some time has passed since the survey. Every effort should be made to avoid interference with any of the normal work routine during the sampling period. Exercise caution to assure that the mere fact that workers are being observed does not cause them to work in an atypical manner. Similarly, avoid making statements that suggest results of the evaluation are already known or, worse, have been predetermined.

Record Keeping. It is not the intent of this chapter to dictate or even describe the record that must be maintained in support of an evaluation; this aspect of occupational hygiene practice is covered in other chapters of this book. The importance of adequate field notes must be recognized and emphasized, however. While samples are being obtained, notes should be made of the time, duration, location, operations underway, and all other factors pertinent to interpretation of the samples, as well as the conditions they are intended to define. Printed forms with labeled spaces for essential data help to avoid the common failure to record important information.

Handling Samples. To be able to rely on the results of measurements, it is necessary that the samples as analyzed are the same as they were when collected and that they are properly identified in the field, in the laboratory and, ultimately, in the report. For example, sample tubes or other containers should be marked *before* the sample is collected in a manner that will withstand the rough handling sometimes encountered. Transit times and temperatures should be within the limits allowed for the type of sample and analysis. A series of documents that establish an accurate "chain of custody" of the samples should exist to ensure that the analytical result is matched with the correct sample. Most analytical laboratories have well-defined chains of custody within their own operations. It is less common for an occupational hygienist who functions primarily as a field investigator to have an established procedure, complete with standardized forms, for assuring proper chain of custody. It would be to the best interests of all practicing occupational hygienists to develop and maintain records supporting chain of custody of all samples collected by them.

Sampling/Analytical Procedures

This section provides an overview of the general considerations that must be made in selecting any measurement method (i.e., the combination of sampling and analysis). Obviously, a measurement method should meet the requirements of the sampling strategy and protocol in order to make decision making easier. Some of the attributes to be considered in selecting a method are discussed here.

Duration of Sampling. When measuring a substance with an eight-hour average time, a single eight-hour sample or several consecutive samples totaling eight hours is best. Short-period or instantaneous (i.e., "grab" samples) are the least satisfactory because of the assumptions and/or extrapolations that are necessary to define full-shift exposures. For a short-term exposure limit (STEL) or ceiling standard, the method should be able to collect enough material to provide adequate sensitivity within the 15-minute, or other short term, time span associated with these criteria.

Sensitivity. The sampling and analytical method together should ideally have a limit of detection much below that representing the exposure limit. However, less sensitive methods are still usable as long as the exposure limit is easily within the range of the method or the lower sensitivity can be accommodated by control of the sampling volume and/or duration.

Freedom from Interferences. To avoid spurious results that can lead to incorrect conclusions, be aware of other substances in the sampled air, or else the result may be too biased and become unusable. Some error due to interferences is acceptable if the outside limits of likely error are known and can be taken into account in developing or using the data.

Time to Result. Evaluation of long-term health hazards, although important, generally is not urgent in terms of time from collection of samples to receipt and interpretation of analytical results. Although timeliness of reporting results is a worthy objective, the criticality of processing the information is less the longer-acting or more chronic the nature of the adverse health effect is. On the other hand, evaluations of some acute hazards, where the adverse effects produced by the substances develop rapidly, require essentially immediate information; direct-reading instrumentation or on-the-spot analytical support would be desirable.

Intrusiveness. Workers are likely to alter their behavior, consciously or subconsciously, when they are being observed or monitored. To the extent that a worker's exposure is related to the worker's actions, this change in behavior can distort the representativeness of the evaluation. Measurement methods that require the close presence of the evaluator (i.e., the person collecting the sample) are more likely to influence the result than samples collected with unobtrusive devices worn by the workers as they go about their normal routine.

Proximity to Breathing Zone

While all exposure measurement methods attempt to sample from the air being breathed by the worker, some are more effective at it than others. A fixed-location sampler—even one positioned near the worker—usually is not measuring exposure of the worker as desired. Even with mobile samples that move with the worker, a few inches difference in distance from the nose and/or mouth, or between left and right shoulders, have been found to make a difference. Proper positioning of the sampling medium, although based on the science of sampling, is influenced often by the art of occupational hygiene accumulated after years of experience in conducting field studies.

Accuracy

Obviously, the more accurate the method, the better the results. However, given the large environmental variability, sampling and analytical imprecision rarely make a significant contribution to the overall error in the final result (the width of the confidence limits of the data). Even highly imprecise methods, such as dust counting, do not add much to the overall variability of results, given that variability exists between workers involved in similar activities and over time. An undetected bias, however, is more serious because it is not dealt with by the statistical analysis and can result in gross, and unknown, error.

Interpretation of Sampling Results

The initial step in interpreting sampling results is to convert all of the raw data into units of concentration or other basis for comparison to reference points of acceptability. When sample periods do not correspond to the averaging time of the exposure limits, some assumptions must be made about the unsampled portions of the work period. It may be necessary to test the impact of various assumptions on the final decision. Next, some test statistics should be calculated and compared with test criteria to make an inference about any hypothesis and to define, at least in general terms, the statistical reliability of the data.

Even when statistical techniques are used to analyze the data collected, use professional judgment to interpret the results of an environmental study. Before an occupational hygienist determines that a worker or group of workers is exposed to a hazard injurious to health, he or she must have the following information: 1) the precise nature of the substance or physical agent involved; 2) the intensity or magnitude of exposure (e.g., the concentration of the contaminant); and 3) reliable knowledge of the duration of exposure. In many cases, adverse effects from exposure to toxic materials or physical agents do not appear until the exposure has continued for several years. An important purpose of occupational exposure limits is to protect against the future appearance of such symptoms.

While statistical tests establish that the results are or are not different from an exposure criterion, the generality of the results must be judged by answering some basic questions: What did the sample(s) represent? May the outcome, which is accepted as covering both sampled and unsampled periods, be extrapolated legitimately into continuing or future exposures? These and similar questions are answered by judgment, based on experience and applied to the observations made at the time of sampling. Answers to the questions are used to interpret the quantitative results.

Time-Weighted Average Exposures

Evaluation of workers' daily time-weighted average (TWA) exposures is best accomplished, when analytical methods permit, by allowing the person to work a full shift with a personal breathing zone sampler attached. The concept of full-shift integrated personal sampling is much preferred to that of short-term or general area sampling if the results are to be compared to standards based on TWA concentrations. The current OSHA permissible exposure limit (PEL) is such a standard. When methods that permit full-shift integrated sampling are not applicable, it may be possible to calculate TWA exposure from alternative short-term or general area sampling methods. Use of these methods is much less precise than personal monitoring, however.

In the absence of personal monitoring data, the first step in calculating the daily, TWA exposure of a worker based on limited, task-specific data is to study the job description for the persons under consideration and determine how much time during the day is spent at various tasks. Such information usually is available from the plant personnel office or foreman. The investigator may have to make time studies himself to obtain the correct information. In fact, even information obtained from plant personnel should be checked by the investigator because, in many situations, job activities as observed by the investigator do not fit the official job descriptions. From this information and the results of the environmental survey, a daily, eight-hour TWA exposure can be calculated, assuming that sufficient sampling has been done at the specific tasks that make up the work day.

Analysis of Short-Term Exposures

Sometimes, when sampling airborne contaminants whose toxicological properties warrant short-term and/or ceiling limit values, it is necessary to use short-term or grab sampling techniques to define peak concentrations and estimate peak excursion durations. For purposes of further comparison, the eight-hour TWA exposure can be calculated using the values obtained by consecutive short-term sampling. Thus, interpretation of short-term sample results usually takes one of two forms: analysis of a series, preferably consecutive, of samples in order to construct an average exposure for a longer period of time; or direct comparison to a standard based on short duration exposures. For the former case, if the sampling and analytical methods confined sampling periods to one hour, a series of eight, consecutive samples representing the worker's exposure could be used to construct an eight-hour TWA exposure. For the second case, if sampling duration is consistent with the time period specified for the short-term limit, a direct interpretation can be made for each sample result. For instance, if the STEL is based on a 15-minute period, a common situation, as long as the sample duration did not exceed 15 minutes it could be compared directly to the standard. Of course, with comparison of a sample result to a "ceiling" limit, the duration of the sampling should be as short as practicable to allow direct comparison with the standard.

Comparing Sampling Results with Appropriate Standards

When the evaluation was performed for purposes of determining compliance with appropriate standards, results must be compared with occupational exposure limits before the level of compliance can be determined or necessary control measures can be recommended. Samples collected for compliance purposes must be collected in a manner that is representative of the worker's daily TWA exposure in order to permit direct interpretation. In fact, in those cases where an occupational exposure limit has not been established, an internal occupational exposure limit should be established, based on relevant information, before the evaluation of exposure is conducted.

For compliance, the most prevalent legal standards are those promulgated by OSHA. Unfortunately, most of the occupational hygiene standards, OSHA's PELs, are in need of update. Although these standards are subject to revision, the process is so cumbersome that revision is a long, drawn-out process and has not been effective in keeping the PELs current. Currently, the exposure standards that are accepted generally as representing current, professional opinion are those recommended by ACGIH as TLVs. In addition, a limited number of consensus standards are published by ANSI.

A state may qualify and be designated to administer and enforce an occupational safety and health program under provisions of the Federal Act. Many states have done this and have adopted the federal health and safety standards. State standards must be "at least as effective" as federal standards, and many states have adopted standards more stringent than the federal or have promulgated standards for conditions not covered by the federal regulations. Consequently, the standards in effect in a particular state must be reviewed by the occupational hygienist. It is important to keep in mind that federal standards are minimum legal requirements.

Comparing Results with Previous Data

Many organizations have developed data for environmental levels of both chemical and physical agents associated with their operations over a period of time. These data may have been generated by sample collection and analysis techniques or by using continuous monitors to maintain a record of concentrations of various gases and vapors in specific areas of the plants. Current evaluation results should be compared to these data, to the extent compatible and comparable. In many cases, the data will not be the same and there are rational explanations for the differences. In some instances the data on exposures maintained by the company will be more complete and detailed than that obtained by the current occupational hygienist; the current evaluation should be examined in light of that historical record.

In addition to data available from company records, other sources of relevant information include the results of previous studies conducted by federal, state and local agencies, some insurance companies, and private consultants. Here again, results of these studies may be different than those obtained by the current occupational hygiene evaluation; use caution in making comparisons.

Other Evaluations Involving the Occupational Hygienist

There are other monitoring activities in occupational health for purposes of decision making that need to be considered and their roles in evaluation understood. The customary evaluation described earlier is the conventional occupational hygiene survey; although it is the most common form, it is only one of several kinds of workplace evaluation that involves the occupational hygienist.

Become familiar with other evaluation methods so as to be able to communicate clearly what objective and methods will be used for each purpose.

Ambient (environment) monitoring is the measurement and assessment of agents at the workplace to evaluate ambient exposure and health risk compared to an appropriate reference. This is the customary method for monitoring air concentration or physical agent intensity when evaluating the occupational environment. Use of the words "ambient" and "environmental" implies that this definition describes general air measurement and interpretation of the results in terms of the general population. However, the general approach to ambient air monitoring is, for all intents, identical with that used for occupational hygiene monitoring. With ambient monitoring, the population at risk is expanded from the workers to all persons in the vicinity, including the general residential population. Ideally, evaluation of health risk in ambient sampling should incorporate estimation of exposure as directly as possible such as by use of personal monitoring. As contrasted with typical occupational hygiene evaluation, the approximate dose in the case of ambient monitoring is total intake, in biological terms.

Biological monitoring is the measurement and assessment of workplace agents or their metabolites, either in tissues, secretions, excretions or bodily fluids, or some combination of these, to evaluate exposure and health risk compared to an appropriate standard. Unfortunately, there are relatively few materials for which meaningful biological parameters have been developed. ACGIH publishes a set of biological exposure indices (BEIs®) as a companion to its annual publication of TLVs; more than 40 substances are covered in that summary.

Health surveillance is the periodic medico-physiological examination of exposed workers with the objective of protecting them and preventing occupationally related diseases. The detection of established disease is outside the scope of this definition; rather, the objective of health surveillance is to detect adverse biological effects that indicate early health problems, but are not yet true symptoms of a clinical disease. Although surveys of this nature are more medical than environmental in nature, the input of an occupational hygienist usually is beneficial.

Biological effect monitoring is the measurement and assessment of early biological effects—of which the relationship to health impairment has not yet been established—to evaluate exposure and/or health risk compared to appropriate references. It is a measurement of effects, often nonspecific, to evaluate exposure and risk, not health status. Again, an occupational hygienist's input usually is helpful.

It is important to understand the distinctions among the various types of survey described above. Ambient or environmental and biological monitoring are agent-related, while health surveillance and biological effect monitoring are worker- related. It is appropriate, therefore, to set exposure criteria for environmental monitoring, equivalent to threshold limit values, and for biological monitoring but not for health surveillance. Conclusions arrived at as a result of environmental, biological, or biological effect monitoring relate to the health risk, those from health surveillance to health status. In the temporal sense of the exposure-effect sequence, environment is most immediate, followed closely by biological monitoring, then biological effect monitoring, then health surveillance. As has been emphasized earlier, keep objectives clear by thinking of the decisions to be made as a result of the measurements.

The evaluation of conditions that exist during an emergency follows the same logic as customary evaluations, but compressed in time. When the release of a chemical might create an emergency in a plant or community, each step must be preplanned in response so that information flow and decision making can be timely. In the simplest plans, worst-case exposure is pre-estimated from the amount of material present, the highest credible release, and the worst possible atmospheric conditions. Evaluation and response are triggered by the event without waiting for air measurements. Control decisions, such as termination of release, cleanup of a spill, and issuance of personal protection are then automatic More complex plans may make use of real-time weather and chemical concentration data (obtained by installed sensors) and process the data quickly via computer, making use of partially presolved dispersion models and terrain constants. However it is done, preplanning is essential to allow decisions to be made quickly and correctly.

In the customary occupational hygiene evaluation, exposure is compared with exposure limit criteria, OSHA PELs, ACGIH TLVs, or NIOSH RELs to arrive at a conclusion regarding safety. In risk assessment, exposure is used to estimate dose, estimate response and, thereby, predict risk.

All of these are not single, absolute quantities; they are represented by probability distributions. Thus, in an exposed population there will be a few individuals with below-average exposure, most will have an average exposure, and a few will have above-average exposure. This distribution of exposure probability usually is normal or perhaps lognormal. Similar distributions exist for conversion of exposure to dose and dose to response. The end result is distribution of the "probability of harm," which is referred to as "risk."

A complication is that although the exposure distribution in a population can be measured, the distribution of the exposure-dose relationship and of the dose-response relationship in a human population is rarely known with any certainty. Thus, there is uncertainty about the risk. Risk assessment and value judgments about the acceptability of risk are most often used to set exposure criteria that are then applied via the occupational hygiene evaluation process.

When the decision has been made to install some exposure reduction control, such as source reduction or exhaust ventilation, there is an expectation of result. Postcontrol measurements are then made to compare with the expected result and with precontrol measurements to determine the effectiveness of control. The performance of the control itself can often be evaluated by short-period or even instantaneous measurements near the source to verify that the emission has been reduced or is being captured. However, since exposure may be the result of multiple sources of which only the largest have been controlled, worker exposure measurements may also be necessary to see if control is adequate.

The occupational hygiene process of anticipation, recognition, evaluation and control can be exercised in discrete episodes called "surveys" or continuously with all phases overlapping as is usually the case when there is an occupational hygiene resident at the plant. The surveys sometimes are called "audits," although an audit is defined more properly as a methodical examination, involving analysis, tests, and confirmations of local procedures and practices leading to a verification of compliance with legal requirements, internal policies, and practices. An audit may be a snapshot of the status of compliance with laws and policy at a point in time that uses occupational hygiene evaluation data as a means of confirming the present status.

Alternatively, an audit can look at the management systems that need to be in place to ensure compliance on a continuing basis if they are to be effective. In this latter case, occupational hygiene data are reviewed to see if the occupational hygiene process is functioning effectively and efficiently. In either case, the audit is not the occupational hygiene process itself, but a management surveillance tool that uses the results of occupational hygiene evaluations.

Summary

The conduct of occupational hygiene surveys and studies is only one phase in the overall effort in determining occupational health hazards. Such surveys are valuable only if all environmental factors relating to the workers' potential exposure are included. In evaluating worker exposure to toxic dusts, fumes, gases, vapors, mists, and physical and/or biological agents, a sufficient number of samples must be collected, or readings made with direct-reading instruments, for the proper duration to permit the assessment of daily TWA exposures and evaluate peak exposure concentrations when needed.

It is essential that the proper instrument be selected for the particular hazard and that it be calibrated periodically to ensure that it is sampling at the correct rate of airflow and, in the case of direct-reading instruments, that they have been calibrated against known concentrations of the contaminant in question. For those samples to be analyzed in the laboratory, a method must be used that is accurate, sensitive, specific, and reproducible for that particular contaminant. Adequate notes taken during the environmental studies are a necessity. An occupational hygienist cannot rely on memory after a study is completed to provide the detailed information necessary for preparation of a report.

Occupational hygiene has often been described as an art *and* a science—meaning that science alone will not lead to the end results but must be augmented with judgment and intuition. This is certainly true in occupational hygiene and in other fields where the goal is not merely to obtain knowledge but to achieve an end result, such as prevention of disease. In some occupational hygiene activities it is possible to follow a fixed set of rules but in others, judgment must guide decisions. There is a danger in that judgment too easily can become highly subjective, leading to differences of opinion among experts that brings into question the value of good work. Judgment is necessary, but it should grow from experience and continually be examined, analyzed, tested, and focused on those points where judgment must be used (i.e., where science is limited in providing the answers). Further, the foundation for, and process of applying, good judgment must be explained more effectively so that it can be understood and accepted by others. Finally, it must be understood that no matter what analytical process is used to develop quantification of data, sound professional judgment always must be exercised by the occupational hygienist, both during all phases of an evaluation: planning, conducting, analyzing, interpreting, drawing conclusions, and recommending the decisions to be made.

Additional Sources

American Conference of Governmental Industrial Hygienists (ACGIH): *Documentation of the Physical Agents Threshold Limit Values.* Cincinnati: ACGIH, 1993, 1996.

American Conference of Governmental Industrial Hygienists (ACGIH): *Documentation of the Threshold Limit Values and Biological Exposure Indices,* 6th Ed. Cincinnati: ACGIH, 1993, 1996.

American Conference of Governmental Industrial Hygienists (ACGIH): *Industrial Ventilation — A Manual of Recommended Practice,* 22nd Ed. Cincinnati: ACGIH, 1994.

American Conference of Governmental Industrial Hygienists (ACGIH): *1996 TLVs® and BEIs®.* Cincinnati: ACGIH, 1996. [Use most recent edition.]

American Conference of Governmental Industrial Hygienists (ACGIH): *Air Sampling Instruments,* 8th Ed. Cincinnati: ACGIH, 1995.

American Industrial Hygiene Association (AIHA): *Field Guide for the Determination of Biological Contaminants in Environmental Samples.* Fairfax, Va.: AIHA, 1996.

American Industrial Hygiene Association (AIHA): *Noise and Hearing Conservation Manual,* 4th Ed. (edited by E.H. Berger, W.D. Ward, J.C. Morrill, and L.H. Royster). Akron, Ohio: AIHA, 1986.

American Society for Heating, Refrigerating and Air-Conditioning Engineers (ASHRAE): *Ventilation for Acceptable Indoor Air Quality (ASHRAE 62*–1989). Atlanta: ASHRAE, 1989.

American Society for Heating, Refrigerating and Air-Conditioning Engineers (ASHRAE): *1997 ASHRAE Handbook – Fundamentals.* Atlanta: ASHRAE, 1997.

American Society for Testing and Materials: *Annual ASTM Book of Standards.* West Conshohocken, Pa.: ASTM, 1997. [Use most recent edition.].

Clayton, G.D., F.E. Clayton, R.L. Harris, L.J. Cralley, L.V. Cralley, and J.S. Bus (eds.): *Patty's Industrial Hygiene and Toxicology* (Complete Set — Volumes I, II, and III). New York: John Wiley & Sons, 1991–1994.

Cralley, L.V., and L.J. Cralley (eds.): *Industrial Hygiene Aspects of Plant Operations.* New York: Macmillan Publishing Co., Inc., 1987.

DiNardi, S.R.: *Calculation Methods in Industrial Hygiene.* New York: Van Nostrand Reinhold, 1995.

Eller, P., and M. Cassinelli (eds.): *NIOSH Manual of Analytical Methods (NMAM),* 4th Ed. (DHHS/NIOSH Pub. No. 94–113). Cincinnati: National Institute for Occupational Safety and Health, 1994.

Gilbert, R.O.: *Statistical Methods for Environmental Pollution Monitoring.* New York: Van Nostrand Reinhold, 1987.

Hawkins, N.C., S.K. Norwood, and J.C. Rock (eds.): *A Strategy for Occupational Exposure Assessment.* Akron, Ohio: American Industrial Hygiene Association, 1991. [Second edition to be published in 1998.]

International Labour Office (ILO): *Encyclopaedia of Occupational Health and Safety,* 4th Ed. Geneva, Switzerland: ILO, 1997.

International Labour Office (ILO): *Occupational Exposure Limits for Airborne Toxic Substances,* 3rd Ed. Geneva, Switzerland: ILO, 1991.

Klaassen, C.D. (ed.): *Casarett and Doull's Toxicology: The Basic Science of Poisons,* 5th Ed. New York: Macmillan Publishing Co., Inc., 1995.

Lewis, R.J., Sr.: *Hawley's Condensed Chemical Dictionary,* 13th Ed. New York: Van Nostrand Reinhold, 1997.

Lewis, R.J., Sr.: *Sax's Dangerous Properties of Industrial Materials,* 9th Ed. New York: Van Nostrand Reinhold, 1996.

Lide, D.R. (ed.): *CRC Handbook of Chemistry and Physics,* 78th Ed. (1997–1998). Boca Raton, Fla.: CRC Press, Inc., 1997.

Lodge, J.P., Jr. (ed.): *Methods of Air Sampling and Analysis,* 3rd Ed. Washington, D.C.: American Public Health Association, 1988.

National Institute for Occupational Safety and Health (NIOSH): *Occupational Exposure Sampling Strategy Manual* by N.A. Leidel, K.A. Busch, and J.R. Lynch (HEW/NIOSH Pub. No. 77–173). Cincinnati: NIOSH, 1977.

National Institute for Occupational Safety and Health (NIOSH): *Laboratory Evaluations and Performance Reports for the Proficiency Analytical Testing (PAT) and Environmental Lead Proficiency Analytical Testing (ELPAT) Programs* (DHHS/NIOSH Pub. No. 95–104) by C.A. Esche, J.H. Groff, P.C. Schlecht, and S.A. Shulman. Cincinnati: NIOSH, 1994.

National Institute for Occupational Safety and Health (NIOSH): *NIOSH Pocket Guide to Chemical Hazards* (DHHS/NIOSH Pub. No. 94–116). Cincinnati: NIOSH, 1994.

National Institute for Occupational Safety and Health: Registry of Toxic Effects of Chemical Substances (RTECS). [Database.]

National Library of Medicine: Hazardous Substances Data Bank (HSDB). Operates off the TOXNET System, Bethesda, Md.

Occupational Safety and Health Administration: *OSHA Analytical Methods Manual,* Vol. 1–4. Salt Lake City: OSHA Technical Center, 1990, 1993.

Occupational Safety and Health Administration: *OSHA Technical Manual,* 4th Ed. Washington, D.C.: U.S. Government Printing Office, 1996.

"Occupational Safety and Health Standards for General Industry," *Code of Federal Regulations* Title 29, Part 1910.

Perry, R.H., and D.W. Green (eds.): *Perry's Chemical Engineers' Handbook,* 7th Ed. New York: McGraw-Hill, 1997.

Powell, C.: "Interpreting Exposure Levels to Chemical Agents." In Patty's *Industrial Hygiene and Toxicology,* Vol. III-B: Theory and Rationale of Industrial Hygiene Practice, 2nd Ed. (edited by L.J. Cralley and L.V. Cralley). New York: John Wiley & Sons, 1985. pp. 333-374.

Sittig, M.: *World-wide Limits for Toxic and Hazardous Chemicals in Air, Water, and Soil.* Park Ridge, N.J.: Noyes Publications, 1994.

U.S. Environmental Protection Agency: "Building Air Quality — A Guide for Building Owners and Facility Managers."

U.S. Environmental Protection Agency: *EPA Compendium of Methods for the Determination of Toxic Organic Compounds in Ambient Air.* Washington, D.C.: EPA, 1984.

World Health Organization: *Environmental and Health Monitoring in Occupational Health* (Technical Report Series No. 535). Geneva, Switzerland: World Health Organization, 1973.

References

1. **American Industrial Hygiene Association:** *"AIHA Mission Statement."* Fairfax, Va.: AIHA, 1994.

Chapter 7

Outcome Competencies

After completing this chapter, the user should be able to:

1. Define underlined terms used in this chapter.
2. Recognize the needs for and uses of systems for preparing known concentrations of air contaminants.
3. Discuss state-of-the-art knowledge of available systems, their applications, advantages, and disadvantages.
4. Apply the calculations involved in the use of these systems.
5. Apply the appropriate system considering one's needs and available resources.
6. Discuss the operating principles of a permeation device.
7. Explain source devices to produce aerosols.

Key Terms

bar • batch mixture • calibration • diffusion system • flow-dilution system • Ideal Gas law • molar gas volume • monodisperse aerosol • Pascal • permeation tube • SI metric units • validated sampling and analysis method

Prerequisite Knowledge

Prior to beginning this chapter, the reader should review *General Chemistry*[1] for general content and S.R. DiNardi, *Calculation Methods for Industrial Hygiene*, Van Nostrand Reinhold, 1995.

Prior to beginning this chapter, the user should review the following chapters:

Chapter Number	Chapter Topic
6	Principles of Evaluating Worker Exposure
8	Principles and Instrumentation for Calibrating Air Sampling Equipment
9	Direct-Reading Instrumental Methods for Gases, Vapors, and Aerosols
10	Sampling for Gases and Vapors
11	Analysis of Gases and Vapors
12	Sampling and Sizing Particles

Key Topics

I. Preparation of Batch Mixtures of Gases and Vapors
   A. Sealed Chambers
   B. Bottles
   C. Plastic Bags
   D. Pressure Cylinders
   E. Calculations of Concentrations in Air

II. Flow-Dilution Systems
   A. Gas-Metering Devices
   B. Construction and Performance of Mixing Systems

III. Source Devices for Gases and Vapors
   A. Vapor Pressure Source Devices
   B. Motor Driven Syringes
   C. Diffusion Source Systems
   D. Porous Plug Source Devices
   E. Permeation Tube Source Devices
   F. Miscellaneous Generation Systems
   G. Illustrative Calculations for Flow-Dilution Systems

IV. Source Devices for Aerosols
   A. Dry Dust Feeders
   B. Nebulizers
   C. Spinning Disc Aerosol Generators
   D. Vaporization and Condensation of Liquids
   E. Miscellaneous Generation Systems

V. Multipurpose Calibration Components and Systems

7

Preparation of Known Concentrations of Air Contaminants

Bernard E. Saltzman, Ph.D., CIH, PE

Introduction

Known low concentrations of air contaminants are required for many purposes: for testing and validation of analytical methods, for calibrating instruments, and for scientific studies.

Because of legal and economic consequences of the measurements to determine regulatory compliance, each regulatory standard should include a validated sampling and analysis method. As part of the Standards Completion Program[2] the National Institute for Occupational Safety and Health validated methods for 386 compounds in 1976. Additional work was subsequently conducted to validate methods for 130 others.[3] An important part of these validation tests was the generation of known concentrations in air in the range of one half to two times the permissible exposure limit (PEL) against which to compare the test results. Many new methods are being developed utilizing charcoal tubes, passive samplers, and direct-reading devices and instruments that require validation.

Numerous electronic instruments are secondary measuring devices that require calibration with accurately known airborne concentrations. The Environmental Protection Agency's quality assurance procedures have been published.[4]

Another use for known concentrations is for toxicological and scientific investigations of the effects of these concentrations. Such work provides the basis for control standards.

Two general types of systems are used for generation of known concentrations. Preparation of a batch mixture has the advantages of simplicity and convenience in some cases. Alternatively a flow-dilution system may be employed. A flow-dilution system requires a metered flow of diluent air and a source for supplying known flows of gases, vapors, or aerosols; these flows are combined in a mixing device. This system has the advantages of compactness and of being capable of providing large volumes at known low concentrations, which can be rapidly changed if desired. Each of these techniques will be described in detail later in this chapter. Many articles have been published on the subject of making known concentrations. Broad coverage is given in articles.[5–18]

Preparation of Batch Mixtures of Gases and Vapors

These methods generally require relatively simple equipment and procedures. However, a serious disadvantage is the fact that only limited quantities of the mixture can be supplied. For reactive compounds erroneously low concentrations may result from appreciable adsorption losses of the test substance on the walls of the chamber. Losses in excess of 50% are common.[19,20] These methods may be used to prepare nominal concentrations of many substances that are not too reactive. They should not be used as primary standards without verification by chemical analyses.

Sealed Chambers

Known concentrations of gases and vapors were first prepared as batch mixtures by introducing and dispersing accurately measured quantities of the test compound into a sealed chamber containing clean air with mixing provided by an electric fan. (Danger! Sparks from brushes may explode some mixtures. When air dilutions of solvents or other combustible materials are prepared, care must be taken to keep the concentrations outside of the explosive limits.) The mixture may be withdrawn from the center of the chamber through a tube. Leakage from sealed exposure chambers, most likely from deterioration of door gaskets, may be evaluated by pressurizing them (to 5 cm H_2O) and determining the rate of pressure drop with time.[21]

Bottles

Figure 7.1 illustrates a simple technique for preparation of vapor mixtures using a 5-gallon glass bottle. A quantity of volatile liquid is pipetted into the bottle onto a piece of paper to assist in its evaporation. The bottle may then be tumbled with aluminum foil inside to facilitate air and vapor mixing. The mixture is withdrawn through a glass tube from the bottom of the bottle rather than from the top to avoid leakage and losses occurring around the stopper.

As the mixture is withdrawn, air enters the top of the bottle to relieve the vacuum. It can be seen that the disadvantage of these techniques is that the concentration decreases during the withdrawal process as clean air enters. Assuming the worst possible case of complete turbulent mixing in the bottle or chamber, the change in concentration is given by Equation 1.

$$C/C_o = e^{W/V} \tag{1}$$

where

C = final concentration in bottle or chamber
C_o = initial concentration
W = volume withdrawn, liters
V = original volume of mixture, liters

Some calculated values are shown in Table 7.1. This table shows the maximum depletion errors produced by withdrawal of the mixture. Smaller errors result if the incoming air does not mix completely with the exiting mixture. Up to 5% can be withdrawn without serious loss. These errors are avoided by use of plastic bags or pressure cylinders, as described below.

Figure 7.2 illustrates a simple commercial assembly for creating a known concentration of hydrocarbon for use in calibrating explosive-gas meters. A sealed glass ampule containing a hydrocarbon such as methane is placed inside a polyethylene bottle and broken by shaking against a steel ball. The mixture is then carefully squeezed into the instrument to be calibrated, taking care not to suck back air. Another commercial system for making a known concentration batch is comprised of a cartridge of isobutane, which is used to fill a small syringe. This is injected into a larger syringe that is then filled with air. The latter syringe serves as a gas holder for the mixture. These devices are relatively simple, convenient, and sufficiently accurate for calibrating explosive-gas meters.

Plastic Bags

A variety of plastic bags have been found to be very useful for preparing known mixtures in the laboratory. Among the materials used have been Mylar®, aluminized Mylar, Scotchpak, Saran®, polyvinyl chloride, Teflon®, Tedlar®, and Kel-F®. Tedlar, Teflon, and Kel-F are considerably more expensive materials preferred for use in photochemical smog studies involving ultraviolet irradiation. Surprisingly, for most applications the less expensive materials give superior performance. Mylar bags are popular because of their strength and inertness.

TABLE 7.1 — Decrease in Concentration Versus Fractional Volume Withdrawn from Chamber Mixture

W/V	0.05	0.10	0.25	0.50
C/C_o	0.951	0.905	0.779	0.607

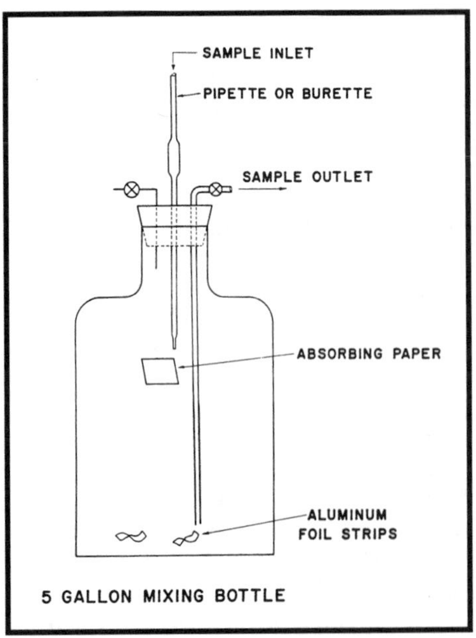

Figure 7.1 — Preparation of vapor mixtures in a glass bottle (Source: Reference 6).

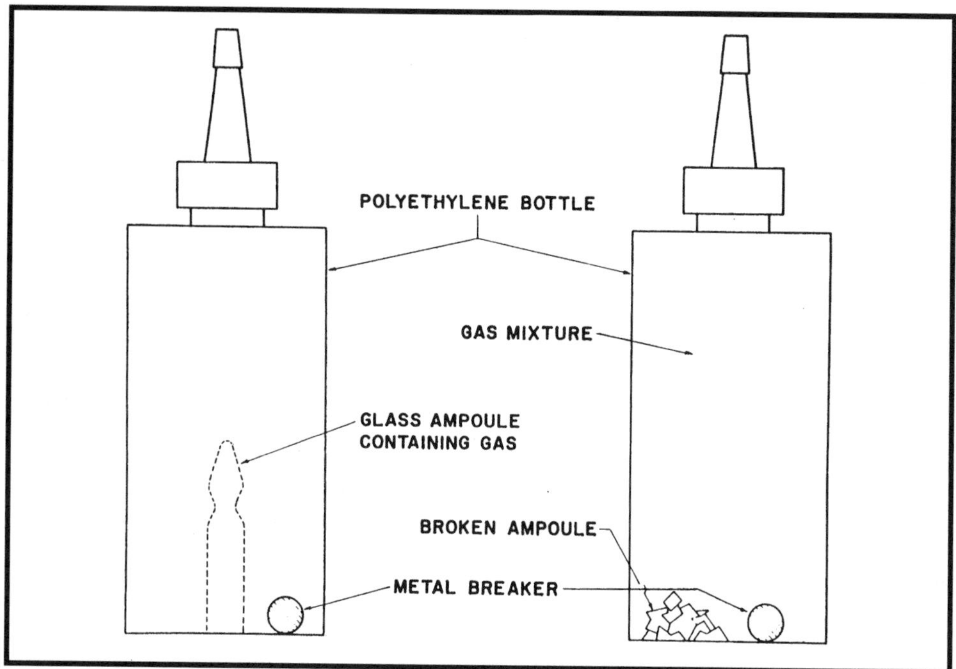

Figure 7.2 — Calibrator for explosive gas meters (Source: Reference 6).

Bags are fabricated from plastic sheets by thermal sealing. Volatile contaminants are baked out of the plastic sheets by some fabricators by keeping them in an oven for a few days. A rigid plastic tube or a valve similar to the type used for tires may be sealed to a side or corner to serve as the inlet, as illustrated in Figure 7.3. The inlet tube to the bag may be closed with a rubber stopper, serum cap, cork, or a valve according to the substance being handled. Bags are available commercially from a number of vendors.[22] The 3 ft × 3 ft or 2 ft × 4 ft size contains over 100 L. Available sizes range from 0.5 L (6 inches × 6 inches) to 450 L (5 ft × 5 ft).

A simple arrangement may be used for preparing a known mixture in a plastic bag. The bag is alternately filled partly with clean air and then completely evacuated several times to flush it out. Then clean air is metered into it through a wet or dry test meter or using a rotameter and stopwatch. The test substance is added to this stream at a tee just above the entrance to the bag. If the test substance is a volatile liquid, it can be injected accurately by syringe through a septum. Sufficient air must subsequently be passed through the tee to completely transfer the injected material. When the desired air volume has been introduced, the bag is disconnected and plugged or capped. The contents may be mixed by gently kneading the bag with the hands.

The major advantage of using flexible bags is that no dilution occurs as the sample is withdrawn. These bags are transportable. Bags should be tested frequently for pinhole leaks. Initial screening tests may be done by filling them with clean air and sealing them. Usually leaks are large enough to produce detectable flattening within 24 hours.

Adsorption and reaction on the walls is no great problem for relatively high concentrations of inert materials such as halogenated solvents and hydrocarbons.[23] However, low concentrations of reactive materials such as sulfur dioxide, nitrogen dioxide, and ozone are partly lost, even after prior conditioning of the bags.[19,20,24,25] Larger bags are preferable to minimize the surface-to-volume ratio. Losses of 5 or 10% frequently occur during the first hour, after which the losses are a small percentage per day. Conditioning of the bags with similar mixtures is essential to reduce these losses. A similar or identical mixture is stored for at least 24 hours in the bag and then evacuated

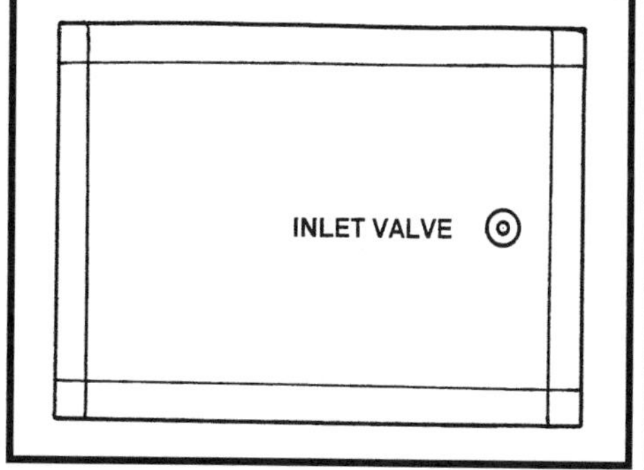

Figure 7.3 — Typical plastic bag for containing air samples.

Certain substances with high positive heats of formation, such as acetylene, can detonate even in the absence of oxygen.

For properly conducted laboratory applications, known mixtures can be prepared very conveniently in plastic bags.

just before use. A tabulation[26] of the uses of these bags described in 12 articles lists the plastic material, the gas or vapor stored, their concentrations, and the researcher comments. Another study[27] focused on occupational hygiene applications of plastic bags. Good stability in Saran bags was found for mixtures containing benzene, dichloromethane, and methyl alcohol; and for Scotchpak bags containing benzene, dichloromethane, and methyl isobutyl ketone. Percentage losses were greater for lower concentrations (i.e., below 50 ppm). Losses greater than 20% were observed in the first 24 hours for Saran bags containing methyl isobutyl ketone vapors; however, concentrations stabilized after two to three days. Concentrations of batch mixtures calculated theoretically were compared with analyses by gas chromatography.[28] Although the theoretical slope was obtained, incorrect intercepts were observed for compounds of low vapor pressure (0.4–8 mm Hg at 25°C) because of significant wall adsorption in the instruments. Losses of carbon monoxide and hydrocarbons in sampling bags of various materials also have been studied.[29] Losses of occupational hygiene analytes with time have been reported.[30] If the mixture is stored for more than a day, permeation into as well as out of the bag must be considered, especially for thin plastic films. Permeation of outside pollutants (hydrocarbons and nitrogen oxides) into Teflon bags has been demonstrated.[31]

It is difficult to draw generalized conclusions from these reports, other than the need for caution in applying plastic bags for low concentrations. Losses should be determined for each material in each type of bag. Even the past history of each individual bag must be considered. For properly conducted laboratory applications, known mixtures can be prepared very conveniently in plastic bags.

Pressure Cylinders

Preparation of certain gas mixtures can be done conveniently in pressurized steel cylinders.[6] This is very useful for mixtures such as hydrocarbons or carbon monoxide in air, which can be stored for years without losses. With other substances, there are losses because of factors such as polymerization, adsorption, or reaction with the cylinder walls. In some cases, as the pressure decreases in the cylinder, material desorbs from the walls, yielding a higher final concentration in the cylinder than was initially present. Concentrations should be low enough to avoid condensation of any component at the high pressure in the cylinder, even at the lowest temperature expected during its use. Care must be taken to use clean regulators made with appropriate materials that will not adsorb or react with the contents of the cylinder. Specially coated aluminum cylinders rather than steel cylinders may be used to assure long-term stability of many reactive gases at ppm concentrations.

A serious safety hazard exists in preparation of compressed gas mixtures. As mentioned previously, there is a possibility of explosion of combustible substances. This may occur because of the heat of compression during a too rapid filling process. Excessive heat also may cause errors in the gas composition. Certain substances with high positive heats of formation, such as acetylene, can detonate even in the absence of oxygen. Also, explosive copper acetylid can be produced if copper is used in the manifolds and connections. Proper equipment, including armor-plate shielding, and experience are required for safe preparation. Because these and accurate pressure gauges are not ordinarily available, it is recommended that such mixtures be purchased from the compressed gas vendors who have professional experience and equipment for such work. These vendors can prepare mixtures either by using accurate pressure gauges to measure the proportions of the components or by actually weighing the cylinders as each component is added. They also can provide an analysis at a reasonable extra charge; however, these analyses are not always reliable.[32]

The Analytical Chemistry Division of the National Institute of Standards and Technology (NIST) supervises gas mixtures offered by certain specialty gas companies as NIST Traceable Reference Material. Mixtures in air include carbon dioxide, carbon monoxide, methane, propane, and nitrogen dioxide. Mixtures in nitrogen (or helium) include benzene, carbon monoxide, carbonyl sulfide, hydrogen sulfide, methyl mercaptan, nitric oxide, and propane.

Calculations of Concentrations in Air

When equations are used for calculations, values in appropriate units must be entered for each term. If there is a constant in the equation, its numerical value corresponds to the units required. Ambient pressures are commonly measured with a mercury barometer and expressed as mm of mercury. Air volumes are commonly measured in liters, and temperatures as degrees celcuis. SI metric units have been agreed on internationally, including length: meter; mass: kg; temperature, °C; pressure: bar (= 10^6 dynes/cm^2), and Pascal (kg/m sec^2). Their relationships to commonly used units are as follows: 1 m^3 = 1000 L = 10^6 mL; 1 atmosphere = 760 mm Hg = 1.0133 bar = 1.0133 × 10^5 Pa.

The calculations for preparation of batch mixtures are based on close adherence to the Ideal Gas law[1] at low partial pressures. Calculations for dilute gas concentrations are based on the simple ratio of the volume of test gas to the volume of mixture, as shown in Equation 2.

$$ppm = 10^6\, v/V \qquad (2)$$

where

ppm = parts per million by volume (definition)
v = volume of test gas in the mixture, L
V = total volume of mixture, L

Changes in ambient temperature and pressure do not change the ppm value because both the numerator and denominator in this equation are changed by the same factor, which cancels out.

In the case of volatile liquids the calculation is based on the ratio of moles of test liquid to moles of gas mixture, Equation 3. The moles of liquid are determined in the numerator by dividing the weight injected by the molecular weight of the liquid compound. The moles of gas mixture are calculated in the denominator by dividing the total volume of mixture by the molar gas volume.

$$ppm = \left(\frac{10^6\, w/MW}{V/V_m}\right) \qquad (3)$$

where

w = weight of volatile test liquid introduced, g
MW = gram molecular weight of test liquid
V_m = gram molecular volume, L, of the mixture under ambient conditions

The molar gas volume is calculated from Equation 4 for the ambient temperature and pressure of the mixture.

$$V_m = 24.45\left(\frac{760}{P}\right)\left(\frac{t + 273.15}{298.15}\right) \qquad (4)$$

where

24.45 = gram molecular volume, L, under standard conditions of 760 mm Hg, 25°C
P = ambient pressure, mm Hg
t = ambient temperature, °C

Calculations for batch mixtures in compressed gas tanks may be made on the basis of the ratio of the partial pressure of the test compound to the total pressure of the mixture, as shown in Equation 5.

$$ppm = 10^6\, p/P_T \qquad (5)$$

where

p = partial pressure of the test compound, and
P_T = total pressure of the mixture.

The same equation may be used for methods relying on the equilibrium vapor pressure of a volatile liquid as a source of a known concentration. In this case p is the equilibrium vapor pressure and P_T is the barometric pressure of the atmosphere.

A more common method of expressing concentrations is in terms of mg/m^3, defined simply as the weight of the test component divided by the volume of the mixture. In this case the ambient pressure and temperature affect the value only of the denominator and thus affect the value of the mg/m^3. For precise work the volume is corrected to standard conditions of 25°C, 760 mm Hg, using another form of Equation 4.

$$V_2 = V_1\left(\frac{P_1}{P_2}\right)\left(\frac{t_2 + 273.15}{t_1 + 273.15}\right) \qquad (4a)$$

where

V_2 = corrected volume (desired at 25°C, 760 mm Hg)
V_1 = ambient volume
P_1 = ambient pressure
P_2 = desired corrected pressure (760 mm Hg)
t_2 = desired corrected temperature (25°C)
t_1 = ambient temperature °C

The standard conditions of 25°C and 760 mm Hg were selected to be close to usual ambient conditions so that the correction is small and is commonly ignored. Results of chemical analyses ordinarily are expressed in terms of milligrams (or micrograms). These may be divided by the sample volume in the same manner to yield mg/m^3 (which is also $\mu g/L$).

The relationship under standard conditions between ppm and mg/m^3 may be derived as follows. In a volume of 24.45 L containing 1 mole of mixture, the ppm is equal to the number of micromoles of the test component. Multiplying ppm by the molecular weight gives the number of micrograms for the numerator.

$$\frac{mg}{m^3} = \frac{\mu g}{L} = \frac{ppm\ MW}{24.45} \quad (6)$$

$$ppm = \frac{24.45\ mg/m^3}{MW} \quad (6a)$$

In Equation 6a if the mg/m^3 are at ambient temperature and pressure, the 24.45 L molar volume is replaced by the ambient value from Equation 4.

These calculations are illustrated by the following examples.

Example 1. A volume of 5 mL of pure carbon monoxide gas is slowly injected into a tube through which 105 L of air are being metered into a plastic bag. What is the concentration (ppm by volume) of the final carbon monoxide mixture?

Answer. Applying Equation 2: ppm = $10^6 \times 0.005\ L/105\ L = 47.6$ ppm.

Example 2. A dish containing 12.7 g of carbon tetrachloride is placed in a sealed cubical chamber with inside dimensions of 2.1 m for each edge. The final temperature is 22.5°C, and the barometer reading is 765 mm Hg. (a) What concentration (ppm) is achieved? (b) What is the concentration in mg/m^3 under ambient conditions? (c) What is the concentration in mg/m^3 under standard conditions?

Answers. (a) The ambient volume in liters is

$V = (2.1)^3 = 9.26\ m^3 \times 1000 = 9260$ L

Applying Equation 4

$$V_m = 24.45 \times \left(\frac{760}{765}\right)\left(\frac{295.65}{298.15}\right) = 24.11L$$

Applying Equation 3

$$ppm = \left(\frac{10^6 \times 12.7/153.84}{9260/24.11}\right) = 214.9\ ppm$$

(b) $mg/m^3 = 12{,}700\ mg/9.26\ m^3 = 1371\ mg/m^3$. (c) There are 1371 mg in a volume of 1 m^3 under ambient conditions. Correcting this volume to standard conditions results in

$$\frac{mg}{m^3} = \frac{1371}{\frac{24.45}{24.11}} = 1351\ mg/m^3$$

Alternatively, Equation 6 could be applied to the ppm, which does not change with ambient pressure and temperature.

$$\frac{mg}{m^3} = \frac{214.9 \times 153.84}{24.45} = 1351\ mg/m^3$$

Flow-Dilution Systems

In flow-dilution systems air and test vapor, gas, or aerosol are continuously and accurately metered and combined in a mixing device. These systems offer the primary advantage of being very compact. Since it is possible to operate them continuously, very large volumes of gas mixture can be provided. In a properly designed system, concentrations also can be changed very rapidly. Because of the relatively small gas volume of this system, the explosive hazard is less than that of batch systems. Any losses by adsorption on surfaces occur only in the initial minutes of operation. After a brief period the surfaces are fully saturated and no further losses occur. Because of these advantages, flow-dilution systems are popular for accurate work with most substances.

Gas-Metering Devices

A variety of devices can be used to measure the air and test substance flows in a flow-dilution system. The accuracy of the final concentration, of course, depends on the accuracy of these measurements. Rotameters are commonly used. Mass flowmeters, orifice meters, and critical orifices are also frequently employed. The calibration equations and techniques for these devices are given in detail in Chapter 8 on instruments for calibration. Because rotameters are very commonly used, a few points of importance to their application in flow-dilution systems will be discussed.

The pulsating flows provided by some diaphragm pumps utilized in many flow-dilution systems may result in serious errors in interpreting meter readings on the basis of a steady flow calibration. The author has observed that rotameter readings may be high by a flow factor of as much as 2, depending on the wave form of the pulsating flow. It is, therefore, essential for accurate measurements to use a pump with a pulsation damper, or to damp out such pulsations by assembling a train comprised of the pump, a surge chamber, and a resistance, such as a partially closed valve or orifice. The error due to the pulsating flow can be determined as the flowmeter is recalibrated by running the pulsating flow through the rotameter and into a soap film meter or wet-test meter, and then comparing results with the steady flow calibration. The reason for this error is that although the mean flow rate of the pulsating flow is proportional to the mean of the first power of the gas velocity, the lifting force on the rotameter ball is proportional to the mean of the gas velocity raised to a power of between 1 and 2. For completely turbulent flows in the rotameter, which are common, the exponent is 2; in this case if the velocity fluctuates in a sine wave, the ball position will correspond to the root mean square value, which is 1.414 times the correct mean value. If the wave form is spiked, even greater deviations can occur. Similar relationships and errors may occur with other types of flowmeters. Wave forms have been measured[33] with small hot-wire anemometers for some popular air sampling pumps not provided with pulsation dampers.

There are two types of corrections of flowmeter calibrations made under standard conditions for readings made under a different ambient pressure and temperature. If the system has substantial pressure drops and the flowmeter is operating under pressure or vacuum, this is added to or subtracted from the barometric pressure to give the absolute pressure, which is used for the correction. The first is the correction because the calibration is dependent on the density, and in some cases the viscosity, of the gas flow, both of which are affected by ambient pressure and temperature. Applying this correction gives the correct ambient flow. The second correction may be applied to convert the actual gas flow rate under ambient conditions to that under standard conditions. The latter correction is made on the basis of the ideal gas relationship using Equation 4a. It should be kept clearly in mind that the first calibration correction depends on the specific device being employed. The two bases, ambient or standard conditions, should not be confused, and the proper one must be employed for the application. Corrections for high altitudes are related to atmospheric pressure.[34]

Construction and Performance of Mixing Systems

A flow-dilution system is comprised of a metered test substance source, a metered clean-air source, and a mixer to dilute the test substance to the low concentration required. The total flow of mixture must be equal to or greater than the flow needed. It is highly desirable to use only glass or Teflon parts for constructing the mixing system. Components preferably should be connected with ball or standard taper ground glass joints. Short lengths of plastic tubing may be used if not exposed in butt-to-butt connections of the glass parts. Some studies have shown that metal and plastic tubing must be conditioned with the dilute mixtures for periods of hours or days before they cease absorbing the test substances.[19,20,35]

Two other factors must be kept in mind in the construction of a mixing system. The first is that the pressure drops must be very small, and the system should preferably be operated at very close to atmospheric pressure. Otherwise, any changes in one part of the system will require troublesome readjustment of the flows of other components. The interactions may require several time-consuming reiterative adjustments. Also, if a monitoring instrument is being calibrated it may not draw the designed sample flow rate from a system not at atmospheric pressure, and the readings may be erroneous. The second factor to consider is that the dead volume of the system must be minimized to achieve a rapid response time. For example, assume that a flow of 0.1 mL/min is metered into a diluent airstream, and that the dead volume of the system to the dilution point is 1 mL. To accomplish one volume change will require 10 minutes. To be certain that this dead volume is completely flushed out, five volume changes are needed, corresponding to a time lag of 50 minutes before the full concentration of test gas reaches the dilution point.

Figure 7.4 illustrates a convenient all-glass system for making gas dilutions. The metered test gas is connected at the extreme right through a ball joint and capillary tube (1). The dilution air is metered into the side arm (2). A trap-like mixing device ensures complete mixing with very little pressure drop. The desired flow can be taken from the side arm (3) and the excess vented through the waste tube (4), which may be connected to a Tygon® tube long enough to prevent entrance of air into the flow system. If necessary, this vent tube should be run to a hood. By clamping down on the vent tube or submerging the outlet (4) in a beaker of water, any desired pressure can be obtained in the delivery system.

The dilution air must be purified according to the needs of the work. Air can be passed through an oil filter and then over a

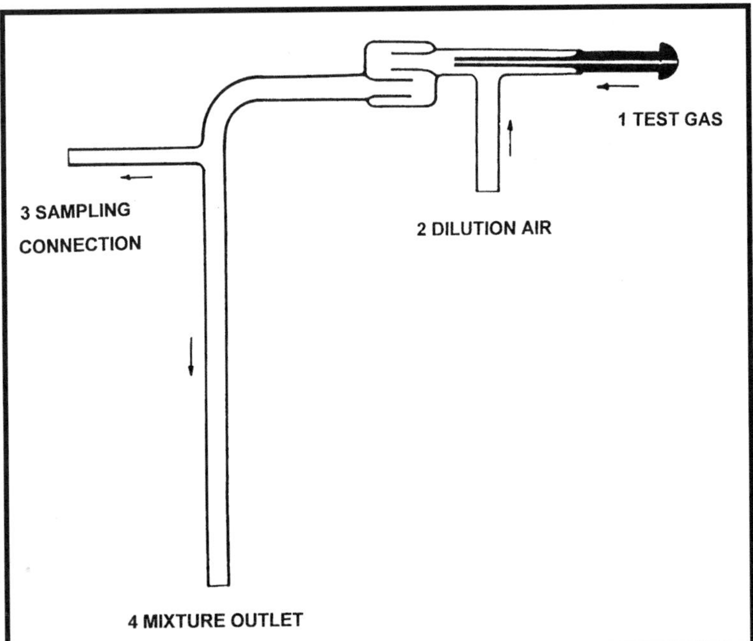

Figure 7.4 — All-glass flow-dilution mixer. 1: Inlet for metered gas; 2: inlet for meterd purified air; 3: samplong connection; 4: waste mixture outlet. (Source: Reference 5)

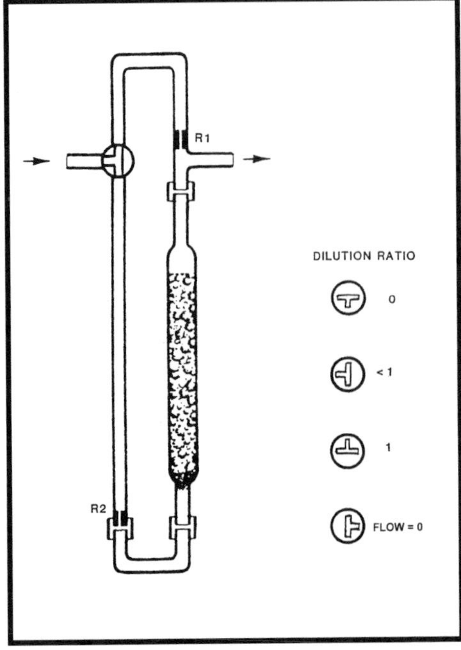

Figure 7.5 — Sel-dilution device. Dilution ratio is the ratio of the exiting to entering concentrations. (Source: Reference 7)

bed of carbon, silica gel, or Ascarite®, or bubbled through a scrubbing mixture of sodium dichromate in concentrated sulfuric acid if necessary. Organic compounds can be destroyed by passing the air over a hot platinum catalyst, or at a lower temperature over a Hopcalite® catalyst. Another convenient method of purification is to pass the air flow through a universal gas mask canister. The purification system must be designed according to the specific needs of the work.

Source Devices for Gases and Vapors

A variety of source devices are described below for providing high concentrations of gases and vapors that can be diluted with pure air to achieve the concentration desired. Each possesses specific advantages and disadvantages. Selection of a source device depends on the needs of the application and the equipment available to the user. Figure 7.5 shows a self-dilution device that can be generally applied to reduce the concentration provided by the source when necessary for work at very low concentrations. The flow of gas or vapor passes through the branches in proportions determined by restrictions R_1 and R_2. An appropriate absorbent such as carbon, soda lime, etc., in the lower branch completely removes the gas or vapor from the stream. Thus, the combined output of the two branches provides the delivered flow at a fractional concentration of the input depending on the relative values of the two flow restrictors. Furthermore, operation of the three-way stopcock also provides either the full concentration or zero concentration or completely cuts off the flow.

Vapor Pressure Source Devices

Figure 7.6 illustrates the method for providing a known concentration corresponding to the equilibrium vapor pressure of a volatile liquid. A flow of inert gas or purified air is bubbled through a container of the pure liquid (1) operated at ambient or elevated temperature. Liquid mixtures are less desirable because the more volatile components will evaporate first; thus, the vapor concentrations change as the evaporation proceeds. Because the output of the first bubbler may commonly be at only 50 to 90% of the saturation vapor pressure, equilibrium concentrations are obtained by then passing the vapor mixture through a second bubbler containing the same liquid, at an accurately controlled lower constant temperature. The excess vapor is condensed, and the final concentration is very close to equilibrium vapor pressure at the cooling bath temperature. A filter must be included to ensure that a liquid fog or mist does not escape. It is desirable to operate the constant temperature bath below ambient temperature so that liquid does not

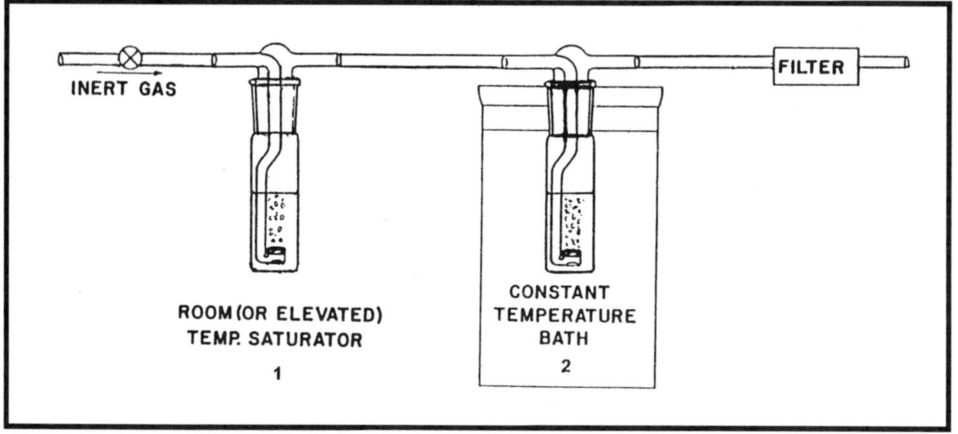

Figure 7.6 — Vapor saturator source system.

condense in the cool portions of the system downstream. The applications of this method to carbon tetrachloride[36] and mercury[37] have been described.

Another version of this arrangement is shown in Figure 7.7. This uses a wick feed from a small bottle containing the additive as a source of the vapor and an ice bath for the constant temperature.

Motor Driven Syringes

Figure 7.8 illustrates a source delivery system using a glass hypodermic syringe (4) that is driven by a motor drive at uniform rates that can be controlled to empty it in periods varying from a few minutes to an hour. A gas cylinder containing the pure test component or mixture is connected at the right side (1). The bubbler (2) is a safety

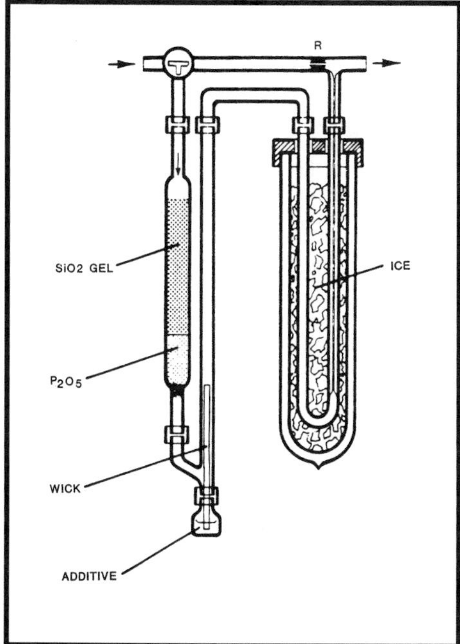

Figure 7.7 — Wick evaporator-condensor source system (Source: Reference 7).

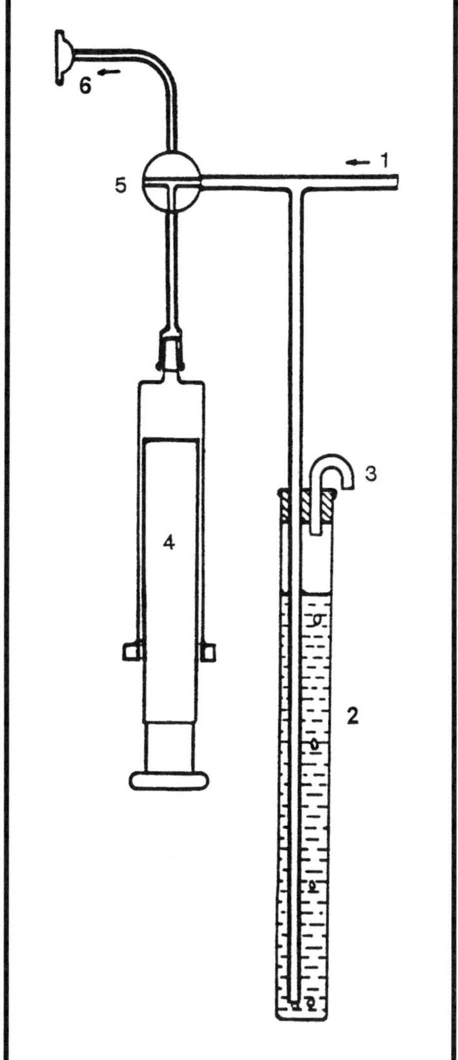

Figure 7.8 — Motor-driven syringe source system. 1: test gas inlet; 2: safety pressure release and flushing vent; 3: waste gas outlet; 4: motor-driven 50-mL glass syringe; 5: stopcock in syringe filling position; 6: metered gas outlet. (Source: Reference 5)

device to protect the glass apparatus from excessive pressures if the tank needle valve is opened too wide. The tank valve is cautiously opened, and a slow stream of gas is vented from the outlet (3) of the bubbler. The syringe is manually filled and emptied several times to flush it with the gas. This is done by turning the three-way stopcock (5) so that on the intake stroke the syringe is connected to the cylinder at (1) and on the discharge stroke to the delivery end (6). After flushing, the syringe is filled from the cylinder and the motor drive is set to discharge it over the desired period of time. This motor drive should include a limit switch to shut off the motor before it breaks the syringe, and a revolution counter for measuring the displacement. From the known gear ratio, the screw pitch, and a measurement of the plunger diameter with a micrometer, the rate of feed can be calculated. An accuracy and reproducibility of parts per thousand can be achieved with an accurate screw and good bearings.

At low delivery rates the back diffusion of air into the syringe from the delivery tip may cause an error. Thus, if the syringe is set to empty over a 1-hour period, toward the end as much as half of the gas mixture contents could be air that has diffused in backward. This error is easily minimized by inserting a loose glass wool plug in the delivery system and using capillary tubing for the delivered flow. The device was designed to connect to the dilution system illustrated in Figure 7.4. Syringes also may be used to meter a volatile liquid into an electrically heated vaporizer connected to an exposure chamber.[38,39] Concentrations can be routinely maintained within 10% of target levels. The possibilities of fire, explosion, or decomposition of a high-boiling compound must be considered.

Diffusion Source Systems

Figure 7.9 illustrates a diffusion system that can provide constant concentrations of a volatile liquid. The test liquid is contained in the bottom of a long thin tube and is kept at a known constant temperature. As the air flow is passed over the tip, vapor diffuses up through the length of the tube at a reproducible rate (assuming the temperature is closely controlled) and mixes with the stream. The rate is determined by the vapor pressure of the liquid, the dimensions of the tube, and the diffusion constants of the vapor and of air. If substantial amounts of a liquid

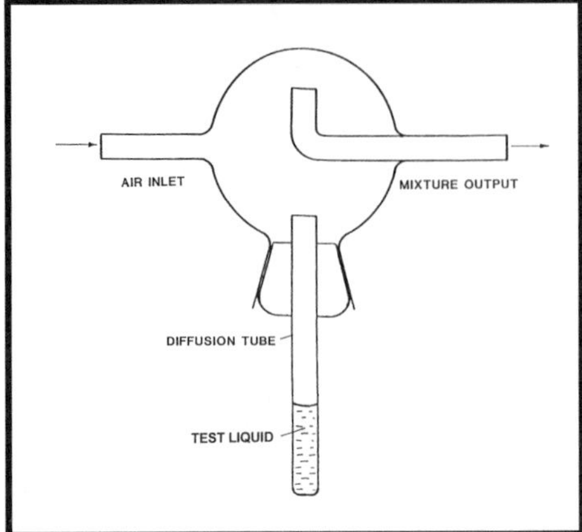

Figure 7.9 — Vapor diffusion source system.

are evaporated and the liquid level drops, the diffusion path length increases slightly. The quantity of liquid evaporated can be determined best from volume markings on the tube or by weighing the tube at the beginning and end of the period of use. It is possible to use a recording balance in some systems. Quantities calculated from the diffusion coefficient and the dimensions of the tube may be less accurate. Experimental values have been tabulated and the limitations of this method described.[40]

Porous Plug Source Devices

Figure 7.10 illustrates a micrometering system[5,41] that both measures and controls small flows of gas in the range of 0.02 to 10 mL/min. This is based on the

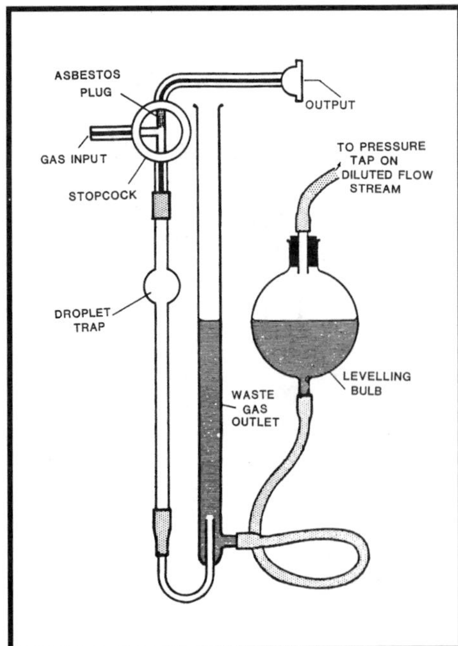

Figure 7.10 — Asbestos plug flowmeter and controller.

principle of diffusion of test gas through an asbestos plug under a controlled pressure difference. The inlet is connected to a cylinder containing pure test gas or a mixture. An acid-washed asbestos plug is contained in one leg of the T bore of a three-way stopcock (which is never turned), as shown in the figure. Any similar inert material may be used if asbestos is not available. The degree of fiber tamping is determined by trial and error to provide the desired delivered flow range. The cylinder needle valve is opened cautiously to provide a few bubbles per minute from the waste outlet in the lower portion of the figure. The height of water or oil above the waste outlet determines the fixed pressure on the lower face of the asbestos plug, which produces a fixed rate of diffusion of the gas through the plug to the capillary delivery tip. The meter is calibrated by connecting the delivery end to a graduated 1-mL pipette with the tip cut off, containing a soap film or drop of water. The motion of the film or drop past the markings is timed with a stop watch. This is repeated for different heights of liquid obtained by adjustment of the leveling bulb. The calibration plot of flow rates in mL/min versus the heights of liquid over the waste outlet in cm is usually a straight line passing through the origin. After the gas cylinder is shut off it should never be disconnected before the liquid pressures equalize; otherwise, the liquid may surge up and wet the asbestos plug. If this occurs, it must be discarded, the bore dried and repacked, and the new plug calibrated.

This device is a very convenient and precise method for metering low flows in the indicated range. The output flow remains constant for weeks, but should be checked occasionally. The delivery tip is connected to the mixer shown in Figure 7.4. For low delivery rates, the dead volume is minimized by using capillary tubing.

The leveling bulb vent is connected to a tap on the diluted gas manifold. This provides a correction for back pressure of the system into which the flow is being delivered. An appreciable back pressure changes the pressure differential across the asbestos plug. The bulb vent connection causes the liquid level to rise in the vent tube. If the vent tube area is small compared with the area of the liquid surface in the bulb, this compensates almost exactly for the back pressure by increasing the upstream pressure on the plug enough to maintain a constant pressure differential.

Permeation Tube Source Devices

Permeation tubes[42,43] are very useful sources for liquefiable gases and volatile liquids. Because of their potential precision they will be discussed in some detail here. NIST certifies sulfur dioxide and nitrogen dioxide permeation tubes. Each is individually calibrated at 25°C to within 1% of the stated permeation rate with 95% confidence. Thus, when maintained in a dry thermostated environment, they can serve as primary standards. Commercial sources[13,22] offer tubes for over 100 compounds and certify their permeation rates as a function of temperature to a precision of ±2%.

In these devices the liquid is sealed under pressure in inert plastic (commonly Teflon) tubes. The vapor pressure may be as high as 10 atmospheres. The gas or vapor permeates out through the walls of the tube at a constant rate (commonly of a few milligrams per day) for periods as long as a year. Figure 7.11 illustrates four commercial types of tubes. In the standard type (1) the liquified gas or volatile liquid is sealed in an inert plastic tube with solid plugs held in place by crimped bands. In some extended lifetime tubes a glass or stainless steel vial containing the liquified gas is attached to the Teflon tube (4). To slow the permeation rate, the low emission

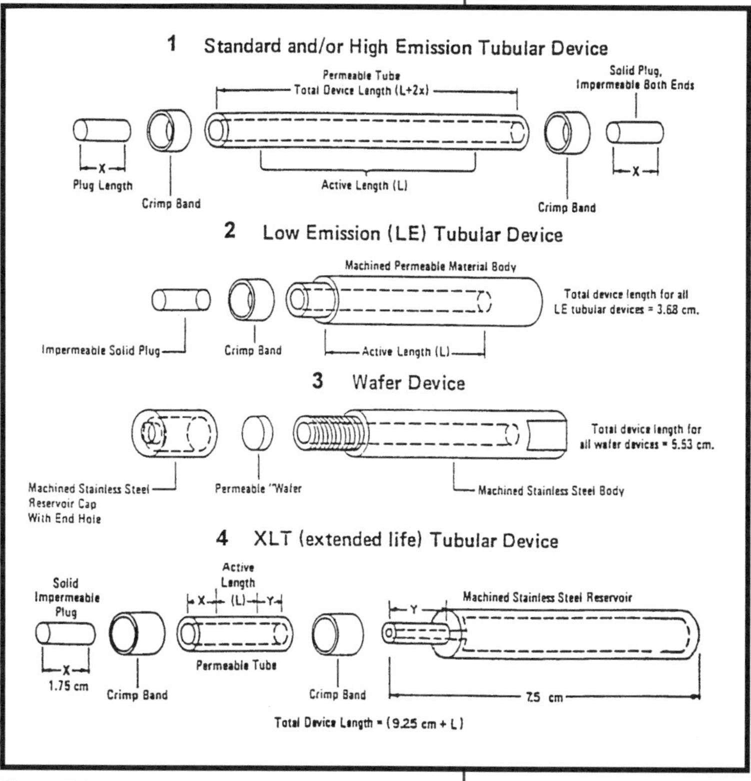

Figure 7.11 — Construction of some commercial permeation tubes (Courtesy of Dynacel® Permeation Devices, VICI Metronics Co.).

devices have thicker permeable walls (2) or wafers connected to an impermeable container (3) holding the liquid. In permeation tubes containing a liquid with a low vapor pressure, such as nitrogen dioxide (boiling point 21.3°C), a sufficiently tight seal may be obtained by pushing the Teflon tube onto the neck of a glass vial and by pushing a glass plug in at the top. For higher vapor pressures, such as in tubes containing propane (vapor pressure 10 atmospheres at 25°C), a stainless steel vial is used. Metals used for bottles or crimping bands must be selected for corrosion resistance, especially if the tubes are calibrated by weighing. In another type of seal, FEP (fluorinated ethylene propylene polymer), Teflon plugs are fused to an FEP Teflon tube by means of heat. Tubes are usually discarded when empty.

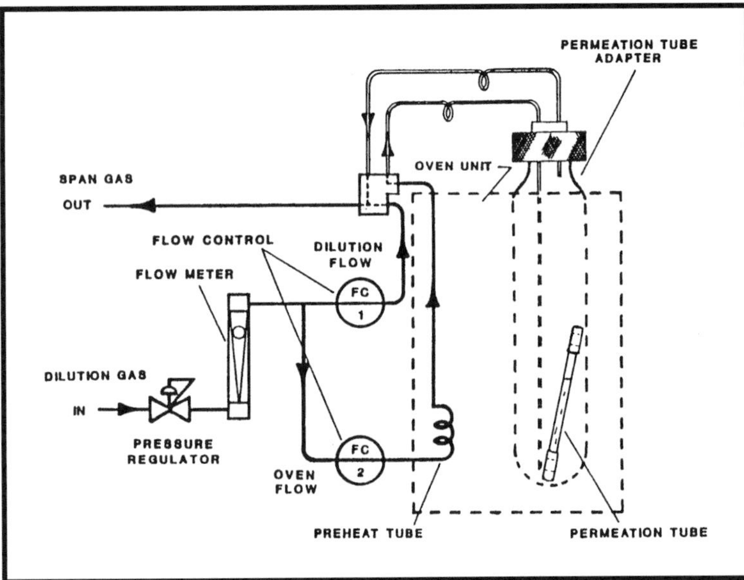

Figure 7.12 — A commercial calibration system using permeation tubes (Model 570C, Precision Gas Standards Generator, courtesy of Kin-Tek Laboratory Inc.).

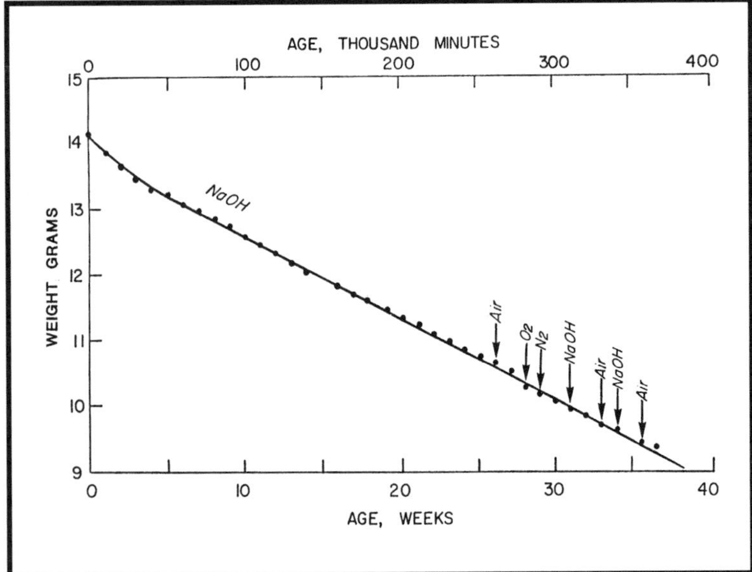

Figure 7.13 — Weight history of a nitrgen dioxide permeation bottle stored over NaOH or in flowing gas streams starting at indicated times (Source: Reference 44).

Figure 7.12 illustrates the typical use of a permeation tube in a commercial precision span gas standard generator for instrument calibration. A small constant flow is passed over the tube in an air oven thermostated to within ±0.05°C and then is diluted with a larger flow to produce the desired concentration and flow rate. In an alternative design that provides an extremely long-lived source, the liquid is contained in a sealed container in the oven. The air flows through the inside of a coiled Teflon tube immersed in the liquid in the container.

All tubes, especially if the contents are under pressure, should be handled with caution since they may present a toxic or explosive hazard. If they have been chilled in dry ice during filling, room for expansion of the liquid on warming to ordinary temperatures should be provided. Tubes should be protected from excessive heat. They should never be manually touched, scratched, bent, or mechanically abused. After a new tube has been prepared, several days or weeks are required before a steady permeation rate is achieved at a thermostated temperature. It was reported[44] that tubes made of FEP Teflon should be annealed at 30°C for a period to relieve extrusion strains and equilibrate the Teflon to achieve a steady permeation rate. Otherwise, a pseudo-stable rate is achieved that is not reproduced after appreciable temperature fluctuations. Tubes made of TFE (tetrafluorinated ethylene polymer) Teflon are more porous than FEP Teflon and may be used to obtain 3 to 10 times higher permeation rates.

Gravimetric calibrations may be made by weighing the permeation tubes at intervals and plotting the weights against time. The slope of the line fitted by the method of least squares to the measured points is the permeation rate. This process may take as long as several weeks with an ordinary balance because of the necessity of waiting to obtain accurately measurable weight differences. However, if a good micro balance is available, the calibration can be shortened to a day. Static charges that develop on some permeation tubes can cause serious weighing errors unless discharged with a polonium strip static eliminator. For a corrosive gas, the balance may be protected from corrosion by inserting the permeation tubes into tared glass-stoppered weighing tubes. The weight history of a nitrogen dioxide tube over a 37-week period[44] is shown in Figure 7.13. The tube was alternately stored in a closed container over sodium hydroxide (to absorb the released gas) and in a container flushed with about 50 mL/min of air (from the laboratory system), with cylinder oxygen, and with cylinder nitrogen. The humidity affected the weights. The most reproducible results were obtained with the cylinder nitrogen.

The environment of some types of permeation tubes must be very carefully controlled. Materials can permeate into the tubes as well as out of them. Thus, nitrogen dioxide tubes exposed to high humidity develop blisters and long-term changes in permeation rates. Even the moisture content of the flowing gas passing over the tube affects the permeation rate. These effects are likely due to the formation of nitric acid within the Teflon walls and/or inside the tube in the liquid nitrogen dioxide. When shipped inside capped iron pipe nipples, the color of the contents changed from brown to green. This was caused by reduction of nitrogen dioxide gas by the iron to nitric oxide, which permeated backward into the tube to combine with nitrogen dioxide to form blue N_2O_3. Exposure to air (or oxygen) for a time restored the brown color by backward permeation of oxygen. A similar problem occurs with hydrogen sulfide tubes, which precipitate colloidal sulfur within the walls of the tube when exposed to oxygen. Depleted sulfur dioxide tubes sometimes contain liquid droplets that may be sulfuric acid from backward permeation of water vapor and oxygen. It is, therefore, desirable never to remove such tubes from their operating environments.

Figure 7.14 illustrates a system that maintains the tube in a constant environment during calibration, use, and storage. A slow stream (50 mL/min) of dry nitrogen (1) from a cylinder is passed through a coil in a thermostated water bath (2). The bath pump (3) circulates water through the coil (4) to maintain the same temperature in an insulated water bath (5), which is mixed by a stirrer (6). The nitrogen stream passes through the stopcock (7) and over the permeation tube (8) to flush away the permeated gases. This stream leaves through the outlet stopcock (9) to the inlet (10) and is blended with a metered pure air flow (11) entering the mixing device to provide known concentrations of the gas at the outlet (12).

The calibration can be made volumetrically[43,44] using a relatively inexpensive Gilmont Warburg compensated syringe manometer (9, 13, 14, 15, 16, 17) supported by a clamp (18). The gas flow from the nitrogen cylinder is temporarily shut

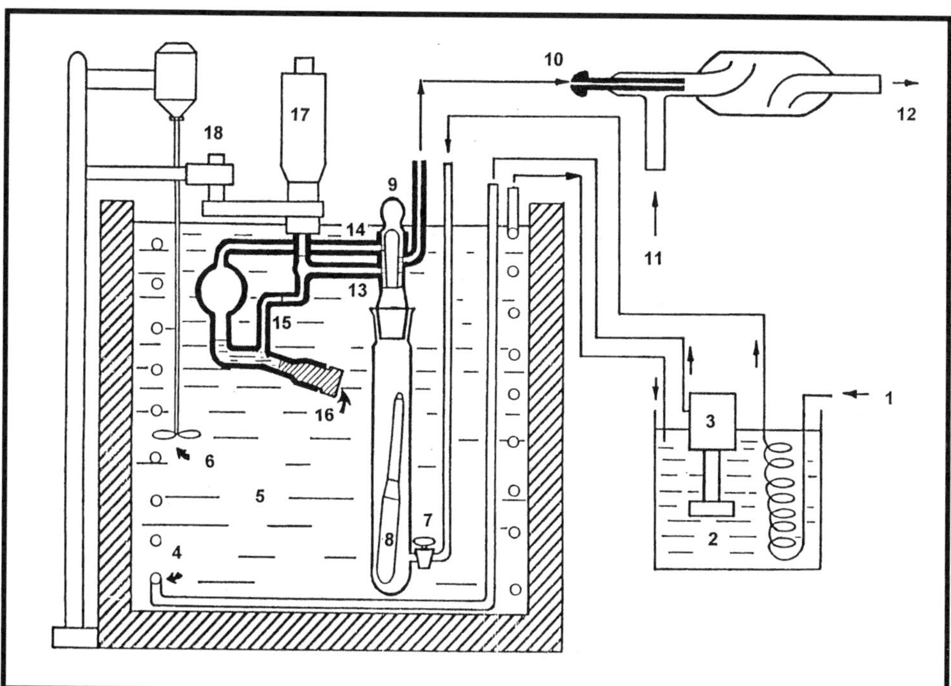

Figure 7.14 — Permeation tube flow system with in situ volumetric calibration. 1: connection for metered dry nitrogen from cylinder; 2: thermostated water bath; 3: circulating water pump; 4: thermostated water coil; 5: insulated water bath; 6: stirrer; 7: gas inlet stopcock; 8: permeation tube; 9: gas outlet stopcock; 10: connection for permeated gas flow; 11: connection for metered dilution air; 12: gas mixture outlet; 13: manometer inlet; 14: manometer outlet; 15: reference line; 16: adjusting screw; 17: 200-µL micrometer syringe; 18: clamp. (Source: Reference 44)

Table 7.2 —
Permeation Rates (ng/min cm) for Various Compounds

Material	*Temp (°C)*	*Wall Thickness (Inches)*	*G (Rate)*	*Life (Months)*
Acetaldahyde	30	.030TFE	235	11
Acetic acid	30	.030TFE	53	67
Acetic anhydride	70	.030TFE	239	15
Acetone	70	.030	270	10
Acrolein	30	.030TFE	164	17
Acrylonitrile	30	.030TFE	90	30
Allyl sulfide	70	.030TFE	48	62
Ammonia	30	.062	210	4.7
	35	.062	300	3.3
Aniline	70	.030TFE	35	99
Benzene	70	.030	260	11
	30	.030TFE	34	88
Bromomethane	30	.030	225	26
Bromine	30	.030TFE	240	42
Bromoform	70	.030TFE	127	77
Butadiene	30	.030	177	12
Butane	30	.030	24	82
1-Butene	30	.030	87	23
cis-Butene	30	.030	57	37
trans-Butene	30	.030	140	14
Butyl amine	70	.030TFE	326	7.7
n-Butyl mercaptan	70	.030TFE	242	11
tert-Butyl mercaptan	70	.030TFE	50	54
Carbon disulfide	30	.030	140	30
Carbon tetrachloride	70	.030	220	24
	30	.030TFE	150	14
Chlorine[A]	30	.062	1250	1.6
Chloroform	70	.030	713	7.1
	30	.030TFE	95	53
Chloroethane	30	.030	56	54
Chloromethane	30	.030	607	5.6
	30	.062	200	16
Cyclohexane	70	.030	20	132
1,2 Diaminoethane	70	.030TFE	448	10
1,2, Dibromoethane	70	.030	115	64
1,1, Dichloroethane	70	.030	265	15
1,2 Dichloroethane	30	.030TFE	69	60
	70	.030	260	16
1,1, Dichloroethene	30	.030	79	52
1,2 cis-Dichloroethene	30	.030TFE	270	16
1,2 trans-Dichloroethene	30	.030	145	29
Dichlorofluoromethane	30	.030	400	5.1
(F21)	30	.062TFE	101	45
Dichloromethane	70	.030	1600	2.8
	30	.030TFE	300	15
	30	.030	60	75
1,2 Dichlorotetra-fluoroethane (F114)	30	.062	670	2.6
Diethyl disulfide	70	.030TFE	58	58
Diethyl sulfide	30	.030TFE	14	202
	70	.030	51	56
Diisopropyl amine	70	.030TFE	157	30
m-Diisopropyl benzene	70	.030TFE	13	219
Dimethyl amine	30	.030TFE	240	9.6
N,N Dimethyl acetamide	70	.030TFE	48	66
Dimethyl disulfide	70	.030TFE	130	28
Dimethyl formamide	70	.030TFE	77	41
1,1 Dimethyl hydrazine	30	.030TFE	550	8.6
	70	.062	320	6.4
Dimethyl sulfate	70	.030TFE	60	75
Dimethyl sulfide	70	.030	300	9.5
	30	70	70	41
	70	.030TFE	1100	2.6
Ethanol	70	.030	49	54
Ethyl acetate	70	.030	296	10
Ethyl amine	30	.030TFE	77	31
Ethyl nitrate	30	.030TFE	224	20

down. Stopcock (7) is closed and stopcock (9) is turned to the position to connect the permeation tube chamber to the manometer inlet (13), and the manometer outlet (14) to the inlet of the mixer (10). The level of the manometer liquid is adjusted to the mark (15) by turning screw (16), and a stopwatch is started. The permeating gas causes the manometer liquid to move away from the mark (15). The pressure is relieved by withdrawing the plunger of the micrometer (17), on which the evolved volume after a timed interval can be read with a sensitivity of 0.2 µL. High precision has been obtained in less than an hour in this manner.

In other systems the permeation tube is suspended by a fiber from a microbalance so that it can be weighed at intervals without removal from a dry inert gas environment. Tubes containing corrosives such as chlorine, nitrogen dioxide, hydrogen fluoride, dimethyl sulfide, and ethyl nitrate should be purged with dry gas while stored.

The quantitative relationships for use of a permeation tube as a source for a flowing gas stream are given by Equation 7.

$$C = G\,L\,K / F \tag{7}$$

where

- C = concentration in the flowing gas stream, ppm
- F = flow rate, mL/min
- G = permeation rate per cm of tube length, ng/min cm
- L = permeation tube length, cm
- K = gas reciprocal density, nL/ng ($= V_m/MW$)

Table 7.2 lists the values of G at 30 and 70°C for various compounds, for commercial tubes of 0.25-inch outside diameter and specified thicknesses. For tubes of other dimensions, the values of G_2 may be calculated from the table value G_1 as follows.

$$G_2 = G_1 \left(\frac{\log(od/id)_1}{\log(od/id)_2} \right) \tag{8}$$

where

- od = outside diameter of tube
- id = inside diameter of tube
- subscripts 1 and 2 refer to values from the table and the new values, respectively

The tube lives listed in Table 7.2 are calculated on the assumption that the tubes are filled with liquid to 90% of their volume and are used at the calibration temperature until only 10% of their volume contains liquid.

The permeation rates have high temperature coefficients. Equation 9 shows the usual relationship in the form of the Arrhenius equation:

$$\log \left(\frac{G_2}{G_1}\right) = \left(\frac{E}{2.303\ R}\right)\left(\frac{1}{T_1} - \frac{1}{T_2}\right) \qquad (9)$$

where

G_1, G_2 = permeation rates at different temperatures
$T_1. T_2$ = corresponding temperatures, K (= °C + 273.15)
E = activation energy of permeation process, cal/g mol
R = gas constant, 1.9885 cal/g mol K

Values of E have been reported[43] ranging from 10 to 16 Kcal/g mol. Thus, the value of the E fraction term on the right may range from 2184 to 3494. To obtain 1% accuracy the temperature must be controlled to at least 0.1°C.

Miscellaneous Generation Systems

Figure 7.15 illustrates an electrolytic generator that was developed[5] as a suitable source of arsine and stibine. The solution was electrolyzed by passing a DC current through the platinum wire electrodes. The lower electrode was the cathode at which hydrogen and small quantities of arsine or stibine were liberated. The stream of purified air bubbled through the fritted tube end near the cathode and flushed the gas mixture into the outlet. The generation of arsine or stibine was not proportional to the DC current but was substantially constant after an initial lag period.

Another system used successfully was an aerated chemical solution mixture.[5] Thus, a 30% w/v solution of potassium cyanide served as a source of hydrogen cyanide. A relatively constant concentration could be obtained for as long as 10 hours. The strength and pH of the solution affected the concentration of hydrogen cyanide produced. The air bubbled through the solution should be free from carbon dioxide, since carbonic acid can displace hydrogen cyanide. In other applications, hydrogen chloride was obtained by aeration of a 1:1 concentrated acid-water mixture, and bromine by aerating saturated bromine

Table 7.2 (continued) — Permeation Rates (ng/min cm) for Various Compounds

Material	Temp (°C)	Wall Thickness (Inches)	G (Rate)	Life (Months)
Ethylene oxide	30	.030	128	23
	30	.030TFE	560	5.3
Ethyleneimine	50	.030	46	61
Ethyl mercaptan	30	.030TFE	64	44
Formaldehyde	70	.030TFE	29	60
	100	.030TFE	320	6
Halothane	30	.030TFE	625	10
Hexane	70	.030	160	14
Hydrazine	70	.030TFE	140	24
Hydrogen cyanide	30	.030	80	54
	30	.030TFE	330	13
Hydrogen fluoride	30	.062	120	12
	45	.062	400	3.7
Hydrogen sulfide	30	.062	240	6
	35	.062	330	4.4
Isopropylamine	70	.030TFE	550	4.3
Isopropyl mercaptan	70	.030	74	37
	70	.030TFE	300	9.2
Methanol	70	.030	216	12
Methyl amine	30	.030	136	17
Methyl bromide	30	.030	225	26
Methyl ethyl ketone	70	.030	100	27
	30	.030TFE	34	80
Methyl ethyl sulfide	70	.030	130	21
Methyl hydrazine	70	.030TFE	198	14
Methyl iodide	30	.030	33	234
	30	.030TFE	200	39
Methyl isocyanate	30	.030TFE	1780	1.8
	30	.062	125	11.3
Methyl mercaptan	30	.030	65	46.7
	30	.030FTE	270	11.2
Nitrogen dioxide[A]	30	.062	1000	2
Perchloro methyl mercaptan	70	.030TFE	125	46
Phosgene	30	.062	400	5
	30	.030	1250	3.8
Propane	30	.030	100	20
	30	.062	37	53
Propanol	70	.030	17	160
Propylene oxide	30	.030TFE	186	15
Styrene	70	.030	72	43
Sulfur dioxide[A]	30	.062	220	9.5
	30	.030	710	6.7
	35	.062	365	5.7
Tetrachloroethene	70	.030	300	18
Tetrahydrothiophene	70	.030TFE	64	52
Thiophene	70	.030	143	25
Toluene	70	.030	120	24
	30	.030TFE	22	133
1,1,1 Trichloroethane	70	.030	113	40
1,1,2 Trichloroethane	70	.030TFE	274	18
Trichloroethene	70	.030	1060	4.7
Trichlorofluoro	30	.062TFE	480	4.4
methane (F11)	30	.030TFE	1700	2.7
Trimethyl amine	30	.030TFE	215	10
Trimethyl phosphite	70	.030TFE	93	38
Vinyl acetate	70	.030	700	4.5
Vinyl chloride	30	.062	120	11
	30	.030	400	7.8
m-xylene	70	.030	44	67
o-xylene	70	.030	40	76

Note: Data from Analytical Instrument Development, Inc. Material is FEP Teflon unless TFE is specified. Rates are for 0.25 inch o.d. tubes.
[A]Also supplied with vial to extend life.

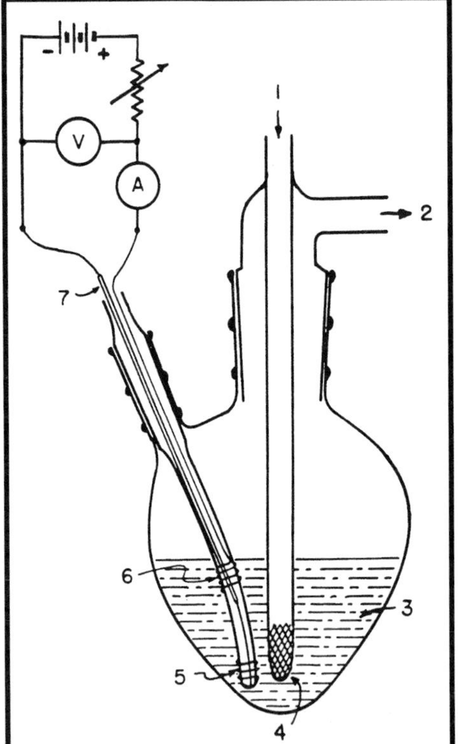

Figure 7.15 — Electrolytic generator for arsine or stibine. 1: inlet for purified airstream; 2: outlet for gas mixture; 3: electrolyte; 4: fritted tube end; 5: cathode; 6: anode; 7: insulating tube. (Source: Reference 5)

Caution: Tubes may explode if heated beyond their design limits.

water in contact with a small amount of liquid bromine.[5] In all of these procedures it is, of course, desirable to thermostat the bubbler to provide constant concentrations.

An interesting technique for preparing highly reactive or unstable mixtures is to utilize chemical conversion reactions. A stable mixture of a suitable compound is passed over a solid catalyst or reactant to produce the desired substance in the airstream. A table of reactions indicated some of these possibilities.[7] Others may be determined from the chemical literature. Multistep conversions also may be used.

The variables involved in flow-dilution systems are the source strength, the desired concentration, and the flow rates.

Illustrative Calculations for Flow-Dilution Systems

The variables involved in flow-dilution systems are the source strength, the desired concentration, and the flow rates. The following examples illustrate some of the calculations.

Example 3. Air is passed over mercury in a heated saturation unit and then equilibrated in a thermostated bath at 20°C at which its vapor pressure is 0.00120 mm Hg. What will the mercury concentration be in ppm and mg/m³ in the airstream warmed to 25°C?

Answer. Applying Equation 5:

$$\text{ppm} = 10^6 \left(\frac{0.00120}{760} \right) = 1.58 \text{ ppm}$$

Applying Equation 6:

$$\text{mg/m}^3 = 1.58 \left(\frac{200.61}{24.45} \right) = 12.9 \text{ mg/m}^3$$

Example 4. A gas stream of 50 mL/min is passed over a sulfur dioxide FEP Teflon permeation tube of 0.25-inch od × 0.030-inch wall thickness × 4.0 cm long, maintained at 30°C. The stream is mixed with a sufficient flow of clean air to total 500 mL/min at 25°C. What is the final concentration of sulfur dioxide in ppm?

Answer. The value of G from Table 7.2 for these dimensions and 30°C is 710 ng/min cm. The molecular weight is 32.06 + 2 × 16.00 = 64.06. The value of $K = V_m/MW = 24.47/64.06 = 0.382$. Applying Equation 7:

$$C = \frac{G\,L\,K}{F} = \frac{710(4.0)(0.382)}{500} = 2.17 \text{ ppm}$$

Example 5. Estimate the concentration that would be produced if the permeation tube temperature in Example 4 were raised to 35°C. Assume that the activation energy, E, is 18.8 Kcal/g mol.

Answer. Applying Equation 9:

$$\log\left(\frac{G_2}{G_1}\right) = \log\left(\frac{C_2}{C^1}\right) =$$

$$\left(\frac{18{,}800}{2.303 \times 1.9885}\right)\left(\frac{1}{303.2} - \frac{1}{308.2}\right) = 0.2197$$

Thus, $C_2/C_1 = 1.66$ and $C_2 = 2.17 \times 1.66 = 3.60$ ppm

Caution: Tubes may explode if heated beyond their design limits.

Source Devices for Aerosols

Preparation of aerosol mixtures is much more complex and difficult than that of gas and vapor mixtures. A major consideration is the size distribution of the particles. Commonly, a lognormal distribution describes the values; this is characterized by a geometric mean and a geometric standard deviation. The usual aerosol source device supplies a range of sizes. However, certain special types supply uniform-sized particles. If the geometric standard deviation is less than 1.1, the particles are considered homogeneous, or monodisperse. There is also a great variety of particle shapes, including spherical, crystalline, irregular, plate-like, spiked, and rod-shaped or fibrous. If the material is a mixture of compounds, the composition may vary with size. Certain substances may be present on the surfaces, which also can be electrostatically charged. All of these properties are affected by the source devices and methods of treatment. In the generation of known concentrations of aerosols, the choices of the operating parameters are determined by the objectives of the study, which may be to duplicate and study a complex aerosol existing naturally or in industry, or to prepare a simple pure aerosol for theoretical examination. A good general treatment of this subject with 257 references is available.[45] Extensive discussions of 28 papers of an aerosol symposium were published.[14] Useful standards for sizing particles by microscopic measurements using latex spheres and pollen of ragweed and mulberry have been suggested.[46]

Another major problem is the proper design of an exposure chamber from which to use or sample the aerosol. Because of settling, impaction, and inertial forces, the spacial distribution of the aerosol or of its different size fractions may not be uniform. One recommended design[47] places the instruments to be calibrated on a turntable in the chamber, and allows them to sample nonisokinetically from a quiescent atmosphere. Alternatively, the aerosol may be produced in a wind tunnel and sampled by the instrument isokinetically. If the aerosol is constituted of different size fractions, the two methods will likely give different results.

Dry Dust Feeders

Methods of producing solid aerosols have been comprehensively described.[48] One of the most convenient and widely used methods is to redisperse a dry powder. Standard test dusts are available, such as road dust, fly ash, silicates, silica, mineral dust, and many pigment powders and chemicals. Because these may tend to agglomerate, the degree of packing of the powder must be controlled and reproducible. A simple method consists of shaking the powder on a screen into the airstream. Mechanical systems attempt to provide a constant feed rate by use of moving belts or troughs, or by rotating turntables, screws, or gears. Because of the erratic behavior of loosely packed dust, the popular Wright dust feed mechanism (Figure 7.16) achieves closer control by compressing the dust in a tubular cup into

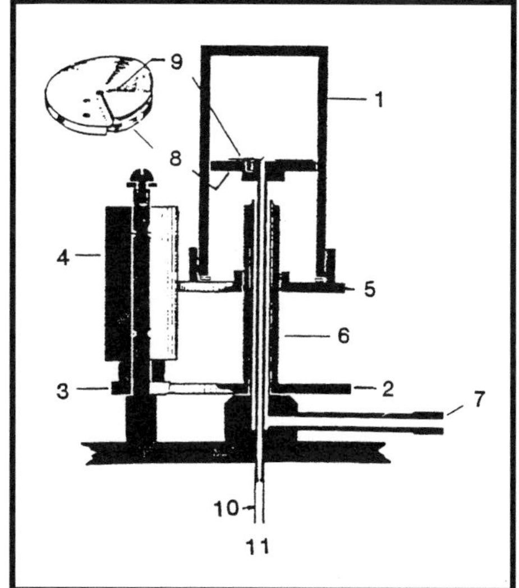

Figure 7.16 — Schematic diagram of Wright dust feeder 1: compressed dust cup; 2, 3, 4, 5: differential gear train that rotates and lowers cup on threaded spindle; 6, 7: compressed air inlet; 8: scraper head; 9: scraper blade; 10: dust outlet; 11: location for impaction plate.

a uniform cake. A rotating scraper advanced by a screw slices off a thin layer of cake continuously. In all of these devices the dust is dispersed by an air jet, which also serves to break up some aggregates. The dusty cloud is passed into a relatively large chamber, which serves to smooth out any rapid fluctuations. Concentrations may fluctuate ±20% over a period of a half hour, because of variations in the packing of the dust or laminations in the cake. Settling chambers, baffles, or cyclones may be added to the system to remove coarse particles, and ion sources to remove electrostatic charges.

Another system utilizes a fluidized bed to disperse a dry powder[49] as illustrated in Figure 7.17. A chain conveyer feeds the dust through an air lock into a chamber containing 100- to 200-µm glass or brass beads. An upward airstream at 9–30 L/min suspends and agitates the beads and blows the dust into the system. Electrostatic charges are produced and are neutralized with an ion source. After 1–3 hours a steady state concentration ±5% is achieved.

Producing fibrous aerosols of desired lengths and diameters is especially difficult. Glass wool fibers have been oriented in a parallel direction in a glycol methacrylate histological embedding medium and sliced into desired lengths with a microtome.[50] The plastic medium is then ashed and the fibers separated from any debris by liquid elutriation. A later study[51] used frozen polyethylene glycol for sectioning on the microtome,

Producing fibrous aerosols of desired lengths and diameters is especially difficult.

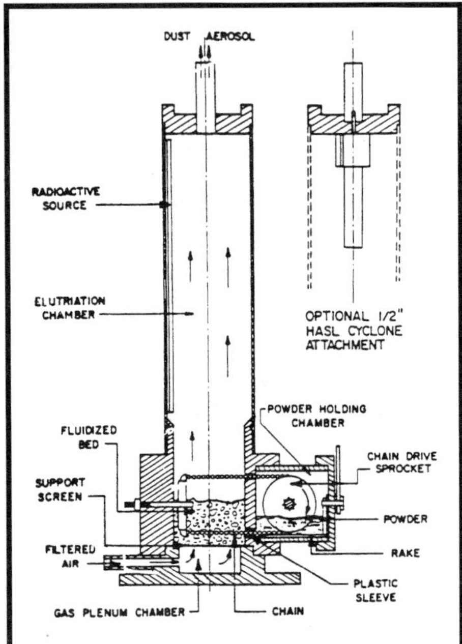

Figure 7.17 — Fluidized bed aerosol generator (Source: Reference 49)

subsequently dissolved in hot water. The fibers are suspended into an airstream from a fluidized bed.

Nebulizers

The compressed air nebulizer, Figure 7.18, is a convenient and useful device to produce aerosols from liquids. The liquid stream is drawn through a capillary tube and shattered into fine droplets by the air jet. The DeVilbiss nebulizer is simple, but holds only about 10 mL of liquid. Modifications can be added[45] such as utilizing a recirculating reservoir system for the liquid (Lauterbach), providing baffles to intercept and return coarse droplets (Collison, Dautrebande), droplet shattering baffles (Lovelace), and nozzle controls. The characteristics of these devices have been described in detail.[52]

Rather coarse sprays are obtained by pumping the liquid mechanically through tangential nozzles, as is done in fuel oil burners. The airflow merely carries off the droplets. Commercial aerosol cans use a mixture of liquid to be atomized and a liquified volatile propellant (such as dichlorodifluoromethane). The rapid evaporation of the propellant from the liquid emerging from the nozzle orifice shatters the stream into droplets having a broad size range. Electrostatic dispersion also has been utilized to break up a liquid stream by electrically charging the orifice. The droplets should be discharged by passage near an ion source soon afterward.

Somewhat different is the popular vibrating orifice generator, Figure 7.19, which uses an intense acoustic field to produce a monodisperse aerosol. In the version illustrated the pressurized liquid is ejected from the orifice as a fine stream, which is disrupted by the vibrations of the piezoelectric ceramic orifice plate into very uniform-sized droplets (coefficient of variation < 1%).

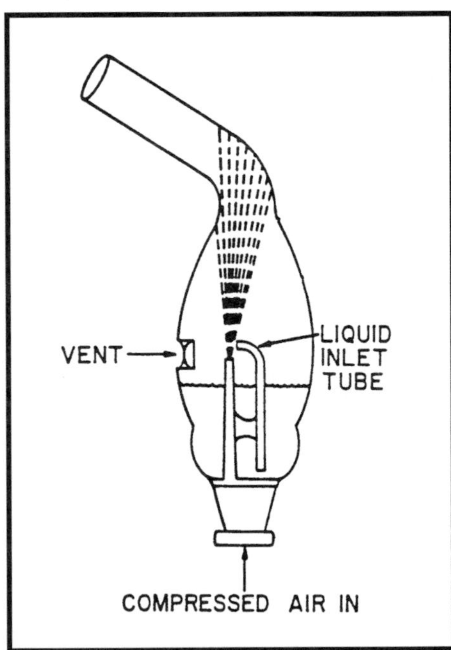

Figure 7.18 — DeVilbiss compressed air nebulizer (Source: Reference 52)

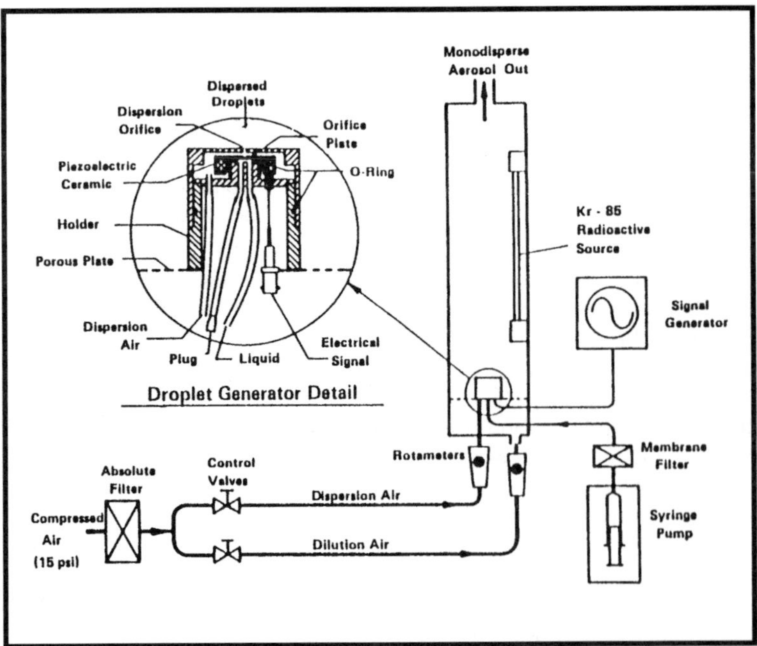

Figure 7.19 — Schematic of vibrating orifice aerosol generator (Source: Reference 14)

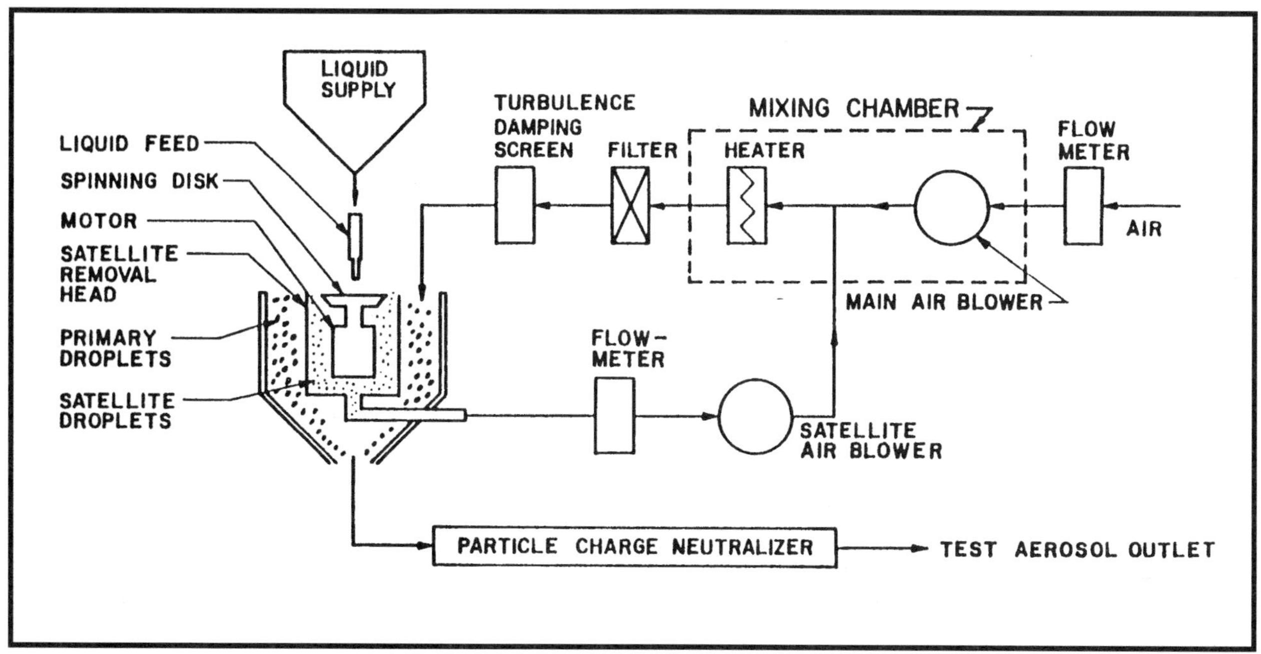

Figure 7.20 — Spinning disc aerosol generator (courtesy of Environmental Research Corp., St. Paul, Minn.)

Spinning Disc Aerosol Generators

A very useful generator for monodisperse aerosols is based on feeding the liquid continuously onto the center of a rapidly spinning disc (60,000 rpm). When the droplet on the edge of the disc grows to a sufficient size, the centrifugal force exceeds that of surface tension and the droplet is thrown off. A commercial version,[22] illustrated in Figure 7.20, produces liquid droplets in the 1 to 10 micron size range. Smaller satellite drops are diverted down by an airstream into a compartment around the disc. The larger particles escape to the outer compartment and are passed around a sealed radioactive ion source to remove the electrostatic charges, then to the outlet.

These liquid sources can be readily applied to supply solid aerosols by dispersing a solution or colloidal suspension. The solvent evaporates from the droplets naturally or on warming, leaving a smaller particle of crystalline solute, or a clump of one or more colloidal particles according to their theoretical probabilities of occurrence in the volume of the droplet. The sizes of the particles are controlled by varying the concentrations. The nature of the materials and of the drying process often affects the nature of the particles, which may exhibit shells or crusts. Passing the particles through a high temperature zone may be employed to decompose them chemically (e.g., production of metal oxides from their salts[53]) or to fuse them into spherical particles.

Vaporization and Condensation of Liquids

The principle of vaporization and condensation was utilized in the Sinclair-LaMer generator for materials such as oleic acid, stearic acid, lubricating oils, menthol, dibutyl phthalate, dioctyl phthalate, and tri-o-cresyl phosphate, as well as for sublimable solids. The system is illustrated in Figure 7.21.[8] Filtered air or nitrogen is bubbled through the hot liquid in the flask

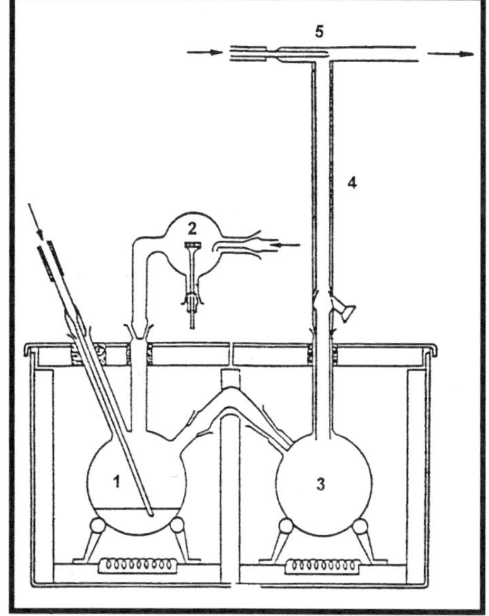

Figure 7.21 — Sinclair-LaMer condensation aerosol generator. 1: vaporizer; 2: condensation nuclei source; 3: superheater; 4: double-walled air cooler; 5: dilution mixer. (Source: Reference 8)

on the left (1). Another portion of the entering air is passed over a heated filament (2) coated with sodium chloride, to provide fine condensation nuclei. The vapor passes into the empty superheater flask (3), in which any droplets are evaporated, and then up the chimney (4) in which it is slowly cooled. The supersaturated vapor condenses on the sodium chloride particles to produce a monodisperse aerosol. Although the condensation nuclei vary in size, they have only a slight effect on the final aerosol droplet size, which is much larger. The aerosol is blended with clean air in a mixer (5). This system has been widely used as a convenient monodisperse source in the 0.02 to 30 micron size range.

Miscellaneous Generation Systems

Many dusts can be produced by means duplicating their natural formation. Thus, hammer or impact mills, ball mills, scraping, brushing, and grinding of materials have been employed. Combustion (e.g., tobacco smoke), high voltage arcing, and gas welding or flame cutting torches can be used. Organic metallic compounds (e.g., lead tetraethyl) may be burned in a gas flame. Metal powders can be fed into a flame or burned spontaneously (thermit and magnesium). Molten metals may be sprayed from metallizing guns. Metal wires can be vaporized by electrical discharges from a bank of condensers. Gaseous reactions also may be employed to produce aerosols, such as reaction of sulfur trioxide with water vapor[54] or of ammonia and hydrogen chloride. Finally, photochemical reactions can be utilized. The natural process for producing oxidative smog has thus been duplicated by irradiating automobile exhaust.

Multipurpose Calibration Components and Systems

A variety of components and systems have been constructed and are commercially available for testing analytical methods and calibrating instruments.[22] Clean air sources for filling respiratory air tanks, as well as for instruments, include oil filters in the air system or use oil-free compressors. Heatless air dryer-scrubbers are available to remove water and contaminants. These utilize two absorption beds alternately. While one is used the other is purged with a portion of the clean air. For ultra pure requirements, a heated catalytic

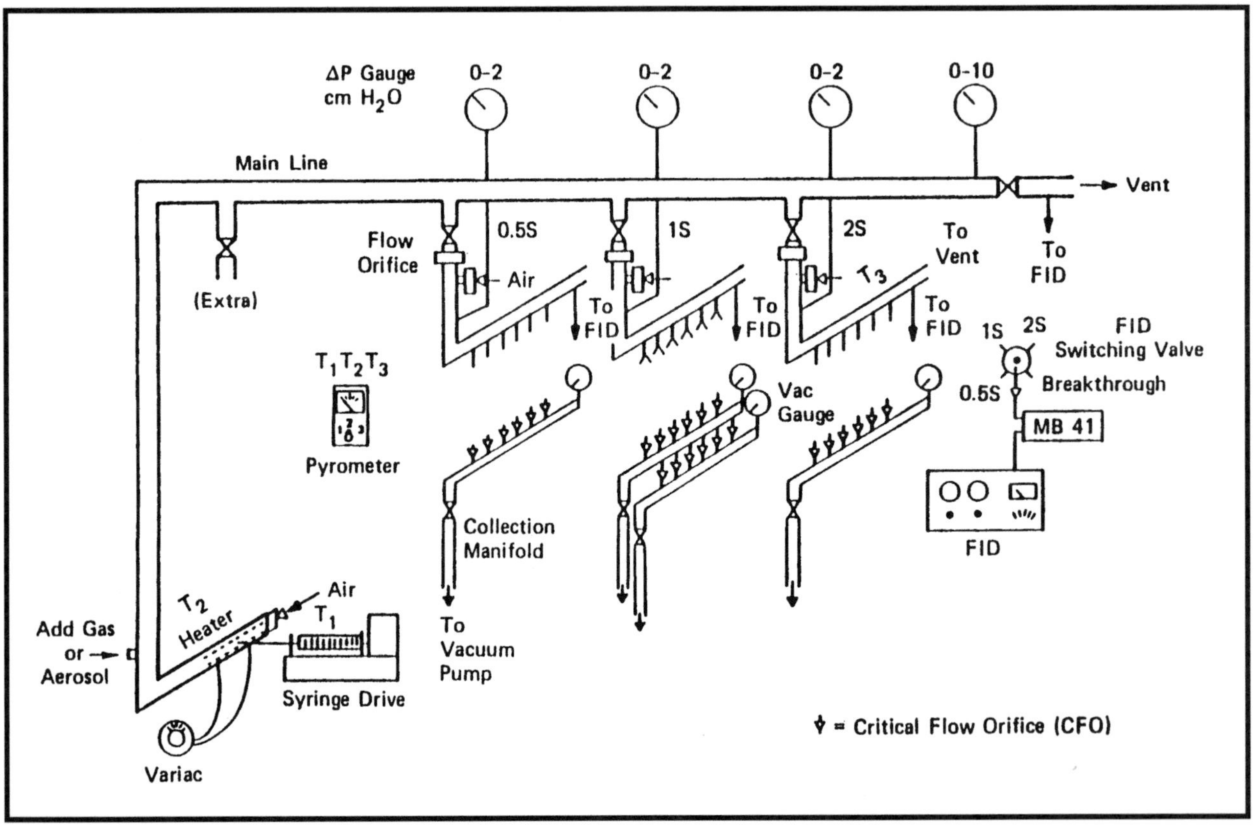

Figure 7.22 — Calibration system for generating multiple concentrations simultaneously. (Source: Reference 3)

oxidizer can be provided to burn off organic impurities. An automated flow-temperature-humidity control system has been described.[55]

In the Standards Completion Program, the 516 methods were evaluated by two National Institute for Occupational Safety and Health contract laboratories. Each set up a multipurpose system[3] for simultaneously producing several test concentrations in the desired range. Figure 7.22 illustrates schematically the one used mostly for testing charcoal tube sampling methods. The test substance was generated at a high concentration and passed through a manifold. Three taps were provided for 0.5, 1, and 2 times the PEL concentrations. To the flow from each the necessary dilution air was added at the throat of a Venturi tube, and the mixed flow was passed to the corresponding sampling manifold having six taps. The flows through the charcoal tubes connected at these taps were controlled by calibrated critical orifices. Thus, the system provided 18 simultaneous samples at three desired concentrations.

The numerous commercial sources of air analyzing instruments[22] commonly provide means for their calibration. Concentrations at the high end of the range (span gas) and at the bottom of the range (zero gas) are used alternately to adjust the electronics of the instrument to provide the proper readings. Quality assurance procedures describe these requirements in detail.[4,10,11] An elaborate computer controlled system also has been described[56] capable of providing mixtures in various desired concentration patterns.

Summary

Preparation of accurately known low concentrations of air contaminants is essential for testing and validating analytical methods associated with control regulations, for calibrating monitoring instruments, and for conducting scientific studies of effects, needed to develop control standards. Batch methods prepare a fixed volume of mixture in an appropriate container, such as a sealed chamber, a glass carboy, a plastic bag, or a compressed gas cylinder. Careful choices and tests must be made to ensure that substantial quantities of the contaminants are not adsorbed or lost on surfaces. If a large volume is prepared there is also a potential danger of toxic exposure from accidental release or of explosion of combustible contaminants. Flow-dilution methods accurately meter and mix a flow of the contaminant and of clean diluent air. They have the advantages of compactness, ability to change concentrations rapidly in a properly designed system, and ability to provide a large quantity of mixture. Any losses of contaminant on the surfaces cease after equilibrium occurs, and a steady state is reached in the operation of the system.

Recommendations for the design of flow-dilution systems include use of inert materials such as glass and Teflon. Connections should preferably be with ball and ground glass joints. Plastic tubing may be used if not exposed to the contaminant in butt-to-butt glass tubing connections. A variety of sources of known flows of contaminants is described, such as devices relying on equilibrium vapor pressures, diffusion, motor-driven syringes, and permeation tubes. Preparation of known aerosol concentrations is a more difficult and complex operation. A number of systems is described.

Theoretical equations are given for calculating concentrations of contaminants in batch and flow dilution systems. Illustrative examples are presented. A few multipurpose test and calibration systems are described. This chapter is designed to give the reader state of the art knowledge of how to design and apply accurate systems for preparation of known concentrations of air contaminants.

References

1. **Hill, J.W. and R.H. Petrucci:** *General Chemistry.* Upper Saddle River, NJ: Prentice Hall, 1996.
2. **Taylor, D.G., R.E. Kupel, and J.M. Bryant:** *Documentation of NIOSH Validation Tests* (DHEW [NIOSH] pub. 77-185). Cincinnati, OH: U.S. Dept. of Health, Education, and Welfare, National Institute for Occupational Safety and Health, 1977.
3. **Gunderson, E.C. and C.C. Anderson:** *Development and Validation of Methods for Sampling and Analysis of Workplace Toxic Substances* (DHHS [NIOSH] pub. 80-133). Cincinnati, OH: U.S. Dept. of Health and Human Services, National Institute for Occupational Safety and Health, 1980.
4. **Environmental Monitoring Systems Laboratory:** *Quality Assurance Handbook for Air Pollution Measurement Systems*, vol. I, *Principles* (EPA 600/9-76-005); vol. II,

Ambient Air Specific Methods (EPA 600/4-77-027a); vol. III, *Stationary Source Specific Methods* (EPA 600/4-77-027b). Washington, DC: U.S. Environmental Protection Agency, 1976–1977. (Superseded in 1990 by the EPA Electronic Bulletin Board System at (301) 589-0046; system questions answered at 202-382-7671.)

5. **Saltzman, B.E.:** Preparation and analysis of calibrated low concentrations of sixteen toxic gases. *Anal. Chem. 33*:1100–1112 (1961).
6. **Cotabish, H.N., P.W. McConnaughey, and H.C. Messer:** Making known concentrations for instrument calibration. *Am. Ind. Hyg. Assoc. J. 22*:392–402 (1961).
7. **Hersch, P.A.:** Controlled addition of experimental pollutants to air. *J. Air Pollut. Control Assoc. 19*:164–172 (1969).
8. **Fuchs, N.A. and A.G. Sutugin:** Generation and use of monodisperse aerosols. In *Aerosol Science,* C.N. Davies (ed.). New York: Academic Press, 1966.
9. **Silverman, L.:** Experimental test methods. In *Air Pollution Handbook,* P.L. Magill, F.R. Holden, C. Ackley, and F.G. Sawyer (eds.). New York: McGraw-Hill Book Co., 1956. pp. 12-1 to 12-41.
10. **American Society for Testing and Materials (ASTM):** *Calibration in Air Monitoring* (ASTM Spec. pub. 598). Philadelphia, PA: ASTM, 1976.
11. **American Society for Testing and Materials (ASTM):** *1987 Annual Book of ASTM Standards,* vol. 11.03. Philadelphia, PA: ASTM, 1987.
12. **Green, H.L.:** *Particulate Clouds: Dusts, Smoke and Mists.* 2nd ed., Chap. 2. Princeton, NJ: Van Nostrand, 1964.
13. **Nelson, G.O.:** *Gas Mixtures—Preparation and Control.* Chelsea, MI: Lewis Publishers, Inc., 1992.
14. **Willeke, K. (ed.):** *Generation of Aerosols and Facilities for Exposure Experiments.* Ann Arbor, MI: Ann Arbor Science Publishers, Inc., 1980.
15. **Woodfin, W.J.:** *Gas and Vapor Generation Systems for Laboratories* (DHHS [NIOSH] pub. 84-113). Cincinnati, OH: National Institute for Occupational Safety and Health, 1984.
16. **Chang, Y.S. and B.T. Chen:** Aerosol sampler calibration. In *Air Sampling Instruments,* 8th ed. Cincinnati, OH: American Conference of Governmental Industrial Hygienists, 1995.
17. **Moss, O.R.:** Calibration of gas and vapor samplers. In *Air Sampling Instruments,* 8th ed. Cincinnati, OH: American Conference of Governmental Industrial Hygienists, 1995.
18. **John, W.:** The characteristics of environmental and laboratory-generated aerosols. In *Aerosol Measurement,* K. Willeke and P. Baron (eds.). New York: Van Nostrand Reinhold, 1993.
19. **Baker, R.A. and R.C. Doerr:** Methods of sampling and storage of air containing vapors and gases. *Int. J. Air Water Pollut. 2*:142–158 (1959).
20. **Wilson, K.W. and H. Buchberg:** Evaluation of materials for controlled air reaction chambers. *Ind. Eng. Chem. 50*:1705–1708 (1958).
21. **Mokler, B.V. and R.K. White:** Quantitative standard for exposure chamber integrity. *Am. Ind. Hyg. Assoc. J. 44*:292–295 (1983).
22. **Cohen, B.S. and S.V. Hering (eds.):** *Air Sampling Instruments,* 8th ed. Cincinnati, OH: American Conference of Governmental Industrial Hygienists, 1995.
23. **Clemons, C.A. and A.P. Altshuller:** Plastic containers for sampling and storage of atmospheric hydrocarbons prior to gas chromatographic analysis. *J. Air Pollut. Control Assoc. 14*:407–408 (1964).
24. **Altshuller, A.P., A.F. Wartburg, I.R. Cohen and S.F. Sleva:** Storage of vapors and gases in plastic bags. *Int. J. Air Water Pollut. 6*:75–81 (1962).
25. **Connor, W.D. and J.S. Nader:** Air sampling with plastic bags. *Am. Ind. Hyg. Assoc. J. 25*:291–297 (1964).
26. **Schuette, F.J.:** Plastic bags for collection of gas samples. *Atmos. Environ. 1*:515–517 (1967).
27. **Smith, B.S. and J.O. Pierce:** The use of plastic bags for industrial air sampling. *Am. Ind. Hyg. Assoc. J. 31*:343–348 (1970).
28. **Samimi, B.S.:** Calibration of MIRAN gas analyzers: extent of vapor loss within a closed loop calibration system. *Am. Ind. Hyg. Assoc. J. 44*:40–45 (1983).
29. **Polasek, J.C. and J.A. Bullin:** Evaluation of bag sequential sampling technique for ambient air analysis. *Environ. Sci. Tech. 12*:708–712 (1978).
30. **Posner, J.C. and W.J. Woodfin:** Sampling with gas bags I: Losses of analyte with time. *Appl. Ind. Hyg. 1*:163–168 (1986).

31. **Kelly, N.A.:** The contamination of fluorocarbon-film bags by hydrocarbons and nitrogen oxides. *J. Air Pollut. Control Assoc. 33*:120–125 (1983).
32. **Saltzman, B.E. and A.F. Wartburg:** Precision flow dilution system for standard low concentrations of nitrogen dioxide. *Anal. Chem 37*: 1261–1264 (1965).
33. **Laviolett, P.A. and P.C. Reist;** Improved pulsation damper for respirable dust mass sampling devices. *Am. Ind. Hyg. Assoc. J. 33*:279–282 (1972).
34. **Treaftis, H.N., T.F. Tomb, and H.F. Carden:** Effect of altitude on personal respirable dust sampler calibration. *Am. Ind. Hyg. Assoc. J. 37*:133–138 (1976).
35. **Altshuller, A.P. and A.F. Wartburg:** Interaction of ozone with plastic and metallic materials in a dynamic flow system. *Int. J. Air Water Pollut. 4*: 70–78 (1961).
36. **Ash, R.M. and J.R. Lynch:** The evaluation of gas detector tube systems: carbon tetrachloride. *Am. Ind. Hyg. Assoc. J. 32*:552–553 (1971).
37. **Scheide, E.P., E.E. Hughes, and J.K. Taylor:** A calibration system for producing known concentrations of mercury vapor in air. *Am. Ind. Hyg. Assoc. J. 40*:180–186 (1979).
38. **Miller, R.R., R.L. Letts, W.J. Potts, and M.J. McKenna:** Improved methodology for generating controlled test atmospheres. *Am. Ind. Hyg. Assoc. J. 41*:844–846 (1980).
39. **Decker, J.R., O.R. Moss, and B.L. Kay:** Controlled-delivery vapor generator for animal exposures. *Am. Ind. Hyg. Assoc. J. 43*:400–402 (1982).
40. **Altshuller, A.P. and I.R. Cohen**: Application of diffusion cells to the production of known concentrations of gaseous hydrocarbons. *Anal. Chem. 32*:802–810 (1960).
41. **Avera, C.B., Jr.:** Simple flow regulator for extremely low gas flows. *Rev. Sci. Instrum. 32*:985–986 (1961).
42. **O'Keeffe, A.E. and G.C. Ortman:** Primary standards for trace gas analysis. *Anal. Chem. 38*:760–763 (1966).
43. **Saltzman, B.E.:** Permeation tubes as primary gaseous standards. In *International Symposium on Identification and Measurement of Environmental Pollutants*, B. Westley (ed.). Ottawa, Ontario, Canada: National Research Council of Canada, 1971. pp. 64–68.
44. **Saltzman, B.E., W.R. Burg, and G. Ramaswami:** Performance of permeation tubes as standard gas sources. *Environ. Sci. Tech. 5*:1121–1128 (1971).
45. **Raabe, O.G.:** Generation and characterization of aerosols. In *Conference on Inhalation Carcinogenesis* (CONF-691001). Gatlinburg, TN: Oak Ridge National Laboratory, 1969.
46. **Fairchild, C.J. and L.D. Wheat:** Calibration and evaluation of a real-time cascade impactor. *Am. Ind. Hyg. Assoc. J. 45*:205–211 (1984).
47. **Marple, V.A. and K.L. Rubow:** Air aerosol chamber for instrument evaluation and calibration. *Am. Ind. Hyg. Assoc. J. 44*:361–367 (1983).
48. **Silverman, L. and C.E. Billings:** Methods of generating solid aerosols. *J. Air Pollut. Control Assoc. 6*:76–83 (1956).
49. **Marple, V.A., B.Y.H Liu, and K.L. Rubow:** A dust generator for laboratory use. *Am. Ind. Hyg. Assoc. J. 39*:26–32 (1978).
50. **Esmen, N.A., R.A. Kahn, D. LaPietra, and E.P. McGovern:** Generation of monodisperse fibrous glass aerosols. *Am. Ind. Hyg. Assoc. J. 41*:175–179 (1980).
51. **Carpenter, R.L., J.A. Pickrell, K.S. Sass, and B.V. Mokler:** Glass fiber aerosols: preparation, aerosol generation, and characterization. *Am. Ind. Hyg. Assoc. J. 44*:170–175 (1983).
52. **Mercer, T.T., M.I. Tillery, and H.Y. Chow:** Operating characteristics of some compressed air nebulizers. *Am. Ind. Hyg. Assoc. J. 29*:66–78 (1968).
53. **Crisp, S., W.A. Hardcastle, J.M. Nunan, and A.F. Smith:** An improved generator for the production of metal oxide fumes. *Am. Ind. Hyg. Assoc. J. 42*:590–595 (1981).
54. **Chang, D.P.Y. and B.K. Tarkington:** Experience with a high output sulfuric acid aerosol generator. *Am. Ind. Hyg. Assoc. J. 38*:493–497 (1977).
55. **Nelson, G.O. and R.D. Taylor:** An automated flow-temperature-humidity control system. *Am. Ind. Hyg. Assoc. J. 41*:769–771 (1980).
56. **Koizumi, A. and M. Ikeda:** A servomechanism for vapor concentration control in experimental exposure chambers. *Am. Ind. Hyg. Assoc. J. 42*:417–425 (1981).

Outcome Competencies

After completing this chapter, the user should be able to:

1. Define underlined terms used in this chapter.
2. Explain the concept of traceability and calibration hierarchy.
3. Recognize the significance of the calibration process for IH measurement.
4. Describe the advantages and disadvantages of the various calibrators used in occupational air sampling.
5. Develop calibration procedures for specific air sampling equipment/methods.
6. Recognize what is needed to establish and maintain a laboratory calibration program.

Prerequisite Knowledge

General college physics

General college algebra

Prior to beginning this chapter, the user should review the following chapters:

Chapter Number	Chapter Topic
9	Direct-Reading Instrumental Methods for Gases, Vapors, and Aerosols
10	Sampling of Gases and Vapors

Key Terms

actual flow rate • bubble meter • bulk meters • by-pass rotameter • calibration • calibration curve • calibration program • critical flow orifices • critical orifices • displacement bottles • dry-gas meters • flask of Mariotte • flow rate standards • flow rate meters • frictionless piston meters • glass piston • graphite piston • heated element anemometer • hierarchical pathway • intermediate standards • laminar-flow meter • mass flowmeters • mercury-sealed piston • meter provers • orifice meters • precision rotameters • primary standards • prover bottle • rotameters • secondary standards • spirometers • standardized flow rate • strapping • thermal meter • traceability • variable-area meter • variable-head meters • velocity pressure • venturi meters • volume syringe • volume meters • wet test gas meter

Key Topics

I. Calibration Process

II. Flow Rate Standards

III. Calibration Hierarchy and Traceability

IV. Primary Standards
- A. Spirometers and Meter Provers
- B. Displacement Bottle
- C. Frictionless Piston Meters

V. Secondary Standards
- A. Wet Test Gas Meter
- B. Dry-Gas Meter
- C. Using Calibrators that Measure Flow Rate

VI. Other Secondary Standards
- A. Variable-Area Meters (Rotameters)
- B. Variable-Head Meters
- C. Critical Flow Orifice
- D. Packed Plug Flowmeter
- E. By-Pass Flow Indicators
- F. Using Point Velocity Meters
- G. Velocity Pressure Meters

VII. Establishing and Maintaining a Calibration Program

8

Principles and Instrumentation for Calibrating Air Sampling Equipment

Peter F. Waldron, MPH, IHIT

Introduction

Air sampling is widely used to measure human exposure and to characterize emission sources. It is often employed within the context of the general survey, investigating a specific complaint, or simply for regulatory compliance. The accuracy and precision of any air sampling procedure can be only as good as the sampling and analytical error that is associated with the method. The difference between the air concentration reported for an air contaminant (on the basis of a meter reading or laboratory analysis) and the true concentration at that time and place represents the overall error of the measurement.

This overall error may be due to a number of smaller component errors rather than a single cause. To minimize the overall error, it is usually necessary to analyze each of the potential components and concentrate one's efforts on reducing the largest component error. For example, it would not be productive to reduce the uncertainty in the analytical procedure from 10 to 1.0% when the error associated with the sample volume measurement is as high as 15 to 20%.

In air sampling the largest portion of the sampling error is frequently due to the flow rate of air and, ultimately, the underestimation or overestimation of the total volume of air that has passed through the sampling device. To define a concentration, the quantity of the contaminant of interest per unit volume of air must be accurately measured. Therefore, to obtain the best estimate of the true concentration to which an employee has been exposed, one must have a thorough understanding of both the setting and maintaining of calibrated flow rates.

The Calibration Process

Before any air sampling device can be relied on as accurate, it must be calibrated. Calibration is defined by the American National Standards Institute as "the set of operations which establishes, under specified conditions, the relationship between values indicated by a measuring instrument or measuring system, and the corresponding standard or known values derived from the standard."[1] In perhaps somewhat simpler terms, the calibration process is a comparison of one instrument's response to that of a reference instrument of known response and known accuracy. Hence, the overall quality of the calibration process can be no better than the quality of the calibrator used or referenced.

The calibrators most frequently used in occupational air sampling can be divided into three basic groups, which are differentiated by the type of measurement being checked. The first and second are volume meters and flow rate meters, both of which respond to the entire airflow of the sampler. The third group is velocity meters, which respond by measuring velocity at a particular point of the flow cross section. There are several different devices within each of these categories that have evolved over the years, primarily from the disciplines of engineering, chemistry, and medicine.

Volume meters include displacement bottles, spirometers/bell-type provers, frictionless meters, wet test meters, dry-gas meters, and positive displacement or roots meters. Flow rate meters are divided into two groups: (1) variable-head meters (which include orifice and venturi meters) and (2) variable-area meters (which include the rotameter). Velocity meters, which are actually a type of flow rate meter, include mass flowmeters, thermo-anemometers, and pitot tubes.

Flow Rate Standards

Flow rate standards would be significantly simplified if the fundamental bases of these measurements were as simple as the "identity standards" for, say, mass and length. These measurements are based on discrete standards, such as the platinum kilogram (known as "K-20") for mass and the platinum meter bar for length.[2] For flow rate measurements there is no off-the-shelf identity standard such as a gallons per minute or a liters per second. To supply a fundamental basis for any flow rate measurement the identity standard must be one that is derived.

The calibrators or standard meters used in flow rate measurements have in the past been classified as either primary, intermediate, or secondary. This classification system is based on the accuracy and ability to directly measure the internal dimensions of the calibrator. As a general rule most texts today limit classifications to either primary or secondary, with all previous intermediate standards now included in the secondary standards category. From a historical perspective, the following definitions for all three classifications of calibrators are offered.[3]

- Primary standards—Devices for which measuring volume can be accurately determined by measurement of internal dimensions alone; the accuracy of this type of meter is ±1% or better.
- Intermediate standards—Devices that are more versatile than primary standards, but for which physical dimensions cannot be easily measured. Intermediate standards are calibrated against primary standards under controlled laboratory conditions. The accuracy of this category of device is usually better than ±2%.
- Secondary standards—Devices for general use that are calibrated against primary or intermediate standards. Typically more portable, rugged, and versatile than devices in the other two categories, these devices generally have accuracies of ±5% or better. The need for recalibration depends on the amount of handling, frequency of use, and type of environment in which they are operated. For example, dust collection within a rotameter may require a recalibration as often as every 3 months.

Calibration Hierarchy and Traceability

The traditional concept of measurement traceability in the United States has focused on an unbroken hierarchical pathway of measurements that leads, ultimately, to a national standard.[4] In air sampling this hierarchical pathway might begin with the need to accurately set the flow rate of a personal air sampling pump at 2 L/min. In this scenario, a precision rotameter (a secondary standard) is used to set this flow rate in the field. The accuracy of the rotameter has been maintained by means of a calibration curve that was generated in the laboratory using an electronic bubble meter (a primary standard). The laboratory checks the accuracy of its primary standard annually. This is done by either obtaining a transfer standard on site, which is directly traceable, or by sending the calibrator to an appropriate laboratory or agency for a direct comparison with a national standard. Virtually all calibration standards currently in use have at some time involved some form of traceability back to an acceptable reference standard of known accuracy. All too often, however, the evidence of traceability is either lost or simply not maintained. When dealing with low permissible exposure levels or low threshold limit values the true accuracy of the calibration must be a consideration. Even a small error in this accuracy when measuring low levels of contaminants may impact the results leading to a potential overexposure.

It is important to note that many federally regulated organizations and their contractors are required to verify that the measurements they make are traceable and that they can support the claim of traceability by auditing records. This regulatory requirement implies the ability to relate individual measurement results through an unbroken chain of calibration to a common source, usually U.S. national standards as maintained by the National Institute of Standards and Technology (NIST).[5]

To establish an audit trail for traceability adequately, NIST recommends that a proper calibration result include (1) the assigned value, (2) a stated uncertainty, (3) identification of the standard used in the calibration, and (4) the specifications of any environmental conditions of the calibration where correction factors should be applied if the standard or equipment were to be used under different environmental conditions.[5]

Volume meters, or bulk meters as they are sometimes called, measure the total volume, V, of a gas that is passed through the meter while it is being operated. The time period, t, of operation as well as the temperature and pressure of the gas as it passes through the meter are also measured. The average flow rate, Q, is derived from the following equation.

$$Q = V/t \tag{1}$$

Primary Standards

Spirometers and Meter Provers

These instruments are examples of integrated volume standards. The term "spirometer" was first applied by the English physician John Hutchinson in 1846 to describe an instrument that measured lung volume.[6] Hutchinson's spirometer required the patient to breathe into a delicately balanced receiver of known volume that was elevated in measurable increments with each expired breath. This work led to the first effective instrument for the detection of early or latent stages of consumption (tuberculosis) and was immediately put to use by the insurance industry.[7]

Today spirometers encompass a group of instruments that include those that measure volume directly, as well as those that measure velocity or pressure differences and convert these indicators through the use of electronics to a volume reading. Only those spirometers referred to as water sealed, dry sealed, or dry rolling sealed actually measure volume directly and would therefore qualify as possible primary volume standards. While spirometers were never designed to be calibrators, it is much easier to adapt the device for this purpose if it can be operated manually. Most dry seal or dry rolling sealed spirometers are electronically operated, leaving the water sealed type as the most useful.

Large water sealed spirometers, also known as respirometers or gasometers, have been manufactured with capacities of 100 to 600 L. The primary difference between these spirometers and Hutchinson's original design is the use of a liquid seal. A typical spirometer, for example, was made of two stainless steel cylinders, each with a sealed end. One cylinder or tank held water while the other, turned upside down like a bell, was suspended in the water by a chain and counterbalance. Volume was calculated by using a bell of known dimensions and measuring how far the bell was lifted when air entering the bell displaced the water. This was usually accomplished by fixing a volume scale to the side of the bell and attaching a pointer to the stationary tank. In this way the total volume entering the system over a given time was measured. Some models also had a built-in mixing fan mounted in the bell and a thermometer to facilitate the measurements needed for gas volume calculations.

It was the size (volume) of these devices that made them so attractive as calibrators; however, both the size, and therefore weight, of these measuring devices eventually made them impractical as lung function testers, and today they are no longer commercially available. What is commercially available is a somewhat more portable survey spirometer that uses guide rods instead of the traditional counterweights and an internal bell that displaces most of the water in the tank, thereby reducing both volume and weight of the liquid (Figure 8.1). While this smaller water sealed spirometer has the advantage of availability, it also has the disadvantage of only a 10-L volume capacity.

Even though the large water sealed spirometers are no longer commercially available, they may still be found in both commercial laboratories and university settings, where they function as primary standard calibrators or possibly training tools. For this instrument to be used as a

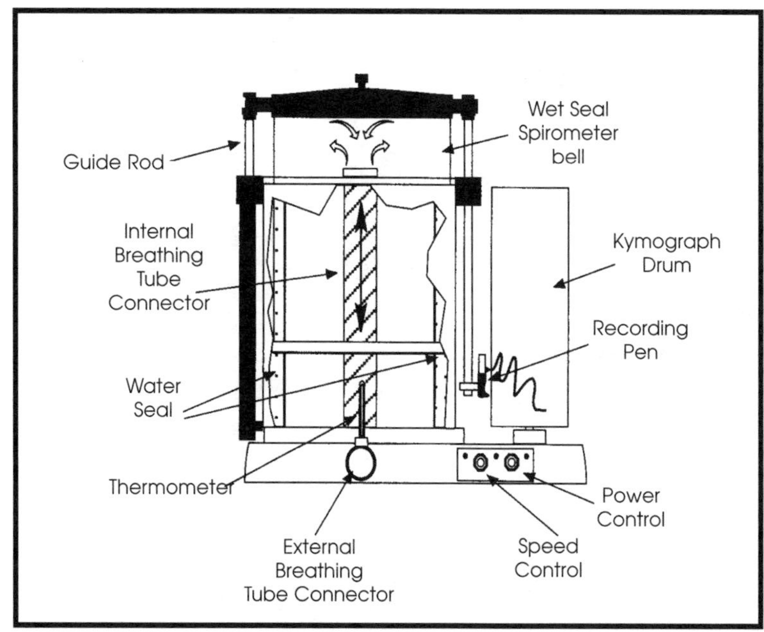

Figure 8.1 — Diagram of a portable survey water sealed spirometer referred to as a respirometer or pulmonary screener with a volume capacity of 10 L (courtesy of Warren E. Collins Inc., Braintree, Mass.).

calibrator, the bell is raised with the gas to be measured while ensuring that the water seal stays intact. At this point the bell can be considered fully charged. As the bell is lowered, the water acts as a piston that drives the gas out of the bell and through the flowmeter under test. The depth of immersion of the bell is a measure of the volume of the gas displaced and can be read off a calibrated scale.

If large volumes are to be measured, the process has to be repeated several times, which can make the operation time-consuming. Before using this spirometer it should be checked for corrosion on the internal surface of the bell and for damage to the shape of the bell that might affect the volume. Allow the water in the spirometer to come to room temperature and the air in the bell to equilibrate to the ambient pressure. The air volume can then be corrected to a standardized condition using the ideal gas laws:

$$V_s = V_{mes}\left(\frac{P_{mes}}{P_{std}}\right)\left(\frac{T_s}{T_{mes}}\right) \quad (2)$$

where

V_s = volume at standard conditions (P_{std} = 760 mmHg at T_{std} = 25°C)
V_{mes} = volume measured at conditions P_{mes} and T_{mes}
T_{mes} = absolute temperature of V_{mes} (usually 298.15 K)
P_{mes} = measured pressure of V_{mes}(mmHg)

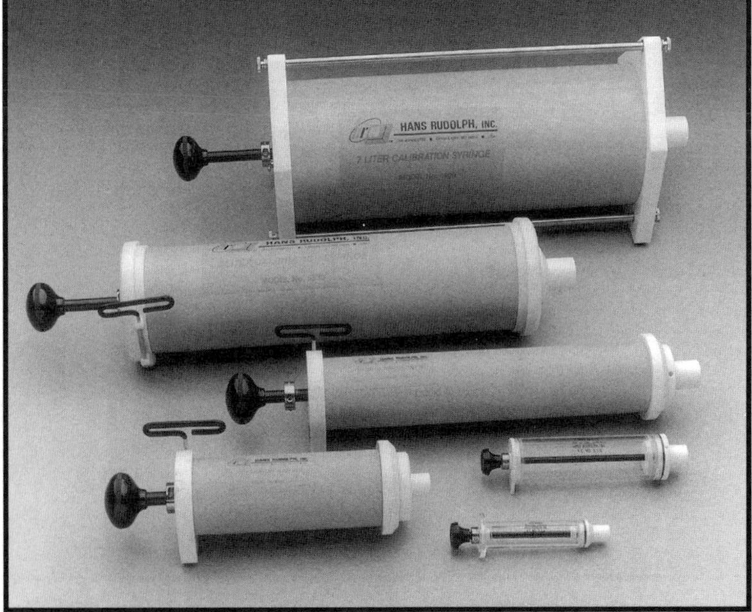

Figure 8.2 — Volume calibration syringes, 0.5 to 7 L (courtesy of Hans Rudolph, Inc., Kansas City, Mo.).

Today most portable dry sealed and water sealed spirometers are calibrated at the factory using a 1- to 3-L volume syringe (for examples see Figure 8.2) that is either certified by or traceable to NIST. While it is possible to ship a large spirometer to NIST for calibration, it is perhaps more practical to use an NIST transfer standard and run the calibration on site. Another method of checking the calibration would be to use a procedure known as "strapping," which consists of measuring the cylinder's dimensions with a steel tape and calculating the volume.[8] For this process to be part of an unbroken calibration hierarchy, the steel tape measure would have to be traceable to a recognized standard.

"Meter provers" refer to a proven tank capacity that is used to check the volumetric accuracy of a gas or liquid that is delivered by a positive-displacement meter. Meter provers, also known as bell provers, are primary volume standards that are actually very similar in both design and function to the large water sealed spirometers. However, unlike the large water sealed spirometers, the bell provers are designed specifically to function as primary volume calibrators, are commercially available with internal volumes 0.18 m^3 (5 ft^3) to 0.71 m^3 (20 ft^3), and are currently in use by NIST. In addition, these instruments employ a low vapor pressure oil instead of water and an internal bell or tank to reduce the overall volume of the liquid used to make a seal (refer to the diagram in Figure 8.1). As a gas flows through the device, it first enters the internal tank and then the gas displaces the bell. The bell is counterbalanced throughout its travel by counterweights suspended on chains, just like a traditional spirometer. By attaching a small-gauge wire to the top of the bell and connecting a linear encoding system, the bell's location can be measured automatically as it is displaced by the gas flowing into the chamber. A computer can collect this information, as well as temperature and pressure measurements, and calculate an instantaneous reading of gas mass flow with an accuracy of 0.5%, according to the manufacturer (see Figure 8.3).

Displacement Bottle

The displacement bottle, sometimes referred to as a prover bottle (see Figure 8.4), is a volume and flow rate calibrator that operates similarly to the bell prover, except that it measures displaced water instead of displaced gas. The operating principle of this instrument was first described in 1686 by the French hydraulician Edme Mariotte and has been referred to as the "flask of Mariotte."[9] Mariotte used a closed cylindrical container filled with water that had a bottom drain to let water out and an inlet tube to let air into the container. The instrument was used primarily to illustrate the weight of the atmosphere but is now commonly employed in the laboratory as a constant-head device.

Displacement bottles usually have a valve at the bottom of the bottle that allows the water to be drained out; this in turn draws air into the bottle in response to the lowered pressure. The volume of the air drawn in is equal to the change in water level multiplied by the cross section at the water surface. A more accurate method used in the laboratory would be to collect the displaced water into a graduated cylinder or volumetric flask. By also measuring the time needed to displace a set volume of water it is possible to calibrate low flow rate meters accurately in the range of 10 to 500 mL/min or less.[8] Measurements for temperature, pressure, and volume will need to be corrected to standard conditions. Wight suggests that when a displacement bottle is used to calibrate another device, there may be substantial negative pressure at the bottle inlet that will need to be a part of the correction process.[3]

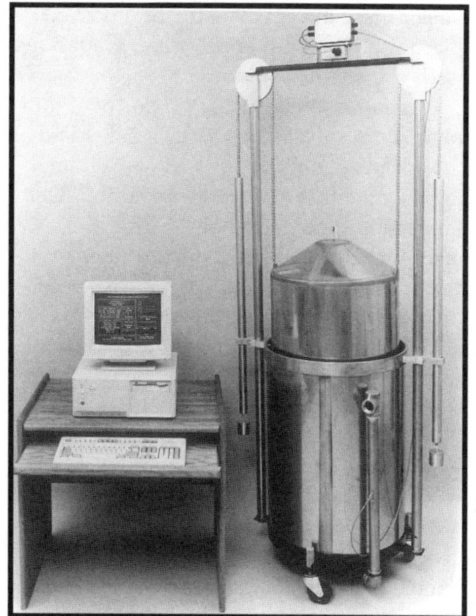

Figure 8.3 — The Sierra Automated Bell Prover is designed and built as a primary standard gas flow calibrator. It is constructed of 300-series stainless steel and fitted with inlet and outlet valving, two counterweights, an optical encoder for measuring bell travel, and automated inputs for both temperature and back pressure measurements (courtesy of Sierra Instruments, Inc., Monterey, Calif.).

"Frictionless" Piston Meters

Frictionless piston meters comprise a group of cylindrical air displacement meters that use nearly frictionless pistons to measure flow rates as primary flow calibrators. The pistons in these instruments are designed to form gas-tight seals of negligible weight and friction and are made from a variety of materials including soap film, mercury, glass, and graphite. Variations in piston material have a direct effect on the meter's cost, accuracy, and portability.

Soap-Film or Bubble Meter

The bubble meter is the most frequently used calibrator in occupational air sampling, primarily because its operation is accurate, economical, and relatively simple. While the first references to a bubble meter go back to the late 1800s, it is difficult to know exactly to whom the honor of invention might belong. In its simplest form the bubble meter consists of a graduated tube, such as a laboratory glass burette, and a soap film. It is also possible to construct

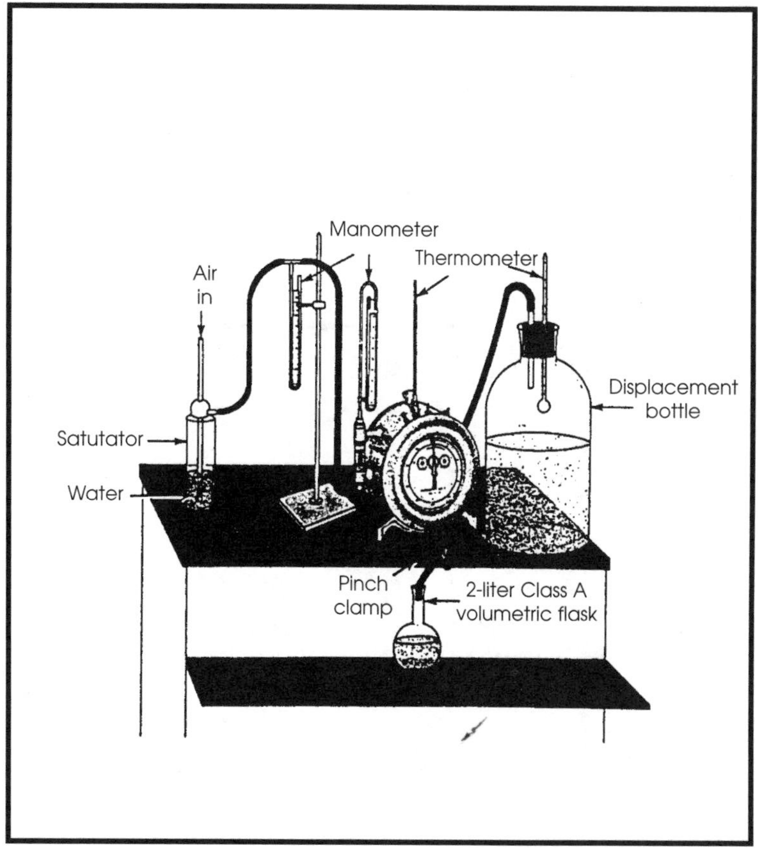

Figure 8.4 — Diagram of a displacement bottle used to calibrate a wet test meter (reprinted with permission from CRC Press Inc.).[3]

one of these meters by using an inverted plastic syringe with appropriate gradations (see Figure 8.5).

A vacuum source, usually a personal sampling pump, is attached to the smaller end of the graduated tube or syringe while immersing the other in a soap/water solution. This immersion should last only as long as it takes for a soap bubble to form and begin to rise in the wet tube. The volume displacement per unit time (i.e., flow rate) can then be determined by measuring the time required for the bubble to pass between two scale markings that enclose a known volume. These meters are generally accurate to within ±1%, although greater accuracy can be achieved under selected conditions.(10)

The simplicity of the soap bubble meter is not without some disadvantages. If the bubble meter is operated in the presence of relatively dry air (e.g., less than 50% relative humidity)(10) and at low flow rates, the air mass that is passing over the soap film may become saturated with water vapor. When this occurs, the volume change being measured will be due to the mixture of water vapor from the solution and the air being measured.(11) The percentage volume due to the water vapor can be obtained from tables of partial pressure of water versus temperature in references such as the *CRC Handbook of Chemistry and Physics*.(12) An appropriate correction factor can then be calculated and applied to any standard

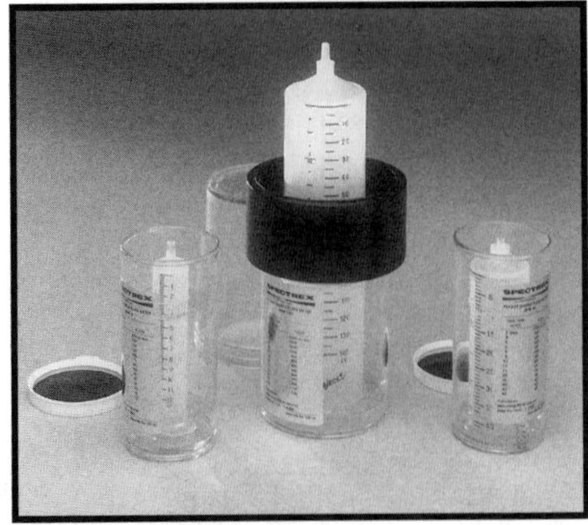

Figure 8.5 — Set of three bubble flowmeters, BFM-10, 40, and 100 (courtesy Spectrex, Redwood City, Calif.).

temperature and pressure corrections that might be needed (see Equation 3).(13)

$$V_{std} = V_{mes}\left(\frac{P_{mes} - P_{wv}}{P_{std}}\right)\left(\frac{T_{std}}{T_{mes}}\right) \quad (3)$$

where P_{wv} = partial pressure of water vapor at the ambient air temperature in mmHg.

An alternative to the time involved in working out the above equation, but not the understanding, is a list of correction factors for reasonable temperatures and elevations that can be expected in a particular region. Such a list could be made for any number of regions taking into account the uniqueness

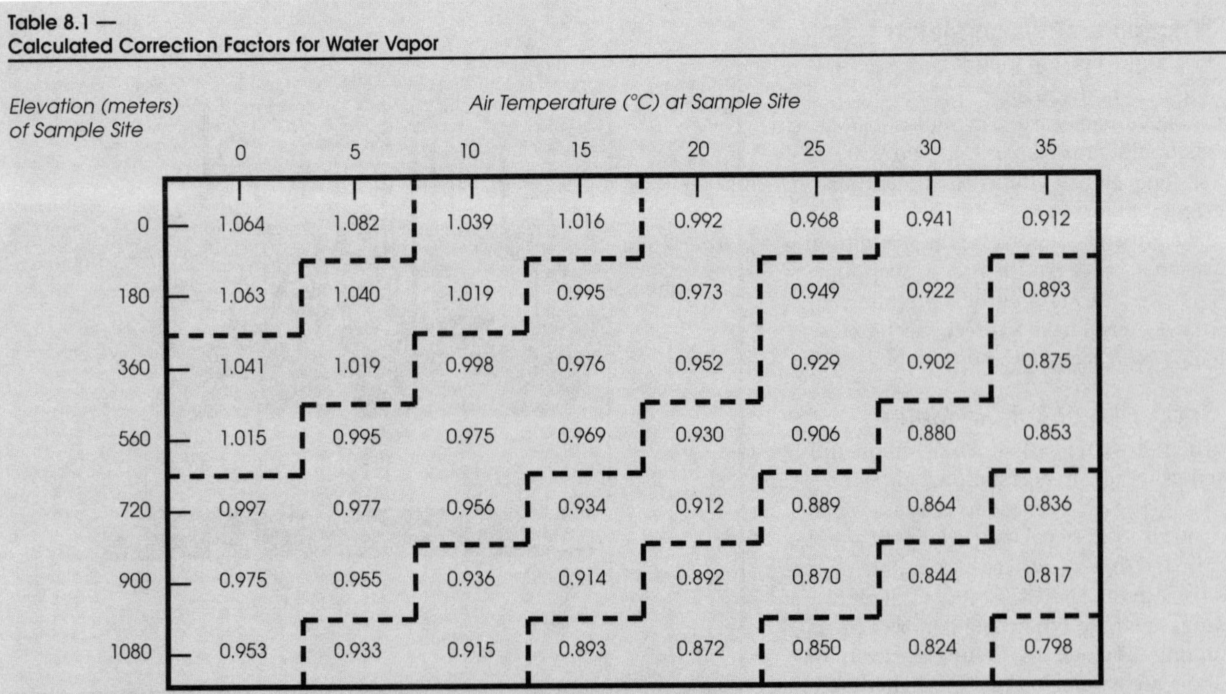

Table 8.1 — Calculated Correction Factors for Water Vapor

Elevation (meters) of Sample Site	Air Temperature (°C) at Sample Site							
	0	5	10	15	20	25	30	35
0	1.064	1.082	1.039	1.016	0.992	0.968	0.941	0.912
180	1.063	1.040	1.019	0.995	0.973	0.949	0.922	0.893
360	1.041	1.019	0.998	0.976	0.952	0.929	0.902	0.875
560	1.015	0.995	0.975	0.969	0.930	0.906	0.880	0.853
720	0.997	0.977	0.956	0.934	0.912	0.889	0.864	0.836
900	0.975	0.955	0.936	0.914	0.892	0.870	0.844	0.817
1080	0.953	0.933	0.915	0.893	0.872	0.850	0.824	0.798

Note: Correction factors for the error associated with the partial pressure of water vapor when using a soap bubble meter (assume low flow rates and dry air) at temperatures and elevations reasonable for a typical cold climate. Corrections are based on 760 mmHg and 25°C. The dashed lines separate increments of volume error by 5%.(14)

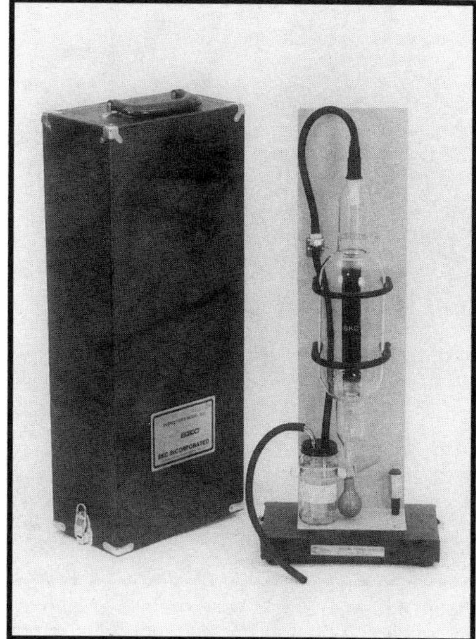

Figure 8.6 — (A) Manual laboratory film flowmeter series 311; (B) manual flowmeter kit model 302 with a range of 100 to 4000 mL/min (figures courtesy SKC Inc., Eighty Four, Pa.).

of both geography and climate.[14] Table 8.1 illustrates the water vapor correction factors that would be appropriate for the temperature and geographical diversity found in a typical cold climate.

Manufactured bubble meters are available from a variety of sources that offer both manual calculation of volume (Figures 8.6A and 8.6B) and electronically determined volumes (Figure 8.7). The electronically determined volumes use a pair of infrared optical triggers that measure the time it takes a piston to travel between two set points. Usually these meters are guaranteed either to meet tolerances that are specified by NIST, or they can be certified for accuracy directly by NIST. Manufacturers usually certify electronic bubble meters for only 1 year, and the meters must be returned to the manufacturer for annual calibration. For nonelectronic bubble meters the need eventually arises through use or misuse for some sort of calibration check of the tube volume. This can be accomplished gravimetrically by inverting the tube and filling it with distilled water between two graduated marks. The water is then drained from the tube, and the weight of the water is measured using a calibrated and traceable gravimetric scale. An accurate volume of the tube can then be determined after applying the appropriate temperature corrections to the liquid.

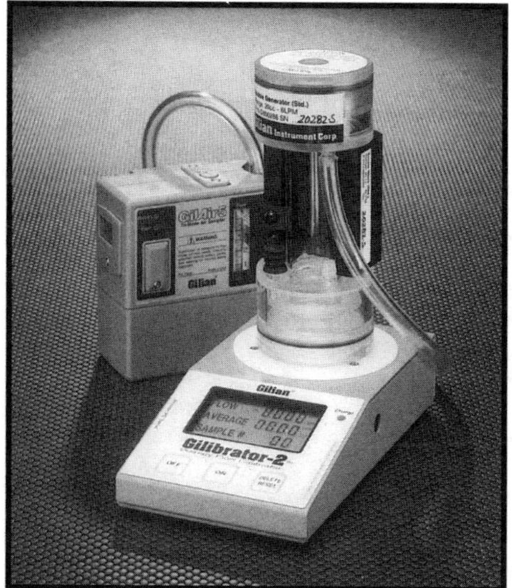

Figure 8.7 — An electronic soap bubble flowmeter, the Gilibrator-2, shown being used to calibrate a personal air sampling pump using one of three possible interchangeable cells (courtesy Sensidyne Inc., Clearwater, Fla.).

Mercury-Sealed Pistons

The mercury-sealed piston meter is another primary reference standard used by NIST. Although these meters are commercially available (see Figure 8.8A), they are extremely expensive and are usually designated for laboratory use only. The mercury-sealed piston meter consists of a precisely bored borosilicate glass cylinder and a close-fitting plate of polyvinyl chloride. The plate is separated from the cylinder wall by an O-ring of liquid mercury that retains its toroidal shape through strong surface tension. The floating seal has a negligible friction loss, but the weight of the piston must be compensated for by calculations. With a manually operated timer the accuracy is said to be ±0.2% for timing intervals of 30 seconds or more.[8]

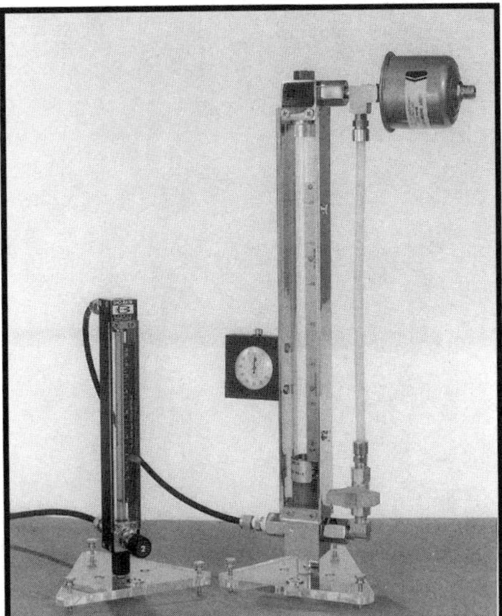

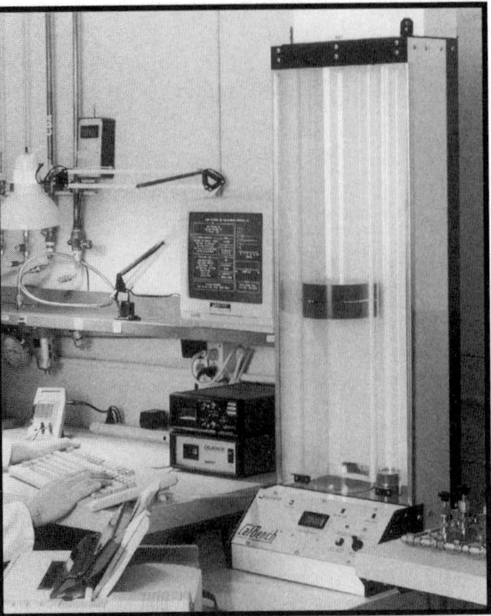

Figure 8.8 — (A) Brooks Vol-u-meter® (shown at right) calibrating a Brooks Sho-rate Purgemeter. Note the activated charcoal filter, designed to remove harmful mercury vapor (courtesy Brooks Instrument Division, Emerson Electric Division, Hatfield, Pa.); (B) the Cal-Bench® model 101, a completely automated primary standard (mercury sealed piston) calibration system (courtesy Sierra Instruments, Inc. Monterey, Calif.).

The mercury piston meter can also be designed to be a completely automated system. The instrument pictured in Figure 8.8B is actually made up of three mercury-sealed piston meters. The tubes are sized to provide a 10:1 ratio in volumetric displacement between the tubes, which allows a calibration flow range of 1 to 50,000 standard cubic centimeters per minute. Instead of using infrared optical triggers, this system employs a specialized sonar transceiver at the top of each tube that emits a pulse of sound energy operating at approximately 50 kHz. As the pulse travels down the tube, it is reflected from the top of the piston and returns to the transceiver. A computer measures the transit time of the energy pulse and calculates the speed of sound based on ambient conditions of pressure, temperature, and relative humidity. By knowing the speed of sound and the transit time of the pulse, the position of the piston is said to be determined with a resolution of 0.152 mL (0.006 inches).

Glass and Graphite Pistons

Solid pistons offer many advantages over liquid seals made of either soap film or mercury. However, unlike the liquid pistons, air leakage can occur between the solid piston and the inside of the tube wall. Figure 8.9 is an example of a gas flowmeter that measures the displacement of a glass piston within a glass cylinder. The manufacturer suggests the use of flow rates that range from 10–200 mL/min with accuracy

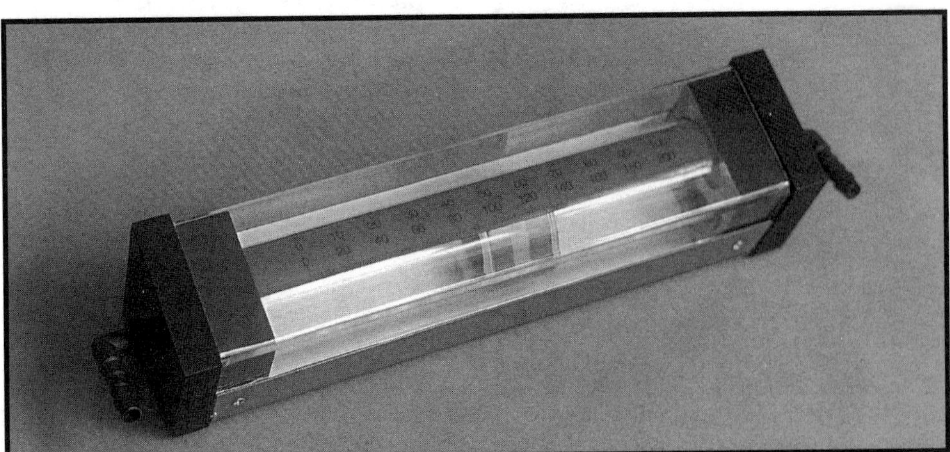

Figure 8.9 — Dry-Flow Flowmeter model 100a uses a glass piston instead of a soap-bubble film (courtesy Spectrex, Redwood City, Calif.).

at 100% full scale of ±2%. By incorporating both electronics and a graphite piston many of the problems associated with soap bubble meters are eliminated (Figure 8.10). The standard cell for this graphite piston meter covers a range of 10 mL to 9.999 L/m. Optional low- and high-flow cells extend the range from 1 mL to 50 L/m. Accuracy is reported by the manufacturer at ±1 to 2% depending on cell size. The variation is due to the clearance between the piston and cell wall, which creates a maximum leakage of approximately 0.005% of full scale.

Secondary Standards

Wet Test Gas Meter

It was not until 1816 that a practical attempt was made to measure commercial coal gas by an automatic and continuous instrument known as Crosley's drum or the wet meter drum.[15] The success of this instrument was in large part responsible for smoothing the way for the popularization of the gas supply industry. The principles used in the Crosley drum are present today in the wet test gas meter, which functions primarily as a laboratory device for calibrating secondary standards. In addition, wet test gas meters are frequently used to meter the flow of other gases directly.

The wet test gas meter (see Figure 8.11) consists of a partitioned drum half submerged in a liquid (usually water) with openings at the center and periphery of each radial chamber. Air or gas enters at the center and flows into an individual compartment causing it to rise, thereby producing rotation. This rotation is indicated by a dial on the face of the instrument. The volume measured depends on the fluid level in the meter, since the liquid is displaced by air. A sight gauge for determining fluid height is provided, and the meter may be leveled by screws and a sight bubble provided for this purpose. Correcting volume measurements to standard conditions is always required. There are several potential errors associated with the use of a wet test meter. The drum and moving parts are subject to corrosion and damage from misuse; there is friction in the bearings and the mechanical counter; and inertia must be overcome at low flows (<1 rpm), while at high flows (>3 rpm), the liquid might surge and break the water seal at the inlet or outlet. In spite of these factors the accuracy of the meter usually is within 1% when used as directed by the manufacturer.

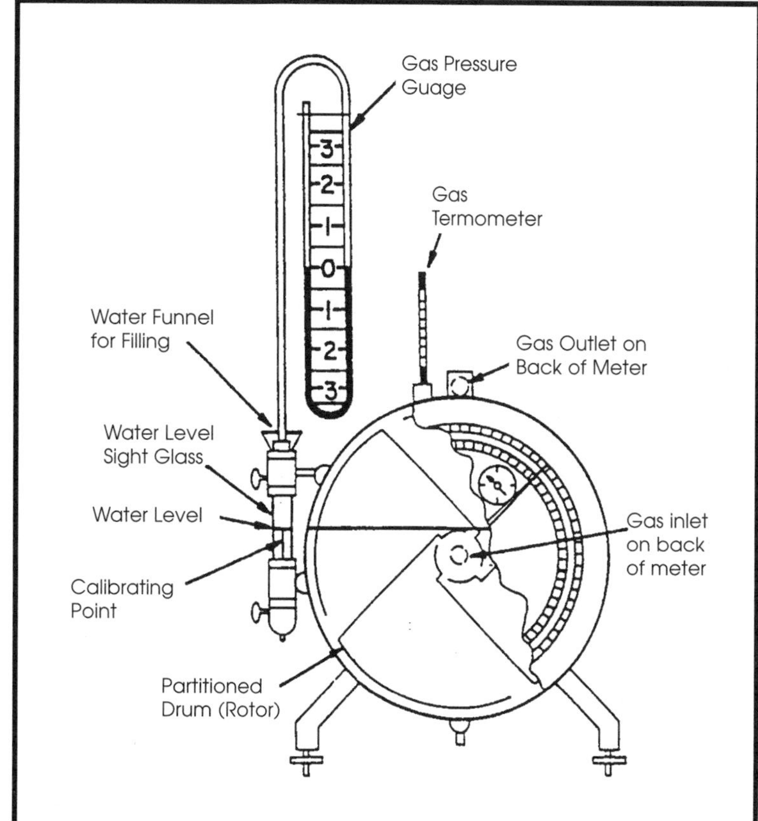

Figure 8.11 — Schematic of a wet test meter.

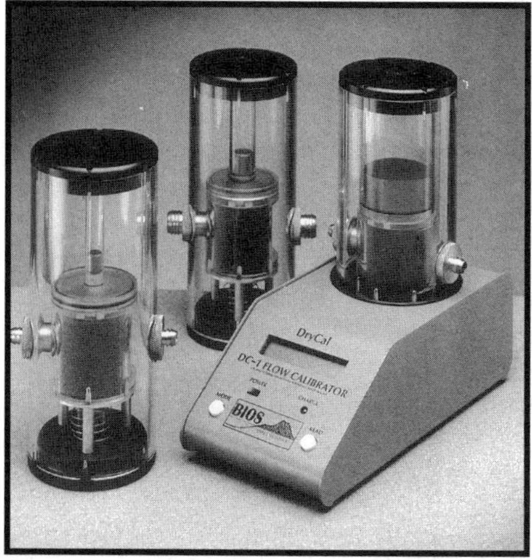

Figure 8.10 — Dry Cal® DC-1 flow calibrator, an example of a graphite piston flowmeter that is completely dry. Three interchangeable cells offer a wide volume range from 1 mL/min to 50 L/min (courtesy of Bois International, Pompton Plains, N.J.).

Before the meter can be checked for accuracy, it must be leveled and filled with water to the calibration point. Then air is run through the meter for several hours to saturate the water and allow the meter to equilibrate.[16] At this point the meter is ready for calibration against a primary standard. While it is possible to use a water sealed spirometer as a primary standard calibrator, it is difficult to find one that has been properly maintained. Manufacturers of wet test gas meters use an NIST-certified meter prover bottle as a primary standard that is capable of providing ±0.05% accuracy. The prover bottle is similar to the displacement bottle except the bottle is precisely manufactured to NIST tolerances. Calibration procedures using a prover bottle are described by the American Society for Testing and Materials.[17]

Dry-Gas Meter

The Crosley drum was used as a consumer's meter from 1816 until the end of the century, by which time it had given way to a dry-gas meter that had come into practical use in 1844.[15] The dry-gas meter shown in Figures 8.12A and 8.12B is probably the second most widely used airflow calibrating device. It consists of two bags interconnected by mechanical valves and a cycle-counting device. The air or gas fills one bag while the other bag empties itself; when the cycle is completed the valves are switched, and the second bag fills while the first one empties. The maximum flow rate of these devices is limited by the volume in cubic feet equivalents to a bagged capacity. Ness explains that a meter with a bag stamped DTM-200 is a dry-test meter that will safely pass a maximum of 200 dry feet per hour at a fi-inch water column differential.[18] Any such device will have the disadvantage of mechanical drag, pressure drop, and leakage; however, the advantage of using the meter under rather high pressures and high volumes often outweighs these errors, which can be determined for a specific set of conditions. Although these meters are not normally used for occupational sampling, they are currently available with flow rates that average around 10–150 L/min (21–318 ft^3/hr). It is also possible to obtain the meter with either a digital or analog readout that measures flow in units of either cubic feet per hour or cubic meters per hour.

The dry-gas meter is calibrated against a primary standard, such as a bell prover, or a recently calibrated secondary standard, such as the wet test gas meter. In the absence of a suitable standard and to maintain a calibration hierarchy, the meter is returned to the manufacturer for annual calibration. Use of a displacement bottle or prover meter bottle is also a possibility following the same procedure employed when calibrating the wet test meter. Accuracy of the meter is corrected by adjusting the meter linkage. In practice, if the meter is within 1% of the known volume, a calibration factor is computed and used to correct all meter values.[3]

Using Calibrators that Measure Flow Rate

Flow rate meters all operate on the principle of the conservation of energy. More specifically, they use Bernoulli's theorem for the exchange of potential energy for kinetic energy and/or frictional heat. Each consists of a flow restriction within a closed conduit. The restriction causes an increase in the fluid velocity and therefore

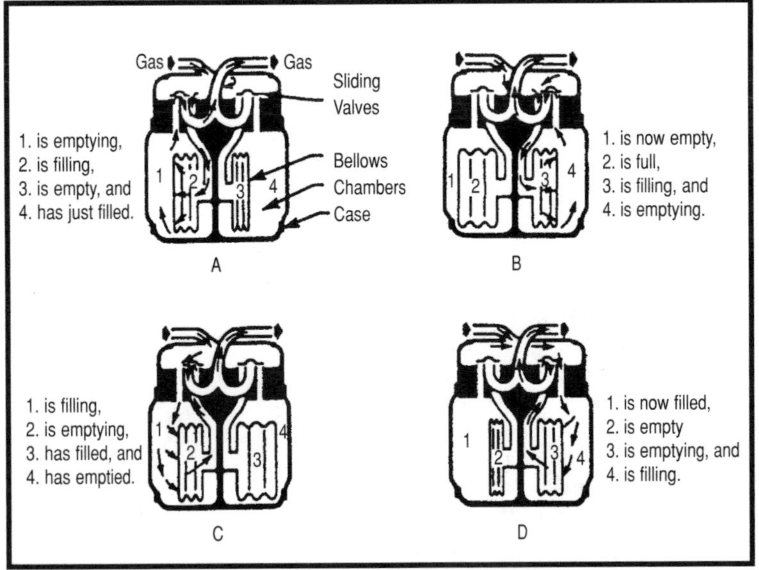

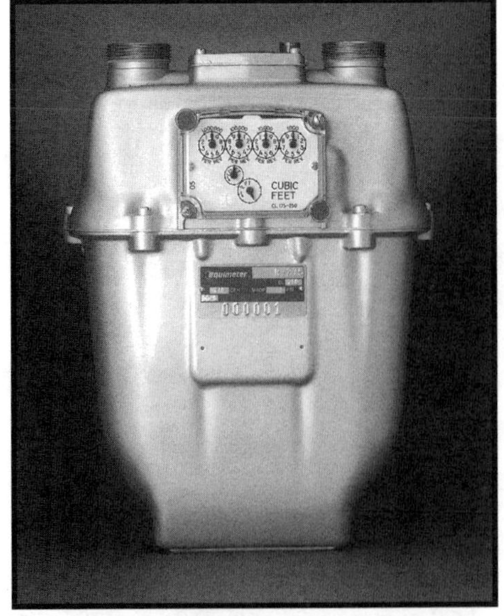

Figure 8.12 — (A) Internal mechanism of a dry-gas meter during operation (reprinted with permission from CRC Press[3]); (B) dry-gas meters are used extensively in stack sampling and pollution monitoring devices (courtesy of Equimeter Inc., DuBois, Pa.).

an increase in kinetic energy, which requires a corresponding decrease in potential energy, i.e., static pressure. The flow rate can be calculated from a knowledge of the pressure drop, the flow cross section at the constriction, the density of the fluid, and the coefficient of discharge. This coefficient is the ratio of actual flow to theoretical flow and makes allowance for stream contraction and frictional effects.

Flowmeters that operate on this principle are divided into two groups. The larger group includes orifice meters, venturi meters, and critical orifices. These have a fixed restriction and are known as variable-head meters because the differential pressure varies with flow. Flowmeters in the other group, which includes rotameters, are variable-area meters because a constant pressure differential is maintained by varying the flow cross section.

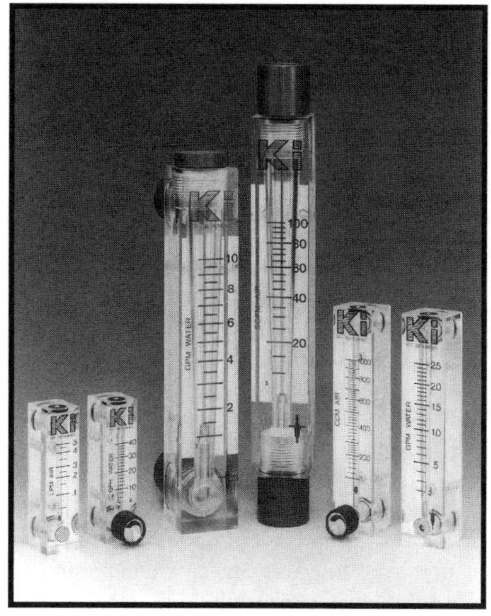

Figure 8.13 — A variety of lightweight (acrylic plastic), single float, field rotameters (courtesy Key Instruments, Trevose, Pa.).

Other Secondary Standards

Variable-Area Meters (Rotameters)

Rotameters are by far the most popular field instruments for flow rate measurements. The device consists of a float that is free to move up and down within a vertical tapered tube that is larger at the top than the bottom (Figure 8.13). Some of the shaped floats achieve stability by having slots that make them rotate. The term rotameter was first used to describe meters that used spinning floats, but now is generally used for all types of tapered metering tubes regardless of whether the float spins. As air flows upward, it causes the float to rise until the pressure drop across the annular area between the float and the tube wall is just sufficient to support the float. The tapered tube is usually made of glass or clear plastic and has a flow rate scale etched directly on one side. The height of the float indicates the flow rate. Floats of various configurations are used, as indicated in Figure 8.14. They are conventionally read at the highest point of maximum diameter, unless otherwise indicated.

Most rotameters have a range of 10:1 between their maximum and minimum flows. The range of a given tube can be extended by using heavier or lighter floats. Tubes are made in sizes from about 0.32 to 15 cm (0.125 to 6 in.) in diameter, covering ranges from a few mL/min to over 28.3 m^3/min (1000 ft^3/min).

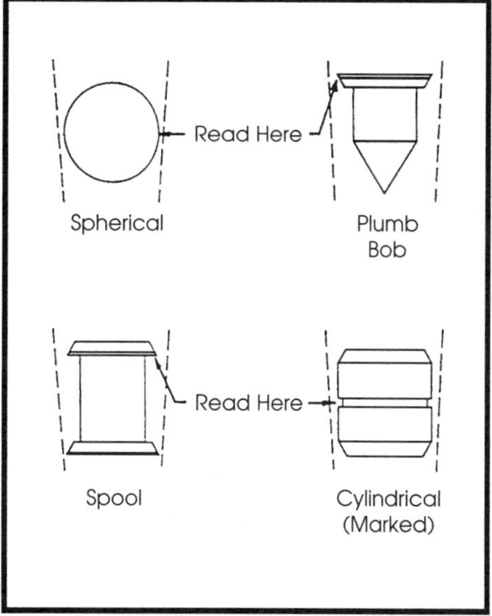

Figure 8.14 — Rotameter floats of various configurations showing reading point of highest maximum diameter for each.

The most widely used material for construction is acrylic plastic. Because of space limitations, the scale lengths are generally small, ranging from 5 to 10 cm (2 to 4 in.). Unless they are individually calibrated, the accuracy is unlikely to be better than ±25%. When individually calibrated, ±5% accuracy can be achieved. It should be noted, however, that with relatively few scale markers on these rotameters, the accuracy of the readings may be a major limiting factor. Precision rotameters are usually larger, 30 cm (12 in), made of glass, and have more accurate numerical scales.

Calibrations of rotameters are performed at an appropriate reference pressure, usually atmospheric. However, since good practice dictates that the flowmeter should precede the sample collector or sensor, the flow is actually measured at a reduced pressure, which may also be a variable pressure if the flow resistance changes with loading. If this resistance is constant, it should be known; if variable, it should be monitored so that the flow rate can be adjusted as needed and appropriate pressure corrections can be made for the flowmeter readings.

The flow rate of a gas through a rotameter is seldom calculated from tube diameters and float dimensions, although the development and use of such equations are available.[19] Instead, curves are generated that relate meter readings to flow rates that are derived from a calibration by using a primary or accurate secondary standard. New rotameters are usually supplied with a vendor or manufacturer's generated calibration curve that should list the conditions of both temperature and pressure of the calibration. All rotameters should be recalibrated at or close to the conditions expected in the field or have appropriate corrections applied for the difference in calibrated and indicated flow rates.

For rotameters with linear flow rate scales, the actual sampling flow will approximately equal the indicated flow rate times the square roots of the ratios of absolute temperatures and pressures of the calibration and field conditions.[20] The ratios increase when the field pressure is less than the pressure in the calibration laboratory or the field temperature is greater than that in the laboratory. Thus, if the flowmeter was accurate at ambient pressure, and the flow resistance of the sampling medium was relatively low (e.g., 30 mmHg), for a flow rate of 11 L/m the flow rate indicated on the rotameter would be $11 \times (730/760)^{0.5} = 10.8$ L/m, a difference of only 1.8%. On the other hand, for a 25 mm diameter AA Millipore with a 3.9 cm^2 filtering area and a sampling rate of 11 L/min, the flow resistance would be approximately 190 mmHg, and the indicated flow rate would be $11 \times (570/760)^{0.5} = 9.5$ L/min, a difference of 14%. It should be noted that any other representative filter calibration should closely indicate a similar difference.

Caplan notes that there is a rather important difference between <u>actual flow rate</u> and <u>standardized flow rate</u>.[21] In the above example a correction is made for differences in pressure resistance that affect the actual flow rate. If a flow rate needs to be corrected to standardized conditions, the correction would have to be inverted. For example, if a rotameter were calibrated at occupational hygiene standard conditions or normal conditions (760 mmHg and 25°C) and used at a different pressure, say 625 mmHg but the same temperature, the difference between actual and standardized flow rates would be 1.103 and 0.907, respectively.

Actual Flow Rate

$$\left(\frac{760 \text{ mmHg}}{625 \text{ mmHg}}\right)^{0.5} = 1.103$$

Standardized Flow Rate

$$\left(\frac{625 \text{ mmHg}}{760 \text{ mmHg}}\right)^{0.5} = 0.907$$

A similar correction will be needed when the sampling is done at atmospheric pressures and/or temperatures that differ substantially from those used for the calibration. For example, at an elevation of 1524 m (5000 ft) above sea level, the atmospheric pressure is only 83% of that at sea level. Thus, the standardized flow rate would be 0.912 or 8.7% less than that indicated on a rotameter scale, based on the altitude correction alone. If the temperature in the field were 35°C while the meter was calibrated at 20°C, the standardized flow rate in the field would be 0.975 or 2.5% less than that indicated. For a summary of the corrections that would be used, refer to the following equation.[13]

$$\text{Q std} = \text{Q ind}\left[\left(\frac{\text{P amb}}{\text{P std}} \times \frac{\text{T std}}{\text{T amb}}\right)\right]^{0.5} \quad (4)$$

where

Q ind = indicated rotameter reading at T amb (ambient temperature in degrees K) and P amb (ambient pressure in mmHg)

Q std = standardized flow based on rotameter calibration at T std (standardized temperature in degrees K) and P std (standardized pressure at 760 mmHg).

In a situation where these kinds of corrections are needed for high altitude and high temperature, the overall correction for the example above would be 0.089 or 10.9%.

The correction in flow generated by the use of this equation is actually a function of the rotameter basic flow equation and not a simple gas-law correction.[21] Most of the terms in the rotameter basic flow equation can be considered constants, thereby allowing the fundamental equations of Boyle and Charles (referred to as the Ideal Gas Law) to be in substitution for density, which will give a quick and reasonably accurate means of correcting flow.

In general, rotameter corrections should only come into play when a significant calibration shift has occurred between the calibration and field conditions. According to both the National Institute for Occupational Safety and Health (NIOSH) and the Occupational Safety and Health Administration (OSHA) a deviation of more than ±5% of the calibration value is considered to be a significant shift. In other words, a volume flow rate will be corrected if measured conditions exceed calibration conditions by 5% or greater using Equation 4.

Once it is determined that a correction for flow is likely to be needed, the specific correction factor can be worked out by either measuring or calculating the sample site air pressure and temperature. When calculating air pressure from elevation information, use a conversion factor of 2.5 mmHg per 30 m elevation increase, or vice versa for an elevation decrease, using Environmental Protection Agency Method 2A—Direct Measurement of Gas Volume Through Pipes and Small Ducts.[22] The actual flow correction would look like the following:

Conversion of elevation to equivalent pressure in mmHg:

$$\left(\frac{1080 \text{ meters}}{30 \text{ meters}}\right)\left(\frac{2.5 \text{ mmHg}}{1}\right) = 670 \text{ mmHg}$$

Rotameter correction factor at occupational hygiene conditions at 760 mmHg and 25°C:

$$\left[\left(\frac{670 \text{ mmHg}}{760 \text{ mmHg}}\right)\left(\frac{298.15 \text{ K}}{303.\ 15 \text{ K}}\right)\right]^{0.5} = 0.931$$

The above equation gives a correction factor value of 0.931, which indicates a 7% error in flow. To correct for this error simply multiply 0.931 by the indicated flow rate from the rotameter calibration curve.

By applying a sufficient range of temperature corrections and pressure corrections (using elevation data) as could be generated

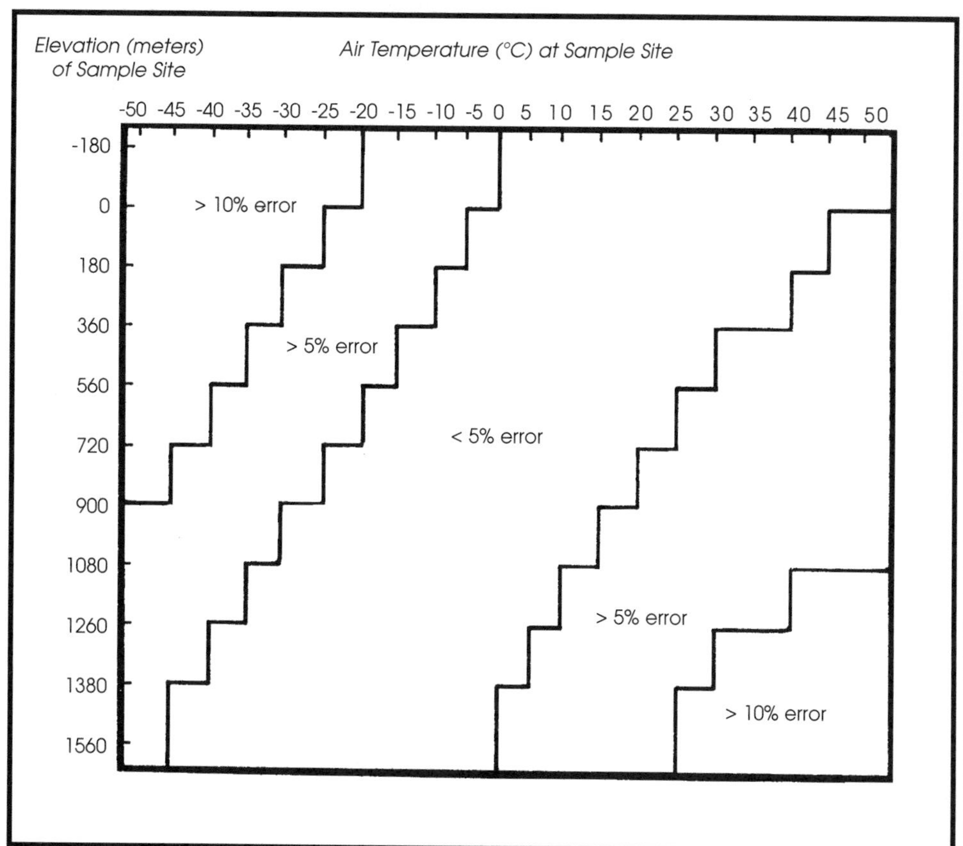

Figure 8.15 — Percent error of indicated flow when using a rotameter at temperatures and elevations that are different from a calibrated state of 760 mmHg and 25°C (normal conditions).[14]

by Equation 4, it is possible to demonstrate exactly under what field conditions a 5% or greater error in flow is likely to occur (see Figure 8.15).[14] If we use the same example as cited above, with an elevation of 1080 m (3543.3 ft) and a warm summer day of 30°C, Figure 8.15 can be used to show that the indicated rotameter flow would be in error by a value greater than 5%. After a detailed examination of Figure 8.15, it becomes evident that there is a rather significant range of temperatures and elevations that would not require a 5% error correction when using a rotameter that was originally calibrated at standard conditions. Whether one chooses to make a rotameter correction at 5% or greater in the long run will depend on one's own professionalism and knowledge of the circumstances in which the error will occur.

Variable-Head Meters

Variable-head meters, or head meters, are devices that produce a differential in pressure caused by a restriction in the airflow stream. This pressure difference was initially studied by Giovanni Battista Venturi in 1774, but it was not until 1894 that an engineer named Clemens Herschel invented the flowmeter, which he named in honor of Venturi.[21] Variations in these head meters include orifice meters, nozzles, and venturi tubes and are discussed in detail in other references.[23–25] These devices are used extensively in chemical processing and other industries but have little impact on occupational air sampling. One notable exception is a type of orifice meter known as a critical flow orifice that is used in the occupational environment to control airstreams in sampling equipment at predetermined constant flow rates.

Critical Flow Orifice

For a given set of upstream conditions the discharge of a gas from a restricted opening will increase with a decrease in the ratio of absolute pressures P_2/P_1, where P_2 is the downstream pressure and P_1 the upstream pressure, until the velocity through the restriction reaches the velocity of sound. The value of P_2/P_1 at which the maximum velocity is just attained is known as the critical pressure ratio. The pressure in the throat will not fall below the pressure at the critical point, even if a much lower downstream pressure exists. For air, this condition is met when P_2 is less than 0.53 P_1, and the ratio of the upstream cross-sectional area to the orifice area is greater than 25.[26,27] Therefore, when the pressure ratio is below the critical value, the rate of flow through the restricted opening depends only on the upstream pressure.

For all differential-producing devices: when $P_2 < 0.53\ P_3$, and $S_1/S_2 > 25$, the mass-flow rate W can be determined by the following equation.

$$W = 0.533\ \frac{C_vS_2P_1}{T_1}\ \text{lbs/sec} \qquad (5)$$

where

C_v = coefficient of discharge (normally ~ 1)
S_1 = duct or pipe cross section in square inches
S_2 = orifice area in square inches
P_1 = upstream absolute pressure in pounds per square inch
T_1 = upstream temperature in °R

Critical flow orifices are widely used in occupational hygiene instruments such as the midget impinger pump for the maintaining and controlling of airstreams at constant flow rates. Systems using critical flow nozzles have also been used in place of bell provers for the calibration of gas meters. They can be purchased but will need to be calibrated, or they can be constructed by using a hypodermic needle. The flowmeter readings are then plotted against the critical flows to yield a calibration curve. The major limitation in their use is that only one critical flow rate is possible from each orifice, and the pressure differential is high. To a large degree this can be overcome by using several orifices in parallel downstream of the flowmeter under calibration. A second limitation is that these orifices can become clogged easily and can erode in time and therefore require frequent examination and/or calibration against other reference meters as part of a regular calibration program.

Packed Plug Flowmeter

One type of variable-head meter that differs significantly from all of the above is the laminar-flow meter. These are seldom discussed in engineering handbooks because they are used only for very low flow rates. Since the flow is laminar, the pressure drop is directly proportional to the flow rate. In orifice meters, venturi meters, and related devices, the flow is turbulent, and flow rate varies with the square root of the pressure differential.

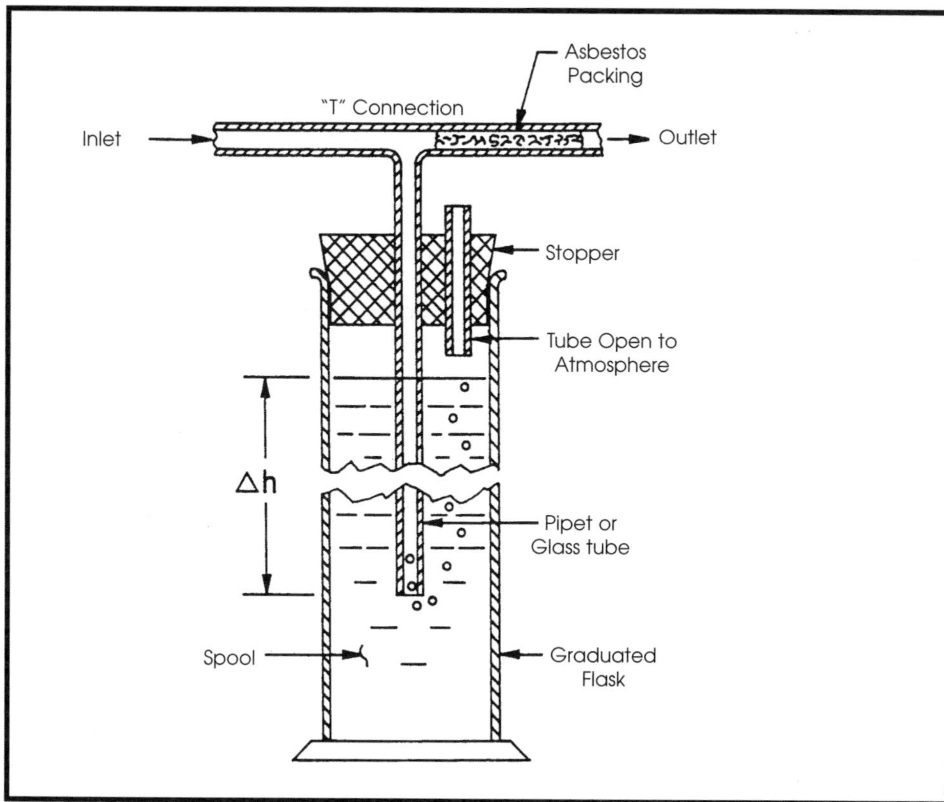

Figure 8.16 — Diagram of a laboratory-assembled packed plug flowmeter used to create laminar flow when calibrating equipment for very low flow rates.

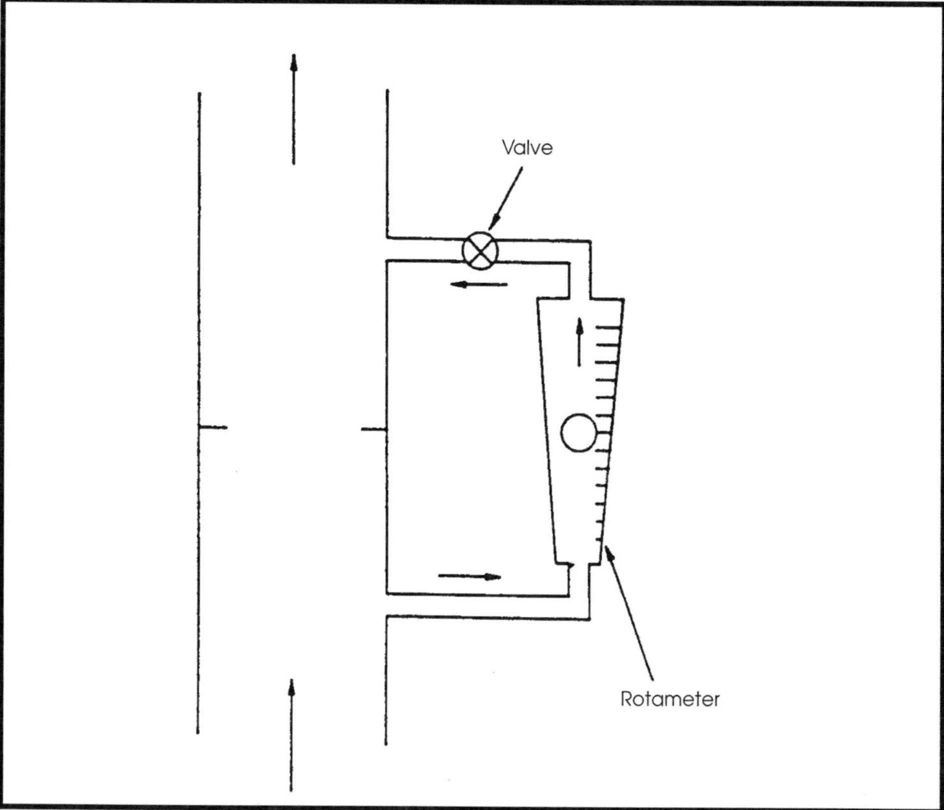

Figure 8.17 — Diagram of a by-pass flow indicator with both the variable-head element (valve) and the variable-area element (rotameter) labeled.

Laminar flow restrictors used in commercial flowmeters consist of egg-crate or tube bundle arrays of parallel channels. Alternatively, a laminar-flow meter can be constructed in the laboratory using a tube packed with beads or fibers as the resistance element. Figure 8.16 illustrates this kind of homemade flowmeter. It consists of a T connection, pipette or glass tubing, cylinder, and packing material. The outlet arm of the T is packed with a suitable material (asbestos has been used in the past), and the leg is attached to a tube or pipette projecting down into the cylinder filled with water or oil. A calibration curve of the depth of the tube outlet below the water level versus the rate of flow should produce a linear curve. Saltzman has used tubes filled with asbestos to regulate and measure flow rates as low as 0.01 cm^3 /min.[28]

By-Pass Flow Indicators

In most high-volume samplers the flow rate is strongly dependent on the flow resistance, and flowmeters with a sufficiently low flow resistance are usually too bulky or expensive. A commonly used metering element for such samplers is the by-pass rotameter, which actually meters only a small fraction of the total flow that is proportional to the total flow. As shown schematically in Figure 8.17, a by-pass flowmeter contains both a variable-head element and a variable-area element. The pressure drop across the fixed orifice or flow restrictor creates a proportionate flow through the parallel path containing the small rotameter. The scale on the rotameter generally reads directly in cubic feet per minute or liters per minute of total flow. In the versions used on portable high-volume samplers there is usually an adjustable bleed valve at the top of the rotameter that should be set initially and periodically readjusted in laboratory calibrations so that the scale markings can indicate overall flow. If the rotameter tube accumulates dirt, or the bleed valve adjustment drifts, the scale readings can depart greatly from the true flows.

Using Point Velocity Meters

Since the flow profile is rarely uniform across the channel, the measured velocity invariably differs from the average velocity. Furthermore, since the shape of the flow profile usually changes with changes in flow rate, the ratio of point-to-average velocity also changes. Thus, when a point velocity is used as an index of flow rate, there is an additional potential source of error, which should be evaluated in laboratory calibrations that simulate the conditions of use. Despite their disadvantages, velocity sensors are sometimes the best indicators available, as, for example, in some electrostatic precipitators where the flow resistance of other types of meters cannot be tolerated.

Velocity Pressure Meters

The basic idea for an instrument that could be used as a reference for measuring the velocity of air was first published by Henri de Pitot in 1732.[9] A standard pitot tube, as it is known, consists of an impact tube with an opening facing axially into the flow and a concentric static pressure tube with eight holes spaced equally around it in a plane that is eight diameters from the impact opening. The difference between the static and impact pressures is the velocity pressure. Bernoulli's theorem applied to a pitot tube in an airstream simplifies to the dimensionless formula:[29]

$$V = 4005(h_v)^{0.5} \quad (6)$$

where

h_v = velocity pressure in inches of water
V = velocity in feet per minute

If the pitot tube is carefully made, it may function as a primary standard. Checking the calibration of a pitot tube, while rarely done, is a simple process that involves comparing the pitot tube with another velocity meter such as an anemometer or current meter that is traceable to a reference standard. There are several serious limitations to pitot tube measurements in most sampling flow calibrations. One is that it may be difficult to obtain or fabricate a small enough probe. Another is that the velocity pressure may be too low to measure at the velocities encountered, even when using an inclined manometer or a low-range Magnehelic® gauge.

Heated Element Anemometers

Any instrument used to measure velocity can be referred to as an anemometer. In a heated element anemometer, the flowing air cools the sensor in proportion to the velocity of the air. Instruments are available with various kinds of heated elements, such as heated thermometers, thermocouples, films, and wires. They are all essentially

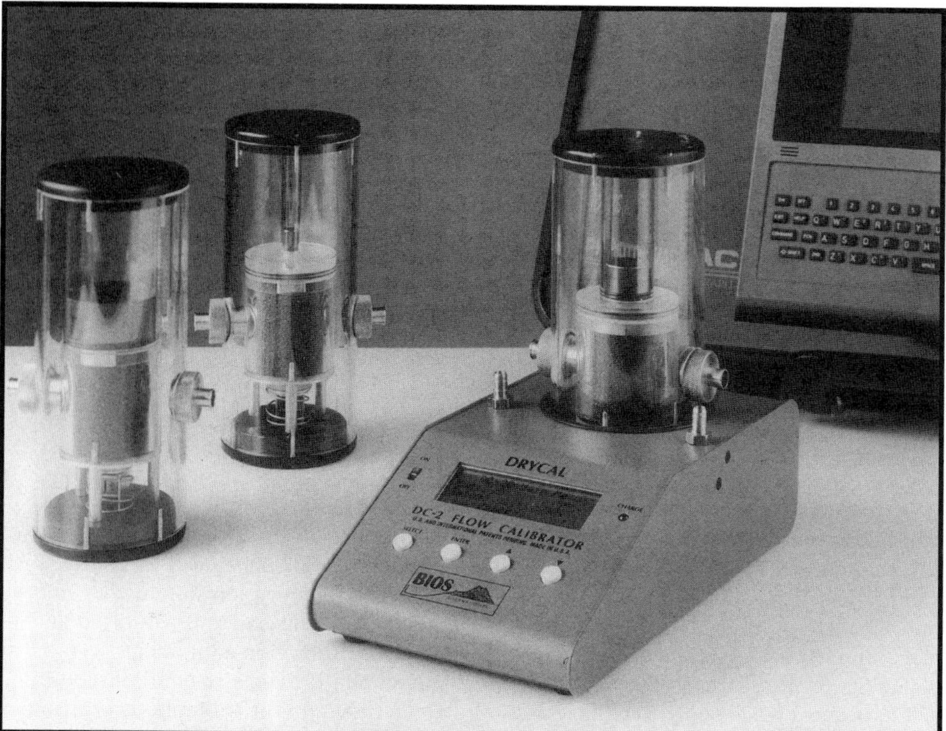

Figure 8.18 — Standardized mass flow calibrator, Dry Cal® DC-2M, in addition to giving true primary volume flow readings, the unit also contains internal barometric pressure and temperature sensors, which allows readings to be corrected to standard conditions (courtesy of Bois International Pompton Plains, N.J.).

nondirectional, i.e., have single element probes, measuring the airspeed but not its direction. They all can accurately measure steady-state airspeed, and those with low mass sensors and appropriate circuits can also accurately measure velocity fluctuations with frequencies above 100,000 Hz. Since the signals produced by the basic sensors depend on ambient temperature as well as air velocity, the probes are usually equipped with a reference element that provides an output that can be used to compensate or correct errors due to temperature variations. Some heated element anemometers can measure velocities as low as 3.1 m/min (10 ft/min) and as high as 2438 m/min (8000 ft/min).

Other Velocity Meters

There are several other ways to use the kinetic energy of a flowing fluid to measure velocity besides the pitot tube. One way is to align a jeweled-bearing turbine wheel axially in the stream and count the number of rotations per unit time. Such devices are generally known as rotating vane anemometers. Some are very small and are used as velocity probes. Others are sized to fit the whole duct and become indicators of total flow rate and sometimes are called turbine flowmeters. These and other instruments such as the velometer or swinging vane anemometer are reviewed in *Industrial Ventilation: A Manual of Recommended Practice.*[(29)]

Thermal Meters

A thermal meter measures mass air or gas flow rate with negligible pressure loss. Known as mass flowmeters, these devices consist of a heating element in a duct section between two points at which the temperature of the air or gas stream is measured. The temperature difference between the two points depends on the mass rate of flow and the heat input. Ness has stated that if these instruments are properly calibrated against a bubble burette their accuracy is within ±3%.[(16)] Figure 8.18 shows a mass flowmeter that combines the primary standard calibrator capability of a frictionless piston meter with internal electronics that make corrections to standardized temperature and pressure. Accuracy of the unit is reported at ±1%.

Establishing and Maintaining a Calibration Program

Each element of the sampling system should be calibrated accurately prior to initial field use. Protocols should also be established for periodic recalibration, since the performance of many transducers and meters will change with the accumulation of dirt, corrosion, leaks, and misalignment due to vibration or shocks in handling, etc. The frequency of such recalibration checks should initially be high, until experience is accumulated to show that it can be reduced safely. The need for calibration and calibration frequency depends on several factors, as outlined by Wight:[3]

- Instrument characteristics—sensitivity and experience with its stability under similar use patterns
- Instrument use—rough handling, moving, heavy usage, and changing environments necessitate frequent calibration
- Instrument users—multiple users and users of various skill and experience

It is important to document the nature and frequency of calibrations and calibration checks to meet legal as well as scientific requirements. Measurements made to document the presence or absence of excessive exposures will be only as reliable as the calibrations on which they are based. Formalized calibration audit procedures established by federal agencies provide a basis for quality assurance where they apply. They can also provide a systematic framework for developing appropriate calibration procedures for situations not governed by specific reporting requirements.

State and local air monitoring networks that are collecting data for compliance purposes must follow calibration procedures and external performance audits as outlined by the Environmental Protection Agency.[30] The OSHA instruction manual has forms to be followed when conducting equipment calibration as well specific requirements for field calibrations.[31] Another governmental source is NIOSH. The *NIOSH Manual of Analytical Methods*[32] recommends specific calibration procedures for both air sampling pumps and analytical equipment.

For any calibration program to be effective it must be performed under a definite, documented, and controlled procedure by competent individuals, in a repeatable manner, and under controlled conditions. It must be reported unambiguously and meet defined traceability requirements. For assurance of safety the calibrations should have an effective quality system. There should also be demonstrated competence in activities that affect reliability, safety, and performance.

The International Organization for Standardization (ISO) series 9000 standard requires that all measurements that affect quality shall be calibrated at prescribed intervals with certified equipment having a known valid relationship to nationally recognized standards. Certification to the ISO 9000 Quality System Standards is primarily in reference to the global business environment but also has an impact on calibration laboratories.[33] The ability to document equipment traceability will no doubt take on greater and greater significance as organizations develop international traceability standards.

The National Conference of Standards Laboratories, with the approval of the American National Standards Institute, has developed the American National Standard for Calibration—Calibration Laboratories and Measuring Test Equipment—General Requirements.[5] The standard outlines the need for inventory, calibration history, location history, and maintenance history to be collected and used for calibration equipment. To satisfy this standard five basic types of reports are necessary: (1) a calibration certificate or report that documents the calibration, (2) a due for calibration report, (3) out of tolerance report, (4) a forward traceability report, and (5) a reverse traceability report.

The following comments summarize the guiding philosophy behind all air sampler calibrations.

- Determine in advance the type of calibrator needed. For example, decide whether a primary or secondary standard is necessary, what level of accuracy is needed, what range of flow will be used, and under what conditions the equipment and calibrator will be used (see Table 8.2).
- Always maintain the record of traceability for every calibration.
- Set up a calibration program. All standard instruments used as calibrators should have available records covering periodic calibration checks as well as annual comparisons with a known standard.
- Be sure to understand the operation of the instrument being calibrated as well as the calibrator before attempting the calibration procedure. Be sure that you are operating within range of the instrument and standard.
- When in doubt about procedures or data, assure their validity before proceeding to the next operation.
- All sampling and calibration train connections should be as short and free of constrictions and resistance as possible. Always check for leaks in

Table 8.2 — Available Methods for the Calibration of Air Sampling Equipment

Standard Meters	*General Flow, Range, or Available Capacity*	*Reported Accuracy*	*Best Usage*
Primary standards			
Water sealed spirometers	100–600 L	±1%	laboratory
Bell provers or meter provers	0.14–0.57 m^3	±0.5%	laboratory
Volume calibration syringe	0.5–7 L	±0.25%	field or lab
Displacement or prover bottle	10–500 mL/min	±0.05 to 0.25%	laboratory
Soap film piston	<1 mL to 30 L/min	±1%	field or lab
Mercury sealed pistons	manual: 30–1200 cc automated: 1–50,000 standard cm^3/min	±0.2% ±0.2%	laboratory laboratory
Glass piston	10–200 mL/min	±2%	field or lab
Graphite piston (accuracy varies with cell size)	1 mL/min to 50 L/min	±1–2%	field or lab
Secondary standards			
Wet test meter	1–480 ft^3/hr (0.5–230 L/min)	±0.5%	laboratory
Dry-gas meter	20–325+ (10–150 L/min)	±1%	field
Rotameter (accuracy depends on calibration)	1 mL/min and up	±25 to 1%	field
Critical flow orifice	depends on orifice diameter	±0.5%	field or lab
Laminar-flow meter	use for very low flow rates	< ±1%	laboratory
Pitot tube traversing	velocities >50 ft/sec (15 m/sec)	±1%	field
Thermo-anemometer (accuracy depends on flow)	10 ft/min (3 m/min) to 8000 ft/min (2439 m/min)	±0.1 to 0.2%	field
Electronic mass flow rate	0–10 mL/min to 0–3000 L/min	±3%	field or lab

Note: Adapted from M. Lippmann, Airflow Calibration.[24]

both the sampling train and the pump. Where appropriate check battery voltage and charging requirements. Whenever battery-operated pumps are used, develop a battery maintenance program to ensure maximum operation life.

- Allow sufficient time for equilibrium to be established, inertia to be overcome, and conditions to stabilize. The time needed for this may vary among equipment and be influenced by environmental conditions or frequency of use.
- Extreme care should be exercised to limit the potential for subjective responses that occur when reading scales, recording times, making adjustments, or even leveling equipment.
- Enough points or different rates of flow should be obtained on a calibration curve to give confidence in the plot obtained. Each point should be made up of more than one (minimum of three) reading(s) whenever practical.
- Do not assume that temperature, pressure, and water vapor corrections do not apply to your particular calibration. First, prove what the correction would be mathematically, and then use professional judgment to determine whether to make the correction.
- Calibration curves should be properly identified as to conditions of calibration, device calibrated and what it was calibrated against, units

involved, range and precision of calibration, date, and who performed the actual procedure. Often it is convenient to indicate where the original data are filed and to attach a tag to the instrument indicating the above information.

- A complete and permanent record of all calibration procedures, data, and results should be maintained and filed. This should include trial runs, known faulty data with appropriate comments, instrument identification, connection sizes, barometric pressure, temperatures, etc.
- Once an instrument has been calibrated it will need to be recalibrated whenever the device has been changed, repaired, received from a manufacturer, subjected to use, mishandled or damaged, and at any time when there is a question as to its accuracy.
- When a recalibration differs from previous records, the cause for this change in calibration should be determined before accepting the new data or repeating the procedure.

Acknowledgments

The author wishes to thank Morton Lippman, Ph.D., CIH, whose original draft of this chapter provided an important framework for a revision and update of sampling equipment as well as an expansion of the subject area. The author is also indebted to Alena F. Chadwick for her skills and assistance as a reference librarian. Lastly, this work benefitted from the dedicated staff of the health sciences, chemistry, and engineering libraries of the University of Iowa.

References

1. **National Conference of Standards Laboratories:** *American National Standard for Calibration—Calibration Laboratories and Measuring and Test Equipment—General Requirements* (ANSI/NCSL Z540-1-1994). Boulder, CO: National Conference of Standards Laboratories, 1994.
2. **Mattingly, G.E.:** Fluid measurement: standards, calibration and traceabilities. In *Heat Transfer Measurements, Analysis and Flow Visualization*, R.K. Shah, ed. New York: American Society of Mechanical Engineers, 1989.
3. **Wight, G.D.:** *Fundamentals of Air Sampling*. Boca Raton, FL: Lewis Publishers, 1994.
4. **Garner, E.L. and S.D. Rasberry:** What's new in traceability. *J. Testing Eval. 21(6)*:505–509 (1993).
5. **National Institute of Standards and Technology (NIST):** *NIST Calibration Services User Guide 250 Appendix Fee Schedule* (NIST special publication). Gaithersburg, MD: NIST, 1994.
6. **Reiser, S.J.:** *Medicine and the Reign of Technology*. Cambridge, U.K.: Cambridge University Press, 1978.
7. **Davis, A.B.:** *Medicine and its Technology*. Westport, CT: Greenwood Press, 1981.
8. **Nelson, G.O.:** *Controlled Test Atmospheres: Principles and Techniques*. Ann Arbor, MI: Ann Arbor Science Publishers, 1971.
9. **Rouse, H. and S. Ince:** *History of Hydraulics*. Iowa City, IA: Iowa Institute of Hydraulic Research, 1980.
10. **Levy, A.:** The accuracy of the bubble meter method for gas measurements. *J. Sci. Instrum. 41(7)*:449–453 (1964).
11. **Baker, W.C. and J.F. Pouchot:** The measurement of gas flow. Part II. *Air Pollut. Control Assoc. J. 33(2)*:156–162 (1983).
12. **Lide, D.R. (ed.):** *CRC Handbook of Chemistry and Physics*, 77th ed. Boca Raton, FL: CRC Press Inc., 1996.
13. **DiNardi, S.R.:** *Calculation Methods for Industrial Hygiene*. New York: Van Nostrand Reinhold, 1995.
14. **Waldron, P.F.:** "Reducing Systematic Error Associated with Asbestos Air Sampling Through the Appropriate Use of Flowmeter Correction Factors." MPH thesis, University of Massachusetts, Amherst, MA, 1991.
15. **Parkinson, B.R:** The history and recent development of the wet gas meter, part 1—history. *Gas J. 252(Oct. 8)*:104–106 (1947).
16. **Precision Scientific Inc.:** *Wet Test Gas Meters* (catalog TS-63111 AT-8). Chicago, IL: Precision Scientific Inc.
17. **American Society for Testing and Materials (ASTM) Committee on Standards:** *Standard Method for Volumetric Measurement of Gaseous Fuel Samples* (D1071-83). Philadelphia, PA: ASTM, 1993. pp. 12–23.
18. **Ness, S.A.:** *Air Monitoring for Toxic Exposures: An Integrated Approach*. New York: Van Nostrand Reinhold, 1991.

19. **Fischer & Porter Co.:** *Variable Area Flowmeter Handbook Volume II: Rotameter Calculations*. Warminster, PA: Fisher & Porter Co., 1982. [Pamphlet]
20. **National Institute of Occupational Safety and Health:** *Occupational Exposure Sampling Strategy Manual*, by N.A. Leidel, K.A. Busch, and J.R. Lynch (DHEW/NIOSH pub. no. 77-173). Washington, DC: Government Printing Office, 1977.
21. **Caplan, K.S.:** Rotameter corrections for gas density. *Am. Ind. Hyg. Assoc. J. 46*:10–16 (1985).
22. "Method 2A-Direct Measurement of Gas Volume Through Pipes and Small Ducts," *Code of Federal Regulations* Title 40, Part 60, App. A. June 1996. pp. 580–583 .
23. **Carvill, J.:** *Famous Names in Engineering*. London: Butterworth, 1981.
24. **Lippmann, M.:** Airflow calibration. In *Air Sampling Instruments for Evaluation of Atmospheric Contaminants*, 8th ed. B.S. Cohen and S.V. Hering, eds. Cincinnati, OH: American Conference of Governmental Industrial Hygienists, 1995. pp. 139–150.
25. **Cusick, C.F.:** *Flow Meter Engineering Handbook*, 3rd ed. Philadelphia, PA: Minneapolis-Honeywell Regulator Co., Brown Instrument Division, 1961.
26. **Bradner, M. and L.P. Emerson:** Flow measurement. In *Fluid Mechanics Source Book*, S.P. Parker, ed. New York: McGraw-Hill, 1987. pp. 203–207.
27. **Perry, J.H. (ed.):** *Chemical Engineering Handbook*, 6th ed. New York: McGraw-Hill, 1984.
28. **Saltzman, B.E.:** Preparation and analysis of calibrated low concentrations of sixteen toxic gases. *Anal. Chem. 33*:1100 (1961).
29. **American Conference of Governmental Industrial Hygienists (ACGIH) Committee on Industrial Ventilation:** *Industrial Ventilation: A Manual of Recommended Practice*, 22nd ed. Lansing, MI: ACGIH, 1995
30. "Standards of Performance of New Stationary Sources," *Code of Federal Regulations* Title 40, Part 60. 1995.
31. **Occupational Safety and Health Administration:** *Technical Manual, TED 1.15*. Washington, DC: Government Printing Office, 1995.
32. **National Institute for Occupational Safety and Health (NIOSH) Division of Physical Sciences and Engineering:** *NIOSH Manual of Analytical Methods*, 3rd ed. (DHHS pub. no. (NIOSH) 84-100). Washington, DC: Government Printing Office, 1994.
33. **Randall, A.:** Calibration and traceability. *Proc. Control Engin. 48(9)*:2 (1995).

Outcome Competencies

After completing this chapter, the user should be able to:

1. Define underlined terms used in this chapter.
2. Explain the principles of operation of selected direct reading survey instruments.
3. Select direct-reading instruments based on fundamental scientific principles to satisfy occupational hygiene objectives.
4. Describe the need for calibration procedures for selected direct reading instruments.
5. Discuss the basic chemical and physical phenomenon underlying the operation of sampling and analysis instruments.

Prerequisite Knowledge

Basic Chemistry

Basic Physics

Prior to beginning this chapter, the user should review the following chapters:

Chapter Number	Chapter Topic
8	Principles and Instrumentation for Calibrating Air Sampling Equipment

Key Terms

aerosol photometers • combustible gas indicators • computed tomography • condensation nucleus counters • detector tube • direct-reading instruments • electrochemical sensors • electron capture detectors • fibrous aerosol monitors • flame ionization detectors • Fourier transform infrared spectrometry • gas chromatograph • infrared gas analyzers • ionization potential • metallic oxide semiconductor (MOS) sensors • multiple particle monitors • optical particle counters • photoacoustic spectroscopy • photoionization detectors • piezoelectric mass sensors • quartz crystal microbalances • real-time monitors • thermal drift • thermal conductivity • Wheatstone bridge • tapered element oscillating microbalance

Key Topics

I. Direct-Reading Gas and Vapor Monitors
   A. Monitoring Single Gases and Vapors
   B. General Survey Monitors for Gases and Vapors
   C. Monitoring Multiple Gases and Vapors

II. Direct-Reading Aerosol Monitors
   A. Optical Techniques for Determinining Aerosol Size and Count
   B. Optical Techniques for Determining Aerosol Mass
   C. Electrical Techniques for Determining Aerosol Size
   D. Electrical Techniques for Determining Aerosol Count
   E. Resonance Techniques for Determining Aerosol Mass
   F. Bata Absorption Techniques for Determining Aerosol Mass

9

Direct-Reading Instrumental Methods for Gases, Vapors, and Aerosols

Lori A. Todd, Ph.D., CIH

Introduction

Direct-reading instruments (real-time monitors) are among the most important tools available to occupational hygienists for detecting and quantifying gases, vapors, and aerosols. These instruments permit real-time or near real-time measurements of contaminant concentrations in the field, thus eliminating the lag time encountered when samples are collected on media and analyzed by a laboratory. Using direct-reading instruments, air contaminants are sampled and analyzed within the instrument in a relatively short time (seconds to minutes). Results are usually indicated on an analog or digital display, a graph, or by a color change that is compared with a calibrated scale. Real-time monitors generally can be used to obtain short-term or continuous measurements. Some monitors have data-logging capabilities that allow digital storage of data. While a data logger does not enhance the accuracy of a measurement, it frees the occupational hygienist from manually recording data and allows for a variety of statistical analyses.

Direct-reading instruments range in size from small personal monitors to hand-held monitors to complex stationary installations with multipoint monitoring capability. Field monitoring instruments are usually lightweight, portable, rugged, weather and temperature sensitive, and are simple to operate and maintain. However, there is no magic black box that can be used to measure all contaminants in air; in addition, instruments used for gases and vapors cannot be used for aerosols, and vice versa.

Direct-reading instruments for gases and vapors are designed to (1) monitor a specific single gas or vapor, (2) monitor specific multiple gases and vapors, or (3) monitor multiple gases and vapors without differentiating among them. All instruments are designed to be used within a designated detection range and should be calibrated before field use. A variety of detection principles are used in direct-reading instruments for gases and vapors including infrared (IR), ultraviolet (UV), flame ionization, photoionization, colorimetric, and electrochemical reaction. Table 9.1 has a partial list of commonly used measurement techniques.

In general, direct-reading instruments for aerosols cannot differentiate between types of aerosols. The type of information provided by the instruments includes particle size distribution, particle count, and total and respirable mass concentrations. However, there is no single instrument that can provide all of these measurements. Direct-reading instruments for aerosols primarily operate using four techniques: (1) optical, (2) electrical, (3) resonance oscillation, and (4) beta absorption. Table 9.2 is a partial list of commonly used measurement techniques for aerosols.

Different aerosol properties are measured by different direct-reading instruments. For example, if instruments measure particle size, the size may be derived from one of the many properties of aerosols such as its optical, aerodynamic, mechanical, or electrical behavior. Thus, direct comparisons of measurements between instruments can be difficult and can give contradictory information.

There is no magic black box that can be used to measure all contaminants in air.

Table 9.1 —
Commonly Used Direct-Reading Instruments for Gases and Vapors

Instrument	*Common Analytes*	*Principle of Operation*	*Range*
Combustible gas detectors	Combustible gases and vapors (nonspecific)	Hot wire—test gas is passed over a heated wire (sometimes in the presence of a catalyst). The test gas burns, changing the temperature of the filament, and the electrical resistance of the filament is measured.	Usually measured in percentage of the lower explosive limit. Some models measure down to 1 ppm.
Colorimetric detectors	Various vapors including formaldehyde, hydrogen sulfide, sulfur dioxide, toluene diisocyanate (specific)	Reaction of the test gas with a chemical reagent (either as a liquid or in some cases an impregnated paper or tape) and measurement of the color produced	Variable
Electrochemical sensors	Carbon monoxide, nitric oxide, nitrogen dioxide, hydrogen sulfide, sulfur dioxide (specific)	Chemical oxidation of test gas	1 to 3000 ppm
Infrared gas analyzers	Organic and inorganic gases and vapors (specific)	Measures infrared absorbance of test gas	Sub-ppm to low percent levels
Metal oxide sensors	Hydrogen sulfide, nitro, amine, alcohol, and halogenated hydrocarbons (specific)	Metal oxide sensor is chemically reduced by the gas, increasing its electrical resistance	1 to 50 ppm
Thermal conductivity sensors	Carbon monoxide, carbon dioxide, nitrogen, oxygen, methane, ethane, propane, and butane	Uses specific heat of combustion of a gas or vapor	Percentage gas
Portable gas chromatographs	Organic and inorganic gases and vapors (specific)	Uses a packed column to separate complex mixtures of gases. Detectors available include flame ionization, electron capture, thermal conductivity, flame photometric, and photoionization	0.1 to 10,000 ppm
Detectors for Gas Chromatographs			
Electron capture detector	Halogenated hydrocarbons, nitrous oxide, and compounds containing cyano or nitro groups	Uses a radioactive source such as ^{63}Ni to supply energy to the detector that monitors the intensity of the electron beam arriving at a collection electrode. When an electron-capturing species passes through the cell the intensity of the electron beam decreases.	0.1 ppb to low ppm
Flame ionization detectors	Organic compounds including aliphatic and aromatic hydrocarbons, ketones, alcohols, and halogenated hydrocarbons	Creates organic ions by passing a hydrogen gas through flame. Measures conductivity of the flame.	0.1 to 100,000 ppm
Photoionization detectors	Most organic compounds, particularly aromatic compounds	Creates ions by exposing test gas to ultraviolet light. Measures conductivity of the gases in the light field.	0.2 to 2000 ppm

Table 9.2 —
Commonly Used Aerosol Monitors

Instrument	*Sample Flow*	*Size Range*	*Concentration Range*
Light-scattering photometers	passive to 100 L/min	0.1 to 20 μm	0.0001 μm/m³ to 200 g/m³
Light-scattering particle counters	0.12 L/min to 28 L/min	0.1 to 8000 m (up to 32,000 μm for drop size analyzers)	1 particle/L to 10^5 particles/cm³
Condensation nucleus counters	0.003 L/min to 4.2 L/min	1.6 nm to 20 nm	0.1 to 10^6 particles/cm³
Single Particle Aerosol Relaxation Time (SPART)	0.5 to 5 L/min	< 0.3 to 10 μm	
Beta attenuation aerosol mass monitors	15 L/min	< 10 μm	10 mg/m³ max
Piezoelectric crystal microbalance	0.24 to 1 L/min	0.05 to 35 μm	100 g/m³ to 100 mg/m³
Tapered Element Oscillating Microbalance (TEOM)	0.5 to 35 μm		5 μg/m³ to 2000 mg/m³
Fibrous aerosol monitor	2 L/min	0.2 μm ∞ 2 to 200 μm	0.0001 to 30 fibers/cm³

Direct-reading instruments provide powerful on-the-spot information and are ideal for situations where the occupational hygienist wants immediate data that are temporally resolved into short time periods. Personal direct-reading instruments, which can be placed on lapels or pockets, can be used for personal exposure monitoring and are available for a limited number of chemicals, particularly for gases that have high acute toxicity, such as carbon monoxide and phosgene.[1] Direct-reading monitors can profile fluctuations in contaminant concentrations that are lost when performing traditional integrated sampling. Thus, measurements can estimate instantaneous exposures, short-term exposures, and time integrated exposures, to compare with ceiling limits, short-term exposure limits (STELs), and time-weighted averages (TWAs), respectively.

Direct-reading instruments can be used as educational and motivational tools in the workplace. The immediate feedback to workers of concentration information can be used to document the impact on exposure of changes in work practices; when combined with the use of video cameras, workers and management can use the information for training and to reduce exposures.[2,3]

In conjunction with traditional integrated sampling methods, direct-reading instruments for gases and vapors can be used in developing personal sampling strategies and for obtaining a comprehensive exposure evaluation. Used to perform an initial survey of the workplace, direct-reading instruments can document the types of contaminants and the range of concentrations in the air. This information would allow a better choice of chemicals to sample, pump flow rates to use, and the representative subset of workers to monitor. When the sample probe of a direct-reading instrument is placed in the breathing zone of a worker, peak exposures to chemicals can be evaluated.

Direct-reading instruments for gases and vapors can be used for evaluating the effectiveness of local exhaust ventilation systems and other controls, detecting leaks, monitoring source emissions, emergency response, and monitoring hazardous waste sites. Carbon monoxide exposures and explosive and oxygen deficient atmospheres create life-and-death situations requiring immediate feedback of concentrations; they cannot wait for laboratory analyses. Some chemicals are difficult to collect on traditional media and for them, real-time instrumentation is the best solution. Direct-reading instruments used in stationary installations are used to provide early warning of high contaminant concentrations resulting from process leaks, spills, failure of ventilation systems, or other catastrophic releases.

Direct-reading instruments for aerosols can be used for surveying the workplace to prioritize sources of exposure to dusts, for respirator fit-testing, for asbestos monitoring, and as an educational tool to evaluate work practices and identify dust generating sources. Aerosol monitoring has been combined with videotaping to evaluate the influence of different operations and activities on exposure.[3]

Which direct-reading instrument to select depends on the application for which it will be used. For monitoring employee exposures to gases and vapors, an instrument would probably be chosen that has high selectivity and can detect and quantify target chemicals in a specific concentration range. For hazardous waste sites or some emergency situations where there are unknown contaminants, a nonspecific general survey instrument, which can respond to a range of contaminants, would be needed. Other factors that impact instrument choice include price, portability, weight, size, battery operation and life, and requirements for personnel training.

While manufacturers have been simplifying operation, direct-reading instruments all require the user to understand the limitations and conditions that can affect performance and calibration, and also to understand maintenance requirements and interpret results. Measurements of gases and vapors can be adversely affected by interferences from other contaminants; therefore, the occupational hygienist needs to be aware of the sampling environment before selecting an instrument. Environmental conditions such as temperature extremes, humidity, elevation above sea level, barometric pressure, presence of particulates, and oxygen concentration can affect instrument performance and accuracy. In addition, electromagnetic fields in the environment can interfere with instrument performance and cause a wide variety of problems including intermittent operation, changes in pump flow rates, illogical displays, and total shut downs.[4]

When direct-reading instruments are used, all potential sources of error should be minimized through proper quality control practices. All instruments require calibration before use. Calibration of gas and vapor monitors includes comparison of instrument readings to known concentrations of gases. Ideally, a multipoint calibration should be

Direct-reading instruments provide powerful on-the-spot information and are ideal for situations where the occupational hygienist wants immediate data that are temporally resolved into short time periods.

performed using specific contaminants that will be measured in the field over the range of expected concentrations. In some cases it may be important to calibrate the instrument for interferences. Interferences can result in false-positive or false-negative results by interfering with either collection, detection, or quantification of contaminants. For nonspecific instruments, calibration is usually performed using a relatively nontoxic gas, such as isobutylene. A few instruments are calibrated electronically and do not require the use of a calibrant gas.

Measurement of aerosols is complicated and, in many ways, more difficult than measurement of gases and vapors. Measurement is affected by many factors such as particle size and shape, particle settling velocity, wind currents, and sampling flow rates. Accurate and defensible measurement of aerosols depends on careful calibration of direct-reading aerosol monitors. While some monitors can be calibrated using fundamental theories of operation, it is more common to establish an empirical relationship using a set of calibrated test aerosols. Calibration may involve testing the system with suspensions of known sizes of nebulized monodisperse polystyrene latex spheres or aerosols generated using a vibrating-orifice generator.[5]

For potentially explosive atmospheres, direct-reading instruments need to be intrinsically safe or explosion-proof. Intrinsically safe means that instruments cannot release thermal or electrical energy that could cause ignition of hazardous chemicals. Explosion-proof means that there is a chamber within the instrument that is designed to contain and withstand any explosion caused by contaminants entering the instrument. Minimum standards for inherent safety in hazardous environments has been defined by the National Fire Protection Association in its National Electrical Code.[6] Testing and certification of equipment is performed by the Underwriters Laboratory or Factory Mutual Research Corp., and other international testing agencies.

This chapter describes the principles of operation, applications, limitations, and calibration procedures for some commonly used direct-reading instruments.

Direct-Reading Gas and Vapor Monitors

Monitoring Single Gases and Vapors

Electrochemical Sensors

A variety of instruments dedicated to monitoring specific single gas and vapor contaminants use electrochemical sensors. Electrochemical sensors are available for up to 50 different individual gases including oxygen, carbon monoxide, nitric oxide, nitrogen dioxide, hydrogen sulfide, hydrogen cyanide, and sulfur dioxide (see Table 9.3).[7, 11] These sensors can be incorporated into a compact design and used as lightweight personal monitors that are small enough to fit into a shirt pocket. They have low power requirements and can operate continuously for up to 4 months without needing replacement batteries.

Carbon monoxide can be monitored using electrochemical cells (the most common detection method), solid-state sensors, and IR methods. Carbon monoxide is produced as a by-product of fuel combustion and is associated with motor vehicle exhaust; therefore, human exposures may be encountered in inadequately ventilated buildings with attached automobile garages; garage workers and traffic police are particularly prone to such exposures. Carbon monoxide poisoning can produce symptoms ranging from headaches and dizziness to unconsciousness and death.

Hydrogen sulfide can be monitored using electrochemical cells and solid-state sensors. It is commonly measured using real-time instruments rather than integrated methods because of its high toxicity, the fact that it causes olfactory fatigue, and its prevalence in confined spaces. Hydrogen sulfide is the primary toxic contaminant that causes death in confined spaces.[8]

Table 9.3 — Gases Detectible by Electrochemical Sensors

Ammonia	Formic acid	Hydrogen sulfide	Phosphine
Arsine	Freon	Nitric acid	Silane
Bromine	Germane	Nitric oxide	Silicon tetrafluoride
Carbon dioxide	Hydrazine	Nitrogen dioxide	Sulfur dioxide
Carbon monoxide	Hydrochloric acid	Nitrogen oxides	Tetrachloroethylene
Chlorine	Hydrogen	Nitrous oxide	Trichloroethylene
Ethylene oxide	Hydrogen chloride	Oxygen	Tungsten hexafluoride
Fluorine	Hydrogen cyanide	Ozone	
Formaldehyde	Hydrogen fluoride	Phosgene	

Sources: References 1 and 7

Oxygen can also be monitored using electrochemical cells; measurements are usually taken in conjunction with combustible gas measurements for confined space entry where the air can be oxygen deficient. The Occupational Safety and Health Administration defines oxygen deficient atmospheres as concentrations of 19.5% or lower; normal air contains 20.9% oxygen.[9,10] It is important to verify adequate oxygen levels to ensure that the combustible sensor is functioning properly. Oxygen deficient atmospheres can be hazardous to health; the hazard can be compounded if the gas displacing the oxygen in the air is toxic. A life-or-death safety hazard may also exist in this atmosphere if the chemical displacing the oxygen is flammable or explosive.

A typical electrochemical sensor (see Figure 9.1) uses a coarse particulate filter, an electrochemical cell with two electrodes, an electrolyte (liquid, gel, or part of a solid matrix), and a porous membrane (normally Teflon). The monitored gas diffuses into the cell, dissolves in the electrolyte, and reacts with a sensing electrode. Charged ions and electrons are produced from this reaction and diffuse across the electrolyte to a counting electrode. This results in a change in the electrochemical potential between the electrodes. The associated electronic circuitry measures, amplifies, and controls the resulting electronic signal and converts it into a meter reading. The electrolytic solutions and composition of electrodes in the sensors can be modified to achieve chemical specificity, sensitivity, and range of detection.

Measurements obtained using electrochemical sensors may be inaccurate due to interferences and contamination. Gases of similar molecular size and chemical reactivity can interfere with measurement by reacting with the electrode and can result in false positive results. This lack of specificity is important to realize when monitoring atmospheres with multiple unknown toxic chemicals. The sensors can be contaminated by acidic or alkaline gases that neutralize the electrolytic solution, which can result in decreased sensitivity. The membrane can become clogged with particulates, and with condensed aerosols, water vapor, and hot gases. A clogged membrane will limit the diffusion of gases into the sensor and will result in underestimation of concentrations.

The electrochemical sensors themselves can be hazardous if they contain a corrosive liquid electrolyte; if the solution leaks, it can cause chemical burns on contact with the skin or eyes.

The electrochemical cells used in oxygen monitors usually contain a sensing electrode of lead or zinc, a counting electrode of gold or platinum, and an alkaline electrolyte of potassium hydroxide and water. The chemical reaction in the cell is proportional to the concentration (partial pressure) of oxygen and is displayed as percentage by volume oxygen within a range of 0 to 25%. Electrochemical cells used to monitor oxygen deteriorate over time, whether they are used or not, because they are exposed to ambient air that contains oxygen. Therefore, replacement is required within a year. Usually, oxygen monitors have an alarm that is activated at 19.5% oxygen.

Oxygen meters must be calibrated with clean air containing 20.9% oxygen at the same altitude and temperature that they will be used. Meters calibrated above sea level (in Denver, for example) and used at a lower elevation (for example, Boston) will overestimate oxygen concentrations. Calibration of oxygen monitors is simple and involves placing the instrument in fresh ambient air and adjusting the potentiometer until it reads the specific oxygen percentage indicated by the manufacturer (20.8% or 20.9%). This should be performed immediately before field use. To check the ability to measure oxygen deficient atmospheres, manufacturers supply gases containing 10 to 15% oxygen.

General Survey Monitors for Gases and Vapors

General survey instruments are capable of measuring a wide range of air contaminants but cannot distinguish among them. They are usually used as area samplers to

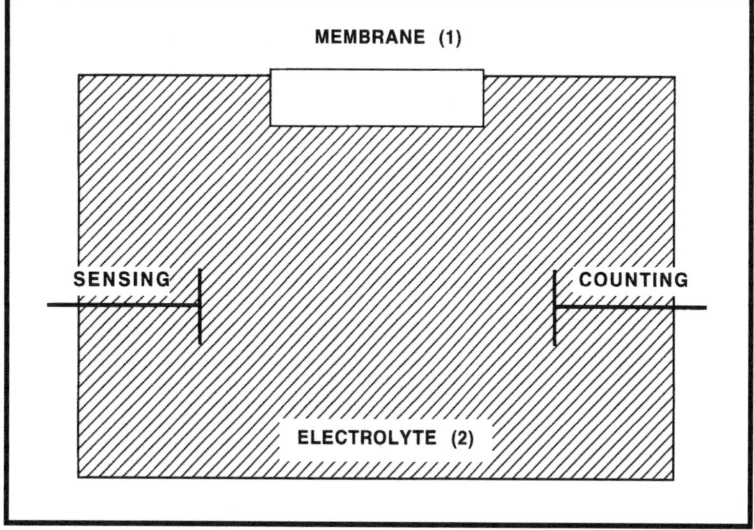

Figure 9.1 — A typical electrochemical sensor.

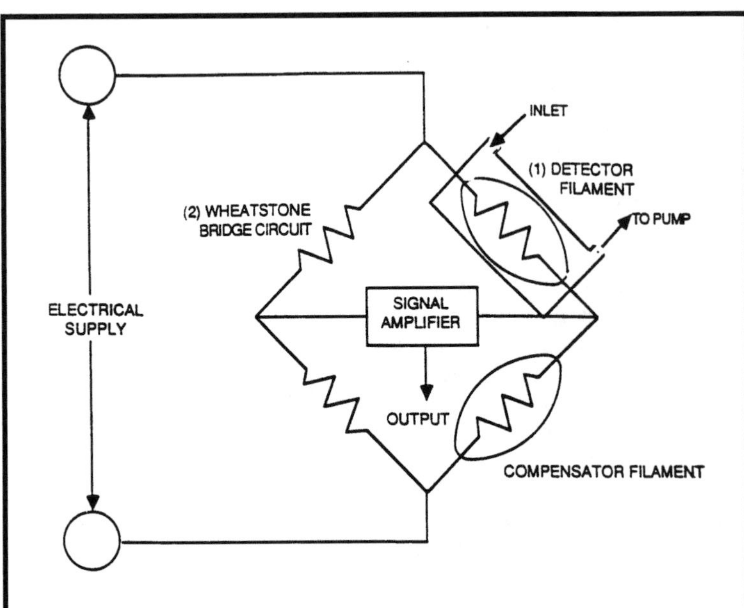

Figure 9.2 — A Wheatstone bridge.

measure concentrations that are immediately dangerous to life and health, and concentrations that are within occupational exposure limits, in the ppm range. Commonly used monitors include combustible gas indicators, photoionization and flame ionization detectors, and solid-state sensors.

Combustible Gas Instruments (CGI)

Combustible gas instruments (CGI) were the first direct-reading instruments to be developed and were used to detect methane in underground mines in Great Britain.(7) CGIs are currently used to measure gases in confined spaces and atmospheres containing combustible gases and vapors (methane and gasoline).(11) These instruments are capable of measuring the presence of flammable gases in percentage of lower explosive limit (LEL) and percentage of gas by volume. A few CGIs are equipped with sensors that allow reliable measurement of low ppm concentrations for a limited number of gases and vapors. The CGI is primarily a safety meter used to detect hazardous concentrations up to 100% of the LEL. When 100% of the LEL is reached, flammable or explosive concentrations are present. In terms of concentrations that relate to occupational exposure limits, a relatively low percentage LEL corresponds to a high concentration. For example, the LEL of methane is 5.3%. Therefore, an LEL of 10% is equivalent to 5300 ppm ($0.10 \times 5.3\% = 0.53\%$ or 5300 ppm). This level is much greater than any threshold limit value or permissible exposure limit. Thus, most CGIs cannot be used to determine compliance with occupational exposure limits.

Virtually all CGIs in use today are based on catalytic combustion. The air containing the contaminant passes over a heated catalytic filament, which is incorporated into a Wheatstone bridge (see Figure 9.2). A Wheatstone bridge is a circuit that measures the differential resistance in an electric current. The catalytic sensor has two filaments; one is coated with a catalyst (usually platinum) to facilitate oxidation or combustion of very low concentrations of a gas. The other filament, the compensating filament, operates at the same voltage as the catalyst, but does not cause oxidation and, therefore, does not increase in temperature.

When the meter is turned on, a current is applied to the Wheatstone bridge and the filaments are heated to a very high temperature. The gas is passed through a coarse metal filter, comes in contact with the heated filaments, and ignites on the filament coated with the catalyst. The burning gas causes the filament to heat; however, there is no temperature change in the compensating filament. The increase in temperature causes an increase in resistance and a decrease in current flow relative to the compensating filament. The heat of combustion changes the temperature in the sensor chamber in proportion to the amount of combustible gas present in the sample stream; this is translated into a meter reading. The greater the change in resistance, the greater the concentration. Catalytic sensors are usually sensitive to concentrations as low as 0.5 to 1% of the LEL.

Thermal conductivity (TC) is another method used for detecting explosive atmospheres that uses the specific heat of combustion of a gas or vapor as a measure of its concentration in air. TC is used in instruments where very high concentrations of flammable gases are expected—greater than 100% of the LEL—and measures percentage of gas as compared with percentage of the LEL. This method is not sensitive to low concentrations of gases. Typical applications include pipeline leaks, tank farms, and landfills. A TC filament is substituted into a Wheatstone bridge. The combustible gases in the sample cool this filament instead of heating it, which decreases the resistance across the Wheatstone bridge and produces a meter reading. The meter reading represents how cool the TC filament is relative to the hot compensating filament. When gases are sampled by this equipment, they will cool the filament even if they are not flammable.

Semiconductor sensors are also used to detect combustible gases and vapors. Sensors consist of a semiconductor, such as silica, covered with a metallic oxide, such as zinc oxide or aluminum oxide. The oxide forms a porous surface that traps oxygen and provides a base level of conduction through the sensor. A conductor is any material that has many free electrons and allows electrons to move through it easily. A heated coil embedded in the semiconductor maintains a constant temperature, controls the number of free electrons, and prevents condensation of water vapor. The sensor is attached to a Wheatstone bridge. When the sensor is exposed to a contaminant, the gas reacts with the trapped oxygen in the sensor, which causes the release of electrons and a decrease in the resistance across the Wheatstone bridge. This results in a meter reading proportional to concentration.

All CGI readings are relative to the gas used to calibrate the equipment in the factory, which may be methane, pentane, propane, or hexane. When the CGI is exposed only to the calibrant, it will respond accurately. When a CGI is calibrated, the instrument response depends on the calibrant gas as well as the type of catalyst employed in the sensor. It is very important to understand the implications of using one calibrant gas over another and the way to interpret the meter readings, because CGIs are used to detect situations where an explosive atmosphere could be present.

Ideally, when the CGI is calibrated, the calibrant gas should have a heat of combustion that is similar to the chemical being monitored. To provide a wide margin of safety, if there are several potential explosive gases in the environment, the instrument should be calibrated for the least sensitive gas. Therefore, a gas should be selected that would require a very high concentration to get the lowest percentage LEL. Thus, all the other gases would overestimate the percentage LEL. For example, methane has a heat of combustion of 55.50 millijoules (mj)/kg and pentane has a heat of combustion of 48.64 mj/kg. Therefore, higher concentrations are required in air for methane to create a potentially explosive atmosphere. If a CGI is calibrated to methane and encounters pentane, it will be more sensitive to pentane and read high. Therefore, the occupational hygienist will be conservative in the interpretation of the measurements and will be provided with a margin of safety. In addition, most CGIs have alarms set for 20 to 25% of the LEL because of the inherent uncertainty when sampling unknown explosion hazards.

When sampling a gas in air that is not the calibrant, the gas may burn more readily on the filament, creating a hotter filament at a lower concentration. This would result in the recording of a higher than actual percentage LEL on the meter. These gases are called hot burning gases. Conversely, when sampling a gas that burns less readily on the filament than the calibrant, this creates a cooler filament at a higher concentration. This would result in the recording of a lower than actual percentage LEL on the meter. These gases are called cool burning gases.

Many manufacturers of CGIs provide instrument response curves to allow conversion of a meter reading to concentration when a specific gas other than the calibrant is being sampled. The response curves are created in the laboratory by using known concentrations of gases and recording the meter readings over the entire LEL range. In the field, if the identity of the vapor to be sampled is known, the response curve can be used to approximate the actual concentration (see Figure 9.3). For example, a 60% LEL reading for the hotter burning gas B would actually represent 40% of the LEL. A 60% LEL reading for the cooler burning gas C would actually represent 75%.

Conversion factors are also used to convert meter readings of specific chemicals

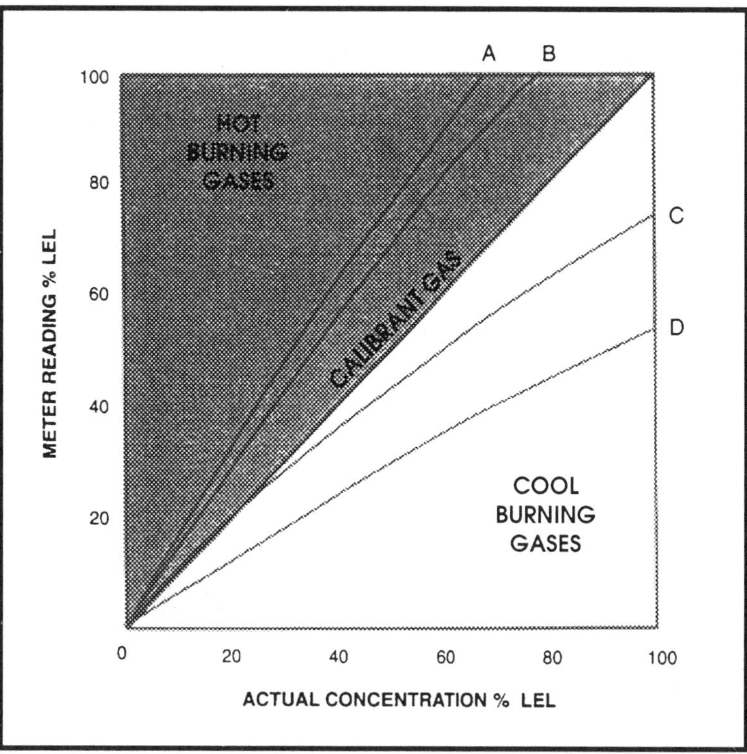

Figure 9.3 — Combustible gas instrument response curve for conversion of a meter reading to concentration.

into concentrations. The conversion factor is a single number provided by the manufacturer that is used to describe the entire response curve; therefore, it is only a conservative approximation. If a conversion factor is 0.9 and the readout is 50% LEL, the estimated concentration is 45% (0.9 × 50%) of the LEL. When response curves or conversion factors are used, an accuracy of ±25% should be factored into the interpretation of the measurement. Manufacturer supplied conversion factors and response curves should only be used for the model and calibrant for which they were generated. Without response curves or conversion factors, the instrument cannot be used to estimate concentrations.

As with all direct-reading instruments, CGIs should be calibrated under the same conditions as those in the field. In the field, the calibration check involves using a single concentration of a gas in the middle of the response range. The response should be within acceptable limits as described by the manufacturer. After the check with the calibrant gas, the user can sample known concentrations of other gases and check the calibration using the response curves or conversion factors.

While CGIs appear relatively simple to use, they have limitations that could result in dire outcomes if not fully understood. Sensors need to be replaced periodically because every time the instrument is used the catalyst deteriorates. It is important for a user to know the response time of an instrument, as response time varies and improper readings may be obtained. CGIs that use a catalytic sensor require the sample air to be oxidized or burned on the filament. Therefore, the occupational hygienist must know the minimum oxygen requirement for oxidation. If the concentration of oxygen in ambient air is too low, the signal obtained will be too low. For confined spaces, in particular, CGIs should always be used in conjunction with oxygen meters. The oxygen concentration should be obtained first, since CGI performance depends on oxygen availability.

Oxygen deficiency may be created when there are concentrations of gases and vapors above the upper explosion limit. In this situation many CGIs will initially peg 100% LEL and then quickly return to zero (see Figure 9.4). Therefore, the meter reading must be observed as soon as measurements are started. For some cool gases, the reading may not peg at 60% rather than 100% before dropping. When concentrations are above the upper explosion limit, the atmosphere is too rich to burn, and oxidation cannot occur. In this case the temperature of both filaments will decrease.

Although CGIs can measure a wide variety of flammable gases and vapors, they cannot measure all materials and can give false-positive or false-negative results. On the other hand, some CGIs have catalysts that enhance the oxidation of chlorinated hydrocarbons, which can result in overestimation of concentrations and meter readings that represent flammable atmospheres when there are none.[7]

High concentrations of some compounds can poison catalysts such as organic heavy metals, silicates, silicones, and silanes. They can coat the catalyst and result in a decreased response of the instrument to other chemicals. Corrosive gases can corrode the sensors, which will ultimately result in depressed readings.

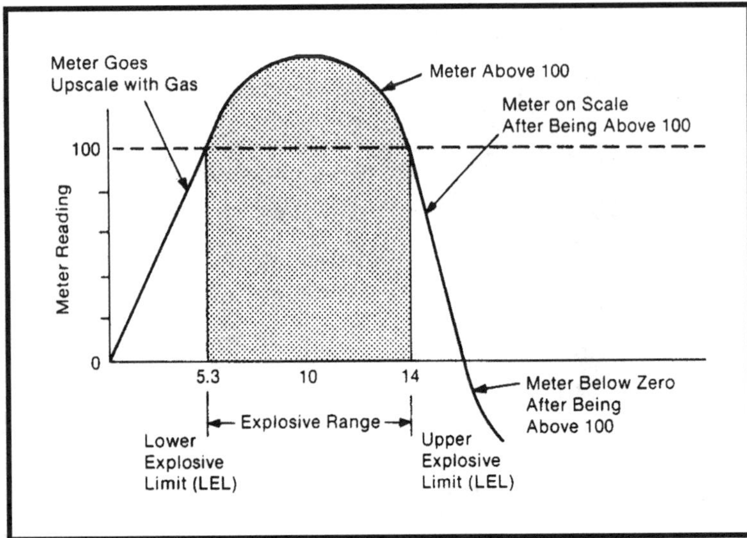

Figure 9.4 — Sensor response range from lower explosive limit to upper explosive limit.

Solid-State Sensors

Solid-state sensors are used to detect ppm and combustible concentrations of gases. These metallic oxide semiconductor (MOS) sensors can be used to detect a variety of compounds including nitro, amine, alcohol, and halogenated hydrocarbons, as well as a limited number of inorganic gases.

When mixtures are present in air, MOS sensors are best used as general survey instruments because they lack specificity and cannot distinguish between chemicals. The sensor will respond to a number of interfering gases and result in inaccurate meter readings. For example, sensors designed to monitor carbon monoxide may be sensitive to sulfur dioxide and hydrogen sulfide as well.

The advantages of MOS sensors include small size, low cost, and simplicity of operation. Disadvantages include lack of specificity, low sensitivity, and low stability. One manufacturer, AIM Safety Co. (Austin, Tex.) has a line of lightweight instruments that are about the size of a large flashlight, have warning lights, alarms, and multiple sensors. Although they are precalibrated for up to 33 chemicals and toxic and combustible gases, they should be used as general survey instruments unless the user is confident that there are no interfering chemicals present in the air.

Instruments with solid-state sensors are initially calibrated in the laboratory; field checks are performed for individual chemicals over several concentrations. Calibration of MOS sensors can be time consuming and more complicated than for other direct-reading instruments. Multiple concentrations must be tested several times over a range of concentrations because MOS sensors are not linear.

Photoionization Detectors (PIDs)

Portable photoionization detectors (PIDs) are general survey instruments used for detecting leaks, surveying plants to identify problem areas, evaluating source emissions, monitoring ventilation efficiency, evaluating work practices, and determining the need for personal protective equipment for hazardous waste site workers. These instruments are nonspecific and are often used to give qualitative information on the amount and class of chemicals present in air. Immediate qualitative results can be obtained for unknowns and mixtures of chemicals. A portable PID can be used around the perimeter of a leaking industrial operation or a hazardous waste site to determine airborne releases of total hydrocarbons into the community. PIDs can be used at hazardous waste sites to determine whether decontamination procedures are effective and for screening of air, water, soil, and drum bulk samples to determine priorities for further sampling.

Quantitative analysis by photoionization is based on the fact that most organic compounds and some inorganic compounds can be ionized when they are bombarded by high-energy UV light. These compounds absorb the energy of the light, which excites the molecule and results in a loss of an electron and the formation of a positively charged ion. The number of ions formed and the ion current produced is directly proportional to mass and concentration. Some compounds are ionized easily while others are not. The amount of energy required to displace an electron is called the ionization potential (IP). Each chemical compound has a unique IP; the higher the IP, the greater the amount of energy required to ionize the material. PIDs operate by collecting the ions on an electrode, amplifying the ion current formed (measured in electron volts), and translating the current into concentration. This process is represented by Equation 1.

$$RH + h\upsilon \rightarrow RH^{+} + e^{-} \quad (1)$$

where

RH = test gas
hv = photon with UV light energy greater than or equal to the IP of the test gas
RH^{+} = positively charged ion
e^{-} = electron

PIDs use a pump or fan to draw the air sample into a UV lamp that is either contained within the housing of the instrument or is located in a separate probe. Lamps of different energies are available; the energy of the lamp determines whether a particular chemical will be ionized. When the UV light energy is greater than the IP of the chemical, ionization will occur. In theory, if the UV light energy is less than the IP of the chemical, there will be no ionization. In practice, a lamp can detect chemicals with IPs that are higher than the rating on the lamp because the rating (in eV) refers to the major emission line of the lamp. However, sensitivity drops off rapidly and is not easily defined or quantified.

The lamps contain a low pressure gas through which a high potential electrical current is passed. By varying the composition of the gas in the lamp, manufacturers can develop lamps with different energy levels. Hydrogen is used to emit 10.2 eV, nitrogen for 10.6 eV, and argon for 11.7 eV. There are only a few lamps available: 9.5 eV, 10.0 eV, 10.2 eV, 10.6 eV, 11.7 eV, and 11.8 eV. When the sample is ionized, the electrical signal is displayed on an analog or digital output. The output does not distinguish between chemicals; it just detects an increase in the ion current.

To use PIDs, the occupational hygienist must know the IP of the chemicals suspected of being present in the air and the IP (eV capacity) of the lamp in the PID. The instrument should not be used to detect compounds with IPs greater than the lamp's capacity. The IPs of most organic compounds are less than 12 eV (see Table 9.4). The IPs of the atmospheric gases nitrogen, oxygen, water, carbon monoxide, and carbon dioxide are greater than 12 eV, and the IP of methane is 12.98 eV. Thus, measurements of contaminants in ambient air will not be affected by these major components of air.

It is possible to use PIDs quantitatively if only one chemical is present in air or if a mixture of chemicals is present and each chemical has the same IP. PIDs are more sensitive to complex compounds than to simple ones. They can detect a range of organic chemicals and some inorganic chemicals, including aromatics, unsaturated chlorinated hydrocarbons, aldehydes, ketones, ethylene oxide, hydrogen sulfide, and glycol ether solvents. Sensitivity increases as carbon number increases and is affected by the functional group (alcohol, amine, halide), structure (straight, branched, cyclical), and type of bond (saturated, unsaturated). In addition to chemical structure, the lamp intensity will also affect the sensitivity of the instrument to a given contaminant. For example, an 11.7 eV lamp is one tenth as sensitive to benzene, which has an IP of 9.25 eV, as a 10.2 eV lamp. Manufacturers usually have charts that list the sensitivity of their lamps to various chemical compounds. The sensitivities are related to a meter response at a known concentration (usually 10 ppm) of the contaminant.

The choice of which UV lamp to use in the instrument depends on the contaminants present in the air as well as the required stability of the lamp. Using lamps with different energies can achieve some degree of selectivity. The 11.7 eV lamp is capable of ionizing the largest number of chemicals and detecting the most chemicals; thus, it has the least selectivity. However, it can be used to selectively detect chemicals with IPs greater than 10.6 eV, such as chlorinated hydrocarbons (methylene chloride, chloroform, and carbon tetrachloride). A 9.5 eV lamp will selectively detect aromatic hydrocarbons such as benzene, xylene, and toluene in atmospheres where there are alkanes and alcohols that have IPs of greater than 10 eV.

In terms of lamp stability, the 11.7 eV and 11.8 eV lamps have the shortest usable lives. The windows in these lamps are made of lithium fluoride, which can transmit high energy, short wavelength UV light; however, lithium fluoride is hygroscopic and is degraded by UV light. Therefore, the windows will absorb water and swell, and light transmission will decrease, even when the instrument is not in use. The more the instrument is used, the greater the damage by UV light. The combination of water vapor and UV damage results in a service life of 3–6 months. Lamps emitting 9.5 eV,

Table 9.4 — Ionization Potentials of Selected Chemicals

Compound	*IP (eV)*
Acetaldehyde	10.21
Acetic acid	10.37
Acetone	9.69
Acrolein	10.10
Allyl alcohol	9.67
Allyl chloride	10.20
Ammonia	10.15
Aniline	7.70
Benzene	9.25
Benzyl chloride	10.16
1,3-Butadiene	9.07
n-Butyl amine	8.71
Carbon disulfide	10.13
Chlorobenzene	9.07
Crotonaldehyde	9.73
Cyclohexane	9.98
Cyclohexanone	9.14
Cyclohexene	8.95
Diborane	11.4
1,1-Dichloroethane	11.06
Dimethyl amine	8.24
Ethyl acetate	10.11
Ethyl amine	8.86
Ethyl benzene	8.76
Ethyl bromide	10.29
Ethyl butyl ketone	9.02
Ethyl chloride	10.98
Ethylene chlorohydrin	10.90
Heptane	10.07
Hydrogen cyanide	13.91
Hydrogen sulfide	10.46
Isoamyl acetate	9.90
Isoamyl alcohol	10.16
Isopropyl acetate	9.99
Isopropyl alcohol	10.16
Isopropyl amine	8.72
Isopropyl ether	9.20
Methanol	10.85
Methyl acetate	10.27
Methyl acrylate	10.72
Methyl ethyl ketone	9.53
Methyl mercaptan	9.44
Morpholine	8.88
Nitrobenzene	9.92
Octane	9.9
Pentane	10.35
2-Pentanone	9.39
Phosphine	9.96
Propane	11.07
n-Propyl acetate	10.04
n-Propyl alcohol	10.20
Propylene dichloride	10.87
Propylene oxide	10.22
Styrene	8.47
Toluene	8.82
Triethylamine	7.50
Vinyl chloride	10.00
Water	12.61
m-Xylene	8.56

10.2 eV, and 10.6 eV are made of magnesium fluoride; the 10.0 eV lamp has windows made of calcium fluoride. These windows are much more stable than lithium fluoride, are not degraded by UV light, and are not hygroscopic.

All PID readings are relative to the factory calibrant gas; the manufacturer adjusts a span setting (potentiometer) so that the PID reads directly for a defined concentration of a known chemical. For chemicals other than the calibrant gas, the meter reading is not the actual concentration. Therefore, meter responses to chemicals other than the calibrant gases should be recorded as ppm-calibrant gas equivalents, not as ppm. PIDs cannot detect all contaminants present in air; therefore, when there is no meter reading it does not mean all contaminants are absent. The typical range of concentrations that can be detected is 0.2 to 2000 ppm for the calibration gas, and readings are usually linear to about 600 ppm. Measurements above 600 ppm are usually not linear and may underestimate actual concentrations. Benzene and isobutylene are most often used to calibrate PIDs.

Factory calibration is performed under optimal conditions; therefore, if the lamp ages or becomes contaminated, the instrument must be recalibrated. The only way to ensure the instrument is working properly is to use the span gas (factory calibrant gas) at multiple concentrations. Some models have microprocessor-controlled calibration procedures. If a single compound is known to be present in the air, it is possible to calibrate the instrument to that chemical to make the readings relevant. Another option is to take the manufacturer's response factor for that chemical and use it to adjust the concentration obtained with the PID. For example, if a chemical has a response factor of 0.5 when using a particular lamp, the user would multiply the meter reading (for example, 50) by 0.5 to get 25 ppm.

PIDs are adversely affected by humidity, particulates, and hot and corrosive atmospheres. Although water vapor has an IP of 12.59, it can cause lamp fogging and deflect, scatter, or absorb light.[12] When UV light is scattered, less light reaches contaminants for ionization, resulting in lower meter readings. It is particularly important to calibrate PIDs in the environment where the equipment will be used (outdoors versus an air-conditioned interior). Dust particles can cover a UV lamp and decrease the amount of UV light transmitted; therefore, most PIDs have an inlet filter to decrease the possibility of particulates entering the ionization chamber. However, if small particles pass through the filter, they can scatter the light and decrease ionization. If the particles are charged, they can act as ions and result in elevated readings or erratic measurements. Hot gases can condense on the lamp or within the instrument and decrease meter readings. Corrosive gases can permanently fog or etch the lamp window or deteriorate the electrodes.

The instruments should be zeroed in the field before obtaining measurements. Clean air always contains some ionizable materials, and readings will always vary depending on what is in the air. However, readings will usually be approximately 0.2 units. If background readings are very high, noncontaminated factory air (zero or hydrocarbon free air) should be used. There are several manufacturers of PIDs including Photovac, MSA, RAE, and HNU.

PIDs from HNU Systems, Inc. (Newton, Mass.) are factory calibrated to benzene, have a separate probe that houses the UV lamp, and a box that contains the readout display and controls (see Figure 9.5). The HNU 101 models have three interchangeable lamps (11.7 eV, 10.2 eV, and 9.5 eV) and three ranges selected by the user (0–20, 0–200, and 0–2000 units). The instruments weigh from 7 to 13 pounds and have a lead-acid gel rechargeable battery that provides at least eight hours of continuous use. The UV probe and readout are factory calibrated together, and it is important that they remain together, or the instrument may not be accurately calibrated.

For most of the units, a small fan inside the probe draws air in, and the response is read on an analog display. The HW-101 model is designed for hazardous waste site monitoring and has a moisture-resistant ion chamber that uses a positive displacement

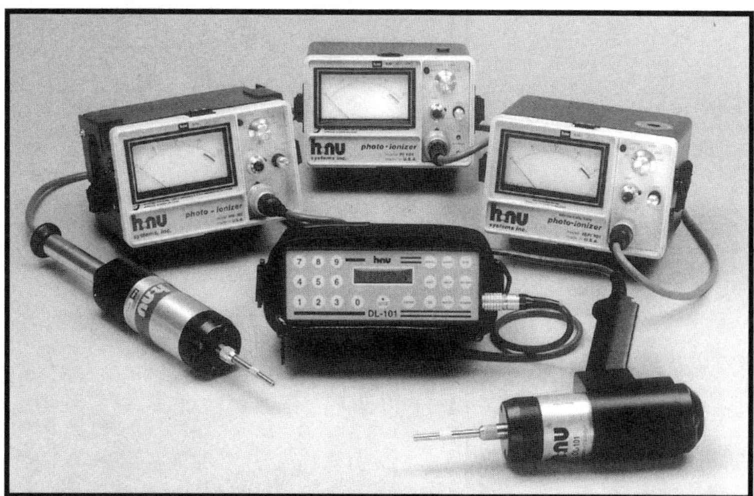

Figure 9.5 — PID: The HNU 101 models.

pump and inlet filter to draw the samples into the probe. The DL-101 model is microprocessor controlled and has data-logging ability.

The Photovac PID (Photovac Monitoring Instruments, Deer Park, N.Y.) does not have a separate probe and meter; it is a single, lightweight unit with a pump, in-line filter, and digital LCD readout. The Photovac 2020 has an advanced microprocessor that can store and receive data for up to 12 hours of sampling (see Figure 9.6). Photovac units offer a range of lamps including 8.5 eV, 9.5 eV, 10.2 eV, 10.6 eV, and 11.7 eV. When a new lamp is installed in the instrument, the unit must be reprogrammed for accurate calibration. Mine Safety Appliances Co. (Pittsburgh, Penn.) (MSA) has a PID unit called the photon gas detector that is similar to the Photovac 2020; however, it is only available with the 10.6 eV lamp. Both units are factory calibrated with isobutylene. Thermo Environmental Instruments, Inc. (Franklin, Mass.) manufactures an OVM Model 580 that can be equipped with a 9.6 eV, 10.0 eV, 10.6 eV, or 11.8 eV lamp and has a variety of data-logging and data retrieval options. It has a field battery life of eight hours.

Passive PID monitors are now available. These units do not use an active sampling pump and are typically much smaller and lighter than other units. RAE Systems (Sunnyvale, Calif.) offers a personal PID that can be mounted in the worker's breathing zone. A new instrument from RAE combines PID with LEL, O_2, and toxic sensors (five-gas monitor).

Figure 9.6 — PID: The Photovac 2020.

Flame Ionization Detectors (FIDs)

Portable flame ionization detectors (FIDs) use a hydrogen flame as the means to produce ions, instead of the UV light used by PIDs. FIDs can be used in similar situations as PIDs, with the exception of landfill sites because of sensitivity to methane, and when inorganic chemicals are expected. In general, FIDs tend to be more difficult to operate than PIDs; however, FIDs are less sensitive to the effects of humidity. FIDs respond to a greater number of organic chemicals than PIDs and are linear over a greater range.

Using a hydrogen flame, the energy is sufficient to ionize materials with IPs of 15.4 eV or less. FIDs respond to virtually all compounds that have carbon-carbon or carbon-hydrogen bonds; therefore, they respond to nearly all organic vapors. The sensitivity of the FID to vapors depends on the energy required to break chemical bonds (the sensitivity of PIDs is related to the IP of the contaminant). Similar to PIDs, the response of the FID will vary depending on the particular chemical. The presence of functional groups will affect sensitivity because they alter the chemical environment of the carbon atom. For example, FIDs respond well to aromatic hydrocarbons, moderately to chlorinated hydrocarbons, poorly to short-chained alcohols, and not at all to formaldehyde (see Table 9.5). Although the detector response is proportional to the number of molecules, the relationship is not linear and is even skewed with organic compounds containing oxygen, nitrogen, sulfur, or chlorine.

The FID has a burner assembly in which hydrogen is mixed with the incoming sample gas stream, and this is fed to the jet where ignition occurs (see Equation 2). FIDs are essentially carbon counters; the organic chemicals present in the gas sample stream form carbon ions. The positively charged carbon ions formed are collected on a negatively charged electrode. The current generated is proportional to the number of ions and concentration.

$$RH + O_2 \rightarrow RH^+ + e^- \rightarrow CO_2 + H_20 \qquad (2)$$

where

Table 9.5 — Relative Response of the OVA (FID Instrument) to Different Chemicals (if Calibrated to Methane)

Compound	*Relative Response (%)*
Acetaldehyde	25
Acetic acid	80
Acetone	60
Acetylene	225
Acrylonitrile	70
Benzene	150
n-Butane	63
1,3-Butadiene	28
Carbon tetrachloride	10
Chloroform	65
Cyclohexane	85
Diethylamine	75
Diethyl ether	18
Ethane	110
Ethanol	25
Ethyl acetate	65
Ethylene	85
Ethylene oxide	70
Hexane	75
Isopropyl alcohol	65
Methane (calibrant)	100
Methanol	12
Methyl ethyl ketone	80
Methylene chloride	90
Octane	80
Phenol	54
Tetrachloroethylene	70
Toluene	110
Trichloroethylene	70
Vinyl chloride	35
o-Xylene	116

RH = test gas
O_2 = is necessary for combustion to occur
RH^+ = positively charged ion
e^- = is an electron.

The FID is insensitive to water, nitrogen, oxygen, and most inorganic compounds, and has a negligible response to carbon monoxide and carbon dioxide. This insensitivity to ambient gases makes the FID extremely useful in the analysis of atmospheric samples.

As with the PID, measurements using the FID are relative to the factory calibrant gas, methane. The meter responds accurately to methane but does not distinguish between methane and other gases. When an FID meter reads 75, it means either that there is 75 ppm of methane in the air or another chemical that is equivalent to 75 ppm of methane. Therefore, readings for all other chemicals should be recorded as ppm-methane equivalents. Methane is used because it is less expensive than most other hydrocarbon gases and is linear over a wide range of concentrations.

It is possible to perform qualitative analysis only when the atmosphere is complex and contains two or more organic compounds. The FID response will not represent the concentrations of specific organic compounds, but rather an estimate of the total concentration of volatile organic compounds. If only one organic contaminant is present, it may be possible to quantify the contaminant if the FID has been calibrated for that specific contaminant.

Before field operation, FIDs should be calibrated with a methane concentration within the concentration range expected in the field. A one-point calibration curve is usually sufficient because the instruments are linear up to 10,000 ppm. However, if concentration ranges larger than 1 to 10,000 ppm are expected, or for instruments with several measurement ranges, a multipoint calibration should be performed. This large dynamic range (up to 10^7) permits field analysis over a wide range of concentrations without need for the sample to be diluted.

To zero FIDs in the field, a background reading is first obtained without the flame being lit. The user zeros the instrument on the most sensitive setting; a stable meter reading denotes proper operation of the collecting electrode in the combustion chamber. After the instrument has been stable for one minute, the flame can be ignited. To minimize contamination, it is important to use fuel that is ultra high purity; for hydrogen this corresponds to 99.999% hydrogen and less than 1% total hydrocarbon contamination. After ignition the meter reading will rise quickly then fall and should stabilize within 15 minutes. The background reading will depend on the location and is usually is 1–5 ppm as methane. This corresponds to approximately 1–1.5 ppm from methane in ambient air and 2–4 ppm from contaminants in the fuel. If clean, ambient air is not available, an activated charcoal filter can be attached to the sample inlet to remove all hydrocarbons except methane and ethane with molecular sizes too small to be trapped. When compared with a PID, the background reading for an FID will be higher because an FID responds to more contaminants.

FIDs usually have inlet particulate filters to prevent large particles from entering the combustion chamber. These filters should be changed frequently, or the particulates can adsorb contaminants and then slowly desorb and release them into the air. This will impact recovery of the instrument after sampling and background measurements. During combustion, fine particulates can be formed in the combustion chamber; over time, these charged particles can build up and attach themselves to the collecting electrode. This will result in erratic meter readings before the flame is ignited.

Some FIDs come equipped with a gas chromatography column so that they can be operated in a general survey mode (FID only) or a gas chromatograph (GC) mode. The GC mode is used for identification and quantification when there are a limited number of known contaminants in the air. The user must first calibrate the GC, as discussed in the section on portable GCs, to establish a calibration curve and peak retention times.

The Foxboro Co. (Foxboro, Mass.) manufactures three portable FIDs called organic vapor analyzers (OVA): the Century OVA 88, Century OVA 108, and Century OVA 128GC (see Figure 9.7). The OVA 88 has a detection range of 1 to 100,000 ppm methane-equivalents displayed on a logarithmic scale. The OVA 108 has a detection range of 1 to 10,000 ppm methane-equivalents displayed on a logarithmic scale and can be equipped with a GC. The OVA 128 GC has a detection range of 1 to 1000 ppm methane-equivalents displayed on a linear scale (the ranges are 1 to 10, 1 to 100, and 1 to 1000 methane-equivalents) and has a GC. Each of the three models has a portable battery that gives approximately eight hours of operation and weighs approximately 12 pounds. The OVA 108 and OVA 128 are currently the only FIDs on the market that are certified for intrinsic safety. The manufacturer does not recommend calibration of this equipment from a gas cylinder. They recommend placing the calibration gas in a sample bag, then withdrawing the sample from the bag into the instrument.

Foxboro carries the TVA 100 vapor analyzer, which has both PID and FID detectors, data-logging capabilities, and can operate for eight hours using a battery. The FID and PID may be used individually or concurrently. The PID uses a 10.6 eV lamp;

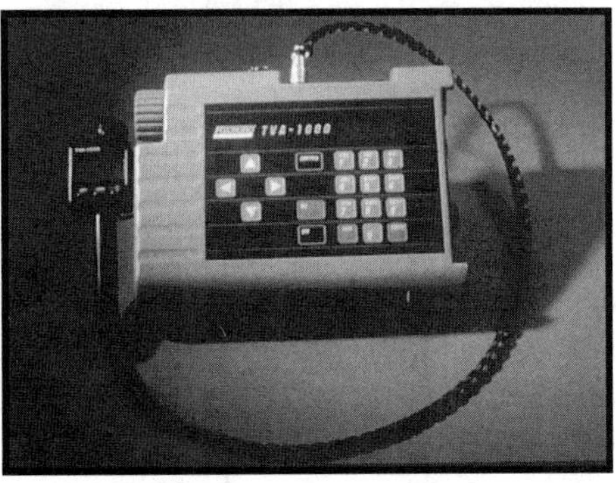

Figure 9.7 — Organic vapor analyzers (OVA): the Century TVA 1000.

Foxboro states that the PID has a range of up to 2000 ppm and the FID has a range up to 50,000 ppm.

Sensidyne, Inc. (Clearwater, Fla.) offers two FID models with GC capabilities; one ranges up to 10,000 and the second up to 1000 ppm methane equivalents. Heath Consultants (Houston, Tex.) manufactures FIDs that are designed for outdoor natural gas pipeline surveys and can measure up to 10,000 ppm methane equivalents.

Monitoring Multiple Gases and Vapors

IR Gas Analyzers

Direct-reading infrared gas analyzers are versatile, can quantify hundreds of chemicals, and are capable of being used for continuous monitoring, short-term sampling, and bag sampling. Advantages of portable IR analyzers are that they can measure a wide variety of compounds, usually as low as 1–10 ppm; and the equipment is easy to use, can be set up quickly, and is relatively stable in the field. These systems have been used to detect leaks of ethylene oxide in hospital sterilization units, and nitrous oxide and other anesthetic gases in hospitals and dentists' offices.[13] IR analyzers are often used in indoor air investigations to measure the buildup of carbon dioxide; carbon dioxide used as a tracer gas can indicate whether the ventilation system is operating efficiently.

While portable IR monitors can be used for personal exposure monitoring, they are better suited to source monitoring because their size usually restricts their use to a fixed position. However, they can be used to supplement personal monitoring when air samples collected in Tedlar bags from the breathing zone of a worker or a source are analyzed with an IR monitor. IR analyzers also have been combined with videotaping to evaluate work practices and to train workers on the impact of work practices on exposure.

While it is possible to identify and quantify unknowns, most instruments are used to quantify known gases and vapors. Identification of unknowns requires a great deal of expertise in interpreting IR spectra, as well as a large IR spectral library of various compounds.

Quantitative analysis by IR spectrometry is based on the principle that compounds selectively absorb energy in the IR region of the electromagnetic spectrum. The mid-IR region (2.5–15.4 microns) is the commonly used wavelength range for quantification. Each compound is comprised of atoms that bend, twist, rotate, and impart a vibrational

Table 9.6 — Specific Infrared Absorption Bands for Hydorcarbons

Chemical Groups	*Absorption Band (microns)*
Alkanes (C-C)	3.35–3.65
Alkenes (C=C)	3.25–3.45
Alkynes (C_C)	3.05–3.25
Aromatic	3.25–3.35
Substituted aromatic	6.15–6.35
Alcohols (-OH)	2.80–3.10
Acids (C-OOH)	5.60–6.00
Aldehydes (COH)	5.60–5.90
Ketones (C=O)	5.60–5.90
Esters (COOR)	5.75–6.00
Chlorinated (C-Cl)	12.80–15.50

Source: Reference 7

energy unique to the compound. Many of the vibrational energy transitions observed in IR spectra are characteristic of certain functional groups (see Table 9.6). When a compound interacts with IR radiation, the pattern of energy absorption, called the absorption spectrum, represents vibrational patterns from the atoms and the functional groups. The characteristic absorption spectrum that is produced can be used to identify the chemical and is considered to be a fingerprint. Figure 9.8 shows an infrared spectrum for the chemical sulfur hexafluoride. The selective absorption of light at specific wavelengths allows the user to focus on defined areas of the absorption spectra, especially where absorption is optimal and where there are no potential interferences from other chemicals.

While a compound can be identified from the spectral pattern of absorption, quantification of a compound from the spectra is possible because the intensity of the IR absorption is proportional to the concentration of the chemical and the distance (path length) that the light travels through the chemical. The longer the path length that the IR light travels before it reaches the detector, the greater the sensitivity and the lower the limit of detection. This is described mathematically by the Bougher-Beer Lambert law as shown in Equation 3.

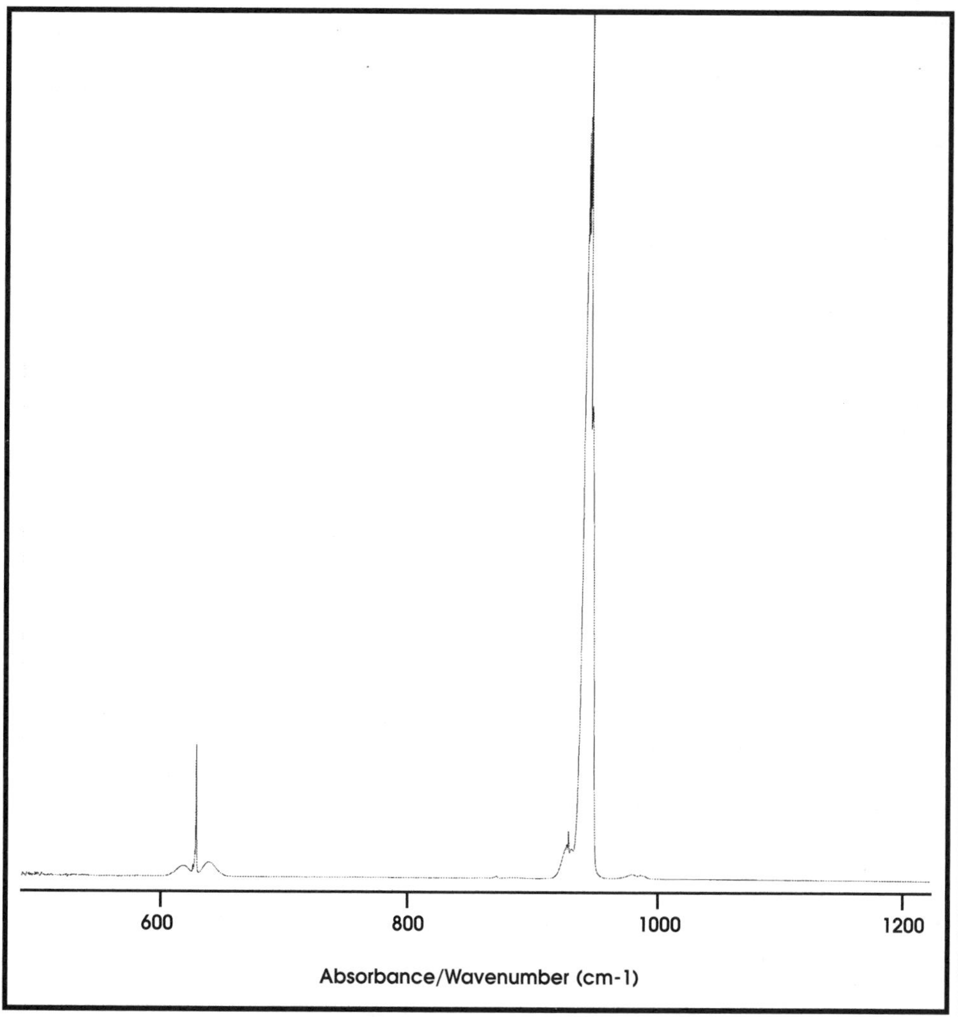

Figure 9.8 — An infared spectrum for the chemical sulfur hexafluoride. Y axis is absorbance unit.

$$I/I_o = \exp^{-\mu CL} \quad (3)$$

where

I = intensity of IR light with molecular absorption
I_o = intensity of IR light without molecular absorption
μ = molecular absorption coefficient of the chemical
C = concentration of the chemical
L = path length of the IR radiation through the chemical

IR spectrometers fall into two categories: dispersive and nondispersive. Dispersive instruments use gratings or prisms to disperse the transmitted beam of IR radiation and are most often used in the laboratory. Nondispersive instruments do not use gratings or prisms, so the entire IR beam passes through the sample. Commercially available portable IR gas analyzers are nondispersive, single beam instruments with a sample inlet, an IR radiation source (usually a Nichrome™ wire resistor), single or multiple filters to select the wavelength of radiation that passes through the sample, a closed sample cell with a fixed or variable path length, and a detector. Path lengths are achieved from 1 to 20 m using mirrors to reflect or fold the IR radiation within the sample cell. This variability in path length is used to compensate for the differences in absorption coefficients of chemicals that affect the detection limits. In the field, the instrument is first zeroed in contaminant-free air before the sample air is drawn into the gas cell by the pump. If contaminant-free air is not available, a charcoal canister, or zeroing charcoal cartridge, can be used to filter the air before it enters the instrument. This cartridge is similar to the cartridges used in respirators. The cartridge must be kept in a plastic bag to avoid contamination. If the zeroing cartridge is saturated with contaminants, and the concentration in air is lower than the concentration through the cartridge, the concentration of contaminant in the sample air may appear negative. IR analyzers are usually equipped with particulate filters to keep dust out of the sample cell, because dust will decrease sensitivity of the instrument by scattering IR light and settling on the mirrors and windows.

IR analyzers are relatively simple to operate and can be dedicated to a single compound or used to detect multiple compounds. For multiple compounds, identification and quantification using portable IR analyzers is usually performed by selecting a limited region in the IR where the compound absorbs strongly and ideally does not interfere with other compounds present in the air. If the compounds in mixtures have overlapping peaks, concentrations will appear falsely elevated. This problem may be overcome by taking a measurement at a secondary peak where the compounds do not overlap. Instruments are usually provided with a chart containing suggested path lengths and wavelengths for specific compounds at specific limits of detection. Therefore, the occupational hygienist must know the nature of both the target compounds and the environment being monitored. With this knowledge it may be possible to select wavelengths with minimal interferences. Fortunately, neither oxygen nor nitrogen interfere with IR measurements; however, water vapor is a major interferer. Measuring a compound in the region where water vapor absorbs infrared light can result in overestimation of concentrations.

IR analyzers need to be calibrated with chemicals when they are first purchased as well as before field use with a multipoint calibration curve. The pump flowmeter also needs to be calibrated when the instrument is first purchased and at regular intervals. The initial calibration ensures that the sample cell and mirrors have not been damaged and the alignment has not changed during shipping. A multipoint calibration curve of absorbance versus concentration (ppm) is created by either flowing a known concentration of a gas through the cell or by generating concentrations by injecting small amounts (microliter) of a known concentration of a gas into the cell using a closed loop system. Small amounts are injected to minimize pressurization of the cell. Calibration curves should be prepared using the same instrument settings that will be used in the field.

Widely used portable IR analyzers, the MIRANs (miniature IR analyzers), are manufactured by the Foxboro Co. The MIRAN Sapphire Analyzers are microprocessor controlled, single beam instruments, that can measure several hundred different gases (see Figure 9.9). An air sample is drawn into the gas cell by a pump at a rate of 25–30 L/min. The sample absorbs IR energy from the Nichrome wire light source, the detector measures the decrease in light intensity, and this absorption is converted to ppm. The MIRAN Sapphire has a variable path length from 0.75 to 20.20 m. Measurement parameters for over 100 chemicals are stored in memory, and the instrument can be programmed for additional gases. These models have built-in libraries with optimum path lengths that can be programmed for measuring several compounds simultaneously and automatically converting absorbance to concentration. The user need only input the gas of interest and the Sapphire automatically sets the path length and the correct wavelength. This unit has an optional gas identification package that automatically compares an entire infrared scan to a library of 400 compounds stored in the computer's memory, and then lists compounds in descending order of matching probability.

The MIRAN 1A has a variable wavelength filter and a long path length cell that can be manually adjusted to detect nearly all organic and many inorganic gases. In contrast, the MIRAN 203 model is a specific vapor analyzer that measures only a single contaminant that is preselected by the user before purchase.

Photoacoustic Analyzers

Quantification of air contaminants by photoacoustic spectroscopy (PAS) involves the use of sound and UV or IR radiation to quantify air contaminants. This type of spectroscopy uses the fact that molecules vibrate at a particular frequency called the resonance frequency. The number and types of atoms determine a chemical's unique resonance frequency. The resonance frequency of most molecules is about 10^{13} Hz or 10^{13} vibrations per second. When molecules are exposed to IR radiation of the same frequency as the resonance frequency, energy is transferred to the molecule, and the molecule gains energy and vibrates more vigorously. The excess energy is then quickly transferred to the surrounding medium, which is usually air or helium, in the form of heat. The transfer of energy causes an increase in the temperature of the medium and produces an increase in pressure; this increase in pressure waves, or sound waves, is detected by a microphone. The intensity of the sound emitted depends on the concentration of the contaminant and the intensity of the incident light.

Similar to analysis by infrared spectroscopy, when a contaminant is irradiated with IR light in the mid-IR region (2.5–15.4 microns), the pattern of energy absorption at specific wavelengths (the fingerprint) can be used to identify the chemical. The intensity of the absorption is proportional to the concentration of contaminants. The fundamental difference between IR spectroscopy and PAS is that in PAS the

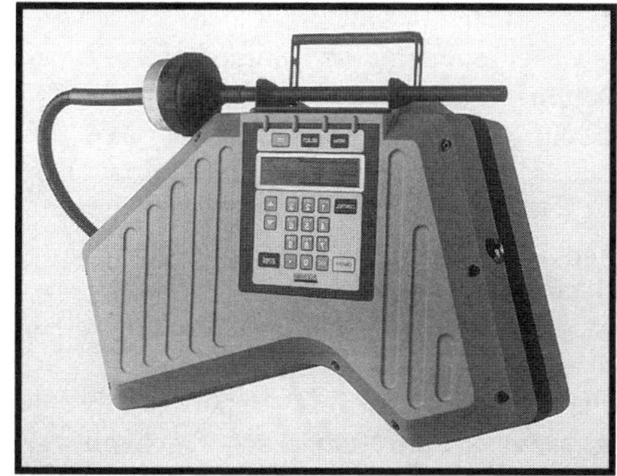

Figure 9.9 — Portable infared analyzer: the MIRAN Sapphire Analyzer.

amount of light energy that is absorbed is measured directly by measuring the sound energy emitted. Interferences by carbon dioxide, water vapor, and other chemicals present in the air, which absorb at wavelengths similar to those used to measure the chemical, will compromise both the limit of detection and the accuracy of the measurement.

A typical photoacoustic analyzer consists of a radiation source that is emitted in pulses, an optical filter or filters to select wavelengths specific for the chemical, a photoacoustic cell filled with air or helium, a microphone, amplifier, recorder, and readout display. The use of optical filters limits the number of chemicals that can be evaluated at one time. Photoacoustic analyzers are usually set up to detect a single chemical or up to five known contaminants. The filters are selected to coincide with a strong absorption band in the IR for the specific gas. Detection limits are chemical-specific and are reported to be between 0.001 and 1 ppm.[(14)]

Both portable and fixed PAS systems are available. Portable systems are frequently used to evaluate indoor air problems by measuring the growth or decay of tracer gases in air. This measurement can be used to calculate air exchange rates and study airflows in a building. Typical tracers include carbon dioxide and sulfur hexafluoride. Tracers can also be used to evaluate the efficiency of local exhaust ventilation systems. Fixed systems can be set up to collect data from multiple points in a workplace and can be used with an alarm to signify elevated concentrations of specific highly toxic contaminants.

Portable GCs

Traditionally, GCs have been laboratory-based instruments; however, portable versions of the GC are increasingly being used as on-site monitoring devices for hazardous waste sites and for environmental and occupational hygiene monitoring. A portable GC can be transported to the field and used under battery power. The use of portable GCs has allowed on-site measurements of vapor and liquid samples; this eliminates the need to collect samples on separate media, ship them to a laboratory, store them over time, and desorb them with other chemicals. Portable GCs are particularly good for identification of specific chemicals in mixtures and unknown chemicals and are best for monitoring volatile compounds.

To date, portable GCs have not been used for compliance monitoring by the Occupational Safety and Health Administration, and generally have not replaced personal sampling methods using collection media. The use of portable GCs for occupational hygiene monitoring is likely to increase in the future; in fact, NIOSH has developed several analytical methods that rely on portable GCs.

In general, a GC consists of an injection system, a GC column, and a detector (see Figure 9.10). An ambient air or liquid sample is usually injected into the system through a septum using a gas-tight syringe, or air is pulled directly into a sampling loop using a vacuum pump. Heated injection ports are used to prevent condensation of chemicals with low volatility and high boiling points, or when ambient temperatures are low.

Compounds are moved by a carrier gas stream from the injection port or sampling loop into the heart of the GC system, the gas chromatography column. The carrier gas can be high-purity air, hydrogen, helium, or nitrogen. The column is filled with packings and coatings that physically interact with the chemicals in the sample and enable separation of complex mixtures into individual compounds. Ideally, each compound will have a different affinity for the column material and will be slowed down to a different extent. The result is that individual compounds will be eluted from the column at different rates and, therefore, reach the detector at different retention times. The retention time of a contaminant depends on the type of column packing material, the length of the column, the flow rate of the carrier gas, and the temperature of the column.

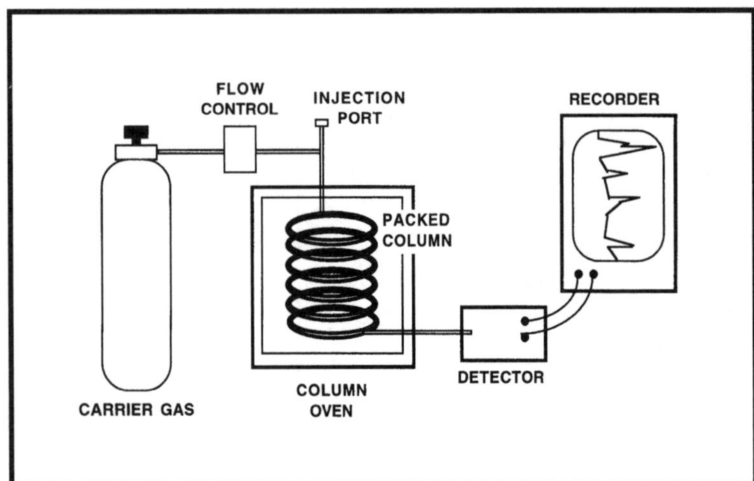

Figure 9.10 — Schematic for a typical gas chromatograph.

There are two different types of columns: packed and capillary. The choice of the appropriate column is essential to adequate resolution of the contaminants. Selection of an improper column can result in inadequate peak separation, broad peaks, tailing, or shouldering. Packed columns consist of a fine, inert, solid material (the support phase) that has a large surface area for holding a thin film coating of a nonvolatile liquid (stationary phase). Some columns have no liquid phase, and the solid material also serves as a solid adsorbent phase. Capillary columns are hollow tubes; the specially treated glass or fused silica walls of the tube serve as the support phase. The liquid stationary phase is coated or bonded to the walls. Columns vary in length from 6 inches up to 30 feet and are usually coiled to allow their length to fit into a portable instrument.

The operating temperature of a column can affect the performance of a GC. Portable GCs with unheated columns are susceptible to changes in ambient temperatures, which can affect retention times and reproducibility of results. A rule of thumb is that column temperatures should be kept at least 5º above ambient temperatures. Lack of temperature control will limit the resolution of compounds eluting from the column, the reproducibility of the results, the range of possible analyses that can be detected, and cause thermal drift. Thermal drift is the change in retention time of a compound over time. To minimize thermal drift, unheated columns should be used only for compounds with short retention times. Heated columns result in stable operating conditions and reproducible results within a large range of ambient temperatures. GCs that have columns enclosed in a temperature-controlled oven offer isothermal operation and enable regulation of temperature in 1ºC increments.

Most field GCs use a technique called backflushing of a column to allow rapid clearing of residual chemicals with long retention times so that the components do not contaminate the column and interfere with subsequent analyses. Two columns are used when backflushing, a short precolumn and a longer analytical column. The precolumn is used to trap heavy, less volatile compounds. After the volatile compounds have cleared the precolumn, the carrier gas flow is reversed, and the remaining contaminants are flushed out away from the detector. Backflushing can significantly shorten the required analytical time.

Most field GCs come with a single detector; the most commonly used detectors are the FID, PID, and electron capture detectors (ECD). The detectors vary in sensitivity, selectivity, and linearity. Table 9.1 lists the characteristics of these detectors. The choice of detector depends on the chemicals for which monitoring is needed, the presence of other contaminants, and the required sensitivity. A strip chart recorder or microprocessor is usually attached to the GC; the separated components detected are represented as peaks. Concentration is then determined automatically using microprocessors that integrate the area under the peak or manually by directly measuring the peak height, calculating the area under the peak using the formula for triangles, counting the number of squares underneath the peak, or cutting out and weighing each peak. In the field, retention time and area of the peaks are then compared with peaks generated during calibration of the instrument. Field operation of GCs requires that they be calibrated with the chemical of interest under the same conditions as the chemical that will be measured in the field.

One limitation of using a portable GC is that it requires a high degree of skill to operate. The operator must calibrate the instrument and know the relative retention times and peak height characteristics of each contaminant of interest. Therefore, known concentrations of individual calibration gases or calibration gas mixtures must be used. Calibration gases can be obtained from specialty gas suppliers and should be introduced into the GC to obtain retention times, sensitivities, and peak heights. Calibration should be repeated when operating conditions are changed.

When evaluating field samples, the operator must be aware that a single peak may represent multiple contaminants that have the same retention time; organic chemicals do not all have unique retention times. Peaks may be detected that cannot be identified without using other analysis methods. The absence of a peak does not ensure that the contaminant is absent; therefore, one of the operating parameters (column, gas flow, temperature) should be changed to confirm the negative results.

If syringes are used to inject low concentration standards or samples, the occupational hygienist should use glass syringes with stainless steel plungers. The sealing ring of the plunger should be Teflon or another material that has little or no affinity for organic compounds. Syringes need to be checked regularly for leaks or blockages and need to be purged with clean air. To avoid cross-contamination, syringes used for injecting calibration gases should not be used for injecting sample gases.

The most important aspect of making accurate GC measurements is repeatability and reproducibility. Therefore, operators should have a thorough quality assurance and quality control program. Proper field operation requires the use of correct carrier and calibration gases and a detector. Some instruments use internal gas cylinders that must be charged before going to the field or external gas cylinders and regulators. Transporting gas cylinders to and from the field can pose logistical problems, particularly when flammable gases such a hydrogen are needed for the FID. Regulations regarding air or land transportation of hazardous compressed gases must be taken into consideration when planning a survey. To reduce contamination of the sample, tubing that is used to connect gas cylinders to GCs should be made of Teflon or stainless steel.

A commonly used portable field GC is the Photovac 10S Plus (see Figure 9.11). This system is relatively lightweight (28 pounds), can use a variety of carrier gases, a photoionization detector, and can be operated for seven hours using an internal rechargeable battery. The system is capable of detecting contaminants as low as 1 ppb. The standard UV lamp used in the instrument is 10.6 eV, and optional 8.4 eV, 9.5 eV, 10.0 eV, and 11.7 eV lamps are also available. An on-board microprocessor allows the user to specify operating parameters, store methods and results, view analyses, and summarize results.

A smaller, hand-held portable GC (8 lbs), the Photovac Snapshot, is a factory-programmed GC designed to detect a limited number of air contaminants. The unit draws air into the instrument using a pump and automatically injects the air into the gas chromatography column; contaminant concentrations are then displayed on a small screen and can be stored in a data

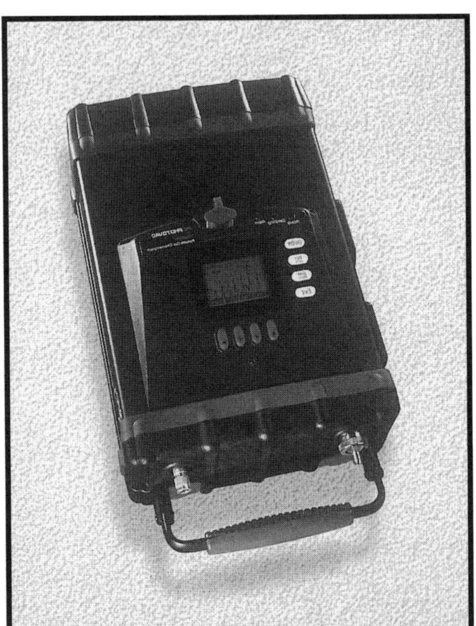

Figure 9.11 — Portable field gas chromatograph: the Photovac Voyager.

logger. The unit maintains isothermal control of column temperatures and uses precolumn backflushing. Interchangeable modules are used in the instrument for evaluating specific industries, such as the rubber industry, or specific chemicals, such as chlorinated hydrocarbons, aromatic hydrocarbons, hydrogen sulfide, and mercaptans. The Snapshot is useful for monitoring environments where chemicals are known or highly suspected to be present; it is not recommended for use in unknown environments. Contaminants that are not present in the selected module will not be identified, and the user will not be alerted to their presence.

The HNU model 301P (HNU Systems Inc., Newton, Mass.) is a portable GC that can be used in the field; however, it is primarily suited for fixed laboratory use. This system weighs around 25 lbs, and a separate battery support pack weighs another 16 lbs. In the field the model can be used for 10 hours. It can use two detectors (PID, FID, or others), which can be operated separately or in series; has a heated injection port; and has an isothermal oven that can hold two packed columns or one capillary column.

The Century OVA model 108 or 128 (The Foxboro Co.) is an FID with a rugged portable field GC option. The GC option includes an exterior gas chromatography column, a manual injection port, and backflush capabilities. In the GC, air contaminants are carried with hydrogen gas through the gas chromatography column. An analog display can be used to count the number of peaks eluting from the column. The operator must know the retention time of known compounds under field operating conditions. The OVA must be calibrated using the compounds of interest with the exact setup planned for use in the field including the column type, field temperature range, and expected concentration range.

Fourier Transform IR (FTIR) Spectrometers

Direct-reading instruments based on Fourier transform infrared spectrometry are on the forefront of monitoring technology for occupational hygiene and environmental applications.[15] These instruments have the potential to monitor a wide range of compounds simultaneously at very low limits of detection (ppb). An FTIR spectrometer can scan the same IR wavelength regions that a conventional filter-based IR spectrometer measures; however, it is much more efficient at collecting and analyzing the radiation, resulting in higher spectral resolution, greater specificity, higher signal-to-noise ratio, and lower limits of detection.[16] Spectral data obtained with the FTIR spectrometer can be used to identify unknown as well as known contaminants and can quantify chemicals in mixtures. Compounds are identified by their fingerprint.

At the heart of the FTIR spectrometer is the interferometer; most models use a Michelson interferometer to simultaneously detect the entire spectral region at once. Most other IR techniques scan the entire wavelength region one wavelength at a time. A simple schematic is provided in Figure 9.12. In its simplest form, the Michelson interferometer consists of a beam splitter and two mirrors, one of which is moveable. For the highest accuracy, the two mirrors should be perpendicular to one another. When a plane monochromatic wave is incident on the beam splitter of the Michelson interferometer, the amplitude of the light is evenly split into two paths toward the mirrors. The light is then reflected by the mirrors back to the beam splitter where they are recombined; interference results from the phase difference in the two IR beams. The phase difference depends on the wavelengths of light present in the beams. The detector records the intensity difference created between the stationary and moving mirrors as a function of the path difference between the two mirrors (the position of the moving mirror); this results in an interferogram (see Figure 9.13). When a Fourier transform is performed on the interferogram, a single beam spectrum is obtained, which is a plot of the intensity versus wave number (cm^{-1}) (see Figure 9.14).

When obtaining data, a clean air background spectrum is first obtained before taking the sample spectrum. A transmission spectrum is created by dividing the single beam of the sample spectrum by the background spectrum obtained in clean air. The clean air spectrum ideally removes spectral features due to chemical interferers and instrument characteristics. An absorbance spectrum is then obtained by taking the negative logarithm of the transmission spectrum. This absorbance spectrum is used for quantitative analysis by comparing it with a reference absorption spectrum for the chemicals of interest at known concentrations. Several techniques are used for quantification, including classical least squares and partial least squares.[17,18]

FTIR systems can be used in extractive or open-path modes. Extractive modes are becoming more commonly used; open-path modes are currently being evaluated and developed. In the extractive mode the FTIR spectrometer is set up similar to IR spectrometers and uses a closed long-path gas cell. The sample is drawn into a cell using a pump, and the cell can use mirrors to create multiple reflections of IR light through the sample to create a long path length. The FTIR measurements represent the concentration of contaminants in the sampled air. Extractive FTIR spectrometers have been used for a variety of industrial and environmental applications including near real-time monitoring of chemicals in a semiconductor industry, auto exhaust emissions, stack gases, chemical processes, and air pollution.[19,20] Extractive systems are calibrated by generating a multipoint calibration curve for known concentrations of calibration gases before field use.

In the open-path (remote sensing) mode, the gas cell is replaced by the open atmosphere. In the bistatic configuration, the IR source is placed at opposite ends of the optical path (as far as 500 m apart) from the detector. In the monostatic configuration, the light source and detector are positioned at the same end of the path; the beam is sent out to a retroreflector placed at the physical end of the path, which reflects the light directly back to the detector. Telescopes can be used to send and receive the beam to and from the interferometer to the retroreflector (monostatic systems) or to receive the light at the detector (bistatic systems); see Figure 9.15.

When pollutants present in the optical path are measured, the concentration is integrated over the entire length of the optical path and is reported as the product of concentration and path length (ppm-meters). The concentration is then divided by the path length to obtain a path-averaged concentration (ppm).

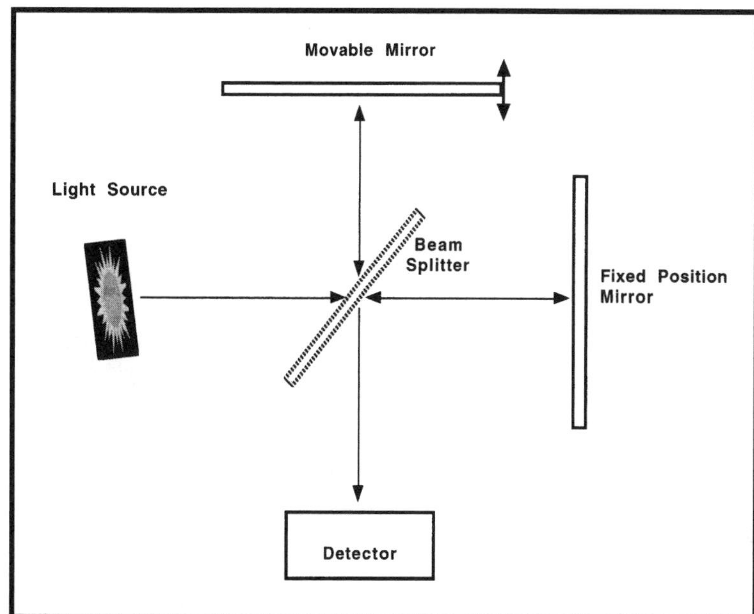

Figure 9.12 — A simple schematic of an FTIR interferometer.

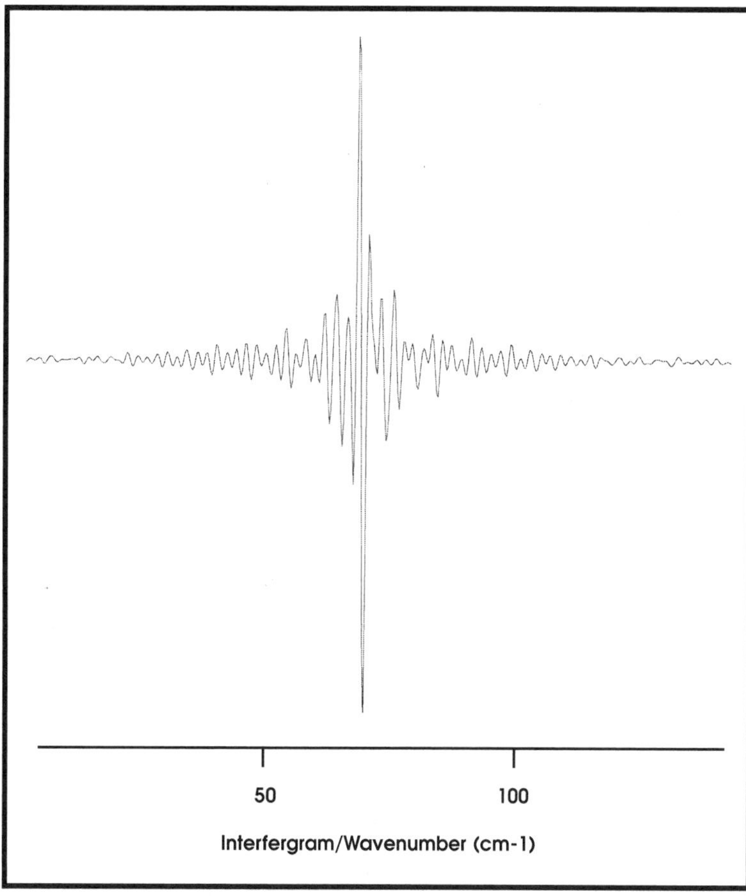

Figure 9.13 — A typical interferogram obtained with an FTIR spectrometer. Y axis is signal strength.

Open-path FTIR spectrometers have

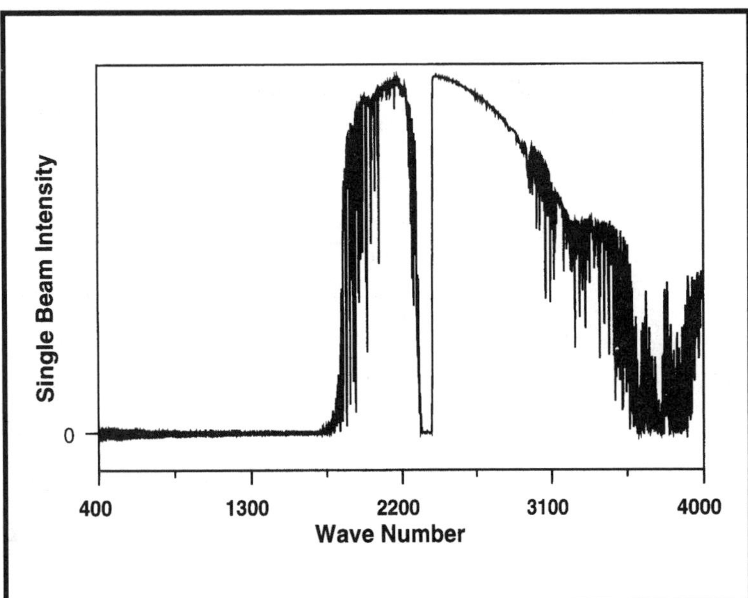

Figure 9.14 — A typical single beam spectrum obtained with an FTIR spectrometer.

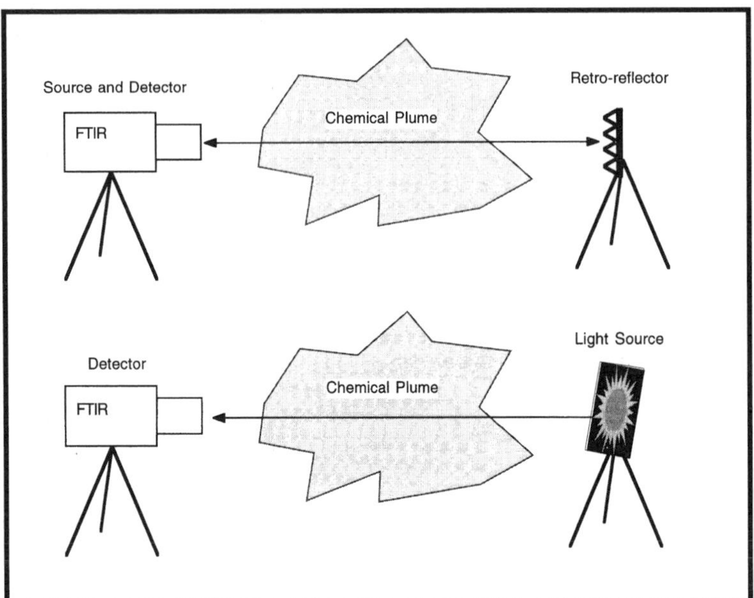

Figure 9.15 — Monostatic (top) and bistatic (bottom) open-path FTIR spectrometer systems.

only recently been used for environmental and occupational sampling. In the environmental arena, open-path FTIR spectrometers can provide real-time monitoring of industrial fence-line and stack emissions and process leaks. In the occupational arena, open-path FTIR spectrometers have the potential to provide the temporal resolution and sensitivity required for measuring short-term exposures to acutely toxic chemicals, and the spatial coverage to detect leaks of toxic chemicals and obtain TWA concentrations of complex mixtures over large areas in a room.[21–24] This technology is attractive because these instruments can simultaneously measure a wide variety of compounds noninvasively, in real-time, and over large areas that would normally require the use of large numbers of point sampling devices. There is no contact of the instrument with the sample, and there are no sampling lines, bags, or canisters that can adsorb compounds and result in unpredictable contaminant losses.

Calibration of open-path FTIR spectrometers poses a challenging problem. Unlike the extractive FTIR, there is no gas cell that can be purged with clean air to obtain contaminant-free background data and filled with known concentrations of contaminants to determine accuracy. A custom designed calibration cell, placed in the open-path, is one of the methods being developed for field calibration and verification.

In the field, a second challenge facing users of remote sensing equipment is the acquisition of a clean background spectrum under the same experimental conditions as those taken for the sample spectrum.[25,26] The clean air spectrum ideally removes spectral features present due to chemical interferers and instrument characteristics. However, in most occupational and environmental applications, it is difficult to find air that is free of target analytes. Even if a clean air background can be obtained at the start of the day, instrument fluctuations and changing environmental conditions (e.g., partial pressure of water vapor) over the course of the day, may require obtaining additional backgrounds for greater accuracy. In the workplace, during a work shift, it may be impossible to obtain a new background spectrum once the chemical processes are underway. The impact on accuracy of using a background taken under different environmental conditions from the sample will depend primarily on the chemical being detected. For example, changes in the partial pressure of water vapor during a day can result in a significant quantification error for toluene because absorption peaks are located in regions where there is considerable water absorption.[24] Research is underway to develop quantification methods that do not require the acquisition of clean background spectra.

Although open-path FTIR spectrometers can provide good temporal resolution of concentrations, spatial resolution is limited because of the long optical path length over which the concentration of chemicals are averaged to provide a path-integrated output of the instrument. However, when a network of intersecting open-path FTIR spectrometers are used in a room, a technique called computed tomography can be applied to the measurements to create both spatially and temporally resolved concentration distribution maps (see Figure 9.16). These maps could be used for exposure assessment, source monitoring, leak detection, and ventilation evaluation. This technique is currently being evaluated and developed for the occupational hygiene application.[27–31]

Detector Tubes

Detector tubes, or colorimetric indicator tubes, are the most widely used direct-reading devices due to their ease of use, minimum training requirements, fast on-site results, and wide range of chemical sensitivities. The first detector tube was developed in 1917 at Harvard University for measuring concentrations of carbon monoxide.[7] Since then detector tubes have been developed for the rapid detection, identification, and quantification of hundreds of chemicals.

A detector tube is a hermetically sealed glass tube containing an inert solid or granular material such as silica gel, alumina, pumice, or ground glass. The inert material is impregnated with one or more reagents that change color when they react with a chemical or group of chemicals. The reagents used in the tubes are selected for specific chemicals or classes of chemicals and may vary by manufacturer. When air is drawn through the tube, the length of the resulting color change (or stain), or the intensity of the color change, which occurs in 1 to 2 min, is compared with a reference to obtain the airborne concentration. There are three methods used to convert color change to concentration: (1) use the calibration scale marked directly on the detector tube, (2) use a separate conversion chart, or (3) use a separate comparison tube. The easiest tubes to use are those with a concentration scale marked directly on the tube (see Figure 9.17).

Most detector tubes contain a glass wool or cotton filter at each end to prevent particulates or aerosols from entering the tubes. Some tubes contain a prelayer that adsorbs interferences, such as water vapor or other chemicals. To use a detector tube, the ends of the tube are broken, and the tube is placed in a bellows, piston, or bulb-type pump. Both the tubes and the pumps are specially designed by each manufacturer; therefore, interchanging equipment from different manufacturers will result in significant measurement errors. When a pump stroke is performed, air is drawn through the tube at a flow rate and volume determined by the manufacturer. A specified number of strokes is used for a given chemical and detection range. Total pump stroke time can range from several seconds to several minutes. A new color detection system from National Draeger (Pittsburgh, Penn.) uses a battery-operated constant mass flow pump and chemical-specific chips that contain 10 capillary tubes. The color change in the capillary tube is read electronically; the concentration is displayed as a digital readout. A limited number of chemicals can be measured with this system at this time (see Figure 9.18).

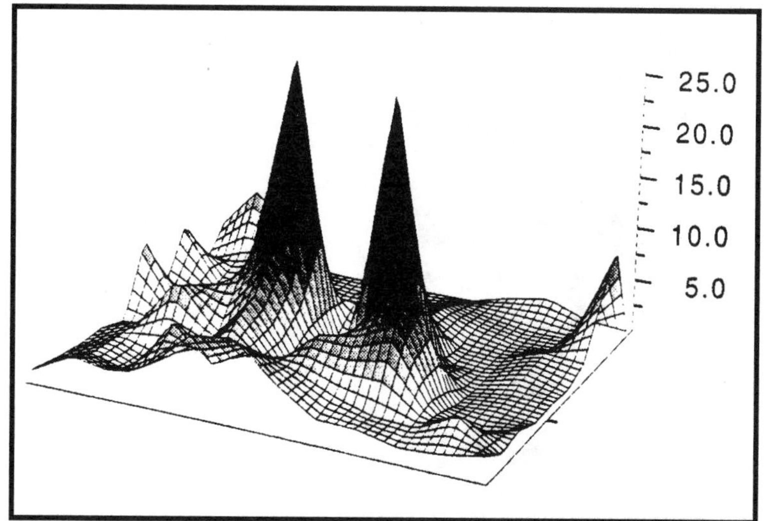

Figure 9.16 — An example of a concentration distribution map created with a computed tomography system. The height of the peaks (z axis) represents the concentration of a chemical in a room. The xy location is the location of the peaks in a room.

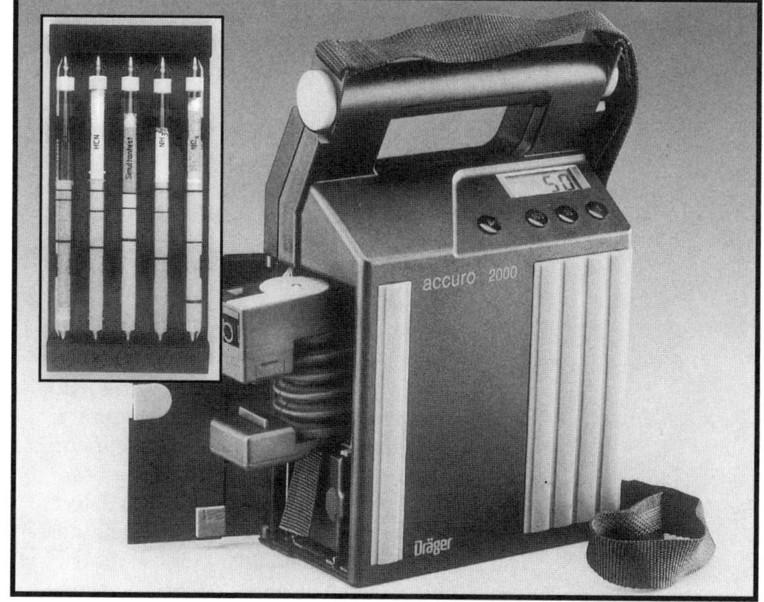

Figure 9.17 — Detector tubes, or colormetric indicator tubes, and pumps from National Draeger.

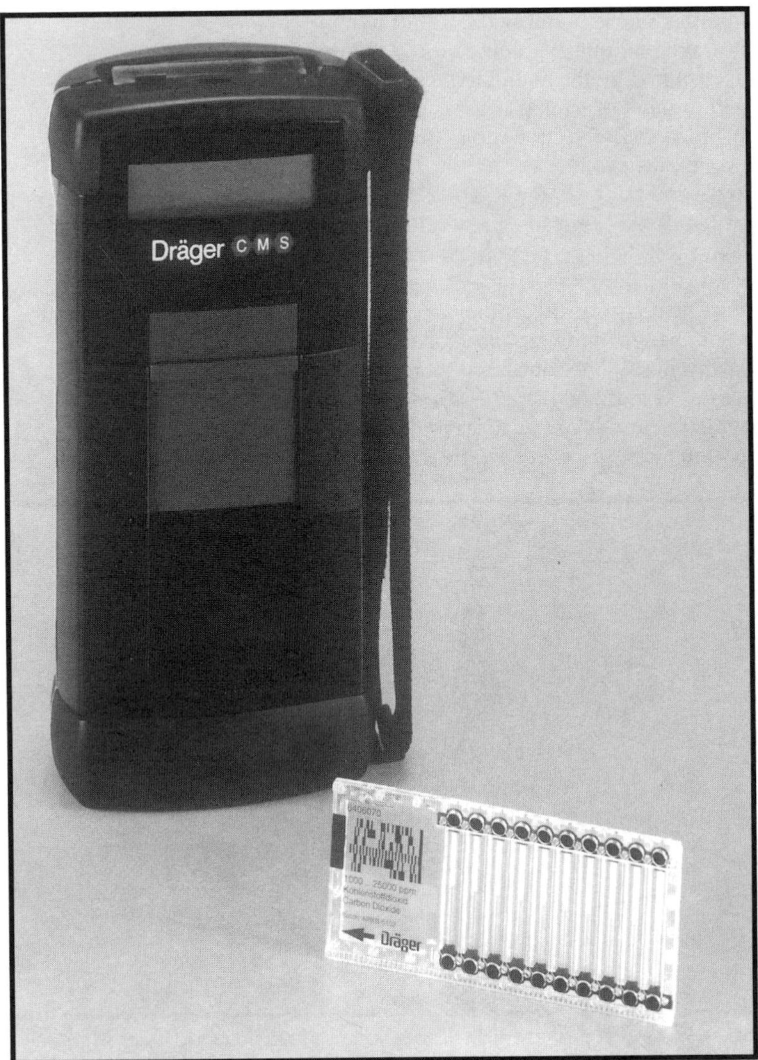

Figure 9.18 — A new generation of tubes and pumps from National Draeger.

Selection of a detector tube depends on the chemical or chemicals to be monitored and the concentration range. While tubes may be used to measure a single contaminant, most tubes react with more than one chemical, usually chemicals that are structurally similar. These interferers are documented by the manufacturer and should be understood by the occupational hygienist when evaluating the results. Usually, for a given chemical there are several tubes available to cover different measurement ranges. For some chemicals there are multiple ranges available in a single tube. All manufacturers offer qualitative indicator tubes that simply record the presence or absence of a contaminant. These tubes, called poly tubes, are designed to detect a large number of organic or inorganic gases and vapors. Qualitative tubes should not be used to estimate airborne concentrations.

Detector tubes are most effective when used for determining the presence in the air of a chemical or class of chemicals. This information can then be used to choose a more accurate and complex sampling method, such as a real-time continuous monitor or personal sampler. In cases where detector tubes are the only sources of information, multiple readings should be taken successively in one location throughout the day to account for temporal variability and over several locations to account for spatial variability of concentrations and contaminant species in air. Detector tubes are not recommended for compliance monitoring of personal exposures unless there are no other alternatives. In this case, detector tubes may be placed in the breathing zone to evaluate ceiling concentrations or STEL concentrations. Detector tubes can be used for source monitoring to evaluate peak exposures; however, many tubes would need to be used to sample adequately for a long enough time period.

Detector tubes are used at hazardous waste sites for measurement of leaks and spills because they offer a wide range of specificities. Inaccessible regions can be sampled using detector tubes by attaching the detector tube to a flexible tube that is then attached to a pump. Several manufacturers have developed Hazmat kits using a variety of tubes to rapidly measure classes of contaminants. These kits have a decision matrix that enables end-users to choose the next appropriate tube or tubes depending on the reading of the previous tubes.

Detector tubes are limited by being sensitive to temperature, humidity, atmospheric pressure, light, time, and the presence of interfering chemicals. The reagents in the detector tubes are chemically reactive and can be degraded over time by temperature and light; therefore, shelf life is limited. Most tubes carry an expiration date of 2 to 3 years, which represents the maximum shelf life when stored under optimal conditions specified by the manufacturer. Chemicals in tubes exposed to sunlight, heat, and ultraviolet light can significantly degrade in a matter of hours.

Generally, detector tubes are recommended to be used in the range of 0 to 40°C; however, most tubes are calibrated at 20 to 25°C, 760 mm Hg, and 50% relative humidity. Sampling under conditions that deviate from the calibration conditions may require corrections or conversions supplied by the manufacturer.

The primary limitation of detector tubes is their susceptibility to interferences, both positive and negative, from other chemicals. Negative interferences will manifest themselves as no color change. Positive interferences can result in overestimation of concentrations, or false positives. Sometimes interferences can result in a different color change. Usually, the manufacturer documents the known interferences. Interferences can be used to the advantage of the occupational hygienist, and several different tubes can be used to screen a class of chemicals to determine what chemicals are in the air.

Some detector tubes have been designed to perform integrated sampling over long monitoring periods of up to eight hours and use low-flow peristaltic or diaphragm sampling pumps (normally at a flow rate of 10–20 cc/min). With these tubes the length of stain is usually calibrated in microliters. Using the sample volume, flow rate, and time, the measurement can be converted to a TWA concentration. Long-term tubes that rely on diffusion are also available with results indicated in ppm-hrs. Dividing by the exposure time will allow for TWA determinations. Long-term results must be corrected for temperature and pressure. Long-term tubes are subject to the same cross-sensitivities as the short-term tubes and may be more affected by environmental conditions because of the long sampling time.

Long-term tubes are effective for investigating sensitive employee concerns that require immediate attention. The tubes are usually selected for a concentration range that includes the PEL or American Conference of Governmental Industrial Hygienists' threshold limit value. The tube can be attached to the clothing of an individual and placed in the breathing zone. Long-term tubes can be used as a screening device and can detect significantly lower concentrations than their short-term counterparts. Lower limits of detection are obtained by sampling for longer periods of time.

Detector tubes and pumps were once certified by the National Institute for Occupational Safety and Health (NIOSH) to meet specific performance standards (described in the *Code of Federal Regulations* Title 42, Part 84), which include specifications for accuracy at four concentrations, standard deviations of the tube readings, and variation of the stain length. A number of tubes are now voluntarily certified by a program administered by the Safety Equipment Institute (SEI). The SEI publishes an annual list of tubes that have been validated by this program (SEI, McLean, Va.). The accuracy of detector tubes has been found to vary among tubes and manufacturers. To be considered acceptable, a detector tube usually must be accurate to within ±25%. Some tubes have not met these specifications and have been found to be accurate to within ±35%.

Detector tube pumps should be calibrated quarterly to ensure proper volume and flow rate measurement. Care must be taken to see that leak-proof valves and connections are maintained. Leak checks should be performed when the pump is first purchased and periodically when the pumps are idle for long periods of time. Leak checks involve inserting an unopened tube into the pump and compressing the bellows or pulling the piston back and locking it (depending on the type of pump). After 2 minutes if the bellows type pump has expanded noticeably, this indicates leakage. In the case of the piston type pump, if the piston on release does not return to a fully closed position, this indicates leakage.

Direct-Reading Aerosol Monitors

Optical Techniques for Determining Aerosol Size and Count

Optical Particle Counter

The most popular direct-reading aerosol monitors are light-scattering devices (also called aerosol photometers). They operate by illuminating an aerosol as it passes through a chamber (sensing volume) and by measuring the light scattered by all the particles at a given scattering angle relative to the incident beam. As the number of particles increases, the light reaching the detector increases. The detector can be a solid-state photodiode or a photomultiplier tube.

Scattering angle has a great influence on aerosol measurements. The smaller the scattering angle, the greater the detection of large particles. A scattering angle of 90º provides the greatest sensitivity for small particles. Scatter depends on the size and shape of the particle, the refractive index of the particle, the wavelength of light, and the angle of scatter. For multiple particles, scatter depends on the particle size distribution, changes in the refractive indices with aerosol composition, dust density, and dust concentration.[32]

Light-scattering monitors are factory and field calibrated. Factory calibration is performed to ensure proper instrument response as compared with similar instruments. Field calibration is performed to improve the specificity of the monitor for particular aerosols and processes. Factory calibration is usually performed by measuring a calibrated aerosol such as Arizona road dust, oil mist, or cement dust. Field calibration involves testing the instrument with a monodisperse test aerosol of known size and refractive index or with a dust with a similar size distribution of the aerosol to be sampled. A side-by-side comparison using a gravimetric method is then performed.

Single-particle techniques based on light scattering cover a size range of 70 nm to more than 100 m and are capable of measuring concentrations of less than 1 particle/L (clean-room monitoring) to about 10^5 particles/cm^3 (aerosol research). Multiple-particle scattering techniques are applicable for concentration measurements of atmospheric aerosols ranging from a few micrograms per cubic meter to several hundred milligrams per cubic meter.

Single-particle, direct-reading instruments, or optical particle counters (OPCs), use monochromatic light such as a laser or light-emitting diode or a broad band light source such as a tungsten filament lamp to illuminate aerosols. In an OPC, aerosol is drawn into the instrument and light scattered by individual particles is measured by a photodetector. From the count rate and pulse height of the photoelectric pulses, the number concentration and size of particles can be determined.

Condensation Nucleus Counter

Condensation nucleus counters (CNCs) can measure very small particles (<1.0 µm) and have been used to study atmospheric aerosols for many years. They are currently used for testing high-efficiency particulate air filters in clean rooms and for quantitatively fit-testing respirators.[33] The CNC has a fast response time, is lightweight and portable, and can be used for real-time measurement. The CNC works by enlarging the particles to a size suitable to measure photometrically. The incoming air stream is first saturated with a vapor such as water or alcohol and is then cooled to induce a super-saturated solution. Vapors then condense on the particles and the particles grow in size. Using CNCs, individual particles can be counted or the intensity of the scattered light can be used to measure concentration (particles/cm^3).

Several types of CNCs have been developed that vary in the method by which the super-saturation is induced and in the method of particle detection. For example, an expansion-type CNC humidifies the aerosol with water vapor at room temperature and then cools by volume expansion or pressure release. The conductive cooling type CNC uses alcohol to saturate the aerosols and then uses conductive cooling for super-saturation; particles are measured by light scattering. The mixing-type CNC combines an aerosol stream at room temperature with a hot air stream of dibutyl-phthalate or dioctyl sebacate. The resultant mixed air is cooled down adiabatically and vapor condenses on particles at a steady-state continuous flow.

Optical Techniques for Determining Aerosol Mass

Multiple Particle Monitors

Multiple particle optical monitors are real-time dust monitors used to measure aerosol concentrations. They have been used in filter testing, atmospheric monitoring, and mine aerosol measurement. The light source used in a photometer can be monochromatic, such as a light-scattering diode or a laser, or a broad-wavelength light source, such as a tungsten filament lamp. The detector can be either a solid-state photodiode or a photomultiplier tube. When particles are drawn into the instrument, the intensity of light scattered into the detector can be used to estimate concentration. As the number of particles increases, the light reaching the detector increases.

The amount of light scattered by a particle into the detector (which determines its sensitivity and the relative response for different types of particles), depends on the size, shape, and refractive index of the particle. These instruments usually use a forward-scattering geometry, and the angle of scattering (theta) is defined relative to the light source. For most instruments, this angle is approximately 90°; however, in some instruments the angle may be as small as 12°.

The advantage of light-scattering methods is that they provide a linear response over a large concentration range. Concentration measurements are not affected by the sampling flow rate; however, sampling rate influences instrument response rate. These instruments actually measure particle count and not mass; therefore, to determine mass the optical properties, density of the dust, and size distribution of the dust must be constant and understood. For quantitative measurements these instruments must be calibrated with an aerosol similar in refractive index and particle size to the one being measured. The instrument must be operated in its linear range, where the number of particles detected is linearly correlated with the photometer signal. This range of linearity is limited at high concentrations by multiple scattering and at low concentrations by the stray-light background in the chamber.

An example of a portable, direct-reading respirable dust photometer designed to detect concentrations of dusts present in workplace air is the tyndallometer TM digital µP (H. Hund GmbH, Wetzlar, Germany). This instrument measures the IR light scattered by airborne dust particles at a mean scattering angle of 70º. The photometer value obtained for dusts can be linearly converted into mass values of respirable dust.

A respirable aerosol monitor (GCA, Bedford, Mass.) allows real-time assessment of respirable dust by light scattering after aerodynamic separation of this fraction by a 10-mm cyclone. These instruments have been used extensively for mineral dust aerosol measurements.

A personal dust monitor, Haz-Dust II, from Environmental Devices (Haverhill, Mass.) can display inhalable, thoracic, and respirable dust levels and collect the specific size fraction by attaching a size-selective device to the sensor. This instrument uses near-forward light scattering of IR radiation to measure the concentration of airborne dust particles; IR light is positioned at a 90° angle from a photodetector. The amount of light received by the photodetector is directly proportional to the aerosol concentration.

The RAM-1 (MIE Inc., Bedford, Mass.) is a portable survey dust monitor that uses a pulsed near-IR light-emitting diode, a silicon detector, and collimating and filtering optics. Light is scattered over an angular range of 45 to 95°. Concentration measurements range from 0.01–2 mg/m^3, 0.01–20 mg/m^3, and 0.01–200 mg/m^3, and the particle size range of the instrument is 0.1–20 µm in diameter. Therefore, the instrument can be used for health hazard evaluations and measuring explosive dust levels. The RAM-1 has been used at hazardous waste sites to monitor aerosols produced when contaminated soils were excavated and moved. Before operation the instrument should be zeroed in clean air.

The MINIRAM (MIE Inc.) is a miniature version of the RAM-1, without an active sampling pump. It is designed to be used for personal sampling of employees and can be used with a data logger (see Figure 9.19). The instrument displays concentrations in milligrams per cubic meter every 10 seconds and is capable of giving short-term measurements and TWAs. Concentration measurements range from 0.01 to 100 mg/m^3.

Electrical Techniques for Determining Aerosol Size

Several instruments combine optical detection techniques with manipulation of particle motion to measure the aerodynamic diameter of a particle. Aerodynamic diameter can then be used to describe the behavior of particles in gravitational settling, filtration, and respiratory deposition.

One instrument, the E-SPART (single particle aerosol relaxation time) (Hosakawa Micron International, Osaka, Japan) determines particle size by subjecting particles to either a superimposed acoustic and DC electric field or only an AC electrical field.[34,35] Aerosol samples are drawn into the instrument using a vacuum pump at a flow rate of 0.5 L/min. In the acoustic/electric mode, particles are subjected to sinusoidal force by an acoustic transducer at a specific frequency that induces an oscillatory velocity component in the horizontal direction. In addition, particles are subjected to a DC electric field that induces a migration velocity component, which

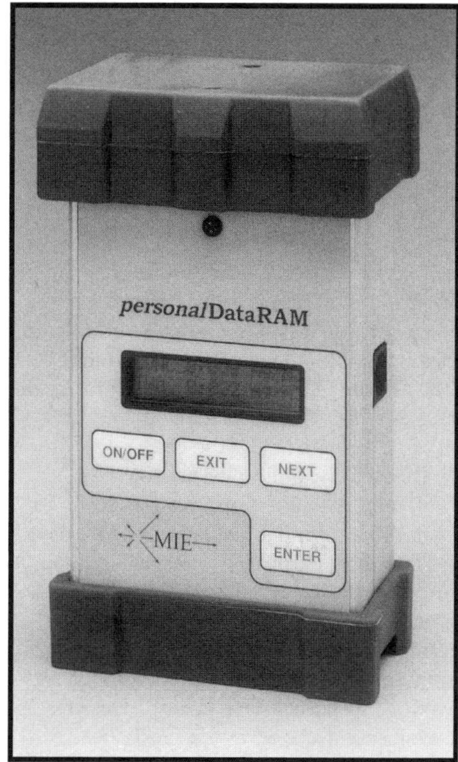

Figure 9.19 — Multiple particle optical monitor: the MINIRAM.

depends on the polarity and the magnitude of the electric charge of the particle and the field strength. A differential laser Doppler velocimeter (LDV) measures the horizontal oscillatory component, and a microphone measures the acoustic field. The motion of the particle, detected by the LDV optics, lags behind the acoustic field by an amount that depends primarily on the particle aerodynamic diameter. Aerodynamic diameters are measured for both electrically charged and uncharged particles using the acoustic mode. Electrical charge is determined from the electrical migration velocity and the aerodynamic diameter. The range of aerodynamic diameters measurable with acceptable sensitivity and resolution is stated by the manufacturer to be 0.2 to 10.0 µm.[36]

In the second mode, particles are only subjected to an AC electric field, and this mode is only applicable to electrically charged particles. A charged particle experiences an oscillatory motion in the electric field, and the oscillatory velocity component will lag behind the electric field by an amount that depends primarily on the particle aerodynamic diameter. The amplitude of the oscillations is directly proportional to the electric charge on the particles. Data generated includes number of particles counted, average charge on the particles, a plot of the size distribution of the particles, and aerodynamic diameter.

Although instrument response can be calculated theoretically, particle size calibration is usually performed using aerosolized monodisperse polystyrene latex spheres (that are less than 10 µm), or using particles greater than 10 µm generated using a vibrating-orifice aerosol generator. Calibration is performed under known conditions of temperature and humidity. Charge calibration is more difficult to perform because it is difficult to generate particles with both a known size and charge.[37] The electric field can be calculated from the distance between the electrodes and the applied voltage across the electrodes.

The Aerodynamic Particle Sizer (APS) (TSI, Inc., St. Paul, Minn.) is a relatively portable instrument that rapidly and precisely sizes most particles by measuring their velocity relative to the air velocity within an acceleration nozzle.[38] When particles are introduced into an accelerated velocity field, small particles follow the motion of the air, while large particles lag behind. This creates an increase in the relative velocity between the air and the particles. This velocity is compared with a calibration curve created using monodisperse spheres.

Aerosols are introduced into the instrument at a flow rate of 5 L/min. Air at the rate of 4 L/min is diverted through a filter and becomes clean sheath air. The remaining 1 L/min aerosol-laden air is fed into a focusing nozzle, recombined with the sheath air, and fed into an acceleration nozzle. At the exit of the acceleration nozzle each particle passes through two laser beams; light scattered from each particle causes two pulses, which are detected by a photomultiplier. The time lag between these pulses is used to calculate particle size.

Calibration of each APS instrument is unique due to variations in the nozzle sizes, laser spacing, and laser beam locations. Biases, which can be on the order of 25%, can be caused by particle density, particle shape factor, gas viscosity, and gas density.

Electrical Techniques for Determining Aerosol Count

Fibrous aerosol monitors (FAMs) (MIE, Inc.) are modified light-scattering monitors that are direct-reading devices designed to measure airborne concentrations of fibrous materials with a length-to-diameter aspect ration greater than three, such as asbestos and fiber glass. Results are reported as a fiber count rather than mass concentration. FAM instruments are extensively used for asbestos removal operations to provide real-time measurement of fiber concentrations to ensure the integrity of asbestos removal engineering control systems. At least two FAMs are used for asbestos monitoring: one inside and one outside the containment area. FAMs are usually equipped with alarms to warn of "containment barrier" failures. Conventional methods for measuring the concentration of airborne asbestos and other fibers involve time-integrated collection on filters and analysis with either a light microscope or an electron microscope.

FAMs operate by drawing air through a tube, using a diaphragm pump, where two pairs of electrodes create electric fields perpendicular to each other and to the axis of the tube. The two electric fields oscillate, which causes fibers to align and oscillate as they travel through the tube. A laser beam shines parallel to the direction of the airflow and illuminates the fibers. The fibers scatter the light in directions perpendicular to their long axes; a photomultiplier tube, positioned in the wall of the flow tube, detects the resulting pulses of scattered light. Fiber count is digitally displayed as fibers per cc and is determined from these pulses. Compact particles do not align in the oscillating field and do not produce pulses synchronous with the oscillating electric field.

One of the limitations of this method is that measurements assume that ideal cylindrical fibers are being detected. However, fibers may not have an ideal cylindrical or even ellipsoidal shape. For example, asbestos fibers may have a noncylindrical cross section, a curvature, or may exist as clumps of fibers. Fibers with curvatures or that have other attached particles may not align at 90° to the detector axis as required for optimum detection geometry. FAMs cannot differentiate among fibers of different materials such as asbestos, fiber glass, cotton, or paper. Another limitation is that the laser beam has a Gaussian illumination of the detection volume; therefore, fibers in the center of the beam scatter more light than fibers at the edge of the beam. Large diameter fibers may settle and go undetected.(39)

Instrument response is calibrated by the manufacturer using a side-by-side comparison with NIOSH Method 7400, which uses a filter to collect amosite aerosol and phase contrast microscopy to count fibers. The manufacturer recommends that this type of calibration be repeated yearly. Field calibration is not possible for testing fiber count accuracy; however, FAMs can be tested for integrity of the electronics, laser alignment, and accurate flow rate.

Resonance Techniques for Determining Aerosol Mass

Piezoelectric Mass Sensors

Piezoelectric quartz crystal microbalances (QCMs) enable quick measurement of aerosol mass, particularly for respirable aerosols, and have been used in occupational hygiene and environmental monitoring for measuring welding and tobacco smoke.(40)

Piezoelectric mass sensors are based on the principle that when crystalline materials are mechanically stressed by compression or tension they produce a voltage proportional to the stress. When these crystals are subjected to an electric current, they oscillate, and the natural vibrational frequency depends on the thickness and density of the crystal. Piezoelectric quartz crystals used in QCMs are usually cut and polished on specific crystallographic planes to produce a single vibrational mode. Most clean quartz crystals vibrate at a natural frequency of 5 or 10 MHz.(41) The crystal is attached to electrodes and is part of an oscillator circuit. When particles are deposited on the crystal, the oscillation frequency decreases in direct proportion to the mass of particles. Concentration is determined from the sample flow rate, sampling time, and mass deposited. Piezoelectric aerosol sensors collect particles on a quartz crystal primarily by electrostatic precipitation or inertial impaction. Commercially available instruments may incorporate size-selective impactors upstream of the sensor to selectively measure respirable aerosols.(42,43) After deposition, most instruments have a mechanism for cleaning the crystal.

Sampling efficiency is affected by both the mass and size of particles. QCM monitors have low sensitivity when particles larger than 10 µm in diameter or large masses of particles are collected. In addition, large, elastic, or spherical particles may bounce or migrate off the crystal surface after deposition. This loss may be eliminated or decreased by applying a specialized sticky coating to the surface of the crystal. Excessive overloading of the instrument by aerosols can cause the vibration of the sensing crystal to cease altogether. Manufactur-ers usually provide procedures to prevent overloading and clean the crystal.

Piezoelectric mass sensors are complex and time consuming to operate. They are initially calibrated by the manufacturer, which sets the display to correspond to a measured aerosol concentration for known changes in the frequency of vibration of the quartz crystal. In the field, instruments should be calibrated by the occupational hygienist using a well-characterized aerosol such as polystyrene latex spheres or, if possible, against the expected challenge aerosol.

Tapered Element Oscillating Microbalance

Another instrument that determines aerosol mass using resonance oscillation is the tapered element oscillating microbalance (TEOM®). In this instrument a specially tapered hollow tube constructed of a glass-like material is used instead of a quartz crystal. The narrow end of the tube supports the collection medium, usually a filter, and is made to oscillate at its natural frequency when driven by an amplifier. Aerosol-laden air is drawn into the instrument, travels first through the filter where the particles are deposited, and then through the hollow tube. As mass is collected on the filter, the natural frequency decreases and is recorded by the oscillation amplifier.

Sampling efficiency is not particularly affected by particle size; particles that are not collected by the filters do not represent a significant mass. Particle size can affect wall losses, which are common to all aerosol sampling devices. Collection of sufficient mass on the filter can conceivably result in particle bounce or migration and damping of oscillations; however, the filters would probably clog before particle loading became too great.[44]

Calibration is performed by adding a known mass to the tapered element and by measuring the change in the oscillating frequency. Rupprecht and Patashnik, (Vorheesville, N.Y.) manufactures a variety of TEOM devices for measuring diesel exhaust, emissions from underground mine fires, and chemicals in ambient air.[45]

Beta Absorption Techniques for Determining Aerosol Mass

Another instrument that measures aerosol mass uses the beta absorption method. This method is well-suited to evaluate fume concentrations and chemicals in indoor air.[46] Beta absorption instruments function by measuring differential absorption of beta radiation before and after collection of particulates on a filter or an impactor. An aerosol is drawn into the instrument, sometimes through a cyclone, and collected on a surface that is positioned between a radioisotope source and an electron counter, which measures intensity. As the mass deposited increases, there is a near exponential decrease in the number of beta particles transmitted through the sample, as shown in Equation 4.

$$I/I_o = \exp^{-\mu x} \tag{4}$$

where

I = transmitted flux of continuous beta particles
I_o = incident flux of beta particles
μ = mass absorption coefficient for beta absorption (cm^2/g)
x = mass thickness of sample (g/cm^2)

Advantages of this technique include simplicity of instrument design, ease of automation, and a dynamic range of sensitivity that is matched to the mass range normally of interest in aerosol monitoring.

Radioisotope sources are selected using beta particle emissions with sufficient half-lives so that frequent source replacements are unnecessary, and sufficient source strengths are available to provide adequate precision. Most beta gauges use ^{14}C or ^{147}Pm isotopes, with half-lives of 5730 and 2.62 years, respectively.[47] Detectors must be sensitive to counting the beta particles at a sufficient rate to perform measurements in a short time interval. Systematic measurement errors may be due to fluctuations in atmospheric density, changes in relative humidity, and long- and short-term drift in the detector.[48]

References

1. **Ness, S.A.:** *Air Monitoring of Toxic Exposures: An Integrated Approach.* New York: Van Nostrand Reinhold, 1991.
2. **Gressel, M.G., W.A. Heitbrink, and P.A. Jensen:** Video exposure monitoring—a means of studying sources of occupational air contaminant exposure, part 1—video exposure monitoring techniques. *Appl. Occup. Environ. Hyg. 8*:334–338 (1993).
3. **Gressel, M.G., W.A. Heitbrink, J.D. McGlothlin, and T.J. Fischbach:** Advantages of real-time data acquisition of exposure assessment. *Appl. Ind. Hyg.* 3:316–320 (1987).
4. **Feldman, R.F.:** Degraded instrument performance due to radio interference: criteria and standards. *Appl. Occup. Environ. Hyg. 8*:351–355 (1993).
5. **Berglund, R.N. and B.Y.H. Liu:** Generation of monodisperse aerosol standards. *Environ. Sci. Technol.* 7:147–153 (1973).
6. **National Fire Protection Agency (NFPA):** *National Electrical Code.* Quincy, MA: NFPA, 1996.
7. **Maslansky, C.J. and S.P. Maslansky:** *Air Monitoring Instrumentation: A Manual for Emergency, Investigatory, and Remedial Responders.* New York: Van Nostrand Reinhold, 1993.
8. **Schivon, G., G. Zotti, R. Tonioloi, and G. Bontempelli:** Electrochemical detection of trace hydrogen sulfide in gaseous samples by porous silver electrodes supported on ion-exchange membranes (solid polymer electrolytes). *Anal. Chem. 67*:318–323 (1995).
9. "Hazardous Waste Operations and Emergency Response" *Code of Federal Regulations*, Title 29, Part 1910.120. 1996.
10. "Permit-required Confined Spaces" *Code of Federal Regulations*, Title 29, Part 1910.146. 1996.
11. **Snee, T.J.:** An instrument for measuring the concentration of combustible airborne material. *Ann. Occup. Hyg.* 29:81–90 (1985).

12. **Barsky, J.B., S.S. Que Hee, and C.S. Clark:** An evaluation of the response of some portable, direct-reading 10.2eV and 11.8eV photoionization detectors, and a flame ionization gas chromatograph for organic vapors in high humidity atmosphere. *Am. Ind. Hyg. Assoc. J. 46*:9–14 (1985).
13. **Elias, J.D., D.N. Wylie, A. Yassi, and N. Tran:** Eliminating worker exposure to ethylene oxide from hospital sterilizers: an evaluation of cost and effectiveness of an isolation system. *Appl. Occup. Environ. Hyg. 8*:687–692 (1993).
14. **Ekberg, L.E.:** Volatile organic compounds in office buildings. *Atmos. Environ. 28*:3571–3575 (1994).
15. **Simonde, M., H. Xiao, and S.P. Levine:** Optical remote sensing for air pollutants—review. *Am. Ind. Hyg. Assoc. J. 55*:953–965 (1994).
16. **Levine, S.P., Y. Li-Shi, C.R. Strong, and X. Hong Kui:** Advantages and disadvantages in the use of Fourier transform infrared (FTIR) and filter infrared (FIR) spectrometers for monitoring airborne gases and vapors of individual hygiene concern. *Appl. Ind. Hyg. 4*:180–187 (1994).
17. **Haaland, D.M.:** Improved sensitivity of infrared spectroscopy by the application of least squares methods. *Appl. Spectrosc. 34*:539–548 (1980).
18. **Haaland, D.M. and R.G. Easterling:** Application of a new least-squares method for the quantitative infrared analysis of multicomponent samples. *Appl. Spectros. 36*:665–672 (1985).
19. **Herget, F.W. and S.P. Levine:** Fourier transform infrared (FTIR) spectroscopy for monitoring semiconductor process gas emissions. *App. Ind. Hyg. 1*:110–112 (1986).
20. **Ying, L.-S., S.P. Levine, C.R. Strang, and W.F. Herget:** Fourier transform infrared (FTIR) spectroscopy for monitoring airborne gases and vapors of industrial hygiene concern. *Am. Ind. Hyg. Assoc. J. 50*:78–84 (1989).
21. **Xiao, H.K., S.P. Levine, W.F. Herger, J.B. D'arcy, et al.:** A transportable remote sensing, infrared air monitoring system. *Am. Ind. Hyg. Assoc. J. 52*:449–457 (1991).
22. **Spellicy, R.L., W.L. Crow, J.A. Draves, W.F. Buchholtz, et al.:** Spectroscopic remote sensing-addressing requirements of the Clean Air Act. *Spectroscopy 8*:24–34 (1991).
23. **Yost, M.G., H.K. Xiao, R.C. Spear, and S.P. Levine:** Comparative testing of an FTIR remote optical sensor with area samplers in a controlled ventilation chamber. *Am. Ind. Hyg. Assoc. J. 53*:611–616 (1992).
24. **Todd, L.A.:** Evaluation of an open-path Fourier transform infrared (OP-FTIR) spectrophotometer using an exposure chamber. *Appl. Occup. Environ. Hyg. 11*:1327–1334 (1996).
25. **Grant, W.B., R.H. Kagan, and W.A. McClenny:** Optical remote measurement of toxic gases. *J. Air Waste Manag. Assoc. 42*:18–30 (1992).
26. **Russwurm, G.M. and J.W. Childers:** *FT-IR Open-Path Monitoring Guidance Document,* 2nd ed. Research Triangle Park, NC: ManTech Environmental Technology, Inc., 1993.
27. **Todd, L. and G. Ramachandran:** Evaluation of algorithms for tomographic reconstruction of chemicals in indoor air. *Am. Ind. Hyg. Assoc. J. 55*:403–417 (1994).
28. **Todd, L. and D. Leith:** Remote sensing and computed tomography in industrial hygiene. *Am. Ind. Hyg. Assoc. J. 51*:224–233 (1990).
29. **A. Samanta and L.A. Todd:** Mapping air toxics indoors using a prototype computed tomography system. *Ann. Occup. Hyg. 40*:675–691 (1996).
30. **R. Bhattacharyya and L.A. Todd:** Spatial and temporal visualization of gases and vapours in air using computed tomography: numerical studies. *Ann. Occup. Hyg. 41*:105–122 (1997).
31. **Yost, M.G., A.J. Gagdil, A.C. Drescher, Y. Zhou, et al.:** Imaging indoor tracer gas concentrations with computed tomography: experimental results with a remote sensing FTIR system. *Am. Ind. Hyg. Assoc. J. 55*:395–402 (1994).
32. **Gebhart, J.:** Optical direct-reading techniques: light intensity system. In *Aerosol Measurement Principles, Techniques, and Applications,* K. Willeke and P.A. Baron, eds. New York: Van Nostrand Reinhold, 1993. pp. 313–344.
33. **Baron, P.A.:** Modern real-time aerosol samplers. *Appl. Ind. Hyg. 3*:97–103 (1988).
34. **Mazumeter, M.K., R.E. Ware, and W.G.Hood:** Simultaneous measurements of aerodynamic diameter and electrostatic charge on single-particle basis. In *Measurements of Suspended Particles by Quasi-elastic Light Scattering,* B. Dahneke, ed. New York: John Wiley & Sons, 1983.

35. **Renninger, R.G., M.K. Mazumder, and M.K. Testerman:** Particle sizing by electrical single particle aerodynamic relaxation time analyzer. *Rev. Sci. Instrum. 52*:242–246 (1981).
36. **Mazumder, M.K., R.E. Ware, J.D. Wilson, R.G. Renniger, et al.:** SPART Analyzer: its application to aerodynamic size distribution measurement. *J. Aerosol Sci. 10*:561–569 (1979).
37. **Mazumder, M.K., R.E. Ware, T. Yokoyama, B.J. Rubin, et al.:** Measurement of particle size and electrostatic charge distributions on toners using E-SPART analyzer. *IEEE Trans. Ind. Appl. 27*:611–619 (1991).
38. **Wilson, J.C. and B.Y.H. Liu:** Aerodynamic particle size measurement by laser-Doppler velocimetry. *J. Aerosol Sci. 11*:139–150 (1980).
39. **Baron, P.A., M.K. Mazumder, and Y.S. Cheng:** Direct-reading techniques using optical particle detection. In *Aerosol Measurement: Principles, Techniques, and Applications,* K. Willeke and P.A. Baron, eds. New York: Van Nostrand Reinhold, 1993. pp. 381–409.
40. **Swift, D.L. and M. Lippmann:** Electrostatic and thermal precipitators. In *Air Sampling Instruments for Evaluation of Atmospheric Contaminants,* 7th ed. S.V. Hering, ed. Cincinnati, American Conference of Governmental Industrial Hygienists, 1989. pp. 387–403.
41. **Williams, K., C. Fairchild, and J. Jaklevic:** Dynamic mass measurement techniques. In *Aerosol Measurement: Principles, Techniques, and Applications,* K. Willeke and P. Baron, eds. New York: Van Nostrand Reinhold, 1993. pp. 296–312.
42. **Samimi, B.:** Laboratory evaluation of RDM-201 respirable dust monitor. *Am. Ind. Hyg. Assoc. J. 47*:354–359 (1985).
43. **Sem, G.J., K. Tsurubayashi and K. Homma:** Performance of the piezoelectric microbalance respirable aerosol sensor. *Am. Ind. Hyg. Assoc. J. 38*:580–588 (1977).
44. **Williams, K.L. and R.P. Vinson:** *Evaluation of the TEOM Dust Monitor.* (Bureau of Mines Information Circular 9119). Washington, DC: U.S. Department of the Interior, Bureau of Mines, 1986.
45. **Shore, P.R. and R.D. Cuthbertson:** *Application of a Tapered Element Oscillating Microbalance to Continuous Diesel Particulate Measurement* (Rep. 850,405). Warrendale, PA: Society of Automotive Engineers, 1985.
46. **Glinsmann P.W. and F.S. Fosenthal:** Evaluation of an aerosol photometer for monitoring welding fume levels in a shipyard. *Am. Ind. Hyg. Assoc. J. 46*:391–395 (1985).
47. **Klein F., C. Ranty, and L. Sowa:** New examinations of the validity of the principle of beta radiation absorption for determinations of ambient air dust concentrations. *J. Aerosol Sci. (U.K.) 15*:391–395 (1984).
48. **Courtney W.J. , R.W. Shaw, and T.C. Dzubay:** Precision and accuracy of beta gauge for aerosol mass determinations. *Environ. Sci. Tech. 16*:236–239 (1982).

Workers exiting the pressure chamber used to treat wood. These systems are used to force preservatives such as creosote and arsenic compounds into the wood. These compounds can not only cause skin problems but are also systemic poisons.

Outcome Competencies

After completing this chapter, the user should be able to:

1. Define underlined terms used in this chapter.
2. Explain sample collection principles to practicing professionals and the public.
3. Select sampling methods for gases and vapors that satisfy occupational hygiene objectives.
4. Select sampling parameters for gases and vapors that satisfy the requirements of the method and the occupational hygiene objectives.
5. Evaluate the applicability of specific media and methods for sampling gases and vapors in defined situations.
6. Perform necessary calculations to ensure the collection of reliable data when sampling gases and vapors.
7. Interpret air sampling data collected.

Key Terms

absorbing medium • accuracy • active sampling • adsorbing medium • analytical methods • area sampling • backup layer • breakthrough volume • cold traps • collection efficiency of solid sorbents • constant flow • derivitization • desorption efficiency • diffusive samplers • electromagnetic susceptibility • exposure • fritted glass bubblers • gas • gas washing bottles • grab sampling • impinger • integrated sampling • intrinsically safe • limit of detection (LOD) • limit of quantitation (LOQ) • lower boundary of working range • monitoring instruments • OSHA compliance • passive sampling • personal sampling • precision • pressure drop • primary standards • sample breakthrough • sampler capacity • sampling media • sampling strategy • secondary standards • solvent extraction • sorbent tube • SUMMA® canister • target concentration • thermal desorption • treated filter • upper measurement limit • vapor • working range

Prerequisite Knowledge

Basic understanding of chemistry and occupational hygiene sampling.

Prior to beginning this chapter, the user should review the following chapters:

Chapter Number	Chapter Topic
8	Principles and Instrumentation for Calibrating Air Sampling Equipment
9	Direct-Reading Instrumental Methods for Gases, Vapors, and Aerosols

Key Topics

I. Sample Collection Principles
   A. Integrated Samples
   B. Grab Samples
   C. Operational Limitations of Sampling and Analysis

II. Sampling Media for Gases and Vapors
   A. Solid Sorbents
   B. Chemically Treated Filters
   C. Liquid Absorbers
   D. Sampling Bags/Partially Evacuated Rigid Containers
   E. Cold Traps

III. Calculations and Interpretation of Data
   A. Conversion from measured analyte to ppm
   B. Comparison to exposure standards

10

Sampling of Gases and Vapors

Deborah F. Dietrich, CIH

Introduction

This chapter introduces the techniques available to the occupational hygienist for evaluating exposures to gases and vapors in the work environment. Over the past decade, rapid improvements have occurred in sampling and analytical methods and in the commercial availability of new, efficient sampling media and monitoring instruments. These advances have enhanced our ability to obtain reliable measurements of workplace exposures. Despite these advances, however, the occupational hygienist must recognize the limitations of the different techniques to ensure that the results are neither misinterpreted nor given undue credibility. In this chapter the applications and limitations of a variety of sampling media and techniques are reviewed.

When developing a particular sampling strategy, the reader should review the specific sampling and analytical methods available for the contaminants of interest. Several organizations, including the National Institute for Occupational Safety and Health (NIOSH), the U.S. Occupational Safety and Health Administration (OSHA), Lewis Publishers/CRC Press Inc., and the American Society for Testing and Materials (ASTM), have compiled and published collections of sampling and analytical methods for gases and vapors. Addresses of these organizations are provided in Table 10.1.

Before beginning a sampling program, it is important to select and consult with a qualified analytical laboratory. The American Industrial Hygiene Association (AIHA) maintains a list of laboratories that have been accredited under the Industrial Hygiene Laboratory Accreditation Program. A qualified laboratory can assist in choosing sampling and analytical methods that meet the sensitivity and specificity criteria appropriate to the environment being evaluated. The laboratory can also help choose a sampling medium and strategy that are compatible with the analytical method selected. Two key factors necessary for the selection of the most appropriate method are (1) a knowledge of the occupational environment and (2) an overall perspective of the limitations of the chemistry of the sampling and analysis.

Some commonly used sampling methods are summarized in Table 10.2. The occupa-

Table 10.1 — Available Publications on Sampling and Analytical Methods for Gases and Vapors

NIOSH Manual of Analytical Methods, 4th ed.
Centers for Disease Control and Prevention
National Institute for Occupational Safety and Health
4676 Columbia Parkway
Cincinnati, OH 45226-1998

Annual Book of ASTM Standards
American Society for Testing and Materials
100 Barr Harbor Dr.
West Conshohocken, PA 19428

OSHA Analytical Methods Manual
Occupational Safety and Health Administration
OSHA Salt Lake Technical Center
P. O. Box 65200
1781 South 300 West
Salt Lake City, UT 84165-0200

Methods of Air Sampling and Analysis
Lewis Publishers/CRC Press Inc.
2000 Corporate Blvd. NW
Boca Raton, FL 33431

Table 10.2 —
OSHA and NIOSH Methods for the Sampling and Analysis of Some Common Gases and Vapors

Gas or Vapor	*Sampling Medium*	*Analytical Method*	*Method Number*
Ammonia	sorbent tube containing H2SO4 impregnated carbon beads	ion chromatography	OSHA ID 188
Benzene	sorbent tube containing coconut shell charcoal	gas chromatography, flame ionization detector	NIOSH 1501
1,3-Butadiene	sorbent tube containing 4-tert-butylcatechol treated charcoal	gas chromatography, flame ionization detector	OSHA 56
Chlorine	fritted glass bubbler containing 0.1% sulfamic acid solution	ion-specific electrode	OSHA ID-101
Formaldehyde	sorbent tube containing 2-(hydroxymethyl) piperidine coated XAD-2	gas chromatography nitrogen selective detector	OSHA 52
Inorganic acids	sorbent tube containing washed silica gel with glass fiber filter plug	ion chromatography	NIOSH 7903
Methyl alcohol	2 sorbent tubes containing Anasorb® 747 carbon beads connected in series	gas chromatography flame ionization detector	OSHA 91
Organophosphorus pesticides	OSHA versatile sampler containing XAD-2 and quartz filter	gas chromatography flame photometric detection	NIOSH 5600
Toluene diisocyanate	glass fiber filter coated with 1-(2-pyridyl) piperazine	high performance liquid chromatography	OSHA 42

tional hygienist should carefully select the sampling and analytical methods most suitable for the specific application. If air sampling is being done for the purposes of OSHA compliance, the choice of a NIOSH or OSHA method may be most appropriate. Note, however, that OSHA does not mandate the sampling method. OSHA standards for individual chemicals will typically list the methods that have been tested by OSHA or NIOSH, but this does not preclude the use of alternative equivalent methods. The employer has the obligation of selecting a method that meets the sampling and analytical accuracy and precision requirements of the standard under its unique field conditions. These requirements typically stipulate measurement at the permissible exposure limit (PEL) within ±25% of the "true" value at a 95% confidence level.[1]

For occupational hygiene purposes a substance is considered a gas if this is its normal physical state at room temperature (25°C) and one-atmosphere pressure. Examples of gases are carbon monoxide, chlorine, oxygen, and nitrogen. If, however, the substance is normally a liquid (or solid) at normal temperature and pressure, then the gaseous component in equilibrium with its liquid (or solid) state is called a vapor. Carbon tetrachloride, formaldehyde, and benzene are examples of compounds that are present in the vapor state.

Sample Collection Principles

There are a number of considerations when designing a sampling plan, such as the location of samples, the number of workers to be sampled, and the duration of samples. Such considerations are reviewed in the exposure assessment chapters of this publication. This discussion will focus on the means of collecting valid samples of gases and vapors. Two basic types of samples are used to evaluate employees' exposures to gases and vapors: integrated and grab. Sampling using continuous monitors will be considered in the chapter on direct-reading instrumental methods.

Integrated sampling for gases and vapors involves the passage of a known volume of air through an absorbing or adsorbing medium to remove the desired contaminants from the air during a specified period of time. With this technique the contaminants of interest are collected and concentrated over a period of time to obtain the average exposure levels during the entire sampling period. On the other hand, grab sampling techniques involve the direct collection of an air-contaminant mixture into a device such as a sampling bag, syringe, or evacuated flask over a short interval of a few seconds or minutes. Thus, grab samples represent the atmospheric concentrations at the sampling site at a given point in time.[2]

Integrated Samples

Integrated sampling, covering the entire period of exposure, is required because airborne contaminant concentrations during a typical work shift vary with time and activity. Instantaneous measurements (grab samples) taken at any given period, therefore, do not reflect the average exposure of the worker for the entire shift and may not capture intermittent high or low exposures. Most integrated sampling is done to determine the 8-hour time-weighted average (TWA) exposure, and results are compared with the OSHA PELs, the threshold limit values (TLV®s) of the American Conference of Governmental Industrial Hygienists (ACGIH), or other applicable limits or guidelines such as the NIOSH recommended exposure limits. Integrated sampling is also required for determining compliance with short-term exposure limits performed over a 15-minute period.

When collecting an integrated sample, it is important that the appropriate sample duration and flow rate are chosen relative to the purpose of sampling, the sensitivity of the analytical method, and the expected concentration of the contaminant of interest. It is also essential that the flow rate and time be accurately measured. The accuracy of any occupational hygiene measurement depends on the precise determination of the mass of contaminant collected as well as the volume of air sampled.

Active Sampling

Most integrated sampling methods published by OSHA or NIOSH use active sampling techniques. Active sampling is defined as the collection of airborne contaminants by means of a forced movement of air by a sampling pump through an appropriate collection device such as a sorbent tube, treated filter, or impinger containing a liquid media. A key element when using active sampling techniques is laboratory or field calibration of the pump's flow rate, thus allowing for an accurate determination of air volume.

Air Sampling Pumps. An integrated air sampling method requires a relatively constant source of suction as an air moving device that can be calibrated to the recommended flow rate.[2] OSHA analytical methods specify that personal sampling pumps be calibrated to within ±5% of the recommended flow rate with the collection media in line.[3] There are a number of lightweight, battery operated pumps available that can be used with a variety of collection devices. Most use a rechargeable nickel-cadmium battery pack that allows for extended sampling over the 8-hour workday or longer. They can be attached to the wearer's belt for personal sampling in the worker's breathing zone, or they can be used as area samplers.

Sampling pumps are available with a variety of flow ranges and automatic features including timers, fault shutdown, constant flow capabilities, and computer compatibility. Most recently developed pumps weigh no more than 2 pounds and are intrinsically safe, allowing safe operation even in potentially explosive environments. Many pumps now also include a mechanism to address electromagnetic susceptibility, as it has been found that degraded instrument performance can result from radio frequency interference or electromagnetic interference from devices such as walkie-talkies, high voltage equipment, and electric motors.[4]

Pumps must be capable of maintaining the desired flow rate over the entire sampling period with the sample collection device in line. This is not normally a problem when sampling gases and vapors with

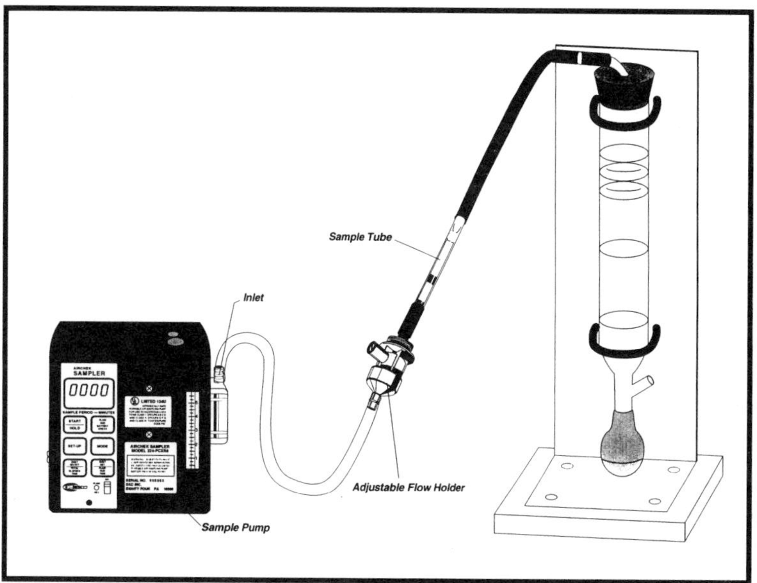

Figure 10.1 — Sampling train connected to calibrator.

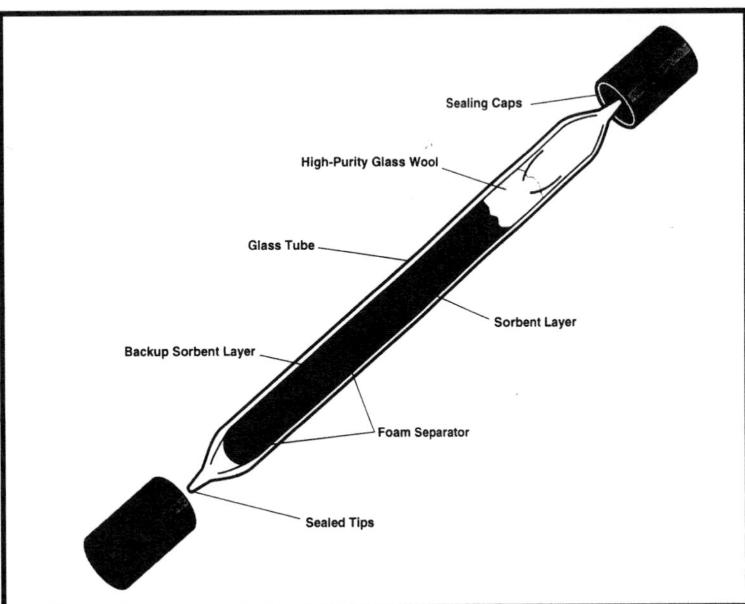

Figure 10.2 — Sorbent sample tube with backup sorbent layer.

solid sorbent tubes. However, when the sorbent material is of a fine mesh size or when sorbent tubes with a relatively high pressure drop are being used to collect short-term samples at flow rates of 1 L/min or greater, some pumps may not be able to handle the pressure drop.[5] Most recently designed personal samplers have a constant flow feature to continuously compensate for varying back pressures. Others have a constant pressure feature to allow for sampling with multiple sorbent tubes each with an independently set flow rate.

Calibration. Pump flow rate must be calibrated with the entire sampling train assembled as it will be used in the field (see Figure 10.1). Good occupational hygiene practice dictates that pumps be calibrated before and after sampling, on the same day, under pressure and temperature conditions similar to those at the sampling site. Calibration should not be done using built-in rotameters found on many sampling pumps. These are not precision devices and will not give a quantitative measure of the rate of airflow.

A wide variety of devices are currently available to measure airflow accurately; these are covered in greater detail in another chapter of this book. Flowmeters that are termed primary standards are based on direct and measurable linear dimensions, such as the length and diameter of a cylinder. These include spirometers, bubble meters, and Mariotte bottles. Secondary standards are flowmeters that trace their calibration to primary standards and maintain their accuracy with reasonable care and handling in operation.[6] Secondary standards include precision rotameters, wet test meters, and dry gas meters. Currently, an electronic readout of the flow rate is available on some pumps. Users are cautioned to read closely the instructions from the manufacturer regarding accuracy of the instrument's reading and the need for calibration using an external flowmeter.

Sample Collection Media. occupational hygienists should consult air sampling methods developed by NIOSH, OSHA, or other recognized testing organizations to determine the appropriate collection media for a specific chemical contaminant. (A more complete discussion of sample collection media is given later in this chapter.) These published methods have been extensively researched and collaboratively tested, and deviations from the requirements are not recommended.

It is important, however, for the occupational hygienist to review the method to determine its applicability relative to the field conditions. Particular attention should be paid to interfering compounds, humidity and temperature effects, and appropriate measuring range. The physical state of the contaminant(s) being sampled is also an important consideration. Air contaminants may often simultaneously exist in multiple-phases, e.g., particulate and vapor phase. It is important to choose the proper sample media to collect all phases of the contaminant of interest.[5]

Advantages and Disadvantages of Active Sampling. Active sampling techniques offer considerable advantages for the measurement of airborne contaminants. Since most methods published by OSHA and NIOSH are active methods, there has been extensive testing and documentation of reliability. If the purpose of sampling is for OSHA compliance, it is possible to select the same active sampling method that would be used by compliance personnel during an inspection.

From a technical standpoint active sampling methods offer the advantage of a calibrated, measured airflow for assurance in accuracy of sample volume. Also, most solid sorbent tubes have a secondary layer of sorbent that serves as a backup layer for the indication of sample breakthrough (see Figure 10.2). If contaminants exist in multiple phases, it is possible to use a series of samplers, such as a prefilter with a sorbent tube, to effectively collect both phases simultaneously from one air sample with one pump.

A disadvantage of active sampling is that the equipment is often cumbersome and may interfere with the job if workers have to wear a pump and sample media throughout the entire workday. Pump calibration can also be time consuming, and technical training is required to perform the necessary tasks. As pumps age, they may become less reliable at maintaining constant flow over the entire sampling period, and more frequent calibration may be necessary.

Passive Samplers

Among the most important developments in air sampling technology in recent years is the development of passive sampling devices. Passive sampling is the collection of airborne gases and vapors at a rate controlled by a physical process such as diffusion through a static air layer or permeation through a membrane without the active movement of air through an air sampler. Most commercially available passive samplers operate on the principle of diffusion. Diffusive samplers rely on the movement of contaminant molecules across a concentration gradient that for steady-state conditions can be defined by Fick's first law of diffusion.[7]

The following equation gives the steady-state relationship for the rate of mass transfer:

$$W = D(A/L)(C_1 - C_o)$$

where:

- W = mass transfer rate, ng/sec
- D = diffusion coefficient, cm^2/sec
- A = cross-sectional area of diffusion path, cm^2
- L = length of diffusion path, cm
- C_1 = ambient concentration of contaminant, ng/cm^3
- C_o = concentration of contaminant at collection surface, ng/m^3

By choosing an effective collection medium, so that C_o remains essentially zero regardless of the value of C_1, the mass transfer or collection rate is proportional to the ambient vapor concentration (C_1). Therefore, the uptake rate of the contaminant is approximately the product of $D(A/L)$ and C_1. It may also be noted that the units of $D(A/L)$ are volume per unit time, the same as for active air-moving devices such as personal sampling pumps.[7]

Each gas or vapor being sampled has a specific diffusion coefficient (D). Therefore, a passive sampler will probably have a different sampling rate for each analyte in the mixture. Diffusion coefficients for various compounds of interest can be determined experimentally or they may be estimated using one of several equations.[8] The diffusion coefficient and the sampler geometry can be used to determine a theoretical uptake rate. In practice, however, uptake rates may vary under various field conditions.[9] To more reliably assess the overall accuracy and precision of a passive sampler, a validation of its performance characteristics can be made using validation protocols published by NIOSH or other testing agencies.[10,11] These protocols stipulate the testing of those aspects unique to passive sampling, as well as those aspects common to both active and passive sampling (see Table 10.3).

Table 10.3 — Performance Characteristics Evaluated in the NIOSH Validation Protocol for Passive Samplers[a]

Analytical recovery	Relative humidity
Sampling rate and capacity	Interferents
Reverse diffusion	Monitor orientation
Storage stability	Temperature
Analyte concentration	Accuracy and precision
Exposure time	Shelf life
Face velocity	Behavior in the field

[a]Data from Cassinelli et al.[10]

Types of Passive Samplers. Many types of passive samplers are commercially available using either solid sorbents or liquid absorbers. Some samplers are designed to collect a broad range of compounds that can be identified by subsequent laboratory analysis, while others, by nature of their sorbent material, will preferentially collect a single chemical or family of chemicals. For example, diffusive samplers containing charcoal are commercially available for the collection of a wide variety of organic vapors. Another passive sampler validated by the OSHA laboratory uses hopcalite® (hydrar®) as the sorbent material to preferentially collect inorganic mercury vapor. Another passive sampler, containing silica gel coated with 2,4-dinitrophenyl hydrazine (DNPH), has proved valuable for assessments of aldehydes (see Figure 10.3). For more information on different sorbent materials, see the section on Types of Sorbent Materials or Table 10.4 later in this chapter.

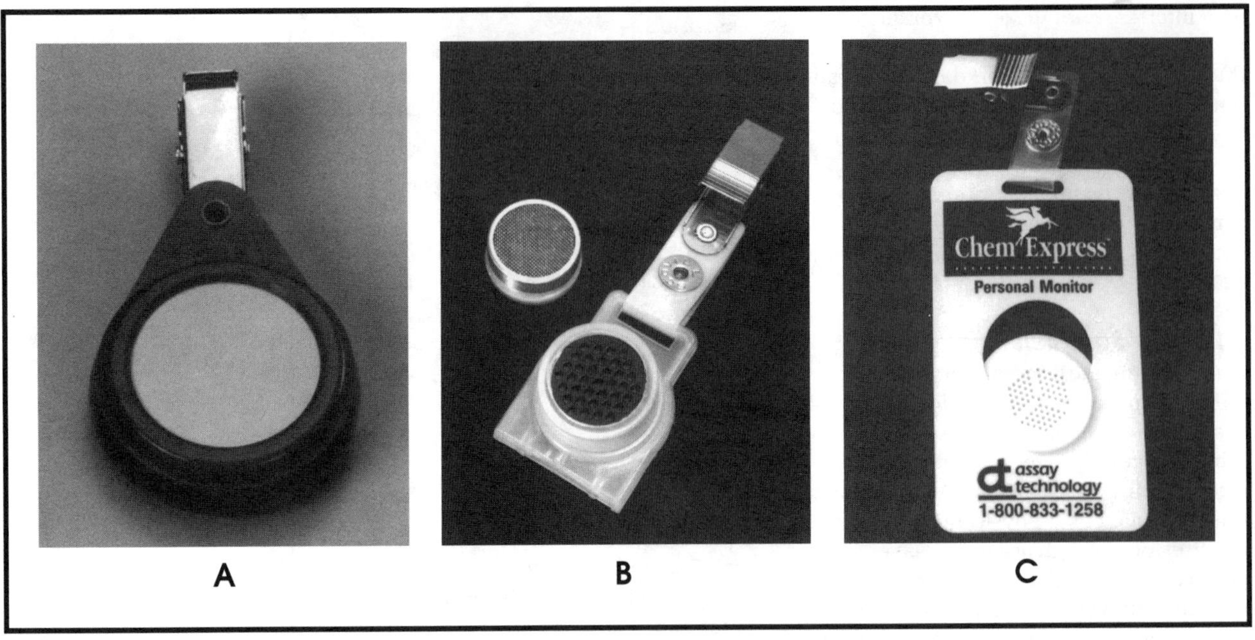

Figure 10.3 — Types of passive samplers: (A) organic vapor sampler (B) inorganic mercury sampler (C) aldehyde sampler.

Table 10.4 —
Properties of Solid Sorbents Commonly Used in Industrial Hygiene Sampling[a]

Sorbent	Specific Surface Area (m^2/g)	Pore Type[b]	Upper Temperature Limit (°C)	Composition
Activated charcoal	>1000	I		coconut shell or petroleum-based charcoal
Silica gel	300–800	I–II		amorphous silica
Tenax® GC	20	III	350–400	polymer of 2,6-diphenyl-p-phenylene oxide
Chromosorb® 102	300–400	II	250	copolymer of styrene and divinyl benzene
104	100–200	III	250	copolymer of acrylonitrile and divinyl benzene
Amberlite® XAD-2	300–400	II	200–250	copolymer of styrene and divinyl benzene
XAD-4	500–850	I–II	200–250	copolymer of styrene and divinyl benzene
XAD-7	325–450	I–II	200–250	acrylate polymer

[a]Data from Crisp,(15) Harper and Purnell,(16) and Stanetzek et al.(17)
[b]Pore type: I is <2 nm; II is 2–50 nm; III is >50 nm

Advantages and Disadvantages of Passive Samplers. Passive samplers offer many advantages to the occupational hygienist. They are very easy to use, allowing samples to be collected by personnel with less technical training. Passive samplers are also less expensive than active sampling when compared to the costs of pumps and flowmeters, and they are less obtrusive to the wearer. Finally, for most occupational hygiene applications the mass of contaminant collected by passive samplers is not significantly affected by temperature or pressure.(12)

Users should weigh these advantages against the possible disadvantages of these devices. In most cases, there are no OSHA or NIOSH methods to reference to ensure the reliability of data when using passive samplers. The sampling rate, if theoretically calculated, may not prove to be valid under field conditions. Reverse diffusion may also be a factor whereby some chemicals diffuse onto the sorbent but are not adequately retained.

Environmental parameters may influence the collection efficiency of passive samplers. Stagnant air (i.e., face velocities less than 25 ft/min) will cause "starvation" of the sampler. As the diffusion zone (L) is lengthened due to the lack of fresh contaminant molecules outside the sampler, the sampling rate is decreased causing a low measurement of the actual concentration. On the other hand, samplers are also affected by high face velocities that cause turbulence to occur in the air gap.(12) (This effect is minimized, however, as most manufacturers use a windscreen to dampen the effect of turbulent air.)

Grab Samples

Grab samples are collected to measure gas and vapor concentrations at a point in time and are used, therefore, to evaluate "peak" exposures for comparison to "ceiling" limits. Grab samples can be used to identify unknown contaminants, to evaluate contaminant sources, or to measure contaminant levels from intermittent processes or other sources. Grab or instantaneous samples of air to be analyzed for its gaseous components are collected using rigid containers, such as syringes or partially evacuated flasks or cans, or in nonrigid containers, such as sampling bags. See a complete discussion of these devices later in this chapter.

Instantaneous (as well as integrated) measurements of gases and vapors may also be performed using detector tubes or direct-reading instruments. These methods are described in another chapter of this book.

An advantage of collecting an air-contaminant mixture directly into a rigid or nonrigid container is that frequently it can be analyzed immediately by direct injection into a gas chromatograph (GC) or by using direct-reading instruments. Thus, quick decisions can be made in the field, such as choosing personal protective equipment, determining the source of leaks, or permitting entry into a vessel. The samples can also be retained for more thorough laboratory analysis at a later time.

A disadvantage of this technique is that for most applications contaminants are collected but not integrated over time. Only some containers will allow for the placement of a metering device, such as a critical orifice on the inlet, that will enable the collection of a sample at near constant flow over a period of time to provide a TWA concentration. Further, if contaminant concentrations are relatively low, analytical instruments may not be sensitive enough to detect them. While grab samples provide information relative to peak exposures, using multiple grab samples to assess full-shift exposures is time consuming and subject to error. This technique requires statistical analysis and will not be discussed in this book.

Operational Limits of Sampling and Analysis

When using any sampling technique, consideration must be given to the inherent limitations of the method, including sampler capacity, limit of detection (LOD) limit of quantitation (LOQ), and the upper measurement limit, which define the useful range for the method.[13] In a given application one or more of these factors will determine the minimum, maximum, or optimum volume of air to be sampled and may determine the confidence that can be placed in the results.

Sampler capacity (Wmax) is a predetermined conservative estimate of the total mass of contaminant that can be collected on the sampling medium without loss or overloading. For gases and vapors, researchers at NIOSH have defined Wmax as two-thirds of the experimental breakthrough capacity of the solid sorbent, i.e., 67% of the mass of contaminant on the sorbent at the breakthrough volume.[13] Breakthrough volume is defined as that volume of an atmosphere containing two times the PEL for the contaminant that can be sampled at the recommended flow rate before the efficiency of the sampler degrades to 95%.

The American Chemical Society (ACS) Committee on Environmental Analytical Chemistry defines the LOD as the lowest concentration level than can be determined to be statistically different from a blank sample. The recommended value of the LOD is the amount of analyte that will give rise to a signal that is three times the standard deviation of the signal derived from the media blank. The LOQ is the concentration level above which quantitative results may be obtained with a certain degree of confidence. The recommended value of the LOQ is the amount of analyte that will give rise to a signal that is 10 times the standard deviation of the signal from a series of media blanks. This corresponds to a relative uncertainty in the measurement of ±30% at the 99% confidence level.[14]

The upper measurement limit (W_u) is the useful limit (mg of analyte per sample) of the analytical instrument. The values of LOD, LOQ, and W_u for a given contaminant and analytical procedure should be obtained from and/or discussed with the analytical laboratory before sampling.

The target concentration (C_t) is a preliminary estimate of the airborne concentration of the contaminant of interest relative to the purpose of testing. This parameter can be used to determine minimum and maximum air volumes. The target concentration may be estimated by using previous sampling data, by using direct-reading instruments, or by relying on the professional judgment of the occupational hygienist.

A minimum sample volume for a quantitative determination (V_{min}) at the target concentration may be calculated as follows:

$$V_{min}\ (L) = \frac{LOQ(mg) \times 10^3 (L/m^3)}{C_t (mg/m^3)}$$

Example: It is desired to collect valid samples of acetaldehyde given the following information: method, NIOSH 3507; range given in method, 2 to 60 mg per sample; target concentration, 200 ppm (360.2 mg/m³). What is the minimum air volume required?

$$V_{min}\ (L) = \frac{2\ mg\ X\ 10^3}{360.2\ mg/m^3}$$

$$= 6\ L$$

The maximum sample volume that may be collected with minimum risk of bias due to breakthrough or sampler overloading may be calculated as follows:

$$V_{max}(L) = \frac{W_{max}(mg) \times 10^3 (L/m^3)}{C_t (mg/m^3)}$$

The working range of a method is the range of contaminant concentration (mg/m³) that may be quantitated at a specified air volume (liters). The lower boundary of the working range is defined by a sample that has a mass of contaminant equal to the LOQ. The upper boundary is defined by sampler capacity. Therefore, the working range may be calculated as follows:

$$\text{lower boundary}(mg/m^3) = \frac{LOQ(mg) \times 10^3 (L/m^3)}{V(L)}$$

$$\text{upper boundary}\ (mg/m^3) = \frac{W_{max}(mg) \times 10^3 (L/m^3)}{V(L)}$$

Sampling Media for Gases and Vapors

Solid Sorbents

The most widely used sampling media for gases and vapors are solid sorbents that adsorb the contaminant onto the surface of the sorbent material. Solid sorbents consist of either small granules or beads. To be effective, a sorbent should

1. Trap and retain all or nearly all of the contaminant from an airstream;
2. Be amenable to desorption of the trapped contaminants from the sorbent;
3. Have sufficient capacity to retain a large enough amount of the contaminant to facilitate analysis without creating too large a pressure drop across the sampling media;
4. Not cause a chemical change of the contaminant except when the analytical method is based on derivitization of the contaminant; and
5. Adsorb the contaminant of interest even in the presence of other contaminants, possibly in higher concentrations than the contaminant of interest.[15]

Properties of commonly used solid sorbents are shown in Table 10.4.

Collection Efficiency of Solid Sorbents

Validated air sampling methods will specify a sorbent material that will effectively trap the contaminant(s) of interest. But several factors may affect the collection efficiency of solid sorbents.[18]

Temperature. Because the adsorption process is exothermic, adsorption efficiency is limited at higher temperatures. In addition, reactivity increases with rising temperatures.

Humidity. As water vapor is adsorbed by a sorbent, the sorbent's adsorption capacity for the contaminant of interest may decrease. This effect is most pronounced for sorbents such as charcoal.

Sampling Rate. Higher flow rates lower the sampling efficiency of solid sorbents. For sorbents whose capacity is significantly reduced by high humidity, reducing the sample flow rate may improve the collection efficiency.

Other Contaminants. The presence of significant concentrations of air contaminants (other than the contaminant of interest) may also reduce the collection efficiency of the sorbent for the target compound. This effect is most pronounced for contaminants within the same chemical family. Some classes of compounds can displace other less tightly adsorbed analytes in mixed atmospheres and cause breakthrough or losses.

Sample Breakthrough. Most commercially available sampling tubes consist of two sections of sorbent separated by glass wool or polyurethane. In charcoal tubes the second or backup section is usually one-third of the total weight of the charcoal. These two sections are desorbed and analyzed separately in the laboratory. As a guideline, if 25% or less of the amount of contaminant collected on the front section is found on the backup section, significant loss of the compound (breakthrough) has probably not occurred. If greater than 25% is detected, breakthrough is evident and results should be reported as "breakthrough, possible sample loss." More specific information can be obtained by a detailed study of the breakthrough profiles for each specific sorbent and chemical.[19]

In some cases, a false indication of breakthrough may be caused by diffusion (migration) of the compounds collected on the front section to the backup section over an extended storage period. Sample migration can be reduced by refrigerating or freezing samples as soon as possible, or it can be eliminated by separating the front and backup sections immediately after sampling. Several methods specify the use of two tubes in series rather than a double-layer tube to avoid this problem (e.g., NIOSH Method 1024 for 1,3-Butadiene). In some cases, field desorption is recommended before transport or storage of the samples.

Desorption of Contaminants

Air contaminants are desorbed from solid sorbents using solvent extraction or thermal desorption techniques. With solvent extraction a few milliliters of a specific solvent are used to extract the contaminants of interest from the adsorbent material. Desorption efficiency is a measure of how much analyte can be recovered from the sorbent tube. It is determined typically at 0.1, 0.5, 1, and 2 times the target concentration based on the recommended air volume, and expressed as a percent of analyte spiked on the sorbent tube.[3] Desorption efficiency should be determined for each lot number of solid sorbent used for sampling and should be done in the concentration range of interest. Solvent desorption, the most frequently used technique, is specified by NIOSH for most occupational hygiene analyses. The most common desorption solvent is carbon disulfide because it has a high desorption efficiency for many organic compounds and produces minimum interference in GC analysis using flame ionization detection.

In some cases, however, the recoveries are poor when using carbon disulfide. For example, higher desorption efficiencies can be attained with mixtures such as methanol and methylene chloride when analyzing for cellosolves®. For specific alcohols, better desorption efficiencies are obtained with a mixture of carbon disulfide and another alcohol.

Thermal desorption works by driving the contaminant off the sorbent by subjecting it to a high temperature. The desorbed contaminant is carried in an inert gas stream such as nitrogen, argon, or helium directly into the analytical instrument, typically a GC or gas chromatograph/mass spectrometer (GC/MS). Because the contaminant is not diluted by a desorption solvent, the entire mass of contaminant collected can be introduced directly into the analytical instrument (rather than an aliquot of a solution). Therefore, this procedure can be used to measure very low levels, often subparts per billion. Additionally, thermal desorption is preferred in GC/MS analyses because solvents used in the desorption process could interfere with the analysis of volatile compounds by masking certain compounds.

Types of Sorbent Material

Activated Charcoal. Charcoal is the most widely used solid sorbent for occupational hygiene sampling. Many of the OSHA and NIOSH sampling procedures for organic vapors are based on sample collection using activated charcoal. Charcoal can be obtained from a variety of sources, including coconut shells, coal, wood, peat, and petroleum.[15] Each form has its own characteristics and uses. However, the most common forms for occupational hygiene sampling are derived from either coconut shells or petroleum.

Activated charcoal has a high adsorptive capacity due to its microporous structure and its high surface area-to-weight ratio. This high adsorptive capacity makes activated charcoal an ideal sorbent for sampling chemically stable compounds over a wide concentration range. High humidity, however, has an adverse effect on the adsorptive capacity of charcoal.[20] If high humidity or other factors that might limit capacity are present in the sampling environment, it may be necessary to use a sampling tube with a larger amount of sorbent material than the one specified in the method.

Despite its wide application, charcoal has some limitations. It is not an efficient collector for very volatile low molecular weight hydrocarbons such as methane and ethane,[21] nor is it effective for low boiling compounds such as ammonia, ethylene, and hydrogen chloride. Due to its oxidizing surface, charcoal is not suitable for the collection of reactive compounds such as mercaptans or aldehydes.[5] Coconut charcoal in particular has been found to be a poor collector for ketones. When ketones are adsorbed, they break down by reactions involving water that are catalyzed by the charcoal surface. Hence, desorption efficiency and storage stability are poor.[22]

Silica Gel. Silica gel is a very useful sorbent for the collection of polar compounds such as alcohols, amines, and phenols. Silica gel is less reactive than charcoal, and it allows for effective analytical recoveries when using nonpolar desorbing solvents such as carbon disulfide.[23] A significant disadvantage of silica gel, however, is that it shows a sharp decrease in capacity with increasing humidity. In fact, silica gel is so hydrophilic that it will preferentially bond water molecules at any humidity causing displacement of less strongly held compounds.[24]

Silica gel has been specified in several OSHA and NIOSH methods for the collection of aliphatic and aromatic amines, aminoethanols, and nitrobenzenes. Several inorganic acids can also be collected using a special prewashed silica gel tube with a glass fiber prefilter inside the tube.

Porous Polymers. Porous polymers are being used extensively in both occupational hygiene and environmental sampling. The wide variety of these materials allows for selectivity to particular applications, and their thermal stability at high temperatures enables them to be thermally desorbed.[15] Commercially available porous polymers include Tenax®, Porapaks®, Chromo-sorbs®, and Amberlite® XAD resins.

Tenax. Tenax is a polymer of 2,6 diphenyl-p-phenylene oxide. It is the sorbent of choice for sampling for low-level contaminants due to its high thermal limit of 375°C, which is amenable to thermal desorption. At very low concentrations, such as those found in environmental samples, the relatively small surface area of Tenax is not as significant a concern. Tenax can be used to collect a broad range of organic compounds, particularly volatile, nonpolar organics having boiling points in the range of 80 to 200°C.[25] It is also used as the sorbent for some explosives, such as trinitrotoluene (TNT) and dinitrotoluene (DNT).

Porapaks. Porapaks comprise a group of porous polymers with a wide range of polarity. The least polar member of the group, Porapak P, has chromatographic properties that facilitate the separation of compounds such as ketones, aldehydes, alcohols, and glycols.[15] Porapak P is specified by NIOSH for the collection of dimethyl sulfate. At the other end of the spectrum, Porapak T is sufficiently polar to separate water and formaldehyde and is useful for the collection of hexachlorocyclopentadiene. Porapaks offer a range of sorbents, but the most polar members have disadvantages because of strong water retention and the greater energy required to remove contaminants for analysis.

Chromosorbs. Chromosorbs have properties similar to those of the Porapaks. Chromosorb 101 is the least polar member of this sorbent group, while Chromosorb 104 is the most polar. Chromosorb 102 is particularly useful as a sorbent for air contaminants because of its large specific surface area. NIOSH methods specify this sorbent for the collection of pesticides including chlordane and endrin. In a 1995 interlaboratory study Chromosorb 106 was found to be the most desirable sorbent for occupational hygiene applications in Europe using thermal desorption as the method of analysis.[26]

Amberlite XAD Resins. These porous polymers include a number of different types, but most air sampling methods using these materials specify XAD-2. NIOSH and OSHA specify XAD-2 for the collection of a wide variety of organophosphorus pesticides, and both agencies have also developed methods for collecting formaldehyde using XAD-2 coated with 2-(hydroxymethyl) piperidine. XAD-7, 8, 9, and 12 are most suitable for polar compounds and XAD-1, 2, and 4 for nonpolar compounds.

Other Solid Sorbents. Some of the more recent OSHA methods specify the use of beaded rather than granular sorbents. These new sorbents are more effective than traditional sorbents in collecting many volatile, polar, and reactive compounds due to improved capacity, desorption, and stability. Anasorb® 747 is a beaded synthetic carbon that has been shown to collect polar compounds effectively, including ketones. It is specified in a number of OSHA methods for the collection of compounds such as methanol, ethanol, propylene oxide, and phosphine. Anasorb CMS is a beaded carbon molecular sieve. Due to the enhanced adsorption potential of this material, it can collect even the most volatile compounds, including anesthetic gases.[24]

Other solid sorbents include molecular sieves, florisil, and polyurethane foam. The molecular sieves have limited application in occupational hygiene, but form useful substrates for liquid media that are suitable for the collection of oxides of nitrogen and for trapping gases such as carbon monoxide and carbon dioxide. The use of florisil is limited to polychlorobiphenyls (PCBs). Polyurethane foam is used primarily for environmental sampling of organochlorine pesticides, PCBs, and polynuclear aromatic hydrocarbons (PNAs).

Multiple Stage Collection Media. In some situations, such as at hazardous waste cleanup sites, occupational hygienists are called on to evaluate employee or community exposures to complex mixtures of organic contaminants. The suitable sorbent for such applications is one that is capable of sampling a wide variety of contaminants with an analytical technique capable of both quantitative and qualitative analyses. Recently developed techniques use multiple sampling tubes to adsorb organic contaminants followed by thermal desorption, cryogenic trapping, and analysis by GC/MS. One such technique, developed specifically for monitoring at hazardous waste sites, consists of two-stage glass sampling tubes (8-mm o.d. and 6-mm i.d., 25 cm long) that are flame-sealed at both ends and contain two sorbent sections retained by glass wool. The two sections include (1) a front section of 100 mg of Tenax-GC, 20/35 mesh; and (2) a back section of 200 mg of Chromosorb 102, 20/40 mesh.

By using a two-stage sorbent tube, the occupational hygienist is able to sample for a greater range of compounds than when using a sampling tube containing only a single sorbent.[27]

Sorbent tubes are also used in combination with filter cassettes to effectively trap multiphase contaminants. For example, the NIOSH method for sampling PNAs (Method 5506) uses a PTFE (polytetrafluoroethylene) filter to collect the particulate fraction, followed by an XAD-2 sorbent tube to collect vapors. Another example is measurement of organotin compounds (NIOSH Method 5504) using a glass fiber filter followed by XAD-2 sorbent tube.

To overcome some of the inconveniences of sampling with a separate tube and prefilter, OSHA researchers developed a tube called the OSHA versatile sampler (OVS). The OVS is a specially designed glass tube that contains a filter to trap aerosols and a two-section sorbent bed to adsorb vapors. OVS tubes are currently available for organophosphorus pesticides, TNT, DNT, and phthalates (see Figure 10.4).

Chemically Treated Filters

Filters have proven to be an effective substrate for various liquid media that can trap a variety of airborne contaminants. Most liquid media used to treat filters chemically derivatize the contaminant(s) of interest producing a more stable compound for storage and analysis. To further enhance the stability of samples, some methods stipulate that filters should be transferred a short time after sampling to glass vials containing a specified liquid.

When compared to wet chemistry methods using bubblers, chemically treated filters are less cumbersome and safer to use in the field and can provide improved collection efficiency for some compounds (e.g., methyl mercaptan). In addition, a front and a back filter can be used in one cassette to determine if sample breakthrough has occurred. Chemically treated filters are specified by OSHA or NIOSH for the collection of aromatic amines, glutaraldehyde, diisocyanates, and fluorides.

Liquid Absorbers

Sampling techniques that use liquid sampling media either include absorption based on solubility or may involve a reaction of the contaminant with the sampling solution. Four basic types of sampling devices have been used in conjunction with liquid absorbers: gas washing bottles, spiral and helical absorbers, fritted bubblers, and glass-bead columns. Of these, only the gas washing bottles and fritted bubblers are

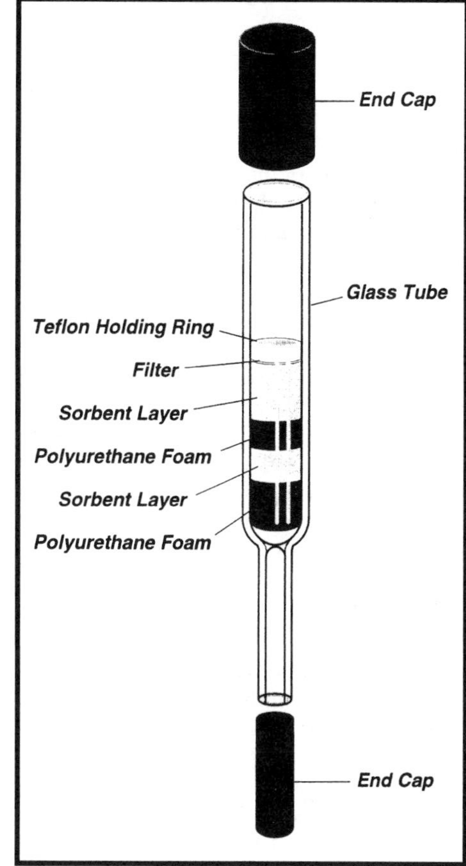

Figure 10.4 — OSHA versatile sampler.

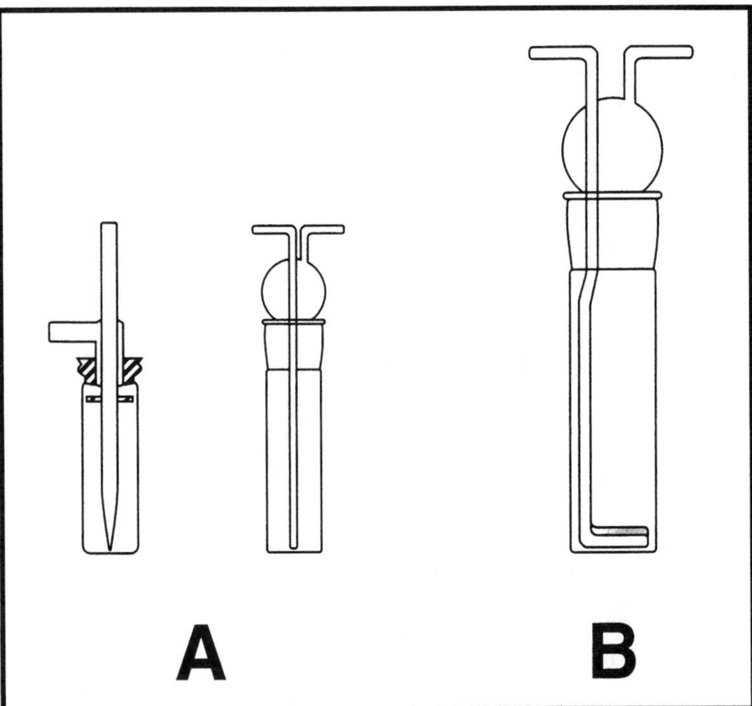

Figure 10.5 — Liquid absorbers: (A) gas washing bottles (B) fritted bubbler.

now used routinely in occupational hygiene sampling (see Figure 10.5).

Gas Washing Bottles

Gas washing bottles include Drechsel types, standard Greenburg-Smith devices, and midget impingers. The air is bubbled through the liquid absorber to secure intimate mixing of air and liquid, and the length of travel of the gas through the collecting medium is equivalent to the height of the absorbing liquid. These scrubbers are suitable for gases and vapors that are readily soluble in the absorbing liquid or react with it. One unit or two units in series may be enough for efficient collection; however, in some cases, several in series may be needed to attain the efficiency of a single fritted glass bubbler. The advantages of these devices include simple construction; ease of rinsing; and, with the exception of Greenburg-Smith impingers, small liquid volume requirements.

Fritted Glass Bubblers

In fritted glass bubblers air passes through formed porous glass plates and enters the liquid in the form of small bubbles. The size of the air bubbles depends on the liquid and the diameter of the orifices from which the bubbles emerge. Frits are classified as fine (25–50 μm), coarse (70–100 μm), or extra coarse (145–175 μm), depending on the pore size. The extra-coarse frit is used when a higher flow rate is desired. The heavier froth generated by some liquids increases the time of contact of gas and liquid. Fritted glass bubblers are more efficient collectors than gas washing bottles and can be used for the majority of gases and vapors that are soluble in the reagent or react rapidly with it. Flow rates between 0.5 and 1.0 L/min are commonly used. Fritted glass is not suitable for sampling contaminants that form a precipitate during sampling that could clog the fritted glass or inhibit quantitative recovery of the sample. In this case, a gas washing bottle should be used. To prevent clogging, it may also be necessary to include a prefilter at the bubbler inlet when sampling air with a high particulate content.

Glass sampling devices are cumbersome, especially for on-worker sampling. Many liquid absorbers contain corrosive or toxic substances, including strong acids, bases, or toxic organics and pose safety and exposure problems for the users. In addition, the evaporation of the liquid media poses problems of sample loss and worker exposure. Thus, the use of liquid absorbers has declined in favor of solid sorbent samplers or chemically treated filters.

Sampling Bags/Partially Evacuated Rigid Containers

In certain situations it may be necessary to collect grab samples of gases and vapors in either flexible sampling bags or partially evacuated rigid containers. Such situations include the following.

- Field applications where samples will be analyzed on site using portable, direct-reading instruments
- Leak, spill, or other emergency situations requiring quick sample collection and analysis so that appropriate control measures can be taken
- Measurements of peak concentrations of contaminants from specific plant processes or worker tasks
- Collection of gases or highly volatile compounds for which adsorption methods are not available or are not efficient

Commercially available sampling bags are listed in Table 10.5. The major drawbacks in the use of the flexible sampling bags are potential instability of the gas or vapor mixture in storage and sample loss through rupture or leakage in transport and handling. Loss of contaminant from a sampling bag may occur through reaction of the contaminant, adsorption of the contaminant onto the bag material, diffusion of the contaminant through the bag material,

Table 10.5 — Commercially Available Sampling Bags[a]

Type	*Chemical Composition*
Five-layer aluminized	inner layer-polyethylene
Halar	copolymer of ethylene and trifluoroethylene
Saran®	polyvinylidene chloride
Tedlar®	polyvinyl fluoride
Teflon®	FEP (fluorinated ethylene, propylene)

[a]Data from Posner and Woodfin[28]

or leakage around the valve stem and seams. Despite these drawbacks, flexible sampling bags are widely used. To minimize errors, the following precautions should be taken:

1. Bags should be purged several times with an inert gas such as nitrogen to remove any trace contaminants before sampling.
2. Sampling bags selected should be relatively impermeable to the contaminant being sampled.
3. Storage of the sample in the bag should not be extended beyond the recommended time for the specific air contaminant and bag material.
4. Bags should not be completely filled, especially if they will be transported by airplane. Changes in pressure or temperature can cause the bag to rupture.
5. If bags are to be reused, they should be cleaned according to the manufacturer's instructions and tested to verify cleanliness before reuse. It may be desirable to leak-test them using a soap solution applied to potential leakage points, such as valve stems and along seams, or by immersion of a filled bag in water.
6. If bags are being used in a heavily contaminated area, it may be desirable to place the bags into an evacuation chamber and fill by negative pressure to avoid contamination of the pump or a degradation of analyte concentration by interaction with the pump (see Figure 10.6).

Evacuated flasks include heavy-walled glass containers (Figure 10.7) and other plastic or metal containers in which 99.97% or more of the air has been removed by a heavy-duty vacuum pump. Sampling is conducted by breaking the seal to permit the air to enter the flask. Commercially available evacuated flasks include glass bulbs and flasks provided by most laboratory supply houses.

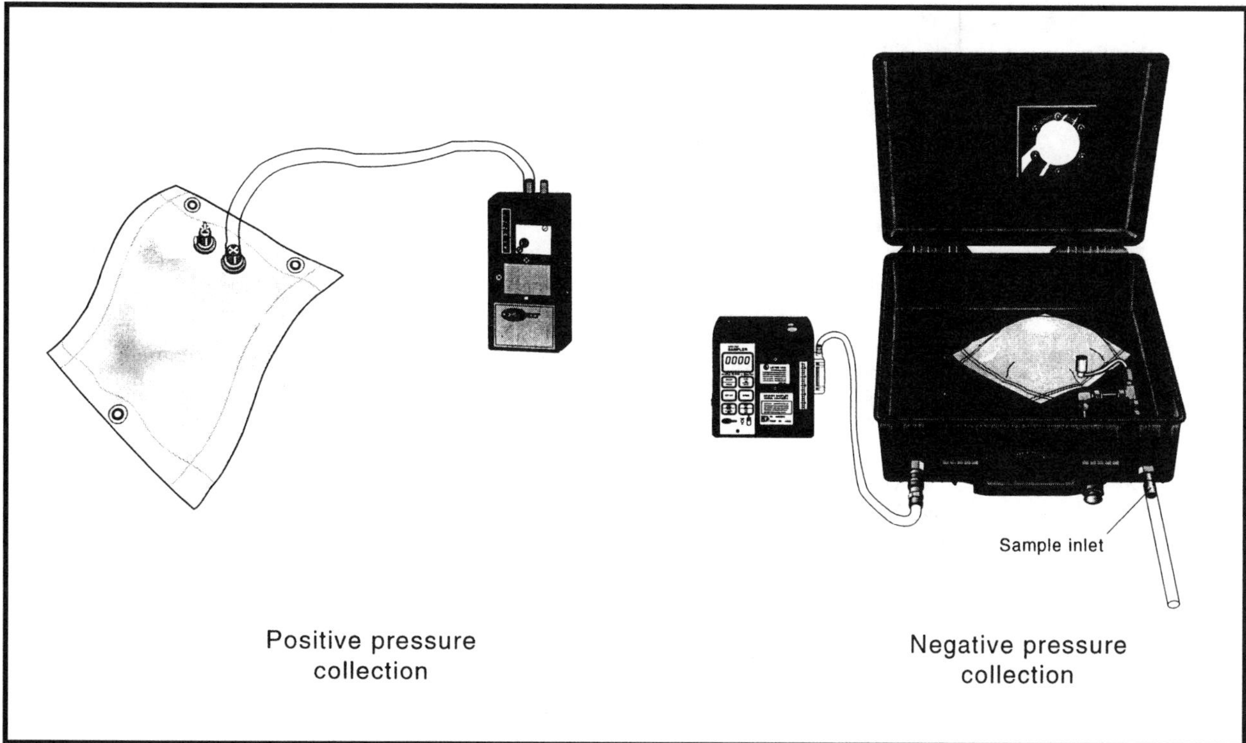

Figure 10.6 — Sample collection using sampling bags.

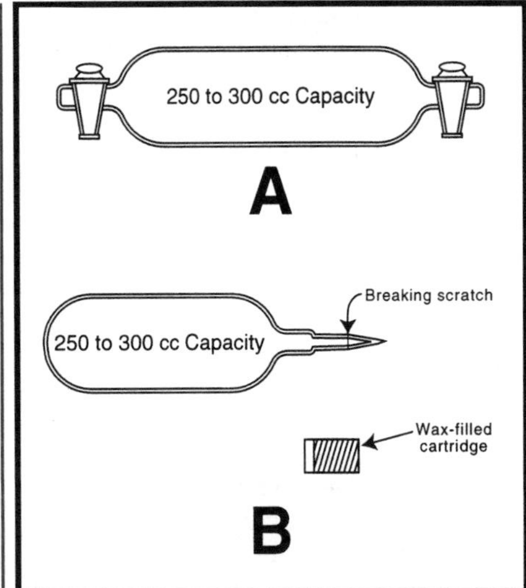

Figure 10.7 — Grab sample bottles: (A) gas or liquid displacement (B) evacuated flask.

Pre-evacuated stainless steel canisters have been widely used as an air collection vessel for contaminants found at extremely low levels for environmental assessments. The interior of the canister is electrochemically passivated by a SUMMA® process that prevents reaction of the sample with the canister, ensuring long-term stability even in the ppb range. Canisters are available in both subatmospheric and pressurized sampling modes. Pressurized sampling requires a pump to provide positive pressure to the sample canister. After sample collection the air is analyzed by GC coupled to one or more appropriate GC detectors.[29]

Cold Traps

Cold traps have been used to collect analytes from a sampled airstream when the compounds are too unstable or reactive or to be collected efficiently by other techniques. Cold traps have also been used to collect unknown contaminants in liquid or solid form for identification purposes. With this technique, vapor is separated from air by passing it through a coil or other vessel that has been immersed in a cooling system such as dry ice, liquid air, or liquid nitrogen.

A variety of traps are available, including the U-tube design. The most efficient traps for field use, however, are the reverse gas washing bottle type in which the analytes are frozen from the airstream by first contacting the outer cold walls of the vessel. The airstream is drawn through the sampling vessels by a constant flow pump. The cold trap technique can also be combined with SUMMA canister sampling in which the sample air is metered into a chilled canister. Sample vapors are transferred from the trap to the analytical instruments by warming, followed by vacuum trapping or selective temperature gas flows using inert gas streams. Cold trap techniques cannot be conveniently used for personal sampling, but apply well to area sampling of indoor or ambient air.

Calculations

The analytical laboratory will determine the total amount of analyte collected on the sampling device and will normally report the results in total mass of the specified contaminant. To determine the airborne concentration, it will be necessary to divide the mass of contaminant by the volume of air sampled. The air volume is calculated by multiplying the average flow rate of the sampler by the sampling time.[30]

Since the results will typically need to be calculated in mg/m^3, it may be necessary to first convert the analytical results from micrograms to milligrams and the air volume from liters to cubic meters (micrograms $\times$ 10^{-3} = milligrams; mg/liter $\times$ 10^3 = mg/m^3).

Many exposure standards for gases and vapors are expressed as the volume fraction of the analyte in the sample atmosphere, usually in ppm. For example, many of the ACGIH TLVs are specified at a normal temperature of 25°C (298K) and a pressure of 760 mm Hg. The following formula, based on the ideal gas law, converts the measured analyte concentration from mg/m^3 to ppm at 25°C and 760 mm Hg:

$$\text{ppm} = \frac{\text{mg}}{\text{m}^3} \times \frac{24.45}{\text{MW}} \times \frac{(T + 273.15)}{298.15} \times \frac{760}{P}$$

where:

mg/m^3 = analytical result, in mg/m^3

24.45 = molar volume (L/mole) at 25°C and 760 mm Hg

MW = molecular weight (g/mole)

T = temperature (°C) of air sampled

273.15 = conversion of °C to K

298.15 = normal temperature, in K

P = pressure (mm Hg) of air sampled

760 = normal barometric pressure (mm Hg)

The conversion factors of 0.001 g/mg, 0.001 m^3/L, and 1,000,000 ppm cancel each other and are not shown in the equation.

The above formula is necessary for an accurate conversion from mg/m³ to ppm when sampling at a high or low pressure, such as at a high altitude or at an extremely hot or cold temperature. Otherwise it is common occupational hygiene practice to assume that sampling was conducted near normal temperature and pressure and to use the following simplified version of the formula to convert the measured analyte concentration from mg/m³ to ppm:

$$\text{ppm} = \text{mg/m}^3 \times 24.45/\text{MW}$$

Once airborne concentrations have been determined, it is necessary to compare those levels to appropriate exposure standards. To determine compliance with full-shift occupational exposure limits, 8-hour TWAs will need to be calculated:

$$\text{8-hour TWA} = \frac{C_1T_1 + C_2T_2 + C_3T_3 ... + C_nT_n}{8}$$

where:

Cx = concentration of an individual sample
Tx = sample time of an individual sample

The number eight is used in the denominator as most standards are based on an 8-hour exposure followed by 16 hours of rest.

Example: If three benzene samples collected on a refinery worker over an 8-hour workday revealed the following exposures—0.5 ppm for a 4.5 hour sample; 1.5 ppm for a 1.0 hour sample; and 0.8 ppm for a 2.5 hour sample—what would be the 8-hour TWA?

$$\text{8-hour TWA} = \frac{(0.5)(4.5) + (1.5)(1) + (0.8)(2.5)}{8} = 0.7 \text{ ppm}$$

If individuals are working extended or unusual shifts in which the exposure time is lengthened or the recovery time is lessened, occupational exposure limits may need to be adjusted. Brief and Scala have developed one such model that has been used to make adjustments to workplace exposure standards for extended or unusual work shifts.[31] This model reduces the allowable limit for both increased exposure time and reduced recovery time.

Additional Sources

Air sampling method manuals by OSHA and NIOSH (Listed in Table 10.1).

Cohen, Beverly S. and Herings, Susanne V. (ed.): *Air Sampling Instruments for Evaluation of Atmosphereic Contaminants.* 8th ed. Cincinnati, OH: ACGIH, 1995.

References

1. *Code of Federal Regulations,* Title 29, Part 1910. 1995. (See standards on individual chemicals, e.g., 1910.1047, 1910.1048, and 1910.1028.)
2. **Soule, R.D.:** Industrial hygiene sampling and analysis. In *Patty's Industrial Hygiene and Toxicology,* vol. 1. New York: John Wiley & Sons. 1991. pp. 73–135.
3. **U.S. Department of Labor:** *OSHA Analytical Methods Manual.* Salt Lake City, Utah: U.S. Department of Labor, Occupational Safety and Health Division Directorate for Technical Support, 1990.
4. **Feldman, R.F.:** Degraded instrument performance due to radio interference: criteria and standards. *Appl. Occup. Environ. Hyg. 8*:351–355 (1993).
5. **McCammon, C.S.:** General considerations for sampling airborne contaminants. In *NIOSH Manual of Analytical Methods,* 4th ed. Cincinnati, Ohio: U.S. Department of Health and Human Services. 1994. pp. 16–31.
6. **Lippmann, M.:** Calibration. In *Patty's Industrial Hygiene and Toxicology,* vol. 1. New York: John Wiley & Sons, 1991. pp. 461–530.
7. **Rose, V.E. and J.L. Perkins:** Passive dosimetry—state of the art review. *Am. Ind. Hyg. Assoc. J. 43*:605–621 (1982).
8. **Lugg, G.A.:** Diffusion coefficients of some organic and other vapors in air. *Anal. Chem. 40*:1072–1077 (1968).
9. **Hickey, J.L.S. and C.C. Bishop:** Field comparison of charcoal tubes and passive vapor monitors with mixed organic vapors. *Am. Ind. Hyg. Assoc. J. 42*:264–267 (1981).
10. **Cassinelli, M.E., R.D. Hull, J.V. Crable, and A.W. Teass:** Protocol for the evaluation of passive monitors. In *Diffusive Sampling: An Alternative Approach to Workplace Air Monitoring.* London: Royal Society of Chemistry, Burlington House, 1987. pp. 190–202.

11. **Brown, R.H., R.P. Harvey, C.J. Purnell, and K.J. Saunders:** A diffusive sampler evaluation protocol. *Am. Ind. Hyg. Assoc. J. 45*:67–75 (1984).
12. **Lautenbeger, W.J., E.V. Kring, and J.A. Morello:** Theory of Passive Monitors. *Ann. of Am. Conf. Gov. Ind. Hyg. 1*:91–99.
13. **Eller, P.M.:** Operational limits of air analysis methods. *Appl. Ind. Hyg.* 2:91–94 (1986).
14. **Keith, L.H., W. Crummett, J. Deegan, R.A. Libby, J.K. Taylor, and G. Wentler:** Principals of environmental analysis. *Anal. Chem. 55*:2210–2218 (1983).
15. **Crisp, S.:** Solid sorbent gas samplers. *Ann. Occup. Hyg. 23*:47–76 (1980).
16. **Harper, M. and C.J. Purnell:** Alkylammonium montmorillonites as adsorbents for organic vapors from air. *Environ. Sci. Technol. 24*:55–62 (1990).
17. **Stanetzek, I., U. Giese, R.H. Schuster, and G. Wunsch:** Chromatographic characterization of adsorbents for selective sampling of organic air pollutants. *Am. Ind. Hyg. J. 57*:128–133 (1996).
18. **Melcher, R.G.:** Laboratory and field validation of solid sorbent samplers. In *Sampling and Calibration for Atmospheric Measurements, ASTM STP 957*. Philadelphia: American Society for Testing and Materials, 1987. pp. 149–165.
19. **Harper, M.:** Evaluation of solid sorbent sampling methods by breakthrough volume studies. *Ann. Occ. Hyg. 37*:65–88 (1993).
20. **Rudling, J. and E. Bjorkholm:** Effect of adsorbed water on solvent desorption of organic vapors collected on activated carbon. *Am. Ind. Hyg. Assoc. J. 47*:615–620 (1986).
21. **Tang, Y.Z., W.K. Cheng, P. Fellin, Q. Tran, and I. Drummond:** Laboratory Evaluation of Sampling Method for C1-C4 Hydrocarbons. *Am. Ind. Hyg. Assoc. J. 57*:245–250 (1996).
22. **Harper, M. and M.L. Kimberland, R.J. Orr, and L.V. Guild:** An evaluation of sorbents for sampling ketones in workplace air. *Appl. Occup. Environ. Hyg. 8*:293–304 (1993).
23. **Guenier, J.P., F. Lhuillier, and J. Muller:** Sampling of gaseous pollutants on silica gel with 2400 mg tubes. *Ann. Occup. Hyg. 30*:103–114 (1986).
24. **Harper, M.:** Novel sorbents for sampling organic vapours. *Analyst 119*:65–69 (1994).
25. **U.S. Environmental Protection Agency:** *Compendium of Method for the Determination of Toxic Organic Compounds in Ambient Air*, by W. T. Winberry, N.T. Murphy, and R.J. Riggan (EPA/600/4-89/017). Research Triangle Park, NC: U.S. Environmental Protection Agency/Office of Research and Development, 1988. Method TO-1.
26. **European Commission:** *Study of Sorbing Agents for the Sampling of Volatile Compounds from Air (EC Contract MAT1-CT92-0038).* Measurements and Testing Programme of the European Commission. Sheffield, UK: Health and Safety Laboratory, Health and Safety Executive, 1995.
27. **Singh, J., K.H. Vora, and A.W. Eissler:** "Onsite Air Monitoring Classification by the Use of a Two-stage Collection Tube." Paper presented at the American Industrial Hygiene Conference, Detroit, MI, May 24, 1984.
28. **Posner, J.C. and W.J. Woodfin:** Sampling with gas bags I: Losses of analyte with time. *Appl. Ind. Hyg.*: 163–168.
29. **U.S. Environmental Protection Agency:** *Compendium of Method for the Determination of Toxic Organic Compounds in Ambient Air*, by W. T. Winberry, N.T. Murphy, and R.J. Riggan (EPA/600/4-89/017). Research Triangle Park, NC: U.S. Environmental Protection Agency/Office of Research and Development, 1988. Method TO-14.
30. **DiNardi, S.R.:** *Calculation Methods for Industrial Hygiene.* New York: Van Nostrand Reinhold, 1995. pp. 186–189.
31. **Brief, R.S. and R.A. Scala:** Occupational Exposure Limits for Novel Work Schedules. *Am. Ind. Hyg. Assoc. J. 36*:467–469 (1975).

Another hazard in the agricultural industry is the application of pesticides. These hazards involve not only application, but mixing as well as re-entry into the fields before the degradation of the pesticide.

Outcome Competencies

After completing this chapter, the user should be able to:

1. Define underlined terms used in this chapter.
2. Distinguish between qualitative and quantitative analytical methods.
3. Recall the operating principles for gas, liquid, and ion chromatograph detectors.
4. Summarize the operating principles for gas, liquid, and ion chromatography.
5. Explain principles of gas and vapor analysis.
6. Select analytical methods to satisfy occupational hygiene objectives.
7. List a minimum of three sources of laboratory analytical methods for gases and vapors.

Key Terms

Beer-Lambert law • electrochemical detector • electron capture detector (ECD) • flame ionization detector (FID) • flame photometric detector (FPD) • fluorescence detector • gas chromatography • gas chromatography/mass spectrometry (GC/MS) • high performance liquid chromatography (HPLC) • ion chromatography (IC) • nitrogen-phosphorus detector • photoionization detector (PID) • retention time • spectrophotometry • thermal conductivity detector • UV absorbance detector

Prerequisite Knowledge

College chemistry, any textbook.

Prior to beginning this chapter, the user should review the following chapters:

Chapter Number	Chapter Topic
21	Nonionizing Radiation
22	Ionizing Radiation
25	Thermal Standards and Measurement Techniques

Key Topics

- I. Chromatographic Methods
  - A. Gas Chromatography
  - B. Gas Chromatography/Mass Spectrometry
  - C. High Performance Liquid Chromatography
  - D. Ion Chromatography
- II. Volumetric Methods
- III. Spectrophotometric Methods
- IV. Quality Assurance

11

Analysis of Gases and Vapors

Patrick N. Breysse, Ph.D., CIH

Peter S.J. Lees, Ph.D., CIH

Introduction

The purpose of this chapter is to introduce the practicing occupational hygienist to the range of analytical techniques used to quantify gases and vapors in the environment. Analytical chemistry techniques allow the occupational hygienist to quantify one or more contaminants in a sample. Quantification of individual air contaminants is accomplished either by the selectivity of the analytical method itself, such as spectrophotometry, or by combining a nonselective analytical method with a separation technique, such as chromatography.

Unlike his or her predecessors the modern occupational hygienist rarely conducts the chemical analysis of samples he or she collects. Due to a general increase in specialization and specifically to the accreditation requirement for laboratories analyzing occupational hygiene samples, virtually all occupational hygiene samples are now analyzed by a relatively small number of regional and national laboratories served by overnight delivery services. Based on a large volume of analyses, these laboratories have the resources to maintain a high degree of analytical proficiency for a wide variety of substances along with rigorous quality control programs that are simply not economically feasible in a laboratory conducting a small number of analyses.

Despite these changes, it is still crucially important for the occupational hygienist to have at least a working understanding of analytical methods and procedures. The most prominent example of the need for such knowledge comes in the selection of a sampling and associated analytical method. The occupational hygienist must be fully cognizant of the requirements and limitations of the analytical method prior to conducting any air sampling survey.

Most important, he or she must be aware of possible interferences that may be present that may complicate or invalidate analysis. A good example of this problem is the collection and analysis of formaldehyde using National Institute for Occupational Safety and Health (NIOSH) Method 3500.[1] In this analytical methodology the presence of phenol in the sample will result in an apparent reduction in the amount of formaldehyde present, a phenomenon called negative interference. If the occupational hygienist is not aware of this methodological limitation, the resulting estimate of worker exposure will be significantly reduced, with possible health and/or regulatory implications. On the other hand, a thorough occupational hygienist will be mindful of these limitations and will select another sampling and analytical method, such as NIOSH Method 2541[2] (which is not subject to this problem), or will inform the laboratory to take steps to eliminate the phenol interference. Compensation for the presence of analytical interferences in the laboratory is not always possible, however, and it is important for the occupational hygienist to maintain communication with the laboratory to prevent such occurrences.

The occupational hygienist must also be very cognizant of the limit of detection of the analytical method selected. Without consideration of analytical limits of detection, sampling efforts can be wasted when nondetected contaminants have a concentration limit of detection in excess of the ACGIH threshold limit value (TLV®) or permissible exposure limit (PEL). Given knowledge of the analytical limit of detection and a defined sampling goal (e.g., to be able to detect a concentration 10% of the TLV or PEL for that substance), a sample volume can be calculated to assure the occupational hygienist that, although the substance was not detected, the concentration is low and not of concern.

Analytical methods are typically associated with a particular sampling procedure and are generally presented in the context of a combined sampling and analytical method. To the extent possible, it is important to use validated standard sampling and analytical methods. Of particular importance for workplace exposure evaluation are the sampling and analytical methods recommended by the NIOSH, the Occupational Safety and Health Administration (OSHA), and to a lesser extent, the U.S. Environmental Protection Agency (EPA). Addresses for obtaining NIOSH, OSHA, and EPA methods as well as sampling and analytical methods recommended by the other organizations can be found in Chapter 10, "Sampling of Gases and Vapors."

Table 11.1 —
List of Analytical Methods and Examples of Common Analytes

Analytical Method	*Examples of Analyte Compounds*
GC/flame ionization detector	PAHs, ketones, halogenated hydrocarbons, alcohols, ethers, aromatic hydrocarbons
GC/photoionization detector	ethylene oxide, tetraethyl lead, tetramethyl lead
GC/nitrogen phosphorus detector	acrolein, nicotine, acetone cyanohydrin, organophosphate pesticides
GC/electron capture detector	butadienes, pentadienes, chlordane, polychorinated benzenes, PCBs
GC/flame photometric detector	mercaptans, carbon disulfide, nitromethane, tributylphosphate
GC/thermal conductivity detector	carbon dioxide
GC/mass spectrometry	aldehyde screening
HPLC/ultraviolet detector	acetaldehyde, anisidine, p-chlorophenol, diethylenetriamine, ethylenediamine, maleic anhydride, p-nitroanaline
HPLC/electrochemical detector	isocyanates
Visible absorption spectrophotometry	acetic anhydride, ammonia, formaldehyde, hydrazine, nitrogen dioxide, phosphine
Ion chromatography	aminoethanol compounds, chloroacetic acid, inorganic acids, iodine, chlorine, hydrogen sulfide, sulfur dioxide

Source: **National Institute for Occupational Safety and Health (NIOSH):** *NIOSH Manual of Analytical Methods*, 4th Ed. (DHHS (NIOSH) Pub. No. 94–113). Cincinnati: NIOSH, 1994.

Standard methods are developed to ensure analytical reproducibility so that results will be comparable across laboratories. Also, standard methods have been extensively evaluated and tested in terms of measurement range, precision, accuracy, and interferences. As a result, occupational hygienists are able to interpret results in a statistically meaningful manner. Nonstandard methods can be used if the hygienist can document that a method is at least equivalent to accepted accuracy and precision guidelines. In this case, it is important that complete documentation of analytical methodology be maintained with the sample results.

As noted previously, it is important to select and consult with a qualified laboratory prior to undertaking any sampling activity. Laboratories accredited by the Industrial Hygiene Laboratory Accreditation Program (IHLAP) of the American Industrial Hygiene Association (AIHA) are listed quarterly in the *American Industrial Hygiene Association Journal* and on the AIHA web site (www.aiha.org). The IHLAP is designed specifically for laboratories analyzing workplace exposure samples. IHLAP accreditation requires an evaluation of (1) laboratory personnel qualifications; (2) laboratory facilities; (3) quality control and equipment; (4) laboratory record keeping; and (5) methods of analyses. Accredited laboratories are also required to participate in a Proficiency Analytical Testing (PAT) program collaboratively sponsored by AIHA and NIOSH.[(3)] The PAT program provides blind reference samples for metals, silica, asbestos/fibers, and six solvents to participating laboratories quarterly. Laboratories are considered proficient if their analysis falls within ±3 standard deviations of the reference value.

In addition to using an accredited laboratory with a rigorous internal quality assurance (QA) program, it is important for the field occupational hygienist to provide blind external QA samples to the laboratory. Such samples include blanks, duplicates, and in some instances, spiked samples or samples of known concentration. Components of a good external QA program are discussed in Chapter 39, "Quality Control for Sampling and Laboratory Analysis."

A wide range of analytical methods is available for the quantification of gases and vapors. A list of analytical methods with examples of analytes taken from the *NIOSH Manual of Analytical Methods* is presented as Table 11.1, and the generic layout of a front page from a NIOSH method is

NAME OF SUBSTANCE **METHOD #**

FORMULA Molecular Weight Chemical Abstracts Service # RTECS #

Method numbers are the same as those in the 3rd edition. Evaluation (Full, Partial, Unrated) is assigned as described on p. 5 of the "blue pages". Issue date reflects current version (August 15, 1993) and previous 3rd edition versions, if any.

OSHA : These exposure limit values are
NIOSH: those in effect at the time of
ACGIH: printing of the method.

PROPERTIES: Boiling/melting points, equilibrium vapor pressure, and density help determine the sample aerosol/vapor composition.

SYNONYMS: Common synonyms for the substance, including Chemical Abstracts Service (CAS) numbers; these are all listed alphabetically in the **Index of Names and Synonyms** ("yellow pages" in this Manual).

SAMPLING

SAMPLER: Brief description of sampling EQUIPMENT

FLOW RATE: Acceptable sampling range, L/min

VOL-MIN: Minimum sample volume (L); corresponds to Limit of Quantitation (LOQ) at OSHA PEL

-MAX: Maximum sample volume (L) to avoid analyte breakthrough or overloading

BLANKS: Each set should have at least 2 field blanks, up to 10% of samples, plus 6 or more media blanks in the case of coated sorbents, impinger solutions, or other

ACCURACY

Data are for experiments in which known atmospheres of the substance were generated and analyzed according to the method. Target accuracy is less than 25% difference from actual concentration at or above the OSHA PEL.

MEASUREMENT

TECHNIQUE: The measurement technique used

ANALYTE: The chemical species actually measured

A summary of the measurement EQUIPMENT, SAMPLE PREPARATION, and MEASUREMENT steps appearing on the second page of the method is given here.

CALIBRATION: Summary of type of standards used

RANGE: Range of calibration standards to be used; from LOQ to upper limit of measurement (Note: More concentrated samples may be diluted in most cases to fall within this calibration range.)

ESTIMATED LOD: Limit of detection (background + 3σ)

PRECISION ($\hat{S}_r$): Experimental precision of spiked samplers

APPLICABILITY: The conditions under which the method is useful, including the working range in mg/m^3 (from the LOQ to the maximum sampler loading) for a stated air volume are given here.

INTERFERENCES: Compounds or conditions which are known to interfere in either sampling or measurement are listed.

OTHER METHODS: Methods from the 2nd edition ("P&CAM" and "S" methods) which are related to this one, as well as similar OSHA and literature methods are keyed to REFERENCES.

Figure 11.1 — Layout of cover page for NIOSH sampling and analytical methods.

presented as Figure 11.1. For discussion purposes, analytical methods for gases and vapors are grouped into chromatographic, volumetric, and optical methods.

Chromatographic Methods

Chromatographic methods are powerful tools for the separation of gaseous contaminants and their subsequent individual analysis. Chromatographic techniques were developed in the early 1900s and were followed in the mid-1940s by the development of modern gas chromatography (GC) elements.(4) The word "chromatography" is taken from the Greek for "color-writing" because early efforts involved separating plant pigments. In general, chromatography is the process of separating the components of a mixture by using a mobile phase and a stationary phase. A diagram of the chromatographic process is presented in Figure 11.2. If the mobile phase is a gas, the separation process is called gas chromatography; if the mobile phase is a liquid, the process is referred to as liquid chromatography. The chromatography column contains a stationary phase, which may be a solid material that reversibly adsorbs the sample components, or a liquid into which the sample components dissolve.

Samples are introduced onto the chromatographic column containing the stationary phase in solution with the mobile phase. The repeated interaction between the solutes in the mobile phase and the stationary phase differentially retards the passage of individual solutes in a mixture, providing separation, given a column of sufficient length. Once separated, the individual analytes can be detected using nonselective detectors to quantify the amount of analyte present.

The output signal of a chromatography detector is normally plotted against time. This plot is called a chromatogram, an example of which is shown in Figure 11.3. The response of the detector to carrier gas produces a constant signal referred to as the baseline. The baseline, therefore, represents a detector signal of zero (i.e., when the detector does not detect anything eluting from the column). A peak is a rise in the plot when one or more chemicals are

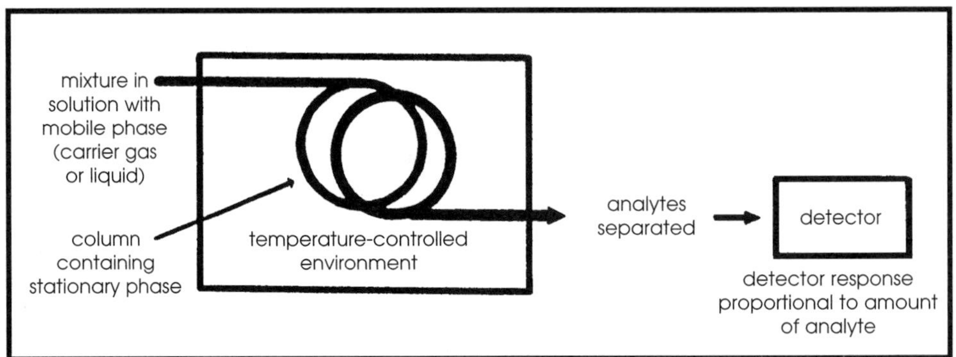

Figure 11.2 — Schematic diagram of generalized chromatographic process.

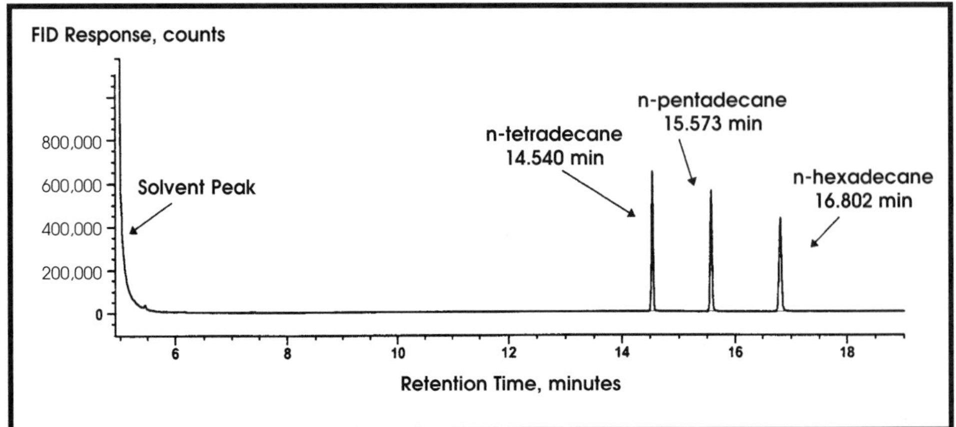

Figure 11.3 — Typical chromatogram produced using a FID and showing separation of n-tetradecane (retention time 14.540 min), n-pentadecane (retention time 15.573 min), and n-hexadecane (retention time 16.802 min).

detected as they elute from the column. The length of time between sample injection and the maximum height of a peak is called the retention time. Retention time is constant for an analyte under a constant set of analytical conditions. Note, however, that retention time does not in and of itself provide definitive identification of a peak; many chemicals may have identical or undetectably similar retention times.

The size (area) of the chromatographic peak corresponding to a given contaminant is directly proportional to the mass of the contaminant injected. The area of the peak is electronically integrated by the gas chromatograph using various algorithms. With proper calibration, the analyst can determine the exact mass of contaminant in an unknown sample. Note, however, that chromatographic detectors do not respond identically to all substances; identical masses of two different contaminants may result in hugely different chromatographic peak areas and vice versa. This difference, known as the response factor, when coupled with the factors that influence retention time, requires individual calibrations for each contaminant of interest under the standard analytical conditions selected for analysis.

The following sections discuss the types of chromatography commonly used for the analysis of gases and vapors in air samples (i.e., GC, high performance liquid chromatography [HPLC], ion chromatography [IC], and GC/mass spectrometry).

Gas Chromatography

The majority of analytical methods for gases and vapors listed in Table 11.1 are gas chromatographic, employing different detectors. GC provides a powerful tool for the analysis of low-concentration air contaminants. GC analysis is applicable to compounds with sufficient vapor pressure and thermal stability to dissolve in the carrier gas and pass through the chromatographic column in sufficient quantity to be detectable. Air samples to be analyzed by GC are typically collected on sorbent tubes and desorbed into a liquid for analysis (see previous chapter).

There are numerous methods that use GC for air sample analysis. NIOSH Method 1500 for hydrocarbons with boiling points between 36°C and 126°C is an example of such a method. This method (which applies to a range of compounds including benzene, toluene, n-hexane, and n-octane) specifies GC/flame ionization detector (FID) analysis.[5]

The basic components of gas chromatograph are a carrier gas system, a sample injector system, a column, a detector, and a recording system. In GC the mobile phase is a gas and is referred to as the carrier gas. The carrier gas system is usually a regulated compressed nonreactive gas (e.g., helium or nitrogen), chosen because it does not interact with the stationary phase.

Samples can be injected onto a GC column as volatile liquid or as a gas. Injection volumes typically range from 0.1 to 50 µL for volatile liquids and from 0.05 to 50 mL for gas samples. The sample must be injected rapidly and evenly to optimize chromatographic separation onto the column. The injector is enclosed in an oven maintained at a higher temperature than the column to prevent condensation of the solvent and/or analyte. Gas or liquid samples can be injected into the column using a syringe with the hypodermic needle inserted through a self-sealing septum contained in the injection port. Laboratories that analyze large numbers of samples use automated sample injection systems that provide for rapid analysis and more precise injection. Gas samples can also be injected using gas sampling valves, although this is not common for routine occupational hygiene analyses.

Recent developments in injection systems involve the use of thermal desorption from sorbent tubes and subsequent gas phase injection of the entire sample. In thermal desorption systems the contaminant is driven from the sorbent at a high temperature into a carrier gas. Since the analytes are not diluted in a desorption solvent, the entire mass of contaminant collected can be introduced directly into the gas chromatograph (rather than an aliquot of a solution). Using this procedure it is possible to quantify lower concentrations of contaminants, often in the subparts per billion range. A limitation of thermal desorption is that there is only one chance for successful analysis since the entire sample is injected. For samples desorbed and diluted in a solvent, the analyst has multiple chances to optimize the analytical conditions, since only a small aliquot of the sample is injected.

After a sample is injected onto the head of a chromatographic column, each component in the sample repeatedly sorbs and desorbs from the mobile phase and the stationary phase. The stationary phase exhibits a different affinity for each component of the sample mixture. Chemical compounds that are more strongly attracted to the stationary phase take longer to be swept through the column by the mobile phase. The end result is that individual compounds elute from the column at different times. Accurate quantification, therefore, depends on the combined abilities of the chromatographic column, carrier gas flow rate, and temperature conditions to separate or "resolve" analytes from other sample components prior to reaching the detector.

GC analysis of air samples uses either packed or capillary columns. Packed columns are typically 0.5- to 2-m long metal or glass tubes in straight, bent, or coiled form, filled with a liquid-coated solid support material. The purpose of the solid support, typically diatomaceous earth, is to provide a large uniform and inert surface area for distributing the liquid coating with which the contaminants interact. The versatility of GC as an analytical tool derives from the wide variety of solvents that can be used as stationary phase coatings in GC columns. The stationary phase polarity, chemical composition, surface area, and film thickness are the major factors influencing the separation process. In general, the composition of the stationary phase chosen will depend on the composition of the analyte. For the most efficient separation, the liquid phase should be chemically similar to the sample being analyzed. For example, nonpolar hydrocarbons are best separated using long chain hydrocarbons, such as paraffin or squalene, as the stationary liquid phase. Polar compounds are better separated using alcohols or amide liquid phases.

In contrast to packed columns, capillary columns are long (30–150 m) open tubes of small diameter (0.25–0.75 mm) with the inside of the tubing coated with the liquid phase. Capillary columns provide better peak resolution because their low resistance to flow allows for longer columns. The tradeoff, however, is that because of their small internal volume, smaller injection volumes must be used.

Most gas chromatographs have large ovens to accommodate a wide variety of columns in a temperature-controlled environment. Temperature control allows chromatographic separation to occur isothermally as well as with programmed temperature changes. Column temperature is adjusted so that it is high enough for the analysis to be accomplished in a reasonable amount of time and low enough that the desired separation is obtained. A rule of thumb is that the retention time of an analyte in a column will double for every 30°C decrease in temperature.[(4)] When there are multiple analytes with a wide range of physical properties in a single sample, temperature programming may be needed to achieve separation in a reasonable time. Temperature programming allows for the separation of analytes with a wide range of boiling points. When a sample is injected onto a relatively cool column, the early peaks result from low boiling compounds that move quickly through the column. Higher boiling compounds, however, will remain immobilized in the stationary phase and will elute very slowly, resulting in long flat peaks. If the temperature in the column is caused to rise through a programmed temperature increase, however, the high boiling point compounds will move more quickly through the column.

Since chromatography is only a separation technique, a detector is required to quantify the amount of each analyte in the column effluent. The response of the detector is typically unspecific and is proportional to the amount of analyte present. To quantify the amount of analyte present, it is necessary to compare the detector responses for the sample components to the detector responses for standards containing known amounts of the compounds of interest. This comparison is referred to as a calibration curve.

Selection of the appropriate detector for the contaminant of interest is essential to realize the full potential of GC analysis. The following detection systems are used most frequently to maximize analytical sensitivity and selectivity.

Flame Ionization Detector (FID). This detector is very sensitive to most organic compounds, including aliphatic and aromatic hydrocarbons, ketones, alcohols, ethers, and halogenated hydrocarbons. It is one of the most widely used gas chromatographic detectors because it has high sensitivity and exhibits a linear response over a wide dynamic range. The FID has a linear range of six to seven orders of magnitude.

FIDs respond only to compounds with oxidizable carbon atoms and therefore will not respond to water vapor, elemental gases, carbon monoxide, carbon dioxide, hydrogen cyanide, formaldehyde, formic acid, water, or to most other inorganic compounds. Importantly, FIDs show little response to carbon disulfide. For this reason, carbon disulfide is frequently used as a charcoal tube desorption solvent for methods using an FID. In an FID, the gas chromatograph column effluent is mixed with hydrogen in excess air prior to passing through a small opening or jet, where it is burned. Ionized combustion products are attracted to a metal "collector" around the flame. The collector is electrically charged at about +200 volts relative to the jet, attracting ions produced by the flame to the collector. Ions reaching the collector create an electrical current, which is amplified and recorded.

Nitrogen-Phosphorus Detector. This detector, also called a thermionic or alkali flame detector, is highly sensitive and selective for nitrogen and phosphorous compounds, including amines and organophosphates. The detector is similar in principle to the FID, except that ionization occurs on the surface of an alkali metal salt, such as cesium bromide, rhobidium silicate, or potassium chloride. The older version of this detector, the alkali flame detector, uses a flame to heat the alkali metal salt. Newer detector designs electrically heat the alkali metal salt to improve stability and reduce response to hydrocarbon interferences.

Flame Photometric Detector (FPD). The FPD is used to measure phosphorus- and sulfur-containing compounds such as organophosphate pesticides and mercaptans. The FPD measures phosphorus- or sulfur-containing compounds by burning the column effluent in a hydrogen-air flame with an excess of hydrogen. Sulfur and phosphorus compounds emit light above the flame; a filter optimized to pass light at 393 nm is used to detect sulfur compounds; a filter optimized to pass light at 535 nm is used to detect phosphorus compounds. A photomultiplier tube is then used to quantify the amount of light passing through the selective filter.

Electron Capture Detector (ECD). The ECD is selective and highly sensitive for halogenated hydrocarbons, nitriles, nitrates, ozone, organo-metallics, sulfur compounds, and many other electron-capturing compounds. Selectivity is based on the absorption of electrons by compounds that have an affinity for free electrons because of an electronegative group or center. In an ECD, electrons are generated using radioactive beta-emitting isotopes, such as nickel 63 or tritium, and captured at a positively charged collector. This electron flux produces a steady current amplified by an electrometer. When an electron-capturing compound is present in the gas chromatograph column effluent, fewer electrons reach the collector. The resulting decrease in current is amplified and inverted so that the output signal from the detector increases as compounds are detected. The loss of current is a measure of the electron affinity of the analytes passing through the detector. Nonchlorinated hydrocarbons have little electron affinity and therefore are not detected using an ECD. Selective sensitivity for halides make this detector particularly useful for pesticides and other halogenated hydrocarbons such as polychlorinated biphenyls (PCBs). A limitation of the ECD is its narrow linear range, which necessitates careful calibration in the range of interest.

Thermal Conductivity Detector. This is the most universal gas chromatographic detector because it can measure most gases and vapors. It has low sensitivity compared with the other detectors, however, and is used primarily for analysis of low molecular weight gases such as carbon monoxide, carbon dioxide, nitrogen, and oxygen. This detector measures the differences in thermal conductivity between the column effluent and a reference gas, made of uncontaminated carrier gas. The most common carrier gas for this detector is helium, because it is inert and has a very low molecular weight. The column effluent and the reference gas pass through separate detector chambers that contain identical electrically heated filaments. Energy is transferred from the filament when analyte molecules strike the heated filament and rebound with increased energy. Heat loss will therefore be directly proportional to the number of collisions per unit time. Differences in thermal conductivity between gases is proportional to the rate of diffusion to and from the filament. Since diffusion is inversely proportional to molecular weight, lighter molecules will have higher thermal conductivities. Compounds with molecular weights greater than the reference gas will conduct more heat away from the filament than the pure, low molecular weight reference gas, thereby reducing the electrical resistance to the filament. The difference in resistance between the two filaments is amplified and recorded.

Photoionization Detector (PID). This detector is sensitive to compounds with low ionization potentials which can be ionized by ultraviolet light. PIDs can be used to selectively detect a wide range of compounds including aromatics, alkenes, ketones, or amines in the presence of aliphatic chromatographic interferences. This detector is similar in principle to the FID except that, instead of using a flame, it uses ultraviolet light to ionize the analyte molecules. Absorption of a photon by a molecule can cause the molecule to lose an electron if the energy of the photon is greater than the ionization potential of the molecule.

Different PID lamps are available to provide different photon energy levels. The lamp photon energy is chosen for selectivity of the analyte over the chromatographic interferences present in the sample. For example, benzene has an ionization potential of 9.2 electron volts (eV), and hexane has an ionization potential of 10.2 eV.

A lamp that emits photons with an energy of 11 eV will ionize both, although it may be more sensitive to benzene. With a 10-eV lamp, the PID will detect only the benzene, even with a large amount of hexane present.

GC/Mass Spectrometry (GC/MS)

In routine practice the occupational hygienist will collect air samples to determine the concentration of a known contaminant or group of contaminants and will request the analysis for these compounds. In this case GC analysis is relatively straightforward as described in the previous section. In other cases, however, the identity of the contaminant is not known and GC analysis alone will be insufficient. Then it is necessary for the GC column effluent to pass through a detector that will provide a qualitative identification of the numerous peaks exiting the column. A gas chromatograph interfaced with a mass spectrometer provides this capability and is a powerful combination for identifying unknown analytes.

In a mass spectrometer the GC column effluent is introduced and ionized, producing parent ions and ion fragments that are accelerated and separated by their mass-to-charge ratio. The mass spectrum is the record of the numbers of each kind of ion. The relative numbers of each ion are characteristic for every compound, including isomers.

The basic parts of a mass spectrometer are (1) the inlet system; (2) the ion source, where the sample is ionized; (3) the accelerating system, which separates the ions by their mass; and (4) the detector system, which measures the number of ions emerging from the accelerating system. The spectrometer is maintained at a high vacuum. Since GCs operate at atmospheric pressures, specialized devices have been developed to inject extremely small samples of the GC effluent into the mass spectrometer to maintain the vacuum.

Ion sources are the primary component of a mass spectrometer. They produce ions without mass discrimination, which are accelerated and passed into the mass analyzer. The most common type of ionization source produces ions by electron impact. As the column effluent from the gas chromatograph enters the mass spectrometer, the low molecular weight carrier gas is removed, and the rest of the column effluent travels through a beam of electrons in a high vacuum. Some of the molecules are fragmented by the electron beam into numerous ions and neutral fragments. The positively charged ions then enter the accelerating system, where rapidly changing electrical and magnetic radiation fields separate the ions according to their mass, allowing ions with each selected mass to pass sequentially through to the detector system. The detector system consists of an ion-multiplier phototube, which measures the number of ions passing through the mass analyzer. As ions strike a collector, the ion-multiplier causes the emission of electrons, which are collected and amplified.

A computer is used to store information on the abundance and associated mass of ions detected by the ion-multiplier. For each scan of the mass analyzer, the set of mass and abundance data pairs is called a mass spectrum, typically displayed as a bar graph or a table showing the abundance of positively charged ions at each mass. The mass spectrum pattern provides information about the structure of the original chemical compound and in many cases is sufficient for identification. An example of a mass spectrogram is presented in Figure 11.4. Some common applications of GC/MS are:

1. Evaluation of complex mixtures, such as mixtures of polynuclear aromatic hydrocarbons (PAHs), or identification of individual component vapors from photocopier and other office machine emissions; GC alone is not capable of positive identification;
2. Identification of pyrolysis and combustion products from fires; and
3. Analysis of insecticides and herbicides—conventional analytical methods frequently cannot resolve or identify the wide variety of industrial pesticides currently in use, but GC/MS can both identify and quantify these compounds.

A specific example of a GC/MS method is NIOSH Method 2539 for aldehyde screening.[6] This method is designed to identify individual aldehydes in an air sample or in a bulk liquid. Another GC/MS method is EPA Method IP–1B, which is designed for the determination of volatile organic compounds (VOCs) in indoor air.[7] This method is based on the collection of VOCs on Tenax® sorbent tubes followed by thermal desorption and capillary GC/MS analysis.

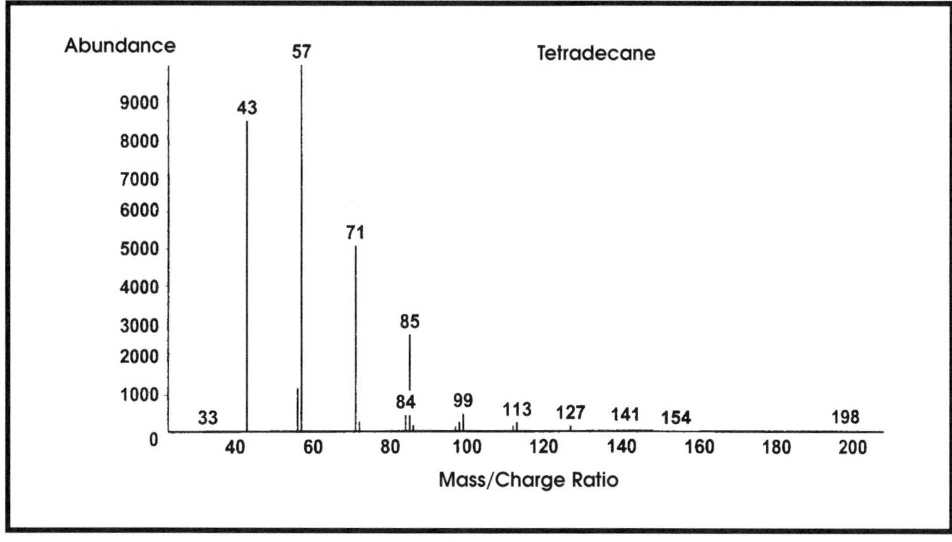

Figure 11.4 — Typical mass spectrum for tetradecane showing the relative abundance of each mass fragment.

High Performance Liquid Chromatography (HPLC)

While GC is used to analyze the vast majority of airborne organic compounds, it is not suitable for compounds that have very high boiling points (low vapor pressures) and chemicals that may be unstable at elevated temperatures. For such chemicals, HPLC is the preferred technique. The design of HPLC instrumentation parallels that of GC and consists of a solvent delivery system, a sample injector, a column, a detector, and a recording device.

In contrast to GC, where the analytes are carried through the column by a gas, liquid chromatography uses a liquid as the carrier or mobile phase. As a result, contaminants need to be only somewhat soluble under the conditions of the experiment. In liquid chromatography, sample components are introduced into a column and partition differentially between the solid adsorbent and liquid phase (solvent) resulting in separation of contaminants as they are swept down the column. In high performance (or pressure) liquid chromatography, high pressure, usually ranging from 500–5000 psi, is required to move the mobile phase through a narrow column containing small sorbent particles.

Similar to GC, HPLC is a separation tool that must be combined with a detector to provide quantitative results. A variety of detectors, such as ultraviolet (UV) and fluorescence detectors, can be used to detect picogram quantities of toxic compounds (e.g., herbicides, insecticides, quinones, phthalates) and many carcinogenic compounds (e.g., nitrosamines, polynuclear aromatics, aflatoxins). HPLC has proved particularly valuable in the determination of PAHs, some of which are known or suspected carcinogens and many of which are ubiquitous in the environment.

Another major use of HPLC has been for the analysis of derivatized airborne organic isocyanates, both because it is more sensitive than the traditional Mercali colorimetric method and because it can separate the individual isocyanates (toluene-2,4-diisocyanate, toluene-2,6-diisocyanate, 1,6-hexamethylene diisocyanate, and methylene diphenyl isocyanate).[8] In addition to air sample analysis, HPLC is suited for the analysis of bulk samples such as oils, tars, and resins. In combination with mass spectrometric techniques, it provides quantitative and qualitative information about occupational hygiene samples.

An example of a specific HPLC method is NIOSH Method 5506 for PAHs.[9] This method is designed for a range of PAHs including benz[a]anthracene, benzo[a]pyrene, and naphthalene using fluorescence (excitation at 240 nm, emission at 425 nm) and UV (254 nm) detectors in series.

Commonly used HPLC detectors include UV, fluorescence, and electrochemical detectors as discussed below.

UV Absorbance Detector. This detector measures the UV light absorbance of the column effluent. It is especially sensitive to aromatic hydrocarbons. The basic components of the UV detector are a UV lamp, a flow-cell with a UV-transparent window for the column effluent, and a photodiode or other light-measuring device. Some UV detectors can be operated at only one wavelength, typically 254 nm, while others can also operate at additional wavelengths using fluorescent waveplates. A variable UV detector can be tuned to any wavelength within its operating range.

Fluorescence Detector. The fluorescence detector measures the emission of light produced by fluorescing eluents and is extremely sensitive to highly conjugated aromatic compounds such as PAHs. Some analytical methods use derivatization reagents to fluoresce the analyte. In these methods a light source raises the fluorescent analyte to an unstable higher energy level, which quickly decays in two or more steps, emitting light at longer wavelengths. The basic detector components are a lamp, a flow cell with windows at a 90º angle for the column effluent, filters or diffraction gratings to select the excitation and emission wavelengths, and a photomultiplier tube or other light-measuring device.

Fluorescence detectors vary in sensitivity and selectivity. Some use a collection of lamps, waveplates, and filters to provide a range of excitation and emission wavelengths. Other fluorescence detectors can be tuned to any wavelength within their range. Some automated models can rapidly change the wavelengths during chromatographic analysis of one sample to optimize the detector for different analytes that elute at different retention times.

Electrochemical Detector. An electrochemical detector responds to compounds (such as phenols, aromatic amines, ketones, aldehydes, and mercaptans) that can be readily oxidized or reduced. Electrode systems use working and reference electrodes to quantify analytes over a range of six orders of magnitude.

Ion Chromatography (IC)

Historically, methods for the analyses of anions have been generally complicated and prone to interferences. Recent advances in IC (which is a form of ion exchange chromatography) have made it the method of choice for anion analysis by the occupational hygiene chemist. Some of the anions most suitable for ion chromatographic analyses include sulfate, nitrate, phosphate, chromate, chloride, cyanamide, isocyanate, sulfite, and thiocyanate. Ion chromatographic methods are also suitable for analyses of cationic species including ions of alkali and alkaline earth metals, inorganic ammonium salts, and salts of various amines. In addition, ion chromatography is used to analyze carcinogens that can be determined as cations, including aromatic amines such as beta-naphthylamine, benzidine, hydrazines, azoarenes, and aziridines.

Major components of an IC include a separation column, background ion suppressor column, various eluents, and a detector. IC systems are commonly divided into suppressed-ion anion and cation chromatography systems and single-column ion chromatography systems. Suppressed-ion chromatography separates ions using an ion-exchange column, with the removal of unwanted electrolytes in a suppressor column prior to detection. Single-column IC can be used if the ion-exchange capacity of the separation column is low and a dilute eluent is used, making ion suppression unnecessary. The most commonly used detector for occupational hygiene sample analysis is the conductivity detector, which measures the electrical conductivity of the column effluent.

It is anticipated that IC will become increasingly relevant to occupational hygiene analysis with the introduction of new types of columns, detectors, and sample derivatization methods.

Volumetric Methods

Analysis using volumetric methods, usually referred to as wet-chemical methods, is performed by measuring the volume of a solution of a known concentration required to react completely with the substance being determined. For the analysis of gases and vapors, titrimetric methods that involve measuring the quantity of a reagent required to react completely with a gaseous contaminant in solution are the principal methods. Detection of an endpoint is based on the observation of some property of the solution that undergoes a characteristic change near the equivalence point. Changes in color, turbidity, electrical conductivity, electrical potential, refractive index, or temperature of the solution are the most commonly used properties to detect reaction endpoint.

For occupational hygiene applications, titrimetric techniques have been used historically to determine airborne concentrations of hydrogen chloride, nitric acid, caustic mist, sulfur dioxide, hydrogen sulfide, and ozone. For the most part, these methods have been replaced by more specific analytical techniques for the contaminant of interest and involve simpler sample collection.

Spectrophotometric Methods

A common technique for many occupational hygiene analyses is the measurement of the absorption of light at a particular wavelength by a solution containing the contaminant or a material that has been quantitatively derived from it. The process is termed "absorption spectrophotometry" or, if visible light is used, "colorimetry." Spectrophotometric methods that use UV and infrared radiation have also been developed. The extent to which light is absorbed by the solution is related to the concentration of the contaminant in solution and the length of the light beam passing through the absorbing solution.

Absorption photometry is described using the Beer-Lambert law, which relates the absorbance of light to the concentration of the absorbing material according to Equation 1.

$$A = \log(I_{in}/I_{out}) = (1/T) = a \times b \times c \qquad (1)$$

where:

A = absorbance
I_{in} = intensity of incident light beam
I_{out} = intensity of beam exiting solution
T = transmittance
a = molar absorptivity constant
b = path length
c = concentration of absorbing material

A plot of absorbance vs. concentration of the absorbing material will yield a straight line. This plot can be used to determine the concentration of that contaminant in solution derived from an air sample. Changes in color intensity have been the basis of many occupational hygiene photometric methods. For example, a classic photometry application is the use of Saltzman's reagent to determine the airborne concentration of nitrogen dioxide.

Although spectrophotometric and colorimetric techniques are still in use in the laboratory, they are rapidly being replaced by more sensitive and convenient instrumental methods. For example, methods of analysis for many organic compounds by UV spectrophotometry have been replaced by GC. UV methods used in occupational hygiene analyses are now generally restricted to direct-reading field instruments, such as those used for measuring mercury vapors and real-time screening of organic compounds. Infrared spectrophotometric techniques are also commonly used to quantify airborne organic compounds. One advantage of the infrared technique is that it can be adapted to direct-reading instruments to measure a variety of gases and vapors. This is discussed further in the chapter on direct-reading instruments.

Quality Assurance

Quality assurance programs required under the IHLAPs are specific and detailed. A QA program includes detailed procedures specific to sampling and analysis. Quality assurance in the occupational hygiene laboratory is described in greater detail in Chapter 39.

Summary

The purpose of this chapter is to present the range of analytical techniques commonly available for the quantification of gases and vapors in occupational environments. In general, the analytical method for a gas or vapor is specified along with a companion sampling method, as discussed in Chapter 10 of this book. Chromatographic, volumetric, and spectrophotometric methods constitute the most common standard techniques for modern occupational hygiene gas and vapor sample analysis.

Of these methods, chromatographic methods are most prevalent. Commonly used chromatographic methods include gas chromatography, high performance liquid chromatography, and ion chromatography to quantify the mass of analyte present in a sample. Of the chromatographic methods, GC is used most often. GC methods can also be linked with mass spectrometry (GC/MS) to identify unknown gases and vapors in samples.

Volumetric methods, also referred to as "wet chemical" methods, have historically been used to quantify gases and vapors in the workplace. These methods, however, have been mostly replaced by other analytic techniques. Methods based on absorption spectrophotometry also have wide application for analysis of air samples for a variety of gases and vapors. These methods, which can be described using the Beer-Lambert law, use visible, ultraviolet, or infrared electromagnetic radiation for the quantification of the mass analyte present in a sample.

Acknowledgments

The authors wish to thank Michael A. Coffman and Jaswant Singh, whose draft version of this chapter provided great assistance.

Additional Sources

Chapman, J.R.: *Practical Organic Mass Spectrometry*. New York: John Wiley & Sons, 1985.

Harris, D.C.: *Quantitative Chemical Analysis*. New York: W.H. Freeman and Co, 1995.

Jennings, W.: *Analytical Gas Chromatography*. New York: Academic Press, 1987.

Johnson, E., and R. Stevenson: *Basic Liquid Chromatography*. Sugarland, TX: Varian Associates, Inc., 1978.

Pryde, A., and M.T. Gilbert: *Applications of High Performance Liquid Chromatography*. London: Chapman and Hall Ltd., 1979. [U.S. distributor: John Wiley & Sons.]

Willard, H., L. Merrit, J. Dean, and F. Settle: *Instrumental Methods of Analysis*. 7th Ed. Belmont, CA: Wadsworth Publishing Co., 1988.

Willet, J.E.: *Gas Chromatography*. New York: John Wiley & Sons, 1987.

References

1. **National Institute for Occupational Safety and Health (NIOSH):** Method 3500. In *NIOSH Manual of Analytical Methods*, 4th Ed. (DHHS [NIOSH] Pub. No. 94–113). Cincinnati: NIOSH, 1994.
2. **National Institute for Occupational Safety and Health (NIOSH):** Method 2541. In *NIOSH Manual of Analytical Methods*, 4th Ed. (DHHS [NIOSH] Pub. No. 94–113). Cincinnati: NIOSH, 1994.
3. **Esche, C.A., and J.H. Groff:** PAT program: background and current status. *Appl. Occup. Environ. Hyg. 11*:522–523 (1996).
4. **Johnson, E., and R. Stevenson:** *Basic Liquid Chromatography*. Sugarland, TX: Varian Associates, Inc., 1978.
5. **National Institute for Occupational Safety and Health (NIOSH):** Method 1500. In *NIOSH Manual of Analytical Methods*, 4th Ed. (DHHS [NIOSH] Pub. No. 94–113). Cincinnati: NIOSH, 1994.
6. **National Institute for Occupational Safety and Health (NIOSH):** Method 2539. In *NIOSH Manual of Analytical Methods*, 4th Ed. (DHHS [NIOSH] Pub. No. 94–113). Cincinnati: NIOSH, 1994.
7. **Winberry, W.T., L. Forehand, N.T. Murphy, A. Ceroli, B. Phinney, and A. Evans:** *Compendium of Methods for the Determination of Air Pollutants in Indoor Air* (EPA/600/4– 90/010). Washington, DC: U.S. Environmental Protection Agency, 1990.
8. **Occupational Safety and Health Administration (OSHA):** *OSHA Analytical Methods Manual*. Salt Lake City, UT: U.S. Department of Labor, OSHA Analytical Laboratory, 1985.
9. **National Institute for Occupational Safety and Health (NIOSH):** Method 5506. In *NIOSH Manual of Analytical Methods*, 4th Ed. (DHHS [NIOSH] Pub. No. 94–113). Cincinnati: NIOSH, 1994.

Taconite mining operation in Hibbing, Minnesota. Taconite is a low-grade iron ore which is mined and processed into steel. (Layne Kennedy, 1995)

Outcome Competencies

After completing this chapter, the user should be able to:

1. Define underlined terms used in this chapter.
2. Explain aerosol morphology.
3. Describe aerosol deposition mechanisms.
4. Relate aerosol deposition to lung physiology.
5. Identify the primary mechanisms of aerosol particle deposition.
6. Describe commonly used techniques for aerosol sampling and size analysis.
7. Select appropriate aerosol sampling devices.
8. Explain the size distribution concept.
9. Calculate critical descriptors of aerosol particle size distribution.

Key Terms

50% cut point size • aerodynamic equivalent diameter • aerosol • aspect ratio • cut size • diffusion • dose • dusts • fibers • fogs • fumes • gravimetric analysis • hygroscopicity • impaction • inhalable fraction • interception • mist • monodisperse • nasopharyngeal (NP) region • particle diffusivity • polydisperse • pulmonary region (P) • respirable fraction • sedimentation • smoke • Stokes diameter • terminal settling velocity • thoracic fraction • tracheobronchial (TB) region

Prerequisite Knowledge

Prior to beginning this chapter, the reader should have an understanding of basic Newtonian physics and mathematics through college algebra. In addition, the reader should review the following chapters:

Chapter Number	Chapter Topic
1	History and Philosophy of Industrial Hygiene
2	Occupational Exposure Limits
4	Environmental and Occupational Toxicology
6	Principles of Evaluating Worker Exposure
9	Direct-Reading Instrumental Methods for Aerosols, Gases, and Vapors

Key Topics

I. Aerosol Morphology

II. Aerosol Deposition Mechanisms
   A. Sedimentation
   B. Inertial Deposition
   C. Diffusion
   D. Interception
   E. Particle Retention

III. Deposition of Inhaled Particles

IV. Aerosol Sampling and Analysis in Occupational Environments
   A. Sampling Theory
   B. Size Selective Sampling
   C. Filtration-Based Techniques
   D. Sedimentation-Based Techniques
   E. Impaction-Based Techniques
   F. Optical Techniques

V. Aerosol Size Distribution Analysis

12

Sampling and Sizing Particles

David Johnson, Ph.D., CIH

David Swift, Ph.D.

Introduction

In common usage the term aerosol is taken to mean the droplet spray produced from an "aerosol can" containing a liquid and a compressed gas propellent, as used for such commercial products as spray deodorants, air fresheners, and spray paints. In more precise terms an aerosol is an assemblage of solid or liquid particles dispersed in a gaseous medium. For occupational hygiene purposes the gaseous medium is almost always air. Familiar aerosol-producing occupational activities include the "dusty trades" such as hard rock and coal mining, in which silica and coal dust aerosols are produced that can lead to chronic obstructive pulmonary disease (silicosis and coal miners' pneumoconiosis, respectively); high-temperature operations such as arc welding, torch cutting, and metal smelting in which extremely fine metal oxide particles are produced that can react with lung tissues to cause potentially fatal chemical pneumonias; and even health care activities in which potentially pathogenic droplet aerosols of infected blood and other body fluids may be produced during surgeries, emergency treatment, and respiratory therapy.

Aerosols may react with or be absorbed through tissues to cause adverse health effects. Depending on the size, shape, and density of the particles, their chemical properties, the airborne concentration and time of exposure, and many other factors, the health effects may range from simple irritation to terminal disease. Occupational aerosol hazards have been recognized since at least the first century A.D., when Pliny (A.D. 23–79) described boat painters who covered their heads with hoods to prevent the inhalation of lead dust.[(1)] A crude form of respiratory protection was also worn by miners at the beginning of the Renaissance, as described by Agricola (1494–1555) in his treatise on European mining practices of the time.[(2)] In the 300 years since Bernardino Ramazzini initiated formal study of occupational diseases, work-related aerosol exposures have been shown to cause numerous diseases of the respiratory tract and other tissues; however, only during the past 100 years have specific mechanisms been identified to explain how these diseases progress. Most recently, advances in aerosol science and inhalation toxicology have extended our understanding of the relationship between aerosol exposure, uptake, and fate, and this improved understanding has in turn promoted the development of refined techniques for aerosol exposure characterization.

It would be a mistake to limit one's thinking about aerosol-related illness to such historically important occupational lung diseases as silicosis, black lung (coal miner's pneumoconiosis), asbestos-related lung cancer, and the like. Improvements in engineering and other control measures promoted by occupational hygienists and other occupational safety and health professionals have significantly reduced the incidence of these historically important diseases, so that other aerosol-related health effects are gaining more attention. For example, diesel combustion aerosol, radon progeny aerosol, environmental tobacco smoke, man-made fibers, and infectious and allergenic bioaerosols are of special interest in occupational hygiene work. Relatively little is known of the health effects of exposure to these aerosols, and the occupational hygienist plays a critical role in exposure characterization, epidemiological study, standards development, and hazard prevention and control related to them. Techniques for aerosol sampling and characterization continue to evolve in a dynamic process driven by advances in technology and our understanding of disease etiology.

In evaluating aerosol exposures and developing appropriate control measures, the occupational hygienist may be required to characterize the aerosol in several ways. No single sampling and analysis technique is appropriate to all needs, so it is important for the occupational hygienist to be familiar with the properties of aerosols in occupational environments and the techniques available for their assessment. This chapter focuses on the sampling and analytical methods and equipment commonly used in assessing aerosols of occupational health concern. It does not address many other areas of aerosol science interest such as ambient atmospheric aerosols, ultrafine particles in clean room environments, or pharmaceutical aerosol preparations, and only incidentally addresses bioaerosols (see Chapter 18 for detailed discussion of biohazards and bioaerosols). Neither does it address techniques and instrumentation not suited to use in the occupational hygiene field. For a discussion of these and other topics related to aerosol generation and measurement the reader is referred to the excellent texts by Willeke and Baron[3] and Cox and Wathes.[4]

Definitions

Aerosols are encountered in several forms in occupational hygiene practice.[5]

Dusts are dry particle aerosols produced by mechanical processes such as breaking, grinding, and pulverizing. Various mining activities (excavation, conveyor transport, pulverizing), materials handling operations (loading, unloading, transport), and dry material preparation and packaging processes (ball milling, sieving, bagging) produce dust particles. These particles are unchanged from the parent material, except that their smaller size and higher specific surface area (surface area per unit mass or volume of material) may enhance their toxicity or explosion potential. As shown in Figure 12.1, dust particle sizes range from less than one micrometer (10^{-6} m or 1 µm) (e.g., ground talc) to 1 mm (e.g., fertilizer and ground limestone) or larger. In general, dust particles tend to be roughly spherical.

Fumes as defined for occupational health purposes are not noxious vapors, but rather are very fine solid aerosol particles produced from vaporized solids. Materials that are normally solid at standard conditions may be first melted and then vaporized if heated to sufficiently high temperature, as occurs during smelting or when metals are bonded in an arc welding operation. Arc welding and oxyacetylene or plasma torch cutting are typical operations producing fume particles in large numbers, along with various gases and vapors. The vaporized metal rises from the process and immediately cools to form spherical molten droplets, which in turn cool to form spherical solid particles. The particles are typically less than 0.1 µm in diameter and are often composed of an oxide of the parent material. These "fresh" oxide particles, which readily penetrate to the deep lung, may be chemically reactive and pose a severe hazard to workers. Zinc oxide fume produced during welding or cutting on galvanized (zinc-coated) metal may cause welders to become nauseous. Fresh cadmium oxide fume, which is produced when cadmium-bearing silver solder is used in brazing together low-melting-point metals, e.g., when joining copper radiator components, is extremely hazardous and can cause potentially fatal pulmonary edema if inhaled. Anticorrosion cadmium coatings on metal bolts are another source of cadmium oxide fume if fasteners are cut apart using an oxyacetylene torch. Metalizing operations, in which worn metal parts such as shafts are "built-up" prior to remachining, involve spraying molten metal onto the worn surface. If not properly controlled using exhaust ventilation, metalizing operations can expose workers to high metal fume concentrations and cause a flu-like immunologic reaction called metal fume fever.

Mists are spherical droplet aerosols produced by mechanical processes such as splashing, bubbling, or spraying. The droplets are unchanged from the parent liquid and range in size from perhaps a few microns to more than 100 µm. Any vigorous process involving liquids has the potential for producing a mist aerosol. Indeed, processes such as spray painting are designed to produce mists and have been the subject of extensive Occupational Safety and Health Administration regulation. In recent years the mist droplet aerosols produced by the coughing or treatment of infected patients have received increased regulatory attention

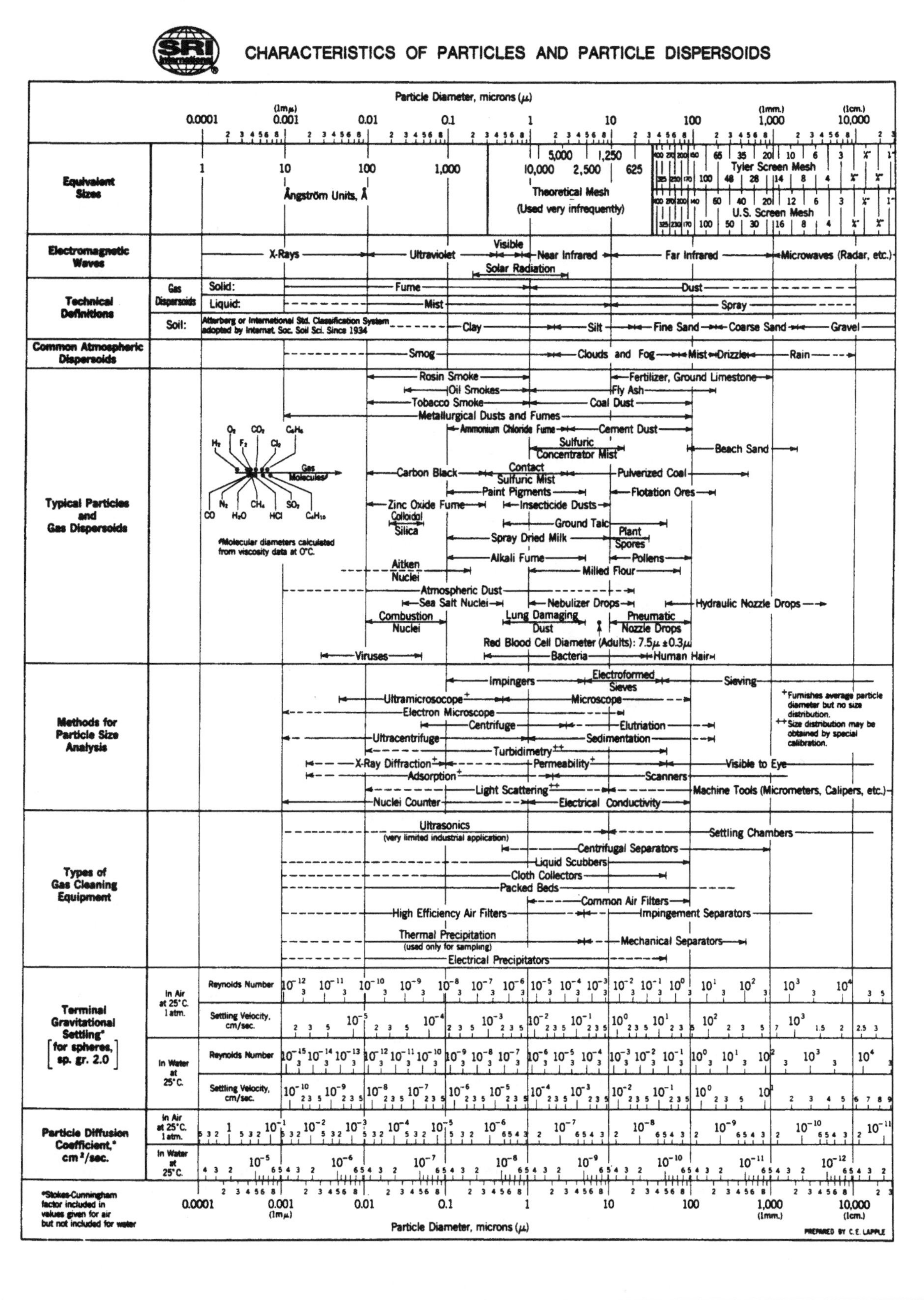

Figure 12.1 — Characteristics of particles and particle dispersoids. (Source: C.E. Lappler, *SRI Journal* 5:94 (1961).)

because of concerns about occupational exposure to multidrug resistant tuberculosis and other pathogens.

Fogs are also droplet aerosols, but are produced by condensation from the vapor phase. The droplets are typically smaller than mist droplets, say on the order of 1 to 10 µm. Whereas mists may visibly settle toward the ground, fogs appear to float in the air indefinitely. Fogs are generally of more interest in atmospheric aerosol work than in occupational health, since few industrial operations produce fogs to which workers may be exposed.

Smokes are complex mixtures of solid and liquid aerosol particles, gases, and vapors resulting from incomplete combustion of carbonaceous materials. Tobacco smoke, for example, contains thousands of chemical species, many of which are toxic or carcinogenic.[6, 7] Smokes from burning plastics, synthetic fabrics, and other petrochemical products may be extremely toxic and are often more dangerous than the flames in a building or aircraft fire. Primary smoke particles are on the order of 0.01 to 1 µm in diameter, but agglomerates containing many particles may be much larger.

Fibers are elongated particles having an aspect ratio, i.e., a ratio of length to width, of greater than 3:1.[8] Fibers may be naturally occurring, such as plant fibers and asbestiform silicate minerals, or synthetic, such as vitreous or graphite fibers. Asbestos-related diseases include asbestosis (a chronic fibrosis of the lung), mesothelioma (cancer of the lining of the lung cavity), and bronchial carcinoma (cancer of the conducting airways of the lung) and have been associated with heavy asbestos exposure in ship builders, pipe fitters, and insulation installers. Increasing use of refractory ceramic fibers, mineral wools, spun glass, and composite materials incorporating reinforcing synthetic fibers has fueled concerns that these fibers may have unsuspected toxic potential.[9] Fibrous aerosol particles display aerodynamic and health effects behaviors different in some respects from spherical or near-spherical particles of the same material and mass, so that aerosol characterization is typically more complex for fibers than for other aerosols.

In virtually all occupational environments the particles making up workplace aerosols exhibit a range of particle sizes, and are said to be polydisperse. Aerosols containing only particles of a single size are monodisperse, but such aerosols are difficult to generate and maintain and are usually only encountered in research and calibration activities.

Aerosol Morphology

Airborne particles may be described in terms of their morphology, or shape and appearance. Particles whose characteristic length dimension is independent of particle orientation are known as isometric. For such particles, it is possible to describe the shape by a single dimension. For the case of spherical particles this dimension is the diameter. Metal fumes and liquid droplets are spherical due to liquid surface tension (during the molten phase for the fumes). Many bioaerosol particles are also spherical or essentially so, including allergenic pollens and dust mite fecal pellets.

Some aerosol particles are not spherical but are essentially isometric. Examples of such particles are dusts produced from grinding or pulverizing processes and some crystalline particles. For the dynamic behavior of such particles, it is suitable to treat them as isometric having a single dimension. One may assume, for simplicity, that such particles are spherical or seek to obtain a "dynamic correction factor" to account for nonsphericity.

Other particles are clearly nonisometric and are treated as ellipsoids of revolution. If the axis of rotation is the minor axis of the ellipse, the resulting particle is known as a prolate spheroid. A fiber is the limiting case of such a particle in which the length exceeds the diameter. As noted above, the conventional definition of a fiber is an elongated particle with a length to diameter ratio ≥ 3. Conversely, if the axis of rotation is the major axis of the ellipse, the particle is an oblate spheroid. A plate-like particle is the limiting case of such a particle in which the particle diameter exceeds its thickness. Fiber particles occur frequently in industrial aerosols. Examples include asbestos, metal whiskers, carbon fibers, and other fiber particles used in composite materials. Examples of plate-like particles include paint pigments, insect parts, mica, and crystalline particles.

The particles discussed above are singlet particles, i.e., they are formed as single particles and remain so for their entire lifetime. When the initial concentration of particles is very high ($>10^6$ particles/cm^3) particles coagulate or flocculate to form aggregate particles. This is most commonly the case when the initial particles are in the ultrafine size range (1 nm–0.1 µm). Such aggregate particles may contain up to hundreds of primary particles. It is very difficult to analyze the dynamic behavior of such particles, as the gross dimensional structure of the aggregates varies widely from one particle to another. Recently, the relationship between the primary and secondary structure of aggregates has been investigated by considering the particles as fractal structures. This general approach was pioneered by Mandelbrot[10] and has been applied to aerosol aggregates by several investigators.[11]

Soot particles, produced by the incomplete combustion of hydrocarbon fuels, are a prime example of aggregates. Compared to primary particles, aggregates display certain important characteristics, chief among which is the very large surface area per unit mass of particle. This large specific surface area may enhance surface condensation of vapors and catalytic reaction of simple vapors to produce new gaseous species.

It must be noted that particle morphology is determined by optical or electron microscopy. When one observes a particle by microscopy, a two dimensional object is seen. To infer the three dimensional structure of the particle, it is necessary to make some assumptions regarding the third dimension. An additional problem is the possibility that particles collected on a substrate for microscopy may suffer deformation due to attractive (or other) forces between the particle and the substrate. Electron microscopy, because it requires a high vacuum, is not feasible for liquids or other volatile substances. This limits the type of particles that may be morphologically analyzed by this method.

Aerosol Deposition Mechanisms

Airborne particles exert their toxic effects on contact with body tissues. The dose of toxicant, i.e., the quantity of physiologically active toxicant in contact with the affected tissue, is influenced to a large extent by the aerosol's physical properties. Aerosol particle size, shape, density, and hygroscopicity (tendency to absorb water vapor) determine to a large extent how the particles behave in the air and how efficiently they are inhaled and deposited in the respiratory tract, and may influence how they interact with body tissues and defense mechanisms. In this section is described the primary deposition mechanisms of inertial deposition, sedimentation, diffusion, and interception. In following sections their influence on aerosol deposition in the body and how they are used in various aerosol sampling and sizing techniques are discussed.

Sedimentation

Sedimentation refers to movement of an aerosol particle through a gaseous medium under the influence of gravity. The rate of settling will depend on the particle's size, shape, mass, and orientation (for nonspherical particles), and on the air density and viscosity. An airborne particle is subjected to a gravitational force and will accelerate according to Newton's Second Law. The accelerating force is given by the product of the particle mass and the gravitational constant g, and for a sphere of diameter d_p and density ρ is

$$F_g = \frac{\pi d_p^3}{6} \rho g \qquad (1)$$

Opposing this force is the drag force on the particle due to gas viscosity. The drag force on a spherical particle moving with velocity U through a medium having a viscosity μ is given by Stokes' Law

$$F_D = 3 \pi \mu d_p U \qquad (2)$$

where $\mu = 1.84 \times 10^{-4}$ g/cm-sec at STP (1 atm, 20°C). Since the particle displaces its own volume of air even at rest, there is also a small buoyant force acting opposite the gravity force, namely

$$F_b = \frac{\pi d_p^3}{6} \rho_f g \qquad (3)$$

where ρ_f is the density of the fluid medium (for air, $\rho_f = 1.29 \times 10^{-3}$ g/cm^3 at STP). The gravitational and buoyant forces are constant, but the drag force increases linearly with particle velocity, so that at some point the sum of the buoyant and fluid drag forces exactly equals the gravitational force, the acceleration becomes zero, and the particle attains a constant velocity (Figure 12.2). This terminal settling velocity is obtained by setting the sum of the forces equal to zero. Normally the buoyant force is negligible compared with the gravitational force. Making this assumption, one obtains an expression for the terminal settling velocity

$$U_T = \frac{d_p^2 \rho}{18 \mu} g = \tau g \qquad (4)$$

where τ is the particle's "relaxation time" and has units of seconds. When the particle diameter is of the order of the gas mean free path, an additional factor must be included in Stokes' Law to take account of the "slip" of the particle. This factor is

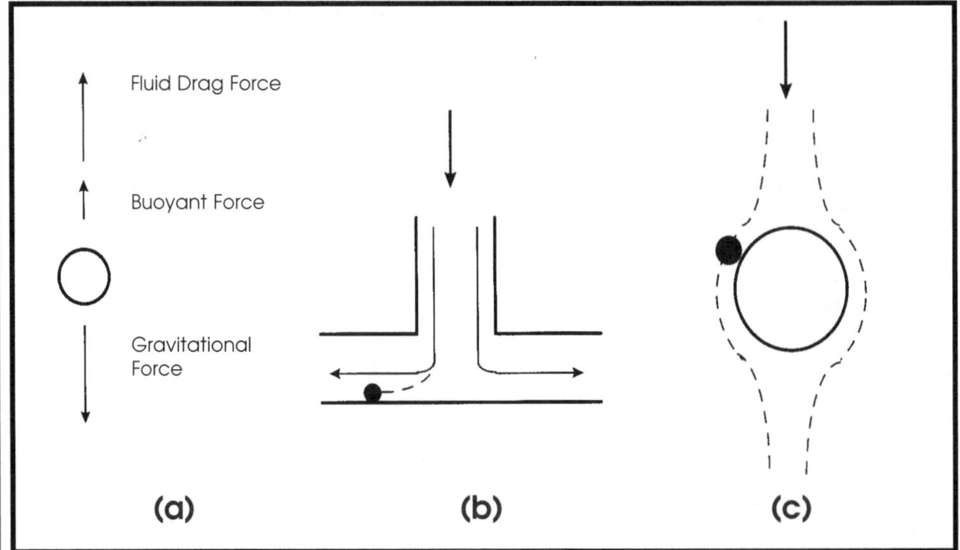

Figure 12.2 — Aerosol particle collection mechanisms: (a) sedimentation—at terminal (maximum) settling velocity, the fluid drag and buoyant forces will exactly offset the particle's weight; (b) inertial impaction—the particle's inertia carries it across airflow streamlines as the air changes direction; and (c) interception—the flow streamline passes the collecting body (such as a filter fiber) within a distance of one-half the particle's diameter.

known as the Cunningham slip correction and is given by the expression

$$C_c = 1 + \frac{0.166}{d_p} \quad (5)$$

where d_p is the particle diameter in micrometers. This expression is appropriate for air at STP for particles down to approximately 0.1 µm.[12] Terminal settling velocities range from approximately 10^{-4} to 10^{-1} cm/sec for 0.1- to 10-µm diameter particles, respectively. The time required for a particle initially at rest to reach U_T is approximately 7τ, which ranges from 10^{-6} to 10^{-3} sec for particles in the range of 0.1 to 10 µm, respectively.

Where the particles are not spherical, Equation 2 may be multiplied by a shape factor with a characteristic particle dimension substituted for sphere diameter. Such a discussion is beyond the scope of this chapter, but may be explored in aerosol science texts including Mercer.[13]

In measuring aerosol sedimentation behavior, the shape, size, and density of particles is often unknown. It is convenient to discuss particle size in terms of the diameter of a spherical particle of the same density that would exhibit the same behavior as the particle in question, or the Stokes diameter, d_{Stk}. The d_{Stk} may be calculated from Equation 4 for a measured U_T and known ρ; however, the usual practice is to extend this analogy to the concept of aerodynamic equivalent diameter, d_{ae}, which is the diameter of a unit density sphere that would exhibit the same settling velocity as the particle in question. Obviously, for unit density aerosols, $d_{Stk} = d_{ae}$. The advantage of using aerodynamic equivalent diameter rather than Stokes diameter is that aerodynamic equivalent diameter normalizes different aerosols to a common basis so that their behaviors may be directly compared.

Inertial Deposition

A particle's inertia, defined as its tendency to resist a change in its motion, is an important deposition mechanism. A particle in motion through a fluid will move in a straight line unless acted on by a net external force, as described by Newton's First Law of Motion. When an aerosol stream is forced to change direction suddenly, the particle's inertia will carry it across flow streamlines some distance before its momentum in that direction is depleted by fluid drag. If a surface is within the distance traveled, the particle may be captured by impaction on the surface (Figure 12.2). In general, the likelihood of impaction increases with the mass and velocity of the particle and with the sharpness of the change in direction. These particle- and system-specific factors are reflected in the Stokes number, Stk, where

$$Stk = \tau \frac{U}{D} C_c \quad (6)$$

in which τ and C_c are as previously defined and U and D are characteristic velocity and dimension values for the system. For example, for a spherical particle in air flowing around a much larger spherical body, D might be the diameter of the larger sphere and U the velocity of the airstream approaching the sphere. Stokes number may take various values depending on the characteristics of the flow system, but for a given system the likelihood of impaction increases with increasing Stokes number.

Diffusion

Aerosol particles in a gaseous medium are bombarded by collisions with individual gas molecules that are in Brownian motion. This causes the particles to undergo random displacements known as diffusion. The particle parameter that describes this process is the particle diffusivity (or diffusion coefficient), D_B. An expression for particle diffusivity is given by the Stokes-Einstein equation:

$$D_B = \frac{k_B\ T}{3\ \pi\ \mu\ d_p}\ C_c \qquad (7)$$

where k_B is Boltzmann's constant (1.38×10^{-23} J/°K) and T is absolute temperature. The diffusion coefficient is inversely proportional to the particle's geometric size and is independent of particle density. The units of D_B are cm^2/sec. Net motion by diffusion occurs when a particle concentration gradient exists. Fick's Law of Diffusion states that the flux of particles, J (particles per unit cross sectional area per unit time), is the product of the diffusivity, D_B, and the concentration gradient, dc/dx. Diffusive transport is favored by small particle diameter, large concentration differences, and short distances over which diffusion occurs.

Interception

Airflow of an aerosol past a collecting surface can produce particle deposition by the process of interception (Figure 12.2). This deposition process does not depend on particle motion across fluid streamlines as is the case for inertial impaction. For fluid streamlines that carry a particle to a distance equal to or less than one-half the particle diameter, the particle will be captured by "touching" the surface. For a collecting object whose characteristic dimension is X_I, the probability of collection by interception is related to the ratio d_p/X_I. The collection mechanism is particularly significant for the case of elongated particles (e.g., fibers) for which d_p is replaced by the particle length, d_l. Fibers are therefore much more likely to be captured by interception than are spherical or near-spherical particles of equivalent mass.

Particle Retention

Particles coming into contact with surfaces may be acted on by various forces, some of which will enhance particle retention on the surface and some of which may tend to cause the particle's release, or resuspension, from the media. Forces that may promote retention are London-van der Waals forces, electrostatic attraction due to charge differences between the particle and surface, and capillary forces arising from the adsorption of a water (or other liquid) film between the particle and the surface.[14,15] Forces tending to dislodge deposited particles include those related to vibration and air currents, as intuition would suggest. Adhesion and resuspension are complex phenomena, but in general it is sufficient to say that smaller particles are more difficult to dislodge than larger particles.

Deposition of Inhaled Particles

Aerosolized materials can affect the skin and eyes, but inhaled aerosols are usually of primary occupational health concern. For a given exposure situation the amount of aerosolized material actually inhaled, the fraction of inhaled aerosol depositing in the different regions of the respiratory tract, and the fate of the deposited material are functions of many factors including the physical and chemical nature of the aerosol (e.g., particle size distribution, hygroscopicity, shape, solubility in water and lipids, chemical reactivity), the exposure conditions (e.g., airborne concentration, temperature, relative humidity, air velocity, physiologic demands on the worker), and characteristics of the individual (e.g., gender, body size, age, overall health, work practices, smoking status). In this section respiratory tract morphology and its effect on particle deposition are discussed in general terms. Detailed discussion of the fate of deposited aerosol particles and the diseases associated with them are found, for example, in Morgan and Lapp [16] and Rosenstock and Cullen.[17]

During breathing, a volume of air is drawn into the nose (or mouth) from the immediate region of the face, i.e., the breathing zone. Most aerosol particles in the zone will be accelerated toward the nose and inhaled, but some larger particles may escape due to their greater inertia. In fact, experimental studies have shown that only half of all particles greater than approximately 50 µm in size are inhaled from still air.[18] Particles in the inhalable fraction, or the fraction of total workplace aerosol actually entering the respiratory tract, must first traverse the moist and convoluted passageways of the nasal turbinates of the nasopharynx (Figure 12.3). Larger particles, on the order of 10 µm and bigger, will have difficulty negotiating the sharp turns at the high velocities in the nasal passages and will deposit by impaction on the nasal mucosa. Further, particles made up of, or coated with, hygroscopic materials may absorb a significant mass of water in this warm and humid nasopharyngeal (NP) region of the respiratory tract, thereby increasing the likelihood that they will be captured by inertial deposition. Inertial impaction is by far the most significant deposition mechanism in the NP region.

Particles not captured in the NP region enter the tracheobronchial (TB) region consisting of the trachea and conducting airways (bronchi and bronchioles). This thoracic fraction contains particles generally smaller than 10 µm. Although various models of lung morphology exist,[19] for discussion purposes the TB region may be described as having a branched structure consisting of 16 generations lined with a mucus-covered, ciliated epithelium.[20] The airways branch at angles of up to 45 degrees,[21,22] with the individual airways inclined from the vertical (in an upright person) at angles of from 0 (trachea) to 60º.[22] The conducting airways are not involved in the gas exchange of respiration; their function is to distribute the inhaled air quickly and evenly to the deeper portions of the lung. Airway diameters become successively smaller with each generation, but the number of airways increases geometrically. The net effect is that air velocities through the conducting airways decrease from approximately 200 cm/sec in the trachea and main bronchi to only a few centimeters per second in the terminal bronchioles, and the time of transit through an airway segment (its residence time in the segment) increases from approximately 0.01 sec in some of the larger bronchi to 0.1 sec in the terminal bronchioles.[23] Higher velocities favor inertial impaction, while lower velocities and higher residence times favor sedimentation and diffusion. Thus, the relative influence of inertial impaction decreases and the influence of sedimentation and diffusion increase from the top to the bottom of the TB region. However, diffusion is not considered a significant deposition mechanism in the conducting airways due to the large diffusion distances involved.

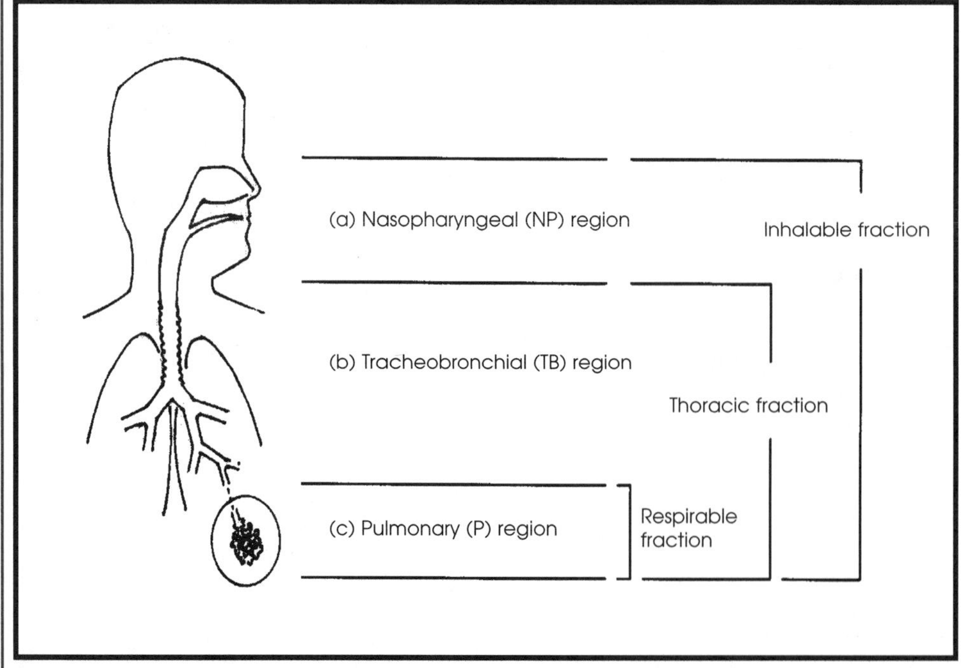

Figure 12.3 — Regions of the respiratory tract: (a) the NP region conditions inhaled air to body temperature and essentially 100% relative humidity and efficiently removes larger particles; (b) the TB region conducts inhaled air quickly and evenly from the mouth and nose to the pulmonary spaces; (c) the P region performs the gas exchange function of respiration.

Particles not captured in the TB region penetrate to the pulmonary (P) region containing the respiratory bronchioles, alveolar ducts, and alveolar sacs across which gas exchange occurs. These particles comprise the respirable fraction of the aerosol. Air velocities in the P region drop from approximately 1 cm/sec in the highest respiratory bronchioles to less than 0.1 cm/sec in the alveolar ducts and sacs, while residence times increase to over 1 sec.[23] Inertial deposition is unimportant at these velocities. Depending on the particle size, either sedimentation or diffusion will be the dominant deposition mechanism in the P region. Particles less than 0.1 µm exhibit significant diffusion and may deposit in large numbers on delicate pulmonary tissues, and, although they may represent a small total mass, may cause severe or fatal injury due to their chemical properties.

It should be noted that mouth breathing due to nasal obstruction or heavy physiologic demand can significantly alter the fraction of inhaled aerosol penetrating to the thoracic region, since the air bypasses the nasal passages completely. Aerosol deposition patterns in a given individual's respiratory tract will be determined by that person's respiratory tract morphology, breathing rate, minute volume, nose versus mouth breathing practices, and numerous other factors. It should also be noted that fiber deposition behavior may be significantly different from that of compact particles of equivalent mass; for example, fibers are more likely to be captured by interception simply due to their shape. Accurate prediction of respiratory tract aerosol deposition patterns is an elusive goal being actively pursued by a number of investigators.

Aerosol Sampling and Analysis in Occupational Environments

Sampling Theory

The general objective of aerosol sampling is to obtain information about aerosol properties at a given location over a specified length of time. One way of classifying sampling methods is to divide them into (1) methods where an aerosol sample is drawn into a sampling device and (2) methods in which the aerosol is sampled in situ. In industrial environments the first class of methods is essentially universally used. This means that the nature of airflow and particle behavior surrounding and within the sampling device becomes a critical issue in assessing the sampling characteristics of the device.

Of the many aerosol properties that can be assessed, the most common are the aerosol mass per unit air volume (mass concentration) and the particle size characteristics. These are important because the toxic properties of aerosols and the associated regulatory standards are most often expressed in terms of one or both of these properties. An example of a class of methods measuring aerosol mass concentration is total dust sampling. Such methods are intended to collect particles of all sizes without separating the sample into size fractions. By contrast, size selective sampling is intended to separate the aerosol into size fractions based on some criterion.

A complete review of sampling theory is given by Vincent,[24] to which the interested reader is referred. Several important conclusions can be drawn from this review of theory: (1) all aerosol sampling is size selective, i.e., what a sampler collects does not correspond to the aerosol as it exists in situ; (2) numerous losses occur in the process of sampling, both external and internal to the sampler; (3) the size selective efficiency of a sampler is a complex function of sampler geometry, sampling rate, flow external to the sampler (e.g., convection, laminar or turbulent flow, sampler orientation to the direction of external flow); and (4) sampler theory can be useful in choosing a sampler for a specific application and for assessing losses during sampling to obtain correction factors. It should be noted that in some instances a sampler exhibits an efficiency > 1.0, i.e., the sampler oversamples some size fraction of the aerosol.

No single aerosol sampling and analysis technique is suited to all possible information needs concerning aerosol exposure and control. Methods and devices are developed to answer specific questions, within the limits of available technology. Advances in both technology and the understanding of aerosol behavior and health effects are continual drivers for further development. In this section aerosol sampling and analysis techniques in common use for workplace evaluation are described. The section does not cover those techniques designed for laboratory or research use, or for characterizing aerosols in ducts, stacks, and so forth; these techniques are reviewed in detail by others (see, for example, Cohen and Hering[25]).

Size Selective Sampling

Aerosol particle size greatly influences where deposition occurs in the respiratory tract, and the site of deposition often determines the degree of hazard represented by the exposure. Aerosol sampling techniques designed to measure aerosol as inhalable, thoracic, and respirable fractions are extremely useful in assessing the hazard potential of occupational aerosol exposures. International criteria have recently been agreed on for the performance of these size selective samplers. A sampler whose performance exactly matched the criteria would collect particles of various sizes with the efficiencies shown in Figure 12.4. A current listing of size-selective samplers for inhalable, thoracic, and respirable sampling is provide by Lippmann.[26] Several of these will be illustrated in the following sections.

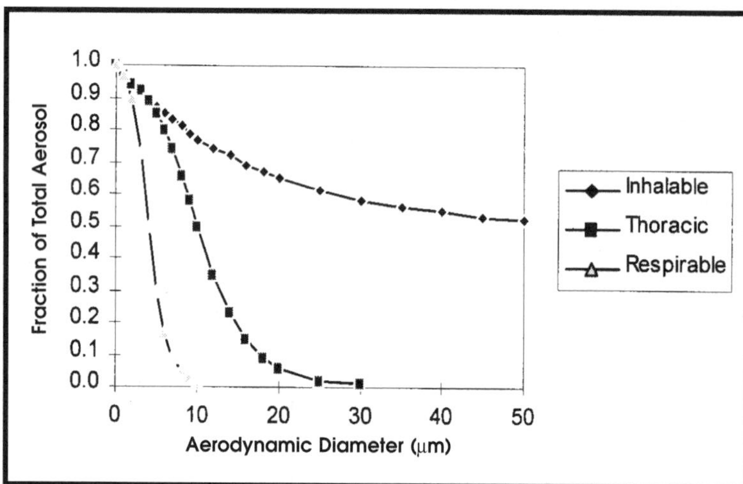

Figure 12.4 — Collection efficiency as a fraction of total aerosol of samplers performing according to the ACGIH/ISO/CEN size selective sampling criteria (as in Reference 11, p. 216).

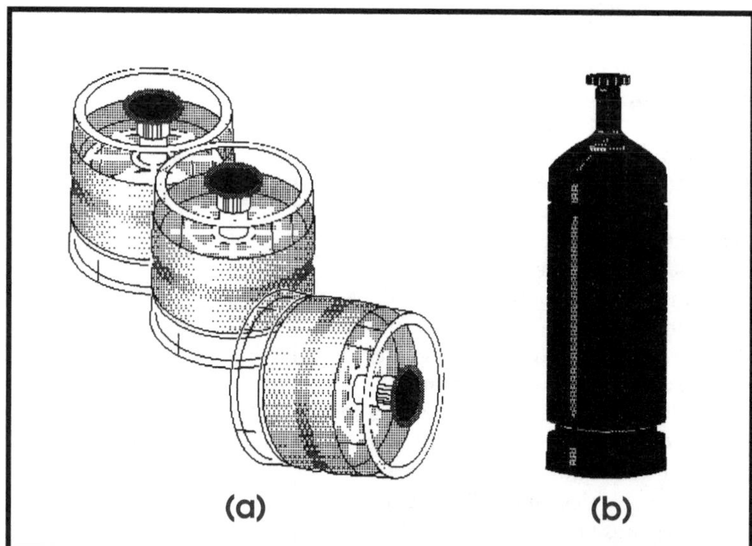

Figure 12.5 — Filter cassettes. Commonly used filter holders include (a) the 37-mm three-piece styrene acrylonitrile cassette used as shown or in open-face mode with one end removed, and (b) the polypropylene 25-mm cassette with cowl, specifically for use in asbestos sampling (end cap shown is removed during sampling) (graphics courtesy SKC, Inc., Eighty Four, Pa.).

Filtration-Based Techniques

Filters have been used for aerosol capture since the ancients first covered their faces with cloth veils to avoid breathing dust and are still used in numerous air cleaning applications including respiratory protection, pollution control, and recovery of valuable fugitive materials. In exposure characterization work, filtration-based techniques allow the study of aerosol mass concentration, number concentration, particle morphology, radioactivity, chemical content, and biohazard potential. Many types of filter media are available, and the choice of media will depend on the aerosol characteristics and the analytical technique to be used. A brief overview of these topics is provided here; for more detailed information the reader is referred to the summaries of filtration theory and methods, filter media, and filter holders by Lee and Ramamurthi[27] and Lippmann.[28]

Gravimetric analysis involves drawing a known volume of aerosol-laden air through a filter of known initial weight, then reweighing the filter to determine the mass captured. The average aerosol concentration is this mass divided by the volume of sampled air, typically expressed as milligrams of aerosol per cubic meter of air sampled (mg/m^3). In personal (breathing zone) sampling, conventional practice is to use a 25-mm, 37-mm, or 47-mm diameter filter cassette containing a filter supported by a screen or backing pad. The cassettes may be used open-faced (with or without an extension hood) or closed-faced. Examples include the three-piece 25-mm cassette with extension cowl used for asbestos sampling, and the three-piece 37-mm polystyrene and two-piece 47-mm stainless steel cassettes used for total aerosol sampling (Figure 12.5). Here "total" is used in the traditional sense, as if all of the aerosol is sampled with equal efficiency; however, research has shown that these devices may significantly undersample even the inhalable fraction of larger particle sizes under typical workplace conditions.[29] The recently introduced British IOM sampler (Figure 12.6) appears to be fairly robust in matching the inhalable fraction sampling performance criterion in a range of sampling conditions.[24,30] This 2-L/min sampling device features a 25-mm filter in a removable capsule and a 15-mm circular inlet and has been shown to sample accurately the inhalable fraction of ambient aerosol for particles up to 80 µm aerodynamic diameter and wind speeds up to 2.6 m/sec.[11]

The widely used filter media may be classified as fiber, porous membrane, or straight-through (capillary) membrane filters.[27] Fiber filters are commonly composed of cellulose, glass, or quartz fibers. They exhibit a low pressure drop at high sampling flow rates, have a high loading capacity, and are relatively inexpensive; however, they may not be adequately efficient for submicron particle sizes, and water absorption may present problems in gravimetric analyses. Since filters must be removed from cassettes for reweighing, the high mechanical strength of fiber filters can be a distinct advantage over other filter types.

Porous membrane filters are produced from gels of cellulose ester, PVC, and other materials that set in a highly porous mesh microstructure, leaving convoluted flow paths. Membrane filter efficiency is reflected in a pore size rating that ranges from approximately 0.1 µm to 10 µm, but pore size rating can be a misleading indicator; for example, mixed cellulose ester membrane filters of 5-µm pore size are typically greater than 98% efficient even for 1-µm and smaller particles, and 8-µm mixed cellulose ester filters are greater than 92% efficient for these particles.[27, 31] These filters, like fiber filters, are considered to be "depth filters" since deposition is throughout the filter matrix, but porous membrane filters have generally higher flow resistance and lower loading capacity than fiber filters.

Straight-through, or capillary, membrane filters are made of polycarbonate or polyester film perforated with straight-through pores of nearly uniform size and distribution. The holes result from neutron bombardment followed by chemical etching to the desired size. Straight-through filters have a high pressure drop and low loading capacity compared with other filters and are susceptible to static charge buildup that can affect particle capture and retention; however, they have distinct advantages for optical or electron microscopic particle analysis. Viewing is excellent since particles larger than the pore size are captured at the filter surface, which is extremely flat and smooth.

Polyester foam media may represent an emerging filtration technology suited to simultaneous sampling of multiple aerosol fractions in a single sampler. A sampler proposed by Vincent et al.[32] utilizes two foam plugs in series with a final high efficiency filter to separate the inhalable, thoracic-less-respirable, and respirable fractions of sampled aerosol. The advantages of such a device are obvious; however, while the technique would be appropriate for gravimetric analysis, there are unresolved questions about its usefulness for other analytical needs and about the feasibility of large-scale manufacture of uniform and well-characterized foams.

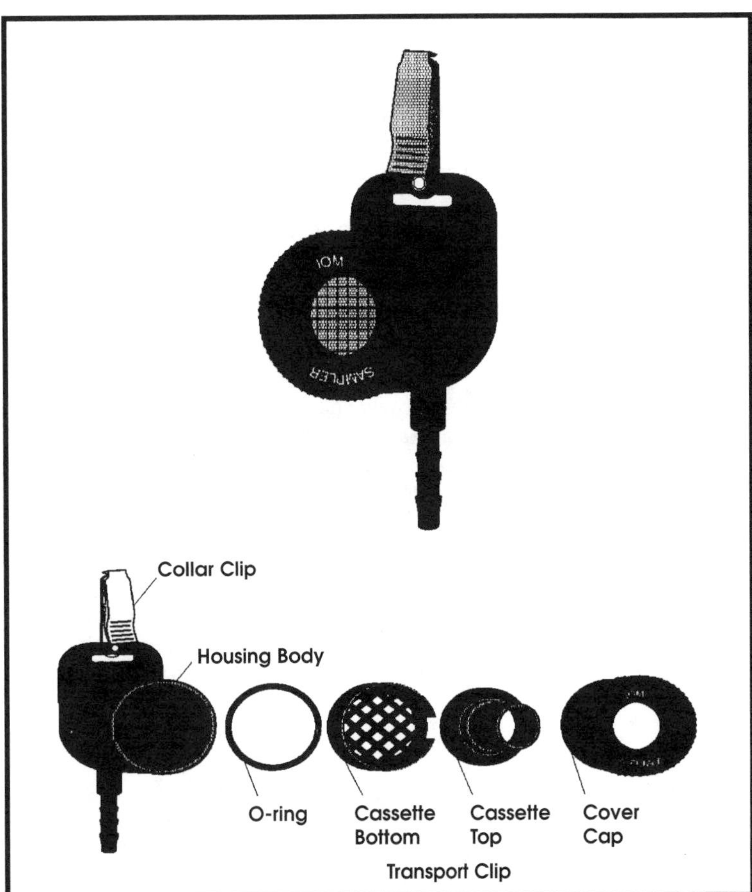

Figure 12.6 — IOM inhalable dust sampler developed by J.H. Vincent and D. Mark at the Institute of Occupational Medicine, Edinburgh, Scotland. For gravimetric analysis the interior cassette, containing the filter, is weighed before and after sampling. The sampler meets international sampling criteria for inhalable particulate matter (graphic courtesy SKC, Inc.).

Eight major suppliers market some 40 varieties and sizes of fiber filters, 58 membrane filters, and 14 straight-through membrane filters,[28] so choice of the appropriate media for a particular application can be a challenge. Where standard methods do not specify the sampling media and method, supplier technical representatives or the supporting analytical laboratory may provide useful guidance.

In any type of filter sampling there is a risk of losing volatile material during collection and of suffering unplanned chemical reactions between collected material and the filter media. Volatiles may be captured on sorbent media in series with and downstream of the filter media, and in some cases chemically treated filters are available that react with and retain the contaminant. Careful choice of sampling media may prevent undesirable species-media reactions.

Sedimentation-Based Techniques

Sedimentation-based aerosol sampling techniques are conceptually simple. In the horizontal elutriator, which is the only type likely to be used in occupational exposure assessment, aerosol passes between two narrowly spaced horizontal plates with an average airflow velocity U as shown in Figure 12.7. An aerosol particle in the flow settles under gravity at its terminal settling velocity U_T, so that in a sufficiently long elutriator the particle will eventually reach the bottom plate and be captured. It may be shown that for a particle entering at the top plate of the elutriator, its deposition distance L from the elutriator entrance is

$$L = H \frac{U}{U_T} \tag{8}$$

where H is the height of the channel. The actual path traveled by the particle is not linear as suggested by this equation; rather, it follows a curved path due to the parabolic velocity profile of the airflow between the plates.[11] Although simple to design and build, horizontal elutriators are not suited to personal sampling and do not provide good resolution of particle sizes; however, they are effective as preseparators for other samplers. For example, the British MRE Type 113A area sampler incorporates a multichannel horizontal elutriator preseparator to remove nonrespirable particles from sampled aerosol prior to filtration.

Other aerosol collection techniques based on sedimentation principles include centrifugation, electrostatic precipitation, and thermal precipitation. In an instrument such as the Stöber spiral duct aerosol spectrometer (centrifuge), aerosol passes through a rectangular channel coiled in a spiral within the spinning centrifuge head, so that particle settling velocities are greatly enhanced by the centrifugal forces. Although these instruments provide excellent size resolution, they are not suited to field use due to their size and cost.

Cyclone separators, like centrifuges, utilize centrifugal forces to effect particle capture. Cyclones cause the sampled air to be spun in a tapering tube as shown conceptually in Figure 12.8. Particles migrate to the tube walls and are removed, and the remaining particles pass out of the device with the air stream. The cut size of the device indicates the particle size captured with 50% efficiency, i.e., the d_{50}. Cyclones are most efficient for large particle sizes, and are used in occupational hygiene primarily as preseparators in respirable aerosol samplers. The d_{50} is controlled by the configuration of the cyclone and the flow conditions, and decreases with decreasing cyclone diameter and increasing flow rate. Commonly used devices operate at flow rates ranging from 1.7 to 2.5 L/min (Figure 12.9). In each device the

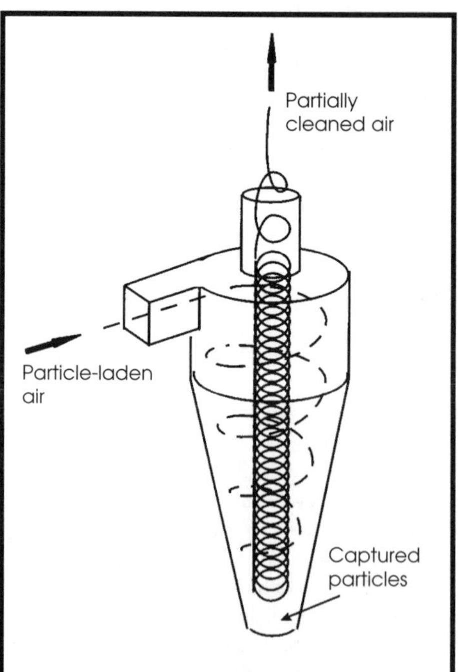

Figure 12.8 — Cyclone separator. Suspended particles are captured by increasing centrifugal forces as the air spirals down the cone of the cyclone; the airflow and uncaptured particles spiral back up the central axis and exit through the top. In a personal respirable aerosol sampler the exiting air, carrying the respirable particle fraction, passes through a cassette filter, where the particles are captured for gravimetric or other analysis.

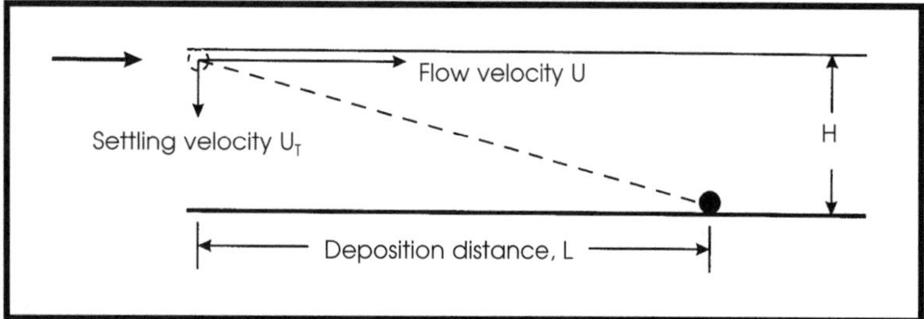

Figure 12.7 — Particle deposition in an idealized horizontal elutriator. In reality, U is not constant because a parabolic flow velocity profile will exist between the top and bottom plates, and the particle will not follow a straight line trajectory; however, the approximation is adequate for instrument design.

respirable aerosol fraction passes through the cyclone to be captured on a filter for subsequent analysis. Collection efficiency performance criteria adopted by the American Conference of Governmental Industrial Hygienists and the International Organization for Standardization/ European Standardization Committee (ISO/CEN) specify that cyclones and other respirable aerosol sampling devices should exhibit a d_{50} of 4 µm for the respirable fraction.[33]

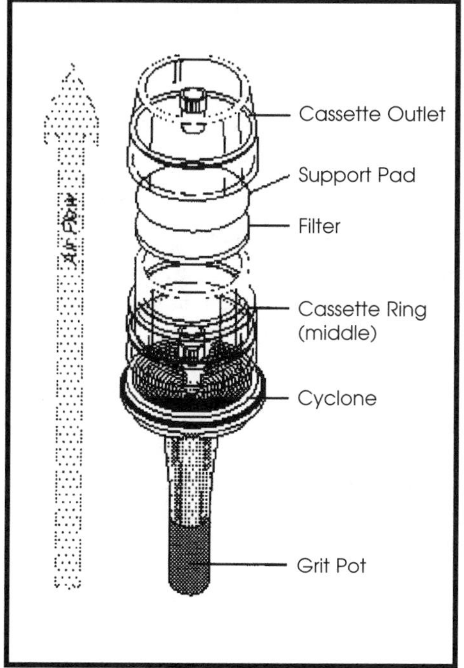

Figure 12.9 — Personal cyclone. Exploded view of a personal respirable dust sampling assembly incorporating an SKC aluminum cyclone and 37-mm three-piece filter cassette. Nonrespirable particles are collected in the grit pot at the base of the cyclone, and the respirable fraction is collected on the filter for subsequent weighing or chemical analysis (graphic courtesy SKC, Inc.).

In electrostatic precipitators, electrically charged aerosol particles migrate to the collection surface because of an applied electric field. These devices are highly efficient for even submicron particles, but require low flow-through velocities and cannot be used in potentially explosive atmospheres because of arcing. Thermal precipitators pass particles between two areas of large temperature difference, as between a heated filament and a cold surface (such as a microscope slide), with the result that air molecules on the warmer side push the aerosol particle toward the colder surface to be captured. These devices also require low flow velocities and are most efficient for submicron particles.

Impaction-Based Techniques

Impaction-based sampling devices are among the most widely used in aerosol characterization. In essence, impactors rely on a particle's inertia to carry it onto a collection surface upon which the sample airstream impinges (Figure 12.10). Jet-and-plate impactors direct a high velocity jet against a collection surface such that the air is forced to make an abrupt 90° turn. Depending on its mass and position in the jet, an entrained particle may or may not strike the plate and be captured. The likelihood of capture increases with increasing Stokes number, as shown in Equation 6, where the characteristic velocity and

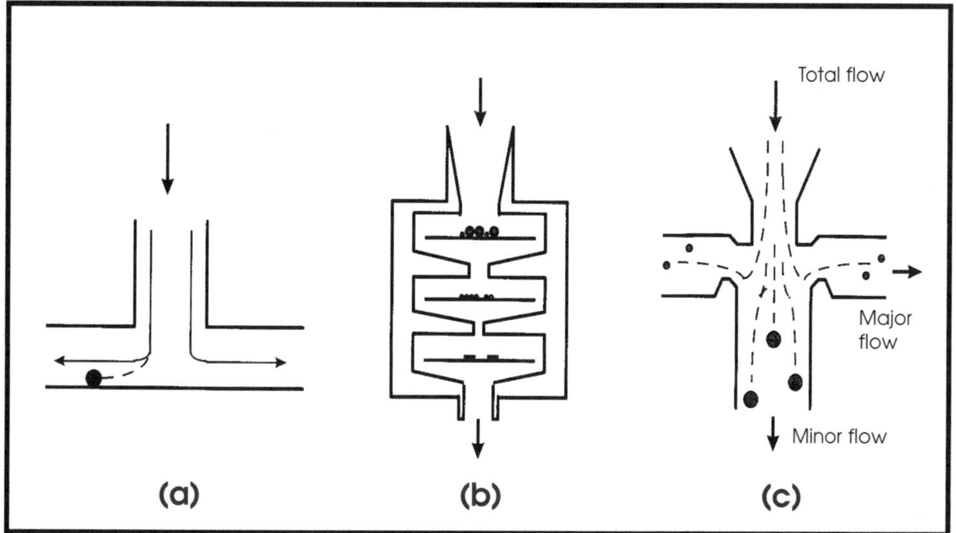

Figure 12.10 — Inertial impactors: (a) conventional jet-to-pate impactor collecting a single size fraction (say all particles over 10 µm); (b) multistage or cascade impactor in which each stage collects a different size fraction; and (c) virtual impactor or dichotomous sampler in which size fractions are separated but not removed from the airstream (as in Reference 34).

dimensions of the system are the jet's average velocity and the slot half-width for a rectangular slot impactor or throat radius for a circular jet impactor. The parameter used to characterize impactor performance is the 50% cut point size, d_{50}, which is the particle size captured by the impactor with 50% efficiency.

Single-stage impactors are most often used as preseparators for other sampling devices as in devices designed to measure only particulate matter less than a specified size, say 2.5 or 10 µm ($PM_{2.5}$ and PM_{10} samplers, respectively). Where the impactor itself is the primary instrument, multistage designs are usually used in a cascade configuration as shown in Figure 12.10. Cascade impactors place impactors of descending cut size in series so that the largest particles are removed in the first stage and serially smaller particles are removed in the subsequent stages. In a well-designed device each stage will collect a narrow range of particle sizes, so that the sampled aerosol is separated into distinct and, ideally, nonoverlapping size fractions (in practice there is always some overlap). The cumulative mass distribution may then be plotted against a characteristic particle size for each stage. Data for aerosols with lognormally distributed particle sizes will plot as a straight line on log-probability paper, so that the distribution's geometric mean particle size and geometric standard deviation may be graphically determined (see, for example, Lodge and Chan[35]). Well-known cascade impactors are the Graseby-Andersen 8-stage area sampler operated at 28.3 L/min and the Graseby-Andersen "Marple" personal cascade impactor operated at 2 L/min.[36] A recent development in this area is the IOM 2-L/min Personal Inhalable Dust Spectrometer (PIDS), which is similar to the Marple personal sampler except that the PIDS utilizes circular rather than slot jets.

Material collected by cascade impaction may be analyzed in various ways, depending on the purpose of the investigation. Gravimetric analysis of the total mass collected on each plate, chemical or radiometric analysis to determine the distribution of species across particle size, and extraction and mutagenic evaluation of different size fractions are some options. These data may be extremely useful in hazard evaluation since more will be known about what potentially hazardous materials are present and where they might deposit in the respiratory tract.

An interesting variation on the plate-type impactor is the virtual impactor (Figure 12.10). In this device there is no collection surface; rather, the aerosol jet exiting the nozzle impinges on the surface of lower-velocity "minor flow" in the collection probe. As the major portion of the flow is forced to reverse direction and exit the device, larger particles fail to negotiate the turn and penetrate into the body of the minor flow. The minor flow thus contains the larger size fraction while the major flow contains the smaller size fraction, and both flows are available for further use or analysis. Such dichotomous samplers have been designed specifically to separate the $PM_{2.5}$ fraction from the PM_{10} fraction in ambient air samples to distinguish between the fine and coarse fractions of PM_{10}.[37]

Another variation of the impactor is the liquid impinger. Liquid impingers project an aerosol jet against a wetted surface (Figure 12.11a), a submerged surface, or a liquid surface (Figure 12.11b). These devices are especially useful when sampling mist aerosols that might evaporate on a dry collection plate or when it is otherwise desirable to react either the collected aerosol or a gas phase component with the collection liquid. The collection liquid may be examined in various ways, including particle counting. Particle counting was used earlier in this century to characterize dust exposure from impinger samples, with dust concentrations expressed in units of millions of particles per cubic foot.[38] Epidemiologists conducting retrospective exposure assessments of the dusty trades must often wrestle with the difficulties of converting data in millions of particles per cubic foot to equivalent exposures in milligrams per cubic meter.

Optical Techniques

Optical measurement of aerosols, as discussed above, may be classified into in situ methods and measurement of a particle sample on a collection medium. In situ methods are rarely used in the occupational setting, being confined to laboratory applications. Discussion of optical methods is thus limited to particles that have been collected on a suitable surface, and the choice of sampling and measurement technique depends on the aerosol properties.

As with other sampling methods, the aerosol particle size characteristics are an important determinant of the optical method chosen. Of primary importance is the particle size with respect to the resolving power of the optical device. In the case

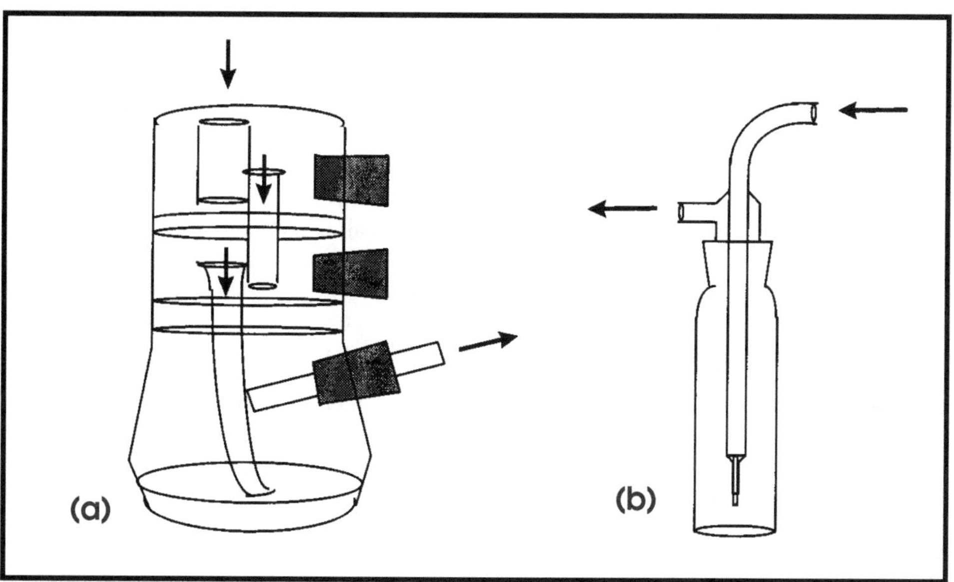

Figure 12.11 — Liquid impingers: (a) multistage impinger in which a jet impinges on a wet surface, and (b) all glass impinger in which a jet impinges on a liquid surface or a submerged jet impinges on the bottom of the impinger.

of an optical microscope, the resolving power is a function of the wave length of the light, which varies from 0.4 µm (violet) to 0.7 µm (red). In ideal conditions the resolving power is approximately 0.2 µm, which is to say that smaller particles cannot be sized by optical microscopy. Currently, the primary application of optical microscopy is to obtain information about the aerosol size distribution.

Electron microscopy requires a special collecting substrate, such as carbon film, and a structure to support the film that will fit the microscope sample holder. The two methods of electron microscopy (EM) are transmission EM and scanning EM. In transmission EM the support structure is a thin metal grid having a diameter of 2 mm. The electron beam is directed perpendicular to the sample grid, and particles are detected by the differing electron transmission of the particles and the substrate. The resolving power of transmission EM is of the order of 0.01 µm. In its simplest form, transmission EM yields a two dimensional image of the particles. It is common to "shadow" the sample by evaporating a metal under vacuum conditions to produce a thin metal film on the particles and substrate. By orienting the metal vapor source oblique to the sample, a shadow is produced behind each particle, which yields some information about the third dimension of the particles.

In scattering EM, the electron beam itself is oriented oblique to the sample, and the scattered electrons are detected to yield images of the particles. This approach is somewhat similar to the shadowing of transmission EM samples in that some three dimensional information is obtained. However, scattering EM differs in that the scattering of electrons by particles is fundamentally different from electron transmission. The contrast between particles and substrate is generally greater for transmission EM. Some particles are difficult to detect in scanning EM because of their low scattering efficiency. Additionally, the resolving power of scattering EM is somewhat less than that for transmission. Thus, the choice between these two methods depends on the particle composition and size.

A major problem with both methods of EM is the question of a representative sample. The area of the collecting substrate for both methods is very small, and it is not feasible to collect an entire sample onto a grid or substrate. Only an extremely small fraction of the total aerosol is collected onto the substrate, and the surface density of particles is quite low. It is thus difficult to obtain a number or mass concentration of an aerosol from an EM analysis. Rather, it is more common to consider the particles as representative of the size distribution and use the results for this purpose.

For additional information on optical and EM microscopy techniques relevant to occupational hygiene practice, the interested reader is referred to the detailed discussions provided by Silverman et al.(39) and Cradle.(40)

Aerosol Size Distribution Analysis

Natural and industrial aerosols are virtually always polydisperse, i.e., they are made up of nonidentical particles. The simplest case might be that of a mist aerosol, such as might be mechanically generated in an industrial process through splashing, spraying, or agitation. Here the particles are morphologically and chemically similar in that they are spherical droplets of the parent material, with the only real difference between particles being in their diameter. As is the case for many quantities occurring in nature, particle sizes in such an aerosol are often approximately log-normally distributed, that is, the logarithms of the particle sizes follow a Gaussian, or normal, frequency distribution. This is fortunate because it allows the entire distribution to be characterized with two descriptors or statistics: geometric mean (or median) size and geometric standard deviation. The distribution is likely to be expressed using either the count median diameter (CMD) and geometric standard deviation (GSD), or the mass median aerodynamic diameter and GSD, depending on how the measurement data were obtained. Count-based frequency distributions result from such measurement devices as optical particle counters, which determine particle size from light scattered by the particle at a specified angle (see, for example, Pui and Swift[41]). Particle counts are grouped in size intervals by such instruments, and the CMD and GSD may be calculated as

$$\log \text{CMD} = \frac{\sum_{i=1}^{N} n_i \log d_i}{\sum_{i=1}^{N} n_i} \quad (9)$$

$$\log \text{GSD} = \sqrt{\frac{\sum_{i=1}^{N} n_i (\log \text{CMD} - \log d_i)^2}{\sum_{i=1}^{N} n_i - 1}} \quad (10)$$

where

n_i = is the number of particles in a given interval,

d_i = is the midpoint size for the interval, and

N = is the total number of particles collected.

These calculations, which are tedious to perform, may give a misleading impression of mathematical certainty since measurement data may be only approximately log-normally distributed. Fortunately, a simple graphical technique is available that is useful for both checking the log-normality of the data and obtaining the CMD and GSD. As shown in Table 12.1, the fraction of the total count in each size interval is as reported by the measurement instrument, the cumulative fraction less than a stated size (e.g., the interval midpoint) is calculated, and the data are graphed as a cumulative frequency distribution in a log-probability plot as shown in Figure 27.12. Note that the top-most data point is not plotted (that corresponding to 100%). If the data are log-normally distributed, then a straight line may be fitted to the data points (the general practice is to perform an "eyeball regression"). CMD is taken as the particle size corresponding to the 50% probability of occurrence, and the GSD is calculated as either the ratio of CMD to the 15.87% particle size or the ratio of the 84.13% particle size to the CMD (both will give the same GSD value).

If a single straight line cannot be fitted to the data, then the distribution is not log-normal, as might be the case for mixtures of aerosols originating from different sources. For example, mine aerosols may be made up of aerosols from diesel engine exhausts, which are largely fine soot particles, mixed with much coarser mine dust. The aerosols taken individually are likely to exhibit log-normally distributed particle sizes, but with much different median diameters and geometric standard deviations. The combined aerosol will then be bimodal, with two modes, or peaks, instead of only one. Measurement data for the aerosol would not fall in a straight line on a log-probability plot; rather, data points on one end of the plot would fall generally along one line while points on the other end would fall along a second line, which would probably have a different slope. Thus, log-probability plots provide more interpretive information than the calculated values from Equations 9 and 10.

Cascade impactors are examples of mass-based measurement instruments that characterize the mass fraction rather than count fraction in specified particle size intervals. In this case the size intervals represent aerodynamic particle size and are usually reported in terms of aerodynamic equivalent diameter, d_{ae}. The data analysis is similar to that performed in Table 12.1 in

Table 12.1 — Analysis of Particle Count Data

Lower Interval Size (µm)	Upper Interval Size (µm)	Midpoint Interval Size d_i (µm)	Number in Interval n_i	Fraction Interval (%)	Cumulative Fraction less Than d_i (%)
0.46	0.54	0.50	3770	0.74	0.74
0.54	0.63	0.585	13,000	2.56	3.30
0.66	0.74	0.685	55,100	10.84	14.14
0.74	0.86	0.80	62,900	12.38	26.52
0.86	1.0	0.93	98,800	19.44	45.96
1.0	1.2	1.1	109,000	21.45	67.41
1.2	1.4	1.3	98,800	19.44	86.85
1.4	1.6	1.5	37,200	7.32	94.17
1.6	1.8	1.7	17,000	3.35	97.52
1.8	2.2	2.0	9710	1.91	99.43
2.2	2.5	2.35	2850	0.56	99.99
2.5	2.9	2.7	52	0.01	100
			508,182		

that the fraction of total collected mass represented by each size interval (collection stage) is calculated, the cumulative mass frequency distribution is determined, a log-probability plot is prepared, and the mass median aerodynamic diameter and GSD are obtained. Note that for a given aerosol the mass median aerodynamic diameter will be larger than the CMD.(14)

Occupational hygienists and epidemiologists may find it necessary to compare occupational exposure data obtained using different measurement techniques. For example, it might be desirable to combine data from cascade impactor sampling with data from optical particle counting when classifying worker exposures. This is by no means a simple task, since such factors as particle shape, density, and refractive index must be well understood if data of one type is to be mathematically converted to another type. An in-depth discussion of the difficulties involved is beyond the scope of this presentation; for additional information the interested reader is referred to Mercer.(13)

Summary

Numerous sampling and analysis techniques are available to characterize potentially hazardous aerosol exposures in occupational environments. Selection of the appropriate technique is a matter of recognizing the potential hazard based on an understanding of process variables and aerosol behavior, defining the information needs required for hazard evaluation, then identifying a sampling and analytical method that will satisfy those needs. No single method can address all possible questions about a given aerosol, nor can any single method answer a given question for all possible aerosols. The choice of which instrument or media to use, or which analysis to perform, may be based on regulatory requirements, current accepted practice, or individual judgment. In all cases an understanding of the capabilities and limitations of the technique used is critical to effective data interpretation and accurate exposure characterization.

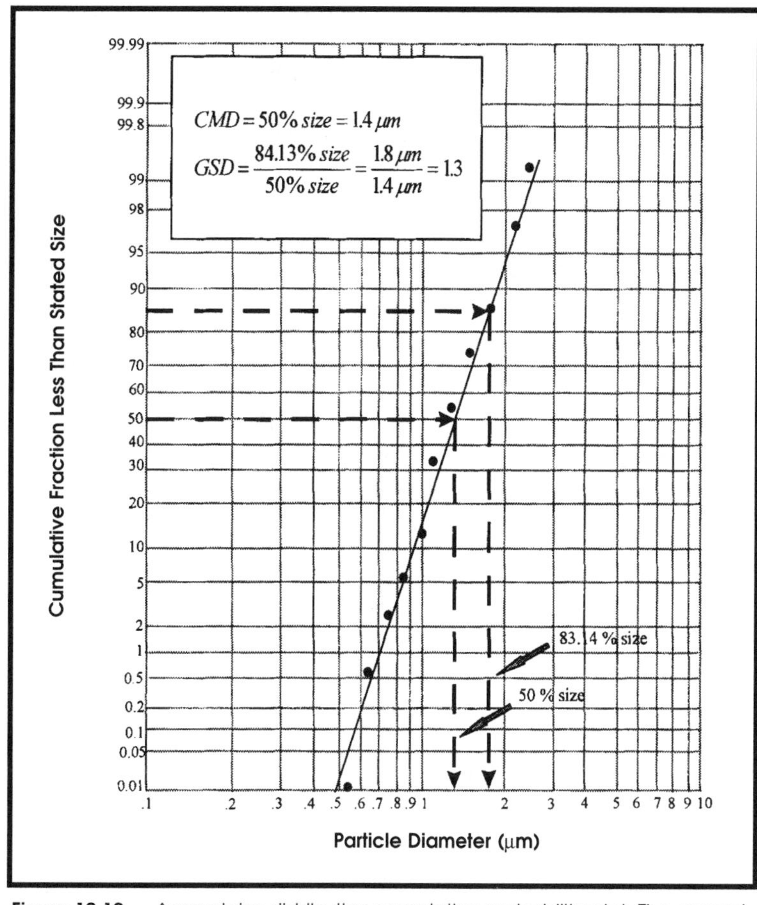

Figure 12.12 — Aerosol size distribution cumulative probability plot. The aerosol particle sizes are approximately lognormally distributed if a straight line provides a good fit to the measurement data. CMD and GSD are determined from the plotted line. Count-based distributions result from optical measurement instruments that estimate particle size by examining particle light scattering behavior.

References

1. **Hunter, D.:** *The Diseases of Occupations*, 6th ed. London: Hodder and Stroughton, 1978.
2. **Agricola, G.:** *De Re Metallica*. New York: Dover Publications, Inc., 1950.
3. **Willeke, K. and P.A. Baron (eds.):** *Aerosol Measurement: Principles, Techniques and Applications*. New York: Van Nostrand Reinhold, 1993.
4. **Cox, C.S. and C.M. Wathes (eds.):** *Bioaerosols Handbook*. Boca Raton, FL: Lewis Publishers, 1995.
5. **Reist, P.C.:** *Aerosol Science and Technology*, 2nd ed. New York: McGraw-Hill, Inc., 1993.
6. **U.S. Department of Health, Education and Welfare (DHEW):** *Smoking and Health: A Report of the Surgeon General* (DHEW pub. [PHS] 79050066). Washington, DC: U.S. Government Printing Office, 1979.
7. **U.S. Department of Health and Human Services (DHHS):** *The Health Consequences of Smoking: Chronic Obstructive Lung Disease. A Report of the Surgeon General* (DHHS [PHS] 84-50205). Washington, DC: U.S. Government Printing Office, 1984.
8. **U.S. Environmental Protection Agency (EPA):** *Workshop Report on Chronic Inhalation Toxicity and Carcinogenicity Testing of Respirable Fibrous Particles* (EPA-748-R-96-001). Washington, DC: U.S. Government Printing Office, 1996.
9. **Donaldson, K., R.C. Brown, and G.M. Brown:** Respirable industrial fibers: mechanisms of pathogenicity. *Br. Assoc. Lung Res. 48*:390–393 (1993).
10. **Mandelbrot, B.B.:** *The Fractal Geometry of Nature*. New York: Freeman, 1983.
11. **Vincent, J.H.:** *Aerosol Science for Industrial Hygienists*. New York: Elsevier Science, Inc., 1995.
12. **Cunningham, E.:** On the velocity of steady fall of spherical particles through fluid medium. *Proc. Royal Soc. A83*:357–365 (1910).
13. **Mercer, T.T.:** *Aerosol Technology in Hazard Evaluation*. New York: Academic Press, 1973.
14. **Hinds, W.C.:** *Aerosol Technology—Properties, Behavior, and Measurement of Airborne Particles*. New York: John Wiley & Sons, 1982.
15. **Esmen, N.A.:** Adhesion and aerodynamic resuspension of fibrous particles. *J. Environ. Eng. 122*:379–383 (1996).
16. **Morgan, W.K.C. and N.L. Lapp:** Diseases of the airways and lungs. In *Occupational Diseases: A Guide to Their Recognition* (rev. ed.), (DHEW [NIOSH] 77-181). Washington, DC: U.S. Government Printing Office, 1977. pp. 103–121.
17. **Rosenstock, L. and M.R. Cullen:** *Textbook of Clinical Occupational and Environmental Medicine*. Philadelphia: W.B. Saunders Co., 1994.
18. **Vincent, J. and L. Armbruster:** On the quantitative inhalability of airborne dust. *Ann. Occup. Hyg. 24*:245–248 (1981).
19. **Yu, C.P. and C.K. Diu:** A comparative study of aerosol deposition in different lung models. *Am. Ind. Hyg. Assoc. J. 43*:54–65 (1980).
20. **Weibel, E.R.:** *Morphometry of the Human Lung*. New York: Academic Press, 1963.
21. **Hansen, J.E. and E.P. Ampaya:** Human air space, shapes, sizes, areas and volumes. *J. Appl. Physiol. 38*:990–995 (1975).
22. **Yeh, H. and G.M. Schum:** Models of human lung airways and their application to inhaled particle deposition. *Bull. Math. Biol. 42*:461–480 (1980).
23. **Harris, R.L. and D.A. Fraser:** A model for deposition of fibers in the human respiratory system. *Am. Ind. Hyg. Assoc. J. 37*:73–89 (1976).
24. **Vincent, J.H.:** *Aerosol Sampling: Science and Practice*. Chichester, UK: John Wiley & Sons, 1989.
25. **Cohen, B.S. and S.V. Hering (eds.):** *Air Sampling Instruments for Evaluation of Atmospheric Contaminants*, 8th ed. Cincinnati, OH: American Conference of Governmental Industrial Hygienists, 1995.
26. **Lippmann, M.:** Size-selective health hazard sampling. In *Air Sampling Instruments for Evaluation of Atmospheric Contaminants*, 8th ed., Cincinnati, OH: American Conference of Governmental Industrial Hygienists, 1995. pp. 81–119.
27. **Lee, K.W. and M. Ramamurthi:** Filter collection. In *Aerosol Measurement: Principles, Techniques and Applications*, K. Willeke and P.A. Baron (eds.). New York: Van Nostrand Reinhold, 1993. pp. 179–205.
28. **Lippmann, M.:** Filters and filter holders. In *Air Sampling Instruments for Evaluation of Atmospheric*

Contaminants, 8th ed., B.S. Cohen and S.V. Hering (eds.). Cincinnati, OH: American Conference of Governmental Industrial Hygienists, 1995. pp. 247–266.

29. **Mark, D., C.P. Lyons, S.L. Upjohn, and L.C. Kenny:** Wind tunnel testing of the sampling efficiency of personable inhalable aerosol samplers. *J. Aerosol Sci. 25(Suppl. 1)*:S339–S340 (1994).
30. **Mark, D. and J.H. Vincent:** A new personal sampler for airborne total dust in workplaces. *Ann. Occup. Hyg. 30*:89–102 (1986).
31. **Liu, B.Y.H., D.Y.H. Pui, and K.L. Rubow:** Characteristics of air sampling filter media. In *Aerosols in the Mining and Industrial Work Environments*, vol. 3. Ann Arbor, MI: Ann Arbor Science, 1983. pp. 989–1038.
32. **Vincent, J.H., R.J. Aitken, and D. Mark:** Porous plastic foam media: penetration characteristics and applications in particle size-selective sampling. *J. Aerosol Sci. 24*:929–944 (1993).
33. **American Conference of Governmental Industrial Hygienists (ACGIH):** *Threshold Limit Values for Chemical Substances and Physical Agents/Biological Exposure Indices.* Cincinnati, OH: ACGIH, 1996.
34. **Marple, V.A., K.L. Rubow, and B.A. Olson:** Inertial, gravitational, centrifugal, and thermal collection techniques. In *Aerosol Measurement: Principles, Techniques and Applications.* New York: Van Nostrand Reinhold, 1993. pp. 206–232.
35. **Lodge, J.P. and T.L. Chan:** *Cascade Impactors: Sampling and Data Analysis.* Akron, OH: American Industrial Hygiene Association, 1986.
36. **Hering, S.V.:** Impactors, cyclones, and other inertial and gravitational collectors. In *Air Sampling Instruments for Evaluation of Atmospheric Contaminants*, 8th ed. Cincinnati, OH: American Conference of Governmental Industrial Hygienists, 1995. pp. 279–305.
37. **Watson, J.G. and J.C. Chow:** Ambient air sampling. In *Aerosol Measurement: Principles, Techniques and Applications.* New York: Van Nostrand Reinhold, 1993. pp. 622–639.
38. **Drinker, P. and T. Hatch:** *Industrial Dust: Hygienic Significance, Measurement and Control.* New York: McGraw-Hill Book Co., 1936.
39. **Silverman, L., C.E. Billings, and M.W. First:** *Particle Size Analysis in Industrial Hygiene.* New York: Academic Press, 1971.
40. **Cradle, R.D.:** *The Measurement of Airborne Particles.* New York: John Wiley & Sons, 1975.
41. **Pui, D.Y.H. and D.L. Swift:** Direct-reading instruments for airborne particles. In *Air Sampling Instruments for Evaluation of Atmospheric Contaminants*, 8th ed. Cincinnati, OH: American Conference of Governmental Industrial Hygienists, 1995. pp. 337–353.

Outcome Competencies

After completing this chapter, the user should be able to:

1. Define underlined terms used in this chapter.
2. Anticipate when biological monitoring is appropriate.
3. Assess the limiting factors to biological monitoring.
4. Provide justification for the concept of biological monitoring.
5. Provide justification for recommendations based on biological monitoring.

Key Terms

absorbed dose • adducts • antibodies • antigens • biochemical epidemiology • biologically effective dose • biological exposure indices (BEIs) • biological monitoring • biomarker • blood • breast milk • chemical • conjugates • determinant • ear wax • elimination • end-exhaled breath (alveolar exhaled breath) • endogenous • enzyme • excretion • feces • flatus • fluids • hair • half-times (pseudo first order) • health surveillance • immune response • macromolecules • markers • medical monitoring • medical surveillance • menses • metabolites • midstream urine • mixed exhaled breath • nail • plasma (blood) • saliva • sebum • semen • serum (blood) • sputum • sweat • uptake • urine

Prerequisite Knowledge

- TLV–TWAs and PELs and the documentations that form their basis.
- Undergraduate biology; human metabolism and physiology; toxicology; undergraduate chemistry; epidemiology; biostatistics; undergraduate kinetics as a part of physical chemistry; undergraduate arithmetic using logarithms, exponentials, differentials, and integration.
- Basic biology; physiology; chemistry; biostatistics; algebra; and epidemology.

Key Topics

I. Hygienists and Biological Monitoring

II. When to Use Biological Monitoring
   A. Federally Mandated Biological Monitoring: The Lead Standard
   B. Biological Monitoring Recommendations
   C. Compounds Without OSHA or ACGIH Recommendations
   D. Routes of Exposure Other Than Inhalation
   E. Presence of PPE
   F. Unanticipated Exposures

III. Factors for Biological Monitoring Data Interpretation Using the BEIs
   A. Vapor Inhalation Predominates as the Route of Exposure
   B. Inhalation of Vapor Present, but Not the Major Route of Exposure

IV. Biological Monitoring Sampling and Quality Assurance/Control
   A. Breath Sampling
   B. Urine Sampling
   C. Other Noninvasive Biological Monitoring Media

Appendix A. Illustrated Examples for Xylene Relative to Workload, the BEIs, and Inhalation Conditions

13

Biological Monitoring

Shane S. Que Hee Ph.D.

Introduction

Biological monitoring is the measurement of chemical markers present in the human body that result from exposure to chemical and physical agents.[1] These markers can be (1) the original exposing chemical; (2) metabolites of a single exposing chemical; (3) conjugates caused by reaction of a single chemical or its metabolites with endogenous biochemical cycles; (4) adducts formed by reaction of a single chemical or its metabolites with endogenous macromolecules, antigens, or antibodies; or (5) an endogenous enzyme or biochemical affected by chemical or physical agents or the product(s) of the immune response of humans. A chemical marker is a specialized biomarker and determinant.

The biological media in which the markers are found can be exhaled breath; flatus; urine; blood; blood serum; blood plasma; sebum; ear wax; semen; the menses; breast milk; sweat; hair; nail; teeth; tears; feces; saliva; fat; skin; sputum; or internal organs.

The markers may be of dose and/or effect. Markers of effect are subdivided into those of adverse effect (medical monitoring), potentially adverse effect or predictive of effect (health surveillance), and markers of susceptibility (genetic markers). Overseeing markers of both dose and effect on workers in a workplace or field setting is termed medical surveillance. Correlating markers within epidemiologic studies is termed biochemical epidemiology or molecular epidemiology.

Both marker types may have to be identified and quantified in samples, which means that the laboratories that can analyze for the marker, as well as the sampling methods, must be known and available. Whereas reference ranges may exist for markers of effect for healthy people, the critical concentration of a marker of dose is generally dependent on an Occupational Safety and Health Administration (OSHA) action level, a permissible exposure limit (PEL), or a recommended occupational air exposure limit such as a threshold limit value (TLV®)-time weighted average. The latter procedure for markers of dose produces equivalent biological monitoring results when inhalation is the only route of exposure and only the exposing chemical is involved.

Biological monitoring results, however, reflect absorption into the body from all routes of exposure: inhalation, oral ingestion, and eye, ear, and skin exposure. Different chemicals and their markers take different times to appear in the same body medium as characterized by different half-times. A single chemical and its markers also take different half-times to appear in different body media or in the same body media after exposure through different routes. The excretion characteristics determine recommended sampling times for markers.

Hygienists and Biological Monitoring

Occupational hygienists lack the legal authority to perform invasive sampling of the human body unless the hygienist is also certified in phlebotomy. Invasive sampling is the province of medical personnel who already have the legal authority to inject medications. Blood sampling in the workplace, even finger sticks, must be performed by licensed medical personnel (although the advent of a home finger-stick blood sampling test for HIV-I in 1996 may set a precedent that will allow a worker to sample blood and give it to a hygienist or an analytical laboratory). Thus, actual medical monitoring, health surveillance, and monitoring for markers of susceptibility and effect are usually outside the practicing scope of hygienists trained in the United States. Hygienists must, however, be able to explain to the worker the relevance of medical surveillance, including blood sampling, during worker training or counseling and must also be able to refer workers to a physician for taking blood samples, medical emergency, disabling injuries, workers' compensation situations, and signs of chronic health problems.

Therefore, generally speaking, the field use of biological monitoring for hygienists is limited to major body media that can be sampled noninvasively, which includes ear wax, exhaled breath, flatus, feces, hair, nails, saliva, sebum, semen, sputum, sweat, tears, and urine. Of these media, only urine and exhaled breath have gained general acceptance. External contamination is a problem with hair, nails, saliva, sebum, and sweat. To assess what is inside or outside of hair or nails is difficult because there is no generally accepted washing technique. Ear wax, feces, flatus, sebum, semen, sputum, sweat, and tears are awkward or inconvenient to sample, and not enough data are available to interpret what the concentrations signify relative to exposure. Also, feces, sebum, semen, and ear wax are complex media requiring skilled and expensive analyses.

An alternative to analytical laboratory analysis is to measure markers directly in excreted body fluids. This aspect of biological monitoring is not well developed, although dipstick technologies now being used for urinalysis and blood analyses in medicine are analogous to detector tube methods of air analysis. Expired breath has higher water vapor and carbon dioxide concentrations than workplace air, and breath biological exposure indices (BEI) are also much lower than the corresponding TLV. Thus, methods developed for TLV air sampling are not directly applicable for breath sampling.

Hygienists may also employ direct reading instruments and devices to measure excretion on external parts of the human body or may sample these parts by wiping or rinsing.[(2)] Differentiating between external exposure and excretion is a major difficulty for breath and skin sampling. Noninvasive detection of substances on the skin is also not very developed at present.[(2)] For example, the use of ionizing radiation methods like X-ray fluorescence for bone lead[(1,3)] may be potentially harmful and can only be used under medical supervision.

When to Use Biological Monitoring

The major uses of biological monitoring occur when

1. Biological monitoring is mandated (for example, when blood lead is keyed to a critical air concentration) or recommended (for example, where BEI documentation exists);
2. Routes of exposure other than inhalation are important (contribute greater than 30% of the exposure);
3. Personal protective equipment (PPE) such as respirators, gloves,[(4)] and protective garments[(4)] are worn; or
4. Unanticipated exposures occur, especially when air monitoring is not performed.

Federally Mandated Biological Monitoring: The Lead Standard

The 1979 OSHA biological monitoring standard for blood lead is a maximum concentration of 50 µg/dL (where dL signifies 100 mL or 0.1 L whole blood) for administrative removal from the workplace of a worker exposed at or above the air action level of 30 µg/m^3.[(5)] Lead is defined in the standard as elemental lead, all inorganic lead compounds, and organic lead soaps. The arithmetic mean of the last three blood sampling tests or the arithmetic mean of all tests conducted over the previous 6 months (whichever time is longer) is at or above 50 µg/dL (unless the most recent blood test is at or below 40 µg/dL). The PEL is 50 µg/m^3. Blood sampling analysis is mandated every 6 months for workers above the air action level for more than 30 days per year; at least every 2 months for blood lead concentrations >40 µg/dL together with at least one annual medical examination; and at least monthly for medically removed workers (at least 2 weeks after the initial blood test) together with a medical examination. The blood lead testing must be done by an analytical laboratory currently licensed by the federal Centers for Disease Control and Prevention, that produces results within 95% confidence and/or within an accuracy of ±15% or ±6 µg/dL, whichever is greater. When the blood lead level exceeds 40 µg/dL employees must be notified of the level by the employer in writing within 5 working days after receipt of biological monitoring results, and that the lead standard requires temporary medical removal with worker benefits for up to 18 months for each removal.

Although medical examinations are the responsibility of a licenced physician, hygienists may have to assist in some of the aspects, and the topic must be covered during worker training supervised by hygienists. The lead standard medical surveillance was the model for subsequent medical surveillance of other chemicals. Medical examinations require detailed work and medical histories with particular attention to past occupational and nonoccupational lead exposures, personal habits (smoking, alcohol consumption, hygiene), and past gastrointestinal, renal, cardiovascular, and neurological symptoms. Pulmonary status must be evaluated if the worker will use respirators. The physical examination must pay particular attention to teeth, gums, and hematologic, gastrointestinal, renal, cardiovascular, and neurologic systems. Blood pressure must be measured and a blood sample taken (with analysis for blood lead, hemoglobin, hematocrit, red cell indices, peripheral smear morphology, zinc protoporphyrin, blood urea nitrogen, and serum creatinine). Routine urinalysis including microscopic examination is also necessary in addition to "any laboratory or other test deemed necessary by sound medical practice."[5-7] (Further information about medical monitoring tests can be found in Chapters 7 through 9 and Chapter 21 of Reference 1.)

If the physical examination indicates lead intoxication (for example, observation of a blue "lead line" at the tooth/gum interface, acute encephalopathy, lead colic, peripheral neuropathy, impotence, or abnormal menstrual cycles), the worker known to be exposed at or above the air action level can be removed immediately. An employee can also request pregnancy testing or laboratory evaluation of male fertility and counseling on the effects of past and current exposure to lead. If the employer selects the first physician, the employee may get a second opinion within 15 days of receipt of the written findings of the first physician, but must signify intent to do this in writing to the employer. A third physician may be designated by the involved physicians in the event of their irreconcilable disagreement. Hygienists make physician recommendations effective, since they are often the key on-site personnel.

The major areas of participation in the lead standard for hygienists are worker training, workplace signs, air sampling, and record keeping. Specific parts of the lead standard must be given to all employees who are potentially exposed to any level of airborne lead. Initial training must be started within the first 180 days of employment but before the time of initial job assignment. Refresher training must be offered at least annually thereafter for all employees potentially or actually exposed above the air action level or for whom the possibility of skin or eye irritation exists. This training involves communicating the content of the lead standard and its appendices; explaining the specific natures of the unit processes that could cause lead exposures at and above the air lead action level; giving information on the purpose, proper selection, fitting, use, limitations, and maintenance of respirators; describing the lead medical surveillance program and its purpose; explaining the lead medical removal criteria; describing the adverse health effects of lead, with particular attention to male and female reproductive effects; giving the engineering controls and work practices associated with the worker's job; communicating the contents of any compliance plan including air monitoring; and instructing employees that any chelating agent like ethylenediaminetetraacetic acid should not be routinely used to remove lead from the body unless this is under the direction of a licensed physician in reaction to lead intoxication.

Relative to air sampling, air monitoring must be at least every 6 months for workers over the action level but under the PEL, whether they wear respirators or not. If the PEL is exceeded, air monitoring must be repeated every 3 months. Air monitoring may be discontinued if 2 samples taken at least 2 weeks apart are below the action level. Any production, process, control, or personnel change triggers air monitoring. Workers may request respirators even if the PEL is not exceeded. Other PPE, change rooms, showers, the requirements for filtered air lunchrooms and housekeeping and engineering controls are beyond the scope of this discussion. Observing hygienists must have had at least the same medical surveillance and PPE program as the employees they are monitoring.

There are four sets of records to be kept:

1. The air sampling record must contain dates, number, duration, locations, and results for each of the samples taken, including a description of the sampling procedure used to determine representative employee exposure

where applicable; a description of the sampling and analytical chemistry methods used and evidence of their accuracy; the type of respirator worn; name, social security number, and job classification of the monitored employee, and of all other employees whose exposure the measurement is intended to represent; and the environmental variables that could affect the measurement of employee exposure.

2. The medical surveillance record similarly must include name, social security number, and description of employee job duties; a copy of the physician's written opinions; results of any air sampling for that employee and representative exposure levels supplied to the physician; and any employee medical complaints about lead exposure. Both the air sampling and the medical surveillance records must be maintained by the employer for at least 40 years (30 years in later OSHA recommendations for different chemicals) or for the duration of employment plus 20 years, whichever is longer.
3. The employer must keep or ensure that the licensed physician keeps a copy of the medical examination results including medical and work histories; a description of any laboratory procedures, and a copy of the standards/guidelines used to interpret the test results or references to that information; and a copy of the results of biological monitoring.
4. The employer must also maintain an accurate record for each employee medically removed. The record must include employee name and social security number; the dates of medical removal and the corresponding date of return; a brief explanation of how each removal was or is being accomplished; and a statement for each removal indicating whether the reason was an elevated blood lead. This record must be maintained at least as long as the worker is employed.

When the air PEL is exceeded in a work area, a sign reading "Warning: Lead Work Area; Poison; No Smoking or Eating" must be posted in addition to signs required by other ordinances, regulations, or statutes.

The above information has been provided as a complete package since air monitoring, biological monitoring, hygienic considerations, record keeping, and preventative measures all must be done simultaneously by the hygienist.

Biological Monitoring Recommendations

OSHA Recommendations

Although OSHA has promulgated only one standard that mandates biological monitoring as the criterion for worker medical removal, the OSHA Medical Records Rule[6] defines "employee exposure record" to include all biological monitoring as "exposure records" by specific OSHA standards. OSHA includes as exposure records all "tests as are needed according to a physician's professional judgment,"[6,7] a common part of all OSHA standards containing medical surveillance. Such tests include those for biological monitoring. Sometimes special tests are mentioned directly in the guidelines.

For example, the 1987 benzene OSHA guidelines[8] (PEL 1 ppm with STEL of 5 ppm and action level of 0.5 ppm) refer to emergency situations after which a spot urine sample is to be analyzed within 72 hours for urinary phenol normalized to a urine specific gravity of 1.024. The critical urinary phenol level is 75 mg/L. At or above this concentration, complete blood counts at monthly intervals for 3 months following the emergency exposure are mandated, triggering further actions if discrepancies in complete blood counts occur (hemoglobin levels or hematocrits are outside the 95% confidence level of reference ranges; the platelet count is at 20% or more below the worker's most recent values or falls below the 95% confidence limit of the lower limit of the reference range; the leukocyte count is $<4000/mm^3$, or there is an abnormal differential leukocyte count relative to reference ranges). Any further persistent (repeatable within 2 weeks) decrease in any of these variables is cause to refer the worker to an internist or hematologist, an act which triggers medical removal of the worker from the areas of benzene exposure, with up to 6 months of benefits.

In its Hazard Communication Final Rule[9] OSHA defines an occupational hygienists as a health professional able to ask manufacturers for the identity of chemicals that are trade secrets for nonemergency health and safety reasons, including biological monitoring purposes. Hygienists must be given trade secret data immediately by manufacturers in an emergency situation.[9] The corresponding state agencies may mandate biological monitoring tests when OSHA or the U.S. Environmental Protection Agency (EPA) do not. Hygienists should always check the regulations of the state in which they reside.

BEIs

The American Conference of Governmental Industrial Hygienists (ACGIH) has issued 36 BEIs as of 1996[10] (Table 13.1), 27 with documentations up to 1991[11] and 9 (plus 6 revised) in the 1996 supplement. Draft and final versions are published in the journal *Applied Occupational and Environmental Hygiene*. Each annual update of the TLVs and BEIs should be consulted as well as their most recent documentations, because they all change. Up to 1996, BEIs have been set using the air TLV and therefore on workplace inhalation exposure only, over 8 hour/day for 5 day/week for the toxic effect on which the TLV is based. This effect is found in the documentation of the latter.[11] BEIs have not yet been issued based on a cancer criterion, though cancer classifications are now established for TLVs. Because the TLV is not designed to protect the health of all workers, hygienists must use professional judgment to assess which TLVs or BEIs are truly protective.

For each exposing compound the BEI documentation is organized into sections: recommended BEI; properties; absorption; elimination; metabolic pathways and biochemical interactions; possible nonoccupational exposure; TLV; and summary. Index subheadings are generally analytical method; sampling and storage; biological levels without occupational exposure; kinetics; factors affecting interpretation of measurements; justification; current database available; recommendations; reference values recommended by other organizations; and other indicators of exposure. There may be more than one BEI for each exposing compound.

Since adverse effects on internal organs after xenobiotic absorption depend on the biologically effective dose exposing them, the absorbed dose is more correlated with the adverse health effects caused by an internal target organ than the external exposure dose. For irritative compounds, the exposure dose is related to the irritative effect directly. Many irritants have ceiling air values rather than TLVs; some amines, aldehydes, and ketones are exceptions. Thus, the documentation of the TLV must be consulted to see whether the critical health effect on which the TLV is based is one on internal target organs. The workload or physical activity at which the TLV was set is then its reference workload condition. Usually the documentation of the TLV for a compound is not explicit about the type of workload, but moderate workload (100 watts) is generally assumed. The compound's BEI documentation is essential for the hygienist to assess whether biological monitoring is feasible or necessary, and to be able to interpret the results. The other piece of necessary documentation is an appropriate National Institute for Occupational Safety and Health (NIOSH) or other sampling and chemical analysis method.

Compounds Without OSHA or ACGIH Recommendations

For those compounds without BEI or OSHA recommendations, other literature sources must be consulted for the critical biological equivalent values. The most useful references are the latest editions of books by Lauwerys and Hoet,[12] Baselt,[13] the textbook by Que Hee,[1] and the biological regulations of other countries like Germany's biological tolerance value (BAT).[14] Other than these, the primary journal literature must be searched through computer and hand searches of Chemical Abstracts, Biological Abstracts, Medline, Index Medicus, the Toxicological Profiles of the Agency for Toxic Substances and Disease Registry, and the Hazardous Substances Data Bank of the National Library of Medicine. Biological monitoring markers and their appropriate analytical methods would be useful additional information in manufacturers' material safety data sheets.

Routes of Exposure Other Than Inhalation

The hygienist must observe whether a worker handles or spills specific solvents, which parts of the skin are exposed, and for how long. Any gloves and garments that are worn when spills occur must be known to be protective against chemical degradation and permeation through consulting standard references.[4]

Hot dusty environments where many chemicals are used may pose an ingestion hazard through the licking of lips, especially when respirators are not in use. Improper storage and cleaning of respirators allowing contaminated dusts to accumulate inside them may lead to both inhalation and ingestion exposures. Food should be forbidden in any process area where toxic chemicals are in use. Workplace smoking may enhance both inhalation and oral exposure of contaminants and their pyrolysis products. Workers should shower before and after work and have clean work clothes available.

Table 13.1 — BEIs Recommended by ACGIH in 1996[10]

Exposing Chemical (BEI Year)	*Marker/Medium*	*Sampling Time*[A]	*BEI*[B]	*Other Notations*[C]
Acetone (1994)	acetone/U	ES	100 mg/L	back, ns
Aniline (1991)	total p-aminophenol/U	ES	50 mg/g CR	ns
	methemoglobin/B	DS/ES	1.5% of Hb	back, ns, sq
Arsenic (As)/soluble	inorganic As/U	EWK	50 µg/g CR	back
As compounds/arsine (1993)				
Benzene (1987)	total phenol/U	ES	50 mg/g CR	back, ns
	benzene/MEA	PNS	0.08 ppm	sq
	benzene/EEA	PNS	0.12 ppm	sq
Cadmium (Cd)/inorg	Cd/U	NC	5 µg/g CR	back
Cd compounds (1993)	Cd/B	NC	5 µg/L	back
Carbon disulfide (1988)	2-thiothiazolidine-4-carboxylic acid/U	ES	5 mg/g CR	
Carbon monoxide (1993)	carboxyhemoglobin/B	ES	3.5% of Hb	back, ns
	carbon monoxide/EEA	ES	20 ppm	back, ns
Chlorobenzene (1992)	total p-chlorocatechol/U	ES	150 mg/g CR	ns
	total p-chlorophenol/U	ES	25 mg/g CR	ns
Chromium (Cr) (VI)	total Cr/U	IS	10 µg/g CR	back
Water-soluble fume (1990)		ES/EWK	30 µg/g CR	back
Cobalt (Co) (1995)	total Co/U	ES/EWK	15 µg/L	back
	Co/B	ES/EWK	1 µg/L	back, sq
N,N-dimethylacetamide (1995)	n-methylacetamide/U	ES/EWK	30 mg/g CR	
N,N-dimethylformamide (1988)	n-methylformamide/U	ES	(40 mg/g CR)	
2-Ethoxyethanol and 2-ethoxyethanolacetate (1994)	2-ethoxyacetic acid/U	ES/EWK	100 mg/g CR	
Ethyl benzene (1986)	mandelic acid/U	ES/EWK	1.5 g/g CR	ns
	ethyl benzene/EEA			sq
Fluorides (1990)	fluorides/U	prior to S	3 mg/g CR	back, ns
		ES	10 mg/g CR	back, ns
Furfural (1991)	total furoic acid/U	ES	200 mg/g CR	back, ns
n-Hexane (1987)	2,5-hexanedione/U	ES	5 mg/g CR	ns
	n-hexane/EEA			sq
Lead (1995)	lead/B	NC	30 µg/dL	back
Mercury (1993)	total inorg Hg/U	preshift	35 µg/g CR	back
	total inorg Hg/B	ES/EWK	15 µg/L	back
Methanol (1995)	methanol/U	ES	15 mg/L	back, ns
Methemoglobin inducers (1990)	methemoglobin/B	DS or ES	1.5%/Hb	back, ns, sq
2-Methoxyethanol and 2-methoxyethyl acetate (1996)	2-methoxyacetic acid/U	ES/EWK		nq
Methyl chloroform (1989)	methyl chloroform/EEA	PLS/EWK	40 ppm	ns, sq
	trichloroacetic acid/U	EW	10 mg/L	ns, sq
	total trichloroethanol/U	ES/EWK	30 mg/L	ns, sq
	total trichloroethanol/B	ES/EWK	1 mg/L	ns
Methyl ethyl ketone (1988)	methyl ethyl ketone/U	ES	2 mg/L	
Methyl isobutyl ketone (1993)	methyl isobutyl ketone/U	ES	2 mg/L	
Nitrobenzene (1991)	total p-nitrophenol/U	ES/EWK	5 mg/g CR	ns
	methemoglobin/B	ES	1.5%/Hb	back, ns, sq
Organophosphorus cholinesterase inhibitors (1989)	cholinesterase/B	discret	70% RBC baseline	back, ns, sq
Parathion (1989)	total p-nitrophenol/U	ES	0.5 mg/g CR	ns
	cholinesterase/B	discret	70% RBC baseline	back, ns, sq
Pentachlorophenol (PCP) (1988)	total PCP/U	PLS/EWK	2 mg/g CR	back
	free PCP/plasma	ES	5 mg/L	back
Perchloroethylene (PER) (1989)	PER/EEA	PLS/EWK	(10 ppm)	
	PER/B	PLS/EWK	(1 mg/L)	
	trichloroacetic acid/U	EWK	(7 mg/L)	ns, sq

Table 13.1 (continued) — BEIs Recommended by ACGIH in 1996[10]

Exposing Chemical (BEI Year)	Marker/Medium	Sampling Time^A	BEI^B	Other Notations^C
Phenol (1987)	total phenol/U	ES	250 mg/g CR	back, ns
Styrene (1986)	mandelic acid/U	ES	800 mg/g CR	ns
		PNS	300 mg/g CR	ns
	phenylglyoxylic acid/U	ES	240 mg/g CR	ns
		PNS	100 mg/g CR	ns
	styrene/venous B	ES	0.55 mg/L	sq
		PNS	0.02 mg/L	sq
Toluene (1986)	hippuric acid/U	ES	(2.5 g/g CR)	back, ns
		last 4-hr S		
	toluene/venous B	ES	(1 mg/L)	sq
	toluene/EEA			sq
Trichloroethylene (TCE) (1986)	trichloroacetic acid/U	EWK	100 mg/g CR	ns
	trichloroacetic acid	ES/EWK	300 mg/g CR	ns
	plus trichloroethanol/U			
	free trichloroethanol/B	ES/EWK	4 mg/L	ns
	TCE/B			sq
	TCE/EEA			sq
Vanadium pentoxide (1995)	total vanadium/U	ES/EWK	50 µg/g CR	sq
Xylenes (1986)	methylhippuric acids/U	ES	1.5 g/g CR	
1996 Intent to Establish or Change BEIs				
Benzene (1996)	s-phenylmercapturic acid/U	ES	25 µg/g CR	
N,N-dimethylformamide (1993)	n-methylformamide/U	ES	20 mg/g CR	
4,4-methylene bis (2- chloroaniline) (MOCA) (1996)	total MOCA/U	ES		nq
Perchloroethylene (PCE) (1993)	PCE/EEA	PLS/EWK	5 ppm	
	PCE/B	PLS/EWK	0.5 mg/L	
	trichloroacetic acid/U	EWK	3.5 mg/L	ns, sq
Tetrahydrofuran (THF) (1996)	THF/U	ES	8 mg/L	
Toluene (1995)	toluene/venous B	PLS/EWK		nq
	hippuric acid/U	ES		back, ns
	o-cresol/U	ES		back, ns
	toluene/U	PLS/EWK		nq

^A B, blood; EEA, end exhaled air; inorg, inorganic; MEA, mixed exhaled air; U, urine
^B Discret, at discretion after a minimum of 2 weeks of exposure; DS, during the shift (within the last 2 hours of exposure); ES, end of shift; EWK, end of workweek; IS, increase during the shift; NC, not critical if prior exposure has occurred for more than 2 weeks; PLS, prior to the last shift (with no exposure overnight); PNS, prior to the next shift (after no overnight exposure); S, shift
^C CR, creatinine; Hb, hemoglobin; ppm (v/v) (one millionth on volume basis for volume of pure vapor in air); RBC, red blood cells; for items in parentheses, see Intent to Establish or Change
^D Back, there is an endogenous background in unexposed individuals; nq, no number is set due to insufficient data; ns, non-specific: other exposing chemicals give the same marker; sq, semiquantitative: there is an ambiguous dose response

A major signal to trigger biological monitoring occurs when workers feel ill when air sampling results are below mandated or recommended concentrations. The worker may be hypersensitive, or other major routes of exposure may exist. The worker should be sent to a physician to check the former condition, and the work practices of the worker observed relative to nearness to exposure sources for the latter. Material safety data sheets must be consulted to identify the specific compounds involved in the exposures and which are likely to pose inhalation, dermal absorption, and oral ingestion potential. The unit processes in the specific workplace must also be known and understood. Toxic interaction information should be obtained.

Those compounds that have a "skin" notation with their TLVs are those for which skin absorption data by liquid, solid, or vapor exposures have been published. ACGIH has a general policy of giving skin notations to those chemicals whose rabbit acute dermal LD_{50} is less than or equal to 1 g/kg body weight, or those causing systemic effects on chronic exposure. The OSHA skin designation is assigned only in cases where dermal exposure has been shown to cause systemic poisoning or where skin exposure leads to an absorbed dose greater than that permitted to be absorbed by inhalation at the PEL. The latter criterion is more conservative than the former. The TLV or PEL is based only on inhalation exposure and does not account for skin or oral exposure over an 8 hour/day-5 day/week exposure regimen for a healthy worker under conditions not immediately dangerous to life and health.

The absence of a skin notation for PELs and TLVs does not mean that the chemical is not absorbed through the skin. Most nonpolar organic and organometallic liquids that are not miscible with water but have water solubilities in the greater-than-ppm (μg/mL) range will potentially pass through the skin, as will solids in solution with such liquids. How relevant potential skin absorption is has to be assessed visually and directly by hygienists. Such evaluations may require skin patches, skin sampling, or tracer studies[2] in addition to biological monitoring.

Presence of PPE

PPE creates a barrier to the hazardous agent so that it is not completely available for body absorption. The final test of effectiveness for PPE is that markers for the exposing compound are not detected in biological media above critical concentrations that correspond to critical air concentrations for air exposure, whether or not engineering controls, PPE, hygienic, and administrative measures have been instituted in the workplace or environment. If marker levels do not decrease after controls are instituted, then either the controls are ineffective, the source of the marker is outside the workplace, or there may be worker connivance.

In addition to appropriate cartridge selection, worker training, and qualitative fit testing, quantitative fit testing is essential to maximize the protective capability of negative pressure respirators. Positive pressure respirators must maintain a protective flow of air across the nose and mouth to prevent entry of exposing chemicals. Gloves and garments must not be degraded or permeated by the exposing chemicals. All PPE must be maintained appropriately.

Unanticipated Exposures

One of the advantages of biological monitoring is that its integrative nature allows warning of unexpected exposures. Thus, while postshift or end-of-shift sampling are generally recommended in BEIs, preshift sampling will warn if carryover from previous work exposures is occurring, or if exposures from outside the workplace are adding to the worker's health risk. Carryover is not likely for compounds of half-times less than 6 hours. However, the excretion of chemicals usually consists of a fast followed by a slower exponential period. Buildup can be addressed by administrative rotation and improving respirator or engineering control protection. Nonoccupational exposures, not being under the control of the worker's employer, are difficult to handle, but the responsibility is the worker's once informed.

Another common workplace situation is when PPE are doffed but the worker unknowingly is still near an emission source of vapor, or puts a bare hand on spilled solvent, for example, on the PPE itself. Biological monitoring may indicate whether a potentially hazardous amount of chemical has been absorbed that is equivalent to a hazardous inhalation exposure as defined in a TLV.

Factors For Biological Monitoring Data Interpretation Using the BEIs

Any marker concentration sampled under BEI conditions that exceeds 0.7 of its BEI concentration reflects a health risk for the worker regardless of exposure route. The appropriate baseline sample (preshift for end of shift; preshift on first workday after weekend for end of week samples) should always be subtracted to obtain the workplace contribution. If this difference still exceeds 0.7 of the BEI, the workplace is definitely the source of overexposure. For effective control, the most important routes of exposure must then be prioritized through a time and motion study relative to job proximity to known sources of the suspected chemical.

If volatile solvents (vapor pressure $>3.0 \times 10^{-3}$ mmHg), gases, or hot processes are present, then inhalation of vapor may be a major exposure route. If hot processes are involved or the environment is dusty, inhalation of particulates may also be major. If there is little respirable dust (<3 mg/m^3 of particulates containing no asbestos and <1% crystalline silica particles not otherwise classified [PNOC] below aerodynamic diameters of 10 μm), or little total dust or inhalable dust (<10 mg/m^3 for PNOC of aerodynamic diameters <100 μm), particulate inhalation exposure should be minor. Aerosols in the thoracic size range (<25 μm aerodynamic diameter) deposit in the lung, with the fraction between 10–25 μm depositing in the upper lung (bronchioles upward to the bronchi) where the ciliary locomotor process can sweep particles up to the trachea and larynx, and then into the stomach by the saliva flow, thus facilitating absorption through ingestion. Particles 25–100 μm aerodynamic diameter are more likely to deposit in the nasal conchae and pharynx, from which some ingestion is expected. If particulate chemicals are soluble in saliva and there is nasal drip, ingestion of part of them will occur. Whole particles will also be ingested during swallowing. Hot and dusty environments may cause dusty faces and lips, and licking of lips and contamination of food and tobacco products may cause oral exposures. Dermal exposures become important when solvents are handled in the absence of PPE, and when PPE are not protective. This route of exposure is especially important for moderately volatile and nonvolatile chemicals of high octanol/water coefficients (K_{ow}) of 10 to 10,000 that contact the skin with water solubilities above ppm values (see Chapter 3 in Reference 1 and the BEI documentations BEI-4 to BEI-11). Hygienic considerations are the key controls for oral and skin exposure in the absence of engineering controls and PPE. They are also key for full compliance with the lead standard.

Vapor Inhalation Predominates as the Route of Exposure

The major parameter to account for when inhalation exposure is dominant (>70% of the total absorbed dose) is physical activity or workload. If a worker is exposed to within 30% of the TLV concentration in a breathing zone personal air sample, a concentration above 0.70 of the BEI of the marker should result if inhalation dominates after correction for the appropriate baseline concentration.

Table 13.2 gives some physiological data that depend on workload.(15) The alveolar ventilation rate is the volume of air per minute inspired into the alveoli, and increases linearly up to 150 watts of work. If the same vapor concentration of a substance is being inspired at these different work rates relative to a moderate workload (using the data of Table 13.2) over the same exposure period, heavy work potentially exposes (38/27) = 1.41 times, light workload (16/27) = 0.59 times, and at rest (5/27) = 0.185 times the mass of a worker at moderate workload. These values represent extremes, since expiration rate generally increases correspondingly but never matches inspiration rate except during hyperventilation. The difference obtained between mass potentially breathed in minus that breathed out, when divided by the mass potentially breathed in, is the uptake for that time period at a specific workload and is usually expressed as a percentage. Uptake is usually determined in bicycle ergometer, treadmill, or step-testing studies where watts of work expended can be measured directly, and biological

Table 13.2 — Some Physiological Parameters Influenced by Physical Activity(15)

Workload (watt)[A]	Oxygen Demand (L/min)	Alveolar Ventilation (L air/min)	Heart Pump Rate (L/min)	Blood Flow (L/min)			Signs
				Muscles	Liver	Brain	
At rest; sedentary (0)	0.3	5.0	6.0	1.0	1.5	0.6	slow regular breaths
Light work (50)	0.8	16.0	9.0	3.6	1.3	0.7	rib cage visibly rises and falls
Moderate work (100)	1.3	27.0	13.0	7.9	1.2	0.7	noisy breath; light sweating
Heavy work (150)	1.8	38.0	19.0	13.0	1.1	0.7	heavy breathing and sweating

[A]Midpoints with ±25-watt range

markers and physiological parameters can also be determined. The literature values of parameters in Table 13.2 for these different regimens vary.

The uptakes of inert solvent vapors such as aliphatic, alicyclic, and aromatic hydrocarbons (for example, hexane, cyclohexane, and benzene, respectively) and their chlorinated analogs have been found[16] to obey similar correlational relationships, but hydrogen-bonding solvents like alcohols (for example, n-butanol) have not. The relationship for inert solvent vapors that is obeyed for % uptake over a 2-hour exposure (work break-to-work break conditions) at a defined concentration near a TLV in a bicycle ergometer study was [16]

$$\% \text{ uptake} = 86 \times AAC_{20-30} / \text{inspired air concentration} + 79.3 \quad (1)$$

where AAC_{20-30} is the alveolar air concentration determined 20 to 30 min after the 2-hour exposure has ended at a given workload.

Appendix A contains examples of the calculations for xylene[15] relative to its BEI at different work loads and exposure conditions. This appendix gives some idea of the strengths and weaknesses of the urinary BEI of metabolite markers relative to exposure conditions at its reference physical activity. Hygienists must ascertain whether the BEI is protective or not for a given exposure scenario rather than relying solely on the urinary concentrations determined by analytical chemistry laboratories. The same may be said for air sampling results. However, hygienists must also understand the strengths and weaknesses of the analytical chemistry methods used, including so-called standard methods.[17–23]

Workers do not work at one physical activity, nor do they inhale the same concentration of a single chemical, as occurs in volunteer studies. Females always have greater uptakes than males for challenges to the same concentration of nonpolar compound vapors at the same physical activity.[1] The xylene BEI documentation[15] also states that ethylbenzene causes up to 20% inhibition of the metabolism of m-xylene, the major xylene isomer in technical xylene. The xylene BEI is valid only for exposure to technical grade xylene containing ethylbenzene and not for exposure to pure xylene isomers. The metabolism of m-xylene is also inhibited up to 50% by alcohol consumption or the ingestion of aspirin. The hygienist must identify whether such additional factors are important for each worker, either directly or through questionnaires. BEI documentation must therefore be consulted to define such questions when a BEI is exceeded.

Methylhippuric acids are also relatively specific urinary markers since xylenes, or methyl alkyl benzenes where the alkyl is a normal (unbranched) aliphatic hydrocarbon chain containing an odd number of carbon atoms (beta-oxidation cuts off 2-carbon units from the side chain), are the only solvent-type precursors that will produce methylhippuric acids on metabolism. Urinary markers can be much less specific. Hippuric acid, the urinary BEI marker for toluene,[15] is also produced by workplace exposure to benzaldehyde, benzoic acid, benzyl halides, and monosubstituted aromatic hydrocarbons with a side chain containing a normal aliphatic side chain of an odd number of carbon atoms (beta-oxidation), and to sodium benzoate preservative in foods, or almond-flavored foods that contain benzaldehyde. Another positive interference is that hippuric acid is produced endogenously as an end product of nitrogen metabolism, with much interindividual variation. For such cases, baseline sampling is essential to discern the workplace contribution.

It is important to realize that a sedentary worker exposed at the TLV is at much lower health risk than a worker exposed at higher workload, because the absorbed dose is lower relative to the absorbed dose at which the TLV was set, as long as an internal target organ effect is the basis of the TLV. This shows that air monitoring alone is not sufficient to judge health risk except when reference conditions are met, and the basis of the TLV thoroughly understood. Every hygienist therefore needs to have a copy of the current documentations of both TLVs and BEIs.

Inhalation of Vapor Present, but Not the Major Route of Exposure

When the personal breathing zone sample cannot account (<70% of TLV) for the concentration of the marker in the biological fluid after baseline correction, workload should again be considered as the first variable to be eliminated. If the breathing zone personal air sample is 0.5 of the TLV, and a worker exposed to xylenes is judged to be doing hard work, the ratio of absorbed mass at an air concentration of 0.5 of the TLV relative to moderate physical activity at the TLV is (ratio of air concentrations) × (ratio of absorbed masses for heavy to moderate workload at the TLV) = 0.5 ×1.24 = 0.62 of that absorbed at moderate workload at the TLV. For example, for xylenes from Appendix A, 2244 × 0.62 = 1391 mg equivalent is absorbed. The critical acute exposure time must be ln[(1391263)/1391 = 0.811] × 3.6/0.693 = 1.1 hour before sampling. This is feasible and not very different from a TLV exposure at moderate work load, and its other exposure scenarios in Appendix A. If the air concentration is 0.1 of the TLV at heavy workload, the critical acute exposure time becomes ln [(0.1 × 2244 × 1.24)263)/(0.1 × 2244 × 1.24) = 0.054] × 3.6/0.693 = 15 hours before sampling, clearly an impossible answer in terms of workplace exposures on the same day. If the air concentrations are 0.2, 0.3, and 0.4 of the TLV, the acute exposure times are, respectively, 4.6, 2.6, and 1.8 hours before sampling. Thus, breathing zone personal air concentrations <0.2 times the TLV must involve exposures in the first part of the work shift to account for >0.7 x BEI concentrations of markers, for xylene exposure, assuming no urination between the exposure and sample taking. This is then a testable hypothesis.

If the half-time is longer, the critical times will lengthen also, so that each marker of an exposing compound is a distinct case, having its own uptake and half-time as major controlling factors. Thus, assuming the same uptake, a half-time of 8 hours will increase these critical times by 2.22 times; that is, all exposures <0.5 of the TLV must have occurred early in the shift for a BEI concentration of marker, with same-day exposures <0.2 TLV being not likely since critical times are longer than the work shift for this longer half-time.

Once heavy workload has been examined and discounted, other modes of exposure should be investigated. If the environment is dusty, the respirable fraction of dusts may contain adsorbed organics. These may be liberated on particle deposition, but they may also be phagocytosed along with the particle by pulmonary alveolar macrophages and eventually transported by the lymph to lymph nodes where they are liberated with lysates spilling into the venous system for further systemic distribution. A personal breathing zone filter sample (Teflon® or glass fiber should be used if possible) for respirable dust can be analyzed for the exposing compound. If this shows negligible exposing compound, or there is little respirable dust, a total dust air sample should also be taken and analyzed similarly to the respirable dust for exposing compound. If workers' faces are dusty, the oral exposure route may be important. If not, skin exposure routes should then be further investigated.

Biological Monitoring Sampling and Quality Assurance/Control

The practicing hygienist should initiate biological monitoring only for those compounds that have BEIs (or foreign equivalents with analogous documentations) so that information on other sources of the markers, baseline values, genetic factors, antagonisms, synergisms, half-times, and the basic metabolic pathways are available. This enables the hygienist to prepare an initial questionnaire for the worker in the event of biological monitoring results that exceed the BEI, that cannot be explained by vapor inhalation, and for biological monitoring interpretation purposes.

The hygienist must think through the objectives of biological monitoring first, and it is useful to prepare a document similar to an informed consent form that sets out these objectives along with what samples are to be taken for what purpose and what will be done with the data. Such material should also be part of a worker training program. Such a generalized form may be required by the employer as a condition of worker employment so that permission for taking noninvasively sampled biological media will always be available. This data will be confidential but open to the worker, who also has the right to medical opinions and counseling.

A licensed physician, certified nurse, or certified phlebotomist must perform BEI blood sampling, for example, for methemoglobin (aniline, methemoglobin inducers, nitrobenzene); cholinesterase; carboxyhemoglobin (carbon monoxide); cadmium, cobalt, free trichloroethanol (trichloroethylene), lead, mercury, perchloroethylene, trichloroethylene, or total trichloroethanol in blood (methyl chloroform); free pentachlorophenol in blood plasma; and styrene or toluene in venous blood. The OSHA lead standard is the model for decision logic once blood concentrations are known.

Since most occupational hygienists are not certified to perform blood sampling, an alternative is to leave all biological monitoring and environmental sampling to physicians and other certified personnel. This is more common in Europe and Japan. However, physicians and others are generally not trained to do air or breath sampling. While hygienists are trained to do air sampling, they are usually not trained for breath sampling. It is far easier for hygienists to learn breath sampling than it is for physicians, since this is related to air sampling. The hygienist's familiarity with NIOSH[17] and OSHA[18] methods is another major advantage.

Regardless of sampling method or personnel, an analytical chemistry laboratory must be consulted for sampling, storage, and transport procedures; whether the relevant sampling containers, appropriate preservatives, and transport containers can be provided; and the time line for sampling and analysis. Making these arrangements before actual sampling will result in lower costs and less invalid sampling and analyses.

The appropriate control sample (usually baseline) must also be provided to aid in chromatography resolution problems and peak assignments, and in interpreting the results relative to workplace exposures. The appropriate blanks (such as an empty container, container filled with isotonic [0.9%] saline or standard reference urine of the same volume as the urine sample, a gas bag filled with pure air or preshift breath, a blood Vacutainer® tube of the same color top filled with the same volume of isotonic saline or standard reference blood as the actual blood sample, an unused blank Vacutainer tube of the same colored top, field blanks) should also be sent.

The hygienist is also urged to send appropriate spiked samples at the BEI level (positive controls) to ensure laboratory quality control/assurance at the user level. Use of standard reference materials from the National Institute for Standards and Technology or from specimen banks with materials characterized through interlaboratory collaboration is also encouraged to prove confidence in the whole sampling and analysis process.

If the BEI is exceeded, in addition to the types of calculations referred to in Appendix A, the sampling should be repeated at least three times to confirm the results before medical action or control measures are initiated. Baseline samples also need to be done.

Breath Sampling

The physiological basis of breath sampling can be consulted in the BEI documentation,[15] BEI- 12 to BEI-14, and in references 24 and 25. References 23 and 26 provide some specific breath sampling methods.

The easiest breath-sampling method is with a clean, nonleaking Tedlar bag. "Clean" means filled and then evacuated three times with pure air from a compressed air cylinder or other certified air source and dried using a hot-air hair dryer during each evacuation (or leaving the bag in an incubator oven at 50°C for 5 minutes). A nonleaking bag is one that is air-filled and does not release bubbles when lightly pressed down upon when immersed in a tub of water, nor does any water enter the bag after such treatment.

The bag is connected to the sidearm of a Pyrex® vacuum flask that has been soaked overnight in 10% nitric acid with copious water rinsing before drying. The bag is attached by a fluorinated ethylene-propylenecopolymer Teflon tube (0.25 inch inner diameter) by leakproof butt-to-butt joints (using Tygon™ collars). The Teflon tube should be cleaned by blowing 50°C pure air through it for at least 30 min after rinsing with a solvent known to solubilize the marker. The flask is rubber-stoppered and the inside of the stopper coated with Teflon tape. The stopper should also contain a clean (10% nitric acid washed) glass tube that almost reaches the flask bottom. The top of the glass tube above the stopper is similarly connected butt-to-butt to similar clean Teflon tubing connected to a Teflon 0.5 to 0.25-inch adapter. Alternatively, the latter Teflon tubing can be worked through a hole in the stopper to be near the flask bottom. The setup must be leak-proof (use soap bubble fluid to detect leaks during sampling), and the transfer tubing must be as short as possible. The function of the flask is to collect any drool, spittle, condensation, or saliva that will inevitably occur. To give a sample, the subject purses the lips and blows through the 0.5-inch part of the adapter tube, and the bag is manipulated to distribute the sample to its distal end. This procedure minimizes the amount of condensed water vapor in the gas bag, especially important if normalization to breath carbon dioxide concentrations is required or intended. The use of desiccants is not recommended, since the intended analyte is often also partially or completely adsorbed or absorbed, resulting in loss.

To sample mixed exhaled breath after breathing in, the worker's nose is closed (held by fingers or a noseclip), and then the worker breathes out through the mouth and into the 0.5-inch adapter and its attachments. The worker repeats this operation until the desired bag volume is obtained (full is best, but more than half full may suffice depending on the compound).

End-exhaled breath is obtained by having the worker breathe out normally into the air and then purse the lips to blow the forced expiration component into the adapter and its attachments. For both modes, the sampling adapter and tubing should be purged by one breath before the sample is collected in the gas bag.

The use of evacuated Pyrex or stainless steel bottles and canisters is discouraged unless they can be evacuated to a known pressure and are equipped with an air-tight on/off valve. If the containers are not evacuated, the volume of breath sampled is uncertain because of dilution with air already in the container (even if a wet-test meter is also connected to record the volume). Another alternative is to replace the Tedlar bag with an appropriate solid sorbent tube connected to a wet-test meter by Tygon tubing. The solid sorbent must be one that can allow water vapor to pass through (such as Tenax™ GC, the Poropak™ series, the Chromosorb™ 101 through 106 series, or the XAD resins). A charcoal tube will be inadequate. However, wet-test meters are heavy and expensive and are not accurate on surfaces that are not level. The solid sorbent can then be analyzed by standard NIOSH, OSHA, or EPA methods, as described later for taking solid sorbent samples from Tedlar gas bag samples.

The hygienist has several options for Tedlar gas bag analysis. If direct reading instruments or devices are available, they can be calibrated using clean Tedlar gas bags in a matrix of mixed or end-exhaled breath generated by the hygienist using the above apparatus. The marker is then added by a 10-μL syringe using the calculated injected volume for the known volume of breath matrix. If detector tubes are used, any relevant interferences should be noted (information is available in the tubes' handbooks) and the appropriate number of strokes used to achieve a reading at the BEI concentration. The latter may not be attainable since detector tubes are designed for the higher concentrations at about the TLV level, in general. Long-term detector tubes developed for sampling at hazardous waste sites may be better. The prior evaluation of a breath control relative to a known Tedlar bag sample at the BEI is recommended before any breath is evaluated with this method. The same applies for measurement with a gas/vapor infrared spectrophotometer of variable wavelength/path length, where the interfering infrared absorption of water vapor and carbon dioxide of the control must be subtracted. The high instrument flow rate is also a major disadvantage.

The hygienist must do prior calculations to define the correct volume gas bag and to assess feasibility. For example, the mixed exhaled BEI for benzene in breath is 0.08 ppm, and the end-exhaled BEI is 0.12 ppm. If 0.10 ppm is chosen to simplify calculations as the first concentration, this is equivalent to 0.32 mg benzene/m^3 at 25°C and 760 mmHg. A 10-L gas bag must contain 320 (μg) × 10 (L)/1000 (L) = 3.2 μg benzene. The liquid density of benzene is 0.879 g/mL or 879,000 μg/1000 μL benzene liquid. Therefore, 3.2 μg benzene is a liquid benzene volume of 3.6 nL. This is an impossible volume for a 10-μL syringe. Multiple dilutions of 10-L gas bags are required. Thus, 1 μL of liquid benzene of mass 879 μg is injected into a gas bag with 10 L of mixed exhaled air. The gas bag is placed in an oven at 50°C or exposed to a hot-air dryer for a few minutes to mix the vapors thoroughly. This creates a benzene concentration of 87.9 μg/L or 27.5 ppm at 25°C and 760 mmHg, and should be checked with a calibrated direct reading instrument for stability and accuracy. A 3.2 μg benzene mass is then contained in 3.2 (μg) 1000 (mL)/87.9 (μg) = 36.4 mL of this 27.5 ppm concentration, which can now be injected into a gas bag filled with 9963.6 mL of mixed exhaled air by a gas-tight syringe and again mixed. Four other concentrations bracketing the 0.10 ppm concentration (usually 0.1, 0.5, 1.0, 1.5, and 2.0 the BEI concentration) are then made up similarly from the first dilution.

The best calibrated direct reading instruments to use for benzene and other aromatics are photoionization detectors (PID) with lamp energies 10.2 eV or 10.6 eV. Aliphatics are best detected by PID detectors with 11.6 eV lamps. These still may not be sensitive enough if ppb (v/v) concentrations of aliphatics are to be measured. If portable capillary column gas chromatography (GC) with PID detection (for example, with a 10.6 eV arc) is available with a gas sampling valve, it can be easily calibrated using these gas bags, and the background resolution of breath peaks can be optimized before the worker breath sample is to be analyzed (along with a bag sample of the worker preshift breath). If there is no gas sampling valve, Tedlar gas bag samples must be injected using a gas-tight syringe of appropriate volume. The quantitation is then by interpolation on the external standards curve generated through these gas bags. This is the basis of NIOSH method 3700,[17] which uses a portable GC/PID of method working range of

0.1–500 ppm, and which is designed for Tedlar bag air samples. The advantage is that answers can be obtained on-site and almost immediately, and injection of samples into the air carrier gas of the column dilutes the effects of humidity and excess carbon dioxide.

If the hygienist does not have the resources to do this type of quantitation, then sampling a known volume of the breath sample from the Tedlar gas bag through a large (200 mg/100 mg) charcoal tube (to minimize breakthrough caused by high humidity) may suffice using a calibrated personal sampling pump. How much volume is necessary? First the NIOSH[17] or OSHA[18] method should be consulted for the marker. For benzene, NIOSH Method 1501 states that the flow rate should be <0.20 L/min for an air volume of 2 (10-min sample) to 30 L for the TLV of 10 ppm or 32 mg/m^3. The meaningful lower limit for this method therefore is a sampled benzene mass of 32,000 (µg) x 2 (L)/1000 (L) = 64 µg benzene. If this is desorbed in 1 mL of carbon disulfide and 5 µL injected, the absolute lower limit on-column is 64,000 (ng) × 5 (µL)/1000 (µL) = 320 ng using flame ionization detection (FID). A 10-L gas bag of 0.1 ppm benzene contains 3.2 µg, so the standard NIOSH carbon disulfide desorption technique is impractical unless the analytical laboratory can be 64/3.2 = 20 times more sensitive or the gas bag is 20 times larger (that is, 200 L), or some combination of larger bags and enhanced laboratory sensitivity. The hygienist must ask the certified analytical laboratory whether this specific degree of enhanced sensitivity is possible before going ahead with this method. If so, nearly the entire bag can be sampled by the pump through the charcoal tube, as long as the volume sampled is accurately known. Such analyses by the standard NIOSH method using carbon disulfide for desorption and GC/FID frees up hygienist time at the expense of timely results. The hygienist also should bear in mind that such tubes may be broken during transport and storage.

The hygienist should also send an accompanying worker control breath sample of the same volume obtained without analyte exposure to facilitate background resolution from analyte peaks, and to confirm that there is no exposure to analyte from outside the workplace. Other blanks include bags or solid sorbent tubes that contain or have sampled an equivalent volume of pure air, as well as unused solid sorbent tubes and field blanks. Positive controls consisting of sampled known air concentrations at the BEI concentration from gas bags should also be sent for analysis. Hygienists should remember that all analyses are relative to the appropriate control, the latter being as essential as the breath sample suspected of containing the marker. The ability to use NIOSH or OSHA air sampling and analysis methods is a very big advantage for breath sampling.

If standard charcoal tube sampling is not possible because of sensitivity problems, a known volume of the 10-L gas bag contents must be sampled onto an appropriate solid sorbent. This then must be analyzed by thermal desorption so that all the sample is placed on-column, using the appropriate GC columns and detectors. For the case of benzene above, the absolute lower limit on-column for the standard GC/FID NIOSH method was 320 ng, well below the contents of the 10-L gas bag containing 0.10 ppm benzene. Therefore, a laboratory specializing in thermal desorption—for example, one that is also EPA-accredited to do purge-and-trap analyses for exposing compounds in ground and surface waters—should be consulted for the solid sorbent of choice. For benzene, this is the solid sorbent used in the purge-and-trap technique of EPA Methods 502.2, 503.1, 524.1, 524.2, 602, 624, 5030, and 8020 (60/80 mesh 2,6-diphenylene oxide polymer commonly called Tenax GC or equivalent [XAD-2™, etc.]). The hygienist should ensure that the solid sorbent tube has the same geometry as that used for the particular laboratory's thermal desorption technique after purge and trapping. Commonly, GC/mass spectrometry, GC/FID, and GC/PID are used for quantitation depending on the desired sensitivity and compound to be analyzed. These are more expensive analyses than standard occupational hygiene analyses. Some investigators have sampled breath directly onto such tubes without a Tedlar bag intermediate step as referred to above. It should be noted that the TLV for benzene had a Notice of Intended Change in 1996 to 0.5 ppm (1.6 mg/m^3), based on its carcinogenicity. This will necessitate eventual changes in the BEIs: the total phenol in urine BEI will be untenable because of endogenous background, with only the breath BEI still tenable, and only by the thermal desorption method of benzene analysis.

The above sampling techniques should suffice for other organic solvent breath samples depending on sensitivity (for example, ethyl benzene, n-hexane, methyl chloroform, perchloroethylene, toluene, and trichloroethylene, all in end-exhaled air). The aromatics, ethyl benzene and toluene, can be sampled and analyzed sensitively by the same analytical techniques as benzene, and also because their breath BEIs are at much higher concentrations. For chlorinated aliphatic analyses, GC/electron capture detection has better sensitivity than FID or PID. GC/FID is best for n-hexane. The BEI breath tests for ethylbenzene, n-hexane, toluene, and trichloroethylene are qualitative or screening tests.

Carbon monoxide in end-exhaled air cannot be analyzed by the above techniques. It can be analyzed using a calibrated portable electrochemical cell involving oxidation to carbon dioxide at a Teflon-bonded diffusion electrode similar to the carbon monoxide ecolyzers familiar to most occupational hygienists. NIOSH Method S340,[17] which is designed for Tedlar gas bag samples over the carbon monoxide range 1–120 ppm, should be used.

Urine Sampling

For urine sampling, a clean wide-mouth container with volume markings (500 mL should suffice for spot samples) should be used. The container should be fitted with a screw-cap lid, which allows fewer spills than other lids. Polyethylene containers should be used for metals. These should be cleaned by overnight soaking in 10% nitric acid with subsequent distilled water rinsing and dried. An acid-washed Pyrex vessel is appropriate for organics, fitted with a Teflon-lined screw-capped lid. The container should also have any indicated preservative. BEI documentation preservative recommendations are copper sulfate for pentachlorophenol; thymol for urinary phenol; hydrochloric acid for mandelic acid (styrene); hydrochloric acid or quinoxalinol or zinc for phenylglyoxylic acid (styrene); thymol or hydrochloric acid for hippuric acid (toluene) and methylhippuric acid (xylenes). The addition of acidic agents like thymol, hydrochloric acid, copper sulfate, and quinoxalinol can cause conjugate hydrolysis.

The worker should be asked to shower and provide a midstream urine sample before putting on clothes, if possible, though this is usually impractical. The hygienist should note the sampling time following that recommended by the BEI documentation (Table 13.1), the urine volume (a container should never be filled entirely), and measure the urine specific gravity with a hygrometer (urinometer). Urine samples should be stored and transported with adequate insulation, frozen in dry ice with no preservatives unless recommended otherwise, and stored in laboratories at -20°C. Such precautions ensure that conjugates are not hydrolyzed, that microorganism growth is prevented, and volatile constituents are not lost. Labels and screw caps should be duct-taped.

Urine samples taken similarly before the work shift (baseline or preshift sample) should also be analyzed at the beginning of a biological monitoring program or whenever a change in process or control occurs in the workplace. Such a sample reflects nonworkplace factors if taken after a nonworking weekend on the first day of the workweek. A baseline should also be taken when the BEI is exceeded to ensure that the exposure source is not external to the workplace, or if there is holdover from any previous exposure. The appropriate baseline sample for an end-of-week BEI is the before-shift spot sample after a weekend.

Creatinine concentrations must also be requested if recommended. The latter accounts for worker fluid intake and urine dilution, but the marker must also be excreted through kidney glomerular filtration rather than through the distal or proximal tubules.

The usual blank samples (empty container; container with preservative; container with preservative with standard reference control urine, preshift urine, or isotonic saline) and positive controls (container with preservative with a standard reference urine containing spiked marker at the BEI concentration; container with preservative with the same standard reference urine as used for spiking) should also be sent for analysis.

To collect urine samples for measurement of excretion rates at the end of a shift, the worker should empty the bladder completely about 3 hours before the end of the shift (the time should be noted) before providing the next void for analysis (the time and urine volume should also be noted and specific gravity measured). The volume divided by the elapsed time should then be provided along with the sample.

The correction for specific gravity should be performed as follows:

$$\text{Corrected specific gravity} = \frac{24 \times \text{observed value}}{\text{last two digits of observed value}} \quad (5)$$

The BEI documentation[15] defines concentrated urine to be either specific gravity >1.03 or creatinine >3 g/L. Dilute urine is defined as specific gravity <1.01, or creatinine <0.5 g/L. Such urines cannot be used and resampling must be done. Hygienists should always request specific gravity and creatinine determinations by the laboratory that analyzes the urine samples. If the samples have not precipitated, the specific gravity measured by the laboratory should be close to that measured by the hygienist when the urine is fresh. If the specific gravity is measured by the hygienist, this will allow resampling if the sample is too dilute or concentrated and avoid the expenses involved with analyzing invalid samples. Reference 27 should be read to supplement the BEI documentation[15] account of urine sampling in BEI-16 to BEI-19.

Sampling methods that involve 24-hour urines involve larger containers that can be kept in a home refrigerator freezer. Such samples may be necessary when testing for pesticides or other compounds of longer half-times than 6 hours, depending on the compound and the BEI. Analytical laboratories often provide these larger containers and include them in the analysis costs.

There are many analytical methods for urine metals, including NIOSH Method 8310, which uses inductively coupled plasma atomic emission spectroscopy for multielemental analyses[17] and Method 8003[17] for lead by atomic absorption spectroscopy. For organic BEI markers that have NIOSH methods for urine, Method 8300 is for hippuric acid; Method 8301 is for hippuric and methylhippuric acids; Method 8303 is for pentachlorophenol; and Method 8305 is for phenol and o-cresol. Non-BEI marker urinary NIOSH methods[17] include those for benzidine (Method 8304 and 8306), fluoride (Method 8308), and 4,4′-methylenebis (2-chloroaniline) or (MBOCA; Method 8302), with the other methods being for markers in blood (delta-aminolevulinic acid dehydratase for lead; 2-butanone, ethanol and toluene; elements; lead; and pentachlorophenol) and serum (polychlorinated biphenyls). The NIOSH methods are preferred, and their sampling directions should be followed closely. Other methods are available.[3,13,19–23]

Other Noninvasive Biological Monitoring Media

Hygienists are advised not to sample other biological media than those detailed here. If it is necessary to do so, sampling and analysis methods are available elsewhere.[28]

Summary

This chapter outlines (1) what a field occupational hygienist should know and understand before implementing a biological monitoring program in the workplace and for worker training purposes; (2) when and how to take noninvasive urine and breath samples; (3) how to conduct user quality control with analytical chemistry laboratories; (4) how to interpret biological monitoring data in conjunction with environmental exposure information that should concurrently be gathered; and (5) major sources of expert information for biological monitoring.

Detailed examples are provided of the entire medical surveillance associated with the OSHA lead standard, and for the ACGIH BEIs for xylenes and benzene.

Appendix A. Illustrated Examples for Xylene Relative to Workload, the BEIs, and Inhalation Conditions

According to Equation (1):

$$\frac{\%\ \text{Uptake} = -86 \times \text{AAC}_{20\text{-}30}}{\text{inspired air concentration} + 79.3} \qquad \text{(A1)}$$

About 63% uptake of 200-ppm of inspired xylene occurs at rest (0 watts) for 2 hours or at light exercise (up to 50 watts) for 90 minutes compared with 50% at 150 watts. An uptake of about 57% might be expected at 100 watts (moderate work), as seen in Table 13.2. Uptake decreases while the absolute absorbed dose increases with increasing physical activity.

Thus, someone working at moderate physical activity exposed at the xylene 1996 TLV of 100 ppm (434 mg/m^3) has an absorbed dose of [434 (mg/m^3) × 0.027 (m^3/min) × 120 (min)) × 0.57 (uptake)] = 802 mg every 2 hours. This must be corrected for the content of ethylbenzene in technical xylene, usually about 10–50% (mean 30%), since ethylbenzene does not produce a methylhippuric acid in urine. Thus, the xylene isomer content absorbed every 2 hours is about (802 × 0.70 = 561) mg. Assuming no change in alveolar ventilation after each 20–30 min break nor in uptake, an 8-hour workday consisting of four 2-hour work periods will cause about 561 × 4 = 2244 mg of xylene isomers to be absorbed. The BEI documentation for xylenes[15] states that pulmonary retention for volunteers is 60–65% of the inhaled amount using mixed exhaled air, independent of exposure concentration or pulmonary ventilation.

The BEI documentation for xylenes[15] further states that about 93–95% (mean 94%) of the absorbed xylenes are metabolized to methyl benzoic acids with 3–6% (mean 4.5%) exhaled (half-times of 1 hour and 20 hours), and <1–2% (mean <1.5%) metabolized to xylenols after 6 hours of inhalation by volunteers at 100 ppm m-xylene. The BEI marker is total methylhippuric acids with pseudo first-order half-times of 3.6 hours and 30.1 hours. Baselt[13] states that the first half-time is about 1.5 hours. The urinary BEI is 1.5 g total methylhippuric acids/g creatinine for a spot urine sample at the end of the shift, or an excretion rate of 2 mg/min for the last 4 hours of a shift.

The molecular weight of methylhippuric acid is 193 compared with 106 for xylene. Thus, 1.5 g of methylhippuric acid is equivalent to 1500 (mg)/193 = 7.77 mmoles of methylhippuric acids, which is the same as the number of moles of methyl benzoic acids if all of the latter are conjugated by glycine. The original absorbed xylene equivalent can now be calculated by correcting for xylenol production and xylene exhalation (7.77/0.94 = 8.27 mmoles). This is a mass of 8.27 × 106 = 876 mg xylene equivalent. Thus, the urine BEI concentration is equivalent to 876 mg absorbed xylene/g creatinine. The volume and creatinine content of the spot urine can be factored in if known. Spot urines are about 300–800 mL in volume (the same as the capacity of the bladder), most being towards the lower volume end, especially if earlier urination has occurred. The daily urine volume is 0.6–2.5 L/day with mean 1.2 L/day.[15] Since the creatinine excretion rate is 1.0–1.6 (mean 1.2) g/day, the mean creatinine urinary concentration is 1.0 g/L.[15] The mean specific gravity of urine is taken to be 1.024, and the volume of urine should be corrected to this specific gravity if different for the spot sample. A 300 mL urine sample corrected to specific gravity 1.024 is therefore expected to contain about 0.3 g creatinine and 0.3 × 876 = 263 mg xylene equivalents at the BEI at 1 g creatinine/L urine. Since 2244 mg of xylene is absorbed over 8 hours, only a fraction of the absorbed xylene equivalents (263/2244 = 0.117) needs to be excreted in the 300 mL spot sample to meet the BEI condition.

Assuming a pseudo first-order process:

$$c_t = c_0 \exp(-kt) = c_o \exp(-0.693\, t/t_{0.5}) \quad \text{(A2)}$$

where c_t and c_o are the equivalent moles/urine volume of untransformed xylene at times t and at the initial exposure time t_0, respectively, and $t_{0.5}$ is the pseudo first-order xylene equivalent half- time relative to the marker, here for methylhippuric acids.

An acute TLV critical exposure (c_0 = 2244 mg) time t corresponding to c_t/c_0 = 1 0.117 = 0.883 must occur in ln 0.883 × 3.6 (hours)/0.693 = 0.65 hours before sampling. All acute TLV exposure times before then cause higher transformed xylene equivalents than the BEI concentration, and the BEI is protective. The other extreme case is acute exposure at the TLV at the beginning of the first shift. The absorbed xylene that remains in 8 hours is c_t = 2244 (mg) exp (-0.693 × 8 (hours)/3.6 (hours)) = 481 mg so that 2244 - 481 = 1763 mg of xylene equivalents is lost. Thus, 1763 x 0.94 (efficiency of metabolism to the BEI marker) = 1657 mg xylene equivalents might be found in the spot sample assuming no other urination. This is 1657/263 = 6.3 times the BEI critical xylene equivalent mass in a 300 mL spot urine sample, assuming no other urination.

TWA exposures are generally not acute single dose. If the workday is 8 a.m. to 12 noon, and 1 p.m. to 5 p.m., and if 0.25 of the total xylene mass is exposed at the midpoint of the first 2-hour exposure period (561 mg xylene absorbed) with sampling at 5 p.m. and no urination in between t = 8 hours:

$$c_t = 561 \exp(0.693 \times 8/3.6) = 120 \text{ mg} \quad \text{(A3)}$$

That is, (561 - 120) = 441 mg of xylene was transformed and 441 × 0.94 = 415 mg of this xylene forms methylhippuric acids. If this same procedure is done for the second 2-hour exposure, then for the two 2-hour exposures after the 1-hour lunch (exposure periods with t = 6, 3, and 1 hour, respectively), the corresponding respective transformed xylene masses are 361, 231, and 92 mg. This type of xylene exposure will cause approximately 415 + 361 + 231 + 92 = 1099 mg absorbed xylene equivalents to have been excreted as methylhippuric acids, assuming no other urination during work. Thus, the BEI is protective for this exposure scenario because the calculated xylene equivalent mass in 300 mL of xylene was 263 mg at the BEI, assuming 1 g creatinine/L urine. Even if urination occurred during lunch time, as long as no urination occurred between the last two work shifts, the minimum excreted at the sampling time is 231 + 92 = 323 mg xylene equivalents, still >263 mg xylene equivalents at the BEI for this 300 mL spot sample. If urination occurred between the last two shifts (3 p.m.), then at t = 6, 4, and 1 hour, the xylene equivalents untransformed from the first three time periods would be 177 + 260

+ 463 = 900 × 0.94 = 846 mg of xylene still available for transformation to methylhippuric acids in the last 2 hours or 846 - 846 exp (-0.693 x 2/3.6) = 270 mg plus the 92 mg from the last exposure of the last shift, making a total of 362 mg xylene equivalents in the final spot urine sample, still above the critical BEI xylene equivalent mass. Thus, to interpret the urine concentrations at a designated sampling time, a urination history must be known as well as duration and intensity of exposures. If the hygienist has direct reading instrument data (for example, from detector tubes or organic vapor analyzers) or can infer exposure times by time-and-motion studies near exposure sources where personal air monitoring data over these times are also known, these exposure conditions can be similarly evaluated to assess whether the BEI is protective using the computational principles above.

The excretion rate of a marker is calculated[15] using the mean urine output of 0.05 L/hour and the urinary concentration corrected to a specific gravity of 1.024. Thus:

$$\text{excretion rate (mg/hour)} = 0.05\ C \qquad \text{(A4)}$$

where C is the concentration adjusted to specific gravity 1.024 in mg marker/L corrected urine volume or for creatinine excretion in mg marker/g creatinine, assuming a mean concentration of 1 g creatinine/L urine. If the BEI concentration is 1500 mg/g creatinine, the excretion rate is (1500 × 0.05/60 = 1.3 mg/min) relative to the actual BEI of 2 mg/min excretion rate over the last 4 hours. The apparent discrepancy arises from the necessity to empty the bladder about 3 hours before the test urine sample, and the logarithmic nature of the metabolism/excretion.

Apart from the case of having high exposure air concentrations near the TLV within 0.65 hr of sampling, the BEI is protective for all the above exposure scenarios.

A problem arises when physical activities lower than the reference occur. The worst case is for workers exposed to the TLV at rest. From Table 13.2, 5/27 = 0.185 the mass is potentially absorbed relative to moderate physical activity, an uptake 63/57 = 1.11 times that at moderate workload. Thus, instead of 2244 mg xylene being absorbed, only (0.185 × 1.11 × 2244 = 461 mg) is absorbed for an exposure at the TLV at rest, or 461/2244 = 0.21 times the absorbed mass at moderate physical activity at the TLV. A 300-mL spot urine sample containing the BEI xylene equivalent mass of 263 mg at a creatinine concentration of 1 g/L is still protective if the TLV is absorbed acutely [-ln (461263)/461 × 3.6/0.693 = 4.4 hours before urine sampling if the worker is sedentary]. Using the above four acute distributed exposure period scenario but with (461/4 = 115) mg xylene per exposure, the expected excreted xylene equivalent mass on urine sampling is now 85 + 74 + 47 + 19 = 225 mg, about 86% of the BEI xylene equivalent mass in a 300-mL spot urine sample. This also justifies an action level below the BEI of about 30%. Note that the effect of previous urination during lunch time and the last break before end-of-shift sampling is more important at low physical activities relative to BEI considerations than at higher physical activities. Working at physical activities greater than the TLV reference will result in higher urine concentrations on exposure to the TLV. Thus, the BEI is still protective.

The above computational example for at-rest workers will aid the hygienist in defining whether a sample that exceeds the BEI is caused solely through enhanced physical activity through substituting the appropriate data related to high physical activity for those used for at-rest physical activity. Thus, the absorbed equivalent mass of xylene at heavy work is (2244 × 1.41 × 50/57 = 2775 mg) xylene equivalents at the TLV, or (2775/2244 = 1.24) times that at moderate workload at the TLV.

If the hygienist has some idea of the temporal frame of the exposures, either through data logging capability of direct reading instruments or through time-and-motion studies relative to the known exposure sources, then such data can be used instead. The observational skill of the hygienist is key to the interpretation of biological monitoring data.

Additional Sources

Lists of laboratories capable of analyzing samples can be found in (1) the annual buyer's guide for *Industrial Hygiene News* under the category Medical Testing Laboratory; (2) the annual directory for *Occupational Hazards* under Laboratory Testing Services for Environmental Management and under Medical Screening and Record-Keeping Services for Occupational Health; (3) the annual purchasing sourcebook of *Occupational Health and Safety* in the Services section under Air Monitoring and Analyses—Exterior; Biological Monitoring; Consultants—Environmental Health Laboratories Analysis; Environmental Health—

Laboratory Analysis Service; Health Monitoring; Laboratory Analysis Service—General; and Screening Services; and (4) the annual laboratory guide edition of *Analytical Chemistry* under the headings Services, Elemental Analysis, Inorganic; Elemental Analyses, Organic; and Forensic Analyses.

Information about state-certified analytical laboratories is available from the list of state OSHA regional offices and state safety and health agencies in the resource finder section of the annual directory of *Occupational Hazards.* The ACGIH BEI Committee and the American Industrial Hygiene Association (AIHA) Biological Monitoring Committee are two other informational resources.

References

1. **Que Hee, S.S. (ed.):** *Biological Monitoring: An Introduction.* New York: Van Nostrand Reinhold, 1993.
2. **Ness, S.A.:** *Surface and Dermal Monitoring for Toxic Exposures.* New York: Van Nostrand Reinhold, 1994.
3. **Ellis, K.J.:** In-vivo monitoring techniques. In *Methods for Biological Monitoring: A Manual for Assessing Human Exposure to Hazardous Substances,* T.J. Kneip and J.V. Crable, eds. Washington, DC: American Public Health Association, 1988. pp. 65–80.
4. **Forsberg, K. and L.H. Keith:** *Chemical Protective Clothing: Permeation and Degradation Compendium.* Boca Raton, FL: Lewis Publishers, 1995.
5. "Lead Standard," *Code of Federal Regulations,* Title 29, Part 1910.1025. 1979.
6. "Access to Employee Exposure and Medical Records; Proposed Modifications; Request for Comments and Notice of Public Hearing," *Code of Federal Regulations,* Title 29, Part 1910. 1982.
7. "Identification, Classification and Regulation of Potential Occupational Carcinogens," *Code of Federal Regulations,* Title 29, Part 1990. 1980.
8. "Benzene, Occupational Safety and Health, Chemicals, Cancer, Health, Risk Assessment," *Code of Federal Regulations,* Title 29, Part 1910. 1987.
9. "Hazard Communication; Final Rule," *Code of Federal Regulations,* Title 29, Part 1910. 1983.
10. **American Conference of Governmental Industrial Hygienists (ACGIH):** *1996 Threshold Limit Values (TLVs) for Chemical Substances and Physical Agents and Biological Exposure Indices (BEIs).* Cincinnati, OH: ACGIH, 1996.
11. **American Conference of Governmental Industrial Hygienists (ACGIH):** *Documentation of the Threshold Limit Values and Biological Exposure Indices,* 6th ed. Cincinnati, OH: ACGIH, 1991. (Supplement 1996).
12. **Lauwerys, R.R. and P. Hoet:** *Industrial Chemical Exposure: Guidelines for Biological Monitoring,* 2nd ed. Boca Raton, FL: Lewis Publishers, 1993.
13. **Baselt, R.C.:** *Biological Monitoring Methods for Industrial Chemicals,* 2nd ed. Littleton, MA: Year Book Medical Publishers, 1988.
14. **Commission for the Investigation of Health Hazards of Chemical Compounds in the Work Area:** *Maximum Concentrations at the Workplace and Biological Tolerance Values for Working Materials.* New York: VCH Publishers, 1991.
15. **American Conference of Governmental Industrial Hygienists (ACGIH):** *Documentation of the Biological Exposure Indices,* 6th ed. Cincinnati, OH: ACGIH, 1991.
16. **Astrand, I., J. Engstrom, and P. Ovrum:** Exposure to xylene and ethyl benzene. I: Uptake, distribution and elimination in man. *Scand. J. Work, Environ. Health 4:*185–194 (1978).
17. **Eller, P. (ed.):** *NIOSH Manual of Analytical Methods,* 3rd ed. Cincinnati, OH: U.S. Department of Health and Human Services, 1987.
18. **U.S. Department of Labor (DOL):** *OSHA Analytical Methods Manual,* 2nd ed. Salt Lake City, UT: DOL, 1990.
19. **Kneip, T.J. and J.V. Crable (eds.):** *Methods for Biological Monitoring: A Manual for Assessing Human Exposure to Hazardous Substances.* Washington, DC: American Public Health Association, 1988.
20. **Sheldon, L., M. Umana, J. Bursey, W. Gutknecht, et al.:** *Biological Monitoring Techniques for Human Exposure to Industrial Chemicals: Analysis of Human Fat, Skin, Nails, Hair, Blood, Urine, and Breath.* Park Ridge, NJ: Noyes Publications, 1986.

21. **Carson, B.L., H.V. Ellis III, and J.L. McCann:** *Toxicology and Biological Monitoring of Metals in Humans: Including Feasibility and Need.* Chelsea, MI: Lewis Publishers, 1986.
22. **U.S. Environmental Protection Agency (EPA):** *Manual of Analytical Methods for the Analysis of Pesticides in Humans and Environmental Samples,* (EPA-600/8-80-038). Research Triangle Park, NC: EPA, 1980.
23. **Angerer, J. and K.H. Schaller (eds.):** Analysis of Hazardous Substances in Biological Materials, vols. 1–5. New York: VCH Publishers, 1997.
24. **Que Hee, S.S. (ed.):** *Biological Monitoring: An Introduction.* New York: Van Nostrand Reinhold, 1993. pp. 148–156.
25. **Kneip, T.J. and J.V. Crable, (eds.):** *Methods for Biological Monitoring: A Manual for Assessing Human Exposure to Hazardous Substances.* Washington, DC: American Public Health Association, 1988. pp. 44–63.
26. **Sheldon, L., M. Umana, J. Bursey, W. Gutknecht, et al.:** *Biological Monitoring Techniques for Human Exposure to Industrial Chemicals: Analysis of Human Fat, Skin, Nails, Hair, Blood, Urine, and Breath.* Park Ridge, NJ: Noyes Publications, 1986. pp. 364–387.
27. **Que Hee, S.S. (ed.):** *Biological Monitoring: An Introduction.* New York: Van Nostrand Reinhold, 1993. pp. 139–148.
28. **Que Hee, S.S. (ed.):** *Biological Monitoring: An Introduction.* New York: Van Nostrand Reinhold, 1993. pp. 124–186.

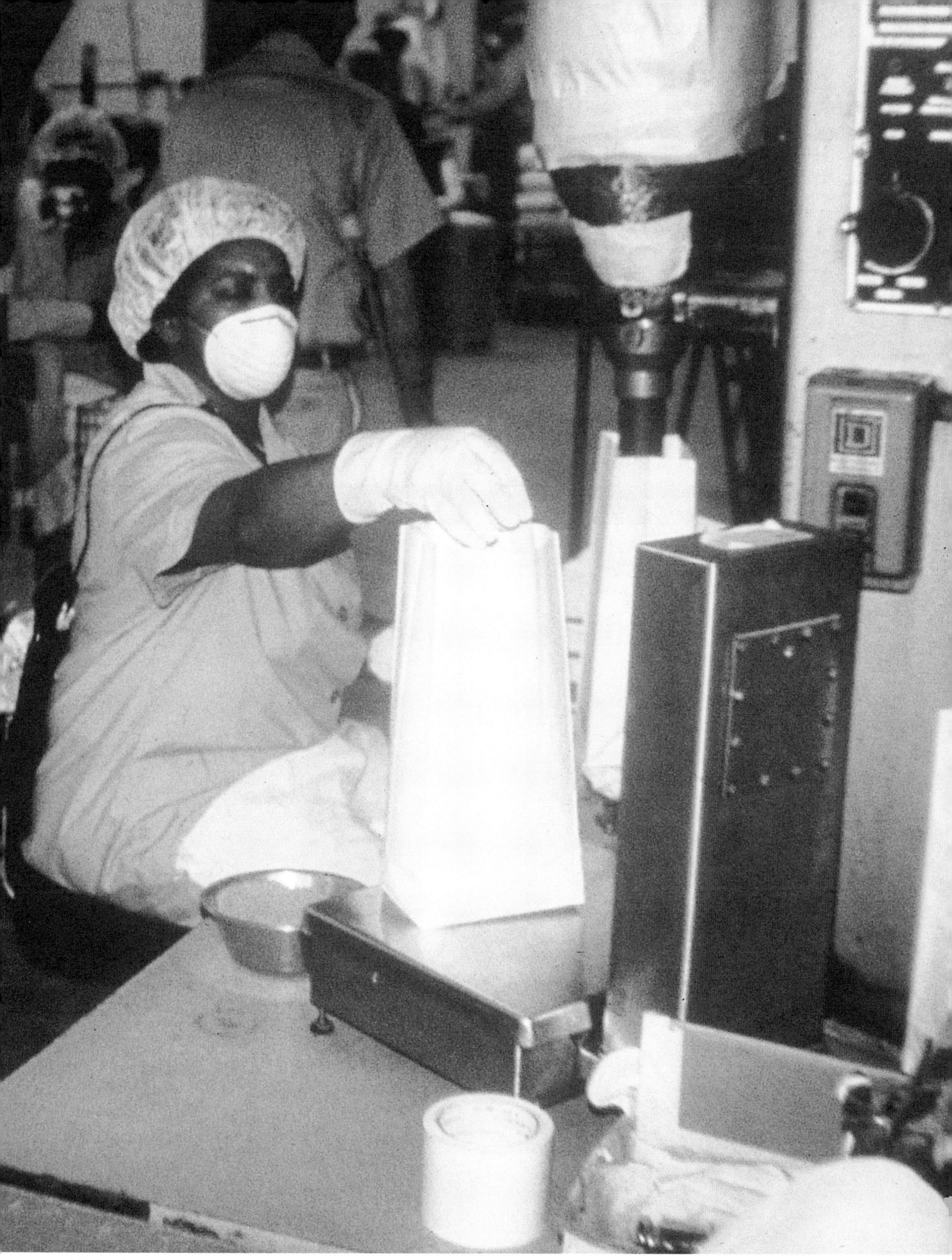

Workers in a spice packaging facility. The primary problem is lung irritation and hypersensitivity. Note the single-use respirator.

Outcome Competencies

After completing this chapter, the user should be able to:

1. Define underlined terms used in this chapter.
2. Describe the importance of the dermal route of exposure.
3. Recognize the dermal absorption process.
4. Summarize approaches for assessing dermal exposures.
5. Describe means of controlling dermal exposures.
6. Discuss the significance of industrial dermatoses.

Key Terms

acne • basal cell carcinomas • biological monitoring • contact dermatitis • critical flux • dermal absorption • dermatitis • flux • lesions • occupational exposure limit • partition coefficients • perfusion • sensitization • skin notation • stratum corneum • systemic effects • urticarial reactions

Prerequisite Knowledge

Anatomy and physiology, differential calculus (basic knowledge only), occupational exposure assessment techniques.

Key Topics

I. Importance of the Dermal Route of Exposure
   A. Significance of Dermatitis and Related Problems
   B. Current Approaches for Assessing Dermal Exposure and Skin Absorption

II. An Overview of the Dermal Absorption Process
   A. Historical Review
   B. A Brief Physicochemical Treatment of Process
   C. Mechanisms of Dermal Absorption
   D. Assessment of Skin Effects: Local and Systemic
   E. Criteria for Determining the Significance of Dermal Absorption

III. Approaches for Occupational Dermal Exposure Assessment and Management
   A. Basic Characterization of Exposure
   B. Dermal Exposure Assessment
   C. Practical Means of Dermal Exposure Monitoring for Exposure Assessment
   D. Medical Evaluation of the Skin
   E. Management and Control of Occupational Dermal Exposures

14

Dermal Exposure and Occupational Dermatoses

J. Thomas Pierce, Ph.D., CIH, DABT

M. Cathy Fehrenbacher, CIH

Lutz W. Weber, Ph.D., DABT

Introduction

Historically, more attention has been given to airborne exposures than dermal exposures. Indeed, in many scenarios it is the airborne exposure potential that dominates the overall picture; however, occupational dermatoses represent one of the most common forms of occupational disease, potentially impairing the health and welfare of millions of workers.[(1)] Skin absorption is the major route of exposure for compounds such as methylene-bis-ortho-chloroaniline, various lipid-soluble pesticides, and the historically important polychlorinated biphenyls (PCBs).[(2)] When a worker's exposure to hazardous materials is primarily from contact with contaminated surfaces, means of dermal monitoring must be considered.[(3)] The American Conference of Governmental Industrial Hygienists (ACGIH) recognizes the importance of dermal exposure by designating some compounds with a "skin" notation attached to the threshold limit value (TLV®).[(4)]

The occupational hygienist should begin to assess dermal exposures to chemical agents when the basic characterization reveals tasks and work practices where skin contact may occur directly or through secondary contact with contaminated tools, work surfaces, or personal protective equipment. Of course, skin contact alone is not enough to create a health risk. A review of available information on the toxicology of the chemical agent and its chemical and physical characteristics should quickly indicate whether or not dermal exposure could present a health risk. Information on the toxicity of the environmental agent, and its chemical (e.g., lipophilicity) and physical (e.g., pH) characteristics, including dermal absorption characteristics, should provide some insight into the target organs that may be affected. Skin contact may be the principal route of exposure for high molecular weight compounds exhibiting low volatility. Another factor that enhances the dermal absorption of a chemical agent is a relatively high degree of lipophilicity for the compound. That is, compounds that are more soluble in oil than water tend to go through the skin more readily. Ironically, the new science of transdermal drug delivery for medicinal substances has fueled skin-related research.[(5)] This development in a seemingly unrelated field has fueled a better understanding of skin function. Still, probably no other occupational health area requires the integration of diverse clinical and exposure assessment skills to address what must be described as a multidimensional dermal absorption and disease problem.[(6-8)]

Importance of the Dermal Route of Exposure

Significance of Dermatitis and Related Problems

Occupational skin disease is one of the leading causes of days lost from the job as a result of work-related illnesses.[(9–11)] The cost of occupational skin disease is high, both for the worker in terms of disability and for industry in terms of days lost and combined medical and compensation dollars spent. Rather than developing an inclusive incidence (attack) rate for all industries, it is more useful to highlight the features of important skin diseases in particular industries.[(12–14)] Ambitious data extrapolations across industry lines or disease classifications are questionable because of the large variation in job classifications and subsequent exposures.

The significance of dermal absorption in terms of its contribution to overall body burden of chemicals is still largely unknown. Only recently have significant efforts been expended, and then only for selected chemicals, to estimate specific body burdens. As a general rule, when low volatility compounds such as polychlorinated hydrocarbons elicit pronounced effects, dermal absorption must be factored into any resulting risk estimate. Until models and assessment methods are refined, it is necessary to work backwards from a verified health effect to absorption parameters because it is unlikely that all of an exposure resulted from inhalation alone.[15]

In 1997 a single incident riveted the attention of safety and health professionals on the seriousness of dermal absorption of toxicants. A distinguished professor of chemistry (48-year-old female) died of mercury intoxication, just days following brief exposure in her laboratory to one or a few drops of dimethyl mercury that leaked across a latex glove.[16] Although not the first lethal incident to result from dermal exposure, it emphasized in a striking fashion the importance of appropriate dermal protection, given the silent latency between exposure and symptoms for this and other compounds. Certainly, strategies for effective intervention must precede the accident investigation stage.

Current Approaches for Assessing Dermal Exposure and Skin Absorption

Standardized protocols to accurately estimate skin absorption have only recently been developed, but have not been applied in occupational and environmental health. While quantitative estimates of dermal absorption are desirable, only a simple semiquantitative or qualitative rating may be necessary to prevent dermatoses.

Biological monitoring is a useful technique for assessing total body dose, especially where skin contact is the major route of exposure. The ACGIH biological exposure indices (BIEs®) were designed to facilitate assessments based on certain dermal absorption measurements.[17] The BEIs represent warning levels of chemicals or their metabolic products in tissues, fluids, or exhaled air of exposed workers, regardless of whether the chemical was inhaled, ingested, or absorbed via skin.

A good example is the use of p-aminoaminophenol in urine to assess exposure to aniline. Biological monitoring strategies are available for several contaminants and are reviewed in the ACGIH monograph *Topics in Biological Monitoring*.[18] Where biological monitoring approaches are not available, other means of determining the dose from skin contact must be considered.

The most commonly used approach in assessing skin exposure uses pads or other dosimeters to determine the amount of contaminant deposited on the skin. This technique is commonly used in assessing skin exposures during pesticide handling and application.[19] The assessment of dermal exposure is generally conservative (i.e., overestimating) because overestimation of risk is preferred to underestimation. However, prior to discussing skin exposure assessments, it must be acknowledged that there are scenarios where the target organs are both internal and external (skin). One example is hydrofluoric acid, which is corrosive to the skin and also absorbed through the skin, thereby depleting total body calcium ion. Another example is dermal exposure to coal tar pitch. Coal tar pitch absorbed into the epidermis can cause transient photosensitization. The occupational hygienist may need to employ a combination of methodologies for assessing and controlling dermal exposures where the target organs are both external and internal.

An Overview of the Dermal Absorption Process

Historical Review

While the nature of occupational dermatologic diseases has occupied clinicians' and researchers' attention for many years, a thorough description of the dermal absorption process has only recently emerged.[7,20] Medical lectures from the 1950s often described skin uptake of any chemical as negligible. In most cases where organic solvents are involved, uptake through the respiratory tract may prevail and obscure a contribution by dermal absorption. It has taken severe industrial accidents with exclusively dermal exposure to make occupational health professionals recognize that the skin is not an impermeable barrier. One classic example would be chloracne.[21,22] The insight that occupational exposure to chlorine-containing materials caused this disease had been gained nearly 100 years ago, but it was only 40 years ago that the true etiology was identified: not chlorine, but chlorinated aromatics like 2,3,7,8-tetra-chloro-di-benzo-p-dioxin (dioxin) are the causative agents; the symptoms result from dermal exposure; and they reflect systemic rather than local absorption.(23) The percutaneous absorption of dioxin itself in human skin is quite poor, about 0.15 to 0.35 percent per hour of an administered dose.(24)

A Brief Physicochemical Treatment of the Process

Depending on resources committed to studying percutaneous absorption in humans, a variety of techniques may be employed. It is safe to say that human percutaneous absorption is a subtle process often necessitating a combination of techniques. Investigators may make use of physicochemical constants or simple modeling methods, while a maximum effort may entail volunteer studies in a manner similar to that of a projected population. Additional data obtained from animal experiments may also be included, given a thorough understanding of species differences.

Dermal absorption in an industrial setting can occur on direct dermal contact with liquid chemicals (or solutions) or on exposure to associated vapors. While vapor absorption is important for high toxicity compounds such as the war gases or in high hazard clean-ups, it represents an extremely complex topic.(8) It is important to remember that the effective surface area for vapors almost invariably exceeds the estimated area for liquid exposure because of the difficulty to provide whole body protection. Condensation of vapors of chemicals with low vapor pressure on the body surface can significantly increase resultant dermal absorption. Dermal absorption of solids (dust, aerosols, etc.) can be facilitated by their dissolution in perspiration.

The dermal penetration rate of dissolved chemicals appears to be directly related to their concentration and indirectly (not inversely) to their solubility in a particular solvent. Prolonged contact with liquids usually results in changes of dermal structure and even permeability (swelling; hydration of the "mortar" or incorporation of solvent into the lipid "bricks" of the stratum corneum [SC]). Thus, dermal absorption is by its nature a dynamic process (see Figure 14.1).

One widely applied technique used to estimate permeation rates involves the use of partition coefficients. Referring to Fick's law of diffusion, from a single vehicle the rate of absorption (J) can be related to the applied concentration as follows:(25)

$$J \propto C_v \quad (1)$$

Here J is the absorption rate per unit area (or flux) and C_v is the concentration of penetrant in the vehicle. Adding the permeability constant, k_p, establishes the equation:

$$J = k_p \times C_v \quad (2)$$

There are limitations in the use of the constant; investigations must be made to ensure that its use is justified over an actual range of concentrations; to see if there are any concentration-dependent changes in the SC itself because of the penetrant; and to examine other deviations from ideality. It is always best to apply these equations as a means of process conceptualization rather than a presumptive means of flux calculation.

Once determined, the permeability constant facilitates comparisons even when different penetrants, vehicles, or experimental conditions are involved. Permeability constants range from 1×10^{-6} (about that of water) to 5×10^{-2} cm/hour for highly lipophilic compounds. The diffusion constant (D_m) indicates the mobility of

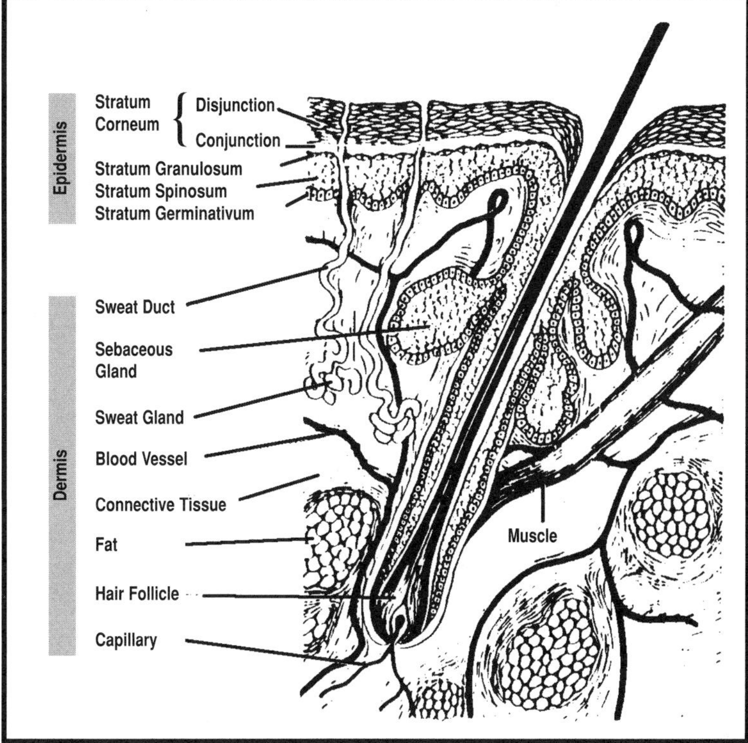

Figure 14.1 — Anatomy of the skin.

Table 14.1 — Useful Data on the Skin Layers of Humans

	Thickness[A] µm	*Weight[A] g*	*Perfusion L/min*
Epidermis	90–1500[B]	100; 90	0
Stratum corneum (dead cells)	5–>1000[C]		
Living cells	160[D]		
Dermis (connective tissue, proteins, blood)	1250	2500; 1700	0.2
Hypodermis (adipose tissue)	1375; 6600	7500; 13000	0.15

[A]Dual values refer to males and females, respectively.
[B]Depends mostly on thickness of the stratum corneum.
[C]Varies with body region: highest in mechanically stressed areas (palms, soles).
[D]Average value: changes somewhat with thickness of the stratum corneum.

the penetrant molecules within the SC, while the partition coefficient (K_m) is the ratio at equilibrium of penetrant concentration in the SC to that in the vehicle, when isolated SC has been immersed in the appropriate liquid.

$$k_p = \frac{K_m \ D_m}{d} \quad (3)$$

(where d is the thickness of skin layer under study). The partition coefficient takes into account the fact that the driving force for net movement of penetrant across the SC is the difference in concentration across the tissue. Therefore, an absorption rate can be expressed as

$$J = \frac{K_m \ Cv \ D_m}{d} \quad (4)$$

The partition coefficient, like the penetrant concentration, is an externally determined factor that helps determine absorption rates. When actual SC: vehicle partition coefficients are not available, substitutes in the form of octanol: water partition coefficients are often used. Effective means have been found to determine these values experimentally as well as to model them. It is a good idea to place confidence intervals around estimates. While these equations are not directly used in exposure measurements, they provide an important conceptual basis to the dermal assessment problem.

Mechanisms of Dermal Absorption

Skin is formed by layers of various cell types which, according to their composition and function, are identified as the three basic layers shown in Table 14.1. The thicknesses of these layers, particularly the SC, vary greatly with body region, resulting in major differences of dermal penetration.(26) Thicknesses of the layers, their weights, and perfusion rates as given in Table 14.1 are for a resting reference human with a cardiac output of 6 L/min. The surface area of the skin can be considered as equal to body surface area, which is about 1.9 m^2 for males (70 kg, 170 cm) and 1.7 m^2 for females (58 kg, 160 cm).(16)

It has been generally assumed that transport of aqueous nonelectrolytes through skin is governed by Fick's law of passive diffusion. The underlying premise has been twofold: (1) that the SC possesses minimal storage capacity; and (2) that the SC represents a relatively uniform or homogeneous diffusion membrane.

There is no doubt that the SC represents the major barrier to percutaneous absorption. Once believed to have properties of a homogeneous barrier to physiological water uptake and loss,(27) it now has been shown to consist of two distinct layers with very different properties, the stratum disjunctum and the stratum compactum. In contrast to all other layers of the skin, the cell-derived structures of the SC are no longer viable. It appears that the disjunctum region acts as a barrier to solute (dissolved substance) uptake, possibly by forming a reservoir.(28,29) The compactum region forms the major water barrier and is currently pictured as a "brick and mortar" structure where the lipid-rich phase represents the brick, and the hydrophilic-proteinaceous phase the mortar.(30,31)

Such findings suggest that the assumption of passive diffusion across the SC applies primarily to nonelectrolytes having relatively small molecular weights and octanol/water partition coefficients greater than approximately 50. Furthermore, the lipid content of SC and its potential for storage of lipophilic solutes cannot always be disregarded in the cases where the objective is to quantify the transdermal absorption of toxic substances.(32) Laboratory studies on percutaneous absorption of industrial chemicals have typically been performed with high solute concentrations, making data somewhat inapplicable to actual scenarios. Recent developments in mathematical modeling have provided a better means of overcoming this problem.(33) Shatkin and Brown(34) uniquely demonstrated that dermal absorption of solutes can be predicted using the kinetic computer model STEL–LA™. They treated the SC as only a barrier and disregarded any storage properties.

Chemicals can penetrate through the SC and diffuse into the dermis, where they then enter capillary blood (see Figure 14.2). Chemicals are carried by the systemic circulation to other tissues, where they may be deposited, metabolized, excreted, or exert biological effects. The rate-limiting step in dermal absorption is either the penetration rate through the SC or the removal rate of the chemical from the dermis. Both SC uptake and dermal absorption are influenced by physicochemical properties of chemicals, such as solubility in biological media and physiological status of the skin. Diseased skin (from whatever cause) behaves much differently than intact, healthy skin.[25] It may be stated here that the ability to metabolize resides only in the living cells of the epidermis and dermis, but certainly not in the SC.

Absorption rates vary with perfusion, which increases with the temperature of the environment and also the physical activity of the worker. Since cardiac output increases in a similar manner, it can be assumed that perfusion of dermis accounts for about 3% of cardiac output and perfusion of the hypodermis for about 2.5%. Skin temperature is usually about 32°C but is dramatically increased through wearing of tight-fitting clothing.[35]

Assessment of Skin Effects: Local and Systemic

While the process of dermal absorption is indeed critical, the key question is whether there are any effects, either local (e.g., dermatologic irritation) or systemic (e.g., other target organs distant from the site of exposure). When industrial chemicals are absorbed, one must evaluate first whether dermatitis occurs, and secondly, whether sufficient quantities have been absorbed to create a significant internal dose. The presence of dermatitis can often be readily determined via visual examination of the skin, but determining whether absorption will create a significant enough internal dose to be of concern is quite challenging.

There have been cases in which there was significant absorption without concurrent dermatitis and vice versa. More predictably, dermatitis itself leads to increased rates of dermal absorption since the SC and other important structures are destroyed or disrupted. The fragility of the SC is demonstrated in the laboratory by the experimental procedure of removing it, layer by layer, through repeated stripping with adhesive tape. The purpose of this procedure is to imitate permeability conditions found with severe dermatitis, but not in all cases does this result in a tremendous increase of dermal absorption. With highly lipophilic molecules (e.g., dioxin) removal of the SC may increase dermal absorption only slightly

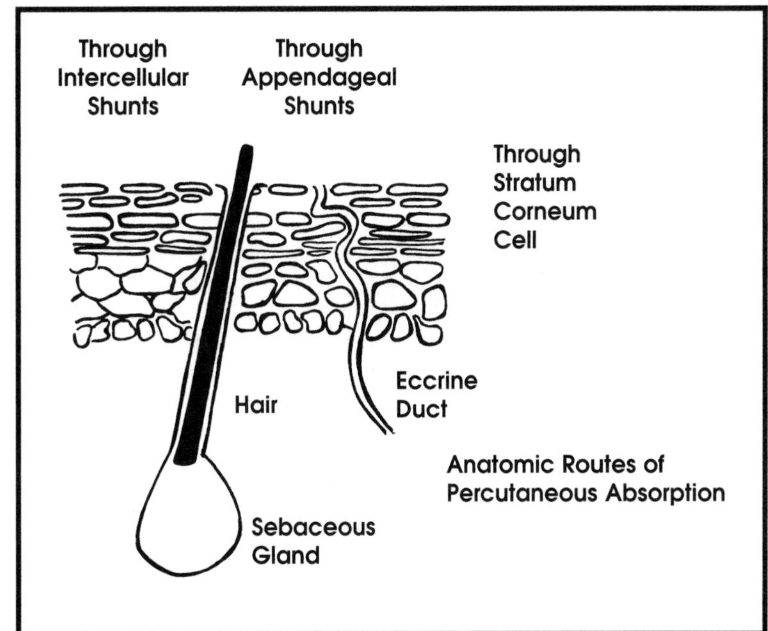

Figure 14.2 — Anatomical routes of percutaneous absorption.

(i.e., two- to fourfold)[24] as compared with 100-fold or more with small water-soluble molecules.

Criteria for Determining the Significance of Dermal Absorption

The criteria for significance of dermal absorption are perhaps best explained in terms of target organ and receptor effects. The fact that a chemical is transferred into the bloodstream does not necessarily make it toxic, and the fact that only small amounts are transported via blood does not necessarily render a chemical harmless.

Criteria for significant dermal absorption potential have been arbitrarily selected, depending on which effect of dermal absorption is being considered. Significance has been attached when the dermally absorbed amount exceeds 30 percent of that from the corresponding respiratory mode. One group of researchers defined a critical flux by comparing the dose resulting from inhalation exposure at the TLV–time-weighted average (TWA), with the dose resulting from the same inhalation exposure, but with additional dermal exposure of a defined area of skin.[35] If this critical flux is found to be equal to a measured (or estimated) actual flux, then by convention the dermal absorption potential is said to be significant. Furthermore, when the critical flux is less than six times the actual flux, then by convention the chemical is said to have dermal toxicity potential—not to be confused with dermatotoxicity. The critical flux is best visualized as a reference value, something made up to compare to an actual flux (e.g., the lower the calculated critical flux) the more likely significant dermal absorption is to occur. However novel these or other calculations may appear, they are merely a means of comparing effective doses via pulmonary and dermal routes, and the potential for dermal toxicity is overestimated when the TLV–TWA is based only on irritation. The TLV skin notation remains a useful guide to prioritizing control efforts. Common sense dictates that for local effects, action should be taken to mitigate contact with the contaminant. For systemic effects, the evaluation and determination of the need for and the mechanism of control is much more complex.

The development of dermal exposure limits is under investigation. Dermal limits could be based on (1) workplace surface contamination levels; (2) skin or clothing contamination levels; or (3) biological measures of exposure or internal dose. It remains to be seen whether or not dermal exposure limits can be obtained using the same approach used for airborne exposure. Skin as an organ differs significantly from lung with respect to metabolic capability (types of metabolic enzymes), capacity to metabolize, and reservoir function. Therefore dermal absorption estimates for airborne materials should be attempted only where basic parameters such as permeability, flux, or total absorbed dose for a given gas or vapor are available.[8]

Approaches for Occupational Dermal Exposure Assessment and Management

Basic Characterization of Exposure

Often the first step in evaluating occupational exposures involves characterizing the work force, work environment, and the contaminants to which workers may be exposed. The Occupational Safety and Health Administration's personal protective equipment standard includes guidance on hazard assessment and personal protective equipment selection that may be helpful in gathering workplace information.[36] When evaluating dermal exposures, one should consider the primary source of contact (e.g., through handling the material) as well as secondary sources of contact (e.g., through handling contaminated equipment or tools). Thorough information on the toxicity of the environmental agent in question is imperative.

Dermatoses can be categorized along the lines of (1) mechanical (including trauma) — friction, pressure and mechanical disruption; (2) chemical — the elements and a wide variety of organic and inorganic compounds; (3) physical — heat, cold, radiation; and (4) biologic — organisms such as viruses, bacteria, fungi, and parasites. Recording of occupational skin disease based on these categories allows particularly hazardous agents to be identified and preventive measures to be taken rapidly. Periodic recording and review of work-related illness at a job site is one of the most important aspects of occupational disease prevention.

Mechanical

Mechanically induced occupational dermatoses constitute the largest category of dermatoses due to trauma. Causes include friction, pressure, and mechanical disruption of the skin. General experience indicates that mechanical causes may account for 75% of occupationally related skin disease as seen in industrial clinics.[11,37] Even a minor injury to the skin in this category may allow more serious dermatoses to occur through compromise to and immunologic dysfunction of the barrier.

Chemical

Chemical dermatoses are caused by elements and compounds that adversely affect dermal structure or metabolism. The harmful effect on the metabolism of cells may be a direct toxic damage to the cells or an indirect injury mediated through immune mechanisms. The more common of these two mechanisms is the direct toxic effect, which is termed "irritant." Typically, irritants injure all the cells in the epidermis. Certain cells may be more sensitive to phenolic chemicals because of the existence of ringed precursors in the synthesis of the pigment melanin.

The list of potential irritants includes many chemicals, ranging from mild soaps to harsh acids and alkalies. Even water sometimes acts as an irritant, because prolonged contact may cause swelling and cell death. Ordinarily aqueous exposure is not a problem because the stratum corneum and intercellular laminar material shield the lower levels of skin from direct contact with water.

Many irritant chemicals may be safely used by limiting the time of exposure and using protective barriers such as gloves, although these approaches are not without problems. Permeability is both a function of the gloves' chemical composition, the wear characteristics, and the nature of the agent of which the worker is to be protected. Because of the variety of chemicals used and the specifics of industrial processes involved, complete elimination is not always feasible. It is important to realize that vinyl, latex or rubber gloves provide comparatively poor protection from small organic molecules, particularly liquids.

Physical and Biological

These causes of occupational dermatoses are well beyond the scope of this chapter. Suffice it to say that the effects of sunlight range from mild erythema to malignant melanoma, and much of the practice of dermatology centers on the bacterial, and fungal origins of disease, but these cannot be covered here.

A brief survey of factors that can increase dermal absorption includes: (1) damage to the SC (by dermatitis, mechanical means, detergents, corrosives); (2) excessive hydration of the SC (extensive hot showering with use of soap; sweating particularly under dense protective clothing; working at high humidity or submerged in water; use of common skin care products that contain moisturizers like urea); (3) exposure to organic solvents that loosen the lipid matrix of the SC and thus impair its barrier function; (4) high skin temperature, be it induced by high ambient temperature, protective clothing, or high skin perfusion due to strenuous work or chemical effect (e.g., alcohol, organic nitro compounds). If decontamination is deemed necessary, circumspect use of soap and cold water will do in most cases.

It may be helpful to use a tiered approach with simple methods of initial priority setting, and progress to more complex approaches for those exposures that cannot initially be eliminated from further consideration. The first tier involves categorizing the magnitude of exposure using descriptive terms such as immersion, routine contact, incidental contact, and negligible contact. A qualitative rating of exposure as a first tier exposure assessment may suffice for purposes of determining whether additional monitoring or further information gathering is needed. These qualitative ratings may be sufficiently descriptive to assign an exposure rating based on the potential for exposure: immersion of hands or frequent contact with other parts of the body (4, poorly controlled exposure); routine contact of the hands or other parts of the body (3, very significant exposure); incidental contact of the hands or other parts of the body (2, significant exposure); and negligible contact of the hands or other parts of the body (1, insignificant exposure).

Recording of occupational skin disease based on mechanical, physical, biological, or chemical categories allows particularly hazardous agents to be identified and preventive measures to be taken rapidly. Periodic recording and review of work-related illness at a job site is one of the most important aspects of occupational disease prevention.

While an occupational hygienist would not be involved in the actual physical examination, health history documentation should include a careful occupational exposure assessment that should be performed by a well-trained individual.[38,39] Indeed, a knowledge of past medical history and the physical examination can be very useful to the exposure assessment process itself.

Skin conditions are often readily apparent to those who have them as well as to their families and co-workers. Since the opportunity for important first observation will probably exist for other than the cognizant physician, various members of the health care team must be well-trained in recognizing fundamental dermatologic exposures and attendant risks.

Dermal Exposure Assessment

The assessment of dermal exposure is often based on limited information relative to the magnitude of skin exposure, the measurements of affected surface, and the dynamics of actual absorption. This generally results in a very conservative estimate of risk.

If the initial basic characterization has not eliminated the exposure from further review, a second tier exposure assessment approach that involves modeling and/or collection of dermal exposure monitoring data, coupled with information and data on dermal absorption potential is recommended.

Determining the amount of skin contact is generally performed using an approach whereby the surface area of the skin potentially exposed to the environmental agent (in units of cm^2), and the amount of environmental agent transferred to the skin during the exposure event (in units of $mg/cm^2/event$) are determined. Generally, it is not possible to determine the actual

Table 14.2 — Surface Area by Region of the Body for Adults 5th to 95th Percentile

Region of the Body	*Men Mean, cm²*	*Min-Max, cm² A*	*n*	*Women Mean, cm²*	*Min-Max, cm² A*	*n*
Head	1180	900–1610	29	1100	953–1270	54
Trunk	5690	3060–8930	20	5420	4370–8670	54
Arms	2280	1090–2920	32	2100	1930–2350	13
Upper arms	1430	1220–1560	6	—	—	—
Forearms	1140	945–1360	6	—	—	—
Hands	840	596–1130	32	746	639–824	12
Legs	6360	2830–8680	32	4880	4230–5850	13
Thighs	1980	1280–4030	32	2580	2580–3600	13
Lower legs	2070	930–2960	32	1940	1650–2290	13
Feet	1120	611–1560	32	975	834–1150	13
Total	19,400	16,600–22,800	48	16,900	14,500–20,900	58

Source: Adapted from *Development of Statistical Distributions or Ranges of Standard Factors Used in Exposure Assessments* (EPA 600/8-85/010), Office of Health and Environmental Assessments, Exposure Assessment Group, Environmental Protection Agency, Washington, DC, 1985.
[A]5th percentile to 95th percentile, respectively.

amount transferred to the skin during the activity. Thus, the amount transferred to the skin (or a surrogate skin such as a glove) is determined at completion of the activity. This amount is influenced by the transfer to, from, and through the skin. Multiplying the skin surface area and the amount of the agent transferred to the skin by the number of exposure events per day provides an estimate of the amount of skin contact per day. The estimated amount absorbed through the skin is then combined with the amount of skin contact to predict the daily potential dermal dose.

The following equation can be used as a screening level estimate of the daily dermal dose rate in units of mg/day, assuming deposition on the skin followed by absorption. Once the predicted dermal absorbed dose rate per day has been determined, it can be compared easily with an occupational exposure level (OEL) expressed as total dose in units of mg/day.

$$DA = S\ Q\ FQ\ ABS\ WF \quad (5)$$

where:

DA = dermal absorbed dose rate per day, mg/day;
S = surface area of skin available for contact, cm^2;
Q = quantity deposited on the skin per event, mg/cm^2/event;
FQ = number of events per day, events/day;
ABS = fraction of applied dose absorbed through the skin during the event; and
WF = weight fraction of the substance in the mixture, unitless.

To determine the surface area of the skin potentially exposed, one needs information on the activities and tasks performed by the worker, and the types and number of skin surfaces (e.g., one hand, two hands, etc.) potentially exposed during the activities. Fortunately, the mean skin surface areas for various parts of the body have been measured for adults and children by several researchers, allowing the development of statistical descriptors for skin surface area.[(40)] Table 14.2 presents measured skin surface area values commonly used by the U.S. Environmental Protection Agency (EPA) as default input values in predicting dermal dose. It is important to note that various parts of the human body have differing rates of absorption.[(8,26,41)]

While determining the skin surface area potentially exposed during various tasks is fairly straightforward, determining the amount of the contaminant deposited on the skin can be challenging. Furthermore, the amount of agent retained on or absorbed into the skin may vary over several orders of magnitude, and is dependent on the specific activities performed by the worker.[(42,43)] Several studies have demonstrated that the hands generally receive the highest exposure, but other parts of the body may also receive significant exposures.[(44)] Cotton, frequently used as a surrogate for skin to assess substance deposition generally results in overestimates. Table 14.3 lists a number of generic

Table 14.3 — Amount of Contaminant Retained on the Skin per Event

Activity	Description	Quantity Transferred to the Skin per Event (mg/cm²/event)[B]	Reference
Manual weighing and handling of powder	Scooping, pouring and mixing dry powder, flakes and granules into liquid; handling of bags	0.03–2.1	42, 43
Handling wet or dried material in a filtration and drying process (powder, slurry)	Removing filter cake; loading wet or dry product onto filter tray; unloading dry material from filter press	0.005–0.63	45
Handling of liquids (solutions, suspensions, pure solvents) without immersion of hands	Opening containers; pouring into mixer; spill cleanup	<0.3–2.1	43
Immersion of hand into a liquid	Immersion of the hand into a liquid followed by partial wipe of the skin with a clean cloth	1.3–10.3	46

[A]These data were obtained under a variety of methods and conditions (see text). For details the reader should consult the original references.

values for substance retention on skin. The data generally represent exposure received by workers wearing no personal protective equipment, and many of the samples were collected using surrogate skin (e.g., gloves). The values were collected using a variety of methods, and many were obtained from single exposure events. Consequently they are fairly uncertain. The relationships to total dermal dose when multiple events are performed per day is also unknown.

Because of the limited amount of data and information available with which to estimate workplace-related dermal exposure, assumptions are generally made when developing estimates. Some common assumptions are that (1) the amount of the agent measured on the absorbent pad is uniformly deposited within a specific area of the body; and (2) the retention rate on absorbent pads or surrogate skin is equivalent to that of the skin. The exposure assessment should also consider the impact of the varying measurement methodologies used by investigators and extrapolation from pesticides mixing and loading data (in agricultural pesticide application operations) to various industrial mixing and loading operations.(44) Nonetheless, the data can be useful in developing screening level estimates of the potential for dermal exposure during some activities. An EPA evaluation of the available literature and field data determined that the screening level estimates developed using deposition modeling of dermal exposure for occupational exposures appear to be reasonable, but may overestimate exposures for some activities.(44)

Practical Means of Dermal Exposure Monitoring for Exposure Assessment

Much of the background of dermal exposure monitoring has been the use of air samples as a surrogate measure of exposure. Biological monitoring results minus air sampling results (both extrapolated back to exposure) then provide only a crude measure of the actual dermal contribution to systemic toxicity. In the absence of skin-specific measurements, air sampling often is complemented by the denotation of skin exposure, often using the TLV skin notation. Other studies may simply focus on the qualitative likelihood of dermal contact. In this circumstance yes/no decision making is acceptable.

Skin Pads and Dosimeters

Direct methods of dermal exposure measurement entail the collection and direct analysis of materials deposited on the skin or clothing of a worker. This is a representation of exposure in the same manner that air sampling represents pulmonary exposure.

While washed exposure pads have been used previously many times, more recently specially washed clothing is exposed and then "laboratory washed," e.g., prepared for analysis. The washed clothing approach in conjunction with videotape methods represents a powerful technique that can predict significance of dermal exposure. Yet such methods suffer from the perception of skin as an anatomical surface area rather than a dynamic organ that is perfused with blood and capable of metabolism. Skin pads or dosimeters are typically used to estimate dermal exposure during insecticide formulation, crop spraying and harvesting, or the aerial application of agricultural chemicals such as herbicides. The Hawthorne effect (outcome improvement produced by studying the situation before implementation of corrective measures) should be recognized when agricultural workers or others are surveyed.

Skin Swabs and Liquid Rinses

Skin swabs and liquid rinses are used to remove and later quantify the types of contaminants on workers' skin. Gauze pads, usually alcohol soaked, can be used to wipe workers' hands. Because contaminants may collect near nail beds, special rinses are applied there to remove contamination. The guiding principle is to use appropriate solvent rinses without creating a dermatitis.

Biological Monitoring Methods

Biological monitoring measures of exposure have been used to integrate the dermal route with other routes of exposure. The summary value of any biological marker can represent exposure from all routes, respiratory, percutaneous, and oral.

Wipe Sampling of Working Surfaces

The use of wipe sampling to assess surface contamination is common practice. Much of its precedent derives from the field of health physics (e.g., leak testing of sealed isotopic sources). While surface wipe sampling may be a routine practice, the methods themselves are not uniform. Although wipe sampling is not routinely used to quantitatively assess occupational and environmental exposures, several studies have been undertaken in which it was an integral part.

Wipe sampling has been used to identify sources of worker exposure to PCBs and other materials following fires involving electrical transformers and capacitors. Compounds that have been collected by wipe sampling of work areas, materials, and tool surfaces include lead and its derivatives, chlorophenols, phenolic antioxidants, 4,4'-methylenedianiline, and many others.

Government agencies such as EPA[19] have published methods for surface or wipe sampling, recommending specific areas to be wiped. Typically, two types of collection media are recommended: glass fiber filters for compounds to be analyzed by high performance liquid chromatography or gas chromatography, and filter paper to collect metals analyzed using absorption or emission spectroscopy. Samples are collected by dry wiping or wetting the collection media. Clearly, surface wiping indicates a potential for exposure but does not necessarily correlate with dose. Careful correlation studies employing biological monitoring as well as surface wiping thus help to denote dose.

Medical Evaluation of the Skin

Despite the absolute desirability of local occupational hygiene support, hazard recognition will likely be made by others, reinforcing the necessity of training. Later in the process the physician may solicit more detailed information from health and safety personnel. Experience indicates that prompt occupational hygiene and clinical consultations must be sought. No situations currently exist to support totally independent or noncollaborative work.

A thorough physical examination of the skin extends from head to toe, including nails and the mucous membranes. Corresponding occupational exposure assessments must parallel such thoroughness. Occupational hygienists should be aware that physicians commonly find occupationally unrelated but significant skin lesions such as basal cell carcinomas or even melanomas of which the patient is totally unaware. The general assessment of the entire skin allows the examiner to determine a pattern of skin problems before focusing on individual lesions. Distribution of a skin problem may follow neural (dermatome) or vascular patterns.

Aging, pigmentation, trauma, nutrition, and hygiene leave critical signs,[48] while color changes can also be related to underlying systemic conditions (e.g., jaundice: hepatobiliary conditions; cyanosis: cardiopulmonary diseases; diffuse hyperpigmentation: Addison's disease; and paleness: anemia). Even though occupational dermatoses are emphasized in training programs, junior occupational hygienists should not become discouraged when findings represent multifactorial problems. Because there are hundreds of dermatoses, a logical process of elimination is required to narrow the possibilities, first to specific groups of diseases and eventually to the exact few. Differential diagnosis may require a biopsy and specialized histopathologic testing. Occupational hygienists should be aware that repeated visits to dermatologic specialists may be required.

Physicians assess where the patient's skin condition first appeared; what it looks like and what symptoms were initially associated with it; how the skin disease may have changed; and what is being done to treat the condition. The careful review of chemical exposures includes occupational chemicals as well as systemic medications, both prescription and over-the-counter. The relationship of the time onset of a skin condition such as a rash to the use of drugs or chemicals is critical.

The history of atopic (allergic) diseases or skin cancer and a careful family history of skin problems may help to alert the physician to genetic (or familial) aspects of the dermatoses. When successive generations of family members work in the same plants, threading through the differential area between family and occupational origins is tedious. Reports of a similar disease or disorder in other exposed workers often helps to denote occupational origins of disease. A distinction is made between clinical impressions and confirmatory laboratory testing and referral. This has proven difficult for quantitative scientists to understand. Some elements of skin disease seem remote in terms of occupational origin, but even psychological stress has been known to exacerbate dermatoses (acne, psoriasis, seborrhea, and atopic eczema).

Contact Dermatitis

While contact dermatitis is the most frequent toxic skin reaction, it is generally confined to the areas actually touched. Clinical signs in affected human subjects may include one or more of erythema (reddening), edema, vesiculation, scaling, and thickening, and may be accompanied by an itch or a burning sensation. When contact dermatitis is suspected, work and hobby history are the keys to identifying pertinent exposures to allergens or irritants. Outdoor workers in particular may experience sun, cold, and heat that can provoke skin reactions: seasonality should always be noted as well as other unusual patterns of occurrence (see Figure 14.3).

Irritant Reactions

Cutaneous irritants do not rely on an immunologic mechanism to produce a local response. Severity depends on (1) the potency of the irritant; (2) the circumstances of contact; and (3) the skin site affected. Irritant response and dermal absorption are somehow related: Factors enhancing one parameter will likely enhance the other, this occurring through largely unknown mechanisms. Since irritation depends on damage to the lower, living layers of the epidermis, those factors that enhance penetration will generally increase the severity of the irritant response. The distribution of the dermatitis may be modified in that sites with a stronger barrier function (e.g., palms and soles) may show little or no dermatitis while other areas are more markedly affected (see Figure 14.4).

Allergic Contact Reactions

Allergic contact dermatitis (or delayed hypersensitivity) occurs as a result of allergy to one or more specific substances (antigens) through cell-mediated immunity. In

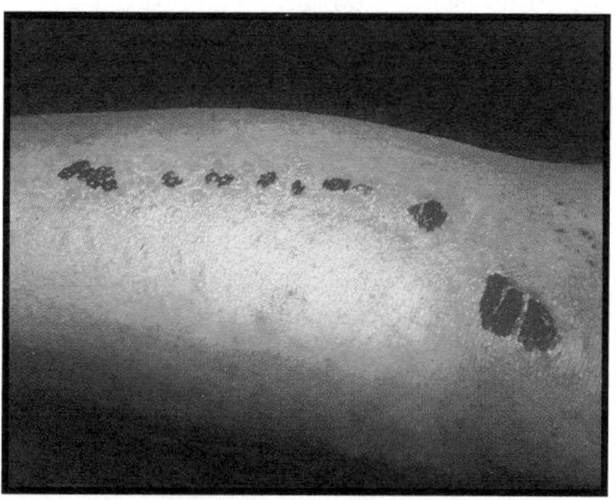

Figure 14.3 — Contact dermatitis. (All rights reserved. Reprinted with permission from the American Academy of Dermatology.)

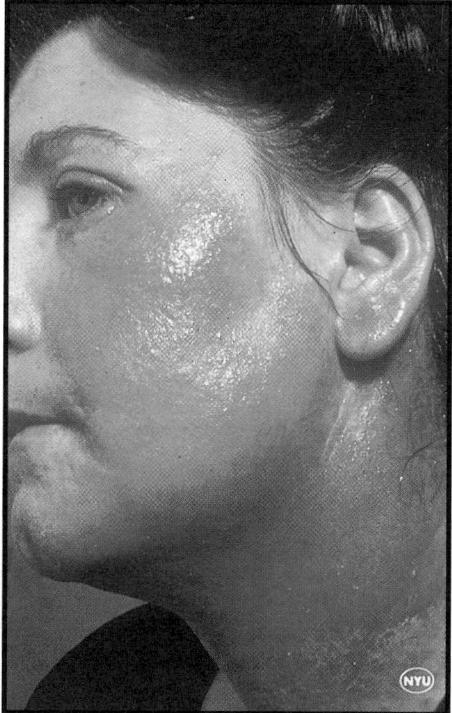

Figure 14.4 — Irritant reactions. (All rights reserved. Reprinted with permission from the American Academy of Dermatology.)

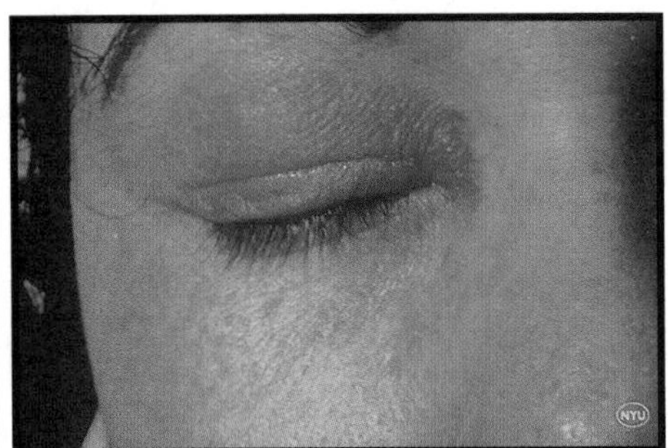

Figure 14.5 — Allergic contact reactions. (All rights reserved. Reprinted with permission from the American Academy of Dermatology.)

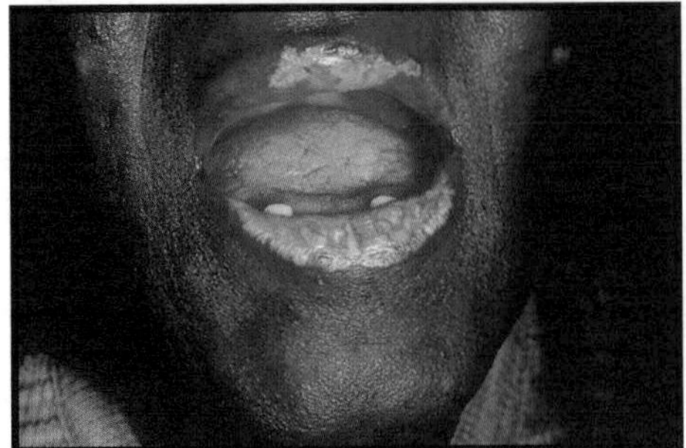

Figure 14.7 — Urticarial reactions. (All rights reserved. Reprinted with permission from the American Academy of Dermatology.)

highly sensitized persons such reactions can be provoked by vanishingly small amounts of the allergen (see Figure 14.5).

There are two main phases in cell-mediated immunity, the sensitization phase (in which the person becomes allergic to the antigen) and an elicitation phase. Sensitization usually takes at least 10 days. When sensitization has been achieved and the individual is then re-exposed, a reaction is obvious after a characteristic delay of about 12–48 hours: hence the name "delayed." While there is some correspondence between skin and respiratory sensitization, it is not exact, and the dermal mode is much more common. This issue remains a dilemma for occupational standard-setters who must deal with both respiratory and dermal routes of exposure.

Photosensitization

Photosensitization is considered to be an abnormal adverse reaction to ultraviolet and/or visible radiation (see Figure 14.6). It may be produced by a number of substances, each having its own action spectrum. Although precise rules do not apply, the action spectrum usually approximates the chemical's absorption spectrum. Most photosensitization reactions appear to yield an action spectrum strongest in the UV-A range (315–400 nm).

Urticarial Reactions

Urticarial reactions (wheal and flare responses) due to chemical contact may be produced as direct reactions or as the result of an immediate hypersensitivity. A number of substances such as histamine directly release vasoactive agents. They include the biogenic polymers released from some plant species and animals such as caterpillars and jellyfish. Other substances produce urticaria as a result of allergic sensitization with the production of immunoglobulin E. Such reactions usually occur within 30 to 60 minutes of contact with the offending agent (see Figure 14.7).

Acne

Greases, oils, coal-tar pitch, and creosote exacerbate acne vulgaris (the usual form of juvenile acne), and may produce the typical lesions of acne, namely comedones (blackheads) and inflammatory folliculitis.[47] Chloracne, on the other hand, is often associated with organochlorine compounds such as the dioxins and various other chlorinated aromatic compounds. It is important not only because of its refractory nature as a skin condition but also because it may signal systemic exposure (see Figure 14.8).[21,22]

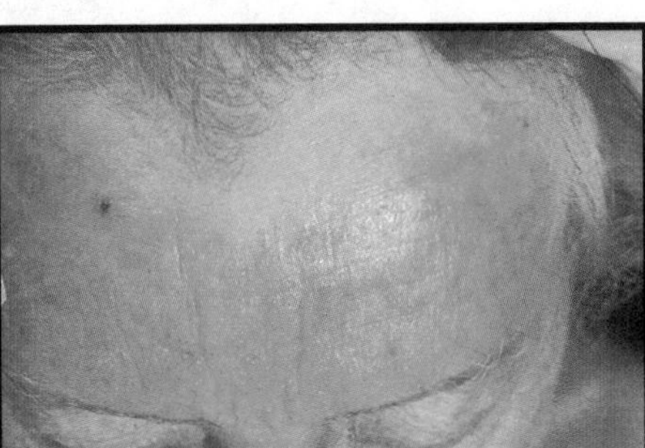

Figure 14.6 — Photosensitization. (All rights reserved. Reprinted with permission from the American Academy of Dermatology.)

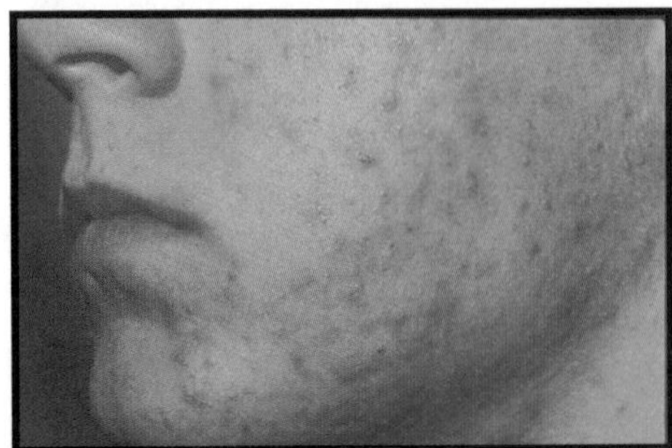

Figure 14.8 — Acne vulgaris. (All rights reserved. Reprinted with permission from the American Academy of Dermatology.)

Management and Control of Occupational Dermal Exposures

Strategies for the control of the occupational dermatoses are quite different from those for classic occupational diseases such as hemangiosarcoma (from vinyl chloride monomer). Whereas the incidence rate for a rare occupational disease may be less than 1 in 10,000, the attack rate for contact dermatitis may be 10 in 100 for a manufacturer. Although one argument holds that the health consequences from industrial dermatitis are slight, the opportunity for effective primary prevention is unparalleled within the occupational health discipline. Additionally, problems in the industrial dermatitis arena can provide a sentinel for noncompliance with respected occupational hygiene practice.

As with any other sampling data or disease reporting system, dermal data must be carefully evaluated before recommendations for protection can be made. The opportunity for effective intervention is balanced against incomplete data. Skin exposure monitoring, the question of dermally absorbed dose, and the dermatoses themselves are complex interrelated problems requiring integration of study disciplines and creativity in problem-solving approaches.

The OSHA personal protective equipment standard requires a written certification of the hazard assessment, requirements for selection of personal protective equipment, and training for employees using eye and face, head, foot, and hand protection. If skin contact is determined to be a critical exposure pathway, it may be helpful to implement procedures to help in evaluating the effectiveness of controls and chemical protective clothing in mitigating exposure. For example, establishing and marking regulated areas, tools, carts, and other equipment will aid in determining whether controls are effective. In addition, skin and surface wipe samples with colorimetric indicators can help in visually determining whether contamination is present. This can indicate the need for reinforcing personal protective equipment procedures, surface decontamination procedures, and other administrative procedures. Frequent surveys to evaluate and improve procedures may be helpful in reinforcing established procedures.

Summary

Direct measurement, biological monitoring, and surface sampling each provide useful insights that exposure assessment scientists such as occupational hygienists can use in assessing a worker's dermal exposure. The greatest challenge is in coupling an understanding of the destructive process of industrial dermatitis with a knowledge of the mechanism of surface contact and ensuing dermal absorption. Mechanisms at each level are only partially understood, but this should not become an impediment to further investigation. Whereas the lung and respiratory tract have been modeled for 30 to 40 years, the skin and its penetration characteristics did not receive much attention until the early 1980s. Exciting new developments will enable exposure assessment scientists to further recognize its potential.

References

1. **Rycroft, R.J.:** Clinical assessment in the workplace: dermatitis. *Occup. Med. 46*:364–366 (1996).
2. **James, R.C., H. Busch, C.H. Tamburro, S.M. Roberts, et al.:** Polychlorinated biphenyl exposure and human disease. *J. Occup. Med. 35*:136–148 (1993).
3. **Ness, S.A.:** *Surface and Dermal Monitoring for Toxic Exposure.* New York: Van Nostrand Reinhold, 1994.
4. **American Conference of Governmental Industrial Hygienists (ACGIH):** *1996 Threshold Limit Values for Chemical Substances and Physical Agents.* Cincinnati, OH: ACGIH, 1996.
5. **Berti, J.J. and J.J. Lipsky:** Transcutaneous drug delivery: a practical review. *Mayo Clin. Proc. 70*:581–586 (1995).
6. **Franklin, C.A., D.A. Somers, and I. Chu:** Use of percutaneous absorption data in risk assessment. *J. Am. Coll. Toxicol. 8*:815–827 (1989).
7. **Wester, R.C. and H.I. Maibach:** Percutaneous absorption of drugs. *Clin. Pharmacokinet. 23*:253–266 (1992).
8. **U.S. Environmental Protection Agency (EPA):** *Dermal Exposure Assessment: Principles and Applications* (preliminary rep. EPA/600/8-91/011B). Washington, DC: EPA, 1992.
9. **Johnson, M.L. and J. Roberts:** Prevalence of dermatologic disease in the United States: a review of the National Health and Nutrition Examination Survey, 1971-1974. *Am. J. Ind. Med. 8*:451–460 (1985).
10. **Percival, L., S.B. Tucker, S.H. Lamm, M.M. Key, et al.:** A case study of dermatitis, based on a collaborative

approach between occupational physicians and industrial hygiene. *Am. Ind. Hyg. Assoc. J. 56*:184–189 (1995).

11. **Hogan, D.J.:** *Skin Lesions and Environmental Exposures* (ATSDR Case Studies in Environmental Medicine). Agency for Toxic Substances and Disease Registry, 1993.
12. **Hostynek, J.J., R.S. Hinz, C.R. Lorence, M. Price, et al.:** Metals and the skin. *Crit. Rev. Toxicol. 23*:171–235 (1993).
13. **Beasley, K.L. and J.W. Burnett:** Common dermatologic manifestations of cutaneous exposure to petroleum and its derivatives. *Cutis 58*:59–62 (1996).
14. **Soni, B.P. and E.F. Sherertz:** Contact dermatitis in the textile industry: a review of 72 patients. *Am. J. Contact Dermatitis 7*:226–230 (1996).
15. **Adams, R.M.:** *Occupational Skin Diseases*, 2nd Ed. Philadelphia, PA: Saunders, 1990.
16. **Anonymous:** An avoidable tragedy. *Occup. Hazards 59*:32 (1997).
17. **American Conference of Governmental Industrial Hygienists (ACGIH):** 1995 *Threshold Limit Values and Biological Exposure Indices Documentation*. Cincinnati, OH: ACGIH, 1995.
18. **American Conference of Governmental Industrial Hygienists (ACGIH):** *Topics in Biological Monitoring*. Cincinnati, OH: ACGIH, 1996.
19. **U.S. Environmental Protection Agency (EPA).** *Pesticide Assessment Guidelines, Subdivision U, Application Exposure Monitoring* (EPA 540/9-87/127). Washington, DC: Office of Pesticide Programs, EPA, 1987.
20. **Stoughton, R.B.:** Percutaneous absorption of drugs. *Ann. Rev. Pharmacol. Toxicol. 29*:55–69 (1989).
21. **Coenraads, P.J., A. Brouwer, K. Olie, and N. Tang:** Chloracne: some recent issues. *Dermatol. Clin. 12*:569–576 (1994).
22. **Zugerman, C.:** Chloracne: clinical manifestations and etiology. *Dermatol. Clin. 8*:209–213 (1990).
23. **Dunagin, W.G.:** Cutaneous signs of systemic toxicity due to dioxins and related chemicals. *J. Am. Acad. Dermatol. 10*:688–700 (1984).
24. **Weber, L.W.D., A. Zesch, and K. Rozman:** Penetration, distribution and kinetics of 2,3,7,8-tetrachlorodibenzo-p-dioxin in human skin *in vitro*. *Arch. Toxicol. 65*:421–428 (1991).
25. **Emmett, E.A.:** Occupational Skin Diseases. In *Occupational Toxicology*, N.H. Stacey, ed. London: Taylor & Francis, 1993.
26. **Feldman, R.J. and H.I. Maibach:** Regional variation in percutaneous penetration of ^{14}C- cortisone in man. *J. Invest. Dermatol. 48*:181–183 (1967).
27. **Scheuplein, R.J. and I.H. Blank:** Permeability of the skin. *Physiol. Rev. 51*:702–747 (1971).
28. **Marzulli, F.N. and H.I. Maibach:** *Dermatotoxicology*. Washington, DC: Taylor & Francis, 1996.
29. **Rice, R.H. and D.E. Cohen:** Toxic responses of the skin. In *Casarett & Doull's Toxicology. The Basic Science of Poisons*, 5th ed. C.D. Klaassen, M.O. Amdur, and J. Doull, eds. New York: McGraw-Hill, 1996. pp. 529–546.
30. **Barry, B.W.:** Penetration enhancers: mode of action in human skin. In *Skin Pharmacokinetics*, Vol. I. B. Shroot and H. Schaefer, eds. Basel, Switzerland: S. Karger, 1987. pp. 121–137.
31. **Elias, P.M., K.R. Feingold, G.K. Menon, S. Grayson, et al.:** The stratum corneum two-compartment model and its functional implications. In *Skin Pharmacokinetics*, vol. I. B. Shroot and H. Schaefer, eds. Basel, Switzerland: S. Karger, 1987. pp. 1–9.
32. **Rougier, A., C. Lotte, and D. Dupuis:** An original predictive method for in vivo percutaneous absorption studies. *J. Soc. Cosmet. Chem. 38*:397–417 (1987).
33. **Fiserova-Bergerova, V.:** Application of toxicokinetic models to establish biological exposure indicators. *Ann. Occup. Hyg. 34*:639–651 (1990).
34. **Shatkin, J.A. and H. Brown:** Pharmacokinetics of the dermal route of exposure to volatile organic chemicals in water: a computer simulation model. *Environ. Res. 56*:90–108 (1991).
35. **Fiserova-Bergerova, V., J.T. Pierce, and P.O. Droz:** Potential of industrial chemicals: criteria for a skin notation. *Am. J. Ind. Med. 17*:617–635 (1990).
36. "Personal Protective Equipment," *Code of Federal Regulations* Title 29, Part 1910.138. 1994.
37. **Hogan, D.J.:** Review of dermatitis for non-dermatologists. *J. Fla. Med. Assoc. 77*:663–666, 1990.
38. **Leung, H.-W. and D.J. Paustenbach:** Techniques for estimating the percutaneous absorption of chemicals due to occupational and environmental

exposure. *Appl. Occup. Environ. Hyg. 9*:187 (1994).

39. **Emmett, E.A.:** Dermatological screening. *J. Occup. Med. 28*:1045–1050 (1986).
40. **U.S. Environmental Protection Agency (EPA):** *Exposure Factors Handbook* (EPA 600/8- 89/043). Washington, DC: EPA, 1989.
41. **VanRooij, J.G.M., J.G.C. de Roos, M.M. Bodelier-Bade, and F.J. Jongeneelen:** Absorption of polycyclic aromatic hydrocarbons through the human skin: differences between anatomical sites and individuals. *J. Toxicol. Environ. Health 38*:355–368 (1993).
42. **Lansink, C.J.M., M.S.C. Beelen, J. Marquart, and J.J. van Hemmen:** *Skin Exposure to Calcium Carbonate in the Paint Industry. Preliminary Modeling of Skin Exposure Levels to powders Based on Field Data* (TNO rep. V 96.064). Rijswijk, The Netherlands: TNO Nutrition and Food Research Institute, 1996.
43. **Health and Welfare Canada, U.S. Environmental Protection Agency, and National Agricultural Chemicals Association:** *Pesticide Handler's Exposure Database*, v. 1.1 [Computer database]. 1992.
44. **U.S. Environmental Protection Agency (EPA):** *Occupational Dermal Exposure Assessment—A Review of Methodologies and Field Data* (Final rep., contract 68-D2-0157, WA No. 2-50). Washington, DC: EPA, 1996.
45. **U.S. Environmental Protection Agency (EPA):** *Exposure and Release Estimations for Filter Press and Tray Dryer Operations Based on Pilot Plant Data.* (EPA 600/R-92/039). Washington, DC: EPA, 1992.
46. **Versar, Inc.:** *Exposure Assessment for Retention of Chemical Liquids on Hands* (Contract 68-01-6271). U.S. Environmental Protection Agency, Exposure Evaluation Division, Washington, D.C.
47. **Harvell, J.D. and H.I. Maibach:** Percutaneous absorption and inflammation in aged skin. *J. Am. Acad. Dermatol. 31*:1015–1021 (1994).
48. **Ancona, A.A.:** Occupational acne. *Occup. Med. 1*:229–243 (1986).

Outcome Competencies

After completing this chapter, the user should be able to:

1. Define underlined terms used in this chapter.
2. Describe the advantages of a comprehensive exposure assessment strategy.
3. Discuss the goals of comprehensive exposure assessment.
4. Outline the major steps in a comprehensive exposure assessment strategy.

Key Terms

authoritative LTA-OELs • basic characterization • compliance strategy • comprehensive exposure assessment• comprehensive strategy • exposure profile • exposure rating • internal OELs • long-term average–OELs (LTA–OEL) • occupational exposure limit (OEL) • regulatory OELs • similar exposure groups (SEGs) • working OELs

Prerequisite Knowledge

Prior to beginning this chapter, the user should review the following chapters:

Chapter Number	Chapter Topic
2	Occupational Exposure Limits
4	Environmental and Occupational Toxicology
6	Principles of Evaluating Worker Exposure
10	Sampling of Gases and Vapors
11	Analysis of Gases and Vapors
12	Sampling and Sizing Particles
16	Modeling Inhalation Exposure
17	Risk Assessment in the Workplace

Key Topics

I. Introduction
   A. Growing Variety of Present and Future Risks
   B. Efficient and Effective Programs

II. Shifting State-of-the-Art: Compliance Monitoring to Comprehensive Exposure Assessment

III. Overview of Exposure Assessment Strategy
   A. Start — Establish the Exposure Assessment Strategy
   B. Basic Characterization
   C. Exposure Assessment
   D. Further Information Gathering
   E. Health Hazard Controls
   F. Reassessment
   G. Communications and Documentation

15

Comprehensive Exposure Assessment

John R. Mulhausen, Ph.D., CIH

Joseph Damiano, CIH, CSP

This chapter contains highlights from the AIHA publication, **A Strategy for Assessing and Managing Occupational Exposures,** *2nd edition. Readers interested in a more comprehensive treatment on this topic are encouraged to refer to this publication.*

Introduction

Occupational hygienists must anticipate, recognize, evaluate, and control health hazards and related risks in the workplace. Central to this effort is the assessment of occupational exposures. Comprehensive exposure assessment is the systematic review of the processes, practices, materials, and division of labor present in a workplace that is used to define and judge all exposures for all workers on all days. Such a rational and methodical approach to exposure recognition and evaluation helps ensure that all health risks are managed, that all related risks are considered, that all organization stakeholders are involved, and that occupational hygiene programs and resources are focused on the most important risks. It also positions the occupational hygiene program to better anticipate and manage future risks.

Growing Variety of Present and Future Risks

Modern workplaces are becoming more complex. The variety of risks associated with workplace exposure to chemical, physical, and biological agents is increasing. While the first priority of the occupational hygienist is to protect the health of workers, health risk is not the only risk he or she is asked to manage. Other risks include the risks posed by noncompliance with regulations, legal risks such as those associated with potential lawsuits or third-party liability, and risks related to the anxiety inherently associated with many people's response to potential exposures.

Occupational hygienists must consider the fact that organizations today are accountable to many more —and more varied—stakeholders than organizations in the past. These new stakeholders include employees, owners, customers, labor unions, regulators, stockholders, the press, and the communities in which the organization operates. Organizations rely on occupational hygienists to satisfy the concerns of these stakeholders on matters related to workplace exposures.

When evaluating risks to employees and the organization, occupational hygienists must also remember that today's programs will be held accountable not only for today's state-of-the-art, but tomorrow's as well. It is not sufficient to limit the question to "Are employee exposures below established exposure limits?" Instead, occupational hygienists must ensure that exposures are characterized well enough, and controlled well enough, to keep present risks within acceptable limits and to position the organization to manage future risks. Questions that must be considered include the following:

- How might this exposure affect the health of employees?
- How good is the exposure limit?
- What additional risks does this exposure present to the organization?

Compliance with current limits is not sufficient. The majority of chemicals do not have an occupational exposure limit (OEL), and the information used to set existing limits is often incomplete. Also, the limits that are set are not always designed to protect all workers or might be out of date.

New toxicological and epidemiological information is gathered each day. That means that new exposure limits will be generated for environmental agents that formerly had none, and that many of the limits currently in place will change. Experience has shown that most exposure limits are lowered when they are changed, and there is no reason to believe that trend will not continue.

Unfortunately, when new limits are set, or old limits are changed, there may be a population of workers who have been exposed to the chemical for some time at concentrations above the new limit. The occupational hygienist today must think about how he or she should position the current occupational hygiene program so it is best able to manage those changes and minimize the future risks they will pose. Having a historical database for all exposures will often allow identification of employees who were exposed above the lowered exposure limit and enable estimation of the extent of their past overexposure. This then allows the formation of a strategy for medical management of the health of those employees.

Efficient and Effective Programs

At the same time that occupational hygienists are being asked to manage a growing variety of risks, their programs are being more carefully scrutinized for efficacy, efficiency, and cost-effectiveness. Economic factors demand that each organizational unit demonstrate its worth and its ability to operate waste-free. Occupational hygiene programs are no exception. The ability to efficiently understand, prioritize, and manage exposures and risks requires a more systematic, better documented approach to occupational hygiene than has typically been practiced in the past.

Occupational Hygiene Program Management

The better the occupational hygienist understands exposures, the better he or she is able to direct and prioritize the occupational hygiene program. This is true whether the goal of the exposure assessment process is regulatory compliance, a comprehensive description of all exposures, or a diagnostic evaluation of health hazard controls. The system for exposure assessment must be integrated with other systems for defining, prioritizing, and managing worker health protection. Assessment results are used to determine the needs and priority for health hazard controls, build exposure histories, and demonstrate regulatory compliance.

Exposure assessment is at the heart of an occupational hygiene program as it supports all of its functional elements (see Figure 15.1). A well-rationalized program relies on a thorough understanding of what is known and not known about exposures. For example, to understand where best to spend precious resources on a monitoring program, the occupational hygienist must

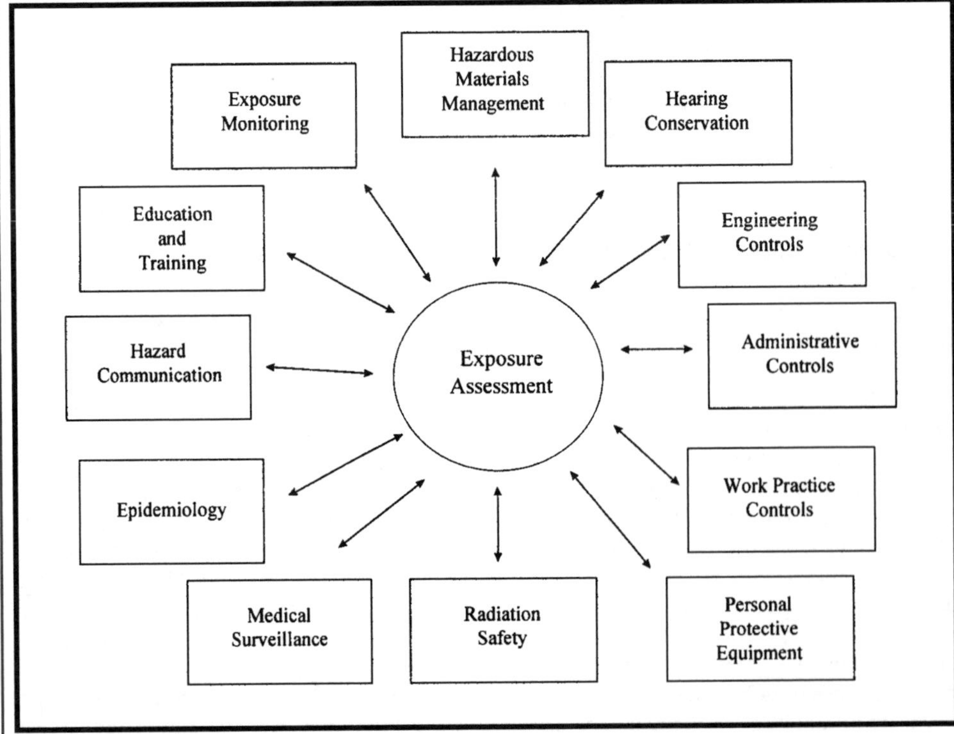

Figure 15.1 — Occupational hygiene program management.

have an understanding of potential exposures that need better characterization or careful routine tracking. A thorough characterization of exposures allows the occupational hygienist to focus worker training programs, better target medical surveillance programs, and define specific requirements for personal protective equipment (PPE).

Better Prioritization of Control Efforts and Expenditures

The better the understanding of exposures and the risks they pose, the more assurance there is that the most important (highest risk) exposures are being controlled first. Control efforts, whether engineering, work practice, or PPE programs, are usually costly to implement and maintain. Therefore, it is critical that those efforts be appropriately prioritized, deployed, and managed.

A thorough understanding of exposures allows prioritization of control efforts to use limited funds wisely. The right combination of control efforts—including short-term, long-term, temporary, and permanent controls—can be implemented based on the prioritized exposure assessments. Plans can be made for improving controls and moving from short-term solutions such as PPE to long-term solutions such as local exhaust ventilation. Management will be assured that money is being spent on the most needed controls first and not wasted on unnecessary control efforts.

Better Understanding of Worker Exposures

A full understanding of exposures, combined with work history, allows better characterization of individual worker exposures and better management of employee medical concerns. Exposure histories, along with health effects information, can indicate the risk that workers have of developing occupational illness and disease. An understanding of exposures allows medical practitioners to target clinical examinations, medical surveillance, or other diagnostic techniques better to detect health effects early. The management of issues related to public health in the community in which the organization operates may be enhanced if there is a well-developed understanding of occupational exposures. When combined with morbidity or mortality data, the comprehensive characterization of exposures greatly improves the power of epidemiologic studies and better positions health care providers to answer questions about an individual's exposures and how they may have affected his or her health.

Shifting State-of-the-Art: Compliance Monitoring to Comprehensive Exposure Assessment

In the past several years the characterization of exposures has received the attention of occupational hygiene professionals and regulatory agencies worldwide.[1-4] The state-of-the-art approach has shifted from compliance monitoring, which focuses on the maximum risk employee to determine whether exposures are above or below established limits, to comprehensive exposure assessment, which emphasizes the characterization of all exposures for all workers on all days.

Regulations in many countries now mandate some periodic review of exposures throughout an organization.[5-7] While current regulations are highly variable in scope and enforcement, the trend is clear, and the reasoning behind the trend indisputable: A comprehensive approach to assessing occupational exposures better positions an organization to understand the risks associated with the exposures and better positions the organization to manage those risks.

No longer is a compliance-based approach to occupational hygiene sufficient. If a broadened definition of risk is accepted, and it is agreed that the customers—workers and the organizations that employ occupational hygienists—are looking to occupational hygiene to help them manage those risks, one comes to the conclusion that the practice of occupational hygiene must

Exposure Assessment vs. Risk Assessment

For the occupational hygienist, exposure assessment and risk assessment are thoroughly mixed into each other; so much so, the reality is that they cannot be reasonably separated.[8-10] Consider the following relationship between health risk and exposure:

$$\text{health risk} = (\text{exposure})\ (\text{toxicity})$$

In the world of occupational hygiene the evaluation of exposure is fully half of the assessment of health risk. The other half is the evaluation of the health effects per unit exposure or the toxicity of the agent to which the worker is exposed. Thus, any exposure in an occupational hygiene sense is only meaningful in its relationship to the health effects that the exposure may cause.

The ultimate goal of occupational hygienists' activities is the reasonable assurance of worker health. In this regard, what one does about risk is called risk management. The control of health hazards can be considered a risk management function. Here again there is an interaction with risk assessment in that good risk management is almost always predicated on good risk assessment, which in turn is driven by the quality of exposure assessments.

progress to embrace a comprehensive and systematic approach to the evaluation of exposures and the risks they pose. Such an approach will include logical systems and strategies for evaluating all exposures, interpreting and assessing the many present and future risks those exposures might pose, and efficiently managing those exposures that present unacceptable risks.

Overview of Exposure Assessment Strategy

An overview of the comprehensive exposure assessment strategy is shown in Figure 15.2. The strategy is cyclic in nature and is most effectively used in an iterative manner that strives for continuous improvement. Early cycles will begin by collecting available information that is relatively easy to obtain. The results of initial exposure assessments based on that information will be used to prioritize follow-up control and information-gathering efforts. Resources should be focused on those exposures that have the highest priority based on the potential health risk they present. As those exposures are better understood and controlled, they will drop in priority and the next cycles through the strategy will focus on the next tier priority exposures.

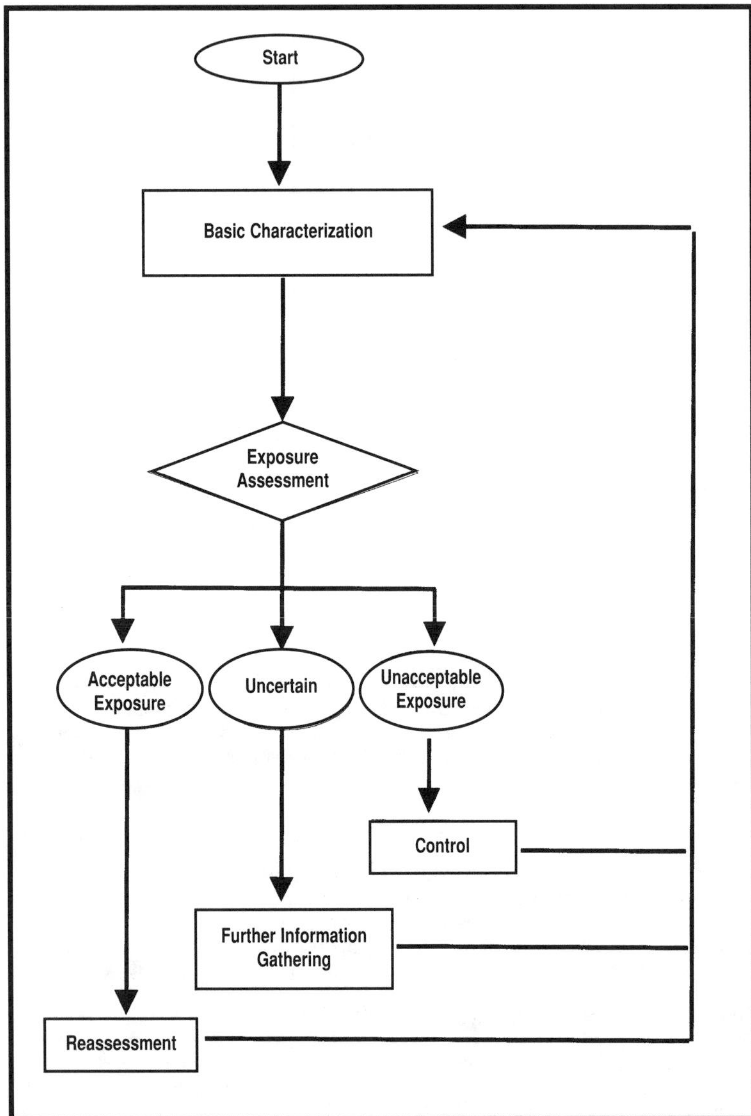

Figure 15.2 — A strategy for assessing and managing occupational exposures.

The major steps in the strategy follow.

1. Start—establish the exposure assessment strategy.
2. Basic characterization—gather information to characterize the workplace, work force, and environmental agents.
3. Exposure assessment—assess exposures in the workplace in view of the information available on the workplace, work force, and environmental agents. The assessment outcomes include groupings of workers having similar exposures, definition of an exposure profile for each group of similarly exposed workers, and judgment about the acceptability of each exposure profile.
4. Further information gathering—implement prioritized exposure monitoring or the collection of more information on health effects so that uncertain exposure judgments can be resolved with higher confidence.
5. Health hazard control—implement prioritized control strategies for unacceptable exposures.
6. Reassessment—periodically perform a comprehensive re-evaluation of exposures. Determine whether routine monitoring is required to verify that acceptable exposures remain so.
7. Communications and documentation—although there is no element in Figure 15.2 for communications and documentation, the communication of exposure assessment findings and the maintenance of exposure assessment data are underlying and essential features in each step of the process.

Start—Establish the Exposure Assessment Strategy

In establishing an organization's exposure assessment strategy, several issues should be carefully addressed, including the role of the occupational hygienist, the exposure assessment goals, and the written exposure assessment program.

Role of the Occupational Hygienist

Exposure assessments should be done by or under the direction of an occupational hygienist. Occupational hygienists have training and experience that makes them uniquely qualified to make judgments about exposure profiles and their acceptability. They are best able to define information gathering needs and strategies. They have an understanding of the control options that would be most effective for a particular situation. They are best equipped to make program modifications and prioritizations that take advantage of the results of the exposure assessment.

The participation of other technically knowledgeable professionals such as engineers, environmental scientists, toxicologists, safety professionals, physicians, nurses, and epidemiologists will facilitate exposure assessment programs and improve the quality of assessments. The interaction of occupational hygienists with colleagues in the occupational health professions will enhance worker protection and help ensure the effective implementation of the exposure assessment and management strategy.

Exposure Assessment Goals

Each organization must define goals for its own exposure assessment program. The goals may include (1) the identification and characterization of actual or potential health hazards; and (2) the development and maintenance of an occupational exposure database. The goals should be clearly articulated and lead to one of two general exposure assessment strategies: compliance strategy or comprehensive strategy.

The compliance strategy usually uses worst-case monitoring with a focus on exposures during the time of the survey. An attempt is made to identify the maximum-exposed workers in a group. One or a few measurements are then taken and simply compared with the occupational exposure limit. If the exposures of the maximum-exposed workers are sufficiently below the OEL, then the situation is acceptable. This strategy provides little insight into the day-to-day variation in exposures levels and is not amenable to the development of exposure histories that accurately reflect exposures and health risk. However, in many organizations with more limited funding, the compliance strategy may be an appropriate first step.

The comprehensive strategy is directed at characterizing and assessing exposure profiles (exposure average and variability) that cover all workers, workdays, and environmental agents. These exposure profiles are used to picture exposures on unmeasured days and for unmeasured workers in the similarly exposed group. In addition to ensuring compliance with OELs, this strategy provides an understanding of the day-to-day distribution of exposures. Exposure assessment findings can be used to address present-day health risks and construct exposure histories. If a historical database is maintained, the exposure assessment data may be used to address future health issues for individual workers and/or groups of workers. In the latter case, the data may be used to support epidemiological studies.

The goals of a system for comprehensive exposure assessment include the following:

- Characterize exposures to all potentially hazardous chemical, physical, and biological agents, including those without formal occupational exposure limits.
- Characterize the exposure intensity and temporal variability faced by all workers.
- Assess the potential risks (e.g., risk of potential harm to employee health, risk of noncompliance with governmental regulations, etc.)
- Prioritize and control exposures that present unacceptable risks.
- Identify exposures that need additional information gathering (e.g., baseline monitoring).
- Document exposures and control efforts, and communicate exposure assessment findings to all affected workers and others who are involved in worker health protection (e.g., management, labor representatives, medical staff, engineering staff, etc.).
- Maintain a historical record of exposures for all workers, so that future health issues can be addressed and managed in view of actual exposure information.
- Accomplish the above with efficient and effective allocation of time and resources.

Because a comprehensive approach to exposure assessment provides a more complete understanding of exposures than the compliance approach, it enables better management of occupational hygiene-related risks. It helps provide assurance to an organization's management, customers, employees, and the communities in which the organization operates that occupational health risks are understood and that the proper steps are being taken to manage the risks.

Written Exposure Assessment Program

A written exposure assessment program is an important reference tool for documenting how an organization will administer occupational exposure assessments. It specifies the strategies, methods, and criteria used in performing the assessments. The written program should address the following:

- Goals of the occupational exposure assessment program.
- Role and responsibilities of the occupational hygienist and other technical support staff.
- Methods for systematized information gathering to form a basic characterization of the workplace, work force, and environmental agents.
- Methods for defining groups of similarly exposed workers and the exposure profile for each group.
- Criteria for making a judgment as to whether the exposure profile for a group is acceptable, unacceptable, or uncertain. Decisions surrounding the selection and application of OELs are crucial.
- Systems for prioritizing and gathering the additional information needed to better characterize uncertain exposure assessments and make a more confident judgment about their acceptability, whether exposure monitoring data or health effects information.
- Exposure thresholds and criteria for conducting quantitative exposure assessments (e.g., baseline monitoring if the initial exposure estimate is greater than 10% of the OEL).
- A system for ensuring that unacceptable exposures are prioritized and controlled.
- Systems for communicating and documenting exposure assessment findings and health hazard control recommendations.
- Systems and criteria for periodic reassessment of workplace exposures, including routine monitoring programs to ensure that acceptable exposures remain acceptable.

Basic Characterization

At the start of the exposure assessment process is the collection and organization of basic information needed to characterize the workplace, work force, and environmental agents. Information is gathered that will be used to understand the tasks being performed, the materials being used, the processes being run, and the controls in place so that a picture of exposure conditions can be made. At a minimum, the information gathered by the occupational hygienist should include an understanding of the operations, processes, and facilities, including:

1. Work force, tasks, and division of labor;
2. Potentially hazardous chemical, physical, and biological agents in the workplace;
3. How and when workers are exposed to the hazardous environmental agents;
4. Exposure controls present in the workplace, including engineering, administrative, and work practice controls and personal protective equipment;
5. Quantities of environmental agents;
6. Chemical and physical properties of the environmental agents; and
7. Potential health effects of the environmental agents, the mechanism of toxicity, and the OELs associated with each agent.

Health effect information about the environmental agents must be gathered, including OELs. Exposure assessments cannot be resolved without an OEL unless there is no exposure. These may be formal OELs such as regulatory OELs (set and enforced by governmental agencies), authoritative OELs (set and recommended by credible organizations, such as the American Conference of Governmental Industrial Hygienists or the American Industrial Hygiene Association), or internal OELs (formally set by an organization for its private use). Or they may be more informal working OELs that have been set by the occupational hygienist based on whatever information might be available to differentiate acceptable from unacceptable exposures. Working OELs are sometimes stated in ranges (e.g., 0.1–1.0 mg/m^3) or incorporate large safety factors to account for uncertainty.

In assessing health effects data and reviewing existing OELs, the occupational hygienist should attempt to answer the following questions:

- Why was the existing OEL set at its particular level?
- What safety factor was used in deriving the OEL?
- What potentially important health effects were not considered?
- How adequate are the health data that support the OEL?

- Should the OEL be adjusted in the presence of a nontraditional work schedule?
- Is there information available on the skin absorption rate? If so, are the data adequate? If not, can the skin absorption rate be predicted from surrogate data for a structurally similar chemical agent?
- Are concomitant exposures present in the workplace that pose additive or synergistic health risks?

The occupational hygienist must also consider the averaging time of the OEL. The averaging time refers to the time span for which an average exposure is estimated. The appropriate averaging time is set by the sponsor of the OEL and in principle can extend over any length of time from seconds or minutes, to a single shift, to multiple shifts, to months and years. To be in compliance with OELs, the occupational hygienist must follow the defined averaging time corresponding to the defined exposure level.

Choosing an appropriate averaging time for an OEL requires knowledge of the uptake, distribution, storage, elimination, and toxic action of the environmental agent. For substances that act quickly to elicit their toxic response, it is important to control the dose across a single shift (8-hour time-weighted average [TWA]) or shorter period of time (e.g., 15 minutes for short-term exposure limits [STELs] and instantaneous for ceiling limits).

If the toxic material acts slowly (through some combination of accumulation of the toxic agent or metabolites or accumulation of bodily damage) then longer averaging times may be appropriate. An exposure profile is deemed acceptable for such chronic agents if the true long-term average exposure is less than an applicable long-term average OEL (LTA–OEL). However, at this time few regulatory or authoritative LTA–OELs have been developed.

In the absence of a regulatory or authoritative LTA–OEL, suggestions for LTA–OELs for chronic disease agents have been proposed at one-third of the 8-hour TWA–OEL(1) and one-tenth to one-fourth of the 8-hour TWA–OEL.(11) The latter suggestion has been endorsed by past chairs of the American Conference of Governmental Industrial Hygienists (ACGIH).(12)

Exposure Assessment

The exposure assessment can be broken into several additional steps (see Figure 15.3), including establishing similar exposure groups (SEGs); defining the exposure profile; and comparing the exposure profile, with its uncertainty, to the OEL, with its uncertainty, to make a judgment that the exposure is acceptable, unacceptable, or uncertain.

Define SEGs

SEGs are groups of workers having the same general exposure profile for the agent(s) being studied because of the similarity and frequency of the tasks that they perform, the materials and processes with which they work, and the similarity of the way that they perform the tasks. SEGs are established using the information gathered during the basic characterization of the workplace, work force, and environmental agents. The occupational hygienist reviews this data and uses his or her training and experience to group employees believed to have similar exposures. SEGs are generally described by process, job, task, and environmental agent.

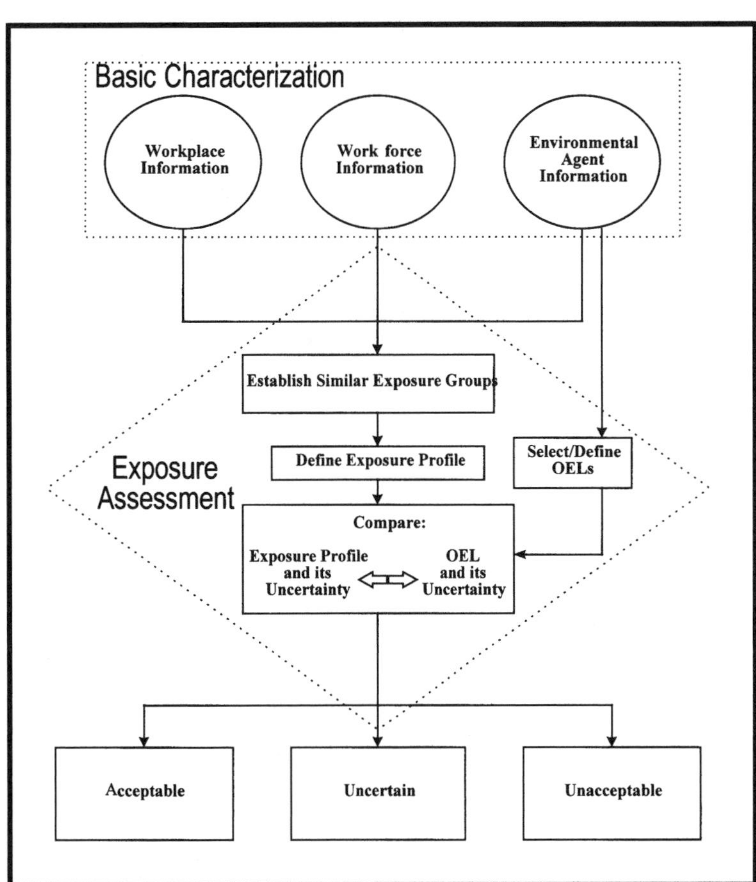

Figure 15.3 — Defining and judging exposure profiles.

Define Exposure Profiles

An exposure profile is an estimate of the exposure intensity and how it varies over time for workers in an SEG. Information used for defining the exposure profile may include qualitative and/or quantitative data. At the start of the exposure assessment process there may be very little quantitative data available, so most early exposure profiles will be based on qualitative information. As such, they may be accompanied by a great deal of uncertainty. As the information gathering and assessing cycle progresses, SEGs may be redefined and their exposure profiles modified and refined based on new information.

A useful tool for beginning to characterize the exposure profile is the exposure rating, particularly during the initial exposure assessments performed when monitoring data may be sparse. An exposure rating is an estimate of exposure level relative to the OEL. Tables 15.1 and 15.2 are examples of categorization schemes for rating exposures. Table 15.1 is based on an estimate of the arithmetic mean of the exposure profile relative to the LTA–OEL, and Table 15.2 is based on an estimate of the upper tail of the exposure profile (the 95th percentile) relative to the OEL. The occupational hygienist's choice of exposure rating schemes will depend on how OELs are defined by regulatory and authoritative standards-setting organizations and on how they are applied in the exposure assessment.

In performing the initial exposure rating, there is value in assuming the absence of personal protective equipment used to control exposures. This will include respirators, hearing protectors, and chemically protective gloves. This approach will allow the occupational hygienist to determine precisely where and to what degree workers depend on personal protective equipment to control a health hazard.

Exposure ratings can be made on the basis of monitoring data (personal monitoring data; screening measurements with easy-to-use instruments such as sound level meters, detector tubes, or other direct reading devices); surrogate data (exposure data from another agent or another operation); and modeling (predictive modeling based on chemical and physical properties or process information).

Table 15.1 — Exposure Rating Categorization: Estimate of the Arithmetic Mean of the Exposure Profile Relative to the LTA–OEL

4	> LTA–OEL
3	50–100% LTA–OEL
2	10–50% LTA–OEL
1	<10% LTA–OEL

Table 15.2 — Exposure Rating Categorization: Estimate of the Exposure Profile 95th Percentile Relative to the OEL

4	> 5% exceedance of the OEL (95th percentile > OEL)
3	> 5% exceedance of 0.5 × OEL (95th percentile between 0.5 × OEL and 1.0 × OEL)
2	> 5% exceedance of 0.1 × OEL (95th percentile between 0.1 × OEL and 0.5 × OEL)
1	Minimal to no exceedance of 0.1 × OEL (95th percentile < 0.1 × OEL)

Make Judgments About the Acceptability of the Exposure Profiles for Each SEG

Based on the SEG exposure profile and information collected about the toxicity of the agent, a judgment is made about the exposure. Specifically, the exposure profile (and the associated uncertainty in the exposure profile) is compared with the OEL (and the associated uncertainty in the OEL) and a judgment made regarding the acceptability of the risk posed by the exposure. Judgment outcomes are that the exposure is acceptable, unacceptable, or uncertain.

Conceptually, uncertainty bands can be placed around the OEL and the exposure profile. If the uncertainty bands do not overlap, the occupational hygienist should be able to resolve the exposure assessment regardless of the level of uncertainty associated with the OEL and the exposure profile.[9] On the other hand, if the uncertainty bands overlap, the occupational hygienist might not be able to judge whether the exposure is acceptable or unacceptable (see Figure 15.4). In that case he or she is forced to reduce the uncertainty associated with the OEL, the exposure profile, or both. The occupational hygienist must decide whether there is enough concern to classify the exposure as unacceptable and initiate a control program, or whether the existing exposure can continue while additional information is gathered.

The exposure judgment is used to prioritize control efforts or the collection of more information based on the environmental agent's estimated level of exposure, severity of potential health effects, and the uncertainty associated with the exposure profile and health effects information. In this system, unacceptable exposures are put on a prioritized list for control, uncertain exposures are put on a prioritized list

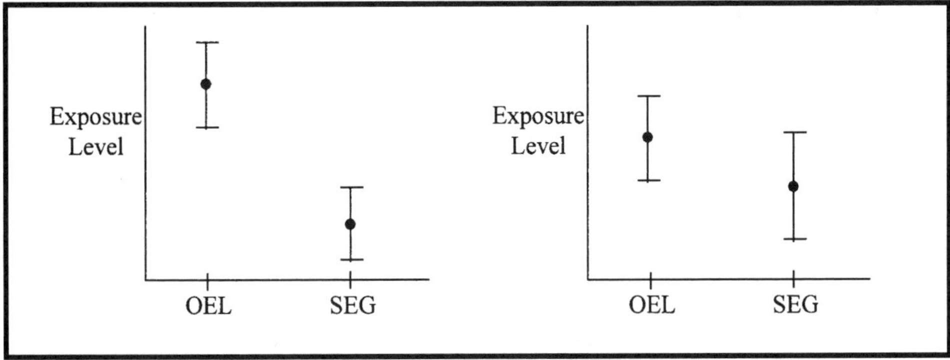

Figure 15.4 — Uncertainty in the exposure estimate and in the OEL.

for further information gathering, and acceptable exposures are documented as such and may be put on a list for periodic routine monitoring to verify that exposures continue to be acceptable.

Further Information Gathering

Exposure profiles that are not well understood, or for which acceptability judgments cannot be made with high confidence, must be further characterized by collecting additional information. Information-gathering efforts should be prioritized. Higher priority should be given to information needs associated with uncertain exposure assessments that involve high exposure rating estimates to highly toxic materials. In some cases, if exposure rating and toxicity estimates are high enough, consideration should be given to the use of personal protective equipment or other interim controls while the information is gathered or generated.

The type of information needed may vary from one SEG to another. The exposure profile for one SEG may be very well understood but there may be little toxicity information available. In that case it will be important to collect, or even generate, toxicological or epidemiological data. Another SEG may have little data or prior knowledge available on which to base an estimate of the exposure profile. In that case it will be important to generate information to better characterize the exposure profile, either through exposure monitoring, modeling, or biological monitoring.

Exposure Monitoring. If an exposure profile is not well characterized, there may be a need for personal monitoring of worker exposures. This may include noise monitoring, air monitoring, skin exposure monitoring, or other environmental measurements.

Exposure Modeling. As tools for using mathematical modeling techniques to predict exposures based on workplace and worker parameters become more sophisticated, they will be used more and more to estimate exposure profiles. Exposure modeling is frequently used to estimate the potential exposures associated with new processes and products. Such models have the advantage of being less expensive and time-consuming than the actual measurement of environmental agents in the workplace.

Biological Monitoring. Biological monitoring may be needed to assess the exposure profile if there are concerns about exposure through skin absorption or inadvertent ingestion. Due to the medical and ethical issues involved, the occupational hygienist should work closely with a physician whenever biological monitoring is considered.

Toxicological Data Generation. If the toxicity of the materials used in the workplace is not understood, then it is difficult to make a judgment about the acceptability of the exposure no matter how well that exposure profile is characterized. In those cases it may be necessary to obtain additional health effects information on the environmental agents of interest by eliciting the aid of toxicologists and other experts.

Epidemiological Data Generation. Epidemiology programs are useful for identifying new relationships between exposures and illness. They can help determine whether illness outbreaks are work-exposure related. They help assure management and workers that if illnesses or diseases are work-related they will be identified and dealt with appropriately. The results of epidemiological studies add to the available toxicological information for an environmental agent and enable better judgments about the acceptability or unacceptability of exposures. One of the biggest weaknesses in current epidemiology practice is the lack of useful exposure data.

Health Hazard Controls

SEG exposure profiles that are judged unacceptable should be put on a prioritized list for control. It is critical that occupational hygiene control programs be deployed and adjusted in view of exposure assessment findings. Control programs balance available resources with control needs to implement and maintain interim (e.g., respirator) or permanent (e.g., local exhaust ventilation) controls. Prioritization of resources for permanent controls can be performed based on the health risk posed by the exposure and the toxicity of the environmental agent. The prioritization factors may also include the uncertainty associated with the judgment, the number of workers exposed, and the frequency of exposure.

Exposure assessment findings can be used to prioritize diagnostic monitoring efforts as well. Diagnostic monitoring is performed to identify the sources of exposure and to understand how the sources, tasks, and other variables (e.g., production rates) contribute to worker exposures. The results help the occupational hygienist devise the most appropriate and efficient control strategies for unacceptable exposures and determine whether the new or modified controls are effective.

Reassessment

Exposures that are judged acceptable may need no further action, other than documentation, until the time comes for reassessment. Or there may be a need to collect further information, such as monitoring data, toxicology data, or epidemiological data to (1) validate the judgment of acceptability; and (2) ensure that the operation does not go out of control.

It is important that exposure profiles and SEGs be kept up-to-date as changes occur in the workplace that affect exposures. This will ensure that exposures continue to be well understood and that the organization's occupational hygiene program continues to respond to changing priorities.

The exposure assessment system should be linked to a management of change program that will help identify changes in the processes, materials, or work force that may significantly alter exposures or the allocation of employees to SEGs. These changes include (1) increased/decreased production rates; (2) increased/decreased production energy; (3) new or untrained workers introduced into the exposure group; (4) changed OELs; (5) new toxicity data; and (6) new or changed material (chemical or physical change).

There should also be a provision for required reassessment at some appropriate interval to account for the fact that many factors that influence changes in exposures are not readily foreseeable.

Communications and Documentation

The entire exposure assessment process, including all exposure assessment findings and follow-up, must be documented, whether or not monitoring data were collected. Moreover, the exposure assessment findings must be communicated in a timely and effective fashion to all workers in the SEG and others who are involved in worker health protection.

Lists of SEGs, their exposure profiles, and the judgments about their acceptability should be permanently stored so that individual exposure histories can be generated. Information on baseline and routine monitoring programs, as well as hazard control plans, must be kept along with evidence that the plans were acted on.

The data documentation efforts required by the comprehensive approach to exposure assessment in many organizations will be difficult to manage without the help of a computerized data management system. In planning an exposure assessment database, the occupational hygienist should carefully consider how exposure data will be used. Records should be established and maintained so that pertinent questions can be answered accurately and within a reasonable period of time. The occupational hygienist should recognize that other disciplines could have an interest in exposure data. This includes workers, management, and perhaps an organization's medical, engineering, and legal staff. Governmental agencies and industry associations may have an interest in occupational exposure data. Exposure records can also be a vitally important component of future epidemiological studies.[13,14]

Summary

Occupational hygienists throughout the world are recognizing that a systematic and comprehensive approach to exposure assessment is an effective mechanism for managing occupational hygiene programs. A thorough understanding of exposures allows the prioritization of the occupational hygiene program—including control efforts—to protect employees and manage exposure-related risks.

It also positions the occupational hygienist for better management of the unpredictable change that will occur both in knowledge of the health effects of environmental agents and in society's tolerance of workplace exposures. Coupled with good work-history information, comprehensive exposure assessments will enable better epidemiology and refinement of the understanding of the relationship between occupational exposures and disease.

The occupational hygienist is strongly encouraged to develop an exposure assessment program that is comprehensive in nature. Such a program will better position organizations to understand and manage ever-broadening occupational health-related risk.

References

1. **Hawkins, N.C., S.K. Norwood, and J.C. Rock:** *A Strategy for Occupational Exposure Assessment.* Akron, OH: American Industrial Hygiene Association, 1991.
2. **American Industrial Hygiene Association:** White paper: A Generic Exposure Assessment Standard. *Am. Ind. Hyg. Assoc. J. 55*:1009–1012.
3. **Organization Resources Counselors, Inc. (ORC):** "A Proposed Generic Workplace Exposure Assessment Standard." Washington, DC: ORC, 1992.
4. **Guest, I.G., J.W. Chessie, R.J. Gardner, and C.D. Money:** *Sampling Strategies for Airborne Contaminants in the Workplace* (British Occupational Hygiene Society Technical Guide 11). Leeds, UK: H and H Scientific Consultants Ltd., 1993.
5. **United Kingdom Health and Safety Executive:** *The Control of Substances Hazardous to Health* (Regulation 1657). London: Her Majesty's Stationery Office Publications Centre, 1988.
6. **Australia National Occupational Health and Safety Commission:** *Control of Workplace Hazardous Substances.* Canberra, Australia: Australian Government Publishing Service, 1993.
7. **Comité Européen de Normalisation (CEN):** *Workplace Atmospheres—Guidance for the Assessment of Exposure by Inhalation of Chemical Agents for Comparison with Limit Values and Measurement Strategy* (European Standard EN 689). London: CEN, 1995. [English version.]
8. **Jayjock, M.A., and N.C Hawkins:** A Proposal for Improving the Role of Exposure Modeling in Risk Assessment. *Am. Ind. Hyg. Assoc. J. 54*:733–741 (1993).
9. **Hawkins, N.C., M.A. Jayjock, and J. Lynch:** A Rationale and Framework for Establishing the Quality of Human Exposure Assessment. *Am. Ind. Hyg. Assoc. J. 53*:34–41 (1992).
10. **Claycamp, H.G.:** Industrial Health Risk Assessment: Industrial Hygiene for Technology Transition. *Am. Ind. Hyg. Assoc. J. 57*:423–427 (1996).
11. **Roach, S.A., and S.M. Rappaport:** But They Are Not Thresholds: A Critical Analysis of the Documentation of Threshold Limit Values. *Am. J. Ind. Med. 17*:727–753 (1990).
12. **Adkins, C.E., A.G. Apol, H.E. Ayer, E.J. Baier, et al.:** Letter to the Editor. *Appl. Occup. Environ. Hyg. 5*:748–750 (1990).
13. **Joint ACGIH–AIHA Task Group on Occupational Exposure Databases:** Data Elements for Occupational Exposure Databases: Guidelines and Recommendations for Airborne Hazards and Noise. *Appl. Occup. Environ. Hyg. 11*:1294–1311 (1996).
14. **Rajan, R., R. Alesbury, B. Carton, M. Gerin, et al.:** European Proposal for Core Information for the Storage and Exchange of Workplace Exposure Measurements on Chemical Agents. *Appl. Occup. Environ. Hyg. 12*:31–39 (1997).

Outcome Competencies

After completing this chapter, the user should be able to:

1. Define underlined terms used in this chapter.
2. Define or recognize their options and responsibilities as occupational hygienists and practicing risk assessors.
3. Describe that true risk is never known but is typically overestimated vis-à-vis the Precautionary Principle.
4. Recall that the degree of risk overestimation is inversely proportional to the resources applied to the estimation.
5. Recognize that the occupational hygienist is a working technologist who uses science.
6. Recall that the heart and soul of this science is in the building, testing, and use of models.
7. Describe how expert judgment is legitimate and valued to the extent that the assumptions and data underlying it can be revealed and explained.
8. Recall that models that are overestimating can be useful, and that they are inexpensive in a tiered approach to risk assessment.
9. Apply equilibrium or steady-state vapor pressure models.
10. Apply the steady-state concentration box model.
11. Apply the steady-state concentration dispersion model.
12. Recall that nonsteady-state concentration modeling is much more complicated.

Prerequisite Knowledge

Basic chemistry, physics, algebra.

Prior to beginning this chapter, the user should review the following chapters:

Chapter Number	Chapter Topic
8	Principles and Instrumentation for Calibrating Air Sampling Equipment
9	Direct-Reading Instrumental Methods for Gases, Vapors, and Aerosols

Key Terms

box model • dispersion model • exposure limits • exposure modeling • exposure assessment • human health risk assessment • steady-state model • vapor pressure • zero ventilation model

Key Topics

I. Why Do Exposure Modeling?

II. Scientific Method and Models
- A. Assumptions in Exposure Models
- B. Tiered Approach to Modeling Exposure

III. Hierarchy of Modeling Estimation Techniques
- A. Tier 1: Saturation or Zero Ventilation Model
- B. Tier 2: General Ventilation Model
- C. Case Study — Modeling Toulene from an Aqueous Product
- D. Tier 2a: Dispersion Model

IV. Determining or Estimating Generation Rates in Other Situations

V. Linked Monitoring and Modeling

VI. Time Element of Exposure

VII. Future of Occupational Hygiene Exposure Modeling

Modeling Inhalation Exposure

16

Michael A. Jayjock, Ph.D., CIH

Introduction

Exposure modeling represents the essence of the science of exposure assessment and should be considered a principal stock in trade of the occupational hygienist. Before delving into the meaning and manner of exposure modeling, it is necessary to understand its context in the overall evaluation of potential impacts to worker health.

The primary function of occupational hygienists is to evaluate the potential risk of exposure to the health of workers. This is called human health risk assessment, and the subject is covered in much greater detail elsewhere in this book; however, for exposure modeling to make sense it is necessary to present the following facts or touchstones of the process.

- Risk is driven equally by the exposure and the health effects per unit exposure.
- The essence of risk assessment is the comparison of exposure to exposure limits.
- When faced with scientific uncertainty the Precautionary Principle(1) advises us to err on the side of safety and thus overestimate risk and/or obtain more information to lower the uncertainty.
- Risk is estimated (typically overestimated) versus the true risk, and the true risk is never known.
- Risk assessment is typically a tiered approach starting with evaluations that generally overestimate risk but are inexpensive, and proceeding to more expensive but less overestimating analytical tools.

It should be reasonably obvious from the above that occupational hygienists measure or otherwise estimate worker exposure, and this exposure has no contextual meaning without a valid exposure limit with which to compare it. The reader is encouraged to review the chapter in this book on risk assessment to understand the full meaning of these important concepts. Other references on this include various American Industrial Hygiene Association (AIHA) books and papers on risk assessment(2,3) and exposure assessment strategies.(4) Suffice it to say here that the following discussion of exposure modeling presents an approach that has as its end the comparison of an estimated exposure to a valid and appropriate exposure limit. Given that relationship it is necessary to understand the origin and basis of the exposure limits used. Anyone who has followed the downward trend in occupational exposure limits will understand that published exposure limits are never written in stone. The working limits, whether derived locally in working establishments or from the regulators or consensus groups such as the American Conference of Governmental Industrial Hygienists (ACGIH), should all be the subject of study, understanding, and judgment regarding their documentation and level of acceptable risk to society.

. . . occupational hygienists measure or otherwise estimate worker exposure, and this exposure has no contextual meaning without a valid exposure limit with which to compare it.

Why Do Exposure Modeling?

Why understand and use exposure models? The simple answer is that practicing occupational hygienists will never be able to monitor every situation everywhere. Also, as a technical expert, an occupational hygienist should have some objective and defensible rationale for why he or she did not do monitoring in the vast majority of exposure scenarios.

Consider a modern plant with hundreds of workers performing perhaps thousands of tasks within it. It is probably safe to say that the majority of tasks are never monitored because the occupational hygienist judges them to be safe. That is, he or she has observed the situation and has concluded that the exposure limit is not exceeded. When asked how that determination was arrived at, the typical answer is that he or she applied expert judgment. That is, the occupational hygienist uses his or her combined experience to make this call. When pressed further the hygienist may say that it is because the system or scenario under consideration is relatively closed, that the vapor pressure is low, the exposure limit is relatively high, etc. These factors combine to tell an experienced occupational hygienist that overexposure will not occur. Some threshold must exist for these skilled estimators where conditions are such that predicted concentrations and exposures approach or exceed the exposure limit. At this point, the occupational hygienist typically moves to action and monitors the situation. The results of that monitoring determines whether controls are implemented.

Undefined expert judgment as described above is the way much of occupational hygiene has been practiced in the past. This manner and technique of working has done a lot of good and protected many workers from overexposure and subsequent adverse health effects. However, it has a number of problems: (1) it is difficult or impossible to explain objectively; (2) it is typically not supported by quantified facts relating specific cause and effect, and (3) as such it is not amenable to technology transfer (i.e., those new to the field find it hard to learn); and (4) it may not be useful or sufficient in the defense against litigation.

The stereotypical "old-time" occupational hygienist who is expert at determining which scenarios do not need monitoring is probably able to do so by running models subliminally or unconsciously in his or her mind while subconsciously comparing the results of this modeled exposure to the exposure limit. This chapter shows how this process can be more open, conscious, objective, and most important, transferable to others. It strives to provide the practitioner with the basic rational and scientific tools and approaches of a technologist doing exposure assessment.

The Scientific Method and Models

Models are the business of science. Think about the basics of science taught in the early years of education and remember the scientific method:

1. State the problem or premise
2. Form a hypothesis
3. Experiment and observe
4. Interpret the data
5. Draw conclusions and make predictions (form new hypotheses or go back to 2)

The hypothesis is the model of what we think the world is like. In the world of occupational hygiene, for example, the hypothesis might be formed (for any number of reasons) that an exposure to an agent in a particular situation or job is not above the exposure limit. To test that hypothesis the occupational hygienist observes the workers and then conducts the experiments (i.e., monitors them). The monitoring results provide objective evidence to accept or reject the hypothesis and to draw conclusions about the risk of this exposure. Perhaps more important, this process feeds the hygienist's internal database (i.e., it increases experience) and makes predictions possible about similar situations in the future. That is, it builds the ability to apply expert judgment.

The lesson here is that there was clearly some model in mind when the hypothesis was formed. There was a mathematical relationship or an algorithm that hypothesized, "given the characteristics of all the causes of exposure in this situation, the resulting exposure will be less than the exposure limit." Often the hygienist is so sure of the hypothesis (the model) that he or she does not monitor. If this important discretionary decision was not made, then everything everywhere would have to be monitored, which, of course, is not possible. Because every exposure possibility cannot be monitored, the occupational hygienist must choose where to spend resources.

Sometimes monitoring proves the hypothesis wrong. This unpleasant reality has probably happened to most of those who have been in the field for some time. However, those who do not learn from their mistakes are doomed to repeat them. Thus, at this point where the model or hypothesis is proven wrong, it obviously needs to be readjusted and refined.

Some choose to call this typical hypothesis forming process of the occupational hygienist "qualitative exposure assessment," when in fact it involves the comparison of a quantitative estimate of the exposure (however unconsciously formulated) with a numerical exposure limit. The point of learning here is that there is a quantitative model present and operating, if only subconsciously, and occupational hygienists need to have a more conscious understanding and explanation of the details of the decision-making process. This conscious understanding will help identify and fix a broken or defective model, and it will allow for the rational explanation to others as to how occupational hygienists operate professionally. Rather than simply invoking a claim of unsubstantiated professional judgment relative to their decisions, by using modeling the occupational hygienist can understand and display the scientific rationale behind those decisions.

It might be helpful at this point in the discussion to think of a relatively simple algebraic model commonly seen in mathematical or scientific training.

$$Y = (a)X_1 + (b)X_2 + (c)X_3 \ldots \quad (1)$$

where:

- Y = dependent variable (exposure)
- X_1 = independent or predictor variable 1 (e.g., vapor pressure)
- X_2 = independent or predictor variable 2 (e.g., surface area)
- X_3 = independent for predictor variable 3 (e.g., temperature)
- a,b,c = predictor strength coefficients

Workers' exposures can be understood only to the extent that the physical world and the entities within it that cause exposures (i.e., the independent or predictor variables) are understandable. Even given a rich database of monitored airborne concentrations (i.e., dependent variable data), one must relate those results to the determinants (i.e., predictors) in the world that produced those exposures.[5] This is necessary to (1) assure the continued validity of the predicted exposures in the future, (2) be reasonably certain that workers are assigned to the appropriate homogeneous exposure group,[4] and (3) allow for prediction of different scenarios.

The construction and use of these models is not mysterious, it is simply science. In the context of modeling inhalation exposure for occupational hygiene, this effort can be thought of as investigating and seeking to understand the determinants of airborne contaminant source generation and control. As the critical variables governing the generation and control of airborne toxicants are discerned, the tools are formed that will build experience, knowledge-base, and confidence to predict actual concentrations and exposures in the real world with simulated scenarios. As such, model development consists of formulating hypotheses about the predictors of exposure and then testing them with data from experiments examining cause and effect; it is the scientific method.

As understanding of why and how physical-chemical models are developed grows, it also becomes clear that the models represent a principal structural basis for the science of exposure assessment. These models, along with the statistical modeling of monitoring data, form the scientific foundation for characterizing worker exposure. Comparing the exposure to the toxicity or the health effect/exposure provides the basis for risk assessment.

Physical-chemical inhalation models are not limited to predicting present exposures. They can be used to estimate historical exposures that cannot easily be re-created and possible future exposures in hypothetical situations or scenarios. By employing a model, an occupational hygienist's insight about possible exposures is enhanced, even if the model is not perfectly accurate. A noted and wise statistician, G.W.E. Box, has been credited with the profound observation that "all models are wrong, but some are useful."

What Dr. Box knew and the rest of us should keep in mind is that all scientific models, including occupational exposure models, are more or less generalized (and therefore crude) representations of reality. Even remarkably elegant and presumably complete and correct basic scientific models such as those devised by Sir Isaac Newton to describe the laws of motion are wrong under certain conditions as described by Einstein. Thus, predictions from physical-chemical inhalation models can be extremely valuable; however, at this point these models are far from being considered elegant or complete. As such, they should be interpreted with caution and the usual judgment and intelligence that an occupational hygienist brings to his or her craft.

Assumptions in Exposure Modeling

Most inhalation exposure models do not estimate human exposure directly. These models estimate the concentration of toxicant in the air and assume that the person is breathing the same air with this concentration. The use of this and other assumptions is important and necessary in exposure assessment. It has been said of many activities that "the devil is in the details." It could also be said of exposure and risk assessment that "the devil is in the assumptions." In order for those who view occupational hygiene to understand it, they need to be able to review the assumptions of the occupational hygienist. Indeed, it is vital to the integrity of the process to sort out and identify each and every assumption used in modeling and estimation of exposure.

Tiered Approach to Modeling Exposure

It is typical to start with relatively simple models that have overestimating assumptions because these models do not require much in the way of resources and thus are simple and quick to run. The downside is that these simple models can dramatically overestimate the exposure potential of the scenario under investigation. Thus, depending on the conclusions of the predicted level of exposure compared with the exposure limit (i.e., the exposure/exposure limit ratio), it may be necessary to try more sophisticated modeling tools. These more complicated models cost more time, effort, and money but they render answers that are less overestimating.

During the tiered process it is not unusual to run out of modeling resource before gaining a definitive answer, and in this case either a better model is developed or representative air monitoring is performed. Unfortunately, for the general development of physical-chemical models the second solution has historically been chosen almost invariably because it is relatively inexpensive and answers the question at hand expeditiously.[3,5]

Hierarchy of Modeling Estimation Techniques

Below is a somewhat consolidated discussion and presentation of inhalation exposure modeling techniques. It is provided to introduce these tools. Some of the fine details are left out, and the reader is encouraged to go to the references for more information. Of particular interest is the appendix on modeling in the AIHA exposure strategies book.[4]

Exposure modeling can become technically complicated very quickly. Indeed, one could easily spend an entire career in this field (although few have). However, the purpose here is to introduce the topic, and thus only a few of the most generally used and useful models are discussed, these being the saturation or zero ventilation model, the box model, and the dispersion model. They are presented here in the form of an example.

For the case study the inhalation exposure potential to toluene from an aqueous solution containing 1 ppm (one weight part per million weight parts, or w/w) of toluene will be considered. In this example the models are shown in order of increasing sophistication and level of information needed to use them appropriately and successfully. Thus, the first model one should think of using is the simple saturation model, followed by the box and dispersion models. More sophistication is brought into the investigation only if the overestimation cannot be tolerated in the evaluation of the risk.

Given minimal information about the use scenario and physical properties of a material, one can estimate the saturation concentration as an estimation of worst case airborne exposure to vapors in a Tier 1 analysis.[6,7]

Tier 1: Saturation or Zero Ventilation Model

This very basic and typically very conservative inhalation exposure model calculates the maximum possible concentration of vapor (i.e., saturation) in air. It is best used for gases and vapors emitted without mist formation when there is no information on ventilation or details of use. For any liquid, saturation will eventually occur in the air above a liquid surface if no ventilation is present and the evaporation rate ultimately overwhelms any removal mechanism such as absorption, adsorption, or chemical transformation.

The equilibrium saturation concentration (C_{sat}) in volume parts of contaminant per million volume parts of air (ppm v/v) will be

$$C_{sat} = \frac{(10^6)(\text{vapor pressure})}{(\text{atmospheric pressure})} \quad (2)$$

Vapor pressure at any ambient temperature is an experimentally determined quantity; however, it can also be estimated from boiling point data for any class of liquids either at atmospheric pressure or under vacuum.[8] The vapor pressure of components within mixtures can also be estimated using established procedures.[9]

This saturation model is usually conservative for the prediction of workroom air concentrations. It has been the author's experience that it overestimates workroom air concentrations of vapor (i.e., nonparticulate) in all but worst case scenarios (e.g., large spills indoors with poor ventilation) by a factor ranging over four orders of magnitude (10–10,000X). This observation is the result of comparing scores of measured concentrations of organic air contaminants in occupational settings with their saturation concentrations calculated from vapor pressure or boiling point data. Worst case scenarios include those in which significant aerosol is released or there is a relatively large area (greater than a few square meters) of evaporating liquid. In these situations the saturation model is often not very overestimating.

This model's value lies in its simplicity as a screen with only a few basic physicochemical properties required as input. As a typically very conservative estimate it represents a good first step in a tiered risk assessment. If exposure levels determined by the model are below the compound's ascribed toxic exposure level (e.g., an occupational exposure limit), a high degree of confidence exists that actual vapor concentrations do not pose an unacceptable risk to worker health via inhalation exposure. Of course, other routes of potential exposure (e.g., dermal or oral) and aerosol generation are not considered in this method.

If one knows very little about the details of the actual exposure scenario, then it can be said that he or she is highly uncertain about the actual level of exposure. However, it is very unlikely that the airborne concentration of toluene in this example in most reasonably conceived scenarios with this product will be higher than its predicted saturation concentration.

In this evaluation one literally assumes that the person breathes the equivalent of headspace concentration of vapors all day. This mythical person would not have the benefit of general dilution ventilation; thus, this is sometimes called the zero ventilation model.

Tier 1 assessment can be done with very little information and at relatively low cost. In the example, the unitless Henry's law constant (HLC) is used, which is defined as

$$HLC = \frac{C_g}{C_l} \tag{3}$$

where:

C_g = equilibrium concentration of toxicant in the air above the liquid (i.e., headspace in weight/volume of air)

C_l = bulk concentration of toxicant in the liquid in weight/volume of liquid

Expressing C_g and C_l in the same units will result in the unitless HLC. For this example, HLC is estimated to be 0.23.[10]

$$HLC = \frac{C_g \text{ (units of mg airborne contaminany/m}^3\text{ air)}}{C_l \text{ (units of mg of contaminant in solution/m}^3\text{ of solution)}} \tag{4}$$

The example product with 1 ppm (w/w) of toluene in aqueous solution has 1000 mg of toluene per cubic meter of water.

$$\left(\frac{1}{10^6}\right)\left(\frac{10^9 \text{ mg of } H_2O}{1 \text{ m}^3 \text{ of } H_2O}\right) = \frac{1000 \text{ mg}}{\text{m}^3} \tag{5}$$

It follows from Equation 3 that the predicted saturation airborne concentration is 230 mg/m^3 (61 ppm v/v). Thus, the estimate of the exposure is 61 ppm v/v toluene, which is above the 1992-1993 ACGIH threshold limit value exposure limit of 50 ppm (v/v). In this example the exposure determined in Tier 1 exceeds the exposure limit, and one needs to go to Tier 2; however, please note that if the product had contained 0.1 ppm (w/w) toluene, then the estimated Tier 1 exposure would have been 6.1 ppm (v/v) toluene, and we would have a fair degree of confidence that an exposure limit of 50 ppm (v/v) would not be exceeded.

Vapor Pressure, Partial Pressure, and Concentration

At this point the relationship between vapor pressure and airborne concentration of gases and vapors should be discussed. The Ideal Gas law from physical chemistry states that the volume of any ideal gas at any particular temperature and pressure is determined entirely by the number of gas molecules. Thus, a mole (6.02×10^{23} molecules) of hydrogen (H_2) at normal temperature and pressure has the same volume (24.4 L) as a mole of butane gas (C_4H_{10}). The Ideal Gas law allows us to express the concentration of a gas (or vapor) as volume parts of the gas (or vapor) per million volume parts of air (ppm v/v). The conversion between ppm v/v and milligrams of gas per cubic meter of air is given as

$$\text{ppm v/v} = \frac{\text{mg}}{\text{m}^3}\left(\frac{24.4}{\text{MW}}\right) \qquad (6)$$

where:

MW = molecular weight of the gas in g/gmole.

An extension of the Ideal Gas law is Dalton's law, which states that in a mixture of gases the total pressure is equal to the sum of the partial pressures of the separate components. This means that at normal atmospheric pressure (or 1 atmosphere of pressure) all of the gases would add up to 1,000,000 ppm. In normal air this usually means that there is approximately 780,840 ppm v/v nitrogen; 209,476 ppm v/v oxygen; 9340 ppm v/v argon; 314 ppm v/v carbon dioxide; 18.2 ppm v/v neon; 8.7 ppm v/v xenon; 5.24 ppm v/v helium; 2 ppm v/v methane; 1.14 ppm v/v krypton; and 0.5 ppm v/v hydrogen.(11) If any gas or vapor is added to this typical air mass with the above concentrations, the partial pressures of these constituents would reduce slightly so that they all still add up to 1 million volume parts per million parts and 1 atmosphere of pressure. For instance, if 61 ppm v/v of toluene from the example is added to this air mass, all of the above values would be proportionately readjusted to allow for this partial pressure of toluene; however, the total pressure would remain at 1 atmosphere.

One meaning of Dalton's law is that one can convert directly between any airborne concentration of gas or vapor and its partial pressure. This is readily seen in Equation 2. In the case of the example with toluene, 61 ppm will exert a partial pressure of 60/1,000,000 atmospheres (6.1×10^{-5} atm). Since 1 atm is equal to 760 torr, the partial pressure of 61 ppm v/v is equal to (6.1×10^{-5} atm) (760 torr/atm) or 4.6×10^{-2} torr.

Tier 2: General Ventilation Model

One of the oldest and most used models in occupational hygiene is the box or general ventilation model. It relies very simply on the concept of the conservation of mass. The model is based on a "black box" of air, into which one cannot go or even look. But as an airborne contaminant is put into the box, any contaminant that subsequently comes out can be constantly measured. The average concentration in the box can be described as

$$\text{Concentration} = \frac{\text{(Amount that went into the Box)-(Amount that came out of the Box)}}{\text{Volume of the Box}} \qquad (7)$$

If the contaminant is going into the box at a steady rate and leaving with the outgoing air at the same rate, then the system is at steady state, and the average concentration in the box is constant.

If we are to believe that the concentration in the box is the same or homogeneous throughout the volume of the box, then we need to make the assumptions that (1) the contaminant remains airborne (is not absorbed onto surfaces), (2) does not change chemically within the box, and (3) on entering the box the contaminant is instantly and completely mixed with the air inside.

Using this simple steady-state model and assumptions, a general ventilation equation for this situation is

$$C_{eq} = \frac{G}{Q} \qquad (8)$$

where:

C_{eq} = steady-state concentration, mg/m^3
G = rate going into the box, mg/hr
Q = ventilation rate of air leaving the box, m^3/hr

Of course, the real world is often much more complicated. The mixing of airborne contaminants is often not complete and instantaneous, and some compounds are removed by nonventilatory mechanisms such as adsorption or chemical reaction. Also, the nonsteady-state situation is significantly more complicated to describe mathematically. A differential equation that attempts to take all of these factors into account can be written for the pollutant concentration within the box for any time.(12)

$$VdC = Gdt - (C)(Q)(m)dt - (C)(k)dt \qquad (9)$$

where:

V = the assumed volume of the box, m^3
t = the time variable, hr
C = the concentration in the box at any given time, mg/m^3
G = the rate of generation of pollutant within the box, mg/hr

Q = the volume flow rate of air exchange in the box, m^3/hr
m = the dimensionless mixing efficiency of ventilation in the assumed box[13]
k = the removal rate by mechanisms other than ventilation and filtration, m^3/hr

Typically, specific information is not available on nonventilatory loss rate (k), the mixing efficiency (m), or the time course of exposure. Thus, values for these factors and for the ventilation (Q) and generation rate (G) are assumed that render a reasonable upper bound estimate of C. Indeed, the steady-state condition is often the default for analysis. The following upper bound physico-chemical model assumptions are typically used for the estimation of airborne concentration (C) from evaporating sources.

Equilibrium Conditions (Time Sufficiently Long That dC = 0). Given a constant source, the maximum airborne concentration will occur at equilibrium. See the discussion on Time Element of Exposure.

G = Environmental Protection Agency (EPA) Algorithm.[14] This is presented in some detail below in the example; however, this algorithm is essentially driven by the mass transfer coefficient (K_t), which in turn is estimated with a relatively simple equation. In the instances where it has been compared with other techniques or measurements, it has significantly overestimated the rate of evaporation. It is used here only as an illustration, and the reader is encouraged to use more sophisticated (less overestimating) methods if needed. A more sophisticated estimate of G from vaporizing sources (HBF model) is also presented later in this chapter.

Q = 0.1 Air Changes/Hour. Many, if not most, residences do not have specific provisions for the circulation of outdoor air. General ventilation often occurs via infiltration of outside air into the house via cracks. Recent data show the air change rates for fresh air in residences in the United States,[15] Finland,[16] and Denmark.[17] The U.S. and Finnish data indicate a median winter value around 0.5 to 0.6/hr. The Danish study reports a median air exchange rate of 0.3/hr. Over 75% of 140 homes in Baltimore, Md., in April had ventilation rates greater than 0.3 air changes/hr.

In the author's experience, typical industrial ventilation rates are higher than 0.3 air changes per hour, and some are very much higher; thus, absent specific information, an assumption of 0.1 air change per hour appears to be a significant overestimating worst case.

m = 0.3. This is based on previous work indoors without fans that reports m values in the range of 0.3 to 0.4.[18] The concept and use of m has limitations. It has been found useful, even necessary, for estimating exposure; however, it violates the basic assumption of conservation of mass and thus has some severe qualifying factors. It might be better thought of as a safety or uncertainty factor associated with concentration hot spots associated with poor mixing around sources.

k = 0 (No Nonventilatory Losses). Some, but relatively few, materials degrade in air quickly enough to significantly affect their airborne concentration.[19] Volatile and semivolatile organic compounds can be deposited onto environmental surfaces where they can accumulate, degrade, or be re-emitted into the air. Short-term sources could have peak concentrations that are significantly lowered by these sink effects even in systems without degradation of the compound.[20] For long-term continuous sources, the equilibrium concentration is ultimately the same but delayed in systems with significant surface deposition. Also, the steady-state airborne concentration could be significantly lower in systems in which degradation is occurring after adsorption onto environmental surfaces.[21]

The following assumptions regarding human activity patterns will maximize the estimation of exposure (the product of airborne concentration ˇ time): The worker is in this exposure field for the entire work shift, and a consumer is in the assumed exposure field for 24 hours/day.

Using these assumptions, the general ventilation model that incorporates the mixing factor and ignores k (i.e., set k = 0) is

$$C_{eq} = \frac{G}{(Q)(m) + k} = \frac{G}{(Q)(m)} \qquad (10)$$

Case Study—Modeling Toluene from an Aqueous Product

In this case study the headspace concentration and the worst case exposure potential associated with it has been estimated. This could be done with only information on the concentration of toluene in the product and its Henry's law constant. To do a more detailed analysis in the case study, it is necessary to get more information about the actual exposure scenario to use this model. This aqueous product is typically used in light industrial settings in which the primary off-gassing and exposure comes from an open container of the product. The open surface area is 100 cm^2, and the workroom is maintained at 25°C. It is also determined that the workers are often in the room but very rarely immediately proximate to the open container. That is, average room concentrations are important rather than near-field concentrations or exposures very close to the open drum. The specific general ventilation rate has not been determined, and there is typically no local exhaust ventilation.

A physical-chemical model to estimate the source strength of an evaporating liquid has been presented by Fleischer[22] and has been used by EPA.[14]

$$G = (10^3)\frac{(K_t)(MW)(AREA)(VPP-VPB)}{(R)(TL)} \quad (11)$$

where:

G = generation rate, mg/hr
K_t = mass transfer rate, m/hr
MW = molecular weight, g/mole
AREA = m^2
VPP = vapor pressure of the substance, atm
VPB = partial pressure of the toxicant in the room air, atm
R = gas constant (8.205×10^{-5} atm m^3/((mole)(deg K)))
TL = temperature of the evaporating liquid, deg K

For this exercise the mass transfer coefficient (K_t) estimated using the following relationship[17] will be used.

$$\text{mass transfer coefficient} = 30\left(\frac{18}{MW}\right)^{1/3} \quad (12)$$

This is done for clarity in the example, because this is a relatively simple relationship; however, because it has been found to overestimate exposure, other more detailed methods exist for this determination.[23–27]

In this example the input values found in Table 16.1 were used.

One assumption in the above model is that the partial pressure of the contaminant in the box air (VPB) is insignificant relative to the vapor pressure (VPP) of the toxicant. This simplifying supposition overestimates the generation rate and thus the resulting concentration. It is useful and justified for relatively small evaporating sources (like the case study example), but it is not appropriate for large sources. The specifics of modeling large evaporating sources are covered elsewhere.[28]

Using these values in Equation 11 results in an estimated off-gassing rate of 40 mg/hour to the workroom.

The ventilation rate Q is equal to the room volume times the air change rate per hour:

$$Q = (V)(\text{air change/hr})$$
$$Q = (50)(0.1) = 5\ m^3/hr$$

Assuming a mixing factor m = 0.3, Equation 10 predicts an average air concentration in the workroom of about 27 mg/m^3 (7 ppm v/v). This is below the current 188 mg/m^3 (50 ppm v/v) exposure limit. It should be noted that most of the predictors in the model have a linear effect on predicted concentration and less than a threefold increase in surface area and vapor pressure (from higher toluene concentration or higher temperatures) would have predicted exposure above this exposure limit. Thus, modeling a significantly larger, warmer, or more concentrated source would have resulted in a prediction of unacceptable upper bound estimate of risk.

Staying with the basic box model, a more sophisticated model to estimate the generation rate can also be used. This work was recently published and represents the

Table 16.1 — Input Values for Example

Value	*Source*
K_t = 17.4 m/hr	Equation 12
MW = 92.1 g/mole	handbook
AREA = 0.01 m^2	100 cm^2 converted to m^2
VPP = 61/1,000,000	61 ppm (v/v) from previous Henry's law calculation converted to atm
VPB = 0	conservative assumption
R = (8.205×10^{-5} atm m^3/((mole)(deg K)))	Universal Gas Constant
T = 298 K	25°C converted to K
V = 50 m^3 (room volume)	conservative assumption
m = 0.3	Reference 18
Air change/hr = 0.1	conservative assumption

method currently used by EPA to estimate inhalation exposures to new chemical substances.[25,26] The Hummel-Braun-Fehrenbacher (HBF) model describes the source rate as:

$$G = \frac{(7.2 \times 10^4)(MW)(VPP)(AREA)}{(R)(TL)} \sqrt{\frac{VxDab}{\pi \Delta x}} \quad (13)$$

where:

Dab = diffusion coefficient, cm^2/sec of a through b (in this case b is air)
Δx = pool length along the direction of the airflow, cm
Vx = air velocity along the x-axis, cm/sec.

The diffusion coefficient (Dab) is further defined as:

$$Dab = \frac{4.09 \times 10^{-5}\,(TL)^{1.9}\left(\frac{1}{29} + \frac{1}{MW}\right)^{0.5}(MW)^{-0.33}}{P_t} \quad (14)$$

where:

Pt = atmospheric pressure, atm.

This model has the added feature (and data requirements) of knowing or estimating the air velocity over the pool (Vx) and length of the pool in the direction of the air movement (Δx). If the pool is assumed to be essentially isometric, then the pool length is reasonably estimated as the square root of the AREA. It has been the author's experience that most indoor environments with reasonably calm air movement will have air moving in the range of 5–20 linear feet per minute in a rather random manner. If the example is continued with the square root of the AREA (converted to cm^2) for Δx and 5 ft/min (converted to cm) for Vx, in this model the estimated off-gassing rate is 15 mg/hr to the workroom. The predicted values are 21 and 30 mg/hr at indoor air velocities of 10 and 20 ft/min, respectively. The original estimate from the relatively crude mass transfer rate estimated using Equation 12 was 40 mg/hr. Thus, the use of this more sophisticated model allowed for a more refined estimate. Indeed, the HBF model predicts that the off-gassing rate will not reach 40 mg/m^3 from this pool until the airflow rate reaches around 50 linear feet per minute. This is a situation that is possible but not very likely in many indoor environments.

Tier 2a: Dispersion Model

The general ventilation model avoids the question of contaminant mixing in the volume. It also ignores near field exposure or sharp gradients of concentration for workers close to the source. A diffusion model has been developed for heat flow[29] and applied to indoor air modeling.[30,31] The equation for a continuous point source is presented in the references to predict concentration at position r and time t.

$$C = \frac{G}{240(\pi)(D)(r)}\left[1 - \mathrm{erf}\left(\frac{r}{\sqrt{4tD}}\right)\right] \quad (15)$$

where:

erf = the error function*
C = concentration, mass/volume, mg/m^3
G = steady-state emission rate, mass/time, mg/hr
r = the distance from the source to the worker, m
D = the effective or eddy diffusivity, area/time, m^2/hr
t = elapsed time, hr

Diffusion of contaminants in workroom air occurs principally because of the turbulent motion of the air.[30] In most industrial environments molecular diffusion is not significant between the emission source and the worker's breathing zone. Instead, the normal turbulence of typical indoor air causes eddys (or packet-like motions) that have the effect of breaking up the contaminant cloud and hastening its mixing with the workroom air. Therefore, applications of diffusion models in industrial environments use experimentally determined diffusion coefficients (D) called eddy or effective diffusivities. These eddy diffusivity coefficients are three to five orders of magnitude larger than molecular diffusivity.

The eddy diffusivity term (D) can be based on experimental measurements at the site being modeled. Some eddy diffusivity values are also available in the literature.[31–33] Measurements of D in indoor industrial environments have ranged from 3 to 690 m^2/hr with 12 m^2/hr being a typical value.

Plotting the predicted airborne concentration (C) at one position r for many values of time t gives an increasing curve of concentration that approaches a steady-state level.

For sources (emitting into a hemisphere) on a surface and at equilibrium, Equation 15 simplifies to

$$C_{eq} = \frac{G}{120(\pi)(D)(r)} \quad (16)$$

Consider the previous example with G = 40 mg/hr. Consider a person working within 1 m of the source (r = 1 m). It is known that the lowest D measured in a very limited database was 3 m^2/hr. If this value is used, the estimated equilibrium airborne concentration of toluene is about 2 mg/m^3. This is less than one-tenth of the amount predicted with the box model; however, the worst case ventilation could be very low, and this could result in very little mixing and a true D value that is much lower than 3 m^2/hr. However, there simply is not enough data to use this model with much confidence. It is useful, however, in that it allows estimation of the effect of distance from the source on worker exposure. It predicts that it is a straight inverse relationship with the exposure going down twofold for every doubling of distance from it from a theoretic point source.

There is little doubt that the eddy diffusivity model could be a very valuable tool that can potentially provide near- and far-field exposure estimations; however, this approach in general suffers because it lacks the reasonable characterization of the primary predictor variable, eddy diffusivity.

Determining or Estimating Generation Rates in Other Situations

This chapter has concentrated on the estimation of generation rate of vapors from evaporation. This was done because vaporizing sources are often important, and the subject readily allows the illustration of physical-chemical modeling. It should be mentioned, however, that there are other sources and other ways of estimating these rates. If it is known or measured that a certain amount of gas is released from a chamber, and the time of release is also known, an average source term or G can be calculated by simply dividing the released mass by the time of release. Sometimes measuring G directly is the best and most expeditious way to attain it for use in a model. For example, painting a glass slide or piece of wood and measuring its loss of weight over time could provide very valuable information for this emission rate. Another example would be to weigh a can of a gaseous product (e.g., refrigerant gas) before and after a measured period of discharge to the air.

Airborne particle concentrations can also be estimated in this manner, but this situation is much more complicated in that relatively large aerodynamic particles will rapidly settle out of the air column. Also, some sprayed aerosol particles are constantly changing their diameter and mass while in the air because of vaporization. Modeling of this situation using indoor-sprayed pesticide as the example has been accomplished with a complicated and very detailed model.(34)

Linked Monitoring and Modeling

The relationship between modeling and monitoring workplace airborne concentrations becomes particularly important when one has some monitoring data to add to the estimates being made by modeling. One can estimate or refine purely model-estimated exposures for an agent by combining the information from monitoring with the theoretical construct of the model.

Simple examples include the predictions of workplace concentrations giving some monitoring data and a change in source or ventilation rate or distance from a point source. Using the above model it can be predicted that reducing the source by half or doubling the ventilation rate will result in a concentration reduced by half. Doubling the distance from a true point source in a random dispersion field will also halve the concentration. These predictions will work for relatively simple situations; however, the more you know about modeling, the more you know where they might not work. Large evaporating sources, for example, will

*The error function is related to the normal or Gaussian distribution. This is the bell-shaped curve described by the function $\phi(x) = \frac{1}{\sqrt{2\pi}} e^{-\frac{x^2}{2}}$. This curve is called the Normal Curve of Error and the area under this curve represents probability integrals such that $\int_0^x \phi(x)dx = \frac{1}{2} erf \frac{x}{\sqrt{2}}$. To evaluate erf(2.3) proceed as follows: Since $\frac{x}{\sqrt{2}} = 2.3$, one finds $x = (2.3)\sqrt{2} = (2.3)(1.1414) = 3.25$. In the normal table entry for area opposite x = 3.25, the value of 0.4994 is given. Thus erf(2.3) = 2(0.4994) = 0.9988. Modern PCs and software (e.g., EXCEL and Mathcad) can do this without effort.

not render half the concentration when the ventilation rate is doubled. The reason and working model for this is presented elsewhere.[28]

Another more detailed example of linking monitoring and modeling is to first run a model on a mixture that predicts a relative exposure to a number of components of that mixture. If there happens to be monitoring data on one of the components in the modeled scenario, the estimates of actual exposure potential can be checked and refined. That is, the monitoring data on the substance can be used to predict the airborne levels of the other components. Using the ratios predicted by the model and multiplying the monitored data by these ratios allows prediction of the concentrations of the other components.

As a tangible illustration of this, assume there is exposure data on substance X (X_{ppm}), but the exposure for substance Y (Y_{ppm}) needs to be estimated. Both are components of the same mixture in a process stream. Their percentages in the stream (%x and %y) and their vapor pressures (VP_x and Vp_y) are known. Assuming their molecular weights are similar, the percentage concentration can be used in place of their mole fractions.

$$Y_{ppm} = \left(\frac{\%_y}{\%_x}\right)\left(\frac{VP_y}{VP_x}\right)X_{ppm} \qquad (17)$$

This modeling approach assumes that the mixture acts as an ideal solution, which may not be true. In any event, the point is that modeled ratios can be combined with monitored data to give more refined estimates of exposure for species that are being monitored.

Time Element of Exposure

Figure 16.1 shows the time course of concentration for nonsteady-state and steady-state models in the case study.

The steady-state Equations 8, 10, and 16 are independent of the room volume (V), and the only thing required about time t is that it be long enough to have attained steady state. The model shown in these equations is thus appropriately applied only to situations that have come to apparent equilibrium. That is, an equilibrium or steady-state model will render some ultimate and unvarying steady-state concentration; however, it says nothing about the time course of the exposure. A steady-state model also says nothing about the actual exposure that occurred at the beginning of the exposure period. If indeed this is the only time a person might be exposed, such as in a batch job that is done at the beginning of an exposure and completed quickly, then this question becomes critical.

The thoughtful reader might ask:

1. How long must a typical operation run before a steady-state concentration is established?
2. How can an exposure that occurs before this apparent equilibrium is established from a source that starts at t=0 be estimated?
3. How can an exposure be estimated that occurs after steady state is established from a source that is then turned off?

Since theoretical steady state is approached but rarely achieved, a practical solution to the problem of gauging the time scale of concentration buildup is to calculate the time required to achieve some percentage of this ultimate concentration. Solving Equation 9 for time (with C=0 @ t=0) yields

$$t = \frac{-V}{(Qm + k)} \, ln[(G - (Qm + k)(C))/G] \qquad (18)$$

At 90% of equilibrium Equation 10 becomes

$$C = \frac{0.90\,(G)}{(Qm + k)} \qquad (19)$$

Combining Equations 18 and 19 gives

$$t\ (@\ 90\%\ \text{of equilibrium}) = 2.303\ V/(Qm + k) \qquad (20)$$

Equation 20 shows that large volumes with low ventilation rates and poor mixing take a relatively long time to reach a

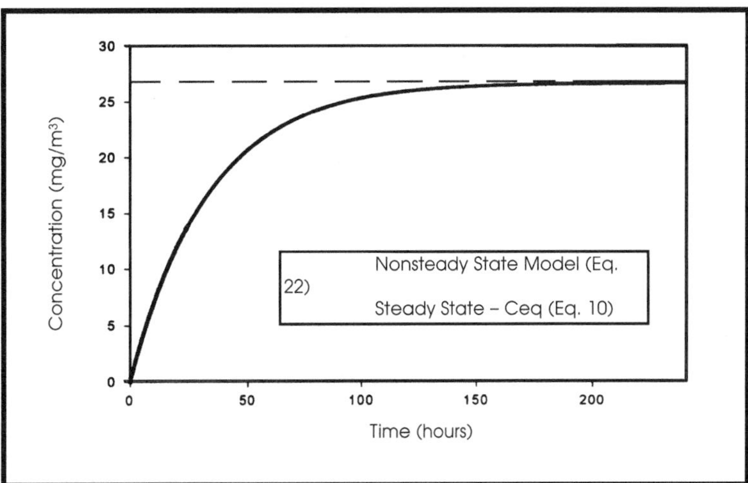

Figure 16.1 — Case study: nonsteady-state and steady-state models.

substantial portion of equilibrium. In the running example, this time (time to attain 90% of equilibrium) is 2.303(50)/((5)(0.3)) or 77 hours. The practical lesson from this is that if one is modeling essentially batch processes where C=0 @ t=0, and the length of time for the exposure is of relatively short duration (less than 1–2 hours), one needs to consider the time-weighted average concentration the worker may be exposed to under non-steady-state conditions.

In the converse situation, there is a workroom air volume containing a concentration of toxicant in which the source has been turned off or removed. Under the same conditions (i.e., large volume, low ventilation rate, poor mixing) it will take a relatively long time to clear this concentration. This scenario is examined and presented below in more detail.

The above analysis and equations of the time variation of airborne concentration assumes that the initial concentration is nil (i.e., C=0 @ t=0), which is the common situation; however, in many industrial situations the initial concentration should not be neglected. For example, in a situation where worker exposure is modeled in the morning and afternoon, and the source is turned off at lunch time (when workers take a 30-minute break), then it might be important for the afternoon estimate to include an initial concentration term based on the decay calculated for the lunch period.

With the source turned off (G=0):

$$C = C_0 e^{-\frac{(Q)(m)(t)}{V}} \tag{21}$$

where:

C_0 = concentration just before lunch and just before source was turned off
C = concentration immediately after the lunch period
t = elapsed time for lunch period

After the lunch break one may need to model the situation as the source is again turned on. This is done by setting k to 0 (unless you have some data on k) and solving Equation 9 for C (with $C=C_0$ @ t=0).

$$C = \frac{G}{(Q)(m)} + C_0 - \frac{G}{(Q)(m)} e^{-\frac{(Q)(m)(t)}{V}} \tag{22}$$

In this situation C_0 in Equation 22 should be set equal to C from Equation 21; that is, the final airborne concentration after the lunch break becomes the initial concentration for the beginning of the afternoon work session.

Please note, if one is modeling a scenario in steady state (i.e., one at apparent equilibrium relative to the airborne concentration of contaminant), initial concentrations need not be accounted for or kept track of, and Equations 10 and 16 should be used.

The question of whether an equilibrium or nonequilibrium model should be used is probably best answered by looking at the scenario of interest and using a reasonable worst case approach. The product of (C_{eq})(time) will always render the highest estimate of exposure for any reasonably constant source. However, if this typically overestimated value is not satisfactory for the evaluation of exposure, one may need to do the more sophisticated analysis of integrating nonequilibrium conditions. Relatively long exposure times and a steady source rate will render essentially the same estimates of exposure regardless of whether one uses a steady-state or non-steady-state model.

Future of Occupational Hygiene Exposure Modeling

As can be seen, the above analyses usually render significantly lower estimates of exposure as they become more sophisticated. These values often remain significant overestimates of the actual exposure. This is because lack of knowledge of basic model details is dealt with by using overestimating, conservative assumptions.

Unfortunately, one can often run out of resources at a relatively low level in Tier 2 and not be able to progress with the modeling approach. If the estimated exposure still exceeds the exposure limit, the assessor needs to go beyond this point to the next stage (Tier 3). He or she has a choice of (1) reducing the uncertainty of the model with research or (2) doing direct monitoring measures of the exposure scenario. The second choice has almost invariably been taken because the first choice is very expensive in the short term. Because direct monitoring databases only look at and provide data on the outcomes of exposure (i.e., the resulting airborne concentrations) and not the determinants that caused the exposure to happen, they have not been useful for model development and validation.

An example of reducing model uncertainty with its concomitant overestimation would be gathering better data on the ventilation rates and conditions extant in different classes of buildings and workplaces. If, for example, one had a probability distribution function of air change rates and eddy diffusivities for rooms in light industrial buildings, he or she could do a Monte Carlo simulation[35,36] showing the range of the predicted exposure in this environment by repeatedly sampling these distributions and using Equation 15 or 16. Of course, this would have to be combined with best estimates of toxicant generation rates.

Because the topic is essentially generic, research in the area of exposure modeling will benefit everyone who does exposure assessment and would be in the best interest of the profession.

References

1. **United Nations:** *United Nations Conference on the Environment and Development (UNCED).* Rio de Janeiro: United Nations, 1992.
2. **Jayjock, M., J. Lynch, and D. Nelson:** *Risk Assessment and the Industrial Hygienist.* [Draft]
3. **Jayjock, M.A. and N.C. Hawkins:** A proposal for improving the role of exposure modeling in risk assessment. *Am. Ind. Hyg. Assoc. J. 54*:733–741 (1993).
4. **Damiano, J., J. Mulhausen, and C. Cole (eds.):** *A Strategy for Assessing and Managing Occupational Exposures.* Fairfax, VA: American Industrial Hygiene Association. [In press]
5. **Jayjock, M.A. and N.C. Hawkins:** Exposure database improvements for indoor air model validation. *Appl. Occup. Environ. Hyg. 10*:379–382 (1995).
6. **Jayjock, M.A.:** Assessment of inhalation exposure potential from vapors in the workplace. *Am. Ind. Hyg. Assoc. J. 49*:380–385 (1988).
7. **Hawkins, N.C.:** "Uncertainy and Residual Risk." Presentation on behalf of the Chemical Manufacturers Association to the National Research Council/National Academy of Science Committee on Risk Assessment of Hazardous Air Pollutants, Washington, DC, November 1991.
8. **Haas, H.B. and R.F. Newton:** Correction of boiling points to standard pressure. In *CRC Handbook of Chemistry and Physics,* 59th ed. Boca Raton, FL: CRC Press, 1978. p. D-228.
9. **Lyman, W.J., W.F. Reehl, and D.H. Rosenblat:** *Handbook of Chemical Property Estimation Methods.* New York: McGraw Hill, 1982. Chapters 11, 14, 15.
10. **Meylan, W.M. and P.H. Howard:** Bond contribution method for estimating Henry's law constants. *Environ. Toxicol. Chem. 10*:1283–1293 (1991).
11. **Lide, D.R.:** *Handbook of Chemistry and Physics,* 76th ed. New York: CRC Press, 1996. Sec. 14, p. 14.
12. **Tichenor, B.A., Z. Guo, J.E. Dunn, L.E. Sparks, et al.:** The interaction of vapor phase organic compounds with indoor sinks. *Indoor Air 1*:1–23 (1991).
13. **Brief, R.S.:** A simple way to determine air contaminants. *Air Eng. 2*:39–41 (1960).
14. **U.S. Environmental Protection Agency (EPA):** *A Manual for the Preparation of Engineering Assessment.* Washington, DC: EPA, September 1984. [Unpublished draft]
15. **Wallace, L.A.:** "EPA/ORD/Monitoring Program." Presentation before the Science Advisory Board Panel on Indoor Air Quality and Total Human Exposure Committee, Washington, DC, March 29, 1989.
16. **Ruotsalainen, R., R. Ronnberg, J. Sateri, A. Majanen, et al.:** Indoor climate and the performance of ventilation in Finnish residences. *Indoor Air 2*:137–145 (1992).
17. **Harving, H., R. Dahl, J. Korsgaard, and S.A. Linde:** The indoor air environment in dwellings: a study of air-exchange, humidity and pollutants in 115 Danish residences. *Indoor Air 2*:121–126 (1992).
18. **Drivas, P.J., P.G. Simmonds, and F.H. Shair:** Experimental characterization of ventilation systems in buildings. *Environ. Sci. Technol. 6*:609–614 (1972).
19. **Tou, J.C. and G.J. Kallos:** Kinetic study of the stabilities of chloromethyl methyl ether and bis(chloromethyl) ether in humid air. *Anal. Chem. 46*:1866–1869 (1974).
20. **Sparks, L.E.:** Modeling indoor concentrations and exposure. *Ann. N.Y. Acad. Sci. 641*:102–111 (1992).
21. **Jayjock, M.A, D.R. Doshi, E.H. Nungesser, and W.D. Shade:** Development and evaluation of a

source/sink model of indoor air concentrations from isothiazolone treated wood used indoors. *Am. Ind. Hyg. Assoc. J. 56*:546–557 (1995).

22. **Fleischer, M.T.:** "An Evaporation/Air Dispersion Model for Chemical Spills On Land." Shell Development Center, Westhollow Research Center, P.O. Box 1380, Houston, Texas, 1980. [Pamphlet]
23. **Schroy, J.M. and J.M. Wu:** Emission from spills. In *Proceedings on Control of Specific Toxic Pollutants.* Gainesville, FL: Air Pollution Control Association, 1979.
24. **Chemical Manufacturers Association (CMA):** *PAVE—Program to Assess Volatile Emissions, Version 2.0.* Washington, DC: CMA, 1992.
25. **Braun, K.O. and K.J. Caplan:** *Evaporation Rate of Volatile Liquids* [EPA/744-R-92-001; NTIS PB 92-232305]. Washington, DC: U.S. Environmental Protection Agency, Office of Pollution Prevention and Toxics, 1989.
26. **Hummel, A.A., K.O. Braum, and M.C. Fehrenbacher:** Evaporation of a liquid in a flowing airstream. *Am. Ind. Hyg. Assoc. J.* 57:519–525 (1996).
27. **Fehrenbacher, M.C. and A.A. Hummel:** Evaluation of the mass balance model used by the Environmental Protection Agency for estimating inhalation exposure to new chemical substances. *Am. Ind. Hyg. Assoc. J. 57*:526–536 (1996).
28. **Jayjock, M.A.:** Back pressure modeling of indoor air concentration from volatilizing sources. *Am. Ind. Hyg. Assoc. J. 55*:230–235 (1994).
29. **Carslaw, H.S. and J.C. Jaeger:** *Conduction of Heat in Solids,* 2nd ed. London: Oxford University Press, 1959. pp. 260–261.
30. **Roach, S.A.:** On the role of turbulent diffusion in ventilation. *Ann. Occup. Hyg. 24*:105–132 (1981).
31. **Wadden, R.A., J.L. Hawkins, P.A. Scheff, and J.E. Franke:** Characterization of emission factors related to source activity for trichloroethylene degreasing and chrome plating processes. *Am. Ind. Hyg. Assoc. J. 52*:349–356 (1991).
32. **Wadden, R.A., P.A. Scheff, and J.E. Franke:** Emission factors from trichloroethylene vapor degreasers. *Am. Ind. Hyg. Assoc. J. 50*:496–500 (1989).
33. **Scheff, P.A., R.L. Friedman, J.E. Franke, L.M. Conroy, et al.:** Source activity modeling of Freon emissions from open-top vapor degreasers. *Appl. Occup. Environ. Hyg. 7*:127–134 (1992).
34. **Matoba, Y., J. Ohnishi, and M. Matsuo:** A simulation of insecticides in indoor aerosol space spraying. *Chemosphere 26*:1167–1186 (1993).
35. **Hawkins, N.C.:** Evaluating conservatism in maximally exposed individual (MEI) predictive exposure assessments: a first-cut analysis. *Reg. Toxicol. Pharmacol. 14*:107–117 (1991).
36. **Evans, J.S., D.W. Cooper, and P.L. Kinney:** On the propagation of error in air pollution measurements. *Environ. Monit. Assess.* 4:139 (1984).

Workers cleaning up hazardous waste sites, especially those involving organic chemicals, are exposed to a wide variety of toxic materials.

Outcome Competencies

After completing this chapter, the user should be able to:

1. Define underlined terms used in this chapter.
2. Explain the purpose and objectives of risk assessment and discuss the increasing reliance on risk assessment in public policy and decision making.
3. Define the role of risk assessment in establishing occupational health exposure limits.
4. Discuss the four steps of the risk assessment paradigm proposed by the National Research Council and modified by EPA.
5. Recall sources of uncertainty in a risk assessment, the impact of uncertainty on outcome, and ways of expressing uncertainty in a risk assessment.
6. Assess the validity of application of classical risk assessment techniques to traditional occupational hygiene.
7. Recall the issues of averaging times, use of IRIS toxicity values, selection of a low-dose extrapolation model, and summation of risks.
8. Explain how iterative risk assessment can optimize the balance between risk assessment and risk management.
9. Describe how the dimensions of risk affect public perception and acceptance of risk.
10. Contrast the purposes and rationales of risk assessment and risk management, and list some of the "nonscientific" factors that may be incorporated into risk management.
11. Given appropriate data, calculate a risk-based cleanup level for water or soil. Adapt this approach to develop a risk-based airborne concentration for the workplace.

Prerequisite Knowledge

Prior to beginning this chapter, the user should review the following chapters:

Chapter Number	Chapter Topic
1	History and Philosophy of Industrial Hygiene
2	Occupational Exposure Limits
4	Environmental and Occupational Toxicology
5	Epidemiological Surveillance
6	Principles of Evaluating Worker Exposure
15	Comprehensive Exposure Assessment
16	Modeling Inhalation Exposure
30	Worker Education and Training
34	Prevention and Mitigation of Accidental Chemical Releases
41	Risk Communication in the Workplace

Key Terms

chronic daily intake • exposure • hazards • hazard index • hazard quotients (HQs) • Integrated Risk Information System (IRIS) • iterative risk assessment • lifetime cancer risk estimate • Monte Carlo analysis • reasonable maximum exposure • reference dose • risk • risk assessment • risk characterization • risk estimate • risk management • toxicity assessment • toxicity value • uncertainty

Key Topics

- I. Introduction
  - A. Purpose
  - B. Definition and Overview of Risk
- II. Development of the Art of Risk Assessment
  - A. Brief History and Background
  - B. Risk Assessment and OSHA
  - C. Risk Assessment in Other Federal Agencies
- III. The Risk Assessment Process
  - A. General
  - B. Occupational Hygiene Risk Assessment
  - C. EPA Superfund Risk Assessment
- IV. Uncertainty and Iterative Risk Assessment
  - A. Sources of Uncertainty in Risk Assessment
  - B. Monte Carlo Analysis
  - C. Tiered Approach to Occupational Hygiene Risk Assessment
- V. Risk Perception, Risk Communication, and the Mismatch Between Risks and Regulations
- VI. Risk Management
  - A. Purpose and Definitions
  - B. Determination of Appropriate Exposure Levels

17

Risk Assessment in the Workplace

Deborah Imel Nelson, Ph.D., CIH

Introduction

Purpose

Although the term "risk assessment" has only recently entered the occupational hygiene vocabulary, the process itself is not new to the profession. In their efforts to provide safe and healthful environments for workers and the general public, occupational hygienists and environmental professionals have always sought answers to questions such as these:

- What is the appropriate workplace exposure level for a chemical not listed in the Occupational Safety and Health Administration (OSHA) permissible exposure limit (PEL) or American Conference of Governmental Industrial Hygienists (ACGIH) threshold limit value (TLV®) tables?
- How much lead can safely remain in contaminated soil at an industrial site?
- Is it acceptable to allow a particular pesticide to be used on apples, a fruit frequently consumed by children?
- Is it necessary to install a ventilation system to protect a few workers during an industrial process that is conducted intermittently?

These questions all involve the subject of risk (that is, the probability and magnitude of harm). The immediate responses are based on risk assessment, which is the process of determining, either quantitatively or qualitatively, the probability and magnitude of an undesired event and estimating the cost to human society or the environment in terms of morbidity, mortality, or economic impact. (For occupational hygienists, the process of risk assessment boils down to asking, "Will any of the workers in this department get sick from working with this chemical?") The ultimate goal of these questions is the efficient management of occupational and environmental risks at acceptable levels. Since we as a society do not possess adequate resources to control all hazards at the zero-risk level (even if such a thing were possible), other factors must be considered in selection of risk management goals and techniques. Economics, technical feasibility, political pressures, and public opinion play large roles in decision making about risk, but accurate risk assessment is the basis for effective distribution of limited human, financial, and organizational resources.

Since the early days of the practice of occupational hygiene, the profession has undergone several changes in its basic motivations and objectives. As described eloquently in J.K. Corn's historical perspective on occupational health hazards,[(1)] changes in the occupational hygiene profession have mirrored the technological advances and political events of the times. Key events in the development of the profession have included the shift from an agricultural to a manufacturing-based economy; the introduction of workers' compensation laws to assist victims and their families; creation of the U.S. Department of Labor and the Office of Industrial Hygiene and Sanitation in the U.S. Public Health Service; the Walsh-Healey Public Contracts Act; provisions in the Social Security Act to aid state efforts in occupational hygiene; founding of ACGIH and the American Industrial Hygiene Association; and the enactment of the

Occupational Safety and Health (OSH) Act of 1970.[1] Events such as the tragedy of Bhopal, India; economic pressures that have led to corporate downsizing; and reinvention of government have continued to affect the practice of occupational hygiene. Corporate decision making is now less influenced by government regulations and more concerned with enhancing profitability and avoiding liability.

These changes call for updated tools in the occupational hygienist's toolbox. It is the thesis of this chapter that risk assessment is a powerful, value-added tool that can assist in decision making, priority setting, and protection of worker safety and health. The major goal of this chapter is, thus, to stimulate the integration of risk assessment into the practice of occupational hygiene. Specific objectives are to introduce risk, risk perception, the development of risk assessment in occupational and environmental health, the role of risk assessment in risk management decision making, and the importance of matching risk assessment efforts to the risk at hand.

Definition and Overview of Risk

In occupational hygiene the source of risk or hazard may include chemicals (the focus of this chapter), noise, ionizing or nonionizing radiation, extremes of temperature, or repetitive motion. A worker's risk from handling a chemical is a function of both the level of exposure (including duration and concentration) and the toxicity of the chemical. Increasing the exposure or the toxicity of the chemical will result in increased risk (a critical concept that is frequently misunderstood by the public). Therefore, risk assessment requires a two-pronged approach: exposure assessment and toxicity assessment. Granted, there are many highly toxic compounds that would be a serious source of risk if exposure occurred; however, if there is little exposure, there is little risk. For purposes of this chapter, risk assessment is thus the science-based evaluation of exposure to a chemical (or process or energy), followed by comparison with some measure of the known human health effects of exposure to that chemical; in other words, predicting the probability of health effects resulting from a given exposure.

One might have the impression that these are extremely risky times. Without question, people are exposed to numerous hazards in today's fast-paced, industrialized society. But despite public perception these are indeed the least risky, safest times in history. In prehistoric times a Neanderthal could expect to live to the age of 29.4, while the average Roman citizen lived to be 32 years old.[2] The average life expectation of Americans began to climb steadily from 33.5 years in 1690,[2] to the current expectation of 76.3 years for babies born in 1995.[3] This level is exceeded only marginally by 19 other countries, primarily in Europe, although Hong Kong is the winner at 80 years. In contrast, the expected lifespan for a child born in 1995 is 36.6 years in Uganda, 39 years in Malawi, and under 49 years in 14 additional countries in Africa, Asia, and the Caribbean.[3]

The cause of death of the vast majority of Americans is not related to unavoidable hazards of the industrial age, but rather to individual lifestyle choices. The top four killers as shown in Table 17.1 — cardiovascular disease, malignancies, chronic obstructive pulmonary diseases, and accidents —are directly related to personal choices of diet, exercise, smoking, alcohol consumption, and use of seat belts. Estimates of the percentage of annual cancer mortality associated with environmental exposures range from 1 to 5% for pollution, 2 to 6% for geophysical factors, 1 to 8% for occupation, and 1 to 2% for consumer products, for a total of 5 to 22%.[2] According to the U.S. Department of Labor, Bureau of Labor Statistics,[4] there were 6210 occupational fatalities related to injuries in 1995, primarily due to transportation accidents and homicides.

The major changes in the leading causes of death in the last quarter of the 20th century reflect the dramatic increase in deaths due to infectious and parasitic diseases, primarily HIV infection, and the increase in death rates due to chronic obstructive pulmonary diseases, for which the major risk factor is cigarette smoking.[5] To quote the cartoon character Pogo, "we have met the enemy and he is us." This is not to downplay occupational and environmental risks, but rather to put them into perspective.

Development of the Art of Risk Assessment

Brief History and Background

People have been exposed to hazards since the beginning of human history, so it is not surprising that risk assessment is not a new concept. Covello and Mumpower[6] described an early group of risk analysts who lived in the Tigris-Euphrates valley about 3200 B.C. These consultants would identify important dimensions of a problem, delineate alternatives, predict outcomes, and recommend the most favorable outcome—a process not dissimilar from modern risk assessment.

Covello and Mumpower listed additional studies conducted during the 16th to the 19th centuries that established the basis for the analysis of environmental risk, including many familiar examples: Agricola's 1556 study linking health effects to mining and metallurgical practices; work by Ramazzini in the early 1700s identifying higher rates of breast cancer in nuns; Sir Percival Pott's study relating scrotal cancer to employment as a chimney sweep; John Snow's removal of the handle from the Broad Street pump in 1854 to prevent the spread of cholera.[6] And what occupational hygiene student has not studied Paracelsus' (1493–1541) famous statement: "All substances are poisons; there is none which is not a poison. The right dose differentiates a poison and a remedy."[7]

Thus, it is difficult to understand why so many Americans today are frightened by the belief that we are surrounded by extremely complex, life-threatening situations that require expert (perhaps governmental) intervention. Also difficult to rationalize is the mismatch of much environmental law and subsequent expenditures on hazards that most experts believe pose little risk to human health (see section on Risk Perception, Risk Communication, and the Mismatch between Risks and Regulations). Plough and Krimsky have suggested that there was a series of events beginning in the late 18th century that led to (1) a transition from folk-centered to expert-centered communication about risk; and (2) if not failure, at best limited success in public risk policy due to difficulty in bridging the gaps between "risk perception guided by human experience" and "probabilistic thinking" by scientists.[8] Their analysis may shed light on the role of the federal government in risk assessment in the United States. According to these authors, the first necessary condition was the rise of the modern state, which has a responsibility to protect the population from physical harm. Second, the development of public health institutions and of modern medicine as the profession that defined health risks and controlled intervention strategies further suggested that the experts should be trusted to control risks. And third, the development of decision analysis during World War II initiated an era of government-sponsored research to develop a rational framework for economic and military decision making. These events led federal regulators, among others, to hope that decision methodologies could be applied to rational analysis of risk factors in public health, medicine, and the environment.

Table 17.1 — Leading Causes of Death in the United States

Cause	*Crude Death Rate/100,000* 1970 (rank)	1993 (rank)
All causes	945.3	879.3
Major cardiovascular diseases, including cerebrovascular diseases	496.0 (1)	366.3 (1)
Malignancies[A]	162.8 (2)	205.8 (2)
Chronic obstructive pulmonary diseases and allied conditions	15.2 (7)	39.2 (3)
Accidents and adverse effects[B]	56.4 (3)	34.4 (4)
Pneumonia and influenza	30.9 (4)	31.7 (5)
Diabetes mellitus	18.9 (5)	21.4 (6)
Other infectious and parasitic diseases[C]	3.4 (11)	17.8 (7)
Suicide	11.6 (8)	12.1 (8)
Homicide and other legal intervention	8.3 (9)	9.9 (9)
Chronic liver disease and cirrhosis	15.5 (6)	9.6 (10)
Nephritis, nephrotic syndrome, and nephrosis	4.4 (10)	9.1 (11)

[A]In 1990 the major cancers for women were lung, breast, and colon-rectal; for men they were lung, stomach, and colon-rectal. (Source: Reference 69, p. 216.)
[B]In 1992 accidents (including motor vehicle accidents) were the leading cause of death (COD) for males and females 1 to 24 years old. For males 25 to 44 the leading COD was human immunodeficiency virus (HIV) infection; for males aged 45 and over, heart disease was the biggest killer. For females aged 25 to 64 malignant neoplasms were the leading COD; after age 65 heart disease was responsible for the most deaths.[3]
[C]When HIV was examined separately, it ranked as the ninth leading COD for males of all ages in the United States in 1992.[3]
Source: U.S. Bureau of the Census. Statistical Abstract of the United States, 1995, 115th ed. Washington, DC: U.S. Bureau of the Census, 1995. Tables 125, 127, and 132.

To this analysis must be added the perspective that with scientific and engineering progress, the nature of many chemical environmental hazards was becoming increasingly complex and difficult for the lay public to comprehend. Most people were forced to rely on interpretations by experts or the media for their information about environmental hazards. So it seems only natural that people tended to rely more heavily on the federal government to provide protection from environmental risks.

Federal regulation of environmental risks began very slowly in the United States with the River and Harbor Act of 1899, which prohibited the discharge of refuse into navigable waters.[2] In 1914 the railroad industry asked the U.S. Public

Health Service to suggest their first drinking water standards to protect passengers from variable water quality along the routes. While there were no regulations passed, it was understood that trains would not stop in towns that did not meet these standards.[9] In 1958 the Food, Drug, and Cosmetics Act was amended by the Delaney clause, which declared as unsafe any additive found to cause cancer in man or animal.

For reasons including those described above, and the emergence of environmentalism as a social movement, the late 1960s and early 1970s saw a virtual explosion in environmental legislation, such as creation of the U.S. Environmental Protection Agency (EPA) and OSHA, the Clean Air Act of 1970, and the Federal Water Pollution Control Act of 1971. These statutes addressed real and present problems: pollution of water and air and hazards in the workplace. Many of these statutes (or subsequent amendments) contained language that, although not specifically requiring risk assessment, danced all around it: "toxic pollutants in toxic amounts"; "unreasonable risk to man or the environment" (see Table 17.2).

The developing art of risk assessment was increasingly used as a means of bridging the gaps between scientific study of the relationships between exposure to toxic chemicals and health effects, and the development of public policy to protect the public and the environment from unreasonable risk. An increased reliance on quantification by policymakers was well evident by the early 1980s.[10]

The second series of environmental laws,[11] which included the Safe Drinking Water Act of 1974 and the Toxic Substances Control Act of 1976, addressed public concerns about pesticides and toxic chemicals. Language became more risk oriented: "no known or anticipated adverse health effects"; "unreasonable risk of injury to health or the environment." Following events such as the hazardous waste disaster at Love Canal, N.Y., the third wave of legislation[11] addressed hazardous waste handling practices and the cleanup of previously contaminated sites. The Comprehensive Environmental Response, Compensation, and Liability Act (Superfund) of 1980 had a tremendous significance for modern environmental risk assessment through its requirement that EPA revise the National Contingency Plan, including addition of methods for evaluating and remedying releases from facilities that pose substantial danger to the public health or the environment. These revisions ultimately led to the frequently cited EPA series on risk assessment for Superfund.[12–14]

Environmental legislation into the 1990s was increasingly directed at specific environmental risks (asbestos in schools, underground storage tanks, lead in drinking water, lead-based paint, etc.), leading indirectly to increased use of risk assessment.[11] National interest in and limited resources for remediation of underground storage tank sites has led to the introduction of risk assessment methodology for establishment of cleanup levels by the American Society for Testing and Materials (ASTM).[15] One piece of legislation from this time frame has had a somewhat delayed impact on risk assessment: the Clean Air Act Amendments (CAAA) of 1990. The amendments established a commission to conduct a review of risk assessment methodology used by EPA to determine carcinogenic risk associated with exposure to hazardous air pollutants and to suggest improvements in such methodology. Published as *Science and Judgment in Risk Assessment* in 1994, the document provided specific recommendations on the practice of risk assessment by EPA: the general retention of the conservative, default-based approach to screening analysis (subject to modifications), an iterative approach to risk assessment, and presentation of sources and magnitude of uncertainty along with point estimates of risk.

Also established by the CAAA was the Risk Assessment and Management Commission, which was charged with making a full investigation of the policy implications and appropriate uses of risk assessment and risk management in regulatory programs under various federal laws to prevent cancer and other chronic human health effects that may result from exposure to hazardous substances.

Risk legislation can be divided into three main categories: (1) risk-based laws designed to reduce risks to zero; (2) balancing laws that balance cost and benefit and provide for some risk above zero, and (3) technology-based laws that require use of a specific technology or that force development of new technology.[1,16] Examples of risk-based regulations are the Safe Drinking Water Act provisions for maximum contaminant level goals established at a level to prevent adverse health effects and the Delaney amendment to prohibit carcinogenic food additives. Examples of risk-benefit balancing include EPA's lead phasedown decision and proposed ban on asbestos-containing materials.[10] The basis of the OSH Act is technology or balancing; that is, it "assures, to the extent feasible . . . that no employee will suffer material impairment of health or functional capacity."[17] The courts have agreed that OSHA can enforce a rule that will require development and/or distribution of new technologies.[10] (Ironically, the cotton dust

Table 17.2 — Selected Major U.S. Statutes Involving Risk Assessment

Law	Date[A]	Agency	Pertinent Language or Impact on Risk Assessment
Delaney amendment to the Food, Drug and Cosmetic Act	1958	FDA	21 USC 348(c)(3): "That no additive shall be deemed to be safe it if is found to induce cancer when ingested by man or animal, or if it is found, after tests which are appropriate for the evaluation of the safety of food additives, to induce cancer in man or animal."
Occupational Safety and Health Act	1970	OSHA	29 USC 655(b)(5): "No employee shall suffer material impairment of health or functional capacity."
Federal Water Pollution Control Act (Clean Water Act)	1971	EPA	33 USC 1251(a)(3): "It is the national policy that the discharge of toxic pollutants in toxic amounts be prohibited."
Federal Insecticide, Fungicide, and Rodenticide Act	1972	EPA	7 USC 136 (bb): "Unreasonable adverse effect on the environment" means any unreasonable risk to man or the environment, taking into account the economic, social, and environmental costs and benefits of the use of any pesticide."
Safe Drinking Water Act	1974	EPA	42 USC 300 g-1(b)(4): "Each maximum contaminant level goal . . . shall be set at the level at which no known or anticipated adverse effects on the health of persons occur and which allows an adequate margin of safety."
Resource Conservation and Recovery Act	1976	EPA	42 USC 6903(5): "'Hazardous waste' means . . . may cause, or significantly contribute to an increase in mortality or an increase in serious irreversible, or incapacitating reversible illness; or pose a substantial present or potential hazard to human health or the environment."
Toxic Substances Control Act	1976	EPA	15 USC 2601(a)(2): "Among the many chemical substances and mixtures . . . there are some whose manufacture, processing, distribution in commerce, use, or disposal may present an unreasonable risk of injury to health or the environment . . . necessitates the regulation of intrastate commerce."
Comprehensive Environmental Response, Compensation and Liability Act (CERCLA, Superfund)	1980	EPA	Established a national program for responding to release of hazardous substances into the environment. Called for revision of the national contingency plan for the removal of oil and hazardous substances, originally required by 33 USC 1321, Federal Water Pollution Control Act. The revision was to include methods for evaluating and remedying releases from facilities that pose substantial danger to the public health or the environment.
Clean Air Act Amendments	1990	EPA	42 USC 7412(f)(1): "The Administrator shall investigate and report . . . methods of calculating the risk to public health remaining . . . after the application of standards, the public health significance of such estimated remaining risk."
			42 USC 7412(f)(2): "If standards . . . do not reduce lifetime excess cancer risks to the individual most exposed to emissions . . . to less than one in one million, the Administrator shall promulgate standards."
			Established the National Academy of Sciences to review risk assessment methodology used by EPA to determine carcinogenic risk associated with exposure to hazardous air pollutants and suggest improvements in methodology. Published as *Science and Judgment in Risk Assessment* by the National Research Council, 1994.
			Established the Risk Assessment and Management Commission, which was to investigate the policy implications and appropriate uses of risk assessment and risk management in regulatory programs under various federal laws to prevent cancer and other chronic human health effects that may result from exposure to hazardous substances. Published as *Risk Assessment and Risk Management in Regulatory Decision-Making—Draft Report for Public Review and Comment* on June 13, 1996.

[A]Most environmental legislation has been amended several times; EPA was not yet in existence when many of the major laws were first enacted.

Sources: Suggested by and adapted from Reference 66, Figure 3. Also, Superintendent of Documents' Home Page on the U.S. Government Printing Office Web Site, http://www.access.gpo.gov./su_docs/; *Selected Environmental Law Statutes, 1994-95*, educational ed. St. Paul, MN: West Publishing Co., 1994; and Reference 2.

standard, which was technology forcing and which industry feared would have negative impacts, resulted in improved competitiveness and productivity.[10] Similarly, the vinyl chloride standard resulted in lower industry costs and improved productivity.[1]) Some laws may have two or three bases. For example, the stationary sources section of the Clean Air Act is risk-based, the vehicle section is technology-forcing, and the fuel section is risk-balancing.[16]

Since the early 1980s there has been a trend toward introduction of more risk-related legislation in Congress.[18] The 1974 and 1978 indices listed no risk-related references, and the 1982 index listed only one. By 1986 the 4-year index included over two pages of risk-related citations, a level that continued through 1990 and 1994. At this writing the U.S. Congress has not passed global risk assessment legislation, although the 104th Congress introduced a large number of proposed bills. These proposals fell into two general categories: (1) how to conduct risk assessment; and (2) when to conduct risk assessment. None of these bills was passed into law, although one was passed by both houses: The Food Quality Protection Act of 1996, which would have required risk assessment before allowing tolerances or exemptions of pesticide chemical residues.[19] The content of risk-related legislation introduced into Congress has become increasingly sophisticated, often with specific language on such issues as selection of low-dose extrapolation, analysis and presentation of uncertainty, use of data from human vs. animal studies, and risk communication.

Risk Assessment and OSHA

When OSHA was created in 1970, the enabling legislation directed it to adopt national consensus standards and existing federal standards. Thus, OSHA adopted some of the American National Standards Institute (ANSI) occupational exposure limits and the 1968 TLVs, which were in effect for employers subject to the Walsh-Healey Public Contracts Act. There were also provisions in the OSH Act for development of new safety and health standards, many of which were challenged in the courts by industry and/or labor organizations. Table 17.3 presents a chronology of OSHA's actions, placing them in context with other events significantly impacting occupational hygiene risk assessment. Highlights of important events are outlined below.[1,16,20,21]

Undoubtedly, the issue with the greatest impact on occupational hygiene risk assessment was the U.S. Supreme Court benzene decision of July 1980. In late 1976, based on epidemiological studies showing elevated rates of leukemia, the National Institute for Occupational Safety and Health (NIOSH) strongly recommended to OSHA that it issue an emergency temporary standard (ETS) to protect workers from benzene exposure. In April 1977 NIOSH forwarded to OSHA the results of epidemiological studies of two Pliofilm plants that showed increased incidence of leukemia. OSHA immediately issued an ETS, lowering the benzene standard from 10 ppm to 1 ppm with a 5-ppm ceiling. In May, the Court of Appeals for the Fifth Circuit issued a temporary restraining order against the ETS. Since ETSs are short-lived (six months), OSHA decided to focus on the permanent standard, and the ETS expired. Based on its "lowest feasible" level policy for exposure to carcinogens, OSHA then issued a new permanent standard (with the same PEL) in 1978. The standard was vacated in October 1978 by the Court of Appeals for the Fifth Circuit, as OSHA had failed to provide an estimate of the benefits and did not evaluate costs and benefits. In July 1980 the Supreme Court, in a divided vote, invalidated the benzene standard, but not, as expected, on the cost-benefits issue. It found that OSHA had failed to show by substantial evidence that the standard would reduce a significant risk in the workplace. The Court stated that a safe workplace was not a risk-free environment and suggested a lifetime occupational cancer of 1 in 1000 (10^{-3}) as a reasonable guideline for identifying significant risk. Seven years later OSHA issued a new benzene standard of 1 ppm (8-hour) and 5-ppm short-term exposure limit (STEL) based on risk assessments. The agency stated that this standard substantially reduced the significant risk of employees exposed at the existing level. The standard was effective December 1987.

The benzene decision was a landmark, both because it established a requirement that the agency determine the existence of a significant risk before it initiated rulemaking and because it provided a benchmark against which to judge occupational risk (i.e., 10^{-3}). It also affirmed the use of conservative assumptions that were supported by a body of reputable scientific thought, and the use of animal studies as a basis for human health risk assessment.

One of the first actions taken by the new agency was to issue an ETS for asbestos in

**Table 17.3 —
A Chronology of Selected Events Impacting Occupational Hygiene Risk Assessment**

Date	Event	Reference
1930s	Industrial hygiene and occupational medicine professionals developed concept of dose-response relationship.	1
1938	Founding of the organization now known as the American Conference of Governmental Industrial Hygienists.	1
1938	First suggested guidelines for asbestos, Dreessen et al., 1938.	1
1939	Founding of the American Industrial Hygiene Association (AIHA).	
1940	Bowditch et al. article on occupational exposure limits.	1
1940s	ACGIH began to develop guidelines for occupational exposure limits, now known as threshold limit values.	1
1947	Enactment of Federal Insecticide, Fungicide, and Rodenticide Act (FIFRA).	10
1958	Passage of Delaney amendment to the Pure Food, Drug, and Cosmetic Act.	16
1961	Mantel and Bryan article on virtually safe doses published in the *Journal of the National Cancer Institute.*	21
1970	Creation of OSHA, NIOSH, and EPA.	
1971	(May) OSHA adopted 1968 ACGIH TLVs as permissible exposure limits.	1
1973	Early FDA regulatory decision utilizing risk assessment based on Mantel and Bryan (1961).	21
1975	Eighth Circuit Court's decision on Reserve Mining Co. v. EPA.	10
1976	EPA proposed the 1976 interim guidelines, the first risk assessment guidelines.	16
1976	In Ethyl Corp v. EPA, the District of Columbia Circuit Court upheld EPA's order reducing lead in gasoline, ruling that actual injury was not required for endangerment to exist, and that risk assessment could form the basis for health-related regulations under the Clean Air Act.	10
1976	Enactment of the Toxic Substances Control Act. Regulation of toxic substances became a major theme of federal environmental statutes, which are characterized by unwillingness to wait for definitive proof.	10
1977	Formation of the Interagency Regulatory Liaison Group (abolished in 1981).	16
1977	OSHA Draft Cancer Policy issued.	16
1977	FDA regulatory decisions used 10^{-6} guideline.	21
1977	(April) OSHA issued ETS for benzene.	20
1977	(May) Court of Appeals for the Fifth Circuit issued a temporary restraining order against the ETS for benzene.	10
1978	OSHA issued new permanent standard for benzene.	1, 20
1978	Congress required publication of annual list of known or suspected carcinogens to which a significant number of people are exposed. National Toxicology Program created to coordinate federal government toxicity testing.	10, 16
1978	(October) Benzene standard vacated by the Court of Appeals for the Fifth Circuit.	20
1979	*Scientific Bases for Identification of Potential Carcinogens and Estimation of Risks* published by the Interagency Regulatory Liaison Group.	1, 16
1979	Court of Appeals for the District of Columbia affirmed the cotton dust standard.	10
1980	Comprehensive Environmental Response, Compensation and Liability Act (also known as Superfund) was enacted.	
1980	OSHA issued the Final Generic Cancer Policy.	16, 20
1980	(July) U.S. Supreme Court invalidated the benzene standard.	10, 20, 21
1981	Executive Order 12291 signed.	1, 10, 20
1981	U.S. Supreme Court decision on OSHA cotton dust standard.	10, 20
1982	EPA standards for pesticide residue in foods based on 10^{-6}, FIFRA risk-benefit analysis.	21
1983	The District of Columbia Circuit Court wrote that the correlation between declining blood lead levels and EPA-mandated reductions in lead-in-gasoline justified EPA's stricter lead limits imposed in 1982.	10

(Continued on p. 336)

**Table 17.3 (continued) —
A Chronology of Selected Events Impacting Occupational Hygiene Risk Assessment**

Date	Event	Reference
1983	National Academy of Sciences published the "Red Book," *Risk Assessment in the Federal Government: Managing the Process.*	
1984	Passage of the Hazardous and Solid Waste Amendments, eventually leading to increased use of risk assessment in management of leaking underground storage tanks.	
1986	Passage of the Superfund Amendments and Reauthorization Act (SARA).	
1986	OSHA announced it would address "existing, significant risks" and that requirements would "significantly" reduce these risks.	20
1986	(November) California voters approved Proposition 65.	10
1986	EPA Cancer Guidelines, and the *Superfund Public Health Exposure Manual.*	
1988	OSHA began the PEL Project, designed to revise and incorporate TLVs as standards.	1, 20
1989	First volume in EPA *Risk Assessment Guidance for Superfund* published.	12
1989	(January) OSHA published the final revised air contaminants standard.	1, 20
1990	Enactment of CAAA, including certain requirements for risk assessments of hazardous air pollutants.	
1992	(February) EPA issued Habicht Memorandum, "Guidance on Risk Characterization for Risk Managers and Risk Assessors."	
1992	(July) Court of Appeals for the Eleventh Circuit ruled that OSHA's actions under the PEL Program were invalid.	20
1993	(March) OSHA announced it would enforce the 1971 PELs.	20
1994	AIHA Risk Assessment Task Group completed the "AIHA Risk Assessment Position Paper" and "AIHA Risk Assessment White Paper."	
1994	As mandated by the CAAA, *Science and Judgment in Risk Assessment* published by the National Research Council.	
1995	The American Society for Testing and Materials published *Risk-Based Corrective Action Applied at Petroleum Release Sites, E 1739-95.*	
1995	(January) AIHA President Jerry Lynch testified at U.S. House of Representatives, Commerce Committee, hearings on H.R. 9, submitting the AIHA Risk Assessment Position Paper and White Paper into the record.	
1995	(March) EPA issued Browner Memorandum, "Policy for Risk Characterization at the U.S. Environmental Protection Agency."	
1996	(June) *Risk Assessment and Risk Management in Regulatory Decision-Making—Draft Report for Public Review and Comment* published by the Commission on Risk Assessment and Risk Management.	
1996	ASTM committee in process of developing RBCA protocol for general chemical exposures.	

May 1971, followed by a final standard of 5 fibers/cc in June 1972. In October 1975 rulemaking was initiated to reduce the PEL to 2 fibers/cc. In November 1983, based on its risk assessment of asbestos exposure, OSHA issued an ETS lowering the PEL to 0.5 fibers/cc; this ETS was held invalid by an appeals court, which stated that risk assessment of controversial issues requires public scrutiny through the notice-and-comment rulemaking procedure. In June 1986 a final asbestos standard that lowered the PEL from 2.0 to 0.2 fibers/cc was issued. Although the Appeals Court for the District of Columbia upheld the standard, it was returned to OSHA for further study. In September 1988 OSHA promulgated a STEL of 1 fiber/cc for 30 minutes. In 1990 OSHA proposed to lower the PEL from 0.2 to 0.1 fiber/cc for all industries; this PEL was issued in August 1994. The standard has been challenged by both union and industry groups, and the appeal is pending.

Cotton dust was the issue on which the cost-benefit issue for occupational safety and health was decided. After years of evidence linking cotton dust and byssinosis, OSHA published a proposed cotton dust standard in December 1976, followed by a final standard in June 1978. It was appealed, and then affirmed by the Court of Appeals for the District of Columbia, which stated that the OSH Act did not require cost-benefit analysis. The Supreme

Court upheld this decision, but also opened the door to the use of cost-benefit analysis, stating that it could be used to adopt the least expensive means to achieve a given level of protection. The Court stated that Congress itself had defined the cost-benefit relationship, placing the benefit of worker health above all considerations except those making this benefit unachievable.

OSHA was an early player in carcinogenic risk assessment and regulation. In 1977 the agency joined the Interagency Regulatory Liaison Group and also issued its draft cancer policy, the purpose of which was to provide a framework for the regulation of carcinogens. The final generic cancer policy was issued in 1980, although it was never used.

Recognizing both the need to update the PELs and the difficulty of regulating substance-by-substance, in 1988 OSHA began the PEL Project, which was designed to revise and incorporate the TLVs as PELs. Shortly thereafter, an article severely criticizing the TLV process was published in the *American Journal of Industrial Medicine*,[22] setting off a firestorm of comments and responses to the journal. In 1989 hearings were held and the final revised air contaminants standard was issued. The standards were challenged, and in July 1992 the Court of Appeals for the Eleventh Circuit ruled that OSHA's actions were invalid. OSHA had failed to show, as required by the benzene decision, that each standard showed a significant hazard in the workplace, or that the PELs were feasible. In March 1993 the agency announced that it would enforce the 1971 PELs.

Other OSHA actions based on risk assessments include the ethylene oxide standard, which was issued in June 1984. In ruling on the challenge to this standard, the Court of Appeals for the District of Columbia basically affirmed the OSHA PEL for ethylene oxide, but remanded the issue to OSHA for further proceedings on the STEL. The ruling granted OSHA leeway in regulating on the frontiers of current knowledge, demanding no more than that OSHA "arrive at a reasonable conclusion based on all the evidence before it."[23] The Court also found that the no-threshold assumption was supported by substantial evidence. The Court of Appeals for the District of Columbia ordered OSHA to issue a final rule on the ethylene oxide STEL by March 1988; a 5-ppm STEL was issued in April 1988.

Formaldehyde is another standard in which risk assessment played a role. In 1987 OSHA issued a formaldehyde standard: 1-ppm PEL (8 hour) and 2-ppm STEL. Both industry and union groups challenged, and the Court of Appeals for the District of Columbia remanded the issue to OSHA, stipulating that (1) if a significant risk remained at 1 ppm, and it were feasible to do so, OSHA must promulgate a lower standard; and (2) medical removal protection must be provided. The agency issued a final rule on formaldehyde, dropping the PEL from 1 to 0.75 ppm and providing medical removal protection.

In January 1997 OSHA issued a proposed standard for methylene chloride. The PEL, based on animal studies demonstrating carcinogenicity, was reduced to 25 ppm, with no ceiling, and a STEL of 125 ppm for 15 minutes.[24]

Risk Assessment in Other Federal Agencies

Other federal agencies that have used and/or developed risk assessment methodologies are the U.S. Food and Drug Administration (FDA), EPA, and the Nuclear Regulatory Commission (NRC). FDA was the first agency to use risk assessment methodologies formally in the regulatory process.[21] In a 1973 proposal on the regulation of food residues containing carcinogenic drugs, they suggested the "virtually safe dose" approach (corresponding to 10^{-8} risk) published by Mantel and Bryan in 1961.[25] Risk assessment was firmly established at the FDA, which later adopted a maximally acceptable lifetime risk of 10^{-6}.

EPA has also had a substantial impact on risk assessment methodology. Since the legislation that the agency enforces is risk-balancing (Federal Insecticide, Fungicide, and Rodenticide Act; Safe Drinking Water Act; Toxic Substances Control Act; Clean Air Act), risk-based (Resource Conservation and Recovery Act, Clean Air Act), or technology-forcing (Clean Air Act), the agency has found itself right in the middle of the risk debate. Its risk assessment activities have been both program-specific and agency-wide and have included development of risk assessment guidelines and methodologies and use of risk assessment in development of standards. Similar to OSHA, EPA has often found itself dealing with risk assessment in the courtroom, with results that impacted risk assessment beyond the agency. For example, in 1976 the District of Columbia Circuit Court upheld EPA's order reducing lead in gasoline, ruling that actual injury was not required for endangerment to exist, and that risk assessment could form the basis for health-related regulations under the Clean Air Act (Ethyl Corp. v. EPA).[10] EPA has also been involved in establishing acceptable risk levels. In 1981 it proposed worker radiation protection rules based on risk assessment (3 deaths per 1000) and comparison with death rates in the safest industries (2.7 in 1000).[16] (NRC made a similar proposal in 1986.) In 1982 the agency issued standards for pesticide residues in foods based on an acceptable risk level of 10^{-6}.

In 1976 EPA issued interim guidelines, that is, the first risk assessment guidelines.[16] In 1984 EPA's *Risk Assessment and Management: Framework for Decision Making* appeared. In September 1986 EPA published the *Risk Assessment Guidelines* in the *Federal Register*.[26] These guidelines were partially in response to the National Academy of Sciences' recommendation[27] and were designed to promote high technical quality and agency-wide consistency. Written by EPA scientists and receiving extensive peer-review, the guidelines covered carcinogenic risk assessment, estimating exposures, mutagenicity risk assessment, health assessment of suspect developmental toxicants, and health risk assessment of chemical mixtures. In addition to use of risk assessment within the agency, in the mid-1980s EPA began issuing guidance for Superfund risk assessors, including the *Superfund Public Health Exposure Manual* (1986) and *Superfund Exposure Assessment Manual* (1988), followed by the *Risk Assessment Guidance for Superfund* series.[12–14] Guidance on risk characterization was provided in the form of memoranda in 1992 and 1995.[28,29] While much of EPA's activity has been in support of human health risk assessment, the agency has recently begun to develop guidance for ecological risk assessment.[30,31] EPA's involvement in risk assessment has been long and varied; only a few highlights have been presented.

The Risk Assessment Process

General

With more federal agencies conducting risk assessments, and vociferous public debate over issues such as saccharin, nitrites in food, formaldehyde in home insulation, etc., consistency in approach was needed. Thus, Congress authorized FDA to perform a study to "strengthen the reliability and objectivity of scientific assessment that forms the basis for federal regulatory policies applicable to carcinogens and other public health hazards."[27] The study was conducted by the National Research Council, which effectively formalized the process with the 1983 publication of the "Red Book," *Risk Assessment in the Federal Government: Managing the Process.*

In the Red Book, a widely quoted four-step paradigm of risk assessment was outlined, along with the recommendation that insofar as possible, these steps were to be guided by the best scientific information available. (In contrast, it would be appropriate for risk management decisions to include political, economic, and social/cultural factors.) These steps, and the questions they ask, are as follows:

1. Hazard identification—Is there a causal link between this chemical and human health effects?
2. Dose-response assessment—What is the relationship between level of exposure and probability of human health effects?
3. Exposure assessment—What is the level of exposure to the chemical?
4. Risk characterization—What is the overall level of risk?

These four questions have been memorably rephrased:[32] Is this stuff toxic? How toxic is it? Who is exposed to this stuff, how long, how often? So what?

This paradigm, which has had broad use, has served to separate risk assessment and risk management and to crystallize thinking about scientific data and inferences (default assumptions). It has provided a structure and a vocabulary for the risk assessment process. Its limitations include the fact that it is a single-chemical rather than holistic approach to risk assessment, and it cannot address adversity of effects, such as cancer vs. teratogenic effects.[32] Despite these limitations, the paradigm provides a useful model for discussing risk assessment. Each of these steps, along with some of the most important issues surrounding them, will be discussed below. It should be noted that there are some differences between the approaches typically taken by occupational hygiene and environmental risk assessors. Also, the various steps of risk assessment may be conducted by different analysts.

Hazard Identification

The first task of risk assessment is to determine whether the chemical under study is linked to human health effects. Sources of data include toxicological, clinical, and epidemiological studies; supporting data may be found in short-term tests, such as those for mutagenicity, and in structure-activity relationships to known carcinogens. Issues to be addressed in toxicological studies include the quality of data, use of laboratory vs. field studies, presence of health effects in single or multiple species, presence of single or multiple tumor sites, discovery of benign or malignant tumors, endpoints other than carcinogenicity, the presence of supporting or conflicting studies, and data gaps.[29] The nature of toxicological mechanisms must be examined to determine whether they are relevant to humans. Clinical and epidemiological studies must be examined to determine the type of study, the method of exposure assessment, and presence of confounding or other causal factors.[29]

With respect to carcinogens, different entities, such as the International Agency for Research on Cancer,[33,34] use varying procedures to evaluate potency, which may result in disparate estimates in risk of exposure. Occupational hygienists will be familiar with the scheme used by ACGIH, which ranks carcinogens: A1, confirmed human carcinogen; A2, suspected human carcinogen; A3, animal carcinogen; A4, not classifiable as a human carcinogen; or A5, not suspected as a human carcinogen.[35] The EPA has used a weight-of-evidence approach to carcinogenic assessment, producing a letter grading scheme. "A" indicates that the chemical is a human carcinogen; "B" and "C" refer to probable and possible human carcinogens. Chemicals that are not classifiable are rated "D"; "E" indicates evidence of noncarcinogenicity. EPA has proposed changes to this classification system.[36]

Depending on the goal of the particular risk assessment, hazard identification may also begin to address exposure assessment through data collection and evaluation. Exposure information tends to be site-specific and includes a description of the chemical(s) of concern, preliminary data on concentrations in relevant media (air, water, food, etc.), and a conceptual model. The source-path-receiver model, which is familiar to occupational hygienists, is an example of a conceptual model, linking the source of the chemical(s), the pathway of movement in the environment, and route(s) of exposure of the various receptors.

Toxicity (Dose Response) Assessment

Both qualitative and quantitative data are developed in a toxicity assessment: (1) a description of the types of health effects that might be expected to occur in humans; and (2) some estimate of toxicity, such as the dose required to cause these health effects. Conceptually, this estimate of toxicity is based on a dose-response curve (that is, a graph depicting the mathematical relationship between dose and response). The outcome of the toxicity assessment may be a slope factor or an exposure limit. Slope factors are based on the slope of the dose-response curve and can be used to develop a probabilistic risk estimate. In contrast, exposure limits, such as occupational exposure limits or maximum contaminant levels established for drinking water, cannot be used to develop probabilistic risk estimates. Rather, as will be described below under Risk Characterization, these limits are used as guidelines or standards against which actual exposures can be compared.

The ideal source of information for a toxicity assessment would be a study in which humans were exposed to chemicals at the same dose levels predicted by the exposure assessment. As appropriate epidemiological studies are usually not available, the toxicity assessor must rely on data from toxicological studies, requiring both biological and numerical extrapolations. Biological extrapolation requires the assumption that adverse health effects in humans will be the same as those observed in animal species, an assumption that may not be warranted due to substantial biochemical, structural, and behavioral differences between humans and laboratory animals. The second assumption, numerical extrapolation, is that equivalent doses of chemicals in humans and animals will result in equivalent rates of cancer or other adverse effects, and that response rates resulting from large doses can be scaled down to predict response rates at lower doses. The issue of equivalent doses has been handled as "dose per body weight" or "dose per surface area." EPA's current proposal[36] is to prorate the dose by a factor equal to the ratio of the human and animal body weights, raised to the 0.75 power. Continuing developments in physiologically based pharmacokinetic models to allow the determination of "dose to target tissue" will further improve the accuracy of this step.[37]

The issue of extrapolating from high to low doses for carcinogens is more problematic. As resources to invest in toxicological studies are limited, relatively large doses of chemicals have been used to initiate statistically significant response rates in relatively small animal populations. The dose-response relationship can thus be accurately characterized in the high-dose range;

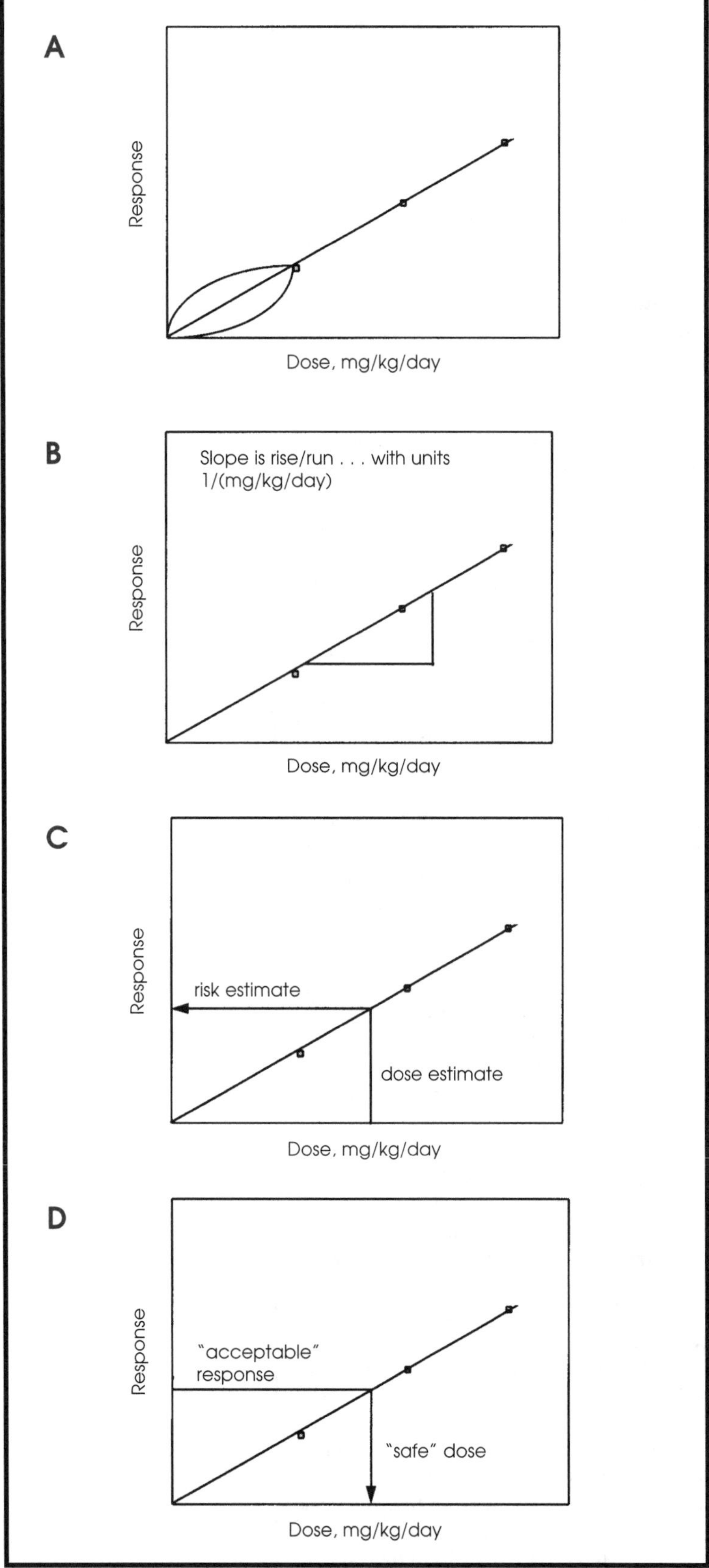

Figure 17.1 — Using dose-response models in carcinogenic risk assessment. A: Dose-response curve for nonthreshold model; B: calculating the slope of the dose-response curve; C: dose estimate determines the risk estimate; D: "acceptable" response determines the "safe" dose.

however, mathematical extrapolation is necessary to extend the curve into the low-dose region (see Figure 17.1). Low-dose extrapolation models can be based on mathematical procedures (e.g., the logistic or probit models) or on assumptions about carcinogenesis (e.g., the one-hit model, the multihit model, the multistage model, etc.). Because of the assumption that there is no threshold dose for carcinogens (that is, any dose will increase the population response rate of cancer) these models force the curve to pass through the origin. Many fit the data equally well in the high-dose region but have widely varying predictions at the lower dose regions.(38) For example, predictions of tumors resulting from saccharin exposure have ranged over six orders of magnitude depending on the extrapolation model and assumptions used.(39) Thus, the selection of the model used to extend the dose-response curve has a tremendous effect on the risk assessment. EPA has elected to use the linear multistage model, as it provides both conservatism and linearity of slope at low doses. Both the need for low-dose linearity and the derivation of the term "slope factor" are easily understood when it is recognized that the slope of the dose-response curve (or rather its 95% upper confidence limit) is the source of this toxicity value. (The x-axis of the dose-response curve is the dose; the y-axis is the response. The slope of the curve will thus be in terms of response/dose, or 1/[mg/kg/day].)

For noncarcinogens the toxicological modeling is less mathematically sophisticated. Determination of the toxicity factor for noncarcinogens begins with the premise that unlike carcinogens, there is a threshold dose (that is, a dose that will not cause adverse health effects). One approach to determining the safe dose for humans (e.g., EPA's reference dose, or RfD) is to translate the threshold dose for animals into the equivalent dose for humans, allowing for an adequate margin of safety. Referring to the dose-response curve (see Figure 17.2), the safe dose for the test animals is ideally represented by the no observable adverse effect level (NOAEL) for the most sensitive species, adjusted to account for differences in surface area between the test animal and humans and for applicable uncertainty factors. These factors, which range from 1 to 10, account for intraspecies and interspecies differences, differences in route of entry, and where necessary, use of the lowest observable adverse effect level (LOAEL). (If human data were available, the interspecies adjustment would not be required.)

Other assumptions that may have to be considered are route-to-route and species-to-species extrapolation. However, despite the difficulties posed by the necessary assumptions, toxicity factors developed from animal studies are generally accepted as indicators of carcinogenic potential for purposes of risk assessment. There is by no means universal acceptance of these procedures, however, and regulatory positions including the exclusion of no-effect thresholds for carcinogens, the equivalency of response in rodents and humans, and the assumption of low-dose linearity have been questioned.(40,41) The risk assessor should carefully examine the data used to develop the dose-response curve, including the selection of animal species (most sensitive or average) or epidemiological data; the model used to develop the dose-response curve, including the rationale, assumptions, and confidence in the results; the methods used to determine exposure in the study; and its relationship to expected exposure in the scenario under analysis.(29)

Exposure Assessment

Exposure assessment is a highly site-dependent process and builds on the source-path-receiver model developed during hazard identification. The level of exposure experienced by members of the worker or general population is a function of the environment, the receptors, and any resulting interactions between the two. Environmental factors that must be considered include the sources of the chemicals and the nature of the exposure. If it is indirect exposure, then the pathways for transfer and the impact of environmental conditions on that transfer must be considered. Chemical, physical, and biological fate processes (such as deposition of chemical on surfaces in the work area) must also be considered. Wide variations in concentration and receptor characteristics may exist, even within a single scenario. Occupational and environmental concentrations tend to be lognormally or sometimes normally distributed, reflecting a range of concentrations with a central tendency that is expressed by the geometric or arithmetic mean. This variance may be handled in different ways in the occupational and environmental settings. The regulator's approach is to characterize the concentration in an occupational setting as the highest concentration (8-hour time-weighted average or ceiling, as appropriate) observed. The approach in environmental risk assessment is to develop an

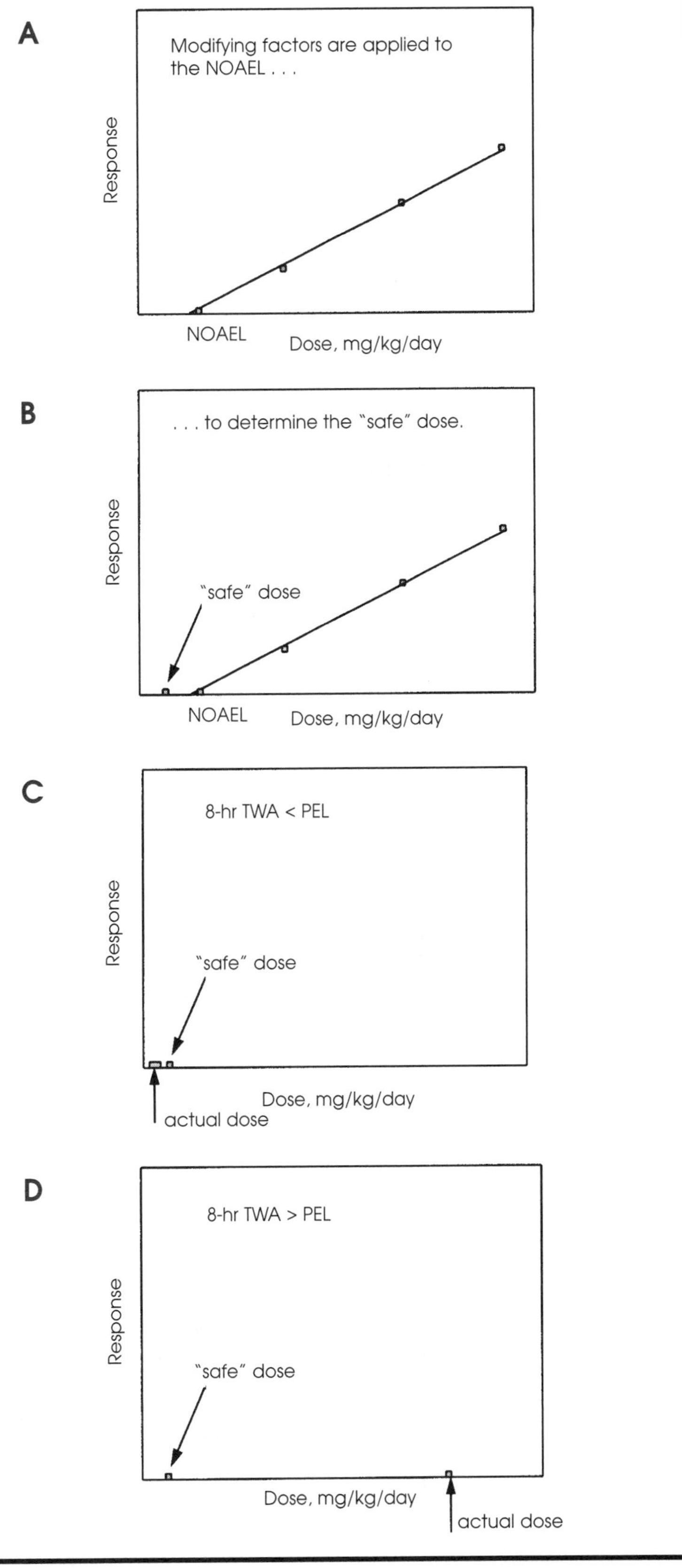

Figure 17.2 — Using dose-response models in noncarcinogenic risk assessment. A: Dose-response curve for threshold model; B: using NOAEL to determine the "safe" dose; C: actual dose is compared with the safe dose (acceptable); D: actual dose is compared with the safe dose (unacceptable).

estimate of the high-end concentration (e.g., upper 95% confidence level) and the mean concentration.

The characteristics (activities, age, size) of the receptors will also influence the level of exposure. For example, occupational exposure to a chemical may be higher than casual recreational exposure to the same chemical in the environment. The interaction of the receptor with the environment will vary and affects exposure. For example, children's exposures tend to exceed those of adults due to higher rates of ingestion and inhalation, as calculated on a body-weight basis. In environmental risk assessment, because of both the range of possible values and the lack of knowledge in a particular situation, default values are often used to describe exposure factors such as breathing rates, ingestion of water, etc.[(42)] A difficult issue, but one that must be considered, is the presence of populations that are susceptible due to increased exposure (e.g., environmental exposure in addition to occupational exposure) or increased sensitivity due to age or health status.

Numerical estimates of exposure can be obtained by monitoring, modeling, or a combination of the two. Occupational hygienists have traditionally relied on collection of airborne samples for analysis of contaminant concentrations, an art that has been advanced by contributions such as the *Occupational Exposure Sampling Strategy Manual*[(43)] and *A Strategy for Occupational Exposure Assessment*.[(44)] The use of modeling to develop exposure assessments has been introduced to the profession by contributions such as those by Jayjock,[(45)] Wadden et al.,[(46)] and Fehrenbacher and Hummel.[(47)] Environmental exposure assessment, which may be equally concerned with current and future exposure scenarios, relies on monitoring to characterize real-time exposures and modeling to predict future exposures. The two approaches may be combined when monitoring data is used to validate and improve model estimates.

Risk Characterization

In the final step, risk characterization, information from the first three steps is assembled, evaluated, and integrated to develop a risk estimate. This is the point in the overall risk assessment process in which exposure is compared with risk/exposure or some other form of toxicity assessment, such as an occupational exposure limit or a drinking water maximum contaminant level.

The specific activities conducted during risk characterization depend on the purpose of the risk assessment and the type of toxicity assessment data available. As described above, the outcome of a toxicity assessment may be a slope factor that can be used to develop a probabilistic risk estimate, or a guideline or standard against which actual exposures may be compared. Figures 17.1 and 17.2 contrast the difference in these two outcomes and the resulting risk characterization. As seen in Figure 17.1A, the data collected from toxicological and/or epidemiological studies is used to develop a dose-response curve. For carcinogens it is assumed that there is no threshold dose (i.e., the curve passes through the origin). Although it is more complicated in actual practice, determining the slope of the line basically results in the slope factor for that chemical (see Figure 17.1B). The curve can be approached from the x-axis, corresponding to dose, and the predicted risk level can be read off the y-axis (Figure 17.1C). Mathematically, this can be done by multiplying the dose by the slope factor to determine a risk estimate. This approach is typically used in EPA Superfund site risk assessment of carcinogens. For example, if the daily dose of benzene, estimated at 1×10^{-4} mg/kg/day, were multiplied by the slope factor for benzene, 0.029 1/(mg/kg/day), the resulting risk estimate would be 3×10^{-6}. This would be correctly interpreted as "the upper bound on the risk estimate is three cases per million persons exposed." Conversely, a tolerable dose or exposure level could be determined by entering the graph from the y-axis at the level of acceptable risk and reading the corresponding dose level along the x-axis (Figure 17.1D). (Note that the use of the term "acceptable level of risk" does not imply that this is a desirable level of risk; perhaps the term "tolerable level of risk" would be more descriptive. For additional discussions of acceptable risk, see Rodricks et al.,[(21)] Percival et al.,[(10)] and Lipton & Gillett.[(48)])

The outcome of the toxicity assessment for noncarcinogens is usually a guideline or standard for exposure levels. For noncarcinogens the dose-response curve typically crosses the x-axis at a point to the right of the origin; i.e., the existence of a threshold dose is assumed (see Figure 17.2A). Recall that in a toxicological study, the NOAEL is modified to obtain an equivalent safe dose for humans—a dose below which adverse health effects should not occur (see Figure 17.2B). The estimated exposure, in units consistent with the dose-response curve, can then be compared with the safe dose. It should be noted that while the terms "exposure" and "dose" have been used somewhat interchangeably in this discussion, the two have slightly different meanings. Dose refers to the amount of contaminant that has been either administered or absorbed by the body and is usually expressed as mg/kg/day. Exposure can be defined as potential contact at the receptor's exchange boundaries, such as lung or skin. In the occupational setting it is customary to describe exposure in terms of airborne concentration of contaminants.

As seen in Figures 17.2C and 17.2D, an exposure below the safe dose would be considered acceptable, while an exposure above the safe dose would not be acceptable. This is not a probabilistic risk estimate (i.e., the ratio of actual and safe levels of exposure cannot be used to predict the number of cases of illness that may occur). Exposures above the safe level should, however, trigger appropriate risk management efforts. This general risk characterization approach is used in occupational settings (e.g., comparison of measured 8-hour time-weighted averages and the PEL), and Superfund site risk assessment of noncarcinogens.

In addition to calculation of risk estimates or comparison of exposures to acceptable exposure levels, risk characterization tasks may involve review of findings of toxicity and exposure assessments (see Table 17.4), quantification of risks from individual chemicals and multiple chemical exposures, combination of risks across exposure pathways, and assessment and presentation of uncertainty. The goal is to produce risk estimates that are accurate, adequately but not overly conservative, and which reflect insofar as possible the degree of risk potentially faced by persons exposed to the processes, chemicals, or energies in question. Complete risk characterization also requires an explanation of any assumptions and models used, along with analysis and discussion of uncertainty. The degree of effort put into a risk characterization may be affected by legal requirements, the availability of resources, the cost of conducting the risk assessment, the cost of risk management, the potential impact of the risk if no mitigation measures are carried out, and the complexity or subtlety of the situation.

Table 17.4 —
Supporting Information in Risk Characterization

Hazard Identification Issues

- key toxicological studies
  - quality of data
  - laboratory vs. field studies
  - single or multiple species
  - single or multiple tumor sites
  - benign or malignant tumors
  - endpoints other than carcinogenicity
  - other supporting studies
  - conflicting studies
  - significant data gaps
- clinical or epidemiological studies
  - type of study
  - exposure assessment
  - confounding factors
  - other causal factors
- toxicological mechanisms
  - relevant studies
  - implications for health effects
- nonpositive data
  - humans
  - animals (laboratory or wildlife)
- summaries of
  - confidence in conclusions
  - alternative conclusions
  - significant data gaps
  - major assumptions

Toxicological Assessment Issues

- data used to develop dose-response curve
  - animal species (most sensitive, average)
  - epidemiological data
- model used to develop dose-response curve
  - rationale
  - noncarcinogenic hazards
    - calculation of RfD/RfC[A]
    - assumptions, uncertainty factors
    - confidence
  - carcinogenic hazards
    - dose-response model
    - basis for its selection
    - other valid models
- exposure
  - route, level
  - same as expected human exposure
  - impact of exposure extrapolations
- adverse effects in wildlife species

Exposure Assessment Issues

- sources of environmental exposure
  - data from different media
  - significant pathways
- populations
  - general
  - highly exposed
  - highly susceptible
- basis for exposure assessment
  - monitoring
  - modeling
  - exposure distributions
- key descriptors
  - average, high end, susceptible populations
  - central tendency
  - high-end estimate
  - highly exposed subgroups
- cumulative exposures
  - ethnic, socioeconomic
- wildlife species
- conclusions
  - different approaches
  - limitations
  - confidence

(Continued on p. 344)

Table 17.4 (continued) — Supporting Information in Risk Characterization

Risk Characterization Issues

- overall picture of risk
- major conclusions and strengths
- major limitations and uncertainties
- science policy options
- qualitative characteristics of risk
  - voluntary vs. involuntary, etc.
  - impact of risk perception
- alternatives to hazard
- risks of hazard alternative(s)
  - comparison to similar risks
  - comparison to similar decisions
  - limitations of comparisons
- community concerns
  - public perception of risk
- other risk assessments on this chemical
- any other useful information

[A]RfC = reference concentration
Source: Reference 29.

Occupational Hygiene Risk Assessment

Risk assessment has been an integral part of the development of occupational exposure limits ever since the benzene decision, in which the Supreme Court found that OSHA must first identify a significant risk of material health impairment and then show that the proposed standard would reduce that risk. As illustrated in Table 17.3, the standards that followed—the 1983 asbestos standard, ethylene oxide, formaldehyde, methylenedianiline, methylene chloride—were based on risk assessments. Risk assessment is also practiced by occupational hygienists in the field whenever exposure measurements are compared with occupational exposure limits. The outcome of this type of risk assessment will not be a probabilistic risk estimate, but a determination that the exposure is less than, equal to, or greater than acceptable levels. (Occupational exposure limits are not typically based on linear dose-response models. Therefore, the degree to which the actual exposure is greater or less than the PEL or TLV is not a linear estimate of risk. While reducing the actual exposure level by half would reduce the risk, one could not say that the resulting risk would be reduced by half.) This information is used in risk management decision making about the type and level of controls that may be necessary to protect worker health.

Thus, the basic concepts of risk assessment are not new to the occupational hygiene profession. While the specific activities may vary between environmental and occupational risk assessment, the objectives of each stage remain the same. Evaluation of any exposure situation—environmental or occupational—requires knowledge of the pathways by which exposure may occur, quantification of the exposure and toxicity of the chemicals involved, and integration of this information (i.e., comparison of the level of exposure with the level of risk per unit of exposure). Table 17.5 compares the familiar model of occupational hygiene practice with the processes of environmental risk assessment, risk management, and risk communication. (It should be noted here that not all occupational hygienists will agree with this mapping of risk assessment and occupational hygiene, maintaining that exposure evaluation is a risk management function.)

While air sampling is the stock-in-trade of occupational hygienists, comparisons of airborne contaminant concentrations to PEL or TLV lists are not the only analytical tools available to the occupational hygienist. A more thorough risk assessment could be performed by investing more resources in the exposure assessment and toxicity assessment processes, potentially resulting in a more accurate comparison of exposure and risk/exposure. A more complete exposure assessment can be performed by the addition of such techniques as:

- The sampling and analysis of surface wipes for presence of heavy metals or pesticides;
- The collection and analysis of blood, urine, or breath samples from exposed workers;
- Incorporation of chemical exposure data from a second job or hobby;
- Evaluation of dermal absorption;
- Evaluation of environmental (food,

Table 17.5 — Comparison of the Classical Occupational Hygiene and Environmental Risk Assessment Models

Classical occupational hygiene	The Environmental Way
Anticipation/recognition	Hazard identification (data collection/evaluation)
Evaluation	Exposure assessment, toxicity assessment, risk characterization
Control	Risk management
Hazard communication	Risk communication

water, soil) as well as occupational exposures to the chemicals in question, along with determination of personal intake factors, such as quantity of contaminated food or water consumed daily, leading to estimation of chronic daily intake (CDI; mg/kg/day); and
- Use of predictive models to estimate exposure to chemicals under varying conditions.

When an expanded exposure assessment has been performed, incorporating data from personal and total environmental sampling, an expanded toxicity assessment can aid in full utilization of this data. Toxicity assessment for occupational hygiene typically begins with looking up the PEL and TLV and reading the latest edition of the *Documentation of the Threshold Limit Values and Biological Exposure Indices*[49] or other readily available reference on industrial toxicology to determine the target organs and types of adverse health effects to be expected. Toxicity assessment can be extended by including analysis of the dose-response relationship or NOAEL/LOAEL values or gathering of quantitative data (from the literature or the laboratory) to develop a dose-response curve. An additional perspective may be gained by examining the traditional methods of environmental risk assessment. The individual worker's total exposure to the chemical in question, including primary and secondary occupational exposure, food, water, soil/dust, and other sources in the home, can be summed and compared with the slope factors and/or RfD toxicity values that will be discussed in the section on environmental risk assessment. A comparison by CAS (Chemical Abstracts Service) number of compounds listed in the 1995-1996 TLV list indicates that nearly 200 chemicals are included in EPA's January 1996 Integrated Risk Information System (IRIS) file list.[50] This is not to say that there are slope factors or RfDs for all (e.g., lead), but the IRIS files do contain extensive toxicological data and serve as an excellent jumping-off point for further data searches, particularly for the more than 300 chemicals that are contained in the IRIS database but do not have TLVs.

The application of environmental risk assessment methodologies to the routine practice of occupational hygiene raises several issues, including (1) the use of averaging times and acute/chronic health effects; (2) situations in which there is a current TLV and/or PEL; and (3) toxicity values that have been developed for ingestion as opposed to inhalation as a route of entry. In addition, there are occupational hygiene risk assessment issues yet to be resolved: selection of mathematical models for low-dose extrapolation, assumptions made when adding risk across chemicals, pathways, populations, etc.

One of the first issues that must be addressed is that of averaging times in relation to regulatory requirements. When dealing with carcinogens, the practice in environmental risk assessment is to determine the reasonable maximum exposure (RME) over a 30-year period (that is, the 95% upper confidence level estimate of the length of a person's residency in a single home). The CDI over this 30-year period is averaged over the anticipated 70-year life span. For noncarcinogens the daily exposure is averaged over the actual exposure time, which will vary for comparison with RfDs based on short-term, subchronic, and chronic exposure periods. This stands in sharp contrast to the usual practice in occupational hygiene, which is to measure actual exposure, typically averaged over an 8-hour day within the context of a 40-hour workweek; some latitude is allowed in the TLVs to account for work schedules longer than 8 hours/day or 40 hours/week.[51] The OSHA interpretation is that any exposure above the 8-hour time-weighted average (after accounting for statistical variation) is a violation of the standards, regardless of whether the worker has any further exposure to the chemical that week or month. While this "daily" approach is reasonable for chemicals that have acute health hazards, it does not make sense for those chemicals associated with chronic health effects and for which the appropriate quantity of concern is the long-term average body burden. A further contrast between occupational and environmental toxicity values is that the acceptable occupational exposure limits are based on intermittent exposures rather than continuous exposures; a higher exposure can be tolerated if frequent recovery periods are built in.

Another issue to be dealt with in application of the IRIS toxicity values in occupational hygiene is the fact that many of them have been developed for ingestion rather than inhalation as the route of entry. Inhalation exposure is more complex due to the dynamics of the respiratory system and interactions with the physicochemical properties of chemicals. In many cases, lead for example, the route of entry is not as critical as is the actual dose absorbed. In the case of respiratory irritants or chemicals that are effectively metabolized by the liver, the route of entry can be critical. A toxicologist should be consulted prior to conversion of an oral slope factor or RfD to an inhalation or dermal toxicity value.

With respect to general risk assessment issues, a point that can be responsible for large differences in risk estimates is the selection of the mathematical model for extrapolation of the low-dose end of the dose-response curve. Similarly, adjustments for differences in animal and human body weights and lifetime equivalences can have a significant impact on the resulting risk estimates. The example for saccharin quoted above illustrates this dramatically with the range of risk estimates spanning six orders of magnitude.[39] It has been said that the range of estimates is like not knowing whether there is enough money in one's pocket to buy a cup of coffee, or pay off the national debt. Obviously, if the public in general and the working population in particular are going to have confidence in the process, they need to be assured of reliability and accuracy in the process.

Finally, assumptions made when adding risks across chemicals, pathways, and populations must be carefully examined. With respect to assumptions of linear additivity of health effects, it must be remembered that for noncarcinogens, the level of concern does not increase linearly as the RfD is approached or exceeded. Errors in risk estimation may result from the combination of hazard quotients (HQs) based on critical effects of varying toxicological significance or varying levels of confidence, uncertainty adjustments and modifying factors, and/or different health effects induced by different mechanisms of action. With respect to summation of carcinogenic risks, questions about linear additivity, weight of evidence in carcinogenic classification, and mechanisms of toxicological action are also relevant. The actual likelihood that any one individual would be exposed to all chemicals/pathways/scenarios also bears careful scrutiny. Is it realistic to assume that the same individual is going to be exposed to a chemical on the job, work with it in a second job or hobby, breathe it in the air at home, ingest it in the food and water, and trespass onto an abandoned landfill to fish in a pond that is contaminated with the same chemical?

EPA Superfund Risk Assessment

While it is not the only federal agency conducting risk assessment, EPA has had a significant impact on the development of environmental risk assessment methodology due to the inclusion of risk assessment as an integral part of the Superfund remediation process, as well as in many other EPA programs. The Superfund methodology is straightforward, although it relies heavily on default assumptions, and allows comparison of dissimilar hazardous waste sites across the nation. The baseline risk assessment process for Superfund sites (see Table 17.3) was based on the National Research Council's four-step paradigm:[27] data collection and evaluation, the simultaneous steps of exposure assessment and toxicity assessment, and the final process of risk characterization, which combines data from the first three steps to develop risk predictions. This section will briefly examine each of these four steps.

Data Collection and Evaluation

The purpose of the EPA baseline risk assessment process is to develop sufficient information to support an informed risk management decision. The outcome of this step is highly site-specific and consists of developing three interconnected parts:

1. A list of chemicals of potential concern that may be present at the site;
2. Reliable data on concentrations of these chemicals in site media; and
3. A site conceptual model (i.e, a source-path-receiver model of the site that hypothesizes ways in which humans may contact the chemicals).

Exposure Assessment

The product of the exposure assessment is data on the number of persons exposed to contaminants, and estimates of the magnitude, duration, and frequency of exposures. This process requires the risk assessor to characterize the physical setting, identify potentially exposed populations, identify potential exposure pathways, estimate exposure concentrations, and then estimate chemical intakes, usually in terms of mg/kg/day. Thorough understanding of environmental parameters, such as climate, soil type, and groundwater hydrology is critical because of their role in fate and transport of contaminants. The general population can be divided into subpopulations characterized by different demographic, geographic, and activity patterns. Because age, activities, relationship to the site, and media intake all influence the expected exposure, the characteristics of potentially exposed persons play an important role in evaluation of exposure. Exposure assessments are usually conducted for subpopulations that share similar characteristics, such as children with residential exposure or adults with occupational exposure.

After identification of subpopulations that may be exposed to contaminants at or from the site, pathway analysis based on the site conceptual model is used to delineate probable locations of environmental contact. This analysis links the sources, locations, and types of environmental releases with population locations and activity patterns to determine the significant pathways of human exposure. An example of a complete exposure pathway is: (1) source—a buried waste drum; (2) release—contaminants leak from the drum and migrate through the soil to a ground water aquifer; (3) exposure point—a drinking water well, used for residential purposes, in the contaminated aquifer; and (4) route of entry—ingestion, dermal absorption, and inhalation during showering.

Contaminant concentrations at the exposure point are site-specific values that are highly dependent on the fate and transport of the chemical through the environment. A chemical released from primary or secondary sources may undergo transport or transformation (physical, chemical, or biological), or it may be accumulated in the environmental compartment. Physical, chemical, and biological processes are strongly affected by properties of the environment and the chemical. Current exposure point concentrations are normally based on actual monitoring of contaminant concentrations in air, surface water and groundwater, soil, and biota. Future exposure point concentrations are more problematic and are typically estimated via computer modeling of environmental fate and transport. There are numerous computer models available, ranging from simple screening level models to complex analytic models with heavy data input requirements. Proper selection and use of environmental models is critical to accurate predictions of exposure point concentrations.

The last step in exposure assessment is to calculate average daily exposure to the contaminants through all relevant media: air, water, soil, or biota. Exposure point concentration of the media, along with media intake, frequency and duration of exposure, body weight, and averaging time are used to calculate the average exposure, expressed as mg/kg/day (see sidebar on this page). Equation 17.1 is generic and must be adapted to calculate exposures for each route of entry: ingestion, inhalation, or dermal absorption. Since in most cases the specific information required in this calculation is not available, generic assumptions or default values are usually substituted (see Table 17.6). These values are based on upper bound estimates of probability distributions of typical exposures.[52] For example, the RME of residents to contaminants in drinking water is based on consumption of 2 L of water each day, for 350 days/year at a home that is occupied for 30 years by a "standard man" who weighs 70 kg. (The reasonable average exposure, based on mean or median rather than upper bound estimates, will be lower.) CDI calculated from Equation 17.1 for water, soil/dust, air, or food products will be in terms of milligrams of contaminant per kilogram of body weight per day, for appropriate comparison with toxicity values developed in

Calculation of Chronic Daily Intake Equation 17.1 —

$$CDI = \frac{C_m \times I_m \times EF \times ED}{BW \times AT}$$

where:

CDI = chronic daily intake (mg/kg/day)
C_m = concentration in affected media (e.g., mg/L)
I_m = intake of affected media (e.g., L/day)
EF = exposure frequency, days/year
ED = exposure duration, years
BW = body weight, kg
AT = averaging time, days

Table 17.6 — Selected Standard Default Exposure Factors

Land Use	Exposure Pathway	Daily Intake Rate	Exposure Frequency	Exposure Duration	Body Weight
Residential	ingestion of portable water	2 L	350 days/year	30 years	70 kg
	ingestion of soil and dust	200 mg (child) 100 mg (adult)	350 days/year	6 years 24 years	15 kg (child) 70 kg (adult)
	inhalation of contaminants	20 m³ (total) 15 m³ (indoor)	350 days/year	30 years	70 kg
Industrial	ingestion of potable water	1 L	250 days/year	25 years	70 kg
	ingestion of soil and dust	50 mg	250 days/year	25 years	70 kg
	inhalation of contaminants	200 m³/workday	250 days/year	25 years	70 kg

Source: **U.S. Environmental Protection Agency (EPA):** *Risk Assessment Guidance for Superfund*, Vol. I, Supplemental Guidance, "Standard Default Exposure Factors" (Pub. 9285.6–03). Washington, DC: EPA, 1991.

the next stage. A possible source of confusion is the concept of averaging time, which is different for carcinogens and noncarcinogens. Since the latency period for carcinogens is frequently measured in decades, the carcinogenic risk is estimated on a per lifetime basis. In contrast, toxicological testing of noncarcinogens is more concerned with health effects occurring during real time (i.e., during the actual exposure time). Therefore, calculation of the CDI for noncarcinogens is averaged over the actual exposure time, which for RME calculations is taken to be 30 years.

Toxicity Assessment

In conducting a Superfund site risk assessment, the toxicity assessment is the least site-dependent of all the stages of risk assessment; therefore, the results should be consistent from site to site. The information needed includes the kinds of health effects (cancer or noncancerous effects) that could occur from exposure and the toxicity values of the chemicals:

1. Slope factors for carcinogens, expressed in terms of 1/(mg/kg/day) based on the slope of the dose-response curve; and
2. RfDs for noncarcinogens, expressed in terms of mg/kg/day based on the NOAEL.

Rather than conducting toxicological or epidemiological studies, the risk assessor typically searches the published literature and databases, beginning with EPA's IRIS database.[50] IRIS contains a wealth of information on approximately 530 chemicals, including solvents, metals, pesticides, and inorganic chemicals. Most available slope factors are based on ingestion, but some are available for inhalation exposure. Fewer still have been determined for dermal absorption. Conversion of toxicity factors determined for one route of entry to another should be done only with the advice and consent of a toxicologist.

Separate toxicity values may thus be available for carcinogenic and noncarcinogenic effects, oral or inhalation exposure, as well as for different exposure periods. In addition, RfDs may reflect the different acceptable exposure levels during short-term, subchronic, or chronic exposure periods.

Risk Characterization

The actual calculation of risk estimates for a Superfund risk assessment is a fairly straightforward, spreadsheet operation, given that correct exposure levels and toxicity values are available and appropriately matched. The type of health effect, the route of exposure, and the length of the exposure period must all be considered. Table 17.7 presents a sample spreadsheet that calculates the carcinogenic risk of adult residential ingestion of arsenic-contaminated soil. The dimensionless risk resulting from exposure to a carcinogen is calculated by multiplying the exposure, in terms of mg/kg/day, by the slope factor in 1/(mg/kg/day) and canceling the units (see Equation 17.2.1 on p. 351). This product is a true probabilistic estimate of risk and can be used to predict the upper bound on the number of cases of cancer per exposed population.

For noncarcinogens, the CDI, expressed as mg/kg/day, is simply compared with the appropriate RfD derived from the modified NOAEL (or LOAEL) and expressed in terms of mg/kg/day (Equation 17.2.3). The resulting ratio of CDI and RfD, which is dimensionless, is the HQ. Unlike the calculations for carcinogens, the HQ is not a true probabilistic estimate of risk of occurrence of adverse health effects and cannot be used to predict the rate of occurrence in a population, the risk of an individual experiencing the adverse health effects, or even the seriousness of any resulting health effects.

Exposure to multiple chemicals is often the case at Superfund sites. While it is not possible to sum carcinogenic and noncarcinogenic risks, it may be appropriate to aggregate risks within each category. In the

Table 17.7 —
Example of Spreadsheet Calculations for Risk Characterization

Carcinogenic Risk Calculations for Ingestion of Arsenic in Soil, Residential Adult Exposure

2	(A)	(B)	(C)	
3	Factor	Value	Units	
4	Intake rate	100	mg/day	
5	Conversion factor	1.00E-06	kg/mg	
6	Exposure frequency	350	days/year	
7	Exposure duration	30	years	
8	Body weight	70	kg	
9	Averaging time	25550	days	
10	Slope factor	1.75	1/(mg/kg/day)	
11				
12	Risk Based on Mean of Arsenic-in-Soil Sample Values			
13	Area	Mean, mg/kg	Intake, mg/kg/day	Risk (unitless)
14	Area A	13.31	7.82E-06[A]	1.37E-05[B]
15	Area B	41.96	2.46E-05	4.31E-05
16				
17	Risk Based on 95% Upper Confidence Limit (UCL) of Arsenic-in-Soil Sample Values			
18	Area	95% UCL, mg/kg	Intake, mg/kg/day	Risk (unitless)
19	Area A	13.94	8.18E-06	1.43E-05
20	Area B	60.04	3.52E-05	6.17E-05

[A]Intake = (B14*B4*B5*B6*B7)/(B8*B9)
[B]Risk = (C14*B10)

absence of specific toxicological information on the effects of multiple chemical exposures, it is usually assumed that the doses (and by extension, the lifetime cancer risk estimate or HQ) are additive. However, the addition of carcinogenic risk estimates or hazards may be excessively conservative. Kodell and Chen[53] have suggested a procedure to reduce conservatism in risk estimation for mixtures of carcinogens.

The risk assessor may also need to consider combining risks across different exposure pathways. This represents a point of departure from typical occupational hygiene practice. Beyond noting a "skin" notation on the TLV list, the occupational hygienist usually does not consider multiple routes of exposure. In an environmental risk assessment it is customary to identify the individual who may face the RME resulting from more than one exposure pathway or scenario: for example, the individual (or group) who has occupational exposure to a chemical, whose drinking water well is contaminated with the same chemical, and who may consume fish from a lake that is also contaminated with the same chemical.

The most thoroughly researched and carefully conducted data evaluation or exposure or toxicity assessment is for nought if the resulting information is not fully considered in forming conclusions about risk and in writing a risk assessment report that is understandable to all stakeholders. EPA has published recent guidance stressing that risk characterizations should use critical information from each stage of the process and communicate this information to risk managers.[28,29] The pertinent messages are that risk characterization should be based on reliable scientific information from many different sources, and that documentation should be clear, transparent, reasonable, and consistent. The risk assessment report should discuss key scientific concepts, data, and methods from each stage of the process. It should also include acknowledgment and analysis of uncertainties (see below) and an estimate of confidence in the risk characterization. The risk assessment should present several types of risk information. A range of exposure scenarios and multiple risk descriptors (central tendency, high end of individual risk, population risk, important subgroups) should be included. Use of several descriptors, rather than a single descriptor (such as the RME individual's risk), enables presentation of a fuller picture of risk. Presentation of a numerical risk estimate is not complete risk characterization; the numbers must be supported by documentation of data, concepts, assumptions, models, and conclusions.

Uncertainty and Iterative Risk Assessment

Sources of Uncertainty in Risk Assessment

Throughout the risk assessment process, the analyst is faced with variability and uncertainty. There is some inconsistency among authors in the use of these terms. The term "uncertainty" often includes both variability and lack of data. Variability is the result of heterogeneity or actual differences among members of a population. It may be more accurately characterized but not reduced with additional data. Uncertainty results from measurement limitations and may be related to study design and analytical techniques or application of data to nonsampled populations. Further measurements will ultimately reduce uncertainty.[54,55] Pending additional data and complete understanding of the underlying physical and biological processes, uncertainty must be resolved through the use of expert judgment.[48] Evans et al.[56] have described a method that incorporates expert judgment, probability distributions, and probability trees in the uncertainty analysis of carcinogenic potency. The probability tree allows components of the risk analysis to be disaggregated, with the judgments of experts weighing more heavily in components related to their expertise. This method encourages differences of opinion and allows all relevant information to be considered in risk characterization.

In lieu of additional data or knowledge, however, risk assessors tend to make conservative assumptions to ensure that any errors in risk estimation are in the direction of public health protection. This may lead to risk estimates that are overly conservative. For example, using conservative estimates for all input factors (e.g., the 95th percentiles of the distributions for inputs that are positively correlated with risk and the 5th percentile of the inputs that are negatively correlated) may lead to risk estimates as high as the 99.99th percentile of the distribution of risk estimates.[57] The risk assessor thus has a responsibility to examine the risk assessment to delineate sources and approximate magnitudes of errors. This analysis has two important purposes: to acknowledge and inform the users of the risk assessment (expert and lay readers alike) of potential sources and directions of error, and to identify those data gaps that can be effectively narrowed through the collection of additional data.

Several methods are commonly used to analyze and present uncertainty and variability in risk assessment, ranging from qualitative to quantitative techniques. The appropriate level of uncertainty analysis depends largely on the level of effort devoted to the risk assessment itself (that is, screening, moderate, or complex). In many situations a qualitative uncertainty analysis will suffice. The risk assessor simply lists the possible sources of uncertainty and variability and indicates the likely direction and magnitude of impact that each source will have on the final risk estimates. If enough data is available, the risk assessor can calculate the risk using the mean of the exposure data (reasonable average exposure) and then compare that with risk associated with the maximum sampling values (RME). The range in the risk values would represent the uncertainty in the estimates (i.e., a semiquantitative uncertainty analysis). Another approach is the sensitivity analysis, in which an input variable is varied by ±10 or ±25% to determine impact on risk estimate.[58]

Quantitative uncertainty analysis, such as first-order Taylor series or Monte Carlo analysis, requires the highest level of effect and amount of input data but also provides the greatest information about uncertainty in the risk assessment.

Monte Carlo Analysis

Monte Carlo analysis has been used for over 50 years to compute difficult multidimensional integrals in physics, chemistry, and other technical disciplines.[55] Its ability to combine probability distributions for several input variables to generate probability distributions for output variables makes it an obvious selection for analysis of uncertainty in risk assessment. Readily available software for personal computers, such as spreadsheet packages (Excel for Windows™) and add-ons (CrystalBall™) has greatly simplified the computational aspects of Monte Carlo analysis.

In a deterministic risk assessment, a point estimate for each input variable (concentration, intake, etc.) is used to calculate a point estimate of risk. As discussed above, the point estimate selected is usually a high-end estimate, leading to overestimates of risk. In contrast, in Monte Carlo analysis a value for each input parameter is randomly selected from a probability distribution value, and risk is then calculated according to the model equation (see Figure 17.3). The process may be repeated thousands of times to produce a probability distribution for the risk estimate, allowing the risk assessor to determine the mean and upper bound of the risk estimate and identify the input variables contributing the most to uncertainty. A comparison conducted by Smith[59] indicated that at the site investigated, the deterministic RME cancer risk estimate was between the 95th percentile and the maximum estimate produced by Monte Carlo analysis, suggesting that the RME risk calculation was protective and the results of probabilistic and deterministic calculations were consistent (see also Thompson et al.[58])

There are, however, limitations to this technique,[59,60] including the requirement for knowledge of probability distributions and covariance (degree of dependence) for input parameters. (Finley et al.[54] have proposed age-specific distributions for a variety of exposure factors, including soil ingestion rates, tapwater consumption, and soil-on-skin adherence.) Exposure factors developed from short-term studies with large populations may not accurately represent long-term conditions in small populations. Also, the tails of Monte Carlo risk distributions, which are of greatest regulatory interest, are very sensitive to the shape of input distributions. For this reason, Burmaster and Anderson[55] have suggested that the risk analyst should run enough iterations (usually 10,000) to demonstrate numerical stability of the tails of the outputs. Other principles of good practice for the use of Monte Carlo analysis developed by these authors include suggestions to show all formulae used, calculate and present point

Deterministic Calculations

$$\frac{A \times B \times C}{D \times E} = F$$

Probabilistic Calculations

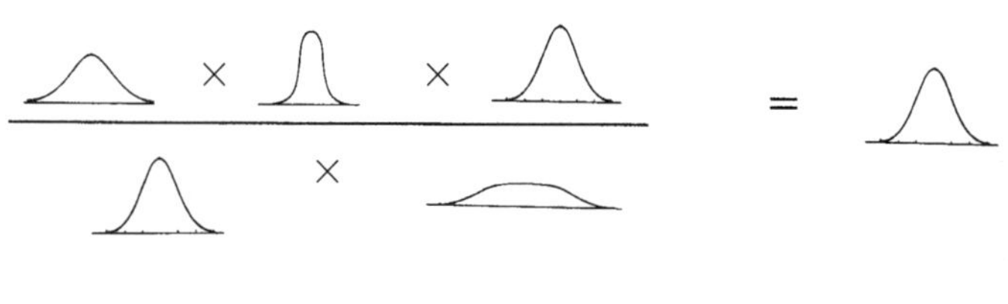

Figure 17.3 — Use of input distributions to generate output distributions in Monte Carlo analysis.

estimates, restrict the use of probabilistic techniques to the pathways and compounds of regulatory interest, and to the extent possible, contrast variability and uncertainty. Whenever possible, use measurement data, present the name and statistical quality of the random number generator used, and discuss the limitations of the methods and interpretation of the results.

Occupational hygienists should familiarize themselves with the Monte Carlo technique. Its ability to provide a range of risk estimates in the form of a probability distribution makes it a potentially valuable tool in risk communication and in selection of risk management options. Several EPA regions have begun to accept Monte Carlo simulations submitted as uncertainty analysis given certain restrictions.[60] Many professional risk assessors have adopted Monte Carlo as a routine analytical technique and believe that it will continue to gain wider acceptance.

Equations for Determination of Cancer Risk and Noncancer Hazard Index

Equation 17.2.1 —
Lifetime Cancer Risk Estimate (low risk, < 0.01)

$$Risk = CDI \times SF$$

where

CDI = chronic daily intake, mg/kg/day
SF = slope factor, 1/(mg/kg/day)
Note: CDI and SF must be expressed in reciprocal units and must represent the same route of entry (ingestion, inhalation, dermal absorption).

Equation 17.2.2 —
Lifetime Cancer Risk Estimate (high risk, ≥ 0.01)

$$Risk = 1 - e^{-(CDI \times SF)}$$

Equation 17.2.3 —
Noncancer Hazard Quotient

$$HQ = CDI / RfD$$

where

HQ = hazard quotient, dimensionless
CDI = chronic daily intake, mg/kg/day
RfD = reference dose, mg/kg/day
Note: CDI and RfD are expressed in the same units and must represent the same exposure period (chronic, subchronic, or shorter-term) and the same route of entry (ingestion, inhalation, or dermal absorption).

Tiered Approach to Occupational Hygiene Risk Assessment

An important objective of this chapter is to suggest that risk assessment efforts must be appropriately matched to the situation at hand. In a deregulated environment the occupational hygienist is under even greater pressure to make risk management decisions that appropriately balance the costs of risk assessment and risk management with the potential impacts of unmitigated risks. The most effective protection of worker health and safety will result from a tiered or iterative approach, which matches the level of evaluation and control with the risk at hand. Basically, this translates to making conservative decisions when faced with lack of data. If the implications of such conservatism are undesirable, additional resources can be devoted to gathering more data, perhaps justifying less conservative decisions. Risk assessment efforts continue until there is acceptable balance among the cost of the risk assessment, the costs of the risk management, and the cost of the unmitigated risk. Additional inputs to selecting the level of risk assessment effort will also include legal requirements, complexity of the situation, time available for analysis, the technical expertise of the analyst, and risk perceptions of workers, management, and the public.

One approach to tiered risk assessments is similar in concept to that developed by ASTM in *Risk-Based Corrective Action Applied at Petroleum Release Sites*, abbreviated here as ASTM–RBCA.[15] The purpose of this framework is the determination of cleanup levels for sites such as those contaminated by leaking underground storage tanks. The first step is to determine whether there are any obvious safety hazards that should be corrected before analysis proceeds to Tier I of ASTM–RBCA. Tier I requires a low level of effort and resources, primarily involving the collection and analysis of site samples. The results are compared to lookup tables that contain suggested concentrations of benzene, toluene, ethyl benzene, xylene, and other petroleum-related compounds in soil, water, and air. These lookup tables are based on on EPA CDI equations and do not take into account any physical, chemical, or biological degradation or transfer. If the cost of achieving Tier I cleanup levels is deemed unnecessary or unfeasible, and continued risk assessment resources are available, the analyst may continue to Tier II. This requires a moderate level of resources, including site-specific data for input to (1) the same risk-based equations used to develop the lookup tables in Tier I; or (2) screening-level models. If degradation or transfer are considered, these cleanup levels will probably be higher than those suggested by the lookup tables in Tier I. If, however, these cleanup levels are also deemed to be unnecessary or unfeasible, and sufficient risk assessment resources exist, the analyst may continue to Tier III. A significant increase in site-specific data and the use of complex modeling characterize this level, but the extra effort may be justified by even higher cleanup levels. Whether or not Tier III cleanup levels are higher, this is considered the last step in the risk assessment. The goal of this tiered approach is not to justify less protection to the public; each tier should achieve a similar level of protection. Rather, the goal of the higher tiers is to allow development of more realistic site-specific cleanup levels. This framework has been adopted by many state underground storage tank programs, and ASTM is in the process of developing an RBCA framework for general chemical exposure.[61]

Collecting additional information or conducting a higher tier of risk assessment may have the effect of shrinking the error bands around the exposure assessment or of shifting the risk estimate to lower (or higher) values. As shown in Figure 17.4, this may allow a less conservative risk estimate. Errors in risk assessment can be of two types: over- or underestimation of risk.

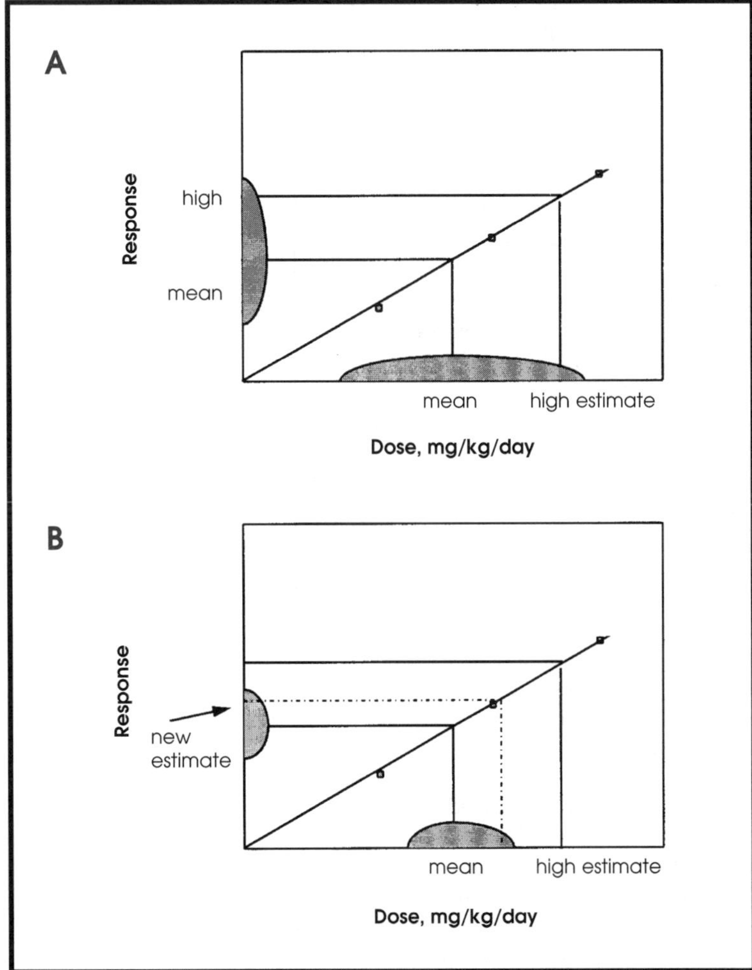

Figure 17.4 — Improved exposure assessment shrinks error bands. A: Dose and risk estimates are conservative; B: shrinking the error bands around the exposure estimate reduces the risk estimate.

In this case the Type I error can be defined as selecting more risk management than is actually needed; the impact is mainly financial. The Type II error is selecting less risk management than is actually needed; the potential impact is more significant, as it involves human health. Higher tier risk assessments should reduce the probability of making either the Type I or Type II error and allow more efficient allocation of risk management resources.

The ASTM–RBCA approach can be adapted for the occupational hygiene profession. Similar to ASTM–RBCA, there are some situations in which risk management is needed immediately; these situations are not appropriate for tiered analysis. In Tier I analysis, judgments are conservative, and the bias is toward making the Type I error. The methodology usually involves collecting air samples and comparing analytical results with the lookup table, which in this case is the PEL/TLV tables. Risk management efforts are conservative, meaning that the preference is to provide too much control rather than too little. Examples would be collecting air samples of welding fumes for comparison with the TLV/time-weighted average, or of H_2S for comparison with the TLV-ceiling.

Tier II represents a higher level of analysis, in which there is a lower probability of making the Type I or II error. Risk characterization is based on exposure assessment of intake from sources (inhalation, dermal absorption, and perhaps ingestion) in the workplace, using air and surface sampling or exposure modeling. Where appropriate, this intake, in terms of mg/kg/day, could be compared with toxicity values expressed as 1/(mg/kg/day) for carcinogens, and mg/kg/day for noncarcinogens. An example would be monitoring air and surface concentrations of cadmium, calculation of intake as mg/kg/day, and comparison with the slope factor and RfD for cadmium.

Tier III is the highest level of risk assessment. Risk management efforts can be closely linked to predicted level of risk, as there is minimum probability of making the Type I or II error (although it should be clearly understood that there is always some probability of making either error). Tools available for this level include extensive modeling, detailed assessment of occupational and environmental exposure to the contaminants of concern, blood/urine analysis, and estimation of personal intake factors. As in Tier II, the results of the exposure assessment may be compared with slope factors and/or RfDs.

When minimum data are collected in an exposure assessment, the confidence bands drawn about the exposure estimates are necessarily wide. To protect worker health and safety adequately, risk must be estimated conservatively (i.e., from the upper bounds of the confidence limits). The occupational hygienist must act conservatively, maximizing the chance of making the Type I error while minimizing the probability of making the more serious Type II error. As risk assessment efforts progress from Tier I to the higher tiers, and more data are collected, the confidence bands about the risk estimate become narrower, perhaps to the point where the upper bound on the risk estimate is below the acceptable risk level. In other words, the additional resources (time, sampling and analytical expenses) expended in risk assessment may be justified by reductions in required risk management efforts. (Recall that risk is a function of both exposure and toxicity; therefore, it is possible that collection of additional toxicity data would also shrink the confidence bands about the risk estimate.) However, collection of additional information may also result in higher estimates of exposure, toxicity, and risk. Thus, a major benefit of higher level risk assessment is reduction of Type I and Type II errors in selection of the risk management strategy.

Risk Perception, Risk Communication, and the Mismatch Between Risks and Regulations

As seen from the brief history of risk regulation provided at the beginning of this chapter, the United States has never had an integrated policy for control of environmental risk. Instead, an ad hoc policy has prevailed, resulting in a patchwork of environmental laws and regulations. Often, legislation has not addressed significant environmental hazards, while the risks that concern experts are left unmanaged. Some familiarity with risk perception will be helpful in explaining this paradox because risk perceptions guide so many of our actions as individuals, as organizations, and as societies. Risk perceptions have a strong influence on levels of risk considered acceptable, and through this route, may even influence congressional and regulatory actions.[62] If a risk is perceived to be significant, then the public will demand, or Congress may perceive a demand for, remedial action. For example, the asbestos-in-schools legislation followed on the heels of public disclosure of the health risks faced by asbestos workers; and the Comprehensive Environmental Response, Compensation and Liability Act of 1980 followed the 1978 disclosure of problems in Love Canal, N.Y. Interestingly, hazardous waste sites, underground storage tanks, and solid waste sites are not considered by many experts to be particularly serious sources of risk to the public.[63,64] In contrast, risks believed by experts to be serious and significant, such as worker exposure to chemicals and radon (which were tied for first place in EPA's *Unfinished Business* report[63]), pesticides, and nonpoint source contamination of drinking water, are not perceived by the public as hazardous. The mismatch between level of risk and allocation of risk management resources may be explained by (1) public perception of risk and resultant pressures on EPA; (2) difficulty of control of some high risk areas with traditional technologies; and (3) the fact that presence of effective control programs may have already reduced risk.[65]

Cohrssen and Covello[66] made an important contribution with the listing of 20 dimensions of risk and the impact of each on risk perception. These dimensions (e.g., increased perception of risks which are unfamiliar, uncontrollable, and which affect children) help in understanding public opinion on asbestos and hazardous waste sites. Morgan[67] further clarified these concepts with the development of a risk matrix. The vertical axis includes the observable constellation of risk dimensions. Location on the continuum depends on such factors as "known to those exposed and to science, effect immediate, old risk" for risks perceived to be low, and "unknown to those exposed and to science, effect delayed, new risk" for risks perceived to be high. Similarly, the horizontal or controllable axis ranges from "not dreaded, not catastrophic, equitable, voluntary, etc." to "dreaded, catastrophic, not equitable, involuntary, etc." With this strategy, nuclear power and recombinant DNA research, which are not observable or controllable, would receive much higher risk scores than alcohol and cigarettes, which are observable and controllable. Without implying that the public's risk perceptions can or should be manipulated, it is certainly ethical and probably necessary to increase the public level of awareness about occupational and environmental risks. Few would argue that increased understanding and encouragement of risk-aversive behavior are unwarranted in the cases of the lifestyle determinants that take the lives of so many Americans. It is equally important that occupational hygienists underscore the importance of lifestyle choices as contributors to individual risks. This is especially important where lifestyle choices and occupational exposures may have synergistic effects, such as smoking and asbestos exposure.

Risk communication is a critical component to successful risk assessment and risk communication. It is a mechanism to ensure stakeholder input, which is extremely helpful in developing consensus. For further reading, see Chapter 41, "Risk Communication in the Workplace," and many excellent references including Covello et al.[68]

Risk Management

Purpose and Definitions

The purpose of this section is to briefly examine the role of risk assessment in the selection of risk management goals and techniques. Although some of the language may differ from traditional "IH talk," the occupational hygienist functions as a risk assessor in the "evaluation" mode, and as a risk manager in the "control" mode. When one person is responsible for conducting risk assessment and making risk management decisions, any distinction between the two processes may be fuzzy and unnecessary. However, the National Research Council envisioned a clear line between risk assessment and risk management:

> *We recommend that regulatory agencies take steps to establish and maintain a clear conceptual distinction between assessment of risks and consideration of risk management alternatives; that is, the scientific findings and policy judgments embodied in risk assessments should be explicitly distinguished from the political, economic, and technical considerations that influence the design and choice of regulatory strategies.*[27]

Risk assessment was to be composed of hazard identification, dose-response assessment, exposure assessment, and risk characterization. Risk management was to consist of development of regulatory options, and evaluation of public health and technical considerations, and the economic, social, and political consequences of regulatory options. In other words, risk assessment was to be a value-neutral scientific process, while risk management could be influenced by nonscientific concerns. Most people willingly agree that the most accurate risk assessment possible is necessary to inform the risk management decision, which may take into account nonrisk factors. But however good this sounds in theory, it is very difficult to put it into actual practice. The reader is well aware of the many points in the risk assessment process where there is uncertainty and variability in the data and a conservative decision is rendered to protect public health: use of toxicological data from the most sensitive species, high-end exposure calculations, selection of the linear multistage model, etc. So it is recognized up front that there is no clear separation between the processes of risk assessment and risk management, a separation made even fuzzier when the risk assessor is the risk manager. To further muddy the waters, EPA has recommended that in the ecological risk assessment process, the risk assessor should meet frequently with the risk manager, so that risk can be evaluated in policy-relevant terms.[31] Risk analysis, as defined by Cohrssen and Covello,[66] goes beyond the scientific process and includes decisions that should be guided by social, cultural, moral, economic, and political factors (i.e., determination of the significance of a risk and the communication of risk information to affected publics). This definition thus embraces risk assessment, risk management, and risk communication, which actually are difficult to separate. With these cautions in mind, several terms can be defined:

- Risk assessment—the estimation of the risk (i.e., the probability and magnitude of harm);
- Risk management—the reduction of risks to an appropriate level;
- R_T—true risk, the actual level of risk;
- R_E—estimated risk, the risk assessor's best estimate of risk; and
- R_A—acceptable risk, the level of risk the public is willing to accept.

An understanding of the relationship between risk and acceptable risk is critical to this discussion because if the risk in a situation exceeds acceptable risk, then risk management efforts must be started. The decision about how much risk management is adequate is based on the gap between risk and acceptable risk; if R_E greatly exceeds R_A, then risk management efforts will probably be significant and immediate. The risk assessor estimates the level of risk based on exposure and effects per exposure, using the risk assessment principles outlined in this chapter. He or she knows that R_T does not equal zero, but its true value might never be known.

Therefore, R_E is selected from what is believed to be the top of the confidence interval about R_T, for example the 95% upper confidence level, because if a mistake is to be made, the bias is toward protection of public health. Society, including the risk assessor, determines whether R_E is acceptable based on social, political, economic, public health, and technical considerations. Sometimes society decides that R_A is too low and can be raised. An excellent example is the recent increase in highway speed limits and the predicted increase in traffic deaths. Despite the fact that this prediction is being proved true, society has not raised a hue and cry to lower the speed limits. In contrast, it has been decided that there are too many deaths from environmental tobacco smoke, as evidenced by the successful efforts to gradually eliminate public smoking.

It has been said that "the best level of risk is zero, if it is free, but it never is."[69] While it seems that society would prefer zero risk, even demanding it at times, different levels of risk are acceptable in different situations. Some levels of risk acceptability have been formalized. In establishing appropriate risk levels for Superfund sites, EPA works in the range of 10^{-6} to 10^{-5}; this is interpreted as "one case per million persons exposed" up to "one case per 100,000 persons exposed." In contrast, OSHA accepts higher risk levels, in the range of 10^{-4} to 10^{-3}. There are reasonable arguments in favor of this differential: workers are generally the healthiest members of society, they are trained in occupational hazards, have some degree of control over their exposure, and are financially compensated for their exposure to risk. The public comprises the young and the old, the healthy and the sick, and persons who may have occupational or environmental exposures to other chemicals. They may not be aware of these exposures, generally have no control over them, and receive no financial compensation. In some cases the level of acceptable risk has been captured in acceptable exposure levels: TLVs and PELs in the occupational setting, National Ambient Air Quality Standards for community air, and certain regional soil cleanup levels for hazardous waste sites. Thus, whether institutionalized or determined individually by the risk manager, the selection of R_A allows determination of corresponding exposure levels to contaminants in air, water, and soil.

Determination of Appropriate Exposure Levels

The acceptable risk level, along with economic, social, political, and technical considerations, is one of the inputs used in selection of (1) acceptable exposure limits and (2) the methods used to achieve these limits. Selection of risk management methods in occupational hygiene is also influenced by the hierarchy of controls: engineering, administrative, and personal protective equipment. If available, knowledge of exposure to other chemicals on the same or different jobs, presence of sensitive subpopulations, and exposures in the home environment might also influence the acceptable level of occupational exposure. Given all these factors, how does the risk manager or occupational hygienist make an appropriate selection? To some extent the determination of acceptable exposure limits for occupational and environmental settings seems to have followed two different trends. As described by Corn,[1] the approach to establishing TLVs was to determine the concentration of material that caused injury and to set threshold values based on the concentrations measured. The risk would then be negligible if exposure levels were kept below certain acceptable levels. (Note that the majority of alterations in TLVs and PELs have been downward, reflecting increasing concern about chronic health effects.) Establishment of acceptable environmental exposures seems to have come from the other direction, as exemplified by the 1958 Delaney clause (or amendment), which prohibited the use in food of any food additive shown to cause cancer in animals or humans at any oral dose.[2,70] The many food additives in use at that time were exempted from the rigorous testing requirements and were listed as GRAS (generally regarded as safe). It was later recognized that long-term use was no guarantee of safety, and several substances, including cyclamates and coal-tar dyes, have been taken off the list. FDA now has four regulatory categories: food additives, GRAS substances, prior-sanctioned substances, and color additives.

The Delaney clause was later subject to legislative change (e.g., saccharin) and interpretation through the GRAS list (e.g., the use of methylene chloride to decaffeinate coffee). Because site-specific cleanup levels for hazardous waste sites have also been based on findings of chronic toxicological or epidemiological studies, there has been a similar upward trend, from zero or background levels to risk-based, or in some cases technology-based levels. This so-called risk-based approach to establishing cleanup levels has been formalized in the EPA documentation,(13) as well as the more recent ASTM *Standard Guide for Risk-Based Corrective Action Applied at Petroleum Release Sites.*(15) These approaches are based on the same equations used to calculate risk (see Equations 17.2.1 and 17.2.2), with the target risk (lifetime cancer risk estimate for carcinogens, or hazard index for noncarcinogens) in this case as a known quantity, and the concentration of the chemical in the media of concern as the dependent variable. See the sidebar below for an example of back-calculation of an acceptable residential water concentration of benzene, using a target risk of 10^{-5}. Similar calculations may be made to determine acceptable risk-based concentrations of contaminants in food, soil, or air.(71) If there is exposure to more than one chemical, or if exposure occurs by more than one route or pathway, the acceptable concentration must be reduced accordingly.

The importance of restricting risk assessment, when possible, to scientific concerns while allowing value judgments to enter into risk management decisions has been stressed. This is a distinction that may cause discomfort to some scientists and engineers, who may execute technically accurate risk assessments and risk management analyses only to see political and social preferences sway decisions to, in their opinions, less appropriate risk management techniques. It is useful to remember that whenever possible risk management decisions should benefit all stakeholders in a situation, for many of whom the economic, social, and political factors are of more immediate importance than the accuracy of the risk assessment itself.

Summary

The purpose of this chapter has been to stimulate the integration of risk assessment into the practice of occupational hygiene. Toward that end, the basic concepts of risk assessment were introduced through discussions of risk, risk perception, regulation of risk, and the risk assessment process. It has been suggested that political, economic, and technological changes force a new approach. "It's required by OSHA" may no longer be a sufficient argument to invest in workplace health and safety controls. Integration of risk assessment and occupational hygiene in an iterative framework can provide effective allocation of resources while minimizing errors, protecting worker health and safety, and meeting the traditional objectives of protecting the bottom line and avoiding liability. Occupational hygienists responsible for worker health and safety must keep in mind the following points:

First, occupational hygienists are professionally and ethically responsible for the outcomes of their evaluations/risk characterizations. Therefore, they must recognize that occupational hygiene risk assessment extends beyond the collection of airborne concentration data and comparison with the occupational exposure limits. New approaches and methods can be adapted from the companion environmental health disciplines.

Calculation of Risk-Based Water Concentration of Benzene

$$TR = \frac{SF_O \times C \times IRW \times EF \times ED}{BW \times AT}$$

$$C = \frac{TR \times BW \times AT}{SF_O \times IR_W \times EF \times ED}$$

$$C = \frac{10^{-5} \times 70\text{ kg} \times 25{,}550\text{ days}}{0.029\text{ mg/kg/day} \times 2\text{ L/day} \times 350\text{ days/year} \times 30\text{ years}}$$

$$= 0.03\text{ mg/L}$$

where

TR = target excess individual lifetime cancer risk, unitless, 10^{-5}
SF_O = oral cancer slope factor, mg/kg/day
C = concentration, mg/L
IRW = daily water ingestion rate, L/day
EF = exposure frequency, 350 days/year
ED = exposure duration, 30 years
BW = body weight, 70 kg
AT = averaging time of 70 years, expressed as 25,550 days

Second, it is the responsibility of occupational hygienists to ensure, insofar as possible, that workers are protected from occupational hazards in a manner that is effective and cost efficient. As the experts in risk assessment for occupational scenarios, occupational hygienists need to continue to explore and develop new methodologies of conducting their profession, especially in the face of a deregulated environment.

If the best of the new methodologies include risk assessment, then the occupational hygiene profession must be about the business of writing a new paradigm for the 21st century.

Acknowledgments

Sincere appreciation is expressed for the generous contributions and thought provoking comments of Dr. Neil Hawkins, Dr. Michael Jayjock, Mr. Jerry Lynch, Dr. Robert Nelson, Mr. Tom Richardson, Dr. Larry Whitehead, and Ms. Becky Ziebro.

References

1. **Corn, J.K.:** *Response to Occupational Health Hazards: A Historical Perspective.* New York: Van Nostrand Reinhold, 1992.
2. **Salvato, J.A.:** *Environmental Engineering and Sanitation,* 4th ed. New York: John Wiley & Sons, Inc., 1992.
3. **U.S. Bureau of the Census:** *Statistical Abstract of the United States,* 115th ed. Washington, DC: U.S. Bureau of the Census, 1995.
4. **U.S. Department of Labor, Bureau of Labor Statistics:** *National Census of Fatal Occupational Injuries.* In http://www.bls.gov/oshcftab.htm/ [Occupational Safety and Health Administration (OSHA) World Wide Web homepage]. Washington, DC: OSHA, June 1997.
5. **Reiser, K.:** General principles of susceptibility. In *Environmental Medicine,* S. Brooks, ed. St. Louis, MO: Mosby Year Book, Inc., 1995.
6. **Covello, V.T., and J. Mumpower:** Risk analysis and risk management: an historical perspective. *Risk Anal.* 5:103–120 (1985).
7. **Williams, P.L., and J.L. Burson:** *Industrial Toxicology: Safety and Health Applications in the Workplace.* New York: Van Nostrand Reinhold, 1985.
8. **Plough, A., and S. Krimsky:** The emergence of risk communication studies: social and political context. In *Readings in Risk,* T.S. Glickman and M. Gough, eds. Washington, DC: Resources for the Future, 1990.
9. **Vesilind, P.A., J.J. Peirce, and R.F. Weiner:** *Environmental Pollution and Control,* 3rd Ed. Boston: Butterworth-Heinemann, 1990.
10. **Percival, R.V., A.S. Miller, C.H. Schroeder, and J.P. Leape:** *Environmental Regulation: Law, Science, and Policy.* Boston: Little, Brown and Co., 1992.
11. **Andrews, R.N.:** Reform or reaction: EPA at a crossroads. *Environ. Sci. Tech. 29*:505A–510A (1995).
12. **U.S. Environmental Protection Agency (EPA):** *Risk Assessment Guidance for Superfund,* vol. I, part A (EPA/540/1-89/002). Washington, DC: EPA, 1989.
13. **U.S. Environmental Protection Agency (EPA):** *Risk Assessment Guidance for Superfund,* vol. I, part B (Pub. 9285.7-01B). Washington, DC: EPA, 1991.
14. **U.S. Environmental Protection Agency (EPA):** *Risk Assessment Guidance for Superfund,* vol. I, part C (Pub. 9285.7-01C). Washington, DC: EPA, 1991.
15. **American Society for Testing and Materials (ASTM):** *Standard Guide for Risk-Based Corrective Action Applied at Petroleum Release Sites* (ASTM E 1739-95). West Conshohocken, PA: ASTM, 1995.
16. **Rushefsky, M.E.:** *Making Cancer Policy.* Albany, NY: State University of New York Press, 1986.
17. "Occupational Safety and Health Act," Pub. Law 91-596, Section 2193. 91st Congress, Dec. 29, 1970; as amended, Pub. Law 101-552, Section 3101, Nov. 5, 1990.
18. **Congressional Information System, Inc.:** *Abstracts of Congressional Publications and Legislative History Citations.* Washington, DC: CIS, Inc. Quadrennial series, including 1974, 1978, 1982, 1986, 1990, and 1994.
19. **U.S. Superintendent of Documents:** "Food Quality Protection Act of 1996." In http://www.access.gpo.gov./su_docs/ [U.S. Government Printing Office World Wide Web homepage] August 1996.
20. **Mintz, B.:** History of the federal Occupational Safety and Health Administration. In *Fundamentals of*

Industrial Hygiene, 4th ed., B.A. Plog, J. Niland, and P.B. Quinlan, eds. Itasca, IL: National Safety Council, 1996.

21. **Rodricks, J.V., S.M. Brett, G.C. Wrenn:** Significant risk decisions in federal regulatory agencies. *Reg. Toxicol. Pharmacol. 7*:307–320 (1987).
22. **Castleman, B.I., and G.E. Ziem:** Corporate influence on threshold limit values. *Am. J. Ind. Med. 13*:531–559 (1988).
23. *Public Citizen Health Research Group v. Tyson*, 796 F2d 1479 (D.C. Cir. 1986).
24. "Methylene chloride." *Federal Register 62* (10 January 1997). p. 1493.
25. **Mantel, N., and W.R. Bryan:** Safety testing of carcinogens. *J. Nat. Cancer Inst. 27*:455–460 (1961).
26. **U.S. Environmental Protection Agency (EPA):** *The Risk Assessment Guidelines of 1986* (EPA/600/8-87/045). Washington, DC: EPA, 1987.
27. **National Research Council, Committee on the Institutional Means for Assessment of Risks to Public Health, Commission on Life Sciences:** *Risk Assessment in the Federal Government: Managing the Process.* Washington, DC: National Academy Press, 1983.
28. **Habicht, F.H.:** "Guidance on Risk Characterization for Risk Managers and Risk Assessors." U.S. Environmental Protection Agency, Washington, DC, February 1992. [Memo.]
29. **Browner, C.:** "Policy for Risk Characterization at the U.S. Environmental Protection Agency." U.S. Environmental Protection Agency, Washington, DC, March 1995. [Memo]
30. **U.S. Environmental Protection Agency (EPA):** *Framework for Ecological Risk Assessment.* Washington, DC: EPA, 1992.
31. **U.S. Environmental Protection Agency (EPA):** "Ecological Risk Assessment Guidance for Superfund: Process for Designing and Conducting Ecological Risk Assessments." (Review draft) Edison, NJ: EPA, 1994.
32. **Barnes, D.G.:** Times are tough—brother, can you paradigm. *Risk Anal. 14(3)*:219–223 (1994).
33. **Vainio, H., M. Coleman, and J. Wilbourn:** Carcinogenicity evaluations and ongoing studies: The IARC databases. *Environ. Health Persp. 96*:5–9 (1991).
34. **Vainio, H., and J. Wilbourn:** Identification of carcinogens within the IARC monograph program. *Scand. J. Work Environ. Health 18(suppl 1)*:64–73 (1992).
35. **American Conference of Governmental Industrial Hygienists (ACGIH):** *1996-1997 Threshold Limit Values (TLVs) for Chemical Substances and Physical Agents and Biological Exposure Indices (BEIs).* Cincinnati, OH: ACGIH, 1996.
36. "Proposed Guidelines for Carcinogen Risk Assessment." *Federal Register 61*:17959–18011 (April 23, 1996).
37. **Gregory, A.R.:** Uncertainty in health risk assessments. *Reg. Toxicol. Pharmacol. 11*:191–200 (1990).
38. **Paustenbach, D.J.:** Health risk assessment and the practice of industrial hygiene. *Am. Ind. Hyg. Assoc. J. 51*:339–351 (1990).
39. **Upton, A.C.:** Epidemiology and risk assessment. In *Epidemiology and Health Risk Assessment*, L. Gordis, ed. New York: Oxford University Press, 1988.
40. **Gori, G.B.:** Cancer risk assessment: The science that is not. *Reg. Toxicol. Pharmacol. 16*:10–20 (1992).
41. **Ames, B.N., and L.S. Gold:** Too many rodent carcinogens: Mitogenesis increases mutagenesis. *Science 249*:970–971 (1990).
42. **U.S. Environmental Protection Agency (EPA):** *Standard Default Exposure Factors* (9285.6-03). Washington, DC: EPA, 1991.
43. **National Institute for Occupational Safety and Health (NIOSH):** *Occupational Exposure Sampling Strategy Manual*, by N.A. Leidel, K.A. Busch, and J.R. Lynch. Cincinnati, OH: NIOSH, 1977.
44. **Hawkins, N.C., S.K. Norwood, and J.C. Rock:** *A Strategy for Occupational Exposure Assessment.* Akron, OH: American Industrial Hygiene Association, 1991.
45. **Jayjock, M.J.:** Back pressure modeling of indoor air concentrations from volatilizing sources. *Am. Ind. Hyg. Assoc. J. 55*:230–235 (1994).
46. **Wadden. R.A., P.A. Scheff, J.E. Franke, L.M. Conroy, et al.:** VOC emission rates and emission factors for a sheetfed offset printing shop. *Am. Ind. Hyg. Assoc. J. 56*:368–376 (1995).
47. **Fehrenbacher, M.C., and A.A. Hummel:** Evaluation of the mass balance model used by the Environmental Protection Agency for

estimating inhalation exposure to new chemical substances. *Am. Ind. Hyg. Assoc. J. 57*:526–536 (1996).

48. **Lipton, J., and J.W. Gillett:** Uncertainty in risk assessment: exceedence frequencies, acceptable risk, and risk-based decision making. *Reg. Toxicol. Pharmacol. 15*:51–61 (1992).
49. **American Conference of Governmental Industrial Hygienists (ACGIH):** *Documentation of the Threshold Limit Values and Biological Exposure Indices*, 6th ed. Cincinnati, OH: ACGIH, 1993.
50. **National Technical Information Service (NTIS):** *EPA's Integrated Risk Information System* (PB 95-591330). Washington, DC: NTIS, January 1996. [Diskettes]
51. **Paustenbach, D.J.:** Occupational exposure limits, pharmacokinetics, and unusual work schedules. In *Patty's Industrial Hygiene and Toxicology*, 3rd ed., vol. 3A. R.L. Harris, L.J. Cralley, and L.V. Cralley, eds. New York: John Wiley & Sons, Inc., 1994. pp. 222–348.
52. **U.S. Environmental Protection Agency (EPA):** *Exposure Factors Handbook* (EPA/600/8- 89/043). Washington, DC: EPA, 1990.
53. **Kodell, R.L., and J.J. Chen:** Reducing conservatism in risk estimation for mixtures of carcinogens. *Risk Anal. 14(3)*:327–332 (1994).
54. **Finley, B., D. Proctor, P. Scott, N. Harrington, D. Paustenbach, and P. Price:** Recommended distributions for exposure factors frequently used in health risk assessment. *Risk Anal. 14*:533–553 (1994).
55. **Burmaster, D.E., and P.D. Anderson:** Principles of good practice for the use of Monte Carlo techniques in human health and ecological risk assessments. *Risk Anal. 14(4)*:477–481 (1994).
56. **Evans, J.S., G.M. Gray, R.L. Sielken, A.E. Smith, C. Valdez-Flores, and J.D. Graham:** Use of probabilistic expert judgment in uncertainty analysis of carcinogenic potency. *Reg. Toxicol. Pharmacol. 20*:15–36 (1994).
57. **Cullen, A.C.:** Measures of compounding conservatism in probabilistic risk assessment. *Risk Anal. 14(4)*:389–393 (1994).
58. **Thompson, K.M., D.E. Burmaster, and E.A.C. Crouch:** Monte Carlo techniques for quantitative uncertainty analysis in public health risk assessments. *Risk Anal. 12(1)*:53–63 (1992).
59. **Smith, R.L.:** Use of Monte Carlo simulation for human exposure assessment at a Superfund site. *Risk Anal. 14(4)*:433–439 (1994).
60. **U.S. Environmental Protection Agency (EPA):** "Use of Monte Carlo Simulation in Risk Assessments" (EPA903-F-94-001). Philadelphia: U.S. Environmental Protection Agency, Region III, 1994.
61. **Begley, R.:** Risk-based remediation guidelines take hold. *Environ. Sci. Technol. 30(10)*:438A–441A (1996).
62. **Canter, L.W., D.I. Nelson, and J.W. Everett:** Public perception of water quality risks—influencing factors and enhancement opportunities. *J. Environ. Sys. 22(2)*:163–187 (1992- 1993).
63. **U.S. Environmental Protection Agency (EPA):** *Unfinished Business: A Comparative Assessment of Environmental Problems* (EPA/230/2-87/025). Washington, DC: EPA, 1987.
64. **U.S. Environmental Protection Agency (EPA):** *Reducing Risk: Setting Priorities and Strategies for Environmental Protection* (SAB-EC-90-021). Washington, DC: EPA, 1990.
65. **Long, J.:** EPA focuses on environmental problems posing least health risk. *Chem. Engin. News*, January 15, 1990. pp. 16–17.
66. **Cohrssen, J.J., and V.T. Covello:** *Risk Analysis: A Guide to Principles and Methods for Analyzing Health and Environmental Risks.* Washington, DC: U.S. Council on Environmental Quality, 1989.
67. **Morgan, M.G.:** Risk analysis and management. *Sci. Am. 269*:32–41 (1993).
68. **Covello, V.T., P.M. Sandman, and P. Slovic:** *Risk Communication, Risk Statistics, and Risk Comparisons: A Manual for Plant Managers.* Washington, DC: Chemical Manufacturers Association, 1988.
69. **Caplan, K.J., and J. Lynch:** A need and an opportunity: AIHA should assume a leadership role in reforming risk assessment. *Am. Ind. Hyg. Assoc. J. 57*:231–237 (1996).
70. **Nadakavukaren, A.:** *Man and Environment: A Health Perspective*, 4th ed. Prospect Heights, IL: Waveland Press, Inc., 1995.
71. **Smith, R.L.:** "Risk-Based Concentration Table, January-June 1995," U.S. Environmental Protection Agency, Region III, Philadelphia, PA, March 1995. [Memo.]

Chapter 18

Outcome Competencies

After completing this chapter the user should be able to:

1. Define underlined terms used in this chapter.
2. Describe the characteristics of biohazardous agents that distinguish them from chemical hazards and hazardous physical agents.
3. Describe the necessary and sufficient conditions for infection to occur.
4. Identify factors to consider in evaluating risk from biological agents.
5. Apply control strategies to biohazards based on risk.
6. Apply containment levels to biohazards.
7. Classify biological agents to hazard categories.
8. Describe limitations of biosafety cabinets.
9. Select effective decontaminants.
10. Use regulations and standards to properly package and ship biological agents.
11. Identify biological agents that are important to industrial processes.
12. Draw process flow diagrams for biotechnology manufacturing.

Prerequisite Knowledge

Prior to beginning this chapter, the user should review the following chapters:

Chapter Number	Chapter Topic
6	Principles of Evaluating Worker Exposure
17	Risk Assessment in the Workplace
19	Indoor Air Quality
48	Laboratory Health and Safety
49	Emergency Planning in the Workplace
50	Anticipating and Evaluating Trends Influencing Occupational Hygiene

Key Terms

biohazards • biohazardous waste • biological safety cabinets • biological agents • biological competition • biosafety • biosafety levels • biotechnology • bloodborne pathogens • chain of infection • colonization • decontaminants • disinfectants • enzymes • etiologic agents • genetic engineering • infection • opportunistic pathogens • personal protective equipment • primary barriers • recombinant DNA • risk assessment • risk communication • secondary barriers • sterilants • zoonotic infection

Key Topics

I. Biohazard Definition

II. Assessment of Biological Risk
   A. Differences Between Biohazards and Chemical or Physical Hazards
   B. Agent, Host, Environment
   C. The Chain of Infection
   D. Risk Assessment Process

III. Working Safely with Biological Agents
   A. Biosafety Program Management
   B. Administrative Controls
   C. Physical Containment- the Biosafety Levels
   D. Decontamination, Disinfection, Sterilization
   E. Treatment and Disposal of Infectious Waste
   F. Accidents, Spills, and Emergency Planning
   G. Transportation, Shipping, and Receiving Biological Agents
   H. Communicating Biological Risk

IV. Implementing Biosafety Programs in Laboratories and in Industry
   A. Biotechnology R&D
   B. Large-Scale Biotech Manufacturing
   C. Intentional Aerosolization Studies Involving Pathogenic Microorganisms
   D. Animal Research
   E. Enzymes in the Detergent Industry

Biohazards in the Work Environment

18

Susan B. Lee, CIH

Barbara Johnson, Ph.D.

Introduction

In the past 10–15 years several events have led to increased awareness of the potential hazards of biological agents. The identification of the biological agent that caused Legionnaires' disease, the recognition of the AIDS (acquired immunodeficiency syndrome) epidemic, and the emergence of recombinant DNA (rDNA) technology as an industry and as a means to create goods and services have all contributed to increased interest and concern about biohazards among the general public and consequently also among occupational hygienists.

Biohazards are as common in certain industries (agriculture, health care, biotechnology, research and clinical labs), as silica, carbon monoxide, and metal fumes are in foundries. Biological agents are inherently different from chemical toxins, carcinogens, or physical agents, and the differences are exploited in assessing risk as well as in selecting and applying control strategies. The model used in occupational hygiene practice—anticipation, recognition, evaluation, and control—can be applied, nevertheless.

The concept of performing a risk assessment and evaluating exposures to biological agents are similar to chemical and physical agents; however, the practices used in evaluation differ. Air sampling and screening for biological samples are not typical methods used to assess potential exposure to biological agents. More often clinical specimens from workers are collected to identify potential exposure to etiologic agents. These results are then compared with bulk and/or wipe sample results for specific organisms. Although techniques for controlling exposure to biohazards are similar, the use of personal protective equipment (PPE) is fundamental protection and not lowest on the hierarchy. Exposure to biohazards by inhalation, direct contact, and breaks in skin are important routes of transmission, and PPE provides crucial primary barrier protection from direct exposure to biological agents.

Although there are specially trained biological safety professionals, their numbers are small,[1] and many organizations may not require full-time specialized expertise. As a result, available health and safety professionals, including occupational hygienists, are increasingly being called on to respond to complaints and to investigate exposure or other issues pertaining to biological agents. The purposes of this chapter are to discuss the nature of biological hazards and the concepts of biological safety, to describe the components of effective biological safety programs, and to demonstrate how these concepts are applied in industry. The role of biological agents in indoor air quality evaluations or environmental monitoring for bioaerosols has purposely been excluded. (The reader should refer to other relevant chapters of this book for further information on these aspects.)

Assessment of Biological Risk

A well thought out risk assessment is a cornerstone in the implementation of a successful biological safety program. Workplace risk assessment is an effort to predict the probability of a given outcome based on quantitative data and estimates from past repeated exposures, and on qualitative valuations of the available information. The objectives are to identify and recommend control strategies that will eliminate the hazard or reduce probability of adverse effects (illness, subclinical infection, harm to plants, animals, or the environment). The goal of effective risk communication should be to convince affected parties (management, workers, community) that the level of risk (probability and severity of adverse outcome) is acceptably low when compared with the level of gain and the preventative measures instituted. Recommendations for action, the result of the risk assessment, may differ from institution to institution, situation to situation, and among individual assessors, but they should always be based on a thorough and systematic process that considers each of the relevant factors.

Initially, it is appealing to consider risk assessment for biological agents in the same way as chemical and radiological materials. Complex quantitative models have been developed and utilized by the U.S. Environmental Protection Agency (EPA),[(2)] International Commission on Radiological Protection,[(3)] National Commission on Radiation Protection,[(4)] and others to evaluate such things as fate and persistence of hazardous chemicals in the environment and the effects of ionizing radiation on human and environmental health. But infectious or allergenic biological materials are intrinsically different from chemicals and radiation, and these differences affect how risks are evaluated. Biological materials typically:

- Have no threshold level of exposure. A single pathogenic organism is capable of replication and in some cases, given the right conditions, can infect a host and cause disease.
- Are ubiquitous in the environment. The concept of permissible exposure limits for biological agents is not appropriate, as pathogens are commonly found in the environment, and it is not solely the concentration that makes them hazardous. There are a multitude of other contributing factors involved (see below).
- Are affected by biological competition. Biological agents do not typically behave synergistically in the same sense that exposure to multiple chemicals can exacerbate symptoms (e.g., methylene chloride exposure can potentiate carbon monoxide exposure). Biological agents compete among each other for dominance in the environment, and after entering the host, they may be biologically inactivated by the immune system or other normal or innate metabolic processes. In the case of immunocompromised individuals, the host's defenses are less effective, and agents that are nonpathogenic in healthy individuals may cause infection, disease, and even death. These are referred to as opportunistic pathogens.

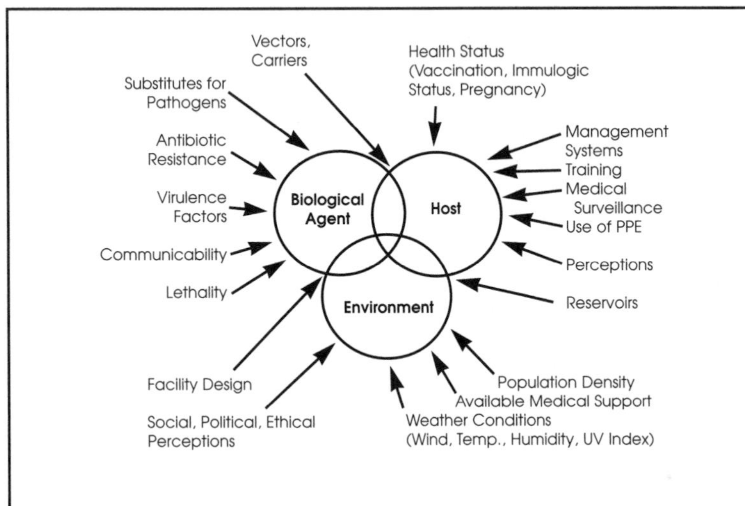

Figure 18.1 — Risk assessment involving biological agents: factors to consider.

To assess risk one must understand and consider the nature of organisms and their products, the nature of the host (i.e., human physiology, including its relationship to host immune responses), and environmental conditions. These interrelationships must be considered within the context of external forces affecting them. Figure 18.1 depicts the interactive nature of the main components of risk assessment for biological hazards.

Agent, Host, and Environment

The interactions among a biological agent, the environment, and a host do not necessarily result in disease or damage. For infection and illness to occur in a new host, there are six necessary and sufficient conditions:

(1) The agent must be pathogenic;
(2) There must be a reservoir of sufficient number for the organism to live and reproduce;
(3) The agent must be able to escape from the reservoir;
(4) The organism must be able to move through the environment (e.g., airborne, contact, carriers such as mosquitoes or ticks);
(5) There must be a portal of entry for the new host (e.g., broken skin,

mucous membrane, inhalation, blood transfer); and

(6) The new host must be susceptible to the agent.

If even one of these conditions is not met, then disease is not an outcome. Each of the biosafety control strategies is intended to break one or more of the links in the chain, thus preventing infection. It follows, therefore, that biosafety can be defined as the art and science of maintaining a broken chain of infection. This chain of infection is depicted in Figure 18.2.[5]

A major aspect of biosafety is the assessment of the risk of infection and disease. The first factor to consider in assessing risk is "how pathogenic is the biological agent?" If work involves using an agent of high virulence, or high mortality, or one for which there is no known vaccine or treatment, then those factors should be considered. Several references provide lists[6–8] that place known biohazardous agents in categories according to their pathogenicity, virulence, and so forth. Least hazardous agents are not known to cause disease in healthy adults. Medium risk agents are associated with human disease, but it is not likely to be fatal, and there usually are treatments. Higher risk agents are indigenous or exotic agents with potential for aerosol transmission, and may produce diseases with serious or lethal consequences. The most hazardous exotic agents pose high risk of life-threatening disease, and there may be no known vaccine or treatment.[9]

Rabies virus is an example of an agent propagating itself in a reservoir (infected bats). Once released, transmission is by aerosol or percutaneous routes, through the environment to the hosts. If the reservoir is not created, if there is no release from the reservoir or transmission through air or skin to the hosts, then there can be no interaction or disease. Procedures such as avoiding the reservoir and using PPE when working in areas where bats are prevalent prevent transmission. In the laboratory setting, use of PPE, procedures to minimize formation of aerosols, and use of containment equipment such as biological safety cabinets prevent release and transmission via the airborne route. Administrative procedures, such as prohibiting the use of sharps, reduce the potential for transmission by percutaneous route.

The last two required conditions for infection are that a host receive an infectious dose through a portal of entry, and that the new host be susceptible to the agent. An infectious dose may be as large as 10^{-6}–10^{-9} organisms, as in the case of enteric pathogens like cholera; or it may require very few organisms (1–10), as in the case of respiratory organisms such as *Mycobacterium tuberculosis* (TB). This wide variation in numbers required for infection is one of several reasons why biosafety professionals have relied on experience and an understanding of the qualitative factors in making meaningful risk assessments.

Routes of entry might be direct contact with mucous membranes or breaks in the skin, by inhalation or ingestion, or by percutaneous injection. Intact skin provides an effective barrier to infection for almost all biohazards via direct contact (*Leptospira*, which can penetrate intact skin, is a notable exception.) The protective nature of skin resides in its characteristic of a dry and slightly acidic barrier (pH 5) and its capacity to support normal flora, which create a hostile and competitive environment for invading microbes. Use of gloves to prevent contact with bloodborne pathogens or respiratory protection against TB via inhalation are examples of barriers to eliminate the portal of entry.

Mucous and other fluids (tears, urine, semen, vaginal secretions, saliva) bathe membranes associated with the urogenital tract, nasal passage, throat and respiratory tract, and eyes act to trap incoming

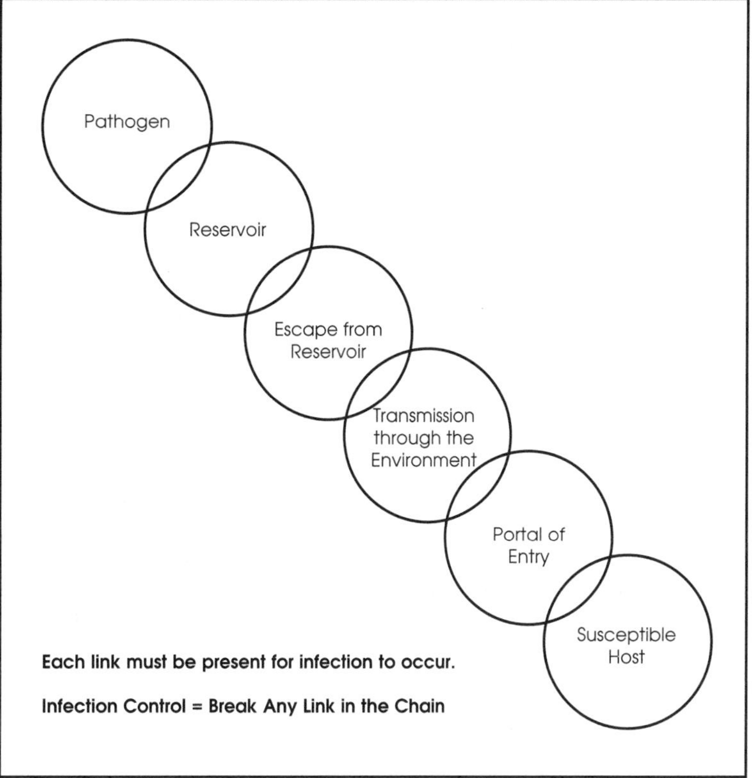

Figure 18.2 — The chain of infection.

invaders and mechanically carry them away from potential sites of adherence. Mucous also functions as a medium for the maintenance of indigenous microflora that prohibit colonization by invading organisms. Without entry, interactions involving the infectious or toxic process such as adherence or receptor binding cannot occur.

Finally, if the biological agent actually enters the host in sufficient numbers, and if the host is susceptible, infection can occur if the agent can overcome biological competition. On ingestion, for example, the biological agent or microorganism is exposed to bile salts, peptidases, trypsin, numerous enzymes, pancreatic and gastric acids, and a variety of indigenous and colonized intestinal microflora. Their role[10] in defense against invading microorganisms is to compete for receptors on cells and for nutrients. They produce proteins that are bactericidal to invading pathogens.[11] Pathogens entering via the respiratory tract encounter cilia on the tracheal rings. The cilia beat toward the mouth, carrying them away from potential target tissues such as the lungs.

Clinical infections are recognized by the display of symptoms and the presence of an immune response. Immunocompromised individuals such as transplant patients in hospitals, neonates, or persons with AIDS-related complex or AIDS, are susceptible to many more infections than most healthy adults. Dust generated in construction projects in hospitals often contains *Aspergillus niger* spores. Exposure to *Aspergillus* spores is not a problem for normal healthy adults, but can result in fatal infection for immunocompromised individuals. The risk assessment considers containment of the project, protection of ventilation systems, and maintaining the health status of patients.

"Infection" does not always mean overt, symptomatic disease. Disease is characterized by a number of overt symptoms exhibited by the infected patient. Colonization is a state in which infection and establishment of an organism within a host has occurred without resulting in subclinical or clinical disease. Examples of colonization include *Staphylococcus salivarius* and *Bacteriodaceae* spp., which colonize the mouth; *S. epidermidis*, which colonizes the skin; and the presence of *Escherichia coli* and *Peptococcus* spp. in the large intestine.[12] In subclinical infections the organism has interacted with the host in such a way that no signs or symptoms are displayed, yet an immune response is mounted (i.e., rise in antibody titer to the organism or its product). Examples are exposures to TB and Cytomegalo virus (CMV).

External Influences on Risk Assessment

Outside influences can be referred to as risk factors. Risk factors alter the natural balance or the perception of actual risk, in either a positive or negative way. One example of a positive influence is the effect of a vaccine, which serves to prime the natural defenses of the host prior to encountering the pathogen. A negative influence is the effect of an immunosuppressive drug on an individual who routinely works with highly infectious materials.

In addition to the agent/host/environment interactions, Figure 18.1 illustrates some external risk factors. Biological agents adapt themselves to compete better in hostile environments. TB has developed multiple antibiotic resistance, and human immunodeficiency virus (HIV) eludes scientists' quest for vaccine or cure, in part because of continual antigenic changes. Biological agents can be genetically altered or simulated in culture, with deletion of hazardous gene sequences, for example, *E. coli* K12 strains that are nonpathogenic. Availability of insect or animal vectors can increase or decrease risk of illness (e.g., mosquitoes carrying yellow fever, deer ticks carrying Lyme disease, bats carrying rabies).

Social and political perceptions of risk can override any purely scientific concern or epidemiologic evidence. A good example is the situation concerning the disposal of medical waste. The scientific data has not shown increased incidence or risk of illness to housekeeping workers or employees of landfills,[13] but the public's perception of the risk (due, in part, to several incidents of needles and syringes washing ashore on recreational beaches) has driven the political process to highly regulate the potentially infectious waste streams from health care facilities.

With respect to the host (i.e., the potentially exposed worker), external influences such as degree of training, perception of risk, motivations to follow precautions, to use PPE, or to take time to treat effluent streams consistently are important to the overall assessment of risk. Management systems (monitoring performance, feedback, rewarding excellent work practices, correcting deficiencies where applicable) should be in place to ensure risks are minimized as much as possible.

Having evaluated all the internal and external factors, the assessor tries to determine (and influence through risk communication) whether that level of risk is acceptable. The question involves personal, social, economic, and scientific values and information.[14] Effective risk communication convinces affected parties (management, workers, community) that the level of risk (probability and severity of adverse outcome) is acceptable when compared with the level of gain and the costs of preventive measures instituted.

Finally, risk assessments should be reviewed and updated periodically. New information, techniques, and protective measures are constantly emerging, and social values are continually evolving such that over time real risk may not change, but the societal level of acceptability of that risk may change. One good example is society's willingness to vaccinate children against childhood diseases.

Until the 1950s, whooping cough (causative agent is *Bordetella pertussis*) was a major lethal disease of children. The administration of the killed DPT vaccine (diphtheria toxin, pertussis killed cells, tetanus toxin) protected against these diseases. Mild side effects were not uncommon, a small population (0.1%) suffered severe side effects, and in more rare instances irreversible brain damage occurred. Since whooping cough was so serious and potentially fatal at that time, parents assessed the low probability of severe side effects to be acceptable when weighed against the protection provided by the vaccine against the disease. Since then, routine vaccination has almost eliminated the disease, and parents have reassessed the relative risks (severe side effects versus the minimal risk of infection). The level of risk associated with the vaccine is the same, but social perspective has changed. Fewer vaccinations are being administered, and the incidence of diseases such as whooping cough are once again on the rise.

The Risk Assessment Process

There is no one sanctioned way to approach assessing biological risk. Quantitative models have been proposed, and certain models currently drive research in this emerging area of concern. One example is research conducted on the aerosol transmission of TB.[15–18] Quantitative modeling for biological risk assessment is necessary research, but applying a theoretical model to a specific situation may require the assessor to make many assumptions that may not necessarily be valid for the case at hand. In addition, the real world situation is likely to involve relatively small numbers of exposed workers, making statistical analysis of limited value. Furthermore, in the workplace the qualitative aspects of risk, such as effectiveness of training, or human perception of risk, and social and political influences are not quantifiable, and they are too important to be set aside. Therefore, the approach recommended here is to evaluate and understand the characteristics of the agent, provide for physical containment, and employ management systems to control work. By effective communication, the assessor may also be able to influence acceptability of risks and perception of safety among employees and in the community.

Each individual should develop a systematic procedure for assessing risk and build into it communication with management, medical advisors, and most importantly, biosafety professionals and occupational hygienists. As a sound, scientifically based assessment procedure is being developed, other groups, including the Institutional Biological Safety Committee, Animal Care and Use Committee, employee representatives, and community officials (i.e., local health providers, a representative from the Department of Health, and elected officials) should be included in discussions related to work conducted at the facility. The following is a recommended approach. Before starting an assessment, determine the mission of the laboratory or facility and the operations to be conducted:

- Identify the agents, gather and evaluate existing data (exposure history, infectious dose, severity of disease, and availability of treatment). Available data are generally from studies of very small populations, and caution must be exercised in attempts to extrapolate the possible effects to a larger or more diverse population.[19]
- Evaluate the effluent streams by which biohazards can escape containment. Biological egress includes infected humans, animals, insect vectors, waste disposal (water, solid, and hazardous waste), and exhaust air.
- Determine the level of personnel proficiency and adequacy of laboratory practices.
- Determine the availability of PPE and equipment designed to minimize or mitigate operational hazards.[20]

- Assess the capability of each facility or room to contain the material during normal and emergency situations (fire, power outage, malfunctioning equipment).
- Determine whether potential exposure is permissible (ethics, policy, law). Notifying local officials about the planned use of biohazardous agents, risk assessments, mitigation practices, and laboratory procedures, and including them in discussions will likely become more commonplace in the future.
- Consider whether the acceptance of risk is voluntary or imposed.
- Recognize factors that give the illusion of the lack of risk (unknown history of the agent, long latency period, lack of models to test the organism).[21]

By weighing all of the factors and defined information (objective) as well as societal, political, and ethical concerns (subjective) a balanced assessment of risk for a specific agent found or used in particular circumstances can be developed, and a rational, proactive, protective biosafety program can be built.

There are numerous references with details beyond the scope of this text. For additional information on assessing risk of biological agents, see References 22–25.

Working Safely with Biological Agents

Biosafety Program Management

The two main objectives for the biosafety program are providing effective containment of hazardous viable organisms and preventing infection. How the biosafety program is organized depends on the classification, use, and concentration of the biological hazards and the risk assessment conducted for the handling of these materials in specific work environments. As an example, the National Institutes of Health Guidelines for Research (NIHG) involving rDNA[26] provides a well-defined model for an effective biosafety program, and compliance is required in government, academic, and industrial laboratories receiving funding from NIH. In addition, NIHG are considered standards of practice in many commercial biotechnology companies.[27] Reference for general laboratory biohazard control programs is found in *Biosafety in Microbiological and Biomedical Laboratories* (BMBL)[28] and in the infection control standards of the Joint Commission on accreditation of Health Care Organizations.[29]

Regardless of the structure of the biosafety program, for it to be effective there are three required elements, and occupational hygienists may have a role in any or all of them. First, effectiveness starts at the top, and management commitment is essential. Management/administration enables the program by providing authority and resources (money, supplies, time, staff, facilities, and equipment), and assigning the roles and responsibilities for implementing the program.

Second, specialized technical expertise is required to provide the scientific knowledge basis for the program. Principal investigators (PIs), a biosafety committee chair, a biosafety officer (BSO), an occupational health physician, occupational hygienists, and/or consultants may be called on to perform risk assessments, provide training, plan for emergencies, and investigate accidents or incidents.

Third, enforcement of rules and procedures is required to provide accountability and responsibility. NIHG assigns this function to the BSO, but many organizations prefer that line supervisors be responsible for their own staffs and enforce rules and procedures through standard techniques of direct supervision and performance evaluations. In health care organizations the infection control practitioner may play roles of both technical expert and enforcement officer.

Given the organizational structure and management commitment and support, what remains in implementing the biosafety program is to:

- Define the scope of the program based on risk;
- Develop policies and procedures;
- Train workers based on job functions, monitor their work, and retrain periodically;
- Provide necessary and proper facilities and equipment; and
- Perform the work safely.

It is at the level of workers performing tasks that the effectiveness of the biosafety program is determined. Each employee has the responsibility to comply with all health and safety rules that apply to his or her own actions and conduct. Employees should participate in the development of health and safety programs, and have measurable goals and objectives that address compliance policies and procedures. They have the responsibility to report hazards to supervision, co-workers, and company health and safety representatives.

NIHG as a System of Administrative Controls

The NIHG[26] describe a systematic approach to preventing both worker exposure and environmental release of genetically engineered (recombinant) organisms. Roles and responsibilities are delineated, and methods for primary and secondary containment, based on risk, are described. For pathogens and other non-recombinant biohazards, the American Industrial Hygiene Association's *Biohazards Reference Manual*[30] and BMBL[28] contain the most current guidelines of biosafety practice. The reader should refer directly to these references for details of the recommendations, and how they might apply to a particular workplace. The following provides a summary.

Biological hazards are best controlled locally by the institution. The institution is responsible for compliance with the NIHG, and should:

- Establish an institutional biosafety committee (IBC);
- Appoint a BSO (required for Biosafety Levels 3 and 4 or at large-scale production facilities);
- Ensure appropriate training regarding laboratory safety and implementation of the NIHG;
- Determine necessity for medical surveillance of personnel involved in rDNA work, and conduct the program if needed;
- Report significant problems or violations of the NIHG, or any significant research-related accidents and illnesses, to NIH.

The IBC is required for rDNA biosafety programs, but its interests need not be limited to rDNA. The primary functions of the IBC are to:

- Review and approve rDNA activities;
- Define physical containment level;
- Assess facilities, procedures, practices, training, and expertise of personnel involved in rDNA research;
- Adopt emergency plans covering spills and personnel contamination;
- Perform other functions as the institution requires, such as implementing the health surveillance program.

Under the NIHG, the BSO is a member of the IBC. His or her role is to inspect laboratories for compliance with the NIHG and report problems and violations to the IBC or PI. The BSO is the technical expert advisor to the IBC, PIs, and laboratory workers on matters of laboratory safety, risk assessment, emergency procedures, and laboratory security. Occupational hygienists are likely to be called on to be members of the IBC. With adequate education, training, and experience in microbiology and biosafety, an occupational hygienist may also serve as a BSO.

The PI registers each rDNA research protocol and submits any significant amendments in writing to the IBC for approval. The protocol registration includes identification of the genetic modifications to be made, the hosts, vectors, and classification of the organisms based on hazard. There must be sufficient information provided, and it must be understandable to all members because there are community representatives and other nonscientists on the IBC. A sample project registration document is shown in Figure 18.3.

Physical Containment

Both the NIHG[26] and BMBL[28] describe containment in terms of primary and secondary barriers. Primary barriers protect the worker and environment in the immediate area of potential exposure. Primary barriers include biological safety cabinets,

rDNA/RNA New Project Registration Document IBC File #

Principal Investigator: ____________________ Phone: ______________

Name of experiment(s) and brief description of the work:

Where will the work be performed? Building and Room #'s

Who else will be working on this set of experiments?

Expected Starting Date:

In order to do a thorough and credible Safety Review and Risk Assessment, the following information is required:

1. Identify the Host Cells
2. Identify the Vectors (map or % of viral elements present, host range)
3. Identify Inserts (source species; nature of gene product, if known; if viral, give % of genome present in a single clone)
4. Maximum anticipated volume of viable material in a single vessel
5. Description of Experimental Procedures which may produce aerosols or significant personal exposures. (centrifugation, sonication, concentraion of virus, etc.)

As Scientist directing this project, I certify that the registration is accurate and complete. I agree to amend it if there is a significant change to the research protocol. I understand that it is my responsibility to assure that all workers on this project comply with the biosafety containment guidelines and other safety rules and procedures defined by the IBC.

P.I. Signature: ______________________________ Date: ____________

Figure 18.3 — Sample project registration document.

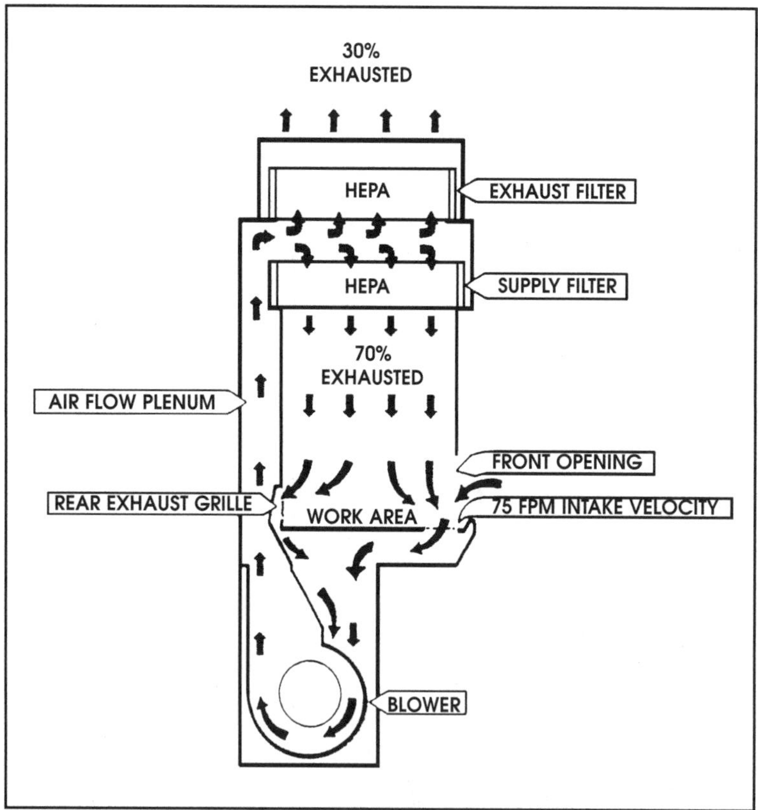

Figure 18.4 — Class II, Type A biological safety cabinet construction and airflow diagram.[31]

sealed centrifuge rotors, glove boxes, high efficiency particulate aerosol (HEPA) filtered animal enclosures, etc. PPE, especially gloves, eye protection, and sometimes respirators, are important primary barriers. Primary containment is also provided by good microbiological technique. Secondary barriers protect the external environment including nonlaboratory work areas and the outside community. Secondary containment is achieved by a combination of facility design (differential pressurization of work areas, HEPA-filtered exhaust ventilation, sterilization of effluent liquids, etc.) and work practices and procedures (see Figure 18.4).

Biohazard potential is described in four degrees of severity. The lowest level (Biosafety Level 1, BSL1) agents are low risk, and not known to cause disease in healthy adult humans. Examples would be asporogenic *Bacillus subtilis*, or *B. licheniformis*, and *E. coli* K12. BSL1 containment is appropriate. It requires only standard microbiological practices and no special equipment for primary containment. Work is performed on open bench tops, there is a sink for secondary protection, windows may open with screens, the furniture is sturdy, and the lab and furniture are easy to clean.

Biosafety Level 2 (BSL2) agents are associated with agents known to cause human disease that can be moderately serious, and

Table 18.1 — Selection and Classification of Biological Safety Cabinets[32,33]

NSF Class	*Min. Intake (ft/min)*	*Origin of HEPA Downflow Air*	*Exhaust Treatment*	*Contamination Ducts/ Plenums*	*Application*
Class I	75	None. Similar to lab fume hood with filtered exhaust.	HEPA to outside.	+ or - pressure	BSL1, BSL2, or BSL3 containment.
Class II Type A (see Figure 3)	75	From common plenum.	70% recirculation within the cabinet, 30% exhausted to lab through HEPA.	+ or - pressure	Low to moderate risk biological agents. BSL1, BSL2, BSL3.
Class II Type B1	100	From separate plenum of uncontaminated recirculated inflow air.	50–70% exhausted to outside through HEPA filter and dedicated duct.	- pressure	Low to moderate risk biological agents. BSL1, BSL2, BSL3. Minute or trace amounts of toxic chemicals and radionuclides.
Class II Type B2	100	From lab air or outside air.	100% exhaust to outside through HEPA; no recirculation within cabinet.	- pressure	Low to moderate risk biological agents. BSL1, BSL2, BSL3; toxic chemicals and radionuclides.
Class II Type B3 (Similar to Type A)	100	From common plenum.	70% recirculated air is exhausted to outside through HEPA filters.	- pressure	Low to moderate risk biological agents, minute or trace amounts of toxic chemicals and radionuclides. BSL1, BSL2, BSL3.
Class III	n/a	Hermetically sealed glovebox enclosure for confining extremely hazardous materials. Used only in maximum containment laboratories.			High risk agents. BL4.

for which preventive or therapeutic interventions are often available. The lists include bacteria (e.g., *Salmonella, Shigella, Legionella, Staphlococcus aureus, Streptococcus*), fungal agents (e.g., *Cladosporium, Cryptococcus neoformans*), parasitic agents (e.g., *Coccidia, Cryptosporidium, Giardia* spp., *Trichinella spiralis*), and viruses (e.g., hepatitis A, B, C, D, and E; Epstein Barr, most poxviruses, and rabies virus). Potential routes of entry are percutaneous injection, ingestion, and direct contact with mucous membranes. Aerosol exposure is rare in nature, but it can occur in laboratory settings. If manipulations cause splash or aerosols, primary containment by means of a biological safety cabinet should be provided (see Table 18.1).

In addition to the BSL1 criteria, additional administrative controls (access control, biosafety manual, medical surveillance, biohazard signs) are required, additional PPE should be used, and an autoclave should be available for waste decontamination.

Biosafety Level 3 (BSL3) agents are indigenous or exotic with potential for infection following aerosol transmission. Agents are associated with serious or lethal human disease for which preventive or therapeutic interventions may be available. These agents pose high individual risk, but low risk to the community. Examples are TB, yellow fever virus, HIV types 1 and 2, human T-lymphotropic virus types 1 and 2, rabies virus, *Coxiella burnetii*, and Venezuelan equine encephalomyelitis virus. BSL3 containment requires special facility design and a relatively large array of special containment equipment and procedures above that required for BSL2 (see Table 18.2).

Biosafety Level 4 (BSL4) organisms are dangerous/exotic agents that pose a high risk of life-threatening disease, and for which preventive or therapeutic interventions are not usually available. Examples are Lassa fever virus, Ebola virus, Marburg virus, and tickborne encephalitis virus complex. BSL4 laboratories are found in only a few places, most of which are related to government or military applications. They can best be described in terms of a room within a room with containment, glove boxes, and/or one-piece supplied-air suits. Neither occupational hygiene comprehensive practitioners nor biosafety professionals should be called on to manage a BSL4 containment facility unless the individual has had extensive work experience inside such a facility. Each of these facilities is unique, and they are used only for the most hazardous work. They are not an appropriate training ground for either the scientist or biosafety personnel. For more detailed information about BSL4, see the previously cited references BMBL, NIHG, *AIHA Biosafety Reference Manual*.

Finally, the concept of a hybrid Biosafety Level 2+ (BSL2+) has evolved in certain locales to allay emergency responders' fears about HIV. BSL2+ refers to facilities and equipment equivalent to BSL2, work practices developed for BSL3, and the use of more PPE by the investigators. This combination provides better primary containment at the lab bench and mitigates the need for a costly BSL3 facility for secondary containment.

Treatment and Disposal of Infectious Waste

Prior to about 1986, disposal of biological waste was not of great public concern nor was it highly regulated. There was no epidemiologic evidence that most hospital waste was any more infective than residential waste, nor evidence that hospital waste had caused disease in the community as the result of improper disposal.[35] The AIDS epidemic, compounded by several incidents of medical wastes (including used needles) washing up on eastern U.S. beaches in the

Table 18.2 — Summary of CDC/NIH Guidelines for Biosafety Level 3 Laboratories[34]

- Plans should incorporate a vestibule with two sets of entry doors, through which all users of the lab must pass when entering the lab from corridor or other public space.
- All wall, ceiling, and floor finishes should be monolithic and sealed (or capable of being sealed) to prevent accidental air leakage out of the space, as well as to allow cleaning and/or decontamination of the space.
- Bench tops should be waterproof as well as resistant to "acids, alkalis, organic solvents, and moderate heat."
- Spaces between casework and equipment should be accessible for cleaning.
- A sink that is foot- or elbow-operated should be located near the entrance to the laboratory.
- All windows in the laboratory should be closed and sealed to prevent air leakage.
- Doors into laboratory and/or into containment modules should be provided with closers.
- An autoclave should be provided within the laboratory for decontamination of waste.
- The heating, ventilating, and air conditioning system should be carefully designed to create a directional airflow from the cleanest area, which should be at the highest positive pressure, to the most contaminated area, which should be at the lowest.
- If the safety cabinets are tested once a year, the cabinet exhaust air may be recirculated within the laboratory. For safety reasons it is more common to design the system to exhaust the air from the biosafety cabinets out of the building.

early and mid-1980s led to a rise in public concern and fear about the disposal of biomedical wastes and a rise in the interest of federal and state agencies in more strictly regulating infectious waste disposal.

Proper (i.e., safe and legal) disposal of biohazardous waste is currently regulated by the individual states. Which waste streams are considered infectious depends on the source, pathogenicity, and the public's perception of risk. For example, health care facilities such as hospitals, nursing homes, walk-in clinics, and clinical laboratories are recognized generators of biohazardous waste because of contamination with human blood and body fluids. Recognizable body parts (pathological waste, regardless of whether fixed in formalin) may also be regulated infectious waste. Other laboratory wastes such as rDNA cultures, laboratory animal carcasses and bedding, and animal biohazardous agents may be regulated infectious waste and require specified treatment(s) before disposal, depending on the local jurisdiction.

EPA drafted a recommendation with 13 categories of infectious wastes,[36] and these form an adequate basis for a definition.

- Isolation wastes: Generated by hospitalized patients who are isolated in separate rooms to protect others from their severe and communicable diseases.
- Cultures and stocks of etiologic agents: All cultures and stocks are included because pathogenic organisms are present in high concentrations.
- Blood and blood products: Included because of the possible presence of the hepatitis agent and other bloodborne pathogens. This set of recommendations recognized the concept of universal precautions for protection from the bloodborne pathogens.
- Pathological wastes: Tissues, organs, body parts, blood, and body fluids from surgery and autopsy. EPA distinguished patients with infectious diseases from others, but called it prudent to handle all pathological wastes as infectious because of the possibility of unknown infection in the patient or corpse.
- Other wastes from surgery and autopsy: Soiled dressings, sponges, drapes, casts, lavage tubes, etc. The American Hospital Association recommends that all surgical dressings from patients should be regarded as potentially infectious because of the possibility of unknown disease.
- Contaminated laboratory waste: Materials that were in contact with pathogens in any type of laboratory work. This would include items such as culture dishes, pipettes, membrane filters, disposable gloves, lab coats and aprons, etc. The NIH and Centers for Disease Control and Prevention (CDC) guidelines for containment require that all wastes from BL3 or BL4 be treated as infectious. There is debate about whether rDNA-contaminated solutions and equipment are infectious. Regardless, common practice is to decontaminate rDNA wastes prior to disposal, and in BL3 and BL4 facilities to autoclave the waste within the facility prior to disposal.[37]
- Sharps (e.g., hypodermic needles, syringes, pasteur pipettes, broken glass, scalpel blades): Those items that present the double hazard of physical injury and infection if the sharp were used in treatment of patients with infectious disease.
- Dialysis unit wastes: Infectious because they are in contact with the blood of patients undergoing hemodialysis. There is a high rate of hepatitis in this patient population.
- Carcasses and body parts of animals exposed to pathogens in research or used in the production of biologicals or in the *in vivo* testing of pharmaceuticals, as well as those that died of known or suspected infectious disease: Biosafety experts recommend that all laboratory animals be considered potentially infectious because of the potential for acquiring zoonotic diseases (diseases transmissible from animals to man).
- Animal bedding and other waste from animal rooms: Similar to the previous paragraph; because pathogens may be secreted or excreted from animals, the waste should be considered infectious.
- Discarded biologicals: Materials produced by pharmaceutical companies for human or veterinary use. These may be discarded because of bad manufacturing lots, out-dating, or recall and removal from the market.
- Food and other products contaminated with etiologic agents: Includes canned foods being recalled because of the danger of intoxication resulting from the ingestion of botulinum, a toxic product of the *Clostridium botulinum* bacterium.

- Contaminated equipment and parts that are contaminated with etiologic agents: These would include laboratory equipment and HEPA filters from biological safety cabinets, if not decontaminated in situ. (National Sanitation Foundation standard NSF-49 dictates that all HEPA filters must be formaldehyde-decontaminated in place prior to removal.)

One could argue that although some of these waste streams are biological materials, they are not infectious, and there is no data to indicate any increased hazard either to the public or to the waste haulers from contact with them. However, it is difficult or impossible to distinguish pathogenic contamination from nonpathogenic contamination by normal senses of sight or smell. Therefore, waste management systems typically consider all-or-none rules for identification and separation of the defined biohazard wastes.

Once the biohazard waste streams are identified, it is necessary to determine how they will be treated and ultimately discarded. There are no federal regulations that apply, although EPA has been considering the authority of the Resource Conservation and Recovery Act (RCRA) for several years, and currently (1996) reconsidering its definition of medical waste. Under RCRA the states can regulate hazardous waste, and several have exercised this authority. Congress passed the Medical Waste Tracking Act of 1988 to determine, on a trial basis, whether infectious waste should be controlled in a similar way to hazardous chemical wastes through a manifesting system. These regulations applied to several states, including New York, New Jersey, Rhode Island, and to Puerto Rico. Ultimately, the trial period passed, and the federal requirements have neither been continued nor expanded to other states.

The nature of the biohazardous waste streams determines the most appropriate collection and storage conditions. For example, collection of sharps from hospital rooms and laboratories should be done in leakproof, puncture resistant, autoclavable containers. There are storage conditions vis-á-vis temperature (ambient, refrigerated, frozen) and time restrictions for pathological wastes. "Double bags," "leakproof containers," and so forth are examples of terminology typically used to describe some of the waste collection and storage requirements for biohazard waste. In the Occupational Safety and Health Administration (OSHA) bloodborne pathogens standard, "red bags" or "BIOHAZARD" labels are required for waste contaminated with human blood or body fluids.[38] In the specific case of biotechnology effluents, validated kill methods are usually employed at the lab bench before disposal as unregulated trash or discharge to drain. Exceptions would be in the case where local ordinances prohibit even treated waste from local landfills, incinerators, or sewers.

Because of the specialized nature of biohazard waste, proper management is more costly than normal solid waste disposal and as such, minimization of the specialized waste streams is desirable. Separation and segregation of the waste streams helps the minimization effort and also helps to ensure each specific waste stream is treated correctly.

Procedures and equipment for transporting biohazard waste within a facility must anticipate the potential for spills as well as the risk of injury from direct contact with the biohazard material, and/or cuts or puncture wounds from sharps. Rigid plastic containers protect workers from sharps, and double containment is recommended for spill prevention. In hospitals, covered carts are often preferred. Work practices and PPE also help to reduce the probability of injury to workers.

Appropriate on-site treatments include autoclaving and/or use of chemical disinfectants. After on-site treatment by effective methods, most biological wastes may be disposed of as nonhazardous solid waste. Local ordinances may require specific treatments, waste bag color, labeling, etc., but the waste becomes the same as any other waste from establishments that are neither laboratories nor health care related. Some of the most common on-site disinfectants are bleach (1:10 or 1:100 dilution of household bleach) to kill liquid solutions, and glutaradehyde or iodophors such as Wescodyne™ for ambient temperature disinfection of objects such as medical devices and reusable pipettes.

Alternatively, biohazard waste may be shipped off-site for incineration. The proper packaging is regulated by U.S. Department of Transportation (DOT) for interstate transportation. Manifests are usually required.

Sterilization by steam autoclave and incineration (either on-site or off-site by contracted vendors) are the two traditional and most widely used methods of infectious waste treatment and disposal. Both have advantages and disadvantages.

Handling of wastes on-site either to autoclave or to package for off-site incineration is a labor-intensive activity. Disposal, either by incineration or removal to a landfill as a special category of waste, is expensive; and health care organizations and research laboratories are not exempt from pressures to be cost effective in managing their businesses. Alternative technologies are being explored and developed to meet the industries' need for less expensive but effective infectious waste treatments. However, none of the new technologies has yet been widely accepted; the well-established methods are difficult to displace.

Decontamination, Disinfection, Sterilization[39–42]

In the context of rendering materials or waste noninfectious, it is necessary to distinguish several terms that are often, and incorrectly, used synonymously:

- Decontamination is the use of physical or chemical means to render materials safe for further handling by reducing the number of organisms present to an acceptable level. What is an acceptable level depends on the situation. Decontamination is used not only to protect personnel and the environment, but also to protect the work area (e.g., operating rooms, clean rooms). "Decontamination" is the generic term that may be applied to both sterilization or disinfection.
- Sterilization is the complete killing of all organisms (including bacterial spores).
- Disinfection kills infectious agents (except bacterial spores) below the level necessary to cause infection. Sanitizers are used on inanimate surfaces; antiseptics are used on skin.

Decontamination may be accomplished by physical means such as heat, steam, and radiation or by chemicals. By definition, sterilants and disinfectants are toxic to viable cells, so it is critically important for users to be familiar with the hazards of the agents they select and to take the necessary precautions to prevent and monitor exposure.[43–46] Ionizing radiation (X- and gamma irradiators) may be used only within the context of federal or state licences. Chemical agents such as ethylene oxide, formaldehyde, glutaraldehyde, and strong acids and bases require special handling techniques. Also, environmental monitoring and medical surveillance may be required, depending on the chemical.

Physical Agents

Wet Heat. High temperatures cause denaturation of enzymes and kill organisms. Boiling (212°F for >30 minutes) and pasteurization (161°F for 15 seconds, or 143°F for 30 minutes) will kill vegetative cells but not bacterial spores.

Dry Heat. In clinical labs, open flames and Bacti-Cinerators™ (an electrical device that dry-heats at 1600°F) are used to heat-sterilize inoculation loops. Hot air ovens (160–180°C for 2 hours) are used for anhydrous materials such as greases or powders. Incinerators destroy infectious wastes.

Steam. Used at 250°F (121°C) under pressure (15–18 psi, depending on altitude) in an autoclave, steam is the most widely used and convenient method of sterilization. Effective steam sterilization requires control of duration of exposure, configuration and size of the load, permeability and dimensions of the containers, and other variables to be assured sterility has been achieved. Indicators (e.g., *Bacillus stearothermophilis* ampules, autoclave indicator tape) are frequently used on each load to validate the sterilization. Biological indicators may be used on a weekly or monthly basis to validate that the autoclave is functioning properly. Chemical indicators and autoclave tape do not give equivalent assurance since they generally change appearance when the temperature set point has been achieved. There is no indication that the proper time was also achieved.

Ionizing Radiation. Ionizing radiation is used for the sterilization of prepackaged medical devices, including operating room supplies such as syringes and catheters.

Ultraviolet (UV) Radiation. In the past ultraviolet radiation has been used as a practical way to inactive viruses, mycoplasma, bacteria, and fungi. It was thought to be effective against airborne microorganisms and within biological safety cabinets to maintain low levels of contamination on exposed surfaces. Its usefulness is limited by its low penetrating power, and shadows and dust on the lamps can also reduce the effectiveness. Other common applications include air locks, animal holding areas, and laboratory rooms during periods of nonoccupancy. However, UV radiation is among the least effective, least perfected methods of sanitization. UV radiation is not a practical disinfectant of liquids.

Filtration. Membrane filters are used to remove bacteria, yeast, and molds from biologic and pharmaceutical solutions. Common pore sizes are 0.22 µm, 0.45 µm, and 0.8 µm.

Chemical Decontaminants

Chemical disinfectants inactivate microorganisms by chemical reaction. The effectiveness of the disinfectant against an infectious agent varies with the nature of the chemical, the concentration, contact duration, temperature, humidity, pH, and the presence of organic matter. Most disinfectant protocols require prewashing with soap and water to remove gross organic material.

Sodium hypochlorite (household bleach) is commonly used as a general disinfectant against a variety of bacteria, viruses, fungi, and TB, including *Staphylococcus aureus* and *Salmonella typhii*. A 1:100 dilution of household bleach to water creates a solution of greater than 500 ppm free available chlorine, which is effective against vegetative bacteria and most viruses. Efficacy is a time dependant function, based on biological load and free chlorine concentration. A 1:10 dilution (5000 ppm free available chlorine) is preferred by some for general applications. A major deficiency of bleach as a disinfectant is that concentrations of approximately 2500 ppm are needed for it to be effective against bacterial spores. Also, these solutions are corrosive and unstable over time. Fresh solutions need to be made daily.

Iodophors (iodine-carrier) possess a wide spectrum of antimicrobial and antiviral activity. They are used for antiseptic and disinfectant purposes at different concentrations. Although not generally sporicidal, Povidone-iodine has been shown to be sporicidal and is important in preventing wound infections. Betadine in alcohol solution is widely used in the United States for disinfection of the hands and operation sites. Iodophors are usually carried in surfactant liquid; they are relatively harmless to man and have a built-in effectiveness indicator: a brown color. Iodophors are inactivated by organic matter, so preliminary surface cleaning may be needed in addition to iodophor disinfection.

Alcohols have rapid bactericidal activity, including acid-fast bacilli, but are not sporicidal and have poor activity against many viruses. Ethanol requires the presence of water for its effectiveness, but concentrations below 30% have little action. The most effective concentration is about 60–70%. Isopropanol and n-propyl alcohol are more effective bactericides than ethanol. However, they are not sporicidal. Seventy-percent alcohol is used for surface decontamination in biological safety cabinets. Solutions of iodine or chlorhexidine in 70% alcohol may be employed for the preoperative disinfection of the skin. All of these alcohols are flammable liquids and may be incompatible with some plastics and rubber.

Quaternary ammonium compounds, as their name implies, are substituted ammonium compounds. They are acceptable as general use disinfectants to control vegetative bacteria and nonlipid-containing viruses; however, they are not active against bacterial spores at typical concentrations (1:750). They have many and varied uses including food hygiene in hospitals, preoperative disinfection of unbroken skin, algal contamination control in swimming pools, and preservatives for eyedrop preparations and contact lens soaking solutions.

Phenolic compounds (0.5%–2%) are recommended for the killing of vegetative bacteria, including TB, fungi, and lipid-containing viruses. They are less effective against spores and nonlipid-containing viruses. They are used for cleaning equipment and floors. Phenolics are derived from the tar obtained as a by-product in the destructive distillation of coal, solublized with soaps. They have been implicated in depigmentation and can cause severe damage to unprotected skin on contact. Lysol™ is a cresol solublized with a soap prepared from linseed oil and potassium hydroxide. It retains the corrosive nature of the phenol.

Two aldehydes are currently of considerable importance as disinfectants: glutaraldehyde and formaldehyde. Glutaraldehyde is used in 2% solution made alkaline before use. It is a highly reactive molecule and works well against bacteria and their spores, mycelial and spore forms of fungi, and various types of viruses. It is considered to be an effective antimycobacterial agent. Dried spores are considerably more resistant to disinfection. Organic matter is considered to have no effect on the antimicrobial activity of the aldehyde, which explains its wide use as a cold sterilant for medical equipment such as cystoscopes and anaesthetic equipment. Glutaraldehyde is used in the veterinary field for the disinfection of utensils and premises. It is an irritant and has potential mutagenic and carcinogenic effects.

Formaldehyde used in solution, 8% formalin in 70% alcohol, is effective against vegetative bacteria, spores, and viruses. Its sporicidal activity is slower than glutaraldehyde, and it is less effective in the presence of protein organic matter. Inhalation of formaldehyde vapor presents potential carcinogenic risk to humans. Paraformaldehyde flakes are vaporized by heating within sealed biological safety cabinets in the typical decontamination procedure. Vaporized paraformaldehyde is used to decontaminate entire rooms, but this procedure is highly hazardous. It requires rendering the room airtight during the procedure and neutralizing the gas prior to release. Personnel conducting this type of procedure must be highly trained, use a buddy system during the process, and conduct the work wearing self-contained breathing apparatus. This technique is the primary option for decontaminating and decommissioning BSL4 facilities, and design criteria should anticipate the need for decontamination activities.

Ethylene oxide (EtO) inactivates all types of microorganisms, including endospores of bacteria and viruses. The action is influenced by the concentration of EtO, temperature, duration of exposure, and water content of the microorganisms. It is used in hospital sterile supply departments and also by commercial suppliers of sterile hospital goods. Pure EtO is explosive and flammable between about 3 and 100% in air. Also, it is mutagenic, potentially carcinogenic in humans, and highly regulated in the United States by OSHA.[43] While the sterilization process itself can be well-controlled with respect to airborne emissions of EtO, the off gassing of the sterile goods, especially plastics, may take several hours. Therefore, there is potential for worker exposure to this highly toxic material in rooms and areas for storage of sterile goods, and in and around clean transport carts.

Accidents, Spills, and Emergency Planning

Considerable information has been accumulated that identifies the major routes of exposure for laboratory acquired infections. The earliest reviews of Sulkin and Pike[47,48] indicate that accidents with needles and syringes gave rise to many laboratory acquired infections. The hazard of sharps is as true today as it was 40 years ago. However, the illnesses we see now are not just brucellosis, Q-fever, or *Salmonella*. The diseases of increasing importance in the 1990s are AIDS, hepatitis, other bloodborne pathogens, enteric pathogens such as *E. coli* strains, *cryptosporidium*, and a reemergence of tuberculosis, including multiple-drug resistant strains.

All accidents and overt and potential exposures involving biological hazards should be immediately cleansed and reported, and the worker should be seen for a medical evaluation. Depending on who in the organization is responsible for the postaccident investigations, record keeping, external reporting, etc., the initial accident report may need to be directed to several individuals including the IBC chairperson. Such reporting permits the monitoring of frequency of accidents and helps in developing procedures or methods of prevention. It also helps to evaluate the types of accidents that result in infection of laboratory workers. In determining the need for external reporting, the nature of the agent involved, the amount of the spills, and the likelihood of exposure that might result in infection are important criteria.

Since biological spills or releases pose unique hazards, emergency responders require special training in hazard evaluation and first aid, as well as decontamination and monitoring procedures. Since few HAZMAT and rescue services are prepared to provide assistance in the event of a biological emergency, it is very important for institutions that use biohazardous agents (pathogens, materials containing pathogens, or toxins) to have in place formal, practiced, internal response and assistance plans to handle potential biological emergencies.

The biological emergency response and assistance plan (BERAP) is an institution's formal guideline for organizing internal response to biological emergencies. Basic information about the institution's mission, the types of work conducted, and an assessment of risks and methods for mitigating anticipated emergencies should be described in an introductory section. In addition, the scope of the BERAP should be defined and addressed by a thorough discussion of the most probable events and worst case scenarios. This information will provide perspective and direction for developing a plan that addresses the needs of the institution and community.

The BERAP defines who is responsible and accountable for specified actions before, during, and after a biological event; how communication and coordination between the institution and local, state, and/or federal authorities will occur; and what procedures will be followed to maintain on-site as well as community safety and environmental integrity.

Optimally, the leader of the emergency response team would be trained in both biological and chemical safety. Under the NIHG, for example, the BSO is charged with developing emergency plans,[49] and this person should function as the emergency response team leader. Primary functions of the technical leader are to assess the scope of the accident, organize the remediation plans, and possibly to communicate with the public or press, although public communication may be reserved for senior management or public affairs representatives.

The core operational emergency response team members are on-site responders who can be trained in topics related to first aid/self aid; spill containment, decontamination procedures, and risks associated with decontaminants (e.g., large-scale sterilization with formaldehyde or sodium hypochlorite); selection and use of proper PPE; mitigation of hazards; procedures for safe entry into and exit from contaminated areas; and signs and symptoms of intoxication or infection.

Emergency response teams within organizations that use BSL3 and BSL4 pathogens should meet quarterly to participate in drills, discuss strategies for hypothetical accidents, or practice donning and removing protective equipment and over-garments (Tyvek® suits), and to continually improve current practices. In some jurisdictions local emergency planning committees require biosafety to be included in the emergency planning required by the Emergency Planning and Community Right to Know Act.[50] Hospital emergency departments plan for community disasters as a part of their accreditation.[29] Radiological disaster scenarios are routinely practiced in areas near nuclear power plants, and depending on the local jurisdiction, hospital emergency departments may wish to prepare for biological emergencies as well.

Communicating Biological Risks

Similar to chemical and physical agent risks, an open and sincere dialogue should help to foster trust and reduce any misconception about biological risk. A strategy for risk communication should be developed based on the content and context of the message and the audience to whom the message will be delivered. Sandman[51–54] and others have described the issues and considerations in detail.[55]

Individuals base perceptions on past experience, values, and previously supplied information. The general public is not likely to have a very high knowledge base with respect to biological hazards, and what they believe is likely to have been influenced by popular press and theater (such as Michael Crichton's *Jurassic Park*, Robin Cook's *Outbreak*, or Richard Preston's *The Hot Zone*). Many people may not feel the risk of death or illness is very high when confronted with a biological exposure concern, but other negative consequences may be very important to them. One example is a potential negative influence on property values in an area near a medical waste incinerator. When the risk of illness or death is perceived as low, these other concerns may become the more important issues to address. Other nonscientific and nonquantifiable risk perceptions include voluntary versus involuntary participation, fair versus unfair treatment, immediate versus delayed health effects, reversible versus irreversible consequences, associated feelings of dread, availability of alternatives to taking the risk, and the reliability or trustworthiness of the source of information.

Consideration should be given to those who are assigned the duty of communicating the risk assessment. There is no one correct person, and the optimal choice depends on the discussion topic, the forum for discussion, and target audience. Biosafety professionals and occupational hygienists would traditionally communicate risk assessment findings and recommendations to employees of an organization. A senior scientist or principal investigator should be able to discuss risk associated with his or her project when addressing the IBC. If the forum is a formal public hearing, organizations might prefer to rely on a laboratory director, public affairs officer, or even corporate counsel to make a public presentation.

Finally, the context in which the risk is presented should be considered. Risk can be communicated in a variety of ways including worst case scenario, reasonably foreseeable event, and most probable event. Clearly, the choice of what to say and how to say it depends on the situation. Communications following unintentional release or an epidemic of symptoms that could be perceived to have arisen from the workplace are made much differently than those in the context of a public hearing or application for a biological materials permit. Biosafety professionals and occupational hygienists need to consult and heed the advice of their public affairs or corporate communications departments about the content and methods of communicating biological risk to the public.

Transportation and Shipping Biological Agents

The main objectives of the hazardous materials shipping standards and regulations are to (1) anticipate and prevent accidents involving hazardous materials in transport and (2) mitigate the potential harm to individuals or the environment when accidents do occur.

These objectives can be accomplished by identifying the biohazardous agents and determining the correct shipping category; using proper packing and packaging materials and procedures; providing warning labels on the packages and documentation in the form of shipping papers; and by thorough training, including emergency response procedures, of persons involved in shipping biological materials. Regulations concerning the transportation of etiologic agents, biological toxins, and diagnostic specimens have changed over the past few years and will continue to evolve to meet safety and security needs.

Shipping Standards

There have been a variety of local and state domestic regulations as well as international standards for shipping.[56–62] Each common carrier would additionally be expected to have their own set of rules. The current need is to harmonize the regulations internationally, and much progress is being made toward that goal. It is not the purpose of this chapter to detail the shipping regulations. They are constantly evolving, and the reader is directed to the references for more information.

To ship biological materials by air that belong to BSL2 through BSL4, diagnostic specimens, live vaccines, genetically modified pathogens, or biological toxins, current industry practice is to follow the requirements of the International Air Transport Association (IATA) Dangerous Goods Regulations. First developed in 1953 by an international board of experts from the member airlines, IATA defines the classes of dangerous materials, identifies which materials comprise these classifications, and specifies requirements for safe packing and packaging materials, documentation, and marking and labeling requirements.

Other standards for shipping biological materials are found in Title 42 of the U.S. *Code of Federal Regulations*, Part 72—Interstate Shipment of Etiologic Agents; U.S. Department of Transportation Hazardous Materials Regulations (49 CFR Parts 171–180); and United States Postal Service regulations. In addition, import permits for infectious agents affecting humans (e.g., as research materials or clinical specimens) must be obtained from CDC for each shipment, prior to bringing the material into the United States. Permits for importation and interstate shipment of certain animal pathogens are required from the U.S. Department of Agriculture. Import permits are required for each shipment. Compliance with IATA standards may not be sufficient to prevent problems in shipping biological materials. Institutions should also consult their shipping carriers prior to shipping any dangerous goods or biological materials.

Receiving Biological Materials

Incoming packages should be packed and labeled to ensure the safety of the recipient as well as the carrier. Packages that are improperly packed or packed using inadequate materials may arrive damaged. If there are delays in transport due to paperwork discrepancies, the material may be unusable, as in the case of frozen material that has thawed.

Packages that are wet, leaking fluid or powder, or otherwise damaged to the extent that the contents of that or another package has been contaminated present an important decision for the receiving organization. Some will not accept any damaged package, and the carrier will need to follow its standard procedures for handling such items. In other cases, such as the expected arrival of a pathogenic or toxic agent, the receiving organization may want to control the potential biohazard immediately. Written procedures should be in place for the following:

- Reporting the problem to the safety office and documenting the problem. It may be advantageous to photograph the site and package, as well as its contents, should questions of liability arise in the future.
- Handling the damaged package. Nonessential personnel should leave the area until it has been adequately decontaminated. The package should not be handled until the properly trained and authorized personnel have donned the proper PPE. The package should be opened inside a biosafety cabinet and decontaminated.
- Inspecting adjacent packages and areas for signs of contamination and decontaminating these areas.
- Medical evaluation of potentially exposed workers.
- Formally reporting the incident to the consignor, shipping company, and recipient company.

A recent regulation[63] went into effect in April 1997. It requires a permit to transfer or receive some 40 different viruses, bacteria, rickettsia, a fungus, and BSL3 and BSL4 biological toxins within the United States.

Training

Depending on the level of involvement in the packaging and shipping activities, both IATA, the International Civil Aviation Organization (ICAO), and DOT require formal training in the regulations and procedures to be followed. IATA/ICAO describes three levels of training: general awareness/familiarization, function specific, and safety and health training. The training is to be provided or verified on employment of a person in a position involving transport of dangerous goods by air. DOT also covers shipping by other vehicles, by rail, sea, etc. Training requirements are similarly categorized, and they allow OSHA and EPA hazardous materials training to qualify for aspects of these requirements. Retraining and certification is required at least every two years.(64)

Implementing Biosafety Programs in Laboratories and Industry

So far in this chapter the real and potential hazards of biological agents as well as control strategies have been discussed. Other texts have described biosafety programs in academic, government, and other biological research laboratories.(65–67) This section will discuss biological hazards and the application of biosafety principles in other industrial settings and operations, specifically biotechnology research and development (R&D), large-scale biotechnology manufacturing, intentional aerosolization of pathogens, animal research, and enzymes in the detergent industry.

Biotechnology R&D:(68,69) Genetically Engineered Microorganisms

The word "biotechnology" describes processes that use organisms and their cellular, subcellular, or molecular components, to provide goods, services, and environmental management. Biotechnology is considered by many to be a new industry, but the application of traditional genetic modification techniques has been used for much of this century to enhance food characteristics (e.g., hybrid corn, selective breeding); to manufacture food (e.g., bread, cheese, yogurt); for bacterial sewage treatment; in medicine (e.g., vaccines, hormones); for pesticides (e.g., *Bacillus thuringensis*) and other uses. What is new and unique about the modern biotechnology industry is the ability to identify and locate specific genes (and their functions) and to manipulate the genetic materials intentionally and precisely within living cells, i.e., genetic engineering.

In pharmaceutical manufacturing the modified cell is the factory. Cells are cultured to produce the desired product, which is then separated from the cells and nutrient broth in which they were cultured (by techniques such as centrifugation, filtration, etc.), purified (by techniques such as column chromatography), formulated, and packaged. In agriculture, plants themselves are genetically altered to add disease or insect resistance or other traits where they are not naturally present. Examples are strawberries that can grow at colder temperatures, tomatoes that do not rot on the grocers' shelves as quickly, and adding nitrogen-fixing ability to rice, corn, and cereal grain crops to reduce the amount of nitrogen fertilizers necessary to produce food. In environmental applications *Pseudomonas* bacteria can break down a range of products including hydrocarbons. DNA from several strains have been added to a single cell to produce a bacterium effective for oil spills. Mixtures of microbes and enzymes have also been designed to digest detergents and paper mill waste, including sulfite liquor.(68)

Biotechnology is a multidisciplinary laboratory science. Molecular and cellular biologists, immunologists, geneticists, protein and peptide chemists, biochemists, and biochemical engineers are most directly exposed to the real and potential hazards of rDNA technology. Other biotechnology workers who have significant exposure to rDNA biohazards would include maintenance and calibration technicians, glasswash and housekeeping department workers, and other support staff. Through a recent survey of health and safety practitioners in the industry, it was found that scientific staff and support staff exposed to genetically engineered materials comprise only about 30%–40% of the total work force in typical commercial biotechnology companies. Other biotechnology workers are found in academic, medical, and government research institutions.(69)

In addition to genetically engineered organisms, biotechnology lab workers are exposed to a wide variety of hazardous and toxic chemicals and radioisotopes, all of which are integral tools of the trade. The products of biotechnology are not necessarily viable organisms, but may present toxic or allergenic hazards similar to their chemically synthesized counterparts. There are ample references and resources pertaining to chemical and physical hazards in the R&D laboratory in other chapters in this text and elsewhere.[38,46,70–73]

Biological hazards, other than exposure to genetically engineered organisms, are common in rDNA labs, and they include human bloodborne pathogens—blood, body fluids, and human derived cell lines and reagents; communicable diseases, such as influenza; laboratory animal dander allergies; and zoonotic illnesses that may be transferred from lab animals to humans. Some of the better known zoonoses are rabies; the nonhuman primate counterpart to HIV (simian immunodeficiency virus, or SIV); Monkey B virus (*Herpesvirus simiae*); anthrax (*Bacillus anthracis*); psittacosis (*Chlamydia psittaci*); brucellosis (*Brucellosis abortus, B. suis, B. melitensis, B.canis*); and Q Fever and Rocky Mountain Spotted Fever (*Coxiella burnetii* and *Rickettsia rickettsii*, respectively). Except for SIV and Monkey B virus, which are hazards for biotechnology research workers in nonhuman primate centers, the most common zoonotic illnesses would be expected in agriculture or food processing facilities and not necessarily in biotechnology laboratories or animal research facilities.

The Evolution of NIH Guidelines for rDNA

With the discovery of the genetic code in 1961–1965 and the first successful DNA cloning experiments in 1972, the science of molecular biology was born, and it advanced very quickly. As early as July 1973 at the Gordon Research Conference on Nucleic Acids, the capability to perform DNA recombinations was apparent, and the scientists called for a worldwide moratorium on certain kinds of experiments to assess the risks and devise appropriate guidelines for biological and ecological hazards. They understood that cloning was an extremely powerful and promising technology, but there were serious risks to consider.[74] Some of the concerns expressed involved the potential for "escape of vectors which could initiate an irreversible process, with a potential for creating problems many times greater than those arising from the multitude of genetic recombinations that occur spontaneously in nature." Also, there were concerns that "microorganisms with transplanted genes may prove hazardous to man or other forms of life. Conceivable harm could result if the altered host has a competitive advantage that would foster its survival in some niche within the ecosystem."[75] It was understood that lab workers would be the "canaries" and some attempt should be made to protect the workers as well as the environment from the unknown and potentially serious hazards.

An international conference sponsored by the National Academy of Sciences and supported by NIH was held in February 1975. The conference report contained the first voluntary guidelines for classification of hazards and matching the level of containment to the level of potential hazard for various types of experiments. Certain experiments were judged to pose such serious potential dangers that the conference recommended against their being conducted at that time.[76]

The first NIH guidelines (NIHG) were published in 1976 with the objective of ensuring that experimental DNA recombination would have no ill effects on those engaged in the work, on the general public, or on the environment. They replaced the 1975 guidelines, which in many instances allowed research to proceed under less strict conditions. The 1976 NIHG were the product of extensive debate within the Recombinant Advisory Committee (RAC). The essence of the NIHG is the subdivision of potential experiments by class, a decision as to which experiments should be permitted at present, and assignment to these of certain procedures for containment of recombinant organisms. No deliberate releases were allowed. Roles and responsibilities of principal investigators, the institution, the NIH and its staff were described, as well as the classification of organisms and containment levels for specific types of experiments.[77]

As scientific knowledge increased, it was expected that the NIHG would need review and revision. The RAC has met quarterly to consider and approve proposals for changes to the NIHG. For example, the NIHG no longer carry a blanket prohibition against deliberate release of genetically engineered organisms. At the present time (1997) the majority of proposals submitted for RAC approval involve specific human gene therapy experiments and corrections and reclassifications of organisms based on hazard potential. One very significant amendment to the NIHG in the past few years was the creation of a new, large-scale containment category known as Good Large-Scale Practice (GLSP). It relaxed the containment requirements for

> *non-pathogenic, non-toxigenic recombinant strains derived from host organisms that have an extended history of safe large scale use, or which have built in*

environmental limitations that permit optimum growth in the large scale setting but limited survival without adverse consequences in the environment.[78]

Products of Biotechnology

In the 1980s the first products of biotechnology emerged in the United States and Europe. Genetically engineered insulin was approved for use in 1982, as was a genetically engineered vaccine against the pig disease, scours.[79] Recombinant bovine somatotropin increases a cow's milk production and increases the weight of beef cattle. Concerns were raised about public health and safety associated with the consumption of these products. Although the NIHG provided a level of protection for research and production workers and for the environment, they do not address product safety or public health issues for consumers of the recombinantly derived products. To assure the public that concerns about product safety were being addressed, in 1985 the U.S. Office of Science and Technology Policy published a comprehensive federal regulatory policy for ensuring the safety of biotechnology research and products.[80] Each agency—Food and Drug Administration, OSHA, EPA, and Department of Agriculture—specified their respective agency policies concerning rDNA within their jurisdictions. Where there were areas of overlapping authority, a lead agency was identified and the framework coordinated by a new Biotechnology Science Coordinating Committee.

The potential risks of biotechnology and its products are a matter of public concern in Europe as well as in the United States. In April 1990 the European Community enacted two directives on the contained use and deliberate release into the environment of genetically modified organisms and microorganisms. Both directives require member states to ensure that all appropriate measures are taken to avoid adverse effects on human health or the environment that might arise from the contained use or deliberate release of genetically modified organisms and microorganisms, in particular by making the user assess all relevant risks in advance. In Germany the Genetic Technology Act was passed in 1990, partially in response to the EC directives, but also to respond to a need for legal authority to construct a trial operation recombinant insulin production facility.[81]

The early fears of creating genetically superior mutant species or toxins have not materialized; however, the debate over the safety of genetic engineering and the products of biotechnology continues.[82] Products are one by one being integrated into modern society, and rDNA technology is being applied in interesting and important areas such as human gene therapy to correct heritable diseases. Recombinant vaccines are well received, and diagnostic kits based on rDNA technology are widely available. DNA fingerprinting is expected forensic evidence in high profile trials, and it can also be used to prove (technically, disprove) paternity. Other potential (and highly controversial) applications include genetic dog tags for military personnel.

Large-Scale Biotech Manufacturing

The vast majority of industrial rDNA large-scale applications use organisms of intrinsically low risk in systems employing physical containment technology that is well-known to industry and has successfully been used to contain pathogenic organisms for years.[80] NIHG Appendix K specifically addresses containment in large-scale (>10 liters) operations, including manufacturing. Like the small-scale guidelines, the levels of physical containment described in Appendix K: Good Large Scale Practice (GLSP), Biosafety Level 1 Large Scale (BSL1-LS), BSL2-LS and BSL3-LS, are based primarily on the hazard potential of the organism. Table 18.3 describes some of the most important industrial microorganisms, and Table 18.4 provides a summary of the large-scale containment guidelines.

It is important for product protection and personal safety from infection and exposure to allergens to identify and adequately contain the rDNA manufacturing process, especially at stages where aerosols might be generated. The unit operations in a typical fermentation process are depicted in Figure 18.5.[83]

Inoculation and primary seed culture are two activities usually performed in dedicated rooms designed to accommodate both the requirement to avoid culture contamination (HEPA-filtered supply air and positive pressure with respect to the anteroom or hall) and within a biological safety cabinet to protect the operator from exposure. The work is typically performed manually.

Transfer of primary seed culture is done aseptically, usually by air pressure into the fermenter (bioreactor). There is potential for leakage and/or aerosolization at any transfer stage if the couplings or gaskets are faulty or not installed correctly, or if the

Table 18.3 — Some Biological Agents Important in Industrial Processes

Aspergillus niger is an asexual fungus commonly found degrading organic matter in nature. It is an opportunistic human pathogen, but has been used safely in the production of citric acid and several enzymes without causing toxic effects in workers.

Aspergillus oryzae is an asexual fungus found in nature and used in the production of soy sauce, miso, and sake without recorded incidents. *A. orzae* does not colonize humans.

Bacillus licheniformis is a spore forming bacterium, readily isolated from the environment where it persists primarily as endospores. It has a history of safe use in the large-scale fermentation production of citric acid and detergent enzymes.

Bacillus subtilis is a spore forming bacterium found naturally in terrestrial environments. It has a history of safe use in large-scale fermentation and is a source of single cell protein for human consumption in Asia.

Chinese Hamster Ovary cells are mammalian cells cultured in a variety of biopharmaceutical applications including the production of recombinant vaccines (e.g., hepatitis B), interferons (α, β, γ), interleukins, coagulation factors, enzymes (e.g., glucose cerebrosidase for treatment of Gaucher's disease), and many genetically engineered mono- and polyclonal antibodies.

Clostridium acetobutylicum is isolated from soils, sediment, well water and from animal and human feces. It is distinguishable from closely related species that are known human pathogens. Used for the production of butanol and acetone from various feedstocks.

Escherichia coli K-12 is a bacterial strain readily distinguishable from close relatives that are human pathogens. It is a debilitated bacterium that does not normally colonize the human intestine.

Penicillium roqueforti is an asexual fungus that decomposes organic matter in nature. ***P. roqueforti's*** long history of use in the production of blue cheese has shown no adverse effects.

Saccharomyces cerevisiae is a yeast that survives well in the environment. It has a history of safe use in the commercial production of many products including beer.

Saccharomyces uvarum is a yeast that has a long history of safe use in the production (by fermentation) of alcoholic beverages and industrial ethanol.

Table 18.4 — Large Scale Guidelines Summary[88]

Criteria	*GLSP*	*Containment Level*[a] *BL1-LS*	*BL2-LS*	*BL3-LS*
Formulate and implement institutional codes of practice for safety of personnel and adequate control of hygiene and safety measures.	X	X	X	X
Provide adequate written instructions and training of personnel to keep the workplace clean and tidy and to keep exposure to biological, chemical, or physical agents at a level that does not adversely affect health and safety of employees.	X	X	X	X
Provide changing and hand-washing facilities as well as protective clothing, appropriate to the risk, to be worn during work.	X	X	X	X
Prohibit eating, drinking, smoking, mouth pipetting, and applying cosmetics in the workplace.	X	X	X	X
Release of aerosols during sampling, addition of materials, transfer of cultivated cells, and removal of material, products, and effluents from the system is				
minimized by work practices and procedures	X			
minimized by engineering controls		X		
prevented by engineering controls			X	X
Internal accident reporting	X	X	X	X
Emergency plans required for handling large losses of cultures	X	X	X	X
Limited access to the workplace		X	X	X
Viable organisms should be handled in a system that physically separates the process from the external environment (closed system or other primary containment).		X	X	X
Culture fluids are not removed from the closed system until organisms are inactivated.		X	X	X
Minimize the release of untreated, viable exhaust gases.		X		
Prevent the release of untreated, viable exhaust gases.			X	X
Closed system is not to be opened until sterilized by a validated procedure.		X	X	X
Inactivate waste solutions and materials with respect to their biohazard potential.	X	X	X	X
Medical surveillance			X	X
Prevent leakage from rotating seals and other penetrations into the closed system.			X	X
Validate the integrity of the closed containment system.			X	X
Closed system shall be permanently identified for record-keeping purposes.			X	X

transfer line ruptures.

Depending on the specifics of the production system, fermentation may occur in batch or continuous processes, and there may be stepwise increases in culture volume with transfers between bioreactors several times. The only break in primary containment is at fermenter sample points where a degree of secondary containment, for example a sample port in a cabinet, may be necessary. Primary containment during fermentation is important for product protection and demands detailed review of vessel design, validated monitoring of seals at all flange joints, continuous welded pipe, absolute filtration of off-gases, etc.

At the completion of the fermentation stage, cell harvest represents a critical stage with respect to containment. Centrifugation can generate large aerosol masses and is a point where operator contact with microorganisms or cells is most likely. Membrane filtration is an alternative separation technique.

Downstream processing techniques do not differ significantly from standard purification steps except where a unique feature has been engineered into the cells. In general, product must be released from the cell (by mechanical, chemical, enzymatic action, osmotic or temperature shock), separated and concentrated (by centrifugation or size specific filtration), purified (by any of several types of chromatography), and prepared into its final form (e.g., formulated, filled, lyophilized, depyrogenated, etc.). Cell breakage and separation of cell debris are the final activities where exposure to the recombinant organism is a risk, and these activities are typically conducted with some means of secondary containment. See figure 18.6.[84] Exposure to endotoxin (a component of the lippopolysaccharide layer of gram negative bacteria) is a recognized hazard during cell disruption activities. Leakage from centrifuges, sonicators, and so forth aerosolizes the cell debris, and inhalation exposures above about 300 ng/m^3 can cause transient allergic type symptoms [85] such as fever, nausea, and headache. The American Conference of Governmental Industrial Hygienists has established a ceiling threshold limit value (TLV®) of 6 ng/m^3 for pure crystalline proteolytic enzymes (subtilisins), but there is no TLV for biologically derived air contaminants such as endotoxin. Currently available evidence does not support a TLV although research continues, especially with respect to assay methods for common aeroallergens. Dose-response observations in experimental and epidemiologic studies (such as that mentioned above) have not been validated, and research in this area is progressing.[86]

Table 18.4 (continued) — Large Scale Guidelines Summary[88]

Criteria	GLSP	Containment Level[a] BL1-LS	BL2-LS	BL3-LS
Universal Biosafety sign to be posted on each closed system.			X	X
Access to workplace is restricted.		X	X	X
Access to the controlled area is restricted.				X
Closed system to be kept at as low pressure as possible to maintain integrity of containment features.				X

[a]GLSP = good large-scale practice; BSL1-LS = large-scale Biosafety Level 1; BSL2-LS = large-scale Biosafety Level 2; BSL3-LS = large-scale Biosafety Level 3

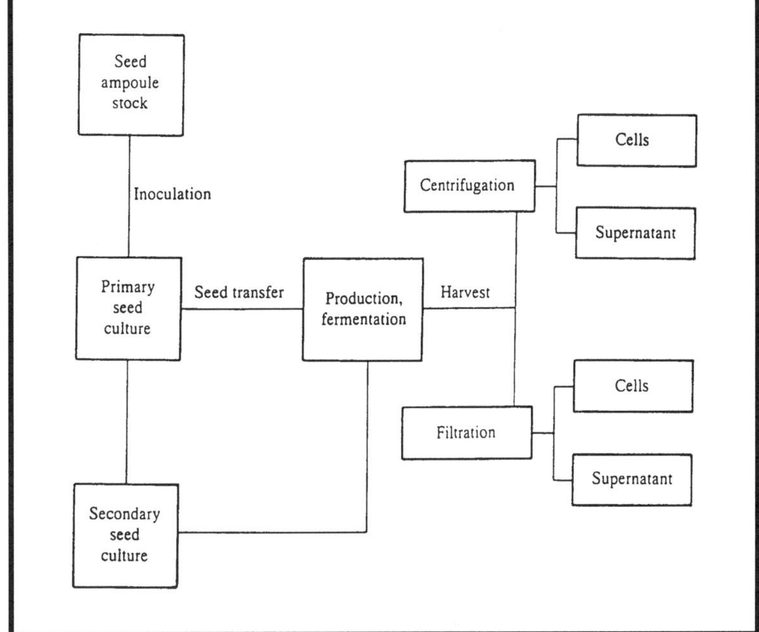

Figure 18.5 — Unit operations for fermentation and separation of recombinant organisms.

Once the product is in a soluble concentrated form, the emphasis turns to product protection from the surrounding environment, and work reverts into a positive pressure, clean-room type of operation. Although any hazard related to the recombinant cells may no longer be at issue, it may still be necessary to control exposure to the product since it may be a highly biologically active molecule in concentrated form.[87]

Intentional Aerosolization Studies with Pathogenic Microorganisms

Particles in the 1–5 µm diameter range are effectively transported to the lung.[89] Because aerosolized droplet nuclei are generally less than 2 µm in diameter they pose

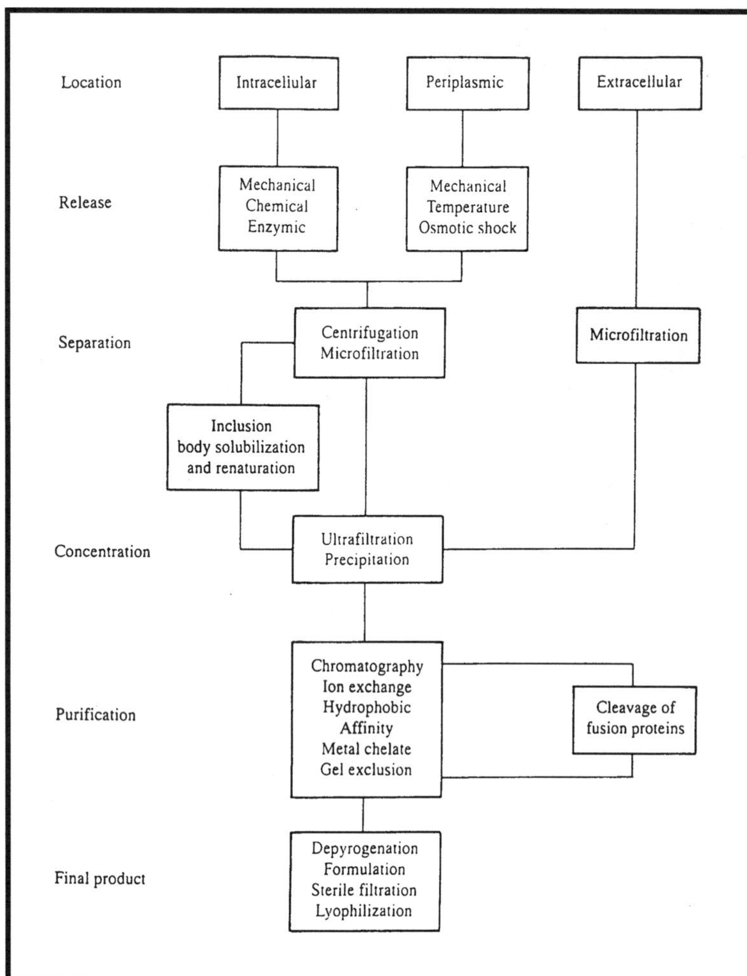

Figure 18.6 — Unit operations for downstream processing of recombinant products.

an inherently greater risk to health by inhalation. Rigorous primary containment is required to prevent release of droplet nuclei containing microorganisms and toxins, and secondary containment and PPE can help prevent transmission in air and access to the respiratory system in numbers sufficient to result in infection or intoxication.

Potential for aerosolization is an important factor in risk assessment and should be assumed for certain types of experimental procedures. For example, animal studies using infectious agents are likely to generate aerosols, as are experimental studies involving shaking, centrifugation, sonication, etc. When aerosol generating procedures are identified, additional precautions can be taken to mitigate the risk from inhalation.

For organisms of relatively low hazard potential (BSL1, BSL2) one approach might be to simply increase the normally assigned biosafety level when aerosol or animal studies are performed. It is helpful to have the scientist perform a dry run to identify problem areas in the methodology, facility safety and operating systems, logistics, practices, and use of PPE.

Work involving intentionally aerosolized toxins or pathogenic microorganisms should be conducted under enhanced BSL3 conditions. Facility and primary barrier designs should be augmented.[90] This is most easily achieved using chambers patterned after a Class III biological safety cabinet. Some considerations in designing such chambers include decontamination ports, dunk tanks, location of glove ports, ergonomics, shatterproof view ports, and HEPA filters prior to the chamber's junction with the building exhaust. Sealed and tested chambers within a secondary airtight chamber that is connected to HEPA-filtered building exhaust may be used to contain the aerosol challenge. The chambers should be successively negative in pressure (0.05-inch differential) with respect to each other and the room. Testing of the chambers should be done with detectable gases (e.g., freon substitutes, halides, or nitrogen) at 2-in. w.g. pressure, and the testing should be repeated until the system is demonstrated to be airtight.[91] Since the containment is patterned after BSL3, there should be a HEPA filtration unit in the building exhaust.

PPE should supplement the containment equipment for studies involving pathogen and toxin aerosolization. Full-facepiece powered air purifying respirators using HEPA filters or, at a minimum, disposable HEPA-filtered respirators and tight-fitting chemical splash goggles to protect the ocular mucosa are recommended. Gloves and back closure/wraparound surgical-style gowns over laboratory clothing should be used. Individuals should be vaccinated against each agent being aerosolized if a vaccine is available. The workers should participate in an immunological surveillance program where a reference serum sample has been taken prior to assignment of work at the facility. The reference sample can be used to determine the need for prophylactic vaccination, and it may be used at a later time should there be an overt or suspected exposure to the pathogen.

Animal Research

Most hazards faced by individuals working with animals or their products in laboratories, veterinary clinics, and breeding facilities are not associated with infectious disease but instead with traditional industrial processes. Handlers and caretakers run increased risks of back injury (working with heavy equipment, supplies, and large animals), animal inflicted injury (bites, kicks, scratches), other physical hazards (slips, trips, falls, electrocution hazards, hearing loss, cuts), and chemical exposure.

Zoonotic infection has been and will continue to be one source of emerging and reemerging disease in humans. In the last 25 years 128 infections were reported by the National Animal Disease Center due to laboratory exposures to infected animals.[92] A well-written review of zoonotic infection has been recently published.[93]

Among the reasons for these infections is that workers fail to recognize the risk of acquiring zoonotic diseases from apparently healthy animals. Only since the advent of human infections with B virus (*Herpesvirus simiae*)[94] (infection only, not disease) and documented aerosol transmission of Ebola strains[95] by airborne route from monkey to man, has the hazard been more widely recognized. In addition, animals can only be screened for those specific etiologic agents suspected to be present in that species and for which diagnostic reagents are available.

Since it is not always possible to know which animals present infectious hazards, to prevent zoonotic illness the occupational hygiene practice model of substitution, engineering, and administrative controls, and personal protection should be universally applied. Hazard controls should always be supported with rigorous training programs, medical surveillance programs, and environmental monitoring.

Substitution

Wherever possible, use pathogen free or specific pathogen free lab animals. This is done in many instances. However, it is costly and in some instances not possible, such as use of nonhuman primates, which are frequently caught in the wild. The quarantine of new animals before moving them into the colony is common practice to protect the colony from disease as well as to protect workers from some infections. However, quarantine is not 100% effective. For example, primates often react negatively to repeated tests for TB throughout the quarantine period only to react positively later. Furthermore, effective quarantine depends on the staff's commitment to either treating or removing infected animals from the facility once they are identified.

Engineering and Administrative Controls

Follow accepted standards of lab animal care. Recommendations for animal biosafety levels are published,[96,97] as are regulations for animal welfare. Adherence to these standards is important to keeping both animals and humans safe.[98,99]

- Follow guidelines and standards with respect to containment; quarantine; separation of species, sexes, and breeding cages; etc.
- Doors and cages should be clearly marked to indicate hazards that may be present and mitigation techniques.
- Facility design should have separate clean and dirty corridors and rooms to prevent spreading infections within the colony.
- Rooms should be designed for ease of cleaning with coved walls, smooth yet slip-resistant floors, pest control via tightly closing doors, screened windows, and well-maintained surfaces (no holes in floors or walls).
- Floor drains should either be sealed or filled with water or an appropriate disinfectant.
- Exhaust air from animal facilities at any level should not be recirculated throughout the building due to the potential for dissemination of zoonotic agents, allergens, and undesirable odors. HEPA-filtered exhaust is required for Animal Biosafety Levels 3 and 4.
- The room should be able to contain animals that have escaped from their enclosures or handlers and prevent entry of unwanted pests and potential vectors.

PPE

Workers should have access to latex or vinyl gloves and dust/mist respirators, and they should be instructed always to wash their hands thoroughly after working with animals, materials containing animal proteins, or dirty equipment. It is optimal to provide workers with a change room, shower, and separate set of clothing, in addition to a lab coat for use in animal areas, to prevent the contamination of street clothing with dander, dusts, hair, and pathogens. Respirators may be provided, in accordance with principles of good occupational hygiene practice and in compliance with OSHA standards.

Training

Training for animal facility workers

should include recognition of potential for the physical hazards of bites and scratches. They should be warned of the potential for developing allergies and contact dermatitis and their symptoms, as well as the risk of zoonotic illnesses. Training should include use and limitations of PPE, laboratory techniques to minimize formation or aerosols, and animal handling techniques to reduce the risk of bites and scratches. It should be emphasized that safe work practices and use of specialized equipment and PPE should be followed at all times, even when working with apparently healthy animals in the absence of perceived risk.

Medical Surveillance

Allergies and contact dermatitis can develop after prolonged work with animals and animal waste. Immunocompromised or allergen sensitive individuals, whether or not they work directly with animals, should not be overlooked as an at-risk population, since a number of zoonotic agents act as opportunistic pathogens. Air, wipe, or bulk sampling[100] for animal hair or dander may be initiated if complaints about odors, unusual illnesses, or allergies are made by individuals who do not normally work in the colony or near animals.

Complaints of illnesses made by animal handlers should be investigated by a team comprised of veterinarians, epidemiologists, health care professionals, biosafety experts, and occupational hygienists. Workers in animal facilities should be vaccinated against moderate to severe disease-causing pathogens in instances where vaccines are available.

Enzymes in the Detergent Industry[101]

Enzymes are complex proteins produced by virtually all living organisms to speed up chemical reactions necessary to maintain life. In detergents they catalyze the breaking down of certain stains into basic components. The breakdown products can be more easily removed by other ingredients in the detergent product. Enzymes have been used in cleaning applications since the early 1900s. They were widely introduced into detergent formulations in the early 1960s, after the industry solved production capacity and alkaline stability problems. By 1969, 80% of all laundry detergent products contained enzymes.

Although enzymes are biologically derived, the hazards and control strategies are quite similar to those for chemical irritants and sensitizers. Routes of exposure are inhalation and direct contact with skin or eyes. If enough enzyme is inhaled, the body will begin to recognize the enzyme, produce allergic antibodies, and become sensitized to the material. Subsequent exposure may trigger allergic symptoms, including occupational asthma. Enzymes have been demonstrated not to be skin sensitizers.[102] Primary irritation is expected, and exposed areas should be protected by the use of gloves and other protective clothing whenever the potential for skin contact exists. Primary irritation of the eye is also expected when detergents or enzymes come in direct contact, but the presence of enzymes in detergents does not necessarily increase the severity of the irritation.

Control of exposure to enzymes in detergent manufacturing operations can be achieved by traditional occupational hygiene methods: developing and using low-dust, encapsulated formulations, implementing capital improvement of the manufacturing equipment, local exhaust ventilation, and measuring and maintaining effectiveness through occupational hygiene monitoring and medical surveillance programs. For more detailed information, see the references.

Summary

Although biological safety grew out of military, disease control, and agricultural research programs in the mid-1900s, it is a relatively new and increasingly more important responsibility for occupational hygienists. The OSHA Bloodborne Pathogens Standard has increased attention to biological hazards by all safety and health practitioners. Some of the industries where biological agents are important and present recognized hazards are health care (hospitals, clinics, continuous care facilities), police and emergency responder services, clinical laboratories, agriculture and veterinary science, in-vitro diagnostic medical device manufacturers, biotechnology, biological R&D, detergent enzymes, and even construction/demolition. Clearly, there are tremendous social and economic benefits derived from work in these occupational settings including public health services, clinical diagnosis and treatments, recombinant vaccines, food supplies, and many more.

The primary goal of the biosafety program is not to prohibit work, but to enable work to be performed safely. In this chapter the objectives were to provide occupational hygiene students and practitioners with an understanding of biological hazards as distinct from chemicals or physical agents, and to present the basic principles and practices of biological safety. A format for comprehensively assessing risks of exposure to biological agents was provided, and the constituents of an effective biological safety program were described. Finally, the chapter showed how biosafety principles are applied in industry.

The field of biological safety, as any other scientific discipline, is constantly changing. In addition to the references cited, the reader is also encouraged to explore the World Wide Web for the latest breaking news and information.

Additional Sources

AIHA Biosafety Committee, Fairfax, Va. (www.aiha.org)

American Biological Safety Association, Mundelein, Ill. (www.orcbs.msu.edu/absa)

American Conference of Governmental Industrial Hygienists, Cincinnati, Ohio (www.acgih.org)
- Infectious Agents Committee
- Agricultural Health and Safety Committee
- Bioaerosols and Air Sampling Committee

American Society of Heating, Refrigerating and Air Conditioning Engineers, Atlanta, Ga. (www.ashrae.org)

American Society of Microbiology, Washington, D.C. (www.asmusa.org)
- Public and Scientific Affairs Board
- Laboratory Practices Committee
- Laboratory Safety Subcommittee

Association of Professionals in Infection Control and Epidemiology, Washington, D.C. (www.apic.org)

CDC, Atlanta, Ga. (www.cdc.org)

Society for Healthcare Epidemiologists of America, Woodbury, N.J.

Society for Industrial Microbiology, Fairfax, Va. (www.simhq.org)

References

1. **American Biological Safety Association (ABSA):** *Membership Directory.* Mundelein, IL: ABSA, 1996.
2. "Guidelines for Carcinogen Risk Assessment," *Federal Register 51*:33992–34003, 1986.
3. Report of committee IV on evaluation of radiation doses to body tissues from internal contamination due to occupational exposure. In *Recommendations of the International Commission on Radiological Protection* (ICRP pub. 10). Oxford: Pergamon Press, 1968.
4. *Review of NCRP Radiation Dose Limit for Embryo and Fetus in Occupationally Exposed Women* (NCRP pub. 53). Washington, DC: National Council on Radiation Protection, 1977.
5. **Nellis, B.F.:** "The Chain of Infection." November 1996. [Unpublished information] Johnson & Johnson, 1999 Lake Avenue, Bldg. 83, Rochester, NY 14650-2209.
6. "Classification of Human Etiologic Agents on the Basis of Hazard," *Federal Register 61*:10004 (1 March 1996). NIH Guidelines Appendix B.
7. "Classification of Biological Agents." Council Directive of the European Communities (90/391/EEC), 1991. Article 18 and Annex.
8. **Centers for Disease Control and Prevention (CDC):** *Biosafety in Microbiological and Biomedical Laboratories*, 3rd ed. J.Y. Richmond and R.W. McKinney, eds. (DHHS pub. no. CDC 93-8395) Washington DC: U.S. Government Printing Office, 1993. Agent summaries.
9. **Centers for Disease Control and Prevention (CDC):** *Biosafety in Microbiological and Biomedical Laboratories*, 3rd ed. J.Y. Richmond and R.W. McKinney, eds. (DHHS pub. no. CDC 93-8395). Washington, DC: U.S. Government Printing Office, 1993. Tables 1 and 2.
10. **Mackowiak, P.A.:** The normal microbial flora. *N. Engl. J. Med.* 307:83–86 (1982).
11. **Smith H.W. and M.B. Huggins:** Further observations on the colicin V plasmid of *Escherichia coli* with pathogenicity and with survival in the alimentary tract. *J. Gen. Microbiol.* 92:335 (1976).
12. **Tramont, E.C.:** Host defense mechanisms. In *Principles and Practices of Infectious Diseases*, 3rd ed. New York: Churchill Livingstone Inc., 1990. pp. 33–40.
13. **Centers for Disease Control and Prevention:** Recommendations for prevention of HIV transmission in health care settings. *Morb. Mort. Wkly. Rep. 36 (Suppl. 1987c)*:2 (1987).

14. **Songer, J.R.:** Management and codification of risks in the laboratory. In *Laboratory Safety Principles and Practice.* Washington, DC: American Society for Microbiology, 1986. pp. 120–122.
15. **Nicas, M.:** Modeling respirator penetration values with the beta distribution: an application to occupational tuberculosis transmission. *Am. Ind. Hyg. Assoc. J. 55*:515–524 (1994).
16. **Nardell, E.A., J. Keegan, S.A. Cheney, et al.:** Theoretical levels of protection achievable by building ventilation. *Am. Rev. Respir. Dis. 144*:302–306 (1991).
17. **Riley, C., C.C. Mills, F. O'Grady, et al.:** Infectiousness of air from a tuberculosis ward. *Am. Rev. Respir. Dis. 85*:511–525 (1962).
18. **Wells, W.F. (ed.):** *Airborne Contagion and Air Hygiene.* Cambridge, MA: Harvard University Press, 1955. pp. 15–17, 121–122.
19. **Griffith J., T. Aldrich, and W. Drane:** Risk assessment. In *Environmental Epidemiology and Risk Assessment.* New York: Van Nostrand, 1993. pp. 212–239.
20. **National Research Council:** *Biosafety in the Laboratory, Evaluation of Hazard.* Washington, DC: National Academy Press, 1989. p. 53.
21. **Rayburn, S.R. (ed.):** *The Foundations of Laboratory Safety: A Guide for the Biomedical Laboratory.* New York: Springer-Verlag, 1990.
22. **Reitman, M. and A.G. Wedum:** Microbiological safety. *Publ. Health Rep. 71*:659–665 (1956).
23. **Johnson, B. and I.G. Resnick:** Safety and containment of microbial bioaerosols. In *Atmospheric Microbial Aerosols: Theory and Application,* New York: Chapman and Hall, 1994. pp. 365–384.
24. "Biological Defense Safety Program," *Code of Federal Regulations* Title 21, Part 626. 1993. pp. 11368–11374.
25. "Biological Defense Safety Program (Technical Safety Requirements)," *Code of Federal Regulations* Title 32, Part 627. 1993. pp. 9424–9449.
26. "Guidelines for Research Involving Recombinant DNA Molecules," *Federal Register 61*:49 (12 March 1996).
27. **Massachusetts Biotechnology Council:** *Biotechnology Regulatory Guide for Communities.* Cambridge, MA: Massachusetts Biotechnology Council, 1995. [Pamphlet]
28. **Centers for Disease Control and Prevention (CDC):** *Biosafety in Microbiological and Biomedical Laboratories,* 3rd ed. J.Y. Richmond and R.W. McKinney, eds. (HHS pub. no. CDC 93-8395) Washington, DC: U.S. Government Printing Office, 1993.
29. **Joint Commission on Accreditation of Healthcare Organizations:** Environment of care standards 1.6, 2.5, 3.1. In *Accreditation Manual for Hospitals.* Chicago: Joint Commission on Accreditation of Healthcare Organizations, 1995.
30. **Heinsohn, P.A., R.R. Jacobs, and B.A. Concoby (eds.):** *Biosafety Reference Manual,* 2nd ed., Fairfax, VA: American Industrial Hygiene Association, 1995.
31. **Stuart, D.G., T.J. Greenier, R.A. Rumery, and J.M. Eagleson:** Survey, use, and performance of biological safety cabinets. *Am. Ind. Hyg. Assoc. J. 44*:265–270 (1982).
32. **Heinsohn, P.A., R.R. Jacobs, and B.A. Concoby (eds.):** *Biosafety Reference Manual,* 2nd ed., Fairfax, VA: American Industrial Hygiene Association, 1995. pp. 64–69.
33. Selection and Classification Chart for Class II Laminar Flow Biological Safety Cabinets. Sanford, ME: The Baker Co., 1983. [Pamphlet]
34. **Bernstein, W.N.:** Designing biosafety cabinets. *Bio/Technology 13*:1068–1070 (1995).
35. **Slavik, N.:** "Statement of the American Hospital Association before the Subcommittee on Regulation and Business Opportunities of the Small Business Committee of the U.S. House of Representatives on the Regulations of Infectious Waste." Washington, DC, August 9, 1988.
36. **Environmental Protection Agency:** "Draft Manual for Infectious Waste Management." September 1982.
37. "Guidelines for Research Involving Recombinant DNA Molecules," *Federal Register 61*:49 (12 March 1996). Appendix G.
38. "Occupational Exposure to Bloodborne Pathogens; Final Rule" *Code of Federal Regulations* Title 29, Part 1910.1030(g)(1).
39. **Plog, B.A., J. Niland, and P.J. Quinlan (eds.):** *Fundamentals of Industrial Hygiene,* 4th ed. Chicago, IL: National Safety Council, 1996. pp. 422–424.
40. **Heinsohn, P.A., R.R. Jacobs, and B.A. Concoby (eds.):** *Biosafety Reference Manual,* 2nd ed. Fairfax, VA: American Industrial Hygiene Association, 1995. pp. 101–110.

41. **Cole, E.C.:** "Environmental Decontamination." Presentation for Practicing Industrial Hygiene within the Biotechnology Industry at the American Industrial Hygiene Conference and Exposition professional development course. Salt Lake City, UT, May 19, 1991.
42. **Russell, A.D., W.B. Hugo, and G.A.J. Ayliffe (eds.):** *Principles and Practice of Disinfection, Preservation, and Sterilization*, 2nd ed. Oxford, U.K.: Blackwell Scientific Publications, 1992. pp. 9–43.
43. "Occupational Exposure to Ethylene Oxide," *Federal Register 49*:122 (22 June 1984).
44. "Occupational Exposure to Formaldehyde," *Federal Register 52*:233 (4 December 1987).
45. "Hazard Communication," *Federal Register 52*:163 (24 August 1987).
46. "Occupational Exposure to Hazardous Chemicals in Laboratories," *Federal Register 55*:21 (31 January 1990). pp. 3300–3335.
47. **Sulkin, S.E. and R.M. Pike:** Survey of laboratory acquired infections. *J. Am. Med. Assoc. 147*:1740–1745 (1951).
48. **Pike, R.M.:** Laboratory associated infections: summary and analysis of 3921 cases. *Health Lab. Sci. 13*:105–114 (1976).
49. "Guidelines for Research Involving Recombinant DNA Molecules," *Federal Register 61*:49 (12 March 1996). Section IV, B-3-c-(3).
50. Title III Superfund Amendments and Reauthorization Act of 1986, Emergency Planning and Community Right to Know Act. Under section 301 of subtitle A, states have established state emergency response commissions, which in turn have appointed local emergency planning committees. *Code of Federal Regulations* Title 40, Parts 350, 355, 370, and 372.
51. **Sandman, P.M.:** *Responding to Community Outrage: Strategies for Effective Risk Communication*. Fairfax, VA: American Industrial Hygiene Association, 1993.
52. **Sandman, P.M.:** *Risk = Hazard + Outrage: A Formula for Effective Risk Communication*. Fairfax, VA: American Industrial Hygiene Association, 1993. [Video]
53. **Sandman, P.M.:** *Quantitative Risk Communication: Explaining the Data*. Fairfax, VA: American Industrial Hygiene Association, 1993. [Video]
54. **Sandman, P.M.:** *Implementing Risk Communication: Overcoming the Barriers*. Fairfax, VA: American Industrial Hygiene Association, 1993. [Video]
55. Center for Environmental Communication (CEC) Publications List, May 1993. A program of the New Jersey Agricultural Experimental Station and the Edward J. Bloustein School of Planning and Public Policy at Cook College, P.O. Box 231, New Brunswick, NJ 08903-0231.
56. "Hazardous Materials Regulations: Editorial Corrections and Classifications; Final Rule," *Federal Register 61*:191 (1 October 1996). pp. 51333–51343. The Department of Transportation Research and Special Programs Administration eliminated over 100 pages of obsolete and duplicative rules, effective Oct. 1, 1996.
57. **International Air Transport Association (IATA):** *Dangerous Goods Regulations*, 36th ed. Montreal: IATA, 1995. IATA Resolution 618, Attachment A.
58. "NAFTA Countries Seek Uniform Code." *Chem. Reg. Rep. 20*. August 9, 1996.
59. "Technical Instructions for the Safe Transport of Dangerous Goods by Air" (Doc. 9284-AN/905). Montreal: International Civil Aviation Organization, 1993–1994.
60. "Interstate Shipment of Etiologic Agents," *Code of Federal Regulations* Title 42, Part 72. 1996.
61. "Hazardous Materials Regulations" *Code of Federal Regulations* Title 49, Parts 171–180. 1996.
62. **U.S. Postal Service (USPS):** *Acceptance of Hazardous, Restricted, or Perishable Matter* (USPS pub. 52). Washington, DC: USPS Headquarters, 1989.
63. "Additional Requirements for Facilities Transferring or Receiving Select Infectious Agents," *Code of Federal Regulations* Title 42, Part 72. 1997.
64. "Training Requirements," *Code of Federal Regulations* Title 49, Part 172.704. 1996.
65. **Department of Health and Human Services:** *Primary Containment for Biohazards: Selection, Installation and Use of Biological Safety Cabinets*. Atlanta, GA: Centers for Disease Control and Prevention, 1995.
66. **Liberman, D.F. and J.G. Gordon (eds.):** *Biohazards Management Handbook*. New York: Marcel Dekker, Inc., 1989.

67. **Fleming, D.O., J.H. Richardson, and J.J. Tulis, and D. Vesley (eds.):** *Laboratory Safety—Principles and Practices*, 2nd ed. Washington, DC: American Society of Microbiology Press, 1995.
68. **Sattelle, D.:** *Biotechnology in Perspective*. Washington, DC: IBA, Hobson Scientific, 1990. p. 4. [Pamphlet]
69. **Lee, S.B. and L.P. Ryan:** Occupational health and safety in the biotechnology industry—a survey of practicing professionals. *Am. Ind. Hyg. Assoc. J. 57*:381–386 (1996).
70. **Furr, A.K. (ed.):** *Handbook of Laboratory Safety*, 3rd ed. Boca Raton, FL: CRC Press, Inc., 1989.
71. **Shapiro, J.:** *Radiation Protection, A Guide for Scientists and Physicians*, 2nd ed. Cambridge, MA: Harvard University Press, 1981.
72. **National Research Council:** *Prudent Practices for Handling Hazardous Chemicals in Laboratories*. Washington, DC: National Academy Press, 1981.
73. **National Fire Protection Association (NFPA):** *Standard on Fire Protection for Laboratories Using Chemicals*. Braintree, MA: NFPA, 1991.
74. **Berg, P., D. Baltimore, H.W. Boyer, S.N. Cohen, et al.:** Potential biohazards of recombinant DNA molecules. *Science 185*:303–305 (1974).
75. "Recombinant DNA Research," *Federal Register 41*:131 (7 July 1976) pp. 27902–27905.
76. "Recombinant DNA Research," *Federal Register 41*:131 (7 July 1976). p. 27903.
77. "Recombinant DNA Research," *Federal Register 41*:131 (7 July 1976). pp. 27903–27906.
78. "Recombinant DNA Research Actions Under the Guidelines," *Federal Register 56*:138 (18 July 1991). Appendix K-I.
79. **Sattelle, D.:** *Biotechnology in Perspective*. Washington, DC: IBA, Hobson Scientific, 1990. pp. 9, 28. [Pamphlet]
80. "Coordinated Framework for Biotechnology Regulation." *Chem. Reg. Rep 101*:0151–0158 (1993).
81. **Reutsch, C.-J. and T.R. Broderick:** New biotechnology legislation in the European Community and Federal Republic of Germany. *Biotechnology*.
82. **Thomas, J.A. and L.A. Myers (eds.):** *Biotechnology and Safety Assessment*. New York: Raven Press, Ltd, 1993.
83. **Collins, C.H. and A.J. Beale (eds.):** *Safety in Industrial Microbiology and Biotechnology*. Oxford, U.K.: Butterworth-Heinemann Ltd., 1992. pp. 165–167.
84. **Collins, C.H. and A.J. Beale (eds.):** *Safety in Industrial Microbiology and Biotechnology*. Oxford, U.K.: Butterworth-Heinemann Ltd., 1992. p. 169.
85. **Balzer, K.:** "Strategies for Developing Biosafety Programs in Biotechnology Facilities." Paper presented to the 3rd National Symposium on Biosafety, Atlanta, GA, March 4, 1994.
86. **American Conference of Governmental Industrial Hygienists (ACGIH):** *1996–1997 Threshold Limit Values for Chemical Substances and Physical Agents, Biological Exposure Indices*. Cincinnati, OH: ACGIH pp. 10–12, 34.
87. **Collins, C.H. and Beale, A.J. (eds.):** *Safety in Industrial Microbiology and Biotechnology*. Oxford, U.K.: Butterworth-Heinemann Ltd., 1992. pp. 167–169.
88. "NIH Guidelines for Research Involving Recombinant DNA Molecules," *Federal Register 61*:42 (1 March 1996). Appendix K, Table 1.
89. **Hatch, T.F.:** Distribution and deposition of inhaled particles in respiratory tract. *Bacteriol. Rev. 25*:237–240 (1961).
90. **Johnson, B. and I.G. Resnick:** Safety and containment of microbial bioaerosols. In *Atmospheric Microbial Aerosols: Theory and Application*. New York: Chapman and Hall, 1994. pp. 365–384.
91. **Department of Health, Education and Welfare (DHEW):** *Laboratory Safety Monograph, A Supplement to the NIH Guidelines for Recombinant DNA Research*. Bethesda, MD: DHEW, 1979. pp. 138–141.
92. **Miller, C.D., J.R. Songer, and J.F. Sullivan:** A 25-year review of laboratory-acquired human infections at the National Animal Disease Center. *Am. Ind. Hyg. Assoc. J. 48*:271–275 (1987).
93. **Fox, J.C. and N.S. Lipman:** Infections transmitted by large and small laboratory animals. *Infect. Dis. Clin. N. Am. 5*:131–163 (1991).
94. **Centers for Disease Control and Prevention:** Guidelines for the prevention of herpesvirus simiae (B virus) infection in monkey handlers. *Morb. Mort. Wkly. Rep. 36*:680–682, 687–689 (1987).
95. **Jaxx, N., P. Jarling, T. Geisbert, J. Geisbert, et al.:** Transmission of Ebola virus (Zaire strain) to uninfected control monkeys in a biocontainment laboratory. *Lancet 346*:1669–1671 (1995).

96. **Richmond, J.Y.:** Hazard reduction in animal research facilities. *Lab. Anim. 20(2)*:23–29 (1991).
97. **Richardson, J.H. and W.E. Barkley (eds.):** *Biosafety in Microbiological and Biomedical Laboratories*, 2nd ed. (HHS pub. no. CDC 88-8395). Washington, DC: U.S. Government Printing Office, 1988.
98. "Laboratory Animal Welfare Regulations," *Code of Federal Regulations* Title 9, Parts 1–3, Subchapter A. 1992.
99. **National Institutes of Health:** *Guide for the Care and Use of Laboratory Animals* (DHHS pub. no. NIH 86-23). Washington, DC: U.S. Government Printing Office, 1985.
100. **Dillon, H.K., P.A. Heinsohn and J. D. Miller (eds.):** *Field Guide for the Determination of Biological Contaminants in Environmental Samples*. Fairfax, VA: American Industrial Hygiene Association, 1996.
101. **Soap and Detergent Association:** *Work Practices for Handling Enzymes in the Detergent Industry*. New York: Soap and Detergent Association, 1995.
102. **Griffith, J.E., H.S. Whitehouse, R.L. Pool, E.A. Newmann, et al.:** Safety evaluation of enzyme detergents. Oral and cutaneous toxicity, irritancy and skin sensitization studies. *Food Cosmet. Toxicol. 7*:581–593 (1969).

Chapter 19

Outcome Competencies

After completing this chapter, the user should be able to:

1. Define underlined terms used in this chapter.
2. Describe indoor air quality impacts on health.
3. Describe the building environment.
4. Describe building HVAC systems.
5. Discuss ASHRAE ventilation standards.
6. Recall how HVAC systems impact air quality.
7. Anticipate and recognize IAQ problems.
8. Describe the planning and conduct of an indoor air quality survey.
9. List pollutant categories.
10. Describe two procedures for sampling indoor air contaminants.

Key Terms

β(1-3)d-glucans • air quality • allergen • building-related disease • endotoxin • hypersensitivity diseases • *Legionella* • multiple chemical sensitivity • mycotoxins • ozone • pollutants • psychosomatic • radon • radon progeny • sick building syndrome • toxicoses • volatile organic compounds

Prerequisite Knowledge

College biology and chemistry.

Prior to beginning this chapter, the user should review the following chapters:

Chapter Number	Chapter Topic
2	Occupational Exposure Limits
4	Environmental and Occupational Toxicology
5	Epidemiological Surveillance
6	Principles of Evaluating Worker Exposure
12	Sampling and Sizing Particles
15	Comprehensive Exposure Assessment
18	Biohazards in the Work Environment
28	Ergonomics
29	Psychology and Occupational Health
30	Worker Education and Training
40	Hazard Communication
41	Risk Communication in the Workplace

Key Topics

I. Health Effects Related to Indoor Air Quality
  A. Overview
  B. Building-Related Disease
  C. Nonspecific Building-Related Symptoms (BRS)

II. The Building Environment
  A. Kinds of Buildings
  B. Relative Humidity (RH)
  C. Thermal Comfort
  D. Odors and Indoor Air Quality

III. Ventilation
  A. Heating, Ventilating, and Air-Conditioning Systems (HVAC)
  B. ASHRAE Ventilation Standards
  C. Indoor/Outdoor Relationships
  D. Filtration
  E. Problems with Maintaining Standard Ventilation
  F. Ventilation System Contamination
  G. Fiber Glass in Ventilation Systems
  H. Assessing Ventilation Problems

IV. Pollutant Categories
  A. Biological Agents
  B. Nonbiological Agents

V. Concluding Remarks
  A. State of the Art
  B. Practical Guidance for Indoor Air Quality Investigations

19

Indoor Air Quality

Harriet A. Burge, Ph.D.

Marion E. Hoyer, Ph.D.

Introduction

The indoor environment has been considered a refuge not only from the weather, wildlife, and unrestricted contact with other people, but also from the air pollution that has become a fact of life in most cities of the world. In fact, Americans, especially in urban environments, spend nearly 90% of their time indoors.[1]

However, the indoor environment is not always safe and comfortable.[2,3] It has been recognized for hundreds of years that infectious disease is more readily transmitted in indoor environments. In addition, modern buildings provide unique sources for some pollutants, and others can accumulate to levels that cause discomfort and disease. The latter is, at least in part, related to energy conservation measures instituted in the late 1960s and early 1970s that led to decreased ventilation rates for most interiors, with increased potential for build-up of internally generated air pollutants. Also, lower levels of potentially harmful pollutants might be expected to have a greater effect indoors because of longer exposure times.[4] It is becoming clear that both comfort and health can be affected by long-term exposure to pollutant levels well below published standards.[5] Finally, new technology is helping to identify and quantify new classes of pollutants heretofore ignored in indoor air quality investigations.

It is important, therefore, to develop standards for air quality that render the indoor environment at least as safe as the outdoors, and within the comfort range for most individuals. ASHRAE (American Society for Heating, Refrigerating and Air-Conditioning Engineers) defines acceptable indoor air quality as "air in which there are no known contaminants at harmful concentrations and with which a substantial majority (usually 80%) of the people exposed do not express dissatisfaction."[6] The problem with this definition is that in nonindustrial environments measurable contaminants are rarely present in levels known to be harmful, even where complaints of discomfort and adverse health effects are considerably in excess of the "acceptable" 20%.[7] Until health risks have been established for chronic low level exposures to both known and currently unrecognized pollutants, one must rely on the second part of this definition for guidance. Indoor air must meet standards that provide for the health and comfort of the majority of occupants.

It should be noted that, in spite of the many publications cited in this chapter, the study of indoor air in other than industrial environments is relatively new, and in a majority of investigations the actual causes of complaints remain elusive. Ongoing research in the epidemiology of building-related complaints and laboratory-based studies on the relationships between exposure and disease for specific agents offer promise for the future.

Health Effects Related to Indoor Air Quality

Overview

Air pollutants, in general, are inhaled and initially impact on the respiratory tract. They may also be absorbed and affect other end organs, and some are stored in body tissues creating the potential for adverse health effects over time even in the absence of acute effects at the time of exposure.[8] In addition, some pollutants affect mucous membranes of the eye, and a few can cause skin rashes and itching.[9,10] With rare exception, menstrual irregularities, difficulties with pregnancy, increased susceptibility to infectious disease, and most cancers have not been shown to be associated with occupancy in nonindustrial indoor environments.[9]

The terminology associated with comfort and health effects of indoor air quality has become confusing. The terms "sick building syndrome (SBS)," "tight building syndrome," "building-related illness," and "building-related disease" have been used interchangeably.[11] For purposes of this discussion, the term building-related disease will be used for cases of infectious, allergic, or toxin-induced disease with objective clinical findings related to building occupancy. Although it is not descriptive, the term sick building syndrome has come into common use for those problems where excessive comfort and health-related symptoms are present that are clearly related to building occupancy, but that are not associated with objective clinical signs.

Sources for agents that result in adverse health effects in buildings include the occupants (contagious diseases, carriage of allergens, and other agents on clothing); building components (volatile organic compounds [VOCs], particles, fibers); contamination of building components (microbial agents, other allergens, pesticides); and outdoor air (chemical air pollutants, microorganisms, other allergens).

Building-Related Disease

Diseases that have been clearly related to building occupancy include hypersensitivity diseases such as hypersensitivity pneumonitis (also called allergic alveolitis), humidifier fever, allergic asthma, and allergic rhinitis; infectious diseases such as legionellosis, tuberculosis, influenza, measles; and toxic syndromes resulting from exposure to (for example) carbon monoxide, pesticides, or (rarely) microbial toxins. Rarely, indoor air pollutants can result in increased risk of lung cancer following long-term (years) exposure.[12,13]

Hypersensitivity Diseases

The hypersensitivity diseases important with respect to indoor air all result from specific immune system responses to environmental challenges.[14] There are two general categories: the IgE-mediated diseases (asthma, allergic rhinitis, or hay fever) and hypersensitivity pneumonitis, which is mediated by IgG and the cellular immune system. All hypersensitivity diseases require an initial series of sensitizing exposures during which the immune system becomes activated. Symptoms occur on subsequent exposures in response to stimulation of the previously activated immune response. Most cases of hypersensitivity disease are caused by proteins or glycoproteins, although some highly reactive chemicals can bind to larger molecules to cause hypersensitivity pneumonitis.

Infections

The most common of the building-related infections are contagious diseases, which are readily spread through indoor air, especially in crowded environments.[15] These diseases include influenza, the common cold, and tuberculosis. The contagious diseases are generally contracted by all those exposed who have not developed specific immunity to the agent (i.e., by either having experienced the disease in the past, or by artificial immunization). The only infectious disease that is commonly spread from environmental reservoirs is Legionnaires' disease. All of the environmental-source infections are opportunistic, requiring some deficit in the immunity of the host before infection can occur. Obviously, all infections are caused by biological pollutants.

Toxicoses

Most of the common indoor air pollutants are toxins that exert their effects in a dose response way, for the most part without regard to host susceptibility. Many are inflammatory agents that act directly on the contacted cells. Some are absorbed and exert effects on organ systems distant from the site of exposure. Some cross the blood-brain barrier to cause central nervous system effects. Most toxicoses associated with indoor air are caused by exposures to chemicals derived from combustion (e.g., carbon monoxide [CO], nitrogen dioxide [NO_2]) or from activities or materials used in the environment (e.g., the VOCs). With the exception of endotoxin and the glucans, exposure to biological toxins in levels sufficient to cause disease is rare in nonindustrial indoor environments.

Nonspecific Building-Related Symptoms (BRS)

SBS

SBS has been defined as a complex of symptoms that includes mucous membrane irritation producing nasal irritation and sinus congestion, eye irritation, nonproductive cough; headaches; fatigue or lethargy; dry skin; dizziness; and nausea.[13,16,17] These symptoms are subjective, rarely associated with objective clinical findings, and are typically present in some occupants of all buildings at some times. To be considered an outbreak of SBS, the attack rate should be at least 20%, there should be some commonality among the symptoms, and a clear temporal association between symptoms and building occupancy should be apparent.[18]

Outbreaks of SBS have been attributed to exposure to VOCs, low relative humidity, endotoxin, some factor within the macromolecular component of house dust, and unknown exposures resulting from inadequate supplies of fresh air.

Psychosomatic Illness

In the case of BRS, where objective clinical findings are not available, it is often difficult to separate environmentally caused symptoms from those created by suggestion or those that are the result of either on-the-job or other stress.[19] Increasing attention by news media to building-related complaints has created heightened occupant concern for air quality, in many cases without apparent cause. For example, clusters of unrelated cancers have led to fears that the work environment is causing the cancer. Such fears can lead to psychosomatic symptoms that can mimic those of SBS. A good initial step where such a situation is suspected is to map the occurrence of cases in space and time. Features of such data to examine are shifting incidence in time and/or space not associated with ventilation system parameters, an epidemic curve consistent with person-to-person rather than common-source transmission, and consistent lack of medical correlates of exposure.[20] Exploring possible causes of widespread worker dissatisfaction is also of help in assessing questionable cases of SBS. On the other hand, poor air quality is a general stressor and can trigger complex psychological reactions including changes in mood, motivation, and problems with interpersonal relations.[19]

Multiple Chemical Sensitivity

Multiple chemical sensitivity (MCS) has been described as resulting from exposure to toxic chemicals that affect the immune system, leading to multiple sensitivities to other chemicals and/or foods.[21] Symptoms of MCS, which may be similar to those of SBS, are often attributed to exposure to trace amounts of chemicals (especially those with perceptible odor) in indoor air. Unfortunately, the methods and theories of practitioners who diagnose and treat this syndrome have not been subjected to rigorous scientific study, and the few patients who have been reevaluated by more traditional methods have, in general, been found to be free of immunologically mediated disease. In many such patients where no physiological mechanism has been found, psychological or psychosocial problems have been predominant.[22] It may not be possible to provide an environment that is clean enough to prevent symptoms in these patients, especially where environmental exposures are not at the root of their problems. Health and safety officers in such situations need to convince co-workers that the environment is safe. Comparing environmental testing results from the patients' environment with similar environments where no complaints exist may be useful, providing the environments meet the highest standards for health and comfort.

The Building Environment

Kinds of Buildings

There are many different kinds of buildings, and a reliable grouping system has yet to be developed. One set of categories that is often used is: residential buildings (divided into single and multiple family dwellings); nonindustrial workplaces (including office buildings, schools, and other "clean" environments); industrial workplaces; agricultural buildings; and health care facilities (hospitals, nursing homes, doctors' offices, outpatient treatment centers, etc.). This review will focus on nonindustrial workplaces and (to a limited extent) residences and health care facilities.

Relative Humidity (RH)

Low RH has been blamed for some of the symptoms of SBS, for increased susceptibility to infectious disease, and for exacerbation of asthma.[23,24] Eye irritation, throat irritation, and cough are often blamed on low RH, but it is becoming clear that most people are comfortable at RH levels that routinely occur even in cold winter climates (<25%), if adequate ventilation is provided to control odors and other air pollutants.

Some evidence indicates that host susceptibility to airborne infectious agents is not affected by RH as was once thought.[25]

Usually, the increased incidence of colds and influenza in winter has been used as evidence for the role of RH in susceptibility to infections. The dry air is considered either to damage mucous membranes or to damage the infectious agents themselves. Damage to mucous membranes at other than extremely low RH (<5%) has not been documented. RH is one of the environmental factors controlling the viability of both airborne and surface microorganisms, but the relation is extremely complex. Some viruses survive well at low and high RH but not at intermediate levels. Bacteria often have a narrow range of RH in which they survive as an aerosol. *Legionella*, for example, was shown in one study to survive best at 65% RH and least well at 55% RH.[26,27] Growth of microorganisms on surfaces and the survival of the house dust mite, a source for potent asthma-inducing allergens, is facilitated by high RH (>60%).[28] Even at lower RH, cool surfaces may lead to condensation or locally much higher RH that can support fungal and/or dust mite amplification.

Thermal Comfort

ASHRAE defines thermal comfort as "that condition of mind which expresses satisfaction with the thermal environment."[29] The perception of thermal comfort is related to metabolic heat production, its transfer to the environment, and resulting body temperature. Personal activity and clothing and the environmental factors of air temperature, air movement, and RH all interactively influence body temperature and therefore thermal comfort. For sedentary workers, air temperature, air movement, and clothing are the most important factors, while RH has little effect on thermal comfort. In fact, it has been shown that sedentary people are unable to perceive changes in RH over a relatively wide range.[30] Lowering RH in the presence of active workers has a cooling effect. In environments where both active and sedentary workers coexist, a slightly warm air temperature at low RH might provide comfort for most, although it is rarely possible to satisfy everyone. Increasing air movement decreases body temperature and causes the temperature to feel lower. However, minimal air movement can lead to complaints of stuffiness and poor air quality.

Odors and Indoor Air Quality

In general, the facet of poor air quality that is most apparent is odor. From the beginning, efforts to establish ventilation rate standards have been directed toward controlling human body odor, and the smell of fresh wood and paint, and even tobacco smoke, were actually considered pleasant. More recently, human body odor has become of less concern because of improvements in both personal hygiene and in mechanical ventilation systems. Perception of almost any odor in a supposedly clean environment will elicit a negative response in some people.[31] Especially problematic are odors that are unexpected in the environment. For example, cooking odors that would be perceived as pleasant in a restaurant or home can cause anxiety or irritation in an office environment, especially where stress or unrelated illness has precipitated concern over environmental conditions. Also, the odor of tobacco smoke causes serious concern in many nonsmokers.[19]

Inadequate ventilation can cause adverse effects in the absence of any perceptible odor. On the other hand, odor can be used as a very rough indication of the concentration of a pollutant in the environment,[31] although minor differences in concentration are beyond olfactory discrimination. For many pollutants (e.g., acrolein, formaldehyde, formic acid, acetic acid, acetone) the odor threshold is either close to or above that for the irritant and/or health effects concentrations, and the perception of the odor is a clear indication of a problem. For other pollutants, the odor threshold is well below that for concentrations causing known health effects, and symptoms tend to be psychologically induced.[19] For most normal, healthy adults, this is probably the case for sporadic, low level environmental tobacco smoke odors as well as odors associated with VOCs released from fungi and bacteria.

Ventilation

SBS apparently occurs most often in mechanically ventilated, air-conditioned buildings in which the amount or distribution of outdoor (fresh) air supplied is inadequate.[32] In buildings lacking sufficient outdoor air supply, effluents from human occupants and their activities, and from building structural and content materials can build up to levels where occupants perceive distinct discomfort.

Heating, Ventilating, and Air-Conditioning (HVAC) Systems

HVAC systems in mechanically ventilated buildings control indoor air quality by providing outdoor air of suitable quality and quantity; filtering, mixing, and distributing both outdoor and indoor air to the occupied space; and providing temperature and humidity control. In addition, ventilation systems may act as sources for specific pollutants including VOCs, fibers (especially fiber glass) and other particles, and biological pollutants.

Major components of HVAC systems include outdoor air intakes, intake fans, and filter banks; heat exchangers to provide heating or cooling; distribution fans and ductwork; diffusers to provide mixing in the occupied space; and a system for returning air to the central system. A well-designed, properly installed and maintained ventilation system should provide air quality that is acceptable to a majority of building occupants. A system that is improperly installed, operated, or maintained may result in degradation of indoor air quality and lead to occupant complaints. Design factors that may adversely impact air quality include insufficient provision for outdoor air; inefficient filtration; inadequate cooling (and dehumidification); improperly designed drip pans; water spray humidification systems; use of porous insulation near water sources; inadequate provision for access to HVAC components that might require maintenance; use of materials that release inappropriate amounts of VOCs or fibers; and inadequate provision for mixing air within the occupied space. Maintenance problems may result in fans that do not operate, clogged filters, and microbial contamination on filters, in condensate pans, and in ductwork.

ASHRAE Ventilation Standards

Most ventilation systems are designed to provide fresh air according to current ASHRAE standards. The current ASHRAE standard for ventilation for acceptable indoor air quality (Standard 62–1989)[6] specifies for most indoor environments that acceptable outdoor air must be supplied to all building occupants at a rate of 15 ft^3/min per person at all times (regardless of the recirculation air treatment provided). Acceptable outdoor air used for building ventilation must meet specific standards, and if it does not, the ventilation system must include air treatment modalities that render outdoor air brought into the building safe and comfortable for building occupants. The standard includes a requirement for adequate mixing throughout the occupied space. Recirculation of indoor air is permitted provided that recirculation does not reduce the amount of outdoor air or reduce air quality below the standard for the occupancy.[6] ASHRAE proposed significant revisions to Standard 62 in August 1996. At the time of publication, these revisions have not been finalized.

Indoor/Outdoor Relationships

The primary source for some indoor air pollutants is the outdoor air.[1,33,34] Fungal spores, some kinds of bacteria, combustion particles and gases, and ozone are all present in outdoor air. For most air pollutants entering a building from outdoors, levels indoors should be lower than those outdoors providing no indoor source exists. Even rough filtration will remove a percentage of all particles from air, and in general, removal rates depend on size of the particle (although charge can have a largely unexplored effect). While many gaseous pollutants are unobstructed by ventilation system and other barriers to entry, many will adsorb to indoor surfaces or react with indoor substances and be removed from the air. Residence times (decay rates) for gaseous pollutants vary with the pollutant, lighting, reactive substrates available, and dilution rate. Outdoor air is, therefore, a necessary and useful control for determining whether a potential pollutant has an indoor or outdoor source.[35]

Filtration

In most large buildings, ventilation system air-handling units bring in outside air, mix it with building air, and circulate the mixture through the building, either through a ducted system or through ceiling plenums. The air mixture is usually passed through filters that are designed to protect the air-handling equipment from accumulations of dirt. Although filter media are available that remove most particles from air, most buildings use filters of minimal efficiency that do not prevent entry of most particles that can be inhaled. As large particles accumulate on the filter media, the efficiency of particle collection increases, and smaller particles begin to be trapped. This effect, combined with particle deposition on duct surfaces, effectively leads to a reduction in indoor levels of most outdoor particles as long as the filters remain undisturbed. This type of filtration is also used in recirculation systems within the occupied space with a similar effect: particles are trapped on the filter and not only do not damage equipment, but are removed from the air.

Handling dust-loaded filters releases clouds of particles into the air. If filters become wet, viable microorganisms (fungal spores and bacteria) can germinate, grow, and produce new spores that are introduced into the airstream either because the fungus organism grows through the filter to the downstream side, or during filter change or other disturbance.[36] Such growth also can lead to the release of volatile compounds and odors characteristic of microbial growth.

Problems with Maintaining Standard Ventilation

Air-handling systems can be designed to deliver a specified amount of outdoor air of acceptable quality and to provide for adequate mixing so that each occupant receives a full measure of fresh air. As long as the system is designed to provide adequate fresh air, built as designed, operated according to design specifications, maintained in top condition, and not subsequently modified, supplying fresh air should not be a problem. Unfortunately, it is rare that all of these conditions are met. Outside air quality may change so that fresh air is no longer adequate (e.g., intensifying traffic may elevate CO levels, or poorly placed cooling towers may become contaminated and produce unacceptable microbial aerosols). In some cases shortcuts may be taken during construction to cut costs. These shortcuts are not always immediately obvious to the investigator trying to solve a ventilation-related problem.

In many cases, especially in the early 1980s, fresh air supply was deliberately reduced and, in some cases, the air intakes were completely blocked to reduce fuel costs. Cost considerations also lead to poor maintenance. Unless all fans are operating and all filters are maintained within pressure drop guidelines, appropriate amounts of fresh air will not be delivered to the occupied space. Contamination of the ventilation system may also result from poor maintenance, leading to degradation of air quality.

Buildings with renovated ventilation systems are often very difficult to evaluate because of the compartmentalized approaches often used in such renovations. Also, changes in space configuration (e.g., temporary walls or dividers) that may be made without concern for ventilation patterns often result in areas with virtually no fresh air supply.

When ventilation system operation parameters are changed in attempts to improve air quality, comfort may be reduced rather than improved, especially if increasing the ventilation rate changes temperature or airflow conditions.[13] If airflow increases or temperature declines, occupants may block air circulation ducts to alleviate their discomfort. In ventilation systems where air is supplied to a pressurized plenum, a common occupant method for environmental control is to remove ceiling tiles in areas where additional cool or fresh air is needed. Unfortunately, removal of a significant number of tiles from one area can result in very low ventilation rates in another.

Ventilation System Contamination

Some of the more serious outbreaks of building-related disease (e.g., hypersensitivity pneumonitis) have involved microbial contamination of the ventilation system downstream of any filtration.[37] Such microbial contamination is always associated with water in the system. Water sources can include cooling coil condensate that drips into and collects in poorly drained pans, humidification and water spray air-cleaning systems, and condensation on cold surfaces including those in ductwork.[38]

While the necessity of draining condensate trays should be obvious, many systems are not designed to drain adequately. Standing water in these trays will become contaminated, especially when mineral scale begins to build up on surfaces. In addition, they provide a continuous moisture source that leads to locally high RH and probably to local condensation.

Humidifiers and water spray air cleaners are not designed to be maintained in a sterile condition, and the benefits of their use should be carefully weighed against the risk of contamination and subsequent disease-causing exposures.[39–42] Water spray systems are especially risky because they not only become contaminated, but also actively aerosolize contaminated droplets. Filters and other surfaces downstream from these units can be constantly wet and support further microbial growth. Evaporative humidifiers are less likely to release particles (including intact microorganisms), although humidifier fever has been shown to result from contamination of evaporative units.[38] As discussed above, the role of RH within the range normally present in buildings for the comfort and health of normal, sedentary individuals is probably minimal (with the possible exception of some contact lens wearers). However, until this finding has been integrated into common building designs, humidifiers will continue to be a part of many building ventilation systems.

Condensation in ventilation system components occurs when cool surfaces are in contact with warm humid air. If ventilation system components are maintained at the same temperature as the airstream, even RH near saturation is unlikely to lead to microbial problems. The accumulation of dirt on ventilation system surfaces tends to slow drying after condensation has occurred, produces an increased surface area for microbial growth, and provides nutrient material for the organisms.

Fiber Glass in Ventilation Systems

Fiber glass is used as sound insulation in ventilation systems and to help prevent condensation. Water evaporates more slowly from fiber glass surfaces than from metal. The rough surfaces of fiber glass may also trap dirt more readily than smooth metal surfaces. Both of these factors may favor microbial growth. Removing extensive microbial growth from fiber glass material is generally not possible, and the material may have to be removed. It is probably not advisable to use unsealed fibrous materials in ventilation system areas where condensation or other wetting cannot be prevented.

Assessing Ventilation Problems

In assessing ventilation problems, then, it is necessary to make sure the central system is properly designed, built, and maintained, and that adequate clean outdoor air is being delivered to the occupied space. Methods for assessing ventilation rates in the occupied space range from indirect assessment (measurement of carbon dioxide [CO_2] and total airborne hydrocarbons) to sophisticated tracer gas techniques. Assuming no internal sources other than the occupants, expected levels of CO_2 in an occupied space can be predicted; any levels exceeding this prediction may indicate inadequate ventilation. In general, if CO_2 in an occupied space exceeds 800 ppm, which is approximately twice ambient levels in outdoor air, occupant complaints are likely to occur.(18) Recently, measurement of total hydrocarbons has been used in a similar way, although prediction of source strength and emission rates is more difficult than for CO_2. Nevertheless, field experience shows that total hydrocarbon levels above 5 mg/m^3 tend to be associated with air quality complaints.(13,43)

Evaluating potential or existing microbial growth in ventilation systems is a more difficult problem, since standards are not available with which to compare sampling data. The best approach may be to rely on visual inspection to identify obvious growth or persistent water problems.

Pollutant Categories

Biological Agents

The Organisms

Viruses are simple organisms composed primarily of a strand of either DNA (deoxyribonucleic acid) or RNA (ribonucleic acid) surrounded by a protein coat. Viruses have no internal mechanism for reproduction. Instead, the virus inserts viral genetic material into the host cell, which sends a message to the cell to make new viral particles. Thus, viruses are always obligate intracellular parasites; they can never grow freely in the environment, although they can survive brief periods in environmental reservoirs, including in air.(44)

Bacteria are relatively simple cellular organisms that lack an organized nucleus and hence are called prokaryotic. The bacterial cell is surrounded by a cell wall, the nature of which allows classification of bacteria into broad groupings: gram-negative bacteria (those readily decolorized during the Gram staining procedure) with a lipopolysaccharide wall; gram-positive bacteria with a peptidoglycan cell wall; and the acid-fast bacteria with a wall rich in mycolic acids. Some bacteria form pseudo-filaments and are called actinomycetes. Within this group are organisms (thermophiles, or heat loving) that grow at high temperatures (56°C) and produce very resistant dry spores. Most bacteria are free-living, although a few pathogenic forms have not been recovered from the environment and may have lost the ability to reproduce outside of a living host. Bacteria are generally small (<1 µm), and can become airborne as intact cells, as smaller fragments, or within water droplets.

Fungi are complex multicellular organisms with an organized nucleus and a complex cell wall composed primarily of acetyl-glucosamine polymers and β(1-3)d-glucans. The fungi are usually filamentous, and produce one to many resistant spore stages during each life cycle. It is these spores that become airborne and result in respiratory exposure and disease. Virtually all fungi are free-living and become airborne from environmental reservoirs. Most fungal spores common in indoor air are dry and hydrophobic (lacking affinity for water); a few are produced in droplets of polysaccharide-rich mucous and are hydrophilic (having a strong affinity for water). Fungal spores range in size from about 2 to >100 µm. In addition, fungal fragments and metabolites can become airborne on much smaller particles.[45]

Protozoa are unicellular animals that are common in soil and water. Many are obligate parasites, but none of these cause disease via the airborne route. The protozoa are relatively large, and the intact animals cannot be considered to form true aerosols; rather, living organisms are carried briefly through the air in large droplets. Protozoa also shed soluble material into water that can become airborne on small particles.[46]

Arthropods are jointed animals with exoskeletons. Cockroaches and dust mites are the arthropods that are considered important from an air quality point of view. Cockroaches are insects and are among the most successful creatures on earth. They are abundant throughout the world, are able to colonize human habitats, and reproduce at an astonishing rate. They colonize environments with available water and a food source. Cockroaches shed proteins as body parts and fecal material that become airborne, although particle sizes remain unknown.[47] Dust mites are acarids, and are related to spiders. Dust mites are abundant in most parts of the world, although they are probably least common in desert environments. They colonize dust in houses when humidity is consistently above 60%. They are small (<100 µm) but live in dust reservoirs from which they rarely become airborne as intact animals. Dust mites shed protein-containing fecal particles that are 10–15 µm in diameter.[28]

Birds and mammals are familiar to all. These vertebrate organisms shed proteins in the form of dander (skin scales), urine, saliva, and blood components into the environment as small particles that can form aerosols.[47,48] Rarely, they also may shed infectious agents that, if they become airborne in high concentrations, may cause human disease. This problem is very rare in occupied buildings, but has been reported in derelict buildings.[15]

Plants are complex multicellular organisms that synthesize carbohydrates from CO_2 and water (mediated by chlorophyll) and have cellulosic cell walls. Plants produce many allergens, the best known of which are borne on pollen grains. Individual pollen grains range from 10–80 µm, with most being in the 15–25 µm range. Although none are abundantly produced indoors, they penetrate interiors and are resistant, residing in dust for long periods.[49] Latex (the sap of the rubber plant), used for many consumer products, also contains important allergens.[50]

Agents of Infection

Nature and Sources. Living viruses, bacteria, fungi, and (rarely) protozoa can grow in other organisms to cause disease. Many of the most common infectious diseases are airborne. This means that the agent becomes airborne in sufficient numbers to result in disease, survives transport through the air, and can cause infection readily through inhalation.

Sources for viral infectious agents are nearly always infected people, and the airborne diseases are usually spread by coughing, sneezing, talking, etc. Sources for airborne bacterial infectious agents are usually infected people or animals, although a few bacterial pathogens are spread primarily from environmental reservoirs (e.g., *Legionella*). The common environmental reservoirs for infectious bacteria are all associated with liquid water and include cooling towers, humidifiers, hot water systems, recirculating water wash systems, etc.[51,52] The fungi and protozoa that cause infectious disease are always released from environmental reservoirs.[53,54] The fungi require gaseous oxygen for growth and are usually found in damp environments or on the edges of liquid water reservoirs. The protozoa are found only in water.

Health Effects. The infectious diseases can be classified into three groups: contagious diseases, virulent environmental-source infections, and opportunistic infections. The contagious diseases are caused by viruses and bacteria and are always transmitted from one person to another. Both the virulent and opportunistic environmental-source infections result from exposure to aerosols produced from environmental reservoirs. Examples of diseases within each of these categories are presented in Table 19.1.

Contagious and virulent diseases can infect all exposed people who are not specifically immunized (either artificially or by having had the infection). Theoretically, a single organism can penetrate to the appropriate site in the respiratory system and initiate the disease process, although exposure to many organisms is probably necessary before this event can occur.[55] The opportunistic infections require that the host's natural immunity be impaired to some degree. Infection

Table 19.1 —
Disease Types, Agents of Infections, and Reservoirs

Type of Disease	*Agent*	*Disease*	*Reservoir(s)*
Contagious disease	influenza viruses	influenza	infected people
	measles virus	measles	infected people
	Mycobacterium tuberculosis	tuberculosis	infected people
Virulent environmental infections	*Histoplasma capsulatum*	histoplasmosis	wet soil enriched with bird droppings
	Coccidioides immitis	coccidioidomycosis	dry soil (deserts)
	Legionella pneumophila	Pontiac fever	hot water systems, cooling towers, etc.
Opportunistic environmental infections	*Legionella pneumophila*	Legionnaires' disease	hot water systems, cooling towers, etc.
	Mycobacterium avium	atypical tuberculosis	natural water reservoirs
	Cryptococcus neoformans	cryptococcosis	dry bird droppings
	Aspergillus fumigatus	aspergillosis	self-heating plant-based organic matter

with *Cryptococcus* and *Legionella* require relatively little immune dysfunction (heavy smoking may be sufficient). Infection with *Mycobacterium avium* or *Aspergillus fumigatus* requires major damage to the immune system, and these agents are hazardous principally for patients with immunosuppresive diseases such as acquired immune deficiency syndrome or some kinds of cancer, or those being treated with immunosuppressants (e.g., to prevent transplant rejection).[(56)]

Monitoring/Sampling. There are no methods readily available to monitor the air for the contagious or virulent infectious agents. The levels that apparently pose a significant risk for infection are very low, and any monitoring procedure would have to be of exquisite sensitivity; detection of the agent would probably follow initiation of an outbreak. Research methods for the detection of specific infectious agents in air include high-volume air sample collection with sample analysis by cell or tissue culture or the detection of specific nucleotide sequences using the polymerase chain reaction.

The opportunistic environmental agents can be sampled using cultural techniques with subsequent identification of the organisms using either traditional morphological and physiological criteria, or using immunological or genetic tracers.

Monitoring is usually not appropriate unless there is some reason to suspect the presence of the organism, or unless an especially sensitive population is likely to be exposed.

Levels Measured. Except for *Aspergillus fumigatus,* levels of infectious agents in indoor environments have not been reported. For the opportunistic infections the host risk factors are far more important than level of exposure. Thus, a relatively normal person will not become infected in the presence of millions of *A. fumigatus* spores, whereas for a severely immunocompromised person, a single spore might be sufficient.

Standards. Standards for levels of *Legionella pneumophila* have been proposed by several groups. One guideline suggests that recovery of culturable *Legionella* in concentrations exceeding 1000/mL of cooling tower water should prompt immediate remediation.[(57)] Lower levels are more difficult to interpret in the absence of reported cases. Miller and Kenepp[(58)] suggest that a dominance of *Legionella* (i.e., >50%) in a water sample should prompt concern.

Control. Control of contagious disease involves isolating either infected or uninfected people by means of respiratory protection. Exposure to aerosols from environmental sources (where the reservoir is known) can also be prevented using respiratory protection. For man-made reservoirs, biocides can be used to kill or at least limit the numbers of residual organisms following thorough cleaning of the reservoir. This approach is well established for *Legionella*.[59] *Cryptococcus*-infested bird droppings have been disinfected using formaldehyde before removal.[60]

Allergens

Nature and Sources. In general, any protein, glycoprotein, or carbohydrate with a molecular weight in excess of 10,000 daltons (unit of molecular weight) can act as an allergen.[61] Most, however, are glycoproteins. A few very highly reactive chemicals can bind to serum proteins following exposure and stimulate antibody production. The major sources for allergens in indoor air are arthropods (cockroaches, dust mites), mammals (cats, dogs, rodents), birds, fungi, and actinomycetes. Accumulated pollen grains may provide a source for indoor allergens. Latex may be abundant in some indoor environments. The arthropod allergens are borne on fecal material or body parts. The mammalian and avian allergens are borne on skin scales, particles of dried urine, saliva, or serum. The fungal allergens are often enzymes that are contained in fungal spores or produced as the spores germinate.[62] All of these organisms occupy indoor environments and shed allergen-containing particles directly into the air or into dust that is subsequently aerosolized. Fungal spores are also abundant in outdoor air, as are pollen grains. As mentioned above, latex allergens are carried on latex gloves and other products, and become airborne on the powder used to facilitate the products' use.

Reservoirs for allergens in the indoor environment include dust (arthropods, fungi, vertebrate dander), water reservoirs from which aerosols are released (e.g., humidifiers), and surface growth of organisms that readily release spores or other kinds of aerosols (e.g., dry-spored fungi).

Health Effects. The hypersensitivity diseases are caused by specific responses of the immune system and require a two-step exposure process: initial exposures that stimulate the immune system, and a second set of exposures that result in mediator release and symptoms. Levels of allergens that induce each of these steps may differ.[14]

Airborne allergens cause diseases such as hypersensitivity pneumonitis, allergic rhinitis, and allergic asthma. Symptoms of hypersensitivity pneumonitis include fever, chills, shortness of breath, malaise, and cough. The disease mimics influenza initially, then pneumonia, but symptoms resolve with cessation of exposure. Long-term exposure can result in permanent lung damage. Attack rates in nonindustrial, nonfarming situations are usually low (around 7%), although much higher attack rates have been reported in some situations.

Symptoms of allergic rhinitis include runny, itchy nose and eyes and sinus congestion, while those of allergic asthma are wheezing and chest tightness resulting from bronchiolar constriction. It is often difficult to separate these hypersensitivity conditions from vasomotor rhinitis and bronchoconstriction related to chemical or cold air exposures.[61] Allergic rhinitis and allergic asthma are controlled by a particular genotype borne by about 20–40% of the U.S. population. Approximately 6% of the U.S. population is sensitive to the nearly ubiquitous house dust mite, 2% to various animal danders, and less than 1% to various microbial allergens.[63]

Monitoring/Sampling. Samples for measurement of allergens are collected either as bulk dust or directly from air. Dust sampling is preferentially used for the arthropod and mammalian allergens, while microorganisms are usually assessed using air sampling. Dust is considered to represent cumulative exposure over time, while air sampling represents only levels at the time of sampling. This distinction is particularly important for the allergens (including dust mite) that are borne on relatively large particles that remain in the air for only short periods after disturbance of reservoirs. The relevance of dust as a surrogate for airborne exposure remains to be clearly documented. At least for the fungi, dust sampling poorly represents simultaneously collected air samples.[64] For both bacteria and fungi, the most reliable method for sampling appears to be visual observation of potential reservoirs, followed by judicious bulk sampling.

Analysis of samples collected for allergen measurement is best performed using specific immunoassays. Such assays are available for major allergens derived from dust mites, cockroaches, cats, and dogs, and others are available as research tools.

Immunoassays are not generally available for the allergen-bearing fungi and bacteria that are common indoors. For the bacteria, culture followed by traditional methods for identification is the only current option. This is a serious shortcoming because many cells may not be culturable, and culturability is not a prerequisite for allergenicity. For the fungi, culture is also the most commonly used analytical method and the only method that ensures that species of fungi can be identified. This step is essential if patient treatment is to include immunotherapy. For total fungi, spore counts overcome the culturability problem, and a few fungi can be identified from spore characteristics alone. A combination of culture and microscopy can provide reasonably accurate quantitation of fungal aerosols.[65,66]

Several analytical chemistry-based methods have been proposed for assessing total fungal load in indoor environments. Measurement of β(1-3)d-glucans or of ergosterol (the principal fungal sterol) have been used.[67] Both correlate well with culturable fungal counts, but provide no information on the kinds of fungi present.

Levels Measured. Levels of the arthropod and mammalian allergens in residences range from essentially undetectable to in excess of 10,000 μg/g of dust. Levels below 2 μg/g are considered low, between 2 and 10 μg/g moderate, and in excess of 10 μg/g sufficiently high that most sensitized people will experience symptoms.

Levels of allergen-bearing bacteria (i.e., actinomycetes) in buildings are usually very low (<1/m^3 of air), so that recovery of any of these organisms should stimulate a search for potential sources. Fungal levels in indoor environments are strongly related to the kind of ventilation, levels outdoors, reservoirs indoors, and disturbance of the reservoirs. Levels in clean, mechanically ventilated office buildings are usually <500 colony forming units (CFU)/m^3, and those in naturally ventilated clean houses <5000 CFU/m^3 when air samples are collected under quiescent conditions. The dominance in indoor air of specific fungi that are not common in outdoor air may signal problems at lower levels.[68]

Standards. There are no published standards for allergens in air or in reservoirs.

Control. Cockroach allergen exposure is best achieved by eradicating cockroaches and by sealing the indoor environment to prevent their reinfestation. Dust mites can be controlled by keeping humidity within reservoirs consistently below 60%, by limiting the use of carpeting and upholstered furniture in humid environments, by encasing mattresses and other soft furnishings in plastic, and by washing bedding in hot water. Mammalian and avian allergens are best controlled by keeping their sources out of the indoor environment. Exposure to allergen-bearing bacteria (actinomycetes) usually results from aerosolization of spores from heated reservoirs. Keeping water out of ventilation systems, limiting the use of water spray humidification, and venting clothes dryers outdoors all will prevent their amplification. Fungi grow primarily on material that is water-soaked or on which condensation is consistently present. Eliminating these water sources will essentially prevent fungal amplification.

Biological Toxins

Many different kinds of organisms produce potent toxins. Notable among these are endotoxins (produced by some kinds of bacteria), -glucans (part of the cell wall of most fungi), and the mycotoxins (secondary metabolites produced by many fungi).

Endotoxin—Nature and Sources. Endotoxin is a lipopolysaccharide that forms the outer cell wall of gram-negative bacteria. The lipid portion of the molecule is responsible for its toxicity.[69] Endotoxin is present in low levels (<1 ng/m^3) in ambient air and is a consistent component of house dust.

Endotoxin—Health Effects. Endotoxin was first recognized as an air quality problem in the cotton processing industry, where it plays a role in diseases such as byssinosis. Exposure to endotoxin can result in acute bronchoconstriction, shortness of breath, cough, fever, and nausea, depending on the level and duration of exposure.[70–73] Disease-causing exposures (>50 ng/m^3) usually occur in association with recirculating water systems that produce aerosols and in agricultural and industrial environments where organic material is handled.[74] Some evidence suggests that endotoxin may also play a role in SBS.[75] Endotoxin also acts as a stimulant to the immune system and may play a role in sensitization to some allergens. There is also some evidence of lowered lung cancer rates in people routinely exposed to high levels of endotoxin.[76,77]

Endotoxin—Monitoring, Levels. Endotoxin can be measured using the *Limulus* amoebocyte bioassay, or by using gas chromatography/mass spectrometry (GC/MS) methods. The *Limulus* method, which is most commonly used, is quite sensitive and measures biologically active endotoxin rather than actual amounts of lipopolysaccharide. GC/MS methods document total lipopolysaccharide and provide information on component structures, but are less sensitive than the *Limulus* method. Both bulk and air samples can be evaluated using these assays. Air samples (minimum volume of about 800 L) are usually collected on filter media. Milton[69] proposed a standard method for sampling and analysis of airborne endotoxin that takes into account the variables associated with its ubiquity in the environment and its ability to adhere firmly to many sample collection media.

Levels of endotoxin in ambient air are usually <1 ng/m^3. Levels as high as 7 μg/m^3 have been measured in cotton mills, industrial settings where recirculating wash water is used, and in agricultural environments. In residential dust, median levels of 1.1 ng/mg have been reported.[69] Reported residential air levels have ranged from undetectable to 18 ng/m^3.[78]

Endotoxin—Standards, Control. There is currently no standard for endotoxin in any environment. Control measures specific to endotoxin usually involve removal of source water or, where this is impossible, use of unrecirculated potable water. In some environments (e.g., machining shops where water-based lubricants/coolants are used) local exhaust ventilation or respiratory protection may be the only alternatives.

β(1-3)d-glucans—Nature and Sources. β(1-3)d-glucans form the major portion of most fungal cell walls and may be chemically bound to chitin (and hence insoluble), or may form a soluble matrix in which the chitin fibrils are embedded.[79] Most fungi that are common in indoor environments contain β(1-3)-glucans. Exceptions are the common black bread mold (*Rhizopus* sp.) and the mirror yeasts (*Sporobolomyces*).

β(1-3)-glucans—Health Effects. β(1-3)d-glucans act as immunostimulants in a similar manner to endotoxin and may be involved in the development of hypersensitivity pneumonitis.[80,81] Their antitumor activities are well-recognized.[83] Soluble glucans appear to have an effect on the lung similar to that of endotoxin.[82]

β(1-3)-glucans—Monitoring, Levels. The β(1-3)d-glucans act similarly to endotoxin in some *Limulus* assays. However, enzyme-linked immunosorbent assays have been recently developed that are sensitive and accurate.[83]

β(1-3)-glucans—Control. Control measures separate from those discussed for the fungi have not been proposed.

Mycotoxins—Nature and Sources. Mycotoxins are secondary products of fungal metabolism.[84] The chemical structures of mycotoxins are quite diverse, ranging from that of moniliformin ($C_4H_2O_3$) to complex polypeptides with molecular weights over 2000. Although only a few have been studied, most fungi probably produce these secondary metabolites, the kinds and amounts depending on the fungal strain and the food source being metabolized. Mycotoxins probably play a role in competition, and their production may depend on the presence of competing organisms.

Exposure to mycotoxins occurs when moldy food is eaten, when contaminated materials are handled so that skin contact occurs, or when spores, mycelial fragments, or materials supporting growth are disturbed and become airborne.

Mycotoxins—Health Effects. The mycotoxins of primary concern with respect to indoor air quality are the potent cytotoxins that cause cell disruption and interfere with essential cellular processes. Some are carcinogenic (e.g., the aflatoxins produced by *Aspergillus flavus* and *A. parasiticus*). Others cause damage to the immune system or specific organs. The macrocyclic trichothecene toxins (produced by *Stachybotrys atra*, *Mycrothecium verrucaria*, and other fungi) fall into this latter class.

Toxicological data on mycotoxins is nearly all derived from ingestion exposures. For some trichothecene toxins, doses well below 1 mg/kg body weight are lethal for experimental animals by ingestion. Creasia et al. report that inhalation exposures are 2 to 20 times more toxic than intraperitoneal injection.[85] Several mycotoxins have been shown to interfere with rat alveolar macrophage function.[86]

Mycotoxins—Monitoring/Sampling. Mycotoxins in indoor environments are assessed using either air or reservoir sampling with cultural or microscopic analysis for known toxigenic fungi. Rarely, analysis for specific mycotoxins is performed. Mycotoxins probably become airborne on particle sizes ranging from approximately 2–10 μm in smallest diameter. Methods for collecting air samples depend on the kind of analysis to be performed. For cultural analysis, culture plate impactors are generally used. It should be noted that this kind of analysis detects only viable (culturable) propagules and may lead to serious underestimates in the potential for mycotoxin exposure. Microslide impactors can be used to sample for fungi such as *Stachybotrys atra* with spores that are readily identifiable microscopically. These have the advantage of reasonable sensitivity with no reliance on viability. Both cultural and slide impactors are generally used as short-term grab samplers and provide only a limited picture of the potential for exposure over time.

For most of the mycotoxins, methods for chemical analysis require more sample volume than is generally available from most air samples, except in the most contaminated of environments. Immunoaffinity methods are available for aflatoxin, while analysis of the trichothecenes depends on either GC/MS or high performance liquid chromatography methods.

Mycotoxins—Standards and Levels Measured. Levels have not been reported for any mycotoxin in indoor air, and risk assessment is usually based on the presence of potentially toxigenic culturable fungi. No standards exist for mycotoxins in indoor air.

Mycotoxins—Control. Approaches for the control of exposure to mycotoxins rely on control of source fungi.

Microbial VOCs. VOCs are produced by all microorganisms. A very wide range of different kinds of compounds are produced, ranging from ethanol to 8 and 9 carbon aldehydes and ketones. Odors of mildew and decay are caused by these compounds. The role of microbially produced volatiles in human health is unknown and merits investigation. It is probable that microbial contamination in indoor environments with inadequate ventilation will at the very least exacerbate irritant symptoms and contribute discomforting odors.[87,88]

Microbial VOCs—Sampling. Microbial VOCs can be monitored in the same way as other VOCs with the provision that the methods used must be sensitive, and samples must usually be collected relatively near sources. Samples have been collected using adsorbents and canisters with analysis using GC/MS methods.

Nonbiological Agents

CO_2

Nature and Sources. Carbon dioxide is an odorless, colorless gas that is formed whenever carbon-containing substances are burned. In nonindustrial occupational settings the primary sources are human respiration and tobacco smoke.

Health Effects. CO_2 is an asphyxiant in that it replaces oxygen, and at high concentrations (>30,000 ppm) it may cause headaches, loss of judgment, dizziness, drowsiness, and rapid breathing.[18] At levels above 800 ppm, air quality complaints begin to increase.[89] It should be remembered that in these cases it is not the CO_2 causing the complaints, but other pollutants accumulating in the environment, probably due to inadequate ventilation. The presence of high CO_2 levels in situations with high measured ventilation rates may be indicative of inadequate mixing in the occupied space.

Monitoring/Sampling. While not usually considered a pollutant, CO_2 is easily measured and is often used as a surrogate for measurement of ventilation rates in indoor air quality investigations. For example, per person ventilation rate can be measured using the following equation:

$$Q_{person} = S(t)/(C_{eq} \ C_o) \qquad (1)$$

where s = emission rate of CO_2 (m^3/hr); C_{eq} = equilibrium level of CO_2; and C_o = CO_2 level in outdoor air. Note that this method works only for spaces where there are enough people to serve as a source, and when they have been present long enough that CO_2 levels have reached equilibrium.

Monitoring for inappropriate levels of CO_2 in indoor air is usually done with colorimetric detector tubes, portable gas chromatographs, or infrared spectroscopy detectors. Detection limits are approximately one order of magnitude below the ambient outdoor level (300 ppm). Air samples may also be collected in bags for subsequent analysis.

Standards and Levels Measured. The U.S. occupational health standard for CO_2 is 5000 ppm, as is the American Conference of Governmental Industrial Hygienists (ACGIH) threshold limit value (TLV®). Japan has specified an indoor standard of 1000 ppm, and a level of 650 ppm is recommended for indoor air in Massachusetts. Although 1000 ppm is calculated to represent a ventilation rate of approximately 15 ft^3/min outdoor air per person in commercial buildings, where the ventilation rate is adequate to dilute human-source air pollutants, CO_2 levels are usually below 800 ppm.[90]

Control. Control of CO_2 in the nonindustrial environment depends almost entirely on the use of fresh air ventilation.

CO

Nature and Sources. CO is an odorless, tasteless, colorless gas. CO exposure usually occurs in a combination of combustion products, many of which have distinctive odors. The most common sources for CO in nonindustrial environments include automobile exhaust from indoor garages or inappropriately placed air intakes, smoking, and in residential environments, unvented combustion appliances.[91]

Health Effects. CO is an asphyxiant that converts hemoglobin to carboxyhemoglobin, thus decreasing the amount of oxygen transported to tissues and resulting in tissue hypoxia. Exposure results in fatigue, shortness of breath, headache, nausea, and at high levels, death.[18] Carboxyhemoglobin levels above 4–5% can exacerbate symptoms of cardiovascular disease, and extreme altitudes may exacerbate the detrimental effects of CO on persons with this disease. Health effects of low-level CO exposure resulting in less than 3% bound hemoglobin are not well established, but probably include effects on the heart and brain. In the general nonsmoking population when CO exposure levels do not exceed 25 ppm, carboxyhemoglobin levels are generally in the range of 0.3–0.7%, while for smokers 2–3% carboxyhemoglobin is considered normal.[13]

Monitoring/Sampling. Active and passive electrochemical CO monitors are available that are sensitive to levels as low as 1 to 5 ppm. A personal exposure monitor is available that monitors and records CO levels continuously, but except in industrial situations, the use of direct reading instruments is usually adequate.[92]

Standards. The current National Ambient Air Quality Standard for CO is 9 ppm maximum for 8-hour average exposure, or 35 ppm maximum for 1-hour average exposure.[93] These levels are intended to provide a margin of safety for people with cardiovascular disease. Mean CO concentrations in nonindustrial indoor environments (excluding garages and other automobile service facilities) range from 1.2–4.2 ppm. The ACGIH time-weighted average (TWA) TLV for CO is 25 ppm.

Control. Control of CO depends on removing sources in the indoor environment and ensuring that outdoor air provided for ventilation meets the average 9 ppm National Ambient Air Quality Standard. In occupational environments, control of indoor sources includes limiting smoking, proper ventilation of indoor garages, and placement of air intakes such that exhaust fumes from loading docks, garages, and adjacent traffic are not entrained in building air. Most of these approaches also apply to the residential environment. In addition, unvented appliances that burn fossil fuel should not be used.

Nitrogen and Sulfur Oxides (NO_2, SO_2)

Nature and Sources. NO_2 is associated with emissions from automobiles, diesel trucks, electrical power-generating stations, and industrial processes. Indoor sources include emissions from unvented appliances that use fossil fuels, including gas stoves and space heaters. These occur primarily in residences and are uncommon in the occupational or commercial environment. Cigarette smoke is also a source of NO_2.[94] While SO_2 is an important pollutant in outdoor air, levels indoors rarely exceed 30% of those outdoors unless an unvented kerosene burner is used with an extremely low grade of fuel.[95]

Health Effects. NO_2 is rarely associated with acute building-related disease except in cases where entrainment from outdoor air occurs and in conjunction with environmental tobacco smoke exposure. Exposure to NO_2 at high concentrations (200 ppm) is associated with acute pulmonary edema and death.[96] Exposure to levels encountered in nonindustrial interiors can lead to a decrease in pulmonary function in asthmatics and may cause effects on lung defenses, biochemistry, function, and structure. Chronic exposure is considered a significant cause for concern, especially for children.[13] In addition, NO_2 is likely to have additive or synergistic effects with other indoor pollutants.[97,98]

SO_2 is an irritant that causes bronchoconstriction in asthmatics, especially during exercise. Levels that cause these effects are in the range of 0.25–0.5 ppm. SO_2 is derived from combustion of sulfur-containing fossil fuels, which are not commonly used indoors. Most indoor exposure results from misuse of equipment, including inadequate venting of oil-burning combustion appliances.[99]

Monitoring/Sampling. Passive monitors using a triethanolamine sorbent are available, as are colorimetric and chemiluminescent real-time detectors. Grab-sample detector tubes are available for measurement of SO_2. It should be emphasized that unless an apparent source for NO_2 or SO_2 is present, it is rarely helpful to monitor for these pollutants in most indoor air quality investigations.

Standards and Levels Measured. NO_2 is regulated in ambient air at 50 ppb, and ACGIH recommends a TWA–TLV of 3 ppm. The U.S. Environmental Protection Agency (EPA) is required to issue a public alert at ambient NO_2 levels of 0.6 ppm, and 1.6 ppm is considered a public emergency. A concentration of 2.0 ppm NO_2 causes significant harm and widespread health effects in the general population. Average outdoor levels range from 0.001 ppm in rural areas to 0.03 ppm in urban air.[94] Indoor levels are almost always below those outdoors unless a significant source is present. In homes with gas-fired stoves, ovens, or heaters, NO_2 levels as high as 0.55 ppm have been measured.

The TLV for SO_2 is 2.0 ppm. Levels reported in houses range from 0.03–0.65 ppm depending on the type(s) of combustion sources present.

Control. Control of NO_2 and SO_2 depends primarily on removing sources, which in the nonindustrial occupational environment involves limiting smoking (NO_2) and limiting the use of kerosene space heaters.

Environmental Tobacco Smoke (ETS)

Nature and Sources. ETS is one of the most frequent causes of complaints in many occupational settings. Burning tobacco releases a complex mixture of chemicals and particles into the air, a mixture that changes as it ages.[13] In addition to CO and particles, nitrogen oxides, CO_2, hydrogen cyanide, and formaldehyde are produced along with a wide variety of other gases and volatile organic compounds.[90] Although always diluted by room air, sidestream smoke contains the same kinds of toxic and carcinogenic substances as inhaled smoke.[100]

Health Effects. Active cigarette smoking has been clearly established as a major and preventable cause of life threatening diseases including heart disease and lung cancer. In addition, exposure to ETS clearly increases risk of lung cancer.[101,102] Exposure of young children to ETS from maternal smoking increases rates of chronic ear infections and middle-ear effusions, respiratory tract infections, and leads to decrements in pulmonary function and predisposes susceptible children to the development of asthma.[11,13] Evidence is accumulating that nonsmoking pregnant women exposed to ETS daily for several hours are at increased risk of producing low-birth-weight babies. Exposure to ETS may also increase heart rate, blood pressure response, and oxygen consumption, especially in nonsmokers, and result in eye and throat irritation, headache, rhinitis, coughing, and bronchoconstriction in asthmatics.[103]

Odor is associated with volatile components of ETS. The relationship between ETS concentration and discomfort of study panels has been well established.[104,105] Also, given the extremely emotional situations that can be generated where smokers and nonsmokers must coexist, it is logical to assume that a variety of psychological symptoms may be traced to ETS, especially in nonsmokers.

Monitoring/Sampling. Because of the extreme complexity of ETS, a tracer must be used for environmental monitoring. A suitable tracer should be unique or nearly unique to tobacco smoke, be present in sufficient quantities to be detected at low smoking rates, have similar emission rates for a variety of tobacco products, and have a fairly consistent ratio to the constituents of interest (e.g., the carcinogenic constituents) over a range of environmental conditions.[13] No single measure meets all of these criteria, and none is recognized as representing ETS exposure. Currently the most commonly used indicator is respirable particulate material. Measures of human exposure ideally use analyses of physiological fluids of exposed individuals rather than relying on indoor air assessment. Nicotine and cotinine are unique to tobacco products and can be measured in saliva, blood, or urine.[100]

Levels Measured. Concentrations of ETS components in air depend on frequency and amount of smoking, outdoor air ventilation rates, presence and efficiency of air cleaning devices, and patterns of air distribution and recirculation. ETS aerosols adsorb readily to surfaces and can discharge volatile constituents long after smokers have left a room.[104] In the presence of ETS, CO levels have been measured from 2 to 35 ppm, and particles in the range of 10–1000 $\mu g/m^3$ have been found. The mean impact of smoking a pack of cigarettes in a fully air-conditioned building with recirculation is to increase particle levels by approximately 41 $\mu g/m^3$. Urinary nicotine measurements in nonsmokers exposed to environmental tobacco smoke have ranged from <10 to >80 µg/hour of exposure.[100]

Standards. Other than outright bans on smoking, there are no standards that delineate acceptable levels of ETS in indoor air. There are standards that cover some specific components not unique to ETS.

Control. The best control option for preventing exposure to ETS is to completely ban smoking, and there has been a trend toward such bans in public buildings. Short of a complete ban, smoking can be confined to areas that are separately ventilated so that recirculation does not spread ETS components into the general environment. Properly designed ventilation systems can lessen the impact of ETS on building occupants, but such design carries a considerable energy penalty, since the exhausted air is generally discharged to the outside. Ventilation rates must be high, and in reality, it is impractical to provide ventilation rates that will reduce odor in an average smoking environment to a level that will satisfy the majority of nonsmokers.[103] Filtration and/or electrostatic precipitation can reduce respirable particle levels to acceptable (nonirritant) levels.[18]

Other Nonbiological Particles

Nature and Sources. Airborne particulate matter includes a broad range of substances that are small enough to remain suspended in air. Particles designated "respirable" are less than 10 μm in diameter and fall into two general categories: larger than 2–3 μm, and smaller than 2–3 μm.[13] In nonindustrial occupational environments, the principal sources of fine particles are cigarette smoke and possibly aerosols from spray air fresheners or cleaning materials.[106] Larger particle aerosols include carpet fragments, dirt carried in from outdoors, and most of the biological particle fraction of the air. Outdoor air could also be a significant source for fine particles in naturally ventilated buildings.

Health Effects. In the indoor environment, health effects related to particle exposures are those discussed for ETS, radon, and biological particles. In addition, respirable particles can be respiratory irritants, especially for asthmatics.[18] Recently, exposure to fine particles in outdoor air at levels below the current National Ambient Air Quality Standard has been associated epidemiologically with acute respiratory distress.

Monitoring/Sampling. Because of the complexity of the respirable particulate component in indoor air, monitors that yield useful data with respect to potential health effects are not available. Commercially available inhaled particle monitors have poor sensitivities and unknown validity at the low levels encountered in nonindustrial environments. In addition, mass concentration is of little help unless something is known about the composition of the aerosol. Obviously, very low levels of a particulate toxin might be greater cause for concern than quite high levels of carpet or other fabric fragments.

Standards. The National Ambient Air Quality Standard for particles is 150 μg/m^3 (24 hour) or 50 μg/m^3 (annual average). The Japanese enforce a similar standard (restricted to particles <3.5 μm) for office building air quality.

Control. Control of fine-particle aerosols in the indoor environment is best achieved by banning smoking or restricting smoking to smoking lounges with dedicated exhaust ventilation.

Fibers

Nature and Sources. Asbestos and fiber glass (synthetic vitreous fibers) are two fibers that have caused serious concern in the nonindustrial environment. Asbestos is covered elsewhere and will not be discussed here. Fiber glass (or fibrous glass) generally includes particles composed of glassy, amorphous material with a length/width ratio greater than 3. Fiber glass is used in textiles and fabrics; as reinforcement in plastics, rubber, and paper; and as thermal insulation material. The insulating properties of fiber glass vary inversely with the particle diameter, so that while most fiber glass fibers are in the range of 4–9 μm in diameter, those with the best insulating properties are 1 μm and less. Fiber glass is used for both thermal insulation and sound insulation in walls, ceilings, and in ventilation system ductwork and is ubiquitous in the work environment.[107]

Health Effects. The National Academy of Sciences Committee on Indoor Air Quality states that no epidemiologic investigation has demonstrated substantial health hazards related to fibrous materials other than asbestos. Lung tumors have been induced in animals by implantation of very small-diameter glass fibers of a sort not common in buildings.[89] Aside from possible chronic health effects, fiber glass has been reported to cause epidemics of rash and itching in office workers. Itching of exposed skin areas, alleviated by showering; associated rashes; and respiratory and eye irritation may all be present. Contact lens wearers may find fibers embedded in soft contact lenses.

Monitoring/Sampling. Air monitoring for fibers is done using filter cassettes with microscopic evaluation as for asbestos (see Chapter 12). However, fiberglass is rarely detected in air in nonindustrial environments. Wipe samples have been a more effective means for documenting exposure.

Standards and Levels Measured. In epidemics of skin rash that have been traced to fiber glass, measures of total suspended particles (presumably including any fiber glass) have been below occupational standards (TLV–TWA 10 mg/m^3) and even below detectable limits. However, the TWA is under review, and proposed changes for

1996 would limit exposure to 1 fiber/cc, or 5 mg/m^3. In most cases where air measurements have been attempted, fiber glass has been undetectable. Wipe samples of surfaces have confirmed the presence of fiber glass, and in most cases sources were discernible.[9]

Control. Control of exposure to airborne fiber glass involves preventing its introduction into the air. This is accomplished by ensuring that fiber glass subject to abrasion is maintained in good condition, and by using care in fiber glass installation and removal operations.

Ozone

Nature and Sources. Ozone is a colorless gas with a characteristic odor that is produced in ambient air during the photochemical oxidation of combustion products such as the nitrogen oxides and hydrocarbons. It can also result from the operation of electrical motors, photocopy machines, and electrostatic air cleaners in occupational environments.

Health Effects. Symptoms associated with exposure to ozone include cough, upper airway irritation, tickle in the throat, chest discomfort including substernal pain or soreness, difficulty or pain in taking a deep breath, shortness of breath, wheezing, headache, fatigue, nasal congestion, and eye irritation. Symptoms usually disappear within 2–4 hours of cessation of exposure. For sedentary people (e.g., office workers) ozone concentrations in excess of 0.1 ppm are likely to result in symptoms such as dryness of upper respiratory tract and throat and nose irritation, and 0.5 ppm will induce decrements in pulmonary function. Symptoms occur at lower ozone levels in exercising individuals.[108]

Monitoring/Sampling. The characteristic odor of ozone is detectable well below the level at which symptoms are induced. Therefore, in the absence of this odor, BRS are unlikely to be due to ozone exposure. Measurement of ozone can be done using chemiluminescence, ultraviolet absorption, or colorimetry. Colorimetric methods actually measure total oxidants, and ultraviolet absorption may be susceptible to interference by organics that absorb the same wavelength.[109]

Standards and Levels Measured. The short-term exposure standard for ozone in ambient air is 235 µg/m^3 or 0.12 ppm (1 hour average), and a ceiling limit of 0.1 ppm has been established by ACGIH. The Occupational Safety and Health Administration permissible exposure limit[110] is 0.1 ppm averaged over 8 hours. Domestic and nonindustrial exposures are usually in the 0.02- to 0.03-ppm range. The odor of ozone can be detected by some people at levels as low as 0.001 ppm, and by most at 0.02 ppm.

Control. Control of ozone in nonindustrial occupational environments depends on adequate ventilation to dilute internal sources and proper maintenance of electrical equipment and other potential sources.

Formaldehyde

Nature and Sources. Formaldehyde (HCHO) is a VOC that, due to its extensive use in a variety of manufacturing processes, is ubiquitous in the modern environment. HCHO is used in bonding/laminating agents, adhesives, paper and textile products, and in foam insulation. It is also used in cosmetics and toiletries as a preservative. In general, products containing large amounts of HCHO are more common in domestic environments (e.g., modular or mobile homes) than in most occupational environments. However, especially in new buildings, HCHO off-gassing from building materials and textiles can be a significant problem.[111] Renovations involving replacement of carpeting can also temporarily raise HCHO levels into the irritant range. While HCHO is a component of sidestream cigarette smoke, levels measured in interiors where smoking occurs but other sources are not present are generally low.

Health Effects. HCHO in excess of 1–3 ppm will cause mucous membrane irritation. The presence of urea formaldehyde foam insulation (in residential insulation) has produced complaints of headache, eye irritation, upper and lower respiratory tract irritation, sore throat, malaise, anorexia, insomnia, and asthmatic reactions. It is not clear whether these symptoms were related to HCHO exposure or to some other toxic substance released from the insulation.[10] HCHO has not been shown to be a respiratory allergen. Chronic, direct skin exposure may induce an immunologically mediated reaction, as can intravenous exposure (e.g., during dialysis). HCHO is a suspected carcinogen, and chronic exposure to levels above 1 ppm should be cause for concern.[7,112]

Monitoring/Sampling. HCHO is generally measured by chemisorption onto a coated solid sorbent, with subsequent colorimetric or fluorimetric analysis after sulfuric acid desorption. Direct reading instruments are available and are cost-effective for high use areas such as medical and veterinary schools.

Standards and Levels Measured. The Occupational Safety and Health Administration 8-hour workplace limit is 1 ppm, while the ACGIH short-term exposure limit is 0.3 ppm. The ASHRAE standard for providing a comfortable indoor environment, and the Swedish limit for new homes, is 0.1 ppm. Levels of HCHO in the range of 0.01–0.03 ppm have been measured in problem office buildings. Mean levels in residences have ranged from 0.078–0.4.[(112)]

Control. Control of HCHO depends on limiting indoor sources, including restriction of smoking and of HCHO-containing building materials and textiles, and provision of adequate ventilation. "Baking-out" of new buildings as described below for other VOCs may also be effective (at least temporarily) in reducing HCHO levels.

Other VOCs

Nature and Sources. Potential sources of VOCs in occupational environments include human bioeffluents, personal care products, cleaning materials, paints, lacquers, varnishes, pesticides, pressed wood products, and insulation, to name a few.[(4)] Microorganisms also release volatiles that are responsible for musty, moldy odors.

Health Effects. A few of the VOCs commonly found in indoor environments are clearly carcinogens (e.g., benzene), although evidence for carcinogenicity is extrapolated from high-level exposures in industrial environments.[(113)] Others (e.g., carbon tetrachloride, chloroform) have produced cancer in laboratory animals, but no direct evidence exists for human effects.[(114)] Evidence is increasing that accumulation of mixtures of VOCs may play a major role in SBS, and in fact, exposure of volunteers in a chamber to mixtures of VOCs has replicated symptoms of SBS.[(11)] Most VOCs are lipid soluble, readily cross the blood-brain barrier, and are easily absorbed through the lungs. Most are neurotoxic and, in levels in excess of occupationally acceptable limits, may cause central nervous system depression, vertigo, visual disorders, and occasionally tremor, fatigue, anorexia, and weakness.[(89,115)] Potential genotoxic effects are still under investigation. Effects of low-level exposures to VOC mixtures over long periods of time are unknown.

Monitoring/Sampling. Tenax®-GC sorbent tubes are generally used to monitor for VOCs. Summa™ canisters may also be an option for VOC sampling. Methods for measuring VOCs have been recently reviewed by Hodgson.[(116)]

Levels Measured. VOC levels in buildings are usually reported either as totals derived from summing all the individual peaks on a chromatogram or as the sum of all the identified peaks.[(117)] Values reported from both residential and nonresidential buildings generally range from <0.15 to 1 mg/m^3.[(118)] Average levels in indoor air for specific VOCs have been reported.[(116)] For example, median levels for carbon tetrachloride, chloroform, and benzene have been reported as 0.6, 0.5, and 10 μg/m^3, respectively.

Standards. For outdoor air 0.16 mg/m^3 of unreactive hydrocarbons is considered unacceptable. Although this standard was designed to prevent irritation from photooxidation products, it may be considered an upper limit for acceptable indoor concentrations of organic pollutants. According to Molhave,[(119)] concentrations below 200 μg/m^3 are within the comfort range, concentrations between 3 and 25 μg/m^3 may cause discomfort, and concentrations above 25 μg/m^3 may be toxic.

Control. Three approaches are of use in reducing indoor VOC exposure: dilution ventilation, air cleaning, and source prevention or removal. Dilution ventilation is often successful in ameliorating some of the symptoms of SBS, providing strongly emitting VOC sources are not present, and providing the ventilation system itself does not constitute a source. Increasing ventilation rates in the presence of strongly emitting sources may increase emission rates by enhancing vaporization. The use of activated charcoal and catalytic oxidation to scrub VOCs from the air is used in submarines, but is not known to be effective in situations where levels are already low. Controlling sources of VOCs involves selecting low emission materials where available, treating high emission sources before installation, proper storage of organic chemicals, and baking out new buildings prior to occupancy.[(13)] ASHRAE suggests heating of an unoccupied building to 120°F for a "short period to drive off gases."[(18)] This method may be effective in some circumstances but has disadvantages that may include damage to building materials and a serious energy penalty.[(120)]

Radon

Nature and Sources. Radon is a noble gas that emits alpha particles with a half-life of 3.8 days. It is a decay product of radium 226, which is a decay product of the uranium 238 series. Radon progeny, two of which emit alpha particles, have half-lives of less than 30 min. Radon equilibrates rapidly with its decay products so that without a replenishing source, significant concentrations cannot be maintained.(13) Prior to about 1983, radon was considered a problem only in such high risk areas as contaminated areas near inactive uranium mines.(12) On the other hand, uranium 238, and hence radium 226, is a constituent of nearly all soils and rocks, although there is considerable geographic variation in concentration. Groundwater in contact with radium-bearing granite can be a source of radon exposure, as can building materials made of such granite, or when contaminated with uranium or radium mill tailings. Elevated radon levels are most likely in below-grade spaces, especially in granitic areas.(121) Residential exposure is almost always higher than that in multistory buildings,(122) and the highest residential levels are in basements, where levels have been measured that carry the same risk of lung cancer as that ascribed to heavy smoking. Nevertheless, commercial buildings supplied with contaminated groundwater or built of radium-containing building materials can also present some risk of radon exposure.(121)

Health Effects. The major risks from exposure to radon decay products include lung cancer, some nasal cancers, and stomach cancer from ingestion of radon-containing water. It is estimated that 5000 to 20,000 deaths per year are due to exposure to radon.(13,123,124) Lung cancer risk is increased in smokers. Radon decay products are present in the environment attached to particles as well as in a free state. Unattached radon decay products are small enough to reach the unciliated bronchiolar or alveolar regions of the lung where, due to long residence times, they cause damage by killing epithelial cells or transforming them into cancer precursors. Particles with attached radon progeny can lodge in the tracheobronchial tree emitting alpha particles that cause similar damage at this site.(125,126) Radon is not an irritant or acutely toxic and is therefore not of concern in acute epidemics of SBS or building-related disease.

Monitoring/Sampling. EPA has published guidelines for the use of continuous monitors (scintillation cell), 3- to 5-day integrating charcoal canisters, alpha-track detectors, and grab-sample scintillation cells that provide reasonable accuracy and precision. A passive radon badge is available that is suitable for monitoring most indoor microenvironments, but must be exposed for several months. The low limit of radon detection is 0.5 pCi/L (pico-curies/liter).

Levels Measured. Average outdoor levels probably occur in the range of 0.1–0.4 pCi/L and vary seasonally.(127) The geometric mean of radon measured in 552 homes was 0.96 pCi/L and ranged as high as 8 pCi/L.(128)

Standards. EPA guidelines suggest that radon levels be maintained below 4 pCi/L.

Control. Scavenging ventilation is the most economical method for removing radon. This technique, combined with the use of radium-free building materials and sealing of below-grade access points to prevent entry of radon from surrounding soil and rock, is effective in minimizing radon exposure in most interiors. If the water supply is contaminated, either changing the supply or removing radon by aeration or absorption with activated charcoal is effective.(124) In buildings where contaminated materials were used for construction, it may be helpful to seal surfaces with epoxy paint.(13)

Concluding Remarks

State of the Art

The nature of indoor air pollution and its related health effects are the subject of intensive ongoing research, and new data is becoming available almost daily. Before embarking on any investigation dealing with indoor air quality issues, the investigator should take the time to review the most current literature for new information. Sources for such data include the proceedings of the most recent scientific meetings focusing on indoor air quality.

Controversial Topics

It should be noted that many issues in the field of indoor air quality remain controversial, and no definitive data are available to guide interpretation of data. These include symptom complexes that remain undefined (MCS, reactive airways disease), the role of mycotoxins and microbial VOCs in building-related complaints, and the role of routine monitoring and control measures in the absence of clearly defined problems, especially with respect to *Legionella*. Reference to the most recent literature on these topics is essential before initiating investigations, interpreting data, or recommending remedial measures.

Practical Guidance for Indoor Air Quality Investigations

No single publication exists that clearly lays out a complete protocol for evaluating indoor air quality. Those that are available generally are strong in specific areas and weaker in others. A few useful documents are the American Industrial Hygiene Association's (AIHA's) *Practitioner's Approach to Indoor Air Quality Investigations* (1989);[129] the ACGIH *Guidelines for the Assessment and Control of Bioaerosols* (1997, in press);[130] the AIHA *Field Guide for Bioaerosols* (1996);[131] and EPA's *Building Air Quality: A Guide for Building Owners and Facility Managers* (1991).[132]

Summary

"The air is the air" is a quotation familiar to many *Star Trek* fans. However, air quality varies widely both outdoors and in. Indoor air is a complex mixture of outdoor air (with its entrained contaminants) and particles and gaseous contaminants released from indoor sources. Common particulate contaminants include infectious microorganisms released from human occupants and (occasionally) from environmental reservoirs, combustion-related particles from cigarette smoke and cooking/heating fuel, aerosol spray particles, allergen-containing particles released from animal occupants (mites, cockroaches, pets, etc.), and microbial toxins associated with bacterial and fungal particles. Gaseous pollutants include carbon dioxide (primarily from human occupants), carbon monoxide, nitrogen oxides, sulfur oxides (from combustion), ozone (from electrical equipment), formaldehyde (from building materials), other volatile organic compounds released from both building materials and microbial growth and radon.

Health effects resulting from exposure to indoor air contaminants include infections, allergic diseases (asthma, allergic rhinitis, hypersensitivity pneumonitis), toxicoses (cancer, asphyxiation, skin rash, mucous membrane irritation), and problems related to perception of the indoor environment as dangerous. Monitoring approaches vary with the pollutant, but usually involve air samples that are either passive or volumetric. For a few pollutants bulk dust or other materials are collected for analysis, and exposure is inferred.

Control of the quality of indoor air depends on ventilation (including adequate filtration) and control of specific sources. For some pollutants there are established exposure limits tied to specific methods of monitoring. For others (especially the biological agents) no standards or guidelines are available, and necessary levels of control rely on common sense assessments of data.

References

1. **Dockery, D.W. and J.D. Spengler:** Indoor-outdoor relationships of respirable sulfates and particles. *Atmos. Environ. 15*:335–343 (1981).
2. **Gammage, R.B. and B.A. Berven:** *Indoor Air and Human Health*, 2nd ed. Boca Raton, FL: CRC/Lewis Publishers, 1996.
3. **Berglund, B., B. Brunekreef, H. Knoppel, T. Lindvall, et al.:** Effects of indoor air pollution on human health. *Indoor Air 2*:2–25 (1992).
4. **Spengler, J.D. and K. Sexton:** Indoor air pollution: a public health perspective. *Science 221*:9–17 (1983).
5. **Samet, J.M. and J.D. Spengler:** *Indoor Air Pollution: A Health Perspective*. Baltimore, MD: Johns Hopkins University Press, 1991.
6. **American Society of Heating, Refrigerating and Air-Conditioning Engineers (ASHRAE):** *Standard 62-1989: Ventilation for Indoor Air Quality*. Atlanta, GA: ASHRAE, 1989.
7. **Gammage, R.R. and S.V. Kaye (eds):** *Indoor Air and Human Health: Proceedings of the Seventh Life Sciences Symposium*. Chelsea, MI: Lewis Publishers, Inc., 1985.
8. **Lebowitz, M.D.:** Health effects of indoor air pollutants. *Annu. Rev. Publ. Health 4*:203–221 (1983).
9. **Kreiss, K. and M.J. Hodgson:** Building associated epidemics. In *Indoor Air Quality*, P.J. Walsh, C.S. Dudney, E.D. Copenhauer, eds. Boca Raton, FL: CRC Press, Inc., 1984. pp. 87–106.
10. **Day, J., R. Lees, R. Clark, and P. Pattee:** Respiratory response to formaldehyde and off-gas of urea formaldehyde foam insulation. *Can. Med. Assoc. J. 131*:1061–1065 (1984).
11. **Samet, J.M., M.C. Marbury, and J.D. Spengler:** Respiratory effects of indoor air pollution. *J. Allergy Clin. Immunol. 79*:685–700 (1987).
12. **Hileman, N.:** Indoor air pollution. *Environ. Sci. Technol. 17*:469A–472A (1983).
13. **U.S. Environmental Protection Agency (EPA):** *EPA Indoor Air Quality*

Implementation Plan (EPA/600/8-87/014). Washington, DC: EPA, 1987. Appendix A. Preliminary indoor air pollution information assessment.

14. **Cookingham, C.E. and W.R. Solomon:** Bioaerosol-induced hypersensitivity diseases. In *Bioaerosols*, H. Burge, ed. Boca Raton, FL: CRC/Lewis Publishers, 1995.
15. **Burge, H.A.:** Airborne contagious disease. In *Bioaerosols*, H.A. Burge, ed. Boca Raton, FL: CRC/Lewis Publishers, 1995.
16. **Finnegan, M.J., C.A.C. Pickering, and P.S. Burge:** The sick building syndrome: prevalence studies. *Br. Med. J. 289*:1573–1575 (1984).
17. **Mendell, M.J.:** Non-specific symptoms of office workers: a review and summary of the epidemiologic literature. *Indoor Air 3*:227–236 (1993).
18. **American Society of Heating, Refrigerating and Air-Conditioning Engineers (ASHRAE):** *ASHRAE Indoor Air Quality Position Paper*. Atlanta, GA: ASHRAE, 1987.
19. **Colligan, M.J.:** The psychological effects of indoor air pollution. *Bull. NY Acad. Med. 57*:1014–1026 (1981).
20. **Guidotti, T.L., R.W. Alexander, and M.J. Fedoruk:** Epidemiologic features that may distinguish between building-associated illness outbreaks due to chemical exposure or psychogenic origin. *J. Occup. Med. 29*:148–150 (1987).
21. **National Research Council:** *Multiple Chemical Sensitivities*. Washington, DC: National Academy Press, 1992.
22. **Terr, A.:** Environmental Illness: a clinical review of 50 cases. *Arch. Intern. Med. 146*:145–149 (1986).
23. **Baetjer, A.M.:** Role of environmental temperature and humidity in susceptibility to disease. *Arch. Environ. Health 16*:565–570 (1968).
24. **Hahn, A., S.D. Anderson, A.R. Morton, J.L. Black, et al.:** A reinterpretation of the effect of temperature and water content of the inspired air in exercise-induced asthma. *Am. Rev. Respir. Dis. 130*:575–579 (1984).
25. **Jennings, L.C. and J.L. Faoagali:** A study of the susceptibility of Antarctic winter-over personnel to viral respiratory illness on their return to New Zealand. *Antarctic Rec. 3*:29–30 (1980).
26. **Hambleton, P., M.G. Broster, P.J. Dennis, R. Henstridge, et al.:** Survival of virulent *Legionella pneumophila* in aerosols. *Journal of Hygiene. 90*:451–460 (1983).
27. **Katz, S.M. and J.M. Hammel:** The effect of drying, heat, and pH on the survival of *Legionella pneumophila. Ann. Clin. Lab. Sci. 17*:150–156 (1987).
28. **Arlian, L.G.:** Humidity as a factor regulating feeding and water balance of the house dust mites *Dermatophagoides farinae* and *D. pteronyssinus. J. Med. Entomol. 14*:484–488 (1977).
29. **American Society of Heating, Refrigerating, and Air-Conditioning Engineers (ASHRAE):** *ANSI/ASHRAE Standard 55-1992, Thermal Environmental Conditions for Human Occupancy*. Atlanta, GA: ASHRAE, 1992.
30. **Andersen, I. and J. Korsgaard:** Asthma and the indoor environment: assessment of the health implications of high indoor air humidity. In *Indoor Air Quality: Papers from the Third International Conference on Indoor Air Quality and Climate*, B. Berglund, R. Berglund, T. Lindvall, J. Spengler, et al., eds. *Environ. Int. 12*:121–127 (1986).
31. **Ruth, J.H.:** Odor thresholds and irritation levels of several chemical substances: a review. *Am. Ind. Hyg. Assoc. J. 47*:A142–A151 (1986).
32. **Melius, J., K. Wallingford, J. Carpenter, and R. Keenlyside:** Indoor air quality: the NIOSH experience (evaluation of environmental office problems). *Ann. Am. Conf. Ind. Hyg. 10*:3–7 (1984).
33. **Thompson, C.R., E.G. Mensel, and G. Kutz:** Outdoor-indoor levels of six air pollutants. *J. Air Pollut. Control Assoc. 10*:881–886 (1973).
34. **Yocum, J.E., W.L. Clink, and W.A. Cote:** Indoor/outdoor air quality relationships. *J. Air Pollut. Control Assoc. 21*:251–259 (1971).
35. **Sexton, K., R. Letz, and J.D. Spengler:** Estimating human exposure to nitrogen dioxide: an indoor/outdoor modeling approach. *Environ. Res. 32*:151–166 (1983).
36. **Elixman, J., W. Jorde, and H. Linskens:** Filters of an air conditioning installation as disseminators of fungal spores. In *Advances in Aerobiology, Proceeding of the 3rd International Conference on Aerobiology*. G. Boehm and P. Leuschner, eds. Basel, Switzerland: Birkhauser Verlan, 1987.
37. **Bernstein, R.S., W.G. Sorenson, D. Garabrant, C. Reaux, et al.:** Exposures to respirable airborne

Penicillium from a contaminated ventilation system: Clinical environmental and epidemiological aspects. *Am. Ind. Hyg. Assoc. J. 44*:161–169 (1983).

38. **Ager, B.P. and J.A. Tickner:** The control of microbiological hazards associated with air- conditioning and ventilation systems. *Ann. Occup Hyg. 27*:341–358 (1983).
39. **Burke, G.W., C.B. Carrington, R. Strauss, J.N. Fink, et al.:** Allergic alveolitis caused by home humidifiers. *J. Am. Med. Assoc. 238*:2705–2708 (1977).
40. **Burge, H.A., W.R. Solomon, and J.R. Boise:** Microbial prevalence in domestic humidifiers. *Appl. Environ. Microbiol. 39*:840–844 (1980).
41. **Robertson, A.S., P.S. Burge, G.A. Wieland, and M.H. Carmalt:** Extrinsic allergic alveolitis caused by a cold water humidifier. *Thorax 42(1)*:32–37 (1987).
42. **Shiue, S.T., H.H. Scherzer, A.C. DeGraff, and S.R. Cole:** Hypersensitivity pneumonitis associated with the use of ultrasonic humidifiers. *N.Y. State J. Med. 90*:263–265 (1990).
43. **Whorton, M.D., S.R. Larson, N.J. Gordon, and R.W. Morgan:** Investigation and work-up of tight building syndrome. *J. Occup. Med. 29*:142–147 (1987).
44. **Knight, V.:** Airborne transmission and pulmonary deposition of respiratory viruses. In *Viral and Mycoplasmal Infections of the Espiratory Tract*. V. Knight, ed. Philadelphia: Lea & Febiger, 1973.
45. **Kendrick, B.:** *The Fifth Kingdom*, 2nd ed. Newburyport, MA: Mycologue Publications, 1992.
46. **Tyndall, R.L. and A.A. Vass:** The potential impact on human health from free-living amoebae in the indoor environment. In *Bioaerosols*, H.A. Burge, ed. Boca Raton, FL: CRC/Lewis Publishers, 1995.
47. **Squillace, S.P.:** Allergens of arthropods and birds. In *Bioaerosols*, H.A. Burge, ed. Boca Raton, FL: CRC/Lewis Publishers, 1995.
48. **Luczynska, C.M.:** Mammalian aeroallergens. In *Bioaerosols*, H.A. Burge, ed. Boca Raton, FL: CRC/Lewis Publishers, 1995.
49. **O'Rourke, M.K., and M.D. Lebowitz:** A comparison of regional atmospheric pollen with pollen collected at and near homes. *Grana 23*:55–64 (1984).
50. **Seaton, A., B. Cherrie, and J. Turnbull:** Rubber glove asthma. *Br. Med. J. 296*:531–532 (1988).
51. **Eichoff, T.C.:** Epidemiology of Legionnaires' disease. *Ann. Intern. Med. 90*:499–502 (1979).
52. **Fraser, D.W.:** Potable water as a source for Legionellosis. *Environ. Health Perspect. 62*:337–341 (1985).
53. **Recht, L.D., S.F. Davies, M.R. Sckman, and G.A. Sarosi:** Blastomycosis in immunosuppressed patients. *Am. Rev. Respir. Dis. 125*:359–362 (1982).
54. **Solomon, W.R., and H.A. Burge:** Allergens and pathogens. In *Indoor Air Quality*, P.J. Walsh, C. Dudney, and E.D. Copenhauer, eds. Boca Raton, FL: CRC Press Inc., 1984. pp. 173–191.
55. **Riley, R.L.:** Indoor airborne infection. *Environ. Int. 8*:317–320 (1982).
56. **Rhame, F.S., A.J. Streifel, K.H. Kersey, and P.B. McGlave:** Extrinsic risk factors for pneumonia in the patient at high risk of infection. *Am. J. Med. 76*:42–52 (1984).
57. **Morris, G.K., and J.C. Feeley:** *Legionella*: impact on indoor air quality and the HVAC industry. In *Proceedings of the American Society of Heating, Refrigerating and Air-Conditioning Engineers' Annual Meeting*. Atlanta, GA: American Society of Heating, Refrigerating and Air- Conditioning Engineers. p. 77. [Abstract]
58. **Miller, R.D., and K.A. Kenepp:** *Legionella* in cooling towers: use of *Legionella*-total bacteria ratios. In *Aerobiology*, M. Muilenberg and H. Burge (eds.). Boca Raton, FL: CRC/Lewis Publishing, 1996. pp. 99–107.
59. **Standards Australia:** Air handling and water systems of buildings, microbial control. (AS3666) Sydney: Standards Australia, 1989.
60. **Emmons, C.W.:** Saprophytic sources of *Cryptococcus neoformans* associated with the pigeon (*Columba livia*). *Am. J. Hyg. 61*:227–232 (1955).
61. **Pope, A.M., R. Patterson, and H. Burge (eds.):** *Indoor Allergens*. Washington DC: National Academy of Sciences Press, 1993.
62. **Sporik, R.B., L.K. Arruda, J. Woodfolk, M.D. Chapman, et al.:** Environmental exposure to *Aspergillus fumigatus* allergen (Asp f I). *Clin. Exper. Allergy 23(4)*:326–331 (1993).

63. **Gergen, P., P. Turkeltaub, and M. Kovar:** The prevalence of allergic skin test reactivity to eight common aeroallergens in the U.S. population: Results from the second National Health and Nutrition Examination Survey. *J. Allergy Clin. Immunol. 80*:669–679 (1987).
64. **Chew, G., M. Muilenberg, D. Gold, and H.A. Burge:** Is dust sampling a good surrogate for exposure to airborne fungi? *J. Allergy Clin. Immunol. 97(1)pt 3*:419 (1996). [Abstract]
65. **Burge, H.A. and W.R. Solomon:** Sampling and analysis of biological aerosols. *Atmos. Environ. 21*:451–456 (1987).
66. **Muilenberg, M.L.:** Aeroallergen assessment by microscopy and culture. *Immunol. Allergy Clin. North Am. 9(2)*:245–268 (1989).
67. **Axelsson, B., A. Saraf, and L. Larsson:** Determination of ergosterol in organic dust by gas chromatography-mass spectrometry, *J. Chromatogr. B 666*:77–84 (1995).
68. **American Conference of Governmental Industrial Hygienists (ACGIH):** *Guidelines for the Assessment of Bioaerosols in the Indoor Environment*. Cincinnati, OH: ACGIH, 1989.
69. **Milton, D.M.:** Endotoxin. In *Bioaerosols*, H.A. Burge, ed. Boca Raton, FL: CRC/Lewis Publishing, 1995.
70. **Castellan, R.M., S.A. Olenchock, K.B. Kinsley, and J.L. Hankinson:** Inhaled endotoxin and decreased spirometric values, an exposure-response relation for cotton dust. *N. Engl. J. Med. 317*:605–610 (1987).
71. **Sandstrom, T., L. Bjermer, and R. Rylander:** Lipopolysaccharide (LPS) inhalation in healthy subjects increases neutrophils, lymphocytes, and fibronectin levels in bronchoalveolar lavage fluid. *Eur. Respir. J. 5(8)*:992–996 (1992).
72. **Kennedy, S.M., D.C. Christiani, E.A. Eisen, D.H. Wegman, et al.:** Cotton dust and endotoxin exposure-response relationships in cotton textile workers. *Am. Rev. Respir. Dis. 135*:194–200 (1987).
73. **Milton, D.K., D. Kriebel, D. Wypij, M. Walters, et al.:** Acute and chronic airflow obstruction and endotoxin exposure. *Am. J. Respir. Crit. Care Med. 149*:A399 (1994). [Abstract]
74. **Milton, D.K., J. Amsel, C.E. Reed, P.L. Enright, et al.:** Cross-sectional follow-up of a flu-like respiratory illness among fiberglass manufacturing employees: endotoxin exposure associated with two distinct sequelae. *Am. J. Ind. Med. 28*:469–488 (1995).
75. **Teeuw, K.B., C.M. Vandenbroucke-Grauls, and J. Verhoef:** Airborne gram-negative bacteria and endotoxin in sick building syndrome. A study in Dutch governmental office buildings. *Arch. Intern. Med. 154*:2339–2345 (1994).
76. **Hodgson, J.T. and R.D. Jones:** Mortality of workers in the British cotton industry in 1968–1984. *Scand. J. Work Environ. Health 16*:113–120 (1990).
77. **Rose, C.S., L.S. Newman, J.W. Martyny, D. Weiner, et al.:** Outbreak of hypersensitivity pneumonitis in an indoor swimming pool: clinical, pathophysiologic, radiographic, pathologic, lavage and environmental findings. *Am. Rev. Respir. Dis. 141*:A315 (1994). [Abstract]
78. **Rylander, R., S. Sorensen, H. Goto, K. Yuasa, et al.:** The importance of endotoxin and glucan for symptoms in sick building. In *Present and Future of Indoor Air Quality in Proceedings of the Brussels Conference*, C.J. Bieva, Y. Courtois, and M. Govaerts, eds. New York: Excerpta Medica, 1989.
79. **Sietsma, J.H. and J.G.H. Wessels:** Solubility of (13)—D/(16)—D-glucan in fungal walls: Importance of presumed linkage between glucan and chitin. *J. Gen. Microbiol. 125*:209–212 (1981).
80. **Fogelmark, B., H. Goto, K. Yuasa, B. Marchat, et al.:** Acute pulmonary toxicity of inhaled beta 1,3-glucan and endotoxin. *Agents Actions 35(1-2)*:50–56 (1992).
81. **Fogelmark, B., M. Sjostrand, and R. Rylander:** Pulmonary inflammation induced by repeated inhalations of beta(1,3)-D-glucan and endotoxin. *Int. J. Exp. Pathol. 75*:85–90 (1994)
82. **Kiho, T., M. Sakushima, S.R. Wang, K. Nagai, et al.:** Polysaccharides in fungi. XXVI. Two branched (1-3)—d-glucans from hot water extract of yu er. *Chem. Pharm. Bull. 39(3)*:798–800 (1991).
83. **Douwes, J., G. Doekes, D. Heederik, B. Brunekreef, et al.:** Measurement of environmental ß(1->3)-glucans by enzyme immunoassay. *Am. J. Respir. Crit. Care Med. 151*:A142 (1995).
84. **Betina, V.:** *Mycotoxins: Chemical, Biological and Environmental Aspects*

(Bioactive molelcules, vol. 9). New York: Elsevier, 1989. [Abstract]
85. **Creasia, D.A., J.D. Thurman, R.W. Wannemacher, and D.L. Bunner:** Acute inhalation toxicity of T-02 mycotoxin in the rat and guinea pig. *Fund. Appl. Toxicol. 14*:45–59 (1990).
86. **Sorenson, W.G., G.F. Gerberick, D.M. Lewis, and V. Castranova:** Toxicity of mycotoxins for the rat pulmonary macrophage *in vitro. Environ. Health Persp. 66*:45–53 (1986).
87. **Rivers, J.C., J.D. Pleil, and R.W. Wiener:** Detection and characterization of volatile organic compounds produced by indoor air bacteria. *J. Exposure Anal. Environ. Epidemiol. 2(suppl. 1)*:177 (1992). [Abstract]
88. **Batterman, S., N. Bartoletta, and H. Burge:** "Fungal Volatiles of Potential Relevance to Indoor Air Quality". Paper presented at the Annual Meeting of the Air and Waste Management Association, Vancouver, Canada, June 1991.
89. **Turiel, I.:** *Indoor Air Quality and Human Health.* Stanford, CA: Stanford University Press, 1985.
90. **Spengler, J.D. and J.M. Samet:** A perspective on indoor and outdoor air pollution. In *Indoor Air Pollution, a Health Perspective.* J.M. Samet and J.D. Spengler, eds. Baltimore, MD: Johns Hopkins University Press, 1991.
91. **Ackland, G., T.D. Hartwell, T.R. Johnson, and R. Whitmore**: Measuring human Exposure to carbon monoxide in Washington, DC and Denver, CO during the winter of 1982-83. *Environ. Sci. Tech. 19*:911–918 (1985).
92. **Coultas, D.B. and W.E. Lambert:** Carbon monoxide. In *Indoor Air Pollution, a Health Perspective,* J.M. Samet and J.D. Spengler, eds. Baltimore, MD: Johns Hopkins University Press, 1991.
93. **U.S. Environmental Protection Agency (EPA):** *Air Quality Criteria for Carbon Monoxide.* Washington DC: EPA, 1979.
94. **Samet, J.M.:** Nitrogen dioxide. In *Indoor Air Pollution, a Health Perspective,* J.M. Samet and J.D. Spengler, eds. Baltimore, MD: Johns Hopkins University Press, 1991.
95. **Leaderer, B.P.:** Air pollutant emissions from kerosene space heaters. *Science 218*:1113–1115 (1982).
96. **Bauer, M.A., M.J. Utell, P.E. Morrow, D.M. Speers, et al.:** Inhalation of 0.30 ppm nitrogen dioxide potentiates exercise-induced bronchospasm in asthmatics. *Am. Rev. Respir. Dis. 1344*:1203–1208 (1986).
97. **Lee, S.D.:** *Nitrogen Oxides and Their Effects on Health.* Ann Arbor, MI: Ann Arbor Science Publishers, 1980.
98. **Ahmed, T., I. Danta, R.L. Dougherty, R. Shreck, et al.:** Effect of NO_2 (0.1 ppm) on specific bronchial reactivity to ragweed antigen in subjects with allergic asthma. *Am. Rev. Respir. Dis. 127*:160 (1983). [Abstract]
99. **U.S. Department of Energy (DOE):** *Indoor Air Quality Environmental Information Handbook: Combustion Sources.* Washington, DC: DOE, Office of Environmental Analysis, 1985.
100. **National Research Council, Committee on Passive Smoking:** *Environmental Tobacco Smoke: Measuring Exposures and Assessing Health Effects.* Washington, DC: National Academy Press, 1986.
101. **Dockery, D.W.:** Environmental tobacco smoke and lung cancer: Environmental smoke screen? In *Indoor Air and Human Health,* 2nd ed., R.B. Gammage and B.A. Berven, eds. Boca Raton, FL: CRC/Lewis Publishers, 1996.
102. **Fontham, E.T.H., P. Correa, P. Reynolds, A. Wu-Williams, et al.:** Environmental tobacco smoke and lung cancer in nonsmoking women; a multicenter study. *J. Am. Med. Assoc. 271*:1752–1759 (1994).
103. **Weber, A. and T. Fischer:** Passive smoking at work. *Int. Arch. Occup. Environ. Health 47*:209–221 (1980).
104. **Cain, W.S., B.P. Leaderer, R. Isseroff, L.G. Berglund, et al.:** Ventilation requirements in buildings—I. Control of occupancy odor and tobacco smoke odor. *Atmos. Environ. 17*:1183–1197 (1983).
105. **Repace, J.L. and A.H. Lowrey:** Indoor air pollution, tobacco smoke, and public health. *Science 208*:464–472 (1980).
106. **Dockery, D.W. and J.D. Spengler:** Personal exposure to respirable particulates and sulfates. *J. Air Pollut. Control Assoc. 31*:153–159 (1981).
107. **Mage, D.T. and R.B. Gammage:** Evaluation of changes in indoor air quality occurring over the past several decades. In *Indoor Air Quality,* P.J. Walsh, C.S. Dudney, and E.D. Copenhauer, eds. Boca Raton, FL: CRC Press, Inc., 1985. pp. 5–36.

108. **Koenig, J.Q.:** Pulmonary reaction to environmental pollutants. *J. Allergy Clin. Immunol. 79*:833–843 (1987).
109. **Hawthorne, A.R., G.M. Thomas, and T. Vo-Dinh:** Measurement techniques. In *Indoor Air Quality*, P.J. Walsh, C.S. Dudney, and E.D. Copenhauer, eds. Boca Raton, FL: CRC Press, Inc., 1984.
110. "Permissible Exposure Limit for Ozone," *Code of Federal Regulations*, Title 29, Section 1910.1000.
111. **Andersen, I., G.R. Lundquist, and L. Molhave:** Indoor air pollution due to chipboard used as a construction material. *Atmos. Environ. 9*:1121–1127 (1975).
112. **Marbury, M.C., and R.A. Krieger:** Formaldehyde. In *Indoor Air Pollution, a Health Perspective*, J.M. Samet and J.D. Spengler, eds. Baltimore, MD: Johns Hopkins University Press, 1991.
113. **Rinsky, R.A., A.B. Smith, R. Hornung, T.G. Filloon, et al.:** Benzene and leukemia: an epidemiologic risk assessment. *N. Engl. J. Med. 316*:1044–1050 (1987).
114. **Rodricks, J.V.:** Assessing carcinogenic risks associated with indoor air pollutants. In *Indoor Air and Human Health*, 2nd ed., R.B. Gammage and B.A. Berven, eds. Boca Raton, FL: CRC/Lewis Publishers, 1996.
115. **Molhave, L.:** Volatile organic compounds as indoor air pollutants. In *Indoor Air and Human Health: Proceedings of the Seventh Life Sciences Symposium*, R.R. Gammage and S.V. Kaye, eds. Chelsea, MI: Lewis Publishers, Inc., 1985.
116. **Hodgson, A.T.:** A review and a limited comparison of methods for measuring total volatile organic compounds in indoor air. *Indoor Air 5(4)*:247–257 (1995).
117. **BSR/American Society of Heating, Refrigerating and Air-Conditioning Engineers (ASHRAE):** 62-1989R, Public Review Draft, August 1996. Atlanta, GA: ASHRAE. Appendix D: D1-D20.
118. **Brown, S., M.R. Sim, M.J. Abramson, and C.N. Gray:** Concentrations of volatile organic compounds in indoor air—a review. *Indoor Air 4*:123–134 (1994).
119. **Molhave, L.:** Volatile organic compounds, indoor air quality and health. *Indoor Air 1*:357–376 (1991).
120. **Girman, J., L. Alevantis, M. Petreas, and L. Webber:** Bake-out of a new office building to reduce volatile organic concentrations. Air & Waste Management Association, Proceedings of 82nd Annual Meeting & Exhibition, Anaheim, CA, June 25–30, 1989.
121. **Auxier, J.A.:** Respiratory exposure in buildings due to radon progeny. *Health Phys. 31*:119–125 (1976).
122. **Abu-Jarad, F. and J.H. Fremlin:** The activity of radon daughters in high-rise buildings. *Health Phys. 43*:75–80 (1982).
123. **Nero, A.:** Indoor concentrations of radon 222 and its daughters: Sources, range and environmental influences. In *Indoor Air and Human Health*, 2nd ed., R.B. Gammage and B.A. Berven, eds. Boca Raton, FL: CRC/Lewis Publishers, 1996. pp. 43–67.
124. **U.S. Environmental Protection Agency (EPA):** *Technical Support Document for the 1992 Citizen's Guide to Radon* (EPA/400-R-92-011). Washington, DC: EPA.
125. **Samet, J.M.:** Radon and lung cancer revisited. In *Indoor Air and Human Health*, 2nd ed., R.B. Gammage and B.A. Berven, eds. Boca Raton, FL: CRC/Lewis Publishers, 1996. pp. 325–340.
126. **Ellet, W. and N. Nelson:** Epidemiology and risk assessment: Testing models for radon- induced lung cancer. In *Indoor Air and Human Health*, 2nd ed., R.B. Gammage and B.A. Berven, eds. Boca Raton, FL: CRC/Lewis Publishers, 1996. pp. 79–129.
127. **Gesell, T.F.:** Background atmospheric 222Rn concentrations outdoors and indoors: a review. *Health Phys. 45*:289–302 (1983).
128. **Alter, W.H. and R.A. Oswald:** Nationwide distribution of indoor radon measurements: a preliminary data base. *J. Air Pollut. Control Assoc. 37*:227–231 (1987).
129. **American Industrial Hygiene Association (AIHA):** *Practitioner's Approach to Indoor Air Quality Investigations*. Fairfax, VA: AIHA, 1989.
130. **American Conference of Governmental Industrial Hygienists (ACGIH):** *Guidelines for the Assessment and Control of Bioaerosols*. Cincinnati, OH: ACGIH, 1997.
131. **American Industrial Hygiene Association (AIHA):** *Field Guide for the Determination of Biological Contaminants in Environmental Samples*. Fairfax, VA: AIHA, 1996.

132. **U.S. Environmental Protection Agency (EPA):** *Building Air Quality: a Guide for Building Owners and Facility Managers.* Washington, DC: EPA, 1991.

Before attempting to perform an indoor air quality survey, the reader is advised to read and understand thoroughly the relevant chapters in the most recent general engineering data section ASHRAE *Handbook of Fundamentals.*

EPA has two publications that also should be read and understood thoroughly before an indoor air quality survey is performed: *Building Air Quality: A Guide for Building Owners and Facility Managers* (EPA/400/1-91/033; DHHS (NIOSH) pub. 91-114); and EPA's Indoor Air Quality Tools for Schools action kit.

The user may create a questionnaire or select one from the EPA materials or the appendixes to this chapter. The user should exercise extreme care in using or creating any questionnaire. The design, construction, validity, administration, and interpretation of questionnaires is a social and behavioral science skill not familiar to most occupational hygienists.

Indoor air quality problems are often emotional issues for building occupants. Not following good IAQ survey practices, as outlined in the above references, can exacerbate the stress placed on the occupants.

Appendix A

Occupant Health and Comfort Questionnaire

1. (Optional) Name: ________________________
   Job Title: ________________________
   Department: ________________________
   Phone: ________________________

2. Area or room where you spend the most time in the building:
   ____
   ____
   ____

3. Do any of your work activities produce dust or odor?
   ❑ Yes ❑ No
   Describe: ________________________
   ____
   ____
   ____

4. Gender ❑ Male ❑ Female
   Age: ❑ Under 25 ❑ 25-34 ❑ 35-44 ❑ 45-54 ❑ 55 and over

5. Do you:
   Smoke . . . .❑ Yes ❑ No
   Have fever/pollen allergies? . . . .❑ Yes ❑ No
   Have skin allergies/dermatitis? . . . .❑ Yes ❑ No
   Have a cold/flu? . . . .❑ Yes ❑ No
   Have sinus problems? . . . .❑ Yes ❑ No
   Have other allergies? . . . .❑ Yes ❑ No
   Wear contact lenses? . . . .❑ Yes ❑ No
   Operate video display terminals? . . . .❑ Yes ❑ No
   Operate photocopiers 10% of the time? . . . .❑ Yes ❑ No
   Use other special office machines? . . . .❑ Yes ❑ No
   Specify: ________________________
   ____

   Take medication currently? . . . .❑ Yes ❑ No
   Reason: ________________________
   ____

6. Office characteristics:
   _____ Number of persons sharing same room/work area
   _____ Number of windows in room/work area
   Do windows open? . . . .❑ Yes ❑ No

   Please rate adequacy of work space per person

Poor		Average		Excellent
1	2	3	4	5

   Please rate room temperature

Poor		Average		Excellent
1	2	3	4	5

   Do others smoke in your work area? . . . . . .❑ Yes ❑ No

7. How long have you worked:
   _____ In this room/area?
   _____ In this building?

Appendix A
Occupant Health and Comfort Questionnaire (continued)

8. Symptoms: Select symptoms you have experienced in this building. This is a random list—not all symptoms listed have been noted in this building.

Symptom	Occasionally	Frequently	Not related to building	Appeared after arrival	Increased after arrival
Difficulty in concentrating					
Aching joints					
Muscle twitching					
Back pain					
Hearing problems					
Dizziness					
Dry, flaking skin					
Discolored skin					
Skin irritation					
Itching					
Heartburn					
Nausea					
Noticeable odors					
Sinus congestion					
Sneezing					
High stress levels					
Chest tightness					
Eye irritation					
Fainting					
Hyperventilation					
Problems with contacts					
Headache					
Fatigue/drowsiness					
Temperature too hot					
Temperature too cold					
Other (specify)					

Have you seen a doctor for any or all of these symptoms?
❑ Yes ❑ No

When do you experience relief from these symptoms?

9. When do these problems usually occur?

TIME OF DAY	Morning Afternoon Evening
DAY OF WEEK	S M T W TH F S
MONTH	J F M A M J J A S O N D
SEASON	Spring Summer Fall Winter

10. Do symptoms disappear?
❑ Yes ❑ No
When? ______________________________

11. In your opinion, what is the cause of perceived indoor air quality problems?

12. Comments: Please take this opportunity to comment on any factors you consider to be important concerning the quality of your work environment.

Thank you very much for your cooperation.

Appendix B
Building Engineering
HVAC Inspection Checklist

Circle all that apply.

1. Is there a mechanical HVAC system?
   Note which floors/rooms are served by each system.
   VAV (variable air volume)
   - VAV with induction
   - VAV with recirculation

   CAV (constant air volume)
   Terminal reheat
   Heat pumps
   Fan coil
   Induction unit
   Unit ventilator
   Ceiling plenum (supply)
   Ceiling plenum (return)

2. If VAV:
   Is there a minimum guaranteed flow?
   What is it? ______________________________

3. Type of HVAC functions:
   Heating
   Cooling
   Dehumidification
   Humidification
   Filtration
   Outdoor air
   Indicate location of functional units

4. Perimeter heating?
   What type?

5. Any recent repairs on HVAC system?

6. HVAC maintenance/operation (note down scheduled times and indicate if done by in-house staff or outside contractor):
   Boiler water treatment, change filters, clean coils, clean ducts, clean drain pans, clean mechanical equipment room, control system maintenance

7. When was the HVAC system last balanced?

8. HVAC controls:
   original controls? modified? economizer? duty cycling? set back?

9. Are system components controlled automatically?
   What are the monitoring/calibration frequencies?

10. Is the amount of makeup (outdoor) air used by the HVAC system the same all year?
    What is the percentage of makeup air used?

11. Is the building run on an economizer cycle?
    What is the maximum percentage of makeup air used?
    What is the minimum percentage of makeup air used?
    What is the makeup air percentage now (today)?

Appendix B
Building Engineering
HVAC Inspection Checklist (continued)

12. How is the HVAC system controlled?
    Where are the thermostats located?
    Do occupants have access to the thermostats?
    When were thermostats last calibrated?
    What is the cycle of the day/night controller, if used?

13. Is the ventilation in the work area(s) decreased or shut off overnight or on weekends?
    If decreased, the system goes down overnight to what percentage of daytime?
    Shutoff/decrease hours are which hours during the week?
    Shutoff/decrease hours on weekends are which hours/days?
    Which hours/days is the building open?

14. Is there a filter system in the outdoor air intake?
    How often are the filters changed?
    What types are the filters?
    What is their theoretical efficiency?

15. Humidification equipment present?
    What type: steam jet, air washer, water spray, pan type, atomizing.

16. Dehumidification system present?
    Central (other than cooling coil), local

17. Heating system type?
    Energy source? (oil, gas, electric, other)
    Primary fluid? (steam, hot water, air, other)
    Supply system? (pipe, duct, plenum, other)
    Distribution system? (radiators, ceiling grilles, other)
    Return system? (pipe, duct, plenum, other)
    Location of heat source?
    Local heating units?

18. Cooling equipment present:
    Chilled water, direct expansion, absorption, cooling tower, air cooled condensing.
    Location of chillers?
    Window units used?

19. Checklist of client documents available:
    Floor plans
    Diagrams
    Design specifications
    Distribution system drawings
    Mechanical room drawings
    Sheet metal drawings
    Plumbing drawings
    Vent specifications
    Equipment schedule
    Previous environmental reports

20. Are HVAC components accessible to maintenance?
    Indicate which areas/components are not accessible ______________________
    ____

21. Are any chemicals, solvents, and/or cleaners stored in the air handling unit/mechanical equipment room (AHU/MER)?

Appendix B
Building Engineering
HVAC Inspection Checklist (continued)

22. Are there any leaks (oil, water, steam) from the HVAC components in the AHU/MER?

23. Are any fans not working?

24. Does the building have an air-conditioning system?
What is its operating schedule?
Is the system operating today?
How often are the condensate trays cleaned?
Is there slime on the condensate trays or the cooling coils?
Are there moldy odors in the system?
Is there water accumulation on the ground?

25. Outdoor air intakes:
Are intakes below third floor level and above a busy street?
Are intakes above entrance to a loading dock?
Are intakes above entrance or exit to a parking garage?
Any other pollution sources near the intakes?
Are there any obstructions lodged in the intake?
Are dampers functioning?
Are intakes within 30 feet of any HVAC, restroom, kitchen, or laboratory exhausts?
Are open doors or windows used for outside air?

26. Filters in AHU:
Does the filter fit so poorly that air bypasses it at the edges (blow-by)?
Are filters matted/dirty/wet?

27. Are the vent ducts or mixing room of the AHU lined with insulation?
Is it external or internal?
What type is it: fiber glass, polystyrene, asbestos
Is the internal lining dirty/wet/shredded/torn

28. Roof: Any flooding or ponding?

29. Operational problems experienced with the HVAC system?

Section 3

Physical Agents

Outcome Competencies

After completing this chapter, the user should be able to:

1. Define underlined terms used in this chapter.
2. Recall the physics of sound.
3. Discuss the physiology of hearing.
4. Explain acceptability criteria.
5. Apply noise measurement techniques.
6. Employ noise abatement technology.
7. Implement a hearing conservation program.
8. Discuss vibration effects, measurement, and control.
9. Discuss ultrasound effects, measurement, and control.

Prerequisite Knowledge

College level algebra, college level physics, introduction to calculus.

Key Terms

A-weighted response • acceleration • accelerometer • acoustic trauma • acoustical absorption • air conduction • amplitude • annoyance • anti-vibration (A/V) • audiogram • audiometer • audiometric testing • bandwidths • bone conduction • C-weighted response • conductive hearing loss • constrained-layer damping • Criteria for Fatigue Decreased Proficiency (FDP) • decibels • directivity • displacement • dissipative mufflers • dosimeter • earmuffs • FFT (fast Fourier transform) spectrum analyzer • flat response • formable earplugs • free-layer damping • frequency • graphic level recorder (GLR) • hand-arm vibration syndrome (HAVS) • hearing protective device (HPD) • lagging • mass loading • noise enclosures • noise-induced hearing loss • noise-induced permanent threshold shift (NIPTS) • noise-induced temporary threshold shift (NITTS) • Noise Reduction Rating (NRR) • occupational vibration • octave bands • periodic vibration • personnel enclosure • premolded earplugs • random vibrations • Raynaud's Phenomenon of Occupational Origin • reduced comfort • resonance • reverberant field • root-mean-square (rms) • semi-inserts • sensorineural hearing loss • sound absorption coefficient • sound intensity analyzers • sound intensity level • sound level meter (SLM) • sound power level • sound pressure level • sound propagation • sound shadow • standing waves • subharmonics • tinnitus • total absorption • transmission loss (TL) • ultrasound • velocity • vibration-induced damage • vibration-induced white finger (VWF) • vibration isolators • vibration transmissibility ratio • wavelength • whole-body vibration

Key Topics

I. Physics of Sound
   - A. Sound Pressure Level
   - B. Sound Intensity Level
   - C. Sound Power Level
   - D. Combining and Averaging Sound Levels
   - E. Frequency
   - F. A- and C-Weighting
   - G. Sound Propagation

II. Physiology of Hearing
   - A. Anatomy
   - B. Auditory Sensitivity
   - C. Audiometry
   - D. Classification of Hearing Loss
   - E. Effects of Excessive Noise Exposure on the Ear

III. Acceptability Criteria
   - A. Hearing Loss Criteria
   - B. Other Considerations

IV. Noise Measurements
   - A. Sound Measuring Devices
   - B. Sound Measurement Techniques

V. Control of Noise Exposure
   - A. Justification for Performing Noise Control
   - B. Administrative Controls
   - C. Engineering Controls

VI. Hearing Conservation Programs
   - A. Sound Survey
   - B. Engineering and Administrative Controls
   - C. Hearing Protection Devices
   - D. Audiometric Testing
   - E. Employee Training and Education
   - F. Record Keeping
   - G. Program Evaluation

VII. Vibration in the Workplace — Measurements and Control
   - A. Effects of Vibration
   - B. Terminology
   - C. Criteria
   - D. Measurement of Vibration
   - E. Conducting the Workplace Measurement Study
   - F. Control of Occupational Vibration

VIII. Ultrasound
   - A. Uses of Ultrasound
   - B. Ultrasound Paths
   - C. Effects of Ultrasound
   - D. Exposure Criteria
   - E. Measurement
   - F. Exposure Controls

20

Noise, Vibration, and Ultrasound

Robert D. Bruce, PE, INCE.Bd.Cert.

Arno S. Bommer, INCE.Bd.Cert.

Charles T. Moritz, INCE.Bd.Cert.

Overview

Excessive exposure to noise, vibration, and ultrasound can be harmful to humans. For at least 100 years, we have known that excessive exposure to noise can result in permanent hearing loss. Since about 1918, it also has been clear that excessive exposure to vibration might cause permanent damage to workers' hands and fingers. And, recently, we concluded likewise that exposure to ultrasound should be limited. In each case, techniques to reduce these exposures have quickly followed recognition of the problem. Still, even though knowledge of and solutions to these problems have been available for some time, workers continue to incur noise-induced hearing loss and vibration-induced damage.

The design profession and the manufacturers and packagers of equipment are entrusted with the responsibility of designing machines, processes, and systems that do not adversely impact the health and safety of the working public. If they are unable to design safe machines, processes, and systems, it is their responsibility to notify the user of the potential danger and to provide information about how the user can be protected. It is, in turn, the occupational hygienist's responsibility to ensure that the sound, vibration, and ultrasound created by machines, processes, and systems comply with health and safety standards. In some companies, occupational hygienists are also responsible for ensuring that new equipment meets the noise limits in the purchase order. As will be discussed later in this chapter, the hygienist's first task often is to quantify the exposure (both magnitude and duration) of the worker to the physical agent. When the exposures exceed the values allowable under applicable standards or regulations, the hygienist is often called on to select and provide the appropriate controls (i.e., administrative controls, engineering controls, or personal protective equipment).

In this chapter, two physical agents are discussed: sound (noise and ultrasound) and vibration. Since noise is the most prevalent complaint of workers in society today, much of this chapter will focus on noise. First, there is a brief discussion of the physics of sound and the physiology of hearing. Next, noise criteria, noise measurements, control of noise, and hearing conservation programs are examined. The chapter closes with a discussion of vibration and ultrasound.

Physics of Sound

Introduction

Sound is a disturbance that propagates as a wave of compressions and rarefactions through an elastic medium. The speed at which the sound propagates (c) is a function of the material's elasticity and density and can be calculated from the following equation:

$$c = \sqrt{\frac{\text{elasticity}}{\text{density}}} \qquad (1)$$

For a solid, the elasticity of a material is represented by Young's Modulus. For a gas, the elasticity is the product of the pressure and the ratio of the specific heat at constant pressure to the specific heat at constant volume. For air at sea level, the equation for the speed of sound reduces to:

$$c = 20.05\ \sqrt{T} \qquad (2)$$

where:

T = absolute temperature in Kelvin (°C + 273.2).

At 21°C, the speed of sound in air is 344 m/sec. In solids and liquids, where the ratio of elasticity to density is higher than in air, sound travels much faster. For example, the speed of sound in water is about 1500 m/sec and in steel it is about 5000 m/sec.

The frequency (f), amplitude, and wavelength (λ) of a sound wave are shown in Figure 20.1. The frequency (measured in hertz or cycles per second and abbreviated Hz), wavelength (in meters), and speed of sound (c, in m/sec) are related in the following equation:

$$\lambda = \frac{c}{f} \qquad (3)$$

The wavelength of sound is important, for example, in designing noise control treatments because sound-absorptive treatments should be at least 1/4 wavelength thick for optimum absorption (though thinner treatments may also be effective depending on the specific material properties). Sound at 125 Hz in 21°C, sea-level air has a wavelength of about 2.75 m, while sound at 2000 Hz has a wavelength of only 0.17 m. For high-temperature air such as in gas turbine exhausts, the speed of sound increases and thus the wavelength of sound becomes longer. At high temperatures, to be effective, sound absorptive treatments need to be very thick, especially for low frequencies.

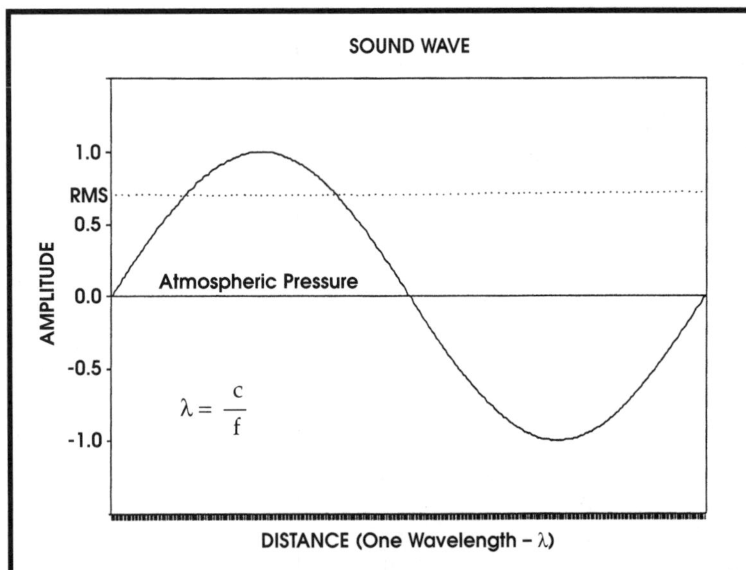

Figure 20.1 — Frequency, amplitude, and wavelength of a sound wave.

Sound Pressure Level

The dynamic range of human hearing sensitivity is large: our ears can detect changes in atmospheric pressure (sound waves) less than 20 μPa (1Pa = $1N/m^2$) in magnitude while still being able to withstand pressure changes greater than 20 Pa in magnitude. Because of this wide range, sound is often described in terms of a sound pressure level which is 10 times the logarithm of the square of the ratio of the instantaneous pressure fluctuations (above and below atmospheric pressure) to the reference pressure. The sound pressure level equation is given below, where P is the instantaneous sound pressure in Pa and P_{ref} is the reference pressure defined as 20 μPa.

$$L_p = 10\log_{10}\left(\frac{P}{P_{ref}}\right)^2 \qquad (4)$$

The units of sound pressure level are decibels, abbreviated dB.

In calculating acoustical levels, logarithms are always base 10. The normal dynamic range of human hearing is from about 20 μPa to greater than 20 Pa; the equivalent range in decibels is from 0 dB (the approximate threshold of hearing) to greater than 120 dB (near the threshold of pain). To understand this marvelous dynamic range, we can consider a postage scale used to weigh packages to be mailed. If we select 6 oz. as the "0" dB level and use 10 log of the square of the ratio of the weight of an object to a 6 oz. letter, the decibel equivalent can be determined. Identified in Figure 20.2 are some of the objects that could be weighted if the postage scale had a dynamic range of 0–150 db.

Since most sounds are made of waves with the intervals above atmospheric pressure roughly equal in number but opposite in magnitude to the intervals below atmospheric pressure, the average pressure is atmospheric pressure and the average fluctuation is 0. To calculate a more meaningful average, the sound-pressure disturbances (positive or negative) are squared (resulting in a positive number), and then the square root is taken of the average of these squared pressure changes. The result is the root-mean-square (rms) sound pressure. To determine the level of the individual sound impulses, the peak sound pressure above atmospheric pressure can be measured instead of the rms average value.

Sound Intensity Level

The sound intensity level (L_i) is similar to the sound pressure level (L_p) except that L_i is a vector quantity, having both magnitude and direction. For example, a sound wave producing an L_p of 90 dB can have an L_i ranging from +90 dB to -90 dB depending on the direction of the sound wave propagation relative to the direction that the sound measurement probe is pointing. In a diffuse, reverberant field with sound waves traveling equally in all directions, the sound intensity will be 0.

Although the L_i does not correspond to the loudness of a sound (the L_p is more comparable to loudness), it is useful as an analytical measurement and is necessary to accurately determine the sound power level as described in the following section.

Sound Power Level

Sound pressure level or sound intensity level data can be used to calculate the sound power level of a sound source. Sound power is analogous to the electrical power rating of a light bulb and is also measured in watts. Whereas the sound pressure, such as the amount of illumination from a light bulb, varies as a function of distance from the source and room conditions, the sound power, like the power rating of a light bulb, is independent of distance and of any room conditions. The sound power level is defined as:

$$L_w = 10\log_{10}\left(\frac{W}{W_{ref}}\right) \quad (5)$$

where:

W = acoustic power in watts; and
W_{ref} = reference acoustic power, 10^{-12} watts.(Prior to about 1960 in the United States, a W_{ref} of 10^{-13} watts was used in some publications.)

Since sound power is equal to the intensity times the area over which the intensity is measured, the L_w of a sound source is calculated by adding the average L_i, measured over an imaginary surface enveloping the source, to 10 times the logarithm of the area (in square meters) of the measurement surface. The imaginary measurement surface must completely surround the sound source and have no sound-absorptive materials within its boundaries. These surfaces are often hemisphere-shaped or conformal. The formula is:

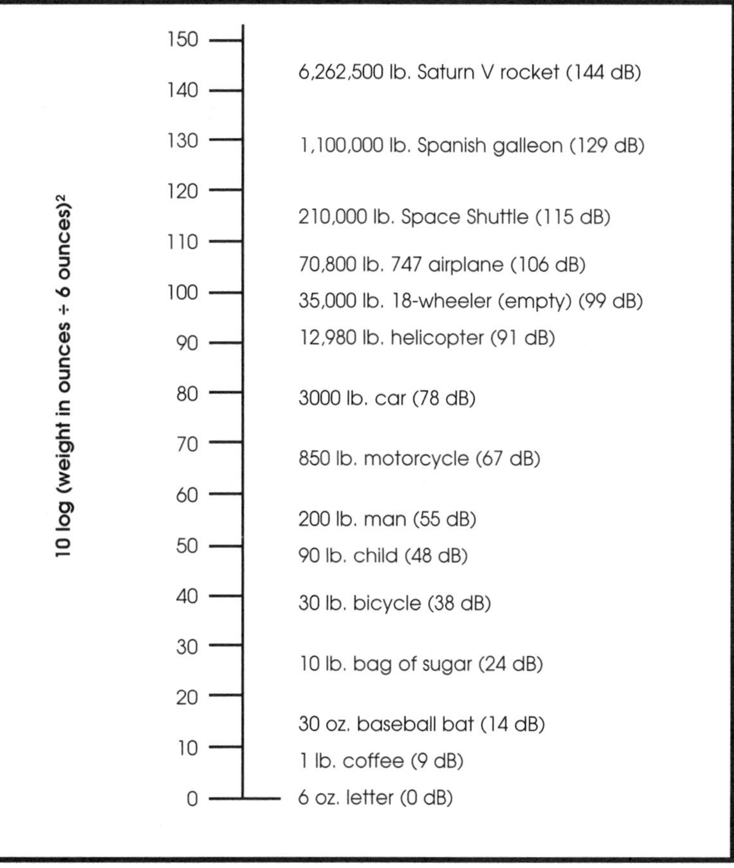

Figure 20.2 — Required dynamic range of postage scale to match capability of the human ear.

$$L_w = L_i + 10\log_{10}(\text{Area}) \quad (6)$$

where:

"Area" = measurement surface area in m^2.

When only L_p data is available, it is either presumed to be equal to the L_i or a correction factor is estimated. The L_p will be much higher than the L_i if the sound measurements are made at a reverberant location.

Combining and Averaging Sound Levels

The decibel scale is logarithmic, which means that the combined sound level of 10 items each producing 70 dB at 10 m will be 80 dB, not 700 dB. The sound pressures add, not the decibel levels. To add sound pressure levels together, sound intensity levels together, or sound power levels together, the following equation should be used:

$$\text{Total L} = 10\log_{10}\left(\sum_{1}^{n} 10^{\frac{L_n}{10}}\right) \quad (7)$$

Although the equation above works well for spreadsheet and other computer applications, it is difficult to use when summing levels quickly without a computer or calculator. A short cut, shown in Table 20.1, can be used to estimate the sum of different sound levels.

For example, if three noise sources each have a sound power level of 100 dB, their total sound power level using the short-cut method is 100 $+_{\log}$ 100 = 103, 103 $+_{\log}$ 100 = 105 dB. [The notation "$+_{\log}$" indicates logarithmic addition.] Using Equation 7, their total sound power level would be 104.8 dB.

Sound pressure levels at different positions around most equipment are likely to vary by at least 3 dB. To calculate the average sound pressure level (in dB), it is necessary first to average the sound pressures (in Pa) and then to determine the sound pressure level of the average sound pressure.

Mathematically, the formula is:

$$\text{Average } L_P = 10\log_{10}\left(\frac{1}{n}\sum_{1}^{n} 10^{\frac{L_{P_n}}{10}}\right) \quad (8)$$

Table 20.1 — Short Cut Decibel Addition

Difference Between Two Levels to be Added	Amount to Add to Higher Level to Determine Sum
0–1 dB	3 dB
2–4 dB	2 dB
5–9 dB	1 dB
10 dB	0 dB

For example, the average of the following eight sound pressure levels (80, 80, 80, 81, 81, 81, 82, and 88 dB) is 83 dB.

Frequency

Most people can hear sounds ranging from very low frequencies below 30 Hz (the rumble of the largest organ pipes) to very high frequencies near 15,000 Hz (the shrill tone of a television). Most sounds have components at many different frequencies. For measurement purposes, this broad range of frequencies is normally divided into nine octave bands. An octave is defined as a range of frequencies extending from one frequency to exactly double that frequency. Each octave band is named for the center frequency (geometric mean) of the band. The center, lower, and upper frequencies for the commonly used octave bands are listed in Table 20.2.

For more detailed acoustical analysis, the octaves can be divided into one-third octave bands. For even more detail, narrow-band analyzers can measure the levels in bandwidths of less than 1 Hz. Narrow-band analysis is especially useful in tracking different tones (relatively high levels at a single frequency) produced by different equipment items.

A- and C-Weighting

Although octave-band sound level data are necessary for designing noise control treatments, a single-number description of the sound is often convenient. Most sound level meters provide the option of quantifying the combined sound at all frequencies with several different "weighting" filters: the "A-weighted" response, the "C-weighted" response, and the "flat" response.

The most common single-number measure is the A-weighted sound level, often

Table 20.2 — Commonly Used Octave Bands

Name of Octave Band (Center Frequency, Hz)	*Defining Frequencies (Hz)*		*A-Weighting of Octave Band (dB)*	*C-Weighting of Octave Band (dB)*
	Lower	*Upper*		
31.5	22.4	45	-39.4	-3.0
63	45	90	-26.2	-0.8
125	90	180	-16.1	-0.2
250	180	355	-8.6	0.0
500	355	710	-3.2	0.0
1000	710	1400	0.0	0.0
2000	1400	2800	+1.2	-0.2
4000	2800	5600	+1.0	-0.8
8000	5600	11,200	-1.1	-3.0

denoted dBA. The A-weighted response simulates the sensitivity of the human ear at moderate sound levels. Low-frequency sounds are significantly reduced in level (-26.2 dB at 63 Hz), and high- frequency sounds are slightly increased (+1.2 dB at 2000 Hz). After this weighting, the levels at all frequencies are summed logarithmically to determine the A-weighted sound level. In addition to being a good estimate of the perceived loudness at moderate levels, there is also a good correlation between the A-weighted sound level and the potential loss of hearing from prolonged noise exposure. Because of this, the A-weighted sound level is used for many applications from community noise ordinances to occupational noise exposure regulations. Table 20.3 presents typical A-weighted sound levels for different sounds.

The C-weighted response simulates the sensitivity of the human ear at high sound levels. It has less weighting applied than the A-weighting filter (for example, only -1 dB at 63 Hz). The flat response looks at the entire audible frequency spectrum without applying any weighting.

The A-weighting and C-weighting values at the standard octave bands are shown in Table 20.2.

Table 20.3 — Typical A-Weighted Sound Levels

Sound Source and Measurement Location	*A-Weighted Sound Pressure Level*
Pneumatic chipper at operator's ear	120 dBA
Accelerating motorcycle at 1 m	110 dBA
Shouting at 1.5 m	100 dBA
Loud lawnmower at operator's ear	90 dBA
School children in noisy cafeteria	80 dBA
Freeway traffic at 50 m	70 dBA
Normal male voice at 1 m	60 dBA
Copying machine at 2 m	50 dBA
Suburban area at night	40 dBA
Air conditioning in auditorium	30 dBA
Quiet natural area with no wind	20 dBA
Anechoic sound testing chamber	10 dBA

Sound Propagation

An ideal point source, in the absence of sound reflections or absorption, will produce sound pressure levels at a distance "r" (in meters) according to the following equation:

$$L_P = L_W + 10\log_{10}\left(\frac{Q}{4\pi r^2}\right) \quad (9)$$

The term $1/4\pi r^2$ is the reciprocal of the spherical area through which sound is radiated. The factor Q can be considered a directivity factor. Q = 1 where there are no reflecting surfaces nearby and sound radiates spherically; Q = 2 where the sound source is near the floor with the sound radiating hemispherically, and the sound source will produce twice the sound energy (+3 dB) at the same distance compared with spherical radiation. If the source is near the intersection of the floor and a wall, the sound radiates in a one-quarter-spherical pattern and Q = 4. If the source is located directly in the corner of a room, it will radiate sound in a one-eighth-spherical pattern and Q = 8. These Q values represent a simplified presentation of directivity, and sound sources may have directional sound radiation patterns of their own without considering reflections.

Alternatively for a measurement surface with a shape that is not a section of a sphere, the reciprocal of the measurement surface area (1/A) can be substituted for $Q/4\pi r^2$.

In most indoor spaces, sound levels near a sound source will decrease with distance from the center of the source as defined in the preceding formula. At moderate distances from the source, multiple sound reflections from room surfaces will increase the sound level above what would be calculated using the formula. At greater distances, sound levels will remain constant regardless of additional distance from the source; this area is defined as the reverberant field. The sound level in the reverberant field depends on the sound power level of the sound source and the amount of sound absorption in the room. The formula for the sound level inside a building is:

$$L_P = L_W + 10\log_{10}\left(\frac{Q}{4\pi r^2} + \frac{4}{TA}\right) \quad (10)$$

where:

TA = total absorption in the room in m^2, calculated by summing the products of each surface's area with its sound absorption coefficient.

Since the sound absorption values of surfaces in rooms vary significantly at different frequencies, reverberation calculations should be made in octave bands, not using the A-weighted sound level. For sound propagation inside rooms with large volumes or over long distances outdoors, there is significant sound reduction in the 4000 and 8000 Hz octave bands due to atmospheric sound absorption. Additional details on the effects of sound absorption are presented in the section of this chapter on the control of noise exposure.

Physiology of Hearing

Introduction

The function of the hearing mechanism is to gather, transmit, and perceive sounds from the environment. The frequency range of the human ear typically extends from 20 Hz to 20,000 Hz. Our hearing sensitivity is greatest in childhood. As we get older, our ability to hear high frequency sound decreases — a condition called presbycusis. From a practical standpoint, many adults have difficulty perceiving sounds above 14,000 Hz.

The translation of acoustical energy into perceptions involves the conversion of sound pressure waves first into mechanical vibrations and then into electrochemical activity in the inner ear. This activity is transmitted by the auditory nerve to the brain for interpretation.

Anatomy

Sound can be perceived through air-conducted or bone-conducted pathways. With air conduction, sound waves pass into the external ear and cause movement of the tympanic membrane. The tympanic membrane excites the bones of the middle ear, which produce signals that are processed in the inner ear. With bone conduction, vibrations travel through the skull to the inner ear or are re-radiated as sound in the external or middle ear. Bone conduction occurs, for example, when you tap your jaw. The sound you perceive is not coming through your ears but through your skull. Under ordinary conditions, bone conduction is much less significant than air conduction in the hearing process. Sound perception via air conduction is the most efficient and common route and is discussed in more detail.

External Ear

A cross section of the ear is shown in Figure 20.3. The ear can be divided into three sections: an external portion (outer ear), an air-filled middle ear, and a fluid-filled inner ear. The external ear consists of the pinna, the external auditory meatus (ear canal), and the tympanic membrane (eardrum). The pinna aids in the localization of sound and increases the sound level at the tympanic membrane at some frequencies.

The external auditory canal extends from near the center of the pinna to the tympanic membrane. The canal is about 26 mm in length and about 7 mm in diameter. The ear canal protects the tympanic membrane and acts as a resonator, providing about 10 dB of gain to the tympanic membrane at around 3300 Hz.

The tympanic membrane separates the external ear from the middle ear. This almost cone-shaped, pearl-gray membrane is about 17.5 mm in diameter. The distance the membrane moves in response to the sound pressure waves is incredibly small, as little as one billionth of a centimeter.[1]

Middle Ear

The middle ear, located between the tympanic membrane and the inner ear, is an air-filled space containing the ossicles, which amplify sound. The ossicles are the three smallest bones in the human body: the malleus (hammer), the incus (anvil), and the stapes (stirrup). The handle of the malleus attaches to the eardrum and articulates with the incus, which is connected to the stapes. The malleus and the incus vibrate as a unit, transmitting the sound waves from the tympanic membrane to the footplate of the stapes, which pushes the oval window in and out.

Not shown in Figure 20.3 are two intratympanic muscles, the tensor tympani and the stapedius. These two muscles help maintain the ossicles in their proper position and help to protect the internal ear from excessive sound levels. When the ear is exposed to sound levels above 80 dB,[2] the tensor tympani and stapedius muscles contract, decreasing the energy transferred to the oval window. This protective reflex, identified as the "aural reflex," does not react fast enough to provide protection against impulsive sounds, nor do the muscles stay contracted long enough to provide protection from long-term steady-sound exposures.

The Eustachian tube connects the anterior wall of the middle ear with the nasopharynx. It is about 38 mm in length and consists of an outer osseous (bony) portion (one-third of the tube which opens into the middle ear) and an inner cartilaginous part (two-thirds of the tube which opens into the throat). The lumen of the osseous portion is permanently opened while that of the cartilaginous portion is closed except during certain periods. The act of swallowing, for example, opens the cartilaginous portion of the Eustachian tube up to the middle ear and equalizes the atmospheric pressure on either side of the tympanic membrane. This equalization of pressure is necessary for optimal hearing.

Inner Ear

The inner ear is a complex system of ducts and sacs that house the three semicircular canals (used for balance) and the

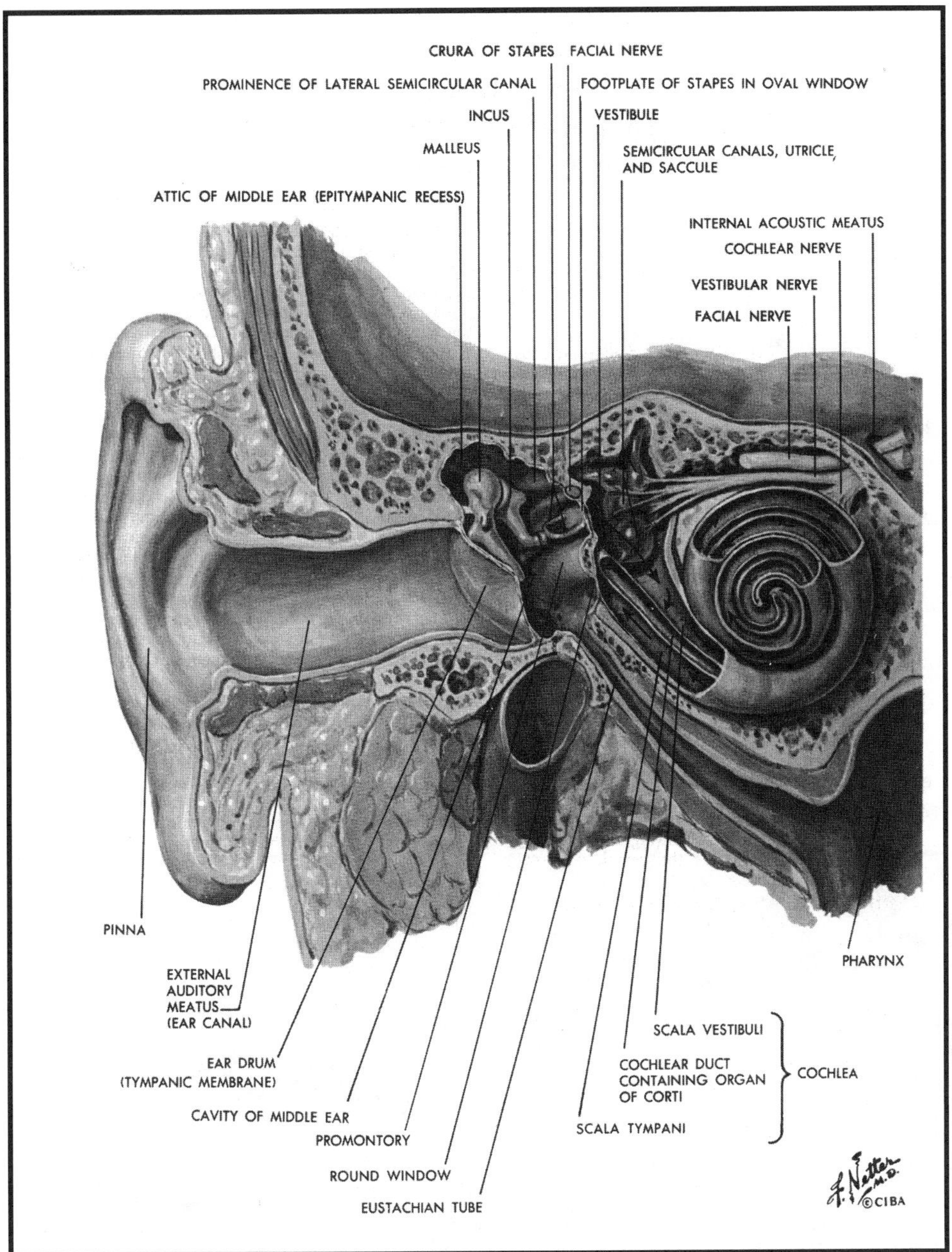

Figure 20.3 — Pathway of sound conduction showing anatomic relationships. (Copyright 1990 by Novartis Medical Education, Summit, N.J. Reproduced with permission from *Atlas of Human Anatomy* by Frank H. Netter, MD. All rights reserved.)

cochlea. The cochlea, shown in Figures 20.3 and 20.4, resembles a snail shell which spirals for about two-and-three-quarter turns around a bony column called the modiolus. Within the cochlea are three canals: the scala vestibuli, the scala tympani, and the scala media or cochlea duct. A bony shelf, the spiral lamina, together with the basilar membrane and the spiral ligament, separate the upper scala vestibuli from the lower scala tympani. The cochlear duct is cut off from the scala vestibuli by Reissner's vestibular membrane.

The scala media is a triangular-shaped duct that contains the organ of hearing, the organ of Corti. The basilar membrane, narrowest and stiffest near the oval window, widest at the apex of the cochlea, helps form the floor of the cochlear duct. On the surface of the basilar membrane are found phalangeal cells that support the critical "hair cells" of the organ of Corti. The hair cells are arranged in a definite pattern with an inner row of about 3500 hair cells and three to five rows of outer hair cells numbering about 12,000. The cilia of the hair cells extend along the entire length of the cochlear duct and are imbedded in the undersurface of the gelatinous overhanging tectorial membrane (see Figure 20.5). In general, the hair cells at the base of the cochlea respond to high-frequency sounds while those at the apex respond to low-frequency sounds.

The movement of the stapedial footplate in and out of the oval window moves the fluid in the scala vestibuli (see Figure 20.6). This fluid pulse travels up the scala vestibuli but causes a downward shift of the cochlear duct with distortion of Reissner's membrane and a displacement of the organ of Corti. The activity is then transmitted through the basilar membrane to the scala tympani. At the end of the cochlea, the round window acts as a relief point and bulges outward when the oval window is pushed inward.

The vibration of the basilar membrane causes a pull, or shearing force, of the hair cells against the tectorial membrane. This "to and fro" bending of the hair cells activates the neural endings so that sound is transformed into an electrochemical response. This response travels through the vestibulocochlear nerve and the brain interprets the signal as sound.

Auditory Sensitivity

To convert acoustical energy in air to a fluidborne signal, the ear must resolve the impedance matching problem that exists. Without such a solution, only about 0.1% of the airborne sound would enter the liquid medium of the inner ear, whereas the other 99.9% would be reflected away from its surface. This represents a 30 dB decrease in the liquid level vs. the airborne level. In other words, the intensity of vibration in the fluid of the inner ear would be 30 decibels less than the intensity present at the external ear.

The middle ear has two arrangements to narrow this potential energy loss. First is the size differential between the comparatively large eardrum and the relatively small footplate of the stapes. The eardrum has an effective area that is 14 times greater than that of the stapedial footplate. This effect increases the force of pressure from the eardrum onto the footplate of the stapes so that there is approximately a 23 dB increase of sound intensity of the fluid of the inner ear. Also, the lever action of the ossicles amplifies the intensity of sound as it traverses the middle ear by

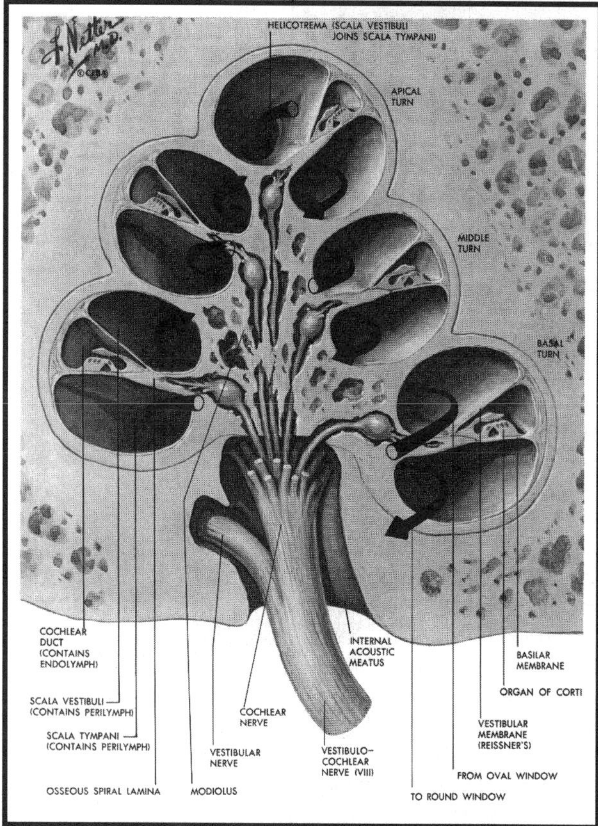

Figure 20.4 — Cross section of the cochlea. (Copyright 1992 by Novartis Medical Education, Summit, N.J. Reproduced with permission from *Clinical Symposia*, Vol. 44, No. 3; illustrated by Frank H. Netter, MD. All rights reserved.)

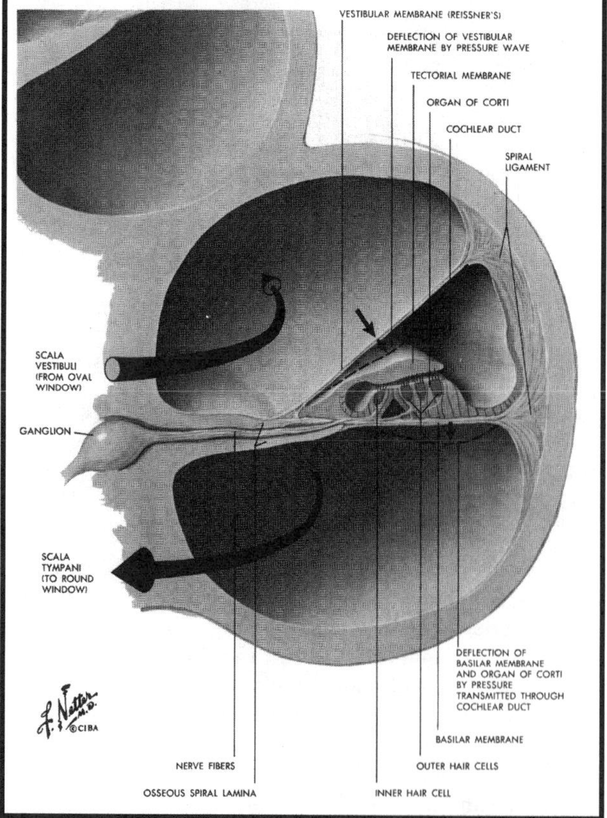

Figure 20.5 — Transmission of sound across the cochlear duct stimulating the hair cells. (Copyright 1992 by Novartis Medical Education, Summit, N.J. Reproduced with permission from *Clinical Symposia*, Vol. 44, No. 3; illustrated by Frank H. Netter, MD. All rights reserved.)

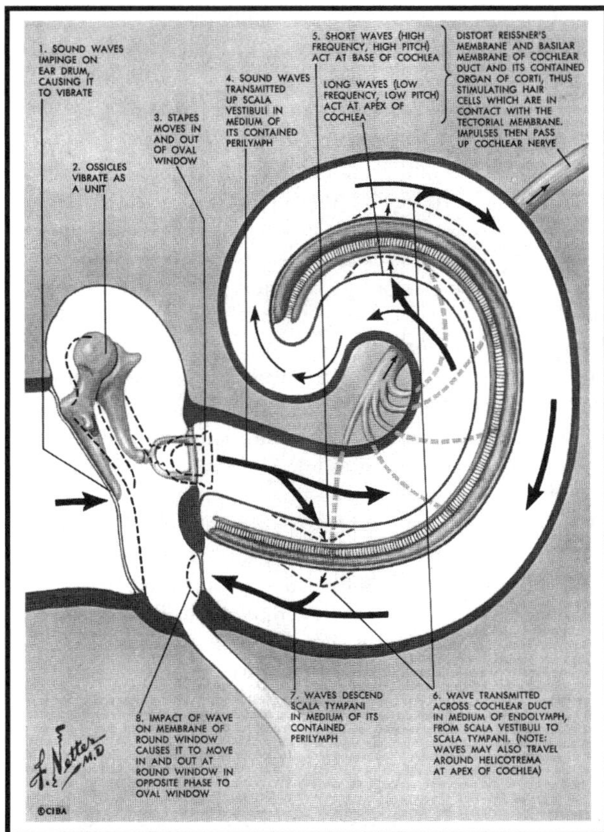

Figure 20.6 — Transmission of vibrations from the eardrum through the cochlea. (Copyright 1970 by Novartis Medical Education, Summit, N.J. Reproduced with permission from *Clinical Symposia*, Vol. 22, No. 4; illustrated by Frank H. Netter, MD. All rights reserved.)

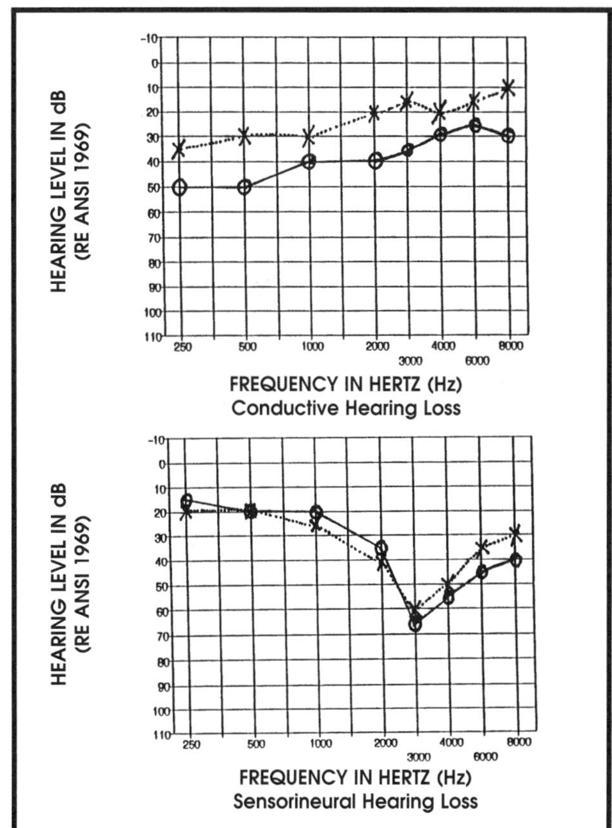

Figure 20.7 — Audiograms showing conductive and sensorineural hearing loss.

about 2.5 dB. Thus, the impedance matching mechanism of the middle ear accounts for a 25.5 dB increase in the intensity of sound pressure at the air-liquid interface.

Audiometry

The pure tone audiometer is the fundamental tool used in industry to evaluate a person's hearing sensitivity. It produces tones at 500, 1000, 2000, 3000, 4000, and 6000 Hz. Many audiometers include a tone at 8000 Hz and some also include tones at 125 Hz and 250 Hz. The intensity output from the audiometer can vary from 0 dB to about 110 dB and is often marked "hearing loss" or "hearing level" on the audiometer.

The zero reference on the audiometer ("0 dB hearing level") is the average level for normal hearing for different pure tones and varies according to the "standard" to which the audiometer is calibrated. Reference levels have been obtained by testing the hearing sensitivity of young healthy adults and averaging the sound level at specific frequencies at which the tones were just perceptible. If a person has a 40 dB hearing loss at 4000 Hz, it means that the intensity of a tone at 4000 Hz must be raised 40 dB above the "standard" to be perceived by that person. Since the threshold of hearing is different for different frequencies, the sound pressure level required to produce 0 dB hearing level will also be different for different frequencies.

The audiogram serves to record the results of the hearing tests. The level of the faintest sound audible is plotted for each test frequency. In conductive hearing losses, the low frequencies show most of the threshold elevation, whereas the high frequencies are most often affected by sensorineural losses. Noise-induced hearing loss often is considered a "dip" at 4000 Hz, but for older individuals this dip might not show due to presbycusis effects at higher frequencies. Examples of audiograms that indicate conductive and sensorineural losses are shown in Figure 20.7.

The recording of an audiogram is deceptively simple, yet for valid test results one must have a properly calibrated audiometer, a proper test environment, and a qualified audiometric technician. When a marked hearing loss is encountered, more sophisticated tests by an audiologist or physician are required to diagnose the site and cause of the hearing loss.

Classification of Hearing Loss

The three main types of hearing loss are conductive hearing loss, sensorineural hearing loss, and a combination thereof.

Conductive Hearing Loss

Any condition that interferes with the transmission of sound to the cochlea is classified as a conductive hearing loss. Pure conductive losses do not damage the organ of Corti or the neural pathways.

A conductive loss can be due to wax in the external auditory canal, a large perforation in the eardrum, blockage of the eustachian tube, interruption of the ossicular chain due to trauma or disease, fluid in the middle ear secondary to infection, or otosclerosis (i.e., fixation of the stapedial footplate). A significant number of conductive hearing losses are amenable to medical or surgical treatment.

Sensorineural Hearing Loss

A sensorineural hearing loss is almost always irreversible. The sensory component of the loss involves the organ of Corti, and the neural component implies degeneration of the neural elements of the auditory nerve.

Exposure to excessive noise causes irreversible sensorineural hearing loss. Damage to the hair cells is of critical importance in the pathophysiology of noise-induced hearing loss. Invariably, degeneration of the nerve fibers accompany severe injury to the hair cells.

Sensorineural hearing loss may be attributed to various causes, including presbycusis, viruses (e.g., mumps), some congenital defects (e.g., heredity), and drug toxicity (e.g., aminoglycosides).

Effects of Excessive Noise Exposure on the Ear

Excessive noise exposure to the ear can cause a noise-induced temporary threshold shift (NITTS), a noise-induced permanent threshold shift (NIPTS), tinnitus, and/or acoustic trauma.

Noise-Induced Temporary Threshold Shift (NITTS)

NITTS refers to a temporary loss in hearing sensitivity. This loss can be a result of the acoustic reflex, short-term exposure to noise, or simply neural fatigue in the inner ear. With NITTS, hearing sensitivity will return to the pre-exposed level in a matter of hours or days (without continued excessive exposure).

Noise-Induced Permanent Threshold Shift (NIPTS)

NIPTS refers to a permanent loss in hearing sensitivity due to the destruction of sensory cells in the inner ear. This damage can be caused by long-term exposure to noise or by acoustic trauma.

Tinnitus

The term "tinnitus" is used to describe the condition in which people complain of a sound in the ear(s). The sound is often described as a hum, buzz, roar, ring, or whistle. This sound is produced by the inner ear or neural system. Tinnitus can be caused by nonacoustic events such as a blow to the head or prolonged use of aspirin. However, the predominant cause is long-term exposure to high sound levels, though it can be caused by short-term exposure to very high sound levels such as firecrackers or gunshots. Many people experience tinnitus during their lives. Often the sensation is only temporary, though it can be permanent and debilitating. Diagnosis and treatment of tinnitus can be difficult because tinnitus is subjective and cannot be measured independent of the subject.

Acoustic Trauma

Acoustic trauma refers to temporary or permanent hearing loss due to a sudden intense acoustic event such as an explosion. The results of acoustic trauma can be a conductive or sensorineural hearing loss. An example of a conductive loss is when the event causes a perforated eardrum or destruction of the middle ear ossicles. An example of a sensorineural loss due to acoustic trauma is when the event causes temporary or permanent damage to the hair cells of the cochlea.

Acceptability Criteria

Introduction

The acceptability of noise can be judged by whether the level of the noise and exposure to the noise can cause hearing loss, annoy people, or interfere with speech communication, hearing or emergency warning signals.

Hearing Loss Criteria

Noise-induced permanent threshold shift affects 10–20 million workers in this country alone.[3,4] If employees are exposed to high noise levels each workday for a working lifetime without adequate hearing protection, they can develop a permanent, irreversible hearing loss.

Figure 20.8 illustrates the impact of noise exposure on hearing. The figure presents the hearing level distribution for four sets of workers.[5] One of the sets is a non-noise-exposed group. The median hearing level as a function of frequency of this non-noise-exposed group is shown in each of the three graphs as the dashed line near the top of each graph. The other three sets of workers (one set for each of the three graphs) were exposed to average daily noise levels of 85 dBA, 90 dBA, and 95 dBA, respectively. The mean age and mean exposure time for these workers are also shown. In addition to the median levels for the noise-exposed workers, the cumulative distribution of the data is also shown. For example, at 4000 Hz, exposure to a noise level of 90 dBA resulted in a median hearing level of about 40 dB, whereas exposure to a noise level of 85 dBA resulted in a median hearing level of only about 30 dB.

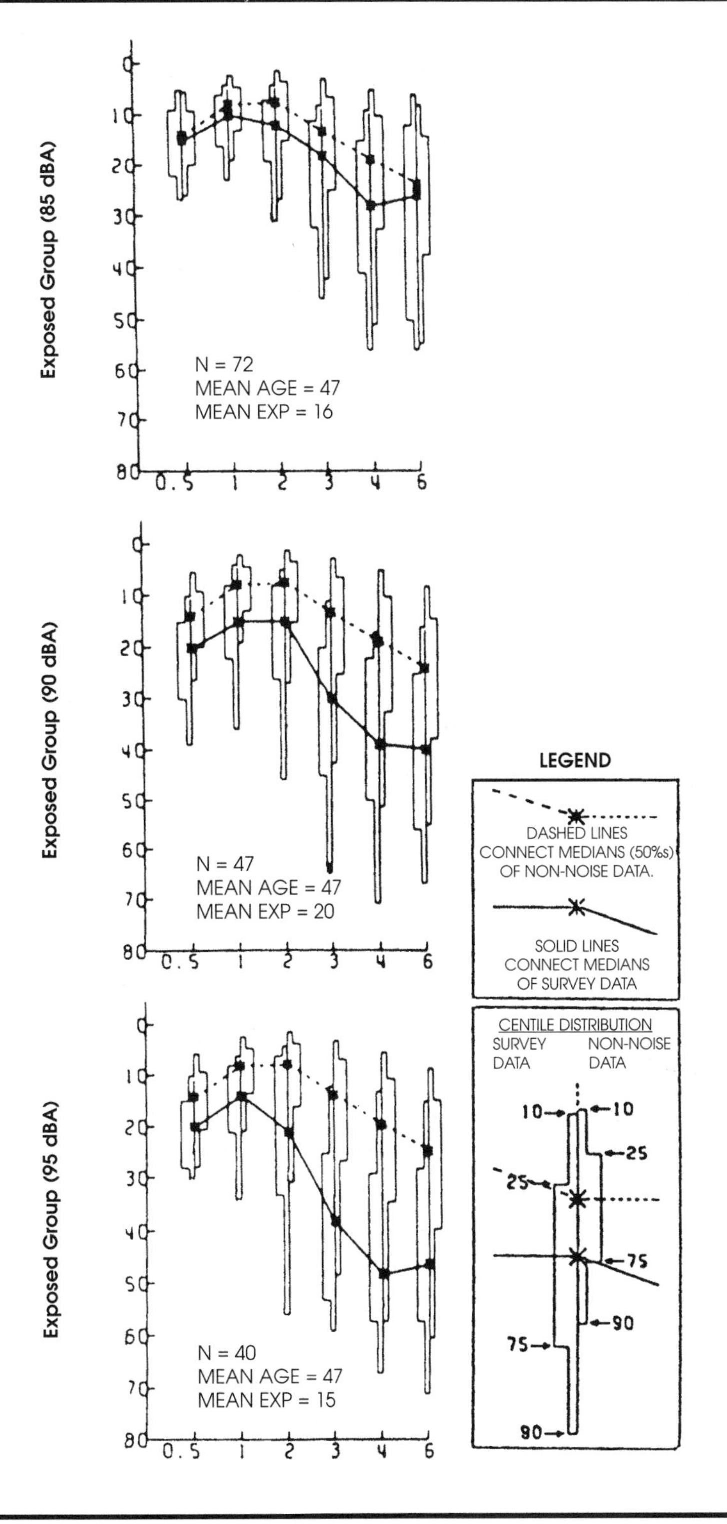

Figure 20.8 — Hearing level distribution for workers aged 43 to 51 exposed to daily average noise levels of 85, 90, and 95 dBA and for workers not exposed. (Adapted from Reference # 5.)

Federal Regulations

The noise exposure of employees in most occupations and industries is regulated by at least one federal agency. In the following paragraphs, the regulations of the following agencies are summarized:

- Occupational Safety and Health Administration (OSHA)
- Mine Safety and Health Administration (MSHA)
- Federal Railroad Administration (FRA)
- U.S. Coast Guard (USCG)
- Federal Highway Administration (FHWA)
- U.S. Department of Defense (DOD).

Occupational Safety and Health Administration. The U.S. Department of Labor enacted the first specific regulation in the United States for protecting the hearing of civilian employees by amending the Walsh-Healey Act to incorporate a limitation on the noise exposure of workers covered by that act. OSHA was created by Congress in the Occupational Safety and Health (OSH) Act of 1970 and authorized to protect workers against material harm. The current OSHA noise exposure regulation, 29 CFR 1910.95, limits the noise exposure of most civilian employees working in the manufacturing, utilities, and service industries to the values given in Table 20.4.[6] It should be noted that the current OSHA regulation is basically the same as the original regulation but with a thorough definition of what OSHA means by a "continuing, effective hearing conservation program." This regulation has an exchange rate or doubling rate of 5 dB. In other

Table 20.4 — OSHA's Noise Exposure Limits

OSHA's Table G-16. Permissible Noise Exposures for Determining the Need for Engineering or Administrative Controls.

Duration (hr/day)	Sound Level (dBA)
8	90
6	92
4	95
3	97
2	100
1 1/2	102
1	105
1/2	110
1/4	115

OSHA's Table G-16a. Permissible Noise Exposure for Determining the Need for Hearing Conservation Program.

A-weighted Sound Level Duration	Reference Hours
80	32
85	16
90	8
95	4
100	2
105	1
110	.50
115	.250
120	.125
125	0.63
130	.031

words, as the sound level is increased by 5 dB, the permissible duration time is decreased by 50%.

Exposure to continuous steady-state noise is limited to a maximum of 115 dBA, while exposure to impulse or impact noise is limited to 140 dB peak sound pressure level.

If workers are exposed to different A-weighted sound levels during the day, the total noise dose (D) is computed with the formula:

$$D = \frac{C_1}{T_1} + \frac{C_2}{T_2} \ldots + \frac{C_n}{T_n} \qquad (11)$$

where:

C_n = employee's exposure time at a particular noise level; and
T_n = total time allowed at that noise level.

A dose of 1.0 (100%) is equivalent to a time-weighted average (TWA) of 90 dBA. The equation for TWA is:

$$TWA = 16.61 * \log_{10}(D) + 90 \text{ dBA} \qquad (12)$$

where:

D = total noise dose.

For determining the need for engineering or administrative procedures, only the exposures to sound levels ≥90 dBA are used in the calculation, as noted in Table 20.4. If the dose is greater than 1.0 (100%) (TWA >90 dBA), feasible administrative or engineering controls must be initiated to reduce the TWA to not greater than 90. For determining the need for hearing conservation programs, all sounds at or above 80 dBA are used in the calculation. If the calculated dose exceeds 0.5 (TWA >85 dBA), employees must be included in a comprehensive hearing conservation program that includes sound monitoring, feasible administrative or engineering controls, audiometric testing, hearing protection, employee training and education, and record keeping. The details of this program are delineated in the current OSHA noise regulation.[6]

Some workers are not covered by the provisions of 29 CFR 1910.95. OSHA provides for the protection of construction workers through 20 CFR 1926.52.[7] The major difference between 1910.95 and 1926.52 is that in the construction industry no protection is required for TWAs between 85 dBA and 90 dBA, and the specific provisions of the hearing conservation program are not listed in detail. As a result, construction workers are more likely to be exposed to noisy operations without proper hearing protection.

Another example of workers not covered by all of the detailed requirements of the hearing conservation portion of CFR 1910.95 are the employees of oil and gas well drilling and servicing operations.[8] Also, civilian employees of other federal agencies are not covered by the OSHA regulations. In accordance with Executive Order 12196, however, they are supposed to be covered by regulations issued by their respective agencies.[9] These regulations were to be as protective as the OSHA regulation. It is unclear which agencies, if any, issued such regulations.

In addition to construction workers, oil and gas well drilling and servicing operations, and most civilian federal employees, a number of other workers are not covered by the OSHA regulation. These include farm workers who are employed by farms with fewer than 10 workers, crew members on all commercial vessels, trainmen, mine workers, and civilian and military employees of the U.S. Department of Defense. Table 20.5 provides a list of these workers, the federal agencies responsible for the administration of their noise limits, and the acoustical aspects of their regulations, including the following considerations:

- Exchange rate, either 3 dB (equal energy) or 5 dB;
- Action level above which a hearing conservation program is required;
- Permissible exposure level and the reference time period;
- Threshold — the lowest sound level that is included in the noise dose calculation;
- Maximum continuous sound, often 115 dBA;
- Maximum impulse noise, often 140 dB; and
- Mandatory hearing protection program, either yes or no.

These regulations are discussed in the following sections.

Mine Safety and Health Administration. There are two categories in the mining industry: coal mining and metal and nonmetal mining. Each of these industries is divided into surface mining and underground mining. The coal mining regulation[10] limits noise exposure to the same values as the 29 CFR 1910.95 regulation. However, the detailed provisions of the OSHA hearing conservation regulation are not mandatory requirements for the mining industry. Recently, MSHA proposed a rule that will require hearing conservation programs whenever the TWA is greater than 85 dBA.[11] The program also requires provision of double hearing protection (earplugs plus earmuffs) when the TWA is greater than 105 dBA.

Federal Railroad Administration. Since trainmen often work 12-hour shifts and since this is the maximum allowable work load, the FRA limits the noise exposure of train personnel to 87 dBA for a 12-hour exposure.[12,13] It uses the 5 dB exchange rate. Hearing conservation program generally are not mandatory; in fact, many railroads have only recently permitted the use of hearing protectors in the locomotive cab.

U.S. Coast Guard. Since crew members on commercial vessels are exposed to sound levels continuously for a 24-hour period, the Coast Guard has developed a criteria of $L_{eff}(24)$ of 82 dBA.[14] This is equivalent to an exposure of 90 dBA for 8 hours and to less than 80 dBA for the remaining 16 hours (since the time at noise

Table 20.5 — Summary of Occupational Noise Regulations in the United States

	Acoustical Requirements								
Agency and Worker Types	*A*	*B*	*C*	*D*	*E*	*F*	*G*	*H*	*I*
OSHA									
Manufacturing, Utilities, and Service Sectors	5	85	90	8	80*	115	140	Yes	3
Construction Workers	5	90	90	8	90	115	140	Yes**	4
Other Agencies' Civilian Employees	5		90	8	80	115	140	No	5
MSHA									
Metal and Nonmetal Mining									
Surface Mining	5	85#	90	8	90	115	140	No#	6,7
Underground Mining	5	85#	90	8	90	115	140	No#	6,7
Coal Mining									
Surface Mining	5	85#	90	8	90	115	140	No#	6,7
Underground Mining	5	85#	90	8	90	115	140	No#	6,7
FRA									
Trainmen	5		87	12		115		No	8,9
DOT									
Crew Members on all Commercial Vessels	5		82	24	80	115	140	No	10
Motor Vehicles	–		90	–	–	–	–	No	11
DOD	3	85	85	8			140	Yes	12

A. Exchange Rate
B. Action Level (dBA)
C. Permissible Exposure Level (dBA)
D. Hours Allowed at Sound Level in B and C
E. Threshold in dBA
F. Maximum Continuous Sound (dBA)
G. Maximum Impulse Noise (dBA)
H. Mandatory Hearing Conservation Program
I. References

* 80 dBA for hearing conservation; 90 dBA for engineering and administrative controls.
**This regulation does not provide any details for a hearing conservation program.
#Proposed Rule requires hearing conservation program.

levels less than 80 dBA is not counted in the calculation). Although hearing protection is encouraged, a mandatory hearing protection program with all of the provisions of the OSHA regulation is currently not required.

Federal Highway Administration. The FHWA limits all vehicle interior noise levels to 90 dBA.[15] In effect, this places a limit on the noise exposure of truck drivers, bus drivers, and other drivers.

U.S. Department of Defense. The different branches of the DOD (principally the Army, Air Force, and Navy) have from time to time used different action levels and different exchange rates for evaluating noise exposure. Currently, the Army, Air Force, and Navy use a 3 dB exchange rate with an 8-hour TWA of 85 dBA as the action level at which civilians and military personnel are included in an audiometry program.[16]

Hearing Damage Risk Recommendations

Other organizations have developed recommendations for limiting noise exposure in order to prevent hearing loss.

U.S. Environmental Protection Agency (EPA). Among other responsibilities, the EPA is charged with protecting the public with an adequate margin of safety under its enabling legislation. The EPA recommended an L_{eq}(8 hours) of 85 dBA and a future regulation of 80 dBA.[17] Inherent in the use of an L_{eq} measure of the noise level is the use of a 3 dB exchange rate.

National Institute for Occupational Safety and Health (NIOSH). In 1972, NIOSH recommended an 85 dBA TWA standard with a 5 dB exchange rate for occupational noise exposures.[18] In 1996, NIOSH released a draft criteria document[19] in which it recommends:

- Limiting noise exposure to 85 dBA TWA for 8 hours;
- Using a 3 dBA exchange rate;
- Reducing the expected performance of hearing protectors (25% for muffs, 50% for formable plugs, and 70% for all other plugs);
- Starting hearing conservation program at an 8-hour TWA of 82 dBA; and
- Using hearing protection for any noise exposure over 85 dBA.

International Institute of Noise Control Engineering (I-INCE). The I-INCE has issued a draft report discussing noise in the workplace. It recommends a TWA of 85 dBA and a 3 dB exchange rate.[20]

American Conference of Governmental Industrial Hygienists (ACGIH). ACGIH recently revised its threshold limit values (TLVs®) for noise.[21] No exposure to continuous, intermittent, or impact noise in excess of a peak C-weighted sound level of 140 dB is allowed. Exposures are based on a 3 dB exchange rate. The recommended TLVs range from 80 dBA for a 24-hour period to 139 dBA for 0.11 seconds. Exposures in excess of these values require personnel to be included in a hearing conservation program.

These recommendations by EPA, NIOSH, I-INCE, and ACGIH — along with the practice of many countries — are likely to influence future corporate policies and government regulations in the United States. Although researchers continue to discuss the merits of these different noise criteria, this is certain: Each individual accumulates hearing loss at his or her own rate, and the higher the noise level and longer the exposure time, the greater the probability of permanent hearing loss.

Other Considerations

Besides hearing loss, there are other concerns about noise such as speech interference and annoyance.

Speech Interference

Face-to-face voice communication and telephone usage are affected by noise levels. For situations in which speech communication is important (e.g., control rooms) the background sound level should be limited to 55 dBA.[22] Although it is possible to communicate at higher background sound levels for short periods, it is very difficult to maintain a high vocal output throughout a work shift.

When exposed to sound levels above about 45–50 dBA, a talker will raise his or her voice by 3–6 dB for each 10 dB increase of noise. In Table 20.6, the column labeled "shouting" represents the maximum normal output from a talker. This communication is just-reliable, which means trained listeners would score 70% on a word list composed of phonetically balanced monosyllabic words.[23] Parenthetically, if you have to shout to be understood at a distance of 1 m, the sound level is high enough to warrant the use of hearing protection.

High noise levels can also interfere with the audibility and clarity of emergency warning devices.

Annoyance

Annoyance is usually associated with community noise concerns; however, employees in occupational environments can also be annoyed by sounds. Methods are available for estimating community responses to sounds,[24] but there are no methods for estimating individual responses. Noises with tonal sounds, aperiodic sounds, screeches, rumbles, and whines can interfere with the performance of tasks requiring concentration: these sounds can also be annoying. In general, when the sound level of an intermittent broad-band noise is about 10 dBA above the background sound levels, it is likely to annoy listeners. Tonal sounds and other "attention-getting" sounds might cause annoyance at lower levels.

Noise can interfere with a person's ability to perform tasks. Conflicting data suggests that noise can disrupt or enhance performance. Obviously, task interference from noise is a multifaceted issue, and the effects of other variables influence the studies.[25]

Noise Measurements

Introduction

The measurement of sound and vibration has been greatly facilitated by advances in the electronics and computer industries. Today, the occupational hygienist can carry instruments for measuring the noise dose, the octave band sound pressure level, the vibration of a surface, and the sound intensity of a source along with a portable computer in several briefcases. The ease of making measurements has resulted in an increase in the number of possible measurements. Whenever sound or vibration is measured, it is imperative to remember the question: Why are these measurements being made? Distractions are numerous and one must not forget this question while making measurements. This section will focus on measurement systems, noise measurements for OSHA compliance, and noise and vibration measurements for evaluating sound sources.

Sound Measuring Devices

For the purpose of evaluating personnel noise exposures, the A-weighted sound level is needed at the ear of the person being monitored. For determining the source of a noise (e.g., part of a machine or system), it may be necessary to measure the sound intensity and/or the vibration of the equipment at different frequencies. This section will discuss the instrumentation that hygienists can use to measure noise and vibration.

Table 20.6 — Speech Communication in Noisy Environments

Distance between Listener and Talker		*Maximum Background Sound Level in dBA for Just-Reliable Communion*
feet	*meters*	*"Shouting"*
1	.3	98
1.6	.5	93
3.3	1	84
6.6	2	78
13.2	4	71
26.4	8	66

Sound Level Meters

The basic instrument for the occupational hygienist or engineer investigating noise levels is the sound level meter (SLM). The SLM consists of a microphone, a preamplifier, an amplifier with an adjustable and calibrated gain, frequency weighting filters, meter response circuits, and an analog meter or digital readout. Also, most meters will have an output jack where the signal can be connected to a tape recorder or a graphic level recorder. There are three levels of precision available, classified by the American National Standards Institute (ANSI) as Types 0, 1, and 2.[26] Type 0 is a laboratory standard; Type 1 is for precision measurements in the field; and Type 2 is for general purpose measurements. A Type 2 meter is the minimum requirement by OSHA for sound measurement equipment and is usually sufficient for surveys by hygienists. For the design of cost-effective noise control treatments, however, the Type 1 meter is preferred.

Most SLMs provide at least two options for frequency weighting: A and C. Some also provide a flat response. Most SLMs have two meter response characteristics: slow and fast. With the fast response, the meter very closely follows the sound level as it changes. Slow response tends to be more sluggish and gives an average of the changing sound level. The OSHA regulation requires measurements to be made with the slow meter response.

Some meters have an "impulse" response or a "peak" response for transient or impulsive sounds. The peak value is the maximum value of the waveform. The impulse response is an integrated measurement and should not be used when measuring peak levels for OSHA compliance. With the impulse response, the meter initially responds very quickly to sounds of short durations, but it decays slowly in order to facilitate reading the value. On some meters with a digital display, the indicated impulse level can remain at the value until the user resets the meter. The

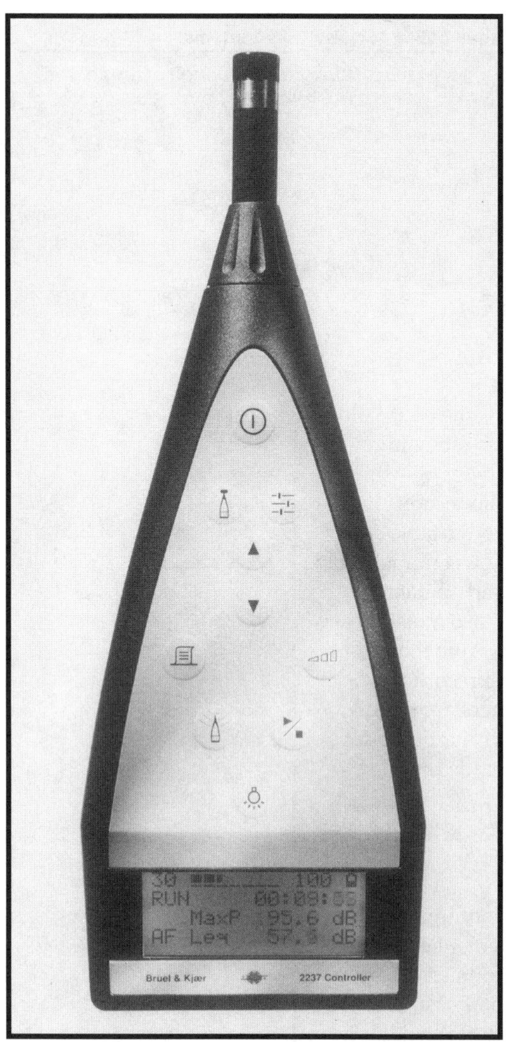

Figure 20.9 — Type 2 sound level meter. (Courtesy Brüel & Kjær.)

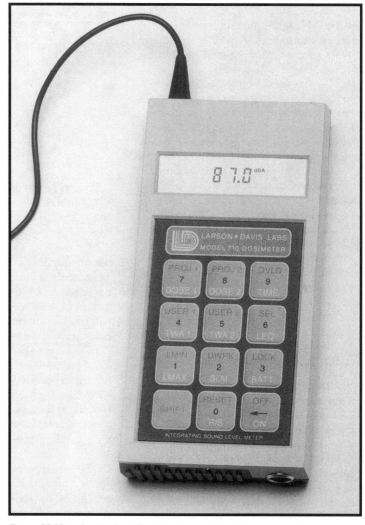

Figure 20.10 — Dosimeter. (Courtesy Larson-Davis.)

peak value can also be displayed until the meter is reset.

In many situations the noise level varies with time such that reading a single number from a meter is difficult. An integrating SLM measures the sound level over a period and calculates the integrated sound level (e.g., L_{eq} [3 dB exchange rate] or L_{osha} [5 dB exchange rate]). An example of an integrating SLM with an A-scale reading and the ability to measure peak values is shown in Figure 20.9.

Dosimeters

Noise dosimeters are a special version of a sound level meter that can measure the noise dose. Many types can be used as a sound level meter to measure the sound level in an area, or as a dosimeter to calculate the daily noise dose based on a full work shift of measurements. Most dosimeters can also calculate a projected dose based on a shorter, exemplary sample. Different noise dose criteria, exchange rates, and thresholds can be selected, such as an 85 dBA criterion, 5 dB exchange rate and an 80 dBA threshold. The dosimeter can be programmed to sample from, say, 8 a.m. to noon, then 1 p.m. to 5 p.m., and can be locked to protect the data from potential tampering by unauthorized personnel. The dosimeter can be interfaced with a computer for data retrieval or it can be connected directly to a printer to print detailed reports of sound level fluctuations over time. Figure 20.10 is an example of a dosimeter.

Sound Intensity Meters

Sound intensity analyzers can be used to identify specific noise sources and to determine compliance with purchase specifications limiting the sound power level. While the sound pressure level indicates the level of the sound (but not the direction from

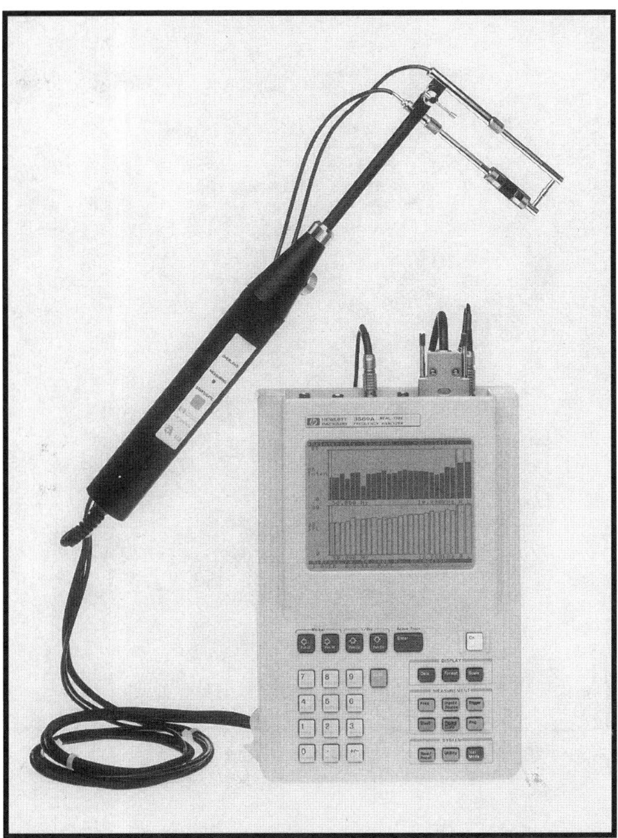

Figure 20.11 — Sound intensity instrumentation system. (Courtesy Hewlett-Packard.)

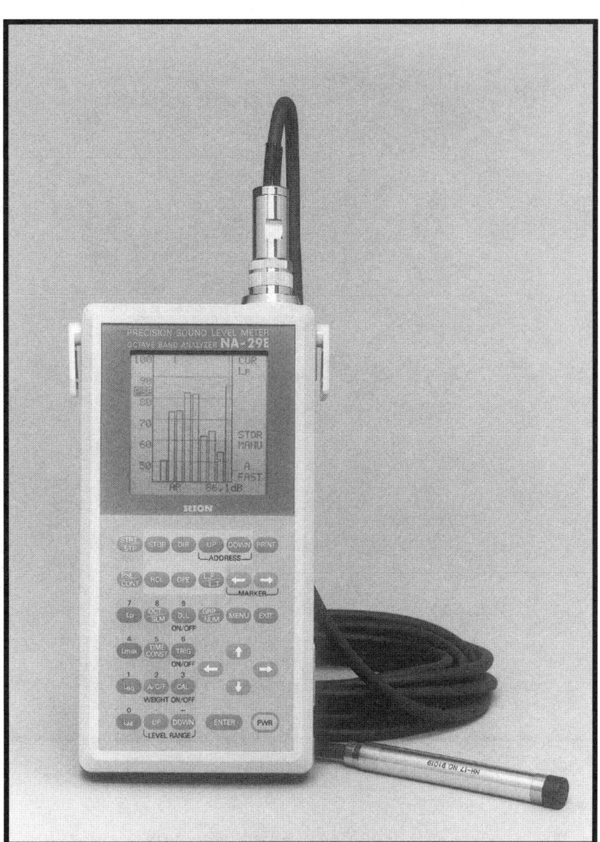

Figure 20.12 — Type 1 integrating sound level meter. (Courtesy Rion.)

which the sound is coming), sound intensity is a measure of both the magnitude and direction of the sound energy. Noise sources can be pinpointed and their sound powers can be rank-ordered using sound intensity, even in occupational environments that are often reverberant. Figure 20.11 shows a battery-powered sound-intensity instrumentation system.

Narrow-Band Analyzers

Measuring the A-weighted sound level is usually sufficient for determining employee noise exposures and determining compliance or violation with some community noise ordinances. However, to analyze sound sources and develop noise control treatments, the frequency content of the sound must be determined. For example, a noise control treatment will have different effects in controlling a low-frequency hum, a mid-frequency whine, or a high-frequency squeal. Usually, the audible frequency spectrum is divided into nine octave bands of frequency ranging from 31.5 Hz to 8000 Hz. For a more detailed analysis, the spectrum is measured in one-third octave bands. Usually a Type 1 (precision) SLM is used for octave and one-third octave analysis. Figure 20.12 shows a Type 1-integrating SLM that can measure levels in octave bands.

Another approach is to use FFT (fast Fourier transform) spectrum analyzer to divide the spectrum into even smaller bands. This type of analysis can be used to identify tones that can be traced to specific pieces of equipment. For example, if a tone appears at 29.2 Hz (1750 rpm), the noise source is probably related to a rotating equipment item such as an electric motor, many of which operate at 1750 rpm. A dual-channel analyzer is part of a sound intensity instrumentation system.

Tape Recorders

Although it is good practice to perform both measurements and analysis in the field, sometimes it is either inconvenient to move the analysis instrumentation to the site or there is too much data to analyze in "real time." For these situations, an instrumentation recorder can be quite helpful. Sound and vibration data can be recorded in the field and played back in the laboratory for analysis. Since the frequency response and the dynamic range of a tape recorder can influence the

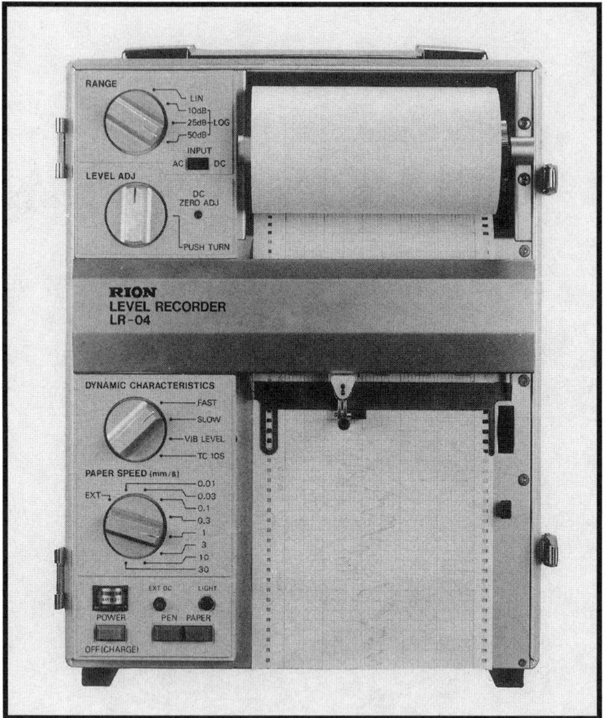

Figure 20.13 — Graphic level recorder. (Courtesy Rion.)

Figure 20.14 — Acoustic calibrator. (Courtesy Larson-Davis.)

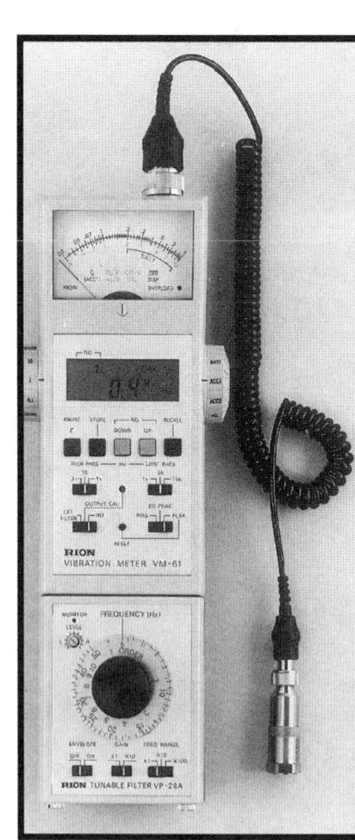

Figure 20.15 — Vibration meter. (Courtesy Rion.)

recording, this should be considered before recording. Various types of recorders are available that can record from one to many channels of data onto different media such as reel-to-reel tape, video cassette, audio cassette, or digital audio tape. These recorders typically have excellent frequency response and a much greater dynamic range than analog recording devices.

Graphic Level Recorders

The graphic level recorder (GLR) is an instrument for providing a written record of the sound levels of particular events as a function of time. Figure 20.13 is an example of one GLR. By connecting the output of a sound level meter or a vibration meter to the GLR, a permanent record on paper of the variation in sound or vibration level over time is written. The response of the writing pen can be set to speeds corresponding to the slow or fast response of the sound level meter.

Calibrators

For accurate measurements, it is necessary to calibrate the SLM before each set of measurements and then check the calibration during the day (approximately every 4 hours) and after the measurements are complete. For dosimetry measurements, calibration before and after a shift is normally sufficient. Figure 20.14 illustrates one type of acoustic calibrator. The calibrator applies a known sound pressure level to the microphone, and the SLM is adjusted to read the proper level. With many calibrators, a correction must be made for the altitude above sea level because this affects the barometric pressure. In general, the correction is needed only at higher altitudes (lower barometric pressures) such as 1500 m. This correction can be determined from the instruction manual or from a barometer included with the calibrator. Only the calibrator designed to calibrate a particular type and size of microphone should be used. In addition to this regular calibration check, sound meters and calibrators should be inspected, calibrated, and certified annually by qualified companies.

Vibration Instruments

Further investigative work can be performed by analyzing the vibrating surfaces of machinery. Vibration is measured by attaching an accelerometer to the surface and connecting it to some type of preamplifier, such as a charge amplifier, a vibration meter, or a sound level meter. The accelerometer can be temporarily attached to the surface by using a thin layer of non-drying, air-conditioning-duct-sealing compound (putty), beeswax, super-glue, or a magnetic mount. Care should be taken not to make measurements at the resonance of the accelerometer or the mounting. For accuracy at high frequencies or for a permanent mounting, a threaded steel stud and a thin film of petroleum jelly should be used. The vibration meter shown in Figure 20.15 allows for measurement of acceleration, velocity, or displacement. By examining the spectrum of the vibration signal, one can determine which surfaces are radiating particular sounds. Additional information about vibration measurements will be presented later in this chapter in the section on vibration.

Sound Measurement Techniques

Sound measurements for the purpose of occupational hearing conservation are focused on determining the duration of employee exposure to various sound levels. Sometimes, occupational hygienists are also called on to investigate and solve community noise problems. Although the criteria will be different, the principles of noise measurements are the same.

OSHA Compliance Survey

An OSHA compliance survey is performed in an industrial facility to identify areas with hazardous levels of noise and to determine the employees who should be included in a hearing conservation program. For example, in the United States, employees covered by 29 CFR 1910.95 with TWAs greater than 85 dBA should be included in a hearing conservation program. Since most measuring dosimeters are accurate to ±2 dB, a conservative approach is to include those persons within 2 dB of the action level.

The sound level meter is the primary instrument used in task-based monitoring and it can provide backup information to verify the noise dose calculated by the dosimeter. It is used to identify areas where sound levels exceed the action level and to identify equipment items or areas that control an employee's noise dose.

Sufficient details on the equipment operating conditions and worker locations must be documented to explain any changes in sound levels or noise doses when the survey is repeated.

The basic plan for determining compliance is to:

1. Tour the facility and develop a detailed understanding of the operation of the facility and its noise sources. Take this tour with someone who is familiar with the plant and its operations. During this tour, speak with knowledgeable personnel about operations and maintenance requirements. Make notes on a floor plan of the area if possible.
2. Identify workers and their locations and estimate the length of time they spend in different areas or how long they use particular tools, etc.
3. Perform an A-weighted sound level survey of the facility, marking the A-weighted sound levels on the floor plan and noting which equipment is on or off.
4. Compare the sound levels in the facility with the locations of personnel, as reported to you and as observed by you in item 1 above.
5. Identify employee groups based on their exposure to noise in common areas. Use a dosimeter to measure the noise exposure of individuals within these different groups.
6. Compare the noise doses as determined by the dosimeter with doses calculated using the measured noise levels and the estimated exposure times.
7. Evaluate any discrepancies and remeasure as necessary.
8. Include individuals with TWAs greater than 85 in a hearing conservation program.
9. Develop noise control treatments where feasible for equipment that causes employees to have a TWA greater than 90 dBA.

This generalized approach can be used to quantify the noise exposures of workers in most industrial facilities.

Before making measurements, the SLM must be calibrated properly. The time, date, location, and other pertinent information should be recorded on a data sheet; Figure 20.16 is an example of a useful data sheet for recording octave band measurements. For OSHA-related measurements, the instrument should be set to measure the A-weighted level with slow meter response. The meter should be held with the microphone at ear height (approximately 1.5 m). The average meter reading should be recorded at each position. If there is significant fluctuation (more than ±3 dB) the minimum and maximum readings, and the central tendency, should be recorded. Care should be taken to ensure that the relevant noise source is operating normally and that other sources that normally do not operate are off.

Noise exposure calculations based on area levels and time estimates are susceptible to many potential inaccuracies, especially if there is high worker mobility, the

CORPORATION NAME
ADDRESS

Sheet ____ of ____

Project Description:

Sketch of Site (with distances to noise sources):

Job Number: *Initials:*

Location:

Date:

Quantity measured:
SPL ACC VEL DISP dB re: _____

Time	Cal, dB	Temp	%RH	Wind Sp	Dir

CONDITIONS:	Time	Pos.	Octave Band Center Frequency, Hz									A–wt
			31.5	63	125	250	500	1000	2000	4000	8000	

NOTES:

Figure 35.16 — Sample data sheet.

sound level fluctuates, or there is significant impulsive sound. Workers can carry noise dosimeters during their normal work shift to document their noise exposure. For OSHA-related measurements, the dosimeter should be calibrated, and the proper settings should be entered for the weighting (A), meter response (slow), threshold (80 dBA), exchange rate (5 dBA), and criterion (90 dBA). The instrument can be worn in a pocket or attached to the user's belt, and the microphone should be attached to the shoulder, pointing upward, approximately 0.15 m (6 in.) from the ear. Detailed operating instructions will be provided by the dosimeter manufacturer.

It is best to conduct individual monitoring on each employee with any noise exposure higher than 85 dBA. If an employee's work schedule varies significantly from day to day, dosimetry should be conducted for several days. If many workers perform essentially the same tasks, it is possible to obtain a representative sample without measuring everyone. Care must be taken in selecting the sample size to ensure that there is high confidence the highest exposures are documented. If there is any doubt on how representative any noise exposure data is, it is best to err on the side of making additional measurements.

Identifying and Locating Noise Sources

In identifying offending noise sources, the hygienist should first use his or her ears. Since all conventional hearing protection devices are most effective at high frequencies, care must be taken not to ignore these less audible sounds when wearing hearing protection.

Ideally, when investigating noise sources (not noise exposures), all pieces of equipment should be turned off except the equipment under investigation, though in occupational situations this often is impossible to do. If there are several sources in a reverberant space, the offending source might be difficult to isolate. In this case, a 1/1- or 1/3-octave band filter set will help to locate the source, since there usually will be excessive sound in a particular band. In some cases, a narrow-band analyzer will be necessary to identify tones that are characteristic of the particular piece of equipment. As noted earlier, another method for analyzing sound sources is to measure sound intensity. Although the equipment is sophisticated and expensive, an intensity analyzer can be used to isolate a source in a reverberant space better than a sound level meter.

Compliance of Equipment with Purchase Noise Specification

Purchase specifications limiting the noise emissions of new or reconditioned equipment can reduce noise exposure in occupational environments. Usually the specification requires that the equipment meet a certain maximum sound level at a distance of 1 m away or at a particular operator position under specified operating conditions. This should be confirmed by the hygienist. The criterion may be A-weighted sound levels, octave-band sound pressure levels, or sound power levels.

Many national and international standards can be referenced in a purchase specification; however, some of these standards do not provide the hygienist with the information needed to determine the noise exposure. The purchase specification must tie the sound level requirements to a specific location, preferably the operator position, and to the normally loaded operating conditions for the equipment.

Control of Noise Exposure

Introduction

Noise exposures must be controlled whenever they exceed government or company noise requirements. Usually the best first step to reduce noise is to develop a written noise control plan. The plan can include the following items:

- Determine current noise exposures of employees.
- Implement a hearing conservation program with audiometric testing and available hearing protection for all employees with time-weighted average noise exposures greater than 85 dBA.
- Reduce noise exposures for employees with TWAs of 90 dBA or above, when feasible, with engineering or administrative controls.
- Determine the cost-effectiveness of different noise control options for reducing noise exposures. Details on noise control treatments are discussed in this section.
- Develop guidelines for the purchase of new and replacement equipment, for the modification of existing facilities, and for the design of new facilities. These guidelines should include the purchase of quiet equipment and the design of building surfaces to absorb sound and to prevent excessive sound transmission. Often, it is much less expensive to prevent the

noise problems from the start than to reduce excessive noise in completed facilities.

Justification for Performing Noise Control

The primary reasons for reducing noise levels in an occupational environment are legal; hearing protection; speech communication; safety; and annoyance.

Legal

In most industrialized countries, hearing conservation is required by regulatory statute. For example, under 29 CFR 1910.95, OSHA requires that all employee exposures to noise be controlled to a TWA of 90 dBA or less. Although the 1983 Hearing Conservation Amendment spells out the detailed requirements of an acceptable hearing conservation program for exposures greater than the equivalent of 85 dBA, it does not indicate that hearing protection is an acceptable form of noise control when the TWA is greater than 90 dBA. Engineering controls (reducing noise levels around people) and administrative controls (removing people from noisy areas and buying quiet equipment) are the only acceptable permanent solutions under this regulation. Although hearing conservation programs can be very effective in reducing the risk of hearing impairment of workers, the only completely effective method is to reduce the noise of all sound sources to acceptable levels. Obviously, this will require a concerted effort on the part of owners and vendors.

Hearing Protection

For more than 100 years, the business community has had access to the knowledge that excessive exposure to high noise levels can cause permanent hearing loss. Frequently this knowledge is not distributed throughout a company and, as a result, noise control is given a low priority. The individuals charged with completing construction projects on time and within budget often do not know efficient ways to reduce noise. Many times their solution is to put some money aside for future measurements and ill-defined retrofit solutions. This is an inefficient way to perform noise control.

High sound levels produce greater risk of dangerous noise exposures than lower noise levels. For example, no permanent hearing loss would be expected for a worker who spends a working lifetime at sound levels less than the equivalent of 80 dBA for an 8-hour day. On the other hand, if the worker were to spend this time at 95 dBA, typical high-frequency hearing losses of greater than 30 dB could occur.

Hearing protective devices provide only about 15 dBA of reliable protection for the majority (about 84%) of the work force. Workers exposed to TWAs above 100 dBA while wearing hearing protection can still incur substantial hearing loss over a lifetime since the protected ears would still be exposed to TWAs higher than 85 dBA.

Speech Communication

Speech communication can be improved dramatically by reducing sound levels. Although it generally is not appreciated, most maintenance personnel and operators depend on speech communication for receiving much of their instructions and information. If two people were attempting to communicate in an 84 dBA environment, they could speak in a normal voice only if they were 1 m (3 ft) or less apart.

In a situation in which communication is crucial (e.g., during an emergency situation or in a control room) the ambient sound level will need to be much lower to achieve marginal speech communication. To compound issues, communication is more difficult in a moderately noisy environment for workers who have incurred some hearing loss.

Safety

Reduced noise levels increase safety by aiding communication and by making it easier to hear alarms or other audible signs of danger. Emergency warning systems must be designed to produce sound levels 10–15 dB above the background sound level. With very high sound levels, high-powered, closely spaced emergency warning speakers must be used.

Annoyance

Lower noise levels means lower annoyance, and in many cases this means better productivity and a safer workplace. Annoyance can also result from wearing hearing protection for extended periods of time. Earmuffs can be very uncomfortable due to the perspiration that develops under the muffs in warm climates or in enclosures or buildings. Lower noise levels can also improve employee concentration.

When residences are located near industrial facilities, the residents may find occupational noise annoying.

Administrative Controls

Administrative noise control measures consist of either relocating people away from noisy areas or redistributing noise exposures such that many employees are given acceptably low exposures instead of a few employees having excessive exposures. These types of measures are conceptually simple but are often difficult to implement and document; thus, they are seldom used.

Clearly, the best administrative program is to eliminate the source of the noise by purchasing quieter equipment and processes. This can also be considered an engineering control as discussed in the following section.

Engineering Controls

When administrative controls cannot be used to reduce noise exposures, engineering noise controls can be used. The simplest (though not necessarily cheapest) option is to replace noisy operations with quieter equipment, quieter processes, or quieter materials. Equipment can also be relocated. Some equipment manufacturers have options for quieter models. Intrinsically quieter equipment is often more efficient, easier to maintain, and less costly than equipment retrofitted with treatments. When purchasing quieter equipment, precise noise specifications should be used to ensure that guaranteed sound levels are for the equipment operating under normal loads and conditions.

When quieter equipment cannot be purchased, noise control treatments can be applied to the equipment. The rest of this section addresses different types of noise control treatments. Before determining which treatment is most appropriate for a particular situation, the source of the noise problem should be analyzed to determine the mechanisms of sound generation and propagation. If there are multiple sound sources, they should be rank-ordered. A range of possible treatments can then be considered. Often, operational and maintenance concerns will eliminate some options. The remaining options can then be evaluated in detail for efficacy and cost-effectiveness. It is often most effective to install simple treatments for the loudest sound sources first. It should then be easier to analyze noise from other sound sources and to develop appropriate additional noise control treatments. Although much can be accomplished by a hygienist with a simple A-weighted sound level meter and a basic understanding of noise control concepts, acoustical consultants are often necessary when complicated measurements or treatment designs are required.

Most occupational noise sources can be separated into two general classifications according to the mechanism by which the sound waves are generated: surface motion of a vibrating solid, and turbulence in a fluid (including air).

Some noise sources are combinations of these two classifications. Noise control for sources with a vibrating surface include reducing 1) the driving force; 2) the response of the surface to the driving forces; and 3) the radiation efficiency of the vibrating surfaces.

Once sound is transmitted to the air, it can be reduced by treating the sound path using the following types of treatments: 1) sound absorption; 2) equipment enclosures; 3) personnel enclosures; 4) shields or barriers; 5) lagging; and 6) mufflers.

Each of these different types of noise control treatments are discussed in detail in the following sections.

Reduce Driving Force

Driving forces in mechanical equipment may be repetitive or nonrepetitive. These forces can be reduced by decreasing the speed, maintaining the dynamic balance, providing vibration isolation, and, in the case of impact operations, increasing the duration of the impact while reducing the force.

The repetitive forces on a machine are often caused by imbalance or eccentricity in a rotating member. These forces strengthen with an increase in rotational speed and this usually results in higher noise levels from the machine. As a result, machines should be selected to operate at slower speeds when this requirement does not conflict with other operational needs.

All rotating mechanical equipment should be in proper dynamic balance to minimize the level of repetitive forces. Proper preventive maintenance of bearings and proper lubrication and alignment of equipment are vital. When machines are balanced improperly or when bearings are worn, it is possible for the noise levels to increase as much as 10 dB above the sound levels of normal operation.

Resilient materials such as rubber, neoprene, and springs can reduce noise by reducing vibration propagation through structures; the design of vibration isolation treatments are discussed in the vibration section of this chapter. Resilient materials installed at impact sites can be used to reduce impact forces. For example, a neoprene-lined bin can be used instead of a bare metal bin to catch small metal parts. This treatment has worked on stock guides, chutes, tumbling barrels, and hoppers and works particularly well when the falling parts do not have sharp edges. The noise levels can be reduced 10–20 dB by these treatments.

Many metal fabricating operations (e.g., forging, riveting, shearing, and punching) use impacts to process material. Due to the short duration of the impact, large forces are required. The noise level is a function of the maximum amplitude of the force that is applied, and a lesser force applied over a longer period will produce less noise.

Reduce Response of Vibrating Surface

Often, the easiest method to reduce the response of a vibrating surface is to increase its damping. A vibrating surface, such as a panel, converts some energy of motion into heat energy as it flexes. The rapidity with which the vibration energy is dissipated is directly related to the damping. A structure with little damping, such as a bell, will continue to vibrate for a long time after it is excited. A highly damped structure, such as a block of rubber, will not ring if struck but will quickly convert the energy to heat. For a given driving force, a highly damped surface will vibrate much less than a lightly damped surface similar in mass and stiffness. Consequently, a highly damped surface will radiate less sound than a surface with low damping.

Before treating a vibrating surface to reduce the noise, the designer needs to determine which surfaces are radiating high levels of noise. This issue is sometimes obvious. Large, flat, unsupported surfaces are likely to be more important radiators of noise than small or stiff surfaces. If a machine is small, it may be expedient to treat all vibrating surfaces. If the machine is large or if the treatment is complicated, this might not be cost-effective, and it may be important to first determine which surfaces make the greatest contributions to the area noise levels.

Increasing the damping of a surface is most effective when the surface is vibrating at or near a resonance frequency. Damping material should be added where the panel has the greatest amplitudes of vibration; damping applied in the center of a vibrating panel attached at its edges is much more effective than damping applied near the edges.

The two types of commonly applied damping treatments are free-layer damping (also called extensional damping) and constrained-layer damping. In free-layer damping, a layer of nonhardening viscoelastic material, usually in the form of tapes, sheets, mastics, or sprays, is adhered to the surface. Constrained-layer damping involves the use of a laminated construction consisting of two or more sheet metal layers, each separated by and bonded to a viscoelastic layer. Because of possible damage to exposed damping surfaces in many occupational environments, the metal exterior surface of constrained-layer damping may be preferable to the less durable exterior surface of free-layer damping.

In addition to reducing the vibration of a surface at resonance frequencies, damping treatments also reduce the transmission of high-frequency sound through the panel.

The temperature of the surface on which the damping material is to be applied must be known since the properties of damping materials are highly dependent on temperature. Resistance to chemicals, oil, and other environmental considerations may also be important in the selection of a damping material. Detailed information on damping materials is available from the individual manufacturers. Liquid mastics and elastomeric damping sheets are the most common damping materials. High-temperature damping materials (up to 320°C [600°F]) are also available for applications such as steel muffler shells, high speed blower housings, steel piping systems, gear boxes, fan housings, and turbine exhausts.

As a basic guideline, free-layer damping treatments should be at least as thick as the surface to which they are applied. The greater the ratio of the thickness of the added layer of damping material to the panel thickness, the greater the noise reduction of the treated panel.

For free-layer damping, viscoelastic materials are usually preferred since the performance of a treatment can be estimated using the manufacturer's product literature. Although the noise reduction provided by a damping treatment is very difficult to predict accurately, the reduction may be estimated from:

$$\text{Noise Reduction} = B * \log_{10}\left(1 + \frac{\eta_c}{\eta_i}\right) \qquad (13)$$

where:

B = 20 for structures vibrating at resonance;
B = 10 for structures with broad-band excitation;

η_c = composite damping loss factor; and
η_i = damping loss factor present initially in the structure.

As shown, the composite damping (η_c) must be considerably greater than the initial damping (η_i) to achieve significant reduction. The composite damping loss factor (η_c) from bonding a viscoelastic layer to a structural plate may be estimated from:

$$\eta_c = \frac{\eta_d}{1 + \dfrac{E * H / (e * h)}{3 + 6h/H + 4(h/H)^2}} \quad (14)$$

where:

E = modulus of elasticity of the plate (N/m^2);
e = modulus of elasticity of the damping material;
H = thickness of the plate (m);
h = thickness of the damping treatment (m); and
η_d = loss factor of the damping material.

The ratio of the moduli of elasticity (e/E) is presumed to be much less than 1 in the derivation of this equation. Representative values of E and η_d are presented in Table 20.7.[27] Since the properties of damping materials vary significantly as a function of frequency and temperature, the values given in this table are presented as general ranges only. The manufacturer's data for specific materials and applications should be obtained before performing final calculations.

The loss factor of a panel before treatment (η_i) is sensitive to construction techniques and edge conditions. If the loss factor of the panel is low before treatment, noise reductions on the order of 10–30 dB can be achieved.

In one application, the lower section of a rubber compounding mill was encased in a metal shell. Functioning as a sounding board, the metal shell radiated the vibrations from the motor, gear, and roll of the mill. The vibration was damped by an application of free-layer damping material to the inner surface of the metal shell. The octave band noise reduction achieved was:

Octave Band Center Frequencies (Hz)							
63	125	250	500	1000	2000	4000	8000
11 dB	8 dB	14 dB	10 dB	6 dB	7 dB	9 dB	11 dB

Table 20.7 — Representative Values of Sound Speed, Density, Modulus of Elasticity, and Loss Factor for Common Building and Damping Materials[27]

Materials	*Speed (m/sec)*	*Density ρ (kg/m³)*	*Modulus of Elasticity E (N/m²)*	*Lost Factor η (dimensionless)*
Aluminum	5150	2700	7.2×10^{10}	0.0001–0.01
Lead	1200	11,000	1.6×10^{10}	0.002–0.015
Steel	5050	7700	20×10^{10}	0.0001–0.01
Plexiglas	1800	1150	$.37 \times 10^{10}$	.0002
Plywood	—	600	.37 ×× 10^{10}	0.01–0.04
Damping Materials	—	—	$7 \times 10^5 - 7 \times 10^8$	0.1–1.0

Design equations for constrained-layer treatments have been derived,[27] but they involve complex frequency and wavelength dependencies and, due to space limitations, are not presented here. Because of the complexities involved, design of constrained-layer treatments often require the assistance of an experienced professional.

In one successful case history, noise was produced by almost continuous impacts between stock material and the tubes that fed the stock to screw machines. A viscoelastic damping material was sandwiched between inner and outer tubes. At 4000 rpm with 13 mm (1/2 in.) stock, the noise measured 0.3 m (1 ft) from the middle of the stock tube was reduced as follows:

Octave Band Center Frequencies (Hz)							
63	125	250	500	1000	2000	4000	8000
12 dB	15 dB	15 dB	14 dB	20 dB	29 dB	34 dB	30 dB

Reduce Radiation Efficiency by Reducing Area of Vibrating Surface

Sound can be radiated from a surface when at least one dimension of the surface is longer than one-fourth of the wavelength of the sound. Thus, low-frequency sound requires larger surfaces for radiation than high-frequency sound.

It sometimes is possible to reduce the sound radiated from a surface by reducing the total area or by dividing a large surface into a number of smaller sections. When possible, perforated or expanded metal can be used in place of solid panels (presuming that the panel itself is not part of an acoustic enclosure).

Use Directivity of Source

Many occupational sources radiate more sound in one direction than in another. This behavior is called the source directivity. Directive sources include

Table 20.8 — Directivity of a Stack (in dB) by Angle Compared with Spherical Radiation[28]

	Octave Band Center Frequencies (Hz)							
Angle	63	125	250	500	1000	2000	4000	8000
45°	+5	+5	+6	+6	+6	+7	+7	+7
60°	+2	+2	+2	+2	+1	0	-1	-2
90°	-3	-4	-6	-8	-10	-12	-14	-16
135°	-4	-6	-8	-10	-13	-16	-18	-20

stacks, intake and exhaust openings on fans and blowers, enclosure openings, and large vibrating metal panels. Occasionally it is possible to orient a source such that a particular location receives less sound than other locations around the source. For stacks, the amount of directivity at different angles compared with a source that radiates spherically are presented in Table 20.8.[28]

The values in this table can be added to the L_p determined by Equation 9 (see page 429) to determine the L_p at different angles to the stack.

In the reverberant sound field inside a building, the directivity of the source cannot normally be used to reduce the sound. The only exception is if the source can be oriented such that the sound radiates directly toward a sound-absorptive material.

Reduce Velocity of Fluid Flow

High-velocity fluid flow can generate noise problems in air-ejection systems, vents, valves, and piping. The noise is generally caused by turbulence in the fluid flow. Although reducing velocities will reduce turbulence and noise, this is often not a practical option.

Consideration should be given to using mechanical part ejectors in place of the air-ejection systems. Quiet nozzles sometimes can be used for air-ejection systems. It also is possible to reduce air velocities in conjunction with more accurate aiming of the nozzles. For exhaust vents on air-operated equipment, small silencers can be installed to reduce the noise.

If the ratio of the absolute pressures upstream and downstream of a valve are 1.9 or greater, the flow through the valve will be choked (sonic). Large pressure drops across valves can be avoided by using diffusers in the line to reduce the pressure upstream of the valve. Valve manufacturers can generally provide sound level data for their valves based on the operating conditions, and some manufacturers have options for quieter valves. Due to their small size and heavy construction, little noise is radiated from the valves themselves; most of the noise is radiated from the piping or vents downstream of the valves. Mufflers can sometimes be installed downstream of valves, or the piping can be lagged as described later in this section.

Increase Sound Absorption

Rooms are neither perfectly reflective nor perfectly absorptive. Sound levels decrease at a rate of 6 dB per doubling of distance at locations close to small sound sources in most rooms (as they do outdoors or in an anechoic chamber). They then approach a constant level at greater distances from sound sources (as they do in reverberation chambers).

To predict the attenuation of sound in a room, it is first necessary to determine the total amount of sound absorption in the room (TA). This is determined by multiplying the surface area of each interior surface material by the material's coefficient of absorption. The coefficient of absorption is the ratio of the sound energy absorbed by a material to the sound energy incident upon the material. An extremely absorptive material such as thick glass fiber will have an absorption coefficient (α) of close to 1.0 (all of the sound is absorbed). An extremely sound reflective surface such as concrete will have a sound-absorption coefficient of close to 0.0 (all of the sound is reflected). Sometimes, manufacturers of sound-absorptive materials will quote α values greater than 1.0; this is an artifact of the test conditions, and a maximum value of 1.0 should be used in equations.

Materials have different absorption coefficients at different frequencies. Absorption coefficients also vary with material thicknesses. Table 20.9 presents octave band sound-absorption coefficients for various materials.[28-32] The formula for calculating the total absorption of a room is:

$$TA = S_1 * \alpha_1 + S_2 * \alpha_2 + ... + S_n * \alpha_n \quad (15)$$

where:

TA = total absorption in metric sabins (1 metric sabin is equal to the sound absorption of 1 m^2 of perfectly absorptive material; $\alpha = 1.0$);

$S_1, S_2, ... S_n$ = areas in m^2 of each interior surface; and

$\alpha_1, \alpha_2, ..._n$ = corresponding coefficients of absorption for these surfaces.

The sound absorption of openings in a room must also be considered when calculating the total absorption. The absorption

Table 20.9 — Sound-Absorption Data for Common Building Materials(28-32)

Material	Sound-Absorption Coefficient 125 Hz	250 Hz	500 Hz	1000 Hz	2000 Hz	4000 Hz
Walls						
Sound-reflecting:						
Brick, unglazed, unpainted	0.03	0.03	0.03	0.04	0.05	0.07
Brick, glazed or painted	0.01	0.01	0.01	0.02	0.02	0.03
Concrete block, painted	0.10	0.05	0.06	0.07	0.09	0.08
Cork on brick or concrete	0.02	0.03	0.03	0.03	0.03	0.02
Glass, typical window	0.35	0.25	0.18	0.12	0.07	0.04
Gypsum board, 1/2-in. paneling	0.29	0.10	0.05	0.04	0.07	0.09
Metal*	0.05	0.02	0.01	0.02	0.02	0.02
Plaster, gypsum or lime, on brick or tile	0.01	0.02	0.02	0.03	0.04	0.05
Plaster, gypsum or lime, on lath	0.14	0.10	0.06	0.05	0.04	0.03
Plywood, 3/8-in. paneling	0.28	0.22	0.17	0.09	0.10	0.11
Wood, 1/4-in. paneling, with air space behind	0.42	0.21	0.10	0.08	0.06	0.06
Sound-absorbing:						
Concrete block, coarse, unpainted	0.36	0.44	0.31	0.29	0.39	0.25
Medium weight drapery, 14 oz/sq. yd., draped to half area	0.07	0.31	0.49	0.75	0.70	0.60
Fiberglass fabric curtain, 8 1/2 oz/sq. yd., draped to half area	0.09	0.32	0.68	0.83	0.39	0.76
Shredded wood fiberboard, 2-in. thick on concrete	0.32	0.37	0.77	0.99	0.79	0.88
Foams: (Acoustical open cell)						
1-in., 2 lb/cu. ft. polyester	.23	.54	.60	.98	.93	.99
2-in., 2 lb/cu. ft. polyester	.17	.38	.94	.96	.99	.91
Glass fiber:						
1-in., 3 lb/cu. ft.	.23	.50	.73	.88	.91	.97
1-in., 6 lb/cu. ft.	.26	.49	.63	.95	.87	.82
Floors						
Sound-reflecting:						
Concrete, terrazzo, marble or glazed tile	0.01	0.01	0.01	0.02	0.02	0.02
Cork, rubber, linoleum, or asphalt tile on concrete	0.02	0.03	0.03	0.03	0.03	0.02
Wood	0.15	0.11	0.10	0.07	0.06	0.07
Wood parquet on concrete	0.04	0.04	0.07	0.06	0.06	0.07
Sound-absorbing:						
Carpet, heavy, on concrete	0.02	0.06	0.14	0.37	0.60	0.65
Carpet, heavy, on foam rubber	0.08	0.24	0.57	0.69	0.71	0.73
Indoor-outdoor carpet	0.01	0.05	0.10	0.20	0.45	0.65
Ceilings						
Sound-reflecting:						
Concrete	0.01	0.01	0.02	0.02	0.02	0.02
Gypsum board, 1/2-in. thick	0.29	0.10	0.05	0.04	0.07	0.09
Plaster, gypsum or lime, on lath	0.14	0.10	0.06	0.05	0.04	0.03
Plywood, 3/8-in. thick	0.28	0.22	0.17	0.09	0.10	0.11
Sound-absorbing:						
Suspended acoustical tile, 3/4-in. thick, 16-in. air space above	0.76	0.93	0.83	0.99	0.99	0.94
Thin, porous sound-absorbing material, 3/4-in. thick (mounted to structure)	0.10	0.60	0.80	0.82	0.78	0.60
Thick, porous sound-absorbing material, 2-in. thick (mounted to structure) or thin material with 1-in. air space behind	0.38	0.60	0.78	0.80	0.78	0.70
Sprayed cellulose fibers, 1-in. thick on concrete	0.08	0.29	0.75	0.98	0.93	0.76
Air Absorption						
Air, per 1000 m^3–@ 50% RH	0	0	0	3.0	7.5	23.6

*Absorption coefficients for metal were estimated by the authors of this chapter. Low frequency absorption coefficients will depend on the metal thickness.

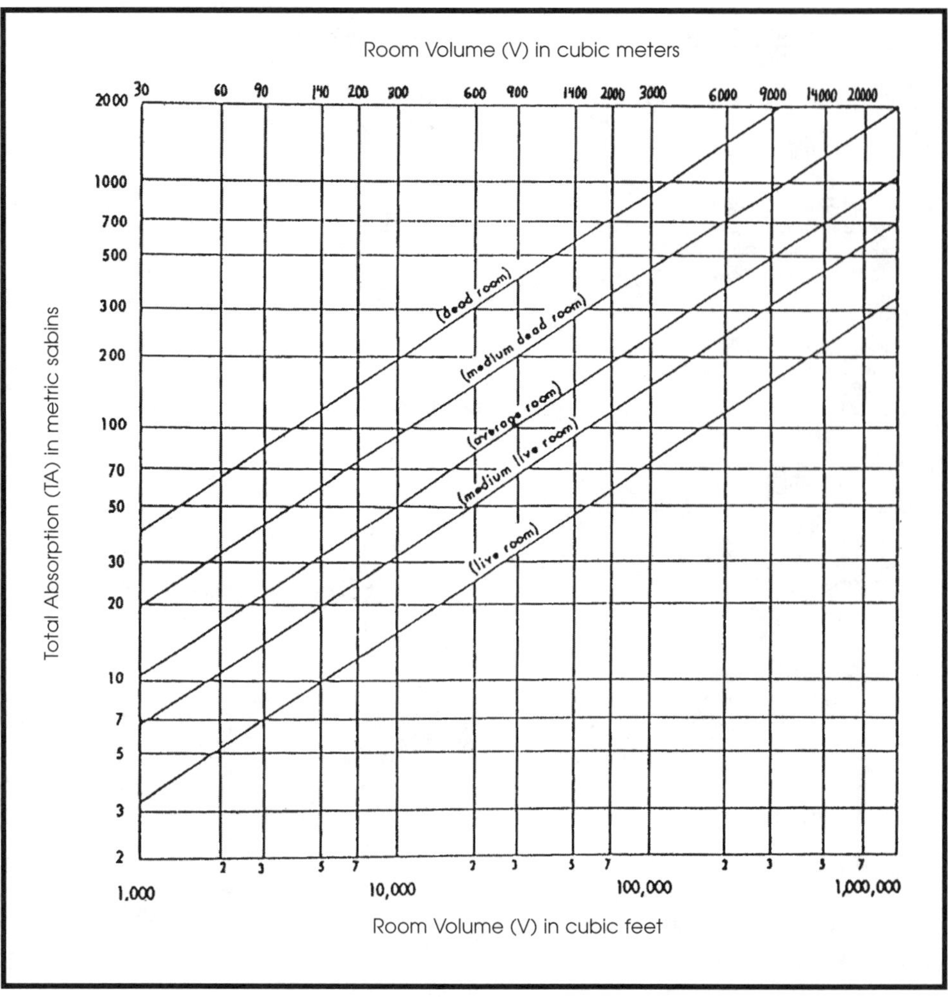

Figure 20.17 — Total room absorption for typical rooms. (Adapted from Reference # 27.)

coefficient of an opening such as an open door or window is 1.0. A 2 m² open window will therefore add 2 metric sabins to the total absorption. Also, the air volume can contribute to the sound absorption. This generally has to be taken into account only with the higher octave band frequencies (2000, 4000, and 8000 Hz) in large rooms.

Example: Calculating Total Absorption at 1000 Hz

The 1000 Hz total absorption for a room with 80 m² of concrete (α = .02), 210 m² of sheet metal (α = .02), 60 m² of windows (α = .12), 40 m² of open windows (α = 1.0), and 150 m² of 25 mm thick acoustical foam (α = .98) is: (80 * .02) + (210 * .02) + (60 * .12) + (40 * 1.0) + (150 * .98) = 2000 metric sabins.

As an alternative to the equation, TA can be approximated by using Figure 20.17 and estimating the relative reverberation of a room.

Example: Calculating Sound Levels in a Reverberant Space

As discussed previously in this chapter, the relationship between the sound pressure level (L_p) and the sound power level (L_w) is given in Equation 6 as:

$$L_w = L_{pa} + 10\log_{10}(A)$$

where:

L_w = sound power level;
L_{pa} = average sound pressure level; and
A = surface area in square meters of the imaginary shell around the equipment where the L_p measurements were made.

This formula is accurate in nonreverberant sound fields such as those found outdoors or in close proximity to the equipment.

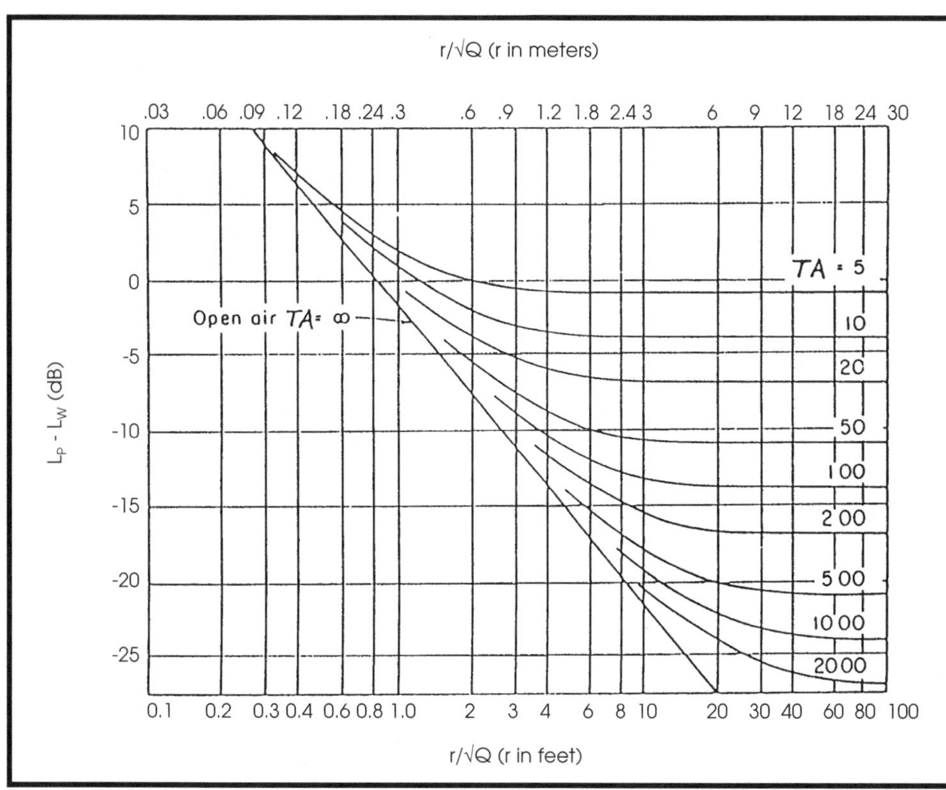

Figure 20.18 — Curves for determining the L_p relative to L_w in a room with total absorption (TA) at a distance (r) from a source of directivity (Q). (Adapted from Reference # 27.)

Inside a reverberant room, sound propagates in accordance with the following formula:

$$L_P = L_W + 10\log_{10}\left(\frac{Q}{4\pi r^2} + \frac{4}{TA}\right) \qquad (16)$$

where:

- Q = directivity factor;
- r = distance from the acoustic center of the source (in meters);
- TA = total absorption in room (in metric sabins); and
- L_w = sound power level.

This equation is presented graphically in Figure 20.18. Note that if the total absorption is very high, as in an anechoic chamber (where 4/TA becomes very small), this equation approaches the earlier simplified equation for outdoor sound radiation: sound levels decrease at a rate of 6 dB per doubling of the distance from the source. Also note that absorption TA has little effect on the sound pressure level at distances close to the sound source. This area is called the acoustic near-field. At greater distances from the source in a reverberant room, the sound level approaches a constant level. This area, called the reverberant field, is the area where sound levels are controlled by the amount of absorption in the room, not the distance from the sound source.

For a location in the reverberant field of a sound source, sound reduction can be achieved by increasing the amount of sound absorption in the room. The purpose of this method of noise control is to reduce the reverberant buildup of sound in a room and, ideally, to achieve the same attenuation of sound over distance that is achieved outdoors (6 dB per doubling of distance). In most cases, the maximum reduction that can be achieved by adding absorption to a room is 3–8 dB. Employees may appreciate this improvement more than an equal decibel improvement of source noise control since a less reverberant space is usually a more pleasant work environment. This noise control technique is only practical when the room is originally reverberant and the sound receiver is located a sufficient distance from the sound source (in the reverberant field). The amount of noise reduction (NR) at a position in the reverberant field that can be achieved by adding sound absorption to a room is determined by the following formula:

$$NR = 10\log_{10}\left(\frac{TA_2}{TA_1}\right) \qquad (17)$$

where:

NR = noise reduction in dB;
TA_1 = TA before treatment; and
TA_2 = TA after treatment.

When increasing room absorption to reduce sound levels, it is important to determine which octave band frequencies need reduction and to choose absorptive materials that are highly absorptive in those frequencies. For example, thick sound absorptive treatments are necessary for reducing low-frequency sound levels.

Example: Room Treatment by Increasing Sound Absorption

Room:	13 m × 14 m × 4 m
Volume:	728 m^3
Ceiling:	Plaster – 182 m^2
Floor:	Concrete – 142 m^2
Floor:	Carpet – 40 m^2
Walls:	Painted Brick – 216 m^2
Frequency of interest:	1000 Hz

The total absorption (TA_1) at 1000 Hz before treatment is equal to:

Ceiling – Plaster	182 × 0.05 =	9.1
Floor – Concrete	142 × 0.02 =	2.8
Floor – Carpet	40 × 0.37 =	14.8
Walls – Brick	216 × 0.02 =	4.3
Air Absorption	.73 × 3.0 =	2.2
Total		**33.2**

The ceiling is to be covered with acoustical tile having an absorption coefficient of 0.80 at 1000 Hz. The total absorption (TA_2) after treatment is equal to:

Ceiling – Tile	182 × 0.80 =	145.6
Floor – Concrete	142 × 0.02 =	2.8
Floor – Carpet	40 × 0.37 =	14.8
Walls – Brick	216 × 0.02 =	4.3
Air Absorption	.73 × 3.0 =	2.2
Total		**169.7**

The noise reduction at a location far from a sound source (in the reverberant field) is calculated as follows:

$$NR = 10\log_{10}\frac{TA_2}{TA_1} = 10\log_{10}\frac{169.7}{33.2} = 7.1 \text{ dB} \quad (18)$$

The results can be quite different for other frequencies; therefore, similar calculations should be repeated for all octave band frequencies of interest. In one case history, 10 automatic wire-cutting machines at a facility were located in an alcove measuring 6 m × 18 m. The operator did not have to tend the machines constantly. The ceiling was wood, three walls were brick, one side was open to a large storage area, and the floor was concrete. The addition of acoustical absorption to this area was recommended because there were multiple noise sources, a low ceiling (2.3 m), very little absorption initially, and the operator was away from the near-field. Acoustical absorption was applied to the ceiling and to one wall. An area of 128 m^2 was covered. After the area was treated, the employees commented that the working conditions had improved considerably. There was no noise reduction in the frequencies below 300 Hz. In the upper octave bands, the L_p was reduced from 4 dB to 12 dB (depending on frequency) near the machines. At 6 m from the center of the machines, the operator location in the reverberant field, the reduction was between 15 dB and 20 dB in the upper octave bands.

Equipment Noise Enclosures

Noise enclosures reduce the noise of a sound source by completely surrounding the source with a barrier material. The sound energy must then propagate through the wall of the enclosure before reaching the area outside. The three primary design concerns for enclosures are use of a good barrier material for the enclosure, application of sound absorption to the interior of the enclosure, and avoidance of enclosure leaks. The ability of the barrier to prevent transmission of sound is quantified as its transmission loss (TL). The amount of sound absorption in an enclosure is quantified as its total absorption, as defined in the previous section. The purpose of the absorption is to prevent the sound level inside the enclosure from building up due to reverberation. The estimated noise reduction at a location outside the enclosure that is achieved by enclosing a noise source is calculated by using the following formula:

$$NR = TL - 10\log_{10}\frac{S}{TA} \quad (19)$$

where:

NR = noise reduction in dB;
TL = transmission loss of enclosure material;
S = exterior surface area of the enclosure in m^2; and
TA = total absorption inside the enclosure in metric sabins.

(S/TA can be considered an absorption correction.)

Sound absorption coefficients are given in Table 20.9, and Table 20.10 lists the measured TL of common building materials.[29,30,33] As shown in Equation 19, the noise reduction of an enclosure depends largely on the TL of the barrier material. This material must normally be constructed of material weighing at least 5 kg/m^2 (1 lb/ft^2) and must be durable enough to withstand any expected abuse. If the interior is completely lined with a perfectly absorptive material, such that the total absorption (TA) equals the exterior surface area (S), then NR = TL. If TA is only half of S (because only half the interior is treated or because the entire interior is treated with a material with an absorption coefficient of 0.5), then NR = TL – 3. If the interior of the enclosure has no sound absorptive treatment, then the enclosure might achieve very little noise reduction. Also, the problems caused by any small leaks in the enclosure will be exacerbated by the increased reverberant sound levels in an untreated enclosure. In many cases, 25–50 mm (1–2 in.) thick glass fiber or acoustical foam is used as the absorptive material. Whenever the enclosed equipment leaks oil, water, or chemicals, it is important to protect the sound absorptive material with thin films (1 mil or 25 µm) of materials such as Mylar® or Tedlar®. In addition to the enclosure barrier TL and the sound absorption inside the enclosure, several other design elements need to be considered:

- No leaks can be allowed in the surface of the enclosure in order to achieve a noise reduction ≥10 dB. Any joints or cracks should be completely sealed. Access and maintenance panels should close and seal tightly.
- The enclosure should be completely vibration-isolated from the enclosed equipment. Any direct structural connection will transmit vibration to the enclosure, which will then radiate sound into the room outside. If a connection is necessary, it should be vibration-isolated (e.g., by using a flexible connection).
- Auxiliary cooling may be required to prevent enclosed equipment from overheating. Any ventilation ducts, with or without fans, should be treated with sound-absorptive materials and should be oriented to prevent a direct line of sight into the enclosed equipment. (See the subsequent section on partial enclosures.)

Example: Design of Full Enclosure

Since transmission loss data and sound absorption coefficient data are often only given for the 125–4000 Hz octave bands, this example only considers those octave bands. Most acoustical problems are in this frequency region, but if a problem arises in the 31.5, 63, or 8000 Hz octave bands, the same formulas apply and TL and sound absorption data can be estimated.

The first step in designing an enclosure is to determine the acoustical requirements. This can be done by subtracting the octave band noise criterion from the measured octave band sound levels at a position of concern. The differences are the required noise reduction of the enclosure in each of the octave bands. If the criterion is only for an A-weighted sound level, it will first be necessary to create an octave band spectrum goal equal to the A-weighted criterion. The following table shows calculations of the required NR.

Octave Band Sound Pressure Level

	125	250	500	1000	2000	4000	dBA
Sound Level	96	91	93	87	86	81	94
Criterion	92	86	81	77	74	72	85
Required NR	4	5	12	10	12	9	—

Next, it is important to determine the size of the enclosure and the amount of absorption that can be applied to the interior of the enclosure. In each octave band, the total absorption (TA) and the absorption correction (10 log(S/TA)), must be calculated. This absorption correction and a safety factor of 3–5 dB should then be added to the required NR to get the required TL. In this example, the exterior surface area of the enclosure is 5 m^2, and the interior is completely treated with a material with the sound absorption coefficients listed in the following calculation table:

Octave Band Center Frequencies

	Surface Area (m^2)	125	250	500	1000	2000	4000
Absorption Coefficient	5	.23	.45	.68	.96	1.00	.98
Total Area and Absorption	5	1.15	2.25	3.4	4.8	5.0	4.9
10 log (S/TA)		6	3	2	0	0	0
Required NR		4	5	12	10	12	9
Safety Factor		5	5	5	5	5	5
Required TL		15	13	19	15	17	14

Table 20.10 — Transmission Loss Data for Common Building Constructions(29,30,33)

Building Construction	125 Hz	250 Hz	500 Hz	1000 Hz	2000 Hz	4000 Hz
			Transmission Loss (dB)			
Walls						
Interior:						
2-in. solid plaster on metal lath (18 psf)	20	22	22	27	36	42
2 × 4 wood studs 16-in. o.c. with 1/2-in.gypsum board both sides (6 psf)	10	28	33	42	47	41
2 × 4 wood studs 16-in. o.c. with 5/8-in. gypsum board both sides – one side screwed to resilient channels. 3-in. glass-fiber batt insulation in cavity (6.5 psf)	32	42	52	58	53	54
6-in. concrete block wall, painted (34 psf)	37	36	42	49	55	58
8-in. concrete block wall with 3/4-in. wood furring, gypsum lath and plaster both sides (67 psf)	43	47	47	55	58	60
2 1/2-in. steel channel studs with 5/8-in. gypsum board both sides (11 psf)	15	24	38	48	40	42
Construction above with glass-fiber insulation in cavity	23	35	44	53	45	43
2 5/8-in. steel channel studs with 2 layers 5/8-in. gypsum board one side, 1 layer other side (8 psf)	22	26	40	51	44	47
Construction above with glass-fiber insulation in cavity	28	38	50	57	50	50
2 5/8-in. steel channel studs with 2 layers 5/8-in. gypsum board both sides (10 psf)	27	34	48	55	50	57
Construction above with glass-fiber insulation in cavity	33	44	55	60	55	60
Exterior:						
4 1/2-in. brick with 1/2-in. plaster each side (55 psf)	34	34	41	50	56	58
9-in. brick with 1/2-in. plaster each side (100 psf)	41	43	49	55	57	60
Two wythes of plastered 4 1/2-in. brick, 2-in. air space with glass-fiber in cavity (90 psf)	43	50	52	61	73	78
2 × 4 wood studs 16-in. o.c. with 1-in. stucco on metal lath on outside and 1/2-in. gypsum board on inside (8 psf)	21	33	41	46	47	51
6-in. concrete with 1/2-in. plaster both sides (80 psf)	39	42	50	58	64	67
Floor–Ceilings						
2 × 10 wood joists 16-in. o.c. with 1/2-in. plywood subfloor, 25/32-in. oak on floor side, and 5/8-in. gypsum board on ceiling side (10 psf)	23	32	36	45	49	56
Construction above with 3-in. glass-fiber batt insulation in cavity	25	36	38	46	51	57
4-in. thick reinforced concrete slab (53 psf)	48	42	45	56	58	66
14-in. precast concrete tees with 2-in. slab, and 2-in. concrete topping (75 psf)	40	45	49	52	60	68
6-in. thick reinforced concrete slab with 3/4-in. T & G wood flooring on 1 1/2-in × 2-in. wooden battens on 1-in. thick glass-wool quilt (83 psf)	38	44	52	55	60	65
18-in. steel joists 16-in. o.c. with 1 5/8-in. concrete on 5/8-in. plywood nailed to joists and heavy carpet on underlay. On ceiling side, 5/8-in. gypsum board nailed to joists (20 psf)	27	37	45	54	60	65
Roofs						
Corrugated steel, 24-gauge with 1 3/8-in. sprayed cellulose insulation on ceiling side (1.8 psf)	17	22	26	30	35	41
2 1/2-in. sand gravel concrete (148 pcf) on 28-gauge corrugated steel supported by 14-in. steel bar joists, with 1/2-in. gypsum plaster on metal lath and 3/4-in. metal furring channels 13 1/2-in. o.c. on ceiling side (41 psf)	32	46	45	50	57	61
Doors						
2 1/2-in. acoustical door with 12-gauge steel facing	43	49	48	55	57	44
1 3/4-in. hollow wood core door, no gaskets or closure, 1/4-in. air gap at sill	14	19	23	18	17	21
Construction above with gaskets and drop seal	19	22	25	19	20	29
1 3/4-in. solid wood core door with gaskets and drop seal (4.3 psf)	29	31	31	31	39	43
1 3/4-in. hollow 16-gauge steel door, glass-fiber filled core with gaskets and drop seal (6.8 psf)	23	28	36	41	39	44

Table 20.10 (continued) — Transmission Loss Data for Common Building Constructions[29,30,33]

Building Construction	125 Hz	250 Hz	500 Hz	1000 Hz	2000 Hz	4000 Hz
	Transmission Loss (dB)					
Glass						
1/8-in. single plate-glass pane	18	21	26	31	33	22
1/4-in. single plate-glass pane with rubber gasket	25	28	30	34	24	35
9/32-in. laminated glass pane (i.e., viscoelastic layer sandwiched between glass layers)	26	29	33	36	35	39
1/4- + 1/8-in. double plate-glass window with 2-in. air space	18	31	35	42	44	44
Construction No. 38 with 4-in. air space	21	32	42	48	48	44
Panels						
lead sheet – 1/16 in.	28	32	33	32	32	33
lead sheet – 1/8 in.	30	31	27	38	44	33
20 g aluminum sheet, stiffened	11	10	10	18	23	25
22 g galvanized sheet steel	8	14	20	23	26	27
20 g galvanized sheet steel	8	14	20	26	32	38
18 g galvanized sheet steel	13	20	24	29	33	39
16 g galvanized sheet steel	14	21	27	32	37	43
18 g fluted steel panels stiffened at edges	30	20	22	30	28	31
1/4-in. plywood	17	15	20	24	28	27
3/4-in. plywood	24	22	27	28	25	27
1/8-in. lead vinyl curtains	22	23	25	31	35	42

After the required TL has been determined, a barrier material and thickness meeting all of the octave band TL requirements can be chosen from Table 20.10. If the criterion is for an A-weighted sound level and not for octave band levels, then an acceptable material will not have to meet every octave band TL requirement as long as the resulting treated octave band sound levels add up to an A-weighted spectrum that meets the criterion.

In one case study at an industrial facility, noise from a steam turbine that was used to drive a boiler feed pump significantly exceeded the exposure criterion and was especially annoying due to its tonal nature. To lower the noise, the turbine was enclosed in 1.6 mm thick (16 gauge) steel, lined with 25 mm (1 in.) thick glass fiber with a density of 48 kg/m^3 (3 lb/ft^3). The noise reduction was 14 dB in the 4000 Hz octave band in which the loudest tones occurred, thus resolving the situation.

Partial Enclosures

Sometimes a complete enclosure around a noise source is not possible. For example, openings in an enclosure are often required for maintenance, controls, drive shafts, ventilation, or process flow. In these cases, a partial enclosure can be built to achieve some noise control. Partial enclosures are constructed similarly to complete enclosures; they have good barrier materials on the outside and sound absorption on the inside, but because of their openings they normally provide less than 10 dB noise reduction.

Since sound will not be attenuated by propagating through direct openings to the outside, paths to the outside should be convoluted and treated with sound-absorptive surfaces when possible. If sound is forced to reflect off sound-absorptive walls before reaching the outside, the sound level can be reduced significantly. Convoluted paths are normally only possible with ventilation openings, but even a straight opening into a partial enclosure (such as an opening around a conveyor belt) can be treated to reduce sound levels by making a long, narrow tubular duct with sound-absorptive walls. The following items should be considered in the design of a partial enclosure:

- Any openings in an enclosure should be directed away from personnel locations; otherwise, there will be little or no sound reduction at the operator positions.
- The number and area of openings in an enclosure should be minimized.
- The interior of the enclosure should be treated with as much sound-absorptive material as possible.
- Partial enclosures work best at reducing high-frequency sounds where the dimensions of the enclosure surfaces are at least several times longer than the wavelength of the sound waves.

The noise reduction provided by a partial enclosure can be estimated by using the following equation:

$$NR = 10\log_{10}\frac{TA}{OS} \quad (20)$$

where:

NR = noise reduction in dB;

OS = area of the openings in square meters; and
TA = total absorption inside the enclosure (including the openings).

To achieve the noise reduction calculated in the formula above, the barrier material of the partial enclosure must have a transmission loss (TL) value at least 10 dB greater than the noise reduction calculated for the partial enclosure. If the opening in the enclosure is pointing toward the employee, the noise reduction at his or her location will probably be at least 3 dB less than that calculated above, depending on the directivity of the noise source.

Personnel Enclosures

When there are multiple noisy sound sources in a room and a low number of operators, it is sometimes useful to enclose the employees instead of the equipment. The equipment room will still be loud, but the operators will be in a relatively quiet enclosure. This personnel enclosure can include equipment controls or just windows for observing operations of the equipment. The noise reduction from the outside of the enclosure to the inside can be calculated with the following equation:

$$NR = TL_C - 10\log_{10}\left(\frac{1}{4} + \frac{S_t}{TA}\right) \quad (21)$$

where:

TL_c = composite transmission loss of the enclosure walls, roof, etc.;
S_t = area of enclosure surfaces between noisy and quiet sides (in m^2); and
TA = total absorption inside the interior of enclosure in metric sabins.

This equation can also be used to determine the NR of a wall separating a noisy room from a relatively quiet room. Like equipment enclosures, personnel enclosures are limited by the TL of the walls and roof and the sound absorption inside the enclosure. The TL of a personnel enclosure is normally more difficult to calculate than the TL of a machine enclosure since it often includes windows, doors, ventilation, and different roof and wall structures. The composite TL of an enclosure consisting of several materials is determined by the following equation:

$$TL_C = 10\log_{10}\left(\frac{S_t}{S_1 \times 10^{(-TL_1/10)} + S_2 \times 10^{(-TL_2/10)} + ...etc.}\right) \quad (22)$$

where:

TL_c = composite TL of enclosure;
TL_1 = TL of wall or roof element 1;
TL_2 = TL of wall or roof element 2, etc.;
S_t = enclosure total exterior surface area;
S_1 = surface area of wall or roof element 1; and
S_2 = surface area of wall or roof element 2, etc.

Since small openings and building elements with low TL values significantly reduce the effectiveness of the entire enclosure, it is important to pay close attention to all components when designing an enclosure. The number and sizes of windows should be minimized and, if possible, laminated or double-layered glass (with as deep an air gap as possible) should be used. Doors should be equipped with seals around all four edges. The walls and roof should also be carefully sealed. Any ventilation ductwork running from the inside of the enclosure to the outside should have an interior lining of sound-absorptive material and should include 90° bends to reduce sound transmission. The interior should be treated with an abundance of sound-absorptive materials. Although carpet is sometimes used on the floor, it is usually effective only at absorbing high-frequency noise (above 2000 Hz). Therefore, the walls and ceiling should be treated with good high- and low-frequency sound-absorptive materials.

Example: Design of a Personnel Enclosure

A personnel enclosure is proposed for a large manufacturing area. The existing sound levels in the area where the enclosure will be located and the desired goal for the interior of the enclosure are listed below:

	Octave Band Center Frequencies						
	125	250	500	1000	2000	4000	dBA
Existing Sound Level	85	89	87	84	78	71	89
Enclosure Design Goal	72	66	61	57	54	52	65
Required NR	13	23	26	27	24	19	

The enclosure will be 2 m wide × 4 m long × 2 m high. The floor will be carpeted, the ceiling will have acoustical tile, and the walls will have some sound-absorptive panels as well as a door and windows. The total absorption (using the absorption coefficients from Table 20.9) and required TL are calculated below:

	Surface Area (m^2)	Octave Band Center Frequencies 125	250	500	1000	2000	4000
Carpet Abs. Coefficient	8	.02	.06	.14	.37	.60	.65
Carpet Absorption		.2	.5	1.1	3.0	4.8	5.2
Ceiling Abs. Coefficient	8	.76	.93	.83	.99	.99	.94
Ceiling Absorption		6.1	7.4	6.6	7.9	7.9	7.5
Plywood Wall Abs. Coefficient	12	.28	.22	.17	.09	.10	.11
Plywood Wall Absorption		3.4	2.6	2.0	1.1	1.2	1.3
Treated Wall Abs. Coefficient	8	.23	.54	.60	.98	.93	.99
Treated Wall Absorption		1.8	4.3	4.8	7.8	7.4	7.9
Window/Door Abs. Coefficient	4	.35	.25	.18	.12	.07	.04
Window and Door Absorption		1.4	1.0	.7	.5	.3	.2
Total Area (S_t) and Abs. (TA)	32	12.9	15.8	15.2	20.3	21.6	22.1
10log(1/4+S_t/TA)		4	4	4	3	2	2
Required NR		13	23	26	27	24	19
Safety Factor		5	5	5	5	5	5
Required Composite TL		22	32	35	35	31	26

In the following table, the composite TL is calculated for the enclosure using TL values from Table 20.10. As shown, this can be fairly complicated to do by hand. One way of reducing the work is to put the calculations on a spreadsheet computer program. The easiest (though least precise) method is to choose roof and wall materials with TL values greater than the required amount for the composite TL and then to try to get windows and doors that do not reduce the composite TL significantly.

	Surface Area (m^2)	Octave Band Center Frequencies 125	250	500	1000	2000	4000
TL of Wall	20	23	35	44	53	45	43
$S \times 10^{(-TL/10)}$		.100	.0063	.0008	.00010	.0006	.001
TL of Roof	8	32	42	52	58	53	54
$S \times 10^{(-TL/10)}$		.005	.0005	.00005	.00001	.00004	.00003
TL of Window	2	25	28	30	34	24	35
$S \times 10^{(-TL/10)}$		.063	.0032	.0020	.00080	.008	.0006
TL of Door	2	23	28	36	41	39	44
$S \times 10^{(-TL/10)}$		.010	.0032	.0005	.00016	.00025	.00008
Total $S \times 10^{(-TL/10)}$		.178	.0132	.0033	.00107	.00889	.00171
Composite TL	32	23	34	40	45	36	43
Required TL		22	32	35	35	31	26

In the above example, the proposed wall is made of 64 mm (2½ in.) steel channel studs with 16 mm (5/8 in.) gypsum board on both sides and glass fiber insulation in the cavity. The roof is constructed similarly except with wooden studs and a resilient channel. The door is a glass-fiber filled 1.6 mm thick (16 gauge) steel door equipped with gaskets and a drop seal. The window is 6 mm (1/4 in.) single plate glass. Note that the weak composite TL in the 2000 Hz octave band is caused by the low TL for the window. This shows that a significant weakness in any building element, even if it is not large in area, can reduce the composite TL of the entire enclosure.

Shields or Barriers

Barriers placed between an employee and noisy equipment can be an effective means of lowering noise exposure. A barrier causes a "sound shadow" at the receiver with a consequent attenuation of L_p. Since sound, like light, diffracts around an edge, the shadow zone is only partially shielded from the noise. Mid- and high-frequency sound is diffracted less than low-frequency sound; therefore, barriers are more effective for shielding from mid- and high-frequency sound sources. The attenuation of a barrier can be calculated as a function of the difference in the length of the sound path imposed by the barrier. The sound path difference is given by:

$$d = \sqrt{S^2 + h_s^2} + \sqrt{R^2 + h_r^2} - \sqrt{(S + R)^2 + (h_r - h_s)^2} \quad (23)$$

where:

- d = difference between the path length from the source to the receiver over a barrier and the straight line-of-sight path between the source and receiver without the barrier;
- S = distance to the barrier from the source (m);
- R = distance to the barrier from the receiver (m);
- h_s = height of the barrier above the source (m); and
- h_r = height of the barrier above the receiver (m).

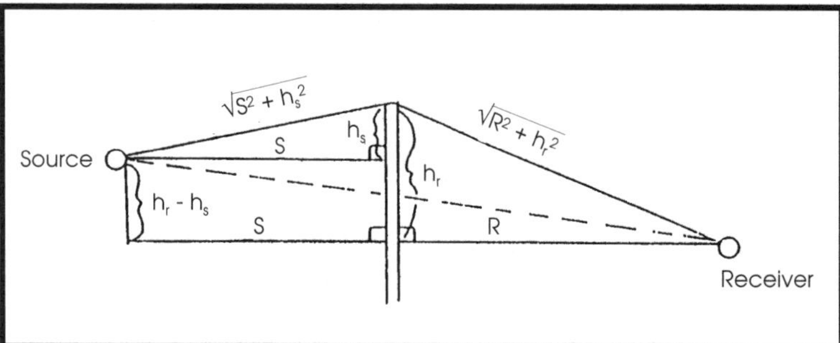

Figure 20.19 — Variables used for calculation of barrier insertion loss.

These variables are illustrated in Figure 20.19. In Table 20.11, an estimate of the attenuation or insertion loss of a barrier is given as a function of the path difference (d) and the frequency.[27] These values are for an ideal barrier with no reflecting surfaces nearby. It can be seen in the table that the insertion loss due to a barrier ranges from 5–24 dB. For the greatest insertion loss, the path difference should be maximized; thus, the barrier should be as tall as possible and placed as near the source or receiver as possible.

If a barrier is used indoors, reflections off ceilings and walls will probably reduce its performance. A rough estimate of the insertion loss of a barrier with a ceiling above the barrier can be derived by considering only the sound path that is reflected off the ceiling to the receiver in the barrier shadow zone. The insertion loss is calculated as:

$$IL = 10\log_{10}\left[\left(\frac{dc}{dl}\right)^2 * \frac{1}{(1-\alpha)}\right] \quad (24)$$

where:

IL = difference between the level of the direct sound path without a barrier and that of the sound path reflected off the ceiling;
dc = path length in meters that the sound travels to the ceiling and down to the receiver position behind the barrier;
dl = line-of-sight distance in meters from the source to the receiver without a barrier; and
α = absorption coefficient of the ceiling.

This equation should be used to calculate the sound level of reflected sound paths off any walls, ceilings, or other surfaces that directly reflect sound from a source to a receiver around a barrier. There may be several such paths. The total expected sound level at the receiver position should be calculated by using Table 20.11 to calculate the sound level at the receiver position from the diffracted sound path and the previous equation to calculate the sound level at the receiver position from each reflected sound path. The resulting diffracted and reflected sound levels

Table 20.11 — Insertion Loss (in dB) of an Ideal Barrier Based on the Path Difference[27]

Path	Octave Band Center Frequency (Hz)								
Difference	31.5	63	125	250	500	1000	2000	4000	8000
3 mm	5	5	5	5	5	5	6	6	7
6 mm	5	5	5	5	5	6	6	7	9
10 mm	5	5	5	5	5	6	7	8	10
15 mm	5	5	5	5	6	6	8	9	12
20 mm	5	5	5	5	6	7	8	10	13
30 mm	5	5	6	6	6	8	9	12	15
60 mm	5	5	6	6	8	9	12	15	18
0.1 m	5	6	6	7	9	11	14	16	19
0.15 m	5	6	7	8	10	13	16	18	21
0.2 m	6	6	8	9	11	14	17	20	23
0.3 m	6	7	9	10	13	16	18	21	24
0.6 m	7	8	11	13	16	18	21	24	24
1 m	8	9	12	14	17	20	23	24	24
1.5 m	9	11	14	16	19	22	24	24	24
2 m	10	13	15	18	21	24	24	24	24
3 m	11	14	16	19	22	24	24	24	24
6 m	14	16	19	22	24	24	24	24	24
9 m	15	18	21	24	24	24	24	24	24

should then be added logarithmically to get the total L_p.

A few other factors should be considered during the design of barriers to be installed in occupational environments:

- The transmission loss of the barrier itself should be at least 10 dB greater than the expected insertion loss of the barrier. This prevents noise transmitted through the barrier from having an appreciable contribution at the receiver position.
- For outdoor barriers, care must be taken to consider reflections off nearby buildings or large equipment. As for the case of indoor barriers, reflected noise will lower the barrier performance to less than that estimated in Table 20.11.
- All diffracted paths should be considered. For example, the sound levels calculated for diffracted paths around the sides of a free-standing barrier should be added to the sound level from the refracted path over the barrier and the reflected paths.
- Table 20.11 is derived with the noise source considered as a point source. For many large sources, such as cooling towers, the position of an appropriate source radiation point is not obvious. One approach is to divide the surface of the large source into many point sources. A conservative and less time-consuming approach is to place the source point at the top of the equipment.
- If two barriers or a barrier and a wall are situated in parallel, noise reflected back and forth between the two surfaces can "walk up" the barrier and lower the insertion loss. This potential can be minimized by choosing a barrier that is acoustically absorptive on the side facing the sound source.

Example: Calculation of Barrier Insertion Loss

A sound source inside a building produces a level of 98 dB in the 1000 Hz octave band at an operator position 9 m (30 ft) distant (horizontally). The sound is radiated from an inlet duct, the top of which is positioned 3.4 m (11 ft) above the floor. The operator is usually standing, and his or her ear height is taken as 1.5 m (5 ft). A barrier is to be erected between the noise source and the operator. Because of access requirements, the barrier will be placed 1.8 m (6 ft) from the source. The barrier height is 3.7 m (12 ft). The ceiling height is 6 m (20 ft) and covered with thermal insulation that has an absorption coefficient of 0.5 at 1000 Hz. Assume that there are no reflected or diffracted sound paths around the side edges of the barrier. Calculate the insertion loss of the diffracted path over the barrier and the loss of the reflected path off the ceiling.

The path difference over the barrier is calculated to be d = 0.15 m (0.5 ft). The insertion loss of the barrier is about 13 dB at 1000 Hz as read from Table 20.11; thus, the L_p would be reduced to 98 – 13 = 85 dB by the barrier, assuming there are no other sound pathways or significant sound sources in the near vicinity. To calculate the ceiling path length, it helps to draw a scale sketch of the source, barrier, receiver, and ceiling in elevation. Choose a ceiling reflection point such that the angle between the ceiling and the sound path from the source is the same as the angle between the ceiling and the sound path to the receiver (a mirror-like reflection) as shown in Figure 20.20. For this example, the total reflected ceiling path length is 11.5 m (37.6 ft). The line-of-sight distance is about 9.3 m (30.6 ft). The reduction along the ceiling pathway from Equation 24 is 5 dB. The contribution from the ceiling path, therefore, is 98 – 5 = 93 dB. The L_p at the receiver is the result of the combination of the barrier and ceiling

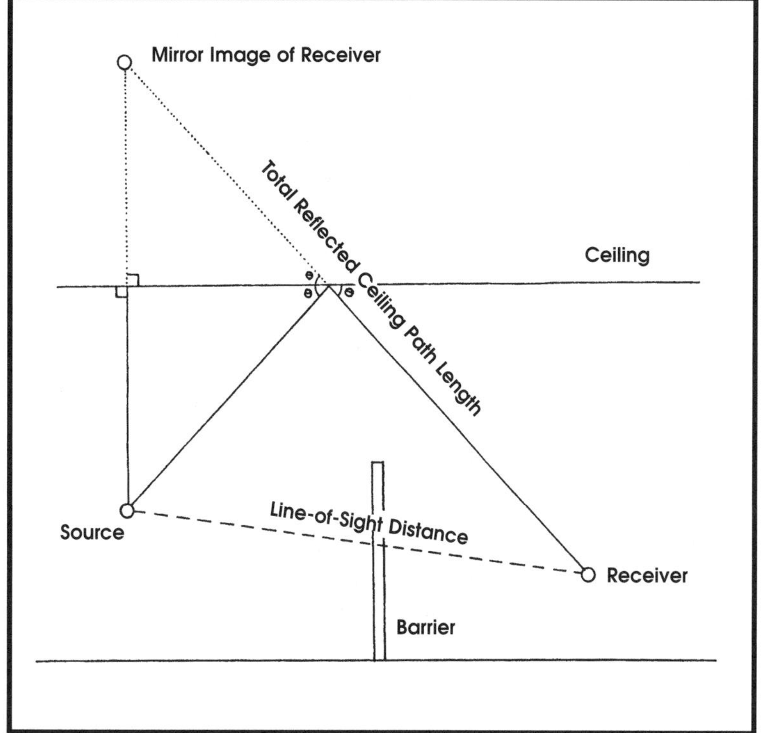

Figure 20.20 — Calculation of sound reflection over a barrier.

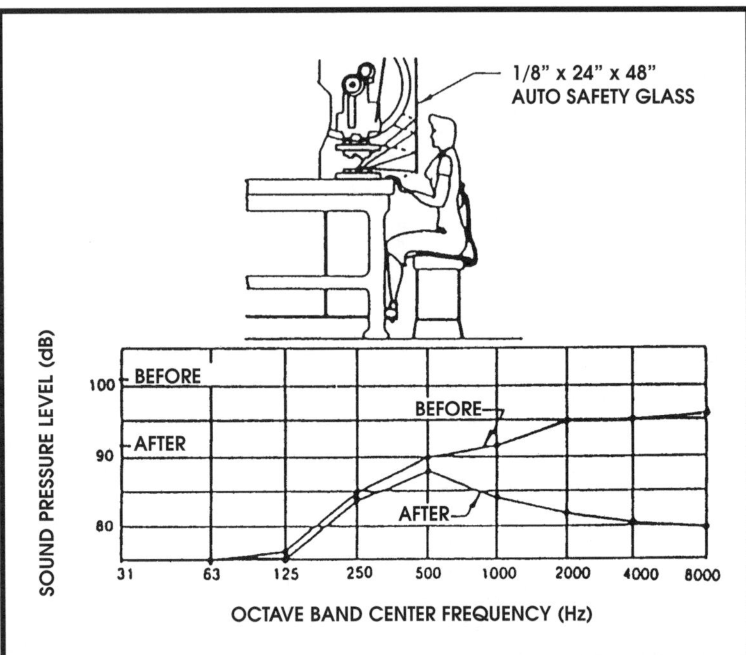

Figure 20.21 — Noise levels before and after installation of a glass shield. (Adapted from Reference # 30.)

contributions of 85 $+_{\log}$ 93 = 94 dB. The barrier reduction of 13 dB is overshadowed by sound reflected off the ceiling.

If the absorption coefficient were increased to 0.98 at 1000 Hz, the reduction reflecting off the ceiling path would be 19 dB for an L_p of 79 dB. The summed level would be 79 $+_{\log}$ 85 = 86 dB. For this latter case, the barrier performance would not be appreciably compromised by the ceiling reflection.

The placement of shields between an employee and a noise source can be an effective treatment, especially if the noise is high frequency and the employee and noise source are close to the shield. A shield used on a punch press is illustrated in Figure 20.21. Safety glass (6 mm [1/4 in.] thick) was installed between the operator and the press. L_ps before and after treatment are presented in the figure. The designer should be aware that in situations in which multiple units are in close proximity, noise from nearby units might make this type of treatment ineffective.

Lagging

It is sometimes impractical to enclose a noise source because of its shape, access requirements for operation and maintenance, or insufficient space. Damping may not be practical if the excitation frequency is different than the panel resonances or if the panel is thick (recall that damping treatments should generally be as thick as the panel to which they are attached). In these situations an enclosure-like treatment attached directly to the vibrating surface of a noisy machine, pressure-reducing valve, or piping can be effective. This treatment is commonly referred to as "lagging."

High-frequency sound radiation can be reduced by wrapping a surface with a sound-absorptive material such as glass fiber or acoustical foam. Sound below about 1000 Hz will be little affected, however. A significant improvement can be attained by covering the absorptive wrapping with an air-tight limp barrier material. This barrier can be asphalt paper, linoleum, neoprene sheeting, lead, loaded vinyl, etc. — the heavier and more limp, the better. A thicker sound-absorptive layer generally also improves the noise reduction. Manufacturers offer lagging materials in the form of foam/vinyl composites as well as other material combinations.

With lagging, the outer layer acts like the walls of an enclosure, and the sound-absorptive layer provides both vibration isolation for the outer layer and also sound absorption for the space between the outer layer and the sound-radiating surface. Lagging treatments will generally be effective above the frequency given by the following equation:[27]

$$f = \frac{63}{\sqrt{wd}} \qquad (25)$$

where:

f = limiting frequency in Hz;
w = limp barrier surface weight in kg/m^2; and
d = thickness of the absorptive layer in meters.

Below the frequency given by this equation, the insertion loss will be essentially zero. In fact, when this treatment is applied to a small pipe, the sound levels in this low-frequency region may increase a few dB over the untreated condition due to the increase in the size of the sound-radiating surface. The insertion loss of a lagging treatment increases above the frequency defined by Equation 25 and may attain 30–45 dB in the highest frequencies, though 15–20 dB is more common.

Lined Ducts and Mufflers

In some situations the noise of HVAC systems becomes an important contribution to the employee noise environment. Ventilation ducts can also be significant noise transmission pathways between noisy and quiet areas. Ducts may also be used as intakes to and discharges from noisy equipment. Noise traveling through ducts can be attenuated by the application of various sound-absorptive duct linings. A simple formula to estimate the attenuation of sound through straight ducts of a regular cross section lined with absorption is:

$$NR = 1.05(\alpha)^{1.4}\frac{P}{S} \quad (26)$$

where:

- NR = noise reduction in dB/m of duct length;
- α = absorption coefficient for the lining material;
- P = lined perimeter of the duct in meters; and
- S = cross-sectional area in m^2.

This formula is most accurate for low frequencies and/or narrow ducts such that $f < 34/w$[27] where "w" is the duct width in meters. At higher frequencies, the attenuation will be less than calculated due to the "beaming" of sound straight down the duct without reflecting off the side walls. This formula can also be applied to parallel baffle sound absorbers with accuracy sufficient for many occupational noise problems. Noise reduction of the parallel baffles generally increases as the distance between the baffles decreases. Of course, muffler performance data from the manufacturer is always preferable.

Blocking the line-of-sight between the source and receiver is an effective element of noise reduction. Bends or staggered absorbers at 90° along a duct will greatly increase the noise reduction, particularly at high frequencies. It might be necessary in these cases to evaluate the resulting increase in back pressure. Figure 20.22 presents sketches and trends in the noise reduction of various arrangements of lined ducts.

Dissipative and reactive mufflers can be used to control noise from exhausts or intakes. Dissipative mufflers have an absorptive lining inside the muffler. In general, these mufflers provide broad-band noise reduction with the maximum attenuation being a function of the thickness of the lining as well as its acoustical properties. Dissipative mufflers are generally used to reduce the noise from fans, blowers, and product entrances and exits from enclosures. A reactive muffler provides attenuation by a series of cavities and side branches with abrupt cross-sectional area changes. The resulting impedance changes reflect the sound back toward the source. These mufflers are likely to have superior performance over a narrow band of frequencies. They are often used on internal combustion engines and fluidborne noise sources.

The design of reactive mufflers is beyond the scope of this chapter. They are usually effective only over a narrow range of frequencies. It might be safer to design an absorptive muffler with a broader frequency response, if possible. However, if the noise is produced in a narrow frequency range and is unlikely to change based on different operations, a reactive muffler may be specified. Since reactive mufflers are complicated to design, they should be purchased with a performance guarantee.

In Figure 20.23, there is a sketch of a combination parallel-baffle dissipative muffler and elbow. This arrangement was attached to the compressor intake of a 5 MW (7000 hp) gas turbine. Six parallel baffles were constructed of 1.2 mm (18 gauge) perforated steel, each 89 mm (3.5 in.) wide and filled with glass fiber. The duct was 2.1 m × 2.4 m (7 in. × 8 in.), and the elbow was constructed of 6 mm (1/4 in.) unlined plate. L_ps measured with and without the muffler installed are presented in the figure.

As an example, the discharge of a filter-bag separator of a pneumatic conveying system handling synthetic fiber fluff produced objectionable noise in the filter area. An absorptive muffler was not desired because of the potential for becoming clogged with fibers from the exhaust

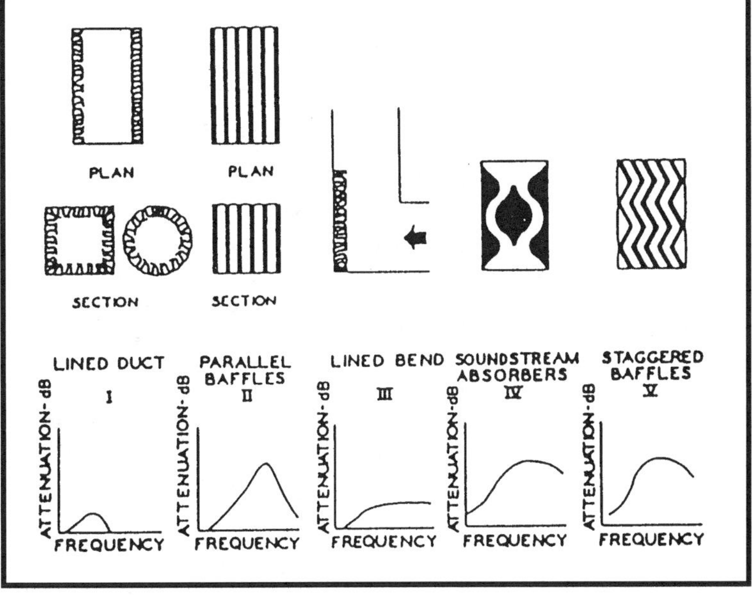

Figure 20.22 — Examples and general acoustic performance of lined ducts and mufflers. (Adapted from Reference # 30.)

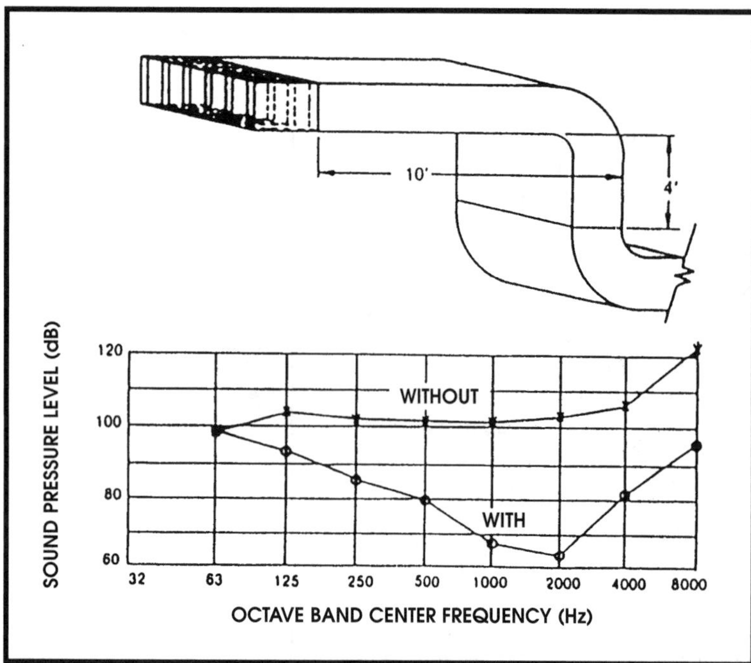

Figure 20.23 — Noise levels with and without muffler. (Adapted from Reference # 30.)

stream. A reactive muffler was installed and provided the following values of noise reduction:

Octave Band Center Frequencies (Hz)				
63	125	250	500	1000
12 dB	23 dB	13 dB	11 dB	10 dB

In this section, a wide variety of noise control treatments were presented with the intent occupational hygienists will become familiar with ways to reduce employee noise exposure.

Hearing Conservation Programs

Introduction

The OSHA hearing conservation amendment provides a good outline of the minimum requirements for an effective hearing conservation program.[34] This includes sound monitoring, feasible administrative or engineering controls, audiometric testing, hearing protection, employee training and education, and record keeping. The true test of an effective program, however, is in the results of the employees' hearing tests.

This section briefly discusses the main requirements for an effective hearing conservation program. Although developing a program might seem straightforward, maintaining an effective program is quite challenging. Many texts have been written to provide assistance.[35-39]

Sound Survey

As noted earlier under the noise measurements section, a sound survey is performed to identify areas with hazardous levels of noise and to determine who should be included in a hearing conservation program. ANSI has requirements for Type I or Type II sound measurement equipment[40-42] and ANSI S12.19–1996 provides specific guidelines for performing noise measurements to determine occupational noise exposure.[43] These measurement procedures were discussed in the measurement section earlier in this chapter.

Engineering and Administrative Controls

Reducing the noise level to below the action level or limiting workers' exposure by reducing their time in noisy areas should be considered the primary methods of protecting workers' hearing,[44,45] although this is often not feasible. Regardless, occupational hygienists should consider using these options to reduce worker noise exposure as the primary means for preventing hearing loss. Engineering and administrative controls combined with hearing protection may be the only way to prevent hearing loss. Also, though a decrease in the sound level of 3–5 dBA might not sound significant, this decrease can double the allowable noise exposure time. Ideally, hearing protection should be considered an interim solution while striving toward long-range goals such as purchasing quieter equipment and using cost-effective noise control solutions. Unfortunately, we have more than 25 years of experience with less-than-ideal situations where feasible solutions have not been available. Consequently, hearing protection remains an important means of conserving employee hearing.

Hearing Protection Devices

A good hearing protective device (HPD) is an effective means to reduce the sound exposure of persons either before engineering or administrative noise controls can be administered or when these controls are not yet feasible. The best HPD for a given situation is the one that is consistently and properly worn by the employee. Selection of an appropriate HPD depends on necessary attenuation; audibility of speech and

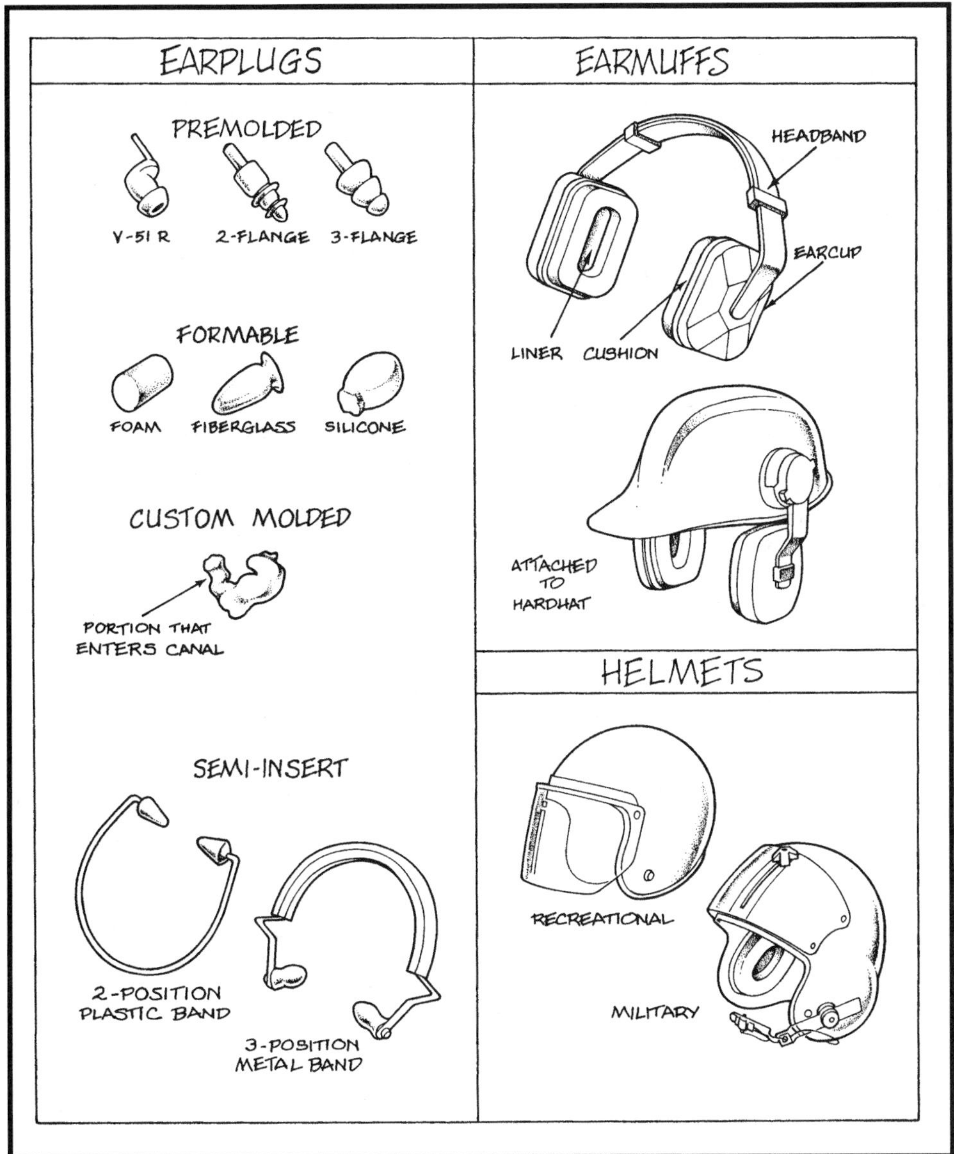

Figure 20.24 — Types of hearing protection. (Adapted from Reference # 46.)

warning signals; compatibility with other safety equipment; care and cleaning requirements; comfort; cost; ease of use; personal preference; temperature and humidity; and visibility (i.e, the HPD is visible so it is easy to monitor compliance with HCP policy).[37]

All of these factors should be evaluated thoroughly before choosing the types of HPDs that will be made available. Workers should be given several choices of appropriate HPDs. Purchasing agents should also be informed of the reasoning behind the selection of the HPDs so they do not inadvertently purchase a lower-cost product they believed was equivalent.[38] The main types of hearing protectors are illustrated in Figure 20.24.[46] Premolded earplugs, formable earplugs, earmuffs, and semi-insert hearing protectors are discussed in the following sections.

Premolded Earplugs

Premolded earplugs, such as the V-51R, E•A•R Express™ Pod Plug, Howard Leight Quiet® plug, or multiflanged plugs are preformed and are simply inserted in the ear. They can be made from soft rubber-like materials or foam. Their main advantages are that they can be inserted and removed easily without touching the portion of the plug that is inserted in the ear canal, and they can be worn easily with other safety equipment such as hard hats and safety glasses. Although laboratory attenuation data can be excellent, good field attenuation can be challenging due to their sensitivity to individual fit. To provide adequate attenuation, they need to seal well and be sized to ensure a proper fit.

Formable Earplugs

Formable earplugs (e.g., foam plugs, fiberglass down, silicone putty, and cotton wax) are formed by the user prior to inserting them into the ear canals. They are less sensitive to individual fit since they can expand or conform to match an individual's anatomy. To provide adequate attenuation, formable earplugs need to seal well and should be sized to ensure a proper fit. Some foam earplugs come in multiple sizes to accommodate different size ear canals. Foam earplugs are among the most comfortable and protective HPDs.

Earmuffs

Although the first earmuffs available did not provide as much attenuation as earplugs, muffs now provide as much attenuation as the best earplugs when worn properly. Proper fitting is still important with earmuffs but is not quite as individualistic as with plugs. Consequently, muffs offer more consistent protection, although performance can decrease when the user wears glasses. Although muffs usually last much longer, they generally are more expensive than plugs, and in hot, humid environments they can be uncomfortable to wear for long periods due to sweat accumulation under the muffs. One notable advantage with muffs is that it is easy to check for employee compliance.

Active noise reduction earmuffs work on the principle of creating a sound wave equal in magnitude but opposite in phase to the noise. When the two sound waves combine, the sound is significantly reduced. Active noise reduction muffs work best for lower-frequency sounds but can be effective up to about 1000 Hz. These systems can be expensive. In environments with extremely high low-frequency sound levels (e.g., in military aircraft) the active noise reduction muff can be used effectively to provide speech communication using the intercom system.

Semi-Inserts

Semi-inserts are a cross between earplugs and earmuffs. They use earplug-like devices attached to the ends of a headband that are pressed into the ear canal. They are moderately comfortable but do not provide as much attenuation as most muffs or plugs. These devices are primarily intended for intermittent use in moderately noisy environments.

HPD Attenuation Ratings

The EPA requires that hearing protectors be labeled with their Noise Reduction Rating (NRR).[47] The NRR is a single-number rating of the hearing protection. The higher the NRR, the higher the attenuation for a specific ideal situation (laboratory-fit of HPD). Unfortunately, this data does not provide a reliable estimate of the attenuation provided to any employee. Consequently, adjustments must be made.

To determine the protected A-weighted sound level at a worker's ear, the effective NRR is subtracted from the C-weighted sound level or the effective NRR is subtracted from the A-weighted sound level plus 7 dB. According to OSHA, the effective NRR is equal to one-half the laboratory NRR.[48] NIOSH recently recommended that instead of reducing the estimated attenuation of all types of hearing protectors by one-half, muffs should be reduced by 25%, formable earplugs by 50%, and premolded earplugs by 70%.[19,49]

Instead of using the NRR, octave band attenuation values from the NIOSH compendium of hearing protection devices can be used.[49] The octave band attenuation values are then subtracted from the measured octave band sound pressure levels at the worker's ear to determine the expected sound level in the ear.

Since all of these procedures are based on the NRR, it should be recognized that the expected sound levels are, at best, good estimates.

Audiometric Testing

An audiometric testing program is an integral part of the hearing conservation program. The audiometric test records provide the only data that can be used to determine whether the program is preventing noise-induced permanent threshold shifts (NIPTS). Audiometric tests should be performed by a licensed or certified audiologist, an otolaryngologist or other physician, or a technician certified by the Council for Accreditation of Occupational Hearing Conservationists (CAOHC). A technician who performs audiometric tests must be responsible to an audiologist, otolaryngologist, or physician. Audiograms should be obtained annually. All employees whose noise exposure exceeds the action level (a TWA of 85 dBA for OSHA compliance), should be included in an audiometric testing program. In some cases it might make sense to include all plant personnel in an audiometric testing program to detect and prevent all causes of hearing loss, not only hearing loss from occupational noise.[39]

Performing audiometric testing after a period of quiet provides an accurate representation of any NIPTS incurred. However, audiometric testing can also be used to measure NITTS as a means of education before any NIPTS is incurred. By testing workers during the work shift or at the end of a work shift and noting any threshold shift, workers who are not wearing their HPDs properly can be identified.[38, 50] Of course, these workers will need to be retested after a period of quiet to ensure that the threshold shift is only temporary.

Employee Training and Education

All persons involved in the hearing conservation program must be educated annually on the effects of noise on hearing, proper use of hearing protection, advantages and disadvantages of different types of hearing protection, purpose of audiometric testing, and their individual audiometric results. This training should include not only the noise-exposed workers but management, supervisors, audiometric technicians, issuers of HPDs, and anyone else involved in the hearing conservation program. Although each group will need a slightly different training focus (management will need to understand any legal responsibilities and supervisors will need to understand how to enforce HPD use), everyone involved needs to understand the risks of noise-induced hearing loss and their particular roles in maintaining an effective program.[51]

Numerous resources are available for use in training programs;[37, 51-54] Training, however, does not have to take the form of videotapes or lectures to be effective. Audiometric testing of workers at the beginning and end of one work day while educating the workers about TTS and its consequences has proved to be very effective. In one case, worker use of hearing protectors increased from 35% to 80% after an education program and the posting of pre-shift and post-shift audiograms.[55]

Record Keeping

Detailed records must be kept of employee's audiometric test records as well as plant noise exposure, types of hearing protection worn by the employee, documentation of employee training, documentation of technician training and certification, audiometer calibration data, and any other medical or audiologic test results or examinations. Also, a record of the employee's recreational noise exposure may be useful. Plant sound survey data and engineering or administrative noise control records likewise need to be maintained. This is not only for review by audiologists and physicians, but also for protection against worker's compensation claims or other legal action. Audiometric records for each worker should be maintained during the worker's employment, but it is recommended that records be kept for the worker's lifetime to protect against possible future legal action.

Program Evaluation

It is important to evaluate the effectiveness of a hearing conservation program. Checklists such as those recommended by NIOSH[56] used in conjunction with audiometric database analysis can ensure that all aspects of a program are monitored. NIOSH recommends that an effective program have a significant threshold shift incidence rate of 5% or less.[19] A significant threshold shift in this case is defined as an increase of 15 dB in hearing threshold, not corrected for age, at any of the audiometric test frequencies (500, 1000, 2000, 3000, 4000, or 6000 Hz) that is repeated for the same ear and frequency in back-to-back tests. OSHA defines a significant threshold shift as a shift of 10 dB or greater, corrected for age, between the most recent audiometric test and the baseline in the average hearing threshold levels at 2000, 3000, and 4000 Hz for either ear, but a recommended maximum incidence rate is not defined. If the average hearing threshold shift exceeds 25 dB, the hearing loss becomes recordable on the OSHA 200 log as a work-related injury or illness. (It is likely OSHA will revise these requirements in conjunction with a new form, OSHA 300.)

Another method of evaluation is through database analysis.[57] By analyzing a database of workers' audiometric test records, hearing trends for different groups can be examined, or individuals can be compared with the reference group. If, for example, workers in a particularly noisy area of a plant show increased hearing loss, this might be an indication that a different hearing protector or engineering noise controls are needed. If one worker shows a significant threshold shift and there are no shifts for all other workers in his or her area, it might be an indication that he or she is using the HPD incorrectly or is receiving excess noise off the job. The draft ANSI standard for evaluating the effectiveness of hearing conservation programs recommends use of audiometric data base analysis for programs having more than 30 participants and provides recommended criteria for various statistical parameters.[58]

Vibration in the Workplace — Measurements and Control

Introduction

Human exposure to vibration is normally divided into whole-body vibration and hand-arm vibration. Although these two different types of vibration exposure usually result in different effects and responses, workers can be exposed to both types simultaneously.[59] For example, when a jack-hammer operator holds the tool away from his body, supporting and guiding it only by his limbs, he is exposed to hand-arm vibration; however, if he leans against the jack hammer with his abdomen, he is exposed to whole-body vibration as well.

In 1974, NIOSH estimated that 8 million workers were exposed to occupational vibration in U.S. industries.[60] Of these, about 6.8 million were exposed to whole-body vibration and 1.2 million to hand-arm vibration. The majority of those exposed to whole-body vibration were truck and bus drivers, heavy equipment operators, and aircraft pilots. Those exposed to hand-arm vibration include operators of gasoline-powered chain saws, string trimmers, and pneumatic tools. Table 20.12 provides a detailed listing of the various occupations and types of vibration exposures.[61,62]

Effects of Vibration

Hand-Arm

Prolonged exposure to hand-arm vibration can lead to a condition known as Raynaud's Phenomenon of Occupational Origin, vibration-induced white finger (VWF), or hand-arm vibration syndrome (HAVS).[63-65] The first symptoms are intermittent tingling and/or numbness of the fingers which does not interfere with work or other activities.[65] The worker may later experience attacks of finger blanching (turning white) usually confined at first to a single fingertip. With added vibration exposure, attacks may extend to the base of the finger. Cold often triggers these attacks, but other factors are involved, such as body temperature, metabolic rate, vascular tone, and emotional state. Attacks usually last 15–60 minutes, but in advanced stages they may last as long as two hours. Recovery from attacks can also be painful. With additional vibration exposure, the symptoms of HAVS become more severe and include increasing stiffness of the finger joints, loss of manipulative skills, and loss of blood circulation, which can lead to gangrene and tissue necrosis. At this point, there are no long-term medical treatments for HAVS.[66]

Table 20.12 — Vibration Exposure in U.S. Industries[61,62]

Industry	*Type of Vibration*	*Common Vibration Sources*
Agriculture	Whole Body	Tractor operation
Automotive Assembly	Hand-Arm	Pneumatic tools
Boiler Making	Hand-Arm	Pneumatic tools
Construction	Whole Body and Hand-Arm	Vehicles, pneumatic tools
Diamond cutting	Hand-Arm	Vibrating hand tools
Forestry	Whole Body and Hand-Arm	Tractors, chainsaws
Foundries	Hand-Arm	Vibrating cleavers, pneumatic tools
Furniture Manufacturing	Hand-Arm	Pneumatic chisels
Grounds Maintenance	Hand-Arm	String trimmers, chainsaws
Iron and Steel	Hand-Arm	Vibrating hand tools
Lumber	Hand-Arm	Chainsaws
Metal Working	Whole Body and Hand-Arm	Drills, sanders, grinders, stand grinding
Mining	Whole Body and Hand-Arm	Vehicles and rock drills
Shipyards	Hand-Arm	Pneumatic hand tools
Stone dressing	Hand-Arm	Pneumatic hand tools
Textile	Hand-Arm	Sewing machines, looms
Transportation	Whole Body	Vehicle operation

Whole-Body

Whole-body vibration can cause both physiological and psychological effects ranging from fatigue and irritation to motion sickness (kinetosis) and to tissue damage. Much of whole-body vibration research is rooted in military settings (e.g., pilots, ship motion studies, tank ride studies). Some work has taken place in the occupational setting, principally in transportation and with on- and off-road vehicles. Since it is potentially dangerous to expose human subjects to high acceleration levels in the laboratory, researchers have simulated high-level exposures using animals and then extrapolated to potential biological effects on humans. Because of the difficulty of obtaining hard epidemiology and medical data, human performance studies have been concerned primarily with performance decrements or comfort reduction at low vibration levels.

The most frequently reported adverse effects of whole-body vibration are lower-back pain, early degeneration of the lumbar spinal system, and herniated lumbar discs. Long-term exposure to whole-body vibration is harmful to the spinal system, but due to the limited numbers of epidemiology studies, firm dose-response relationships are undetermined.[67]

Terminology

Periodic Vibration

Vibration is considered periodic if the motion of a particle repeats itself considerably over time. The simplest form of periodic vibration is called simple harmonic motion, which can be represented by a sinusoidal curve similar to the sound wave shown in Figure 20.1 earlier in this chapter. The motion of any vibrating particle can be characterized at any time by its:

- Displacement from the equilibrium position;
- Velocity, or rate of change of displacement; or
- Acceleration, or rate of change of velocity.

For simple harmonic motion, these three characteristics of motion are related mathematically.

Displacement

The displacement (X) of a particle from its reference position under influence of harmonic motion can be described mathematically as:

$$X = X_{peak}\sin(2\pi ft) = X_{peak}\sin(\varpi t) \qquad (27)$$

where:

X_{peak} = maximum displacement;
f = frequency in Hz;
t = time in seconds; and
ϖ = angular frequency (equal to $2\pi f$).

Displacement is most significant in the study of the deformation and bending of structures.

Velocity

As a particle displaces up and down, it moves with a characteristic velocity (V). The velocity is the time rate of change of displacement and can be determined mathematically for simple harmonic motion by taking the derivative of Equation 27.

$$V = \varpi X_{peak}\cos(\varpi t) = V_{peak}\cos(\varpi t) \qquad (28)$$

Velocity is often measured in preventative maintenance programs for rotating machinery.

Acceleration

The acceleration of a particle (A) is the time rate of change of the velocity. The acceleration is determined mathematically for simple harmonic motion by taking the derivative of Equation 28.

$$A = -(\varpi)^2 X_{peak}\sin(\varpi t) = A_{peak}\sin(\varpi t) \qquad (29)$$

The minus sign indicates that the acceleration is a one-half cycle (180°) out of phase with reference to the displacement. For occupational vibration, the acceleration is the most important quantity since it is proportional to the forces applied to the hand, arm, or whole body, and it is believed that the forces are the source of damage. Also, for periodic vibrations the velocity and displacement can be calculated from the acceleration through integration.

RMS

Peak values are useful for pure sinusoidal vibration; however, because workplace vibration is complex and contains many vibration frequencies, the root-mean-square value is used. The rms average is given by Equation 30 and is calculated by summing the squares of the acceleration values measured over time, dividing by the measuring time, and then taking the square root of the resulting value.

$$A_{rms} = \sqrt{\frac{1}{T}\int_0^T A^2(t)dt} \qquad (30)$$

The rms value of the acceleration is directly related to the energy content of the vibration being measured.

Resonance

Resonance is a condition in which the movement of the human body (or any other mechanical system) acts in concert with an externally generated vibration force, resulting in an amplification of the resulting vibration movement. In other words, parts of the human body like to vibrate more at certain frequencies than at other frequencies. At these resonant frequencies the body amplifies the vibration and exacerbates the occupational vibration problem. Resonance frequencies for the whole-body occur in the 3–6 Hz range and the 10–14 Hz range. Resonance frequencies for the hand-arm system tend to be higher in frequency, in the 150–300 Hz frequency range. In structures, resonant vibration can cause large displacements that can cause severe damage.

Random Vibration

Random vibrations occur quite frequently in nature and may be defined as motion in which the vibrating particles undergo irregular motion cycles that never exactly repeat themselves. Because of their nonperiodicity, random vibrations result in energy over a broad-frequency spectrum. A mechanical shock pulse is an example of random vibration. Note that for a pulse or a random signal the measured acceleration or vibration signal cannot be integrated to obtain the velocity or displacement as with a periodic signal.

Vibration Transmissibility Ratio

In many cases, it is desirable to know how much vibration is transmitted through structures (e.g., vehicle seats or gloves). In other words, how well does the driver's seat remove, transmit, or amplify the vibration coming up from the floor of the vehicle to the driver's buttocks? The vibration transmissibility ratio is the ratio of the vibration output (vibration at the seat) to the vibration input (vibration at the floor of the vehicle). A transmissibility ratio of 1 means there is no change in the level of vibration between the input and output; a ratio >1 indicates an amplification of the original vibration; a ratio <1 indicates a reduction or attenuation of the original vibration.

Criteria

The acceptability of hand-arm or whole-body vibration is judged on whether the level of vibration exposure can potentially cause damage to the body or decrease performance. Guidelines for hand-arm vibration are generally concerned with potential damage to the body, whereas guidelines for whole-body vibration are more concerned with performance degradation. Because the effects of vibration depend on a number of factors in addition to the magnitude, and because of the tentative nature of the current guidelines, the user must thoroughly understand the standard or guide before comparing criteria with measured data. Failure to do so might result in inaccurate interpretation and results.

Hand-Arm

The etiology of hand-arm vibration disorders is not well-understood. The frequency spectrum, amplitude, and exposure duration are important in evaluating hand-arm vibration, but so are worker posture, particular part of the hand exposed, the temperature, the noise level, and any other conditions that could affect the worker's circulatory system (e.g., disease, smoking, chemical exposure).

Three hand-arm vibration consensus standards or guides are currently in use in the United States:

- ISO 5349–1986: *Guidelines for the Measurement and Assessment of Human Exposure to Hand-Transmitted Vibration;*[68]
- ANSI S3.34–1986: *Guide for the Measurement and Evaluation of Human Exposure to Vibration Transmitted to the Hand;*[69] and
- ACGIH's "Hand-Arm (Segmental) Vibration Syndrome (HAVS)" section in its TLVs handbook.[70]

Also, the NIOSH hand-arm criteria document[71] provides a work practices and medical monitoring approach without including a numerical exposure recommendation. OSHA has not adopted any of these standards.

When analyzing sound for hearing conservation, the concern generally is with the sound level in decibels at frequencies between 20 Hz and 10,000 Hz. In hand-arm vibration, the concerned is mostly with the rms acceleration in meters per second squared at frequencies between 5 Hz and 1500 Hz.

The three hand-arm vibration standards are discussed in the following sections.

ISO 5349–1986: Because the etiology of hand-arm vibration disorders are not well-understood, in part because of a limited amount of data, this standard by the

International Organization for Standardization (ISO) presents recommended methods for measuring, analyzing, and reporting hard-arm vibration data. It does not, however, include an exposure criterion as part of the standard. Guidance for acceptable hand-arm vibration exposure is provided in Annex A, where the time before the onset of finger blanching can be compared with the weighted vibration level for different percentiles of a population (as shown here in Figure 20.25). This figure assumes that a person is exposed to the vibration regularly from one tool or industrial process.

Guidelines for procedures to prevent HAVS are presented in Annex B of the ISO standard. These preventative measures include physical examinations, use of low-vibration tools, careful maintenance, worker training and education, use of gloves, and having the worker grip the tool with the least amount of force possible.

ANSI S3.34–1986: The ANSI standard is similar to the ISO standard in that it presents recommended methods for measuring, analyzing, and reporting hard-arm vibration data, but it does not include an exposure criterion as part of the standard. Guidance for acceptable hand-arm vibration exposure is provided "for information only" in the standard's appendices. In addition to analyzing the weighted vibration level, exposure curves are provided for one-third octave band levels (as shown here in Figure 20.26). If the actual acceleration levels and exposures exceed the allowable exposure time and if there is good coupling between the worker's hands and the vibration source, the worker might be at risk for developing HAVS.

Hand-Arm (Segmental) Vibration Syndrome: The ACGIH document specifies that measurements shall be made according to the procedures of ANSI S3.34–1986 or ISO 5349. Unlike the other two standards, ACGIH provides threshold limit values for hand-arm vibration. The TLVs for vibration in the X_h, Y_h, or Z_h axis (the subscript h signifies hand-arm vibration) are shown here in Table 20.13.

ACGIH also recommends use of antivibration tools, antivibration gloves, proper work practices (such as keeping the worker's hands warm), and a medical surveillance program.

Whole-Body

The main criteria document for whole-body vibration is ISO 2631/1–1985.[72] The American National Standards Institute

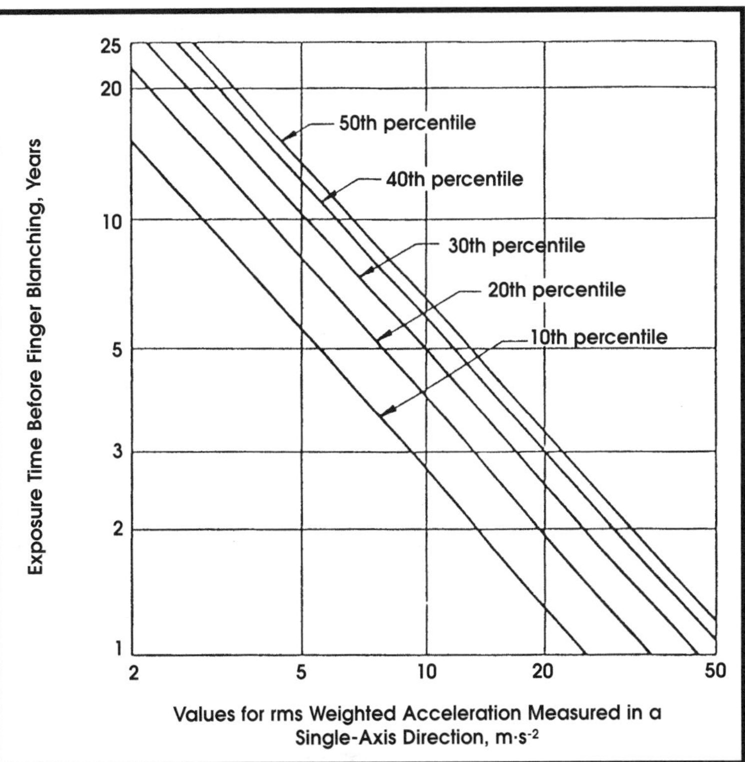

Figure 20.25 — Exposure time for different percentiles of a population group exposed to vibrations in three coordinate axes. (Adapted from Reference # 68.)

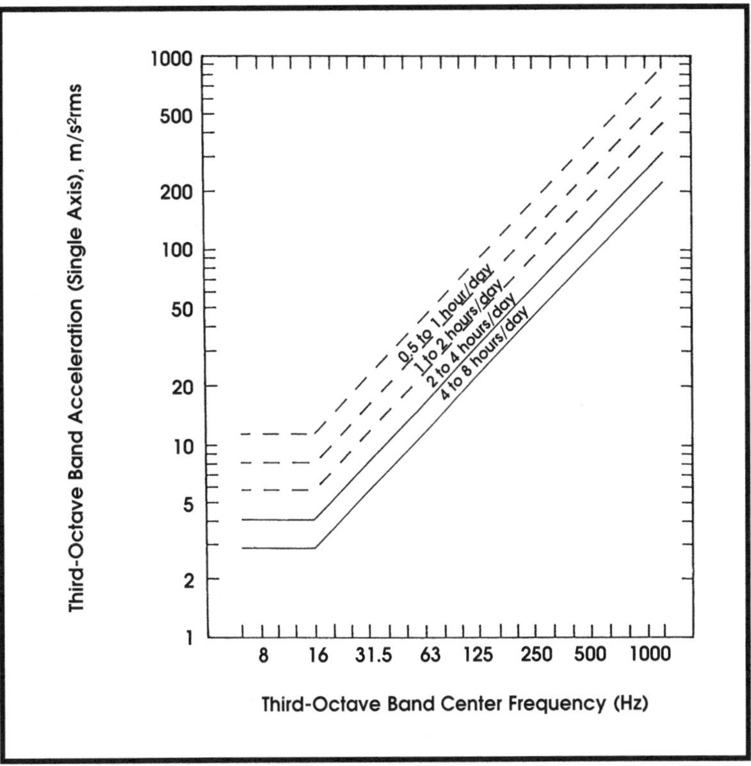

Figure 20.26 — Vibration exposure zones for the assessment of hand-transmitted vibration. (The zones of daily exposure time are for rms accelerations of discrete frequency vibration and for narrow-band or broad-band vibration analyzed as third-octave band rms acceleration. The values are for the dominant single-axis vibration generating compression of the flesh of the hand. The values are for regular daily exposure and for good coupling of the hand to the vibration source.) (Adapted from Reference # 69.)

Table 20.13 — ACGIH Threshold Limit Values (TLVs) for Exposure of the Hand to Vibration in the X_h, Y_h, or Z_h Axis[21]

Total Daily Exposure Duration	Maximum Frequency-Weighted, RMS, Component Acceleration (which shall not be exceeded in any axis)
4 hr–<8 hr	4 m/sec^2
2 hr–<4 hr	6 m/sec^2
1 hr–<2 hr	8 m/sec^2
<1 hr	12 m/sec^2

issued the ISO 2631 standard as ANSI S3.18–1979,[73] and the ACGIH TLVs for whole-body vibration[74] are also drawn from this document. With whole-body vibration, measurements are made in the back-to-chest (a_x) side-to-side (a_y) and foot-to-head (a_z) directions. Criteria for Fatigue Decreased Proficiency (FDP), reduced comfort (RC), and exposure limits (EL) are provided in Figure 20.27 (a_z axis) and Figure 20.28 (a_x and a_y axis).

The FDP boundaries represent the ability of a person to work at task(s) under vibration exposure without the vibration interfering with the worker's ability to perform. The RC boundaries are concerned with preservation of comfort during vibration exposure. The EL attempts to preserve health and safety of workers during vibration exposure.

Measurement of Vibration

A variety of component systems consisting of mechanical, electrical, and optical elements are available to measure vibration. The most common system uses a vibration transducer to transform the mechanical motion into an electrical signal, an amplifier to enlarge the signal, an analyzer to measure the vibration in specific frequency ranges, and a metering device calibrated in vibrational amplitude units.

Vibration Transducers

Accelerometers are the most common type of vibration transducer, and they produce an output voltage signal proportional

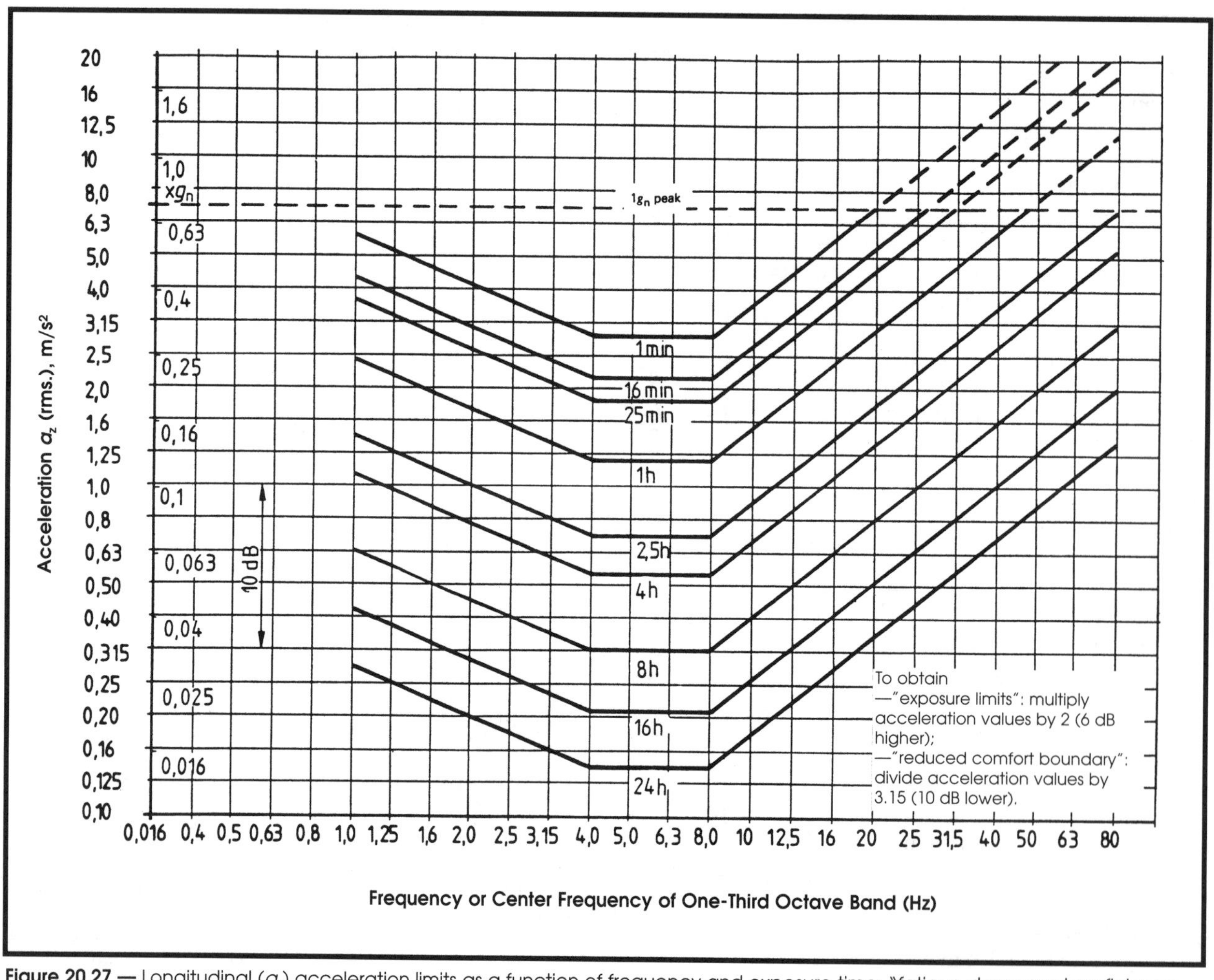

Figure 20.27 — Longitudinal (a_z) acceleration limits as a function of frequency and exposure time: "fatigue-decreased proficiency boundary." (Adapted from Reference # 72.)

to the acceleration. The most common type of accelerometer is the piezoelectric type. Two piezoelectric discs produce a voltage on their surfaces due to the mechanical strain on the asymmetric crystals composing the discs. The strain is in the form of vibrational inertia from a moving mass atop the discs. The output voltage is proportional to the acceleration, and thus to the vibration signal. The upper limit of the accelerometer's useful frequency range is determined by the resonant frequency of the mass and the stiffness of the whole accelerometer system. The lower limit of the frequency range varies with the cable length and the properties of the connected amplifiers. The accelerometer's sensitivity and the magnitude of the voltage developed across the output terminals depends on the properties of the materials used in the piezoelectric discs and the weight of the mass. The mechanical size of the accelerometer, therefore, determines the sensitivity of the system; the smaller the accelerometer, the lower the sensitivity. In contrast, a decrease in size results in an increase in frequency of the accelerometer's resonance and, thus, a wider useful range. Other factors to consider in the selection of a suitable accelerometer include the transverse sensitivity (which is the sensitivity to accelerations in a plane perpendicular to the plane of the discs) and the environmental conditions during the accelerometer's operation (primarily temperature, humidity, and varying ambient pressure).

Often, two types of sensitivities are stated by the manufacturer: voltage sensitivity and charge sensitivity. Voltage sensitivity is important when the accelerometer is used in conjunction with voltage measuring electronics. Charge sensitivity, an indication of the charge accumulated on the discs for a given acceleration, is important when the accelerometer is used with charge-measuring electronics.

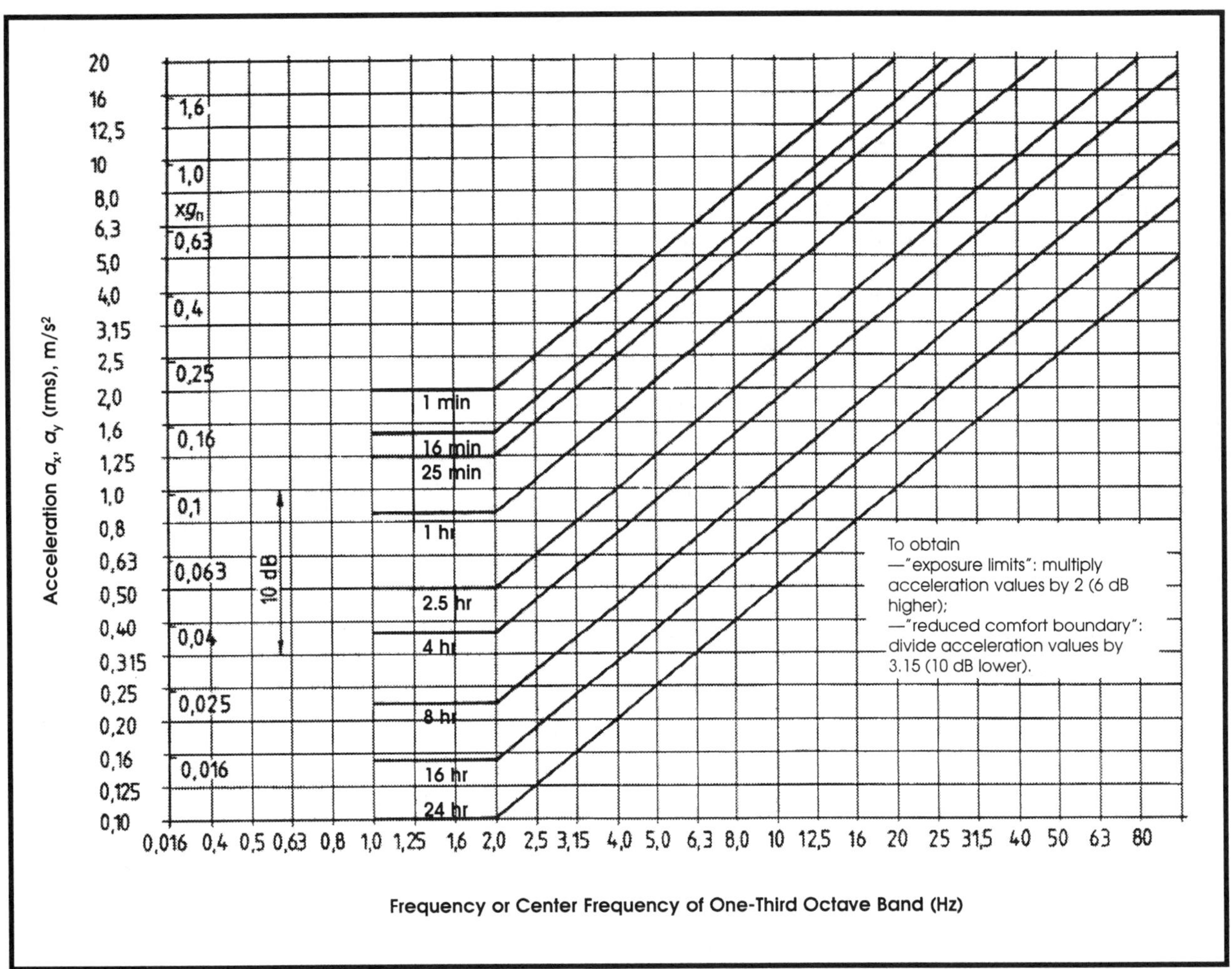

Figure 20.28 — Transverse (a_x, a_y) acceleration limits as a function of frequency and exposure time: "fatigue-decreased proficiency boundary." (Adapted from Reference # 72.)

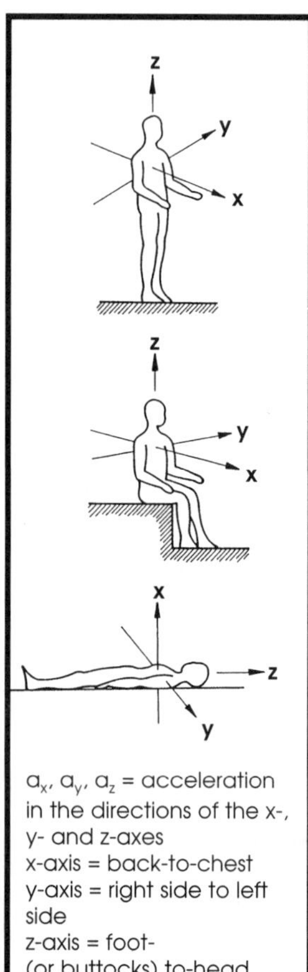

Figure 20.29 — Basicentric coordinate system for whole-body vibration. (Adapted from Reference # 72.)

Since the acceleration, velocity, and displacement for nonrandom vibration are all interrelated by differential operations, all of the variables can be measured with an accelerometer.

Preamplifier

The preamplifier is introduced in the measurement circuit for two reasons: 1) to amplify the weak output signal from the accelerometer; and 2) to transform the high output impedance of the accelerometer to a lower, acceptable value. It is possible to design the preamplifier in either of two ways: one in which the preamplifier output voltage is directly related to the input voltage (a voltage amplifier); and one in which the output voltage is proportional to the input charge (a charge amplifier).

The major difference between the two types of amplifiers is in their performance characteristics. When a voltage amplifier is used, the overall system is very sensitive to changes in the cable length between the accelerometer and the preamplifier, whereas changes in cable length produce negligible effects on a charge amplifier. The input resistance of a voltage amplifier will also affect the low-frequency response of a system.

Measurement

Since vibration is a vector quantity and therefore has magnitude and direction, it is mandatory when specifying or taking vibration measurements, comparing data, etc., to define both of these quantities. Vibration can appear in six directions at any one given measurement point: three linear perpendicular vector components (x, y, and z) and three rotational components (pitch, yaw, and roll). For human vibration work, only the linear motion components are measured.

Measurement coordinate systems are defined and standardized for both hand-arm and whole-body vibration. The coordinate systems are shown in Figures 20.29 and 20.30. To avoid confusion, whole-body measurement components are denoted as a_x, a_y, and a_z, whereas hand-arm measurement components are denoted as X_h, Y_h, and Z_h. Hand-arm vibration measurements can be made using one of two coordinate systems depending on the accelerometer mounting: a basicentric system if measurements are made on the handle of the tool nearest the position where the worker grips the tool, or a biodynamic system if the measurements are made on the third metacarpal.

The particular work situation and measurement requirements determine which type of accelerometer to use. The weight or mass of the accelerometer must be as small as possible in relation to the structure to be measured. An accelerometer that is too heavy will weigh down the surface and give inaccurate results, which is called mass loading. To avoid mass loading, the general rule is that the accelerometer's mass should be no more than one-tenth of the effective mass of the surface to which it is mounted.

The dynamic amplitude range of the device must accommodate the maximum acceleration level anticipated. For example, a 10 g (98 m/sec^2) accelerometer should not be used to measure a 100 g (980 m/sec^2) acceleration source, since the accelerometer would be destroyed very quickly. Similarly, a 100 g accelerometer is not used to measure a 1 g signal because the accelerometer would be insensitive to small signals and, thus, small changes would not be detected. Also, it should be noted that the frequency response of the accelerometer must match the overall frequency spectrum that is measured. The output of the accelerometer should be monitored during the measurements to ensure that overloads do not cause direct current (D.C.) shifts.

Z_h Z_h X_h Y_h

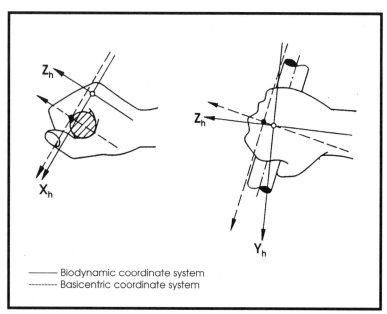

——— Biodynamic coordinate system
--------- Basicentric coordinate system

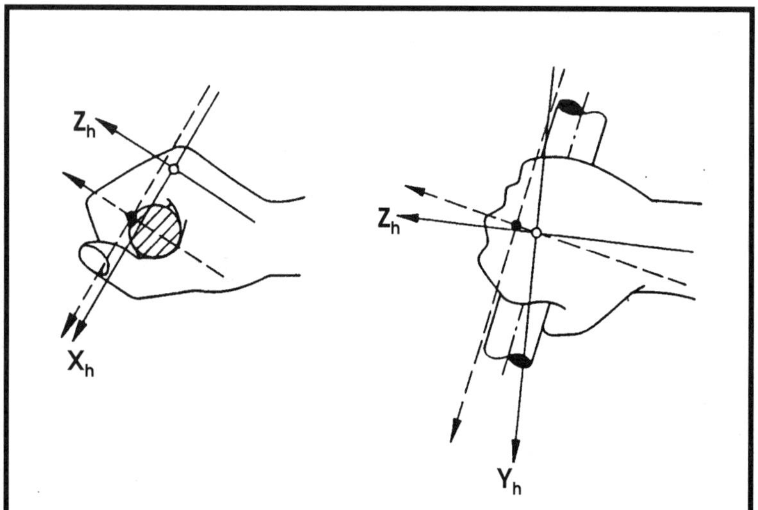

Figure 20.30 — Biodynamic and basicentric coordinate systems for hand-arm vibration. (Adapted from References # 68 and # 69.)

Accelerometer Mounting

Proper mounting of an accelerometer to a structure means the device is rigidly affixed to the vibrating source in line with the direction of vibration. If the accelerometer is not rigidly affixed, it will produce a false

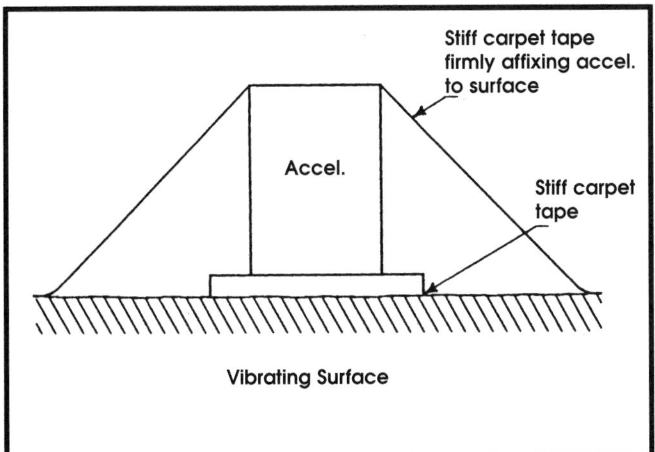

Figure 20.31 — Accelerometer mounting technique for the measurement of vibration impinging on the surface of the human body. (Adapted from Reference # 59.)

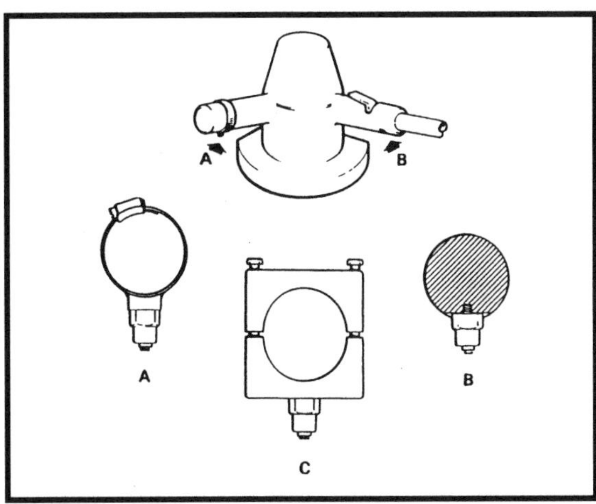

Figure 20.32 — Mounting techniques for measuring the vibration on the handle of a pneumatic grinder. (Adapted from Reference # 75.)

output signal. With poor mounting, the tendency is to lower the natural frequency of the system, thereby reducing the effective bandwidth (window) of the accelerometer. If the device is not in line with the vibration, a component of the vibration will not be measured due to misalignment.

Many accelerometers come with threaded studs that can be screwed directly into the structure. Another mounting method is to use a thin layer of beeswax between the accelerometer and the test structure. Epoxy or cyanoacrylate cement can also be used; however, soft glue should not be used because the soft material acts as a mechanical low-pass filter.

In some instances, the accelerometer can be mounted using a strong permanent magnet that is affixed to a metal test structure. A probe-type fixture, where the accelerometer is simply held against the surface, is not recommended for human vibration analysis.

To mount an accelerometer to the hand, the skin surface must be thoroughly cleaned. The accelerometer is then taped to the hand using stiff carpet tape as shown in Figure 20.31.[59]

When measuring vibrating pneumatic hand-tools, or whenever the acceleration level is expected to be high, other mounting techniques need to be used. Figure 20.32 shows three mounting techniques for measuring the vibration on the handle of a pneumatic grinder.[75] The accelerometer is mounted to an automotive hose clamp or removable bracket that is firmly attached to the tool handle where the worker grasps the tool. Alternatively, a section of the handle can be milled flat, a mounting screw tapped, and the accelerometer screwed into the handle. Also available are special adapters that are placed between a worker's hand and the tool or between a worker's buttocks and his seat.

Conducting the Workplace Measurement Study

Before performing any vibration measurements, it is important to understand the role vibration plays in the work process. It is important to walk through the entire process and spend time observing how workers work, especially with a vibrating process. It is also important to meet with the appropriate participants (workers, management, labor, occupational hygienists) and describe what will be done, how it will be done, and how much disruption will be caused by the measurements.

If possible, a camera should be used to record each step of the overall work process and particularly the actual work process involving vibration. In vibration processes, there is usually a repetition in work cycles that can be broken down into many steps. Depending on the variability of the levels, average readings might be required. It is recommended that observations be made of experienced workers rather than inexperienced workers. New workers on the job may have had little training and may not have developed good work habits.

Control of Occupational Vibration

Occupational vibration can be controlled by modifying work practices, such as limiting exposure time, gripping a tool with less force, purchasing low-vibration equipment, wearing antivibration (A/V) gloves, and by using engineering controls such as damping or isolation. It is often necessary to use several techniques to minimize the effects of harmful vibration.

Modifying Work Practices

Because the effects of vibration depend on factors such as worker posture, ambient temperature, and exposure level and time, the effects of hand-arm or whole-body vibration can be reduced significantly by modifying work practices. The following antivibration work practices for hand-arm vibration can be used:[76, 77]

- Use both A/V tools and A/V gloves.
- Use work breaks (e.g., 10 minutes per continuous exposure hour) to avoid constant vibration exposure.
- Measure and monitor the vibration of tools so that tools with increased acceleration levels due to age and wear can be replaced or repaired.
- Screen all workers regularly for HAVS through a specialized medical exam.
- Advise workers to have several pairs of warm antivibration gloves available and require that the gloves and warm clothing be worn.
- Advise workers to reduce smoking while using vibrating hand tools to avoid the vasoconstrictive effects of nicotine.
- Advise workers to grasp the tool as lightly as possible to reduce mechanical coupling to the hands.
- Have a physician examine workers if symptoms of tingling, numbness, or signs of white or blue fingers appear.

The following work practices should help minimize the effects of whole-body vibration:[78]

- Limit the time spent by workers on a vibrating surface to no more than is absolutely necessary to perform the job safely.
- Have machine controls moved off vibrating surfaces (whenever possible).
- Isolate vibrating equipment from surfaces where workers are stationed.
- Have work breaks to avoid constant vibration exposure (e.g., 10 min/hr).
- Maintain vibrating machinery to prevent development of excessive vibration.
- Have a physician consult with workers who have a history of, or who are currently suffering from, musculoskeletal problems (especially the lower spine) before they are exposed to whole-body vibration.

Training courses on the effects of occupational vibration and on proper use and maintenance of equipment are important in heightening the awareness of the problem by workers, management, labor, occupational hygienists, and medical personnel.

Low-Vibration Tools

As in noise control, the most effective place to control vibration exposure is at the source. For hand-arm vibration, A/V tools can be purchased. Since the early 1970s, for example, gasoline-powered chain saw manufacturers have been aware of HAVS and have designed a series of A/V chain saws. Where A/V saws were used extensively, there was a lower prevalence of HAVS. Today, nearly all gas-powered chain saw manufacturers have at least one A/V saw in their product line. In some cases, acceleration levels have fallen from the 20–30 g range to <1 g.

These saws must be maintained and their vibration isolators replaced periodically to keep the vibration levels low. This same expertise was applied by many manufacturers to other professional tools such as gas-powered string trimmers, edgers, and other tools used in gardening and grounds keeping. In the pneumatic tool industry, many manufacturers now offer A/V tools or tools with A/V grips.

Keep in mind that because most pneumatic A/V designs are new, their long-term effectiveness in reducing HAVS are not yet known.

A/V Gloves

A/V gloves use special vibration isolation materials such as Sorbothane® brand elastomer or Viscolas® brand elastomer. The amount of protection afforded by A/V gloves varies as a function of frequency and gripping force. Although workers should be encouraged to wear A/V gloves and to keep their hands warm and dry, the amount of protection gloves provide is difficult to predict. In one study using chipping hammers, researchers measured a 41%–67% decrease in acceleration levels at the workers' hands.[79] In another study, increased vibration levels were measured in some frequencies.[80] ISO's 10819–1996 standard now provides guidelines with which gloves marketed in Europe must comply to be considered "antivibration" gloves.[81]

Vehicle Seats

For whole-body vibration, special seats can help isolate the worker from the vibration of the vehicle. For many years, manufacturers of heavy equipment, farm vehicles, and over-the-road trucks and buses offered and installed reduced-vibration or suspended seats for operators; in some cases, these suspended seat designs are offered as options in lieu of a cheaper, more rigid seat. As a rule, these special seats do help in removing undesirable vibration from the operator's lower trunk and they should be used when available. There are three difficulties with these seats, however: 1) the operator must adjust the seat to his or her body weight, and many workers do not understand the need to do so; 2) these seats must be serviced and maintained according to the manufacturer's specifications; and 3) these seats have limited isolation and damping at low frequencies.

Engineering Control of Occupational Vibration

From an engineering standpoint, vibration can be diminished by reducing the driving force, reducing the response of the vibrating surface, or isolating the vibrating source. Each of these techniques and their limitations are discussed in the following sections.

Reduce Driving Force

As with sound, vibration can be reduced by decreasing the driving force. These forces can be reduced by decreasing the speed, maintaining the dynamic balance, or in the case of impact operations increasing the duration of the impact while reducing the forces. Sometimes these options are only practical in designing new equipment; however, mechanical vibration can often be reduced by proper balancing of rotating machinery and proper maintenance of machinery. For pneumatic hand tools, the driving force can be reduced by using the smallest tool possible for the job.

Reduce the Response of the Vibrating Surface

The most common method to reduce the response of a vibrating surface is to increase its damping. Damping reduces the response to resonant vibrations in the structure but does not reduce the response to a driven force such as a hammer blow. The amount of vibration reduction from damping depends on the thickness of the damping material, the method in which the damping material is applied, and the surface temperature. Since the driving forces are most often the concern in occupational vibration problems, damping alone is rarely a solution.

Vibration Isolation

Vibration isolators decouple structures, such as the human hand from a pneumatic tool or a driver from a vehicle, by using "soft" connections. Vibration energy is absorbed by the isolator instead of being transmitted to the user. The transmissibility ratio (the ratio of the vibration output to the vibration input) varies depending on the resonant frequency of the isolator and the frequency being isolated. Figure 20.33 shows the transmissibility vs. frequency ratio (driving frequency divided by natural frequency) for different amounts of damping in a system. To be effective, the resonant frequency of the vibration isolators must be less than one-half the lowest frequency to be isolated. Forces at driving frequencies near the isolator resonance are amplified by an amount dependent on the damping.

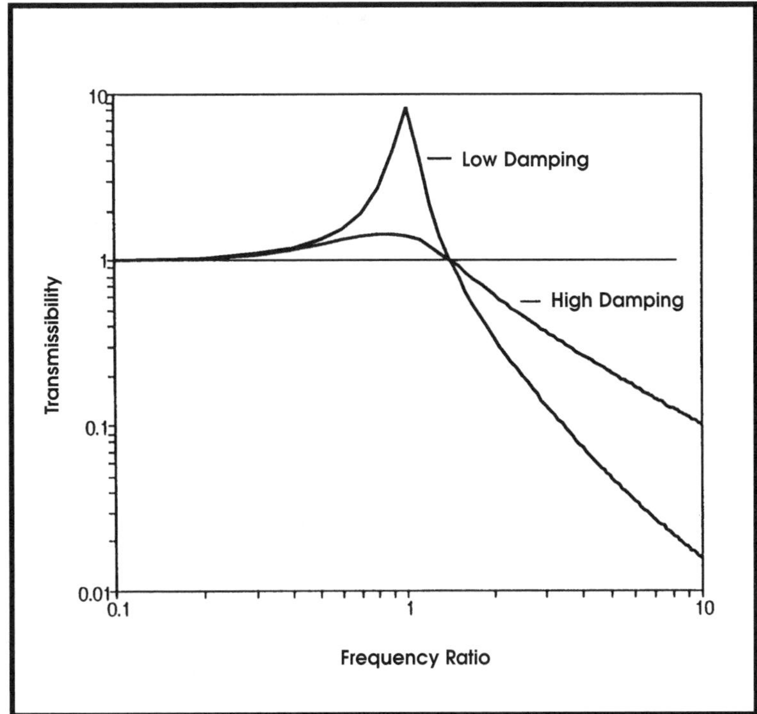

Figure 20.33 — Transmissibility vs. frequency ratio for different amounts of damping.

The resonant frequency of the isolator (f_n) can be calculated using Equation 31.

$$f_n \approx 0.5\sqrt{\frac{1}{\delta_{st}}} \qquad (31)$$

where:

δ_{st} = static deflection in meters under load, such as the weight of a driver sitting on a seat. A reduc tion in transmissibility can take place only by allowing the isolator to deflect by motion. Thus, certain space clearances must be provided for the isolated equipment.

Ultrasound

Introduction

Ultrasound is commonly defined as sound with frequencies above the range audible to the human ear. Although the audible range varies from person to person, the division between audible sound and ultrasound is often placed at 10 kHz. The upper frequency limit for ultrasound ranges from 50 kHz to above 100 kHz, but most effects on hearing are at the lower ultrasonic frequencies. In addition to ultrasonic waves in the air, ultrasonic energy in liquids (liquidborne ultrasound) can also be a hazard.

Uses of Ultrasound

In the past several decades, uses of ultrasound have increased rapidly in industry, the military, and the home. Both commercial and industrial uses may produce potentially harmful levels of airborne ultrasound, and both also use liquid propagation media. Table 20.14 is a listing of established and promising ultrasound applications.

Commercial uses of airborne ultrasound include automatic door openers, automatic photographic and video devices, dog whistles, security systems, sirens, etc. Commercial uses of liquidborne ultrasound include cleaning (e.g., jewelry and metal), air treatment (e.g., humidifiers and cleaners), and medical applications. Ultrasound is also used in both destructive and nondestructive testing of materials.

Ultrasound Paths

Airborne Ultrasound

Like sound at audible frequencies, the velocity of ultrasound in air is about 344 m/sec. Unlike low-frequency (below 500 Hz) sound that can curve around corners with very little loss in amplitude, high-frequency (above 2 kHz) sound propagates in a very directional manner. Indeed, the higher the frequency, the more directional the sound propagation. Because of this, high-frequency sounds

Table 20.14 — Ultrasonics in Industry — Established Uses, Promising Applications, and Recent Developments[82,83]

Application/Frequencies	*Description of Process*	*Intensity Range*
Cleaning and degreasing/20–50 kHz	Cavitated cleaning solution scrubs parts immersed in solution	<1–6 W/cm^2
Soldering and braising	Displacement of oxide film for bonding without flux	1–50 W/cm^2
Plastic welding/about 20 kHz	Welding soft and rigid plastic	about 100 W/cm^2
Metal welding/10–60 kHz	Welding similar and dissimilar metals	about 2000 W/cm^2
Machining/about 20 kHz	Drilling; rotary machining; impact grinding using slurry	
Extraction/about 20 kHz	Extracting juices and chemicals from flowers, fruits, and plants	about 500 W/cm2
Atomization/20–300 kHz	Fuel atomization to improve combustion efficiency	
Emulsification and dispersion	Mixing and homogenizing liquids, slurries, and cream	
Defoaming and degassing	Separation of foam and gas from liquid	
Foaming of beverages	Displacing air by foam in containers prior to capping	
Electroplating	Increasing plating rates and producing more uniform deposits	
Erosion	Cavitation erosion tests, deburring, and stripping	
Drying	Drying powders, foodstuff, pharmaceuticals, paper, and plastic	
Control and measurements	Interruption or deflection of beam; Doppler effect	
Drilling and braiding	Slurry used between sonically vibrated tool and workpiece	
Nondestructive testing	Pulse-echo exploration for flaws and gauging thickness	
Agglomeration and precipitation	Separating solids from gases producing larger particles	
Impregnation of porous materials	Increased density, absence of gas inclusions	
Degassing of melts (metal glass)	Improvement of material density; refinement of grain structure	
Mixing of slurry (pulp)	Improvement of consistency	
Agitation of chemicals	Maintaining uniform solutions and concentration	
Accelerating chemical reactions	Aging of liquors; tanning of hides; extractions	
Food treatment	Destroying molds and bacteria; tenderizing, removing starch	
Metal insertion into solid plastic	Vibrating metal softens plastic as it is inserted	
Measuring fluid flow particle size	Noncontacting measuring method	

are effectively blocked by barriers resulting in sound shadows. Only a small portion of high-frequency sound will enter the ear canal unless the sound is transmitted directly toward its entrance. Standing waves (areas of closely spaced high and low levels near a sound-reflective surface) are also common for high-frequency sounds.

Liquidborne Ultrasound

The velocity of sound in water (and the human body) is approximately 1500 m/sec. Physical characteristics of solvents and other materials can be found in physics and chemistry handbooks. Ultrasound within liquids is generally diffuse and uniform in level when the liquid is contained within sound-reflective walls.

Effects of Ultrasound

Physiological Effects of Airborne Exposure

Although a wide range of biological and chemical effects have been reported for ultrasound exposures, it is possible to cover only the most common effects in this chapter. Most exposures to ultrasound in the workplace are the result of ultrasound processes radiating incidental ultrasound energy into the surrounding rooms where employees are located. Table 20.15 lists examples of physiological effects of ultrasound exposures.

It might be impossible to develop a single meaningful hearing conservation guideline for all ultrasound exposures. Three reasons contribute to this difficulty. First, the sensitivity of the ear significantly varies over the wide range of ultrasound frequencies. Second, the measurement of exposure and the assessment of the effects of airborne ultrasound are difficult. Finally, ultrasound is used in liquid as well as air media, so there is potential danger from contact with either the source (transducer) or the liquid medium.

Von Gierke(84) found that subharmonics of incident sounds (3.5–23 kHz) formed on the eardrums of animals. If this phenomenon were also to occur in human ears, hearing damage from ultrasound exposures could result in the inner ear at audible-frequency subharmonics (integer fractions [e.g., 1/2] of the ultrasonic frequency) without these subharmonics existing (and being measurable) in the work area. Parrack supported this theory

Table 20.15 — Physiological Effects of Ultrasound(83)

HUMAN		SMALL ANIMALS
Death (calculated)(85)	— 180 dB —	
	— 170 dB —	
Loss of equilibrium(86)		Death (rabbits)(86)
Dizziness(86)		
	— 160 dB —	
Mild warming (body surface)(87)		Body temperature rise (hairless mice)(88)
	— 150 dB —	
Mild heating (skin clefts)(83)		Death (mice, rats, guinea pigs)(83)
	— 140 dB —	Body temperature rise (haired mice)(88)
	— 130 dB —	
	— 120 dB —	Mild biological changes (rats, rabbits)(89)
No physiological changes(90)		
	— 110 dB —	
No hearing loss occupational exposure(91)	— 100 dB —	

by reporting temporary threshold shifts for humans as a result of 1/2-order subharmonics generated within the ear for a wide range of frequencies.[85]

Airborne ultrasound exposures, by localized heating of the tissue, also affect parts of the body other than the ear. Fortunately, the acoustic impedance of air is significantly different from that of the body and, as a result, only a small fraction of incident airborne ultrasound is absorbed by the body; the rest is reflected. As for nonauditory effects of airborne ultrasound, airborne levels must be extremely high to cause damage to inner body parts.

In summary, the potential of permanent hearing impairment from exposures to ultrasound is not known definitely. Noise-induced hearing impairment is much less likely to result from ultrasound than from audible sound exposures. Subharmonics of ultrasound are most likely to cause noise-induced hearing impairment or other problems, but these subharmonic levels might not be detectable in work-area noise measurements.

Annoyance of Airborne Exposure

The adverse effects of airborne ultrasound include headaches, sore throats, dizziness, and nausea. These responses, sometimes called ultrasonic sickness, are highly variable among individuals. Exposed individuals may experience these symptoms while the sound levels are low or inaudible. High-level ultrasound may cause changes in vestibular function, which may explain the dizziness. These reactions are considered temporary and are often considered psychological; however, they should be taken seriously and exposure levels should be reduced. Annoyance, distraction, or stress may also be caused by sources other than ultrasound, and many of the symptoms listed above may result from these other sources.

Direct Contact with Liquidborne Ultrasound

Because acoustic impedances are better matched between liquids and the body than between air and the body, the human body can absorb a much higher percentage of incident ultrasound when in direct contact with a transducer or a liquid medium than in air. Such exposure results in localized tissue heating; therefore, workers must be prohibited from touching transducers or immersing any part of their body in a liquid while an immersed transducer is operating. Ultrasound can decrease the surface tension of the solvents, which could result in higher concentrations of vapors. The air around these operations must be monitored regularly until it is determined that the exposure is safe.

Exposure Criteria

Table 20.16 shows ultrasound exposure limits proposed in the United States and other countries. The objectives of the different limits vary. Some recommendations intend to prevent objective physiological effects (primarily hearing impairment) while others intend to limit subjective effects. With the exception of the U.S. Air Force's 85 dB limit across all frequency bands, most of the exposure limits allow higher levels at higher frequencies (20 kHz) and lower levels at lower frequencies (<20 kHz).

The first known document to set high-frequency exposure limits (above 10 kHz) in the United States was Air Force Regulation 161–35 (1973).[92] This document, "Hazardous Noise Exposure," deals with whole-body effects. Hearing protection is required when the sound level in

Table 20.16 — Various Occupational Exposure Limits (in dB)

	Sound Pressure Levels in One-Third Octave Bands Proposed By (*)							
Frequency in kHz	*1*	*2*	*3*	*4*	*5*	*6*	*7*	*8*
8	90	75	—	—	—	—	—	—
10	90	75	—	—	—	80	—	—
12.5	90	75	75	85	—	80	—	—
16	90	75	85	85	—	80	—	75
20	110	75	110	85	105	105	75	75
25	110	110	110	85	110	110	110	110
31.5	110	110	110	85	115	115	110	110
40	110	110	110	85	115	115	110	110
50	110	—	110	—	115	115	110	110

* Legend: 1–Japan (1971), 2–Acton (1975), 3–USSR (1975), 4–USAF (1976), 5–Sweden (1971), 6–USA ACGIH (1988), 7–INTL IRPA (1984), 8–Canada (1989)

any one-third octave band above 10 kHz exceeds 85 dB. A limit of 30 minutes each working day is set for levels at or above 150 dB in the same bands. Although this regulation is still in effect, the required use of hearing protection for exposures above 85 dB is very conservative; therefore, the USAF may adopt the levels proposed first by Parrack[85] and later by ACGIH.[93]

ACGIH first published its recommended TLVs for ultrasound in 1988.[93] Another valuable guideline is Canadian Safety Code–24, *Guidelines for the Safe Use of Ultrasound, Part II — Industrial and Commercial Applications*.[83]

There are several reasons why ultrasound exposures were not included in many of the existing health and safety regulations:

- Most of the current rules, regulations, and guidelines on noise exposures were written when it was assumed that the most harmful spectral components of noise were in the frequency range from 125 Hz to 8 kHz. Many authorities still agree with this opinion.
- Adding the ultrasound frequency range to regulations would have delayed achieving consensus and publishing these documents.
- Relatively few persons were exposed to high level ultrasound when the first noise regulations were written.
- The effects of ultrasound were not well-understood, and susceptibility to ultrasound was generally considered to be highly variable.
- Hearing threshold levels were not normally measured above 8 kHz.

As the number of ultrasound sources continues to increase, further knowledge should be gained on the effects of ultrasound exposures (especially if uniform records are kept on exposures and hearing levels). ACGIH's current TLVs should be used as guidelines until there is a regulation.

Measurement

When measuring ultrasound (even more than audible sound) one must carefully consider the objective of the measurements before performing the measurements. Measurement at ultrasonic frequencies presents several problems that require special equipment and procedures that differ from standard noise measurements. Many instruments are not intended for use above 10 kHz, and only a few sound level meters and analyzers can measure sound above 20 kHz. Measurements taken in liquid media involve further complications. Hence, care must be taken to obtain accurate data.

Since ultrasound exposure limits (whether for hearing conservation or psychologically related problems) are often given in one-third octave bands, appropriate instrumentation should be used rather than dosimeters which provide A-weighted data. Objects the size of a human head can shield a microphone from ultrasound; therefore, measurements should be made either at the location of the worker's ears without the worker present or on both sides of the worker's head.

To evaluate exposures related to physiological effects other than hearing impairment, instrumentation and procedures must be capable of determining sound levels in frequency bands on different areas of the body.

Instrumentation

One of the most vital and yet restrictive components of any sound measurement system is the microphone. Although many rules-of-thumb were developed for microphone selection, nothing replaces the manufacturer's instruction books and the advice of those who use the instruments.

Microphones and hydrophones must be selected for the specified frequency ranges, and several different transducers may be required to cover the full frequency range. In general, 6 mm (1/4 in.) or smaller diameter microphones must be used in order to obtain a relatively flat frequency response at ultrasonic frequencies. These microphones have lower sensitivity and a poorer signal-to-noise ratio at lower frequencies than larger diameter microphones. Besides frequency, the following considerations may be important in choosing ultrasound microphones and hydrophones:

- Level of sensitivity to obtain an acceptable signal-to-noise ratio;
- Directional characteristics (free-field, pressure-type, random-response, etc.);
- Rugged construction to cope with environmental factors (heat, rough handling, electromagnetic fields, corrosive vapors, liquids, etc.); and
- Adequate dynamic range.

Since instrumentation for ultrasound measurement is generally more sophisticated, more expensive, and less familiar than comparable instrumentation for audible sound, more time should be allowed for the selection of ultrasound equipment. Sound level meters that meet OSHA's minimum specifications (Type II) should not be used for ultrasound measurements because there are no tolerance limits defined above 10 kHz.

Techniques

Because ultrasound is much more directional than audible sounds, measurement samples should be made with microphones oriented in different directions to account for the microphones' directional characteristics. Generally, the required number of samples increases as the frequency increases. For each orientation, many samples might be required to account for the standing waves common in some spaces with high-frequency noises. More samples might also be required to account for localized areas in "ultrasound shadows." To determine the exposure dose, samples can be averaged over a period of time. In general, samples should be repeated at different orientations and locations until the data trend toward a stable average value.

A good rule-of-thumb is to use a 6 mm free-field microphone to take samples at different orientations at each location. At the highest levels, the microphone will point toward the loudest sources, where additional measurements might be needed.

Some ultrasound sources such as sewing or welding with ultrasound can be difficult to measure because they use pulsing operations with different peak levels, pulse lengths, and pulse repetition rates. Since the available damage-risk criteria do not include these parameters, the value of this exposure information is difficult to evaluate.

Different body locations (such as the crevices between fingers) are more susceptible to heating from ultrasound.[83,94-97] Table 20.14 lists some intensity levels expected from several ultrasound applications.

In summary, ultrasound measurements may require special frequency bandwidths, numbers of samples, microphones, and microphone orientations. Considerable care must be taken in placing and orienting the microphone. Locations of reflecting surfaces and barriers must be considered and recorded as a part of the measurement data. Many samples must be taken in most cases to determine a representative value because of the extreme changes in ultrasound levels that might be caused by acoustic shadows and standing waves.

High-Frequency Hearing Measurement

ANSI has no standard test procedures for measuring hearing levels at frequencies between 8 kHz and 20 kHz, and OSHA has no requirements for such tests. Although it is more difficult to conduct tests at these frequencies than at the normal audible frequencies, it is possible with specialized commercially available audiometers. Using special techniques and calibration procedures, reliable data can be obtained.

Exposure Controls

Administrative Controls

It is sometimes possible to reduce employee exposure to ultrasound by relocating work areas or ultrasound equipment to another room. Because of the high directivity of ultrasound, it is possible to reduce exposure significantly by changing locations even within a room.

Engineering Controls

Engineering controls for ultrasound are often effective because of its high directivity and the high transmission loss characteristics of most materials at ultrasound frequencies. High-frequency sounds are blocked more effectively by barriers than are lower frequencies, thereby creating sound shadows. Barriers can be used to isolate either the noise source or the worker. When simple barriers are insufficient due to very high levels of ultrasound, complete enclosures can be built around the ultrasound source or the worker.

If the dimensions of a barrier are large compared with the wavelength of a sound, the barrier will generally be effective in blocking the sound. Since ultrasound wavelengths are very small, any solid object with dimensions greater than 0.3 m (1 ft) will be an effective noise barrier.

Most acoustical data for noise control treatments extend to a maximum of 8 kHz. In general, transmission loss (TL) at ultrasound frequencies can be presumed to be at least as high as the TL for the highest given frequencies. Similar presumptions cannot be made for sound absorption data or silencer insertion loss data; however, unfaced porous materials such as glass fiber and acoustical foam generally provide good absorption at ultrasonic frequencies.

Standing waves are often found around high-frequency noise sources. Standing waves can be reduced or eliminated by applying sound-absorptive treatments to the surfaces reflecting the ultrasound. Sound-absorptive values at ultrasound frequencies vary significantly depending on the frequency and the physical properties of the materials. Even a very thin film (1/1000 in. or 25 μm) covering sound-absorptive materials will significantly reduce sound absorption at high frequencies by reflecting the ultrasound.

Other treatments discussed in this chapter (such as damping and lagging) may also be appropriate for ultrasound.

Personal Protection

Little information is available on the protection levels offered by personal hearing protectors for ultrasound exposures. Methods for measurement of hearing protector attenuation characteristics generally cover only the frequency range from 125 Hz to 8 kHz. For frequencies 8–32 kHz, the attenuation values for 8 kHz can be used.[98] With ultrasound, extra care should be taken in inserting earplugs or putting on earmuffs to avoid any gaps or leaks through which ultrasound can pass.

Summary and Acknowledgments

The problems of overexposure to two physical agents — sound (noise and ultrasound) and vibration — have been discussed in this chapter. For each of these agents, criteria have been reviewed either in the form of governmental regulations, international standards, or recommendations from distinguished professional organizations concerned with worker health and safety. Recommendations have been presented for quantifying worker exposure to these physical agents by noise dosimetry, sound pressure level measurements, sound intensity measurements, vibration measurements, and ultrasound measurements. Also, different methods of reducing worker exposure to various physical agents have been presented in the form of engineering controls, administrative controls, and personal protective equipment and programs.

The authors are indebted to the many individuals who have contributed to this investigation of noise, vibration, and ultrasound in the occupational environment. Unfortunately, the records available to us at this point are not sufficient enough to thank everyone involved, but we do wish to recognize the efforts of those who participated in preparing chapters in earlier versions of AIHA documents on noise and in the first edition of *The Industrial Environment—Its Evaluation & Control*, published by NIOSH in 1973, in addition to the following people who assisted us in researching this latest chapter:

Joseph R. Anticaglia, MD
Elliott H. Berger
Thomas B. Bonney
James H. Botsford
John F. Brower
R. Douglas Bruce, MD
George J. Butler
Allen L. Cudworth, Sc.D.
Irving H. Davis
W. Gregory Deskins
Leo G. Doerfler, Ph.D.
John J. Earshen
Margaret F. Edmond
Julius H. Fanney, Jr.
Kathy A. Foltner, CCC-A
Lewis S. Goodfriend, PE
Lee D. Hager
Vaughn H. Hill
William C. Janes
Herbert H. Jones
Stanley H. Judd
Walter H. Koon
Joseph A. Lamonica
Paul L. Michael, Ph.D.
Jeffrey C. Morrill
Arthur Nielson
Paul Ostergaard, PE
Stanley E. Phil
Julia D. Royster, Ph.D.
Larry R. Royster, Ph.D.
Joseph Sataloff, MD
Edward J. Schneider
Robert D. Soule, PE
C. Grissom Steele, PE
Alice H. Suter, Ph.D.
Edwin H. Toothman
Floyd A. Van Atta, Ph.D.
W. Dixon Ward, Ph.D.
Donald E. Wasserman
Alonzo M. Webbe

References

1. **von Bekesy, G.:** The Ear. *Scientific American 197*:66 (August 1957).
2. **Gulik, W.L., G.A. Gescheider, and R.D. Frisina:** *Hearing: Physiological Acoustics, Neural Coding, and Psychoacoustics.* New York: Oxford University Press, 1989. p. 83.
3. **Simpson, M.A., and R.D. Bruce:** *Noise in America: The Extent of the Noise Problem* (EPA Report No. 550/9–81–101). Washington, D.C.: U.S. Environmental Protection Agency/Office of Noise Abatement and Control, 1981.
4. **Suter, A.H.:** The Development of Federal Noise Standards and Damage Risk Criteria. In *Hearing Conservation in Industry, Schools, and the Military* (edited by David M. Lipscomb). Boston: College Hill Press, 1988.

5. **National Institute for Occupational Safety and Health:** *Occupational Noise and Hearing: 1968–1972* (NIOSH Pub. No. 74–116). Cincinnati: National Institute for Occupational Safety and Health, 1974.
6. "Occupational Noise Exposure," *Code of Federal Regulations* Title 29, Part 1910.95. 1971. [39 FR 23502, June 27, 1974, as amended at 46 FR 4161, Jan. 16, 1981; 46 FR 62845, Dec. 29, 1981; 48 FR 9776, Mar. 8, 1983; 48 FR 29687, June 28, 1983; 54 FR 24333, June 7, 1989.]
7. "Occupational Noise Exposure," *Code of Federal Regulations* Title 20, Part 1926.52.
8. "Occupational Noise Exposure," *Code of Federal Regulations* Title 29, Part 1910.95—subparagraph (o). 1971.
9. "Occupational Safety and Health Programs for Federal Employees" (Executive Order 12196; as amended by Executive Order 12223). Feb. 26, 1980.
10. *Code of Federal Regulations* Title 30, Chapter I. Subchapter N, Parts 56.5050, 57.5050, 70.500, 71.800.
11. "Health Standards for Occupational Noise Exposure in Coal, Metal and Nonmetal Mines," *Code of Federal Regulations* Title 30, Parts 56, 57, 62, 70, and 71. 1996. [Proposed rule published in *Federal Register,* Dec. 17, 1996—Part II, U.S. Department of Labor/Mine Safety and Health Administration.]
12. "Locomotive Cab Noise," *Code of Federal Regulations* Title 49, Chapter II, Part 229, Subpart C, Paragraph 229.121. 1980.
13. "Requirements of the Hours Service Act: Statement of Agency Policy and Interpretation," *Code of Federal Regulations* Title 49, Chapter II, Part 228, Appendix A. 1977. [42 FR 27596, May 31, 1977, as amended at 43 FR 30804, July 18, 1978; 53 FR 28601, July 28, 1988.]
14. **U.S. Coast Guard:** "Navigation and Vessel Inspection Circular No. 12–82: Recommendations on Control of Recessive Noise." Washington, D.C.: U.S. Department of Transportation/U.S. Coast Guard, June 1982.
15. "Vehicle Interior Noise Levels," *Code of Federal Regulations* Title 49, Chapter III, Part 393.94. 1973. [38 FR 30881, Nov. 8, 1973, as amended at 40 FR 32336, Aug. 1, 1975; 41 FR 28268, July 9, 1976.]
16. **U.S. Department of Defense:** "DOD Hearing Conservation Program" (DOD Instruction 6055.12). Washington, D.C.: U.S. Department of Defense, April 22, 1996.
17. **U.S. Environmental Protection Agency:** "Occupational Noise Exposure Regulation; Request for Review and Report." Washington, D.C.: U.S. Environmental Protection Agency, 1974. [39 FR 43802–43809.]
18. **National Institute for Occupational Safety and Health:** *Criteria for a Recommended Standard ... Occupational Exposure to Noise* (NIOSH Pub. No. HSM 73–11001). Cincinnati: National Institute for Occupational Safety and Health, 1972.
19. **National Institute for Occupational Safety and Health:** *Criteria for a Recommended Standard ... Occupational Noise Exposure; Revised Criteria.* Cincinnati: National Institute for Occupational Safety and Health, 1996. [Draft.]
20. **International Institute of Noise Control Engineering:** "Technical Assessment of Upper Limits on Noise in the Workplace" (I-INCE Pub. No. 94–1). *Noise/News Intl.,* Dec. 1994. pp. 227-237. [Draft.] (The current unpublished draft has been extensively changed, but the essential features have been retained.)
21. **American Conference of Governmental Industrial Hygienists:** Noise. In *1995–1996 Threshold Limit Values (TLVs) for Chemical Substances and Physical Agents and Biological Exposure Indices (BEIs).* Cincinnati: American Conference of Governmental Industrial Hygienists, 1995. pp. 108-110.
22. **Bruce, R.D., and E. McKinney:** "Noise Criteria for Ships and Offshore Platforms." Proceedings of Noise-Con 94, Progress in Noise Control for Industry, Fort Lauderdale, Fla., 1994. pp. 995–1000
23. **Levitt, H., and J.C. Webster:** Effects of Noise and Reverberation on Speech. In *Handbook of Acoustical Measurements and Noise Control* (edited by C.M. Harris). New York: McGraw-Hill Inc., 1991. p. 16.10.
24. **Hoover, R.M., and R.D. Bruce:** Powerplant Noise and Its Control. In *Standard Handbook of Powerplant Engineering* (edited by Thomas C. Elliott). New York: McGraw-Hill Inc., 1989.

25. **Suter, A.H.:** *Communication and Job Performance in Noise: A Review* (ASHA Monographs No. 28). Rockville, Md.: American Speech-Language-Hearing Association, November 1992.
26. **Beranek, L.L.:** "Acoustical Measurements, Revised Ed." Acoustical Society of America, American Institute of Physics, 1988. p. 803.
27. **Beranek, L.L. (ed.):** *Noise & Vibration Control.* New York: McGraw-Hill Inc., 1971.
28. **Miller, L.N.:** *Noise Control for Buildings and Manufacturing Plants.* Cambridge, Mass.: Bolt Beranek and Newman Inc., 1981.
29. **Egan, M.D.:** *Concepts in Architectural Acoustics.* New York: McGraw-Hill Inc., 1972.
30. **Bruce, R.D., and E.H. Toothman:** Engineering Controls. In *Noise and Hearing Conservation Manual,* 4th Ed. (edited by E.H. Berger, W.D. Ward, J.C. Morrill, and L.H. Royster). Akron, Ohio: American Industrial Hygiene Association, 1986. pp. 417–521.
31. **National Institute for Occupational Safety and Health:** *Compendium of Materials for Noise Control* (DHEW Pub. No. 75–165) Cincinnati: National Institute for Occupational Safety and Health, 1975.
32. **Acoustical and Insulatory Materials Association:** *Bulletin of the Acoustical and Insulatory Materials Association.* Park Ridge, Ill.: AIMA, 1974.
33. **Bies, D.A., and C.H. Hansen:** *Engineering Noise Control, Theory and Practice.* London: Unwin Hyman, 1988.
34. "Hearing Conservation Program," *Code of Federal Regulations* Title 29, Section 1910.95, Part C. 1971.
35. **Berger, E.H., W.D. Ward, J.C. Morrill, and L.H. Royster (eds.):** *Noise and Hearing Conservation Manual,* 4th Ed., Akron, American Industrial Hygiene Association, 1988.
36. **Lipscomb, D.M. (ed.):** *Hearing Conservation in Industry, Schools, and the Military.* Boston: College Hill Press, 1988.
37. **Gasaway, D.C. (ed.):** *Hearing Conservation: A Practical Manual and Guide.* Englewood, N.J.: Prentice-Hall, Inc., 1985.
38. **Royster, L.H., and J.D. Royster:** Hearing Conservation Programs. In *Handbook of Acoustical Measurements and Noise Control,* 3rd Ed. New York: McGraw-Hill Inc., 1991.
39. **Sataloff, R.T., and J. Sataloff:** *Occupational Hearing Loss,* 2nd Ed. New York: Marcel Dekker Inc., 1993.
40. **American National Standards Institute:** *Specifications for Sound Level Meters* (ANSI S1.4–1983 [R1994]). New York: American National Standards Institute, 1995.
41. **American National Standards Institute:** *Specifications for Sound Level Meters (Supplement to ANSI S1.4–1983)* (S1.4a–1985). New York: American National Standards Institute, 1985.
42. **American National Standards Institute:** *Specification for Personal Noise Dosimeters* (ANSI S1.25–1991). New York: American National Standards Institute, 1991.
43. **American National Standards Institute:** *Measurement of Occupational Noise Exposure* (ANSI S12.19–1996). New York: American National Standards Institute, 1996.
44. **Erdreich, J.:** Alternatives for Hearing Loss Risk Assessment. *Sound and Vibration 19(5):*22- 23 (1985).
45. **Suter, A.H.:** Noise Control: Why Bother? *Sound and Vibration 18(10):*5 (1984).
46. **Nixon, C.W., and E.H. Berger:** Hearing Protective Devices. In *Handbook of Acoustical Measurements and Noise Control,* 3rd Ed. New York, McGraw-Hill Inc., 1991.
47. "Product Noise Labeling," *Code of Federal Regulations* Title 40, Part 211, Subpart B — Hearing Protective Devices. 1979.
48. "Guidelines for Noise Enforcement," *Code of Federal Regulations* Title 29, Part 1910.95(6)(1)—Instruction CPL 2–2.35. 1983.
49. **Franks, J.R., C.L. Themann, and C. Sherris:** *The NIOSH Compendium of Hearing Protection Devices* (NIOSH Pub. No. 94–130) Cincinnati: National Institute for Occupational Safety and Health, 1994.
50. **Suter, A.H.:** Hearing Conservation. In *Noise and Hearing Conservation Manual,* 4th Ed. (edited by E.H. Berger, W.D. Ward, J.C. Morrill, and L.H. Royster). Akron, Ohio: American Industrial Hygiene Association, 1986. pp. 1-18.
51. **Royster, L.H., and J.D. Royster:** Education and Motivation. In *Noise and Hearing Conservation Manual,* 4th Ed. (edited by E.H. Berger, W.D. Ward, J.C. Morrill, and L.H. Royster). Akron, Ohio: American Industrial

Hygiene Association, 1986. pp. 383–416.

52. **Feldman, A.S., and C.T. Grimes:** Employee Training Programs in Occupational Hearing Conservation. In *Hearing Conservation in Industry.* Baltimore: Williams & Wilkins, 1985. pp. 156-163.
53. **Berger, E.H., et al.:** "An Earful of Sound Advice about Hearing Protection." Indianapolis: E•A•R Division of Cabot Corporation, 1988.
54. **Day, N.R.:** *Hearing Conservation: A Guide to Preventing Hearing Loss.* Daly City, Calif.: Krames Communications, 1983.
55. **Zohar, D., A. Cohen, and N. Azar:** Promoting Increased Use of Ear Protectors in Noise Through Information Feedback. *Human Factors 22:*69-79 (1980).
56. **National Institute for Occupational Safety and Health:** *Preventing Occupational Hearing Loss: A Practical Guide* (NIOSH Pub. No. 96–110). Cincinnati: National Institute for Occupational Safety and Health, 1996.
57. **Royster, J.D., and L.H. Royster:** Audiometric Data Base Analysis. In *Noise and Hearing Conservation Manual,* 4th Ed. (edited by E.H. Berger, W.D. Ward, J.C. Morrill, and L.H. Royster). Akron, Ohio: American Industrial Hygiene Association, 1986. pp. 293–318.
58. **American National Standards Institute:** *Evaluating the Effectiveness of Hearing Conservation Programs* (draft ANSI S12.13–1991). New York: American National Standards Institute, 1991.
59. **Wasserman, D.E.:** *Human Aspects of Occupational Vibration.* Amsterdam, The Netherlands: Elsevier, 1987.
60. **Wasserman, D.E., D.W. Badger, T. Doyle, and L. Margolies:** Occupational Vibration — An Overview. *J. Am. Soc. Safety Eng. 19:* (1974).
61. **Wasserman, D.E.:** *Occupational Diseases: A Guide to Their Recognition* (NIOSH Pub. No. 77–181). Cincinnati: National Institute for Occupational Safety and Health, 1977.
62. **Radwin, R.G., T.J. Armstrong, and E. Vanbergeijk:** Vibration Exposure for Selected Power Hand Tools Used in Automobile Assembly. *Am. Ind. Hyg. Assoc. J. 51(9):*510-518 (1990).
63. **Taylor, W.:** *The Vibration Syndrome.* London: Academic Press, 1974.
64. **Pelmear, P.L., W. Taylor, and D.E. Wasserman:** *Hand-Arm Vibration — A Comprehensive Guide for Occupational Health Professionals.* New York: Van Nostrand Reinhold, 1992.
65. **Taylor, W., and P. Pelmear (eds.):** *Vibration White Finger in Industry.* London: Academic Press, 1975.
66. **Taylor, W., and D. Wasserman:** Occupational Vibration. In *Occupational Medicine* (edited by C. Zens). Chicago: Yearbook Medical Publishers, 1988.
67. **Hulshof, C., and B.V. Van Zanten:** Whole Body Vibration and Low-Back Pain — A Review of Epidemiological Studies. *Intl. Arch. Occup. Environ. Health 59:* (1987).
68. **International Organization for Standardization:** *Guidelines for the Measurement and Assessment of Human Exposure to Hand-Transmitted Vibration* (ISO 5349–1986). Geneva, Switzerland: International Organization for Standardization, 1986.
69. **American National Standards Institute:** *Guide for the Measurement and Evaluation of Human Exposure to Vibration Transmitted to the Hand* (ANSI S3.34–1986). New York: American National Standards Institute, 1986.
70. **American Conference of Governmental Industrial Hygienists:** Hand-Arm (Segmental) Vibration Syndrome (HAVS). In *1995–1996 Threshold Limit Values (TLVs) for Chemical Substances and Physical Agents and Biological Exposure Indices (BEIs).* Cincinnati: American Conference of Governmental Industrial Hygienists, 1995. pp. 84-87.
71. **Henschel, A., and V. Behrens:** *Criteria for a Recommended Standard ... Occupational Exposure to Hand-Arm Vibration* (NIOSH Pub. No. 89–106). Cincinnati: National Institute for Occupational Safety and Health, 1989.
72. **International Organization for Standardization:** *Evaluation of Human Exposure to Whole- Body Vibration — Part 1: General Requirements* (ISO 2631/1–1985). Geneva, Switzerland: International Organization for Standardization, 1985.
73. **American National Standards Institute:** *Guide for the Evaluation of Human Exposure to Whole-Body Vibration* (ANSI S3.18–1979). New York: American National Standards Institute, 1979.

74. **American Conference of Governmental Industrial Hygienists:** Whole-Body Vibration. In *1995–1996 Threshold Limit Values (TLVs) for Chemical Substances and Physical Agents and Biological Exposure Indices (BEIs).* Cincinnati: American Conference of Governmental Industrial Hygienists, 1995. pp. 124-131.
75. *Test Procedure for the Measurement of Vibration from Hand Held (Portable) Power Driven Ginding Machines* (PNEUROP 6610/1983). London, England.
76. **National Institute for Occupational Safety and Health:** *Current Intelligence Bulletin #38: Vibration Syndrome* (NIOSH Pub. No. 83–110). Cincinnati: National Institute for Occupational Safety and Health, 1983.
77. **Radwin, R.G.:** "Neuromuscular Effects of Vibrating Hand Tools on Grip Extensions, Tactility, Discomfort, and Fatigue." 1986. [University of Michigan, Industrial and Operations Engineering Dept./DHHS (NIOSH) Grant Reports 5–RO3–OH01852–02 and 1–T15–OH07207–01.]
78. **Wasserman, D.:** *Occupational Vibration in the Foundry* (NIOSH Pub. No. 81–114). Proceedings of the Symposium on Occupational Health Hazard Control Technology in the Foundry and Secondary Non-Ferrous Smelting Industries. Cincinnati: National Institute for Occupational Safety and Health, 1981.
79. **Goel, V.K., and K. Rim:** Role of Gloves in Reducing Vibration: An Analysis for Pneumatic Chipping Hammer. *Am. Ind. Hyg. Assoc. J. 48(1):*9-14 (1987).
80. **Gurram, R., S. Rakheja, and G.J. Gouw:** Vibration Transmission Characteristics of the Human Hand-Arm and Gloves. *Intl. J. Ind. Ergon. 13:*217-234 (1994).
81. **International Organization for Standardization:** *Method for the Measurement and Evaluation of the Vibration Transmissibility of Gloves at the Palm of the Hand* (ISO 10819–1996). Geneva, Switzerland: International Organization for Standardization, 1996.
82. **Steinberg, E.B.:** Ultrasonics in Industry. *Proceed. IEEE 53(10):*1292-1304 (1965).
83. **Bly, S.H.P., and D.A. Benwell:** *Guidelines for the Safe Use of Ultrasound: Part II — Industrial and Commercial Applications.* Ottawa, Canada: Bureau of Radiation and Medical Devices/Non-Ionizing Radiation Section, 1989.
84. **von Gierke, H.E.:** Subharmonics Generated in Human and Animal Ears by Intense Sound. *J. Acoust. Soc. Am. 22:*475 (1950).
85. **Parrack, H.O.:** Effects of Air-borne Ultrasound on Humans. *Intl. Audio. 5:*294-308 (1966).
86. **Allen, C.H., H. Frings, and I. Rudnick:** Some Biological Effects of Intense High Frequency Airborne Sound. *J. Acoust. Soc. Am. 20:*62-65 (1948).
87. **Ravizza, R., H. Heffner, and B. Masterton:** Hearing in Primitive Mammals. I: Opossum (Didelphis virginianus). *J. Audio. Res. 9:*1-7 (1969).
88. **Danner, P.A., E. Ackerman, and H.W. Frings:** Heating in Haired and Hairless Mice in High-Intensity Sound Fields from 6–22,000 hertz. *J. Acoust. Soc. Am. 26:*731 (1954).
89. **Acton, W.I.:** The Effects of Industrial Airborne Ultrasound on Humans. *Ultrasonics:*124-268 (1974).
90. **Grigor'eva, V.M.:** Effect of Ultrasonic Vibrations on Personnel Working with Ultrasonic Equipment. *Soviet Phys.-Acoust. 11:*426-427 (1966).
91. **Acton, W.I., and M.B. Carson:** Auditory and Subjective Effects of Airborne Noise from Industrial Ultrasonic Sources. *Br. J. Ind. Med. 24:*297-304 (1967).
92. **U.S. Air Force:** "Hazardous Noise Exposure" (USAF Regulation 161–35). 1973.
93. **American Conference of Governmental Industrial Hygienists:** *Threshold Limit Values and Biological Exposure Indices for 1988-1989.* Cincinnati: American Conference of Governmental Industrial Hygienists, 1988. p. 107.
94. **Kelly, E. (ed.):** Ultrasonic Energy. In *Biological Investigations and Medical Applications.* Urbana, Ill.: University of Illinois Press, 1965.
95. **Kelly, E. (ed.):** *Ultrasound in Biology and Medicine* (Pub. No. 3). Washington, D.C.: American Institute of Biological Sciences, 1957.
96. **Knight, J.J.:** Effects of Airborne Ultrasound on Man. *Ultrasonics 6:*39-42 (1968).

97. **Reid, J.M., and M.R. Sikov (eds.):** *Interaction of Ultrasound and Biological Tissues.* Proceedings of a Workshop at Battelle Seattle Research Center, Nov. 8-11, 1971. [DHEW (FDA) Pub. No. 73–8008 (BRH/DBE 73–1). 1972.]

98. **Berger, E.H.:** "Attenuation of Hearing Protectors at the Frequency Extremes." Proceedings of the Eleventh International Congress on Acoustics, Paris, France, 1983. pp. 289-292.

Maintenance worker climbs a 600-foot tower in Tuxedo, Texas. (Layne Kennedy, 1997)

Outcome Competencies

After completing this chapter, the user should be able to:

1. Define underlined terms in this chapter.
2. Describe nonionizing electromagnetic radiation.
3. Review biological/health effects for ultraviolet, laser, and radio-frequency radiation, and extremely low frequency fields.
4. Calculate hazard distances for laser radiation and some sources of radio-frequency radiation.
5. Review measurement fundamentals for nonionizing radiation.
6. Describe elementary control measures.

Prerequisite Knowledge

Introductory college physics (any text).

Key Terms

absolute gain • aphakes • aversion response • broadband • cancellation • Class 1 • Class 2 • Class 2a • Class 3a • Class 3b • Class 4 • contact currents • continuous wave • current density • cutaneous malignant melanoma (CMM) • diffuse reflection • divergence • duty cycle • effective irradiance • electric fields • electric-field strength • electromagnetic spectrum • electromagnetic radiation • extremely low frequency (ELF) • far field • flash blindness • frequency • geometrical resonance • hazard distance • illuminance • induced currents • infrared (IR) • intrabeam viewing • irradiance • Lambertian surface • laser radiation • Laser Safety Officer (LSO) • luminance • magnetic fields • magnetic flux density • magnetic-field strength • microwave radiation • minimum erythemal dose (MED) • near field • nominal hazard zone (NHZ) • nonbeam hazards • nonionizing radiation • nonmelanoma skin cancer (NMSC) • optical density (OD) • optical radiation • permeability • permittivity • perturbation • phosphenes • photoconjunctivitis • photokeratitis • photon energy • photosensitivity • plane waves • power density • prudent avoidance • radiance • radiant exposure • radio-frequency (RF) • radiometer • radiowaves • retinal hazard region • reversible behavior disruption • specific absorption (SA) • specific absorption rate (SAR) • shielding effectiveness (SE) • spatial averaging • specular reflection • ultraviolet (UV) • UV-A • UV-B • UV-C • visible • wavelength

Key Topics

I. Introduction
   A. Electromagnetic Radiation
   B. Electromagnetic Spectrum
   C. Nonionizing Radiation

II. Optical Radiation
   A. Anticipation
   B. Recognition
   C. Evaluation
   D. Control Measures

III. Laser Radiation
   A. Anticipation
   B. Recognition
   C. Evaluation
   D. Control Measures

IV. Radio-Frequency and Microwave Radiation
   A. Anticipation
   B. Recognition
   C. Evaluation
   D. Control Measures

V. Extremely Low Frequency Fields
   A. Anticipation
   B. Recognition
   C. Evaluation
   D. Control Measures

VI. Nonionizing Radiation Control Program
   A. Responsibility
   B. Medical Monitoring
   C. Information and Training
   D. Hazard Communication
   E. Self-Checks and Audits

21

Nonionizing Radiation

R. Timothy Hitchcock, CIH

William E. Murray

Robert M. Patterson, Sc.D., CIH

R. James Rockwell, Jr.

Introduction

In the past 75 years, there has been a remarkable increase in the number of man-made sources and applications of nonionizing radiation. These include a myriad of uses of laser and radio-frequency radiation in industrial, scientific, military, consumer, and medical applications. During much of the same period, research has focused on the possible health effects associated with nonionizing radiation (NIR), both man-made and naturally occurring. The body of information generated in those studies demonstrate that not only can overexposure to nonionizing radiation produce a number of serious health effects, but that there are thresholds between safe exposures and over exposures. Hence, it is possible to use nonionizing radiation for beneficial applications in the occupational environment without subjecting workers to harmful exposures. To achieve this goal, however, it is necessary to recognize and evaluate sources of nonionizing radiation in the workplace and provide expertise in appropriate controls to the worker. This is the role of the occupational hygienist.

The intent of this chapter is to assist the occupational hygienist in anticipating, recognizing, evaluating and controlling nonionizing radiation. Other useful references on nonionizing radiation are available.[1-4]

Electromagnetic Radiation

Because all nonionizing radiation is electromagnetic radiation (or electric and magnetic fields), it is necessary to have a basic understanding of electromagnetic radiation before delving into the nonionizing radiation spectral regions in more detail.

Electromagnetic radiation is the propagation, or transfer, of energy through space and matter by time-varying electric and magnetic fields, as shown in Figure 21.1. In the case of electromagnetic energy, the fields are composed of vector quantities. A vector field is any physical quantity that takes on different values of magnitude and direction at different points in space.

As depicted in Figure 21.1, the electric and magnetic fields are in time phase and space quadrature. Space quadrature means the electric and magnetic vector fields are mutually orthogonal. Time phase means that the waves have reached the same stage in their periodic oscillation with respect to time.

Electric fields are produced by electric charges, while magnetic fields are produced by moving charges, or a current.

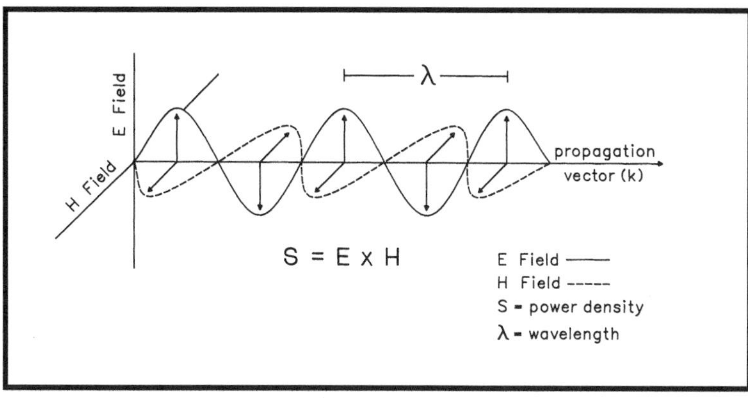

Figure 21.1 — The waves represent the electric and magnetic fields. The electric field vector (solid line) is vibrating up and down in the plane of the paper, while the magnetic field vector (dashed line) is vibrating in and out of the plane of the paper. The direction the radiation is moving is defined by a third vector — the propagation vector, k. Electromagnetic fields are transverse to the direction of propagation and contained within the envelope formed by the axis of propagation and the sinusoidal waves.

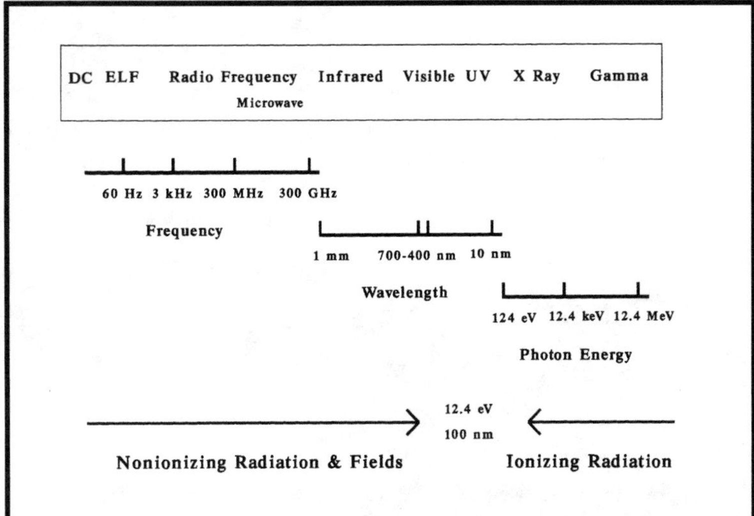

Figure 21.2 — The electromagnetic spectrum is a continuum that spans high-energy gamma radiation to non-time-varying fields (direct current). The spectrum is often divided into two regions: ionizing radiation and nonionizing radiation. The boundary between these regions is a photon energy of 12.4 eV.

Electric and magnetic fields in turn will exert a force on electric charges, and this is the basis for interactions with matter. Electric fields will act on a charge regardless of its motion, while the charge must be in motion relative to the magnetic field, or vice versa, before an interaction occurs between the magnetic field and the charge.

Electromagnetic Spectrum

The electromagnetic spectrum is subdivided into named regions (see Figure 21.2). Any location on the spectrum may be characterized by wavelength, frequency, and photon energy.

Wavelength, designated lambda (γ), is the distance between the ends of one complete cycle of a wave. Frequency, designated by the letter "f" or the Greek letter nu (v), is the number of complete wave cycles that pass a point in space in 1 second. Frequency and wavelength are related by the speed of light, c, where $f = c/\lambda$.

Photon energy describes the energy possessed by electromagnetic energy when characterized as discrete bundles, as described by quantum theory. The unit of photon energy is the joule (J) or the electron volt (eV).

Nonionizing Radiation

Photons with energies less than 12.4 eV are considered to have insufficient energy to ionize matter, and are nonionizing in nature, as shown in Figure 21.2. Quantities and units often used in the practice of nonionizing radiation safety are compiled in Table 21.1. Characteristics of the nonionizing radiation spectral region are shown in Table 21.2, and discussed elsewhere.[5-8]

The nonionizing spectral region includes the ultraviolet (UV), visible, infrared (IR), radio-frequency (RF), and extremely low frequency (ELF) spectral regions. As indicated in Table 21.3, wavelength is the descriptor used for UV, visible, and IR radiation and frequency is used for RF and ELF. Because these spectral regions have differences in absorption, biological effects, exposure guidelines, measurement instrumentation, and control measures; each is discussed separately in the following text.

Optical Radiation*

Anticipation

UV, visible, and IR radiation are called optical radiation because they behave according to the laws and principles of geometric optics. The optical wavelengths can be divided into spectral bands, as shown in Table 21.4. The boundaries between the bands have no physical basis and serve only as a framework for addressing biological effects.

The UV divisions are named regions. UV-A is the blacklight region, UV-B the erythema region, and the UV-C the germicidal region. The 100–180 nm region is known as the vacuum UV region, because these wavelengths are readily absorbed in air. UV-B and UV-C regions are often called actinic UV, because they are capable of causing chemical reactions.

Quantities and Units

Two systems of quantities and units, radiometric and photometric, are used for optical radiation. Radiometric quantities constitute a physical system that can be used to describe the emission from, or exposure to, any optical radiation. The photometric system is a subset of the radiometric system but is based on the response of the human eye to optical radiation and is used only for visible radiation. The radiometric system is used mainly for assessing optical radiation hazards, whereas the photometric system is used for specifying exposure limits for visible radiation and lighting requirements.[2]

Radiometric and photometric quantities and units are shown in Table 21.1. Similar quantities from both systems include irradiance-illuminance, radiance-luminance,

* This section was written by William E. Murray and R. Timothy Hitchcock

radiant intensity-luminous intensity, and radiant exposure-light exposure. For example, radiance is radiometric brightness (all optical wavelengths), while luminance is brightness as perceived by the eye (just visible wavelengths). Conversions between the two systems are possible but require spectral data and the use of wavelength-dependent conversion factors (called luminous efficacy), since the eye does not respond equally to all wavelengths. Spectral data are usually provided by the quantity spectral irradiance (E_γ), which has units of $W/m^2/nm$ or W/m^3. Spectral irradiance is the irradiance within a narrow spectral band.

Radiance is the radiometric brightness of an extended source. It is the radiant power per solid angle per unit area, where the solid angle is usually defined as a cone (area projected from the surface of a source) perpendicular to the direction of the flux.[9-11] Irradiance is the dose rate in photobiology. It describes the radiant flux that crosses a given surface (receiver), divided by the area of that surface. Radiant exposure is the photobiological dose of radiant energy. Hence, the dose in optical radiation is in joules/centimeter squared (J/cm^2), while the dose rate is in watts/centimeter squared (W/cm^2). (Remember: $1 W = 1 J/s$ or $P = Q/t$).

Luminance describes the brightness of an object, or the luminous intensity per unit area of surface. Illuminance is the luminous flux crossing a surface of a given area. The photometric quantities shown in Table 21.1 use SI units. Useful conversions to the English units are:

luminance – $1 cd/m^2 = 10^{-4} cd/cm^2 =$ 0.2919 footlamberts (ftL)

illuminance – $1 lm/m^2 = 1 lx =$ 0.929 footcandles (ftcd).

Table 21.1 — Important Quantities and Units

Quantity	Symbol	Unit Name	Unit Abbreviation
Beam size at lens	b_o	length	cm
Conductivity	σ	siemen/meter	S/m
Contact current	I_c	milliampere	mA
Current density	J	ampere/meter square	A/m^2, mA/cm^2
Divergence	ϕ	radian	rad, mrad
Duration of single pulse	t	time	s, sec
E-field strength	E	volt/meter	V/m, V^2/m^2, kV/m
Energy	Q	joule	J
Exit beam diameter	a	length	cm, mm
Focal length	f_o	length	cm
Frequency	υ,f	hertz	Hz (GHz, MHz, kHz)
H-field strength	H	ampere/meter	A/m, A^2/m^2
Induced currents	I_i	milliamperes	mA
Irradiance	E	watt/meter square	W/m^2, mW/cm^2
Magnetic flux density	B	tesla	T (mT, µT)
Maximum viewing angle	α_{max}	radian	rad, mrad
Minimum viewing angle	α_{min}	radian	rad, mrad
Mode field diameter	ω_o	length	µm
Numerical aperture	NA	dimensionless	—
Optical density	OD,D_γ	dimensionless	—
Permeability	μ	henry/meter	H/m
Permittivity	ε	farad/meter	F/m
Photon energy	Q_p	joule, electron volt	J, eV
Power	P,Φ	watt	W
Power density	W,S	watt/meter square	W/m^2, mW/cm^2
Radiant exposure	H	joule/meter square	J/m^2, mJ/cm^2
Reflectivity	ρ	dimensionless	—
Specific absorption rate	SAR	watts/kilogram	W/kg
Specific absorption	SA	joules/kilogram	J/kg
Total exposure duration	T	time	s, sec
Viewing angle	Θ	radian	rad, mrad
Wave impedance	Z	ohm	Ω
Wavelength	λ	meter	m (nm, µm, mm)

Table 21.2 — Composite Characteristics of NIR

Wavelength:	100 nm to 300,000 km
Frequency:	3.0 PHz to 1 Hz
Photon energy:	1.987×10^{-18} J to 6.6×10^{-34} J

Table 21.3 — Fundamental Characteristics

Region	Wavelength	Frequency
Ultraviolet	100–400 nm	—
Visible	400–770 nm	—
Infrared	770 nm–1 mm	—
Radio-frequency	—	300 GHz–3 kHz
Extremely low frequency	—	3 kHz–3 Hz

Table 21.4 — Spectral Bands for Optical Radiation

Region	Band	Wavelength
Ultraviolet	UV-C	100-280 nm
	UV-B	280–315 nm
	UV-A	315–400 nm
Visible		400–770 nm
Infrared	IR-A	770–1400 nm
	IR-B	1.4–3.0 µm
	IR-C	3.0 µm–1 mm

Biological and Health Effects

It is important to remember that a biological effect is not necessarily a health effect. For example, vision is a biological effect that results from exposure to light, a form of electromagnetic energy, but it is not a health effect. On the other hand, photokeratitis, which results from overexposure to UV radiation, is an example of a biological effect that leads to an adverse health effect.

Ultraviolet Radiation

Interaction with Matter. Interactions of optical radiation with matter primarily include transmission, propagation, absorption, reflection, and refraction. When optical radiation is incident on a surface, it may be reflected or transmitted into the material. Reflection may be specular (mirrorlike) or diffuse (non-mirrorlike), depending on the relative smoothness of the surface. Refraction, which is a change in the direction of propagation, occurs at an interface between two transmitting media that have different indices of refraction.

The target organs for UV radiation are the skin, eyes, and immune system. Skin effects that may be of importance from occupational exposures include erythema, photosensitivity, aging, and cancer. Ocular effects are photokeratoconjunctivitis, cataracts, and retinal effects.

Skin. The most common adverse response of the skin to UV is erythema or sunburn. The maximum erythemal effect is at 260 nm, with a secondary peak at 300 nm.[12] However, the International Commission on Illumination erythema reference action spectrum suggests that the skin is most sensitive (lowest effective doses) from 250 nm to about 300 nm.[13,14] The erythemal response (skin reddening) is a vascular reaction involving vasodilation and increased blood volume, and ranges from a slight reddening to severe blistering. The changes observed depend on the dose and spectral content of the incident radiation, the pigmentation and exposure history of the exposed skin, and the thickness of the stratum corneum. The erythemal response begins at 2 to 4 hr postexposure, peaks at 14 to 20 hr, and lasts for up to 48 hr. Parrish et al. determined that the minimum erythemal dose (MED) from 250 to 304 nm ranged from 14 to 47 mJ/cm^2 and increased dramatically above 313 nm.[15] This value of 15–50 mJ/cm^2 is also referred to as the threshold dose for erythema and applies to lightly pigmented skin not recently exposed to UV radiation (not conditioned). The defense measures taken by the skin on exposure, melanogenesis and skin thickening subsequent to hyperplasia, will increase the threshold required to produce the same degree of reddening.[16]

Certain chemicals found in medications, plants, or industrial operations are photosensitizing agents. Photosensitivity includes two types of reactions: phototoxicity and photoallergy. Phototoxicity is more common and affects all individuals if the UV dose or the dose of the photosensitizer is high enough. Photoallergy is an acquired altered reactivity in the exposed skin resulting from an immunologic response.[17]

Various lists of photosensitizing agents are available.[18,19] Agents include many medications, some sunscreen agents, plants (e.g., figs, parsley, limes, parsnips, and pinkrot celery), and industrial photosensitizers including coal tar, pitch, anthracene, naphthalene, phenanthrene, thiophene, and many phenolic agents.

The carcinogenic effects of UV radiation have been studied extensively, and literature reviews are available.[20-25] Evaluating the carcinogenicity of solar and UV radiation, an International Agency for Research on Cancer (IARC) working group found that solar radiation is carcinogenic to humans and all UV regions are probably carcinogenic to humans.[23]

Three types of skin cancers are of concern: squamous cell carcinomas (SCC) and basal cell carcinomas (BCC) [referred to jointly as nonmelanoma skin cancer (NMSC)] and cutaneous malignant melanoma (CMM). The American Cancer Society estimates that about 800,000 new cases of NMSC are diagnosed annually.[26] Four variables have been implicated in NMSC: (1) lifetime sun exposure; (2) the intensity and duration of the UV-B component in sunlight; (3) genetic predisposition; and (4) other factors unrelated to sunlight, such as exposure to ionizing radiation and polycyclic aromatic hydrocarbons.[17,27] A study that examined nonsolar sources of UV found no increased risk of BCC and SCC associated with sources including welding, mercury vapor lamps, and black lights.[28] The carcinogenic action spectrum for humans is known to include the UV-B region but may extend throughout the UV-A region.

The incidence of cutaneous malignant melanoma has increased rapidly in the United States and worldwide in the past 30 to 40 years. In the United States, about 40,000 new cases of CMM occur each year, with more than 7000 deaths anticipated.[26] From 1973 to 1988, the death rate for malignant melanoma among U.S. males increased 50%, a greater increase than that for any other cancer. For females, the death rate rose 21%. Although several factors, such as the presence of moles and genetic predisposition, have been associated with this disease, increased solar exposure has been implicated as a contributing factor,[29] as has intermittent exposure, that is, "short bursts of intense exposure."[30] An action spectrum for CMM in a hybrid fish suggests the involvement of not only UV-B, but also UV-A and blue light.[31,32]

Recent research suggests that UVR exposure also depresses both the systemic and local immunologic response.[33] Changes have been observed in Langerhans cells, and the density of these cells is reduced after exposure.[34] Since Langerhans cells are located above the basal membrane of the epidermis, they are not protected by the melanocytes and this effect may be independent of skin pigmentation.[35]

UV exposure does produce some beneficial or therapeutic effects,[36] including vitamin D_3 synthesis and therapy for skin conditions including vitiligo, psoriasis, mycosis fungoides, acne, eczema, and pityriasis rosea. In some treatments, photosensitizing agents are used to enhance the effect of UV radiation.[37]

Eye. The cornea and conjunctiva are the primary absorbers of UV-B and UV-C. Corneal transmission ranges from 60 to 83% in the UV-A band, with much of the energy absorbed by the lens. Photokeratitis and photoconjunctivitis result from acute, high-intensity exposure to UV-B and UV-C. Commonly referred to as "arc eye" or "welder's flash" by workers, this injury results from exposure of the unprotected eye to a welding arc or other artificial sources rich in UV-B and UV-C. Sunlight exposure produces these sequelae only in environments where highly reflective materials are present such as snow ("snow blindness") or sand.

Symptoms include lacrimation, blepharospasm, and photophobia accompanied by a sensation of sand in the eye and severe pain. These symptoms become apparent at 2–12 hr postexposure, and persist for up to 48 hr, usually without residual damage. Photochemical denaturation and coagulation of protein structures are the basic mechanisms of cell damage.[38] No increasing tolerance to subsequent irradiation develops.[16]

Threshold doses for humans range from 4 to 14 mJ/cm² between 220 and 310 nm.[38,39] An action spectrum has been established, and the wavelength of minimum threshold occurs at 270 nm. Reciprocity holds for exposure durations up to several hours; thus, the tissue damage depends on the total dose, and not the dose rate.

The only lenticular effect linked with UVR exposure, primarily solar radiation, is the cataract.[40,41] In the United States, estimates of the number of cataract operations range up to 1 million annually and cataracts are the third leading cause of blindness.[42] Predisposing factors include age, gender, family history, nutritional status, and certain medical conditions and medications.

An action spectrum has been established for lenticular opacities in rabbits. For exposures of less than one day, threshold radiant exposures from 290 to 320 nm ranged from 0.15 to 12.60 J/cm² for transient opacities, and 0.5 to 15.5 J/cm² for permanent opacities. In both cases, the minimum threshold was 300 nm. From 325 to 395 nm cataracts could not be produced even at very high exposures. Primates were exposed only at 300 nm; the threshold radiant exposure was 0.12 J/cm².[38] The mechanism for UV-induced cataracts may be photochemical in the 295 to 320 nm range and thermal at longer wavelengths.[43]

Epidemiological studies of human populations in six countries on four continents have attempted to discern if an association exists between sunlight exposure and the occurrence of cataracts, specifically senile cataracts. Although individual solar or UV radiation exposures were not measured, diverse surrogates such as UV-B counts, hours of sunlight exposure, and geographic location were used to characterize average daily exposures. Relative risk or odds ratios ranged from 1.3 to 5.8, demonstrating a consistent association between UV exposure and cataracts.[43,44] In reviewing the literature, Waxler concluded that the threshold for cataracts is 2500 MEDs yearly, corresponding to daily exposures of 7 to 90 mJ/cm² for UV-B and 0.4 to 98 J/cm² for UV-A.[45] Also, there is some evidence that systemic photosensitizers, primarily drugs, may play a role in cataractogenesis.[46,47]

Exposure to UV-A and UV-B, primarily from solar radiation, has been implicated as an etiologic agent in solar retinitis, cystoid macular edema (CME), and senile macular degeneration (SMD).[48] In adults, transmission of these wavelengths through the ocular media is minimal except in an aphake (absence of the lens). Although originally thought to be a thermal injury, solar retinitis (also known as eclipse blindness) is a photochemical lesion resulting from absorption of UV-A and blue wavelengths. Experimental studies have determined that retinal damage occurs at lower exposures at 325 nm than at 441 nm (5.0 vs. 30 J/cm² respectively). The threshold for broad-band UV-A exposure is 0.09 J/cm².[45]

Visible Radiation

Skin. Above 400 nm, high-intensity exposure to visible wavelengths can lead to thermocoagulation of skin similar to that produced by electrical or thermal burns. Several variables influence the threshold for and the amount of damage: absorption and scattering by the skin, the incident intensity, exposure duration, area of skin exposed, and degree of vascularization of the irradiated tissue.[49] Individuals with certain diseases or genetic deficiencies are sensitive to light, but the mechanisms and action spectra are not well documented.[17] Some photosensitizing agents may have action spectra that reach into the visible region. As mentioned earlier, the action spectrum of malignant melanoma in hybrid fish extended through the UV spectral region and included some blue-light wavelengths.[31,32]

Eye. Since the human eye is relatively transparent to light, it is transmitted through the ocular media to the retina, with minimal absorption. These photons initiate a photochemical chain reaction in light-sensitive absorbers in the retina resulting in the sensation of vision.[16] The light intensity at the retina is a factor of 10^4 to 10^6 greater than that incident on the pupil, which is called optical gain.

Retinal effects occur through four interaction mechanisms: (1) thermal; (2) photochemical; (3) elastic or thermoacoustic transient pressure waves; and (4) nonlinear phenomena. The latter two mechanisms are seen primarily with lasers. The absorption of light by tissue produces heat. At sufficient intensities, the resulting rapid rise in temperature can denature proteins and inactivate enzymes. The thermal mechanism is the primary damage mechanism for 1 ms to 10 sec exposure durations.[9] Long-term (>10 to 100 sec) exposure to light levels above normal environment levels may produce photochemical damage in the retina. The action spectrum mimics the spectral absorption of melanin through the lower end of the visible (500 to 400 nm) region down into the UV-A region. Photochemical damage can be enhanced by the thermal mechanism that predominates in the 500 to 1400 nm range, although there is no sharp cutoff point between the two.[50]

Exposure to intense light, such as the sun, carbon arc, or welder's arc, without proper protection may produce temporary or permanent retinal scotomas (blind spots). Reports of injuries from observing solar eclipses date back as far as Hippocrates (460 to 370 B.C.), and sporadic references to this have been made throughout the centuries.[51] Factors affecting the degree of retinal hazard include the size, type, and spectral intensity of the source, the pupil size, retinal image quality, and the spectral transmittance of the ocular media, the spectral absorption of the retina and choroid, and the exposure duration.

Thresholds for retinal damage have been determined for both incoherent and coherent optical radiation sources under a wide variety of exposure conditions. The threshold radiant exposure for retinal damage is 30 J/cm^2 for a 1000 sec exposure at 441 nm for retinal photochemical damage, corresponding to a retinal irradiance of 30 mW/cm^2. The damage may be cumulative over 2 or more days. At 633 nm, the threshold increases to 950 mW/cm^2, for the same exposure duration. Threshold retinal irradiances for broad-band sources as low as 0.93 μW/cm^2 have been reported for all-day exposures. Visual impairment has been shown in monkeys exposed repeatedly to 6 and 100 mJ/cm^2 at 514 nm over many months.[45] Because of the lack of information on the human eye, exposure limits are based on data obtained primarily on rabbits and monkeys, which are extrapolated to the human eye.

Glare may produce visual discomfort, often due to squinting in an effort to screen light. If glare is substantial or frequently induced, it may result in tiredness, irritability, possibly headache, and a decrease in work efficiency.[51]

Glare can be differentiated into veiling glare, discomfort glare, or blinding glare (flash blindness). Although all three are present in the case of high-intensity light, the effects of the first two are primarily evident only when the source is present. Blinding glare is especially significant where it produces afterimages that persist long after the light itself has vanished. Regardless of whether the glare source is specularly or diffusely reflected, it can cause discomfort or affect visual performance, or both. The visual discomfort or annoyance from glare is well understood and has been confirmed by many experiments. Certain experimental studies have found that people sometimes become more physically tense and restless under glare conditions. Although the cause is physical, the discomfort brought about by glare is often subjective. The evaluation of discomfort, then, must make use of subjective responses as criteria. Involved in such procedures is the concept of the borderline between comfort and discomfort.[52]

Flash blindness, a relatively new problem occurring only since the development of intense light sources brighter than the sun, is a temporary effect in which visual sensitivity and function is decreased severely in a very short time period. This normal visual response, whose complex mechanism is not well understood, is related to the eye's adaptive state and the size, spectral distribution, and luminance of the light source.[9]

Infrared Radiation

Skin. The damage to skin from IR exposure results from a temperature increase in the absorbing tissue. The increase depends on the wavelength, the parameters involved in heat conduction and dissipation, the intensity of the exposure, and the exposure duration. The most prominent effects of near-IR include acute skin burn, increased vasodilation of the capillary beds, and an increased pigmentation that can persist for long periods of time. With continuous exposure to high-intensity IR, the erythematous appearance due to vasodilation may become permanent.[53] Many factors mediate the ability to produce a skin burn. It is evident that the rate at which the temperature of the skin is permitted to increase is of prime importance. High levels of far-IR, often referred to as radiant heat, are encountered in glassblowing, foundries, and furnaces. This IR exposure can be a significant contributor to thermal stress.

Eye. IR produces thermal effects to the eye. The cornea is highly transparent to IR-A, has water absorption bands at 1.43 and 1.96 µm, and becomes opaque to IR above 2.5 µm.[54] In rabbits the near-IR threshold radiant exposure decreased from 5.5 to 1.25 kJ/cm^2 as the source irradiance was increased from 2.93 to 4.55 W/cm^2. The corneal damage ranged from epithelial haze to corneal erosion with a focused beam incident on a miotic (contracted) pupil. Complete recovery occurred, usually in 24 hr. The threshold in primates was 8.0 kJ/cm^2 at an irradiance of 4.23 W/cm^2[55]. Chronic corneal exposure to subthreshold doses may lead to "dryeye," characterized by conjunctivitis and decreased lacrimation.[54]

Moderate IR doses can result in constriction of the pupil (miosis), hyperemia, and the formation of aqueous flares. More severe exposures may lead to muscle paralysis, congestion with hemorrhage, thrombosis and stromal inflammation. Within a few days, necrosis of the iris may cause bleached atrophic areas to form. Pigmentation loss at the border of the iris follows in 2 to 4 days.[57] A stromal haze was observed in the rabbit iris at the threshold radiant exposure ranging from 4.0-1.25 kJ/cm^2 as the irradiance increased from 2.93 to 4.55 W/cm^2. In primates, the threshold radiant exposure was 8.0 kJ/cm^2 at an irradiance of 4.23 W/cm^2.[58]

Damage to the lens of the eye from IR has been investigated for many years. The term "glassworker's cataract" has become generic for lenticular opacities found in workers exposed to processes hot enough to be luminous.[16] The occupational groups at risk include glassblowers, foundry and forge workers, cooks, and laundry workers, as well as those who worked in sunlight.[54] Robinson reported that lens opacities on the posterior surface found in glassworkers were different in appearance from senile cataracts.[56] Dunn found no evidence of any ocular disturbance among glassworkers,[57] and Keating et al. found no posterior cortical changes in iron rolling mill workers.[58] The incidence of posterior capsular opacities was higher, however, this differs from what is classified as a cataract.[58]

The mechanism for IR cataractogenesis has long been debated. Some results appear to support the hypothesis that the iris must be involved in IR- or heat-induced cataractogenesis, at least with regard to acute exposure conditions.

The retina is susceptible to near-IR since the ocular media are relatively transparent to these wavelengths. For extended exposures, the corneal irradiance required to produce a minimal retinal lesion at 1064 nm is almost three orders of magnitude greater than that at 442 nm.[54] Absorption of IR by the retinal pigment epithelium produces heat. If the heat is not dissipated rapidly and the temperature of the tissue rises about 20°C, irreversible thermal damage will result from the denaturation of protein and other macromolecules.[50] The size or area of the image on the retinal choroid apparatus and the absorbed retinal irradiance (dose rate) are two of the most critical factors in the production of thermal injury.

Recognition

Generation and Sources

A simple scheme for categorizing optical radiation sources is the following: (1) sunlight; (2) lamps; (3) incandescent (warm-body) sources; and (4) lasers.[9] (Lasers are discussed later in this chapter.) Table 21.5 lists some occupations in which workers are at risk of exposure to optical radiation.

The sun provides a large component of worker exposure to optical radiation since as much as one-third of the work force is exposed occupationally to sunlight. The exposure varies greatly with climate, geography, altitude, and work activities. The solar irradiance outside the atmosphere is

Table 21.5 — Common Exposures to Optical Radiation Sources[A]

Sources	Spectral Regions of Concern	Potential Exposures
sunlight	UV, visible, near-IR	farming, construction, landscaping, life guards, other outdoor workers
arc lamps	UV, visible, near-IR	photoreproduction, optical laboratories, entertainment
germicidal lamps	actinic UV	hospitals, laboratories, medical clinics, maintenance
Hg-HID lamps	UV-A, blue light	maintenance, industry, warehouses, gymnasiums
(broken envelope)	actinic UV	
carbon arcs	UV, blue light light activities	laboratories, search
industrial IR sources	IR	steel mills, foundries, glassmaking, drying equipment
metal halide UV-A lamps	near-UV, visible	printing plants, maintenance, integrated circuit manufacturing
sunlamps	UV, blue light	tanning parlors, beauty salons, fitness parlors
welding arcs	UV, blue light	construction, repair and maintenance, bypassers

[A]Adapted from Reference 10.

about 135 mW/cm^2 while at sea level, the maximum is about 100 mW/cm^2.[(10)] The solar spectrum peaks at about 450 nm and drops off rapidly in the UV and more gradually in the IR region.

UV and visible radiation are generated by electronic transitions in excited atomic systems. IR radiation is produced by changes in the vibrational, rotational, or translational energy states in molecules.

UV Radiation. Uses of UV radiation include curing materials, suntanning, photoluminescence, chemical manufacturing, treating skin disorders, germicidal applications, UV spectroscopy, and photomicrolithography. UV is an unwanted by-product in processes such as welding, metal cutting, and glass manufacturing. A listing of selected occupations that may be associated with UV exposure is available.[(59)]

The main UV source is the sun. Global UV radiation has two components: the sun's beam and sky radiation. Sky radiation is diffuse and caused by scattering in the atmosphere. Global UV-A is a stable component of global radiation.[(60,61)] Solar UV-B wavelengths less than about 295 nm are attenuated by stratospheric ozone, although wavelengths as low as 286 nm have been detected at the earth's surface.[(62)]

UV-B is sensitive to the angle of the sun, being greatest when the sun is overhead, from about 10 a.m. to 2 p.m., solar time,[(63)] and the time of year.[(64)] Dosimetric studies with mannequins have shown that the greatest sunlight exposures are received by horizontal surfaces such as the tops of the feet, shoulders and ears, back of the neck, and the area around the nose, cheekbones, and lower lip.[(65-67)] Although the UV irradiance is often sufficient to produce skin and eye effects for relatively brief exposure times, ocular effects generally do not develop because of shielding of the eye by the orbit, eyebrows, and eyelashes, and squinting.[(68,69)]

A number of outdoor occupations and activities have been evaluated for UV exposure. These include fishermen,[(70)] construction workers,[(70)] farmers and gardeners,[(71-73)] landscape workers,[(70)] physical education teachers,[(71)] outdoor sports athletes,[(74,75)] lifeguards,[(71)] and airline pilots.[(76)]

Of the man-made sources, welding and various lamps are the major sources of occupational exposure. Another UV source that is gaining popularity in materials processing and medical and laboratory applications is the family of excimer lasers.

Some of the more hazardous welding processes include argon-shielded gas-metal arc welding (GMAW) with aluminum or ferrous alloys and helium-shielded gas tungsten arc welding (GTAW) with ferrous alloys.[(9)] Parameters important in determining the UV irradiance include the arc current, shielding gas, electrode wire material, base metal, and joint geometry. Weld geometry and the degree to which the arc is buried in the weld also influence UV levels.[(59)] Welding and some surface applications with carbon dioxide lasers may generate potentially hazardous actinic UV levels because of the interaction of the energetic far-infrared laser beam with metallic materials.

Lamps that may emit relatively high levels of UV include metal halide, high-pressure xenon, mercury vapor, and other high-intensity discharge lamps.[(9,77-79)] The spectral output of mercury vapor lamps varies with the gas pressure. Low pressure lamps emit line spectra characteristic of mercury (mostly 253.7 nm) and are used primarily for germicidal applications. Overexposure[(80)] and potential overexposures to users have been reported.[(81,82)] Medium to high-pressure lamps are used for curing and area lighting, such as in gymnasiums and parking lots. These lamps produce a bluish light and may be hazardous if viewed at close distances. Mercury lamps used for area lighting have two envelopes. The inner envelope is quartz (transparent to UV), while the outer envelope is made of materials that are largely opaque to UV. The output may produce hazardous exposures some distance from the lamp if the outer envelope is broken and the lamp continues to operate, which can occur in lamps that are not self-extinguishing. UV curing processes irradiate UV-curable materials and monomeric photosensitizers with 300 to 400 nm (predominantly 365 nm) UV to produce a hard, durable film.

UV-A, -B, and -C have been found from fluorescent lamps used in open fixtures, but UV-B and UV-C were not found when an acrylic diffuser was used with the fixture. The highest UV-B and UV-C irradiances were associated with high-output (HO) and super high-output (SHO) lamps.[(83)]

Medical uses include phototherapy, photochemotherapy, tanning, and disinfection and sterilization. The spectral output from some lamps used in photomedicine has been reported,[(84,85)] and some exposures may be in excess of exposure limits.[(86)] Phototherapy workers have a greater risk of dying from skin cancer than medical workers who do not work with UV, but their risk is much less than medical workers who use ionizing radiation.[(87)]

Dentists may use UV and blue-light-curable materials. Generally, handheld applicators that are properly designed, maintained, and used will minimize exposure to the dentist. However, improper use or excessive leaks[(88)] may result in mouth burns to the patient.

Visible Radiation. The sun is the major source, along with various lamps, projection systems, welding arcs, and lasers (see Table 21.5). The luminance of the noon-day sun is 1.6×10^5 cd/cm^2,[(9)] while the time necessary for blue-light injury is about 90 seconds.[(69)] Solar blue-light exposures on the flight deck of a commercial aircraft were higher than those experienced by the general population, but within exposure guidelines.[(89)]

Luminance from welding arcs can be hundreds to ten thousands of candelas per centimeter square.[(9)] Blue light from various types of welding ranges from 0.16 to 135 W/cm^2 • sr.[(9)]

Xenon short-arc lamps emit relatively high levels of blue light, while low-pressure fluorescent lamps emit relatively low levels. Luminance levels for xenon short-arc lamps were around 10^4 to 10^5 cd/cm^2.[(9)] For potential blue-light exposures to photoflood lamps used in TV studios and theaters, acceptable exposure duration varied from 1 min to 3 hr at distances between 2.25 and 10m. Calculated values of luminance were between 30 and 1900 cd/cm^2.[(90)] Quartz linear lamps used for space heating are reported to emit negligible levels below 600 nm, when fitted with a ruby sleeve.[(91)]

Workers from NIOSH evaluated luminance levels from thermal arc spraying,[(92)] laboratory furnaces,[(93)] aluminum reduction potrooms,[(94)] and glass furnaces[(95)] finding maximum levels of 16.7, 1.9, 0.3, and 0.93 cd/cm^2, respectively.

Infrared Radiation. Most of the sources discussed under visible radiation also emit IR radiation. Also, a number of IR lasers and optical wireless communications systems use IR light-emitting diodes (LEDs).[(96,97)]

In the NIOSH studies cited earlier, levels of IR radiation were thermal arc spraying (170 mW/cm^2),[(92)] laboratory furnaces (130 mW/cm^2),[(93)] aluminum reduction potrooms (190 mW/cm^2),[(94)] and glass furnaces (173 mW/cm^2).[(95)] Peak IR-A irradiance values in the Swedish iron and steel industry were 140 mW/cm^2 while hotrolling ingots and 130 mW/cm^2 in an electrosteel plant.[(98)]

Quartz linear lamps with ruby sleeves were reported to pose no hazard for normal use, but the evaluators recommended that they should be switched off for maintenance tasks at distances closer than 1 m.[(91)] An evaluation of IR illuminators used with closed-circuit television (CCTV) surveillance systems found a value of irradiance of 10 mW/cm^2 at 5 m, with increasing levels as one approached the device. Hence, there is a potential hazard for individuals involved with installing and maintaining these systems.[(99)] Spot welding produced IR (800 to 1800 nm) levels between 3 and 2500 mW/cm^2.[(100)]

Evaluation

Exposure Guidelines

Both exposure and emission standards are employed to protect people from hazardous optical radiation. Exposure standards are used to limit the intensity of optical radiation to which a worker or the general public can be exposed, regardless of the source. Emission standards apply only to specific optical radiation sources, such as high-intensity mercury vapor lamps, or a type of source such as the laser. Such standards and guidelines are available from a number of sources, and examples of each are highlighted here. This overview is not a compendium of available standards; the reader is encouraged to seek out further sources of such information.

Ultraviolet Radiation.The American Conference of Governmental Industrial Hygienists (ACGIH),[101] the National Institute for Occupational Safety and Health (NIOSH),[102,103] and the International Commission on Nonionizing Radiation Protection (ICNIRP, which used to be part of the International Radiation Protection Association)[104] recommend occupational exposure limits for UV radiation.

The ACGIH threshold limit values (TLVs®) for UVR incident on skin or eye are similar to the exposure limits (ELs) established by ICNIRP. The TLVs apply to UVR in the spectral band from 180 to 400 nm and to exposure of eye and skin to solar radiation and all artificial UVR sources except lasers. These TLVs should not be used for photosensitive individuals or to persons concomitantly exposed to systemic or topical photosensitizing agents, or ocular exposure of aphakes. They are applicable for exposure durations from 0.1 sec to the whole working day.

These exposure guidelines are based primarily on studies of acute exposures in humans and animals, keratoconjunctivitis, erythema, and cataracts. They do not take into account the effects of long-term exposure on eye and skin.

Light and Near-Infrared Radiation. ACGIH has established a TLV for ocular exposure in any 8-hr workday in the spectral region from 400 to 1400 nm.[69,101] The TLVs protect against retinal thermal injury in the 400 to 1400 nm range and retinal photochemical injury from chronic blue-light exposure between 400 and 700 nm.

Workers who have had their lens(es) removed (aphakes) in cataract surgery have an increased risk of retinal photochemical injury. Wavelengths in the 300 to 400 nm range can be transmitted to the retina if these workers do not have a UV-absorbing intraocular lens implant(s). The latter TLV is modified in the 300 to 700 nm region using an appropriate hazard function.

The TLV for near-IR (IR-A, 770 to 1400 nm) protects against delayed effects on the lens (cataracts) by limiting exposure to 10 mW/cm^2.[101] Sources used as heat lamps or lacking a strong visual stimulus require a more complex analysis.

Emission (Product) Standards

The U.S. Food and Drug Administration (FDA) has product standards for sources of optical radiation including high-intensity discharge mercury vapor lamps, sunlamps, and lasers. (See the section on laser safety for a discussion of the standard on products containing lasers.)

The standard on high intensity mercury vapor discharge lamps (21 CFR Part 1040.3) requires lamp manufacturers to provide a label to inform the user if the lamp will self-extinguish when the outer glass envelope is broken. The sunlamp standard (21 CFR Part 1040.2) requires manufacturers to affix a warning (danger) label on each product, provide the consumer with instructions for using the sunlamp, incorporate a timer and a manual control to terminate the exposure, and include protective eyewear with each sunlamp.

Instruments

The radiometer used by most occupational hygienists is a broad-band instrument that integrates spectral irradiance over a wide range of wavelengths (see Figure 21.3). In general, radiometers are used to measure radiant power and have calibrated output in irradiance. Typically, these instruments also have an integration function that allows radiant exposure to be measured over time. The output of some radiometers used to measure actinic UV is in effective irradiance (E_{eff}) in W/cm^2, which is the irradiance of a broad-band source weighted against the peak of the spectral effectiveness curve, which is 270 nm.

Scanning radiometers (spectroradiometers) measure irradiance in a narrow bandwidth. They may have their output in current (amperes), which must be compared to wavelength-specific calibration factors to determine the spectral irradiance.

The optical sampling system may include input optics, band filters/monochromators, detectors, and the electronics package. Useful UV detectors for continuous-wave (CW) sources include photoelectric detectors (photodiodes and photomultiplier tubes) and thermal detectors (thermopiles and pyroelectric detectors). Ideally, detectors should display a wavelength-independent (flat) response within the spectral region of interest with a rapid decrease outside of this band. Some detectors have been noted to deviate from linearity at high UV levels.[105]

Filters[106,107] include interference, neutral-density, and special filters. A bandpass

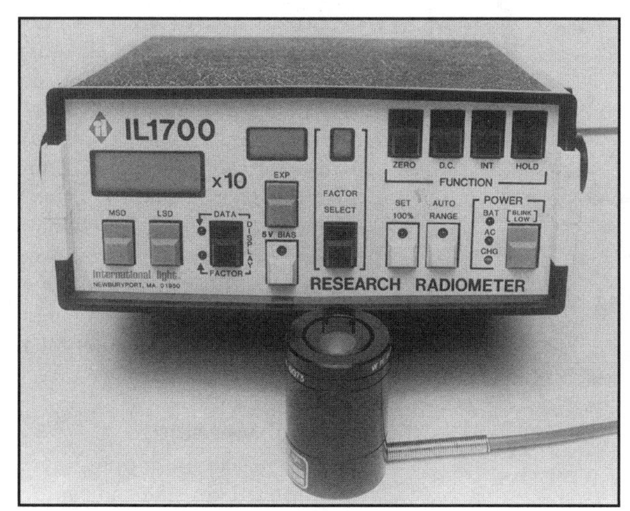

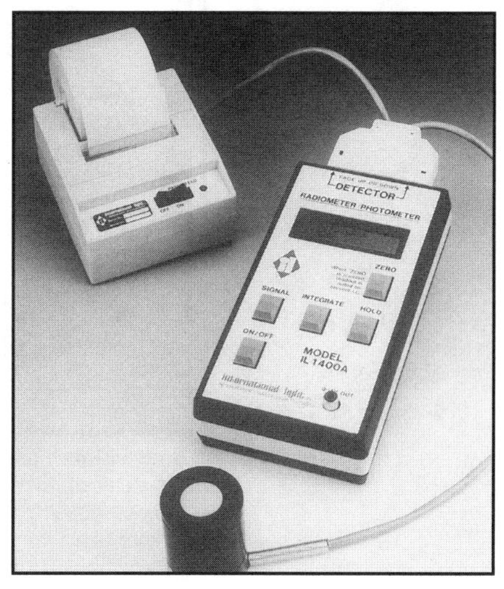

Figure 21.3 — Broad-band radiometers: (a) A portable research radiometer with an optical sampling system. The sampling probe incorporates three elements: diffuser, filter, and detector; (b) A hand-held radiometer and optical sampling probe is shown with a small printer. (Courtesy International Light, Inc.)

filter used with portable radiometers provides spectral weighting so that transmission closely matches the exposure criteria. Some detectors may respond to visible or IR radiation, or both, and require a filter to absorb these wavelengths.

Diffusers provide uniform illumination of the detector, make the detector-source alignment less critical, and provide a cosine response at the detector.

Monochromators are components of spectroradiometers and are used to measure a narrow spectral region without the use of filters. The monochromator includes an inlet slit, diffraction grating/prism, alignment linkage, internal mirrors, and an exit slit. The monochromator is connected to the detection system. The bandwidth or bandpass of a monochromator is the width of the transmitted spectral distribution. Bandwidth depends on the slit width and grating. Generally, smaller slit openings produce smaller bandwidth, up to the limiting bandwidth or instrument resolution.

There are a number of sources of error in radiometry,(108-111) the most common being susceptibility to stray source and ambient light. Ambient lighting and broad-band source emissions may exceed the spectral irradiance one is attempting to measure. Unless the instrument can reject this light, considerable error may be associated with the measurement data.(59)

Dosimeters, composed of a photosensitive material, have been used for UV measurements.(86,112-116) The erythemal dose determined with dosimeters made from the plastic CR-39 (allyl diglycol carbonate) and from polysulphone film have been compared, with mixed results.(117-119)

Exposure Assessment

Prior to measurement, the evaluator should determine source characteristics (e.g., spectral distribution of a lamp or current of a welding arc) operator-source interaction, and the location of the source relative to reflective objects such as aluminum surfaces. Make sure that the instrument is calibrated and the spectral distribution of the source is within the calibrated response spectrum of the instrument. Allow the instrument to warm up, if required, zero the output indication, then check for zero-point drift. The evaluator should pay attention to personal safety and take any personal protective equipment that may be necessary.

When measuring, the effective irradiance or radiance is read directly as the output or is calculated from the output quantity (see Figure 21.4). The effective UV irradiance is calculated for instruments that have their output in spectral irradiance (E_λ), which is a wavelength-dependent quantity, as:

$$E_{eff} = \Sigma E_\lambda S_\lambda \Delta_\lambda \quad (1)$$

Here, E_λ is the measured spectral irradiance in a narrow bandwidth, S_λ is the relative spectral effectiveness at a given wavelength within that band, and Δ_λ is the actual bandwidth in nanometers. S_λ is read from a tabular compilation of the latest booklet of TLVs for UV radiation, and the bandwidth is a function of the instrument.

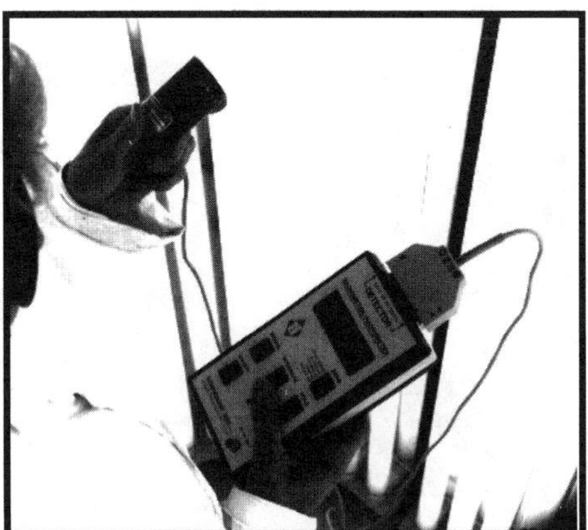

Figure 21.4 — Use of a hand-held radiometer. The optical sampling probe is directed toward the source and the irradiance (or radiant exposure) is read from the display. (Courtesy International Light, Inc.)

For welding arcs, the effective UV irradiance at a given distance can be estimated prior to measurement if the arc current is known. It has been found that values of measured effective irradiance do not exceed:

$$E_{eff} = ((4 \times 10^{-4} \text{ watts/ampere}^2) \times I^2)/r^2 \quad (2)$$

where I is the arc current in amperes and r is distance from the source in centimeters. This equation may be used to estimate the upper limit of the effective irradiance for arc currents from 10–20 A to greater than 700 A.[9]

For broad-band measurement data in the UV-B and UV-C, the allowable exposure duration may be calculated as the quotient of the radiant exposure at 270 nm (30 J/m^2 = 3 mJ/cm^2 = 0.003 J/cm^2) and the measured value of effective irradiance, as:

$$t\ (s) = \frac{30\ J/m^2}{E_{eff}\ (W/m^2)} \quad (3)$$

If the effective irradiance (E_{eff}) at a specific wavelength or bandwidth is known, the allowable exposure duration may be determined by using the wavelength-specific values of radiant exposure (e.g., H = 6 mJ/cm^2 at 254 nm) and irradiance, where t = H/E. (Equation 3 should be applied with caution if the measured quantity is not relatively constant.)

At times, collecting data at a large number of locations or distances from the source may not be practical. Here, the inverse square law may be used to estimate the radial increase or decrease of irradiance of uncollimated, extended sources as:

$$E_1 = E_2 r_2^2 / r_1^2 \quad (4)$$

Use of the inverse square law may provide a conservative estimate for levels of actinic UV that are readily absorbed by certain molecular species including atmospheric molecules and water vapor.[109] However, the absorption by air molecules at wavelengths greater than 200 nm is insignificant for distances of up to several hundred yards.[109] Conversely, an underestimation of the irradiance may occur if the source geometry produces a high gain, that is, if the optical power is contained in a smaller cross-sectional area or the radiation is collimated.[120]

Control Measures

Control measures for optical radiation must address both skin and eye. Successful engineering controls include isolation or enclosure of the process and safety subsystems such as interlocks, shutters, and alarms. Personal protection includes clothing, sunscreens and sun blocks, and the use of protective eyewear.

Engineering Controls

Barriers or enclosures may be made from a variety of efficient attenuating materials including metals, plastics, and glasses. Transmission curves of various filter materials and UV inhibitors are available.[106] Some useful UV filter materials are polyester films, cellulose acetate, and acrylics such as methyl methacrylate based polymers.[42,106,121]

Most common materials transmit little UV, although this is a function of wavelength, material thickness, and angle of incidence. Window glass (2.4 mm thick) transmits almost no UV below 300 nm; around 330 nm the transmission increases to about 50%. Data indicate that little radiation at 308 nm is reflected from the surface of glass at angles of incidence less than 60. For angles greater than this, the degree of reflection increases exponentially.[122]

Commercially available welding curtains may be made of materials that are either opaque or transparent to visible wavelengths. Opaque materials include canvas duck, asbestos substitutes, and polymer laminates. Transparent welding curtains may allow visual contact with welders, reduce arc glare, and increase general illumination levels.[123] Generally, curtains should attenuate UV and blue light to minimize potential hazards while transmitting near the peak of visual sensitivity, around 550 nm. In testing 25 transparent, colored welding curtains, NIOSH found that wavelengths less than 340 nm were highly attenuated.[123]

Many plastics, glasses, and some dyes may degrade with long-term exposure to UV radiation. Solarization, resulting from photochemical reactions, is a decrease in UV transmissivity in quartz and glass and may lead to a perceptible color change in the glass.

Doors or access panels to barriers or enclosures that contain UV radiation in excess of the applicable exposure limit should be interlocked. Interlocks can be connected to power supplies or shutter mechanisms and, in some cases, may also be connected to the exhaust ventilation systems for welding effluent or ozone.

Some lamp enclosures may be interlocked, such as self-extinguishing mercury vapor lamps. These lamps stop operating within 15 min of breakage of the outer of the two envelopes. The FDA requires that self-extinguishing mercury vapor lamps be marked with the letter "T." If the lamp is not self-extinguishing, the FDA requires that such lamps be marked with the letter "R."

Visual or audible alarms, or both, should be used where workers may access a noninterlocked enclosure that contains hazardous levels of UV. This is especially important if the source does not emit visible radiation that may indicate potential exposure.

Skin Protection

Sun Blocks and Sunscreens. In the past decade, the availability of, and emphasis on the need for, sunblocks and sunscreens that are applied to the skin to reduce cutaneous UV exposure from solar radiation have increased considerably. A sunblock reduces exposure by reflecting and scattering the incident radiation. Zinc oxide and titanium dioxide are very effective sunblocks, reflecting up to 99% of the radiation both in the UV and visible regions, possibly into the IR region also. Although very effective, the sunblocks are messy and some people think they present an unacceptable appearance.

Sunscreens, on the other hand, absorb UV usually over a limited wavelength range in the UV-B and UV-A regions, and are considered to be drugs by the FDA.[(124)] The agents include para- aminobenzoic acid (PABA) and its esters, benzophenones, salicylates, cinnamates, and anthranilates.[(17)] Usually in a gel or cream, the preparations have good substantivity, allowing them to bind with the stratum corneum and resist removal by perspiration during heavy activity or by water when swimming. All provide protection in the UV-B range; only the benzophenones and anthranilates provide limited protection in the UV-A region. PABA preparations can discolor clothing and can cause contact-type, eczematous dermatitis. The body's natural photoprotector, melanin, absorbs strongly in the UV-A.[(125)]

Sunscreens are designed to protect against UV-induced erythema and sunburn, although their efficacy in reducing the risk of cancer[(126,127)] and immunologic effects[(34,128)] has been questioned. The sun protection factor is the measure of effectiveness of these agents; SPF equals the ratio of the UV-B dose required to produce erythema with protection to that required without protection. Although many commercial products claim very high SPFs, protection above an SPF of 15 has not been documented.[(129)] To achieve a product's SPF rating, it is important to apply the proper amount of sunscreen to the skin, a fact often overlooked in use. It has also been reported that sunscreens provide less protection in the outdoor environment than suggested by label values. This was attributed to the changing ratio of UV-B to UV-A with changes in the angle of the sun throughout the day.[(130)]

Quick-tanning preparations containing β-carotene or carthoxanthine produce a "tan" by coloring the skin but offer no protection against UV.[(17)] Hawk et al. tested 55 commercial products available in the United Kingdom in 300 human volunteers.[(131)] Sunscreens provided good protection against UV-B but only a few products were effective in the UV-A region, mostly at the short wavelengths. Sunblocks reflected the radiation throughout the UV region far into the visible range (up to 650 nm).

Since their effectiveness in preventing melanoma has not been established, other protective measures such as hats and protective clothing, and avoidance of exposure during peak hours (10 a.m.–2 p.m. sun time) should be used.[(132)]

Clothing. Factors affecting UV protection of fabrics include weave, color, stretch, weight, quality, and water content.[(133,134)] Evaluation of transmission in one study showed that UV leaks through openings in the mesh structure of woven fabrics. Fabrics made of continuous films exhibited good attenuation. Transmission of materials ranged from 1% for a FR-8 treated breezetone to 0.0003% for a number of fabrics shown in Table 21.6. Generally, laundering of fabrics increased the attenuation.[(134)]

Twenty commonly used textiles were evaluated at 313, 365, and 436 nm (blue light). The transmission of dark blue denim was 0.06% while women's beige tights transmitted 75% of the radiation.[135]

Researchers at the Australian Radiation Laboratory have tested the UV transmission between 280 and 400 nm of more than 2000 samples of fabrics and materials. From this research, UV protection factors have been derived (see Table 21.7).[133] The UV protection factor (UPF) is defined as the ratio of the effective UV dose for unprotected skin to the effective dose for protected skin. The range of UPF values for cotton and polyester-cotton fabrics was broad, while all samples of lycra exceed a UPF value of 40.[133]

Most polymer gloves used for chemical protection will provide some UV protection. The degree of UV reduction can be ascertained by field measurement of the incident and transmitted UV.

Eye Protection

Optical Density. Optical density (OD) is the quantity used to specify the ability of protective eyewear to attenuate optical radiation, where:

$$OD = \log_{10} (ML/EL) \quad (5)$$

Here, OD is the level of attenuation necessary to reduce the measured (or calculated) level (ML) to the exposure limit (EL). For optical radiation, ML and EL will usually be in terms of irradiance or radiant exposure, but both numerator and denominator must be in the same terms. OD values are shown in Table 21.8.

Table 21.6 — Selected Fabrics with 0.003% Transmission of Actinic UV

Welder's leather
FR–7A (R) treated sateen
FR–8 (R) treated twill
FR–8 (R) treated denim
1006 aluminized rayon
1019 aluminized rayon
Kevlar (R) blend - 22 oz
11501–1 unaluminized Perox (R)
1000–1 aluminized Perox (R)
Cotton interlock knit coated with nitrile rubber latex
Leather-stained sheepskin - lightweight
Leather-chrome tanned side split cowhide

Adapted from Reference 134.

Table 21.7 — UV Protection Factors (UPF) for Clothing

UPF	Absorption (%)
10	90
20	95
30	96.7
40	97.5
50	98
100	99
200	99.5

Adapted from Reference 133.

Table 21.8 — Optical Density (OD)

OD Value	Attenuation
1	10
2	100
3	1000
4	10,000
5	100,000
6	1,000,000
7	10,000,000
8	100,000,000

Field tests of eyewear can be made to determine OD. This requires measuring the level at the position of the worker's eye, with and without the eyewear covering the detector. If the measured quantity were irradiance at the eye, OD would be determined from

$$OD = \log_{10} (E_o/E) \quad (6)$$

Here, E_o is the measured effective irradiance (W/m^2) without eyewear, and E is the measured irradiance (W/m^2) with the eyewear in place.

Recommended levels of attenuation can be found in the technical literature. For example, Parrish et al.[136] state that UV-A-opaque eyewear for environmental use (solar UV) should transmit no more than 0.1% (OD = 3) to the central part of the pupil.

The quantity used to specify the attenuation of protective eyewear for welders is the shade number. Shade number (S#) is related to optical density by:

$$S\# = (7/3)OD + 1 \quad (7)$$

where a shade number of 10 equates to an OD of 3.857. It should be noted that OD in Equation 7 reflects broad-band attenuation of optical radiation, and is not specific to a particular spectral region. Generally, the UV attenuation of commercially available filter shades is effectively eliminated,[9,137] although UV entering from the rear of welding helmets may produce unacceptable exposures.[138,139]

Absorbing Material. Materials commonly used in the construction of protective eyewear are glass, polycarbonate, and CR-39 plastic. Glass and polycarbonate absorb actinic UV well, while glass, polycarbonate and CR-39 may not provide sufficient attenuation in the near UV at wavelengths above about 350 nm. Hence, always ensure that the spectral transmission of the protective eyewear is compatible with the spectral distribution and intensity of the source. This information should be available from equipment and eyewear manufacturers.

Limited test data of UV-protective eyewear are available in the literature[140] as well as information on prescription eyewear,[141-143] contact lenses,[144-146] children's sunglasses,[147] and sunglasses.[141,148,149] Some studies of sunglasses have reported finding UV transmission windows.[150,151] It has been suggested that if sunglasses with UV transmission windows reduce visible light levels to comfortable levels of illumination, the wearer may view potentially hazardous UV-A sources for relatively lengthy periods of time, which would increase the ocular dose of UV-A.[152]

Some sources of optical radiation may produce broad-band emissions requiring filtration in more than one spectral region. Under these conditions, protective eyewear must be capable of attenuating all potentially hazardous wavelengths to acceptable levels. This usually requires the use of a suitable base material plus coatings or lens additives. For example, some arc lamps may be rich in actinic UV and blue light. Polycarbonate could be used to absorb the actinic UV, while amber or yellow pigments could be added to absorb the blue light. The use of pigments may reduce the luminous transmittance of the eyewear, however, which is troublesome in some environments.

Other Factors. In addition to transmission data, the comfort, frame design, and fit are important factors. Leakage around frames may pose functional and measurement problems, because UV leakage as high as 25% has been found for poorly fitting eyewear.[136]

Laser Radiation*

Anticipation

Laser radiation is optical radiation that propagates in the form of a beam and has some special properties including low divergence, monochromaticity, and coherence. Optical radiation emitted from conventional sources usually is broad-band regarding the emitted wavelengths (polychromatic), and spreads out in all directions as it propagates away from the source. In contrast, most lasers emit light in a very narrow bandwidth, which is described as monochromatic (i.e., a single wavelength or color). Laser radiation also propagates in a highly directional manner and is characterized by a low level of divergence, or beam spread. Coherence means that the wavelengths of laser radiation are in phase both in space and time.

Temporal Characteristics

Lasers may operate continuously (called continuous wave or CW) or be pulsed (normal pulse, Q-switched, or mode-locked). CW lasers emit a "temporally constant power of laser light."[153] Pulsed lasers may emit single pulses or repetitive pulses.

For a frame of reference, consider that the innate human aversion response time to bright light, including visible laser light, is assumed to be within 0.25 sec. Note that the aversion response does not occur with exposure to invisible radiation such as UV and IR. With this in mind, if a laser emits radiation for a time greater than or equal to 0.25 sec, it is defined as a CW laser. It follows that pulses are noncontinuous emissions in which the duration of each pulse is shorter than 0.25 sec.

Pulse widths, however, may be much shorter than the aversion response time. These are realized with Q-switched or mode-locked pulses. Q-switching produces pulses on the order of a few nanosecond to microseconds, while mode-locked pulses are even shorter, in the picosecond domain. Special pulsing techniques can produce pulses on the order of femtoseconds.[153]

In some applications, such as information management with bar-code scanners, the laser beam may be scanned. In scanning, the output is spatially distributed (often in some form of a linear pattern) in a manner that makes it physically impossible for the entire output to enter the eye. This reduces the retinal irradiance compared to a raw (unscanned) beam.

Direct Beam vs. Reflection

Exposures may be direct (primary beam or intrabeam viewing) or indirect (scattered or diffusely reflected). Most intrabeam exposures will be to a small beam or point source of light, called small source viewing. Most laser beams fall into this category. The direct beam may also be an extended source—one that can be resolved into a geometrical image by the eye.[11]

Reflections may be specular or diffuse, or a combination of the two. Specular reflection, which may be regarded as a type of direct viewing, occurs when the beam is incident on a mirrorlike surface. This depends on the wavelength and the

* This section was written by R. Timothy Hitchcock and R. James Rockwell, Jr.

dimension of the surface irregularities of the reflector (i.e., it is a wavelength-dependent phenomenon). Specular reflections occur when the size of the surface irregularities of the reflecting surface are smaller than the incident wavelength.

In practice, most slightly rough, nonglossy surfaces act as diffusing surfaces to incident laser beams. A diffusing surface acts as a plane of very small scattering sites that reflect the beam in a radially symmetric manner. The roughness of the surface is such that the surface irregularities are larger than the laser wavelength, which randomly scatters the incident beam. The reflected radiant intensity depends on the incident intensity (I_o) and the cosine of the viewing angle (Θ) where:

$$I(r,\Theta) = I_o \cos \Theta \quad (8)$$

This relationship is known as Lambert's Cosine Law and a surface behaving in this manner is referred to as a Lambertian surface. This relationship defines an ideal plane diffuse reflector.

It should be stressed that "rough" surfaces do not always act as diffuse reflectors at all wavelengths. For example, brushed aluminum, which is partially diffuse for visible wavelengths, is a good specular reflector for far infrared wavelengths, such as those emitted by a CO_2 laser.

Also, beam reflections from so-called semispecular reflectors may include both a specular and a diffuse component. In this case, the reflection has features of a diffuse reflection, but in a preferred direction. Examples of semispecular surfaces include epoxy paint over plywood, aluminum painted plywood, and ice.(154)

Quantities and Units

Laser radiation is described by irradiance and radiant exposure. The unit of irradiance usually applied to exposure considerations is the watt per centimeter square (W/cm^2); the joule per centimeter square (J/cm^2) is used for radiant exposure. By convention, irradiance is usually applied to exposures to CW lasers, and radiant exposure is used for exposure to pulsed laser beams.

The spectral character of laser radiation is described by the wavelength. Although the nanometer is applicable to the output from UV, visible light, and most IR lasers, the unit used in American National Standard Z136.1 is the micrometer (µm).(155)

Biological and Health Effects

The target organs for laser radiation are the skin and eyes. Generally, due to the larger surface area of the skin, it may be at greater risk of exposure than the eye. However, the eye is more vulnerable to laser radiation.

Scientific understanding of the hazards of laser radiation to these target organs is derived primarily from studies of laboratory animals. These studies, and studies of broad-band optical radiation, have shown that exposure to laser radiation or radiation at laser wavelengths produces threshold effects, with the exception of some stochastic effects associated with UV radiation.

Laser-induced biological effects depend primarily on wavelength, irradiance, and exposure duration.(156-158) The penetration depth into the skin and eyes is wavelength dependent. In general, actinic UV wavelengths and IR-B and IR-C are absorbed topically, primarily by the epidermis or cornea, whereas visible and IR-A wavelengths are transmitted through the ocular media to the retina. IR-A wavelengths also penetrate relatively deeply into the skin.

Interactions with Matter and Damage Mechanisms. Damage mechanisms may be photomechanical, thermal, or photochemical. Photomechanical effects, sometimes called acoustic effects, occur when brief pulses are incident on tissues. Such pulses may be generated by Q-switched or mode-locked lasers. Since $P = Q/t$, where t is pulse duration, which is generally less than 100 ns for Q-switched lasers, exposure produces very high peak irradiance for even modest values of pulse energy. Such exposure has been shown to produce a plasma at the site of absorption within the eye. The plasma expands rapidly, often in the form of shock waves or sonic transients, disrupting tissues as it propagates. After a short distance, the pressure waves may terminate in cavitation, as the vapor bubble collapses.(159) Photomechanical effects can destroy tissues directly and may cause hemorrhaging, with blood collecting in the vitreous. A number of accidents involving photomechanical effects to the eye have occurred, and a relatively large number of these involve neodymium:YAG (Nd:YAG) lasers.(160-168)

Thermal effects affect the eye or skin, and occur in exposure times from microseconds to seconds, although microsecond pulses may produce both photomechanical and thermal effects. Thermal effects involve absorption of radiant energy by a chromophore (absorbing structure) such as melanin or hemoglobin.(156,157,169) Absorption increases the random molecular motion and, hence, the total energy of the tissues. This is manifest in the tissues as heat, which produces damage by denaturing protein and inactivating enzymes.[170] For threshold retinal thermal lesions, the damage is greatest at the center of the lesion, and diminishes toward the border of the lesion.[156,157] All types of laser radiation can produce thermal effects, although it may not be the predominate mechanism for tissue effects for lasers emitting UV radiation.

Photochemical effects to the eye and skin primarily involve UV radiation, while retinal effects also involve visible light up to 550 nm (blue and green laser light). Photochemical damage is attributed to the photoproducts of light-induced chemical reactions or changes. An example is the UV-activated dimerization of adjacent pyrimidine molecules in a deoxyribonucleic acid (DNA) molecule, which may interfere with DNA synthesis and replication. Cellular studies with a type of UV laser, the excimer laser, determined that a wavelength of 248 nm (KrF laser) was most effective in damaging DNA, producing mutagenic and cytotoxic effects. The relative effectiveness of DNA damage decreased at 308 nm (XeCl laser), and again at 193 nm (ArF laser).[171] Studies with test animals have established that photochemical effects occur from relatively lengthy (>10s of seconds) exposure of the eye (retina) to blue and green laser radiation.[156,172-175] Photochemical threshold lesions of the retina exhibit relatively uniform damage across the lesion, in contrast to thermal lesions,[156,157] as discussed earlier. Photochemical effects to the skin from exposure to blue and green laser radiation have not been reported.

Eye. The major ocular structures at risk are the cornea, iris, lens, retina, and optic nerve. The target tissue is determined by penetration depth, which is wavelength dependent. For example, overexposure to UV laser light can cause photokeratitis of corneal tissues, while exposure to laser wavelengths in the UV-B and UV-A spectral regions can contribute to cataracts.

The spectral region where the retina is at greatest risk is called the retinal hazard region, encompassing visible and IR-A wavelengths between 400 and 1400 nm. A number of studies have examined the issue of retinal sensitivity as a function of wavelength, finding that the retina is more sensitive to short visible wavelengths than to IR-A radiation.[156,172-174]

Early in their history, it was recognized that lasers had a great potential for causing retinal injury. The reason is that viewing a laser beam can produce small retinal images with intensities that are orders of magnitude greater than conventional light sources; in fact, they are brighter than the sun. An important consideration when viewing a laser beam is its size relative to the diameter of the pupil. Most laser beams are small sources, where the diameter of the beam is smaller than the pupil size.* In this case, the eye, like any optical system, has a limitation (called the diffraction limit) on the smallest image size it may resolve and focus. To determine the effect of a source imaged on the retina, it is necessary to know the retinal image size, which is a function of beam divergence and lens focal length. For a first approximation, one can show that a small laser beam with a divergence of 1 mrad will produce a diffraction-limited retinal spot diameter of approximately 17 μm. When a small (diffraction-limited) beam is incident on the retina, the optical gain (retinal irradiance/corneal irradiance) of the eye is on the order of 100,000 times, for a 7-mm pupil. Hence, a corneal irradiance of 1 W/cm^2 translates to a retinal irradiance of about 10^5 W/cm^2.

Viewing the diffuse reflection of a large laser beam or an array of laser diodes produces a larger retinal image size. This is referred to as extended (large) source viewing. This provides, at first consideration, some degree of protection since the retinal irradiance will be significantly lower due to the larger spot size. Some lasers are, however, of sufficient power (Class- 4, as discussed later) to be extended-source (diffuse-reflection) hazards. In this case, the degree of retinal damage would be significant because of the larger retinal spot sizes associated with a typical extended source viewing condition. Also, larger image sizes (typically 100 μm or greater) do not dissipate heat as rapidly as smaller image sizes. Consequently, the retinal irradiance that produces a minimal burn on the retina is about 10 to 100 times lower for large image sizes than for small (20 μm) image sizes. Hence, different exposure criteria are needed for the two exposure conditions.

Exposure of the retina may produce a spectrum of effects, including behavioral effects (aversion response), temporary impairment (glare, afterimage, flash blindness), reversible effects that temporarily impair vision (hemorrhaging or local swelling), and scarring that results in permanent visual impairment (scotoma or

* In laser safety, a pupil diameter of 7 mm (0.4 cm^2) is used as the so-called limiting aperture in the retinal hazard region. This value assumes a dark-adapted, dilated pupil.

blind spot). Retinal effects are most severe if the location of damage is along the visual axis of the eye, in the fovea centralis, the region responsible for visual acuity and color vision. Exposure in the peripheral retina, away from the visual axis, may result in little functional impairment.

A transition zone between retinal effects and effects on the anterior structures of the eye begins at the far end of the visible spectrum and extends into the IR-A region. In the IR-B spectral region, damage is observed to both the lens and the cornea. The ocular medium becomes opaque to radiation in the IR-C region because the absorption by water is high in this region. In the IR-C region, as in the UV-A and UV-B regions, the threshold for damage to the cornea is comparable to that of the skin. Damage to the cornea, however, is much more disabling and of much greater concern.

A number of specific references are available on wavelength-specific exposures,(156,172-177) beam-spot size,(178-180) pulsed exposures,(181-185) subnanosecond pulses,(186-189) and subthreshold effects.(190) Results have also been published from studies with many types of lasers including carbon dioxide,(181-183,191-193) Nd:YAG,(170,177,184,194) titanium:sapphire,(195) helium-neon,(170,194,196,197) diode,(198,199) argon,(170,175,194,197,200) and excimer.(201,202)

Skin. In general, the skin is considered the secondary target tissue, since the eye is more vulnerable to laser-induced damage. Laser radiation may be reflected from or transmitted into the skin. The structural inhomogeneities of the skin, skin structures such as hair follicles and glands, irregular shape, and layering can cause absorption and scattering of optical radiation in tissues.(203) As a result, there are multiple internal reflections in addition to absorption and transmission of the incident laser beam.

The magnitude of the absorption of laser light by tissues is strongly related to the wavelength of the laser and is expressed by the linear absorption coefficient. For the common laser sources in the 300 to 1000 nm range, almost 99% of the radiation penetrating the skin will be absorbed in at least the outer 4 mm of tissue. For example, the absorption coefficient at the argon-ion laser wavelength of 514 nm is 41 cm^{-1}. This means that 99% of the incident beam is absorbed in approximately 1.12 mm of tissue. By comparison, the beam from a Nd:YAG laser (1064 nm) has an absorption coefficient of 10 cm^{-1}, and hence the 99% absorption depth will be approximately 4.6 mm. The 99% absorption depths for most other common lasers occur in tissue thicknesses less than 4 mm.

UV laser radiation predominately produces photochemical effects to the skin,(204-206) and these are the same as those discussed previously for broad-band UV sources. Also, there is the potential for photosensitization from a combined exposure to a chemical agent and UV laser radiation.

For wavelengths greater than 400 nm, the reaction of the skin to absorbed optical radiation is essentially that of a thermal effect. This may involve the sensation of warmth or pain, or result in whitening, or a burn (nonspecific coagulation necrosis).(9,207-209) Pulsed lasers may produce a "pinprick" sensation.(209) The basic mechanisms of thermally induced tissue destruction result from denaturation of cell protein, interference with basic cell metabolism, and secondary effects such as interference with vascular blood supply. Healing of laser-induced skin lesions is similar to any localized thermal wound, and should be medically treated in a similar fashion.

Thermal reaction of absorbed laser radiation in tissues depends strongly on wavelength, exposure duration, and to some degree, the location of the exposure. Research studies have shown that skin can withstand brief temperature rises, even approaching 100°C for very short exposure times. The response appears to be logarithmic as the exposure times become shorter.

For example, a 21°C rise above body temperature (37°C) to 58°C produces cell destruction for exposure times longer than 10 seconds. Tissues, however, can withstand a temperature rise up to 70°C if the exposure time is less than 1 second.

Nonbeam Hazards. The hazards discussed so far deal with human access to the beam. Other potential hazards, however, are associated with lasers where human exposure is not to the beam, but to other hazardous agents generated as a result of the laser application. These nonbeam hazards include electricity, airborne contaminants, plasma radiation, fires, and explosions.

Electrical hazards, shock and electrocution, are primary hazards of laser systems.(210,211) Most deaths attributed to laser systems have been electrical fatalities.(167)

Airborne contaminants(212) may be generated when a powerful laser beam interacts with matter. Studies have examined this issue for metals,(213-215) concrete,(213) plastics,(216-220) composites,(221) glasses,(222,223) wood,(224) and tissues.(225-232) Airborne contaminants may also occur from the normal operation of gas lasers,(233-235) whereas the skin, respiratory system, and other organs or systems may be targeted during preparation or use of organic dye lasers that involve handling of organic dyes and solvents.[236-238]

Lasers may generate other wavelengths by the normal operation of the laser system. This is called collateral radiation, and includes a plethora of nonlaser wavelengths. For example, some laser power supplies can produce X-rays,[239] and X-rays may be used to preionize gases in some excimer lasers. X-rays are usually shielded by the laser enclosure, but they may be of concern during service activities when the housing is removed and the laser energized. Collateral radiation may also include radio-frequency radiation,[240] and power-frequency fields.[241]

Plasma radiation occurs at the site of interaction of an energetic laser beam with a metal. A few studies of plasma radiation have demonstrated that the interaction of a focused beam from a CO_2 laser (10.6 µm) with various metals produces potentially hazardous levels of actinic UV and blue light near the site of interaction.[242,243] X-rays may be generated when high-energy, rapidly pulsed laser beams interact with plasmas in laboratory experiments using lasers.[244-246]

The optical power contained in a Class 4 laser beam (Table 21.9) is a potential fire hazard if it interacts with combustible materials, such as clothing[266] or surgical materials,[248,249] or other target materials. Hence, this may be an issue in the selection of shielding materials. Also, components of lasers[239] and systems are potential explosions hazards,[250] such as some pumping lamps and dirty optical components.

Accidents. Relative to the tremendous growth in laser applications and uses, relatively few accidents have occurred, even considering that accidents are probably highly underreported.[251,252] Lasers cited in specific accident reports include ruby,[163,253-256] Nd:YAG,[161-167,257,258] doubled-YAG,[256] CO_2,[259] argon,[161-163,260] krypton,[161] dye lasers,[161-163,256] HeCd,[163] and HeNe.[261]

Accidents have included exposure to the direct beam,[162,163,165,252,256] reflected beam,[162,163,166,168,252,255,256,261] and a number of nonbeam accidents/incidents including fires,[166,251,259,262-264] and electrocution and shock.[9,167,262] Accident and incident reports may be grouped into research and development,[162,165,254-256] education,[253] medical,[166,263-265] industrial,[161,259] military,[258] and entertainment.[260]

Reviews of accidents and accident data are available.[161,163,164,168,252,258,262,265,266] Summary findings in these reviews include (1) Q-switched Nd:YAG lasers have been involved in a large number of the reported ocular accidents; (2) most skin injuries and fires are associated with CO_2 lasers; (3) the only deaths attributed to laser systems were from electrocution and embolism (following medical procedures); (4) performing beam alignment is a relatively high-risk task; and (5) a relatively high number of ocular accidents have occurred when eyewear was available but not worn.

Classification

The United States legally requires that lasers and products containing lasers be certified to the Federal Laser Product Performance Standard, in 21 CFR Subchapter J, Part 1040.10.[267] The first step in certification is classification, where laser products are classified according to their output power and potential hazard during

Table 21.9 — Federal Laser Product Classification Scheme

Laser Class[A]	Allowable Power (watts)[B]	Emission Duration (sec)	Relative Output Power	Relative Hazard
1[C]	0.39×10^{-6} [D]	> 10,000	extremely low	none known
2a[E]	0.39×10^{-6} [D]	> 1000	very, very low	very low
2[E]	≤ 0.001	> 0.25	very low	low
3a[E]	≤ 0.005	> 0.00038	low	low-moderate
3b[C]	≤ 0.5	> 0.25	moderate	moderate-high
4[C]	> 0.5	——	high	high

[A]Currently, the U.S. Food and Drug Administration uses Roman numerals in its classification scheme. However, the FDA plans to amend its rules to use Arabic numbers as do ANSI and the International Electrotechnical Commission.
[B]The FDA calls the allowable power for a given exposure duration the accessible emission limit (AEL). AELs are used for classification of laser products, and are not exposure limits. For example, a laser product would be Class 3a if the AEL is >0.001 W, but ≤ 0.005 W. In general, the AELs in this table have been determined for the visible spectral region.
[C]This class applies to all spectral regions.
[D]This power level is for visible wavelengths.
[E]This class applies just to visible wavelengths in the FDA regulations. Note that the FDA plans to amend the rules to delete Class 2a and to allow the inclusion of low-power nonvisible lasers in Class 3a.

normal (intended) use. This is a numerical system with four broad classes (1 to 4), as shown in Table 21.9.

Class 1 lasers cannot emit laser radiation at known hazard levels. Users of Class 1 laser products are generally exempt from radiation hazard controls during operation. It is important to remember that a laser product with a classification indicating a relatively low hazard level, such as Class 1, may contain a higher class (power) laser embedded within its protective housing. When this occurs, the laser radiation emitted from the product will not be a hazard for its intended use, since the system is Class 1, but may be a significant hazard if the product is modified or when the device is serviced.

Class 2a is a special designation based on a 1000-second (16.7min) exposure and applies only to lasers that are not intended for viewing such as supermarket laser scanners. The emission from a Class 2a laser is defined such that the emission does not exceed the Class 1 limit for an emission duration of 1000 seconds (see Note E, Table 21.9).

Class 2 lasers are low-power devices that emit visible radiation above Class 1 levels but do not exceed 1 mW. Since the laser light is very bright, the human aversion response (blink reflex) to bright light controls exposure. Hence, Class 2 lasers are a potential hazard to the eye if the exposure exceeds 0.25 seconds. Consider that, if a Class 2 laser beam is transmitted through a 7-mm aperture (the size of a fully dilated pupil), an irradiance of 2.5 mW/cm^2 (1 mW ÷ 0.4 cm^2) results. An irradiance of 2.5 mW/cm^2 is therefore incapable of damaging the retina for exposure durations less than or equal to the human aversion response time. Only limited controls are specified for this class.

Class 3a lasers have intermediate power levels (CW = 1–5 mW). Class 3a emissions may be "low" or "high" irradiance. If the beam size is relatively large and the irradiance is less than or equal to 2.5 mW/cm^2 (low irradiance), directly viewing the beam is the same as viewing a Class 2 laser beam, and the aversion response limits the exposure. However, this is not the case if a low-irradiance beam is viewed with collecting optics that concentrate the beam, such as binoculars or telescopes. The second case, high irradiance, occurs if the beam size is small enough that the irradiance exceeds 2.5 mW/cm^2. Here, there is an increased probability of retinal damage if the beam is viewed directly, even for times less than 0.25 seconds, (i.e., the aversion response may not be protective). Some limited controls are usually recommended for Class 3a lasers, especially those in the high-irradiance category.

Class 3b includes moderately powerful lasers (e.g., CW = 5–500 mW). In general, Class 3b lasers are not a fire hazard, nor are they generally capable of producing a hazardous diffuse reflection except for conditions of staring at distances close to the diffuser. Specific control measures are recommended.

Class 4 lasers are high-power devices (e.g., CW >500 mW). These are hazardous to view under any condition (direct or scattered radiation) and are potential fire and skin hazards. Significant control measures are required in Class 4 laser facilities.

Generally, the laser classification scheme is consistent universally, but there are some differences among the schemes used by organizations such as the Center for Devices and Radiological Health (CDRH) of the U.S. Food and Drug Administration, ANSI, and the International Electrotechnical Commission (IEC). The FDA reserves Classes 2a, 2, and 3a for lasers with visible emissions, for example, whereas ANSI allows lasers with invisible emissions to be Class 3a (see Note E, Table 21.9). This chapter uses the FDA classification scheme.[267]

Recognition

Generation

Lasers have three major components: optical cavity, energy source, and lasing medium. The optical cavity is terminated at either end by a mirror—one highly reflective and one partially reflective. The partially reflective mirror, called the output coupler mirror, forms the aperture where the laser radiation is emitted from the optical cavity. The energy source may be electrical, chemical, or electromagnetic. Its purpose is to energize the lasing medium, which is contained within the optical cavity. The lasing medium may be gas, liquid, solid, or semiconductor.

Energy absorption excites electrons in the lasing medium, which de-excite with the liberation of photons. Commonly, excited electrons emit absorbed energy by returning to the ground state, in a process called spontaneous emission. In the generation of laser radiation, however, a large number of electrons reside at an energy level higher than normal, a metastable state called a population inversion. Photons propagating down the axis of the optical cavity interact with these electrons, stimulating them to emit photons of the same wavelength and phase, and move in the same direction. Thus, stimulated emission causes an immense amplification in the number of photons propagating between the mirrors, and these photons are emitted as a beam of coherent radiation. Hence, the process by which laser radiation is generated is "light amplification by the stimulated emission of radiation: laser."

Sources

Most lasers derive their name from the lasing medium. Some common names are listed in Table 21.10. Some lasers are named for the process that generates the laser radiation, such as excimer lasers, where the laser light is generated by a chemical reaction involving a small polymer, or dimer.

A useful feature of some lasers is tunability, where the wavelength can be adjusted within a given spectral region. The first type of tunable laser was the dye laser. The lasing medium is a powder dissolved in an organic solvent, which is then pumped into the optical cavity. Vibronic lasers and color-center lasers are examples of tunable solid-state lasers.

Laser applications have grown tremendously since 1960 when the first functional laser was developed, and are integrated into many areas of modern life. The most notable areas are medicine,(268-274) dentistry,(275-277) science,(278-284) industry,(285-294) communications,(295-298) construction,(284,299) commerce,(300,301) education,(302-304) entertainment,(305) criminal justice,(306,307) and military applications.(308-310)

In industry, one of the major applications of laser radiation is material processing, which uses a powerful laser beam to weld, cut, drill, mark, surface treat, or remove material from a variety of substances such as plastics, metals, ceramics, glasses, semiconductors, wood, and cloth. The major types of lasers used for material processing include CO_2, Nd:YAG, and excimer, along with Nd:glass, ruby, and chemical oxygen-iodine-lasers (COIL).

Some other uses are semiconductor diode lasers in optical fiber communication systems, and krypton and argon lasers in entertainment. HeNe and diode lasers are used to scan information contained in universal product codes in industry and commerce, and for alignment and positioning in industry, construction, and agriculture. Excimer, argon, krypton, HeNe, Nd:YAG (fundamental frequency and harmonics), and CO_2 lasers are used in the health care industry.

Table 21.10 — Some Common Lasers and Their Wavelengths[A]

Lasing Medium	*Abbreviation*	*Type*	*Wavelength (nm)*	*Spectral Region*
Excimer[B]		gas		
Argon Fluoride	ArF		193	UV-C
Krypton chloride	KCL		222	UV-C
Krypton fluoride	KrF		248	UV-C
Xenon chloride	XeCl		308	UV-B
Xenon fluoride	XeFl		351	UV-A
Nitrogen	N	gas	337	UV-A
Helium-cadmium	HeCd	gas	325	UV-A
			441.6	blue
Argon	Ar	gas	351	UV-A
			488	blue
			514.5	green
Krypton	Kr	gas	568	yellow
			647	red
Rhodamine 6G[C]	—	dye	570-650	yellow-red
Helium-neon	HeNe	gas	543	green
			632.8	red
			1152.6	IR-A
			3392	IR-C
Ruby	—	solid	694.3	red
Titanium:sapphire[D]	Ti:sapphire	solid	660–1180	red - IR-A
Alexandrite[D]	—	solid	700–800	red - IR-A
Gallium aluminum arsenide	GaAlAs	diode	670–830	red - IR-A
Gallium arsenide	GaAs	diode	850	IR-A
Neodymium:YAG	Nd:YAG	solid	1064	IR-A
second harmonic[E]			532	green
third harmonic			355	UV-A
Chemical oxygen iodine laser	COIL	chemical	1315	IR-A
Holium:YAG	Ho:YAG	solid	2100	IR-B
Hydrogen fluoride	HF	chemical	2.6–3.6	IR-B–IR-C
Carbon monoxide	CO	gas	5500	IR-C
Carbon dioxide	CO_2	gas	10,600	IR-C

[A]Although X-ray lasers are being developed, the majority of lasers in use today operate at wavelengths between 180 nm and 1 mm.
[B]Excimer = excited dimer
[C]Tunable organic dye laser also known as Rhodamine 590.
[D]Tunable solid state lasers, also called vibronic lasers.
[E]The second harmonic is sometimes called a frequency-doubled YAG, or a KTP laser. KTP (potassium titanyl phosphate) is a crystalline material used to change the wavelength from 1064 nm to 532 nm.

Evaluation

Exposure Guidelines and Standards

Exposure Guidelines. Guidelines used in the United States have been published by ANSI,(155,311,312) ACGIH,(101,313) ICNIRP,(314-316) and IEC.(317,318) In addition to these general guidelines, there are also guidelines for specific environments including health care[319,320] and optical fiber communications.[321,322]

The guidelines include exposure limits for the eye and the skin. In general, the limits for the eye are more conservative. The limits are wavelength dependent over a range between 180 nm and 1 mm.

The exposure limit in ANSI Z136.1–1993 is called the maximum permissible exposure (MPE). It should be emphasized that viewing a beam at MPE levels would be annoying and/or uncomfortable. It would not be hazardous, however, because MPEs are below known hazard levels. As with all laser exposure limits, MPEs were established in studies using animals. In the assessment, the data were often displayed using a regression line on probability graph, as shown in Figure 21.5. The data as collected are expressed in terms of probability. MPEs are based on the MRD_{50}, or laser dose level at which effects are seen in one-half of the exposures, with a tenfold safety factor.

As noted earlier, the exposure limits depend on wavelength and exposure duration. Since laser beams should not be purposely viewed except when viewing diffuse reflections where exposure is less than the MPE, exposure duration usually represents the anticipated time of an accident. Because actual exposure conditions may be difficult to determine, standard recommendations for exposure times are often used.[155] These are compiled in Table 21.11.

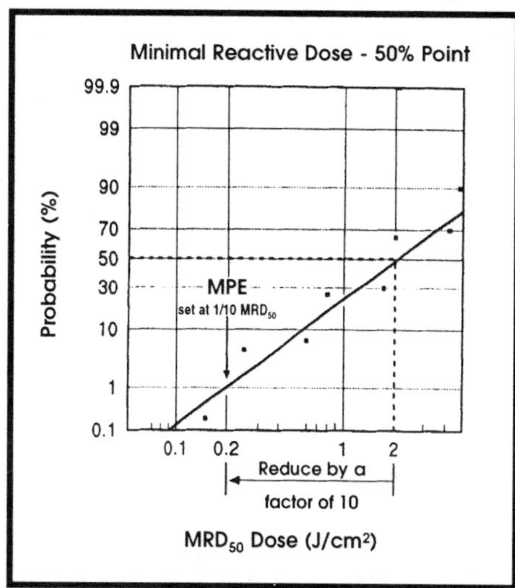

Figure 21.5 — Typical regression plot converting MRD_{50} data into MPE data. Note that the MPE is set a factor of 10 below the 50% probability point for ocular damage.

Table 21.11 — Suggested Viewing Times

Application	*Suggested Exposure Duration*
All wavelengths, single-pulse	Pulse width
Visible beams, unintended viewing, aversion response time	0.25 sec
Infrared viewing time	10 sec
Long-term viewing of a diffuse reflection	600 sec
Occupational exposure (entire workday)	30,000 sec
Cumulative daily exposure time (worst-case)	24 hr

The guidelines include a number of wavelength-dependent correction factors and parameters. The factors and their biological basis are compiled in Table 21.12. In use, when the exposure limit is determined, it will include reference to any correction factors that must be used.

As an example of determination of an exposure limit, one of the most commonly used limits is for lasers with visible emissions where the duration of exposure is the human aversion response time (0.25 sec). This is:

$$1.8\ t^{3/4} \times 10^{-3}\ J/cm^2 \qquad (9)$$

Note that there are no applicable correction factors. For 0.25 seconds, this equals a radiant exposure of 6.4×10^{-4} J/cm². If the potential exposure is to a CW laser, this limit is divided by the exposure duration, yielding an exposure limit as a value of irradiance of 2.5 mW/cm².

If the exposure is to a near-IR laser beam, the exposure limit is:

$$1.8\ C_A\ t^{3/4} \times 10^{-3}\ J/cm^2 \qquad (10)$$

where C_A is a frequency-dependent correction factor, defined in Table 21.12. At a wavelength of 1000 nm, C_A is equal to 4, and the MPE for a 10-sec exposure (IR wavelengths) is 0.04 J/cm² for a pulsed emission and 0.004 W/cm² for a CW laser.

Laser safety practitioners are often puzzled over the use of correction factor C_E. Before using this correction factor, one must determine whether the exposure scenario will result in viewing the laser beam as a small source or as an extended source. This is simplified somewhat if the evaluator imagines that when viewing a large object at close distance, an extended image forms on the retina. Here, the viewing angle (α) is determined by D_L/r, where D_L is the diameter of the laser beam and r is the viewing distance. Now, as the observer moves away from the source, the viewing angle changes, resulting in a smaller image on the retina. Ultimately, the viewing angle is sufficiently small that small-source viewing dominates, and a minimal image is formed on the retina. It follows that there must be some distance from the source where there is a "crossover" between the two types of viewing. The maximum distance at which one still sees an extended object is called r_{max}. The viewing angle at r_{max} is called α_{min}, or the limiting angular subtense; α_{min} is the apparent visual angle that divides small-source viewing from extended source viewing, where:

$$\alpha_{min} = D_L / r_{max} \quad (11)$$

The minimum value for the retinal image size for pulsed sources has been determined to be about 24 µm, which corresponds to a visual angle of 1.5 mrad. This factor and biological studies of eye movements have shown that a small source will not remain in one spot of the retina for long time periods unless the eyes have been anesthetized. When exposure times are greater than about 10 sec, these eye movements will cause the image size to cover a much larger area, about 11 mrad.

There are differences in the retinal heat flow (e.g., the ability of the tissues to dissipate heat) associated with small-source and extended source viewing. This has necessitated two exposure limits. The exposure limit for small source (intrabeam) viewing is used at distances beyond r_{max}. The exposure limit for extended-source viewing is the product of the exposure limit for small-source viewing and the correction factor C_E, where the viewing distance does not exceed r_{max}.

In addition to the Federal Laser Product Performance Standard, the Federal Aviation Administration (FAA) has published procedures for handling airspace matters associated with outdoor laser light shows (FAA 7400.2D Chapter 34). In federal workplace safety standards, the Occupational Safety and Health Administration (OSHA) has addressed the use of lasers in the construction industry standard. This includes the general use of lasers (29 CFR 1926.54) and eye and face protection (29 CFR 1926.102). Other applicable OSHA standards have been reviewed.(323) Also, OSHA has made a number of written interpretations concerning lasers and the construction industry standard. These are available from the OSHA homepage (http://www.osha.gov).

Some states have requirements for lasers.(324,325) In general, these lasers must be registered with the state agency, so they should be inventoried and tracked to ensure compliance. State agencies may also request information concerning the Laser Safety Officer (LSO), and require a minimum level of competence for them. The ANSI standard provides guidance on training of LSOs.(155)

Table 21.12 — Correction Factors Defined in Exposure Guidelines

Correction Factors and Parameters	*Applicable Wavelengths (nm)*	*Biological Basis*
C_A	700–1400	Retinal absorption by melanin where the IR limits are increased because of higher retinal absorption of visible wavelengths, i.e. retinal absorption is lower in the IR-A
C_B	400–700	Blue-light and photochemical effects for long-term exposures where the limits for orange and red wavelengths are increased because of the decreased likelihood of photochemical effects
T_1	550–700	Time to replacement of exposure limits based on thermal effects with limits based on photochemical effects, i.e., C_B is used when the duration $\geq T_1$
C_C	1050–1400	Pre-retinal absorption of IR wavelengths where the limits are increased for IR-A wavelengths between 1150 and 1400 nm because less of this radiation reaches the retina
C_E	400–1400	Correction for extended source laser beams where the limits for small source laser beams are increased because exposure is to a large beam that forms an image on the retina
C_P	400–1,000,000	Correction for repetitive pulses which decreases the limit for repetitive pulses because of a greater potential for injury (repetitively pulsed lasers have lower damage threshold levels than CW radiation of comparable average power)
2.5	280–400	Reduction factor for UV exposure limits for exposure on consecutive days, where anticipated exposure approaches the exposure limit.

Information adapted from ANSI Z136.1–1993, "Safe Use of Lasers".

Exposure Assessment

In contrast to the classical occupational hygiene approach where the agent under investigation is measured, field measurement of laser radiation is usually not necessary because of the complexity of making meaningful measurements. For example, measurement requires alignment between the laser beam and a small aperture that covers the detector (e.g., a 7-mm aperture is used when measuring laser radiation within the retinal hazard region). As mentioned earlier, alignment is a process that involves an increased risk of accidental contact with the beam.

Instead of measurement, hazard evaluation uses numerical methods to analyze potential laser hazards. ANSI Z136.1 provides guidance on hazard evaluation, including a number of solved problems in

Appendix B of the standard.[155] Also, a number of useful reviews of laser hazard analytical techniques are available.[326-337]

As a first step in this process, the evaluator should perform a survey so the numerical analysis can be meaningfully related to the workplace. This should include an assessment of possible nonbeam hazards, as well as an appraisal of beam hazards. At this time, some of the operational parameters required for the numerical analysis (see Table 21.13) may be obtained or reviewed. Sources of this information include labels on the laser product, specifications in operations and service manuals, and the manufacturer.

The next step in the hazard evaluation is determining the exposure limit, which depends on wavelength and exposure duration. The exposure limit is then used to determine the nominal hazard zone (NHZ), the fundamental tool of laser hazard evaluation where there is access to an open beam. Conceptually, the NHZ is the zone around the laser where the beam intensity exceeds the exposure limit, as demonstrated in Figure 21.6. As a laser beam propagates in space, it spreads out due to divergence, and it may undergo atmospheric attenuation if used outdoors. As the beam diverges, the irradiance decreases until, ultimately, it is reduced to a value equal to the exposure limit. Since this occurs at the boundary of the NHZ, persons outside the NHZ boundary are exposed below the exposure limit and considered to be in a "safe" location.

Table 21.13 — Important Laser Operational Parameters

- Wavelength
- Exposure duration
- Optical power
- Beam divergence
- Exit beam diameter
- Pulsed laser
  - energy per pulse
  - pulse repetition frequency
  - pulse width
  - average power
- Laser with lens
  - focal length
  - size of beam incident on lens
- Fiber optic output
  - mode of fiber
  - numerical aperture of fiber (multimode fiber)
  - mode field diameter (singlemode fiber)

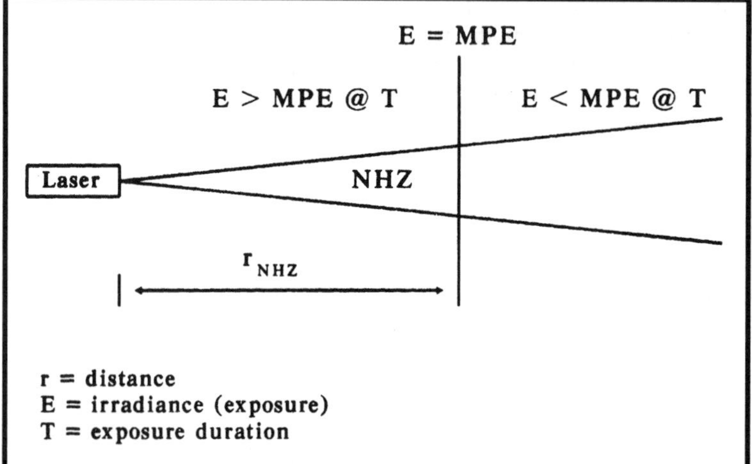

Figure 21.6 — The nominal hazard zone (NHZ) indicates the area or volume in space where there is the potential for overexposure.

Nominal Hazard Zones. The purpose of an NHZ evaluation is to define that region where control measures are required.[338] Also, knowledge of the NHZ is useful in preinstallation planning meetings and laser safety training sessions. ANSI Z136.1 requires that the LSO determine the NHZ for Class 3b and 4 lasers with limited open-beam paths and open-beam paths,[155] but it may also be useful to do so for certain high-irradiance Class 3a lasers.

Four NHZ models have been suggested in ANSI Z136.1: intrabeam, lens-on-laser, fiber optics, and diffuse reflection.[155] These are compiled in Table 21.14. When these equations are used, it is important that the units are consistent, since this is a common source of error. In this regard, note that beam divergence is usually given in terms of milliradians, but all calculations must be performed in radians.

For a laser with constant output characteristics, the largest (worst-case) NHZ will occur for intrabeam (small-source) viewing (see Figure 21.7). In this case, the irradiance is limited only by beam divergence, unless used outdoors at distances greater than 300 m,[9] where atmospheric absorption may become significant. In general, the intrabeam NHZ may extend to relatively large distances.

Most occupational uses of lasers incorporate a lens as the final component in the beam path. Typically, a converging lens focuses the beam with the effect of increasing the irradiance, which reaches a maximum in the focal plane of the lens. However, this also causes the beam to spread with an angle usually many times larger than the inherent laser beam divergence in the space beyond the focal plane. Subsequently, the irradiance decreases at relatively short distances from the focal plane, which reduces the extent of the NHZ in comparison to the intrabeam NHZ.

Generally, laser radiation spreads out rapidly when emitted from the end of an optical fiber, although beamlike qualities

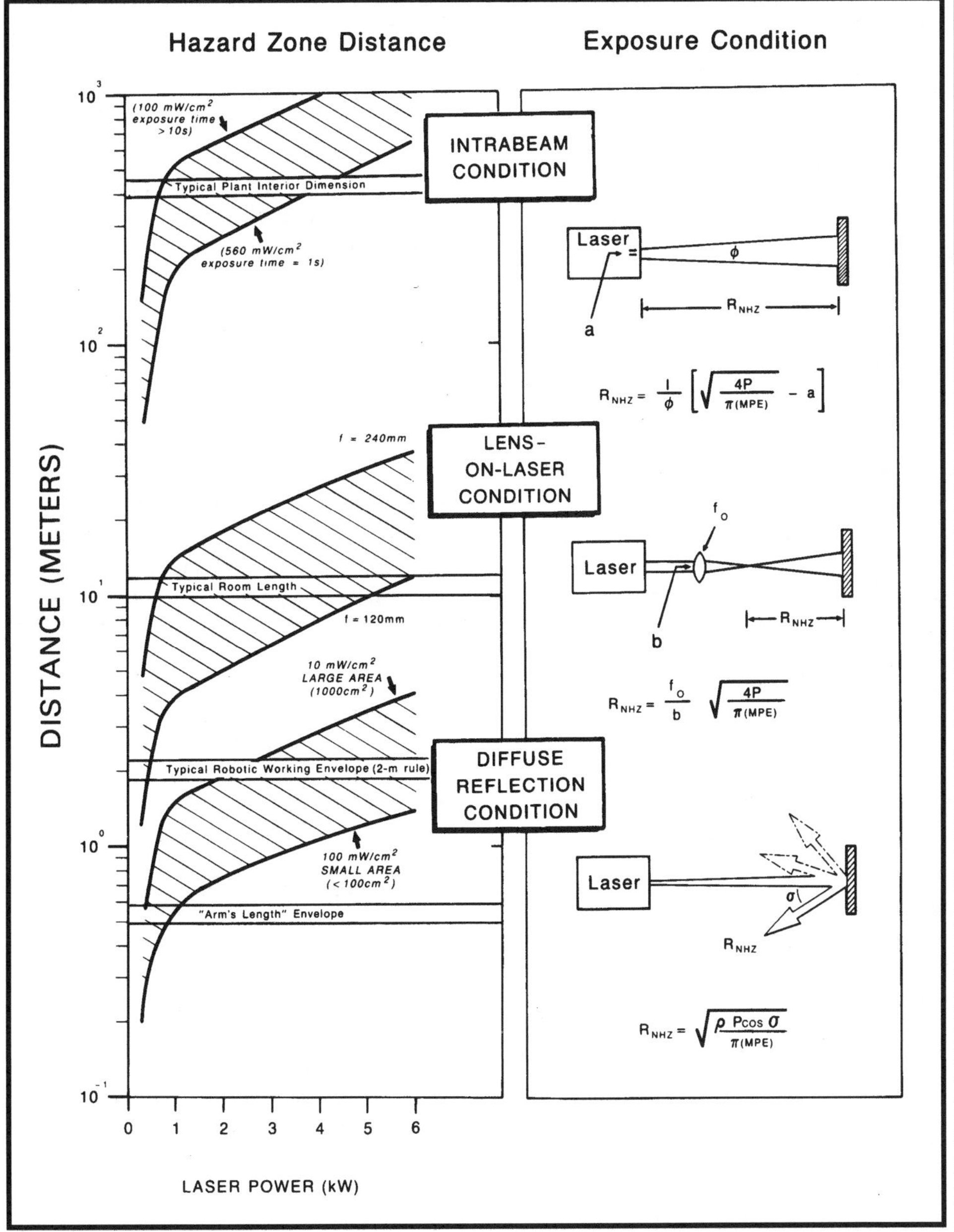

Figure 21.7 — Comparison of the extent of three nominal hazard zone models (intrabeam, lens-on-laser, and diffuse reflection) for a high power carbon dioxide laser in an industrial setting. (Note that the equation for the intrabeam NHZ has been reformulated as shown in Table 21.12.)

have been reported from fractured fibers used in the operating room.[340] Hence, the optical fiber is a beam-expanding element that is optically equivalent to a short-focus lens placed in the beam path. The effect of the optical fiber is that it shrinks the range of the NHZ, again, compared to the intrabeam NHZ.

The fourth model is for diffuse reflection of the beam. The reflected irradiance or radiant exposure from a Lambertian surface at some distant point is inversely related to the square of the distance (r) from the surface. This describes diffuse reflections from a point source. For a CW laser, this is:

$$E = \frac{\rho\, P\, \cos\theta}{\pi\, r^2} \quad (12)$$

When Equation 12 is rearranged to solve for distance and the exposure limit is substituted for irradiance (or radiant exposure), the equation for diffuse reflection in Table 21.14 results. In general, worst-case conditions are often used where the viewing angle is 0° (where cos Θ = 1) and the reflectivity ρ is 100%. As might be expected, diffuse reflection reduces the NHZ to relatively short distances because the beam is randomly scattered.

Table 21.14 — Nominal Hazard Zone Equations[A,B]

Equation	Application
$r = \frac{1}{\Phi}\left[\left(\frac{4p}{\Pi\ MPE}\right) - a^2\right]^{1/2}$	Viewing the direct beam (intrabeam)[C]
$r = \left(\frac{f_o}{b_o}\right)\left(\frac{4p}{\Pi\ MPE}\right)^{1/2}$	Lens on laser
$r = \left(\frac{1.7}{NA}\right)\left(\frac{P}{\Pi\ MPE}\right)^{1/2}$	Multimode optical fiber
$r = \left(\frac{\omega_o}{\lambda}\right)\left(\frac{\Pi P}{2\ MPE}\right)^{1/2}$	Singlemode optical fiber
$r = \left(\frac{\rho P \cos\theta}{\Pi\ MPE}\right)^{1/2}$	Diffuse reflection

[A]The ANSI Z136.1 exposure limit is the maximum permissible exposure (MPE).
[B]NHZs may be determined for the eyes or skin, since exposure limits exist for both target tissues. The NHZ for the eyes is called the nominal ocular hazard distance (NOHD).
[C]This equation was reformulated in ANSI Z136.1–1993 from

$$r = \frac{1}{\Phi}\left[\left(\frac{4p}{\Pi\ MPE}\right)^{1/2} - a\right]$$ (See Reference # 339.)

Table 21.15 — Nominal Hazard Zones for Select Lasers

		NHZ (m)		
Laser Type	Exposure Duration	Direct Beam	Lens-on-Laser	Diffuse Reflection
Argon[A]	0.25 s	1000	—	0.5
	8 hr	50,500	—	25.2
Nd:YAG[B]	10 s	2680	134	1.3
	8 hr	7280	364	3.6
CO_2[C]	10 s & 8 hr	564	5.6	0.6

[A]λ = 488 nm, CW, Φ = 20 W, a = 1 mm, ϕ = 1 mrad, Θ = 0°, ρ = 100%, MPE (0.25 s) = 0.0025 W/cm², MPE (8 hr) = 1×10^{-6} W/cm²

[B]λ = 1064 nm, pulsed, Φ = 5 W, Q = 50 mJ, a = 1 mm, ϕ = 1 mrad, Θ = 0°, ρ = 100%, F = 100 Hz, t = 30 ns (Q-switched), b_o = 5 mm, f_o = 25 cm, MPE (10 s) = 8.89×10^{-5} W/cm², MPE (8 h) = 1.2×10^{-5} W/cm²

[C]λ = 10.6 µm, CW, Φ = 1 kW, a = 20 mm, ϕ = 2 mrad, b_o = 2.54 cm, f_o = 12.7 cm, Θ = 0°, ρ = 100%, MPE (10 s & 8 hr) = 0.1 W/cm²

It is important to remember that since exposure limits are time dependent, it is possible to calculate more than one hazard zone (distance) for a given laser. Table 21.15 includes estimates of the NHZ for some typical laser conditions.

Control Measures

Control Measures are required if there is potential for human access at levels above the exposure limits. The priority should be to use engineering controls over administrative and procedural controls. In general, control measures are necessary for normal operation and maintenance of Class 3b and 4 lasers that have limited open-beam paths and open-beam paths. Accordingly, it is a good practice to use engineering controls, such as beam enclosures, whenever possible, to minimize the need for administrative and procedural control measures. Obviously, if the beam path is totally enclosed, then the potential for hazardous exposure may exist only during service, if there is an embedded Class 3b or 4 laser. If open-beam systems are necessary, the ANSI Z136.1 standard is a useful reference since it lists control measures as a function of the laser class.[155]

Laser Safety Officer (LSO)

The LSO, as described in the ANSI standard, is one who administers the overall laser safety program. Importantly, this standard defines the LSO as one with "the authority and responsibility to monitor and enforce the control of laser hazards and effect the knowledgeable evaluation and control of laser hazards."[155] Also, the LSO should receive training commensurate with the level of hazard posed by the lasers in use.

Engineering Controls

It is sometimes necessary to view the beam in industrial applications. This can be done using special window filter materials or remotely with closed-circuit TV or video cameras.[341] For benchtop work, beams should not be delivered at eye level, for standing or seated individuals, nor should they be directed at doorways or windows. Windows should be made of laser absorptive materials when the NHZ extends beyond them. An attempt should be made to preclude the use of shiny or glossy materials that may produce a specular reflection. Unauthorized personnel should be excluded from the area of the beam by walls, curtains, or even rope barriers.[342,343] In industrial flowthrough systems, the equipment may be designed so that the parts become part of the protective housing of the laser tool.[344]

Laser Controlled Area. When the entire beam path from a Class 3b or 4 laser is not sufficiently enclosed and there is the potential for overexposure, a laser-controlled area is required. Class 3b and 4 laser controlled areas must be posted with a warning sign, operated by qualified and authorized personnel, and transmission outdoors separated from indoor-controlled areas. A number of other requirements apply to Class 4 laser controlled areas.[155]

An important feature of a Class 4 laser-controlled area is the entryway control. The ANSI standard allows the LSO to select from three control options: nondefeatable, defeatable, and procedural entryway controls. Nondefeatable entryway controls use nondefeatable controls, such as a magnetic interlock, to switch the laser off or enable a shutter in the beam path, when the entryway door is opened with the laser energized. This is the simplest method, but is often unworkable since individuals who are approved to work with the laser must have access to the area while the laser is working. The other two methods allow access to an area where there is an open beam, which requires authorization, training, additional engineering controls, and the proper use of personal protective equipment at the entrance. Defeatable entryway controls have an override feature that allows an entryway interlock to be temporarily bypassed if it is evident that there is no hazard at the point of entry. Procedural entryway controls use additional engineering controls such as a light baffle (curtain or barrier) and a warning light or sign (see Figure 21.8) or audible signal at the entryway, and requires laser workers with a high level of training.

Establishing a temporary laser-controlled area is also frequently required. This usually occurs where service activities require access to the beam from a Class 4 laser embedded within a Class 1 application. Often, barrier curtains[342] are used to exclude unauthorized individuals from a temporary-laser controlled area, as shown in Figure 21.9.

Administrative and Procedural Controls

The LSO should be notified concerning the purchase of any Class 3b or 4 laser, or any Class 3b or 4 embedded laser that will require on-site service. The ANSI standard requires the use of hazard warning signs, with the signal word "Danger," when working with Class 3b and 4 levels of laser radiation (see Figure 21.8). Make sure that all lighted signs, emission indicators, and displays are visible through the prescribed eyewear. The ANSI standard requires written standard operating procedures (SOP) for Class 4 lasers and for alignment practices. Suggested criteria for a laser SOP have been published.[242,345]

Medical Approval and Training. Individuals who work with Class 3b and 4 lasers must have medical approval. Medical approval includes an ocular history and an evaluation of visual acuity, color vision, and macular function. If individuals work with UV lasers, the medical examination should address the skin and the possibility of photosensitization. Preplacement exams and exams subsequent to potentially hazardous exposures must be included. Periodic exams are not required but may be prudent for individuals using UV lasers, especially if they are taking prescription medications. Exit (termination) exams are not required by the ANSI standard.[155]

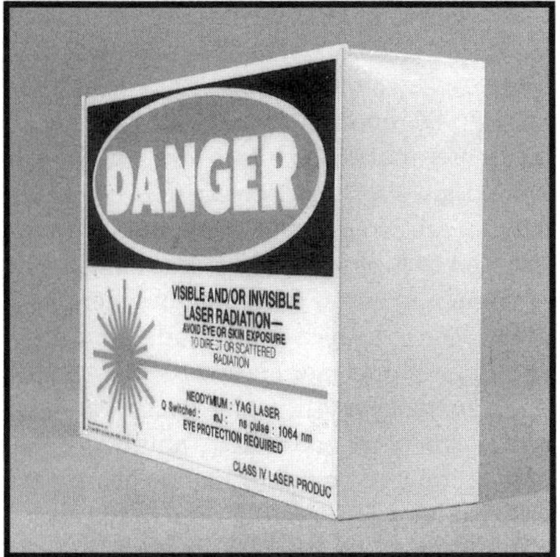

Figure 21.8 — A lighted entryway warning sign with the signal word "Danger" and the warning message for a Class 4 laser (Avoid Eye or Skin Exposure to Direct of Scattered Radiation). (Courtesy Rockwell Laser Industries, Inc.)

Figure 21.9 — A temporary laser controlled area is usually established when servicing a Class 4 laser that is embedded in a Class 1 laser product. Here, the perimeter of the controlled area is a barrier curtain that incorporates a "sign pocket" where the appropriate warning or notice sign may be displayed. (Courtesy Rockwell Laser Industries, Inc.)

In the United States, the ANSI standard requires training for users of Class 3b and 4 lasers,[155] while the IEC standard requires training for operators of Class 3a, 3b, and 4 lasers.[317] OSHA requires training for users of lasers in the construction industry in 29 CFR 1925.54.

Training may be general laser safety or application based. Training formats include classroom,[346] video-based,[347] interactive video-based,[348] computer-based,[349] and a combination of these formats. Regardless of format, training should be documented. Although not required by the various standards, awareness training for users of low-power lasers may be advisable in some environments.[350]

Personal Protective Equipment. Skin protection is best achieved through engineering controls. If there is the potential for damaging skin exposure, then skin covering must be used. Sunscreens may prove effective for diffuse exposure to lasers that emit UV-B and UV-A radiation. However, careful consideration should be given to the active ingredient in a particular sunscreen to assure effectiveness at the laser wavelength. With this in mind, it should be noted that sunscreens are formulated for use with environmental UV (about 295–400 nm), and may not be adequate for use with exposure to actinic UV radiation, such as that emitted by some excimer lasers.

Most gloves provide some protection from laser radiation. Tightly woven fabrics and opaque gloves provide the best protection. A laboratory jacket or coat can provide some protection for the arms. However, these may be a fire hazard if directly exposed to Class 4 laser beams. A good practice is to use flame-resistant materials around Class 4 beams. This applies not only to clothing, but to curtains and drapes. In the medical environment, surgical drapes are often saturated with a saline solution.[351]

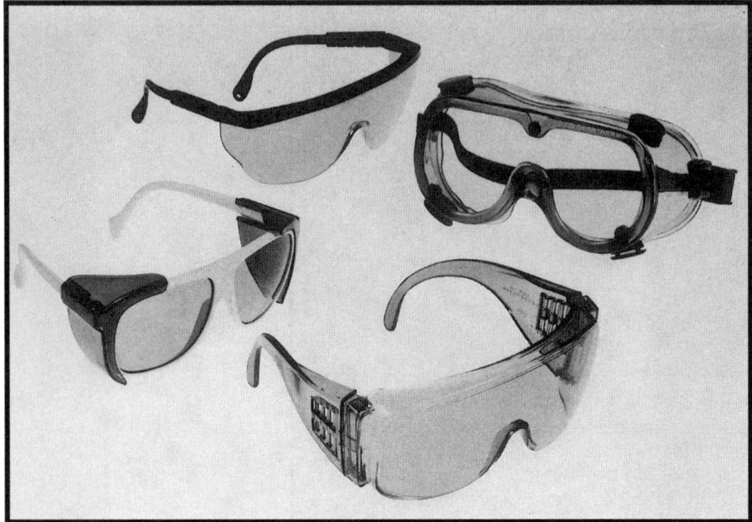

Figure 21.10 — Various styles of laser protective eyewear. (Courtesy Glendale Protective Technologies.)

There are many types of laser protective eyewear (see Figure 21.10) because of the large number of laser types and output powers. It is therefore extremely important to make sure that the proper eyewear is prescribed and used. Useful guides to the selection of laser protective eyewear have been published.[352,353]

Eyewear frame styles typically used in industry are spectacles or goggles, while other special designs include "flip-up" frames for multiple wavelength environments.[354] Faceshields may be used to protect the face and eyes from scattered UV radiation[355] or possibly from reflected CO_2 laser radiation.[356] Lenses may be glass, plastic, or a combination. In some cases, glass lenses may be stacked[357] or laminated with plastic to protect the eyes in case the lens fractures on exposure. In general, glass can withstand higher irradiance levels, may be used for corrective lenses, and are scratch-resistant but heavy. Reflective (dielectric) coatings may be used with glass, and this provides a relatively narrow bandpass with good transmission of visible wavelengths necessary to see (visual transmittance).[358] Presently, most plastics are absorptive filters that require the incorporation of dyes within the plastic matrix, usually polycarbonate. In general, this reduces the visual transmittance.

Laser-protective eyewear is characterized by optical density (OD; see Equations 5 and 6) and wavelength.[357,359] For example, an OD of 6 represents a reduction of the incident radiation by a factor of 1 million. It is possible for the eyewear to be damaged or fail (crack, melt, etc.) when exposed to high-intensity laser beams.[360-362] Hence, eyewear is used only when other control measures are inadequate. Although it does not necessarily physically damage the eyewear, a potentially hazardous phenomenon called "nonlinear optical transmittance" (dynamic bleaching) has also been observed in a small number of filter types. This occurs when the laser radiation saturates an absorber with a resulting increase in the amount of laser radiation transmitted.[363,364] Also, other researchers observed that some filter materials exhibited a spike in the initial values of power transmittance, which rapidly decayed to a steady-state value.[365]

In the United States, the ANSI standard requires that eyewear must be marked with OD and wavelength so users can select the proper eyewear.[155] Obviously, this becomes an important issue when more than one type of eyewear is used in an area with more than one type of laser. Eyewear should be used as directed, and stored properly. It should be inspected routinely to determine if there is pitting, crazing, or solarization of the lenses and if goggle straps or spectacle sideshields are in good condition.

Alignment Practices. An area of particular concern in laser safety is beam alignment. In general, this should be done with the beam power as low as practical, or with an alignment laser, which is usually a low-power laser with a visible beam that is coaxial with the beam to be aligned.

Only diffuse reflections of visible beams should be viewed. Consider that the inverse square relationship with distance is valid as long as the distance is much greater than the laser spot diameter. Consequently, a diffuse surface acts as a distance-dependent attenuator that permits indirect viewing of low-powered laser beams when the reflecting spot is small. An example of this might be a point source-diffuse reflection of a relatively low-powered beam from a nonglossy business card held at arm's length. However, it should be stressed that diffuse viewing should be performed only when wearing the appropriate type of laser protective eyewear. So-called alignment eyewear has a relatively low value of OD because it is for safe viewing of a diffuse reflection. Some eyewear manufacturers label such eyewear "DVO," which stands for diffuse viewing only.

The following example illustrates how to determine the necessary OD for an alignment using a 5-W argon laser, where a diffusing surface with a reflectivity of 80% is viewed at an angle of 0° for 600 sec at arm's length (assume $r = 0.5$ m). From Equation 12, the irradiance at 0.5 m is calculated to be 0.5 mW/cm^2. The MPE for 600 sec in the visible spectrum is 16.7 µW/cm^2, so the necessary optical density is 1.5, which is an attenuation of about 30 times.

Invisible beams can be viewed using various phosphor image converter materials. These are often in the form of a "viewing card." The cards absorb the invisible optical radiation, then emit at wavelengths that are visible through protective eyewear. Since there may be significant reflections from viewing cards that are laminated with high-gloss plastics, it is best to purchase viewing cards with diffusing finishes, or to place matte-finish tape over the laminate.[366] IR beams can be observed with infrared viewers that produce a visible spot that can be seen through IR-protective eyewear.

Radio-Frequency and Microwave Radiation*

Anticipation

The RF spectral region is 300 GHz to 3 kHz. Usually, microwave radiation is considered a subset of RF radiation, although an alternative convention treats radiowaves and microwaves as two spectral regions. In the latter context, microwaves occupy the spectral region between 300 GHz and 300 MHz, while radiowaves include 300 MHz to 3 kHz.

Various order-of-magnitude band designations (see Table 21.16) have been assigned to the RF and sub-RF portion of the spectrum. Frequencies in the various bands are allocated for uses including navigation, aeronautical radio, broadcasting, and citizens' radio. In addition to band designations, specific frequencies are designated for industrial, scientific, and medical (ISM) uses. ISM frequencies are 13.56, 27.12, 40.68, 915, 2450, 5800, and 24,125 MHz. Designations for frequency bands used with radar are shown in Table 21.17.

Quantities and Units

Seven quantities may be used to characterize exposure to RF fields: the specific absorption rate (SAR), specific absorption (SA), electric-field strength (E), magnetic-field strength (H), power density (W or S), induced currents (I_i), and contact currents (I_c). The SAR and SA are quantities that represent exposure within tissues. E, H, and W are measures of the intensity of the

* This section was written by R. Timothy Hitchcock

Table 21.16 — Nomenclature of Band Designations

Frequency Range	Designation	Abbreviation
* <30 Hz	sub-Extremely Low Frequency	sub-ELF
* 30–300 Hz	Extremely Low Frequency	ELF
* 300–3000 Hz	Voice Frequency	VF
3–30 kHz	Very Low Frequency	VLF
30–300 kHz	Low Frequency	LF
300–3000 kHz	Medium Frequency	MF
3–30 MHz	High Frequency	HF
30–300 MHz	Very High Frequency	VHF
300–3000 MHz	Ultra High Frequency	UHF
3–30 GHz	Super High Frequency	SHF
30–300 GHz	Extremely High Frequency	EHF

* The IEEE definition of band designations does not include VF, and defines ELF as 3–3000 Hz, and <3 Hz as ultralow frequency (ULF).

Table 21.17 — Radar Letter-Band Designations

Letter Designation	Nominal Frequency Range (GHz)	Specific Radar Bands (GHz)
HF	0.003–0.03	
VHF	0.03–0.3	0.138–0.144
		0.216–0.225
UHF	0.3–1.0	0.420–0.450
		0.890–0.940
L	1–2	1.215–1.400
S	2–4	2.300–2.500
		2.700–3.700
C	4–8	5.250–5.925
X	8–12	8.500–10.680
Ku	12–18	13.4–14.0
		15.7–17.7
K	18–27	24.05–24.25
Ka	27–40	33.4–36.0
V	40–75	59–64
W	75–110	76–81
		92–100
mm	110–300	126–142
		144–149
		231–235
		238–248

fields in space, while I_i and I_c are measures of RF-induced electric currents.

The SAR is the rate at which energy is absorbed per unit mass, or dose rate. It is the fundamental quantity of the exposure criteria, and the dosimetric quantity of choice in studies of biological effects. It is generally expressed in units of watts per kilogram (W/kg), representing the power deposited in a unit mass. The unit, W/kg, is also used for metabolic rate, where the resting metabolic rate of an adult human being is about 1 W/kg. The SAR depends on the electric-field strength in tissues (E_i), the electrical conductivity (σ) of tissues, and the density of tissues (ρ) as in:

$$SAR = \sigma E_i^2/\rho \qquad (13)$$

The SA is the time integral of the SAR and, as such, is the RF dose. It is the energy absorbed per unit mass of tissue, with units of joules per kilogram (J/kg).

In general, the determination of SAR and SA does not lend itself to field evaluations of exposure but is limited to carefully controlled studies in the laboratory or numerical model. Because of this, exposure guidelines are written in terms of related quantities, called derived limits, that are more easily evaluated when determining exposure. These include E, H, and W for electromagnetic fields, which may be viewed as surrogate measures for the SAR. Hence, the exposure guidelines are in terms of frequency-dependent derived limits that will maintain the SAR at an acceptable level.

Electric-field strength and magnetic-field strength are vector quantities, but only the magnitude is reported in safety evaluations. E fields are generated by electric charges and are measured in terms of the electric potential (V) over some distance. The unit is the volt per meter (V/m). Since power is related to the square of the voltage, E^2 (V^2/m^2) is often used in exposure guidelines or in describing the output of measurement instruments.

H fields are generated by moving electric charges, such as a current (I) moving through a long, thin wire. The flux density (B) of the magnetic field at some distance (r) from the wire is

$$\mathbf{B} = I/2\pi r. \qquad (14)$$

H is used for RF magnetic fields and has units of amperes per meter (A/m). Since power deposition in tissues is related to I^2, H^2 (A^2/m^2) may also be used. (To convert between **B** and **H**, use $\mu = \mathbf{B}/\mathbf{H}$, where μ is the magnetic permeability. For fields in air, μ is the value of the magnetic permeability of free space, μ_o, 1.257×10^{-6} henry/meter.)

Power density (W) represents the time-averaged energy flow across a surface, and typically is used when measuring microwave radiation. The unit of power density is watts per meter squared (W/m^2), although the use of milliwatts per centimeter squared (mW/cm^2) in hazard evaluation is common. (Conversion: $1\ W/m^2 = 0.1\ mW/cm^2$.) W is related to E and H by:

$$W = E\approx/120\pi = 120\pi H\approx. \qquad (15)$$

RF fields can induce currents within exposed tissues. These induced currents (I_i) flow through the body to ground, with a common path through the foot, which is called the "short-circuit" or foot current. The unit of electric current is the ampere (A), although the milliampere (mA) is the magnitude usually addressed in safety evaluations.

In an environment where there are RF fields, it is possible for contact with conductive objects to result in currents that flow into the body at the point of contact, which is usually the hand. If this occurs, exposure guidelines may require evaluation of contact currents (I_c) in mA.

Physical Characteristics

Understanding certain properties of an RF field is necessary for the satisfactory performance of a hazard evaluation. These include near and far fields, plane waves, impedance, polarization, modulation, gain, and duty cycle.

Near and Far Fields. The antenna is the circuit element that causes the RF energy to be radiated. The transition of RF energy from the conduction state on the surface of an antenna to the radiation state in free space is not immediate but passes through the near field then into the far field (see Figure 21.11).

In the space immediately surrounding the antenna is the reactive near field. Here, the E- and H-field components exist in a complex temporal and spatial pattern, and the energy is nonpropagating or "stored." A short distance from the antenna is the radiating near field, characterized by both energy storage and propagation (radiation). Here, the spatial pattern of the beam intensities is complex and may increase or decrease with distance or may remain unchanged. Beyond the near field is a transition zone, then the far field. In the far field, there is propagation, and the power density follows the $1/r^2$ law for distance.(367-370)

Plane Waves. To understand plane waves, consider an ideal point-source antenna which emits radiation in an isotropic pattern. The radiation pattern would be spherical (uniform in all directions), so near this antenna a receiver detects curvature in the approaching field. However, if removed sufficiently far from the source, some distance into the far field, a receiver would sample only a very small area of an immense curved wavefront. In the local region of space occupied by the receiver, it would detect a flat, or planar front; hence, the name, plane waves.

Two important characteristics of plane waves are space quadrature and time phase (see Figure 21.1). Space quadrature exists when the E- and H-field vectors are at right angles to one another and to the direction of propagation. The time phase of the fields is such that the **E** and **H** vectors reach their maximum and minimum values simultaneously.

Free-Space Impedance. The impedance, Z, of free space is the quotient of **E**/**H**. In the near-field impedance must be determined. However, when plane-wave conditions exist, the free-space wave impedance is a constant value, 377 (120 π) ohms (Ω), at a given point in space.

Combining the expressions for power density and impedance for a plane wave yields:

$$W\ (\text{mW/cm}^2) = E^2/3770 \qquad (16)$$

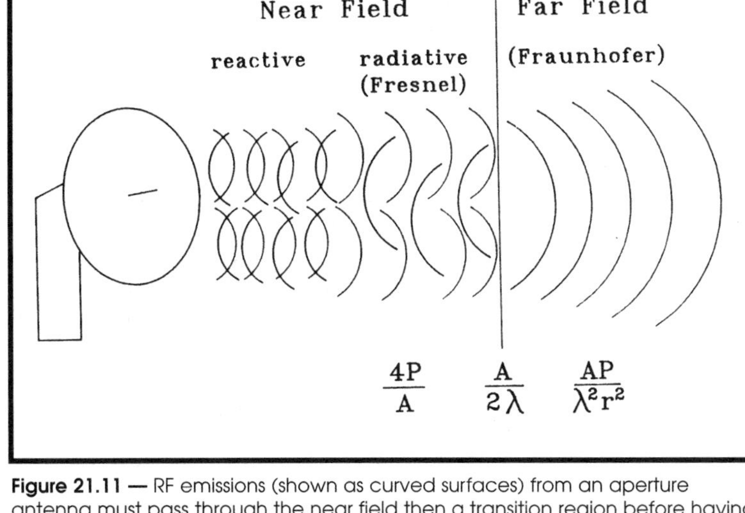

Figure 21.11 — RF emissions (shown as curved surfaces) from an aperture antenna must pass through the near field then a transition region before having the characteristics of radiation. The power density in the near field and far field as well as the distance to the boundary between the two zones may be determined from the equations. Note that the power density in the far field is dependent on distance, while the power density in the near field is not.

and

$$W\ (\text{mW/cm}^2) = 37.7H^2 \qquad (17)$$

which are useful formulas for conversions involving field strength and power density.

Polarization. Polarization is that property of an electromagnetic field describing the time-varying direction and amplitude of the electric-field vector. A field may be polarized linearly, circularly, or elliptically, or it may be unpolarized. Specifically, the type of polarization is described as the figure traced as a function of time by the amplitude of the E-field vector at a fixed location in space. An observer would view this trace along the direction of propagation.

Modulation. Modulation is the process by which some characteristic of an unmodulated carrier wave (a carrier of information) is varied by a modulating wave (called a signal). The modulating signal, which is lower in frequency than the carrier, is used to superimpose information on the carrier wave. If the amplitude of the carrier is varied, the wave is amplitude modulated (AM). If the frequency or phase of the carrier is varied, then the wave is frequency modulated (FM) or phase modulated (PM), respectively. AM and FM are used in broadcasting, while FM and PM are used in communications. Digital modulations such as time-division multiple-access (TDMA), and code-division multiple-access (CDMA) are also used in communications.

Some industrial and medical RF sources may be opportunistically amplitude modulated, with the modulating signal in the ELF spectral region. This occurs because the electric circuitry allows the imposition of the fundamental or a higher harmonic of the power frequency (50 or 60 Hz) on the RF carrier.[371,372]

Gain. Gain is a measure of the directional properties of an antenna, representing an increase in power output of the system in relation to an ideal isotropic emitter. For a point-source antenna, gain equals 1. If a reflector is placed near this antenna, it changes the radiation pattern due to collimation or focusing, and the gain is greater than 1.

Absolute gain (G) is a ratio of the actual transmitted power density in the main beam to the power density transmitted from an isotropic radiator. If the absolute gain of an antenna with area (A) operating at a given wavelength is unknown, a conservative estimate of absolute gain can be calculated from:

$$G = 4\pi A/\lambda^2 \qquad (18)$$

When gain is used in calculations, values of absolute gain must be used:

$$G = 10^{(g/10)} = \text{antilog}\ (g/10) \qquad (19)$$

where:

g = gain in decibels.

Duty Cycle. Some sources may exhibit a cyclic or intermittent operation, where RF is emitted for only a fraction of the total time of operation, the "on-time," in an operational cycle. The ratio of on-time to the total time of operation (on-time plus off-time) is called the duty factor or duty cycle (DC). DC is also the product of the pulse repetition frequency (PRF) in pulses per second (Hz) and the pulse width or duration (PW) in seconds (PRF × PW).

Duty cycle is important, because it allows the determination of average power (where average power = peak power × DC), and most exposure guidelines are in terms of average power. Sources that typically have a duty cycle less than 1 include radar, dielectric sealers, induction heaters, RF welding units, some communication devices, and medical diathermy units.

Biological and Health Effects

Human data are limited and present no clear trends, so scientists have relied on animal models to establish biological effects. These effects have been extrapolated to humans and used in setting exposure limits. Animal studies have found effects in most major systems including nervous, neuroendocrine, reproductive, immune, and sensory.[2,4,369,370,373-380] The following review addresses effects on behavior, reproduction and development, and the eyes and cancer.

Interaction with Tissues. Tissues are classified as a lossy dielectric material, that is, a material that interacts with the field and absorbs considerable RF energy. RF energies are more highly absorbed in tissues of high water content (e.g., muscle) than in tissues with low water content (e.g., fat). RF interaction with tissues may be complex, with standing waves formed at the interface of tissues with different dielectric properties, such as skin-fat or fat-muscle.

In general, short-wavelength RF has a relatively shallow penetration depth into tissues, such that at frequencies of more than a few gigahertz, absorption is in the skin. Longer wavelengths may penetrate more deeply, and the body is relatively transparent to long-wavelength magnetic fields. Obviously, penetration depth may affect what organs or systems are at risk.

Numerous studies have clearly shown that absorption of RF energies produces thermal effects, although some athermal and nonthermal effects have been reported.[381-386]

RF exhibits three modes of tissue interaction at the molecular level: polar molecule alignment, molecular rotation and vibration, and the transfer of kinetic energy to free electrons and ions.[387] Alignment of polar molecules with the field is an ubiquitous mechanism that results in frictional heating.

Behavior and Nervous System Effects. Exposure guidelines are based on a few well-established effects observed in test animals. One of these is reversible behavior disruption in short-term studies, a sensitive measure of RF exposure.[388-392] In general, behavioral changes are thermal effects attributed to significant increases in body temperature due to absorbed RF energy.[391,392]

There are East European and Russian reports of RF workers with certain nonspecific symptoms (also found in workers not exposed to RF and members of the public) associated with the nervous system. These effects may have clinical signs extending to the cardiovascular system, called "radiowave illness" or "microwave sickness."[393-397] Some similar symptoms have been reported in the United States medical literature in two cases of apparently high, acute overexposure to microwaves.[398,399]

In addition to behavioral effects, combined interactions of RF fields with neuroactive drugs and chemicals have been reported in test animals.[400,401]

Reproduction and Development. Teratogenic effects have been reproducibly demonstrated at 27.12 MHz with rats, when the whole-body average (WBA) SAR was relatively high, around 10 to 11 W/kg.[402-404] These appear to be thermal effects, as a dose-response effect was seen with high rectal temperatures.[439] Developmental abnormalities were observed in rodents at 2450 MHz, with high values of WBA SAR.[405,406]

Epidemiological studies of reproductive endpoints have demonstrated no trends.[407-4415] Two studies reported effects on semen quality,[416,417] but the small sample size and lack of exposure data make interpretation difficult.

Ocular Effects. Effects have been reported to the cornea, iris vasculature, lens, and retina.[418,419] Effects to the corneal endothelium of monkeys were reported at 2450 MHz by one research group,[420,421] but not by another.[422]

Historically, cataracts have been of concern because the avascular nature of the lens increases its susceptibility to heat-induced change. The most effective frequencies in test animals are 1 to 10 GHz. Acute thresholds have been determined for restrained rabbits receiving ocular exposure in the near field of a 2.45-GHz applicator.[423,424] No cataracts were observed with far-field exposure of unrestrained animals,[425] even if exposures were almost lethal.[426] Studies of cataracts in people with purported RF exposure do not support a causal link,[396,427-433] although one study suggested an aging effect on the lens.[430]

Cancer. Limited data suggest that RF may be a possible tumor promotor in animals,[434,435] while other studies demonstrate no significant differences in RF-exposed groups and controls.[436-438] One study found significant differences in primary malignancies in RF-exposed rats when the tumor data for all tissue types were combined. However, no other health effects were observed in the more than 100 other endpoints evaluated.[439,440] In sum, the animal data on cancer are inconclusive.

Epidemiological studies have provided some suggestive evidence that RF energies are carcinogenic to human beings.[4,395] No differences were observed in cancer mortality in two studies of military personnel categorized as RF exposed,[441,442] but a third study found a link between exposure of Polish military personnel to pulsed RF and an increased incidence of cancers of the alimentary canal, brain, and hematopoietic and lymphatic systems.[443] Another study categorized possible RF exposure of Air Force personnel on the basis of job title. A modest increase (39%) in brain cancer was observed, which was marginally statistically significant.[444]

No differences were observed between the staff at the Radiation Laboratory at the Massachusetts Institute of Technology (radar development during World War II) and the general population. A significant difference in cancer of the gall bladder and bile ducts was reported when MIT personnel were compared with a cohort of physician specialists.[445]

Although cancer was not the only endpoint, findings in a study of overall mortality in cellular telephone customers were not remarkable.[446] In an hypothesis-generating study an increase of testicular cancer was found in a group of policemen in Washington State who used radar guns.[447]

Although not occupational studies, studies examining the proximity to radio and TV towers and cancer provide information of interest to broadcast workers. A study in Great Britain reported statistically significant results for adult leukemia, skin cancer, and bladder cancer for people living within 10 km of 1 FM radio and TV antenna.[448] However, this result was not replicated in a study of all high-power transmitters in Great Britain, although some evidence was found of a decrease in leukemia with distance from the antennas.[449] In Australia, Hocking and colleagues determined a statistically significant increase in the incidence of and mortality from childhood leukemia.[450] This hypothesis-generating study needs further evaluation with a more rigorous study design.

The human studies to date have a number of limitations. One major limitation is the absence of information on doses, dose-response, or duration of exposure. Others include the classification of possible exposures on the basis of job title, a small number of cases or small sample size, little control for confounding factors, and no consistency in findings in the studies reporting statistically significant positive associations. However, because of some suggestive findings, further study is needed.

Accidents, Incidents, and Other Concerns. A number of accidents and incidents have been reported. They may be categorized as ocular,(451-455) nervous system,(398,399,456-461) reproductive,(453,462,463) skin,(453,455-458,461,464-470) and cancer.(471-474) Reviews of accidents are available.(370,380)

Other concerns with exposure to radiofrequency fields include possible perturbation of the field by metallic implants (e.g., metal staples, cochlear implants)(476-479) or metallic objects that are worn (e.g., jewelry, watches, metal-framed spectacles).(480-482) RF fields may also interfere with electronic devices such as sampling devices,(483,484) medical electronics (e.g. cardiac pacemakers),(485-488) and electroexplosive devices. In some chemical sample preparation methods, microwaves have been reported to superheat solutions, which may lead to melting of reaction vessels or explosions.(489)

Table 21.18 — Information on Some Important RF Sources

Source	Frequencies (MHz)	Uses	System Components
Dielectric heater	10–70 many at 27.12	Heat, seal, weld, emboss, mold, or cure dielectric materials	Power supply, RF generator, tuning circuitry, press, electrodes (die)
Induction heater	0.250–0.488	Heat conductive materials via electromagnetic induction	Power supply, RF generator, transmission line, induction coil
Plasma processors	0.1–27 many at 13.56	Chemical milling; nitriding of steel; synthesis of polymers; modifying polymeric surfaces; deposition (sputtering) and hardening of coatings and films; and etching, cleaning, or stripping photoresist	RF generator, transmission line, reactor vessel, RF tuning and control module, vacuum pump, gas storage, and gas delivery system
Radar	EHF, SHF, UHF	Detection, tracking, ranging	Transmitter, waveguide, antenna, receiver, display
Broadcasting	see Table 21.19	Radio, TV transmission lines, tower, antennas	Transmitter
Communications	HF, VHF, UHF	Fixed position and mobile systems used for voice/data transmission	Transmitter, receiver, transmission lines, antennas
Video display terminals	VLF, LF	Visual imaging component of information processing systems	Cathode-ray-tube VDTs have been the focus of most evaluations
Diathermy	Microwave (915 or 2450) or shortwave (13.56 or 27.12)	Heat therapy	Generator, control console, transmission line, applicators
Electrosurgical units	0.5–100	Cauterizing or coagulating tissues	Generator, transmission line, surgical probe, current return cable

Recognition

Generation

RF energy is generated by the acceleration of charge in oscillatory circuits. Generators include RF vacuum tubes (triode, tetrode, and pentodes), microwave vacuum tubes (klystrons, magnetrons, and traveling-wave tubes), and solid-state devices (semiconductor diodes and triodes).(369-370)

Sources

Sources are natural (sun, galaxies, lighting, human body) and man made.(370,490-496) Man-made sources are radiators, leakage sources, or a combination. Radiators (e.g., antennas) intentionally emit RF energy into the environment. Leakage may be the consequence of poor design, lack of maintenance, or improper maintenance. Information on selected sources is contained in Table 21.18 and discussed in the following text.

Because of the low operational frequencies, evaluations of dielectric heaters must include E and H fields, I_i, and possibly I_c. Cyclic operation requires correction for duty cycle. Evaluations have shown that work with unshielded or improperly shielded heaters may result in overexposure. In general, overexposures are to free fields (primarily electric),(497-499) and induced currents.(500,501)

Induction heaters are also low-frequency sources. Typically, measurements of induction heaters require duty-cycle correction. H-field values in excess of exposure guidelines have been reported near these devices.(502)

Measurement of E, H, I_i, and possibly I_c is required for plasma processing units, too. Generally, leakage is low, but unshielded viewing ports are a potential leakage source. Also, units configured with twin reactor vessels may leak from an open vessel while the second vessel is in use,(503) and equipment attached to plasma units can act as antennas and emit RF energy.(504)

The beam emitted from radar antennas is highly directional and, in some cases, a very narrow beam is produced. Some antennas move horizontally (scan) or vertically (elevation), and most units are pulsed at relatively high peak powers. The combination of scanning, elevation, and pulsing (duty cycle <1) usually results in brief exposures for individuals irradiated by the beam. Evaluations of commercial radar (airport surveillance and approach traffic control),[505] aircraft,[506,507] marine,[508-510] and police traffic-control radars[511-514] have not demonstrated over exposures during normal operation. However, exposure to airport radar[505] and aircraft units[506,507] may be of concern during maintenance activities. Reports have characterized overexposure from a portable military unit[398] and a military aircraft unit.[399] Evaluation of other military radars have demonstrated high RF levels when a rotating beam was stopped, which required an interlock to be disconnected.[515]

Broadcasting types and frequencies are in Table 21.19. Broadcast towers may support a single antenna or multiple, stacked antennas (FM and TV), or be the antenna (AM radio). If an antenna occupying a high position on the tower must be serviced, it is possible that personnel may be exposed to fields from energized antennas located lower on the tower.[516] The hands and feet of climbing personnel may also receive high exposures, especially if the transmission line is located near the ladder. When maintenance work is performed on energized AM towers, it is possible body current may exceed the exposure limits.[517] Also, workers may experience spark discharge and sustained contact currents.

Evaluations of fixed communications systems have not demonstrated overexposures of operational personnel.[492,518-520] It is possible that relatively high levels of RF may be encountered by maintenance personnel who work on satellite communications (SATCOM) systems or near energized cell-site antennas.[521] Mobile[522-524] and portable communication systems may produce relatively high RF levels very near the antennas. Hence, when the antenna is located very near the head, exposure may result in relatively high local values of SAR.[525,526] RF currents on metallic parts of the radio case may also produce relatively high levels of exposure to the face.[527,528]

Numerous evaluations of low-frequency E and H fields from cathode-ray-tube type video display terminals (VDTs)[529-533] and televisions[4,534] have not demonstrated the potential for overexposure. RF fields may be emitted from any surface of these devices, but the magnitude decreases rapidly with distance. In general, most of the RF energy emitted by VDTs/TVS resides in the VLF and/or LF band designations.

Table 21.19 — Broadcast Frequency Allocation

Type	*Carrier Frequency (MHz)*
AM radio	0.535–1.605
FM radio	88–108
Low-band VHF-TV	54–72, 76–88
High-band VHF-TV	174–216
UHF-TV	470–806

Diathermy units may operate continuously or be pulsed, and some may be amplitude modulated or have a ripple at ELF frequencies.[371,372] The leakage field of the applicator depends on the type of applicator. Relatively high levels may be found in the vicinity of the cables.[535,536] Adjustment of the equipment during operation may result in exposure to the hands.[537,538]

Evaluations of solid-state and spark-gap electrosurgical units (ESUs) demonstrated that field strengths increased with increasing output power and levels were higher for solid-state units.[539] Levels near the probe and unshielded leads may exceed exposure criteria.[537,539,540]

Evaluation

Exposure Guidelines

RF guidelines for occupational exposure are recommended by groups including the Institute of Electrical and Electronics Engineers (IEEE),[541] the ACGIH,[101] the ICNIRP,[542] the National Council on Radiation Protection and Measurements (NCRP),[543] and the Federal Communications Commission (FCC).[544,545] Discussions of other occupational guidelines are available.[370]

The NCRP, ICNIRP, and FCC also have limits for members of the general public, while the IEEE recommends limits for controlled and uncontrolled environments. A major difference between these two environments is that individuals in the controlled environment (CE) are aware of the potential for exposure, whereas individuals in the uncontrolled environment (UE) have no knowledge or control over their exposure.[541] Since members of the general public or workers may be in either environment, this must be determined during the evaluation so the proper exposure limits are used.

The exposure limits are derived from a fundamental quantity of exposure called the basic limit. Current density is the basic

Table 21.20 — Specific Absorption Rates (W/kg)

	Whole-Body Average (W/kg)	Spatial Peak (W/kg/g)	Extremities (W/kg/10 g)[A]
Controlled Environment[B]	0.4	8.0	20
Uncontrolled Environment[B]	0.08	1.6	4
Occupational[C]	0.4	8.0	20
General Population[D]	0.08	—	—

[A]Local SAR based on absorption in 10 g of tissue.
[B]From reference 541.
[C]From reference 101.
[D]From reference 543.

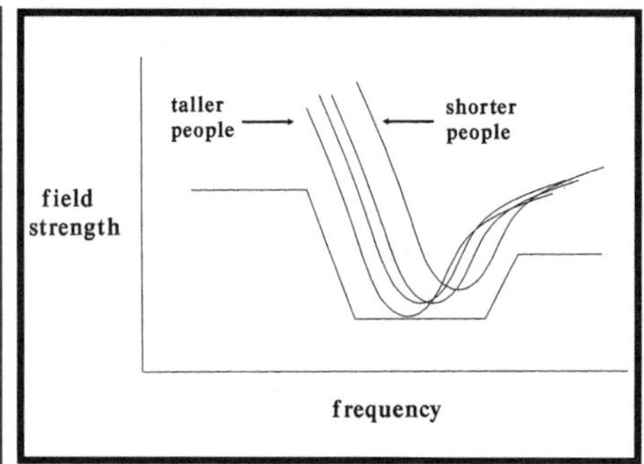

Figure 21.12 — The derived limits form an envelope around the family of RF absorption spectra for individuals of different heights. The envelope is formed by three regions where the limits plateau, and two transition (sloping) regions where the limits must be calculated.

limit at the lowest frequencies, less than 100 kHz, and SAR is the basic limit throughout much of the remainder of the RF region.[(541)] SARs have been recommended for whole-body average exposure, spatial-peak exposure, and exposure to the extremities, as shown in Table 21.20.

The exposure (derived) limits form an envelope around the human whole-body absorption spectra, as shown in Figure 21.12. The relationship between frequency and body height in this figure demonstrates that maximum absorption occurs at a given frequency for a given body height, and this concept is called geometrical resonance. For any individual, resonance (maximum absorption rate) is established when body height is about 40% of the incident wavelength, and the E-field vector is parallel with the long axis of the body.[(546)] The guidelines address resonance for grounded or ungrounded (free space condition) bodies. These conditions have been integrated to produce the resonance range that typically spans 30–300 MHz, the region where the exposure limits are lowest.

Because the body absorbs RF more poorly at frequencies outside the resonance region, the exposure limits increase via the transition regions, to the other plateau regions. In the low-frequency region, the exposure limits are the highest because the body is highly sub-resonant and RF energy is poorly absorbed. In the high-frequency

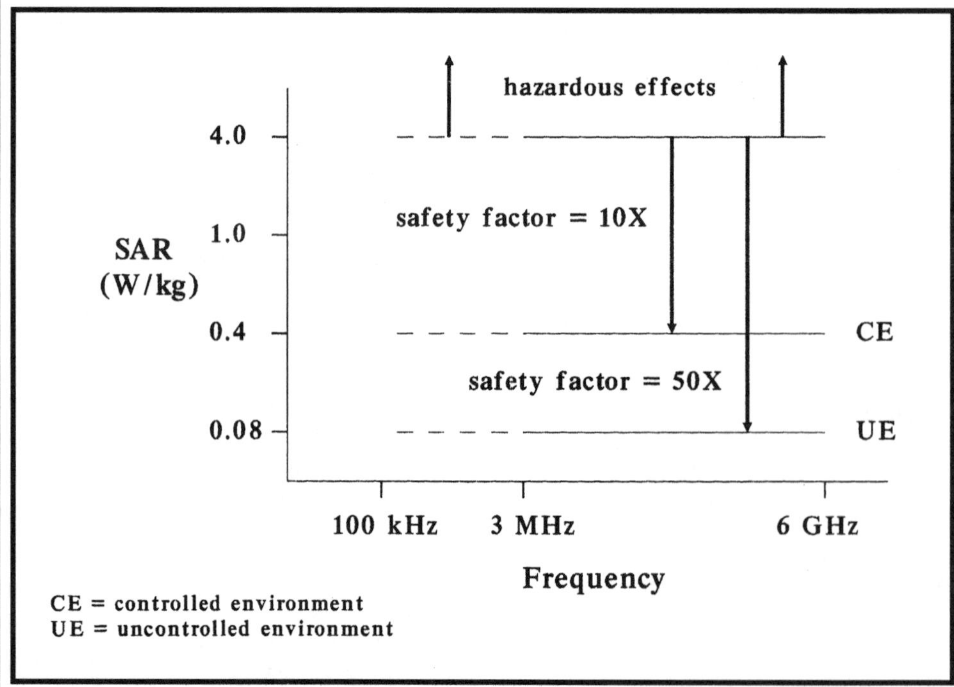

Figure 21.13 — The whole-body average SAR was determined from studies with test animals. Potentially hazardous effects occur at SARs in excess of 4 W/kg. As discussed in the text, the SARs for the controlled and uncontrolled environments were derived by the application of tenfold and fiftyfold safety factors, respectively.

region, RF energy absorption and hence, the limits, are at levels between the other regions. This is called the quasi-optical region since wavelength is decreasing and interaction begins to approximate that of optical wavelengths.

Note that the SAR does not vary with frequency as do the derived limits. This is because a single threshold dose rate for unfavorable biological effects (4 W/kg for reversible behavior disruption in test animals) was determined from a review of the biological database. As shown in Figure 21.13, a tenfold safety factor was applied to this to produce a whole-body average SAR of 0.4 W/kg for occupational exposures and the controlled environment. The WBA-SAR for the UE or the general population (0.08 W/kg) includes an additional fivefold safety factor. Hence, this lower SAR is not based on a lower threshold dose rate for potentially hazardous effects.

Additional criteria for induced and contact currents exist for exposure at low frequencies, typically less than 100 MHz. At these low frequencies, the body is highly conductive and RF E-fields in space induce currents within the body. It has been demonstrated that these currents may produce locally high values of SAR, so guidelines have been developed to control this type of exposure. RF energies may also couple to conductive objects, such as a vehicle body, metal fence, or metal roof, and be conducted into the body when contact is made. Since this may result in RF shock or burns, organizations such as IEEE[541] and ACGIH[101] recommend a guideline for grasping contact with the hand.

In general, the averaging time for occupational and general population limits for free fields is 6 and 30 minutes, respectively. Six minutes applies to the CE except at frequencies between 15 and 300 GHz where it decreases to 10 seconds at 300 GHz, and must be calculated.[541] In general, 30 minutes is used for the extended resonance range, where limits for the UE are less than limits for the CE. The averaging time for induced and contact currents is presently 1 second. (IEEE SCC28 is considering revising the averaging time for induced currents to 6 minutes, to be consistent with the free-field limits, since both types of exposure are believed to produce similar effects.)

Although not an exposure guideline, the FDA has a performance standard for microwave ovens in 21 CFR Part 1030.[267] This includes requirements for manufacturers of these ovens, and allowable microwave leakage at the time of manufacture and throughout the service life of the oven.

Instruments

Instruments are available to measure RF fields, body currents, and contact currents. In general, these instruments are broad-band, meaning they have a wide calibrated frequency range and a frequency-independent response within this range. Some instruments, typically used for microwave oven measurements, have been designed and calibrated for 915 or 2450 MHZ, and are not broad-band instruments. All instruments should be calibrated at least annually and, if broad-band, at a number of frequencies across the frequency range.

Densitometry. The measurement of RF field strength and power density is called densitometry.[373] Densitometric measurements of fields in space are made with analog or digital survey instruments that have three components: probe, connective cable or optical fiber, and metering instrumentation, as shown in Figures 21.14 and 21.15. The probe includes antennas to couple the field into detectors. The output from the detectors is directed to the metering instrumentation where it is processed and displayed and/or logged. Output units may include V/m, A/m, V^2/m^2, A^2/m^2, $|FSU|^2$ (field-strength units squared = V^2/m^2 or

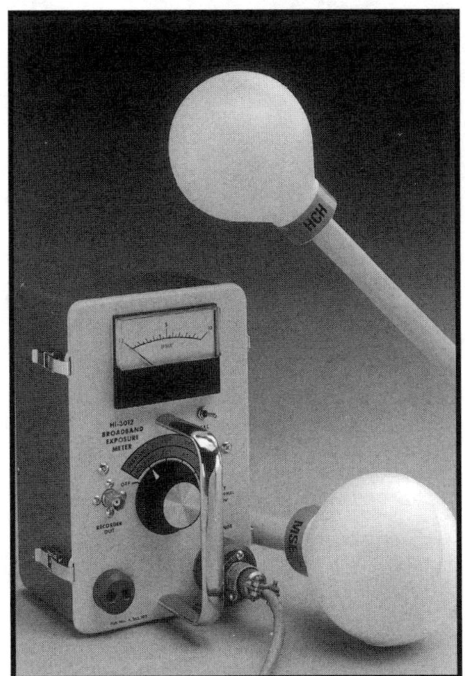

Figure 21.14 — A broad-band survey meter with E- and H-field probes. The antennas and detectors (diodes) are enclosed in the polystyrene insulator. The tape at the base of the insulator is labeled for the type of probe (E field = MSE; H field = HCH) and color-coded to match one of two scales located above the indicator knob on the metering instrumentation. (Courtesy Holaday Industries, Inc.)

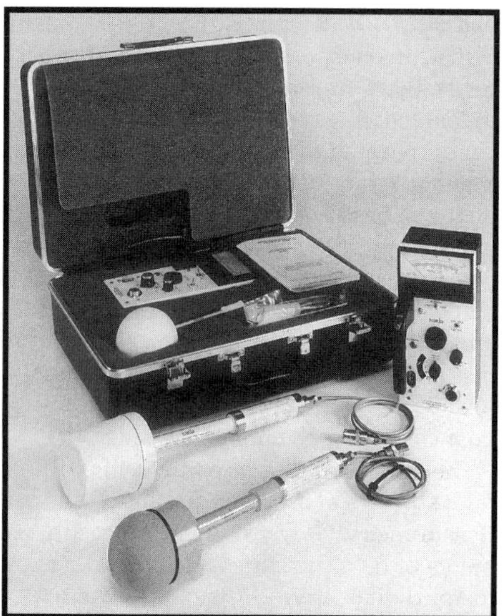

Figure 21.15 — A broad-band survey meter with probes, carrying-case, and operations manual. The antennas and detectors (thermocouples) are located within the RF-transparent plastic or polystyrene. Calibration factors are included under clear plastic that covers the handle of the probe. (Courtesy Loral-Narda.)

Figure 21.16 — A displacement current sensor that may be used for monitoring low-frequency E and H fields such as those emitted by video display terminals. (Courtesy Holaday Industries, Inc.)

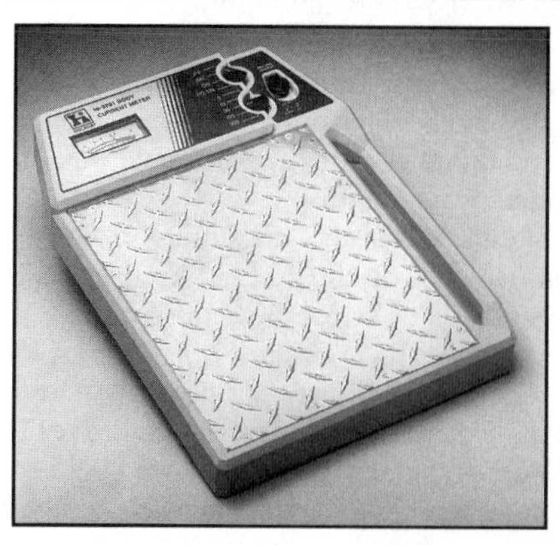

A

B

Figure 21.17 — (a) A stand-on meter used to measure the short-circuit current flowing out of the foot; (b) A clamp-on current transformer measuring currents flowing in the ankle. The unit is connected, via optical fiber, to a system readout device that is located on the user's waist. (Courtesy Holaday Industries, Inc.)

A^2/m^2), W/m^2, mW/cm^2, pJ/m^3, or percentage of standard.

For the most part, linear antennas (monopoles or dipoles) are used for E-fields,[547-549] and loop or coil antennas are used for H fields.[550,551] Commonly, detectors are thermocouples or diodes,[549,552] and are encased in an RF-transparent material (polystyrene, plastic) along with the antennas. Most instruments have interchangeable broad-band probes, which are color-coded or labeled to differentiate between E- and H-field response and to link the probe with a color-coordinated scale, at least on instruments with analog meters. Some newer instruments "handshake" with the probe and automatically select the proper output parameters.

Displacement current sensors have been developed for use in monitoring low-frequency E fields, such as those associated with video display terminals (see Figure 21.16). This instrument is a type of parallel-plate capacitor with two small, conductive plates electrically connected by a wire. Immersed in an RF field, a current flows between the plates and is measured by an operational amplifier.(553,554) Also, this instrument has a loop antenna for magnetic fields, located around the perimeter of the conductive plates.

Current Monitors. Two types of induced current monitors are available: stand-on and clamp-on devices (see Figure 21.17). Stand-on meters use a parallel-plate capacitor design to measure foot currents flowing to ground.(555,556) The bottom plate must make contact with ground, which may be difficult in some environments, especially outdoors. Both analog and digital devices are available. A human body model antenna that may be coupled to a stand-on meter is also available. A review of the performance of two manufacturers' makes of stand-on meters is available.(557)

The clamp-on device (see Figure 21.17b) is a current transformer that determines the current flowing through an appendage such as the ankle or wrist, or through the torso.(517,558,559) In this configuration, the body is the primary circuit and the measurement device includes the secondary circuit and a detector. The readout module may be on the transformer or may be remote. In the latter case, an optical fiber connects to the readout, a logging device, or both. Output is in mA or percent of standard.

A contact current monitors currents flowing through a handheld probe, which is connected to a ground plate with a connective cable (see Figure 21.18). The instrument has the impedance equivalent of the human body when making grasping contact with the hand, and output is in mA. The performance of this instrument has also been reviewed.(557)

Personal Monitors. At present, RF dosimeters are not in general use, because of difficulties in making meaningful measurements. This is partly due to an RF shadowing effect that shields a measurement device worn near the body. The shadow develops on the side of the body facing away from the source,(560,561) which introduces bias into the measurement data. Although not dosimeters, personal monitors with audible alarms that are activated when the intensity reaches some threshold value are available. In some units, the alarm threshold is preset, but in others it may be selected.

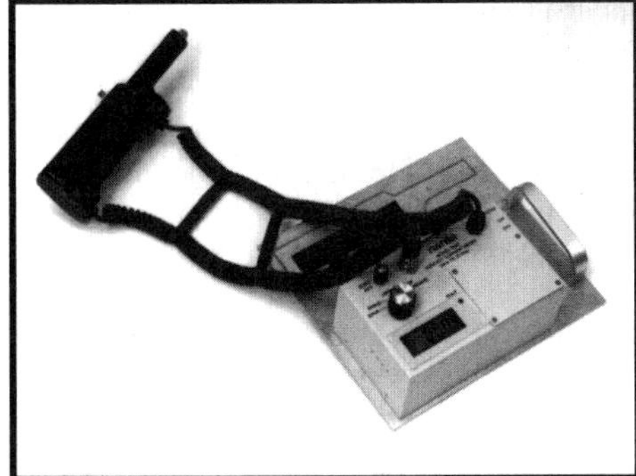

Figure 21.18 — A contact current meter showing the sampling probe, connective cable, meter, and baseplate. (Courtesy Lora-Narda.)

Frequency Counters. These are portable, handheld instruments that allow the user to determine or confirm the most intense single frequency associated with a source. These instruments offer broad-band operation, and high sensitivity, and they are relatively inexpensive. Care must be taken not to overload the instrument by using it too near powerful RF sources.

Exposure Assessment

When preparing to evaluate exposure, one should be aware of potential sources of measurement errors (see Table 21.21), should determine source characteristics (see Table 22.22), and know what measurements are required (see Table 21.23). A useful guide to help prepare for an evaluation is available from the NCRP.(496)

Table 21.21 — Sources of Error

- No knowledge of operational and/or calibrated frequencies.
- Instrument responses to amplitude modulated fields or pulsed emissions.
- Perturbation of the fields to be measured by evaluator and/or instrument.
- Standing wave formation and multipath effects.
- Probe interaction with the source and reflections.
- Interaction of the leads and metering instrumentation with the field.
- Zero drift and field zeroing problems.
- Probe burn-out in intense fields.

Table 21.22 — Source Characteristics

Primary Characteristics

- Frequency (fundamental and harmonic content)
- Modulation (FM, AM, other)
- Polarization
- Continuous wave (CW) or pulsed
- Characteristics of pulsed sources
  - pulse repetition frequency (prf)
  - pulse width or duration
  - duty cycle (factor)
- Antenna characteristics
  - power (effective radiated power (ERP), output)
  - effective area
  - diameter of aperture antenna
  - gain
  - beam width
  - scan rate (angle)

Table 21.23 — Types of RF Measurements Required by C95.1–1991

Frequency	*Measurement*
300 GHz–300 MHz	E or H or W: spatial average
300 MHz–100 MHz	E & H: spatial average
100 MHz–3 kHz	E & H: spatial average[A] Induced currents Contact currents[B]

[A]Consideration should be given to remote monitoring of low-frequency E fields, since the evaluator's body may significantly perturb these fields.
[B]Only required if there are conductive objects that may contain/store RF energy as an electrical current.

Because all RF instruments respond properly to a finite frequency range, the operational frequency of the source must be within the calibrated frequency range of the instrument. If it is not, use an instrument that has the proper frequency range, or contact the instrument manufacturer about extending the calibrated frequency range.

Prior to collecting measurement data or making calculations, the evaluator should be familiar with the source and tasks related to RF exposure, preferably by observation. In general, the evaluation is done under worst-case conditions, since all other conditions of exposure for a given task will be acceptable if the worst-case does not result in overexposure.

Field Survey Procedures. Before making measurements, allow the instrument to equilibrate with the temperature in the survey area, then check the instrument zero in a RF-free zone. First, evaluate source emissions (look for "hot spots" and areas of worst-case leakage), then assess exposure.

To evaluate emissions, extend the probe in the direction of the source, then proceed toward the source, slowly immersing the probe in the field, so that it "sees" the field first. Move the probe in the horizontal direction and then the vertical direction across a relatively small area, probing for hot spots. A thorough evaluation will probably involve placing the probe at locations not normally occupied, and may result in determining relatively intense fields. Remember that, although source characterization may produce useful data on emissions, it should not be used to assess potential overexposure, which is the next step in the evaluation. However, if emissions are low (e.g., <1/5 the occupational limits) or not detectable (limits of detection should be <1/5 the occupational limits), it is reasonable to document the results and not perform further evaluation.[370]

Now, from the evaluation of emissions and the observations of the job, determine the possible worst-case exposure condition(s). Exposure should be evaluated without the operator being present, the operator-absent condition, since the presence of the operator's body may perturb (influence) the RF fields and their measurement. One way of doing this is spatial averaging.[541,552] This is a method of estimating average exposure of the whole body by collecting densitometric data across the vertical dimension of a simple linear model of the body. A typical model is a 200-cm tall dielectric "stickman," made from materials of low relative permittivity such as PVC or CPVC pipe, which is placed at the operator position.[370] If the instrument does not have a spatial-average function, a minimum of 10 equally spaced (e.g., 20-cm spacing) measurements must be made between 0 and 200 cm, then arithmetically averaged. When averaging field-strength units, remember to use the square of the units, since the SAR is power dependent. For power density, squaring the measured values is not necessary; just sum them and divide by the number of observations.

When collecting data, be aware of the requirements for the measurement distance, since this varies with different standards. For example, C95.1 specifies 20 cm,[541] or three probe diameters, from the source, while the ACGIH recommends 5 cm.[101] Because of spatial variability of the field and the rapid change of field strength with distance, large differences may occur in measurement data collected at 5 and 20 cm.

The exposure guidelines specify that exposure data must be time-averaged. This is done manually, or with a data logger, or time-averaging module. A 6-min averaging time usually applies to occupational exposures at frequencies less than 15 GHz. Between 15 and 300 GHz, averaging time is inversely related to frequency, and must be calculated.[101,541]

If exposure is relatively uniform and continuous for the averaging time, it is not necessary to sample for the entire period. However, if the intensity of exposure varies, as might occur if the worker moves in and out of the field (e.g., when supplying parts or removing product from a manufacturing process), it will be necessary to determine a time-weighted average (TWA), where:

$$\text{TWA} = \frac{(ML_1 \times t_1) + (ML_2 \times t_2) + \ldots + (ML_n \times t_n)}{t_a} \quad (20)$$

Here, ML is the measured level (spatial average) for a given exposure duration, t, and t_a is the averaging time, where the sum of the values of t equal t_a.

Hazard Calculations for Intentional Radiators. These provide an estimate of either the power density at a given distance from an antenna or the hazard distance.[367,370,492] Hazard distance (range) is the linear distance from the antenna at which the field intensity is reduced to the exposure limit.

To calculate the maximum power density on the beam axis (W) in the near field of an aperture antenna, use:

$$W = 16P\eta/\pi D^2 = 4P\eta/A \quad (21)$$

Here, P is the antenna power (watts), η is the antenna efficiency (values ≤ 1), D is the diameter of the antenna, and A is the cross-sectional area of the antenna.

Since the near-field equations are independent of distance, it is presumed that the calculated power density exists between the source and the boundary between the near and far fields, R, which is estimated by:

$$R = \frac{2D^2}{\lambda} \quad (22)$$

Now, to calculate the power density in the far field, use:

$$W = GP/4\pi r^2 \quad (23)$$

or $W = GP/\pi r^2$, which assumes 100% ground reflection and produces a slightly higher estimate of power density.

To estimate the hazard distance in the far field of aperture antennas, substitute the appropriate exposure limit (EL) for W, in Equation 23, then rearrange to solve for distance.

$$r = (GP/4\pi EL)^{1/2} \quad (24)$$

Again, when 100% ground reflection is assumed, the hazard distance equation becomes $r = (GP/EL)^{1/2}$. Gain, G, may be calculated as shown in Equation 18.

For FM and TV antennas, hazard distance formulas without and with ground reflections are, respectively:

$$r = (30\ ERP)^{1/2} / EL \quad (25)$$

$$r = (48\ ERP)^{1/2} / EL \quad (26)$$

ERP is the effective radiated power, in watts, which is the product of gain and transmitter power. Although Equations 25 and 26 can be derived in terms of power density, they are normally expressed in terms of electric-field strength for broadcast sources. So, EL is in V/m and r is in meters.

Measurement of Induced and Contact Currents. Induced and contact current must be measured for exposures at frequencies between 3 and 100 MHz. Note, this does not include FM broadcast frequencies between 100 and 108 MHz (see Table 21.23). (IEEE Standard Coordinating Committee 28 is considering this as a possible revision to the C95.1 standard.) It has been shown that values of induced current may exceed the exposure limits, even when the strength of the E- and H-fields do not exceed the limits for free fields.[562]

To measure induced currents, stand-on instruments should be located at the position where the operator normally stands or sits. Connect the remote readout device, if one is used. Have the operator stand on the instrument and perform his or her normal function while you monitor the output. If the source is pulsed, be sure to monitor a sufficient number of work cycles to determine operator exposure. For current transformers, clamp the device around the appendage (usually the ankle, as shown in Figure 21.17b). Locate the data logger on the belt and secure the optical fiber so that it is out of the way. Switch the units on, and determine exposure. (Note that the measured value of the induced current varies, even if the source has a relatively constant RF field, if the operator moves around, momentarily contacts conductive objects, or is joined by another individual at the work location.) Presently, the averaging time for induced currents is 1 second. However, this may be changed to 6 minutes in the future.

Exposure guidelines require that contact currents be measured if there are conductive surfaces or objects that can store RF electric energy, and release this on contact. To make this determination, one must be familiar with objects in the workplace and how these objects are grounded, or their lack of grounding. (Note: electrical grounding does not ensure that the machine is properly grounded for RF currents.) Conductive objects to consider include ladders, metal tables, metal handles or pulls, vehicles, fences, reinforcing bar, chains, guy wires, frames, metal roofs, and shipboard components.

Currently, only one contact current instrument is commercially available. To use this instrument, connect the sampling probe to the instrument. Select the highest range if the magnitude of the current is unknown, and locate the meter at ground potential, usually on the floor. The path of the current is through the handheld sample probe and to the meter via the connective cable. Touch the sample probe to the surface and depress the probe switch to determine the current. Remember that these numbers represent human impedance where the worker is barefoot and grasping the object with his hand.

Control Measures

Engineering Controls

Engineering controls include interlocking, shielding, bonding, grounding, filtering, and waveguides below cutoff. Good design practices include fail-safe interlocks, built-in leakage detectors, visual "power-on" indicators, and visual and audible alarms. Interlocks are used at system access points, such as transmitter access panels. Because of the potential for faulty relays and switches, it may be prudent to use redundant relays in circuits for alarms and power-off conditions.[563] A maintenance program may be established to evaluate safety system performance periodically, including interlock checks. Lock-out/tag-out procedures for working with sources of hazardous energies must be implemented and followed. If flammable materials are used, fire/smoke detectors, alarms, and extinguishers should also be included in system specifications.[564]

Shielding and Enclosures. Incident RF fields induce currents in the surface of conductive materials such as those typically used for shielding. Shields use reflection, absorption, and internal reflection[565,566] to manage these currents and reduce the incident fields that penetrate the conductive material.

Shielding effectiveness (SE) is used to describe the capability of a material as a shield (see Table 21.24). SE is a function of losses resulting from absorption, reflection, and internal reflection. SE, in decibels (dB), is calculated by:

$$SE = 10 \log_{10} (W_i/W_t) \text{ dB} \qquad (27)$$

where W_i and W_t are, respectively, the incident and transmitted power density.

Shields are constructed from a single layer of a suitable material or layers of materials. Shielding materials exhibit a frequency-dependent nature in SE. Plane waves are effectively attenuated by materials that have high conductivity, low permeability, and low characteristic impedance. Generally, materials suitable for E fields produce high reflective losses. Metals that are most effective for E fields and plane waves include silver, copper, gold, aluminum, brass, bronze, tin, lead, chromium, and zinc.[567] Polymers and polymer blends containing conductive materials are also used.[568,569]

H-field shielding materials include iron, some stainless steels (430), steel (SAE 1045), grain-oriented silicon-iron, nickel-iron alloys such as Mu-Metal, and cobalt-iron alloys such as Vaccoflux.[567,570] Composites, made of multiple layers of materials such as copper-stainless steel-copper, copper-alloy-copper,[571] and copper-plywood,[572] may also be used.

RF-shielded enclosures use shielding materials to reduce leakage and penetration of RF fields. In designing an enclosure, attention is given to the selection of the base shielding material and to seams, panels, flanges, cover plates, doors, ventilation openings, cable penetrations, and grounding. Seams should be overlapped[573] and welded.[574] Panels and cover plates should be attached with conductive gaskets such as those made of Monel, neoprene, silicone, and tin-plated beryllium-copper.[575,576] Gasket materials may be fabricated into various forms including metal meshes, metal meshes on elastomer cores, knitted wire mesh, spring-finger stock, foam-backed metal composites, hollow tube (conductive neoprene), and various specialty shapes.[576,577] Periodic maintenance of gaskets will be necessary, and the frequency of maintenance depends on the environment and the type of gasket used.

Waveguide Below Cutoff. A waveguide is often a hollow metal tube (circular, rectangular, or square) used to confine and guide electromagnetic waves with minimum transmission loss. Conversely, the purpose of a waveguide below cutoff is to attenuate

Table 21.24 — Interpretation of Field Reduction Values

Attenuation (dB)	*Qualitative Description*
0–10	Negligible shielding
10–30	Minimum significant shielding
30–60	Average shielding
60–90	Above average shielding
90–120	Excellent shielding
>120	Superior shielding

electromagnetic waves. If there are openings in shielded enclosures for ventilation or conveyor openings in RF process equipment, RF leakage through these openings will decrease the SE of the enclosure. The use of waveguide below cutoff sleeves in the openings will help maintain the SE of the enclosure. An example is a honeycomb filter that is used to allow air movement but attenuates electromagnetic waves.[578,579] Waveguides below cutoff have also been used successfully to control leakage around conveyors in industrial microwave dryers and RF sealers.[580]

Resonant Frequency Shift. If the source operates near the whole-body, grounded-resonance frequencies, around 10 to 40 MHz, the WBA-SAR may be reduced by separating the body from ground by a small distance. The SAR reduction is achieved by effectively shifting the body into the free-space resonance condition, while exposure is actually at grounded-resonance frequencies. This reduces the body's ability to absorb low-frequency RF energy. This has been demonstrated by simulating an air gap between human subjects and ground with expanded polystyrene and hydrocarbon resin foam, both of which are electrically insulating materials.[581] Other insulating materials may also prove successful in reducing the SAR, as long as they are relatively transparent to RF, as indicated by low values of relative permittivity (measure of the effect of a material on the E field relative to free space), and the separation distance is sufficiently large.[500,581,582]

Location of Equipment and Personnel. Studies have indicated that placement of a lossy dielectric, such as a person, near a flat, RF-reflective surface, can enhance the SAR. This effect was further enhanced if the person was located in a reflective corner.[546,583,584] The materials of building construction and the placement of RF devices within the building are therefore important considerations in minimizing exposures.

Administrative and Procedural Controls

Administrative controls include footwear, protective clothing and gloves, prepurchase review of sources, controlling the duration of exposure, increasing the distance between the source and workers (although this may be achieved by engineering modifications), restricting access, and placing warning signs.[370]

Footwear, Protective Clothing and Gloves. Footwear may modify absorption by volunteers exposed on a ground plane at frequencies between 0.63 and 50 MHz.[500,555,559,581,585] This is accomplished by reducing the grounding effect and shifting the body toward the free-space resonance condition, as discussed earlier. The level of reduction depends on the type of shoes and socks worn and the RF frequency. For example, the use of nylon or wool socks reduced absorption to the same extent as an air gap of comparable thickness,[581] and rubber-soled shoes[500] and clogs with thick wooden soles[559] had a similar effect.

RF-protective suits are made of a base material (e.g., wool, polyester, or nylon) that is impregnated with a highly conductive metal, such as silver, or is woven with metallic stainless-steel thread.[586-590] If the metallic fibers are oriented in one direction, they may demonstrate polarization sensitivity. A mesh design, where the fibers occupy vertical and horizontal positions, is not sensitive to field polarization.[591]

Openings in RF-protective suits for the hands or feet, zippers, or facepiece may introduce the potential for RF leakage into the suit. In addition to RF penetration, other concerns include suit ignition, arcing, and standing-wave formation. If RF-protective clothing is used, it is important to ensure that adequate test data indicate the necessary attenuation will be achieved at the specified frequency or frequencies, the suit is not flammable, and no electrical safety hazard exists. Users should be trained on the proper use of the suit and to understand and recognize use limitations in their workplace.

Protective gloves may provide protection from RF-induced contact currents and spark discharge. Rubber insulated gloves can be effective in reducing low-frequency contact currents. OSHA inspectors have documented two cases in which gloves were effective in controlling spark discharge to the hands of longshoremen. In one report, workers wore fleece-lined rubber gloves where the open-circuit voltage was 100 to 330 V and the short-circuit current was 180 to 1900 mA.[592] In the second case, workers wore two pairs of gloves. The outer pair was leather, while the inner pair was rubber. The open-circuit voltage and the short-circuit current were 300 V and 300–1000 mA, respectively.[593]

Tell reported current reduction factors for gloved hands vs. bare hands at AM radio frequencies of 4 to 31 times. The magnitude of the reduction depended on the RF frequency, glove type, and hand pressure. Reduction factors varied inversely with frequency, and heavy duty leather work gloves had the greatest reduction factors, while a rubber glove had the lowest.[594]

Distance. Increasing the distance between the person and the source is probably the most frequently used control measure, but it is also the control measure that can be most easily circumvented. Workers should be informed of the extent of the zone of exclusion and the supportive rationale. Zone limits are often delineated by clearly visible methods, such as tapes, paints, or signs, or by physical delimiters such as rope, fencing, or traffic cones. Both horizontal and/or vertical distance from a source may be limited. Controlling vertical distance may be necessary to minimize exposure of maintenance personnel on radio and MW towers.

Duration of Exposure. Many exposure guidelines allow an RF energy dose (SA) of 144 J/kg for the applicable averaging time, where SAR $\times$ t_a = 144 J/kg (e.g., 0.4 W/kg $\times$ 360 sec = 0.08 W/kg $\times$ 1800 sec = 144 J/kg). Reciprocity allows the SAR to be higher if the exposure duration (T) is less than t_a, as long as the product of the two (SAR $\times$ T) does not exceed 144 J/kg. Since the exposure limits are derived from the SAR, it follows that if the product of exposure duration and exposure intensity (e.g., measured level, ML) does not exceed some constant value (e.g., T $\times$ ML $\leq$ EL $\times$ t_a), the exposure is within acceptable limits. For a given exposure (ML), the allowable exposure duration (t*) can be calculated from:

$$t^* = (EL \times t_a) / ML \tag{28}$$

where EL is the exposure limit and ML is the measured (or calculated) level. Remember, this control measure should be used only when ML exceeds EL and the exposure duration is less than t_a.

Warning Signs. The ANSI C95.2 subcommittee recommends the design and color scheme of a warning symbol to be used for RF radiation between 300 kHz and 100 GHz. Recommendations for including this symbol in a sign are made by ANSI.[595] (SCC28 is revising this standard to include "no touch" and contact current warning symbols and signal words.)

Work Practices. Good work practices include reducing the source operating power during "hot" maintenance and service to levels that do not result in potential overexposure. Following maintenance activities, it is important to ensure that all fasteners have been replaced and that all areas of potential leakage, such as waveguide flanges and access panels, are secure. It is important to verify that the area is clear of personnel before switching on sources that generate RF beams. Maps of areas of potential over exposures may be generated and used with area demarcation.

Extremely Low Frequency (ELF) Fields*

Anticipation

When the source-receptor distance for electromagnetic radiation is large compared with the wavelength, the electric and magnetic fields are linked and must be considered together (although measurement of either will suffice). This is referred to as the "far" or "radiation" zone, where there is an electromagnetic field. A television signal with a frequency on the order of 10^8 Hz and a wavelength of about 3 m broadcasts, or radiates, as an electromagnetic quantity.

When the source-receptor distance is small with respect to the wavelength, the electric and magnetic fields are not linked. This is the "near" or "static" zone, where the fields are independent and can be considered as separate entities. ELF fields have wavelengths on the order of 1000 km, so one is always in the near or static zone. ELF fields are considered as separate, independent, nonradiating electric and magnetic fields at any conceivable observation point. This is referred to as the "quasi-static approximation."

Quantities and Units

Electric fields are created by electric charges. The electric field, **E**, is defined by the magnitude and direction of the force it exerts on a static unit charge

$$\mathbf{F} = q\mathbf{E} \tag{29}$$

where:

F = force (in newton);
q = electric charge (coulomb); and
E = electric field (V/m). [Vector quantities are in bold type.]

The magnitude or intensity of **E** describes the voltage gradient, or the difference in voltage between two points in

* This section was written by Robert M. Patterson and R. Timothy Hitchcock

the field. Electric field intensities near alternating current (ac) high-voltage transmission lines are in the range of kilovolts per meter (kV/m).

The magnetic flux density, **B**, characterizes the magnetic field strength, **H**, as **E** characterizes the electric field and is linearly related to the magnetic field through:

$$\mathbf{B} = \mu\mathbf{H} \quad (30)$$

where:

B = magnetic flux density (tesla [T]);
μ = magnetic permeability; and
H = magnetic field strength (A/m).

The magnetic permeability depends on the medium; the magnetic permeability of a vacuum is designated μ_0. Air and biological matter have permeabilities essentially equal to μ_0. This means that the magnetic flux density is unchanged by these materials.

Basic Physics

Magnetic fields are created by moving charges, or currents. This applies to all fields, whether they are from magnets, power lines, or the earth. Just as the electric field is defined by the force on a unit charge, the magnetic field is defined by the magnitude and direction of the force exerted on a moving charge or current

$$\mathbf{F} = q(\mathbf{v} \times \mathbf{B}) \quad (31)$$

where:

v = velocity (m/sec).

The magnetic flux density, which may be represented by lines of induction per unit area, has units of tesla, T. One tesla is 10,000 gauss (G), a frequently used engineering unit. The microtesla (μT) is more convenient for environmental levels. Convenient conversions are 1 μT equals 10 mG, and 1 mT equals 10 G. The static magnetic flux density of the earth is about 50 μT.

The CGS system of units, which preceded the present SI system, used units of centimeters, grams, and seconds. In the CGS system, the permeability, μ_0, is dimensionless and equal to 1. As a result, **B** numerically approximates **H** in the CGS system, and the two came to be used interchangeably. The value of μ_0 in the SI system is 4×10^{-7} H/m. This factor can be used to convert true magnetic field strength (A/m) to magnetic flux density (T) in air by the expression $\mathbf{B} = \mu_0\mathbf{H}$.

Interaction with Matter/People. Electric fields interact with humans through the outer surface of the body, inducing fields and currents within the body. Hair vibration or other sensory stimuli may occur in fields greater than 10 kV/m. A safety issue arises from currents induced in metal structures, which may produce shocks when humans contact the structure and provide a path to ground.

An electric field causes currents to flow in the body, as expressed by Ohm's law:

$$\mathbf{J} = \sigma\mathbf{E} \quad (32)$$

where:

J = induced current density (A/m^2); and
σ = tissue conductivity in siemens per meter (S/m).

A grounded person in an electric field experiences a short-circuit current (in this case, current flowing out of the foot to earth) of approximately:[596]

$$I_{sc} = 1.5 \times 10^{-7}\, f\, W^{2/3}\, E_o \quad (33)$$

where:

I_{sc} = short-circuit current (μA);
f = frequency (Hz);
W = weight (g); and
E_o = external electric field strength (V/m).

Thus, a person weighing 70 kg would have a total short-circuit current of about 153 μA in a 10 kV/m field. Deno,[597] and Kaune and Phillips,[598] investigated current flow in models of humans and laboratory animals exposed to 60-Hz electric fields. Their data indicate that current densities induced in a grounded, erect person exposed to a 10 kV/m vertical electric field are 0.55 $\mu A/cm^2$ through the neck and 2 $\mu A/cm^2$ through the ankles.

Time-varying magnetic fields induce electric fields, which in turn induce currents, in tissue in direct proportion to the magnetic flux density, the frequency of oscillation, and the radius of the current loop. The electric field and current flow are perpendicular to the flux density. A vertically directed flux density causes current to flow in standing humans in loops whose plane is perpendicular to the vertical axis. For a sinusoidally varying flux density and a circular current flow, the current density can be expressed as:

$$J = \sigma\pi frB \qquad (34)$$

where:

f = frequency (Hz); and
r = loop radius in meters (around 0.15–0.2 m).

With average tissue conductivity equal to 0.2 S/m, the current density at the perimeter of the torso of an adult can be approximated as:

$$J = 0.1\ f\ B \qquad (35)$$

The maximum current density induced in the normal residential environment is of the order of $\mu A/m^2$. For electric arc welders it may be of the order of mA/m^2.

With photon energy directly proportional to frequency, it is readily apparent that ELF fields will cause neither ionization nor heating. Because the body is a relatively good conductor, the highest internal field that can be induced by an E-field strength in air is about 1 V/m, which leads to a mass-normalized rate of energy transfer of 10^{-4} W/kg, 4 orders of magnitude less than the resting metabolic rate.[(599)] Pulsed magnetic fields can produce higher internal electric fields, but they are still too small to produce measurable tissue heating. Any interactions of ELF fields in air with humans are thus nonthermal.

Biological and Health Effects

The search for biological and health effects of EMF exposure has been going on at an accelerating pace for nearly three decades. Results have been mixed, and their interpretation has been controversial, with little consensus on biological effects and virtually none on health effects that might arise from fields at the levels found in occupational or general community environments.

The next few paragraphs discuss so-called robust biological effects; that is, effects that are reproducible and generally accepted among researchers. This is followed by a discussion of the more controversial nonrobust biological effects, which are found by some investigators but not by others who attempt to replicate their results. The section closes with an overview of the epidemiological studies into possible health effects of EMF.

The previous section described how the effect of fields exerts force on charged entities, creating currents, and how the relations between fields and induced currents are quantified through the conductivity of the medium, for example, tissue. Organizations have recommended exposure limits for ELF-EMF based on the following correspondence between current density and biological effects:[(600)]

- 1–10 mA/m^2; minor biological effects have been reported;
- 10–100 mA/m^2; well established effects occur, including effects on the visual and nervous systems;
- 100–1000 mA/m^2; stimulation of excitable tissue occurs, causing possible health hazards;
- 1000 mA/m^2 and above; extrasystoles and ventricular fibrillation can occur.

Recalling the approximation, $J \approx 6$ B (at 60 Hz), magnetic fields of the order of 1 mT would lead to current densities in the range of 6 mA/m^2.

Phosphenes. Perhaps the most commonly mentioned robust biological effect attributed to EMF is the sensation of flashes of light within the eye, called phosphenes. These are termed electrophosphenes and magnetophosphenes, depending on the field that causes them, and they originate when the induced current density is of the order of 10 mA/m^2 or more. For example, experiments show that the threshold for production of magnetophosphenes is between 2 and 10 mT for magnetic fields in the frequency range of 10–100 Hz.

Calcium Efflux. The common paradigm guiding exposure limits, such as TLVs is that more is worse, or "the dose makes the poison." This is the basis for present EMF occupational exposure limits, but it is also the subject of considerable discussion and debate. This controversy has arisen because of a number of experiments that are reported to have shown effects occurring in windows of field frequency and power. A chemical exposure analogy would be as if a little benzene were harmful but a greater amount were not, and a still greater amount were again harmful. The idea of effects of electromagnetic energy occurring in windows of different parameters does have some foundation: consider that we see in a frequency window at about 10^{15} Hz (or more commonly expressed in wavelengths from about 400–700 nm). One immediate difference, however, is that this is a single window rather than multiple windows. The studies that suggest window effects are the so-called calcium efflux experiments. These are in vitro effects and are summarized next.

In the first of these studies,[(601)] fresh chick brain tissue was spiked with the radioactive isotope, $^{45}Ca^{2+}$. Half of a brain was exposed to a 147-MHZ field that was amplitude modulated at 16 Hz; the other half served as a control. Compared to the unexposed half, the exposed half had about a 20% increase in calcium exchange with the physiologic solution in which it was immersed. Other modulation frequencies were used, but none produced as large an effect. Plotted against modulation frequency, the relative amount of $^{45}Ca^{2+}$ exchanged followed a curve indicating a resonant response. Without the 147-MHZ carrier frequency, a 16-Hz electric field had the opposite effect, namely, the exchange from the exposed half was decreased relative to the unexposed half.

A follow-up study in another laboratory found an increase rather than a decrease in $^{45}Ca^{2+}$ exchange in chick brains exposed to combined 15-Hz electric and magnetic fields. Later work suggested that the effect depended on the relative orientation and strength of the earth's magnetic field, and that it occurred in "windows" of field frequency and intensity and even temperature of the experimental preparation.[602-605]

Liboff suggested that the resonant response was due to a cyclotron resonance phenomenon operating on K^+ ions, which in turn were involved in transmembrane exchange with the $^{45}Ca^{2+}$ ions.[606] Studies of rat operant behavior by Thomas and colleagues[607], of diatom motility by Smith et al.,[608] and of lymphocyte incorporation of $^{45}Ca^{2+}$ by Liboff et al.[609] supported the proposed resonance mechanism. However, later studies of turtle colon by Liboff and coworkers,[610] cells by Parkinson and Hanks,[611] lymphocytes by Prasad et al.,[612] and rat operant behavior by Stern and Laties were all negative.[613] The last two studies are noteworthy as they attempted to replicate the earlier work of Liboff[609] and Thomas.[607]

The suggestion that the movement of $^{45}Ca^{2+}$ in these studies is due to a cyclotron resonance mechanism driving the ions through cell membrane pores has been discounted on a number of grounds (e.g., the ions are hydrated and the radius of the ion's orbit is too large). However, other theoretical descriptions have been proposed,[614,615] using classical physics and quantum physics descriptions of the modification of energy levels of a charged oscillator in a magnetic field. The modification of energy levels changes the binding rate of ions to proteins, which in turn alters their biological activity. Theoretical problems also exist with these descriptions,[616] and initial laboratory studies have been unable to confirm them.[617]

Neurite Outgrowth. More recently, neurite outgrowth from cells has been attributed to EMF exposures in vitro, and the causal mechanism has been described by an ion parametric resonance model.[618] However, the significant neurite outgrowth response to EMF exposure seen in one laboratory[618] was not replicated in another.[619]

Genetic Effects. Genetic toxicology studies have found no reliable effects, but there are scattered positive findings.[620,621]

Reproduction and Development. Graves et al. exposed chicken embryos to 60-Hz, sinusoidal electric fields with strengths up to 100 kV/m. No effects were found on mortality, deformity, birth weight, or postnatal development.[622]

A large-scale study of chicken embryos was conducted in six laboratories in North America and Europe.[623] The exposures were to magnetic fields with a 1 µT flux density and rapidly pulsed at 100 Hz. Two laboratories found a significant increase in abnormal embryos in the exposed groups. Pooled data from all six laboratories were also significant. However, the interaction between the incidence of abnormalities and the laboratory doing the experiment was also significant, that is, the effect of exposure differed significantly among laboratories.

In a multigenerational study, female Hanford miniature swine were exposed to 60-Hz electric fields of 35 kV/m for 20 hr/day. When these and an unexposed control group were bred with unexposed males, the control group had a greater rate of fetal malformations than the exposed group. This result reversed in a second breeding. Breeding of the offspring produced similarly conflicting results.[624]

Because of the ambiguities in the swine study, a study using a similar protocol was performed with rats.[625] Electric field exposures included 0, or control, and 10, 65, and 130 kV/m. There were no significant increases in litters with malformations among exposed animals.

Magnetic field exposures were used in a developmental toxicology study in rats. Flux densities were 0.09 µT (sham exposure), 0.61 µT, and 1000 µT. There were no significant differences in fetal body weight or the incidence of malformations among the exposure groups.[626]

Effect on Melatonin. The possible effects of ELF fields on levels of the pineal hormone melatonin have been investigated. Night-time suppression of melatonin was found in rats exposed to 60-Hz E fields.[627] Since then, some studies have reported effects from E fields,[628] B fields,[629] and no effects.[630-633] Other findings were that E-field exposure decreased serum melatonin but not pineal melatonin levels;[634] pulsed static B fields affected melatonin if exposure was during the mid- or late-dark phase, but not when exposure was during the day time or early dark phase;[635] short-term B-field exposure affected pineal and nocturnal serum melatonin, while long-term exposure affected just nocturnal serum melatonin;[636] serum melatonin was not affected by day-time E- and B-field exposure with a slow onset,[637] but it was reduced by variable exposure with a rapid onset/offset.[638]

A possible role of ELF fields in carcinogenesis was proposed by Stevens who hypothesized that reducted melatonin, which has been suggested to have tumor-suppression properties, may lead to greater tumor growth.[639] Research using the mammary tumor inducer dimethyl benzanthracene, or DMBA, followed by a 60-Hz, 40-kV/m electric field exposure showed no significant difference between exposed and sham-exposed groups of rats.[640,641] Stevens and Davis have published a review of findings relative to the "melatonin hypothesis" and breast cancer.[642] Exposures of leukemia-implanted mice to 60-Hz magnetic fields of up to 500 μT did not change their survival compared with that of unexposed mice.[643]

Other Studies. Animal research on possible effects of EMF exposure continues, including additional cancer studies and studies of the immune system. Although some evidence suggests a promotional effect, a robust response has yet to emerge, and cause and effect has not been established.[644-647]

Studies of Human Volunteers. Studies of human volunteers exposed to combined electric and magnetic fields (60 Hz, 9 kV/m, and 20 μT) have found a statistically significant slower heart rate and a change in brain evoked potential. No significant responses occurred at exposures of 12 kV/m and 30 μT. The observed responses appeared to be associated with changes in field conditions, that is, with intermittent exposure, rather than exposure strength or duration.[648,649] Exposure to a continuous, 200-mG field produced no effect.[650] None of these results has been independently replicated. However, other researchers observed no effects on circadian rhythm of hematologic and immunologic functions in young men exposed to continuous and intermittent magnetic fields (50 Hz and 10 μT).[651]

Male volunteers were exposed to intermittent 60-Hz magnetic fields during a single night, while hourly samples for plasma melatonin were collected. No differences were observed between sham controls and the two exposure groups (10 mG and 200 mG). However, subjects with pre-existing low levels of melatonin who were exposed to 200-mG B fields and bright light demonstrated further melatonin reductions that were statistically significant. A second study evaluated the potential for effects on men with preexisting low levels of melatonin and exposed to a 200-mG B field. As in the first experiment, there was no overall effect on serum melatonin levels. In contrast to the initial findings, there was no significant decrease in serum melatonin levels in subjects with pre-existing low levels.[652]

Epidemiological Studies. Occupational epidemiological studies of ELF exposures and adverse health effects have focused on cancer. Designs have ranged from analyses of cohort death records to case-control studies of occupational groups. A number of useful review articles are available.[644,653-658]

An increased risk of various forms of cancer among those employed in "electrical occupations" such as electricians, electronics engineers, and radio repairmen has been identified in statistical analyses of death records.[659-661] These studies commonly employ statistics known as the standardized mortality ratio or the proportionate mortality ratio to identify increased risk. Deficiencies included a lack of consistency in designating occupational classification, no consideration of mobility between occupations, and lack of consideration of confounding exposures to other physical and chemical agents.[8]

A review article covering 11 separate studies of workers in electrical occupations concluded that the most consistent finding was a small increase in the risk of leukemia. By combining the results from all studies, a relative risk of 1.18 was calculated, with a 95% confidence interval (CI) of 1.09–1.29. The authors warned that the results are equivocal with respect to cause, because electrical workers are also exposed to agents other than fields, some of which may be leukemogenic.[662]

In contrast to these results for leukemia, a case-control study of deaths due to primary brain cancer or leukemia found no increased risk of leukemia among those in electrical occupations (odds ratio of 1.0, 95% CI 0.8–1.2).[663] There was an increased risk of brain cancer (odds ratio of 1.4, 95% CI 1.1–1.7), with electrical engineers and technicians, telephone workers, electric power workers, and electrical workers in manufacturing industries all showing elevated risk. Among these groups, the odds ratios for leukemia were from 1.1 to 1.5, but none was significantly different from 1.

Welders are among the workers most highly exposed to power-frequency magnetic fields. Analyzing published data, Stern found the relative risk of leukemia among welders to be essentially equal to 1.[664] There were no statistically significant findings in a study of cancer among railway engine-drivers and conductors by individual occupation. However, when data were combined for both occupations, the relative risk of lymphocytic leukemia was significantly elevated (2.3 95% CI 1.3–3.2).[665]

Tynes and colleagues found significantly increased risk among a cohort of 37,945 Norwegian electrical workers (evaluated by job title) for cancers of the colon, pancreas, larynx, breast, and bladder. For leukemia, acute and chronic myeloid leukemia were significantly elevated (SIR 1.56 and 1.97, 95% CI 1.06–2.26 and 1.10–3.26, respectively), but not other types.[666] However, a Danish study, which also used job title, found no increase in the risk of breast cancer or brain tumor, but the incidence of leukemia was significant among men (but not in women), with a value of 1.64 (95% CI 1.20–2.24).[667] Also, in contrast to the results of Tynes,[666] Floderus and coworkers reported an increased risk of chronic lymphocytic leukemia but not acute myeloid leukemia in a case-control study in Sweden.[668]

A number of studies have examined the risk of breast cancer and ELF-EMF exposure. Guenel classified exposure by industry and occupation and as continuous or intermittent. No significant increase in risk of breast cancer was observed in men or women.[669] Coogan et al. classified exposure on the basis of "usual occupation." Odds ratios for the high exposure category were 1.43 (95% CI 0.99–2.09) for all subjects, 1.98 (95% CI 1.04–3.78) for premenopausal women, and 1.33 (95% CI 0.82–2.17) for postmenopausal women. Computer equipment operators had an odds ratio of 1.79 (95% CI 1.03–3.11) while precision inspectors, testers, and relative workers were 7.99 (95% CI 1.69–37.84).[670] In a mortality study, Loomis and colleagues reported a statistically significant increase in the risk of breast cancer for telephone installers, repairers, and line workers (odds ratio 2.17, 95% CI 1.17–4.02). Compared with the findings of Coogan et al.,[670] the odds ratio for computer equipment operators was not elevated.[671]

Three large occupational epidemiological studies of electric utility workers were completed recently. Sahl et al. found no increased risk of cancer among the group of utility workers that they studied.[672] Theriault and colleagues used a nested case-control study to examine morbidity in 250,000 workers in electric utilities in France and Canada. Results suggested an increased risk of a form of leukemia.[673] Savitz et al. found an increased risk of brain cancer but not leukemia among the utility workers in their study.[674] The studies' results are summarized in Table 21.25.

In a further study of Ontario utility workers, researchers reported a statistically significant increase in leukemia (odds ratio = 4.45 95% CI 1.01–19.7) for cumulative electric-field exposure, but not for magnetic-field exposure. For interaction between E and B fields, risk for all leukemia was significantly increased, but the estimates were imprecise, probably because of the small number of cases.[675]

Recognition

Generation

Electric fields are generated by electric charge, while magnetic fields are produced by the motion of electric charge. In the workplace, these fields are produced by the generation, transmission, and use of electricity, and anything in this path is a potential exposure source, from the generator, to the power lines, to an electric drill or clock.

Sources

Electrical appliances may have flux densities of about 10 μT in their very immediate vicinity. Some occupational environments, such as those where induction heaters are in use, may have flux densities of 10–100 mT, and electric furnaces may generate fields on the order of 1–100 mT.[644] Useful reviews of ELF magnetic fields are available.[644,676-683] Others have examined exposure from specific sources or environments including residential,[684-693] appliances,[695-699] schools,[699]

Table 21.25 — Summary of Three Power Industry Studies

Study	*Sahl et al. (1993)*	*Theriault et al. (1994)*	*Savitz et al. (1995)*
Exposure measure	*Above mean (25 μT-years)*	*Above 90th percentile (15.7 μT-years)*	*Above 90th percentile 4.3 μT-work years or 18.8 μT-years*
Brain cancer RR (95% CI)	0.81 (0.48–1.36)	1.95 (0.76–5.00)	2.29 (1.15–4.56)
Leukemia RR (95% CI)	1.07 (0.80–1.45)	1.75 (0.77–3.96)	1.11 (0.57–2.14)
AML RR (95% CI)		2.68 (0.50–14.50)	1.62 (0.51–5.12)

office,[700-707] medical,[708-710] transportation,[711-714] farming,[715] communications and broadcasting,[716,717] industry,[718-724] and utilities.[725-733]

The strongest electric fields generally occur in power plants, where maximum levels of 15–29 kV/m have been recorded, or near transmission lines and are typically about 10 kV/m directly under the wires and 1 or 2 kV/m at the edge of the right-of-way.[644] Fields from building wiring, power tools, and other electrical appliances typically range only up to 100 V/m.[732]

Occupational sources and source characteristics are as varied as the occupational environments in which they are found. However, two basic facts may be noted: sources of strong electric fields are associated with high electrical charge, such as around high-voltage equipment, and sources of high magnetic fields are generally characterized by high currents, such as around high-amperage equipment or locations of high current flow. These sources need not be associated with heavy industry; electrical transformers and wiring located in office building vaults are nonindustrial sources of high fields.

Stuchly and Lecuyer measured fields at the worker's position for 22 electric arc welding machines, which use high amounts of current. Electric fields were generally low, about 1 V/m (maximum = 300 V/m; mean = 47 V/m). Magnetic flux densities ranged from about 1 μT to a few hundred μT, and averaged 136 μT. The highest measurement was 1 mT at a worker's hand. The frequency of highest flux density was usually 60 Hz, although for some sources it was 120 or 180 Hz.[721]

Rosenthal and Abdollahzadeh measured flux densities in microelectronics fabrication rooms. In the aisles of the rooms, values ranged from 0.01 to 0.7 μT and averaged 0.07 μT. Higher levels were found near specific pieces of equipment. The authors estimated an 8-hr time-weighted-average exposure for a worker using the furnace to be 1.8 μT, and 0.6 μT for one using the sputterer.[722]

Very high flux densities were measured at the worker's position at welding machines and steel furnaces. Seventy mT was measured at an induction heater, and 10 mT was found near a spot welder.[723]

Bowman et al. sampled 114 work sites. "Electrical worker" environments had geometric mean electric fields of 4.6 V/m and flux densities of 0.5 μT. Secretaries had values of 2–5 V/m, 0.31 μT if they used a VDT, and 0.11 μT if they did not. For power line workers, the overhead line environment yielded geometric means of 160 V/m and 4.2 μT. A value of 5.7 μT was determined for underground lines. Other findings included 298 V/m and 3.9 μT at a transmission substation, and 72 V/m and 2.9 μT at a distribution substation. Radio and television repair shops yielded 45 V/m, while ac welding produced 4.1 μT.[241]

In a petroleum refinery, low-voltage electrical distribution workers had higher 8-hr geometric mean exposure than did high-voltage workers, 0.53 compared with 0.46 μT. These exposures are not remarkable, but high exposures occurred in the latter group during one particular operation, with values ranging from 0.2 to 1.8 mT. The geometric mean exposure was 71 μT, and the mean peak was 870 μT.[724]

Personal exposure data have been collected for work, nonwork, and sleep periods in a study of 36 Canadians — 20 utility workers and 16 office workers. The time-weighted average of 1-week's data yielded a geometric mean electric field of 10 V/m. The utility workers' geometric mean magnetic field exposure was 0.31 μT. It was 0.19 μT for the office workers. Both groups had a level of 0.15 μT while sleeping. At work, the utility workers' exposures averaged 48.3 V/m and 1.66 μT. Office workers were exposed to a geometric mean level of 4.9 V/m and 0.16 μT.[734]

Evaluation of exposure to workers grouped into 28 occupational categories at 5 United States power companies found the highest time-averaged B-field personal exposures in electricians, linemen, and cable splicers. Also, power-plant operators, instrumentation and control technicians, and machinists may also receive relatively high B-field exposures.[733]

Evaluation

Exposure Guidelines

Ideally, exposure guidelines and standards are established on the basis of an accepted mechanism of interaction, dose-response studies in animals, and epidemiological evidence of similar effects in humans. None of this has occurred for ELF fields. No accepted, biologically plausible mechanism has been advanced to explain how fields interact with biological systems to yield observed in vitro responses, much less disease in an organism. In fact, the field parameter or parameters to be measured, because of their possible biological significance, are unknown. The traditional approaches that "more is worse" and that time-weighted-average or even peak exposures are the quantities of interest have been called into question by studies suggesting that effects occur in windows of frequency and power. The choice of exposure metric is thus made doubly difficult by the lack of knowledge about interaction mechanisms and the lack of a clearly defined, exposure-associated health effect.

Exposure guidance, based on the current understanding of ELF-induced biological effects, has been developed by a number of countries and organizations. The ACGIH,[(101,735,736)] the National Radiological Protection Board (NRPB) in the United Kingdom,[(737,738)] the German government,[(600)] and the ICNIRP[(739)] have all promulgated standards or guidelines for exposures in the ELF range. The ACGIH guideline extends into part of the RF spectral regions, which that organization calls subradiofrequency.[(101)]

The rationale for ELF exposure limits is based on induced body currents. Both the ACGIH and ICNIRP recommend limiting induced current densities in the body to those levels that occur normally, that is, up to about 10 mA/m^2 (higher current densities can also occur naturally in the heart). They acknowledged that biological effects have been demonstrated in laboratory studies at field strengths below those permitted by the exposure guidelines, but both concluded that there was no convincing evidence that occupational exposure to these field levels leads to adverse health effects.[(735,736,739)]

ICNIRP recommends power-frequency E-field exposure limits of 10 kV/m for a whole working day, with a short-term limit of 30 kV/m. Interim times and field strengths are related by the formula, $t = 80/E$, where t is in hours and E is in kV/m. The limits for magnetic flux density for occupational exposure are 0.5 mT for the entire workday, 5 mT for exposures of 2-hr or less, and 25 mT for exposure to limbs.[(739)]

The ACGIH TLV for occupational exposure to ELF electric fields states that exposure should not exceed 25 kV/m for frequencies from 0 (DC) to 100 Hz. For frequencies in the range of 100 Hz to 4 kHz, the TLV is given by

$$E_{TLV} = 2.5 \times 10^6 / f \ (V/m) \qquad (36)$$

where:

E has units of V/m (rms value); and

f = frequency in Hz.

A proviso is added for workers with cardiac pacemakers, limiting power-frequency exposures to 1 kV/m. Electromagnetic interference with pacemaker function may occur in some models at power-frequency electric fields as low as 2 kV/m.

The TLV for magnetic fields from 1 to 300 Hz limits routine occupational (rms) exposure to ceiling values determined by

$$B_{TLV} = 60 / f \ (mT) \qquad (37)$$

where f is the frequency in Hz. At frequencies below 1 Hz, the TLV is 60 mT. From 300 Hz to 30 kHz the ceiling value is a limit of 0.2 mT. For workers with cardiac pacemakers, the limit is 0.1 mT at power frequencies.

The NRPB issued a Consultative Document in 1986.[(737)] In 1993, they published a basic restriction and an investigation level. The basic restriction is the fundamental criterion of the exposure limit, and the investigation level is for "investigating whether compliance with basic restrictions is achieved."[(738)]

The German DIN VDE Standard 0848 calls for a limit for "permanent exposure" of 40 kV/m for frequencies from 0 up to 10 Hz, and a functional dependence on frequency of $102{,}840/f^{0.4101}$ for 10 Hz to 30 kHz, with "about 50% higher field strength ... permitted in the occupational areas of electrotechnical staff."[(600)] This leads to occupational exposure limits of about 20 kV/m at power frequencies, falling to 1.5 kV/m at 30 kHz. Magnetic field exposures are limited to 16,000 A/m, rms, for 0 to 2 Hz (equivalent magnetic flux density 20 mT), and $21{,}593/f^{0.4325}$ A/m for frequencies of 2 Hz to 30 kHz. At 60 Hz, the limit would be 3,675 A/m (equivalent flux density 4.6 mT). Five-minute exposures of the extremities can be 10 times higher. As explained by Bernhardt and colleagues, the standard is based on there being "no indications that there exists a direct specific field effect other than caused by induced body currents," and that the limiting current densities are those avoiding biological effects on excitable cells.[(600)]

Instruments

Discussions of ELF EMF measurement instrumentation are available.[(644,740-744)] Instrument types include survey instruments and personal exposure monitors, as well as waveform capture instruments that provide spectral data. In general, instruments are broad-band and dedicated to either E-field or B-field measurement,

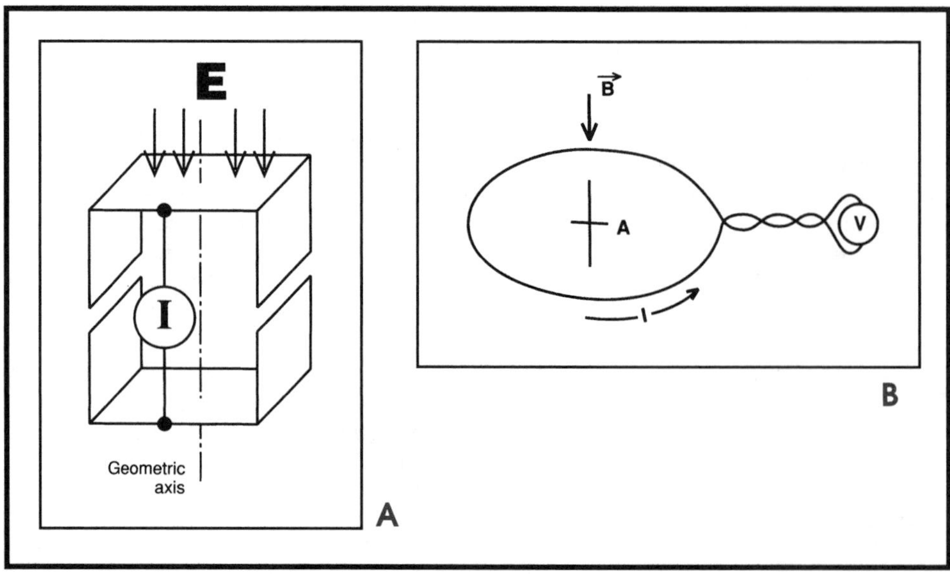

Figure 21.19 — (a) Schematic showing the principle of operation of a free-body electric-field meter. A current flows between the two halves in proportion to the electric-field strength; (b) Schematic of the principle of operation of a magnetic-field coil. The voltage induced in the coil is proportional to the magnetic-flux density, its frequency, the area of the coil, and the number of loops.

although some instruments measure E and B fields. Antennas are a single-axis or tri-axis design. Instruments incorporating single-axis antennas are sensitive to the orientation of the antenna relative to the incident field; the tri-axis designs are isotropic.

Electric Fields. E fields are usually measured with a device shown schematically in Figure 21.19a. Called a free-body dipole probe, it measures the induced current between two halves of an isolated conducting body. In an electric field that is parallel to the axis of the box, the current flowing between the two halves is proportional to the field strength and frequency: I = KfE, where I is the current (A), f is the frequency (e.g., 60 Hz), E is the electric field strength (V/m), and K is a proportionality factor. Electric fields are modified or "perturbed" by conducting objects, such as persons making measurements. For this reason the device is often extended at the end of a nonconducting pole.

Magnetic Fields. Instruments commonly used to measure power-frequency magnetic fields (or flux densities) are based on Faraday's law of voltage induction in a conducting coil (see Figure 21.19b). A conducting loop in a time-varying magnetic field will have a voltage (that is, an electromotive force, or emf) induced in it that is proportional to the time rate of change of the field and the area of the loop. In a sinusoidally varying field, the induced emf equals 2fnAB, where f is the frequency, n is the number of turns in the loop, A is the area of the loop, and B is the magnetic flux density (tesla) perpendicular to the plane of the loop. The device can be made more sensitive by increasing the area or the number of loops, or both. A standard measurement procedure is to orient the coil within the field to obtain a maximum reading.

Alternatively, the magnitude of the resultant flux density can be calculated from three measurements taken at right angles to each other:

$$B = (B_x^2 + B_y^2 + B_z^2)^{1/2} \quad (38)$$

However, it is difficult for the surveyor to rotate the probe precisely around the center of the coil, leading to measurement

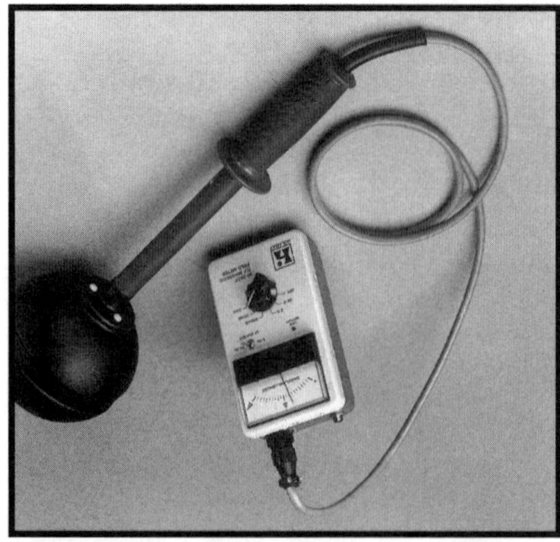

Figure 21.20 — A broad-band ELF magnetic-field instrument that includes a tri-axis antenna array contained within the black, plastic sphere. (Courtesy Holaday Industries, Inc.)

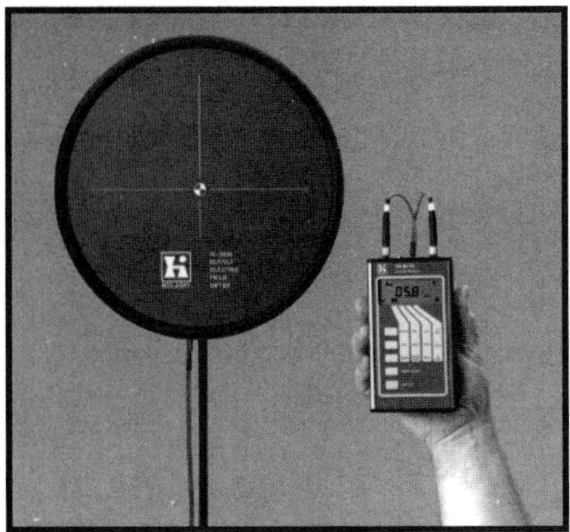

Figure 21.21 — A broad-band E-field antenna connected to a datalogger with a fiber optic connection. (Courtesy Holaday Industries, Inc.)

error. More sophisticated instruments incorporate three orthogonal coils and electronic circuitry to find the resultant (i.e., they do not require orientation). Such an instrument is shown in Figure 21.20.

Many companies manufacture monitors that use the principles described here. Their products differ in how they record measurements. Some, useful mainly for surveys, indicate only the instantaneous field intensity. When connected to a device such as a chart recorder or data logger (see Figure 21.21), these instruments can collect data over time or space for later analysis.

Personal Exposure Monitors. A number of personal exposure monitors that are worn on the body have been introduced. These are most useful for magnetic fields since the human body does not perturb low-frequency magnetic fields (i.e., the body appears to be largely transparent to these fields), as it does electric fields. In general, these instruments include on-board microprocessors to control data recording and recovery. Comparisons and specifications of some personal monitors have been reviewed in the literature.(745-752)

Exposure Assessment

Field surveys include source identification and characterization, observation of the interaction of the workers with the source, followed by collection of measurement data. Various organizations and workers have suggested measurement strategies and exposure assessment. These suggestions include evaluations of the workplace,(644,753-756) schools,(699) office environment,(706,707) utilities,(727,733,757-759) transportation,(754) and residence.(683,758,760-763)

Control Measures

Engineering Control

Methods of reducing field levels and exposure levels are receiving increasing attention. This area of inquiry and its application are generally referred to as field management. Three broadly described methods are available for reducing field levels: shielding, separation, and cancellation.

Shielding is probably the simplest way to reduce electric fields.(764) For 60-Hz fields, air has a conductivity of about 10^{-9} S/m, while the conductivity of metals is above 10^{+7} S/m. The factor of 10^{+16} greater conductivity of metals causes charge to be conducted to a grounded metal surface, terminating the field.(765) Shielding of magnetic fields can be accomplished near isolated equipment, such as transformers, but not easily. Special materials and designs are required.

Separation is effective and can be useful when adequate distances can be maintained between the source of the field and the person exposed. The strength of a magnetic field from an electric current is expressed by the Law of Biot and Savart:

$$dB = K\,I\,dl \times \mathbf{r} / r^3 \qquad (39)$$

where:

K = a constant;
I = current in a conductor of length dl; and
r = distance from the conductor.

Applying the Biot-Savart Law with I in amperes and r in feet, the magnetic field due to current in an effectively infinitely long wire is given by $B = 6.56\ I/r$, and for two closely spaced, parallel wires, the field is $B = 6.56\ Id/r^2$, with d being the spacing of the wires. Finally, if the current flows in a loop, the field is $B = 10.31\ Ia^2/r^3$, where a is the area of the loop.(765) Thus, magnetic field intensities decline with distance for a single wire, with the square of the distance from parallel wires such as power lines or house wiring, and with the cube of the distance from wire loops such as those in motors or on electric stoves.

Cancellation can be applied on scales from power lines to electric blanket wiring. The general principle involves having wires in proximity carry current, and produce fields, that are opposite in phase. The field from one wire then effectively cancels the field from the other at locations removed from the two.

Table 21.26 — Elements of a Nonionizing Radiation Protection Program

- Program responsibility
- Inventory of sources
- Pre-purchase approval of sources
- Hazard assessment
- Accident/incident investigation
- Control measures
- Information and training
- Hazard communication
- Medical surveillance
- Instrument calibration
- Self-checks and audits
- Documentation
- Record keeping

Administrative and Procedural Control

Administrative control measures may be limited because the exposure guidelines (e.g., TLV) are expressed as a ceiling value, and not as a time-weighted average. Thus, time limitation of exposures may not be effective.

An alternative is the administrative approach of prudent avoidance to limit exposures.(644) Prudent avoidance is an approach in which one chooses a low-cost, easily accomplished method to reduce exposure but makes no concerted effort in this regard. An example is the choice between walking in the vicinity of a high-strength field source or taking an equally effective, alternative route farther away from the source.

Nonionizing Radiation Control Program

The goal of the radiation control program is to ensure that sources of nonionizing radiation minimize the risk of adverse health effects to the users. An operational nonionizing radiation control program, which will help meet this goal, should incorporate the elements compiled in Table 21.26. A few of the key features included in this table are discussed next.

Responsibility

Management must recognize the necessity of having a nonionizing radiation protection program and take ownership. The authority and responsibility to implement the program should reside with a knowledgeable (competent) person, such as the Nonionizing Radiation Safety Coordinator (NRSC).(766) It is possible that the individual responsible for the ionizing radiation program or the laser safety program (the RPO or LSO, respectively) may be assigned the responsibility for the complete nonionizing radiation safety program. The competent person should have or receive technical training commensurate with the degree of hazard presented by the workplace sources.(370)

In organizations with a variety of NIR sources, it may be prudent to establish a Nonionizing Radiation Safety (NRS) committee.(766) The responsibilities of the committee include, but are not limited to, recommending policies and procedures, providing technical advice, and reviewing qualifications of users, equipment requests, and exposure reports. In practice, there are few NRS committees, but a number of facilities do have laser safety committees, as recommended in ANSI Z136.1.(155)

Medical Monitoring

In the United States, there are few consensus recommendations for medical monitoring for nonionizing radiation workers. ANSI Z136.1 recommends medical surveillance prior to work with Class 3b and Class 4 lasers and following accidental exposure. Workers are classified as either incidental or laser personnel. Incidental personnel should be evaluated for visual acuity, while laser personnel should recieve a more comprehensive ocular examination. Skin examinations are suggested if individuals have a history of photosensitivity and work with UV lasers.(155)

NIOSH made recommendations for workers who are exposed to UV radiation. These include a review of the medical history with respect to sunlight-related conditions, and a suggestion to provide information to workers to raise their level of awareness of possible UV-related conditions.(102)

OSHA and NIOSH have supported medical monitoring for individuals who work with RF sealers and heaters.(767) The World Health Organization (WHO) recommends medical surveillance if RF exposure "would significantly exceed the [IRPA] general population limits."(380) Some organizations, primarily military and defense contractors, perform ocular surveillance exams of RF workers. These may include preplacement, termination, and accident exams.(380,768-770)

Regarding medical examinations for ELF-exposed individuals, according to WHO, "In view of the fact that there is no health effect that could be attributed specifically to ELF exposure, it is not practicable to recommend any specific medical examinations, apart from those that may be appropriate for electrical fitters and linemen in general."(8)

Information and Training

Employees should be provided with information concerning NIR sources in their workplaces, potential exposures, and operating procedures. Some organizations recommend that individuals who have a high likelihood of being exposed at or above the applicable exposure limits should receive education and training commensurate with the level of risk. For example, U.S. Department of Defense Instruction 6055.11 requires RF safety training for "personnel who routinely work directly with equipment that emits RF levels in excess" of the exposure limits.[770] Also, ANSI Z136.1 requires laser safety training of users who are at greatest risk of overexposure (i.e., users of Class 3b and Class 4 lasers). It may be prudent, however, to provide brief information sessions to individuals who have low-level exposures or where there is a low risk of exposure.

Informational and training requirements are often reviewed during the prepurchase review meeting. DOE recommends that workers be trained "prior to assignment to a job involving potential exposure to NIR and at least annually thereafter."[771]

NIR safety training should provide the exposed individuals with an understanding of the sources and exposure levels in their workplaces. This information will help place the potential exposures in perspective. Other information that might be in the training program includes company information (policy, operating procedures, medical surveillance, intracompany contacts, etc.) and information on instrumentation, as appropriate.

Hazard Communication

Although OSHA did not extend the hazard communication law to physical agents, the concept of effective risk communication is also integral to programs dealing with physical agents. As a minimum, the nonionizing radiation safety program should familiarize exposed workers with the potential hazards of nonionizing radiation. To maximize effectiveness, the program will also proactively share information about exposure levels and the resulting risks with the workers. Such an approach helps simplify and explain otherwise complex and often confusing issues dealing with real and purported biologic effects of nonionizing radiation. Documentation of the results of any hazard evaluations should be shared with employees who use the source, and their supervisors.

The conclusions of accident investigations should be made available to employees, including the outcome of field evaluations and clinical examinations.

Self-Checks and Audits

Periodic self-checks are often performed by individuals responsible for significant sources of nonionizing radiation. These include checks of hardware and administrative controls. Periodic audits may be used by the competent person to evaluate the effectiveness of the self-check program.

Summary

Nonionizing radiation includes UV, visible, IR, RF, and ELF spectral regions and laser radiation. Sources of NIR are both naturally occurring and man-made. Because the number of man-made sources of NIR is increasing, and a larger number of these may generate more than one type of NIR, the potential for exposure to NIR is increasing.

Overexposure to NIR electromagnetic radiation and fields may cause a variety of biological and health effects. These span the spectrum from benign effects, such as constriction of the pupillary opening on exposure to light, to serious diseases, such as UV-induced skin cancer. In general, most effects are nonstochastic, with the intensity of the response varying with the magnitude of the dose or dose rate. Some stochastic effects (e.g., skin cancer) arise from UV exposure, and effects windows have been reported in in-vitro studies of ELF fields.

Exposure guidelines have been recommended from 180 nm in the UV-C spectral region through the ELF and sub-ELF region to fields generated by direct current (non-time-varying electric and magnetic fields). The guidelines for each spectral region are discrete, whereas the limits for laser radiation cover all optical radiation (UV, visible, and IR). The UV and RF guidelines are examples of envelope guidelines, and the RF and ELF guidelines (including sub-RF) are derived from the basic limits, SAR and current density, respectively. The guidelines are dynamic, and are revised when there is consensus concerning the available scientific evidence.

Broad-band field survey instruments are available for most NIR regions, except for some infrared and microwave wavelengths. Dosimeters or personal exposure monitors are available in the UV and ELF spectral regions, although NIR dosimetry is still a developing area at this time. Numerical modeling to predict potential exposure is useful for some sources, especially when the radiation is in the shape of a beam, as it is in lasers and some collimated microwaves.

Engineering and administrative/procedural control measures apply throughout the NIR spectrum. Shielding is an effective engineering control throughout, but is more complicated for low-frequency magnetic fields. Distance is also useful throughout, since the intensity of NIR decreases with radial distance from the source. Personal protective equipment is most useful for optical radiation.

A nonionizing radiation protection program needs management support, the involvement of a competent technical staff, and workers who are aware of the potential hazards of NIR sources with which they work. Successful implementation of such a program will minimize the risk of NIR-related health effects, meet regulatory requirements, help control insurance costs, and reduce the impact of possible negative publicity.

References

1. **Wilkening, G.M.:** Nonionizing Radiation. In *Patty's Industrial Hygiene and Toxicology General Principles,* 4th Ed. New York: John Wiley & Sons, 1991. pp. 657-742.
2. **Murray, W.E., R.T. Hitchcock, R.M. Patterson, and S.M. Michaelson:** Nonionizing Electromagnetic Energies. In *Patty's Industrial Hygiene & Toxicology,* Vol. 3, Part B, 3rd. Ed. New York: John Wiley & Sons, 1995. pp. 623-727.
3. **Miller, G.:** Nonionizing Radiation. In *Fundamentals of Industrial Hygiene,* 4th Ed. Itasca, IL: National Safety Council, 1995. pp. 273-317.
4. **Patterson, R.M., and R.T. Hitchcock:** Nonionizing Radiation and Fields. In *Health and Safety Beyond the Workplace.* New York: John Wiley & Sons, 1990. pp. 143-176.
5. **International Radiation Protection Association:** Review of Concepts, Quantities, Units and Terminology for Non- Ionizing Radiation Protection. *Health Phys. 49:*1329-1362 (1985).
6. **Sutter, E.:** Quantities and Units of Optical Radiation and their Measurement. In *Dosimetry of Laser Radiation in Medicine and Biology* (SPIE Vol. IS 5). Bellingham, WA: SPIE Optical Engineering Press, 1989. pp. 38-79.
7. **National Council on Radiation Protection and Measurements (NCRP):** *Radiofrequency Electromagnetic Fields — Properties, Quantities and Units, Biophysical Interaction, and Measurements.* Washington, D.C.: NCRP, 1981.
8. **World Health Organization:** *Extremely Low Frequency (ELF) Fields* (Environmental Health Criteria 35). Geneva: WHO, 1984.
9. **Sliney, D., and M. Wolbarsht:** *Safety with Lasers and Other Optical Sources.* New York: Plenum Press, 1980.
10. **World Health Organization:** *Lasers and Optical Radiation* (Environmental Health Criteria 23). Geneva: WHO, 1982. pp. 42-45.
11. **Sutter, E.:** Extended Sources—Concepts and Potential Hazards. *Optics Laser Technol. 27:*5-13 (1995).
12. **Everett, M.A., R.M. Sayre, and R.L. Olsen:** Physiologic Response of Human Skin to Ultraviolet Light. In *The Biological Effects of Ultraviolet Radiation (With Emphasis on the Skin).* New York: Pergamon Press, 1969. pp. 181-186.
13. **McKinlay, A.F., and B.L. Diffey:** A Reference Action Spectrum for Ultraviolet Induced Erythema in Human Skin. *CIE J. 66:*17-22 (1987).
14. **Diffey, B.L.:** Observed and Predicted Minimal Erythema Doses: A Comparative Study. *Photochem. Photobiol. 60:*380-382 (1994).
15. **Parrish, J.A., K.F. Jaenicke, and R.R. Anderson:** Erythema and Melanogenesis of Normal Human Skin. *Photochem. Photobiol. 36:* 187-191 (1982).
16. **Matelsky, I.:** Non-ionizing Radiations. In *Industrial Hygiene Highlights, Vol I.* Pittsburgh: Industrial Hygiene Foundation of America, Inc, 1968. pp. 140-179.
17. **Epstein, J.H.:** Photomedicine. In *The Science of Photobiology,* 2nd Ed. New York and London: Plenum Press, 1989. pp. 155-192.
18. **U.S. Food and Drug Administration:** *Medications that Increase Sensitivity to Light: A 1990 Listing* by J.I. Levine (DHHS [FDA] Pub. No. 91–8280). Rockville, MD: U.S. Department of Health and Human Services, 1990.
19. **Hawk, J.L.M.:** Photosensitizing Agents in the United Kingdom. *Clin. Exptl. Dermatol. 9:*300-302 (1984).

20. **Blum, H.F.:** *Carcinogenesis by Ultraviolet Light.* Princeton, NJ: Princeton University Press, 1959.
21. **Epstein, J.H.:** Photocarcinogenesis, Skin Cancer and Aging. *J. Amer. Acad. Dermatol. 9*:487-506 (1983).
22. **van der Leun, J.C.:** Yearly Review: UV-Carcinogenesis. *Photochem. Photobiol.*:861-868 (1984).
23. **International Agency for Research on Cancer (IARC):** *Solar and Ultraviolet Radiation* (IARC Monographs on the Evaluation of Carcinogenic Risks to Humans, Vol. 55). Lyon, France: IARC, 1992.
24. **Lefell, D.J., and D.E. Brash:** Sunlight and Cancer. *Scient. Am. 275(1)*:52-59 (1996).
25. **Kusewitt, D.F., and R.D. Ley:** Animal Models of Melanoma. *Cancer Surveys 26*:35-70 (1996).
26. **American Cancer Society:** *Cancer Facts and Figures – 1996.* Atlanta: American Cancer Society, 1996.
27. **National Institutes of Health (NIH):** *Consensus Statement: Sunlight, Ultraviolet Radiation and the Skin 7.* Bethesda, MD: NIH, 1989.
28. **Bajdik, C.D., R.P. Gallagher, G. Astrakianaskis, G.B. Hill, S. Fincham, and D.I. McLean:** Non-Solar Ultraviolet Radiation and the Risk of Basal and Squamous Cell Skin Cancer. *Brit. J. Cancer 73*:1612-1614 (1996).
29. **Centers for Disease Control and Prevention (CDC):** Death Rates of Malignant Melanoma Among White Men – United States, 1973-1988. *MMWR 41*:20-21, 27 (1992).
30. **Nelemans, P.J., H. Groenendal, L.A.L.M. Kiemeney, F.H.J. Rampen, D.J. Ruiter, and A.L.M. Verbeek:** Effect of Intermittent Exposure to Sunlight on Malignant Melanoma Risk Among Indoor Workers and Sun-Sensitive Individuals. *Environ. Health Perspect. 101*:252-255 (1993).
31. **Setlow, R.B., E. Girst, K. Thompson, and A.D. Woodhead:** Wavelengths Effective in Induction of Malignant Melanoma. *Proc. Natl. Acad. Sci. 90*:6666-6670 (1993).
32. **Setlow, R.B.:** Relevance of *In Vivo* Models in Melanoma Skin Cancer. *Photochem. Photobiol. 63*:410-412 (1996).
33. **National Radiological Protection Board (NRPB):** *Cellular and Molecular Effects of UVA and UVB* by N.A. Cridland and R.D. Saunders (NRPB-R269). Chilcon, Didcot, Oxon, U.K.: NRPB, 1994.
34. **Council on Scientific Affairs:** Harmful Effects of Ultraviolet Radiation. *J. Amer. Med. Assoc. 262*:380-384 (1989).
35. **Morison, W.L.:** Effects of Ultraviolet Radiations on the Immune System in Humans. *Photochem. Photobiol. 50*:515-524 (1989).
36. **Green, C., B.L. Diffey, and J.L.M. Hawk:** Ultraviolet Radiation in the Treatment of Skin Disease. *Phys. Med. Biol. 37*:1-20 (1992).
37. **Epstein, J.H.:** Phototherapy and Photochemotherapy. *N. Engl. J. Med. 322*:1149-1151 (1990).
38. **National Institute for Occupational Safety and Health:** *Ocular Ultraviolet Effects from 295 nm to 400 nm in the Rabbit Eye* by D.G. Pitts, A.P. Cullen, P.D. Hacker, and W.H. Parr (HEW [NIOSH] Pub No. 77–175). Cincinnati: U.S. Department of Health, Education and Welfare, 1977.
39. **Pitts, D.G., and T.J. Tredici:** The Effects of Ultraviolet Radiation on the Eye. *Amer. Ind. Hyg. Assoc. J. 32*:235-246 (1971).
40. **Hollows, F., and D. Moran:** Cataract — The Ultraviolet Risk Factor. *Lancet II–1981*:1249-1250 (1981).
41. **West, S.K., and C.T. Valmadrid:** Epidemiology of Risk Factors for Age-Related Cataract. *Survey Ophthalmol. 39*:323-334 (1995).
42. **Taylor, H.R., S.K. West, F.S. Rosenthal, B. Munoz, H.S. Newland, H. Abbey, and E.A. Emmett:** Effect of Ultraviolet Radiation on Cataract Formation. *N. Engl. J. Med. 319*:1429-1433 (1988).
43. **Pitts, D.G., L.L. Cameron, J.G. Jule, and S. Lerman:** Optical Radiation and Cataracts. In *Optical Radiation and Visual Health.* Boca Raton, FL: CRC Press, Inc., 1986. pp. 5-41.
44. **Zigman, S.:** Recent Research on Near-UV Radiation and the Eye. In *The Biological Effects of UVA Radiation.* New York: Praeger Publishers, 1986. pp. 252-262.
45. **Waxler, M.:** Long-Term Visual Health Problems: Optical Radiation Risks. In *Optical Radiation and Visual Health.* Boca Raton, FL: CRC Press, Inc., 1986. pp. 183-204.
46. **Dayhaw-Barker, P., D. Forbes, D. Fox, and S. Lerman:** Drug Phototoxicity and Visual Health. In *Optical Radiation and Visual Health.* Boca Raton, FL: CRC Press, Inc., 1986. pp. 147-175.

47. **Lerman, S.:** Effect of UVA Radiation on Tissues of the Eye. In *The Biological Effects of UVA Radiation.* New York: Praeger Publishers, 1986. pp. 231-251.
48. **Marshall, J.:** Ultraviolet Radiation and the Eye. In *Human Exposure to Ultraviolet Radiation: Risks and Regulations.* Amsterdam: Elsevier Science Publishers B.V. (Biomedical Division), 1987. pp. 125-42.
49. **Goldman, L., S.M. Michaelson, R.J. Rockwell, D.H. Sliney, B.M. Tengroth, and M.L. Wolbarsht:** Optical Radiation, with Particular Reference to Lasers. In *Nonionizing Radiation Protection,* 2nd Ed. Copenhagen: World Health Organization, 1989. pp. 49-83.
50. **Ham, W.T., Jr., R.G. Allen, L. Feeney-Burns, and M.F. Marmor:** The Involvement of the Retinal Pigment Epithelium (RPE). In *Optical Radiation and Visual Health.* Boca Raton, FL: CRC Press, Inc., 1986. pp. 43-67.
51. **Geeraets, W.J.:** Radiation Effects on the Eye. *Ind. Med. 39*:441-450 (1970).
52. **Luckiesh, M., and S.K. Guth:** Brightness in Visual Field at Borderline Between Comfort and Discomfort. *Illum. Eng. 44*:650-670 (1949).
53. **Moss, C.E., R.J. Ellis, W.E. Murray, and W.H. Parr:** Infrared Radiation. In *Nonionizing Radiation Protection,* 2nd Ed. Copenhagen: World Health Organization, 1989. pp. 85-115.
54. **National Institute for Occupational Health and Safety:** *Biological Effects of Infrared Radiation* by C.E. Moss, R.J. Ellis, W.H. Parr, and W.E. Murray (DHHS [NIOSH] Pub. No. 82–109). Cincinnati: U.S. Department of Health and Human Services, 1982.
55. **National Institute for Occupational Health and Safety:** *Determination of Ocular Threshold Levels for Infrared Radiation Cataractogenesis* by D.G. Pitts, A.P. Cullen, and P. Dayhaw-Barker (DHHS [NIOSH] Pub. No. 80–121). Cincinnati: U.S. Department of Health and Human Services, 1980.
56. **Robinson, W.:** On Bottle-Maker's Cataract. *Brit. Med. J.* 2:381-384 (1907).
57. **Dunn, K.L.:** Cataract from Infrared Rays."Glass Workers' Cataract:' A Preliminary Study. *Arch. Ind. Hyg. Occup. Med. 1*:166-180 (1950).
58. **Keating, G.F., J. Pearson, J.P. Simons, and E.E. White:** Radiation Cataract in Industry. *Arch. Ind. Health 11*:305- 315 1955.
59. **Hitchcock, R.T.:** *Ultraviolet Radiation* (Nonionizing Radiation Guide Series). Fairfax, VA: American Industrial Hygiene Association, 1991.
60. **Thorington, L.:** Spectral, Irradiance and Temporal Aspects of Natural and Artificial Light. *Ann. N.Y. Acad. Sci. 453*:28-54 (1985).
61. **Frederick, J.E. and A.D. Alberts:** The Natural UV-A Radiation Environment. In *Biological Responses to Ultraviolet A Radiation.* Overland Park, KS: Valdenmar Publishing Company, 1992. pp. 7-18.
62. **Garrison, L.M., L.E. Murray, and A.E.S. Green:** Ultraviolet Limit of Solar Radiation at the Earth's Surface with a Photon Counting Monochromator. *Appl. Opt. 17*:683-684 (1978).
63. **Scotto, J., T.R. Fears, and G.B. Gori:** Ultraviolet Exposure Patterns. *Env. Res. 12*:228-237 (1976).
64. **Driscoll, C.M.H.:** Solar UVR Measurements. *Radiat. Prot. Dosim. 64*:179-188 (1996).
65. **Diffey, B.L., M. Kerwin, and A. Davis:** The Anatomical Distribution of Sunlight. *Brit. J. Dermatol. 97*:407-410 (1977).
66. **Urbach, F.:** Geographic Pathology of Skin Cancer. In *The Biological Effects of Ultraviolet Radiation.* New York, Pergamon Press, 1969. pp. 635-650.
67. **Diffey, B.L., T.J. Tate, and A. Davis:** Solar Dosimetry of the Face: The Relationship of Natural Ultraviolet Radiation Exposure to Basal Cell Carcinoma Localisation. *Phys. Med. Biol. 24*:931-939 (1979).
68. **Sliney, D.H.:** The Merits of an Envelope Action Spectrum for Ultraviolet Radiation Exposure Criteria. *Am. Ind. Hyg. Assoc. J. 33*:644-653 (1972).
69. **Sliney, D.H.:** Eye Protective Techniques for Bright Light. *Ophthal. 90*:937-944 (1983).
70. **Rosenthal, F.S., C. Phoon, A.E. Bakalian, and H.R. Taylor:** The Ocular Dose of Ultraviolet Radiation to Outdoor Workers. *Investig. Ophthalmol. Vis. Sci. 29*:649-656 (1988).
71. **Gies, H.P., C.R. Roy, S. Toomey, R. MacLennan, and M. Watson:** Solar UVR Exposures of Three Groups of Outdoor Workers on the Sunshine Coast, Queensland. *Photochem. Photobiol. 62*:1015-1021 (1995).

72. **Larko, O., and B.L. Diffey:** Natural UV-B Radiation Received by People with Outdoor, Indoor, and Mixed Occupations and UV-B Treatment of Psoriasis. *Clin. Exp. Dermatol. 8*:279-285 (1983).
73. **Challoner, A.V.J., D. Corless, A. Davis, G.H.W. Deane, B.L. Diffey, S.P. Gupta, and I.A. Magnus:** Personnel Monitoring of Exposure to Ultraviolet Radiation. *Clin. Exp. Dermatol. 1*:175-179 (1976).
74. **Igawa, S., H. Kibamoto, H. Takahaski, and S. Arai:** A Study on Exposure to Ultraviolet Rays During Outdoor Sports Activity. *J. Therm. Biol. 18*:583-586 (1993).
75. **Diffey, B.L., O. Larko, and G. Swanbeck:** UV-B Doses Received During Different Outdoor Activities and UV-B Treatment of Psoriasis. *Brit. J. Dermatol. 106*:33-41 (1982).
76. **Diffey, B.L., and A.H. Roscoe:** Exposure to Solar Ultraviolet Radiation in Flight. *Aviat. Space Environ. Med. 61*:1032-1035 (1990).
77. **McKinlay, A.F.:** Artificial Sources of UVA Radiation: Uses and Emission Characteristics. In *Biological Responses to Ultraviolet A Radiation.* Overland Park, KS: Valdenmar Publishing Company, 1992. pp. 19-38.
78. **Bergman, R.S., T.G. Parham, and T.K. McGowan:** UV Emission from General Lighting Lamps. *J. Illum. Eng. Soc. 24*: 13-24 (1995).
79. **Moseley, H.:** Ultraviolet and Laser Radiation Safety. *Phys. Med. Biol. 39*:1765-1799 (1994).
80. **Gulvady, N.U.:** UV Keratoconjunctivitis vs. Established Dose Effect Relationships. *J. Occup. Med. 18*:573 (1976).
81. **Murray, W.E.:** Ultraviolet Radiation Exposures in A Mycobacteriology Laboratory. *Health Phys. 58*:507-510 (1990).
82. **Boettrich, E.P.:** "Hazards Associated with Ultraviolet Radiation in Academic and Clinical Laboratories." Masters thesis, Department of Radiation Biology and Biophysics, University of Rochester, Rochester, NY, 1985.
83. **Cole, C., P.D. Forbes, R.E. Davies, and F. Urbach:** Effect of Indoor Lighting on Normal Skin. *Ann. N.Y. Acad. Sci. 453*:305-316 (1985).
84. **Diffey, B.L., and A.F. McKinlay:** The UVB Content of `UVA Fluorescent Lamps' and Its Erythemal Effectiveness in Human Skin. *Phys. Med. Biol. 28*:351-358 (1983).
85. **Fischer, T., J. Alsins, and B. Berne:** Ultraviolet-Action Spectrum and Evaluation of Ultraviolet Lamps for Psoriasis Healing. *Int. J. Dermatol. 23*:633-637 (1984).
86. **Larko, O., and B.L. Diffey:** Occupational Exposure to Ultraviolet Radiation in Dermatology Departments. *Brit. J. Dermatol. 114*:479-484 (1986).
87. **Diffey, B.L.:** The Risk of Skin Cancer from Occupational Exposure to Ultraviolet Radiation in Hospitals. *Phys. Med. Biol. 33*:1187-1193 (1988).
88. **Lerman, S.:** Human Ultraviolet Radiation Cataracts: *Ophthalmic. Res. 12*:303-314 (1980).
89. **Roscoe, A.H., and B.L. Diffey:** A Preliminary Study of Blue Light on an Aircraft Flight Deck. *Health Phys. 66*:565- 567 (1994).
90. **Hietanen, M.T.K., and M.J. Hoikkala:** Ultraviolet Radiation and Blue Light from Photofloods in Television Studios and Theaters. *Health Phys. 59*:193-198 (1990).
91. **McIntyre, D.A., W.N. Charman, and I.J. Murray:** Visual Safety of Quartz Linear Lamps. *Ann. Occup. Hyg. 37*:191- 200 (1993).
92. **National Institute for Occupational Safety and Health:** *Health Hazard Evaluation Report Miller Thermal Technologies, Inc., Appleton, Wisconsin* by C.E. Moss and R.L. Tubbs (Rpt. No. HETA-88-136-1945). Springfield, VA: National Technical Information Service (PB89-188031), 1989.
93. **National Institute for Occupational Safety and Health:** *Health Hazard Evaluation Report HETA 91–095–2142, Cone Geochemical, Inc., Lakewood, Colorado* by C.E. Moss. Springfield, VA: National Technical Information Service (PB92- 133214), 1992.
94. **National Institute for Occupational Safety and Health:** *Health Hazard Evaluation Report HETA 88–229–1985, Ormet Corporation, Hannibal, Ohio* by C.E. Moss and R.L. Stephenson. Springfield, VA: National Technical Information Service (PB90-180704), 1988.
95. **National Institute for Occupational Safety and Health:** *Health Hazard Evaluation Report HETA 88–299–2028, Louie Glass Factory, Weston, West Virginia* by C.E. Moss, R.L. Tubbs, L.L. Cameron, and E. Freund, Jr. Springfield, VA: National Technical Information Service (PB91-115311), 1990.

96. **Chu, T.S., and M.J. Gans:** High Speed Infrared Local Wireless Communication. *IEEE Comm. Mag. 25(8)*:4-10 (1987).
97. **Smythe, P.P., D. Wood, S. Ritchie, and S. Cassidy:** Optical Wireless: New Enabling Transmitter Technologies. In *IEEE International Conference on Communications '93, Technical Program, Conference* (Vol. 1/3). New York: IEEE, 1993. pp. 562-566.
98. **Lydahl, E., A. Glansholm, and M. Levin:** Ocular Exposure to Infrared Radiation in the Swedish Iron and Steel Industry. *Health Phys. 46*:529-536 (1984).
99. **Devereux, H., and M Smalley:** Are Infra Red Illuminators Eye Safe? In *Proceedings of the Institute of Electrical and Electronics Engineers 29th Annual 1995 International Carnahan Conference on Security Technology* (IEEE Cat. No. 95CH3578–8). New York: IEEE, 1995. pp. 480-481.
100. **Chou, B.R., and A.P. Cullen:** Ocular Hazards of Industrial Spot Welding. *Optom. Vis. Sci. 73*:424-427 (1996).
101. **American Conference of Governmental Industrial Hygienists:** *1996 TLVs and BEIs.* Cincinnati: American Conference of Governmental Industrial Hygienists, 1996.
102. **National Institute for Occupational Safety and Health (NIOSH):** *Occupational Exposure to Ultraviolet Radiation* (HSM 73–11009). Cincinnati: NIOSH, 1972.
103. **National Institute for Occupational Safety and Health:** *Recommended Standard for Occupational Exposure to Ultraviolet Radiation.* Springfield, VA: National Technical Information Service (PB93–213122), 1993.
104. **International Radiation Protection Association:** Guidelines on Limits of Exposure to Ultraviolet Radiation of Wavelengths between 180 nm and 400 nm (Incoherent Optical Radiation). *Health Phys. 49*:331-340 (1985).
105. **Gies, H.P., C.R. Roy, S. Toomey, and D. Tomlinson:** The ARL Solar UVR Measurement Network: Calibration and Results. *SPIE 2282*:274-284 (1994).
106. **Klein, R.M.:** Cut-off Filters for the Near Ultraviolet. *Photochem. Photobiol. 29*:1053-1054 (1979).
107. **Racz, M. I. Reti, and S. Ferenczi:** Measurement of UV-A and UV-B Irradiance with Glass Filtered Detectors. *SPIE 2022*:192-195 (1993).
108. **Landry, R.J., and F.A. Andersen:** Optical Radiation Measurements: Instrumentation and Sources of Error. *J. Natl. Cancer Inst. 69*:155-161 (1982).
109. **Fanney, J.H., and C.H. Powell:** Field Measurement of Ultraviolet, Infrared, and Microwave Energies. *Am. Ind. Hyg. Assoc. J. 28*:335-342 (1967).
110. **Tug, H., and E.M. Baumann:** Problems of UV-B Radiation Measurements in Biological Research: Critical Remarks on Current Techniques and Suggestions for Improvements. *Geophys. Res. Lett. 21*:689-692 (1994).
111. **McKenzie, R.L., and P.V. Johnston:** Comment on "Problems of UV-B Radiation Measurements in Biological Research: Critical Remarks on Current Techniques and Suggestions for Improvements" by H. Tug and M.E.M. Baumann. *Geophys. Res. Lett 22*:1157-1158 (1995).
112. **Rosenthal, F.S., M. Safran, and H. R. Taylor:** The Ocular Dose of Ultraviolet Radiation from Sunlight Exposure. *Photochem. Photobiol. 42*:163-171 (1985).
113. **Diffey, B.L., O. Larko, B. Meding, H.G. Edeland, and U. Wester:** Personal Monitoring of Exposure to Ultraviolet Radiation in the Car Manufacturing Industry. *Ann. Occup. Hyg. 30*:163-170 (1986).
114. **Diffey, B.L.:** A Comparison of Dosimeters Used for Solar Ultraviolet Radiometry. *Photochem. Photobiol. 46*:55-60 (1987).
115. **Young, A.R., A.V.J. Challoner, I.A. Magnus, and A. Davis:** UVR Radiometry of Solar Simulated Radiation in Experimental Photocarcinogenesis Studies. *Brit. J. Dermatol. 106*:43-52 (1982).
116. **Wong, C.F., R. Fleming, and S.J. Carter:** A New Dosimeter for Ultraviolet-B Radiation. *Photochem. Photobiol. 50*:611-615 (1989).
117. **Wong, C.F., R.A. Fleming, S.J. Carter, I.T. Ring, and D. Vishvakarman:** Measurement of Human Exposure to Ultraviolet-B Solar Radiation Using A CR-39 Dosimeter. *Health Phys. 63*:457-461 (1992).
118. **Wong, C.F., S. Toomey, R.A. Fleming, and B.W. Thomas:** UV-B Radiometry and Dosimetry for Solar Measurements. *Health Phys. 68*:175-184 (1995).

119. **Sydenham, M.M., M.J. Collins, and L.W. Hirst:** The Effectiveness of Poly(Allyl Diglycol Carbonate) (CR-39) for Low-Dose Solar Ultraviolet Dosimetry. *Photochem. Photobiol. 59*:58-65 (1994).
120. **Sliney, D.H., F.C. Bason, and B.C. Freasier:** Instrumentation and Measurement of Ultraviolet, Visible, and Infrared Radiation. *Am. Ind. Hyg. Assoc. J. 32*:415-431 (1971).
121. **Gies, H.P., and C.R. Roy:** Bilirubin Phototherapy and Potential UVR Hazards. *Health Phys. 58*:313-320 (1990).
122. **Koller, L.R.:** *Ultraviolet Radiation*, 2d Ed. New York: John Wiley & Sons, 1965. pp. 153-160.
123. **Sliney, D.H., C.E. Moss, C.G. Miller, and J.B. Stephens:** Semitransparent Curtains for Control of Optical Radiatiion Hazards. *Appl. Opt. 20*:2352-2366 (1981).
124. **Wuest, J.R., and T.A. Gossel:** Update on Sunscreens: Part I. *Kentucky Pharm. 55(6)*:185-188 (1992).
125. **Kollias, N., and A.H. Baqer:** Photoprotection by the Natural Pigment: Melanin. In *Human Exposure to Ultraviolet Radiation: Risks and Regulations*. Amsterdam: Elsevier Publishers B.V. (Biomedical Division), 1987. pp. 121-124.
126. **Wolf, P., C.K. Donawho, and M.L. Kripke:** Effect of Sunscreens on UV Radiation-Induced Enhancement of Melanoma Growth in Mice. *J. Nat. Cancer Inst. 86*:99-105 (1994).
127. **Armstrong, B.K., and A. Kricker:** Epidemiology of Sun Exposure and Skin Cancer. *Cancer Surveys 26*:133-153 (1996).
128. **Hersey, P., M. MacDonald, C. Burns, and S. Schibeci:** Analysis of the Effect of A Sunscreen Agent on The Suppression of Natural Killer Cell Activity Induced in Human Subjects Radiation from Solarium Lamps. *J. Invest. Dermatol. 88*:271-276 (1987).
129. **Wuest, J.R., and T.A. Gossel:** Update on Sunscreens: Part II. *Kentucky Pharm. 55(7)*:213-216 (1992).
130. **Sayre, R.M., N. Kollias, R.D. Ley, and A.H. Baqer:** Changing the Risk Spectrum of Injury and the Performance of Sunscreen Products throughout the Day. *Photodermatol. Photoimmunol. Photomed. 10*:148-153 (1994).
131. **Hawk, J.L.M., A.V.J. Challoner, and L. Chaddock:** The Efficacy of Sunscreening Agents: Protection Factors and Transmission Spectra. *Clin. Exptl. Dermatol. 7*:21-31 (1982).
132. **Koh, H.K., and R.A. Lew:** Sunscreen and Melanoma: Implications for Prevention. *J. Nat. Cancer Inst. 86*:78-79 (1994).
133. **Gies, H.P., C.R. Roy, C.R. Elliott, and W. Zongli:** Ultraviolet Radiation Protection Factors for Clothing. *Health Phys. 67*:131-139 (1994).
134. **Sliney, D.H., R.E. Benton, H.M. Cole, S.G. Epstein, and C.J. Morin:** Transmission of Potentially Hazardous Actinic Ultraviolet Radiation through Fabrics. *Appl.Ind. Hyg. 2*:36-44 (1987).
135. **Berne, B., and T. Fischer:** Protective Effects of Various Types of Clothes against UV Radiation. *Acta Derm. Venereol. 60*:459-460 (1980).
136. **Parrish, J.A., R.R. Anderson, F. Urbach, and D.Pitts:** *UV-A Biological Effects of Ultraviolet Radiation with Emphasis on Human Responses to Longwave Ultraviolet*. New York: Plenum Press, 1978. pp. 248-252.
137. **Sliney, D.H., and B.C. Freasier:** Evaluation of Optical Radiation Hazards. *Appl. Opt. 12*:1-24 (1973).
138. **Tenkate, T.S.D., and M.J. Collins:** Angles of Entry of Ultraviolet Radiation into Welding Helmets. *Am. Ind. Hyg. Assoc. J. 58*:54-56 (1997).
139. **Tenkate, T.S.D., and M.J. Collins:** Personal Ultraviolet Radiation Exposure of Workers in a Welding Environment. *Am. Ind. Hyg. Assoc. J. 58*:33-38 (1997).
140. **Moseley, H.:** Ultraviolet and Visible Radiation Transmission Properties of Some Types of Protective Eyewear. *Phys. Med. Biol. 30*:177-181 (1985).
141. **Rosenthal, F.S., A.E. Bakalian, and H.R. Taylor:** The Effect of Prescription Eyewear on Ocular Exposure to Ultraviolet Radiation. *Am. J. Pub. Health 76*:1216-1220 (1986).
142. **Wojno, T., D. Singer, and R.O. Schultz:** Ultraviolet Light, Cataracts, and Spectacle Wear. *Ann. Ophthalmol. 15*:729-732 (1983).
143. **van Kuijk, F.J.G.M.:** Effects of Ultraviolet Light on the Eye: Role of Protective Glasses. *Environ. Health Perspect. 96*:177-184 (1991).
144. **Harris, M.G., M. Dang, S. Garrod, and W. Wong:** Ultraviolet Transmittance of Contact Lenses. *Opt. Vis. Sci. 71*:1-5 (1994).
145. **Nilsson, S.E.G., P. Lovsund, P.A. Oberg, and L.-E. Flordahl:** The

Transmittance and Absorption Properties of Contact lenses. *Scand. J. Work, Environ. Health 5*:262-270 (1979).

146. **Quensel, N.-M., and P. Simonet:** Spectral Transmittance of UV-Absorbing Soft and Rigid Gas Permeable Contact Lenses. *Optom. Vis. Sci. 72*:2-10 (1995).
147. **Werner, J.S.:** Children's Sunglasses: Caveat Emptor. *Opt. Vis. Sci. 68*:318-320 (1991).
148. **Davey, J.B., B.L. Diffey, and J.A. Miller:** Eye Protection in Psoralen Photochemotherapy. *Brit. J. Dermatol. 104*:295-300 (1981).
149. **Rosenthal, F.S., A.E. Bakalian, C. Lou, and H.R. Taylor:** The Effect of Sunglasses on Ocular Exposure to Ultraviolet Radiation. *Am. J. Pub. Health 78*:72-74 (1988).
150. **Anderson, W.J., and R.K.H. Gebel:** Ultraviolet Windows in Commercial Sunglasses. *Appl. Opt. 16*:515-517 (1977).
151. **Magnante, D.B.O., and D. Miller:** Ultraviolet Absorption of Commonly Used Clip-On Sunglasses. *Ann. Ophthalmol. 17*:614-616 (1985).
152. **Segre, G., R. Reccia, B. Pignalosa, and G. Pappalardo:** The Efficiency of Ordinary Sunglasses as a Protection from Ultraviolet Radiation. *Ophthalmic Res. 13*:180-197 (1981).
153. **Birngruber, R.:** Laser Output Characteristics. *Health Phys. 56*:605-611 (1989).
154. **Franks, J.K.:** Potential Ocular Hazards from Reflective Surfaces. In *Proceedings of the 1992 International Laser Safety Conference*. Orlando, FL: Laser Institute of America, 1993. pp. 7I-55–7I-60.
155. **American National Standards Institute:** *American National Standard for the Safe Use of Lasers* (ANSI Z136.1–1993). Orlando, FL: Laser Institute of America, 1993.
156. **Ham, W.T., Jr., H.A. Mueller, J.J. Ruffolo, Jr., and A.M. Clarke:** Sensitivity of the Retina to Radiation Damage as a Function of Wavelength. *Photochem. Photobiol. 29*:735-743 (1979).
157. **Ham, W.T., Jr., J.J.Ruffolo, Jr., H.A. Mueller, and D. Guerry:** The Nature of Retinal Radiation Damage: Dependence on Wavelength, Power Level and Exposure Time. *Vision Res. 20*:1105-1111 (1980).
158. **Lund, D.J., P.R. Edsall, D.R. Fuller, and S.W. Hoxie:** Ocular Hazards of Tunable Continuous-Wave Near Infrared Laser Sources. *Proc. SPIE 2674*:53-61 (1996).
159. **Puliafito, C.A., and R.F. Steinert:** Short-Pulsed Nd:YAG Laser Microsurgery of the Eye: Biophysical Considerations. *IEEE J. Quantum Electon. QE–20*:1441-1448 (1984).
160. **Decker, C.D.:** Accident Victim's View. *Laser Focus 13(8)*:6 (1977).
161. **Boldrey, E.E., H.L. Little, L., M. Flocks, and A. Vassiliadis:** Retinal Injury to Industrial Laser Burns. *Ophthalmol. 88*:101-107 (1981).
162. **Pleven, C.:** A Description of Fourteen Accidents Caused by Lasers in a Research Environment. In *Lasers et Normes de Protection (First International Symposium on Laser Biological Effects and Exposure Limits)*. Paris: Centre de Recherches du Service de Sante des Armees, 1986. pp. 406-417.
163. **Haifeng, L., G. Guanghuang, W. Dechang, X. Guidao, S. Liangshun, X. Jiemin, and W. Haibiao:** Ocular Injuries from Accidental Laser Exposure. *Health Phys. 56*:711-716 (1989).
164. **Wolfe, J.A.:** Laser Retinal Injury. *Military Med. 150*:177-185 (1985).
165. **Henkes, H.E., and H. Zuidema:** Accidental Laser Coagulation of the Central Fovea. *Ophthalmologica 171*:15-25 (1975).
166. **Rockwell, R.J., Jr.:** Learning from Case Studies: How to Avoid Laser Accidents. In *Expert Strategies for Practical and Profitable Management*. Atlanta: American Health Consultants, Inc., 1985. pp. 57-66.
167. **Rockwell, R.J., Jr.:** Laser Accidents: Are They All Reported and What Can Be Learned from Them? *J. Laser Appl. 1(4)*:53-57 (1989).
168. **Sliney, D.H.:** Ocular Injuries from Laser Accidents. *Proc. SPIE 2674*:25-33 (1996).
169. **Marshall, J.:** Eye Hazards Associated with Lasers. *Ann. Occup. Hyg. 21*:69-77 (1978).
170. **Birngruber, R., F. Hillenkamp, and V.-P. Gabel:** Theoretical Investigations of Laser Thermal Retinal Injury. *Health Phys. 48*:781-796 (1985).
171. **Kochevar, I.E.:** Biological Effects of Excimer Laser Radiation. *Proc. IEEE 80*:833-837 (1992).
172. **Ham, W.T., Jr., H.A. Mueller, and D.H. Sliney:** Retinal Sensitivity to Damage from Short Wavelength Light. *Nature 260*:153-155 (1976)

173. **Ham, W.T., Jr., J.J. Ruffolo, Jr., H.A. Mueller, and D. Guerry III:** The Nature of Retina Radiation Damage: Dependence on Wavelength, Power Level and Exposure Time. *Vision Res. 20*:1105-1111 (1980).
174. **Ham, W.T., Jr., and H.A. Mueller:** The Photopathology and Nature of the Blue Light and Near-UV Retinal Lesions Produced by Lasers and Other Optical Sources. In *Laser Applications in Medicine and Biology* (Vol. 4). New York: Plenum Press, 1989. pp. 191-246.
175. **Zwick, H., and E.S. Beatrice:** Long-Term Changes in Spectral Sensitivity after Low-Level Laser (514 nm) Exposure. *Mod. Probl. Ophthal. 19*:319-325 (1978).
176. **Frisch, G.D., E.S. Beatrice, and R.C. Holsen:** Comparative Study of Argon and Ruby Retinal Damage Thresholds. *Invest. Ophthal. 10*:911-919 (1971).
177. **Zuclich, J.A., S. Schuschereba, H. Zwick, F. Cheney, and B.E. Stuck:** Comparing Laser-Induced Retinal Damage from IR Wavelengths to that from Visible Wavelengths. *Proc. SPIE 2674*:66-79 (1996).
178. **Courant, D., L. Court, B. Abadie, B. Brouillet, J. Garcia, J. L. Paradis, and J. C. Perot:** Experimental Determination of a Laser Retinal Lesion Threshold Produced by a Single Pulse in the Visible Spectrum. In *Lasers et Normes de Protection (First International Symposium on Laser Biological Effects and Exposure Limits)*, Paris: Centre de Recherches du Service de Sante des Armees, 1986. pp. 157-174.
179. **Courant, D., L. Court, and D. H. Sliney:** Spot-Size Dependence of Laser Retinal Dosimetry. In *Dosimetry of Laser Radiation in Medicine and Biology* (vol IS 5). Bellingham, WA: SPIE Optical Engineering Press, 1989. pp. 156-165.
180. **Courant, D., L. Court, and D. H. Sliney:** Research Relative to Safety Formulations for Retinal Damage from Extended Sources and Large Retinal Images. In *Proceedings of the International Laser Safety Conference*, Orlando, FL: Laser Institute of America, 1991. pp. 4-25–4-33.
181. **Zuclich, J.A., M.F. Blankenstein, S.J. Thomas, and R.F. Harrison:** Corneal Damage Induced by Pulsed CO_2 Laser Radiation. *Health Phys. 47*:829-835 (1984).
182. **Zuclich, J.A., and M.F. Blankenstein:** Comments on "Pulsed CO_2-Laser Corneal Injury Thresholds." *Health Phys. 50*:552 (1986).
183. **Ham, W.T., Jr., and H.A. Mueller:** Pulsed CO_2-Laser Corneal Injury Thresholds. *Health Phys. 50*:551-552 (1986).
184. **Allen, R.G., S.J. Thomas, R.F. Harrison, J.A. Zuclich, and M.F. Blenkenstein:** Ocular Effects of Pulsed Nd Laser Radiation: Variation of Threshold with Pulsewidth. *Health Phys. 49*:685-692 (1985).
185. **Sliney, D.H., and W.J. Marshall:** Bioeffects of Repetitive Pulsed Lasers. In *Proceedings of the International Laser Safety Conference*, Orlando, FL: Laser Institute of America, 1991. pp. 4-15–4-23.
186. **Birngruber, R., C. A. Puliafito, A. Gawande, W.-Z. Lin, R. W. Schoenlein, and J. G. Fujimoto:** Femtosecond Laser-Tissue Interactions: Retinal Injury Studies. *IEEE J. Quantum. Electron. QE-23*:1836-1844 (1987).
187. **Farrer, D.N.:** Bioeffects of Ultrashort Pulses: An Overview. In *Proceedings of the International Laser Safety Conference*, Orlando, FL: Laser Institute of America, 1991. pp. 4-9–4-12.
188. **Roach, W.P., C.D. DiCarlo, G.D. Noojin, D.J. Stolarski, R. Amnotte, V. Caruthers, B.A. Rockwell, and C.P. Cain:** Sub-nanosecond, Single Laser Pulse Minimum Visible Lesion Studies in the Near Infrared. *Proc. SPIE 2681*:366- 374 (1996).
189. **Toth, C.A., D.G. Narayan, C. Osborne, B.A. Rockwell, C.D. Stein, R.E. Amonette, C. DiCarol, W. Roach, G. Noojin, and C. Cain:** Histopathology of Ultrashort Laser Pulse Retinal Damage. *Proc. SPIE 2681*:375-381 (1996).
190. **Belkin, M.:** Low-Energy Laser Effects on the Eye. *Proc. SPIE 1883*:31-41 (1993).
191. **Campbell, C.J., M.C. Rittler, H. BredeMeier, and R.A. Wallace:** Ocular Effects Produced by Experimental Lasers. II. Carbon Dioxide Laser. *Am. J. Ophthalmol. 66*:604-614 (1968).
192. **Leibowitz, H.M., and G.R. Peacock:** Corneal Injury Produced by Carbon Dioxide Laser Radiation. *Arch. Ophthalmol. 81*:713-721 (1969).
193. **Borland, R.G., D.H. Brennan, and A.N. Nicholson:** Threshold Levels

for Damage of the Cornea following Irradiation by a Continuous Wave Carbon Dioxide (10.6 μm) Laser. *Nature 234*:151-152 (1971).

194. **Birngruber, R., V.-P. Gabel, and F. Hillenkamp:** Experimental Studies of Laser Thermal Retinal Injury. *Health Phys. 44*:519-531 (1983).
195. **Lund, D.J., P.R. Edsall, D.R. Fuller, and S.W. Hoxie:** Ocular Hazards of Tunable Continuous-Wave Near-Infrared Laser Sources. *Proc. SPIE 2674*:53-61 (1996).
196. **Ham, W.T., Jr., W.J. Geeraets, H.A. Mueller, R.C. Williams, A.M. Clarke, and S.F. Cleary:** Retinal Burn Thresholds for the Helium-Neon Laser in the Rhesus Monkey. *Arch. Ophthalmol. 84*:797-809 (1970).
197. **Ham, W.T., Jr., H.A. Mueller, M.L. Wolbarsht, and D.H. Sliney:** Evaluation of Retinal Exposures from Repetitively Pulsed and Scanning Lasers. *Health Phys. 54*:337-344 (1988).
198. **Ham, W.T., H.A. Mueller, J.J. Ruffolo, Jr., R.K. Guerry, and A.M. Clarke:** Ocular Effects of GaAs Lasers and Near Infrared Radiation. *Appl. Opt. 23*:2181-2186 (1984).
199. **Ham, W.T., Jr., and H.A. Mueller:** Ocular Effects of Laser Infrared Radiation. *J. Laser Appl. 3(3)*:19-21 (1991).
200. **Tso, M.O.M., and B.S. Fine:** Repair and Late Degeneration of the Primate Foveola after Injury by Argon Laser. *Invest. Ophthalmol. Visual Sci. 18*:447-461 (1979).
201. **Sliney, D.H., R.R. Krueger, S.L. Trokel, and K.D. Rappaport:** Photokeratitis from 193 nm Argon-Fluoride Laser Radiation. *Photochem. Photobiol. 53*:739-744 (1991).
202. **Schmidt, R.E. and J.A. Zuclich:** Retinal Lesions due to Ultraviolet Laser Exposure. *Invest. Ophthalmol. Vis. Sci. 19*:1166-1175 (1980).
203. **Van Gemert, M.J.C., S.L. Jacques, H.J.C.M. Sterenborg, and W.M. Star:** Skin Optics. *IEEE Trans. Biomed. Eng. 36*:1146-1154 (1989).
204. **Zhao-zhang, L., W. Jia-nu, G. Bao-kong, and Z. Yan:** Ultraviolet Erythema of Laser Radiation. *Lasers Life Sci. 2(2)*:91-101 (1988).
205. **Zhao-zhang, L., W. Jia-nu, and G. Bao-kong:** Damage Thresholds of Skin Irradiated by Ultraviolet Lasers. *Health Phys. 56*:683-686 (1989).
206. **Sterenborg, H.J.C.M., F.R. de Gruijl, G. Kelfkens, and J.C. van der Leun:** Evaluation of Skin Cancer Risk Resulting from Long Term Occupational Exposure to Radiation from Ultraviolet Lasers in the Range from 190 to 400 nm. *Photochem. Photobiol. 54*:775-780 (1991).
207. **Goldman, L.:** The Skin. *Arch. Environ. Health 18*:434-436 (1969).
208. **Brownell, A. S., W. H. Parr, and D. K. Hysell:** Skin and Carbon Dioxide Laser Radiation. *Arch. Environ. Health 18*:437-442 (1969).
209. **Hruza, G.J., J.S. Dover, T.J. Flotte, M. Goetschkes, S. Watanabe, and R.R. Anderson:** Q-Switched Ruby Laser Irradiation of Normal Human Skin. *Arch. Dermatol. 127*:1799-1805 (1991).
210. **Varanelli, A.G.:** Electrical Hazards Associated with Lasers. *J. Laser Appl.7*:62-64 (1995).
211. **Thomas, D.K.:** Often Overlooked Electrical Hazards Common in Many Lasers. In *Proceedings of the 1992 International Laser Safety Conference.* Orlando, FL: Laser Institute of America, 1993. pp. 4I-41–4I-43.
212. **Kokosa, J.M.:** Hazardous Chemicals Produced by Laser Materials Processing. *J. Laser Appl. 6*:195-201 (1994).
213. **Tarroni, G., C. Melandri, T. De Zaiacomo, C. Lombardi, C., and M. Formignani:** Characterization of Aerosols Produced in Cutting Steel Components and Concrete Structures by Means of a Laser Beam. *J. Aerosol Sci. 17*:587-591 (1986).
214. **Hardaway, G.A.:** Lasers in Metalworking and Electronic Industries. *Arch. Environ. Health 20*:188-192 (1970).
215. **Hietanen, M., A. Honkasalo, H. Laitinen, L. Lindross, I. Welling, and P. von Nandelstadh:** Evaluation of Hazards in CO_2 Laser Welding and Related Processes. *Ann. Occup. Hyg. 36*:183-188 (1992).
216. **Haferkamp, H., M. Goede, K. Engel, and J.-S. Witebecker:** Hazardous Emissions: Characterization of CO_2 Laser Material Processing. *J. Laser Appl. 7*:83-88 (1995).
217. **National Institute for Occupational Safety and Health:** *Occupational Hazards of Laser Material Processing* by R.J. Rockwell, Jr., R.M. Wilson, S. Jander, and R. Dreffer (PB89–186530). Springfield, VA: National Technical Information Service, 1976.
218. **Ball, R.D., B. Kulik, and S.L. Tan:** The Assessment and Control of

Hazardous By-Products from Materials Processing with CO_2 Lasers. In *Industrial Laser Handbook*, Tulsa, OK: Penn Well Books, 1988. pp 3-13.

219. **National Institute for Occupational Safety and Health:** *Occupational Hazards of Laser Material Processing* by C.E. Moss and T. Seitz (HETA–90–102–L2075). Springfield, VA: National Technical Information Service (PB91–146233), 1976.
220. **Doyle, D.J.:** Spectroscopic Evaluation of Toxic By-Products Produced During Industrial Laser Processing. In *Proceedings of the International Laser Safety Conference*, Orlando, FL: Laser Institute of America, 1991. pp. 3-109–3-114.
221. **Kwan, J.K.:** Tocicological Characterization of Chemicals Produced from Laser Irradiation of Graphite Composite Materials. In *Proceedings of the International Laser Safety Conference*, Orlando, FL: Laser Institute of America, 1991. pp. 3-69–3-96.
222. **National Institute for Occupational Safety and Health:** *Health Hazard Evaluation Report HETA 89–331–2078, Photon Dynamics Ltd., Inc. Longwood, Florida*, by C.E. Moss and A. Fleeger (PB91–188946). Springfield, VA: National Technical Information Service, 1990.
223. **Fleeger, A., and C.E. Moss:** Airborne Emissions Produced by The Interaction of A Carbon Dioxide Laser with Glass, Metals, and Plastics. In *Proceedings of the International Laser Safety Conference*, Orlando, FL: Laser Institute of America, 1991. pp. 3-23–3-31.
224. **Engel, K., A. Hampe, D. Seebaum, I. Vinke, and J. Wittbecker:** Review of Scientific Research in the Federal Republic of Germany Concerning Emissions in Laser Material Processing. In *Proceedings of the International Laser Safety Conference*, Orlando, FL: Laser Institute of America, 1991. pp. 5-29–5-52.
225. **Kokosa, J.M. and J. Eugene:** Chemical Composition of Laser-Tissue Interaction Smoke Plume. *J. Laser Appl. 1(3)*:59-63 (1989).
226. **National Institute for Occupational Safety and Health:** *Health Hazard Evaluation: Report No. HETA 88–101–2008, University of Utah Health Sciences Center, Salt Lake City, Utah*, by C.E. Moss, C. Bryant, J. Stewart, W.-Z. Whong, A. Fleeger, and B.J. Gunter (PB91-107789). Springfield, VA: National Technical Information Service, 1990.
227. **Weber, L., and T. Meier:** Concepts of Risk Assessment of Complex Chemical Mixtures in Laser Pyrolysis Fumes. *Proc. SPIE 2624*:259-269 (1996).
228. **Waesche, W., and H.J. Albrecht:** Investigation of the Distribution of Aerosols and VOC in Plume Produced during Laser Treatment under OR Conditions. *Proc. SPIE 2624*:270-275 (1996).
229. **Albrecht, H., and W. Wasche:** Evaluation of Potential Health Hazards Caused by Laser and RF-Surgery. *Proc. SPIE 2624*:200-204 (1996).
230. **Ferenczy, A., C. Bergeron, and R. M. Richart:** Human Papillomavirus DNA in CO_2 Laser-Generated Plume of Smoke and its Consequences to the Surgeon. *Obstet. Gynecol. 75*:114-118 (1990).
231. **Matchette, L.S., T.J. Vegella, and R.W. Faaland:** Viable Bacteriophage in CO_2 Laser Plume: Aerodynamic Size Distribution. *Lasers Surg. Med. 13*:18-22 (1993).
232. **Sawchuk, W.S., and R.P. Felten:** Infectious Potential of Aerosolized Particles. *Arch. Dermatol. 125*:1689-1692 (1989).
233. **Sliney, D.H., and T.N. Clapham:** Safety with Medical Excimer Lasers with An Emphasis on Compressed Gases. *J. Laser Appl. 3(3)*:59-62 (1991).
234. **Benoit, H., J. Clark, and W.J. Keon:** Installation of A Commercial Excimer Laser in The Operating Room. *J. Laser Appl. 1(3)*:45-50 (1989).
235. **Lorenz, A.K.:** Gas Handling Safety for Laser Makers and Users. *J. Lasers Appl. 6(3)*:69-73 (1987).
236. **Kues, H., and G. Lutty:** Dyes Can Be Deadly. *Laser Focus 11(4)*:59-60 (1975).
237. **Miller, G.:** Industrial Hygiene Concerns of Laser Dyes. In *Proceedings of the International Laser Safety Conference*, Orlando, FL: Laser Institute of America, Orlando, 1991. pp. 3-97–3-105.
238. **Wuebbles, B.J.Y., and J.S. Felton:** Evaluation of Laser Dye Mutagenicity Using the Ames/Salmonella Microsome Test. *Environ. Mutagen. 7*:511-522 (1985).

239. **Bos, A.J.J., and M.J. De Haas:** On the Safe Use of A High Power Ultraviolet Laser. In *Human Exposure to Ultraviolet Radiation: Risks and Regulations,* New York: Elsevier Science Publishers. 1987. pp. 377-382.
240. **Seitz, T.A., and C.E. Moss:** RF-Excited Carbon Dioxide Lasers: Concerns of RF Occupational Exposures. In *Proceedings of the International Laser Safety Conference.* Orlando, FL: Laser Institute of Ameica, 1991. pp. 3-35–3-40.
241. **Bowman, J.D., D.H. Garabrant, E. Sobel, and J.M. Peters:** Exposures to Extremely Low Frequency (ELF) Electromagnetic Fields in Occupations with Elevated Leukemia Rates. *Appl. Ind. Hyg. 3*:189-194 (1988).
242. **Rockwell, R.J., Jr., and C.E. Moss:** Optical Radiation Hazards of Laser Welding Processes. Part II: CO_2 Laser, *Am. Ind. Hyg. Assoc. J. 50*:419-427 (1989).
243. **Hietanen, M., and P. Von Nandelstadh:** Scattered and Plasma-Related Optical Radiations Associated with Industrial Laser Processes. In *Proceedings of the International Laser Safety Conference,* Orlando, FL: Laser Institute of Ameica, 1991. pp. 3-105–3-108.
244. **O'Neill, F., C.E. Turcu, D. Xenakis, and M.H.R. Hutchinson:** X-ray Emission from Plasmas Generated by An XeCl Laser Picosecond Pulse Train. *Appl. Phys. Lett. 55*:2603-2604 (1989).
245. **Kuhnle, G., F.P. Schafer, S. Szatmari, and G.D. Tsakiris:** X-Ray Production by Irradiation of Solid Targets with Sub-Picosecond Excimer Laser Pulses. *Appl. Phys B 47*:361-366 (1988).
246. **Chen, H., Y.-H. Chuang, J.A. Delettrez, S. Uchida, and D.D. Meyerhofer:** Study of X-ray Emission from Picosecond Laser-Plasma Interaction. *Proc. SPIE 1413*:112-119 (1991).
247. **Hietanen, M., A. Honkasalo, H. Laitinen, L. Lindross, I. Welling, and P. von Nandelstadh:** Evaluation of Hazards in CO_2 Laser Welding and Related Processes. *Ann. Occup. Hyg. 36:*183-188 1992.
248. **Domin, M.A.:** Ignition Potential of Surgical Appliances and Materials. In *Proceedings of the 1992 International Laser Safety Conference.* Orlando, FL: Laser Institute of America, 1993. pp. 4M-5–4M-12.
249. **Sosis, M.B., and F.X. Dillon:** Comparison of CO_2 Laser Ignition of the Xomed Plastic and Rubber Tracheal Tubes. In *Proceedings of the 1992 International Laser Safety Conference.* Orlando, FL: Laser Institute of America, 1993. pp. 4M-13–4M-16.
250. **Engel, D.:** Laser Generated Metal Dust Explosive Potential. In *Proceedings of the 1992 International Laser Safety Conference.* Orlando, FL: Laser Institute of America, 1993. pp. 4I-21–4I-23.
251. **Bauman, N.:** Laser Accidents: Why Only 10% Get Reported. *Laser Med. Surg. News Advances,* August:1-7 (1988).
252. **Bandle, A.M., and B. Holyoak:** Laser Incidents. In *Medical Laser Safety—Report #48.* London, England: U.K. Institute of Physical Sciences, 1988. pp. 47-55.
253. **Rathkey, A.S.:** Accidental Laser Burn of the Macula. *Arch. Ophthal. 74*:346-348 (1965).
254. **Zweng, H.C.:** Accidental Q-Switched Laser Lesion of Human Macula. *Arch. Ophthal. 78*:596-599 (1967).
255. **Curtin, T.L., and D.G. Boyden:** Reflected Laser Beam Causing Accidental Burn of Retina. *Am. J. Ophthal 65*:188- 189 (1968).
256. **Gabel, V.-P., R. Birngruber, and B. Lorenz:** Clinical Observations of Six Cases of Laser Injury to the Eye. *Health Phys. 56*:705-710 (1989).
257. **Asano, T.:** Accidental YAG Laser Burn. *Am. J. Ophthalmol. 98*:116-117 (1984)
258. **Stuck, B.E., H. Zwick, J.W. Molchany, D.J. Lund, and D.A. Gagliano:** Accidental Human Laser Retinal Injuries from Military Laser Systems. *Proc. SPIE 2674*:7-20 (1996).
259. **Laitinen, H., and T. Jarvinen:** Accident Risks and the Effect of Performance Feedback with Industrial CO_2 Lasers. *Opt. Laser Technol. 27*:25-30 (1995).
260. **Makhov, G.O.:** Accidents and Incidents in the Laser Entertainment Industry. In *Proceedings of the 1992 International Laser Safety Conference.* Orlando, FL: Laser Institute of America, 1993. pp. 7I-39–7I-43.
261. **Armstrong, C.E.:** Eye Injuries in Some Modern Radiation Environments. *J. Am. Opt. Assoc. 41*:55-62 (1970).

262. **Rockwell, R.J., Jr.:** Laser Accidents: Reviewing Thirty Years of Incidents: What Are the Concerns—Old and New? *J. Laser Appl. 6*:203-211 (1994).
263. **Cozine, K., L.M. Rosenbaum, J. Skanazi, and S.H. Rosenbaum:** Laser-Induced Endotracheal Tube Fire. *Anesthesiology 55*:583-585 (1981).
264. **Pashayan, A.G., J.S. Gravenstein, N.J. Cassisi, and G. McLaughlin:** The Helium Protocol for Laryngotracheal Operations with CO_2 Laser: A Retrospective Review of 523 Cases. *Anesthesiology 68*:801-804 (1988).
265. **Felten, R.P.:** A Review of Laser Injuries as Reported through the MDR System. In *Proceedings of the International Laser Safety Conference*, Orlando, FL: Laser Institute of America, 1993. pp. 7M-9–7M-12.
266. **Wolbarsht, M.L.:** Permanent Blindness from Laser Exposures in Laboratory and Industrial Accidents. *Proc. SPIE 2674*:21-24 (1996).
267. **U.S. Food and Drug Administration:** Subchapter J — Radiological Health (21 CFR). Washington, D.C.: U.S. Government Printing Office, 1971.
268. **Spicer, M.S., and D.J. Goldberg:** Lasers in Dermatology. *J. Am. Acad. Dermatol. 34*:1-25 (1996).
269. **Fitzpatrick, R.E.:** Lasers in Dermatology and Plastic Surgery. *Opt. Photonics 6(11)*:23-31 (1995).
270. **Pettit, G.H.:** Lasers Take Up Residence in the Surgical Suite. *IEEE Circuits Devices 8(3)*:18-25 1992.
271. **Ren, Q., R.H. Keates, R.A. Hill, and M.W. Berns:** Laser Refractive Surgery: A Review and Current Status. *Opt. Eng. 34*:642-660 (1995).
272. **Grossweiner, L.I.:** Photodynamic Therapy. *J. Laser Appl. 7(1)*:51-57 (1995).
273. **Wolbarsht, M.L.:** Low-Level Laser Therapy (LLLT) and Safety Considerations. *J. Laser Appl. 6(3)*:170-172 (1994).
274. **Mommsen, J., and M. Strumer:** Use of Excimer Lasers in Medicine: Applications, Problems and Dangers. *Proc. SPIE 1503*:348-354 (1991).
275. **Wigdo, H.:** Laser Safety in Dentistry. In *International Laser Safety Conference*, Orlando, FL: Laser Institute of America, 1993. pp. 6M-25–6M-28.
276. **Miserendino, L.:** Recommendations for Safe Use in Dentistry. *J. Laser Appl. 4(3)*:16-17 (1992).
277. **Balin, V.N., A.S. Guk, S.P. Kropotov, D.Y. Maday, T.A. Kuzovkova, V.A. Serebryakov, and S.V. Frolov:** Experimental Caries Treatment Using the Pulsed Erbium Laser. *Proc. SPIE 2672*:103-105 (1996).
278. **Richter, P.I.:** Air Pollution Monitoring with Lidar. *Trends Anal. Chem. 13*:263-266 (1994).
279. **Patonay, G., M.D. Antoine, and A.E. Boyer:** Semiconductor Lasers in Analytical Chemistry. *Proc. SPIE 1435*:52- 63 (1991).
280. **Ventzas, D.E., and A. Angelopoulos:** Laser Particle Size Analysis. *Model. Meas. Control 54(4)*:17-26 (1994).
281. **Avila, V.:** Laser Polarimeters: Recent Developments, Design and Application. *J. Laser Appl. 8(1)*:43-53 (1996).
282. **Bescos, B., and A.G. Urena:** Laser Chemical Analysis of Metallic Elements in Aluminum Samples. *J. Laser Appl. 7(1)*:47-50 (1995).
283. **Mooradian, A.:** Raman Spectroscopy of Solids. In *Laser Handbook*, Vol. 2. New York: North-Holland Publishing Co., 1972. pp. 1411-1456.
284. **Mauldin, J.H.:** *Light, Lasers and Optics*. Blue Ridge Summit, PA: Tab Books Inc., 1988. pp. 153-157.
285. **Betz, A., M. Retzbach, G. Alber, and W. Prange:** Automated Laser Systems for High Volume Production Sheet Metal Application. *Proc. SPIE 1990*:627-636 (1993).
286. **O'Conner, L.:** Making Light Work of Fabric Cutting, *Mechanical Eng. 115(11)*:64-66 (1993).
287. **Belforte, D. A.:** High-Power Nd:YAG Lasers Shine in Industrial Processing. *Laser Focus World 28(9)*:69-76 (1992).
288. **Ready, J.F.:** *Industrial Applications of Lasers*, New York: Academic Press, 1978.
289. **Folkes, J.A.:** Developments in Laser Surface Modification and Coating. *Surface Coatings Technol. 63*:65-71 (1994).
290. **Fly, D.E., J.T. Black, and B. Singleton:** Low Power Laser Heat Treatment to Improve Fatigue Life of Low Carbon Steel. *J. Laser Appl. 8(2)*:89-93 (1996).
291. **Chen, X., W.T. Lotshaw, A.L. Ortiz, P.R. Staver, C.E. Erikson, M.H. McLaughlin, and T.J. Rockstroh:** Laser Drilling of Advanced Materials: Effects of Peak Power, Pulse Format and Wavelength. *J. Laser Appl. 8(5)*:233-239 (1996).

292. **Larsson, L.-O and P. Hedenborn:** Laser Scanners as a Measuring Device and Its Application in Arc Welding. *Int. J. Joining Materials 5*:14-18 (1994).
293. **Rebhan, U., H. Endert, and G. Zaal:** Micromanufacturing Benefits from Excimer-Laser Development. *Laser Focus World 30(11)*:91-96 (1994).
294. **Woodin, R.L., D.S. Bomse, and G.W. Rice:** Lasers in Chemical Processing. *Chem. Eng. News 68(51)*:20-31 (1990).
295. **Higgins, T.V.:** Light Speeds Communications. *Laser Focus World 31(8)*:67-74 (1995).
296. **Jaeger, J.L., and R.T. Carlson:** Laser Communication for Covert Links. *Proc. SPIE 1866*:95-106 (1993).
297. **Mazzaferro, J.:** Laser-Based Communications Systems Emit Radiation, An Eye Hazard. *Occupat. Health Safety 64(2)*:59-61 (1995).
298. **Milonni, P.W., and J.H. Eberly:** *Lasers*. New York: John Wiley & Sons, 1988. pp. 609-618.
299. **Gorham, B.J.:** Safety Aspects in the Use of Outdoor and Surveying Lasers. *Opt. Laser Technol. 27*:19-24 (1995).
300. **Gibbs, R.:** Printing Benefits from New Technologies. *Laser Focus World 31(11)*:77-81 (1995).
301. **Health and Safety Executive:** *Laser Safety in Printing*. London: HMSO Publications Centre, 1990.
302. **Seeber, F.P., and D. Haes:** Laser Practices and Problems in Educational Institutions. In *International Laser Safety Conference*, Orlando, FL: Laser Institute of America, 1993. pp. 5I-25–5I-29.
303. **Sweet, J.:** Effective Laser Applications in High School Physics; *J. Laser Appl. 6(4)*:248-251 (1994).
304. **Brill, M.L.:** Taking Advantage of Laser Properties to Enhance Demonstrations and Student Laboratories. *J. Laser Appl. 8(5)*:259-263 (1996).
305. **Makhov, G.:** Safety Concerns for Laser Displays and Entertainment Installations. In *Proceedings of the International Laser Safety Conference*, Orlando, FL: Laser Institute of Ameica, 1991. pp. 5-65–5-69.
306. **Menzel, E.R.:** Laser Applications in Criminalistics. *J. Laser Appl. 3(2)*:39-44 (1991).
307. **Menzel, E.R.:** *Fingerprint Detection with Lasers*. New York: Marcel Dekker, Inc., 1980.
308. **Franks, J.:** Military Laser Range Safety. In *Proceedings of the International Laser Safety Conference,*. Orlando, FL: Laser Institute of Ameica, 1991. pp. 9-17–9-23.
309. **Ashley, S.:** Searching for Land Mines. *Mech. Eng. 118(4)*:62-67 (1996).
310. **Lamberson, S.E.:** Airborne Laser. Proc. *SPIE 2702*:208-213 (1996).
311. **Sliney, D.H.:** Infrared Laser Effects on the Eye: Implications for Safety and Medical Applications. *Proc. SPIE 2097*:36-43 (1993).
312. **Sliney, D.H.:** Laser Effects on Vision and Ocular Exposure Limits. *Appl. Occup. Environ. Hyg. 11*:313-319 (1996).
313. **Vos, J.J.:** Certainities and Uncertainities about Safe Laser Exposure Limits. *Proc. SPIE 2052*:467-474 (1993).
314. **International Radiation Protection Association:** Guidelines on Limits of Exposure to Laser Radiation of Wavelengths between 180 nm and 1 mm. *Health Phys. 49*:341-359 (1985).
315. **International Radiation Protection Association:** Recommendations for Minor Updates to the IRPA 1985 Guidelines on Limits of Exposure to Laser Radiation. *Health Phys. 54*:573-574 (1988).
316. **Sliney, D.H.:** The IRPA/INIRC Guidelines on Limits of Exposure to Laser Radiation. In *Light, Lasers and Synchrotron Radiation*. New York: Plenum Press, 1990. pp. 341-346.
317. **International Electrotechnical Commission:** *Safety of Laser Products Part 1 — Equipment Classification, Requirements and User's Guide* (IEC 825-1). Geneva: IEC, 1993.
318. **Boucouvalas, A.C.:** IEC 825-1 Eye Safety Classification of Some Consumer Electronic Products. *IEE Colloquim on Optical Free Space Communications Links* (Ref. No. 1996/032). London: IEE, 1996. pp. 13/1–13/6.
319. **American National Standards Institute:** *American National Standard for Safe Use of Lasers in Health Care Facilities* (Z136.3–1993). Orlando, FL: Laser Institute of America, 1993.
320. **Hattin, H.C.:** Laser Safety in Health Care Facilities: A National Standard of Canada. *J. Clin. Eng. 19*:218-221 (1996).
321. **American National Standards Institute:** *American National Standard for the Safe Use of Optical Fiber Communications Systems Utilizing*

Laser Diodes and LED Sources (Z136.2–1997). Orlando, FL: Laser Institute of America, 1997.

322. **International Electrotechnical Commission (IEC):** *Safety of Laser Products Part 2 — Safety of Optical Fibre Communications Systems* (IEC 825–2). Geneva, Switzerland: IEC, 1993.
323. **Curtis, R.A.:** OSHA Standards Related to Laser Safety. In *Proceedings of the International Laser Safety Conference*, Orlando, FL: Laser Institute of Ameica, 1991. pp. 1-35–1-39.
324. **Handren, R.T., Jr.:** An Overview of State Laser Regulations. In *Proceedings of the International Laser Safety Conference*, Orlando, FL: Laser Institute of Ameica, 1991. pp. 1-51–1-57.
325. **Occupational Safety and Health Administration:** *Guidelines for Laser Safety and Hazard Assessment* (OSHA Instruction Pub. No. 8–1.7). Washington, D.C.: U.S. Government Printing Office, 1991.
326. **Marshall, W.J., and P.W. Conner:** Field Laser Hazard Calculations. *Health Phys. 52*:27-37 (1987).
327. **Marshall, W.J., and W.P. Van DeMerwe:** Hazardous Ranges of Laser Beams and Their Reflections from Targets. *Appl. Optics 25*: 605-611 (1986).
328. **Marshall, W.J.:** Laser Reflections from Relatively Flat Specular Surfaces. *Health Phys. 56*:753-757 (1989).
329. **Marshall, W.J.:** Determining Hazard Distances from Non-Gaussian Lasers. *Appl. Optics 30*:696-698 (1991).
330. **Marshall, W.J.:** Comparative Hazard Evaluation of Near-Infrared Diode Lasers. *Health Phys. 66*:532-539 (1994).
331. **Marshall, W.J.:** Understanding Laser Hazard Evaluation. *J. Laser Appl. 7*:99-105 (1995).
332. **Marshall, W.J., R.C. Aldrich, and S.A. Zimmerman:** Laser Hazard Evaluation Method for Middle Infrared Laser Systems. *J. Laser Appl. 8*:211-216 (1996).
333. **Lyon, T.L.:** Hazard Analysis Technique for Multiple Wavelength Lasers. *Health Phys. 49*:221-226 (1985).
334. **Lyon, T.L.:** Laser Measurement Techniques Guide for Hazard Evaluation (Tutorial Guide — Part 1). *J. Laser Appl. 5(1)*:53-58 (1993).
335. **Marshall, W.J.:** Focused Laser Beam Hazard Calculations. In *Proceedings of the International Laser Safety Conference*, Orlando, FL: Laser Institute of America, 1991. pp. 9-33–9-39.
336. **Lyon, T.L.:** Laser Measurement Techniques Guide for Hazard Evaluation (Tutorial Guide — Part 2). *J. Laser Appl. 5(2&3)*:37-42 (1993).
337. **Rockwell, R.J., Jr., and C.E. Moss:** Hazard Zones and Eye Protection Requirements for a Frosted Surgical Probe Used with an Nd:YAG Laser. *Lasers Surg. Med 9*:45-49 (1989).
338. **Rockwell, R.J., Jr.:** Utilization of the Nominal Hazard Zone in Control Measure Selection. In *Proceedings of the International Laser Safety Conference*, Orlando, FL: Laser Institute of America, 1991. pp. 7-25–7-42.
339. **Marshall, W.J.:** Modified Range Equation. In *Proceedings of the International Laser Safety Conference*, Orlando, FL: Laser Institute of America, 1993. pp. 7I-45–7I-53.
340. **Labo, J.A., and M.E. Rogers:** Can Broken Fiber Optics Produce Hazardous Laser Beams. *Proc. SPIE 1892*:176-187 (1993).
341. **Kestenbaum, A., R.J. Coyle, and P.P. Sloan:** Safe Laser System Design for Production. *J. Laser Appl. 7*:31-37 (1995).
342. **Wilson, D.F.:** Barrier Curtains for Laser Hazard Control. In *Proceedings of the International Laser Safety Conference*, Orlando, FL: Laser Institute of America, 1991. pp. 5-71–5-75.
343. **Gorham, B.J.:** Safety Aspects in the Use of Outdoor and Surveying Lasers. *Opt. Laser Technol. 27*:19-24 (1995).
344. **Rockwell, R.J., Jr., J.F. Smith, and W.J. Ertle:** Playing It Safe with Industrial Lasers. *Photonics Spectra 29(4)*:118- 124 (1995).
345. **Barat, K.:** A Guide to Developing a Laser Standard Operating Procedure. *J. Laser Appl. 8*:255-257 (1996).
346. **Stocum, W.E.:** Laser Safety Training Programs for a Large and Diverse Research and Development Laboratory. In *Proceedings of the International Laser Safety Conference*, Orlando, FL: Laser Institute of America, 1993. pp. 5I-13–5I-17.
347. **Johnson, K.L.:** Laser Safety Training Video for Use in an R&D and Industrial Setting. In *Proceedings of the International Laser Safety Conference*, Orlando, FL: Laser Institute of America, 1991. pp. 7-87–7-89.

348. **Hitchcock, R.T., J.F. Smith, J.R. Johnson, and M. Siegel:** Development of an Interactive Video Laser Safety Training Program. In *Proceedings of the International Laser Safety Conference*, Orlando, FL: Laser Institute of America, 1993. pp. 5I-1–5I-8.
349. **Moskal, P.:** Laser Safety Training. *Proc. SPIE CR58*:149-163 (1995).
350. **Smith, J.F.:** Class 2 Laser Safety Training Programs. In *Proceedings of the International Laser Safety Conference*, Orlando, FL: Laser Institute of America, 1991. pp. 7-83–7-86.
351. **Coleman, J.A., and R.H. Ossoff:** Laser Safety in Otolaryngology Head and Neck Surgery. In *Proceedings of the International Laser Safety Conference*, Orlando, FL: Laser Institute of America, 1991. pp. 6-27–6-37.
352. **ECRI:** Laser Safety Eyewear. *Health Devices 22(4)*:159-207 (1993).
353. **Sliney, D.H.:** *Guide for the Selection of Laser Eye Protection*, 3rd Ed. Orlando, FL: Laser Institute of America, 1993.
354. **Spaeth, D.J.:** Options in Choosing Laser Protective Eyewear. In *Proceedings of the International Laser Safety Conference*, Orlando, FL: Laser Institute of America, 1991. pp. 8-19–8-29.
355. **Bowker, K.W.:** Hazards Associated with Sources of Ultra-Violet Radiation Used in a Research Environment. In *Human Exposure to Ultraviolet Radiation: Risks and Regulations*. New York: Elsevier Science Publishers, 1987. pp. 371-375.
356. **Sliney, D.H.:** Laser Safety. *Lasers Surg. Med. 16*:215-225 (1995).
357. **Swope, C.H.:** The Eye — Protection. *Arch. Environ. Health 18*:428-435 (1969).
358. **Sutter, E.:** The DIN and CEN Laser Eyewear Standards: Review and Testing Requirements. In *Proceedings of the International Laser Safety Conference*, Orlando, FL: Laser Institute of America, 1991. pp. 2-23–2-40.
359. **Sliney, D.H.:** Laser Eye Protectors. *J. Laser Appl. 2(2)*:9- 13 (1990).
360. **Hack, H., and N. Neuroth:** Resistance of Optical and Colored Glasses to 3-nsec Laser Pulses. *Appl. Optics 21*:3239-3248 (1982).
361. **Swearengen, P.M., W.F. Vance, and D.L. Counts:** A Study of Burn-Through Times for Laser Protective Eyewear. *Am. Ind. Hyg. Assoc. J. 49*:608-612 (1988).
362. **Tucker, R.J.:** Damage Testing of Polycarbonate Laser Protective Eyewear. In *Proceedings of the International Laser Safety Conference*, Orlando, FL: Laser Institute of America, 1991. pp. 9-25–9-27.
363. **Lyon, T.L., and W.J. Marshall:** Nonlinear Properties of Optical Filters—Implications for Laser Safety. *Health Phys. 51*:95-96 (1986).
364. **Mayo, M.W., W.P. Roach, C.M. Bramlette, and M.D. Gavornik:** Nonlinear Transmittance through Laser Protective Material: Reverse Photosaturation of 1-Phenylazo 2-Napthalenol in Polymer Host. *Proc. SPIE 1864*:86-95 (1993).
365. **Scott, T.R., R.J. Rockwell, Jr., and P. Batra:** Optical Density Measurements of Laser Eye Protection Materials. In *Proceedings of the International Laser Safety Conference*, Orlando, FL: Laser Institute of America, 1993. pp. 8-7–8-16.
366. **Adams, A.J., and J.E. Skipper:** Safe Use of Infrared Viewing Cards. *Health Phys. 59*:225-228 (1990).
367. **Mumford, W.W.:** Some Technical Aspects of Microwave Radiation Hazards. *Proc. IRE 49*:427-447 (1961).
368. **Hicks, C.W.:** "Communication on Antennae, Dosimetry, and Waveguides." February 1986. [Telephone Conversation]. U.S. Army Environmental Hygiene Agency, Aberdeen Proving Grounds, MD 21010.
369. **Hitchcock, R.T.:** *Radio-Frequency and Microwave Radiation*, 2nd Ed. (Nonionizing Radiation Guide Series). Fairfax, VA: American Industrial Hygiene Association, 1994.
370. **Hitchcock, R.T., and R.M. Patterson:** *Radio-Frequency and ELF Electromagnetic Energies — A Handbook for Health Professionals*. New York: Van Nostrand Reinhold, 1995.
371. **Eriksson A., and K. Hansson Mild:** Radiofrequency Electromagnetic Leakage Fields from Plastic Welding Machines. *J. Microwave Power 20*:95-107 (1985).
372. **Martin, C.J., H.M. McCallum, and B. Heaton:** An Evaluation of Radiofrequency Exposure from Therapeutic Diathermy Equipment in the Light of Current Recommendations. *Clin. Phys. Physiol. Meas. 11*:53-63 (1990).

373. **U.S. Environmental Protection Agency:** *Biological Effects of Radiofrequency Radiation*, edited by J.A. Elder and D.F. Cahill (Report No. EPA–600/8–83–026F). Springfield, VA: National Technical Information Service, 1984.
374. **Polk, C., and E. Postow (eds.):** *Handbook of Biological Effects of Electromagnetic Fields*, 2nd Ed. Boca Raton, FL: CRC Press, 1996.
375. **U.S. Air Force:** *Critique of the Literature on Bioeffects of Radiofrequency Radiation: A Comprehensive Review Pertinent to Air Force Operations* by L. Heynick (Report No. USAFSAM–TR–87–3). Brooks Air Force Base, Texas: USAF School of Aerospace Medicine, 1987.
376. **Beers, G.J.:** Biological Effects of Weak Electromagnetic Fields from 0 Hz to 200 MHz: A Survey of The Literature with Special Emphasis on Possible Magnetic Resonance Effects. *Mag. Res. Imag. 7*:309-331 (1989).
377. **National Radiological Protection Board:** *Biological Effects of Exposure to Non-Ionising Electromagnetic Fields and Radiation* by R.D. Saunders, Z.J. Sienkiewicz, and C.I. Kowalczuk (NRPB–R240). Chilton, Didcot, Oxon, U.K.: NRPB, 1991.
378. **Gandhi, O.P. (ed.):** *Biological Effects and Medical Applications of Electromagnetic Energy*. Englewood Cliffs, NJ: Prentice Hall, 1991.
379. **National Radiological Protection Board:** *Human Health and Exposure to Electromagnetic Radiation* by J.A. Dennis, C.R. Muirhead and J.R. Ennis (NRPB–R241). Chilton, Didcot, Oxon, U.K.: NRPB, 1992.
380. **World Health Organization:** *Electromagnetic Fields (300 Hz to 300 GHz)* (Environmental Health Criteria 137). Geneva, Switzerland: WHO, 1993.
381. **Teixeira-Pinto, A.A., L.L. Nejelski, Jr., J.L. Cutler, and J.H. Heller:** The Behavior of Unicellular Organisms in An Electromagnetic Field. *Exp. Cell Res. 20*:548-564 (1960).
382. **Saito, M., and H.P. Schwan:** The Time Constants of Pearl-Chain Formation. In *Proceedings of the Fourth Annual Tri-Service Conference on the Biological Effects of Microwave Radiation*, Vol. 1. New York: Plenum Press, 1961. pp. 85-97.
383. **Schwan, H.P.:** Nonthermal Cellular Effects of Electromagnetic Fields: AC-Field Induced Ponderomotoric Forces. *Brit. J. Cancer 45(Supp. V)*:220-224 (1982).
384. **Schwan, H.P.:** *Biological Effects of Non-Ionizing Radiations: Cellular Properties and Interactions* (Lecture No. 10). Bethesda, MD: National Council on Radiation Protection and Measurements, 1987.
385. **Michaelson, S.M.:** Subtle Effects of Radiofrequency Energy Absorption and Their Physiological Implications. In *Interactions between Electromagnetic Fields and Cells*. New York: Plenum Press, 1985. pp. 581-601.
386. **Elder, J.:** Radiofrequency Radiation Activities and Issues: A 1986 Perspective. *Health Phys. 53*:607-611 (1987).
387. **U.S. Air Force:** *Radiofrequency Radiation Dosimetry Handbook*, 2nd Ed., by C.H. Durney, C.C. Johnson, P.W. Barber, H. Massoudi, M.F. Iskander, J.L. Lords, D.K. Ryser, S.J. Allen, and J.C. Mitchell (School of Aerospace Medicine Report SAM–TR–78–22). Brooks Air Force Base, TX, 1978. pp. 31-32.
388. **D'Andrea, J., O. Gandhi, and J.L Lords:** Behavioral and Thermal Effects of Microwave Radiation at Resonant and Nonresonant Wavelengths. *Radio Sci. 12(6S)*:251-256 (1977).
389. **de Lorge, J.O.:** The Effects of Microwave Radiation on Behavior and Temperature in Rhesus Monkeys. In *Biological Effects of Electromagnetic Waves* (HEW [FDA] Pub. No. 77–8010). Rockville, MD: Bureau of Radiological Health, 1976. pp. 158-174.
390. **de Lorge, J.O.:** Disruption of Behavior in Mammals of Three Different Sizes Exposed to Microwaves: Extrapolation to Larger Mammals. In *Proceedings of the 1978 Symposium on Electromagnetic Fields in Biological Systems*, Edmonton, Canada: International Microwave Power Institute, 1978. pp. 215-228.
391. **de Lorge, J.O.:** The Thermal Basis for Disruption of Operant Behavior by Microwaves in Three Animal Species. In *Microwaves and Thermoregulation*. New York: Academic Press, 1983. pp. 379-399.
392. **D'Andrea, J.A.:** Microwave Radiation Absorption: Behavioral Effects. *Health Phys. 61*:29-40 (1991).
393. **Dodge, C.H.:** Clinical and Hygienic Aspects of Exposure to Electromagnetic Fields. In *Biological*

Effects and Health Implications of Microwave Radiation. Washington, D.C.: U.S. Government Printing Office, 1969. pp. 140-149.

394. **Silverman, C.:** Nervous and Behavioral Effects of Microwave Radiation in Humans. *Am. J. Epidemiol. 97*:219-224 (1973).
395. **Silverman, C.:** Epidemiology of Microwave Radiation Effects in Humans. In *Epidemiology and Quantitation of Environmental Risk in Humans from Radiation and Other Agents*. New York: Plenum Press, 1985. pp. 433-458.
396. **Sadcikova, M.N.:** Clinical Manifestations of Reactions to Microwave Irradiation in Various Occupational Groups. In *Biological Effects and Health Hazards of Microwave Radiation*, Warsaw, Poland: Polish Medical Publishers, 1974. pp. 261-267.
397. **Baranski, S., and P. Czerski:** *Biological Effects of Microwaves*. Stroudsburg, PA: Dowden, Hutchinson & Ross, Inc., 1976.
398. **Forman, S.A., C.K. Holmes, T.V. McManamon and W.R. Wedding:** Psychological Symptoms and Intermittent Hypertension Following Acute Microwave Exposure. *J. Occup. Med. 24*:932-934 (1982).
399. **Williams, R.A., and T.S. Webb:** Exposure to Radio-Frequency Radiation from An Aircraft Radar Unit. *Aviat. Space Environ. Med. 51*:1243-1244 (1980).
400. **Lai, H., A. Horita, C.K. Chou, and A.W. Guy:** A Review of Microwave Irradiation and Actions of Psychoactive Drugs. *IEEE Engineering in Medicine and Biology Magazine 6(1)*:31-36 (1987).
401. **Frey, A.H., and L.S. Wesler:** Interaction of Psychoactive Drugs with Exposure to Electromagnetic Fields. *J. Bioelect. 9*:187-196 (1990).
402. **Lary, J.M., D.L. Conover, E.D. Foley, and P.L. Hanser:** Teratogenic Effects of 27.12 MHz Radiofrequency Radiation in Rats. *Teratology 26*:299-309 (1982).
403. **Lary, J.M., D.L. Conover, P.H. Johnson and J.R. Burg:** Teratogenicity of 27.12-MHz Radiation in Rats is Related to Duration of Hyperthermic Exposure. *Bioelectromagnetics 4*:249-255 (1983).
404. **Lary, J.M., D.L. Conover, P.H. Johnson and R.W. Hornung:** Dose-Response Relationship and Birth Defects in Radiofrequency—Irradiated Rats. *Bioelectromagnetics* 7:141-149 (1986).
405. **Berman, E., H. Carter, and D. House:** Observations of Syrian Hamster Fetuses After Exposure to 2450-MHz Microwaves. *J. Microwave Power 17*:107-112 (1982).
406. **Berman, E., H. Carter, and D. House:** Reduced Weight in Mice Offspring After in utero Exposure to 2450-MHz (CW) Microwaves. *Bioelectromagnetics 3*: 285-291 (1982).
407. **Cohen, B.H., A.M. Lilienfeld, S. Kramer amd L.C. Hyman:** Parental Factors in Down's Syndrome: Results of The Second Baltimore Case-Control Study. In *Population Cytogenetics—Studies in Humans*. New York: Academic Press, 1977. pp. 301-352.
408. **Kallen, B., G. Malmquist, and U. Moritz:** Delivery Outcome Among Physiotherapists in Sweden: Is Non-Ionizing Radiation A Fetal Hazard? *Arch. Environ. Health 37*:81-84 (1982).
409. **Kolmodin-Hedman, B., K.H. Mild, M. Hagberg. E. Jonsson, M.-C. Andersson, and A. Eriksson:** Health Problems Among Operators of Plastic Welding Machines and Exposure to Radiofrequency Electromagnetic Fields. *Int. Arch. Occup. Environ. Health 60*:243-247 (1988).
410. **Taskinen, H., P. Kyyronen, and K. Hemminki:** Effects of Ultrasound, Shortwaves, and Physical Exertion on Pregnancy Outcome in Physiotherapists. *J. Epidemiol. Comm. Health 44*:196-201 (1990).
411. **Schnorr, T.M., B.A. Grajewski, R.W. Hornung, M.J. Thun, G.M. Egeland, W.E. Murray, D.L. Conover and W.E. Halperin:** Video Display Terminals and The Risk of Spontaneous Abortion. *New Engl. J. Med. 324*:727-733 (1991).
412. **Nielsen, C.V., and L. Brandt:** Spontaneous Abortion Among Women Using Video Display Terminals. *Scand. J. Work Environ. Health 16*:323-328 (1990).
413. **Brandt, L., and C.V. Nielsen:** Congenital Malformations Among Children of Women Working with Video Display Terminals. *Scand. J. Work Environ. Health 16*:329-333 (1990).
414. **Larsen, A.I., J. Olsen, and O. Svane:** Gender-Specific Reproductive Outcome and Exposure to High-Frequency Electromagnetic Radiation

Among Physiotherapists. *Scand. J. Work Environ. Health 17*:324-329 (1991).

415. **James, W.H.:** Sex Ratio of Offspring of Female Physiotherapists Exposed to Low-Level High-Frequency Electromagnetic Radiation. *Scand. J. Work. Environ. Health 21*:68-69 (1995).
416. **Lancranjan, I. M. Maicanescu, E. Rafaila, I. Klepsch and H.I. Popescu:** Gonadic Function in Workmen with Long-Term Exposure to Microwaves. *Health Phys. 29*:381-383 (1975).
417. **Weyandt, T.B.:** *Evaluation of Biological and Male Reproductive Function Responses to Potential Lead Exposures in 155 MM Howitzer Crewmen.* Springfield, VA: National Technical Information Service (AD-A247 384), 1992.
418. **Paulsson, L.-E., Y. Hamnerius, H.-A. Hansson, and J Sjostrand:** Retinal Damage Experimentally Induced by Microwave Radiation at 55 mW/cm^2. *Acta Ophthalmol. 57*:183-197 (1979).
419. **Kues, H.A., and J.C. Monahan:** Microwave-Induced Changes to the Primate Eye. *Johns Hopkins APL Tech. Dig. 13*:244-254 (1992).
420. **Kues, H.A. L.W. Hirst, G.A. Lutty, S.A. D'Anna, and G.R. Dunkelberger:** Effects of 2.45-GHz Microwaves on Primate Corneal Endothelium. *Bioelectromagnetics 6*:177-188 (1985).
421. **Kues, H.A., and S. D'Anna:** Changes in the Monkey Eye Following Pulsed 2.45-GHz Microwave Exposure. *Proceedings of the Ninth Annual Conference of the IEEE Engineering in Medicine and Biology Society* (Vol. 2; IEEE Cat. No. 87CH2513-0), New York: Institute of Electrical and Electronic Engineers, 1987. pp. 698-700.
422. **Kamimura, Y., K.I. Saito, T. Saiga, and Y. Amemiya:** Experiment of 2.45 GHz Microwave Irradiation to Monkey's Eye. In *1994 International Symposium on Electromagnetic Compatibility* (IEEE Cat. No. 94TH0680-9). New York: Institute of Electrical and Electronics Engineers, 1994. pp. 429-432.
423. **Carpenter, R., and C. Van Ummersen:** The Action of Microwave Radiation on The Eye. *J. Microwave Power 3*:3-19 (1968).
424. **Guy, A.W., J.C. Lin, P.O. Kramar and A.F. Emery:** Effect of 2450-MHz Radiation on The Rabbit Eye. *IEEE Trans. Microwave Theory Tech. MTT-23*:492-498 (1975).
425. **Michaelson, S.M., J.W. Howland, and W.B. Deichmann:** Response of The Dog to 24,000 and 1285 MHz Microwave Exposure. *Ind. Med. 40*:18-23 (1971).
426. **Appleton, B., S.E. Hirsch, and P.V.K. Brown:** Investigation of Single-Exposure Microwave Ocular Effects at 3000 MHz. *Ann. N.Y. Acad. Sci. 247*:125-134 (1975).
427. **Appleton, B., and G.C. McCrossan:** Microwave Lens Effects in Humans. *Arch. Ophthalmol. 88*:259-262 (1972).
428. **Zydecki, S.:** Assessment of Lens Translucency in Juveniles, Microwave Workers and Age-Matched Groups. In *Biologic Effects and Health Hazards of Microwave Radiation*. Warsaw, Poland: Polish Medical Publishers, 1974. pp. 273-280.
429. **Cleary, S.F., B.S. Pasternack, and G.W. Beebe:** Cataract Incidence in Radar Workers. *Arch. Environ. Health 11*:179-182 (1965).
430. **Cleary, S.F., and B.S. Pasternack:** Lenticular Changes in Microwave Workers. *Arch. Environ. Health 12*:23-29 (1966).
431. **Majewska, K.:** Investigations on The Effect of Microwaves on The Eye. *Pol. Med. J. 38*:989-994 (1968).
432. **Siekierzynski, M., P. Czerski, A. Gidynski, S. Zydecki, C. Czarnecki, E. Dziuk and W. Jedrzejczak:** Health Surveillance of Personnel Occupationally Exposed to Microwaves. III. Lens Translucency. *Aerospace Med. 45*:1146- 1148 (1974).
433. **Bonomi, L., and R. Bellucci:** Considerations of The Ocular Pathology in 30,000 Personnel of The Italian Telephone Company (SIP) Using VDTs. *Bollettion Di Oculistica 68 (S7)*:85-98 (1989).
434. **Szydzinski, A., A. Pietraszek, M. Janiak and J. Wrembel, M. Kalczak and S. Szmigielski:** Acceleration of the Development of Benzopyrene-Induced Skin Cancer in Mice by Microwave Radiation. *Dermatol. Res. 274*:303-312 (1982).
435. **Szmigielski, S.A., A. Szudzinski, A. Pietraszek, M. Bielec, M. Janiak, and J. Wrembel:** Accelerated Development of Spontaneous and Benzopyrene-Induced Skin Cancer in Mice Exposed to 2450 MHz Microwave Radiation. *Bioelectromagnetics 3*:171-191 (1982).
436. **Santini, R., M. Hosni, P. Deschaux and H. Pacheco:** B16 Melanoma

Development in Black Mice Exposed to Low-Level Microwave Radiation. *Bioelectromagnetics 9*:105-107 (1988).

437. **Wu R.Y., H. Chiang, B.J. Shao, N.G. Li, Y.D. Fu:** Effects of 2.45-GHz Microwave Radiation and Phorbol Ester 12-O-Tetradecanoylphorbol-13-acetate on Dimethylhydrazine Colon Cancer in Mice. *Bioelectromagnetics 15*: 531- 538(1994).
438. **Svedenstal B.M., and B. Holmberg:** Lymphoma Development Among Mice Exposed to X-rays and Pulsed Magnetic Fields. *Int. J. Radiat. Biol. 64*:119-125 (1993).
439. **United States Air Force:** *Effects of Long-Term Low-Level Radiofrequency Radiation Exposure on Rats Volume 9. Summary* by A.W. Guy, C.-K. Chou, L.L. Kunz, J. Crowley, and J. Krupp (School of Aerospace Medicine Report USAFSAM-TR-85-64). Brooks Air Force Base, TX., 1985. pp. 16-19.
440. **Chou, C.-K., A.W. Guy, L.L. Kunz, R.B. Johnson, J.J. Crowley and J.H. Krupp:** Long-Term, Low-Level Microwave Irradiation of Rats. *Bioelectromagnetics 13*:469-496 (1992).
441. **Robinette, D., C. Silverman, and S. Jablon:** Effects upon Health of Occupational Exposure to Microwave Radiation (Radar). *Am. J. Epidemiol. 112*:39-53 (1980).
442. **Garland F.C., E. Shaw, E.D. Gorham, C.F. Garland, M.R. White, P.J. Sinsheimer:** Incidence of Leukemia in Occupations with Potential Electromagnetic Field Exposure in United States Navy Personnel. *Am. J. Epidemiol.* 132:293-303 (1990).
443. **Szmigielski, S.:** Cancer Morbidity in Subjects Occupationally Exposed to High Frequency (Radiofrequency and Microwave) Electromagnetic Radiation. *Sci. Total Env. 180*:9-17 (1996).
444. **Grayson, J.K.:** Radiation Exposure, Socioeconomic Status, and Brain Tumor Risk in the US Air Force: A Nested Case-Control Study. *Am. J. Epidemiol. 143*:480-486 (1996).
445. **Hill, D.G.:** "A Longitudinal Study of a Cohort with Past Exposure to Radar: The MIT Radiation Laboratory Follow-up Study." Ph.D. diss., Johns Hopkins University, (Available from Ann Arbor, Michigan: University Microfilms International) 1988.
446. **Rothman, K.J., J.E. Loughlin, D.P. Funch, and N.A. Dreyer:** Overall Mortality of Cellular Telephone Customers. *Epidemiology 7*:303-305 (1996).
447. **Davis, R.L., and F.K. Mostofi:** Cluster of Testicular Cancer in Police Officers Exposed to Hand-Held Radar. *Am. J. Ind. Med. 24*:231-233 (1993).
448. **Dolk, H., G. Shaddick, P. Walls, C. Grundy, B. Thakrar, I. Kleinschmidt, and P. Elliott:** Cancer Incidence near Radio and Television Transmitters in Great Britain I. Sutton Coldfield Transmitter. *Am. J. Epidemiol. 145*:1-9 (1997).
449. **Dolk, H., P. Elliott, G. Shaddick, P. Walls, and B. Thakrar:** Cancer Incidence near Radio and Television Transmitters in Great Britain I. All High Power Transmitters. *Am. J. Epidemiol. 145*:10-17 (1997).
450. **Hocking, B., I.R. Gordon, H.L. Grain, and G.E. Hatfield:** Cancer Incidence and Mortality and Proximity to TV Towers. *Med. J. Australia 165*:601-605 (1996).
451. **Hirsch, F.G., and J.T. Parker:** Bilateral Lenticular Opacities Ocurring in a Technician Operating a Microwave Generator. *A.M.A. Arch. Ind. Hyg. Occup. Med. 6*:512-517 (1952).
452. **Dougherty, J.D., J.C. Caldwell, W.M. Howe, and W.B. Clark:** Evaluation of an Alleged Case of Radiation Induced Cataract at a Radar Site. *Aerospace Med. 36*:466-471 (1965).
453. **Rose, V.E., G.A. Gellin, C.H. Powell, and H.G. Bourne:** Evaluation and Control of Exposures in Repairing Microwave Ovens. *Am. Ind. Hyg. Assoc. J. 30*:137-142 (1969).
454. **Joyner, K.H.:** Microwave Cataract and Litigation: A Case Study. *Health Phys. 57*:545-549 (1989).
455. **Lim, J.I., S.L. Fine, H.A. Kues, and M.A. Johnson:** Visual Abnormalities Associated with High-Energy Microwave Exposure. *Retina 13*:230-233 (1993).
456. **Fleck, H.:** Microwave Oven Burn. *Bull. N.Y. Acad. Med. 59*:313-317 (1983).
457. **Dickason, W.L., and J.P. Barutt:** Investigation of an Acute Microwave-Oven Hand Injury. *J. Hand Surg. 9A*:132- 135 (1984).
458. **Ciano, M., J.R. Burlin, R. Pardoe, R.L. Mills, and V.R. Hentz:** High-Frequency Electromagnetic Radiation Injury to the Upper Extremity: Local and Systemic Effects. *Ann. Plast. Surg. 7*:128-135 (1981).

459. **Tintinalli, J.E., G. Krause, and E. Gursel:** Microwave Radiation Injury. *Ann. Emerg. Med. 12*:645-647 (1983).
460. **Hocking, B., K. Joyner, and R. Fleming:** Health Aspects of Radio-Frequency Radiation Accidents. Part I: Assessment of Health after a Radio-Frequency Radiation Accident. *J. Microw. Power Electromag. Energy* 23:67-74 (1988).
461. **Schilling, C.J.:** Effects of Acute Exposure to Ultrahigh Radiofrequency Radiation on Three Antenna Engineers. *Occup. Environ. Med 54*:281-284 (1997).
462. **Rosenthal, D.S., and S.C. Beering:** Hypogonadoism after Microwave Radiation. *JAMA 205*:105-108 (1968).
463. **Rubin, A., and W.J. Erdman:** Microwave Exposure of the Human Female Pelvis during Early Pregnancy and prior to Conception. *Am. J. Phys. Med. 38*:219-220 (1959).
464. **Brodkin, R.H., and J. Bleiberg:** Cutaneous Microwave Injury. *Acta Dermatol. 53*:50-52 (1973).
465. **Heins, A.P.:** "RF Investigation Involving the *Maersk Constellation*" (letter to M. Grueber). Salt Lake City, UT: U.S. Department of Labor, Health Response Team, 1990.
466. **Shepich, T.J.:** "Safety Hazard Information Bulletin on Radiofrequency Radiation-Caused Burns" (memorandum to regional administrators). Washington, D.C.: U.S. Department of Labor, Occupational Safety and Health Administration, 1990.
467. **Hocking, B., K.J. Joyner, H.N. Newman, and R.J. Allred:** Radiofrequency Electric Shock and Burn. *Med. J. Australia 161*:683-685 (1994).
468. **McLaughlin, J.T.:** Tissue Destruction and Death from Microwave Radiation (Radar). *Calif. Med. 86*:336-339 (1957).
469. **Ely, T.S.:** Microwave Death. *JAMA 6*:1394 (1971).
470. **Ely, T.S.:** Science and Standards — A Letter to the Editor. *J. Microw. Power 20*:137 (1985).
471. **Archimbaud, E., C. Charrin, D. Guyotat, and J.-J. Viala:** Acute Myelogenous Leukemia Following Exposure to Microwaves. *Br. J. Haematol. 73*:272-273 (1989).
472. **Jauchem, J.R.: Correspondence:** Leukaemia following Exposure to Microwaves: Analysis of a Case Report. *Br. J. Haematol. 76*:312 (1990).
473. **Archimbaud, E.:** Correspondence. *Br. J. Haematol. 73*:313 (1990).
474. **Hocking, B., and M. Garson:** Correspondence. *Br. J. Haematol. 76*:313-314 (1990).
475. **Budd, R.:** Burns Associated with the Use of Microwave Ovens. *J. Microw. Power Electromag. Energy 27*:160-163 (1992)
476. **Feucht, B.L., A.W. Richardson, and H.M. Hines:** Effects of Implanted Metals on Tissue Hyperthermia Produced by Microwaves. *Arch. Phys. Med. 30*:164-169 (1949).
477. **Hepfner, S.T.:** Radio-Frequency Interference in Cochlear Implants. *New Engl. J. Med. 313*:387 (1985).
478. **Chou, C.K., J.A. McDougall, and K.W. Chan:** Absence of Radiofrequency Heating from Auditory Implants during Magnetic Resonance Imaging. *Bioelectromagnetics 16*:307-316 (1995).
479. **Hocking, B., K.H. Joyner, and A.H.J. Fleming:** Implanted Medical Devices in Workers Exposed to Radio-Frequency Radiation. *Scand. J. Work Environ. Health 17*:1-6 (1991).
480. **Moseley, H., S. Johnston, and A. Allen:** The Influence of Microwave Radiation on Transdermal Delivery Systems. *Brit. J. Dermatol. 122*:361-363 (1990).
481. **Murray, K.B.:** Hazard of Microwave Ovens to Transdermal Delivery System. *New Engl. J. Med. 310*:721 (1984).
482. **Davias, N., and D.W. Griffin:** Effect of Metal-Framed Spectacles on Microwave Radiation Hazards to the Eyes of Humans. *Med. Biol. Eng. Comput. 27*:191-197 (1989).
483. **Cook, C.F., and P.A. Huggins:** Effect of Radio Frequency Interference on Common Industrial Hygiene Monitoring Instruments. *Am. Ind. Hyg. Assoc. J. 45*:740-744 (1984).
484. **Shackford, H., and B.E. Bjarngard:** Disturbance of Diode Dosimetry by Radiofrequency Radiation. *Med. Phys.* 22:807 (1995).
485. **Carrillo, R., O. Garay, Q. Balzano, and M. Pickels:** Electromagnetic Near Field Interference with Implantable Medical Devices. In *IEEE International Symposium on Electromagnetic Compatibility*. New York: IEEE, 1995. pp. 1-3.
486. **Tan. K.-S., and I. Hanberg:** Radiofrequency Susceptibility Tests on Medical Equipment. In *Proceedings of the 16th Annual*

International Conference of the IEEE Engineering in Medicine and Biology Society (Vol. 2). New York: IEEE, 1994. pp. 998-999.

487. **Coray, R., and H. Schaer:** Immunity of Cardiac Pacemakers and Risk Potential of Pacemaker Patients, with Special Regard to High-Power Medium- and Short-Wave Transmitters. In *Ninth International Conference on Electromagnetic Compatibility* (Conf. Pub. No. 396). Manchester, UK: Institution of Electrical Engineers, 1994. pp. 6-12.
488. **Boivin, W.S., S.M. Boyd, J.A. Coletta, and L.M. Neunaber:** Measurement of Radiofrequency Electromagnetic Fields in and around Ambulances. *Biomed. Instrum. Technol. 31*:145-154 (1997).
489. **Kingston, H.M., P.J. Walter, W.G. Engelhart, and P.J. Parsons:** Chemical Laboratory Microwave Safety. In *Microwave Enhanced Chemistry: Fundamentals, Sample Preparation, and Applications* (ACS Professional Reference Book Series). Washington, D.C.: American Chemical Society, 1997.
490. **Stuchly, M.A., M.H. Repacholi, D. Lecuyer, and R. Mann:** Sources and Applications of Radiofrequency (RF) and Microwave Energy. In *Biological Effects and Dosimetry of Nonionizing Radiation Radiofrequency and Microwave Energies* (edited by M. Grandolfo, S. Michaelson, and A. Rindi). New York: Plenum Press, 1981. pp. 19-41.
491. **Tell, R.A., and E.D. Mantiply:** Population Exposure to VHF and UHF Broadcast Radiation in the United States. *Proc. IEEE 68*:6-12 (1980).
492. **Environmental Protection Agency:** *The Radiofrequency Radiation Environment: Environmental Exposure Levels and RF Radiation Emitting Sources* by N.N. Hankin (EPA 520/1-85-014). Washington, D.C.: U.S. Government Printing Office, 1986.
493. **Stuchly, M.A.:** Potentially Hazardous Microwave Radiation Sources - A Review. *J. Microwave Power* 12:369-381 (1977).
494. **Mild, K.H., and K.G. Lovstrand:** Environmental and Professionally Encountered Electromagnetic Fields. In *Biological Effects and Medical Applications of Electromagnetic Energy*. Englewood Cliffs, NJ: Prentice Hall, 1990. pp. 48- 74.
495. **Environmental Protection Agency:** A Review of Radiofrequency Electric and Magnetic Fields in the General and Work Environment: 10 kHz to 100 GHz by E.D. Mantiply, S.W. Poppell, and J.A. James. In *Summary and Results of the April 26-27, 1993 Radiofrequency Radiation Conference* (Vol. 2: Papers) (402-R-95-011). Washington, D.C.: U.S. Government Printing Office, 1995. pp. 1-23.
496. **National Council on Radiation Protection and Measurements:** *A Practical Guide to The Determination of Human Exposure to Radiofrequency Fields* (NCRP Report No. 119). Bethesda, MD: NCRP, 1993.
497. **Conover, D.L., W.E. Murray, E.D. Foley, J.M. Lary and W.H. Parr:** Measurement of Electric- and Magnetic-Field Strengths from Industrial Radio-Frequency (6-38 MHz) Plastic Sealers. *Proc. IEEE 68*:17-20 (1980).
498. **Stuchly, M.A., M.H. Repacholi, D. Lecuyer and R. Mann:** Radiation Survey of Dielectric (RF) Heaters in Canada. *J. Microwave Power 15*:113-121 (1980).
499. **Cox, C., W.E. Murray, and E.P. Foley:** Occupational Exposures to Radiofrequency Radiation (18-31 MHz) from RF Dielectric Heat Sealers. *Am. Ind. Hyg. Assoc. J. 43*:149-153 (1982).
500. **Gandhi, O., J.-Y. Chen, and A. Riazi:** Currents Induced in Human Beings for Plane-Wave Exposure Conditions 0-50 MHz and for RF Sealers. *IEEE Trans. Bio-Med. Eng. BME-33*:757-767 (1986).
501. **Conover, D.L., C.E. Moss, W.E. Murray, R.M. Edwards, C.Cox, B. Grajewski, D.M. Werren, and J.M. Smith:** Foot Currents and Ankle SARs Induced by Dielectric Heaters. *Bioelectromagnetics 13*:103-110 (1992).
502. **Stuchly, M.A., and D.W. Lecuyer:** Induction Heating and Operator Exposure to Electromagnetic Fields. *Health Phys. 49*:693-700 (1985).
503. **Robert Desrosiers:** "Radio Frequency (RF) Using Tools and Some Observations on RF Emissions." September 30, 1986 [Personal Communication]. IBM Corporation, Burlington, VT 05452.
504. **Ungers, L.J., J.H. Jones, and G.J. Mihlan:** "Emission of Radio-Frequency Radiation from Plasma-Etching Operations." Paper presented at the 1985 American Industrial

Hygiene Conference, Las Vegas, NV, May 19-24, 1985.

505. **Joyner, K.H., and M.J. Bangay:** Exposure Survey of Civilian Airport Radar Workers in Australia. *J. Microwave Power 21*:209-219 (1986).

506. **Tell, R.A., and J.C. Nelson:** Microwave Hazard Measurements Near Various Aircraft Radars. *Rad. Data Rep. 15*:161-179 (1974).

507. **Tell, R.A., N.N. Hankin, and D.E. Janes:** Aircraft Radar Measurements in the Near Field. In *Operational Health Physics, Proceedings of the Ninth Midyear Topical Symposium of the Health Physics Society*. McLean, VA: Health Physics Society, 1976.

508. **Food and Drug Administration:** *Measurement of Power Density from Marine Radar* by D.W. Peak, D.L. Conover, W.A. Herman and R.E. Shuping (HEW [FDA] Pub. No. 76-8004). Washington, D.C.: U.S. Government Printing Office, 1975.

509. **National Institute for Occupational Safety and Health:** *Hazard Evaluation and Technical Assistance Report HETA 89-284-L2029 Technical Assistance to the Federal Employees Occupational Health Seattle, Washington* by C.E. Moss (NTIS # PB91-107920). Springfield, VA: National Technical Information Service, 1990.

510. **National Institute for Occupational Safety and Health:** *Hazard Evaluation Report HETA 93-002-22829 United States Coast Guard Governors Island, New York* by C.E. Moss and A.T. Zimmer (PB93-215051). Springfield, VA: National Technical Information Service, 1993.

511. **Fisher, P.D.:** Microwave Exposure Levels Encountered by Police Traffic Radar Operators. *IEEE Trans. Electromag. Compat. 35*:36-45 (1993).

512. **Ontario Ministry of Labour:** *Microwave Emissions and Operator Exposures from Traffic Radars used in Ontario* by M.E. Bitran, D.E. Charron and J.M. Nishio. Weston, Ontario: Ministry of Labour, 1992.

513. **National Institute for Occupational Safety and Health:** *Health Hazard Evaluation Report HETA 92-0224-2379 Norfolk Police Department, Norfolk, Virginia* by R. Malkin (NTIS PB 94-183456). Springfield, VA: National Technical Information Service, 1994.

514. **Balzano, Q., J.A. Bergeron, J. Cohen, J.M. Osepchuk, R.C. Petersen, and L.M. Roszyk:** Measurement of Equivalent Power Density and RF Energy Deposition in the Immediate Vicinity of a 24-GHz Traffic Radar Antenna. *IEEE Trans. Electromag. Compat. 37*:183-191 (1995).

515. **United States Air Force:** *Radio Frequency Radiation Hazard Survey 141 Tactical Control System Ramey PR* by N.D. Montgomery (AFOEHL Report 90-088RC00679ERA). Springfield, VA: National Technical Information Service (AD-A225-343), 1990.

516. **Curtis, R.A.:** Occupational Exposures to Radiofrequency Radiation from FM Radio and TV Antennas. In *Nonionizing Radiation—Proceedings from a Topical Symposium*. Cincinnati: American Conference of Governmental Industrial Hygienists, 1980. pp. 211-222.

517. **Tell, R.A.:** *Induced Body Currents and Hot AM Tower Climbing Assessing Human Exposure in Relation to the ANSI Radiofrequency Protection Guide.* Report prepared for the Federal Communications Commission. Springfield, VA: National Technical Information Service (PB92-125186), 1991.

518. **Petersen, R.C.:** Electromagnetic Radiation from Selected Telecommunications Systems. *Proc. IEEE 69*:21-24 (1980).

519. **Joyner, K.H., and M.J. Bangay:** Exposure Survey of Civilian Airport Radar Workers in Australia. *J. Microwave Power 21*:209-219 (1986).

520. **Environmental Protection Agency:** *An Evaluation of Selected Satellite Communication Systems as Sources of Environmental Microwave Radiation* by N.N. Hankin (EPA-520/2-74-008). Washington, D.C.: EPA, 1974.

521. **Petersen, R.C., and P.A. Testagrossa:** Radio-Frequency Electromagnetic Fields Associated with Cellular-Radio Cell-Site Antennas. *Bioelectromagnetics 13*:527-542 (1992).

522. **Environmental Protection Agency:** *An Investigation of Energy Densities in the Vicinity of Vehicles with Mobile Communications Equipment and Near a Hand-Held Walkie-Talkie* by D.L. Lambdin (Report No. ORP/EAD 79-2). Las Vegas, NV: EPA, 1979.

523. **Food and Drug Administration:** *Measurements of Electromagnetic Fields in the Close Proximity of CB Antennas* by P.S. Ruggera (HEW [FDA] Pub. No. 79-8080). Washington, D.C.: Government Printing Office, 1979.

524. **Guy, A.W. amd C.-K. Chou:** Specific Absorption Rates of Energy in Man

Models Exposed to Cellular UHF Mobile-Antenna fields. *IEEE Trans. Microwave Theory Tech. MTT-34*:671-680 (1986).

525. **Cleveland, R.F., and T.W. Athey:** Specific Absorption Rate (SAR) in Models of the Human Head Exposed to Hand-Held UHF Portable Radios. *Bioelectromagnetics 10*:173-186 (1989).
526. **Dimbylow, P.J., and S.M. Mann:** SAR Calculations in an Anatomically Realistic Model of the Head for Mobile Communication Transceivers at 900 MHz and 1.8 GHz. *Phys. Med. Biol. 39*:1537-1553 (1994).
527. **Kuster, N., and Q. Balzano:** Energy Absorption Mechanisms by Biological Bodies in the Near Field of Dipole Antennas. *IEEE Trans. on Vehicular Technol. VT-41*:17-23 (1992).
528. **Rothman, K.J., C.-K. Chou, R. Morgan, Q. Balzano, A.W. Guy, D.P. Funch, S. Preston-Martin, J. Mandel, R. Steffens, and G. Carlo:** Assessment of Cellular Telephone and Other Radio Frequency Exposure for Epidemiologic Research. *Epidemiology 7*:291-298 (1996).
529. **Center for Devices and Radiological Health:** *An Evaluation of Radiation Emission from Video Display Terminals* (DHHS [FDA] Pub. No. 81-8153). Rockville, MD: CDRH, 1981.
530. **Australian Radiation Laboratory:** *Electromagnetic Emissions from Video Display Terminals (VDTs)* by K.H. Joyner, C.R. Roy, G. Elliott, M.J. Bangay, H.P. Gies and D.W. Tomlinson (ARL/TR067). Yallambie, Victoria: Australian Radiation Laboratory, 1984.
531. **Wolbarsht, M.L., F.A. O'Foghludha, D.H. Sliney, A.W. Guy, A.A. Smith and G.A. Johnson:** Electromagnetic Emission from Visual Display Units: A Non-Hazard. *Proc. SPIE 229*:187-195 (1980).
532. **National Institute for Occupational Safety and Health (NIOSH):** *Potential Health Hazards of Video Display Terminals* by Murray, W.E., C.E. Moss, W.H. Parr, M.J. Smith, B.G.F. Cohen, L.W. Stammerjohn and A. Happ (DHHS [NIOSH] Pub. No. 81-129). Cincinnati: NIOSH, 1981.
533. **Weiss, M. M., and R. C. Petersen:** Electromagnetic Radiation Emitted from Video Computer Terminals. *Am. Ind. Hyg. Assoc. J. 40*:300-309 (1979).
534. **Boivin, W. S.:** RF Electric Fields: VDTs vs. TV Receivers. In *Proceedings of the International Scientific Conference: Work with Display Units.* Stockholm, Sweden, May 12-15, 1986. Amsterdam: Elsevier Scientific Publishers, 1986.
535. **Food and Drug Administration:** *Measurements of Emission Levels During Microwave and Shortwave Diathermy Treatments* by P.S. Ruggera (DHHS [FDA] Pub. No. 80-8119). Washington, D.C.: U.S. Government Printing Office, 1980.
536. **Stuchly, M.A., M.H. Repacholi, D.W. Lecuyer and R.D. Mann:** Exposure to The Operator and Patient During Short Wave Diathermy Treatments. *Health Phys. 42*:341-366 (1981).
537. **Hansson Mild, K.:** Occupational Exposure to Radio-Frequency Electromagnetic Fields. *Proc. IEEE 68*:12-17 (1980).
538. **Kalliomaki, P.L., M. Hietanen, K. Kalliomaki, O. Koistinen and E. Valtonen:** Measurements of Electric and Magnetic Stray Fields Produced by Various Electrodes of 27-MHz Diathermy Equipment. *Radio Sci. 17 (5S)*:29S-34S (1982).
539. **Ruggera, P.S.:** Near-Field Measurements of RF fields. In *Symposium on Biological Effects and Measurement of Radio Frequency/Microwaves* (HEW [FDA] Pub. No. 77-8026). Washington, D.C.: Government Printing Office, 1977. pp. 104- 116.
540. **Paz, J.D.:** Potential Ocular Damage from Microwave Exposure During Electrosurgery Dosimetric Survey. *J. Occup. Med.* 29:580-583 (1987).
541. **Institute of Electrical and Electronics Engineers:** *IEEE Standard for Safety Levels with Respect to Human Exposure to Radio Frequency Electromagnetic Fields, 3 kHz to 300 GHz* (IEEE Std. C95.1-1991). New York: IEEE, 1992.
542. **International Radiation Protection Association:** Guidelines on Limits of Exposure to Radiofrequency Electromagnetic Fields in the Frequency Range from 100 kHz to 300 GHz. *Health Phys. 54*:115-123 (1988).
543. **National Council on Radiation Protection and Measurements:** *Biological Effects and Exposure Criteria for Radiofrequency Electromagnetic Fields* (NCRP Report No. 86). Bethesda, MD: NCRP, 1986.
544. "Guidelines for Evaluating the Environmental Effects of Radiofrequency Radiation," *Federal*

Register 61:153 (7 August 1996), pp. 41006-41019.

545. "Guidelines for Evaluating the Environmental Effects of Radiofrequency Radiation," *Federal Register 62*:14 (22 January 1997), pp. 3232-3240.
546. **Gandhi, O.P.:** State of Knowledge for Electromagnetic Absorbed Dose in Man and Animals. *Proc. IEEE 68*:24-32 (1980).
547. **Aslan, E.:** Broad-band Isotropic Electromagnetic Radiation Monitor. *IEEE Trans. Instrum. Meas. IM-21*:421-424 (1972).
548. **Larsen, E.B., and F.X. Ries:** *Design and Calibration of the NBS Isotropic Electric-Field Monitor (EFM-5), 0.2 to 1000 MHz* (NBS Technical Note 1033). Washington, D.C.: U.S. Government Printing Office, 1981.
549. **Tell, R.A.:** Instrumentation for Measurement of Electromagnetic Fields: Equipment, Calibrations and Selected Applications Part I — Radiofrequency Fields. In *Biological Effects and Dosimetry of Nonionizing Radiation Radio-frequency and Microwave Energies* (edited by M. Grandolfo, S. Michaelson, and A. Rindi). New York: Plenum Press, 1981. pp. 95-144.
550. **Aslan, E.:** A Low Frequency H-Field Radiation Monitor. In *Biological Effects of Electromagnetic Waves* (HEW [FDA] Pub. No. 77-8011). Washington, D.C.: Government Printing Office, 1976. pp. 229-238.
551. **National Institute for Occupational Safety and Health (NIOSH):** *Development of Magnetic Near-Field Probes* by F.M. Greene (HEW [NIOSH] Pub. No. 75-127). Cincinnati: NIOSH, 1975.
552. **Institute of Electrical and Electronics Engineers:** *IEEE Recommended Practice for the Measurement of Potentially Hazardous Electromagnetic Fields—RF and Microwave* (IEEE Std. C95.3-1991). New York: IEEE, 1992.
553. **Conti, R.:** Instrumentation for Measurement of Power Frequency Electromagnetic Fields. In *Biological Effects and Dosimetry of Static and ELF Electromagnetic Fields*, edited by M. Grandolfo, S.M. Michaelson, and A. Rindi. New York: Plenum Press, 1985. pp. 187-210.
554. **Baron, D.:** Measuring EMF Emissions from Video Display Terminals. *Compliance Eng. Mag.* Fall:1-5 (1991).
555. **Gandhi, O.P., I. Chatterjee, D. Wu and Y.-G. Gu:** Likelihood of High Rates of Energy Deposition in the Human Legs at the ANSI Recommended 3-30 MHz RF Safety Levels. *Proc. IEEE 73*: 1145-1147 (1985).
556. **Gandhi, O.P., J.-Y. Chen and A. Riazzi:** Currents Induced in a Human Being for Plane-Wave Exposure Conditions 0-50 MHz in RF Sealers. *IEEE Trans. Biomed. Eng. 33*:757-767 (1986).
557. **Federal Communications Commission:** *Engineering Services for Measurement and Analysis of Radiofrequency (RF) Fields* by Richard A. Tell (FCC Report No. OET/RTA 95-01). Washington, D.C.: U.S. Government Printing Office, 1995.
558. **Lubinas, V., and K.H. Joyner:** *Measurement of Induced Current Flows in the Ankles of Humans Exposed to Radiofrequency Fields* (Report 8000). Clayton, Victoria, Australia: Telecom Research Laboratories, 1991.
559. **National Institute of Occupational Health:** *Guidelines for the Measurement of RF Welders* by P. Williams and K.H. Mild (Undersokningsrapport 1991:8; Rapportkod: ISRN AI/UND-91-8-SE). Umea, Sweden: National Institute of Occupational Health, 1991.
560. **Beischer, D.E., and V.R. Reno:** Microwave Reflection and Diffraction by Man. In *Biologic Effects and Health Hazards of Microwave Radiation.* Warsaw, Poland: Polish Medical Publishers, 1974. pp. 254-259.
561. **Beischer, D.E., and V.R. Reno:** Microwave Energy Distribution Measurements in Proximity to Man and Their Practical Application. *Ann. N.Y. Acad. Sci. 247*:473-479 (1975).
562. **Tofani, S., G. d'Amore, G. Fiandino, A. Benedetto, O.P. Gandhi, and J.Y. Chen:** Induced Foot-Currents in Humans Exposed to VHF Radio-Frequency EM Fields. *IEEE Trans. Electromag. Compat. 37*:96-99 (1995).
563. **Bassen, H., and J. Bing:** An EM Radiation Safety Controller. *J. Microwave Power 14*:45-48 (1979).
564. **Eure, J.A., J.W. Nicolls and R.L. Elder:** Radiation Exposure from Industrial Microwave Applications. *Am. J. Pub. Health 62*:1573-1577 (1972).
565. **Keiser, B.E.:** *Principles of Electromagnetic Compatibility.* Dedham, MA: ARTECH HOUSE, INC, 1979.

566. **Ott, H.W.:** *Noise Reduction Techniques in Electronic Systems*. New York: John Wiley & Sons, 1976. pp. 137-171.
567. **Yasufuku, S.:** Technical Progress of EMI Shielding Materials in Japan. *IEEE Elect. Insulat. Mag. 6(6)*:21-30 (1990).
568. **Naishadham, K.:** Shielding Effectiveness of Conductive Polymers. *IEEE Trans. Electromag. Compat. 34*:47-50 (1992).
569. **Colaneri, N.F., and L.W. Shacklett:** EMI Shielding Measurements of Conductive Polymer Blends. *IEEE Trans. Instrum. Measure. 41*:291-297 (1992).
570. **Brailsford, F.:** *Magnetic Materials*, 3rd Ed. New York: John Wiley & Sons, 1960. pp. 111-136.
571. **Trenkler, Y., and L.E. McBride:** Characterization of Metals as EMC Shields. In *IEEE Instrumentation and Measurement Conference* (IEEE Cat. No. 86CH2271-5). New York: IEEE, 1986. pp. 65-69.
572. **Hoeft, L.O.:** Measured Magnetic Field Reduction of Copper-Sprayed Wood Panels. *IEEE 1985 International Symposium on Electromagnetic Compatibility* (IEEE Cat. No. 85CH2116-2). New York: IEEE, 1985. pp. 34-37.
573. **Jonnada, R.K.R., and K.A. Peebles:** Effects of Shield Discontinuities on The Performance of Shielded Enclosures. *1965 Symposium Digest Seventh National Symposium on Electromagnetic Compatibility*. New York: Institute of Electrical and Electronics Engineers, 1965.
574. **Honig, Jr. E.M:** Electromagnetic Shielding Effectiveness of Steel Sheets with Partly Welded Seams. *IEEE Trans. Electromag. Compat. EMC-19*:377-382 (1977).
575. **Soltys, J.J.:** Maintaining EMI/RFI Shielding Integrity of Equipment Enclosures with Conductive Gasketing. In *IEEE International Symposium on Electromagnetic Compatibility* (IEEE Cat. No. CH1304-5). New York: IEEE, 1978. pp. 333-338.
576. **White, D.R.J., and M. Mardiguian:** Gasket Types and Materials: A Basic Selection Guide. *EMC Technol. & Interference Control News 8(1)*:59-64 (1989).
577. **Molyneux-Child, J.W.:** Knitted Gasketing for Enclosure Shielding. *New Electronics 19(14)*:48,51 (1986).
578. **Bereuter, W.A., and D.C. Chang:** Shielding Effectiveness of Metallic Honeycombs. *IEEE Trans. Electromag. Compat. EMC-24*:58-61 (1982).
579. **Kunkel, G.M.:** Shielding Characteristics of Honeycomb Filters. In *1986 IEEE International Symposium on Electromagnetic Compatibility* (IEEE No. 86CH2294-7). New York: IEEE, 1986. pp. 299-303.
580. **Bureau of Radiological Health:** *Concepts and Approaches for Minimizing Excessive Exposure to Electromagnetic Radiation from RF Sealers* by P.S. Ruggera and D.H. Schaubert (DHHS [FDA] 82-8192). Washington, D.C.: Government Printing Office, 1982.
581. **Hill, D.A.:** Effect of Separation from Ground on Whole-Body Absorption Rates. *IEEE Trans. Microwave Theory Tech. MTT-32*:772-778 (1984).
582. **National Institute for Occupational Safety and Health:** *Health Hazard Evaluation Report HETA 90-389-2272 Dometic Corporation, La Grange, Indiana* by T.A. Seitz, C.E. Moss, and R. Shults. Springfield, VA: National Technical Information Service (PB93-215028), 1992.
583. **Gandhi, O.P., and E.L. Hunt:** Corner-Reflector Applicators for Multilateral Exposure of Animals in Bioeffect Experiments. *Proc. IEEE 68*:160-162 (1980).
584. **Gandhi, O.P., E.L. Hunt, and J.A. D'Andrea:** Deposition of Electromagnetic Energy in Animals And in Models of Man With And Without Grounding and Reflector Effects. *Radio Sci. 6(S)*:39-47 (1977).
585. **Chen, J.-Y., and O.P. Gandhi:** Electromagnetic Deposition in An Anatomically Based Model of Man for Leakage Fields of A Parallel-Plate Dielectric Heater. *IEEE Trans. Microwave Theory Tech. 37*:174-180 (1989).
586. **Reynolds, M.R.:** Development of A Garment for Protection of Personnel Working in High-Power RF Environments. In *Proceedings of the Fourth Annual Tri-Service Conference on the Biological Effects of Microwave Radiation* (Vol. 1). New York: Plenum Press, 1961. pp. 71-81.
587. **Klascius, A.F.:** Microwave Radiation Protective Suit. *Am. Ind. Hyg. Assoc. J. 32*:771-774 (1971).
588. **Chou, C.-K., A.W. Guy, and J.A. McDougall:** Shielding Effectiveness of Improved Microwave-Protective Suits. *IEEE Trans. Microwave Theory Tech. MTT-35*:995-1001 (1987).

589. **Joyner, K.H., P.R. Copeland, and I.P. MacFarlane:** An Evaluation of A Radiofrequency Protective Suit and Electrically Conductive Fabrics. *IEEE Trans. Electromag. Compat. 31*:129-137 (1989).
590. **Amato, J.A.:** Use Protective Clothing for Safety in RF Fields. *Mobile Radio Technol. 12(4)*:40,42,44 (1994).
591. **De Bruyne, R., and W. Van Loock:** New Class of Microwave Shielding Materials. *J. Microwave Power 12*:145-154 (1977).
592. **Heins, A.P., and R. Curtis:** "RF Shock Hazard Inspection Terminal 5, Port of Seattle" (letter to D. Beeston). Salt Lake City, UT: USDOL OSHA Health Response Team, 1990.
593. **Heins, A.P.:** "RF Investigation Involving The Maersk Constellation" (letter to M. Grueber). Salt Lake City, UT: USDOL OSHA Health Response Team, 1990.
594. **Tell, R.A.:** *RF Current Reduction Provided by Work Gloves at AM Radio Broadcast Frequencies* (FCC/OET Contract Report No. RTA 93-01). Springfield, VA: National Technical Information Service (PB94-117041), (1993).
595. **American National Standards Institute:** *Radio Frequency Radiation Hazard Warning Symbol* (ANSI C95.2-1982). New York: IEEE, 1982.
596. **Tenforde, T.S., and W.T. Kaune:** Interaction of Extremely Low Frequency Electric and Magnetic Fields with Humans. *Health Phys. 53*:585-606 (1987).
597. **Deno, D.W.:** Currents Induced in the Human Body by High Voltage Transmission Line Electric Field — Measurement and Calculation of Distribution and Dose. *IEEE Trans. Power Appar. Syst. PAS-96*:1517-1527 (1977).
598. **Kaune, W.T., and R.D. Phillips:** Comparison of the Coupling of Grounded Humans, Swine, and Rats to Vertical, 60-Hz Electric Fields. *Bioelectromagnetics 1*:117-130 (1980).
599. **Tenforde, T.S.:** Biological Interactions of Extremely-Low-Frequency Electric and Magnetic Fields. *Biochem. Bioenergetics 25*:1-17 (1991).
600. **Bernhardt, J.H., H.J. Haubrich, G. Newi, N. Krause, and K.H. Schneider:** Limits for Electric and Magnetic Fields in DIN VDE Standards: Considerations for the Range 0 to 10 kHz. *CIGRE, International Conference on Large High Voltage Electric Systems.* 112 Boulevard Haussmann, 75008 Paris, August 27th-September 4th, 1986.
601. **Bawin, S.M., L.K. Kaczmarek, and W.R. Adey:** Effects of Modulated VHF Fields on the Central Nervous System. *Ann. N.Y. Acad. Sci. 247*:74-81 (1975).
602. **Blackman, C.F., S.G. Benane, L.S. Kinney, W.T. Joines, and D.E. House:** Effects of ELF Fields on Calcium-Ion Efflux from Brain Tissue in vitro. *Radiat. Res. 92*:510-520 (1982).
603. **Blackman, C.F., S.G. Benane, J.R. Rabinowitz, D.E. House, W.T. Joines:** A Role for the Magnetic Field in the Radiation-Induced Efflux of Calcium Ions from Brain Tissue in vitro. *Bioelectromagnetics 6*:327-337 (1985).
604. **Blackman, C.F., L.S. Kinney, D.E. House, and W.T. Joines:** Multiple Power-Density Windows and Their Possible Origin. *Bioelectromagnetics 10*:115-128 (1989).
605. **Blackman, C.F., S.G. Benane, D.E. House:** The Influence of Temperature During Electric- and Magnetic-Field- Induced Alteration of Calcium-Ion Release from in vitro Brain Tissue. *Bioelectromagnetics 12*:173-182 (1991).
606. **Liboff, A.R.:** Cyclotron Resonance in Membrane Transport. In *Interactions Between Electromagnetic Fields and Cells.* New York: Plenum Publishing Company, 1985. pp. 281-296.
607. **Thomas, J.R., J. Schrot, and A.R. Liboff:** Low-Intensity Magnetic Fields Alter Operant Behavior in Rats. *Bioelectromagnetics 7*:349-357 (1986).
608. **Smith, S.D., B.R. McLeod, A.R. Liboff, and K. Cooksey:** Calcium Cyclotron Resonance and Diatom Motility. *Bioelectromagnetics 8*:215-227 (1987).
609. **Liboff, A.R., R.J. Rozek, M.L. Sherman, B.R. McLeod, and S.D. Smith:** Ca^{2+}-45 Cyclotron Resonance in Human Lymphocytes. *J. Bioelectricity 6*:13-22 (1987).
610. **Liboff, A.R., W.C. Parkinson, and D.C. Dawson:** Ion Cyclotron Resonance Study in Turtle Colon. In *Abstracts, Tenth Annual Meeting of the Bioelectromagnetics Society*, Stamford, CT, 1988. p. 32.
611. **Parkinson, W.C., and C.T. Hanks:** Search for Cyclotron Resonance in Cells in vitro. *Bioelectromagnetics 10*:129- 145 (1989).

612. **Prasad, A.V., M.W. Miller, E.L. Carstensen, Ch. Cox, M. Azadniv, and A.A. Brayman:** Failure to Reproduce Increased Calcium Uptake in Human Lymphocytes at Purported Cyclotron Resonance Exposure Conditions. *Radiat. Environ. Biophys. 30*:305-320 (1991).
613. **Stern, S., and V.G. Laties:** Magnetic Fields and Behavior. In *Project Abstracts of The Annual Review of Research on Biological Effects of Electric and Magnetic Fields From the Generation, Delivery & Use of Electricity*, p. A-35, San Diego, CA, November 8-12, 1992. (Available from W/L Associates, Ltd., 120 W. Church Street, Frederick, MD 21701.)
614. **Male, J.C., and D.T. Edmonds:** Ion Vibrational Precession: A Model for Resonant Biological Interactions with ELF Fields. In *Abstracts, Twelfth Annual Meeting of the Bioelectromagnetics Society*, San Antonio, TX, 1990. p. 98.
615. **Lednev, V.V.:** Possible Mechanism for The Influence of Weak Magnetic Fields on Biological Systems. *Bioelectromagnetics 12*:71-76 (1991)
616. **Adair, R.K.:** Criticism of Lednev's Mechanism for The Influence of Weak Magnetic Fields on Biological Systems. *Bioelectromagnetics 13*:231-235 (1992).
617. **Bruckner-Lea, C., C.H. Durney, J. Janata, C. Rappaport, and M. Kaminski:** Calcium Binding to Metallochromic Dyes and Calmodulin in The Presence of Combined ac-dc Magnetic Fields. *Bioelectromagnetics 13*:147-162 (1992).
618. **Blackman, C.F., J.P. Blanchard, S.G. Benane, D.E. House, and J.A. Elder:** Blind, Double-Blind and Triple-Blind Tests of the IPR Model. In *Project Abstracts. The Annual Review of Research on Biological Effects of Electric and Magnetic Fields from the Generation, Delivery & Use of Electricity*. Palm Springs, CA, November 12-16, 1995.
619. **Lee, W., D.B. Lyle, T.A. Fuchs, and M. Swicord:** Proliferation/ Differentiation and Neurite Outgrowth of PC-12 Cells Exposed to Extremely Low Frequency (ELF) Magnetic Fields. In *Project Abstracts. The Annual Review of Research on Biological Effects of Electric and Magnetic Fields from the Generation, Delivery & Use of Electricity*. Palm Springs, CA, November 12-16, 1995.
620. **Murphy, J.C., D.A. Kaden, J. Warren, and A. Sivak:** Power Frequency Electric and Magnetic Fields: A Review of Genetic Toxicology. *Mutation Res. 296*:221-240 (1993).
621. **McCann, J., F. Dietrich, C. Rafferty, and A.O. Martin:** A Critical Review of the Genotoxic Potential of Electric and Magnetic Fields. *Mutation Res. 297*:61-95 (1993).
622. **Graves, H.B., and T.J. Reed:** *Effects of 60-Hz Fields on Chick Embryo and Chick Development, Growth, and Behavior* (Technical Report RP-1064). Electric Power Research Institute, Palo Alto, CA, 1985.
623. **Berman, E., L. Chacon, D. House, B.A. Koch, W.E. Koch, J. Leal, S. Lovtrup, E. Mantiply, A.H. Martin, G.I. Martucci, K.H. Mild, J.C. Monahan, M. Sandstrom, K. Shamsaifar, R. Tell, M.A. Trillo, A. Ubeda, and P. Wagner:** Development of Chicken Embryos in A Pulsed Magnetic Field. *Bioelectromagnetics 11*:169-187 (1990).
624. **Sikov, M.R., D.N. Rommereim, J.L. Beamer, R.L. Buschbom, W.T. Kaune, and R.D. Phillips:** Developmental Studies of Hanford Miniature Swine Exposed to 60-Hz Electric Fields. *Bioelectromagnetics 8*:229-242 (1987).
625. **Rommereim, D.N., R.L. Rommereim, L.E. Anderson, and M.R. Sikov:** Reproductive and Teratologic Evaluation in Rats Chronically Exposed at Multiple Strengths of 60-Hz Electric Fields. In *Abstracts, Tenth Annual Meeting of the Bioelectromagnetics Society*, Stamford, CT, 1988.
626. **Rommereim, D.N., R.L. Rommereim, D.L. Miller, R.L. Buschbom, and L.E. Anderson:** Developmental Toxicology Evaluation of 60-Hz, Horizontal Magnetic Fields in Rats. *Thirtieth Hanford Symposium on Health and the Environment*, October 29-November 1, 1991, Richland, WA, 1991.
627. **Wilson, B.W., L.E. Anderson, D.I. Hilton, and R.D. Phillips: Chronic Exposure to 60-Hz Electric Fields:** Effects on Pineal Function in the Rat. *Bioelectromagnetics 2*:371-380 (1981).
628. **Wilson, B.W., E.K. Chess, and L.E. Anderson:** 60-Hz Electric-Field Effects on Pineal Melatonin Rhythms: Time Course for Onset and Recovery. *Bioelectromagnetics 7*:239-242 (1986).

629. **Kato, M., K. Honma, T. Shigemitsu, and Y. Shiga:** Effects of Exposure to a Circularly Polarized 50-Hz Magnetic Field on Plasma and Pineal Melatonin Levels in Rats. *Bioelectromagnetics 14*:97-106 (1993).

630. **Lee, J.M., F. Stormshak, J.M. Thompson, P. Thinesen, L.J. Painter, E.G. Olenchek, D.L. Hess, R. Forbes, and D.L. Foster:** Melatonin Secretion and Puberty in Female Lambs Exposed to Environmental Electric and Magnetic Fields. *Biol. Reprod. 49*:857-864 (1993).

631. **Lee, J.M., F. Stormshak, J.M. Thompson, D.L. Hess, and D.L. Foster:** Melatonin and Puberty in Female Lambs Exposed to EMF: A Replicate Study. *Bioelectromagnetics 16*:119-123 (1995).

632. **Truong, H., and S.M. Yellow:** Continuous or Intermittent 60 Hz Magnetic Field Exposure Fails to Affect the Nighttime Rise in Melatonin in the Adult Djungarian Hamster. *Biol. Reprod. 52*:72 (1995).

633. **Levine, R.L., J.K. Dooley, and T.D. Bluni:** Magnetic-Field Effects on Spatial Discrimination and Melatonin Levels in Mice. *Physiol. Behav. 58*:535-537 (1995).

634. **Grota, L.J., R.J. Reiter, P. Keng, and S. Michaelson:** Electric Field Exposure Alters Serum Melatonin but Not Pineal Melatonin Synthesis in Male Rats. *Bioelectromagnetics 15*:427-438 (1994).

635. **Yaga, K. R.J. Reiter, L.C. Manchester, H. Nieves, J.H. Sun, and L.D. Chen:** Pineal Sensitivity to Pulsed Static Magnetic-Fields Changes during the Photoperiod. *Brain Res. Bull. 30*:153-156 (1993).

636. **Selmaoui, B., and Y. Touitou:** Sinusoidal 50-Hz Magnetic-Fields Depress Rat Pineal NAT Activity and Serum Melatonin. Role of Duration and Intensity of Exposure. *Life Sci. 57*:1351-1358 (1995).

637. **Rogers, W.R., R.J. Reiter, L. Barlow-Walden, H.D. Smith, and J.L. Orr:** Regularly Scheduled, Day-Time, Slow Onset 60 Hz Electric and Magnetic Field Exposure Does Not Depress Serum Melatonin Concentration in Nonhuman Primates. *Bioelectromagnetics (Suppl.) 3*:111-118 (1995).

638. **Rogers, W.R., R.J. Reiter, H.D. Smith, and L. Barlow-Walden:** Rapid-Onset/Offset, Variably Scheduled 60 Hz Electric and Magnetic Field Exposure Reduces Nocturnal Serum Melatonin Concentration in Nonhuman Primates. *Bioelectromagnetics (suppl) 3*:119-122 (1995).

639. **Stevens, R.G.:** Electric Power Use and Breast Cancer: A Hypothesis. *Am. J. Epidemiol. 125*:556-561 (1987).

640. **Leung, F.C., D.N. Rommereim, R.G. Stevens, and L.E. Anderson:** Effects of Electric Fields on Rat Mammary Tumor Development Induced by 7,12 Dimethylbenz(a)anthracene (DMBA). In *Abstracts, Ninth Annual Meeting of the Bioelectromagnetics Society.* Portland, OR, 1987. p. 41.

641. **Leung, F.C., D.N. Rommereim, R.G. Stevens, B.W. Wilson, R.L. Buschbom, and L.E. Anderson:** Effects of Electric Fields on Rat Mammary Tumor Development Induced by 7,12 Dimethylbenz(a)anthracene (DMBA). In *Abstracts, Tenth Annual Meeting of the Bioelectromagnetics Society,* Stamford, CT, 1988. p. 2.

642. **Stevens, R.G., and S. Davis:** The Melatonin Hypothesis: Electric Power and Breast Cancer. *Environ. Health Perspect. 104(Suppl. 1)*:135-140 (1996).

643. **Thomson, R.A.E., S.M. Michaelson, and Q.A. Nguyen:** Influence of 60-Hertz Magnetic Fields on Leukemia. *Bioelectromagnetics 9*:149-157 (1988).

644. **Hitchcock, R.T., S. McMahan, and G.C. Miller:** *Extremely Low Frequency (ELF) Electric and Magnetic Fields.* Fairfax, VA: American Industrial Hygiene Association, 1995.

645. **Loscher, W., and M. Mevissen:** Animal Studies on the Role of 50/60-Hertz Magnetic Fields in Carcinogenesis. *Life Sci. 54*:1531-1543 (1994).

646. **Holmberg, B., and A. Rannug:** Magnetic Fields and Cancer Development in Animal Models. *Radio Sci. 30*:223- 231 (1995).

647. **Liburdy, R.P., and W. Loscher:** Laboratory Studies on Extremely Low Frequency (50/60-Hz) Magnetic Fields and Carcinogenesis. In *The Melatonin Hypothesis Breast Cancer and Use of Electric Power* edited by R.G. Stevens, B.W. Wilson, and L.E. Anderson. Columbus, OH: Battelle Press, 1997.

648. **Graham, C.M., M.R. Cook, and H.D. Cohen:** *Immunological and Biochemical Effects of 60-Hz Electric and Magnetic Fields in Humans* (Report No.

DOE/CE/76246-T1). U.S. Department of Energy Office of Scientific & Technical Information, P.O. Box 62, Oak Ridge, TN 37830, 1990.

649. **Cook, M.R., C. Graham, C.D. Cohen, and M.M. Gerkovich:** A Replication Study of Human Exposure to 60-Hz Fields: Effects on Neurobehavioral Measures. *Bioelectromagnetics 13*:261-285 (1992).

650. **Sastre, A., C. Graham, M.R. Cook, S.J. Hoffman, and M.M. Gerkovich:** Human Heart Rate Variability in Magnetic Fields: Continuous Versus Intermittent Exposure. In *Project Abstracts of The Annual Review of Research on Biological Effects of Electric and Magnetic Fields From the Generation, Delivery & Use of Electricity*. Palm Springs, CA, November 12-16, 1995.

651. **Selmaoui, B., A. Bogdan, A. Auzeby, J. Lambrozo, and Y. Touitou:** Acute Exposure to 50 Hz Magnetic Field Does Not Affect Hematologic or Immunologic Functions in Healthy Young Men: A Circadian Study. *Bioelectromagnetics 17*:364-372 (1996).

652. **Graham, C., M.R. Cook, D.W. Riffle, M.M. Gerkovich, and H.D. Cohen:** Nocturnal Melatonin Levels in Human Volunteers Exposed to Intermittent 60 Hz Magnetic Fields. *Bioelectromagnetics 17*:263-273 (1996).

653. **Heath, C.W.:** Electromagnetic Field Exposure and Cancer: A Review of Epidemiologic Evidence. *Ca. Cancer J. Clin. 65*:29-44 (1996).

654. **Kheifets, L.I., A.A. Afifi, P.A. Buffler, and Z.W. Zhang:** Occupational Electric and Magnetic Field Exposure and Brain Cancer: A Meta-Analysis. *J. Occupat. Environ. Med. 37*:1327-1341 (1995).

655. **Inskip, P.D., M.S. Linet, and E.F. Heineman:** Etiology of Brain Tumors in Adults. *Epidemiologic Rev. 17*:382-414 (1995).

656. **Shaw, G.M., and L.A. Croen:** Human Adverse Reproductive Outcomes and Electromagnetic Field Exposures: Review of Epidemiologic Studies. *Environ. Health Perspect. 101(Suppl. 4)*:107-119 (1993).

657. **Carstensen, E.L.:** Magnetic Fields and Cancer. *IEEE Eng. Med. Biol. Mag. 14(4)*:362-369 (1995).

658. **Feychting, M.:** Occupational Exposure to Electromagnetic Fields and Adult Leukemia: A Review of the Epidemiological Evidence. *Radiat. Environ. Biophys. 35*:237-242 (1996).

659. **Milham, S.:** Mortality from Leukemia in Workers Exposed to Electrical and Magnetic Fields. *New Engl. J. Med. 307*:249 (1982).

660. **Coleman, M., J. Bell, and R. Skeet:** Leukemia Incidence in Electrical Workers. *Lancet i*:982-983 (1983).

661. **McDowall, M.E.:** Leukemia Mortality in Electrical Workers in England and Wales. *Lancet 8318*:246 (1983).

662. **Coleman, M., and V. Beral:** A Review of Epidemiological Studies of the Health Effects of Living Near or Working with Electricity Generation and Transmission Equipment. *Int. J. Epidemiol. 17*:1-13 (1988).

663. **Loomis, D.P., and D.A. Savitz:** Mortality from Brain Cancer and Leukaemia among Electrical Workers. *Brit. J. Ind. Med. 47*:633-638 (1990).

664. **Stern, R.M.:** Cancer Incidence among Welders: Possible Effects of Exposure to Extremely Low Frequency Electromagnetic Radiation (ELF) and to Welding Fumes. *Environ. Health Perspect. 76*:221-229 (1987).

665. **Alfredsson, L., N. Hammar, and S. Karlehagen:** Cancer Incidence Among Male Railway Engine-Drivers and Conductors in Sweden, 1976-90. *Cancer Causes Control 7*:377-381 (1996).

666. **Tynes, T., A. Anderson, and F. Langmark:** Incidence of Cancer in Norwegian Workers Potentially Exposed to Electromagnetic Fields. *Am. J. Epidemiol. 136*:81-88 (1992).

667. **Guenel, P., P. Rashmark, J.A. Anderson, and E. Lynge:** Cancer Incidence in Danish Persons Who Have Been Exposed to Magnetic Fields at Work. *Cancer Registry of the Danish Cancer Control Agency*. Copenhagen, 1992.

668. **Floderus, B., T. Persson, C. Stenlund, G. Linder, C. Johansson, J. Kiviranta, H. Parsman, M. Lindblom, B. Knave, A. Wennberg, and A. Ost:** *Occupational Exposure to Electromagnetic Fields in Relation to Leukemia and Brain Tumors. A Case-Control Study*. National Institute of Occupational Health, Solna, Sweden. Printed by LITOHUSET, Stockholm, Sweden, 1992

669. **Guenel, P., P. Raskmark, J.B. Andersen, and E. Lynge:** Incidence of Cancer in Persons with Occupational Exposure to Electromagnetic Fields in Denmark. *Brit. J. Ind. Med. 50*:758-764 (1993).

670. **Coogan, P.F., R.W. Clapp, P.A. Newcomb, T.B. Wenzl, G. Bogdan, R. Mittendorf, J.A. Baron, and M.P. Longnecker:** Occupational Exposure to 60-Hz Magnetic Fields and Risk of Breast Cancer in Women. *Epidemiol. 7*:459- 464 (1996).
671. **Loomis, D.P., D.A. Savitz, and C.V. Ananth:** Breast Cancer Mortality among Female Electrical Workers in the United States. *J. Natl. Cancer Inst. 86*:921-925 (1994).
672. **Sahl, J.D., M.A. Kelsh, and S. Greenland:** Cohort and Nested Case-Control Studies of Hematopoietic Cancers and Brain Cancers among Electric Utility Workers. *Epidemiology 4*:104-14 (1993).
673. **Theriault, G., M. Goldberg, A.B. Miller, B. Armstrong, P. Guenel, J. Deadman, E. Imbernon, T. To, A. Chevaller, D. Cyr, and C. Wall:** Cancer Risks Associated with Occupational Exposure to Magnetic Fields among Electric Utility Workers in Ontario and Quebec, Canada, and France: 1970-1989. *Am. J. Epidemiol. 139*:550-72 (1994).
674. **Savitz, D.A., and D. P. Loomis:** Magnetic Field Exposure in Relation to Leukemia and Brain Cancer Mortality among Electric Utility Workers. *Am. J. Epidemiol. 141*:123-34 (1995).
675. **Miller, A.B., T. To, D.A. Agnew, C. Wall, and L.M. Green:** Leukemia following Occupational Exposure to 60-Hz Electric and Magnetic Fields among Ontario Electric Utility Workers. *Am. J. Epidemiol. 144*:150-160 (1996).
676. **Grandolfo, M., and P. Vecchia:** Natural and Man-Made Environmental Exposures to Static and ELF Electromagnetic Fields. In *Biological Effects and Dosimetry of Static and ELF Electromagnetic Fields* (edited by M. Grandolfo, S. Michaelson and A. Rindi). New York: Plenum Press, 1985. pp. 49-70.
677. **Kaune, W.T.:** Assessing Human Exposure to Power-Frequency Electric and Magnetic Fields. *Env. Health Perspect. 101 (suppl. 4)*: 121-133 (1993).
678. **Bonneville Power Authority:** *Electrical and Biological Effects of Transmission Lines* (DOE/BP-945). Portland, OR: Bonneville Power Authority, 1993.
679. **Skotte, J.H.:** Exposure to Power-Frequency Electromagnetic Fields in Denmark. *Scand. J. Work Environ. Health 20*:132-138 (1994).
680. **Tenforde, T.S.:** Spectrum and Intensity of Environmental Electromagnetic Fields from Natural and Man-Made Sources. In *Electromagnetic Fields Biological Interactions and Mechanisms* (Advances in Chemistry Series 250). Washington, D.C.: American Chemical Society, 1995. pp. 13-35.
681. **National Institute for Occupational Safety and Health (NIOSH):** Occupational Exposure Assessment for Electric and Magnetic Fields in the 10-1000 Hz Frequency Range by T.D. Bracken (DHHS [NIOSH] Pub. No. 91–111). In *Proceedings of the Scientific Workshop on the Health Effects of Electric and Magnetic Fields on Workers*. Cincinnati: NIOSH, 1991.
682. **Randa, J., D. Gilliland, W. Gjertson, W. Lauber, and M. McInerney:** Catalogue of Electromagnetic Environment Measurements, 30-300 Hz. *IEEE Trans. Electromag. Compat. 37*:26-33 (1995).
683. **Maruvada, P.S.:** Characterization of Power Frequency Magnetic Fields in Different Environments. *IEEE Trans. Power Del. 8*:598-606 (1993).
684. **Merchant, C.J., D.C. Renew, and J. Swanson:** Exposures to Power-Frequency Magnetic Fields in the Home. *J. Radiol. Prot. 14*:77-87 (1994).
685. **Kaune, W.T., R.G. Stevens, N.J. Callahan, R.K. Severson and D.B. Thomas:** Residential Magnetic and Electric Fields. *Bioelectromagnetics 8*:315-335 (1987).
686. **Juutilainen, J., J. Eskelinen and K. Saali:** Residential Exposure to ELF Magnetic Fields. In *Proceedings of the Ninth Annual Conference of the IEEE Engineering in Medicine and Biology Society*. [Cat. No. 87CH2513-0.] New York: Institute of Electrical and Electronics Engineers, 1987. pp. 1177-1178.
687. **Friedman, D.R., E.E. Hatch, R. Tarone, W.T. Kaune, R.A. Kleinerman, S. Wacholder, J.D. Boice, Jr., and M.S. Linet:** Childhood Exposure to Magnetic Fields: Residential Area Measurements Compared to Personal Dosimetry. *Epidemiology 7*:151-155 (1996).
688. **Silva, M. N. Hummon, D. Rutter and C. Hooper:** Power Frequency Magnetic Fields in the Home. *IEEE Trans. Power Del. 4*: 465-478 (1989).

689. **Rauch, G.B., G. Johnson, P. Johnson, A. Stamm, S. Tomita, and J. Swanson:** A Comparison of International Residential Grounding Practices and Associated Magnetic Fields. In *Proceedings of the 1991 IEEE Power Engineering Society Transmission and Distribution Conference* [Cat. No, 91CH3070-0.] New York: Institute of Electrical and Electronics Engineers, 1991. pp. 743-748.
690. **Valjus, J., M. Hongisto, P. Verkasalo, P. Jarvinen, K. Heikkila, and M. Koskenvuo:** Residential Exposure to Magnetic Fields Generated by 110-400 kV Power Lines in Finland. *Bioelectromagnetics 16*:365-376 (1995).
691. **Mild, K.H., M. Sandstrom, and A. Johnsson:** Measured 50-Hz Electric and Magnetic Fields in Swedish and Norwegian Residential Buildings. *IEEE Trans. Instrum. Meas. 45*:710-714 (1996).
692. **Kavet, R., J.M. Silva and D. Thornton:** Magnetic Field Exposure Assessment for Adult Residents of Maine Who Live Near and Far Away from Overhead Transmission Lines. *Bioelectromagnetics 13*:35-55 (1992).
693. **Hartwell, F.:** Magnetic Fields from Water Pipes. *Elec. Const. Main. 92(3)*:63-70 (1993).
694. **Florig, H.K., and J.F. Hoburg:** Power-Frequency Magnetic Fields from Electric Blankets. *Health Phys. 58*:493-502 (1990).
695. **Gauger, J.R.:** Household Appliance Magnetic Field Survey. *IEEE Trans. Power Apparat. Systems PAS-104*:2436-2444 (1985).
696. **Mader, D.L., and S.B. Peralta:** Residential Exposure to 60-Hz Magnetic Fields from Appliances. *Bioelectromagnetics 13*:287-301 (1992).
697. **Tofani, S., P. Ossola, G. D'Amore, and O.P. Gandhi:** Electric Field and Current Density Distributions Induced in an Anatomically-Based Model of the Human Head by Magnetic Fields from a Hair Dryer. *Health Phys. 68*:71-79 (1995).
698. **Preece, A.W., W. Kaune, P. Grainger, S. Preece, and J. Golding:** Magnetic Fields from Domestic Applicances in the UK. *Phys. Med. Biol. 42*:67-76 (1997).
699. **Sun, W.Q., P. Heroux, T. Clifford, V. Sadilek, and F. Hamade:** Characterization of the 60-Hz Magnetic Fields in Schools of the Carleton Board of Education. *Am. Ind. Hyg. Assoc. J. 56*:1215-1224 (1995).
700. **Breysse, P., P.S.J. Lees, M.A. McDiarmid and B. Curbow:** ELF Magnetic Field Exposures in an Office Environment. *Am. J. Ind. Med. 25*:177-185 (1994).
701. **Stuchly, M.A., D.W. Lecuyer and R.D. Mann:** Extremely Low Frequency Electromagnetic Emissions from Video Display Terminals and Other Devices. *Health Phys. 43*:713-722 (1983).
702. **Juutilainen, J., and K. Saali:** Measurements of Extremely Low-Frequency Magnetic Fields around Video Display Terminals. *Scand. J. Work Environ. Health 12*:609-613 (1986).
703. **Walsh, M.L., S.M. Harvey, R.A. Facey, and R.R. Mallette:** Hazard Assessment of Video Display Units. *Am. Ind. Hyg. Assoc. J. 58*:324-331 (1991).
704. **Sapashe, D., and J.R. Ashley:** Video Display Terminal Fringing Magnetic Field Measurements. *IEEE Trans. Instrum. Meas. 41*:178-184 (1992).
705. **Tell, R.A.:** *An Investigation of Electric and Magnetic Fields and Operator Exposure Produced by VDTs: NIOSH VDT Epidemiology Study.* Srpingfield, VA: National Technical Information Service (PB91-130500/XAB), 1990.
706. **National Institute for Occupational Safety and Health:** *Health Hazard Evaluation Report HETA 93-0734-2401 Santa Clara County Administrative Office Building, San Jose, California*, by A. Tepper, C.E. Moss, and D. Booher. Springfield, VA: National Technical Information Service (PB94-193711), 1994.
707. **Ontario Ministry of Labour:** *Background ELF Magnetic Fields in Ontario Offices* by M. Bitran, D. Charron, and J. Nishio. Toronto, Ontario: Ministry of Labour, 1995.
708. **Phillips, K.L.:** "Characterization of Occupational Exposures to Extremely Low Frequency Magnetic Fields in a Health-Care Setting." Master's thesis, School of Public Health, University of Texas, Houston, TX, 1993.
709. **National Institute for Occupational Safety and Health:** *Electric and Magnetic Fields in a Magnetic Resonance Imaging Facility: Measurements and Exposure Assessment* Procedures by T.D. Bracken. Springfield, VA: National Technical Information Service (PB94-174489), 1994.

710. **Paul, M., K. Hammond, and S. Abdollahzadeh:** Power Frequency Magnetic Field Exposures Among Nurses in a Neonatal Intensive Care Unit and a Normal Newborn Nursery. *Bioelectromagnetics 15*:519-529 (1994).
711. **Minder, Ch.E., and D.F. Pfluger:** Extremely Low Frequency Electromagnetic Field Measurements (ELF-EMF) in Swiss Railway Engines. *Radiat. Prot. Dosimetry 48*:351-354 (1993).
712. **Department of Transportation:** *Safety of High Speed Guided Ground Transportation Systems. Magnetic and Electric Field Testing of the Amtrak Northeast Corridor and New Jersey Transit/North Jersey Coast Line Rail Systems, Volume I: Analysis* by F.M. Dietrich, W.E. Ferro, P.N. Papas, and G.A. Steiner (DOT/FRA/ORD-93/01.I). Springfield, VA: National Technical Information Service (PB93-219434), 1993.
713. **Department of Transportation:** *Safety of High Speed Guided Ground Transportation Systems. Magnetic and Electric Field Testing of the Amtrak Northeast Corridor and New Jersey Transit/North Jersey Coast Line Rail Systems, Volume II: Appendices*, by F.M. Dietrich, D.C. Robertson, and G.A. Steiner (DOT/FRA/ORD-93/01.II). Springfield, VA: National Technical Information Service (PB93-219442), 1993.
714. **Fisher, R.B.:** Electric and Magnetic Fields and Electric Transit Systems. *Trans. Res. Rec. 1503*:69-76 (1995).
715. **Silva, M., and D. Huber:** Exposure to Transmission Line Electric Fields During Farming Operations. *IEEE Trans. Power Apparat. Syst. PAS-104*:2632-2640 (1985).
716. **Enk, J.O., and M.M. Abromavage:** Exposure to Low-Level Extremely Low Frequency Electromagnetic Fields Near Naval Communications Facilities. *IEEE J. Oceanic Eng. OE-9*:136-142 (1984).
717. **National Institute for Occupational Safety and Health:** *HETA 93-0424-2486 Chicago Television Stations Chicago, Illinois* by R. Malkin and C.E. Moss. Springfield, VA: National Technical Information Service (PB95-241121), 1995.
718. **Barroetavena, M.C., R. Ross, and K. Teschke:** Electric and Magnetic Fields at Three Pulp and Paper Mills. *Am. Ind. Hyg. Assoc. J. 55*:358-363.
719. **Vogt, D.R., and J.P. Reynders:** An Experimental Investigation of Magnetic Fields in Deep Level Gold Mines. In *Seventh International Symposium on High Voltage Engineering 1991*. Dresden, Germany: Dresden University of Technology, 1991. pp. 63-66.
720. **National Institute for Occupational Safety and Health:** *HETA 93-1038-2432 L-S Electro-Galvanizing Company Cleveland, Ohio* by C.E. Moss and D. Mattorano. Springfield, VA: National Technical Information Service, 1995.
721. **Stuchly, M.A., and D.W. Lecuyer:** Exposure to Electromagnetic Fields in Arc Welding. *Health Phys. 56*:297-302 (1989).
722. **Rosenthal F.S., and S. Abdollahzadeh:** Assessment of Extremely Low Frequency (ELF) Electric and Magnetic Fields in Microelectronics Fabrication Rooms. *Appl. Occupat. Environ. Hyg. 6*:777-784 (1991).
723. **Lovsund P., P.A. Oberg, and S.E.G. Nilsson:** ELF Magnetic Fields in Electrosteel and Welding Industries. *Radio Sci. 17(5S)*:35S-38S (1982).
724. **Cartwright, C.E, P.N. Breysse, and L. Booher:** Magnetic Field Exposures in a Petroleum Refinery. *Appl. Occup. Environ. Hyg. 8*: 587-592 (1993).
725. **Hayashi, N., K. Isaka and Y. Yokoi:** ELF Electromagnetic Environment in Power Substations. *Bioelectromagnetics 10*:51-64 (1989).
726. **Heroux, P.:** 60-Hz Electric and Magnetic Fields Generated by a Distribution Network. *Bioelectromagnetics 8*:135-148 (1987).
727. **Vinh, T., T.L. Jones, and C.H. Shih:** Magnetic Fields Near Overhead Distribution Lines — Measurements and Estimating Technique. *IEEE Trans. Power Del. 6*:912-921 (1991).
728. **Chartier, V.L., T.D. Bracken and A.S. Capon:** BPA Study of Occupational Exposure to 60-Hz Electric Fields. *IEEE Trans. Power Apparat. Systems PAS-104*:733-744 (1985).
729. **Gamberale, F., B.A. Olson, P. Eneroth, T. Lindh and A. Wennberg:** Acute Effects of ELF Electromagnetic Fields: A Field Study of Linesmen Working with 400 kV Power Lines. *Brit. J. Ind. Med. 46*:729-737 (1989).
730. **Sahl, J.D., M.A. Kelsh, R.W. Smith, and D.A. Aseltine:** Exposure to 60 Hz Magnetic Fields in the Electric Utility Work Environment. *Bioelectromagnetics 15*:21-32 (1994).

731. **Bracken, T.D., R.F. Rankin, R.S. Senior, J.R. Alldredge, and S.S. Sussman:** Magnetic Field Exposure Among Utility Workers. *Bioelectromagnetics 16*:216-226 (1995).
732. **Miller, D.A.:** Electrical and Magnetic Fields Produced by Commercial Power Systems. In *Biological and Clinical Effects of Low-Frequency Magnetic and Electric Fields*. Springfield, IL: Charles C. Thomas, 1974. pp. 62-70.
733. **Kromhout, H., D.P. Loomis, G.J. Mihlan, L.A. Peipins, R.C. Kleckner, R. Iriye, and D.A. Savitz:** Assessment and Grouping of Occupational Magnetic Field Exposure in Five Electric Utility Companies. *Scand. J. Work Environ. Health 21*:43-50 (1995).
734. **Deadman, J.E., M. Camus, B.G. Armstrong, P. Heroux, D. Cyr, M. Plante, and G. Theriault:** Occupational and Residential 60-Hz Electromagnetic Fields and High-Frequency Electric Transients: Exposure Assessment Using A New Dosimeter. *Am. Ind. Hyg. Assoc. J. 49*:409-19 (1988).
735. **American Conference of Governmental Industrial Hygienists:** Notice of Intended Change — Sub- Radiofrequency (30 kHz and below) and Static Electric Fields. *Appl. Occupat. Environ. Hyg. 5*:734-737 (1990).
736. **American Conference of Governmental Industrial Hygienists:** Notice of Intended Change — Sub- Radiofrequency (1 Hz to 30 kHz) Magnetic Fields. *Appl. Occupat. Environ. Hyg. 5*:884-892 (1990).
737. **National Radiological Protection Board:** *Advice on the Protection of Workers and Members of the Public from the Possible Hazards of Electric and Magnetic Fields with Frequencies below 300 GHz: A Consultative Document*. NRPB, Chilton, Didcot, Oxfordshire, OX11 0RQ, U.K., 1986. p. 2.
738. **National Radiological Protection Board:** *Board Statement on Restrictions on Human Exposure to Static and Time Varying Electromagnetic Fields and Radiation*. NRPB, Chilton, Didcot, Oxfordshire, OX11 0RQ, U.K., 1993.
739. **International Non-ionizing Radiation Committee of the International Radiation Protection Association:** Interim Guidelines on Limits of Exposure to 50/60 Hz Electric and Magnetic Fields. *Health Phys. 58*:113-122 (1990).
740. **IEEE Magnetic Fields Task Force:** An Evaluation of Instrumentation Used to Measure AC Power System Magnetic Fields. *IEEE Trans. Power Del. 6*:373-383 (1991).
741. **Environmental Protection Agency:** *Final Report Laboratory Testing of Commercially Available Power Frequency Magnetic Field Survey Meters* by SC&A, Inc. and Science Applications International Corporation (EPA 400R-92-010). Washington, D.C.: EPA, 1992.
742. **Zipse, D.W.:** Electric and Magnetic Fields: Equipment and Methodology Used for Obtaining Measurements. *IEEE Trans. Ind. Appl. 30*:262-268 (1994).
743. **Bartington, G.:** Sensors for Low Level Low Frequency Magnetic Fields. In *IEE Colloquium 'Low Level Low Frequency Magnetic Fields'* (Digest No. 1994/096). London, UK: IEE, 1994. pp. 2/1-2/9.
744. **Sussman, S.S.:** Exposure Assessment at Extremely Low Frequencies: Issues, Instrumentation, Modeling, and Data. *Radio Sci. 30*:151-159 (1995).
745. **Deno, D.W., and M. Silva:** Method of Evaluating Human Exposure to 60 Hz Electric Fields. *IEEE Trans. Power Apparat. Sys. PAS-103*:1699-1705 (1984).
746. **Hayashi, N., K. Isaka, S. Yura, and Y. Yokoi:** Development of Magnetic Field Dosimeter for Exposure Application. In *7th International Symposium on High Voltage Engineering* (Paper 93.09, Vol. 9). Dresden, Germany: Lechnische Universitat, 1991. pp. 83-86.
747. **Dlugosz, L., K. Belanger, P. Johnson, and M.B. Bracken:** Human Exposure to Magnetic Fields: A Comparative Assessment of Two Dosimeters. *Bioelectromagnetics 15*:593-597 (1994).
748. **Lo, C.C., T.Y. Fujita, A.B. Geyer, and T.S. Tenforde:** A Wide Dynamic Range Portable 60-Hz Magnetic Dosimeter with Data Acquisition Capabilities. *IEEE Trans. Nuc. Sci 33*:643-646 (1986).
749. **Lindh, T., and L.-I. Andersson:** Comparison between Two Power-Frequency Electric-Field Dosimeters. *Scand. J. Work Environ. Health 14(suppl 1)*:43-45 (1988).
750. **Heroux, P.:** A Dosimeter for Assessment of Exposures to ELF Fields. *Bioelectromagnetics 12*:241-257 (1991).

751. **Douglas, J.:** Taking the Measure of Magnetic Fields. *EPRI J. 17*:16-17 (1993).
752. **Kaune, W.T., J.C. Niple, M.J. Liu, and J.M. Silva:** Small Integrating Meter for Assessing Long-Term Exposure to Magnetic Fields. *Bioelectromagnetics 13*:413-427 (1992).
753. **Patterson, R.M.:** Exposure Assessment for Electric and Magnetic Fields. *J. Exp. Analy. Environ. Epidemiol. 2*:159- 176 (1992).
754. **Zaffanella, L.E.:** Magnetic Field Exposure Characterization During Environmental Field Surveys for the EMF RAPID Program. In *Proceedings of the American Power Conference* (Vol. 58-1), edited by A.E. McBride. Chicago, IL: Illinois Institute of Technology, 1996. pp. 263-268.
755. **National Institute for Occupational Safety and Health (NIOSH):** *Health Hazard Evaluation Report HETA 92-0009-2362 New York Telephone Company White Plains, New York* by R. Malkin, C.E. Moss, N.C. Burton, and D. Booher. Cincinnati: NIOSH, 1993.
756. **National Institute for Occupational Safety and Health (NIOSH):** Occupational Exposure Assessment for Electric and Magnetic Fields in the 10-1000 Hz Frequency Range by T.D. Bracken. In *Proceedings of the Scientific Workshop on the Health Effects of Electric and Magnetic Fields on Workers* edited by P.J. Bierbaum and J.M. Peters. Cincinnati: NIOSH, 1991.
757. **Institute of Electrical and Electronic Engineers:** *IEEE Standard Procedures for Measurement of Power Frequency Electric and Magnetic Fields from AC Power Lines* (IEEE Standard 644-1987). New York: IEEE, 1987.
758. **IEEE Magnetic Fields Task Force:** Measurements of Power Frequency Magnetic Fields away from Power Lines. *IEEE Trans. Power Del. 6*:901-911 (1991).
759. **Rauch, G.B., G. Chang, M. Keller, and T.D. Bracken:** Protocol for Measurement of Transmission and Distribution Line Workplace Magnetic Fields by Waveform Capture. In *Proceedings of the 1994 IEEE Power Engineering Society Transmission and Distribution Conference* (IEEE Cat. No. 94CH3428-0). New York: IEEE, 1994. p. 504.
760. **IEEE Magnetic Fields Task Force:** A Protocol for Spot Measurements of Residential Power Frequency Magnetic Fields. *IEEE Trans. Power Del. 8*:1386-1394 (1993).
761. **Yost, M.G., G. M. Lee, D. Duane, J. Fisch, and R. R. Neutra:** California Protocol for Measuring 60 Hz Magnetic Fields in Residences. *Appl. Occ. Environ. Hyg. 7*: 772-777 (1992).
762. **Kaune, W.T.:** Assessing Human Exposure to Power-Frequency Electric and Magnetic Fields. *Env. Health Perspect. Suppl. 101 (s4)*:121-133 (1993).
763. **Kaune, W.T., S.D. Darby, S.N. Gardner, Z. Hrubec, R.N. Iriye, and M.S. Linet:** Development of a Protocol for Assessing Time-Weighted-Average Exposures of Young Children to Power-Frequency Magnetic Fields. *Bioelectromagnetics 15*: 33-51 (1994).
764. **Cotten, W.L., K.C.K. Ramsing, and C. Cai:** Design Guidelines for Reducing Electromagnetic Field Effects from 60-Hz Electrical Power Systems. *IEEE Trans. Ind. Appl. 30*:1462-1471 (1994).
765. **Feero, W.E.:** Electric and Magnetic Field Management. *Am. Ind. Hyg. Assoc. J. 54*:205-210 (1993).
766. **Glaser, Z.R.:** Organization and Management of A Nonionizing Radiation Safety Program. In *CRC Handbook of Management of Radiation Protection Programs* (2nd Ed.), Boca Raton, FL: CRC Press, 1992. pp. 43-52.
767. **National Institute for Occupational Safety and Health (NIOSH):** *Radiofrequency (RF) Sealers and Heaters: Potential Health Hazards and Their Prevention* (Current Intelligence Bulletin 33, #80-107). Cincinnati: NIOSH, 1979.
768. **U.S. Air Force:** *Exposure to Radiofrequency Radiation* (AFOSH Standard 161-9). Washington, D.C.: Department of the Air Force, 1987.
769. **Australian Standard:** *Radiofrequency Radiation Part 1: Maximum Exposure Levels—100 kHz to 300 GHz* (AS 2772.1- 1990). Crescent, Homebush: Standards Australia.
770. **U.S. Department of Defense:** *Protection of DoD Personnel from Exposure to Radiofrequency Radiation and Military Exempt Lasers* (DoD Instruction 6055.11). Washington, D.C.: Department of Defense, 1995.
771. **U.S. Department of Energy:** *Industrial Hygiene Standard for Non-Ionizing Radiation* (draft 11/16/92). Washington, D.C.: U.S. Department of Energy, 1992.

Outcome Competencies

After completing this chapter, the user should be able to:

1. Define underlined terms used in this chapter.
2. Describe the physics of radiation production.
3. List categories of ionizing radiation.
4. Explain dosimetry and attenuation.
5. List radiation detection devices.
6. Recall the operating principles for radiation detection devices.
7. Summarize the sources of exposure to ionizing radiation.
8. Describe radiation production equipment.
9. Describe the biological consequences of exposure to ionizing radiation.
10. Recall agencies that regulate ionizing radiation.

Key Terms

absorption • alpha • atomic mass • atomic mass unit • atomic number • atomic weight • attenuation • beta • bremsstrahlung • Compton effect • contamination • critical target • densitometer • gamma rays • Geiger counter • hot lab • ionization chamber • ionizing chamber • ionizing radiation • isomers • isotope • LD_{37} • LD_{50} • metastable • Naturally Occurring Radioactive Material (NARM) • neutrino • neutron • nucleon • nuclear force • nuclides • pair production • photoelectric effect • positrons • proportional counter • radioactivity • radionuclide • radon progeny • sensitive volume • sensitometer • specific activity • standard man • survey • thermoluminescent dosimetry • tissue equivalent • TLD chips • total absorption process

Prerequisite Knowledge

Basic college physics, college chemistry.

Prior to beginning this chapter, the user should review the following chapters:

Chapter Number	Chapter Topic
2	Occupational Exposure Limits
4	Environmental and Occupational Toxicology
21	Nonionizing Radiation

Key Topics

I. Radiation
   A. Ionizing Radiation—Photons and Particles

II. Physical Aspects of Production of Ionizing Radiation
   A. Electromagnetic Radiation (X-Rays, Gamma Rays, Bremsstrahlungen, Cosmic Rays)
   B. Radioactivity

III. Dosimetry
   A. Attenuation Processes
   B. Attenuation of Photons

IV. Radiation Detection Devices

V. Sources of Exposure from Ionizing Radiation
   A. Natural Background Radiation
   B. Artificial and Technologically Enhanced Radiation Exposure
   C. radon

VI. Radiation Production Equipment
   A. Nuclear Reactors
   B. Accelerators

VII. Biological Consequences of Ionizing Radiation
   A. Dose Response Curve

VIII. Agencies

IX. Radiation Protection: Time, Distance, Shielding

22

Ionizing Radiation

Margaret E. McCarthy

Radiation

Ionizing Radiation—Photons and Particles

Ionizing radiation is a general term applied to both electromagnetic waves and/or particulate radiation capable of producing ions by interaction with matter. Radiation dose units are limited to ionizing radiation photons and ionizing particles (Tables 22.1, 22.2, and 22.3). Typically, ionizing radiation photons are defined in terms of energy expressed in multiples of the electron volt, keV or MeV, rather than wavelength (m) or frequency (Hz). Wavelength units typically apply to the part of the electromagnetic spectrum in the optical, infrared, and microwave region and frequency units to the lower energy portion of radio and television regions. Though physically the same in any unit, measurement in the electromagnetic spectrum historically is described in the unit in which the component can be measured. For biological measurement that compares ionizing photons with ionizing particles, measurement in energy units is required. It is appropriate to state units of ionizing photons in keV or MeV. Heretofore, the term radiation will assume ionizing radiation. All the radiation rays used in occupational hygiene are listed in Table 22.5. The common symbols for particles are noted in Table 22.6.

Nomenclature pertaining to electromagnetic waves refers to their origin. Photons produced from nuclear transition emission are called gamma rays, typically MeV. Those photons produced from electron

Table 22.1 — Radiation Units

Dose Unit	*SI Unit*	*Comment*
Exposure dose	Roentgen	used only for X-ray machine; not SI, but quantity
Activity	Becquerel	limited to radionuclides only; quantity
Absorbed dose	Gray	any classification of radiation dose; quality
Dose equivalent	Sievert	only for human dose monitoring; integrated quality
Linear energy transfer	LET	energy transfer per path length (keV/μm)
Absorbed dose	Kerma	dose for indirectly ionizing radiation, quality

Table 22.2 — Scaling Factors for Ionizing Radiation

Relative biological effectiveness (RBE)	ratio of a standard cell effect to test effect
Quality factor (QF)	factor to convert Gy to Sv (rad to rem)
f-factor	factor to convert R to Gy (energy dependent)

Table 22.3 — International System (SI) Radiation Units

SI Unit (mks[A])	*Definition*	*cgs[B] Unit*	*Conversion*
Becquerel (Bq)	1 disintegration per sec	Curie (Ci)	1 Bq = 2.7×10^{-11} Ci
Gray (Gy)	1 Joule per kg	rad	1 Gy = 100 rad
Sievert (Sv)	1 Gy times quality factor	rem	1 Sv = 100 rem
Roentgen (R)	2.58×10^{-4} Coulomb per kg		X-ray machines only; limited to photons <3 MeV

[A]mks = meter/kilogram/second system of measurement
[B]cgs = centimeter/gram/second system of measurement

Table 22.4 — Quality Factor

Type of Radiation	*QF (One Significant Figure)*
Photons, electrons, betas	1
Thermal neutrons (≤10 keV)	3
Neutrons (>10 keV)	10
Protons	10
Alpha	20
Heavy recoil nuclei	20

Source: *Code of Federal Regulations* Title 10, Part 835.

Table 22.5 — Category of Ionizing Rays

Common Name	*Nature*	*AMU*	*Charge*	*Origin*	*QF*
Gamma	photon	0	0	nucleus	low
X-ray	photon	0	0	atomic	low
Bremsstrahlung	photon	0	0	atomic	low
Cosmic (EM)	photon	0	0	extraterrestrial	low
Cosmic particle	particle	*	**	extraterrestrial	high
Alpha	particle	4	2+	nucleus	high
Beta	particle	~0	1-	nucleus	low
Positive beta	particle	~0	1+	nucleus	low
Positron	particle	~0	1+	atomic	low
Electron	particle	~0	1-	atomic	low
Neutron	particle	1	0	nucleus	middle/high
Proton	particle	1	1+	nucleus	middle/high
Nucleon (proton/neutron)	particle	1	1+/0	nucleus	high

* has mass.
** has charge.

Table 22.6 — Shorthand Notation of Common Nuclides

Symbol	*Name*	*Equivalent Symbol*	*Other Equivalents*
p	proton	$^{1}_{1}H$	hydrogen ion
α	alpha	$^{2}_{4}H$	doubly charged helium ion
d	deuteron	$^{2}_{1}H$	heavy hydrogen
t	triton	$^{3}_{1}H$	radiohydrogen

Notes: 1 amu = 1 proton = 931 MeV; 1 beta particle = 1 electron = 0.511 MeV. Water has the chemical formula H_2O. If one of the hydrogens is the deuteron, then it is called deuterium and abbreviated as D_20. If one of the hydrogens in the water molecule is triton, then the water is called tritium, which happens to be a weak beta emitter and is used as a tracer.

transitions in high atomic mass nuclides are called characteristic X- rays.

Photons produced from free electron interactions, typically keV, are called X-rays or *bremsstrahlungen*. A bremsstrahlung (plural bremsstrahlungen) is an X-ray created when a high energy particle is negatively accelerated over a very short distance and the kinetic energy is transformed into an ionizing photon. Depending on the rate of deceleration, the energy of the photon ranges between the lower limit of ionizing radiation and the maximum kinetic energy of the particle. The particle can be any charged particle such as a beta from radioactive decay, a proton in a cyclotron, or electrons in an X-ray machine.

Unwanted bremsstrahlung radiation is of concern for protection against radioactivity.

Photons whose origins are extraterrestrial are called cosmic electromagnetic rays. There are also cosmic particles. Characteristic X-rays, bremsstrahlungen, X-rays in general, gamma rays, and cosmic electromagnetic rays are collectively called photons, independent of origin and of energy. Direct reference made to the energy (wavelength, frequency) of the photon or groups of photons is termed "quality." Reference to the number of photons is termed "quantity." Graphical displays of quantity vs. quality are two-dimensional spectra.

Physics units are limited to ionizing radiation with specific applications. Radiation quality or quantity does not necessarily imply a predictable radiation dose response. There are dose equivalent units for that conversion. See Table 22.4 for the application of radiation and Table 22.6 for the definition of the units. The scaling factors in Table 22.2 are used to convert a dose unit to a radiobiological response unit. These factors are energy-dependent, and correct usage involves table conversions at a specific energy.

Physical Aspects of Production of Ionizing Radiation

Electromagnetic Radiation (X-rays, Gamma Rays, Bremsstrahlungen, Cosmic Rays)

Any photon produced as a result of a transition from a higher energy level to a lower energy level in a bound atomic structure has an energy characteristic of the energy difference in the shell of that atom. Optical energy transitions produce scintillation spectra with characteristic wavelengths. When the transition energy of the characteristic photons occurs in the ionizing energy region, then the photons are ionizing and are termed characteristic X-rays. Characteristic X-ray production occurs in materials of high Z and competes with another process called the Auger process, which predominates in materials of low Z. The letter "Z" refers to the number of protons and is defined in the next section. The Auger process is initiated in the same manner with the removal of a lower energy electron in an atom; however, the potential energy differential is not produced as a photon but imparted to an atomic electron passing in the vicinity of the transitional electron. At ionizing energy this unsuspecting electron is ejected from its shell in the atom with the ionizing kinetic energy of the transition minus the binding energy of the particular shell. The kinetic energy of the Auger electron is characteristic of the atom and can be statistically predicted.

The nomenclature in Table 22.5 evolved over the recent history of physics. The first column gives the common name of the radiation quality; the second whether the radiation is a particle or a photon; the third gives the atomic mass unit (AMU) one significant figure. The origin column shows where each item is found in the atom: (1) the nucleus, the nuclear component governed by the nuclear force, which includes protons and neutrons among others, or (2) the atomic component, which comprises the electrons and its electron shell structure. The quality factor (QF) of low, medium, or high is listed so that the relative importance of biological interaction can be compared.

There are items above that are physically the same but whose origin and common name are different. X-rays, gamma rays, and bremsstrahlungen are ionizing photons with a specified energy. Electrons and beta particles are particles with a rest mass of 0.511 MeV with a negative charge; positive betas and positrons are positive electrons with a rest mass of 0.511 MeV bearing a positive charge. The kinetic energy must be specified to compare them. The term nucleon refers to either a proton or a neutron.

Radioactivity

Ionizing radiation is defined as the spontaneous emission of ionizing particles and/or ionizing photons (or gamma) from a nucleus. The nucleus consists only of protons, neutrons, and a pion, the exchange particle, which works in relation to the nuclear force. Charge is considered to be integral to the proton as the fundamental stable building block. The nucleus of an atom contains a specific combination of nucleons (proton or neutron) attracted by the nuclear force operating through the pion. The attractive nuclear force is the strongest force in nature, overcoming the Coulomb repulsion of the positive protons, and is charge independent, attracting equally either the proton or neutron. This force extends over a finite short range ($\sim 10^{-15}$ m = 1 fm). The nucleons obey the Pauli exclusion principle, allowing two protons of opposite spin and two neutrons of opposite spin to fill one energy level.

To define a specific nucleus, the number of protons, the number of neutrons, and the energy levels of the nucleons must be specified. Stable nuclei have protons and neutrons in the ground state. The organization of nuclides by number of protons and number of neutrons is represented on the chart of the nuclides and contains currently known nuclei, stable and radioactive. The chart of nuclides can be found in the book, *Atoms Radiation and Radiation Protection*, 2nd Ed. (see Additional Sources).

The basis for the AMU on the chart of the nuclides is the 12-carbon nucleus. To account for binding energy, 1 AMU is defined as one-twelfth of the mass of a 12-carbon nucleus, so that to one significant number the proton and the neutron each has a mass equal to one. The neutron/proton (n/p) ratio of 12-C is 6/6 = 1; the nucleus is even/even, binding energy per nucleon is ~7.5 MeV, and the shell level is closed.

The atomic number, is called the Z of the nuclide. The Z can also be represented by the chemical symbol with the atomic number written in the lower left corner. The range of atomic numbers from Z=1 to Z=109 can be broken roughly into three groups: low Z, where Z ranges from 1 to 20; mid-Z in the range 21–82; and high Z, greater than 82. The mass number of the nuclide is represented by the summation of the number of protons and number of neutrons and written in the upper left corner of the chemical symbol or the Z. A note here on nomenclature: to refer to one specific nucleus is to refer to a nuclide, not an isotope. An isotope is one nuclide of two or more nuclides with the same atomic number but different mass. If the entity is radioactive, then reference is made to the radionuclide. Each nuclide has its own mass number (A) represented by integers. Atomic weight refers to the number with four decimal places on the periodic table that is the weighted average of naturally occurring masses (isotopes) of a particular element.

The nucleons, similar to the electrons in the atomic structure, are bound with a typical nuclear binding energy per nucleon of about 8 MeV. The larger the binding energy per nucleon, the more stable the nucleus. Stable nuclei tend to be symmetrical in number and parity. For stability the ratio of neutrons to protons for low Z nuclei is one and the number of protons and neutrons is an even number. As the Z number increases, the size of the nucleus increases, and the trend for stability is to have more neutrons than protons, the Coulomb repulsion becomes significant at large numbers (n/p ~1.5 for stability). There is a point at which the nucleus is too large for stability (Z > 83 and A > 150), and those nuclides are radioactive.

Each of the radioactive decay processes below results in a lower energy nuclide, which may or may not be stable. The measure of radioactivity, the rate of decay, is the Becquerel (1 disintegration per second) independent of the energy of the decay product. The time to decay to one-half of the initial activity is the physical half-life, typical of the original nucleus. Short half-lives are typical of highly unstable nuclei; long half-lives, of quasi-stable nuclei. The energy of the decay product is characteristic of the specific nuclide and is found in published documents. Decay may or may not end in a stable nuclide. High Z nuclides may decay through a series of decay radionuclides to stable nuclides.

If the physical half-life of the initial nuclide (parent) is longer than that of a radioactive decay nuclide (progeny), then an equilibrium condition can exist. If the half-life of the initial decay nuclide is very much greater than the decay nuclide (e.g., 226-Ra of 1622 years to 222-Rn of 3 days), a condition called secular equilibrium exists. For two nuclides with approximately equal half-lives, transient equilibrium exists (e.g., 212-Pb of 10.6 hours to 212-Bi of 1 hour). These equilibria occur under closed conditions. If the half-life of the progeny exceeds that of the parent, no equilibrium is possible.

Gamma Decay

Gamma ray emission occurs from natural background and man-made sources. The gamma ray is a photon (no mass and no charge, and spin = 1) emitted from a nucleus when a nucleon makes a transition from a high energy level to a lower one. After emission of a gamma (γ), the energy of the nucleus is less.

Pure gamma emission occurs when only a gamma is emitted; atomic number and mass remain the same.

$$^{91}_{43}\mathrm{Nb} \rightarrow {}^{91}_{43}\mathrm{Nb} + \gamma \quad (1)$$

Isomeric transition occurs if the nucleus stays in an excited state longer than 10^{-6} sec and less than 24 hours. The two nuclides are isomers of each other, each being a different nuclide. The mass number of the initial radionuclide is accompanied by an "m" for metastable and always must be expressed. A metastable nuclide at the higher energy will always decay by isomeric transition, emitting a gamma. The product may or may not be stable.

$$^{99m}_{43}\mathrm{Tc} \rightarrow {}^{99}_{43}\mathrm{Tc} + \gamma \quad (2)$$

Internal conversion occurs when a nucleon has a higher than normal energy state. The nucleon makes a transition to a lower energy state by giving off the energy to an electron (rather than emitting a gamma). Most probably the K-shell electron receives this energy and is ejected from the atom with the energy of transition from the nucleus minus the K-shell electron binding energy. The electron is monoenergetic and is termed the conversion electron. An excited nucleus can branch between gamma decay and internal conversion. This nuclear process is analogous to the branching between characteristic X-rays and the Auger process in the photoelectric effect. The electron behaves identically to a beta particle and warrants similar health physics controls.

$$^{131}_{54}\mathrm{Xe}^* \rightarrow {}^{131}_{54}\mathrm{Xe} + e \quad (3)$$

Beta/gamma emission occurs almost simultaneously when a beta is emitted from a nucleus. See the section on Isobaric Transitions.

Excited state emission occurs when a nucleon is momentarily in a higher energy level and decays in less than 10^{-6} sec to the ground state. The excited state is indicated by an asterisk to the upper right of the Z.

$$^{14}_{6}\mathrm{C}^* \rightarrow {}^{14}_{6}\mathrm{C} + \gamma \quad (4)$$

The difference between the excited state, isomeric state, and gamma emission state is the length of the half-life. Normal use of radionuclides relies on the half-lives being long enough to be measurable after the creation of the radionuclide. Knowledge of excited state radiation is important for developing shielding requirements near cyclotrons and accelerators but such radiation occurs so rapidly that it is not important in the use of sealed sources. The isomeric transition of 99m-Tc is the workhorse of nuclear medicine departments for medical imaging of radionuclides. The half-life of 6 hours is long enough to perform a medical procedure and short enough to decay rapidly to reduce body burden.

Isobaric Transitions

A beta (β or β+) particle is a high kinetic energy electron of nuclear origin possessing a charge of plus or minus 1, the mass equivalent of an electron (0.511 MeV), and a spin of one half. Emission of a beta particle makes the transition isobaric because the mass of the radioactive product is essentially the same as the mass of the radioactive nuclide that emitted the beta. A beta particle emitted in an isobaric transition is not monoenergetic but ranges between zero and a maximum value. Its kinetic energy plus that of its companion, the neutrino, maintains the conservation of energy. The neutrino is a massless particle that travels at the speed of light, is almost undetectable, and is not of interest for purposes of radiation protection but is presented here to balance the energy equations. Beta decay changes the n/p ratio toward stability and can result in a stable nucleus, which occurs most frequently in beta decay of low Z. The beta/gamma decay predominates over other decay modes with nuclides in the middle Z range.

(Negative) beta decay is a process initiated by an imbalance in the nucleus, which has too many neutrons in comparison with the number of protons. The n/p ratio is too large (>1 for low Z) for stability. Effectively, a neutron is converted into a proton, reducing the ratio toward stability. This process of beta decay occurs in naturally occurring radionuclides and fission products.

The following decay process occurs whenever a neutron is converted to a proton and also when the neutron is free from the nuclear force.

$$^{1}_{0}n \rightarrow {}^{1}_{1}p + {}^{0}_{-1}\beta + \nu \qquad (5)$$

An example of pure beta emission is as follows.

$$^{32}_{15}P \rightarrow {}^{32}_{16}S + {}^{0}_{-1}\beta + \nu \qquad (6)$$

A neutron in the nucleus is converted into a proton, a beta particle, and an antineutrino. The beta particle and the antineutrino are emitted, leaving the nucleus with the same mass (isobaric transition).

If the decay creates a nuclide in an excited state, the subsequent very rapid decay of the progeny nucleus can emit one gamma or multiple gammas. Since the transition occurs in a immeasurable time, the transition is written as one equation.

$$^{60}_{27}Co \rightarrow {}^{60}_{28}Ni + {}^{0}_{-1}\beta + \nu + \gamma_1 + \gamma_2 \qquad (7)$$

The beta/gamma transitions predominate over pure beta in the mid-Z range.

Positive beta decay is a process initiated by an imbalance in the nucleus caused by too many protons in comparison with the number of neutrons. The n/p ratio is too small for stability and is increased toward stability when a proton is converted into a neutron. Positive beta radionuclide emitters do not normally occur in nature and must be produced on cyclotrons or accelerators. Positive beta radionuclides are used in medicine for diagnostic applications.

The net process that occurs in the nucleus is shown in the following equation.

$$^{1}_{1}p + \rightarrow {}^{1}_{0}n + {}^{0}_{+1}\beta^{+} + \nu \qquad (8)$$

An example of a pure positron emitter would be

$$^{8}_{5}B \rightarrow {}^{8}_{4}Be + {}^{0}_{+1}\beta^{+} + \nu$$

Similar to negative beta/gamma decay, in positive beta/gamma decay a positive beta is emitted from the nucleus followed by the gamma emission.

$$^{68}_{31}Ga \rightarrow {}^{68}_{30}Zn + {}^{0}_{+1}\beta^{+} + \nu + \gamma_1 + \nu \qquad (9)$$

With the emission of a positive beta particle there is always a secondary radiation reaction called annihilation radiation. The positive beta encounters an electron in the medium. The two annihilate one another by converting the rest mass of the two particles into two photons of 0.511 MeV each, traveling in opposite directions (180° or π radian).

$$^{0}_{+1}\beta^{+} + {}^{0}_{-1}e^{-} \rightarrow \gamma + \gamma \qquad (10)$$

This is the same annihilation radiation as in the attenuation process of pair production.

Electron capture competes with positive beta decay. The effect of the decay process is the same; the proton is converted into a neutron, rendering the transition isobaric with a net decrease in the Z number by one. The excited nucleus captures an electron from the atomic structure, usually a K-shell electron. The electron at nuclear force range combines with the proton to form a neutron. There is, of course, a "hole" in the lower levels of the atomic structure whereby secondary radiation can occur.

$$^{1}_{1}p + {}^{0}_{-1}e \rightarrow {}^{1}_{0}n + \nu \qquad (11a)$$

$$^{67}_{31}Ga + {}^{0}_{-1}e \rightarrow {}^{67}_{30}Ni + \nu \qquad (11b)$$

Alpha Decay

The alpha (α) particle consists of one shell level of a nucleus, two protons, and two neutrons at nuclear distances. For a high Z radionuclide less energy is required to dump an entire shell of four nucleons than to emit a single particle. Alpha decay results in a nuclide with an atomic number less by two AMU and a mass less by four AMU than the original nucleus. The n/p ratio after emission usually does not result in forming a stable nucleus. The alpha is monoenergetic, with energy in the MeV range, and occurs only in heavy nuclei, $Z > 82$ and $A > 150$. After emission the alpha has a limited range in matter because of its large charge and large mass. (In equations, the mass and Z of the alpha are not normally written.)

$$^{238}_{92}U \rightarrow {}^{234}_{90}Th + {}^{4}_{2}\alpha \qquad (12)$$

Alpha emitters are never used in the medical area. There are some industrial product uses in moisture gauges, static electrometers and smoke detectors. It is now problematic to dispose of the smoke detectors. It is suggested that any product that contains an alpha emitter be stored until the disposal procedures are checked. Even though initially unlicensed, some of these items must now have disposal records traceable to an approved site. Table 22.7 provides a profile of the characteristics of ionizing rays.

Dosimetry

Attenuation Processes

For both particles and photons the incident ray can interact elastically or inelastically. The word "scattering" is misused. In the strictest sense it applies to interactions of either photon or particle origin in which an incident ray is changed in direction with or without an energy loss. In inelastic scattering, energy is imparted from the ray to the medium. Scattering does not apply to secondary processes such as characteristic X-ray production or annihilation radiation. Photons are attenuated exponentially; particles are attenuated to a finite depth.

Table 22.7 — Physical Dosimetry

Photon	attenuation is exponential monoenergetic gamma from radionuclides bremsstrahlung or X-rays from machines can compare energy of different photons by HVL radionuclides emitting gamma have specific gamma factor
Alpha particle	always monoenergetic MeV energy range displays Bragg curve of specific ionization limited range in tissue
Beta (positive & negative)	not monoenergetic energy of disintegration shared by positive beta and neutrino or negative beta and antineutrino average energy approximated by 1/3 maximum energy does not display Bragg curve finite range in matter (for electrons in tissue range in cm = 1/2 E in MeV) energy deposited in a medium by beta is due to inelastic collisions that are directly proportional to the Z of the medium and to the energy of the particle (to shield from beta use low Z medium to lower danger from bremsstrahlung-produced during collisions)
Neutrino	not important for dosimetry

Note: HVL = half value layer. (For more information, see http:\\www.nrc.gov)

Attenuation of Photons

The processes described below are organized into two major groups—interactions without energy deposition and interactions with energy deposition. The conservation laws of physics governing all interactions are not explicitly stated. Processes in which a reduction in intensity of a beam, particles, and/or photons is known but energy transfer is unknown is termed attenuation. For those processes relating to energy transfer from the beam to the medium, then the word absorption can be applied. For the majority of radiation dose processes applicable to humans (the photoelectric and the Compton effect from medical sources and background) the photon interacts with an electron within the body. It is sufficient to state that the photon must interact with an electron attached to an atom, ion, or radical and not with a free electron. A free electron is one not bound in an atomic structure and that can have energy or momentum equal to zero.

Elastic, Classical, or Unmodified Scatter for Photons

The energy of the photon after scattering is the same as before the scattering. Therefore, the frequency and the wavelength remain the same. The only difference is that the direction of travel probably changes. At the atomic level, the photon is absorbed and re-emitted at the same energy. There is a chance that it can be emitted in the same general direction, called forward scatter. Elastic scattering is less probable in the ionizing range than in the nonionizing region of the electromagnetic spectrum, and its probability diminishes with an increase in photon energy. See the discussion of the Compton effect later in this section for inelastic or modified scattering.

Absorption Process for Photons: Photoelectric, Compton, Pair Production, and Photo Disintegration

For an ionizing photon ($E > 10$ keV) the most probable process of interaction in matter depends on its energy. Low energy photons are those whose energies lie between 10 and 125 keV; medium energy photons those with energies from ~100 keV to ~10 MeV; and high energy photons are those with threshold energies beginning at 1.02 MeV, but predominate at energies greater than 10 MeV. The electrons released or created in these interactions deposit energy, which is the ionizing radiation dose. The specific names for these interactions apply to photon interactions only and not to nuclear interactions.

For photons whose energies begin at the ionization energy to about 125 keV, the photoelectric effect is the predominant absorption process. The photoelectric effect, also called a total absorption process, occurs when an inner shell electron of an atom absorbs the energy of an incident photon. The energy of the photon must exceed the binding energy of the electron. The electron is ejected from its shell with the kinetic energy of the initial photon minus the binding energy of the particular shell. The electron deposits its energy within a local region of where it was released. This energy is the dose deposited from the primary interaction and is specifically called the photoelectron. After the initial removal of an inner shell electron to become the photoelectron, an outer shell electron quantum mechanically moves to fill in the space of the ejected photoelectron. This transitional electron must release excess energy in the process of moving to a lower energy level. The release of its energy is considered a secondary process. The original atom is now an ion and will eventually collect an electron to balance charge after a secondary process has occurred.

There can be two secondary processes from the primary photoelectric effect. These two processes statistically compete with each other depending on the atomic number of the target atom. The fluorescent yield is the fraction of secondary photons (characteristic X-ray production) compared with the secondary electrons (Auger process) created in the secondary process subsequent to the photoelectric effect.

For atoms of low atomic number the Auger electron emission predominates. The Auger electron is one of the atomic electrons selected to be given the energy from the transitional electron. Statistically, the Auger electron originates from the same atomic shell as the photoelectron. Given the shell level, this energy is predictable and is considered monoenergetic. Most detectors measure photons and not electrons and would not measure the energy of this electron, but this secondary electron contributes to human dose.

For atoms with high atomic number the excess energy from the transitional electron appears as a photon (or photons) equal in energy to the shell(s) of transition. In high Z material such as tungsten, copper, materials used in shielding, and target material, these photons have energies in the ionizing region and are termed characteristic X-rays. They are characteristic energy of the atomic shell, and, therefore, of the element (Z) creating them and are X-rays from creation by electrons. These X-rays are attenuated as any other ionizing photon. Any condition removing an inner shell electron from an atom will induce a similar secondary effect of characteristic X-ray(s) or the Auger electron process. Auger electrons are produced in some radioactive decay processes and become important for hand and eye dosimetry in the use of radioactive materials.

The photoelectric effect is the principle attenuating process underlying use of diagnostic X-rays and dental films. The Z-cubed dependence allows a differential absorption of the incident photons. With the proper detection on film or tape the fine differences between normal and abnormal medical conditions can be detected. With low energy, Z-dependent, incident photons, personnel protection includes the use of lead gloves, lead thyroid blockers, lead aprons, and lead glass to absorb the incident radiation photons and ejected photoelectrons.

The Compton effect, also called inelastic scattering, predominates in attenuating photons in the energy region between 100 κeV and 10 MeV. The incident photon interacts with any outer shell electron. Outer shell electrons are loosely bound at typically low electronvolt (eV) binding energies. The incident photon is absorbed and re-emitted from the interaction site as a scattered photon with less energy. The energy imparted to the electron releases it, the Compton electron, at an angle obeying conservation laws. The scattered photon is reduced in energy but still carries a large fraction of the initial energy. The atom is now an ion and will eventually collect an electron. There is no secondary process for the Compton effect.

The Compton process is considered to be Z-independent for equal numbers of electrons; interaction occurs with outer-shelled electrons. The medical field of radiation therapy is predicated on use of the Compton effect in which a tumor (Z ~ 7) can be irradiated through nonhomogeneous material and through relatively high Z biological material such as bone (Z ~ 13). The

Compton effect is electron density dependent. Water has the highest electron density. Oxygenated tumors with high electron density respond to radiation therapy. Radiation protection from photons that interact in this region necessitates thick shielding such as concrete walls rather than personnel protection such as lead aprons.

For photons with a minimum energy of 1.02 MeV the attenuation process of pair production can occur but predominates at high energies above 10 MeV. The threshold requirement of 1.02 MeV is predicated on the mass equivalent of 0.511 MeV for the mass equivalent of one electron. A high energy photon passes in the local region near the nucleus of an atom. The threshold energy of 1.02 MeV of the photon is converted into the mass of two electrons (1.02 MeV total) with the remaining energy transferred to the electrons as kinetic energy. To obey conservation of momentum and charge, two electrons must be created with opposite charges. The (negative) electron continues in the medium as a high energy electron until its kinetic energy is dissipated. The positive electron, the positron, is an antiparticle, which cannot exist very long in matter. The positron loses its kinetic energy until its speed is zero relative to a newly found electron and then necessarily undergoes the secondary process of annihilation radiation.

Somewhere in the vicinity of the positron is an electron. Opposite charges attract. On collision, the mass of the positron and the mass of the electron annihilate each other by changing the energy from the rest masses into photons. These two photons have 0.511 MeV of energy each and travel in opposite (180º or π radian) directions from one another.

Any two particles with the same mass but opposite charge and angular momentum are termed antiparticles of each other. Examples include antiprotons and antineutrons. Annihilation radiation is a process that necessarily follows any particle-antiparticle annihilation process, releasing photons with the mass-equivalent energies. The most common one is the 0.511 MeV positron-electron interaction. Annihilation radiation is used in some nuclear medicine procedures. Rather than use a positron to interact with an electron, a positive beta is created by selecting the appropriate radioactive material to yield a positive beta. Positive betas and positrons differ only in nomenclature relating to their origin. The 0.511 MeV photons are measured with coincidence counters.

Photodisintegration completes the attenuation processes for photons. An incident photon with a threshold energy ~ 8 MeV or greater interacts with the nucleus of an atom emitting a nucleon or a combination of nucleons. One exception is 9-Be with a threshold energy of 1.66 MeV for a gamma-in, neutron-out reaction. With a high threshold and low probability of interaction this process is limited to high energy machines in research.

Attenuation of Particles

There are similarities between the attenuation of free particles and attenuation of photons in matter. Both have an associated wavelength, the deBroglie wavelength, and are characterized by energy. However, the differences are greater than the similarities. Photons are exponentially attenuated; particles deliver dose in a finite, short range. Particles created on cyclotrons or accelerators, of course, will have a long range. Photons carry no charge; most particles are charged. Even the uncharged neutron free from the nucleus will decay into two charged particles.

Charged Particles

A charged particle loses energy in a medium proportional to charge mass, and inversely proportional to its velocity squared. If the particle is created by a radioactive decay process, its maximum energy is predicable.

Monoenergetic charged particles display the Bragg peak when dissipating energy in matter. The most common example is the alpha particle. As the particle slows down and spends more time per path length, the curve increases in ionizing density and terminates at a predicable depth dependent on the density of the attenuator and kinetic energy of the alpha. Due to its large mass and double negative charge, an alpha particle is limited to a very short range. In air the typical 5 MeV alpha has a range of 5 cm and in skin is attenuated by the dead layer. Internally, alpha particles are a serious biological hazard particularly to the lung and other organs and tissues to which the radionuclides are biochemically transported. See the section on radon later in this chapter.

The beta (negative) or positive beta is emitted in radioactive decay with its companion neutrino sharing the total energy and is not monoenergetic. The kinetic energy of the beta varies between zero and a maximum value. The neutrino, the energy of which varies between the maximum value and zero, is not considered interactive with matter and is ignored for radiation protection purposes. Beta particles created from radioactive decay are listed according to an average energy E_{av}, approximately one-third the maximum energy. Beta particles and free ionizing electrons deposit energy in tissue to a depth of 0.5 cm for each MeV. A 10 MeV electron (or beta) will penetrate the skin to about 5 cm.

Noncharged Particles

Neutrons

The topic of neutrons is added for completion. With no net charge a neutron can interact with target nuclei by being captured by the target nucleus to create, in most cases, an unstable nuclide. This process is known as neutron activation and serves as the beginning of artificially produced radionuclides. The activated isotopic nucleus provides a beta/gamma emitting product. If in the process of capture a particle is ejected, the newly formed nuclide is transmuted into a new element. Neutrons are part of the process that provides fission of a uranium nucleus and are an important radiation protection consideration.

Free neutrons occur in nuclear facilities where fission and neutron activation are of interest. There are no pure neutron emitters; 252-californium is a close approximation with some gamma emission. Free neutrons are considered radioactive with a half-life of about 10.4 minutes. Their interaction can be either elastic scattering (fast neutrons with energy > 100 keV) or inelastic, in which a small fraction of the neutron energy is transferred to the target, resulting in gamma emission (medium energy of 100 eV–100 keV; slow 0.025–100 eV). It is the thermal neutron with kinetic energy of the order of room temperature ~ 0.025 eV that undergoes capture.

The objective in neutron radiation protection is to reduce the energy of the high energy neutrons by inelastic scattering and then capture them. Materials to thermalize neutrons are made from low Z materials such as water, plastic, or paraffin. Capture is then accomplished by boron or cadmium. Protection from secondary radiation is provided by specially selected materials impregnated in the protective shielding. Neutron dosimetry is complex, and measurement involves special neutron detectors.

A summary of relative attenuation due to energy and atomic number of the attenuator is listed in Table 22.8. The energies of the attenuation processes are in Table 22.9. Radiation dosimetry is radiation energy — type specific. For specific calculations, see Additional Sources at the end of this chapter (MIRD 89 & 91).

Radiation Detection Devices

Radiation detection devices detect and/or measure the radiation dose from a field of radiation. Selection of the appropriate detector is critical in the evaluation of the radiation dose or dose rate. The sensitive volume is that part of the detector where interaction between the radiation and the detection medium actually occurs. Survey devices or equipment consist of portable, multiscaled radiation detectors to evaluate radiation dose over a relatively short period of time (usually dose per minute). Survey instruments can operate on batteries, be light enough for an adult to carry in one hand, and respond in multiple ways to radiation fields. Many have audio speakers in addition to multistaged scaling factors, and some can measure particles and/or photons.

Monitoring devices or equipment evaluates a radiation field over a relatively long period of time (week or month). Personnel monitors are generally evaluated on a monthly schedule. Personnel who require shorter time evaluation wear special monitors. Room monitoring equipment may be active continuously with threshold warning devices. Table 22.10 provides an overview of most of the radiation detectors in health physics application; Table 22.11 gives more detail on the common survey methodologies.

Table 22.8 — Summary of Attenuation Processes

Process	*Z*	*Energy*	*Threshold*
Elastic scattering	—	1/E	entire EM spectrum
Photoelectric effect	Z^3	$1/E^3$	K binding energy of Z
Compton effect	independent	1/E	—
Pair production	Z or Z^2	E	incident photon of 1.02 MeV
Photodisintegration			nuclear binding energy

Table 22.9 — Process by Which Ionizing Radiation is Absorbed in Soft Tissue

Energy Range	*Attenuation Process*
10–50 keV	photoelectric much more than Compton
60–90 keV	photoelectric and Compton equally
100–200 keV	Compton more than photoelectric
200 keV–2 MeV	Compton alone
2–5 MeV	Compton and some pair production
5–50 MeV	pair production begins to predominate
>50 MeV	pair production most important

Note: Water Z ~ 7.4.

The ideal ionizing chamber displays an increase in ion pair production (quantity) with an increase in output voltage (quality) applied across the chamber. The response of the ideal chamber is not linear but has various regions. As the output voltage increases from zero, a threshold value of voltage is reached to produce an immediate rise in ion pair production at relatively low voltage to a plateau called the ionizing region. As the voltage increases beyond the plateau, the number of ion pairs produced is proportional to the applied voltage over that range. This region is called the proportional region and terminates in the limited proportion region. In this region the response of the chamber is relatively linear in shape and is energy dependent. The response of different energy particles and photons can be separated. At relatively high voltage the response of the ion pairs plateaus. The response in this region, the Geiger region, is fairly flat and terminates with a rapid rise in voltage, which will destroy the tube, the discharge region. See Figure 22.1 for the response curve.

A radiation detector designed to operate in one and only one of these regions is called an ionization chamber, proportional counter, or Geiger counter, respectively. Selection of the instrument depends on application. Ionization chambers and Geiger counters are usually portable and used for surveys. Proportional counters measure mixed radiation fields such as might be found in environmental monitoring.

A radiation survey is performed to measure the radiation field; measure contamination levels, fixed or removable; and/or measure airborne radioactivity. Removable radiation from liquid sources or fragments of radioactive particles is called contamination. If the lower level of the survey equipment lies within the measurability of the small amounts of radiation, then the equipment is capable of locating the contamination. Radioactive solid sources can flake, crack, or chip to release radioactive particles to the environment. Liquid sources can leak, drip, or spread with poor transfer technique to the work environment. Control of unwanted

Table 22.10 — Classification of Ionizing Radiation Detecting Equipment

Classification	*Sensitive Volume*	*Energy Range (Rad)*	*Health Physics Function*
Electrical			
Ionization	gas	all energies	survey meter
Proportional counter	gas	all energies	environmental monitor
GM	gas	0.001–100	survey GM
Solid state material	Ge(Li)/Si(Li)	all	tissue equivalent survey
Chemical			
Film	(AgX)	$0.01–10^4$	diagnostic and personnel
Chemical (Fricke)	ferrous sulfate	$5–10^8$	absolute calibration
Light			
Scintillation	crystal NaI	all photon	Photomultiplier (PM) tube
Scintillation	liquid	N/A[A]	low beta detection (wipe test)
Cherenkov	liquid	blue light	not for personnel
Thermoluminescence			
TLD	LiF	$0.01–10^5$	tissue equivalent monitors
TLD	CaF_2		high energy monitors
Heat			
Calorimeter	thermistor	l–105	absolute dosimeter

Note: An absolute dosimeter is one in which all the primary and secondary ionizing events are measured.
[A]N/A=not applicable

Table 22.11 — Electrical Detection Systems

Detector	*Radiation*	*Energy Range*	*Efficiency*
Ionization	alpha	all energies for counting/spect.	high
	beta	all energy < 3 MeV	moderate
	gamma	all energy	<0.1%
	neutrons	thermal with moderator	moderate
	X-rays	diagnostic range	window thickness dependent
Proportional counters	alpha	all energies	window thickness, high
	beta/electron	all < 3 MeV	moderate
	gamma	all	< 1 %
	neutrons	thermal with BF_3	moderate
	X-rays	N/A[A]	
GM counters	alpha	insensitive to energy	moderate
	beta/electron	insensitive to energy < 3 MeV	moderate
	gamma	all through a secondary process	< 1%
	neutrons	N/A	
	X-rays	common survey equipment	window thickness dependent

Note: The ideal ionization chamber exhibits a threshold response and then displays the following regions: ionization, proportional, GM, discharge. See Figure 22.1. The detectors above are used in the specific part of the response curve.
[A]N/A=not applicable.

radioactive contamination is an additional consideration in the protection of personnel. Radioactive materials require NRC licenses or state registration for use. Each radionuclide is governed by regulation for its proper compliance. Laboratories are required to survey periodically (daily or weekly) with instruments to detect the radionuclide. If the energy is too low (tritium; 14-C) then wipe testing is required. The wipe test involves swabbing a 100-cm^2 area with a small swab (~2 cm^2) and then measuring the activity with instruments that can detect the specific radionuclide. If, however, contamination occurs in a high radiation field such as a radiopharmacy (hot lab) where locked storage and current NRC postings are required, then the radiation surveyor wipes a defined area (10 cm × 10 cm) with an absorbent disposable cloth to measure that activity on an instrument outside the radiation field. Wipe tests and records to document the procedure are performed routinely as per license requirements. The swab may be wetted with a chelating agent. Certain radionuclides with low beta energy (14-C and 3-H) require wipe tests with special instrumentation.

The most widely used detector for personnel monitoring and medical diagnosis is X-ray film. The sensitive volume is the silver ion of either silver bromide or silver iodide. The percentage of AgI is about 10% to allow the atomic number of I (Z=53) to attenuate radiation by the photoelectric effect. Care must be taken to handle and store unexposed film as it is highly dependent on processing controls, pressure, humidity, storage on edge and not flat, and artifacts by air pockets. Processing of exposed film must be done using standard conditions.

When properly developed, film displays an optical density increment with an increase of radiation exposure called the Hurter-Driffield curve. The curve displays a linear central portion, called the gradient or film gamma, which gives information on film contrast and film latitudes. Optical density (OD) is defined as the logarithm of base 10 of the ratio of the intensity of incident light divided by the intensity of transmitted light. There is always a minimum OD such that the curve does not go through the origin due to base fog, somewhere around 0.2 density units for film well-stored and within the shelf-life date. Good diagnostic quality lies in a range of OD 0.25–2.5 in increments of 0.02. The density of the film is read by a densitometer, which emits light through a processed film. A sensitometer exposes undeveloped film to known intensity of a known photon energy. The film is then processed to produce a calibration strip of known density.

The advantages of film use are numerous: radiation field alignment; light field coincidence with radiation field; field flatness; symmetry; radiation distribution patterns. It is good for electron dosimetry and fine for photons with a limit on ±3% isodose curve accuracy. Properly processed and stored, it provides a permanent record over the years. A limitation of film is its dependence on the energy of the exposure photons and dose rate of the photons. To evaluate a mixed field of radiation the monitors are equipped with filters of different densities encased in the plastic mold. The radiation doses can be determined.

The alternative personnel monitor is one with the sensitive volume of a thermoluminescent tissue equivalent material, LiF. Thermoluminescent dosimetry (TLD) is a radiation detection and measurement procedure in which a TLD material is exposed to radiation. The electrons accrue radiation-induced energy by being elevated to a higher energy level and remaining there. In TLD material the electron remains in a trap; electron energy information is regained by heating the TLD material in a calibrated oven with a photodetector. As the electron returns to ground state, an optical photon is emitted. The intensity of light is proportional to energy stored. The TLD material is annealed for reuse by resetting the electrons in the ground state. These TLD chips are reusable, small in size, and convenient to use. A disadvantage is that lost information cannot be regained.

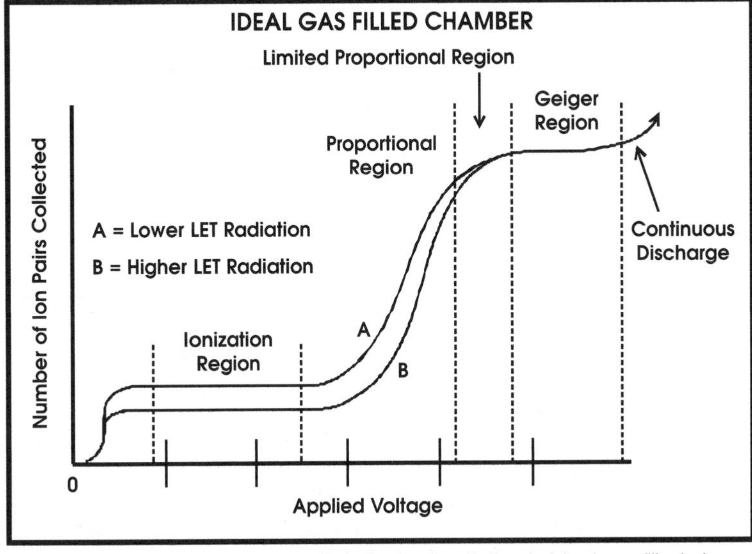

Figure 22.1 — Response of Ion pairs to ionized radiation in ideal gas filled chamber: A = lower LET radiation; B = higher LET radiation.

The choice between a film and a TLD personnel dosimeter depends on the radiation field and physical working conditions. As with any detector, the choice of the detector depends on the use, conditions, and mandates of any radionuclide license. See Tables 22.10 and 22.11.

Sources of Exposure from Ionizing Radiation

A series of reports to advise the U.S. government on health consequences of radiation exposures are authored by the National Research Council's committees on the Biological Effects of Ionizing Radiation (BEIR). The result of these reports (see Additional Sources section) is the analysis of the bulk of ionizing radiation data to produce a comprehensive updated review of ionizing radiation health effects. The areas of these committees are divided as follows: heritable genetic effects; cellular radiobiology and carcinogenic mechanisms; radiation carcinogenesis; radiation effects on the fetus; and radiation epidemiology and risk modeling. The radiation values are then used by EPA to set guidelines and promulgate radiation safety regulations.

Natural Background Radiation

Most radiation is natural in origin and ubiquitous. Terrestrial radiation is emitted by the long-lived radionuclides in rocks and soils (Table 22.12), some of which become incorporated into the human body through the food and water supply. Soils formed from igneous rock contain large amounts of uranium and, therefore, radon. Most soils contain some uranium, making radon ubiquitous. The U.S. Environmental Protection Agency (EPA) estimates that radon contributes two-thirds of the natural dose to the average American. Indoor radon exposure is reducible, and low concentration levels are mandated. This topic is given more consideration later in this chapter.

Cosmic radiation (photons and particles) originates in outer space and contributes to external radiation dose (around 8% of total) especially at high altitudes closer to the propagating sources without the benefit of air attenuation. Internal dose (about 11% of total) results from radionuclides (14-C) created in the upper atmosphere distributed by meteorological conditions. Long-lived 40-potassium contributes to internal dose via food consumption.

Table 22.12 — Natural Radioactivity

Series Name	*Mass*	*Initial n*	*Final n*	*Initial*	*T*[A]	*Final Stable*
Thorium	4n + 0	58	52	$^{232}_{90}Th$	10^{10}	$^{208}_{82}Pb$
Neptunium	4n + 1	60	52	$^{237}_{93}Np$	10^{6}	$^{209}_{83}Bi$
Uranium	4n + 2	59	51	$^{238}_{92}U$	10^{9}	$^{206}_{82}Pb$
Actinium	4n + 3	58	51	$^{235}_{92}U$	10^{9}	$^{207}_{82}Pb$
^{40}K	$^{40}Ca\ ^{+\beta+v}$				10^{9}	
^{14}C	$^{14}N\ ^{+\beta+v}$				10^{4}	

Note: All nuclides with $Z \geq 84$ are naturally radioactive; some with $Z < 84$ are naturally radioactive. In particular, 40-K and 14-C are important. There are three naturally occurring series of radionuclide. The fourth series, the neptunium series, has a half-life so short that any naturally occurring members of the series have decayed. It can be produced in the laboratory.
[A] T = physical half-life in year

Artificial and Technologically Enhanced Radiation Exposure

Human activity raises radiation exposure due to discharges from coal fired (14-C) and nuclear plants (a large spectrum of short-lived radionuclides and low levels of long-lived radionuclides). Radioactive discharges from nuclear power plants are limited to specific discharge levels above environmental background. Radiocarbon and ash from the coal fired plants are not regulated. According to National Council on Radiation Protection and Measurements (NCRP) Reports 92 and 95 (see Additional Sources section), population exposure from operation of a 1000-MW coal fired plant is 100 times that for a nuclear plant (490 person-rem/yr compared to 4.8) and releases 5.2 tons of uranium and 12.8 tons of thorium to the atmosphere each year. Air from underground uranium mines and phosphorous plants contain 222-Rn and are regulated to maximum concentration effluent limits.

Artificial radiation exposure to the average American may or may not be regulated. Medical facilities can discharge low level, nonbiologically contaminated medical radionuclides into the sewer system. Patient excreta with radionuclides is exempt from regulation. Medical exposures are not regulated, and the largest radiation from medical causes is due to diagnostic X-rays, including dental X-rays, that contribute about 11% of the total annual dose to the average American. Any licensed practicing professional (e.g., physician, chiropractor, dentist) using diagnostic X-rays, photons for therapy, or radiopharmaceuticals can prescribe what is considered state of the art doses to deliver the level of care required for the specific instance. Radiation protection occurs by limiting duplicate procedures. Other contributors include fallout, nuclear fuel cycle, occupational dose, and consumer products (gas mantles, Fiesta™ dishes).

Radon

Radon, a noble gas, is classified as a lung carcinogen in humans, with sufficient evidence to support its classification. It is regulated in air and in water by EPA under various acts. Exposure to the naturally occurring radioactive element radon and its progeny is regarded as the largest contributor of radiation exposure from natural sources to humans. EPA projects that 5000–20,000 lung cancers per year are associated with radon exposure in air. The National Academy of Sciences derives an estimate of 13,300 death per year.

Radon in drinking water poses the risk of stomach cancer from ingested water, based on 1 L per day intake. EPA uses a 10^4 to 1 transfer coefficient of radon in water to radon in air (10,000 pCi/L of radon in water contributes 1 pCi/L of radon in air) and estimates that about 5% of radon in the home originates in drinking water. Exposure to radon imbibed in drinking water is about 20% of the risk from inhaling the radon.

The origin of the radon (222-radon) in air and in water derives from the naturally occurring decay chain of 238-uranium. Each of the intermediate decay products in this chain has a characteristic physical half-life before terminating in the stable nuclide of 206-lead. There are two other isotopes of radon (219-radon and 220-radon) produced in natural soils from two independent natural decay schemes. Since the predominant decay series of the four original series is the 238-uranium series with the 222-radon daughter, each is termed, respectively, uranium and radon (the terms used by the regulatory agencies) without the mass numbers. Uranium other than 238-U or radon other than 222-Rn are given the individual mass number and contribute trivial radiation to humans.

The decay series of 222-radon begins with 238-uranium, emits alpha, beta, and gamma radiation at specific nuclides, and terminates at stable 206-lead. Radium in rock, soils, or concrete decays into the noble gas radon with the emission of a 4.78-MeV alpha particle. The recoil momentum can release the radon nucleus from its surroundings and allow it to migrate. Gaseous radon with a half-life of 3.82 days diffuses and evolves from soils and then on decay begins a solid phase series of radon daughters.

As a monatomic noble gas, radon is relatively insoluble, is exhaled rather than absorbed, and will not be deposited on surfaces. It emits alpha radiation of high energy, which, if it decays during residence in the lung, produces damage and radioactive daughter products. Radon becomes a biological hazard when it remains long enough in the lung to decay into its daughter products. It is the progeny of radon, not radon per se, that contribute to lung dose. The daughter products exist as unattached ions, atoms, condensation nuclei, or attached to particles, all of which emit high energy alpha radiation that cause lung damage. The term radon daughters refers to any of the decay products from radon to the stable nucleus of 206-lead. The term radon progeny applies to the first four decay products of 222-radon: 218-Po, 214-Pb, 214-Bi, and 214-Po. These four radon progeny have short enough half-lives to decay in the lung before being physiologically removed and are responsible for the radiation dose delivered to the lung. The target cell in the lung from radon exposure is the bronchial epithelium, which is the site of the majority of lung cancers considered to be caused by radon. The fifth decay step is 210-Pb, which has a long enough half-life (22.3 years) to be removed from the lung by physiological processes before it decays.

The radon concentration in air over typical soil depends on meteorological conditions. When confined to a defined volume, the radon progeny decay by secular equilibrium to 210-Pb in approximately 4 hours. The equilibrium fraction is the fraction of decaying daughters divided by the number of radon atoms. A fraction of 1 is the value when the number of decay products from progeny equals the number of decay products from the 226-Ra. If a fraction of the radon is removed from the defined volume, the fraction of progeny decreases. The values of radon progeny outdoors are at about 70% of the equilibrium value. EPA estimates that the average outdoor concentrations are 8 Bq/m^3 or 200 pCi/L or 0.001 WL (working level) for someone working outdoors continuously.

The WL (J/m^3) represents a derived standard based on human epidemiological studies of working miners to relate the risk from radon exposure to the incidence of lung cancer. The working level month (WLM) is defined in terms of energy deposition over a working month. The dose equivalent unit of the Sievert is not used.

The progeny of radon are solid phase, ionized nuclei, which have a high likelihood of becoming electrostatically attached to aerosols. Because they are ionized, progeny can attach to an aerosol or remain unattached. It has a characteristically high surface area to volume ratio with a geometric diameter range between 0.001 mm–100 mm.

Neutral particles adsorb on aerosols. Then, when inhaled, the aerosols with radon deposit on the interior surfaces of the lung. The aerosol factors include the size distribution of the particle, the fraction unattached, and the equilibrium factor. Outdoors, the unattached fraction is below 10%. Unattached fractions consist of free ions, micrometer-size agglomerates with water molecules, which plate out with about 100% efficiency in the lung. Thus, the dose (per WLM) from one unit of an unattached fraction will be greater than the dose (per WLM) from the same unit of an attached fraction.

The radon progeny preferentially attach to aerosol particles. The dosimetry becomes complex when the radon becomes attached to particles of different solubilities and particle size. This distribution of the fraction of the attached/unattached groups is critical to dose evaluation since the dose is proportional to the unattached fraction. For example, nasal breathing influences more unattached radon deposition, which is cleared from the nasal passages on exhale. Particle size attachment determines in which part of the airway the radon will be deposited. Radiation dose to the bronchial epithelium depends on the amount of radon, the aerosol mixture, and the physiology of the lung.

When confined to a constant volume, radon remains in equilibrium with its progeny. Measurement of radon in air requires knowing or estimating the equilibrium constant. When radon is subject to air convection currents or ventilation, the equilibrium value is reduced from a value of 1 for equilibrium conditions. When trapped indoors, the fraction of radon in equilibrium with its progeny is dependent on the ventilation rate for the structure. The equilibrium fraction used by BEIR IV and the International Commission on Radiological Protection (ICRP) for the risk evaluation from radon exposure is 0.5 for existing housing and is the fraction required for EPA evaluation. The rationale of radon mitigation is that the equilibrium fraction is disrupted and reduced in value to 0.5 by venting the radon to an outside air space, where the decay process continues away from the living space. Measurement of radon in air in a dwelling is predicated on a constant 0.5 fraction and is defined as "closed house conditions."

It may be prudent for the occupational hygienist to evaluate potentially elevated radon concentrations or reevaluate existing mitigation conditions. Radon in water and radon in air in the home and public buildings in excess of a mandated level are required to be mitigated. Radon in the ambient air is usually mitigated by depressuration and radon removal in drinking water by aeration. The mandates govern public water systems, which are defined as having 15 or

Table 22.13 — EPA Action Levels for Indoor Radon in Air

Concentration (pCi/L)	*EPA Protocol*
<0.25	not practical; considered de minimis
1–4	recommend reduction
4 (0.02 WL)	action level as annual average
4–20	action required within year
20–200	action required within months
>200	action required weeks

Note: Outdoor radon standards are limited to uranium and phosphorus mine vents. (For more information, see http:\\www.epa.gov)

Table 22.14 — Instrumentation Approved for Indoor Radon and Decay Measurement

Measurement Device	Closed House	Minimum Time	Outdoor Condition Unsuitability
Continuous radon monitors	yes	48 hours	equilibrium conditions required for diffusion chamber
Alpha track detectors	no	90 days	uncertainty at low concentration
Electret ion chambers	no	—	temperature dependent ±10°F
Activated charcoal adsorption	no	2–7 days	not integrate uniformly; moisture sensitive
Charcoal liquid scintillation cells	no	2–7 days	not integrate uniformly; moisture sensitive
Grab radon sampling	strict yes		filters required; damage/saturate
scintillation cells		4 hours	high variability; best 3 days
activated charcoal		1 hour	moisture sensitive
pump-collapsible bag		4–24 hours	
Pump/collapsible bag devices	yes	24 hours	required radon equilibrium
Unfiltered track detectors	yes	uncertainty at low counts	not sensitive radiation <5 MeV; sensitive to pressure changes, e.g., ±600 feet
Continuous working level monitors		48 hours	radon equilibrium with daughters
Radon progeny integrating			
sampling units	no	48	concentrations per hour or per day not achievable
thermoluminescent dosimeter		48	
alpha track			uncertainty at low concentration
electret		48	temperature dependent ±10°F
Grab sampling–working level	yes	5 min	measurement of flow through a filter short duration; not monitoring technique

more service connections or regularly serving at least 25 persons for 60 or more days per year. Normally, a public water supply does not have elevated levels of radon but may have other regulated radionuclides.

Radon mitigation regulations cover indoor air and drinking water but do not account for the environmental fate of the radon in outdoor air. The U.S. Geological Survey and EPA developed the Map of Radon Zones to pinpoint communities of high radon concentration. There are EPA-approved methods of measuring radon in the environment for compliance with EPA sensitivity and accuracy. These procedures for measurement are lengthy (days or months) and require sensitive instrumentation. Assistance in assessing radon levels is provided by EPA. Measurement of radon is presently time-consuming and requires instruments specific to its measurement. See Table 22.13 for a detailed list and 22.14 for relative measurement methods. (For more information, see http:\\www.nrc.gov)

Radiation Production Equipment

Nuclear Reactors

A nuclear reactor operates on the physical principle that uranium fissions easily with the addition of a thermal neutron. With the absorption of the neutron, the uranium atom disintegrates into two fission nuclei, gamma, and (on the average) two free neutrons. A nucleus is said to be fissile if absorption of a thermal neutron will cause a split into two nuclei. Fissile material contained in a fuel rod is covered with cladding to contain the uranium fission by-products and transfer heat. Control rods such as boron or cadmium made of low Z material with a high cross section for neutrons reduce the kinetic energy of free neutrons for a capture and control of subsequent fission.

Most reactors are intended for electric power generation; there are a few for research. Any reactor can be used as a source of neutrons for research. A pure nonradioactive material is lowered through a specialized port into the reactor pool for exposure to free neutrons. After neutron capture, a radioactive uranium isotope rich in excess neutrons is produced. The disadvantage of neutron activation is that the contained product is a mixture of radioactive and nonradioactive isotopes, yielding a relatively low specific activity product. The time of irradiation can be lengthened to increase specific activity, but the radioactive component decays, prohibiting very short-life radionuclides.

After the uranium fuel rod is spent and no longer of use in the reactor, the fission products can be chemically separated from one another, yielding radioactive products that have high specific activity. The disadvantage of this process over neutron activation is that the product can be contaminated with materials of similar chemical properties. Radionuclides with excess neutrons decay by the beta/gamma process. Radionuclides that are produced from neutron activation or from fission by products are licensed by the Nuclear Regulatory Commission (NRC).

Accelerators

Accelerators are machines that operate at high voltage or in high intensity electric fields to accelerate charged particles. The system is maintained under high vacuum and may utilize special focusing devices on the charged particles to confine them to a small volume or cross sectional area before slamming them into the target. Each type of accelerator uses a specific charged particle for acceleration and production of photons and/or particles.

A Van de Graaff generator collects electrons by friction from a moving belt and stores the electrons on a large metal sphere. The high potential energy electrons are later released to form a uniform highly precise beam of electrons. Without a target the beam of electrons is small but uniform; with a target, a small uniform beam of photon (bremsstrahlung) is produced. The Van de Graaff is one of the original machines to be used in radiation therapy for head and neck treatments. One is located at MIT, Cambridge, MA, with a maximum potential of 12 MeV. Due to the precision of the beam, a Van de Graaff accelerator can be used as a preaccelerator for very large research accelerators.

The cyclotron is an accelerator that accelerates a charged particle (except an electron) in a circle in a vacuum system. Power to the beam is supplied by a radio frequency power tube. The free particle is driven by a changing electric field and maintained between two large magnets called dees. The energy of the particle is highly controllable. A cyclotron can produce radiopharmaceuticals and specific radionuclides used in research such as positive beta emitters.

The betatron is a cyclotron that accelerates electrons in a circle with no radio frequency. The electrons are highly controllable, and the betatron has the capability of delivering variable energy photons or an electron beam. This machine was developed exclusively to be used in radiation therapy for photon and electron treatments. The disadvantage of the betatron is that the beam is small in area. It is used in the energy range of 4–10 MeV with 30 MeV as a maximum. The output is pulsed.

The linear accelerator, or linac, accelerates charged particles along a wave guide using radio frequency power. Powering devices are magnetrons for 12 MeV or less or klystrons for large accelerators. Charged particles accelerate on the radio frequency wave to very high energies. The LAMPF machine at Los Alamos, NM, can attain energies to 800 MeV; SLAC in Stanford, CA, attains energies to 25 GeV. Medical linac energies are typically 4 or 6 MeV. High energy linacs are for research only. Special radiation survey instruments are used near radio frequency fields.

A conventional X-ray machine accelerates electrons in the keV energy range. The X-ray tube is an evacuated tube that accelerates electrons to a maximum voltage, predicting the maximum energy of the X-ray, which is the stated value of the spectrum. When the electrons are rapidly slowed down, X-rays are produced by the conversion of kinetic energy to electromagnetic energy. The majority of X-rays produced are emitted from a target at a back angle. Most of the X-rays produced in X-ray machines produce a spectrum of a continuous distribution of photons from a minimum energy determined by filtration to the peak voltage energy. The energy produced is stated in terms of kVp, the "p" referring to peak voltage set on the controls. The bremsstrahlung curve is called the continuous X-ray distribution curve. The window of an X-ray machine is made with low Z material to allow penetration of the lower energy X-rays. Application of medical X-ray machines can be dental or radiographic X-ray or some therapy. The average energy of an X-ray spectrum is determined by a process called the half value layer. An output radiation rate for each machine is specified with the peak voltage value and the half value layer. All dental, diagnostic radiology, and some superficial radiation therapy machines fall into this group. Research X-ray machines in this range require personnel protection, particularly for the hands and eyes.

Radionuclides produced by accelerators are called Naturally Occurring Radioactive Material (NARM) and are licensed or registered by each state rather than by the NRC. Machines for medical use fall into special categories, and their operation and compliance requirements fall within state jurisdiction. Radiopharmaceuticals for medical and veterinary purposes fall under a jurisdiction of the NRC. Current applications of such machines are listed in Table 22.15.

Biological Consequences of Ionizing Radiation

Chromosome morphology and normal recombination events can be altered by radiation. This alteration can lead to death of the cell, can be carried on as a mutation, or perhaps repaired. Background radiation is of the order of 2-3 mGy per year and is not considered a significant influence for risk from ionizing radiation. That ionizing radiation at significant doses above background radiation can cause measurable cellular damage has been known since shortly after its discovery. The initiation of the field of what is now known as radiobiology began when Pierre Curie asked two French biologists, Bergonie and Tribondeau, around the year 1906 to study the effect of radiation on the development of frog eggs. The theses of their work has been generalized into a law named after them: that radiation sensitivity is directly proportional to mitotic activity and inversely proportional to the degree of differentiation of the cell. This statement can been applied to most mammalian cells (one exception is the lymphocyte).

The critical target for ionizing radiation damage in any cell is the deoxyribonucleic acid (DNA). Damage is caused by an interaction or the initiation of free radicals at

Table 22.15 — Machines to Produce Radioactive Rays

Machine	Particle	Direction	Product	Use
Van de Graaff	electron	linear	photons/electrons	RT; preaccelerator electrons
Cyclotron	any charged	circle	radionuclides	radiopharmaceuticals research (except electrons)
Betatron	electron	circle	photons/electrons	medical
Linac	any charged	linear	radionuclides	research
Medical linac	electron	linear	photons/electron	medical
X-ray	electron	linear	photons	medical
Reactor	neutron	any	fission products	energy source; free neutrons

the point of interaction between the ionizing radiation and the cell target. These lesions, as they are called, are classified as either the result of a direct or an indirect hit. A direct hit is one in which the interaction from the ionizing radiation actually occurred at the DNA site. The particle or photon deposits all or part of its energy at the site and causes the immediate death of the cell. Death of a cell is defined as the loss of reproductive integrity and not necessarily loss of viability. Cell survival is defined as the ability of a cell to produce daughter cells.

The direct radiation effect on the DNA is the disruption of the DNA strand, either a single strand break or a double strand break, causing a major discontinuity in the primary structure. If a direct effect of ionizing radiation on cells causes a double strand break, the event is usually lethal to the cell within a very short time, defined as a period shorter than the cycling time of the cell. Repaired double strand breaks are the most important lesions giving rise to chromosomal aberrations; however, most cells with double strand breaks do not have the capacity to repair, and therefore die. They are selectively removed from the statistical pool of those cells that are the precursors of alterations in the cell. High LET (linear energy transfer; see Table 22.1) and high RBE (relative biological effectiveness; see Table 22.2) ionizing radiation, such as alpha and low-energy beta particles, tend to cause more direct hit deaths; however, the few survivors from this type of radiation induce more chromosomal aberrations than the more plentiful survivors from low LET and low RBE damage. Sublethal damage caused by high LET radiation is not repaired to the same extent as that from exposure to low LET radiation.

Indirect hits result from interaction with a molecule other than DNA, which then releases a free radical or radicals. Damage from the indirect effect requires more time to manifest the effects of cell damage. Since cell content is mostly water, the majority of free radicals are free water radicals. Indirect interaction can cause multiple effects, one of which is chromosome damage. The free radical then interacts with the DNA to cause the lesion. The amino acid bases of irradiated DNA display a relative radio sensitivity: thymine (most sensitivity), cytosine, adenine, and guanine (least sensitive). The lesions may or may not end in the death of the cell but may produce altered cellular metabolism.

The majority of surviving cells from a radiation interaction can be injured with sublethal damage caused by an indirect hit. Low LET and low RBE radiation and high energy beta rays are not efficient in the production of double strand breaks and usually produce single strand breaks. The intermediaries of the free radicals in the cell cause the majority of DNA damage. This damage is not usually lethal to the cell because some of it is repaired. However, repair mechanisms themselves are subject to misrepair, and cells are altered with repeated injury.

A note: in the nomenclature of toxicology, the terms direct and indirect as applied to the effects of chemicals on DNA have different definitions from those used in radiation physics. The direct effect in toxicology means that the DNA has a chemical additive, an adduct, attached to the DNA. That attachment is the putative cause of disruption of normal functioning, which may lead to cancer. The indirect chemical effect, an epigenetic effect, requires chemical modification by a metabolic process to initiate the damage process. The activated form of the molecule is then the putative cause of a change in the DNA and lesions that can cause cancer or terata (birth defects).

Membrane integrity is required for normal cell metabolism. Damage to the cell membrane is considered one of the major factors causing cell death. Membranes can be degraded in one of two ways, either by activation of membrane phospholipases in an ischemic condition or by free radical attack. Free radicals are generated in normal cell metabolism, and all free radical formation is highly lipophilic, causing damage to the cell membrane. In particular, unsaturated lipids are susceptible to peroxidation by ionizing radiation. Radiation injury, like chemical and ultraviolet radiation exposure to membranes, increases cell free radical formation and cellular damage.

Exogenous chemicals can enhance or diminish cellular damage. Certain chemicals in cigarette smoke, for instance, are highly associated with lung cancer. In uranium mine worker epidemiological studies, the nature of the dust is important in the evaluation of lung cancer risk. If the factors influencing membrane composition are altered, as with chemical application or excessive cell regeneration (hyperplasia), the sensitivity to radiation can be modified, increased, or even decreased.

In summary, ionizing radiation particles in comparison with photons have higher RBE and higher LET and tend to kill the cells directly with double strand DNA breaks. There are relatively few survivors to this damage, but those that survive will

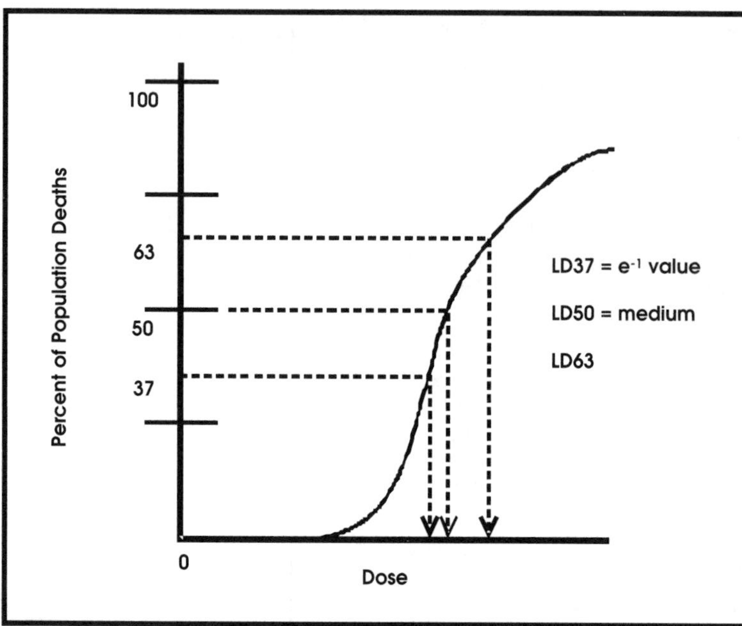

Figure 22.2 — Typical dose response curve.

tend to have greater chromosomal damage even after repair. Low energy photons with a low RBE and low LET tend to kill and injure the cells indirectly by free radical attack and cause single strand DNA breaks. This type of damage occurs more frequently and can possibly be handled by cell repair mechanisms.

Dose Response Curve

The dose response curve is a mathematical relationship between the severity of the radiation effect and the total dose (see Figure 22.2). Dose response curves are used to characterize living systems. The respondents can be individual cells in a cell line or a complete animal such as the standard test animals used by the National Toxicology Program. Even with the standard rabbit or laboratory inbred rat lines, individual variation to a toxin necessitates the establishment of a response curve under controlled conditions to represent the median response. Epidemiology studies are used to estimate the dose response of humans.

The value of the dose at the median response is the LD_{50} (lethal dose to 50% of the population). Human population radiation doses are expressed as LD_{50}/LD_{30}, indicating death to 50% of the population within 30 days. The overall shape of this curve is sigmoid. Deviations from the characteristic shape can provide information about hypothesized mechanisms. The slope of the linear portion from ionizing dose will depend heavily on the dose rate. Investigation of the low dose portion of the curve and its mathematical approach to the x-axis can yield information on the occurrence of a threshold response. The ICRP adopted a univariate logistic regression linear nonthreshold model for radiation protection in 1977 based on the consensus of the scientific community. Now the existence for the possibility of a threshold is being debated.

In radiation studies, radiation dose rate is one of the most important variables that must remain constant throughout the experiment. For low LET radiation the linear quadratic model for the dose response curve governs the shape. A reduction in the dose rate reduces the quadratic coefficient, which changes the slope of the linear portion of the curve. If the slope rises too rapidly, a small error in the dose delivered will be reflected in a large change in cell death. Relatively high dose rates introduce a much higher rate of double strand breaks. Likewise, a shallow slope will be displayed with a large change in dose with a relatively small change in cell death, typically found at low dose rates. Ideally, to find the LD50 with accuracy, the slope should be of the linear portion of the curve around the value of 1 [$(LD_{63}–LD_{37})/(D_{63}–D_{37})$ approximately equals 1]. The LD_{37} is also labeled the 1/e value for radiobiological work; LD_{63} is defined as 1-D_{37}, and both represent the outer points for the linear portion of the dose response curve.

At low LET there is a finite probability that the radiation will not interact within the two strands separating the critical biological target, the DNA. Strands are separated at about 3 nm. Radiation can cause single or double strand breaks, the double strand break being more serious. At a high spatial rate of energy deposition (high LET) the probability of interaction between the radiation and the DNA increases and approaches 1. The DNA is highly likely to receive a double strand break from the high LET radiation. The frequency of double strand breaks is associated with chromosomal aberrations. Somatic chromosome abnormalities are found at a low rate in the general population but are omnipresent in cancer cells.

In 1949 the NCRP formulated the first "standard man," a human model to specify the masses of important critical organs and tissue to be used for radiation dose evaluation. Refinement of the model in 1959 included distribution of the elements in the total body, effective radii of organs,

Table 22.16 — NRC Maximum Permissible Dose Equivalent Limits

Occupational Limit	MPD[A] Annual (mSv)
Whole body	50
Individual organ or tissue except lens of eye	500
Lens of the eye	150
Skin or extremity[B]	500
Embryo/Fetus (declared pregnancy)	5 mSv/9 months
Nonoccupational	*MPD Annual (mSv)*
Continuous	1
Special application	5

[A]MPD = Maximum permissible dose
[B] Extremity means hand, elbow, arm below the elbow, foot, knee, or leg below the knee. Monitoring badges must include the whole body. Higher limits such as a finger dose require separate badging.

and intake and excretion rates. The ICRP model is now called Reference Man and describes an occupational radiation worker who is a 70-kg, 1.7 m tall, 20–30 year old Caucasian male, who lives and therefore breathes in a temperate climate of 10–20°C and has Western European or North American habits. A partial list of current regulations is listed in Table 22.16. A complete, up-to-date list can be found on the web page of the NRC at http:\\www.nrc.gov.

Agencies

Radioactive materials are governed by federal and state regulation regarding the use of radionuclides and how each one is listed for labeling purposes. The radionuclide with its specific chemical symbol, energy level above ground state, and mass must be specified (in addition to chemical form and activity) when designating a radionuclide, for example, $^{99m}T_c^{(+7)}$ 100mBq.

There are agencies that regulate various aspects of ionizing radiation. The NRC regulates special nuclear material (plutonium and uranium) and by-product material, which is chemically or physically separated fission product material. Radioactive material produced by other machines such as cyclotrons or linear accelerators is not regulated by the NRC but by most states. These radionuclides are NARM (naturally occurring radioactive material). Some states regulate these materials and some simply require registration of the source material. An "agreement state" is one that has an agreement with the NRC to enforce both NRC and the respective state rules.

The NRC regulations are listed under Title 10 in the *Code of Federal Regulations* (CFR). The sections of importance to radiation protection are Part 20, "Standards for Protection Against Radiation," and Part 35, "Medical Use of By-Product Material." Transportation is covered in Title 49 of CFR, and EPA regulates the release into the environment. The regulations are now available at the NRC web site at http:\\www.nrc.gov.

Radiation Protection: Time, Distance, Shielding

Radiation protection principles cover the concepts of time, distance, and shielding. For persons who work with radioactive materials the understanding of contamination control is also important.

Time

Radiation exposure increases linearly with an increase in the amount of time spent. The dose received, of course, depends on the quality factors and dose rate dependence. The dose rate for the same boundary conditions determines the dose received. Dose rates from diagnostic level radionuclides are much lower than from X-ray machines and sealed calibration sources. The rule is to reduce the time of exposure to a minimum period. Practice of techniques with dummy sources not only reduces the time spent but also minimizes mistakes that can lead to contamination.

Distance

The standard rule for radiation protection regarding distance from a source is that the dose, or more appropriately dose rate, is inversely proportional to the distance squared from the source. The assumptions of the inverse square law are that the source radiates isotropically from a point source, includes only gamma emission, and that there is no air absorption between the source and the point of calculation with no scatter into or out of the field of interest. That relationship is

$$I_1 = I_2(D_2/D_1)^2 \quad (13)$$

where

I_1 = is the intensity at a distance D_1, and I_2 = is the intensity at a distance I_2.

This formula applies to effective point sources but fails when the distances are too close to the source of radiation, where particle radiation may be included for radioactive sources. Since sources are labeled with activity and date of that activity, the exposure rate for gamma emitters can be calculated at any distance by use of the specific gamma constant defined as follows.

$$\text{Exposure rate (R/hr)} = \Gamma A/d^2 \quad (14)$$

where

Γ = specific gamma ray constant listed in texts in units of R cm^2/mCi hour
A = activity of radionuclide in mCi, and
d = distance from a point source of radioactivity in cm.

Gamma constants can be found in any standard text or requested from the gamma emitter manufacturer.

Diagnostic and therapeutic X-ray machines are calibrated with dose rate exposures at a given distance and field size. Focal sizes are known and the formula applied. The inverse square law is not applied for multiple sources as in radioactive laboratories.

Shielding

Shielding of photons includes the discussions on half value layer and the linear and the mass attenuation coefficients. In general, high Z material is used to shield against photons, e.g., lead and concrete. Shielding for beta emission and neutrons requires low Z material to reduce the initial energy by inelastic scattering and prevent bremsstrahlung production.

Personnel protection for contamination requires use of latex gloves, eye shields, and protective clothing. Radiation monitors for the whole body are required as well as finger badges for those materials that involve hand transfer techniques. Radiation dose from contamination is increased at close distances because low energy beta emission and Auger emission, which normally are absorbed in air, are deposited on the skin or gloves and increase hot particles. The use of shield layers such as eye shields and gloves not only prevent the particles from reaching the body but also shield by absorption.

In addition to the correct shielding of the container, airborne radioactive contamination requires use of hoods with the proper filters and air volume flow.

Summary

Human exposure to ionizing radiation is part of coexistence with the planet. Background radiation includes exposure from long-lived radioactive nuclides in the earth, water, and air. With the advent of modern medicine, exposure from dental and medical X-rays, nuclear medicine studies, and radiation therapy add to the accumulated exposure for both the patient and the occupational worker. Industrial and consumer products add to the body burden. Radiation limitation levels are set for the occupational worker, with more limited measures taken for the general public. The accumulated knowledge of the biological effects of radiation, the collection of which spans more than a century, is reviewed by government agencies and is eventually translated into radiation protection for the worker and the general public.

Additional Sources

Cember, H. *Introduction to Health Physics.* New York: Pergamon Press, 1987.

Fundamentals of Nuclear Medicine, 2nd ed. Maryland Heights, MO: Mathews Medical Book Co., 1991.

International Commission on Radiation Units and Measurements (ICRU). Title? (Rep. 33). Place: Publisher, 1980. Defines radiation output in Systeme International de Unite (SI) units.

International Commission on Radiological Protection (ICRP). *Report of the Task Group on Reference Man* (ICRP pub. 23). Oxford: Pergamon Press, 1975.

International Commission on Radiological Protection (ICRP). *Biological Effects of Inhaled Radionuclides* (ICRP pub. 31). Oxford: Pergamon Press, 1979.

International Commission on Radiological Protection (ICRP). *Limits for Intakes of Radionuclides by Workers* (ICRP pub. 30, pt. 1). Oxford: Pergamon Press, 1979.

International Commission on Radiological Protection (ICRP). *Nonstochastic Effects of Ionizing Radiation* (ICRP pub. 41). Oxford: Pergamon Press, 1984.

International Commission on Radiological Protection (ICRP). *Radiation Protection of Workers in Mines* (ICRP pub. 47). Oxford: Pergamon Press, 1985.

International Commission on Radiological Protection (ICRP). *Lung Cancer Risk From Indoor Exposures to Radon Daughters* (ICRP pub. 50). Oxford: Pergamon Press, 1987.

International Commission on Radiological Protection (ICRP). *Principles for Limiting Exposure of the Public to Natural Sources of Radiation* (ICRP pub. 39). Oxford: Pergamon Press, 1983.

International Commission on Radiological Protection (ICRP). *Report of Committee II on Permissible Dose for Internal Radiation* (ICRP pub. 2). Oxford: Pergamon Press.

MIRD. *Primer for Absorbed Dose Calculations*, rev. ed. Maryland Heights, MO: Mathews Medical Book Co., 1991.

MIRD. *Radionuclide Data and Decay Schemes*. Maryland Heights, MO: Mathews Medical Book Co., 1989.

National Council on Radiation Protection and Measurements (NCRP). *Evaluation of Occupational and Environmental Exposure to Radon and Radon Daughters in the United States* (NCRP rep. 78). Bethesda, MD: NCRP, 1984.

National Council on Radiation Protection and Measurements (NCRP). *Exposure from the Uranium Series with Emphasis on Radon and its Daughters* (NCRP rep. 77). Bethesda, MD: NCRP, 1984.

National Council on Radiation Protection and Measurements (NCRP). *Ionizing Radiation Exposure of the Population of the United States* (NCRP rep. 93). Bethesda, MD: NCRP, 1987.

National Council on Radiation Protection and Measurements (NCRP). *Measurement of Radon and Radon Decay Daughters in Air* (NCRP rep. 97). Bethesda, MD: NCRP, 1988.

National Council on Radiation Protection and Measurements (NCRP). *Radon Exposure of the U.S. Population—Status of the Problem* (NCRP commentary 6). Bethesda, MD: NCRP, 1991.

National Council on Radiation Protection and Measurements (NCRP). *Public Radiation Exposure from Nuclear Power Generation in the United States* (NCRP rep. 92). Bethesda, MD: NCRP, 1987.

National Council on Radiation Protection and Measurements (NCRP). Radon. In *Proceedings of the Twenty-Fourth Annual Meeting*. Bethesda, MD: NCRP, 1988.

National Council on Radiation Protection and Measurements (NCRP). *Control of Radon in Houses* (NCRP rep. 103). Bethesda, MD: NCRP.

"National Primary Drinking Water Regulations: Radionuclides." *Code of Federal Regulations* Title 40, Parts 141, 142. 1991.

National Research Council, Committee on Biological Effects of Ionizing Radiation. *The Effects on Populations of Exposure to Low Levels of Ionizing Radiation*. Washington, DC: National Academy Press, 1980.

National Research Council, Committee on the Biological Effects of Ionizing Radiations. *Health Risks of Radon and Other Internally Deposited Alpha-emitters*. Washington, DC: National Academy Press, 1988.

National Research Council, Committee on the Biological Effects of Ionizing Radiation. *Health Effects of Exposure to Low Levels of Ionizing Radiation*. Washington, DC: National Academy Press, 1990.

National Research Council. *Comparative Dosimetry of Radon in Mines and Homes*. Washington, DC: National Academy Press, 1991.

Radiological Health Handbook, rev. 1970. Washington, DC: U.S. Government Printing Office, 1970.

Till, J.E., and H.R. Meyer. *Radiological Assessment, a Textbook on Environmental Dose Analysis* (NRC FIN B0766). Washington, DC: Office of Nuclear Reactor Regulation, U.S. Nuclear Regulatory Commission, 1983.

United Nations Scientific Committee on the Effect of Atomic Radiation. *Sources, Effects, and Risks of Ionizing Radiation*. New York: United Nations, 1988.

United Nations Scientific Committee on the Effect of Atomic Radiation. *Sources, Effects and Risks of Ionizing*

Radiation. New York: United Nations, 1993.

U.S. Environmental Protection Agency (EPA). *Guidelines on Air Quality Maintenance Planning and Analysis*, vol. 10R (EPA 450/4-77-001). Washington, DC: EPA, 1977.

U.S. Environmental Protection Agency, Office of Radiation Programs (EPA). *Radon Reference Manual* (EPA 520/1-87-20). Washington, DC: EPA, 1987.

U.S. Environmental Protection Agency, Office of Radiation Programs (EPA). *Indoor Radon and Radon Decay Product Measurement Protocols* (EPA 520/1-86-04). Washington, DC: EPA, 1989.

U.S. Environmental Protection Agency, Office of Radiation Programs (EPA). *Risk Assessments Methodology. Environmental Impact Statement. NESHAPS for Radionuclides. Background Information Document*, vol. 1 (EPA 520/1-89-005). Washington, DC: EPA, 1989.

U.S. Environmental Protection Agency, Office of Radiation Programs (EPA). *Risk Assessments Methodology. Environmental Impact Statement. NESHAPS for Radionuclides. Background Information Document*, vol. 2 (EPA 520/1-89-006-1). Washington, DC: EPA, 1989.

U.S. Environmental Protection Agency, Office of Radiation Programs (EPA). *Risk Assessments Methodology. Environmental Impact Statement. NESHAPS for Radionuclides. Background Information Document*, vol. 2 (EPA 520/1-89-006-2). Washington, DC: EPA, 1989. Appendixes.

U.S. Environmental Protection Agency (EPA). *Occurrence and Exposure Assessment for Radon in Public Drinking Water Supplies*, Wade Miller and Associates, Inc. Washington, DC: EPA, 1990.

U.S. Environmental Protection Agency (EPA). *Protocols for Radon and Radon Decay Product Measurements in Homes* (EPA 402-R-92-003). Washington, DC: EPA, 1992.

U.S. Environmental Protection Agency (EPA). *Radon Mitigation Standards* (EPA 402-R-078). Washington, DC: EPA, 1993; rev. April 1994.

U.S. Department of Health and Human Services, Public Health Service (PHS). *Seventh Annual Report on Carcinogens. Summary 1994*. Research Triangle Park, NC: PHS, 1994.

World Wide Web

Radon and related sites. www.epa.gov

Nuclear Regulatory Commission Regulations and References. www.nrc.gov

Bibliography of radiation protection and evaluation textbooks, journals, and societies: NRCP and WHO listings. www.medphysics.wisc.edu/~cameron

Technical resources for nuclear data. Dept. of Advanced Technology at Brookhaven National Laboratory. www.dna.bnl.gov

Rural Kansas oil field at sunset. (Layne Kennedy, 1992)

Outcome Competencies

After completing this chapter, the user should be able to:

1. Define underlined terms used in this chapter.
2. Calculate the pressure at any habitable depth under water or altitude above sea level.
3. Calculate the molar fraction or partial pressure of oxygen, nitrogen, carbon dioxide, etc., in air at any of the above pressures.
4. Anticipate the effects and estimate the incidence of hypoxia and benign acute mountain sickness.
5. Propose some potential control schemes for hypoxia and benign acute mountain sickness.
6. Anticipate the effects of nitrogen narcosis, oxygen and carbon dioxide toxicities and the conditions at which they might occur.
7. Explain the principles behind changing the composition of the air used in NITROX and saturation diving.
8. Calculate the change in trapped gas volume resulting from a change in depth or altitude.
9. Anticipate the magnitude of change associated with barotrauma.
10. Discuss the cause and forms of decompression sicknesses and describe the control approaches used to mitigate decompression sickness.

Key Terms

acclimatization • airtight caisson • barotrauma • benign acute mountain sickness • bottom time • Boyle's law • carbon dioxide toxicity • Dalton's law • decompression schedules • decompression sickness • dysbaric osteonecrosis • dysbarism • hematocrit • hemoglobin • Henry's law • high altitude cerebral edema • high altitude pulmonary edema • high pressure nervous syndrome • hyperbaric • hypobaric • hypoxia • NITROX • oxygen toxicity • partial pressure • pressure • saturation diving • solubility coefficient • time of useful consciousness • Valsalva maneuver

Prerequisite Knowledge

Prior to beginning this chapter, the user should review the following chapters:

Chapter Number	Chapter Topic
4	Environmental and Occupational Toxicology
8	Principles and Instrumentation for Calibrating Air Sampling Equipment

In addition, a college-level knowledge of physics, chemistry, and mathematics is helpful.

Key Topics

I. Physical Principles
   A. Boyle's Law
   B. Dalton's Law
   C. Henry's Law

II. Hypobaric Hazards
   A. Recognition of Hypobaric Hazards
   B. Control of Hypobaric Hazards

III. Hyperbaric Hazards
   A. Recognition and Control of Hyperbaric Hazards

IV. Changing Pressure Effects
   A. Recognition of Changing Pressure Hazards
   B. Control of Pressure Changes

23

Barometric Hazards

William Popendorf, Ph.D., CIH

Introduction

Several books and numerous book chapters are available on health hazards associated with abnormal atmospheric pressure. The focus of most of these references is on the physiological and medical responses to pressure rather than the environmental elements and work practices that hygienists might be trying to control. Although this chapter will discuss health effects, its focus will be primarily on anticipating and recognizing the physical conditions constituting a barometric hazard; secondary focus will be on their control.

From an occupational hygiene perspective, barometric hazards can be categorized as 1) hypobaric (low pressure) hazards, 2) hyperbaric (high pressure) hazards, and 3) hazards from changes in pressure, predominantly, but not exclusively, decreases in pressure.

Hypobaric conditions produce adverse health effects due to a lack of oxygen, specifically the low absolute partial pressure of oxygen (PO_2). In normal air (20.9% oxygen), these effects do not begin to be detectable until at least 2000 m (6000 ft) above sea level (ASL); however, the same range of effects can occur at or near sea level if the fraction of oxygen is reduced, as might be present in a confined space. Health effects include both direct symptoms of hypoxia and a group of indirect symptoms that have come to be known as benign acute mountain sickness (benign AMS) and their life threatening cousins high altitude pulmonary edema (HAPE) and high altitude cerebral edema (HACE).

Hyperbaric conditions can produce narcotic-like effects from high inert gas pressure (especially nitrogen, although high pressure helium also has neurologic effects) and toxic effects from high oxygen or carbon dioxide pressure. Nitrogen and helium are also responsible for decompression sickness following a rapid decrease in total pressure during ascent.

Changes in pressure can cause adverse health effects via at least two mechanisms: 1) pain and traumatic injury from the expansion or contraction of trapped gas as the pressure changes; and 2) the formation of inert gas bubbles within supersaturated tissues that can produce a range of decompression sicknesses (DCS). DCS can arise following a rapid decrease either from a hyperbaric pressure to normal pressure (typical of diving, underwater construction and work in pressurized caissons or tunnels) or from near sea level pressure to a hypobaric pressure (typical of flight crews).

The following sections will discuss the physical principles and physiological mechanisms underlying all of these hazards. Then each of the above three categories of barometric hazards will be discussed individually. The general sequence will cover conditions defining recognized hazards, the nature of the health effects, and viable controls of these hazards. This chapter will not stress either medical diagnosis and treatment or the toxicological mechanisms underlying these hazards; the interested reader is directed toward other texts on high altitudes or on high pressure.

Table 23.1—Common Units of Pressure Equivalent to One Standard Atmosphere

14.696 pounds per square inch (lb/in^2 or psi)
29.920 inches of mercury (in Hg)
406.14 inches of fresh water (in H_2O)
33.08 feet of sea water (fsw)
101325 Newtons per square meter (N/m^2)
101.325 kiloPascals (kPa)
760 millimeters of mercury (mmHg)
1.01325 bars (B)

Physical Principles

Dysbarism is a generic term applicable to any adverse health effect due to a difference between ambient pressure and the total gas pressure in tissues, fluids, or cavities of the body. To understand the physical and physiological effects of pressure, an occupational hygienist should have a thorough understanding of three physical gas laws: Boyle's law, Dalton's law, and Henry's law, which are discussed in subsequent sections. These three laws underlie most of the hazards associated with dysbarisms. All three of these laws relate to ambient pressure that changes with altitude above sea level and with depth below the surface of water. Pressure is the force per unit surface area exerted by the molecules of a fluid in contact with a body. Barometric hazards to man are most easily referenced to differences from the normal living environment of one standard atmosphere, which can be expressed in several common units as listed in Table 23.1. Pressure measured relative to the local atmosphere is sometimes called gauge pressure with units like psig. Hypobaric hazards are always reported in absolute pressure of some form as does the diving literature, but conditions in compressed air construction work are usually stated in gauge pressure.

Boyle's Law

Boyle's law, postulated in 1662, states that the volume of a gas at constant temperature is inversely proportional to its pressure. As a historical footnote, Robert Boyle hired a then young Robert Hooke to make an air pump with which he not only studied the physical behavior of gases but also observed animal responses to pressure. Boyle's law can be formulated as Equation 1 or as the more general Universal (or Ideal) Gas law, Equation 2.

$$P\,V = \text{constant} \qquad (1)$$

$$P\,V = nRT \qquad (2)$$

where:

P = total pressure, atmospheres;
V = gas volume, liters;
n = moles of gas; its mass in grams divided by its molecular weight;
R = universal gas constant, 0.08205 L × atm /K/mole; and
T = absolute temperature in degrees Kelvin, °K = °C + 273.15

Boyle's law applies to the expansion and contraction of gases within the body due to external pressure changes. Expanding trapped gas within the lung, middle ear, sinuses, or stomach (gastrointestinal [GI] tract) can cause pain, and rapidly expanding gas can actually cause traumatic injury called a barotrauma. One use of the Universal Gas law familiar to occupational hygienists is to find the molar volume of any gas at normal temperature (T = 25C = 298.15 K) and pressure (P = 1 atm):

$$\frac{V}{n} = \frac{RT}{P} = \frac{0.08205\ \text{L} \times \text{atm/K/mole} \times 298.15\ \text{K}}{1\ \text{atm}} = 24.45\ \text{L/mole} \qquad (3)$$

Another useful application is finding the density (ρ) of a known gas. For instance, knowing the molecular weight (MW) the density of air is 28.96 g/mole.

$$\rho = \frac{\text{mass}}{V} = \frac{n\,MW}{V} = \frac{(28.96\ \text{g/mole})}{24.45\ \text{L/mole}} = 1.184\ \text{g/L} \qquad (4)$$

The pressure depends on the height and density of the column of air or water above it. Thus, pressure decreases with altitude above sea level and increases with depth below the surface of water. Changes in absolute pressure with depth are easy to anticipate because water is practically incompressible; thus, pressure increases linearly with depth. However, water density does differ between fresh water at 1 kg/L (62.4 lb/ft^3) and sea water 1.025 kg/L (64.6 lb/ft^3). It is important to remember that the pressure at the water's surface is always 1 local atmosphere (at sea level this is the nominal 1 atmosphere). Thus, the pressure in absolute total atmospheres (ATA) at any depth in terms of either feet or meters may be found using Equation 5:

$$P_{ATA} = P_{local} + (\text{depth}/\kappa) \approx 1 + (\text{depth}/\kappa) \qquad (5)$$

where P_{local} is either 1 atm or a lower pressure at altitude given by Equation 6 and κ is chosen from Table 23.2 based on the density of the water and units of depth

Table 23.2—Values of κ Coefficients in the Units of Depth Below the Surface of Water

	Depth in Feet	Depth in Meters
Fresh water	33.8	10.3
Sea water	33.1	10.1

Example 1: Find the total pressure while repairing an oil rig at a depth of 185 feet under the Gulf of Mexico.

Use κ = 33 in Equation 5 to find the total pressure at a depth in sea water given in feet (denoted as "fsw" for feet of sea water). Because the surface is at sea level, $P_{ATA} = 1$ atm + 185/33 = 1 atm + 5.6 = 6.6 ATA.

Changes in pressure with altitude are slightly more complex because air is compressible. Its density varies according to Boyle's law inversely with pressure that itself varies with the height of the atmosphere above it. If the air temperature were constant, this change in pressure with altitude would be an exact exponential relationship of the form in Equation 6.

$$P_{ATA} = P_{\text{at sea level}} \times e^{(\text{-altitude}/\kappa)} \qquad (6)$$

where κ is chosen from Table 23.3 based on the units of altitude.

While air temperature does change with altitude, it turns out that this change is sufficiently uniform that atmospheric pressure can still be approximated by the above exponential formula.(1) Under standard conditions, temperature drops 2°C per 1000 ft (technically 1.9803°C = 3.5645°F) up to 36,000 ft (~ 11,000 m) where the constant temperature stratosphere begins. The coefficients in Table 23.3 were optimized to predict P to within ±1% for most terrestrially accessible altitudes up to 20,000 ft (6100 m). However, these coefficients will overestimate P by >10% above 25,000 ft. Table 23.3 also includes coefficients for powers of 2, which some readers may find more intuitive (similar to a half-life). Thus, the atmospheric pressure at 18,000 feet is approximately one half that at sea level.

Example 2: Find the local barometric pressure at Logan, Utah (altitude 4455 ft or 1358 m ASL) on a normal day.

Table 23.3—Values of κ for Use in Equation 6 (or Power of 2) to Anticipate the Normal Pressure at Altitudes Above Sea Level (ASL)

	≤20,000 ft	≤6100 m
For Equation 6	25,970	7915
For Power of 2	18,000	5485

$P_{\text{at sea level}}$ = normal pressure = 1 atm = 760 mmHg

$P_{\text{at Logan}} = 760 \times e^{(-1358/7915)} = 760 \times 0.842 =$ 640 mmHg

$P_{\text{at Logan}} = 760 \times 2^{(-4455/18{,}000)} = 760 \times 0.842 =$ 640 mmHg

The above example predicts that normal atmospheric pressure at that location measured by a barometer will be 640 mmHg. Note however, that weather bureaus and airports always adjust their readings for their local altitude and would still "report" a pressure of 760 mmHg or 29.92 inches Hg. Changes in the equivalent sea level pressure caused by weather fronts are normally within ±25 mmHg (or ±1 inch Hg). Thus, on a nonstandard day, a hygienist could specify the local pressure either by inserting the pressure reported by a local weather bureau or airport into Equation 6 with errors within ±1%, by assuming the day is standard and inserting 760 or 29.92 with errors of ±3% (25/760 or 1/29.92), or by finding a working barometer.

Dalton's Law

Dalton's law involves a term called "partial pressure." The partial pressure of substance i (P_i) is simply the force per unit surface area exerted by molecules of one specific chemical in contact with a body. John Dalton conducted extensive research in physical chemistry and formulated the modern atomic theory (for which the unit of atomic mass was given his name). Dalton's law (1801), sometimes called the Law of Partial Pressures, states that the total pressure, P, of a mixture of gases is equal to the sum of its independent partial pressures, Equation 7.

$$P_{ATA} = \Sigma P_i = \Sigma[Y_i\, P_{ATA}] = Y_1 P_{ATA} + Y_2 P_{ATA} + \ldots + Y_n P_{ATA} \qquad (7)$$

where Y_i = the molar fraction of gas i in the total mixture = P_i/P_{ATA}

In other words, the partial pressure exerted by each component is proportional to its molecular concentration in the mixture. Thus, partial pressure, P_i, is but one measure of airborne concentration. Equation 8 relates P_i to the more familiar occupational hygiene concentration term of ppm or molecules of a contaminant per million molecules of air.

$$ppm_i = \frac{P_i \times 10^6}{P_{ATA}} = Y_i \times 10^6 \qquad (8)$$

Dalton's law can be used to determine how much oxygen is available in the ambient air, in the lung, or in the alveoli either at altitude when the total P is low or when high concentrations of other gases displace oxygen even at sea level. The molecular composition of air is quite constant with altitude. Table 23.4 lists the United States and internationally agreed standard composition applicable to all humanly habitable altitudes. The U.S. standard atmosphere is identical to those adopted by the ISO (International Organization for Standardization) and ICAO (International Civil Aviation Organization) through 11 km. This table also calculates the MW of standard dry air using Equation 9;[1] humidity can reduce the molecular weight of air by 0.1 to 0.2 grams.

$$MW_{mixture} = \frac{\Sigma(Y_i \times MW_i)}{\Sigma(Y_i)} \qquad (9)$$

Henry's Law

Henry's law, proposed by William Henry in 1803, states that the equilibrium concentration of a gas dissolved into a liquid will equal the product of the partial pressure of the gas times its solubility in the liquid. The gas solubility in a given liquid (shown in Equation 10 as S_i) is usually called "Henry's constant."

$$C_{i\ in\ solution} = S_i \times P_i \qquad (10)$$

where

Ci = concentration of gas i dissolved in solution, cmΔ/mL = cc/mL
Pi = partial pressure of gas i, atm
Si = solubility coefficient of gas i in a given solute, cc/mL/atm

Outside the body, Henry's law has been successfully used to relate known solvent concentrations in water or other solutes to its vapor pressure and, therefore, to its rate of evaporation from or absorption into that media.[2–5] Physiologically, Henry's law can predict the body's absorption of most gases from the alveoli of the lung, its rate of transport via the blood, and the amount of gas that can be stored in tissue where it can eventually have an adverse effect. It is important to realize that because blood is mostly water, the rates of gas absorption, transportation to tissues, and eventual desorption from tissues are all primarily dependent on the gas's water solubility (with the exception of oxygen because of hemoglobin). On the other hand the mass of gases stored in lipid tissues, such as myelinated neurons and collagen at joints, is determined by the gas's lipid solubility. Henry's law predicts that the greater the ratio of a gas's lipid to water solubility, the more slowly these gases can be carried back out of lipid tissues on leaving high pressure, which is especially a problem for poorly perfused tissues like collagen within joints. It also turns out that the anesthetic quality of a gas is highly correlated to its lipid or oil solubility. Values of Henry's constants for some physiologically important gases are listed in Table 23.5, rank ordered by their lipid/water solubility ratios.

Example 3: Assuming that a carbonated beverage is initially bottled in equilibrium with carbon dioxide at 1 atm (i.e., 100% CO_2), how much CO_2 gas is dissolved in a 12-oz (0.355-L) bottle?

Use Equation 10 to find the concentration of gas in the bottled liquid, then the volume of gas trapped in the bottle.

CCO_2 = 1 atm × 0.5797 cc/mL/atm = 580 cc/L

$VCO_2 = V_{liquid} \times CCO_2$ = 355 mL × 0.580 cc/mL = 206 cc

Using YCO_2 in normal atmosphere from Table 23.4 in Equation 8, the ppm of CO_2 in normal air is only about 314 ppm. Using

Table 23.4—Chemical Composition of Standard Dry Air[1]

Chemical Component	*Molecular Weight*	*Y_i (%)*	*$\frac{\Sigma(Y_i \times MW_i)}{\Sigma(Y_i)}$*
Nitrogen (N_2)	28.0134	78.084	21.8740
Oxygen (O_2)	31.9988	20.948	6.7031
Argon (A)	39.948	0.934	0.3731
Carbon dioxide (CO_2)	44.0099	0.0314	0.0138
Neon (Ne)	20.183	0.00182	0.0004
Helium (He)	4.0026	0.00052	0.0000

Sum of molar fractions = 99.9997
Molecular weight via Equation 9 = 28.96440

Table 23.5—Solubility Parameters (Henry's Coefficients) of Some Gases of Physiological Interest

Gas	*S in Water (cc/mL/atm)*	*S in Lipid (cc/mL/atm)*	*S Lipid / S Water*
Cyclopropane	0.204	11.2	55.0
Argon	0.0262	0.1395	5.3
Nitrogen	0.01206	0.0609	5.0
Oxygen	0.0238	0.112	4.7
Nitrous oxide	0.435	1.4	3.2
Helium	0.0087	0.0148	1.7
Carbon dioxide	0.5797	0.88	1.5
Ethyl ether	15.6	15.2	1.0

Equation 10 again in proportions (as shown below), one can see that when that bottle is first opened, there is 3185 times more CO_2 in the beverage than there will be when it comes back into equilibrium with ambient air.

$$\frac{C_{\text{in a fresh beverage}}}{C_{\text{in an old beverage}}} =$$

$$\frac{SCO_2 \times (1{,}000{,}000 \text{ ppm})}{SCO_2 \times (314 \text{ ppm})} = 3185$$

This ratio is sufficient to cause bubbles to form rapidly within the beverage on opening— bubbles that can comprise as much as 58% of its liquid volume. Only if the pressure is released slowly and the evolved gas dissipates, is it possible for such a beverage to lose its fizz without forming bubbles.

Hypobaric Hazards

Occupational examples of hypobaric conditions include high-altitude construction, mining, and aviation, especially aircrews or passengers under rapid loss of pressurization conditions. The number of hygienists actively involved in these settings is probably less than warranted by the range of hazards and number of people exposed. Effects of hypobaric health hazards include the following:

- Hypoxia due to insufficient oxygen. Symptoms range from barely detectable to completely disabling depending on the severity of the cellular oxygen depletion. Normal increases in respiratory ventilation cannot prevent some decrease in a person's ability to perform extended strenuous work. In cases of rapid decompression or removal of supplemental oxygen, one's ability to perform lifesaving responses can be limited to a potentially very short time of useful consciousness.
- Benign acute mountain sickness (benign AMS) is a constellation of symptoms (highlighted by frontal headaches) that can range from discomforting to incapacitating and is precipitated by a rapid ascent but will generally resolve spontaneously within 3 to 5 days.
- Acute mountain sickness without the above benign qualifier refers to high altitude pulmonary edema (HAPE) and/or high altitude cerebral edema (HACE). Their symptoms are more objective than benign AMS and can rapidly become life threatening if not treated by a rapid descent to a lower altitude.
- Chronic mountain sickness or Monge's disease is a rare response to prolonged stays at elevation. Its symptoms include those of benign AMS (perhaps also of HACE but not HAPE) but occur only after several years of exposure. This condition is hypothesized to be the cascading result of very high increases in hematocrit. Because the time of response of chronic mountain sickness is so delayed relative to industrial personnel transfers, it is considered herein to be outside the occupational hygienist's realm. The interested reader is referred to Heath and Williams[(6)] or Ward et al.[(7)]
- Decompression sickness (DCS) at high altitude is identical in form to that following underwater diving but usually less severe. Because it can occur in both hypobaric and normal pressure environments, the discussion of DCS is deferred to the Changing Pressure Effects section.

Recognition of Hypobaric Hazards

As discussed above, ambient total pressure changes with altitude can be predicted using Equation 6. If total pressure decreases but the mixture of gases stays the same, the partial pressures of oxygen and nitrogen will decrease in parallel with the total pressure. Thus, a quantitative prediction of these ambient partial pressures at any altitude can be made by applying the molar composition YN_2 and YO_2 from Table 23.4 and the change in total pressure from Equation 6:

$$P_{\text{ambient N2}} = 0.78084 \times 760 \times 2^{(\text{-feet}/18{,}000)} \quad (11a)$$

$$P_{\text{ambient O2}} = 0.20948 \times 760 \times 2^{(\text{-feet}/18{,}000)} \quad (11b)$$

In other words, as long as there are no local sources of emission, absorption, or consumption of either gas, the molar ratio of nitrogen to oxygen in ambient air will always be 0.78084 to 0.20948, or about 3.73 to 1. Now using Dalton's law (Equation 7):

$$0.99032\ P_{ATA} = PN_2 + PO_2 = 3.76\ PO_2 + PO_2 = 4.76\ PO_2 \quad (11c)$$

Therefore in ambient air, $PO_2 \approx P_{ATA} / 4.76$. Physiologically, the situation becomes more complicated. The composition of gases changes as air enters the respiratory tract, as summarized in Table 23.6.[8,9] One of the first changes to occur is the complete humidification of the air before reaching the alveoli. The lung's concentration of water vapor is nominally always 47 mmHg, equal to the vapor pressure of water at the body's core temperature of 37.2°C (or 99°F). This constant PH_2O of 47 mmHg is a small molar fraction when the total P is 760 mmHg at sea level, but it becomes an increasing fraction as the total pressure drops with altitude. The body can easily exhale 1–2 L/day, continually humidifying typically dry mountain air and contributing an additional risk of dehydration at altitude.[7]

The next change is the simultaneous absorption of oxygen and release of carbon dioxide within the alveoli. Alveolar PO_2 is decreased below what would otherwise be expected because some oxygen is absorbed into the blood for distribution to the body, listed in Table 23.6 as ΔPO_2. ΔPO_2 is ~38 mmHg at sea level and decreases with altitude in a nonlinear fashion in response to decreasing oxygen initially within the alveoli and increasing respiratory minute volume, which varies with the degree of acclimatization. Increased respiration decreases the amount of oxygen absorbed per breath (ΔPO_2), thereby increasing the average alveolar oxygen and the oxygen saturation within the blood. Meanwhile, the PCO_2 released from blood into the alveoli is about 40 mmHg at sea level and decreases at higher altitudes to a plateau of about 24 mmHg at 24,000 feet (7300 m). Normal ambient PCO_2 is so much smaller than physiological levels at any altitude that it may be disregarded. (Because ambient YCO_2 is only 310–320 ppm [see Table 23.4], ambient PCO_2 found using Equation 7 is only 0.2 mmHg at sea level and decreases with altitude, similar to Equation 11.)

Dalton's law can be used again to approximate the physiological dynamics of respiration at increased altitudes shown in Table 23.6. The effect of a potential inert gas is inserted here for completeness because the same hypoxic effects caused by a low total pressure of air at altitude can also occur at sea level if an inert gas displaces air. Inert gas concentrations are normally negligible except in confined spaces (discussed in Chapter 32). Applying Equations 7 and 11c:

$$P_{ambient\ total} = \Sigma P_i = PN_2 + PO_2 + (P_{inert\ gas}) \quad (12a)$$

$$P_{ambient\ total} = 4.76PO_2 + (P_{inert\ gas}) \quad (12b)$$

Accounting for the presence of water vapor in the lung and for the liberation of physiological PCO_2 yields a new distribution of gases, and in particular, a reduced concentration of oxygen reaching the lung:

$$P_{ATA} = 4.76PO_2 \text{ in lung} + PH_2O + PCO_2 + (P_{inert\ gas}) \quad (13a)$$

$$PO_2 \text{ in lung} = [P_{ATA} - PH_2O - PCO_2 - (P_{inert\ gas})]/4.76 \quad (13b)$$

From this a certain amount of oxygen will be absorbed within the alveoli (PO_2) to yield Equation 14:

$$PO_2 \text{ in alveoli} = [P_{ATA} - PH_2O - PCO_2 - (P_{inert\ gas})]/4.76 - \Delta PO_2 \quad (14)$$

Example 4: Find the alveolar oxygen partial pressure in an unacclimatized person at 30,000 feet, the approximate height of the Mount Everest summit.

Equations 6 and 12 through 14 can be used in sequence:

Table 23.6—Physiological Partial Pressures (mmHg) when Breathing Normal Air

Altitude	*Ambient Air*		*Physiological Pressures[A,B]*			*Alveolar[B]*
(ft ASL)	*total P*	*PO_2*	*PH_2O*	*PCO_2*	*ΔPO_2*	*PO_2*
0	760	159	47	40 (40)	−38 (−38)	104 (104)
10,000	523	110	47	36 (23)	−26 (−19)	67 (77)
20,000	349	73	47	24 (10)	−19 (−9)	40 (53)
30,000	226	47	47	24 (7)	−15 (−6)	18 (30)
40,000	141	29	47	not humanly tolerable		
50,000	87	18	47	not humanly tolerable		
62,800	47	10	47	water at body temperature boils		

[A]The partial pressures (mmHg) of alveolar CO_2 exhaled and O_2 absorbed decreases at higher altitudes due to increasing respiratory minute volume.
[B]The first set of PO_2 are unacclimatized values; the second are for acclimatized persons.[8,9]

local $P_{ATA} = 760\ e^{(-30,000/24,540)} = 224$ mmHg (versus 226 from NOAA[1])

lung $PH_2O = 47$ mmHg (at body core temperature)

alveolar $PCO_2 = 24$ mmHg (known by experiment)

lung $PO_2 = (224\text{-}47\text{-}24)/4.76 = 153/4.76 = 32$ mmHg

$\Delta PO_2 = 15$ mmHg (also known by experiment)

alveolar $PO_2 = (32 - 15) = 17$ mmHg (versus 18 mmHg from Table 23.6)

Hemoglobin's affinity for oxygen is a major contributor of physiological tolerance to hypobaric conditions. Hemoglobin in red blood cells binds about 50 times more oxygen than is dissolved in blood plasma. One can calculate using Dalton's law, Henry's law, and sea level data from Tables 23.5 and 23.6, that only about 0.2 cc O_2 can be dissolved into 100 mL of blood plasma acting as water vs. about 20 cc O_2/100 mL (often called "20 volume percent") contained in normally oxygenated blood with hemoglobin. Moreover, rather than a linear relationship with the partial pressure of oxygen predicted by Henry's law, hemoglobin binds with oxygen in a beneficially nonlinear way. As shown by the center line in Figure 23.1 depicting hemoglobin at a normal blood pH of 7.4 (corresponding to an alveolar PCO_2 of 40 mmHg), blood is at least 95% oxygen saturated at an alveolar oxygen partial pressure as low as 85 mmHg. The body's response to less oxygen in the blood is to increase its respiration rate, driving off CO_2, increasing the blood's pH, and further increasing the carrying capacity of hemoglobin.[9,10] Thus, hemoglobin gives the body a very robust tolerance to modest altitudes, as summarized in Table 23.7.

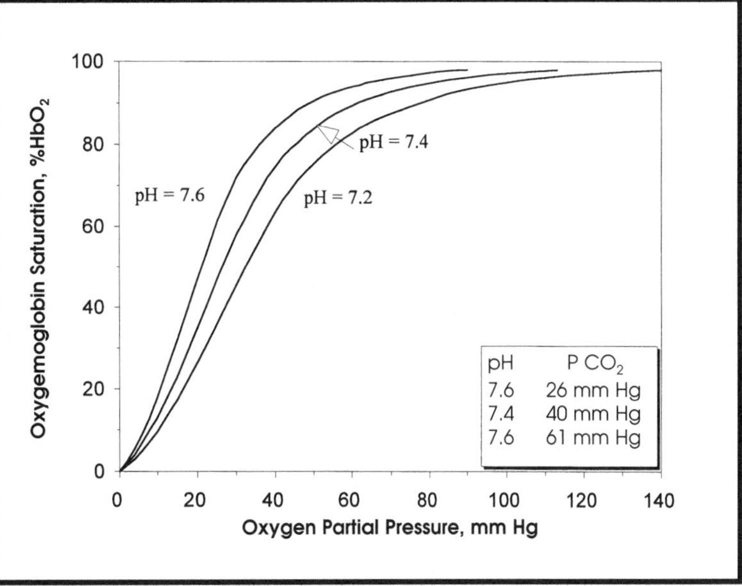

Figure 23.1—The oxyhemoglobin dissociation curves for human blood at 37°C and pH of 7.6, 7.4 (normal), and 7.2.[10]

One of the first physiological symptoms of hypoxia is shortness of breath on exertion. The unacclimatized person's initial physiological response of increasing respiration is somewhat thwarted by the secondary effect of hyperventilation to decrease the blood's carbon dioxide concentration (see PCO_2 in Table 23.6) and pH, which tends to lower respiration. The body will acclimatize to altitude in 2-5 days, facilitating hyperventilation (see further discussion in the Control of Hypobaric Hazards section). The combined benefits of hemoglobin's natural affinity for oxygen,

Table 23.7—Summary of Direct Physiological Responses to Hypobaric Pressures[11-14]

Altitude (ft)	Ambient PO_2 mmHg	Alveoli PO_2 mmHg	Blood O_2 % sat.	Health Effects	Eqv. Sea Level YO_2	Eqv. Sea Level Y_{inert}
< 6000	>127	>82	>95	none except on maximum exertion	17%	21%
12,000	101	65	90-95	decreased night vision and AMS symptoms	13%	37%
18,000	79	44	75-85	euphoria, loss of coordination	10%	51%
>18,000 limited by the time of useful consciousness (TUC)						
20,000	73	40	74-82	TUC = 10-20 minutes	9.6%	55%
25,000	59	25	45-55	TUC = 3-5 minutes	8%	63%
30,000	47	21	30-40	TUC = 1-2 minutes[A]	6%	71%
35,000	37	12	15-20	TUC = 30-60 seconds[A]	5%	77%
40,000	29	12	10-15	TUC = 15-20 seconds[A]	4%	82%

Note: The concentrations of inert gas sufficient to create the equivalent levels of hypoxia at sea level were calculated from ambient PO_2 using Equation 12.

[A]Complete loss of consciousness will result above 30,000 feet.

an initial increase in cardiac output, and even modest increases in respiration are so effective that very few physiological effects of altitude can be detected below 6000 feet (1800 m) except that an oxygen debt can develop more rapidly if near maximum exertion. This was perhaps most vividly demonstrated in the 1968 Summer Olympics in Mexico City (7546 ft, 2300 m) in which no world records were established in events lasting longer than 2.5 minutes.[15] Mental performance is also not affected below a PO_2 equivalent to 6000 ft.[12]

From the perspective of inert gases in confined spaces, it is noted in Table 23.7 that 6000 feet is equivalent to 17% oxygen at sea level. Figure 23.1 shows that the oxyhemoglobin will still be >95% saturated under this condition. This observation implies that the OSHA requirement to ventilate any time the oxygen content is less than 19.5% (29 CFR 1910.94(d)(9)(vi)) has no real basis in health. The ACGIH threshold limit value recommendation of a YO_2 of 18% or an equivalent PO_2 of 135 mmHg, is a similarly conservative health hazard at sea level that if applied literally (using Equation 11) would ban all work over 3750 feet above sea level. Such guidance might best be described as a good practice standard: Given that providing fresh air to a workplace is generally cheap (although perhaps time-consuming), abundant oxygen should be available unless it is consumed, an unreasonable amount of some other gas or vapor is allowed in the workplace air (that is likely to be toxic or explosive well before it creates an oxygen deficient health hazard), or ventilation is marginal. On the other hand, more severe displacements of oxygen at sea level (shown on the right side of Table 23.7) are capable of causing the full range of hypoxic symptoms listed.

The decrease in night vision acuity in the next group of symptoms in Table 23.7 manifests itself in lower sensitivity to stimuli and decreased peripheral vision and contrast discrimination.[11,13] The percentage increase in the light intensity necessary to maintain an equivalent retinal response may be estimated using Equation 15, determined by regressing the data summarized by Gagge and Shaw.[8]

$$\frac{\%\ \text{increase in}}{\text{light intensity}} = 80 \times \ln(1 - \text{altitude in feet}/19{,}400) \qquad (15)$$

This effect suggests that special precautions should be taken to avoid working in poor lighting conditions at high altitude work sites. Symptoms of acute mountain sickness (AMS) that can occur at about 12,000 feet are discussed below.

Except for the aviation and recreation industries, work above 18,000 is quite rare. High altitude hazard recognition training for pilots and flight crews includes their exposure to the early symptoms of hypoxia (lightheadedness and peripheral tingling) that usually precede euphoria, incoordination, and the loss of the ability to take corrective steps to ensure one's survival. The special case of responding to rapid decompression at altitudes above 20,000 feet emphasizes the limited time of useful consciousness (TUC),[8] sometimes called the effective performance time.[11] It is incumbent on the flight crew and beneficial to passengers to put on masks providing 100% oxygen within the sometimes very short times listed in Table 23.7.

In contrast to the direct effects of hypoxia described above, two important groups of indirect and slightly delayed responses to altitude have been identified. What was initially called AMS has now been subdivided into benign AMS and what some call malignant AMS.[16]

Benign acute mountain sickness constitutes an array of symptoms that may begin to develop in travelers 6–12 hours after arriving at altitudes above 8000 ft (2500 m). Symptoms include headache (very common and nearly always in the frontal region), difficulty sleeping (next most common), lightheadedness or dizziness, nausea or vomiting, and fatigue or weakness. Symptom severity ranges from mild (discomforting) to severe (incapacitating). Physical examination of those with symptoms has revealed that about 25% exhibited chest crackles or peripheral pulmonary edema.[17] Symptoms more severe than a headache will normally increase gradually, peak on the second or third day, and resolve themselves by the fourth or fifth day. Thus, the term "benign" was appended to AMS and has been widely used to differentiate this pattern from the more life threatening manifestations of acute mountain sickness.[6,7] The incidence of benign AMS can be anticipated from prior studies as summarized in Table 23.8, although the subjective nature of benign AMS makes its diagnosis a variable.[6,7]

Symptoms of benign AMS will subside spontaneously (without treatment) and will not necessarily affect the same traveler repeatedly or with the same severity. Treatment of symptoms with ibuprofin may be better at relieving symptoms of headache than aspirin, but Ward et al.[7] advocates voluntary hyperventilation, which also promotes acclimatization. Acetazolamide (Diamox®, 250 mg twice daily) may be used either as a prophylaxis beginning 24 to 48 hours before ascending or to relieve symptoms.[7] Prevention by avoiding rapid ascents is widely touted,[6,7] but the recommended schedule of one to two days per 1000 feet above 9000 feet is not consistent with the fast pace of some industrial or military temporary assignments.

It is important to be able to differentiate benign AMS symptoms from the less common but more severe and life threatening forms of AMS that may develop. Dickinson[16] proposed the term "malignant AMS" to encompass high altitude pulmonary edema and high altitude cerebral edema, although this categorization is not as widely accepted as benign AMS.[6] The edema in HAPE is characterized by the release of large quantities of a high protein fluid into the lung. Differential symptoms, which are often denied by the patient, include severe breathlessness (in 84% of cases) and chest pain (in 66%), with or without the above symptoms of benign AMS. Symptoms of patients with HAPE will rapidly progress to a dry cough, production of a foamy pink sputum, audible bubbling and gurgling sounds while breathing, and cyanosis of the lips and extremities. Early recognition of these symptoms, conservative field diagnosis, and prompt action are essential to prevent further progression into a coma followed by death within 12 hours. Oxygen should be administered if available, and the patient should be taken immediately to a lower altitude. Recovery without complications is normally quite rapid, and although the recovered patient should be cautious, he or she may later return to high altitude without further trouble.[6,7]

The incidence of HAPE is uncertain. One study reported rates of 0.9% in residents returning to 10,000 feet ASL after short visits to a lower altitude.[21] Heath and Williams summarized the incidence among studies of mixed populations at altitudes between 10,000 and 20,000 ft (2800–6195 m) as 0.5 to 1.5%.[6] They also cited studies reporting rates of subclinical pulmonary edema diagnosed radiologically ranging from 12 to 66%. HAPE is slightly more prevalent among the young and apparently healthy.

The mechanism(s) of HAPE is unclear. It may or may not be related to those causing benign AMS, but the most prevalent theory imputes pulmonary vasoconstriction due to the accumulation of water in extravascular spaces. Preventive guidelines are broadly similar to those for benign AMS with the added cautions against overexertion the first few days after rapidly traveling or returning to altitudes above 9000 feet (2700 m). Acetazolamide (Diamox) is not preventive against HAPE.

Table 23.8—Reported Incidence of Benign Acute Mountain Sickness

Altitude (ft)	Altitude (meters)	Incidence	Data Source
6200-9600	1900-2940	25%	ref. 18
9350	2850	9%	ref. 19
10,000	3050	13%	ref. 19
11,975	3650	34%	ref. 19
14,250	4343	43%[A]	ref. 17
13,910	4240	53%	ref. 20

[A]60% if flown to 9186 ft; 31% if hiking from 3940 ft

HACE is even less understood than HAPE. Ward et al.[7] believe that HACE is a direct progression of benign AMS to include cerebral edema, while Heath and Williams[6] believe that thrombosis also plays a part. The symptoms of HACE include many benign AMS symptoms but are differentiated by disturbed consciousness (irrationality, disorientation, and even hallucinations), abnormal reflex and muscle control (ataxia, bladder disfunction, and even convulsions), and/or perhaps most characteristically papilloedema (swelling of the optic disc). HACE is more rare than HAPE although symptoms of mixed HACE and HAPE frequently occur. As with HAPE, early recognition and action are essential to prevent a fatal HACE outcome. Fortunately, both conditions require the same treatment with oxygen and evacuation to a lower altitude. Taken together, knowledge of HAPE and HACE are vital components of a hazard communication program for supervisors and workers at high altitude.

Control of Hypobaric Hazards

The full paradigm of occupational controls is applicable to the hypobaric workplace. The classically preferred option of source control is only practical in aircraft. Engineering control can be instituted by increasing the total pressure, i.e., building a pressurized cockpit or cabin. Modern commercial, turbine-powered aircraft maintain a maximum interior-to-exterior pressure differential of 8.6 psi, which will maintain a cabin altitude of no more than 8000 feet. General aviation operations are restricted to cabin altitudes of 12,500 ASL without personal protection.

Personal protective equipment (similar to "supplied air respirators") is available to increase YO_2 in the breathing air. The maximum option of providing 100% oxygen will extend the no-effect zone to about 35,000 feet.[9,11] An annoying, sometimes painful, but usually resolvable problem is the tendency for the body to absorb the high concentrations of oxygen from the middle ear overnight.[22] If the eustachian tube does not open spontaneously to relieve the pressure difference, the "Valsalva maneuver," described in the Changing Pressure Effects section, should be performed. Supplemental oxygen is generally limited to short-term use in aircraft systems or mountain climbing expeditions. However, Ward et al.[7] suggest adding 5% oxygen indoors to relieve symptoms of hypoxia via the use of electrically powered oxygen concentrators.

Acclimatization is a remarkably effective long-term control for habitable high altitudes. Acclimatization changes the balance between two respiratory control mechanisms. On initial exposure to low oxygen, the reaction of peripheral chemoreceptors (PO_2 sensors in the carotid and aortic bodies) is to increase the respiratory minute volume; however, increased respiration decreases the blood PCO_2 and increases its pH, which decreases the stimulation of the "respiratory center" within the brain. This natural balance initially limits the body's ability to increase respiration in response to a feeling of breathlessness. For example, the work capacity of a new arrival at 17,000 feet would be expected to be reduced by 50%.[9] This would be aggravated if the work required a respirator.

The first adaptation over 2 to 5 days at altitude is a reduction in the blood's bicarbonate ion concentration (HCO_3^-), decreasing the negative sensitivity of the respiratory center to increased ventilation. Thereafter, the peripheral chemoreceptors can more easily increase respiratory minute volumes four- to fivefold, increasing one's work capacity back toward normal. This is also the time period over which symptoms of benign AMS, should they occur, will generally subside. The length of a corresponding administrative restriction to the intensity of a new arrival's work schedule is shorter than but roughly analogous to the 1 week often recommended for heat stress acclimatization.

For more extended stays at altitude, longer term physiological changes will further benefit one's working capacity. After a period of 2–3 weeks, the body's hematocrit and blood volume will begin to increase 50 to 90% above normal, and the initial increase in cardiac output will begin to return to normal. Hematocrit is the percent of cellular matter in a volume of whole blood—normally 42% (15 g Hb/100 mL) for males and 38% (13.5 g Hb/100 mL) for females.[9] Following the initial drop to 50% of one's sea level capacity at 17,000 feet, these changes can be expected to raise one's work capacity to about 70% within 2 or 3 months.[9] Other changes in cardiovascular circulation occur even more slowly but are most pronounced in persons born and raised at high altitude.

Selection criteria for temporary work at high altitude are not particularly restrictive. Among the factors not considered detrimental to high altitude are increased age, postmyocardial infarction if symptom-free for several months, controlled hypertension, asthma, and well-controlled diabetes.[6] Travel to high altitudes is not recommended for those with effort angina, a recent myocardial infarction, chronic bronchitis, emphysema, and interstitial lung disease.[6] Hard data on reproductive hazards to pregnant females have not been developed, but high altitude travel while pregnant is generally not advised due to fetal oxygen requirements.[6,7]

Hyperbaric Hazards

The most common occupation associated with hyperbaric conditions is underwater diving.[23] Occupational diving is expanding into new frontiers like fish farming.[24] Compressed air work in construction is a less common occupation. Pressure supplied to an airtight caisson used to be a common technique to reduce the flow of water or mud while digging bridge pilings (see Figure 23.2). As workers removed the undersurface mud and sand, the caisson would settle until reaching a stratum where a stable structural foundation could be formed. Similar air pressure has also been applied to tunnels and mines to control water intrusion during construction. The Occupational Safety and Health Administration (OSHA) limits compressed air workers' maximum pressures to the equivalent of 112 fsw to protect them not only from the direct hazards of hyperbaric conditions described in this section, but also from the indirect hazards resulting after return to normal pressures (described in the Changing Pressure Effects section). The National Institute for Occupational Safety and Health (NIOSH) estimated there were about 5000 professional divers and caisson workers in the United States exposed to hyperbaric hazards in the 1970s.[26] Hygienists are often involved in construction projects but rarely have direct responsibilities for diving operations. The material covered in this section and the Changing Pressure Effects section should provide the technical bases to enhance support functions to specialized and highly trained supervisory staff.

The array of hazards associated with hyperbaric conditions includes:

- Gas narcosis caused by nitrogen in normal air during dives of more than 120 feet (35 m). Helium, substituted for nitrogen in "mixed gas diving," can cause an effect called high pressure nervous syndrome beyond 500 fsw.
- Gas toxicities caused by oxygen and carbon dioxide. The damage of oxygen to the lung and brain (central nervous system [CNS]) will vary with time of exposure and depth. While a carbon dioxide partial pressure of 15–40 mmHg will stimulate the central respiratory sensor, concentrations > 80 mmHg suppress respiration.
- Another group of effects can occur after leaving hyperbaric conditions too rapidly. Because they do not occur while staying in one barometric condition, Decompression sickness and dysbaric osteonecrosis are discussed in the Changing Pressure Effects section.

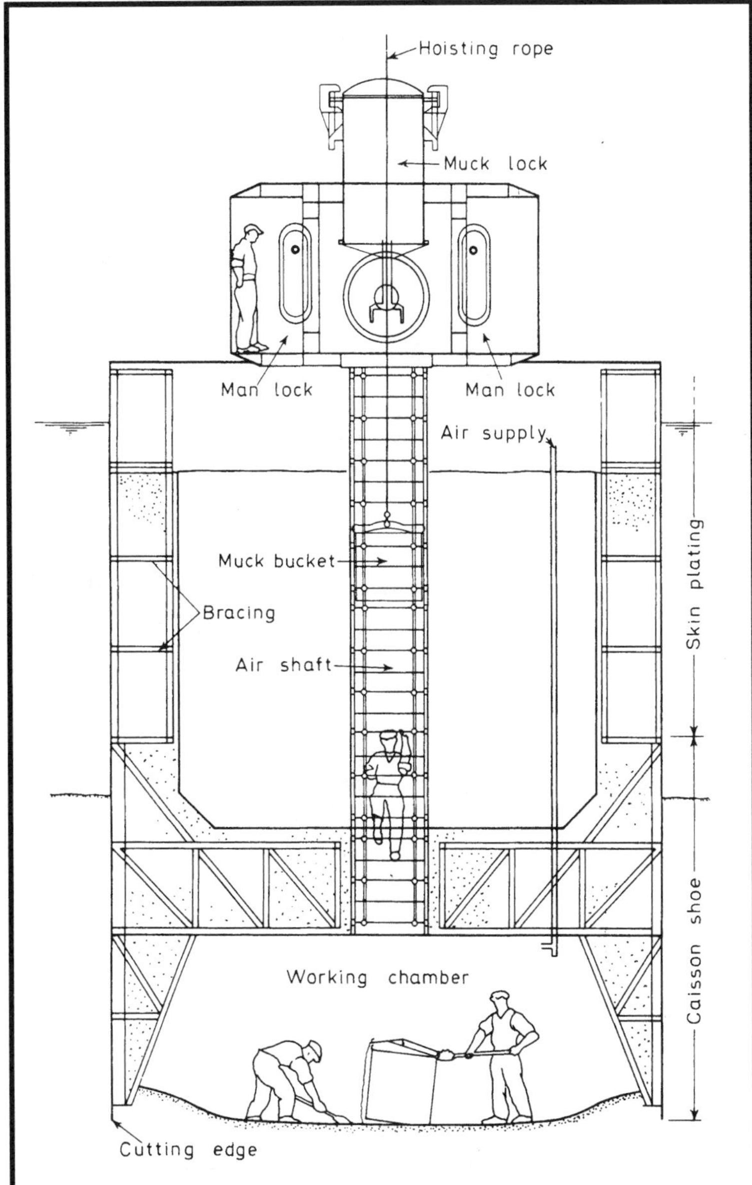

Figure 23.2—A compressed air caisson with separate air locks for personnel and bottom muck.[25]

Recognition and Control of Hyperbaric Hazards

The first of these hazards is simply the result of the narcotic effect of any gas absorbed into neural tissues. As a good approximation, the relative narcotic potential of gases is related to their solubility in the lipid layers surrounding neural tissue (the Meyer-Overton rule for anesthetic gases). Henry's constants for selected anesthetic gases (cyclopropane, nitrous oxide, and ethyl ether) are provided in Table 23.5 as useful points of reference. The amount of a gas that will dissolve in lipid tissue increases with its oil solubility and with its partial pressure, in accordance with Henry's law (Equation 10). Pressure increases with depth underwater, as described by Equation 5. Each component of the breathing air maintains its own partial pressure as a fraction of the increasing total pressure in accordance with Dalton's law (Equation 7). Thus, at any depth (or known pressure created by other means), the pressure and potential lipid concentration of each gas can be determined.

Example 5: Find the N_2 partial pressure in air and the potential concentration of nitrogen in saturated tissues while repairing an oil rig 185 feet under the Gulf of Mexico.

Starting with a total pressure of 6.6 ATA from Example 1, Dalton's law as expressed in Equation 7 can be used to find the fraction of the total pressure contributed by nitrogen:

$$PN_2 = YN_2 \times P_{ATA} = 0.7808 \times 6.6 = 5.2 \text{ atm}$$

The concentration of nitrogen in solution can then be determined from Henry's law as expressed in Equation 10 and data from

Table 23.5:
CN_2 in water = SN_2 in water × PN_2 = 0.01206 cc/mL/atm × 5.2 atm = 0.062 cc/mL

CN_2 in lipid = SN_2 in lipid × PN_2 = 0.0609 cc/mL/atm × 5.2 atm = 0.314 cc/mL

One can see that at saturation the concentration of N_2 in lipid tissues is much more than in the blood. While it takes time for sufficient nitrogen to be transported to saturate the whole body, neurologic tissue is so perfused by blood that symptoms of nitrogen narcosis can be quite rapid. Because the severity of symptoms listed in Table 23.9 depends on the gas concentration in neural lipids, severity depends on depth; however, severity also depends strongly on personal susceptibility, experience, training, rate of descent, and level of exertion.[22,27–29]

Administratively limiting depth has been the most common control for nitrogen narcosis. For instance, the deepest routine air supplied dive recommended in the *U.S. Navy Diving Manual* is 190 fsw, which is limited to a bottom time of 40 minutes (bottom time is actually measured from the time the diver leaves the surface until beginning the ascent).[22] Diving time limits are based more on practical considerations than on gas physiology. Reducing or removing nitrogen as the source can be a cost-effective control in certain conditions. Reducing the nitrogen/oxygen ratio by using enriched oxygen mixtures (called NITROX) can speed the ascent rate thus decrease the total diving time but is limited to a shallower depth than air diving because of oxygen's own toxicity at pressures of more than 1 atmosphere. A separate decompression schedule developed for 68% N_2, 32% O_2 NITROX diving limits depth to 130 fsw.[29]

Substituting helium for all or most of the nitrogen (called "mixed gas diving") is a cost-effective control for deeper dives. Helium has a higher molecular diffusivity and a lower lipid/water solubility ratio than nitrogen, causing it to saturate the body's tissues more quickly during a dive. It is also less stable in solution, requiring its decompression schedule to have more stops than nitrogen to prevent bubbles from forming in tissues, e.g., supersaturation limited to 1.7 × ambient, cf. 3× for nitrogen. Schedules for surface supplied He/O_2 dives to 380 fsw are available.[30] Dives beyond that are only practical by keeping the diver under pressure for several days (called "saturation diving"). The rate of compression must be kept slow to avoid symptoms of high pressure nervous syndrome (HPNS) such as nausea, fine tremors, and incoordination that can begin to appear at about 500 fsw.[22,30,31] Dives deeper than 1000 fsw have been made using a trimix of nitrogen, helium, and oxygen; physiological research has found that the narcotic potential of a small amount of nitrogen can be used to balance the stimulatory effect of helium at high pressure. Helium presents other problems. Its high thermal diffusivity combined with the high gas density and specific heat at depth cause more rapid heat exchange rates requiring careful protection from hypothermia in the typically cold underwater temperatures.[28,30,32,33] Helium's low molecular weight causes a high-pitched distortion of human speech (a "Donald Duck" effect) that eventually requires electronic processing to become intelligible.[30,34]

The second group of hyperbaric hazards is due to the toxicity of common air constituents such as oxygen or carbon dioxide at high pressures. The hazards of oxygen were first explored as a result of World War II attempts to dive with pure oxygen to avoid nitrogen hazards and to avoid creating detectable bubbles by using a closed-circuit self-contained breathing apparatus (rebreathers).[30,35,36] Most symptoms of oxygen toxicity can be categorized as either pulmonary (coughing, substernal soreness, and pulmonary edema) or CNS (including body soreness, nausea, muscular twitching, and convulsions). The toxic mechanism is believed to be related to the increase in oxygen free radicals.[9,37] Both symptoms and severity vary inversely with pressure and time of exposure, as shown in Figure 23.3.[27,28,36,38,39] CNS hazards predominate for exposures in the time frame of a working day, while pulmonary effects are more of a concern during saturation diving or recompression therapy for DCS. The pulmonary curve in Figure 23.3 corresponds to about a 12% change in vital capacity.[40] The CNS curve is more judgmental. While the curve implies an asymptote near 1.5 ATA,[22,39] a plan should be in place to deal

Table 23.9—Increasing Severity of Nitrogen Narcosis Symptoms with Depth in Feet and Pressure in ATA[9,22,29]

Depth (ft)	P_{ATA} (atm)	PN_2 (atm)	Symptoms
100	4.0	3.0	reasoning measurably slowed.
150	5.5	4.3	joviality; reflexes slowed; idea fixation.
200	7.1	5.5	euphoria; impaired concentration; drowsiness.
250	8.3	6.4	mental confusion; inaccurate observations.
300	10.	7.9	stupefaction; loss of perceptual faculties.

with convulsions any time the oxygen partial pressure exceeds 1.0 atm.[28] It is notable that the onset of life-threatening convulsions is not necessarily preceded by the less severe symptoms.[22,35]

The simplest prevention of oxygen toxicity can be achieved by administratively limiting the time of exposure above one atmosphere (see Table 23.10). However, Dalton's law suggests that controlling the partial pressure of oxygen by reducing its molar fraction (YO_2) is a natural extension of mixed gas diving using helium.[30] In fact, dives to depths of more than 1000 ft (300 m) use only around 1% oxygen to keep PO_2 to less than 0.5 ATA.[31] For saturation diving the Navy manual recommends an oxygen partial pressure of 0.21 ATA (equivalent to normal air at sea level), between 0.44 and 0.48 ATA during depth changes, and a maximum of 1.25 ATA for short intervals.[30]

Carbon dioxide becomes toxic when it suppresses respiration. Normally an increase in PCO_2 decreases blood pH, which acts to increase the respiratory minute volume. However, at $PCO_2 > 80$ mmHg (about twice the IDLH [immediately dangerous to life and health] value), the respiratory control center becomes depressed and will soon cease to function.[41] Thus, carbon dioxide is not toxic at exhaled air concentrations at sea level (40 mmHg in Table 23.6). Nor is it toxic if normal ambient air (314 ppm CO_2 in Table 23.4) is compressed over 90 times to 3000 fsw yielding $PCO_2 = 24$ mmHg. However, the combination of the accumulation of exhaled carbon dioxide at increased pressure (either in the breathing system's dead space or due to a malfunction) can rapidly cause toxic effects.[42] Using Henry's law and Table 23.5, the concentration of carbon dioxide in lipids at 80 mmHg is less than half that of nitrogen at 3.6 ATA, supporting the suggestion that the toxic effect of carbon dioxide may be due to a different mechanism than that of oxygen and nitrogen.[43] On the other hand, there is likely to be an interaction between the early response to carbon dioxide causing an increase in respiration and an episode of CNS oxygen toxicity.[30] Prevention of carbon dioxide toxicity is simply a matter of good system design and maintenance. OSHA regulations for commercial diving operations (29 CFR 1910.430 and 1926.1090) limit CO_2 to 1000 ppm in supply air and to 0.02 ATA within the mask, usually by assuring that the flow of surface supplied air to masks and helmets is at least 4.5 actual cubic feet per minute at any depth at which they are operated. The 0.02 ATA is equivalent to 1000 ppm at 20 atm (or 627 fsw).

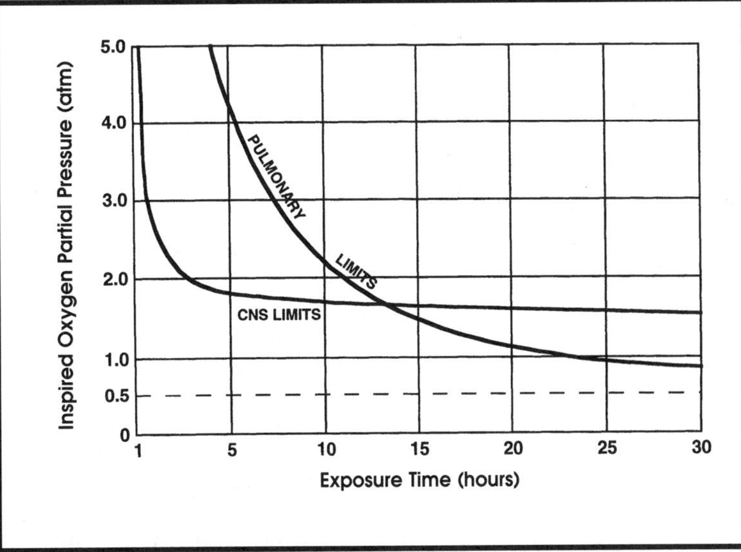

Figure 23.3—Recommended limits of exposure to inspired oxygen.[28]

Table 23.10—Maximum Bottom Time (Minutes) for Specified Levels of Oxygen Exposure

PO_2, ATA	Surface Supplied HeO_2 Diving[30]	Standard Air Diving[22]	Emergency Air Diving[22]
1.30	unlimited	60	480
1.40	50	40	
1.50	40		360
1.60	30		
1.70	20		
1.80	15		240

Other long-term hyperbaric effects, such as those summarized by Farmer and Moon[31] are neither well established nor otherwise discussed here. Similarly, divers and (more specifically) compressed air workers can face other nonbarometric risks including microbes and parasites,[28,29,44–46] noise,[28,29,47–49] silica,[50] radon,[51] fire,[23,29,52] and chemicals during underwater cleanup operations.[29] Thus, the recognized acute and chronic barometric effects covered here are only a portion of the total health risks faced by these workers.

Changing Pressure Effects

The recognized adverse health effects of changing pressure manifest themselves in two acute symptoms and one chronic symptom. The following effects can occur in either hypobaric or hyperbaric conditions and were therefore grouped separately.

- Pain due to expanding or contracting trapped gases, potentially leading to barotrauma. This acute symptom and potential damage can occur during either ascent or descent but are potentially most severe when gases are expanding.

- Decompression sickness (DCS) due to the evolution of inert gas bubbles in the body. Acute symptoms of DCS can occur during a decrease in pressure but most commonly occur soon after the ascent has been completed.
- Dysbaric osteonecrosis causing detectable lesions most commonly on the body's long bones. Although its etiology is unknown, this chronic disease may be related to the evolution of gas bubbles that may or may not be diagnosed as decompression sickness.

Recognition of Changing Pressure Hazards

Pain and barotrauma from expanding or contracting gases while transiting between pressure zones are the most direct effects predictable from Boyle's law. The most common sites of pain from trapped gases are teeth, the GI tract, sinuses, middle ear, and lungs (the latter particularly during ascent).[53–55] The expansion of trapped gas caused by dental decay can actually cause a tooth to crack or a dental filling to become dislodged during ascent; good dental care will prevent this problem. Divers and flyers should anticipate and not attempt to suppress the release of natural gases of digestion that expand during ascent.

The sinuses are membrane-lined, hollow spaces within the skull bones connected to the nasal cavity by narrow passages. Blockage of these passages due to nasal congestion or a head cold can cause pain during either ascent or descent. Sinus pain during descent is called "sinus squeeze." Divers should be trained to detect blocked sinuses and should not dive with a cold or an allergic inflammation.

The most common source of pain on descent is from the contraction of air in the middle ear if the eustachian tubes are inflamed or blocked. The eustachian tubes normally relieve outwardly (during ascent) at a small pressure difference (ΔP) of only a couple of mmHg. However, it usually requires more, at least 15 mmHg, to relieve inwardly (during descent). If not relieved, pain can begin to occur at 50–100 mmHg, and the eardrum will rupture at 100–500 mmHg. Equation 5 can be used to find the change in depth for any pressure. Some examples are given in Table 23.11. However, because pressure is not linear with altitude above sea level, the change in altitude to achieve a similar fixed ΔP will vary with the starting altitude above sea level and the direction (ascending or descending) as given by Equation 16. To achieve an air pressure difference of 500 mmHg is rare because it requires, for instance, a descent to sea level starting at a pressure altitude of at least 27,000 feet ASL (8200 m).

$$\Delta\text{altitude descending} = \kappa \times \ln[1-(\Delta P/P)] \qquad (16a)$$

$$\Delta\text{altitude ascending} = \kappa \times \ln[(\Delta P/P) + 1] \qquad (16b)$$

where

κ = the altitude coefficient for Equation 6 taken from Table 23.3
ΔP = the change in pressure in the same units as P, below
P = the initial pressure found using Equation 6

For most people, opening the eustachian tubes during descent requires some conscious action like yawning or swallowing. The Valsalva maneuver is a more active technique used by flyers and some divers to force air up their eustachian tube by closing their mouth, holding their nose, and trying to exhale. This technique may also clear slightly blocked sinuses. However, external forces on the eustachian tube at a P of 90 mmHg will usually prevent it from opening, even with the help of the Valsalva maneuver.[54] Thus, Farmer and Moon[31] recommend that divers clear their ears every 2 feet (corresponding to 50 mmHg in Table 23.11). Should a blockage occur, divers should be trained to stop and rise back up a few feet before attempting to clear and proceed.[22]

The most severe outcome of expanding gases is pulmonary barotrauma. An increase in gas volume of 20 to 30% can cause an initially full lung to rupture. A trapped gas volume expands in proportion

Table 23.11—Change in Seawater Depth Corresponding to a Selected Change in Absolute Pressure (ΔP)

ΔP		ΔDepth	
mmHg	atm	fsw	m
15	0.020	0.662	0.202
50	0.066	2.18	0.667
100	0.132	4.37	1.33
500	0.658	21.78	6.65

to the change in relative pressure, as predicted by Boyle's law. In contrast with changes in absolute pressure described above, changes in relative pressure are not constant with depth. Use Equation 17 (derived from Equation 5) to find the change in depth necessary to create a given relative change in pressure. Note that since depth increases downward, an expansion occurs when Δdepth is negative.

$$\Delta\text{depth} = -([\text{initial depth}] + \kappa) \times \left[1 - \frac{V_{initial}}{V_{final}}\right] \quad (17)$$

where κ = coefficient from Table 23.2 depending on the water density and units of depth and V = the gas volumes. During ascent the initial volume is smaller than the final (larger) volume; this ratio is also the final (lower) pressure to the initial (higher) pressure.

The examples below use Equation 17 to show that equal relative changes in pressure occur over smaller distances at shallow depths than when starting from deeper depths. This implies an important lesson to be learned in training experiences: that the risk of pulmonary barotrauma from a breathhold ascent is actually greater for a given change in shallow depth than for the same change at a greater depth. Divers must also be trained to exhale during rapid ascent.

Example 6: Find the change in depth necessary for a gas volume to expand by 25% starting at initial depths of 10, 100, and 500 feet of sea water.

From Table 23.2 for feet of sea water, κ = 33.1, and for this problem, $V_{initial}/V_{final}$ = 1/1.25 = 0.80. Using Equation 17 for the initial depths given:

10 feet:
Δdepth = -(10 + 33.1) × (1 - 0.80) = -8.6 feet

100 feet:
Δdepth = -(100 + 33.1) × (1 - 0.80) = -27 feet

500 feet:
Δdepth = -(500 + 33.1) × (1 - 0.80) = -107 feet

Decompression sickness (DCS) is the most commonly known of the many dysbarisms. It is sometimes referred to as "evolved gas dysbarism," "compressed air sickness," "caisson worker's syndrome," or various common names listed in Table 23.12. DCS is completely different from the preceding direct effects. DCS is caused indirectly by the formation of inert gas bubbles at one or more locations within the body.

Example 5 can be extended to a human analogy of the "pop bottle" in Example 3. The nominal distribution of a 70 kg human body is 58% water (or 40.6 L), 20% fats, lipids, and oils (or ~14 L), and 22% solids (mainly bone). Thus, the volumes of gas (V) can be determined in each compartment:

VN_2 in body water =
0.062 cc/mL × 40.6 L = 2.5 L N_2

VN_2 in body lipid =
0.314 cc/mL × 14 L = 4.4 L N_2

A total volume of 7 L nitrogen is about the size of a basketball. As pointed out in the Physical Principles section when discussing high lipid to water solubility ratios, the rate at which inert gases like N_2 are transported by blood is slow compared to the capacity of lipid tissue to absorb them. During descent this difference creates a beneficial time lag that allows short dives to be made without any decompression, but during ascent from longer dives the reverse process can be dangerous. Should they desire, divers can easily decompress to lower pressures at rates much faster than the stored gases can be resorbed back into the blood and exhaled out of the body. The desorption rate from any location in the body is determined by:

- The difference between tissue and blood gas concentrations (a closely

Table 23.12—Distribution of Initial DCS Symptoms by Professional Divers and Tunnel Workers Reported

Location of Bubbles	*Symptom(s)*	*DCS Type*	*Common Term*	*Professional Divers*(58)	*Tunnel Workers*(58)	*Recreational Divers*(31,57)
Joints	pain on flexure	I	bends	70-90%	55-90%	41%
Skin	altered skin sensation, itching, or rash	I		1-15%	0-10%	20%
Brain-spine	dizziness, headache, loss of coordination, weakness	II	staggers	10-35%	8-25%	35%
Chest	cough, dyspnea, pain on breathing	II	chokes	2-8%	1-7%	3%

related term called "supersaturation" is the ratio of the gas concentration within tissue or a liquid to that in the atmosphere or lung), which depends on "bottom time," and

- The perfusion of tissue(s) by blood into which the inert gas must dissolve (in general, skeletal lipid tissues are perfused less thoroughly than are muscle, CNS, or other organs).

Note that the tissue has to have a higher partial pressure compared with the blood in order for gas to be removed from the tissue. However, if the pressure ratio is too large, bubble formation and DCS occurs.

Symptoms of DCS can range from irritating to severe. The common names given to DCS depend on its symptoms; its symptoms in turn depend on the location of the gas bubbles (Table 23.12). The location of the bubbles largely determines the seriousness of the sickness. Beyond the descriptors in Table 23.12, a simple medical classification of DCS has evolved. Type I DCS symptoms involve only skin, lymphatic, or joint pain. Type II DCS involves respiratory symptoms, neurologic or auditory-vestibular symptoms, and symptoms of shock or barotrauma. Type II is potentially life threatening. Of course, nothing is completely simple. For instance, Arthur and Margulies[56] point out that skin marbling (from intradermal bubbles) is indicative of impending systemic involvement and should be treated as Type II DCS. Elliott and Moon[57] report that recreational divers suffering DCS are initially more likely to have Type I symptoms, but most divers eventually progress to Type II.

The incidence of DCS is largely unknown for various reasons. Literally thousands of cases of DCS have been reported among divers,[31,58,59] but the frequency of diving or even the number of divers from which rates could be assessed, is unknown. The distribution of symptoms in Table 23.12 is only among those cases reported to the respective databases. The incidence following three sets of hyperbaric chamber dives is summarized in Table 23.13. Farmer and Moon[31] cite reports of DCS risk of 0.1 to 0.2% in commercial diving operations, while another report claims 31% of divers have experienced DCS at least once.[59] These two rates would be compatible statistically after 185 to 370 dives. The incidence of DCS among compressed air workers has been reported around 0.5% in two large groups[60,61] and 0.07% in another group.[62] Differences in rates may be due to different decompression schedules used (both between and within divers and compressed air workers), the consistent adherence to those schedules (a function of training and supervision), or detection and reporting protocols (differences in day-to-day versus periodic medical supervision, and in clustered large-scale versus individual working groups).

Table 23.13—Incidence of DCS Among Chamber Dive Trials in the United States and Canada[63]

Number of Dives	Depth Mean (range) Meters	Bottom Time Mean (range) Minutes	Breathing Gas	DCS Incidence
1041	45 (15-88)	22 (5-120)	air	3.0%
647	66 (36-100)	32 (10-100)	$He\text{-}O_2$	4.2%
261	92 (43-123)	33 (15-90)	$He\text{-}O_2$	12.%

The same DCS phenomenon can occur in flight crews in unpressurized aircraft, in someone flown from near sea level to a high mountain facility, in those in hypobaric chambers used for training, and in someone who flies soon after diving. Incidence rates among hypobaric chamber technicians generally have been reported from about 0.35%[64,65] to 0.62%[66] and one report of 0.25% at pressure altitudes ranging from 25,000 to 45,000 ft (7620–13,106 m).[67] Both physiological and epidemiologic studies show that DCS is likely to occur at about a 15% incidence rate when underwater diving is followed by flying—going from hyperbaric to hypobaric conditions.[68] The *U.S. Navy Diving Manual* recommends not flying above 2300 feet ASL (700 m) for 12 hours after a dive requiring decompression and for 2 hours following a no-decompression dive.[22]

Dysbaric osteonecrosis is perhaps the least known barometric pressure hazard, both technically and publicly. Although it was first recognized among caisson workers early in the 20th century by Bornstein and Plate,[69] it is now known to also affect divers.[31,59,70] Still it is not widely recognized, adequately researched, or effectively controlled by current practices.[71,72] Dysbaric osteonecrosis (also called aseptic bone necrosis) manifests itself as regions of bone and marrow necrosis, especially of the humerus, femur, or tibia. The lesions are indistinguishable histologically from necrosis from other causes. The condition is generally asymptomatic, with detection relying on differential diagnosis of high quality radiographs and by excluding other causes.[71,73] In two British reports, the prevalence of detectable bone lesions was reported as 24% among compressed air workers[73] and 6.2% among divers.[59] Most of these lesions were in the head, neck, or shaft of the long bones where they are generally benign. However, 3.7% of compressed air workers had lesions adjacent to articulating surfaces, where they can cause degenerative changes.(73) "Juxtaarticular" lesions were found in 1.2% of divers, with at least 15% of these divers (0.2% overall) actually experiencing joint damage (in shoulders of divers; in shoulders and hips of compressed air workers).(59) There are strong positive associations between lesions and length of diving experience (but not age), the maximum depth dived (none were found in those who had never dived below 30 m [100 ft]), and a history of at least one prior DCS, although it can also occur without any known prior acute DCS symptoms.(59)

There is a tendency to assume that bone necrosis is caused by evolved gas bubbles that did not necessarily cause acute symptoms, but there is no direct evidence for any clear etiology.(59) It is important to understand that existing decompression schedules have been defined and refined experimentally based on symptoms rather than on preventing bubbles *per se* or by maintaining and applying good epidemiologic health surveillance.(28) It has been shown that bubbles can be detected electronically before symptoms are detected.(74–76) In the absence of other pathological etiologies, it is plausible that these asymptomatic bubbles could account for the prevalence of osteonecrosis in divers without a history of DCS.(59) The prevalence of dysbaric osteonecrosis is significant and perhaps still being underestimated by the occupational health establishment.

Control of Pressure Changes

The risk of DCS is controlled by administratively limiting the pressure ratio during ascent (the inverse of the volume ratio used in Equation 17). Early experimental research by Haldane recommended a maximum ratio of 2:1 for saturated tissues, the Haldane rule.(77) However, the majority of the recommended initial standard air decompression ratios are close to but exceed this ratio, as denoted by the gray area of Figure 23.4. Only a small portion of the long dives to depths between 35 and 60 feet complies with this 2:1 guidance.(22) The background level of DCS even when decompression guidelines are followed, the ragged pattern of the exceedance zone, and the detection of bubbles in blood by Eckenhoff et al.(75) and Ikeda et al.(76) after saturation dives to depths of only 25 feet suggest the limited degree of control afforded by these guidelines.

The substitution of helium for nitrogen (discussed in the Hyperbaric Hazards section) changes the dynamics of gas absorption and desorption but does not remove the bubble hazard. The use of 1-atmosphere suits is a recent development that has some promise if issues of functional flexibility can be overcome.(31) However, the costs and availability of new technologies keep the vast majority of divers using conventional administrative controls that rely on decompression schedules.

The *U.S.Navy Diving Manual* has four basic decompression schedules, along with various options, exemptions, and response contingencies.(22,30)

1) Standard Air Decompression (USN Table 7.5;(22) see Example 7). A schedule applicable to either scuba or surface air supplied divers who completely decompress either in the water or in a diving bell or chamber before reaching surface pressures. The maximum recommended air dive is 190 feet for 40 minutes with emergency dives to 300 feet for 180 minutes.

Example 7: U.S. Navy schedule for in water ("standard") decompression from depth = 100 fsw (30.4 m; 4 ATA), bottom time = 120 minutes.(22)

Stop, ft (m)	Stop Time	Total Ascent Time
30 (9.1)	12	12
20 (6)	41	53
10 (3)	78	131 = 2:11 hours

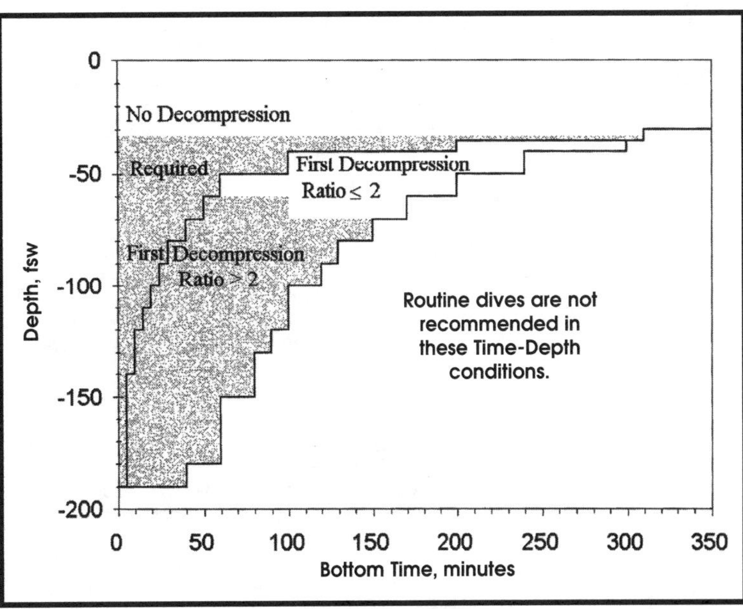

Figure 23.4—Diving conditions above the top solid line require no decompression. Conditions below the bottom solid line require exceptional approval. The first decompression stop for conditions in gray is at a pressure ratio greater than 2. (22)

2) Surface Decompression using either oxygen (USN Table 7.6;[22] see Example 8) or air (USN Table 7.7;[22] see Example 9). These schedules provide a more convenient, often safer, and in the case of using oxygen, a shorter decompression schedule that requires the diver to transfer to a recompression chamber within 5 minutes of reaching the surface. The maximum recommended dive is still 190 feet.

Example 8: U.S. Navy schedule for surface decompression using oxygen. Depth = 100 fsw (30.4 m; 4 ATA), bottom time = 120 minutes.[22]

Stop, ft (m)	*Stop Time*	*Total Ascent Time*
30 (9.1)	3	3
40 foot equiv.	53	56 = 1:09 hours

Example 9: U.S. Navy schedule for surface decompression using air. Depth = 100 fsw (30.4 m; 4 ATA), bottom time = 120 minutes.[22]

Stop, ft (m)	*Stop Time*	*Total Ascent time*
30 (9.1)	12	12
20 (6)	41	53
20 foot equiv.	41	94
10 foot equiv.	78	172 = 3:00 hours

3) Surface Supplied Helium-Oxygen Decompression (USN Table 11-4;[30] see Example 10). A schedule that can involve stops breathing either the bottom supplied mixture, a 40% oxygen mixture, or pure oxygen. The maximum dive on this schedule is 380 feet for 120 minutes.

Example 10: U.S. Navy surface supplied helium oxygen decompression schedule for depth = 100 ft, bottom time = 120 minutes.[30]

Stop, ft (m)	*Stop Time*	*Total Ascent Time*
40 (12.1)	87	87 on 100% O_2 87 = 1:27 hours

4) Helium Oxygen Saturation Diving (USN Table 12-7[30]): A decompression schedule specifying a set rate of feet per hour. For example, an ascent from saturation diving at 340 feet would require 120 hours (or 5 days). On the other hand, saturation diving allows more working time per day, greatly reducing the total time for long jobs in addition to avoiding the hazards of multiple compressions and decompressions.

One important exception is that short dives may be made with no decompression time. These limits are depicted as the times above the top heavy line in Figure 23.4. For deeper and/or longer dives, decompression time requirements are a cost burden on employers and a potentially boring time for employees, an inviting incentive for both parties to cut corners, resulting in a higher incidence of DCS and potentially of osteonecrosis.[58] Motivational training and close supervision are essential components of a successful diving program.

In the United States, OSHA regulates compressed air work, land based diving, and diving from vessels not subject to Coast Guard inspection.[1] Such diving is governed by either General Industry Standards Subpart T (29 CFR 1910.401–441) or Construction Standards Subpart Y. Other diving is governed by U.S. Coast Guard and U.S. Department of Transportation (DOT) Marine Occupational Safety and Health Standards (46 CFR 197.200–488). Construction Standards Subpart S includes 29 CFR 1926.801 governing caissons and 29 CFR 1926.803 governing compressed air work (in addition, 29 CFR 1926.804 contains Definitions Applicable to all of Subpart S). Tables 23.14 and 23.15 provide a quick overview of these OSHA regulations, which are basically work practice standards that refer to a diving manual and by implication to a diving and decompression schedule, plus certain other specified requirements that parallel the principles and mechanisms outlined above and the schedules and guidelines in the *U.S. Navy Diving Manual*.[22,30]

Table 23.14—Overview of OSHA 29 CFR 1910.401–441: Subpart T: Commercial Diving Operations (Parallels 29 CFR 1926.1071–1092: Subpart Y: Construction Diving)

1910.401 –	Scope and application
1910.402 –	Definitions (a glossary of terms)
1910.410 –	Qualifications of dive team (covers training requirements)
1910.420 –	Safe practices manual (a written procedures manual shall be developed and maintained)
1910.421 –	Pre-dive procedures (covers emergency planning)
1910.422 –	Procedures during dive (covers communication, decompression tables, and the dive depth-time record to be maintained)
1910.423 –	Post-dive procedures (covers instructions to diver, provision of recompression chamber (required to be on site if the dive is outside the "no-decompression limits" and deeper than 100 fsw), and recompression requirements if needed)
1910.424 –	SCUBA diving (limited to ≤ 130 fsw and specifies certain procedures)
1910.425 –	Surface-supplied air diving (limited to ≤ 190 fsw (with 30 minutes to 220 fsw excepted) and specifies certain procedures)
1910.426 –	Mixed gas diving (specifies certain constraints and procedures)
1910.427 –	Lifeboating (puts certain constraints on air supplied or mixed gas diving while the support vessel is underway)
1910.430 –	Equipment (various specifications including supplied air quality limits of 20 ppm CO and 1000 ppm CO_2, hoses, lines, masks, helmets, decompression chamber, etc.)
1910.440 –	Recordkeeping requirements (retention of most records by employer for 5 years except records of nonincident dives for only 1 year, and all 5-year records to be forwarded to NIOSH)

The Divers Alert Network (DAN) at Duke University, Durham, N.C., provides a 24-hour emergency hotline (919-684-8111) for diving accidents or emergency medical information. Other emergency facilities are listed by Thalmann:[28]

United States Navy Experimental Diving Unit (904-230-3100); United States Naval Medical Research Institute (301-295-1839); Canada Diving Division of the Defense and Civil Institute of Environmental Medicine (416-635-2000); England Institute of Naval Medicine (011 44 1705 722-351).

Nonemergency access to DAN during business hours is available at (919-684-2948).

Information may also be obtained from the Undersea and Hyperbaric Medical Society (301-942-2980); or the National Board of Diving and Hyperbaric Medical Technology (504-366-8871).

Table 23.15—Overview of OSHA 29 CFR 1926.800 - 804: Subpart S: Underground Construction, Caissons, Cofferdams, and Compressed Air

1926.800 –	Underground Construction: defines general program requirements such as air quality monitoring by a "competent person"
1926.801 –	Caissons (specifies certain fall safety and pressure testing requirements)
1926.802 –	Cofferdams (e.g., specifies escape provisions in case of flooding)
1926.803 –	Compressed Air (describes onsite supervision, annual medical certification of each employee, provision of a "medical lock" (decompression chamber), medical emergency identification badges, e.g., bracelets, for all compressed air workers, posting of decompression schedules, a maximum working pressure of 50 psig, air supply ventilation, sanitation, and fire prevention requirements)
1926.804 –	Definitions, e.g., "decanting" when a person is rapidly brought to atmospheric pressure then recompressed immediately (to be undertaken only under medical direction)
Appendix A to Subpart S –	Decompression tables that differ from diving table schedules. These schedules cover much longer working times than are recommended in the *U.S. Navy (Air) Diving Manual*,[22] e.g., working times of more than 8 hours for ≤ 46 psig (equivalent to ~ 100 fsw). The decompression schedule calls for continuous slow decompression vs. stops at multiple stages and somewhat longer times than Navy diving recommendations.[22]

Conclusion

Currently, there are no OSHA regulations governing work at altitude. Hygienists have had little direct influence on managing diving hazards. However, hygienists should be able to contribute to improving the control of hazards in each of these environments. This chapter should provide both the incentive and the basic tools to anticipate, recognize, and control barometric hazards in some of industry's more novel workplaces.

Hypobaric hazards include hypoxia, benign AMS, and the more life threatening forms of AMS—HAPE and HACE. Decompression sickness is also possible at very high altitudes, although symptoms are less severe than those subsequent to underwater diving or compressed air work. Engineering, personal protection, and administrative controls may all be beneficial.

Hyperbaric hazards include narcosis from a PN_2 above 4 atm, HPNS from PHe above about 15 atm, toxic effects from oxygen above 1 atm or carbon dioxide above 80 mmHg, DCS caused by inert gas bubble formation, and osteonecrosis perhaps from a similar cause. Administrative control of depth to ≤ 190 fsw is recommended for air diving, decreasing the nitrogen content of air (NITROX) is useful in shortening decompression time for shallower dives, and replacing nitrogen with helium will allow dives in excess of 1000 fsw. DCS and osteonecrosis remain a lingering problem with all decompression schedules.

This chapter also explains the physical laws developed by Boyle, Dalton, and Henry that underlie and help to quantify barometric hazards. An extensive bibliography should facilitate further investigations into many of these areas. With this information, hygienists should be better prepared to contribute to improving the health of workers facing barometric hazards.

Additional Sources

Rom, W.N.: High altitude environments. In Environmental and Occupational Medicine, 2nd ed. Boston, MA: Little, Brown and Company, 1992. pp. 1143–1151.

Tredici, T.J.: Ophthalmology in Aerospace Medicine. In *Fundamentals of Aerospace Medicine*, ed. by R.L DeHart; Philadelphia: Lea and Febiger, 1985. pp. 465-510.

References

1. **National Oceanic and Atmospheric Administration, National Aeronautics and Space Admininstration, United States Air Force (NOAA):** *U.S. Standard Atmosphere, 1976* (NOAA-S/T 76-1562). Washington, DC: NOAA, Oct. 1976.
2. **Mackay, D. and A.T.K. Yeun:** Mass transfer coefficient correlations for volatilization of organic solutes from water. *Environ. Sci. Technol. 17(4)*:211–217 (1983).

3. **Stiver, W. and D. Mackay:** Evaporation rate of spills of hydrocarbons and petroleum mixtures. *Environ. Sci. Technol. 18(11)*:834–840 (1984).
4. **Fthenakis, V.M., and V. Zakkay:** A theoretical study of absorption of toxic gases by spraying. *J. Loss Prev. Process Indr. 3(2)*:197–206 (1990).
5. **Siegrist, R.L. and P.D. Jenssen:** Evaluation of sampling method effects on volatile organic compound measurements in contaminated soils. *Environ. Sci. Technol. 24(9)*:1387–1392 (1990).
6. **Heath, D. and D.R. Williams:** *High Altitude Medicine and Pathology.* Oxford: Oxford University Press, 1995.
7. **Ward, M.P., J.S. Milledge, and J.B. West:** *High Altitude Medicine and Physiology*, 2nd ed. London: Chapman & Hall, 1995.
8. **Gagge, A.P. and R.S. Shaw:** Aviation medicine. In *Medical Physics*, O. Glasser, ed. Chicago, IL: The Year Book Publ., 1985.
9. **Guyton, A.C.:** Aviation, high altitude, and space physiology. In *Textbook of Medical Physiology*, 8th ed. Philadelphia, PA: Saunders Co., 1991. pp. 464–470.
10. **Fox, S.I.:** *Human Physiology*, 5th ed. Dubuque, IA: Wm. C. Brown, 1996. pp. 458–498.
11. **Sheffield, P.J. and R.D. Heimbach:** Respiratory physiology. In *Fundamentals of Aerospace Medicine*, R.L DeHart, ed. Philadelphia, PA: Lea and Febiger, 1985. pp. 72–109.
12. **Knight, D.R., C.L. Schlichting, C.S. Fulco, and A. Cymerman:** Mental performance during submaximal exercise in 13 and 17% oxygen. *Undersea Biomed. Res. 17(3)*:223–230 (1990).
13. **McFarland, R.A. and J.N. Evans:** Alterations in dark adaptations under reduced oxygen tensions. *Am. J. Physiol. 127*:37–50 (1939).
14. **Henderson, Y. and H.W. Haggard:** *Noxious Gases*, New York: Van Nostrand Reinhold, 1943.
15. **McArdle, W.D., F.I. Katch, and V.L. Katch:** *Exercise Physiology: Energy, Nutrition, and Human Performance.* Philadelphia, PA: Lea and Febiger, 1991.
16. **Dickinson, J.G.:** Terminology and classification of acute mountain sickness. *Br. Med. J. 285*:720–721 (1982).
17. **Hackett, P.H. and D. Rennie:** Rales, peripheral edema, retinal hemorrhage and acute mountain sickness. *Am. J. Med. 67*:214–218 (1979).
18. **Honigman, B., M.K. Thesis, J. Koziol-McLain, et al.:** Acute mountain sickness in a general tourist population at moderate altitude. *Ann. Intern. Med. 118*:587-592 (1993).
19. **Maggiorini, M., B. Buhler, M. Walter, and O. Oelz:** Prevalence of acute mountain sickness in the Swiss Alps. *Br. Med. J. 301*:853–855 (1990).
20. **Hackett, P.H., D. Rennie, and H.D. Levine:** The incidence, importance, and prophylaxis of Acute Mountain Sickness. *Lancet II*:1149–1154 (1976).
21. **Scoggin, C.H., T.M. Hyers, J.T. Reeves, and R.F. Grover:** High-altitude pulmonary edema in the children and young adults of Leadville, Colorado. *N. Engl. J. Med. 297*:1269–1272 (1977).
22. *U.S. Navy Diving Manual Volume 1 (Air Diving).* NAVSEA 0994-LP-001-9110. Washington, DC. 1993.
23. **Elliott, D. and A.M. Grieve:** The offshore oil and gas industry. In *Recent Advances in Occupational Health*, J.M. Harrington, ed. Edinburgh: Churchill Livingstone, 1987. pp. 21–36.
24. **Douglas, J.D.M. and A.H. Milne:** Decompression sickness in fish farm workers: a new occupational hazard. *Br. Med. J. 302*:1244–1245 (1991).
25. **Blake, L.S. (ed.):** *Civil Engineering Reference Book*, 3rd ed. London: Butterworth & Co., 1985. pp. 16–29 to 16-31.
26. **Gillen, H.W.:** Proposal for an osteonecrosis registry in the United States. *Proceedings of a Symposium on Dysbaric Osteonecrosis*, E. L. Beckman and D. H. Elliott, eds. (HEW [NIOSH] #75-153). Cincinnati, OH: National Institute for Occupational Safety and Health, U.S. Department of Health, Education, and Welfare, 1974. pp. 221–226.
27. **Bennett, P.B. and D.H. Elliott:** *The Physiology and Medicine of Diving.* London: W.B. Saunders, 1993.
28. **Thalmann, E.D.:** Diving hazards. In *Safety and Health in Agriculture, Forestry, and Fisheries.* R.L. Langley, R.L. McLymore, W.J. Meggs, and G.T. Roberson, eds. Rockville, MD: Government Institutes, 1997. pp. 617–641.
29. **National Oceanic and Atmospheric Administration (NOAA):** *NOAA Diving Manual.* Washington, DC: U.S. Department of Commerce, Oct. 1991.
30. *U.S. Navy Diving Manual Volume 2 (Mixed-Gas Diving).* NAVSEA 0994-LP-001-9020. Washington, DC. 1991.

31. **Farmer, J.C. and R. Moon:** Occupational injuries of divers and compressed air workers. In *Occupational Injuries: Evaluation, Management, and Prevention,* T.N. Herington and L.H. Morse, eds. St. Louis, MO: Mosby – Year Book, Inc., 1995. pp. 423–445.
32. **Keatinge, W.R., M.G. Hayward, and N.K.I. McIver:** Hypothermia during saturation diving in the North Sea. *Br. Med. J. 290*:291 (1980).
33. **Timbal, J., H. Vieillefond, H. Guenard, and P. Varene:** Metabolism and heat losses of resting man in a hyperbaric helium atmosphere. *J. Appl. Physiol. 36(4)*:444–448 (1974).
34. **Maitland, G. and A. Findling:** Structural analysis of hyperbaric speech under three gases. *Aerospace Med. 45(4)*:380–385 (1974).
35. **Donald, K.W.:** Oxygen poisoning in man. *Br. Med. J. 1*:712–717 (1947).
36. **Clark, J.M. and C.J. Lambertsen:** Pulmonary oxygen toxicity: a review. *Pharmacol. Rev. 23(4)*:37–133 (1971).
37. **Jacobson, J.M., J.R. Michael, R.A. Meyers, M.B. Bradley, et al.:** Hyperbaric oxygen toxicity: role of thromboxane. *J. Appl. Physiol. 72(2)*:416–422 (1992).
38. **Piantadosi, C.A.:** Physiologic effects of altered barometric pressure. In *Patty's Industrial Hygiene and Toxicology*, 4th ed., vol. 1A. New York: Wiley & Sons, 1982. pp. 329–359.
39. **Lipsett, M., D. Shusterman, and R.R. Beard:** Inorganic compounds of oxygen, nitrogen and carbon. In *Patty's Industrial Hygiene and Toxicology*, 4th ed., vol. 2F. New York: Wiley & Sons, 1994. pp. 4523–4643.
40. **Harabin, L.D. Homer, P.K. Weathersby, and E.T. Flynn:** An analysis of decrements in vital capacity as an index of pulmonay toxicity. *J. Appl. Physiol. 63(3)*:1130–1135 (1987).
41. **Case, E.M. and J.B.S. Haldane:** Human physiology under high pressure I: Effects of nitrogen, carbon dioxide, and cold. *J. Hygiene. 41(3)*:225–249 (1941).
42. **Warkander, D.E., W.T. Norfleet, G.K. Nagasawa, and C.E.G. Lundgren:** CO2 retention with minimal symptoms but severe dysfunction during wet simulated dives to 6.8 atm abs. *Undersea Biomed. Res. 17(6)*:515–523 (1990).
43. **Hesser, C.M., L. Fagraeus, and J. Adolfson:** Roles of nitrogen, oxygen, and carbon dioxide in compressed-air narcosis. *Undersea Biomed. Res. 5(4)*:391–400 (1978).
44. **Anonymous:** Divers' ear. *Br. Med. J.* 2:1104–1105 (1977).
45. **Schane, W.:** Prevention of skin problems in saturation diving. *Undersea Biomed. Res. 18(3)*:205–207 (1991).
46. **Victorov, A.N., V.K. Ilyin, N.A. Policarpov, M.P. Bragina, et al.:** Microbiologic hazards for inhabitants of deep diving hyperbaric complexes. *Undersea Biomed. Res. 19(3)*:209–213 (1992).
47. **Summitt, J.K. and S.D. Reimers:** Noise — a hazard to divers and hyperbaric chamber personnel. *Aerospace Med. 42(11)*:1173–1177 (1971).
48. **Brady, J.I., Jr., J.K. Summitt, and T.E. Berghage:** An audiometric survey of navy divers. *Undersea Biomed. Res. 3(1)*:41–47 (1976).
49. **Tobias, J.V.:** Effects of underwater noise on human hearing. *Polish J. Occup. Med. Environ. Health 5(2)*:153–157 (1992).
50. **Ng, T.P., K.H. Yeung, and F.J. O'Kelly:** Silica hazard of caisson construction in Hong Kong. *J. Soc. Occup. Med. 37(2)*:62–65 (1987).
51. **Lam, W.K., T.W. Tsin, and T.P. Ng:** Radon hazard from caisson and tunnel construction in Hong Kong. *Ann. Occup. Hygiene 32(3)*:317–323 (1988).
52. **Dorr, V.A.:** Fire studies in oxygen-enriched atmospheres. *J. Fire Flammability* 1:91–106 (1970).
53. **Garges, L.M.:** Maxillary sinus barotrauma — case report and review. *Aviat. Space Environ. Med. 56(8)*:796–802 (1985).
54. **Melamed, Y., A. Shupak, and H. Bitterman:** Medical problems associated with underwater diving. *N. Eng. J. Med. 326(1)*:30–35 (1992).
55. **Molenat, F.A. and A.H. Boussuges:** Rupture of the stomach complicating diving accidents. *Undersea Hyperbaric Med. 22(1)*:87–96 (1995).
56. **Arthur, D.C. and R.A. Margulies:** The pathophysiology, presentation, and triage of altitude-related decompression sickness associated with hypobaric chamber operation. *Aviat. Space Environ. Med. 53(5)*:489–494 (1982).
57. **Elliott, D.H. and R.E. Moon:** Manifestations of the decompression disorders. In *The Physiology and*

Medicine of Diving. London: W.B. Saunders, 1993. pp. 481–505.

58. **Rivera, J.C.:** Decompression sickness among divers: an analysis of 935 cases. *Mil. Med. 129(4)*:314–334 (1964).
59. **Decompression sickness central registry and radiological panel:** asceptic bone necrosis in commercial divers. *Lancet II*:384–388 (1981).
60. **Golding, F.C., P. Griffiths, H.V. Hempleman, W.D.M. Paton, et al.:** Decompression Sickness during construction of the Dartford Tunnel. *Br. J. Ind. Med. 17*:167–180 (1960).
61. **Lo, W.K. and F.J. O'Kelly:** Health experience of compressed air workers during construction of the mass transit railway in Hong Kong. *J. Soc. Occup. Med. 37(2)*:48–51 (1987).
62. **Lee, H.S., O.Y. Chan, and W.H. Phoon:** Occupational health experience in the construction of phase I of the mass rapid transit system in Singapore. *J. Soc. Occup. Med. 38(1-2)*:3–8 (1988).
63. **Tikuisis, P., P.K. Weathersby, and R.Y. Nishi:** Maximum likelihood analysis of air and HeO_2 dives. *Aviat. Space Environ. Med. 62(5)*:425–431 (1991).
64. **Bason, R., H. Pheeny, and F. Dully:** Incidence of decompression ssickness in Navy low pressure chambers. *Aviat. Space Environ. Med. 47*:995–997 (1976).
65. **Crowell, L.B.:** A five-year survey of hypobaric chamber physiological incidents in the Canadian forces. *Aviat. Space Environ. Med. 54(11)*:1034–1036 (1983).
66. **Piwinski, S., R. Cassingham, J. Mills, A. Sippo, et al.:** Decompression sickness incidence over 63 months of hypobaric chamber operation. *Aviat. Space Environ. Med. 57(11)*:1097–1101 (1986).
67. **Bason, R. and D. Yacavone:** Decompression sickness: U.S. Navy Altitude Chamber Experience 1 October 1981– 30 September 1988. *Aviat. Space Environ. Med. 62(12)*:1180–1184 (1991).
68. **Conkin, J. and H.D. Van Liew:** Failure of the straight-line DCS boundary when extrapolated to the hypobaric realm. *Aviat. Space Environ. Med. 63(11)*:965–970 (1992).
69. **Bornstein, A. and E. Plate:** Uber chronische Gelekveranderungen Enstanden Durch Presslugterkrankung. Fortchr. Feb. *Roetgenstrahlem. 18*:197–206 (1911-1912).
70. **Harrison, J.A.B.:** Aseptic bone necrosis in Naval clearance divers: radiographic findings. *Proc. Roy. Soc. Med. 64*:1276–1278 (1971).
71. **Lee, T.C. and K. Neville:** Barometric medicine. In *Environmental and Occupational Medicine*, 2nd ed., W.N. Rom, ed. Boston, MA: Little, Brown and Company, 1992. pp. 1133–1142.
72. **Downs, G.J. and E.P. Kindwall:** Aseptic necrosis in caisson workers: a new set of decompression tables. *Aviat. Space Environ. Med. 57*:569–574 (1986).
73. **Gregg, P.J. and D.N. Walder:** Caisson disease of bone. *Clin. Orthop. 210(5)*:43–54 (1986).
74. **Spencer, M.P.:** Decompression limits for compressed air determined by ultrasonically detected blood bubbles. *J. Appl. Physiol. 40(2)*:229–235 (1976).
75. **Eckenhoff, R.G., S.F. Osborne, J.W. Parker, and K.R. Bondi:** Direct ascent from shallow air saturation exposures. *Undersea Biomed. Res. 13(3)*:305–316 (1986).
76. **Ikeda, T., Y. Okamoto, and A. Hashimoto:** Bubble formation and decompression sickness on direct ascent from shallow air saturation diving. *Aviat. Space Environ. Med. 64(2)*:121–125 (1993).
77. **Boycott, A.E., G.C.C. Damant, and J.S. Haldane:** The prevention of compressed-air illness. *J. Hyg. 8(3)*:342–443 (1908).

The foundry industry involves the casting of molten metal into various shapes and subsequent treatment of the product. Pouring of molten metal results in the release of metal fume into the air.

Chapter 24

Outcome Competencies

After completing this chapter, the user should be able to:

1. Define underlined terms used in this chapter.
2. Evaluate the role of major physiological systems relating to heat strain
3. Evaluate thermal injuries and prescribe first aid treatment
4. Evaluate the role of cold in cumulative trauma disorders
5. Evaluate environments that may precipitate heat/cold stress problems
6. Analyze heat/cold stress problems to identify the most effective and practical solutions
7. Describe administrative, engineering, and job controls
8. Develop appropriate mechanisms to identify heat/cold problems arising from new situations
9. Develop appropriate mechanisms to prevent new heat/cold problems
10. Describe the major physiological systems that respond to heat and cold stress
11. Evaluate the worker with respect to potential heat/cold strain problems
12. Develop practical methods for controlling heat/cold stress

Prerequsite Knowledge

Basic biology, basic chemistry, and basic physics.

Key Terms

acclimatization • axillary • circadian effects • conduction • convection • countercurrent heat exchange • electrolytes • frostbite • heat strain • heat stress • hidromeiosis • hunter reflex • hyponatremia • hypothermia • hypotonic • im/clo ratio • internal body • internal body temperature • microenvironment • natural wet bulb • outer shell • psychometric • pyrogenic • radiation • shock • strain • stressors • vasoactive • vasoconstrict • vasodilation • vortex coolers • water intoxication • wet-bulb globe temperature

Key Topics

I. Introduction
   A. Personal Protective Clothing and Equipment
   B. Thermal Strain Disorders
   C. Heat Stroke
   D. First Aid for Heat Stroke
   E. Heat Exhaustion and Cramps
   F. Heat Rash and Other Problems
   G. Cold Strain Disorders
   H. Heat and Cold Injuries

II. Physiology of Thermoregulation
   A. Heat Production

III. Mechanisms of Thermal Exchange
   A. Evaporation
   B. Physiology

IV. Assessing Environmental and Microenvironmental Stress
   A. Assessing Cold

V. Assessing Environmental and Microenvironmental Strain
   A. Anatomical Aspects of Heat Distribution

VI. Measuring Temperatures of Workers
   A. Internal Body Temperature
   B. Other Methods of Estimating Internal Body Temperature
   C. Skin Temperature

VII. Factors Affecting Thermal Strain
   A. Controllable Factors Affecting Strain

VIII. Uncontrollable Factors Affecting Strain
   A. Body Size
   B. Age
   C. Worker Health
   D. Heat Stress and Reproduction
   E. Gender Differences in Thermal Tolerance

IX. Controlling Thermal Exposure
   A. Administrative Controls
   B. Engineering Controls
   C. Environmental Controls
   D. Microenvironmental Control
   E. Microclimate Cooling

X. Prediction of Thermal Work Tolerance
   A. Generalized Prediction of Thermal Tolerance
   B. Individualized Prediction of Thermal Tolerance in Protective Clothing

XI. Future Directions for Work in Hot and Cold Environments and Protective Clothing

24

Applied Physiology of Thermoregulation and Exposure Control

Phillip A. Bishop, Ed.D.

Introduction

Purpose

The purpose of this chapter is to provide readers with the necessary background information to permit evaluation of the thermal characteristics of the working environment, determine the need for intervention, and devise appropriate, practical intervention methods. Hot and cold environments are thermal stressors, and workers experience a physiological strain as a result. It would be impossible to anticipate every possible situation that could arise and every potential solution that could be implemented. Therefore, the purpose of this chapter is to elucidate the basic principles necessary to enable occupational hygienists to solve the specific problems in their workplaces.

For example, an occupational hygienist might be called on to participate in the planning of a new plant, a new expansion of production or space, or a new process. Anticipating potential worker heat stress in the planning phase can improve safety and permit relatively low-cost improvements in environmental control systems that may preclude later health and productivity problems. In another scenario, the hygienist may need to plan for field work involving protective clothing. The biophysics of the microenvironment inside the clothing may greatly affect manpower needs or work-hour requirements. Understanding the microenvironment may save a company money and, more importantly, may prevent an injury or death. A hygienist also may be called on to help with some emergency in extremely hot or cold conditions. A basic knowledge of the biophysics of the thermal environment and how to minimize the dangers to workers while maximizing productivity is essential in all these situations. Similarly, knowledge of the advantages and disadvantages of different thermal remediation techniques may lead to more rapid and efficient solutions.

Essentially, hot and cold environments may reduce both safety and productivity. Besides the obvious dangers from frostbite or heatstroke, even a milder thermal strain can be problematic. Environments do not have to be life-threatening to cause problems. A cold environment, for example, may increase the risk of cumulative trauma disorders.[1,2] Performance deteriorates during cold stress before physiological limits have been reached.[3] Even if the internal body is warm, hand dexterity begins to decrease when skin temperatures fall to 15–20°C (59–68°F),[3,4] and muscle strength declines when muscle temperature falls below 28°C (82°F).[5] Likewise, a warm environment can compromise the concentration, steadiness, or vigilance of workers.[6] Even moderately warm environments may require interruption of work with extensive rest breaks, especially when protective clothing is worn (Figure 24.1).[7]

Across most of the United States, and in many other places in the world, the outdoor environment can pose problems for workers. For example, the average wet-bulb globe temperature (WBGT) in July for much of North and South America, Europe, and Asia, is above 29°C (84°F).[8] (WBGT is most commonly calculated as the sum of 0.7 natural wet bulb + 0.2 black globe + 0.1 dry bulb.) Temperatures do not have to be close to record-breaking to

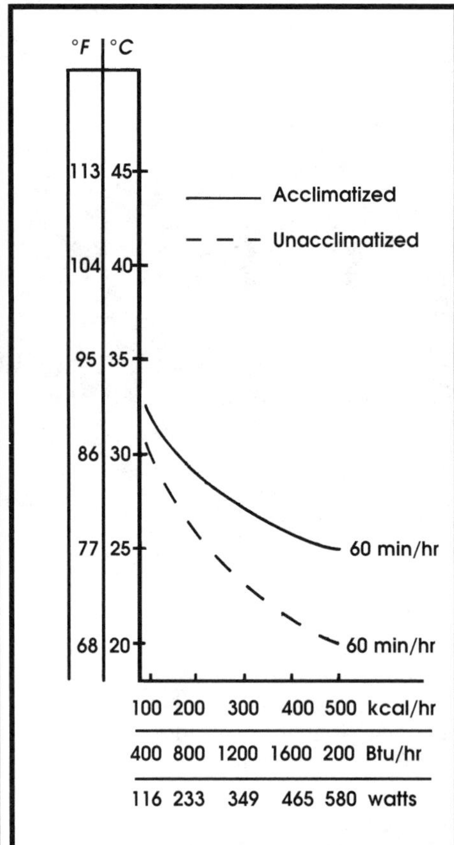

Figure 24.1 — ACGIH threshold limit values for heat acclimatized and unacclimatized workers.

produce problems. In the United States, for example, the first moderately warm days of spring tend to result in more heat-stress medical emergencies than the hotter days of summer, because of the lack of heat acclimatization. Likewise, cold exposure problems can be a serious threat to utility employees responding to an ice storm even if temperatures are not extremely cold.

Some potential hot areas are easily anticipated, such as foundry operations. Many occupational and protective clothing environments result in WBGTs as high as 43°C (109°F).[8] The heat stress is less obvious in other occupations and becomes apparent only through experience. For example, heating and air conditioning contractors working in the attic of a private dwelling can encounter dry bulb temperatures higher than 49°C (120° F), even when the weather is moderate. The situation is more complex when a worker must vary his or her work rate extensively while wearing semipermeable chemical protective clothing in cold climates. During the periods of heavy work, the worker may become overheated; yet during periods of very light work, the wetness accruing from previous sweating can result in hypothermia.

Personal Protective Clothing and Equipment

Protective clothing and equipment of all types may compound thermal problems for workers. Welders, for example, may suffer from the heat in warm environments and from the cold in cool environments because their protective clothing retains heat in the summer, but the design of welding masks and gloves hampers adding adequate insulation in winter. Class A, B, C, or D chemical protective clothing can create a microenvironment under the clothing that may be uncomfortably warm or hot during work even in otherwise comfortable ambient temperatures.

Thermal Strain Disorders

Human deep-body temperature is maintained within a few degrees of 37°C (98.6°F). Extreme variations from this internal body temperature can adversely affect important chemical reactions and the structure and function of proteins. The most common thermo-physiological threats to worker safety and productivity in industry are heat stress disorders. Of these, the most serious risk is heatstroke. If heatstroke continues unchecked, it will result in blood clots, tissue death, cerebral (brain) hypoxia, general central nervous system dysfunction, and finally death.[9] Death from heat strain occurs when internal body temperature approaches 43°C (109°F). Likewise, hypothermia, or low internal body temperature, can be dangerous. If the body temperature begins to drop, shivering will commence at 34°C (93°F), at 26.5°C (80°F) workers can become unresponsive, and death due to cold occurs rapidly at a rectal temperature of about 25°C (77°F).

Heat Stroke

There are two types of heatstroke. Classic heatstroke usually is a threat to the elderly, infirm, and very young during extremely hot weather. The type most likely encountered in industry is exertional heatstroke. In both types of heatstroke the body temperature rises, and brain damage may result.

Unacclimatized workers exposed to moderately warm temperatures or acclimatized workers exposed to hot temperatures should be continuously monitored to detect signs of potential heat injury. Typically, workers in the early stages of heatstroke evidence hot, dry skin (significant because normally the person would be sweating profusely); however, in exertional heatstroke the skin can also be wet. In both types the skin is typically red but may be mottled or pale blue-gray (indicating very low oxygen delivery), and internal body temperature will be very high.[10] The victim may evidence mental confusion or lose consciousness. Breathing may be faster and deeper than normal. All workers who may be exposed to heat stress, or who supervise others exposed to heat stress, should be reminded periodically of these symptoms and first aid procedures. Heat stroke is a life-threatening medical emergency, and trained medical help should be summoned immediately, even if the diagnosis is uncertain.

First Aid for Heat Stroke

While waiting for medical help or while transporting potential heatstroke victims, first aid should be initiated. For all heat injuries, remove the victim from the heat source. Stricken workers should be cooled as rapidly as possible, as danger to the victim increases the longer cooling is delayed. Cooling is best accomplished by maximizing airflow across the body by fanning, removing clothing, and whatever additional means are available. Supplementing this by applying ambient-temperature water to the victim enhances evaporative heat dissipation. Using whole-body ice or cold water immersion can decrease blood flow to the skin, thereby reducing and actually hindering body cooling. However, medical personnel at a major road-running race have reported very rapid lowering of internal body temperature by packing the torso in ice, while leaving the limbs at ambient temperature.[11] The American Red Cross recommends the use of ice packs on the victim's wrists, ankles, groin, each armpit, and the neck to cool the blood. In either case, cooling must be stopped once the victim's temperature is lowered to about 39°C (102°F) to prevent hypothermia. Conscious victims can be given half a glass (4 ounces) of water every 15 minutes (drinking too fast can cause vomiting).[12] Anyone giving first aid should observe universal precautions against bloodborne pathogens (i.e., avoid direct contact with body fluids).

Victims of heat injury can go into shock, or circulatory collapse, which can be life-threatening itself. Shock is characterized by rapid breathing and a rapid, irregular pulse. It should be treated by placing the victim in a comfortable horizontal position. If shock is present, and there are no head, neck, or back injuries, elevate the feet[12] and continue cooling the person, but do everything possible to secure immediate medical treatment. Administration of intravenous fluids by qualified personnel is usually needed, and getting the victim to medical care should not be delayed even if the person regains consciousness. Death typically occurs when internal body temperature exceeds 43°C (109°F) and is a result of cardiovascular failure, although brain damage occurs earlier in the process.[13]

Heat Exhaustion and Cramps

Less risky but more common heat injuries are heat exhaustion and heat cramps. Although these disorders do not put life at risk, they may be intermediate steps on the way to heatstroke. Workers who experience undue fatigue or muscle cramps while working in the heat are likely to be suffering from heat exhaustion and should be forced to sit or lie down in a cooler environment. Heat exhaustion victims often have a headache and may feel nauseous as well. Their skin is usually pale, and they may feel faint. They should be kept well hydrated and observed closely. If sweating stops suddenly, or the worker loses consciousness or becomes disoriented, he or she should be treated as a heatstroke victim and medical help summoned immediately. If heat cramps persist, an intravenous infusion, and hence skilled medical care, will be required. Any worker who has experienced a previous heat injury is more susceptible to a subsequent injury and should be afforded more protection, such as reduced heat exposure, more frequent or longer rest periods, etc.[14]

Since it may be difficult to distinguish between heat exhaustion and exertional heatstroke, it is wise to observe all heat exhaustion victims closely. If they show any confusion, bizarre behavior, or loss of consciousness, they should be treated as heatstroke victims, and emergency medical assistance should be called immediately.

Heat Rash and Other Problems

The final common form of heat injury is heat-related rashes. Kerslake,[15] in his classic text on heat strain, provides a detailed explanation of how sweat glands can become clogged. Sweat glands often become infected, causing discomfort that may reduce worker productivity. If large areas of skin become affected, sweat production can be compromised, reducing heat tolerance. Professional medical treatment will be required.

Other complications of work in hot situations include sweat in the eyes and dripping sweat, which may damage sensitive equipment and cause electrical hazards. In impermeable clothing 100 mL or more of sweat can collect in boots and gloves. Thermal stress can distract workers, reduce concentration, and lead to early fatigue, all of which are likely to reduce worker safety. Ramsey and Kwon[16] have reported deterioration of other performance variables during heat stress situations.

Cold Strain Disorders

Cold strain injuries are also potentially dangerous. The most dangerous cold threat is hypothermia (abnormally low internal body temperature), which fortunately is rare in industry. However, anyone who could be exposed to near-freezing temperatures for prolonged periods should be trained in the prevention and treatment of hypothermia. This could include workers such as highway maintenance personnel and search and rescue personnel (including volunteers) working in cold climates.

Similar to the way in which high temperatures can cause heat strain, extremely low temperatures can interfere with vital biochemical processes. The best data suggests that humans with internal body temperatures of 25°C (77°F) or below would be expected to die. Symptoms of hypothermia include uncontrollable shivering and intense feelings of cold, falling blood pressure, and irregular heartbeat. Victims become incoherent and disoriented and may be very drowsy. Once hypothermia begins, the blood vessels near the skin may dilate, causing further heat loss,[10] which may result in further reduction in internal body temperature and cause death even after a person is moved to a warmer environment. Any person who cannot be warmed at the first onset of hypothermia symptoms should receive medical treatment as soon as possible. First aid consists of warming the person as rapidly as possible and protecting against shock by keeping the feet elevated and the trunk warm. Care should be taken to gently warm hypothermia victims to prevent cardiovascular problems. Since circulatory and ventilatory function may be compromised in hypothermia, cardiopulmonary resuscitation may be needed. The victim's pulse and breathing should be checked periodically.

Frostbite is more common than hypothermia. Frostbite is a result of freezing of the extracellular fluid in the skin, which can permanently damage the tissue. It usually occurs on the extremities, such as the tips of the fingers, ears, and nose and manifests itself with initial pain at the afflicted site, which subsides as nerves are damaged. The tissue becomes white or grayish. Since the face is often less protected than other body parts, the victim may be unaware of the first signs of frostbite there because he or she cannot see the discoloration. Though not life-threatening, frostbite damage can be severe and permanent, so prompt medical attention is required.

Rapid heat transfer by conduction can occur if a body part comes in contact with a very cold object, even if the ambient air temperature is mild. Workers should be trained to avoid contact with cold metal or with liquids of low vapor pressure such as alcohol or cleaning fluids, all of which can increase the possibility of frostbite.[3] Workers should try to avoid direct contact with metallic surfaces below 0°C (32°F).[4] Workers exposed to extreme cold (-25°C (-13°F) or below) or moderate cold (15°C [5°F] or below) with high wind (25 mph or greater, i.e., windchill effective temperatures of -51°C (-60°F) and below) are in danger of frostbite and hypothermia. However, frostbite can occur in even warmer temperatures if workers are not properly clothed.[10] Table 24.1 shows the effects of wind velocity on the freezing point for exposed flesh.

Problems can arise before serious cold injury occurs. Cold exposure can reduce dexterity and strength.[3] For example, even if the internal body temperature is normal, dexterity begins to decline when hand skin temperature falls to 15–20°C (59–68°F), which is a very common temperature for work in cold environments.

Heat and Cold Injuries

Protecting against heat strain disorders is primarily a matter of anticipating problems and trying to prevent even mild disorders from developing. Gradually acclimatizing workers, keeping them well-hydrated (discussed in more detail later), and detecting symptoms early are the most important defensive measures. Workers showing initial signs of any heat or cold disorder should be

Table 24.1 — Windchill Index[3,4,8,10,12,17,18]

There is moderate danger of freezing exposed flesh if the ambient temperature is less than:
-7°C (20°F) if wind velocity is over 20 mph
-12°C (10°F) if wind velocity is over 15 mph
-18°C (0°F) if wind velocity is over 10 mph
-23°C (-10°F) if wind velocity is over 5 mph
-29°C (-20°F) if wind velocity is over 0 mph

There is great danger of freezing exposed flesh if the ambient temperature is less than:
-29°C (-20°F) if wind velocity is over 20 mph
-23°C (-10°F) if wind velocity is over 25 mph

removed from exposure and given fluids for rehydration. It is especially important to note that the use of protective clothing may hamper both the worker's temperature control and rehydration. It is difficult to evaluate problems when workers are entirely clad in coveralls, respiratory protective masks, and other equipment. The need for decontamination before drinking also may greatly hamper adequate hydration.

Self-monitoring is often difficult, hence it is imperative that both workers and work site supervisors appreciate the importance of the cues the body may provide. Workers should be paired with others for work in hot or cold environments, with instructions to watch for signs of thermal strain in each other.[4] It is also important for workers and supervisors to realize that there is tremendous variability among workers in thermal tolerance; what one worker may find merely uncomfortable could result in a dangerous level of heat strain for other workers doing the same work in the same environment. Often it is the most industrious "get the job done" employees who place themselves at greatest risk for heat or cold injury.

Physiology of Thermoregulation

Thermoregulation in humans is very complicated and is not fully understood. Fortunately, the level of understanding needed to protect workers and maximize productivity is relatively basic. When the body uses carbohydrates, fats, and proteins for energy, about 70–80% of the energy available is released as heat.[17] This heat energy always moves from hotter to cooler locations, and in a closed ideal system would achieve equilibrium, or uniform distribution of heat energy. Ultimately, the uniform distribution of heat is inevitable. That is, no matter how good your thermos bottle may be, without the application of external heat, eventually your coffee will become the same temperature as the ambient environment.

Heat Production

Working muscles will always generate heat, whether it is the muscles of the stomach and small intestines or the biceps muscle in the arm. As heat is produced, muscle temperature rises, and heat transfer occurs from the muscle to the cooler blood. As the blood flows through the body, it encounters other cooler areas and transfers heat to them. So the heat generated in the muscles is carried to the inactive tissues and to the skin, which in most cases is much cooler than the blood. For simplicity then, in terms of thermal strain, the worker's body should be thought of as having two principal components: an internal body consisting of the brain and viscera and the outer shell comprised of the skin and muscles. The temperature in the shell may vary considerably, whereas ideally the internal body temperature is maintained within about 3°C (5°F) of its mean temperature (37°C, 98.6°F).

How much heat is produced is determined by the rate of muscular activity. A person resting quietly produces about 100 W (1.4 kcal/min) depending on body size. In general, the bigger the person the higher the resting metabolic heat production. Some normal bodily functions, such as food digestion, increase heat production even further.

During work, metabolic heat production increases. Among well-trained workers the metabolic rate can be sustained as high as 700 W (600 kcal/hour) for 2 or 3 hours. A sample of the metabolic requirements for various tasks is shown in Table 24.2. These metabolic rates were generally determined from a very small sample, and actual energy requirements for a given task will vary according to the mechanical efficiency of the task and the worker (efficiency tends to increase to an upper limit in relation to task experience), body or limb weight, loads carried, speed of activity, and clothing worn. Clothing may interfere with performance of a task because the bulk of clothes can reduce range of motion, gloves decrease dexterity, and the weight of clothing can contribute to energy costs and fatigue.

Mechanisms of Thermal Exchange

Evaporation

In humans the primary means for heat dissipation is the evaporation of sweat. Sweat, a hypotonic fluid (i.e., more dilute than most body fluids), is secreted from thousands of glands per square inch all across the body. The evaporation of sweat under optimal conditions requires the absorption of about 0.58 kcal/mL at normal body temperature. The author has noted sustained sweat rates as high as 10 mL/min for almost 8 hours. Sustained sweat rates as high as 1 L/hour have been reported,[18] and even higher rates have been reported for shorter durations.[15,19,20] Thus, if 1 L of

Table 24.2 — Approximate Metabolic Requirements of Some Representative Work Activities

Activity	*kcal/hour*
Keyboarding, quiet standing	108
Seated writing	120
Driving tractor	150
Sewing with a machine	190
Cooking	198
Machining	198
Sheet metal work	198
Carpet sweeping, wallpapering	200
General carpentry	210
Lathe operation	210
Welding	210
Raking	222
Electrical work	234
Mopping floors	252
Scraping and painting	260
Vacuuming	264
Feeding animals	264
Tapping and drilling	265
Laundry	270
Planting seedlings	288
Using a chain saw	305
Plastering	320
Walking—smooth road	325
Shoveling grain	348
Erecting mine supports	360
Forestry hoeing	372
Tipping molds	378
Drilling coal, rock	384
Walking at 4 mph	396
Steel forging	408
Shoveling coal	438
Mowing	456
Tending steel mill furnace	516
Trimming trees	528
Felling trees	540
Barn cleaning	552
Rapid marching	582
Digging trenches	594
Fast ax chopping	1212

Note: Adapted from McArdle et al.,[17] pp. 769–780. These approximate requirements are based on 68 kg body weight. Adjust by proportionate body weight (e.g., a worker weighing one-third more will roughly use one-third more energy). Values are approximate; see text for additional considerations.

sweat were totally evaporated from the skin in 1 hour, 580 kcal of heat would be lost, which would be more than adequate for even the most demanding work activities. For maximal effectiveness evaporation must occur from the skin, not from the clothing, although evaporation from the clothing may provide some cooling.[21] Regrettably, in most industrial situations, although 1 L of sweat may be produced in an hour, seldom can this much sweat be evaporated from the skin.

Thus, the body has a tremendous ability to dissipate heat under normal circumstances. With regard to heat balance, it is important to realize that in many industrial situations, the only means for workers to dissipate heat is by evaporation of sweat. Therefore, during heat stress, any time sweat evaporation is hampered, the potential for overheating becomes very serious.

Heat loss through sweat evaporation can be hampered in many industrial situations. Sweat evaporation is reduced when the relative humidity level nears 100%, and ambient temperature is near skin temperature. Humidity alone, which is a function of air temperature, is not the controlling factor in evaporation, but rather the gradient between the vapor pressure at skin temperature and the water vapor pressure of the air.[22] The cessation of evaporation occurs when the ambient water vapor pressure equals that of the skin. This is more likely to occur under any protective clothing and equipment such as welding aprons or encapsulating chemical protective coveralls. Under vapor-barrier clothing, such as coated Tyvek®, the microenvironment (i.e., the environment lying between clothing and skin) may become saturated by evaporated sweat and reach near 100% relative humidity after only a few minutes of work.

Similarly, any failure to produce adequate sweat will result in rapid overheating. Lack of sweat can occur due to extreme dehydration and in heatstroke when the sweat glands simply shut down.

One physiological sweating response of interest is hidromeiosis, or a small reduction in sweat production once the skin surface is wetted with sweat. This effect could possibly be used to advantage by prewetting the skin of workers who will be exposed to a hot humid macroenvironment or microenvironment (e.g., protective clothing use). The hidromeiosis effect could then be used to reduce excess sweating, or sweating at such a high rate that it cannot be evaporated to provide additional cooling, but merely drips away and contributes to dehydration. Dripped sweat carries away much less heat than sweat evaporated from the skin.

Convection

Convection results in heat loss by two avenues: (1) within the body by blood circulation, and (2) by movement of water or air across the skin. The transfer of heat by the flow of some liquid or gas is termed convection and is vital to thermoregulation in the body. This process controls the heat flow between the shell and internal body through the movement of blood. There are several situations in which convection may become important. If the skin is cooler than the internal body, heat can be moved via convection, with heat ultimately passed to the environment. However, if the skin is the same temperature or hotter than the blood, either no transfer will occur or the blood can actually gain heat, and the internal body can become still hotter.

In most cases the skin is cooler than the blood because the body takes steps to ensure this. In fact, the skin and surface muscles act dynamically to protect the stability of the internal body where most vital organs lie. In hot situations some amount of heat can be stored in this shell (periphery) before the internal body temperature starts to rise. Likewise, this shell tissue can cool significantly before the internal body temperature begins to drop.[23] Since convective heat flow between the internal body and the skin can be controlled by adjusting blood flow, a good system exists for controlling internal body temperature.

Blood flow in humans may produce some countercurrent heat exchange, which occurs when arteries, which carry oxygen-rich blood away from the heart, lie in very close proximity to the veins that bring blood back to the heart. In thermal extremes this countercurrent can be somewhat effective in helping control heat balance, with the rate of heat exchange dependent on the thermal gradient and proximity of the arteries and veins. For example, in very cold environments the blood leaving the hand will be relatively cold and the blood entering it relatively warm. When these vessels come together in the wrist, there will be a heat transfer such to warm the venous blood exiting the hand and cool the arterial blood entering the hand with the net effect of conserving internal body heat. Many cold-dwelling animals such as the penguin have highly effective countercurrent heat exchange anatomies to conserve heat.

The role of convective heat transfer by blood flow within the body has been tested as a marker for heat tolerance.[24] It was hypothesized that when skin temperature was the same as internal body temperature, humans could not continue to work in the heat. Eastman Kodak[4] also agreed the convergence of skin and internal body temperature marks the end of a person's capability to transfer heat out of the body. Although this is a theoretically sound argument and potentially useful in signaling a boundary to heat tolerance, Nunneley et al.[25] showed that skin temperature can actually exceed rectal temperature for some time without any significant change in the rate of rise of rectal temperature or heart rate. Workers were able to continue work and dissipate heat for substantial periods of time.

One explanation for this apparent paradox is that the mean skin temperature can exceed internal body temperature for some body surface areas (e.g., over working muscles), whereas the underlying tissue and the tissue in other parts of the body can be cooler than the internal body. Those parts of the body that are cooler than the internal body then provide avenues for some heat transfer.

Heat also is transferred by the motion of cooler air across the skin. Convective heat loss can be experienced as the dangerous windchill in cold environments, or the pleasant cool breeze in warm environments. The importance of convective heat loss is that equilibrium will occur eventually in any situation. When the air next to the skin is still, and there is no air convection, the air nearest the skin will achieve equilibrium of temperature by conduction and also may become saturated with water vapor from the skin. When this occurs, neither heat transfer by convection nor by evaporation of sweat can occur. This illustrates the need for air movement, especially if humidity is relatively high. If heat loss is desired, increasing the movement of even humid air across the skin will help to increase the evaporation rate.

In the cold, convective loss of heat is usually undesirable. If cool air is continuously brought into contact with the skin, that air will continue to absorb heat as it tries to reach equilibrium, and heat will be continuously lost. By the same token, in the absence of convection the air nearest the skin surface will gradually become saturated with water vapor, and evaporative cooling will be reduced. Likewise, it will move towards a temperature equilibrium with the skin, and conduction to the air will be reduced. Thus, when trying to protect from the cold, a layer of dead (nonconvective) air is needed to prevent continuous convective heat loss.

This principle is illustrated by the skin diver's wet suit for cold water activities. In the wet suit a layer of ambient water is brought in contact with the diver's skin, and quickly warmed to body temperature by conduction. As long as that water is held against the skin, little heat transfer occurs due to convection; however, conductive loss continues, as discussed next.

Conduction

Conductive heat transfer occurs when there is direct contact between a hotter and a colder substance. In most industrial situations conduction of heat to substances other than air is not usually important because there is relatively little contact between the skin and other materials. However, when workers are required to handle cold materials, particularly those with high thermal conductivity, conduction can lead to localized frostbite, which can be very serious. When large areas of the body are in contact with a surface of high conductivity (e.g., workers in prone postures lying on metal plating), the gain or loss of heat can be significant. Likewise, conduction through machinery and equipment and through tools can lead to serious burns if there is a substantial heat source and good conductivity.

The best conductors of heat are very dense materials, such as metals; the worst conductors are of low density, such as a vacuum or still (nonconvective) air. Even the insulating value of fur is primarily attributed to the air it traps.(15)

One can keep warm in a cold environment by layering clothing to trap nonmoving air between the multiple layers. If, however, the outer layer is snug because of the bulk of clothing underneath, the clothing may actually lose dead air space and become a conductor itself. For example, wearing three or four pairs of socks underneath boots that were intended to fit well with only a single pair, probably will result in a situation wherein the insulating layer of air is displaced by the sock, and the sock fabric may form a conductive link between the foot and the outside environment. Similarly, wearing a heavy coat over a down vest will compress the down and reduce the vest's insulating ability.

Conduction is also an important factor in wet clothing, in which the water will displace air. Since water is denser than air, it is 23 times more conductive,(13) which will result in a rapid loss of heat.

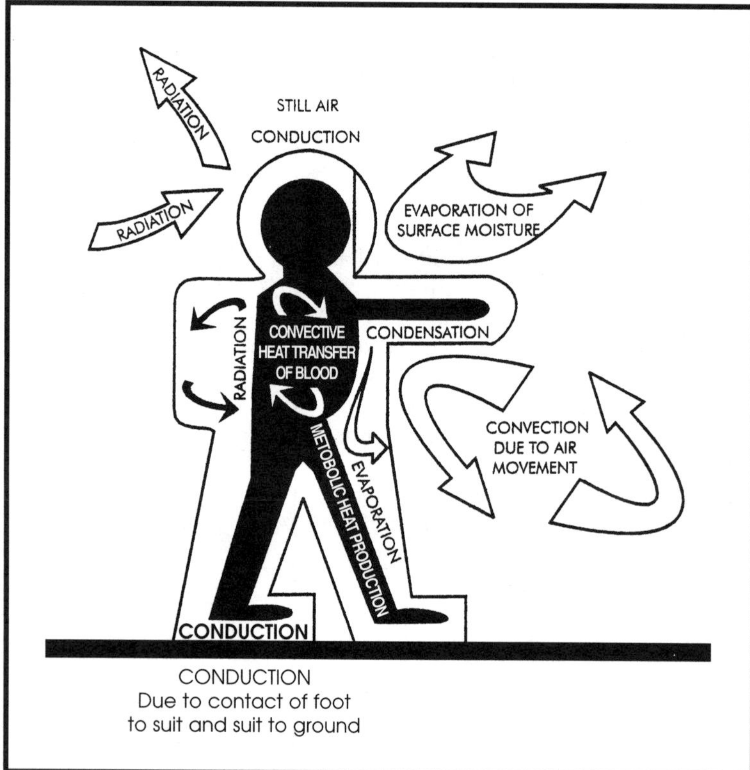

Figure 24.2 — Thermal factors affecting workers in clothing. Portrayed is a worker in Class A protective clothing. Heat is produced by the working muscles and metabolic processes. It is carried to the skin through blood flow (convection). From the skin it may be radiated, evaporated, conducted, and lost through convection (or in some cases gained) depending on the situation inside the suit (the microenvironment). The worker plus suit then reacts with the external (macro) environment to gain or lose heat depending on the conditions. The integration of both the micro and macro heat exchanges will determine whether the worker's average body temperature continuously rises or establishes an equilibrium at some higher temperature. (Drawing by John Kelley).

Radiation

Radiation is the electromagnetic transfer of heat energy without direct contact. Radiant heating from the sun provides the best illustration. Despite the vacuum of space, sunlight strikes the earth's surface and is both absorbed and reflected, producing heat. Workers in hot environments exposed to high radiant loads will benefit from shielding. This, of course, explains the appeal of shade to those laboring in the sun. It is important to recognize that all objects radiate to other objects, thus the total thermal radiation to which a worker is exposed is the sum of all direct and indirect (reflected) radiation, minus the worker's radiation to cooler objects. For simplicity, when the mean radiant temperature is above about 35°C (95°F) (a common mean skin temperature during work in warm environments), the body will gain heat, whereas below 35°C (95°F), the body loses heat through radiation. Again, both the macro- and the microenvironments must be taken into account.

Physiology

The integration of all the biophysical factors determines the physiological status of the worker. Figure 24.2 illustrates the composite problem. At the onset of work, blood temperature increases due to metabolic heat production from the working muscles. The hypothalamus, the brain's thermostat, integrates the signals from the body's internal sensors plus those from thermal sensors in the skin. The central processor signals appropriate responses at the skin level. If the clothing characteristics, the skin temperature, and the metabolic rate are such to cause an increase in internal body temperature, the response is to try to return to the normal internal body temperature (assuming no fever). To dissipate more heat, the body will dilate the skin blood vessels to permit convective heat transfer to the skin. Skin temperature will rise, allow convective heat loss if there is air movement, and conductive heat loss where there is contact with cooler materials. In addition, the skin will radiate to surrounding cooler objects. Heart rate will increase to provide this extra blood flow. Sweat glands will be signaled to start releasing sweat to the skin surface where it can be evaporated. Sweat rate will increase to a maximal level depending on the individual and the degree of heat acclimatization. If the sweat is not evaporated, some hidromeiosis will occur and feedback will reduce the sweat rate.

All of this will occur in the microenvironment between the clothing and the skin. Since the volume of air in the microenvironment is relatively small, it can be greatly affected by the heat and moisture delivered to it by the body. If it is relatively tightly sealed, the microenvironment may become very different from the external macroenvironment. However, if the microenvironment has good exchange with the external environment (clothing is porous, highly permeable to water vapor, loose fitting, with generous openings at the cuffs and neck), the two environments will be much more similar.

When heat dissipation equals heat production, a new internal body temperature equilibrium will be obtained. Once metabolic rate is decreased, the heat dissipation mechanisms will slowly reduce to return to that individual's normal internal body temperature of approximately 37°C (98.6°F).

If the worker's metabolic rate and clothing are such that internal body temperature begins to fall, shivering and voluntary movements will commence to generate more metabolic heat to try to maintain internal body temperature. If possible, the worker will add more clothing or seek a less drafty, warmer, work location.

Assessing Environmental and Microenvironmental Stress

For the occupational hygienist it is important to recognize that most thermoregulation problems deal with two environments, the macroenvironment of the office, tool room, warehouse, etc. and the microenvironment that lies underneath the clothing and protective equipment that a worker wears. The macroenvironment will influence the microenvironment, but the two are usually considerably different.

Totally encapsulating chemical protective clothing provides a very stark example of micro- and macroenvironmental differences. Consider a situation in which an individual is working hard while wearing an impermeable protective coverall, hood, gloves, and respirator. If the encapsulating suit is impermeable to the transmission of water vapor, predictably the suit will soon become saturated with sweat. If the environment is hot and the work rate is high, this may happen quickly, depending on the volume of air inside the suit. Even if the environment is relatively cool, and the worker is at rest, the continuous insensible water loss from the skin will eventually saturate the microenvironment. Likewise, the temperature inside the suit will become several degrees warmer than the ambient environment.

In other clothing the same will occur at a slower rate and to a lesser extent. In fact, in protecting against cold exposure, the microenvironment is what is most important to maintain. Table 24.3 shows the upward adjustments to the WBGT according to the clothing worn. Laboratory tests of the full Tyvek® suit adjustment (+10°C) found it to be to be fairly accurate.[26]

The American Conference of Governmental Industrial Hygienists (ACGIH) uses WBGT as a means for quantifying the hot environment. This system, invented by Yaglou and Minard in 1957[27]

Table 24.3 —
Addition (°C) to WBGT Needed to Account for Clothing

Clothing	Adjustment[A] (°C)	(°F)	Source
Shorts/seminude	-2	-4	Ramsey (1978)[B]
Summer work uniform	0	0	ACGIH (1990, 1991)[C,D]
Cotton coveralls	+2	+4	ACGIH (1990, 1991)[C,D]
Cotton coveralls	+3.6	+6.3	Kenney (1987)[E]
Impermeable jacket/body armor	+2	+4	Ramsey (1978)[B]
Winter work uniform	+4	+7	ACGIH (1990, 1991)[C,D]
Completely enclosed suits	+5	+9	Ramsey (1978)[B]
Water barrier, permeable	+6	+11	ACGIH (1991)[D]
Gortex	+6	+11	ACGIH (1990)[C]
Tyvek full suit	+10	+18	ACGIH (1990)[C]
Vapor barrier suit	+10.6	+19	Kenney (1987)[E]

[A]Ambient WBGT (°C) should be adjusted by the numbers shown to get the worker's WBGT for all heat stress/strain work schedule calculations.
[B]J.D. Ramsey. Abbreviated guidelines for heat stress exposure. *Am. Ind. Hyg. Assoc. J. 39*:491–495 (1978).
[C]American Conference of Governmental Industrial Hygienists (ACGIH). *Threshold Limit Values for 1990-1991*. Cincinnati, OH: ACGIH, 1990.
[D]American Conference of Governmental Industrial Hygienists (ACGIH). *Threshold Limit Values for 1991-1992*. Cincinnati, OH: ACGIH, 1991.
[E]W.L. Kenney. WBGT adjustments for protective clothing. *Am. Ind. Hyg. Assoc. J. 48*:576–577 (1987). Data adjusted to show summer work uniform as 0°C.

for military work in open areas, relies on three different thermometers to assess the convective, conductive, radiant, and evaporative heat stress: a dry bulb thermometer to measure air temperature, a wet-bulb thermometer with a wetted wick covering the bulb and exposed to the air to measure humidity, and a 15-cm diameter flat black globe to estimate the radiant load. There are two types of wet bulb measurement: natural wet bulb relies on ambient air motion, or the psychometric (artificially ventilated) wet bulb. For WBGT the natural wet bulb should be used when air motion is minimal, because natural wind motion affects how fast the air in contact with the skin is replaced by cooler, drier air.[3,15] The natural wet bulb relies on the ambient air movement as opposed to a maximal airflow in the psychometric wet bulb. Measuring both the natural and psychometric wet bulb values provides the best assessment for corrective action. The wet-bulb temperature should always be measured in the environment experienced by the workers. If there is air motion, the wet bulb temperature will be lower, but if no air motion is present, the natural wet bulb will best relate to the heat strain experienced by the workers. The readings from the three thermometers are weighted and summed to give the WBGT.

The readings of each of these thermometers are based on their relative impact on human thermal strain. Since simple conductive cooling has little impact on the body's cooling ability, dry-bulb temperature is usually weighted as 10% of the WBGT. However, because the evaporation of sweat is very important in body temperature control, the natural wet bulb is the most important measure. The wet bulb is weighted as 70% of the WBGT.

Finally, the globe thermometer is used to gauge radiant stress. On a cloudy day the potential for heat stress will be lower, as will the globe temperature. This temperature is weighted as 20% of the WBGT. Indoors, without a radiant load, the weightings are 70% for the wet bulb and 30% for the globe or dry-bulb temperature (which should be identical in the absence of significant radiant energy). In any indoor situation with a radiant heating source (such as foundry operations) the radiant load can be high, and the globe temperature cannot be ignored, so the globe temperature should be incorporated into the WBGT.

Antunano et al.[28] tested a Heat-Humidity Index, which consisted of equal weightings of wet-bulb and dry-bulb temperatures (i.e., weightings of 0.5 and 0.5 for the wet and globe temperatures, as opposed to the normal inside WBGT, which is the sum of 0.7 natural wet bulb + 0.3 black globe), for use with heavy semipermeable chemical protective clothing (U.S. military chemical defense ensemble that is very thick). They found that the new weightings were better related to heat strain than the WBGT. They suggest that other weightings might be better for thin, impermeable clothing. This is logical in that the microenvironment under the protective clothing quickly becomes saturated and minimizes the impact of environmental humidity, although this has not been established definitively.

Assessing Cold

The windchill index is a combination of dry-bulb temperature and wind speed. Windchill has become a popular measure of cold stress because convective loss of heat from the skin by the movement of air is extremely important in the cold. Windchill tables that are calculated as equivalent temperatures based on wind speed and ambient dry-bulb temperature can be found in variety of sources.[3,4,10,12,17,18] Table 24.1 provides a summary of the windchill index.

Most people are aware that ambient humidity plays a role in comfort in the cold. This is not because of the impact on sweat evaporation but rather because higher humidity seems to sensitize the cold sensory nerves. This suggests that whereas workers in more humid environments may sense the cold more, they are not at an increased risk for frostbite or hypothermia relative to drier cold and might in fact be protected by the increased discomfort.

Assessing Environmental and Microenvironmental Strain

The first and simplest step in assessing the thermal environment is to recognize that many jobs are inherently thermally stressful. High temperatures are frequently encountered in smelting, boiler cleaning or maintenance, plastics extrusion or molding, asphalt paving or roofing in the summer, and fire fighting. Low temperature jobs include emergency work in inclement weather (including unseasonable but otherwise mildly cool weather), and any task in which workers may inadvertently become wet in cold environments.

Since work involves the production of energy through metabolism, workers are constantly producing heat. The higher the work rate, the higher the rate of heat production. Whether a worker is cooling off or heating up depends on the balance of heat production through metabolism combined with heat gains or losses caused by interaction with the environment. This is typically illustrated in the heat balance equation discussed later. The body uses metabolic production to help control body temperature. Hence, shivering is an involuntary mechanism that causes the muscles to raise their metabolic rate to produce more heat. Lethargy reduces the metabolic rate and results in less heat production.

Since at any given moment different parts of the body are individually generating various amounts of heat and experiencing different interactions with the external environments, the heat balance equation is very complex. This partly explains why simple equations cannot be used to predict accurately how workers will respond to a given environment. However, understanding the heat balance equation is very useful in predicting thermal problems for workers and devising solutions.

A qualitative assessment of the heat balance equation can be very useful in assessing the potential for heat or cold problems. Consider the example of a worker laboring at an accident site with a hazardous chemical spill on an interstate highway on a warm spring day. If the air temperature is 18°C (64°F), the humidity level is 50%, the wind is still, the sun is out, and the worker is performing moderately hard work in Class A protective clothing (i.e., totally enclosed suit with a self-contained breathing system that offers the greatest protection from chemical hazards), the suit's microenvironment creates a very good potential for heat problems despite the relatively mild temperatures. The microenvironment will quickly become saturated with moisture resulting in minimal evaporative cooling. Convective cooling by outside air is also prevented by the suit. If the metabolic heat production is high, the lack of evaporative and convective cooling can lead to considerable heat strain.

Although a very quantitative approach to evaluating thermal strain is useful, sole reliance on quantitative evaluation is usually ill-advised. Attempts to predict any individual worker's response to a given thermal environment is very difficult, and there is much data to support great variability among workers in response to the environment.(19,29,30) Exactly what causes this variability is not fully understood. Undoubtedly, there are many differences among workers in size, state of acclimatization, amount and distribution of fat (which is insulative), distribution and function of the cardiovascular and sweat production systems, and the maximal sustainable output of blood from the heart. Theoretically, a given quantity of heat will raise the temperature of a smaller worker more rapidly than that of a larger worker. However, this simple observation is complicated by the fact that the surface area-to-mass ratio is higher in smaller workers than in bigger workers. More will be said about the contribution of individual factors later in this chapter.

Since humans are so highly variable, the only way to determine the true thermal strain is to measure the individual's response. In hot environments internal body temperature should be monitored as the best single gauge of heat strain; but in the cold, skin temperature is also important, because frostbite can easily occur while internal body temperature is normal.

Safe prediction of the heat strain experienced by workers is extremely difficult.(31) At present, the simplest approach is to monitor environmental conditions and take appropriate precautions in hot or cold situations. It must be kept in mind that some workers may experience heat intolerance problems even in very mild conditions. Ideally, every worker's internal body temperature would be monitored, and protective and rehabilitative procedures would be instituted whenever internal body temperatures reached a certain predetermined threshold. Unfortunately, that is not practical at present. Recently there have been some attempts to measure skin or ear canal temperature to predict internal body temperature, but these procedures have not been found valid.(32,34)

The *Occupational Safety and Health Guidance Manual for Hazardous Waste Site Activities*(14) says workers should be monitored when ambient work temperature is above 21°C (70°F). Monitoring workers consists of measuring heart rate, oral temperature, and body weight (which reflects sweat losses). If heart rate exceeds 110 beats/min or oral temperature exceeds 37.6°C (99.7°F), work periods must be shortened and rest periods lengthened to reduce the physiological strain. In addition, the National Institute for Occupational Safety and Health (NIOSH)(14) recommends that in temperatures of 32°C (90°F) (with no sun) or higher, when workers are wearing impermeable clothing, physiological monitoring of acclimatized workers should be repeated every 15 minutes. Konz also recommends(3) multiple measures rather than a single measure of heat strain. This is to preclude a dangerous assumption based on a single faulty measurement. Multiple measures might include heart rate plus temperature, when practical.

Obviously, for workers in contaminated environments physiological monitoring would likely prove impractical. However, this does highlight the need for constant vigilance and extreme caution when workers wearing protective clothing are exposed to warm temperatures.

Anatomical Aspects of Heat Distribution

Heat is not evenly distributed throughout the body. Gradients of one sort exist continuously. The most obvious heat gradient exists between the internal body and the shell. The deepest internal body tissues generally are maintained at about 37°C (99°F), controlled by the hypothalamus in the brain. Variation in internal body temperature among people is well known. It is not uncommon for healthy individuals to have a normal internal body temperature of 38°C (100°F) or higher.

Whatever an individual's normal internal body temperature may be, it is typically much higher than the skin or shell temperature. The skin temperature is variable, being greatly influenced by the environment. Skin temperatures in the low 30s centigrade (mid-80s Fahrenheit) are reported frequently, and temperatures as high as 38°C (100°F) can be seen as well, particularly under protective clothing.[(24,25)]

In hot conditions the blood flow rising to the skin may produce a flushed appearance in some workers. While this increased blood flow to the skin increases the convective heat flow, it also requires extra work by the heart and may compromise blood flow to the viscera. Many workers sense this increased demand and reduce their work rate to lower the strain on the heart. When workers cannot or will not reduce their work rate, the added strain may be enough to result in heat exhaustion, heatstroke, or a heart attack (in those with existing heart disease).

All workers in hot conditions experience an elevated heart rate relative to the rate they would have doing the same work in a cooler environment. There is a close relationship between heart rate and rectal temperature in an individual working at a steady rate. Anyone supervising workers laboring in hot environments must be made aware of the increased risk of heart attack.

Since the extra blood flow to the skin raises the heart rate, some have proposed using heart rate as a gauge of heat strain. At rest or in steady-state work this could be very accurate. Unfortunately, many factors affect heart rate, with work rate being only one of the major influences. Heart rate can also be influenced by posture, hydration status, changes in level of physical fitness, use of stimulants including caffeine, and others. Bernard and Kenney[(32)] have proposed a sliding average of heart rate that may be useful.

Measuring Temperatures of Workers

Internal Body Temperature

The internal body temperature is the temperature of the body's internal organs. Internal body temperature is considered to be the best single measure of heat strain. Unfortunately, this measure is very difficult to obtain and can only be accurately measured during physical work by inserting a thermometer to make contact with the body's central deep internal tissue. This can be done in the following three ways: (1) inserting a thermometer 8–12 cm into the rectum, (2) inserting a wire thermometer through the nose or mouth and down the esophagus to approximately heart level, or (3) having the subject swallow a radiotelethermometer that can transmit temperature data to the outside. Understandably, most workers are not enthusiastic about taking part in any of these options.

Several alternatives have been suggested to overcome the difficulties of inserting thermometers deep into the body. Oral temperature and axillary (armpit) temperature are used clinically to measure pyrogenic (fever) changes in body temperature. A distinction must be made between work-induced and pyrogenic-induced elevations in body temperature. Work-induced temperature increases result from the difference between the greatly increased heat production in the muscles and heat dissipation rate from the skin. Pyrogenic temperature rise is due to a resetting of the thermostat in the hypothalamus of the brain, which lowers heat dissipation with only a small rise in heat production. Consequently, workers with a fever may feel cold, even though they have an elevated internal body temperature. Also, in work-induced temperature increases, the blood flow is much higher than in pyrogenic temperature increases.[(33)] Differences between pyrogenic and work-induced temperature increases are important because methods typically acceptable for measuring temperature increases during pyrogenic sickness have not proven accurate during work.[34–36]

For example, oral temperature is typically used by physicians to determine the presence of a fever. However, oral temperatures during physical work can be very unreliable. Some marathon runners, for example, have had highly elevated rectal temperatures, yet will show near-normal oral temperatures. Likewise, some physicians now use ear canal or tympanic (ear drum) measurements of body temperature. Studies during physical work show these measures to be quite variable for measuring internal body temperature during work, depending on the particular conditions.[34–36]

Other Methods of Estimating Internal Body Temperature

There have been some creative attempts to use other measures to estimate internal body temperature. Skin temperature has been proposed and tested.[32] Since the heart rate reflects both work strain and heat strain on the body, Bernard and Kenney[32] proposed combining heart rate and chest temperature information to monitor the overall strain on workers. However, tests of a device that measured heart rate and chest temperature, using chemical protective clothing, have not been reassuring.[37]

Another commercially available device uses ear canal temperature to predict internal body temperature. One of these systems recognizes that there are discrepancies between ear canal and internal body temperature and uses prework oral temperature to provide a correction to ear canal temperature as an estimator of internal body temperature. Although reports on the accuracy of this device are lacking, research suggests that this approach may also be inaccurate.[38] Practical measurements or estimates of internal body temperature would contribute greatly to worker safety. Hopefully, technological advances will permit an accurate noninvasive measurement of internal body temperature in the next few years.

Skin Temperature

Skin temperature is very easy to measure at any particular site on the body, but varies from site to site on any given individual according to both environment and activity. At rest in a cool climate, the skin temperature in body extremities will be relatively low and may vary by several degrees across the body. In hot environments skin temperatures become more uniform, but variation remains. Workers exposed to high radiant heat will obviously experience high skin temperatures on the exposed side and much lower temperatures on the shaded side. In the cold the lowest skin temperatures are found in feet, followed by hands, with highest skin temperatures at the head.[39]

Skin temperature is important because it usually determines comfort level. Sensors in the skin alert us to potential damage. The pain threshold varies from 45 to 77°C (113–171°F) and the first-degree burn threshold from 60 to 138°C (140–280°F) depending on the conductivity of the material in contact with the skin of the finger (temperatures based on a 1-second contact).

Skin temperature is usually spoken of as mean (average) skin temperature, which is usually a weighted average of the individual temperatures of several sites.[40,41] In some industrial situations, the skin temperature at a specific site may well be more important than the mean temperature (e.g., workers with a body part exposed to a high heat source). Skin temperature at an individual body surface location can be measured by simply taping on a skin thermistor or thermocouple, which are commercially available as small, flat, wired disks. If the temperature of bare skin is desirable, you should be careful not to apply too much tape or other insulation, as this will reduce heat loss and raise the skin temperature artificially. Small amounts of tape will do this, but the result will be only a slightly more conservative reading in the heat and a slightly less conservative reading in the cold.

Factors Affecting Thermal Strain

Heat stress is the external heat load placed on the body due to the characteristics of the environment, and heat strain is the body's response. Workers show a large variation in heat strain, even though all may be working at the same work rate in the same environment. This is because many controllable and uncontrollable factors affect heat strain, as summarized in Table 24.4.

Controllable Factors Affecting Strain

Clothing

In protecting against cold, clothing insulating values are measured in clo units. One clo equals the insulating value needed

Table 24.4 — Summary of Variables That May Influence Work in Thermally Stressful Environments

Variable	*Impact*
Controllable variables	
Work task	
Rate	Metabolic rate influences heat storage rate
Type	Mobility influences ability and type of cooling possible; psychomotor function may be affected
Workers (may be controlled through selection/training)	
Physical fitness	Improved fitness increases thermal tolerance
Training	Increases safety
Acclimatization	Increases tolerance; with impermeable clothing, sweat effects may be mitigated
Size	Both mass/area and absolute mass may influence tolerance
Body fat content	Theoretically may influence heat loss
Hydration	Dehydration increases heat injury risk; repeated workdays may affect prework hydration and electrolytes
Electrolyte levels	Will influence rehydration and physiological function
Health	Fever, other illness, or medications may affect tolerance
Genetics	Large interindividual variability
Gender	Generally unstudied; thermoregulation shows a sex difference, but not in overall response, except that females may be less cold-tolerant
Age	Does not affect thermal tolerance except to the degree it affects physical fitness
Clothing (including protective clothing) required	
Insulating value	As insulation increases, potential heat loss decreases
Permeability	As permeability decreases, less opportunity for sweat evaporation; this may retard cooling or increase clothing wetness
Weight	Increases in weight increase metabolic requirements
Stiffness	Increases in stiffness raise metabolic costs of movement
Glove/mitten	Effects dexterity and hand/arm type fatigue
Gas mask	Reduces field of vision, may fog, impedes communication, raises metabolic costs
Uncontrollable factors	
Work task	
Type	Mobility will influence ability and type of cooling possible; psychomotor function may be affected
Workers	
Size	Both mass/area and absolute mass may influence tolerance
Body fat content	Theoretically may influence heat loss
Genetics	Large interindividual variability
Gender	Thermoregulation shows particular sex differences, but not in overall response
Age	Does not affect thermal tolerance except to the degree it affects physical fitness
Environment	
Temperature	Increases in temperature increase heat storage
Humidity	Major impact on heat tolerance, but in protective clothing the role of humidity is less
Radiant load	Can be a major heat source
Wind velocity	Can play a major role in heat loss, in both hot and cold environments

for someone to be comfortable sitting in a typical office environment of 21°C (70°F), 50% relative humidity, and air speed of 10 cm/sec (20 ft/min). Clo is generally determined empirically (by observation of experimental conditions) in the laboratory. A value of 3 clo would then represent three such layers of clothing. Clo values are readily obtainable from many sources.[3,4] Practically speaking, clo values are often more useful for comparing clothing than for trying to calculate any one individual's specific responses to the environment. The clo needed is a function of the metabolic heat production. Figure 24.3 is an illustration of the relationship between activity, temperature, and clothing needed.

Clothing keeps us warm because it creates a microenvironment that results in a comfortable balance between heat production and heat loss. In other words, properly chosen clothing reduces conduction, convection, evaporation, and radiation in such a way that the metabolic heat production will maintain internal body temperature and skin temperature. The introduction of a strong draft (i.e., convective heat loss) will alter the clo level needed for comfort, as will an increase in metabolic rate.

Sweat evaporation is important both to minimize condensation that would wet the clothing and to prevent saturation of the microenvironment. For this reason, the ratio of clo to the impermeability-to-water-vapor is an important characteristic of clothing, known as the im/clo ratio. The higher the clo, the warmer the clothing. The higher the im/clo ratio, the greater the problems in evaporating sweat.

Acclimatization

Among controllable factors, the most important is acclimatization. When workers are exposed to hot environments, particularly when they perform physical labor in hot environments, their bodies gradually adapt in several ways.

Heat acclimatization.—In heat acclimatization, the human body adaptations include

- Increase in the amount of sweat, which increases evaporative cooling potential
- Earlier onset of sweating, which reduces heat storage prior to activation of evaporative cooling
- More dilute sweat (lower salt concentration), which reduces electrolyte losses

- Increased skin blood flow, which provides greater convective heat transfer between internal body and skin
- Reduction in heart rate at the same work rate, which lowers cardiovascular strain and the oxygen requirements of the heart
- Greater use of fat catabolism during heavy work, which conserves carbohydrates that are useful when very high rates of energy production are needed.
- Reduction in skin and internal body temperature at the same work rate, which maintains a larger heat storage reserve and permits the worker to work at a higher maximal rate.[17]

These adaptations all function in concert to lower the strain at a given work rate and hot environment. Increased sweating makes increased evaporative cooling possible, and earlier sweating reduces the amount of heat stored in the body. The fewer electrolytes in sweat, the less body electrolyte deficit incurred in prolonged sweating. The increased skin blood flow enhances convective heat losses, and the reduced heart rate lowers the energy requirements of the heart, which gives the heart more rest. The increased fat catabolism conserves the body's store of carbohydrates that are essential for nervous system metabolism. All these work together to reduce the internal body and skin temperatures (i.e., heat strain) for a given amount of work, providing a greater reserve for emergency or prolonged work requirements.

Acclimatization can make a large impact on work tolerance in the heat for the lightly clothed worker. Wyndham[43] reported that well-acclimatized South African miners could maintain productivity up to a wet-bulb temperature of about 29°C (84°F), but above this temperature productivity rapidly fell to approximately 50% at a wet-bulb temperature of about 33°C (91°F). Therefore, despite a high degree of acclimatization, when environmental conditions are extreme, productivity will inevitably fall, and this reduced productivity should be anticipated. However, if workers wear impermeable encapsulating chemical protective clothing, some of the value of acclimatization may be mitigated. The increased sweating under impermeable clothing may actually hasten dehydration. The other changes would be positive, and heat acclimatization most likely enhances work tolerance and certainly reduces cardiovascular strain in the heat in protective clothing.

Heat acclimatization occurs very rapidly, with substantial adaptation apparent after only 2 hours of heat exposure per day[18] for 8 consecutive days. Additional acclimatization will continue to occur with additional exposure. Figure 24.4 illustrates the change in heart rate and rectal temperature during a 9-day heat exposure. Note the gradual change in physiological strain evidenced by the lower temperature and heart rate for the same work rate. This suggests that, if the seasonal changes are gradual, people working outside will make a natural adaptation to either heat or cold. However, sudden changes in weather may result in dangerous strain levels, particularly of heat strain. Obviously, workers just beginning to work around foundries and similar high heat-stress jobs will not have a natural acclimatization, so supervisors need to be diligent to ensure that workers have enough time to acclimatize. Acclimatization to one heat level may only partially acclimate the individual to higher heat exposures. Likewise, acclimatization may be temporarily lost after a long weekend or a vacation. The more time an individual spends away from the heat, the longer time is required for readaptation.

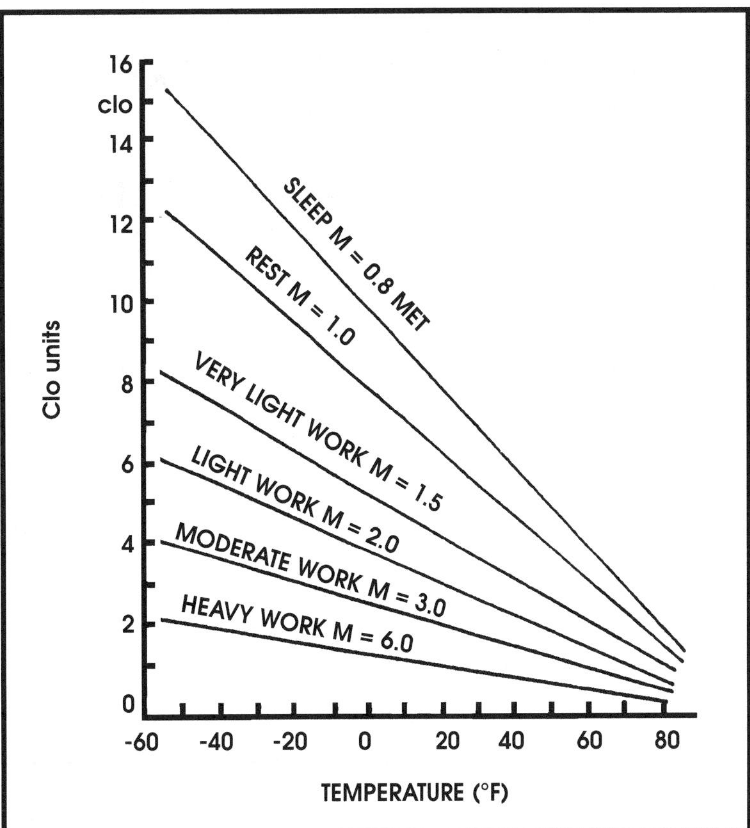

Figure 24.3 — Insulation of clothing and air required at different metabolic rates to maintain thermal comfort. The unit clo is defined in text. One MET equals the resting metabolic rate of approximately 100 W (3.6 mL/kg of body weight/min of oxygen consumption, or an average of 200–250 mL/min). (From Burton & Edholm,[42] with permission).

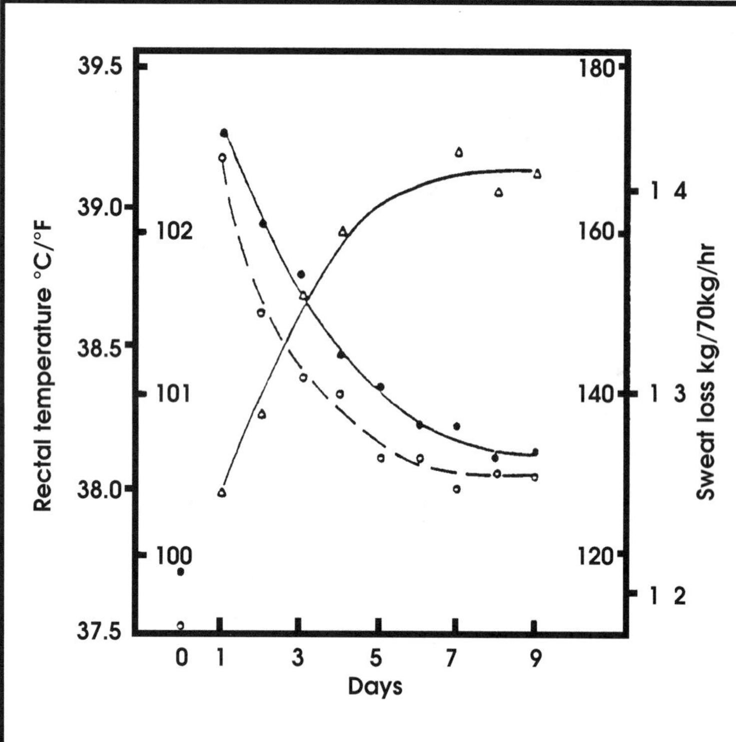

Figure 24.4 — Typical average rectal temperatures (•), pulse rates (°), and sweat losses (Δ) of a group of men during the development of acclimatization to heat. On Day 0 the men worked for 100 min at an energy expenditure of 349 W (300 kcal/hour) in a cool climate; the exposure was repeated on Days 1 to 9, but in a hot climate with dry- and wet-bulb temperatures of 48.9 and 26.7°C (120 and 80°F). (Used with permission from C.S. Leithead and A.R. Lind, *Heat Stress and Heat Disorders*. London: Cassell, 1964).

Cold Acclimatization.—Cold acclimatization, which is much less profound than heat acclimatization, produces a lowered internal body temperature and an increased blood flow through the exposed extremities.[17] These changes help conserve heat by reducing the heat loss gradient.

Table 24.5 —
Fluid Replacement Guidelines

- Workers should be careful to consume a well-balanced diet and drink plenty of nonalcoholic beverages in the day preceding severe heat exposure.
- Workers should avoid diuretic drinks immediately prior to work and drink as much as a half liter prior to commencement of work.
- During work, workers should try to drink as much and as frequently as possible.
- Workers should be provided cool drinks that appeal to them. Fluids can contain 40–80 g/L of sugar and 0.5 to 0.7 g/L of sodium.
- Workers should be encouraged to rehydrate between work shifts.
- Body weight should be monitored at the start and end of each shift to ensure that progressive dehydration is not occurring.

Note: These guidelines were adapted in part from McArdle et al.,[17] pp. 814–818.

One physiological aspect of cold tolerance that is not a part of acclimatization is the hunter or Lewis reflex. When extremities are very cold, the blood vessels will vasoconstrict (become smaller in diameter, which reduces blood flow) to conserve heat for the internal body organs. When the hunter reflex occurs, the finger tips, palms, toes, sole of foot, ear lobe, and parts of the face react to the cold exposure by occasional vasodilation (increases in vessel diameter, which increases blood flow) that periodically rewarms peripheral tissues without the loss of excess heat that would occur if higher temperatures were maintained constantly in these areas.[13] This periodic rewarming delays frostbite and minimizes heat loss from the internal body. This reflex is seen to varying degrees in humans, being well developed in some and virtually absent in others.

Physical Fitness

As might be expected, the greater the physical fitness level of the worker, the more the worker will adapt to, and tolerate, both the heat and the cold. Physical fitness leads to better thermal tolerance chiefly because fitness leads to increased blood volume and cardiovascular capabilities. Aerobic fitness is known to increase blood volume, cardiac stroke volume, maximal cardiac output, and capillarization of the muscles. These changes would lower the cardiovascular strain for any given work rate, as well as increase the reserves. The increased blood volume, for example, becomes important when blood must simultaneously supply the muscles with oxygen at the same time it must transport heat to the skin for heat dissipation.

Hydration

The most easily controlled factor in heat tolerance is the hydration level of workers. Central venous blood pressure, stroke volume of the heart (amount of blood pumped with each beat), and central blood volume are often lower in the heat. Research by Coyle and Montain[44] has shown that fluid replacement lowers body temperature and cardiovascular responses during work in the heat, both of which are linked to heatstroke probability. Providing copious amounts of potable fluids to workers is considered to be one of the most important precautions that can be taken to maintain the highest blood volume. It is a good idea to monitor body weight very carefully to detect chronic dehydration. Coyle and Montain[44] suggest workers try to drink enough to make up for about 80% of the working sweat loss (100% replacement is virtually unattainable in hard-working laborers). For example, if workers lose 1 kg (2.2 lbs) of sweat, they need to replace the lost sweat with 800 mL of fluid intake. Between shifts the remaining deficit should be recovered so that workers start each shift fully hydrated.

Workers chronically exposed to hot environments should have their body weights monitored. NIOSH[14] suggests that body weight loss in a workday should not exceed 1.5%. It is important to recognize that it takes a great deal of effort for workers to maintain body weight within 1.5% over the course of a work shift when engaged in heavy work resulting in heavy sweating. Regardless, complete rehydration should be achieved before the start of the next shift. Rehydration guidelines that may be helpful are included in Table 24.5.

When hydration is maintained, workers seem to be able to undergo several successive days of intense heat exposure without obvious cumulative adverse effects. Solomon et al.[20] measured the stability of tolerance for protective clothing within subjects across days. Six individuals wearing military protective clothing exercised at an external work rate of 38 watts (33 kcal/hour), in a 23°C (73°F) WBGT room, 4 hours per day, for 4 consecutive days. Subjects varied somewhat from day to day in work tolerance, work/rest ratio, sweat production, and perception of effort, but there were no clear trends in the data suggesting a cumulative increase or decrease in work tolerance.

It must be recognized that in the American culture, most workers tend to be reluctant to sacrifice convenience for safety. It is best if the employer maximizes the availability of appealing fluids. This may take some extra planning in situations requiring respirator use, since the respirator greatly inconveniences drinking. Coyle and Montain[44] suggest that the ability to tolerate large volumes of fluid while engaged in heavy work is a learned ability. Workers chronically exposed to situations eliciting heavy sweating should be made aware of this and encouraged to develop the ability to increase fluid intake.

Electrolytes

Sweat is a mixture of water, electrolytes, and lactic acid. Electrolytes are essentially salts, such as sodium, potassium, and calcium, that are vital to the normal functioning of the body. Sweat of acclimatized workers contains 1–4 g/L of electrolytes, and the sweat of unacclimatized workers contains even more. One study has shown that during heavy sweating, it may be very difficult to maintain the body's electrolyte levels. Armstrong et al.[45] reported fluid and electrolyte losses in a 20°C (68°F) WBGT (dry-bulb temperature = 30 and wet-bulb temperature = 18°C; 86 and 64°F, respectively) environment while wearing protective clothing. Their findings showed that the losses of the electrolytes sodium and calcium resulting from 6 hours of work can exceed the normal daily intake. This suggests that workers exposed to repeated days of protective clothing use in the heat or other situations inducing profuse sweating may incur electrolyte deficits that can pose a serious health risk. It was also shown that total body fluid losses and electrolyte losses in sweat were consistently greater than in urine. These findings suggest that monitoring electrolytes only in urine to protect workers is inadequate, and that attention should be given to replacing electrolytes in these workers.

Several investigators have suggested that adding large amounts of electrolytes to workers' drinks may slow down fluid absorption, thereby hindering rehydration. Consequently, manufacturers of some of the popular fluid replacement drinks have lowered the electrolyte content of these drinks. In contrast, Nadel et al.[46] argue that rehydration cannot be complete until all the electrolytes lost in sweat are restored such that all the fluid compartments are returned to preexposure status. They also point out that the thirst drive is reduced too quickly by drinking low-concentrate drinks and recommend that the drinks be as high in electrolytes as possible.

In further support of the need for electrolyte replacement is the occurrence of water intoxication or hyponatremia (low blood sodium levels), which can result from uncompensated electrolyte loss combined with ingestion of large quantities of water. In this situation, body electrolytes fall so low that coma can ensue. Regardless of what your workers' beverage of preference might be on the work site, they should be encouraged to drink as often and as much as they can tolerate. Those who sweat a great deal should be encouraged to eat foods high in electrolytes (e.g., cantaloupe, bananas) as well as drink electrolyte replacement drinks. The loss of 1–4 g of electrolytes per liter of sweat means a person losing ½ L of sweat per hour for 4 hours would still need up to 8 g of electrolyte replacement to return to the pre-work physiological status. Conversely, workers with diagnosed electrolyte problems or who are on sodium-restricted diets should not work in the heat. The typical American diet contains common foods that are high in electrolytes and can be taken in increased volumes when needed. This obviates the need for salt tablets, which are not generally recommended because too much salt can result in reduced work capacity, electrocardiogram changes, increased urine production, and reduced sweat production.

Medications

Many therapeutic and social drugs can have an impact on a worker's tolerance for heat or cold. A weekend of social drinking or a big party can leave a worker dangerously dehydrated. Some therapeutic drugs, such as heart-rate controlling (beta-blocking) drugs, will compromise work ability in jobs with high heart strain, such as moderate or hard labor while wearing protective clothing. Vasoactive (affecting blood vessel size) drugs can influence heat loss and blood supply and thereby contribute to hypothermia or frostbite. The *Physician's Desk Reference Guide to Drug Interactions, Side Effects and Indications* lists 15 different medications that may affect heat tolerance and many more that affect hydration levels in some manner.[47] Any worker who is taking any medication should receive medical clearance before being exposed to hot or cold conditions.

Uncontrollable Factors Affecting Strain

Body Size

There are also some uncontrollable factors that influence heat and cold tolerance. For example, body size and fat content will influence tolerance. With regard to heat, size can be both an asset and a liability. On one hand, the temperature rise in response to a given amount of heat energy is inversely proportionate to mass. That is, the bigger the person, the less increase in total body temperature per unit of heat stored, hence the less heat strain experienced. On the other hand, the larger the person, the greater the energy required to perform a task (and hence the higher the metabolic heat production), particularly for weight-supported activities such as walking. Also, the bigger the person, the lower the surface area-to-mass ratio, so the person's ability to dissipate heat is reduced, and it will take longer for the person to cool off after heat exposure. In the cold, size is generally an advantage because typically more heat is generated in the body, and the reduced surface area-to-mass ratio keeps the worker warmer.

Fat is a good insulator, which means that the fatter the person is, the less heat tolerant and more cold tolerant they should be. However, for some people, more fat may be located internally, while for others, more fat may be located under the skin (subcutaneously). Since fat located subcutaneously will have a greater impact in both hot and cold environments, those people will be more affected. Wherever the fat is located, the extra weight will raise the energy costs and metabolism for workers who must support and transport their body weight as part of the job (e.g., if the worker must squat repeatedly to pick up things). There is much individual variation in the influence of this factor on heat and cold tolerance.

Age

Although not well studied, thermal tolerance to both cold and heat tends to be generally unaffected by age,[48] although some earlier studies suggest that heat tolerance decreases in older people.[18] Any observed declines in thermal tolerance with age may be related more to decreased physical activity than aging as such; however, there is a definite decline in maximal work capacity with age. The fall in maximal cardiac output capacity with age probably contributes as well.

Worker Health

Sick workers are at special risk in stressful work environments. Fever is produced by the effects of circulating pyrogens on the central nervous system, and body temperature is regulated at a higher temperature than normal. This means that the same amount of work will produce the same heat storage, but at a higher, more dangerous temperature. A worker who ordinarily tolerates the heat well will thus be impaired.

Any disease that may influence cardiovascular or renal function or hydration state (e.g., diarrhea results in dehydration) may impact on thermal tolerance. Generally speaking, it is dangerous for the ill to work in hot environments.[49] Workers and supervisors should be trained to screen themselves and each other to avoid unsafe hot and cold exposures.

Heat Stress and Reproduction

The effects on reproduction of acute and chronic exposure to physical labor in combination with heat are not well known. The effects on pregnant women are potentially very serious. Exposure of pregnant women to high internal body temperatures during the first trimester (3 months) of pregnancy may result in fetus malformation.[3,17] This is vital information for workers, since it is possible that a female worker might not realize she is pregnant until well into the first trimester. It is well known that intense work can elevate rectal temperature as high as 40°C (104°F) even in moderate environments. Therefore, risks of heat exposure during pregnancy are quite serious. Efforts should be made to inform workers and supervisors fully regarding this risk. Likewise, female workers who might be pregnant should be protected from heat exposure.

In men, the potential impact of heat on reproduction would be chiefly manifest in terms of its effects on fertility. Bagatell and Bremner[50] found no effects for heavy exercise in heat on sperm count, sperm morphology, mean testosterone level, free testosterone, and other reproductive hormones. Hackney et al.[51] found testosterone levels to be lower in highly trained endurance athletes compared with controls; however, the control subjects were significantly heavier and fatter than the athletes. Griffith et al.[52] reported that a 2-week bout of intense exercise significantly decreased testosterone but not sperm count in endurance trained athletes. Based on the available research, it does not appear that heat has a major effect on reproduction in men, but further research may lead to other conclusions.

Gender Differences in Thermal Tolerance

The earliest detailed studies of gender differences in heat tolerance found that women's ability to thermoregulate was decidedly inferior to men's.[53–55] These observations supported the earlier studies of Hertig,[56] Wyndham,[55] and Fox,[57] which reported lower sweat production rates for women relative to men. Subsequent studies that matched the cardiorespiratory fitness levels and size of male and female subjects found that women's heat tolerance was at least as good and occasionally more favorable than that of men.[58–62] It appears then, that conflicting opinions on the gender difference in thermal tolerance may be attributed mainly to differences in fitness levels rather than differences in physiological responses.

Pandolf et al.[63] and Kenney[64] have verified these findings of minimal gender difference in heat tolerance. They found that males and females were similar in tolerance for work in heat, response to hypohydration, and response to a heat acclimation protocol. Kenney,[64] in a review, concluded that women might be slightly more efficient thermoregulators in hot humid environments because the frequently observed lower sweat rate for women relative to fitness-matched men[58,59,65,66] was an advantage. Lower sweat production rate slows dehydration and presents no disadvantage when the humidity level reduces the effectiveness of evaporative cooling to the point that more sweat is produced than evaporated. In work settings requiring semipermeable or impermeable protective clothing, this reduced sweat rate might also be advantageous in that the usefulness of evaporative cooling is minimal, and the ability to rehydrate would be compromised. Unfortunately, most studies of protective clothing have used male subjects, so gender differences in tolerance to protective clothing remain speculative.

With regard to cold environments, the larger proportion of insulating fat would theoretically offer a gender-related advantage to women. McArdle et al.[67] attributed an observed lower tolerance to cold to the greater surface area-to-mass ratio of females even at the same level of fatness. Another explanation offered was that women possess a less sensitive thermoregulatory system. This hypothesis is supported by the aforementioned lower sweat rates in women relative to men.

In a second study McArdle et al.[68] found that females at rest in water (20, 24, and 28°C; 68, 75, 82°F) were less tolerant of cold than males. Body fatness did correlate with the fall in rectal temperature of the males, but not the females. They also found in this study that exercise in water and air at 20, 24, and 28°C effectively attenuated the gender difference in cold tolerance seen in the previous resting study. Fat distribution advantage for females seemingly offset the previously mentioned area/weight disadvantage and also offset any difference in thermoregulation sensitivity. Pandolf et al.[69] concluded that ambient temperature, body size, and work type and rate were more important factors in cold tolerance than surface area-to-mass ratio and gender. It should be noted that the most cold-tolerant group ever studied is Korean and Japanese pearl divers, who are women.

There are some fundamental physiological differences between sexes in thermal tolerance; however, these differences tend to combine during work in such a way as to minimize the difference in overall response. Additionally, studies suggest no gender difference in heat acclimation.[69–71]

In summary, it appears that for very low work rates, such as inspection or supervision tasks, gender differences in cold tolerance are important. For work rates eliciting substantial amounts of metabolic heat production, gender responses are somewhat different, but the net effect is a similar overall response, regardless of gender. For work tasks requiring more than minimal energy expenditure, considerations other than gender, such as size, acclimation state, and physical work capacity, should be used in assigning workers to tasks involving thermal tolerance. Additionally, individual differences in response to heat loads in particular tend to be very great.[30,72–76] Gender differences in response to a broad range of work and environmental conditions need to be studied.

Controlling Thermal Exposure

Administrative Controls

Worker Selection

There are several ways of controlling the physical strains of hot and cold environments. As always, administrative controls can be used to advantage. One obvious though complex administrative approach is worker selection, which raises ethical and moral issues. For example, excluding women from some hot jobs may be unethical and illegal sex discrimination, but failure to exclude pregnant women from jobs that threaten heat strain is certainly unethical. Keeping workers with heart conditions from performing certain jobs may be highly ethical in some situations and unethical in others, depending on the circumstances. Ethical issues must be considered on a case-by-case basis.

Workers may be selected based on the nature of the work and a number of other criteria. Selecting workers based on obvious factors seems reasonable. For example, an acclimatized, fit, lean worker generally would be expected to tolerate greater heat stress than a fat, unfit, unacclimatized worker. Although this is generally true, the only way to determine worker tolerance is to observe workers over a period of time to see who is most tolerant of a given work load and environmental combination. Measuring core temperature response to work would be desirable but is generally impractical.

Worker Training

Worker training is always a good idea. Teaching workers to recognize potential hot/cold problems and training them to deal with these should improve both safety and productivity. For example, training workers to maintain good levels of hydration and to select protective clothing that provides the maximum protection with the minimum heat strain improves safety and productivity. Research has shown that if workers select coated Tyvek protective coveralls when only regular (uncoated) Tyvek is needed, the heat strain is much higher than necessary, and worker productivity and safety are reduced.[77] Workers and supervisors should be taught that therapeutic and recreational drugs may alter hot/cold tolerance, and that workers who report a fever should be protected from heat exposure.

Scheduling

Most industries do not have a great degree of scheduling flexibility, but annual planning and careful scheduling to minimize stressful exposure to heat or cold when possible would improve safety and increase productivity. Factors to be considered in scheduling include time of day, season, and locale. Time of day, especially for hot outside work, can have a major impact on heat stress. When possible, work in very hot outside environments should be scheduled for night, or for early and late in the day. Outside work requiring continual protective clothing use should, as far as practical, be scheduled for the coolest months.

But when such scheduling is possible, it must be done with attention to thermal physiology. For example, workers in protective clothing will not be greatly affected by humidity but will probably be affected by radiant heat exposure. Additionally, temperatures in attic-like areas often fall greatly at night. Circadian effects (the routine alterations in physiology related to time of day) can alter temperature as much as 1°C (2°F), which will influence both heat and cold tolerance.[33]

One industry with which many readers are familiar, American football, illustrates the failure to consider scheduling. The fact that professional and collegiate football exposes players to the heat of August (consider that they also wear heavy, insulating, protective clothing) is much more a case of tradition than necessity. It is a testimony to the power of tradition that the death of players each year has not resulted in a great outcry for rescheduling.

Work-Rest Intervals

Work-rest intervals are the means used by ACGIH to set controls for the environmental exposure of workers. A look at the threshold limit values for heat as presented in Figure 24.1 shows that as heat strain increases, the ratio of work to rest must fall. Safe and wise scheduling of work and rest is not as simple as it might at first appear. For example, as can be seen in Figure 24.5, it was discovered that in protective clothing in very hot environments when rest must occur while wearing the clothing, it was better to have workers work continuously rather than work and rest, because they were unable to cool off during rest.[78] Resting in the protective clothing in this situation would only result in a longer duration of heat exposure without an increase in productivity. The balance between work and rest must consider both safety and the thermal physiology, because workers resting in protective clothing may not cool much; however, doffing and donning protective clothing would increase worker risk of toxic chemical exposure.

ACGIH guidelines provide specific information on work/rest ratio requirements under specific environments. There is also useful information on work-rest intervals in other sources.[79,80]

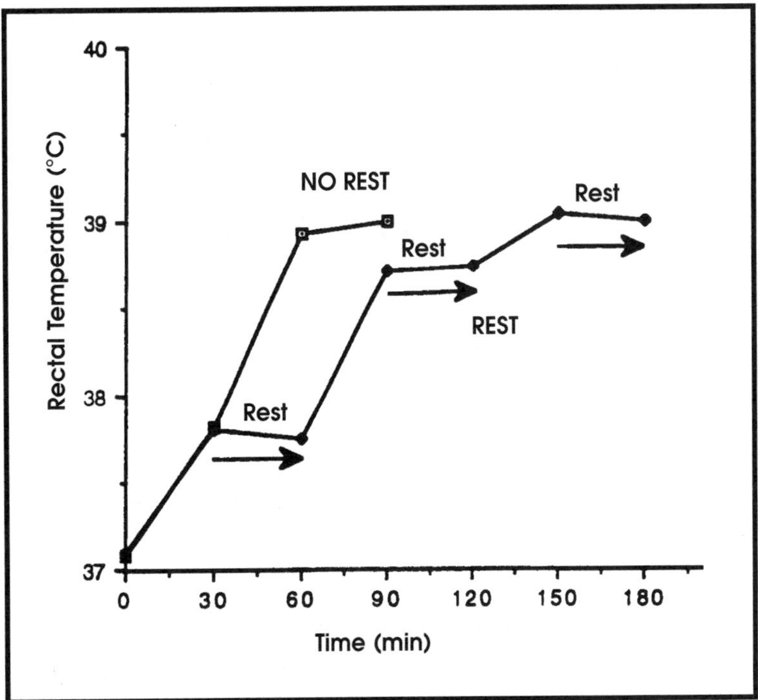

Figure 24.5 — Mean rectal temperatures for intermittent work in protective clothing with and without rest. Adapted with permission from Bishop et al.[78]

Engineering Controls

Engineering controls can also be employed to improve safety and productivity in hot and cold environments. For example, reducing the heat loss from kilns by improving seals and insulation will save money and lower the heat strain in the surrounding areas. Installing radiant heat shields may be very cost effective in many situations because of the resulting increase in worker productivity. Job redesign may be necessary to lower work rates (i.e., lower metabolic heat production) and reduce heat strain, which can increase productivity.

Eastman Kodak[4] lists several steps that can be taken to minimize thermal stress in hot environments, such as

- macro cooling (i.e., cooling the general work environment)
- reduction of ambient humidity
- increase in air velocity
- lowering metabolic rate by implementation of slower work rate or provision of mechanical assistance
- adjustment of clothing or addition of microenvironmental cooling
- provision of radiant shielding
- provision of a cool location, or at least spot-cooling, for rest breaks
- provision of plenty of palatable fluids of the workers' preference for rehydration

For cold environments, they recommend[4]

- reduced air velocity
- placement of windbreaks as needed
- minimization of drafts
- balancing of work rate so that periods of intense work are not followed by low work rates
- increasing the clothing insulation
- use of windproof clothing as appropriate
- providing opportunities and equipment to dry clothing that is wet or damp
- increasing radiant heat with micro- or spot heaters

Environmental Controls

Macroenvironmental change to reduce thermal problems is the most obvious but typically the least practical change. When these macroenvironments cannot be changed, the microenvironments can sometimes be improved. One simple and well-studied example is the microenvironment inside protective clothing. As previously discussed, this microenvironment can quickly become very warm and humid. Such a microenvironment can be cooled in a variety of ways.

Microenvironmental Control

The simplest means of controlling the microenvironment is to control the clothing worn. In hot jobs, adjust clothing to protect

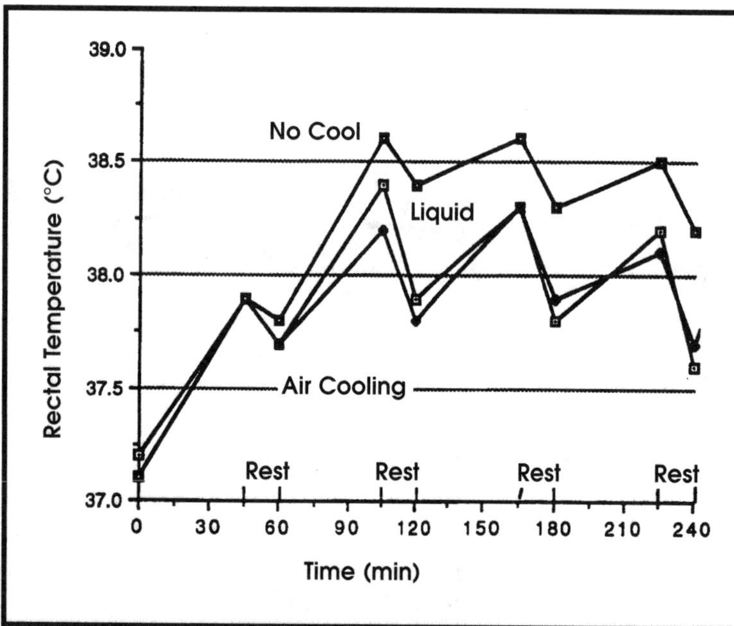

Figure 24.6 — Mean rectal temperatures for intermittent work in protective clothing (metabolic rate of 430 W or 370 kcal/hr) at WBGT = 25°C (77°F) with no cooling (n = 14), air and liquid cooling (n = 13) during rest. Number of subjects completing 240 min was: no cool = 5, liquid cool = 6, air cool = 11. Adapted with permission from Bishop et al.[78]

against radiant loads, but maximize air movement and evaporation of sweat. Sometimes this can be accomplished simply by minimizing clothing. If protective clothing must be worn, use a fabric or design that maximizes sweat evaporation and air movement.[78]

In the cold, microenvironmental control means using outer clothing that minimizes airflow across the skin and maximizes the insulating, or clo, value. There is a practical limit to how much clo can be raised, because of both the bulk and the increase in surface area. Increase in clo is negated, at some point, by the increase in surface area. This is readily illustrated in the problem of insulating small cylinders, such as the fingers. Adding a small amount of 4.7 clo/inch insulation actually increases the heat loss because of the resulting increase in diameter of the finger-plus-insulation, which greatly raises the surface area of the insulation-covered cylinder.[81] Thus, to some extent, increased insulation will be offset by increased surface area.

The microclimate of the head is also especially important, since at -4°C (25°F), the heat loss from the head on average is about 50% of resting metabolism. Adding 2.4 clo units of insulation to the head reduces heat loss equivalent to adding 4 clo to the rest of the body.[3]

Microclimate Cooling

As illustrated in Figure 24.6, all microcooling is most effective in hot temperatures; however, it declines in impact as the temperature falls. Likewise, good results have been observed in some situations by cooling workers only during rest. For example, a worker in protective clothing who must be mobile on a hazardous waste site might plan to spend 30–60 minutes working and plan to finish near a support site, where he or she could be attached to either a clean liquid or air cooling system for a rest break. It should be recognized that the worker's subjective comfort may be as important, and sometimes more important, than physiologic benefits.

Phase-Change Vests

The simplest cooling system that can be used under protective clothing or with regular work clothing is the ice or phase-change vest. This equipment basically consists of a vest and several ice packs that can be frozen and carried to the job site. These ice packs may contain only water but more often contain a mild antifreeze that lowers the freezing point slightly and thus can absorb more heat. Recently a cooling system has been developed using a substance that undergoes phase change at about 18°C (65°F). The advantage of this system is that the cooling packs can be refrozen at higher temperatures than water or water/ antifreeze. This permits refreezing in an ice chest.

These systems are among the least expensive. The main disadvantage to such systems is that the cooling packs tend to be heavy, which will add to the metabolic costs and heat production when the workers must move around. In addition, the cooling packs generally do not last very long; therefore, long jobs require many ice packs. Furthermore, if the cooling vests are worn under the protective clothing, replenishing the cooling is problematic. If they are worn outside the protective clothing, their effectiveness is reduced and vests are contaminated. The only temperature control in this system is how often the ice packs are replenished. Some users may experience unpleasant coldness when first donning the vest. Published reports are few, but most agree that this approach is one of the cheapest and is reasonably effective.

Liquid Cooling Systems

Another microcooling system commercially available is the circulated liquid system. This system has a closed-loop heat exchanger, usually in the form of long underwear with small-gauge tubing woven into it. Both torso and lower-body clothing are available. The clothing is worn next to the skin, and cool fluid is circulated by an external cooler, which may be as simple as a reservoir with a coolant pack and pump. The advantages of this system are that it can be worn under protective clothing, and the coolant packs can be replenished without removing the protective clothing and so with minimal interruption of work. Most systems offer the user control of the amount of cooling provided, and the system is fairly efficient. Numerous studies have found these liquid systems to be quite effective.

The main disadvantages of liquid cooling are that the suit adds a layer of insulation; the coolant packs have to be frequently replenished; and since the fluid is moved via a small pump, battery power must be replenished as well. Also, the coolant pack and pump tend to be heavy, which adds to the energy costs of the work. They are also subject to snagging in tight quarters. One caution would be that leaks from this system can pose shock and slip (fall) hazards and could damage some equipment.[81]

Air Cooling

Another common microcooling system is the air system. This is an open-loop system in which cool air is supplied via some type of manifold to a vest or other distribution system. The cool air can be provided from a conventional air cooler such as an air conditioner or by vortex coolers, which generate both a hot and cold stream of air from a compressed air source. Advantages of this system are that it is one of the most comfortable because the flow provides for evaporation of some sweat, and in most systems flow control and temperature can be regulated to the user's taste. The disadvantage is that these systems usually require a tether to provide a continuous supply of cool air. Since it is an open cooling system, it is simpler than the liquid system; however, the efficiency is lower because there is typically only a single pass of air across the body, and regardless of any remaining cooling capacity, it is then exhausted to the atmosphere. An open-loop air system can work in some protective clothing situations because the positive pressure created will protect the worker, but it is obviously unusable in a Class A (totally closed) protective clothing situation. Since this is an open cooling system, leaks are in most cases not problematic. In a toxic environment, extreme care must be taken to ensure that supply air is clean. Air cooling systems have been well studied and seem to work as well as liquid systems, and in at least one study worked better.[81]

Microenvironmental Heating

Conversely, in cold situations the use of space heaters, heated boots, socks, or gloves may alter the microenvironments to improve safety and cost-effectiveness. However, improperly used heaters can be dangerous due to fire and burn potential and carbon monoxide production. Since heat generation requires a lot of energy, the monetary cost of low productivity must be balanced against heating costs. Phase change, solid and liquid fuel, circulating fluid, and thermoelectric heating have been used to provide microenvironmental heating, but the physiological and cost-effectiveness of these methods have not been reported.

Prediction of Thermal Work Tolerance

Because hot environments (micro- and macro-) particularly limit work tolerance and pose a significant health and safety threat, there is a strong need for accurate prediction of worker tolerance in hot and cold environments. Prediction generally takes two forms: (1) a generalized prediction equation that attempts to predict for groups of people, and (2) an individualized approach that tries to use individual information to predict for a single worker.[31] Because of the great variability among workers in their heat/cold responses, one of the greatest needs in the prevention of heat strain is a practical means of monitoring or predicting heat strain within individuals rather than generally.

Generalized Prediction of Thermal Tolerance

Several attempts have been made at creating generalized models of the thermal environment that would predict worker tolerance based on work rate and environmental conditions. The U.S. Army has developed a predictive model for use with soldiers.[82] Another very complex model has been developed[83] and has been tested by the U.S. Air Force. Although often not good predictors of individual responses, generalized models can be made more specific by inserting personal data, such as acclimatization and fitness status of individual workers.

Also, the Navy's physiological heat exposure limits, which have been used to predict tolerance times under a variety of conditions,[80] may be useful in anticipating group mean responses to a given situation.

Individualized Prediction of Thermal Tolerance in Protective Clothing

Previous attempts to predict individual responses have not been accurate for a broad range of workers in a broad range of situations. Some type of personalized prediction is likely to be more accurate for individual workers than a generic prediction in which one tolerance time is predicted for all workers in a given situation.

Several studies have used field tests to predict individual's heat tolerance in protective clothing.[84–86] In contrast to the findings of Armstrong et al.,[45] heart rate response to exercise has not predicted tolerance.[86]

Unfortunately, a recently completed study of prediction based on an 8-hour test using a walk-plus-arm-work protocol proved much less successful. It appears that this prediction approach at present could only be used as a screening tool rather than a true predictor of performance.[87] Should this approach eventually prove accurate and applicable to other work rates and environments, safe and accurate prediction of work tolerance in the heat would be feasible.

Future Directions for Work in Hot and Cold Environments and Protective Clothing

There are numerous health, safety, and productivity problems associated with work in hot and cold that are not well studied. Field of vision restriction, loss of tactility, hindrances to communication, altered center of gravity, increased need to avoid equipment damage (e.g., tears in protective clothing, etc.), and loss of mobility raise the risk of injury and lower productivity of workers in heavy winter clothing or protective clothing. The economic costs to employers of adherence to current recommendations for work in very hot environments and the potential benefits resulting from improving workers' safety and productivity require that resources be devoted to improving protective clothing work practices and equipment.

A number of questions remain to be answered with respect to maximizing safety and work productivity. How can cold and hot weather clothing and protective clothing be redesigned to reduce ergonomic limitations and control heat transfer? Under what work and environmental conditions does microenvironmental heating and cooling become metabolically and economically cost-effective? What work-rest schedule is optimal with, and without, heating and cooling? What guidance should be provided for industry regarding microclimate cooling of protective clothing? How can workers be trained or conditioned to increase safety and productivity in hot and cold environments? Does acclimatization effectively improve work tolerance in protective clothing since sweat evaporation is restricted? These and other key questions await further investigation.

Summary

Protecting the worker against the stresses of the thermal environment requires recognition of potential problems, prevention of heat or cold strain, and determination of the most practical and effective solutions. This can best be achieved by teaching workers and supervisors to understand the fundamentals of the biophysics of worker thermoregulation and exposure control. Undoubtedly, there will be major advances in materials, equipment, and techniques in the future. Evaluating and selecting the most appropriate innovations in individual situations is again a function of grasping the underlying principles.

Work in hot environments, or warm environments in protective clothing, shortens work tolerance even for the most hardy workers, and work-rest scheduling may not extend work time effectively in hot, hazardous environments.[88] In previous military studies, workers in protective clothing have suffered from physical and heat exhaustion after relatively short work duration and even in cool temperatures (WBGT = 9°C, 48°F), which emphasizes the need for constant efforts to ensure worker safety even in cool ambient conditions when protective clothing is used. Worker safety in protective clothing is not strictly an issue of work rates. Workers in some military studies have performed relatively light duty tasks in protective clothing, such as patient medical care, and consequently suffered heat problems[89]

Microenvironmental cooling effectively increases work tolerance and worker comfort. Liquid, air, and phase-change cooling have all proved effective. Phase-change vest cooling presently appears the most feasible for short-term cooling of very mobile workers.

In addition to the serious diminution of physical work capacity, work in winter clothing and protective clothing can also result in possible loss of dexterity, strength, work capacity, and perceptual motor performance. This loss results in a higher safety risk as well as decreased productivity.

The mitigation of work capacity presents occupational hygienists and ergonomists with a very challenging problem. Currently, guidance for employers of personnel in hot environments is very restrictive of productivity and is thus problematic for industry. Continued research and cooperation is needed to maximize productivity while simultaneously minimizing health risks for workers in extremely hot and cold environments. A checklist to help improve worker safety and productivity in extreme environments is included as Tables 24.6 and 24.7.

Acknowledgment

The author appreciates the contributions of the late Alexander R. Lind, whose draft version of this chapter provided great assistance, and Stuart M.C. Lee for his assistance.

Additional Sources

Alpaugh, E.L.: Temperature extremes. In Fundamentals of Industrial Hygiene, 3rd ed., B.A. Plog, ed. Washington, DC: National Safety Council, 1988.

American Red Cross: American Red Cross Community First Aid and Safety. St. Louis, MO: Mosby Lifeline, 1993. pp. 221, 146–147.

Bureau of Medicine and Surgery, Navy Dept.: Manual of Naval Preventative Medicine, NAVMED-P-5010-3. Palo Alto, CA: Research Reports Center, 1988.

Table 24.6 — Checklist for Heat Exposures

- ❑ Are adequate supplies of palatable cool drinks available?
- ❑ What is the major source of heat stress and how can it be mitigated (e.g., protective clothing requires particular strategies)?
- ❑ If radiant shielding (including shade) is possible, is it in the most strategic location?
- ❑ Is temperature-monitoring equipment available at the work site?
- ❑ Are work guidelines that are appropriate to the situation available to workers and supervisors?
- ❑ Are first aid supplies available that are appropriate to heat/cold emergencies?
- ❑ Has an appropriate work rate been determined, and is there sufficient manpower to stay on schedule despite a slower work pace?
- ❑ Have supervisors been instructed to remove workers at the first sign of problems?
- ❑ Have workers been properly and thoroughly acclimatized (or reacclimatized after time away from the stressing environment)?
- ❑ Is a cool recovery/rest area available?
- ❑ Are workers and supervisors trained in recognizing the symptoms, and providing first-aid treatment of heat injury?
- ❑ Is there a means of calling emergency medical support? Do workers know how and where to call emergency medical support?
- ❑ Is the clothing appropriate (minimal obstruction of sweat evaporation and maximal protection from radiant heat; i.e., use the lightest, most permeable clothing that provides adequate safety)?
- ❑ Is air velocity as high as practical?
- ❑ Are workers well hydrated at the beginning of work?
- ❑ Is spot cooling available?
- ❑ Is microclimate cooling available as needed?
- ❑ Have workers who might be pregnant, or those with cardiovascular problems, previous heat injuries, on problematic medications, and who have fever, been protected from elevated internal body temperatures?
- ❑ Have workers been reminded of appropriate safety precautions?

Table 24.7 — Checklist for Cold Exposures

- ❑ Are workers and supervisors trained in recognizing the symptoms and providing first-aid treatment of frostbite and hypothermia?
- ❑ Is there a means of calling emergency medical support? Do workers know how and where to call emergency medical support?
- ❑ Are appropriate clothing and replacements for wet items available?
- ❑ Is emergency warming available?
- ❑ Are there facilities available for drying clothing items that become damp or wet?
- ❑ Are windbreaks erected in the most beneficial locations?
- ❑ Is a windchill chart available?
- ❑ Have supervisors been instructed to remove workers at the first sign of problems?
- ❑ Are hand/foot warmers available?
- ❑ Has the work rate been modified as much as possible to avoid following very high work rates with very low ones (i.e., avoid causing workers to sweat, followed by very low work rates that might cause them to become hypothermic)?
- ❑ Is spot warming available?
- ❑ Are drinks available? (Avoid drinks high in caffeine since caffeine is a vasodilator.[3])

Eastman Kodak Co.: Ergonomic Design for People at Work, vol. 1. Workplace, equipment and environmental design and information transfer, S.H. Rodgers, ed. London: Lifetime Learning Publications, 1983.

Electrical Power Research Institute: Heat Stress Management Program for Nuclear Power Plants, NP-4453. Palo Alto, CA, 1986.

Konz, S.: Work Design, 4th ed. Scottsdale, AZ: Publishing Horizons, 1995.

National Institute for Occupational Safety and Health (NIOSH), Occupational Safety and Health Administration (OSHA), U.S. Coast Guard (USCG), U.S. Environmental Protection Agency (EPA): Occupational Safety and Health Guidance Manual for Hazardous Waste Site Activities (DHHS/NIOSH pub. 85-115). Washington, DC: NIOSH, OSHA, USCG, EPA, 1985. pp. 8–21.

Pandolf, K.B., M.N. Sawka, and R.R. Gonzalez (eds.): Human Performance Physiology and Environmental Medicine at Terrestrial Extremes. Indianapolis, IN: Benchmark Press, 1988.

References

1. **Kroemer, K.H.:** Avoiding cumulative trauma disorders in shops and offices. *Am. Ind. Hyg. Assoc. J. 53*:596–604 (1992).
2. **Frederick, L.J.:** Cumulative trauma disorders an overview. *Am. Assoc. Occup. Health Nurses J. 40(3)*:113–116 (1992).
3. **Konz, S.:** *Work Design*, 4th ed. Scottsdale, AZ: Publishing Horizons, 1995. pp. 378, 381–383, 389, 390.
4. **Eastman Kodak Co.:** *Ergonomic Design for People at Work*, vol. 1. S.H. Rodgers, ed. London: Lifetime Learning Publications, 1983. pp. 253, 267, 271.
5. **Simonson, E. and A.R. Lind:** Fatigue in static effort. In *Physiology of Work Capacity and Fatigue*, E. Simonson, ed. Springfield, IL: C.C. Thomas Publishing, 1971.
6. **Robinson, M.C. and Bishop, P.A.:** Influence of thermal stress and cooling on fine motor and decoding skills. *Int. J. Sports Med. 9*:148 (1988). [Abstract]
7. **American Conference of Governmental Industrial Hygienists (ACGIH):** *1993-1994 Threshold Limit Values and Biological Exposure Limits.* Cincinnati, OH: ACGIH, 1993.
8. **Sawka, M.N., C.B. Wenger, and K.B. Pandolf:** Thermoregulatory responses to acute exercise-heat stress and heat acclimation. In *The Handbook of Physiology*, sec. 4, vol. I. M.J. Fregly and C.M. Blatteis, eds. New York: Oxford University Press, 1996, pp. 157–161.
9. **Werner, J.:** Temperature regulation during exercise: an overview. In *Perspectives in Exercise Science and Sports Medicine*, vol. 6. C.V. Gisolfi, D.R. Lamb, and E.R. Nadel, eds. Dubuque, IA: Brown and Benchmark, 1993. pp. 63–68.
10. **Alpaugh, E.L.:** Temperature extremes. In *Fundamentals of Industrial Hygiene*, 3rd ed. B.A. Plog, ed. Washington, DC: National Safety Council, 1988.
11. **Armstrong, L.E.:** "Treatment of Heat Stroke." May 1996. [Personal communication] University of Connecticut, Dept. of Exercise Science, Box U-110, 2095 Hillsdale Road, Storrs, CT 06269-1110.
12. **American Red Cross:** *American Red Cross Community First Aid and Safety.* St. Louis, MO: Mosby Lifeline, 1993. pp. 146, 147, 221.
13. **Folk, G.E.:** *Textbook of Environmental Physiology*, 2nd ed. Philadelphia, PA: Lea & Febiger, 1974. pp. 105, 154–155, 244.
14. **National Institute for Occupational Safety and Health, Occupational Safety and Health Administration, U.S. Coast Guard, and the U.S. Environmental Protection Agency:** *Occupational Safety and Health Guidance Manual for Hazardous Waste Site Activities* (DHHS [NIOSH] pub. 85-115). Washington, DC: U.S. Government Printing Office, 1985. pp. 8–21.
15. **Kerslake, D.M.:** *The Stress of Hot Environments.* Cambridge, UK: Cambridge University Press, 1972. pp. 95, 96, 134, 147, 238.
16. **Ramsey, J.D. and Y.G. Kwon:** Recommended alert limits for perceptual motor loss in hot environments. *Int. J. Ind. Ergonom. 9*:245–257 (1992).
17. **McArdle, W.D., F.I. Katch, and V.L. Katch:** *Exercise Physiology*, 4th ed. Baltimore, MD: Williams and Wilkins, 1996. pp. 157–158, 169, 512, 521.
18. **Brief, R.S.:** *Basic Industrial Hygiene, A Training Manual.* New York: Exxon Corp., 1975. pp. 189, 191, 192.
19. **Bishop, P.A., S.A. Nunneley, and S.H. Constable:** Comparisons of air

and liquid personal cooling for intermittent heavy work in moderate temperatures. *Am. Ind. Hyg. Assoc. J. 52*:393–397 (1991).

20. **Solomon, J., P. Bishop, J. Beaird, and J. Kime:** Responses to repeated days of light work at moderate temperature in protective clothing. *Am. Ind. Hyg. Assoc. J. 55*:16–19 (1994).
21. **Robinson, S., S.D. Gerking, and L.H. Newburgh:** Interim Report #27 to the CMR, Jul 1945. Cited in *The Physiology of Heat Regulation and the Science of Clothing*, L.H. Newburgh, ed. Philadelphia, PA: W.B. Saunders Co., 1949. p. 351.
22. **Goldman, R.F.:** Prediction of human heat tolerance. In *Environmental Stress: Individual Adaptations*, L.F. Folinsbee, J.A. Wagner, J.F. Borgai, B.L.Drinkwater, et al., eds. New York: Academic Press, 1978. pp. 57.
23. **Bazett, H.C.:** The regulation of body temperatures. In *The Physiology of Heat Regulation and the Science of Clothing*, L.H. Newburgh, ed. Philadelphia, PA: W.B. Saunders Co., 1949. p. 110.
24. **Pandolf, K.B., and R.F. Goldman:** Convergence of skin and rectal temperatures as a criterion for heat tolerance. *Aviat. Space Environ Med. 49(9)*:1095–1101 (1978).
25. **Nunneley, S.A., M.J. Antunano, and S.H. Bomalaski:** Thermal convergence fails to predict heat tolerance limits. *Aviat. Space Environ Med. 63*:886–890 (1992).
26. **Reneau, P. and P. Bishop:** A review of the suggested WBGT adjustment for encapsulating protective clothing. *Am. Ind. Hyg. Assoc. J. 57*:58–61 (1996).
27. **Yaglou, C.C. and D. Minard:** Control of heat casualties at military training camps. *Arch. Ind. Health 16*:302–316 (1957).
28. **Antunano, M.J. and S.A. Nunneley:** Heat stress in protective clothing: validation of a computer model and the heat-humidity index (HHI). *Aviat. Space Environ Med. 63*:1087–1092 (1992).
29. **Bishop, P.A., R.E. Pieroni, J.F. Smith, and S.H. Constable:** Limitation to heavy work at 21ºC of personnel wearing the U.S. military chemical defense ensemble. *Aviat. Space Environ. Med. 62(3)*:216–220 (1991).
30. **Mar'yanovich, A.T., V.S. Balandin, A.K. Bekuzalov, and G.M. Lapikov:** Individual features of responses to a combination of heat and physical exertion. *Hum. Physiol. 10*:49–55 (1984).
31. **Bishop, P.A.:** A new approach to predicting response to work in hot environments. In *Advances in Ergonomics and Safety II*, B. Das, ed. New York: Taylor & Francis, 1990. pp. 913–918.
32. **Bernard, T.E. and W.L. Kenney:** Rationale for a personal monitor for heat strain. *Am. Ind. Hyg. Assoc. J. 55*:505–514 (1994).
33. **Stitt, J.T.:** Central regulation of body temperature. In *Perspectives in Exercise Science and Sports Medicine*, vol. 6. C.V. Gisolfi, D.R. Lamb, and E.R. Nadel, eds. Dubuque, IA: Brown and Benchmark, 1993. p. 4.
34. **Morgans, L.F, S.A. Nunneley, and R.F. Stribley:** Influence of ambient and core temperatures on auditory canal temperature. *Aviat. Space Environ Med. 52(5)*:291–293 (1981).
35. **Armstrong, L.E., A.E. Crago, R. Adams, J.M. Senk, et al.:** Use of the infrared temperature scanner during triage of hyperthermic runners. *Sports Med. Train. Rehab. 5*:243–245 (1994).
36. **McCafferey, T.V., R.D. McCook, and R.D. Wurster:** Effect of head skin temperature on tympanic and oral temperature in man. *J. Appl. Physiol. 39(1)*:114–118 (1975).
37. **Reneau, P. and P. Bishop:** Validation of a personal heat stress monitor. *Am. Ind. Hyg. Assoc. J. 57*:650–657 (1996).
38. **Green, J.M., T. Clapp, D. Gu, and P.A. Bishop:** Validity of QUEST-TEMP II Personal Heat Strain Monitor. *Am. Ind. Hyg. Assoc. J.* [Submitted]
39. **Day, R.:** Regional heat loss. In *The Physiology of Heat Regulation and the Science of Clothing*, L.H. Newburgh, ed. Philadelphia, PA: W.B. Saunders Co., 1949. pp. 245–247.
40. **Burton, A.C.:** Human calorimetry II. The average temperature at the tissues of the body. *J. Nutr. 9*:261–280 (1935).
41. **Ramanathan, N.L.:** A new weighting system for mean surface temperature of the human body. *J. Appl. Physiol. 19(3)*:531–533 (1964).
42. **Burton, A.C. and O.G. Edholm:** *Man in a cold environment*. London: Edward Arnold Publishers, 1955. p. 61.
43. **Wyndham, C.H., N.B. Strydom, H.M. Cooke, and J.S. Maritz:** "Studies on the Effects of Heat on Performance of Work" (Applied

Physiology Laboratory Report 1-3/59). Johannesburg, S. Africa: Transvaal and Orange Free State Chamber of Mines, 1959.

44. **Coyle, E.F. and S.J. Montain:** Thermal and cardiovascular responses to fluid replacement during exercise. In *Perspectives in Exercise Science and Sports Medicine*, vol. 6. C.V. Gisolfi, D.R. Lamb, and E.R. Nadel, eds. Dubuque, IA: Brown and Benchmark, 1993. pp. 183–187.
45. **Armstrong, L.E., P.C. Szlyk, J.P. De Luca, I.V. Sils, et al.:** Fluid electrolyte losses in uniforms during prolonged exercise at 30ºC. *Aviat. Space Environ. Med. 63*:351–355 (1992).
46. **Nadel, E.R., G.W. Mack, and A. Takamata:** Thermoregulation, exercise, and thirst: interrelationships in humans. In *Perspectives in Exercise Science and Sports Medicine*, vol. 6. C.V. Gisolfi, D.R. Lamb, and E.R. Nadel, eds. Dubuque, IA: Brown and Benchmark, 1993. p. 248.
47. **Mehta, M. (ed.):** *Physician's Desk Reference Guide to Drug Interactions, Side Effects and Indications*, 50th ed. Montvale, NJ: Medical Economics, 1996. pp. 1263, 1268, 1317.
48. **Seales, D.R.:** Influence of aging on autonomic-circulatory control at rest and during exercise in humans. In *Perspectives in Exercise Science and Sports Medicine*, vol. 6. C.V. Gisolfi, D.R. Lamb, and E.R. Nadel, eds. Dubuque, IA: Brown and Benchmark, 1993. pp. 291–293.
49. **MacPherson, R.K.:** The effect of fever on body temperature regulation in man. *Clin. Sci. 18*:281–287 (1959).
50. **Bagatell, C.J. and W.J. Bremner:** Sperm counts and reproductive hormones in male marathoners and lean controls. *Fertil. Steril. 53(4)*:688–692 (1990).
51. **Hackney, A.C., W.E. Sinning, and B.C. Bruot:** Reproductive hormonal profiles of endurance-trained and untrained males. *Med. Sci. Sports Exer. 20(1)*:60–65 (1988).
52. **Griffith, R.O., R.H. Dressendorfer, C.D. Fullbright, and C.E. Wade:** Testicular function during exhaustive endurance training. *Phys. Sports Med. 5(18)*:54–64 (1990).
53. **Dill, D.B., M.K. Yousef, and J.D. Nelson:** Responses of men and women to two-hour walks in desert heat. *J. Appl. Physiol. 35*:231–235 (1973).
54. **Shoenfield, Y., Udassin, R. Shapiro, Y., Ohri, A., et al.:** Age and sex difference in response to short exposure to extreme dry heat. *J. Appl. Physiol. 44*:1–4 (1978).
55. **Wyndham, C.H., J.F. Morrison, and C.G. Williams:** Heat reactions of male and female Caucasians. *J. Appl. Physiol. 49*:1–8 (1965).
56. **Hertig, B.A., H.S. Belding, K.K. Kraning, D.L. Batterton, et al.:** Artificial acclimatization of women to heat. *J. Appl. Physiol. 18*:383–386 (1963).
57. **Fox, R.H., B.E. Lofstedt, P.M. Woodward, F. Eriksson, et al.:** Comparison of thermoregulatory functions in men and women, *J. Appl. Physiol. 26*:444–453 (1969).
58. **Frye, A.J. and E. Kamon:** Responses to dry heat of men and women with similar aerobic capacities. *J. Appl. Physiol. 50*:65–70 (1981).
59. **Avellini, B.A., E. Kamon, and J.T. Krajewski:** Physiological responses of physically fit men and women to acclimation to humid heat. *J. Appl. Physiol. 44*:254–261 (1980).
60. **Kamon, E., B.A. Avellini, and J. Krajewski:** Physiological and biophysical limits to work in the heat for clothed men and women. *J. Appl. Physiol. 41*:71–76 (1976).
61. **Paolone, A.M., C.L. Wells, and G.T. Kelly:** Sexual variations in thermoregulation during heat stress. *Aviat. Space Environ. Med. 49*:715–719 (1978).
62. **Drinkwater, B.L., J.E. Denton, P.B. Raven, and S.M. Horvath:** Thermoregulatory responses of women to intermittent work in the heat. *J. Appl. Physiol. 41*:57–61 (1976).
63. **Pandolf, K.B., M.N. Sawka, and Y. Shapiro:** *Physiological Differences Between Men and Women in Exercise-Heat Tolerance and Heat Acclimatization* (NTIS, AD-A152 048). Washington, DC: U.S. Department of Commerce, 1985. pp. 1–25.
64. **Kenney, W.L.:** Physiological correlates of heat intolerance. *Sports Med. 2*:279–286 (1985).
65. **Shapiro, Y., K.B. Pandolf, B.A. Avellini, N.A. Pimental, et al.:** Physiological responses of men and women to humid and dry heat. *J. Appl. Physiol. 49*:1–8 (1980).
66. **Hori, S., M. Mayuzumi, N. Tanaka, and J. Tsujita:** Oxygen uptake of men and women during exercise and recovery in hot environment and a comfortable environment. In *Environmental Stress: Individual Adaptations*, L.F. Folinsbee, J.A.

Wagner, J.F. Borgai, B.L.Drinkwater, et al., eds. New York: Academic Press, 1978. pp. 51.

67. **McArdle, W.D., J.R. Magel, T.J. Gergley, R.J. Spina, et al.:** Thermal adjustment to cold-water exposure in resting men and women. *J. Appl. Physiol. 56(6)*:1565–1571 (1984).
68. **McArdle, W.D., J.R. Magel, T.J. Gergley, R.J. Spina, et al.:** Thermal adjustment to cold-water exposure in exercising men and women, *J. Appl. Physiol. 56(6)*:1572–1577 (1984).
69. **Pandolf K.B., M.M. Toner, W.D. McArdle, J.R. Magel, et al.:** Influence of body mass, morphology, and gender on thermal responses during immersion in cold water. In *Proceedings of the 9th International Symposium of Underwater and Hyperbaric Physiology Undersea and Hyperbaric Medical Society*. Bethesda, MD: Underwater and Hyperbaric Physiology Undersea and Hyperbaric Medical Society, 1987. pp. 145–152.
70. **Sawka, M.N., M.M. Toner, R.P. Francesconi, and K.B. Pandolf:** Hypohydration and exercise: effects of heat acclimation, gender, and environment. *J. Appl. Physiol. 55*:1147–1153 (1983).
71. **Horstman, D.H. and E. Christensen:** Acclimatization to dry heat: active men vs. active women. *J. Appl. Physiol. 52*:825–831 (1982).
72. **Wyndham, C.H., N.B. Strydom, A.J. van Rensburg, A.J.A. Benade, et al.:** Relation between VO_2 max and body temperature in hot humid air conditions. *J. Appl. Physiol. 29(1)*:45–50 (1970).
73. **Bell, C.R. and J.D. Walters:** Reactions of men working in hot and humid conditions. *J. Appl. Physiol. 27*:684–686 (1969).
74. **Craig F.N., H.W. Garren, H. Frankel, and W.V. Blevins:** Heat load and voluntary tolerance time. *J. Physiol. 6*:634–644 (1954).
75. **Krajewski, J.T., E. Kamon, and B. Avellini:** Scheduling rest for consecutive light and heavy work loads under hot ambient conditions. *Ergonomics 22*:975–987 (1979).
76. **Vogt, J.J., J.P. Libert, V. Candas, F. Duall, et al.:** Heart rate and spontaneous work-rest cycles during exposure to heat. *Ergonomics 26*:1173–1185 (1983).
77. **Reneau, P. and P. Bishop:** A comparison of two vapor barrier suits across two thermal environments. *Am. Ind. Hyg. Assoc. J.* [In press]
78. **Bishop, P., P. Ray, and P. Reneau:** A review of the ergonomics of work in the U.S. military chemical protective clothing. *Int. J. Ind. Ergonom. 15(4)*:278–283 (1995).
79. **Electrical Power Research Institute (EPRI):** *Heat Stress Management Program for Nuclear Power Plants* (NP-4453). Palo Alto, CA: EPRI, 1986.
80. **Bureau of Medicine and Surgery, Navy Dept.:** *Manual of Naval Preventative Medicine* (NAVMED-P-5010-3). Palo Alto, CA: Research Reports Center, 1988.
81. **Bishop, P.A., S.A. Nunneley, and S.H. Constable:** Comparisons of air and liquid personal cooling for intermittent heavy work in moderate temperatures. *Am. Ind. Hyg. Assoc. J. 52*:393–397 (1991).
82. **Pandolf, K.B., L.A. Stroschen, L.L. Drolet, R.A. Gonzalez, et al.:** Prediction modeling of physiological responses and human performance in the heat. *Comput. Bio. Med. 16*:319–329 (1986).
83. **Wissler, E.H.:** Simulation of fluid cooled or heated garments that allow man to function in hostile environments. *Chem. Eng. Sci. 41*:1689–1698 (1986).
84. **Shvartz, E., S. Shibolet, A. Meroz, A. Magazanik, et al.:** Prediction of heat tolerance from heart rate and rectal temperature in a temperate environment. *J. Appl. Physiol. 43(4)*:684–688 (1977).
85. **Kenney, W.L., D.A. Lewis, R.K. Anderson, and E. Kamon:** A simple test for the prediction of relative heat tolerance. *Am. Ind. Hyg. Assoc. J. 47*:203–206 (1986).
86. **Bishop, P.A., G. Smith, P. Ray, J. Beaird, et al.:** Empirical prediction of physiological response to prolonged work in encapsulating protective clothing. *Ergonomics 37*:1503–1512 (1994).
87. **Bishop, P., P. Reneau, P. Ray, and M. Wang:** Empirical prediction of physiological response to prolonged work in encapsulating protective clothing. In *Proceedings of the 7th International Conference on Environmental Ergonomics*, London: Freund Publishing House, Ltd., 1996.

88. **Carter, B.J. and M. Cammermeyer:** Biopsychological responses of medical unit personnel wearing chemical defense ensemble in a simulated chemical warfare environment. *Mil. Med. 150(5)*:239–249 (1985).

89. **Carter, B.J. and M. Cammermeyer:** Emergence of real casualties during simulated chemical warfare training under high heat conditions. *Mil. Med. 150(12)*:657–663 (1985).

Workers construct a 500-foot tower in Ropesville, Texas for a local FM radio station. (Layne Kennedy, 1997)

Outcome Competencies

After completing this chapter, the user should be able to:

1. Define underlined terms used in this chapter.
2. Recognize instruments and methods for measuring environments.
3. Evaluate heat and cold exposure limits.
4. Demonstrate the process for making decisions concerning thermal environments.

Prerequsite Knowledge

Basic chemistry, physics, and general occupational hygiene background.

Prior to beginning this chapter, the user should review the following chapters:

Chapter Number	Chapter Topic
1	History and Philosophy of Industrial Hygiene
2	Occupational Exposure Limits
24	Applied Physiology of Thermoregulation and Exposure Control

Key Terms

air velocity • air temperature • anemometer • apparent temperature • Botsball • clothing insulation • corrected effective temperature • dew point temperature • dry bulb temperature • effective temperature • effects on safety behavior • equivalent chill temperature • globe thermometer • globe temperature • heat stress index • hygrometer • mean radiant temperature • metabolic heat • natural wet bulb temperature • new effective temperature • perceptual-motor performance • permissible exposure limit • permissible heat exposure threshold limit values • physiological heat exposure limit • psychrometer • psychrometric chart • psychrometric wet bulb temperature • radiometers • recommended exposure limits • recommended alert limits • relative humidity • required clothing insulation • required sweat rate • thermal balance • thermometer • vapor pressure • wet bulb globe temperature • wet kata thermometer • wet globe temperature • windchill index

Key Topics

I. Definitions
   A. Psychrometric Chart

II. Instruments and Mmethods for Measuring Thermal Components
   A. Temperature Measurements
   B. Humidity Measurements
   C. Air Velocity Measurements
   D. Radiant Heat Measurement

III. Heat Stress Indices
   A. Thermal Balance
   B. Metabolic Heat
   C. Heat Stress Index (HSI)
   D. Effective Temperature (ET)
   E. Wet Bulb Globe Temperature (WBGT)
   F. Wet Globe Temperature (WGT)
   G. Thermal Stress and the Required Sweat Rate
   H. Cooling Power of Air
   I. Other Heat Indices
   J. Correlation Between Heat Indices

IV. Heat Exposure Limits
   A. WBGT Recommendations
   B. Stay Times
   C. Recommend Limits in Other Indices

V. Cold Exposure Limits
   A. Windchill Index
   B. Required Clothing Insulation
   C. ACGIH Recommendations

VI. Thermal Effects on Performance and Safety
   A. Effects of Heat
   B. Effects of Cold
   C. Effects on Accidents and Safety Behavior

VII. Evaluating the Hot Workplace: An Example

VIII. Evaluating the Cold Workplace: An Example

25

Thermal Standards and Measurement Techniques

Jerry D. Ramsey, Ph.D., PE

Mohamed Y. Beshir, Ph.D.

Introduction

This chapter deals with the standards and measurement techniques for natural and artificial thermal environments. It discusses definitions of thermal components, instruments and methods for measuring thermal components, heat stress indices, heat and cold exposure limits, and thermal effects on performance and safety. The chapter also presents examples to demonstrate how to use this knowledge in making decisions concerning environmental exposures. The units and symbols used in this chapter are those proposed by the International Organization for Standardization (ISO), i.e., units of the International System (SI units).[1]

Definitions

The thermal environment can be assessed by measuring its thermal components: dry bulb (air) temperature, psychrometric wet bulb temperature, natural wet bulb temperature, relative humidity, vapor pressure, dew point temperature, air velocity, globe temperature, and mean radiant temperature.

The dry bulb temperature is measured with a thermometer, commonly a liquid-in-glass thermometer. Temperature units are expressed in degrees Celsius (°C), Kelvin (K) (K = Celsius + 273), or degrees Fahrenheit (°F) (F = 9/5 Celsius + 32). Celsius and Kelvin units are proposed for temperature in SI units.[1] Equivalent °F units will be included for convenience of the reader. The term air temperature is synonymous with dry bulb temperature.

The psychrometric wet bulb temperature, commonly called wet bulb temperature, is measured by a thermometer on which the sensor is covered by a wetted cotton wick that is exposed to forced movement of the air. Accuracy of wet bulb temperature measurements requires using a clean wick, distilled water, and proper shielding to prevent radiant heat gain.

The natural wet bulb temperature is the temperature measured when the wetted wick covering the sensor is exposed only to naturally occurring air movements.

Nomenclature

°C = degrees Celsius
°F = degrees Fahrenheit
°K = degrees Kelvin
C = convective heat
E = evaporative heat
e_a = the saturation vapor pressure at Ta, mm Hg
E_{max} = maximum evaporative capacity of the climate
E_{req} = evaporative heat required
$e_{S(ET^*)}$ = the saturation vapor pressure at TET*, mm Hg
h_c = the convective heat transfer coefficient
h_r = the linear radiation exchange coefficient
H_{res} = respiratory heat loss (convective and evaporative)
i = the ratio of the transfer coefficients for sensible heat.
I_{cl} = insulation, clothing
K = conductive heat
M = metabolic heat
R = radiant heat
S = heat storage
T_a = air temperature
t_{ch} = equivalent chill temperature
t_{cl} = dry heat loss from clothing to environment
T_{ET^*} = the new effective temperature, i.e., ET*
T_g = globe temperature
T_o = the operative temperature
T_r = the mean radiant temperature
t_{sk} = heat conducted from the skin
v = air velocity
V_{ar} = relative air velocity, m/sec
w = the skin wetness
W = external work rate
WCI = windchill index

The relative humidity is the ratio of the actual amount of moisture in the air to the amount of moisture that the air could hold if saturated at the same temperature. The percentage value of this ratio is usually used to describe relative humidity.

The vapor pressure is the pressure at which a vapor can accumulate above its liquid (water), if the vapor is confined over its liquid and the temperature is held constant. Normal units for vapor pressure are mm Hg, torr, or kPa, where 1 mm Hg = 1 torr, and 7.5 torr = 1 kPa.

The dew point temperature is the temperature at which condensation of water vapor in a space begins, for a given state of humidity and pressure, as the vapor temperature is reduced. There is a unique dew point temperature associated with each combination of dry and wet bulb temperatures.

The velocity of the air movement is also called wind speed. Units for air velocity are meters per second (m/sec), feet per min (ft/min), or miles per hour (mph).

The globe temperature is a measure of radiant heat. It is obtained by placing the sensor of a thermometer in the center of a hollow copper sphere, usually 15 cm (6 in) in diameter and painted a matte black to absorb the incident infrared radiation. Globe temperature can also be obtained from electronic instruments that use black spheres with diameters about 5 cm (2 in.).

The mean radiant temperature is the temperature of an imaginary black enclosure, of uniform wall temperature, that provides the same radiant heat loss or gain as the environment measured. It can be approximated from readings of globe temperature, dry bulb temperature, and air velocity.

Psychrometric Chart

The psychrometric chart (Figure 25.1) is the graphic representation of the relationship between the dry bulb temperature, wet bulb temperature, relative humidity, vapor pressure, and dew point temperature. If any two of these thermal components of the environment are known, the other three can be obtained from the chart. For example, if the dry bulb temperature and wet bulb temperature of an environment are 43 and 24°C, respectively, it can be determined from Figure 25.1 that the relative humidity, vapor pressure, and dew point temperature of this environment are 20%, 13 mm Hg, and 15°C, respectively. Note that at 100% relative humidity, dry bulb temperature = wet bulb temperature = dew point temperature.

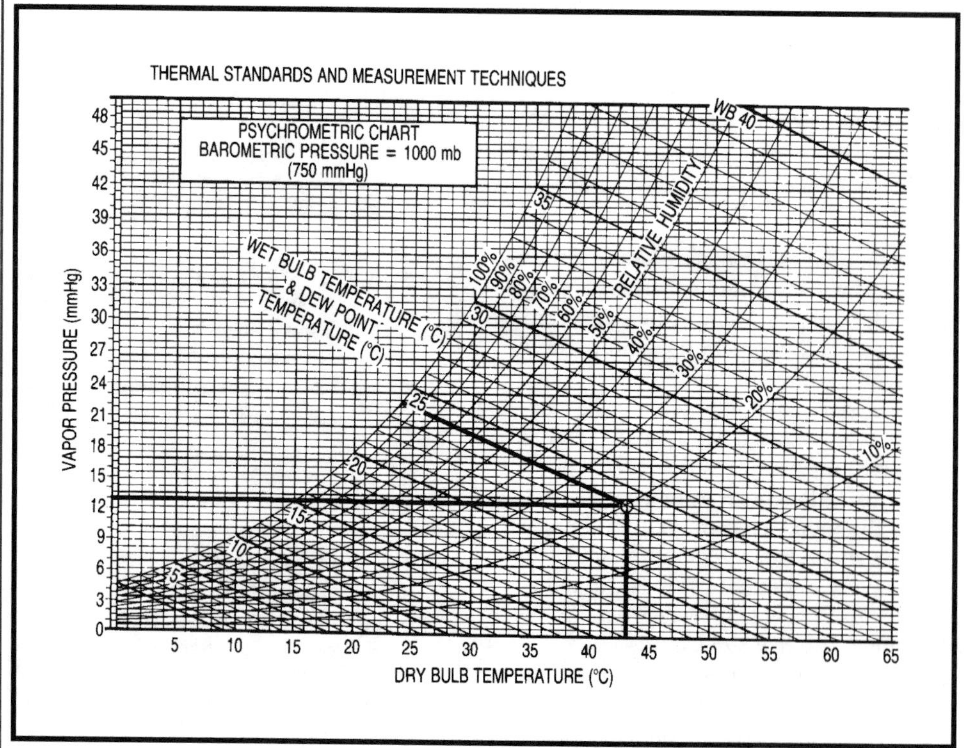

Figure 25.1 — The psychrometric chart.(100)

Instruments and Methods for Measuring Thermal Components

Temperature Measurements

Any instrument that measures temperature is called a thermometer. The term is most commonly used, however, for liquid-in-glass thermometers. Thermometers can be classified according to the nature, properties, characteristics, and materials of the sensing element. The main types are liquid-in-glass, bimetallic, and resistance thermometers, and thermocouples. Depending on the methods and setup used for taking the measurements, the same type of thermometer can be used to measure dry bulb, psychrometric wet bulb, natural wet bulb, or globe temperature. Following are some hints for obtaining accurate temperature measurements.

- The time allowed for measurement must be longer than the time required for thermometer stabilization.
- The measured temperature must be within range of the thermometer.
- Under radiant conditions (i.e., temperatures of the surrounding surfaces are different from the air temperature), the accuracy of air temperature measurement can be improved either by shielding the sensing element or by accelerating the movement of the air. Under high radiant conditions, both shielding the sensor and accelerating the air may be required.
- The sensing element must be in contact with, or as close as possible to, the area of interest.

Liquid-in-Glass Thermometers

The liquid-in-glass thermometer, first developed in 1706, is the most widely used and familiar type of thermometer. The more commonly used liquids are mercury and alcohol. Mercury-in-glass thermometers have wider application, but one exception is the measurement of extreme cold, since the freezing point of alcohol (114°C) is considerably below that of mercury (40°C).

The liquid-in-glass thermometer consists of a bulb and a stem of glass. The bulb serves as the liquid container, and the stem has a capillary tube into which the liquid will expand from the bulb when heated. The length of the liquid in the capillary tube depends on the temperature to which the thermometer is exposed. The calibration of the temperature scale on the thermometer stem depends on the coefficient of expansion of both the glass and the liquid.

Two major types of liquid-in-glass thermometers are available. The total immersion thermometer is calibrated by immersing the whole thermometer in a thermostatically controlled medium. It should be used when the whole thermometer is exposed to the measured temperature, such as when measuring the air temperature. The partial immersion thermometer is calibrated by immersing only the bulb of the thermometer in a thermostatically controlled medium. It should be used when only the sensing element (the bulb) is exposed to the measured temperature (as in measuring wet bulb and globe temperatures).

Bimetallic Thermometers

The bimetallic thermometer uses the principle that each metal has a certain linear coefficient of expansion. This thermometer consists of two strips of different metals of equal length welded or soldered together on one end, with the free ends connected to an indicator. When the thermometer is exposed to a certain temperature, the two strips change length according to their coefficients of expansion, which produces calibrated movement of the indicator.

Bimetallic thermometers are usually in a dial form. By changing the material and length of the strips, the temperature range and the sensitivity of the thermometer will be changed. The longer the strip, the more sensitive the thermometer will be.

Resistance Thermometers

The resistance thermometer implements the principle that changing temperature creates changes in electric resistance. A resistance thermometer consists of a resistor (the sensing element) and a wheatstone bridge, galvanometer, or other means for measuring the resistance change that results from the temperature change. The thermometer may be calibrated to give a direct temperature reading. The resistor is either a metal wire or a semiconductor. The resistance increases as the temperature increases for the metal wire, whereas the resistance decreases as the temperature increases for the semiconductor. When a semiconductor is used as a sensor, the resistance thermometer is called a thermistor.

Thermocouples

The thermocouple thermometer uses the principle that when a junction of two wires of dissimilar metals is formed, an electromotive force is generated. Its potential depends on the junction temperature. A thermocouple is simply a junction of two wires of different metals, formed by soldering, welding, or merely twisting the two wires together. A thermocouple circuit is formed when the free ends of the wires are also joined together. If one junction is kept at a constant reference temperature, the temperature of the second junction can be determined by measuring the existing electromotive force or the induced electric current with a potentiometer or a millivoltmeter. The temperature of the second junction is determined from an appropriate calibration table or curve.

Humidity Measurements

Humidity is the amount of water vapor within a given space. It is commonly measured as the relative humidity. It has an important effect on human thermal exchange with the environment: A lower humidity level allows faster sweat evaporation from the skin and larger amounts of heat removal from the body.

Two types of instruments are used for measuring relative humidity. A psychrometer gives an indirect measurement of the relative humidity, and a hygrometer gives a direct measurement.

Psychrometers

A psychrometer consists of two mercury-in-glass thermometers. One thermometer measures dry bulb temperature, and the other measures wet bulb temperature; in the latter case, the bulb is covered with a cotton wick wetted with distilled water. The thermometers are mounted parallel to each other on the frame of the psychrometer. The relative humidity and water vapor pressure can be determined from the dry and wet bulb values, using a psychrometric chart as shown in Figure 25.1. Often this is presented as a table or nomogram drawn on the psychrometer frame. The wet bulb thermometer must be read only after the application of forced air movement. Forced air movements can be obtained manually with a sling psychrometer or mechanically with a motor-driven or an aspirated psychrometer.

The sling psychrometer is usually whirled by a handle for approximately one minute. Accurate measurement is obtained when both thermometers stabilize (i.e., no temperature changes between repeat readings). On the aspirated psychrometer, a battery or spring-operated fan blows air across the wick. To ensure appropriate fan speed, the battery must be checked before using the psychrometer. Air blown across the wet wick must not pass over the dry bulb thermometer or errors in air temperature readings may result. The sensors of both the sling psychrometer and the aspirated psychrometer must be shielded from radiation.

Hygrometers

Any instrument that measures humidity is a hygrometer; however, the term is commonly used for instruments that provide a direct reading of the relative humidity.

Organic hygrometers do not provide a high degree of accuracy, but because of their low cost they are widely used. Organic hygrometers operate on the principle that organic materials change their length according to their moisture content, which depends on the humidity of the ambient air. The change in length of the organic material can be transferred as a direct readout of relative humidity on a percentage scale.

Other types of hygrometers for laboratory and more precise applications include dew point, electrolytic, electronic, and chemical hygrometers. Dew point hygrometers measure the dew point temperature, and the relative humidity is determined from a conversion chart. The principle of electrolytic hygrometers is that the electric resistance of a salt film depends on the humidity of the exposed atmosphere. On electronic hygrometers, a sulfonated polystyrene strip is the sensing element whose conductance varies as it absorbs water vapor. Chemical hygrometers (e.g., lithium chloride) measure the humidity directly by extracting and weighing the water vapor from a known sample.

Air Velocity Measurements

The movement of air affects the mechanism of exchange of convective and evaporative heat between the human body and the environment. In a warm, dry environment, an increase in air velocity reduces the effects of the environment on the body because the resultant heat loss through evaporation is greater than the heat gain from convective heat. However, in a hot, humid environment, in which air temperature exceeds skin temperature and the water vapor pressure of ambient air exceeds the water vapor pressure of the skin, an increase in air velocity increases the heat-loading effects of the environment on the body. In any cold environment, an increase in air velocity increases the rate of heat loss by convection.

Any instrument that measures air velocity or wind speed is an anemometer, either a vane anemometer (velometer) or a thermal anemometer (thermoanemometer). Accurate determination of an air velocity contour map in a work area is very difficult, or often impossible, due to the variability in air movement both with time and in space. Also, most of the available anemometers are insensitive to low air velocity measurement.

If an anemometer is not available for accurate indoor air velocity measurement, air velocity (v) in meters per second can be estimated as follows:

- No sensation of air movement (closed room with no air source): $v < 0.2$ m/sec
- Sensing light breezes (perception of slight air movement): $0.2 \leq v \leq 1.0$ m/sec
- Sensing moderate breezes (at a few meters from a fan, definite perception of air movement, causing tousling of hair and movement of paper): $1.0 < v \leq 1.5$ m/sec
- Sensing heavy breezes (located close to a fan, air causes marked movement of clothing): $v > 1.5$ m/sec

Sensitivity to air movement is increased when the skin is wet and/or when body movements generate airflow across the skin.

Vane Anemometers

Measuring air velocity with vane or cup anemometers involves measuring either the rotation of a fan (rotating vane anemometer) or the deflection of an internal vane (velometer) by placing the anemometer in the airstream. The vane anemometer usually has an air inlet and an air outlet. The vane or fan is placed in the pathway of, and perpendicular to, the air direction, so that movement of the air causes vane deflection or causes the fan wheels to rotate, producing a direct readout of air velocity.

Thermal Anemometers

The basic idea of the thermal anemometer is to determine air velocity by measuring the cooling effects of air movements on an electrically heated thermometer. Two types of thermal anemometers are commonly used: hot-wire anemometers, which use resistance thermometers, and heated thermocouple anemometers.

One technique for measuring the cooling effects of air movement is to heat the wire by applying an electric current of specified value and then determine the air velocity from a calibration chart based on air velocity and the resultant wire resistance (hot-wire) or the resultant electromotive force (heated thermocouple). A second technique is to bring the resistance, or the electromotive force, to a specified level and then measure the current required to maintain this level. The air velocity is obtained from a calibration chart relating air velocity to the required electric current. In both cases correction is made for air temperature.

Radiant Heat Measurement

Determining radiant heat exchanges is necessary to define the thermal environment. A variety of radiometers have been used to measure radiant flux in surface pyrometry and meteorological applications.[2] The net radiometer has been used to measure the radiant energy balance of human subjects.[3] A radiometer with a sensor consisting of a reflective polished disk and a black absorbent disk can also be used. These laboratory and special purpose instruments, however, are not commonly used in occupational heat measurements.

In the occupational environment, the Vernon globe thermometer[4] is the most commonly used device for estimating radiant heat. The thermometer recommended by the National Institute for Occupational Health and Safety (NIOSH),[5] and described above under the definition of globe temperature, has an emissivity of 0.95. The globe temperature is an integrated measure that has a time lag, but it does provide a means for approximating the mean radiant temperature (T_r) from air temperature (T_a), globe temperature (T_g), and air velocity (v), according to the following equation.

$$T_r = T_g + (1.8v^{0.8})(T_g - T_a)^{\circ}C \qquad (1)$$

The globe thermometer exchanges heat with the environment by radiation, convection, and conduction. It stabilizes when the heat exchange by radiation is equivalent to that by convection and conduction: normally 15 to 20 min is required for a globe that is 15 cm in diameter. It has been demonstrated that the globe diameter affects heat exchange and stabilization time, but that appropriate conversion equations can be applied so that black globes of different diameters will yield functionally equivalent results. There are small globe thermometers (4.2 cm) that reduce the response time to one-quarter that required for the standard Vernon globe thermometer.[6]

The precise evaluation of mean radiant temperature for use in research or critical occupational applications is discussed elsewhere.[7]

Heat Stress Indices

Heat stress is the aggregate of environmental and physical work factors that constitute the total heat load imposed on the body. The environmental components of heat stress are air temperature, water vapor pressure, radiant heat, and air movement. Physical work contributes to total heat stress by producing metabolic heat in the body in proportion to the work intensity.

A heat stress index is a composite measure used for the quantitative assessment of heat stress. Over the years, various indices have been developed to integrate into a single number the components of thermal environment and/or the physical and personal factors that influence heat transfer between the person and the environment. Heat indices can be classified as those based on physical factors of the thermal environment, thermal comfort assessment, rational heat balance equations, and physiological strain.[8]

Thermal Balance

A major criterion for evaluating the usefulness of a heat stress index is its correlation with the changes that occur in human physiological response to heat strain. The major readily measured physiological responses to heat stress are increases in body temperature, which are measured by oral, tympanic, esophageal, or rectal temperature; heart rate; and sweat production.[9,10] Oxygen consumption is affected less by heat exposure than by the physical work load.

In addition to the environmental and physical factors mentioned above, age, physical fitness, health status, clothing, and acclimatization are also major factors contributing to heat strain. Unfortunately, an index that integrates all these parameters and hence correlates them precisely to one or more physiological responses has not yet been developed.[11–13] However, there are several indices for measuring heat stress, each with special advantages that make it more suitable for use in a particular environment.

Several thermal indices use the basic construct of thermal balance or heat exchange between the human body and the environment, as represented by the following equation.

$$M \pm R \pm C \pm K \ E = \pm S \qquad (2)$$

where

M = metabolic heat
R = radiant heat
C = convective heat
K = conductive heat
E = evaporative heat
S = heat storage

In this equation heat storage equals zero if the body is in heat balance. The metabolic heat is positive as a heat gain, while the evaporative heat is negative as a heat loss. Other components may be positive or negative based on their influence in the thermal exchange. The conductive heat is very small and is usually neglected. Heat production resulting from external work is sometimes identified separately from metabolic heat. Similarly, respiratory heat loss by evaporation and convection may be considered separate components in the heat balance equation.[14] These components have a relatively small impact on heat balance and normally are neglected unless metabolic heat is to be accurately measured. Heat exchange is measured in terms of the watt or kilocalories per hour (kcal/hour). One watt equals 0.8606 kcal/hour.

Metabolic Heat

The metabolic heat generated by work or activity represents a major component in human heat balance and is more difficult to measure than are the environmental components. Metabolic heat can be measured directly or estimated indirectly from physiological measurements such as oxygen consumption. In the evaluation of occupational environments, however, metabolic heat is usually estimated by means of tabulated descriptions of energy cost for typical work activity, as shown in Table 25.1,[15] or from tables that specify incremental metabolic heat resulting from activity or movement of different body parts, such as arm work, leg work, standing, and walking.[16] The metabolic heat can then be estimated by summing the component metabolic heat values based on the actual body movements of the worker.

ISO[17] has recommended determining the metabolic rate analytically by adding values of basal metabolic rate, metabolic rate for body position or body motion, metabolic rate for type of work, and metabolic rate related to work speed. The basal metabolic rate is a function of age, body weight, and height and equals 44 watts/m^2 for a standard man. Metabolic rate values for body position, body motion, and type of work are specified in Table 25.2. For example, the metabolic rate estimate for a standard man sitting and performing average handwork is 84 watts/m^2; 44 (basal), plus 10 (sitting), plus 30 (average handwork). Assuming a surface area of 1.8 m^2 (standard man) results in an estimated metabolic rate of 1.8 m^2 × 84 watts/m^2 = 151 watts. Tables for assessing metabolic rate related to work speed are also presented in the ISO standard.[17]

Several tables, charts, and equations relating to metabolic rate determination have been summarized in a procedure for systematic work load estimation.[18] Figure 25.2 shows this system, which can be used with an activity log to determine metabolic heat as a function of stationary, walking, or extra effort work for one hand, two hands, or whole body work. The numbers assigned to each row represent the range of intensity for the activity, with the lower numbers indicating slow or light activity and the higher numbers indicating fast or heavy activity. For example, a worker standing and doing heavy two-arm work would be coded S-9, which shows a metabolic heat of 310 watts.

Table 25.1 — Some Selected Types of Work Classes According to Work Load Level[15]

Work Load	*Energy Expenditure Range (M)*
Level 1, resting	Less than 117 watts (100 kcal/hr)
Level 2, light	117–232 watts (100–199 kcal/hr)
sitting at ease: light handwork (writing, typing, drafting, sewing, bookkeeping); hand and arm work (small bench tools, inspecting, assembly or sorting of light materials); arm and leg work (driving car under average conditions, operating foot switch or pedal). Standing: drip press (small parts); milling machine (small parts); coil taping; small armature winding; machining with light power tools; casual walking (up to 0.9 m/sec, i.e., 2 mph). Lifting: 4.5 kg (10 lb) < 8 lifts/min; 11 kg (25 lb) < 4 lifts/min	
Level 3, moderate	233–348 watts (200–299 kcal/hr)
hand and arm work (nailing, filing); arm and leg work (off-road operation of trucks, tractors, or construction equipment); arm and trunk work (air hammer operation, tractor assembly, plastering, intermittent handling of moderately heavy materials, weeding, hoeing, picking fruits or vegetables); pushing or pulling lightweight carts or wheelbarrows; walking 0.9–1.3 m/sec (2–3 mph). Lifting: 4.5 kg (10 lb), 10 lifts/min; 11 kg (25 lb), 6 lifts/min	
Level 4, heavy	349–465 watts (300–400 kcal/hr)
heavy arm and trunk work; transferring heavy materials, shoveling; sledge hammer work; sawing, planting, or chiseling hardwood; hand mowing, digging, walking 1.8 m/sec (4 mph), pushing or pulling loaded hand carts or wheelbarrows; chipping castings; concrete block laying. Lifting: 4.5 kg (10 lb), 14 lifts/min; 11 kg (25 lb), 10 lifts/min	
Level 5, very heavy	above 465 watts (400 kcal/hr)
heavy activity at fast to maximum pace; ax work; heavy shoveling or digging; climbing stairs, ramps, or ladders; jogging, running, or walking faster than 1.8 m/sec (4 mph). Lifting: 4.5 kg (10 lb) > 18 lifts/min; 11 kg (25 lb) > 13 lifts/min	

Heat Stress Index (HSI)

Belding and Hatch developed the heat stress index (HSI) to express thermal stress of a hot climate as the ratio of the evaporative heat required (E_{req}) to maintain the body in thermal equilibrium (i.e., stationary equals zero) to the maximum evaporative capacity of the climate (E_{max}).[11] Therefore

$$HSI = E_{req}/E_{max} \times 100 \qquad (3)$$

The index assumes individuals of average build (weight, 70 kg; height, 1.7 m; body surface area, 1.8 m^2), dressed in shorts and gym shoes, experiencing a skin temperature of 35°C (95°F), and uniformly wetted with sweat. There is no storage of heat in the body at the beginning of heat exposure, and thermal exchanges by conduction and respiration can be ignored. From the equation of heat exchange above:

$$(E_{req}) = M + R + C \qquad (4)$$

where M = metabolic heat, R = radiant heat, and C = convective heat. The equations and coefficients needed to compute the values of R, C, E_{max}, E_{req}, and HSI for various combinations of clothing are available.[8,19]

The approximations and assumptions introduced in the HSI construction, mainly for simplicity and practicality, have resulted in areas of reduced accuracy. For example, the HSI does not correctly differentiate between heat stress resulting from a hot, dry climate and that resulting from a warm, damp climate. Similarly, a work change that results in a 100-watt increase in metabolic heat would have a greater physiological impact than an environmental change that results in a 100-watt increase in radiant or convective heat, even though in the HSI, the radiant-plus-convective heat value has the same weighting as the value of metabolic heat. Approximations and limitations notwithstanding, the HSI has been used widely and successfully as a tool for evaluating hot work environments. Nomograms are also available as an aid to determining E_{req} and E_{max}.[20]

Kamon and Ryan[21] developed an effective heat strain index based on the equations for heat exchange through radiation, convection, and evaporation, with the addition of the expected physiological responses involving metabolism and sweating. The effective heat strain index is readily determined by the value of the wetness of the skin (w), that is, the fraction of the total body surface covered by sweat, and is calculated as[22]

$$w = HSI/100 \qquad (5)$$

Table 25.2 — Metabolic Rate for Body Posture, Type of Work, and Body Motion Related to Work Speed[17]

Body Posture	*Metabolic Rate (watts/m²)*	
Sitting	10	
Kneeling	20	
Crouching	20	
Standing	25	
Standing stooped	30	
	Metabolic Rate (watts/m²)	
Type of Work	*Mean Value*	*Range*
Handwork		
light	15	< 20
average	30	20–35
heavy	40	> 35
One-arm work		
light	35	< 45
average	55	45–65
heavy	75	> 65
Two-arm work		
light	65	< 75
average	85	75–95
heavy	105	> 95
Trunk work		
light	125	< 155
average	190	155–230
heavy	280	230–330
very heavy	390	> 330
Work Speed Related to Distance	*Metabolic Rate Related to Work Speed (watts/m⁻²)*	
Walking, 2 to 5 km/hr	110	
Walking uphill, 2 to 5 km/hr		
inclination, 5°	210	
inclination, 10°	360	
Walking downhill, 5 km/h		
declination, 5°	60	
declination, 10°	50	
Walking with load on back, 4 km/hr		
10 kg load	125	
30 kg load	185	
50 kg load	285	
Work Speed Related to Height		
Walking upstairs	1725	
Walking downstairs	480	
Mounting inclined ladder		
without load	1660	
10 kg load	1870	
50 kg load	3320	
Mounting vertical ladder		
without load	2030	
10 kg load	2335	
50 kg load	4750	

Note: Values exclude basal metabolism (44 watts/m²)

The effective heat strain index assumes that the mean skin temperature is 36°C, while the HSI assumes a value of 35°C. Other investigators have also concluded that 36°C is a more accurate estimate of mean skin temperature for use in heat transfer equations.[23,24]

Effective Temperature (ET)

The effective temperature (ET) scale was developed by Houghten and Yaglou in 1923.[25] It was the first heat index and, in revised form, is still one of the widely used indices for evaluating thermal environments. The scale was based on equivalent subjective estimates of the thermal environments with different combinations of air temperature, air velocity, and humidity.

Two environmental chambers were used. Subjects moved between a reference chamber and a second chamber while adjustments were made until the subjects felt they had equivalent thermal sensations. All conditions that had the same effects were grouped together under the same ET as the still, saturated conditions of the reference chamber.

Two ET scales were developed: the normal scale for men wearing ordinary, indoor summer clothing and the basic scale for men stripped to the waist. Measuring the ET of a room requires dry bulb temperature, wet bulb temperature, and air velocity values. An example of ET, using the normal scale, obtained from the nomogram in Figure 25.3 is: if the dry and wet bulb temperatures and the air velocity of an environment are 30°C, 20°C, and 2 m/sec, respectively, the corresponding ET value for this environment is 23°C. This means that the standard man wearing ordinary, summer clothing will sense a thermal environment of 30°C dry bulb temperature, 20°C wet bulb temperature, and 2 m/sec air velocity as equivalent to a 23°C dry bulb temperature environment of still, saturated air (i.e., zero air velocity and 100% relative humidity).

Because radiation heat measuring devices were not available, the radiation parameter was not included in the ET scale. Later, after the introduction of the Vernon globe temperature,[4] Bedford (1946)[26] amended the ET scale to include allowances for radiation. This scale, called corrected effective temperature (CET), uses the globe temperature instead of the dry bulb temperature. CET can also be determined from Figure 25.3.

Although the ET and CET scales have been widely used, they make limited allowance for the effects of clothing worn and no allowance for the level of physical activity. Under severe conditions approaching the limits of tolerance, wet bulb temperature becomes the major determinate of heat strain, and the ET scale may underestimate the severity of these conditions.[27]

A committee on atmospheric comfort modified CET by replacing the wet bulb temperature with the dew point and called their index the equivalent ET corrected for radiation.[28] This modification yields only minor differences from CET for most thermal conditions.

The new effective temperature (ET*) developed by Gagge et al.[29] is similar to the ET scale, but it uses as its basis an environment at 50% relative humidity instead of 100% relative humidity. Gagge and Gonzalez[22] claimed that the ET* is more comparable with everyday experience than the ET scale. Quantitatively, ET* is the solution, by iteration, of the following equation.[30,31]

$$e_a - 0.5e_{S(ET^*)} = -(i/w)(T_o - T_{ET^*}) \quad (6)$$

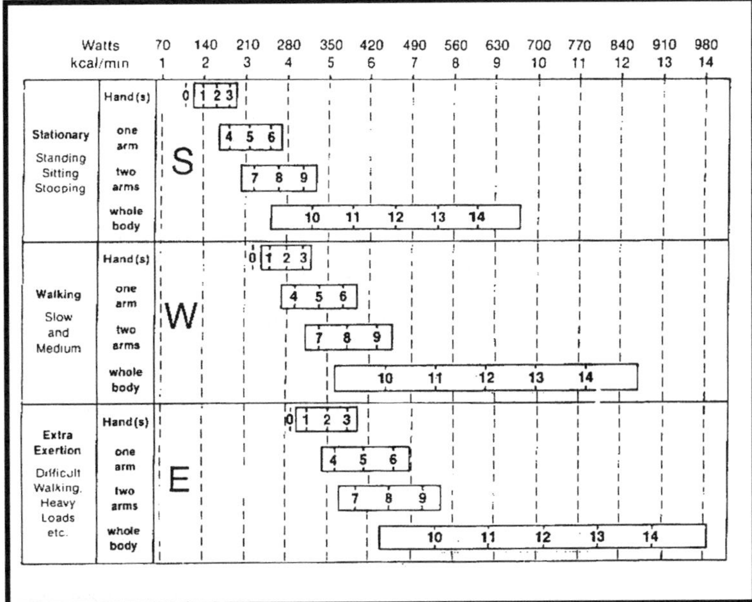

Figure 25.2 — Systematic work load estimation (SWE) schema.[18]

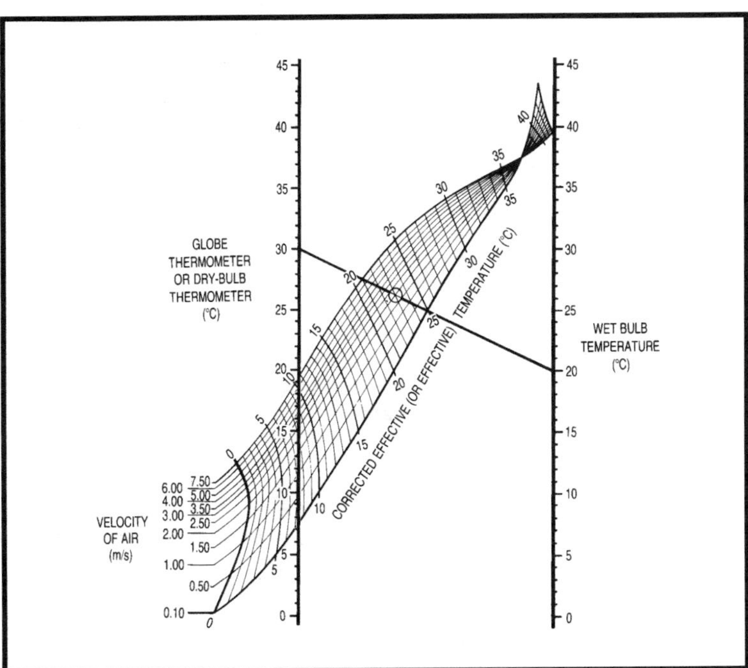

Figure 25.3 — Nomogram for the CET, or ET, normal scale.[100]

or

$$0.5e_{S(ET^*)} + (i/w)T_{ET^*} = e_a + (i/w)T_o \quad (7)$$

where:

e_a = the saturation vapor pressure at T_a, mm Hg

$e_{S(ET^*)}$ = the saturation vapor pressure at T_{ET^*}, mm Hg

w = the skin wetness and ranges from 0.06 to unity

T_o = the operative temperature

defined as: $To = (h_rT_r + h_cT_a)/(h_r + h_c)$, °C

T_r = the mean radiant temperature, °C

T_a = the air temperature, °C

h_r = the linear radiation exchange coefficient. Its value, rather constant at normal temperature range, is approximately 4.7 W/m² °C

h_c = the convective heat transfer coefficient is a function of air velocity and may be estimated by the equation: $h_c = 8.6v^{0.53}$, W/m² °C

i = the ratio of the transfer coefficients for sensible heat $(h_r + h_c)$ and insensible heat (h_e): $i = (h_r + h_c)/h_e$, kPa/K.

T_{ET^*} = the new effective temperature, i.e., ET*, °C

The ratio i, a unique transfer characteristic of the total environmental heat exchange for a human subject, is a function of air movement, clothing worn, and barometric pressure. It has a value of 0.11 kPa/K for a normally clothed, sedentary subject in still air and decreases with clothing insulation and/or increasing air movement or barometric pressure.[30]

The ET* scale is a theoretically accurate estimate of human heat transfer and is finding increased use in assessing the thermal environment. Computer and calculator programs have been developed for determination of the ET* value.[31,32]

Wet Bulb Globe Temperature (WBGT)

The wet bulb globe temperature (WBGT) developed by Yaglou and Minard (1957)[33] was not based on analysis of a new set of prime data. Rather, it was derived from, and was a means for estimating, the CET. It was originally developed for use in controlling heat casualties at military training centers.

The WBGT combines the effect of the four main thermal components affecting heat stress: air temperature, humidity, air velocity, and radiation, as measured by the dry bulb (T_{db}), natural wet bulb (T_{nwb}), and globe (T_g) temperatures. For indoor conditions (or outdoors without solar load, e.g., cloudy or shaded), WBGT values can be determined from the following equation.

$$WBGT = 0.7\ T_{nwb} + 0.3\ T_g \qquad (8)$$

For conditions with solar radiation, the equation becomes

$$WBGT = 0.7\ T_{nwb} + 0.2\ T_g + 0.1\ T_{db} \qquad (9)$$

Different forms of these equations have been proposed to define WBGT estimates under different conditions, but these two are the most widely used and accepted.

In 1969 a panel of international experts from the World Health Organization (WHO) reviewed several thermal indices including CET and HSI. The panel determined that none of the indices specifically predicted the physiological strain to be expected from exposure to heat.[34]

NIOSH later established five principal criteria for a standard heat stress index for industrial use.[5] Based on these criteria it recommended the WBGT index as the standard heat stress index. The Occupational Safety and Health Administration (OSHA) Advisory Committee,[35] American Conference of Governmental Industrial Hygienists (ACGIH),[16] and ISO[36] have also recommended the WBGT as the primary heat stress index.

NIOSH (1972)[5] suggested the use of a tripod with thermometers measuring air, natural wet bulb, and globe temperature as the standard means for determining the WBGT. Integrated electronic instruments for measuring WBGT, such as those shown in Figure 25.4, are also commercially available. Such instruments provide a direct or digital readout of WBGT and, in some models, air velocity and dry bulb, wet bulb, and globe temperatures also. The stabilization time required for the WBGT standard tree is at least 20 min, since it includes a globe thermometer that is 15 cm (6 in.) in diameter. The stabilization time required for the integrated electronic instruments is usually about 5 min, since all the sensors are resistance thermometers, and the globe has a small diameter of approximately 4 cm (< 2 in.).

Electronic instruments for personal heat stress monitoring are also available to register heart rate and/or body temperature via earplug or chest-mounted sensors. These are useful for highly stressful situations where it is important to have information on an individual's physiological responses. Typically, however, if the work allows it, a worker trained to recognize heat illness can recognize limiting conditions and make the necessary modifications.

Wet Globe Temperature (WGT)

Botsford (1971)[37] developed the wet globe temperature (WGT), which combines air temperature, humidity, air velocity, and radiation into a single reading. The WGT is

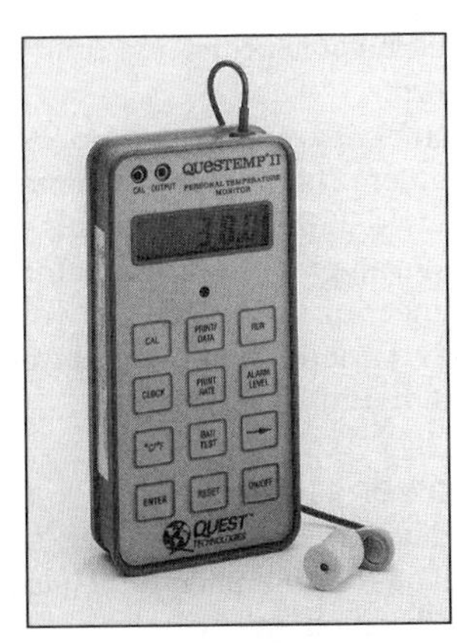

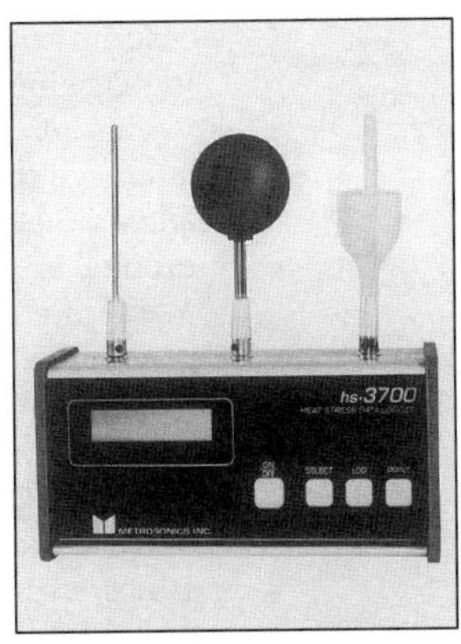

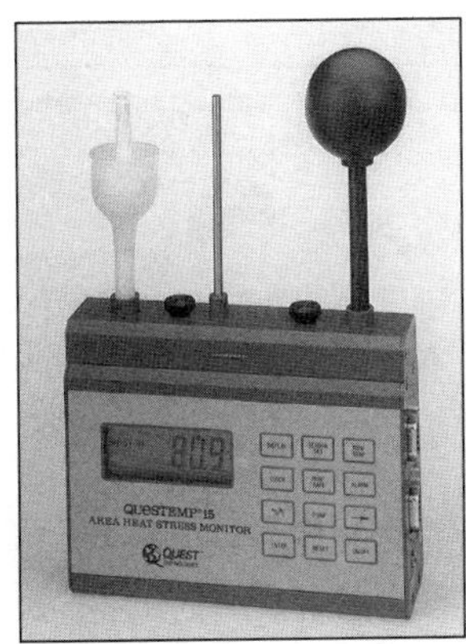

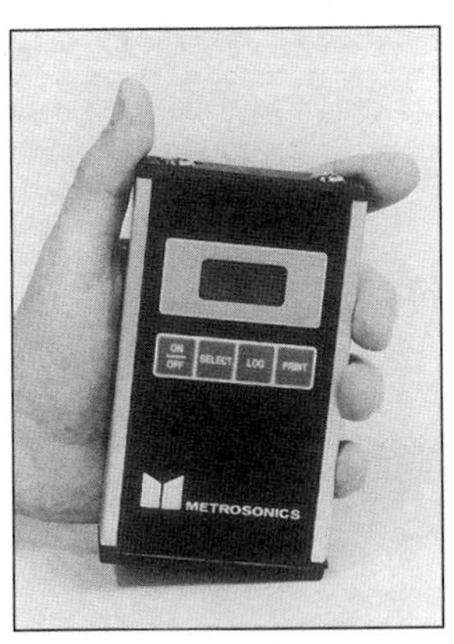

Figure 25.4 — Typical electronic instruments for measuring WBGT.

measured by the Botsball thermometer, which consists of a metal probe with its heat sensor in the center of a 6 cm (2³/₈ in.) diameter hollow, copper sphere. The sphere is painted black and covered with a double layer of black cloth that can be saturated with water. The Botsball has a water reservoir of approximately 7 cc.

The WGT considers the effect of the four main thermal components, and several researchers have developed predictive equations with high coefficients of correlation for the relationship between the WBGT and the WGT.[38–42] Results of these studies were the basis for proposed equations, WBGT = WGT + 3(°C), or WBGT = 1.1 WGT, as means for approximating this relationship for indoor or nonsolar load environments.[43] The equation WBGT = WGT + 1(°C) has been suggested for approximating these index differences for outdoor conditions with solar radiation.[44,45]

A NIOSH workshop on recommended heat stress standards[46] concluded that the WGT was a reasonable alternative to WBGT or ET for monitoring the industrial environment. The WGT was suggested as a basic index because of its simplicity and relatively low cost. Disadvantages of the Botsball relate to the lower sensitivity and accuracy of the thermometer and to the risk of obtaining erroneous values due to the condition and wetness of the cloth-covered ball. The black cloth can fade in the sun, become contaminated with mineral deposits from the water, or deteriorate from fumes or dusts in the operating area. Maintaining a clean, wet, black cover is critical to use of this device. If the Botsball is to be used for long periods, risk of dryness can be minimized with a siphon to a nearby bottle or other auxiliary water source. The stabilization time for the Botsball is approximately 5 min when the temperature differential between subsequent readings is small (5 to 10°C), but may require 15 min when this differential is larger (>15°C).[47]

Thermal Stress and the Required Sweat Rate

The international standard, ISO 7933, incorporates the latest scientific information for computation and interpretation of thermal balance and the sweat rate required to maintain this balance, i.e., the required sweat rate.[14]

This rational approach to assessing hot environments requires measurements of the thermal environment as well as a determination of the metabolic heat production and the clothing being worn. These data serve as input to the previously discussed basic heat balance equation, and calculations of each heat balance component are made using the equations shown in Table 25.3. This standard also includes equations for the respiratory heat loss and mechanized power, but they are excluded here since they typically have less effect on thermal balance than is represented by the imprecision in determining metabolic heat.

Comparison of the required sweat rate and skin wetness with the maximum limiting values for skin wetness and sweat rate provides an indication of limiting thermal severity of the work/heat. These limits are expressed for both acclimatized and unacclimatized workers. Heat storage and resulting increases in body temperature occur when thermal equilibrium is exceeded. Two levels of heat limits (warning and danger) are presented in terms of

Table 25.3 — Components of Thermal Balance[14]

E_{req} = Evaporation rate required (watts/m²) for the maintenance of the thermal equilibrium = M - C - R

M = Heat flow by metabolism (watts/m²)

C = Heat flow by convection at the skin surface (watts/m²) = $h_c \times F_{cl} (\bar{t}_{sk} - \bar{t}_a)$

R = Heat flow by radiation at the skin surface (watts/m²) = $h_r F_{cl} (\bar{t}_{sk} - \bar{t}_r)$

E_{MAX} = Maximum evaporation rate (watts/m²) if skin is completely wetted = $(P_{sk,s} - pa)/RT$

w_{req} = Required skin wetness (dimensionless) = E_{req}/E_{max}

SW_{req} = Required sweat rate (SW_{req}, watts/m²) = E_{req}/r_{req}

where:

h_c = the convective heat transfer coefficient, watts/m² K
F_{cl} = the reduction factor for sensible heat exchange due to the wearing of clothes (dimensionless)
$\bar{t}_{sk}$ = the mean skin temperature, °C
$\bar{t}_a$ = the air temperature, °C
h_r = the radiative heat transfer coefficient, watts/m², K
$_r$ = the mean radiant temperature, °C
$p_{sk,s}$ = the saturated vapor pressure at the skin temperature, kilopascals (kPa)
p_a = the partial water vapor pressure in the working environment, kPa
R_T = the total evaporative resistance of the limiting layer of air and clothing, m² kPa/watt
r_{req} = the evaporative efficiency of sweating (dimensionless), which corresponds to the required skin wetness

Table 25.4 — Reference Values for the Different Criteria of Thermal Stress and Strain[14]

Criteria	*Nonacclimatized Subjects* Warning	Danger	*Acclimatized Subjects* Warning	Danger
Maximum skin wetness				
(w_{max})	0.85	0.85	1.0	1.0
Maximum sweat rate				
Rest: (M < 65 watts/m²)^A				
SW_{max}, watts/m²	100	150	200	300
SW_{max}, g/hr	260	390	520	780
Work: (M > 65 watts/m²)^A				
SW_{max}, watts/m²	200	250	300	400
SW_{max}, g/hr	520	650	780	1040
Maximum heat storage				
Q_{max}, watts hr/m²	50	60	50	60
Maximum water loss				
D_{max}, watts hr/m²	1000	1250	1500	2000
D_{max}, g	2600	3250	3900	5200

^A M = metabolic heat

heat storage and maximum water loss. These limiting values for thermal stress and strain are shown in Table 25.4.

Allowable exposure time, or duration of limiting exposure, can also be calculated when thermal balance is not achieved. These thermal balance relationships are based on a 36°C body temperature associated with the onset of sweating. They also assume no impermeable protective clothing is being worn.

This ISO document also includes a BASIC language computer program for calculating all parameters used in the standard. This rational method also provides useful information for engineering controls in the workplace, since the contribution of each thermal component is individually determined.

Cooling Power of Air

This index, developed for use in hot mines, uses a kata thermometer, which is a liquid-in-glass thermometer with two marks etched on the stem.[48] The wet kata thermometer (the bulb covered by a wetted wick) is heated in water and then allowed to cool in the environment being measured. The time required for the liquid in the wet kata to cool between the two marks can be measured with a stop watch, and this time can be used with a calibration sheet to determine the cooling power of the air.

Due to the fragile nature of the kata thermometer, and the stable relationship between the cooling power, the wet bulb temperature, and air velocity, the equation below is also commonly used in lieu of wet kata readings to determine cooling power in the hot mines of South Africa.[49]

$$K = (0.7 + v^{0.5})(36.5 - T_{wb}) \qquad (10)$$

where:

K = wet kata cooling power, mcal/cm²/sec
v = air velocity (m/sec)
T_{wb} = ambient wet bulb temperature (°C)

Other Heat Indices

Other indices that have been reported in the literature include predicted 4-hour sweat rate;[50] wet bulb-dry bulb index;[51] temperature humidity index;[52] index of physiological effect;[53] index of thermal stress;[54] relative strain index;[55] reference index;[56] and others.[8]

Since 1985 the U.S. National Weather Service has been using an abbreviated apparent temperature (called a heat index) as an index of heat discomfort during the summer months. This index includes the amplifying effect of increasing humidity on the discomfort level and is frequently reported by the news media to complement the report of daily summertime air temperatures. As shown in Table 25.5, an air temperature of 38°C/100°F and a 50% relative humidity "feels like" a temperature of 49°C/120°F. Apparent temperature

Table 25.5 —
Apparent Temperatures, °C/°F, and Heat Syndromes (adapted from National Oceanic and Atmospheric Administration)[57]

Air Temperature	*Apparent Temperatures*^{A}									
52/125°	51/123	61/141								
49/120°	47/116	54/130	64/148							
46/115°	44/111	49/120	57/135	66/151						
43/110°	40/105	44/112	51/123	58/137	66/150					
40/105°	38/100	40/105	45/113	51/123	57/135	65/149				
38/100°	35/95	37/99	40/104	43/110	49/120	56/132	62/144			
35/95°	32/90	34/93	36/96	38/101	42/107	45/114	51/124	58/136		
32/90°	29/85	31/87	32/90	34/93	36/96	38/100	41/106	45/113	50/122	
29/85°			29/84	30/86	31/88	32/90	34/93	36/97	39/102	42/108
27/80°				26/79	27/81	28/82	29/85	30/86	31/88	33/91
24/75°					24/75	24/76	25/77	26/78	26/79	27/80
	10%	20%	30%	40%	50%	60%	70%	80%	90%	100%
	Percent Humidity									

^{A}130°F or more (extremely hot), heatstroke or sunstroke is imminent; 105–130°F (very hot), sunstroke, heat cramps, and heat exhaustion likely, and heatstroke possible with prolonged exposure and physical activity; 90–105°F (hot), sunstroke, heat cramps, and heat exhaustion possible with prolonged exposure and physical activity; 80–90°F (very warm), fatigue possible with prolonged exposure and physical activity.

ranges associated with different heat syndromes/ risks are also presented. The original apparent temperature model was developed in the field of meteorology.[57]

Correlation Between Heat Indices

A heat stress index is a numerical evaluation of the hot environment. Therefore, if the level of one or more of the thermal components increases, the numerical value of the heat index increases. Logically, strong correlation should exist among most of the heat indices, and this has been reported as shown in Table 25.6 from several laboratory and field studies.[38,39,41,56,58,59] The WBGT is highly correlated to the other indices (0.913 to 0.977), and a linear relationship of similar slope (0.727 to 0.934) exists among the thermal indices.

Heat Exposure Limits

The heat stress indices incorporate important thermal variables into a single number or scale with different degrees of success. The major application for a single index number is to define limits for exposure to hot occupational environments. Such limits may correspond to the upper, midrange, or lower levels of physiological impact and risk. Threshold values of thermal indices represent exposures above which a worker will be at some risk to heat illness. Upper, tolerance, or ceiling limits represent absolute maximum exposures above which work should not continue because the risk of heat illness is high. The term permissible exposure limit has been used in the literature alternately to represent threshold and ceiling limits as well as the range in between, and this has generated confusion at times.

Heat exposure limit values are typically based on some set of assumed physiological, personal, and/or environmental conditions. A number of factors, however, create variability and imprecision in

Table 25.6 —
Correlation of WBGT and Other Indices

Source	*Number of Observations*	*Regression Equation (°C)*^{A}	*Coefficient of Correlation*
Beshir et al.[38]	13,489	WGT = 0.905 WBGT - 0.909	0.956
Brief and Confer[39]	34	ET = 0.823 WBGT + 4.131	0.913
Brief and Confer[39]	34	ETCR = 0.786 WBGT + 6.029	0.966
Jensen and Heins[58]	200	CET = 0.727 WBGT + 7.380	0.964
Ramsey and Beshir[59]	6825	T_{nwb} = 0.934 WBGT - 1.903	0.977

^{A}WGT, wet globe temperature; ET, effective temperature; ETCR, effective temperature corrected for radiation; CET, corrected effective temperature; T_{nwb}, natural wet bulb temperature the application of such limits: individual differences, worker populations, age, acclimatization, and health status, or errors in measuring or estimating metabolic work load or the environment.[46,60]

Since threshold limits are not associated with the high risk of heat illness, they may be defined in terms of levels at which nearly all, or some specific percentile of workers (e.g., 95%) may be repeatedly exposed without adverse effects. Upper or tolerance limits, however, are commonly based on an individual worker or work group in a work environment rather than on the general population of workers.

WBGT Recommendations

The most commonly used index for expressing limiting thermal exposure levels in occupational environments is the WBGT.

One fundamental assumption used in threshold heat stress limits is that a worker's deep body temperature should not exceed 38°C. This was recommended by a WHO panel of physiologists[34] and is similar to the strain criteria of the Belding and Hatch HSI,[61] which suggest that deep body temperature should not increase by more than 1°C.

Another major support for both the WHO recommendation and the development of the WBGT limits are Lind's prescriptive zone studies.[51,62] To use these studies to develop WBGT threshold limits, it was necessary to account for the effects of metabolic heat gain caused by work and to convert studies reported in ET into WBGT units. Lind's studies used ET measures with subjects who were unacclimatized and wore only shorts, whereas an industrial worker normally would be acclimatized and wear some form of worker uniform. The effect of clothing, which lowers the limits, and the effect of acclimatization, which raises them, were assumed to be approximately equal in magnitude and opposite in direction, and thus would cancel each other with no further modification. Figure 25.5 charts this development procedure and the resulting permissible exposure limit (PEL).[63]

Lind's data using the basic ET scale for partially clothed men was first converted to the normal ET scale for clothed men and then to WBGT, using the relationships suggested by Minard (1964).[64] The permissible exposure threshold limits were designed to provide a work temperature and WBGT combination so that 95% of the workers would not have a deep body temperature exceeding 38°C. The 38°C limit is widely accepted as an average limit for prolonged daily exposure to heavy work; however, it is considered conservative as a limit for many individuals and for short exposure periods. Persons who are medically screened and under experimental surveillance commonly exceed 39°C without noticeable ill effects.[45] One way to circumvent debate on the 38°C/39°C limits is to reference rate of changes in deep body temperature to specific physiological heat tolerance and exposure limits.[65]

Time-Weighted Average

Because any workday or work period can consist of varying levels of both thermal exposure (WBGT) and metabolic heat, it is necessary to calculate time-weighted average (TWA) values to determine the total heat or work impact. To calculate the time-weighted average of the variable x

$$TWA(x) = (x_1t_1 + x_2t_2 + \ldots - x_nt_n)/(t_1 + t_2 + \ldots + t_n) \quad (11)$$

where:

x = value of either WBGT or metabolic workload
x_1 = value during period 1 of WBGT or metabolic workload
t_1 = time in minutes of period 1

NIOSH Recommendations

A criteria document for a recommended standard concerning occupational exposure to hot environments was developed by NIOSH.[5] This document, a work practices standard, recommended that one or more work practices be instituted when "exposure of an employee is continuous for one hour or intermittent for a period of two hours at a time-weighted average WBGT that exceeds 79°F (26.1°C) for men or 76°F (24.4°C) for women." The suggested limits represent the point of WBGT intersection with 400 kcal/hour (465 watts) work load on the WBGT chart (Figure 25.5). This work load is the maximum anticipated as an hourly TWA in American industry. The lower value for women was intended to compensate for differences from men. Subsequent research concerning male/female differences has established that such differences more commonly represent differences in physical work capacity or heat acclimatization, and that different limits would be better based on these factors than on gender.[45]

A revised criteria document was issued in 1986 that changed exposure limit calculations to a 1-hour TWA and also removed the differential related to gender.[66] Also included in this revision were recommended alert limits for unacclimatized workers, recommended exposure limits (REL) for acclimatized workers, and ceiling limits beyond which appropriate and adequate heat protective clothing and equipment should be required. A summary of the recommended limits for acclimatized workers is shown in Figure 25.6.

ACGIH Threshold Limit Values

ACGIH[16] has adopted permissible heat exposure threshold limit values (TLVs) that are basically the same as the NIOSH criteria documents. The work load categories are light work < 200 kcal/hour (233 watts), moderate work 200–350 kcal/hour (233–407 watts), and heavy work 350–500 kcal/hour (407–581 watts). Each limiting temperature (TLV) listed is the most conservative WBGT number for the corresponding work load range. Many feel that the 500 kcal/hour (581 watts) as an hourly TWA work load exceeds what will be encountered in American industry, since workers tend to adjust their work intensity and rest pauses during work with high metabolic demand. ACGIH also adopted WBGT limits for different work-rest regimens. It should be noted, though, that these limits relate only to the listed subset of different work-rest time combinations, and to conditions where rest breaks are taken in the same temperature as the work.

OSHA Advisory Committee Recommendations

OSHA established in 1973 the Standards Advisory Committee on Heat Stress to evaluate the proposals from NIOSH, conduct hearings, receive public exhibits and comments, and prepare recommendations for a heat stress standard. Their recommendation was similar to the one proposed by NIOSH; the main difference was the provision of a higher set of limiting WBGT values if the work was performed in high air velocity. The OSHA recommendation divided the work practices into three sets:

(1) those practices required in all instances where environmental conditions and work load stresses exceeded listed limits and included potable water, written company policy on acclimatization procedures, first aid training for workers, and, in the case of workers at heavy work load, medical examination;

(2) additional work practices of a special nature that can be used to ameliorate stress in general work; and

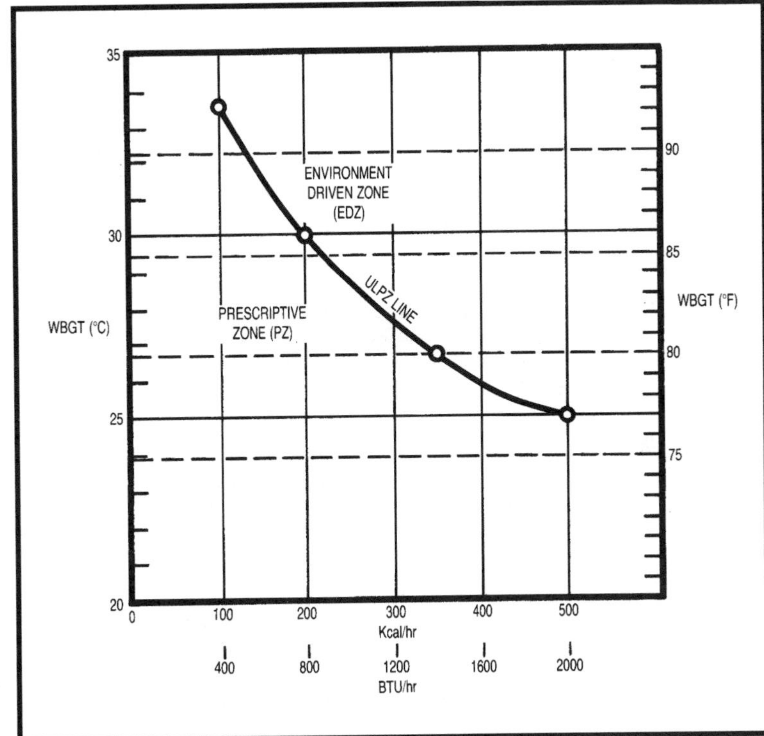

Figure 25.5 — Permissible exposure limit chart.[63]

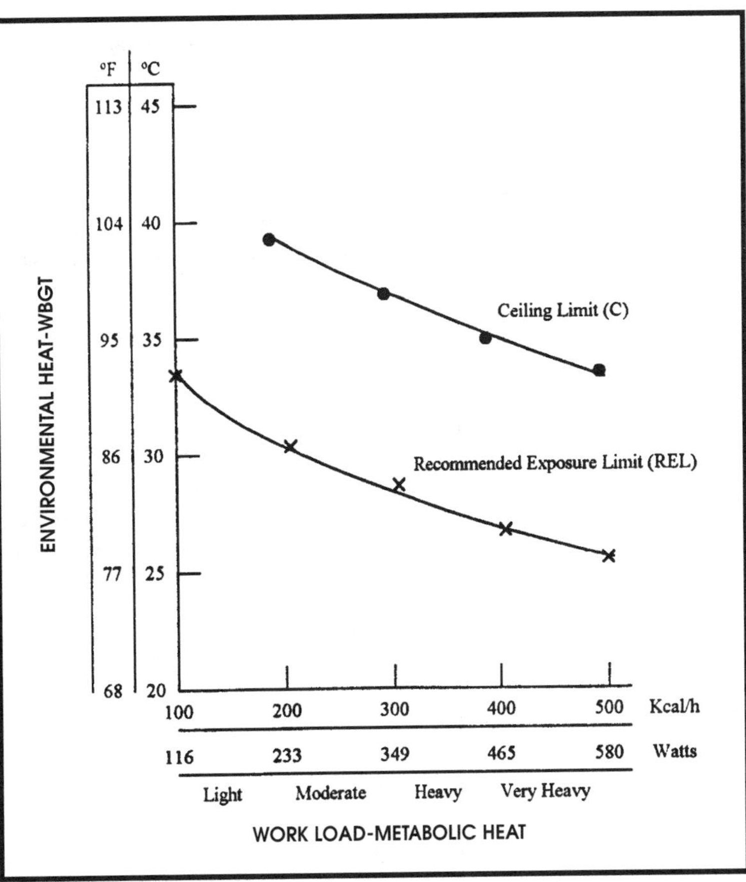

Figure 25.6 — Recommended heat stress exposure limit and ceiling limit for acclimatized workers.[66]

Table 25.7 —
Use of the WBGT Index in Control of Physical Activity[68]

WBGT Index	Precautions Indicated
78°F (26°C)	Extremely intense physical exertion may precipitate heat exhaustion or heat stroke.
82°F (28°C)	Discretion should be used in planning heavy exercise for unseasoned personnel.
85°F (29°C)	Strenuous exercise such as marching at standard cadence should be suspended in unseasoned personnel during their first three weeks of training.
>85°F (29°C)	Outdoor classes in the sun should be avoided.
88°F (31°C)	Strenuous exercise should be curtailed for all recruits and other trainees with less than 12 weeks training in hot weather. Hardened personnel, after having been acclimatized each season, can carry on limited activity at WBGT of 88°F to 90°F (31°C–32°C) for periods not exceeding six hours a day.
90°F (32°C)	Physical training and strenuous exercise should be suspended for all personnel (excluding essential operational commitments not for training purposes, where the risk of heat casualties may be warranted).

(3) work practices required for extreme heat exposure where intensity is severe and work duration is normally short.

These recommendations were submitted on majority vote to OSHA, but no standard for heat stress has been issued. A complete description of the advisory committee actions can be found in Ramsey (1976).[67]

U.S. Military Recommendations

The U.S. military has issued joint criteria for using the WBGT index in control of physical activity.[68] This military standard (see Table 25.7) cautions that measurements should be taken in the specific environments to which personnel are exposed.

The U.S. Navy developed for use in its fleet operations a set of physiological heat exposure limit curves.[68] Each curve represents a different level of work intensity or metabolic cost, so the maximum hours per shift associated with different levels of TWA WBGT can be directly determined. Since the physiological heat exposure limit addresses only hot operations, the WBGT values do not go below 82°F (27.8°C). Note also that the limits apply to military personnel who represent a relatively young, healthy, and physically fit working population. These limits represent upper exposure limits above which the risk of heat illness becomes unacceptably high for exposed individuals.

ISO Recommendations

ISO[36] has also adopted a heat stress standard based on the WBGT index. Limits are based on 1-hour time-weighted WBGT values for workers who are normally clothed, physically fit, of standard size with surface area of 1.8 m^2, and in good health. Included are five metabolic rate classes that range from resting to a very high metabolic rate and an adjustment for acclimatization. Limits for most unacclimatized workers are 1 to 3°C lower than for acclimatized workers. In the case of heavy and very heavy metabolic work load, a different WBGT is indicated where there is sensible air movement compared with no sensible air movement. These limits are shown in the summary information of Table 25.8.

AIHA Recommendations

Industrial heat exposure limits are also suggested by the American Industrial Hygiene Association[28] as shown in Table

Table 25.8 —
A Comparison of Proposed WBGT Threshold Values for Acclimatized Workers

Work Load	ACGIH[16]	AIHA[28]	OSHA[35]	ISO[36]	NIOSH[66]
Resting		32.2°C 100 kcal/hr (117 watts)		33°C ≤100 kcal/hr (117 watts)	
Light	30°C 100–200 kcal/hr (117–233 watts)	30°C 200 kcal/hr (233 watts)	30.0°C^A, 32.2°C^B <200 kcal/h (233 watts)	30°C 100–201 kcal/hr (117–234 watts)	30°C < 200 kcal/hr (233 watts)
Moderate	26.7°C 201–350 kcal/hr (234–407 watts)	26.7°C 300 kcal/hr (349 watts)	27.8°C^A, 30.6°C^B 201–300 kcal/hr (234–349 watts)	28°C 201–310 kcal/hr (234–360 watts)	28°C 201–300 kcal/hr (234–349 watts)
Heavy			26.1°C^A, 28.9°C^B > 301 kcal/hr (350 watts)	25°C^A, 26°C^B 310–403 kcal/hr (360–468 watts)	26°C 301–400 kcal/hr (350–465 watts)
Very Heavy	25°C 350–500 kcal/hr (407–581 watts)			23°C^A, 25°C^B > 403 kcal/hr (468 watts)	25°C 401–500 kcal/hr (466–580 watts)

^{A}Low velocity
^{B}High velocity

25.8. These limits, in WBGT units, were developed at Exxon for acclimatized workers engaged in either continuous work or intermittent work alternated with 1-hour rest periods under cooler conditions. The continuous work limits assume eight hours per day with 10 min rest each hour.

Comparison of WBGT Recommendations

Table 25.8 summarizes and compares several of the proposed WBGT threshold limit values where work load levels have been specified to correspond to certain WBGT limits. The specified AIHA work load levels are "light, moderate, and heavy,"[28] but the corresponding metabolic costs are consistent with the categories of "resting, light, and moderate" used by other authors and groups. Similarly, the ACGIH category "heavy"[16] is more consistent with the metabolic cost range of "very heavy." When the different metabolic heat assumptions are considered and threshold limit proposals are compared, a strong pattern of consistency is observed: resting, 32–33°C; light, 30°C; moderate, 27–28°C; heavy, 25–26°C; and very heavy, 23–25°C. Given the imprecision attendant on estimating metabolic work load and the variability of metabolic costs during the day and from day to day, the WBGT threshold values in Table 25.8 are all basically equivalent. None of the limits differ by more than 2°C, and the error in estimating metabolic work load can create differences this large or larger.[19]

Adjustments to WBGT Limits

Different levels of thermal conditions and metabolic heat requirements generate different WBGT limiting values. The variables of clothing and acclimatization also affect thermal stress and strain, and methods for adjustment of WBGT limits based on these factors have been developed.[16, 69–71] Table 25.9 is a summary of these modification factors, but it should be noted that these factors are approximate and should be used for general guidance only.[72]

Threshold RELs shown in Figure 25.6 basically refer to the worker who is acclimatized and dressed in typical work attire of trousers and a cotton work shirt. Limits for the unacclimatized worker would need to be adjusted, however, as shown in Table 25.9; the heavier the work load, the greater the adjustment needs to be. For example, the unacclimatized person engaged in a moderate work load task should have the REL lowered 2°C (3.6°F) WBGT.

Table 25.9 — WBGT Modification Factors for Acclimatization and Clothing[72]

Unacclimatized		
Low work load	117–232 watts (100–199 kcal/hr)	1°C (1.8°F)
Moderate work load	233-348 watts (200-299 kcal/hr)	-2°C (-3.6°F)
Heavy work load	349-465 watts (300-400 kcal/hr)	-3°C (-5.4°F)
Clothing		
Light work clothing		0°C (0°F)
Cotton coverall, jacket		-2°C (-3.6°F)
Winter work clothing, double cloth coveralls, water barrier		-4°C (-7.2°F)
Lightweight vapor barrier suits		-6°C (-10.8°F)
Fully enclosed suit with hood and gloves		-10°C (-18°F)

Note: The recommended exposure limits of Figure 6 should be reduced by the amounts shown. These are approximate values; use for general guidance only.

Similarly, the WBGT limits should be lowered with increasing insulation or impermeability of the clothing worn by the worker. For example, the REL for a worker in coveralls and jacket should be reduced by 2°C (3.6°F), and if the worker is in a light vapor-barrier suit the limits should be reduced by 6°C (10.8°F). These recommended modification factors for clothing are also shown in Table 25.9.

Stay Times

Work/rest schedules associated with threshold limits have been proposed.[5,16] Such limits relate to the onset of increased risk for heat strain during 1-hour work periods. However, some jobs require excursion into very hot conditions where ceiling limits are more appropriate than threshold limits and the proper limiting factor is the time of exposure. Typical jobs would be boiler or cupola maintenance, inspection, or repair in steam tunnels or confined hot spaces, emergency work, or response, etc.

The physiological heat exposure limit represents an early and successful effort in developing stay times for U.S. military personnel based on the metabolic heat generated by the work.[68] More recently a comprehensive set of stay times have been developed for use in the electric power industry.[73] These guidelines include not only the metabolism factor, but also the effects of different levels of clothing. Table 25.10 is a summary of the above stay time recommendations.[74] Presented are a range of ceiling limit stay times, where the lower number refers to a basic, more conservative stay time, and the larger number refers to highly fit and selected workers. The modification factors of Table 25.9 can be used to adjust the stay times based on different levels of clothing or acclimatization. It is important

Table 25.10 — Range of Stay Times (Min) for Ceiling Limits (WBGT)

°F	°C	WBGT Metabolic Heat Workload (M)[A] Low	Moderate	Heavy
122	50	15-45	0-30	0-12
115	46	20-65	5-40	0-15
111	44	30-75	10-45	0-20
108	42	45-110	15-65	5-25
104	40	60-120	15-80	10-30
100	38	90-150	20-110	15-45
97	36	120-240	30-150	15-70
93	34	180-480	45-200	20-90
90	32	360-480	90-250	30-120
86	30	480	120-360	60-175
82	28	480	480	120-240
79	26	480	480	240-480

Note: Assumes worker who is acclimatized, healthy, physically fit, wearing lightweight work clothing, trained in heat stress, takes normal breaks, can rest in cool areas, can self-limit the heat exposure, and has instituted appropriate protective practices and controls.
[A]Range shows "basic time" (short time) and "highly selected personnel" (longer time).[74]

to note that these stay times are general guidelines only, which should be used cautiously and only as part of a well-developed, understood, and practiced heat management program.

Recommended Limits in Other Indices

CET has also been used as an index for tolerance limits to heat stress. The WHO scientific panel recommended for the acclimatized worker a set of CET tolerance limits as follows: light work, 32°C; moderate work, 30°C; and heavy work, 28.5°C. The acclimatized worker limits are 2°C higher than those for the unacclimatized worker. These limits are based on the studies of Lind[75] and refer to a worker stripped to the waist.

Exposure limits in WGT units have also received attention due to the simplicity of the Botsball instrument and the reasonable correlation with WBGT. WGT limits are approximately 3°C below the corresponding WBGT limits if thermal conditions are not extreme, i.e., less than 38°C/101°F WGT.[43]

The dial thermometer of the Botsball also lends itself to color-coding of the dial face to indicate action levels without the necessity of converting to temperature numbers. Outdoor action limits for acclimatized military troops are based on temperature conditions reflected on the Botsball as a given color.[45] For example, the red condition measures a WGT reading between 30 and 31.1°C (86 and 88°F) and specifies an appropriate water intake of 1.5 to 2.0 L or quarts per hour and a schedule of 30 min rest, 30 min work per hour. The green condition ranges between 26.7 and 28.3°C (80–83°F) and specifies 0.5 to 1.0 L of water and a schedule of 50 min work, 10 min rest per hour.

The revised NIOSH criteria document states that if environmental heat exposure is assessed by ET, CET, or WGT, then it is appropriate to convert it to equivalent WBGT for comparison with WBGT limits.

Cold Exposure Limits

Limits for exposure to cold environments are difficult to specify since the amount of clothing and its insulating characteristics have such a dominant effect on the extent of cold exposure. The effects of exposure to cold environments also vary dramatically based on the amount of metabolic heat being generated by any work being performed. A sleeping bag can provide enough insulation to withstand extreme cold conditions, even though the contribution of metabolic heat is low. The worker dressed in heavy clothing may be comfortable at light work, sweating inside these clothes during heavy work periods, then very cold during a subsequent rest break with the damp clothing and low metabolic heat. Thus, the worker can go through zones of comfort, heat stress, and cold stress with the same clothing and thermal exposure.

Windchill Index

The windchill index represents the most universally accepted scale for describing the combined cooling effects of air temperature and wind velocity. This index provides a more accurate description of cold thermal conditions than air temperature alone since, at constant temperature, the risk of tissue freezing increases with air movement. Although the index was developed using the freezing rate of a container of water, it does provide a useful means of estimating those combinations of temperature and wind speed likely to freeze human flesh.

Windchill reflects the cooling power of wind on exposed flesh and is commonly expressed as an equivalent chill temperature. Flesh may freeze within 30 sec in the great danger area, and within 1 min in the increasing danger area. Temperature and wind speed combinations in the little danger area pose potential cold injury problems unless the skin is completely dry and exposure time is less than 5 hr. The windchill index and the equivalent chill temperature (t_{ch}) can be calculated as follows.[77]

$$WCI = 1.16\,(10.45 + 10\,v_{ar}^{1/2} - v_{ar})\,(33 - T_a),\ W/m^2$$

$$t_{ch} = (33 - WCI)/25.5,\ ^{\circ}C \qquad (12)$$

where:

WCI = windchill index
v_{ar} = relative air velocity, m/sec
T_a = air temperature, °C
(t_{ch}) = equivalent chill temperature

When equivalent chill temperature is below freezing, cold injury can occur. Frostbite or frost nip occur when tissue fluid freezes, and immersion foot (trenchfoot) occurs when reduced blood flow from chilling causes nerve damage. The windchill index does not recognize the amount of clothing being worn, but relates instead to bare skin such as the face and hands. It does provide a comparative scale for cooling power, but greatly exaggerates the importance of wind for the person dressed in heavy clothing and having face/hand protection.(77)

Required Clothing Insulation (IREQ)

An index of required clothing insulation (IREQ) is presented in ISO/TR 11079.(78) This index uses the same general concepts of thermal equilibrium as previously discussed for hot environments. Originally proposed by Holmér(79) IREQ assumes a minimal cold tolerance level for skin temperature of 30°C and skin wetness of 0.06 in a stationary standing man. With these assumptions the amount of insulation required to obtain thermal balance can be calculated. Equilibrium in the cold occurs when the heat conducted from the skin (t_{sk}) to the clothing surface equals the dry heat loss from the clothing surface to the environment (t_{cl}). Thus, IREQ can be expressed as

$$IREQ = (t_{sk} - t_{cl})/M - W - H_{res} - E \qquad (13)$$

where:

$$M - W - H_{res} - E = R + C$$

and

t_{sk} = heat conducted from the skin
t_{cl} = dry heat loss from clothing to environment
M = metabolic rate
W = external work rate
H_{res} = respiratory heat loss (convective and evaporative)
E = evaporative heat loss
R = radiative heat loss
C = convective heat loss

Calculation of IREQ requires an iterative procedure since t_{cl} is not a measured value. Computer programs are available for this and related calculations.(78)

The low skin temperature (30°C) and wetness (0.06) relate to uncomfortably cool conditions that will not produce sweating and the interference from moisture absorbing into the clothing, refreezing, and producing complications in determining the insulation requirements. Figure 25.7 depicts typical relationships of IREQ and metabolic rate, where increased metabolic rate allows reduction in insulating requirements at any given temperature.

To determine IREQ for a specific environment requires a measurement of the various temperatures, air velocity, and an estimate of metabolic activity using methods or tables previously discussed. Then the basic thermal balance equations are used to determine the IREQ, which is not exactly the same as the clothing insulation normally sited with standard clothing. In most cases a higher clothing insulation value should be selected since differences in motion and wind penetration affect it and reduce the effectiveness of the clothing. These factors are included in the IREQ.

ACGIH Recommendations

ACGIH recommends that dry insulating clothing be provided to maintain worker core temperature above 36°C (96.8°F), for exposures to air temperatures below 4°C (40°F).(16) This represents a body temperature

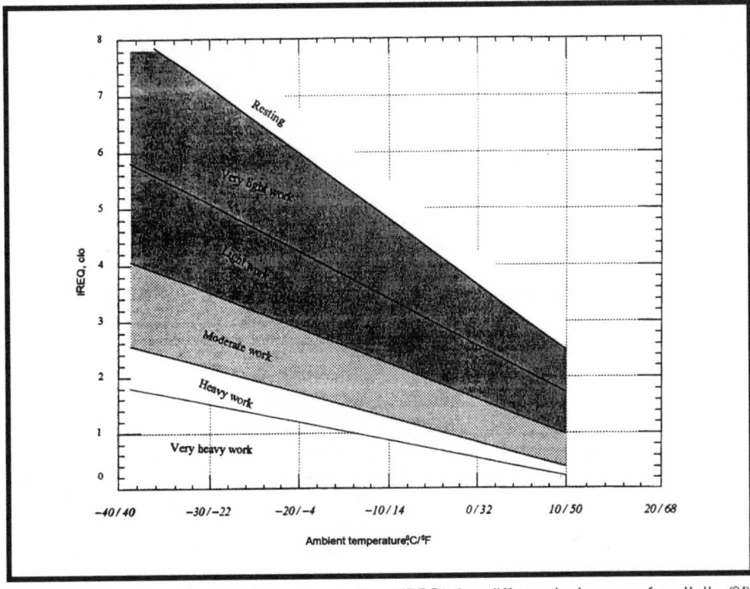

Figure 21.7 — Required clothing insulation (IREQ) for different classes of activity.(94)

above the point of intense shivering, and more serious health effects. The uncontrollable physiological response of shivering provides a good warning signal to a worker that the combined exposure effects of temperature, clothing, and metabolic heat are going beyond discomfort and into potential health-related problems.

Work/rest schedule guidelines for worker exposure to extreme cold (below -26°C/15°F) conditions have also been proposed as shown in Table 25.11. The suggested threshold limiting values of this table are for workers in dry clothing. They must also have the opportunity for periodic breaks in a warm location. These data do not specify a level of clothing (clothing insulation value) or a warm-up temperature associated with the schedules, so they should be used only for general guidance. It is also recommended that bare skin should not be exposed to equivalent windchill temperatures below 32°C (25.6°F).[16]

Thermal Effects on Performance and Safety

Effects of Heat

Effects on Physical Performance

The effects of hot environments on physical work output and on manual task performance are reasonably well understood and documented.[80,81] Physical work in a hot environment generates physiological responses that are readily measurable and physical fatigue that is readily noticeable.

Thermal exposure of such intensity and duration that it contributes to localized or general fatigue will negatively affect physical work capacity. Loss of performance capacity during physical activity in the heat can be measured in reduced production output, speed, quality, or repetitions per unit time. Deterioration in work output due to heat, as measured by military marching rates, has been observed.[82] It has been concluded that hot environments significantly reduce the work load of manual handling tasks. A temperature increase from 17 to 27°C WBGT has reduced work load by 20% for lifting, 16% for pushing, and 11% for carrying tasks.[83] Scheduling interim rest cycles during physical work under hot, ambient conditions has been suggested as a means to increase work efficiency by reducing physiological cost.[84]

Effects on Perceptual-Motor Performance

Although decreases in physical work capacity in the heat are relatively predictable, human performance is much more unpredictable when it involves sedentary work, low metabolic costs, perceptual-motor skills, or mental activities. Research concerning the effect of heat on perceptual-motor performance has been extensive and often contradictory.

Many studies have attempted to isolate the effects of heat on a sedentary or standing person performing light work-load tasks involving perceptual motor and mental

Table 25.11 —
Threshold Limit Values; Work/Warm-Up Schedule for 4-Hour Shift[16]

Air Temperature (Sunny Sky)		*No Noticeable Wind*		*5 mph Wind*		*10 mph Wind*		*15 mph Wind*		*20 mph Wind*	
°C (approx.)	*°F (approx.)*	*Max. Work Period*	*No. of Breaks*	*Max. Work Period*	*No. of Breaks*	*Max. Work Period*	*No. of Breaks*	*Max. Work Period*	*No. of Breaks*	*Max. Work Period*	*No. of Breaks*
-27	-17	(norm. breaks)	1	(norm. breaks)	1	75 min	2	55 min	3	40 min	4
-30	-22	(norm. breaks)	1	75 min	2	55 min	3	40 min	4	30 min	5
-33	-27	75 min	2	55 min	3	40 min	4	30 min	5	nonemergency work should cease	
-36	-32	55 min	3	40 min	4	30 min	5	nonemergency work should cease		↓	
-38	-37	40 min	4	30 min	5	nonemergency work should cease		↓		↓	
-41	-42	30 min	5	nonemergency work should cease		↓		↓		↓	
<-43	<-45	nonemergency work should cease		↓		↓		↓		↓	

Note: Schedule applies to any 4-hr (480 min) work period with moderate to heavy work activity, with warm-up periods and an extended break (e.g., lunch) at the end of the 4-hr work period in a warm location. For light-to-moderate work (limited physical movement), apply the schedule one step lower. For example, at -35°C (-30°F) with no noticeable wind (Step 4), a worker at a job with little physical movement should have a maximum work period of 40 min with four breaks in a 4-hr period (Step 5).

tasks. Several authors have summarized or generalized from these individual investigations.[85–89] The data summarized by Wing[85] became the basis for the upper limits of exposure for unimpaired mental performance suggested by NIOSH.[5] However, these recommendations were omitted from the revised criteria document.[66]

Continuing research has indicated that the degree of performance decrement on such work tasks in the heat is dominantly affected by the type of task being performed. Heat appears to have a very small effect on mental or very simple tasks and, indeed, brief exposures to the heat may even enhance task performance. This does not imply that work on a simple task will not deteriorate in performance over time, but simply that it will not deteriorate differentially based on the presence of heat. Other perceptual motor tasks such as tracking, vigilance, eye-hand coordination, and combinations of these tasks do tend to show losses in performance due to the heat. These tasks depict a pattern of onset of performance decrement in the 30–33°C (86–91°F) WBGT temperature range, and the decrement appears to be relatively independent of exposure time.[90] This is the same temperature range that is associated with the onset of physiological heat stress for the worker performing sedentary or light work, as shown in the REL of Figure 25.6.

In summary, performance on a specific task, under a specific set of thermal conditions, is difficult to predict precisely due to the wide range of individual personal characteristics that relate to the handling of thermal loads and to the person's skill in performing the specific task. General relationships that may be useful, however, in designing work tasks in hot environments include the following:[89]

- Performance in a hot environment and for exposure periods that yield general fatigue will also yield general performance decrements.
- Acclimatization will aid physiological adjustment and reduce the strain of a task. With perceptual-motor tasks, the effect is less pronounced, although there is evidence that acclimatization reduces performance variability and improves performance.
- Performing mental or simple perceptual-motor tasks during brief exposure to high temperature levels results in only minor decrement, or even enhancement of performance because of the arousal effects of the heat. Most other tasks, however, show onset of performance decrement around 30–33°C (86–91°F) WBGT.
- Hot temperatures seem to affect skilled or trained personnel differently, depending on the level of mental or physical load at nonheat levels. If the work task does not load the operator, the addition of heat causes an arousal effect that enhances or minimally affects the performance. If the work task has already created an overload, the addition of heat will tend to degrade performance.
- Perceptual-motor performance within the comfort zone is generally best at the cooler end of the zone.
- A commonly reported relationship between performance of perceptual-motor tasks and thermal conditions is the inverted U, where performance is highest at midrange temperatures and decreases with positive or negative changes in the temperature.

Effects of Cold

The most significant effect of cold exposure is on manual dexterity. Many studies of performance in cold have emphasized the effects of cold on the skillful use of the hands. An individual required to work in the cold normally has protective, insulating layers of clothing consistent with the severity of the cold. Thus, the internal microclimate of the body may be relatively stable under normal cold conditions. Adequately protecting the hands, however, is difficult because manual dexterity and general handwork are adversely affected by increased thermal insulation and protection. Dusek[91] reported that at finger temperatures below 15.6°C (60°F), significant decrease in manual performance occurred. Below 10°C (50°F), there is onset of pain and extensive loss of manual abilities. Below 4.4°C (40°F), tactual discrimination is lost, as is the ability to perform fine manipulative movements. Significant correlations were observed between manual task performance and the skin temperatures of the hand, the finger, and the upper arm, mean skin temperature and mean body temperature during cold exposure.[92] The skin temperature of the hand and the mean skin temperature showed highest correlations with performance. After a 1-hour exposure to cold, manual performance of subjects wearing cold-protective clothing with cotton working gloves showed continuing deterioration as the environmental temperature level lowered to -20°C (-4°F).[92]

Motor skill loss in the cold occurs more rapidly for fine motor skills than for gross motor skills. The National Association of Building Contractors considers guidelines for reduction in motor skills as 10%, 20%, and 25%, for gross motor skills and 60%, 80%, and 90% for fine motor skills, at -18°C/0°F, -23°C/-10°F, and -29°C/-20°F ET, respectively. These guidelines assume that the worker has the proper clothing to provide protection.

The consensus is that loss of perceptual-motor performance in the cold is primarily a deterioration of motor capabilities rather than mental deterioration. The opportunity to warm the hands or remove them from cold exposure periodically can help reduce loss of manipulative motor skills. Hand protection is recommended, if manual dexterity is not critical, for temperatures below 16°C (60.8°F) for sedentary work or below -7°C (19.4°F) for moderate work.(16) Mittens offer more protection than gloves since air does not have contact with each finger. A layered ensemble of gloves covered with mittens offers the highest degree of cold protection. Performance losses as a function of lower body temperature have also been reported.

Cognitive or mental tasks are much less affected by the cold than are motor tasks. However, it is often difficult to separate these effects because results of mental activity are generally manifested in some motor movement that is more likely to be affected by cold. Cold exposure affects visual activity as a result of ocular discomfort, including tearing and sensations of dryness, accompanied by frequent blinking.(93) The general psychological responses to prolonged work in the cold can be significant. Elements of isolation and confinement due to cold work locations as well as discomfort from the cold have been noted to result in behavioral responses. These include changes in arousal, perception, mood, personality, apathy, etc.(77)

Prolonged contact with surfaces having temperatures below 15°C (59°F) may impact dexterity; below 7°C (45°F) may induce numbness; and below 0°C (32°F) may induce frost nip or frostbite.(94) When surfaces are below -7°C (-19.4°F), warnings should be given to prevent inadvertent contact by bare skin, and hand protection should be provided if the air temperature is below 17.5°C (0°F).(16)

Effects on Accidents and Safety Behavior

There have been only a few extensive studies of thermal effects on safety. One report describes a study of over 2000 accidents in four types of industrial workshops.(95) Air temperature had a significant effect on accidents in two of the observed plants; the observed temperatures were in the 15–24°C (59–75°F) range and higher accident rates occurred at the lower temperatures. In the third plant, there were no temperature effects on accidents. The fourth plant was analyzed differently because of the small number of accidents.

As previously mentioned, NIOSH recommendations for upper limits of thermal exposure for unimpaired mental performance imply that a thermal environment that impairs mental performance may also negatively affect safety and accidents. Results of a 4-year study of accidents in a steel mill show that peak accident rates occur during peak air and dew point temperatures.(96)

Surry(97) states that "the conditions for thermal comfort are essentially the same for peak work efficiency and for minimum accident rates," and other studies have confirmed similar relationships. A summary of over 30 independent studies on thermal comfort reports the preferred air temperature in the majority of these studies to be between 17°C (63°F) and 26°C (79°F) dry bulb,(98) with the highest preference being 23–25°C.(99) Studies included data for both males and females, from offices, schools, laboratories, and light industry. It was further noted that as the monthly mean outdoor temperatures increase, the preferred temperature for comfort tends to increase.

Comfort is also affected by metabolic work load and clothing. In general, the higher the metabolic work load and/or the clothing insulation, the lower the temperature required for comfort. By the addition or removal of clothing, the individual can establish his or her comfort temperature.(100)

Thermal effects on safety behavior have been reported in a 14-month study conducted in a metal products manufacturing plant and a foundry. A wide variety of industrial work tasks and workstations were observed.(101) Conclusions from this investigation support the relationship between unsafe work behavior and ambient temperatures as a U-shaped curve, with the minimum unsafe behavior rate occurring in the preferred temperature zone of 17°C (63°F) to 23°C (73°F) WBGT, and with the unsafe behavior rate increasing when ambient temperatures increase or decrease from this range. Grouping work tasks according to metabolic work load and different job risk groups yielded higher unsafe work behavior for both heavy work and high risk jobs, but the basic U-shaped relationship was consistently portrayed. A generalized curve demonstrating this relationship between unsafe behavior rate and temperature is shown in Figure 25.8.(101)

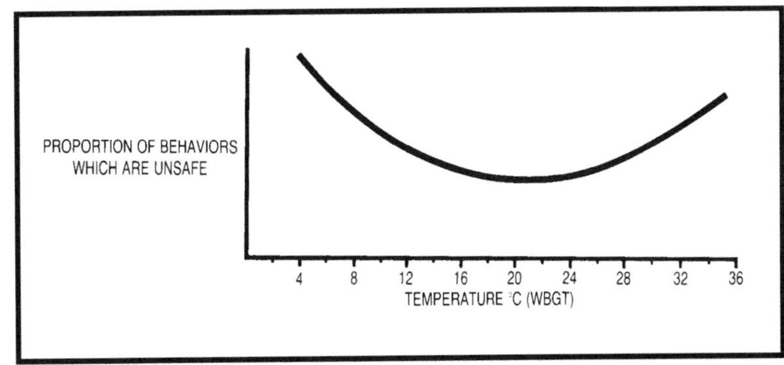

Figure 25.8 — Effects of workplace thermal conditions on safe work behavior.(101)

Evaluating the Hot Workplace: An Example

The potentially hot workplace will usually become apparent; it will be obvious, or the workers will make the hot conditions known. The important objective, however, is to determine if the workplace is excessively hot according to heat standards or heat illness criteria. Such a determination can be made using the steps described below. Example calculations in this section will use SI units, i.e., °C, watts, and m/sec.(102)

(1) Preliminary Review. The hot work site should be observed as to the nature of the work, the thermal characteristics, type of clothing worn, and other relevant job and worker information. Records of prior heat-related injury or illness should be reviewed. This information will be helpful in deciding what kind of heat stress assessment is to be made.

(2) Select a Heat Index. A heat stress index should be selected that best represents the hot work in question. Rational indices based on calculations of thermal balance have been discussed. For most applications, the WBGT is the preferred index. It is relatively easy to measure and to interpret the results, and it is also the most widely used and accepted index.

(3) Select Instruments. WBGT can be determined by measuring the thermal components of globe temperature, air temperature, and natural wet bulb temperature, and then calculating its value. Or, it can be obtained directly using electronic instruments (Figure 25.4), or approximately using a Botsball. If a Botsball is used, its reading will be approximately 3°C (5.4°F) below the corresponding WBGT.

(4) Measure the Thermal Environment. Measurement of the thermal environment requires care. The instruments should be placed in the actual working locations to ensure the readings are representative of the heat exposure. Instruments should be allowed time to stabilize to the environment, air temperature should be shielded from sun or radiant sources, and wicks on the wet bulb thermometers should be clean and fully wetted. Repeated readings may be desirable if conditions can change during the period for which a TWA is to be calculated. The temperatures and the amount of time spent at each location (work or rest) must be determined.

(5) Calculate the WBGT. If direct reading instruments are not used, the WBGT can be easily calculated. Assume the temperature readings obtained at an indoor workplace were

$$T_a = 35C;\ T_{nwb} = 30C;\ T_g = 39C \qquad (14)$$

where

T_a = air temperature
T_{nwb} = natural wet bulb temperature
T_g = globe temperature

The WBGT is calculated as

$$WBGT = 0.7\ (30°C) + 0.3\ (39°C) = 21.0 + 11.7 = 32.7\ WBGT \qquad (15)$$

(6) Calculate TWA for WBGT. Frequently the worker's thermal exposure varies during the hottest hour of the day due to working at different locations or due to changing thermal conditions. If so, it will be necessary to calculate a $WBGT_{TWA}$. Assume the measurements at the work site show 30 min exposure at 30°C WBGT, 20 min at 28°C WBGT, and 10 min at 38°C WBGT. Then

$$WBGT_{TWA} = (30\ min\ (30°C) + 20\ min\ (28°C) + 10\ min\ (38°C))/60\ min = (900 + 560 + 380)/60 = 30.7°C \quad (16)$$

(7) Estimate Metabolic Heat. The work effort and intensity required by the job being performed in the heat must be well understood, since it makes such a major contribution to total heat load on the worker. Thermal balance is affected by the cumulative heat load over the work hour, so it is necessary to consider all tasks and rest periods in the assessment of metabolic heat. SI, Table 25.2, and Figure 25.2 represent some of the common approaches to relating the work to its energy cost, or metabolic heat. For example, assume the worker is carrying and stacking boxes, which appears from Figure 25.4 to represent a W-8 work load (i.e., 390 watts). If the heaviness of this work requires a 5-min rest break (S-0 workload, or 115 watts) after each 15 min of work, the best estimate of metabolic heat would again be the hourly TWA

$$M_{TWA} = (45\ min\ (390\ watts) + 15\ min\ (115\ watts))/60\ min = (17550 + 1725)/60 = 321\ watts \quad (17)$$

(8) Evaluate Recommended Threshold Limits. Various proposals for heat stress thresholds are available and reasonable, but most are very similar. Those presented by ACGIH(16) or NIOSH(66) are most commonly recommended. From the above calculated example, metabolic heat = 321 watts, and using the RELs from Figure 25.6, a WBGT limit of 27°F is indicated. The calculated WBGT was 30.7°F, which is above the REL and below the ceiling limit of 37°F. The REL represents a threshold value of heat stress and the onset of increased risk of heat illness.

(9) Adjustments of WBGT. Table 25.9 depicts numerical corrections to WBGT limits based on acclimatization and levels of clothing worn. If the worker of the above example was unacclimatized, instead of a WBGT limit equal to 27°C, the limit would be 25°C (i.e., -2°C at a moderate work load). Or, if this worker was acclimatized but wearing winter work clothing, the WBGT limit would be 23°C (i.e., 27 - 4°C).

(10) Determine Thermal Components. Although WBGT provides a good composite index of heat load on the body, knowledge of the specific thermal components can provide diagnostic information concerning control of the heat. High globe temperature readings support the use of radiant shielding, low velocity speaks for increased ventilation, and high natural wet bulb temperature relates to high moisture and the call for moisture control or refrigeration. The use of the equations of Table 25.3 and measurement of the individual thermal components can provide quantification of the convective, radiative, and evaporative heat flow useful in making engineering modifications at the workplace.

(11) Modification of Hot Work Exposure. When a job or work site has been identified as above the REL, a broad array of practices can be considered for protecting the worker. Actions related to the worker can involve acclimatization, clothing, personal protection, fluid replacement, health issues, and self-determination during exposure. It is especially important that individual workers are knowledgeable about the symptoms and control of heat illness. Many engineering controls are also available and can be selected based on the specific thermal components in a work environment. These include ventilation, spot cooling, refrigeration, fans, radiant shielding, moisture reduction, and isolation. Relevant administrative practices for consideration are training about heat stress, reducing metabolic heat through mechanization or work scheduling, providing rest breaks, and use of a buddy system. Thermal environment controls are discussed more extensively in the preceding chapter.

(12) Re-evaluate Heat Limits. After modifications of the hot work have been made, the work should be re-evaluated using the previously described methods. Note that these threshold limits do not represent upper limits for work in the heat. They imply that risk for heat stress is increasing, and appropriate work practices and other controls must be implemented. As with most occupational hygiene controls, the use of engineering modifications provides a more complete and permanent solution. However, in the

case of heat stress, humans have their own built-in sensors of heat strain, so the understanding and use of all practices should be a requirement for anyone who works in the heat.

Evaluating the Cold Workplace: An Example

Occupational exposure to the cold can occur indoors or outdoors. Indoor cold work in refrigerated areas is typically easier to evaluate and control since the conditions of temperature, wind speed, and total working environment are consistent. Also, indoor thermal conditions can often be modified (e.g., changing wind speed) and the work clothing and work schedule can be finally selected for the environment. The larger variability in temperature and wind found with outdoor cold work or with "indoor" work sites dominated by outdoor temperatures, makes it more difficult to establish a safe and healthful workplace.

There are considerably fewer guidelines concerning cold work from standards, recommendations, and even the research literature than are found for hot work. The steps in evaluation of a cold outdoor work site are discussed in the example below. Again, the example will be presented in SI units.

(1) Preliminary Review. The cold work site should be observed to determine the nature of the work, the characteristics of both the work and warm-up areas, clothing worn, and other relevant job and worker information. Climatic records and injury/illness records may also be useful in determining the type of analysis to be made.

(2) Decide Type of Analysis. An initial decision involves selecting either a simple and general analysis or a more detailed, rational analysis. The general approach involves less measurement, more estimates, and the use of tables for decision making. The basic equations of thermal balance can be used along with additional measurements to evaluate rationally the effects of environment, work, and clothing parameters. A general analysis is recommended as appropriate for most cold work decisions.

(3) Measure the Thermal Environment. The primary measurements are air temperature (t_a) and wind speed (v). If air velocity instruments are not available, wind speed can be estimated: 2 m/sec (5 mph)—light flag moves; 4 m/sec (10 mph)—light flag fully extended; 7 m/sec (15 mph)—raises newspaper sheet; 9 m/sec (20 mph)—blowing and drifting snow.[16] It may be proper in some instances to use temperature information from the National Weather Service as a substitute for actual readings. Assume in the example T_a = -18°C and v = 7 m/sec.

If rational analysis is to be used, measures of mean radiant temperature and humidity must also be obtained.

Surface temperatures should be measured when bare skin/hands can contact cold surfaces. Assume a surface temperature of -9°C (16°F) is measured using a surface thermistor. This surface would be considered uncomfortably cold and hand protection would be recommended.

(4) Determine Windchill Index. The WCI could be calculated directly and converted to an equivalent chill temperature or this information could be obtained from tables, if available. For this example, air temperature = -18°C and air velocity = 7 m/sec, an equivalent chill temperature of -36°C. This represents a condition where exposed skin should be covered, since flesh is likely to freeze within a few minutes.

(5) Estimate Metabolic Heat. The metabolic heat can be estimated using the information found in Tables 25.1 and 25.2, or Figure 25.2. For example, assuming the work involves arm and trunk work of intermittent handling of boxes, Table 25.1 would suggest this to be moderate work.

(6) Determine Insulation Required. The IREQ of clothing adequate for thermal balance in an environment can be calculated if all variables have been measured. Generally an adequate approximation of clothing requirements can be obtained from Figure 25.8. For the example air temperature = 18°C and moderate work, Figure 25.8 shows an IREQ between 1.7 clo and 2.7 clo, where 1 clo = 0.155 meter square °C per watt. Clothing for this work and temperature combination should provide insulation of this level.

(7) Modification of Cold Work Exposure. There are a large number of practices that may be useful in alleviation of cold stress. Of major importance is assurance that the worker is adequately educated and trained concerning all aspects of controlling and responding to cold environments. The selection and use of clothing and of

personal protection for the head and extremities is critical. Engineering controls may also be applicable including the proper selection of tools, equipment, and machinery; the use of shielding or heated enclosures, local or personal heaters; mechanization of the work, etc. Administrative controls should also be considered as they relate to careful planning and preparation for the work, work scheduling, work sharing, work breaks, and the use of buddy systems. An extensive discussion of cold stress preventive measures can be found in Holmér[94] and ACGIH.[16] If modifications of the work or environment are substantial, it may be useful to re-evaluate the situation using the previously described methods.

Summary

Adverse thermal environments represent a commonly encountered occupational exposure, and the risks to humans are real and recognizable.

The examples cited above demonstrate some of the methods and procedures discussed in this chapter, and they provide guidance on evaluating and making decisions for hot and cold occupational exposures where conditions are simple and straightforward. Most often, however, the making of thermal environment decisions in the real workplace involves knowledge and application of the broader principles covered throughout this chapter.

References

1. **International Organization for Standardization:** Proposed standard system of symbols for thermal physiology. *J. Appl. Physiol. 27*:439–446 (1969).
2. **Gagge, A.P.:** Effective radiant flux, an independent variable that describes thermal radiation on man. In *Physiological and Behavioral Temperature Regulation*, J.D. Hardy, A.P. Gagge, and J.A.J. Stolwijk, eds. Springfield, IL: Charles C. Thomas, 1970.
3. **Cena, K. and J.A. Clark:** Physics, physiology, and psychology. In *Bioengineering, Thermal Physiology and Comfort*, K. Cena and J.A. Clark, eds. New York: Elsevier Scientific Publishing Co., 1981.
4. **Vernon, H.:** The measurement of radiant heat in relation to human comfort. *J. Ind. Hyg. 14*:95 (1932).
5. **National Institute for Occupational Safety and Health:** *Criteria for a Recommended Standard. Occupational Exposure to Hot Environments.* (HSM 72-10629) Washington, DC: U.S. Government Printing Office, 1972.
6. **Kuehn, L.A. and L.E. Machattie:** A fast responding and direct-reading WBGT index meter. *Am. Ind. Hyg. Assoc. J. 36*:325–331 (1975).
7. **International Organization for Standardization (ISO):** *Thermal Environments: Instruments and Methods for Measuring Physical Quantities.* (ISO/DIS 7726). Geneva: ISO, 1996.
8. **Witherspoon, J.M. and R.F. Goldman:** Indices of thermal stress. *ASHRAE Bull. LO-73-8*:5–13, 1974.
9. **Belding, H.S.:** Strains of exposure to heat. Standards for occupational exposure to hot environments. In *Standards for Occupational Exposure to Hot Environments, Proceedings of Symposium*, S.M. Horvath and R.C. Jensen, eds. (DHEW/NIOSH pub. 76-100). Washington, DC: U.S. Government Printing Office, 1976.
10. **Wyndham, C.H. and A.J. Heyns:** The accuracy of the prediction of human strain from heat stress indices. *Arch. Sci. Physiol. 27*:295–301 (1973).
11. **Belding, H.S.:** The search for a universal heat stress index. In *Physiological and Behavioral Temperature Regulation.* J.D. Hardy, A.P. Gagge, and J.A.J. Stolwijk, eds. Springfield, IL: Charles C. Thomas, 1970.
12. **Dukes-Dobos, F.N.:** Rationale and provisions of the work practices standard for work in hot environments as recommended by NIOSH. In *Standards for Occupational Exposure to Hot Environments, Proceedings of Symposium*, S.M. Horvath and R.C. Jensen, eds. (DHEW/NIOSH pub. 76-100). Washington, DC: U.S. Government Printing Office, 1976.
13. **Kuhlemeier, K.V. and T.B. Wood:** Laboratory evaluation of permissible exposure limits for men in hot environments. *Am. Ind. Hyg. Assoc. J. 40*:1097–1103 (1979).
14. **International Organization for Standardization (ISO):** *Hot Environments—Analytical Determination and Interpretation of Thermal Stress Using Calculation of Required Sweat Rate* (ISO 7933). Geneva: ISO, 1989.

15. **Smith, J.L. and J.D. Ramsey:** Designing physically demanding tasks to minimize levels of worker stress. *Ind. Eng. 14*:44–50 (1982).
16. **American Conference of Governmental Industrial Hygienists (ACGIH):** *TLVs: Threshold Limit Values for Chemical Substances and Physical Agents and Biological Exposure Indices*. Cincinnati, Ohio: ACGIH, 1992.
17. **International Organization for Standardization (ISO):** *Determination of Metabolic Rate* (ISO 8996). Geneva: ISO, 1990.
18. **Tayyari, F., C.L. Burford, and J.D. Ramsey:** Guidelines for the use of systematic workload estimation. *Int. J. Ind. Ergonom. 4*:61–66 (1989).
19. **Ramsey, J.D. and C.P. Chai:** Inherent variability in heat-stress decision rules. *Ergonomics 26*:495–504 (1983).
20. **McKarns, J.S. and R.S. Brief:** Nomographs give refined estimate of heat stress index. *Heat./Piping/Air Cond. 38*:113–116 (1966).
21. **Kamon, E. and C. Ryan:** Effective heat strain index using pocket computer. *Am. Ind. Hyg. Assoc. J. 42*:611–615 (1981).
22. **Gagge, A.P. and R.R. Gonzalez:** Physiological bases of warm discomfort for sedentary man. *Arch. Sci. Physiol. 27*:409 (1973).
23. **Stewart, J.M. and C.H. Wyndham:** Suggested thermal stress limits for safe physiological strain in underground environments. *J. S. Afr. Inst. Min. Metal 76*:334 (1975).
24. **Vogt, J.J., V. Candas, and J.P. Libert:** Graphical determination of heat tolerance limits. *Ergonomics 25*:285–294 (1982).
25. **Houghten, F.C. and C.P. Yaglou:** Determining lines of equal comfort. *J. Am. Soc. Heat. Vent. Eng. 29*:163–176 (1923).
26. **Bedford, T.:** *Environmental Warmth and Its Measurement*. Medical Research Council, War Memo No. 17. London: His Majesty's Safety Office, 1946.
27. **MacPherson, R.K.:** *Physiological Responses to Hot Environments*. Med. Res. Coun. Special Report Serial No. 298, London: His Majesty's Safety Office, 1960.
28. **American Industrial Hygiene Association (AIHA):** *Heating and Cooling for Man in Industry*. Akron, OH: AIHA, 1975.
29. **Gagge, A.P., J.A.J. Stolwijk, and Y. Nishi:** An effective temperature scale based on a simple model of human physiological regulatory response. *ASHRAE Trans. 77*:247 (1971).
30. **Gagge, A.P.:** Rational temperature indices of thermal comfort. In *Bioengineering, Thermal Physiology and Comfort*, K. Cena and J.A. Clark, eds. New York: Elsevier Scientific Publishing Co., 1981.
31. **Nishi, Y.:** Field assessment of thermal characteristics of man and his environment by using a programmable pocket computer. *ASHRAE Trans. 83(I)*:103 (1977).
32. **Chai, C.P.:** "The Assessment of the Ability of Various Heat Stress Indices to Predict Safe Work Behavior." Master's thesis, Texas Tech University, Lubbock, TX, 1981.
33. **Yaglou, C.P. and D. Minard:** Control of heat stress casualties at military training centers. *Arch. Ind. Health 16*:302 (1957).
34. **World Health Organization (WHO):** "Health Factors Involved in Working Under Conditions of Heat Stress." (WHO tech. rep. ser. 412) Geneva: WHO, 1969.
35. **Ramsey, J.D.:** Heat stress standard: OSHA's advisory committee recommendations. *National Safety News 6*:89–95 (1975).
36. **International Organization for Standardization (ISO):** "Hot Environments—Estimation of the Heat Stress on Working Man, Based on the WBGT Index (Wet Bulb Globe Temperature)" (ISO 7243). Geneva: ISO, 1989.
37. **Botsford, J.H.:** A wet globe thermometer for environmental heat measurement. *Am. Ind. Hyg. Assoc. J. 32*:1–10 (1971).
38. **Beshir, M.Y., J.D. Ramsey, and C.L. Burford:** Threshold values for the Botsball: A field study of occupational heat. *Ergonomics 25*:247–254 (1982).
39. **Brief, R.S. and R.G. Confer:** Comparison of heat stress indices. *Am. Ind. Hyg. Assoc. J. 32*:11–16 (1971).
40. **Ciriello, V.M. and S.H. Snook:** The prediction of WBGT from the Botsball. *Am. Ind. Hyg. Assoc. J. 38*:264–271 (1977).
41. **Mutchler, J.E. and J.L. Vecchio:** Empirical relationships among heat stress indices in 14 hot industries. *Am. Ind. Hyg. Assoc. J. 38*:253–263 (1977).

42. **Sundin, D., F.N. Dukes-Dobos, P. Jensen, and C. Humphreys:** "Comparison of the ACGIH TLV for heat stress with other heat stress indices." Paper presented at the American Industrial Hygiene Conference, San Francisco, CA, 1972.
43. **Ramsey, J.D.:** Practical evaluation of hot working areas. *Prof. Saf. Feb*: 42–48 (1987).
44. **Onkaram, B., L.A. Stroschein, and R.F. Goldman:** Three instruments for assessment of WBGT and a comparison with WGT (Botsball). *Am. Ind. Hyg. Assoc. J. 41*:634–641 (1980).
45. **Goldman, R.A.:** Prediction of heat strain, revisited 1979–1980. In *Proceedings of a NIOSH Workshop on Recommended Heat Stress Standards* (DHHS [NIOSH] pub. 81-108), F.N. Dukes-Dobos and A. Henshel, eds. Washington, DC: U.S. Government Printing Office, 1980.
46. **National Institute for Occupational Safety and Health (NIOSH):** *Proceedings of a NIOSH Workshop on Recommended Heat Stress Standards* (DHHS [NIOSH] pub. 81-108), F.N. Dukes-Dobos and A. Henschel, eds. Washington, DC: U.S. Government Printing Office, 1980.
47. **Beshir, M.Y.:** A comprehensive comparison between WBGT and Botsball. *Am. Ind. Hyg. Assoc. J. 42*:81–87 (1981).
48. **Hill, L.:** The science of ventilation and open air treatment, part 1. *Med. Res. Coun. Special Report No. 32*. London: Medical Research Council, 1919.
49. **Young, P.A., W.L. Potts, and A.C. Mandal:** Kata thermometry in relation to the specific cooling power of a mine environment. *J. Mine Vent. Soc. S. Afr. 31*:136–137 (1978).
50. **McCardle, B., W. Dunham, H.E. Holling, W.S.S. Ladell, et al.:** The prediction of the physiological effects of warm and hot environments: The P4SR index. *Med. Res. Coun. R.N.P. Report 47*:391 (1947). London.
51. **Lind, A.R.:** Tolerable limits for prolonged and intermittent exposures to heat. In *Temperature: Its Measurement and Control in Science and Industry*, vol. 3, J.D. Hardy, ed. New York: Van Nostrand Reinhold, 1963.
52. **Thom, E.C.:** A new concept for cooling degree-days. *Air Cond., Heat. Vent. 54*:73 (1957).
53. **Robinson, S., E.S. Turrell, and S.D. Gerking:** Physiologically equivalent conditions of air temperature and humidity. *Am. J. Physiol. 143*:21–32 (1945).
54. **Givoni, B.:** The influence of work and environmental conditions on the physiological responses and thermal equilibrium of man. In *Proceedings of a UNESCO Symposium on Arid Zone Physiology and Psychology*. Lucknow, India: UNESCO, 1962.
55. **Lee, D.H. and A.H. Henschel:** *Evaluation of Thermal Environment in Shelters* (Public Health Service pub. TR-8). Washington, DC: U.S. Department of Health, Education, and Welfare, 1963.
56. **Pulket, C., A. Henschel, W.R. Burg, and B.E. Saltzman:** A comparison of heat stress indices in a hot- humid environment. *Am. Ind. Hyg. Assoc. J. 41*:442–444 (1980).
57. **Steadman, R.G.:** The assessment of sultriness, part I: A temperature humidity index based on human physiology and clothing science. *J. Climate Appl. Meteorol. 18*:861–873 (1979).
58. **Jensen, R.C. and D.A. Heins:** "Relationship Between Several Prominent Heat Stress Indices." (DHEW [NIOSH] pub. 77-109) Washington, DC: U.S. Government Printing Office, 1976.
59. **Ramsey, J.D. and M.Y. Beshir:** "How Good is NWB as a Simple Indicator of Heat Stress?" Paper presented at the American Industrial Hygiene Conference, Portland, OR, 1981.
60. **Ramsey, J.D.:** Threshold limits for workers in hot environments. In *Proceedings of the International Mine Ventilation Congress*, Johannesburg, South Africa, 1975. pp. 249–253.
61. **Belding, H.S. and T.F. Hatch:** Index for evaluating heat stress in terms of resulting physiological strains. *Heat./Piping/Air Cond. 27*:129–135 (1955).
62. **Lind, A.R., P.W. Humphreys, K.J. Collins, K. Foster, et al.:** Influence of age and daily duration of exposure on response of men to work in heat. *J. Appl. Physiol. 28*:50–56 (1970).
63. **Dukes-Dobos, F.N. and A. Henschel:** Development of permissible heat exposure limits for occupational work. *ASHRAE J. 15*:57–62 (1973).
64. **Minard, D.:** "Effective Temperature Scale and its Modifications," (NMRI research rep. 6). Bethesda, MD: Naval Medical Research Institute, 1964.
65. **Dasler, A.R.:** *Heat Stress, Work Function and Physiological Heat*

Exposure Limits in Man (special pub. 491). Washington, DC: National Bureau of Standards, Commerce Department, 1977. pp. 65–92.
66. **National Institute for Occupational Safety and Health (NIOSH):** *Criteria for a Recommended Standard. Occupational Exposure to Hot Environment* (DHHS [NIOSH] 86-113). Washington, DC: U.S. Government Printing Office, 1986.
67. **Ramsey, J.D.:** Standards Advisory Committee on heat stress–recommended standard for work in hot environments. In *Standards for Occupational Exposure to Hot Environments, Proceedings of a Symposium* (DHEW [NIOSH] pub. 76-100). S.M. Horvath and R.C. Jensen, eds. Washington, DC: U.S. Government Printing Office, 1976. pp. 191–294.
68. **Triservice Publication:** Occupational and Environmental Health: Prevention, Treatment and Control of Heat Injury (TB MED 507/NAVMED P-5052-5/AFP 160-1). Washington, DC: Departments of the Army, the Navy, and the Air Force, 1991.
69. **Ramsey, J.D.:** Abbreviated guidelines for heat stress exposure. *Am. Ind. Hyg. Assoc. J. 39*:491–495 (1978).
70. **Kenney, W.L., D.A. Lewis, C.G. Armstrong, D.E. Hyde, et al.:** Psychometric limits to prolonged work in protective clothing ensembles. *Am. Ind. Hyg. Assoc. J. 49*:390–395 (1988).
71. **Bishop, P.A., G. Smith, P. Ray, J. Beard, et al.:** Empirical prediction of physiological response to prolonged works in encapsulating protective clothing. *Ergonomics 37(9)*:1503 (1994).
72. **Ramsey, J.D., T.E. Bernard, and F.N. Dukes-Dobos:** Evaluation and control of hot working environments: part I—guidelines for the practitioner. *Int. J. Ind. Ergonom. 14*:119–127 (1994).
73. **Bernard, T.E., W.L. Kenney, L.F. Hanes, J.F. O'Brien:** "Heat Stress Management Program for Power Plants," (NP4453L). Palo Alto, CA: Electric Power Research Institute, 1991.
74. **Ramsey, J.D.:** Heat and clothing effects on stay times. *Int. J. Ind. Ergonom. 13*:157–168 (1994).
75. **Lind, A.R.:** A physiological criterion for setting thermal environmental limits for everyday work. *J. Appl. Physiol. 18*:51–56 (1963).
76. **International Organization for Standardization (ISO):** *Ergonomics of the Thermal Environment—Principles and Applications of International Standards* (ISO/DIS 11399). Geneva: ISO, 1994.
77. **Parsons, K.C.:** *Human Thermal Environments*. London: Taylor and Francis, Ltd., 1993.
78. **International Organization for Standardization (ISO):** *Evaluation of Cold Environments—Determination of Required Clothing Insulation (IREQ)* (ISO/TR 11079). Geneva: ISO, 1993.
79. **Holmér, I.:** Required clothing insulation (IREQ) as an analytical index of cold stress. *ASHRAE Trans. 90*:1116–1128 (1984).
80. **Mackworth, N.H.:** High incentives versus hot and humid atmosphere in a physical effort task. *Br. J. Psychol. 38*:90–102 (1947).
81. **Pepler, R.D.:** *Performance and Well-Being in Heat, Temperature, Its Measurement and Control in Science and Industry*, vol. 3. New York: Reinhold, 1963. p. 319.
82. **Soule, R.G., K.B. Pandolph, and R.F. Goldman:** Voluntary march rate as a measure of work output in the heat. *Ergonomics 21*:455–462 (1978).
83. **Snook, S.H. and V.M. Ciriello:** The effects of heat stress on manual handling tasks. *Am. Ind. Hyg. Assoc. J. 35*:681–685 (1974).
84. **Kamon, E., J. Benson, and K. Soto:** Scheduling work and rest for the hot ambient conditions with radiant heat source. *Ergonomics 26*:181–192 (1983).
85. **Wing, J.F.:** *A Review of the Effects of High Ambient Temperature on Mental Performance* (AMRL-TR- 65-102). Wright-Patterson Air Force Base, OH: U.S. Department of the Air Force, Aerospace Medical Research Laboratory, 1965.
86. **Grether, W.F.:** Human performance at elevated environmental temperatures. *Aerosp. Med. 44*:747–755 (1973).
87. **Ramsey, J.D. and S.J. Morrissey:** Isodecrement curves for task performance in hot environments. *Appl. Ergonom. 9*:66–72 (1978).
88. **Kobrick, J.L. and B.J. Fine:** Climate and human performance. In *The Physical Environmental at Work*, D.J. Osborne and M.M. Gruneberg, eds. Chichester, UK: John Wiley & Sons, 1983. pp. 69–107.
89. **Ramsey, J.D.:** Task performance in heat: a review. *Ergonomics 38*:154–163 (1995).

90. **Ramsey, J.D. and Y.G. Kwon:** Recommended alert limits for perceptual motor loss in hot environments. *Int. J. Ind. Ergonom. 9*:245–257 (1992).
91. **Dusek, E.R.:** *Effect of Temperature on Mental Performance. Protection and Functioning of the Hand in Cold Climates.* Washington, DC: National Academic Science, Natural Resource Council, 1957.
92. **Tanaka, M., M. Tochihara, S. Yamazaki, S. Ohnaka, et al.:** Thermal reaction and manual performance during cold exposure while wearing cold-protective clothing. *Ergonomics 26*:141–150 (1983).
93. **Kobrick, J.L.:** Effects of exposure to low ambient temperature and wind on visual acuity. *J. Eng. Psychol. 4*:92–98 (1965).
94. **Holmér, I.:** Cold stress: part I—Guidelines for the practitioner. *Int. J. Ind. Ergonom. 14*:139–149 (1994).
95. **Powell, P.I., M. Hale, J. Martin, and M. Simon:** "2000 Accidents, A Shop Floor Study of Their Causes." (Rep. 21). London: National Institute of Industrial Psychology, 1971.
96. **Belding, H.S., R.F. Hatch, B.A. Hertig, and M.L. Riedsel:** Recent developments in understanding of effects of exposure to heat. In *Proceedings of the 13th International Congress of Occupational Health*. New York: International Congress of Occupational Health, 1960.
97. **Surry, J.:** *Industrial Accident Research: A Human Engineering Approach.* Toronto, Canada: University of Toronto, Department of Industrial Engineering, 1968.
98. **Humphreys, M.A.:** The dependence of comfortable temperatures upon indoor and outdoor climates. In *Bioengineering, Thermal Physiology and Comfort*, K. Cena and J.A. Clark, eds. New York: Elsevier Scientific Publishing Co., 1981.
99. **Reddy, S.P. and J.D. Ramsey:** Thermostat variations and sedentary job performance. *ASHRAE J. 18*:32–36 (1976).
100. **Fanger, P.O.:** *Thermal Comfort: Analysis and Applications in Environmental Engineering*. New York: McGraw-Hill, 1972.
101. **Ramsey, J.D., C.L. Burford, M.Y. Beshir, and R.C. Jensen:** Effects of workplace thermal conditions on safe work behavior. *J. Safety Res. 14*:105–114 (1983).
102. **Ellis, F.P., F.E. Smith, and J.D. Walters:** Measurement of environmental warmth in SI units. *Br. J. Ind. Med. 29*:361–377 (1972).

Section 4

The Human Environment at Work

Outcome Competencies

After completing this chapter, the user should be able to:

1. Define underlined terms used in this chapter.
2. Explain basic biomechanical principles involving forces.
3. Describe how muscles generate tension.
4. Describe lever systems in the human body.
5. Discuss the basic concepts of anthropometry.
6. List the basic experimental techniques used in biomechanics.
7. Demonstrate basic types of biomechanical analyses.

Key Terms

accelerometers • antagonists • anthropometry • biomechanics • concentric • contraction • eccentric • electrogoniometer • electromechanical delay • electromyogram • electromyography • equilibrium • ergonomics • fixators • force couple • force platform • force • force-velocity relationship • free-body diagram • general motion • human factors • innervation ratio • inverse dynamics • isometric • kinesiology • kinetics • length-tension relationship • levers • mass • mechanical advantages • moment arm • moment of inertia • moment of force • motor • motor unit • neuron • net force • prime movers • radius of gyration • rotation • static • synergists • torque • translation • videography • volume

Prerequisite Knowledge

Prior to beginning this chapter, the user should review the following references for general content.

Vector Mechanics for Engineers, by F.P. Beer and E.R. Johnson. New York: McGraw-Hill Book Co., 1984.

Occupational Biomechanics, 2nd ed., by D.B. Chaffin and G.B.J. Andersson. New York: John Wiley & Sons, 1991.

College physics and algebra.

Prior to beginning this chapter, the user should review the following chapters:

Chapter Number	Chapter Topic
27	Work Physiology
28	Ergonomics

Key Topics

I. Basic Principles
   - A. Definitions
   - B. Biomechanical Principles
   - C. Muscle Function
   - D. Levers

II. Anthropometry
   - A. Segment Dimensions
   - B. Inertial Properties
   - C. Measurement of Segmental Characteristics

III. Methodology
   - A. Kinematics
   - B. Kinetics
   - C. Electromyography

IV. Types of Analyses
   - A. Static Analysis
   - B. Dynamic Analysis

Chapter 26

26

Biomechanics

Joseph Hamill, Ph.D.

Elizabeth C. Hardin

Basic Principles

Definitions

Whereas people may use the terms kinesiology, biomechanics, ergonomics, and human factors interchangeably, that use is not necessarily appropriate. Each has a unique definition, unique areas of study, and unique methodologies. The terms are not synonymous and should be used to refer to specific questions in health and safety. They all refer in some manner, however, to human performance. "Kinesiology" is a misunderstood term that broadly refers to the study of movement. Unfortunately, this term also has been used as a synonym for all of the other terms. In addition, the terms "ergonomics" and "human factors" are also considered synonymous. Ergonomics involves such disciplines as anatomy, physiology, psychology, occupational hygiene, and design engineering. Human factors, on the other hand, has its basis more in the areas of experimental psychology and engineering. These disciplines concern themselves with the interface of humans with their environment. For example, the accommodation of the worker in the workplace by correct placement of controls, or the mental stress of the worker while performing a task, are viable areas of study in both ergonomics and human factors.

Biomechanics is a discipline that deals with the mechanical aspects of body motion. This term has been defined by the American Society of Biomechanics as "the application of the principles of mechanics to the study of biological systems." Biomechanics thus utilizes the knowledge base of anatomy, physiology, and mechanical engineering. The studies of muscle activity and the forces on and within the body are points of interest to the biomechanist.

Biomechanical Principles

The purpose of a biomechanical analysis is to investigate the effect of both internal and external forces on the body while it accomplishes a given task. This section will discuss the nature of forces and the movement produced by these forces and will present Newton's three laws of motion, which are the basis for all biomechanical investigations.

Forces and Torques

For the movement of a segment or a body to occur, a force must be applied. A force is an interaction of two objects that produces a change in the state of motion of an object. A force may cause an object to move, accelerate or decelerate it, change its direction, or stop it from moving. In most situations multiple forces act concurrently on either a segment or the total body. In these cases, the concept of net force is important. A net force is the sum of all concurrent forces.

A force has four characteristics, two of which are defined because a force is a vector. These characteristics are magnitude and direction. The remaining two characteristics are the line of action and the point of application of the force. The point of application of a force is the location at which the force acts on the body. The line of action is the line passing through a force's point of application in the direction of the force's action. Figure 26.1 illustrates the characteristics of the biceps muscle force. The unit of force in the Systeme Internationale d'Unites

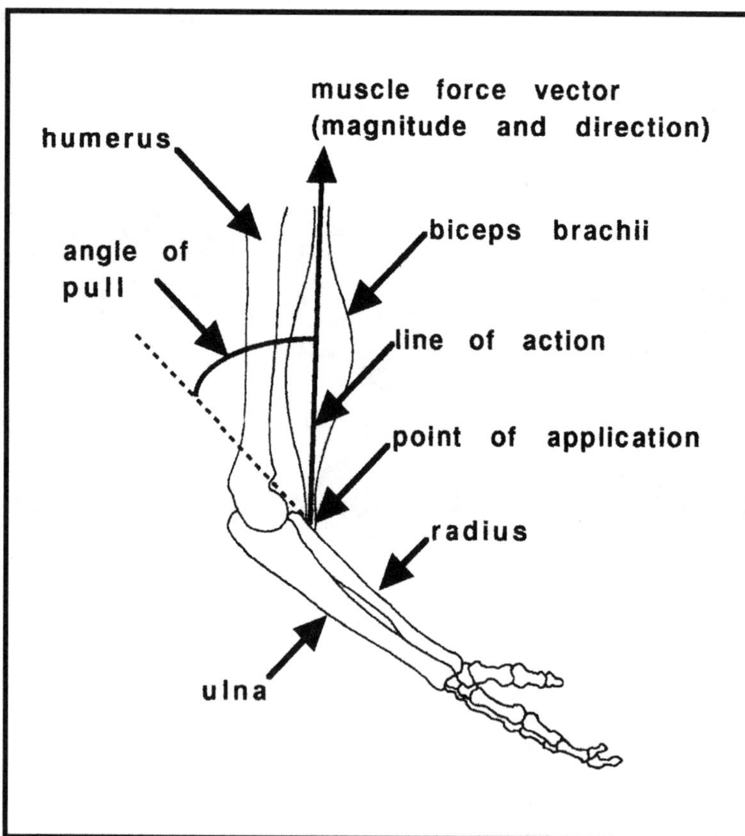

Figure 26.1 — Charcteristics of a force.

(SI), the metric system, is the Newton (N). Use of English units in the scientific literature is discouraged.

Depending on how forces are applied, they may cause specific types of motion referred to as (1) pure translation, or straight line motion; (2) pure rotation, or angular motion; or (3) general motion, a combination of both translation and rotation. When a force is applied such that its line of action is directly through the center of mass of an object, the resulting motion is pure translation. When pure translation occurs, all points of the body move through the same distance in the same time interval.

For a pure rotation to take place, two equal and opposite forces must act at a distance from an axis of rotation and not through the center of mass of the body. The product of a force and the perpendicular distance from the line of action of the force to the axis of rotation is called a moment of force or a torque. The unit of torque in the SI system is the Newton-meter (N × m). Each of the two forces results in a torque about the same axis of rotation with each causing translation and rotation. Since the forces are equal in magnitude and act in opposite directions, the translation from each force cancels out the other. Pure rotation, however, results because each torque produces a rotation in the same direction. The pair of forces arranged to produce pure rotation is referred to as a force couple.

In causing both translation and rotation to occur, a force must be applied such that it results in a moment of force or a torque. A single force causing a torque results in both a rotation and a translation in the direction of the force application or general motion of the body. In most instances, the forces that act on the human body and result in human movement are of this type.

Newton's Laws of Motion

The three laws of motion promulgated by Sir Isaac Newton (1642–1727) describe the interaction of forces on a body that result in movement. These laws are known as (1) the law of inertia, (2) the law of acceleration, and (3) the law of action-reaction.

The law of inertia states that a body continues in its state of rest, or of uniform motion in a straight line, unless a force acts on it. Mathematically, this law can be expressed as follows:

$$\text{If } \sum F = 0 \text{ then } v = \text{constant} \qquad (1)$$

where F = force and v = velocity. That is, if the sum of the forces acting on a body is zero, then the velocity will not change. To produce motion in an object that is at rest, a force must be applied. Likewise, to stop or alter an object's motion, a force must be applied to the object. The inertia of an object describes the object's resistance to motion and is directly related to the amount of matter (kilograms) of the object.

The second law, the law of acceleration, states that a body acted on by an external force moves such that the force is equal to the time rate of change of linear momentum. This gives the expression

$$F = \frac{d(mv)}{dt} \qquad (2)$$

where:

F = the net force,
mv = the linear momentum, and
t = time.

However, this law is probably more commonly expressed as "a force applied to an object causes an acceleration of the object that is proportional to the force and inversely proportional to the mass of the object." This statement results in the well-known expression

$$F = ma \quad (3)$$

where:

F = the net force acting on the object,
m = the mass of the object, and
a is the resulting acceleration of the object.

This statement provides a cause-effect relationship. The force, F, can be thought of as the cause, while the result can be thought of as the acceleration of the mass the force acts on, ma.

The third of Newton's laws states that for every action there is an equal and opposite reaction. This law can be expressed mathematically as:

$$F_{AB} = -F_{BA} \quad (4)$$

When Objects A and B interact, Object A produces an effect on B. In turn, the second object, B, produces an equal and opposite effect on A. This law illustrates that forces never act in isolation and, in fact, act in pairs.

Muscle Function

Muscle tissue accounts for approximately 40% of the total body mass and converts chemical energy into mechanical energy. There are three types of muscle in the human body: skeletal, cardiac (found only in the heart), and smooth (found only in the viscera). However, for this discussion, we will refer only to skeletal muscle. A muscle is composed of cells called muscle fibers. Muscle fibers are composed of myofibrils, actin, and myosin, that lie parallel to each other. Myofibrils are arranged in a repeating pattern of structures called sarcomeres that are the functional contractile units. Muscle fibers are grouped to form fascicles, and fascicles are grouped to form a muscle. Each fascicle is surrounded by a fascia and other connective tissue as is the muscle itself. Muscles are connected to bone via tendons. The connective tissue surrounding the muscle and the tendons have elastic properties that can be utilized to do work.

The purpose of the skeletal muscular system is to provide an engine for the body to accomplish motion. Muscles cause motion by producing force. In this section, we will not concentrate on the physiological process by which muscles produce force, but we will emphasize the outcome of this force production. That is, we will discuss the effects of force production on the skeletal system.

Skeletal muscles are complex organs that are primarily responsible for initiating, stopping, stabilizing, and controlling the motion of body segments. Muscles can only generate tension, and thus act in pairs to move a segment and then return the segment to its starting position. To accomplish the motion of a segment, muscles must cross a joint, that is, a point of intersection of two bones. Only at the extremes of the range of motion of joint movement do other structures such as ligaments and bony structures produce restraining forces. It is unusual that a single muscle is responsible for a given movement. In fact, there is a great deal of redundancy in the system such that movements result from the coordinated effort of several muscles. Thus, muscles are categorized as prime movers, antagonists, synergists, or fixators. A prime mover is responsible for the primary action at the joint while the antagonist acts to brake or decelerate the limb. Synergist and fixator muscles generally stabilize and control the bony structures that constitute the joint.

For a muscle to reach its active state it must receive a signal from the nervous system via a nerve referred to as a motor neuron, at which time the muscle exerts tension on its skeletal attachments. These signals from the nervous system may be either voluntary or reflex. A single motor neuron and the muscle fibers that it innervates are referred to as a motor unit. A muscle may have many motor units, ranging from a few hundred to up to a thousand per muscle. This difference in the number of motor units within a muscle enables the muscle to meet a variety of needs.[1]

Although a single muscle fiber is innervated by a single motor neuron, each motor neuron may innervate more than one muscle fiber. The number of muscle fibers innervated by a single motor neuron is called the innervation ratio, which varies from 1:1900 (1 neuron per 1900 muscle fibers) as in the gastrocnemius to 1:15 (1 neuron per 15 muscle fibers) as in the extraocular muscles. The lower the innervation ratio, the finer the control of the muscle force. Motor units can be differentiated from each other based on a number of physiological factors such as the speed of contraction, the magnitude of force development, and the resistance to fatigue. Motor units can be classified into three groups.[2] These are (1) slow-contracting, fatigue-resistant; (2) fast-contracting, fatigue-resistant; and (3) fast-contracting, fast-to-fatigue. The first of these produces the least force while the last produces the greatest force.

When a muscle is stimulated to its active state, it is said to undergo a contraction. An active muscle may undergo an isometric contraction (muscle remains at the same length while developing tension), a concentric contraction (muscle shortens while developing tension) or an eccentric contraction (muscle lengthens while developing tension). Depending on the relationship between the external load on the body segment and the force exerted by the muscle, the segment either rotates or remains still. Thus, the outcome of these types of contractions depends on whether the muscle force exerted is equal to, greater than, or less than the load it is acting against.

Muscle has two general mechanical properties that relate to force production. The first of these is the length-tension relationship. This relationship describes the maximal force that a muscle, muscle fiber, or sarcomere can exert and its length. Gordon, Huxley, and Julian[3] reported findings on the length-tension relationship that supported cross-bridge theory of muscle contraction and confirmed this theory as the primary explanation of muscle force production. Figure 26.2 illustrates the length-tension relationship. This curve was generated under isometric conditions at lengths greater than and less than resting length. Resting length refers to the length of an isolated muscle and has no significance to muscles *in vivo*. Total tension is the sum of the contractile tension and the passive tension or tension from noncontractile structures. Figure 26.2 shows that at shorter muscle lengths, all of the force production is generated by the contractile components of the muscle, while at longer muscle lengths the total tension also includes the passive structures as well as the contractile components.

For practical purposes, the length-tension relationship can be simulated using an angle-torque curve. As the joint changes position, the moment arm of the muscle changes as a function of joint angle. The practical implication of this relationship is that the muscle generates a different torque or moment of force (force × moment arm) throughout the range of motion, with less torque when the joint is flexed or extended fully. It is also evident that there is an optimal joint angle that results in a maximum torque.

The second relationship that influences muscle force production is referred to as the force-velocity relationship. This relationship describes the capacity of a muscle to exert tension while changing its length or changing its length against an external load. Hill[4] fitted an equation to the concentric portion of this curve as follows.

$$(P + a)(V + b) = (P_o + a)b \qquad (5)$$

where:

P = the tension in the muscle
P_o = the isometric tension
V = the velocity of shortening
a and b are constants

Figure 26.3 illustrates the force-velocity relationship. It can be seen that the velocity of muscular contraction varies inversely as

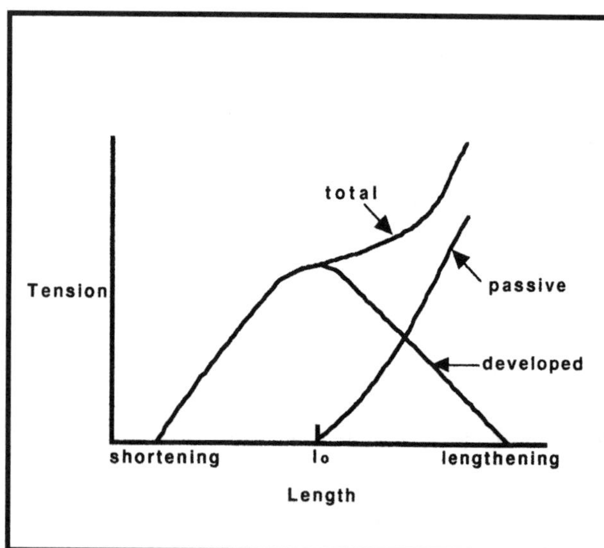

Figure 26.2 — Length-tension relationship.

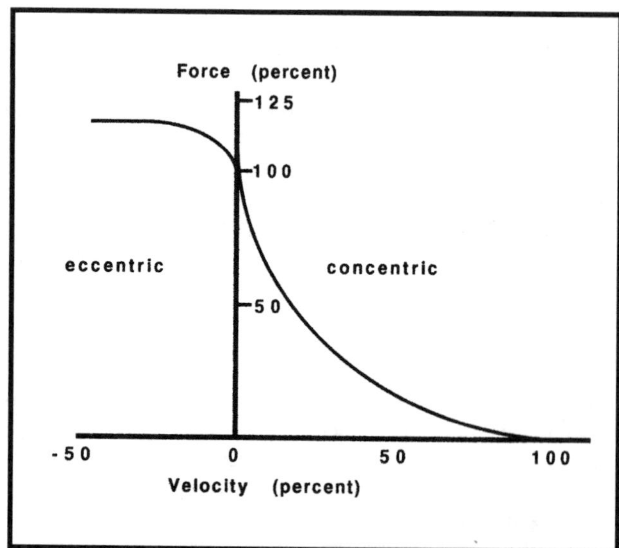

Figure 26.3 — Force-velocity relationship.

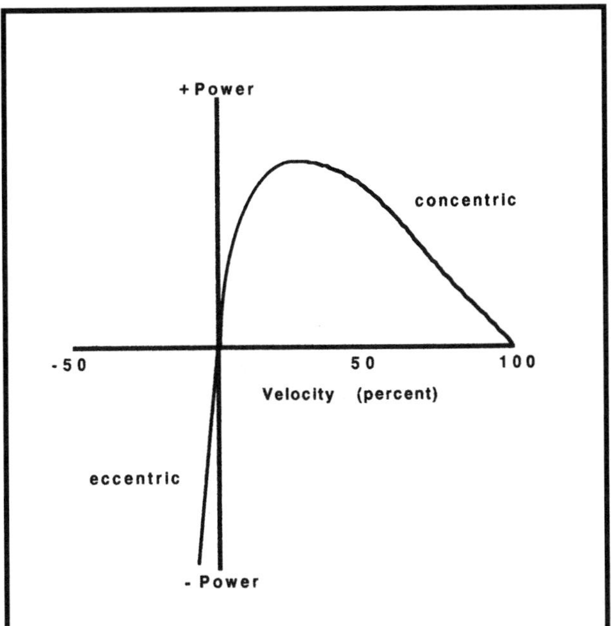

Figure 26.4 — Mechanical power-velocity relationship.

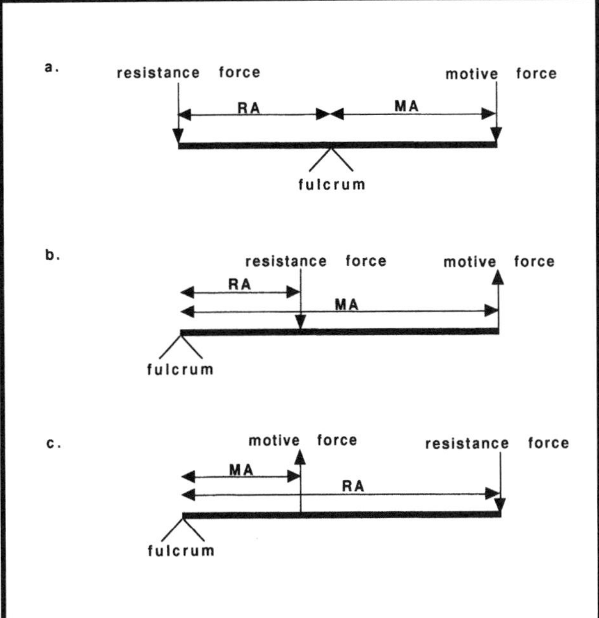

Figure 26.5 — Types of levers.

the tension developed by the muscles. This equation does not hold for the eccentric case, however. Eccentric tensions are substantially greater than are isometric tensions.

Mechanical power output (the product of force and velocity) can be determined from the force-velocity relationship (Figure 26.4). Mechanical power describes the rate of doing work and can be calculated as the product of force and velocity. For muscle, it reaches a maximum when the force and speed are between a third and a quarter of their maximal values. This indicates that the speed of shortening is related to the rate at which mechanical energy leaves the muscle.[5] In the eccentric case, mechanical power is negative, indicating that the muscle is absorbing energy from an external source. That is, the muscle is doing work against an external force.

Levers

The musculoskeletal system is a system of levers in which the bones are rigid bodies and the muscles are the force actuators. Anatomical levers are much like those in applied mechanics, with the same arrangement, movement, and function. Knowledge of anatomical levers is necessary to perform a biomechanical analysis and subsequent evaluation of a task.

Levers consist of a resisting load, a fulcrum (axis of rotation), and a motive force (applied by muscle). The distance from the fulcrum to each force is referred to as a moment arm. For example, the resistance moment arm is the distance from the resistance force to the fulcrum, and the motive moment arm is the distance from the motive force to the fulcrum. The ratio of motive moment arm to resistance moment arm is referred to as the mechanical advantage (MA). MA may be greater than 1.0, less than 1.0, or equal to 1.0. Each situation results in the lever arrangement providing a particular mechanical advantage.

There are three classes of levers in which these components vary in arrangement (Figure 26.5). In a first class lever the fulcrum is located between the resisting and motive forces that act in the same direction (Figure 26.5a). This arrangement provides a great deal of versatility in that it can magnify the motive force, thereby enabling great loads to be lifted (MA > 1.0); it can magnify the speed of movement of a limb (MA < 1.0); or it can simply change the direction of the motive force (MA = 1.0). In the human body all three functions are accomplished, although some are more prominent than others. An anatomical example of a first class lever involves head movement when looking up and down. The atlanto-occipital joint acts as the fulcrum, the muscles of the neck are the motive force, and the weight of the head is the resisting force.

A second class lever has the fulcrum located at one end, the motive force at the other end (in the same direction as the supporting part of the fulcrum), and the resisting load between the fulcrum and the motive force (in a direction opposite to that of the motive force). Figure 26.5b illustrates a second class lever. In this arrangement MA > 1.0, with the result that this lever magnifies the motive force increasing the capability of lifting a greater load. An anatomical example of a second class lever occurs at the ankle joint during locomotion. At the end of each ground contact phase of each step, the individual pushes off the ground by strongly plantar flexing the ankle. The fulcrum is at the metatarsal joint; the resisting force is the total body weight roughly passing close to the ankle joint; and the motive force is provided by the triceps surae muscle via the Achilles tendon. While movements resulting from second class levers are very powerful, they are also not precise.

The majority of levers in the limbs are third class levers. This lever system has the motive force between the fulcrum and the resistance force; thus, MA < 1.0, and the motive force acts in the opposite direction to the resistance force (Figure 26.5c). This type of lever system magnifies the range of motion of the lever. In the extremities the joint acts as the fulcrum while the long bone acts as the rigid lever. The motive force is the muscle, and the resistance force is the weight of the limb plus any external load opposing rotation of the limb. An anatomical example of a third class lever is exemplified by the flexion action of the elbow. The biceps brachii muscle is the motive force, the elbow joint is the fulcrum, and the resistance force is a combination of the weight of the forearm and some external weight.

Anthropometry

Anthropometry is the science of measurement of the body's mass, size, shape, and inertial properties. It is an empirical science that has developed a number of techniques to measure various quantities. A number of factors such as age, sex, race, and body type can influence anthropometric parameters. To develop or to use biomechanical models, it is necessary to understand and measure a number of anthropometric parameters. Some of the more important anthropometric parameters in biomechanical analyses are segment dimensions, mass (both of individual segments and the total body), and individual segment moments of inertia.

Segment Dimensions

Segment dimensions include length, volume, mass, location of the mass center, and the radius of gyration. Anatomical segments are defined by a proximal and distal joint center about which rotation occurs (with the exception of the foot and the hand.) The approximate segment length is the distance between these two joints. Dempster[6] determined segment lengths based on measurements taken from X-rays of living subjects. Lengths may be obtained from data such as Dempster's or measured directly. Direct methods of measuring segment length will be discussed later.

Segment mass and volume are important in determining the influence of gravity on movement performance. Measurement of the relationship of segment mass to volume may be determined by immersing the segment in water. Segment mass is proportional to volume and density as given in the formula:

$$M = V \times D \tag{6}$$

where:

M = the segment mass (g),
V = the volume of water displaced by the segment (cm^3),
D = the average segment density ($g \times cm^3$), and water density = 1.0.

Density values are available from cadaver data, and segment volume may be measured directly.[7]

Location of the segment center of mass allows determination of the point at which the gravitational force acts on the segment. Methods to determine this parameter include taking the segment from a cadaver and suspending it from a pin while mounted in a frame. The pin is moved along the frame until the point of balance is determined. A common method of this measurement in living subjects is to have the person position themselves while lying on a board mounted on a force platform. Assuming that the sum of moments around a point is equal to zero (static equilibrium) and knowing segment weights, one may solve for the center of mass.[8] Data on segment mass center locations are generally presented as a proportion of segment length.

Inertial Properties

The data necessary to calculate joint moments and forces are the segment mass center locations, the segment masses, and the segment lengths. However, since movements are generally dynamic, that is, the segment rotates, the rotational inertial property of the segment is important to consider. This property is the measure of resistance to rotational change and is referred to as the moment of inertia.

$$I = \sum_{i=1}^{n} m_i \times r_i^2 \quad (7)$$

where:

I = the moment of inertia (kg × cm²)
m = the mass of the segment in discrete increments (kg)
r = the distance of each mass increment from the axis of rotation (cm)

Methods of measuring the moment of inertia have been developed using cadavers and living subjects.(9–12) In particular, Plagenhoef(9) has calculated these as a proportion of segment length.

Measurement of Segmental Characteristics

Segment length data exist as a proportion of body stature and have been derived by researchers using living subjects and anatomical landmarks. Figure 26.6 illustrates one such set of data.(7)

Tables of segment mass data also have been derived and are expressed as a proportion of body mass. Dempster's(6) values of segment center of mass locations, based on cadaver data, are often used in biomechanical calculations. This is because of the difficulty of measuring center of mass locations of segments in living subjects. Measuring segmental moments of inertia and radius of gyration is also difficult in living subjects. The values generally used are also based on Dempster(6) with other researchers deriving additional values.(13)

In recent years, computer models have been used to develop anthropometric parameters for use in biomechanics.(14,15) These models require input in the form of segment lengths and girths and body mass and calculate body parameters by modeling the human body segments as geometric solids. Using data from either cadaver studies or from computer models is appropriate. However, computer models have the advantage of using measurements taken directly from the subject.

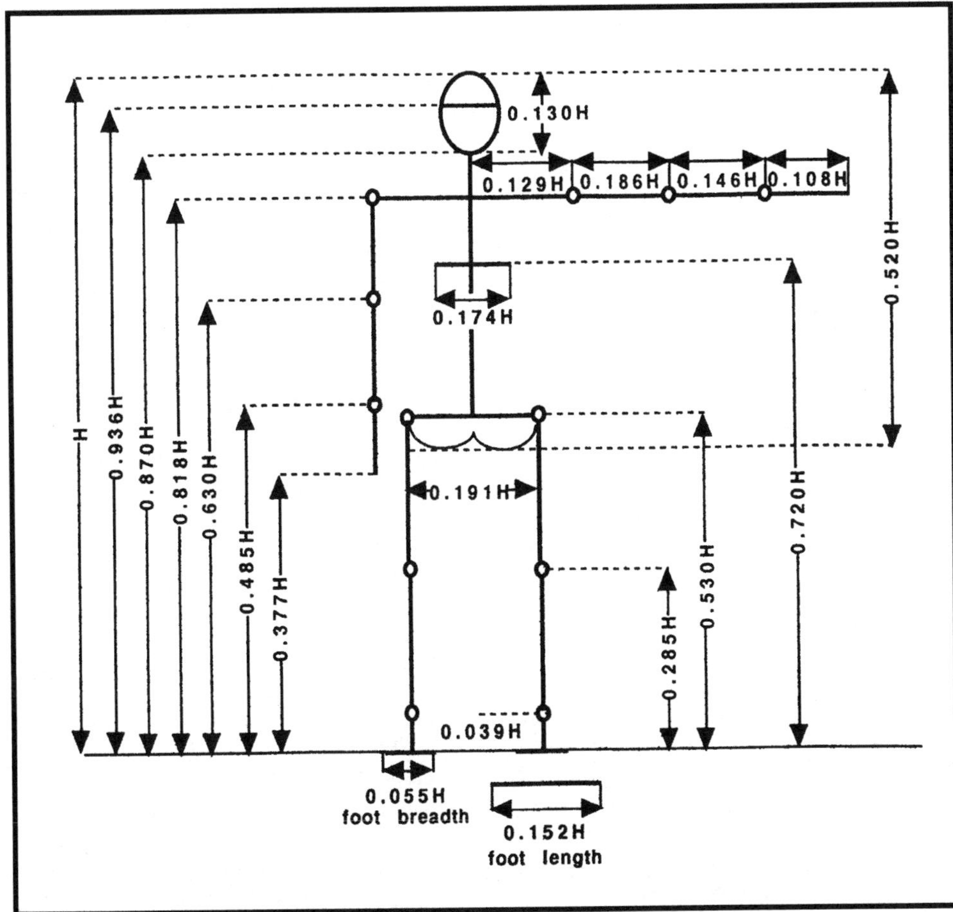

Figure 26.6 — Anthropometric dimensions.

Methodology

A number of methodologies are used in a biomechanical analysis, and for a detailed recounting of the full range of these techniques one should read Dainty and Norman,(16) Chaffin and Andersson,(17) Nigg and Herzog,(18) or Winter.(8) The most common types of measurements employed in biomechanics include instrumentation that records kinematics, kinetics, and electromyography.

Kinematics

Kinematics is the branch of mechanics that concerns the description of a body's motion in space without an explanation of the cause of the observed motion. Kinematic parameters that are often measured include time (temporal analysis), position (either in two-dimensional or three-dimensional space) and linear and angular displacement, velocity, and acceleration. The measurement of kinematic parameters provides an excellent starting place for the understanding and evaluation of a movement. This type of analysis, however, should not be thought of as a finishing point.

The most common tools used to conduct a kinematic analysis include high-speed videography, opto-electric monitoring systems, accelerometers, and electrogoniometers. Both the videographic and opto-electric systems use the digitization of key anatomical landmarks into Cartesian coordinates that are recorded on a storage device. The sampling frequency of these systems vary from 30 Hz to 1 KHz, although the general trend is to sample at 60 Hz. Each sampling instance, therefore, contains the positions of the landmarks being observed at a particular instant in time. The positions can then be used to calculate displacements and further, in conjunction with time, to calculate velocities and accelerations. There are also conventions suggested by the Canadian Society of Biomechanics and International Society of Biomechanics for the calculations of joint and segment angles. From the angle-time data, angular velocity and angular acceleration may be calculated.

Videographic and opto-electric techniques provide a great deal of flexibility in a workplace setting. In the case of videography, the event in question can be replayed quickly, and a permanent record is obtained. Although markers or light-emitting diodes are placed on the anatomical landmarks of interest, the subject's normal movement is generally not constrained. In addition, the accuracy of the measures is quite good if proper experimental protocol is observed. In many instances, a two-dimensional analysis may be satisfactory, but a three-dimensional analysis, while considerably more involved, is more advantageous.

Accelerometers are devices that measure accelerations, or changes in velocity, directly. These instruments generally have a mass of less than 4 g including the mounting device. These devices may be uniaxial or triaxial depending on the question being asked. Accelerometers may be sampled at a relatively high sampling rate and, if proper experimental protocol is used, produce very accurate results. Because of their very high frequency response, they are most often used to measure impacts or vibrations. This implies that the accelerometer must be mounted on a rigid surface preferably attached directly to the skeletal system of the subject. Light et al.(19) mounted an accelerometer on a pin that was screwed directly into the subject's bone. While there are distinct advantages to this procedure, it is impractical in other than a laboratory setting. Therefore, most researchers mount the accelerometer on a bony prominence with little superficial soft tissue.(20) Such anatomic locations include the distal medial tibia, the tibial tuberosity, the frontal bone of the skull, or on a bite bar held in the teeth. It should be noted that the acceleration measured does not represent bone acceleration but rather the acceleration of a mass element on the surface of the bone.(21)

While an accelerometer provides interesting information, this device also has some drawbacks. First, the attachment of the accelerometer to the subject, even without using bone pins, can be uncomfortable. Secondly, these devices are only useful for a single segment, so they may be inappropriate for certain motions. In addition, the absolute motion of the accelerometer is difficult to determine.

The last device used to measure certain kinematic parameters is the electrogoniometer. This device measures joint angles. An electrogoniometer is a potentiometer that is placed at the joint center and two extensions that are attached to the limbs that intersect at the joint. The electrogoniometer is generally interfaced to a computer via an analog-to-digital converter and may be sampled at a very high rate. These devices can be designed to measure rotations about one, two, or three axes at a joint.

However, because the device must be directly attached to the subject, it may constrain the natural motion of the subject during the test period. In addition, possible errors might arise when the axis of rotation of the electrogoniometer does not coincide with that of the joint. Another major disadvantage is that the electrogoniometer gives information relative to the body and not to the absolute motion of joints.[22]

Kinetics

Kinetics is a branch of mechanics that concerns the underlying causes of movement rather than the result of the movement. This involves the measurement or calculation of force (and torque) and related constructs such as work, energy, and power. This information provides an understanding of human motion at a much more complex level than does kinematics. A kinetic analysis "allows the researcher to evaluate the causes of motion, to present a diagnosis and to formulate pertinent recommendations with regard to training."[23] A number of methods are used in biomechanics to study the kinetics of movement, including direct measurement of force, the calculation of force using indirect techniques, and the estimation of force using mathematical models. In most cases the anthropometric measures of the body and its individual segments (addressed in a previous section of this chapter) are necessary.

The direct measurement of force involves both the measurement of external loads or the direct measurement of muscle forces. External forces can be measured using transducers, force platforms, or pressure platforms. A transducer is a device that provides a force output calibrated to a detected change in an electrical current. Several types of transducers are used in biomechanics, from strain gauges to load cells.

A force platform measures the ground reaction force (GRF) applied to the body by the surface of the platform. They have been used for a number of years to quantify the external forces during human locomotion. The GRF is a force vector consisting of three components: a vertical component and two shear components, the anterio-posterior (A/P) and medio-lateral (M/L) components. The shear components act parallel to the force platform surface. In addition, three moments about the corresponding axes are also obtained. The force and moment values can be used to calculate the center of pressure and the free moment. Using the time histories of the vertical, A/P, and M/L components, a number of GRF parameters have been derived to evaluate shoe function during locomotion.[24] A description of the path of the center of pressure has also been used in the evaluation of athletic footwear.[25] The free moment has been used as a measure of rotational friction and to predict pronation/supination actions of the foot.[26]

A third technique of force measurement is to quantify pressure distribution. Pressure is defined as the force perpendicular to a surface per unit area of the surface. Generally, the surface is a sensor of a known area that measures the force applied. Researchers have designed platforms that are made up of a matrix of sensors, each of which quantifies the pressure on a certain contact area. Recordings from each sensor then produce a pressure profile over the whole contact area. The force considered is normal to the surface of the sensor, and the shear forces tangential to the surface are ignored. This technique has been used quite often in locomotion studies.[27]

The direct measurement of internal forces has been accomplished by placing a transducer on the tendon of the muscle or on a ligament, which requires a surgical procedure. Komi et al.[28] and Komi[29] placed a "buckle" transducer on the Achilles tendon of subjects and recorded forces up to 12.5 time body weight during running. This technique, however, is not well developed for long term studies.

Electromyography

The study of the electrical signal associated with a muscle contraction is referred to as electromyography, and the detected signal is called an electromyogram (EMG). Since a muscle has a number of motor units and a tension increase can result from either the recruitment of a number of motor units or an increase in the stimulation of rate for a given motor unit, an EMG signal is often the sum of the activity of several motor units or several muscles. Voluntary muscle activity results in an EMG that generally increases with tension, but a number of other factors influence the signal including the velocity of contraction, rate of tension buildup, fatigue, and reflex activity.

All living cells, including muscle cells, are surrounded by a membrane that is selectively permeable to various ions. When activated, the muscle fiber membrane is depolarized propagating in both directions along the muscle fiber. The depolarization of the muscle fiber creates a small electrical potential. Because all of the fibers of a motor unit do not contract at the same time, the motor unit potential is complex, resulting from the overlapping of several muscle fibers.

The detection surfaces, referred to as electrodes, may be placed directly into the muscle (in-dwelling or intramuscular electrode) or on the skin (surface electrode). For a detailed analysis of an individual muscle, in-dwelling electrodes must be used. In-dwelling electrodes may be either needle electrodes or fine-wire electrodes. Surface electrodes are placed on the skin directly above the muscle in question, and thus, are good for studying superficial muscles. This type of electrode is most often used when a general recording of muscle activity is sufficient. Careful application, signal detection, amplification, recording, and processing must be employed. These techniques and protocols are outlined in a report by an ad hoc committee of the International Society of Electrophysiological Kinesiologists.[16]

A number of techniques have been used to process EMG signals. EMG signals may be collected as raw signals and subsequently processed or electronically processed as they are collected. Regardless, the signal may be evaluated in a number of ways, including (1) full wave rectification (FWR), the absolute value of the raw signal; (2) linear envelope, FWR followed by a low pass filter; (3) integration of the FWR over the entire muscle contraction; (4) integration of the FWR over a fixed time interval; or (5) integration of the FWR to a preset level. These are illustrated in Figure 26.7. Analysis of the EMG in the frequency domain using Fourier transforms to calculate the power spectrum density of the signal has also been accomplished.[30,31] The most important parameters for analyzing the power spectrum density of the EMG signal are the mean and median frequencies of the signal.

Chaffin and Andersson[17] stated that the primary reason for studying EMG signals is to predict muscle tension. Part of the difficulty in studying this relationship involves the problem of isolating individual muscles in humans. In addition, there is a temporal dissociation of the EMG signal and the force signal. An electromechanical delay occurs between the onset of the EMG signal and the following onset of the corresponding force signal. However, researchers have studied the EMG-muscle tension relationship under isometric conditions.[32] Both linear and nonlinear relationships have been reported between EMG amplitude and muscle tension.[33,34] With the difficulty in associating EMG and muscle tension in isometric conditions noted, description of this relationship in a dynamically contracting muscle is problematic. Many of these experiments have been done on constant velocity dynamometers.

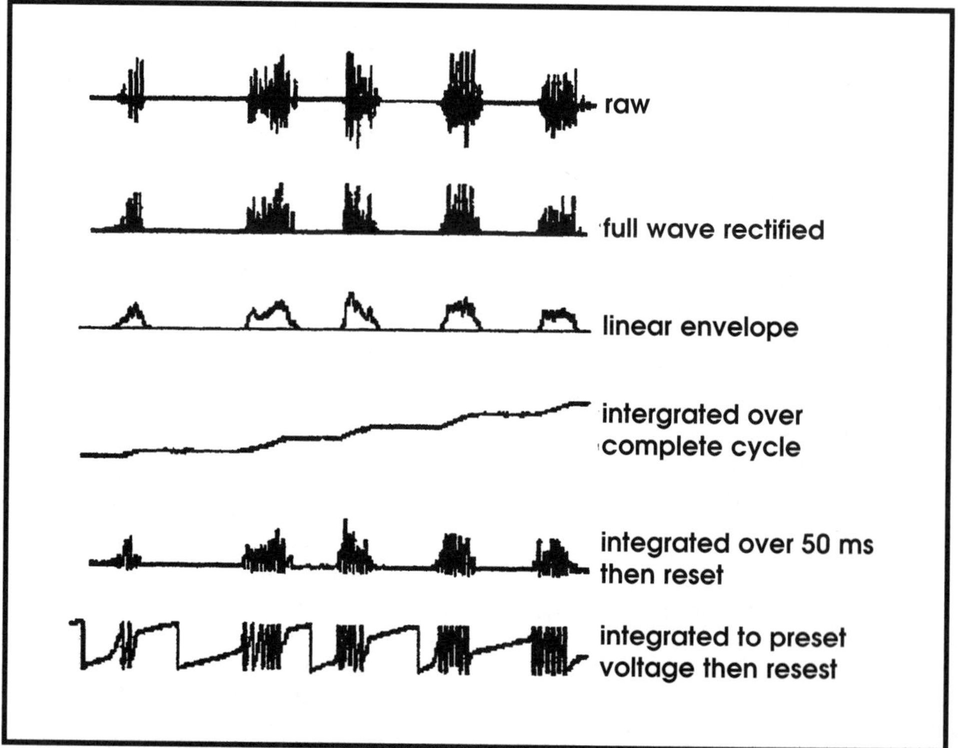

Figure 26.7 — Forms of processed EMG signals.

Relating a constant change in joint angle to a constant change in muscle length is not possible. Several studies, however, have related EMG and joint moments during certain unrestrained movements.[35,36]

Types of Analyses

The previous sections of this chapter have described the background for conducting a biomechanical analysis. While there are a number of data sets that provide information for many biomechanical parameters, it is also important to conduct a full biomechanical analysis. It is only through a full biomechanical analysis that we can truly understand the complexity of the system under study. For example, by means of a biomechanical analysis both the internal and external forces acting on different structures can be estimated. These data may then be used to predict maximum loads that may be accomplished in the particular task. For example, in a lifting task, a particular load may be reasonable in a certain posture, but in another posture the load may be dangerous. Biomechanical analyses can provide this type of information by yielding an insight that cannot be gained from any other type of information.

Static Analysis

In a biomechanical analysis there are several methods of computation that may be used to determine the kinetics of a task. In a kinetic evaluation one determines the external forces that act on a mass to cause motion. However, two types of kinetic evaluations may be undertaken: (1) static and (2) dynamic. A static evaluation considers the situation in which there is either no motion or motion at a constant velocity. If there is no change in velocity, there is no acceleration; or, more correctly, the acceleration is zero. Since the latter case is not a reasonable situation in tasks of interest to this audience, the former case is more likely. When there is no motion it does not mean that there are no forces. It simply means that the forces counteract each other resulting in no motion. The following equations describe a planar (two-dimensional) system.

$$\Sigma F_x = 0$$
$$\Sigma F_y = 0 \qquad (8)$$
$$\Sigma M_z = 0$$

where:

x = the horizontal axis
y = the vertical axis
z = the third orthogonal axis

For example, in the horizontal direction there may be two equal and opposite forces acting in which the sum total of the forces is zero. Thus, no motion in the x-direction can occur. Similarly, no motion can occur in the y-direction. The third equation indicates that there are moments of force that sum to zero, indicating that no rotation about the z-axis can occur. A static analysis generally begins with a free-body diagram. A free-body diagram is a conceptual drawing of the forces and moments acting on the system. Figure 26.8 illustrates a free-body diagram of a static situation. The system in this figure is said to be in equilibrium, since a static situation is described.

A static biomechanical model may involve a single segment or multiple segments. For example, holding an object that has been lifted may involve a single limb model or it may involve a multisegment model. Chaffin and Andersson[17] presented a six-segment, two-dimensional lifting model. There are, however, many more complicated static models that include either more segments or a three-dimensional analysis. In this chapter, however, a static analysis will be illustrated using only a single segment, planar (two-dimensional) model as described in Figure 26.8. The mass of the arm and the distance of the center of mass of the arm from the elbow joint are derived from anthropometric data. Using the known forces and the distances

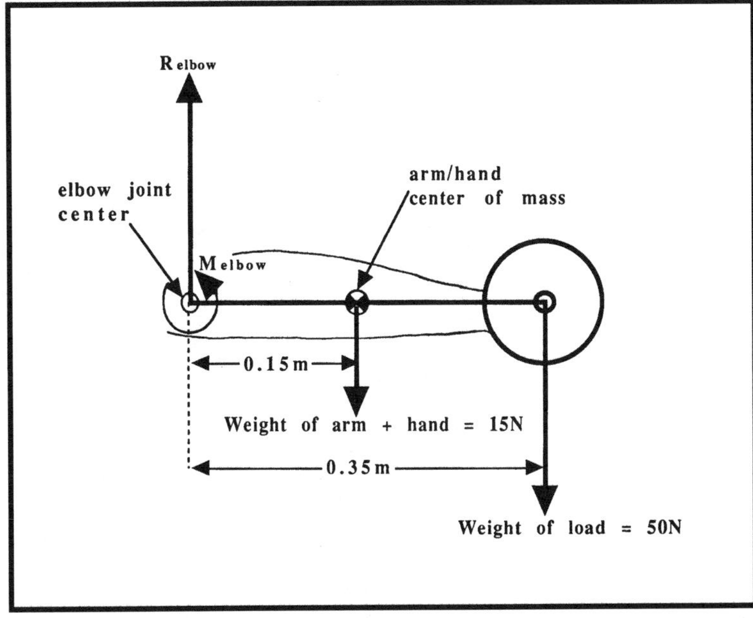

Figure 26.8 — Free-body diagram for a static analysis.

to the axis of rotation, the magnitude of the unknown forces and moments can be calculated as follows. The first condition of equilibrium is that the sum of the forces must equal zero. First, the forces in the horizontal direction must sum to zero. Therefore:

$$\Sigma F_x = 0 \quad (9)$$

Since there are no horizontal forces, this condition needs no further calculation. However, there are three forces in the vertical direction, the sum of which will equal zero. These forces include the weight of the load in the hand (50 N acting downward), the weight of the arm and hand (15 N acting downward), and the heretofore unknown reaction force acting at the elbow that must balance the other two forces (acting upward). Thus:

$$\Sigma F_x = 0$$
$$50\text{ N} - 15\text{ N} + R_{elbow} = 0 \quad (10)$$
$$R_{elbow} = 65\text{ N}$$

The reaction force at the elbow is therefore 65 N. This value is indicative of the magnitude of the tensile forces in the ligaments and muscles holding the joint together and the shearing and compressive forces acting on the joint contact surfaces.[17]

The second condition of equilibrium is that the sum of the moments must equal zero. Moments are calculated via the product of force and the moment arm, that is, the perpendicular distance to the axis of rotation. In this position the moment arm is simply the distance from the line of action of the force to the elbow joint center. There are three moments acting on this system: (1) the moment due to the weight of the load in the hand (acting clockwise); (2) the moment due to the weight of the arm and hand (acting clockwise); and (3) the reaction moment at the elbow (acting counter-clockwise). Thus:

$$\Sigma M_z = 0$$
$$-(15\text{ N} \times 0.15\text{ m}) - (50\text{ N} \times 0.35\text{ m}) + M_{elbow} = 0$$
$$M_{elbow} = 2.25\text{ N-m} + 17.5\text{ N-m}$$
$$M_{elbow} = 19.75\text{ N-m} \quad (11)$$

This reactive moment at the joint represents the strength of specific muscle actions to maintain posture.[17] It should be noted that this moment is a net moment and does not represent any specific muscle.

One example in which static analyses are used is in the measurement of muscular strength. Muscular insufficiency is the lack of muscular strength and is a critical factor in musculoskeletal injuries. Muscle strength is defined as the maximum force that a muscle can develop in a given situation. To measure muscle strength, a static situation is developed to test isometric strength (also referred to as static strength). In an isometric muscle contraction, the muscle generates tension but there is no shortening of the muscle. In the force-velocity relationship, isometric contractions occur at zero velocity (i.e., no motion) and result in force production that is greater than concentric contractions. In measuring isometric strength, the muscle force exerted and the resisting external load are in equilibrium, resulting in no movement of either the segment or the external load. The resisting external load generally is much greater than the individual can exert and is calibrated to give a reading of the force applied by the individual.

A number of other factors affect muscle strength that must be taken into consideration during strength measurement. Muscle strength may be influenced by posture, age, gender, and anthropometry. For example, posture will influence strength measures due to the change in moment arm length as the joint rotates. An angle-moment curve is therefore frequently used to describe strength because strength depends on joint position (Figure 26.9). Posture also influences strength because the force-length relationship of the muscle under which maximal force production is dependent on the muscle length.

Safe, reliable, and practical static strength measures may be collected following the guidelines of Chaffin.[37] These procedures dictate the use of a force measurement device that can collect peak force, and a 3-second average force. This device should be standardized but adjustable between individuals to accommodate anthropometric differences and should not cause subject discomfort. The exertion duration is recommended to be 4 to 6 seconds, with rest time between exertions of 30 seconds to 2 minutes. Standardized instructions to subjects should inform them of the inherent risks of the test, allow them to modify the pace of the test or length of the rest period, inform them of future use of the data, and eliminate the confounder of coercion. Environmental distractions should be minimized during testing, and competition between subjects avoided. Reports of strength test results must include characterization of the subjects, conditions during the test, and statistical results.

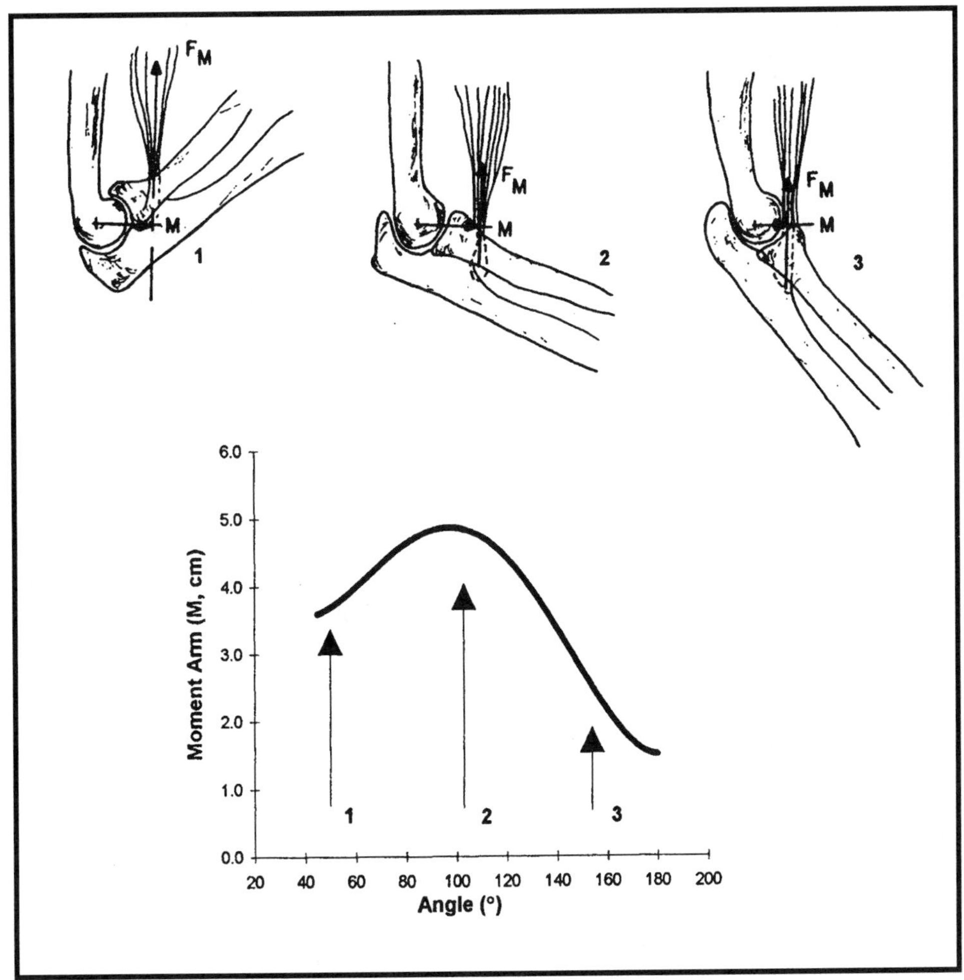

Figure 26.9 — Elbow joint and the biceps brachii moment arm.

Many static models exist. One of the simplest models available using computer software is the Two-Dimensional Static Strength Model developed at the University of Michigan Center for Ergonomics. This model also provides conservative estimates for low-back risk factors. Using this model in a task analysis, one would select the postures used in completing the task. For example, unloading crates from a truck incorporates the initial, the carrying, and the placement postures. The back compression design limit (BCDL) is one parameter that can be calculated with this model. Epidemiological, physiological, and psychological evidence suggests that jobs exceeding the BCDL provide an increased injury risk to some workers. The National Institute for Occupational Safety and Health (NIOSH) *Work Practices Guide*[38] suggests that over 75% of the females and 99% of the males could lift loads at the BCDL. The back compression upper limit (BCUL) is another parameter that can be calculated with this model. Data indicate that musculoskeletal injury rate and severity of injury are significantly greater for workers placed in jobs exceeding the BCUL. The *Work Practices Guide*[38] suggests that only about 25% of males and 1% of females have the muscle strength to perform tasks above the BCUL.

The model described above uses formulae developed by NIOSH.

$$BCDL = 392\,(15/H)\,(1-(0.0041\,|V-75|))(0.7 + (7.4/D)(1-F/F_{max})$$

$$BCUL = 3\,BCDL \qquad (12)$$

where:

H = the horizontal distance of the object from the midpoint between the ankles (cm)
V = the vertical starting point of the load center above the floor 175 cm maximum
D = the vertical travel distance of the lift 25 cm $< D < 200 - V$ (cm)

F = the average frequency of lifting (lifts/min)
F_{max} = the maximum frequency of lifts/min allowed for different postures and lifting periods, as shown in Table 26.1.

The compressive force on L5/S1 (lumbar 5-sacral 1) may be computed with the following formula.

$$F_{comp} = F_{abd} + F_{muscle} + BW_a \sin\theta + F_{hand} \sin\theta \quad (13)$$

where:

F_{comp} = force at L5/S1
F_{abd} = force produced in the abdominal cavity
F_{muscle} = back muscle force
BW_a = body weight above L1
θ = trunk angle

Although these formulae allow one to calculate the compressive force on the spine, it is of the utmost importance for biomechanists to identify the factors that influence this force.

This model may be used to define job conditions. Job conditions above the BCUL should be unacceptable; conditions between the BCUL and the BCDL require tighter administrative or engineering modifications, while conditions below the BCDL would represent nominal risk to workers. Modifications might include decreasing the vertical height of the unloading platform by adding a second intermediate platform.

Static analyses of muscle forces and moments have been used extensively in ergonomics. For example, static analyses have been used as pre-employment screening,(39,40) to test endurance in manual materials handling tasks,(41) and for static strength evaluation.(42) In a study that compared static and dynamic lifting tasks, McGill and Norman(43) reported that peak dynamic moments at L4/L5 were 19% greater than a static analysis predicted. Although static measures do not appear to be representative of a dynamic task, the assumption has been made by some researchers that the dynamic task is performed slowly, and thus, static measures are representative. However, in most instances, this assumption is tenuous.

Table 26.1 — Maximum Frequency of Lifts per Minute for Different Postures

Duration of Lift Period	*V > 75 cm (Standing)*	*V ≤ 75 cm (Stooping)*
1 hour	18	15
8 hours	15	12

Dynamic Analysis

In a dynamic analysis, there are significant changes in the kinematics of the body. That is, there are significant changes in both linear and angular velocities of both the individual segment and the total body. The kinetic forces and moments of the system must be taken into consideration. During movement, complex inertial forces are generated. Additionally, the inertial parameters for a segment or a body must be determined, usually from anthropometric data. Methods for obtaining kinematic data (videography), kinetic data (force measuring devices), and anthropometric data have been presented earlier in this chapter.

The conditions for a two-dimensional, planar dynamic analysis are more complicated than for the static case, although the number of equations does not change. The first condition states that:

$$\Sigma F = ma \quad (14)$$

when m is the mass of a body and a is the resulting acceleration. This equation is actually two equations describing motion in the horizontal and vertical directions.

$$\Sigma F_x = ma_x \text{ (horizontal direction)} \quad (15)$$
$$\Sigma F_y = ma_y \text{ (vertical direction)} \quad (16)$$

Forces acting on the body may be considered to be muscular, gravitational, contact, or inertial. The gravitational forces are the weights of the segments. The contact forces can be the reaction forces with another segment, the ground, or an external object. The inertial forces are ma_x and ma_y.

The second condition of a dynamic analysis states that

$$\Sigma M_z = I_z\alpha_z \quad (17)$$

where:

ΣM_z = the sum of the moments about a joint center,
I = the moment of inertia of the segment, and α is the angular acceleration of the segment.

ΣM_z represents all of the moments acting about a joint center resulting from the forces acting on the body. $I_z\alpha_z$ represents the inertial torque.

Since all of the forces and moments that cause the moment are calculated by evaluating the resulting motion itself, the technique generally used for this calculation is known as an inverse dynamics approach. To calculate muscle forces, kinematic, kinetic, and anthropometric data are combined in a process referred to as a linked segment or Newton-Euler model. This model requires the following assumptions concerning each body segment in the calculation: (1) each segment has a fixed center of mass; (2) the location of the center of mass remains fixed; (3) the joints are hinge joints; (4) the mass moment of inertia of the segment is constant; and (5) the length of the segment remains constant. The forces acting on the segment model are (1) gravitational forces, (2) ground reaction or external forces, and (3) muscle and ligament forces. A free-body diagram of a segment model of a two-dimensional dynamic analysis is presented in Figure 26.10. Known values include the kinematics of the segment, the anthropometry of the segment, and the reaction forces at the distal end of the segment. The dynamic equations can then be written to calculate the proximal reaction forces and the net muscle moment. The net muscle moment represents the net effect of the muscle activity at a joint. The analysis begins at the most distal segment and proceeds up the chain to the next segment and so on. The force generated by individual muscles that cross a joint cannot be ascertained by this method. However, there are models that, with additional information, can calculate individual muscle forces.

This type of analysis has been used in locomotion studies[44] and in the evaluation of the lower back during lifting.[43] It is certainly beyond the scope of this chapter to do a dynamic analysis of even a single segment. It should be obvious, however, that a dynamic analysis is certainly more complicated than a static analysis. In addition, a three-dimensional dynamic model is much more complicated than a two-dimensional dynamic model.

To relate dynamic strength measures, for example, to static strength measures we must again note that a dynamic analysis is complicated by motion. Initiating any motion requires that segments accelerate to overcome inertia, and this requires a large muscle force. Once motion has commenced, the force-velocity relationship must be considered. As muscle velocity changes, so does muscle force. Using static strength measures to predict performance in a dynamic task will typically provide an overestimate of about 20%.[45]

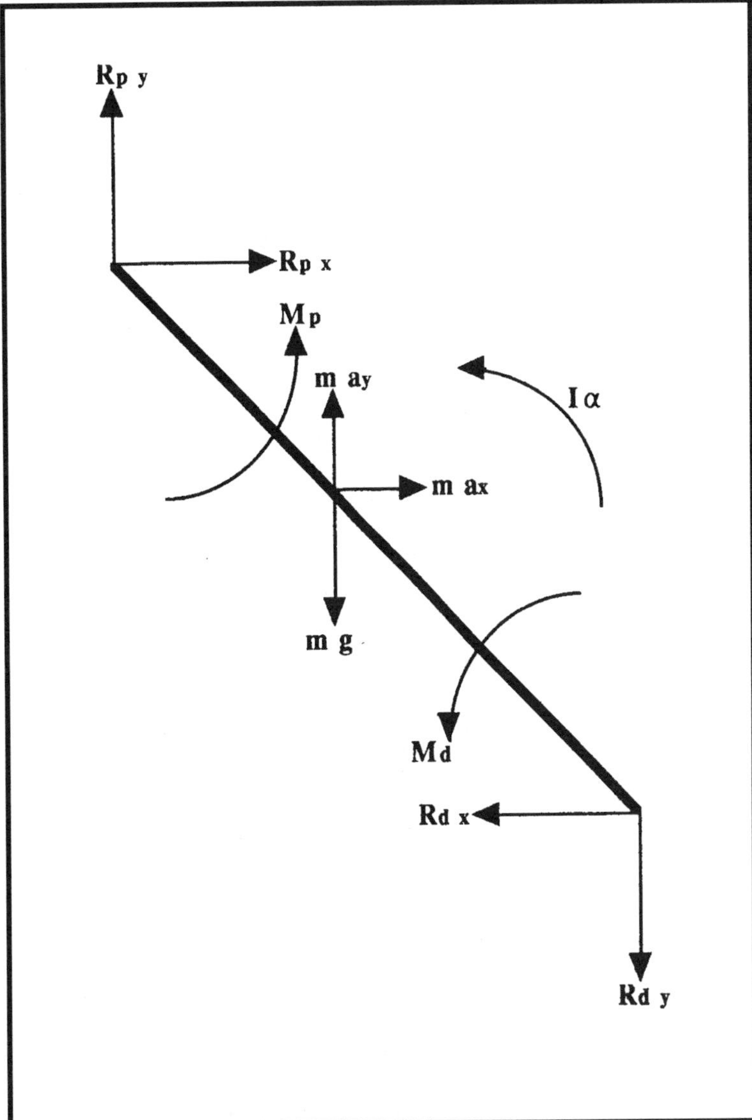

Figure 26.10 — Free-body diagram for a dynamic analysis.

Dynamic analyses are certainly more accurate than static analyses, but they are also more complicated and more difficult to conduct in other than a laboratory setting. However, dynamic analyses have been undertaken to evaluate, for example, lifting tasks[46,47] and asymmetric load carrying.[48] Plagenhoef[49] stated that there is a need for dynamic biomechanical models. He indicated that the ability to interpret the forces and moments at each joint brings about a more critical analysis of motion. He suggested that the timing of the relative accelerations of each body part and the magnitudes of the moments of force indicate the motion's efficiency. Finally, he indicated that the contribution of each body segment to the whole motion may also be determined.

Summary

The purpose of this chapter was to present basic biomechanical principles that may be used in a workplace analysis. Basic muscle mechanics and how muscles work with the skeletal system to produce movement were presented. The two principal relationships in terms of muscle force are the length-tension and force-velocity relationships. In the former relationship, muscle force decreases with either increasing or decreasing muscle length. In the latter relationship, muscle force is inversely proportional to the velocity of shortening. Muscles act on the skeletal system in the form of levers. There are three types of lever, each of which magnifies muscle force or the range of motion of a segment.

In many types of biomechanical analyses anthropometric data are necessary. Body segment dimensions such as segment length, segment mass, location of the segment center of mass, and segment moment of inertia were discussed, and sources for these data were presented. The anthropometric parameters used in biomechanical analyses generally come from two sources, cadaver studies and mathematical modeling. Either of these methods is a satisfactory source.

Biomechanical data collection procedures such as the measurement of kinematic, kinetic, and electromyographic data were presented. Kinematic data, the position, velocity, and acceleration of a body, are collected by means of high-speed video, goniometry, or accelerometry. Kinetic data, the forces that act on a body, are collected via force measuring devices. Electromyography, the study of the electrical signals from muscles, was presented as a further biomechanical tool.

With these considerations as a background, general types of biomechanical analyses such as static and dynamic evaluations were discussed. It should be emphasized that both of these types of analyses are generally conducted in a laboratory setting. A static analysis refers to a biomechanical analysis conducted on a body that is stationary or is moving at a constant velocity. There are two conditions that must be met for a static analysis to be done: (1) the sum of the forces acting on the system must equal zero and (2) the sum of the moments acting on the system must equal zero. Static strength measurements were given as an example of a static analysis. A dynamic analysis is conducted when the body or body segments are in motion. This type of analysis is certainly more complicated than the static analysis. In both cases these analyses give information about the joint forces and the moments generated across the joint.

References

1. **McComas, A.J.:** Motor unit estimation: methods, results and present status. *Muscle and Nerve 14*:585–597 (1991).
2. **Burke, R.E.:** Motor units: anatomy, physiology and functional organization. In *Handbook of Physiology*, V. B. Brooks, ed. Bethesda, MD: American Physiological Society, 1981. pp. 345–422.
3. **Gordon, A.M., A.F. Huxley, and F.J. Julian:** Tension development in highly stretched vertebrate muscle. *J. Gen. Physiol. 184*:143–169 (1966).
4. **Hill, A.V.:** The heat of shortening and the dynamic constraints of muscle. *Proc. Roy. Soc. London, B, 126*:136–195 (1938).
5. **McMahon, T.A.:** *Muscles, Reflexes and Locomotion*. Princeton, NJ: Princeton University Press, 1984. pp. 1–26.
6. **Dempster, W.T.:** *Space Requirements of the Seated Operator* (WADC-TR-55-159). Wright-Patterson Air Force Base, Dayton, OH: Aerospace Medical Research Laboratories, 1955.
7. **Drillis, R. and R. Contini:** *Body Segment Parameters* (BP174-945, Tech. Rep. 1166.03). New York: School of Engineering and Science, New York University, 1966.
8. **Winter, D.:** *The Biomechanics and Motor Control of Human Movement*, 2nd ed. New York: John Wiley & Sons, 1990. pp. 51–74.
9. **Plagenhoef, S.C.:** Methods for obtaining kinetic data to analyze human motions. *Res. Q. Am. Assoc. Health Phys. Ed. 37*:103–112 (1966).
10. **Roebuck, J.A., K.H.E. Kroemer, and W.G. Thomson:** *Engineering Anthropometry Methods*. New York: Wiley-Interscience, 1975. pp. 77–128.
11. **Miller, D.I. and R.C. Nelson:** *Biomechanics of Sport*. Philadelphia, PA: Lea and Febiger, 1976. pp. 28–32.
12. **Peyton, A.J.:** Determination of the moment of inertia of limb segments by a simple method. *J. Biomech. 19*:405–410 (1986).
13. **Becker, E.B.:** "Measurement of Mass Distribution Parameters of Anatomical Segments" (SAE Paper 720964). Detroit, MI: Society of Auto Engineers, 1972.

14. **Hanavan, E.P.:** *A Mathematical Model of the Human Body* (AMRL-TR-64-102). Wright-Patterson Air Force Base, OH: Aerospace Medical Research Laboratories, 1973.
15. **Hatze, H.:** A mathematical model for the computational determination of parameter values of anthropometric segments. *J. Biomech. 13*:833–843 (1980).
16. **Dainty, D.A. and R.W. Norman (eds.):** *Standardizing Biomechanical Testing in Sport*. Champaign, IL: Human Kinetics Publishers, 1987.
17. **Chaffin, D.B. and G.B.J. Andersson:** *Occupational Biomechanics*, 2nd ed. New York: John Wiley & Sons, 1991.
18. **Nigg, B.M. and W. Herzog (eds.):** *Biomechanics of the Musculo-skeletal System*. New York: John Wiley & Sons, 1994.
19. **Light, L.H., G.E. McLellan, and L. Klenerman:** Skeletal transients on heel strike in normal walking with different footwear. *J. Biomech. 13*:477–480 (1980).
20. **Shorten, M.R. and D.S. Winslow:** Spectral analysis of impact shock during running. *Int. J. Sports Biomech. 8*:288–304 (1992).
21. **Nigg, B.M.:** Acceleration. In *Biomechanics of the Musculo-skeletal System*, B.M. Nigg and W. Herzog, eds. New York: John Wiley & Sons, 1994. pp. 237–253.
22. **Peat, M., R.E. Grahame, R. Fulford, and A.O. Quanbury:** An electrogoniometer for the measurement of single plane motions. *J. Biomech. 9*:423–424 (1976).
23. **Gagnon, M., D.G.E. Robertson, and R.W. Norman:** Kinetics. In *Standardizing Biomechanical Testing in Sport*, D.A. Dainty and R.W. Norman, eds. Champaign, IL: Human Kinetics Publishers, 1987. pp. 21–57.
24. **Bates, B.T., S.L. James, L.R. Osternig, and J.A. Sawhill:** An assessment of subject variability, subject-shoe interaction and evaluation of running shoes using ground reaction force data. *J. Biomech. 16*:181–191 (1983).
25. **Cavanagh, P.R. and M.A. Lafortune:** Ground reaction forces in distance running. *J. Biomech. 13*:397–406 (1980).
26. **Holden, J.P. and P.R. Cavanagh:** The free moment of ground reaction in distance running and its changes with pronation. *J. Biomech. 24*:887–897 (1991).
27. **Hennig E.M., P.R. Cavanagh, H.T. Albert, and N.H. MacMillan:** A piezoelectric method of measuring the vertical contact stress beneath the human foot. *J. Biomed. Eng. 4*:213–222 (1982).
28. **Komi, P.V., M. Salonen, M. Jarvinen, and O. Kokko:** In vivo registration of achilles tendon forces in man. I. Methodological development. *Int. J. Sports Med. 8*:3–8 (1987).
29. **Komi, P.V.:** Relevance of in vivo force measurements to human biomechanics. *J. Biomech. 23*:23–34 (1990).
30. **Bigland-Ritchie, B., E.F. Donavan, and C.S. Roussos:** Conduction velocity and EMG power spectrum changes in fatigue of sustained maximal efforts. *J. Appl. Physiol. 51*:1300–1305 (1981).
31. **DeLuca, C.J.:** Myoelectrical manifestations of localized muscular fatigue in humans. *Crit. Rev. Biomed. Eng. 11*:251–279 (1983).
32. **Bouisset, S.:** EMG and muscle force in normal motor activities. In *New Developments in Electromyography and Clinical Neurophysiology*, vol. 1, J.E. Desmedt, ed. Basel, Switzerland: Karger, 1973.
33. **Lippold, O.C.J.:** The relation between integrated action potential in human muscle and its isometric tension. *J. Physiol. 117*:492–499 (1952).
34. **Zuniga, E.N. and D.G. Simmons:** Non-linear relationship between averaged electromyogram potential and muscle tension in normal subjects. *Arch. Phys. Med. 50*:613–620 (1969).
35. **Hof, A.L. and J. van der Berg:** EMG to force processing: 2. Estimation of model parameters for the human triceps surae muscle and assessment of accuracy by means of a torque plate. *J. Biomech. 14*:771–785 (1981).
36. **Olney, S.J. and D.A. Winter:** Prediction of knee and ankle moments of force in walking from EMG and kinematic data. *J. Biomech. 18*:9–20 (1985).
37. **Chaffin, D.:** Ergonomics guide for the assessment of human strength. *Am. Ind. Hyg. Assoc. J. 36*:505–510 (1975).
38. **National Institute for Occupational Safety and Health:** *A Work Practices Guide for Manual Lifting* (Tech. rep. 81-122). Cincinnati, OH: U.S. Department of Health and Human Services, 1981.
39. **Pytel, J.L. and E. Kamon:** Dynamic strength test as a predictor for maximal and acceptable lifting. *Ergonomics 24*:663–672 (1981).

40. **Kamon, E., D. Kiser, and J. Pytel:** Dynamic and static lifting capacity and muscular strength of steelmill workers. *Am. Ind. Hyg. Assoc. J. 43*:853–857 (1982).
41. **Lind, A.R., R. Burse, R.H. Rochelle, J.S. Rinehart, et al.:** Influence of posture on isometric fatigue. *J. Appl. Physiol. 45*:270–274 (1978).
42. **Garg, A. and D.B. Chaffin:** A biomechanical computerized simulation of human strength. *Trans. Am. Inst. Ind. Eng. 7*:1–15 (1975).
43. **McGill, S.M. and R.W. Norman:** Dynamically and statically determined low back moments during lifting. *J. Biomech. 18*:877–885 (1985).
44. **Winter, D.A. and D.G.E. Robertson:** Joint torque and energy patterns in human gait. *Biol. Cybern. 29*:137–142 (1978).
45. **Garg, A., A. Mital, and S.S. Asfour:** A comparison of isometric strength and dynamic lifting capability. *Ergonomics 23*:13–27 (1980).
46. **Fisher, B.O.:** Analysis of spinal stresses during lifting. M.S. thesis, Industrial Engineering, University of Michigan, Ann Arbor, 1967.
47. **Kromodihardjo, S. and A. Mital:** Kinetic analysis of manual lifting activities: Part I—Development of a three-dimensional computer model. *Int. J. Ind. Ergonom. 1*:77–90 (1986).
48. **DeVita, P., D.M. Hong, and J. Hamill:** Effects of asymmetric load carrying on the biomechanics of walking. *J. Biomech. 24*:1119–1129 (1991).
49. **Plagenhoef, S.:** *Patterns of Human Motion.* Englewood Cliffs, NJ: Prentice-Hall, 1971.

A worker in the forestry industry cutting down a tree. Hazards include noise, hand vibration, injury due to falling branches and trees, as well as dermatitis (poison ivy and oak) and possibly even bites of animals such as snakes.

Outcome Competencies

After completing this chapter, the user should be able to:

1. Define underlined terms used in this chapter.
2. Describe cardiovascular responses to exercise.
3. Recognize limits to exercise.
4. Describe cumulative trauma problem.
5. Recommend solutions to cumulative trauma.
6. Evaluate fatigue problems.
7. Recommend approaches to reducing fatigue.

Key Terms

administrative solutions • automation • boredom • cardiovascular system • carpal tunnel syndrome • cumulative trauma disorders • damaging wrist motion • day sleeping • dynamic load • engineering solution • force • job enlargement • job rotation • joint deviation • maximum work rate • mechanization • metabolism • neutral (handshake) position • overload • part-time work • pulmonary system • rating of perceived exertion • recovery/work ratio • repetition/duration • risk factors • shift work guidelines • static load • systemic system • working rest

Prerequisite Knowledge

The reader is assumed to have a university education in science or engineering, college level biology and chemistry.

Prior to beginning this chapter, the user should review the following chapters:

Chapter Number	Chapter Topic
4	Environmental and Occupational Toxicology
24	Applied Physiology of Thermoregulation and Exposure Control

Key Topics

I. Cardiovascular Responses to Exercise
   A. Anatomy
   B. Response to Exercise
   C. Cardiovascular Limits
   D. Decreasing Cardiovascular Stress

II. Cumulative Trauma
   A. Concept

III. Risk Factors
   A. Hand/Wrist
   B. Shoulder/Neck/Elbow
   C. Back

IV. Fatigue
   A. Cardiovascular Fatigue
   B. Musculoskeletal Fatigue
   C. Brain Fatigue
   D. Shift Work

27

Work Physiology*

Stephan A. Konz, Ph.D.

Cardiovascular Responses to Exercise

This section will describe system anatomy, the responses to exercise, limits, and how to decrease cardiovascular stress.

Anatomy

There are two connected subsystems in the cardiovascular system: the pulmonary system and the systemic system. In the pulmonary system, blood from the right ventricle is pumped to the lungs, where carbon dioxide is removed and oxygen is added before the blood is returned to the left side of the heart.

In the systemic system, blood from the left ventricle is pumped to the body arteries. Oxygen and nutrients are removed and carbon dioxide and metabolic waste products are added before the blood is returned in the veins to the right side of the heart. Nutrients are added from the intestines. Fat-soluble wastes are biotransformed in the liver into water-soluble wastes and put back into the blood. The kidney then eliminates water-soluble wastes through urine. Some wastes are eliminated in the intestines (feces).

Figure 27.1 shows an engineer's view of the cardiovascular system.

Response to Exercise

The cardiovascular system has five responses to exercise: (1) heart rate, (2) stroke volume, (3) artery-vein differential, (4) blood distribution, and (5) going into debt. The burning of fuel is called metabolism (see the sidebar on page 714 for an in-depth discussion).

Heart Rate. For light and medium work loads, heart rate is a good predictor of metabolic rate. Two exceptions to this rule are when heart rate is increased by (1) emotions or (2) by vasodilation (a reaction of unacclimatized people in heat stress).

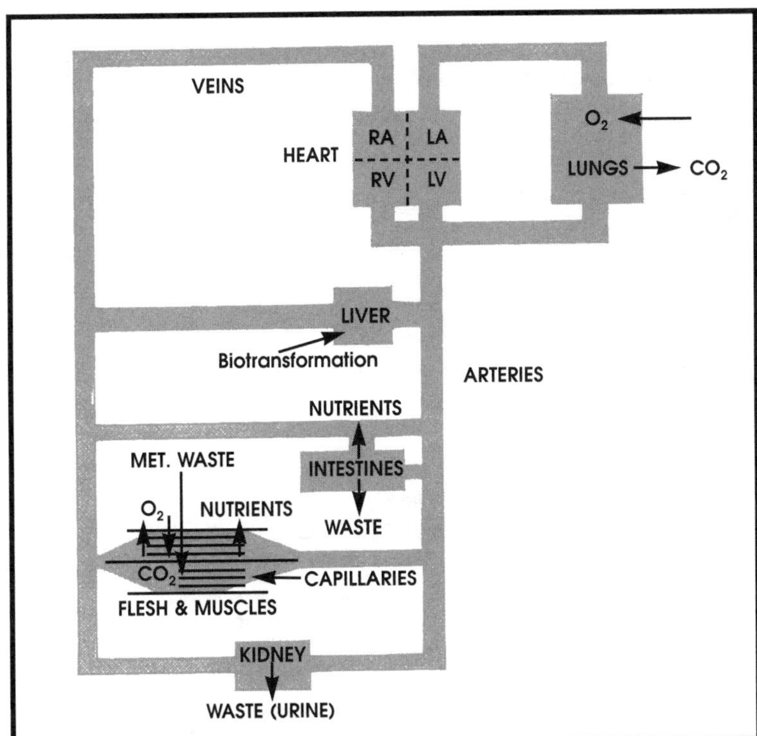

Figure 27.1 — Engineer's view of the cardiovascular system, showing a pulmonary circuit and a systemic circuit. The heart has two pumps. In pulmonary circulation the right ventricle sends blood to the lungs, where O_2 is added and CO_2 is removed. The blood returns to the left atrium. In systemic circulation the left ventricle sends blood out to the arteries. After passing through the various organs and capillaries, the blood returns in the veins. In the "marsh" of the capillaries, the blood gives up O_2 and nutrients to the flesh and obtains CO_2 and metabolic wastes. Nutrients are added to the blood from the intestines. Wastes are biotransformed in the liver and reinserted into the blood to be removed by the kidney (urine) and intestines (feces). (**Source:** Reference 1.)

**Part 1 and 2 of this chapter is a modification of Chapter 11, Work Physiology, and Chapter 16, Cumulative Trauma, in Stephan Konz,* Work Design: Industrial Ergonomics, *4th ed., Publishing Horizons, Scottsdale, AZ, 1995.*

Metabolism

Metabolic rate is divided into three parts: basal, activity, and digestion.

TOTMET = BSLMET + ACTMET + DIGMET

where:

TOTMET = total metabolic rate
BSLMET = basal metabolic rate
ACTMET = activity metabolic rate
DIGMET = digestion metabolic rate

In ergonomics, metabolic rate typically is expressed in watts, abbreviated W. The energy content of food is given in kcal (commonly spoken of as calories rather than the technically correct kilocalories). 1 W = .859 kcal/h.

Basal metabolism maintains body temperature, body functions, and blood circulation.

BSLMET = BSMET (WT)

where:

BSLMET = basal metabolic rate, W
BSMET = 1.28 W/kg for males, 1.16 W/kg for females
WT = body weight, kg

BSMET is lower for females because females tend to have a higher percentage of fat than males, and fat has a limited metabolism.

Activity metabolism, not surprisingly, provides the energy for activities. In laboratories, metabolism can be estimated from oxygen consumption, but in industrial applications it usually is estimated from tables or formulas (see Table 27.1). When using tables or formulas, remember that most work is not continuous because people take microbreaks. Estimate the time of each activity from a videotape of the job or from occurrence sampling.

Pandolf et al.[2] gave the following as the cost of walking without a load:

$$\text{WLKMET} = C\,(2.7 + 3.2\,(v\ .7)^{1.65})$$

where:

WLKMET = walking metabolism (total), W/kg of body weight
C = terrain coefficient
= 1.0 for treadmill, blacktop road
= 1.1 for dirt road
= 1.8 for swamp
= 2.2 for sand
v = velocity, m/sec (for v > .7 m/sec (2.5 km/hour, 1.56 miles/hour))

Pandolf et al.[3] give the cost of standing or walking slowly is

$$\text{WLKMETT} = 1.5\ \text{WT} + 2\ (\text{WT} + \text{WTL})(\text{WTL}/\text{WT})^2 + C(\text{WT} + \text{WTL})(1.5\ v^2 + .35\ vG)$$

where:

WLKMETT = walking (slowly), total, W
WT = weight of body, kg
WTL = weight of load on shoulders, kg
v = velocity of walking, m/sec (v < 1 m/sec)
C = terrain coefficient (see above)
G = grade, %

The first term (1.5 WG) is the metabolic cost of standing without a load. The second term is the cost of load bearing, while standing. Walking on the level is C (WT + WTL)1.5 v^2. Cost for climbing a grade is C (WT + WTL)(.35 vG).

Digestion metabolism is the energy cost of digesting food. For the typical carbohydrate/fat/protein mixture of the U.S. diet and disregarding time since the meal

DIGMET = .1 (BSLMET + ACTMET)

Table 27.1 — Activity Cost for Various Activities

W/kg	Activity
.4	Sitting quietly, writing
.6	Standing relaxed, playing cards
.7	Standing office work
.8	Dressing and undressing
1.0	Driving a car
1.2	Keying rapidly
1.7	Painting furniture with a brush
3.1	Cleaning with upright vacuum sweeper

Source: Reference 1.

Although heart rate can be measured with electronics, for many jobs it usually can be estimated with sufficient accuracy simply by estimating the perceived exertion. A worker could be shown Table 27.2 and, using the words for guidance, would vote their rating of perceived exertion (RPE) with a number. The worker's heart rate would be approximately 10 × RPE.

Stroke Volume. The second method the body uses to adjust oxygen supply is stroke volume, the amount of blood pumped by the left ventricle, as shown in the following equation.

$$\text{CO} = \text{HR} \times \text{SV} \tag{1}$$

where:

CO = cardiac output, L/min
HR = heart rate, beats/min
SV = stroke volume, L/beat

Table 27.2 — Rating of Perceived Exertion (RPE)

Vote[A]	Subjective Description
6	No exertion at all
7	Extremely light
8	
9	Very light
10	
11	Light
12	
13	Somewhat hard
14	
15	Hard
16	
17	Very hard
18	
19	Extremely hard
20	Maximal exertion

Note: This scale was developed by Borg[(4)] to predict heart rate. Use the words for guidance, then vote with a number.
[A]Heart rate = 10 × vote.

Arterial-Venous (AV) Differential. The third method of adjusting oxygen supply to the body is AV oxygen differential. When resting, the arterial oxygen content is 19 mL/100 mL of blood; the venous oxygen content is 15 mL/100 mL. That is, for every 100 mL of blood passing through the muscles, the muscles get 4 mL of oxygen. However, in an emergency (fleeing from a tiger, for example), the veins drop to 6 mL and the AV differential becomes 13 mL/100 mL.

Blood Redistribution. Redistribution also adjusts oxygen supply. As muscle blood flow increases, the kidneys and intestines use less blood; if a person has eaten recently, he or she may get stomach cramps.

Debt. Finally, if the oxygen supply is not sufficient, the body can go into debt—that is, it uses anaerobic oxygen stored as compounds in the blood (see Figure 27.2). Although the anaerobic oxygen is replaced, the process of replacement uses oxygen, leading to a "debt + interest" situation.

Cardiovascular Limits

The fact that people can work hard does not mean they should. High-metabolic rate jobs should be considered prime candidates for mechanization, because a motor is a far more efficient source of power than muscles. Although scientists may be interested in physiology, engineers should be interested in productivity.

For most of the population, work should be limited to about 350 W and 100–120 beats/min for the working day. During more intensive work periods, 130 beats/min should be the maximum.

Deceasing Cardiovascular Stress

Both engineering and administrative solutions should be considered to reduce cardiovascular stress. The primary engineering solution is to use a motor. For materials handling, the use of fork lifts, hoists, and powered conveyors may be necessary. Workers should slide rather than lift objects (horizontal transfer rather than vertical transfer); lower objects instead of lifting them; use wheeled transfer, such as carts or dollies, instead of carrying heavy loads; and use powered hand tools. Balancers and manipulators can reduce static load.

Two administrative solutions are job rotation and part-time work. In job-rotation, people shift jobs periodically during the day. In part-time work, the job is split among several people, each of whom works part of the shift. For example, Joe works for four hours in the morning and Pete for four hours in the afternoon. Another alternative is to hire enough people so that the entire job is done as part of an eight-hour shift (for example, package handling for delivery services often is done in two to three hours).

Cumulative Trauma

This section will describe the cumulative trauma concept and then discuss, in more detail, the hand/wrist, the shoulder/neck/elbow, and the back.

Concept

The first section of this chapter briefly described overload to the cardiovascular

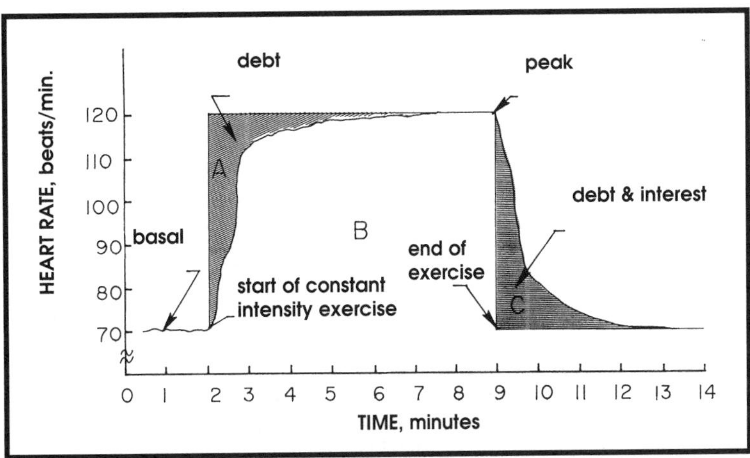

Figure 27.2 — Aerobic response (and therefore heart rate) lags the start of exercise. The deficit (Area A) is replaced by anaerobic oxygen. Aerobic refers to reactions using oxygen from the lungs; anaerobic (without oxygen) reactions use oxygen stored as compounds in the blood. During recovery (Area C), the anaerobic oxygen stored in the blood is replaced but the replacement process itself uses oxygen ("interest"), so Area C is larger than Area A. For task cost, use Area B + C. (**Source:** Reference 1.)

Recovery/Work Ratio

Both the amount of recovery (rest) and the distribution are important.

Amount. Recovery (repair, rest) time can be calculated as a ratio of exposure time; that is, a recovery/work ratio. For example, in a 24-hour day, if a specific joint is used for 8 hours, then there are 16 hours available for recovery. That is, there is 16/8 = 2 hours of recovery for every 1 hour of exposure. If the joint is used for only 4 hours per day (say by alternating with the other arm), then there is 20/4 = 5 hours of recovery for every 1 hour of exposure. Overtime or long shifts can cause considerable reduction of the ratio; 12 hours of work gives 12/12 = 1 hour of recovery for 1 hour of exposure.

Working rest (job variety) allows rest for part of the body while another part works. The total body may recover during at-work breaks (coffee, lunch) and evening/nights. In addition, weekends, holidays, and vacations permit additional recovery.

Guideline: Increase recovery time.

Distribution. The above assumes that each minute of recovery (repair, rest) time is equally effective; recovery rate versus time is a horizontal line. But recovery rate declines exponentially with time (see Figures 27.3 and 27.4). That is, the amount of recovery for minutes 6–10 is much less than for minutes 0–5. Thus, for the same total length of break, many short breaks are better than occasional long breaks. A single break of 15 minutes is not as effective as three breaks of 5 minutes. An 8-hour break is not as effective as four breaks of 2 hours. On a practical basis, this means that it is better to rotate jobs within days rather than between days. For example, have Joe work on Job A in the morning and have Pete work on it in the afternoon. This is better than having Joe working for 8 hours on Job A on Monday and then on Job B on Tuesday. Waersted and Westgaard[5] suggest rotation should be after 1 to 2 hours rather than after 4 hours.

Janaro and Bechtold[6] point out there is a time penalty for breaks (for example, when an operator turns off equipment, goes to rest area, returns from rest area, powers up equipment); thus, very short breaks may not be cost-effective. However the microbreak concept allows the operator to take a break (perhaps 20 seconds) at the workstation and leave the equipment on, thus eliminating the travel cost and the on/off cost.

Guideline: Frequent short breaks are better than occasional long breaks.

Source: Reference 1.

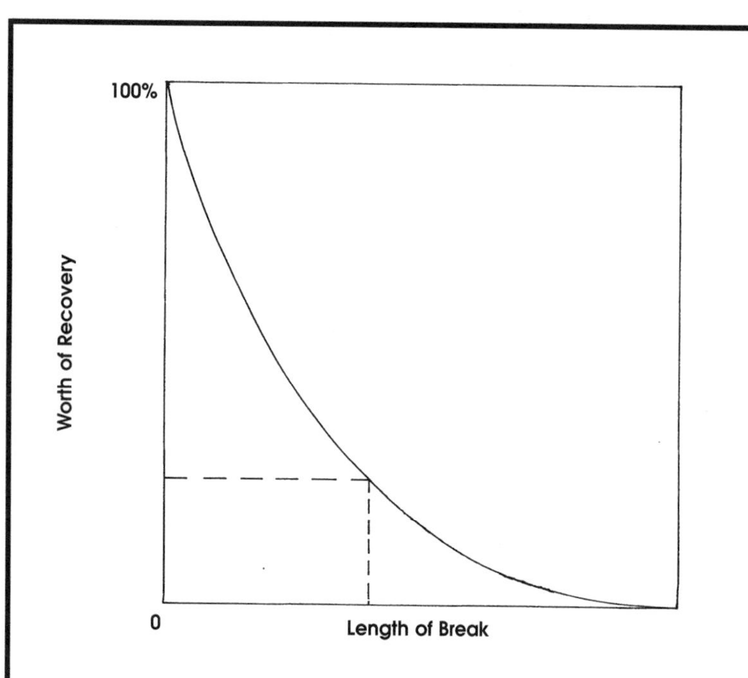

Figure 27.3 — Since recovery from fatigue is exponential, when time is low, the percentage change in concentration is greater than the percentage change in time. When time is high, the change in concentration is less than the change in time. That is, there is more recovery in the first part of the break than in the latter (see also Figure 27.4). Fatigue (while working) increases exponentially (curve is reversed). When time is low, the change in concentration is less than the change in time; when time is high, the change in concentration is greater than the change in time. Thus, three breaks of 5 minutes have more benefit than one break of 15 minutes as (1) recovery is better and (2) fatigue has not increased as much.

system. However, the nature of work has changed over the years, and many overloads now are localized to smaller areas. These local overloads have been called many things: repetitive strain; occupational overuse syndrome; work-related musculoskeletal disorders; or, most commonly, cumulative trauma disorders (CTD).

Cumulative trauma concerns intermediate-term (spanning months or years) effects of body activity on the nerves, muscles, joints, and ligaments. The primary problems occur in the hand/wrist, the shoulder/neck/elbow, and the back.

Risk Factors

There are many risk factors for CTDs, both occupational and nonoccupational. The primary occupational risk factors are repetition/duration, joint deviation, and force. (In some cases, cold and vibration are risk factors, probably because of reduced blood flow and thus slower repair time.) Unfortunately, at the present state of knowledge, it is not possible to quantify the risk factors. Lower values of force (say, from 4 kg to 2 kg) are preferred, but exactly what impact this will have on a specific individual is unknown. The relationship between tobacco use and cancer is analogous. Although we may not be able to predict the

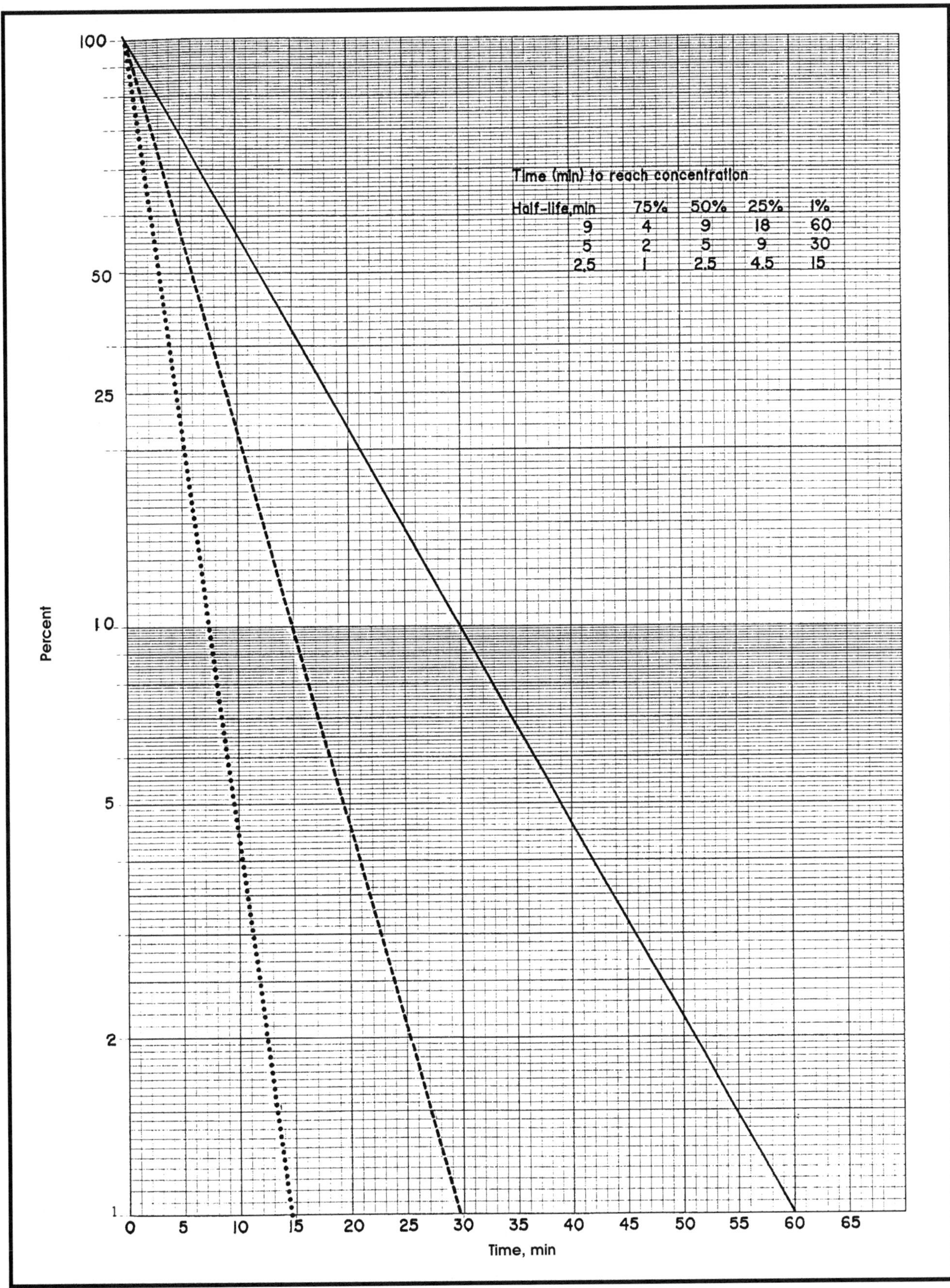

Figure 27.4 — An exponential curve (such as shown in Figure 27.3) can be approximated by a straight line on semi-log paper. In theory, an exponential curve approaches 0 concentration asymptotically (i.e., never reaches zero). An approximation of the end point is the time at which concentration reaches 1%. If the concentration is 1% at 60 minutes, the 50% level was reached in 9 minutes. The concentration took only 4 minutes to drop from 100% to 75% but took 42 minutes to drop from 25% to 1%.

probability of lung cancer in Joe, a 35-year-old male who has smoked a pack a day of Brand X for 17 years, there is little doubt Joe should reduce the number of cigarettes he smokes. Similarly, a certain amount of force added to a certain amount of joint deviation, for example, may cause cumulative trauma; just how much is too much is not yet known.

Repetition/Duration. A hand/wrist task is considered repetitive if it has a cycle time of less than 30 seconds. Back or shoulder movements (i.e., more stressful movements) might be repetitive with an interval of several minutes. However, duration during the shift also is important. Short duration will be defined as less than one hour per day, moderate as one to two hours, and long as greater than two hours. Thus, repetition really concerns the number of repetitions per shift.

However, there is another complication: the body is (to some extent) self-repairing. This repair can occur during work (e.g., arm extension muscles rest while arm rotation muscles work, and vice versa; see the sidebar on page 716 for a discussion of recovery:work ratio). In addition to this working rest, recovery can occur during nonwork time, such as coffee breaks, lunch breaks, sleeping at home, etc. A further complication is the duration of the work over months and years; a problem occurs if the daily repair is not sufficient to return to normal, and the trauma starts from a higher baseline.

There is also great individual variation in physical condition; that is, strong bodies usually can repair themselves better than weak bodies (hence, many firms try to encourage strong bodies with wellness programs). If the trauma is sufficient, however, even strong bodies fail eventually.

Joint Deviation. Ideally, joints should operate at the neutral position (zero joint deviation) to prevent injury.

Force. Reducing the magnitude and duration of external forces and their torques can help avoid cumulative trauma.

Nonoccupational Risk Factors. Trauma outside of work can be caused by such activities as sewing or keying musical instruments; by sports such as lifting weights or bowling; or by an injury. A person also can have anatomical imperfections (small carpal tunnel, weak back muscles) or physiological imperfections (diabetes, insufficient hormones) that lead to trauma.

Solutions to Cumulative Trauma. Engineering and administrative approaches to dealing with CTDs have been mentioned already. However, when cumulative trauma is already apparent, medical/rehabilitation solutions such as physical treatment, medication, or surgery must be tried. The sidebar on page 720 gives more information on all of these approaches.

Hand/Wrist

CTDs can occur in the tendons, the muscles, or the nerves of the hands and wrists. One of the many CTDs of the hand/wrist is carpal tunnel syndrome (see Figure 27.5). Engineering solutions focus on repetition/duration, joint deviation, and force.

Repetition/Duration. Reducing lifetime use of a joint helps reduce injury. For example, a foot- operated control instead of a hand-operated control could be used. Instead of 5000 repetitions per day for the right wrist, there could be 2500 for the right wrist and 2500 for the left wrist.

Joint Deviation. Since the goal is to avoid bending the wrist, it should be kept in the neutral (handshake) position either by changing the job or changing the tool. Changing the job may require a change in worker posture (from sitting to standing) or a change in the work orientation (by tilting the table or fixtures on the table). Changing a tool involves bending the tool, not the wrist. But finding one perfect angle is unlikely, and experimentation may be necessary. For example, Jegerlehner(12) reported reducing wrist deviations on the spindle lever of a tapper by using a ball handle on the end of the lever.

Force. Reducing force and duration helps prevent wrist injuries. Wick(13) defines a damaging wrist motion as a bent wrist involving a force. McCarty et al.(14) give 1000 damaging wrist motions per hour as an upper ergonomic limit. In addition, all pinch grips requiring more than eight pounds of pressure are considered dangerous.

In looking for solutions, consider whether a robot or clamp could be used in place of the worker's hand. Can force be reduced? For example, a ballpoint pen requires about 180 g of force, pencils and fountain pens about 120 g, and felt tips about 100 g. Can the tool (pliers, scissors, clippers) be returned with a spring instead of requiring a muscle action? For cutting with hand tools, use motors instead of human muscle. For triggers, reduce force/finger by using a trigger strip instead of a trigger button. Vibrating tools require more grip force than nonvibrating tools; therefore, reducing the vibration reduces the force.

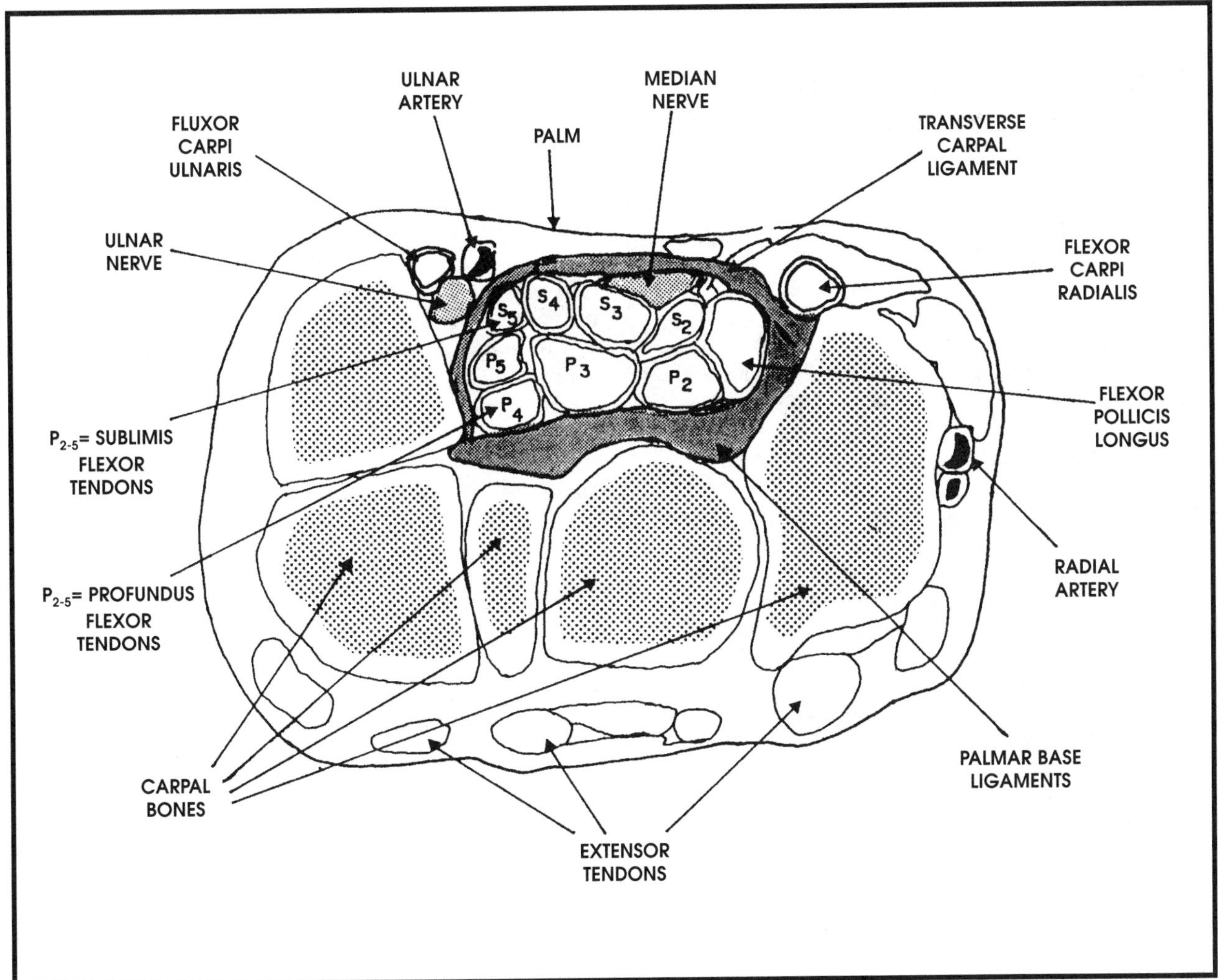

Figure 27.5 — Section of the right wrist (looking toward fingers). Carpal tunnel syndrome (CTS) is a condition in which the median nerve is pinched at the wrist. (The median nerve can be pinched at a number of locations in the arm and shoulder; thus not all pinched median nerves are CTS.) Since a pinched nerve conducts more slowly, a common clinical test for CTS is nerve conduction velocity. Johnson and Evans[11] strongly question the criterion of 4.3 msec as indicating CTS, as there are large variances both within and among individuals. (**Source:** Reference 10.)

Shoulder/Neck/Elbow

As with the hand and wrist, engineering solutions for the shoulder, neck, and elbow focus on repetition/duration, joint deviation, and force.

Repetition/Duration. Reduce lifetime use of the joint through automation or mechanization. Minimize one-sided work by using both the left and right hands. Elbow problems tend to occur when frequent use is accompanied by extreme deviations or considerable force (tennis elbow) or by rotating the forearm while the wrist is bent (golfer's elbow).

Joint Deviation. A key guideline is "Don't lift your elbow." That is, the upper arm should be vertical. When the work forces a joint into a deviated position, either the work or the shoulder should be moved rather than the upper arm. Adjusting the work implies adjustable-height work surfaces. Adjusting the shoulder while seated probably means an adjustable height chair. Slanting (tilted) work surfaces may give better elbow as well as shoulder orientation. When used with computers, bifocals tend to cause unnatural neck postures, so bifocal-wearers may wish to use single-vision lenses with a .6 m focal length.

Force. Use balancers for hand tools. Even when not being used to hold something, a person's arm weighs about five percent of his or her body weight.

In a situation where force is being used to move an object from one surface to another, consider whether the two surfaces can be made the same height, so that sliding replaces lifting.

Solutions for Cumulative Trauma

Three approaches to dealing with cumulative trauma include engineering solutions, administrative solutions, and medical/rehabilitation approaches.

Engineering Solutions

Automation. Eliminate the need for a person to do the job. For example, use a robot to load boxes, use a palletizer to place cartons on a pallet.

Mechanization. A machine does part of the job but an operator is still present. For example, a checkout clerk could use a bar code scanner instead of keying information. This reduces task difficulty.

Job Enlargement. Increase the total job content for a specific person. Since the total task is split among fewer people, each performs a greater variety of tasks, and the job becomes less repetitive.

Reduced Joint Deviation. For example, use keyboard design and location to reduce wrist deviation for keyboarders. Neck deviation can be alleviated by replacing bifocal glasses with computer glasses (single-vision glasses with a focal length of about .6 m).

Reduced Force (Amount and Duration). For example, keeping knives sharp allows the worker to expend less effort. Force commonly is associated with manual materials handling.

Administrative Solutions

Some approaches (job rotation, part-time workers) seek to reduce exposure. Other approaches (exercise, stress reduction, supports) are designed to increase the person's ability to endure the stress.

Job Rotation. By employing the concept of working rest, a specific part of the body rests but the person is still working. The concept works best when the alternate tasks differ from one another. For example, a person picking boxes off a conveyor could rotate with a person doing paperwork. A less satisfactory solution would be doing checkout using a left-hand station instead of a right-hand station or shifting from packing large bags of snack food to packing small bags. In addition to stress-reduction benefits, job rotation has managerial advantages in that people are cross-trained (able to do more than one job), which allows more flexible scheduling, and there is a perceived fairness because good and bad jobs are shared. Job rotation requires broad job descriptions rather than narrow ones. For example, "office worker" rather than "secretary 1" and "secretary 2."

Part-Time Worker. When there is no other job to rotate to, several people can be hired part-time. For example, hire Joe to key for four hours in the morning and Pat to key for four hours in the afternoon. In addition to reducing cumulative trauma, advantages of part-time work include lower labor cost per hour (wages + fringes), better fit to fluctuating demand (a 25 percent increase in demand requires each person to work 5 hours per day vs. 10 hours for a full-time worker), and (possibly) higher quality workers. (It may be difficult to hire full-time high-quality people for repetitive low-paying jobs.) Disadvantages include less time on the job (and thus less learning), possible moonlighting for other employers (so workers are tired when working for you), and a high fixed cost per employee for hiring and training.

Exercise. For cumulative trauma reduction, exercises should emphasize flexibility and strength rather than cardiovascular endurance. Note that stress can be caused by static loading as well as dynamic loads. For dynamic loads, relax and stretch muscles; for static loads, move the muscles.

Stress Reduction. Social factors, both on and off the job, can cause stress. On-the-job stressors include work load, deadlines, and interpersonal relationships. Off-the-job stressors include domestic and financial problems. Some treatments are relaxation (learning to "let go"), rest (especially microbreaks and working rest), exercises (warm up before work, dynamic exercises to counteract static loads), and medicine to minimize sleep disturbances.

Supports. An appealing idea is to support the body. Four commercially available possibilities are wrist braces, elbow braces, back braces (back belts, lifting belts), and lower-arm supports. However, in general it seems that using such supports actually weakens the body. Workers should therefore weigh the pros and cons of using supports before they begin wearing them.

Using a wrist splint off the job is relatively noncontroversial but wearing the splint during work may impair recovery. Although back belts are popular, there seems to be little scientific evidence that they provide benefits.[7] In tests, arm supports did not provide worthwhile benefits during welding or light assembly.[8]

Medical/Rehabilitation Approaches

Once a person has a CTD, medical personnel enter the battle. Medical personnel can provide specific recommendations for treatment, which can include physical therapy, medication, and surgery.

Physical treatments include ice (to reduce inflammation), heat (to increase blood flow), exercise, massage, and splints.

Medication can be used to reduce swelling, promote healing, etc.

Surgery is a last resort. Note that surgery does not always prevent the problem from recurring. For example, Owen[9] reports that "80 percent of carpal tunnel syndrome surgery is highly successful." Naturally, surgery cannot help if the nerve has been permanently damaged.

Back

Back problems result from both overload (generally manual materials handling) and underload (lack of movement when sitting). For overload, risk factors are categorized as personal and job-related. Personal factors include age, gender, body size, physical fitness and training, lumbar mobility, strength, medical history, years of employment, smoking, psychosocial factors, and structural abnormalities. Job risk factors include heavy physical work, lifting, bending, stretching and reaching, twisting, pushing and pulling, and accidents. For underload, job risk factors include prolonged sitting and standing and whole-body vibration. This long list of risk factors emphasizes that back problems are not simple.

Underuse of the Back. When people sit for prolonged periods, the intervertebral discs are not properly nourished, so prolonged sitting should be avoided. In addition, people in sedentary jobs should be encouraged to be active in their activities outside work. The problems seem to be worse when sitting is combined with whole-body vibration (for example, truck driving). Possible solutions include reducing vibration transmitted to the driver, good posture, and variability in posture.

Overuse of the Back. Examination of the risk factors shows that the problem results primarily from manual materials handling.

Fatigue

This section will discuss cardiovascular fatigue, musculoskeletal fatigue, and brain fatigue. Fatigue is most often related to long work hours—especially if there is a lack of sleep. Performance effects of fatigue are more likely to be reflected in errors than in lower productivity.

Cardiovascular Fatigue

The most common occupational task stressing the cardiovascular system is manual materials handling. Mital[15,16] determined that male materials handlers could sustain, without overexertion, 29 percent of their maximum oxygen uptake for eight-hour shifts but only 23 percent for 12-hour shifts. Mital[17] found workers in an air-cargo firm's package handling area, hired only for a two-hour shift, working at 43–50 percent of their maximum oxygen uptake. Thus, maximum work rate depends on the hours worked per shift.

Musculoskeletal Fatigue

Muscle (local) fatigue is divided into static and dynamic. When muscles are loaded but do not move, it is a static load; when loaded muscles move, it is dynamic.

Static Load

Static load primarily results from posture. There are great differences in the fatigue resistance of various muscles. For example, Birgland-Richie[18] reported endurance times of the soleus seven times greater than that of the quadriceps. Although recovery is rapid for low stress, full recovery from maximal exertion may take several days.

Dynamic Load

Dynamic load is divided into video display terminal (VDT) work and non-VDT work. There have been some studies on rest for VDT tasks.

Work and Break Lengths. Misawa et al.[19] suggest VDT work should be 60 minutes or less before a break. Yoshimura and Tomada[20] studied a work period of 50 minutes with rest periods of 5, 10, 15, 20, and 25 minutes; they recommended the 15-minute rest. Kopardekar and Mital[21] recommended 60 minutes of VDT work followed by 10 minutes of rest. The German standard is 10 minutes of nonworking rest after 60 minutes of continuous computer work.

Microbreaks. Short breaks (perhaps 20 seconds) are called "microbreaks." Henning et al.[22] concluded that both regimented breaks and operator-controlled breaks are useful, but operator-controlled breaks had fewer task interruptions. Therefore, feedback on the amount and type of short breaks taken is beneficial.[23]

Active vs. Passive Rest. What is done during the break? Passive rest is sitting at the workplace; active rest varies from going to another work area to doing gymnastics. The little research available tends to support active rest (although not necessarily gymnastics).

Non-VDT Work. Bhatia and Murrell[24] reported that industrial workers preferred six breaks of 10 minutes each over four breaks of 15 minutes each. Several other studies have reported that frequent short breaks are better than occasional long breaks.

Brain Fatigue

The category of work-related stress classed as brain fatigue can be divided into four main divisions: (1) optimum stimulation, (2) concentration and attention, (3) sleep/biological clock, and (4) shift work.

Optimum Stimulation

Fatigue is not all physiological; there is a strong psychological (lack of motivation) component. For example, Finkleman[25] analyzed 3700 people who reported fatigue in their work. Workers reported less fatigue from physically demanding jobs than from jobs with low physical demand. An important predictor of fatigue is too much information (overload) or too little information (boredom).

Stimulation (information) can come from the task or the environment. If the task is high stimulation, try lowering the stimulation from the environment. If the task is low stimulation, add stimulation to the task (such as using movement) or the environment (conversation with others, windows with a view, varied environmental temperature, and so on).

Concentration and Attention

Sedentary tasks often require considerable mental activity. Overload examples include simultaneous translation and education; underload examples include control-room monitoring and driving vehicles.

Sleep/Biological Clock

Sleep restores the functions of the brain, rather than those of the body. For most people, maximum sleepiness is around 3 a.m., with a secondary dip coming around 2 p.m. (postlunch dip). The postlunch dip is not related to food consumption, since it does not occur after breakfast or dinner.

Fatigue is probably more affected by lack of sleep than by the number of hours spent working. The regulations all assume that the hours of the day are equal and interchangeable, but this is not true. Employees who work nights need to make adjustments in their sleeping habits (see Table 27.3 for day sleeping tips).

The most common drug to decrease sleepiness and increase alertness is caffeine; amphetamines work also, but should be used with caution.

Alcohol or some of the older antihistamines have the effect of increasing sleepiness and decreasing alertness. Alcohol-related vehicle accidents at night may result as much from decreased alertness as

Table 27.3 — Tips for Day Sleeping

Develop a Good Sleeping Environment

- It is important to have darkness (e.g., opaque curtains).
- Quiet is also important because it is difficult to go back to sleep when daytime sleep is interrupted. Consider using earplugs, unplugging bedroom phones, turning down phone volume in other rooms, and reducing TV volume in other rooms. Explain to your children that you need quiet to sleep.
- If the sleeper is part of an "augmented crew" (truck, aircraft), good mattresses, minimum noise and light, good thermal conditions, etc., should be provided.

Plan Your Sleeping Time

- Tell others your schedule to minimize interruptions.
- Consider sleeping in two periods (5–6 hours during the day and 1–2 hours in the late evening before returning to work). Less daytime sleep and more evening sleep not only makes it easier to sleep but also may give a better fit with family and social activities.
- Morning to noon bedtimes are the most unsuitable times for sleep.

Develop Good Eating Habits

- Have a light (not zero or heavy) meal before sleep.
- Liquid consumption increases the need to urinate (which wakes you up).
- Avoid caffeine.
- A warm drink before your bedtime (perhaps with family members starting their day) may also help with social needs.
- Avoid foods that upset your stomach and thus wake you up.

Learn to Relax

- If under emotional stress, try to relax before going to bed.
- Light exercise, reading, and listening to soothing music are all possible stress reducers.

Source: Adapted from Reference 1.

Table 27.4 — Shift System Design Recommendations

- Permanent night work does not seem to be advisable for the majority of shift workers. Few people are able to make a complete physiological switch to night work. Even permanent night workers have problems because weekends, holidays, and vacations usually require readapting to a day cycle. If shifts rotate, rapid rotation is preferable to slow (weekly) rotation.
- Shift durations of 12 hours have advantages and disadvantages. Some potential problems are fatigue, covering absences, overtime, limitation of toxic exposure, and possible moonlighting when workers have large blocks of leisure time.
- Avoid an early (before 7 a.m.) start for the morning shift.
- Distribution of leisure time is important. Workers should have sufficient time to sleep between shifts (e.g., during shift changeovers). Limit the number of consecutive working days to no more than 5 to 7 days. For each shift system, have some non-working weekends with at least 2 successive full days off.
- Rotate shifts forward (day, then evening, then night).
- Keep the schedule simple and predictable. People want to be able to plan their personal lives. Make work schedules understandable. Publicly post them in advance so employees can plan; 30 days in advance is a good policy.

Source: Reference 27

from impaired motor function or reaction time.[26] Sleepiness, even without the presence of alcohol, is itself a major culprit in traffic accidents.

Shift Work

Table 27.4 gives general shift work guidelines.

Summary

This chapter briefly covers three major topics: (1) cardiovascular responses to exercise, (2) cumulative trauma (hand/wrist, shoulder/neck/elbow, back), and (3) fatigue (cardiovascular, musculoskeletal, and brain).

References

1. **Konz, S.:** *Work Design: Industrial Ergonomics*, 4th ed. Scottsdale, AZ: Publishing Horizons, 1995.
2. **Pandolf, K., M. Haisman, and R. Goldman:** Metabolic energy expenditure and terrain coefficients for walking on snow. *Ergonomics 19*:683–690 (1976).
3. **Pandolf, K., B. Givoni, and R. Goldman:** Predicting energy expenditure with loads while standing or walking very slowly. *Appl. Physiol. Resp. Environ. 43*:577–581 (1977).
4. **Borg, G.:** Psychophysical scaling with applications in physical work and the perception of exertion. *Scand. J. Work, Environ. Health 16(suppl. 1)*:55–58 (1990).
5. **Waersted, M. and R. Westgaard:** Working hours as a risk factor in the development of musculoskeletal complaints. *Ergonomics 34*:265–276 (1991).
6. **Janero, R. and S. Bechtold:** A study of the reduction of fatigue impact on productivity through optimal rest break scheduling. *Hum. Factors 27*:459–466 (1985).
7. **Rys, M. and S. Konz:** Lifting belts—a review. *Int. J. Occup. Saf. Ergonom. 1*:294–303 (1995).
8. **Jarvolm, U., G. Palmerud, R. Kedefors, and P. Herbets:** The effect of arm support on supraspinatus muscle load during simulated assembly work and welding. *Ergonomics 34*:57–66 (1991).
9. **Owen, R.:** Carpal tunnel syndrome: a products liability perspective. *Ergonomics 37*:449–476 (1994).
10. **Tanaka, S. and J. McGloughlin:** A conceptual quantitative model for prevention of work-related carpal tunnel syndrome (CTS). *Int. J. Ind. Ergonom. 11*:181–193 (1993).
11. **Johnson, S. and B. Evans:** Tracking median nerve conduction as a method of early detection of carpal tunnel syndrome. In *Proceedings of the Human Factors and Ergonomics Society*. Santa Monica, CA: Human Factors and Ergonomics Society, 1993. pp. 759–763.
12. **Jegerlehner, J.:** Ergonomic analyses of problem jobs using computer spreadsheets. In *Advances in Industrial Ergonomics and Safety III*, W. Karwowski, and J. Yates, eds. London: Taylor & Francis, 1991. pp. 865–871.
13. **Wick, J.:** Force and frequency: How much is too much? In *Advances in Industrial Ergonomics and Safety VI*, F. Aghazadeh, ed. London: Taylor & Francis, 1994. pp. 521–525.
14. **McCarthy, M., C. Thayer, and J. Wick:** Reducing risk of repetitive motion injuries in wafer processing. In *Advances in Industrial Ergonomics III*, W. Karwowski, and J. Yates, eds. London: Taylor & Francis, 1991. pp. 83–86.
15. **Mital, A.:** Maximum weights of lift acceptable to male and female industrial workers for regular 8-hour shifts. *Ergonomics 27*:1127–1138 (1984).
16. **Mital, A.:** Maximum weights of lift acceptable to male and female industrial workers for extended work shifts. *Ergonomics 27*:1115–1126 (1984).
17. **Mital, A., F. Hamid, and M. Brown:** Physical fatigue in high and very high frequency manual materials handling: perceived exertion and physiological factors. *Hum. Factors 36*:219–231 (1994).
18. **Bigland-Ritchie, B., F. Furbush, and J. Woods:** Fatigue of intermittent submaximal voluntary contractions: central and peripheral factors. *J. Appl. Physiol. 61*:421–429 (1986).
19. **Mishawa, T., K. Yoshina, and S. Shigeta:** An experimental study of the duration of a single spell of work on VDT performance. *Jpn. J. Ind. Health 26*:296–302 (1984).
20. **Yoshimura, I. and Y. Tomoda:** A study on fatigue grade estimation for VDT work—investigation of the rest pause. *Jpn. J. Ergonom. 31*:215–223 (1995).

21. **Koparderkar, P. and A. Mital:** The effect of different work-rest schedules on fatigue and performance of a simulated directory assistance operator's task. *Ergonomics 37*:1697–1707 (1994).
22. **Henning, R., G. Kissel, and D. Maynard:** Compensatory rest breaks for VDT operators. *Int. J. Ind. Ergonom. 14*:243–249 (1994).
23. **Henning, R., E. Callaghan, A. Ortega, G. Kissel, et al:** Continuous feedback to promote self-management of rest breaks during computer use. *Int. J. Ind. Ergonom. 18*:71–82 (1996).
24. **Bhatia, N. and K. Murrell:** An industrial experiment in organized rest pauses. *Hum. Factors 11*:167–174 (1969).
25. **Finkleman, J.:** A large database study of the factors associated with work-induced fatigue. *Hum. Factors 36*:232–243 (1994).
26. **Walsh, J., T. Humm, M. Muehlback, J. Sugerman, et al:** Sedative effects of alcohol at night. *J. Stud. Alcohol 52*:597–600 (1991).
27. **Knauth, P.:** The design of shift systems. *Ergonomics 36*:15–28 (1993).

A step in a foundry process usually involves grinding on the product to remove imperfections. This can involve the release of metal particles as well as sand particles which can be inhaled and cause internal damage. This process also results in high noise levels.

Outcome Competencies

After completing this chapter, the user should be able to:

1. Define underlined terms used in this chapter.
2. Understand the role ergonomics plays in overall worker health and safety.
3. Identify common work-related musculoskeletal disorders (WMSDs) and cumulative trauma disorders.
4. Identify risk factors associated with WMSDs.
5. Select controls that mitigate or eliminate work-related health and safety risks.
6. Apply anthropometric solutions to work environments to best fit the needs of workers.
7. Perform a work analysis using either the work-methods study technique or checklists.
8. Understand the basic elements of an ergonomic control program.

Key Terms

abduction • anthropometry • carpal tunnel • center of gravity • checklists • cumulative trauma disorders • epicondylitis • ergonomics • flexion • ligaments • methods study • physical work capacity • popliteal • pronation • supination • tendinitis • tenosynovitis • work • work-related musculoskeletal disorders

Prerequisite Knowledge

College level biology and chemistry.

Prior to beginning this chapter, the user should review the following chapters:

Chapter Number	Chapter Topic
1	History and Philosophy of Industrial Hygiene
3	Legal Aspects of the Occupational Environment
5	Epidemiological Surveillance
26	Biomechanics
27	Work Physiology
29	Psychology and Occupational Health
30	Worker Education and Training
37	Program Management
40	Hazard Communication
41	Risk Communication in the Workplace

Key Topics

I. Brief History of the Scientific Application of Ergonomics

II. Point of Connection Between Occupational Hygiene and Ergonomics

III. Work-Related Musculoskeletal Disorders (WMSDs)
- A. Why are WMSDs a Problem?
- B. Occupational Risk Factors Associated with the Upper Extremities
- C. Occupational Risk Factors Associated with the Lower Back

IV. Work Methods Evaluation
- A. Analyzing Jobs
- B. Risk Factor Assessment Techniques
- C. Psychophysical
- D. Anthropometry: Designing the Workplace to Fit the Worker

V. Putting It All Together: Case Study Using Ergonomic Principles and Applications
- A. Work Force and Physical Plant
- B. Methods
- C. Results
- D. Musculoskeletal Disorders
- E. Discussion

VI. Epidemiologic Evidence of the Work-Relatedness of Musculoskeletal Disorders

VII. Basic Elements of an Ergonomic Control Program

28

Ergonomics

Salvatore R. DiNardi, Ph.D., CIH

Introduction

Ergonomics is the science of fitting workplace conditions and job demands to the capabilities of the working population. Effective and successful "fits" assure high productivity, avoidance of illness and injury risks, and increased satisfaction among the work force. Although the scope of ergonomics is much broader (see below), the term in this chapter refers to assessing those work-related factors that may pose a risk of musculoskeletal disorders and recommendations to alleviate them.[1]

The American Industrial Hygiene Association Ergonomics Committee defines ergonomics as *"a multidisciplinary science that applies principles based on the physical and psychological capabilities of people to the design or modification of jobs, equipment, products, and workplaces. The goals of ergonomics are to decrease risk of injuries and illnesses (especially those related to the musculoskeletal system), to improve worker performance, to decrease worker discomfort and to improve the quality of work life. The benefits of well-designed jobs, equipment, products, work methods and workplaces include: enhanced safety and health performance; improved quality and productivity; reductions in errors; heightened employee morale; reduced compensation and operating costs; and accommodation of diverse populations, including those with disabilities. Although ergonomics is an evolving science, proper application of its principles can achieve benefits that are significant and immediate.*

Industrial hygienists are concerned with the anticipation, recognition, evaluation, and control of all hazards in the work environment, including hazards related to musculoskeletal disorders. Ergonomics principles should be utilized by industrial hygienists to ensure that the physical and psychological demands of jobs match workers' capabilities. Appropriately trained industrial hygienists should apply the science of ergonomics to ensure that the workplace is as free of recognized hazards as possible."[2]

The National Institute for Occupational Safety and Health (NIOSH) draws a distinction between ergonomics and musculoskeletal disorders. NIOSH defines ergonomics as "the science of fitting workplace conditions and job demands to the capabilities of the working population. Ergonomics is an approach or solution to deal with a number of problems — among them are work-related musculoskeletal disorders."

Musculoskeletal disorders include a group of conditions that involve the nerves, tendons, muscles, and supporting structures (such as intervertebral discs). They represent a wide range of disorders, which can differ in severity from mild periodic conditions to those that are severe, chronic, and debilitating. Some musculoskeletal disorders have specific diagnostic criteria and pathological mechanisms (hand-arm vibration syndrome). Others are defined primarily by the location of pain and have a more variable or less clearly defined pathophysiology (back disorders). Musculoskeletal disorders of the upper extremities include carpal tunnel syndrome, wrist tendinitis, epicondylitis and rotator cuff tendinitis. Both nonoccupational and occupational factors contribute to the development and exacerbation of these disorders.

This chapter outlines the approach most commonly recommended for identifying and correcting work-related musculoskeletal disorders and offers practical information for applying ergonomic applications in

Thanks to James D. McGlothlin, Ph.D., CPE, AIHA Ergonomics Committee Chair (1995–1996), for his assistance in writing this chapter.

workplaces. Additional information about the techniques, instruments, and methods mentioned in this chapter, along with references to other resources, can be found in the appendices. This chapter is limited to health professionals who need to apply ergonomics solutions to the workplace to ensure safe and healthful work conditions.

To accomplish the broad goal of making work safe and humane, to increasing human efficiency, to promoting human well-being, ergonomics can be applied by several scientific disciplines and principles, methods, and data as shown in Table 28.1.[3] Also, to accomplish this broad goal, the disciplines applied to solve ergonomic issues include psychologists, cognitive scientists, physiologists, biomechanicists, applied physical anthropologists, occupational hygienists, safety specialists, and occupational and systems engineers. The approach used by these scientists could range from the application of ergonomic principles in designing a simple tool to evaluating complex technological systems. This chapter focuses on ergonomics from a much simpler and narrower platform — anticipation, recognition, evaluation, and control of work-related musculoskeletal disorders (WMSDs) of the upper extremities and back in the workplace. The reader is encouraged to seek additional information on the myriad applications of ergonomics, including cognitive aspects, in the ergonomics journals and publications listed in the Appendices at the end of this chapter. The reader is also referred to other chapters in this book (physiology, biomechanics, psychology) that touch on the subject of ergonomics.

Brief History of the Scientific Application of Ergonomics

The term ergonomics is derived from the Greek roots *ergon*, meaning work, and *nomos*, meaning law. Literally translated ergonomics means "The Laws of Work." Ergonomics was first coined in 1950 by scientists in the United Kingdom to describe the interdisciplinary efforts to design equipment and work tasks to fit the person. These scientists were from the physical, biological, and psychological disciplines. Ergonomics is also known by the term "human factors engineering." While both "ergonomics" and "human factors" are the same in meaning, the difference is that the term human factors has an American origin, while ergonomics has a European origin. In 1957, American engineers, behavioral scientists, and anthropometrists, who performed similar work formed the Human Factors Society, and in 1992 renamed the organization the Human Factors and Ergonomics Society.

Even though the word "ergonomics" was coined in 1950, the observation that musculoskeletal disorders were an occupational hazard dates back to the 1700s when the physician Ramazzini quoted the following: *"Manifold is the harvest of diseases reaped by craftsman... As the...cause I assign certain violent and irregular motions and unnatural postures... by which... the natural structure of the living machine is so impaired that serious diseases gradually develop therefrom."*[4] Not much attention was paid to human musculoskeletal injury and illness because it was generally theorized that during the industrial revolution manual labor was plentiful and inexpensive. If someone experienced an occupational injury or illness, he or she was replaced. Indeed, it was not until the 1970s when the Occupational Safety and Health Act was created that, under the General Duty Clause, musculoskeletal injuries and illnesses were identified as problems to be prevented. This was further supported by one of the first successful cases filed by an employee against a company in a court of law. The court determined that *"An injury which develops gradually over time as a result of the performance of the injured worker's job-related duties is compensable."*[5] Following this historic case, industry had an economic incentive to consider ergonomics.

Table 28.1 — Origins and Applications of Ergonomics and Human Factors (Adapted from **Kroemer, K.H.E.**: Ergonomics (Chapter 13). In *Fundamentals of Industrial Hygiene*, 4th Ed. 1996. p. 347).

	Ergonomics	
Anatomy ——>	Anthropometry	Industrial engineering
Orthopedics —>	Biomechanics	Bio-engineering
Physiology ——>	Work physiology	Systems engineering
Medicine ——>	Industrial hygiene	Safety engineering
Physiology ——>	Management	Military engineering
Sociology ——>	Labor relations	Computer-aided design
	Human Factors	

Point of Connection Between Occupational Hygiene and Ergonomics

The scope of occupational hygiene includes a range of chemical, biological, and physical hazards. However, in recent years the application of ergonomics to solve work-related musculoskeletal disorders has become more prominent because of loss of productivity, workers' compensation costs, and human suffering. As a result, some of the health and safety responsibilities for controlling musculoskeletal injuries and illnesses have fallen on the shoulders of the occupational hygienist. However, not all occupational hygienists have adequate training in ergonomics. Part of the reason is that there are only a handful of universities in the United States that formally teach ergonomics as part of the occupational hygiene curriculum. For example, several graduate programs (based in environmental and occupational health in schools of public health) traditionally teach programs that focus on controlling airborne contaminants in the workplace, while other graduate programs (based in occupational engineering in schools of engineering) focus on ergonomic applications to control work-related physical stressors, such as repetition, force, posture, and lack of recovery time. Fortunately, the same basic skills taught in occupational engineering for ergonomic applications can be also transferred to the occupational hygiene curriculum to expand the scope of their responsibilities. The occupational hygienist can also apply these skills to simultaneously reduce chemical, biological, and physical hazards. For example, walk-through evaluations of the workplace by occupational hygienists enable them to see first-hand the hazards in the workplace, conduct informal interviews with workers, take measurements, and make recommendations through engineering and administrative controls. The expanded role of occupational hygienists allows them the same measurement techniques, except that these measurements can be for musculoskeletal as well as chemical hazard controls. In such instances, the occupational hygienist can evaluate and control hazardous exposures when the individual's work practices and/or job processes are the cause of higher or lower chemical exposures. The key is to apply the principles of job analysis where work can be broken into its fundamental elements and then to link those elements to work risk factors. Occupational hygienists, by virtue of their fundamental training in assessing workplace exposures, can apply these skills to connect the work being performed with other exposures, such as respiratory hazards from chemical exposure, and thereby applying more effective controls. Finally, the role of the occupational hygienist is expanding to include ergonomics as part of the job responsibility. By evaluating the workplace from the perspective of reducing job-related musculoskeletal injuries and illnesses, occupational hygienists can be more comprehensive in applying their skills to workplace controls.

Work-Related Musculoskeletal Disorders (WMSDs)

According to a recent NIOSH publication, the general term "musculoskeletal disorders" describes the following: 1) disorders of the muscles, nerves, tendons, ligaments, joints, cartilage, or spinal discs; 2) disorders that are not typically the result of any instantaneous or acute event, such as a slip, trip, or fall, but reflect a more gradual or chronic development; 3) disorders diagnosed by a medical history, physical examination, or other medical tests that can range in severity from mild and intermittent to debilitating and chronic; and 4) disorders with several distinct features, such as carpal tunnel syndrome, as well as disorders defined primarily by the location of the pain (i.e., low back pain).(1)

However, the specific term "work-related musculoskeletal disorders" refers to 1) musculoskeletal disorders to which the work environment and the performance of work contribute significantly; or 2) musculoskeletal disorders that are made worse or longer lasting by work conditions. These workplace risk factors, along with personal characteristics (e.g., physical limitations or existing health problems) and societal factors, are thought to contribute to the development of WMSDs.(6) They also reduce productivity or cause dissatisfaction. Common examples are jobs requiring repetitive, forceful, or prolonged exertions of the hands; frequent or heavy lifting, pushing, pulling, or carrying of heavy objects; and prolonged awkward postures. Vibration and cold may add risk to these work conditions. Jobs or working conditions presenting multiple risk factors will have a higher probability of causing a musculoskeletal problem. The level of risk depends on the intensity, frequency, and duration of the exposure to these conditions and the individual's capacity to meet the force or other job demands that might be involved. These conditions are more correctly called "ergonomic risk factors for musculoskeletal disorders" rather than "ergonomic hazards" or "ergonomic problems." But like the term "safety hazard," these terms have become popular.

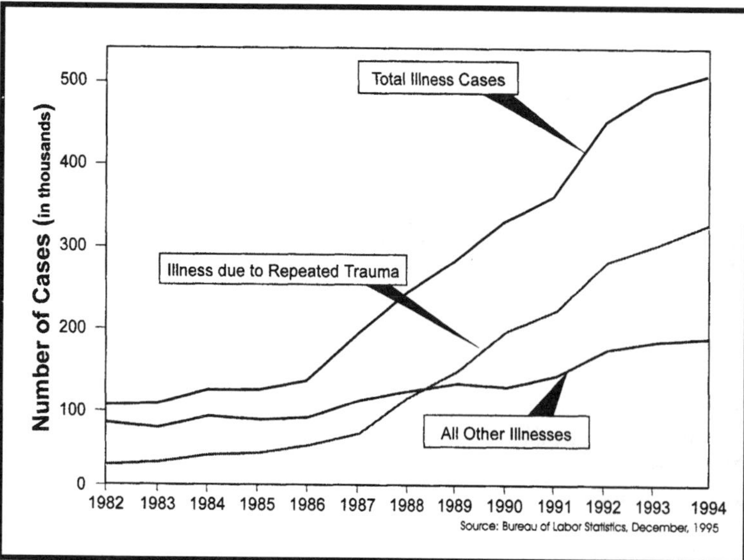

Figure 28.1 — Number of new occupational illnesses in private industry from 1982–1994.

Scientific research has provided important insights into the etiology and prevention of these disorders.

Musculoskeletal disorders of the neck and upper extremities resulting from work factors affect employees in every type of workplace and include such diverse workers as food processors, automobile and electronics assemblers, carpenters, office data entry workers, grocery store cashiers, and garment workers. The highest rates of these disorders occur in the industries with a substantial amount of repetitive, forceful work. Musculoskeletal disorders affect the soft tissues of the neck, shoulder, elbow, hand, wrist, and fingers. These include the nerves (e.g., carpal tunnel syndrome, tendons (e.g., tenosynovitis, peritendinitis, epicondylitis), and muscles (e.g., tension neck syndrome).(14)

In 1994, 332,000 musculoskeletal disorders from repeated trauma were reported in U.S. workplaces. This figure represents nearly 65 percent of all illness cases reported to the U.S. Bureau of Labor Statistics — an increase of nearly 10% compared with 1993 figures and more than 15 percent relative to 1992 figures. (Figure 28.1 shows the number of new occupational illnesses for private industry, 1982–1994.) The rapid rise in illnesses from repeated trauma began in 1987.(14)

The most frequently reported upper-extremity musculoskeletal disorders affect the hand/wrist region. In 1993, carpal tunnel syndrome, the most widely recognized condition, occurred at a rate of 5.2 per 10,000 full-time workers. This syndrome required the longest recuperation period of all conditions resulting in lost workdays, and a median of 30 days away from work. Table 28.2 lists the common cumulative trauma disorders of the upper extremities.

WMSDs of the Low Back. Low back musculoskeletal disorders are common and costly. Although the causes of low back disorders are complex, substantial scientific evidence identifies some work activities and awkward postures that significantly contribute to the problem. In the United States, back disorders account for 27% of all nonfatal occupational injuries and illnesses involving days away from work.(14) Prevention activities should be undertaken based on current knowledge, but important new research efforts are needed to assure that work-related low back disorders are successfully prevented and treated. For some occupations and tasks, there is a pressing need for more information about safe levels of exposure and for further validation of promising

Why are WMSDs a Problem?

Many reasons exist for considering WMSDs a problem, including the following: 1) WMSDs are among the most prevalent lost-time injuries and illnesses in almost every industry;(7–9) 2) WMSDs, specifically those involving the back, are among the most costly occupational problems;(8–12) 3) Job activities that may cause WMSDs span diverse workplaces and job operations (see Table 28.2)(3); 3) WMSDs may cause a great deal of pain and suffering among afflicted workers; and 4) WMSDs may decrease productivity and the quality of products and services. Workers experiencing aches and pains on the job may not be able to do quality work.

Because musculoskeletal disorders can be associated with nonwork activities (e.g., sports) and medical conditions (e.g., renal disease, rheumatoid arthritis), it is difficult to determine the proportion due solely to occupation. For example, in the general population, nonoccupational causes of low back pain are probably more common than workplace causes.(13) However, even in these cases, the musculoskeletal disorders may be aggravated by workplace factors.

WMSDs of the Upper Extremities. Musculoskeletal disorders of the upper extremities (such as carpal tunnel syndrome and rotator cuff tendinitis), resulting from work factors, are common and occur in nearly all sectors of our economy. More than $2 billion in workers' compensation costs are spent annually on these work-related problems.(14) Workers' compensation costs undoubtedly underestimate the actual magnitude of these disorders.

Table 28.2 — List of Common Cumulative Trauma Disorders of the Upper Extremities
(Adapted from **Kroemer, K.H.E.:** Ergonomics, (Chapter 13). *Fundamentals of Industrial Hygiene*, 4th Ed. 1996. p. 396-397).

Disorder	*Description*	*Typical Job Activities*
Carpal tunnel syndrome (writer's cramp, neuritis, median neuritis)(N)	The result of compression of the median nerve in the carpal tunnel of the wrist. This tunnel is an opening under the carpal ligament on the palmar side of the carpal bones. Through this tunnel pass the median nerve, the finger flexor tendons, and blood vessels. Swelling of the tendon sheaths reduces the size of the opening of the tunnel and pinches the median nerve and possibly blood vessels. The tunnel opening is also reduced if the wrist is flexed or extended or ulnarly or radially pivoted.	Buffing, grinding, polishing, sanding, assembly work, typing, keying, cashiering, playing musical instruments, surgery, packing, housekeeping, cooking, butchering, hand washing, scrubbing, hammering.
Cubital tunnel syndrome (N)	Compression of the ulnar nerve below the notch of the elbow. Tingling, numbness, or pain radiating into ring or little fingers.	Resting forearm near elbow on a hard surface or sharp edge or reaching over obstruction.
deQuervain's syndrome (or disease) (T)	A special case of tendo synovitis that occurs in the abductor and extensor tendons of the thumb where they share a common sheath. This condition often results from combined forceful gripping and hand twisting, as in wringing cloths.	Buffing, grinding, polishing, sanding, pushing, pressing, sawing, cutting, surgery, butchering, use of pliers, "turning" control such as on motorcycle, inserting screws in holes, forceful hand wringing.
Epicondylitis ("tennis elbow") (T)	Tendons attaching to the epicondyle (the lateral protrusion at the distal end of the humerus bone) become irritated. This condition is often the result of impacting or jerky throwing motions, repeated supination and pronation of the forearm, and forceful wrist extension movements. The condition is well-known among tennis players, pitchers, bowlers, and people hammering. A similar irritation of the tendon attachments on the inside of the elbow is called medical epicondylitis, also known as "golfer's elbow."	Turning screws, small parts assembly, hammering, meat cutting, playing musical instruments, playing tennis, pitching, bowling.
Ganglion (T)	A tendon sheath swelling that is filled with synovial fluid, or cystic tumor at the tendon sheath or a joint membrane. The affected area swells up and causes a bump under the skin, often on the dorsal or radial side of the wrist. (Because it was in the past occasionally smashed by striking with a bible or heavy book, it was also called a "bible bump.")	Buffing, grinding, polishing, sanding, pushing, pressing, sawing, cutting, surgery, butchering, use of pliers, "turning" control such as on motorcycle, inserting screws in holes, forceful hand wringing.
Neck tension syndrome (M)	An irritation of the levator scapulae and trapezius group of muscles of the neck, commonly occurring after repeated or sustained overhead work.	Belt conveyor assembly, typing, keying, small parts assembly, packing, load carrying in hand or on shoulder.
Pronator (teres) syndrome (N)	Result of compression of the median nerve in the distal third of the forearm, often where it passes through the two heads of the pronator teres muscle in the forearm; common with strenuous flexion of elbow and wrist.	Soldering, buffing, grinding, polishing, sanding.
Shoulder tendinitis (rotator cuff syndrome or tendinitis, supraspinatus tendinitis, subacromial bursitis, subdeltoid bursitis, partial tear of the rotator cuff) (T)	This is a shoulder disorder located at the rotator cuff. The cuff consists of four tendons that fuse over the shoulder joint, where they pronate and supinate the arm and help to abduct it. The rotator cuff tendons must pass through a small bony passage between the humerus and the acromion, with a bursa as cushion. Irritation and swelling of the tendon or of the bursa are often caused by continuous muscle and tendon effort to keep the arm elevated.	Punch press operations, overhead assembly, overhead welding, overhead painting, overhead auto repair, belt conveyor assembly work, packing, storing, construction work, postal letter carrying, reaching, lifting, carrying load on shoulder.

Table 28.2 — List of Common Cumulative Trauma Disorders of the Upper Extremities (continued)

Disorder	*Description*	*Typical Job Activities*
Tendinitis (tendinitis) (T)	An inflammation of a tendon. Often associated with repeated tension, motion, bending, being in contact with a hard surface, vibration. The tendon becomes thickened, bumpy, and irregular in its surface. Tendon fibers may be frayed or torn apart. In tendons without sheaths, such as within elbow and shoulder, the injured area may calcify.	Punch press operation, assembly work, wiring, packaging, core making, use of pliers.
Tendo synovitis (tenosynovitis, tendovaginitis) (T)	This disorder occurs to tendons inside synovial sheaths. The sheath swells. Consequently, movement of the tendon with the sheath is impeded and painful. The tendon surfaces can become irritated, rough, and bumpy. If the inflamed sheath presses progressively onto the tendon, the condition is called stenosing tendo synovitis. DeQuervain's syndrome is a special case occurring in the thumb, while the trigger finger condition occurs in flexors of the fingers.	Buffing, grinding, polishing, sanding, punch press operation, sawing, cutting, surgery, butchering, use of plier, "turning' control such as on a motor cycle, inserting screws in holes, forceful hand wringing.
Thoracic outlet syndrome (neurovascular compression syndrome, cervicobrachial disorder, brachial plexus neuritis, costoclavicular syndrome, hyperabduction syndrome) (V,N)	A disorder resulting from compression of nerves and blood vessels between clavicle and first and second ribs at the brachial plexus. If this neurovascular bundle is compressed by the pectoralis minor muscle, blood flow to and from the arm is reduced. This ischemic condition makes the arm numb and limits muscular activities.	Buffing, grinding, polishing, sanding, overhead assembly, overhead welding, overhead painting, overhead auto repair, typing, keying, cashiering, wiring, playing musical instruments, surgery, truck driving stacking, material handling, postal letter carrying, carrying heavy loads with extended arms.
Trigger finger (or thumb) (T)	A special case of tendo synovitis where the tendon becomes nearly locked so that its forced movement is not smooth but snaps or jerks. This is a special case of stenosing tendo synovitis crepitans, a condition usually found with digit flexors at the A1 ligament.	Operating finger trigger, using hand tools that have sharp edges pressing into the tissue or whose handles are too far apart for the user's hand so that the end segments of the fingers are flexed while the middle segments are straight.
Ulnar artery aneurysm (V,N)	Weakening of a section of the wall of the ulnar artery as it passes through the Guyon tunnel in the wrist; often from pounding or pushing with heel of the hand. The resulting "bubble" presses of the ulnar nerve in the Guyon tunnel.	Assembly work.
Ulnar nerve entrapment (Guyon tunnel syndrome) (N)	Results from the entrapment of the ulnar nerve as it passes through the Guyon tunnel in the wrist. It can occur from prolonged flexion and extension of the wrist and repeated pressure on the hypothenar eminence of the palm.	Playing musical instruments, carpentering, brick laying, use of pliers, soldering, hammering.
White finger ("dead finger," Raynaud's syndrome, vibration syndrome) (V)	Stems from insufficient blood supply bringing about noticeable blanching. Finger turns cold, numb, tingles, and sensation and control of finger movement may be lost. The condition is due to closure of the digit's arteries caused by vasospasm triggered by vibrations. A common cause is continued forceful gripping of vibrating tools, particularly in a cold environment.	Chain sawing, jack hammering, use of vibrating tool that is too small for the hand, often in a cold environment.

N=nerve disorder; T= tendon disorder; M=muscle disorder; V= vessel disorder.

intervention approaches, such as mechanical lifting devices for nursing aids.

The economic costs of low back disorders are staggering. In a recent study, the average costs of worker's compensation claim for low back disorder was $8300, which was more than twice the average cost of $4075 for all compensable claims combined. Estimates of the total costs of low back pain to society in 1990 were between $50 billion and $100 billion per year, with a significant share (about $11 billion) borne by the workers' compensation system. Moreover, as many as 30 percent of American workers are employed in jobs that routinely require them to perform activities that may increase their risk of developing low back disorders.(14)

Despite the overwhelming statistics on the magnitude of the problem, more complete information is needed to assess how changes implemented to reduce the physical demands of jobs will affect workplace safety and productivity in the future. A tremendous opportunity exists for prevention efforts to reduce the prevalence and costs of low back disorders, since a significant number of occupationally related low back disorders are associated with certain high-risk activities. For example, female nursing aids and licensed practical nurses were about 2fi times more likely to experience a work-related low back disorder than all other female workers. Male construction laborers, carpenters, and truck and tractor operators were nearly two times more likely to experience a low back disorder than all other male workers.

Occupational Risk Factors Associated with the Upper Extremities

Several work-related factors have been identified as occupational risk factors that cause musculoskeletal injuries and disorders in the upper extremities. These risk factors have been identified by several researchers, as well as what actions should be taken.[15–21] The work risk problems and solutions are briefly summarized below.

Repetition

Controls: Work enrichment; increase rest allowances; worker rotation. Repetition, a series of motions having little variation and performed every few seconds, may produce fatigue and muscle-tendon strain. If adequate recovery time is not allowed for these effects to diminish, or if the motions also involve awkward postures or forceful exertions, the risk of actual tissue damage and other musculoskeletal problems will probably increase. A task cycle time of less than 30 seconds has been considered as "repetitive."

Estimates vary as to repetition rates that may pose a hazard, because other factors, such as force and posture, also affect these determinations. One proposal for defining high risk repetition rates for different body parts is shown in Table 28.3.[22]

Work repetition and its relationship to carpal tunnel syndrome has been shown in studies of workers in a frozen food processing plant,[23] in automotive plants,[24] in a pork processing plant,[25] and among supermarket workers.[26] It should be noted that in the context of evaluating jobs for work risk factors one of the first aspects of the job to be analyzed is to characterize it by the content of the job and how long it takes to do the job. As a result, cycle time, which can also be a measure of job repetition, is one of the most commonly cited job risk factors in the literature. Other risk factors such as posture and force usually follow repetition by the nature of how jobs are analyzed and by the degree of difficulty in analysis (i.e., it is harder and less precise in measuring posture and force from video analysis than repetition). Work repetition alone can be a significant risk factor, but others risk factors such as posture and force—which are not as easily measured—can be just as significant, but are not mentioned as often due to lack of quantitative measures by the investigators.

Repetition can be reduced by increasing the variety of tasks performed and by a corresponding increase in work time or worker rotation. Additional tasks must be compatible with the original workstation and should not involve the same types of work stresses as the original job.[27] The feasibility of worker rotation obviously depends on the level of skill required to perform a given job and on the existence and detail of labor contracts.

Force

Controls: Reduce weight of tools and part; handle smaller quantities; control balance of objects; use quality control program to maintain fit tolerances; use or develop mechanical aids. Changes in weight, size, shape, and balance of hand-held objects can reduce the force of the exertion required to perform a task. In its simplest application, workers can pick up fewer objects at a time or lift with two hands instead of one, to reduce weight-induced stress. They need less strength to grasp an object with a power grip, fingers wrapped around the object(s), than with a pinch grip, finger tips supporting the load.[28–29] Grasping an object at its center of gravity is easier than elsewhere because the weight of the object does not tend to twist it out of the worker's hands. If the worker cannot change

Table 28.3 — High Risk Repetition Rates by Different Body Parts

Body Part	*Repetitions Per Minute*
Shoulder	More than 2fi
Upper Arm/Elbow	More than 10
Forearm/Wrist	More than 10
Finger	More than 200

(From **Kilbom, Å:** Repetitive Work of the Upper Extremity. Part II: The Scientific Basis for the Guide. *Int. J. Ind. Erg. 14*:59–86 (1994).)

hand location easily, reducing, shifting, or adding weight in order of preference will shift the center of gravity. For example, attachment of an airline to a tool at a right angle can reduce the torque caused by the airline. Another factor adding to the force is the torque required to hold the tool. External torque control devices and different kinds of connectors are effective in controlling torque; many times these same devices can serve an additional purpose of retracting and holding the tool when not in use, thereby reducing the time for getting the tool.

Although there have been many studies investigating the association between force and musculoskeletal injuries and illnesses, only a few have quantitative assessments.[17,19,21] Based on these studies it was concluded that grip force should not exceed 40%–50% of the maximum grip strength. Hand tools should not weigh more than five pounds; preferably, the weight should be just under four pounds. Pinch grips should be avoided when applying force because they require up to five times more force than power grips. To reduce force the following are recommended: the tool handle should be 50–60 mm thick for power grip, and 8–13 mm thick for precision grip; the tool length should be 120 mm long for power grip, and 100 mm for precision grip (125 mm when wearing gloves).[30]

Posture

Controls: Work location; work orientation; tool design. Work posture is a function of the location and orientation of the work and the design of tools used in performing the work. It affects the ability of workers to reach, hold, and use equipment and influences how long they can perform their jobs without adverse health effects.[31] Posture can be controlled through the location of work and design of the tool.[32, 29] The controls range from the obvious to the obscure: it is easier to reach a case stacked at heart height. Without stock and storage problems, the solution is apparent and simple. As the movements become more precise, the approach becomes more sophisticated. If a jig can be adjusted to minimize ulnar deviation, it may not be necessary to investigate alternative tool design. If both options exist, management has a choice; it can shop. Both workers and management benefit.

Extreme postures can lead to discomfort and joint stresses, reduced blood flow, high muscle forces, fatigue, reduced endurance time, acute shoulder and neck pain, shoulder tendinitis, and carpal tunnel syndrome. Figure 28.2 shows the standard anatomical position with classical terminology for major body movements.[5] Figure 28.3 shows the various body positions and corresponding terminology.[33]

Figure 28.4 shows the hand and wrist postures and corresponding terminology.[17] Figure 28.5 shows the various types of hand grip postures and corresponding terminology.[17] Figure 28.6 shows the postures commonly associated with cumulative trauma disorders.[34] Figure 28.7 shows grip strength as a function of degree of wrist deviation as measured in the neutral position.[17] As these figures show, to reduce awkward postures, the hands should not work above mid-chest height; the shoulders should not be elevated; shoulder abduction should not be more than 30°; shoulder flexion should be minimized; overhead reaches should be minimized; elbow/forearm pronation and supination should be minimized; and wrists should not deviate more than 20° so that wrist flexion/extension are minimized.[30]

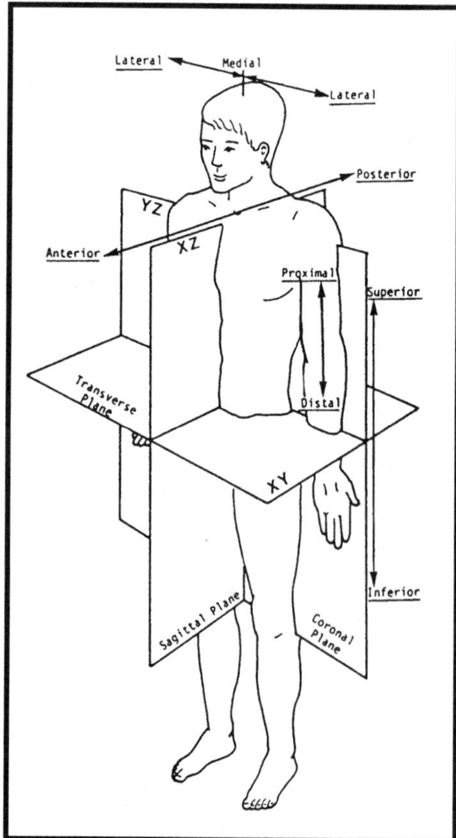

Figure 28.2 — Terminology used to define position and location on the body. (Reprinted with permission from **Annis and McConville:** Anthropometry (Chapter 1). In *Occupational Ergonomics: Theory and Applications* (edited by Bhattacharya and McGlothlin). 1996. pg. 3.)

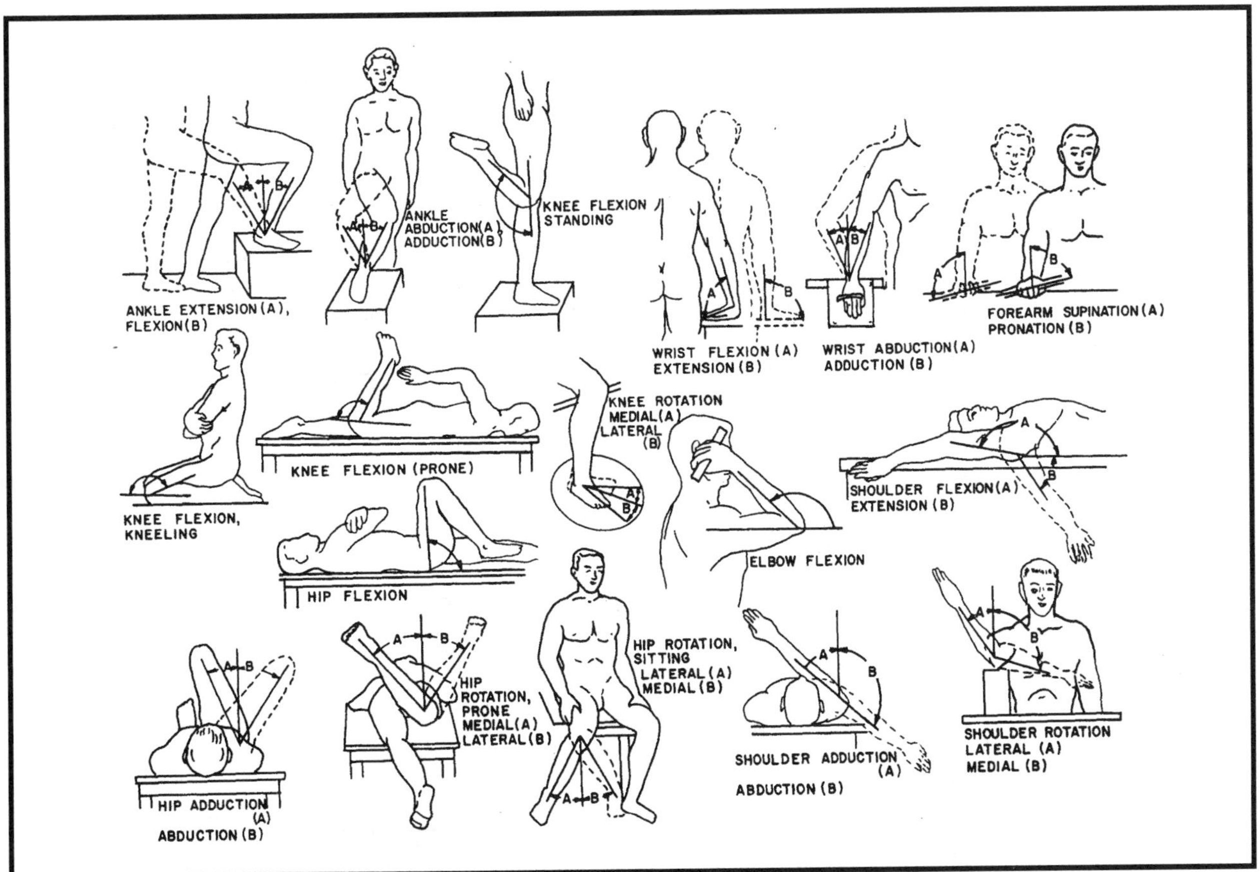

Figure 28.3 — Various body postures and corresponding terminology. (**Chaffin and Anderson:** *Occupational Biomechanics, Joint Motion — Methods and Data.* New York: John Wiley & Sons, Inc., 1984. p. 86. Reprinted with permission.)

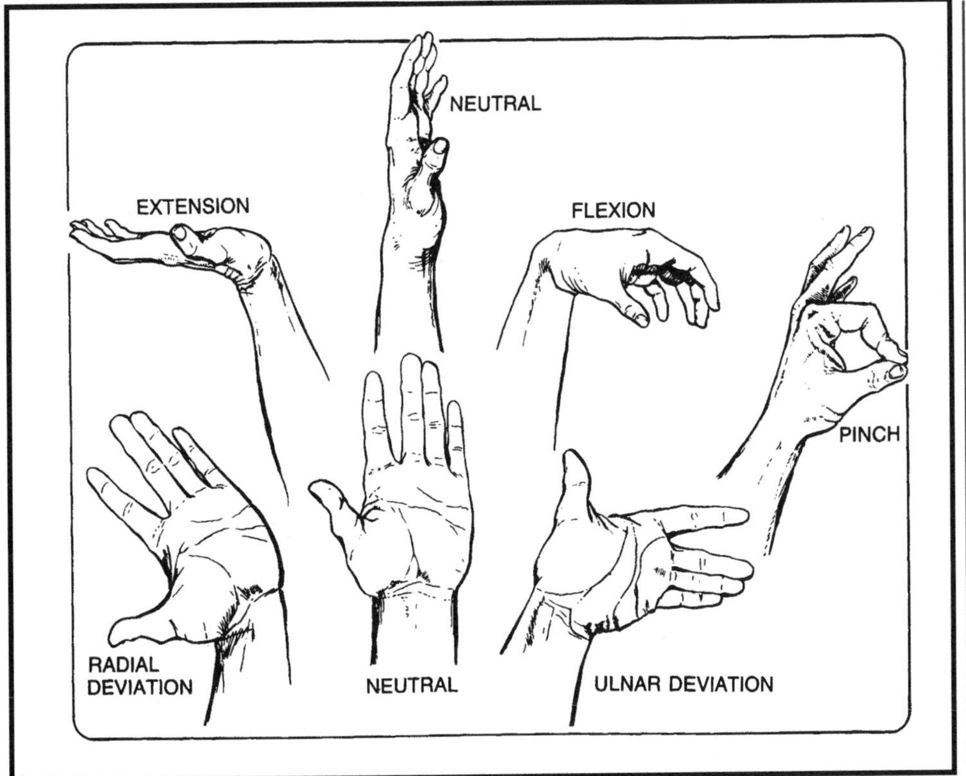

Figure 28.4 — Hand and Wrist postures and corresponding terminology (Reprinted with permission from "Analyzing Jobs." In **Putz-Anderson, V. (ed.):** *Cumulative Trauma Disorders: A Manual for Musculoskeletal Diseases of the Upper Limbs.* Philadelphia: Taylor & Francis, 1988. p. 54.)

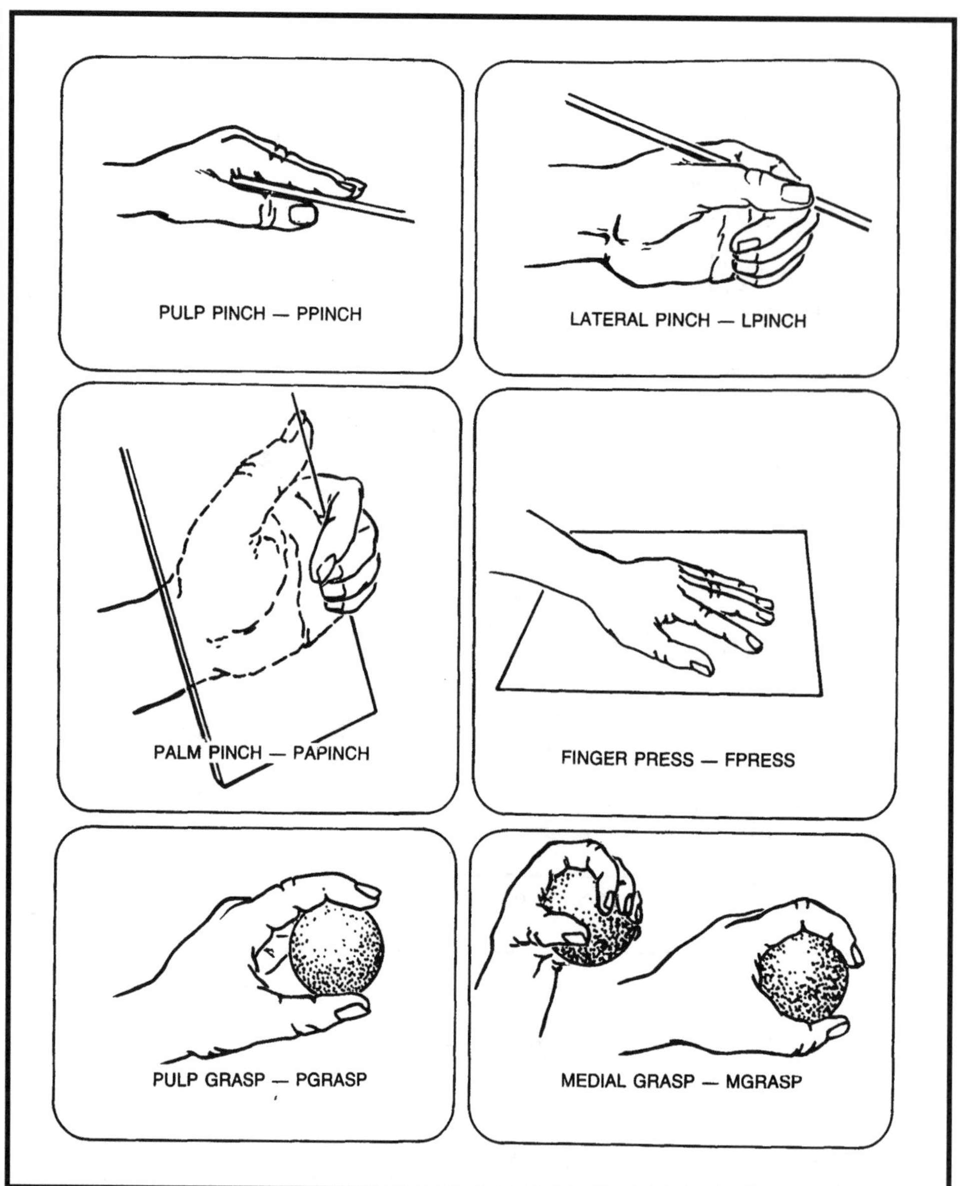

Figure 28.5 — Various types of hand grip postures and corresponding terminology. (Reprinted with permission from "Analyzing Jobs." In **Putz-Anderson, V. (ed.):** *Cumulative Trauma Disorders: A Manual for Musculoskeletal Diseases of the Upper Limbs.* Philadelphia: Taylor & Francis, 1988. p. 56.)

Static Loads

Controls: Change posture; bring load closer to body; use mechanical jigs. When the body is in the same posture for an extended period of time, the metabolic energy requirements are high, but blood circulation is low. As such, the muscles do not receive the needed oxygen, and the muscle easily becomes fatigued.[35] To avoid this, design the job so that the body does not remain in a fixed position for prolonged periods of time. Have a place to set the hand tool after performing a job instead of holding it to work on another piece; fashion a mechanical jig to hold parts rather than using the hand as a "bioclamp" while the other holds a tool.

Mechanical Stress

Controls: Handle size; handle shape; handle materials. One of the more common mechanical stresses is hand contact with hard, sharp edges of an object. The average male can exert 100–130 pounds of hand grip force; the average female can exert 60 to 70 pounds of grip force.[36] Repeated exertion of these forces results in reaction stresses transmitted through the hand to the underlying tendons. For example, hand-held industrial scissors to cut fabric for clothing can cause stenosing tenosynovitis crepetans. Commonly known as "trigger finger," stenosing tenosynovitis crepetans is often associated with tools that have hard or sharp edges on their handles.[37,38]

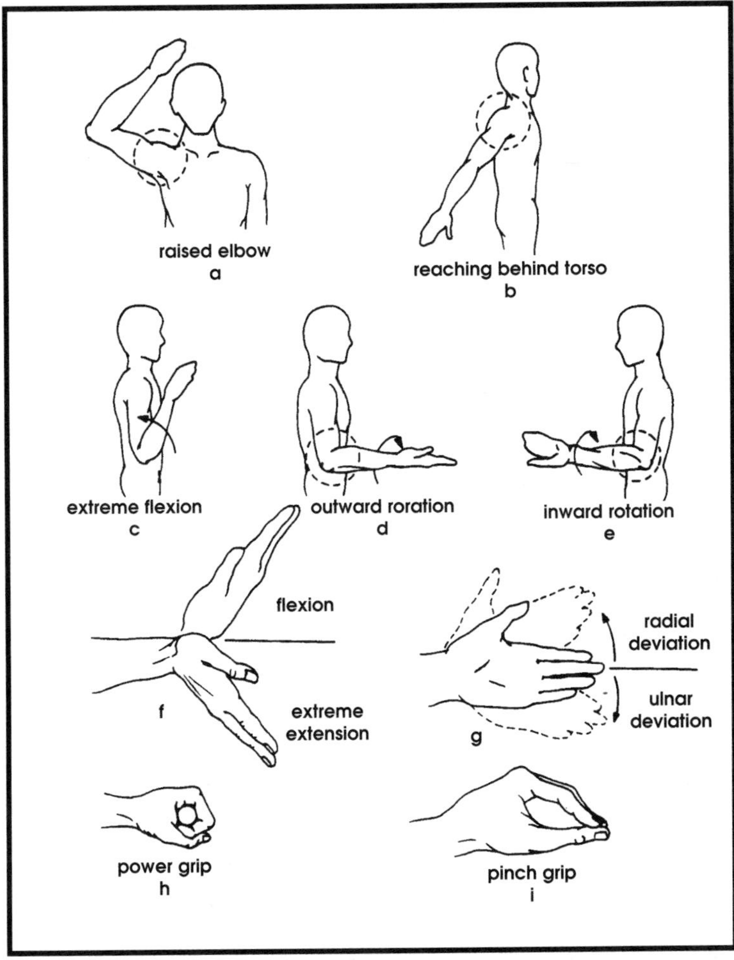

Figure 28.6 — Postures commonly associated with cumulative trauma disorders. (Reprinted with permission from **Armstrong, T.J.:** Upper-Extremity Posture: Definition, Measurement, and Control (Chapter 6). In *The Ergonomics of Working Postures* (edited by N. Corlett, J. Wilson, and I. Manenica) (Proceedings of the 1st International Ergonomics Symposium, Zadar, Yugoslavia, April 1985). Philadelphia: Taylor & Francis, 1985. p. 61.)

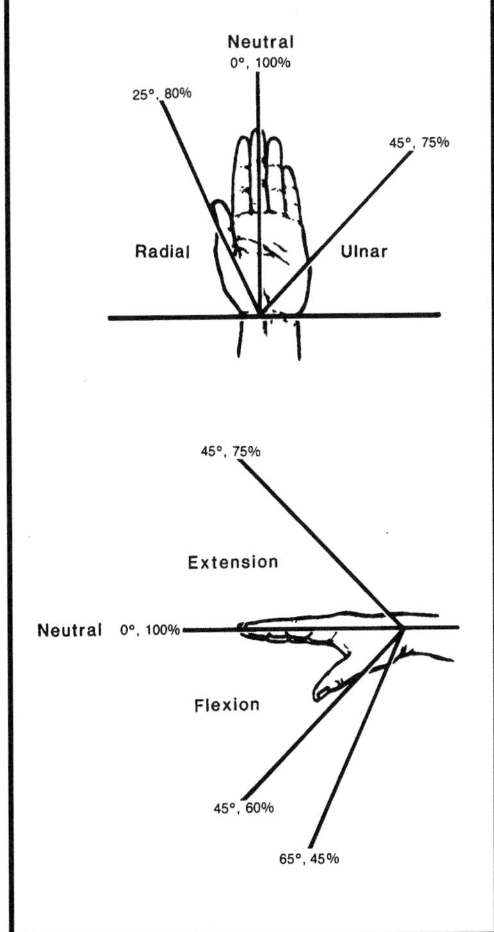

Figure 28.7 — Grip strength as a function of the degree of wrist deviation expressed as a percentage of power grip as measured in the neutral position. (Reprinted with permission from "Analyzing Jobs." In **Putz-Anderson, V. (ed.):** *Cumulative Trauma Disorders: A Manual for Musculoskeletal Diseases of the Upper Limbs.* Philadelphia: Taylor & Francis, 1988. p. 57.)

To decrease the potential for mechanical stress related disorders, handles should have well-rounded corners, and their diameters should be at least 1.5 inches. Pliant rubber or plastic should cover metal handles. Ideally, the small joints of the hand should be near midflexion when applying power to tools to provide a combination of tool retention and partially stretched muscles. This is especially useful when using such tools as wirecutters and pliers.

Also, local contact stresses that are pressure concentration points, resulting from manipulating external objects (pressure caused by uneven surface projections), can cause injury to nerves, blood vessels, and skin. To avoid or reduce contact stresses it is suggested that the contact area be enlarged; the pressure be reduced; and the pressure on the elbows, wrists, and hands be reduced by not leaning on them.

Low Temperatures

Controls: Personal protective equipment; material temperature; environmental temperature; air exhausts. Workers' hands often encounter temperatures below 65°F. Typical cold environment exposures occur in meat packing plants, near air exhaust systems of power tools, or in the handling of cold objects (e.g., metal hand tools). Obviously, in unheated work or storage environments the problem can be more severe in the winter. Chronic exposure to low temperature environments or low temperature objects can contribute to numbness, decreased blood flow, and diminished sensory feedback.[39] As a result, workers may unconsciously exert grip forces that are much greater than needed, causing unnecessary strain on tendons and possible injury. Also, a worker's hands when in contact with tools, materials, and environmental air should not be colder than 20°C for a prolonged period of time.[40–43] As before, safeguards are conceptually simple: keep the body warm and use comfortable gloves; minimize localized cold air supplies with attachments that direct the air away from the worker; provide rubber- or plastic-handled tools to add insulation to the stress relief benefit discussed earlier. Also, working in cold climates can either cause or exacerbate work-related cumulative trauma disorders. Cold temperatures reduce blood flow to the upper extremities, especially the hands, which have an abundance of motor and sensory nerves. The cold weather may reduce the sensory feedback from the hands and cause workers to have greater exertion of force to perform their jobs. Wearing gloves may also alter sensory feedback and cause the worker to grip the tool more than needed.

Vibration

Controls: Change process; change tool; change operating parameters. Workers exposed to vibration of the upper limbs often complain of numbness of fingers, wrist pain, sensitivity to cold, and circulatory disturbances in the fingers.[44] There are many factors to consider in examining vibration exposure, including frequency and magnitude of vibration, duration of exposure, temporal exposure pattern and work method, posture of hand, and type of tool. Vibration induced vascular diseases can result from frequencies between 20–1000 Hz, the range of most power hand tools.[45,27] Minimizing vibrations decreases the potential for vibration related disorders. When the process cannot be changed, change the tool. There are many hand tools available today with effective vibration dampening devices. Routine maintenance, in addition to increasing lubrication of a power tool, retards the development of wear-induced vibration. In some cases, rubber and/or foam sleeves placed between the tool and hand will minimize vibration.

Epidemiological studies have shown that workers who are exposed to hand-arm vibration have a greater risk of injury and musculoskeletal disorders than those who are not. Vibration causes over-gripping of the tool and leads to higher forearm muscle activation and higher muscle loads. When this is combined with repetitive work, the vibration may exacerbate musculoskeletal disorders. Generally, it is recommended that segmental vibration entering the hands be avoided below 1000 Hz, specifically hand-arm vibrations in the 2–200 Hz range.[46]

Gloves

Controls: Process change; variable size gloves; glove replacement. Gloves protect the hands from environmental temperature extremes, chemical contaminants, and mechanical insult. They also enhance friction and in some tasks reduce strength requirements.[47] When a process cannot be changed, gloves are one of the most logical and cost-effective control measure to protect the hands. In manual-intensive industries the primary benefit from using gloves is to diminish mechanical insult from repetitive job tasks. A drawback is that gloves can interfere with movements of the hand and may attenuate strength by up to 30% depending on glove material and fit.[48-50] Also, task completion time may be delayed by up to 37% with certain types of gloves, which may impede performance in emergency situations such as hazardous chemical releases.[51] Getting the proper glove fit is difficult because of the variations in hand size; management's tendency to purchase only a few sizes creates problems for the workers. If gloves are needed, the best solution is to ensure that a variety of sizes is available, and that access to new, well-fitting gloves is easy and encouraged.

Fit and Reach

Controls: Adjustable workstation; assortment of hand tool sizes; bring work closer to body. The ability of individuals to fit in the workplace, to reach objects and hand tools, and to see without obstruction can force them to adopt awkward postures and have static loads that can cause musculoskeletal injuries or aggravate existing disorders. The solution is that work should accommodate the work population, from 5% females to 95% males.[52] The work, objects, and tools should be located so that the worker does not have to lean forward and flex the back, neck, shoulder, or extend arms and hands beyond their "functional" reach (approximately 12 inches from the front of the torso).

Work Organization

Controls: Balance work between workers; adjust line speed so workers do not "chase" parts; allow adequate recovery time between tasks. Machine pacing can cause problems because workers cannot control work frequency. Also, the type of job workers are doing may not allow them to recover from the previous task. Self pacing versus machine pacing is preferable because it allows workers to control the frequency of the work being performed and allows them to recover from previous tasks.[18]

Multiple Risk Factor Exposures

Controls: Systems analysis of work; mechanize; worker and management input for work changes. Though each of the risk factors above can lead to work-related cumulative trauma disorder, multiple risk factors may be more problematic not only in terms of increasing risk but also in terms of deriving solutions. For example, repetitive jobs may also have some forceful motions and awkward postures associated with them. If the job is machine paced as well, there is the additional risk factor of minimal recovery time. The combination of repetition, force, awkward postures, and insufficient recovery time may lead to an increased incidence and severity of work-related cumulative trauma disorders above that resulting from an individual exposure to one of these risk factors. Therefore, the most effective solution is to reduce risk factors by carefully examining what is being done, and then determine which solution will most benefit the worker and company.[53]

Personal Protective Equipment

Wrist splints. Wrist splints or braces used to keep the wrist straight during work are not recommended, unless prescribed by a physician for rehabilitation. Keeping the wrist straight during work helps to optimize the biomechanical leverage of the individual to perform work. However, using a splint to achieve the same end may cause more harm than good since the work orientation may require workers to bend their wrists. If workers are wearing wrist splints, they may have to use more force to work against the brace. This is not only inefficient, it may actually increase the pressure in the carpal tunnel area, causing more damage to the hand and wrist.[54] The proper remedy for this situation is to orient the workpiece or adjust the workstation so that the worker can perform in a neutral posture with optimum biomechanical leverage.

Occupational Risk Factors Associated with the Lower Back

Low back pain and injuries attributed to manual lifting activities continue as one of the leading occupational health and safety issues facing the workplace today. As such, several work risk factors have been identified as leading to overexertion and musculoskeletal injuries during manual material handling. Some of these risk factors are outlined below.

Posture

Controls: Avoid extreme range of motion when lifting; Keep loads to be lifted between knee and heart height; avoid twisting when lifting loads; keep loads close to body when lifting. Body posture changes force requirements and may cause work to become very strenuous. Often the activity forces the body to assume different postures. For example, stoop postures can be advantageous when the load is lifted repeatedly. The squat posture is desirable when the load can fit between the knees and must be lifted occasionally. Loads that cannot fit between the knees and must be lifted repetitively should be handled by two individuals or must be moved with the help of a mechanical assist device.

Generally, to reduce back injury risk during manual material handling, one should avoid an extreme range of movements, moving loads from the floor, turning and twisting, "jerking motions," fixed postures, lifting loads above heart height, and pushing or pulling heavy loads. Manual material handling should be done between knee and heart height. Use caution during material handling when the size of the load is too large to fit between the knees and a stooped lift is needed. The key to remember, especially when lifting, is "posture" and "load"— the greater the load, the more good posture plays a role in back injury prevention. Pushing force should be exerted in an erect posture, and the load should have handles. For heavy and awkward loads, the load should be lifted by using the squat posture, the weight of the load should be less than the sum of the capacities of the individuals, and the workers engaged in lifting (i.e., buddy lift) should be similar in height. However, when the width of the load is too wide to fit between the knees, a squat lift may be inappropriate. In this case, the focus of the lift should be on getting the load as close to the body as possible, thereby reducing the likelihood for back injury. Specific recommendations for load stability include the following: The load should be rigid and symmetrical in shape; the weight should be distributed uniformly; if the load is not uniform, the heavier end should be closer to the body; the center of gravity for the load should be along the line joining the two hands; the load dimension in the sagittal plane should not exceed 50 cm; the load dimension between the hands should be minimized. The load height should be determined by practical considerations, such as body size and ability to clearly view obstructions in the path, and the maximum load should not exceed 51 pounds. The limit is based on the revised NIOSH lifting guidelines.[55] More details about the NIOSH guidelines are presented in the chapter on biomechanics. However, additional aspects for manual material handling are outlined below.

Frequency/Repetitive Handling

Controls: Mechanize task; use "buddy" system; reduce line speed. Materials handling activities that require frequent handling either should be redesigned to reduce the frequency or should depend on mechanical equipment to aid material handling. The revised NIOSH guidelines do not recommend load handling frequencies of more than 10 minutes if the work is done for eight hours. The handling frequency can increase to 12 minutes if the working duration is reduced to two hours. Therefore, if higher frequencies are encountered, the work load should be reduced.[30]

Static Work

Controls: Redesign work station/work area to allow freedom of movement; have adjustable work station to allow changes in posture; remove physical barriers so worker can stand closer to work. Nearly all activities involving materials handling contain both a static and dynamic component. Tasks such as repetitive lifting have a dominating dynamic component; tasks such as load holding have a dominating static component. The static work effort is characterized by contraction of muscles over extended periods of time (e.g., adopting a posture for extended periods of time).[53]

Static work should be avoided as much as possible. As with the upper extremities, static work has high metabolic demand; however, the blood supply is reduced because of restricted circulation. Consequently, little oxygen gets to the starved muscles and fatigue quickly sets in. Therefore, it is better to design jobs so that they allow the workers to freely move their bodies during manual material handling tasks.

Handles/Couplings

Controls: All manual material handling containers should have handles; Material handled should be slip resistant; handles should be designed for "power grip" handling. Good handles or couplings are essential to provide load and postural stability during materials handling. Cut-out handles should be 115 mm long, 25–38 mm wide (or diameter, in case cylindrical handles are provided); cylindrical handles should have 30–50 mm clearance all around; handles should have a pivot angle of 70° from the horizontal axis of the box; handles should be located at diagonally opposite ends to provide both vertical and horizontal stability for the load. According to the NIOSH lifting model, a reduction of up to 10% should be made if the containers or objects being handled do not have handles.[30]

Asymmetrical Handling

Controls: Use hoists where possible to lift loads; avoid twisting when lifting loads asymmetrically; balance the load when carrying. Asymmetric handling of loads is common in industry. However, when this is done it leads to reduced load handling capabilities and strength, increased intra-abdominal and intra-discal (shear) pressures, and increased muscle activity of the lower back muscles. The materials handler is advised not to keep his or her feet in a locked position. When the worker pivots the feet in the direction of the load, the task is less stressful. The reduction in load lifting capability in such cases is expected to be no more than 15% for a 90° turn. For example, if a person can lift 50 lb without twisting, that person can only lift 42.5 lb when turning 90°.[30]

Space Confinement

Controls: Clear path or work area before manual material handling; Use overhead hoists for manual materials handling; avoid putting materials on ground or above heart height. Performing load handling activities within the confines of spatial restraint is common in manual material handling tasks in industry. For loads that are to be placed on shelves, the shelf opening clearance for inserting boxes by hands should be approximately 30 mm (1.2 inches). If the workplace layout does not allow erect posture (e.g., because of limited height to stand erect) the load should be reduced significantly from the 51-pound maximum load during trunk flexion.

Personal Protective Equipment

Personal protective equipment, including shoes, gloves, vests, trousers, goggles, respirators, aprons, overalls, and masks, should permit free movement, be easily removable, and allow for personal cooling to prevent body metabolic heat build-up.

Gloves should fit and fit the task; they should allow maintenance of dexterity. Shoes should be non-slip, comfortable, and water resistant.

Back Belts. The use of back belts to prevent back injury has not been shown to be a preventative measure according to a NIOSH study.[56] Specifically, noted in the document the NIOSH Back Belt working group concluded the following:

- *There are insufficient data indicating that typical industrial back belts significantly reduce the biomechanical loading of the trunk during manual lifting.*
- *There is insufficient scientific evidence to conclude that wearing back belts reduces risk of injury to the back based on changes in intra-abdominal pressure (IAP) and trunk muscle electromyography (EMG).*
- *The use of back belts may produce temporary strain on the cardiovascular system.*
- *There are insufficient data to demonstrate a relationship between the prevalence of back injury in healthy workers and the discontinuation of back belt use.*

The NIOSH working group recommends:

- *Caution in interpreting the results of studies that evaluated the effects of belt use on predictions of biomechanical loading of the spine.*
- *Caution in interpreting the results of epidemiological studies; the experience with these studies should be used to develop better designed epidemiological research.*
- *Future research should be designed to evaluate the efficacy of wearing back belts to prevent work-related back injury.*

Work Methods Evaluation

Analyzing Jobs

Finding out as much about the job as possible is a critical element in the recognition, evaluation, and control of musculoskeletal disorders. This procedure, known as job analysis, is one of the most tedious aspects of mastering ergonomics. Basically, there are two ways to perform a job analysis: 1) by conducting a methods study, which is a common method used by occupational engineers; and 2) by using checklists. Both methods are briefly described below.[57–60]

Work-Methods Study

Work-methods study is the systematic recording and critical examination of existing, as well as proposed ways of doing work. It is a means of developing and applying safe, easy, and effective solutions and reducing overall job costs. This method can be applied at two levels: 1) recording work sequence using a flow process chart; and 2) recording work sequence using the techniques of micro motion study. The second method of recording workplace movements is commonly used for very short cycle jobs and requires filming the job. See Appendix A for tips on properly videotaping jobs for analyses.

Jobs that have a very short cycle time, usually in seconds, and are repeated hundreds of times per day, such as in assembly jobs, are subjected to micro motion study. This is a process in which all job elements are divided into fundamental motions known as "therbligs." Therbligs were coined after Frank and Lilian Gilbreth (Therbligs are Gilbreths spelled backwards) who were known in the early twentieth century for their research on work efficiency.[18] There are 18 fundamental motions in therblig analysis. Table 28.4 shows the basic work elements developed by the Gilbreths. Such detail can be obtained by analyzing film or videotape in slow motion and stop action; hence, the tedious nature of job analysis mentioned earlier. Such detail, however, can yield much information about the job and solutions to control musculoskeletal disorders.

Describing the job in a series of elements can help the researcher determine which ones may be job risk factors for

Table 28.4 — Table of Work Elements Developed by the Gilbreths

Element	*Description*
Search	Looking for something with the eyes or hand.
Select	Locating one object (that is) mixed with others.
Grasp	Touching or gripping an object with the hand.
Reach	Moving of the hand to some object or location.
Move	Movement of some object from one location to another.
Hold	Exerting force to hold an object on a fixed location.
Position	Moving an object in a desired orientation.
Inspect	Examining an object by sight, sound, touch, etc.
Assemble	Joining together two or more objects.
Disassemble	Separating two or more objects.
Use	Manipulating a tool or device with the hand.
Unavoidable	Delay Interrupting work activity because of some factor beyond the worker's control.
Avoidable	Delay Interrupting work activity because of some factor under the worker's control.
Plan	Performing mental process that precedes movement.
Rest to overcome fatigue	Interrupting work activity to overcome the effects of repeated exertions or movements.

musculoskeletal disorders. Elements can be grouped into tasks that are usually performed in a similar sequence. Each task should be timed by a stop watch or by calculating the elapsed time on a videotape (it is good practice to verify the timing element on the videotape with a stop watch). Examples of a task include reaching for a part, grasping the part, moving the part to an assembly line, positioning the part on a jig, etc. Table 28.5 shows how such work tasks can be defined for the right and left hand for an assembly job. After this breakdown of tasks, the job can be analyzed for risk factors that can cause musculoskeletal disorders. Elements that can increase the probability of a musculoskeletal disorder are extracted from the previous table and shown in Table 28.6 with recommended solutions. Such information can be useful when complex jobs are broken down into defined units of work activity. It is also useful when trying to establish an ergonomics control program since this approach not only documents specific exposure problems but also lists solutions. The solutions can be used in ergonomics task force meetings where the pros and cons of the solution can be fleshed out and action can be taken.

As the tables 28.5 and 28.6 show, breaking the job into its parts can be time-consuming and involved. However, this systematic approach yields highly specific results in terms of identifying aspects of the job that may cause musculoskeletal disorders. Additionally, it gives the evaluator a mechanism for thinking about specific ergonomic solutions.[18] With practice, the evaluator can become proficient in this method. The key point here is several job risk factors can be identified that may seem small risks, but when several of these risk factors are reduced or eliminated by ergonomic solutions, there can be a large reduction in the overall physical stressors of the job. This alleviation of physical stressors may lead, in turn, to a reduction in work-related musculoskeletal disorders.

Table 28.5 — Tasks for One Job Cycle for an Automotive Assembler

Left Hand	*Right Hand*
1. Reach to get dashboard	1. Reach to get dashboard
2. Grasp dashboard	2. Grasp dashboard
3. Move dashboard to line	3. Move dashboard to line
4. Pre-position dashboard on line	4. Pre-position dashboard on line
5. Position dashboard on line	5. Position dashboard on line
6. Release dashboard on jig	6. Release dashboard on jig
7. Reach to dashboard support	7. Idle (hand at rest)
8. Grasp dashboard support	8. Idle (hand at rest)
9. Move support to dashboard	9. Reach for tool
10. Position support on dashboard	10. Grasp screw driver
11. Hold support on dashboard	11. Move screw driver to dashboard
12. Hold support on dashboard	12. Position screw driver on dashboard
13. Hold support on dashboard	13. Use screw driver on dashboard
14. Move to load screw driver	14. Move screw driver to dashboard
15. Position screw on screw driver	15. Move screw driver to dashboard
16. Hold dashboard	16. Use screw driver on dashboard
17. Idle (hand at rest)	17. Move screw driver to bench
18. Idle (hand at rest)	18. Position screw driver on bench
19. Idle (hand at rest)	19. Release screw driver on bench
20. Idle (hand at rest)	20. Release jig posts
21. Idle (hand at rest)	21. Grasp jig posts
22. Idle (hand at rest)	22. Move (turn) jig posts
23. Idle (hand at rest)	23. Release jig posts
24. Idle (hand at rest)	24. Idle (hand at rest)
25. Idle (hand at rest)	25. Reaching for jig posts
26. Idle (hand at rest)	26. Grasp jig posts
27. Idle (hand at rest)	27. Move (turn) jig posts
28. Idle (hand at rest)	28. Release jig posts

Table 28.6 — Summary of Risk Factors and Recommendations for Automotive Assembler

Work Element	*Work Risk Factor*	*Recommendation*
5-6 10-23	Sharp edges from work bench causes uncomfortable contact pressure on operator's waist.	Pad front of bench with rounded 2 in. rubberized strip with foam core.
Right hand 12,13	Pistol grip screw driver used to drive screw in vertical position causing high wrist flexion	Provide in-line screw driver to drive screw to keep wrist posture neutral during this task.
1-27	The use of thick work gloves may decrease the tactile sensitivity of the hands during work	Use flexible light weight gloves that fit the worker's hands, to aid in sensory feedback.
Left hand 7,8	Worker reaches back (arm abduction) to get dashboard part from bin.	Move dashboard support cart next to work bench so the worker does not have to reach back for the part.
Right hand 21	Grasping jig sprues requires repetitive finger pinch, with high finger forces.	Provide larger grip for jig handle to improve biomechanical leverage by using the entire hand.

Checklists

Checklists provide an alternative approach to methods study. On the plus side, checklists are easy to use because they are a qualitative method that quickly evaluates job risk factors. Depending on how they are structured, they can offer a "relative ranking" of jobs—from those with the highest number of job risk factors to those with the least. On the negative side, checklists are not quantitative, and in some instances may not tell the evaluator what the real risk factors are. Perhaps that is why there are as many checklists as there are industries. Checklists come in all sizes from simple one page lists, having yes/no responses, to multiple page lists, which try to semiquantify various risk factors. Two frequently used check lists are shown in Figures 28.8 and 28.9.[61,62] The first was developed by the University of Michigan's Center for Ergonomics to evaluate risk factors for upper extremities; the other checklist was developed by NIOSH to qualitatively assess lifting hazards. A third checklist, (see Figure 28.10) developed by Dr. Sue Rogers, an ergonomics consultant, attempts to quantify and categorize such jobs by assessing a raw score, then totaling the score for each job evaluated.[63] Again, the objective is to rank order each job. The advantage of Rogers' checklist is its inherent attempt to look at the synergistic nature of job risk factors — that is, to evaluate the job risk factors of repetition, force, posture, and recovery time as one value.

When using or developing checklists, one should use caution to make sure the questions are relevant, as the collection and analysis of the data can be time-consuming. More important, when working within the framework of a checklist, evaluators have to make sure the problems they are trying to solve are within that framework, that is, validating the checklist for giving enough detail to derive effective solutions. Additional examples of simple yes/no proactive checklists are shown in Appendix B.

Yes/no checklists are useful for walk-through surveys and can be used to rank order jobs into high, medium, and low risk. To do this, the evaluator can add up the number of yes or no responses for each job. Jobs with the highest scores for these job risk factors would be ranked high, and so on.

Risk Factor Assessment Techniques

The techniques to assess risk factors associated with musculoskeletal injuries and illnesses are broadly classified as 1) biomechanical; 2) physiological; and 3)

Risk Factors	no	yes
1. Physical Stress		
1.1 Can the job be done without hand/wrist contact with sharp edges?	____	____
1.2 Is the tool operating without vibration?	____	____
1.3 Are the worker's hands exposed to temperature > 21C (70F)	____	____
1.4 Can the job be done without using gloves?	____	____
2. Force		
2.1 Does the job require exerting less than 4.5 Kg (10 lbs) of force?	____	____
2.2 Can the job be done without using finger pinch grip?	____	____
3. Posture		
3.1 Can the job be done without flexion or extension of the wrist?	____	____
3.2 Can the tool be used without flexion or extension of the wrist?	____	____
3.3 Can the job be done without deviating the wrist from side to side?	____	____
3.4 Can the tool be used without deviating the wrist from side to side?	____	____
3.5 Can the worker be seated while performing the job?	____	____
3.6 Can the job be done without "clothes wringing" motion?	____	____
4. Workstation hardware		
4.1 Can the orientation of the work surface be adjusted?	____	____
4.2 Can the height of the work surface be adjusted?	____	____
4.3 Can the location of the tool be adjusted?	____	____
5. Repetitiveness		
5.1 Is the cycle time longer than 30 seconds?	____	____
6. Tool Design		
6.1 Are the thumb and finger slightly overlapped in a closed grip?	____	____
6.2 Is the span of the tool's handle between 5 and 7 cm (2-2 3/4")?	____	____
6.3 Is the handle of the tool made from material other than metal?	____	____
6.4 Is the weight of the tool below 4 kg (9 lbs)?	____	____
6.5 Is the tool suspended?	____	____

["No" responses are indicative of conditions associated with the risk of Work-Related CTDs.] ***Lifshitz, Y., and T. Armstrong,** 1986, A Design Checklist for Control and Prediction of Cumulative Trauma Disorders in Hand Intensive Manual Jobs. *Proceedings of the 30th Annual Meeting of Human Factors Society,* pp. 837-841.

Figure 28.8 — Michigan's checklist for upper extremity cumulative trauma disorders.*

Risk Factors	yes	no
General		
1.1 Does the load handled exceed 50 lb?	____	____
1.2 Is the object difficult to bring close to the body because of its size, bulk, or shape?	____	____
1.3 Is the load hard to handle because it lacks handles or cutouts for handles, or does it have slippery surfaces or sharp edges?	____	____
1.4 Is the footing unsafe? For example, are the floors slippery, inclined, or uneven?	____	____
1.5 Does the task require fast movement, such as throwing, swinging, or rapid walking?	____	____
1.6 Does the task require stressful body postures, such as stooping to the floor, twisting, reaching overhead, or excessive lateral bending?	____	____
1.7 Is most of the load handled by only one hand, arm, or shoulder?	____	____
1.8 Does the task require working in environmental hazards, such as extreme temperatures, noise, vibration, lighting, or airborne contaminants?	____	____
1.9 Does the task require working in a confined area?	____	____
Specific		
2.1 Does lifting frequency exceed 5 lifts per minute?	____	____
2.2 Does the vertical lifting distance exceed 3 feet?	____	____
2.3 Do carries last longer than 1 minute?	____	____
2.4 Do tasks which require large sustained pushing or pulling forces exceed 30 seconds duration?	____	____
2.5 Do extended reach static holding tasks exceed 1 minute?	____	____

Comment: "Yes" responses are indicative of conditions that pose a risk of developing low back pain. The larger the percentage of "yes" responses, the greater the risk. **Note:** Checklist was developed by NIOSH researchers, including Dr. Tom Waters.

Figure 28.9 — Hazard evaluation checklist for lifting, carrying, pushing, or pulling.

psychophysical. These techniques can be qualitative or quantitative. Listed below are a few examples of the techniques that can be used for risk factor assessment.

Biomechanical

Spinal stresses: Spinal stresses (compressive and shear) provide an indication of the hazard the body can be subjected to during manual material handling. Several biomechanical models have been developed to assess these stresses. Some perform static and two-dimensional (2-D) analysis while others perform dynamic and three-dimensional (3-D) analysis. Even though several biomechanical models are available, most are complex and not user friendly.

Before using a model, it is essential to understand the limitations of these and other existing biomechanical models. Static models consider the motion of the human body as a series of static postures and perform static analysis in each posture to determine musculoskeletal stresses. Dynamic models are more complex and can account for inertial effects of the body segments and the loading because of acceleration. However, while the static model is easy to use, some quantitative data are lost. The dynamic model gives more quantitative data and, as some argue, more accurate risk assessment for back injury, but the data collection and analysis are much more complex and time consuming. As with checklists, the utility of biomechanical models is greater for comparative task analysis than for stand-alone task analysis to determine absolute load values. Table 28.7 shows the relationship between static vs. dynamic model evaluation parameters.[64]

The static biomechanical model (2-D or 3-D) is available from the University of Michigan's Center for Ergonomics. These models are basically static strength models and predict populations capable of performing each task from a variety of inputs: body posture, object, force needed to oppose the object (average, maximum), and location of the hands.[65]

Along the same line as the static biomechanical model mentioned above is the development of several software programs that calculate the 1991 Revised NIOSH Lifting Equation (discussed later). These programs simply take the input data from the lifting equation and automatically calculate the NIOSH Recommended Weight Limit (RWL)and the Lifting Index (LI). The LI is the ratio of the weight lifted (by the worker), divided by the RWL. For example,

Body Part		Effort Level	Continuous Effort Time	Efforts/ Minute	Priority
Neck/Shoulders	R	___	___	___	___
	L	___	___	___	___
Back		___	___	___	___
Arms/Elbows	R	___	___	___	___
	L	___	___	___	___
Wrists/Hands/	R	___	___	___	___
Fingers	L	___	___	___	___
Legs/Knees	R	___	___	___	___
	L	___	___	___	___
Ankles/Feet/	R	___	___	___	___
Toes	L	___	___	___	___

Effort Categories
1 = Light
2 = Moderate
3 = Heavy

Continuous Effort Time Categories
1 = < 6 secs
2 = 6 to 20 secs
3 = > 20 secs

Efforts/Minute Categories
1 = < 1/min
2 = 1 to 5/min
3 = > 5 -15/min

Priority for Change

Moderate = 1 2 3, 1 3 2, 2 1 3, 2 2 2, 2 3 1, 2 3 2, 3 1 2

High = 2 2 3, 3 1 3, 3 2 1, 3 2 2

Very High = 3 2 3, 3 3 1, 3 3 2

JobTitle: ___________

Specific Task: ___________

Job Number: ___________

Department: ___________

Location: ___________

Contact Person(s): ___________
Phone: ___________

Analyst: ___________
Phone: ___________

Date of Analysis: ___________

Figure 28.10 — Ergonomic job analysis. (Reprinted with permission from: Sue Rogers, Ph.D., consultant.)

if the worker lifts a box weighing 75 lb, and the RWL is 25 lb, then, the LI is 3.0. The LI may be used to establish priorities for evaluating jobs. Because of limited space, all details of the NIOSH formula cannot be presented here, and the reader is referred to the NIOSH publication entitled: *Applications Manual for the Revised NIOSH Lifting Equation by Tom Waters, Vern Putz-Anderson, and Arun Garg.*[66] However, the information given below, as well as the example of a chemical bag handler (Figure 28.11), highlight the utility of the revised NIOSH lifting equation. Additional examples of NIOSH RWL and LI output are shown in the case study near the end of this chapter.

The revised NIOSH Lifting Equation is defined by the following equation:

$$RWL = LC \times HM \times VM \times DM \times AM \times FM \times CM$$

Table 28.7 — Dependent and Independent Variables in the Measurement of Muscle Strength

	Isometric (Static)		*Isovelocity (Dynamic)*		*Isoacceleration (Dynamic)*		*Isojerk (Dynamic)*		*Isoforce (Static or Dynamic)*		*Isoinertia (Static or Dynamic)*		*Free Dynamic*	
Variables	*I.D*	*D.*	*I.D*	*D.*	*I.D*	*D.*	*I.D*	*D.*	*I.D*	*D.*	*I.D*	*D.*	*I.D*	*D.*
Displacement, linear/angular	Constant* (zero)		C or X		C or X		C or X		C or X		C or X		C or X	
Velocity, linear/angular	0		Constant* (zero)		C or X		C or X		C or X		C or X			X
Acceleration, linear/angular	0		0		Constant * (zero)		C or X		C or X		C or X			X
Jerk, linear/angular	0		0		0		Constant* (zero)		C or X		C or X			X
Force, torque	C		C or X		C or X		C or X		Constant* (zero)		C or X			X
Mass, moment of inertia	C		C or X		C		C		C or X		Constant* (zero)			X
Repetition	C		C or X		C or X		C or X		C or X		C or X		Constant * (zero)	

Legend: I.D = independent; D.= Dependent; C=Variable can. be controlled; * = set to zero; 0 = variable is not present (zero); X= can be dependent variable; The boxed constant variable provides the descriptive name. (Adapted from Kroemer, Marras, McGlothlin, et al., 1990.)

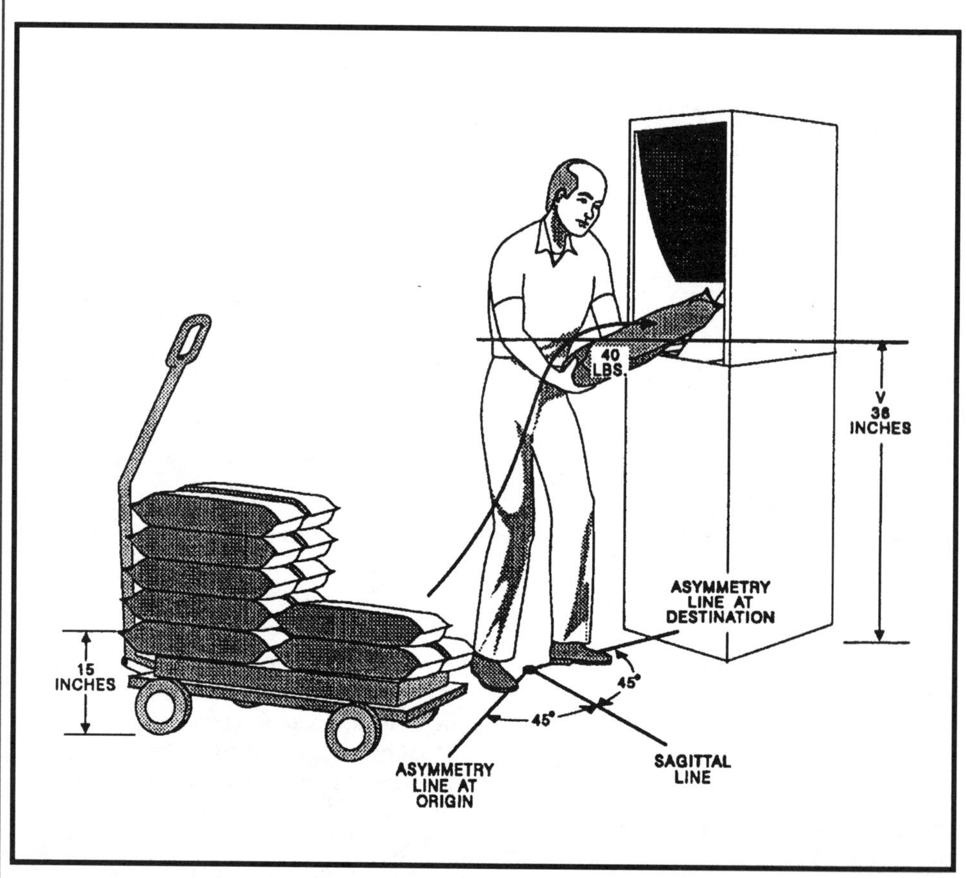

Figure 28.11 — Example of worker loading bags of powder into hopper. (From **Waters, T.R., V. Putz-Anderson, and A. Garg:** *Applications Manual for the Revised NIOSH Lifting Equation* (NIOSH Pub. No. 94–110). 1994. p. 66.)

where RWL = Load Constant (LC) * Horizontal Multiplier (HM) * Vertical Multiplier (VM) * Distance Multiplier (DM) * Asymmetric Multiplier (AM) * Frequency Multiplier (FM) * Coupling Multiplier (CM) (* indicates multiplication). Each of these variables is defined below with the formulas (in metric and U.S. Customary) for each variable shown in Table 28.8. Table 28.9 serves as a useful reference for determining the multiplier values for each of the six multipliers used in the equation. The tabular format of Table 28.9 shows these multiplier values in inches and centimeters.[66]

Figure 28.11 consists of a worker lifting bags of chemicals off a hand truck and to a mixing hopper. The worker positions himself midway between the hand truck and the mixing hopper. Without moving his feet, he twists to the right and picks up a bag off the hand truck. He then twists to his left to place the bag on the rim of the hopper. A sharp-edged blade within the hopper cuts open the bag to allow the chemical contents to fall into the hopper. This task is done infrequently (i.e., 1–12 times per shift) with large recovery periods between lifts (i.e., 1.2 recovery time/work time ratio). Significant control is not required at the destination, but the worker twists at the origin and destination of the lift. Although several bags are stacked on the hand truck, the highest risk of overexertion injury is associated with the bag on the bottom of the stack; therefore, only the lifting of the bottom bag will be examined. Note, however, that the frequency multiplier is based on the overall frequency of lifting for all of the bags. Figure 28.12 shows the job analysis worksheet calculations that were derived from Table 28.9.[66] While the formula looks somewhat daunting to the uninitiated, it is relatively easy to use with some practice. Here, the user simply has to locate the values on the table, find the associated multiplier (no value exceeds 1.0 for each multiplier), and derive the RWL and LI. Note that if the exact number is not given, for example, 13.5 inches for horizontal distance, then the user should extrapolate the value. In this case the horizontal multiplier is 0.74 (i.e., between 0.77 at 13 in. and 0.71 at 14 in.).

Physiological Techniques

Physiological techniques are useful for repetitive and whole body work, such as manual materials handling. The two human body responses that indicate the extent of hazard and respond to various risk factors are oxygen consumption and

Table 28.8 — Revised NIOSH Equation for the Design and Evaluation of Manual Lifting Tasks

Component	*Metric*	*U.S. Customary*
LC = Load Constant	23 kg	51 lb
HM = Horizontal Multiplier	(25/H)	(10/H)
VM = Vertical Multiplier	(1-(.003 |V-75|))	(1-(.0075 |V-30|))
DM = Distance	(.82+(4.5/D))	(.82+(1.8/D))
AM = Asymmetric	(1-(.0032A))	(1-(.0032A))
FM = Frequency	(see Table 28.9)	(see Table 28.9)
CM = Coupling	(see Table 28.9)	(see Table 28.9)

Where:

H = Horizontal location of hands from midpoint between the ankles. Measure at the origin and the destination of the lift (cm or in).

V = Vertical location of the hands from the floor. Measure at the origin and destination of the lift (cm or in).

D = Vertical travel distance between the origin and the destination of the lift (cm or in).

A = Angle of asymmetry — angular displacement of the load from the sagittal plane. Measure at the origin and destination of the lift (degrees).

F = Average frequency rate of lifting measured in lifts/min. Duration is defined to be: ≤ 1 hour; ≤ 2 hours; or ≤ 8 hours assuming appropriate recovery allowances.

Table 28.9 — Tabular Presentation of Six Multipliers (Horizontal, Vertical, Frequency, Distance, Asymmetric, and Coupling) Showing the Multiplier Values in Inches and Centimeters for the NIOSH Revised Lifting Equation

Horizontal Multiplier

H	HM	H	HM
	in	cm	
≤10	1.00	≤25	1.00
11	.91	28	.89
12	.83	30	.83
13	.77	32	.78
14	.71	34	.74
15	.67	36	.69
16	.63	38	.66
17	.59	40	.63
18	.56	42	.60
19	.53	44	.57
20	.50	46	.54
21	.48	48	.52
22	.46	50	.50
23	.44	52	.48
24	.42	54	.46
25	.40	56	.45
>25	.00	58	.43
		60	.42
		63	.40
		>63	.00

Vertical Multiplier

V	VM	V	VM
in		cm	
0	.78	0	.78
5	.81	10	.81
10	.85	20	.84
15	.89	30	.87
20	.93	40	.90
25	.96	50	.93
30	1.00	60	.96
35	.96	70	.99
40	.93	80	.99
45	.89	90	.96
50	.85	100	.93
55	.81	110	.90
60	.78	120	.87
65	.74	130	.84
70	.70	140	.81
>70	.00	150	.78
		160	.75
		170	.72
		175	.70
		>175	.00

Frequency Multiplier

F lifts/min	DURATION					
	< 1 hour		1-2 hours		2-8 hours	
	V< 30 in	V ≥ 30 in	V< 30 in	V ≥ 30 in	V< 30 in	V ≥ 30 in
≤2	1.00	1.00	.95	.95	.85	.85
.5	.97	.97	.92	.92	.81	.81
1	.94	.94	.88	.88	.75	.75
2	.91	.91	.84	.84	.65	.65
3	.88	.88	.79	.79	.55	.55
4	.84	.84	.72	.72	.45	.45
5	.80	.80	.60	.60	.35	.35
6	.75	.75	.50	.50	.27	.27
7	.70	.70	.42	.42	.22	.22
8	.60	.60	.35	.35	.18	.18
9	.52	.52	.30	.30	.00	.15
10	.45	.45	.26	.26	.00	.13
11	.41	.41	.00	.23	.00	.00
12	.37	.37	.00	.21	.00	.00
13	.00	.34	.00	.00	.00	.00
14	.00	.31	.00	.00	.00	.00
15	.00	.28	.00	.00	.00	.00
>15	.00	.00	.00	.00	.00	.00

Distance Multiplier

D	DM	D	DM
in		cm	
≤10	1.00	≤25	1.00
15	.94	40	.93
20	.91	55	.90
25	.89	70	.88
30	.88	85	.87
35	.87	100	.87
40	.87	115	.86
45	.86	130	.86
50	.86	145	.85
55	.85	160	.85
60	.85	175	.85
70	.85	>175	.00
>70	.00		

Asymmetric Multiplier

A	AM
deg	
0	1.00
15	.95
30	.90
45	.86
60	.81
75	.76
90	.71
105	.66
120	.62
135	.57
>135	.00

Coupling Multiplier

COUPLING TYPE	CM	
	V < 30 in	V ≥ 30 in
GOOD	1.00	1.00
FAIR	.95	1.00
POOR	.90	.90

JOB ANALYSIS WORKSHEET

DEPARTMENT	Manufacturing	JOB DESCRIPTION	Dumping bags into mixing hopper
JOB TITLE	Batch Processor		Example 3
ANALYST'S NAME			
DATE			

STEP 1. Measure and record task variables

Object Weight (lbs)		Hand Location (in)				Vertical Distance (in)	Asymmetric Angle (degrees)		Frequency Rate lifts/min	Duration (HRS)	Object Coupling
		Origin		Dest.			Origin	Destination			
L (AVG.)	L (Max.)	H	V	H	V	D	A	A	F		C
40	40	18	15	10	36	21	45	45	<.2	<1	Fair

STEP 2. Determine the multipliers and compute the RWL's

RWL = LC × HM × VM × DM × AM × FM × CM

ORIGIN RWL = 51 × .56 × .89 × .91 × .86 × 1.0 × .95 = 18.9 Lbs

DESTINATION RWL = 51 × ☐ × ☐ × ☐ × ☐ × ☐ × ☐ = ☐ Lbs

STEP 3. Compute the LIFTING INDEX

ORIGIN LIFTING INDEX = OBJECT WEIGHT (L) / RWL = 40 / 18.9 = 2.1

DESTINATION LIFTING INDEX = OBJECT WEIGHT (L) / RWL = ___ = ☐

Figure 28.12 — Job analysis worksheet for dumping bags of chemical powder into mixing hopper. (From **Waters, T.R., V. Putz-Anderson, and A. Garg:** *Applications Manual for the Revised NIOSH Lifting Equation.* (NIOSH Pub. No. 94–110). 1994. p. 68.)

heart rate. These two methods are briefly described below.

Oxygen Consumption. Oxygen consumed by an individual is influenced by the intensity of the task he or she performs. Oxygen consumption is compared with the physical work capacity (PWC), also known as aerobic capacity, maximum aerobic power, and maximum oxygen uptake ($V0_2$ max). PWC is the maximum amount of oxygen that an individual can consume per minute. The higher the percentage of PWC a task requires, the higher the resulting physical stress, fatigue, and possibility of injury.

PWC can be measured by maximal or submaximal methods.[67] In the maximal method, the worker is stressed to the maximum, and oxygen consumption at that level is recorded. This method is not recommended for occupational practitioners. In submaximal methods, individuals are required to perform at least three workloads on either a bicycle ergometer or treadmill, or—for a specific task like arm strength— an arm crank ergometer. The treadmill is recommended when trying to assess maximal aerobic capacity. The technique is to increase the treadmill pace from 3 mph, 4 mph, and 5 mph at a 10 percent slope. When the heart rate and oxygen consumption have stabilized, the values are recorded. These values are plotted on a x-y graph (e.g., heart rate on x-axis, and oxygen uptake on y-axis) and a straight line is drawn through the plotted points. The value of the maximum heart rate for the individual is determined and plotted in the x-axis. One of the more accurate formulas for an individual's maximum heart rate is:

Maximum heart rate =
214 - 0.71 × Age in years.

A vertical line is projected from the maximum heart rate point. A horizontal line is projected from the point of intersection between this vertical line and the straight line joining the three plotted points. The value given by the intersection of the horizontal line and the y-axis is the PWC of the individual.

When the PWC is determined and compared with the steady state oxygen consumption of a worker on a job, an indication of the physical stress of the job is obtained. If a job requires more than 21%–23% of PWC for eight-hour shifts, it is likely to lead to overexertion and, possibly, musculoskeletal disorders.[68]

Measurement of the volume of oxygen consumed can provide an overall index of energy consumption and will show the energy demands of work. For example, use of 1 liter of oxygen will yield approximately 5 kilocalories (kcal). The kilocalories can be added to determine metabolic energy expenditure for several activities as shown in Table 28.10.[69] For example, the average man in his 20s has an average maximal capacity of 3–3.5 L/min., while the "average" woman of the same age has an average capacity of 2.3–2.8 L/min. This translates to approximately 1020 kcal/hr for the male,

Table 28.10 — Metabolic Energy Costs of Several Activities

Activity	*Kcal/hr*
Resting, prone	80-90
Resting, seated	95-100
Standing, at ease	100-110
Drafting	105
Light assembly (bench work)	105
Medium assembly	160
Driving automobile	170
Walking, casual	175-225
Sheet metal work	180
Machining	185
Rock drilling	225-550
Mixing cement	275
Walking on job	290-400
Pushing wheelbarrow	300-400
Shoveling	235-525
Chopping with axe	400-1400
Climbing stairs	450-775
Slag removal	630-750

Values are for male worker of 70 kg (154 lb). (Reprinted with permission from Ergonomics Guide to Assessment of Metabolic and Cardiac Costs of Physical Work. *Am. Ind. Hyg. Assoc. J. 32*:560-564 (1971).)

and 750 kcal/hr for the female. It must be kept in mind that oxygen capacity diminishes with age. For example, for a 60-year-old male, the average maximal capacity is 2.2–2.5 L/min. and the 60-year-old female has 1.8–2.0 L/min. maximal capacity.

Oxygen consumption can be measured by several commercially available devices, such as the MRM-1 Oxygen Consumption Computer (Waters Instruments, Inc., Rochester, Minnesota) and the Morgan Oxylog (Ambulatory Monitoring, Inc., Ardsley, New York). These devices require the worker to put on a mask, breath room air, and exhale into the mask. The exhaled air is analyzed for oxygen content and compared with oxygen content in the room's air to determine average minute oxygen uptake. The 0_2 measurements should be taken when the worker has achieved a "steady state" when sampling.

Heart Rate. Heart rate is a frequently measured indicator of physical stress. The easiest direct method is to take the pulse of the worker once steady state has been reached. The pulse rate should be recorded for a full minute to avoid the effect of sinus arrhythmia. If the pulse rate is being recorded at the end of the work period, a 15-second reading is sufficient; readings of longer duration tend to carry into the recovery period. A three-minute average at the steady state should be recorded. The maximum working heart rate should not exceed 130 beats/min.

The other method for recording heart rate is somewhat similar to recording electrocardiogram (EKG) readings in that active and passive electrodes are used. Two electrodes are placed on the rib cage, about 10 inches apart, and the ground is placed on a bony landmark. The skin is lightly abraded to get better conduction if passive electrodes are used; such electrodes are covered with an adhesive collar filled with electrode gel.

Heart rate recording devices vary from simple and relatively inexpensive, such as the Polar and Quinton Heart Rate Monitors, which only provide a rate, to physiographs and data graphs, which can be used to record EKG. Physiographs and data graphs can be relatively expensive and may not be necessary for occupational use. Figure 28.13 shows the application of the heart rate monitor on a beverage delivery person where the heart rate data are overlayed on a video monitor to show cardiovascular demands as a function of task demands.[70] Real-time monitoring of cardiovascular demands while work is videotaped can be useful when determining the job demands, as well as the impact of engineering controls and work practices to lower such demands. Table 28.11 shows the relationship between heart rate, oxygen uptake, and total energy expenditure.[67]

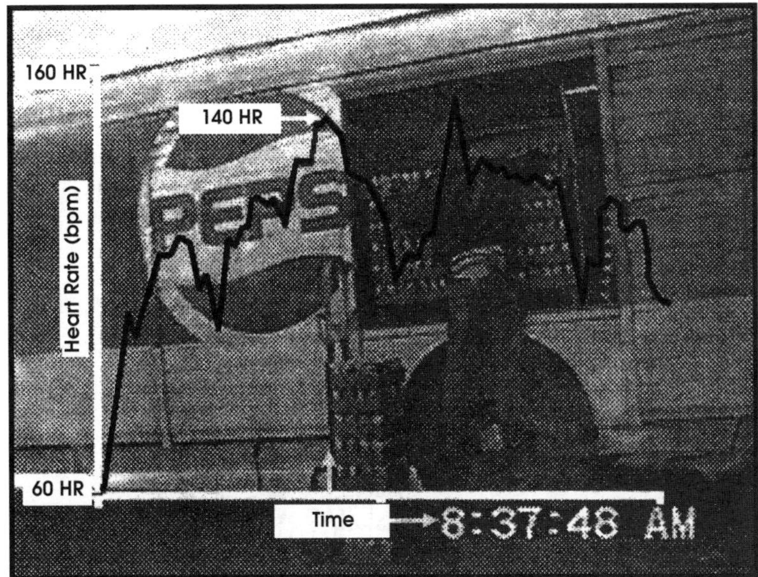

Figure 28.13 — Heart rate overlay (chart) on videograph of worker getting soft drinks from top shelf in truck. (Arrow points to driver-sales-worker's current heart rate.) (From **McGlothlin, J.D.:** *Ergonomic Interventions for the Soft Drink Beverage Delivery Industry* (NIOSH Pub. No. 96–109). 1996. p. 36.)

Table 28.11 — The Relationship Between Heart Rate, Oxygen Uptake and Total Energy Expenditure

Classification	*HeartRate beats/min*	*Oxygen uptake L/min*	*Total Energy Expenditure (Kcal/min)*
Light work (e.g., typing)	90 or less	up to 0.5	2.5
Medium work	100	0.5–1.0	5.0
Heavy Work	120	1.0–1.5	7.5
Very heavy work	140	1.5–2.0	10.0
Extremely heavy work (e.g., ditch digging)	160	over 2.0	15.0

Adapted from **Astrand, P., and K. Rodahl:** *Textbook of Work Physiology: Physiological Bases of Exercise*, 3rd Ed. New York: McGraw-Hill, 1986. p. 502.)

Psychophysical

Psychophysical techniques are suitable for repetitive as well as nonrepetitive or infrequently performed tasks. These techniques are also inexpensive since little or no equipment is needed. The techniques described in this section can be used for both lower back and upper extremities and include measurements of postural discomfort, static and dynamic work (perceived exertion), and fatigue.

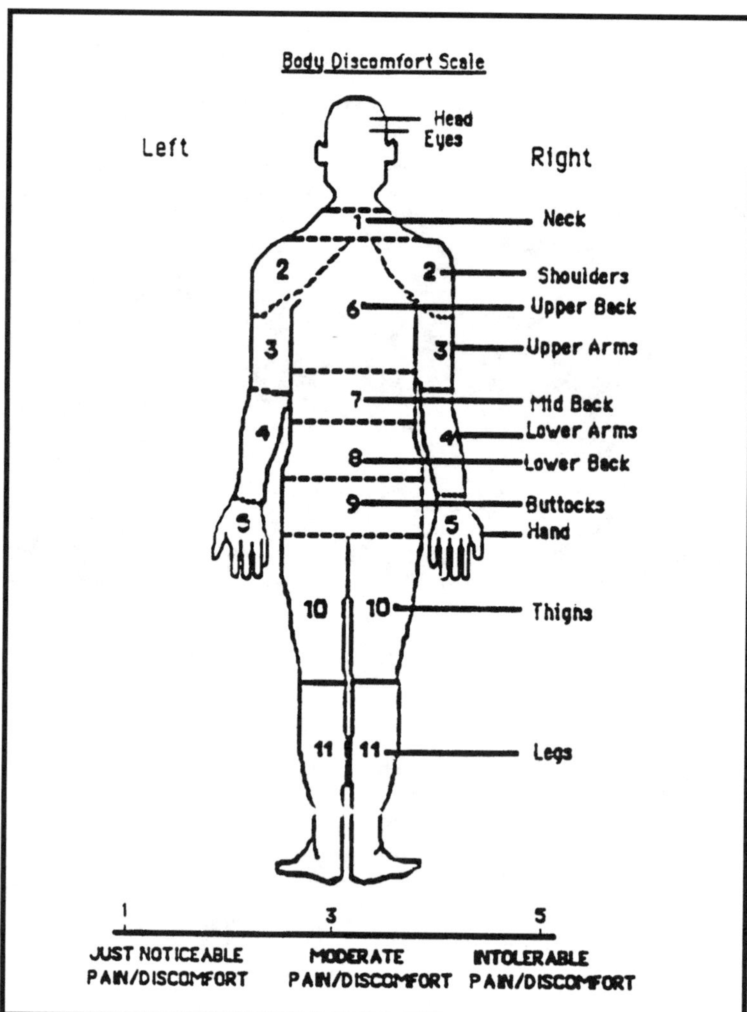

Figure 28.14 — Body part discomfort form and rating scale. (Reprinted with permission from **Corlett, E.N., and R.P. Bishop:** A Technique for Assessing Postural Discomfort. *Ergonomics 19*:175–182. (1976).)

A

6	NO EXERTION AT ALL
7	
	EXTREMELY LIGHT
8	
9	VERY LIGHT
10	
11	VERY LIGHT
12	
13	SOMEWHAT HARD
14	
15	HARD (HEAVY)
16	
17	VERY HARD
18	
19	EXTREMELY HARD
20	MAXIMAL EXERTION

B

0	NOTHING AT ALL
0.5	EXTREMELY WEAK
1	VERY WEAK
2	WEAK
3	MODERATE
4	SOMEWHAT STRONG
5	STRONG (HEAVY)
6	
7	VERY STRONG
8	
9	
10	EXTREMELY STRONG (ALMOST MAX)

Figure 28.15 — (A) Borg's rating of perceived exertion (RPE) scale of physical tasks; (B) Borg's rating of perceived exertion (RPE) scale for large muscles. (Reprinted with permission from **Borg, G.:** *An Introduction to Borg's RPE-Scale.* Ithaca, NY: Movement Publications. 1985.)

Postural discomfort. Extreme postures or postures that are maintained for prolonged periods of time can be uncomfortable and lead to fatigue and pain. It has been determined that the level of discomfort is linearly related to the force exertion time. Maintaining the same posture is physiologically equivalent to applying a force. A body map and a five-point scale are used to determine the body part discomfort.(71) Figure 28.14 shows both the body map and the scale.(71) The worker experiencing discomfort in the task being evaluated specifies the location of discomfort on the chart and rates it using the scale. The worker rates body parts at regular intervals ranging from 30 minutes to three hours. Some ergonomists suggest that the discomfort rating be taken just before a break to get more reliable readings.

Rating of perceived exertion (RPE). There is a curve linear relationship between the intensity of a range of stimuli and workers' perception of their intensity. These perceived exertions can be rated on a Borg Scale shown in Figure 28.15A and 15B.(72)

Figure 28.15A shows the Borg rating of perceived exertion for physical tasks. Note that in addition to the overall perception of exertion, these ratings are influenced by previous experience and motivation. In general, highly motivated subjects tend to underestimate their exertion.

The scale in Figure 28.15B is valid only for large muscle groups and should not be used for work performed by fingers and hands. As shown in this figure, each scale has a point value. Scores for all statements are added to determine the worker's opinion.

Modern technology also has allowed such information to be entered into a computer by the workers using a light pen. If the questionnaire is easily understood by the worker and data input is designed as yes/no type responses, This technology is very useful and can save time. Figure 28.16A and 16B shows the questionnaire data shown on a computer screen and filled out by a worker in the field using a light pen. This technology proved useful in a recent NIOSH study of beverage delivery workers in terms of efficient data management and analysis.(70)

Anthropometry: Designing the Workplace to Fit the Worker

Anthropometry literally translated means *measuring the human,* primarily for body size and shape. The systematic measurement of physical properties of the

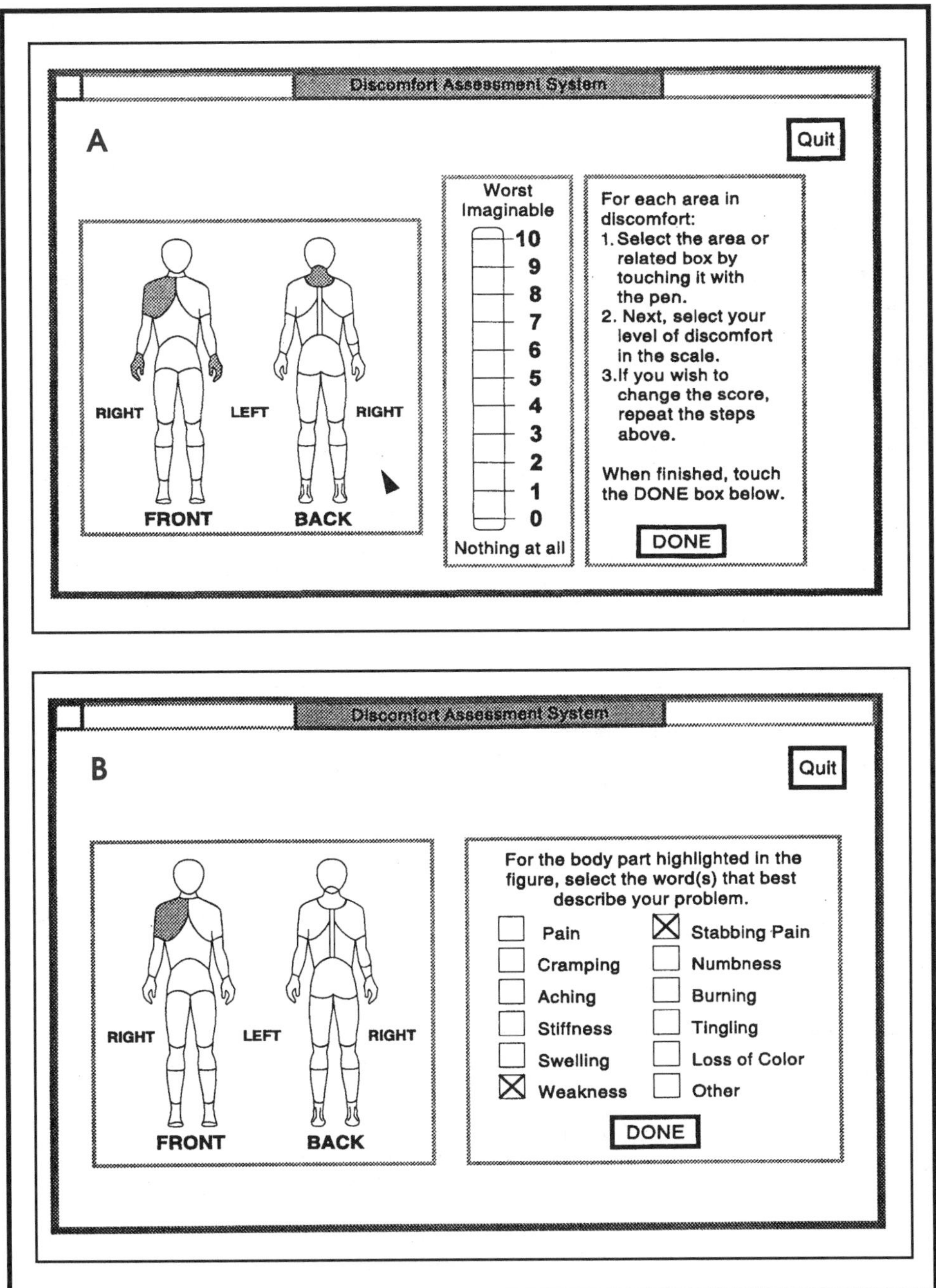

Figure 28.16 — (A) Discomfort Scores Screen (Body Figures) shown on computer and activated by light pen. (B) Discomfort Descriptors Screen shown on computer screen and activated by light pen. (From **McGlothlin, J.D.:** *Ergonomic Interventions for the Soft Drink Beverage Delivery Industry* (NIOSH Pub. No. 96–109. 1996. p. 36.). **Note:** Software developed by Norka Saldana, Ph.D., University of Michigan, Center for Ergonomics.

human body is the application of anthropometry. Measurement of humans has been done for hundreds of years, such as fitting soldiers with battle dress, but only within the last 50 years have the human dimensions been used to improve the design and sizing of everyday things.[52] It is when anthropometry is applied to the design and construction of things from personal protective equipment to modern office furniture that this discipline enters the realm of ergonomics. The purpose is to improve the effectiveness and efficiency of the person wearing and/or using the equipment. Figure 28.17 shows the selected skeletal landmarks used to define traditional anthropometric measurements.[52] These landmarks are traditional dimensional descriptors of body size. The measurements provide the distance between two points obtained under static conditions. The basic categories of static dimensions include lengths, depths, breadths, and distances for body size. This information can be useful when applied to equipment purchases, such as office furniture. For example, if the occupational hygienist was given the task of buying ergonomic office chairs, the knowledge and application of anthropometry might be useful. Table 28.12 shows body dimensions of a U.S. civilian adult sampling.[3] If the task was to purchase chairs to fit a population range from the 5th percentile female to the 95 percentile male, the occupational hygienist would want to select chairs that could be height adjusted from 35.1 inches to 42.9 inches. This is the sitting popliteal height (the distance from the floor to the joint in back of the knee) for the 5th percentile female to 95th percentile male. Therefore, when the chair can be adjusted for this range, the small female or large male can sit in a comfortable position with the upper and lower leg approximately 90° to each. While the example above seems deceptively simple, the applications can become very complex, and the occupational hygienist is encouraged to seek the advice from experts in this field when designing workplaces.

Hand Tools

Proper hand tool design, selection, installation, and use are crucial to the prevention of cumulative trauma disorders of the upper extremities. The high incidence of musculoskeletal injuries in manufacturing has prompted concern that workers be protected from excess physical stress arising from exposure to repeated and sustained motions and exertions, work requiring awkward postures, contact stress, cold, and vibration. It has been recognized by management and labor that the interface between the human operator and the tools frequently used for repetitive manual work can affect a large number of jobs. Consequently, manufacturers of both nonpowered and powered hand tools are being challenged by customer demands for products that will minimize physical stress. The physical stressors associated with hand tool operation include awkward postures, forceful exertions, repetitive motion, contact stress, and vibration. Use of hand tools in manufacturing may show that the physical stress associated with chronic, work-related musculoskeletal disorders may be smaller in magnitude than any single occurrence and seems harmless, but repeated exposure over time may lead to work-related cumulative trauma disorders. The objective when selecting, installing, and using hand tools is to minimize worker exposure to each physical stress factor. The principal physical factors to consider when using tools include posture, force, repetitiveness, contact stresses, and vibration, exposure time, and temperature.[73]

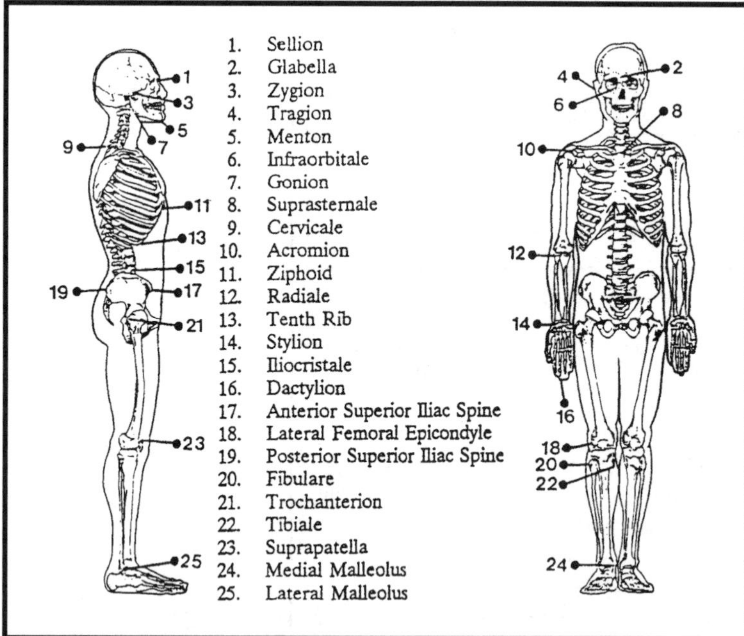

Figure 28.17 — Selected skeletal landmarks used to define traditional anthropometric measurements. (Reprinted with permission from **Annis and McConville:** Anthropometry (Chapter 1). In *Occupational Ergonomics: Theory and Applications* (edited by Bhattacharya and McGlothlin). 1996. pg. 4.)

Tool Selection

Tool selection should be made within the context of the job. By considering the ergonomic aspect of tool application for a specific job, the adverse effects of using the wrong tool can be prevented. The appropriate hand tool for the job should 1) maximize work performance; 2) enhance work quality; 3) minimize worker stress; and 4) prevent the onset of fatigue. Also, there are process engineering requirements for tools that also need to be addressed. For example, power hand tools must be capable of performing specific tasks in terms of speed, dimensions, torque, feed force, power, weight, trigger activation, spindle and chuck diameter, noise level, air pressure, precision and tolerance, bits, blades, abrasives, and power source. More important is the consideration of workers and how they will use the tools. For example, when a hand tool is selected, the worker characteristics to be considered are strength, anthropometry, manual dexterity, and motor capabilities.

Table 28.12 — Body Dimensions of U.S. Civilian Adults (female/male in cm)

	Dimensions	Percentile 5th	50th	95th	Standard Deviation
	Heights, Standing				
99	Stature ("height")	152.8/164.7	162.94/175.58	173.7/186.6	6.36/6.68
D19	Eye	141.5/152.8	151.61/163.39	162.1/174.3	6.25/6.57
2	Shoulder (acromion)	124.1/134.2	133.36/144.25	143.2/154.6	5.79/6.20
D16	Elbow	92.6/99.5	99.79/107.25	107.4/115.3	4.48/4.81
127	Wrist	72.8/77.8	79.03/84.65	85.5/91.5	3.86/4.15
38	Crotch	70.0/76.4	77.14/83.72	84.6/91.6	4.41/4.62
84	Overhead fingertip reach (on toes)	200.6/216.7	215.34/32.80	231.3/249.4	9.50/9.99
	Heights, Sitting				
93	Sitting height	79.5/85.5	85.20/91.39	91.0/97.2	3.49/3.56
49	Eye	68.5/73.5	73.87/79.02	79.4/84.8	3.32/3.42
3	Shoulder (acromion)	50.9/54.9	55.55/59.78	60.4/64.6	2.86/2.96
48	Elbow rest	17.6/18.4	22.05/23.06	27.1/27.4	2.68/2.72
73	Knee	47.4/51.4	51.54/55.88	56.0/60.6	2.63/2.79
86	Popliteal	35.1/39.5	38.94/43.41	42.9/47.6	2.37/2.49
104	Thigh clearance	14.0/14.9	15.89/16.82	18.0/19.0	1.21/1.26
	Depths				
36	Chest	20.9/21.0	23.94/24.32	27.8/28.0	2.11/2.15
187	Elbow–fingertip	40.6/44.8	44.35/48.40	48.3/52.5	2.36/2.33
26	Buttock–knee sitting	54.2/56.9	58.89/61.64	64.0/66.7	2.96/2.99
27	Buttock–popliteal sitting	44.0/45.8	48.17/50.04	52.8/54.6	2.66/2.66
106	Thumbtip reach	67.7/73.9	73.46/80.08	79.7/86.7	3.64/3.92
	Breadths				
53	Forearm–forearm	41.5/47.7	46.85/54.61	52.8/62.1	3.47/4.36
66	Hip, sitting	34.3/32.9	38.45/36.68	43.2/41.2	2.72/2.52
	Head Dimensions				
62	Length	17.6/18.5	18.72/19.71	19.8/20.9	0.64/0.71
60	Breadth	13.7/14.3	14.44/15.17	15.3/16.1	0.49/0.54
61	Circumference	52.3/54.3	54.62/56.77	57.1/59.4	1.46/1.54
68	Interpupillary breadth	5.7/5.9	6.23/6.47	6.9/7.1	0.36/0.37
	Hand Dimensions				
65	Wrist circumference	14.1/16.2	15.14/17.43	16.3/18.8	0.69/0.82
59	Length, stylion to tip 3	16.5/17.09	18.07/19.41	19.8/21.1	0.98/0.99
63	Breadth, metacarpal	7.4/8.4	7.95/9.04	8.6/9.8	0.38/0.42
60	Circumference, metacarpal	17.3/19.8	18.65/21.39	20.1/23.1	0.86/0.98
4	Digit 1: breadth, distal joint	1.9/2.2	2.06/2.40	2.3/2.6	0.13/0.13
1	Length	5.6/6.2	6.35/6.97	7.2/7.8	0.48/0.48
15	Digit 2: breadth, distal joint	1.5/1.8	1.73/2.01	1.9/2.3	0.12/0.15
10	Length	6.2/6.7	6.96/7.53	7.7/8.4	0.46/0.49
27	Digit 3: breadth, distal joint	1.5/1.7	1.71/1.98	1.9/2.2	0.11/0.14
22	Length	6.9/7.5	7.72/8.38	8.6/9.3	0.51/0.54
39	Digit 4: breadth, distal joint	1.4/1.6	1.58/1.85	1.8/2.1	0.11/0.14
34	Length	6.4/7.1	7.22/7.92	8.1/8.8	0.50/0.52
51	Digit 5: breadth, distal joint	1.3/1.5	1.47/1.74	1.7/2.0	0.11/0.13
46	Length	5.1/5.7	5.83/6.47	6.6/7.3	0.46/0.49
	Foot Dimensions				
51	Length	22.4/24.9	24.44/26.97	26.5/29.2	1.22/1.31
50	Breadth	8.2/9.2	8.97/10.06	9.8/11.0	0.49/0.53
75	Lateral malleolus height	5.2/5.8	6.06/6.71	7.0/7.6	0.53/0.55
124	Weight (kg) U.S. Army	49.6/61.6	62.01/78.49	77.0/98.1	8.35/11.10
	Weight (kg) civilians*	39/58*	62.0/78.5*	85/99*	13.8/12.6*

* Estimated (from Kroemer, 1981).
Note that all values (except for civilians' weight) are based on measured, not estimated, data that may be slightly different from values calculated from average plus or minus 1.65 standard deviation.
(Excerpted from Gordon, Churchill, Clauser, et al., 1989; Greiner, 1991.)

Worker strength is affected by body position, the direction of exerted force relative to the body, the type of grip used for holding the tool, and the handle size. Grip strength depends on the type of grip used for holding a handle. The power grip is preferred because of the large muscles recruited to perform the job. To optimize grip strength, handle shape is important. It is important to know handle span, handle circumference, as well as handle shape. The thumb, index, and middle fingers are the strongest fingers and should be used for producing the most grip force. By increasing the surface area of contact between the handle and the hand the amount of torque can be increased. However, grip strength has been shown to decrease when handles with diameters greater than 50 mm are used.[74]

Worker anthropometry can affect the posture a worker assumes when operating a hand tool. The location, orientation, and tool design should all be considered together along with the stature of the worker.[27] In some situations where the work location and orientation cannot be adjusted, it may be possible to select another tool or to change the location or orientation of work.

A tool that is too large may be difficult to use for performance or precision work. The type of grip suitable for a manual tool operation is often limited by the size of the tool. For example, a power grip can provide greater strength than a pinch grip; however, a pinch grip provides greater control of precision movements. As a result, tool selection should optimize the proportion of strength that an operator must exert with the ability to make necessary movements with speed and precision. There is a tradeoff between the increased mechanical advantage provided by a long tool, versus its weight and size, and the ability for an operator to manipulate and handle it.[50]

Table 28.13 — Tool Properties and Tool Parameters Affecting Tool Properties

Tool Properties	*Tool Parameters Affecting Tool Properties*
Load	Center of gravity location Tool mass Use of counterbalance or articulating arm Power line installation
Handle size	Type of grip needed (power, pinch)
Handle shape	Type of grip needed (power, pinch)
Handle orientation	Type of grip needed (power, pinch) Distribution of load Work location
Feed force	Type of fastener head Type of fastener tip or drill bit Work material
Sound level	Tool speed Power Work material Tool location
Reaction torque	Spindle torque Tool and handle length Stiffness of joint Torque reaction bar
Vibration	Tool weight Work material Abrasive material Tool speed Tool power Handle location Moment of inertia

Workstation and Task Factors

The main power hand tool characteristics and operational requirements for hand tools may include the following: 1) tool weight and load distribution; 2) triggers; 3) feed and reaction force; 4) handles; 5) work location and orientation; 6) tool accessories; 7) vibration; and 8) noise. The following Table 28.13 shows the power hand tool mechanical properties affected by certain tool parameters.[73]

Engineering Control and Design Consideration

The following list of areas are recognized for consideration, analysis, and implementation of controls for work-related musculoskeletal disorders: posture, exertion/force, contact pressure, handle friction, gloves, center of gravity, tool location, tool activation and throttle, reaction torque, balancer and suspension, and vibration. Table 28.14 shows examples of specific tool properties that are affected through design.[73]

The proper design and selection of hand tools is a complex process of analyzing the work station, materials, and methods; of accounting for worker characteristics; and of considering all of the variations that may result from the process or the individual. To accomplish this it is necessary to identify and have an adequate understanding of the interrelationship of most of the process variables. This information must be integrated to critically evaluate the physical characteristics of the tool and the location of its application.

Putting It All Together: Case Study Using Ergonomic Principles and Applications

This case study shows one of the many methods to evaluate work-related musculoskeletal disorders in the workplace, and how to evaluate the effectiveness of the interventions.

Case Study — An Ergonomics Evaluation of a Flywheel Milling Department at a Motorcycle Manufacturing Plant: An Intervention Study[75]

Background. In the winter of 1990, NIOSH received a joint request to evaluate musculoskeletal disorders of the upper limbs and back from the union and management of a vehicle-manufacturing facility employing approximately 650 workers. Particular concern was expressed about the flywheel milling areas.

The trigger for the NIOSH request began in 1989 when worker's compensation costs increased because of a growing number of injuries. This company had experienced an economic decline during the 1970s and early 1980s resulting in worker layoffs. In the mid-1980s, after an economic recovery, experienced workers on layoff were recalled to work. Eventually, as the recovery continued, the company began hiring new employees. About this same time, the company hired a new nurse for the health unit who has subsequently become the safety director. She began to more rigorously maintain the OSHA 200 logs and to educate workers and management concerning work-related musculoskeletal disorders. Because of a combination of these two factors, around 1989 the company experienced a large increase in workers' compensation cases.

Based in part on the initial NIOSH report,[76] several ergonomic interventions were developed and implemented by the company, which resulted in a reduction (from approximately 38% in 1989 to approximately 18% in 1992) in the number of reported musculoskeletal injuries in this department. This reduction was documented in three follow-back visits, which were conducted by NIOSH representatives in May 1992, October 1992, and August 1993. These visits resulted in information and recommendations that provided guidance to the company to help improve medical surveillance of musculoskeletal injuries.

Work Force and Physical Plant

Preintervention Evaluation. In January 1990, approximately 253 motorcycle engines and 170 motorcycle transmissions were fabricated and assembled each day at this facility. Production was 24 hours per day, and two to three employees worked in the flywheel milling area per eight-hour shift. The flywheels came in the following two sizes: large (FL) (approximately 19 lb premilled), and small (XL) (approximately 16 lb premilled). In another area in this department, there were five full-time employees assembling, truing, and balancing the flywheels. Employees rotate through the truing task every two to four hours. Occasionally, these employees would work 10 to 12 hour days to keep pace with production demands. There were 38 full-time workers in this department.

Table 28.14 — Physical Factors for Hand Tools and Design Objectives

Physical Factor	*Design Objectives*
Tool load	Use of light weight and composite materials Optimum load distribution Optimum handle location
Handle size/shape	Optimum size handle Optimum handle shape Adjustable handle size
Handle orientation	Optimum angle Adjustable orientation
Work location	Optimum location for tool load, handle size, and handle orientation
Sound level	Motor Housing and suspension Mufflers
Fasteners	Fastener head design
Torque	Shutoff mechanism
Vibration	Mortor mounting Tool balance and load distribution

Postintervention Evaluation. In August 1993, approximately 340 motorcycle engines and 254 motorcycle transmissions were fabricated and assembled each day at this facility. Production was 24 hours per day, and two to three employees worked in the flywheel milling area per 8-hour shift. The flywheels came in the following sizes: large (FL) (approximately 17.5 lb premilled), and small (XL) (approximately 13.5 lb premilled). In another area in this department, there were seven full-time employees assembling, truing, and balancing flywheels. Because of the changes to the truing area, employees did not need to be rotated as mentioned above. Also, improved forging specifications (discussed in detail later) of the FL and XL flywheels reduced the premilled weight. There were 50 full-time employees in this department.

Process Description

The NIOSH evaluation of this motorcycle manufacturing plant focused on the flywheel milling and assembly department, where milling, assembly and truing, and balancing flywheels were done.

Milling of the flywheels consists of a series of steps to complete the job cycle. The basic steps include the following: manually getting the forged flywheel from a supply cart, drilling and machine milling it, grinding off metal burrs, inspecting, measuring, and placing the finished flywheel in a receiving cart. Each milling "cell" contains three to four milling machines, a drill press, and two to three worktables. Approximately two to three pounds of metal are cut from each flywheel during the milling process.

After milling, the next phase is assembly of the flywheel unit. This consists of the gear and sprocket side of the flywheel, two connecting rods, bearings, and a crank pen. These components are assembled and put together by a "marriage press." After this, the unit is taken to the truing area for straightening and centering.

Truing of the flywheel is done by manually mounting it on a fixture on top of a table, and manually rotating the flywheel to determine misalignment (a centering gauge is viewed by the operator to determine misalignment). In the initial NIOSH evaluation, when the misalignment area was found, a 5-lb brass-head hammer, which was held by the employee, was repeatedly struck against the flywheel to straighten and center (true) the unit. In the follow-back evaluations, a 40-ton press performed the truing operation, and the hammers were eliminated. After the flywheel unit is trued, it is manually lifted and placed onto a cart that is moved to the balancing area.

A flywheel unit is balanced by picking it up from a cart and placing it in a cradle in the balancing machine. The flywheel connecting rods are attached to balancing arms that rotate the unit at high speeds. A computer determines where holes are to be drilled to provide balance when the flywheel unit is operating at high speeds. Following this procedure, the flywheel is manually moved from the balancing machine to a cart. The weights of fully assembled post-milled flywheels (sprocket and gear) and their components (bearings, crank pin, and two connecting rods) are large, (FL) 32.5 lb (34 lb maximum weight); and small, (XL) 25.6 lb (26 lb maximum weight). The finished flywheel units are then moved from this department to the engine assembly department.

Methods

NIOSH researchers conducted an initial visit (January 25, 1990), and three (May 19, 1992, October 28, 1992, August 30, 1993) follow-back health hazard evaluations. The evaluation of the flywheel milling department consisted of a walk-through survey of motorcycle engine fabrication and assembly, a review of OSHA 200 logs and company medical compensation data, informal interviews with employees, and an ergonomic evaluation of jobs in the flywheel milling area.

Ergonomic Evaluation

NIOSH Initial Evaluation. An in-depth ergonomic evaluation of the flywheel milling area was conducted during the initial survey; follow-back surveys consisted of assessing ergonomic changes in the flywheel milling jobs, as recommended in the in-depth report sent to the company.

The initial ergonomic evaluation consisted of 1) discussions with flywheel milling employees regarding musculoskeletal hazards associated with their job; 2) videotape of the flywheel milling process; 3) a biomechanical evaluation of musculoskeletal stress during manual handling of the flywheels; and 4) a recording of workstation dimensions. Two flywheel milling cells were evaluated.

Videotapes of the jobs were analyzed at regular speed to determine job cycle time, slow-motion to determine musculoskeletal hazards of the upper limbs during manual material handling tasks, and stop-action to sequence job steps and perform biomechanical evaluations of working postures. All video analysis procedures were used to document potential musculoskeletal hazards in performing the job.

Time and motion study techniques were used for the first phase of job analysis. *Work methods analysis* was used to determine the work content of the job. The second phase of job analysis was to review the job for recognized occupational risk factors for WMSDs. These WMSDs risk factors include repetition, force, posture, contact stress, low temperature, and vibration. Also, biomechanical evaluation of forces, which are exerted on the upper limbs, back, and lower limbs of the worker while performing the task, also were performed. This two-phase approach for job analysis and quantification of forces, which act on the body during materials handling, forms the basis for proposed engineering and administrative control procedures aimed at reducing the risk for musculoskeletal stress and injury.

After receipt of the initial NIOSH report in October 1990, the company conducted several meetings over a one- to two-year period to engineer out specific job hazards in the flywheel milling and assembly department. The meetings led to the systematic selection of equipment based on more than 20 performance criteria. Some of these criteria included the following: reduction or elimination of the specific hazard (vibration from hand tools); user friendly controls; noise reduction; easy access for maintenance personnel; parts availability; cycle time; machine guarding; and machine durability.

NIOSH Follow-Back Evaluations. Three follow-back evaluations were conducted between May 1992 and August 1993. During these evaluations NIOSH researchers spoke with the safety director, as well as the operators, managers, and engineers involved in the redesign of the work processes in the flywheel department. An evaluation was also done on the changes made since the initial evaluation. Specific NIOSH activities during these follow-back evaluations included 1) discussions with employees regarding changes in their job for musculoskeletal hazards; 2) videotaping the flywheel milling, truing, and balancing process; 3) reviewing company ergonomic committee activities on reducing job hazards in this department; 4) presenting education and training sessions on ergonomics to plant supervisors, engineers, and workers; and 5) reviewing OSHA 200 logs. Recommendations were made on site during these follow-back evaluations to encourage continuous improvement of jobs to reduce musculoskeletal disorders.

Rates of Musculoskeletal Disorders. OSHA 200 logs for the years 1987 to 1992 were obtained and coded. All musculoskeletal problems, including such conditions as sprains, strains, tendinitis, and carpal tunnel syndrome involving the upper extremities and back, were included in the analysis. Since it is extremely difficult to determine from the OSHA 200 logs whether a musculoskeletal sprain or strain is due to acute or chronic trauma, all of these events were included. Musculoskeletal contusions, which are likely to be more acute events, were not included. Information on the number of employees for each year was obtained and used to develop incidence rates. Although the change in the truing machine was completed in 1991, many of the other changes in the flywheel department were not in place until the end of 1993. Therefore, these data may not show the full effect of all the design changes.

Results

Table 28.15 summarizes the initial ergonomic recommendations for the flywheel milling area made by NIOSH researchers in January 1990, and the actions completed by the company on the last follow-back evaluation in September 1993.

This table shows that several actions were taken to address concerns about musculoskeletal injuries in the flywheel milling cell. Pre- and post-ergonomic intervention activities for the milling, assembly and truing, and balancing area are summarized below.

Table 28.15 — Ergonomic Changes

Engineering Controls	
Initial Recommendation (January 1990)	*Result (September 1993)*
Reduce the weight of the flywheels by improving forging specifications. This will reduce milling time and the amount of weight handled over the workday.	Weight of fly wheels were reduced by improving forging specifications by nearly 2 lbs. In addition, only one type of flywheel forging (for the gear and sprocket sides) is shipped to plant, and is milled to specifications. This simplifies the milling process, reduces waste, and multiple handling of flywheels.
Reduce or eliminate exposure to vibration from powered hand grinder. Twenty percent of the work cycle time consists of vibration exposure from this tool.	Customized metal deburring machines were purchased to eliminate over 90% of the exposure from the hand grinding operations. The hand grinder is used less than 1% of the work cycle time (for minor touch up of fly wheel).
Layout of the flywheel milling job is inefficient from a production and material handling perspective. Consider movable flywheel carts and/or gravity conveyors between milling work stations to reduce musculoskeletal stress.	The flywheel milling cell was reorganized into 2 work cells, reducing the number of machines per cell, and the amount of material handled per worker. For the FL flywheel, this resulted in a 38% reduction in material handling from 28,175 lbs to 17,472 per shift, and a 43% reduction in the number of times the operator needed to handle the flywheel during the milling process. Similar results were documented for the XL flywheel milling process.
Reduce the size of the metal pan that is built around the base of the indexing machine and round the corners to reduce the reach distance to attach the flywheels to the machine.	The indexing machine has been eliminated, and replaced by the another more efficient machine. Physical barriers were considered and designed out of this machine before it was put into operation.
Install durable rubberized floor matting around flywheel milling cells to reduce lower limb fatigue of workers.	Several types of rubberized floor matting were evaluated for durability, slip-resistance, and comfort by the operators in this department. A selection of rubberized mats were made available for the operators.
Remove all physical barriers that may cause workers to overreach, such as limited toe and leg space where the worker has to reach over barriers to manually position flywheels for processing.	Most physical barriers were eliminated because the worker was part of the workstation redesign process. Toe and leg space were considered when the work cells were redesigned. Machines, such as the drill press, were adjusted up to chest height of worker. This reduced stooping to position the flywheels in the machines.

Table 28.15 — Ergonomic Changes (continued)

Work Practices	
Initial Recommendation (January 1990)	*Result (September 1993)*
Recommend workers use the "power grip" rather than the "pinch grip." when handling the flywheel. The "pinch grip" requires handling of the flywheel by the fingertips and thumb, resulting in high musculoskeletal forces and fatigue. Use of two hands is also recommended when handling parts to reduce asymmetric biomechanical loading of the limbs and back.	All of the workers in this department received ergonomics training on material handling techniques. When the flywheels were handled at the wheel end, both hands were used, especially when positioning flywheels in or out of the milling machines.
When wheel carts are brought into the flywheel milling cell, they should be brought in with the cart bumper facing away from the traffic area to avoid contact with the worker's shins.	The wheel carts bumpers were retrofitted with tubular steel to reduce mechanical contact with the worker's shins. Several of the wheel carts were also fitted with hinged bumpers that can be manually rotated in the vertical position and out of the worker's way. Workers position the wheel carts close to their work area to reduce distance and material handling.
Operators should avoid overreaching while handling flywheels during milling. Overreaching may result in excess musculoskeletal stress and possibly injury, especially later in the work shift when the worker may become fatigued.	On-site training of workers about biomechanical aspects of work may have increased their awareness to reduce overreaching while performing their job. Redesign of the workstation also helped reduce overreaching by providing leg and toe clearance, and adjusting the height of the workstations to fit the worker.

Milling

Preintervention Evaluation. Milling of the large (FL) flywheel (average weight 19.0) and the small (XL) flywheel (average weight 16.0) consisted of 37 steps for the FL flywheel and 25 steps for the XL flywheel. It was estimated that 28,175 pounds of flywheels were manually handled for the FL flywheel and 18,980 pounds for the XL flywheel per eight-hour day. These total weights were derived by multiplying the average weight of the milled flywheel (17.5 lb FL, and 14.5 lb XL) times the average number of times the flywheel was picked up (23 and 17, respectively), times the average number of flywheels milled per day (70).

Postintervention Evaluation. Milling of flywheels was divided into two jobs. Instead of two flywheel castings for the left and right half (gear and sprocket sides), there was one master cast flywheel, weighing 17.5 lb for the FL flywheel, and 13.5 lb for the XL flywheel. In the first modified flywheel milling job, the number of steps required to complete the work cycle was 13. The number of flywheels milled per eight-hour day was approximately 84; this represents 17,472 pounds handled per day for the FL flywheel and 13,759 pounds for the XL flywheel. In the second flywheel milling job, the number of steps to complete the work cycle was nine. The number of flywheels milled per day also was approximately 84, representing 10,202 pounds for the FL, and 9526 pounds for the XL flywheel handled per day. Because of the short cycle time, the worker on this job also worked on the flywheel balancing job. Table 28.16 summarizes the material handling results of the flywheel milling job before and after ergonomic interventions.

Preintervention Hand-Arm Vibration Exposure. In addition to the potential overexertion injuries for manual handling of the flywheels in the milling cells, another concern was excess hand-arm vibration exposure from a hand-held grinder that removed metal burrs from the flywheel. It was determined that approximately 20 percent of the job cycle was used for removing metal burrs. As noted in Table 28.15, recommendations were provided to reduce vibration exposure and improve job efficiency.

Postintervention Hand-Arm Vibration Exposure. Vibration exposure was virtually eliminated with the purchase of a customized metal deburring machine. This machine was designed according to specifications from engineers and workers performing the job and would automatically remove burrs with grinding media (stones) inside the unit. The installation of this unit in the flywheel milling cell resulted in over a 90% reduction in hand-held grinders and a reduction from 20% of the work cycle to less than one percent (occasional touch-up) for hand-held grinding operations. The deburring machine allowed the worker to move on to other work elements while this job was done, thus making the job more efficient and reducing potential hazardous vibration exposure.

The cost of the deburring machine was more than $200,000. To justify the costs over hand-held grinding, the company established an evaluation program that incorporated the goal of sound engineering and production principles with ergonomic design. Table 28.17 lists the steps in which decisions and actions of plant personnel accomplished their goal as applied to the deburring machine.

Truing (Assembly and Centering)

Preintervention Truing Evaluation. After milling, the flywheels were assembled together with connecting rods, bearings, and a crank pen. A marriage press was used to sandwich the parts into one unit. The flywheel unit was then "trued." After mounting the flywheel on a fixture, the flywheel unit was manually rotated, using a centering gauge to detect misalignment. The unit was struck, using a five lb brass-head hammer held by the worker. Depending on the amount of straightening necessary, the initial impact of the brass-head hammer could be as high as 92,000 pounds. Impact forces were reduced as the flywheel was straightened to specifications. The repeated forces needed to straighten the unit were somewhat traumatizing to the workers, and they needed to

Table 28.16 — Comparison of Pre- and Post-Interventions of Manual Handling of Flywheel Milling

Pre-intervention (January 1990)	*FI-Flywheel*	*XI-Flywheel*
Premilled Flywheel Weight	19.0	16.0
Avg. Weight	17.5	14.5
Avg. Cycle Time	5 minutes	4 minutes
# Flywheels/8-hr day	70	75
# Steps Moving Flywheel	23	17
# lbs. Moved/8-hr shift	28,175	18,980
Post-Intervention First Flywheel Cell[1] (September 1993)		
Premilled Flywheel Weight	17.5	13.5
Avg. Flywheel Weight	16.0	12.6
Avg. Cycle Time	4 minutes	4 minutes
# Flywheels/8-hour day	84	84
# Steps Moving Flywheel	13	13
# lbs. Moved/8-hour shift	17,472	13,759
Post-Intervention Second Flywheel Cell[2]		
Average Wt. Flywheel	16.0	12.6
Average cycle time	1.5 minutes	1.5 minutes
# Flywheels/8-hour day	84[3]	84
# Steps Moving Flywheel	9	9
Avg. Weight/8-hour day	12,096	9,526

[1] Flywheel milling completed by another worker in adjacent cell.
[2] Second flywheel cell completes the milling process.
[3] Up to 280 flywheels can be milled per 8-hour day. However, this worker also does flywheel balancing job and only keeps pace with the first flywheel milling cell.

Table 28.17 — Steps in the Metal Deburring Machine Purchase

Steps	*Activity*	*Comments*
1.	NIOSH report (January, 1990), observes potential problem from hand-arm vibration exposure from hand-held grinder	Recommends several options to reduce exposure, including a metal finishing machine to remove burrs.
2.	Problem-solving team formed by the company	Team participants: Manufacturing engineer -1, operators -2, maintenance machine repairman -1, supervisor -1, tool designer -1, medical -1, purchasing -1 , and facility -1.
3.	Mission statement formed	"Deburr flywheels and connecting rods in a manner to decrease musculoskeletal injuries from hand grinders, while improving quality and reducing variability."
4.	Overview of method	1. three vendors quoted project; 2. team ran trials with all three and rated results on a matrix; 3. one vendor, received highest quality matrix rating plus received a consensus favorable rating from the team.
5.	Definition of priorities	1. "What the customer wants." (safety, quality, ergonomics); 2. How the company can meet these requirements: a. reduce hand grinding > 90%, b. machine construction, c. ease of load and unload.
6.	Analysis methods	1. trials and analysis; 2. interview of vendors; 3. discussion and review of machine and process details.
7.	Justification	1. Safety, (a). eliminate flywheel grinding by 90%, (b). estimated savings from prevented lost-time accidents $53,679 (1987-1991). 2. Quality, same or improved — no loose burrs, reduced variability, complexity, increase throughput. 3. Ergonomic, (a). easy to load and unload, (b) no forward bending, especially with weight out in front of body, (c). both hands available to handle flywheels. 4. Housekeeping and environmental, (a). noise cover, (b). eliminates flying metal.
8.	Delivery and payback impacts	(a). headcount - meets planned requirements for future layout and schedule production increases, (b). meets capacity effect with less increased manual time, (c). cycle time less than 2 minutes/flywheel, 20 seconds to load and unload, (d). labor cost savings - none except cost increase avoidance savings with increases in schedule, (e). flexibility - increased, (f). set-up less than 10 minutes, (g). in-process inventory, (h). floor space - more than hand grinders; same for alternatives, (I). over head - increased. annual usage cost saved (grinders and bits $1730, new process annual costs $7848)
9.	Employee modification recommendations	(a). insulation covers (noise reduction, (b). load arms presented to operator (to reduce bending over), (c). rinse cycles.
10.	Costs	$229,616
11.	Timetable	Delivery (12-23-92), Installation (1-11-93), Implementation (6-14-93).

be rotated from this job every two to four hours. Engineers and workers were developing ways to reduce exposure to this job when NIOSH researchers arrived during the initial visit in January 1990. NIOSH researchers agreed that the job needed to be changed to reduce the impact force stressors to the upper limbs.

Postintervention Truing Evaluation. Recommendations to reduce exposure to this job (called the "hammer slammer" job by workers) resulted in the use of a 40-ton press that was modified, based on plant ergonomic committee input, for truing the flywheels. The press completely eliminated the need for the brass hammers, thus eliminating mechanical trauma to the upper limbs from this task.

Balancing

Preintervention Balancing Evaluation. After the flywheel unit was trued, the next step was balancing. This process involved manually picking up the flywheel from a cart and placing it in a cradle in the balancing machine. The flywheel connecting rods were attached to balancing arms, which rotated the flywheel at high speeds. Balance sensors relayed a profile of the unit's balance characteristics to a computer, which determined where the holes were to be drilled. After the holes were drilled, the flywheel unit was rotated once more for a final balance check. The flywheel was manually picked up from the balancing machine and placed in a cart. The process was then repeated. Using the 1991 NIOSH lifting formula, it was determined that workers performing this job were occasionally at risk for back injury when manually handling the FL flywheel unit. It was determined that when the flywheel unit was picked up, a safe weight was approximately 30 pounds, and when it was placed in the balancing cradle, the safe weight was approximately 21 pounds. The difference in safe lifting weights is mainly attributable to the location of the load from the body when it is placed in the balancing cradle. Figure 28.18 summarizes the information used to determine the NIOSH recommended weight limit (RWL) of 30 and 21 pounds.

Postintervention Balancing Evaluation. A similar procedure for balancing the flywheels was performed using the sensors and computer. However, because the flywheel unit was heavy, an overhead hoist mounted on an x-y trolley (gantry hoist) was used to lift the unit and place it in the balancing unit cradle. Balancing was performed by the computer, and the gantry hoist was used once more to put the finished part back in the cart. The hoist is an excellent engineering control to address this material handling problem.

Musculoskeletal Disorders

Table 28.18 shows the yearly rates of musculoskeletal disorders for the entire production facility, as well as specific rates for the most commonly affected body parts: shoulder, hand/arm, and back.

Table 28.18 also shows a breakdown of the disorders into those affecting the upper extremities and back injuries during that five-year period for the flywheel department. Upper extremity disorders have steadily been decreasing since the intervention program began in 1990, while back injuries have decreased in a more erratic pattern. Since many of the interventions that would have decreased the amount of manual material handling were only implemented in late 1992, the more modest decrease in back injuries is not surprising.

Table 28.19 shows the rate of musculoskeletal disorders for the flywheel department and for all other departments combined between 1987 and 1993. This table shows that the rate of injuries increased dramatically in 1989 in the flywheel department and has subsequently been decreasing. In the other departments there has been a modest, but continued increase over the same period.

Figure 28.19 shows a graph of the incidence rates as well as the change in

Table IV. Calculations using 1991 NIOSH lifting formula for flywheel Assembly lift for balancing job.

Job Analysis Worksheet

Department: 909
Job Title: Balancer
Date: August 30, 1993

Job Description: Load Flywheel Unit into Balancing Machine

Step 1. Measure and record task variables

Object Weight (lbs)		Hand Location (in)				Vertical Distance (in)	Asymmetric Angle (degrees)		Freq. Rate	Duration	Object Coupling
		Origin		Dest.			Origin	Dest.	lifts /min	Hours	
L(avg)	L(Max.)	H	V	H	V	D	A	A	F		C
32	34	15	40	20	35	5	0	30	< .2	< 2	Fair

Step 2. Determine the multipliers and compute the Recommended Weight Limits (RWL's)

RWL = LC x HM x VM x DM x AM x FM x CM

ORIGIN RWL = 51 x .67 x .93 x 1.0 x 1.0 x .95 x 1.0 = 30 lbs

DESTINATION RWL = 51 x .50 x .96 x 1.0 x .90 x .95 x 1.0 = 21 lbs

Step 3. Compute the LIFTING INDEX

ORIGIN Lifting index = Object Weight / RWL = 34/30 = 1.13

DESTINATION Lifting index = Object Weight / RWL = 34/21 = 1.62

Formulas for calculating Recommended Weight Limit: Load Constant = 51 lb; Horizontal Multiplier (HZ) = (10/H); Vertical Multiplier (VM) = 1-(.0075)|V-30|); Distance Multiplier (DM) = .82 + (1.8/D); Asymmetric Multiplier (AM) = 1-(0032A); Frequency Multiplier (FM) = from Appendix A; Coupling Multiplier (CM) = from Appendix A

Figure 28.18 — Case study. Calculations using 1991 NIOSH lifting formula for flywheel assembly lift for balancing job.

employment over the past decade. This graph illustrates the effects of the economic recovery described at the beginning of this case study. After an initial period of increasing production where experienced workers were recalled to work, in about 1988 or 1989 new and inexperienced workers were hired. This rapidly increasing rate of production, combined with the hiring of a new nurse and safety coordinator who increased awareness of work-related musculoskeletal disorders, resulted in the initial increase in rates from 1989 to 1990.

Discussion

The goal of any effective ergonomic intervention effort is to eliminate the job hazards and to use reduced morbidity as a measure of success. However, if the process of how this is done has been recorded and that process follows a systematic procedure that benefits the workers and the company, then the process can be effectively repeated again and again and has a high probability of success. This case study presents a process where there was commitment from top management to provide resources to manufacture flywheels better, from company engineers to select the most cost-effective equipment available, and from workers to be involved in every aspect of the equipment from selection to custom design. The goal is a better product, made in a cost-effective manner by a healthy work force.

In this motorcycle manufacturing plant there was a strong commitment by both the management and workers to improve the ergonomic design of their equipment. Although there has been a downward trend in musculoskeletal injuries particularly affecting the upper extremities, the full effect has only begun to be seen since many design changes have only recently been implemented.

The problem-solving approach used by this company was effective because it used a team of employees, engineers, managers, and medical personnel. This resulted in a participation approach in which all parties in the flywheel milling and assembly department contributed with their knowledge and experience. It was noteworthy that after the committee selected this press, management supported the committee's decision. This fostered productive meetings and timely approval of other purchases to reduce and eliminate job hazards associated with musculoskeletal injuries on the job. The examples given below highlight these points.

Table 28.18 — Entire Production Facility Incidence Rates of Musculoskeletal Disorders from OSHA 200 Logs

Year	*Total Rate (% of workers)*	*Shoulder Rate (% of workers)*	*Hand/Arm Rate (% of workers)*	*Back Rate (% of workers)*
1987	9	3	2	3
1988	10	1	4	4
1989	14	1	5	6
1990	17	5	5	7
1991	17	4	3	7
1992	16	3	4	6
1993	13	3	3	4

Table 28.19 — Flywheel Department: Trends in Work-Related Musculoskeletal Disorders 1987–1993

		WRMD[1] Incidence Rate (% of Workers)		*Lost/Restricted Workdays*	
Year	*Workers*	*Total*	*Lost/Restricted Workday Cases*	*Number of 100 Workers*	*Median # per case[2]*
1987	34	17.6	11.8	110	10
1988	34	11.8	8.9	130	13
1989	36	38.9	27.6	610	13
1990	44	20.5	11.5	390	33
1991	43	27.9	18.7	480	21
1992	45	17.8	13.4	560	12
1993	48	20.8	12.5	190	11

[1] Work-related musculoskeletal disorders—Includes all neck, upper extremity and back cases.
[2] This includes only those cases that had some lost or restricted workdays.

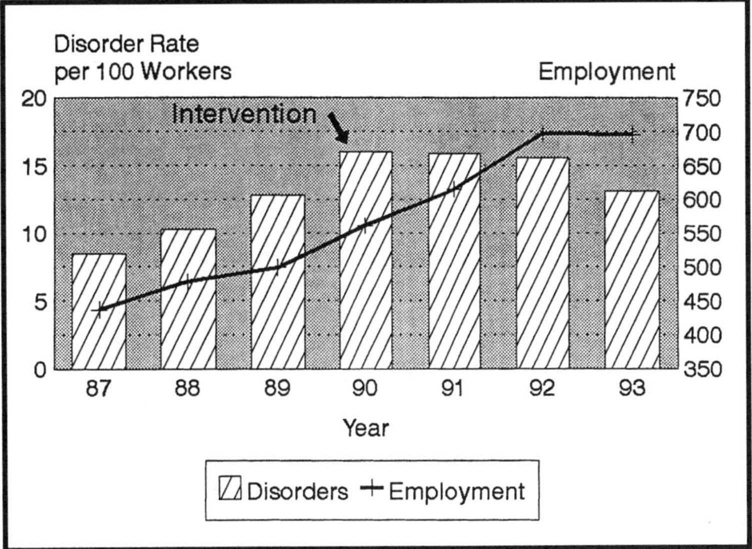

Figure 28.19 — Case study. Results of employment growth and rates of musculoskeletal disorders at a motorcycle manufacturing plant.

A committee was formed to resolve the problem of high morbidity from the flywheel truing job. The committee included two supervisors, two truing operators, and six engineers. They first established the four most important criteria for the new process and then identified two potential vendors with machines that could meet these criteria. The selection was a 40-ton press. The next step was to send a sampling of flywheels to the press manufacturer for truing by the manufacturer's personnel. Subsequent evaluation by engineers and workers in the company's flywheel department was performed. When the quality issues were satisfied, workers from the flywheel department were sent to the press manufacturing plant to test the machine. Based on employee input, the equipment was modified prior to purchase. The new equipment was phased in as the old equipment was phased out, over a 2-month period. Although some workers had some reservations about the new press because they were highly skilled in their jobs, the department supervisor worked with them to allow a phase-in period where they performed their jobs six hours a day using the brass hammers and two hours using the new press. The workers were encouraged to comment on the press and how it could be improved. Gradually the workers gave up the hammer job and used the press.

The new press cost the facility about $58,000, whereas the annual costs of brass hammers had been $40,000. Also, during the previous six years, there had been 10 injuries from the old truing process. These injuries cost more than $20,000 in medical costs alone, plus additional expenses from lost work time.

A similar change was made for the flywheel balancing job. Using the revised NIOSH lifting formula, analysis of this task showed that manually lifting the flywheel unit from the cart, twisting the body, and placing the flywheel in the cradle of the balancing machine could pose a musculoskeletal hazard. Initially, an articulating arm having a base pod was used to perform this job. This was done without input from the committee. When the unit was installed, it proved to be too cumbersome and took too much time. The unit was soon pushed to the side, and the workers performed the job as usual. When the committee, including the worker who performed this job, thought about how to achieve the goals of reducing material handling, saving time, and keeping the unit out of the way during the balancing process, they conceived of a gantry hoist system that would do the job. The result was a suspended gantry hoist, which was easily controlled by the worker in three directions.

As these examples show, the company and its employees learned several lessons about what it takes to sustain a successful ergonomics program. The first lesson was that problem solving usually includes a series of steps rather than one leap from the problem to a solution. Depending on the training and resources of the company, this process can be immediate or take months. Also, resources needed to do the job can be nominal or very costly. Examples of the two extremes in this study are raising the drill press to eliminate stooping while loading flywheels into its fixture and purchasing a customized spindle deburring machine.

Another lesson is that successful ergonomic programs need to be sustained because of the dynamic nature of today's business and production environment. A variety of approaches can be used to achieve this. However, to sustain a successful ergonomics program, providing stimulation to the process by hiring competent outside experts brings about more ergonomic changes. Often management may become complacent and react to problems as they arise, rather than forming a plan to engineer ergonomics into the machines and processes they control. The outside expert can be most effective on the planning side of the equation, so that ergonomic factors can be engineered into the machines and processes prior to operation.

Finally, the importance of the front-line supervisor who serves as the communicator between management and the production worker is very important. The front-line supervisor can make or break an ergonomics program. The supervisor provides a supportive environment for workers' ideas, developing their concepts into practical applications that use sound engineering principles. The front-line supervisor also needs to effectively communicate with upper management, to present needs in a systematic way, and to secure resources to get the job done right. Ergonomics training for the front-line supervisor is important, but just as important is skillful communication of the needs of the workers and the goals of the company.

Epidemiologic Evidence of the Work-Relatedness of Musculoskeletal Disorders

The question of whether musculoskeletal disorders are work-related was recently addressed in a NIOSH study titled *Musculoskeletal Disorders and Workplace Factors: A Critical Review of Epidemiologic Evidence for Work-Related Musculoskeletal Disorders of the Neck, Upper Extremity, and Low Back.*[53] While musculoskeletal disorders were recognized as having occupational etiologic factors as early as the 18th century, it was not until the 1970s that occupational factors were examined in a systematic way using epidemiologic methods to associate the work-relatedness of these conditions. In this document, NIOSH concluded that "a large body of credible epidemiological research exists that shows a consistent relationship between MSDs and certain physical factors, especially at higher exposure levels... and in particular when workers are exposed to several risk factors simultaneously (Table 28.20).[53]

Because the link between MSDs and work risk factors has been more clearly established through this body of work, then logic would dictate that reducing and preventing such risk factors from occurring in the work setting should reduce the incidence and severity of such disorders as well. In order to do this effectively, ergonomic control programs may be one of the more effective mechanisms to achieve this end. In addition to the NIOSH science document mentioned above, a companion document entitled: *Elements of Ergonomics Programs: A Primer Based on Workplace Evaluations of Musculoskeletal Disorders*[1] was developed to guide industry and labor representatives in methods to systematically evaluate, reduce, and eliminate work risk factors that may cause MSDs. The basic elements are outlined in the next section of this chapter.

Table 28.20 — Evidence for Causal Relationship Between Physical Work Factors and MSDs

Body Part Risk Factor	*Strong Evidence* (+++)	*Evidence* (++)	*Insufficient Evidence* (+/0)
Neck and Neck/Shoulder			
Repetition		X	
Force		X	
Posture	X		
Vibration			X
Shoulder			
Posture		X	
Force			X
Repetition		X	
Vibration			X
Elbow			
Repetition			X
Force		X	
Posture			X
Combination	X		
Hand/Wrist			
Carpal tunnel syndrome			
Repetition		X	
Force		X	
Posture			X
Vibration		X	
Combination	X		
Tendinitis			
Repetition		X	
Force		X	
Posture		X	
Combination	X		
Hand-Arm Vibration Syndrome			
Vibration	X		
Back			
Lifting/forceful movement	X		
Awkward posture		X	
Heavy physical work		X	
Whole body vibration	X		
Static work posture			X

Basic Elements of an Ergonomic Control Program

The first step in forming an ergonomics team is to make sure all personnel resources in the plant are represented, including management, labor, engineering, medical, and safety personnel. The team establishes a training schedule in which an outside expert, familiar with the plant operations, teaches ergonomics principles to management and workers.

Over time, medical surveillance is used to determine the effectiveness of the ergonomic interventions. Medical surveillance can be active or passive. Active surveillance is usually conducted by administering standardized questionnaires to workers in problem and nonproblem jobs. Passive surveillance is conducted by examining medical injury or illness records, such as OSHA 200 Logs, workers' compensation reports, and attendance records for absenteeism. Analysis is done on both approaches to identify patterns and changes over time.

Decreases in the incidence and severity of musculoskeletal disease and injury serve as one measure of success. Increases in productivity and product quality serve as another. In many instances, workers' awareness of their musculoskeletal disease and injuries will show an increase in incidence rates early in the ergonomics program. However, as the program matures, both incidence and severity usually decrease. The length of time required to observe such effects can be a function of the company resources, worker participation, company size, corporate culture, and type of product produced. On average, it takes two to three years before "real" effects are seen. Two important lessons can be learned from successful ergonomics programs: 1) The program should not be created as an entity separate from the mission of the plant. Rather it should be woven into existing programs, such as safety and medical programs; and 2) The ergonomics program must be sustained, as it is an iterative process that incorporates the philosophy of continuous improvement, transfer of technologies from one department to another, and documentation of ergonomic success and failures.

As referenced above, *Elements of Ergonomics Programs: A Primer Based on Workplace Evaluations of Musculoskeletal Disorders* spells out seven basic training elements for developing an in-house ergonomics program. They are:

1. *Determine if training is needed:* If the evidence gathered from checking health records and results of the job analysis indicates a need to control ergonomic risk factors, then employees must be provided with the training necessary for them to gain the knowledge to implement control measures.
2. *Identify training needs:* Different categories of employees will require different kinds of ergonomics instruction.
3. *Identify goals and objectives:* Define the objectives of training in clear, directly observable, action-oriented terms.
4. *Develop learning activities:* Whatever the mode of training—live lectures, demonstrations, interactive-video programs, or varied instructional aids—learning activities should be developed that will help employees demonstrate that they have acquired the desired knowledge or skill.
5. *Conduct training:* Training should take into account the language and educational level of the employees involved. Trainees should be encouraged to ask questions that address their particular job concerns, and hands-on learning opportunities should be encouraged.
6. *Evaluate training effectiveness:* A common tool for training evaluations is to ask questions about whether workers found the instruction interesting and useful to their jobs and if they would recommend it to others. More important, however, are measures of the knowledge gained or improvements in skills, as may be specified in the course objectives. Knowledge quizzes, performance tests, and behavioral observations are also evaluation tools. One exercise recommended here is that the class propose improvements in workplace conditions, based on information learned in class, and that they present these to management for their review. This relates to another level of evaluation, which is whether the training produces some overall change at the workplace. The latter measure is complicated by the time required before results are apparent, and the training may have been only one of several factors responsible for such results.
7. *Improving the program:* If the evaluations indicate that the objectives of the training were not achieved, a review of the training plan elements would be in order and revisions should be made to correct shortcomings.

Although the above-mentioned steps can help employers develop ergonomics training activities without having to hire outside help, much depends on the existing capabilities of the staff. If in-house expertise in ergonomics is limited, start-up activities could necessitate the use of consultants or outside special training for those employees who would ultimately assume responsibility for ergonomic activities within the workplace.

For more information on ergonomics and its applications refer to Appendix C which lists 1) journals that commonly contain articles related to ergonomics; 2) electronic sources of information; and 3) glossary of terms.

Acknowledgments

Thanks are extended to Robert Radwin, Ph.D., of the University of Wisconsin for his contribution to the section on hand tools; to Anil Mital, Ph.D., of the University of Cincinnati for his contribution to the section on Risk Factor Assessment Techniques; to Alex Cohen, Ph.D., consultant, and to Karl H.E. Kroemer, Ph.D., from Virginia Tech, Don Chaffin, Ph.D., from the University of Michigan, and Jim Annis and John McConville of Anthropology Research Project, Yellow Springs, Ohio, for their consent to use selected tables and figures. Thanks also to Sue Rogers, Ph.D., and Dave Ridyard for their contributions to the section on checklists.

References

1. **National Institute for Occupational Safety and Health (NIOSH):** *Elements of Ergonomics Programs. A Primer Based on Workplace Evaluations of Musculoskeletal Disorders* (pub. no. 97-117). Cincinnati, OH: U.S. Department of Health and Human Services, Public Health Service, Centers for Disease Control and Prevention, NIOSH, 1997.
2. **American Industrial Hygiene Association (AIHA) Ergonomics Committee:** Position statement on ergonomics, approved by AIHA board, May 1997.
3. **National Safety Council:** *Fundamentals of Industrial Hygiene*, 4th ed. Barbara A. Plog, Jill Niland, and Patricia J. Quinlan, eds. Chicago, Il: National Safety Council, 1996. pp. 347–401.
4. **Tichauer, E.:** *The Biomechanical Basis of Ergonomics*. New York: John Wiley & Sons, 1978.
5. **Bhattaharya, A. and James D. McGlothlin (eds.):** *Occupational Ergonomics: Theory and Applications*. New York: Marcel Dekker, Inc., 1996. pp. 685.
6. **Armstrong, T.J., P. Buckle, L.J. Fine, M. Hadberg, et al.:** A conceptual model for work-related neck and upper-limb musculoskeletal disorders. *Scand. J. Work, Environ. Health 19(2)*:73–84 (1993).
7. **Bureau of Labor Statistics:** *Workplace Injuries and Illnesses by Selected Characteristics*. Washington, DC: U.S. Department of Labor, Bureau of Labor Statistics, 1993. pp. 95–142.
8. **Bureau of Labor Statistics:** *Characteristics of Injuries and Illnesses Resulting in Absences from Work*. Washington, DC: U.S. Department of Labor, Bureau of Labor Statistics, 1994. pp. 96–163.
9. **Tanaka, S., D. Wild, P. Seligman, W. Halperin, et al.:** Prevalence and work-relatedness of self-reported carpal tunnel syndrome among U.S. workers: analysis of the occupational health supplement data to the 1988 National Health Interview Survey. *Am. J. Ind. Med. 27(4)*:451–470 (1995).
10. **Webster, B.S. and S.H. Snook:** The cost of 1989 workers' compensation low back pain claims. *Spine 10*:1111–1115 (1994).
11. **Guo, H., S. Tanaka, L. Cameron, P. Seligman, et al.:** Back pain among workers in the United States: national estimates and workers at high risks. *Am. J. Ind. Med. 28(5)*:591–602 (1995).
12. **Frymoyer, J.W. and W.L. Cats-Baril:** An overview of the incidence and costs of low back pain. *Orthop. Clin. N. Am. 22*:262–271 (1991).
13. **Liira, J.P., H.S. Shannon, L.W. Chambers, and T.A. Haines:** Long term back problems and physical work exposures in the 1990 Ontario health survey. *Am. J. Pub. Health 86(3)*:382–387 (1996).
14. **National Institute for Occupational Safety and Health (NIOSH):** *National Occupational Research Agenda* (pub. no. 96-115). Cincinnati, OH: U.S. Department of Health and Human Services, Public Health Service, Centers for Disease Control and Prevention, NIOSH, 1996.
15. **Tichauer, E.R. and H. Gage:** Ergonomic principles basic to hand tool design. *Am. Ind. Hyg. Assoc. J. 38*:622–634 (1977).
16. **Armstrong, T.J.:** *An Ergonomic Guide to Carpal Tunnel Syndrome*. Akron, OH: American Industrial Hygiene Association, 1983.
17. **Putz-Anderson, V. (ed.):** *Cumulative Trauma Disorders: A Manual for Musculoskeletal Diseases of the Upper Limbs*. London: Taylor & Francis, 1988.
18. **McGlothlin, J.D.:** "An Ergonomic Program to Control Work-Related Cumulative Trauma Disorders of the Upper Extremities." Ph.D. diss., University of Michigan, Ann Arbor, Mich., 1988.
19. **Mital, A. and A. Kilbom:** Design, selection, and use of hand tools to alleviate cumulative trauma of the upper extremities: Parts I and II—Guidelines for the practitioner. *Int. J. Ind. Ergonom. 10(1-2)*:1–6 (1992a,b).
20. **Westgaard, R.H. and C. Jensen, and K. Hansen:** Individual and work-related risk factors associated with symptoms of musculoskeletal complaints. *Int. Arch. Occup. Environ. Health 64*:405–413, 1993.
21. **Kuorinka, I. and F. Forcier (eds.):** *Work Related Musculoskeletal Disorders (WMSDs): A Reference Book for Prevention*. London: Taylor & Francis, 1995.
22. **Kilbom, A.:** Repetitive work of the upper extremity: Part II—The scientific basis (knowledge base) for the guide. *Int. J. Ind. Ergonom. 14(1-2)*:59–86 (1994b).

23. **Chiang, H, Y. Ko, S. Chen, H. Yu, et al.:** Prevalence of shoulder and upper-limb disorders among workers in the fish-processing industry. *Scand. J. Work, Environ. Health 19(2)*:126–131 (1993).
24. **Silverstein, B.A.:** "The Prevalence of Upper Extremity Cumulative Trauma Disorders in Industry." Diss., University of Michigan, Ann Arbor, Mich., 1985.
25. **Moore, J.S. and A. Garg:** Determination of the operational characteristics of ergonomic exposure assessments for prediction of disorders of the upper extremities and back. In *Proceedings of the 11th Congress of the International Ergonomics Association*. London: Taylor & Francis, 1991. pp. 144–146.
26. **Osorio, A.M., R.G. Ames, J. Jones, J. Castorina, et al.:** Carpal tunnel syndrome among grocery store workers. *Am. J. Ind. Med. 25(2)*:229–245 (1994).
27. **Armstrong, T.J., P. Buckle, L.J. Fine, M. Hagberg, et al.:** A conceptual model for work- related neck and upper-limb musculoskeletal disorders. *Scand. J. Work, Environ. Health 19(2)*:73–84 (1993).
28. **Swansen, A., I. Matev, and G. Groot:** The strength of the hand. *Bull. Prosthet. Res. 10-14*:145–153 (1970).
29. **Armstrong, T.J., J. Foulke, J. Bradley, and S. Goldstein:** Investigation of cumulative trauma disorders in a poultry processing plant. *Am. Ind. Hyg. Assoc. J. 43*:103–116 (1982).
30. **Mital, A.:** "Recognition of Musculoskeletal Injury Hazards for the Upper Extremity and Lower Back" (final report, contract no. CDC-94071VID). Cincinnati, OH: U.S. Department of Health and Human Services, Public Health Service, Centers for Disease Control and Prevention, National Institute for Occupational Safety and Health, 1996.
31. **Marras, W. and R.W. Schoenmarklin:** Wrist motions in industry. *Ergonomics 36(4)*:341–351 (1993).
32. **Tichauer, E.R.:** *Occupational Biomechanics. An Introduction to the Anatomy of Function of Man at Work.* New York: Institute of Rehabilitation Medicine, New York University Medical Center, 1975.
33. **Chaffin, D.B. and G.B.J. Andersson:** *Occupational Biomechanics*, 2nd ed. New York: Wiley Interscience, 1991.
34. **Corlett, N., J. Wilson, and I. Manenica (eds.):** The ergonomics of working postures. *Proceedings of the First International Occupational Ergonomics Symposium, Zadar, Yugoslavia.*
35. **Sommerich, C.M., J.D. McGlothlin, and W.S. Marras:** Occupational risk factors associated with soft tissue disorders of the shoulder: a review of recent investigations in the literature. *Ergonomics 36(6)*:697–717 (1993).
36. **Armstrong, T. and D. Chaffin:** Carpal tunnel syndrome and selected personal attributes. *J. Occup. Med. 21*:481–486 (1979).
37. **Tichauer, E.R.:** Biomechanics sustains occupational safety and health. *Ind. Eng. 8*:46–56 (1976).
38. **Boiano, J., A. Watanabe, D. Habes:** *Armco Composits* (DHHS/NIOSH pub. no. HETA 81-143-1041). Cincinnati, OH: Department of Health and Human Services, Public Health Service, Centers for Disease Control and Prevention, National Institute for Occupational Safety and Health, 1982.
39. **Schiefer, R.E., R. Kok, M.I. Lewis, and G.B. Meese:** Finger skin temperature and manual dexterity 0—some inter-group differences. *Appl. Ergon. 15(2)*:135–141 (1984).
40. **Fox, W.F.:** Human performance in the cold. *Hum. Factors 9(3)*:203–220 (1967).
41. **Pelmear, P.L., D. Leong, I. Taraschuk, and L. Wong:** Hand-arm vibration syndrome in foundrymen and hard rock miners. *J. Low Freq. Noise Vib. 5(4)*:163–167 (1986).
42. **Olsen, N., S.L. Nielsen, and P. Voss:** Cold response of digital arteries in chain saw operators. *Br. J. Ind. Med. 38*:82–88 (1981).
43. **Williamson, D.K., F.A. Chrenko, and E.J. Hamley:** A study of exposure to cold in cold stores. *Appl. Ergon. 15*:25–30 (1984).
44. **Taylor, W. and P.L. Pelmear (eds.):** *Vibration White Finger in Industry.* New York: Academic Press, 1975.
45. **Brammer, A.J., W. Taylor, and J.E. Piercy:** Assessing the severity of the neurological component of the hand-arm vibration syndrome. *Scand. J. Work, Environ. Health 12(4)*:428–431 (1986).
46. **National Institute for Occupational Safety and Health (NIOSH):** *Criteria for a Recommended Standard:*

Occupational Exposure to Hand-Arm Vibration (DHHS/NIOSH pub. no. 89-106). Cincinnati, OH: NIOSH, 1989.

47. **Riley, M.W., D.J. Cochran, and C.A. Schanbacher:** Force capability differences due to gloves. *Ergonomics 28*:441–447 (1985).
48. **Hertzberg, H.T.E.:** Some contributions of applied physical anthropology to human engineering. *Ann. NY Acad. Sci. 63*:616–629 (1995).
49. **Lyman, J.:** The effects of equipment design on manual performance in protection and functioning of the hands in cold climates. In *Production and Functioning of the Hands in Cold Climates*, R.R. Fisher, ed. Washington, DC: National Academy of Sciences, National Research Council, 1957. pp. 86–101.
50. **Sperling, L. R. Kadefors, and A. Kilbom:** Tools and hand function: the cube model—a method for analysis of the handling of tools. In *Designing for Everyone*, Y. Queinnec and F. Daniellou, eds. London: Taylor & Francis, 1991. pp. 176–178.
51. **Plummer, R., T. Stobbe, R. Ronk, W. Myers, et al.:** Manual dexterity evaluation of gloves used in handling hazardous materials. *Proceedings of the Human Factors Society, 29th Annual Meeting*. Santa Monica, CA: Human Factors Society, 1985. pp. 819–823.
52. **Bhattaharya, A. and J.D. McGlothlin (eds.):** *Occupational Ergonomics: Theory and Applications*. New York: Marcel Dekker, Inc., 1996. p. 4.
53. **National Institute for Occupational Safety and Health (NIOSH):** *Musculoskeletal Disorders and Workplace Factors: A Critical Review of Epidemiologic Evidence for Work-Related Musculoskeletal Disorders of the Neck, Upper Extremity, and Low Back* (pub. no. 97- 141) B. Bernard, ed. Cincinnati, OH: U.S. Department of Health and Human Services, Public Health Service, Centers for Disease Control and Prevention, NIOSH, 1997.
54. **National Institute for Occupational Safety and Health (NIOSH):** *Investigation of Occupational Wrist Injuries in Women. Terminal Progress Report*. Cincinnati, OH: U.S. Department of Health and Human Services, Public Health Service, Centers for Disease Control and Prevention, NIOSH, 1991.
55. **Waters, T.R., V. Putz-Anderson, A. Garg, and L.J. Fine:** Revised lifting equation for the design and evaluation of lifting tasks. *Ergonomics 36(7)*:749–776 (1993).
56. **National Institute for Occupational Safety and Health (NIOSH):** *Workplace Use of Back Belts; Review and Recommendations* (pub. no. 94–127). Cincinnati, OH: U.S. Department of Health and Human Services, Public Health Service, Centers for Disease Control and Prevention, NIOSH, 1994.
57. **Barnes, R.M.:** Motion and time study. In *Design and Measurement of Work*, 6th ed. New York: John Wiley & Sons, 1968.
58. **Karger, D. and W. Hancock:** *Advanced Work Measurement*. New York: Industrial Press, 1982.
59. **Niebel, B.W.:** *Motion and Time Study*, 8th ed. Homewood, IL: Irwin, 1988.
60. **Konz, S.:** *Work Design: Industrial Ergonomics*, 3rd ed. Worthington, OH: Publishing Horizons, Inc., 1990.
61. **Lifshitz, Y. and T.J. Armstrong:** A design checklist for control and prediction of cumulative trauma disorder in intensive manual jobs. In *Proceedings of the Human Factors Society 30th Annual Meeting*, Dayton, Ohio, September 29–October 3, 1986. vol. 2, pp. 837–841. Santa Monica, CA: Human Factors Society, 1986.
62. **Waters, T.:** A checklist for manual materials handling. Cincinnati, OH: National Institute for Occupational Safety and Health, Division of Biomedical and Behavioral Sciences, 1989.
63. **Rodgers, S.H.:** A functional job analysis technique. In *Occupational Medicine: State of the Art Reviews*, Moore, J.S. and A. Gary, eds. Philadelphia, PA: Hanley & Belfuss Publishers, 1992.
64. **Marras, W.S., J.D. McGlothlin, D.R. McIntyre, M. Nordin, et al.:** *Dynamic Measures of Low Back Performance, An Ergonomics Guide*. Akron, OH: American Industrial Hygiene Association, 1993.
65. **Chaffin, D.B. and G.B.J. Andersson:** *Occupational Biomechanics*. New York: John Wiley & Sons, 1984. pp. 166–178.
66. **National Institute for Occupational Safety and Health (NIOSH):** *Revised Guide to Manual Lifting*. Cincinnati, OH: Robert A. Taft Laboratories, DHHS/NIOSH, 1991.

67. **Astrand, P.O. and K. Rodhal:** *Textbook of Work Physiology*, 3rd ed. New York: McGraw- Hill, 1986.
68. **Mital, A., A. Garg, W. Karwowski, S. Kumar, et al.:** Status in human strength research and application. *Trans. Inst. Ind. Eng. 25(6)*:57–69 (1993b).
69. **Gordon, C.C., T. Churchill, C.E. Clauser, B. Bradtmiller, et al.:** *Anthropometric Survey of U.S. Army Personnel* (TR-89-027). Natick, MA: U.S. Army Natick Research, Development, and Engineering Center, 1989.
70. **McGlothlin, J.D.:** Ergonomic interventions for the soft drink beverage delivery industry. In *Proceedings of the International Ergonomics Association World Conference*, A. de Moraes and S. Marino, eds. Rio de Janeiro, Brazil: International Ergonomics Association, 1995. pp. 16–20.
71. **Corlett, E.N. and R.P. Bishop:** A technique for assessing postural discomfort. *Ergonomics 19(2)*:175–182 (1976).
72. **Borg, G.A.V.:** Psychophysical bases of perceived exertion. *Med. Sci. Sports Exerc. 14*:377–381 (1982).
73. **National Institute for Occupational Safety and Health (NIOSH):** *Proceedings of a NIOSH Workshop: A Strategy for Industrial Power Hand Tool Ergonomic Research—Design, Selection, Installation, and Use in Automotive Manufacturing* (DHHS/NIOSH pub. no. 95-114). Cincinnati, OH: U.S. Department of Health and Human Services, Public Health Service, Centers for Disease Control and Prevention, NIOSH, 1995.
74. **Pheasant, S. and D. O'Neil:** Performance in gripping and turning—a study in hand/handle effectiveness. *Appl. Ergonom. 6(4)*:20–208 (1975).
75. **McGlothlin, J.D. and S. Baron:** *Harley-Davidson Incorporated, Milwaukee, Wisconsin* (DHHS/NIOSH pub. no. HETA 91-0208-2422). Cincinnati, OH: Department of Health and Human Services, Public Health Service, Centers for Disease Control and Prevention, National Institute for Occupational Safety and Health, 1994.
76. **McGlothlin, J.D., R.A. Rinsky, and L.J. Fine:** *Harley-Davidson Incorporated, Milwaukee, Wisconsin* (DHHS/NIOSH pub. no. HETA 90-134-2064). Department of Health and Human Services, Public Health Service, Centers for Disease Control and Prevention, National Institute for Occupational Safety and Health, 1990.

Appendix A. Tips for Properly Videotaping Jobs for Analysis

(Developed by Dan Habes, NIOSH)

Protocol for Videotaping Jobs for Risk Factors

The following is a guide to preparing a videotape and related task information for facilitating job analyses and assessments of risk factors for work-related musculoskeletal disorders.

Materials needed:

Video camera and blank tapes
Spare batteries (at least two) and battery charger
Clipboard, pens, paper, blank checklists
Stopwatch, strain gauge (optional) for weighing objects

Videotaping Procedures:

1. To verify the accuracy of the video camera to record in real time, videotape a worker or job with a stopwatch running in the field of view for at least 1 min. The play-back of the tape should correspond to the lapsed time on the stopwatch.
2. Announce the name of the job on the voice channel of the video camera before the taping of any job. Restrict running time comments to the facts. Make no editorial comments.
3. Tape each job long enough to observe all aspects of the task. Tape five to 10 min for all jobs, including at least 10 complete cycles. Fewer cycles may be needed if all aspects of the job are recorded at least three to four times.
4. Hold the camera still, using a tripod if available. Don't walk unless absolutely necessary.
5. Begin taping each task with a whole-body shot of the worker. Include the seat/chair and the surface the worker is standing on. Hold this for two to three cycles, then zoom in on the hands/arms or other body parts which may be under stress due to the job task.
6. It is best to tape several workers to determine if workers of varying body size adopt different postures or are affected in other ways. If possible, try to tape the best- and worst-case situations in terms of worker "fit" to the job.

   The following suspected upper body problems suggest focusing on the parts indicated:
   - wrist problems/complaints. . . . . hands/wrists/forearms.
   - elbow problems/complaints. . . . . .arms/elbows.
   - shoulder problems/complaints. . . . . . .arms/shoulders.

   For back and lower limb problems, the focus would be on movements of the trunk of the body and leg, knee, and foot areas under stress due to task loads or other requirements

7. Video from whatever angles are needed to capture the body part(s) under stress.
8. Briefly tape the jobs performed before and after the one under actual study to see how the targeted job fits into the total department process.
9. For each taped task, obtain the following information to the maximum extent possible:
   - ❏ if the task is continuous or sporadic;
   - ❏ if the worker performs the work for the entire shift, or if there is rotation with other workers;
   - ❏ measures of work surface heights and chair heights and whether adjustable;
   - ❏ weight, size and shape of handles and textures for tools in use; indications of vibration in power tool usage;
   - ❏ use of handwear;
   - ❏ weight of objects lifted, pushed, pulled or carried; and
   - ❏ nature of environment in which work is performed — (too cold or too hot?).

Appendix B. Proactive Ergonomics Program Checklists

General Workstation Design Principles*

1. Make the workstation adjustable, enabling both large and small persons to fit comfortably and reach materials easily.
2. Locate all materials and tools in front of the worker to reduce twisting motions. Provide sufficient work space for the whole body to turn.
3. Avoid static loads, fixed work postures, and job requirements in which operators must frequently or for long periods:
   - ❑ lean to the front or the side;
   - ❑ hold a limb in a bent or extended position;
   - ❑ tilt the head forward more than 15°; or
   - ❑ support the body's weight with one leg.
4. Set the work surface above elbow height for tasks involving fine visual details and below elbow height for tasks requiring downward forces and heavy physical effort.
5. Provide adjustable, properly designed chairs.
6. Allow the workers, at their discretion, to alternate between sitting and standing. Provide floor mats or padded surfaces for prolonged standing.
7. Support the limbs; provide elbow, wrist, arm, foot, and back rests as needed and feasible.
8. Use gravity to move materials.
9. Design the workstation so that arm movements are continuous and curved. Avoid straight-line, jerking arm motions.
10. Design so arm movements pivot about the elbow rather than around the shoulder to avoid stress on shoulder, neck, and upper back.
11. Design the primary work area so that arm movements or extensions of more than 15 inches are minimized.
12. Provide dials and displays that are simple, logical, and easy to read, reach, and operate.
13. Eliminate or minimize the effects of undesirable environmental conditions such as excessive noise, heat, humidity, cold, and poor illumination.

Design Principles for Repetitive Hand and Wrist Tasks*

1. Reduce the number of repetitions per shift. Where possible, substitute full or semi-automated systems.
2. Maintain neutral (handshake) wrist positions:
   - Design jobs and select tools to reduce extreme flexion or deviation of the wrist.
   - Avoid inward and outward rotation of the forearm when the wrist is bent to minimize elbow disorders (i.e., tennis elbow).
3. Reduce the force or pressure on the wrists and hands:
   - Wherever possible, reduce the weight and size of objects that must be handled repeatedly.
   - Avoid tools that create pressure on the base of the palm which can obstruct blood flow and nerve function.
   - Avoid repeated pounding with the base of the palm.
   - Avoid repetitive, forceful pressing with the finger tips.
4. Design tasks so that a power rather than a finger pinch grip can be used to grasp materials. Note that a pinch grip is five times more stressful than a power grip.
5. Avoid reaching more than 15 inches in front of the body for materials:
   Avoid reaching above shoulder height, below waist level, or behind the body to minimize shoulder disorders.
   Avoid repetitive work that requires full are extension (i.e., the elbow held straight and the arm extended).
6. Provide support devices where awkward body postures (elevated hands or elbows and extended arms) must be maintained. Use fixtures to relieve stressful hand/arm positions.
7. Select power tools and equipment with features designed to control or limit vibration transmissions to the hands, or alternatively design work methods to reduce time or need to hold vibrating tools.
8. Provide for protection of the hands of working in a cold environment. Furnish a selection of glove sizes and sensitize users to problems of forceful over gripping when worn.
9. Select and use properly designed hand tools (e.g., grip size of tool handles should accommodate majority of workers).

*Adapted from design checklists developed by Dave Ridyard, CPE, CIH, CSP, Applied Ergonomics Technology, 270 Mather Road, Jenkintown, PA 19046-3129

Hand Tool Use and Selection Principle*

1. Maintain straight wrists. Avoid bending or rotating the wrists. Remember, bend the tool, not the wrist. A variety of bent-handle tools are commercially available.
2. Avoid static muscle loading. Reduce both the weight and size of the tool. Do not raise or extend elbows when working with heavy tools. Provide counter-balanced support devices for larger, heavier tools.
3. Avoid stress on soft tissues. Stress concentrations result form poorly designed tools that exert pressure on the palms or fingers. Examples include short-handled pliers and tools with finger grooves that do not fit the worker's hand.
4. Reduce grip force requirements. The greater the effort to maintain control of a hand tool, the higher the potential for injury. A compressible gripping surface rather than hard plastic may alleviate this problem.
5. Whenever possible, select tools that use a full-hand power grip rather than a precision finger grip.
6. Maintain optimal grip span. Optimum grip spans for pliers, scissors, or tongs, measured from the fingers to the base of the thumb, range from 6 to 9 cm. The recommended handle diameters for circular-handle tools such as screwdrivers are 3 to 5 cm when a power grip is required, and 0.75 to 1.5 cm when a precision finger grip is needed.
7. Avoid sharp edges and pinch points. Select tools that will not cut or pinch the hands even when gloves are not worn.
8. Avoid repetitive trigger-finger actions. Select tools with switches that can be operated with all four fingers. Proximity switches are the most desirable triggering mechanism.
9. Isolate hands from heat, cold and vibration. Heat and cold can cause loss of manual dexterity and increased grip strength requirements. Excessive vibration can cause reduced blood circulation in the hands causing a painful condition known as white-finger syndrome.
10. Wear gloves that fit. Gloves reduce both strength and dexterity. Tight-fitting gloves can put pressure on the hands, while loose-fitting gloves reduce grip strength and pose other safety hazards (e.g., snagging).

Design Principles for Lifting and Lowering Tasks*

1. Optimize material flow through the workplace by:
   - ❏ reducing manual lifting of materials to a minimum;
   - ❏ establishing adequate receiving, storage, and shipping facilities; and
   - ❏ maintaining adequate clearances in aisle and access areas.
2. Eliminate the need to lift or lower manually by:
   - ❏ increasing the weight to a point where it must be mechanically handled;
   - ❏ palletizing handling of raw materials and products; and
   - ❏ using unit load concept (bulk handling in large bins or containers).
3. Reduce the weight of the object by:
   - ❏ reducing the weight and capacity of the container;
   - ❏ reducing the load in the container; and
   - ❏ limiting the quantity per container to suppliers.
4. Reduce the hand distance from the body by:
   - ❏ changing the shape of the object or container so that it can be held closer to the body; and
   - ❏ providing grips or handles for enabling the load to be held closer to the body.
5. Convert load lifting, carrying, and lowering movements to a push or pull by providing:
   - ❏ conveyors;
   - ❏ ball caster tables;
   - ❏ hand trucks; and
   - ❏ four-wheel carts.

Design Principles for Pushing and Pulling Tasks*

1. Eliminate the need to push or pull by using the following mechanical aids, when applicable:
   - Conveyors (powered and nonpowered);
   - Powered trucks;
   - Lift tables; and
   - Slides or chutes.
2. Reduce the force required to push or pull by:
   - ❏ reducing side and/or weight of load;
   - ❏ using four-wheel trucks or dollies;
   - ❏ using nonpowered conveyors;

*Adapted from design checklists developed by Dave Ridyard, CPE, CIH, CSP, Applied Ergonomics Technology, 270 Mather Road, Jenkintown, PA 19046-3129

- ❑ requiring that wheels and casters on hand-trucks or dollies have (1) periodic lubrication of bearings, (2) adequate maintenance, and (3) proper sizing (provide larger diameter wheels and casters);
- ❑ maintaining the floors to eliminate holes and bumps; and
- ❑ requiring surface treatment of floors to reduce friction.

3. Reduce the distance of the push or pull by:
   - ❑ moving receiving, storage, production, or shipping areas closer to work production areas; and
   - ❑ improving the production process to eliminate unnecessary materials handling steps.
4. Optimize the technique of the push or pull by:
   - ❑ providing variable-height handles so that both short and tall employees can maintain an elbow bend of 80° to 100°;
   - ❑ replacing a pull with a push whenever possible; and
   - ❑ using ramps with a slope of less than 10%.

Design Principles for Carrying Tasks*

1. Eliminate the need to carry by rearranging the workplace to eliminate unnecessary materials movement and using the following mechanical aids, when applicable:
   - Conveyors (all kinds);
   - Lift trucks and hand trucks;
   - Tables or slides between workstations;
   - Four-wheel carts or dollies; and
   - Air or gravity press ejection systems.
2. Reduce the weight that is carried by:
   - ❑ reducing the weight of the object;
   - ❑ reducing the weight of the container;
   - ❑ reducing the load in the container; and
   - ❑ reducing the quantity per container to suppliers.
3. Reduce the bulk of the materials that are carried by:
   - ❑ reducing the size or shape of the object or container;
   - ❑ providing handles or hand-grips that allow materials to be held close to the body; and
   - ❑ assigning the job to two or more persons.
4. Reduce the carrying distance by:
   - ❑ moving receiving, storage, or shipping areas closer to production areas; and
   - ❑ using powered and nonpowered conveyors.
5. Convert carry to push or pull by:
   - ❑ using nonpowered conveyors; and
   - ❑ using hand trucks and push carts.

*Adapted from design checklists developed by Dave Ridyard, CPE, CIH, CSP, Applied Ergonomics Technology, 270 Mather Road, Jenkintown, PA 19046-3129

Appendix C.

1) Journals in which Articles Related to Ergonomics are Likely to Appear.
2) Electronic Sources of Information.

Journals in which Articles Related to Ergonomics are Likely to Appear

American Industrial Hygiene Association Journal
American Industrial Hygiene Association, 2700 Prosperity Ave., Ste. 250, Fairfax, VA 22031-4307, Tel 703-849-8888, Fax 703-207-3561

American Journal of Industrial Medicine
John Wiley & Sons, Inc., Journals, 605 Third Ave., New York, NY 10158, Tel 212-850-6000, Fax 212-850-6088

Applied Ergonomics
Butterworth-Heinemann, Part of the Reed Elsevier group, Linacre House, Jordan Hill, Oxford OX2 8DP, England, Tel 0865-310366, Fax 0865-310898

Applied Occupation & Environmental Hygiene
Elsevier Science Inc., 655 Avenue of the Americas, New York, NY 10010

Clinical Biomechanics
Butterworth-Heinemann, Part of the Reed Elsevier group, Linacre House, Jordan Hill, Oxford OX2 8DP, England, Tel-0865-310366, Fax 0865-310898

Ergonomics
Taylor & Francis Ltd., Rankine Road., Basingstoke, Hants, RG24 8PR, England. Tel 4-1256-840366, Fax 44-1256-479438

Human Factors
Human Factors and Ergonomics Society, Box 1369, Santa Monica, CA 90406-1369, Tel 310-394-1811, Fax 310-394-2410

International Journal of Industrial Ergonomics
Elsevier Science B.V., P.O. Box 211, 1000 AE Amsterdam, Netherlands, Tel 31-20-4853911, Fax 31-20-4853598

Journal of Biomechanics
Elsevier Science Ltd., Pergamon, P.O. Box 800, Kidington, Oxford OX5 IDX, England, Tel 44-1865-843000, Fax 44-1865-843010

Journal of Hand Surgery: American Volume
Churchill Livingston International, 650 Ave. of the Americas, New York, NY 10011, Tel 212-206-5040, Fax 212-727-7808

Journal of Neurosurgery
American Association of Neurological Surgeons, 1224 W. Main St., Ste. 450, Charlottesville, VA 22903, Tel 804-924-5503, Fax 804-924-2702

Muscle and Nerve
John Wiley & Sons, Inc., Journals, 605 Third Ave., New York, NY 10158, Tel 212-850-6645, Fax 212-850-6021

Neurology
Advanstar Communications, Inc., 7500 Old Oak Blvd., Cleveland, OH 44130, Tel 216-826-2839, Fax 216-891-2726

Occupational Ergonomics
Chapman and Hall, Inc., 29 W. 35th St., New York, NY 10001, Tel 212-244-3336, Fax 212-563-2269

Occupational Medicine
Butterworth-Heinemann, Part of the Reed Elsevier group, Linacre House, Jordan Hill, Oxford OX2 8DP, England, Tel 0865-310366, Fax 0865-310898

Orthopedics
Slack, Inc. 6900 Grove Road, Thorofare, NJ 08086, Tel 609-848-1000, Fax 609-853-5991

Rheumatology
S. Karger AG, Allschwilerstr., 10. P.O. Box, CH-4009 Basel, Switzerland, Tel 061-3061111, Fax 061-3061234

Scandinavian Journal of Work, Environment and Health
Topeliuksenkatu 41 aA, FIN-00250 Helsinki, Finland, Tel (+358 0)474-7694 and (+358 31) 260-8644, Fax (+358 0) 878-3326

Spine
(Philadelphia, 1986); state of the art reviews, 1986. 3/yr. $90 (foreign $99).
Hanley & Belfus, Inc., 210 S. 13th St., Philadelphia, PA 19107, Tel 215-546-7293, Fax 215-790-9330

Work
Elsevier Science Ireland Ltd., P.O. Box 85, Limerick, Ireland, Tel 353-61-471944, Fax 353-61-472144

Electronic Sources of Information

BIOMCH-L

BIOMCH-L is a listserve that provides a forum for discussion among experts/practitioners in the fields of biomechanics, bioengineering, and ergonomics. The wide variety of membership generates a multidisciplinary platform for discussing relevant topics of interest. To subscribe to the list, send e-mail to:

Bitnet users: listserv@hearn.bitnet

Other users: listserv@nic.surfnet.nl

Mail the following one-line message: subscribe biomch-l YourFirstName YourLastName (YourAffiliation). Follow the instructions sent to you by the listserver to complete the process. World Wide Web (WWW) archives are located at http://www.kin.ucalgary.ca/isd/biomch_l.html

BIOMECHANICS YELLOW PAGES

This is an electronic database of products and services related to the field maintained by Pierre Baudin at the following WWW address:

http://www.orst.edu/~bowenk/byp.html

ERGOWEB

This is a WWW site for useful information and services related the field. It provides a comprehensive source of information, software, references, and simple software programs for practicing ergonomists and biomechanists:

http://ergoweb.mech.utah.edu/

BIOMECHANICS WORLDWIDE (BWW)

Another comprehensive and updated source maintained by Pierre Baudin for navigating the WWW sites dealing with biomechanics and ergonomics:

http://dragon.acadiau.ca/~pbaudin/biomch.html

COMPUTER SYSTEMS TECHNICAL GROUP

This is a forum for discussing topics related to computer applications in human factors. Send a one-line e-mail message to listserv@vtvml.cc.vt.edu. The body of the mail should contain the message:

subscribe CSTG-L "Your Full Name in Quotes"

HUMAN FACTORS TECHNICAL GROUP ON AGING

This is a forum for discussing topics related to human factors and aging. Send a one-line e-mail message to listserv@UCFIVM.cc.ucf.edu. The body of the mail should contain the message:

subscribe HFTGA "Your Full Name in Quotes"

HUMAN FACTORS COMMUNICATIONS TECHNICAL GROUP

This is a forum for discussing topics related to technical discussions in communications in human factors. Send a one-line e-mail message to listserv@VTVMl.cc.vt.edu. The body of the mail should contain the message:

subscribe COMMS-L "Your Full Name in Quotes"

BIOMECHANICS-ERGONOMICS RESOURCE TANK

This is a source of information related to biomechanics and ergonomics, with useful links to WWW sites serving a parallel purpose; maintained by the authors:

http://www.uc.edu/~bhattatt/welcome.html

GOVERNMENT AND ORGANIZATIONAL ELECTRONIC SOURCES OF ERGONOMICS INFORMATION

http://www.cdc.gov/niosh/homepage.html

NIOSH home page

http://www.osha.gov/index.html

Occupational Safety & Health Administration

http://thomas.loc.gov

Thomas legislative information on the Internet

http://www.ccohs.ca/

Canadian Centre for Occupational Health & Safety

http://www.who.ch/

World Health Organization (WHO) home page

http://www.osha-slc.gov/

OSHA Salt Lake City home page

http://ergo.human.cornell.edu

Cornell Ergonomics

http://www2.ncsu.edu/CIL/NCERC/index.html

North Carolina Ergonomics Resource Center

http://www.os.dhhs.gov/

U.S. Department of Health and Human Services home page

Owner of a small oil drilling company in Kansas sits in the "doghouse," a small trailer for workers on site, while the last of many drill samples are taken to determine if oil production will be a financially sound investment. (Layne Kennedy, 1992)

Outcome Competencies

After completing this chapter, the user should be able to:

1. Define underlined terms used in this chapter.
2. Recall the Safety Triad.
3. Describe the approaches to changes in people.
4. Recognize the complexity of people.
5. Discuss stress and distress.
6. Explain behavior-based approach.
7. Select activators vs. consequence to promote safe behavior.
8. Define four types of consequences.
9. Describe the relevance of person-based psychology.
10. Explain the social dynamics of occupational health and safety.

Key Terms

actively caring • applied behavior • attitudes • authority • behavior-based coaching • behavior-based observation • belonging • commitment • counter control • distress • feedback • human dynamics • incentive/reward program • ingratiation • interdependency • interpersonal trust • motivation and emotion • optimism • personal control • consistency principle • psychology • reciprocity • recognition process • scarcity principle • self-efficacy • self-esteem • social psychology • social thought and behavior • stress • values

Prerequisite Knowledge

General college psychology course.

Key Topics

I. The Safety Triad
   A. Behavior-Based vs. Person-Based Approaches

II. The Complexity of People
   A. BASIC ID
   B. Stress vs. Distress

III. Fundamentals of the Behavior-Based Approach
   A. The DO IT Process
   B. Intervening with Activators
   C. Four Types of Consequences

IV. Relevance of Person-Based Psychology
   A. Self-Esteem
   B. Empowerment
   C. Enhancing the Actively Caring Person-States

V. The Social Dynamics of Industrial Health and Safety
   A. We Try to be Consistent in Thought and Deed
   B. We Reciprocate to Return the Favor
   C. The Power of Authority
   D. We Actively Care for People We Like
   E. Value Increases with Scarcity

29

Psychology and Occupational Health

E. Scott Geller, Ph.D.

Introduction

Psychology influences and can be used to benefit almost every aspect of our lives — including our health and safety. But what is "psychology" anyway? The *American Heritage Dictionary* defines psychology as "1. The science of mental processes and behavior, 2. The emotional and behavioral characteristics of an individual, group, or activity." Similarly, the first two definitions in the *New Merriam-Webster Dictionary* are "1. The science of mind and behavior, 2. The mental and behavioral characteristics of an individual or group."

Two features of both first definitions of "psychology" are noteworthy. Each uses the term "science," and each refers to both behavioral and mental processes. Behaviors are the outside, objective and observable aspects of people, whereas mental or mind refers to the inside, subjective and unobservable characteristics of people. Science implies the objective and systematic analysis and interpretation of reliable observations of natural or experimental phenomena.

A psychology of safety must be based on rigorous research, not common sense or intuition. That's what science is all about. Much of the psychology in self-help paperbacks, audiotapes, and motivational speeches is not founded on scientific investigation, but is presented because it sounds good and will "sell." The psychology in this chapter was not selected on the basis of armchair hunches but rather from the relevant research literature.

The Safety Triad

Generally, occupational health and safety requires continual attention to three domains: 1) environment factors (including equipment, tools, physical layout, procedures, standards, and temperature); 2) person factors (including people's attitudes, beliefs, and personalities); and 3) behavior factors (including safe and at-risk work practices, as well as going beyond the call of duty to intervene on behalf of another person's safety). This triangle of safety-related factors has been termed "The Safety Triad"[1,2] and is illustrated in Figure 29.1.

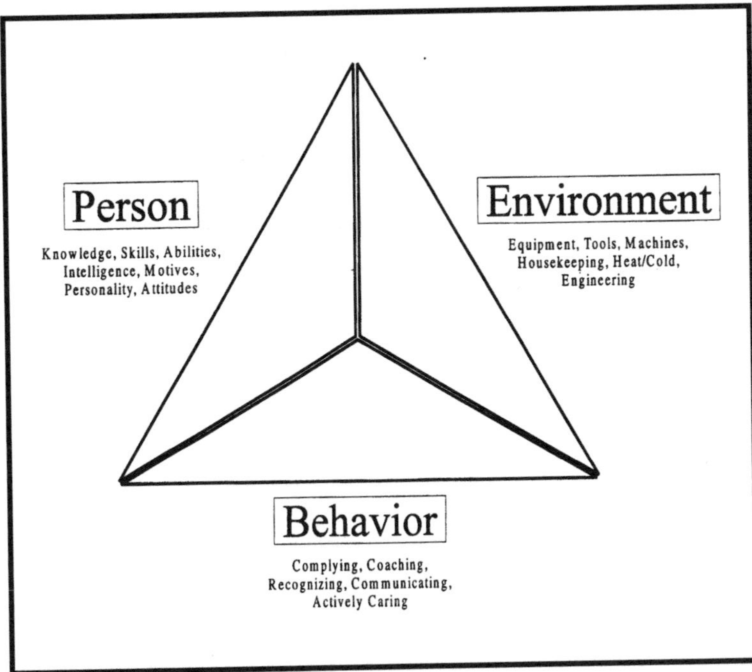

Figure 29.1 — The Safety Triad depicting three domains of factors determining occupational health and safety.

The domains in Figure 29.1 are dynamic and interactive. A change in one factor eventually impacts the other two. For example, behaviors that reduce the probability of injury often involve environmental change, and lead to attitudes consistent with the safe behaviors, especially if the behaviors are viewed as voluntary. In other words, when people choose to act safely, they act themselves into safe thinking. Such behaviors often result in some environmental change. The behavior-based and person-based domains represent the human dynamics of occupational safety and need to be addressed in order to improve industrial health and safety.

Behavior-Based vs. Person-Based Approaches

Most of the myriad of opinions and recommendations given to address the psychology of occupational health and safety can be classified into one of two basic approaches to produce beneficial change in people — a person-based approach and a behavior-based approach. In fact, most of the numerous psychotherapies available to treat developmental disabilities and psychological disorders (from neurosis to psychosis) can be classified as essentially person-based or behavior-based. That is, most psychotherapies focus on changing people either from the inside out (as in "thinking people into acting differently") or from the outside in (as in "acting people into thinking differently"). In other words, person-based approaches attack individual attitudes or thinking processes directly (e.g., by teaching clients new thinking strategies or giving them insight into the origin of their abnormal or unhealthy thoughts, attitudes, or feelings). In contrast, behavior-based approaches attack the clients' behaviors directly (e.g., by changing relationships between behaviors and their consequences).

Many clinical psychologists use both person-based and behavior-based approaches with their clients, depending on the nature of the problem. Sometimes the same client is treated with both person-based and behavior-based intervention strategies. This chapter integrates relevant principles from these two psychological approaches in order to continually improve industrial health and safety.

The Person-Based Approach. Imagine observing two employees pushing each other in a parking lot as a crowd gathers to watch. Is this aggressive behavior, horseplay, or mutual instruction for self-defense? Are the employees physically attacking each other to inflict harm, or does their physical contact indicate a special friendship and mutual understanding of the line between aggression and play? Perhaps longer observation of this interaction, with attention to verbal behavior, will help determine whether this physical contact between individuals is aggression, horseplay, or a teaching/learning demonstration. However, a truly accurate account of this event might require an assessment of each individual's intentions or feelings. It is possible, in fact, that one person was aggressing while the other was just having fun; or the interaction started as horseplay and progressed to aggression (from the perspective of personal feelings, attitudes, or intentions of the two individuals).

This scenario illustrates a basic premise of the person-based approach. Focusing only on observable behavior does not explain enough. People are much more than their behaviors. Concepts like intention, creativity, intrinsic motivation, subjective interpretation, self-esteem, and mental attitude are essential to understanding and appreciating the human element of a problem. Thus, a person-based approach applies surveys, interviews and focus-group discussions to find out how individuals feel about certain situations, conditions, behaviors, or personal interactions.

A wide range of therapeutic approaches fall within the general framework of person-based, from the psychoanalytic techniques of Sigmund Freud, Alfred Adler, and Carl Jung to the client-centered humanism developed and practiced by Carl Rogers, Abraham Maslow, and Viktor Frankl.[(3)] Humanism is the most popular person-based approach today, as evidenced by the current market of "pop psychology" videotapes, audiotapes, and self-help books. Some current popular industrial psychology tools (e.g., the Myers-Briggs Type Indicator and other trait measures of personality, motivation, or risk-taking propensity) stem from psychoanalytic theory and practice.

The key principles of humanism included in most "pop psychology" approaches to increasing personal achievement are: 1) everyone is unique in numerous ways and the special characteristics of individuals cannot be understood or appreciated by applying general principles or concepts; 2) individuals have far more potential to achieve than they typically realize and should not feel hampered by past experiences or present liabilities; 3) the present state of an individual in terms of feeling, thinking, and believing is a critical determinant of personal success; 4) one's self-concept influences mental and physical health, as well as personal effectiveness and achievement; 5) ineffectiveness and abnormal thinking and behavior result from large discrepancies between one's real self ("who I am") and ideal self ("who I would like to be"); and 6) individual motives vary widely and come from within a person. Readers familiar with the writings of W. Edwards Deming[4,5] and Stephen R. Covey[6,7] will recognize that these eminent industrial consultants would be classified as humanists (or advocates of a person-based approach).

The Behavior-Based Approach. The behavior-based approach to applied psychology is founded on behavioral science as conceptualized and researched by B.F. Skinner.[8,9] In his experimental analysis of behavior, Skinner rejected for scientific study unobservable inferred constructs such as self-esteem, cognition, intentions, and attitudes. He researched only observable behavior and its social, environmental, and physiological determinants. Therefore, the behavior-based approach starts with an identification of observable behaviors to change and environmental conditions or response consequences that can be manipulated to influence the target behavior(s) in desired directions.

This approach has been used effectively to solve environmental, safety, and health problems in organizations and throughout entire communities. First the problem is defined in terms of relevant observable behavior. Then an intervention process is designed and implemented to decrease behaviors causing the problem and/or increase behaviors that can alleviate the problem.[10–13] The behavior-based approach to occupational safety is reflected in the research and scholarship of several safety consultants and is becoming increasingly popular for industrial applications.[14–17]

Considering Cost-Effectiveness. When people act in certain ways they usually adjust their mental attitude and self-talk to be consistent with their actions;[18] and when people change their attitudes, values, or thinking strategies, certain behaviors change as a result.[19] Thus, person-based and behavior-based approaches to changing people can influence both attitudes and behaviors, either directly or indirectly. Furthermore, most parents, teachers, first-line supervisors, and safety captains have used both of these approaches in their attempts to change other persons' knowledge, skills, attitudes, or behaviors. When we lecture, counsel, or educate others in a one-on-one or group situation, we are essentially using a person-based approach. When we recognize, correct or discipline others for what they have done, we are operating from a behavior-based perspective. Unfortunately, we are not always effective with our person-based or behavior-based change techniques, and often we do not know whether our intervention worked as intended.

In order to apply person-based approaches to psychotherapy, clinical psychologists receive specialized therapy or counseling training for four years or more, followed by an internship of at least a year. Such intensive experiential training is necessary because tapping into an individual's perceptions, attitudes, and thinking styles is a demanding and complex process. Also, these internal dimensions of people are extremely difficult to measure reliably, making it cumbersome to assess therapeutic progress and obtain straightforward feedback regarding one's therapy skills. Consequently, the person-based therapy process can be very time consuming, requiring numerous one-on-one sessions between professional therapist and client.

In contrast, the behavior-based approach to psychotherapy was designed for administration by individuals with minimal professional training. From the start, the idea behind the behavior-based approach was to reach people in the settings where their problems occur (e.g., the home, school, rehabilitation institute, workplace) and teach the managers or leaders in these settings (e.g., parents, teachers, supervisors, friends, or co-workers) the behavior-change techniques most likely to work under the circumstances.[20] More than three decades of research have shown convincingly that this on-site approach is cost effective, primarily because behavior-change techniques are straightforward and relatively easy to administer, and because intervention progress can be readily monitored by the ongoing observation of target behaviors. Thus, intervention agents can obtain objective feedback regarding the impact of their intervention techniques, and accordingly refine or alter components of a behavior-based process.

A Need for Integration. A common perspective, even among psychologists, is that humanists (as in person-based) and behaviorists (as in behavior-based) represent opposite poles of an intervention continuum.[21,3] Behaviorists are considered cold, objective, and mechanistic, operating with minimal concern for people's feelings; whereas humanists are viewed as warm, subjective and caring, with limited concern for directly changing another person's behaviors or attitudes. In fact, the basic humanistic approach to therapeutic intervention is termed "nondirective" or "client-centered." This refers to the principle that therapists, counselors, or coaches do not directly change their clients, but rather provide empathy and a caring and supportive environment for enabling clients to change themselves (i.e., from the inside out).

Given the conceptual foundations of humanism and behaviorism, it is easy to build barriers between person-based and behavior-based perspectives, and assume one must follow either one or the other approach when designing an intervention process. In fact, many consultants in the safety management field market themselves as using one or the other approach, but not both. It is my firm belief that an integration of these approaches is not only possible but necessary to truly understand the psychology of safety and health.

The Complexity of People

Statements like "All injuries are preventable," "It's human nature to work safely," "Safety is just common sense," and "Safety is a condition of employment" could send the message that working safely is easy or natural. Nothing could be further from the truth. In fact, it is often more convenient, more comfortable, more expedient, and more common to work at risk than safely. And, our past experiences in the situation usually support our decisions to choose the unsafe or at-risk behavior, whether we're working, traveling, or playing. Thus, the promotion of industrial health and safety is often a continuous fight with human nature.

BASIC ID

The complexity and uncontrollability of human nature can be appreciated by considering the acronym BASIC ID. Each letter represents one of seven human dimensions of an individual. Many clinical psychologists use a similar acronym as a reminder that helping people to improve their psychological state requires attention to each of these seven domains.[22,23] The relevance of these people factors in understanding human nature on the job is demonstrated in the following simple scenario of a near miss.

Dave, an experienced and skilled craftsman, worked rapidly to make an equipment adjustment, while the machinery continued to operate. As he worked, the production-line employees watched and waited to resume their work. Dave realized all too well that the sooner he completed his task, the sooner would his co-workers resume quality production. Therefore, he had not shut-down and locked-out the equipment power. After all, he had adjusted this equipment numerous times before without locking out the power and never experienced an injury.

A morning argument with his teenage daughter pervaded Dave's thoughts as he worked, and suddenly he experienced a near miss. His late timing nearly resulted in his hand being crushed in a "pinch point." After removing his hand just in time, Dave felt weak in his knees and began to perspire. This stress reaction was accompanied by a vivid image of a crushed right hand.

After gathering his composure, Dave walked to the switch panel, shut-down and locked-out the power; and then lit up a cigarette. He thought about this scary event for the rest of the day, and during his breaks, he related his near miss to fellow workers.

This brief episode illustrates each of the psychological dimensions represented by BASIC ID and demonstrates the complexity of human activity. *Behavior* is illustrated by such overt and observable actions as adjusting equipment, pulling a hand away from the moving machinery, lighting up a cigarette, and talking to co-workers.

Dave's *attitude* about work was fairly neutral at the start of the day, but immediately following his near miss he felt a rush of emotion or affect. His attitude toward "energy control and power lock out" (ECPL) changed dramatically, and his commitment to locking out power when necessary increased after relating his near miss to his friends.

Sensation was evidenced by Dave's dependence on visual acuity, hand-eye coordination, and a keen sense of timing when adjusting the machinery. His ability to react quickly to the dangerous situation prevented severe pain and potential loss of valuable touch sensation.

Imagery occurred after the near miss when Dave visualized a crushed hand in his "mind's eye," and this contributed to the significance and distress of the incident. Dave will probably experience this mental image periodically for some time to come. And this will likely enhance his motivation to perform appropriate ECPL procedures, at least for the immediate future.

Cognition or "mental speech" about the morning argument with his daughter may have contributed to the timing error that resulted in the near miss. Dave will probably remind himself of this episode in the future, and these cognitions may help to activate ECPL behavior.

Interpersonal refers to the other people in Dave's life who contributed to his near miss, and will be influential in determining whether Dave will initiate and maintain appropriate ECPL practices. For example, it was the interpersonal discussion with his daughter that occupied his thoughts or cognitions before the near miss, and the presence of production-line workers influenced Dave (through subtle peer pressure) to attempt a quick adjustment of equipment without ECPL practices. These on-lookers may have distracted Dave from the task, or they could have motivated him to show-off his efficient adjustment skills. After Dave's near miss, his interpersonal discussions were "therapeutic," helping him relieve his distress and increase his personal commitment to occupational safety.

Drugs in the form of caffeine (from morning coffee) may have contributed to Dave's timing error. The extra cigarettes Dave smoked as a "natural" reaction to distress also had physiological consequences, which could have been reflected in Dave's subsequent behavior, attitude, sensation, or cognition.

The lesson in this brief scenario and interpretation is people are complex. This complexity of human nature interacts with environmental factors to cause at-risk work practices, near misses, and sometimes personal injuries. It's relatively easy to control the environmental factors contributing to work injuries. And, as explained later, it's feasible to measure and control the behavioral factors contributing to work injuries. However, the complex person factors, implied by the BASIC ID acronym, are quite illusive. They are difficult to define, to measure objectively, and to manage efficiently. This leads to the occurrence of some work injuries that could not have been prevented. However, the field of psychology provides insights for understanding the individual worker, and this information can be used to increase the acceptance and beneficial impact of occupational safety and health programs.

Stress vs. Distress

Many people use the terms stress and distress interchangeably to mean something bad. The first definition of stress in the *American Heritage Dictionary* is "importance, significance, or emphasis placed on something." Similarly, the *New Merriam-Webster Dictionary* defines stress as "a factor that induces bodily or mental tension... a state induced by such a stress . . . urgency, emphasis." In contrast, distress is defined as "anxiety or suffering . . . severe strain resulting from exhaustion or an accident" (*American Heritage Dictionary*), or "suffering of body or mind: pain, anguish: trouble, misfortune . . . a condition of desperate need (*New Merriam-Webster Dictionary*).

Not only are the dictionary definitions of stress and distress quite different, results from psychological research indicate value in distinguishing between "stress" and "distress." Stress can be positive, resulting in heightened awareness, augmented mental alertness, and an increased readiness to perform. In fact, some psychological theories presume that some stress is necessary for people to perform.[(24)] With this notion, stressors motivate us, and for a number of other reasons, the results can be positive (stress) or negative (distress). The individual who asserts, "I work best under pressure (or deadlines) understands the motivational power of stress.

Where are the stressors? Stress or distress can be provoked by a wide range of situations and circumstances. Some stressors are acute and sudden life events, such as death or injury to a loved one, marriage, marital separation or divorce, birth of a baby, failure in school or at work, and a job promotion or relocation. Other stressors include the all-too-frequent minor hassles of everyday life, from long lines and excessive traffic thwarting our goals, lean and mean work conditions making unreasonable demands on our time, and never-ending expenses causing worry about personal finances.[(25)]

Since most adults spend more active time at work than in any other situation, our jobs or careers are a prime source of stressors. Some workplace stressors are obvious, while others might not be as evident but are just as powerful. Work overload can obviously become a stressor and provoke either stress or distress. In addition, work underload — being asked to do too little — can produce profound feelings of boredom, which can also lead to distress. Performance appraisals are stressors which can motivate constructive performance if they are perceived as objective and fair, or they can contribute to distress and inferior performance if they are viewed as subjective and unfair.

Other work-related factors that can be perceived as stressors and lead to distress and eventual burnout include: 1) role conflict or ambiguity (uncertainty about one's job responsibilities; 2) responsibility for others; 3) a crowded, noisy, smelly, or dirty work environment; 4) lack of involvement or participation in decision making; 5) interpersonal conflict with other employees; and 6) insufficient support from co-workers.[26]

Coping with Stressors. Understanding the multiple causes of conflict, frustration, overload, boredom, and other potential stressors in our lives can sometimes lead to effective ways of reducing distress or turning distress into positive stress. Sometimes our schedules can be revised to reduce environmental barriers like traffic and shopping lines that thwart our daily goals. Sometimes we can say "no" to a request that will overload us. Sometimes we can find time to truly relax and recuperate from tension and fatigue. Sometimes we can communicate effectively with others to clarify work duties, reduce interpersonal conflict, gain support from co-workers, or feel more comfortable about an additional job responsibility. And, sometimes we can get reassigned to a task that better fits our present talents and aspirations. However, it is often impossible to remove sudden (acute) or continuous (chronic) stressors in our lives, and we need to deal with these head-on. In fact, believing you can handle the stressors in a constructive manner is the first step toward experiencing stress rather than distress, and acting constructively rather than destructively.

Person Factors. Some people are more resistant to distress because they have certain personality characteristics. More specifically, research has shown that individuals who believe they are in control of their destinies and have an optimistic outlook toward life (they generally expect the best) are more likely to gain control of their stressors and experience positive stress rather than distress.[27–29] It's important to realize that these person factors — an expectancy of high personal control and optimism — are not permanent inborn traits of people. Rather, they are changeable states or expectancies developed from personal experience. Hence, it's possible to provide people with experiences that increase feelings of being "in control" of their stressors and facilitate expectations of good outcomes from their attempts to turn stress into constructive action.

Fit for Stressors. We can increase our sense of personal control and optimism by improving our level of fitness. And, being physically fit increases our body's ability to cope with the fight-or-flight syndrome discussed earlier. It's likely you've already heard the basic guidelines for improving fitness, which include: 1) stop smoking; 2) reduce or eliminate alcohol consumption; 3) exercise regularly (at least three times a week for about 30 minutes per session); 4) eat balanced meals with decreased fat, salt, and sugar, and don't skip breakfast; and 5) obtain enough sleep (usually 7 or 8 hours per 24-hour period for most people). For some of us, following each of these guidelines over the long term is easier said than done. Often we need the support and encouragement of others when attempting to break a smoking or drinking habit or maintain a regular exercise routine.

Social Factors. A support system of friends, family, and coworkers can do wonders at helping us reduce distress in our lives.[30–32] As mentioned above, social support can motivate us to do what it takes to stay physically fit. And, the people around us can make a boring task bearable and even satisfying, or they can make a stimulating job dull and tedious. Other people can motivate us to participate in group meetings and increase our commitment to overcome our stressors in constructive ways. On the other hand, the communication between people can set the stage for conflict, frustration, hostility, a win/lose perspective, and distress.

We can learn how to handle stressors from other people who take effective control of stressful situations and expect the best. Or, we can listen to the complaining, backstabbing, cynicism of others, and thereby add fuel to our own potential for distress. In other words, we can manage our stressors by choosing to interact with other people who can help us build resistance against distress and help us feel better about situations with potential stressors. We can also set the right example, and be the kind of social support to others we want others to be for us. The good feelings of personal control and optimism you experience from reaching out to help others with their stressors can do wonders in helping you cope with your own stressors. And, this actively caring stance builds your own support system which you might need if your own stressors get too overwhelming to handle yourself.

Fundamentals of the Behavior-Based Approach

Whether treating clinical problems (such as drug abuse, sexual dysfunction, depression, anxiety, pain, hypertension, and child or spouse abuse) or preventing any number of health, social, or environmental problems (from developing healthy and safe lifestyles to improving education and protecting the environment), overt behavior is the focus of behavior-based intervention. Treatment or prevention is approached by asking three basic questions: 1) What behaviors need to be increased or decreased to treat or prevent the problem?; 2) What environmental conditions (including interpersonal relationships) are currently supporting the undesirable behaviors or inhibiting desirable behaviors?; and 3) What environmental or social conditions can be changed to decrease undesirable behaviors and increase desirable behaviors? Thus, behavior change is the desired outcome of treatment or prevention, as well as the means to solving the identified problem.

The DO IT Process

Applications of the behavior-based approach to industrial health and safety can be summarized with the acronym DO IT. The process is continuous and involves the following four steps: D = Define critical target behavior(s) to increase or decrease; O = Observe the target behavior(s) during a preintervention baseline period to set behavior-change goals, and perhaps to understand natural environmental or social factors influencing the target behavior(s); I = Intervene to change the target behavior(s) in desired directions; T = Test the impact of the intervention procedure by continuing to observe and record the target behavior(s) during the intervention program. From this data the cost-effectiveness of the intervention can be evaluated and an informed decision made regarding whether to continue the program, implement another intervention strategy, or define another behavior for the DO IT process.

The intervention phase of DO IT involves the application of one or more behavior-change techniques. When designing an intervention program, the simple ABC model depicted in Figure 29.2 is followed. Activators direct behavior (as when the ringing of a telephone or doorbell signals certain behaviors from residents), and consequences motivate behavior (as when residents answer or don't answer the telephone or door depending on current motives or expectations developed from prior experience at telephone or door answering).

The most motivating consequences are soon, sizable, and certain. In other words, we work diligently for immediate, probable, and large positive consequences (positive reinforcers or rewards), and we work hard to escape or avoid soon, certain, and sizable negative consequences (negative reinforcers or punishers). Safe behaviors are usually not reinforced by soon, sizable, and certain consequences. In fact, safe behaviors are often punished by soon and certain negative consequences (including inconvenience, discomfort, and slower goal attainment).

Competing with the lack of soon, certain, and sizable consequences for safe behaviors, are soon, certain (and sometimes sizable) consequences for at-risk behaviors. Specifically, at-risk behaviors avoid the discomfort and inconvenience of most safe behaviors, and they often allow people to achieve their production and quality goals faster and easier. In their attempts to achieve more production, supervisors sometimes activate and reward at-risk behaviors (unintentionally, of course).

Because activators and consequences are naturally available throughout our everyday existence to support at-risk behaviors in

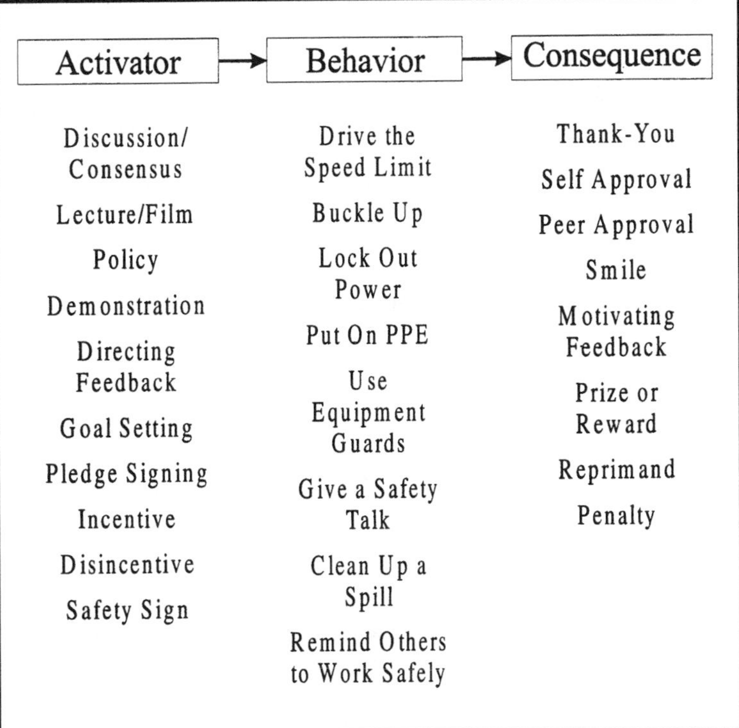

Figure 29.2 — The basic ABC model used to develop behavior-change interventions.

Perspectives on Psychology Issues in the Workplace: Job Design to Improve Working Conditions

Literature on occupational health and worker stress issues identifies many working conditions, both physical and psychosocial, that impact on the psychological well-being of workers. The term psychosocial generally refers to the social environment at work and the organizational and content aspects of a job. In an article on preventing work-related psychological disorders, researchers with the National Institute for Occupational Safety and Health recommended the guidelines below for controlling psychosocial risk factors at work. Essentially, the recommendations offer methods for improving or enhancing employee morale. Although not necessarily an occupational health professional's job, awareness of these motivational factors can assist in getting the job done.

Work load and work pace. Workloads need to match up with the worker's physical and mental capabilities. Work underload is as damaging as work overload. Interestingly, heavy work load or work pace are not the issue in worker morale. Increased control in the decision making by the individual worker is the most important aspect.

Work schedules. Again, the issue is worker control over how to handle the work demands and responsibilities. Flextime, compressed work weeks, and job sharing are positive steps that enhance worker morale and productivity.

Work roles. Job duties—roles and responsibilities—need to be clearly defined. *Job future.* Matters regarding job security, opportunities for career development, and even benefits, need to be clear and in no way ambiguous.

Social environment. Workers should have the opportunity for a certain amount of personal interaction both for emotional support and actual job assistance while on the job.

Content. The design of jobs should provide meaning, stimulation, and an opportunity to use skills. Many jobs that were originally narrow, fragmented activities (i.e., assembly lines) have been changed to allow for job rotation and an increase in the scope of the individual worker's activity and responsibility.

Participation and control. Giving individuals the opportunity to have input on decisions and actions relevant to them and their jobs has a direct positive effect on their physical and psychosocial health on the job.

lieu of safe behaviors, safety can be considered a continuous fight with human nature. Hence, the development and maintenance of safe work practices often requires the implementation of intervention strategies to keep people safe. These intervention strategies serve as activators or consequences, or they combine both approaches. They focus on increasing safe behaviors and/or decreasing at-risk behaviors.

Intervening with Activators

Activators are generally much easier and less expensive to implement than consequences. Therefore activators are used much more frequently than consequences to promote safe behavior. Posters or signs are perhaps the most popular activators for safety. Some signs bear only a general message ("Safety is a Condition of Employment"), whereas others refer to a specific behavior ("Hard Hat Required in this Area"). Some signs request the occurrence of a specific behavior ("Walk," "Wear Ear Plugs in This Area"), whereas other behavior-based signs request the avoidance of a certain behavior ("Don't Walk," "No Smoking Area"). Some behavior-specific signs request a relatively convenient response ("Buckle-Up"), other signs prompt behaviors which are relatively inconvenient to perform ("Lock Out All Energy Sources Before Repairing Equipment"). Some signs imply consequences ("Use Eye Protection — Vision is Precious"); others do not ("Wear Safety Goggles"). Some signs remind us of a general purpose ("Actively Care for a Total Safety Culture"), others remind us of a challenging goal ("Zero Lost Time Injuries is Our Goal"). Which of these types of signs are more effective? How could we make some of these messages more effective? The following principles will help you answer these and other questions regarding the design of activators.

Specify Behavior. Behavioral research has demonstrated that signs with general messages and no specification of a desired behavior to perform or an undesirable behavior to avoid, have very little impact on actual behavior. On the other hand, signs that refer to a specific behavior can be beneficial.[33–35]

Maintain Salience with Novelty. All of the field research demonstrating the impact of response-specific signs were relatively short term. None lasted more than a couple of months. Thus, the activators were salient because they were different (or novel). It is perfectly natural for activators like sign message to lose their impact over time. This process is called habituation, and is considered by some psychologists to be the simplest form of learning.[36] In other words, through habituation we learn not to respond to an event that occurs repeatedly. Hence, staying attentive to safety activators is a continuous fight with human nature. In this case the human nature is habituation.

Vary the Message. The principle of habituation tells us we need to find ways to vary activators. When an activator changes it becomes more salient and noticeable. A variety of techniques are available for changing the message on safety signs. Some signs have removable slats on which different messages can be placed. Other signs or billboards allow for the interchange of letters, thus allowing maximum flexibility in word display. And computer generated signs have an infinite amount of convenient versatility to vary safety messages. Some plants have video screens throughout their facility (in break rooms, lunch rooms, visitor lounges, and hallways) which can portray all kinds of safety messages, as monitored from a control location.

The people who should attend to the message and follow its specific behavioral advice should have as much input as possible in defining message content. These are usually the operators or line workers of an organization — those most at risk when ignoring the behavioral advice of an activator. Many organizations will get suggestions for safety messages just by asking. However, in some cultures, people may not be used to giving safety suggestions. Therefore, it may be necessary to add a consequence to motivate the suggestion of safety messages.

Involve the Participants. This principle of activator design is obvious by now, and relevant for the development and implementation of any behavior-change intervention. When people contribute to a safety effort, their ownership and commitment to safety increases. Of course, this principle works both ways. When individuals feel a greater sense of ownership and commitment, their involvement in safety achievement is more likely to continue. Thus, involvement feeds ownership and commitment, and vice versa.

Implicate Consequences. Field research has shown that activators which do not implicate consequences can influence behavior when they are salient and implemented in close proximity to an opportunity to perform the specified target behavior.[11] It's important to realize, though, that the target behaviors in these studies were all relatively convenient to perform. We're talking about depositing handbills in a particular receptacle, choosing certain products, using available safety glasses and safety belts. There is plenty of evidence that activators alone won't succeed when target behaviors require more than a little effort or inconvenience.[37–39]

Activators that specifically signal the availability of a consequence are either incentives or disincentives. An incentive announces to an individual or group, in written or oral form, the availability of a reward. This pleasant consequence follows the occurrence of a certain behavior or an outcome of one or more behaviors. In contrast, a disincentive is an activator announcing or signaling the possibility of receiving a penalty. This unpleasant consequence is contingent on the occurrence of a particular undesirable behavior.

Research has shown quite convincingly that the impact of a legal mandate varies directly with the amount of media promotion or disincentive.[40] Similarly, the success of an incentive program depends on making the target population aware of the possible rewards. In other words, marketing positive or negative consequences with activators (incentives or disincentives) is critical for the motivating success of a consequence intervention.

Four Types of Consequences

Table 29.1 summarizes the different types of motivating consequences. Relative to a task or activity, consequences can be natural or extra. Natural consequences, produced by the target behavior, are usually immediate and certain. In contrast, extra consequences are added to the situation and are often delayed and may be uncertain. Extra consequences are necessary when the natural consequences are insufficient to motivate the desired behavior, as is often the case with safety-related activities.[41]

Relative to the person performing the task, consequences can be considered external or internal. External consequences are observable by others, and thus can be studied objectively. Internal consequences are subjective and biased by the performer's perceptions. It's difficult to know objectively the exact nature of the internal consequences influencing an individual's performance. But we know from personal experience that internal consequences accompany performance and dramatically influence motivation and subsequent performance.

Table 29.1 — Behavior is Motivated by Four Different Types of Consequences (adapted from Geller)[39]

Source of Consequence Relative to Person		Source of Consequence Relative to Task: Natural (immediate)	Extra (often delayed)
	Internal	Reading for Pleasure Watching Television Listening to Music	Reading for Work Monitoring Instruments Reading for Homework
	External	Painting a Picture Doing Crossword Puzzles Playing Recreational Sports	Doing Math Homework Playing Professional Sports Working on an Assembly Line

The term "intrinsic" has been eliminated from this classification scheme, because of the different uses of this word. Note, however, that "natural" is synonymous with "intrinsic"" from a behavior-based perspective,[42,43] while "internal" is the same as "intrinsic" from a humanistic (or person-based) perspective.[44,45]

Table 29.1 classifies various activities according to the type of consequence relative to the task (natural versus extra) and the performer (internal versus external). While these activities illustrate particular types of consequences available to motivate performance, the categorizations are neither mutually exclusive nor inclusive. Even the most straightforward task classifications, for example, can overlap with other categories, according to perceptions of the performer.

For example, if you play a musical instrument, complete a crossword puzzle, plant a garden, or participate in recreational sports, natural and external consequences are immediately available. You've performed well, done a good job, or maybe you're not pleased with the results. Add to this the fact that you might compare your results to past results, or the accomplishments of others. This is adding a personal evaluation bias to the natural feedback — internalizing the external consequences. Now you've created internal consequences to accompany your activity.

Let's take it a step further. Perhaps another person adds an extra consequence by commending or condemning your performance. This could dramatically influence your motivation. And what if you got paid for gardening, or playing the piano? Your motivation could be further influenced.

As we've discussed, some activities or behaviors are not readily motivated by certain types of consequences, thus requiring extra support. Table 29.1 can be used to identify these tasks, and guide approaches for intervention with extra consequences. Since safe behavior competes with at-risk behavior that is supported by external and natural consequences, it is often necessary to support safe behavior with extra consequences. There is a right and wrong way to do this, and for health and safety promotion, the wrong way is more commonly used.

Using Safety Rewards Incorrectly. The definitions of incentive and reward imply that a specific target behavior must be specified. However, most incentive/reward programs for occupational safety do not specify behavior. That is, the most common application of incentives for safety management specifies a certain reward which employees can receive by avoiding a work injury (or by achieving a certain number of "safe work days"). Many of these nonbehavioral, outcome-based incentive programs implicate substantial peer pressure because they use a group-based contingency. That is, if anyone in the company or work group is injured, everyone loses their reward.

So what behavior is motivated by such an outcome-based, group-contingency incentive/reward program? Obviously, if workers link certain safe behaviors directly with a high probability of avoiding an injury, then an outcome-based incentive program can have a beneficial impact. However, the injury-avoidance advantages of safe behavior are seldom observable, and at-risk behavior is rarely followed by an injury or even a near miss. Thus, the most likely behavior to be influenced by an outcome-based incentive/reward program is injury reporting.

If having an injury loses one's reward (or worse, the reward for an entire work group), there is pressure to avoid reporting that injury, if possible. For example, it's not uncommon for a co-worker to cover for an injured employee in order to keep accumulating "safe days" and not lose their reward possibility. Hence, these incentive programs might decrease the numbers (the injury rate), at least over the short term, but corporate safety is obviously not improved.

Outcome-based incentive programs often lead to a detrimental attitude of apathy or helplessness regarding safety achievement. In other words, employees develop the perspective that they can't really control their injury record, but must cheat or beat the system to celebrate the "achievement" of an injury reduction goal. In one situation an employee was disqualified from an outcome-based lottery when he was burned on the leg by a welding torch. But, the employee whose at-risk use of the torch caused the burn remained eligible for the raffle prize.[46]

Using Safety Rewards Correctly. This discussion leads logically to seven basic guidelines for establishing an effective incentive/reward program for managing the human element of industrial health and safety.

1. The behaviors required to achieve a safety reward should be specified and perceived as achievable by the participants.
2. Everyone who meets the behavioral criteria should be rewarded.
3. It's better for many participants to receive small rewards than for one person to receive a big reward.

4. The rewards should be displayable and represent safety achievement (for example, coffee mugs, hats, shirts, sweaters, blankets, or jackets with a safety message), rather than rewards that will be hidden, used, or spent.
5. Contests should not reward one group at the expense of another.
6. Groups should not be penalized (lose their rewards) for failure by an individual.
7. Progress toward achieving a safety reward should be systematically monitored and publicly posted for all participants.

Guideline 2 recommends against the popular lottery or raffle drawing. A lottery results in one "lucky" winner being selected and a large number of "unlucky" losers. Thus, the announcement of a raffle drawing (the incentive) might get many people excited. And if lottery tickets are dispensed for specific safe behaviors, there would be some motivational benefit. Eventually, however, the valuable reward would be received by only a lucky few. Also, there is a disadvantage in linking chance with safety. It's bad enough we use the word "accident" in the context of safety processes. The first definition of "accident" in the *New Merriam-Webster Dictionary* is "an event occurring by chance or unintentionally." We shouldn't add to this inference that injuries are chance occurrences.

Volk[47] interviewed a number of safety directors who implemented a lottery incentive program for safety and vowed they would never do it again. The big raffle prize (like a snowmobile, pick-up truck or television set) was displayed in a prominent location. Everyone got excited (temporarily) about the possibility of winning. Thus, everyone's attention was directed to the big prize instead of the real purpose of the program — to keep everyone safe. The material rewards in an incentive program should not be perceived as the major payoff. The incentives are only reminders to do the right thing, and the rewards serve as feedback and a statement of appreciation for doing the right thing.

When the rewards include a safety logo or message (Guideline 4), they become activators for safety when they are displayed. Also, if the safety message or logo was designed by representatives from the target population, the reward takes on special meaning. Special items like these cannot be purchased anywhere, and from the perspective of internal consequences, they are more valuable than money.

More important than the type of external reward is the way it is delivered. Rewards should not be perceived as a means of controlling behavior, but as a declaration of sincere gratitude for making a contribution. The more people who receive this recognition, the more deposits made in the emotional bank accounts of potential participants in a safety improvement process. That's why it's better to reward many than few (Guideline 3).

Contests that pit one group against another can lead to an undesirable win-lose situation (Guideline 5). Safety needs to be perceived as win-win. This means developing a contract of sorts between employees, which makes everyone a stockholder in improving safety and health. Thus, everyone in the organization is on the same team. Team performance within departments or work groups can be motivated by providing team rewards or bonuses for team achievement. Every team that meets the "bonus" criteria should be eligible for the reward. In other words, Guideline 2 should be applied to teams when developing a safety incentive program to motivate team performance.

Penalizing groups for individual failure (Guideline 6) reflects a problem inherent in many outcome-based incentive programs. It is certainly easy to administer a contingency that simply withdraws reward potential from everyone whenever one person makes a mistake. However, this type of incentive plan can do more harm than good. As discussed earlier, it can promote unhealthy group pressure and develop feelings of helplessness or lack of personal control. Displaying the results of such a program only precipitates these undesirable perceptions and expectations. It motivates a reluctance to report injuries, as well as near misses, and this severely undermines the essence of injury prevention — the identification of hazards and at-risk behaviors to change.

On the other hand, when the safety incentive program is behavior-based and perceived as equitable and fair, it is advantageous to display progress toward reaching individual, team, or company goals (Guideline 7). When people see their efforts transferred to a feedback chart, their motivation and sense of personal control is increased (or at least maintained). Obviously, the development and administration of an effective incentive/reward program for safety requires a substantial amount of dedicated effort. There is no quick fix. But it is worth doing, if you take the time to do it right. As Daniels[48] wisely stated, "If you think this is easy, you are doing it wrong."

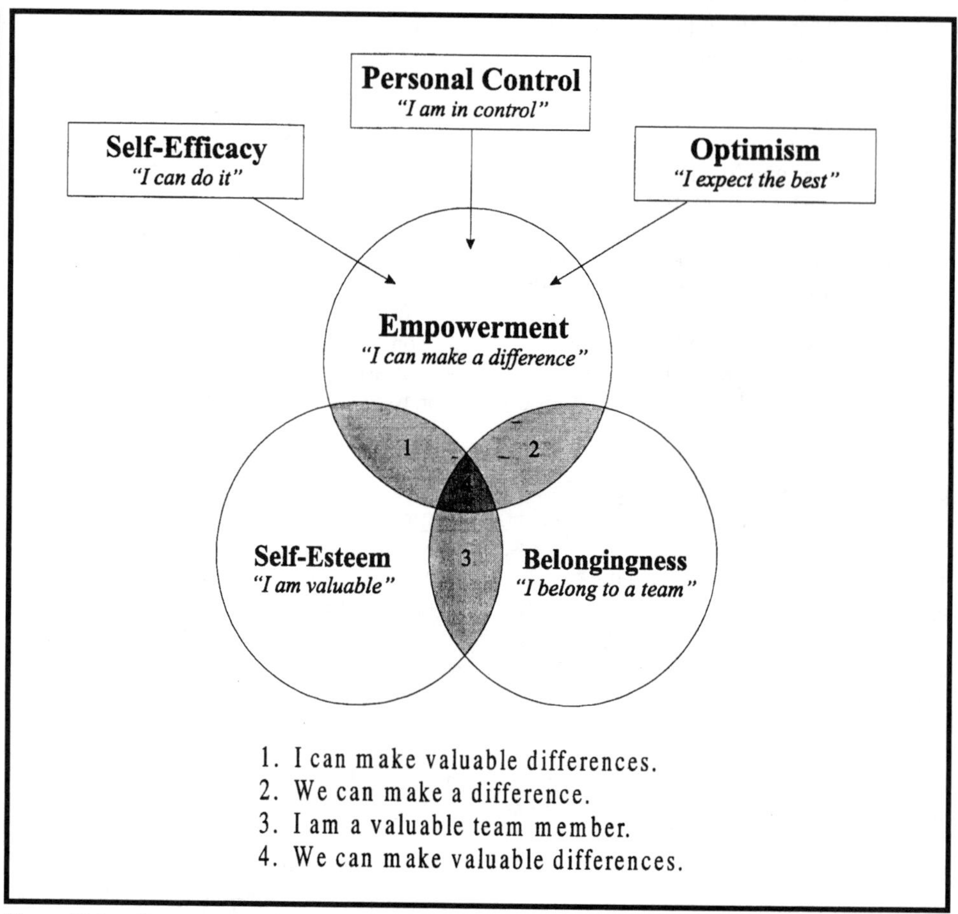

Figure 29.3 — Five person-states that increase people's willingness to actively care.

Relevance of Person-Based Psychology

The behavior-based tools discussed above for improving industrial health and safety, including observation and feedback through a DO IT process and behavior-based intervention, require people to implement them. And, the more people who help implement the behavior-based techniques, the greater the impact on safety and health. In other words, having the right tools to improve health and safety is not sufficient. The people of an organization need to actively care with regard to health-related goals, action plans, and supportive activators and consequences. They need to go beyond their normal routine to remove health and safety hazards from the environment and to influence their co-workers' health and safety-related behaviors in desirable directions.

Research indicates that certain psychological states or expectancies affect the propensity for individuals to actively care for the safety or health of others. Furthermore, certain conditions or establishing operations (including activators and consequences) can influence these psychological states.[49,50] These states are illustrated in Figure 29.3 — a model useful to stimulate discussions among industry employees of specific situations, operations, or incidents that influence their willingness to actively care for industrial/health and safety.

Self-Esteem ("I am Valuable")

How do you feel about yourself? Generally good or generally bad? The extent to which you generally feel good about yourself is your level of self-esteem. If we don't feel good about ourselves, it's unlikely we'll care about making a difference in the lives of others. Thus, the principle here is that the better we feel about ourselves, the more willing we are to actively care for the safety and health of other people.

One's self-concept (or relative feeling of worth) is the central theme of most humanistic therapies.[51,52] According to Carl Rogers and his followers, we have a real and ideal self-concept. That is, we have notions (dreams) of what we would like to be (our ideal self) and what we think we are (our real self). The greater the gap between our real and ideal self-concepts, the lower our self-esteem. Thus, a prime goal of many humanistic therapies is to help people reduce this gap between their ideal and actual self concept. This can be done by raising people's perceptions of their real self-concept ("Count your strengths and blessings and you'll see that you're much better than that"), or by lowering one's aspirations or ideal self-concept ("You expect too much; no one is perfect; take life one-step at a time and you'll eventually get there").

It is important to maintain a healthy level of self-esteem, and to help others raise their self-esteem. Research has shown, for example, that persons with high levels of self-esteem report fewer negative emotions and less depression than people with depressed self-esteem.(53) Furthermore, those with higher self-esteem handle stressors better than persons with lower self-esteem.(54) In other words, higher self-esteem enables the conversion of stressors to positive stress rather than negative distress.

Researchers have also found that individuals who score higher on measures of self-esteem are less susceptible to outside influence techniques,(55) are more confident of achieving their personal goals,(56) and make more favorable impressions on others in social situations.(57) And supporting the model depicted in Figure 29.3, persons with higher self-esteem have been found to help others more frequently than those scoring lower on a self-esteem scale.(58)

Empowerment ("I Can Make a Difference")

In the management literature, empowerment typically refers to delegating authority or responsibility, or sharing decision making.(59) In contrast, the person-based perspective of empowerment focuses on the reaction of the recipient to increased power or responsibility. From a psychological perspective, empowerment is not giving or receiving more responsibility, but rather empowerment is feeling more responsible. In other words, this view of empowerment requires the personal belief that "I can make a difference." This belief is strengthened with perceptions of personal control,(60) self-efficacy,(61) and optimism.(62,63) Such an empowerment state is presumed to increase motivation (or effort) to "make a difference" (perhaps by going beyond the call of duty). Behavioral science research supports this intuitive hypothesis.(64–67)

Self-Efficacy ("I Can Do It"). Self-efficacy (or self-effectiveness) is a key factor in social learning theory, determining whether a therapeutic intervention will be successful over the long term.(61) It refers to self-confidence that a person can successfully accomplish what he or she sets out to do. Dozens of studies have demonstrated that subjects who score relatively high on a measure of self-efficacy perform better at a wide range of tasks. They show more commitment to a goal and work harder to pursue it. They demonstrate greater ability and motivation to solve complex problems at work. They have better health and safety habits; and they are more apt to react to stressors with positive stress than negative distress.(27,68,69)

Self-efficacy contributes to self-esteem, and vice versa; but these constructs are different. Simply put, self-esteem refers to a general sense of self-worth; whereas self-efficacy refers to feeling successful or effective at a particular task. In other words, self-efficacy is task specific and can vary markedly from one situation to another, while one's level of self-esteem remains rather constant across situations.

Personal Control ("I am in Control"). Rotter(60) used the term locus of control to refer to a general expectancy regarding the location of forces controlling a person's life — internal or external. Those with an internal locus of control believe they usually have direct personal control over significant life events as a result of their knowledge, skills, and abilities. They believe they are captain of their lives' ships. In contrast, persons with an external locus of control believe factors like chance, luck, or fate play important roles in their lives. In a sense, externals believe they are victims (or sometimes beneficiaries) of circumstances beyond their direct personal control.(60,70) Thus, internals generally expect to have more personal control over the positive and negative consequences in their lives than do externals.

This personal control dimension has been one of the most researched individual difference dimensions in psychology. Since Rotter developed the first measure of this construct in 1966, more than 2,000 studies have investigated the relationship between perceptions of personal control and other variables.(71) Internals are more achievement-oriented and health conscious than externals. They are less prone to distress, and more likely to seek medical treatment when they need it.(72,73) In addition, having an internal locus of control helps reduce chronic pain, facilitates psychological and physical adjustment to illness and surgery, and hastens recovery from some diseases.[74] Internals perform better at jobs which allow them to set their own pace, whereas externals work better when a machine controls the pace.[75,76].

Optimism ("I Expect the Best"). Optimism is the learned expectation that life events, including personal actions, will turn out well.[62,63] Optimism relates positively to achievement. Seligman[63] reported, for example, that Olympic-level swimmers who scored high on a measure of optimism recovered from defeat and swam even faster compared to those swimmers scoring low. Following defeat, the pessimistic swimmers swam slower.

Compared to pessimists, optimists maintain a sense of humor, perceive problems or challenges in a positive light, and plan for a successful future. They focus on what they can do rather than on how they feel.[77,78] As a result they handle stressors constructively and experience positive stress rather than negative distress.[79] Optimists essentially expect to be successful at whatever they do, and therefore they work harder than pessimists to reach their goals. As a result, optimists are beneficiaries of the self-fulfilling prophecy.[80].

Belonging ("I Belong to a Team"). Peck[81] challenges us to experience a sense of true community with others. People need to develop feelings of belonging with one another regardless of political preferences, cultural backgrounds and religious doctrine. We need to transcend our differences, overcome our defenses and prejudices, and develop a deep respect for diversity. Peck proclaims we must develop a sense of community or interconnectedness with one another if we are to accomplish our best and ensure our survival as human beings.

It seems intuitive that building a sense of community or belonging among our co-workers will improve industrial health. Safety and health improvement requires interpersonal observation and feedback, and for this to happen people need to adopt a collective win/win perspective instead of the individualistic win/lose orientation common in many workplace settings. A sense of belonging (or inter-connectedness) leads to interpersonal trust and caring, which sets the stage for actively caring behavior.

Enhancing the Actively Caring Person-States

Some people express concern that the actively caring person-state model may not be practical. "The concepts are too soft or subjective," is a typical reaction from these individuals. They accept the behavior-based approach because it is straightforward, objective, and clearly applicable in an organizational setting. On the other hand, person-based concepts like self-esteem, empowerment, and belonging appear ambiguous and difficult to deal with. "The concepts sound good and certainly seem important, but how can we get our arms around these concepts and use them to promote health and safety?"

In a workshop setting, after introducing the actively caring person-states, participants can be divided into discussion groups and asked to discuss practical implications of a particular person-state. Group members are asked to define events, situations, or contingencies that decrease and increase the person-state assigned to their group. This verifies personal understanding of the concept and leads to practical ramifications of such understanding. The groups are then asked to derive simple and feasible action plans to increase the person-state assigned them. Common suggestions are summarized below; many other ways to enhance these person-states are detailed in Geller.[14]

Self-Esteem. Factors consistently listed as determinants of self-esteem include communication strategies, reinforcement and punishment contingencies, and leadership styles. Participants have suggested a number of ways to build self-esteem, including: a) provide opportunities for personal learning and peer mentoring, b) increase recognition for desirable behaviors and personal accomplishments, and c) solicit and follow up a person's suggestions.

Self-Efficacy. A "can do" attitude requires the perception of having sufficient resources, training, and opportunities to carry out task assignments. Again, personal perception is the key. A supervisor, parent, or teacher might believe he or she has provided everything needed to complete a task successfully. However, the employee, child, or student might not think so. Hence, it is necessary to check with the recipient of an assignment with regard to feelings of self-efficacy. This is easier said than done, however, because people are often reluctant to admit their incompetence. We frequently deny or conceal perceptions of "can't do it," in order to maintain the appearance of self-efficacy.

Personal Control. Most people find a healthy balance between internal (personal) and external control. They essentially believe, "I'm responsible for the good things that happen to me, but bad luck or uncontrollable factors are responsible for the bad things."[82,83] Thus, people are apt to attribute injuries to rotten luck and beyond their control, especially when they happen to them. This is the prime reason industrial safety must focus on process achievements rather than failure (injury) outcomes.

There are a number of ways to increase perceptions of personal control, including: a) setting short-term goals and tracking achievements; b) offering frequent rewarding and correcting feedback for process activities rather than only for outcomes; c) providing opportunities to set personal goals, teach peers, and chart "small wins";[84] d) teaching employees basic behavior-change intervention strategies (especially feedback and recognition procedures); e) providing workers time and resources to develop, implement and evaluate intervention programs; f) showing employees how to graph daily records of baseline, intervention, and follow-up data; and g) posting response feedback graphs of group performance.

Optimism. Optimism results from thinking positively, avoiding negative thoughts, and expecting the best to happen. Thus, anything that increases our self-efficacy should increase optimism. Also, if our personal control is increased we perceive more influence over our consequences. And we have more reason to expect the best when we have adequate self-efficacy. Thus, the person-states of self-efficacy, personal control and optimism are clearly intertwined. A change in one will likely influence the other two.

Belonging. Common proposals for increasing an atmosphere of belonging among employees include: a) decreasing the frequency of top-down directives and "quick-fix" programs; b) increasing team-building discussions, group goal-setting and feedback, as well as group celebrations for both process and outcome achievements; and c) using self-managed (or self-directed) work teams.

When groups are given control over important matters (like developing a safety observation and feedback process or a behavior-based incentive program), both their empowerment and belonging can be enhanced. When resources, opportunities, and talents are available among team members to enable feelings of group empowerment ("We can make a difference"), feelings of belonging occur naturally. This leads to synergy, whereby the group achieves more than could be possible from the group members working independently.

Social psychologists have identified barriers to group belonging and synergy, and they have researched ways to increase interdependency in organizational cultures. This research is summarized next with reference to six basic principles of social dynamics. Each of these principles influences people's willingness to actively care for the health and safety of others.

The Social Dynamics of Occupational Health and Safety

Safety improvement requires a transformation from dependency and independency to interdependency. In other words, it's not enough for people to rely on the organization to keep them safe through engineering interventions. Nor is it sufficient for people to count on only their own individual efforts to keep them free from injury. Rather, people need others to remove environmental hazards they don't notice and to provide corrective feedback for at-risk behavior they might not realize they're performing. Such interdependency requires effective personal interaction. The social dynamics of the situation determine whether such exchanges are likely to occur and whether the impact of such interaction is beneficial.

The social dynamics of an organization both reflect and influence the culture at the same time. That is, certain aspects of a work setting affect social dynamics, and these social dynamics in turn influence the culture. The six social influence principles explained here reflect basic social dynamics that inhibit or facilitate industrial health and safety.

We Try to be Consistent in Thought and Deed

Many psychologists consider the consistency principle a weapon of influence lying deep within us and directing our everyday actions. It reflects our motivation to be (and appear) consistent. Simply put, when we make a choice or take a stand, we encounter personal and social pressures to perform consistently with our commitment. We obtain this pressure to be consistent from three basic sources: 1) society values consistency within people; 2) consistent conduct is beneficial to daily existence; and 3) a consistent orientation allows for shortcuts in information processing and decision making.[85] Instead of considering all relevant information in a certain situation, people need only remember their commitment or decision and respond consistently. This principle explains people's resistance to change, while also suggesting particular ways to motivate lasting improvement in both behavior and attitude.

Public and Voluntary Commitment. When people sign their name to a petition or pledge card they are making a commitment to behave in a certain way. Later, they behave in this way to be consistent with their commitment. Safety professionals can use this variation of the consistency principle to increase various safety-related behaviors.(86) After a discussion about a particular work procedure, for example, the audience could be asked to make a commitment to perform the desired behavior. What kind of commitment should be requested?

Commitments are most effective (or influential) when they are public, require effort, and are perceived as voluntary or not coerced. Thus, it would be more beneficial to have employees make a public rather than private commitment to perform a certain safe behavior. And, it would be better to have them sign their name to a card or public declaration display than to merely raise their hands. In addition, it is very important for those pledging to follow a certain work practice to believe they made the commitment voluntarily. The reality might be that decisions to make a public commitment are dramatically influenced by external forces like peer pressure. However, if people write an internal script that they made a personal choice, consistency is most likely to follow the commitment. Thus, the promoter of a commitment strategy needs to realize the influence of personal choice and make statements that allow participants to believe the commitment is not coerced but is up to them.

Start Small and Build. Sometimes referred to as the "foot-in-the-door" technique,(87) this influence strategy also follows directly from the consistency principle. To be consistent, a person who follows a small request is likely to comply with a larger request later. Thus, after agreeing to serve on a "safety steering committee," an individual is more willing to give a safety presentation at a plant-wide safety and health meeting. Research has found this commitment strategy to be successful in boosting product sales, monetary contributions to charities, and blood donations.(88)

The pledge-card commitment technique referred to above uses this principle. More specifically, after people sign a pledge card that commits them to perform a certain behavior for a specified period of time (such as "Buckle vehicle safety belts for one month," "Use particular personal protective equipment for two months," "Walk behind yellow lines for the rest of the year"), they are more likely to actually do the safe behavior.

The "foot-in-the-door" technique only works to increase safe behaviors when people comply with the initial small request. In fact, if a person says "no" to the first request, this individual might find it even easier to refuse a subsequent, more important request. Thus, if the circumstances suggest a "no" to your request, you didn't start small enough. In this case, you should be prepared to retreat to a less demanding request. This technique is described later under the reciprocity principle.

Raise the Stakes Later. This technique occurs when a person is persuaded to make a decision or commitment (for example, to serve on the safety steering committee) because of the relatively low stakes associated with the decision (the monthly safety meetings will not require too much time and effort). Then, when the individual gets committed to the decision (attends the first two safety meetings), the stakes are raised (more meetings are requested for a special safety effort). Because of the consistency principle, the individual will likely stick with the original decision (remain an active member of the committee).

Almost 20 years ago Cialdini and colleagues(89) demonstrated the powerful influence of this technique when attempting to get college students to sign-up for an early 7:00 a.m. experiment on "thinking processes." During the solicitation phone calls, the 7:00 a.m. start time was mentioned up-front for half the subjects. Only 24 percent of these individuals agreed to participate. For the other subjects, the caller first asked if they wanted to participate in the study. Then, after 56 percent agreed, the caller said the experiment started at 7:00 a.m. The caller gave subjects a chance to change their minds, but none did. Furthermore, 95 percent of these individuals actually showed up at the 7:00 a.m. appointment time. After making an initial commitment to participate, practically all of the subjects showed consistency and kept their commitment — in spite of the higher stakes.

This procedure is similar to the "foot-in-the-door" technique in that a larger request occurs after the target person agrees with a smaller request. A key difference, however, is there is only one basic decision in this procedure, with the costs or stakes raised after initial commitment. This compliance tactic is common among car dealers. Once a customer has agreed to purchase a car at a special price (for example, $800 below all other competitors), the price is raised for a number of reasons. The salesperson's boss might have refused to approve the deal, certain options had not been included in the special price, or the dealership manager may have decreased the value of the customer's trade-in. Customers who have agreed to the special price will usually not change their minds with a price increase, because reneging on a purchase decision may suggest a lack of consistency or indicate failure to fulfill an obligation (even though the obligation is only imaginary). Often customers will develop a set of new reasons to justify their initial choice and the additional costs.

Don't Stifle Trust. The stakes-raising strategy raises a critical issue with regard to using certain techniques to increase safe behavior. Even though this influence strategy is used rather frequently to increase compliance, how would you feel about a change agent using this technique if you knew it was used intentionally to get more money, commitment or safe behavior from you? For example, do you trust the waiter who brings you an expensive wine list only after you've been seated and made selections from the food menu? Your answer probably depends on whether you believe this sequence of events was done intentionally to get you to buy more.

Similarly, you might not dislike or mistrust the car salesman who adds cost to a vehicle's advertised purchase price unless you are suspicious the price differential was fabricated deliberately to increase revenue. In other words, our trust, appreciation, or respect for people might decrease considerably if we believe they intentionally used a particular influence technique to trick or deceive us into modifying our attitude or behavior. Of course, there may be no harm done if the result is clearly for our own good (as for our health or safety) and we realize this.

Which first — attitude or behavior? Because of the consistency principle it doesn't matter whether attitude or behavior changes first. The issue is whether a technique is available to influence one or the other. The three influence techniques discussed under the consistency principle were introduced as targeting behavior. It could be argued, however, that internal (attitudinal) dimensions were intertwined throughout each technique. For example, the pledge-card procedure works best when the person believes (internally) that the commitment was voluntary. Following successive compliance with escalating demands, internal commitment is developed, until eventually an "attitude" results. Moreover, the stakes-raising technique depends on target individuals developing an internal justification for their initial decision, which then strengthens personal commitment and leads to more behavior following the additional cost or effort required. Consequently, the key lesson is that people attempt to keep their internal dimensions (like attitude) and external actions (behavior) consistent. Thus, whether attitude or behavior is influenced first, if the person does not feel coerced, the other will likely follow.

We Reciprocate to Return the Favor

Some psychologists, sociologists, and anthropologists consider reciprocity a universal norm that motivates much of interpersonal behavior. It can be used to increase people's involvement in a safety-improvement process and to cultivate interdependence. Simply put, the reciprocity principle is reflected in the slogan, "Do for me and I'll do for you." In other words, if you are nice to people, they will feel obligated to return the favor. And research has shown that the favor might be returned to someone other than the original source.[90]

It's important to realize that how we react to people after doing them a favor can either stifle or mobilize reciprocity. When a person thanks you for actively caring, you should not demean the favor by saying things like "No problem," or "It was really nothing." Anything that makes the actively caring seem insignificant or trivial will reduce the impetus for reciprocity. However, adding words to make the actively caring appear more significant or meaningful can be quite awkward and create an uncomfortable or embarrassing situation.

To maintain a comfortable verbal exchange that does not belittle the actively caring and stifle reciprocity, you could react to someone's "Thank you for actively caring" with something like, "Thank you for appreciating my effort to promote health and safety, but I know you'd do the same for me." This reaction shows admiration for the thank-you, and thus increases the likelihood that more thanks will be given. Plus, it activates the reciprocity principle in a way that will be perceived as genuine and valid.

Gifts Aren't Free. In a field experiment, Isen and Levin[91] found that 84 percent of those individuals who found a dime in the coin-return slot of a public phone (placed there by researchers) helped an accomplice pick up papers he dropped in the subject's vicinity. In contrast, only 4 percent of those who did not find a dime helped the man pick up his papers. Similarly, students given a cookie while studying at a university library were more likely than those not given a cookie to agree to help another person by participating in a psychology experiment.[91]

Does this justify the distribution of free safety gifts, such as pens, tee-shirts, caps, cups, and other trinkets? Yes, to some extent, but the amount of reciprocity activated depends on the recipient's perceptions. How special is the gift? Was the gift given to a select group of people, or was it distributed to everyone? Does the gift or its delivery represent significant sacrifice in money, time, or effort? Can the gift be purchased elsewhere, or does the safety slogan on it make it special?

The bottom line here is that the more "special" the safety gift — as perceived by the recipient — the more reciprocity activated. Remember also, the way a safety gift is presented can make all the difference in the world. The labels and slogans linked with the gift can influence the amount and kind of reciprocity implicated. If the gift is presented to represent the actively caring safety leadership expected from a "special" group, a certain type of reciprocity is activated. People will tell themselves they are considered safety leaders and they need to justify this label by actively caring for the safety of others.

Start Big and Retreat. Suppose the plant safety director pulls you aside and asks you to chair the safety steering committee for the next two years. Let's assume you perceive this request as outrageous, given your other commitments and the fact you have never even served on a safety steering committee before — let alone chaired one. You say "Thanks for asking, but no." The safety director says he understands, and then asks whether you'd be willing to serve on the committee. According to social influence research, because the safety director "backed down" from his initial request, you will feel subtle pressure to make a similar concession — to reciprocate — and agree to the second, less demanding assignment.

Cialdini and associates[92] were among the first to demonstrate the influence power of this "door-in-the-face" technique. Posing as representatives of the "County Youth Counseling Program," they approached college students walking on campus and requested their volunteer service to chaperon a group of juvenile delinquents on a day trip to the zoo. When this was the first and only request, only 17% of those approached volunteered to help. However, three times more students volunteered when the researchers first asked for a much larger favor. Specifically, they asked whether the student would be willing to counsel juvenile delinquents for two hours a week over a two-year period. All subjects refused this request, but then half of them agreed to serve as unpaid chaperons for a day at the zoo. Apparently, the researchers' willingness to retreat from their initial request influenced several college students to reciprocate and comply with a smaller favor.

Have you ever wondered why lawyers ask for such outlandish amounts of money at the start of a civil trial? Or, why labor negotiators start with extreme demands? It's likely these influence agents do not expect to receive their initial request. But, perhaps they've learned they are more apt to gain a second request after retreating from the first. They've learned the power of this reciprocity-based technique through real-world negotiation.

We Follow the Crowd. In research conducted by Solomon Asch and associates in the mid-1950s, more than 33% of intelligent and well-intentioned college students were willing to publicly deny reality in order to be consistent with the obviously inaccurate judgments of their peers. Asch's classic studies of conformity[93] involved six to nine individuals sitting around a table and verbalizing their judgment of which of three comparison lines was the same length as the standard. All but the last individual to voice a decision in sequence were research associates posing as subjects. On some trials, these confederates uniformly gave obviously incorrect judgments. The last decision in the sequence was from the real subject of the experiment.

The correct answer was always obvious, and on several trials everyone gave the correct answer. On some judgment trials, however, an obviously incorrect comparison line was selected by each research associate according to plan, and the critical question was whether the subject would deny the obvious truth in order to conform with the group.

This and similar procedures were used in numerous social psychology experiments to study social dynamic factors that influence amount of conformity. For example, a subject's willingness to deny reality in order to go along with the group consensus was facilitated by increasing group size,[94] apparent competence or status of group members,[95] and anonymity (i.e., by having group members wear disguises). On the other hand, the presence of one dissenter (or nonconformist) in the group was enough to significantly decrease conformity by increasing a subject's willingness to chose the correct line even when the previous decisions of only 1 out of 15 individuals reflected this correct choice.[96].

The phenomenon of social conformity is certainly not new to any reader. We see examples of conformity every day, from the types of clothes people wear to their particular styles of communication in both written correspondence and verbal presentation. Thus, the role of conformity as an outside factor influencing at-risk behavior cannot be overlooked. Plus, we need to realize that group pressure to conform, even with at-risk behavior, is greater when the group is larger and the group members are perceived as relatively competent or experienced. Also important is the fact that one dissenter (e.g., a leader willing to ignore group pressure and do the right thing) is enough to prevent another person from succumbing to social conformity to work at risk. And when the actively caring dissenter has authority in the situation, many others will likely follow suit. This is explained by the next basic principle of social influence.

The Power of Authority

In research conducted by Stanley Milgram and associates in the 1960s, 65% of intelligent and well-meaning college students followed orders to administer 450-volt electric shocks to a screaming peer. Imagine you are one of nearly 1000 participants in one of Milgram's 20 obedience studies at Yale University. You and another individual are led to a laboratory to participate in a human learning experiment. First, you draw slips of paper out of a hat to determine randomly who will be the "teacher" and who will be the "learner." You get to be the teacher.

The learner is taken to an adjacent room and strapped to a chair wired through the wall to an electric shock machine containing 30 switches with labels ranging from "15 volts — light shock" to "450 volts — severe shock." You are seated behind this shock generator and instructed to punish the learner for errors in the learning task by delivering brief electric shocks, starting with the 15-volt switch and moving up to the next higher voltage with each of the learner's errors.

If you comply with the experimenter's instructions, you hear the learner moan when you flick the third, fourth, and fifth switches. When flicking the eighth switch (labeled "120 volts"), the learner screams "these shocks are painful;" and when the tenth switch is activated the learner shouts, "Experimenter, get me out of here!" At this point, you may show some concern, but the experimenter prompts you to continue with words like, "Please continue — the experiment requires that you continue." If you continue to increase shock intensity with each of the learner's errors, you reach the 330-volt level. At this point you hear shrieks of pain from the learner. At the 330-volt level, the learner pounds on the wall and then becomes silent. Nevertheless, the experimenter urges you to flick the 450-volt switch when the learner fails to respond to the next question.

At what point do you think you would refuse to obey the experimenter's instructions? If you believe you would stop playing this sadistic game soon after the learner indicated the shock was painful, your prediction would be the same as people who Milgram surveyed before conducting the experiment, including 40 psychiatrists. Thus, Milgram[97] was surprised that 65 percent of his actual subjects, ranging in age from 20 to 50, complied fully with the experimenter's requests — right up to the last 450-volt switch.

Did the subjects figure out the learner was a confederate of the experimenter and was not really receiving shocks? Did they realize they were being deceived in order to test their obedience to authority? No, all subjects displayed genuine concern and stress when giving the shocks. They sweated, trembled, bit their lips, and some laughed nervously. Some openly questioned the experimenter's instructions, but with an additional directive from the person in charge (the experimenter) most obeyed.

As with the conformity research, Milgram and associates studied the influence of various situational factors on amount of obedience. Full obedience exceeded 65% (with as many as 93% flicking the highest shock switch) when: 1) the authority figure (the one giving the orders) was in the room with the subject; 2) the authority figure was supported by a prestigious institution (Yale

University); 3) the shocks were given by a group of "teachers" in disguise (remaining anonymous); 4) there was no evidence of deviance (i.e., no other subject was observed disobeying the experimenter); and 5) the victim was depersonalized or distanced from the subject (i.e., in another room).

According to Milgram himself, an important lesson from this research is that "ordinary people, simply doing their jobs, and without any particular hostility on their part, can become agents in a terrible destructive process."[98] In a similar vein, people might perform unsafe acts or overlook obvious safety hazards and put themselves and others at risk as a result of social obedience or social conformity.

The statement, "I was just following orders" reflects this obedience phenomenon, and "everyone else does it" implies conformity or peer pressure. As health and safety professionals, we need to realize the powerful impact of both conformity and authority, and plan interventions to overcome their potential negative influence. Remember that a leader who deviates from the norm and sets a safe example can decrease destructive conformity and obedience. And, when a critical mass of individuals actively care for the safety and health of others, constructive conformity and obedience occur to cultivate a total safety culture.

We Actively Care for People We Like

For whom would you most likely actively care — someone you dislike or someone you like? The answer is obvious, of course, and reflects the value of increasing our personal appeal to others when cultivating interdependency and a total safety culture. Social psychology research has verified a number of intuitive techniques to facilitate ingratiation, including: 1) demonstrate agreement with target persons on other matters before making a request; 2) offer genuine praise, recognition, or rewarding feedback; 3) use "name dropping" to show association with people respected by target persons; 4) radiate positive nonverbal cues (e.g., smiles, friendly gestures) that show appreciation and interest in the target person; and 5) modify personal appearance to be more acceptable and appealing to the target person. This might mean, for example, donning or removing a tie in order to appear more similar to a target audience.

Guidelines for giving and receiving recognition are presented elsewhere[88,99,100] and won't be repeated here, except to point out the special value of recognition in increasing ingratiation and spreading a sense of belonging and interdependency throughout a work culture. When you recognize a person appropriately, you do more than support that person's desirable behavior and build his or her self-efficacy and self-esteem. You influence that person to like and appreciate you more. This increases your ability to exert positive social influence on that person.

And, there are other benefits from an effective recognition process. Not only does liking and appreciation increase from the person being recognized to the person doing the recognizing, but the reverse also happens. After all, when we find behavior to recognize we learn to appreciate the person performing the desired behavior. This mutual ingratiation increases feelings of belonging and interdependency. And through the power of reciprocity, one appropriate recognition interaction will likely lead to more interpersonal recognition. This in turn affects more ingratiation and interdependency throughout a work culture.

In industrial safety, we frequently experience the opposite of ingratiation. Safety management is too often a confrontation between an authoritative individual and a subordinate who is not following a rule. Or, a request to serve on a safety steering committee or special safety task force is given in an impersonal memo or in a computer e-mail message. Although the principle of ingratiation is known and understood through personal experience, it's often not followed when making a safety request.

In fact, safety is too often presented as a top-down condition of employment, and this leads to the perception that we follow safety rules because we "have to" not because we "want to." This can limit individual perceptions of freedom and choice when it comes to safety. As discussed under the final social influence principle, this reduced sense of personal freedom or control can lead to counterproductive behaviors and attitudes.

Value Increases with Scarcity

Have you ever gone out of your way to purchase front-row tickets to a sports event or music concert? Have you ever jammed into a department store in order to take advantage of "special, limited-time" bargains? Or, perhaps you're a collector of rare coins, stamps, baseball cards, or antique furniture? Have you ever participated at an auction where one-of-a-kind items are displayed and sold to the highest bidder? Scarce items become even more valuable when other people want them. Have you ever paid more than a fair price at an auction in order to win a bidding competition? Each of these situations illustrates how the value of something increases with perceived scarcity.

The Forbidden Fruit Phenomenon. Why do teenagers consume drugs, including alcohol, cigarettes, and marijuana? The principle of social conformity (or peer pressure) discussed above is certainly a factor, but why is the purchase and consumption of these drugs considered valuable? They don't taste good, at least at first, and everyone knows they are not good for you. For years teens hear the slogan "Just say no to drugs." Why is it so difficult for some teenagers (and adults) to say no?

Consider that the redundant message "Say No" can make the drug seem desirable. It's the "forbidden fruit" phenomena, as told in the Biblical story of Adam and Eve in the Garden of Eden. Forbidding something makes that something seem valuable, and if we can beat the system to partake of the "forbidden fruit" we might experience an extra rush of pleasure because we asserted our freedom in a top-down situation perceived as stifling our individuality and creativity.

Items or opportunities that appear scarce in the eyes of the beholder seem more valuable. Thus, drugs seem desirable to some people because they are scarce, and consuming drugs can feel more pleasurable because the behavior itself represents a scarce freedom. I'm not advocating the legalization of drugs or the lowering of the legal age for alcohol consumption. But, understanding the scarcity principle can help us appreciate why certain illegal acts occur and why increased enforcement of laws might not help in the long run.

Reacting with Counter Control. When individuality or perceived personal control is made scarce with top-down control, some people will exert contrary behavior in an attempt to assert their freedom. This is referred to as psychological reactance by social psychologists[101] and counter control by behavior analysts.[102] Whatever you call it, the results can be devastating to industrial safety and health.

For example, in one company, a worker wore safety glasses without lenses. That's right, not safety glasses, only safety frames. He had removed the lenses. When a supervisor walked up the aisle, this employee looked right at him and waved. His co-workers gave him a lot of peer approval for thumbing his nose at the system. Safety was perceived as a top-down mandate that restricted individual freedom. This worker increased his status in that culture by pushing against the system and demonstrating his individuality and independence.

Thus situations that appear to take away personal freedom or control can activate behaviors opposite to what we want. When people perceive a system as restrictive they might attempt to beat it. However, when the threat of punitive consequences is severe and the probability of getting caught high, most people will comply. But compliance will typically be limited to situations in which the rule is reinforced. Don't expect people to internalize the rule and follow it when they don't have to. The more control you exert on the outside of people, the less self-control people develop on the inside.

External Control and Internal Values. When we perform in certain ways because of external controls or threats, we say we are doing it because we "have to" not because we "want to." Under such circumstances we feel no obligation to adjust our inner self (including our beliefs and values) to conform with our outer self (our behavior). We can live with the inconsistency between what we do (follow the rules) and our belief (that the rule is silly and unreasonable). Then, when we are in situations where enforcement of a rule is difficult, we are apt to break the rule. Hence, most workers use their personal protective equipment (PPE) when it's called for in the workplace, but many don't use appropriate PPE when doing even riskier jobs at home. What percentage of your neighbors wear ear plugs, safety glasses, and steel-toed shoes while mowing their lawns?

Our outside self (behavior) influences our inside self (attitudes, beliefs, values) when we perceive our behavior was our idea (at least to some extent). That's why it's so important for people to volunteer for safety promotion efforts. When people choose to get involved in a safety process, their behaviors (from developing components of the process to teaching it to others) help to develop or support an internal value or belief system that drives the appropriate behavior in the absence of external controls. This is the consistency principle in action, as discussed earlier.

Consequently, values-driven safety will not come from increasing external controls over behavior. Rather, it is cultivated in situations that: a) provide a clear purpose or mission (the value), b) promote a variety of straightforward methods for accomplishing the mission, and c) allow people to choose and customize procedures for their particular setting. Of course, this is easier said than done, and it's often tempting to exert external controls when things don't go our way.

We need to understand and believe that values cannot be dictated. Values can be developed from outside sources that encourage voluntary participation in activities representing or supporting a particular value. We internalize the principles and lessons we choose to experience, but we are apt to resist those principles and lessons we feel were forced upon us. Obviously, it's a difficult but important challenge to provide just enough outside influence to make a program or process appear worthwhile and inviting without inhibiting feelings or personal choice and control. In this regard, please remember another psychological principle: we perceive more choice and personal control when working to achieve positive consequences (rewards) than when working to avoid negative consequences (penalties).

Summary

As a science of mind and behavior, psychology is actually a vast field of numerous subdisciplines. Areas covered in a standard college course in introductory psychology, for example, include research methods, physiological foundations, sensation and perception, language and thinking, consciousness and memory, learning, motivation and emotion, human development, intelligence, personality, psychological disorders, treatment of mental disorders, social thought and behavior, industrial/organizational psychology and human factors engineering, and environmental psychology. This chapter clearly does not cover all of these areas of psychology, not even from an introductory perspective. Rather, this coverage of psychological principles and procedures focuses on only those research-based aspects of psychology directly relevant to reducing industrial injury and illness.

Even with the narrow health and safety focus of this chapter, however, many relevant topics are not considered, and those topics addressed are only given introductory coverage. Nevertheless, this chapter is quite comprehensive (including relevant topics from clinical psychology, applied behavior analysis, organizational psychology, social psychology, learning, motivation, and personality), and offers a broad overview of the psychology of occupational safety and health. Most important, the topics covered represent the current state-of-the-art in applying what we know about human dynamics to the challenge of involving more people in safety and health improvement.

As reviewed here, behavior-based psychology holds many of the tools for increasing safe behavior and reducing at-risk behavior on a large-scale. However, the availability of techniques to conduct behavior-based observation and feedback, behavior-based incentive/reward programs, behavior-based coaching and injury investigation is not sufficient. We need to develop the type of organizational culture that both enables and facilitates the actively caring application of the various behavior-based tools. Such a culture needs to support the social influence principles of consistency, reciprocity, conformity, authority, and ingratiation in ways that cultivate interpersonal trust, interdependency and the five person states that increase an individual's propensity to actively care, self-esteem, self-efficacy, personal control, optimism, and belongingness. Some ways to start the cultivation of this kind of ideal culture were introduced here, but numerous other techniques (perhaps more effective) are yet to be discovered. And even the procedures reviewed here need to be customized for your particular setting and circumstances.

References

1. **Geller, E.S.:** Managing occupational safety in the auto industry. *J. Organ. Behav. Manage. 10(1)*:181–185 (1989).
2. **Geller, E.S., G.R. Lehman, and M.R. Kalsher:** *Behavior Analysis Training for Occupational Safety*. Newport, VA: Make-A-Difference, Inc., 1989.
3. **Wandersman, A., P. Popper, and D. Ricks:** *Humanism and Behaviorism: Dialogue and Growth*. New York: Pergamon Press, 1976.
4. **Deming, W.E.:** *Out of the Crisis*. Cambridge, MA: Massachusetts Institute of Technology, Center for Advanced Engineering Study, 1982.
5. **Deming, W.E.:** *The New Economics for Industry, Government, Education*. Cambridge, MA: Massachusetts Institute of Technology, Center for Advanced Engineering Study, 1993.

6. **Covey, S. R.:** *The Seven Habits of Highly Effective People.* New York: Simon and Schuster, 1989.
7. **Covey, S.R.:** *Principle-Centered Leadership.* New York: Simon and Schuster, 1990.
8. **Skinner, B.F.:** *The Behavior of Organisms.* Acton, MA: Copley Publishing Group, 1938.
9. **Skinner, B.F.:** *About Behaviorism.* New York: Alfred A. Knopf, 1974.
10. **Elder, J.P., E.S. Geller, M.F. Hovell, and J.A. Mayer:** *Motivating Health Behavior.* New York: Delmar Publishers, Inc., 1994.
11. **Geller, E.S., R.A. Winett, and P.B. Everett:** *Preserving the Environment; New Strategies for Behavior Change.* Elmsford, NY: Pergamon Press, 1982.
12. **Goldstein, A.P. and L. Krasner:** *Modern Applied Psychology.* New York: Pergamon Press, 1987.
13. **Greene, B.F., R.A. Winett, R. Van Houten, E.S. Geller, et al.:** *Behavior Analysis in the Community: Readings From the Journal of Applied Behavior Analysis.* Lawrence, KS: University of Kansas Press, 1987.
14. **Geller, E.S.:** *The Psychology of Safety: How to Improve Behaviors and Attitudes on the Job.* Radnor, PA: Chilton Book Company, 1996b.
15. **Krause, T.R., J.H. Hidley, and S.J. Hodson:** *The Behavior-Based Safety Process,* 2nd ed. New York: Van Nostrand Reinhold, 1996.
16. **McSween, T.E.:** *The Values-Based Safety Process.* New York: Van Nostrand Reinhold, 1995.
17. **Petersen, D.:** *Safe Behavior Reinforcement.* New York: Aloray, Inc., 1989.
18. **Festinger, L.:** *A Theory of Cognitive Dissonance.* Stanford, CA: Stanford University Press, 1957.
19. **Fishbein, M. and I. Ajzen:** *Belief, Attitude, Intention, and Behavior: An Introduction to Theory and Research.* Reading, MA: Addison-Wesley, 1975.
20. **Ullman, L.P. and L. Krasner (eds.):** *Case Studies in Behavior Modification.* New York: Holt, Rinehart, & Winston, 1965.
21. **Newman, B.:** *The Reluctant Alliance: Behaviorism and Humanism.* Buffalo, NY: Prometheus Books, 1992.
22. **Lazarus, A.A.:** *Behavior Therapy and Beyond.* New York: McGraw-Hill, 1971.
23. **Lazarus, A.A.:** *Multimodal Behavior Therapy.* New York: Springer Publishing Company, Inc., 1976.
24. **Yerkes, R.M. and J.D. Dodson:** The relation of strength of stimulus to rapidity of habit formation. *J. Comp. Neurol. Psychol. 18*:459–482 (1908).
25. **Holmes, T.H. and M. Masuda:** Life change and illness susceptibility. In *Stressful Life Events: Their Nature and Effects,* B.S. Dohrenwend and B.P. Dohrenwend, eds. New York: Wiley, 1974.
26. **Maslach, C.:** *Burnout: The Cost of Caring.* Englewood Cliffs, NJ: Prentice-Hall, 1982.
27. **Bandura, A.:** Self-efficacy mechanism in human agency. *Am. Psychol. 37*:122–147 (1982).
28. **Scheier, M.F. and C.S. Carver:** *Perspectives on Personality.* Boston, MA: Allyn and Bacon, 1988.
29. **Scheier, M.F. and C.S. Carver:** Effects of optimism on psychological and physical well-being: theoretical overview and empirical update. *Cognit. Therap. Res. 16*:201–229 (1992).
30. **Coyne, J.C. and G. Downey:** Social factors and psychopathology: Stress, social support, and coping processes. *Ann. Rev. Psychol. 42*:401–425 (1991).
31. **Janis, I.L.:** The role of social support in adherence to stressful decisions. *Am. Psychol. 38*:143–160 (1983).
32. **Lieberman, M.A.:** The effects of social support on response to stress. In *Handbook of Stress Management,* L. Goldbert and D.S. Breznitz, eds. New York: Free Press, 1983.
33. **Delprata, D.J.:** Prompting electrical energy conservation in commercial users. *Environ. Behav. 9*:433–440 (1977).
34. **Geller, E.S., C.D. Bruff, and J.G. Nimmer:** "Flash for Life": Community-based prompting for safety belt promotion. *J. Appl. Behav. Anal. 18*:145–149 (1985).
35. **Thyer, B.A. and E.S. Geller:** The "buckle-up" dashboard sticker: an effective environmental intervention for safety belt promotion. *Environ. Behav. 19*:484–494 (1987).
36. **Carlson, N.R.:** *Psychology: The Science of Behavior,* 4th ed. Needham Heights, MA: Allyn and Bacon, 1993.
37. **Geller, E.S., J.B. Erickson, and B.A. Buttram:** Attempts to promote residential water conservation with educational, behavioral, and engineering strategies. *Popul.Environ. 6*:96–112 (1983).
38. **Hayes, S.C. and J.D. Cone:** Reducing residential electrical use: payments, information, and feedback. *J. Appl. Behav. Anal. 14*:81–88 (1977).

39. **Witmer, J.F. and E.S. Geller:** Facilitating paper recycling: effects of prompts, raffles, and contests. *J. Appl. Behav. Anal. 9*:315–322 (1976).
40. **Ross, H.L.:** *Deterring the Drinking Driver: Legal Policy and Social Control.* Lexington, MA: Lexington Books, 1982.
41. **Sulzer-Azaroff, B.:** Is back to nature always best? In *The Educational Crisis: Issues, Perspectives, Solutions,* E.S. Geller, ed. Lawrence, KS: Society for the Experimental Analysis of Behavior, Inc., 1992. pp. 68–69.
42. **Skinner, B.F.:** *Verbal Behavior.* New York: AppletonCentury-Crofts, 1957.
43. **Vaughan, M.E. and J. Michael:** Automatic reinforcement: an important but ignored concept. *Behaviorism 10*:217–227 (1982).
44. **Deci, E.L.:** *Intrinsic Motivation.* New York: Plenum, 1975.
45. **Kohn, A.:** *Punished by Rewards: The Trouble with Gold Stars, Incentive Plans, A's, Praise, and Other Bribes.* Boston, MA: Houghton Mifflin, 1993.
46. **Geller, E.S.:** The truth about safety incentives. *Prof. Saf. 41(10)*:34–39 (1996c).
47. **Volk, D.:** Learn the do's and don'ts of safety incentives. *Saf. Health (March) 153*:54–57 (1994).
48. **Daniels, A.C.:** *Bringing Out the Best in People.* New York: McGraw-Hill, Inc., 1994
49. **Geller, E.S.:** If only more would actively care. *J. Appl. Behav. Anal. 24*:607–612 (1991).
50. **Geller, E.S.:** Ten principles for achieving a total safety culture. *Prof. Saf. 39(9)*:18–24 (1994).
51. **Rogers, C.:** The necessary and sufficient conditions of therapeutic personality change. *J. Counsult. Psychol. 21*:95–103 (1957).
52. **Rogers, C.:** *Carl Rogers on Personal Power: Inner Strength and Its Revolutionary Impact.* New York: Delacorte, 1977.
53. **Straumann, T.J. and E.G. Higgins:** Self-discrepancies as predictors of vulnerability to distinct syndromes of chronic emotional distress. *J. Pers. 56*:685–707 (1988).
54. **Brown, J. D. and K.L. McGill:** The cost of good fortune: when positive life events produce negative health consequences. *J. Pers. Soc. Psychol. 57*:1103–1110 (1989).
55. **Wylie, R.:** *The Self-Concept,* vol. 1. Lincoln: University of Nebraska Press, 1974.
56. **Wells, L.E. and G. Marwell:** *Self-Esteem.* Beverly Hills, CA: Sage, 1976.
57. **Baron, R.A. and D. Byrne:** *Social Psychology: Understanding Human Interaction,* 7th ed. Boston, MA: Allyn and Bacon, 1994.
58. **Batson, C.D., M.H. Bolen, J.A. Cross, and H.E. Neuringer-Benefiel:** Where is altruism in the altruistic personality? *J. Pers. Soc. Psychol. 1*:212–220 (1986).
59. **Conger, J.A. and R.N. Kanungo:** The empowerment process: integrating theory and practice. *Acad. Manage. Rev. 13*:471–482 (1988).
60. **Rotter, J.B.:** Generalized expectancies for internal versus external control of reinforcement. *Psychol. Monogr. Vol. 80,* #609 (1966).
61. **Bandura, A:** *Self-Efficacy: The Exercise of Control.* New York: W.H. Freeman and Company, 1997.
62. **Scheier, M.F. and C.S. Carver:** Optimism, coping and health: assessment and implications of generalized outcome expectancies. *Health Psychol. 4*:219–247 (1985).
63. **Seligman, M.E.P.:** *Learned Optimism.* New York: Alfred A. Knoff, 1991.
64. **Bandura, A.:** *Social Foundations of Thought and Action: A Social Cognitive Theory.* Englewood Cliffs, NJ: Prentice-Hall, 1986.
65. **Barling, J. and R. Beattie:** Self-efficacy beliefs and sales performance. *J. Organ. Behav. Manage. 5*:41–51 (1983).
66. **Ozer, E.M. and A. Bandura:** Mechanisms governing empowerment effects: a self-efficacy analysis. *J. Pers. Soc. Psychol. 58*:472–486 (1990).
67. **Phares, E.J.:** *Locus of Control in Personality.* Morristown, NJ: General Learning Press, 1976.
68. **Betz, N.E. and G. Hackett:** Applications of self-efficacy theory to understanding career choice behavior. *J. Soc. Clin. Psychol. 4*:279–289 (1986).
69. **Hackett, G., N.E. Betz, J.M. Casas, and I.A. Rocha-Singh:** Gender, ethnicity, and social cognitive factors predicting the academic achievement of students in engineering. *J. Couns. Psychol. 39*:527–538 (1992).
70. **Rushton, J.P.:** *Altruism, Socialization, and Society.* Englewood Cliffs, NJ: Prentice Hall, Inc., 1908.
71. **Hunt, M.M.:** *The Story of Psychology.* New York: Doubleday, 1993.
72. **Nowicki, S. and B.R. Strickland:** A locus of control scale for children. *J. Consult. Psychol. 40*:148–154 (1973).

73. **Strickland, B.R.:** Internal-external control expectancies: from contingency to creativity. *Am. Psychol. 44*:1–12 (1989).
74. **Taylor, S.E.:** *Health Psychology*, 2nd ed. New York: McGraw-Hill, 1991.
75. **Eskew, R.T. and C.V. Ricke:** Pacing and locus of control in quality control inspection. *Hum. Factors* 24:411–415 (1982).
76. **Phares, E.J.:** *Introduction to Personality*, 3rd ed. New York: Harper Collins, 1991.
77. **Carver, C.S., M.F. Scheier, and J.K. Weintraub:** Assessing coping strategies: a theoretically based approach. *J. Pers. Soc. Psychol. 56*:267–283 (1989).
78. **Peterson, C. and L.C. Barrett:** Explanatory style and academic performance among university freshmen. *J. Pers. Soc. Psychol. 53*:603–607 (1987).
79. **Scheier, M.F., J.K. Weintraub, and C.S. Carver:** Coping with stress: divergent strategies of optimists and pessimists. *J. Pers. Soc. Psychol. 51*:1257–1264 (1986).
80. **Tavris, C. and C. Wade:** *Psychology in Perspective*. New York: Harper Collins College Publishers, 1995.
81. **Peck, M.S.:** *The Different Drum: Community Making and Peace*. New York: Simon and Schuster, 1979.
82. **Beck, A.T.:** Cognitive therapy: a 30-year retrospective. *Am. Psychol. 46*:368–375 (1991).
83. **Taylor, S.E.:** *Positive Illusions: Creative Self-Deception and the Healthy Mind.* New York: Basic Books, 1989.
84. **Weick, K. E.:** Small wins: Redefining the scale of societal problems. *Am. Psychol. 39*:40–44 (1984).
85. **Cialdini, R.B.:** *Influence: Science and Practice*, 3rd ed. New York: Harper Collins College Publishers, 1993.
86. **Geller, E.S. and G.R. Lehman:** The buckle-up promise card: a versatile intervention for large-scale behavior change. *J. Appl. Behav. Anal. 24*:91–94 (1991).
87. **Freedman, J.L. and S.C. Fraser:** Compliance without pressure: The foot-in-the-door technique. *J. Pers. Soc. Psychol. 4*:195–203 (1966).
88. **Geller, E.S.:** How to give quality recognition. *Ind. Saf. Hyg. News 30(12)*:12 (1996a).
89. **Cialdini, R. B., Cacioppo, J. T., Bassett, R., & Miller, J. A.** (1978). Low-ball procedure for producing compliance: Commitment then cost. Journal of Applied Social Psychology, 15, 492-500.
90. **Berkowitz, L. and L.R. Daniels:** Responsibility and dependency. *J. Abnorm. Soc. Psychol. 66*:429–436 (1963).
91 **Isen, A.M. and P.F. Levin:** Effect of feeling good on helping: cookies and kindness. *J. Pers. Soc. Psychol. 21*:384–388 (1972).
92. **Cialdini, R.B., J.E. Vincent, S.K. Lewis, J. Catalan, et al.:** Reciprocal concessions procedure for inducing compliance; the door-in-the-face technique. *J. Pers. Soc. Psychol. 1*:206–215 (1975).
93. **Asch, S.E.:** Effects of group pressure upon the modification and distortion of judgments. In *Groups, Leadership, and Men*, H. Guetzkow, ed. Pittsburgh, PA: Carnegie Press, 1951.
94. **Asch, S.E.:** Opinions and social pressure. *Sci. Am. 193*:31–35 (1955).
95. **Crutchfield, R.S.:** Conformity and character. *Am. Psychol. 10*:191–198 (1955).
96. **Nemeth, C.:** Differential contribution of majority and minority influence. *Psychol. Rev. 93*:23–32, 1986.
97. **Milgram, S.:** Behavioral studies of obedience. *J. Abnorm. Soc. Psychol. 67*:371–378 (1963).
98. **Milgram, S.:** *Obedience to Authority.* New York: Harper Collins, 1974.
99. **Geller, E.S.:** Hey, I'm trying to give you a compliment. *Ind. Saf. Hyg. News 31(1)*:12 (1997a).
100. **Geller, E.S.:** How to celebrate safety success. *Ind. Saf. Hyg. News 31(2)*:12–13 (1997b).
101. **Brehm, J.W.:** *A Theory of Psychological Reactance*. New York: Academic Press, 1966.
102. **Skinner, B.F.:** *Beyond Freedom and Dignity*. New York: Knopf, 1971.

Appendix A

Perspectives on Psychology in the Workplace: The National Institute for Occupational Safety and Health and Behavioral Science Research in the Workplace*

The National Institute for Occupational Safety and Health (NIOSH) has the longest standing behavioral science research program in occupational health in the United States. For 25 years the Institute has conducted and sponsored laboratory, field, and epidemiological studies that have helped to define the role of work organization factors in occupational safety and health. Work organization broadly refers to the way work processes are structured and managed and addresses elements such as the scheduling of work, job design, interpersonal aspects of work, career concerns, management style, and organizational characteristics.

Overview of NIOSH Research

Psychosocial research of NIOSH has focused on the health effects of specific job conditions, such as machine-paced work, shift work, and worker control or autonomy; [1-3] on occupational stressors in specific occupations, such as coal miners, police officers, postal workers, nurses, and office workers; [4-9] and on occupational differences in the incidence of stressors and stress-related disorders.[10] For example, NIOSH sponsored and helped conduct the classic study, *Job Demands and Worker Health: Main Effects and Occupational Differences*, which provided the first comparative analysis of psychological job demands and health for more than 2,000 workers in 23 occupations.[11]

Keeping abreast of the changing work environment has been an ongoing challenge for NIOSH. NIOSH researchers responded to the growth of office work and the introduction of computer technology with pioneering studies on health complaints among office workers. The first indications of a link between the psychosocial environment and musculoskeletal disorders among computer users emerged in a study conducted at NIOSH.[12] Subsequent research confirmed the initial finding that psychosocial factors, particularly increased work pressure and job insecurity or uncertainty, predicted objectively defined and measured musculoskeletal problems.[13]

As office environments changed, NIOSH also saw a dramatic increase in the number of requests to evaluate indoor air quality. Although deficiencies in building ventilation systems are the most common problems found, NIOSH investigators also have found that job-related psychosocial stressors have contributed to the symptoms workers experience.[14] NIOSH research on air quality and computer use demonstrates the complex interaction of psychological and physical workplace factors and their impact on health. Identifying the mechanisms underlying this connection remains an important task for NIOSH and for occupational health psychology.

The Role of Intervention Research at NIOSH

NIOSH is mandated to conduct research that develops practical strategies and techniques to reduce or prevent workplace illnesses and injuries. Therefore, much of NIOSH research on work organization has an intervention component. Several NIOSH studies have evaluated the merits of workplace-based individually oriented stress management techniques on stress responses, such as anxiety and elevated blood pressure. This research led to recommendations for a strategy for employers interested in establishing worksite stress control programs that include not only prescriptive individually oriented strategies but also target the job and the organization.[15] Consistent with this emphasis, in collaboration with occupational health and safety professionals, in 1990 NIOSH developed a national strategy for the prevention of work-related psychological disorders.[16]

Through its intramural and sponsored research, NIOSH is also testing specific work organization interventions. These include (a) the introduction of rest breaks and job redesign to prevent or at least ameliorate musculoskeletal disorders in computerized work[17] and (b) intervention research to improve adherence to safe work practices among health care workers.[18]

In recent years NIOSH also has responded to the impact of large scale economic changes. The trend toward corporate downsizing in order to reduce costs and improve competitiveness in a global economy began in the late 1980s and increased substantially in the 1990s, affecting both blue collar and white collar workers. At the request of two Fortune 500 companies, NIOSH has begun to evaluate organizational characteristics and interventions that will minimize stress and loss of productivity associated with corporate downsizing. This NIOSH study will address three key aspects of downsizing or reorganization: (a) the purpose of downsizing; (b) the process of downsizing; and (c) the provision of assistance to employees who lose their jobs and those who do not. NIOSH will develop practice guidelines for companies to help reduce the negative health and performance consequences of downsizing.

In another key study, NIOSH researchers are hoping to identify the characteristics of "healthy" work organizations—those that are profitable and competitive and also promote good employee mental and physical health.

*Excerpted from: **Rosenstock, L.:** Work Organization Research at the National Institute for Occupational Safety and Health. *J. Occup. Health Psychol. 2(1)*:7–10 (1997).

Appendix A References

1. **Hurrell, J.J. Jr. and M.M. Colligan:** Machine pacing and shift work: evidence of job stress. *J. Organ. Behav. Manage. 8*:159–175 (1986).
2. **National Institute for Occupational Safety and Health (NIOSH):** *Health Consequences of Shift Work* (DHEW/NIOSH pub. no. 78-154). Cincinnati, OH: NIOSH, 1978.
3. **Sauter, S.L., J.J. Hurrell Jr., and C.L. Cooper:** *Job Control and Worker Health.* Chichester, England: Wiley, 1989.
4. **Cohen, B.F.G. (ed.):** *The Human Aspect of Office Automation.* Amsterdam: Elsevier Science, 1984.
5. **Hurrell, J.J. Jr. and M.L. Smith:** Machine paced work and type A behavior pattern. *J. Occup. Psychol. 58*:15–25 (1995).
6. **Hurrell, J.J. Jr. and M.L. Smith:** Sources of stress among machine-paced letter sorting- machine operators. In *Machine Pacing and Occupational Stress*, G. Salvendy and M. Smith, eds. London: Taylor & Francis, 1981. pp. 253–260.
7. **McLaney, M.A. and J.J. Hurrell Jr.:** Control, stress and job satisfaction in Canadian nurses. *Work Stress* 2:217–224 (1988).
8. **National Institute for Occupational Safety and Health (NIOSH):** *Job Stress and the Police Officer: Identifying Stress Reduction Techniques* (DHEW/NIOSH pub. no. 76-186). Cincinnati, OH: NIOSH, 1976.
9. **National Institute for Occupational Safety and Health (NIOSH):** *An Analysis of Job Stress in Coal Mining* (DHEW/NIOSH pub. no. 77-217). Cincinnati, OH: NIOSH, 1977.
10. **Colligan, J.J., M.J. Smith, and J.J. Hurrell Jr.:** Occupational incidence rates of mental health disorders. *J. Hum. Stress 3*:34–39 (1977).
11. **Caplan, R.D., S. Cobb, J.R. French, R. Van Harrison, et al.:** *Job Demands and Worker Health: Main Effects and Occupational Differences.* Washington, DC: U.S. Department of Health, Education, and Welfare, 1975.
12. **Smith, M.J., B. Cohen, L.W. Stammerjohn Jr., and A. Happ:** An investigation of health complaints and job stress in video display terminal operations. *Hum. Factors 23*:387–400 (1981).
13. **Hales, T.R., S.L. Sauter, M.R. Peterson, L.F. Fine, et al.:** Musculoskeletal disorders among visual display terminal users in a telecommunications company. *Ergonomics 37*:1603–1621 (1994).
14. **National Institute for Occupational Safety and Health (NIOSH):** *Hazard Evaluation and Technical Assistance Report, Library of Congress* (NIOSH rep. no. HETA 88-364-2104). Cincinnati, OH: NIOSH, 1991.
15. **Murphy, L.R. and T.F. Shoenborn (eds.):** *Stress Management in Work Settings.* New York: Praeger Publishers, 1989.
16. **Sauter, S.L., L.R. Murphy, and J.J. Hurrell Jr.:** Prevention of work-related psychological disorders: a national strategy proposed by the National Institute for Occupational Safety and Health (NIOSH). *Am. Psychol. 45*:1146–1158 (1990).
17. **Sauter, S.L. and N.G. Swanson:** The relationship between workplace psychosocial factors and musculoskeletal disorders in office work: suggested mechanisms and evidence. In *Repetitive Motion Disorders of the Upper Extremities*, S.L. Gordon, S.J. Blair, and L.J. Fine, eds. Rosemont, IL: American Academy of Orthopedic Surgeons.
18. **DeJoy, D., L.R. Murphy, and R.M. Gershon:** The influence of employee, job/task and organizational factors on adherence to universal precautions among nurses. *Int. J. Ind. Ergonom. 16*:43–55 (1995).

Appendix B

Perspectives on Psychology in the Workplace: The Growth of Intervention and the Variety of Approaches

During the past decade there have been an increasing number of publications addressing interventions aimed at preventing work-related illness and injury and promoting employee health. The focus of these publications has included (a) the development of conceptual models to guide worksite stress and social health programs;[1–7] (b) an examination of approaches for prevention of work-related psychological disorders[7–8] as well as approaches for addressing health at work more broadly,[9] (c) a review of stress management interventions, (d) an assessment of studies aimed at preventing and reducing stress at work,[10–13] and (e) an examination of occupational safety and health interventions.[14]

While the overall aim of this literature is similar, to gain increased knowledge and understanding of effective interventions for promoting employee health and preventing work-related illness and injury, the diversity of perspectives and fields has for the most part evolved independently and with little overlap. For example, the target of change (e.g., individual and organization), the intervention strategy (e.g., job redesign and stress management), and the proposed outcomes (e.g., reduced stress and improved safety) differ considerably across these fields. Thus, it is not uncommon within a single organization to separate interventions developed by staff from different disciplinary backgrounds, for example, industrial-organizational psychology, occupational health and safety, health promotion-education, and clinical psychology and health psychology. The result has been the proliferation of programs that are fragmented from one another that may, on the one hand, focus solely on individual behavior change (e.g., use of protective clothing and psychological counseling) with little consideration of the broader organizational context or focus on changing the organization with little attention paid to individual employee differences.

References

1. **Baker, E.A., B.A. Israel, and S.J. Schurman:** The integrated model: implications for worksite health promotion and occupational health and safety practice. *Health Educ. Q. 23*:175–190 (1996).
2. **Gottlieb, N.H. and K.R. McLeroy:** *Social health. In Health Promotion in the Workplace*, 2nd ed. M.P. O'Donnell and J.S. Harris, eds. Albany, NY: Delmar, 1994.
3. **Heaney, C.A. and M. van Ryn:** Broadening the scope of worksite stress programs: a guiding framework. *Am. J. Health Promot. 4*:413–420 (1990).
4. **Israel, B.A. and S.J. Schurman:** Social support, control, and the stress process. In *Health Behavior and Health Education: Theory, Research and Practice*, K. Glanz, F. Lewis, and B. Rimer, eds. San Francisco, CA: Jossey-Bass, 1990.
5. **Ivancevich, J.J., M.T. Matteson, S.M. Freedman, and S.J. Phillips:** Worksite stress management interventions. *Am. Psychol. 45*:252–261 (1990).
6. **Karasek, R.A.:** Stress prevention through work reorganization: a summary of 19 international case studies. Conditions at Work Digest: *Preventing Stress at Work 11*:23–42 (1992).
7. **Quick, J.C., L.R. Murphy, and J.J. Hurrell Jr., and D. Orman:** The value of work, the risk of distress, and the poser of prevention. In *Stress and Well-Being at Work: Assessments and Interventions for Occupational Mental Health*, J.C. Quick, L.R. Murphy, and J.J. Hurrell, eds. Washington, DC: American Psychological Association, 1992.
8. **Sauter, S.L., L.R. Murphy, and J.J. Hurrell:** Prevention of work-related psychological disorders: a national strategy proposed by the National Institute for Occupational Safety and Health. *Am Psychol. 45*:1146–1158 (1981).
9. **Ilgen, D.R.:** Health issues at work: opportunities for industrial/organizational psychology. *Am. Psychol. 45*:273–283 (1990).
10. **DiMartino, V. (ed.):** *Conditions of Work Digest: Preventing Stress at Work*. Geneva: International Labour Office, 1992.
11. **Landsbergis, P.A. and J. Cahill:** Labor union programs to reduce or prevent occupational stress in the United States. *Int. J. Health Serv.* pp. 105–129 (1994).
12. **Landsbergis, P.A., S.J. Schurman, B.A. Israel, M. Schnall, et al.:** Job stress and heart disease: evidence and strategies for prevention. *New Solutions 3*:42–58 (1993).
13. **Schurman, S.J., and B.A. Israel:** redesigning work systems to reduce stress: a participatory action research approach to creating change. In *Job Stress Interventions*, L.R. Murphy, J.J. Hurrell Jr., S.L. Sauter, and G.P. Keita, eds. Washington, DC: American Psychological Association, 1995. pp. 235–263.
14. **Goldenhar, L.M. and P.A. Schulte:** Intervention research in occupational health and safety. *J. Occup. Med. 36(7)*:763–775 (1994).

More than 100 professional development courses were available at the 1997 American Industrial Hygiene Conference & Exposition held in Dallas, Texas.

chapter 30

Outcome Competencies

After completing this chapter, the user should be able to:

1. Define underlined terms used in the chapter.
2. Recall the relation mundating training.
3. Recognize the importance of program evaluation.
4. Prepare learning objectives and competencies.
5. Outline the steps to develop training programs.
6. Select appropriate training methods and media, given situational constraints.

Key Terms

formative evaluation • instructional objectives • instructional systems design • instructional technology • performance measures • summative evaluation • training needs assessment

Prerequisite Knowledge

Prior to beginning this chapter, the user should review the following chapters:

Chapter Number	Chapter Topic
29	Psychology and Occupational Health
40	Hazard Communication
41	Risk Communication in the Workplace

A selection of related readings are listed in the Background Reading Material section at the end of the chapter.

Key Topics

I. Why Do We Need to Train?
   A. Scenario 1. Is Training Needed?
   B. Importance of Effective Training and Education
   C. Regulatory Drivers
   D. Internal Drivers

II. Steps in the Development of Effective Training Programs — How Do We Train?
   A. Instructional Systems Design
   B. Training Needs Assessments
   C. Scenario 2. An Integrated Health, Safety, and Environmental Training Approach
   D. Defining Performance Objectives
   E. Selecting Appropriate Training Methods and Media
   F. Evaluating the Effectiveness of Existing Programs
   G. Planning and Presenting Training
   H. Evaluating Process and Product Outcomes
   I. Improving Programs Continuously
   J. Scenario 3. Following the Steps: A Typical Workplace Situation

30

Worker Education and Training

Margaret C. Samways

Why Do We Need to Train?

Scenario 1. Is Training Needed?*

Susan Brown, occupational hygienist, was contacted by the plant manager, Mr. Cummings, concerning a recent accident involving a worker's overexposure to radio frequency (RF) radiation. The employee's workstation was beside a high-power RF drying oven, where he placed textile packages on a conveyor that conveyed them into the drying chamber of the oven. His position was very close to the drying chamber RF source, which was not well shielded. Power density levels at the workstation were measured at 300 mw/cm². Power densities just inside the drying chamber opening were over 1500 mw/cm².

Recently, because of increased production demands, more textile packages were being placed on the conveyor, power levels were elevated, and spacing between packages had been reduced. The worker was placing a 35-pound package on the conveyor when he noticed that several packages about to enter the chamber were spaced too closely for proper drying. He reached over to rearrange the spacing on the conveyor, and his hands and forearms entered the chamber momentarily. He received first degree burns on his hands and minor erythema on his forearms from the high power densities just inside the drying chamber.

Mr. Cummings wants a cost-effective way to fix the problem and maintain his production quotas.

Cummings: "Susan, I want you to put a training session together for the drying oven workers so we don't have another accident. Show them a video and make sure you tell them to keep their hands out of the oven entrance! I don't want this to happen again."

Susan: "I don't think training these workers is the answer to preventing this kind of accident—there's a simpler one-time fix that won't require constant enforcement measures and periodic refresher training."

Cummings: "And how much would that cost?"

Susan: "I don't have exact costs, but the solution to the problem requires that we eliminate the hazard by changing the design of the workstation and the drying oven itself. If we install a sheet metal, tunnel-like extension or barrier over the conveyor belt, we will move the workstation 5 feet away from the drying oven entrance and high power densities. Even if a worker has to stick his hands into the opening occasionally, the tunnel will prevent him from reaching the hazardous RF levels. His whole body exposures will be reduced at the same time."

Cummings: " Supposing you're right—will this interfere with production?

Susan: "No, and we won't have to conduct training to remind workers of the risks of sticking their hands in the oven, or have to spend time enforcing work procedures to overcome the deficiencies of the existing workstation setup. I can guarantee that the one-time cost of the sheet metal for the tunnel and installation will be much less than the costs of setting up training, the man-hours in initial and periodic training, lost production, and other hidden training costs."

Cummings: "Sounds logical. Let's talk to plant engineering and come up with the sort of solution that you're suggesting."

Importance of Effective Training and Education

More often than we realize, training is requested in an attempt to address occupa-

*This scenario was developed with the assistance of Mark R. Francis, CIH, CSP.

tional hygiene problems that cannot be corrected by a training intervention. This chapter will discuss the appropriate use of training programs, proper techniques for training, and the positive results that can be achieved. Scenario 1 outlines an example where training is the wrong approach for ensuring protection of the health and safety of employees, although some brief training may be required to supplement the new engineering control.

In spite of the potential misapplication of training described above, training is recognized as an essential element of an effective occupational hygiene program. In the following sections, the importance of training is discussed, as well as some of the major forces that drive its use.

Training and education designed to protect the health and safety of workers may frequently be ineffective, as evidenced by the high number of citations issued by Occupational Safety and Health Administration (OSHA) inspectors. Of all OSHA citations, those for noncompliance with the Hazard Communication Standard have been the most numerous for many years.[1] Worker training is often the final point of delivery for all the occupational hygienist's professional activities in the workplace; engineering, administrative, and other controls can be implemented successfully if training is effective, but can be sabotaged when information and instructions have been poorly communicated or forgotten.

Although considerable sums of money are spent on training, both in terms of direct training costs and time dedicated to training, discrepancies between desired and actual levels of worker behavior continue to present a problem to the occupational hygienist. Very large costs can be incurred when ineffective training is the identified cause of lost time accidents and fatalities.

It should be noted that in this chapter the terms educational technology and instructional technology are used interchangeably. As described by Seels and Richey, "instructional technology is the theory and practice of design, development, utilization, management and evaluation of processes and resources for learning."[2] The purpose of the chapter is to show how the sound and tested principles of instructional technology can be used effectively by the occupational hygienist in the workplace. This systematic approach to instruction has been endorsed by OSHA in the Voluntary Training Guidelines,[3] which were first issued in 1984 and were most recently reissued in 1995. The recommended approach produces a learning environment that shapes behavior to satisfy stated objectives.

Failure to achieve desired training outcomes can result from a number of factors.

- Training is not the solution; that is, poor performance is not caused by a lack of knowledge or skills.
- Training is too generic; that is, not targeted toward the specific occupational hygiene needs of the workplace or situation.
- The special needs and preferences of adult learners, as represented by the work force, are not taken into account.
- Resources are inadequate, or are not used in a cost-effective manner.
- Inappropriate training methods and media are selected, or appropriate methods are not selected.
- There is a lack of transfer of learning from the training setting to the workplace.
- Training is a one-shot event, with no followup.
- No evaluation of training effectiveness is made.
- Improvements, based on evaluation results, are not implemented.
- Adequate documentation of worker training history is not maintained.

Sometimes, effective training seems difficult for the occupational hygienist to achieve, since he or she wears many hats and lacks time and resources. However, positive changes can be made in the working environment if the training issue is approached systematically, as described later in this chapter.

Regulatory Drivers

Training is driven by some external or internal need. It is generally recognized that productivity is enhanced by safe work practices, and that a well-trained work force performs better than one that has received insufficient training. The occupational hygienist is often made responsible for the identification and prioritization of health and safety training needs. The need for training is easy to determine when it is dictated by a regulation. The federal government has generally recognized the importance of worker education and training and has established requirements concerning various industrial hazards. States with OSHA-approved programs have had to meet or exceed the federal requirements; for example, several states have mandated annual retraining when OSHA has not.

OSHA, following the passage of the

Occupational Safety and Health Act[4] in 1970 (OSH Act), has taken the lead in the design of effective worker training. OSHA has also established rules concerning specific chemicals, personal protection devices and procedures, and hazard communication. Many of the early OSHA rules were derived from national consensus standards and do not have the specificity of later rules, such as the training required under the Hazardous Waste Operations and Emergency Response Standard.[5] Training also appears as a requirement under other statutes, such as Title 10, Section 19.12 of the Nuclear Regulatory Commission, although OSHA is preeminent in the area of worker health and safety training. A brief summary of some major regulatory requirements follows.

OSHA General Duty Clause

The General Duty Clause of the OSH Act requires that each employer

> *shall furnish to each of his employees employment and a place of employment free from recognized hazards that are causing or are likely to cause death or serious physical harm to his employees, and shall comply with occupational safety and health standards promulgated under this Act.*[4]

It is generally understood that the requirement to provide training and information is implicit in these statements. More specific implications can be found in the description of the responsibilities of the employee, who "shall comply with occupational safety and health standards and all rules, regulations and orders pursuant to this Act, which are applicable to his own actions and conduct."[4] Obviously, compliance with rules and safe work actions can be achieved only through effective training.

Specific Vertical Standards

In contrast to the general injunction to train that is implicit in the General Duty Clause, there have been a number of "vertical" standards issued by OSHA that apply to specific chemicals or operations and contain evaluation and control provisions for that chemical or operation. Examples are the standards issued for benzene and lead. All such standards issued in recent years contain training requirements. These requirements can be found at the end of each standard, since training is perceived as the last function to be performed once all other controls are in place. Key rights of the employee include the "right to be made aware of the hazards found in the workplace." For example, what hazardous physical or chemical agents are used or stored in the workplace? What health problems are associated with exposure to such agents? What methods can be employed to minimize exposure? Workers are also entitled to know the magnitude and severity of any or all of their exposures to physical or chemical agents and may, if they so desire, obtain copies of medical evaluations made for the purpose of managing such exposures. This knowledge is best handled by means of systematic education and training.

The implications for training of the General Duty Clause and the standards that have been promulgated are immense. The hazard communication standard[6] and the state/local right-to-know laws, which are discussed separately, have served to focus even more specifically on the essential role of worker communication and training.

Environmental Protection Agency (EPA)

The Clean Air Act was originally passed and the EPA established in 1970; this had the effect of consolidating all federal pollution control agencies into one organization. In 1976 the Toxic Substances Control Act (TSCA) gave the EPA administrative and enforcement powers over new and existing chemicals entering the environment. Section 8(e) of TSCA requires that workers be informed annually of their right to report to the EPA any information that reasonably supports a conclusion that a chemical substance or mixture presents a substantial risk of injury to health or the environment. Similarly, under the 1976 Resource Conservation and Recovery Act (RCRA), workers should be informed of their right to report to EPA concerns regarding the handling and disposal of hazardous and other wastes. These worker-informing requirements of TSCA and RCRA are less onerous to the trainer than are the actual training requirements mandated in OSHA standards. They can be met through simple informational means, such as posters, safety meeting talks with handouts, or inexpensive programs available from vendors.

An interesting recent development has been the promulgation by EPA of the Accidental Release Prevention: Risk Management Program[7] requirements under the Clean Air Act Amendments. Effective on August 19, 1996, this section specifically requires worker training. Workers must be "trained and competent" in operating procedures and safe work practices, must be trained about health and safety hazards, and must be able to participate in emergency operations during shutdown. The section cross references OSHA's Process Safety Management rule,[8] so that it is possible to satisfy both EPA and OSHA with one combined training program.

Department of Transportation (DOT)

Several sections of *Code of Federal Regulations* Title 49 relate to the shipment of hazardous materials, which requires training. Section 173.1(6) states that "[it] is the duty of each person who offers hazardous materials for transportation to train each of his officers, agents and employees having any responsibility for preparing hazardous materials for shipment as to the applicable regulations."

Training is also required of carriers, and the sections for rail, air, and water (Sections 174, 175, and 176, respectively) all state that "it is the duty of each carrier to make the prescribed regulations effective and to thoroughly instruct each of his officers, agents and employees in relation thereto." These training requirements have prompted organizations such as the American Trucking Association to offer seminars across the country, and some instructional packages are also available.

Hazard Communication Standard

The OSHA hazard communication standard, which applies also to the construction industry,[9] has been termed the most significant landmark in the history of OSHA regulatory activity. The standard is unique because it is a performance rule; that is, industry has great latitude in the manner of compliance, as long as the intent of the standard is met. Since the intent is to communicate hazards to workers in an effective manner, training is the key requirement that validates the performance of all the other requirements, such as hazard determination, labeling, and material safety data sheet (MSDS) distribution. This standard alone has caused great visibility of the training function and has had a positive impact on the availability of budgetary and other resources to the occupational hygienist-trainer.

The standard makes it clear that general topics, such as the meaning of warning phrases, can be dealt with generically. However, much of the training must be workplace-specific. This means that purchased generic programs will not meet all the requirements, although they may be useful as part of the overall training plan. The job of developing, implementing, and documenting training will therefore fall on the field or location coordinator, who is the only person fully acquainted with the specific hazards in each workplace. Although this person may not be the occupational hygienist, particularly if the occupational hygienist operates at the corporate or regional level, his or her help will be needed to guide and support the local trainer.

The compliance directives, issued in August 1995, have provided further indications of what constitutes compliance in the training area, and guidance has been given by OSHA in the training guidelines mentioned previously. OSHA also proposes to produce a model training program in 1997 to further assist trainers faced with thorny issues such as having to provide site-specific training before a worker starts a new job, and finding ways to explain difficult toxicological concepts.

The state and local right-to-know laws and ordinances are generally the same in intent as the federal standard, but they are not performance oriented and are often more concerned with the disclosure of hazardous chemical information by industry. Also, they reach beyond the workplace to the community, an area not touched by OSHA. The lack of a performance orientation is significant, and there is no requirement, either explicit or implicit, that training has to be effective.

Legal Liability Issues

Since 1973, when the last edition of this book was published, there has been a revolution in the legal arena that has forced industry to focus its attention on the legal issues associated with toxic substances. The case viewed as the landmark was *Borel v. Fibreboard Paper Products Corporation et al.*,[10] in which Clarence Borel, an industrial insulation worker, sued the manufacturers of insulation materials for failure to warn of the dangers involved in handling asbestos. As a result of this and subsequent cases, personal injury suits known as toxic torts have been initiated by a manufacturer's own employees. They have also been brought by the employees of other companies to whom a company may supply raw materials or other products, consumers, contractors, and persons who may be exposed to emissions or wastes generated during manufacturing operations. From 1908, when the first workers' compensation law that successfully passed the test of constitutionality was adopted by the federal government, the concept of compensation for any disease or injury occurring at work or caused by the work environment became universal. Generally, acceptance by the worker of workers' compensation payments precludes recovery of additional compensation by a legal suit. However, most state laws contain exceptions that will allow employees, under certain circumstances, to sue their employers directly. The most common basis for such suits is the employer's duty to warn workers if the employer knows or has reason to believe that products or materials may be toxic at concentrations at which exposures may occur. The scientific data regarding the toxicity of a substance need not be conclusive for a duty to warn to exist. An obligation to warn arises if enough evidence exists that a reasonable person would want to be warned in order to avoid exposure. As was stated in *Borel*, "an insulation worker, no less than any other product user, has a right to decide whether to expose himself to the risk."[10]

The duty to warn not only is significant to the lawyers, but raises some practical questions for the occupational hygienist, such as what information must be transmitted? How must it be transmitted to warn adequately? And how can one tell if the transmission of information was adequate? These questions often fall into the lap of the occupational hygienist, who must plan and budget for training programs using the most effective methods and media. Careful documentation of the training also becomes of paramount importance. Even though toxic tort law is unsettled, it is clear that effective and documented training and education, particularly where chemical hazards are concerned, can play a leading role in preventing a significant legal liability. This may, in turn, have a major impact on the organization's balance sheet.

Internal Drivers

While the regulatory and legal liability drivers are common to everyone, the internal factors that drive the occupational health and safety training effort are specific to each organization's products and processes, existing data collection and analysis systems, and management climate. Although the most obvious internal driver is company policy, policy does not occur in a vacuum. It can be triggered by many factors, such as OSHA fines or an increase in injury incidence or severity rates. When costs mount, the tendency to use training as the solution to many different kinds of problems may increase the pressure on the occupational hygienist to supply a quick training fix. For many of these problems training is not only an inappropriate solution, but may even mask the true need. This issue will be discussed further in the section on Training Needs Assessment.

Unlike the training required by regulations, training that is driven by internal factors must be prioritized based on the perceived risks involved. Because of his or her professional expertise in the area of human health, the occupational hygienist may be involved in assisting the company's risk managers to determine these priorities for each specific work site.

In a survey of 82 chemical manufacturing companies conducted by the Chemical Manufacturers Association (CMA) in 1984, it was found that the primary drivers for training were statutory requirements, which have been discussed, and health and safety audits. Next came corporate policy, and last came hazard identification and health and safety statistics. In the OSHA Voluntary Training Guidelines, two methods of identifying employee populations in greatest need of health and safety training are suggested. One is to pinpoint hazardous occupations, and the second is to examine the incidence of accidents and injuries, both within the company and within the industry. The methodology for both of these can be found in other chapters in this book.

Examples of some of the different kinds of data that might commonly indicate the existence of a training need are given in the following paragraphs.

Health and Safety Audit Results. Health and safety audits are usually conducted by trained teams of auditors, using protocols that have been developed internally or by an industry group. Often, the team contains both labor and management representatives, as well as company experts or consultants. Because so many companies have downsized and have limited expertise at the local plant level, audits have become an important vehicle for identifying problems and maintaining consistent standards throughout the organization. Modern audits do more than take a snapshot of plant conditions. They also review programs and programmatic trends that are predictors of possible future problems.

Morbidity and Mortality Data. Company or industry studies identify groups of workers who have been or may be at a greater risk of disease than the general population because of the workers' exposures to specific hazardous materials.

Results are reported in terms of statistical probability. From the trainer's viewpoint, this is probably the least useful of all the internal indicators because of its lack of immediacy and specificity.

Medical Surveillance Data. Several OSHA standards require regular medical surveillance of workers exposed to identified physical, chemical, or biological hazards, or in the case of hazardous waste site workers, the possibility of encountering unknown hazards. The occupational physician may also note an unexplained increase in visits to the medical department for skin or eye irritation. Abnormal findings may indicate a need for changes in engineering, administrative, or personal protection controls. Training, generally considered an administrative control, may be involved in any or all of these actions.

Occupational Hygiene Monitoring Data. The data generated by the occupational hygienist can trigger immediate actions. For example, noise measurements may show that a hearing conservation program, including training, is necessary. Changed levels of contaminants may indicate the need for a more comprehensive respiratory protection program, which will require training. Since making occupational hygiene monitoring data available to affected employees is required by *Code of Federal Regulations* Title 29, Section 1910.20, a training video on the role of the occupational hygienist and the interpretation of findings would be timely and useful, both for management's and workers' information. The American Industrial Hygiene Association (AIHA) has developed such a video program, and it can be purchased quite inexpensively.

Sickness and Absenteeism Costs. These costs, as with epidemiologic data, tend to be tracked over a long period of time and are of little immediate use to the occupational hygienist/trainer. However, they may reveal the need for training and information about off-the-job hazards. Off-the-job accidents are at least four times more frequent than on-the-job accidents and can result in great, and sometimes hidden, losses to the company. Any reduction in such losses by helping workers transfer concepts of health and safety from the workplace to the home will have a positive effect on the company's bottom line. The occupational hygienist also may be asked by management to organize a workplace health promotion program covering such topics as stress management or protection from exposures to hazardous substances while pursuing hobbies. Resources are available in most communities to advise or present programs of this sort.

Company Policies and Trade Association Initiatives. Most large companies and governmental agencies have well-established health and safety policies backed by their compliance audits. It has already been noted that audit findings often are a powerful driver for expanded and more effective training programs. Several trade associations have issued guidance to their members. For example, CMA has its Responsible Care® program, which contains specific guidance on all facets of worker safety and health. Many of the policies and guidelines developed by the associations within the last 5 years contain helpful sections focusing specifically on worker training. The organizations are usually happy to make them available to nonmembers.

Steps in the Development of Effective Training Programs—How Do We Train?

Instructional Systems Design

The assumption that an instructional need does exist should never be taken lightly. Training programs are bound to fail if they are offered as temporary solutions to problems that could be solved more effectively with engineering or other controls. Training programs are successful only when they are designed to meet goals that meet instructional needs. Management and/or other training is often needed to orient supervisory personnel to the appropriate applications of training, so that unnecessary instruction can be precluded. The problems that can be addressed successfully by training are as follows.

- A specified level of knowledge is needed to perform a job safely, and employees do not currently have this knowledge.
- A specified skill is needed, and employees do not now have this skill.
- Employees have the requisite knowledge and/or skills, but the job will soon require new or different knowledge or skills.

All possible workplace problems fall into these categories. For example, a new worker lacks both knowledge and skills. An experienced worker may be assigned to a new job, or a new process or piece of equipment may be introduced into the workplace. The process of retraining is simply one of bringing knowledge and skills up to the requisite level. A fourth category, motivation or attitudes toward health and safety, is sometimes considered to be a training problem. Attitudes, however, can be more effectively addressed by management commitment to safety and health, and supervisory acknowledgment or workers' efforts to conform to safety rules and to maintain a healthy and safe workplace. As discussed elsewhere by the author, "the goal of training is best conceptualized by imagining a physical gap between actual and desired levels of performance and recognizing that, by giving people the knowledge and skills they need to enhance this performance, this gap can be closed."[(11)]

The concept of the gap also encourages the occupational hygienist to focus on issues that are important to the development of effective training. Once a gap has been identified, the occupational hygienist must determine where this particular need stands in relation to other needs, what specific instructional objectives should be satisfied, when training should be delivered, what worker knowledge base exists, what resources are available, and other concerns. These issues are addressed comprehensively by practitioners of instructional technology.

Instructional technology has been described in many different ways, and there are numerous models to illustrate instructional systems design. Despite some variations, the systems approach to instruction always emphasizes the importance of a training needs assessment; the specification of instructional objectives; precisely controlled learning experiences to achieve these objectives; criteria for performance; and evaluative information based on performance measures. Other characteristics of instructional technology include the use of feedback to continually modify and improve the instructional process and recognition of the complex interactions between the components of the system. As Goldstein states in his book, "from this perspective, training programs are never finished products; they are continually adaptive to information that indicates whether the program is meeting its stated objectives."[(12)]

The simplest model, and also the most appropriate for the occupational hygienist, is the one used by OSHA in the Voluntary Training Guidelines.[(3)] They provide employers with a blueprint for designing, conducting, evaluating, and revising training programs in any or all areas of occupational health and safety. The model is based on and consistent with models used by professional instructional designers, which can be found in the extensive body of literature on human learning. The guidelines are in fact the distillation of years of study and experience. They afford the employer great flexibility in the selection of training content, methods, and media, and encourage a personalized approach to the training programs at individual work sites. Thus, using the systems approach to training recommended in the guidelines, an occupational hygienist is able to conduct informational and training programs that are consistently effective and, at the same time, are customized to meet site-specific needs.

The model consists of seven steps that are simple to follow. The steps are depicted graphically in Figure 30.1. The first step is to determine if training is needed, an issue that was discussed in Scenario 1. Examples of problems for which the occupational hygienist might be asked to supply a training solution are:

- Complaints of headaches or other symptoms caused by poor ventilation;
- Indications, through sound level measurements, that there is excessive noise in one area;
- Failure to contain a leak or a spill because appropriate materials are unavailable; and
- Repetitive pattern of accidents in one plant operation.

The first three examples are problems that clearly require engineering or administrative solutions. The last may or may not be a training problem; more information would be necessary to make a decision. The decision will be easier to reach if the occupational hygienist has previously oriented management to the functions and limitations of training, either by means of a formal training session or through a series of discussions.

A. Training is the solution
B. Training needs assessment
C. Specify instructional objectives
D. Develop content
E. Conduct training
F. Evaluate the training
G. Improve

From the OSHA Voluntary Training Guidelines, July 27, 1984

Figure 30.1 — Systems approach to training.

More than 100 professional development courses were available at the 1997 American Industrial Hygiene Conference & Exposition held in Dallas, Texas.

Training Needs Assessments

The determination of an instructional need, triggered either by external or internal factors, leads to an analysis of the training task or training needs assessment. Sometimes called a problem analysis, the training needs assessment "includes identifying needs, determining to what extent the problem can be classified as instructional in nature, identifying constraints, resources and learner characteristics, and determining goals and priorities."(13) Experience has shown that, because of the time and expense involved, there is always pressure to begin training without a thorough analysis of the needs. However, a program initiated without a complete assessment of tasks, learner characteristics, and resources is comparable to constructing a building without first studying the needs and constraints of the client.

Identifying Relevant Learning Domains

The term knowledge embraces several different levels of complexity of learning. In the field of instructional technology, terms have been established to identify the various levels and types of learning, called learning domains. There is obviously a difference, for example, between having to memorize a list of facts and having a general understanding of what those facts mean. On a more superficial level, it may be necessary only to know where the facts are kept and how to gain access to them.

In the field of occupational health and safety, the facts may represent some very complex concepts that are difficult for a nonprofessional to understand. A good example is the language used in some sections of an MSDS intended for the treating physician or nurse. Although the hazard communication standard requires that workers be trained concerning the chemical hazards in their workplace, it is unproductive to try to teach workers to memorize the meanings of terms such as nephrotoxin. It is more efficient to communicate the risks in everyday terms and provide a glossary for reference purposes. The decision as to what kind or level of knowledge is needed will, of course, have an impact on the cost of training and also on the selection of methods and media.

Goldstein(12) differentiates knowledge, skills, and abilities as follows.

- Knowledge—an organized body of knowledge, usually factual or procedural, which if applied makes adequate job performance possible.
- Skills—capability to perform job operations with ease and precision. Implies a performance standard that is required for effective job operations.
- Abilities—cognitive abilities necessary to perform a job function, requiring the application of some knowledge base.

An example of knowledge is that a worker knows where to go for first aid assistance. However, knowledge operates on several different levels of complexity. There is obviously a difference between writing a description of the position of the first aid station and simply pointing out its location to a supervisor. The first, more complex activity is unnecessary. In fact, a worker's inability to write clearly might interfere with the objective, which is simply to locate the station. Examples of different levels or types of knowledge and skills are given below.(14)

Recognition. A good example of a situation in which recognition is an appropriate level of knowledge is the interpretation of warning signs. It would be absurd and counterproductive to require a worker to memorize the words on all warning signs and be able to reproduce them, correctly spelled, in a quiz. Nor would we expect a worker to be able to tell us what the dimensions of the signs are, or of what material they are made. It is sufficient if the worker recognizes a sign saying "DANGER" or "NOISY AREA: Hearing protection must be worn" and takes the appropriate action. The signs present images through the use of shapes, colors, and signal words that are meant to trigger certain actions, such as donning safety glasses or hearing protection. Training for recognition is simple to conduct once the purpose of the training is clearly identified. For training about signs, a discussion using color slides or a pamphlet containing pictures and explanations probably will be adequate.

Discrimination. Discrimination requires a worker to know what a thing is or is not in comparison with other objects or situations. For example, most work facilities have a system of emergency whistles or claxons that indicate instructions ranging from "gather at a designated place" to "emergency evacuation." Each signal is meaningful only in the context of the others. Similarly, discrimination is required in selecting the correct respirator for different kinds of hazards. An occupational hygienist or supervisor needs to know all possible choices before deciding that in a suspected hydrogen sulfide environment, self-contained breathing apparatus will be necessary. The training in these instances should present the range of choices and give practice at comparing and contrasting them. Job aids such as charts or tables in which the characteristics or pros and cons are listed are useful tools for teaching discrimination.

Understanding. The term "understand" is so frequently misused that it has to be approached warily. A training objective that states that at the end of training "workers must understand the material safety data sheet" could mean many different things. For example, it could mean that the worker should have an understanding of terminology equal to that of a professional occupational hygienist, or that he or she should understand the origin and history of the MSDS, or should be able to reproduce one by rote. Such levels could require a training effort far beyond the budget of the trainer. They also would be counterproductive to the usefulness of the MSDS as an informational tool. What is needed in this instance is training that will enable the worker to interpret important health and safety protection information. A reference guide that supports the training will be useful in this context.

In other contexts, and for other tasks, the need to understand might require a totally different kind of training activity. Using the MSDS to identify the correct types of extinguishing agents to use on different types of fires might, on closer examination, turn out to be a discrimination task such as those described in the previous section. When the desired outcomes of training are clearly defined, the appropriate training methods will also become clear.

Skills. Skills training involves the worker doing something rather than knowing something, although all skills training involves elements of knowledge. Since effective training simulates the real-world task as closely as possible, it follows that skills training involves the actual performance, with guidance, of the target skill. The task is facilitated if it is broken down into a series of steps, so that each step is mastered before the entire task is attempted. Skills are acquired most effectively if:

- The trainer gives an overview of the entire task, explaining essential nomenclature;
- The trainer performs the task, describing the sequence as it progresses;
- The worker performs the task, with guidance and reminders;
- The worker practices alone, or with another worker; and
- Recognition is given for skill mastery.

Abilities. The abilities needed to perform a job function, as defined by Goldstein,[12] usually are based on clusters of knowledge and/or skills. For example, an occupational hygienist might need the ability to organize facts and figures for presentations to management, or to pass on information to workers verbally so that the information is understood. For the purposes of this chapter, however, concentration will be on the training of knowledge and skills that are necessary to instruct workers; the synthesis of many higher level knowledge and skill clusters necessary to acquire an ability is beyond the scope of the chapter.

The occupational hygienist may be asked by management to develop or contract for training designed to have an impact on worker attitudes. Poor safety performance is often attributed by management to poor worker attitudes, and training is frequently prescribed as a solution. It is important to remember that training is never a solution for problems in which no training deficit exists. Other factors, usually not accessible to the occupational hygienist, may lie at the root of the undesirable safety behaviors, which seem to reflect poor attitudes. Another chapter in this book deals with factors that foster employee involvement in health and safety efforts through worker ownership of the program.

Assessing Task, Population, and Resource Characteristics

The regulatory and internal drivers that trigger the need for training often overlap and must be identified and sorted, as will be discussed later in this chapter. The occupational hygienist may find that training is not the answer to problems identified by management, as was discussed in the first scenario. Some training programs may take priority over others. For example, a requirement to train all new employees before they can start the job will take priority over annually scheduled refresher training. Before training, some resolution will be made to differentiate between knowledge and skills training, or some combination.

Generally, training will be more effective and less costly if a targeted rather than a shotgun approach is used. Some major factors to be considered when evaluating the goal or target of training are:

- Identification of appropriate types of training for different work groups within the employee population, including engineers and management;

- Turnover rate, because higher turnover increases the frequency of training for individual new hires;
- Literacy levels of trainees;
- Different languages of workers;
- The impact of shift work;
- Logistical problems, such as the number of people available for training from each work operation;
- Availability of audiovisual equipment and aids;
- Availability of trained trainers; and
- Budget.

The Communication and Training Methods Committee of AIHA has developed a one-page publication, *Needs Assessment Guidelines*, obtainable from AIHA, which summarizes the key factors of audience characteristics, content requirements (knowledge and skills), and administrative considerations.

Identifying the Key Goals of Regulatory Training

The training needs most often identified by the occupational hygienist are, of course, those that are driven by regulations. The various requirements appear confusing, since they have been issued by many different agencies over a long period of time. Early training requirements used terms such as "competent to perform," without specifying the level of competence to be achieved. In the standards that were issued later, it is possible to identify overlapping or duplicate requirements. As a useful aid to untangling these issues, the American Petroleum Institute (API) and CMA developed an easy reference document for its members. This was later adopted by API for its members. This exhaustive document, *Federally Mandated Training and Information,* is available from API.[15]

Scenario 2 offers a typical situation an occupational hygienist might face in addressing overlapping requirements.

Scenario 2. An Integrated Health, Safety, and Environmental Training Approach*

Regulations today can be overwhelming, particularly when they overlap. EPA, OSHA, and other regulating bodies all have training requirements, some of which can simultaneously apply to one chemical or product. This is true within OSHA rules as well as between different agencies. The challenge to the occupational hygienist is to train comprehensively, but also efficaciously.

How should overlapping or multiple training requirements be handled? To demonstrate one approach, consider the chemical toluene as an example and follow it through its life cycle at Pretty Paints, Inc. For brevity, the regulating bodies will be limited to OSHA, EPA, and DOT.

Assume that toluene is an ingredient in Pretty Paints' commercial products line, and that the following very simplified unit operations are involved: receiving and storage of raw ingredients; plant processing into many different commercial paints; product packaging and shipment to customers; and waste handling associated with in-plant production, accumulation, and off-site disposal.

Within these unit operations, consider the matrix of what workers are required to know and implement in some of the areas involved, as shown in Table 30.1. Although there are many more requirements applicable to these unit operations, you can see in this example that each regulatory agency has requirements designed to meet special needs. For labeling, for instance, OSHA labels communicate occupational health hazards, EPA labels communicate waste hazards, and DOT labels present transportation hazards.

To further complicate the issue, most manufacturers have different people responsible for OSHA, EPA, and DOT compliance. The worker may have to go through several training sessions, delivered by various individuals, to perform the job properly. An integrated training approach would consider elements common to all regulations for handling toluene and other chemicals.

In this scenario, the OSHA, EPA, and DOT team representatives agree to create a baseline chemical hazards training session. Regulatory compliance is made as much a part the job training as product mixing and weighing techniques. Furthermore, the team has agreed to discuss all of the regulatory requirements as they apply to the sequence of the production process, rather than instructing separately on each of the regulations. Each of the specialists will take turns leading the discussions for unit operations based on the heaviest regulatory input. In this way, the required concepts will be woven into the instructional design without compromising the process flow training. Although these sessions may take longer to conduct than the traditional "this-is-the-regulation and this-is-what-you-need-to-know" approach, the total amount of employee time off the job may be less because major parts of all three sessions (EPA, OSHA, and DOT) will be conducted simultaneously. As an additional bonus, workers will begin to think of their regulatory responsibilities in the same way as any other job function. Eventually, ideas such as

*This scenario was developed with the assistance of Mark R. Francis, CIH, CSP.

Table 30.1 — Scenario 2: Required Worker Knowledge and Activities

Activity	Regulatory Training Requirements OSHA 29 CFR	EPA 40 CFR	DOT 49 CFR
Receiving	assure MSDS is available for product	none	none
Processing	understand hazards as required by HazCom; understand PPE requirements for handling toluene	understand proper waste handling within the plant; assure wastes are properly accumulated by waste stream type	none
Product packaging	assure correct labeling is placed on each container	none	assure proper shipping containers are used and proper product shipping labeling is placed on each container
Wastes	hazardous waste handlers must understand hazards	assure correct waste labeling is placed on each container; comply with time limits for waste accumulation	assure proper waste shipping labeling is placed on each container

waste minimization methods or substitution of less hazardous materials will surface in a similar manner as suggestions presently generated for improving individual job functions.

By designing integrated training to be within the context of the workers' requirements, health, safety, and environmental needs can be accepted more readily as part of the job. In this example the challenge was met by stepping back and viewing the desired results from a practical standpoint—that of the people who have to make it happen every day.

Defining Performance Objectives

In the world of civil engineering, it would be inconceivable to attempt to build a structure without first planning its purpose, location, and other details. Huge amounts of money would be at risk if this detailed planning step was omitted. Concern with specific objectives applies equally in the world of training.

According to the OSHA Voluntary Training Guidelines, clear and measurable objectives that are thought out before the training begins can do much to ensure effective training. A well-defined instructional objective describing what the trainee should know or be able to do after the training session has three elements.

1. The desired performance must be observable. Individuals must demonstrate what they have learned.
2. Performance must be measurable. The objective should define what constitutes acceptable performance.
3. Conditions under which performance is to occur must be stated; that is, objectives should describe the important conditions under which the individual will demonstrate competence.

The following is an example of an objective that contains all three elements: "Given a label (conditions), workers will point (observable) with 100% accuracy (measurable) to the signal word."

The objective should describe the desired practice or skill and its observable behavior in sufficient detail to allow other qualified persons or trainers to recognize when the desired behavior has been exhibited. This point is significant in industrial situations, where it is often important to present standardized training of equivalent quality to all workers in many different locations. Consistency will be lost unless all trainers work with the same learning objectives.

Instructional objectives provide a road map for the development of training content and, to some extent, what training format should be used. They also act as a means of measurement that allows the occupational hygienist to evaluate whether performance does, or does not, reach the desired goal. The more detailed the objectives, the easier it is for the occupational hygienist to fine-tune those portions of the training that are not effective. Thus,"expensive retrofitting of entire training programs can be avoided, and a consistent approach can be adopted by many different trainers."[11]

Selecting Appropriate Training Methods and Media

The identification and selection of instructional methods and media becomes much simpler once the objectives and kinds of learning have been defined. Because of the rapid growth in technology, the occupational hygienist now has many options, ranging from traditional didactic (lecture) presentations to simulations involving virtual reality. In an article written in 1992, Hannafin sees the advent of "increasingly realistic stimuli, quick access of large quantities of information, rapidly linking information and media, and removing the barriers of distance between instructor and learners and between learners."[16] Seels and Richey state that "as instructional projects become more sophisticated, the demarcations between [learning] domains blur, and the activities of one domain are inescapably dependent on the activities of another."[2] Certainly the new technologies provide avenues of development that allow the trainer to adapt instruction to unique situations, devise new approaches, and address nontraditional learning environments found in many workplaces.

Over and above specific methods and media considerations, several strategies should be considered, such as in-house development of training materials, bringing in an outside trainer and/or materials, sharing resources with other company facilities, training trainers to take care of ongoing training needs at scattered operations, and incorporating new topics in existing safety or tailgate meetings.

Also consider the possibility of participation in distance education, via satellite, or self-instructional computer-managed programs where worker turnover is high or classes cannot easily be convened. As stated in the OSHA guidelines,

> *the determination of methods and materials for the learning activity can be as varied as the employer's imagination and available resources will allow. The employer may want to use charts, diagrams, manuals, slides, films, viewgraphs, videotapes, audiotapes, or simply blackboard and chalk, or any combination of these and other instructional aids. Whatever the method of instruction, the learning activities should be developed in such a way that the employees can clearly demonstrate that they have acquired the desired skills or knowledge.*[3]

Some general selection criteria will assist the occupational hygienist to navigate through the maze of choices. First, the learning situation should simulate the real-life job as closely as possible. The closer the simulation, the easier it is for the worker to transfer knowledge and skills from the classroom to the job. Examples of various training approaches, in rough order of descending effectiveness,[11] are as follows.

- The real thing (e.g., handling a real MSDS);
- A simulation (e.g., practicing the handling of simulated chemical spills using water or other harmless agents);
- An interactive compact disk that enables students to interact with simulations of dangerous situations
- Passive viewing of an audiovisual representation (e.g., a videotape of a spill being handled);
- Visuals to accompany a lecture (e.g., pictures of appropriate respirators);
- Lectures; and
- Handouts.

Other considerations also come into play. For example, a simulation may be greatly preferable to the real thing for reasons of safety, control, repeatability for groups of students, or other factors. Handouts appear at the bottom of the list because there is no assurance that they will be studied or even opened. As supplements to other methods, however, handouts can be valuable for many reasons, including their potential to involve the worker's family in protecting the health and safety of the worker. Availability and cost are additional factors that should be weighed against the effectiveness of methods that most closely approximate the actual work situation.

Generally, an effective training approach incorporates several different training methods and media, since selection depends on the skills and/or knowledge to be demonstrated as learning outcomes. Tasks that require group interaction or team response on the job require group-oriented learning activities such as role-playing, small group problem-solving sessions, and team practice. Tasks requiring the individual acquisition of knowledge, such as learning specific lockout/tagout procedures, can be taught with self-paced instruction, such as that provided by computer-assisted instruction or interactive video.

Training methods and media selection considerations involve many complex issues for the occupational hygienist anxious to provide high-quality and effective training programs. The term "resources" covers many factors that are not always considered, including the most important resource, management support. Only if management is truly supportive can the necessary funds, time, personnel, and facilities be allocated. Without such support the training program may be doomed to failure, either through lack of start-up funds or poor implementation and follow-up. Note, however, that money alone does not ensure effective training. There have been many cases where companies have thrown money at a training problem, only to end up with expensive hardware and software that missed the training target and quickly became outdated.

Many effective low-cost training programs have been designed that were preceded by a careful needs assessment, were conducted with a clear definition of objectives, and used available resources effectively. For example, by following some simple rules of scripting, photography, and/or videotaping, it is possible to develop a very professional-looking slide or video program in-house, showing real workers in specific and relevant activities and situations. Group discussions following the presentation of such a video can be interesting, lively, and effective.

The selection of specific methods and media is influenced by additional factors. Many selections, for example, are clearly suitable or unsuitable in a given set of conditions.

Evaluating the Effectiveness of Existing Programs

It is not always necessary for the occupational hygienist to start from scratch in developing effective programs; programs may exist in-house or in the catalogs of vendors that will meet the need or at least supplement the training. Part of the needs assessment, then, is an evaluation of what is currently available that meets the specific instructional objectives. It is a good idea to get a rough idea of the applicability of the existing programs by constructing a matrix. Along the top can be listed the categories of performance outcomes, such as "list correct actions to take in the event of a spill" (knowledge) or "perform the correct actions in the event of a spill" (skills). Down the left side can be listed all the manuals, guidelines, and training programs available. If any of them satisfy one or more of the outcomes listed along the top, put a checkmark in the matrix where demand and supply intersect. In this manner, the remaining empty squares can be identified as presenting a need for purchased or developed programs or services.

The next step is to eliminate from all available programs those that will be useless because of hardware demands or other factors. Selection among the remaining programs should be made on the basis of answers to the following questions:

- Does this program meet precisely the needs identified?
- If not, is there some built-in flexibility that will allow the appropriate modifications?
- Could this program be used to supplement what is already available or for annual retraining?
- Are there data that indicate that this program has been effective with groups of workers similar to those at this specific workplace?
- Is the vendor willing to give the names of previous clients, so that the performance of the program in similar operating environments can be ascertained?
- If it is a purchase, can all or part of the program be previewed?
- Is there a leader's guide that will help in administering the program?
- Are there tests that will allow documentation of workers' understanding of the content of the program?
- Are there student handouts or job aids that will be useful to workers as continuing reminders when they get back on the job?
- Are appropriate examples used?

Additional questions may be appropriate, depending on the special characteristics of each workplace. For example, it may be important to find out whether a Spanish version of the program is available.

Planning and Presenting Training

A well-known training cartoon shows two instructors walking into the classroom, where the trainees are assembled. Says Trainer No. 1 to Trainer No. 2, "Tell the class to wait while I prepare my lessons!" This situation is not unusual; many trainers bypass all the previous steps discussed in this chapter and start planning the training in the fifth step: conducting the training. Conducting the training is a process that begins long before the class actually meets. Preparation for training is essential. This includes the general preparation involved in making decisions about such things as the physical layout of the training room and the elimination of distractions, to the preparation of equipment and written materials.

General Preparation

The following general preparation factors should be considered for training that will take place in a classroom setting.

Physical Layout. The training room should be large enough to accommodate comfortably the number of participants expected plus the needed equipment and furniture. The effectiveness of instruction can be reduced by overcrowding. The shape of the room also can affect the nature

Training Room Checklist

Physical layout
- seating arrangement
- acoustics
- illumination control
- doors and exits
- other

Furniture
- chairs
- tables
- lectern
- displays
- other

Projection equipment
- overhead or slide projector
- remote controls
- extra carousels
- screens
- video playback
- spare bulbs
- window shading
- other

Display writing surfaces
- easels
- pads
- chalkboard
- wall space
- blank transparencies for overhead projection

Supplies
- note pads
- pens
- pointer
- paper clips
- masking tape
- chalk
- markers for transparencies
- other

Training materials
- lesson plans
- handouts
- copies of quizzes
- participant roster
- training certificates
- other

Prefunction check
- equipment set up and operational
- wiring covered and taped
- help resources such as electrician identified
- room key obtained
- other

of group interaction, the acoustics, and visibility. Low ceilings with obstructions such as chandeliers should be avoided; high ceilings (12 feet) are preferred, particularly when images are to be projected on a screen.

Physical Conditions. Extremes of heat or cold and poor ventilation interfere with learning efficiency. High temperatures and humidity cause drowsiness; cold causes discomfort. The instructor should know where the controls are and how to operate them. Similarly, a poorly illuminated facility interferes with the readability of display surfaces and printed training materials and can cause stress and fatigue. The training facility should also be free of sources of disruption and distractions that could impede the learning process. Frequently, the occupational hygienist has to make do with the cafeteria or other public room. In this case it would be preferable to rent an outside room. Noise sources outside the facility, such as public address systems or activities in adjoining rooms are particularly irksome. The instructor may be able to deal with such distractions by changing rooms, increasing the level of sound within the training room, or by changing the planned activities.

Seating Arrangements. The choice of seating arrangements will depend on the type of training session, the number of participants, the space available, and the learning objectives to be achieved. The way participants are seated will influence the degree of control the instructor has over the group, the level of interaction between the instructor and individual participants, and interaction among the various participants. For example, a U-shape, V-shape or half-circle will encourage group interactions because participants have eye contact with each other. Preplanning the classroom layout and seating arrangements will enhance the instructor's overall effectiveness in managing the instructional process and in dealing with individual differences. Generally, it is best to limit a class to 25 participants; when numbers are higher, the less aggressive members of the group tend to withdraw, and it is difficult to get the whole group to participate actively.

Specific Preparation

Every instructor has had the experience of having to cope with burned-out bulbs, shortage of electrical outlets, and other training glitches. These experiences can be minimized by the use of a good checklist.

Conducting Training

How training is conducted will depend, of course, on the initial assessment of the needs and the audience and resource factors that have been identified.

Most occupational hygienist deal with relatively small groups of employees in a changing environment of new health and safety regulations and new processes and hazards. He or she must interpret the regulations for specific work processes and must also respond promptly to workplace incidents. Realistically, then, most training will be addressed to issues specific to each workplace and will be conducted at the local level with small groups of employees. Time is usually available only in chunks of not more than 40 minutes, or the time normally devoted to the regular safety meeting. The following basic steps apply equally to the most common situation, where the occupational hygienist has limited time and resources, as well as to situations where high technology training solutions are available. The four major activities involved in conducting training are overview, presentation, application, and practice or review.

Overview. The overview is important to the adult learner. It should be designed to present the big picture into which the training fits, or to present needed information to bring all trainees up to speed. For example, as a preamble to a detailed session on the MSDS, the trainer might spend a few minutes describing the purpose and function of the MSDS as a key component of the hazard communication program.

The overview should also emphasize the importance and relevance of the material to be discussed. It should place the session in relation to past and future sessions and state clearly the expected learning outcomes. Examples related to the job experience of the trainees will reinforce the message that the training is job related and important. Approximately 10% of the time available for short training sessions should be devoted to the overview. Training conducted over several days would not require as great a percentage of the time spent on the overview.

Presentation. In presenting material, it is best to proceed from known information to unknown, or from simple to complex material, in small steps. Each step in the process should be related to the whole picture as well as to other steps. Key points should be emphasized and demonstrations and visual aids used as appropriate. Active participation by the trainees is an important part of the process. As stated in the OSHA Voluntary Training Guidelines, "employees can become involved in the training process by participating in discussions, asking questions, contributing their knowledge and expertise, learning through hands-on experience, and through role-playing exercises."[3] Using a variety of methods and media will also help trainees retain information.

Application. The application step can be most clearly defined in skills learning. It allows the learner the opportunity to perform the skills that are being taught in a supervised and nonthreatening situation. For example, in a session on respiratory protection, the instructor's live demonstration of donning a respirator would be followed by the opportunity for trainees to practice donning respirators. Another type of application exercise is to divide the class into small groups and to assign workplace-related problems for the groups to solve. Solutions to the problems should be based on the pre-existing knowledge and experience of the trainees as well as on the material presented in class. Errors can be checked and corrected and positive responses developed in this context. Encouraging and positive comments by the instructor, such as "you have almost completed the whole sequence—there are only two more steps to master," can spur the employees on to proficiency. Application activities, by their very nature, relate the classroom instruction to the job and actively involve the trainees in the acquisition of new knowledge and skills, as recommended by the OSHA training guidelines.

Review and Practice. Most occupational hygienists have experienced courses where they are bombarded with interesting facts, with little opportunity to integrate and assimilate the material. In a college setting, students have the luxury of time to review and make sense of their notes. This kind of voluntary review is not possible in the workplace; the review must be built in throughout the instructional sequence. If there are 10 major instructional objectives, for example, review is needed following each of the 10 course segments, as well as at the end. Cumulative review is necessary if the objectives represent sequential building blocks toward the desired new knowledge or skills. For example, if concepts of adverse health effects resulting from overexposure to hazardous substances are built on knowledge of routes of entry and acute versus chronic effects, review items should not be independent of each other, but should include the earlier concepts and show how they interrelate.

Review enables trainees to cement new knowledge and to relate it to previously learned material. There are many different ways to perform review. For example, workers can be paired and given the opportunity to observe and critique each other's performance at skill tasks. If videotapes are used, the tape can be stopped at important points for purposes of review and discussion. Computer managed programs contain built-in review, often geared to the performance of each individual. The instructor may choose to ask questions of the trainees, or to employ a review game. All methods of review give trainees the opportunity to practice and to receive feedback, strengthening the desired knowledge and skills and increasing the probability of transfer to the workplace.

Evaluating Process and Product Outcomes

Evaluation is an essential step in the development and maintenance of an effective training program. As described by Goldstein, it "is the systematic collection of descriptive and judgmental information necessary to make effective training decisions related to the selection, adoption, value and modification of various instructional activities."[12] The information gathered through evaluation can be used by the occupational hygienist to revise instructional programs to better achieve the stated objectives. As a simple example of an easy-to-measure objective, either employees will know where the emergency shower is or they will not. Similarly, either employees will follow correct lockout/tagout procedures or they will not, and either they will be able to answer questions posed by an OSHA inspector or they will not. Without an evaluation, the occupational hygienist will not know whether the training is successful or completely off target, nor will he or she know how to correct any problems. Questions answered by the evaluation include the following:

- Have I achieved the objectives of the training program, in terms of the immediate acquisition of knowledge and/or skills?
- Do observations on the job indicate that the desired changes in knowledge and/or skills have occurred?
- Can the changes or lack of change be attributed to the instructional program?
- Is it likely that similar changes will occur for other employees taking this program in the future?

- What can I do to improve the training program and to make it more effective?

There are five commonly accepted types of evaluations.

Participant Satisfaction (such as opinion ratings of the training). The participant satisfaction type of evaluation is subjective. It is the easiest to conduct, but does not include an assessment of learning. It can, however, yield important information. Questions on the appropriateness of training materials or the adequacy of the time taken to cover certain points often reveal unanticipated problems. For example, if participants rate a presentation on waste handling as "too fast" or "too complicated," another approach that emphasizes only the key points might need to be considered.

Learning Outcomes (such as true/false tests and other quizzes). Measures of learning outcomes, administered immediately following the training, reveal how well participants have learned principles, skills, information, and techniques. The test items are based on the instructional objectives that were initially defined. For example, the objective "Given an MSDS, worker will point with 100% accuracy to the location of the precautions for safe handling" would be turned into the following test item: "Point to the safe handling section of this MSDS." Learning outcome methods include paper and pencil or oral test items, job simulations, or any other activity that directly reflects the instructional objectives.

Attitude Changes (as observed in interviews and on the job). The occupational hygienist should approach the topic of worker attitude with great caution. Too frequently, management believes that workers will not learn nor practice safe workplace behavior because of bad attitudes. Poor attitudes often turn out to be the result of factors such as poor communication, discouragement of worker involvement in the health and safety program, or other external factors. Experienced hygienists have found that when the program is cooperative and encouraging, most workers are eager to learn about the hazards of their jobs and ways to protect themselves against these hazards. There is, however, a legitimate reason to attempt to measure attitudes. Employee attitudes—opinions about the job, their supervisor, and their workplace—directly affect behavior and can have a significant effect on performance. In turn, positive changes in behavior, such as the outcomes of effective instruction on working safely, will result in more positive attitudes. Questions, surveys, interviews, and observation are the usual means to measure attitudes before and after training.

Behavior or Job Performance Changes (measured directly or indirectly). Evaluations of behavior changes focus on measuring the changes that training has produced in performance on the job. This kind of evaluation attempts to determine whether the worker has transferred new knowledge and/or skills acquired in the classroom to the work setting. This is particularly important in the case of health and safety, where it is essential to have zero errors on the job. Measurement of behavior change requires hard data about actual performance. Methods include observations of work practices by trained observers; analysis of usage of required equipment, such as PPE; measurable results of specific training objectives; review of safety records before and after training; and comments from supervisors and co-workers describing significant changes in safe work behaviors.

Accomplishment of Organizational Goals and Improvements (as reflected in higher productivity, for example). Although the primary goal of all health and safety training is protection of the worker, it is generally recognized that the organization also benefits greatly when employees are knowledgeable about potential hazards in the workplace and have learned to work safely. Since every illness and injury has a direct impact on productivity and profitability, a reduction in illnesses and injuries will have a significant effect. For example, the financial consequences of an acute overexposure to hazardous substances, or of chronic ill health resulting from long-term repeated exposures, are very severe. These organizational effects apply to all phases of the occupational hygienist's job, but since training is the key to successful implementation of any program, it is also seen as a significant factor in achieving quantifiable organizational goals.

These five types of evaluations can be assigned to one or both of two major categories, formative evaluation and summative evaluation. For example, participant satisfaction and the measurement of learning outcomes are generally used to evaluate effectiveness of training while it is still being developed (formative), while accomplishment of organizational goals is measured some time after the training (summative). Seels and Richey[2] describe these terms as follows: "Formative evaluation involves gathering information on adequacy and using this information as a basis for further development. Summative evaluation involves gathering information on adequacy and using this information to make decisions about utilization." Stated simply, "when the cook tastes the soup—that's formative. When the guests taste the soup—that's summative."(2) Formative evaluation stresses tryout and revision processes, enabling the occupational hygienist to identify and correct weak features of the training, while summative evaluation uses outcome criteria to assess the effectiveness of the program. The formative evaluation should be completed and judged satisfactory before training on a large scale is begun; the trainer will find it helpful to pilot the training on sample groups of employees to identify major problems and to make modifications. Both types of evaluation, however, can lead to program improvements.

The results of an evaluation must be documented so that improvements may be made. Although OSHA does not require documentation of training, it is good to remember the adage "if it wasn't documented, it didn't happen." Documentation of attendance at a course is not enough to demonstrate to the OSHA inspector that learning has occurred. The course outline, the objectives, and the results of the evaluation should be included in the documentation. Fortunately, there are now inexpensive and user-friendly software programs available from distributors like the National Safety Council to assist in all documentation needs.

Improving Programs Continuously

The final step identified in the OSHA Voluntary Training Guidelines, Improving the Program, flows from the evaluation. If the evaluation shows that the desired knowledge and skills have been acquired, no further improvement is necessary. If, however, some of the instructional objectives have not been satisfied, the program must be revised/improved until they are met.

A common mistake is to add more elaborate explanations where weak spots have been identified, in an attempt to correct deficiencies. Often, the deficiency is due to a lack of clarity rather than a shortage of explanation. Instead of adding more details and possibly increasing confusion, the solution may be to restructure and simplify the materials. Some initial assumptions may need to be re-examined. For example, the objective of teaching employees terms such as toxicology may be redundant with the goal of training them to avoid contact with certain hazardous chemicals. Another common error is to retain material in the course that is already known to the employees and could therefore be omitted. If the steps in the training process have been followed systematically, it is relatively easy to retrace the route to identify where improvements are needed.

It is tempting to consider the training program as finished once the evaluation shows that it is effective. This is particularly true when the regulations have no specific requirement for retraining, or when annual retraining appears to be all that is required. However, if the occupational hygienist considers all possible changes that can occur in workplace processes and practices, as well as the natural process of forgetting, it becomes obvious that training has to be an ongoing and dynamic process rather than a one-shot deal. Ideally, training should be reinforced daily with reminders by supervisory personnel and with posters, hard-hat decals, handouts, brief reviews during regularly scheduled safety meetings, and other methods. The behaviors learned in the classroom should become habitual components of standard work practices. When behaviors such as emergency response are unlikely to be called on very often, drills should be held periodically. The continuous improvement of training is an important part of the occupational hygienist's job, and time should be assigned for this activity.

Scenario 3. Following the Steps: A Typical Workplace Situation

The Situation

Jim Hughes is an occupational hygienist with multiple tasks and responsibilities. He has done a good job of organizing the training activities at his plant, and he uses an inexpensive documentation software package to keep careful records of courses, retraining requirements, data from evaluations, and other details. Occasionally, an unexpected training need surfaces. In this instance, 17 workers have been assigned to work on a new process. They will be handling a variety of chemicals, some of which are potentially hazardous. The workers are experienced, trained in safe handling of chemicals and using PPE, and already have been given an overview of the company's hazard communication program. However, Jim needs to present an in-depth discussion of the health and safety information found in the MSDS of the chemicals specific to the new process ***(Step A: Training is the solution)****.*

The Particulars

Jim has only one week to prepare for this training session, which for administrative reasons is limited to two hours. He has a copy of the inventory of chemicals that will be used in this workplace and has noted that there are two suspected carcinogens, some common acids and caustics, and some flammables. He has records of prior training of these employees in his database. The facility supervisor who is handling logistics at the location has copies of the MSDS for each substance in the inventory. There is a quiet training room available that seats 20 comfortably. It has large windows on two walls. He also checks to see if the windows have shades or curtains to block the light ***(Step B: Training needs assessment)****.*

Based on the major goals of communicating the information in the MSDS most pertinent to health and safety and teaching the meaning of some important terms, Jim writes 10 instructional objectives ***(Step C: Specify instructional objectives)****, and then outlines his presentation. Jim is thinking about an advance handout (homework). He has reviewed each MSDS and sees that there are one or two unique hazards, while the rest of the substances belong to some major classes of chemicals. He may discuss some of these as members of a generic group, such as acids and caustics. He sees that almost all of the protective measures needed for the new process are familiar to the employees; however, they should be reviewed in the context of the unfamiliar hazards. He thinks about audiovisual aids, and checks on the availability of relevant videos and slides. Remembering the importance of employee participation, he plans time for discussion and questions. He prepares a short quiz, based on his instructional objectives, so that he has some data on the effectiveness of the program* ***(Step D: Develop content)****.*

Jim presents the training and receives interesting comments from the employees, which he plans to weave into any future training sessions ***(Step E: Conduct training)****. He administers the quiz and enters the results into his records. He also realizes that these workers are going to be receiving many types of training before transferring to their new jobs and plans to follow up at the work site when the process comes on line* ***(Step F: Evaluate the training)****. He will continue to monitor this work group to see if the knowledge has been transferred to the job and will make further improvements in the program as necessary* ***(Step G: Improve)****.*

Summary

In this chapter the reasons for training, including some of the major regulatory drivers, have been reviewed, and the instructional systems design steps that are necessary to ensure effective training have been discussed. The role of effective training as a key component of an occupational hygiene program has been emphasized throughout. In two scenarios the following important points were made: training is often not an answer to health and safety deficiencies in the workplace, and an integrated approach is desirable when there are many different training demands. A third scenario showed an occupational hygienist applying the instructional systems design process in a typical workplace situation. By following the steps outlined in this chapter, the occupational hygienist will be able to continuously improve the training offered to employees. An important point to remember is that all levels of management need training, from the CEO on down. If everyone is aware of potential hazards and/or controls, training becomes more easily accepted and also more effective.

Additional Sources

Instructional Technology and Training Basics

American Petroleum Institute (API): *Training Competencies* (Pub. 1210). Washington, DC: API, 1994.

American Society for Training and Development (ASTD): *Basic Training for Trainers*, vol. 1–3. Baltimore, MD: ASTD Publishing Services, 1996. Collected from the ASTD *INFO-LINE* series.

American Society for Training and Development (ASTD): *Training and Development Handbook*, 4th ed. R.L. Craig (ed.): Baltimore, MD: ASTD Publishing Services, and New York: McGraw-Hill, 1996.

Anglin, G.C. (ed.): *Instructional Technology: Past, Present, and Future*, 2nd ed. Englewood, CO: Libraries Unlimited, 1995.

Bullard R., M.J. Brewer, N. Gaubas, A. Gibson, et al.: *The Occasional Trainer's Handbook*. Troy, MI: Educational Technology Publications, 1994.

Gagne, R. and K.L. Medsker: *The Conditions of Learning: Training Applications*. Fort Worth, TX: Harcourt Brace College Publishers, 1996.

Head, G.E.: *Training Cost Analysis: A How-To Guide for Trainers and Managers*. Baltimore, MD: ASTD Publishing Services, 1993.

Rothwell, W.J. and H.C. Kazanas: *Mastering the Instructional Design Process: A Systematic Approach*, San Francisco: Jossey-Bass, 1992.

Methods and Media

American Society for Training and Development (ASTD): *Toolkit Series: Lesson Plans,* Baltimore, MD: ASTD Publishing Services, 1995.

Barron, A.E. and G. Orwig: *Multimedia Technologies for Training*. Englewood, CO: Libraries Unlimited, 1995.

Kirby, A.: *The Encyclopedia of Games for Trainers*. Amhurst, MA: HRD Press, 1992.

National Audiovisual Center: *1994 Reference List of Audiovisual Materials*. Washington, DC: National Audiovisual Center, 1994.

Occupational Safety and Health Administration (OSHA): OSHA publications and audiovisual programs. OSHA 2019, OSHA Publications Distribution Office, U.S. Department of Labor, Room N-3101, Washington, DC.

Piskurich, G.M. (ed.): *Selected Readings on Instructional Technology*. Baltimore, MD: American Society for Testing and Development, 1987.

Willis, B. (ed.): *Distance Education Strategies and Tools*. Troy, MI: Educational Technology Publications, 1994.

Background Reading Material

Prior to beginning this chapter, the reader should review the following references for general content.

Instructional Technology

I.L. Goldstein. *Training in Organizations; Needs Assessment, Development, and Evaluation*. Pacific Grove, CA: Brooks/Cole Publishing Co., 1993

B. Seels and Z. Glasgow. *Exercises in Instructional Design*. Columbus, OH: Merrill Publishing, 1990 (2nd ed. due 1997).

R. Gagne and K.L. Medsker. *The Conditions of Learning: Training Applications*. Fort Worth, TX: Harcourt Brace College Publishers, 1996.

B. Seels and R.C. Richey. *Instructional Technology: The Definition and Domains of the Field*. Washington, DC: Association for Educational Communications and Technology, 1994.

G.C. Anglin (ed.). *Instructional Technology: Past, Present, and Future,* 2nd ed. Englewood, CO: Libraries Unlimited, 1995.

Multimedia Technologies

A.E. Barron and G. Orwig. *Multimedia Technologies for Training*. Englewood, CO: Libraries Unlimited, 1995.

Occupational Safety & Health

M.C. Samways. How to select training methods and media. In *Handbook of Occupational Safety and Health,* L. Slote (ed.). New York: John Wiley & Sons, 1987.

M.C. Samways. Training. In *Hazard Communication Compliance Manual,* M.B. Kent and J.C. Silk (eds.). Washington, DC: BNA Books, 1995.

References

1. **National Advisory Committee on Occupational Safety and Health Hazard Communication Workgroup:** *Report to OSHA on Hazard Communication*. Washington, DC: Occupational Safety and Health Administration, 1996.
2. **Seels, B. and R.C. Richey:** *Instructional Technology: The Definition and Domains of the Field*. Washington, DC: Association for Educational Communications and Technology, 1994.
3. "Voluntary Training Guidelines." *Federal Register 49*:30290.
4. "Occupational Safety and Health Act," Pub. Law 91-596, Section 2193. 91st Congress, Dec. 29, 1970; as amended, Pub. Law 101-552, Section 3101, Nov. 5, 1990.
5. "Hazardous Waste Operations and Emergency Response Standard." *Code of Federal Regulations,* Title 29, Section 1910.120.
6. "Hazard Communication Standard." *Code of Federal Regulations,* Title 29, Section 1910.1200.
7. "Accidental Release Prevention: Risk Management Program." *Code of Federal Regulations,* Title 40, Section 68 112(r)(7). 1996.
8. "Process Safety Management." *Code of Federal Regulations,* Title 29, Section 1910.119.
9. *Code of Federal Regulations,* Title 29, Section 1926.58.
10. *Borel v. Fibreboard Paper Products Corp. et al.,* 93 F.2d 1076 (1973).
11. **Samways, M.C.:** Training. In *Hazard Communication Compliance Manual,* M.B. Kent and J.C. Silk (eds.). Washington, DC: BNA Books, 1995.
12. **Goldstein, I.L.:** *Training in Organizations: Needs Assessment, Development, and Evaluation*. Pacific Grove, CA: Brooks/Cole Publishing Co., 1993.

13. **Seels, B. and Z. Glasgow:** *Exercises in Instructional Design*. Columbus, OH: Merrill Publishing, 1990.
14. **Samways, M.C.:** How to select training methods and media. In *Handbook of Occupational Safety and Health*, L. Slote (ed.). New York: John Wiley & Sons, 1987.
15. **American Petroleum Institute (API):** *Federally Mandated Training and Information* (API/CMA pub. 1200). Washington, DC: API, 1994.
16. **Hannafin, M.J.:** Emerging technologies, ISD, and learning environments. *Educ. Technol. Res. Dev. 40*:49–63 (1992).

Section 5

Controlling the Occupational Environment

Outcome Competencies

After completing this chapter, the user should be able to:

1. Define underlined terms used in this chapter.
2. Identify examples of workplaces and operations that require controls.
3. Recall and use occupational exposure and flammable/explosive limits.
4. List and define exposure control methods.
5. Identify goals of ventilation including local and general exhaust.
6. Recognize limitations of various control approaches.
7. List sources of make-up air.
8. Recognize components of a ventilation system.
9. Derive and use dilution ventilation models.
10. Use dilution ventilation models when designing ventilation controls.

Prerequisite Knowledge

In conjunction with this chapter, the user should read or review the following chapters.

Chapter Number	Chapter Topic
6	Principles of Evaluating Worker Exposure
17	Risk Assessment in the Workplace
18	Biohazards in the Work Environment
19	Indoor Air Quality
36	Respiratory Protection

Key Terms

administrative controls • control • density correction factor • dilution ventilation • emission • exposures • engineering controls • local exhaust ventilation (LEV) • PPE controls • prevention • problem characterization

Key Topics

I. The Occupational Hygienist's Role

II. Fundamental Control Approaches

III. General Types and Approaches to Control
   A. Prevention
   B. Control Begins at the Design Stage
   C. Applications of Controls
   D. Problem Characterization

IV. Emission Source Control in Nonindustrial Environments

V. Emission and Exposure Control in Industrial Environments
   A. Aerosol Emission Source Behavior
   B. Vapor and Gas Emission Source Behavior
   C. Open and Closed Industrial Processes
   D. Substitution a of Less Hazardous Material
   E. Process Changes
   F. Isolation and Enclosure
   G. Ventilation
   H. Administrative Controls
   I. Maintenance Approaches
   J. Respirators

VI. Cost and Energy Considerations

VII. Successful Controls

31

General Methods for the Control of Airborne Hazards

D. Jeff Burton, CIH, CSP, PE

Introduction

This chapter deals with the general subject of emission control of airborne contaminants to minimize employee exposures. Noise and other hazards are discussed in the chapters dealing with those specific subjects. Subsequent chapters in this section cover local exhaust ventilation (LEV) and other specific emission and control measures.

Control of emissions and employee exposures may be accomplished by many methods. The occupational hygienist has a responsibility to choose controls that are compatible with (and acceptable to) the employee, the process or task, and are cost-effective. The type of control chosen also depends on regulatory requirements (some operations require specific controls, e.g., welding stainless steel indoors requires LEV), the nature of the hazard, and the way it affects the employee (e.g., the route of entry to the body for chemical hazards.)

The selection of controls must also consider the control's effect on operations and maintenance; that is, controls should not materially interfere with normal operations, and maintenance should be possible with little effect on ease of access and safety.)

Some of the major control techniques introduced in this chapter (e.g., local exhaust ventilation, personal protection equipment [PPE], respirators) are covered in more detail in other chapters of this book.

The Occupational Hygienist's Role

The industrial or occupational hygienist's traditional role was to anticipate and/or recognize, evaluate, and control occupational health hazards. In recent years the actual control of emissions and hazardous conditions has often become the first-line responsibility of engineering departments, maintenance departments, or others. Nevertheless, today's occupational hygienist must continue to be an active team player as a consultant, reviewer, information provider, conceptual designer, and final authority for employee protection.

In most cases the engineering/maintenance staff should rely on the occupational hygienist for guidance, calculations, and even design assistance. The occupational hygienist should insist on being involved in the review and commissioning of all emission and exposure controls during design, installation, operation, and maintenance. Otherwise, employee exposure protection may be compromised.

Fundamental Control Assumptions

There are five fundamental assumptions that the occupational hygienist should recognize:

1. All hazards can be controlled to some degree and by some method.
2. There are alternative approaches to control.
3. More than one control may be useful or required.
4. Some control methods are more cost-effective than others.
5. Controls may not completely control the hazard.

Controls chosen based on these assumptions will be realistic and cost-effective.

Table 31.1 — Approaches and Examples of the Major Control Approaches

Type of Control	*Approaches and Examples*
Administrative controls	Management involvement, training of employees, rotation of employees, air sampling, biological sampling, medical surveillance
Engineering controls	Process change, substitution, isolation, ventilation, source modification
PPE	Gloves, aprons, rubberized clothing, hard hats
Source modification	Changing a hazard source to make it less hazardous (e.g, wetting dust particles or lowering the temperature of liquids to reduce off-gassing and vaporization)
Substitution	Substituting a less hazardous material, equipment, or process for a more hazardous one (e.g., use of soap and water in place of solvents, use of automated instead of manually operated equipment)
Process change	Changing a process to make it less hazardous (e.g., paint dipping in place of paint spraying)
Isolation	Separating employees from hazardous operations, processes, equipment, or environments (e.g., use of control rooms, physically separating employees and equipment, barriers placed between employees and hazardous operations)
Ventilation	Two fundamental approaches: general exhaust (dilution of air contaminants) and local exhaust (of air contaminants)
Process controls	Continuous processes typically are less hazardous than intermittent processes
Isolating techniques	Storage of hazardous materials (e.g., use of ventilated storage cabinets for chemicals, size of storage container) isolating equipment (e.g., physical isolation of valves and pump seals, barriers around equipment) isolating employees (e.g., use of closed control rooms, isolation booths, supplied-air islands)

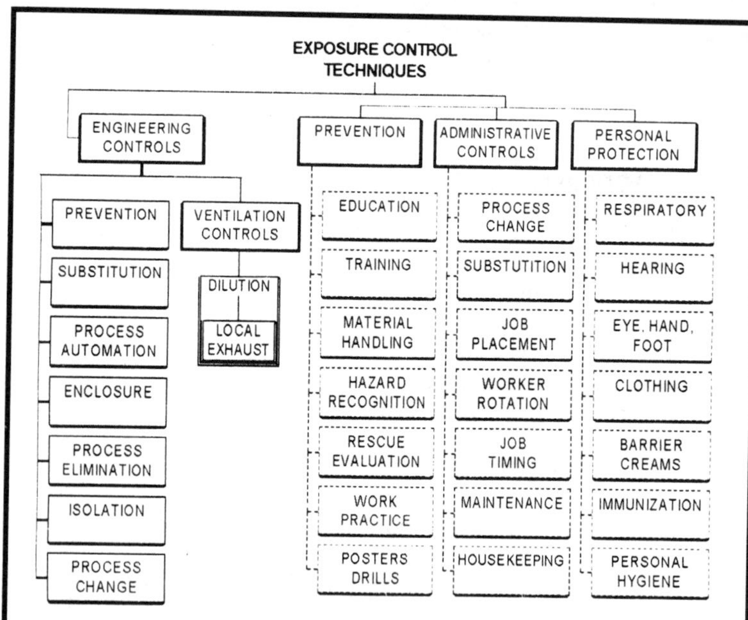

Figure 31.1 — Exposure control techniques. (Source: S. DiNardi.)

General Types and Approaches to Control

Administrative actions, engineering controls, and PPE are the three traditional approaches used to control emissions and employee exposures. Each of these is discussed in more detail in this and other chapters of the book. In-depth coverage may be found in the references listed at the end of this chapter. Samples of these control approaches are shown in Table 31.1 and Figure 31.1.

Prevention

The use of prevention is important to the overall application of controls. The occupational hygienist should cultivate an attitude of prevention in every situation where employees risk exposure to air contaminants. Every control evaluation should begin with the question, "Is it possible to prevent the condition needing control?"

Case Study 1. Part-time welders in a plant were exposed to welding smoke when welding parts for a machine manufactured in the plant. The occupational hygienist was asked to research emission and exposure control options. Reviewing the literature, the occupational hygienist found that traditional exposure control approaches included LEV, dilution ventilation, and respirators.

Outcome. Prevention suggested finding an alternative method of bonding the metals. Would chemical bonding be feasible? What about riveting, bolting, or automated welding in a closed system? A report was prepared listing all options. A management team eventually selected bolting of parts as the most cost effective control.

Control Begins at the Design Stage

It is more difficult and expensive to apply emission or exposure controls after an operation or process has already been installed or constructed. The occupational hygienist should be involved from the very beginning of any project where emissions and exposures are likely to occur, especially during design. Design criteria found in good occupational health programs are shown in Table 31.2.

Application of Controls

Once a potential airborne hazard has been anticipated or recognized, the application of controls usually follows the implementation steps shown in Table 31.3. A team consisting of the occupational hygienist and others

Table 31.2 — Occupational Hygiene Design Criteria for Projects with Emission/Exposure Potential

(1) meet all applicable codes, regulations, and standards
(2) obtain occupational hygiene input
(3) design processes to be as enclosed as possible, or to be remote from employees
(4) design to maintain employee exposures as low as possible—below the action level if possible, and never to exceed the permissible exposure limit
(5) use low-hazard materials and processes
(6) design with employee safety and health as a major objective
(7) make process equipment as maintenance-free as possible
(8) automate systems as much as possible
(9) obtain occupational hygiene review of proposed plans and specifications
(10) include the occupational hygienist in the commissioning process

Table 31.3 — Approach to the Implementation and Application of Controls

(1) identify/characterize hazard
(2) identify emission/exposure sources
(3) characterize sources
(4) characterize worker involvement with sources
(5) characterize air movement
(6) identify all alternative controls available
(7) choose the most effective control(s) considering compliance requirements, costs, and ethics
(8) implement controls
(9) follow up with testing and maintenance of control systems

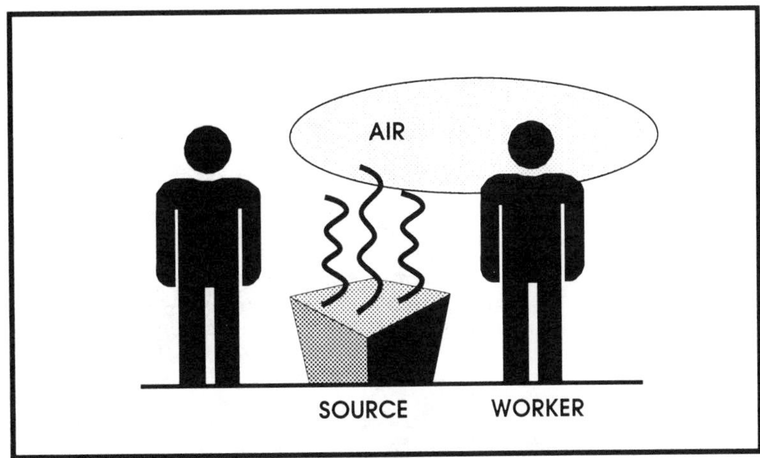

Figure 31.2 — Relationship between emission source, air, and employee.

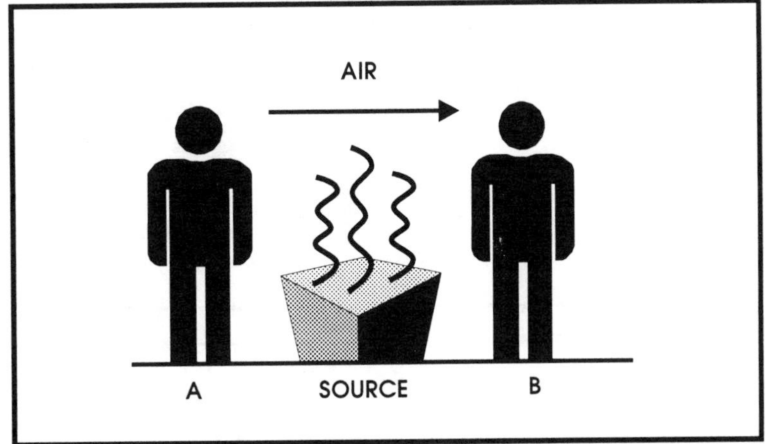

Figure 31.3 — Evaluation of source, employee, and air assists in strategy development.

(e.g., facilities engineering, designers, maintenance supervisors, fire marshal, safety professional, operations) will generate a database and an application plan.

Problem Characterization

The first five steps of Table 31.3 have traditionally been called problem characterization. Without these steps, the chances of finding and implementing cost-effective controls is minimal. Problem characterization is one of the most important functions of, and has been the exclusive domain of, the occupational hygienist. Who else has the training, expertise, or the interest to perform this important work?

As controls are being considered, the team must know everything about the behavior of the emission source, the worker, the air, and their relationships to each other. Figure 31.2 shows the relationship that must exist in any exposure problem.

In Figure 31.3, which employee looks more at risk? Based on the proximity of the worker to the emission source and the direction of air movement, it appears that Employee B may be at higher risk of exposure. Obviously, employee exposures do not occur unless a hazard source is located somewhere in the workplace. And air must move the contaminant or hazard into the occupied space of the worker (or the worker must move into the contaminated or hazardous area.) Table 31.4 lists the questions that must be answered to evaluate employee risk. Some of these cannot be answered with much accuracy, but the occupational health professional should develop an estimate for each, even if the estimate only borders on the quantitative.

Case Study 2. Figure 31.4 shows a simple problem characterization. The top figure shows a plan view (looking down on the floor plan) of a grain handling system in a food manufacturing plant. Grains are dried and then stored in two bins. Two employees work in the building.

Outcome. Note that all three aspects of the problem are presented on three more plan views. Emission sources are identified

Table 31.4 — Information Needs for Problem Characterization

Emission source behavior
- Where are emission sources, or potential emission sources?
- Which emission sources actually contribute to exposure?
- What is the relative contribution of each source to exposure?
- Characterize each contributor (e.g., chemical composition, temperature, rate of emission, direction of emission, initial emission velocity, continuous or intermittent, time intervals of emission).

Air behavior
- How does the air move (e.g., direction, velocity)?
- Characterize the air (e.g., air temperature, mixing potential, supply and return flow conditions, air changes per hour, effects of wind speed and direction, effects of weather and season).

Worker behavior
- How do workers interact with emission sources?
- Characterize employee involvement (e.g., worker location, work practices, worker education and training, cooperation).

by arrows — direction and magnitude are suggested by the arrows. Worker locations show proximity to emission sources. Airflow is indicated by arrows and volume flow rates. For example, air flows out of the building through a roof fan, as shown in Figure 31.4.

To summarize, exposure is directly related to the amount of airborne contaminant that reaches the breathing zone (BZ) of the worker and the time duration of exposure. Controls then must attempt to reduce the transfer of contaminants into the BZ and/or the time of exposure. Problem characterization is the primary technique used to identify and evaluate these parameters.

The remainder of this chapter is devoted to the application of various control approaches. The first section deals with source control in nonindustrial environments. Subsequent sections present control strategies found primarily in industrial environments, but the general principles apply to all control projects.

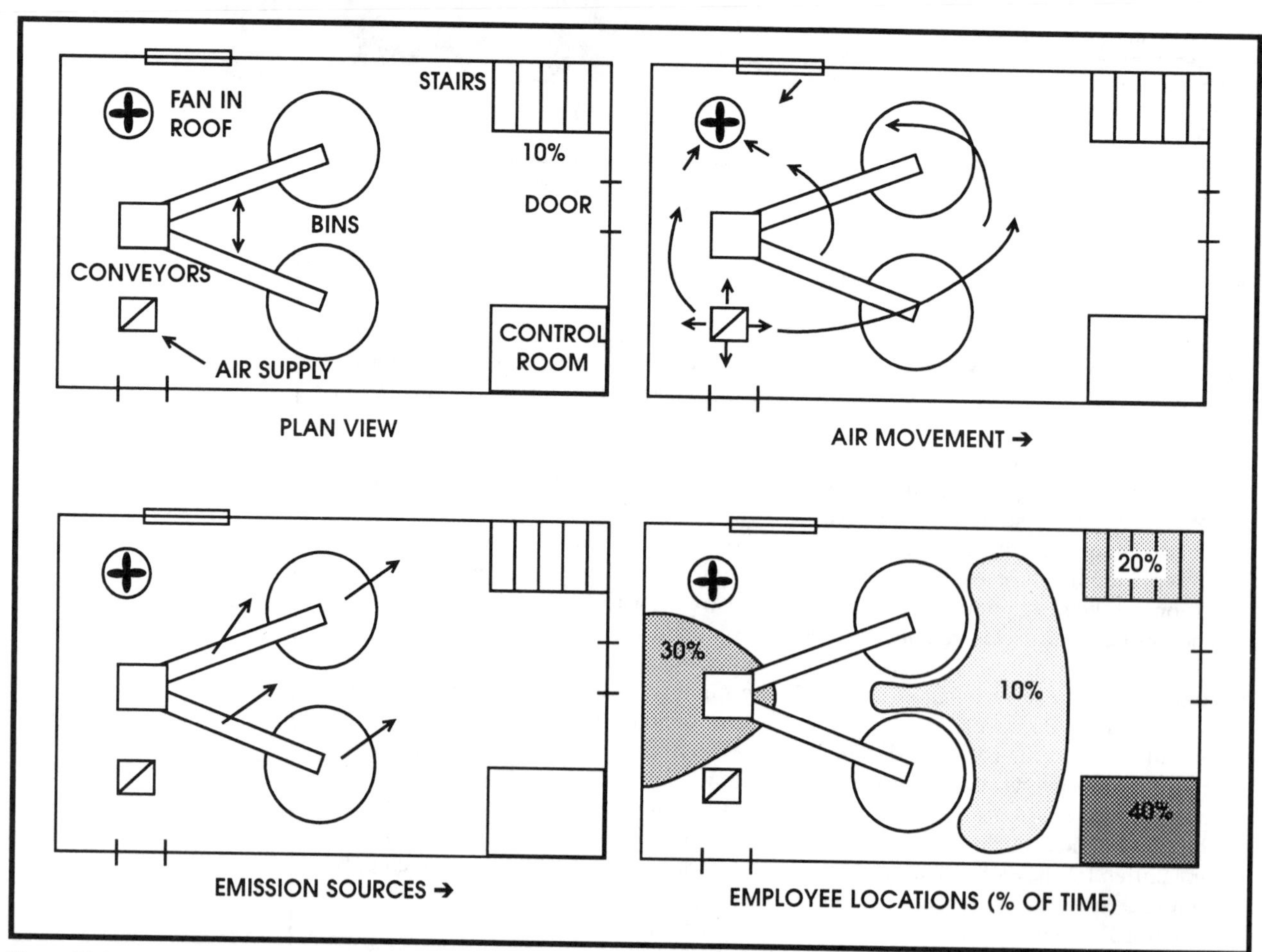

Figure 31.4 — Problem characterization of a grain-handling operation.

Emission Source Control in Nonindustrial Environments

Employee groups have traditionally been divided into two categories. Blue-collar workers work in industrial and production-type plants. These employees are often exposed to airborne chemical contaminants related to plant processes and materials. White-collar workers are those who work in commercial establishments like stores and warehouses and those who work in office environments.

In its infancy, occupational hygiene concentrated on the control of emissions and exposures to blue-collar workers (e.g., control of exposures to silica and coal dust in mines, solvents and acid mists in manufacturing facilities, and heavy metals in smelters and foundries.) In recent decades it has been recognized that white-collar workers are also exposed to airborne hazards (e.g., tobacco smoke, ozone, formaldehyde, mold spores). In today's occupational hygiene world, airborne hazards control in the commercial/office environment has become as important as traditional work in industrial environments.

In the nonindustrial environment two primary controls are used: emission source control and dilution ventilation. This section discusses source control. Dilution ventilation is covered in a later section.

There are two basic approaches to source control in office buildings, commercial establishments, and other employee environments where exposure problems exist: (1) use, specify, and/or install equipment, materials, and furnishings with low emissions; and (2) limit or minimize emissions once equipment, materials, or furnishings are in place and the building has been occupied.

Types of Emissions. Emissions can be separated into two families: the wet emitters and the dry emitters. Table 31.5 shows the various types. When selecting or specifying building materials, equipment, or furnishings, follow the useful approaches found in Table 31.6.

Table 31.5 — Types of Emission Sources Found in the Nonindustrial Environment

Type of Emission	Description
Wet materials emitting into the building air during construction	These include solvents, paints, adhesives, and sealers applied during the construction or remodeling of a building. Most of the emissions occur during the first few hours or days after application, but some may continue to emit at low rates for days, weeks, or months after. Exposure may be controlled by applying materials well before the building or space is (re)occupied.
Wet materials emitting into the building air during occupancy	These include such materials as cleaners, waxes, disinfectants, and biocides. Because the building may be occupied during application, the potential for exposure and complaints is high. Another source is the use of boiler steam, which contains water treatment chemicals, for humidification.
Dry materials emitting into the building air	These include materials such as the SBR latex backing found on many carpets, styrene monomer from plastics, or formaldehyde from pressed wood. The offending emission is usually a material that was added during the manufacturing process. When the product is unpacked at the site, a large emission may occur, decaying quickly as time passes. Others may emit for months or years (e.g., formaldehyde from wood products.)
Dry materials emitting adsorbed chemicals (sink materials)	Surfaces can absorb/adsorb and desorb chemicals from the air adjacent to the material (sinks). The rate of take-up and desorption is a function of the sink material, the surface area, the temperature, the chemical, and the chemical's concentration in the air. Materials with rough or fleecy surfaces (such as cloth or paper) adsorb more readily than smooth materials (such as glass or steel). Sinks can become loaded with chemicals during high air concentrations (i.e., during construction) and may off-load slowly during occupancy.

Emission and Exposure Control in Industrial Environments

The majority of life- and health-threatening employee exposures to airborne chemicals occur in the industrial environment. However, the emission sources and their controls as described in the remainder of this chapter can be applied to both the industrial and nonindustrial occupational environments.

Aerosol Emission Source Behavior

Particle air contaminants become airborne by a number of mechanisms in industrial environments. This emission activity is called pulvation. Table 31.7 and Figure 31.5 show typical mechanisms of pulvation. Control of particle emissions must recognize these characteristics and attempt to minimize the pulvating energy (e.g, reduce wind, vibration, and heat.)

Vapor and Gas Emission Source Behavior

Vapor contaminants evaporate from a liquid surface or are emitted from gas-containing vessels. The emission rate from evaporation is related to the vapor pressure

Table 31.6 — Approaches to the Emission Control of Building Materials and Furnishings

Emissions	Select materials with lowest emission rates and toxicity, least potential for creating irritation, and highest odor detection threshold (zero emissions is ideal). Where not available, choose materials with low long-term emission rates, or materials with high emission decay factors. Some materials may emit low quantities initially but may emit over a long term.
Environment	Choose materials not adversely affected by temperature or humidity. Materials that significantly increase emissions with increases in temperature or humidity will likely create indoor air quality (IAQ) episodes eventually.
Sink potential	Choose materials that are least likely to become sinks, and as such, secondary emission sources. Factors affecting sink adsorption, absorption, and desorption include the following. *roughness:* Smooth materials are less likely to become sinks than fleecy or rough materials. *temperature:* Affects rate of uptake and desorption, diffusion, and rates of evaporation. *humidity:* Affects rates of emission of formaldehyde from pressed wood and other carriers. *air movement:* Affects concentrations of chemicals in air; affects size of boundary layer of air (a tiny layer of quiet air on the surface that controls the rate of gas-phase mass transfer at the sink).
Maintenance	Maintenance-free materials are less likely to become emitters. Materials that can be cleaned with water are less likely to become sources of IAQ complaints.
Supplier	Choose suppliers and manufacturers who have information and IAQ programs. Manufacturers who have conducted emission testing and can provide the above information—along with testing results—should be supported. Without good information from suppliers and manufacturers, it is difficult to provide good IAQ through source control.
Aging	Many specifiers require newly manufactured equipment and furnishings to be "aired out" or "aged" before installation or before building occupation. Aging creates an opportunity for the initial burst of emissions to occur remote from the building and its occupants. For example, carpet can be rolled out for several days or weeks in a warehouse before installation. "Baking out" has not been shown be cost-effective.

(vp) of the material. Vapor pressures vary with temperature (T). Higher temperatures create higher vapor pressures. For example, the vapor pressure of water at sea level and T = 20°C is about vp = 20 mmHg. It increases to vp = 760 mmHg at 100°C. (At the boiling point, the vapor pressure equals the atmospheric pressure.)

The maximum concentration a vapor can attain in air (saturated condition) is given by the ratio of the vapor pressure

Table 31.7 — Pulvation Mechanisms and Resulting Airborne Particles

Mechanism	*Particle*
Agitated liquids	mist
Liquids in which bubbles rise to the surface and break	mist
Heated viscous liquids	vapor, then solidified/ condensed particles
Heated solids or metals	fume
Sprayed materials	particles, condensed particles
Conveyor-transferred dry materials	dust
Tumbled, abraded, or agitated dry materials	dust
Materials through which air passes	dust
Materials over which air passes	dust
Materials that fall from one level to another	dust
Vibrated materials	dust

Source: British Occupational Hygiene Society, Controlling Airborne Contaminants in the Workplace. London: Science Reviews Ltd., 1987.

(sometimes called the partial pressure) to atmospheric pressure. For water at standard conditions the maximum concentration of water vapor in air is 20/760 = 2.5% by volume. When additional water vapor is added to air, existing water vapor will condense somewhere in the space to maintain the vapor equilibrium.

$$CR = \text{partial pressure} = vp/BP \text{ (consistent units)} \quad (1)$$

where CR is concentration ratio and BP is boiling point. (Note: Multiply CR by 100 to obtain percentage, by 10^6 to obtain ppm.)

Gases generally emit to the environment from closed systems (e.g., vessels, pipes, process equipment) or from processes that generate gases (e.g., carbon monoxide from combustion). One approach suggested by the British Occupational Hygiene Society for estimating the relative hazard potentials of vaporous materials is to compare the ratio of the concentration (C) at saturated vapor pressure with the threshold limit value (TLV®; both in ppm).

$$\text{vapor hazard index} = C_{\text{vp-saturated}}/TLV \quad (2)$$

Case Study 3. Two chemicals were being considered in an open cleaning operation. The two chemicals' properties were:

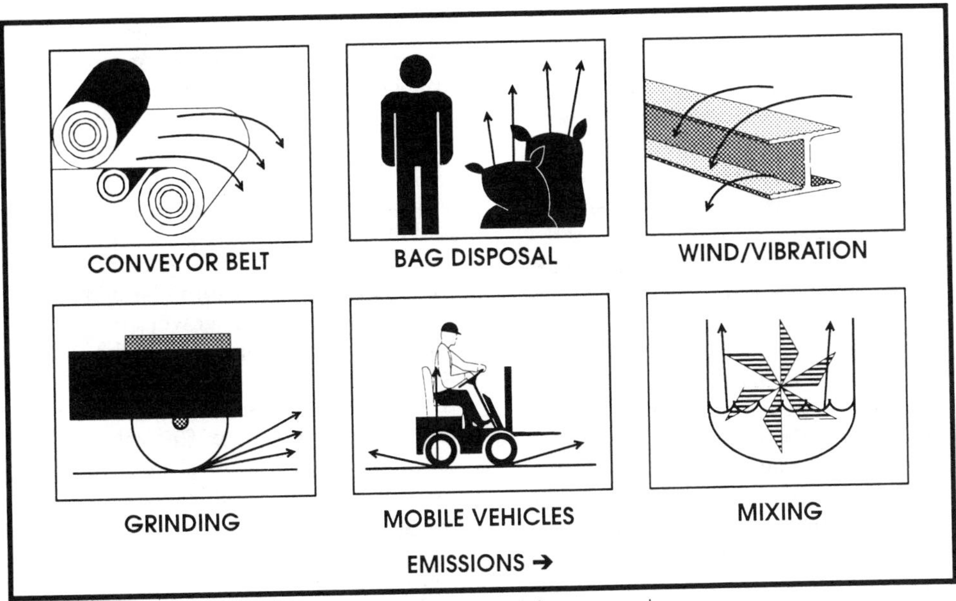

Figure 31.5 — Particle emitting process found in industrial occupancies.

Chemical A: vp = 3 mmHg, TLV = 100 ppm; Chemical B: vp = 6 mmHg, TLV = 350 ppm. The occupational hygienist was asked to comment on which chemical would be more appropriate.

Outcome. The atmospheric pressure at the plant averages BP = 650 mmHg. Concentration ratio at saturation, Chemical A = vp/BP = 30/650 = 0.00462 (4620 ppm); concentration ratio at saturation, Chemical B = vp/BP = 60/650 = 0.00923 (9230 ppm). The indexes were calculated as follows.

vapor hazard index A =
$C_{saturated}/TLV = 4620/100 = 46.2$

vapor hazard index B =
$C_{saturated}/TLV = 9230/350 = 26.3$

Based on this outcome and other considerations, the occupational hygienist recommended Chemical B.

The vapor hazard index is an interesting approach, but must not be the sole criteria for decision making. Other factors may include work practices, cost of materials, difficulty of separation at the air cleaner, permit requirements, routes of entry, usage, open or closed process, specific gravity of the materials, and so forth. The occupational hygienist must use all such criteria when making a final decision regarding emission and exposure control.

Settling. Many gases and vapors are heavier than air in their pure states. A saturated air/vapor mixture (such as generated on the top of a tank of solvent) will often be heavier than air and will settle or flow downward in response to gravity. However, its rapid mixing with air and the quick dilution of such saturated air/vapor mixtures retards settling almost as soon as the air/vapor cloud leaves the source. Settling is a function of the relative density of the air/vapor mixture to air. See the last column of Table 31.8. A good rule of thumb is: In percentage quantities, an air/vapor mixture will settle; in ppm quantities, the air/vapor mixture will behave like the air.

Table 31.8 —
Densities of Air/Vapor Mixtures

Chemical	*VP (20°C)*	*MW/MW_{air}*	*SG (Re air)*	*Relative Density at 5000 ppm (Re: Air)*
Acetone	185	2.0	2.7	1.005
Toluene	22	3.2	1.06	1.011
n-hexane	150	3.0	1.4	1.010
Xylene	10	3.7	1.04	1.014

Note: VP = vapor pressure in mmHg; MW = molecular weight, unitless; SG = specific gravity, unitless

Source: British Occupational Hygiene Society, *Controlling Airborne Contaminants in the Workplace*. London: Science Reviews Ltd., 1987.

Open and Closed Industrial Processes

According to McFee and Garrison,[1] all industrial processes are either open or closed. Open systems are more apt to emit contaminants routinely to the air shared by employees. Indeed, closing an open system can be a primary emission control approach. (For example, an open surface tank containing sulfuric acid is totally enclosed.) On the other hand, enclosing a chemical emission source in a building may concentrate airborne chemicals and create a higher-exposure experience for workers in the building. Additionally, a closed system may give a false sense of security because a closed system invariably leaks, or creates fugitive emissions, which must be evaluated and controlled.

Open systems include such processes or conditions as (1) stockpiles of chemicals or materials that can be pulvated (or evaporated) by vibration, wind, or other mechanical action; (2) open conveyor belts that emit because of the vibration of the equipment and the movement of material; (3) tanks and kettles that contain evaporation chemicals; (4) open containers; (5) spray painting; and (6) hand welding.

The occupational hygienist should evaluate each real or potential exposure problem (and its control) by asking the following questions.

- Is the process or equipment open or closed?
- Is there an opportunity for the material within the process or equipment to be emitted to the air shared by employees?
- Will employees be (or are they) exposed?
- Can the process, equipment, or system be closed (or opened, if the employee is required to be within the enclosure)?
- What effect on exposure will the closure (or opening) have?

Case Study 4. An employee spray paints small metal parts with an air-powered spray gun. The painter's complaints and air monitoring suggest that the painter is exposed to solvent vapors and paint particles at about 50% of the permissible exposure limit (PEL). The occupational hygienist has been asked to evaluate the situation and make recommendations for reducing exposures to below 10% of the PEL.

Outcome. The occupational hygienist answers the suggested questions above: The process is open; there is an opportunity for the material to be emitted into air; employees will be exposed; and the process can be partially enclosed in a small spray booth. Because the booth will be enclosed on five sides (top, bottom, two sides, and back), there is a very good chance that emissions and exposures can be controlled if the ventilation is adequate and the painter uses good work practices. The amount of reduction can be mathematically estimated using formulas described in References 2–8. Proper work practices must be used as well to control the behavior of the worker (e.g., the painter must remain entirely outside the enclosure and in the upwind position). This will help assure that the enclosure is effective as an exposure control. (Paint booths are designed to contain paint overspray. To be effective as employee protection, other measures must be taken.)

Continuous Operations. Closed systems are often used in chemical and refining industries where production processes are continuous (as opposed to batch processes). Closed production systems are becoming increasingly popular in the semiconductor industry, as well.

According to Lynch and Lipton,[1] health hazards may arise when a closed system fails to maintain containment. In such cases secondary control measures are required. Such secondary systems may include the following:

- Secondary LEV (e.g., around flanges, pumps, gas cylinders, or other closed systems);
- Dilution ventilation;
- Isolation of the employee in a controlled atmosphere (a separately ventilated control room, for example);
- Personal protective devices (e.g., emergency respirators worn on the belt); and
- Warning systems to alert employees (an air monitoring system with alarm, for example).

No closed system can be assumed to provide perfect control. Accidents, poor design, and/or faulty equipment will invariably create fugitive emissions. The occupational hygienist must anticipate such events (through modeling) and prepare to prevent unnecessary exposures.

Case Study 5. The labor union safety committee expressed concern that employees would be exposed to silane if a particular flange in a piping system failed. The occupational hygienist was asked to investigate and provide recommendations for control of potential fugitive emissions.

Outcome. On investigation the occupational hygienist found that according to the manufacturer, the flange had been shown to fail about 10% of the time within 10,000 hours when used with silane. The flange had never failed catastrophically; it was most likely to leak up to 0.001 lbs/hour when used at the pressure found in the piping system. Control options included using a different, more expensive flange, secondary ventilation around the valve, air monitoring systems to warn when the flange failed, shutoff valves operating upstream of the flange in the event of a failure, and employee respirators. The occupational hygienist mathematically modeled the situation (determining how much air was required under the worst case accident to maintain air concentrations in the exhaust air to below 1 ppm, for example) and provided information on cost-effectiveness of each available control option (see References 2 and 3 for detailed approaches.) A report was written, and management made a control choice.

According to Raterman,[9] the occupational hygienist may institute controls at the source, the air path, and/or at the worker. Many of these control strategies are described in more detail in Figure 31.6.

Substitution of a Less Hazardous Material

Eliminating a toxic emitter is generally the best approach to control. It comes close to the ideal of prevention. Many toxic materials have suitable substitutes with lower toxicity ratings or evaporation rates. Examples are legion: walnut shells for sand in blasting, hydrochlorofluorcarbons for fluorinated or chlorinated hydrocarbons in solvent usage, toluene for benzene in paint thinners, titanium dioxide for white lead in paint, calcium silicates for asbestos, oil for mercury in gauges, and so forth.

When substituting a less toxic material, always consider the resulting hazard. Substituting acetone for toluene, for example, while providing a less toxic material, will increase emissions (acetone has a lower boiling point) and increase the fire hazard (acetone has a lower flash point). In this case the hazard may actually increase. Sometimes a material's properties may be changed to reduce emissions (e.g., lowering the temperature of a solvent, wetting a dry powder, pelletizing dusty powders). The occupational hygienist should evaluate potential substitutes during the planning phase of control application and consider costs and suitability in the process.

Case Study 6. Perchloroethylene was being used to clean metal parts in a parts cleaner. The commercial cleaner had been intended for use with Stoddard Solvent, but PCE was found to be more effective. Unfortunately, emissions had increased and employee exposures were close to the PEL. The occupational hygienist was called.

Outcome. The occupational hygienist investigated the situation and tried several different cleaning materials and strategies. In the end, plain soap and water proved to be as effective in cleaning the parts when one step in the manufacturing process was altered to eliminate coating the part with oil. In this case the successful control included both substitution and process change, not an uncommon occurrence.

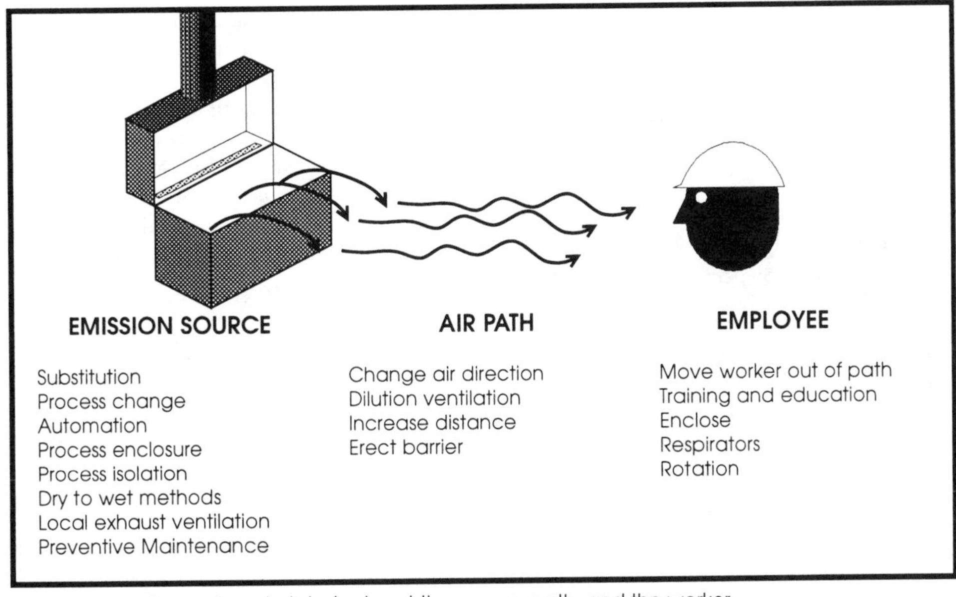

Figure 31.6 — General control strategies at the source, path, and the worker.

Process Changes

Process change, like substitution, attempts to replace more hazardous equipment or processes with less hazardous approaches. Historic examples include paint dipping for paint spraying, polyester resin sealants for lead solder in automobile body seams, low-speed oscillating grinders for high-speed rotary grinders, bolting for welding, cooling-coil vapor degreasing tanks for hand-washing of parts, continuous closed processes for open batch processes, and so forth.

Again, the occupational hygienist should evaluate the proposed change with regard to the final hazard being created by the change and its effects on productivity and product quality. Fortunately, process or equipment changes used for hazard control often result in increased productivity and product quality.

Case Study 7. Lead-contaminated dust was collected from a baghouse and placed in drums for shipment to an approved landfill. Exposures to employees handling the dust exceeded the PEL for lead, which required the constant use of respirators. The occupational hygienist was asked to evaluate the process and recommend changes.

Outcome. After thorough study, the occupational hygienist recommended that a small sintering furnace be installed at the discharge of the baghouse. This would agglomerate the dust into less-dusty chunks of material. The exhaust for the furnace was routed into the baghouse inlet ductwork, eliminating any emissions from the furnace. Exposures to employees handling the resulting chunky materials were reduced below the PEL.

Table 31.9 — Ventilation Types

Type	Usage/Approach
HVAC	Mechanically provides clean, tempered, fresh air for comfort and health. Traditionally known also as "general ventilation."
General exhaust or dilution	Controls the air environment by removing and replacing contaminated air before chemical concentrations reach unacceptable levels.
Local exhaust	Removes chemical contaminants at or near the emission source.
Local supply	Provides clean, fresh air for cooling, exclusion of contaminated air, or to replace exhausted air. Also called supplied-air island, or SAI.
Natural ventilation	Uses wind or temperature differences to induce airflow through the building, creating dilution of air contaminants.
Make-up	Replaces exhausted air.

Isolation and Enclosure

The principle of isolation is found in almost every production plant. It may be as simple as erecting a wall between a process and an employee (e.g., separating office from production employees by a wall), or as complicated as automated processes using robotic techniques (e.g., used to spray paint automobile body frames). Enclosure, closely related to isolation, usually takes the form of building an airtight enclosure around the process or the employee. Examples of isolation and enclosure include building enclosures around piles or open storage of dusty materials; enclosing asbestos in airtight containers; enclosing a sandblasting operation in an airtight housing; enclosing a chemical operation in a totally enclosed system; enclosing crane operators in an air-supplied cab; consolidation of chemical pipes and lines into an enclosed and exhausted area; pneumatic conveying of material; enclosing and exhausting sampling and view ports; and so forth.

The principle of isolation includes separation by time as well as space. For example, foundry shakeout can be scheduled during the evening when most employees have left the workplace, or maintenance can be performed during regular shutdown periods when production employees are not present.

Case Study 8. Two employees were overexposed to lead dust in a ball mill. Their jobs consisted of periodically inspecting each mill during the operation. The occupational hygienist was asked to recommend control strategies.

Outcome. The most cost-effective control turned out to be isolating the employees in a control room adjacent to the ball mill operation and providing closed-circuit TV cameras at the mouth of each ball mill. The control room was provided with a dedicated air handling system isolated from the air in the ball mill area. During subsequent infrequent in-person visits to the ball mill area, employees wore PPE. Exposures were reduced by over 500%.

Ventilation

Ventilation is a widely used and time-tested approach to emission and exposure control. Almost every building in the western world employs some type of ventilation to promote or protect the comfort or health of occupants. Within the category of ventilation one finds a number of approaches. These are listed in Table 31.9 and covered in more depth in subsequent sections and chapters. See references and additional sources for more detailed information on all ventilation topics.

LEV

LEV systems are designed to contain, control, or capture emissions at or near their sources. Typical LEV systems consist of control structures (hoods, exhausted enclosures), ductwork, air cleaners, fans, and stacks (see Figure 31.7).

The LEV approach attempts to eliminate emissions to the workroom air, as opposed to dilution ventilation systems (see next section), which allow emissions to occur and dilute contaminants to some acceptable concentration before the contaminated air reaches the breathing zone.

LEV systems are covered in detail in Chapters 32 and 33.

Dilution (General Exhaust) Ventilation

Emissions and subsequent employee exposures are often controlled with fresh dilution air in both industrial and nonindustrial environments. For a constant emission source, no sinks, perfect mixing, and a constant airflow, the equation below describes the resulting equilibrium concentration in a ventilated space (e.g., chamber, room, building, space).

$$C = E/Q \text{ or } Q = E/C \quad (3)$$

where:

E = emission rate;
C = concentration; and
Q = ventilation rate in the same units as E.

Figure 31.8 shows the situation for five different emission rates, all other conditions equal and at steady state conditions. The numbers suggest ratios of emission rates based on typical emission factors of 0.1 to 5 mg/m^2 per hour. Note that background concentrations increase significantly at air exchange rates less than N = 0.5 air changes/hour. Also, increases above 1–2 air changes/hour do not offer equivalent reductions in concentrations.

Contaminated air must be diluted to some acceptable level. This safe or comfortable level of exposure is called the acceptable concentration, CA (e.g., carbon dioxide at CA = 1000 ppm, carbon monoxide at 9 ppm). The number is often chosen by the occupational hygienist (in consultation with toxicologists and medical personnel.)

Table 31.10 shows typical selection criteria for dilution ventilation. When conditions in the second category of the table prevail,

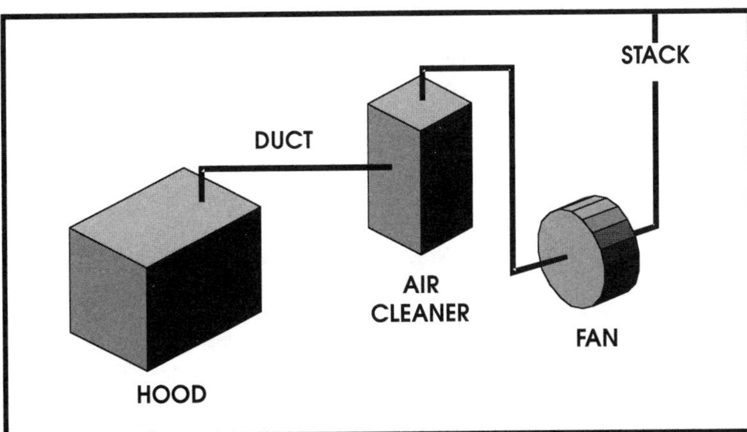

Figure 31.7 — Components of an LEV system.

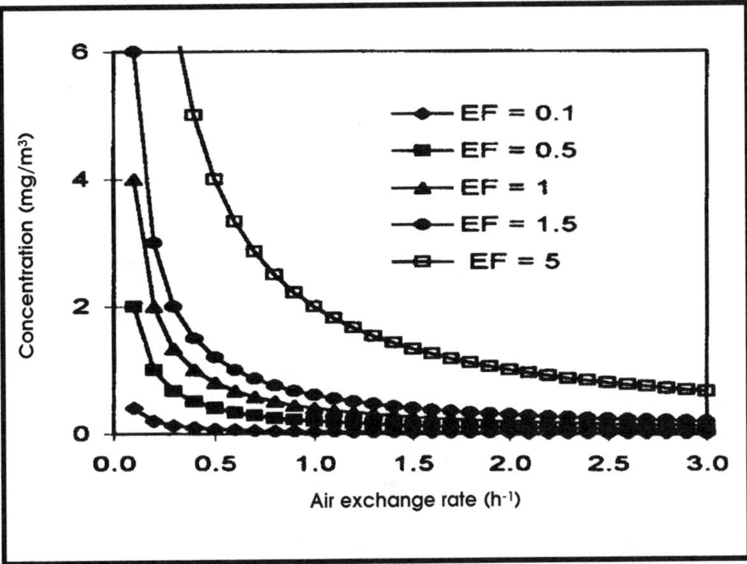

Figure 31.8 — Concentration vs. air exchange rate. Source: Hal Levin.

Table 31.10 — Selection Criteria for Dilution Ventilation

Conditions that lend themselves to dilution with outdoor air

- Major air contaminants are of relatively low toxicity.
- Contaminant concentrations are not hazardous.
- Smoking is not allowed in the occupied space.
- Emission sources are difficult or expensive to remove.
- Emissions occur uniformly in time.
- Emission sources are widely dispersed.
- Emissions do not occur close to the breathing zone of people.
- Moderate climatic conditions prevail.
- The outside air is less contaminated than the inside air.
- The HVAC system is capable of conditioning the dilution air.

Dilution ventilation is less effective and more expensive in the following conditions

- Air contaminants are highly toxic materials.
- Contaminant concentrations are hazardous.
- Smoking is allowed in the occupied space.
- Emission sources are easy to remove.
- Emissions vary with time.
- Emission sources consist of large point sources.
- People's breathing zones are in the immediate vicinity of emission sources (i.e., 2–3 feet).
- The building is located in severe climates.
- The outside air is more contaminated than the inside air.
- The existing HVAC system is not capable of treating the air.

costs will be high. If the outside air is more contaminated, dilution will not work, period. Source removal, LEV, emission control, substitution, or other forms of control should be considered (possibly in conjunction with dilution ventilation).

In the majority of cases, the outside air will be less contaminated than indoor air. Researchers report, however, that outdoor concentrations of the reactive gases sulfur dioxide and ozone are usually greater than indoor concentrations. Another exception occurs when there is an outdoor source near the air intake, such as a truck idling next to a building air intake, in which case the outdoor concentrations of carbon monoxide, nitrogen oxides, and smoke will be higher than those in indoor air. Also, summertime concentrations of mold spores and pollen are often greater outdoors.

Carbon monoxide, nitrogen oxides, and respirable dust concentrations are usually higher indoors. But this may be reversed, again if significant outdoor sources exist. For example, when the wind blows in a dry climate, outdoor levels of dust may exceed indoor concentrations. Or, if a parking garage is attached to a building, outside air concentrations of carbon monoxide may be higher at times. Total volatile organic compounds, carbon dioxide, formaldehyde, radon, and environmental tobacco smoke concentrations are almost always greater indoors than outdoors.

Case Study 9. Office employees were complaining about "bad air." Specific complaints included headache, watery eyes, irritation of the nose and throat, and dry, itchy skin. The occupational hygienist was asked to investigate and determine if dilution ventilation was appropriate.

Outcome. The occupational hygienist found that at the end of the day carbon dioxide totalled about 1200 ppm in the return air plenum. No specific chemical or biogenic sources, indoors or outdoors, were immediately obvious. However, the building was only six weeks old. The amount of carbon dioxide concentration suggested insufficient outside air. The fact that this was a new building suggested the problem of off-gassing. Fortunately, however, newer heating, ventilating, and air conditioning (HVAC) systems are capable of providing more air and better distribution. The occupational hygienist recommended that during the first 4-6 months of occupancy, the system be set to deliver 100% fresh air.

Case Study 10. A Los Angeles-based computer components firm operated 15 small dip tanks in a room adjacent to a set of offices. Although the dip tanks were enclosed and exhausted, office employees were concerned about potential exposure because the building's air handling unit (AHU) served both the office and the production area. An occasional "chemical" odor infiltrated the office area from the tank room. Office employees were asking for more dilution air. The occupational hygienist performed an emission/exposure characterization study.

Outcome. Odors suggested fugitive emission from the tanks. Complaints were alleviated by additional dilution ventilation. Other solutions could have included isolating the office HVAC system from the tank area AHU, isolating the tank area from the rest of the building, and moving the tank operation or the offices to another building.

Approach to Implementing Dilution Ventilation

If preliminary indications point to dilution, the occupational hygienist should develop the following information: (1) a profile of the contaminant sources; (2) an estimate of emission rates; (3) a description of the space (e.g, volume); (4) an acceptable concentration for exposure; and (5) an appropriate air dilution flow rate.

Sources. A basic understanding of emission sources should be obtained, if possible. Any information is better than none. Intelligent professional judgement is better than no estimate. This information gathering will include descriptions of emission sources; chemical description of emissions to include chemical composition, size and shape, and temperature; current airborne concentrations; and rates of emission or evaporation over time.

Steady State Emission Rates. The most intimidating of these items is the last one. Estimating emission rates for contaminants requires ingenuity, detective work, and skill. Often emission factors must be used.

Once you have an estimate of emission rate, it is possible to estimate an evaporation or emission rate as shown below.

$$q \approx \frac{387 \times \text{lbs evaporated}}{MW \times t \times d} \left(= \frac{0.0244 \times \text{grams}}{MW \times t} \right) \quad (4)$$

where:

q = volume of vapor generated in ft^3/min at American Conference of Governmental Industrial Hygienists (ACGIH)-ventilation standard temperature and conditions, d = 1 (Systeme Internationale d'Unites [SI]: m^3/sec)

MW = molecular weight or molecular mass
t = approximate time, minutes (SI: seconds)
d = density correction factor
lbs = amount evaporated, in lbs (SI: grams)

Accuracy. Note that "≈" is used instead of "=" in most cases. The formulas can provide only approximations of the real world. Always take such estimates with a grain of salt and use good judgment about the answers obtained.

The Space. Obtain all of the physical parameters of the building, its equipment and processes, and its occupants. Gather data on the occupied space—width, height, length, barriers, obstructions, and so forth. Building plans are often helpful.

HVAC and Other Existing Ventilation. Information on existing ventilation is also important—general ventilation, space heaters, open doors and windows, freestanding fans, LEV systems, and other dilution systems already in place. You should also study prevailing air movement in the space—directions and velocities. Ask about other times of the day and other operations (for instance, what happens during very cold weather or when the local exhaust system is turned off).

Occupants. Obtain a thorough understanding of the physical locations of people; the time considerations (such as how much time a person spends in a particular location and how they interact with emission sources); and training, education, and cooperativeness.

Acceptable Concentration. The occupational hygienist must determine the acceptable level of exposure. Usually it is some fraction of existing exposure standards (e.g., 10% of the PEL). Without this concentration, modeling is impossible, and predicting the benefits of dilution is more difficult.

Dilution Rates. Dilution air volumes should be based on the information developed above. The volume flow rate of air required to dilute a constant volume flow rate of emitted vapor is approximated as follows.

$$Q_{OA} \approx \frac{q \times K_{eff} \times 10^6}{Ca\ (ppm)} \qquad (5)$$

where:

Qoa = volume flow rate of dilution air, scfm (m^3/sec)
q = volume flow rate of vapor, scfm (m^3/sec)
Ca = the acceptable exposure concentration, ppm
$K_{(eff)}$ = a mixing factor to account for incomplete or poor delivery of dilution air (OA) to occupants

Air Mixing. The mixing of air is sometimes called ventilation efficiency or ventilation effectiveness. Mathematically, it can be stated as:

K_{eff} = Actual Q_{OA} required to provide minimum OA to occupied zone ÷ Ideal Q_{OA} required to provide minimum OA to occupied zone.

The value of k_{eff} ranges from 1.0 to 3 in most cases. If it seems likely that $k_{eff} > 2$, think first of improving mixing or of going to another form of control. Dilution is likely to be too expensive, and the uncertainties are too high. Typical values for k_{eff} are shown in Table 31.11.

Dilution will be more effective, and lower values of the mixing factor k_{eff} can be selected when (1) dilution air is routed through the occupied zone; (2) supply air is distributed where it will be most effective (e.g., be sure a supply register serves every office); (3) returns are located as close to contaminant sources as possible; and (4) auxiliary or freestanding fans are used to enhance mixing.

A measured estimate of K_{eff} can be made using tracer gases, but the procedure is beyond the scope of this book.

Case Study 11. Toluene will evaporate from a process. The occupational hygienist is asked to evaluate and determine the requirements for adequate dilution ventilation.

Outcome. Investigation produced the following estimates: Four pounds of toluene will be uniformly evaporated from the process during an 8-hour period. The volume flow rate (Q_{OA}) required for dilution

Table 31.11 — Typical Mixing Factors

K_{eff}	Typical Conditions
1.0	Wide open office spaces with good supply (makeup) and return air locations, all ventilation equipment function adequately; no point sources of emission
1.1	Conditions not ideal, but wide use of freestanding fans to create mixing; no point sources of emission
1.2–1.5	Poor placement of supply and return registers; partitioned cubicles or work locations with barriers; generally adequate distribution of supply and return locations; discernable but small point sources near employees
1.5–3.0	Crowded spaces with tight partitions/walls/barriers and poor supply and return locations; point sources of emission

to Ca =10 ppm is shown below (k_{eff} = 1.0, STP [ACGIH ventilation], MW = 92.1).

$$q \approx \frac{387 \times \text{lbs evaporated}}{\text{MW} \times \text{minutes}}$$

$$q \approx \frac{387 \times 4 \text{ lbs evaporated}}{92.1 \times 480 \text{ minutes}} \approx 0.0350 \text{ scfm}$$

$$Q_{OA} \approx \frac{q \times k_{eff} \times 10^6}{\text{Ca (ppm)}}$$

$$Q_{OA} \approx \frac{0.035 \times 1 \times 10^6}{10 \text{ (ppm)}} \approx 3500 \text{ scfm}$$

Case Study 12. New particleboard wood shelving is to be installed in a large office bay. The occupational hygienist was asked to estimate the volume flow rate of formaldehyde (HCHO) vapor formed and the dilution volume flow rate required to keep background concentrations of HCHO at or below 0.05 ppm. Assume: STP, MW ≈ 30,250 m^2 of particleboard, Ca ≈ 0.05 ppm, $k_{eff} \approx 1.25$.

Outcome. After searching product literature the occupational hygienist found that for particleboard, 2000–25,000 µg/m^2 per day might be emitted. Converting milligrams to pounds yields the following equations.

$$\text{lbs/day} \approx \frac{25{,}000 \ \mu g/m^2 \times 250 \ m^2}{1{,}000{,}000 \ \mu g \times 454 \text{ lb}}$$

$$q \approx \frac{387 \times \text{lbs emitted}}{\text{MW} \times \text{minutes}}$$

$$q \approx \frac{387 \times 0.0138}{30 \times 1440} \approx 0.000123 \text{ scfm}$$

and

$$Q_{OA} \approx \frac{q \times k_{(eff)} \times 10^6}{\text{Ca (ppm)}}$$

$$Q_{OA} \approx \frac{0.000123 \times 1.25 \times 10^6}{0.05 \text{ (ppm)}} \approx 3000 \text{ scfm (rounded)}$$

Administrative Controls

Engineered controls that eliminate or reduce the hazard are generally preferable to administrative or PPE controls.

Administrative controls place at least part of the burden of responsibility for protection on the worker. However, administrative controls are often effectively used in conjunction with or in addition to engineering controls. Indeed, very few airborne hazards are controlled by one control measure alone, and some types of administrative controls are always required (e.g., training and education).

A successful control strategy of a welding operation might include LEV (which automatically provides some dilution), substitution of a less toxic material, good housekeeping of the welding area to prevent secondary pulvation of dust, preventive maintenance of welding equipment, and training and education of the welders so they can work with minimum exposure. (Something as simple as adjustments of head position can impact exposure concentrations.)

Traditional administrative controls include rotation, reduction of exposure times, instituting exposure-reducing work practices, developing preventive maintenance programs, good housekeeping programs, training and education of employees, and teaching and requiring good personal hygiene.

Employee Rotation and Reduction of Exposure Times

Exposure is generally reported as a time-weighted average, or TWA (for compliance purposes).

$$\text{TWA} = f(C, t) \tag{6}$$

where C = average exposure concentration and t = time. The TWA concentration can be reduced by lowering the concentration in air. This chapter has concentrated on reducing emissions and subsequent exposures.

TWA exposures can also be lowered by reducing the time of exposure. See the evaluation in Table 31.12. The downside to this approach is that more employees are exposed. This goes against the traditional occupational hygiene philosophy of reducing exposures as much as possible

Table 31.12 — Employee Rotation Reduces Individual Exposure

Airborne Concentration	Time of Exposure	8-Hour TWA
100 ppm	8 hours	100 ppm
100 ppm	4 hours	50 ppm
100 ppm	2 hours	25 ppm
100 ppm	1 hour	13 ppm

for as many as possible. For example, using the evaluation in Table 31.12, eight employees must be exposed to achieve an average TWA exposure of 13 ppm. Additionally, eight people must be trained to do the job safely, eight sets of PPE must be supplied, eight medical exams must be given, and so forth. Cost-effectiveness considerations alone may disqualify the use of rotation.

Case Study 13. In spite of excellent LEV, workers soldering battery terminals were exposed to lead fume concentrations near the TWA PEL of 50 mg/m^3. Although workers were being adequately protected through the additional use of respirators, medical exams, good personal hygiene, and so forth, the plant desired to reduce TWA air exposures to below the action level, 30 mg/m^3. The occupational hygienist was asked to provide suggestions.

Outcome. In the end, the most cost-effective approach involved worker rotation. A similar work crew in a plastics fabricating section of the plant was trained to perform the soldering operation safely. Eventually both crews worked a half-day in each department, and resulting TWA exposures averaged 25 mg/m^3.

Housekeeping

Emitted particulate materials must go somewhere: out the stack, out the window, or settle on horizontal surfaces in the plant. As such, settled materials become secondary emission sources (often with as much or more impact on exposures as the primary source). For example, sweeping—although a small source compared, say, to a vibratory conveyor—can be a significant source of exposure because the BZ of the sweeper is only a few feet away from the emission point. All buildings vibrate and such vibration is the energy source—along with wind and traffic—to pulvate settled dust into the air.

Some Occupational Safety and Health Administration (OSHA) standards require good housekeeping (e.g., the standard for inorganic lead). All occupational hygiene guidelines for good practice mention housekeeping as an important control.

Dust may be removed and controlled by wet cleaning methods and vacuum systems. Sweeping and blowing simply create unwanted emissions. Spilled liquids should be cleaned up before evaporation causes a problem. Rags and absorbent materials should be disposed of following local fire and hazardous waste requirements.

Wet Methods

Wetting a dusty material reduces pulvation (but must be compatible with the process.) Adding humidity to a material is also effective because it may create agglomeration among small particles and add weight.

Examples of wetting include fogging sprays at conveyor belt transfer points, water washing of dusty surfaces, wetting of sand in sand blasting, and adding humidity to dry air to reduce indoor air quality complaints.

Personal Hygiene

Personal hygiene is important to the overall protection of the worker but can also impact the inhalation exposure. For example, if a worker smokes without washing his or her hands, hazardous chemicals may be inhaled through contaminated cigarettes. Similarly, if dirty clothes are taken home, additional exposure may result during clothes handling and washing.

Maintenance Approaches

As noted previously, emissions may be planned or unplanned and continuous or intermittent. All are impacted by the quality of maintenance applied to the emitting process or equipment.

Insufficient maintenance increases the likelihood of both catastrophic emissions (e.g., rupture failure of pipe flange) and fugitive emissions (e.g., slowly increasing rates of leaks from flanges, seals, joints, access doors). Poor maintenance practices may increase emissions when normal emission controls are detached or impaired (e.g., during filter replacement, fan seal repair, or equipment repair).

Maintenance itself may lead to elevated emissions in the vicinity of maintenance personnel. In these cases, training and PPE are required to minimize exposures to maintenance personnel.

Preventive maintenance (PM) of emission controls is an approach that evaluates the potential for increased emissions over time (e.g., the seal fails within three months of installation). PM attempts to repair, replace, or correct potential deficiencies before the problem occurs (e.g., the seal will be replaced every two months).

Because of its great potential to impact emission sources, for good or ill, maintenance should be a primary concern of the occupational hygienist.

Respirators

Part of the third tier of controls (PPE, after engineering and administrative controls), appropriate respirators are often used to provide employee exposure protection. However, respirator use should not substitute for engineering and administrative controls because (1) it places the responsibility of protection on the worker and (2) the hazard is not actually reduced in the workplace (the preferred approach).

The four generally accepted occasions for respirator use are (1) during the time other control measures are being installed or instituted; (2) during maintenance and repair work; (3) when other controls are not feasible; and (4) during emergencies.

There are two basic environments in which respirators are used—those atmospheres in which atmospheric conditions (e.g., chemicals concentrations) are immediately dangerous to life or health, and those that are not.

There are two basic types of respirators—those that supply safe, clean air (atmosphere supplying) and those that purify the air (air purifying). Air purifying respirators have a limited life that depends on the concentration of the contaminant in air, the type of canister, the ambient temperature, and the wearer's breathing rate.

All respirators sold in the United States for industrial use are approved/certified under National Institute for Occupational Safety and Health (NIOSH) testing methods. Testing traditionally has been conducted at NIOSH's Testing and Certification Laboratory in Morgantown, W.V. Under a new federal rule (*Code of Federal Regulations* [CFR] Title 42, Part 84) others will be enabled to conduct testing, as well.

OSHA requires that respirators used for employee protection be certified or approved (29 CFR 1910.134.). OSHA also requires respirator users to be fit-tested using one of two new test methods, one quantitative and one qualitative.

The subject of respirators is covered in depth in Chapter 36.

Cost and Energy Considerations

Costs are related to (1) types of controls applied and (2) the level of control desired. One can control emissions, for example, such that resulting employee exposures are at the TLV/PEL, the action level (usually 50% of the PEL or TLV), 25% of the PEL, 10% of the PEL, and so forth. All of the above target levels are used by practicing occupational hygienists, although the most common target is probably 10% of the TLV. Much of the semiconductor industry has adopted standards that call for target exposures at 1% of the TLV/PEL. Other philosophical approaches to control include the following:

- *Maximum Achievable Control.* These approaches are expensive and do not include cost-effectiveness as a primary concern. Such approaches regularly include after-installation measures such as air monitoring, reporting and record keeping procedures, medical exams, and so forth. This is the most effective (in terms of exposure control) but also potentially the most costly.
- *Maximum Financially Feasible Achievable Control.* This approach takes the attitude of achieving the lowest exposures possible within financial feasibility. This is a popular approach but makes control strategies difficult to model mathematically because no set exposure level is targeted. Establishments using this approach often attain exposure levels well below the PEL.
- *Reasonably Achievable Control.* This approach says, "We'll do the best we can, but within reason," where "reason" is defined by the user. This approach is probably less desirable than selecting a target exposure concentration.

Successful Controls

According to Lynch and Lipton,[1] emission and exposure controls must demonstrate the following characteristics to be considered successful:

- Protection must be reliable and consistent throughout its life.
- The effectiveness of control must be measurable throughout its life.
- The control must minimize the need for worker responsibility for his or her own protection.
- The control must not create additional uncontrolled health hazards.
- The control must protect against all routes of entry (e.g., skin and eye, gastrointestinal tract, and lung).

To those might be added the following:

- The control must be able to achieve the target exposure concentration.
- It must be accepted by the employee and be compatible with his or her work.
- It should be cost effective and financially feasible for the organization.

Summary

The fundamental assumptions considered in this chapter were that (1) all hazards can be controlled to some degree; (2) there are alternative approaches to control; (3) more than one control may be useful or required; and (4) some control methods are more cost-effective than others. Types of controls are administrative, engineering, and personal protective devices.

Administrative controls include management involvement, training of employees, rotation of employees, air sampling, biological sampling, and medical surveillance. Engineering controls include process change, substitution, isolation, ventilation, and source modification. PPE controls include personal protective equipment, clothing, and devices.

Various methods can be used to control hazards, such as source control, which is changing a hazard source to make it less hazardous; substituting a less hazardous material, equipment, or process for a more hazardous one; changing a process to make it less hazardous; separating employees through isolation from hazardous operations, processes, equipment, or environments. Process controls might also be used; continuous processes typically are less hazardous than intermittent processes. Finally, ventilation can dilute (general exhaust) or remove (local exhaust) contaminants.

Control implementation methods include:

1. Identification/characterization of the hazard;
2. Identifying emission/exposure sources;
3. Characterizing sources;
4. Characterizing worker involvement with sources;
5. Characterizing air movement;
6. Identifying all alternative controls available;
7. Choosing the most effective control(s) considering compliance requirements, costs, and ethics;
8. Implementing controls; and
9. Testing and maintenance of control systems.

Additional Sources

American Society of Heating, Refrigerating and Air Conditioning Engineers (ASHRAE): *ASHRAE 55-1992—Thermal Environmental Conditions for Human Occupancy* [Standard]. Atlanta, GA: ASHRAE, 1992.

American Society of Heating, Refrigerating and Air Conditioning Engineers (ASHRAE): *ASHRAE 62-1989 and 62R—Ventilation for Acceptable Indoor Air Quality* [Standard]. Atlanta, GA: ASHRAE, 1989.

American Society of Heating, Refrigerating and Air Conditioning Engineers (ASHRAE): *Guideline for the Commissioning of HVAC Systems* (ASHRAE guideline 1-1989). Atlanta, GA: ASHRAE, 1989.

American Society of Heating, Refrigerating and Air Conditioning Engineers (ASHRAE): *Handbook of Fundamentals*. Atlanta, GA: ASHRAE, 1993.

American Society of Heating, Refrigerating and Air Conditioning Engineers (ASHRAE): *Preparation of O&M Documentation for Building Systems* (ASHRAE guideline 4-1993). Atlanta, GA: ASHRAE, 1993.

British Occupational Hygiene Society: *Controlling Airborne Contaminants in the Workplace*. London: Science Reviews Ltd., 1987.

Burton, D.J.: *IAQ and HVAC Workbook*. Salt Lake City, UT: IVE, Inc., 1997.

Institut de Recherche en Santé et en Sécurité du Travail du Québec (IRSST): *Guide for the Prevention of Microbial Growth in Ventilation Systems*. Montreal, PQ, Canada, 1995.

Jorgensen, R.: *Fan Engineering*, 8th ed. Buffalo, NY: Buffalo Forge Co., 1983.

Levin, H.: *Indoor Air BULLETIN*, 2548 Empire Grade, Santa Cruz, CA 95060.

National Institute for Occupational Safety and Health and Environmental Protection Agency: *Building Air Quality—A Guide for Building Owners and Facility Managers* (EPA/400/1-91-033). Washington DC: Government Printing Office, 1991.

Talty, J.T. (ed.): General methods of control. *Ind. Hyg. Eng.* pp. 70–77 (1988).

References

1. **Cralley, L.V. and L.J. Cralley (eds.):** *In-Plant Practices for Job Related Health Hazards Control*, vol. 2. New York: Wiley Interscience, 1989.
2. **Burton, D.J.:** *Industrial Ventilation Workbook*. Salt Lake City, UT: IVE, Inc., 1997.
3. **American Conference of Governmental Industrial Hygienists**

(ACGIH): *Industrial Ventilation—A Manual of Recommended Practice*, 22nd ed. Cincinnati, OH: ACGIH, 1994.

4. **American National Standards Institute (ANSI):** *Exhaust Systems—Open-Surface Tanks—Ventilation and Operation* (ANSI Z9.1–1991). New York: ANSI, 1991.
5. **American National Standards Institute (ANSI):** *Fundamentals Governing the Design and Operation of Local Exhaust Systems* (ANSI Z9.2–1979 [R1991]). New York: ANSI, 1979.
6. **American National Standards Institute:** *Spray Finishing Operations—Safety Code for Design, Construction, and Ventilation* (ANSI/AIHA Z9.3–1994). Fairfax, VA: American Industrial Hygiene Association, 1994.
7. **American National Standards Institute (ANSI):** *Ventilation and Safe Practices of Abrasive Blasting Operations* (ANSI Z9.4–1985). New York: ANSI, 1985.
8. **American National Standards Institute:** *Laboratory Ventilation* (ANSI/AIHA Z9.5–1992). Fairfax, VA: American Industrial Hygiene Association, 1992.
9. **Plog, B.A. (ed.):** *Fundamentals of Industrial Hygiene*, 4th ed. Chicago, IL: National Safety Council, 1996. Chapter 18.

During metal working processes, chrome and nickel plating produce airborne hexavalent chrome and nickel. The airborne carcinogens can be inhaled and retained in the body.

Outcome Competencies

After completing this chapter, the user should be able to:

1. Define underlined terms used in this chapter.
2. Describe collection hoods for dusts, gases, and vapors.
3. Review basic principles of ventilation design.
4. Calculate airflow requirements for specific process hood applications.
5. Calculate fan static pressure requirements for a simple local exhaust system.
6. Review the parameters for determining the most appropriate fan size.

Key Terms

air cleaner • baffle • breathing zone • capture velocity (V_c) • compound hoods • continuity equation • ducts • enclosing hood • exterior hood • fan • flange • hood face velocity (V_f) • LEV energy requirements • local exhaust ventilation (LEV) • minimum duct transport velocity • plenum • process hood • receiving hood • slot hood • stack • static pressure (SP) • static pressure losses • total pressure (TP) • velocity pressure (VP)

Prerequisite Knowledge

Prior to beginning this chapter, the user should review the following references for general content:

Subject	Resource
College Physics	Serway, Prentice Hall

Prior to beginning this chapter, the user should review the following chapters:

Chapter Number	Chapter Topic
6	Principles of Evaluating Worker Exposure
31	General Methods for the Control of Airborne Hazards
33	Evaluating Ventilation Systems

Key Topics

I. Mechanical Ventilation for Contaminant Control
  A. Exposure Control Methods
  B. Choosing LEV for Contaminant Control

II. LEV Design Guidance
  A. Basic Components of an LEV System
  B. Local Exhaust Hoods
  C. Hood Selection
  D. Control Velocity for Enclosing Hoods
  E. Airflow Requirements for Enclosing Hoods
  F. Control Velocity for Exterior Hoods
  G. Airflow Requirements for Exterior Hoods
  H. Exterior Hood Design Summary and Examples
  I. Effect of Flanging an Exterior Hood
  J. Using Published Hood Design Criteria
  K. Design of Slot Hoods, Plenums, and Baffles
  L. Summary of LEV Hood Design

III. LEV System Energy Requirements

IV. Pressure Relationships in an LEV System
  A. Measurement of LEV System Pressures
  B. Quantifying the Energy Lost at the Hood

V. Transporting Contaminant Through Ductwork

VI. Quantifying Energy Losses in Ductwork

VII. Fan Static Pressure Requirements

VIII. A Bookkeeping Method

IX. Trial and Error Approach to Ventilation Design

32

An Introduction to the Design of Local Exhaust Ventilation Systems

Dennis K. George, Ph.D., CIH

Salvatore R. DiNardi, Ph.D., CIH

Introduction

Many professionals enter the field of occupational hygiene with previous experience in the field of mechanical engineering or a similar engineering-related discipline. Such individuals have a distinct advantage in the design of mechanical control systems such as local exhaust ventilation (LEV), frequently finding the LEV design process to be a natural extension of general ventilation or heating, ventilation, and air-conditioning (HVAC) system design. However, after several years of teaching LEV design in both the academic and continuing education environments, the authors have observed that a number of occupational hygienists lack a strong engineering background. These individuals often become frustrated and perhaps even a bit intimidated when asked to make comments or render decisions on LEV systems. It is to this latter group of professionals that this introductory chapter is addressed.

A basic knowledge of the fundamentals of LEV design is an important tool for the occupational hygienist. Although an occupational hygienist might never be called on to design a system from scratch, most will occasionally find it necessary to recommend the installation of LEV systems as exposure control measures; consult with mechanical engineers regarding LEV requirements; review drawings and specifications for proposed LEV systems; periodically test, troubleshoot, and evaluate the performance of LEV systems; and explain why existing LEV systems are ineffective at controlling employee exposures to potentially hazardous airborne contaminants. For these reasons, it is worthwhile to spend the time necessary to become familiar with at least the basic principles of LEV design.

The authors acknowledge the virtual impossibility of thoroughly presenting the many intricacies of the LEV design process in a relatively brief book chapter. Hopefully, this chapter will serve as a starting point for those not familiar with local exhaust design or those wishing to review the fundamental concepts. The reader must be aware that in the interest of space, thorough treatment of many very important topics relating to LEV will not be possible. References will be made to other more detailed sources where appropriate.

At the outset of this chapter, a comment on units is in order. Because calculations and units are used extensively throughout this chapter, the authors have determined to use a single consistent set of units in lieu of presenting both English (IP) and Le Systeme International d'Unites (SI) units. The authors hope that this will make for a cleaner, more readable chapter. To be consistent with most of the more widely used references in the field, the English system of units has been selected. Please note, however, that in the authors' opinion, the field of industrial ventilation design should seek to move to the SI system as a goal for the future.

Mechanical Ventilation For Contaminant Control

LEV is a primary means of controlling employee exposure to gases, vapors, and particles in traditional workplaces such as foundries, machine shops, assembly lines, and refineries. These exhaust systems are termed "local" in the sense that the source of exhaust or suction is located adjacent to the source of contamination. If properly designed, such an arrangement removes a contaminant directly from its source before it has an opportunity to disperse into the general workplace atmosphere, where it could subsequently be inhaled by a worker. Capturing and removing a contaminant at its source is the principle objective of LEV systems.

Please note that local exhaust is not the only strategy for reducing employee exposures that uses mechanical ventilation. For example, general or dilution ventilation systems rely on the supply and exhaust of air with respect to an area, room, or building rather than on a localized exhaust source to control airborne contaminant concentrations.(1) However, controlling employee exposure via LEV is quite distinct from the use of dilution ventilation for exposure control. A dilution ventilation control strategy actually allows contaminants to be released into general workplace air under the assumption that they will be immediately and completely mixed with a predetermined quantity of uncontaminated air sufficient to dilute contaminant concentration to some acceptable level.

Most simple dilution ventilation models are derived from a fundamental material balance equation in which the rate of contaminant accumulation is a function of the difference between the rate at which the contaminant is being generated and the rate at which it is being removed (i.e., the ventilation rate). The target concentration is determined *a priori* and usually depends on whether the control objective relates to controlling potential airborne health hazards, fire and explosive conditions, odors, or nuisance type contaminants.(1) Generally speaking, if the primary objective is contaminant control and worker protection, LEV is more effective than dilution ventilation.

Local exhaust systems also differ significantly from HVAC systems. The latter systems provide thermal comfort and conditioned air to residential and commercial spaces and to nontraditional workplaces such as offices, hospitals, shopping malls, and schools. HVAC systems also serve to supply the fresh air needed to control the myriad air pollutants such as human respiratory byproducts (carbon dioxide), body odor, and other common pollutants (volatile organic compounds, ozone, particles, etc.) found in the indoor environment. LEV system design, while theoretically similar, differs from HVAC system design in component selection, specific calculation processes, and other practical details.

When applied to local exhaust systems, HVAC system design methods may yield energy inefficient systems that do not provide adequate employee protection in traditional workplaces. Although some contaminant dilution may occur, the quantity of clean air delivered via the typical HVAC system is only a fraction of that needed in industrial applications for adequate reduction of contaminant concentration. Also, most HVAC systems recirculate the tempered or conditioned air with a minimal addition of fresh outside air. Since these systems are designed primarily to provide thermal comfort, provisions for contaminant removal from this recirculated air are generally absent.

The major goal of this chapter is to provide an introduction to energy efficient LEV systems that protect employees from adverse exposures to workplace hazards (gases, vapors, and particles) that are the result of manufacturing, production, or maintenance operations. This material is intended for use by occupational hygienists with a basic science background and will build on techniques commonly used in occupational hygiene. Application of dilution ventilation models is discussed in Chapter 16 in the context of modeling inhalation exposure. HVAC systems and their role in the indoor environmental quality of nontraditional workplaces, a topic of enormous interest and continuing concern, is covered in Chapter 19.

Exposure Control Methods

Before delving into the LEV design process, it may be useful to briefly review the overall concept of occupational hygiene controls. The fundamental definition of occupational hygiene is the anticipation, recognition, evaluation, and control of hazards arising in and of the workplace. These hazards may be of a chemical, physical, biological, or ergonomic nature. Other chapters in this book describe these factors in great detail. The major exposure routes for workplace chemical hazards are inhalation, ingestion, and dermal absorption, with inhalation being the principal route for those airborne chemical hazards (e.g., gases, vapors, dusts, fumes, mists, and bioaerosols).

Protecting an employee from inhalation exposure is achieved by protecting the worker's breathing zone. In general terms the breathing zone is defined as the volume surrounding a worker's nose and mouth from which he or she draws breathing air over the course of a work period. This zone can be pictured by inscribing a sphere with a radius of about 10 inches centered at the worker's nose. Exposures occur when there is contact at the air boundary layer between the worker's breathing zone and airborne contaminants in the workplace environment. If this contact is prevented or minimized by proper control methods, significant exposures will not occur. Such is the goal of occupational hygiene controls for airborne chemical and biological agents.

Although this chapter focuses on LEV as a means to control chemical exposure, it should be emphasized that LEV is but one of many control methodologies available to the knowledgeable occupational hygienist. In fact, LEV is frequently not the most desirable option. Basic techniques available to control or minimize a worker's adverse exposure include prevention, engineering controls, administrative controls, and personal protection (see Figure 31.1 from Chapter 31). Occupational Safety and Health Administration (OSHA) regulations and occupational hygiene textbooks discuss these topics in great detail (see also Chapter 31).

Before applying LEV a skilled occupational hygienist must review the process and consider whether exposure to the hazardous material can be prevented. Prevention is accomplished through techniques such as elimination, substitution, isolation, or process change. If prevention cannot be achieved, other engineering control techniques such as LEV may be implemented to secure control of inhalation exposures. However, one must always bear in mind that local exhaust systems cannot be relied on completely to protect the workers without fully considering other exposure control methods.

An occupational hygienist should always remember that successful engineering controls are passive or transparent to the worker, requiring minimal interaction between the worker and the system. User friendly systems are essential if they are to provide proactive control, thereby protecting a worker's breathing zone from adverse exposure. The more thought or activity that a worker must invest in the implementation of a control, the less likely that he or she will take the trouble to use it.

Choosing LEV for Contaminant Control

When determining whether LEV is the best option for a particular application, several aspects of the contaminant-generating process must be considered. Various authors have previously identified factors or conditions that may indicate the need for local exhaust as a means for contaminant control. For example, Burton states that a local exhaust system is generally appropriate when employees are potentially overexposed to airborne chemical contaminants and when one or more of the following conditions is present:[2,3]

- More cost-effective controls are not available or feasible;
- The contaminant of concern is relatively toxic (e.g., contaminant has a low occupational exposure limit [OEL]);
- The employee is in the immediate vicinity of the emission (note that dilution ventilation is inappropriate in this situation since adequate dilution may not occur before the contaminant enters the breathing zone);
- Contaminant emission rate varies with time;
- Sources of contaminant emission are large and few as opposed to small, many, and dispersed (in the latter case, dilution may be a more feasible option);
- Contaminant emission sources are fixed as opposed to mobile; and
- LEV systems are required or suggested by federal, state, local, building, or consensus standards.

Consult References 2 and 3 for additional details on the general applicability of LEV as a means of industrial contaminant control.

LEV Design Guidance

If a LEV system is selected as the most appropriate control option, the professional called on to specify the system must prepare and design the most effective and energy efficient system possible for the given operation. Fortunately, many excellent references are available to guide the interested reader through a complete discussion of LEV design.[1–8] The most widely used and often cited resource is *Industrial Ventilation: A Manual of Recommended Practice* developed by the Committee on Industrial Ventilation of the American Conference of Governmental Industrial Hygienists (ACGIH). Regardless of any additional resources that might be available to the budding industrial ventilation designer, use of the *Industrial Ventilation Manual* could almost be classified as a must. This reference combines many years of collective professional experience beginning with that of pioneers in the field and continuing through that of present day researchers. The depth and breadth of the information collected in this manual in the form of charts, tables, figures, design drawings, equations, etc., is essentially unavailable from any other single source.

Basic Components of an LEV System

As stated, the basic design objective for LEV systems is to control employee exposures to gases, vapors, and particles by capturing and removing these contaminants at the source of generation before they reach the breathing zone. When this objective is satisfied and concentrations of contaminants in the breathing zone are kept below applicable OELs, the major goal of protecting the employee from adverse chemical exposure is achieved.

Figure 32.1 pictures a typical LEV system. Each component makes a unique contribution toward meeting the primary design objective and must be specifically selected, designed, and integrated into the system.

The terms used in Figure 32.1 to describe LEV systems may be new to some readers. They are briefly defined as follows and will be more fully explained in subsequent sections.

A process hood is an entry into a duct from the workroom or process. It is essentially a device designed to capture, enclose, or receive contaminants from a process. As discussed in subsequent sections of this chapter, energy must be supplied to the system to induce the air to move into the hood and to overcome the turbulence created at the air entry. A well-designed hood is "transparent" to the worker; otherwise the worker will remove the hood or bypass it.

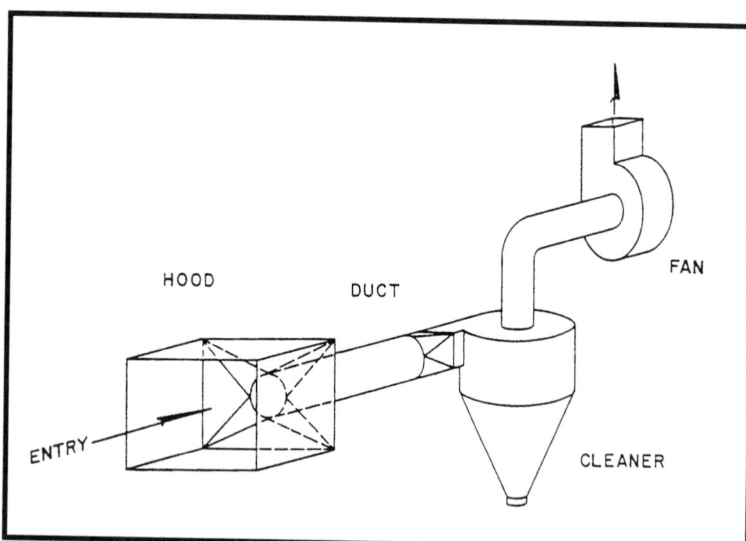

Figure 32.1 — Typical local exhaust system components.

Ducts carry contaminants from the hood, through the system, out of the workroom, through an air-cleaning device, and into the ambient environment. The air velocity in the duct must be sufficient to convey the contaminant and prevent the settling of particle contaminants. For air to move through a duct, sufficient energy must be provided to overcome the friction between the constituents of air and the duct's surface. Air moving through a duct does not change direction unless an elbow or branch entry intervenes. Additional energy must be added to overcome the friction loss created by these occurrences. Consequently, energy efficient designs require smooth, hard, round ducts with a minimum number of fittings (elbows and branches).

An air cleaner is a device used for removing a chemical hazard from an airstream before discharge into the ambient air. The design and selection of air cleaners are beyond the scope of this chapter. Consider them a black box requiring energy to operate at a specific air volume. Chapter 4 of the *Industrial Ventilation Manual* is dedicated to the selection and design of air cleaning devices.

The fan provides the energy required by a specific design to move air through the system. The selection of fans is based on specific calculations introduced in this chapter and detailed in Chapter 6 of the *Industrial Ventilation Manual*. Use only Air Movement and Control Association (AMCA)-certified fans for industrial exhaust ventilation systems. There is no assurance that an uncertified fan will perform at the manufacturer's stated static pressure and air volume.

Finally, the stack (not pictured in Figure 32.1) is a device used to discharge air into the ambient environment and away from the building wake. Location of exhaust stacks relative to fresh air intakes is a critical issue. Stack design and location errors may result in re-entrainment of contaminated local exhaust emissions into the building's fresh air supply system, possibly causing odor complaints and unexpected exposures to hazardous materials.

Each component must be designed to specific, clearly stated criteria. These criteria must uniquely fit the emissions for each process to meet the goal of protecting the worker from adverse exposure. Careful selection of the various components of the system has the added benefit of ensuring energy conservation.

Local Exhaust Hoods

The process of LEV design begins with decisions made regarding the hood. It is probably safe to say that the hood is the most important part of the entire LEV system. Proper selection and design of the hood will go a long way toward achieving the goal of capturing and removing the contaminant at its source before it has a chance to reach the worker's breathing zone. In fact, this is precisely what hoods are designed to do: "capture and remove process emissions prior to their escape into the workplace environment."[1] A hood accomplishes this mission by creating an airflow at the point of contaminant generation that is sufficient to overcome any competing air motion and induce the movement of contaminant into the system.[1] As will be seen, the pattern of airflow created in front of or within the hood and the ability of this airflow to capture and/or contain process emissions are determined by the following factors:

- The hood's configuration or shape;
- The extent to which the hood encloses the contaminant source; and
- The quantity of air flowing into the hood.

Simply stated, whenever an airborne contaminant is being generated by some process in the workplace, it is desirable to have it removed and transported away from the worker's breathing zone via an LEV system. The LEV hood is selected and designed to generate the airflow pattern necessary to accomplish this goal.

As formally defined, "the local exhaust hood is the point of entry into the exhaust system and is defined to include all suction openings regardless of their physical configuration."[1] Indeed, there are hoods in many different sizes, shapes, and geometries. The type of hood selected will, of course, depend on the process and nature of the emissions the hood is expected to control. For this discussion it will be useful to follow the typical convention of dividing hoods into two broad categories: enclosing hoods and exterior hoods.

An enclosing hood is a hood that "completely or partially encloses the process or contaminant generation point."[1] It essentially surrounds the contaminant source, thereby isolating the ventilation process from the worker and the workspace. Thus, when the contaminant is emitted from the process, it is already either totally or at least partially inside the hood. The contaminant is contained inside the enclosure by an inward flow of air through the hood opening(s) and is thereby prohibited from escaping into the workroom air. Designs for enclosing hoods range from those with minimal openings that provide essentially complete enclosure (e.g., Figures 32.2 or 32.3) to those with one side of the hood either partially or completely open (e.g., Figure 32.4).

An exterior hood is defined as a hood that is "located adjacent to an emission source without enclosing it."[1] The movement of air flowing into an exterior hood is designed to reach out and capture contaminants in the workroom at the source of contaminant production and induce them to flow into the hood along with the moving air. In contrast to the enclosing hood the exterior hood does not surround the contaminant source and thus the contaminant is released outside rather than inside the hood. Hoods used to remove contaminants emitted from open surface tanks (e.g., Figure 32.5) are common examples of exterior hoods.

A subcategory of the exterior hood is the receiving hood (i.e., a hood designed and positioned to take advantage of any initial velocity or motion imparted to the contaminant by the generating process). Examples include grinding wheel hoods that take advantage of the momentum of particles thrown from the grinding wheel/workpiece and the canopy hood, which is positioned to benefit from the upward movement of contaminants emitted by hot processes.

Hood Selection

The cardinal rule in LEV hood design is to select or design a hood that encloses the contaminant-generating process as much as possible. Control is greatly simplified when the contaminant is released inside an enclosure rather than into the general workplace atmosphere. Exterior hoods have an inherent disadvantage in that they must overcome any external influence that may be competing against the hood's ability to entrain the contaminant.

The most notable of these external influences is the cross draft, "extraneous air movements which can disrupt the airflow patterns between the point of contaminant release and the exhaust hood."[4] The *Industrial Ventilation Manual* identifies several sources of this external air motion, including thermal air currents, motion of machinery, material motion, operator movement, room air currents (nominally

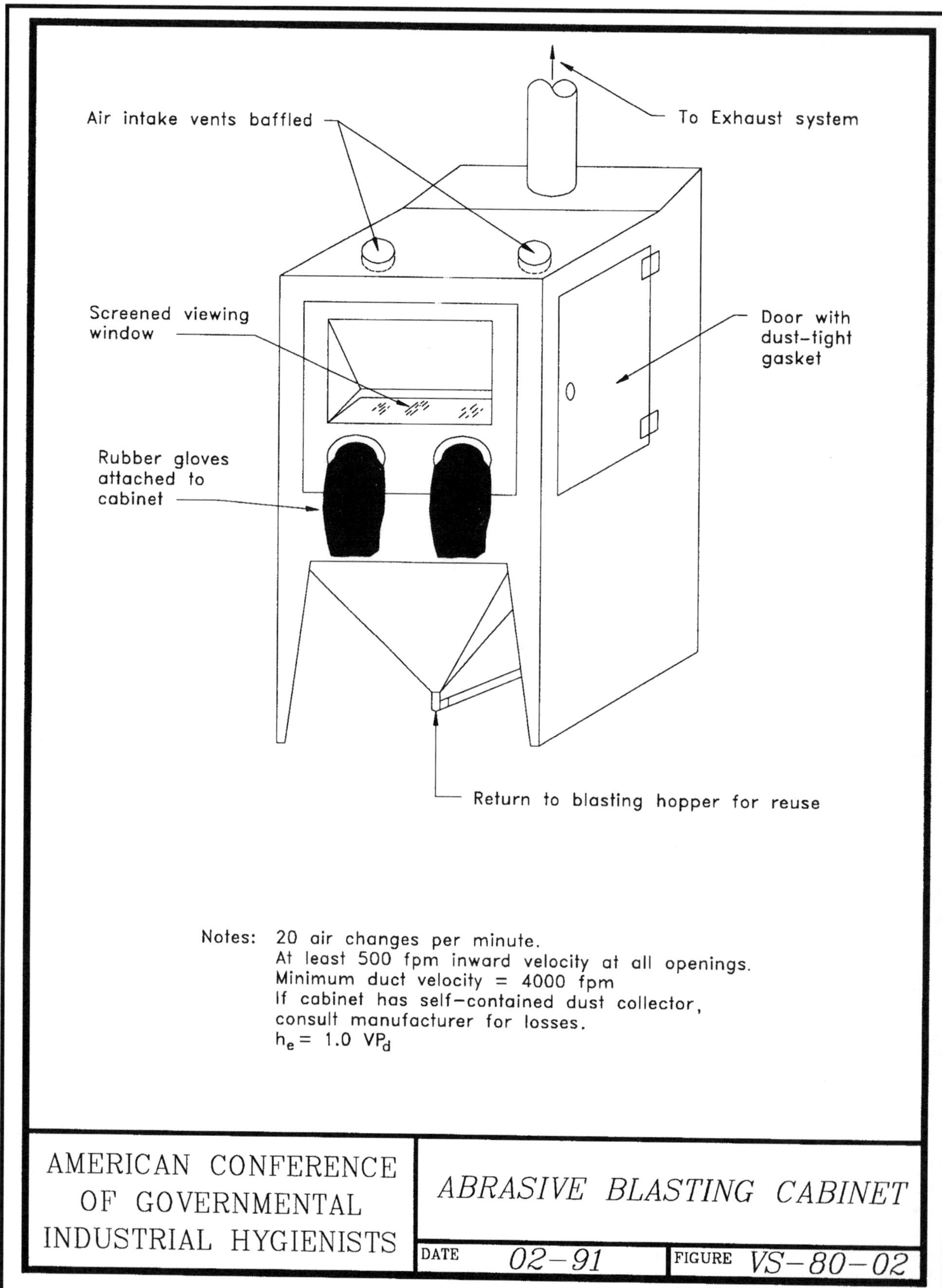

Figure 32.2 — Abrasive blasting hood. (From **American Conference of Governmental Industrial Hygienists (ACGIH®):** *Industrial Ventilation: A Manual of Recommended Practice*, 22nd Ed. Copyright 1995, Cincinnati, Ohio. Reprinted with permission.)

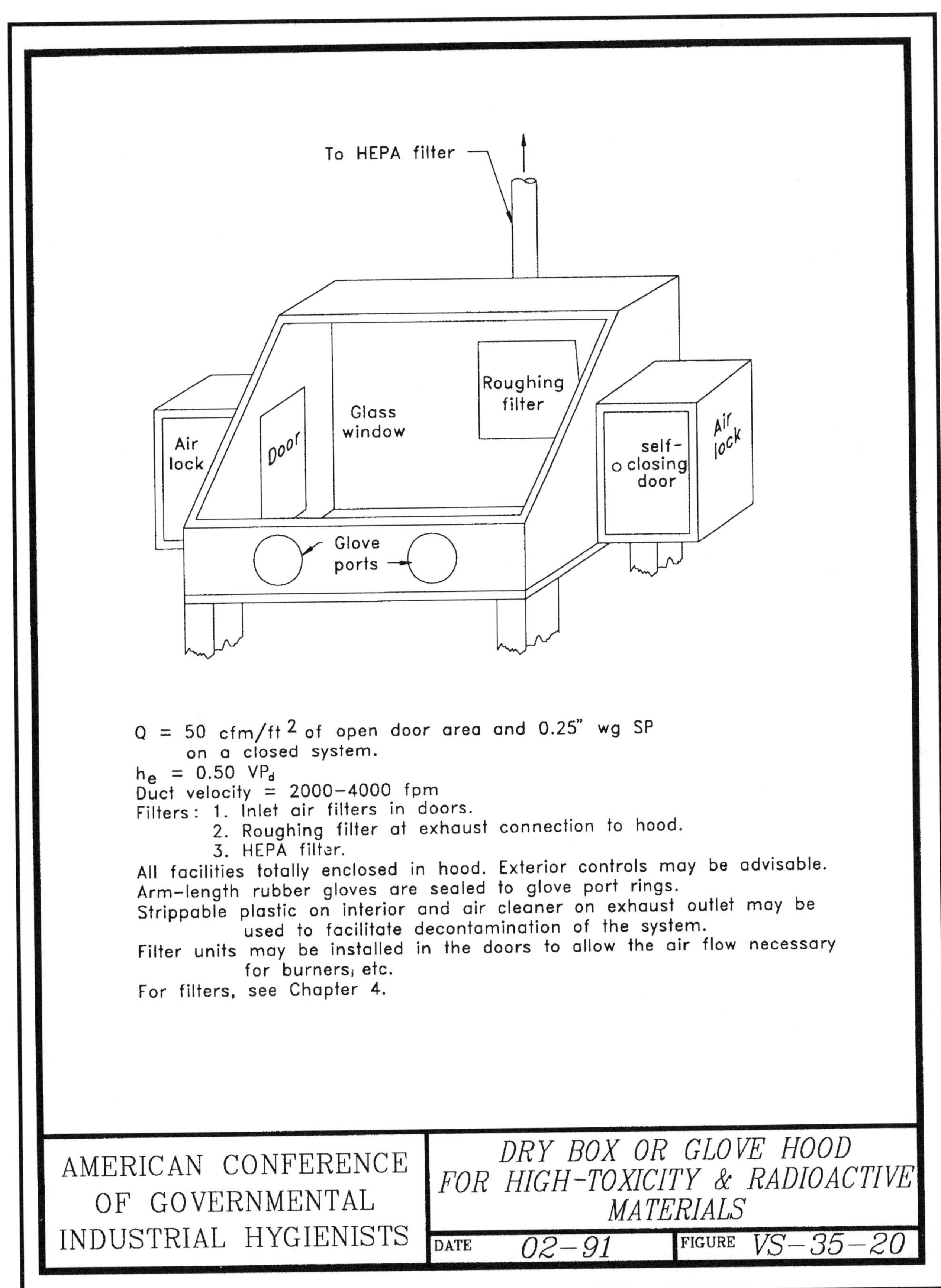

To HEPA filter
Roughing filter
Glass window
Air lock
Door
self-closing door
Air lock
Glove ports
Q = 50 cfm/ft 2 of open door area and 0.25" wg SP on a closed system.
he = 0.50 VPd
Duct velocity = 2000–4000 fpm
Filters: 1. Inlet air filters in doors.
2. Roughing filter at exhaust connection to hood.
3. HEPA filter.
All facilities totally enclosed in hood. Exterior controls may be advisable.
Arm-length rubber gloves are sealed to glove port rings.
Strippable plastic on interior and air cleaner on exhaust outlet may be used to facilitate decontamination of the system.
Filter units may be installed in the doors to allow the air flow necessary for burners, etc.
For filters, see Chapter 4.
AMERICAN CONFERENCE OF GOVERNMENTAL INDUSTRIAL HYGIENISTS
DRY BOX OR GLOVE HOOD FOR HIGH-TOXICITY & RADIOACTIVE MATERIALS
DATE 02-91
FIGURE VS-35-20

Figure 32.3 — Dry box/glove hood. (From **American Conference of Governmental Industrial Hygienists (ACGIH®):** *Industrial Ventilation: A Manual of Recommended Practice*, 22nd Ed. Copyright 1995, Cincinnati, Ohio. Reprinted with permission.)

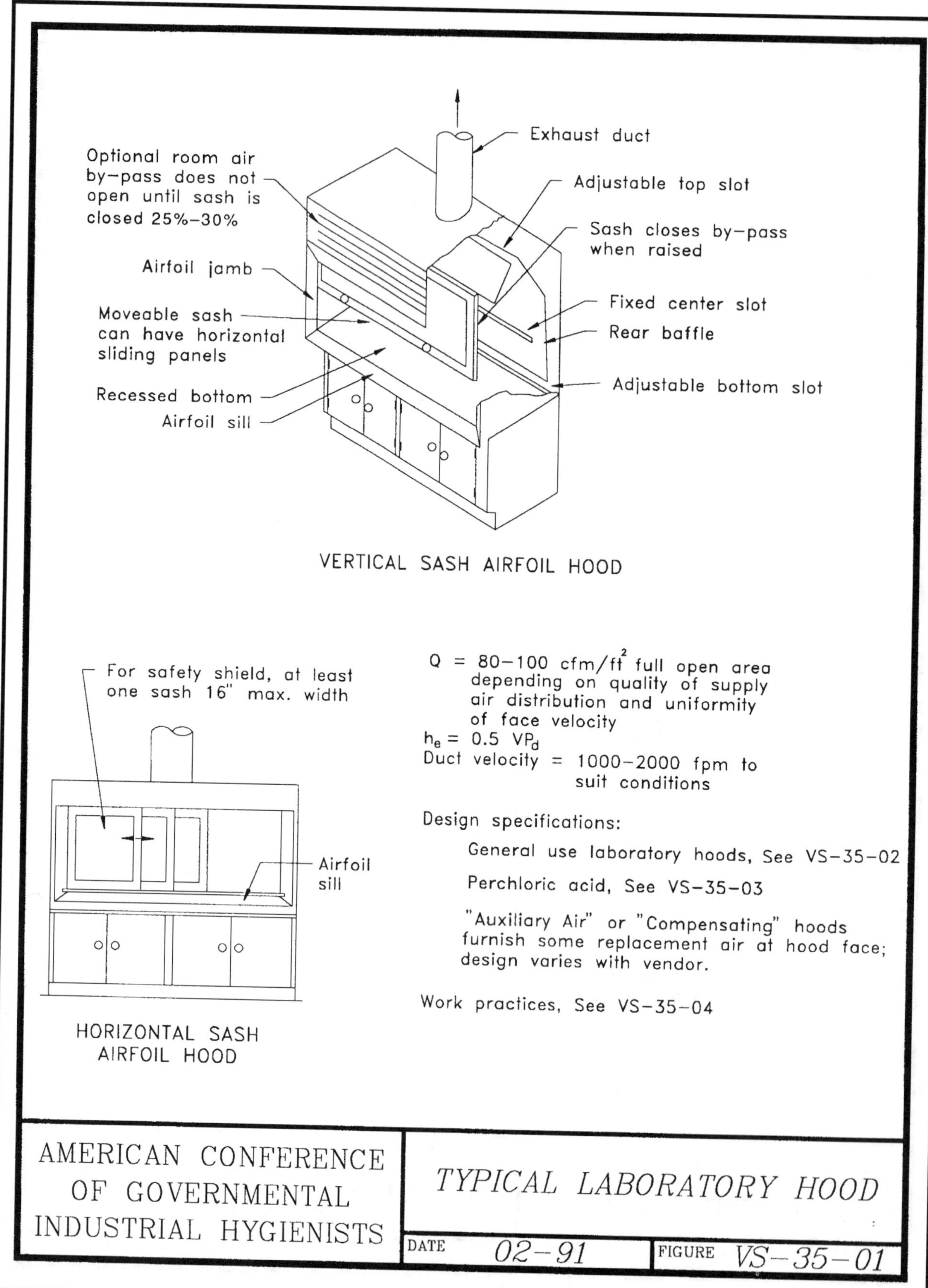

Figure 32.4 — Laboratory hood. (From **American Conference of Governmental Industrial Hygienists (ACGIH®):** *Industrial Ventilation: A Manual of Recommended Practice*, 22nd Ed. Copyright 1995, Cincinnati, Ohio. Reprinted with permission.)

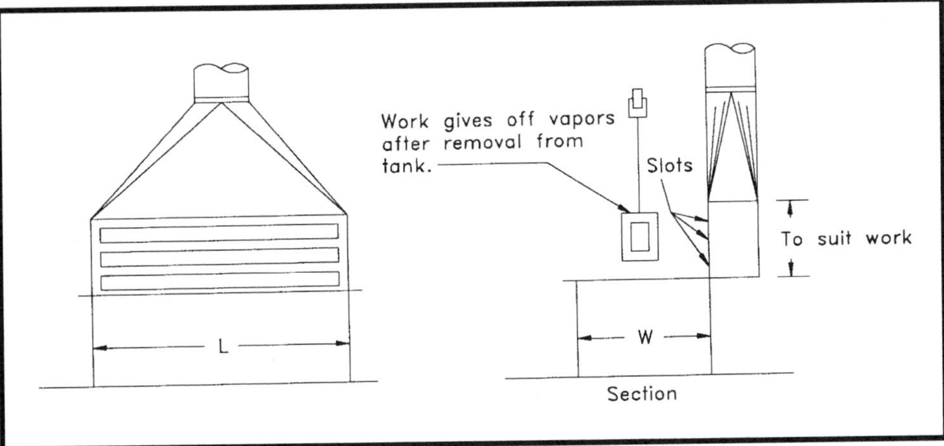

Figure 32.5 — Open surface tank hood. (From **American Conference of Governmental Industrial Hygienists (ACGIH®):** *Industrial Ventilation: A Manual of Recommended Practice*, 22nd Ed. Copyright 1995, Cincinnati, Ohio. Reprinted with permission.)

assumed to be 50 ft/min), and rapid air movement caused by spot cooling and heating equipment.[1] The air velocity generated by the exterior hood at the point of contaminant generation must be sufficient to overcome these conditions if the hood is to entrain the contaminant. The contaminant control objective may be defeated if cross drafts disrupt hood performance to the extent that a significant portion of the contaminant escapes the influence of the hood and remains in the workplace air. Burgess et al. provide an extensive discussion of the effects of cross drafts on hood performance and how they are handled in the design of exterior hoods.[4]

Another potential problem in the use of exterior hoods is the contamination of a worker's breathing zone with process emissions before they are drawn into the hood. Obviously, it is little consolation that a hood captures and removes 100% of a process emission if the contaminant is pulled through the worker's breathing zone as it makes its way from the process to the hood. This is a particularly troublesome occurrence for the canopy hood (see Figure 32.6), a design that typically allows easy worker access to the area between the process emission and the hood face.

For these reasons, it is desirable to enclose the contaminant emission point with the hood to the maximum extent feasible as illustrated in Figure 32.7. In fact, when conceiving a hood configuration, most experienced LEV designers recommend visualizing the process as totally enclosed by the hood and then removing only as much as is necessary for adequate access to the process.[3] Since the contaminant will be emitted inside the hood, this "enclose to the maximum" approach will minimize the disruptive effect of cross drafts, reduce the potential for contaminants to be drawn through the breathing zone, and will in general provide much more effective contaminant control. Exterior

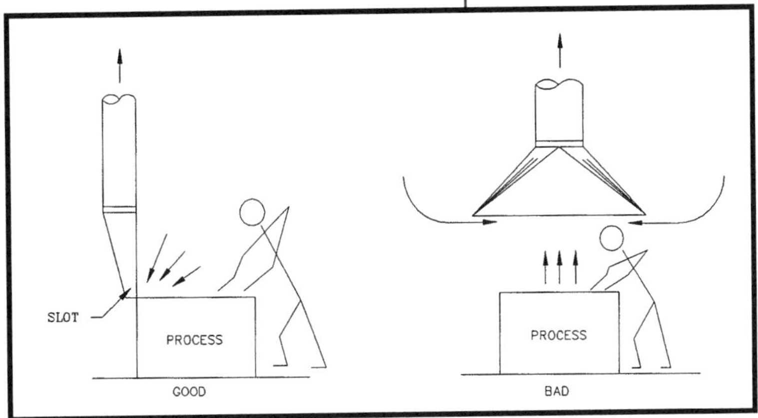

Figure 32.6 — Canopy hood. (From **American Conference of Governmental Industrial Hygienists (ACGIH®):** *Industrial Ventilation: A Manual of Recommended Practice*, 22nd Ed. Copyright 1995, Cincinnati, Ohio. Reprinted with permission.)

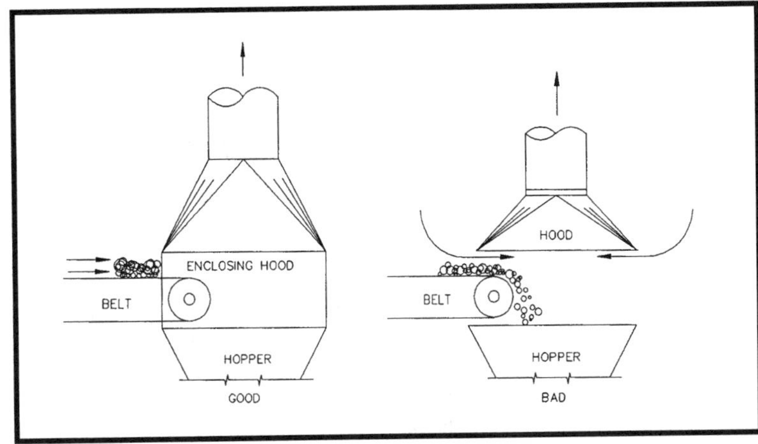

Figure 32.7 — Use of enclosure hood. (From **American Conference of Governmental Industrial Hygienists (ACGIH®):** *Industrial Ventilation: A Manual of Recommended Practice*, 22nd Ed. Copyright 1995, Cincinnati, Ohio. Reprinted with permission.)

Table 32.1 — Minimum Maintained Velocities into Spray Booths

Operating conditions for objects completely inside booth	Cross draft, fpm	Airflow velocities, fpm Design	Range
Electrostatic and automatic airless operation contained In booth without operator	Negligible	50 large booth	50–75
		100 small booth	75–125
Air operated guns, manual or automatic	Up to 50	100 large booth	75–125
		150 small booth	125–175
Air operated guns, manual or automatic	Up to 100	150 large booth	125–175
		200 small booth	150–250

Notes:
(1) Attention is invited to the fact that the effectiveness of the spray booth depends on the relationship of the depth of the booth to its height and width.
(2) Cross drafts can be eliminated through proper design and such design should be sought. Cross drafts in excess of 100 fpm (feet per minute) should not be permitted.
(3) Excessive air pressures result in loss of both efficiency and material waste in addition to creating a backlash that may carry overspray and fumes into adjacent work areas.
(4) Booths should be designed with velocities shown in the column headed "Design." However, booths operating with velocities shown in the column headed "Range" are in compliance with this standard.

hoods should be selected only if dictated by the configuration of the process.

Control Velocity for Enclosing Hoods

If the process allows an enclosing hood design (i.e., if the designer can completely or partially surround the source such that the contaminant is generated inside the hood) contaminant control can be achieved by the inward flow of air through the hood openings. When this inward flow of air is sufficient, contaminants cannot escape. By definition, the air velocity at the hood opening is called the hood face velocity (V_f). Thus, if source enclosure is possible in a given application, the first question addressed by the LEV designer is simply what face velocity is required for a given enclosing hood to ensure adequate contaminant control.

The answer to this question depends to a great extent on the nature of the process being ventilated. Although a thorough discussion is beyond the scope of this chapter, the basic information considered by a designer when specifying face velocity for an enclosing hood can be illustrated with an example of a commonly encountered enclosing hood — the spray paint booth. Table 32.1 appears as Table G-10 in the OSHA General Industry Standards (Title 29 *Code of Federal Regulations*, Part 1910.94).

Note from Table 32.1 that the face velocity required for a paint booth may vary between 50 and 250 ft/min, depending on such factors as the size of the booth, the magnitude of cross drafts, and the nature of the spray-painting process occurring inside the booth (e.g., electrostatic, airless, air-operated, etc.). Each of these parameters must be considered before specifying a face velocity.

In addition to the air velocity, another important consideration for a booth-type enclosure is the necessity for an even distribution of air across the face of the booth. In other words, the velocity of air at all points across the hood face should be approximately equal. Even though the average face velocity may meet some minimal specification, hood performance might be inadequate if most of the air flows through the center of the booth face (the path of least resistance) with little air moving, and consequently low face velocities, near the edges of the booth face.[4] Various sources discuss design practices that can be initiated to ensure the even distribution of exhaust air.[1,4] These steps include making the booth deeper, using a baffle, and/or using filters or other air-cleaning devices to distribute the incoming air more evenly.

The designer should also note that "the higher the better" is not necessarily a valid maxim in regard to face velocity, particularly in applications where the inward flow of air may be disturbed by a person working at the hood. This occurrence may cause air turbulence or reverse flow, which can actually pull contaminant out of the enclosure. This is a particular concern with another commonly encountered enclosing hood, the laboratory fume hood, where face velocities of greater than 150 ft/min have created serious problems in this regard.[4] In the *Industrial Ventilation Manual* there are several pages of design guidance relating to laboratory hoods and recommendations for a face velocity range of 60 to 100 ft/min depending on a variety of factors such as room air currents (as affected by traffic past the hood and the terminal throw velocity of supply air jets) and the location of the equipment in the hood relative to the hood face. Other factors cited as affecting lab hood performance include hood aerodynamic entry characteristics, sash design position(s), and the degree to which the velocity is uniform across the hood face.

Consideration of the spray paint booth and the laboratory hood illustrates some of the many factors that come into play when determining the face velocity required to contain emissions within an enclosing hood. Specification of face velocity is so dependent on various process-related variables that rules of thumb are difficult to establish. In general, Burgess et al.[4] write that a face velocity of 100 to 200 ft/min is generally sufficient for most applications. However, these investigators also warn that, because of the possibility of contaminant carry-out from turbulence at the worker/airflow interface, excessive face velocity can cause just as many exposure problems as insufficient face velocity.

For more specific direction on hood design and face velocity specifications for the more common industrial processes, the reader is encouraged to consult Chapter 10 of the *Industrial Ventilation Manual*. This chapter provides a number of enclosing hood illustrations for several of the more typical industrial operations. These prints contain hood design guidance, including a range of face velocity recommendations "taken from designs used in actual installations of successful local exhaust ventilation systems."[1]

Even with this information, however, professional judgment is required since the hood face velocity requirement for a given application will vary depending on the presence of cross drafts and the toxicity of materials being emitted. Disruptive cross drafts and/or the handling of materials of high toxicity will typically require a face velocity toward the higher end of the specified range.

Airflow Requirements for Enclosing Hoods

As stated, the first step in designing an LEV system with an enclosing hood is the determination of the velocity in ft/min through the hood face (i.e., V_f) that is adequate to contain emissions for the particular application at hand. However, to proceed with the design of the system, the designer must understand that face velocity is not a typical LEV design parameter. That is, LEV systems are not specified nor are fans selected on the basis of providing a certain linear velocity at the hood openings. At this point the designer must convert face velocity in ft/min to a set of units that can be used in specifying a system and selecting a fan. Rather than linear velocity, airflow requirements in LEV systems are expressed in units of volumetric flow rate, or the volume of air flowing past a given location during some specified time interval. Volumetric flow rate is typically expressed as cubic feet per minute (ft^3/min or cfm) or, in the SI system, cubic meters per second (m^3/sec or cms).

Conceptually, volumetric flow rate is a product of the velocity of air and the cross-sectional area through which the air flows. This relationship, sometimes referred to as the continuity equation, is one of the most fundamental in LEV design. Mathematically, it can be expressed as:

$$Q = VA \qquad (1)$$

where:

Q = the volumetric flow rate of air (ft^3/min or cfm);
V = the velocity of air through area, A (ft/min or fpm); and
A = the cross-sectional area through which the air flows (ft^2).

In the case of enclosing hoods, the velocity of air, V, is the specified face velocity, V_f, and the area, A, is the total cross-sectional area of the hood openings. Calculation of the required volumetric flow rate necessary to generate a specific face velocity through a hood opening can be illustrated with an example.

Example 1. Consider the enclosure hood design for a swing grinder in Figure 32.8. If the face opening for this swing grinder installation is to be 3 feet high and 5 feet wide, what volumetric flow rate, Q, is required to generate a face velocity of 150 ft/min?

Solution. For this configuration, the manual recommends a face velocity of 150 ft/min for large openings (4 feet to 6 feet wide). Therefore, V_f = 150 ft/min; A = 5 ft × 3 ft = 15 ft^2; $Q = V_fA$; Q = 150 ft/min × 15 ft^2; and Q = 2250 ft^3/min.

So if 2250 ft^3/min of air is moved through the 3- × 5-foot opening to the hood enclosing this swing grinder, the average velocity at the face of that opening will be the desired 150 ft/min. Thus, the design parameter for this hood is a volumetric flow rate, Q, of 2250 ft^3/min.

In addition to minimizing the competition from cross drafts and providing more effective control, one can now begin to appreciate another reason to enclose the process as much as possible. Specifically, minimizing the area of the openings into a hood conserves the volumetric flow rate of air needed to achieve the desired face

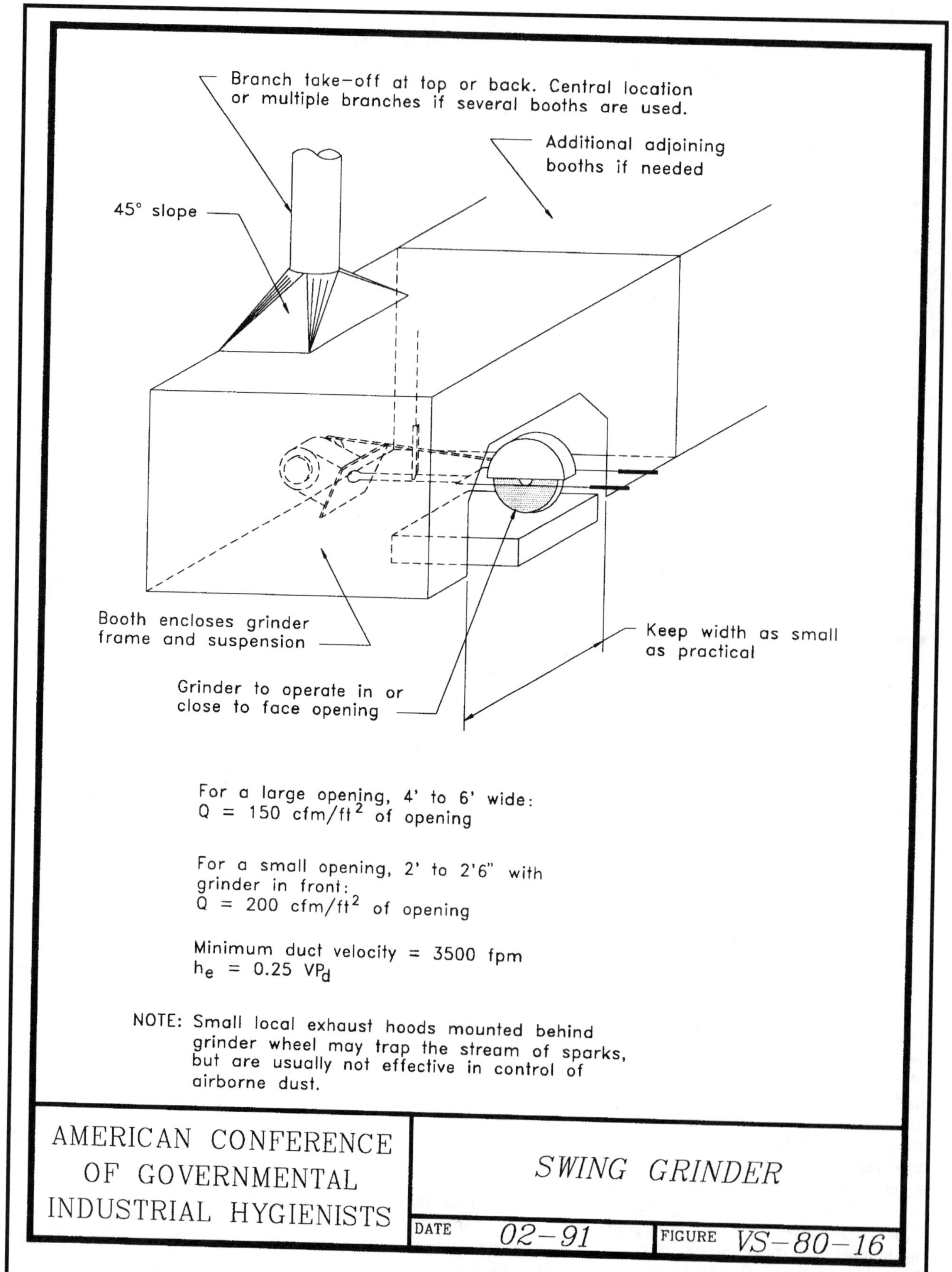
Branch take-off at top or back. Central location or multiple branches if several booths are used.
Additional adjoining booths if needed
45° slope
Booth encloses grinder frame and suspension
Keep width as small as practical
Grinder to operate in or close to face opening
For a large opening, 4' to 6' wide:
Q = 150 cfm/ft² of opening
For a small opening, 2' to 2'6" with grinder in front:
Q = 200 cfm/ft² of opening
Minimum duct velocity = 3500 fpm
he = 0.25 VPd
NOTE: Small local exhaust hoods mounted behind grinder wheel may trap the stream of sparks, but are usually not effective in control of airborne dust.
AMERICAN CONFERENCE OF GOVERNMENTAL INDUSTRIAL HYGIENISTS
SWING GRINDER
DATE 02-91
FIGURE VS-80-16

Figure 32.8 — Swing grinder hood. (From **American Conference of Governmental Industrial Hygienists (ACGIH®):** *Industrial Ventilation: A Manual of Recommended Practice*, 22nd Ed. Copyright 1995, Cincinnati, Ohio. Reprinted with permission.)

velocity. For example, if the opening to the swing grinder in Example 1 had been 4 × 6 feet, the face area would be 24 ft^2, which would require a volumetric flow rate of 150 ft/min × 24 ft^2 = 3600 ft^3/min to achieve the same level of control (i.e., 150 ft/min face velocity at the opening). This is a significant increase in the required flow rate of air. Larger volumetric flow rates translate into higher initial expenditures for fans as well as higher energy costs to operate the system and heat and/or cool the additional makeup air. Designing the hood enclosure as completely as the process will allow is indeed an important practice in LEV hood design for many reasons.

A perusal of the design prints in the *Industrial Ventilation Manual* indicates that most of the face velocity recommendations provided for various applications are given in terms of ft^3/min/ft^2 (i.e., the number of cubic feet per minute of air for every square foot of hood opening). For example, in the swing grinder problem above the manual recommends Q = 150 ft^3/min/ft^2 for large openings. This is another way of using the continuity equation to show that the recommended face velocity for this installation is 150 ft/min (i.e., ft^3/min divided by ft^2 gives ft/min). Specifying the airflow requirements in this manner permits direct calculation of the design parameter, Q, by multiplying the ft^3/min/ft^2 specification by the hood face area.

Some enclosing hood designs have more than one opening. For example, a tunnel is considered to be a hood with an opening at each end. It is important to note that the calculation of Q is based on total hood opening area. Thus, the cross-sectional areas of each opening must be summed to a total face area before multiplying by the desired face velocity to obtain the total volumetric flow rate required.

Control Velocity for Exterior Hoods

Unfortunately, the layout of certain processes and the requirement for convenient access prohibit the selection of an enclosing hood in many industrial situations. In these cases an exterior hood must be used. By definition, an exterior hood is a hood designed to control a contaminant emitted outside the physical confines of the hood. When the process dictates an exterior hood, the designer must be prepared to compete with all of those external factors that work to disperse the contaminant beyond the influence of the hood. To understand the design considerations for exterior hoods it is essential to examine more closely the concepts relating to the capture of contaminants by air flowing into a hood.

Consider a contaminant being emitted into workroom air by some source at a given distance, X, in front of an exterior hood (see Figure 32.9). Obviously, the designer wants this contaminant to flow into the hood and be removed before it escapes into the general workplace atmosphere. As mentioned previously, an important consideration here is that other forces (e.g., cross drafts or motion imparted to the contaminant by the process, etc.) may be acting on the contaminant simultaneously in such a way as to oppose its entrainment by the air flowing into the hood. If the contaminant is to be controlled, the velocity of the air flowing past the contaminant source must be adequate to overcome these opposing influences and to capture or entrain the contaminant at that point with the moving air.[1] This velocity, which is the important issue for exterior hoods, is referred to as the capture velocity (V_c). Thus, the design of an exterior hood begins with the determination of the required capture velocity in ft/min.

The question is how an LEV designer determines the velocity at the point of contaminant generation in such a way as to overcome these opposing forces and cause the contaminant to flow into the hood along with the moving air. Intuitively, one would expect that capture velocity will depend on such factors as the velocity of the opposing air currents and the magnitude and direction of the force imparted to the contaminant by the process. But, how does an exterior hood designer use this information to quantify an adequate capture velocity in ft/min.?

As with the estimates of face velocities for enclosing hoods, the answer to this question is essentially empirical/experience-based and relates primarily to the

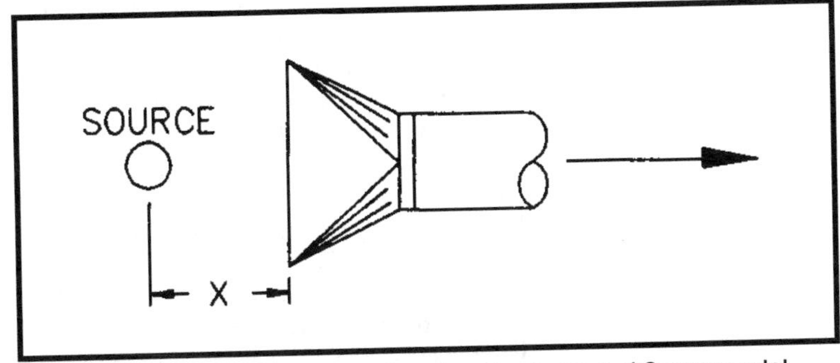

Figure 32.9 — Exterior hood. (From **American Conference of Governmental Industrial Hygienists (ACGIH®):** *Industrial Ventilation: A Manual of Recommended Practice*, 22nd Ed. Copyright 1995, Cincinnati, Ohio. Reprinted with permission.)

Table 32.2 — Range of Capture Velocities

(From **American Conference of Governmental Industrial Hygienists (ACGIH®):** *Industrial Ventilation: A Manual of Recommended Practice*, 22nd Ed. Copyright 1995, Cincinnati, Ohio. Reprinted with permission.)

Condition of Dispersion of Contaminant	*Example*	*Capture Velocity, fpm*
Released with practically no velocity into quiet air.	Evaporation from tanks; degreasing, etc.	50–100
Released at low velocity into moderately still air.	Spray booths; intermittent container filling; low speed Conveyor transfers; welding; plating; pickling.	100–200
Active generation into zone of rapid air motion.	Spray painting in shallow booths; barrel filling; Conveyor loading; crushers.	200–500
Released at high initial velocity into zone at very rapid air motion.	Grinding; abrasive blasting; tumbling	500–2000

In each category above, a range of capture velocity is shown. The proper choice of values depends on several factors:

Lower End of Range
1. Room air currents minimal or favorable to capture.
2. Contaminants of low toxicity or of nuisance value only.
3. Intermittent, low production.
4. Large hood (large air mass in motion).

Upper End of Range
1. Disturbing room air currents.
2. Contaminants of high toxicity.
3. High production, heavy use.
4. Small hood (local control only).

nature of the contaminant-generating process. One of the most common methods of estimating a minimum capture velocity in a particular situation involves tapping into the collective judgment and experience of many LEV hood designers who have successfully designed and implemented exterior hoods in a variety of industrial applications. Based on this experience, tabulated values of capture velocities as a function of various process-related variables have appeared in engineering literature dating back at least 50 years. This information, which has presumably withstood the test of time, appears in the *Industrial Ventilation Manual* and is reproduced here as Table 32.2.

Note from this table that the minimum capture velocity for a given situation is specified as a range of appropriate velocities categorized by various conditions of contaminant dispersion (i.e., the velocity with which the contaminant is released from the process and the motion of the air into which the contaminant is released). As one might expect, a contaminant released with practically no velocity into quiet air requires a lower velocity to achieve capture (50–100 ft/min) than one released at high initial velocity into a zone of very rapid air motion (500–2000 ft/min).

Obviously, quite a wide range of capture velocities exist within a given category. As explained in the table, selection of a specific velocity within the given range depends on several additional factors, such as the nature of the room air currents, the toxicity of the contaminant, the rate of production, and the size of the air mass in motion as reflected by the size of the hood. Thus armed with a thorough knowledge of the process, the contaminants being generated, and the behavior of the room air currents, the LEV designer is able to consult Table 32.2 and thereby quantify the minimum velocity necessary at the point of contaminant generation to capture and convey the contaminant into the hood.

Admittedly, this somewhat arbitrary process of determining capture velocity is not particularly satisfying for those accustomed to more objective design criteria. One can easily see that, given exactly the same situation, two different designers might very well come up with two quite different values for a required minimum capture velocity. As this determination is such a critical issue for external hoods, many individuals new to LEV design become thoroughly disheartened at the subjectivity of this procedure. Whatever might be said about the inherent disadvantages of making such an important determination as capture velocity in such a subjective fashion, at least it emphasizes the importance of a thorough working knowledge of the process and the interrelated variables when attempting to design effective LEV systems. It also serves as a reminder that occupational hygiene is an art as well as a science and that professional judgment is absolutely critical in the LEV design process, as it is in so many aspects of occupational hygiene. Other references contain a more complete discussion of the pitfalls and precautions related to capture velocity determination.[3,4]

Airflow Requirements for Exterior Hoods

If an exterior hood configuration is dictated by a given process, the next step, as demonstrated above, is the determination of the velocity required at the point of contaminant generation to achieve the capture and removal of that contaminant (i.e., the capture velocity). Capture velocity is ascertained from Table 32.2 by considering various process and contaminant-related issues such as the nature of contaminant generation, room air currents, contaminant toxicity, process production rate, and air mass moved by the hood. However, capture velocity, like face velocity for enclosure hoods, is not a typical LEV system design parameter (that is, LEV systems are not designed, nor are fans selected based on a required capture velocity at a certain point in front of the hood). As with face velocity, it is necessary to convert capture velocity in ft/min into the more common design parameter, volumetric flow rate of air in ft^3/min entering the hood. The designer must calculate the quantity of air, Q, in ft^3/min that must enter a hood opening to generate a certain capture velocity at some point, X, in front of the hood, where X is the distance from the point of contaminant generation to the hood opening. The question is how much Q is necessary to generate a certain V_c.

To answer the question regarding the relationship between Q and V_c it becomes necessary to return to the continuity equation (Q = VA) and discuss it in a bit more detail. This equation states that the volumetric flow rate of air in a system is equal to the product of the linear velocity of the air flowing through the system and the cross-sectional area through which the air flows. Assuming that the flow of air is incompressible (a typical assumption in LEV design), and if one subscribes to the concept that matter can neither be created nor destroyed, we can apply this fundamental equation to air flowing anywhere in a ventilation system, including that flowing into the hood from the ambient environment. This leads to another important ramification of the continuity equation: Once a given quantity of air is defined and moving into or through a ventilation system, the volumetric flow rate through any cross-sectional area of the system will be identical to that flowing through any other cross-sectional area of the system. Note from Figure 32.10 that regardless of whether one is concerned with air moving into the system from outside the hood (Q_1), air moving through the hood opening (Q_2), or air moving through the ductwork leading to the hood (Q_3), the volumetric flow rate of air is continuous (thus the name continuity equation). Remember, matter cannot be created or destroyed. It is physically impossible to have one quantity of air flowing at a given point in the system and a different quantity flowing at some other point, assuming that no other branches or entries into the system are bringing in additional air. Thus, in Figure 32.10, $Q_1 = Q_2 = Q_3$.

Since the volumetric flow rate is the same at each location in a ventilation system, it is clear from the continuity equation that an inverse relationship exists between V and A. For volumetric flow rate (Q) to remain continuous within a system, the velocity (V) must increase as the cross-sectional area (A) through which it is flowing decreases and vice versa. That is, in Figure 32.10, if $Q_2 = V_2A_2$, $Q_3 = V_3A_3$, and $Q_2 = Q_3$, then $V_2A_2 = V_3A_3$. Thus, if $A_2 > A_3$, then V_2 must be proportionally $< V_3$ to maintain the equality. This can be easily illustrated with an example.

Example 2. Consider the exterior hood in Figure 32.10 (plain round opening). Assume that the hood opening is 12 inches (1 ft) in diameter and that the duct leading to the hood is also round with a diameter of 6 inches (0.5 ft). If 1000 ft^3 of air per minute is entering the hood, what is the velocity of air at the hood face and in the duct leading to the hood?

Solution.

$$Q_{face} = V_{face}A_{face}$$

$$Q_{face} = 1000\ ft^3/min$$

$$A_{face} = \frac{\pi d^2_{face}}{4} = \frac{(\pi)(1 ft^2)}{4} = 0.7854\ ft^2$$

$$V_{face} = \frac{Q_{face}}{A_{face}} = \frac{1000\ ft^3/min}{0.7854\ ft_2} = 1273\ ft/min \quad (2)$$

So, as the 1000 ft^3/min of air flows through the 0.7854 ft^2 opening of the hood, it is traveling at a velocity of 1273 ft/min. However, as this air flows into the duct leading to the hood, the cross-sectional area through which the air is flowing has decreased. Since air (matter) can neither be

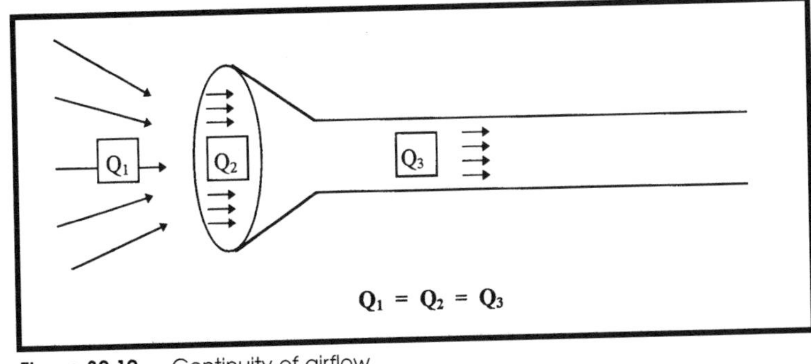

Figure 32.10 — Continuity of airflow.

created nor destroyed, Q_{face} must equal Q_{duct}. Therefore, to maintain continuity, a proportional increase in the velocity of air flowing through the duct is necessary.

$$Q_{duct} = V_{duct}A_{duct}$$

$$Q_{duct} = 1000 \text{ ft}^3/\text{min}$$

$$A_{duct} = \frac{\pi d2_{duct}}{4} = \frac{(\pi)(0.5 \text{ feet}^2)}{4} = 0.1963 \text{ ft}^2$$

$$V_{duct} = \frac{Q_{duct}}{A_{duct}} = \frac{1000 \text{ ft}^3/\text{min}}{0.1963 \text{ ft}^2} = 5093 \text{ ft/min} \quad (3)$$

Once inside the 0.1963 ft^2 duct, this 1000 ft^3/min of air accelerates to a velocity of 5093 ft/min. The fourfold decrease in cross-sectional area (0.7854 ft^2/0.1963 ft^2 = 4) through which the 1000 ft^3/min of air is flowing causes a proportional fourfold increase (5093 ft/min/1273 ft/min = 4) in the velocity of the air.

Returning to the original questions of this section, just how does the continuity equation relate to what is occurring in front of the hood, and what is the relationship between the volumetric flow rate of air entering the hood and the velocity of air at certain distances from the hood? In other words, in the Q = VA equation, what is the relationship between Q, the volumetric flow rate of air entering the hood, and V, which in this situation is the capture velocity (V_c) desired at the point of contaminant generation some distance, X, from the hood? Furthermore, in applying the continuity equation in front of the hood, what is considered to be the cross-sectional area, A, through which the air is flowing? The answers to each of these questions can be best illustrated by the theoretical case of a point suction source freely suspended in space as is illustrated in Figure 32.11.

Assume that the hood depicted in Figure 32.11 is an infinitely small point source of suction suspended freely in space such that air flows toward and into the hood equally from all directions. If a point is chosen at some distance, X, from the hood's face, a sphere of radius, X, and surface area, $A = 4\pi X^2$, surrounding the hood can be identified. Once a quantity of air is defined and flowing into the hood, all of the air must pass through this surface area with equal velocity at all points on the sphere's surface. In terms of the continuity equation, Q = VA, Q is the volumetric flow rate of the air moving into the point hood, and A is the cross-sectional area through which this air moves as defined by the surface area of the sphere of radius, X. Thus, the continuity equation defines the relationship

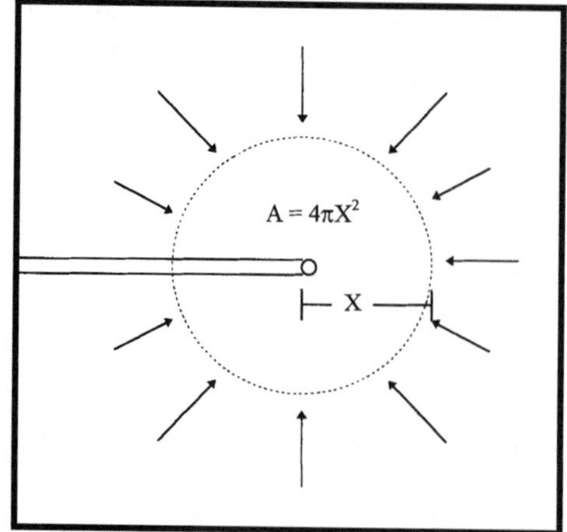

Figure 32.11 — Point source suction.

between volumetric flow rate and the velocity of air at a certain distance, X, from the point of suction as follows.

$$Q = VA$$

$$A = 4\pi X^2$$

$$Q = (V)(4\pi X^2) \quad (4)$$

Considering this point source of suction to be an exterior hood and the velocity at a distance, X, to be the capture velocity necessary to capture and remove the contaminant being generated at distance, X (as determined from Table 32.2), this equation can be used to calculate the volumetric flow rate necessary to achieve the desired capture velocity. So, as might have been expected, the volumetric flow rate (Q) needed to provide a certain capture velocity (V_c) at the point of contaminant generation (X) is a function of the velocity required and the distance from the hood to the source of contaminant generation.

$$Q = (V_c)(4\pi X^2) \quad (5)$$

Example 3. Suppose that 1000 ft^3/min is flowing into the point source suction hood illustrated in Figure 32.11. What is the velocity of the air at a distance of 12 inches (1 ft) from the hood?

Solution.

$$Q = (V_c)(4\pi X^2)$$

$$V_c = \frac{Q}{4\pi X^2} = \frac{1000 \text{ ft}^3/\text{min}}{(4)(\pi)(1 \text{ ft})^2}$$

$$V_c = 80 \text{ ft/min} \quad (6)$$

So, 1000 ft^3/min flowing into a point source suction hood generates a velocity of 80 ft/min at a distance of 12 inches.

Of course, an infinite number of concentric spherical surface areas exist around the point source of suction through which the air entering the hood must flow. In accordance with the continuity equation, the volumetric flow rate will be equivalent through each of them. However, note that as the air moves closer to the source (i.e., as X decreases) the surface area of the sphere — and thus the cross-sectional area through which the air is flowing — decreases as a function of X^2. As the cross-sectional area decreases, the velocity increases proportionally. This illustrates a very important point regarding capture velocity: it increases exponentially as the air moves closer to the hood.

Example 4. Using the information from Example 3 above, what will the velocity of air be at a distance of 6 inches (0.5 ft) from the hood?

Solution.

$$Q = (V_c)(4\pi X^2)$$

$$V_c = \frac{Q}{4\pi X^2} = \frac{1000\ ft^3/min}{(4)(\pi)(0.5\ ft)^2}$$

$$V_c = 320\ ft/min \qquad (7)$$

At 6 inches, this same volumetric flow rate of 1000 ft^3/min generates a velocity 320 ft/min. Decreasing the distance to the hood by a factor of 2 (i.e., 12 inches/6 inches) causes a decrease in surface area of 4 (i.e., 2^2), which causes a corresponding fourfold increase in velocity (320 ft/min/80 ft/min = 4).

To keep matters in perspective, remember that the main point of this discussion is that a fundamental relationship exists, based on the continuity equation, between the volumetric flow rate of air (Q) flowing into an exhaust opening (i.e., exterior hood) and the velocity of air (V_c) at some distance (X) away from that opening. For the theoretical case of a point source of suction, that relationship is expressed as $Q = (V_c)(4\pi X^2)$.

Obviously, the point source suction hood is a theoretical construct used to illustrate this simple relationship. For an actual hood the suction area is not limited to a point in space but is applied over the entire face area of the hood. Also, the physical presence of an actual hood cannot be ignored as can that of a theoretical point hood. To be useful to LEV hood designers, the relationship must be modified and extended to cover more practical exterior hood geometries such as round and rectangular openings, slotted openings, etc.

Fortunately, over the years a variety of equations have developed such that the relationship between air velocity at a distance, X, as a function of volumetric flow rate and hood dimensions can be estimated for most simple exterior hood geometries. Many of these equations have evolved as extensions of the point source suction geometrical model described above. In this approach the designer attempts to visualize the three-dimensional shape through which the air enters the given hood and relate the velocity at a given point, X, to the volumetric flow through the surface area inscribed by that particular shape. The application of this approach, including a table of formulae (reproduced as Table Figure 32.12) and several example problems is described in greater detail by Burton.[3]

Other, more exact theoretical approaches relating the velocity at some point in front of the hood to the flow rate of air into the hood have appeared in recent literature. The reader is encouraged to consult Chapter 5 of Burgess et al. for a more thorough discussion of this important research, including an excellent summary table of exterior hood velocity models and a thorough bibliography of relatively current research.[4]

Perhaps the most commonly used set of equations relating capture velocity to volumetric flow rate for simple hood geometries are those appearing in Chapter 3 of the *Industrial Ventilation Manual* and reproduced as Table 32.3. These equations primarily describe the volumetric flow rate of air going into the hood as a

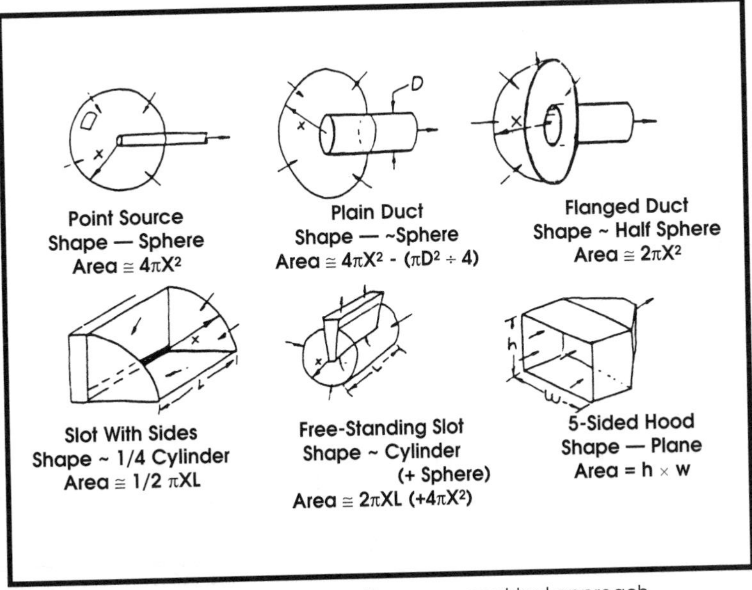

Figure 32.12 — Capture velocity equations — geometrical approach.

Table 32.3 — Hood Types

(From **American Conference of Governmental Industrial Hygienists (ACGIH®):** *Industrial Ventilation: A Manual of Recommended Practice*, 22nd Ed. Copyright 1995, Cincinnati, Ohio. Reprinted with permission.)

HOOD TYPE	DESCRIPTION	ASPECT RATIO, W/L	AIR FLOW
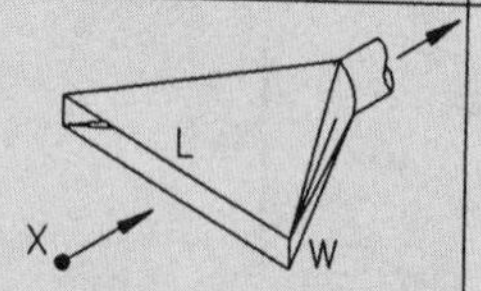 X, L, W	SLOT	0.2 OR LESS	$Q = 3.7\ LVX$
 X	FLANGED SLOT	0.2 OR LESS	$Q = 2.6\ LVX$
 W, L, X; A = WL (sq.ft.)	PLAIN OPENING	0.2 OR GREATER AND ROUND	$Q = V(10X^2 + A)$
 X	FLANGED OPENING	0.2 OR GREATER AND ROUND	$Q = 0.75V(10X^2 + A)$
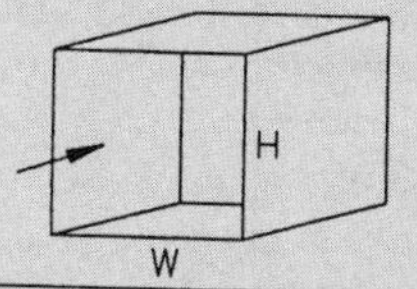 H, W	BOOTH	TO SUIT WORK	$Q = VA = VWH$
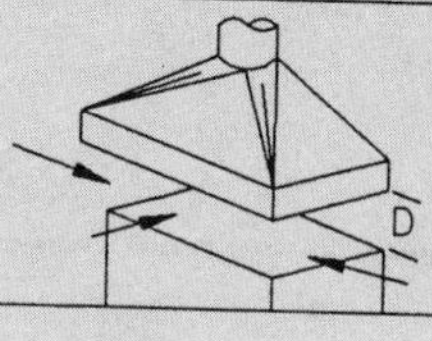 D	CANOPY	TO SUIT WORK	$Q = 1.4\ PVD$ SEE VS-99-03 P = PERIMETER D = HEIGHT ABOVE WORK
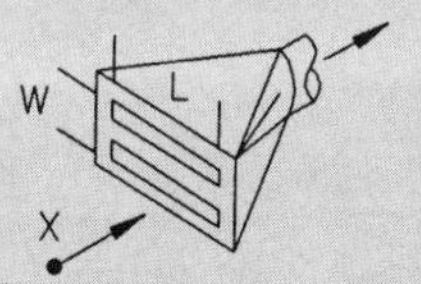 W, L, X	PLAIN MULTIPLE SLOT OPENING 2 OR MORE SLOTS	0.2 OR GREATER	$Q = V(10X^2 + A)$
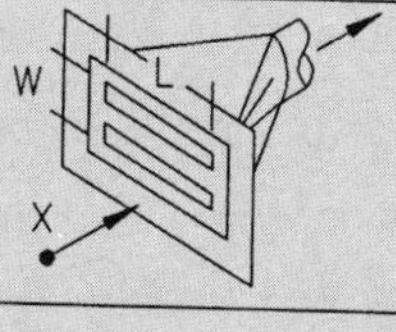 W, L, X	FLANGED MULTIPLE SLOT OPENING 2 OR MORE SLOTS	0.2 OR GREATER	$Q = 0.75V(10X^2 + A)$

AMERICAN CONFERENCE OF GOVERNMENTAL INDUSTRIAL HYGIENISTS

HOOD TYPES

DATE 1-88 FIGURE 3-11

function of the air velocity at various distances along the hood centerline (line from the center of the hood face extending perpendicularly outward) and are considered to be reasonably accurate out to distances of X equivalent to about 1.5 times the diameter of the duct leading to the hood. The origin of these relationships dates back to a series of empirical experiments conducted by J.M. Dalla Valle and Leslie Silverman at Harvard University in the 1930s and 1940s. It is interesting that the work of these industrial ventilation pioneers still forms the basis for the design of modern exterior exhaust hoods. Again, the reader is encouraged to consult Burgess et al. for a discussion of important extensions of this work that have emerged in recent years.

Many readers may be interested in a more rigorous treatment of the continuity equation, including a discussion of basic fluid mechanics as related to the fundamentals of airflow and conservation of mass. These subjects are described in more detail in various sources listed in the bibliography.[1,4-9]

Exterior Hood Design Summary and Examples

At this point in the chapter it is appropriate to review what has been presented thus far regarding exterior capture hoods and work through an illustrative example. Remember, the objective is to capture and remove a contaminant that is being generated by some process before that contaminant reaches a worker's breathing zone. This assumes that the contaminant-generating process, due to its nature and physical configuration, cannot be either partially or completely enclosed by an enclosing hood (which typically provides better control), and therefore the designer is left with specifying an exterior hood. As stated earlier, the air flowing into an exterior hood must generate sufficient air velocity at the point of contaminant generation to overcome opposing air currents and draw the contaminant into the hood (capture velocity, V_c). Thus, the design process for exterior hoods requires the designer to address two questions:

1. What air velocity, V_c, is required at the point of contaminant generation to overcome opposing air currents, capture the contaminated air at that point, and cause it to flow into the hood?
2. What volumetric flow of air, Q, into the hood is required to generate the necessary capture velocity?

Question 1 may be answered by studying the process and the conditions of contaminant dispersion and then consulting Table 32.2. From this table and its associated descriptive parameters a desired capture velocity can be estimated. Once the designer has determined an adequate capture velocity, the equations appearing in Table 32.3 (or other similar equations[3,4]) allow the calculation of the volumetric flow rate of air into the hood necessary to achieve the desired capture velocity at the point of contaminant generation.

Example 5. An occupational hygienist has been asked to design an LEV system for a process in which a worker, using a fairly stiff, 1-inch brush, brushes a large stained-glass plate covered with finely divided lead dust on a production basis. Occasional disturbing air currents may be present, and the worker works with her face very near the source. The LEV designer is requested to determine the following:

- Type of hood (enclosure or exterior), hood dimensions, and hood location relative to the contaminant source;
- The air velocity at the point of contaminant generation necessary to adequately capture the contaminant (i.e., V_c); and
- The volumetric flow rate of air, Q, necessary to produce the desired capture velocity.

Solution. The first step in the design process is to determine the hood configuration that would achieve the most effective control. Assume that on investigation of the process the designer concludes that because of the required worker access to the workpiece and the size of the glass plate an enclosure hood is impractical for this operation. Therefore, an exterior hood must be designed. Also assume that on further examination it is determined that the process configuration allows location of the hood face no closer than 9 inches from the contaminant (i.e., lead dust) generation point. Finally, after considering this information and noting the glass plate surface area over which the lead dust is generated, the designer consults Table 32.3 and decides on a hood with a flanged rectangular 8- × 5-inch opening. The physical dimensions of the hood are illustrated in Figure 32.13.

The next step is the determination of capture velocity from Table 32.2. As mentioned previously, this is the most subjective part of the design procedure and requires a good deal of thoughtful consideration regarding the nature of the contaminant-generating process and of the properties of

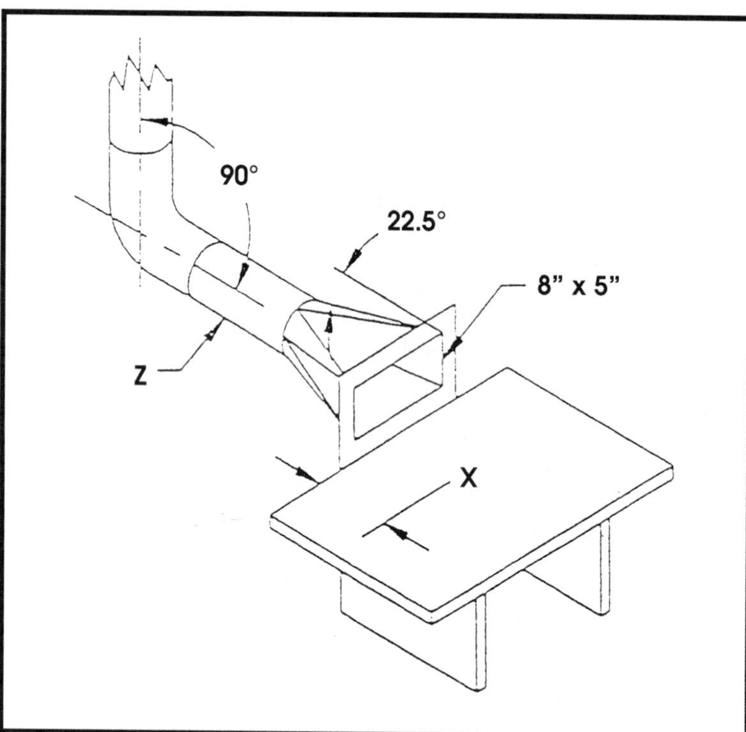

Figure 32.13 — Hood configuration for lead dust problem.

the contaminant itself. From the process investigation, the designer notes these points:

- The worker uses a stiff brush to brush finely divided lead dust, indicating that the contaminant generation may be moderately active with at least some (albeit probably low) velocity imparted to the dust by the brushing action.
- The process is conducted on a production basis and it is therefore likely that the contaminant is generated fairly continuously throughout the work shift.
- Competing air currents (i.e., air currents unfavorable to capture) may be present, although the magnitude of these currents is unquantified.
- The contaminant (lead dust) may be considered of relatively high toxicity.
- The hood is small, producing local control only.

These process parameters do not precisely match any category listed in Table 32.2. One might argue that the category "Released at low velocity into moderately still air" is the "Condition of Dispersion of Contaminant" that most closely relates to the process in question. The capture velocity range specified for this category is 100–200 ft/min. Considering that disturbing room air currents are present, the contaminant is of high toxicity, the process is production oriented, and the hood is small and provides local control only, a capture velocity at the high end of this range might be selected (e.g., 200 ft/min.) However, because the contaminant in this case is lead dust, which is not only highly toxic but also very dense and heavy, the designer might choose to be even more conservative and design for a capture velocity of 250 ft/min or perhaps even higher. This is a decision that the designer must make after careful investigation of the process and observation of the manner in which the work is conducted. For the sake of this example, assume that a capture velocity of 250 ft/min is selected.

Note again that, without doubt, another designer assessing this same process could quite possibly obtain a different value of capture velocity from Table 32.2 that would perhaps be an equally valid estimate of what constitutes an adequate capture velocity for this situation. Remember, too, that this is the weak link in the exterior hood design process and depends to a large extent on the experience and professional judgment of the designer and on the individual's personal bent toward conservatism in LEV design. Once again, it is crucial that the designer give conscientious consideration to this decision so that he or she is able to justify, at least in his or her own mind, the capture velocity selected.

Now that the configuration and dimensions of the hood have been determined and an adequate capture velocity estimated, the next step is to determine the volumetric flow rate, Q, required to generate this velocity at the specified distance for this particular hood. Remember, Q is the critical design parameter for LEV systems and its calculation is of utmost importance.

Q can be calculated by referring to Table 32.3, the equations that relate the required capture velocity to volumetric flow rate. For a flanged opening, the equation for Q as a function of V_c is as follows:

$$Q = 0.75V_c(10X^2 + A) \tag{8}$$

In this equation, Q is the volumetric flow rate necessary to achieve an air velocity of V_c (in ft/min) at a distance, X (in ft), along the centerline of a flanged rectangular hood of face area, A (in ft^2). For the hood in question V_c is 250 ft/min as determined from Table 32.2, X is 9 in. or 0.75 ft as dictated by the process, and A is 8- × 5 inches = 40 in.^{2} = 0.28 ft^2, the hood face area.

Q is then calculated as follows:

$$Q = (0.75)(250 \text{ ft/min})(10(0.75 \text{ ft})^2 + 0.28 \text{ ft}^2) = 1107 \text{ ft}^3/\text{min} \quad (9)$$

Thus, to generate an air velocity of 250 ft/min at a distance of 9 inches along the centerline of an 8- × 5-inch flanged rectangular hood, a volumetric flow rate of approximately 1107 ft^3/min must be entering that hood. This 1107 ft^3/min is the important design parameter for this exterior hood.

Note: In Figure 32.12, the hood is depicted as resting on the work surface. In reality, this configuration would reduce the area from which the hood draws exhaust air and thus actually reduce the quantity of air necessary to provide the specified capture velocity. Therefore, the 1107 ft^3/min will likely be, in this case, a conservative overestimation of the required flow rate.

Several additional points are worthy of note regarding the above example. The first has to do with the effect the distance from the source to the hood has on the volumetric flow rate necessary to generate the required control velocity. The distance, X, from the hood face to the point of contaminant generation is usually dictated by process-related factors such as the physical dimensions of the workpiece, the requirements for employee access to the work area, layout of the work space, and so forth. However, since Q is proportional to X^2, it is extremely important to keep this distance as small as possible, consistent with process restrictions, to minimize airflow requirements. Even small changes in the specified location of the hood relative to the contaminant source can make large differences in volumetric flow rate. For example, in the above problem the designer determined, based on process access constraints and workpiece size, that the hood could be placed no closer than 9 inches from the point of lead dust generation. At this distance, 1107 ft^3/min was required to generate the capture velocity of 250 ft/min. How would the flow rate requirements change if the designer had decided to locate the hood 12 inches from the contaminant generation point?

$$Q = 0.75V_c(10X^2 + A)$$

$$Q = (0.75)(250 \text{ ft/min})(10(1.0 \text{ ft})^2 + 0.28 \text{ ft}^2) = 1928 \text{ ft}^3/\text{min} \quad (10)$$

Thus, increasing the distance from the hood face to the contaminant generation point by 3 inches in this design requires an additional 825 ft^3/min (about 75% more) volumetric flow rate to achieve the required 250 ft/min capture velocity. This is quite a significant increase. In general, since velocity decreases inversely with the square of the distance from the hood, a doubling of distance requires a 2^2 or fourfold increase in required airflow to generate the same velocity.

More important, what happens if the worker is not properly instructed in how to work with the new ventilation system? Specifically, what if he or she is not told that for the exhaust system to be effective the brushing activity should be conducted no farther than 9 inches away from the hood face and along the hood centerline? Assume that the worker conducts the brushing activity 12 inches away from the hood face instead of 9 inches. At the specified exhaust flow rate of 1107 ft^3/min, what effect will this have on the velocity at the contaminant generation point?

$$V_c = \frac{Q}{0.75(10X^2 + A)} = \frac{1107 \text{ ft}^3/\text{min}}{0.75(10(1 \text{ ft})^2 + 0.28)} = 144 \text{ ft/min} \quad (11)$$

At a distance of 12 inches, the centerline velocity has fallen from the specified 250 ft/min to less than 150 ft/min. Such a dramatic drop in velocity would have a significant impact on the effectiveness of this hood in capturing and removing the lead dust. Obviously, it is absolutely crucial for workers to be informed of the proper positioning of work relative to the hood as well as other important aspects of LEV system operation in order to achieve the objective of a protected employee breathing zone. This is particularly true for exterior hoods.

With all the uncertainty related to the estimation of an adequate capture velocity, the calculation of volumetric flow rate necessary to achieve that capture velocity, and the extreme sensitivity of control effectiveness to hood position, it is not surprising that many designers tend to be conservative in the design of exterior hoods.

Effect of Flanging an Exterior Hood

The previous example problem also introduced the concept of adding a flange to an exterior hood. A flange is defined as "a surface at and parallel to the hood face which provides a barrier to unwanted air flow from behind the hood."(1) Note the difference between the flanged opening and the plain opening in Table 32.3 (page

866). The effect of the flange can be illustrated by comparing Figure 32.14 with Figure 32.15. Figure 32.14 depicts the air streamline pattern into a plain round opening. Note that air is pulled from behind as well as from in front of the hood opening. Drawing air from behind the hood is, in reality, undesirable in that this air is not useful in entraining the contaminant being generated in front of the hood. In a sense, this portion of Q is being wasted. If a flange is added around this opening (see Figure 32.15), the flow from behind the hood is essentially blocked, thus reducing the flow area and effectively extending the exhaust air streamlines farther from the hood face where the contaminant is being generated. Use of a flange can achieve either of two positive effects: (1) decreasing the volumetric flow rate required to achieve a given capture velocity at a given point; or (2) increasing the velocity achieved by a given flow rate at a given point. Thus, in the design process less air may be designated to achieve the same capture velocity if a flange is specified. Also, the capture velocity of an existing hood may be significantly improved by the simple act of adding a flange.

Comparing the equations from Table 32.3 for a plain opening and a flanged opening, one can see that the flanged opening can achieve the same capture velocity with about 75% of the flow rate required for a plain opening. Similarly, the flanged slot requires only about 70% (i.e., 2.6/3.7) of the flow necessary for a plain slot to achieve the same velocity at a given distance, X. Obviously, flanging is a simple yet effective means of increasing the effectiveness of a an exterior LEV hood. Note that in the *Industrial Ventilation Manual* it is written that for most applications, the flange width should be equal to the square root of the hood face area (i.e., flange width = $\sqrt{A}$).

Using Published Hood Design Criteria

As was the case for enclosure hoods, Chapter 10 of the *Industrial Ventilation Manual* provides numerous practical examples of exterior hood designs for common industrial operations and processes requiring more complex hood configurations than those illustrated in Table 32.3. These prints typically allow direct calculation of volumetric flow rate as a function of various process-related parameters. For example, a simple exterior welding hood design is given as Figure 32.16.

Volumetric flow rate requirements can be determined from Table 32.3 as a function of distance from the hood to the source and whether the hood has a plain, flanged, or cone-shaped opening. Other useful notes

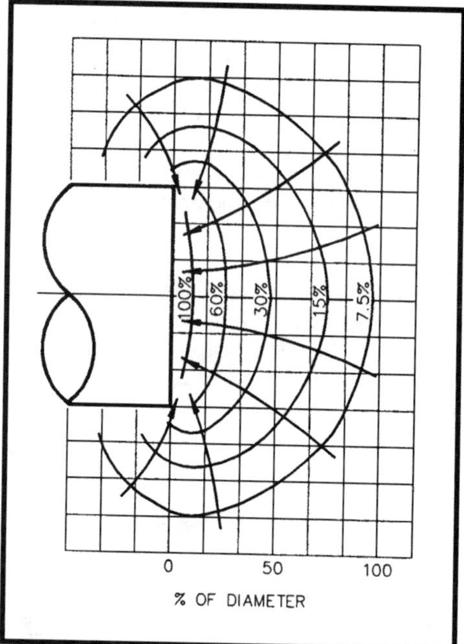

Figure 32.14 — Velocity contours: plain circular opening. (From **American Conference of Governmental Industrial Hygienists (ACGIH®)**: *Industrial Ventilation: A Manual of Recommended Practice*, 22nd Ed. Copyright 1995, Cincinnati, Ohio. Reprinted with permission.)

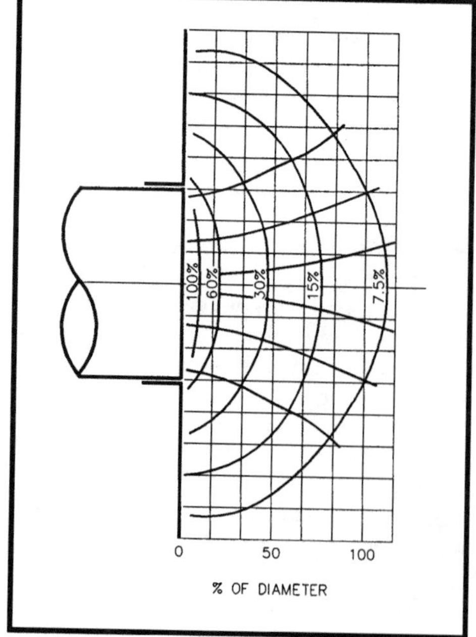

Figure 32.15 — Velocity contours: flanged circular opening. (From **American Conference of Governmental Industrial Hygienists (ACGIH®)**: *Industrial Ventilation: A Manual of Recommended Practice*, 22nd Ed. Copyright 1995, Cincinnati, Ohio. Reprinted with permission.)

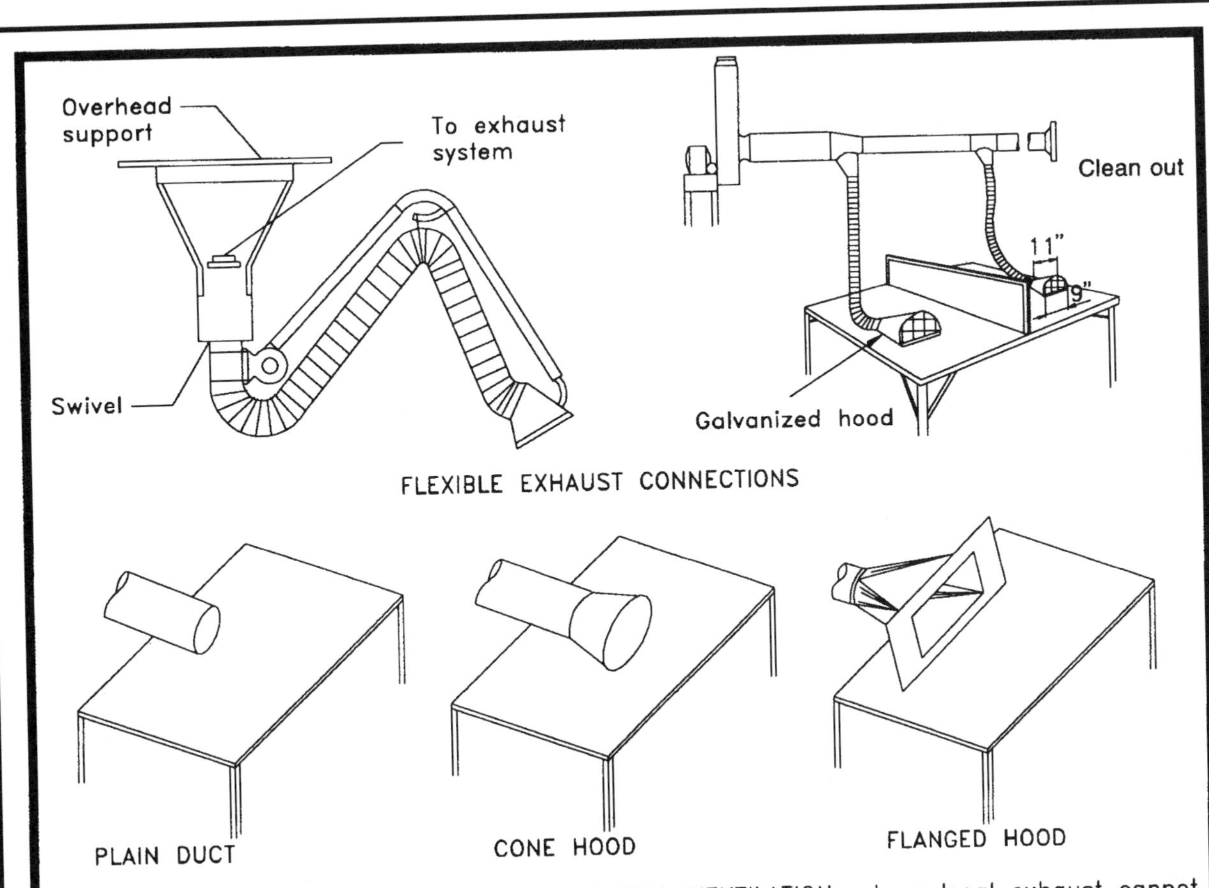

RATE OF EXHAUST

X,	Plain duct, cfm	Flange or cone, cfm
Up to 6	335	250
6–9	755	560
9–12	1335	1000

Face velocity = 1500 fpm
Minimum duct velocity = 3000 fpm
Plain duct entry loss = 0.93 VP_d
Flange or cone entry loss = 0.25 VP_d

Notes:

1. Locate work as close as possible to hood.
2. Hoods perform best when located to the side of the work.
3. Ventilation rates may be inadequate for toxic materials.
4. Velocities above 100–200 fpm may disturb shield gas.

GENERAL VENTILATION, where local exhaust cannot be used :

Rod, diam.	cfm/welder
5/32	1000
3/16	1500
1/4	3500
3/8	4500

OR

A. For open areas where welding fume can rise away from the breathing zone:
cfm required = 800 x lb/hour rod used

B. For enclosed areas or positions where fume does not readily escape breathing zone:
cfm required = 1600 x lb/hour rod used

For toxic materials, higher air flows are necessary and operator should use respiratory protection equipment.

Other types of hoods:
Bench, see VS-90-01
Booth, for design see VS-90-30
Q = 100 cfm/ft^2 of face opening

AMERICAN CONFERENCE OF GOVERNMENTAL INDUSTRIAL HYGIENISTS

WELDING VENTILATION MOVABLE EXHAUST HOODS

DATE *1–91* FIGURE *VS–90–02*

Figure 32.16 — Welding ventilation flexible hoods. (From **American Conference of Governmental Industrial Hygienists (ACGIH®):** *Industrial Ventilation: A Manual of Recommended Practice*, 22nd Ed. Copyright 1995, Cincinnati, Ohio. Reprinted with permission.)

on the application of this hood to the welding process are also offered to the designer.

Presumably, design guidance such as this already incorporates considerations regarding the conditions of contaminant dispersion and thus relieves the designer of the stress of capture velocity estimation from Table 32.2. Be aware, however, that these flow rate specifications often relate to "typical" environmental and material conditions. Specific application of a given hood in an environment where conditions may not be conducive to capture or where highly toxic contaminants are generated may require adjustment of calculated values for Q. The designer would do well to heed the advice offered by the ACGIH Industrial Ventilation Committee in the opening paragraphs of Chapter 10 of the *Industrial Ventilation Manual*:

> *The following illustrations of hoods for specific operations are intended as guides for design purposes and apply to usual or typical operations.... All conditions of operation cannot be categorized; because of special conditions (i.e., cross-drafts, motion, differences in temperature, or use of other means of contaminant suppression), modifications may be in order.*
>
> *Unless it is specifically stated, the design data are not to be applied indiscriminately to materials of high toxicity, i.e., beryllium and radioactive materials. Thus, the designer may require higher or lower air flow rates or other modifications because of the peculiarities of the process in order to adequately control the air contaminant.*[1]

Design of Slot Hoods, Plenums, and Baffles

The discussion of exterior hoods would not be complete without an introduction to some additional terminology that frequently arises in the discussion of LEV system design: slots, plenums, and baffles. These terms are illustrated in Figure 32.17 and described in the paragraphs below.

The exterior hoods discussed thus far in this chapter have been of the plain or flanged opening variety. This type of hood is designed to remove contaminants generated over a relatively small area, such as might be produced by a localized or welding operation. However, there are many industrial processes that permit contaminants to emanate from very large surface areas. For example, processes using degreasing tanks, plating tanks, fluidized beds, or welding tables (see Figure 32.17) may give off contaminants across the entire open surface area of the tank or bed. This necessitates a hood design that will distribute the airflow evenly across the surface and provide an adequate capture velocity at all points over the contaminant-generating area. Slot hoods with plenums are frequently used for this function. A slot hood is defined as a hood with an opening (i.e., slot) that has a width to length (W/L) ratio of 0.2 or less.[1] The function of the slot is solely to provide uniform air distribution."[1]

A slot hood can evenly distribute exhaust air when it is used in conjunction with a plenum. Note from Figure 32.17 that a plenum is essentially a large air compartment or chamber connected to the slot that functions to distribute the suction evenly across the slot area. The distribution of suction is accomplished by sizing the slot width and plenum depth such that the velocity of the slot is much higher than that in the plenum. This subjects all points across the slot to approximately equal suction resulting in essentially uniform slot velocity.[1] (Please note that the term suction is actually a specific type of pressure that exists in a ventilation called static pressure. Static pressure and its significance in an LEV system will be discussed in subsequent sections of this chapter.)

Obviously, it would be of little use to design a slot hood to control emissions from an open surface tank if virtually all of the air flows through one section of the slot with very little flowing through other sections. For example, consider Figure 32.18A and 32.18B.

Since there is little opportunity for the suction from the duct to be distributed across the slot in Figure 32.18a, most of the air will flow into the hood from the center of the slot with very little air flowing at either end. If this hood were being used to control emissions from a tank, it would likely be ineffective at capturing contaminant given off at the tank edges. However, in Figure 32.18B, the plenum allows this suction to be distributed more evenly, providing uniform suction through all portions of the slot. Thus, the objective of a plenum is to spread out or distribute the air evenly such that the face velocity is equal across the slot and, more important, that the capture velocity generated at the far corners of the tank is every bit as effective as that in the middle.

Many people mistakenly assume that slots are designed for the purpose of decreasing the area of the hood opening, thereby increasing the hood face velocity (i.e., slot velocity) and ultimately improving

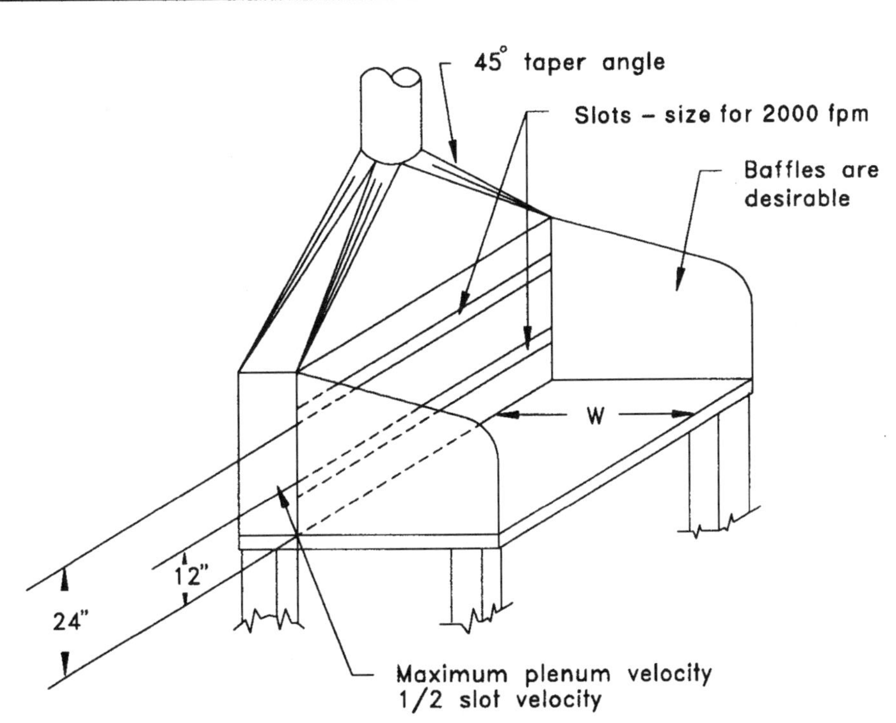

Q = 350 cfm/ft of hood length
Hood length = required working space
W = 24" maximum, if W>24", see Chapter 3
Minimum duct velocity = 2000 fpm
$h_e = 1.78\ VP_s + 0.25\ VP_d$

General ventilation, where local exhaust cannot be used:

Rod, diam.	cfm/welder
5/32	1000
3/16	1500
1/4	3500
3/8	4500

or

A. For open areas, where welding fume can rise away from the breathing zone: cfm required = 800 x lb/hour rod used

B. For enclosed areas or positions where fume does not readily escape breathing zone: cfm required = 1600 x lb/hour rod used

For toxic materials, higher air flows are necessary, and operator may require respiratory protection equipment.

Other types of hoods:
Local exhaust: See VS-90-02
Booth: For design, see VS-90-30
Q = 100 cfm/ft^2 of face opening
MIG welding may require precise air flow control.

AMERICAN CONFERENCE OF GOVERNMENTAL INDUSTRIAL HYGIENISTS	WELDING VENTILATION BENCH HOOD	
	DATE 1-91	FIGURE VS-90-01

Figure 32.17 — Welding ventilation bench hood. (From **American Conference of Governmental Industrial Hygienists (ACGIH®)**: *Industrial Ventilation: A Manual of Recommended Practice*, 22nd Ed. Copyright 1995, Cincinnati, Ohio. Reprinted with permission.)

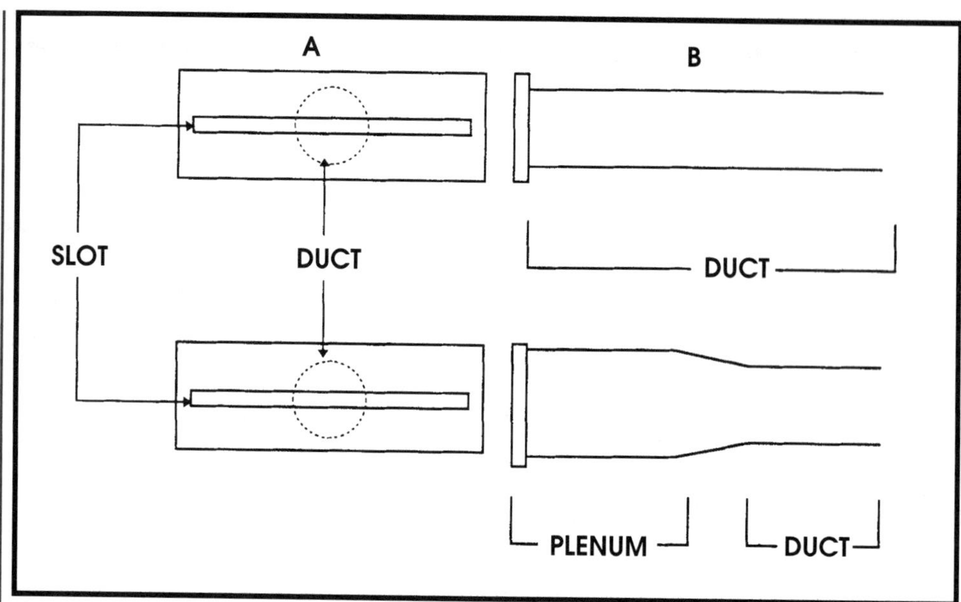

Figure 32.18 — (A) Slotted hood without plenum; (B) slotted hood with plenum.

the hood's ability to reach out and capture contaminants. It is tempting to assume that higher slot velocities translate into higher capture velocities at points in front of the hood. Although adding a slot to an open-faced hood does indeed decrease face area and increase the face velocity in accordance with the continuity equation, this result does not increase the velocity at points in front of the hood. As can be seen from the equations in Table 32.3 (page 866), the capture velocity at distance X is a function of slot dimensions and volumetric flow rate, not slot velocity. Therefore, a higher slot velocity does not contribute to a higher capture velocity.

Since it is the uniform distribution of the volumetric flow across the slot that improves capture effectiveness and not the magnitude of the slot velocity, an important design question is simply, "What is a sensible slot velocity?" In the *Industrial Ventilation Manual* it is written that 2000 ft/min is a reasonable slot velocity for most hoods with a corresponding plenum velocity of 1000 ft/min. These values provide a relatively uniform flow for most slot hoods.

The size of the slot(s) is also an important specification in the design. Since the slot should extend across the entire length of the tank for most operations, the slot length is typically matched to the length of the tank. The width of the slot(s) is determined from the continuity equation in accordance with the design slot velocity (usually 2000 ft/min) and the volumetric flow rate necessary to provide adequate capture velocity. An example should clarify this process.

Example 6. For a fluidized bed installation with a slot/plenum combination, the *Industrial Ventilation Manual* gives the following guidance: Q = 150 $ft^3/min/ft^2$ of bed (150 LW); slot velocity = 2000 ft/min. Calculate the following assuming that L = 6 ft and W = 2.5 ft: (a) required volumetric flow rate, Q; (b) slot dimensions (design for two slots); (c) plenum dimensions.

Solution. The first step in proceeding with this slot hood design for a fluidized bed is to use the design guidance to calculate the volumetric flow rate required. In this example, Q is calculated as a function of the total bed area in square feet in accordance with the formula, Q = 150 $ft^3/min/ft^2$ of bed. In this installation the total bed area is L × W = (6 ft)(2.5 ft) = 15 ft^2. Thus, the required Q is (150 $ft^3/min/ft^2$)(15 ft^2) = 2250 ft^3/min.

Remember when using the *Industrial Ventilation Manual* design guidance that the volumetric flow rate calculated (in this case 2250 ft^3/min for this fluidized bed) has been demonstrated in actual industrial installations to be sufficient to effectively capture contaminants generated from the given process. But, again, the designer must assure that atypical factors such as cross drafts, motion, differences in temperature, or high toxicity contaminants do not render the design Q insufficient for the situation at hand.

The next step is to determine the slot dimensions, both length and width. Perhaps the easiest way to proceed is to first determine the total slot area, A, by using the calculated volumetric flow rate and the specified slot velocity in the continuity equation. In Part (a) the flow rate was determined to be 2250 ft³/min. The design specified a slot velocity of 2000 ft/min. Thus, the total slot area can be calculated as follows:

$$Q = VA$$

$$2250\ \text{ft}^3/\text{min} = (2000\ \text{ft/min})(A)$$

$$A = \frac{2250\ \text{ft}^3/\text{min}}{2000\ \text{ft/min}} = 1.125\ \text{ft}^2 \qquad (12)$$

As previously mentioned, the slot width is usually determined by the width of the tank. In this case the fluidized bed length is given as 6 feet. Therefore, the slot is also designed to be 6 feet in length. Since area is simply length times width, the total slot width is given by the following.

$$A = L \times W$$

$$1.125\ \text{ft}^2 = (6\ \text{ft})(W)$$

$$W = \frac{1.125\ \text{ft}^2}{6\ \text{ft}} = 0.188\ \text{ft} = 2.25\ \text{ft} \qquad (13)$$

Note that this 2.25 inches is the total slot width. Two slots were specified in the design; therefore, the width of each slot is one-half of 2.25 inches, or 1.125 inches.

Typically, plenum velocity is designed to be one-half of the magnitude of the slot velocity. In accordance with the continuity equation, this would occur if the plenum cross-sectional area were designed to be twice that of the total slot area. Since the plenum length and the slot length are approximately the same, the velocity through the plenum would be approximately one-half of that through the slot if the plenum depth were twice the total slot width. Thus, the design plenum depth should be twice 2.25 inches or 4.5 inches. The plenum are would therefore be (4.5 inches)(72 inches) = 324 inches^2 = 2.25 ft^2, and the plenum velocity would be 1000 ft/min as specified.

$$Q = VA$$

$$2250\ \text{ft}^3/\text{min} = (V)(2.25\ \text{ft}^2)$$

$$V = 1000\ \text{ft/min} \qquad (14)$$

Note the presence of side baffles in Figure 32.17. A baffle is defined as "a surface which provides a barrier to unwanted air flow from the front or sides of the hood."[1] It functions in essentially the same manner as a flange in that it reduces the area through which the air flow is drawn and focuses the exhaust streamlines out in front of the hood. Recall the continuity equation, Q = VA. If the cross-sectional area through which the air flows is reduced, air velocity must correspondingly increase. Thus, the net result of baffling a hood is an increase in the air velocity at a given point in front of the hood over that at the same point for the hood without baffles. The precise magnitude of this effect is difficult to quantify and depends on the baffle location and size.[1]

Often the effectiveness of an existing exterior hood can be greatly increased by adding baffles to the side and perhaps even across the top if allowed by the process. Of course, this essentially serves to convert an exterior hood into an enclosing hood if the contaminant is generated inside the confines of the baffles. Not only does this improve the hood's contaminant-capturing ability it also serves to protect the contaminant source against the undesirable effects of cross drafts that might otherwise disperse the contaminant beyond the reach of the hood.

The reader is encouraged to consult the section on open surface tank design in Chapter 10 of the *Industrial Ventilation Manual* for a more thorough discussion of this important aspect of ventilation design. Calculation of volumetric flow rate for the fluidized bed example above was a fairly straightforward exercise; however, there are a variety of open surface tank configurations, sizes, and types existing in the industrial environment. These tanks contain contaminants of varying degrees of toxicity with differing contaminant evolution characteristics. Each of these attributes must be considered in determining a minimum control capture and subsequent minimum volumetric flow rate for these tanks. An in-depth discussion of open surface tank design considerations and a detailed example of flow rate calculations is too lengthy to attempt in this chapter. However, several pages are dedicated to this topic in the *Industrial Ventilation Manual*, and because these tanks are quite common in industrial settings, it is well worth the time to carefully read this material and work the example exercises.

Summary of LEV Hood Design

Although the preceding discussion of LEV hood design is far from exhaustive, hopefully the reader has at least a basic understanding of how hoods function in a LEV system and why they are often considered to be the most important system component. It is at the hood where the system interfaces with the contaminant as it emanates from some process and begins to mix with workroom air. If the hood is poorly designed, the system might not generate the air movement necessary to capture and maintain control of the contaminant until it reaches the hood. If significant quantities of contaminant escape the hood, the overall design objective (i.e., to protect the employee's breathing zone) will be threatened, and the employee's health may ultimately be compromised. The importance of proper hood selection and design cannot be overemphasized.

In addition to the *Industrial Ventilation Manual*, several excellent, more detailed references can be consulted for in-depth discussion of hood design.(3–5) For example, Burgess et al. devote an entire chapter to hood design for specific operations in which the authors provide several pages of explanatory text and descriptive material on hood design for various applications.

A summary of important hood design considerations appeared in the original edition of this book and is offered here as a conclusion to this section on hoods.(6)

- An attempt should be made to minimize or eliminate all air motion in the area of the contaminant source. This will reduce the amount of air needed to be exhausted and subsequently reduce system power and equipment requirements.
- Air currents that necessarily exist should be utilized by the hood whenever possible.
- The hood should enclose the process as much as possible without endangering workers' safety.
- When enclosure is impractical, the hood should be located as close to the contaminant source as possible. The air velocity created by an exhaust hood varies inversely with the square of the distance for all but long, slot-type hoods.
- The hood should be located so that the contaminant is removed away from the breathing zone of the worker.
- The use of flanges and baffles should be considered. Flanges can increase hood effectiveness and may reduce air requirements by 25%.
- Use of a hood larger than required should be considered. Large hoods can reduce the danger of spills by diluting them rapidly to safe levels. It has also been shown that small hoods require higher capture velocities to be as effective as large hoods.

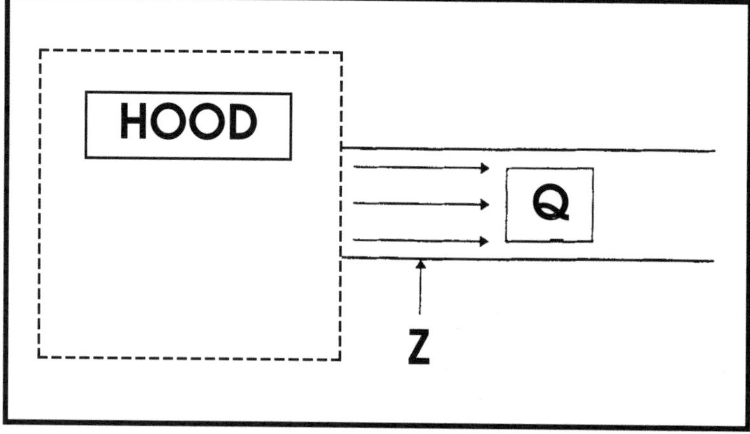

Figure 32.19 — Hood configuration and Q are known as Point Z.

LEV System Energy Requirements

At this point in the LEV design process the hood has either been designed from scratch to fit a particular process or has been adapted from a print in the *Industrial Ventilation Manual* or a similar source. The designer is at point Z as illustrated in Figure 32.19, with a volumetric flow rate of air, Q, that will capture and/or control the contaminant being generated.

Accurate determination of the airflow rate, Q, that must be moved into a hood for adequate contaminant control is indeed a critical component of LEV design. This parameter is essential for selecting a fan capable of moving the sufficient volumetric flow of air. In fact, Q is one of the two main parameters used to specify a fan. Once this quantity has been properly determined, a fan of adequate capacity can be selected. However, calculating Q is only half the story of LEV design and fan selection. The other important design parameter relates to the LEV energy requirements necessary to achieve the design airflow. Not only must the designer determine the essential airflow, Q, but he or she must also quantify the energy that must be continually delivered to the specified system to overcome the inertia of still air in front of the hood, move it into the LEV system, through the hood and ductwork, across the air cleaner, and eventually out the stack. This quantity, along with Q, is a critical parameter in LEV design and fan selection.

As mentioned, the fan is the device that furnishes the energy needed to drive a ventilation system. The magnitude of energy supplied to the system by the fan must be calculated accurately by the designer. If the energy requirement is incorrectly determined, an inadequate fan may be selected, preventing the system from performing as expected. Ultimately, the resulting LEV system may not move the designed volumetric flow rate of air and thus may fail to meet the objective of removing contaminants before they reach the worker's breathing zone.

Fundamentally, determining the overall energy delivery requirement of the fan is nothing more than a systematic summation of all the individual components of this energy. In other words, once the designer has selected the hood and calculated the necessary airflow, he or she can proceed in a logical, step-by-step fashion to quantify the energy requirements necessary to move this air through the specified system. The quantification procedure consists of sequentially focusing on each energy "consumer" in the system, calculating the magnitude of energy necessary to satisfy each consumer, and finally, summing all of these individual energy components. The sum of these components represents the overall energy required to move the specified quantity of air through the given system. Knowing the total quantity of air to be moved and the energy necessary to move it, the designer can select an appropriate fan for the given system. A brief analysis of LEV system energy consumers and the common methods of quantification will be the primary topic of discussion of the remainder of this chapter.

Introduction to the Energy Requirements in an LEV System

In discussing energy requirements for LEV systems, a logical place to begin is with the still air out in front the hood. This air, into which the contaminant is being generated, must be induced to flow into the LEV system in accordance with the calculated airflow requirements. To fulfill the objective of LEV, this air must be "energized"; in other words, enough energy must be supplied to the system by the fan to overcome the inertia of this still air so that the specified volumetric flow rate will move into the hood, hopefully entraining and transporting the contaminant along with it. Thus, the first component of energy overcomes the inertia of still air and accelerates it to the desired system velocity. This energy component represents the useful work accomplished by the system.

Once the proper flow rate of air is in motion, it must remain in motion until such time as it is discharged from the workplace through the properly designed stack. Unfortunately, as the air makes its way through the system, additional energy is consumed or lost due to factors such as turbulence and friction. Although this energy is lost in the sense that it contributes nothing toward the useful work of accelerating air, its magnitude must be tallied into the overall energy requirements of the system. The following sections of this chapter will be spent discussing several of these energy losses as well as the methods by which they are quantified.

Generally speaking, LEV system energy losses can be broken down into the following categories:[3]

- Turbulence losses as the air enters the hood;
- Friction losses as the air moves through ductwork;
- Losses in fittings (elbows, contractions, expansions, orifices, blast gates, valves, exhaust caps, etc.);
- Entry losses for branch entries;
- System effect losses at the fan; and
- Losses across air-cleaning devices.

Energy losses occur throughout the system between the hood face and the exhaust stack. Burton states that it is not uncommon for losses to outweigh useful work by 10 or 20 to 1.[3] Each component of this lost energy must be accurately quantified and summed with the other components to determine the total system energy requirements. This total energy must be constantly delivered to the system by the fan to ensure that the specified volumetric flow rate of air continually moves through the system. Otherwise, the system will not operate properly. If the designer fails to account for significant energy consumers and thus underestimates energy requirements in the system design phase, a fan with inadequate energy capacity may be selected. When this system is installed and operated, the fan will deliver an insufficient amount of energy to the system. Thus, after all the energy consumers between the exhaust stack and the hood face are satisfied, not enough energy will remain to accelerate the required quantity of air. Such a system risks failure to accomplish the basic goal of LEV.

To understand properly how energy requirements are calculated, one must first understand that the energy delivered to an LEV system by the fan is not typically described in familiar energy units but rather in terms of the different types of pressures that are encountered within the system. Thus, a description of the calculation processes used to determine system energy requirements necessitates a discussion of the various LEV system pressure relationships.

Pressure Relationships in an LEV System

A virtually universal principle of air movement is simply this: air moves from one point to another because of a difference in pressure between those two points. Regardless of whether one is considering an LEV system or global weather patterns, air moves from an area of higher pressure to an area of lower pressure. Therefore, if the goal is to induce workplace air to move into an LEV system, the pressure inside the system must be less than the pressure outside the system (typically the ambient, atmospheric pressure existing in the workplace). In an LEV system, this pressure differential is generated by the fan.

To understand how the fan accomplishes this mission, consider it to be analogous to a "bucket brigade" with the fan blades as the buckets. As the blades turn they scoop air from one side of the fan and deposit it on the other side. (Note some common LEV terminology: The side of the fan from which the air is scooped is commonly called the suction or downstream side of the fan. The side where the air is deposited is typically called the pressure or upstream side of the fan. See Figure 32.20.) The process of scooping up a blade (bucket) full of air creates a partial vacuum or area of low pressure on the suction side of the fan. Adjacent air moves over to fill in this area of lower pressure. This movement of air proceeds on down the duct in a manner similar to the domino effect as the air continues to flow toward the fan to fill in adjacent areas of lower pressure. At the hood face, the point at which the LEV system interfaces with ambient pressure, a pressure differential is established between the outside atmospheric pressure and the pressure inside the system. In this case, the pressure at the hood face (which is on the suction side of the fan) is less than atmospheric (that is, it is negative with respect to atmospheric pressure). Thus, the outside air, which is at atmospheric pressure, flows into the LEV system through the hood.

Of course, just the opposite occurs on the pressure side of the fan. When the air is deposited by the fan blade, an area of higher pressure is created that proceeds up the duct toward the exhaust stack, where the system once again interfaces with ambient pressure. However, on the pressure side of the fan the pressure inside the system is greater than (i.e., positive with respect to) the ambient pressure. Thus, the air flows out of the stack and into the ambient environment.

In summary, the rotating fan blades establish a condition in which the pressure on the suction side of the fan is continuously less than atmospheric pressure, causing air to flow into the LEV system. On the pressure side of the fan, a pressure that is greater than atmospheric is created, causing the air to flow out of the system.

This fan-generated differential pressure between the inside of the LEV system and the outside ambient pressure is called static pressure (SP). Static pressure, like volumetric flow rate, is one of the most fundamental quantities of LEV design. The static pressure represents the energy that is continually delivered to the system by the fan. It is analogous in concept to potential energy in that it indicates the capability of the system to do work (i.e., overcome the inertia of still air, accelerate it to the desired system velocity, and convey it through the system until it is finally discharged into the ambient environment).

Referring to Figure 32.20, note again that static pressure is a differential pressure, not an absolute pressure. The static pressure measured at any point in an LEV system is

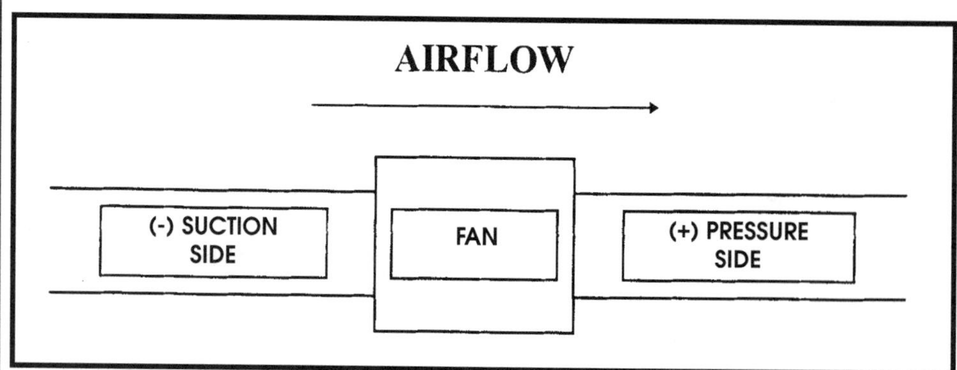

Figure 32.20 — Illustration of suction side and pressure side of fan.

the difference between the atmospheric pressure and the pressure inside the system at that point.

The units used for static pressure in LEV design are not the familiar pressure units such as pounds per square foot or pounds per square inch (psi). These units are not very useful in LEV work because of the small pressure differentials involved. Instead, it is more convenient to express static pressure as the height of a liquid column (usually mercury or water) that would be supported by that pressure. As a point of reference, atmospheric pressure at sea level is typically given as 14.7 psi. This pressure would support a water column approximately 406 inches high. Thus, atmospheric pressure can be reported as 406 in. wg (where wg stands for water gage). If the static pressure existing in an LEV system at a given point would support a 2-inch column of water, the static pressure at that point is said to be 2 inches of water or 2 in. wg. If this pressure is being measured on the suction side of the fan, the algebraic sign of the static pressure would be negative, indicating that it is less than atmospheric at this point in the LEV system. It would be reported as 2 in. wg.

Static pressure acts in all directions in an LEV system. On the suction side of the fan, the static pressure tends to collapse the duct, much like a straw in a thick milkshake collapses when lower pressure is created by someone drinking the milkshake. On the fan's pressure side, the static pressure acts to expand the duct in a manner similar to blowing up a balloon (in fact, it is static pressure that keeps a balloon inflated). When LEV designers speak of energy delivered to the system by the fan, they are referring to potential energy in the form of static pressure. This energy represents this system's ability to do work.

The fan must be capable of generating a sufficient magnitude of static pressure (in in. wg) at the hood opening so that the desired quantity of still air will be accelerated into the hood and up to the proper system velocity. The important question is how much static pressure the fan must deliver at the hood opening to accomplish this.

To answer this question, another important pressure existing in an LEV system must be introduced—velocity pressure (VP). As the fan blades turn and continually generate static pressure (potential energy), the pressure differential created begins to induce the still air to move as discussed previously. As the fan blades begin spinning, a portion of the static pressure generated is immediately converted into pressure exerted by moving air. In other words, some of the potential energy created by the fan is converted into kinetic energy of air in motion. The pressure exerted by this moving air is referred to as velocity pressure. In an operating LEV system, negative static pressure at a hood face directs room air into the system and accelerates it to the desired system velocity. Thus, a portion of the fan-generated static pressure available at the hood face is continuously converted into velocity pressure as the air is accelerated to the desired system velocity.

Essentially, velocity pressure is the kinetic energy generated in a ventilation system as the result of air movement. The wind you feel on a windy day, or when you hold your hand out of the window of a moving car is due to velocity pressure. Unlike static pressure, which acts in all directions, velocity pressure acts only in the direction of flow. Velocity pressure is also typically measured in units of in. wg. In this case it is the height of a water column that would be held up by the pressure exerted by the moving air. Unlike static pressure, velocity pressure is always positive in an LEV system, regardless of the side of the fan on which it is measured. It is the pressure exerted by moving air and cannot be negative. If no air is moving, the velocity pressure is 0.

(Note that up to this point in the chapter, the phrase "desired system velocity" has been used several times without explanation. Stating that the static pressure must be sufficient to accelerate still air to some desired system velocity implies that the LEV designer has in mind some target velocity at which he or she desires the air to be moving as it makes its way through the LEV system. This is, in fact, the case. As the designer calculates the static pressure required to accelerate still air and move it into the LEV system, this calculation is made with a specific system (i.e., duct) air velocity in mind. The procedure used by the LEV designer to estimate the desired duct velocity will be discussed in subsequent paragraphs.)

Since it is the total fan static pressure requirement that the LEV designer is calculating, there must be some relationship between the static pressure at the hood opening and the desired velocity of air in the duct leading to that hood. To understand the relationship between static pressure and velocity one must first understand the relationship between velocity and velocity pressure. Velocity pressure is defined in the *Industrial Ventilation Manual* as "that pressure required to accelerate air from zero velocity to some velocity (V) and is proportional to the kinetic energy of the air stream."[1] Obviously, some relationship exists between the velocity pressure (in in. wg) exerted by moving air and the velocity at which the air is moving. Mathematically, this relationship can be expressed quite simply as follows:

$$V = 1096\sqrt{\frac{VP}{\rho}} \qquad (15)$$

where:

V = the velocity of the moving air (ft/min);
VP = the velocity pressure exerted by the moving air (in. wg); and
ρ = the density of air (lbs/ft^3).

For standard air (ρ = 0.075 lbs/ft^3), the equation can be expressed as follows.

$$V = 4005\sqrt{VP} \qquad (16)$$

This relationship between velocity and velocity pressure was described many years ago by Bernoulli. Its actual derivation, while beyond the scope of this chapter, is treated in some detail in various other texts.[1,4-9] (Obviously, the equation in the above form is not properly "dimensional" in the sense that all units do not cancel properly. In this equation the constant, 4005, accounts for dimensionality by incorporating the various conversion factors between velocity pressure in in. wg and velocity in ft/min. It is essential to use this constant and to have velocity pressure in units of in. wg to calculate the velocity in ft/min.)

This equation is extremely useful to the LEV designer. Once the desired system velocity is determined, the velocity pressure exerted by air moving at this velocity can be determined. For example, consider Figure 32.21.

Assume that in this particular system, the LEV designer desires an air velocity of 3500 ft/min in the duct leading to the hood. In other words, the still air in front of the hood must be accelerated to a velocity of 3500 ft/min in the duct leading to the hood. (The reason that 3500 ft/min was selected will be discussed later.) The question is, what velocity pressure will be exerted by air moving at a velocity of 3500 ft/min?

$$V = 4005\sqrt{VP}$$

$$VP = \left(\frac{V}{4005}\right)^2 = \left(\frac{3500}{4005}\right)^2 = 0.76\text{in. wg} \qquad (17)$$

So, following the definition of velocity pressure given in the *Industrial Ventilation Manual,* to accelerate still air to a duct velocity of 3500 ft/min, 0.76 in. wg of velocity pressure is required. The question is: Where does this 0.76 in. wg of velocity pressure come from? The answer: It is continuously generated from static pressure produced by the fan.

This relationship between velocity and velocity pressure enables the LEV designer to address the question raised previously regarding the amount of static pressure that must be present at the hood opening to accelerate the desired quantity of still air into the hood and up to the proper system velocity. As stated, the static (differential) pressure existing at the hood face induces the air to move into the LEV system. In this process the static pressure (potential energy) is converted to velocity pressure (kinetic energy) as the air begins to move. In the system under consideration a sufficient quantity of static pressure must be continuously available at the hood face and converted to velocity pressure to accelerate the air to 3500 ft/min in the duct leading to the hood. The 3500 ft/min corresponds to 0.76 in. wg of velocity pressure. This velocity pressure comes from static pressure generated by the fan. Therefore, 0.76 in. wg of static pressure

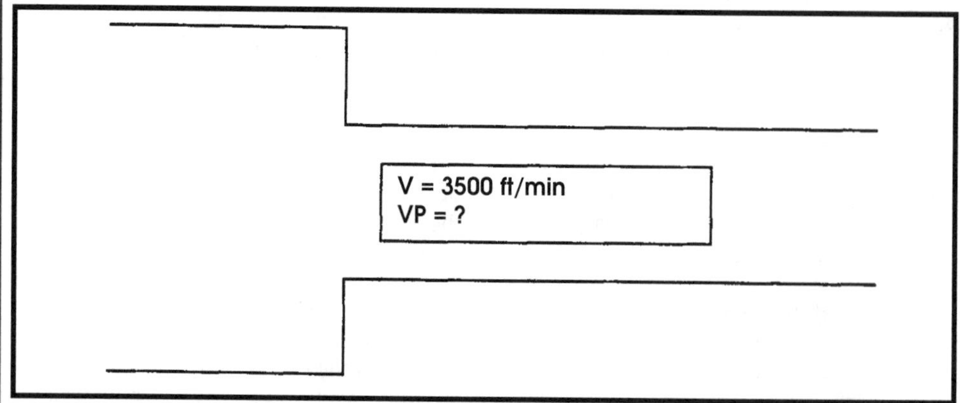

Figure 32.21 — Relationship between velocity and velocity pressure.

must be continuously generated by the fan and continuously converted to 0.76 in. wg of velocity pressure to achieve air acceleration to 3500 ft/min. Thus, the fan must continuously generate one velocity pressure's worth of static pressure to accelerate still air up to the desired duct velocity. In the LEV design process in which the designer must sum up the total static pressure to be generated by the fan, he or she must start with one velocity pressure's worth of static pressure that will be used to accelerate the still air up to the desired duct velocity. Note that as this static pressure is present on the suction side of the fan, the algebraic sign would be negative (i.e., -0.76 in. wg).

Example 7. Assume that a velocity of 4500 ft/min is desired in the duct leading to the hood in Figure 32.21. How much static pressure must the fan generate to achieve the acceleration of still air out in front of the hood to this velocity?

Solution. The LEV designer knows that one velocity pressure's worth of static pressure must be generated by the fan in any system to achieve the desired duct velocity. In this system, in which a duct velocity of 4500 ft/min is desired, what is the value of one velocity pressure's worth of static pressure?

$$VP = \left(\frac{V}{4005}\right)^2 = \left(\frac{4500}{4005}\right)^2 = 1.26 \text{ in. wg} \quad (18)$$

In this case, the velocity pressure corresponding to the desired duct velocity of 4500 ft/min is 1.26 in. wg. Therefore, the LEV designer must start the static pressure summation process with -1.26 in. wg, which must be continuously generated by the fan to accelerate still air out in from of this hood to a velocity of 4500 ft/min in the duct leading to the hood.

As mentioned earlier, accelerating the air to the desired duct velocity represents the useful work accomplished by the fan. If no other energy consumers existed in the system, the LEV designer's job would be a simple one indeed. All he or she would need to do is determine the desired duct velocity, calculate the velocity pressure exerted by air moving at this velocity, and then specify a fan that continuously provided that velocity pressure's worth of static pressure to the system. Note from Figure 32.22 that in the absence of any other energy losses, one velocity pressure's worth of static pressure provided by the fan at Point C would be sufficient to properly accelerate the air at Point A and move it all the way through the system. Unfortunately, many energy consumers exist between points A and C that must be quantified in terms of the static pressure consumed. Each of these must be added to that original one velocity pressure's worth of static pressure needed to accelerate the air to accurately determine the total static pressure that must be generated by the fan at Point C to ensure proper airflow and velocity.

These energy consumers or LEV system energy losses have been listed previously. For example, between points A and B in Figure 32.22, some energy is lost as turbulence as the air enters the hood. This energy loss must be calculated and summed into the total fan energy requirement. As air moves from Point B to Point C it encounters friction from the walls of the ductwork. Additional energy must also be supplied to the system to overcome these friction losses as well. Although this figure shows a straight run of ductwork between points B and C, many systems have elbows, contractions, expansions, and so forth in the ductwork. Any time the air changes direction, expands, or contracts, additional energy is lost as turbulence, friction, and/or inertial changes. This energy, too, must be added to the total system energy requirements. During the design phase, each of these losses must be accounted for and quantified in terms of static pressure. This static pressure, which will ultimately be lost in the system, must be added to the original one velocity pressure's worth of static pressure that will perform the useful work of accelerating the air. The summation of the energy losses and the energy used to accelerate the air determines the total static pressure that must be added to the system by the fan to have a properly functioning system.

Note that because of these energy losses, the static pressure in an LEV system increases in absolute value (i.e., grows more negative) as the system proceeds from the

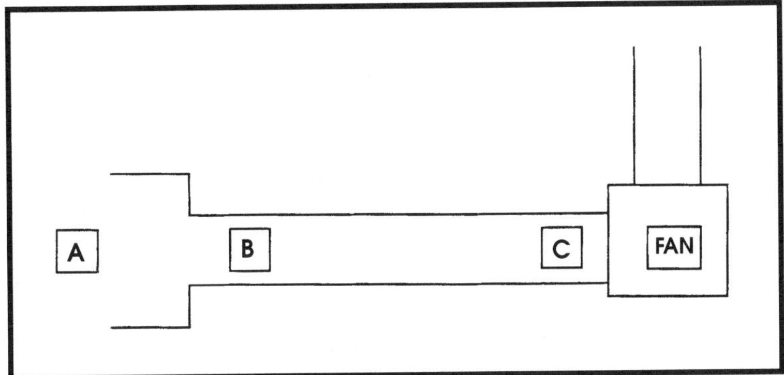

Figure 32.22 — Illustration of simple system.

hood toward the fan. Obviously, in Figure 32.22, the static pressure at Point B must be less than (i.e., more negative than) the static pressure at Point A or the air will not flow into the system. As has been shown, this difference in static pressure between A and B must be at least one duct velocity pressure's worth to accelerate the still air to the desired duct velocity. However, the total difference in static pressure between A and B is one velocity pressure's worth plus what is lost between A and B because of hood turbulence. Similarly, static pressure is greater in absolute value (more negative) at Point C that at Point B; however, at Point C we need all of that static pressure required at B to accelerate air and overcome hood losses plus whatever additional losses are incurred between B and C in the form of friction, elbows, expansions, and so forth. Static pressure measured at any point in the system downstream of the fan represents the total energy required to move the air to that point. This includes the energy needed to accelerate the air plus any energy that is lost downstream of that point because of turbulence, friction, and so forth. In the LEV design phase, the designer starts at Point A and sums static pressure requirements as the system proceeds from the hood to the fan. The goal is to arrive at the fan with the total static pressure needed by the specified system to accelerate the desired amount of air and overcome all of the static pressure losses.

Another way to picture what's happening in a LEV system is to consider the manner in which static pressure is consumed in an operating LEV system. Between any two points on the suction side of the fan, the static pressure will be greater in absolute value at the point closest to the fan. In fact, static pressure is greatest (most negative) immediately downstream of the fan (Point C in Figure 32.22). This is the maximum static pressure in the system. From Point C to Point A, the absolute value of static pressure decreases as it is used to overcome energy losses. In other words, between the fan and the hood the amount of static pressure available will progressively become less and less. The LEV designer must ensure that enough static pressure is generated at Point C so that as portions are used up from C (fan) to A (hood opening), enough is left over to accelerate the air to the specified velocity.

A concept that emerges from the above discussion is that static pressure and velocity pressure are mutually convertible (i.e., static pressure can be converted into velocity pressure and vice versa). This is analogous to the mutual convertibility of potential and kinetic energies. Just as the potential energy and the kinetic energy can be summed to give the total energy in a system, the static pressure and the velocity pressure can be summed to give the total pressure of an LEV system. Algebraically, this can be expressed as follows:

$$TP = SP + VP \tag{19}$$

Referring again to Figure 32.22, note that the total pressure at Point C (TP_c) is the sum of the static pressure and the velocity pressure at Point C.

$$TP_c = SP_c + VP_c \tag{20}$$

This equation can be rearranged as:

$$-TP_c = -SP_c + VP_c \tag{21}$$

This illustrates that the static pressure measured at Point C is comprised of essentially two components: that which has accomplished the useful work of air acceleration (which is represented mathematically by the velocity pressure) and that which has been lost in the system up to Point C (which is represented mathematically by the total pressure). In effect, total pressure at any point in the system represents the amount of static pressure consumed up to that point in overcoming friction, turbulence, and so forth.

Earlier in this chapter the concept of conservation of mass was discussed in the sense that once a quantity of air is defined and moving into or through a ventilation system, the volumetric flow rate through any cross-sectional area of the system will be identical to that flowing through any other cross-sectional area of the system. The relationship between total pressure, static pressure, and velocity pressure illustrates another basic principle of fluid mechanics that governs the flow of air in LEV systems: the conservation of energy. Energy, like mass, must be conserved in an industrial ventilation system. "Conservation of energy means that all energy changes must be accounted for as air flows from one point to another."[1] In terms of Figure 32.22 and the pressures discussed previously, this principle can be expressed as follows:

$$TP_c = TP_B + h_L$$

or (22)

$$SP_c + VP_c = SP_B + VP_B + h_L$$

where:

h_L = the energy losses encountered by the air as it flows from Point B to Point C.[1]

For an example of the manner in which total pressure, static pressure, and velocity pressure are related in a typical ventilation system, please note Figure 32.23.

Measurement of LEV System Pressures

It was stated previously that pressures in LEV systems are measured in inches of water (in. wg). Several sophisticated pieces of equipment are available to conduct the measurements; however, the simplest and most straightforward device for measuring these pressures is a simple tube filled with a suitable liquid (e.g., water or mercury) as illustrated in Figures 32.24 through 32.26. This instrument is called a liquid manometer.

Figure 32.24 illustrates the measurement of static pressure on both the suction and pressure sides of the fan. Recall that static pressure acts in all directions. It tends to collapse the duct on the suction side of the fan and expand the duct on the pressure side of the fan. If a hole is drilled into a duct and one end of a manometer tube inserted, static pressure would be registered by the manometer liquid regardless of the orientation of the tube (e.g., perpendicular to flow, parallel to flow, or any orientation in between). However, to eliminate the impact of the moving air on the manometer liquid (that is, to prevent the manometer liquid from being deflected by velocity pressure as well as static pressure), static pressure is measured perpendicular to flow as illustrated in Figure 32.24. Note that on the suction side of the fan, the ambient pressure, which is greater than the pressure inside the system, presses down on the liquid through the open end of the manometer causing the negative deflection as shown. On the pressure side of the fan, where the pressure inside the system is greater than ambient pressure, the opposite liquid deflection is noted.

Total pressure measurement is illustrated in Figure 32.25. Here, the manometer tube inside the duct is oriented parallel and into the direction of flow. In this orientation the manometer liquid is deflected by both the velocity pressure (which is the pressure exerted by the moving air) and the static pressure (which, as stated, acts in all directions).

If a measurement of velocity pressure is desired, a manometer arrangement such as the one illustrated in Figure 32.26 is required. In this setup, both ends of the manometer tube are open to the inside of the system. However, one end of the tube is positioned perpendicular to the direction of flow, measuring only static pressure. The other end of the tube is oriented parallel and into the direction of flow, measuring both velocity pressure and static pressure (i.e., total pressure). Thus, static pressure,

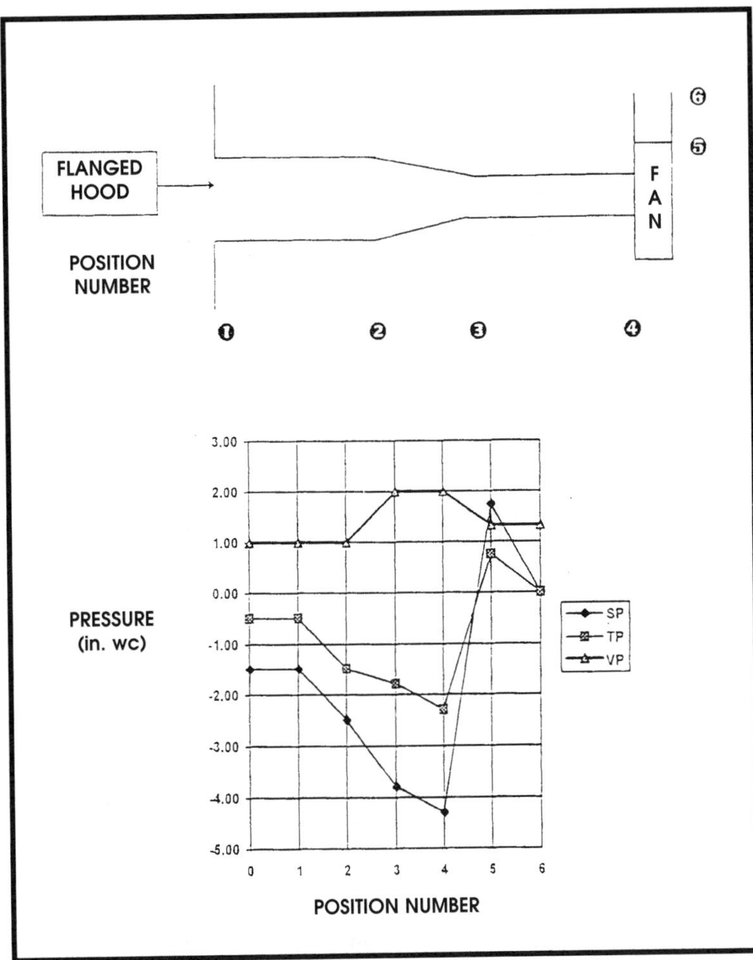

Figure 32.23 — Behavior of TP, SP, and VP in a typical system.

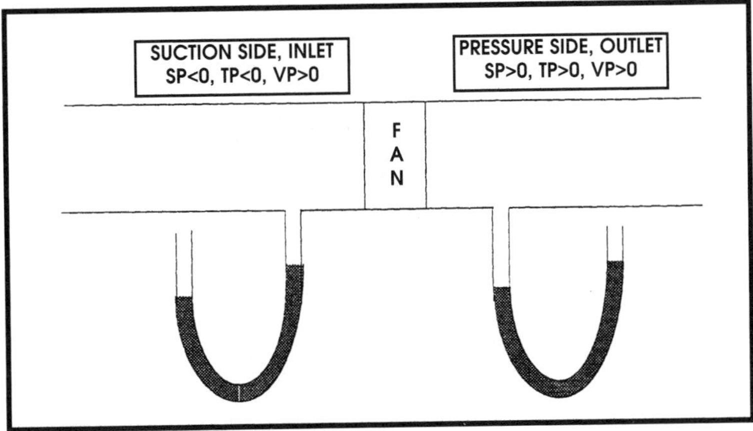

Figure 32.24 — Static pressure measurement.

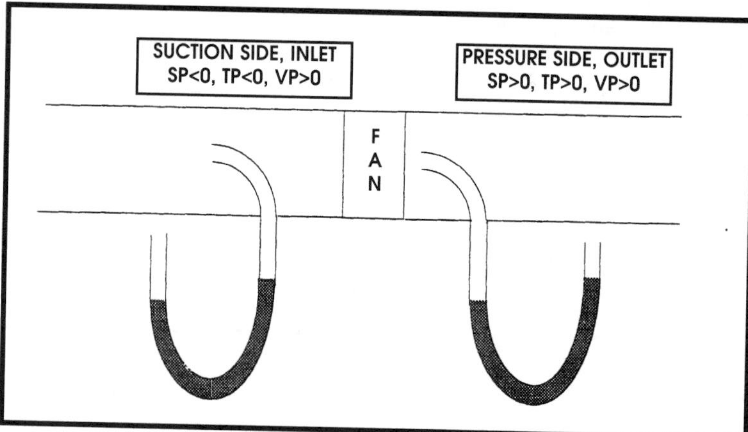

Figure 32.25 — Total pressure measurement.

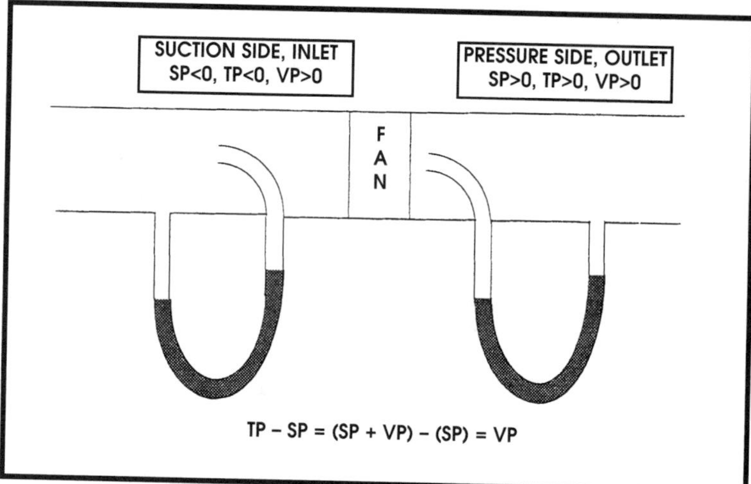

Figure 32.26 — Velocity pressure measurement.

which acts in all directions, is being registered by both ends of the manometer tube. In this manner, the effect of static pressure is essentially canceled out, leaving only the deflection caused by the impact of the moving air (i.e., velocity pressure) to be measured by the manometer. Algebraically, the manometer is performing the following operation:

$$TP - SP = (SP + VP) - (SP) = VP \quad (23)$$

A summary of the properties of total pressure, static pressure, and velocity pressure is given as Table 32.4.

Quantifying the Energy Lost at the Hood

As stated previously, when quantifying LEV system energy (i.e., static pressure) requirements, the goal is to arrive at the fan with the total static pressure needed by the specified system to both accelerate the air to the desired duct transport velocity and to overcome all of the static pressure losses incurred. The amount of static pressure needed for air acceleration has been identified as one duct velocity pressure's worth. That is, if the duct transport velocity desired is 4500 ft/min (as in Example 7), it will take 1.26 in. wg of static pressure to achieve this acceleration, since standard air moving at 4500 ft/min exerts a velocity pressure of 1.26 in. wg.

Thus far in the chapter, static pressure losses have been mentioned and discussed in a rather general way. Getting more specific, Burgess et al. point out that "as a fluid moves through a system, it will encounter resistance to flow."[(4)] This resistance decreases static pressure throughout the system that must be overcome by the fan.

> *This resistance can arise from two general mechanisms: (1) friction associated with shearing stresses and turbulence within the duct, and (2) shock from sudden velocity (speed or direction) changes or flow separation at elbows, branches, and transitions (i.e., expansions or contractions). Frictional pressure losses result from rubbing the fluid against the walls of the conduit as well as against itself.*[(4)]

Shock losses result "from turbulence or shock in the fluid stream, which causes a violent mixing of the fluid and subsequent eddy formation. These disturbances are usually associated with redirection of flow or sudden changes in the duct size that consequently cause drastic velocity changes."[(4)]

The next several sections will describe the procedures for quantifying these frictional and shock losses at various points in the system in terms of the amount of static pressure consumed. A logical place to begin is with those losses occurring at the hood.

Consider Figure 32.27, which depicts the streamlines of air flowing into a plain hood opening.

> *As air is pulled into the duct opening from the area outside the duct, the individual airstreams converge to a higher velocity air stream that does not completely fill the duct. The point where the airstream diameter is the smallest is called the vena contracta (which means a contraction of the airstream diameter).*[(5)]

Since the actual diameter of the airstream (and thus the cross-sectional area through which the air is moving) is smaller at the vena contracta than at other points in the duct, the velocity at the vena contracta is higher than the average duct velocity in the rest of the system in accordance with the continuity equation (Q = VA). Therefore, at the vena contracta, the air

Table 32.4 — Properties of SP, VP, and TP

Pressure Type	Measurement Technique	This Type of Pressure Will	Normal Sign Suction Side	Normal Sign Pressure Side
STATIC PRESSURE (SP)	Perpendicular to the duct	SUCTION SIDE Tend to collapse a duct on the suction side (-) PRESSURE SIDE Tend to expand a duct on the pressure side (+) Overcome friction in ducts, air cleaners, coils, etc. Accelerate the air to the required velocity.	NEGATIVE	POSITIVE
VELOCITY PRESSURE (VP)	Difference between TP - SP = VP	Maintain the airflow.	POSITIVE	POSITIVE
TOTAL PRESSURE (TP)	Parallel to axis of flow TP = SP + VP	Air flows from a zone of higher total pressure to a zone of lower pressure	NEGATIVE	POSITIVE

velocity has increased and, correspondingly, the velocity pressure also increases. This velocity pressure increase, of course, can only come at the expense of static pressure, which decreases at the vena contracta as depicted in Figure 32.27. Thus, a small energy loss occurs in this conversion of static pressure to velocity pressure; however, as shown in Figure 32.27, as the air passes through the vena contracta, it slows down as the airstream diameter enlarges to fill the duct. In this process, velocity (and thus velocity pressure) decreases. Interestingly, at this point some velocity pressure actually converts to static pressure. However, this uncontrolled deceleration of the air from vena contracta velocity to the downstream duct velocity results in significant shock and friction losses. "The more pronounced the vena contracta, the greater will be the energy loss."(1) This energy loss is called hood entry loss (h_{ed}) and represents additional static pressure that must be quantified by the LEV system designer added to the system by the fan. Overcoming this loss is a continual demand placed on an operating LEV system. This is energy that is continually lost to heat of friction, shock, turbulence, and so forth as the air moves from Point A to Point B in Figure 32.22.

The next logical question is how hood entry loss is quantified in terms of static pressure. As noted above, the magnitude of the energy (i.e., static pressure) loss at a hood is directly related to the size of the vena contracta. The smaller the diameter of the airstream at the vena contracta, the larger the energy loss. Therefore, factors that affect the size of the vena contracta will also affect the amount of static pressure lost (and thus the amount of static pressure that the LEV designer must add to the overall specification for the fan). Any factor that causes the airstream diameter at the vena contracta to be smaller will increase the static pressure lost at a hood, and any factor that causes the airstream diameter to be larger at that point will have the opposite effect.

Two major factors affect the size of the vena contracta: (1) the ultimate velocity to which the air is being accelerated; and (2) the shape or geometry of the hood. The first of these factors may be somewhat intuitive. The higher the velocity of air

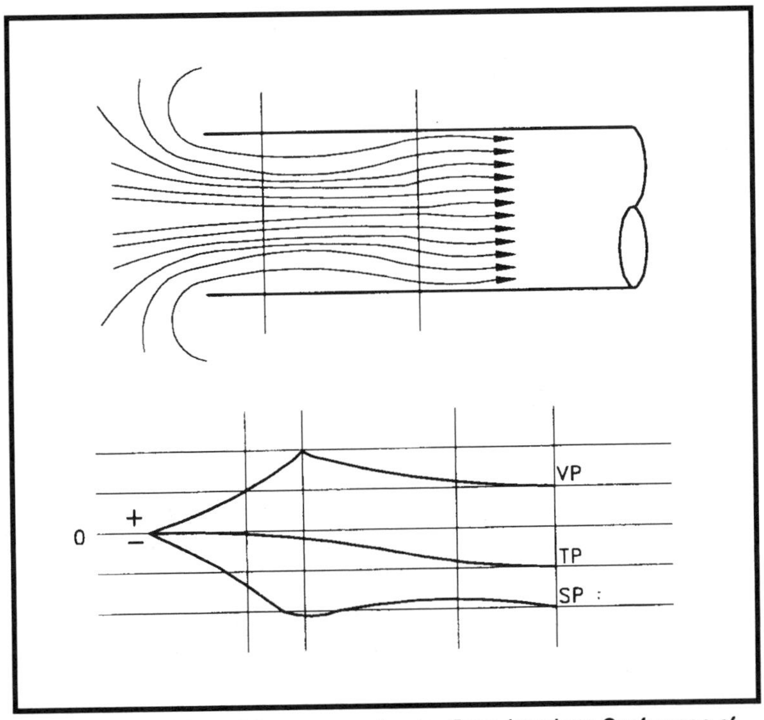

Figure 32.27 — Airflow at the vena contracta. (From **American Conference of Governmental Industrial Hygienists (ACGIH®):** *Industrial Ventilation: A Manual of Recommended Practice*, 22nd Ed. Copyright 1995, Cincinnati, Ohio. Reprinted with permission.)

entering a given hood, the greater the inertia and thus the more pronounced or "scrunched" the airstream becomes as the air enters the hood. Indeed, for a given hood, the hood entry loss increases in direct proportion to the ultimate velocity (or velocity pressure) to which the air is accelerated. (Note that this increasing loss with increasing duct velocity pressure is completely separate and distinct from the additional static pressure needed to perform the useful work of accelerating the air to some higher velocity pressure.)

The shape of the hood is also critical in determining the size of the vena contracta. As shown in Figure 32.27, the vena contracta is formed by the airstream lines as they enter the hood. By virtue of their geometry, certain hood configurations allow a smoother entry of airstream lines into the duct. The less bending, twisting, and contorting to which the stream lines of the air entering the hood are subjected, the less pronounced the vena contracta and the lower the corresponding shock and friction losses.

Compare a bell mouth inlet hood to a plain opening hood as shown in Figure 32.28. The shape of the bell mouth inlet more naturally matches the geometry of the airstream lines as they flow into the hood. Note that the stream lines are not forced to bend, turn, or twist but rather flow very smoothly into the system. Thus, they do not converge into an area that is much smaller than that of the duct leading to the hood (i.e., the stream lines do not form a tight vena contracta). However, consider the airstream lines entering the plain hood opening. Here the significant changes in airstream direction and the accompanying turbulence create a much tighter vena contracta, leading to increased shock and friction losses. More energy, in the form of static pressure, must be invested in the plain hood entry than in the bell mouth entry to pull room air into the system. A bell mouth entry, into which the stream lines of air can flow very smoothly, is said to be a more efficient hood than a plain round opening because less static pressure is wasted as the air enters the system.

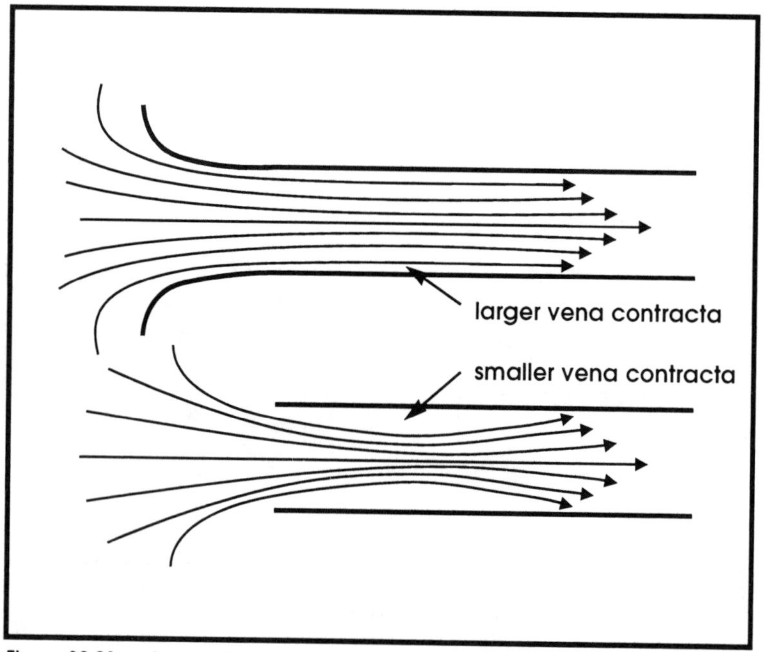

Figure 32.28 — Comparison of vena contracta formation for different hoods.

The degree to which a given hood shape can accomplish a smooth transition of air streamlines into the duct gives rise to an extremely important hood parameter, the hood entry loss factor (F_H). Essentially, the lower the factor, the smoother (i.e., less turbulent, twisted, or contorted) the airflow into the hood and, thus, the less pronounced the vena contracta. Of course, this ultimately translates into less static pressure being wasted as the air enters the hood.

So the hood entry loss (in in. wg of static pressure) for a given hood is ultimately determined by the velocity to which the air is being accelerated, combined with the ability of the given hood geometry to affect a smooth transition of this air into the duct (as reflected by the hood entry loss factor). These two factors can be used by the LEV designer in the following relationship to estimate the magnitude of static pressure lost at the hood.

$$h_{ed} = F_H \times VP_d \qquad (24)$$

where:

h_{ed} = the hood entry loss (in in. wg);
F_H = the hood entry loss factor for the given hood (dimensionless); and
VP_d = the velocity pressure in the duct leading to the hood (in in. wg).

In this expression h_{ed} is the amount of static pressure (in in. wg) that is lost or consumed as the air makes its way from the room into the duct. Note that it is determined by the velocity (velocity pressure) in the duct combined with the shape of the hood.

It is important to understand that the hood entry loss factor is determined solely by the hood geometry and is therefore a constant for a given hood. This factor does not change with velocity pressure. However, as noted from the above equation, the actual magnitude of static pressure lost at the hood does, in fact, change as the velocity pressure to which the air is ultimately accelerated increases. Thus, for a given hood, more static pressure is lost (and therefore must be accounted for in total fan static pressure requirements) as the duct velocity increases.

Obviously, before an LEV designer can determine the amount of static pressure that must be added to the system by the fan to overcome hood loss he or she must know the hood entry loss factor for the hood that will be used in the system. Actually, every hood in existence has its own unique hood entry loss factor. Although the method is too lengthy to discuss in this chapter, the actual procedure for hood entry loss factor determination is quite straightforward for an operational hood in the field. (See Burton for a more complete description of hood entry loss factor measurement.[3]) However, being able to measure this factor in the field does not help the designer who is attempting to quantify static pressure requirements in the design phase (i.e., before the hood has even been constructed).

In reality, similar hood shapes have similar values for F_H. These values are tabulated for common hood geometries in the *Industrial Ventilation Manual* and other sources.[1-5] Figure 32.29 illustrates F_H values for common hoods.

Note from Figure 32.29 that F_H for the bell mouth inlet is only 0.04 while the plain opening F_H is 0.93. This, of course, is consistent with the previous discussion, since the bell mouth entry is a much more efficient hood than the plain opening.

Example 8. Calculate the hood entry loss in in. wg for (a) a bell mouth inlet hood and (b) a plain opening hood if the desired duct velocity is 3450 ft/min.

Solution. The first step is the calculation of the velocity pressure exerted by air traveling at a velocity of 3450 ft/min is as follows.

$$V = 4005\sqrt{VP}$$

$$VP = \left(\frac{V}{4005}\right)^2 = \left(\frac{3450}{4005}\right)^2 = 0.74 \text{ in. wg} \quad (25)$$

(a) For the bell mouth inlet for which $F_H = 0.04$

$$h_{ed} = F_H \times VP_d$$

$$h_{ed} = F_H \times VP_d = 0.04 \times 0.74 \text{ in. wg} = 0.03 \text{ in. wg} \quad (26)$$

(a) For the plain opening for which $F_H = 0.93$

$$h_{ed} = F_H \times VP_d$$

$$h_{ed} = F_H \times VP_d = 0.93 \times 0.74 \text{ in. wg} = 0.03 \text{ in. wg} \quad (27)$$

This example demonstrates the significant difference between these two common hood geometries in terms of the static pressure lost in the process of moving air from the room into the system. At a duct velocity of 3450 ft/min, only 0.03 in. wg is consumed by the more efficient bell mouth inlet compared with the 0.69 in. wg consumed by the plain opening. Remember that this is extra energy that must be continuously supplied to the system by the fan and therefore summed into the total energy (static pressure) requirements. Remember, too, that the difference becomes even more pronounced as the duct velocity increases.

Obviously, hood geometries considerably more complex than those illustrated in Figure 32.28 exist in the real world. In Chapter 10 of the *Industrial Ventilation*, hood entry loss factors are given for more than 100 typical industrial applications.

For certain types of hoods, shock and friction losses can occur at more than one point between the air in the room and the air in the duct downstream of the duct vena contracta. These hoods, called <u>compound hoods</u>, are hoods that have two or more points of significant energy loss. "Common examples of hoods having double entry losses are slot type hoods and multiple opening, lateral draft hoods commonly used on plating, paint dipping and degreasing tanks, and foundry side-draft shakeout ventilation."[1] For example, consider the slot/plenum arrangement illustrated in Figure 32.30. An energy loss occurs as a vena contracta is formed by the stream lines downstream of the slot. The air then continues through the plenum and converges into the duct, forming the second vena contracta and thus the second point of energy loss. The energy loss occurring at the slot as the air is accelerated to the slot velocity pressure is referred to as the slot entry loss (h_{es}) as distinguished from the duct entry loss (h_{ed}) that occurs as the air is accelerated to the (usually) higher duct velocity pressure.[1]

In calculating the overall static pressure loss for the compound hood, each point of significant energy loss must be independently calculated and added together.

$$h_e = h_{ed} + h_{es} = (F_H \times VP_d) + (F_S \times VP_S) \quad (28)$$

where:

HOOD TYPE	DESCRIPTION	COEFFICIENT OF ENTRY, CE	HOOD ENTRY LOSS (F_N)
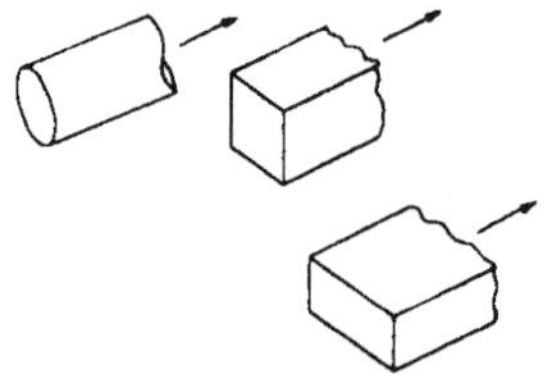	PLAIN OPENING	0.72	0.93
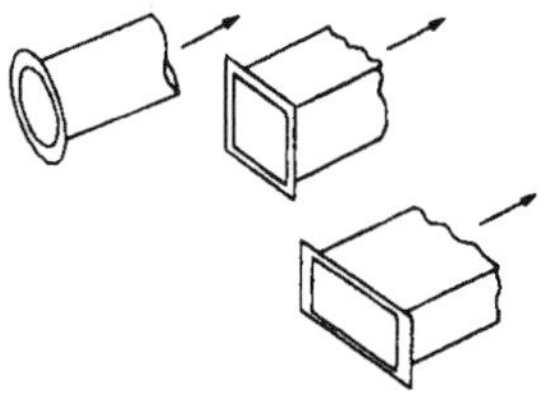	FLANGED OPENING	0.82	0.49
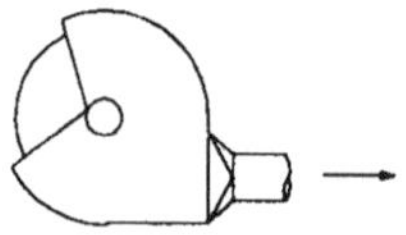	TYPICAL GRINDING HOOD	0.78	(STRAIGHT TAKEOFF) 0.65
		0.85	(TAPERED TAKEOFF) 0.40
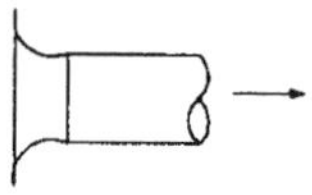	BELL MOUTH INLET	0.98	0.04
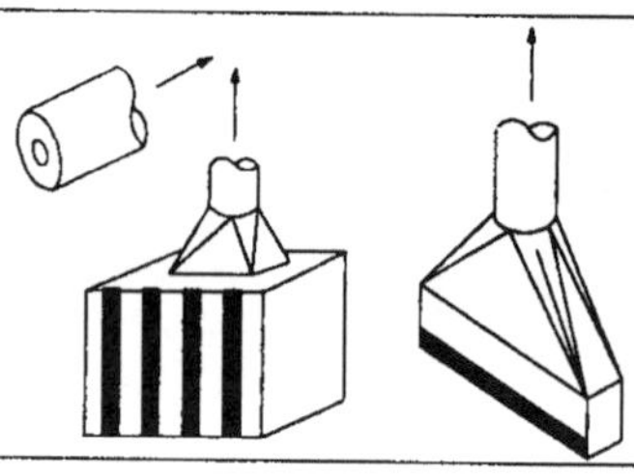	ORIFICE	SEE FIGURE 5-12	
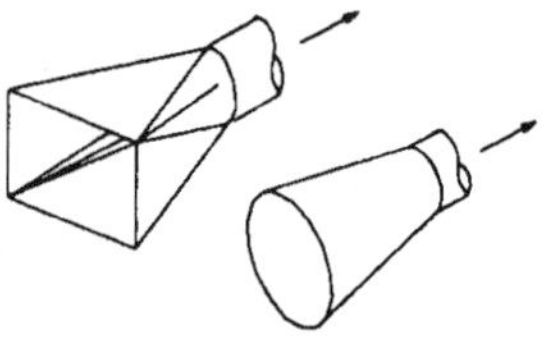	TAPER OR CONE HOOD	SEE FIGURE 5-12	

AMERICAN CONFERENCE OF GOVERNMENTAL INDUSTRIAL HYGIENISTS

HOOD LOSS FACTORS

DATE 4-94 FIGURE 3-16

Figure 32.29 — Hood loss factors. (From **American Conference of Governmental Industrial Hygienists (ACGIH®):** *Industrial Ventilation: A Manual of Recommended Practice*, 22nd Ed. Copyright 1995, Cincinnati, Ohio. Reprinted with permission.)

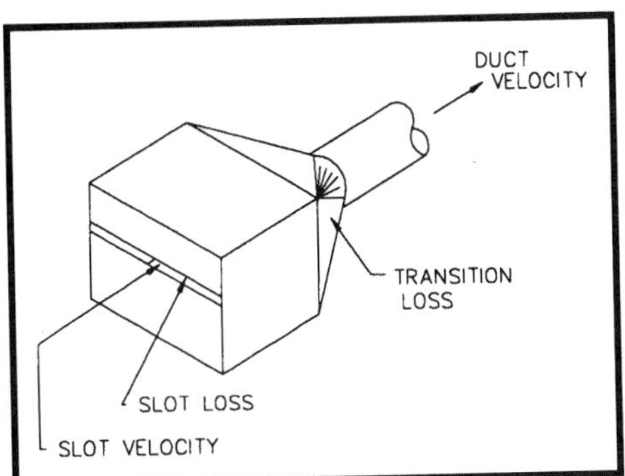

Figure 32.30 — A compound hood. (From **American Conference of Governmental Industrial Hygienists (ACGIH®):** *Industrial Ventilation: A Manual of Recommended Practice*, 22nd Ed. Copyright 1995, Cincinnati, Ohio. Reprinted with permission.)

h_e = total hood entry loss (in. wg);
h_{ed} = hood entry loss of air entering duct (in. wg);
h_{es} = hood entry loss of air entering slot (in. wg);
F_S = slot entry loss factor (dimensionless);
F_H = hood entry loss factor (dimensionless);
VP_S = slot velocity pressure (in. wg); and
VP_d = duct velocity pressure (in. wg).

The slot entry loss factor, F_S, has a value typically in the range of 1.00 to 1.78.(1)

Example 8. Calculate the total hood entry loss for a slot/plenum hood given the following information: slot velocity = 2000 ft/min, duct velocity = 4000 ft/min, slot entry loss factor = 1.78, and hood entry loss factor = 0.50.

Solution.

$$VP = \left(\frac{V_S}{4005}\right)^2 = \left(\frac{2000}{4005}\right)^2 = 0.25 \text{ in. wg}$$

$$VP = \left(\frac{V_d}{4005}\right)^2 = \left(\frac{4000}{4005}\right)^2 = 1.00 \text{ in. wg}$$

$$h_e = h_{ed} + h_{es} = (F_H \times VP_d) + (F_S \times VP_S)$$

$$h_e = (0.50 \times 1.00) + (1.78 \times 0.25) = 0.95 \text{ in. wg} \quad (29)$$

It is essential to keep in mind that hood entry loss is exactly that: it is lost. It is energy in the form of static pressure that must be continuously put into the system by the fan just to overcome the resistance to airflow imposed by the hood. In the previous example of the compound hood, the fan must supply 0.95 in. wg of static pressure just to surmount this loss. Therefore, the designer must account for this loss in the design phase by adding 0.95 in. wg to the total static pressure requirements of the fan.

At this point, look back to Figure 32.22 on page 881. In terms of the total static pressure requirements at Point B, two distinct components have been discussed thus far. The first component is that required to accelerate still room air up to the desired duct velocity. This necessitates one duct velocity pressure's worth of static pressure and is the useful work accomplished by the system. The second component is that energy or static pressure that is lost as the air enters the hood. This component is determined by the ultimate duct velocity pressure as well as the shape of the given hood (as reflected in the hood entry loss factor). Thus, the total static pressure required to accelerate still room air up the desired duct velocity and move it to Point B is given by the following:

$$SP_B = VP_d + (F_H \times VP_d) \quad (30)$$

where:

SP_B = the static pressure measured at Point B (in in. wg).

The static pressure measured at Point B (i.e., just upstream of the hood) is an important parameter for a ventilation system. It is typically referred to as the hood static pressure, SP_H.

$$SP_H = VP_d + (F_H \times VP_d) \quad (31)$$

The hood static pressure represents the total static pressure required to accelerate room air up to the desired duct velocity (VP_d) plus the additional static pressure needed to overcome friction and shock losses at the hood ($F_H \times VP_d$). The duct velocity pressure, VP_d, is often factored out of the above equation to give the following expression:

$$SP_H = (1 + F_H)\, VP_d \quad (32)$$

For the compound hood, the traditional equation for SP_H is given by the following.

$$SP_H = h_{ed} + h_{es} + VP_d = (F_H \times VP_d) + (F_S \times VP_S) + VP_d \quad (33)$$

However, the above equation implies that the acceleration of air achieved in the slot (i.e., VP_S) can be completely carried on into the duct. In other words, none of the air's momentum is lost in the plenum. Burton states that this conservation of slot velocity "occurs only rarely, e.g., when a carefully designed duct takeoff lies directly behind the slot. Most of the time, the acceleration of the slot is lost in the plenum.[3] To be conservative, it is desirable to go ahead and add the slot velocity pressure to the static pressure requirements for a compound hood, giving the following equation:

$$SP_H = h_{ed} + h_{es} + VP_d = (F_H \times VP_d) + (F_S \times VP_S) + VP_d + VP_S \quad (34)$$

Example 9. Calculate the hood static pressure for the slot/plenum hood in Example 8.

Solution.

$$SP_H = (F_H \times VP_d) + (F_S \times VP_S) + VP_d + VP_S$$

$$SP_H = (0.50 \times 1.00) + (1.78 \times 0.25) + 1.00 + 0.25 = 2.20 \text{ in. wg} \quad (35)$$

Thus, 2.20 in. wg of static pressure is required to accelerate still air through the slot/plenum, into the duct, and up to 4000 ft/min. In this example, 1.00 in. wg + 0.25 in. wg = 1.25 in. wg of this static pressure does the useful work of accelerating the air through the slot and again into the duct. The other 0.95 in. wg is spent overcoming the losses in the compound hood. If this hood configuration were used in Figure 32.23 and the static pressure measured at Point B (i.e., hood static pressure), the liquid deflection would read approximately -2.20 in. wg. Remember, static pressure measured at any point in an LEV system on the suction side of the fan represents the total static pressure required to get the air to that point and includes that component needed to accelerate air as well as that required to overcome all the losses to that point.

Transporting Contaminant Through Ductwork

At this point in the LEV design process, the hood has been designed, the volumetric flow rate necessary to control the contaminant has been determined, and the energy required to get this air moving and through the hood has been calculated. The design has now progressed to Point B in Figure 32.22. The next challenge faced by the LEV designer is the determination of the static pressure necessary to overcome the various sources of resistance between points B and C in Figure 32.22 and convey the design Q forward from the hood in the direction of the fan.

The transportation function of a ventilation system is accomplished by the ducts. Ducts carry contaminants from the hood, through the system, out of the workroom, and to the air cleaner where it is separated from the moving air. Ducts then transport the contaminant-free air to its point of discharge into the ambient environment. Ducts also provide the channel through which static pressure is distributed throughout the system from the fan to the hood face.

Although ducts have fewer design considerations than hoods, several important decisions on ductwork must be made by the designer to ensure effective and efficient operation of the system. Before proceeding to shock and friction losses occurring in ductwork, a few general comments regarding LEV duct specifications are in order.

For most LEV applications, round ducts are typically considered superior to nonround (e.g., rectangular, square, oval, etc.) ducts for the following reasons:[1,3]

- Round ducts provide for lower friction loss;
- Round ducts have higher structural integrity allowing lighter gauge materials and fewer reinforcing members;
- Round ducts resist collapsing better than other shapes; and
- Round ducts provide better aerosol transport conditions.

LEV ductwork can be constructed of a variety of materials. The two most common materials are black iron (heavier than 18 gauge, which has been welded, flanged, and gasketed) and welded galvanized steel sheet (recommended for temperatures less than 400°F).[1] Other construction materials include PVC plastic, ABS plastic, and reinforced fiber glass.[3] According to the *Industrial Ventilation Manual*, "To minimize friction loss and turbulence, the interior of all ducts should be smooth and free from obstructions, especially at joints."[1]

If the contaminant gas stream being conveyed contains corrosive gases, vapor, or mist, corrosive-resistant materials (metals, plastics, or coatings) may be required. The *Industrial Ventilation Manual* provides a chart as a guide in selecting construction materials for various corrosive atmospheres. However, "It is recommended that a specialist be consulted for the selection of materials best suited for applications when corrosive atmospheres are anticipated."[1] Extensive guidance on duct specification and layout can be found in the *Industrial*

Ventilation Manual. In general, "the location and construction of the ductwork must provide sufficient protection against external damage, corrosion, and erosion, to provide a long, useful life for the LEV system."[(6)]

The phrase "desired duct velocity" has been used several times in this discussion on LEV system energy requirements. Just what is an appropriate velocity for air moving through ductwork in an LEV system? Is there some target duct velocity for which the LEV designer is aiming? If so, what factors are considered by the designer in determining this velocity? The fact is, the appropriate velocity or velocity range for the air moving through LEV ductwork is determined by the type of material transported in the duct. The important consideration, especially when particulate matter is being handled by the system, is the prevention of contaminant settling and plugging within the duct. For systems handling particulate, this gives rise to the concept of a minimum duct transport velocity. According to the *Industrial Ventilation Manual*, "When solid material is present in the airstream, the duct velocity must be equal to or greater than the minimum air velocity required to move the particles in the air stream."[(1)]

How does the LEV designer know the minimum duct design velocity to target for a specific application? As with so many other parameters in the field of LEV design, knowledge of acceptable minimum duct velocities for various types of particulate matter has accumulated over the years from experience and experimentation. The values are tabulated in several resources.[(1,3)] For example, Table 32.5 has been reproduced from the *Industrial Ventilation Manual* and provides minimum design velocities as a function of the nature of the contaminant.

Note that for gases and vapors, which essentially move along with the airstream, any desired duct velocity is theoretically acceptable. However, for various economic considerations (e.g., economic balance between ductwork costs and fan, motor, and power costs), the optimum velocity is usually specified to fall between 1000 and 2000 ft/min. Also be aware that several of the design prints in Chapter 10 of the *Industrial Ventilation Manual* also specify a minimum duct velocity to be used in the given application.

Once an appropriate duct velocity has been selected based on the material being conveyed, how does the designer achieve that velocity in the system? The LEV designer can ensure that the air moving through the LEV system ductwork will be traveling at the target velocity by selecting an appropriate diameter for the duct leading to the hood. The target duct transport velocity, V, and the design volumetric flow rate, Q, are used in the continuity equation to determine the duct size.

Table 32.5 — Minimum Duct Transport Velocity

(From **American Conference of Governmental Industrial Hygienists (ACGIH®):** *Industrial Ventilation: A Manual of Recommended Practice*, 22nd Ed. Copyright 1995, Cincinnati, Ohio. Reprinted with permission.)

Nature of Contaminant	*Examples*	*Design Velocity*
Vapors, bases, smoke	All vapors, gases, and smoke	Any desired velocity (economic optimum velocity usually 1000-2000 fpm)
Fumes	Welding	2000–2500
Very fine light dust	Cotton lint, wood flour, litho powder	2500–3000
Dry dusts & powders	Fine rubber dust, Bakelite molding powder dust, jute lint, cotton dust, shavings (light), soap dust, leather shavings	3000–4000
Average industrial dust	Grinding dust, buffing line (dry), wool jute dust (shaker waste), coffee beans, shoe dust, granite dust, silica flour, general material handling, brick cutting, clay dust, foundry (general), limestone dust, packaging and weighing asbestos dust in textile industries	3500–4000
Heavy dusts	Sawdust (heavy and wet), metal turnings, foundry tumbling barrels and shake-out, sand blast dust, wood blocks, hog waste, brass turnings, cast iron boring dust, lead dust	4000–4500
Heavy or moist	Lead dusts with small chips, moist cement dust, asbestos chunks from transite pipe cutting machines, buffing lint (sticky), quick-lime dust	4500 and up

Consider again the continuity equation ($Q = VA$). At this point in the design process, Q has already been determined. It is the quantity of air needed with the given hood configuration and process contaminant generation characteristics to capture the contaminants and move them into the system. In accordance with the continuity equation, this Q, once defined and started in motion, will remain constant throughout the system (unless other branches are added). Therefore, the same Q that enters the hood will also be moving through the ductwork leading from the hood. Also at this point, the duct velocity, V, has been established. This is the transport velocity that must be maintained to ensure that the contaminant remains entrained in the airstream. So, since both Q and V are essentially fixed, the only variable left to the designer is A, the cross-sectional area through which the air is flowing. In this case, A refers to the cross-sectional area of the duct leading to the hood.

For a round duct, the area is given by the following equation:

$$A = \frac{\pi D^2}{4} \qquad (36)$$

where:

D = the duct diameter (in square units, typically ft^2).

Thus, with a given volumetric flow rate, Q, a LEV system designer can ensure that the duct velocity, V, is adequate by adjusting his or her specification for D, the duct diameter (which determines the cross-sectional area, A, through which the air flows).

Example 10. Consider the fluidized bed hood discussed in Example 6. The volumetric flow rate, Q, for this hood was calculated to be 2250 ft^3/min based on a bed area of 15 ft^2. The *Industrial Ventilation Manual* describing this design also specifies a minimum duct velocity of 3500 ft/min. Given this Q and V, what diameter should be specified for the duct leading to this hood?

Solution. First, set up the continuity equation for this problem.

$$Q_{duct} = V_{duct} A_{duct}$$

$$A_{duct} = \frac{Q_{duct}}{V_{duct}} = \frac{2250\ ft^3/min}{3500\ ft/min} = 0.6429\ ft^2 \qquad (37)$$

So to ensure a duct velocity of 3500 ft/min in a system through which the fan is moving 2250 ft^3/min, a duct with a cross-sectional area of 0.6429 ft^2 should be specified. This translates into a diameter of

$$A = \frac{\pi D^2}{4}$$

$$D = \sqrt{\frac{4A}{\pi}} = \sqrt{\frac{(4)(0.6429\ ft^2)}{\pi}} =$$

$$0.9047\ ft = 10.86\ in \qquad (38)$$

So, if the designer specified a duct with a diameter of 10.86 inches (which has a cross-sectional area of 0.6429 ft^2), the 2250 ft^3/min of air would travel through the duct at a velocity of 3500 ft/min in accordance with the continuity equation.

There is, however, a problem with specifying a duct with a diameter of 10.86 inches. This is not a standard size duct diameter. Duct fabricators and suppliers distribute ducts in a wide range of standard sizes (that is, 3.0 inches to more than 100 inches). Table 32.6 lists generally available duct diameters and the corresponding calculated areas. Notice that 10.86 inches does not appear on the table. Generally, ducts are available in half-inch sizes up through 9 inches. Beyond this, standard sizes increase in 1-inch increments. An LEV designer could conceivably specify a diameter of 10.86 inches, but such a special order to a sheet metal shop would be very (and needlessly) costly, indeed.

Fortunately, it is not necessary to achieve a velocity of exactly 3500 ft/min in this problem. Remember, the velocities given in Table 32.6, ventilation system prints, and similar sources are minimum velocities needed to prevent settling and plugging. It is acceptable to exceed these velocities slightly to select a standard sized duct. In this example, a duct with either a diameter of 10 inches (area = 0.5454 ft^2) or 11 inches (area = 0.6600 ft^2) would generally be available from a supplier. However, if the 11-inch duct were chosen, the actual velocity in the duct would be:

$$V_{duct} = \frac{Q_{duct}}{A_{duct}} = \frac{2250\ ft^3/min}{0.6600\ ft^2} =$$

$$3409\ ft/min \qquad (39)$$

which drops below the minimum transport velocity of 3500 ft/min. Therefore, the smaller diameter (and thus smaller area) duct should be selected to ensure that the actual velocity in the duct exceeds the recommended minimum velocity of 3500 ft/min. In this case, if the 10-inch duct were selected, the actual velocity of air moving through the duct would be

$$V_{duct} = \frac{Q_{duct}}{A_{duct}} = \frac{2250 \text{ ft}^3/\text{min}}{0.545 \text{ ft}^2} =$$

$$4125 \text{ ft/min} \qquad (40)$$

This corresponds to an actual velocity pressure in the duct of

$$VP = \left(\frac{V}{4005}\right)^2 = \left(\frac{4125}{4005}\right)^2 = 1.06 \text{ in. wg} \qquad (41)$$

It is extremely important to recognize that it is the actual duct velocity pressure, not the velocity pressure corresponding to the recommended minimum duct velocity, that is used in the determination of static pressure requirements.

Example 11. Calculate the hood static pressure for the fluidized bed system in Example 6.

Solution. For this problem Q = 2250 ft³/min, D = 10 inches, A = 0.5454 ft², V_{duct} = 4125 ft/min (the actual velocity in the duct leading to the hood), and VP_{duct} = 1.06 in. wg.

From Figure 32.16, V_{slot} = 2000 ft/min, VP_{slot} = 0.71 in. wg, and $h_e = 1.78\ VP_{slot} + 0.25\ VP_{duct}$.

$$SP_H - h_{es} + h_{ed} + VP_d = (F_S \times VP_S) + (F_H \times VP_d) + VP_d$$

$$SP_H = (1.78 \times 0.71 \text{ in. wg}) + (0.25 \times 1.05 \text{ in. wg}) + 1.06 \text{ in. wg} = 2.59 \text{ in. wg} \qquad (42)$$

So, for the fluidized bed hood, 2.59 in. wg of static pressure would be required to continuously accelerate 2250 ft³/min of still air through the slot/plenum compound hood and up to 4125 ft/min in the duct leading to the hood. Of this 2.59 in. wg of static pressure, only 1.06 in. wg (i.e., one duct velocity pressure's worth) is doing the useful work of accelerating the air. The rest is spent in overcoming the shock and friction losses of the compound hood. Thus, in the design phase, the LEV designer must ensure that 2.59 in. wg of static pressure remains at this point to accomplish the desired goal. Remember also that if this system were installed and operating in accordance with the specifications, a manometer measuring hood static pressure at this point would read -2.59 in. wg.

Quantifying Energy Losses in Ductwork

At this point, the LEV designer is ready to proceed to the quantification of the static pressure consumers between points B and C in Figure 32.22. As discussed, it takes energy to get the air moving and overcome turbulence at the hood. Now that this air is in motion, through the hood, and moving at the desired duct velocity, additional energy must be added to the system to overcome energy losses related to the ducting system. This includes losses incurred because of the friction and turbulence of the air as it moves through the duct, as well as losses at elbows, expansions, contractions, entries, and other fittings.

The first loss to be considered is that resulting from the friction of the air "rubbing" against the sides of the ductwork as it makes its way toward the fan. In this situation, static pressure is lost in the sense that it is being irretrievably converted into heat, vibration, and/or noise. The actual static pressure lost to duct friction is affected by several variables.[3]

- Length of the duct, in a linear relationship (i.e., if length doubles, friction loss doubles);
- Roughness of the wall (i.e., a material with a smoother duct surface will give rise to less friction loss than a material with a higher surface roughness);
- Velocity of air, in a squared relationship (i.e., if the velocity doubles, the friction loss will increase by four);
- Duct area, in an inverse squared relationship (i.e., if the duct area doubles, the friction loss will decrease by four)
- Duct diameter, in an inverse relationship to the fifth power when Q remains constant (i.e., if duct diameter

Table 32.6 — Duct Diameters and Corresponding Areas

Diameter (inches)	Area (ft²)	Diameter (inches)	Area (ft²)
3.0	0.0491	22.0	2.6398
3.5	0.0668	24.0	3.1416
4.0	0.0873	26.0	3.6870
4.5	0.1104	28.0	4.2761
5.0	0.1364	30.0	4.9087
5.5	0.1650	32.0	5.5851
6.0	0.1963	34.0	6.3050
6.5	0.2304	36.0	7.0686
7.0	0.2673	38.0	7.8758
7.5	0.3068	40.0	8.7266
8.0	0.3491	42.0	9.6211
8.5	0.3941	44.0	10.5592
9.0	0.4418	46.0	11.5410
10.0	0.5454	48.0	12.5664
11.0	0.6600	50.0	13.6354
12.0	0.7854	52.0	14.7480
13.0	0.9218	54.0	15.9043
14.0	1.0690	56.0	17.1042
15.0	1.2272	60.0	19.6350
16.0	1.3963	70.0	26.7254
17.0	1.5763	80.0	34.9066
18.0	1.7671	90.0	44.1786
19.0	1.9689	100.0	54.5415
20.0	2.1817	120.0	78.5398

is doubled, friction loss will be reduced to 1/32 of the original when Q remains constant); and

- Duct diameter, in an inverse relationship to the 1.4 power when velocity remains constant (i.e., if duct diameter is doubled, friction loss will be reduced by about 65% when the velocity remains constant).

Fortunately, as a result of the work of various researchers over the past 50 or more years, use of these factors in the quantification of static pressure loss in ductwork can be accomplished through the use of a single factor, the friction loss factor, H_f.[10-13] The simplified version of the H_f equation is as follows:

$$H_f = \frac{aV^b}{Q^c} \quad (43)$$

where:

H_f = the friction loss factor (in duct velocity pressures per foot of duct);
V = duct velocity (in ft/min);
Q = volumetric flow rate in duct (ft³/min); and
a,b,c = constants that vary as a function of duct material as shown in Table 32.7.

This simplified expression provides reasonably good accuracy (less than 5% error). See the *Industrial Ventilation Manual* for details of the development of the friction loss factor.

Note that this factor represents static pressure in terms of velocity pressures per foot of straight duct. In other words, to calculate the total static pressure lost in a run of duct of a given length, this factor must be multiplied by the length of the run in feet as well as by the duct velocity pressure.

$$SP_{duct\ friction} = H_f \times \text{duct length} \times VP_{duct} \quad (44)$$

Example 12. Assume that in Figure 32.22, 100 feet of straight galvanized sheet duct lies between points B and C. Calculate the additional static pressure that must be added to the system by the fan to overcome the duct friction between points B and C if these points are connected by a 10-inch diameter duct through which 2000 ft³/min of air is moving.

Solution. The area of a 10-inch duct is 0.5454 ft². Thus, this 2000 ft³/min will be traveling at a velocity of

$$V = \frac{Q}{A} = \frac{2000\ \text{ft}^3/\text{min}}{0.5454\ \text{ft}^2} = 3667\ \text{ft/min} \quad (45)$$

which corresponds to a velocity pressure of

$$VP = \left(\frac{V}{4005}\right)^2 = \left(\frac{3667}{4005}\right)^2 = 0.84\ \text{in. wg} \quad (46)$$

Calculating the duct friction factor, we note the following from given information and Table 32.7 for galvanized sheet duct: Q = 2000 ft³/min; V = 3667 ft/min; a = 0.0307; b = 0.533; c = 0.612.

$$H_f = \frac{aV^b}{Q^c} = \frac{(0.0307)(3667\ \text{ft/min})^{0.533}}{(2000\ \text{ft}^3/\text{min})^{0.612}} =$$

$$0.023\ \text{VPs per foot} \quad (47)$$

Using the friction loss factor and the duct length, the static pressure lost to duct friction can be calculated as follows:

$$SP_{duct\ friction} = H_f \times \text{duct length} \times VP_{duct} = 0.023\ \text{VPs/ft} \times 100\ \text{ft} \times 0.84\ \text{in. wg} = 1.95\ \text{in. wg} \quad (48)$$

Thus, 1.95 in. wg of static pressure will be lost to heat, vibration, and/or noise as 2000 ft³/min of air is conveyed through a 10-inch duct between points B and C. The LEV designer must ensure that this much additional static pressure is put into the system by the fan to overcome this loss. In an operating system the static pressure measured at Point C will be 1.95 in. wg higher (i.e., 1.95 in. wg more negative) than at Point B because of that consumed by this 100 feet of straight duct in between.

Note that there are various charts and nomographs that allow the friction factor to be estimated from the parameters Q, V, and D. For example, Figure 32.31 is taken from the *Industrial Ventilation Manual* and applies to galvanized sheet metal ducts. Knowing any two of Q, V, and D allows a point to be fixed in the body of the graph. From this point the friction loss factor can be estimated by projecting to either the top or bottom chart axis. Although this and similar charts and nomographs are useful, the exponent functions available on most modern calculators make use of the above equation quite simple.

Table 32.7 — Friction Loss Factor Calculation

	Duct Material	a	b	c
$H_f = a\frac{V^b}{Q^c}$	Galvanized	0.0307	0.533	0.612
	Black iron, AL, PVC, stainless steel	0.0425	0.465	0.602

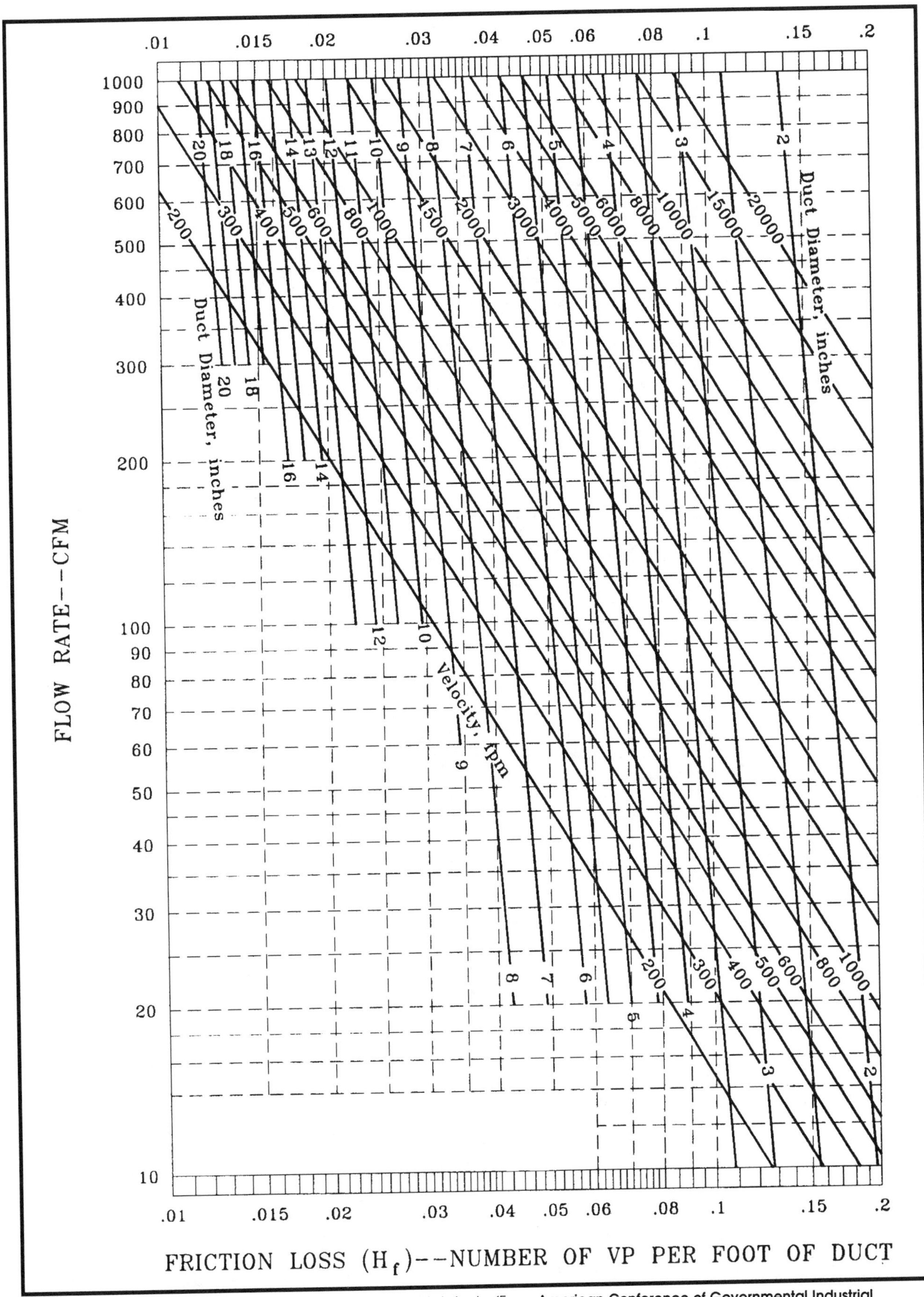

Figure 32.31 — Friction loss factors for galvanized sheet metal ducts. (From **American Conference of Governmental Industrial Hygienists (ACGIH®):** *Industrial Ventilation: A Manual of Recommended Practice*, 22nd Ed. Copyright 1995, Cincinnati, Ohio. Reprinted with permission.)

For straight runs of duct, the frictional pressure loss that results from air turbulence and from the rubbing of the air against the walls of the duct is the principal static pressure consumer. As shown, the quantification of this static pressure loss depends on the friction factor (which incorporates Q, V, D, and surface roughness), the duct length, and the velocity pressure. However, any time the flow changes direction (such as at a turn or elbow) or undergoes a sudden change in the duct size that causes drastic velocity changes, the violent air mixing and eddy formation also give rise to system energy losses. That is to say, if elbows, expansions, contractions, etc., exist between points B and C in Figure 32.23, shock losses occurring in these fittings will consume static pressure over and above that lost to duct friction. These losses, like all such sources of energy loss, must be quantified in terms of static pressure and added to the overall energy requirements of the fan.

Fortunately, "most of the losses encountered in other parts of a ventilation system (e.g., expansions, contractions, elbows, hood entries, branch entries) are also proportional to the velocity pressure."[(4)] This is consistent with what has been discussed thus far. The faster the air moves through the system, the more violent and turbulent the mixing that would occur at an elbow or transition. Thus, if there are elbows, expansions, or contractions between points B and C in Figure 32.22, the quantity of static pressure required to overcome the losses incurred by these fittings can be calculated as some factor times the velocity pressure. The value of the factor depends on the type of fitting encountered.

For elbows, the loss factor depends on three variables: (1) the number of sheet metal joints or pieces making up the elbow; (2) the magnitude of the direction change (in degrees) the air must make; and (3) the ratio of the centerline radius of curvature of the turn to the diameter of the duct. Charts of these factors are tabulated in various sources such as Figure 32.32, which appears in the *Industrial Ventilation Manual*. These factors are for 90° elbows. Loss factors for smaller elbows are presumed to be proportional (i.e., the loss factor for a 60° elbow is 2/3 that of the 90° elbow, etc.).

$$SP_{elbow} = F_{elbow}VP \tag{49}$$

where:

SP_{elbow} = static pressure consumed in the elbow (in. wg);
F_{elbow} = elbow loss factor (dimensionless); and
VP = duct velocity pressure (in. wg).

In practice, most LEV designers simply use an average loss factor for elbows that depends on the radius of curvature to duct diameter ratio and magnitude of the air direction change, as illustrated in Figure 32.33.

Example 13. Calculate the additional static pressure loss that would be incurred if a 90° elbow and a 60° elbow, both with R/D = 2.0, were added to the 100 feet of straight duct in Example 12.

Solution. In Example 12 the velocity pressure of the moving air in the 10-inch duct is 0.84 in. wg. From Figure 32.33 note that the loss factor for a 90° elbow with R/D = 2.0 is 0.27. The factor for a 60° elbow is 2/3 of this value or 0.18. Calculate the static pressure loss as a result of the elbows as in the following:

$$SP_{elbow} = (0.27)(0.84 \text{ in. wg}) + (0.18)(0.84 \text{ in. wg}) = 0.38 \text{ in. wg} \tag{50}$$

Thus, an additional 0.38 in. wg of static pressure must be continuously supplied to the system by the fan to overcome the mixing and turbulence losses occurring at these elbows.

Losses occurring at contractions are handled similarly in that these losses are also proportional to the velocity pressure. However, another phenomenon occurs at a contraction that impacts the total static pressure requirements of a fan. Specifically, at a duct contraction the cross-sectional area, A, through which the air is traveling decreases. Since Q remains constant in accordance with the conservation of mass, the velocity, V, at which the air is traveling must proportionally increase as would be predicted by the continuity equation. This acceleration of air and the accompanying increase in velocity pressure must be accounted for by a corresponding decrease in static pressure. Remember, at any point in the system where the air is accelerated, velocity pressure increases. In accordance with the conservation of energy, an increase in velocity pressure cannot be obtained without an equal price being paid in terms of static pressure. So, if the velocity pressure increases from, say, 1.25 in. wg to 1.50 in. wg as a result of a contraction, this extra 0.25 in. wg of velocity pressure must be "bought" with an additional 0.25

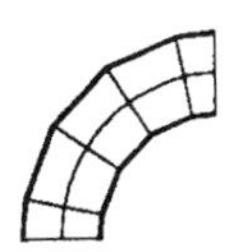

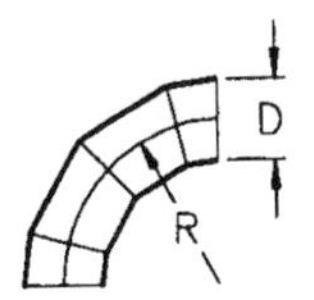

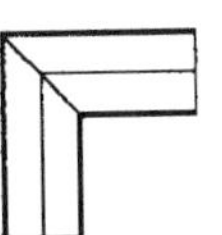

	R/D					
	0.5	0.75	1.00	1.50	2.00	2.50
Stamped	0.71	0.33	0.22	0.15	0.13	0.12
5-piece	–	0.46	0.33	0.24	0.19	0.17*
4-piece	–	0.50	0.37	0.27	0.24	0.23*
3-piece	0.90	0.54	0.42	0.34	0.33	0.33*

* extrapolated from published data

OTHER ELBOW LOSS COEFFICIENTS

Mitered, no vanes	1.2	
Mitered, turning vanes	0.6	
Flatback (R/D = 2.5)	0.05	(see Figure 5-23)

NOTE: Loss factors are assumed to be for elbows of "zero length." Friction losses should be included to the intersection of centerlines.

ROUND ELBOW LOSS COEFFICIENTS

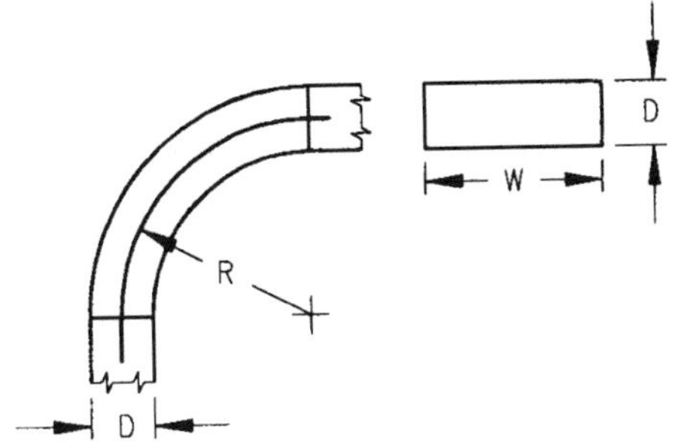

R/D	Aspect Ratio, W/D					
	0.25	0.5	1.0	2.0	3.0	4.0
0.0(Mitre)	1.50	1.32	1.15	1.04	0.92	0.86
0.5	1.36	1.21	1.05	0.95	0.84	0.79
1.0	0.45	0.28	0.21	0.21	0.20	0.19
1.5	0.28	0.18	0.13	0.13	0.12	0.12
2.0	0.24	0.15	0.11	0.11	0.10	0.10
3.0	0.24	0.15	0.11	0.11	0.10	0.10

SQUARE & RECTANGULAR ELBOW LOSS COEFFICIENTS

AMERICAN CONFERENCE OF GOVERNMENTAL INDUSTRIAL HYGIENISTS

DUCT DESIGN DATA
ELBOW LOSSES

DATE *1-95* FIGURE *5-13*

Figure 32.32 — Elbow loss coefficients. (From **American Conference of Governmental Industrial Hygienists (ACGIH®):** *Industrial Ventilation: A Manual of Recommended Practice*, 22nd Ed. Copyright 1995, Cincinnati, Ohio. Reprinted with permission.)

90° ROUND ELBOW LOSS FACTORS

C.L. R/D	Factor
Mitered	1.25
1.5	0.39
2.0	0.27
2.5	0.22

60° elbow = 2/3 loss
45° elbow = 1/2 loss

C.L. = duct centerline
R = radius of duct
D = diameter of duct

Figure 32.33 — Elbow loss coefficients.

in. wg of static pressure supplied by the fan. This, technically, is not a loss in the sense that it is, in fact, accomplishing the useful work of air acceleration. But the designer must account for it in summing up static pressure requirements for the fan.

Along with the air acceleration taking place at a contraction comes additional turbulence and air mixing, which give rise to actual static pressure losses. Contractions can be either gradual or abrupt, the former being more efficient in terms of lower static pressure losses. As with all the other losses discussed, static pressure losses for contractions can be quantified in terms of a contraction loss factor multiplied by the velocity pressure in the smaller diameter duct. For tapered contractions the factor increases with increasing taper angle. For abrupt contractions, the factor is a function of ratio of the larger to smaller duct area. The factors are tabulated in various sources such as Figure 32.34 adapted from the *Industrial Ventilation Manual.*

The equation quantifying static pressure and acceleration losses at a contraction is usually of the following form:

$$SP_{contraction} = (VP_2 - VP_1) + L(VP_2 - VP_1) \quad (51)$$

for tapered contractions and

$$SP_{contraction} = (VP_2 - VP_1) + K(VP_2) \quad (52)$$

for abrupt contractions.[1]

In these equations, the variables with a subscript of 1 indicate conditions in the larger duct; those with a subscript of 2 indicate conditions in the smaller duct (i.e., upstream of the contraction). The term $VP_2 - VP_1$ indicates the additional static pressure that is required to accelerate the air. L is the loss factor for tapered contractions; K is the loss factor of abrupt contractions.

Example 14. Consider the abrupt contraction pictured in Figure 32.35. Calculate the total static required to move air through this contraction, including that used to accelerate the air as well as that lost to turbulence. The following information is provided: Q = 2000 ft³/min; D_1 = 14 inches = 1.1667 feet; D_2 = 10 inches = 0.8333 feet.

Solution.

$$A_1 = \frac{(\pi)(1.1667 \text{ ft})^2}{4} = 1.069 \text{ ft}^2$$

$$A_2 = \frac{(\pi)(0.8333 \text{ ft})^2}{4} = 0.5454 \text{ ft}^2$$

$$V_1 = \frac{2000 \text{ ft}^3/\text{min}}{1.069 \text{ ft}^2} = 1871 \text{ ft/min}$$

$$V_2 = \frac{2000 \text{ ft}^3/\text{min}}{0.5454 \text{ ft}^2} = 3667 \text{ ft/min}$$

$$VP_1 = \left(\frac{1871 \text{ ft/min}}{4005}\right)^2 = 0.22 \text{ in. wg}$$

$$VP_2 = \left(\frac{3667 \text{ ft/min}}{4005}\right)^2 = 0.84 \text{ in. wg}$$

$$\frac{A_2}{A_1} = \frac{0.5454}{1.069} = 0.51 \quad (53)$$

Note from Figure 32.34 that for an abrupt contraction with $A_2/A_3 = 0.5$, the contraction loss factor is 0.32. Calculating the total static pressure needed for the abrupt contraction

$$SP_{contraction} = (VP_2 - VP_1) + K(VP_2)$$

$$SP_{contraction} = (VP_2 - VP_1) + K(VP_2)$$

$$SP_{contraction} = (0.84 \text{ in. wg} - 0.22 \text{ in. wg}) + 0.32(0.84 \text{ in. wg})$$

$$SP_{contraction} = 0.62 \text{ in. wg} + 0.27 \text{ in. wg} = 0.89 \text{ in. wg} \quad (54)$$

Thus, to move the air through this contraction requires a total of 0.89 in. wg of static pressure to be supplied by the fan. Of this, 0.62 in. wg of static pressure is required to accelerate the air from a velocity pressure of 0.22 in. wg up to a velocity pressure of 0.84 in. wg while 0.27 in. wg of static pressure is lost to turbulence.

STATIC PRESSURE REGAINS FOR EXPANSIONS

Within duct

Regain (R), fraction of VP difference					
Taper angle degrees	Diameter ratios D_2/D_1				
	1.25:1	1.5:1	1.75:1	2:1	2.5:1
3 1/2	0.92	0.88	0.84	0.81	0.75
5	0.88	0.84	0.80	0.76	0.68
10	0.85	0.76	0.70	0.63	0.53
15	0.83	0.70	0.62	0.55	0.43
20	0.81	0.67	0.57	0.48	0.43
25	0.80	0.65	0.53	0.44	0.28
30	0.79	0.63	0.51	0.41	0.25
Abrupt 90	0.77	0.62	0.50	0.40	0.25
Where: $SP_2 = SP_1 + R(VP_1 - VP_2)$					

At end of duct

Regain (R), fraction of inlet VP						
Taper length to inlet diam L/D	Diameter ratios D_2/D_1					
	1.2:1	1.3:1	1.4:1	1.5:1	1.6:1	1.7:1
1.0:1	0.37	0.39	0.38	0.35	0.31	0.27
1.5:1	0.39	0.46	0.47	0.46	0.44	0.41
2.0:1	0.42	0.49	0.52	0.52	0.51	0.49
3.0:1	0.44	0.52	0.57	0.59	0.60	0.59
4.0:1	0.45	0.55	0.60	0.63	0.63	0.64
5.0:1	0.47	0.56	0.62	0.65	0.66	0.68
7.5:1	0.48	0.58	0.64	0.68	0.70	0.72
Where: $SP_1 = SP_2 - R(VP_1)$*						

*When $SP_2 = 0$ (atmosphere) SP_1 will be (−)

The regain (R) will only be 70% of value shown above when expansion follows a disturbance or elbow (including a fan) by less than 5 duct diameters.

STATIC PRESSURE LOSSES FOR CONTRACTIONS

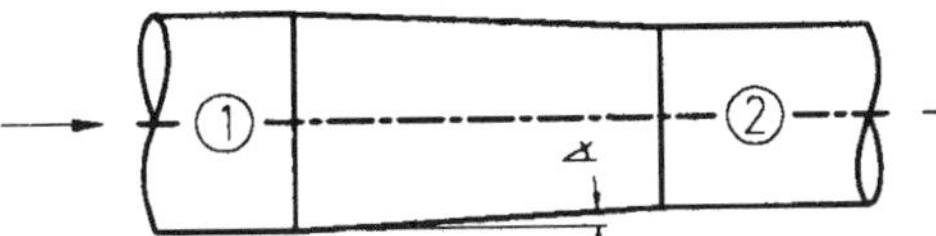

Tapered contraction
$SP_2 = SP_1 - (VP_2 - VP_1) - L(VP_2 - VP_1)$

Taper angle degrees	L(loss)
5	0.05
10	0.06
15	0.08
20	0.10
25	0.11
30	0.13
45	0.20
60	0.30
over 60	Abrupt contraction

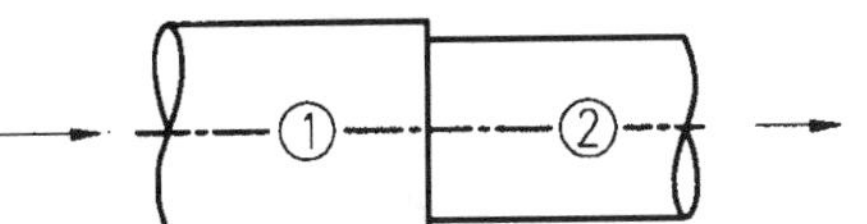

Abrupt contraction
$SP_2 = SP_1 - (VP_2 - VP_1) - K(VP_2)$

Ratio A_2/A_1	K
0.1	0.48
0.2	0.46
0.3	0.42
0.4	0.37
0.4	0.32
0.6	0.26
0.7	0.20

A= duct area, ft^2

Note:
In calculating SP for expansion or contraction use algebraic signs: VP is (+), and usually SP is (+) in discharge duct from fan, and SP is (−) in inlet duct to fan.

AMERICAN CONFERENCE OF GOVERNMENTAL INDUSTRIAL HYGIENISTS	*DUCT DESIGN DATA*	
	DATE *1–95*	FIGURE *5–17*

Figure 32.34 — Static pressure regain/loss factors for expansions/contractions. (From **American Conference of Governmental Industrial Hygienists (ACGIH®):** *Industrial Ventilation: A Manual of Recommended Practice*, 22nd Ed. Copyright 1995, Cincinnati, Ohio. Reprinted with permission.)

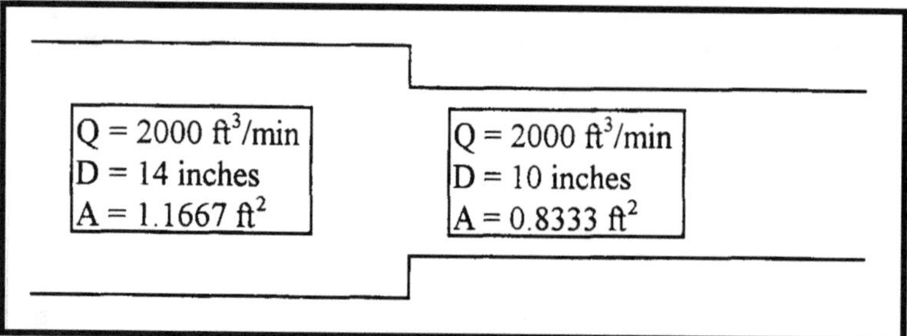

Figure 32.35 — Abrupt contraction calculation.

In terms of the mutual convertibility of velocity pressure and static pressure, an effect opposite that occurring at a contraction takes place at a duct expansion. In this case, as the duct cross-sectional area increases the air slows down and the velocity pressure decreases, causing a corresponding increase in static pressure (sometimes called a static pressure regain). The static pressure regain can actually be claimed as credit (that is, subtracted from the total static pressure requirements of the fan). However, because of the flow separation and turbulence, some energy is lost at an expansion. This means that only a certain portion of the static pressure can be reclaimed, the rest is lost. The magnitude of the regain can be calculated as a factor multiplied by the velocity pressure differences.

$$SP_{regain} = R(VP_1 - VP_2) \quad (55)$$

Here, VP_1 is the velocity pressure in the smaller diameter duct; VP_2 the velocity pressure in the larger diameter duct. The factor, R, is a function of the expansion taper angle and the ratio of the two duct diameters as shown in Figure 32.34 from the *Industrial Ventilation Manual*.

Fan Static Pressure Requirements

Note that between points B and C in Figure 32.22 any number of static pressure consumers may exist. These include such items as duct friction, elbows, expansions, and contractions, plus many other that have not been discussed. However, regardless of how many sources of energy loss are present, the fundamental approach is the same. The static pressure loss for each one must be quantified and summed with all the other losses to determine the total static pressure requirements for the fan. This brings the designer to Point C, the entry into the fan. The question is, what happens at the fan?

As stated, the fan is the source of suction in an LEV system. This suction or static pressure differential is what drives the system, causing contaminated air to be drawn into the hood. However, this air must also be discharged (after the contaminant has been removed by some appropriate air cleaner) back into the environment at atmospheric pressure. Thus, on the pressure side of the fan, where the static pressure is greater than atmospheric, the air proceeds along the ductwork from points D to E in Figure 32.22. The summation of static pressure requirements proceeds from the hood face all the way to the fan and then resumes on the other side and continues all the way to the stack, where the air is discharged once again into atmospheric pressure.

Static pressure is also lost on the pressure side of the fan just as it is on the suction side. Duct friction, elbows, expansions/contractions, etc., also cost static pressure on the pressure side. This must be tabulated and summed to determine the total energy requirements of the fan. Of course, on the pressure side of the fan, the algebraic sign of the static pressure is positive: "The static pressure at the fan outlet must be higher than atmospheric by an amount equal to the losses in the discharge duct."[1] This static pressure is lost as the air moves from Point D to Point E. Refer again to Figure 32.23, which illustrates the behavior of total pressure, static pressure, and velocity pressure in a hypothetical system.

Once all of these losses on both sides of the fan have been tabulated and the energy needed to accelerate the air at various points in the system has been accounted for, the total fan static pressure requirement can be calculated. When an LEV designer selects a fan, he or she does so on the basis of volumetric flow rate, Q, and the parameter that reflects this total energy or static pressure requirement. This energy parameter may be either fan total pressure (FTP) or fan static pressure (FSP). Some fan manufacturers publish their fan ratings catalogs in FTP, some in FSP. The designer must examine the rating tables to determine which parameter is being used. These quantities can be calculated with the following formulae:

$$FTP = TP_{outlet} - TP_{inlet}$$

$$FTP = (SP_{outlet} + VP_{outlet}) - (SP_{inlet} + VP_{inlet})$$

$$FSP = FTP - VPoutlet$$

$$FSP = SP_{outlet} - SP_{inlet} - VP_{inlet} \quad (56)$$

When using the above formulae, it is critical to use the proper algebraic sign for TP, SP, and VP.

Example 15. In a particular LEV system, a 24-inch diameter duct delivers 11,400 ft^3/min to a fan. Leaving the fan is a 26-inch diameter duct carrying this same volumetric flow rate. Given that the static pressure at the fan inlet is 7.36 in. wg and the outlet static pressure is 0.41 in. wg, calculate both the FTP and the FSP.

Solution.

$$D_{inlet} = 24 \text{ in.} = 2 \text{ ft}$$

$$A_{inlet} = 3.142 \text{ ft}^2$$

$$V_{inlet} = \left(\frac{Q}{A_{inlet}}\right) = \left(\frac{11{,}400 \text{ ft}^3/\text{min}}{3.142 \text{ ft}^2}\right)$$
$$= 3629 \text{ ft/min}$$

$$VP_{inlet} = \left(\frac{V_{inlet}}{4005}\right)^2 = \left(\frac{3629 \text{ ft/min}}{4005}\right)^2$$
$$= 0.82 \text{ in. wg}$$

$$D_{outlet} = 26 \text{ in.} = 2.167 \text{ ft}$$

$$A_{outlet} = 3.687 \text{ ft}^2$$

$$V_{outlet} = \left(\frac{Q}{A_{outlet}}\right) = \left(\frac{11{,}400 \text{ ft}^3/\text{min}}{3.687 \text{ ft}^2}\right)$$
$$= 3092 \text{ ft/min}$$

$$VP_{outlet} = \left(\frac{V_{outlet}}{4005}\right)^2 = \left(\frac{3092 \text{ ft/min}}{4005}\right)^2$$
$$= 0.60 \text{ in. wg}$$

$$FTP = (SP_{outlet} + VP_{outlet}) - (SP_{inlet} + VP_{inlet})$$

$$FTP = (0.41 \text{ in. wg} + 0.60 \text{ in. wg}) - (-7.36 \text{ in. wg} + 0.82 \text{ in. wg})$$

$$FTP = 1.01 \text{ in. wg} - (-6.54 \text{ in. wg}) = 7.55 \text{ in. wg}$$

$$FSP = SP_{outlet} - SP_{inlet} - VP_{inlet}$$

$$FSP = 0.41 \text{ in. wg} - (-7.36 \text{ in. wg}) - 0.82 \text{ in. wg} = 6.95 \text{ in. wg} \quad (57)$$

A Bookkeeping Method

At this point in the chapter, most of the fundamental calculations in LEV design have been discussed. Boiling the chapter down into simple terms, the LEV designer must make two very important determinations: (1) what quantity, Q, of air in cubic feet per minute is necessary to achieve proper contaminant control; and (2) how much energy, in the form of static pressure, is necessary to accelerate this air and move it through the ventilation system. The first part of this chapter dealt with selecting the most effective hood design for the given application and determining the proper Q for contaminant control. The latter portion of the chapter introduced the various consumers of static pressure in an operating LEV system, which include the energy needed to accelerate air as well as that required to overcome losses due to hood entries, duct friction, and in fittings (e.g., elbows, expansions, contractions, etc.). As many readers have probably noted, the quantity of static pressure needed to satisfy each of these consumers (with the exception of expansions and contractions that must also account for static pressure and velocity pressure conversions) can be calculated as some factor times the duct velocity pressure. Thus, an expression for tabulating static pressures up to this point could be given as follows:

$$SP = F_H VP_{duct} + (H_f \times \text{length})VP_{duct} + F_{elbow}VP_{duct} + VP_{duct} \quad (58)$$

Note that all loss factors, which are nothing more than multiples of velocity pressure, can be factored out and summed before multiplying by the duct velocity pressure.

$$SP = [F_H + (H_f \times \text{length}) + F_{elbow} + 1]VP_{duct} \quad (59)$$

In effect, all of the factors summed together indicate how many velocity pressures' worth of static pressure are required to overcome energy losses in the system. These must, of course, be added to the one velocity pressure's worth of static pressure that is required to accelerate the still air to duct velocity pressure (i.e., the +1 in the above equation) to obtain the total static pressure requirements for which the LEV designer must account.

Although this chapter has dealt only with a single run system (i.e., a system with only one hood and ductwork leading to the fan), one can imagine how difficult the task of keeping up with each of the factors, velocity pressures, volumetric flow rates, and so forth might become in a system with several ducts/hoods branching off from the main ductwork. Fortunately, a bookkeeping system or worksheet has been developed that makes the design task more manageable. The version of the worksheet appearing in the current (22nd) edition of the *Industrial Ventilation Manual* is illustrated as Figure 32.36.

Velocity Pressure Method Calculation Sheet

Plant Name: PROBLEM 2 **Elevation:** ______ **Date:** 2/94

Location: ______ **Temperature:** ______ **Drawing No.:** Fig. 5-8

Department: ______ **Factor:** ______ **Designer:** ______

No.			Item	Units						
1			Duct Segment Identification		1-A	2-B	3-B	B-A	1-A	A-C
2			Volumetric Flowrate	cfm	9600	960	700	1743	9600	11413
3			Minimum Transport Velocity	fpm	3500	3500	3500	3500	3500	3500
4			Duct Diameter	inches	22	7	6	9	18	24
5			Duct Area	sq. ft.	2.64	.2673	.1963	.4428	1.767	3.142
6			Actual Duct Velocity	fpm	3637	3592	3565	3945	5432	3633
7			Duct Velocity Pressure	"wg	0.82	0.80	0.79	0.97	1.84	0.82
8	H		Slot Area	sq. ft.	10				10	
9	O	S	Slot Velocity	fpm	960				960	
10	O	L	Slot Velocity Pressure	"wg	0.06				0.06	
11	D	O	Slot Loss Factor — Fig. 5-12 or Chap. 10		1.78				1.78	
12		T	Acceleration Factor	0 or 1	0				0	
13	S	S	Slot Loss per VP — Items 11 + 12		1.78				1.78	
14	U		Slot SP	"wg	0.10				0.10	
15	C		Duct Entry Loss Factor — Fig. 5-12 or Chap. 10		0.25	0.25	0.25		0.25	
16	T		Acceleration Factor	1 or 0	1	1	1		1	
17	I		Duct Entry Loss per VP — Items 15 + 16		1.25	1.25	1.25		1.25	
18	O		Duct Entry Loss — Items 7 x 17	"wg	1.03	1.01	0.99		1.03	
19	N		Other Loss	"wg				0.13		
20			Hood Static Pressure — Items 14 + 18 + 19	"wg	1.03	1.01	0.99	0.13	1.03	
21			Straight Duct Length	feet	13	3	4	18	13	34
22			Friction Factor (H_f)		.0089	.0361	.0436	.0263	.0110	.0080
23			Friction Loss per VP — Items 21 x 22		0.12	0.67	0.17	0.47	0.14	0.27
24			No. of 90° Elbows		1	0.67	1.67	2	1	
25			Elbow Loss per VP — Item 24 x Loss Factor		0.27	0.18	0.45	0.54	0.27	
26			No. Entries			1	1	1		
27			Entry Loss per VP — Item 26 x Loss Factor			0.18	0.18	0.18		
28			Special Fittings Loss Factors							
29			Duct Loss per VP — Items 23 + 25 + 27 + 28		0.39	0.47	0.81	1.19	0.41	0.27
30			Duct Loss — Items 7 x 29		0.32	0.38	0.64	1.16	0.76	0.22
31			Duct SP Loss — Items 20 + 30		1.45	1.38	1.63	1.29	3.16	0.22
32			Cumulative Static Pressure	"wg	-1.45	-1.38	-1.63	-2.92	-3.16	-3.38
33			Governing Static Pressure	"wg		-1.63		-3.16		
34			Corrected Volumetric Flowrate	"wg		1043		1813		
35			Resultant Velocity Pressure	"wg		0.84	0.84	1.71	1.71	0.82

PERTINENT EQUATIONS:

$$Q_{corr} = Q_{design}\sqrt{\frac{SP_{gov.}}{SP_{duct}}}$$

$$VP_r = \frac{Q_1}{Q_3}VP_1 + \frac{Q_2}{Q_3}VP_2$$

$$H_f = a\frac{V^b}{Q^c}$$

(See Chapter 1)

Duct Material	a	b	c
Galvanized	0.0307	0.533	0.612
Black iron, AL, PVC, Stainless steel	0.0425	0.465	0.602

$$FSP = SP_{outlet} - SP_{inlet} - VP_{inlet}$$

90° ROUND ELBOW LOSS FACTORS

C.L. R/D	Factor
Mitered	1.25
1.5	0.39
2.0	0.27
2.5	0.22

60° elbow = 2/3 loss
45° elbow = 1/2 loss

BRANCH ENTRY LOSS FACTORS

Angle	Factor
15°	0.09
30°	0.18
45°	0.28
60°	0.44
90°	1.00

Figure 32.36 — Velocity pressure design method calculation worksheet. (From **American Conference of Governmental Industrial Hygienists (ACGIH®):** *Industrial Ventilation: A Manual of Recommended Practice*, 22nd Ed. Copyright 1995, Cincinnati, Ohio. Reprinted with permission.)

An explanation of each of the row entries is provided as follows.

Row 1: Duct Segment Identification

This row allows the designer to assign a unique identification to each segment of the system. This becomes increasingly important as the system becomes more complicated (i.e., has several branches). Generally speaking, a single run system will have two basic segments: (1) from the hood face to the fan; and (2) from the fan to the discharge. For example, in Figure 32.22 the designer might choose to identify the first segment as A-C and the second segment as D-E. If A-C contains an expansion, contraction, air cleaner, and so forth such that the duct diameter beyond the fitting/air cleaner is different from that before, the designer must separate the run into two different segments because the velocity pressure will change in accordance with the continuity equation. In this case, he or she may choose to identify the portion of the system from the hood face to the fitting/air cleaner as the first segment, and the portion from the fitting/air cleaner to the fan as a separate segment. (Of course, the additional pressure drop imposed by the fitting/air cleaner, etc., must not be neglected.) If another branch entered the system between A and C, the first segment would extend from the hood face to the point of the new branch entry. Beyond that point, volumetric flow rate, duct size, and velocity pressure will change (because of the additional airflow brought in by the new segment), requiring the ductwork beyond the entry to be identified as a new segment.

Row 2: Volumetric Flow Rate, Q (in ft^3/min)

The volumetric flow rate, Q, in ft^3/min, being conveyed through this segment is entered in this row. In the simple unbranched systems illustrated in this chapter, this is simply the flow rate required with the given hood design to provide adequate contaminant control as discussed in the first sections of this chapter.

Row 3: Minimum Transport Velocity, $V_{transport}$ (in ft/min)

This is the minimum velocity, $V_{transport}$, in ft/min that is recommended to prevent contaminant settling and plugging in the duct. This information is obtained from Table 32.5 and will, as discussed, be different for different materials. The important thing to remember is that this will not be the actual duct velocity. The minimum velocity entered in this row will be used in the continuity equation, along with volumetric flow rate in Row 2, to determine the duct diameter needed to achieve this minimum flow.

Row 4: Duct Diameter, D (inches)

Once the minimum transport velocity has been obtained from Table 32.5 (Row 3), the designer can use this value along with the volumetric flow rate (Row 2) in the continuity equation ($A = Q/V$) to determine the area, A, and ultimately the diameter, D, of the duct in this particular segment. Remember, however, that the duct area obtained by using the minimum transport velocity will likely not give a duct with a standard diameter. To maintain the transport velocity, especially for particulate contaminant, it is usually necessary to specify the next smaller standard diameter duct for entry in this row. Refer to Table 32.6 and Example 10.

Row 5: Duct Area, A_d (ft^2)

After a standard duct diameter for the LEV system segment has been specified (Row 4), the area of this duct, A_d, can be calculated and entered into this row.

Row 6: Actual Duct Velocity, V_d (ft/min)

Using the volumetric flow rate (Row 2) and the actual duct area (Row 5) in the continuity equation ($V = Q/A$), the designer can now calculate the actual velocity, V_d, in this segment of the system.

Row 7: Duct Velocity Pressure, VP_d (in. wg)

Once the actual duct velocity has been determined (Row 6), the duct velocity pressure, VP_d, in this segment can be calculated ($VP = (V/4005)^2$) and entered here.

Note that Rows 8 through 14 are used for duct segments that include a compound hood as discussed in the section on "Quantifying the Energy Lost at the Hood." If the segment in question does not start with a hood (e.g., a segment between a contraction and the fan) or if the hood is a simple hood in that is has only one point of significant energy loss, these rows may be ignored. Refer to the previous section that discusses compound hoods.

Row 8: Slot Area, A_s (ft^2)

This is the area of the slot, A_s, or initial entry point for the compound hood. Note, if the hood design consists of multiple slots, the total area should be entered in this row.

Row 9: Slot Velocity, V_S (ft/min)

Using the volumetric flow rate (Row 2) and the slot area (Row 8) in the continuity equation, the LEV designer can calculate the velocity of the air as it moves through the slot, V_S.

Row 10: Slot Velocity Pressure, VP_S (in. wg)

Slot velocity pressure is calculated from slot velocity (Row 9): $VP_S = (V_S/4005)^2$.

Row 11: Slot Loss Factor, F_S (Dimensionless)

Recall that the slot entry loss factor, F_S, is the factor that when multiplied by the slot velocity pressure determines the quantity of static pressure lost to shock losses and turbulence at the vena contracta formed downstream of the slot. This value can be obtained from prints in the *Industrial Ventilation Manual* for specific hood applications. In the absence of specific guidance, a value of 1.78 is commonly used for the slot entry loss factor for many slot/plenum hoods.

Row 12: Acceleration Factor (Dimensionless)

Recall that static pressure requirements at the hood are composed of two components. One component is that which is lost to shock and turbulence due to the vena contracta formation. The other component is that which accomplishes the useful work of accelerating the air. One velocity pressure's worth of static pressure is required to accelerate still air out in front of the hood up to the desired system velocity. The static pressure required to accelerate air need only be accounted for at one point in the hood: either at the slot or in the duct, whichever has the higher velocity pressure. In other words, if the velocity pressure of the air in the duct leading to the compound hood is higher than the velocity pressure of the air going through the slot, which is normally the case, the one velocity pressure's worth of static pressure is calculated based on the duct velocity pressure. In that case, a 0 would be inserted in this row (and a 1 goes in Row 16 below). However, if the slot velocity pressure is greater than the duct velocity pressure, which would be the case if the slot area were smaller than the duct area, the one velocity pressure's worth of static pressure would be based on the slot velocity pressure. In that case, a 1 would go in this row (and a 0 in Row 16).

Row 13: Slot Loss per VP (Dimensionless)

This is the total number of slot velocity pressures' worth of static pressure required to move the air through the slot and into the plenum. This includes that lost at the slot as indicated by the slot loss factor (Row 11) plus that needed for acceleration (Row 12, if the slot velocity pressure is greater than the duct velocity pressure).

Row 14: Plenum SP (in. wg)

When the number of slot velocity pressures' worth of static pressure (Row 13) is multiplied by the slot velocity pressure (Row 10), this gives the quantity of static pressure in in. wg that is required to move the air into the plenum. As stated, this included that lost at the slot and that required to accelerate still air if slot velocity is larger than duct velocity.

Rows 15 through 20 also deal with system segments that have hood entries. These hoods may be simple hoods, in which case Rows 8 through 14 above will be blank, or they may be compound hoods where Rows 15 through 20 function as a continuation after the slot and plenum losses have been dealt with in rows 8 through 14. Just as in rows 8 through 14, rows 15 through 20 will essentially add the hood entry loss factor (Row 15) to the 1 velocity pressure needed to accelerate air (Row 16) to obtain a total factor that represents the total number of duct velocity pressures' worth of static pressure. When this total loss factor is multiplied by the duct velocity pressure, the magnitude of static pressure needed to accelerate air and move it into the hood is obtained (Row 17).

Row 15: Hood Entry Loss Factor, F_H (Dimensionless)

The hood entry loss factor, F_H, as discussed in the section on duct losses, is inserted in this row.

Row 16: Acceleration Factor (Dimensionless)

If the segment in question is a simple hood, a 1 will always go in this row, indicating the 1 duct velocity pressure's worth of static pressure required to accelerate air up to duct velocity pressure. Even if the hood is compound, a 1 will still normally go here since duct velocity pressure is usually higher than slot velocity pressure.

Row 17: Duct Entry Loss Per VP (Dimensionless)

This is the sum of the hood entry loss factor (number of velocity pressures lost due to vena contracta) and the acceleration factor (1 velocity pressure to accelerate the air).

Row 18: Duct Entry Loss (in. wg)

Multiplying the duct velocity pressure (Row 7) times the total number of velocity pressures' worth of static pressure consumed in the hood (Row 17, the duct entry loss per velocity pressure) gives the total static pressure in in. wg needed to accelerate air to the duct velocity pressure and overcome hood entry losses (Point B in Figure 32.23). If this hood is a simple hood and no other losses are present up to this point, this value will represent the hood static pressure, SP_H.

Row 19: Other Loss (in. wg)

If any other source of significant static pressure loss exists before Point B in Figure 32.23, it would go into this row. Also, this row is often used as a convenient spot on the worksheet to account for special losses, such as an air-cleaning device, in duct segments that may not even have a hood (e.g., a duct segment that extends from an expansion to the fan).

Row 20: Hood Static Pressure, SP_H (in. wg)

This row contains the hood static pressure, SP_H, which is a sum of all static pressure losses up to Point B in Figure 32.23 including slot/plenum losses (Row 14, for compound hoods), hood entry and acceleration losses (Row 18), and any other source of static pressure loss up to this point (Row 19).

In Rows 21 through 28 the various loss factors for duct friction, elbows, entries, and special fittings are entered and subsequently summed in Row 29. Row 29 thus represents the total number of velocity pressures' worth of static pressure lost to these fittings. When this total factor is multiplied by the duct velocity pressure (Row 7), the total static pressure in in. wg lost in this duct segment is obtained (Row 30).

Row 21: Straight Duct Length, L (ft)

This row contains the total length, L, of straight duct in this segment. Recall that the static pressure lost to duct friction is a function of the length of the duct in the given segment.

Row 22: Friction Factor, H_f (Dimensionless)

The duct friction factor is entered into this row. The friction factor can be obtained from a chart (e.g., Figure 32.30) or calculated from the following equation.

$$H_f = \frac{aV^b}{Q^c} \qquad (60)$$

Row 23: Friction Loss per VP (Dimensionless)

Recall that when the friction factor (Row 22) is multiplied by the duct length (Row 21), the number of velocity pressures' worth of static pressure lost to duct friction is obtained. This value goes in this row.

Row 24: Number of 90° Elbows (Dimensionless)

The total number of 90° or 90° equivalent elbows is entered in this row. Recall that elbow loss factors are given in terms of 90° elbows with losses occurring in elbows of lower magnitude being considered proportional (e.g., a 45° elbow carries one-half the loss of a 90° elbow). So, for example, if the duct segment in question has two 90° elbows, two 60° elbows, and two 45° elbows, the total 90° elbow equivalent of 4 1/3 would be entered in this row.

Row 25: Elbow Loss per VP, F_{elbow} (Dimensionless)

When the total number of 90° elbows (Row 24) is multiplied by the loss factor per 90° elbow (from Figure 32.33), the total number of duct velocity pressures' worth of static pressure lost in the elbows in this segment is obtained.

Row 26: Number of Entries (Dimensionless)

Although losses occurring in branch entries have not been discussed in this chapter, these losses are also accounted for with a loss factor (which depends on the angle of entry) times the number of entries. This row contains the number of branch entries.

Row 27: Entry Loss per VP (Dimensionless)

In this row the number of entries is multiplied by the entry loss factor (see bottom of sheet) to give the total number of duct velocity pressures' worth of static pressure that will be consumed by the entry.

Row 28: Special Fittings Loss Factors (Dimensionless)

If there is any other special fitting in this particular duct segment, the loss factor can be entered into this row. Note, however, that this is a loss factor (that is, a factor denoting how many duct velocity pressures' worth of static pressure are lost due to this fitting). It is not a static pressure loss in in. wg. If, for example, the static pressure loss occurring in an air-cleaning device or some special fitting were listed in terms of in. wg, it would not be appropriate to tabulate this value in this row.

Row 29: Duct Loss per VP (Dimensionless)

This row now proceeds to tabulate all the loss factors existing in this duct segment other than those related to the hood (which were dealt with in Rows 9–20). This includes the duct friction loss factor, the elbow loss factor, the entry loss factor, and any special fitting loss factor. Therefore, in this row goes the sum of Rows 23, 25, 27, and 28. This gives the total number of duct velocity pressures' worth of static pressure lost in this segment (excluding the hood).

Row 30: Duct Loss (in. wg)

When the total number of duct velocity pressures' worth of static pressure (Row 29) is multiplied by the duct velocity pressure (Row 7), the total quantity of static pressure lost in the ductwork of this segment is obtained in in. wg. This value goes in this row.

Row 31: Duct SP Loss (in. wg)

At this point, the static pressure lost in the hood (Row 20) is added to the static pressure lost in the ductwork (Row 30) the total static pressure, in in. wg, required to move the design Q (Row 2) through this segment is obtained. Note that the 1 velocity pressure's worth of static pressure used to accelerate the air is included in the tally in Row 20.

Rows 31–35 are used with more complex systems (i.e., those having more than one branch) and have not been discussed in this chapter.

Further Study

This chapter has presented the basic concepts involved in designing efficient LEV systems for the purpose of removing industrial contaminants at their source and thereby protecting the breathing zone of employees. Although only simple, one-run systems have been introduced, the same basic design principles are followed for more complex systems with several branches and multiple hoods. Many of the more advanced design topics not discussed in this introductory chapter are discussed in great detail in the *Industrial Ventilation Manual* and other references listed at the end of the chapter.

For example, further information on the following topics can be found in the specified chapters of the *Industrial Ventilation Manual*.

- Fans, fan curves, fan selection, fan laws, system effects, etc.—Chapter 6;
- Balancing static pressures at junctions—Chapter 5;
- Exhaust stack outlets and other stack considerations—Chapter 5; and
- Air-cleaning devices—Chapter 4.

Note, however, that even though all of these issues may make the design process more complicated, they do not change the basic concepts of calculating the volumetric flow rate necessary for adequate contaminant control and then summing up the energy requirements necessary to achieve that airflow.

As a final point in this regard, please note that this chapter has introduced the basic, fundamental concepts of industrial LEV system design for dry air at standard conditions (70°F and sea level). It is important to realize that the density of air changes with changes in temperature and pressure in accordance with the ideal gas law. According to Burton:[(3)]

> *Air's weight-to-volume ratio (lbs/ft³), or density, varies with temperature and pressure. For example, when we heat air, it expands and its density decreases; when we go to a higher altitude, the barometric pressure drops, the air expands and the density drops.*
>
> *All other conditions equal, changes in air density vary linearly with changes in absolute temperature. Similarly, air density varies linearly with the change in air pressure (e.g., if the pressure increases by 10%, the density will increase by 10%). These two relationships are combined into an equation called the Ideal Gas Law.*
>
> *Using the Ideal Gas Law, we can calculate a density correction factor, d. This factor can be used to correct measured air velocities to actual velocities, actual volume flow rates to standard volume flow rates, and so forth. At standard*

conditions, d = 1.0. At higher temperatures and altitudes, d is less than one; at lower temperatures and altitudes below sea level, d increases above one."

The density correction factor can be used to covert between actual and standard conditions as follows:

Air Density (actual) =
Air Density (standard) × d (61)

As stated, d is calculated from the ideal gas law using the following equation:

$$d = \left(\frac{530}{°F + 460}\right)\left(\frac{BP}{29.92}\right) \quad (62)$$

where:

d = the density correction factor;
°F = temperature (degrees Fahrenheit); and
BP = barometric pressure (in. Hg).

Since standard air was assumed for each of the problems presented in this chapter, d always equaled 1. However, remember that when working at temperatures and/or pressures that are markedly different from standard, the density correction is necessary. According to the *Industrial Ventilation Manual*, corrections are seldom required for temperatures between 40°F and 100°F and/or elevations between 1000 feet and +1000 feet.[(1)] Refer to Chapter 5 of the *Industrial Ventilation Manual* for an in-depth discussion of corrections for nonstandard density.

Trial and Error Approach to Ventilation Design

As a fitting conclusion to this chapter on proper LEV design techniques, it is worthwhile to briefly mention several common mistakes made when attempting to use ventilation to control worker exposure. Unfortunately, uninformed designers who do not follow appropriate design procedures may cause untoward discomfort or long-term adverse health outcomes.

Designers and/or workers will occasionally attempt to supplement existing controls with improper and ineffective techniques. For example, large propeller (axial) fans are frequently installed in an attempt to remove contaminated air or to provide clean outside air to dilute the contaminant. Unfortunately, this might serve to merely distribute contaminants throughout the space. Another common practice is the placement of shields or baffles around the source (partially enclosing the process) to minimize the release of fugitive emissions but failing to provide local exhaust. This does nothing to remove the contaminant or reduce worker exposure and might lead to additional complications such as ground level emissions leaching into the surrounding environment.

Improper positioning of exhaust stacks relative to air inlets is another all too frequent occurrence. Locating the stack of a local exhaust system adjacent to a fresh air inlet can easily cause contaminated air to be drawn back into the building. This can, at the very least, lead to complaints of fugitive odors and might create more serious indoor air quality problems. A detailed discussion of adequate stack height and location relative to intakes including calculations and examples can be found in other references. Consultation of these resources is imperative for anyone engaging in local exhaust design.

Finally, adding to an existing system without a critical analysis of the effects this addition may have on the performance of the system is a serious error. The usual and unintended result is that neither the original system nor the addition will function as intended. The calculation methods necessary to correctly add to an existing system are also discussed elsewhere and must be consulted when the necessity of system modification arises.

In summary, trial and error and other uninformed approaches to ventilation design are a colossal waste of time and energy and will not protect the worker. It is critically important to spend the time and effort necessary to master the proper techniques of designing local exhaust ventilation systems.

References

1. **American Conference of Governmental Industrial Hygienists (ACGIH):** *Industrial Ventilation: A Manual of Recommended Practice*, 22nd Ed. Cincinnati: ACGIH, 1995.
2. **Burton, D.J.:** Local Exhaust Ventilation of Industrial Occupancies. In *Fundamentals of Industrial Hygiene*, 4th Ed. (edited by B.A. Plog, J. Niland, and P.J. Quinlan). Itasca, IL: National Safety Council, 1996.
3. **Burton, D.J.:** *Industrial Ventilation Workbook*, 3rd Ed. Bountiful, UT: IVE, Inc., 1994.

4. **Burgess, W.A., M.J. Ellenbecker, and R.D. Treitman:** *Ventilation for Control of the Work Environment.* New York: John Wiley & Sons, 1989.
5. **McDermott, H.J.:** *Handbook of Ventilation for Contaminant Control.* Ann Arbor, MI: Ann Arbor Science Publishers, 1976.
6. **Mutchler, J.E.:** Principles of Ventilation. In *The Industrial Environment—Its Evaluation and Control.* Cincinnati: National Institute for Occupational Safety and Health, 1973.
7. **Mutchler, J.E.:** Local Exhaust Systems. In *The Industrial Environment—Its Evaluation and Control.* Cincinnati: National Institute for Occupational Safety and Health, 1973.
8. **Engineering Staff of George D. Clayton & Associates:** Design of Ventilation Systems. In *The Industrial Environment—Its Evaluation and Control.* Cincinnati: National Institute for Occupational Safety and Health, 1973.
9. **Soule, R.D.:** Industrial Hygiene Engineering Controls. In *Patty's Industrial Hygiene and Toxicology,* Vol. I, Part B, 4th Ed. (edited by G.D. Clayton and F.E. Clayton). New York: John Wiley & Sons, 1991.
10. **Alden, J.L., and J.M. Kane:** *Design of Industrial Ventilation Systems,* 5th Ed. New York: Industrial Press, 1982.
11. **Talty, J.T. (ed.):** *Industrial Hygiene Engineering: Recognition, Measurement, Evaluation, and Control,* 2nd Ed. Park Ridge, NJ: Noyes Publications, 1988.
12. **Heinsohn, R.J.:** *Industrial Ventilation—Engineering Principles.* New York: Wiley-Interscience, 1991.
13. **Hemeon, W.C.L.:** *Plant and Process Ventilation.* New York: Industrial Press, Inc., 1954.

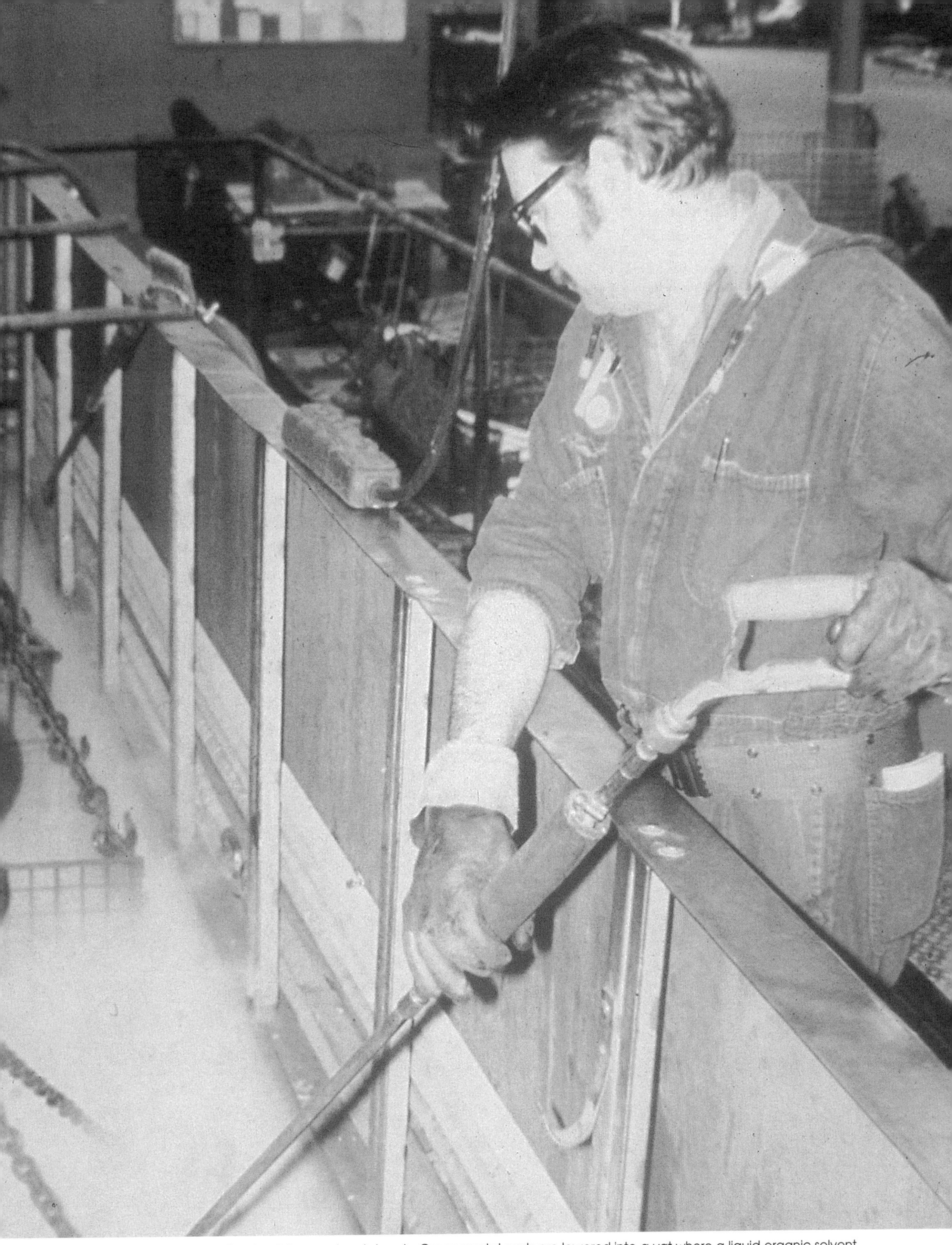

A metal processing operation involving used metal parts. Greasy metal parts are lowered into a vat where a liquid organic solvent is heated.

Outcome Competencies

After completing this chapter, the user should be able to complete the following items during evaluation of an industrial exhaust ventilation system:

1. Define underlined terms used in this chapter.
2. Define acceptable performance criteria.
3. Develop a work plan.
4. Collect and evaluate field data.
5. Assess the ventilation system's performance.
6. Troubleshoot existing ventilation systems.
7. Document the results.
8. Establish a plan to retest systems.

Key Terms

communications plan • field data • field data sheets • OHS plan • process flow and instrumentation diagram (PF&ID) • volumetric flow rate • work plan

Prerequisite Knowledge

Prior to beginning this chapter, the user should review the ACGIH publication *Industrial Ventilation — A Manual of Recommended Practice* for general content. The user should also review the following chapters:

Chapter Number	Chapter Topic
31	General Methods for the Control of Airborne Hazards
32	An Introduction to the Design of Local Exhaust Ventilation Systems
44	Report Writing

Key Topics

I. Establish the Objectives of the Evaluation
II. Define Acceptable Performance Criteria
III. Define Requirements for Data Accuracy and Precision
IV. Identify Existing System Components
V. Write a Work Plan
VI. Write an Occupational Health and Safety Plan
VII. Select Field Instrumentation
VIII. Collect the Field Data
IX. Analyze and Interpret the Data
X. Write the Report
XI. Have the Report Peer-Reviewed
XII. Retest the System

Chapter 33

33

Evaluating Ventilation Systems

A. David Scarchilli, PE, D.E.E.

A practical guide to performing field measurements on an existing industrial exhaust ventilation system is provided in this chapter. This information should be used in conjunction with ACGIH's **Industrial Ventilation — A Manual of Recommended Practice** *and related publications.*

Introduction

This chapter provides a practical guide for evaluation of industrial exhaust ventilation systems. It provides a generic methodology that can be adapted to most industrial ventilation systems that contain a fan, hoods, connecting ducts, and a pollution control device. A good reference for use with this chapter is the "Measurement Techniques and Instrumentation" chapter in the American Conference of Governmental Industrial Hygienists' publication *Industrial Ventilation — A Manual of Recommended Practice.*

The evaluation of creature comfort or process HVAC systems and stack sampling methodologies (i.e., qualifying and quantifying what is in the exhaust gas by either physical or chemical analysis) are better described in other references. The American Society of Heating, Refrigerating and Air-Conditioning Engineers (ASHRAE) has several excellent references for HVAC system evaluations, and the U.S. Environmental Protection Agency (EPA) has published references for stack sampling procedures.

Establish the Objectives of the Evaluation

The most important part of any project is to establish clear objectives. Reasons for evaluating an industrial ventilation system include:

- To establish the baseline performance and operating conditions;
- To determine if additional exhaust hoods can be added to an existing system;
- To evaluate initial system acceptance;
- To verify compliance with National Fire Protection Association (NFPA) design guidelines for special hazards;
- To determine if adequate transport velocities are being maintained;
- To provide performance data for stack sampling or air pollution emissions calculations;
- To demonstrate compliance with Occupational Safety and Health Administration (OSHA) regulations or with federal, state, or local environmental air pollution control regulations and air permit provisos;
- To balance the airflow of a system after modifications have been installed; and
- To provide engineering data supporting recommendations to improve system performance.

Define Acceptable Performance Criteria

Once the project objectives have been clearly stated and accepted, establish and document acceptable performance criteria. Be sure to include both measurable performance criteria (e.g., hood capture velocity) and definitions of subjective criteria (e.g., elimination of workplace odors).

The most important measure of success is the acceptance of the ventilation system by the workstation operator. If the operator is not satisfied with the performance of the ventilation system, the system will probably not be used as intended and the design has failed.

One sure way of achieving success is to interview the operator before conducting the field evaluation. This ensures operator participation and brings up operator concerns early in the evaluation. Always be candid with the operator. If a ventilation system will increase the noise level at the operator's workstation, informing that person early of this change will reduce the trauma of accepting a new environment in which to work. For multiple workstations, consider designing a prototype workstation that includes ventilation performance with ergonomic factors (e.g., lighting, noise, proper posturing), productivity enhancements, and quality control. This will help with acceptance of the redesigned workstation by all stakeholders.

Define Requirements for Ensuring Data Accuracy and Precision

Next, determine the required accuracy and precision of the data. For example, is the data to be submitted to a regulatory agency for review? Will it be used for legal purposes? Or is the data to be used in house for operational purposes? The accuracy must be adequate for the intended use. This determination of accuracy will define the specification for field instrumentation and dictate the skill level required of the field technician or engineer who will collect the data and supervise the project. Prepare a quality control procedure that meets the minimum requirements for factory calibration, pre- and postmeasurement function checks, and documentation of the results of each check.

Identify Existing System Components

Identifying all of the existing ventilation system components may not be a trivial task for complex systems, systems modified without the supervision of an engineer, or systems that do not have "as-built drawings." A good starting point is a process flow and instrumentation diagram (PF&ID). If an accurate PF&ID is not available, a field sketch may suffice. An as-built drawing may be required for critical systems. Figure 33.1 is an example of a simple system PF&ID.

Determine the original design objectives of the system (e.g., the removal of vapors, dust, heat, mist, or moisture). Determine the original system limitations (e.g., temperature, pressure, corrosivity, flow) on all system hoods, branch lines, submains, and

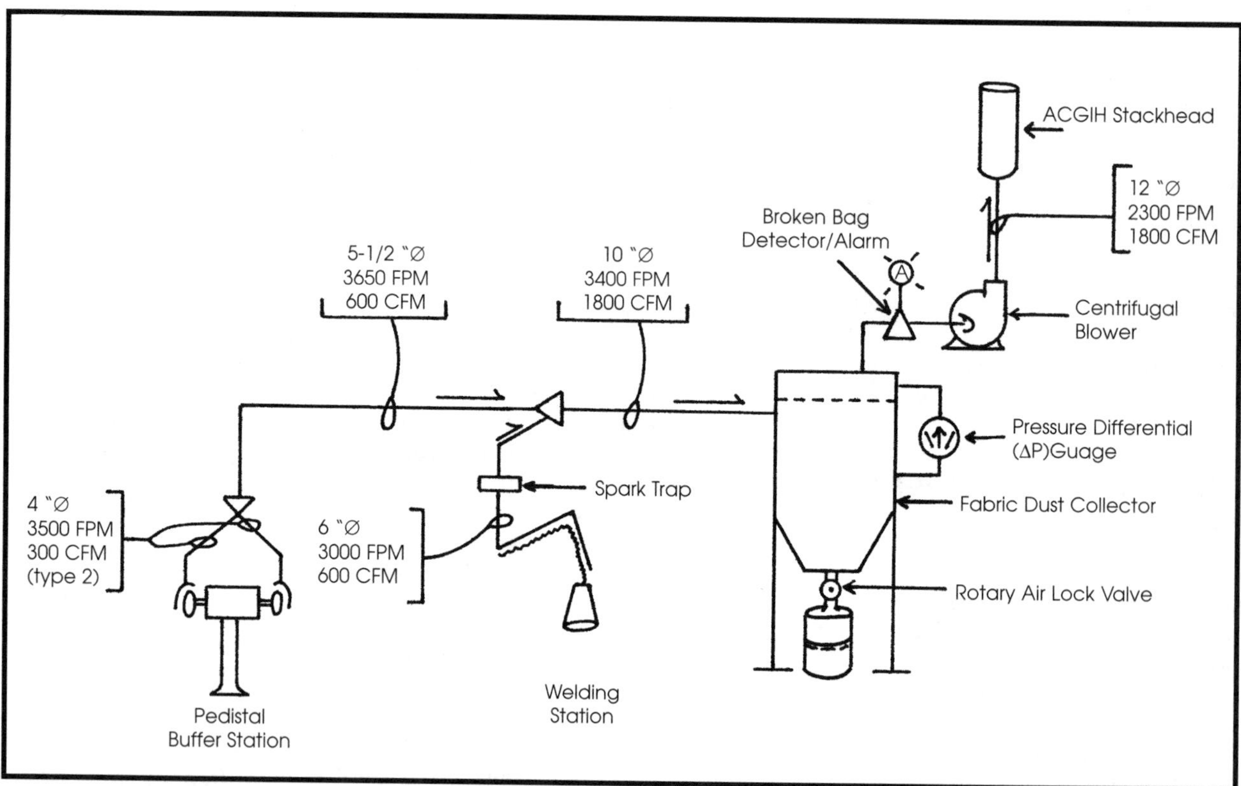

Figure 33.1 — Process flow and instrumentation diagram of a local exhaust ventilation system.

mains. Determine all system mechanical components such as fans, air pollution control devices, filters, pressure gauges, fire dampers, manual and automatic dampers, hood inlet screens, condensate drain lines, and system instrumentation (e.g., particle sensors, pressure gauges, flow gauges). Determine, if available, the design make-up air requirements of the system and identify the associated system components. Document the design flow rates, temperature and pressure ranges, hours of operation, variability of process throughput, and operation of all system components.

Document the operation of each piece of equipment identified in the ventilation system. This information can generally be found in the facility engineering files or from the equipment manufacturer. Compile the equipment serial numbers, model numbers, and the vendor technical support phone numbers for future reference. Obtain equipment specification sheets, operation and maintenance manuals, engineering performance tables, fan performance tables, and operating curves. Air pollution control equipment specifications should include the filter-cleaning mechanism and acceptable pressure drops.

Conduct a system inspection and walk-through with employees familiar with the system to verify that the PF&ID is accurate. Interview each operator to determine the system's start and stop sequence.

Write a Work Plan

A work plan should be developed to ensure that all required data are efficiently collected and properly documented. The work plan should include field data sheets, a marked-up floor plan or system schematic to show where the field data is to be collected, the methodology for collecting the data, the individual responsible for collecting the data, and a communications plan.

Field Data Sheets. The use of field data sheets ensures that the data is collected in a concise manner, that all of the data required to meet the project objectives is recorded, and that future data is collected in a similar manner for trend analysis and correlation purposes.

Examples of field data sheets are included at the end of this chapter. Sheets specific to individual equipment components should address, at a minimum, the following:

- Stack gas parameters (velocity, flow, pressure, temperature);
- Fan operation (rotation from motor side, RPM, belt tension, number of belts, motor amperage and voltage, external static pressure);
- Filters (removal efficiency, low and high [i.e., clean and dirty], pressure drops, condition of seals);
- Dry fabric dust collectors (pressure drop, operation of cleaning cycle, bin indicator, presence of material in hopper, operation of screw conveyor or rotary air lock valve);
- Clean air plenum test for breaches by fluorescence powder or stack sampling;
- Mist eliminators ("P-Trap," moisture in clean air plenum, pressure drop);
- Scrubbers (water flow and pressure, nozzle spray pattern, condition of mist eliminator, controls operational);
- Electrostatic precipitators (voltage, arcing, inlet filter pressure drop, electrode condition); and
- Cyclones (pressure drop, material in hopper).

Where to Collect the Data. Develop a clearly marked diagram showing where to collect the data. Another technique is to mark the locations in the field using duct tape, permanent marker, or chalk.

Methodology for Collecting the Data. Develop specific procedures for collecting the data. Identify the number of vertical and horizontal velocity traverse readings, determine hood or booth face velocity traverse locations or spacing, etc. There are several references at the end of this chapter that define procedures for collecting accurate field data.

Communications Plan. The communications plan identifies the individuals who must be informed during the evaluation. It is intended to comply with all of the facility security and notification procedures. It includes the team roster that identifies the team members by name and the tasks each is to perform.

Document sign-in policies, parking and unloading areas, and alarms and emergency procedures. Define the notification procedures if there is a possibility for "uncontrolled releases" to the environment. Distribute a copy of the facility contractor health and safety policy. Inform facility engineering and maintenance if power is to be interrupted. Inform production supervisors and operators if interference with production operations is expected and issue a project schedule.

Write an Occupational Health and Safety (OHS) Plan

Many hazards are involved with evaluation of an industrial ventilation system, including explosion hazards, corrosive materials, and hot surfaces. Define the basic elements of the OHS plan by having a qualified occupational hygienist (preferably a certified industrial hygienist [CIH]) complete a job hazard analysis. The hazard assessment includes an evaluation of material safety data sheets (MSDSs), and the work practices outlined in the work plan.

As a minimum, the OHS plan should include specific instructions on personal protective equipment (PPE); lockout/tagout (LO/TO); confined space entry (CSE); fall protection, security; spills and contingency plans; emergency procedures; phone numbers of the facility emergency coordinator, police, fire, ambulance and nearest hospital; locations of telephones, first aid kits; eye washes and showers; and the facility contractor OHS policy.

A safety meeting with all field personnel at the start of each workday is recommended. Document attendance with sign-in sheets and file them in an appropriate place.

Select Field Instrumentation

Select field instrumentation based on the work plan objectives, experience in evaluating similar systems, and the anticipated range of measurement values. Another important aspect of selecting the appropriate field instrumentation is the required sensitivity to meet the work plan objectives. For example, if a temperature reading is required to be rounded to the nearest tenth, be sure that the field instrumentation can be read accurately to the nearest one-hundredth.

For instruments requiring periodic calibrations, check the interval since the last calibration, compare it with the manufacturer's recommendation, and recalibrate if necessary. If field calibration using reference gases will be required, ensure that gases in the appropriate concentrations are available.

Table 33.1 — Field Checklist

Equipment Description	*Availability*
personal protective equipment	
drill, key, bits	
grounded electrical extension cords (GFI protected)	
extra batteries	
LO/TO equipment	
plastic sheets	
broom/dust pan	
ladders	
clip board	
field data sheets	
confined space entry equipment/permit	
level	
measuring tape	
duct tape	
sheet metal plugs	
tools	
construction cones/barrier tape	
instruments (i.e., velometer, tachometer, manometer, etc.)	
tool bag	
other MSDSs	

Collect the Field Data

Conduct a walk-through of the system to be evaluated with the field crew before the actual day of sampling. Locate all system components to be evaluated, inspect measurement points for safe access, review the field data sheets, and discuss the elements of the OHS plan. Request that the field crew provide proper training and medical records consistent with the OHS plan. Confirm the schedule with production supervisors to ensure that any changes in the operation of the system are documented and approved before system checkout.

Ensure that each piece of instrumentation is properly prepared, charged, and calibrated. Use of a checklist ensures that all required equipment and supplies are ready. Table 33.1 is an example of a field checklist.

Analyze and Interpret the Data

Properly document the calculations on which report findings and recommendations are based. An important parameter commonly calculated from field data is the volumetric flow rate, which may require corrections for temperature and pressure (i.e., air density) when not at standard conditions. If the report is to be in metric units, establish the conversion factors at the beginning of the calculations. Many velometers can store individual data points and display average readings. If it is important to record all of the velocity traverse readings, record them in the calculations. A typical set of calculations is shown in the following example.

Example: Calculate the Exhaust Flow Rate from a Heated Source.

Field conditions: T_{stack} = 233°F

Duct diameter = 12 in.; Area (A) = 0.785 $ft.^2$

Velometer accuracy = ± 2.0% full-scale reading

$V_{average}$ = 1443 fpm based on two 12-point traverses

Q_{acfm} = VA = (1443 fpm)(0.785 ft.^{2}) = 1133 acfm

Q_{scfm} = ([70°F + 460°F]/[233°F + 460°F]) (1133 acfm) = 870 scfm

Correcting for instrument accuracy: (±2.0%)(1500 fpm full-scale reading) = ±30 fpm

Corrected flow rate: 840 < Q_{scfm} < 900

Summarize the data in tables, graphs, and figures as appropriate. Use tables that show both acceptable performance and the results derived from the evaluation to make the comparison easy to understand. Conclusions should be made based on the operating parameters and review of the fan curve. See Table 33.2 (troubleshooting guide) for assistance in developing recommendations.

Write the Report

The final engineering deliverable is a report that documents the survey's findings, conclusions, and recommendations. The report should be easy to read and contain all of the information and assumptions used during the survey. A sample outline follows. (See Chapter 44 for more details on report writing.)

Executive Summary: A brief summary of the survey's most important elements presented in order of decreasing priority.

Introduction: Include the contractual arrangements for completing the work; a reference to the agreed-on scope of work; date the field data was collected; who was responsible for collecting and interpreting the data; and the name of the ventilation system analyzed.

Objective(s): Concise statement that contains the survey's objective(s).

Work Plan Summary: Summary of the work plan and any field changes necessary to complete the work.

Instrumentation: Complete list of all of the instruments used, including the results of any field calibrations, the date of the last factory calibration, manufacturer's name, and the system's model and serial number.

Operation Conditions: Detailed summary of the manufacturing parameters and ventilation component settings in effect during collection of the field data, including process flow rates, temperatures, equipment loads, position of blast gates, filter pressure differential settings and any unusual conditions observed during the field data collection.

Table 33.2 — Troubleshooting Guide

Problem	Probable Cause	What to Do
$Q_{actual} < Q_{design}$ or $V_{actual} < V_{design}$	• Incorrect fan wheel rotation	• Reverse electrical leads
	• Duct or hood obstruction; material deposited in elbows or transitions	• Remove obstruction
	• Fan belts slipping/low speed	• Tighten or replace belts; reset adjustable sleeve
	• Filters plugged or blinded	• Clean or replace filters
	• Back draft damper obstructed in closed position	• Repair back draft damper
	• Too many hoods	• Increase fan speed/motor size and rebalance system
	• Poor fan inlet conditions	• Replace with straight inlet pipe
	• Low system pressure	• Compare system pressure requirements with fan curve; adjust accordingly
System noisy	• High velocity	• Install silencer
	• Fan bearings	• Repair or replace bearings
	• Fan vibration	• Install vibration isolators or balance fan wheel
Poor hood capture efficiency	• Hood is improperly designed or positioned with respect to the point of contaminant generation	• Redesign/reposition hood and train operators
	• Turbulence (tool or room)	• Redirect air flow into hood
	• Branch flow and velocity low	• Increase system pressure at hood
	• Obstructions in duct	• Remove obstruction

Results: Results of the field data and engineering calculations should be presented in a summary table. The results should be complemented with a brief narrative discussion of the results highlighting any unusually high or low readings, and an explanation for any outliers.

Conclusions: Conclusions derived from the results should be listed in order of importance to the outcome of the survey. These conclusions should also be stated in the report's Executive Summary.

Recommendations: List prioritized recommendations appropriate to the objective and scope of work. These recommendations should be listed in the Executive Summary.

Assumptions: Include all applicable assumptions used to draw conclusions.

References: A list of references used to collect data or to complete the calculations should be included.

Limitations: Limitations on the use of the data collected and the source of information provided by others.

Appendix A: Field Data Sheets

Appendix B: Engineering Calculations

Appendix C: OHS Plan and Minutes of the Safety Meeting

Have the Report Peer-Reviewed

Issue the report in draft format and have the complete report peer-reviewed. Ensure that the review includes checking all engineering calculations and assumptions, the conclusions drawn from the survey, and all recommendations. This quality assurance procedure should become standard practice for all engineering reports.

Retest the System

In many instances industrial ventilation systems are not an integral part of the process in terms of production or product throughput. Therefore, running periodic checks on their performance is essential. This is especially true for systems that may be affected by accumulations of the material they are conveying or controlling. Therefore, the final recommendation should always be to retest the performance of the system periodically, even if no recommendations for corrective action are made. The timing of the retests should be stated in the recommendations contained in the report.

Summary

This chapter sets up a program that when adopted in conjunction with guidance from the "Measurement Techniques and Instrumentation" chapter in ACGIH's *Industrial Ventilation — A Manual of Recommended Practice* can be used to establish the evaluation of industrial exhaust ventilation systems.

Acknowledgments

The author would like to express his gratitude to several individuals who volunteered time to help in the preparation of this chapter: Greg Lynch, OccuHealth, Inc.; Virginia (Ginny) Peck-Viau, Document Design; Salvatore R. DiNardi, University of Massachusetts; Richard D. Fulwiler (author of this chapter from the previous edition of this book); the staff at OccuHealth, Inc., and to my family.

Additional Sources

The following references are updated periodically. It is recommended that the latest edition of each be used as a reference to reflect changes in the science.

American Conference of Governmental Industrial Hygienists: *Industrial Ventilation — A Manual of Recommended Practice.* Cincinnati: American Conference of Governmental Industrial Hygienists.

American Industrial Hygiene Association: *American National Standard for Laboratory Ventilation* (ANSI/AIHA Z9.5–1992). Fairfax, Va.: American Industrial Hygiene Association, 1993.

American National Standards Institute: *Fundamentals Governing the Design and Operation of Local Exhaust Systems* (ANSI Z9.2). New York: American National Standards Institute.

American Society of Heating, Refrigerating and Air Conditioning Engineers: *ASHRAE Handbooks (i.e., Fundamentals, HVAC Applications, Refrigeration and HVAC Systems and Equipment).*

Field Data Sheet (Example)
Fabric Filter Dust Collector

Company Name:
Department:
Location:
Manufacturer/Model:

Date:
Prepared by:

1.0 Initial Field Measurements/Calculations

ΔP Primary Filters = ________"spwg
ΔP Secondary Filters = _______"spwg
$V_{main\ duct}$ = __________fpm

Temp. Gas =_________°F
Duct Diam. = _________in.
Fan Speed = _________rpm

$Q_{main\ duct} = V_{average\ main\ duct}$ ______ × Cross-sectional Area _______(ft^2) = __________acfm

2.0 Visual Inspection

Primary Filters	____ Satisfactory	_____ Unsatisfactory
Secondary Filters	____ Satisfactory	_____ Unsatisfactory
Hopper	____ Satisfactory	_____ Unsatisfactory

3.0 Filter Integrity Test (fluorescent powder)

_____ Satisfactory _____ Unsatisfactory

4.0 Cleaning Cycle

Solenoid Valves	_____ Operational	_____ Not Operational
Diaphragm Valves	_____ Operational	_____ Not Operational
Photohelic Gage	_____ Operational	_____ Not Operational
Mechanical Shaker	_____ Operational	_____ Not Operational

5.0 Fan Information

Rotation	____ Satisfactory	_____ Unsatisfactory
Belt Tension and Condition	____ Satisfactory	_____ Unsatisfactory
Bearings	____ Satisfactory	_____ Unsatisfactory
Pulley Alignment	____ Satisfactory	_____ Unsatisfactory

6.0 Notes:

Field Data Sheet (Example)
Mist Collector

Company Name:
Department:
Location:
Manufacturer/Model:

Date:
Prepared by:

1.0 Initial Field Measurements/Calculations:

P Primary Filters = ________"spwg
P Secondary Filters = _______"spwg
$V_{average\ main\ duct}$ = __________fpm

Temp. Gas = _________°F
Duct Diam. = _________in.
Fan Speed = _________rpm

$Q_{main\ duct} = V_{average\ main\ duct}$ ______ × Cross-sectional Area _______(ft^2) = __________acfm

2.0 Visual Observation

Coalescing Filter	____ Satisfactory	_____ Unsatisfactory
Accumulated Liquid	____ Satisfactory	_____ Unsatisfactory
Clean Air Plenum	____ Satisfactory	_____ Unsatisfactory
Filters	____ Satisfactory	_____ Unsatisfactory

3.0 Fan Information

Rotation	____ Satisfactory	_____ Unsatisfactory

4.0 Notes

Hazmat team tests leaking valve at transportation test center in Pueblo, Colorado. (Layne Kennedy, 1994)

Outcome Competencies

After completing this chapter the user should be able to:

1. Define underlined terms used in this chapter.
2. Explain risk reduction methods.
3. Explain the framework needed to prevent accidental releases.
4. Recall regulations related to chemical releases.
6. Describe the difference between hazard evaluation and hazard identification.
7. List the important models to evaluate chemical releases.
8. Summarize disasters of historical importance.

Prerequisite Knowledge

Prior to beginning this chapter, the user should review the following chapters:

Chapter Number	Chapter Topic
2	Occupational Exposure Limits
3	Legal Aspects of the Occupational Environment
6	Principles of Evaluating Worker Exposure
30	Worker Education and Training
47	Hazardous Waste Management
48	Laboratory Health and Safety
49	Emergency Planning in the Workplace

Key Terms

disasters • emergency response • exposure assessment • extrinsic safety • hazard and operability (HAZOP) surveys • hazard analysis (HAZAN) • hazard evaluation • hazard identification • human factors • intrinsic safety • risk analysis • risk management • toxic chemical releases

Key Topics

I. Historical Background
   A. Characteristics of Disasters
   B. Evolution of Planning, Prevention, and Response

II. Purpose
   A. Framework
   B. Sources of Information
   C. Roles

III. Principles of Prevention
   A. Hazard Management
   B. Community Interaction

IV. Legislation
   A. Hazardous Materials Transportation Act
   B. Superfund
   C. Emergency Planning and Community Right to Know Act of 1986
   D. OSHA
   E. EPA
   F. Other U.S. Legislation
   G. Seveso Directive
   H. Control of Industrial Major Accident Hazards (CIMAH)

V. Hazard Identification
   A. Data Collection
   B. Hazard Indices
   C. Dow and MOND Indexes
   D. HAZOP
   E. HAZAN
   F. Fault Tree Analysis
   G. Event Tree Analysis
   H. Cause-Consequence Analysis
   I. Human Error Analysis
   J. Environmental Hazards

VI. Hazard Evaluation
   A. Release Characteristics
   B. Dispersion Models
   C. Biological Effect Prediction
   D. Integrated Systems

VII. Risk Reduction
   A. Intrinsic Safety
   B. Extrinsic Safety
   C. Human Factors

VIII. Planning and Response
   A. Scope
   B. Planning Process
   C. Health Planning Considerations
   D. Casualty Handling
   E. Exposure Assessment
   F. Organizational Communication
   G. Drills

34

Prevention and Mitigation of Accidental Chemical Releases

Jeremiah R. Lynch, CIH, CSP, PE

Historical Background

Characteristics of Disasters

Natural disasters such as earthquakes and floods have been occurring since before recorded history, but the advent of man-made disasters, other than war, required the development of cities, factories, and transportation systems. Some significant natural and man-made disasters are listed in Table 34.1.[1] In each case the disaster and the subsequent public reaction resulted in improved means of detection and prevention. This also is likely to happen in the aftermath of recent examples such as the Exxon Valdez oil spill on March 24, 1989, and the explosion and fire at the Phillips Petroleum Co. plant in Pasadena, Texas, on October 23, 1989. While it is natural to tighten precautions following a disaster, it would be better to recognize and evaluate the potential for a disaster before

Table 34.1 — Disasters and Results

Type	*Location and Date*	*Total Deaths*	*Results*
Earthquake	Chicago, IL; 10/9/1871	250	building codes prohibiting wooden structures; water reserve
Flood	Johnstown, PA; 5/31/1889	2209	inspections of dams
Tidal wave	Galveston, TX; 9/8/1900	6000	sea wall built
Fire	Iroquois Theater, Chicago, IL 12/30/03	575	stricter safety standards
Marine fire	*General Slocum*, East River, NY 6/15/04	1021	stricter ship inspections; revision of statutes (life preservers, experienced crew, fire extinguishers)
Earthquake	San Francisco, CA; 4/18/06	452	widened streets; limited heights of buildings; steel frame and fire-resistant buildings
Mine	Monongah, WV; 12/6/07	361	creation of Federal Bureau of Mines; stricter mine inspections
Fire	N. Collinwood School; Cleveland, OH; 3/8/08	176	need realized for fire drills and planning of school structures
Fire	Triangle Shirt Waist Co. New York, NY; 3/25/11	145	strengthening of laws concerning alarm signals, sprinklers, fire escapes, fire drills
Marine	*Titanic* struck iceberg, Atlantic Ocean; 4/15/12	1517	regulation regarding number of lifeboats; all passenger ships equipped for around-the-clock radio watch; International Ice Patrol
Explosion	New London School, TX; 3/18/37	294	odorants put in natural gas
Fire	"Coconut Grove," Boston, MA; 11/28/42	492	ordinances regulating aisle space, electrical wiring, flameproofing of decorations
Explosion	Caprolactam factory; Flixborough, England, 6/1/74	28	increased emphasis on hazard analysis
Toxic release	Union Carbide plant, Bhopal, India, (methyl isocyanate); 12/3/84	2500	tighter laws and regulations to prevent toxic chemical releases
Nuclear	Chernobyl, Russia (nuclear plant burns, releases radiation over Russia and Europe); 4/86	31+	re-evaluation of nuclear power safety

Source: Reference 1.

it happens and thus prevent it and others like it. The development of modern disaster prevention technology has as its focus 1) methods of hazard analysis that allow the prediction of the potential for a disaster in the absence of the event; and 2) the application of preventive measures not learned by experience.

Evolution of Planning, Prevention, and Response

As recently as the space shuttle *Challenger* disaster it was apparent that even sophisticated agencies such as the National Aeronautics and Space Administration were not uniformly applying the most advanced risk analysis technology. While the many acknowledged uncertainties in the input data limit the precision of all risk analysis techniques, it is possible to calculate the bounds on the probability of an event with useful levels of certainty. These techniques include fault tree analyses, failure mode and effect analysis, what-if studies, and the Baysean statistics of decision analysis, which will be briefly discussed later in this chapter. Other more advanced methods are described in the books and journals listed in the references. In the future, man-made disasters may not be considered unavoidable accidents unless the probability of their occurrence has been shown to be vanishingly small by a pre-event risk analysis.

Advances in control technology are not so much scientific breakthroughs as the application of many small measures the value of which has long been recognized. These measures are discussed under the headings Intrinsic Safety, Extrinsic Safety, and Human Factors. The important advances in prevention are the use of such techniques as hazard and operability (HAZOP) surveys, coupled with risk analysis, to determine and prioritize what needs to be done to control risk in each situation.

Prevention is obviously preferable to response, but prudence requires that workable response plans exist wherever a disaster could occur. Spurred by experience and legislation, plans are evolving to become multisite/community-based designs that are rationalized by exacting studies of response experience and tested and refined by realistic drills.

Purpose

Framework

The prevention and mitigation of the nonroutine chemical releases that may result in a disaster is an exercise in risk management. In occupational hygiene, the risk management process has traditionally been divided into the categories of recognition, evaluation, and control. This same framework is applicable to the present topic except that the real recognition and evaluation phases are often combined, and control activities are divided into those measures taken to prevent the event and those taken to mitigate or minimize the consequences of the event. In this chapter the categories of hazard identification, hazard evaluation, risk reduction, mitigation, and response will be used.

Identification begins with a data collection activity that asks the question, "what chemicals are present in sufficient quantities to create a disaster if the entire amount was released?" HAZOP, hazard analysis (HAZAN), and various failure mode analyses are done next to develop release scenarios for credible releases. There are techniques to quantitatively estimate probabilities, but these are used less often. In the evaluation phase the possible consequence is estimated from the scenario assumptions, dispersion models, and biological effect criteria. Interaction between the identification and evaluation phases is necessary to be sure that the most serious consequence scenarios are identified.

A value judgment regarding risks applied to the results of the hazard evaluations may lead to the conclusion that risk reduction is necessary. The best way to reduce risk is to build plants so that disastrous events cannot happen. Where such events are still possible, equipment (scrubbers, alarms) and procedures are added to reduce risk. In some high hazard industries such as explosives manufacturing, isolation and distance are used to reduce the consequences of events.

If an event does occur, the consequence will depend on what actions are taken to respond to the emergency. Emergency forces must be available and properly used. People at risk must be evacuated or protected. Casualties must be properly moved and attended. All of this requires intelligent planning tested by frequent realistic drills.

Sources of Information

The body of knowledge on disaster prevention and response is enormous. The literature is filled with anecdotes from emergencies, detailed analyses of disasters and their causes, results of experimental chemical releases, mathematical dispersion models, risk analysis techniques, engineering means of prevention, and guides to emergency planning. Consequently, this chapter will not be an exhaustive coverage of the subject but an overview and source guide. Some of the major elements of technology in disaster prevention, mitigation, and response will be described. Organizational sources of assistance are listed in the conclusion, and the references provide additional help. Many of these are secondary sources that can in turn provide long lists of primary resources. Taken as a whole, this body of information can provide access to essentially everything published on the subject as of the time of writing and to the main organizations and research groups working in this field. While such a broad view is useful, it is likely that there also will be certain customary traditional approaches taken by specific industries (chemical manufacturing, transportation).

All of the assembled and codified experience of others is not enough. Each manager of a facility or operation that has the potential to cause a disaster has to create a unique understanding of that potential and how to control it. Much of the disaster prevention and mitigation literature deals with process tools by which the probabilities of unlikely events are examined and how to reduce those probabilities. However, no two sets of circumstances that might lead to disaster are alike, and thus, each manager must become expert in his or her own problem. This is a continuing effort, since circumstances change because of new laws, plant aging, engineering or process changes, new knowledge of chemical toxicity, and the evolving expectations of the public. The manager must be sensitive to these changes and to improvements in the understanding of how accidents happen and to the near-miss events that provide performance experience without major loss. The literature is helpful, but only learning by evaluation will lead to safety.

Roles

The manager of any process or operation has the actual and legal responsibility for what happens. Management will delegate, depending on the size of the organization, the responsibility for planning and coordinating disaster prevention, mitigation, and response activities. These coordinators will rely on internal and external experts to advise them and to perform certain tasks. Engineers will need to describe in detail how the equipment operates so that HAZOP analysts can ask "what if." The network of firefighters, first responders, and hospitals will need to consider what they would do in a disaster. In this multidisciplinary effort, the health professional (physician/occupational hygienist) is not usually in the lead role but performs a key specialist role, particularly in toxic disasters. Some of these contributions are as follows:

- Expert advice on the potency and acute and chronic health effects of chemicals that might be released;
- Response strategies (evacuation versus "button-up," or sealing off those who may be exposed) for various release scenarios;
- Planning and drilling for casualty handling involving all available resources;
- Technical health and environmental hazard information to help inform authorities and the public;
- Exposure assessment before, during, and after a release event by modeling and/or measurement; and
- Planning, auditing, and drilling the health and environmental components of the overall response plan.

These may be new roles for some health professionals, but they derive from their basic skills and from increasing concern for toxic chemical releases. This chapter addresses the full range of roles in disaster prevention, mitigation, and response but does so from the point of view of the health professional with the background needed to ease integration into the overall disaster planning team.

Principles of Prevention

Hazard Management

A systematic approach to chemical process hazard management requires that the means for preventing catastrophic release, fire, and explosion are understood, and that the necessary preventive measures and lines of defense are installed and maintained.[2,3] A schematic representation of this management system is shown in Figure 34.1.[4] While the process is logical and technically correct, the successful management of chemical process hazards will not occur without management commitment. This commitment takes the form of a requirement to know the magnitude of the hazards, a determination to follow through on necessary changes to reduce risk, and a reasonable and prudent criterion for what is acceptable. This last is the most difficult

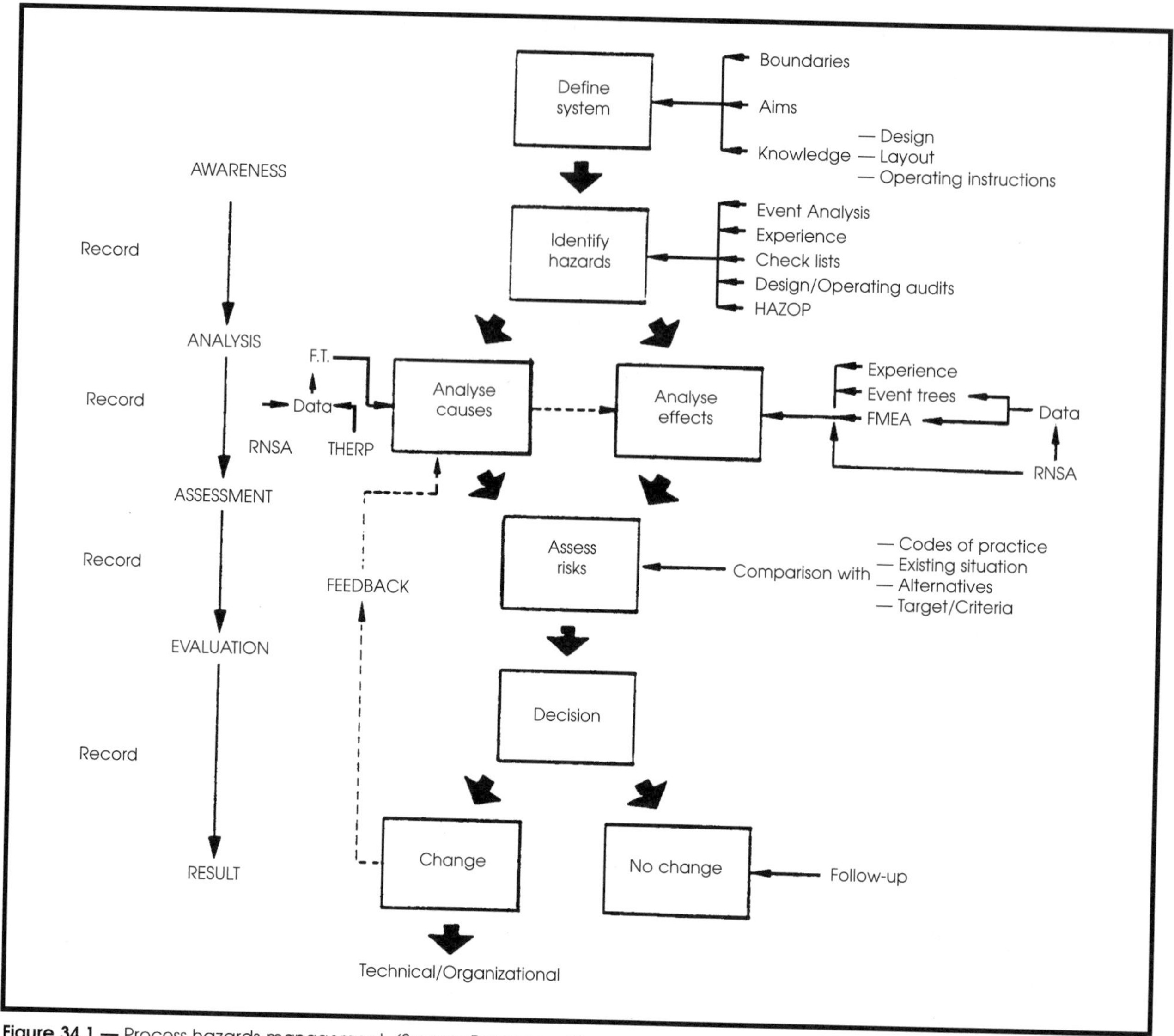

Figure 34.1 — Process hazards management. (Source: Reference 4.)

principle on which to achieve agreement. Historically, the threshold of acceptability was defined and reduced by public reaction to catastrophes. These criteria were not explicitly stated but could be implied from the laws, regulations, standards, codes, and practices accepted and applied. As the criteria of acceptability changes, old codes, standards, and practices need to be re-examined to see whether they achieve the currently required level of risk reduction. This is a continuing process and one that requires new risk management and analysis skills not widely used in the past.

Community Interaction

Decision making in risk management has been broadened beyond the facility manager and owners to include the community at risk.[5-7] As a practical matter, the community will be involved in decisions on new facilities and even on existing facilities either because the law requires it, or because intense public pressure would substantially hinder any action that did not have public acceptance. To gain this acceptance, the community needs to be involved from the earliest phase of the planning process, and the facility management needs to be completely open and accurate as the project development evolves. The local facility management needs to work closely with community leaders to understand and resolve their concerns. The principles of risk communication teach that it is not only the magnitude of the risk that determines acceptability but also how decisions about the imposition of risks and benefits are made. Failure to observe these principles can result in the failure of a project that would otherwise have met pure objective risk acceptability criteria.

Even within the enterprise there can be problems of risk communication. For example, the design of a new or modified facility will be in the hands of engineers who are in the habit of dealing with concepts that can be quantified with some accuracy. The risks to be considered can rarely be well quantified, and the health risks in particular have an added level of uncertainty. Beyond this, there is the uncertain question of public reaction. There is no fixed formula for managing the engineer-health professional-public interaction. The universal principle of openness, respect, and sensitivity are the best guides.

Legislation

Hazardous Materials Transportation Act

The U.S. Department of Transportation (DOT) is responsible for administering the Hazardous Materials Transportation Act[8] as well as the Natural Gas Pipeline Safety Act[9] and the Hazard Liquid Pipeline Safety Act.[10,11] All of these acts are intended to adequately protect the public against the risk to life and property that are inherent in the transportation of hazardous materials in commerce. They give the secretary of transportation the authority to regulate any safety aspect of the transportation of hazardous materials including packing, handling, labeling, marking, placarding, and routing. Regulations on packing cover the manufacture, marking, maintenance, repairing, and testing of packages. Regulations on handling cover number of personnel, level of training, frequency of inspection, and safety assurance procedures for handling of hazardous materials. The requirements are promulgated in a series of regulations in *Code of Federal Regulations* (CFR) Title 49, Parts 171–199. They incorporate by reference a long series of guides and standards developed by some of the organizations listed in Table 34.2.

In 49 CFR 172 are listed hazardous materials subject to labeling and other requirements of law. This list gives the hazard class, the United Nations or North American identification numbers, labeling and packaging requirements, and quantity restrictions. It is supplemented by the Reportable Quantities list for hazardous substances designated under the Comprehensive Environmental Response, Compensation and Liability Act of 1980 (CERCLA) as published by the U.S. Environmental Protection Agency (EPA). To aid those who must comply with these complex laws and regulations, DOT publishes a series of information bulletins for shippers, carriers, freight handlers, and container manufacturers. These bulletins cover duties of employees, shipping documents, labels, container specifications, quality control, violation reporting, and incident investigations. The main thrust of DOT's activity is accident prevention rather than emergency response.

In addition to DOT, the Nuclear Regulatory Commission regulates radioactive materials transport (10 CFR 71), the Federal Aviation Administration regulates aircraft carriage enforcement procedures (14 CFR 13), and the Coast Guard regulates hazardous cargo on board ships and in ports (33 and 46 CFR).

Superfund

CERCLA,[12] more commonly known as Superfund, was enacted "to provide for liability, compensation, cleanup and emergency response for hazardous substances released into the environment and the cleanup of inactive hazardous waste disposal sites." While Superfund is best known for its efforts to deal with hazardous waste sites, there are several provisions relevant to nonroutine chemical releases.[13,14]

Response. Section 104 provides the authority for the removal of any hazardous substance that may be released into the environment and the authority to respond to a release as necessary to "protect the public health or welfare of the environment."

Planning. Section 105 calls for the revision of the National Contingency Plan[15] to include the requirements of CERCLA.

Liability. Section 107 establishes a standard of strict liability for incidents involving the release of hazardous substances and provides for fines of up to $50,000,000 plus response costs plus triple punitive

Table 34.2 — Organizations Involved in Hazardous Materials Transportation Safety

American Society of Mechanical Engineers (ASME)
American National Standards Institute (ANSI)
American Society for Testing and Materials (ASTM)
Compressed Gas Association (CGA)
The Chlorine Institute (CI)
American Iron and Steel Institute (AISI)
Chemical Manufacturers Association (CMA)
National Fire Protection Association (NFPA)
Institute of Makers of Explosives (IME)
The Fertilizer Institute (TFI)
International Maritime Organization (IMO)
Society of the Plastics Industry (SPI)
International Organization for Standardization (ISO)

damages. Of interest to first responders to hazardous materials incidents is a provision stating that

> *no person shall be held liable . . . for damages as a result of actions taken or omitted in the course of rendering care, assistance, or advice in accordance with the National Contingency Plan or at the direction of an on-scene coordinator appointed under such plan with respect to an incident creating a danger to public health or welfare or the environment as a result of any release of a hazardous substance or the threat thereof unless there was gross negligence.*

CERCLA was amended by the Superfund Amendments and Reauthorization Act (SARA), which is discussed below.

Emergency Planning and Community Right to Know Act of 1986

This broad piece of legislation, known as Title III of SARA,[16] establishes a number of emergency planning, notification, reporting, and training requirements. The law requires that states establish emergency response commissions, emergency planning districts, and local emergency planning committees (LEPCs) consisting of representatives of government, police and fire departments, hospitals, the media, community groups, and industry. Each LEPC writes an emergency plan covering hazardous substance facilities, emergency response and coordination, notification, release warning and dispersion, emergency equipment and facilities, evacuation, training, and drills. The LEPC is guided by publications from the national response team established by the National Contingency Plan (Section 105 of CERCLA). Local personnel are to be trained in hazard mitigation, emergency preparedness, fire prevention and control, disaster response, and other emergency training topics.

The act also requires facilities to notify the community emergency coordinator and the LEPC in the event of a release of certain hazardous chemicals in excess of specified amounts. The notification must identify the substance; the amount released; time and duration of the release; media (air, water); anticipated health risks; medical advice; precautions; and a telephone contact. To avoid releases of extremely hazardous substances, EPA will review systems for monitoring, detecting, and preventing such releases, including perimeter alert systems. Facilities are also required to provide material safety data sheets (MSDS) or lists of MSDSs to LEPCs.

Other major sections of this act provide for the reporting of routine nonemergency releases of certain toxic chemicals (Section 313). SARA provisions also require (Section 126) that the Occupational Safety and Health Administration (OSHA) issue standards for the heath and safety of employees in hazardous waste operations (HAZWOPER).

OSHA

In 1985 OSHA initiated a special emphasis program for the chemical industry (ChemSEP). As part of this program OSHA inspected facilities to study current practices for the prevention of catastrophic incidents and the mitigation of releases. At the same time, EPA conducted a review of emergency systems for monitoring, detecting, and preventing releases of extremely hazardous substances as required by Section 305 of SARA. These two efforts led OSHA and EPA to cooperate on the development of standards for the prevention of chemical release disasters. Since workers at a facility are usually the population most immediately and seriously impacted by a release, and since OSHA has more experience in inspecting safety hazards at industrial facilities, the coordinated regulations will be issued and enforced by OSHA.

On June 1, 1992, OSHA issued a final rule, "Process Safety Management of Highly Hazardous Chemicals."[2] This rule has its antecedents in OSHA's ChemSEP effort, EPA's SARA review, and the Chemical Manufacturers Association's (CMA's) Chemical Awareness and Emergency Response (CAER) program.[17] Input from the Organization Resources Counselors,[18] the American Petroleum Institute, the Oil Chemical and Atomic Workers, and the United Steelworkers of America was also considered. The rule is performance oriented and requires that process safety management (PSM) systems be put in place in covered workplaces.

In its Appendix A the rule includes a list of highly hazardous chemicals with threshold quantity amounts that trigger the provisions of the rule. In addition, the rule applies to flammable liquids or gases (except certain fuels) in excess of 10,000 lb. Appendix C of the rule is a compliance guideline for PSM, and Appendix E provides further sources of information.

EPA

On June 20, 1996, EPA issued a final rule, "Accidental Release Prevention Requirements," under the authority of Section 112(r)(7) of the Clean Air Act.[3] The purpose of the regulation is to prevent accidental releases of regulated substances and reduce the severity of those releases that do occur. The rule requires a risk management plan and one of three levels of management system programs depending on the degree of risk as determined by a hazard assessment. This rule covers stationary sources, which includes those establishments covered by OSHA's PSM rule. The highest (most stringent) level of risk management program required by the EPA rule is fundamentally identical to OSHA PSM requirements; so any establishment that meets PSM will meet all three EPA levels.

The chemicals regulated under this rule are listed in "List of Regulated Substances and Threshold Quantities for Accidental Release Prevention."[19] The list gives the name of the chemical, the Chemical Abstract Service number, the threshold quantity, and the basis for the listing.

Other U.S. Legislation

Several other U.S. laws have provisions that relate to nonroutine chemical releases.

Resource Conservation and Recovery Act. This law lists certain hazardous wastes and requirements for their treatment, disposal, and storage.

Clean Water Act. The Clean Water Act sets limits on the release of hazardous substances into water.

Toxic Substance Control Act. Under certain circumstances, EPA may require health and environmental properties testing of new and existing chemicals. In addition, if releases may present substantial risk to health or the environment, there may be instances where such releases are reportable to the EPA administrator.

Seveso Directive

The Council of the European Communities adopted the Seveso directive, named after an accidental release of dioxin in Italy in 1976, on June 24, 1982.[20,21] This directive sets up an approach to the prevention of major chemical release disasters in Europe and covers many of the same requirements as the U.S. regulations.[22]

Control of Industrial Major Accident Hazards (CIMAH)

Following the Seveso disaster the European Community issued in 1983 a directive on major accident hazards of certain industrial activities. Each member country was required to implement this directive by means of national laws. The United Kingdom issued the CIMAH regulations in 1984[23] to control fire, explosion, and toxic hazards. The regulations require safety study, notification, emergency plans, and public information where listed chemicals are held in excess of stated quantities.

Hazard Identification

Hazard identification is the first step in the risk assessment and management process. It starts with data collection to define the system and proceeds through the identification of hazards and the analysis of causes and effects. The output of the hazard identification step leads to hazard evaluation and risk reduction, as needed. Many formal techniques have been developed for the systematic analysis of complex systems. Some of the most commonly used methods (HAZOP and HAZAN) and some specialized methods like human error analysis are discussed later in this chapter. Some others are listed in the references.[24–33] They all attempt to consider all reasonable possibilities and all suffer from the drawback that the probability of future events can only be guessed.

Data Collection

A necessary first step in all hazard identification and analysis procedures is the collection of accurate data about the substances present, the process and equipment, and the layout, topography, meteorology, and demographics of the plant and surrounding community. A complete inventory of substances present would include raw materials, additives, catalysts, intermediates, products, and wastes. The inventory should specify quantities, locations, use, and marking and type of storage. Chemicals used in utilities (hydrazine, chlorine) should also be listed. For each substance on the inventory, the substance specific information listed in Table 34.3 should be obtained. No single source (e.g., MSDS) will have all of this information, and measurements may need to be made to collect some of the data.

Plant and process equipment data collection starts with an up-to-date set of process and instrumentation drawings (P&ID). As plants age and are modified, the engineering details of plant changes are not always carried over to the P&ID, so it becomes out of date. This can be a serious problem when conducting the hazard analyses described below if the plant is not, in fact, what the analysts think it is. In addition to

Table 34.3 — Substance Specific Information

Health effects
Exposure limits—TLV, IDLH, ERPG, Community Exposure Guides Effects—irritancy, systems organ, cancer reproduction effect, neurotoxicity Toxicity—oral, dermal LD50 Odor—threshold, persistence
Physical/chemical properties
Physical—solubility, vapor pressure, melting/boiling point, density Chemical—formula, structure, molecular weight, corrosivity
Fire and explosion properties
Flammability—flash point, autoignition temperature Explosivity—upper and lower explosive limits, detonation, explosive decomposition, pyrophoric Reactivity—auto reaction, incompatibilities, effect of impurities
Environmental
Fate—biodegradation, bioconcentration, hydrolysis, photolysis, volatility Effects—terrestrial, aquatic

accurate P&IDs, additional detail should be collected on the age and condition of equipment, the state of maintenance, and the actual process operating procedures and practices used. What is known about the degree of corrosion of pipes and vessels? Are battery limit block valves operable? Have safety valves been tested, and are rupture disks properly installed? Have prescribed operating procedures been modified by short cuts? Will remote sensors work?

To predict the consequences of a release, it is necessary to know how the land lies, which way the wind blows, and where the people are. Simple dispersion models make worst case assumptions and need little data, while complex models require a lot of data, as discussed below. Demographic data, transportation routes, and the location of fire stations and hospitals are also needed for casualty care planning.

The collection and continuous updating of this assembly of information is a difficult task and a burdensome effort. Companies need to assure by means of their audit or review process that the job is being done wherever it is needed.

Hazard Indices

A complete inventory will list a few substances that have the potential to cause a catastrophic incident and many that do not. A necessary step in going from the inventory to the development of event scenarios is a ranking of the list by hazard potential. Several hazard indices have been developed for this purpose.[34,35] The vapor hazard index[36] is a ratio of the concentration of saturated vapor divided by the threshold limit value (TLV®) times 1000. However, the TLVs are based on a variety of end points rather than the acute toxicity of primary concern in chemical release incidents. The substance hazard index[18] is the ratio of the vapor pressure of a substance to its acute toxicity concentration, which is calculated from either the American Industrial Hygiene Association (AIHA) Emergency Response Planning Guideline Level Three (ERPG–3; see below), the EPA Levels of Concern, or the Acute Toxic Concentrations developed by the State of New Jersey. Substances with substance hazard indexes over 5000 are considered dangerous. Both of these indexes consider only intrinsic human toxicity. Other ranking systems have been developed[37] that consider both human and environmental end points.

EPA has proposed a screening index in its rulemaking under SARA. The index is equal to a "level of concern" value (usually the National Institute for Occupational Safety and Health [NIOSH] immediately dangerous to life or health [IDLH] value in milligrams per liter) divided by V the vapor fraction (1 for gases). EPA uses this index to determine threshold quantities for the planning and notification provisions of their rules. An obvious weakness of this index is that it depends on IDLH values that may not be consistent and accurate measures of intrinsic hazard.

Dow and MOND Indexes

The Dow Chemical Co. Fire and Explosion Index[38] was originally intended as an aid to fire protection by calculating a fire and explosion index. It also calculates the maximum probable property damage and the maximum probable days outage. It also considers general and special process hazards such as toxicity hazards and arrives at a toxicity index. A simplified flow diagram is shown in Figure 34.2.

The American Institute of Chemical Engineers Center for Chemical Plant Safety[39] has summarized this relative ranking procedure as follows:

1. Identify on the plot plan those process units that would have the greatest effect or contribute the most to a fire, explosion, or release of toxic material.
2. Determine the material factor for each unit.
3. Evaluate the appropriate contributing hazard factors, considering fire, explosion, and toxicity.

4. Calculate the unit hazard factor and damage factor for each unit.
5. Determine the fire and explosion index and area of exposure for each unit.
6. Calculate maximum probable property damage.
7. Evaluate maximum probable days outage and business interruption costs.

Based on the nature and severity of the hazard, various preventive and protective features are selected and the process repeated to calculate new indexes.

The MOND fire, explosion, and toxicity index, which is a variation of the Dow Index and was developed by the MOND division of Imperial Chemical Industries (ICI) in the United Kingdom., has been extended to consider more operations and equipment and to evaluate the hazards from materials and reactions more extensively.

Both methods can be used during the design stage of a project or on an existing unit. A team of specialists in engineering, fire, safety, and health will best be able to conduct them.

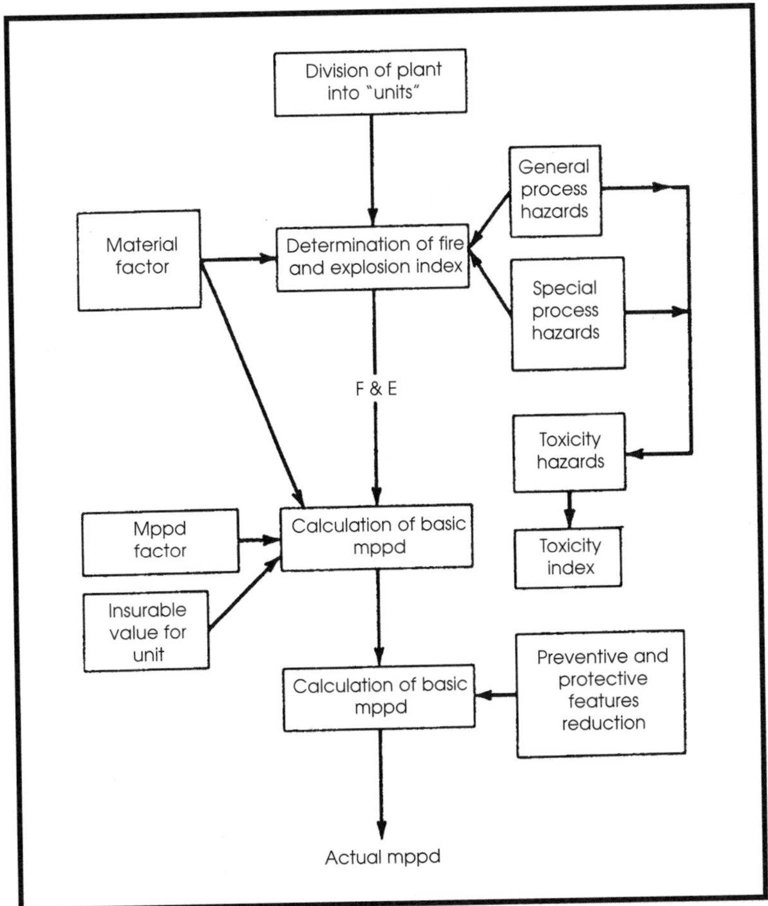

Figure 34.2 — Dow index procedure. (Source: Reference 4.)

HAZOP

A HAZOP survey is one of the most common and widely accepted methods of systematic qualitative hazard analysis. It is used for both new and existing facilities and can be applied to a whole plant, a production unit, or a piece of equipment. It uses as its database the usual sort of plant and process information and relies on the judgment of engineering and safety experts in the areas with which they are most familiar. The end result is reliable, therefore, in terms of engineering and operational expectations, but it is not quantitative and may not consider the consequences of complex sequences of human errors.

The objectives of a HAZOP study can be summarized as follows:[40]

- Identify areas of the design that may possess a significant hazard potential.
- Identify and study features of the design that influence the probability of a hazardous incident occurring.
- Familiarize the study team with the design information available.
- Ensure that a systematic study is made of the areas of significant hazard potential.
- Identify pertinent design information not currently available to the team.
- Provide a mechanism for feedback to the client of the study team's detailed comments.

A HAZOP study is conducted in the following steps.

Specify the Purpose, Objective, and Scope of the Study. The purpose may be the analysis of a plant yet to be built or a review of the risk of an existing unit. Given the purpose and the circumstances of the study, the above objectives can be made more specific. The scope of the study includes the boundaries of the physical unit and also the range of events and variables considered. For example, at one time HAZOPs were mainly focused on fire and explosion end points, but now the scope usually includes toxic release, offensive odor, and environmental end points. The initial establishment of purpose, objectives, and scope is very important and should be precisely set down so that it will be clear, now and in the future, what was and was not included in the study. These decisions need to be made by an appropriate level of responsible management.

Select the HAZOP Study Team. The team leader should be skilled in HAZOP and in the interpersonal techniques to facilitate successful group interaction. As many other experts as needed should be included in the team to cover all aspects of design, operation, process chemistry, and safety. The team leader should instruct the team in the HAZOP procedure and should emphasize that the end objective of a HAZOP survey is hazard identification; solutions to problems are a separate effort.

Collect Data. Theodore[39] has listed the following materials that are usually needed: 1) process description; 2) process flow sheets; 3) data on the chemical, physical, and toxicological properties of all raw materials, intermediates, and products; 4) P&IDs; 5) equipment, piping, and instrument specifications; 6) process control logic diagrams; 7) layout drawings; 8) operating procedures; 9) maintenance procedures; 10) emergency response procedures; and 11) safety and training manuals.

Conduct the Study. Using the information collected, the unit is divided into study nodes, and the sequence diagramed in Figure 34.3 is followed for each node. Nodes are points in the process where process parameters (i.e., pressure, temperature, and composition) have known and intended values. These values change between nodes as a result of the operation of various pieces of equipment such as distillation columns, heat exchangers, or pumps. Various forms and worksheets have been developed to help organize the node process parameters and control logic information.

When the nodes and parameters are identified, each node is studied by applying the specialized guide words to each parameter. These guide words and their meanings are key elements of the HAZOP procedure. They are listed in Table 34.4.[38]

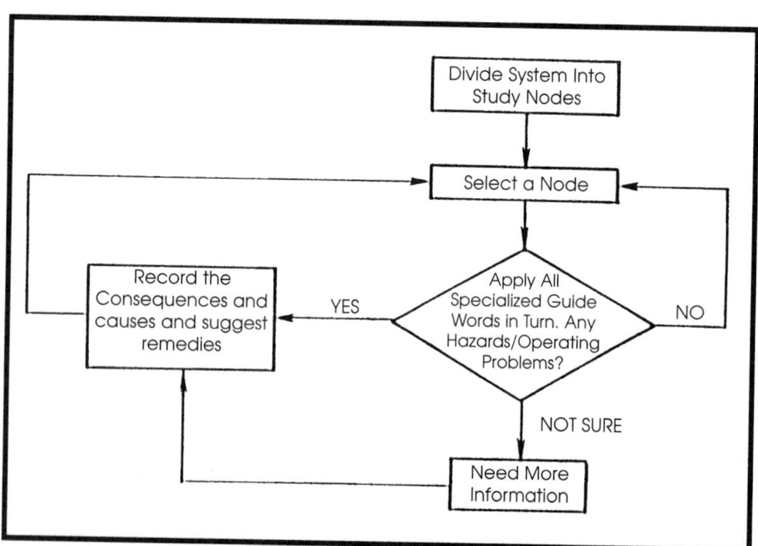

Figure 34.3 — HAZOP method flow diagram. (Source: Reference 38.)

Table 34.4 — HAZOP Guide Words and Meanings

Guide Word	Meaning
No	negation of design intent
Less	quantitative decrease
More	quantitative increase
Part of	qualitative decrease
As well as	qualitative increase
Reverse	logical opposite of the intent
Other than	complete substitution

Repeated cycling through this process, which considers how and why each parameter might vary from the intended and what the consequence would be, is the substance of the HAZOP study.

Write the Report. As much detail about events and their consequences as is uncovered by the study should be recorded. Obviously, if the HAZOP identifies a credible sequence of events that would result in a disaster, appropriate follow-up action is needed. Thus, although risk reduction action is not a part of the HAZOP per se, the HAZOP may trigger the need for such action.

HAZOP studies are time-consuming and expensive. Just getting the P&IDs up to date on an older plant may be a major engineering effort. Still, for processes with significant risk, they are cost-effective when balanced against the potential loss of life, property, business, and even the future of the enterprise that may result from a major release.

HAZAN

The acronym HAZAN is a generic term for a variety of quantitative hazard and risk analysis methods. A logical sequence for the systematic examination of a facility is to use hazard identification and ranking techniques first. If there are possible scenarios that could lead to unacceptable consequences, then qualitative techniques (HAZOP) can be applied next. Where a qualitative hazard exists, HAZAN methods can be used to estimate the quantitative probability of adverse events. How much analysis is worthwhile is a function of the consequence of the adverse event and the difficulty in preventing it.

Fault Tree Analysis

Fault tree analysis is a logical method of analyzing how and why a disaster could occur. It is a graphical technique that starts with the end event—the accident or disaster, such as a nuclear fuel meltdown—and works backward to find the initiating event or combination of events that could lead to the final event. If the probabilities of each potential initiating event are known or can be estimated, then the probability of the end or top event can be calculated.

The following example of fault tree analysis is taken from *Accident and Emergency Management*.[40] A water pumping system consists of two pumps, A and B, where A is the pump ordinarily operating and B is a standby pump that automatically takes over if A fails. Flow of water through the pump is regulated by a control value in both cases. Suppose that the top event is no water flow, resulting from the following initiating events: failure of Pump A and failure of Pump B, or failure of the control valve. The fault tree diagram for this system is shown in Figure 34.4. This diagram shows the logical relationship of the events and allows the calculation of the probability of the top event if the probabilities of the initiating events are known. Let A, B, and C represent the failure of Pump A, the failure of Pump B, and the failure of the control valve, respectively. Then it can be seen from the diagram that the top event T (no water flow) occurs if both A and B occur or if C occurs. A, B, and C can be assumed to be independent events unless they have a common cause or are otherwise related. From the basic theories of probability, we can write:

$$P(T) = P(AB) + P(C) - P(ABC)$$
$$\text{and } P(AB) = P(A)P(B)$$
$$\text{and } P(ABC) = P(A)P(B)P(C)$$

$$\text{Therefore: } P(T) = P(A)P(B) + P(C) - P(A)P(B)P(C).$$

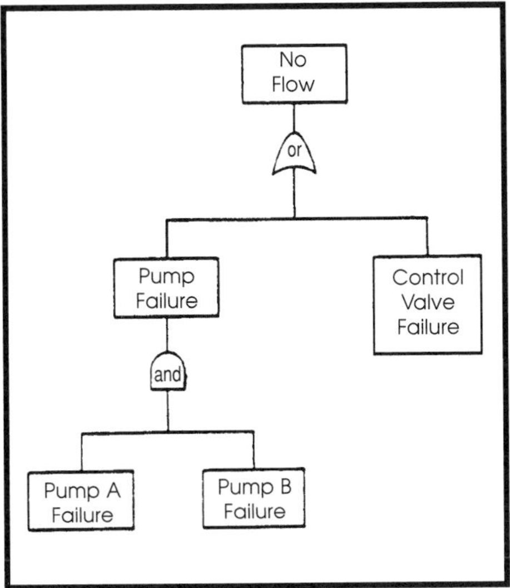

Figure 34.4 — Fault tree diagram. (Source: Reference 39.)

If the individual initiating event probabilities P(A), P(B), and P(C) are known, then the probability of P(T), the top or end event, can be calculated.

The output of fault tree analysis (and of several other methods of reliability analysis described below) is a list of the minimum combinations of events sufficient to cause the outcome. This list of events is called the minimal cut set and can be ranked in terms of likelihood and consequence as a means of prioritizing safety improvement programs.

Event Tree Analysis

Another technique for applying the basic concepts of probability to the evaluation of reliability of actual systems that have the potential to cause disasters is event tree analysis.[32] This technique is similar to fault tree analysis in that it examines, in a probabilistic manner, the consequences of a series of logically connected events. Event tree analysis, however, differs from fault tree analysis in that it starts with the initiating event rather than working backward from the end event. As in fault tree analysis, a diagram is constructed showing the logical relationship between events and outcomes. Event tree analysis may be qualitative or, if reasonably accurate probabilities can be assigned to events, a quantitative estimate of reliability can be obtained. In practice, the analyst reads off the diagram the sequences of events that result in disaster and then describes and examines the events in these sequences to describe how an accident may occur. As with fault tree analysis, the result is useful both in the design of safe new plants and to improve the safety of existing facilities.

Cause-Consequence Analysis

Fault tree analysis and event tree analysis have been combined into cause-consequence analysis so that sequences can be worked out either from the event to the outcome or from the outcome back to the initiating events. The result can be either qualitative or quantitative depending on the nature of the input data. Because the analysis, while in some ways more complex, is more intuitively obvious, the display and description of sequences better serves the purposes of communication and of training those who can influence or control events. The output is usually a ranking of accident sequences based on their consequences and a ranking of minimal cut sets to evaluate important initiating events.

Human Error Analysis

The preceding methods of analysis are easiest to do if the initiating events can be accurately assumed to be independent. Indeed, this assumption is often implicit in analyses, as they are usually performed even if the independence of initiating events is not stated. Yet it is obvious that equipment failure events often have common causes (a storm, a power failure) and thus are not independent. Another potential common cause is when operating personnel have an erroneous assessment of the situation or a mind set about what is happening, sometimes persisting in the face of evidence to the contrary. This human error then becomes the common cause of a series of initiating events that, if looked at as independent events, would seem to have a very low probability of occurring simultaneously. At Three Mile Island, the operator's misunderstanding of the cooling water status led to several actions or inactions that formed a part of a sufficient minimal cut set to result in the accident. In the cause of the Ocean Ranger oil rig disaster,[41] the operators did not appreciate the damage that sea water had done to the ballast control panel. Because of confusing and conflicting information presented by the system, their attempts to operate the ballast controls became a common cause in a series of unlikely events that resulted in the capsizing of the rig and the loss of 84 lives.

The techniques described above can be used to analyze the reliability of complex systems even where common causes and man-machine interactions are involved.[42–44] Human error analysis is a specialized subdiscipline of hazard evaluation that adds considerations of human performance to the hazard evaluation process. It is specialized and complex because of the difficulty in understanding the multitude of factors that influence human error rates and their variability, the high degree of interaction between human and system failure (lack of independence), and the nature of the options for altering human error rates.[45] Yet, human behavior must be factored into reliability analysis for it to be accurate in the increasing number of systems where errors in man-machine interaction can be an initiating event in a disaster.

The most commonly used quantitative method for the measurement and assessment of personnel-induced errors is the technique for human error prediction (THERP).[4] This procedure involves 1) identification of human activities that create a hazard, 2) estimation of failure rates; and 3) effect of human failures on the system.

The output of THERP is an input to fault tree or other methods of hazard analysis. While THERP can estimate failure rates for the routine performance of tasks, it cannot cope with error in human decisions and has difficulty with task error rates altered by stress, as in an emergency.

Environmental Hazards

In addition to fire, explosion, or the sudden release of a toxic gas, an emergency may be precipitated by the sudden release of an environmentally hazardous material.[46] A large oil spill can threaten aquatic organisms, birds, and those who depend on them for food. Chemicals released into rivers can threaten fish and drinking water supplies. Spills on land can reach and contaminate streams and aquifers and be transported to human or animal receptors or be taken up by plants and enter the food chain.

The identification of the potential for environmental disasters is similar in some respects to that of other kinds of disasters. Critical facilities that could potentially release environmentally toxic substances should be identified. Release scenarios and the pathways by which the substance would reach environmentally sensitive locations can be described and the consequences of each release-pathway combination analyzed. The engineering risk analysis methods described above for releases of substances toxic to humans can be applied to these environmental release scenarios.

The variety of transport media and the chemical and biological changes that may occur, as illustrated in Figure 34.5, complicate prediction of human or environmental receptor exposure and consequent harm. These complex processes can result in a delay of years before the consequences of a release are manifest. Also, because of the many uncertainties associated with each leg of a pathway, it is very difficult to predict consequence with any accuracy.

Human health consequences are predicted based on exposure of the human receptor compared with certain accepted exposure limits for food or drinking water. For risk management purposes, these human health consequences are valued in the same way as human health consequences resulting from air contaminants. For environmental damage, the situation is more complex. There is a substantial lack of data on the effects of a wide variety of chemi-

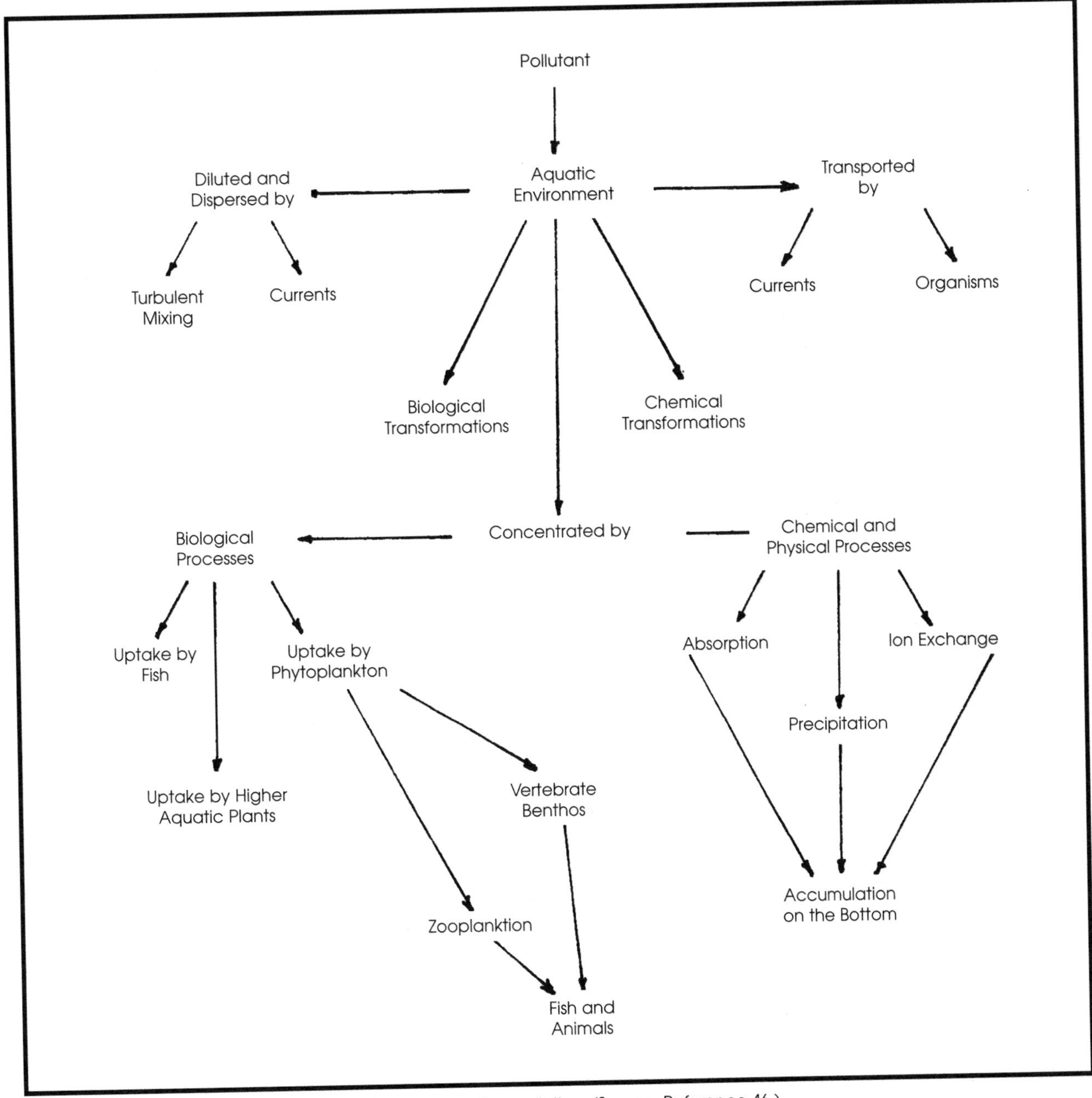

Figure 34.5 — Environmental distribution processes affecting pollution. (Source: Reference 46.)

cals on an even wider variety of organisms. Further, there is no accepted system for making environmental value judgments across different species.

Hazard Evaluation

Hazard evaluation, in this context, means estimation of the harm that might be done by a toxic release. The preceding section discussed how to identify hazards and analyze the likelihood of release events. Based on this analysis, various release scenarios can be developed. This section will discuss the methods used to estimate the consequences of a release. Basically, there are five steps: 1) scenario definition (part of hazard identification), 2) release rate amount and physical state estimation, 3) transport and dispersion modeling, 4) biological effects prediction, and 5) consequence estimation. Each of these steps is based on a combination of scientific theory, empirical data, and judgment. As a result, there is considerable uncertainty associated with the final conclusion(s).

Release Characteristics

In the section above, a number of methods for identifying scenarios that could result in a disaster were discussed. These events would result in a release of a hazardous material. The release would have certain physical, chemical, and biological characteristics that need to be known or modified to predict the consequence of the events and thus develop appropriate mitigation in response plans. Specifically, some event will result in a loss of containment of a substance that will then be released at some rate for some time. The release may be a gas, a liquid, or a two-phase mixture. All of the characteristics of the release are taken into account in source models to predict the source terms that are inputs to various dispersion models.

Loss of containment is any event that allows substances normally contained in the tanks and piping of a facility to be released into the environment. In emergencies the concern is not so much about losses of containment that result in a slow release, such as the leaks that produce fugitive emissions (although these are important in the prevention of air pollution). The loss of containment events that produce disastrous releases are those capable of releasing large amounts of material or energy in a short time.

In general, properly designed, intact, undisturbed pipes and vessels do not spontaneously fail. Failure and release occur as a result of improper design, as in the case of the flexible coupling between reactors used at Flixborough;[47] or because of some unintended event. Pipes may fail as a consequence of unintended stress brought on by weather or physical forces compounded by corrosion, fatigue, or embrittlement. Vessels fail in fires or because there is either too much or too little pressure. In the transportation of chemicals by road or rail, vehicle accidents can cause loss of containment. Storms, groundings, or collisions of tankers can cause the release of their cargoes.

The manner in which containment is lost will determine the release mechanism. Figure 34.6 illustrates several ways in which containment could be lost. The type and location of rupture and the phase,

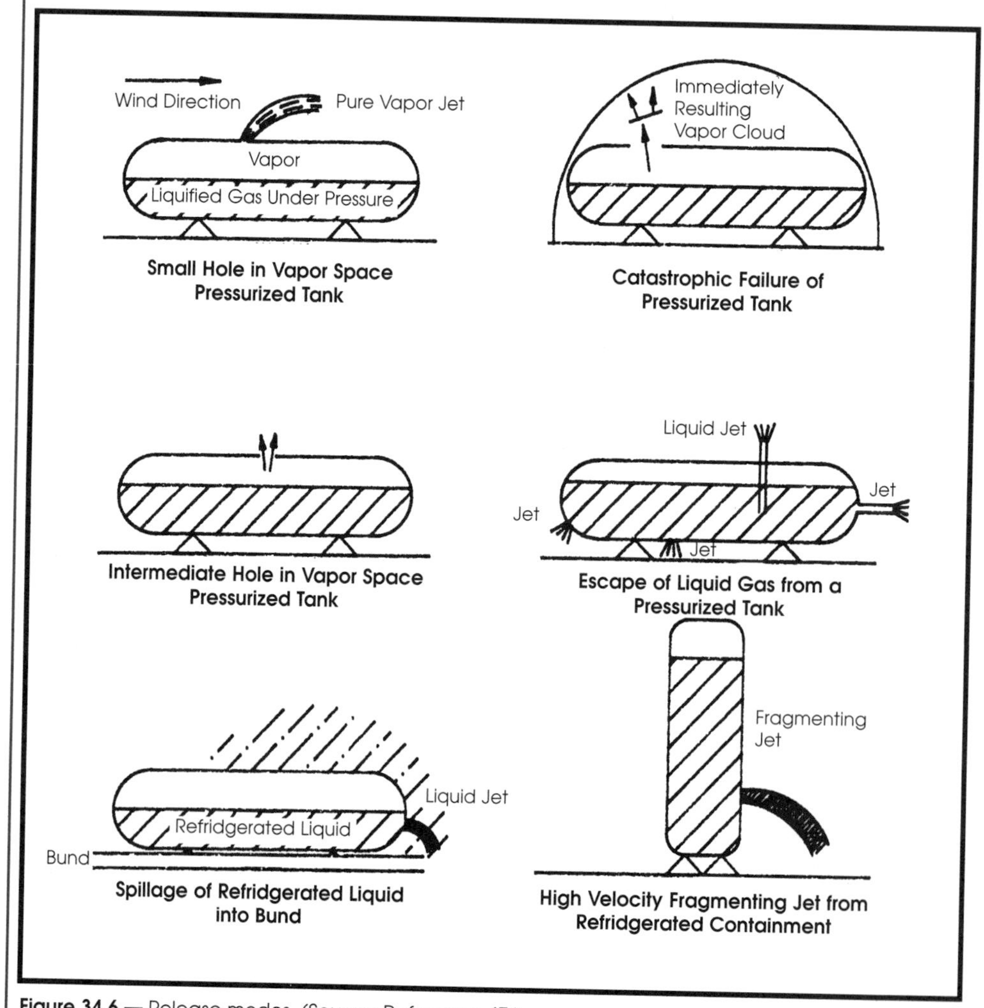

Figure 34.6 — Release modes. (Source: Reference 47.)

pressure, and temperature of the material determine the rate, period, and velocity of the release. The release could be a pressurized jet resulting in a momentum plume of gas or a rapidly expanding vapor cloud of a positively or negatively buoyant gas. Two-phase releases can be a mix of a gas and a bulk liquid that may subsequently evaporate or a liquid aerosol in a gas stream that is dispersed with or without complete evaporation.[48]

Loss of containment can also result in a spill of a substance in water, such as an oil spill, or on the ground where soil or groundwater contamination may be a problem unless appropriate remedial action is taken.

Depending on all of these factors, releases may be divided into four general time/area classes.[49]

1. Continuous point release—a prolonged, small area release; a hole or broken pipe on a large vessel.
2. Term point release—a brief period release from sudden and complete loss of a container of a gas. When the release duration is less than one-tenth of a minute, it is considered to be spontaneous or a "puff."
3. Continuous area release—a prolonged release over some area, as from a spill of a volatile liquid on the ground or water.
4. Term area release—a rapid release over some area, as from the evaporation of a pool of spilled refrigerated liquid.

Modeling of the source terms for a release requires knowledge of the class of release, the geometry of the release aperture and surroundings (contaminant wall, pond), the physical chemical properties of the material at the time of the release, and the environmental conditions. All of these data are brought together as input to a source emission model, as shown in Figure 34.7.

A large number of models have been developed to estimate release rates and source terms as inputs to subsequent dispersion models. Release of a gas, liquid, or two-phase mixture from a hole in a pressurized container can be modeled from first principles and from fluid dynamics, assuming the size and shape of the hole.[50] For buoyant gases these models yield dispersion model input source terms directly. For heavy gases and liquids the situation is more complicated, and empirical models based on test data developed at Edgewood Arsenal and Thorney Island are used.[51] A number of evaporation models have been developed by the states, the U.S. government, and by other governments and private concerns. These models deal adequately with the most frequently expected accident conditions. However, additional work is needed for such unique conditions as chemical reaction after release, evaporation of mixtures, vadose layer release, and terrain effects. Toxic releases resulting from fires are a special case of loss of containment. The release and subsequent ignition of a flammable liquid or gas can result in the formation of a toxic gas or aerosol that can then be dispersed. Release can be modeled as in the noncombustible gases, but the products of combustion are difficult to predict quantitatively, particularly when they will vary with temperature and the availability of oxygen. Fires can also cause the destruction of toxic materials and can in those cases reduce the hazard to the general population. Another concern for fires in plants where radioactive materials are used for gauges, radiography, or other purposes is that the fire will destroy the radioactive material shielding and possibly disperse the material. This needs to be considered in facility design[52] and in disaster planning.

Dispersion Models

When a toxic substance is released into the atmosphere, it moves or spreads as carried by its own momentum, by gravity, and by the wind. This dispersion of the gas cloud will both dilute it and cause it to cover or involve a wider area. It is necessary to predict the concentration of the toxic substance that may occur at any location and time in order to predict the consequences to people and the protective measures necessary to avoid those consequences.[53–55] While long-range transport of clouds may be best described by box or compartment models, the intermediate range (100–1000 m) movement of a toxic substance is usually dealt with by dispersion modeling.

Methods of modeling the dispersion of a release of gas in the atmosphere are based on the following premises.[14,47]

- The gas is not lost or losses are taken into account (conservation of matter).
- Through the action of turbulence in the atmosphere, the gas is distributed horizontally and vertically according to a Gaussian (normal) distribution, the parameters of which are a function of distance from the source.
- The parameters of the distribution are derived from experiments and are described by empirical equations.

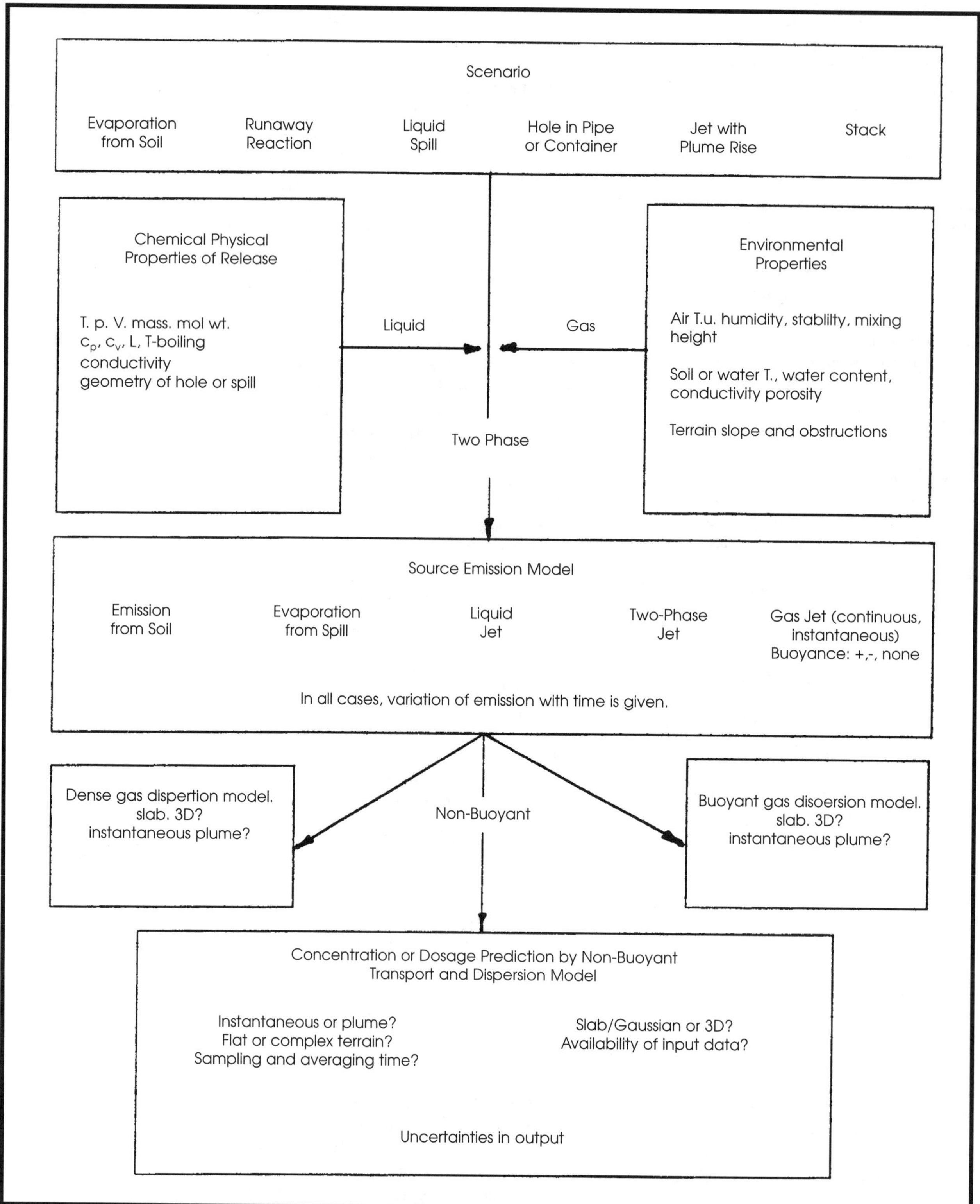

Figure 34.7 — Generalized vapor cloud model logic sequence. (Source: Reference 56.)

These premises lead to the idealized Gaussian plume model for a continuous source shown in Figure 34.8. Since this is an idealized model, it has a number of limitations. The model is valid only for a continuous source over flat open ground. It assumes that meteorological conditions are constant in time and space, that there is some wind, and that the gas is the same density as air.

Hanna[56] noted that there were over 100 hazardous materials transport and

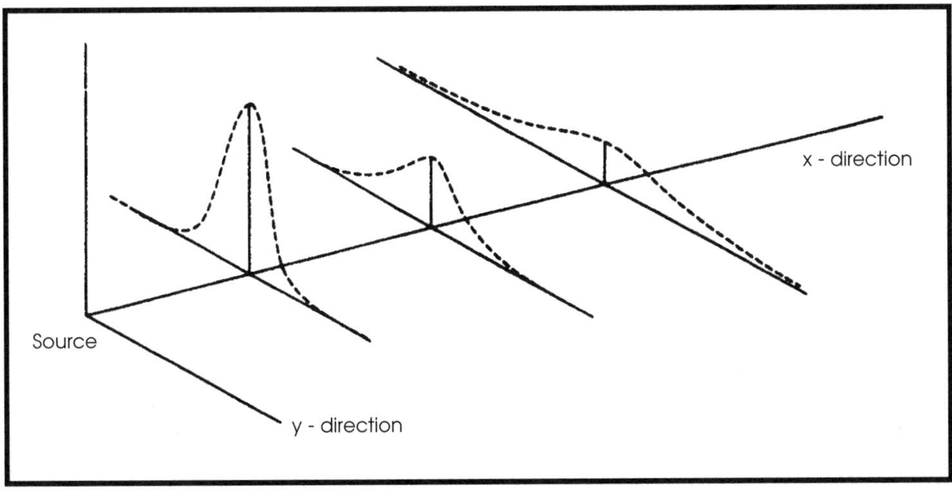

Figure 34.8 — Gaussian distribution. (Source: Reference 50.)

Table 34.5 —
Models for Hazardous Gas Transport and Dispersion

Model	Reference[A]	Instantaneous Puff	Continuous Plume	Dense Gas	Slab Similarity	Numerical (Grid)	Winds Vary
AIRTOX	Paine et al. (1986)	X	X	X	X		
AVACTA II	Zannetti et al. (1986)	X	X		X		X
Britter	Britter (1979, 1980)		X	X	X		
CARE	Verholek (1986)	X	X	X	X		X
CHARM	Radian (1986); Balentine & Eltgroth (1985)	X	X	X	X	X	X
Chatwin	Chatwin (1983s)	X		X	X		X
CIGALE 2	Crabol et al. (1986)	X		X	X		
COBRA III	Oliverio et se. (1986)	X	X	X	X		
CRUNCH	Fryer (1980); Jagger (1983)		X	X	X		
3D-MERCURE	Riou & Saab (1985)	X	X	X		X	X
D2PC	Whitacre et al.	X	X		X		
DEGADIS	Havens & Spicer (1985)	X	X	X	X		
DENS20	Meroney (1984)	X	X	X	X		
DENZ	Fryer & Kaiser (1979)	X		X	X		
Fay & Zemba	Fay & Zemba (1985, 1986)	X	X	X	X		
FEM3	Chan (1983)	X	X	X		X	X
HAZARD	Drivas et al. (1983)	X	X	X	X	X	X
HEAVY GAS	Deaves (1985)	X	X	X		X	X
HEAVYPUFF	Jensen (1983)	X		X	X		
HEGADAS	Colenbrander & Puttock (1983)	X	X	X	X		
HEGDAS	Morrow		X	X	X		
Hoot et al.	Hoot et al. (1973)		X	X	X		
INPUFF 2.0	Peterson & Lavdas (1986)	X			X		X
MADICT	Ludwig (1985)	X			X	X	X
MESOPUFF 11	Scire and Lurmann (1983)	X			X		X
MIDAS	Woodward (1987)	X	X	X	X		X
NILU	Eidsvik (1980)	X		X	X		
Ooms et al.	Ooms et al. (1974)		X	X	X		
Port Comp. System	MOE (1983, 1986)	X	X	X	X		
RIMPUFF	Mikkelsen et al. (1984)	X	X		X		X
SAFEMODS	Raj (1981, 1986)	X	X	X	X		
SAFER	Personal correspondence	X	X	X	X	X	X
SIGMET	England et al. (1978)	X	X	X		X	X
SLAB	Morgan et al. (1983)	X	X	X	X		
SPILLS	Fleischer (1980)	X	X		X		
TOXGAS	McCready et al. (1986)	X			X		
Van Ulden	Van Ulden (1974)	X		X	X		
VAPID	Jensen (1983)	X	X	X	X		
Webber	Webber & Brighton (1986)	X		X	X		
Wilson	Wilson (1981)	X	X		X		
Zeman	Zeman (1982)	X	X	X	X		

Source: Reference 46.
[A]References refer to the source article (Hanna 1987).

dispersion models, that the number was growing at the rate of about 10 per year, and that only a few have been adequately evaluated with field data. He listed the characteristics of 40 of these (Table 34.5) and characterized the models in terms of the phenomena they treat. Four models are discussed here to illustrate the basic principles important for the transport and dispersion of hazardous gases.

AIRTOX

The AIRTOX model for the release and dispersion of toxic air contaminants can account for single-phase denser-than-air releases. It is a hybrid model that incorporates various theoretical models, and it uses a slab model approach to calculate air dispersion. It assumes a pseudoinstantaneous dense gas cloud to calculate instantaneous sources. Continuous sources are also modeled, and the passive plume is assumed to disperse according to standard Gaussian formulas with a wind shear component. The model has been shown to compare fairly well with field data.

DEGADIS

The dense gas dispersion (DEGADIS) model assumes that the crosswind distribution is uniform in the middle of the cloud and Gaussian at the edges, and that the vertical distribution follows a modified power law. The model deals with the dilution of a dense gas cloud by entrainment as the cloud moves under the influence of wind and gravity. It can calculate the initial cloud behavior near its source (the source blanket) and can simulate both steady-state and transient spills. Transient spills are modeled as a series of pseudo steady-state releases with a coordinate system that moves with the wind over the transient gas source. This moving observer concept and the source blanket concept are unique to DEGADIS, but there are several aspects of the model that need further evaluation.

FEM3

The FEM3 model is representative of three-dimensional models. While simpler two-dimensional models provide great insight into the physical processes important for the transport and dispersion of hazardous gases and agree fairly well with the limited databases, the physical event is three dimensional; therefore, much of what is happening cannot be modeled by two-dimensional models. FEM3, like most three-dimensional models, is extremely complex, and its computer code requires a long time to execute even on a supercomputer. Yet, as complex as it is, it is necessary to neglect certain processes, such as interphase heat exchange, to keep the run time within reason. Obviously, FEM3 is not suitable for real-time simulations. Its best application is as a means of understanding what is happening.

INPUFF 2.0

The INPUFF model is typical of a group of puff trajectory models for nondense emissions. The model cannot deal with the negative buoyancy of a dense gas cloud close to the source. It can model plume rise from positively buoyant plumes using standard formulas found in EPA UNAMAP models. INPUFF uses a Gaussian puff to calculate the distribution of concentrations in each puff. Dispersion coefficients are based on extensive experimental evidence, and the rate of dispersion can change with time. Multiple sources can be handled by summing the contribution from individual puffs at the receptor.

Model Uncertainty

Uncertainty in model predictions is an important consideration when models are used to predict the consequences of toxic chemical releases. The causes of uncertainty can be divided into three components.

Model Assumptions. As noted above in the discussion of FEM3, all models that attempt to approximate reality can take a very long time to run on a supercomputer; consequently, useful models contain some simplifying assumptions and therefore are only approximations of reality. How much error these assumptions introduce depends on how far they are from true and how robust the model is. One reason why there are so many models is that each model is a better or worse fit of an event to be modeled. Therefore, it is incumbent on the risk assessor to select the best model for the release scenario.

Input Errors. All models require the input of values for certain parameters that describe the physical, chemical, and meteorological conditions. These input parameters may not be single fixed values but may be variables that are functions of time and space. Often, they are not well known and must be approximated or assumed, which introduces error. There is a tradeoff between model accuracy and input error. The more realistic the model, the greater the number of less accurately known input parameters it will require. At some point short of maximum possible reality, an optimum combination of model detail and input error results in minimum overall error. When models are used for planning purposes to predict the consequences of a hypothetical future event, it is of course necessary to assume all of the input parameters. Often, worst case assumptions are made so that the predicted consequence will be conservative. Unfortunately, the selection of a series of worst case values for the input parameters can result in an excessive but unknown degree of conservatism.

Random Error. Turbulence of a fluid is an inherently chaotic process and thus cannot be usefully predicted. This stochastic uncertainty is dealt with as a random variable with a variance roughly equal to the square of the mean of the predicted concentration. It is usually the smallest of the three components of error.

Validation

Comparing predicted results with actual data from field tests validates models.(57,58) The number of models has increased much more rapidly than the quantity of field-test data, so most models are not validated. Where a complex model is validated using a carefully assembled mass of accurate input data, the agreement between predicted and actual can be quite good—less than a factor of two difference. Where assumptions are substituted for data, the agreement is not as good, and the best that can be said is that the predicted concentration is not inconsistent with the actual concentration given the anticipated level of error. In general, models that predict long-term averages are more accurate than dynamic models of instantaneous concentrations, since errors in parameters tend to be averaged out.

Biological Effect Prediction

The modeling technology and techniques described above can predict with some error the concentration time pattern of the released chemical as it moves out from a point or area of release. People at locations where this cloud passes may be exposed and as a result may suffer adverse effects. The exposure may be by inhalation or by skin contact with or without absorption through the skin. The degree of adverse effect will depend on the dose, which in turn depends on the concentration and duration of the exposure and on whether the person is out of doors, in a car, or inside a house.

The prediction of the effects of exposure is a classic question of toxicology. Where little information is available on the toxicity of the chemical or only gross estimates of exposure concentrations are available, it may be adequate to classify chemicals according to relative toxicity using some estimate such as the median lethal dose (LD_{50}). A number of such schemes are shown in Table 34.6. The LD_{50} is the dose at which half of the animals in an experiment would be expected to die as a result of a short-term exposure. It is a common reference point for acute toxicity and may be the only data point available for some less common chemicals. In some cases, the best adverse effects prediction that can be done will be based on a judgment of concentration from the amount released and some

Table 34.6 — Classification Systems for Acute Toxicity

Textbook (Hodge & Sterner)	*Industrial Chemicals (EEC)*	*Poison Law (Japan)*	*Pesticide Act (US)*	*Transport of Goods (United Nations)*	*Pesticide (WHO)*
Extremely toxic (< 1)	very toxic (< 25)	designated poison (< 15)	category I (< 50)	group I (< 5)	extremely hazardous < 5
Highly toxic (> 1–50)	toxic (< 25–200)	poisonous (< 30)	category II (< 5–500)	group II (< 5–50)	highly hazardous (> 5–50)
Moderately toxic (> 50–500)	harmful (< 200–2000)	deleterious substance (< 300)	category III (> 500–5000)	group III solids (> 50–500)	moderately hazardous (> 50–500)
Slightly toxic (> 500–5000)			category IV (> 5000)	liquids (> 50–2000)	slightly hazardous (> 500)
Practically nontoxic (> 5000–15,000)					
Relatively harmless (> 15,000)					

Note: Based on LD50 (mg/kg body weight)
Source: Institution of Chemical Engineers (ICE). Risk Analysis in the Process Industries. (Pub. series 45). London: ICE, 1985.

simple box model applied to a lethality class for the chemical to arrive at a consequence prediction suitable for emergency response.

The time-concentration relationship to adverse effect may be divided into three domains. At the most distant fringe of the area touched by the cloud the concentration may be below an actual or practical biological threshold such that no effect occurs regardless of the duration of the exposure. In this region, effect is predicted by whether the concentration alone is over or under some threshold. Closer in, where the concentration is somewhat higher, the effect may depend on the product of the time and the concentration. Still closer to the source, where the concentration is highest, and the adverse effects of some substances like chlorine or ammonia(59) may include death, the adverse effect is best predicted by the square of the concentration times the time. This more complex biological behavior, which occurs in the zone of exposure of greatest concern in disasters, requires that both the modeling and the effects prediction guidelines include a time component.

A simple way of relating exposure concentration-time-effect is shown in Table 34.7. This table divides the exposure concentration ranges into hazard bands and gives some indication of the proportion of people affected (e.g., lethal for 50%). This type of data presentation based on direct human experience is only possible for a few chemicals such as chlorine. For most chemicals, where only animal data are available, some process of extrapolation must be used, and only the most serious end points are estimated.

Various organizations have set emergency exposure limits or guidelines for use in hazard evaluation, remediation, and emergency response planning. To be useful, such guidelines should be developed both for substances where human experience is known and for those where only animal data are available. They should have uniform effect criteria (e.g., threshold of lethality) and should specify the exposure duration. It is also helpful if the guidelines are set by a process of consensus among experienced scientists so they have broad credibility.

The Committee on Toxicology of the National Research Council set out to develop emergency exposure limits in the early 1960s. These levels, later called Emergency Exposure Guidance Levels (EEGLs) were defined as

> *a ceiling limit for an unpredicted single exposure, usually lasting 60 minutes or less and never more than 24 hours, whose occurrence is expected to be rare in the lifetime of any person. It reflects an acceptance of the statistical likelihood of the occurrence of a nonincapacitating, reversible effect in an exposed population. It is designed to avoid substantial decrements in performance during emergencies and might contain no uncertainty factor.*(60)

The National Research Council's (NRC's) EEGL effort was sponsored by the military, and the exposed population was healthy adults. Recognizing that the EEGLs were not applicable to the general population, the NRC introduced the concept of Short-Term Public Emergency Guidance Levels. These guidelines are intended for public emergencies, but the NRC has recommended few of them.

During the 1970s NIOSH developed a set of IDLH exposure levels as part of an effort known as the Standards Completion Program. The concept and the term IDLH came from respiratory protection regulations that made a distinction between

Table 34.7 — Hazardous Bands for Chlorine Gas

Hazard Band	*Exposure (ppm)*	*Effect*
Distress (3–15 ppm)	3–6	Causes stinging or burning sensation but tolerated without undue ill effect for up to 1 hour
	10	Exposure for less than 1 min causes coughing
Danger (15–150 ppm)	10–20	Dangerous for 1/2–1 hour exposure Immediate irritation to eyes, nose, and throat, with cough and lachrymation
	100–150	5–10 min exposure fatal for some vulnerable victims
Fatal (> 150 ppm)	300–400	A predicted average lethal concentration for 50% of active healthy people for 30 min exposure
	1000	Fatal after brief exposure (few breaths)

Source: Reference 53.

respiratory protection devices that are suitable for life-threatening environments and those that are not. The IDLHs were occupational exposure levels based on a 30-minute exposure time. Since they were never intended as general public exposure limits, and since some of them were set by simply taking 100 times the TLV in effect at the time, they are not suitable as community emergency guidelines.

Increased concern over accidental chemical releases following the Bhopal disaster prompted AIHA to establish the Emergency Response Planning Committee. This committee was charged with the development of a set of levels that could be used as boundaries between different degrees of an emergency. They are not recommended exposure levels in any sense, because the existence of exposure over any of them is an unwanted event. As indicated by the name, they are intended for response planning purposes but are also useful in prioritizing disaster prevention measures. They are developed according to a protocol initially developed by the Organizational Resource Counselors and are defined as follows.[61]

Emergency Response Planning Guideline 3 (ERPG–3) is the

> *maximum airborne concentration below which it is believed that nearly all individuals could be exposed for up to one hour without experiencing or developing life-threatening health effects. The ERPG-3 level is a worst-case planning level above which there is the possibility that some members of the community may develop life-threatening health effects. This guidance level could be used to determine if a potentially releasable quantity of a chemical could reach this level in the community, thus demonstrating the need for steps to mitigate the potential for such a release.*

ERPG–2 is defined as the "maximum airborne concentration below which it is believed that nearly all individuals could be exposed for up to one hour without experiencing or developing irreversible or other serious health effects or symptoms which could impair an individual's ability to take protective action." Above ERPG–2, for some members of the community, there may be significant adverse health effects or symptoms that could impair an individual's ability to take protective action. These symptoms might include severe eye or respiratory irritation or muscular weakness.

ERPG–1 is defined as the "maximum airborne concentration below which it is believed that nearly all individuals could be exposed for up to one hour without experiencing other than mild, transient adverse health effects or without perceiving a clearly defined objectionable odor." The ERPG–1 identifies a level that does not pose a health risk to the community but that may be noticeable due to slight odor or mild irritation. In the event that a small, nonthreatening release has occurred, the community could be notified that they might notice an odor or slight irritation but that concentrations are below those that could cause health effects. For some materials, because of their properties, there may not be an ERPG–1. Such cases would include substances for which sensory perception levels are higher than the ERPG–2 level. In such cases no ERPG–1 level would be recommended.

Human experience takes precedence in developing ERPGs, but they can also be developed from animal data. The documentation recognizes that human response is not precise and that there are uncertainties in the data on which the limits are based. While limits could be set for a range of exposure times, the committee chose the single time of 1 hour based on the available toxicology information and typical exposure scenarios.

Guidance on acutely hazardous chemicals has also been provided by other organizations. The World Bank and the European Economic Community have used Lethal Concentration for 50% (LC_{50}) to arrive at acutely hazardous quantities rather than concentrations.[62] The National Fire Protection Association (NFPA) has developed a health hazard rating scale for use by emergency service personnel. The U.K. Health and Safety Executive sets specified level of toxicity exposure concentrations as an aid in calculating individual risk from major hazards.[63]

Finally, the Clean Air Act Amendments of 1990 required EPA to promulgate an initial list of 100 substances that, in the case of an accidental release, are known to cause death, injury, or serious adverse effects to the environment, and to set threshold quantities for these substances. This list includes chlorine, ammonia, methyl chloride, ethylene oxide, vinyl chloride, methyl isocyanate, hydrogen cyanide, hydrogen sulfide, toluene diisocyanate, phosgene, bromine, hydrogen chloride, hydrogen fluoride, sulfur dioxide, and sulfur trioxide. The threshold quantities are based on the toxicity, reactivity, volatility, dispersibility, combustibility, or flammability of the substance.

Integrated Systems

The risk manager needs to have the whole risk assessment process summarized so that prevention and response options will be clear. The crisis center director needs a system that takes all the available data on the conditions and circumstances at the time of a release and predicts the consequences fast enough and clearly enough for proper response decisions to be made. A number of integrated risk assessment systems have been developed for these purposes.

Manual Systems. The few most likely scenarios can be modeled for several weather conditions or a worst case set of conditions. The predicted concentration time patterns can be plotted as isopleths at the ERPG or other effect boundary concentrations. These plots can be on separate pages for different wind directions or as movable transparent overlays that can be oriented on a map for the site of release and the current wind direction. These consequence prediction systems have the advantage of being easy to understand and use and are not dependent on the availability of a computer and a source of electricity. They have the disadvantage of being very approximate and of covering only a limited number of possibilities. For these reasons a number of rather sophisticated computer systems have been developed, but even when a computer-based integrated risk assessment system is used, it is prudent to have a manual backup available.

Computer Systems. Several integrated risk assessment commercial computer programs have been developed (Table 34.8). While the capabilities of the various systems are different, and newer versions generally add capabilities, the following attributes are found in at least one system.

- Accepts substance, circumstance, and meteorology input;
- Provides substance hazard data;
- Includes source term/initial dispersion modeling;
- Does air dispersion modeling;
- Provides models for site-specific terrain;
- Analyzes sensor data;
- Provides a real-time graphical display; and
- Assesses risk.

Computer systems have been used for risk assessment, planning, training, drills, accident critiques, and response to emergencies. While the systems are capable of real-time operation, they are most commonly used to provide simulations for planning and training purposes.

Table 34.8 — Integrated Risk Assessment Computer Systems

System	Source
CARE	Environmental Systems Corp.
CAMEO	National Safety Council
EPI Code	Homann Associates, Inc.
HASTE	Environmental Research and Technology Inc.
SAFER	SAFER Emergency Systems, Inc.

Risk Reduction

The material presented above describes systems that allow the prediction of the likelihood and consequence of various chemical releases of various event scenarios. These predictions can be made for either an existing plant or a projected new plant during the design phase. Given the uncertainties of the modeling and estimation methods used, qualitative descriptions of probability and consequence are usually appropriate. Thus, the probability will vary from frequent (meaning likely to occur repeatedly) to very unlikely (meaning not likely to occur in hundreds or thousands of years). The consequences range from minor odor or discomfort events to very serious events that may cause multiple deaths. These probability consequence combinations for each release scenario can be displayed on a graph or matrix that can serve as a graphical aid to mitigation priority setting. Obviously, an event that is predicted to occur frequently with very serious consequences is an unacceptable situation, and a design or operation that results in such a prediction must be modified to reduce the risk before proceeding. For existing facilities, probabilities are more likely to be infrequent and consequences moderate. Mitigation effort priorities in this zone may be set based on a judgmental estimate of the degree of unacceptability of each probability consequence combination. At some low probability/low consequence point, it will be decided based on criteria derived either from policy or law that nothing further needs to be done. This scheme for deciding whether something needs to be done leads logically to consideration of the mitigation options available.[64,65]

Intrinsic Safety

Kletz[47] has analyzed plant safety in terms of attributes that make the plant "friendly," or forgiving of error. These plants are designed so that departures from normal tend to be self-correcting or at most lead to minor events rather than major disasters. Plants designed to be forgiving and self-correcting are inherently safer than plants where equipment has been added to control hazards or where operators are expected to control them. Some characteristics of friendly plants are as follows.

- Low inventory—smaller equipment and vessels and less intermediate product in storage means less hazardous material on site.
- Substitution—hazardous materials can sometimes be replaced with less hazardous materials.
- Attenuation—if hazardous materials must be handled, they can be used in their least hazardous forms.
- Simplification—operators are less likely to make mistakes if the plant and its instrumentation are easy to understand.
- Domino effects—an event should self-terminate rather than initiating other events.
- Incorrect assembly impossible—equipment design can make assembly mistakes impossible.
- Obvious status—an operator should be able to look at a piece of equipment and know if it is correctly installed or if a valve is open or shut.
- Tolerance—small mistakes need not create big problems.
- Low leak rates—leaks should be small and easy to control.

The application of these principles is summarized in Table 34.9.

Extrinsic Safety

Friendly plants as described by Kletz are designed to avoid major problems, and this approach to plant safety should be taken wherever possible. Where it is not, external features of the plant should be arranged to minimize the undesirable consequences of a release. The most obvious means of reducing the exposure of community populations is to locate high hazard facilities remote from communities. In general, siting is a major design consideration both within a facility as well as in the relation of the facility to population centers. LeVine[66] provides some guidelines for the layout of plants storing highly toxic hazardous materials (HTHM).

High-integrity storage tank design is essential for HTHM. Consideration should be given to overdesign or the use of double integrity storage tanks, for example, with a concrete outer tank or berm.

If the liquid storage tank for HTHM is diked, it is good practice that the same dike does not contain inventories of other materials that are flammable, reactive, or incompatible with the HTHM. If located next to a diked area containing flammable and/or reactive chemicals, the toxic storage may have to be provided with fixed fire protection for cooling. Factors to be considered include drainage in the diked areas, response time, adequacy of the site fire brigade or public fire department that would respond, and the flammability and reactivity of the nearby materials. The impounding area should be designed to minimize the vaporization rate of volatile HTHM. Installed means of evaporation suppression should be considered (e.g., foam).

HTHM should not be stored in proximity to explosive/flammable/reactive materials that might affect the storage tank. Diking for materials that are flammable or reactive should be sloped or drained to carry spilled material to a separate impoundment tank or area and away from any nearby HTHM tank. For HTHM the impoundment system should be designed to minimize the surface area and move material away from other hazard areas.

The concept and design of a designated dump tank should be investigated at the earliest stages of site layout. Depending on the potential hazard and the time required to empty the HTHM storage tank, it may be desirable to keep some storage vessels empty, or at low levels, for such emergency transfers. The possible need to empty a leaking tank is one important reason for providing a submerged pump or other transfer method.

The momentum of the tank contents might overflow a traditionally designed dike wall if the tank were to burst. A dike wall of about one-half tank height is required for containment of this scenario. Full-height, close-in dikes or berms have been provided for some cryogenic ammonia (and liquefied natural gas) storage tanks. The dike must be sized to contain at least the volume of the largest tank within the dike.

In case all means of containment fail, releases should be controlled through safety valves or other pressure relief systems and directed by vent headers to a safe release point or to an emergency relief treatment system. These end-of-pipe systems act as release countermeasures by means of secondary containment or by destroying the released materials. They include scrubbers, flares, incinerators,

Table 34.9 —
Characteristics of Friendly Plants

Characteristic	Examples	
	Friendliness	*Hostility*
Low inventory		
Heat transfer	miniaturized	large
Intermediate storage	small or nil	large
Reaction	vapor phase	liquid phase
	tubular reactor	pot reactor
Substitution		
Heat transfer	media nonflammable	flammable
Solvents	nonflammable	flammable
Carbaryl production	Israeli process	Bhopal process
Attenuation		
Liquefied gases	refrigerated	under pressure
Explosive powders	slurried	dry
Runaway reactants	diluted	neat
Any material	vapor	liquid
Simplification		
	hazards avoided	hazards controlled by added equipment
	single stream	multistream with many crossovers
	dedicated plant	multipurpose plant
	one big plant	many small plants
Domino Effects		
	open construction	closed buildings
	fire breaks	no fire breaks
Tank roof	weak seam	strong seam
Horizontal cylinder	pointing away from other equipment	pointing at other equipment
Incorrect assembly impossible		
Compressor valves	noninterchangeable	interchangeable
Device for adding water to oil	cannot point upstream	can point upstream
Obvious status	rising spindle valve or ball valve with fixed handle	nonrising spindle valve
	figure-8 plate	spade
Tolerant of maloperation or poor maintenance		
	continuous plant	batch plant
	spiral-wound gasket	fiber gasket
	expansion loop	bellows
	fixed pipe	hose
	articulated arm	hose
	bolted arm	quick-release coupling
	metal	glass, plastic
Low leak rate		
	spiral-wound gasket	fiber gasket
	tubular reactor	pot reactor
	vapor phase reactor	liquid phase reactor

Source: Adapted from Reference 47.

absorbers, condensers, and any other device placed in the vent stream to prevent release to the atmosphere. Even when release to the atmosphere has occurred, some mitigation is possible by means of such vapor release countermeasures as water sprays or curtains, steam or air curtains, or deliberate ignition.

The operation of secondary containment and vapor release countermeasures will be triggered by some release warning system. This warning may come from observation of process parameters as displayed in the control room; from the detection of odor or the sighting of a cloud of smoke, fume, or colored gas; or from a signal from an installed vapor detection sensor. The early warning that can be provided by an installed detection system can trigger a quick response that may prevent most or all of a release. The detectors sense and measure gases and vapors by combustion, catalytic, electrochemical, and solid-state devices. They may alarm locally or remotely in the control room or be part of a network connected to a computer that analyzes the information from a number of locations and presents the results in a way that highlights response options. For early warning vapor detection systems to be useful tools they must be reliable and well maintained. Also, they must be specific enough for the chemical of concern and sufficiently sensitive to differentiate between routine releases (fugitives) and emergencies.

Human Factors

The best-designed plant can fail if improperly operated. Case histories of plant disasters[67] have shown that human error was often involved in the chain of events that lead to the outcome. At Three Mile Island, it was a persistent misunderstanding by the operators of the status of the cooling system that led them to overriding of automatic safety systems.

The possibility of human error cannot be eliminated, so plants should be designed to be forgiving of operator mistakes. Operator training and motivation can substantially reduce human error. OSHA has listed operating procedures necessary to control plants that handle highly hazardous chemicals.[2] These operating procedures should be detailed in readily available operating manuals that are kept evergreen as plant equipment or procedures change. Operators should be thoroughly trained in the operating procedures before they assume responsibility for plant operations, and supplemental and refresher training should be provided as needed—typically annually. Some form of testing or qualification should be set up to ensure that each operator has learned the procedures. Periodically, management should compare prescribed and actual operating procedures. Where there are differences, they should be analyzed to determine whether it is because of an employee lapse or because the written procedures are impractical or could be improved.

Contractors present a special concern in the control of human factors related to chemical releases. Teams of contractor employees often work on site under the supervision of their own company with only indirect control by the site owner. The training of contractor employees is the responsibility of the contractor and is complicated by relatively high turnover in some jobs. Where contractors hire skilled trades from a hiring hall, the contractor may not know who will be working on a job until the day of the assignment. Even when all of a contractor's regular employees are trained and aware of safe operating procedures, the contractor may subcontract specific tasks and thus bring on site people who are not trained and aware. Frequently, there are multiple contractors on a site doing jobs that may affect each other's safety. Some major disasters appear to be related to a lack of a clear division of responsibilities and close coordination between the owner and the several contractors.[47,68] While the legal consequences of owner-contractor interaction complicate the matter, at least one jurisdiction[63] has stated that "where . . . a special risk is likely to arise due to the nature of the work performed (and the owner of the premises has special knowledge of it), the owner must retain sufficient control of the operator to ensure that contractors' employees are properly protected against the risk." In the United States an owner has the responsibility under OSHA to ensure that the contractor is informed of hazards and that the contractors' employees are trained.

A typical program addressing the contractor issue might include the following:

- The contract must include the site safety and health manual and the operating procedures to be followed.
- The contractor must attend a work review before the job starts, at which specifications, safety, process hazards, and employee training is detailed.
- The contractor must be provided with MSDSs for all hazardous materials to be encountered.
- The contractor must train his or her employees and provide the owner with a list of trained employees. The contractor must understand that only trained employees will be admitted to the site.
- The contractor must obtain any necessary work permits (hot work, confined space entry) before starting such work.
- The contractor must keep accident records and inform the owner of any accidents.
- The contractor is subject to safety reviews and inspections.
- Any violation of safety rules can be cause for termination of the contract.

The above list is not intended to cover all requirements and may not fully apply to certain contractors who have little involvement with hazardous processes, but it gives an example of the level of interaction some owners have found necessary to ensure adequate control of contractors.

Planning and Response

Even when all reasonable measures to prevent a chemical release have been taken, the risk or probability of a release occurring is not zero. A prudent management practice, therefore, is to anticipate that an emergency may occur eventually and to plan for it. A well-developed and implemented emergency plan can minimize loss and help protect people, property, and the environment. To accomplish this requires planned procedures, clearly understood responsibilities, designated and accepted authority and accountability, and trained and drilled people. When an emergency occurs, it will be too late to plan, so it must be done ahead of time based on the kinds of emergencies that might occur.[39,69–71]

Scope

An emergency response plan must be tailored to fit the facility, community, and the types of emergencies that can occur. Michael[72] has divided emergency response functions to consider in planning into the following five categories: 1) accident assessment, which includes detecting abnormal conditions, assessing the potential consequences, and immediately taking appropriate measures to mitigate the situation; 2) notification and communication, which includes the physical and administrative means whereby plant operators can rapidly notify plant management, off-site emergency response agencies, and the public; 3) the command and coordination function, which clearly establishes who is in charge of the emergency response in the plant and in off-site communities; 4) protective actions, which are those taken to protect the health and safety of plant personnel and the public, such as release mitigation, sheltering, and evacuation; and 5) support actions, include fire fighting, emergency medical care, social services, and law enforcement.

Plan evacuators should expect to find the attributes listed in Table 34.10 considered in the plan.[73] Not all of these attributes will be required in every plan, but it is useful to consider all of them to ensure that the plan is complete.

Planning Process

Successful planning requires the cooperation and support of all involved. For this reason, planning should be under the direction and overview of a committee that includes those groups who have a significant interest or stake in the successful implementation of the plan in the event of an emergency. The CAER program developed by CMA recognizes the importance of community involvement and seeks to achieve two goals:[70] 1) development of a community outreach program and means to provide the public with information on chemicals manufactured or used at local chemical plants; and 2) improvement of local emergency response planning by combining chemical plant emergency plans with other local planning to achieve an integrated community emergency response plan.

In the United States, SARA established LEPCs for each planning district. These committees facilitate the preparation and implementation of emergency plans and must include representatives from key public and private groups or organizations.

Regardless of the composition of the planning committee, it must have a commitment to work together on a common goal to be successful. The group must have the ability, authority, and resources to do the job. It must have access to all the industrial, community, transportation, and planning skills needed. Most importantly, the group must agree on its purpose and be able to work together cooperatively.

Health Planning Considerations

The planning process proceeds through various situations and contingency and

Table 34.10 — Plan Evaluation Attributes

Scope and provenance
Emergency response organization
Emergency alarm
Operational communications
Control centers
Evacuation sheltering
Accounting for people
First aid
Transportation
Security
Fire fighting
Chemical release detection and mitigation
Community notification and information
Shutdown of operations
Continuity of utilities and services
Mutual (industry) aid organizations
Outside agency notification and coordination
Public affairs
Governmental relationships
Restoration to normal operations
Clean up and rehabilitation
Legal issues
Training
Drills
Detailed action plans and standard operating procedures
Critique and plan revision

resource assessment steps to arrive at a set of options that is incorporated into a plan that can be tested and refined. Each of the members of the planning committee will have special considerations to input to the planning process. Facility medical and occupational hygiene members of the planning committee have specialized considerations involving casualty handling, exposure assessment, and external communications.

Casualty Handling

Adequate provision must be made in the plan to ensure that optimum treatment is provided or made available for all individuals who experience either injury, illness, or psychological distress as a consequence of an emergency.[74,75] A well-developed plan for handling medical emergencies should be integrated into the facility emergency plan and, where appropriate, the local community plan.

Provisions for handling casualties will vary significantly; depending on such factors as the nature of the operation, associated hazards, size of the work force, and proximity to major hospital facilities. Staffing, training, equipment, facilities, and transportation needs must be considered in a comprehensive plan to minimize the effects of illness and injury resulting from an emergency.

It is unlikely that a plant will equip itself with staffing and facilities capable of definitively treating all illness or injury arising from a major incident. Appropriate use of outside specialized medical units is, in general, most likely to result in optimum clinical care for ill or injured employees. The proximity of these specialized facilities, which will determine the levels of initial treatment or stabilization required on site, can be described as follows.

1. Patient can be delivered to a hospital within 10 minutes. Only triage, resuscitation, first aid treatment of serious injuries or illnesses, basic stabilization, and definitive treatment of minor casualties are needed on site.
2. Patient can be delivered to a hospital within 60 minutes. Formal stabilization may be required and, in some cases, treatment may need to be initiated on site.
3. Patient cannot be delivered to a hospital within 60 minutes. On-site facility must be equipped to provide more definitive treatment before patient is transferred.

Treatment Locations. Treatment for injuries may need to be provided at various locations. The casualty handling provisions of the medical plan need to identify these locations and the level of care available at each.

- Incident site—because the casualty should be moved to a safe place of treatment at the earliest possible opportunity, only lifesaving procedures are normally conducted at the incident site.
- Site facility—this is the first medical unit where equipment and conditions are conductive to assessment, stabilization, and/or initial treatment of casualties.
- Safe facility—it may be necessary to designate a location outside the site for use if the incident or gas release requires evacuation of the site medical facility.
- Minor injury facility—in the event of multiple casualties with minor injuries, transfer to a local satellite center may be appropriate.
- General hospital—the usual location of surgical teams and facilities for continuing patient care.
- Specialized medical units—definitive care centers for burns, plastic surgery, and neurosurgery are often available.

Transportation. It is unlikely that the site will be able to provide all the vehicles necessary to remove casualties to the hospital or even transport multiple casualties within the site. Arrangements should be made with external agencies to provide multiple ambulances and to move casualties in such a way that therapy can continue during transit. For rapid transport from remote sites consideration should be given to helicopter evacuation of casualties. Transport of equipment may be necessary in the event the site medical facility must be evacuated. Minor casualties can be discharged to home or to other treatment facilities by car or taxi. In all transport cases, consideration should be given to providing escort by a fellow worker or supervisor.

Close liaison with the ambulance services must be maintained, and it is advisable that there should be a shared communications network. Accurate records of all patients in transit must be kept and the need for escorts taken into account when determining emergency response staffing. Signs to direct ambulances to the treatment center/triage area, escorts for ambulances on site, and police escorts for ambulances in the community may also be required.

Staffing. Staffing requirements for emergencies will depend on the potential number of casualties from each credible event scenario and the probable nature and severity of the illnesses or injuries. The goal is to ensure that the sum of in-house and external staff is enough to deal with credible emergencies at each stage of the incident. Plans that maximize the likelihood of stabilizing and maintaining the patient until transfer can be effected must be developed. The proximity of specialized medical units and ease of transportation will influence these plans. Identification of appropriate external specialized medical units should be made on the basis of anticipated types of illness and injury. Staff must have an understanding of these facilities and their medical capabilities, and in turn the external medical units need to be made familiar with the operations and hazards at the facility. Both staff and the medical facilities also need to be equipped with adequate toxicity and treatment data on relevant materials.

In addition to trauma and burn specialists, consideration should be given to the provision of neurosurgeons for head and spinal injuries, plastic surgeons, and other physicians to handle patients affected by the inhalation of chemicals or combustion products. In the absence of ready availability of such medical specialists, consideration should be given to the training needs of site and community medical staff.

Equipment. Decisions on the appropriate level of emergency medical equipment also depend on the evaluation of the scenarios that may follow a major incident. The nature of the materials handled, anticipated injury or illness outcome, number of potential casualties, on-site staffing levels, and proximity to local hospitals and specialized medical units are factors to be considered.

Record Keeping. Adequate record keeping is essential in any clinical situation, and it is no less important when there may be a significant number of casualties arriving in a short space of time and being rapidly transferred for further treatment. Patients must, where possible, be adequately identified and records kept of their arrival and departure times at the treatment center, together with details of their clinical condition. This information is necessary to monitor the developing staffing requirements; control the flow of patients through the center; provide adequate information to hospitals of anticipated number and extent of injuries; provide transport groups with anticipated requirements; answer requests from distraught relatives; and provide management with an overview for internal and media information.

Exposure Assessment

A significant contribution to disaster response planning and execution can be made by the assessment of the magnitude and impact of exposures resulting from the release of chemicals into the environment or from the toxic products of fires. Exposure assessment may be defined as the estimation or measurement of the exposure of a population in terms of concentration and duration, evaluation of that exposure by comparison with health effects criteria, and estimation of the health consequences. This assessment should be made for both toxic chemical releases and for toxic decomposition products generated by fires. It may be needed for both on-site and off-site populations including rescue and emergency personnel. Effective emergency response requires that exposure for all involved populations be quickly and accurately assessed, interpreted, and communicated so that it can form a basis for management decision making.

The focus of emergency planning for toxic chemical release is on those releases that could cause serious acute effects or death. Other less serious releases can, however, have a major impact. Experience has shown that major odor incidents, while not harmful to health, can cause serious aggravation and hostility in local communities. Releases of chemicals that are chronic health hazards or carcinogens can also have serious consequences even when the release is barely detected by local communities and is of such short duration that no health effect is anticipated.

There are several distinct situations and purposes that may need either a predicted, contemporary, or retrospective assessment of exposure by either dispersion modeling or air contaminant concentration measurement.

Prevention. As discussed above, exposure patterns for various scenarios are predicted by dispersion modeling for the purpose of prioritizing risk reduction measures.

Emergency Action. In the event of a toxic release into a community, people may need to be advised to evacuate or button-up (seal themselves indoors) as appropriate.[76,77] Since this decision must be made quickly (usually in minutes), exposure assessment for all plausible scenarios is usually estimated in advance by modeling. Affected areas, predicted by map overlays or similar means, are typically communicated to local authorities that advise the public.

Declaring the Emergency Over. Exposure assessment is needed to decide when people can return to an affected area or "unbutton." Since the time constraints on the emergency decision no longer apply, and since modeling is not much use in predicting where pockets of gas may remain, actual measurement is often the best way to make the decision that it is safe to return.

Event Record. Documentation of the time course of exposure for various population segments can serve several purposes. In the event of claims for injury, this record can be used to show where injurious concentrations did and did not occur. It is also useful as an aid to understanding any symptoms seen and as part of the critique of emergency prevention and response actions. This record is based on model predictions supplemented by measurements.

Public Reassurance. Releases of odorous or small quantities of toxic materials not requiring emergency action may still need exposure assessment for public reassurance. A public perception of harm can sometimes be as much of a problem as actual harm.

Emergency and rescue crews operating near a release automatically use self-contained breathing apparatus, so no measurement is needed at that time. At some distance away, contamination will be low enough so that other respirators could be used or that no respirators are needed. These areas could be better defined by measurements if necessary. For re-entry and other post emergency purposes, the objective is to be sure that no pockets remain. A wide net of measurements that takes into account weather conditions is needed. Familiarity with the modeling results and the map overlay predictions will help decide where to measure. Measurement teams will need to be provided with and trained in the use of personal protective equipment.[78]

The direct-reading instruments and detector tubes necessary for near-term measurements should be on hand before an emergency arises, and these instruments must be kept in a state of readiness. Batteries go dead and detector tubes expire. Long-term measurements may be made with sorbent tubes or filters sent to an outside laboratory for analysis. In some circumstances even gas bags may be useful. Supplies should be kept in reserve and a fast, accurate laboratory identified.

Since the measurement data collected may be needed later, careful records should be kept. A form designed to be easily completed by measurement crews working under the severe time constraints of the field would be useful.

Organizational Communication

Fast, reliable, accurate operational communication is the hallmark of a good emergency plan. It is also very difficult to achieve if left until an emergency actually occurs. Things happen so fast, and there are so many people who have or need information, that careful preparation, training, and discipline is needed for the communication system to work effectively.

Initial Notification. It is essential that there be early communication with the medical group concerning the occurrence of an emergency situation, even when there are no immediate casualties. This notice allows decisions to be made about appropriate short-term staffing requirements and the alerting of other agencies of the potential need for support.

Lines of Communication. The lines of flow for information and instruction are derived from the organization and command structure of the emergency response organization. Open communication channels are needed with the incident manager, transport organization, hospitals, and public relations manager. Reliable, redundant links are needed for information to flow out. Since the disaster organization is usually different from the routine organization, many of these communications links are not routinely staffed. It is necessary for response team members to get to their posts, report in, and test primary and backup communication systems as quickly as possible.

Communication Roles. Since members of the emergency response team may not be on site at the time of an emergency, it is necessary to assign responsibility for calling in medical/health personnel needed at the appropriate times. Names and telephone numbers should be kept up to date. Once the plan is in operation, it may be necessary to designate individuals whose only role is to ensure that all messages are immediately handled without tying up the communications network or distracting other people from their duties. Everyone who needs to communicate during an emergency will need to be trained and drilled in the operation of the equipment and in the strict communication discipline necessary to avoid system overload.

Since critical operational communications will often be to and from external resources (hospitals, ambulance services, fire departments, government offices, etc.), the external resources should be contacted in advance to determine their communication capabilities and expectations.

Frequently, these organizations have back-up systems of communication available that are not widely known. In the case of a natural disaster involving both the community and the plant, the external resources will also be operating in an emergency mode and may not have the same resources and communications ability as in normal times.

Public Information. Serious incidents involving employee or public exposure to toxic chemicals are likely to receive intensive mass media coverage in many countries. Public concern over safety has increased, so it is important that the facts of an emergency are properly communicated to the media. For purposes of communication, an emergency can be divided into three time phases:

1. Initial response—because an emergency may occur at night or on a weekend, there could be an initial period (30 minutes to 1 hour) when no senior managers will be present. Decisions and communications during this phase must be planned.
2. Management control—when senior managers and the designated spokesperson arrive (usually within an hour), decisions and communications can be matched more exactly to the situation and fixed planned procedures modified.
3. Expert input—within 8 to 24 hours of the emergency it should be possible to enlist outside experts to help with complex technical or scientific issues that may arise.

It is important that spokespersons be aware of the toxic hazards present on site and have access to information concerning their health effects. They should have at least MSDS-type information available. Some facilities have developed briefer, more focused fact sheets for emergency use. Health professionals should become involved in developing this communication reference material and in helping the designated spokesperson understand the data and be able to anticipate media questions. Particular attention needs to be given to technical terminology that may be unfamiliar to the spokesperson and to the media.

Drills

Realistic tests of plan execution are necessary both to evaluate the plan and preparations to implement the plan and as a means of training.[79,80] Experience has shown that only the simplest of plans are likely to work the first time they are tested. Consequently, a facility should not assume that it is prepared to deal with an emergency unless the emergency response plan has been realistically exercised. Drills can be broad or specialized and can vary in depth of involvement of the facility and community. Some specific types of drills include:

- Tabletop, a step-by-step read-through of the emergency response plan by key controllers;
- Notification test, an exercise for communication facilities, notification procedures, call-lists, and backups;
- Organization test, which covers assembly, team makeup, understanding responsibilities, and response time;
- Rolling equipment test, involving emergency equipment call-up and movement, security and access, off-site responders, mobile communications; and
- Full scale test, involving casualty simulation, off-site response, community involvement, and media.

By one test drill or another, command and control, communications, casualty handling, mediation, and all other significant aspects of the plan should be tested. The drills should be carefully planned so that they are realistic within their scope. Drill evaluators who can input later to a comprehensive critique should observe them. Drills often detect some common emergency response plan failures such as:

- Designated individuals who do not have the training, time, or resources to carry out their duties;
- Delayed notification because of inadequate on-site authority at the time of the event;
- Dommunications overloads that block access to local authorities;
- Security breakdowns that inhibit access by responders and/or allow unauthorized access;
- Difficult decisions (e.g., sacrifice property to avoid a toxic release) not being made quickly enough;
- Incorrect casualty handling decisions being made because of confusion and inadequate communication among emergency crews, first responders, and hospitals;
- Control rooms becoming isolated and providing inadequate protection;
- Media/external communication needs interfering with operational communications;
- Chemical exposure effects and treatment information not being communicated to the hospital that receives the chemical casualties; and

- Support function (transportation, utilities, etc.) responsibilities listed in site plan that are not reflected and detailed by function plans.

The objective of drilling is to find these potential defects by realistic exercises so that the plan can be revised and training improved to maximize the likelihood that an actual emergency can be responded to without any critical errors.

Transportation Emergencies. Planning for successful response to chemical release emergencies that occur while a chemical is in transit involves some special considerations.[81–83] The first responders will not necessarily know what chemical is involved, and it may take some time at the critical early stage of the emergency to obtain that information. When the required information is on hand, it may be that special skills or equipment are needed to deal with the situation, so arrangements must be made to bring these to the scene. For these reasons, planning for transportation emergencies is more complex, and response often takes longer than for site emergencies.

The first item of information that may be available to firefighters, police, rescue workers, or other emergency services is the DOT placard on the side of the vehicle. A four-digit number on the placard or on an orange panel is the identification number for a material in the DOT guidebook.[84] This guidebook also provides a series of generic guides for each material that describe potential hazards and initial emergency actions.

The DOT identification number and the name of the material will also be shown on the shipping papers that should be on the truck, train, ship, or aircraft. In the case of a tank truck, the driver should also have a copy of the MSDS that gives more specific hazard, precaution, and emergency information for the material.

Additional information and assistance is available from the Chemical Transportation Emergency Center (CHEMTREC), a public service of CMA. CHEMTREC provides immediate advice by telephone and also contacts the shipper of the hazardous material for detailed assistance and appropriate response. CHEMTREC needs to know either the identification number or name of the material and the nature of the accident. CHEMTREC operates continuously and should be called as soon as immediate needs have been met and the facts ascertained.

Backing up CHEMTREC is another CMA service, the Chemical Emergency Network or CHEMNET. This network consists of about 240 response teams representing about 100 chemical companies. These CHEMNET teams can provide on-site experts to assist local emergency services with advice on the best way to handle a specific chemical accident. A similar network in Canada is called the Transportation Emergency Assistance Plan and is operated out of 11 regional response centers.

Summary

Sources of Information. The body of information and technology relating to the subject of this chapter is very large, far larger than could be covered in the space available. Consequently, this chapter does not attempt to instruct the reader on how to do the many tasks involved in disaster prevention and response planning. Instead, this chapter should be regarded as an overview of the subject that identifies the sources the reader may consult to expand on each detailed topic. Many of these sources are referenced in the text. In addition, there are some institutions that have a continuing involvement in this field and can be looked to for the latest information and technology.

DOT. The DOT Office of Hazardous Materials Transportation has developed an *Emergency Response Guidebook* (ERG) (DOT P 5800.4), which is a manual primarily for initial response to hazardous materials incidents that occur in the transport of hazardous materials. The ERG, which is updated periodically, gives the name of the material corresponding to the identification number displayed on the placard on the vehicle and listed on the shipping paper. Each identification/substance listing refers the user to a numbered guide that briefly describes potential hazards (fire, explosion, health hazards) and emergency action (fire fighting, dealing with spills or leaks, first aid). The ERG is a means of quickly providing critical information to first responders. The guide advises the responder to immediately contact CHEMTREC for additional assistance beyond the initial phase of the incident. In addition to the ERG, DOT is the lead agency in transportation safety research, guidance, and regulation.

NFPA. The NFPA is primarily concerned with fire protection and suppression systems but is also an authority on certain aspects of hazardous materials. Their *Fire Protection Guide on Hazardous Materials*[52] contains the complete texts of the four NFPA documents that classify most common hazardous materials: NFPA 325M: "Fire Hazard Properties of Flammable Liquids, Gases and Volatile Solids"; NFPA 49: "Hazardous Chemicals Data"; NFPA 491M: "Manual of Hazardous Chemical Reactions"; and NFPA 704: "Identification of the Fire Hazards of Materials."

This consolidated guide includes information on fire and explosion hazards, life hazards, personal protection, fire fighting, and other facts needed to make informed decisions in an emergency.

American Institute of Chemical Engineers Center for Chemical Process Safety (CCPS). A number of manuals developed by the CCPS(39,66,85,86) were used as a basis for this chapter. The CCPS and its committees maintain an awareness of the technology, and their handbooks reflect the state of the art.

CMA. The CMA CHEMTREC and CAER programs have been described above. These programs and other activities are the responsibility of CMA staff and committees who stay current on developments related to chemical emergencies. They are available to assist member chemical companies and the general public as appropriate.(17,73,87)

University of Delaware Disaster Research Center (DRC). This center provides a unique resource that "engages in a variety of sociological and social science research on group and organizational preparations for, responses to, and recovery from community-wide emergencies, particularly natural and technological disasters."(88,89) The DRC conducts studies, sends teams to disaster sites, and maintains a publication collection of over 20,000 items related to the social and behavioral aspects of disasters.

References. The list of publications at the end of this chapter contains a number of general and specialized references to the modern literature on this subject.

Future Trends. Presently proposed governmental regulations will require process hazard risk assessment for many facilities that have not done it in the past. Because of these regulations and as a consequence of expansion of the risk assessment field, the process will become more formal and detailed. New practitioners will enter the field, and it will be increasingly important for the facility manager to be sure of the qualifications of those who would provide this service. Some 5 or more years ahead, as the results of a number of risk assessments become known, it is probable that new regulations, at least in the United States, will establish expectations for the results of the risk assessment process and will try to define more precisely what risk is acceptable.

The technology of process hazard risk assessment, in addition to being more formal and detailed as mentioned above, is likely also to become more probabilistic. The analogy to health risk assessment that claims to calculate the likelihood of an adverse consequence will be persuasive. The lack of credible input data (e.g., mean time to failure) will be partly relieved by expanded databases developed in the course of increased regulatory activity. A need will be seen for a measure of the actual or limiting probability on chemical releases of various magnitudes so that there will be a number to compare with acceptable risk criteria or bright-lines. A challenge to all involved in process hazard analysis will be to try to ensure that the databases support the outcome both in quantity and quality. Further, it must be assured that the assumptions of independence of causes typically made are carefully considered, and that the consequences of multiple human errors be included in any analyses.

Community involvement in disaster prevention and response planning is already a fact in the United States and is likely to spread. In the future the public will expect to be even more involved and to participate in decision making. As a consequence, some situations that were considered to constitute an acceptable level of risk will need to be revisited and new consensus criteria of acceptability applied. Risk assessors and the public will need to educate each other to achieve common understandings and expectations.

References

1. **Thygerson, A.L.:** *Accidents and Disasters: Causes and Countermeasures.* Englewood Cliffs, NJ: Prentice-Hall, 1977.
2. "Process Safety Management of Highly Hazardous Chemicals." *Code of Federal Regulations* Title 29, Part 1910.119. 1992.
3. "Accidental Release Prevention Requirements." *Code of Federal Regulations* Title 40, Part 68. 1996.
4. **Concawe:** *Concawe Methodologies for Hazard Analysis and Risk Assessment in the Petroleum Refining and Storage Industry.* The Hague: Concawe, 1982.
5. **O'Reilly, J.T.:** *Emergency Response to Chemical Accidents. Planning and*

Coordinating Solutions. New York: McGraw-Hill, 1987.

6. **Wilson, E.A.:** Selected annotated bibliography and guide to sources of information on planning for and response to chemical emergencies. *J. Hazard. Mater. 4* (1981).
7. **Elkins, C.L. and J.L. Makris:** Emergency planning and community right-to-know. *J. Air. Pollut. Control Assoc. 38* (1988).
8. "Hazardous Materials Transportation Act." Pub. L. 93-633.
9. "Natural Gas Pipeline Safety Act." Pub. L. 90-481.
10. "Hazardous Liquid Pipeline Safety Act." Pub. L. 96-129.
11. **Solomon, C.H.:** The Exxon chemicals method of identifying potential process hazards. London: *Inst. Chem. Eng. Loss Preven. Bull. 52* (1983).
12. Comprehensive Environmental Response, Compensation and Liability Act of 1980, Pub. L. 96-510 (December 1980).
13. **DePol, D.R. and P.N. Cheremisinoff:** *Emergency Response to Hazardous Materials Incidents*. Lancaster, PA: Technomic Publishing Co., 1984.
14. **Environmental Protection Agency (EPA):** *Chemical Emergency Preparedness Program: Interim Guideline*. Washington, DC: EPA, 1985.
15. **Federal Emergency Management Agency (FEMA):** *Planning Guide and Checklist for Hazardous Materials Contingency Plan* (FEMA-10). Washington, DC: U.S. Government Printing Office, 1981.
16. "Superfund Amendments and Reauthorization Act." Pub. L. 99-499.
17. **Chemical Manufacturers Association (CMA):** *Community Awareness and Emergency Response Program Handbook*. Washington, DC: CMA, 1985.
18. **Organization Resources Counselors (ORC):** Recommendations for Process Hazards Management of Substance with Catastrophic Potential. Washington, DC: ORC, 1988.
19. "List of Regulated Substances and Thresholds for Accidental Release Prevention." *Code of Federal Regulations* Title 40, Part 68.130. 1994.
20. **European Economic Communities (EEC):** "Council Directive of 24 June 1982 on the Major Hazards of Certain Industrial Activities (Seveso Directive)" (85/501/EEC). EEC, 1982.
21. **European Communities (EC):** "Council Common Position on the Control of Major Accident Hazards Involving Dangerous Substances." (96/C120/03/EC). EC, 1996.
22. **Lykke, Erick:** *Avoiding and Managing Environmental Damage from Hazardous Industrial Accidents*. Pittsburgh, PA: Air and Waste Management Association, 1986.
23. **Health and Safety Executive (HSE):** *A Guide to the Control of Industrial Major Accident Hazards Regulations of 1984*. [HS(R)21] London: HSE, 1985.
24. **Air Pollution Control Association:** *Avoiding and Managing Environmental Damage from Major Industrial Accidents*. Pittsburgh, PA: AWMA (1986).
25. **Apostolakis, G.:** The concept of probability in safety assessments of technological systems. *Science 250* (1990).
26. **Clifton, J.J.:** Risk analysis and predictive techniques. *Major Hazards (UK)* July 1987.
27. **Frankel, E.G.:** Systems reliability and risk analysis. Amsterdam, The Netherlands: Martinus Nijhoff, 1984.
28. **Freeman, R.A.:** Use of risk assessment in the chemical industry. *Plant Oper. Prog. 4* (1985).
29. **Gressel, M.G. and J.A. Gideon:** An overview of process hazard evaluation techniques. *Am. Ind. Hyg. Assoc. J.* 52:158–163 (1991).
30. **Institution of Chemical Engineers (ICE):** *Risk Analysis in the Process Industries*. (Pub. Series No. 45). London: Institution of Chemical Engineers, 1985.
31. **Institution of Chemical Engineers (ICE):** *The Assessment and Control of Major Hazards*. (EFCE Pub. Services No. 42). London: Institution of Chemical Engineers, 1985.
32. **Lees, F.A.:** *Loss Prevention in the Process Industries*, vol. 2. London: Butterworth, 1980.
33. **Lees, F.A.:** Some aspects of hazard survey and assessment. *The Chemical Engineer (UK)* Dec. 1980.
34. **Coppock, R.:** *Regulating Chemical Hazards in Japan, West Germany, France, the United Kingdom and the European Community: a Comparative Examination*. Washington, DC: National Academy Press, 1986.
35. **Harris, N.C. and A.M. Moses:** *The Use of Acute Toxicity Data in the Risk Assessment of Accidental Releases of Toxic Gases* (Symp. series 80, 136).

London: Institute of Chemical Engineers, 1983.
36. **Pitt, M.J.A.:** Vapour hazard index for volatile chemicals. *Chem. Ind.* (1982).
37. **Institution of Chemical Engineers:** Bhopal: The company report. *Loss. Prevent. Bull. 63-1* (1985).
38. **Dow Chemical Company:** *Fire and Explosion Index Hazard Classification Guide*, 5th ed. Midland, MI: Dow Chemical Co., 1981.
39. **American Institute of Chemical Engineers (AIChE):** *Guidelines for Hazard Evaluation Procedures*. New York: AIChE, 1985.
40. **Theodore, L., J.P. Reynolds, and F.B. Taylor:** *Accident and Emergency Management*. New York: John Wiley & Sons, 1989.
41. **Royal Commission on the Ocean Ranger Marine Disaster:** "Report One: The Loss of the Semisubmersible Drill Rig *Ocean Ranger* and its Crew" (Pub. ZI-1982/1-IE). Toronto: Canadian Government Publishing Center, 1984.
42. **Heising, C.D., and W.S. Grengebach:** The *Ocean Ranger* oil rig disaster: A risk analysis. *Risk Anal. 9* (1989).
43. **Bell, B.J. and A.D. Swain:** A procedure for conducting human reliability analyses (NUREG/CR-2254). Washington, DC: U.S. Nuclear Regulatory Commission, 1983.
44. **McCormick, N.J.:** *Reliability and Risk Analysis*. New York: Academic Press, 1981.
45. **Rasmussen, J.:** Approaches to the control of the effects of human error in chemical plant safety. (Riso-M-2638). Roskilde, Denmark: Riso National Laboratory, 1987.
46. **Hart, G.S.:** Avoiding and managing environmental damage from major industrial accidents. Executive summary of the international conference. *J. Air Pollut. Control Assoc. 36* (1986).
47. **Kletz, T.A.:** Friendly plants. *Chem. Eng. Prog.* (July 1989).
48. **Ramskill, P.M.:** *Discharge Rate Calculation Methods for Use in Plant Safety Assessment* (SRD R 352). Culcheth, UK: Safety and Reliability Directorate, 1986.
49. **Beddows, N.A.:** Emergency prediction information. *Prof. Safety* (Oct. 1990).
50. **TNO (Netherlands Organization for Applied Scientific Research):** "Methods for the calculation of the physical effects of the escape of dangerous material" (Pub. 9092). Delft, The Netherlands: Directorate-General of Labour, 1980.
51. **McNaughton, D.J., C.G. Wirly, and P.M. Bodner:** Evaluating emergency response models for the chemical industry. *Chem. Eng. Prog.* (Jan. 1987).
52. **National Fire Protection Association (NFPA):** *Fire Protection Guide on Hazardous Materials*. Quincy, MA: NFPA, 1986.
53. **Baxter, P.J., P.C. Davies, and V. Murray:** Medical planning for toxic release into the community: the example of chlorine gas. *Brit. J. Ind. Med. 46* (1989).
54. **Coleman, R.J.:** Evacuation planning: crisis or control. *Fire Chief Magazine* (Feb. 1983).
55. **Cooney, W.D.C.:** The role of public services—off-site arrangements. *Major Hazards* (UK) (July 1987).
56. **Hanna, S.R. and P.J. Drivas:** *Guidelines for the Use of Vapor Cloud Dispersion Models*. New York: American Institute of Chemical Engineers, 1987.
57. **McQuaid (ed.):** Heavy gas dispersion trials at Thorney Island—2. *J. Hazard. Mater. 16* (1987).
58. **Nielson, M. and N.O. Jensen:** Research on continuous and instantaneous gas clouds. *J. Hazard. Mater. 21* (1989).
59. **Pedersen, F. and R.S. Seleg:** Predicting the consequences of short-term exposure to high concentrations of gaseous ammonia. *J. Hazard. Mater. 21* (1989).
60. **National Research Council:** *Emergency and Continuous Exposure Guidance Levels for Selected Airborne Contaminants*. Washington, DC: National Academy Press, 1985.
61. **American Industrial Hygiene Association (AIHA):** *Emergency Response Planning Guides*. Akron, OH: AIHA, 1987.
62. **World Health Organization:** *Rehabilitation Following Chemical Accidents: A Guide for Public Officials*. Copenhagen, Denmark: FADL, 1989.
63. **Health and Safety Executive:** *Annual Report of the Chief Inspector of Factories for* 1974. London: Her Majesty's Stationary Office, 1975.
64. **Edmundson, J.N.:** The use of risk results in planning and design. *Major Hazards (UK)* (July 1987).
65. **Prugh, R.W. and R.W. Johnson:** *Guidelines for Vapor Release Mitigation.*

New York: American Institute of Chemical Engineers, 1987.

66. **LeVine, R.:** *Guidelines for Safe Storage and Handling of High Toxic Hazard Materials*. New York: American Institute of Chemical Engineers, 1987.
67. **Kletz, T.A.:** *What Went Wrong?* London: Gulf, 1985.
68. **Suro, R.:** Plastics plant explodes; many missing. *New York Times* (October 24, 1989).
69. **Brown, G.N.:** Disaster planning for industry. *Occup. Health (UK)* (Sept. 1977).
70. **Eberlein, J.:** Emergency plans—the industry approach. *Chem. Ind. 82* (1987).
71. **Krikorian, M.:** *Disaster and Emergency Planning*. Loganville, AL, 1982.
72. **Michael, E.J., O.W. Bell, J.W. Wilson, and G.W. McBride:** Emergency planning considerations for chemical plants. *Environ. Prog.* (Feb. 1988).
73. **Chemical Manufacturers Associations (CMA):** *Site Emergency Response Planning*. Washington, DC: CMA, 1986.
74. **Fulde, G.W.O. (ed.):** *Emergency Medicine, the Principles of Practice*. Sydney, Australia: Williams and Williams, 1988.
75. **Laing, G.S.:** *Accidental and Emergency Medicine*. New York: Springer Verlag, 1988.
76. **Duclos, P., S. Binder, and R. Riester:** Community evacuation following the Spencer Metal Processing plant fire, Nanticoke, Pennsylvania. *J. Hazard. Mater. 22* (1989).
77. **Rogers, G.O. and J.H. Sorensen:** Warning and response in two hazardous materials transportation accidents in the U.S. *J. Hazard. Mater. 22* (1990).
78. **American Conference of Governmental Industrial Hygienists (ACGIH):** *Guidelines for the Selection of Chemical Protective Clothing*, 3rd ed. Cincinnati: ACGIH, 1987.
79. **Merriman, M.:** Not a drill. *Occup. Health Safety* (March 1990).
80. **Michael, E.J. and R.E. Vanesse:** *Planning and Implementation of Emergency Preparedness Exercises Including Scenario Preparation*. Boston, MA: Stone and Webster, 1985.
81. **Bennett, G.F., F.S. Feates, and J. Welder:** *Hazardous Materials Spills Handbook*. New York: McGraw-Hi11, 1982.
82. **Maloney, D.M., A.J. Policastro, L. Coke, and W. Dunn:** The development of initial isolation and protective action distances for U.S. DOT publication 1990 *Emergency Response Guidebook*. In *Proceedings of HAZMAT CENTRAL*, 1990.
83. **Student, P.J. (ed.):** *Emergency Handling of Hazardous Materials in Surface Transportation*. Washington, DC: Association of American Railroads, 1981.
84. **U.S. Department of Transportation (DOT):** *Emergency Response Guidebook* (DOT P 5800.4). Washington, DC: DOT, 1987.
85. **American Institute of Chemical Engineers (AIChE):** *Guidelines for the Technical Management of Chemical Process Safety*. New York: AIChE, 1989.
86. **American Institute of Chemical Engineers (AIChE):** *Plant Guidelines for the Technical Management of Chemical Process Safety*. New York: AIChE, 1992.
87. **Chemical Manufactures Association (CMA):** *Evaluating Process Safety in the Chemical Industry*. New York: CMA, 1992.
88. **Disaster Research Center (DRC):** "Publication List." [Booklet] Newark, DE: University of Delaware, 1989.
89. **Hughes, M.A.:** A selected annotated bibliography of social science research on planning for and responding to hazardous material disasters. *J. Hazard. Mater. 27* (1991).

Chapter 35

Outcome Competencies

After completing this chapter, the user should be able to:

1. Define underlined terms used in this chapter.
2. Identify the significance of dermal hazards and their control in the workplace.
3. Summarize the types of protective clothing by hazard category.
4. Explain the processes of degradation, penetration, and permeation of protective clothing.
5. Summarize the test methods used to determine the effectiveness of protective clothing.
6. Explain the exposure assessment process used for selection of protective clothing.
7. Organize criteria for the selection of appropriate protective clothing.
8. Describe the process of contamination and summarize example methods of decontamination.
9. Summarize the causes, effects, and control of heat stress caused by protective clothing.
10. Identify the components of a protective clothing program.

Key Terms

breakthrough time • degradation • penetration • permeation • permeation rate

Prerequisite Knowledge

Prior to beginning this chapter, the user should review the following chapters:

Chapter Number	Chapter Topic
2	Occupational Exposure Limits
13	Biological Monitoring
14	Dermal Exposure and Occupational Dermatoses
17	Risk Assessment in the Workplace
18	Biohazards in the Work Environment
22	Ionizing Radiation
24	Applied Physiology of Thermoregulation and Exposure Control
25	Thermal Standards and Measurement Techniques
31	General Methods for the Control of Airborne Hazards
32	An Introduction to the Design of Local Exhaust Ventilation Systems
34	Prevention and Mitigation of Accidental Chemical Releases
45	Occupational Safety

Key Topics

I. An Overview of Dermal Hazards
   A. Chemical Hazards
   B. Physical Hazards
   C. Biological Hazards

II. Types of Protective Clothing
   A. Gloves
   B. Boots
   C. Garments

III. Ergonomics of Protective Clothing

IV. Selection of Protective Clothing
   A. Determining the Hazards and Their Potential Effects
   B. Other Control Options
   C. Determining Performance Characteristics
   D. Determining the Need for Decontamination
   E. Determining the Ergonomic Constraints and Cost

V. Maintenance, Inspection, and Repair

VI. Worker Education and Training

VII. Developing and Managing a Protective Clothing Program

35

Personal Protective Clothing

S. Zack Mansdorf, Ph.D., CIH, CSP, QEP

Introduction

Dermatological disorders are one of the National Institute for Occupational Safety and Health's (NIOSH) top 10 leading occupational health problems.[1] These disorders are primarily a result of unprotected exposures to harmful chemical, biological, and physical agents. Most of the injuries and disease risks from dermatological disorders can be prevented or reduced through the appropriate selection and use of protective clothing or other control methods.[2]

While protective clothing can be an effective control method for occupational hazards, its effectiveness depends on proper use by the wearer, and it should be used only after careful consideration of other more effective and less user-dependent control measures. Failure of the protective clothing or its improper selection and use frequently can result in injury or illness.

In this chapter the categories of dermal hazards, types of protective clothing, ergonomic considerations, selection, maintenance issues, worker education and training, and management of a protective clothing program will be reviewed.

An Overview of Dermal Hazards

There are several general categories of bodily hazards for which specialized clothing can provide protection. These general categories include chemical, physical, and biological hazards (see Table 35.1 for a summary).

Chemical Hazards

Many chemicals may present more than one type of dermal risk (for example, a substance such as benzene is both toxic and flammable). There are at least three key factors to be considered when assessing the dermal risk that chemical hazards pose: (1) the likely routes of exposure (e.g., inhalation, ingestion, dermal, injection); (2) potential adverse effects of unprotected exposure; and (3) the exposure potential (likely dose) associated with the work assignment. Of the three factors, the adverse dermal or systemic effects through skin permeation are the most important consideration for determining the need for protective clothing.[3,4] Some exposure situations simply present a cleanliness issue (e.g., oil and grease) while other situations (e.g., skin contact with anhydrous hydrofluoric acid) could present a situation that is immediately dangerous to life and health (see Figure 35.1).

As shown in Table 35.1, adverse effects of skin contact with chemicals can be generally categorized as causing irritation, an allergic response, corrosion (chemical burns), skin toxicity, systemic toxicity (permeation through the skin), and promotion of cancer of the skin or other body cancers. Nicotine, an example of a chemical that normally presents the greatest risk by the dermal route, has significant toxicity due to excellent skin permeability but is not generally an inhalation hazard (except when self-administered). This is only one of many instances where the dermal route offers a much more significant risk than the other routes of entry.[5] Chemicals with the potential to contribute significantly to a worker's overall dose by the dermal route are identified by a "skin" notation in the Occupational Safety and Health Administration's

Figure 35.1 — Worker handling chemicals.

**Table 35.1 —
Examples of Dermal Hazard Categories**

Hazard	*Examples*
Chemical	irritants allergens corrosives dermal toxins systemic toxins cancer causing agents
Physical	trauma producing thermal hazards (hot/cold) fire vibration radiation
Biological	human pathogens animal pathogens environmental pathogens

permissible exposure limits and in the American Conference of Governmental Industrial Hygienists (ACGIH) threshold limit values (TLVs®).[(6)] However, many other substances that do not normally present inhalation hazards can have significant adverse effects on unprotected skin.[(7)] For example, the inorganic acids have low vapor pressures and are hazardous to the skin because of their corrosive nature. As a worst case example, a single unprotected skin exposure to anhydrous hydrofluoric acid (above 70% concentration) can be fatal. As little as a 5% burn can result in death from both the corrosion and the effects of the fluoride ion.[(8)] On the opposite end of the spectrum, inorganic lead is an example of a material that is highly toxic to humans but has little skin toxicity. In this case the concern is contamination that could later lead to ingestion or inhalation, since the vast majority of particulates will not permeate intact skin.

Chemical Permeation of Barriers

Research showing the diffusion of solvent through "liquid-proof" protective clothing barriers has been published for almost three decades.[(9,10)] Acetone, for example, has been shown to go through neoprene rubber (of typical glove thickness) within 30 minutes of direct liquid contact on the normal outside surface of the barrier.[(11)] This is called permeation (movement of a chemical through a protective clothing barrier that has no visible holes). The permeation process consists of the diffusion of chemicals on a molecular level through protective clothing. Permeation, since it is essentially diffusion, is best described by Fick's first law.[(12)] Without going into detail on Fickian diffusion, the significance is that the rate increases with temperature and is inversely related to thickness. The permeation process occurs in three steps: (1) absorption of the chemical at the barrier surface; (2) diffusion through the barrier; and (3) desorption of the chemical on the normal inside surface of the barrier (see Figure 35.2). It should also be noted that for permeation to occur, continuous contact with the challenge material is not required. Intermittent exposures can also result in significant permeation, as has been shown by a testing method used to model these types of exposure situations.[(13)]

The time elapsed from the initial contact of the chemical on the outside surface until detection on the inside surface is called the breakthrough time. The permeation rate is the rate of movement (mass flux) of the chemical through the barrier. The permeation rate is normally reported in mass per unit area per unit time (e.g., $\mu g/cm^2/min$) after equilibrium is reached and may be normalized for thickness. Most current testing done for permeation resistance is for periods of up to eight hours to reflect normal work shifts.[(14)] However, these tests are conducted under conditions of direct liquid or gaseous contact that typically do not occur in the work environment. Therefore, it might be assumed that there would be a significant safety factor built into the test. While this point has some validity, the permeation test is static, while the work environment is dynamic (flexing of materials, pressures generated from gripping or movement, prior physical damage to the glove or garment, etc.).

Given the lack of published skin permeability and dermal toxicity data, the approach taken by most safety and health professionals is to select a barrier with no breakthrough for the duration of the job or task (usually 8 hours), which is essentially a no dose concept.[3] This is an appropriate conservative approach; however, it is important to note that there is no protective barrier currently available that provides permeation resistance to all chemicals. For situations where the breakthrough times are short, the safety and health professional should select the barrier(s) with the best performance (i.e., longest breakthrough time and/or lowest permeation rate) as well as considering other control measures (such as a clothing change).

Permeation Testing

A standard test method for determining permeation of protective clothing is F739–91, "Test Method for Resistance of Protective Clothing Materials to Permeation by Liquids or Gases Under Conditions of Continuous Contact," published by the American Society for Testing and Materials.[15] In essence, the test consists of placing the barrier material between a reservoir of the challenge chemical (liquid or gas) and a collection cell connected to an analytical detector (see Figure 35.3). Hence, the actual reported breakthrough time is also related to the sensitivity of both the analytical method and system (i.e., collection system). Variations on this method can also be performed, such as a determination of permeation resulting from intermittent contact or a procedure for permeation of volatiles based on weight loss from evaporation through the barrier.[13,16]

Aside from the permeation process just described, there are two other chemical resistance properties of concern to the safety and health professional. These are degradation and penetration. Degradation is a deleterious change in one or more physical properties of a protective material caused by contact with a chemical. For example, the polymer polyvinyl alcohol is a very good barrier to most organic solvents but swells and is degraded by water. Latex rubber, which is widely used for medical gloves, is readily soluble in toluene and hexane, as another example. Therefore, latex or natural rubber gloves would be ineffective for protection against these chemical solvents. In many cases

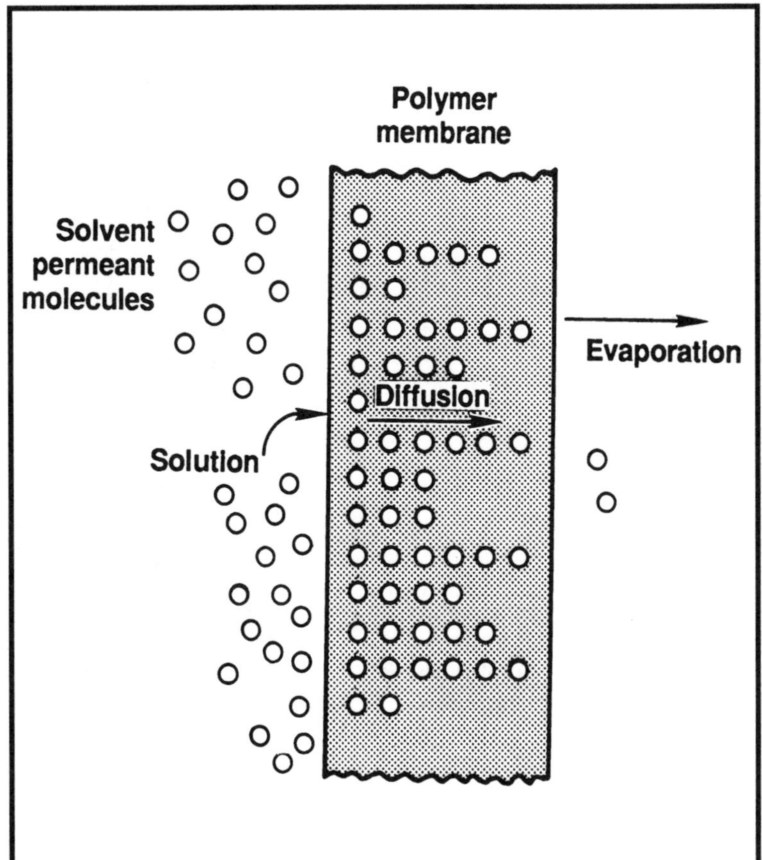

Figure 35.2 — The permeation process.

degradation can be assumed if the barrier swells or has a change in physical appearance (wrinkles, burns, color changes, etc.). Degradation may not always be visible. Solvents can wash out the plasticizer in some polymers causing them to become brittle but not change the appearance of the barrier.[17] Data showing degradation may be used to disqualify a clothing material from consideration but not used solely for selection, since permeation may occur without evidence of degradation.

Penetration is the flow of a chemical through zippers, weak seams, pinholes, cuts, or imperfections in the protective clothing on a nonmolecular level. Even the best protective barriers will be rendered ineffective if punctured or torn. Penetration protection is important when the exposure is unlikely or infrequent and the toxicity or hazard minimal. Penetration is usually a concern in garments used for splash protection and for some applications of breathable garments that have liquid resistance.

Chemical Resistance Data and Complicating Factors

Several guides have been published listing chemical resistance data for many of

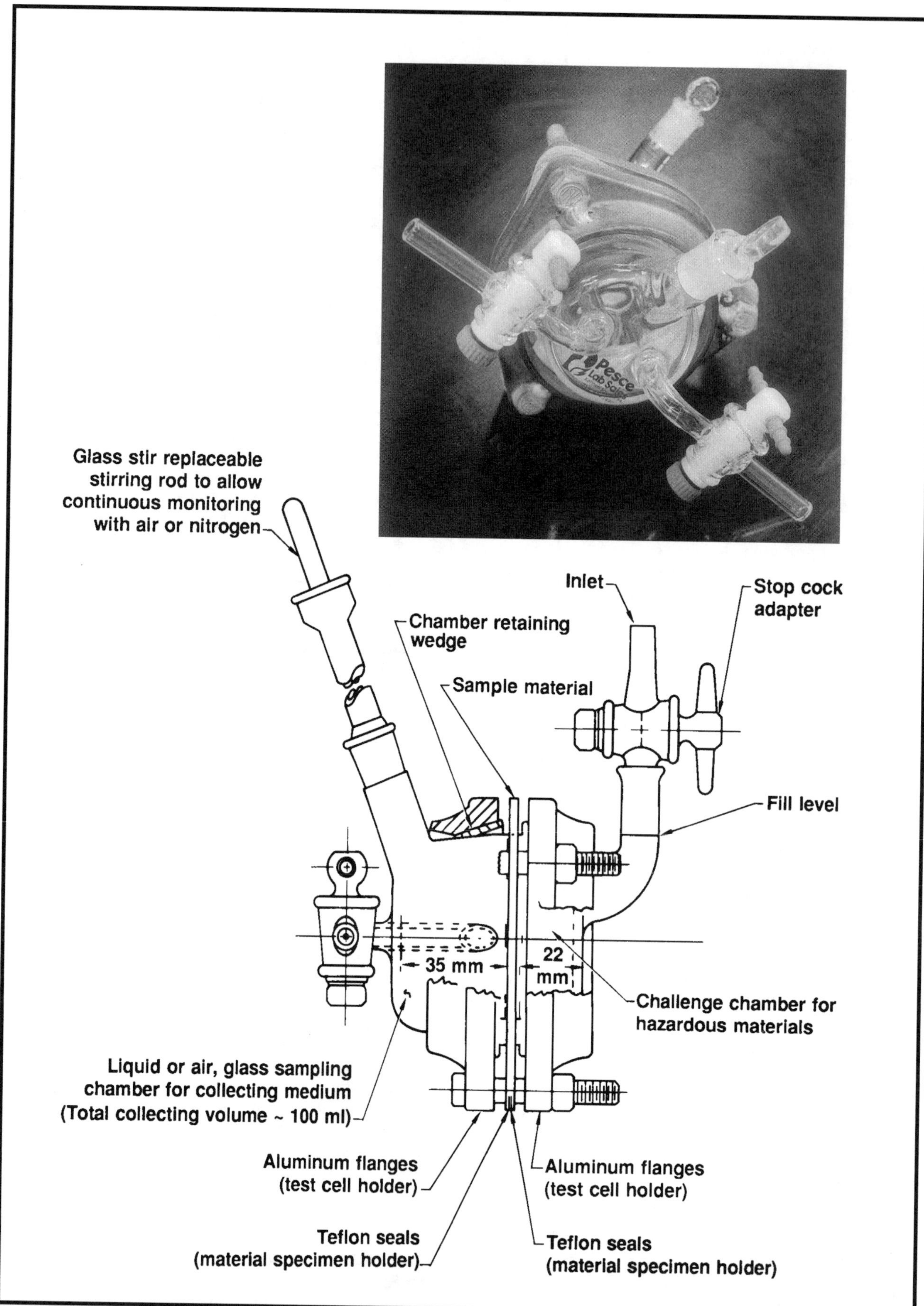
Glass stir replaceable stirring rod to allow continuous monitoring with air or nitrogen
Inlet
Stop cock adapter
Chamber retaining wedge
Sample material
Fill level
35 mm
22 mm
Challenge chamber for hazardous materials
Liquid or air, glass sampling chamber for collecting medium (Total collecting volume ~ 100 ml)
Aluminum flanges (test cell holder)
Aluminum flanges (test cell holder)
Teflon seals (material specimen holder)
Teflon seals (material specimen holder)

Figure 35.3 — ASTM permeation test cell.

the most widely used industrial chemicals; many are also available in an electronic format.[18–20] In addition to these guides, most manufacturers in the industrially developed countries also publish, in printed and electronic form, current chemical and physical resistance data for their products.

In general, the permeation of protective clothing can also be estimated using solubility parameters and other factors such as molecular size.[12] From this model, chemicals that are most closely related can be expected to behave most alike in terms of permeation. That is to say, "like dissolves like." For example, the ketones would be expected to have similar permeation characteristics since they have similar chemical structures. Generally, this holds true, and testing supports this approximation in many instances. Hence, a barrier with good resistance to permeation by ethyl alcohol would be expected to behave in a like fashion when exposed to methyl alcohol. Nevertheless, while a good approximation, it is not universal. Also, most of the rubber compounds and some of the polymers contain varying mixtures of additives that can have significant effects on permeation. Neoprene rubber barriers from different manufacturers and different lots, for example, can perform quite differently.[21] Finally, and to further complicate the issue, chemical mixtures can behave much differently than neat or pure compounds. For some mixtures, the breakthrough times can be much shorter than any of the individual component times.[22] This is probably because of the effects of the interactions among the components and the potential for one or more of the components to act as a vehicle. Very little published permeation data exists for mixtures. In these situations it is important to test the mixture against the barrier of choice.

Physical Hazards

As noted in Table 35.1, physical hazards can be categorized as those that are trauma producing (produce injuries such as cuts), those from thermal effects, and those produced by vibration or radiation. Trauma to the skin from physical hazards (cuts, abrasions, etc.) is common to many occupations, with construction and meat cutting as two examples. Thermal hazards to the skin include the adverse effects of extreme cold and heat such as those from molten metals or handling cryogenic liquids. The protective attribute of clothing for these hazards is related to the insulation provided (generally increasing with thickness), whereas protective clothing for flash fire and electric arc exposures requires flame resistance properties.[23] Firefighters and others with similar exposures to both heat and flame require protection that is both fire resistant and insulating.[24] They typically use a multilayered garment that also includes a vapor barrier to protect them from steam generated from the heat of a fire.

Protection from some forms of both ionizing and nonionizing radiation can be achieved using protective clothing. In general, protective clothing for ionizing radiation is based on the principle of shielding (e.g., lead-lined aprons and gloves) or preventing particulate radionuclides and liquids from direct contact with the skin. Clothing for electromagnetic radiation, such as microwave, is based on grounding, whereas protection from light (ultraviolet, visible, and infrared) depends on the wavelength of the radiation.[25] For example, infrared protection is usually afforded by reflective clothing such as aluminumized coverings (Figure 35.4 illustrates an example of thermal hazards.)

Excessive vibration can have several adverse effects on body parts, primarily the hands. Occupations such as mining (handheld drills) and road repair (pneumatic hammers or chisels) are two examples where excessive hand vibration can lead to bone degeneration and loss of circulation in the hands (Raynaud's phenomenon).[26] Specialized protective clothing (e.g., urethane or foam containing gloves) can help to damp the vibration received by the body and to keep the hands warm and dry.

Biological Hazards

Biological hazards include infection from agents and diseases common to humans, those common to animals, and those common to the work environment that can have adverse effects on humans. Biological hazards common to humans have received great attention with the increasing spread of bloodborne pathogens such as AIDS (acquired immune deficiency syndrome) and hepatitis. Hence, health care work that might involve exposure to blood or body fluids usually requires some type of liquid-resistant garment and gloves (see Figure 35.4).[27] OSHA requires that appropriate protective clothing be used for these exposure situations under its bloodborne pathogens regulation (29 CFR 1910.1030). Diseases transmitted from animals through handling (e.g., anthrax) have a long history of recognition and require protective measures similar to those used for handling

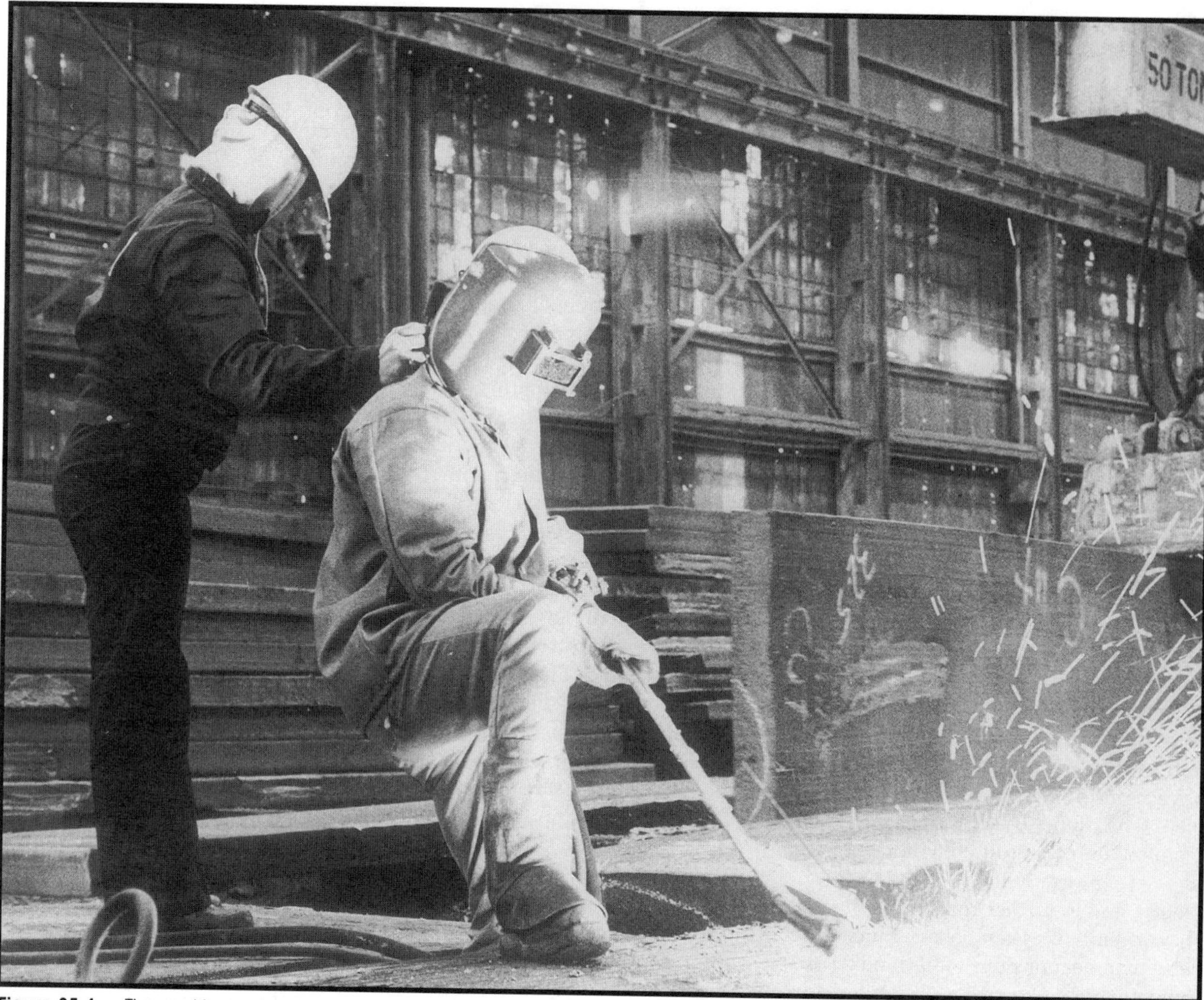

Figure 35.4 — Thermal hazards.

bloodborne pathogens from humans.[28] Work environments that can present a hazard from biological agents include sewage treatment plants, composting facilities, and clinical and microbiological laboratories as well as other special work environments.

Types of Protective Clothing

Protective clothing in a generic sense includes all elements of a protective ensemble (e.g., garments, gloves, boots, etc.). Thus, protective clothing can include everything from a finger cot providing protection against paper cuts to a fully encapsulating suit with self-contained breathing apparatus used for an emergency response to a hazardous chemical spill.

Protective clothing can be made of natural materials (e.g., cotton, wool, leather), man-made fibers (e.g., nylon, rayon), or various polymers (plastics and rubbers such as butyl rubber, polyvinyl chloride, chlorinated polyethylene, etc.). Table 35.2 summarizes by hazard the typical physical, chemical, and biologic performance requirements and common protective materials used.

Protective clothing materials and configurations vary greatly depending on the intended use. However, normal components are analogous to personal clothing (e.g., pants, jacket, hood, boots, gloves) for most physical hazards. Special use items for applications such as flame resistance in the molten metals industries can include chaps, armlets, and aprons constructed of both treated and untreated natural (including asbestos weaves) and synthetic fibers and materials. Specially treated or inherently flame-resistant porous fabrics and materials are commonly used for flash fire and electric arc (flash over) protection (e.g., petrochemical industry and electric utility industries) but usually do not provide protection from any long-term heat or flame exposure. It should be noted that

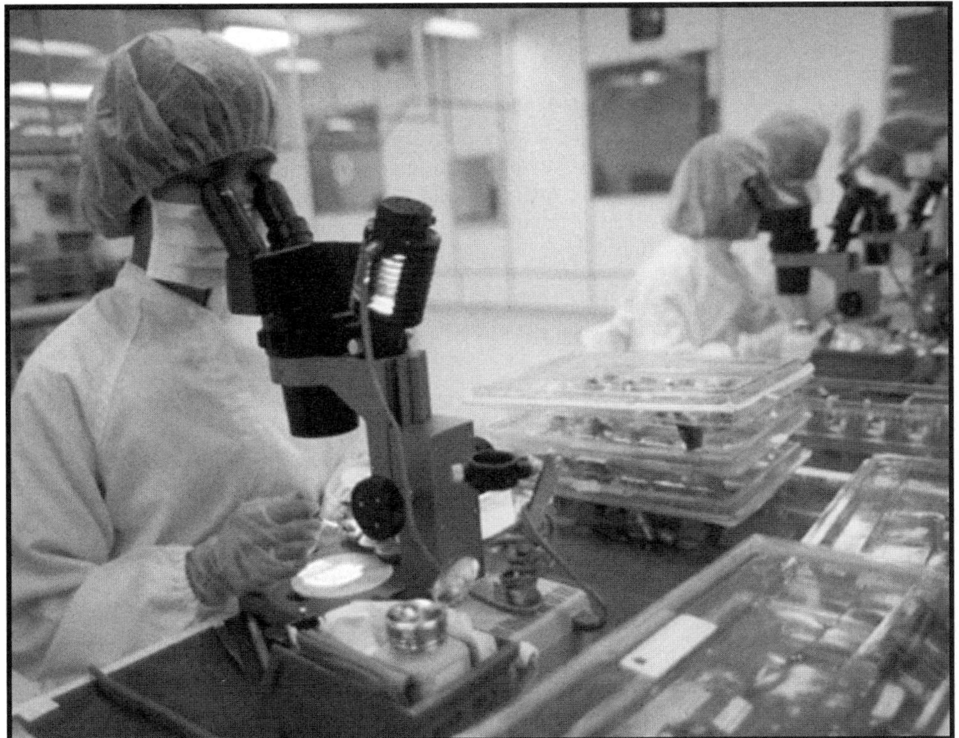

Figure 35.5 — Biohazards.

firefighting requires specialized clothing that provides flame (burning) resistance, a liquid barrier, and thermal insulation (protection from heat), while entry into fuel fires (with their strong infrared component) also requires reflective clothing (e.g., aluminumized cover). Clothing for cut protection can range from garments of aramid fibers to chain mail gloves of metal construction (including titanium) or special fiber batting for protection against chain saws. Protection from cold extremes usually includes multiple components of high insulating values that allow for the wicking and/or evaporation of perspiration. Chemical protective clothing can be even more exotic in terms of construction.

Gloves

Protective gloves are made from a wide variety of natural and synthetic materials. Light-duty cotton gloves and heavy-duty leather gloves are used to protect against a variety of physical hazards. Synthetic gloves can be used to protect against both physical and chemical hazards. They fall into four major construction categories: knit, unsupported, supported, and laminated.

Table 35.2 —
Common Physical, Chemical, and Biological Performance Requirements

Hazard	*Performance Characteristic Required*	*Common Protective Clothing Materials*
Thermal	insulation value	heavy cotton or other natural fabrics
Fire	insulation and flame resistance	aluminized gloves; fire retardant; aramid fiber and other special fabrics
Mechanical abrasion	abrasion resistance; tensile strength	heavy fabrics; leather with metal studding
Cuts	cut resistance	metal mesh; aromatic polyamide fiber and other special fabrics
Punctures	puncture resistance	leather; fabric coated with filled plastic; thick elastomers
Vibration	damping	natural or polymeric gloves with elastomeric linings
Chemical/toxicologic	permeation resistance	polymeric materials; elastomeric materials
Biological	"fluid-proof;" puncture resistant	latex or polymer
Radiologic	usually fluid resistant or particle	polymer gloves; lead lined resistant for radioneuclides or shield containing

Knit synthetic gloves are typically used for cut resistance. They can be made of high strength synthetic yarns or fiber-wrapped steel yarns. Gloves of chain mail and steel-stapled leather are also available.

Chemically protective gloves (see Figure 35.6) are usually available in a wide variety of polymers and combinations such as cotton gloves coated by the polymer of interest (manufactured using a dipping process). Some of the new foil and multilaminate gloves are only two dimensional (flat) and hence have some ergonomic constraints but are highly chemical resistant. These gloves typically work best when a form-fitting outer polymer glove is worn over the top of the inner flat glove (called double gloving) to conform to the shape of the hands. Polymer gloves are available in a wide variety of thicknesses ranging from very light weight (<2 mm) to heavy weight (>5 mm) with and without inner liners or substrates (called scrims). Gloves are also commonly available in a variety of lengths ranging from approximately 25 cm for hand protection to gauntlets of approximately 80 cm extending from the workers shoulder to the tip of the hand. The correct choice of length depends on the extent of protection required; however, the length should normally extend at least to the worker's wrist to prevent drainage into the glove.

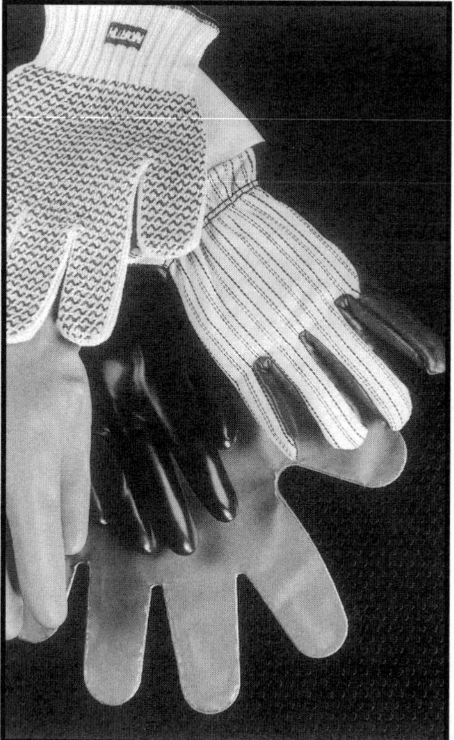

Figure 35.6 — Glove examples. (Photo courtesy of North Safety Products.)

Boots

Boots are available in a wide variety of lengths ranging from hip length to those that only cover the bottom of the foot. Chemical protective boots are available in only a limited number of polymers since the boot heal and sole require a high degree of abrasion resistance. Common polymers and rubbers used in chemically resistant boot construction include PVC (polyvinyl chloride), butyl rubber, nitrile, and neoprene rubber. Specially constructed laminate boots using other polymers can also be obtained but are quite expensive and limited in polymer choices at the present time. Integrated foot coverings (part of the suit) of the same protective barrier as the protective suits are available to extend the chemical resistance properties of the barrier to the feet. They are intended as foot coverings to be used inside a boot, although some fully encapsulating suits do have integrated boots intended for contact with the ground.

Garments

Chemical protective garments can be obtained as one-piece fully encapsulating (gas-tight) garments with attached gloves and boots or as multiple components (e.g., pants, jacket, hoods, etc.) (see Figure 35.7). Some protective materials used for construction of ensembles have multiple layers or laminants. Layered materials are generally required for polymers that do not have good inherent physical integrity and abrasion resistance properties (e.g., butyl rubber versus Teflon™) to permit manufacture and use as a garment or glove. Common support fabrics are nylon, polyester, aramides, and fiber glass. These substrates are coated or laminated by polymers such as PVC, Teflon, polyurethane, polyethylene, and other proprietary materials. Some suits use layering of different polymers to improve the range of chemical resistance (e.g., a layer of neoprene, nylon for support, and butyl rubber).

Over the past decade there has been enormous growth in the use of nonwoven polyethylene- and polypropylene-based materials for disposable garment construction. These garments, sometimes incorrectly called paper suits, are made using a special process where the fibers are bonded together rather than knitted or woven.(29) These nonwoven fabric garments are low in cost, very lightweight, and have good applications for protection against particulates, but are not normally chemical nor liquid resistant.(30) Nonwoven fabric-based

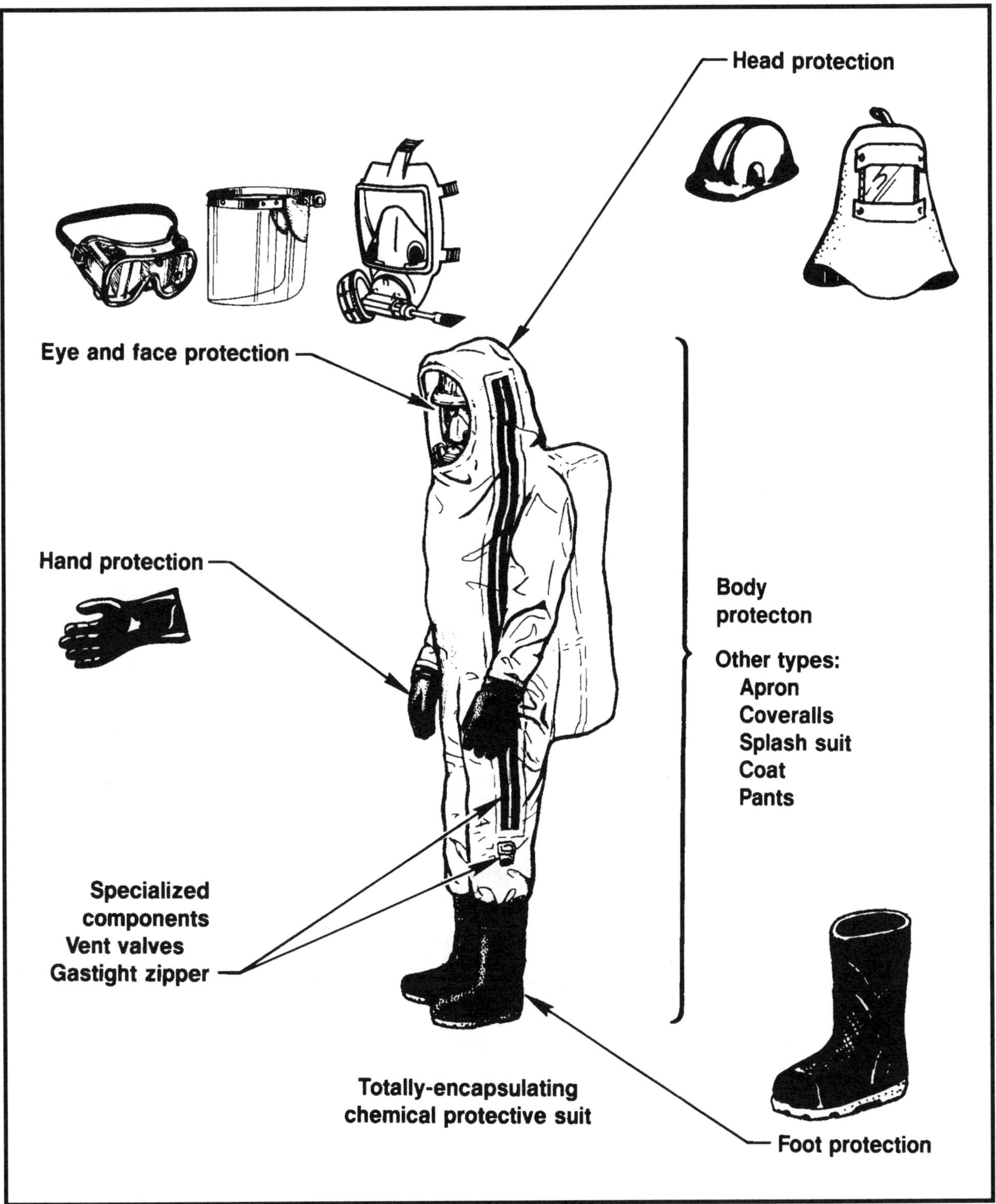

Figure 35.7 — Types of chemical protective clothing.

garments are also available with various coatings or films such as polyethylene, Saran™, and other polymers. Depending on the coating or lamination characteristics, these garments offer good chemical resistance to many common substances. Garments are also available that include construction with microporous films. These products are often considered breathable because they allow some water vapor transmission. This feature may make them less prone to producing heat stress under some conditions. However, microporous films still allow vapors to penetrate and may not always prevent liquid penetration. It should also be noted that this type of protective clothing is available for clean room applications; however, this is for protection of the product rather than the wearer. The concern here is debris from the human and lint from street clothing.

The fully encapsulating gas-tight suit of one-piece construction provides the highest level of protection available from chemical protective clothing. In the majority of these configurations, the respiratory protection device (airline or self-contained breathing apparatus [SCBA]) is also protected by the suit since it is worn within the suit. Protection factors for suits of this type are typically higher than those for the respiratory protection (protection factor of 10,000) provided the suit has appropriate chemical resistance to the challenge.(31)

The U.S. Environmental Protection Agency, in cooperation with a number of other agencies such as NIOSH, has devised a scheme to describe levels of protection for protective ensembles used on hazardous waste sites.(32) Their scheme consists of four levels designated by the letters A through D.

- Level A: SCBA or positive pressure airline system with escape SCBA and totally encapsulating chemical protective suit, gloves (double layer), chemically resistant boots, plus other safety equipment.
- Level B: Same respirator as Level A and hooded chemical-resistant suit, gloves (double layer), chemically resistant boots, plus other safety equipment.
- Level C: Full-face or half-mask air-purifying respirator and protective clothing as in Level B, plus other safety equipment.
- Level D: No respiratory protection; coveralls with an option for gloves, boot coverings, and other related safety equipment.

It should be recognized that Levels B and C require the same splash suit. Most industrial situations require a much more specific protective clothing ensemble selection beyond the generic requirement of a splash suit and gloves. The selection of the appropriate barriers for the industrial situation is usually easier to complete since the challenge chemicals and work requirements will usually be known.

Ergonomics of Protective Clothing

In all but a few cases the addition of protective clothing and equipment will decrease productivity and increase worker discomfort.(33) Some exceptions might be cold environments as related to improved gripping power with some gloves for certain objects. The use of protective clothing may also lead to decreased work quality since error rates increase with the use of protective clothing.

For chemical protective and some fire resistant clothing there are some general guidelines that need to be considered concerning the inherent conflicts between worker comfort, efficiency, and protection. First, the thicker the barrier the better (increases time to breakthrough or provides greater insulation for thermal protection); however, the thicker the barrier the more likely it will decrease ease of movement and user comfort. Thicker barriers also increase the potential for heat stress. Second, barriers with excellent chemical resistance tend to increase the level of worker discomfort and heat stress. This is because the barrier normally will also deter water vapor transmission (i.e., evaporative cooling through sweating). Third, the higher the overall protection of the clothing, the more time the job will take to accomplish and the more likely there will be work errors. There are also a few jobs or tasks where use of protective clothing could increase risk (e.g., around moving machinery, where risk of heat stress is greater than the chemical hazard, etc.). While this situation is rare, it must be considered.

Other issues relate to the physical limitations imposed by using protective clothing. For example, a worker issued a thick pair of gloves will not be able to perform tasks easily that require a high degree of dexterity and repetitive motions. As another example, a spray painter in a totally encapsulating suit usually will not be able to look to the side, up, or down since typically a respirator facepiece and suit visor restrict the field of vision. These are only some examples of the ergonomic restrictions created when wearing protective clothing and equipment.

One of the primary ergonomic constraints for chemically resistant suits and ensembles is the issue of heat stress. Once the worker dons the suit, the microclimate quickly approaches 100% humidity, and the body core temperature rises since the suits have good insulating properties and present a barrier to the evaporation of perspiration. The amount of heat retained within the suit and the level of perspiration will depend on the metabolic work load. Nevertheless, almost all suits that are chemically resistant and cover all or most of the body present heat stress challenges. Solutions to the problem include providing protection from radiant heat (shielding), cooling outside environments (air conditioning or early/late work), replacing fluids frequently, and using rest breaks. Additional controls can include the use of ice-containing body vests, portable and integrated cooling systems (most use circulating cool water in a body garment), and venturi (air-based) coolers where applicable. Whatever measures are used, workers in heat stress situations should be closely monitored.

The work situation must always be considered in the selection of the protective clothing for the job. The optimum solution is to select the minimum level of protective clothing and equipment that is necessary to do the job safely.

Selection of Protective Clothing

In the United States, OSHA regulations (29 CFR 1910.132) require that a hazard assessment for personal protective equipment be conducted before assignment.(34) In summary, the regulation requires the employer to assess the workplace to determine whether hazards are present or likely to be present, and if so, to select the appropriate protective equipment for the specific hazard(s) involved. It also requires the selection decision to be communicated to the affected employee, for the equipment to fit properly, and training of employees in the use and limitations of the equipment provided.

The overall approach to the selection of protective clothing for most situations can be illustrated using an eight-step process that incorporates the required OSHA hazard assessment. These steps are as follows:

1. Determine the type of hazard(s) most likely to occur.
2. Determine the adverse effects of unprotected exposure.
3. Determine whether other control options can be used instead of protective clothing.
4. Determine performance characteristics needed for protection.
5. Determine the need for decontamination (as applicable).
6. Determine the ergonomic constraints presented.
7. Determine the cost of the various options.
8. Make the selection.

Determining the Hazards and Their Potential Adverse Effects

The risk assessment process is fundamental to the practice of occupational hygiene. It is covered in detail in Chapter 17. Therefore, the discussion of this topic will be limited to how it might be applied to protective clothing. In this regard, chemical hazards will be used as the model. The process should begin with a determination of the chemical hazards of the process or work task, likely exposures and routes of exposure, and extent of exposure.

The best method to determine the likely worker exposures and potential routes of exposure is to actually inventory the chemicals used and observe the task or work assignment. A second but less informative approach is to have the work process described or to evaluate it from a written description. Coupled with knowledge of the chemicals used in the process such as physical characteristics (e.g., vapor pressure, physical state, etc.) from the material safety data sheet (MSDS), an assessment of the most likely routes of exposure can be made. Likewise, an assessment of the dermal hazards presented by the chemicals used can be obtained from the MSDS and other reference sources.

Once the evaluation of the likely exposures, route(s) of entry, and toxicity of the materials has been completed, an assessment of the extent of potential exposure needs to be determined. That is, what is the nature and extent of worker contact? For liquid chemicals, as an example, is the nature of potential contact simply from an inadvertent splash, or do workers become wet from contact with the chemical? For those scenarios where the material is highly hazardous (e.g., liquid sodium cyanide) although the likelihood of contact is remote, the worker must obviously be provided with the highest level of protection available. For situations where the exposure represents a very minimal risk (e.g., a nurse applying rubbing alcohol to a patient), the level of protection does not need to be absolute. This selection logic is based essentially on an estimate of the adverse effects of the material combined with an estimate of the likelihood of exposure.

Other Control Options

The first consideration should always be whether the job or task can be safely done without the use of protective clothing. It has long been the philosophy of NIOSH that personal protective clothing should not be used before consideration of other control options.[35] These include substitution of a less hazardous material, use of automated or mechanical means for accomplishing the task, and use of engineering controls (e.g., ventilation) or nonengineering controls (e.g., administrative controls such as working in the mornings or evenings for hot work). As stated earlier in the chapter, this is because protective clothing is the last line of defense and depends heavily on proper use by the worker.

Determining Performance Characteristics

The physical and chemical hazards of the job will define the performance characteristics of the protective clothing. Often these performance characteristics will require compromise, since no single selection will meet all of the criteria. For example, there is no commercially available glove that provides both good chemical and thermal or fire resistance. In this case other approaches such as the wearing of a sacrificial insulating outer glove over a chemically resistant glove would need to be considered. This approach would require disposal of the outer glove if contaminated since the fiber-containing gloves would tend to retain the chemical contamination. As another example, the patient would not want a brain surgeon to select a glove with excellent cut and puncture resistance for protection against bloodborne pathogens if it also offered poor dexterity. In these cases the performance characteristics needed must be rank-ordered so that the greatest risk is resolved before lesser risks. This approach may also be needed for selection of chemical protective clothing where the challenge material is a mixture that has not been specifically tested. That is to say, select the barrier that provides the best protection against the most hazardous component of the mixture and consider other options such as double gloving using two different barriers.

Determining the Need for Decontamination

For protective clothing that is used with hazardous chemicals, decontamination must be considered even if the clothing is for a single use. This is because cross-contamination can occur with doffing of the protective clothing. Not all situations will require decontamination; however, there are many case histories of workers needlessly contaminated because decontamination was not considered. This is an especially important consideration if the work clothing is stored with street clothing, reused, taken home, or laundered with regular clothing.[36]

Determining the Ergonomic Constraints and Cost

There are two final factors before a selection can be made. These are determining the ergonomic constraints presented by the job or task and evaluating the costs for the protective clothing options that meet the needs of the work to be performed. In some cases it may be necessary to redesign the work tasks to better fit the ergonomic requirements of performance while protective clothing is worn. For some tasks it may even be concluded that the work cannot be efficiently or safely performed using protective clothing. In these cases a different approach to accomplishing the task must be developed.

The final factor is the cost of the options available. The financial impact of reuse (with consideration of decontamination measures as applicable) versus single use should be considered along with the other aspects of the costs of administering an effective protective clothing program and of determining the overall life cycle costs of the clothing (or cost per use). Finally, the selection can be made using all of the information gathered. From this section of the chapter, it should be fairly clear that the selection choice will usually be the result of the consideration of many factors and not simply a choice based on product literature or a guidebook.

Maintenance, Inspection, and Repair

The proper storage, inspection, cleaning, and repair of protective clothing is important to the overall protection provided by the products to the wearer.

Some protective clothing will have storage limitations such as a prescribed shelf-life as well as requiring protection from ultraviolet radiation (sunlight), ozone, moisture, temperature extremes, and product folding. For example, natural rubber products are usually susceptible to all of these environmental factors. Many of the encapsulating polymer suits can be damaged if folded rather than allowed to hang upright. The manufacturer or distributor should be consulted for any specific storage limitations its products may have.

Inspection of protective clothing should be performed frequently by the user (e.g., on receipt, prior to use, and after each use). Inspection by co-workers is another technique that may be used to involve the wearer in ensuring the integrity of the selected protective clothing. As a management policy, it is also advisable to have the worker's supervisor periodically inspect protective clothing items that are reused routinely (e.g., weekly). Inspection criteria will depend on the intended use of the protective item; however, it would normally include examination for any obvious defects such as tears, holes, imperfections, and degradation. As one example of an inspection technique, clean polymer gloves used for protection against liquids can be blown up with exhaled air to check for integrity.

Cleaning of protective clothing for reuse must be performed with care. Natural fabrics can be cleaned with normal washing procedures provided they are not contaminated with toxic materials. Synthetic fibers and materials commonly have cleaning procedure limitations. For example, some flame-resistant treatments will lose their effectiveness if the garments are not properly cleaned.[37] Clothing used for protection against chemicals that are not water soluble cannot usually be decontaminated by simple water and soap washing. Tests performed on protective clothing used by pesticide applicators indicate that normal washing procedures are not effective for many pesticides.[38] Dry cleaning of some protective clothing is not recommended since it is commonly ineffective and can degrade or contaminate the product. It is important to consult the manufacturer or distributor of the product before attempting cleaning procedures that are not specifically recommended.

Most protective clothing is not repairable. Repairs can be made on some limited items such as fully encapsulating polymer suits. However, the manufacturer should be consulted for the proper repair procedures.

Worker Education and Training

Adequate education and training for users of protective clothing is essential. Training and education should cover the following.

- The nature and extent of the hazard(s);
- When protective clothing should be worn;
- What protective clothing is necessary;
- Use and limitations of the protective clothing to be assigned;
- How to properly inspect, don, doff, adjust, and wear the protective clothing;
- Decontamination procedures, if necessary;
- Signs and symptoms of overexposure or clothing failure;
- First aid and emergency procedures; and
- The proper storage, useful life, care, and disposal of protective clothing.

Pertinent training areas not already provided to the worker through other programs should also be included. For those topical areas already provided to the worker, refresher training should be given to the clothing user. For example, if the signs and symptoms of overexposure have already been provided as part of training for working with chemicals, symptoms that are a result of significant dermal exposures versus inhalation should be reemphasized. Finally, the workers should have an opportunity to try out the protective clothing before a final selection decision is made and significant quantities ordered.

Knowledge of the hazards and limitations of the protective clothing not only reduces the risk to the worker but also provides the health and safety professional with a person capable of providing feedback on the effectiveness of the protective equipment.

Developing and Managing a Protective Clothing Program

A written protective clothing program can reduce the chance for error, increase worker protection, and establish a consistent approach to the selection and use of protective clothing. A model program could contain the following elements.

- An organization scheme and administrative plan;
- A risk assessment methodology;
- An evaluation of other control options to protect the worker;
- Performance criteria for the protective clothing;
- Selection criteria and procedures to determine the optimum choice;
- Purchasing specifications for the protective clothing;
- A validation plan for the selection with medical surveillance, as appropriate;
- Decontamination and reuse criteria, as applicable;
- A user training program ; and

- An auditing plan to assure that procedures are consistently followed.

Several examples of misuse of protective clothing are commonly seen in industry. Misuse is usually the result of a lack of understanding of the limitations of protective clothing by either management, workers, or both. An example is the use of protective clothing not resistant to flame for workers handling flammable solvents or in situations where open flames, burning coals, or molten metals are present. Some protective clothing made of polymeric materials such as polyethylene will support combustion and actually melt, causing a more severe burn.

A second common example is the reuse of protective clothing (including gloves) where the chemical has contaminated the inside, resulting in increased worker exposure with each subsequent use. It is very common to see another variation of this problem where workers use natural fiber gloves (e.g., leather or cotton) or personal (leather) shoes to work with liquid chemicals. If the chemicals are spilled on or contact the natural fibers, they may be retained for long periods of time and migrate to areas where they may contact the skin. A variation of this problem is where contaminated work clothing is taken home for cleaning, which can result in the exposure of an entire family.[36] The problem is common because the clothing typically is cleaned with articles of family clothing. Since many chemicals are not water soluble, they can be spread to other clothes simply by mechanical actions. There have been several examples of this occurring, especially in industries that manufacture or process pesticides and heavy metals (e.g., poisoning of the families of workers handling mercury and lead).

These are only a few of the more prominent examples of the misuse of protective clothing. Many of these problems can be overcome by simply making sure the wearer understands the proper use and limitations of the protective clothing.

Summary

Dermal hazards are a leading cause of occupational injuries and illness. They may be categorized as chemical, physical, and biological. Chemical hazards result from direct contact or the more subtle permeation, mass penetration, or degradation of protective barriers. Physical hazards include those that can produce direct trauma to the skin (e.g., cuts) and those of a thermal nature, vibration, and radiation. Biological hazards include human, animal, and environmental pathogens. Performance requirements for protective clothing will be determined by the nature and degree of the hazard.

Protective clothing can be obtained in a variety of configurations. Most typically these include gloves, boots, and garments. These range from fully encapsulating gas-tight suits to simple hand coverings. Protective clothing of all types typically present ergonomic constraints and may lead to increased stresses, such as heat stress and lower productivity.

Selection of protective clothing should be based on:

1. A determination of the hazard;
2. Determination of the adverse effects of exposure;
3. Examination of other applicable control measures;
4. Determination of the performance requirements for the protective clothing;
5. Determination of the need for decontamination;
6. Determination of the ergonomic constraints presented; and
7. Determination of the cost benefit of selection options.

All protective clothing requires a program for maintenance, inspection, and repair. Worker education is especially important to ensure that the protective clothing selected is properly used and the hazards of exposure are understood.

Development of a comprehensive program for the selection and use of protective clothing can enhance the level of protection afforded to workers. A model approach could contain the following key elements:

1. An organization scheme and administrative plan;
2. A risk assessment methodology;
3. An evaluation of other control methods to protect the worker;
4. Performance criteria for the protective clothing;
5. Selection criteria and procedures to determine the optimum choice;
6. Purchasing specifications for the protective clothing;
7. A validation plan for the selection with medical surveillance, as appropriate;
8. Decontamination and rescue criteria, as applicable;
9. A user training program; and
10. An auditing plan to ensure procedures are consistently followed.

References

1. **National Institute for Occupational Safety and Health (NIOSH):** *Proposed National Strategies for the Prevention of Leading Work-Related Diseases and Injuries*. Atlanta, GA: NIOSH, 1988.
2. **Mansdorf, S.Z.:** Risk assessment of chemical exposure hazards in the use of protective clothing—an overview. In *Performance of Protective Clothing* (ASTM STP 900). Philadelphia, PA: American Society for Testing and Materials, 1986. pp. 207–213.
3. **Perkins, J.L.:** *Chemical Protective Clothing*, vol. I. *J. Appl. Ind. Hyg.* 2:222–230 (1987).
4. **Mansdorf, S.Z.:** Industrial hygiene assessment for the use of protective gloves. In *Protective Gloves for Occupational Use*. Boca Raton, FL: CRC Press, 1994.
5. **Grandjean, P:** *Skin Penetration: Hazardous Chemicals at Work*. New York: Taylor & Francis, 1990.
6. **American Conference of Governmental Industrial Hygienists (ACGIH):** Dermal absorption. In *Documentation of Threshold Limit Values and Biological Exposure Indices*. Cincinnati, OH: ACGIH, 1992.
7. **Forsberg, K. and S.Z. Mansdorf:** *Quick Selection Guide to Chemical Protective Clothing*, 2nd ed. New York: Van Nostrand Reinhold, 1993.
8. **Mansdorf, S.Z.:** Anhydrous hydrofluoric acid. *Am. Ind. Hyg. Assoc. J. 48*:7 (1987).
9. **Linch, L.L:** Protective clothing. In *The CRC Handbook of Laboratory Safety*. Boca Raton, FL: CRC Press, 1971. pp. 124–137.
10. **Sansone, E.B. and Y.B. Tewari:** The permeability of laboratory gloves to selected solvents. *Am. Ind. Hyg. Assoc. J. 39*:169–174 (1978).
11. **Johnson, J. and K. Anderson (eds.):** *Chemical Protective Clothing*, vol. II. Fairfax, VA: American Industrial Hygiene Association, 1990.
12. **Perkins, J.L.:** Solvent-polymer interactions. In *Chemical Protective Clothing*, vol. I. Fairfax, VA: American Industrial Hygiene Association, 1990.
13. **American Society for Testing and Materials (ASTM):** "Test Method for Resistance of Protective Clothing Materials to Permeation by Liquids or Gases Under Conditions of Intermittent Contact (Method F1383)." West Conshohocken, PA: ASTM, 1992.
14. **Henry, N. and N. Schlatter:** Development of a standard method for evaluating chemical protective clothing to permeation of hazardous liquids. *Am. Ind. Hyg. Assoc. J. 42*:202–207 (1981).
15. **American Society for Testing and Materials (ASTM):** "Test Method for Resistance of Protective Clothing Materials to Permeation by Liquids or Gases Under Conditions of Continuous Contact (Method F739)." West Conshohocken, PA: ASTM, 1991.
16. **American Society for Testing and Materials (ASTM):** "Test Method for Resistance of Protective Clothing Materials to Liquid Permeation—Permeation Cup Method (Method F1407)." West Conshohocken, PA: ASTM, 1995.
17. **Coletta, G.C., S.Z. Mansdorf, and S.P. Berardinelli:** Chemical protective clothing test method development: Part II. Degradation test method. *Am. Ind. Hyg. Assoc. J. 41*:26–33 (1980).
18. **Forsberg, K. and S.Z. Mansdorf:** *Quick Selection Guide to Chemical Protective Clothing*. New York: Van Nostrand Reinhold, 1997.
19. **Johnson, J., A. Schwope, R. Goydan, and D. Herman:** *Guidelines for the Selection of Chemical Protective Clothing, 1991 Update*. Springfield, VA: National Technical Information Service, 1992.
20. **Forsberg, K. and L. Keith:** *Chemical Protective Clothing Performance Index Book*, 2nd ed. New York: John Wiley & Sons, 1997.
21. **Mickelsen, R.L. and R. Hall:** A breakthrough time comparison of nitrile and neoprene glove materials produced by different manufactures. *Am. Ind. Hyg. Assoc. J. 48*:941–947 (1985).
22. **Mickelsen, R.L., M. Roder, and S.P. Berardinelli:** Permeation of chemical protective clothing by three binary solvent mixtures. *Am. Ind. Hyg. Assoc. J. 47*:189–194 (1986).
23. **Stull, J., M. Connor, and C. Heath:** Development of a combination thermal and chemical protective ensemble for U.S. Navy fire fighting applications. In *Performance of Protective Clothing*, vol. 5 (STP 1237). West Conshohocken, PA: American Society for Testing and Materials, 1996. pp. 408–427.

24. **Day, M.:** A comparative evaluation of test methods and materials for thermal protective performance. In *Performance of Protective Clothing* (ASTM STP 989). West Conshohocken, PA: American Society for Testing and Materials, 1988. pp. 108–120.
25. **Davies, J.:** Conductive clothing and materials. In *Performance of Protective Clothing* (ASTM STP 989). West Conshohocken, PA: American Society for Testing and Materials, 1988. pp. 813–831.
26. **Plog, B.A. and J.B. Olishifski:** Overview of industrial hygiene. In *Fundamentals of Industrial Hygiene,* 3rd. ed. Chicago: National Safety Council, 1988. pp. 15–16.
27. **Brown, P.L.:** Protective clothing for health care workers: liquid proofness versus microbiological resistance. In *Performance of Protective Clothing* (STP 1133). West Conshohocken, PA: American Society for Testing and Materials, 1992. pp. 65–82.
28. **Miller, A. and C. Volk:** Biological hazards. In *Fundamentals of Industrial Hygiene,* 3rd. ed. Chicago: National Safety Council, 1988. pp. 335–355.
29. **Noonan, E.:** Spunbonding in the 1990s: a technology on the move. *Nonwovens Ind.* March 1991, pp. 58–63.
30. **Mansdorf, S.Z.:** "The Role and Future of Nonwovens in Personal Protective Equipment." Paper presented at the International Nonwovens Symposium, Munich, Germany, 1992.
31. **Johnson, J.S. and J. Stull:** "Measuring the Integrity of Totally Encapsulating Chemical Protective Suits." In *Performance of Protective Clothing* (ASTM STP 989). West Conshohocken, PA: American Society for Testing and Materials, 1988. pp. 525–534.
32. **National Institute for Occupational Safety and Health (NIOSH), Occupational Safety and Health Administration, U.S. Coast Guard, and Environmental Protection Agency:** *Occupational Safety and Health Guidance Manual for Hazardous Waste Site Activities* (NIOSH pub. 85-115). Cincinnati, OH: NIOSH, 1985. Chapter 10.
33. **Slater, K.:** Comfort or protection: the clothing dilemma. In *Performance of Protective Clothing,* vol. 5 (STP 1237). West Conshohocken, PA: American Society for Testing and Materials, 1996. pp. 486–497.
34. "General Industry Standards," *Code of Federal Regulations* Title 29, Part 1910.132.
35. **Roder, M.:** *A Guide for Evaluating the Performance of Chemical Protective Clothing (CPC)* (NIOSH pub. 90-109). Cincinnati, OH: National Institute for Occupational Safety and Health, 1990.
36. **National Institute for Occupational Safety and Health (NIOSH):** *Report to Congress on Workers' Home Contamination Conducted Under the Workers Family Protection Act.* Cincinnati, OH: NIOSH, 1995.
37. **Makinen, H.:** The effect of wear and laundering on flame-retardant fabrics. In *Performance of Protective Clothing* (STP 1133). West Conshohocken, PA: American Society for Testing and Materials, 1992. pp. 754–765.
38. **Laughlin, L. and C. Nelson:** Decontaminating personal protective equipment of applicators: A synthesis of research results. In *Proceedings of Quality and Usage of Protective Clothing,* Kittila, Finland: NOKOBETEF IV, 1992. pp. 179–185.

Workers wearing yellow tyvek and blue SCBAs prepare for a HAZMAT training exercise. ©1993 Paul Watson, HAZ-120.

Outcome Competencies

After completing this chapter, the user should be able to:

1. Define underlined terms in this chapter.
2. Describe operating principles of the various types of respirators.
3. Select appropriate respiratory protection.
4. Design, implement, and evaluate a management program for respirator use.
5. Explain respirator capabilities and limitations.
6. Recall respirator selection for nonroutine uses.
7. Explain the training needs for respirator users.
8. Recognize the importance of fit-testing.
9. Recognize that different countries have different test standards.

Key Terms

air-line respirator • assigned protection factor (APF) • bed residence time • chemisorption • closed-circuit SCBA • continuous flow respirators • demand respirators • diffusion • end-of-service-life indicator (ESLI) • filter efficiency degradation • fit check • fit factor • fit-tested • full-facepiece • half-mask • hazard ratio (HR) • IAA fit-test • immediately dangerous to life or health (IDLH) • impaction • interception • irritant smoke fit-test • maximum use concentration (MUC) • mouth bit respirator • negative pressure device • open-circuit SCBA • positive pressure device • powered air-purifying respirator (PAPR) • pressure demand respirators • program administrator • qualitative fit-test • quantitative fit-test • quarter-mask • saccharin fit-test • saturated vapor concentration (SVC) • self-contained breathing apparatus (SCBA) • sorbents • total airborne concentration (TAC) • wear time

Prerequisite Knowledge

Prior to beginning this chapter, the user should review the following chapters:

Chapter Number	Chapter Topic
15	Comprehensive Exposure Assessment
16	Modeling Inhalation Exposure
42	Confined Spaces

Key Topics

I. Introduction
   A. Respiratory Protection Program
   B. Respiratory Hazards

II. Types of Respirators
   A. Air-Purifying Respirators
   B. Atmosphere-Supplying Respirators
   C. Combination Air-Purifying and Atmosphere-Supplying Respirators

III. Respirator Selection
   A. Routine Use
   B. Nonroutine Use

IV. Training
   A. Training Requirements
   B. Wear Time

V. Respirator Fit-Testing
   A. Quantitative Fit-Tests (QNFTs)
   B. Qualitative Fit-Tests (QLFTs)
   C. Test Exercises
   D. Respirator Sealing Problems

VI. Standards
   A. Respirator Test Standards
   B. Respirator Use Standards

36

Respiratory Protection

Craig E. Colton, CIH

Thomas J. Nelson, CIH

Introduction

A primary objective of an occupational health program is the prevention of adverse health effects. When a material becomes airborne, or dangerous concentrations of gases and vapors are present, the primary means of preventing injury is through the use of engineering and work practice control measures. When these measures are not feasible, or while they are being installed or implemented, appropriate respirators should be used during emergencies and as a preventive measure.

If respirators are to function as designed, they must be properly selected, fit-tested, maintained, and used by trained employees. Because of medical or psychological conditions everyone may not be able to wear a respirator. Respirators may interfere with vision or voice communications. They can be hot and cumbersome to wear. These difficulties make respirators the least satisfactory method of control.

Despite these difficulties, respirators in many cases are the only feasible form of protection. To use a respirator and have it perform effectively requires a well-managed, complete, and systematic program. Even with the best program and most efficient respirator, inhalation exposures are only reduced, not completely eliminated.

Respiratory Protection Program

Government agencies,[1] consensus groups[2,3] and technical committees[4] have recognized that when respirators are used, a formal, effective, and complete program is required. The safe use of a respirator requires at a minimum that the following program elements be addressed:

- Program administration;
- Standard operating procedures;
- Exposure assessment;
- Medical evaluation of respirator wearers;
- Proper selection of respiratory protective equipment;
- Training;
- Respirator fitting; and
- Cleaning, inspection, maintenance, and storage.

The Occupational Safety and Health Administration (OSHA) respiratory protection standard requires that these elements be addressed within a respiratory protection program.[1] The American national standard for respiratory protection (ANSI Z88.2–1992) is a voluntary consensus standard containing similar program recommendations.[2] Both the OSHA and ANSI standards should be consulted in developing an effective and current program.

The most important person in a respiratory protection program is the <u>program administrator</u>. Program administration should be assigned to a single individual who is given the authority and responsibility for the overall program and who reports to the facility manager. Others may assist, but final responsibility remains with the single individual, thereby ensuring that there is coordination and direction for the program.

The complexity of the administrator's task in supervising the program will vary depending on the respiratory hazards present, types of respirators used, workers, and workplace. The administrator must have sufficient knowledge and must keep abreast of current issues, technological advances, and regulatory changes pertaining to respiratory protection.

The program administrator's responsibilities include:

- Conducting an exposure assessment by measuring, estimating, or reviewing information on the concentrations of airborne contaminants in the work area. This is done prior to respirator selection and periodically during respirator use to ensure that the proper type of respirator is being used.
- Selecting the appropriate type of respirator that will provide adequate protection for all contaminants present or anticipated.
- Maintaining records and written procedures in a manner that documents the respirator program and allows evaluation of the program's effectiveness.
- Evaluating the respiratory protection program's effectiveness through ongoing surveillance of the program and respirator use.

In addition to ongoing surveillance, the program must be audited periodically to ensure that the written procedures are being followed and that the requirements of applicable regulations and industry accepted standards are still being met.

To aid objectivity, the audit should be conducted by a knowledgeable person not directly associated with the program rather than by the respiratory protection program administrator. An audit checklist should be prepared and updated as necessary. Any defects or shortcomings found during the audit should be documented, including plans to correct problem areas and target dates for completion.

When a respirator program is implemented, there will be other required administrative information. This will include policies, procedures, and forms for recording the results of program activities.

Several policies that will need to be addressed include when respirators should be used, voluntary use of respirators, employee-purchased respirators, medical disqualification, and facial hair prohibition. These policies will require that groups such as the company's human resources, medical, safety, hygiene, legal, and management personnel agree on specific program elements.

Written procedures must document the entire program. These specify the activity that will occur, where it will be done, who will perform the activity, and how often. For example, a maintenance procedure for a half-mask respirator will specify whether the employee or a designated person is responsible for cleaning the respirator, how the employee or maintenance person is trained, the specific cleaning and inspection procedure, how often the respirator is to be cleaned, and where replacement parts are obtained.

Forms should be used to document the completion of required activities and their outcomes. The AIHA book *Respiratory Protection: A Manual and Guideline* contains examples of forms and audit checklists that may be required.[4]

Respiratory Hazards

Respiratory hazards may be excessive airborne concentrations of gases, vapors, or aerosols, or the reduced concentration or partial pressure of oxygen. The degree of the respiratory hazard present may be classified either as immediately dangerous to life or health (IDLH) or not IDLH. The degree and type of respiratory hazard are important factors for choosing the appropriate respirator type.

Types of Respirators

Respirators are designed to cover the entrances to the respiratory system—the nose and mouth. They vary widely in design and function and can be described in terms of the design of the respiratory inlet covering and by the mechanism used to provide protection. Respirators are tested in the United States by the National Institute for Occupational Safety and Health (NIOSH). Respirators that pass NIOSH test criteria published in the *Code of Federal Regulations* (CFR) are referred to as "approved" respirators.[5] Approval systems that exist in other parts of the world are mentioned later (see the "Respirator Test Standards" section later in this chapter). The NIOSH classification and limitations will be used as the examples for specific respirator descriptions given in this chapter.

Respiratory inlet coverings provide a barrier between the environment and the wearer's respiratory system. The inlet coverings are also used to hold the pieces that make a functioning respirator and are classified as either tight-fitting or loose-fitting.

Tight-fitting respiratory inlet coverings take the following forms:

Mouth Bit Respirator. This respirator has a short tube that is designed to fit into the mouth and a nose clip to seal the nostrils. When the lips are closed over the mouth bit and the nose clip is used, the respiratory system is sealed. Mouth bits are used exclusively in escape-only type respirators.

Quarter-Mask. This type covers the area of the face from above the nose to just

Figure 36.1 — Quarter-mask respiratory inlet covering (Courtesy Aearo, Indianapolis, IN).

under the lips and seals tight to the face. This type of respirator is not widely used (see Figure 36.1).

Half-Mask. This respirator (see Figures 36.2A, B, and C) covers the area from above the nose to underneath the chin.

Full-Facepiece. This respirator (see Figure 36.3) covers from above the eyes to below the chin. Both the half-mask and full-facepiece fit tight to the face. A full-facepiece provides protection from eye irritants.

There are four types of loose-fitting respiratory inlet coverings:

- A hood covers the head and neck and may cover portions of the shoulders (see Figure 36.4A).
- A helmet is a hood that offers head protection against impact and penetration (see Figure 36.4B) and has a bib that may cover the area down to the chest.
- A loose-fitting facepiece (Figures 36.5A and B) covers the entire face but forms only a partial seal with the face. It does not cover the neck and shoulders and may not offer head protection against impact and penetration.
- A suit covers most of the body (e.g., from the head to the waist) or the entire body. It differs from a splash suit in that breathing air is supplied directly into the air suit. Splash suits and other body coverings are intended to be used with other types of respirators or no respirator at all.

For each type of inlet covering there are many variations in the materials of construction and individual design features. A half-mask respiratory inlet covering may be made of a variety of elastomers (e.g., rubber, silicone, thermoplastic; see Figures 36.2A and B) or made entirely of a filtering material (see Figure 36.2C). Full-facepieces are also made of a variety of materials.

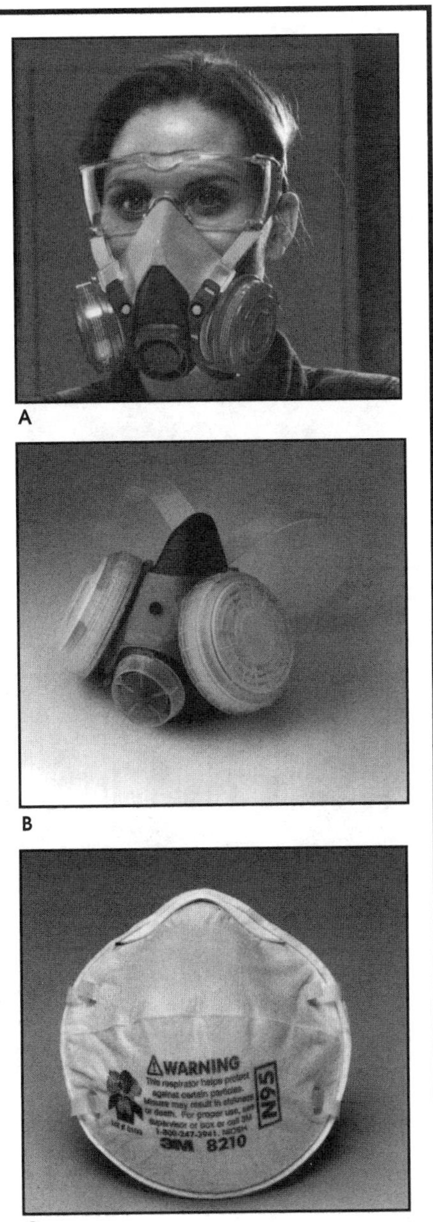

Figure 36.2 — Various half-mask respiratory inlet coverings. A, B: elastomeric dual cartridge half-masks made of two different materials; C: N-95 filtering facepiece half-mask (Courtesy 3M, St. Paul, MN).

Figure 36.3 — Full-facepiece respiratory inlet covering. (Scott Health and Safety, Monroe NC).

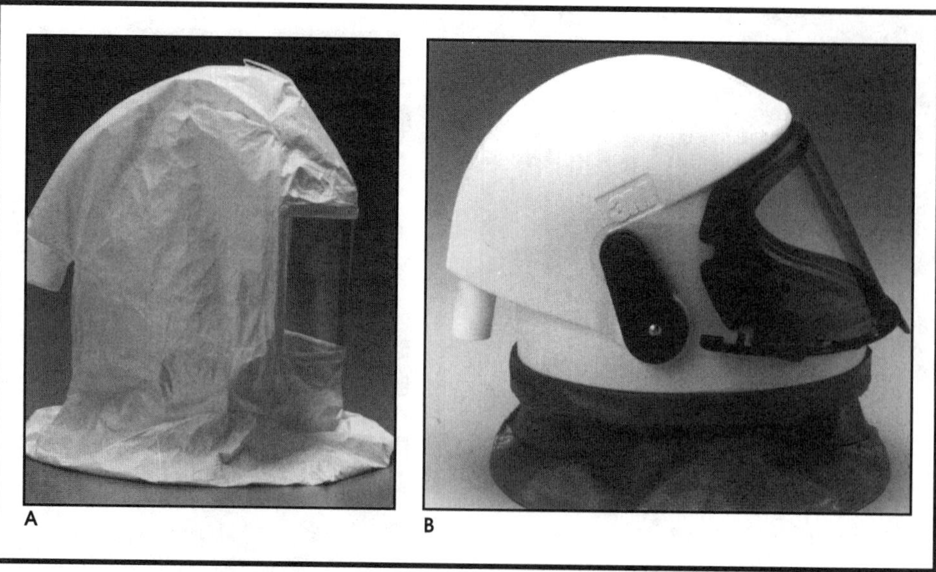

Figure 36.4 — Loose-fitting respiratory inlet coverings. A: hood; B: helmet (Courtesy 3M, St. Paul, MN).

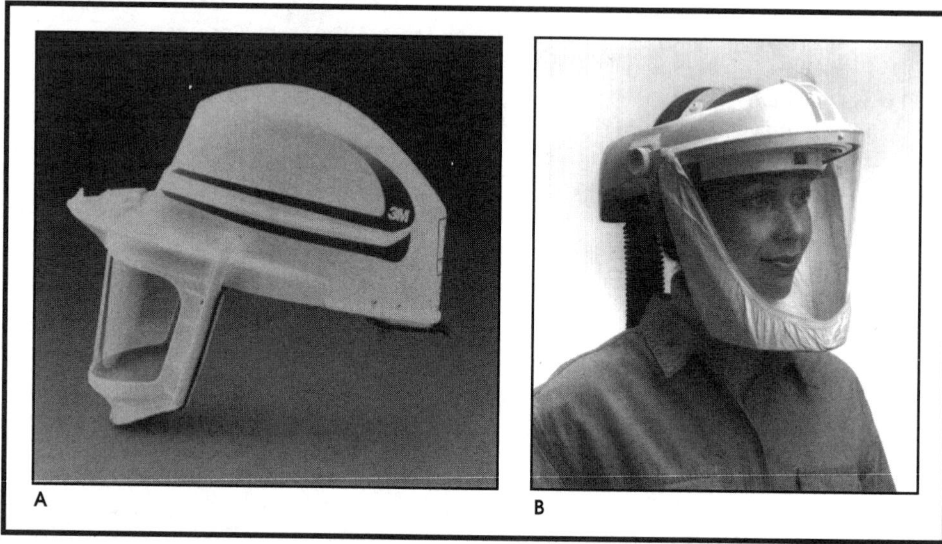

Figure 36.5 — Loose-fitting respiratory inlet coverings. A: loose-fitting facepiece with rubber side shields (Courtesy 3M, St. Paul, MN); B: loose-fitting facepiece with fabric side shields (Courtesy Racal, Frederick, MD).

Hoods and helmets may have wide viewing areas, provide eye protection (see Figure 36.4), have single or double bibs, and may be made out of a variety of materials (e.g., coated fabrics, plastics). Loose-fitting facepieces may include head protection (see Figure 36.5A) and have rubber or fabric side-seals (see Figure 36.5B).

The device providing respiratory protection can be described as either an air-purifying, atmosphere-supplying, or combination air-purifying and atmosphere-supplying respirator. Each will be discussed in turn.

Air-Purifying Respirators

The air-purifying respirator cleanses the contaminated atmosphere. Ambient air passes through an air-purifying element that can remove specific gases and vapors, aerosols, or a combination of these contaminants. Particulate filters (commonly referred to as filters) are used to cleanse aerosols. Chemical filters (commonly referred to as chemical cartridges or canisters) are used to remove gases and vapors. Air-purifying respirators are classified as either powered or nonpowered. The nonpowered air-purifying respirator uses the person's breathing (inhalation) to draw air through the air-purifying element. The powered air-purifying respirator (PAPR) uses a blower to force air through the air-purifying element. Air-purifying respirators are limited to those environments in which there is sufficient oxygen to support life. The useful life of an air-purifying element is limited by the concentration of the air contaminants, the breathing rate of the wearer, temperature and humidity levels in the workplace, and the contaminant removal capacity of the air-purifying medium.

To be certified as a PAPR by NIOSH, the blower must provide at least 115 L/min (4 ft^3/min) of air to a tight-fitting facepiece (i.e., half-mask or full facepiece) and at least 170 L/min (6 ft^3/min) to a loose-fitting facepiece, helmet, or hood.(5) The respirator manufacturer provides a device for verifying that the airflow exceeds these levels. These devices include flow meters, flow plates, and pressure gauges that can be attached to the respirator. The great advantage of the PAPR is that it usually supplies air at positive pressure, reducing inward leakage of contaminants when compared with the negative pressure (non-powered) respirators. This is why PAPRs are assigned a higher level of protection than their negative pressure counterpart.(2) It is possible, however, at high work rates to create a negative pressure in the facepiece, thereby increasing facepiece leakage. This concern can be reduced by fit-testing tight-fitting PAPRs.

Aerosol Removing Respirators

Air-purifying respirators using filters provide respiratory protection against aerosols such as dusts, mists, fumes, and other particles, but do not protect against gases, vapors, or oxygen deficiency. Filters may be made of randomly laid nonwoven fiber materials, compressed natural wool, or synthetic fiber felt, or fibrous glass that may be loosely packed in a filter container or made into a flat sheet of filter material that is pleated and placed in a filter container. Pleating increases filter surface area, which can improve filter loading and efficiency and lower breathing resistance.

Removal Mechanisms. Aerosols can be filtered by various mechanisms as they are drawn through the filter.(6) These filtration mechanisms include particle interception, sedimentation, impaction, and diffusion. In addition, some filters also use electrostatic attraction. As air moves through the filter, suspended particles flow with airstreams. As the airstreams approach a fiber lying perpendicular to their path, they split and compress to flow around the fiber. The airstreams rejoin on the other side of the fiber (see Figure 36.6).

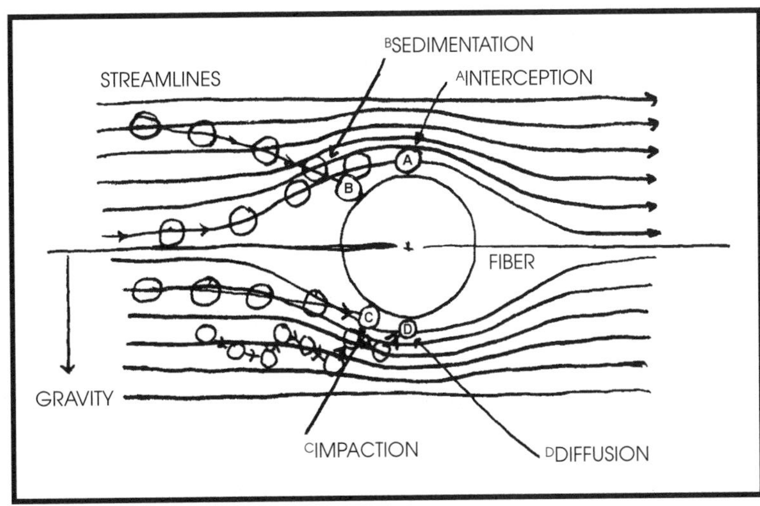

Figure 36.6 — Airflow pattern around a filter fiber showing four filtration mechanisms. A: interception; B: sedimentation; C: impaction; D: diffusion.

In interception capture, the particles follow their original airstream. If the particle center comes within one particle radius of the fiber, it contacts the fiber surface and is captured. As particle size or airflow increases, the probability of interception increases.

Sedimentation capture works through the effect of gravity on the particle; therefore, the flow rate through the filter must be low. Particle settling in the filter results in contact with the fiber. It is most significant for large particles; for example, those larger than 3 μm.

Particles with sufficient inertia cannot change direction sufficiently to avoid the fiber and therefore impact on it. As the airstreams split and change direction suddenly to go around the fiber, these particles are captured because of impaction on the surface of the fiber. A particle's size, density, speed, and shape determine its inertia.

Diffusion capture results from particle movement caused by air molecule bombardment (Brownian motion) and is important only for smaller particles. Particles randomly cross the airstreams and encounter a filter fiber. This random motion depends on particle size and temperature. For example, as particle size decreases, diffusive activity of the particle increases, which increases the chance of capture. A lower flow rate through the filter also increases the chance of capture because the particle spends more time in the area of the fiber.

Every respirator filter uses all of these filtration mechanisms to some degree, as the filter manufacturer attempts to make an efficient filter with low breathing resistance. The exact contribution of each mechanism depends on flow rate, filter solidity, fiber diameter, and particle size and density.

In addition to mechanical removal, attractive forces such as electrostatic attraction may augment filtration. Filters using both mechanical and electrical removal mechanisms are often referred to as electrostatic filters. In electrostatic capture the charged particles are attracted to filter fibers or regions of the filter fiber having the opposite charge. Uncharged particles may also be attracted depending on the level of charge imparted on the filter fiber. This removal process aids the other removal mechanisms, especially interception and diffusion. The advantage of using fibers with an electrical charge is that the filtration efficiency can be enhanced without making any contribution to airflow resistance. Two types of electrostatic materials used in respirator filters are referred to as resin wool and electrets.

The older version of the electrostatic filter is the resin wool filter. Resin wool typically consists of wool and synthetic fibers combined with resin particles. Friction generated between the resin and wool during the filter carding process results in negatively charged resin particles and positively charged wool fibers. Due to this separation, local nonuniform electrostatic fields develop throughout the filter. This type of filter is not as commonly available as it once was.

Electret fibers are a recent development in filtration technology. These fibers have a strong electrostatic charge permanently embedded into their surface during processing. They maintain a positive charge on one side of the fiber and a negative charge of equal magnitude on the opposite side. Both charged and uncharged particles will be attracted to electret fibers. Charged particles are attracted to the parts of the fiber that have an opposite charge. Uncharged particles have equal internal positive and negative charges. The strong electrostatic forces of the electret fibers polarize these charges, inducing a dipole within the particle, and the particle is then attracted to the fiber by a polarization force. Long-term environmental testing of resin wool electrostatic filters has indicated they are susceptible to filter efficiency degradation at high humidity and elevated temperature conditions.(7) Some electret filters are not affected by exposure to these same conditions.

Generally, large heavy particles are removed by impaction and interception, and large light particles are removed by diffusion and interception. Diffusion removes very small particles. When a single fiber is joined by other filter fibers to create a filter maze of certain average porosity and thickness, the different filtration mechanisms will combine at different particle sizes to affect total filtration efficiency and breathing resistance. The capture mechanisms of sedimentation, interception, and inertial impaction combine effectively to remove nearly all particles larger than 0.6 μm. Also the low flow rates through respirator filters of only a few centimeters per second let diffusion play its part very effectively for particles below 0.1 μm.

However, in the region between these two particle sizes (i.e., 0.1–0.6 μm), diffusion and impaction are not as effective, and a minimum filtration efficiency exists, as shown in Figure 36.7. The lowest point on this curve is called the most penetrating particle size and can be determined empirically in the laboratory. The most penetrating size range can vary slightly with filter design and flow rate.(8) The addition of an electrostatic charge to the fibers can greatly improve the filtering ability for small particles.(9) Most respirator filters have a most penetrating particle size between 0.2 μm and 0.4 μm. This is the basis for the widely used dioctyl phthalate and sodium chloride filter tests using a 0.3-μm mass median aerodynamic diameter particle.

Filter Efficiency and Degradation. NIOSH and certification regulations of other countries allow for various levels of filter efficiency ranging from 80% to 99.97%. For example, NIOSH certifies three levels of efficiency: 95%, 99%, and 99.97%.(5) The efficiency of the filter in a given workplace will depend on the conditions in that workplace. Factors such as the particle size distribution, the nature of the aerosol, and work rate affect efficiency. Since NIOSH tests with the most penetrating particle size (as discussed above), the actual filter efficiency will be greater than that indicated by the approval. Because a respirator filter has measurable penetration of particles in the 0.2–0.4 μm range, it is easy to forget that anywhere else, filtration efficiency is essentially 100%. It is especially important when considering the relatively large particle size in the workplace.(10) For the filter tested in Figure 36.8, the penetration at the most penetrating particle size is around 10%, while at 1 μm the penetration is only about 1.5%. It is the reduction of the entire actual work environment particulate challenge in particle number or mass that is important to protecting the worker. The filters NIOSH certifies are listed in Table 36.1.

Filter efficiency degradation is defined as a lowering of filter efficiency or a reduction in the ability of the filter to remove particles as a result of workplace exposure.

Generally, solids and water-based aerosols do not reduce filter efficiency. Certain oils, such as dioctyl phthalate, have been shown to reduce filter efficiency of some filters for small particles. It may be due to the wetting of the filter fiber by the oil, masking the electrostatic charge. As a result, NIOSH has established the N-series of filters for nonoil aerosols and the R- and P-filter series for oil aerosols.

NIOSH test requirements assure that filter efficiency does not fall below the stated level with up to 200 mg of oil loaded on the respirator for R-series filters. Since the test is stopped at this point, it is not known whether the filter efficiency will decrease beyond the 200-mg load. It is recommended that R-series filters be limited to 8 hours of use or discarded when the respirator has collected 200 mg of oil (100 mg/filter for dual filter respirators).(11)

The NIOSH test requirements for P-series filters measures the filter efficiency until it stabilizes or is no longer decreasing. If filter efficiency is decreasing when the 200-mg point is reached, filter loading continues. When it stabilizes, the filter efficiency is determined. This ensures that the filter efficiency never goes below the stated efficiency level. NIOSH requires that the filter manufacturer state a time-use limitation recommendation for P-series filters when oil aerosols are present. P-series filters should be used and reused in accordance with the manufacturer's time-use limitation. For nonoil aerosols there is no time restriction on either N-, R-, or P-series filters because the nonoil aerosols are relatively nondegrading. Any decrease in filter efficiency can be offset by continual loading of the filter.(9)

Aerosol filters generally become more efficient as particles are collected and plug the spaces between the filter fibers. Filters should be changed when users find it difficult to breathe through them. Also, the filters should be changed when they are damaged or for hygiene reasons.

Gas/Vapor Removing Respirators

These air-purifying respirators protect against certain gases and vapors by using various chemical filters to purify the inhaled air. They differ from aerosol filters in that they use either cartridges or canisters containing sorbents that remove harmful gases and vapors. Sorbents are granular porous materials that interact with the gas or vapor molecule to clean the air. The cartridges and canisters may be replaceable, or the entire respirator may be disposable.

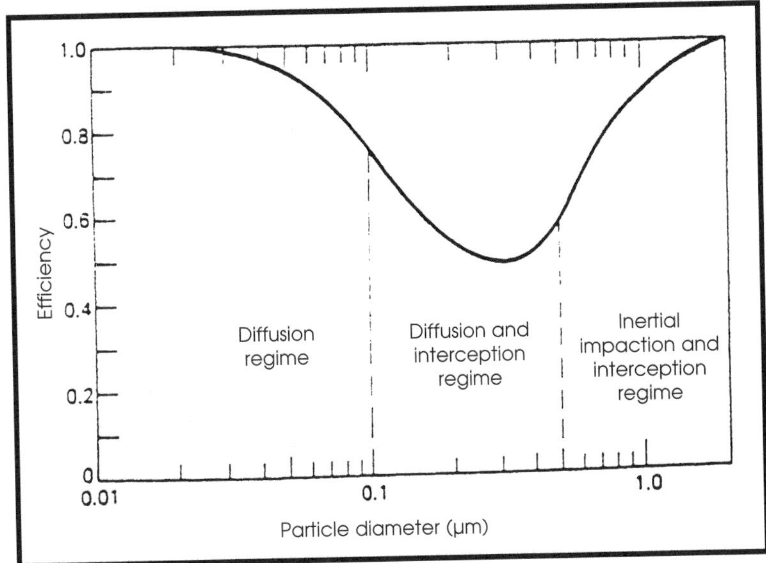

Figure 36.7 — Filter efficiency vs. particle size schematic illustrating the different filtration regimes (Source: **Lee, K.W., and B.Y.H. Liu:** On the Minimum Efficiency and the Most Penetrating Particle Size for Fibrous Filters. *J. Air Pollution Control Assoc. 30(4)*:377–381 (1980).)

Removal mechanisms are generally classified as adsorption, absorption, chemisorption, or catalysis.(12) In contrast to aerosol filters, which are effective to some degree no matter what the particle, cartridges and canisters are designed for protection against specific contaminants (e.g., ammonia gas or mercury vapor) or classes of contaminants (e.g., organic vapors). Adsorption is the adherence of gas or vapor molecules to the

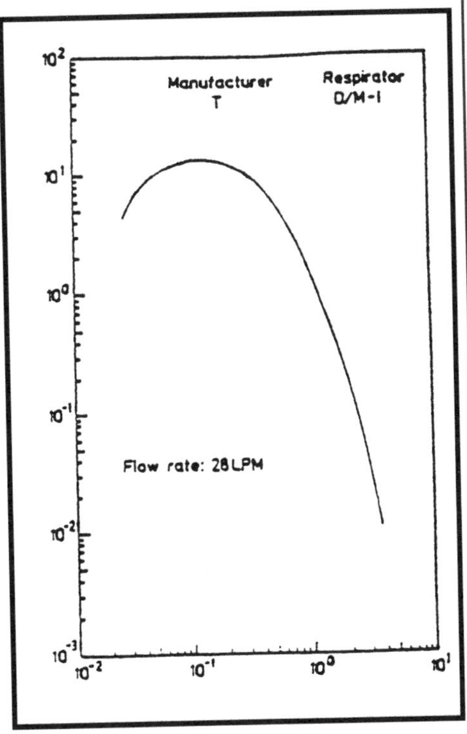

Figure 36.8 — Actual filter efficiency curve for a dust/mist filter approved under 30 CFR 11.

Table 36.1 —
Description of Filter Classes Certified under 42 CFR 84

Filter Class (Test Agent)	*Efficiency (%)*	*Test Maximum Loading (mg)*	*Use*
N-series (NaCl[A])		200	nonoil aerosols
N-100	99.97		
N-99	99		
N-95	95		
R-series (DOP oil[B])		200	oil and nonoil aerosols (time use restriction may apply[C])
R-100	99.97		
R-99	99		
R-95	95		
P-series (DOP oil[B])		stabilized efficiency	oil and nonoil aerosols
P-100[D]	99.97		
P-99	99		
P-95	95		

Source: Reference #5.
[A]NaCl = sodium chloride.
[B]DOP oil = dioctyl phthalate.
[C]In the presence of oil aerosols, service time may be limited to 8 hours of use or up until the total mass loading is less than 200 mg (100 mg/filter for dual filter respirators).
[D]The P-100 filter must be colored magenta.

surface of another substance called the adsorbent or sorbent. The attractive force between the sorbent and the adsorbate molecule is relatively small, on the order of Van der Waals' forces. Generally, the more sorbent surface available, the more adsorption that occurs. Since only weak physical forces are involved, the process can be reversed.

Activated carbon is commonly used for removal of organic vapors via adsorption. For use in respirators, the surface of carbon is greatly enhanced or "activated" using heat and steam. The most common starting material for activated carbon is coconut or coal. Activated carbon has an extensive network of molecular-size internal pores and, consequently, large internal surface areas. The typical range of surface area is 1000–2000 m^2/g of carbon. In general, this material has a greater affinity for less volatile materials; the less volatile the solvent, the greater the amount of solvent adsorbed.(12) Generally, organic vapors of molecular weight greater than 50 or boiling points (BPs) greater than 70°C are effectively adsorbed by activated charcoal.(13) However, attempts to predict breakthrough time by relating it to a single property of a vapor have proved to be unrealistic.(12)

For gases and vapors that would otherwise be weakly adsorbed, sorbents (usually activated carbon) can be impregnated with chemical reagents to make them more selective. Chemical interaction occurring at the interface through the use of treated carbons is referred to as chemisorption.(12) Chemisorption is usually irreversible. Examples are activated charcoal impregnated with iodine to remove mercury vapor or with metal salts like nickel chloride to remove ammonia gas. If the chemical does not remain on the surface but enters into the solid or its impregnant and reacts chemically with it, changing both, it is referred to as absorption.(12) The sorbent can also act as a catalyst (e.g., Hopcalite for converting carbon monoxide to carbon dioxide), which decomposes and detoxifies the contaminant by formation of relatively innocuous substances.(12)

These removal mechanisms are essentially 100% efficient until the sorbent's capacity is exhausted or the catalyst is poisoned, thus preventing its operation. At this point, breakthrough occurs as the contaminant passes through the cartridge or canister and into the respirator.

Canisters are very similar to cartridges. The basic difference is the volume of sorbent rather than function. Canisters have the larger sorbent volume. They do not provide greater protection than cartridges; they just last longer than a chemical cartridge under identical conditions. "Gas mask" is a term often used for a gas- or vapor-removing respirator that uses a canister. While gas masks are limited by contaminant concentration for routine use, they can be used for escape only from IDLH atmospheres that contain adequate oxygen to support life (19.5% oxygen). They must never be used for entry into an IDLH atmosphere.

Service life of cartridges and canisters depends on the following factors:

- Quality and amount of sorbent;
- Packing uniformity and density;
- Exposure conditions, including breathing rate of the wearer;
- Relative humidity;
- Temperature;
- Contaminant concentration;
- Affinity of the gas or vapor for the sorbent; and
- Presence of other gases and vapors.

Generally, high concentrations, high breathing rate, and humid conditions adversely affect service life. Because exposure conditions are subject to wide variation, it is difficult to estimate the service life of canisters and cartridges, even when other conditions (e.g., temperature and relative humidity) are constant. Table 36.2 shows various chemical cartridge breakthrough times for different organic gases

Table 36.2 — Selected Chemical and Physical Properties of Organic Chemicals Mentioned in this Chapter

Chemical	Mol. Wt.	Boiling Point °C	Vapor Pressure torr (20°C)	IDLH[14] (ppm)	Threshold Limit Value[15] (ppm)	Geometric Mean Odor Threshold[16] (ppm)	Range Odor Threshold[16] (ppm)[A,B]	Laboratory Breakthrough Time[17] (min.)[C]
Benzene	78.11	80.1	75	3000	10	61	34–119	88.6
Toluene	92.1	110.6	22	2000	50	1.6	0.16–37	114
m-Xylene	100.6	139.1	9	10,000	100	20	20	116
Methanol	32.04	64.7	97	25,000	200	160	4.2–5960	3.2
Ethanol	46.07	78.5	44	—	1000	180	49–716	45.3
Isopropanol	60.09	82.5	33	20,000	400	43	37–610	81.8
n-Butanol	74.12	117	6	8000	C 50	1.2	0.12–11	141
3-methyl-1-butanol	88.15	131.4	2.8	8000	100	121		
Methylene chloride	84.94	40.1	349	5000	50	160	160	15.8
Chloroform	119.4	61.3	160	1000	10	192	133–276	52.4
Methyl chloroform	133.4	74.1	100	1000	350	390	390	58.9
Trichloroethylene	131.4	87	58	1000	50	82	82	83
Carbon tetrachloride	153.8	76.75	91	300	5	252	140–584	90
Acetone	58.1	56.2	180	20,000	750	62	3.6–653	46
2-Butanone	72.1	79.6	77.5	3000	200	16	2-85	94.4
2-Pentanone	86.1	102.2	27	5000	200	7.7	7.7	12
4-Methyl-2-pentanone	100.2	115.8	16	—	50	0.88	0.1–7.8	
Pentane	72.15	36	426	5000	600	—	119–1147	71.3
n-Hexane	86.18	68.7	124	5000	50	—	65–248	64.6
Heptane	100.2	98.43	40	4250	400	230	230	89.8
Cylclohexane	84.16	80.74	78	10,000	300	780	780	82.3
Methyl acetate	74.08	56.9	173	10,000	200	180	46.5	
Ethyl acetate	88.12	77.06	73	10,000	400	18	6.4–50	84.7
n-Propyl acetate	102.13	101.6	25	8000	200	4.1	0.5–34	85.6
Butyl acetate	116.16	126.5	10	10,000	150	0.31	0.063–7.4	96.9
Isopentylacetate	130.19	145.6	4	3000	100	0.22	0.0034–209	88.3

[A]Range of acceptable values based on an evaluation of the studies.
[B]Single acceptable study, geometric mean is the single value.
[C]Tested at 22°C, 50% RH, at an airflow of 53.3 L/min. Test concentration: 1000 ppm; breakthrough concentration: 10 ppm.

and vapors. While the cartridge is approved for organic vapors by testing against carbon tetrachloride, it may last longer (e.g., butanol) or much shorter (e.g., methanol) when compared with the test agent. Hence, an organic vapor cartridge may be recommended for use against butanol, but not for methanol (molecular weight < 50; boiling point < 70°C) even though both compounds are classified as organic vapors.

Most data reported for chemical cartridge breakthrough times have been for single contaminants. The evaluation of cartridges against mixtures of contaminants has received very little attention.[12] Cartridge breakthrough may occur earlier in the presence of mixtures than would have been predicted from data for a single chemical. Unfortunately, cartridge use for mixtures is more reflective of the real world.

Another concern when using cartridges or canisters is the desorption of adsorbed compounds during respirator use. A well-adsorbed vapor at a high concentration can cause early breakthrough of a poorly adsorbed vapor when both are present in the workplace.[18] In general, compounds with low boiling points have increased desorption properties. Another concern is whether desorption will take place if the cartridge is partially used and then reused after a short period (hours) without use (e.g., overnight). Organic vapors adsorbed on a used cartridge can also migrate through the carbon bed without airflow. The user will then detect breakthrough when the respirator is used next. This is most significant for the most volatile and poorly retained organic vapors (e.g., BP < 65°C), especially low-boiling oxygenated compounds and low-boiling fluoro- and chlorofluorocarbons.

It has been recommended, especially for agents that desorb readily, that even if a respirator is specially designed for an agent, the cartridge should not be worn intermittently because of the desorption that can lead to hazardous concentrations during subsequent use, even in an uncontaminated work area.(12)

Combination Aerosol Filter/Gas or Vapor Removing Respirators

These respirators use aerosol-removing filters with a chemical cartridge or canister for exposure to multiple contaminants in different physical forms or a single chemical in more than one physical form (e.g., mist and vapor). The filter is generally a permanent part of a canister but can be either permanent or replaceable on the chemical cartridge. Replaceable filters are sometimes used because the filter and chemical cartridge are not exhausted at the same time. This allows for disposal of only the part that needs changing. Filters used in combination with cartridges must always be located on the inlet side of the cartridge. This way, any gas or vapor adsorbed onto a filtered particle is captured by the sorbent as it evaporates or desorbs from the particle.

Table 36.3 — Air Quality for Atmosphere-Supplying Respirators

Limiting Characteristics	*Allowable Maxima*
Percent O_2 balance predominantly N_2[A]	atm/19.5%–23.5%
Water, ppm (v/v)[B]	
Dew Point, °F[B]	
Oil (condensed) (mg/m³ at NTP[C])	5 ppm
Carbon monoxide	10 ppm
Odor	D
Carbon dioxide	1000 ppm

Source: **Compressed Gas Association (CGA):** *Commodity Specification for Air.* Arlington, VA: CGA, 1989.
Note: Grade D air (quality verification Level D)
[A]The term "atm" (atmospheric) denotes the oxygen content normally present in atmospheric air; the numerical values denote the oxygen limits for synthesized air.
[B]The water content of compressed air required for any particular verification level may vary from saturated to very dry. For breathing air in conjunction with SCBA in extreme cold, where moisture can condense and freeze causing the breathing apparatus to malfunction, a dew point not to exceed -50°F (63 ppm v/v) or 10° lower than the coldest temperature expected in the area is required. If a specific water limit is required, it should be specified as a limiting concentration in ppm (v/v) or dew point. Dew point is expressed in °F at one atmosphere pressure absolute, 101 kPa abs. (760 mmHg).
[C]NTP = normal temperature and pressure
[D]Specific measurment of odor in gaseous air is impractical. Air normally may have a slight odor. The presence of a pronounced odor should render the air unsatisfactory.

Atmosphere-Supplying Respirators

Atmosphere-supplying devices are the class of respirators that provide a respirable atmosphere to the wearer, independent of the ambient air. The breathing atmosphere is supplied from an uncontaminated source, which must conform to certain purity levels. In the United States, Grade D air is required as specified in the Compressed Gas Association standard, "Commodity Specification for Air."(19) Table 36.3 lists the air quality requirements for Grade D breathing air. Atmosphere-supplying respirators fall into three groups: air-line respirators, self-contained breathing apparatus (SCBA), and combination air-line and SCBA.

Air-Line Respirators

Air-line respirators, sometimes referred to as Type C or CE supplied-air respirators, deliver breathing air from either a compressor or compressed air cylinders through a supply hose connected to the respiratory inlet covering. A flow control valve, regulator, or orifice is provided to govern the rate of airflow to the worker. Depending on the NIOSH certification, up to 300 feet of air supply hose is allowable.(5) Hose supplied by the respirator manufacturer along with recommended hose lengths and operating pressures must be used. The maximum permissible inlet pressure is 125 pounds per square inch.(5) The approved pressure range and hose length is noted on the certification label or operating instructions provided with each approved device.

Air-line respirators should be used only in non-IDLH atmospheres—or in other words, atmospheres in which the wearer can escape without the use of a respirator. This limitation is necessary because the air-line respirator depends entirely on an air supply that is not carried by the wearer of the respirator. If this air supply fails, the wearer may have to remove the respirator to escape from the area. Another limitation is that the air hose limits the wearer to a fixed distance from the air supply source.

Air-line respirators operate in three modes: demand, pressure demand, and continuous flow. The respirators are equipped with half facepieces, full facepieces, helmets, hoods, or loose-fitting facepieces. Some of these respiratory inlet coverings may provide eye and head protection.

Versions of these respirators may be designed for welding or abrasive blasting. Respiratory protective equipment designed for abrasive blasting (Type CE) is equipped to protect the wearer from impact of the rebounding abrasive material. A special hood or shroud may be used to protect the wearer's head and neck, and shielding material may be used to protect the viewing windows of the head enclosures.

Demand. These air-line respirators are equipped with either half or full facepieces. The design of the regulator allows for airflow only on inhalation. These respirators are negative-pressure devices, meaning that a negative pressure with respect to the outside of the respirator is created in the facepiece on inhalation. While these respirators can still be found in work sites, they are not recommended; the pressure demand air-line respirator is much more protective, and the cost differential between the two is negligible.

Pressure Demand. These respirators are very similar to the demand type but are designed so that the pressure inside the respirator is generally positive with respect to the air pressure outside the respirator during both inhalation and exhalation. This positive pressure means that when a leak develops in the face seal due to head movement, for example, the leakage of air would be outward.

The regulator generally has a small spring that forces the airflow valve open. In the facepiece the exhalation valve also has a small spring or another mechanism that forces the exhalation valve closed. Air flows into the facepiece until the air pressure inside the facepiece and the regulator are equal. This results in a small positive pressure inside the facepiece that remains positive during inhalation, provided airflow through the regulator is greater than the breathing rate.

Thus, pressure demand respirators provide a higher degree of protection to the user compared with demand respirators. Pressure demand respirators are available only with half and full facepieces. These respirators are normally used when the air supply is restricted to high-pressure compressed air cylinders. Continuous flow respirators will generally exhaust a bottled air supply too quickly to be practical.

Continuous Flow. A continuous flow unit has a regulated amount of air delivered to the respiratory inlet covering and is normally used where there is an ample air supply, such as that provided by an air compressor. These devices may be equipped with either tight-fitting or loose-fitting respiratory inlet coverings.

NIOSH establishes airflow requirements.[(5)] Units equipped with tight-fitting enclosures, (i.e., a half or full facepiece) must provide at least 115 L/min (4 ft^3/min) measured at the facepiece. When loose-fitting helmets, hoods, or facepieces are used, the minimum amount of air to be delivered is 170 L/min (6 ft^3/min). In either case, the maximum flow is not to exceed 450 L/min (15 ft^3/min). Also, the pressure requirements of the respirator must be compatible with the pressure ranges of the air source. Operating these respirators using the proper pressure, number of air hose sections, and lengths of air hose as identified in the operating instructions will ensure proper airflow is obtained.

SCBA

SCBA provides respiratory protection against gases, vapors, particles, and an oxygen-deficient atmosphere. The wearer is more mobile than with an air-line respirator and independent of the surrounding atmosphere as the breathing gas is carried by the wearer. SCBA may be used in IDLH and oxygen-deficient atmospheres either as escape-only devices or for entry into and escape from these atmospheres.[(5)] A full facepiece is most commonly used with SCBAs, though half facepieces, hoods, and mouth bits are available on some units. There are two major types of SCBAs: closed-circuit and open-circuit.

Closed-Circuit SCBA. In closed-circuit SCBA all or a percentage of the exhaled gas is scrubbed and rebreathed. Closed-circuit units have the advantage of lower weight for the same use duration as open-circuit devices. Service life for the units ranges from 15 minutes to 4 hours. Disadvantages include increased complexity (e.g., a carbon dioxide scrubber is required in many of the units) and cost. In many of the devices the air supply can become quite warm because the exhaled air is breathed again. Closed-circuit SCBA are available as both negative and positive pressure devices. The positive pressure devices with service lives greater than or equal to 30 minutes are recommended for entry into and escape from IDLH atmospheres.[(2)] They may be designed as stored oxygen or oxygen-generating systems.

Stored oxygen systems use compressed oxygen from cylinders or oxygen carried as a liquid. Oxygen is admitted to a breathing bag either in a continuous flow or controlled by a regulator governed by the pressure or degree of inflation of the bag. The wearer inhales from the bag and exhales into it. Exhaled breath is scrubbed of carbon dioxide by a chemical bed, usually a caustic such as sodium hydroxide.

Oxygen-generating systems rely on chemical reactions to provide the needed oxygen. Water vapor and carbon dioxide from the exhaled breath react with a solid chemical, usually potassium superoxide, in a canister-size container that releases oxygen. This reaction in the canister removes the carbon dioxide from the exhaled breath.

Open-Circuit SCBA. In open-circuit SCBA, the exhaled breath is released to the surrounding environment after use rather than being recirculated. The breathing gas is generally compressed air. They are typically designed to provide 30 to 60 minutes of service and are available in both demand (negative pressure) and pressure demand (positive pressure) styles. Because of the greater protection provided by pressure demand devices, they are recommended over negative pressure systems. Only the positive pressure devices with 30 minutes or longer of service life are recommended for entry into and escape from IDLH atmospheres.(2)

Escape SCBA. Some SCBA are designed for escape only. They are similar in design to the types described above, but the use duration tends to be shorter, typically 5, 7, 10, or 15 minutes. Units approved as escape-only may not be used to enter a hazardous atmosphere.

Combination SCBA and Air-Line Respirators

These units are air-line respirators with an auxiliary self-contained air supply that can be used if the primary air supply fails. Since they have backup or escape provisions, these devices are usable in IDLH and oxygen-deficient atmospheres.

An advantage of these devices is they can be used in IDLH situations requiring extended work periods where an SCBA alone does not provide sufficient time. Also, since the SCBA needs only supply enough air for escape, the cylinder can be smaller compared with a 30-minute SCBA. The smaller SCBA cylinder on many of these units makes them particularly convenient for use in confined spaces. Operation in the air-line mode allows for longer service times than a fixed-duration SCBA.

The auxiliary SCBA may be NIOSH-approved either in the 3-, 5-, or 10-minute service time category or the 15 minutes or longer category.(5) If the SCBA portion is rated for a service life of 3, 5, or 10 minutes, the wearer must use the air-line during entry into a hazardous atmosphere, and the SCBA portion is used for emergency egress only. When the SCBA is rated for service of 15 minutes or longer, the SCBA may be used for emergency entry into a hazardous atmosphere (e.g., to connect the air-line) when not more than 20% of the air supply's rated capacity is used during entry. This allows for enough air for egress when the warning device indicates a low air supply. These units must also have a gauge visible to the wearer and a low-air warning device.

The combination SCBA/air-line respirator may operate in demand, pressure demand, or continuous flow modes. These devices use the same principles as the respective air-line respirator. Demand mode is not recommended.

Combination Air-Purifying and Atmosphere-Supplying Respirators

Another type of respirator is a combination of an air-line respirator with an auxiliary air-purifying element attached, which if properly selected provides protection in the event the air supply fails. NIOSH has approved combination air-line and air-purifying respirators with the air-line operating in either continuous flow or pressure demand mode. These respirators can be used in either an air-purifying or atmosphere-supplying mode. The most popular versions have high efficiency particulate air (HEPA) filters such as an N-, R-, or P-100, but devices are available with complete arrays of chemical cartridges as well.

These respirators have additional limitations. They are not for use in IDLH atmospheres or in atmospheres containing less than 19.5% oxygen; only the hose lengths and pressure ranges specified on the approval label may be used; and they may be used only in atmospheres for which the air-purifying element is approved.

The approval label must be consulted for proper use of the respirator in the air-purifying mode. The restrictions can vary from manufacturer to manufacturer depending on the respirator design.

Respirator Selection

Routine Use

Selection of a particular respirator will depend on a number of factors. The selection process includes an analysis of workplace hazards, the physical characteristics of the work area, the physical demands of the work, and the capabilities and limitations of the various types of respirators.

Workplace Hazards

An exposure assessment is one component of hazard analysis in the workplace. Other components include determining characteristics of the hazard such as the physical state of the contaminant, skin absorption and eye irritation potential, oxygen concentration, and warning properties for gases and vapors.

For respirators, an exposure assessment is usually limited to airborne hazards. Chapters 6 and 15 contain discussions of the information needed and methods used to conduct an assessment. For respirator selection, the exposure assessment may be based on data collected by air sampling, by estimating or modeling exposures, or on the potential for exposure.

Chapter 16 has a discussion of methods for modeling or estimating the magnitude of exposures. Respirators are selected for routine use in two situations: documented overexposure or positive exposure assessment and potential exposure. For selection based on potential exposure, it is presumed that an exposure will occur even though prior experience indicates otherwise.

For example, in the chemical industry the use of a respirator is generally required for the first break into a closed system such as a pipe. Work practices and prior experience at such tasks may have shown that exposures do not occur. Yet, it is standard practice to require a full-facepiece pressure demand air-line respirator in many chemical plants on the chance that material may still be in the line being opened and a release may occur.

Another important step in evaluating the workplace hazard is determining the physical state of the airborne contaminant. Care is required in selecting air-purifying respirators for materials that exist as both particles and vapors. In the past, generalizations about the contaminant phase have been based on the listing of the ACGIH threshold limit value (TLV®).[20] Materials with a TLV listed only in mg/m^3 were expected to exist in the particle phase, requiring a particulate filter to be selected. Materials with a TLV listed in both ppm and mg/m^3 are generally found in the vapor phase, requiring a chemical cartridge to be selected. However, it is possible that some materials with TLVs listed only in mg/m^3 may also occur in the vapor phase. Some chemicals that have TLVs in mg/m^3 only, such as acrylamide, have vapor pressures high enough that vapor can be released from the particles in the air or trapped on the filter. Likewise, some materials expected to exist primarily as vapors can have a significant portion of their airborne mass as particles. In atmospheres where the vapor of some of these materials coexists with particles that may adsorb or solubilize the vapor, a significant fraction may exist in the particle phase. For example, a significant percentage of ethylene glycol would be expected to exist in the particle phase when the relative humidity is greater than 60% and the total airborne concentration is equal to 127 mg/m^3.[21] Respirators with both aerosol- and vapor-removing capabilities would be required.

General guidelines have been developed for estimating the airborne phase for air sampling and can be applied to respirator selection.[21] These guidelines require three items of information: (1) identity of the major constituents; (2) the saturated vapor concentrations of the constituents; and (3) an estimate of the total airborne mass concentration.

A chemical's saturated vapor concentration (SVC) compared with the estimate of its total airborne concentration (TAC) can help predict the phase of the chemical. A substance with an SVC much lower than its TAC will always be found predominantly in the particle phase (e.g., lead, SVC = zero at 20°C, for which only a particulate filter would be recommended). A substance with an SVC much higher than its TAC will always be found predominantly in the vapor phase. In the case of acrylamide, presuming a total airborne concentration of 0.6 mg/m^3 (2 x TLV), the SVC of 27 mg/m^3 at 20°C indicates its presence in the vapor phase should be considered despite the TLV being listed in mg/m^3 only. A chemical cartridge and a filter may be recommended for acrylamide instead of a filter alone. For butyl acetate, on the other hand, only the vapor phase may have been considered because the TLV is listed in both ppm and mg/m^3. Presuming a total airborne concentration of 710 mg/m^3 of butyl acetate, the SVC/TAC ratio is 9 at 20°C. Although only the vapor phase may be predicted, it may exist in both vapor and particle phase in an atmosphere with other particles present. It is possible for a significant fraction of the vapor to dissolve into or be adsorbed onto particles of another substance. Again, a chemical cartridge and filter would be recommended.

Physical Characteristics of the Work Area

The area where the respirator will be used and its environmental conditions can affect respirator selection. The distance a person will have to travel to a location having respirable air also needs to be considered in selection. This includes planning for the escape of workers if an emergency occurs, for the entry of workers to perform maintenance duties, and for rescue operations. For example, an air-line respirator generally requires that the person enter and exit the work area the same way because the air hose limits the direction and distance a person can travel. In areas with moving equipment, such as a warehouse, hoses may get in the way or be run over.

Environmental conditions such as temperature can affect which respirator can be used. In hot humid conditions, use of lightweight respirators is one means to lessen the burden of respirator wear. Air-line respirators can be equipped with vortex heaters or coolers to condition the air. An air-line respirator will also provide cooling because of the flow of air and adiabatic expansion of the compressed air.

Physical Demands of the Work

The work an employee will perform and other required personal protective equipment used while wearing a respirator needs to be considered. Factors such as work rate, work movements, noise levels, communication and vision requirements, and length of time that the respirator will be worn affect selection. For jobs that demand physical effort, selecting the lightest and easier breathing respirator may reduce the burden. Respirators that facilitate communication or are compatible with communication devices may need to be selected so workers can talk clearly with each other. Full-facepiece respirators may interfere with field of vision. (Glasses with temple bars cannot be worn with full-facepiece respirators.)

Table 36.4 — NIOSH and ANSI Assigned Protection Factors

Type of Respirator	NIOSH Respirator Decision Logic[22]	ANSI Z88.2–1992[2]
Air-purifying		
Single use or quarter-mask	5	10
Half-mask	10[A]	10
Full-facepiece, D/M or D/F/M	10	100
Full-facepiece, all other media	50	100
Powered air-purifying		
Half-mask	50	50
Full-facepiece	50	1000[B]
Loose-fitting facepiece	25	25
Hood or helmet	25	1000[B]
Air-line		
Half-mask		
demand	10	10
continuous flow	50	50
pressure demand	1000	50
Full-facepiece		
demand	50	100
continuous flow	50	1000
pressure demand	2000	1000
Loose-fitting facepiece	25	25
Hood or helmet	25	1000
SCBA		
Demand	50	100
Pressure demand	10,000	10,000[C]

[A]Includes disposable particulate respirators if QNFT is used.
[B]100 if dust/mist filter is used.
[C]For emergency planning where concentrations can be estimated.

Respirator Capabilities and Limitations

In addition to the capabilities and limitations discussed in the description of respirator types, other limitations include the level of performance for a specific type of respirator (the assigned protection factor), the need for an independent supply of air, the efficiency of filters, the capacity of the cartridges and canisters, and any regulatory maximum use concentration.

Assigned Protection Factors. An assigned protection factor (APF) is the expected workplace level of respiratory protection that would be provided by a properly functioning respirator or a class of respirators to users who are properly fitted and trained.[2]

In selecting a specific respirator for use, the APF must be greater than the expected air contaminant concentration (C_{air}) divided by its exposure limit (TLV).

$$APF \geq C_{air} / TLV \quad (1)$$

For example, if the expected concentration is 30 ppm and the exposure limit is 2 ppm, a respirator with an APF of at least 15 must be used. This is also called the hazard ratio (HR) or APF needed, where the HR equals the concentration divided by the exposure limit. A similar selection concept used in many regulations is the maximum use concentration (MUC). The respirator MUC for a given chemical is the APF x the exposure limit. Other factors affect the MUC, as discussed below.

Many groups have developed lists of APFs. In the United States the two main lists of APFs are from ANSI[2] and NIOSH.[22] The APFs listed by these groups are not the same in all cases because they were developed at different times using different types of respirator performance information and use restrictions. Table 36.4

compares the APFs developed by each. OSHA lists APFs in several substance-specific standards (e.g., lead). The ANSI APFs, developed more recently than those from NIOSH, are based primarily on workplace performance information, whereas the NIOSH APFs are based primarily on laboratory performance information gathered on respirators available in the mid-1970s. Therefore, the ANSI APFs represent the best estimates of respirator performance available today. The APFs listed by OSHA substance-specific standards must be used when they are less than the ANSI value. The authors recommend that the ANSI APF be used in place of the OSHA APF when the ANSI value is lower.

The maximum use concentration of a respirator is based on several factors including the APF, IDLH concentrations, and regulatory limits. A respirator may not be used at a concentration greater than the exposure limit times the APF for the respirator. In addition, air-purifying and air-line respirators may not be used at concentrations greater than the IDLH. Finally, respirator-use regulations may preclude the use of a respirator above specific limits. For example, in the formaldehyde standard OSHA limits the use of a full-facepiece pressure demand air-line respirator to no more than 100 times the PEL.[23]

Filter Selection. Filter selection is based on classification of the aerosol. For NIOSH-approved filters, if the aerosol is an oil mist or includes oil particles, only R- or P-series filters can be considered.[5,11] The decision to use an R- or P-series filter can be made after consideration of desired service time, applicable time use restrictions, breathing resistance, and cost differences. If no oil is present, then an N-series filter can be selected. The filter efficiency needed will depend on which inlet covering is used and the anticipated concentration of the aerosol. If a half-mask is being used, a 95% filter will normally be adequate. For any PAPR or full-facepiece negative pressure respirator, a filter efficiency greater than 95% may be needed based on the expected concentration. For example, use of a full-facepiece PAPR at a concentration 100 times the exposure limit may require a 99.97% efficiency filter.

Cartridges/Canisters. Unlike filters, cartridges have high removal efficiency and at some point become less efficient and lose the ability to trap the contaminant completely. Therefore, selection of an air-purifying respirator for gases and vapors must consider the service life of the cartridge or canister. To protect workers, the chemical cartridge or canister must be changed before significant breakthrough occurs. There are several ways to set a replacement schedule before service life is exhausted. These include chemical warning properties, an end-of-service-life indicator (not widely available), or a replacement schedule that changes the cartridges or canisters before breakthrough occurs.

The most frequently used method for identifying the time to change cartridges or canisters is by monitoring the chemical's warning properties. Warning properties of gases and vapors refer to an odor, taste, or irritation indicating the chemical's presence in the environment. Warning properties can be classified as adequate or poor. Poor warning properties are defined as an odor, taste, or irritation effect that is not detectable or persistent at concentrations at or below the occupational exposure limit (includes those with no warning properties).[2] Odor thresholds are the most commonly used warnings; irritation levels are second. Nose and throat irritation will work for all respirators, but eye irritation is only useful for full-facepiece respirators.

The organic vapor respirator is not approved for organic vapors with poor warning properties.[5] When the odor or irritation threshold of the substance is greater than the PEL or TLV, and the substance is considered to have poor warning properties, only atmosphere-supplying respirators should be used.

Odor thresholds are determined by having a test panel of people smell an airstream with a known concentration of a test chemical. The panel notes when the material can be smelled. By reporting this for several concentrations, a concentration-response curve can be established. Two types of odor threshold tests can be performed: (1) detection, which is noting that a material is present; and (2) recognition, or being able to identify the material. Normally the concentration responded to by 50% of the panel is reported as the odor threshold. For replacement of the cartridge or canister it is not necessary to be able to identify the chemical by its odor, hence detection threshold is acceptable.

Warning properties can provide some indication to the wearer that the cartridge or canister is reaching the end of its useful service life. Since they rely on human senses, they are not foolproof. Amoore and Hatula reported that "the ability of members of the population to detect a given odor is strongly influenced by the innate variability of different persons' olfactory powers, their prior experience with that odor, and by the degree of attention they accord to the matter."[24] Another concern is that the threshold values reported are typically median values. Therefore, 50% of the respirator wearers can detect a substance's odor only at levels that must exceed the reported threshold.

Odor thresholds can be found in various publications, on material safety data sheets, and in information supplied by various respirator manufacturers. Compilations of chemical lists with odor thresholds also exist.[16,25] Table 36.2 lists odor thresholds for some common chemicals.

It should be noted that odor threshold data reported in the literature differ considerably; commonly the values may range over four orders of magnitude for the same chemical.[16] Selecting the upper end of the odor threshold range provides the greatest chance of classifying the chemical as having poor warning properties. If the range of values is not accurate, however, this approach may not be appropriate. In an attempt to reduce this variability and provide better estimates of the odor threshold, a review and critique of odor threshold data has been published.[16] Because of limitations of odor thresholds, NIOSH has recommended screening tests for workers who wear gas- or vapor-removing respirators to determine their ability to detect the odor below the contaminant's exposure limit.[22]

An end-of-service-life indicator (ESLI) is a system that warns the user when the chemical cartridge or canister is nearing the end of its service life. Currently, the ESLIs in use are passive devices, so the respirator user must periodically check the ESLI status to know when to replace the cartridge or canister. Most ESLIs use a color change indicator. When the indicator color matches the reference color corresponding to exhausted service life, the cartridge or canister must be changed. Typically, ESLIs are used on cartridges and canisters for removal of contaminants that do not provide adequate warning of impending breakthrough to the respirator user. Examples include mercury vapor, ethylene oxide, and carbon monoxide gas.

When a respirator with an ESLI is not available, or the chemical has poor warning properties and an atmosphere-supplying respirator may be impractical because of lack of feasible air supply or need for worker mobility, the only method available for replacing cartridges or canisters before breakthrough is via a cartridge replacement or change-out schedule. This schedule must be based on cartridge service life data, desorption studies, expected concentrations, patterns of use, and duration of exposure.[2] If the cartridges are replaced daily or more frequently, desorption studies may not be necessary.

Given the variability among people with respect to detection of odors and differences in measuring odor thresholds, etc., the best practice is to establish cartridge change-out schedules even for chemicals with adequate warning properties. The warning properties in these cases should be used as a secondary indicator for cartridge change-out.

Cartridge service life is a function of sorbent characteristics, chemical concentration, environmental conditions, and work rate. Service life or the change-out time can be estimated using general guidelines, determined by testing cartridges in the laboratory or in the field, estimated from surrogate respirator carbon tubes, or calculated using breakthrough equations.

If the material has adequate warning properties, estimates of service life can be used to set a cartridge change-out schedule. High-BP materials are collected more efficiently than low-BP materials. Following are guidelines for estimating change-out for organic vapor cartridges:

- If the organic vapor's boiling point is greater than 70°C and its concentration less than 200 ppm, the organic vapor cartridge should last 8 hours at a normal work rate (assuming normal breathing rate).
- Service life is inversely proportional to flow rate.
- If the concentration is reduced by a factor of 10, the service life will only increase by 5.
- Humidity greater than 85% generally reduces service life by 50%.[13]

Using these guidelines, one would expect an organic vapor cartridge to last at least 8 hours for n-butanol (BP > 70°C) at a concentration of 200 ppm, 50% relative humidity, and a normal work rate (30 L/min). At 20 ppm of n-butanol it would be expected to last at least 40 hours (8 hours at 200 ppm × 5 = 40 hours at 20 ppm). At 200 ppm n-butanol, relative humidity at 90%, and a breathing rate of 60 L/min, the guidelines would predict a service life of 2 hours (8 hours × 0.5 [50% reduction at 90% RH] ÷ 2 = 2 hours at 90% RH and work rate of 60 L/min).

If the material does not have adequate warning properties, estimates of service life can only be used to determine whether the material may be an acceptable candidate for further testing. This testing may be done using the actual cartridge in laboratory or field tests or by the use of respirator carbon tubes.

In a laboratory test the time to breakthrough is measured at various concentrations and humidity chosen to represent the workplace. Many commercial labs are equipped to run tests for a fee. Service life studies have been reported in the literature.[17, 26, 27] Table 36.2 lists breakthrough data for several chemicals. Although this information is for single chemicals, it is possible to test more than one chemical simultaneously at concentrations and relative humidity that mimic the work environment. When the testing conditions are not representative, it is more difficult to extrapolate to the workplace.

In a field test the time to breakthrough is measured directly in the workplace. The cartridge or canister is connected to a high-flow pump to simulate breathing (~30 L/min is light work, ~60 L/min is moderate), and samples are collected downstream to determine the time to breakthrough. Two advantages of this testing are that no equipment is needed to reproduce the work environment, and the results are easier to extrapolate to the workplace. The disadvantages include the awkwardness and space limitations of setting up laboratory equipment (e.g., high flow pumps, detectors) in the workplace.

A respirator carbon tube is a small glass tube filled with the charcoal from the respirator cartridge or canister.[28] A breakthrough time for the respirator carbon tube is determined using a standard personal sampling pump to draw workplace air through the tube until the contaminant is detected downstream. Contaminant breakthrough can be determined by using either direct-reading instruments or charcoal sampling tubes. The charcoal tubes are analyzed by a laboratory to establish breakthrough. The breakthrough time is used in conjunction with a bed residence model to predict the cartridge service life. Bed residence time is defined as the time taken by a molecule of air to pass through a packed adsorbent bed.[29] A linear relationship exists between breakthrough time and bed residence time. The bed residence time of the respirator carbon tube and the respirator cartridge can both be calculated, and then the breakthrough time of the carbon tube is used to predict the cartridge breakthrough time. The advantages include a service life estimate under workplace conditions, since these tests can be conducted in the workplace and typical industrial hygiene sampling pumps can be used. This equipment is much more portable than that used in the field tests described above. A disadvantage is having to manually pack the respirator carbon tubes.

Procedures have been developed to estimate breakthrough time for organic vapor cartridges and canisters.[30] The calculation of breakthrough time depends on solvent variables, carbon variables, and ambient conditions. The model currently used is for liquids only, is limited by humidity, and is restricted to single chemicals.

In summary, cartridges and canisters should be replaced earlier than the change-out schedule indicates in any of the following situations:

- If the ESLI shows the specified color change;
- If breakthrough is detected by smell or taste, or by eye, nose, or throat irritation;
- If the shelf life is exceeded; and
- If an OSHA regulation specifies a disposal frequency (e.g., formaldehyde).

If a person is wearing a cartridge or canister that needs replacement, he or she should return to fresh air as quickly as possible. In addition, if uncomfortable heat in the inhaled air is detected or the wearer has a feeling of nausea, dizziness, or ill-health, it is imperative to return to fresh air. (A properly operating cartridge or canister may become warm on exposure to certain gases or vapors, but a device that becomes extremely hot indicates that concentrations greater than the device's limits have been reached.)

Nonroutine Use

Three types of hazardous atmospheres require careful consideration for respirator selection because of the unusual nature of the hazard. These are entries into confined spaces, oxygen-deficient atmospheres, and emergencies. Confined spaces and oxygen-deficient atmospheres may be IDLH.

IDLH

Numerous definitions have been presented for IDLH atmospheres. ANSI defines any atmosphere that poses an immediate hazard to life or poses immediate, irreversible debilitating effects on health as being IDLH.[2] The common theme in all the definitions is that IDLH atmospheres will affect the worker acutely as opposed to chronically. Thus, if the concentration is above IDLH levels, only highly reliable respiratory protective equipment is allowed. The only two devices that meet this requirement and provide escape provisions for the wearer are (1) pressure demand or other positive pressure SCBA rated for 30 minutes or longer; or (2) combination type, pressure demand air-line respirators with auxiliary self-contained air supply.

With the exception of substance-specific standards, OSHA has not established IDLH limits. The most complete set of IDLH limits have been established by NIOSH in its *NIOSH Pocket Guide to Chemical Hazards* for the purpose of respirator selection.[14] Two factors have been considered when establishing IDLH concentrations: (1) The worker must be able to escape within 30 minutes without losing his or her life or suffering permanent health damage; and (2) the worker must be able to escape without severe eye or respiratory irritation or other reactions that could inhibit escape.

A location is considered IDLH when an atmosphere is known or suspected to have chemical concentrations above the IDLH level. A confined space containing less than the normal 20.9% oxygen should also be considered IDLH, unless the reason for the reduced oxygen level is known.[2] Otherwise, according to NIOSH, oxygen levels of less than 19.5% are IDLH.[5] When there is doubt about the oxygen content, the contaminants present, or their airborne levels, the situation should be treated as IDLH. If an error in respirator selection is made, it should be on the side of safety. Thus, in emergency situations, such as a spill where the chemical or its airborne concentration is unknown, one of the above two respirators must be selected.

Concentrations in excess of the lower explosive limit (LEL) are considered to be IDLH. Generally, entry into atmospheres exceeding the LEL is not recommended except for lifesaving rescues. For concentrations at or above the LEL, respirators must provide maximum protection.

Since fire fighting is also considered IDLH by NIOSH, the only practical device providing adequate protection for this activity is pressure demand SCBA.[22] In addition to being NIOSH-approved, the SCBA used for fire fighting should comply with the most current edition of the National Fire Protection Association (NFPA) standard, NFPA 1981.[31] The NFPA requirements exceed the NIOSH requirements in many areas such as flow rate, harness assembly, and fire and heat resistance.

Emergency Procedures

In some industries, where the possibility of a chemical release, process upset, or the need to rescue someone during an emergency may require advance planning, a wide variety of possible conditions requiring the emergency or rescue use of respirators can be envisioned. An adequate emergency and rescue respirator response capability can be achieved through a serious effort to plan for the worst foreseeable consequences of particular malfunctions or mishaps.

Written procedures for the emergency and rescue uses of respirators can be developed by anticipating the likely processes, tasks, and materials that may lead to an emergency. The possible consequences of equipment or power failures, uncontrolled chemical reactions, fire, explosion, or human error should be evaluated. Past occurrences requiring emergency or rescue uses of respirators, as well as conditions that resulted in such respirator applications, can be used to help determine what may be necessary. The procedure should be reviewed by a person who is thoroughly familiar with the particular process or operation, ensuring that appropriate types of respirators are selected; an adequate number should be provided where they may be needed for emergency or rescue use; and the respirators should be stored, maintained, and regularly inspected so that they are readily accessible and operational when needed.

Training

Training is an important aspect of a respirator program. For the safe use of any respirator it is essential that the user be properly instructed in its operation. Supervisors as well as the person issuing respirators must be instructed by a qualified person who has knowledge of respiratory protection and workplace contaminants. Emergency and rescue teams must be given adequate training to ensure proper respirator use as well as covering other issues such as health hazards, work practices, use of other equipment on site, and medical surveillance requirements. All workers need to be trained on initial assignment of a respirator and kept current with annual training as a minimum.

Training Requirements

Each respirator wearer should be given training that includes the following:

- An explanation of the need for the

respirator, including an explanation of the respiratory hazard and what happens if the respirator is not used properly;

- Instructions to inform their supervisor of any problems related to respirator use;
- A discussion of what engineering and administrative controls are being used and why respirators are still needed for protection;
- An explanation of why a particular type of respirator has been selected;
- A discussion of the function, capabilities, and limitations of the selected respirator;
- Instruction in how to don the respirator and check its fit and operation;
- Successful completion of either a qualitative or quantitative fit-test;
- Instruction in respirator maintenance; and
- Instruction in emergency procedures and the use of emergency escape devices and regulations concerning respirator use.

Training on donning the respirator must include an opportunity to handle the respirator, and it should provide instructions for each wearer in the proper fitting of the respirator, including demonstrations and practice in how the respirator must be worn, how to adjust it, and how to determine whether it fits properly. Respirator manufacturers can provide training materials that tell and show how the respirator is to be adjusted, donned, and worn. The training session must also allow time to practice. Hence, a lecture or just showing a videotape is not sufficient unless it is followed up with actual hands-on time. Close, frequent supervision can be useful to ensure that the workers continue to use the respirator in the correct manner. Supervisory personnel should periodically monitor the use of respirators to ensure they are worn properly.

As a minimum, written records of the names of those trained and the dates when the training occurred, who conducted the training, and what was covered must be kept.

Wear Time

Wear time is the percentage of time the respirator is worn during the time it is needed to prevent inhalation of a contaminant. Not wearing a respirator for short periods while it is needed can have a profound effect on overall protection. While a respirator is not worn, the protection factor it provides is 1; that is, the individual is exposed to the ambient contaminant concentration.

The effect of nonwear time can be calculated from the following equation.

Effective Protection Factor =
Work shift time in minutes ÷
(1/APF)(Wear time in minutes) +
Nonwear time in minutes (2)

NOTE: The exposure during wear time can be reduced by the APF or any assumed level of protection.

For example, if a person removes his or her respirator for 1 minute to talk during a task that takes 1 hour, the wear time is 98%. If the person uses a respirator with a level of protection of 1000, the effective level of protection actually achieved is 56, including this 1 minute nonwear time. In training it is important that people understand the effect of nonwear time on the level of performance that can be achieved. As nonwear time increases for any respirator, the protection levels for all respirators approach 1. When poor wear habits are practiced, the effective protection levels of an SCBA and a half-mask may be identical (see Figure 36.9).

Respirator Fit-Testing

Each respirator wearer of tight-fitting devices must be provided with a respirator that fits. To find the respirator that fits, the worker must be fit-tested. There are both qualitative and quantitative fit-test methods.

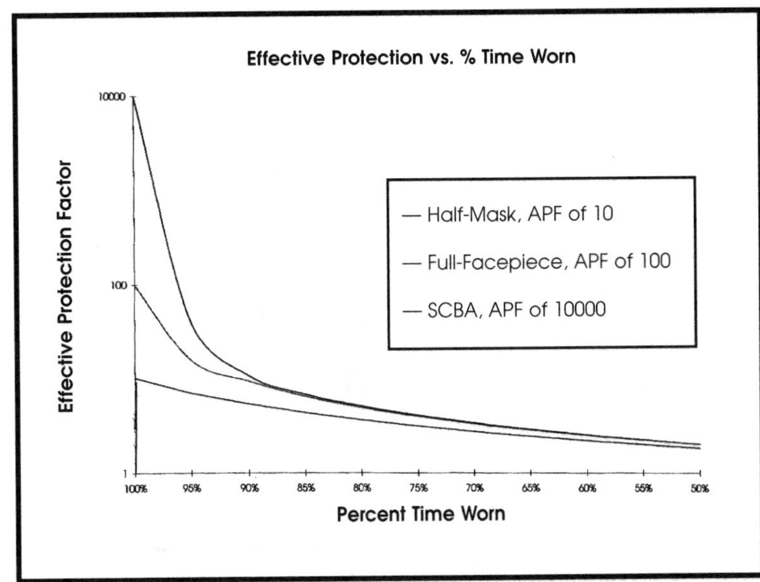

Figure 36.9 — Relationship of respirator effective protection factor to time worn.

In addition, each respirator wearer must be required to check the seal of the respirator by appropriate means before entering a harmful atmosphere. Each respirator manufacturer provides instructions on how to perform these fit checks. A fit check is a test conducted by the wearer to determine whether the respirator is properly adjusted to the face. The procedures may vary slightly from one respirator to another because of differences in construction and design. In either case the employee is either checking for pressure or flow of air around the sealing surface.[32] Fit checks are not substitutes for qualitative or quantitative fit-tests. Care must be taken in conducting fit checks. Respirator wearers must be given thorough training in carrying out these tests and in how to recognize the test end point.

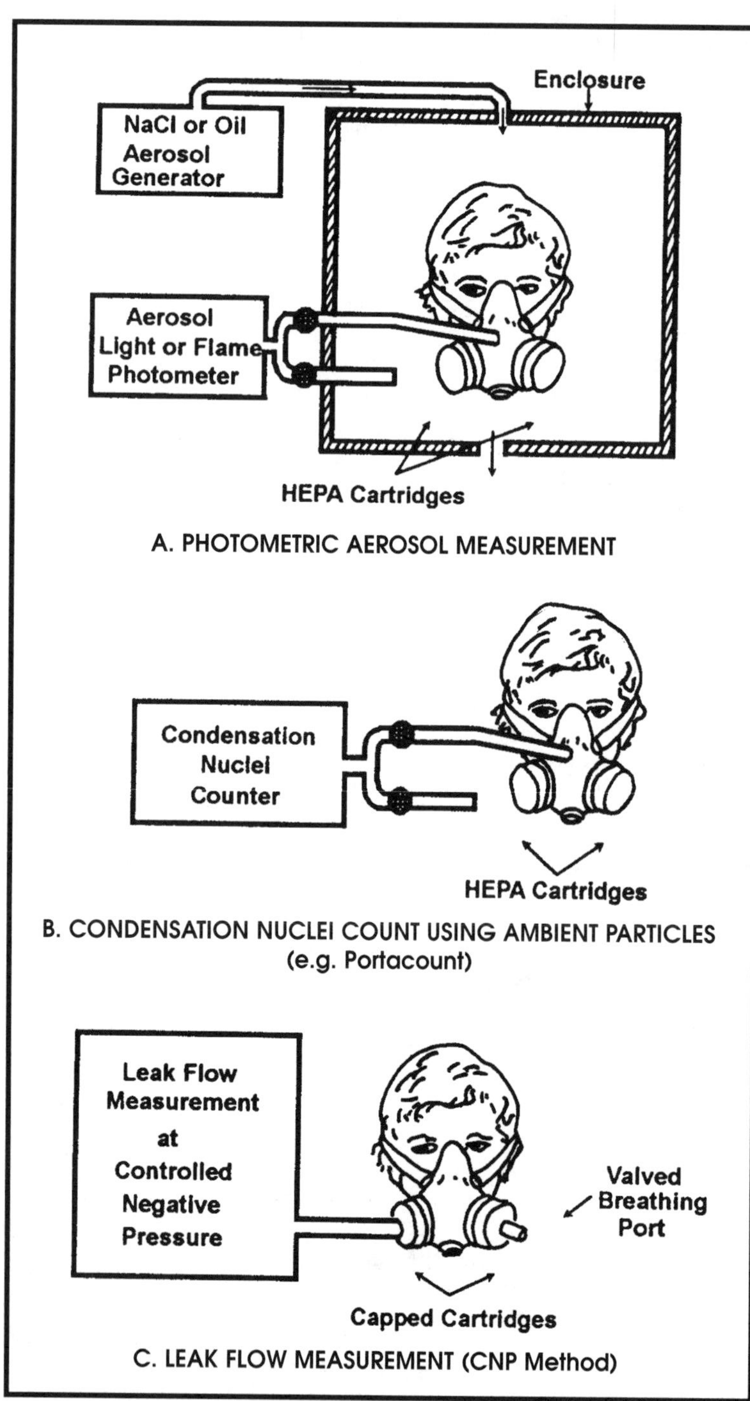

Figure 36.10 — Schematic representation of commercially available QNFTs. (Adapted from Reference 37).

Respirator fit-testing is conducted to determine whether a particular model and size of mask fits an individual's face. Prior to the test the person should be trained on proper respirator use and donning, and the purpose and procedures for the fit-test. Since respirator fit-testing is done to evaluate face-seal leakage only, the respirator must be equipped with the appropriate filters or cartridges to prevent penetration through the air-purifying element. Either qualitative or quantitative fit-tests can be used;[2] each has its advantages and disadvantages.

Quantitative Fit-Tests (QNFTs)

In a quantitative fit-test, either air leakage is determined (known as the controlled negative pressure technique [33]) or a test substance is measured both inside and outside the respirator while it is being worn. The test result can be reported as percentage leakage or the ratio of test agent concentration outside the respirator (C_o) to the concentration inside (C_i). The ratio of C_o to C_i is called a fit factor. A number of methods are available commercially that use oil mists,[34] salt mists,[35] or ambient particles[36] as test substances. Figure 36.10 shows schematics of the commercially available QNFT methods.

There are advantages and disadvantages associated with a quantitative fit-test. The advantages include a direct and objective measurement of fit and, with some methods, real-time documentation. The disadvantages include the cost of equipment needed (typically $4000–$10,000), maintenance, training in the operation of the sophisticated test equipment, and the need for specially probed respirators or adapters. Methods using test aerosols require HEPA filters (N-, R-, or P-100).

The minimum fit factor required to pass a quantitative fit-test is a value 10 times the APF of the respirator.[2] For a half-mask respirator with an APF of 10, the minimum fit factor required is 100. The factor of 10 is used to account for differences in performance that occur when a respirator is actually used vs. the measured performance during the fit-test.

Qualitative Fit-Tests (QLFTs)

In a qualitative fit-test, a face-seal leakage evaluation similar to QNFT is made, but no direct measurements of the test and leak concentrations are made. The commonly performed QLFTs use a person's sense of taste, smell, or irritation to determine whether facepiece leakage occurs. A test procedure is used to control the concentration outside the respirator.

QLFTs are generally simple to perform, require only a few pieces of inexpensive equipment, and can be done in almost any location. The tests do not give a numerical indication of fit. The person being tested must be honest when he or she indicates whether the test material is detected.

Examples of QLFT agents used today include odorous vapor tests (e.g., isoamyl acetate [IAA]), taste tests (e.g., saccharin), and irritant smoke (e.g., titanium and stannic chloride).(37) Early IAA QLFTs used a stencil brush, cloth, or small tube containing IAA waved around the respirator. If no odor was detected, the fit was presumed to be adequate. In the irritant smoke test, the fumes from a ventilation smoke tube were directed at the person wearing the respirator. If significant leakage occurred, the fume would cause coughing or irritation.

These early tests did not contain well-documented procedures for their performance. There was no guidance indicating how close the stencil brush should be to the respirator, the effect of room ventilation on the challenge agent concentration, and assurance the test subject could detect the low levels of the challenge agent.

As a result of new research, a number of new fit-test protocols were developed that carefully controlled the test variables. The instructions were specific for determining the ability of a person to detect the test material and for the generation of the test atmosphere. For half-mask respirators, qualitative and quantitative fit-tests are expected to provide approximately the same level of protection, since the pass/fail levels for the qualitative fit-test protocols discussed below are based on the same pass/fail fit factor of 100 used for a quantitative fit-test.

IAA Fit-Test

The IAA fit-test protocol requires three jars, water, IAA, and a large plastic bag (55-gallon drum liner).(38) The respirator used must be equipped with organic vapor cartridges. Since the respirator is not probed, a worker's own respirator can be used.

The sense of smell is used as the indicator of fit, so careful controls are necessary to prevent odor fatigue and to determine whether the test subject can smell low levels of the IAA. To determine that a test subject can smell IAA at low concentrations, a sensitivity test using dilute IAA is conducted. A stock is made by adding 1 cc of IAA to 800 mL of water. From the stock solution, 0.4 cc are added to 500 mL of water in a 1-L jar. This produces a 1-ppm concentration of IAA in the head space of a jar. The person is asked to choose between two jars: one with water and IAA, the other with only water. If the jar with IAA is chosen, it is presumed the person can smell IAA at 1 ppm. If the water jar is chosen, the person cannot be fit-tested by this method.

A test enclosure is made from a large plastic bag (55-gallon drum liner). By placing 0.75 mL of IAA on a piece of paper towel suspended inside the enclosure, a concentration of 150 ppm is produced. Theoretically, people with fit factors less than 150 will fail this fit-test. It has been shown that 95% of the people passing would have fit factors greater than 100 when compared with QNFT.(39)

Saccharin Fit-Test

A similar QLFT protocol has been developed using a saccharin mist.(40) This test requires two solutions of sodium saccharin, one that is essentially a saturated solution in water (83 g saccharin in 100 cc of water), the other a 1:100 dilution of the saturated solution. The saccharin fit-test can be used for any respirator equipped with any type particulate filter. A worker's own respirator can be used with no modification other than the addition of a prefilter for those respirators not equipped with filters.

A person's ability to taste saccharin is used as the criteria to determine whether the fit is acceptable. This ability is tested by having the person place a small test hood over his or her head while not wearing a respirator. The dilute (sensitivity) solution of saccharin is sprayed into the hood with an inhalation nebulizer. The number of squeezes (either 10, 20, or 30) of the nebulizer bulb required before the person tastes the saccharin is noted. If the person cannot taste the saccharin after 30 bulb squeezes, another type of fit-test is needed.

Next, while wearing the respirator to be tested, the person places the hood over his or her head and is exposed to a solution of saccharin that is 100 times stronger than the taste sensitivity solution. The same number of bulb squeezes noted in the taste sensitivity portion of the test are used. Thus, the saccharin test has a theoretical passing fit factor of 100. When compared with QNFT, at least 95% of those passing achieve a fit factor of 100.[41]

Irritant Smoke Fit-Test

The irritant smoke fit-test uses a ventilation smoke tube containing stannic chloride on pumice. When exposed to air, the metal chloride reacts with moisture to form a mixture of hydration products and hydrochloric acid. This fume is irritating and will cause people to sneeze, cough, or react in some fashion to the irritation.

To perform the test, the person being tested is exposed to a slight concentration of the irritant to determine whether he or she is irritated by it.[38] Some people have been found not to react to the material. Next, while wearing the respirator, seated in front of a laboratory hood, they are exposed to a concentration of the fume from approximately 1 m. The distance is decreased until the end of the smoke tube is approximately 6 in. from the sealing surface of the respirator. A pump delivering 200 mL/min is connected to the tube to generate a steady stream of smoke.[38] Few data on the concentrations generated during the testing or sensitivity check phase of the test have been published.

Bitter Aerosol Fit-Test

Another QLFT has been developed using denatonium benzoate, commonly known as Bitrex™.[42] Bitrex has been used as a taste aversion agent in household liquids (to keep children from drinking them). The fit-test is based on the saccharin fit-test protocol.

A dilute solution of Bitrex in a 5% sodium chloride solution is used to predict whether people can detect the bitter taste. A stronger solution is used for the testing phase. The stronger solution is not 100 times more concentrated than the sensitivity solution because the response to the taste of Bitrex appears to be exponential. Empirical studies conducted to evaluate the taste response effects at various Bitrex concentrations indicated that small increases in solution concentration caused greater increases in taste response than might be expected.[43] A comparison study showed that the Bitrex fit-test performs equally with the saccharin fit-test and can be used to assess fit factors of 100.[42]

Test Exercises

To mimic the movements of a person in the workplace, a series of test exercises is normally performed during a fit-test. The following six exercises are recommended: normal breathing, deep breathing, moving head side to side, moving head up and down, talking, and normal breathing. Each exercise is performed for 30–60 seconds while standing still. Some OSHA substance-specific standards (e.g., formaldehyde) may require additional exercises.[23]

Respirator Sealing Problems

A respirator equipped with a facepiece (tight- or loose-fitting) must not be worn if facial hair comes between the sealing periphery of the facepiece and the face or if facial hair interferes with valve function. Both negative and positive pressure tight-fitting respirator performance has been shown to be reduced when the respirators are worn over facial hair.[44] Only respirators equipped with hoods or helmets are acceptable with facial hair.[2]

In some industrial environments safety glasses or goggles must be worn to protect the eyes from flying objects. With a half-mask respirator, safety glasses, regular eyeglasses used to improve vision, or goggles can interfere with the fit of the respirator where it sits on the bridge of the nose. Wearing glasses with temple bars that pass through the sealing surface of a full facepiece respirator is not acceptable.

For half-mask respirators, a fit-test is used to determine whether the respirator fits properly. The subject of the fit-test should wear any glasses, goggles, or other protective equipment normally used while wearing a respirator that may interfere with how the respirator fits. The fit-test is used to demonstrate that the glasses or other equipment do not interfere with the function of the respirator.

For full-facepiece respirators, the options are to use contact lenses or special eyeglasses that mount inside the facepiece. Most manufacturers supply a special spectacle kit that can be used. Research has shown that people can use contact lenses with respirators without difficulty.[45]

Standards

Respirator regulations fall into two categories: respirator testing or approval standards, and respirator use standards. While they appear to address two different concerns, both types of standards affect the use of respirators and have implications for respiratory protection programs.

Respirator Test Standards

The primary objective of a respiratory protective equipment test standard or certification program is to assure users that a minimum level of performance is achieved by the respirator. This is important since this equipment is used to guard health or protect life. A second objective of the certification program is to supply the user with respirator use and limitation information. Many of the selection restrictions placed on the various respiratory protective devices are a result of performance levels established by the certification tests.

Many countries and standards organizations have established respirator testing standards. In addition, countries that do not have their own testing standards may recognize certified or approved respirators developed elsewhere. Table 36.5 lists some of the countries with respiratory protective device certification programs.

Table 36.5 —
Countries with Respirator Testing Standards

Country	*Standard Name*	*Location*
Australia	AS/NZS 1716:1994	Standards Australia, 1 The Crescent, Homebush NSW 2140 Australia
Brazil	Several standards covering many types of respiratory protective devices	ABNT, Av. Treze de Maio, 13-28.o andar, CEP20003 – Caixa Postal 1680, Rio de Janerio – RJ Brazil
CEN[A]	Several standards covering all respiratory protective devices	CEN, Cenral secretariat, rue de Stassart,36 B-1050 Brussels, Belgium. For English: British Standards Institute, Linford Wood, Milton Keynes, MK14 6LE Tel:44-(0)1908 221166
China	Various standards	Mr. Lu Haiyan or Mr. Xu Shaoduo, Division Chief, Bureau of Occupational Safety and Health Ministry of Labour, 12, Heipingli Zhongjie, Beijing 100716, P.R.C.
Japan	Standards for Dust Respirators Notification No. 19 of Ministry of Labor (March 30, Showa 63-nen (1988)) Standards for Gas Masks Notification No.68 of MOL (September 26, Heisei 2-nen (1990)) Notification No. 1 of MOL (January 8, Heisei 8-nen (1996)) Revised	Ministry of Labor 2-2, Kasumigaseki 1-chome Chiyoda-ku Tokyo 100, Japan
	Various standards	Japanese Standards Association, 1-24, Akasaka 4-chome, Minato-ku, Tokyo 107 Japan
Korea	Standard for Dust Respirators Notification No. 90–71 of Ministry Corporation), (Revised Dec. 28, 1990) Standard for Gas Masks Notification No.91–82 of Ministry of Labor (Dec. 7, 1991) Standards for Supplied Air Respirators Notification No. 95–12 of Ministry of Labor (Jun. 17, 1995)	KISCO (Korea Industrial Safety Corp.), 34-4, Gusan-dong, Bupyong-gu, Incheon city, Zipcode: 403-120
Mexico	NOM–116–STPS–1994	STPS, Secretaria "B", Direccion General de Seguridad e Higiene en el Trabajo, Direccion de Normalizacion e Investigacion, Avenida Azcapotzalco la Villa no. 209, Barrio de Santo Tomas, Delegacion Azcapotzalco, C.p. 02020 Mexico D.F.
New Zealand	AS/NZS 1716:1994	Standards New Zealand, Standards House, 155 The Terrace, Wellington 6020, New Zealand
Russia	Various standards	Gosstandart of Russia, c/o Ms. N. Tinofeeva, 3, Electrichesky per., Moscow, 123856, Russia
South Africa	SABS 1455:1988 Standard Specification Part I-IV	South African Bureau of Standards, Private Bag X191, Pretoria, South Africa 0001
United States	42 CFR 84	NIOSH, Certification and Quality Assurance Branch, Division of Safety Research, 944 Chestnut Ridge Road, Morgantown, WV 26505

[A]Standards cannot be bought from CEN; they can only be purchased from the member associations in the language of the country. Committee for European Normalization members: Austria, Belgium, Denmark, Finland, France, Germany, Greece, Iceland, Ireland, Italy, Luxembourg, Netherlands, Norway, Portugal, Spain, Sweden, Switzerland, and United Kingdom.

Table 36.6 — Countries with Respirator Use Standards

Country	Standard Name	Location
Australia	AS/NZS 1715:1994	Standards Australia, 1 The Crescent, Homebush NSW 2140 Australia
Brazil	Respiratory Protection Program—Selection and Use of Respirators	Fundacentro, Rua Capote Valente, 710, São, Paulo, Brazil
Canada	Selection, Use and Care of Respirators, Z94.4–93, Compressed Breathing Air and Systems, CAN3–Z180.1–M85	Canadian Standards Association, 178 Rexdale Boulevard, Rexdale (Toronto), Ontario, Canada M9W 1R3,
New Zealand	AS/NZS 1715:1994	Standards New Zealand, Standards House, 155 The Terrace, Wellington 6020, New Zealand
South Africa	SABA 0220:1988 Code of Practice, The Use, Care and Maintenance of Respiratory Protective Equipment	South African Bureau of Standards, Private Bag X191, Pretoria, South Africa 0001
United States	29 CFR 1910.134	OSHA, 200 Constitution Ave. NW, Washington, DC 20210
	ANSI Z88.2–1992	ANSI, 11 West 42nd Street, New York, NY 10036
Venezuela	Covenin 1056/I-91	Norma Venezolana Covenin, Comision Venezolana de Normas Industriales, Ministerio de Fomento, Av. Andres Bello Edif. Torre Fondo Comun Piso 11, Caracas, Venezuela

Respirator Use Standards

Various standards, regulations, and voluntary use standards exist for proper use of respirators. Most voluntary standards are consensus standards where a group of respirator experts meet to establish a standard of practice. Table 36.6 lists some of the organizations with use standards. These organizations may also have established standards relating to other areas of respiratory protection. For example, the Canadian Standards Association (CSA) has developed a standard for compressed breathing air and systems,[(46)] and ANSI has established standards for medical evaluations of respirator wearers.[(47)]

In the United States various government agencies have established respirator use regulations, including OSHA,[(1)] the Mine Safety and Health Administration (MSHA),[(48–50)] and the Nuclear Regulatory Commission (NRC).[(51)]

In addition to these general regulations, OSHA has established substance-specific standards for such contaminants as asbestos, lead, benzene, and cadmium, which contain specific respiratory protection provisions for these substances.

Summary

The proper selection and use of respiratory protection is essential to controlling exposures with respirators. A complete respiratory protection program must be established and implemented that is based on knowledge of the performance limitations and capabilities of the respirators and the appropriate standards and regulations for their use.

The information presented here is not intended to be all-inclusive in content or scope. A description of common respirators explains basic modes of operation. This information is key to proper selection of respirators. The information provided in Table 36.2 can be used to illustrate many of the selection principles discussed. A complete summary of the OSHA regulations was not attempted as they are not followed worldwide. For additional information the reader should consult the references and regulations and standard sources listed.

References

1. "Respiratory Protection," *Code of Federal Regulations* Title 29, Part 1910.134. 1994. pp. 413–417.
2. **American National Standards Institute:** *American National Standard for Respiratory Protection* (ANSI Z88.2). New York: American National Standards Institute, 1992.
3. **Canadian Standards Association:** *Z94.4–93, Selection, Use, and Care of Respirators.* Rexdale, Ontario, Canada: Canadian Standards Association, 1993.
4. **Colton, C.E., L.R. Birkner, and L.M. Brosseau (eds.):** *Respiratory Protection: A Manual and Guideline.* 2nd Ed. Fairfax, VA: American Industrial Hygiene Association, 1991.

5. "Approval of Respiratory Protective Devices," *Code of Federal Regulations* Title 42, Part 84. 1996. pp. 528–593.
6. **Japuntich, D.A.:** Respiratory particulate filtration. *J. Int. Resp. Prot. 2*:137–169 (1984).
7. **Ackley, M.W.:** Degradation of electrostatic filters at elevated temperature and humidity. In *World Filtration Congress III*. Croydon, England: Upland Press, 1982. pp. 169–176.
8. **Stevens, G.A., and E.S. Moyer:** "Worst case" aerosol testing parameters: I. Sodium chloride and dioctyl phthalate aerosol filter efficiency as a function of particle size and flow rate. *Am. Ind. Hyg. Assoc. J. 50*:257–264 (1989).
9. **Brown, R.C.:** *Air Filtration*. Oxford, England: Pergamon Press, 1993.
10. **Hinds, W.C., and P. Bellin:** Effect of facial-seal leaks on protection provided by half-mask respirators. *Appl. Ind. Hyg. 3*:158–164 (1988).
11. **National Institute for Occupational Safety and Health (NIOSH):** *NIOSH Guide to the Selection and Use of Particulate Respirators Certified Under 42 CFR 84* (DHHS/NIOSH Pub. No. 96–101). Washington, DC: U.S. Department of Health and Human Services/NIOSH, 1996.
12. **Moyer, E.S.:** Review of influential factors affecting the performance of organic vapor air-purifying respirator cartridges. *Am. Ind. Hyg. Assoc. J. 44*:46–51 (1983).
13. **Nelson, G.O.:** " Rules of Thumb for Cartridge Service Life." July 29, 1996. [Private communication]. Miller-Nelson Research, 8 Harris Court, Suite C-6, Monterey, CA 93940.
14. **National Institute for Occupational Safety and Health (NIOSH):** *NIOSH Pocket Guide to Chemical Hazards* (DHHS/NIOSH Pub. No. 90–117). Washington, DC: U.S. Department of Health and Human Services/NIOSH, 1990.
15. **American Conference of Governmental Industrial Hygienists (ACGIH):** *Threshold Limit Values (TLVs) for Chemical Substances and Physical Agents and Biological Exposure Indicies (BEIs)*. Cincinnati: ACGIH, 1995.
16. **American Industrial Hygiene Association (AIHA):** *Odor Thresholds for Chemicals with Established Occupational Health Standards*. Fairfax, VA: AIHA, 1989.
17. **Nelson, G.O., and C.A. Harder:** Respirator cartridge efficiency studies: VI. Effect of concentration. *Am. Ind. Hyg. Assoc. J. 37*:205–216 (1976).
18. **Yoon, Y.H., J.H. Nelson, J. Lara, C. Kamel, and D. Fregeau:** Effect of solvent mixtures on service life. *Am. Ind. Hyg. Assoc. J. 52*:65–74 (1991).
19. **Compressed Gas Association:** *Commodity Specification for Air* (ANSI/CGA G7.1). Arlington, VA: Compressed Gas Association, Inc., 1989.
20. **Perez, C., and S.C. Soderholm:** Some chemicals requiring special consideration when deciding whether to sample the particle, vapor, or both phases of an atmosphere. *Appl. Occup. Environ. Hyg. 6*:859–864 (1991).
21. **Soderholm, S.C.:** Particle and Gas Phase Interactions in Air Sampling. In *Air Sampling Instruments for Evaluation of Atmospheric Contaminants*, 8th Ed. Cincinnati: American Conference of Governmental Industrial Hygienists, 1995. pp. 67–80.
22. **National Institute for Occupational Safety and Health (NIOSH):** *Respirator Decision Logic* (DHHS/NIOSH Pub. No. 87–108). Washington, DC: U.S. Department of Health and Human Services/NIOSH, 1987.
23. "Formaldehyde," *Code of Federal Regulations* Title 29, Part 1910.1048. 1994. pp. 411–443.
24. **Amoore, J.E., and E. Hautala:** Odor as an aid to chemical safety: odor thresholds compared with threshold limit values and volatilities for 214 industrial chemicals in air and water dilution. *J. Appl. Toxicol. 3*:272–290 (1983).
25. **Fazzalari, F.A. (ed.):** *Compilation of Odor and Taste Threshold Value Data*. Philadelphia: American Society for Testing and Materials, 1978.
26. **Yoon, Y.H., and J.H. Nelson:** Breakthrough time and adsorption capacity of respirator cartridges. *Am. Ind. Hyg. Assoc. J. 53*:303–316 (1992).
27. **Henry, N.W., III and R.S. Wilhelme:** An evaluation of respirator canisters to acrylonitrile vapors. *Am. Ind. Hyg. Assoc. J. 40*:1017–1022 (1979).
28. **Cohen, H.J., and R.P. Garrison:** Development of a field method for evaluating the service life of organic vapor cartridges: results of laboratory testing using carbon tetrachloride. *Am. Ind. Hyg. Assoc. J. 50*:486–495 (1989).

29. **Ackley, M.W.:** Residence time model for respirator sorbent beds. *Am. Ind. Hyg. Assoc. J. 46*:679–689 (1985).
30. **Wood, G.O.:** Estimating service lives of organic vapor cartridges. *Am. Ind. Hyg. Assoc. J. 55*:11–15 (1994).
31. **National Fire Protection Association (NFPA):** *Open-Circuit Self-Contained Breathing Apparatus for Firefighters* (NFPA 1981). Quincy, MA: NFPA, 1992.
32. **Myers, W.R., M. Jaraiedi, and L. Hendricks:** Effectiveness of fit check methods on half mask respirators. *Appl. Occup. Environ. Hyg. 10*:934–942 (1995).
33. **Cruthchfield, C.D., M.P. Eroh, and M.D. VanErt:** A feasibility study of quantitative respiratory fit testing by controlled negative pressure. *Am. Ind. Hyg. Assoc. J. 52*:172–176 (1991).
34. **Hyatt, E.C., J.A. Pritchard, and C.P. Richards:** Respirator efficiency measurement using quantitative DOP man tests. *Am. Ind. Hyg. Assoc. J. 33*:635–643 (1972).
35. **Hounam, R.F., D.J. Morgan, D.T. O'Connor, and R. J. Sherwood:** The evaluation of protection provided by respirators. *Ann. Occup. Hyg. 7*:353–363 (1964).
36. **Willeke, K., H.E. Ayer, and J.D. Blanchard:** Methods for quantitative respirator fit testing with aerosols. *Am. Ind. Hyg. Assoc. J. 42*:121–125 (1981).
37. **Han, D.H., K. Willeke, and C.E. Colton:** Quantitative fit testing techniques and regulations for tight-fitting respirators: current methods measuring aerosol or air leakage, and new developments. *Am. Ind. Hyg. Assoc. J. 58*:219–228 (1997).
38. "Lead," *Code of Federal Regulations* Title 29, Part 1910.1025. 1994. pp. 139–175.
39. **Nelson T.J., O.T. Skredtvedt, J.L. Loschiavo, and S.W. Dixon:** Development of an improved qualitative fit test using isoamyl acetate. *J. Int. Soc. Respir. Prot. 2*:225–248 (1984).
40. "Qualitative Fit Testing Protocols Under the Lead Standard" (Posthearing Data Submission, Docket No. H–049A). 3M, Box 33275, Building 275-6W-01, St. Paul, MN 55144, Oct 23, 1981.
41. **Marsh J.L.:** Evaluation of irritant smoke qualitative fitting test for respirators. *Am. Ind. Hyg. Assoc. J. 45*:371–376 (1984).
42. **Mullins, H.E., S.G. Danisch, and A.R. Johnston:** Development of a new qualitative test for fit testing respirators. *Am. Ind. Hyg. Assoc. J. 56*:1068–1073 (1995).
43. **Mullins, H.E.:** "Taste Response of Human Taste Panels to Bitrex." April 2, 1996. [Private communication.] 3M OH&ESD, Building 260-3A-07, St. Paul, MN 55144.
44. **Stobbe T.J., R.A. daRoza, and M.A. Watkins:** Facial hair and respirator fit: a review of the literature. *Am. Ind. Hyg. Assoc. J. 49*:199–2.4 (1988).
45. **Lawrence Livermore National Laboratory:** *Is it Safe to Wear Contact Lenses with a Full Facepiece Respirator?* by R.A. daRoza and C.S. Wearver (UCRL–53653). Livermore, CA: Lawrence Livermore National Laboratory, 1986.
46. **Canadian Standards Association:** *Compressed Breathing Air and Systems* (CAN3–Z180.1–M85). Rexdale, Ontario, Canada: Canadian Standards Association, 1985.
47. **American National Standards Institute:** *American National Standard for Respiratory Protection—Respirator Use—Physical Qualifications for Personnel* (ANSI Z88.6). New York: American National Standards Institute, 1984.
48. "Control of Exposure to Airborne Contaminants," *Code of Federal Regulations* Title 30, Part 56.5005. 1993. pp. 338.
49. "Control of Exposure to Airborne Contaminants," *Code of Federal Regulations* Title 30, Part 57.5005. 1993. pp. 409–410.
50. "Respiratory Equipment," *Code of Federal Regulations* Title 30, Subpart D. 1993. pp. 493–494.
51. "Use of Individual Respiratory Protection Equipment," *Code of Federal Regulations* Title 10, Part 20.1703. 1993. pp. 385–386.

Section 6

Program Management

Chapter 37

Outcome Competencies

After completing this chapter, the user should be able to:

1. Define underlined terms used in this chapter.
2. Explain total quality management (TQM).
3. List general components of TQM.
4. Discuss goal setting in occupational and environment health.
5. Discuss the rules of accountability and measurability.
6. List total quality rules for communication.
7. Discuss kaizen or continuous improvement.

Prerequisite Knowledge

None

Key Terms

activity-based costing (ABC) • benchmark • brainstorming • catchball • continuous improvement • customer • cycle time • failure modes and effects analysis • fault tree analysis • flowchart • House of Quality • inspiring • integrated product development (IPD) • interpreting • kaizen • mental models • milestones • mobilizing • operating plan • pareto analysis • personal mastery • policy deployment • process • quality function deployment (QFD) • queue time • reflective listening • rework • robust design • root cause • shaping • shared vision • strategic plan • supplier • systems thinking • Taguchi experiments • team learning • total quality • total quality control (TQC) • total quality environmental management (TQEM) • total quality management (TQM) • vision statement • what gets measured gets managed

Key Topics

I. Introduction
  A. Integration into Organizational Priorities
  B. Program Planning
  C. Rationale for EHS Program Elements
  D. The Role of Management Skills in Occupational Hygiene
  E. Management from a Support Position
  F. Career Planning

II. Management Theory
  A. Organizational Structure
  B. Leadership

III. Modern Management Practices and Application to Occupational Hygiene
  A. Total Quality
  B. Integrated Product Development
  C. Quality Function Deployment
  D. Benchmarking
  E. Resource Management
  F. Cost Accounting
  G. Policy Deployment
  H. Goal Setting
  I. Negotiation Skills
  J. Team Dynamics

37

Program Management

Alan Leibowitz, CIH, CSP

Introduction

Occupational hygiene is, at its core, a technical profession. Given this fundamental orientation, occupational hygiene has historically tended to focus on the sampling and analytical aspects of the profession. While appropriate for a profession charged with using environmental evaluations to protect life and environment, this technical focus has left many practitioners ill-prepared for a future where interpersonal and managerial skills are as highly valued as technical prowess; a future where possession and transmission of technical information is not sufficient to ensure that concerns and recommendations are acted on.

If occupational hygiene is to progress as a profession, its practitioners must prepare for a future where team contributions and performance will eclipse the value of the individual technical contributor. To participate successfully in this environment, it is increasingly important that occupational hygienists, and more broadly environmental health and safety (EHS) professionals, be capable of communication on many levels and to various audiences.

To ensure that their recommendations are appropriately received and evaluated, occupational hygienists must first understand their working environment. Whether they practice as outside consultants, educators, or in industry, they must understand the fundamental rituals observed by their customers. Many programs have failed because of an inability to recognize how decisions are made in an organization and how change is initiated and managed.

Change comes only when the benefit to an organization can be clearly articulated and understood. The pace and magnitude of change in business today is increasing at a rapid rate. To participate in this process, successful hygienists must function as "agents of change" to influence their customers and organizations toward the proper course in the protection of health and the environment. As Tom Peters states in his book *Thriving On Chaos*,[1] "Today, loving change, tumult, even chaos is a prerequisite for survival, let alone success."

Integration into Organizational Priorities

A key attribute of a successful EHS program is a high level of distribution of responsibility for program actions among functional departments. When EHS professionals are solely responsible for the success or failure of all program elements it is much more difficult, if not impossible, to succeed. Every organization should have a clearly stated EHS policy that articulates the organization's commitment to EHS and clearly distributes responsibilities to appropriate functional areas within the organization. An example corporate EHS policy is provided in Appendix A. To ensure that all departments understand and accept their EHS responsibilities, it is important that EHS personnel become a part of the company culture. The occupational hygienist must speak the language of the organization and participate in the important rituals that establish internal priorities and goals.

Each aspect of an EHS program requires integration with appropriate functional departments. Figure 37.1 illustrates the relationship between the essential tools of EHS and the critical processes they support. For each tool a potential supporting

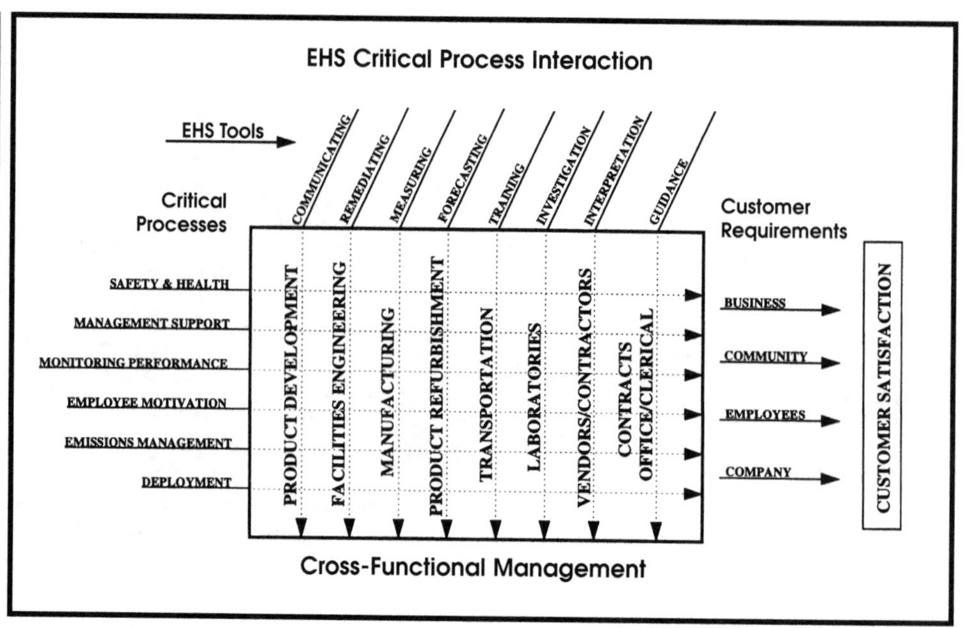

Figure 37.1 — Example cross-functional EHS program support relationship analysis (adapted from a matrix developed by D. Polzo, ITT Avionics).

function is identified. Involvement of all functional aspects of an organization are critical to an effective EHS program as measured by customer satisfaction.

An occupational hygienist's first goal should be to understand the larger organization. The professional must identify previously established priorities and how EHS issues fit into them. For example, if the organization's current goals include growth in international sales, the EHS department must have a goal to understand regulations in the markets to be explored. If this goal also includes an increase in production requirements, then the EHS department must be prepared to review production plans for new equipment requirements, facility modifications, and training and sampling needs. Often the most critical aspect of production changes, from an EHS perspective, is the evaluation of new regulatory requirements that may be introduced. In some cases obtaining a permit can take many months, and if the need is not properly anticipated, this aspect can delay important programs. This does not endear the EHS department to the rest of the organization. Even development of appropriate personal protective equipment strategies can take time if sampling is required on a phased approach to determine what exposures may be anticipated before full production is initiated.

Program Planning

To avoid these unacceptable delays and interference with global goals, the occupational hygienist must be aware of the plans of the organization both in general terms and, where IH issues might exist, in specific detail. Later in this chapter a discussion of integrated product development will describe how IH issues can be included in specific process design planning. Before the specific detail of a particular process is established, the occupational hygienist should participate in the goal-setting process for the organization. Most often this takes the form of long-range or strategic planning and near-term or operations planning.

An organization's strategic plan generally describes where the organization would like to be in the next three to five years. As a general rule, strategic plans are likely to change each year as new information becomes available and as the marketplace evolves. The dynamic nature of the organizational planning process requires constant vigilance to ensure that EHS issues are included and are kept up-to-date. This part of the planning process has a longer time horizon than development of an operating plan, which generally looks one to two years out. The longer viewpoint requires management to assess potential modifications to their current strategy in terms of opportunities or threats that might arise in the future. This planning includes an evaluation of future resource needs, particularly if there is a requirement to move the organization into new areas. The occupational hygienist can be most valuable when the organization is evaluating entry into previously unexplored areas. EHS should provide input to the strategic plan based on goals and current expectations of impacts on the organization from regulatory and technical trends. Some organizations include a section in their written strategic plans that focuses on EHS trends and assumptions. Others simply include EHS issues as part of the body of their plans where principle strategic issues have EHS elements.

Operating plans establish the organization's specific short-term goals. These plans are much more fixed than strategic plans, and considerable resources are expended to ensure that the goals established have a strong chance of success. A company's operating plan usually focuses on significant opportunities and challenges faced in the near term. Resource-intensive or high-profile EHS activities should be included in the written plan.

The deployment of both strategic and operating plan objectives throughout an organization is critical to its success. There are many different routes to help ensure that top-level goals are aligned throughout an organization.[2] In larger organizations the procedure is often highly structured. Initial planning is first accomplished on the corporate level, then at business units, and then in functional departments within the unit. At the corporate level, broad initiatives are established that provide a direction for the rest of the organization. Business units must then look at what areas of their operations can potentially support the larger organization's goals. At the business unit level, detailed financial goals are established and action plans are developed delineating specific functional responsibilities. Finally, each functional area must look at its plans and determine how its intended actions support the broader goals of the organization. On this bottom tier of the planning process, the nature and sequence of specific required actions are established.

While not every action of the EHS department must fit the global plan, it is important that the link between EHS and broad operational goals is clearly established throughout the organization. This linkage must be articulated in a manner that supports acceptance of EHS integration for all of management including members of the EHS department. EHS managers must participate in planning and deployment to ensure that EHS issues are considered at all stages of the process. Since these documents become the bibles of the organization, it is also very helpful if specific EHS actions are included in the plans. This reminds the organization that EHS is part of its responsibilities to the business as a whole and is not an added responsibility to be managed when time permits. A strategic plan should include an evaluation of anticipated impacts for the plan period. These impacts might, for example, include new regulations or enforcement initiatives. Operating plans can by their nature include more specific detail on what actions are to be taken during the life of the plan. It is appropriate to require that an action plan, such as the one shown in Figure 37.2, be included for major items in each business unit's portion of the plan.

Rationale for EHS Program Elements

Early in the development of the program every EHS manager must provide management with a rationale for each of the areas of program focus. Maintaining a successful EHS program is expensive, and it is important that management understand where resources are being applied and why. The first step in establishing focus areas is an evaluation of the internal and external EHS issues that face the organization. Areas to

ACTION REGISTER						
LOCATION/PROGRAM: Plant C - East			**SUBJECT AREA:** Industrial Hygiene/Safety			**DATE:** 10/7/97
ITEM NO.	**ISSUE**	**DATE ASSIGNED**	**DATE DUE**	**STATUS**	**ASSIGNED TO**	**REMARKS/ACTION**
AJL-1.3	Wave solder pot cleaning IH sampling.	4/7/97	5/30/97	Complete	A. Samuel/ M. Sue	Above standard. Ventilation improvement required.
AJL-1.3 V1	Wave solder ventilation upgrade.	8/10/97	4/9/98	Duct work being fabricated.	J. Abel/ A. Samuel	
FM-95.1	Toxic gas handling training required.	9/18/97	5/10/98	Wafer lab complete.	M. Bezar	Additional SCBA on order.
APL - 1e	No sprinkler - main aisle b15 - b23.	8/23/97	2/21/98	Design under development.	A. Polin/ J. Abel	

Figure 37.2 — Example operating plan action register.

explore include regulatory compliance history, marketing priorities, environmental exposure potential, and internal initiatives.

Regulatory compliance history and challenges are important because they have impact on the magnitude of exposure to the organization and responsible management. An organization with a historically poor compliance record faces a higher level of fines for repeat violations. Previous compliance history also has a bearing on the level of criminal liability under federal sentencing guidelines. When an organization is in a heavily regulated area, pharmaceuticals for example, the potential for episodes of noncompliance are magnified by the sheer number of compliance issues. In such cases regulatory compliance will be a priority for the organization.

Marketing priorities have an impact based on the organization's intended customers and shareholder interests. As an example, a supplier of consumer goods, such as health and safety products, has a much greater potential impact on the bottom line from publicity regarding EHS issues than does a manufacturer of heavy machinery for industrial use. The benefits of good publicity and the consequences of bad will be magnified in a company where the customer chooses a product based on a notion of the "goodness" of the organization from an ethical viewpoint. Shareholder initiatives may also provide incentive for participation in activities such as ISO 14000 international environmental management systems standards or other external, structured initiatives that might not otherwise be required in well-managed EHS programs. Given that such initiatives generally require substantial resources, including them as a priority must have sound business advantages.

The potential for environmental exposure to the organization based on its mode of operation or history can also have a bearing on priorities. If an organization is in an acquisition or divestiture mode, considerable resources will be required for due diligence activities, which evaluate the risks associated with the transaction. These risks can include noncompliance issues, improper hazardous materials use, and possible contamination concerns.

The nature and location of operating locations is also an issue. If there are many older facilities, evaluation of the properties and remediation of historic issues will be a priority. Operating many smaller locations will require resources to support nonprofessional EHS coordinators at these locations. Foreign operations also can require different priorities, as other regulations must be evaluated, and transaction times for all aspects of the program will be increased due to language and cultural barriers.

Internal commitments developed by the organization must also be considered in establishing program priorities. If the organization has leadership aspirations, resources must be dedicated to the programs that support this goal. For example, senior management might commit to elimination of a particular material, such as beryllium, from their processes. This effort may require considerable resources from EHS to evaluate process changes and potential substitute materials. Too often, in a rush to meet commitments new materials are rushed into use without proper scrutiny. These substitutes may present a greater employee hazard that could go unrecognized because users may feel that, since this is a replacement for a "bad" material, they have nothing to be concerned about.

When establishing the rationale for the EHS program for management each of these areas must be evaluated and inclusion of resources to address them clearly articulated. Since many companies today are data driven, it may be helpful to use a methodology that evaluates each aspect of the EHS function semiquantitatively and calculates the needed resources in a relatively objective fashion. Appendix B contains an example of a tool developed by one company to assist in determining the level of resources required at each location.

Except in consulting, an organization's EHS is rarely a profit center. Since most accomplishments are injury- or cost avoidance-based, it is difficult to quantify the benefits of program improvements. Using the resource evaluation tool provides a starting point for resource discussions with management. These discussions should follow an agenda, established prior to the meeting, which includes senior management support for approved initiatives. A detailed proposal should be presented and specific actions and resource needs addressed. Once finalized, the program plan should serve as a guide to future actions and should be updated on a regular (at least annual) basis.

The Role of Management Skills in Occupational Hygiene

As a general rule, occupational hygienists have limited real authority to accomplish their job tasks. Their role is typically focused on obtaining the data necessary to determine what changes, if any, are required in a company's processes to ensure the protection of workers and other potentially exposed individuals. That data is then used to develop recommendations that can be conveyed to the individual(s) responsible for the process in question. The success or failure of a facility's IH program depends on the ability to convince those individuals able to change a process, of the value of the occupational hygienist's advice and the benefits of the required action. Effective IH management must consider not only presentations but also the added value represented by the recommendations to the business. For example, a dust collector installed to protect employees may also collect valuable product that can be recycled or reclaimed.

Most occupational hygienists practice not among their scientifically inclined peers, but in the greater business world where leaders are more likely to have a business education. To communicate with this decision-making population, occupational hygienists must be able to convey necessary information in a manner that is comfortable for their audience. This requires an understanding of management philosophy, nomenclature, and presentation norms. For example, most organizations have a method for presenting statistical data that is typically used for internal presentations. That style should be adopted for presenting IH data.

It is also important that as their stature advances within the organization occupational hygienists be comfortable with making business presentations. As senior EHS managers advance, they will more frequently be asked to participate in projects and meetings that are not specifically focused on EHS issues. Their participation will be of value only if they can understand the issues under discussion in sufficient detail to identify where EHS concerns might arise.

Management from a Support Position

Occupational hygienists typically function from a support position. This means that though they may exercise considerable influence over decisions, they do not have the final say. In his book *Getting Things Done When You are Not in Charge*,[3] Geoffrey Bellman identifies several characteristics of support positions that occupational hygienists may recognize:

- More expertise than authority;
- Important influence but not the final say;
- Free access to the organization;
- Regular interaction with people of greater authority;
- Future planning is difficult;
- Accused of not appreciating "the big picture" or the "bottom line";
- Important customers do not appreciate or understand the function;
- Seen as a cost center to be at best tolerated;
- Predecessor reputation may have an adverse effect; and
- Battles with other support professionals vying for limited resources.

Given these characteristics, the support role brings some daunting challenges that are best addressed by focusing on those areas where change is required and convincing those in authority to take required actions. This often requires the creation of a vision of a future state where the issues of concern have been addressed, and the development of a plan to reach that goal. A vision statement identifies where you want to be in the future. In broad terms it establishes the end point for an improvement process and allows the development of a plan to move from an unacceptable present to a beneficial future.

The key to convincing others that your recommendations provide the most appropriate path for your organization is understanding customer motivations. In this case the customer is anyone who may be affected by the information developed though sampling or other analysis of the work environment. These customers may be internal or external. Examples of IH customers are regulatory agencies, employees, purchasers of services/products, and company management.

Career Planning

In management, as in all other aspects of life, the adage "to thine own self be true" applies. If managers are to continue to be productive contributors to their organizations they must continue to develop and grow. Like most aspects of management behavior, this growth occurs most efficiently when a clear plan is developed and followed.

Career planning for EHS professionals is not unlike that for most technical specialties except that the breadth of the profession offers a greater range of options than most. Generally, unless the individual is part of a large organization focused on EHS, career options are not obvious, and development of a career plan requires careful deliberation. These plans are usually specific to the individual and only a few organizations provide assistance with their preparation. The following lists some basic steps to be considered when developing a career plan:

- Develop a vision based on various career stages. What would you like to do next year, in five years, or in 10 years? Try not to limit options to what appears obvious. Seek assistance from associates, mentors, and outside resources.
- Evaluate each option in terms of requirements to meet the goal and the way each development step supports future goals. Asking someone you admire who has achieved a level of success to tell you what was required in the process can help clarify the steps you need to take.
- Explore other functional areas or aspects of EHS when the opportunity presents itself. This will help you evaluate what others do and avoid the "grass is always greener" phenomenon.
- Write down your plan and share it with others for comment as appropriate. Sharing the plan with your boss, if possible, can have significant advantages in ensuring that your ambitions are understood.
- Readjust your plan on a regular basis.

Opportunities should be considered for growth both within and outside of the organization. Most companies recognize the need to develop their employees and to enhance the skills they bring to their jobs. A good company will continue to educate employees even when their education and experience exceed the needs of their present jobs. While this presents a possibility that the employee may leave for a more advanced position, smart organizations know that having the best person for the job entails some risk. In EHS, overqualified rather than underqualified is the preferred situation from a corporate perspective. If EHS managers do not have the education their positions require, serious consequences could arise while they are learning on the job.

If the organization does not encourage individual growth and development, employees may feel they have nowhere to go. Morale and productivity will be reduced. Harvey Mackay[4] suggests exploring the following areas to avoid career stagnation:

- *Readjust your pace:* Ensure that your expectations are realistic. See what other organizations offer.
- *Analyze your competition:* How do you compare with others in the same functional area in or out of the company? Get to know your value and options.
- *Look for sideways growth:* If you have reached your limit in EHS, explore other functions such as engineering, marketing, or product safety.
- *Study the corporate culture:* Compare how you present yourself and your ideas to the presentations of those who succeed in the organization.
- *Take a look at your personal risk factor:* There are always ways to advance, but more drastic measures carry greater risk. Before you propose bold new ideas or take on difficult projects, make sure you can accept the possibility of failure. Often the risk is worth it, but acceptance of risk should be a conscious decision, not a surprise.

A career is a lifetime pursuit made up of a series of jobs. Employees control their own careers by the choices they make, and only one thing is certain: mistakes will be made along the way. The true test of whether it is a career or a job is the ability to recover from the bumps in the road.

Management Theory

Ever since people organized into groups with common goals there has been a need to establish methods to manage the group to minimize duplication of effort and to ensure completion of all tasks necessary for the group's survival. Management of these groups was often determined by leaders chosen for physical characteristics such as strength or heredity. While history is replete with great civilizations flourishing under effective management, there are also many examples of failure when inappropriate leaders were selected or developed. Where success occurred, the techniques employed by the leaders were often not recorded and were lost to those who might later try to emulate those methods. In more recent times organizational and management theorists have tried to quantify what makes successful leaders and organizations.

Organizational Structure

Organizations use many different structures in their attempts to ensure that processes are conducted in the most efficient manner possible. Some of the theories describing such structures have been designated by letters. They are Theory X, Theory Y, and Theory Z.

Theory X directs a work-centered approach to organization. This theory, as presented by Douglas McGregor in the early 1960s,[5] rests on four assumptions: (1) work is a chore; (2) people dislike work and will avoid it if possible; (3) people must be coerced or otherwise forced to put forth maximum effort; and (4) people prefer to be led and to avoid responsibility.

Given these factors, there is little self-motivation possible, and an organization requires strong leaders and a rigid hierarchy to meet its goals. In a Theory X organization all decisions are made at the top, where leaders use their authority to align the workers. In this application authority may be defined as the power to exact obedience. The organization is structured around a strict chain of command, with decision making at each level of the organization dependent on approval from the next higher level. This centralized structure makes communication very difficult, as access to the appropriate individual, either higher or lower in the organization, often requires passing through several intervening functions or individuals. In this type of structure it is most important for the EHS function to be represented at the highest level of the organization. If not, the often difficult information that must be communicated in a good EHS program could be suppressed. Such circumstances can force the professional to balance career security against ethical conduct.

At the same time he discussed Theory X, McGregor presented an alternative, designated Theory Y. People are at the center of a Theory Y organization. The assumptions of Theory Y are (1) work is a normal part of life and is not inherently disliked; (2) people can exercise self control and do not necessarily require threats to do work effectively; (3) people commit to objectives based on their perception of potential rewards; (4) people seek responsibility under appropriate circumstances; (5) most people can be innovative when given the opportunity; and (6) most people use a fraction of their intellectual potential in most jobs.

Taking these factors into account leads the Theory Y organization to seek a participative structure. In this arrangement, decision making is distributed throughout the organization. Objectives are achieved by persuading the group that a given direction is appropriate. Employees are organized into groups with significant interaction among them.

Both Theory X and Y are limited by the extremes of their perception of the organization. Theory Z attempts to address this deficiency by moving beyond simply considering work and people to evaluation of such factors as organizational size, degree of interaction, personalities, alignment of goals, level of decision making, and the state of the system. This type of analysis focuses on an organization's culture and how it would best reflect selected goals. The focus is on clear and open communication at all levels and a sense of collective responsibility for all members of the organization. This type of organization depends on team performance to meet its objectives and reflects the general trend in business today to drive decisions and responsibility down into the organization where the actual work is performed. Decision making is participative, but an individual is responsible for the final decision. Even with the team focus so popular today, there is still a need for individuals to manage the process and make decisions. Team dynamics are discussed later in this chapter.

In his book *Reengineering Management*,[6] James Champy recognizes the continued need for managers despite ongoing flattening of organizations. He sees managers at four levels:

1. Self-managers, who answer only for the quality of their work whether as individuals or as part of a team.
2. Process and people managers, who answer for the work of others. In current team thinking, this position often rotates as the project develops and the team requires different skills.
3. Expertise managers, whose responsibility is for identifying future trends and needs.
4. Enterprise managers, with profit and loss responsibilities.

While occupational hygienists might fall in any of these categories, they are most often expertise or self-managers. In most organizations they are valued for their abilities to help the organization avoid trouble and to act without specific direction.

All of these elements acting in concert allow a decentralized organization to work. This structure with distributed responsibility greatly facilitates the function of an EHS program. When provided with a fair and open forum in which to make a case for EHS improvement, few proposals will be ignored. Whether proposals are approved or denied, decisions will be quicker and opportunities to revisit areas of concern will continue to exist.

Most current business publications exhort organizations to develop the ability to respond quickly to the rapidly changing global marketplace. Like Theories Y and Z, their advice leads to more decentralized organizations. In his book *The Fifth Discipline*,[7] Peter Senge discusses the need for an organization to learn as a group. He sees this fundamental requirement as dependent on five "component technologies" or disciplines:

1. The ability to see all processes as a system of interrelated events. This systems thinking allows identification and correction of unacceptable patterns.
2. Employees' commitment to their own learning, as demonstrated by a desire to focus their energies by continually aligning their personal growth activities with the needs of the organization. This personal mastery addresses the organization's need, on group and individual levels, to have a spiritual commitment to learning.
3. The ability to identify basic mental models or paradigms that limit the organization's ability to move forward. These ingrained assumptions about how the world works may limit the team's ability to move past obsolete practices and procedures.
4. Developing a shared vision of an appropriate future for the organization. This is a level beyond simply signing onto the personal vision of a charismatic leader. It is the ability to translate such a vision throughout the organization in a manner that fosters a genuine commitment.
5. Finally, and most important, is the concept of team learning where teams develop the ability to think as a group. As is discussed later in this chapter, teams are the fundamental units in many organizations today, but all too often patterns of defensive behavior limit their ability to achieve the collective insights that are not available to individuals. Teams must develop an atmosphere of open communication to allow these achievements to occur.

An organizational structure that focuses on learning, with an emphasis on teams, is ideal if change in the EHS program is required. EHS deficiencies are rarely the result of conscious decisions to violate the law or adversely affect employees. Rather, these concerns arise from a lack of awareness of responsibilities and the basis for policy requirements. In a learning organization with EHS representatives participating on teams, corrective education and training comes naturally. Concerns that arise can be dealt with in the normal course of business, rather than as special events that cause resentment when they interrupt production processes.

Leadership

Leadership is the ability to mobilize available resources toward achieving the organization's vision and goals. These resources may be employees and their ideas, or the organization's physical assets. Leadership is most important when the organization is undergoing, or in need of, change. In their role of agents of change occupational hygienists are often called on to exercise their leadership abilities.

Many of the opportunities to exercise leadership in the workplace today come in the role of team leader. As is described later in this chapter, teams have become a critical part of the developmental process in industry today. The evolution of leadership applications is described in Table 37.1.[8]

Traditionally it was believed that leadership was not a learned ability but rather some inherent quality of individuals that made them stand out among the group. Current thinking indicates that leadership is more related to a given situation than an accident of birth. Each employee possesses a set of personality traits and abilities. Different situations require different skills in leaders. Leadership needs in the relatively stable past were quite different from those required today with fierce competition and a bewildering array of external requirements. Training and education can develop further skills that can be applied to leadership situations.

Table 37.1 — Evolution of Leadership Applications

Supervisory Leadership	*Participative Leadership*	*Team Leadership*
Direct people Explain decisions Train individuals	Involve people Get input for decisions Develop individual performance	Build trust and inspire teamwork Facilitate and support team decisions Expand team capabilities
Manage one-on-one Contain conflict React to change	Coordinate group effort Resolve conflict Implement change	Create a team identity Make the most of team differences Foresee and influence change

The Forum Institute[9] categorizes the tools necessary for leadership as abilities in four areas:

1. Interpreting: Data is available from many sources. Communication with external customers, peer organizations, and information from regulatory sources can provide substantial data. Another source of information is inside the organization. Every group has information regarding its goals and the role of each function in meeting them. It is sometimes necessary to dig beneath surface policies and procedures to find the real expectations of the group. On a fundamental level, information and observations of the motivation and capabilities of the IH work group provides valuable information. Finally, introspection is necessary to ensure understanding of personal motivation and capabilities to identify their effect on relationships and work. A leader must be able to obtain and use this information to identify the potential impact on the organization. Interpretation of information from many different sources is a critical leadership skill. By using active communication skills, listening, and asking questions, a leader can select the information most important to the organization's future.
2. Shaping: With this information leaders shape a vision for the future that reflects the organization's values and goals. With the current pace of change many groups feel that it is impossible to keep up with the present, let alone plan for the future. Good leaders can develop a strategy that seems both appropriate and achievable. They can elicit the support of their organizations by creating a meaningful picture of the future and the group's role in achieving it. Each step along the path is identified and measurable action plans developed.
3. Mobilizing: Achieving the organization's vision requires mobilizing available resources toward an identified goal. Leaders can focus the efforts of a group comprised of individuals with varying motivations, abilities, and ideas. This takes an ability to communicate clearly to all members of the group in a manner that ensures their understanding of the specific tasks they are expected to complete. This clear communication allows the leader to trust the team and fosters the development of a supportive and cooperative work environment where employees can contribute to the best of their abilities.
4. Inspiring: Finally, leaders inspire their team to achieve success. By sharing authority and acknowledging the contributions of others, leaders enable others to feel and act like leaders. This sense of ownership by team members helps the group to achieve peak performance. Under strong leaders, team members are encouraged to develop their talents and to accept increasing levels of responsibility. Feeling that they have a stake in the results of their actions and that their efforts will be recognized encourages the entire work group to respond to the peaks and valleys of events in the workplace with acceptance of the present and enthusiasm for future improvement.

Modern Management Practices and Application to Occupational Hygiene

Total Quality

A major theme in business today is the concept of total quality. Total quality programs are often called total quality management (TQM), total quality control (TQC), or even total quality environmental management (TQEM). Total quality is a management methodology that emphasizes the improvement of the processes by which businesses operate and products are produced. The process includes all activities that produce an output for a customer. The provider of that output is the supplier. All aspects of business focus on supplier-to-customer relationships. The suppliers and customers may be internal or external to the organization. For example, customers of the EHS department might be regulators, management, or employees. Suppliers might be regulators, consultants, or internal departments such as purchasing.

Total quality initiatives involve all suppliers and customers and thus all employees in an organization. This type of management embraces the use of data and quantitative analysis over subjective processes such as the use of intuition or brute force. The methods used to improve processes center on the reduction of variability and the elimination of wasteful activities that often accompany processes developed without structured tools.

The philosophy of total quality requires a paradigm shift, away from a reluctance to tamper with processes that appear to be working to one stating that all processes are imperfect and an organization must strive for continuous improvement. This philosophy of continuous improvement or kaizen ensures that the organization will never be satisfied with less than optimal performance in any of its processes. Kaizen indicates that every process can and should be continually evaluated and improved, in terms of time required, resources used, resultant quality, and other aspects relevant to the process.

Many tools have been developed to facilitate the pursuit of total quality. Most of these tools focus on identifying, measuring, controlling, predicting, and reducing waste. One of the fundamental tools used in this process is the Deming Wheel, named after W. Edwards Deming, who introduced statistical process control to the Japanese in the 1950s. While the concepts he adapted from work by Shewhart and others at Bell Laboratories were largely ignored in his native United States, they were eagerly embraced by the Japanese. The unprecedented growth in Japanese economic power can largely be attributed to this single impetus.

The Deming Wheel (see Figure 37.3) is used to help solve problems by identifying weaknesses in processes. The circuit consists of Plan, Do, Check, and Act. The use of this tool, and others like it, usually entails the establishment of a team to evaluate a process or complete a project. The team members are trained in problem solving methodologies and team building. They are also provided technical support as required.

Since many of these evaluations are conducted on existing processes, they would begin with the Check phase, which looks at the past. This step is used to determine what has or has not been done. It most often begins by development of a flowchart that breaks the process down into discrete steps (see Figure 37.4). A properly constructed flowchart provides a pictorial representation of all steps in a given process.

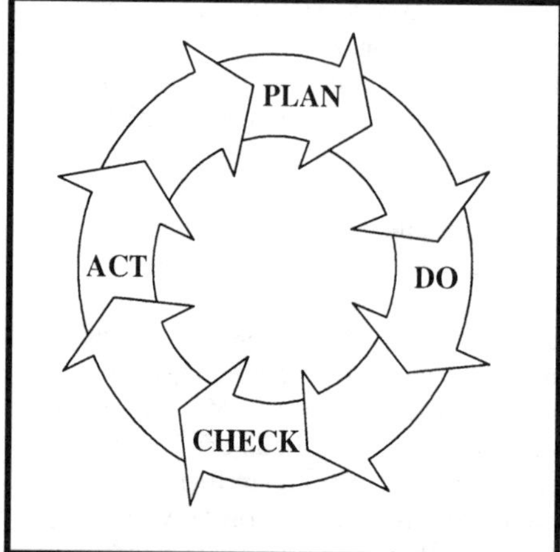

Figure 37.3 — The Deming Wheel.

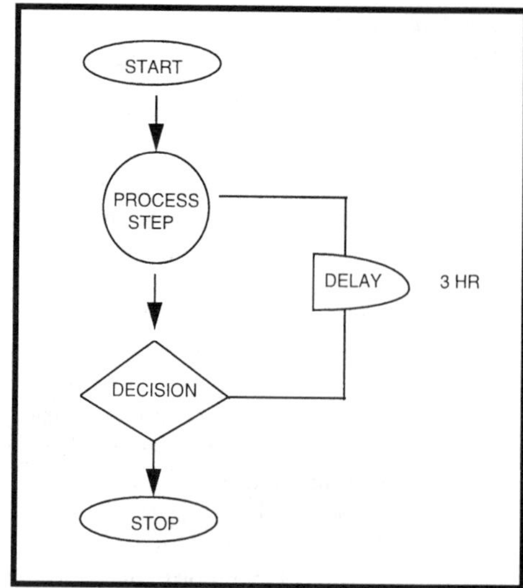

Figure 37.4 — The flowchart.

A set of standard symbols is used to identify the type of action taking place in each step (see Figure 37.5), and the amount of time for each step and time between steps, or queue time, is clearly identified.

Each step in the process is then evaluated to determine if it adds value or is wasteful. Evaluation of the flowchart is also used to determine where bottlenecks exist and what actions are constraining optimal performance of the process. Data is gathered detailing performance in key areas of constraint or bottleneck and is then evaluated using process analysis tools such as pareto analysis. This tool was developed by Wifredo Pareto at the turn of the century. He looked at the distribution of wealth in the Italian states and discovered that 80% of the wealth was controlled by 20% of the population. This same ratio has been found in many other systems. Often 80% of the problems in a process are the result of only 20% of the potential causes. This 80/20 ratio is referred to in some areas of safety management as "the principle of the critical/vital few."[10] Pareto analysis is used to isolate and identify areas of significant concern from a group of many potential concerns. Proper use of pareto diagrams can help the team avoid using its limited resources to address a less important aspect of the problem.

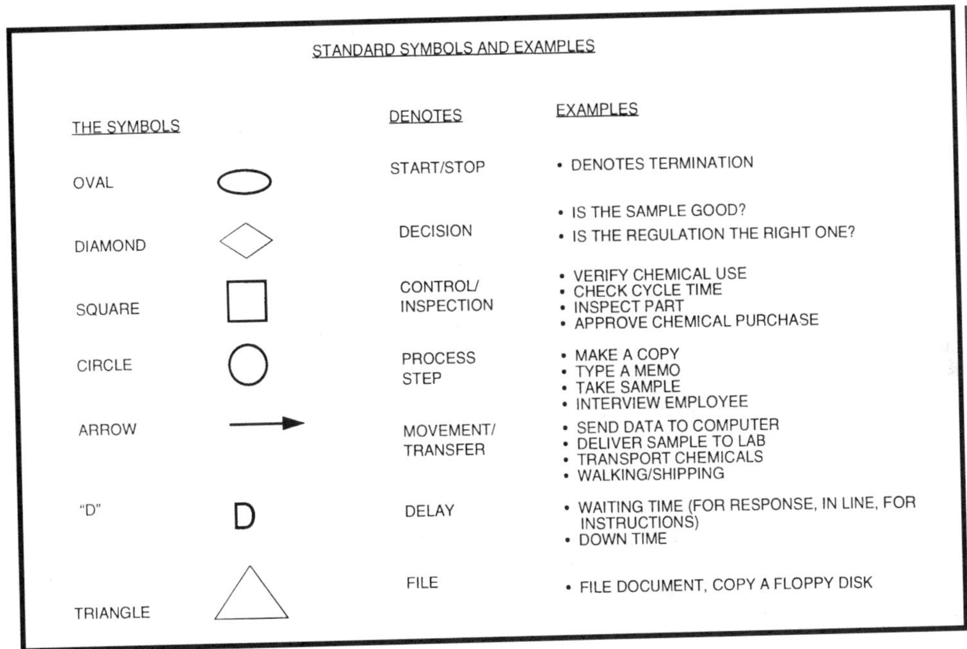

Figure 37.5 — Flowchart symbols.

As part of the Act step, pareto charts developed earlier are analyzed in an effort to establish cause-and-effect relationships for identified areas of concern. Reactions to the cause are then sought, often using brainstorming techniques. In brainstorming, groups are encouraged to think creatively about potential solutions. This process usually begins with unfiltered collection of potential solutions, which are later analyzed to select the most likely answers. Potential quick solutions identified during this step are then implemented in an effort to contain the problem, while more targeted solutions to the root cause of the problem are identified, using more formal problem-solving tools. The root cause is the activity that if corrected or eliminated will solve the identified problem. Root causes might be identified using more advanced tools such as additional sublevel paretos, Taguchi experiments, and fault tree analyses.

Taguchi experiments provide a method for evaluating several different elements of a process at the same time, as opposed to classic experimental design, which focuses on time- and resource-consuming technical analysis of one factor at a time. This tool was designed by Genichi Taguchi as a cost-effective way to improve processes under industrial conditions and limitations. Taguchi experiments use standard statistical tools to evaluate how much each selected cause contributes to the variation of the product. Using the Taguchi method insignificant factors are quickly eliminated, allowing focused attention on the manipulation of those factors that increase the product's or process's robustness against variation. Robust design is a technique for making the utility of the final product insensitive to variations in the manufacturing process. In EHS terms this might mean that no matter how different the input chemicals are the final product will have no increased exposure potential from one batch to the next.

Fault tree analysis is a graphical method of performing a failure modes and effects analysis. These techniques focus on relationships between data and not its quantification. They are useful in foolproofing a process or design and in problem solving.

The Plan step focuses on the development of a plan to eliminate the root cause of the problem. Central to any such plan is establishing accountability and a method to track the progress of necessary corrective actions. An action plan is often developed for this purpose that identifies the problem and chosen corrective action, along with a schedule for completion and the person responsible for the action.

The Do portion of the wheel is the measurement step. Here the effect of the chosen action is measured to evaluate its impact on the root cause. Again, statistical measurement tools such as Pareto and trend charts along with histograms may be employed. Elimination of each root cause is analyzed in the context of its contribution to the overall problem. Most operational or EHS concerns have many root causes, each of which has a different potential to address the initial concern. Finally, the Do step ends with an analysis to ensure that a permanent solution to the problem has been implemented. Again, statistical process evaluation tools are employed in this process.

Then the wheel is "spun" again if the chosen solutions do not appear to have achieved final correction of the problem. Remember, continuous improvement demands that these processes continue until optimal solutions are found. In reality, if at this stage the process has been significantly improved to an extent that other concerns have a higher priority for attention, it is preferable to address these greater concerns and then return to the initial problem when time and resources allow.

Integrated Product Development

Integrated product development (IPD) is a systematic approach to the multifunctional, concurrent design of products and their related processes (see Figure 37.6). It includes manufacturing and support of the products through their life cycle. This multifunctional team approach is intended to decrease product development and deployment cycle time by addressing all elements of the product life cycle from conception through disposal, from the outset of the process. The multifunctional team includes representatives from all internal functional areas and external members from customers and suppliers. Team members participate during appropriate phases of the project and are provided with the authority, responsibility, and accountability necessary to meet customer requirements. An important aspect of continuous improvement is the reduction of the time it takes to bring a new product to market. This cycle time includes all elapsed time from the start of any process to its conclusion.

Process cycle time is reduced though open and rapid communication among all participants in the process. This helps ensure that major issues are not overlooked, and problems are quickly surfaced and resolved. In a typical IPD process each phase of the project is managed by a team whose leaders and members vary as appropriate to the task at hand. Each phase has a thorough checklist that broadly identifies the major elements that must dealt with at that time.

Process cycle time (see Figure 37.7) has become a critical indicator of a successful enterprise. Under current conditions, where the first-to-market company often has the greatest success, any wasted effort can have a significant impact on profit and survival. Every process consists of many cycle time elements. The cycle time is the length of time each discrete element of a process takes from initiation through completion. Individual elements can be added together to give a total process cycle time.

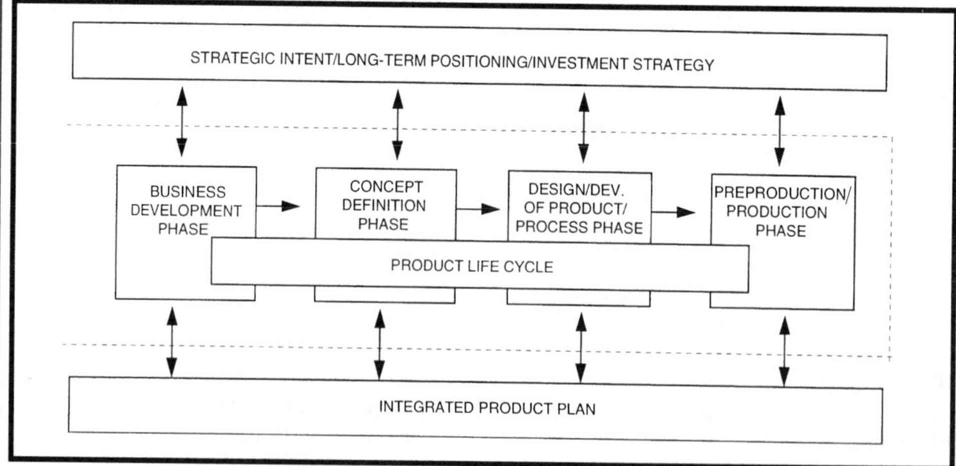

Figure 37.6 — IPD process overview.

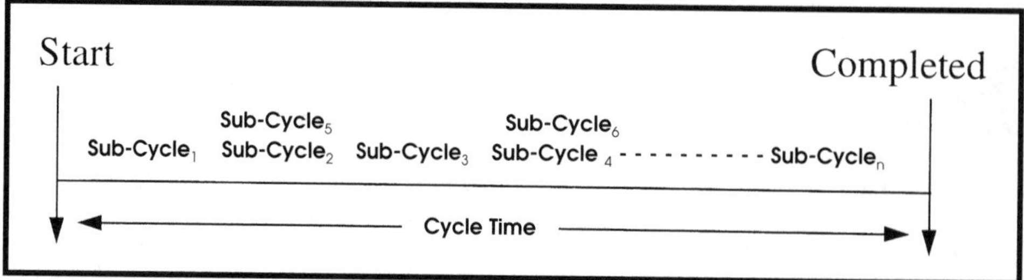

Figure 37.7 — Chart showing total cycle time (the sum of the cycle times of all process subelements).

A typical method of measuring cycle time is to follow one example of the process's output from beginning to end. Issues that must be considered when measuring cycle time include whether or not to count time when people are not working, differences in the lengths of units of measure such as months, and how to include rework. Rework is that portion of the output that must undergo additional processes to correct any unacceptable characteristics introduced during initial production.

Occupational hygienists can use the IPD process to ensure that EHS issues are raised earlier in the design process, so that the overall program will not be delayed by uncovering IH concerns too late. Like other functional experts, IH professionals would typically not be a part of the core IPD team. Rather, a top-level view of their issues would be included in the various checklists and they would be called in to participate with the team when issues in their area arose. Successful implementation of the IPD process and inclusion of EHS issues can help overcome one of the principal frustrations occupational hygienists face: getting other departments to identify IH issues in a timely fashion. The IPD process allows these issues to be properly addressed without the conflict that inappropriate process delays can create.

The IPD process consists of the four phases below. In each phase EHS issues are considered where appropriate.

1. Business development: This phase begins with identification of opportunities and ends with a decision to pursue or not. Effort is focused on determining the customer's actual requirements, the risks involved, and the capability of the organization and/or partners to achieve the desired result. IH issues at this stage might include new chemicals or processes to be introduced.
2. Concept definition: This involves further definition of requirements as part of the construction of a strategy to win the business. EHS input is necessary to ensure that internal EHS advantages are included and that all areas requiring additional EHS activities are identified.
3. Design/development: In this phase there is completion of a detailed design along with validation of production readiness. EHS personnel must ensure that product safety issues have been addressed and any required protective equipment has been included, and that necessary training has been identified. This phase is the most important from an accident prevention standpoint.
4. Preproduction/production: This phase involves qualification of the process and production implementation. Once the actual manufacturing activity begins, a thorough EHS review is required to ensure that all hazards are as described during the development process and that they have been adequately addressed. This phase also includes product life cycle maintenance through disposal.

Quality Function Deployment

Quality function deployment (QFD) is a systematic means of ensuring that the demands of the customer are accurately translated into action within the supplier organization. The use of this tool must be carefully managed, since its output can be used in the development of operating and strategic plans. To use QFD effectively,

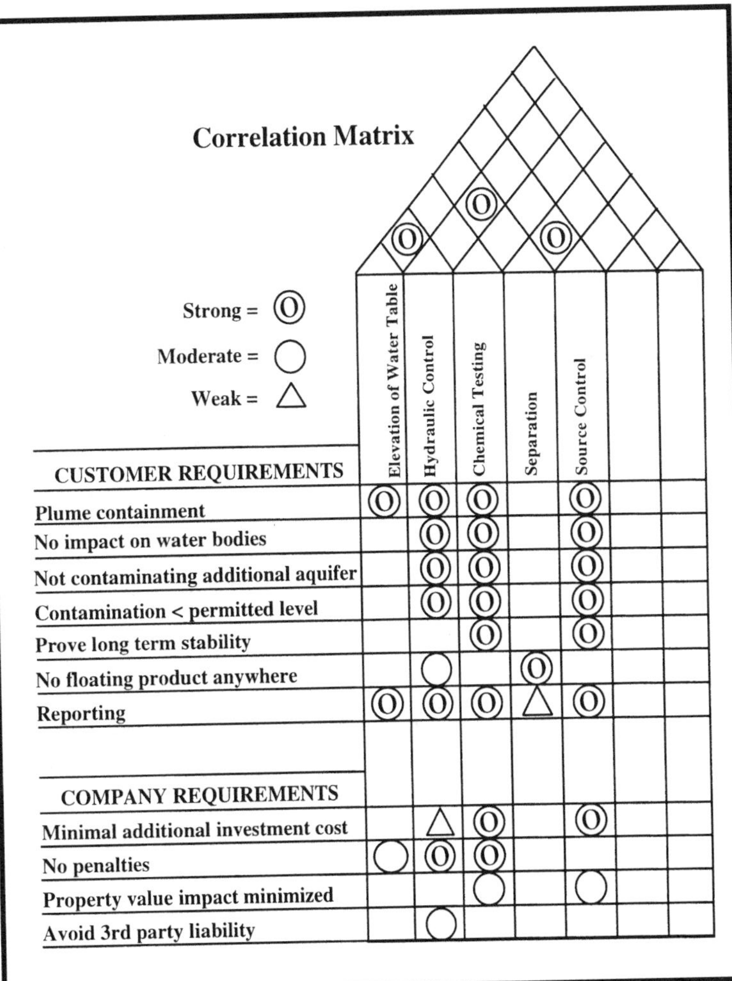

Figure 37.8 — Environmental remediation House of Quality (D. Polzo and M. Kamenir, ITT Avionics).

substantial information on customer requirements is necessary. From customer data, product and process requirements can be developed, including appropriate performance measurement tools.

A useful tool in correlating customer needs with process design requirements is known as "The House of Quality." As is seen in Figure 37.8, this matrix evaluates the "whats" of customer needs in terms of the "hows" or capabilities of the operation. This structure helps identify what priorities exist based on the level of correlation between the hows and the whats.

This House of Quality correlates customer and company requirements against possible approaches. In this example, these requirements are the whats, or what the company wants to accomplish. They are correlated in the body of the house with potential remedial approaches, the hows, or specific actions that could be taken. For example, in Figure 37.8, plume containment is strongly correlated with hydraulic control. The triangular correlation matrix, or roof, represents the correlation among the hows. In this example, elevation of the water table is strongly correlated with hydraulic control. Such a tool can assist the user to determine quickly what actions will most satisfy the customer. In an IH context this approach might be used to evaluate regulatory and aesthetic requirements prior to an asbestos removal project.

Benchmarking

A benchmark is a standard of excellence or achievement against which other similar processes can be measured or judged. Use of benchmarks allows an occupational hygienist to compare a specific program to other world-class operations to determine where areas of potential improvement exist.

Benchmark facilities can also be used as a source of program ideas and for obtaining program examples to avoid reinventing what others have already accomplished. EHS is typically thought of as a noncompetitive area, and most companies will freely share or trade their programs. Table 37.2 shows a 10-step model for benchmarking.

Benchmarking is not simply a matter of comparing internal processes to those of other organizations. It involves evaluating another organization to select those methodologies that provide a level of performance worth emulating. Selecting appropriate organizations to benchmark requires thought and consideration of the specific objectives of the project at hand. Rarely does a benchmark begin without a detailed plan of what will be evaluated and the anticipated results.

Table 37.2 — 10-Step Model for Benchmarking

Action	*Explanation*
Identify benchmarking subject.	Select area of substantial impact where improvement would have real meaning to customer satisfaction.
Identify benchmarking partners.	Look for the most productive examples. Do not restrict selections to your own industry.
Determine data collection method; collect data.	Develop a formal process and follow it.
Determine current performance gap.	Look for differences between your processes and those used by benchmark partner.
Project future performance.	Evaluate future impact of integration of identified improvements.
Communicate findings; gain agreement.	Present results/analysis of improvements that could be adopted.
Establish functional goals.	Develop specific goals to meet selected improvements.
Develop action plan.	Have measurable milestones and identify leaders to meet selected goals.
Implement plans; monitor progress.	Conduct periodic progress evaluations.
Recalibrate benchmark.	Determine if completed actions meet expectations; look for further improvements (may include new round of benchmarking).

Most world-class organizations have only a few areas where they are truly exceptional. These areas are where benchmarking efforts should be focused. It is also important not to limit yourself to only peer organizations. For example, even light manufacturing operations can often gain substantial, useful information by benchmarking a major chemical manufacturer's chemical handling procedures.

Benchmarking is most effective when carried out by the employees who are responsible for the benchmarked activity within their own operation. Participation by the affected employees and departments ensures their support when changes are required. Once the team sees what could be accomplished, its members will be more willing to accept challenging performance targets.

Resource Management

Every EHS program must support its organization's financial, as well as operational, goals. To evaluate performance against financial goals, all EHS-related costs must first be identified. For those costs directly associated with the function of the EHS department, a separate budget should be established that accounts for all anticipated expenditures to manage the program. All costs to other departments should be determined and conveyed to the affected departments prior to budget preparation. While it is all too common for unexpected costs to arise in EHS, it is unacceptable not to budget funds for known activities. Too often, EHS items will be left off initial budgets on the theory they will have to be addressed anyway and the funds will be found somewhere. Omitting EHS items from the budgeting process is a classic example of a lack of control in company EHS management.

Currently most companies assign EHS costs to overhead and do not associate them with particular processes or products. This method assures that the costs are identified but does not encourage accountability from those areas generating the costs. As an example, suppose an operation has a goal to reduce the amount of solvent waste generated in its manufacturing process. There is little incentive for local supervision to meet that goal if all waste disposal costs are paid out of the EHS department. If, however, they were required to budget for their own waste costs, manufacturing departments would be more likely to pay better attention to how they manage their wastes.

Cost Accounting

In an effort to address this issue, several organizations have begun to use activity-based costing (ABC) to identify and allocate costs based on measurable activities.[11] ABC can be conducted on any aspect of the EHS program and begins by identifying all costs associated with that portion of the program. There are two elements of ABC: activity analysis and cost object analysis. Activity analysis for all of EHS would be difficult, given its broad scope. For IH, the following are some examples of activities that should be included in this process:

- *Sampling:* planning, setup, on-site work, employee training, results review, and presentation to employees;
- *Regulatory:* interaction with agencies, review of publications, communication to management and employees, and any fines associated with noncompliance;
- *Chemical issues:* material safety data sheet review, evaluation of alternative products and processes, employee training, and communication with customers and suppliers.

Four types of resources may be expended on each identified activity. These resources are payroll costs, operating costs, capital expenditures, and current and future liability costs.[12] The next phase of ABC extracts the costs in each resource area for each activity. The first three are relatively easy to identify. Future liability costs, however, are difficult to quantify and can only be an estimate developed by experts in the given area.

Cost object analysis focuses on the activities required by products. For example, if a product requires the use of a toxic solvent in its production, costs for monitoring and control might be assigned on a per-product basis. This is an example of unit-based activity. Other types of events that drive activities are batch-based, product-sustaining, and facility-sustaining. Each case identifies specific, required activities and the resources they consume. In this way the costs are distributed where they are actually used.

ABC provides a tool to identify what drives activities requiring corporate resources, and where those resources are used. Given this information, a management program could be developed to eliminate or reduce activities that are not value-added. The distribution of costs also places accountability where it belongs, with the processes and departments that require the expenditure of resources. These departments are also much more able to manage and potentially reduce these costs, since they control the processes involved.

Policy Deployment

Every EHS program has many policies and procedures designed to ensure that the organization understands the requirements that apply to the operations. These policies must be deployed to all affected employees in a manner that ensures that they understand and accept their responsibilities.

The basic principles of policy deployment require a company to develop policies, including improvement targets, and deploy them throughout the organization. This deployment is performed in a manner that permits the operating organizations to establish supporting goals and targets, along with a method to measure performance. In the policy deployment process senior management evaluates the organization's performance in a given area and develops policies designed to correct any deficiencies. This policy moves through the organization, gaining supporting actions and associated measurements. The results of these measurements are fed back to senior management and the process begins anew. An important aspect of this process is continuing two-way communication at all levels of the deployment process, to evaluate lower-level plans and the continuing utility of the original policies. The two-way communication required by this process is often referred to as catchball.

This same process is appropriate to deployment of EHS policies. The principal difficulty in ensuring that all appropriate individuals are aware of, and are following, EHS policies is the lack of feedback on how the process is working. If, when a policy is created, time is taken to work with the affected locations and functions to ensure their understanding and to agree on appropriate measurement and reporting frequencies, there will be little doubt whether the process is working. For example, if the EHS department determines that a policy to eliminate the use of all known human carcinogens from the process is appropriate, it would first develop a clear policy, working with the affected functions in its development. Affected individuals would then receive training in the application of the policy. Their adherence to the policy might then be measured by first requiring that an inventory of all human carcinogens be completed and specific action plans for their elimination developed. Performance against these plans could be measured on a monthly basis, with percent reductions displayed on bar charts for ease of understanding.

Goal Setting

Most statistical measures of performance, or metrics, in EHS have zero as their ultimate goal. Depending on how the metrics for hazardous waste, employee exposures, accidents, and the like are constructed, they generally improve as they decrease. The issue of goal setting arises in the time period between the identification of an area of concern and its elimination. Unless the concern can be eliminated in a short period, interim goals or milestones will be required to evaluate the success of the program. Beating or achieving the goal on an ongoing basis is an indication that the process is functioning acceptably, assuming the goal was properly chosen in the first place. It is important to remember that what gets measured gets managed. If the measurement criteria is not properly aligned with the goal, employees may find methods to meet the goal without addressing the problem. For example, if emergency response time has been a problem, a location might choose "average minutes to respond" as one of their metrics. If EHS starts sending inadequately trained personnel to the emergency because they are more available, they will have reduced response time but will not have addressed the root cause issue.

In establishing functional goals, it is important to keep some basic principles in mind. First limit the number of goals to the manageable, given the available resources, unless the consequences of such a limitation are unacceptable. If, for example, 10 situations presenting an immediate threat to employees are uncovered, the organization cannot choose to address only eight because the full project is not included in the budget. In most circumstances, however, a few appropriate goals that can be achieved are better than dozens of goals that scatter resources and make it unlikely that any will be achieved satisfactorily.

Second, goals should be aligned with the organization's long-term plans. As is discussed earlier in this chapter, functional goals should flow from a company's strategic and operating plans. For example, if the organization has an aggressive acquisition strategy, the EHS department might have a goal to develop and implement a comprehensive due diligence policy and procedure.

Finally, at least some of the organization's goals should be very ambitious. As Champy writes, "What companies actually require to thrive in today's revolutionary climate is rarely anything so 'reasonable' and 'realistic' as a 10 percent improvement in some performance measure or other. What we actually require is more often something like a 50 percent improvement or a 75 percent improvement."[6] Setting challenging goals encourages the organization to use its resources to their maximum potential. There is a risk, however, that if unreasonable goals are established, employees may become discouraged at continued failure and operate at less than their best if they feel that even their best is not sufficient.

Negotiation Skills

EHS management has sometimes been called making the unpleasant palatable, if not preferable. While many of the programs developed to protect health and environment have obvious positive results, many of these results are often avoidance of a problem, rather than specific contribution to a company's bottom line. While these efforts may have a real effect on profitability, it is often difficult to convince management to commit what may be considerable initial resources to projects with long-term benefits. Unfortunately, business today works on an ever-decreasing horizon. What was once an inadequate 2-year planning cycle has continued to erode to the point where many operations focus on quarter-by-quarter, if not month-by-month, results. This can present a substantial obstacle to the occupational hygienist proposing a program to avoid chronic effects that might not appear for 10 to 20 years. The fact that such programs are usually initiated is usually the result of successful negotiations.

Negotiation skills are an essential tool for the EHS professional. The skills required are generally the same as those required by any salesperson. A set of eight appropriate points for selling your point of view are described by Chester Karass in his book *Give and Take*.[13]

1. *Talk less and listen more:* One key point in negotiations is to ensure that there is not already initial agreement. Too often, difficult negotiations end when the parties involved realize they just misunderstood each other's position.
2. *Don't interrupt:* Interruptions block communication and can irritate the other party.
3. *Don't be belligerent:* Raising the level of hostility in a negotiation is rarely a good tactic. Despite the success that may occur by forcing a decision through intimidation, it is usually short-lived. In most cases EHS negotiations will occur among people with long-term relationships that are likely to involve a series of such events. If a negotiation ends in a confrontation, the result is likely to affect all future discussions adversely.
4. *Don't be in a hurry to bring up your points:* If you listen to the other side's points, you may find many areas of common ground on which to build a final agreement.
5. *Restate the other party's position and objectives as you understand them:* Many experts in communication suggest that practicing reflective listening, where you summarize and repeat the position of the other party, can aid in clear communication. Knowing that you will have to repeat the other party's position will also force you to listen better.
6. *Identify the key point and stick to it:* Through focused discussion each element of the proposal can be evaluated. If there are several options involved in the proposal, make it clear what can be modified and what must be implemented exactly as proposed.
7. *Don't digress from the key point and keep the other party from digressing:* There are three ways to avoid digression: agree to some nonessential point, agree to discuss it later, or identify the intrusion as being somewhat off the point.
8. *Be "for" a point of view, not "against:"* Accentuating the benefits of a proposal rather than deriding others viewpoints is usually the best tactic. While an opinion is often required, make it clear that you respect the other party's efforts and viewpoint. There is rarely gratuitous opposition; the other side has a reason for its position.

Negotiations will not always succeed, and when they do not the importance of the proposal must be evaluated in deciding what to do next. If the issue does not require immediate action to prevent irreparable harm, a decision to delay or cancel the project can be accepted. In this case the proposal might be re-evaluated and presented in a modified form at a later date when more data becomes available or conditions change. If, however, there is an imminent danger or other ethically unacceptable concern, further action must be taken. This action could consist of gathering more data and presenting the proposal again, elevation of the issue to a higher level in the organization, or involving external assistance. Each of these options is progressively more difficult and potentially costly to the professional. When taking these more drastic steps, facts must be checked and rechecked, and evaluation of the issue by another professional is appropriate. These "reality checks" are necessary to ensure that the action required is worth the risk involved.

Each of the EHS professions has it own code of ethics. The "Canon of Ethical Conduct" from the "Code of Ethics for the Practice of Industrial Hygiene" reads:

Occupational hygienists shall:

1. Practice their profession following recognized scientific principles with the realization that the lives, health and well-being of people may depend on their professional judgment and that they are obligated to protect the health and well-being of people.
2. Counsel affected parties factually regarding potential health risks and precautions necessary to avoid adverse health effects.
3. Keep confidential personal and business information obtained during the exercise of IH activities, except when required by law or overriding health and safety considerations.
4. Avoid circumstances where a compromise of professional judgment or conflict of interest may arise.
5. Perform services only in their areas of competence.
6. Act responsibly to uphold the integrity of the profession.

Such codes leave little room to feel that one's duty has been discharged by simply notifying management or the client of a serious problem. Technical professionals must ensure that proper action is taken to correct situations of potential serious harm. In most cases those potentially affected will be otherwise unaware of the risk they face.

Team Dynamics

The era of functional areas acting independently in an organization is dwindling. There is currently a recognition that to work efficiently all parts of an organization must work together. Where formality is appropriate, this is often accomplished through the use of teams. Teamwork requires shared responsibility and accountability. Each participant of the team comes to rely on the other members to help ensure the team's success. There is diminished emphasis on individual contributions, developed and presented in isolation.

Team members are selected based on their expertise and their potential contribution to solving the problem at hand. Each member's contribution is critical to team success, and there must be a commitment on the part of the individual and management to seeing the process to completion. If a team is to be successful, all members must feel empowered to participate on an equal level with all other members, without regard to their rank outside the team. Team dynamics will determine the success or failure of the team, as all products should be developed by consensus. The most useful decisions are developed by exploring all views and options and then reaching agreement.

An essay from Zenger Miller Inc.[8] describes the required attributes of a successful team:

- Clear, articulated vision and sense of mission;
- Strong, visible support from senior management;
- Long-term, organization-wide commitment
- Sharing of information;
- Transfer of real authority and decision-making power;
- Skills training to support the assumption of new duties and new ways of interacting;
- Continued coaching of teams and feedback from team leaders; and
- Systems that support teams.

With the evolution of the workplace from a top-down-directed structure to one where responsibility is distributed, teams have become essential. Teams can succeed only when they clearly understand the boundaries of their authority and responsibility. This information can best be conveyed when the team is established. Senior management should work with the team to ensure that the limits the team faces and its mission are clearly understood. At the same time, training should be provided to the team members, and particularly the team leaders, to prepare them for the interaction they can expect as members of a team.

Summary

Key elements of management science and its application to effective control of the EHS process have been delineated in this chapter. By using the tools and approaches described, EHS professionals can help ensure that their programs become part of the standard management approach used by their organizations. EHS management often suffers from a belief that the issues faced are too unique to be managed using standard techniques. Use of common tools such as TQM and benchmarking, however, can enable the technical professional to speak the language of the organization and be part of the process rather than an afterthought.

References

1. **Peter, T.:** *Thriving on Chaos; Handbook for a Management Revolution.* New York: Alfred A. Knopf, Inc., 1988.
2. **Lorange, P.:** *Strategic Planning and Control: Issues in the Strategy Process.* Oxford, U.K.: Blackwell Publishers, 1993.
3. **Bellman, G.M.:** *Getting Things Done When You are Not in Charge: How to Succeed from a Support Position,* San Francisco, CA.: Berrett-Koehler Publishers, 1992.
4. **Mackay, H.:** *Beware of the Naked Man Who Offers You His Shirt.* New York: William Morrow & Co., 1990.
5. **McGregor, D.:** *The Human Side of the Enterprise.* New York: McGraw-Hill, 1960.
6. **Champy, J.:** *Reengineering Management: The Mandate for New Leadership.* New York: HarperCollins, 1995.
7. **Senge, P.M.:** *The Fifth Discipline: The Art and Practice of the Learning Organization.* New York: Doubleday Currency, 1990.
8. **Berrey, C., A. Klausner, and D. Russ-Eft:** "Highly Competitive Teams: The Key to Competitive Advantage." San Jose, CA.: Zenger Miller, 1993. [Essay]
9. **Forum Institute:** *Individual Leadership.* Chicago, IL.: Forum Corp., 1995.
10. **Bird, F.E.:** *Loss Control Leadership,* Loganville, GA.: International Loss Control Institute, 1985.
11. **Heller, M.:** Pay attention now or pay dollars later, *Logistics Spectrum 30(2)*:7–13 (April 1996).
12. **Gordon, A.W. and R.W. Michaud (eds.):** *Principles of Environmental Health and Safety Management.* Rockville, MD: Government Institutes, 1995.
13. **Karrass, C.L.:** *Give and Take: The Complete Guide to Negotiating Strategies and Tactics.* New York: Thomas Y. Crowel, 1974.

Appendix A

Example Corporate EHS Policy

Policy

The health and safety of our employees is a primary consideration in all company operating decisions. The success of the company EHS program is directly related to the level of management support it receives. Therefore, it is the firm and continuing policy of the company to:

- Conduct all operations in a manner that complies with applicable laws designed to ensure the protection of our employees, facilities, local community, and the environment; and
- Hold management responsible for the compliance of their operations with applicable laws and requirements concerning the workplace health and safety of employees.

Responsibilities

The president and chief executive officer shall be responsible for:

- Ensuring that each unit is aware of its responsibilities under this program;
- Including EHS program compliance in unit performance reviews;
- Approving company EHS policies; and
- Ensuring that sufficient funds are provided to allow proper operation of this program.

The unit general manager or a delegate shall be responsible for:

- Including EHS considerations in operating decisions and performance reviews;
- Selecting a qualified EHS program coordinator who has an appropriate background to implement and manage an effective EHS program;
- Allocating sufficient resources to maintain an effective program;
- Approving unit EHS policies; and
- Notifying the company director, environmental health and safety, during the same business day of an incident that involves a fatality or environmental, safety, or health circumstance likely to subject the company to adverse consequences.

The company director, environmental health and safety, shall be responsible for:

- Guiding management in the establishment of a sound EHS program and for coordinating defense efforts in achieving stated EHS goals;
- Visiting company facilities to review EHS performance;
- Advising company management of the compliance status and occurrences relating to the environmental health and safety program;
- Providing consulting services to company units;
- Receiving and reviewing unit inputs and providing executive summaries for company management;
- Preparing and maintaining company EHS policies and reporting forms;
- Administering ongoing EHS programs; and
- Reviewing facility lease, acquisition, or divestiture plans and other capital requests to ensure that EHS issues have been properly considered.

Appendix B

XYZ Company's EHS Resources Analysis Worksheet Instructions

Section 1: Task Details

1. In Column 2 (Estimated Annual Workdays) indicate the estimated total number of annual workdays required to complete the task indicated in Column 1. Use blanks in Column 1 to describe tasks unique to your location or otherwise not listed on the form.
2. In Column 3 (Cost Excluding Personnel) indicate the estimated cost of meeting the requirements of the task indicated in Column 1.

   Note: Where data is available it should be used to make these estimates as accurate as possible. However, it is important to recognize that this process will only provide an estimate of a location's required resources. The results of this exercise should be used as a data point in discussions of needed resources.

3. Sum each column as indicated and transfer the data to the XYZ Company's EHS Annual Workdays Estimate sheet. (**Note:** This transfer and subsequent calculations will be performed automatically when using the computer spreadsheet.)

Section 2: XYZ Company's EHS Annual Workdays Estimate

4. The summation data from each section should be entered in the appropriate 75% block.
5. The 75% data is then extrapolated to 100% using the following formula:

$$\frac{75\%\ \text{sum}}{3} \times 4 = 100\%\ \text{estimate}$$

Example:

$$\frac{210\ \text{safety days}}{3} \times 4 = 280\ \text{days at}\ 100\%$$

Note: This assumes that only 75% of an individual's time can be identified with specific tasks.

6. These days are then converted to individual employees using the following formula:

$$\frac{100\%\ \text{days}}{100} = \text{number of required full-time equivalents}$$

Example:

$$\frac{280\ \text{days}}{250} = 1.12\ \text{full-time safety equivalents}$$

Full-time equivalent = personnel resources equal to a full-time employee.

Section 3: Level Analysis

7. Complete the questionnaire by placing the indicated weighting score (Column 2) in each box in Column 3 (Score) that applies to your facility.

8. Determine the minimum level of personnel your location requires by summing the score column and comparing your unit total with the ranges in Table 37.1B. See Table 37.2B for examples of skill categories.

Table 37.1B —
Scores and Qualification Levels

Total Score	*Qualification Level*[A]
0–3	locally developed EHS coordinator
4–6	technician level
7 or greater	professional level

[A]Minimum level for at least one of the site EHS personnel

Table 37.2B —
EHS Skill Competencies Required to Meet XYZ Company's Commitment

Professional Level Skill Set
(required by units with significant EHS exposure)

1. Knowledgeable of all federal, state, and local regulations.
2. Knowledgeable of all XYZ Company's policies and requirements.
3. Able to function as a creditable resource to comment on pending legislation and be an adviser within appropriate industry groups.
4. Familiar with upcoming trends in environmental legislation.
5. Knowledgeable about the various solutions to complex environmental problems.
6. Able to direct company policy to establish leading-edge programs.
7. Key member of company's senior management team.
8. Able to work effectively with line managers and plant employees at all levels.
9. Professional degree or equivalent experience.

Technician Level Skill Set
(required by units with moderate EHS exposure)

9. Active role in the EHS management of a particular company.
10. Primary responsibility is EHS (little shared responsibility).
11. Strong commitment to XYZ Company's EHS program.
12. Able to work with line managers and plant employees.
13. Working knowledge of federal, state, and local regulations and technical requirements.
14. Technician level degree or equivalent experience.

Locally Developed EHS Coordinator
(required by units with minimal EHS exposure)

15. Familiar with XYZ Company's environmental and audit protocols.
16. Possible shared responsibility with EHS and other disciplines outside the environmental field.
17. Committed to XYZ Company's EHS program.
18. Able to work with line managers and plant employees.
19. Familiar with applicable regulatory requirements for a localized business.

Task Details Worksheet

General EHS	Estimated Annual Workdays[A]	Cost (Excluding Personnel)
Reports		
XYZ Company's EHS self-evaluation		
Company headquarters		
Metrics		
Other		
Inspections		
Qualitative audit program		
Consultant		
Company staff		
Insurance		
Program		
State/local		
Subcontractor/vendor		
Other		
Training/Meeting		
Coordinator education		
Emergency response		
Hazard communications		
Supervisory		
XYZ Company		
Trade/professional associations		
Other		
Programs		
Department meetings		
Management presentations		
Procedure/process review		
Contract/proposal input		
Community/regulator relations		
Integrated product development		
Industry/professional groups		
Budgeting		
Due diligence		
Other		
Total General EHS (sum of columns)		

Task Details Worksheet

Industrial Hygiene	Estimated Annual Workdays[A]	Cost (Excluding Personnel)
Reports		
Chemical inventory		
Regulatory		
Other		
Inspections		
Regulatory		
Other		
Training		
Hearing conservation		
Respirator		
Indoor air quality		
Other		
Programs		
Asbestos		
Labeling		
Lasers		
MSDS review/management		
Radiation		
Ventilation		
Other		
Monitoring		
Atmospheric		
Noise		
Radiation		
Other		
Total Industrial Hygiene (sum of columns)		

Task Details Worksheet

Environmental	Estimated Annual Workdays[A]	Cost (Excluding Personnel)
Reports		
Waste water discharge		
Hazardous waste generator		
Hazardous waste manifests		
Emissions (i.e., SARA)		
Chemical management (i.e., TSCA)		
Permitting/authorizations		
Air		
Polychlorinated biphenyls (PCBs)		
Sewer		
Hazardous solid waste		
Nonhazardous solid waste		
Storage vessels (i.e., tanks)		
Groundwater/surface water		
Other		
Inspections		
Regulatory agencies		
Subcontractor		
Waste disposal facilities		
Other		
Training		
Waste management/handling		
Spill prevention		
Chemical handling		
Waste minimization		
Other		
Programs		
Remediation		
Hazardous/special waste disposal		
Medical waste disposal		
Other		
Monitoring		
Air		
Water		
Total Environmental (sum of columns)		

Task Details Worksheet

Safety	Estimated Annual Workdays[A]	Cost (Excluding Personnel)
Reports		
Accident		
Emergency preparedness		
Loss prevention/loss control		
Liability		
Workers' compensation		
Other		
Inspections		
Regulatory		
Fire systems		
Departmental		
Other		
Training		
Accident investigation		
Job safety analysis		
Specific operation		
Ergonomics		
Driver		
Fire prevention		
Other		
Programs		
Department meetings		
Purchase review		
Safety committees		
Product safety		
Fire prevention		
Other		
Permitting		
Building		
Hotwork		
Radiological		
Confined space		
Other		
Total Safety (sum of columns)		

Task Details Worksheet

Medical	**Estimated Annual Workdays**[A]	**Cost (Excluding Personnel)**
Reports		
Accident/near miss		
Records management		
Absence paperwork		
Workers' compensation		
Other		
Training		
Bloodborne pathogens		
Medical counseling		
Health education		
Other		
Programs		
Consultations		
Examinations		
Emergencies		
Pre-/postplacement evaluation		
Other		
Total Medical (sum of columns)		
Total EHS (sum of all disciplines)		

[A]Data is for scheduled activities that account for approximately 75% of workload; does not account for clerical assistance.

EHS Annual Workdays Estimate	
General EHS workdays 75%	
Extrapolated to 100%	
Number of general EHS employees	
Safety workdays 75%	
Extrapolated to 100%	
Number of safety employees	
Industrial hygiene 75%	
Extrapolated to 100%	
Number of IH employees	
Environmental 75%	
Extrapolated to 100%	
Number of environmental employees	
Medical 75%	
Extrapolated to 100%	
Number of medical employees	
Combined EHS 75%	
Extrapolated to 100%	
Number of EHS employees	
Current staffing	
Full-time equivalents required	

Level Analysis	**Weighted Score**	**Score**
Population		
Fewer than 100 employees	0	
100–300 employees	1	
300–500 employees	3	
500 or more employees	5	
Hazardous chemical use[A]		
Extremely hazardous substances	1	
Greater than 10,000 pounds per year	1	
Greater than 100,000 pounds per year	1	
Remedial activities[A]		
Water	1	
Soil	1	
Processes[A]		
Plating	2	
Machine shop	1	
Spray coating	1	
Laser user	1	
Ionizing radiation (X-ray, radioactive materials)	2	
Radiofrequency sources	1	
Heavy equipment use	1	
Fleet operations	1	
Construction	2	
Confined spaces	1	
Cleaning operations	1	
Other		
Human factors[A]		
Ergonomic issues	1	
Total Score (total facility weighting)		

[A]Select all that apply.

Outcome Competencies

After completing this chapter, the user should be able to:

1. Define underlined terms used in this chapter.
2. Differentiate baseline and comprehensive surveys.
3. Differentiate surveys from audits.
4. List the elements that comprise an industrial hygiene audit.
5. Contrast the industrial hygiene survey with the industrial hygiene audit in terms of goals and outcomes.

Key Terms

baseline survey • comprehensive survey • full-shift sampling • grab sampling • occupational hygiene audit • occupational hygiene survey

Prerequisite Knowledge

Basic chemistry, biology, and general occupational hygiene background.

Prior to beginning this chapter, the user should review the following chapters:

Chapter Number	Chapter Topic
1	History and Philosophy of Industrial Hygiene
3	Legal Aspects of the Occupational Environment
5	Epidemiological Surveillance
6	Principles of Evaluating Worker Exposure
10	Sampling of Gases and Vapors
12	Sampling and Sizing Particles
13	Biological Monitoring
15	Comprehensive Exposure Assessment
16	Modeling Inhalation Exposure
17	Risk Assessment in the Workplace
31	General Methods for the Control of Airborne Hazards
37	Program Management

Key Topics

I. Occupational Hygiene Surveys
   A. Survey Types
   B. Survey Methods
   C. Survey Quality
   D. Survey Forms

II. Occupational Hygiene Audits
   A. Management Commitment
   B. Audit Philosophy
   C. Audit Scope
   D. Internal vs. External Audits
   E. Preparing for an Audit
   F. The Audit Team
   G. Audit Elements
   H. Audit Logistics
   I. Audit Report
   J. Postaudit Actions
   K. Legal Concerns

38

Surveys and Audits

Frank J. Labato, CIH, CSP

Occupational Hygiene Surveys

The Occupational Safety and Health Act of 1970 requires safe and healthful working conditions for every working person in this country. One of the primary means of following through on this mandate is to conduct inspections and surveys of the workplace to ensure that standards are being met, and that the work environment is generally free from recognized health hazards that are likely to cause death or serious physical impairments. An occupational hygiene survey is an activity carried out by a qualified individual to measure and evaluate various biological, chemical, ergonomic, and physical parameters in the workplace or environment to determine their standings in relation to established health and safety standards. By utilizing this proactive mechanism to identify and quantify existing or potential health and safety hazards, appropriate measures can be taken to upgrade the work environment and reduce the risk of injury and illness. The goal of an occupational hygiene survey is to prevent or control foreseeable workplace hazards through a systematic identification and evaluation process. Surveys should be conducted periodically to track known potential hazards and also to detect new or missed hazards.

Whenever any type of survey is conducted, it is important to involve the workers in the process and to use the time as an added means of communicating to employees the nature and extent of the hazard, the measures they can take to protect themselves, and the reasons for continued monitoring. These discussions can help the employees understand the hazards and realize their own responsibilities in minimizing potential overexposures through proper work practices and correctly wearing assigned personal protective equipment (PPE).

Surveys are useful to detect not only outlying conditions, i.e., levels of contaminants that do not conform to regulatory standards or to internal facility requirements, but also to discover unsafe practices that may adversely affect employee health and safety.(1) Surveys may be done with a prescribed regularity and placed on a checklist as part of a routine monitoring program. They may also follow an irregular schedule and be conducted on an as-needed basis (e.g., new machine start-up, seldom-run operations, or one-time events, including spills). Though the initiating reasons may vary, the primary purpose of surveys—to evaluate and reduce risks—remains the same. Often this takes the form of a quantitative assessment with some instrumental device, ranging from a sampling pump with the appropriate sorbent collector to direct-reading instrumentation.

The importance of employing only qualified individuals to perform occupational hygiene surveys cannot be overemphasized, since the prevention, recognition, evaluation, and control of workplace health and safety hazards can only be as effective as the experience and aptitude of the person conducting the evaluation. It is easy to imagine two individuals walking through the same area and one identifying several areas of occupational risk that merit follow-up testing, while the other sees only a few of the very obvious hazards. If an individual is well-versed primarily in airborne chemical assessments, and thus selectively sees only these types of problems, other problems such as noise, electrical, and machine guarding hazards may remain undetected.

Survey Types

A baseline survey refers to the initial evaluation of a health and safety parameter that will be used as the basis for all future comparisons. A baseline survey may be the very first evaluation of a process at start-up, or it may be the first recorded measurement of a long-standing process. The scope may focus on a single airborne contaminant or physical agent, or be expanded to include several factors that have not previously been adequately examined and documented. Baseline surveys may evaluate the following agents individually or in combination: airborne levels of particulates including asbestos, silica, metal fumes, wood dusts, or a variety of nuisance-type dusts; any number of chemical vapors and gases for which there are occupational exposure standards; ambient temperature levels for heat and cold stress; noise; light; nonionizing radiation; vibration; lasers; and ionizing radiation.

Obviously, it is preferable to characterize a process or an operation at its inception to obtain an accurate description of a given activity on day number one. However, if an initial assessment was not conducted at start-up, the first evaluation becomes the surrogate baseline for assessing the current level of risk of worker overexposure or injury. Also, the effectiveness of engineering controls to mitigate airborne contaminant or physical agent levels can be determined. From either a physical measurement or visual evaluation standpoint, the baseline survey represents the beginning or reference point against which future readings will be gauged.

The primary reason to conduct any type of survey must be worker protection through the assessment of the risk of injury or illness from working in a defined environment. Another important reason to perform a survey is to determine the level of compliance with existing Occupational Safety and Health Administration (OSHA) standards or internal facility standards that may be some percentage of the OSHA standard. Surveys may also be conducted to ascertain standings compared with a particular national consensus standard (e.g., the American National Standards Institute, the American Society of Heating, Refrigerating and Air-Conditioning Engineers, the American Conference of Governmental Industrial Hygienists [ACGIH], etc.). However, only the OSHA standards have the force of law behind them; the latter standards may have legally binding aspects only if they are incorporated into federal or state statutes.

A comprehensive survey is a multifaceted examination of all recognized health and safety hazards in a defined work environment. It encompasses a wall-to-wall evaluation of the health and safety factors that may place an employee at increased risk of developing an adverse effect from working in a specific setting. It is apt to include an evaluation of a wide range of chemical and physical agents as a result of a "walkaround" inspection of the workplace for apparent and potential hazards. For example, a comprehensive occupational hygiene survey of a machine shop could include measurements of noise levels, airborne oil mist levels, degreasing solvents used, metal fumes from welding operations and metal dusts from grinding operations, lighting levels, and exhaust ventilation capture velocities at contaminant control points. It should evaluate the presence of machine guarding and required clearances; shields to contain ultraviolet light from welding operations; the secure strapping of gas cylinders to prevent falling; storage areas for full and empty gas cylinders; job-required eye, respiratory, hearing, face, foot, skin, hand, and head protection; identification and hazard labels on chemical containers; a master list of all hazardous chemical products; the presence of material safety data sheets for these products; adequate hand-washing facilities; unobstructed general and emergency exits; unobstructed and labeled electrical panels; and general housekeeping.

Survey Methods

Whether one is conducting a baseline survey or a comprehensive evaluation, the measuring techniques used may determine the validity and applicability of the test results. Grab sampling employs a short measurement period for quantifying the level of airborne contaminant or physical agent present. It is often associated with a direct-reading instrument or colorimetric detector tube. While the grab sample approach takes the least time to conduct, it also carries the heaviest burden when an attempt is made to characterize the entire day based on several seconds' or minutes' worth of monitoring data. In the strictest sense a grab sample result is a snapshot of the ambient concentration at that specific moment in time, and any inferences about full-shift exposures must be qualified by stating certain assumptions, e.g., the constancy of a process. For example, if one looks solely at the numerical value of an OSHA 8-hour permissible exposure limit (PEL), it is easy to see how a single grab sample result could mislead a person into concluding that the full shift's average airborne concentration falls within a regulatory limit. In reality, the actual 8-hour time-weighted average and 15-minute short-term exposure may have exceeded the PEL or short-term exposure limit (STEL), respectively, depending on the time interval during which the grab sample was obtained. If the process or operation is very steady and continuous, then the ability to extrapolate the grab sample results to an 8-hour average concentration is strengthened. However, there are very obvious pitfalls if the operation is highly variable and measurements are made either at a peak concentration/level or during a low point. Therefore, the potential exists to report a significantly erroneous result, and the resulting strategy may not adequately protect the workers' health, or it may be an unnecessary expense based on an incorrect presumption that the measured levels are representative of the entire day.

Full-shift sampling collects survey information over the course of an 8-hour day to determine the actual 8-hour time-weighted average exposure concentration. It is the preferred method to use to accurately characterize an operation for the gathering of baseline and routine monitoring data and to document regulatory compliance. Even though OSHA has adopted STELs for many chemicals, compliance determinations for hundreds of agents are routinely based on personal sample results that cover the entire work shift and provide an averaging of the minute-to-minute fluctuations that normally occur during the shift.

Survey Quality

The accuracy of occupational hygiene exposure monitoring depends on several factors, including the use of proper equipment; calibration of the equipment before and after sampling; specifying the correct analytical method; comparing results to the proper standard; issuing appropriate recommendations to minimize employee exposures and risk; and accurate and complete record keeping of all pertinent survey information. It is especially important to calibrate and operate equipment according to the manufacturer's instructions and to use only supplies and repair centers that are approved by the equipment maker to maintain operational validity.

Flow rate is critical when conducting air sampling with a personal or area pump, since airborne concentrations and the regulatory status will depend heavily on the determination of the air volume sampled. The main function of an analytical laboratory is to report results in terms of mass, e.g., milligrams or micrograms of material present. This mass is then divided by the volume of air sampled to yield the concentration of the contaminant in air in terms of milligrams per cubic meter or parts per million. Therefore, if a pump stopped or its flow rate changed during the sampling process, either due to an operational problem or employee tampering, the resulting concentration could vary greatly from the actual average airborne level. Another source of possible error relates to the effects of temperature and pressure on the volume of air sampled. The OSHA PELs and the ACGIH threshold limit values (TLV®) are based on normal temperature and pressure (NTP), which is defined as 25°C and 760 mmHg. Charles's, Boyle's, and Gay-Lussac's laws describe the effects of temperature and pressure on air volumes. When these laws are taken into account, significant error can result if field air volumes are not corrected back to NTP. Once corrected, the resulting airborne concentration can be legitimately compared with the appropriate PEL or TLV.(2)

Another important aspect to recognize is the dynamic nature of the workplace. Whether one is conducting sampling to assess the extent of a chemical concentration in air or to measure the level of a physical agent like noise or heat, the reading or result is not a constant but will vary the next hour, day, week, etc. Variability is an inherent aspect in the measurement process that must be examined in light of random occurrences, process or operational fluctuations, and seasonal effects. The concept of statistical distributions, confidence intervals, and confidence levels can be seen as playing an important role in the quantification and assessment of worker exposures to the agents of interest. It is much more meaningful to state that the measured level is 14 ppm with a standard deviation of ± 0.7 at the 95% confidence level, than to simply say that the level is 14 ppm.

Survey Forms

Survey forms are quite useful as a type of checklist to ensure that various pieces of

Table 38.1 — Sample Survey Information Categories

Employee name
Social security or employee no.
Job title
Length of time at current assignment
Shift
Survey date
Engineering controls in use
Sample shipment date
Personal protective equipment worn
Priority: normal or rush
Work area location
Name of facility
Address of facility
Facility phone and fax no.
Agent being sampled or evaluated
Work activity or process
Adverse symptoms
Start and end time of sampling or evaluation
Sampling or evaluation duration (minutes)
Flow rate of pump (cc/min, L/min)
Volume sampled (liters)
Pump serial number and calibration date
Collection medium
Sample number
Total number of samples and blanks
Direct-reading instrument used
Calibration date
Temperatures, wet and dry bulb
Humidity
Analytical method
Laboratory name and address
Barometric pressure
Airflow/wind velocity and direction
Surveyor's name and title, etc.
Result (ppm, mg/m^3, °F, °C, dB(A),%, ft/min, etc.)
Allowable level (PEL, TLV, %LEL, °F, °C, etc.)
Sampling, operation and analytical comments

information are collected for analytical, concentration calculation, and record keeping purposes. The actual categories of information vary according to the kind of survey being conducted, but generally include those listed in Table 38.1.

Occupational Hygiene Audits

An occupational hygiene audit is a process used to evaluate periodically the existence and the effectiveness of biological, chemical, ergonomic, and physical health and safety program elements present in a workplace. An audit can prove to be a key tool in determining the adequacy of the practices, procedures, and management controls necessary to achieve internal or external goals.[3] A primary goal may be directed toward regulatory compliance alone, or in combination with a firm's own internal standards for maintaining a healthy work force and acceptable levels of environmental effluents.

If all necessary practices, procedures, and management controls are in place for a given workplace, then the purpose of an audit becomes an adherence issue. If the minimum internal controls are not in place, then the audit functions to identify those areas that are not adequately addressed to meet the facility's desired goals, including maintaining a healthy work force, low turnover of personnel, a decrease in workers' sick days and compensation claims, regulatory compliance, avoidance of monetary fines, third-party torts, and adverse publicity.

An audit will usually determine if the facility's means of addressing routine or unanticipated events follow established facility procedures and whether those procedures are adequate to satisfactorily handle those events. The audit can also serve a very useful function in determining if there is a prudent distribution of resources within a facility and in bringing important issues to the forefront for prioritizing.[3]

Management Commitment

The most important phase of an audit starts well in advance of the actual auditing function. It begins with a management commitment to excellence. The top-level leadership in an organization must buy-in to the process, the advantages of determining the status of key issues in advance of regulatory discovery and, of prime importance, to implementing all necessary corrective actions once deficiencies are identified. Thinking that all problems found by an audit must be rectified overnight would be a nightmarish burden. Rather, the whole purpose of an audit revolves around risk assessment, risk management, and risk reduction, which, in turn, seek to preserve the ultimate valuable resource: an experienced and skilled work force.

The audit process represents a systematic approach to the comprehensive examination of functions that must be in place and the strategy or game plan that details how to reach that desired end. At the conclusion of a successful audit the organization should have a neatly compiled collection of items with differing degrees of deficiencies and levels of seriousness from the standpoint of protection of human health and the environment. Once this list is compiled into categories based on priority and ease of remedy, the next phase reverts back to the organization's commitment to excellence. The most serious pitfall a facility can encounter is the undertaking of an audit and the acceptance of the audit findings with no strategy, or worse no intention, in mind as to correction. An audit report stashed in the back of a file drawer due to inadequate financial or personnel resources becomes a smoking gun for the organization and a regulatory compliance whip for the agencies. The precept that cannot be overstated is that no audit should be undertaken without a strong commitment to correct the identified regulatory inadequacies and to follow through with the compliance-based recommendations.

Audit Philosophy

An audit can be a tremendous learning experience for the facility embarking on this exhaustive probe.(4) Often there is no clear definition of responsibilities within an organization. In the aftermath of an incident it is sometimes heard that a certain worker thought that another worker was supposed to perform a specified function. Therein lies the reason why some things fall through the cracks. Therefore, a major accomplishment of an audit can include the delineation of areas of responsibility, which results in an effective mechanism for covering the wide range of regulatory requirements. Actions to be taken at the conclusion of an audit can be aptly paraphrased using the five Ws: who, what, when, where, and why; or in slightly different order, what are we supposed to do, why do we need to do it, when (or how often) do we need to do it, who is responsible for doing it, and where is the reference location for this required action. A product of the audit may include an itemized listing of the above that can be used as a checklist to ensure future compliance. Placing this checklist on a computerized database will provide additional assurance of timeliness of mandated tasks.

An audit should not be viewed as a one-shot venture. Rather, the concept should be incorporated into a program that is repeated periodically to ensure continuous improvement. No audit is capable of detecting all of the deficiencies within the wide spectrum of occupational and environmental health and safety. Therefore, different approaches may be undertaken to systematically review this continually expanding arena. One such approach may be to focus on a few major compliance issues at a time, such as OSHA's standards on hazard communication, PPE, respiratory protection, hearing conservation, bloodborne pathogens, confined space, and lockout/tagout. These areas account for the majority of citations issued by the agency. Once these areas are examined and adequately addressed, attention can be turned to other key topics, including injury and illness record keeping, trenching and excavation, machine guarding, electrical safety, and required OSHA postings. The choice of areas to examine in the first round of visits will be determined in large part by the business nature of the site. However, after a few audit rounds have been completed, many major risk areas will have been systematically identified and addressed. The process should commence with the issues or hazards that have the greatest propensity to impact worker health and safety and to generate large fines and then proceed to those lower in priority.

An audit should be thought of more as a financial investment than as an expenditure, and as one forward step in a series of efforts toward continuous quality improvement. These persistent strides begin the process of gaining control over the issues likely to result in a fatality, serious injury, crippling environmental fine, or negative publicity, all of which can negatively impact potential customers' opinions and the bottom line. The systematic elimination of these items will facilitate efforts to concentrate on the remaining mandates considered less serious by the regulating agency.

Audit Scope

There is a plethora of regulations today in the combined health, safety, environment, transportation, and numerous national consensus standards that may either be incorporated into regulations by reference or become de facto regulations in the absence of specific laws in a particular area. Thus, the tendency may be to request a compliance audit for all pertinent regulations governing a facility's operations. While this is an admirable intention, it may result in bid proposals ranging in the hundreds of thousands of dollars and taking several months to complete. A more realistic approach is to focus on issues that represent the greatest risk in terms of worker health and safety and environmental damage. An examination of an agency's most numerous citations over the past few years combined with a consultant's input can help direct the audit undertaking toward significant issues with the greatest potential for harm. While ensuring that grinding machines have a written operating procedure is undoubtedly an important concern, perhaps confined space or hazardous waste handling procedures may outweigh machine guarding issues as potential hazards at a given facility. Therefore, to maximize the beneficial output of an audit, limited resources need to be directed to serious issues with an increased likelihood of occurrence and adverse effects.

If initial resources will allow, a more comprehensive audit can be undertaken that includes a wide range of issues encompassing worker health and safety (OSHA); the preservation of the environment in light of hazardous waste regulations (U.S. Environmental Protection Agency, EPA: Resource Conservation and Recovery Act, RCRA); hazardous material transportation (U.S. Department of Transportation, DOT); and radioactive material usage (U.S. Nuclear Regulatory Commission, NRC). An extensive audit of this scope will most likely require several specialists to determine programmatic adequacy, may extend over several months and could cost up to hundreds of thousands of dollars. However, even high audit costs may seem low in comparison to the actual costs associated with bringing the organization into an acceptable level of regulatory compliance, and also low in comparison to litigation costs if unsafe conditions lead to workers' compensation claims and third-party personal injury suits.

Internal vs. External Audits

Internal Audits

Ideally, an audit should be conducted by someone capable of making an unbiased and independent assessment of the status of a program.[(4)] It is for this reason that many firms hire an outside consultant to examine these issues. The reasoning is based on the presumption that a fresh set of eyes can, in many instances, see inadequacies that may be missed by personnel internal to the organization. Independent reviewers hired on a fee-for-service basis can also ask probing questions with little fear of future retaliation and impact on career advancement.

This in no way infers that an internally conducted audit serves no useful purpose. The success of an internal audit is highly dependent on the experience, motivation, and upper administrative support of an introspective self-evaluation. Choosing the members of the audit team is a particularly important task. In fact, the team's composition can be a predictor of the final audit quality. Therefore, the innate nature of an audit must be matched to the personality of the prospective auditor(s). An internal audit will generally assemble several individuals who are experienced within their respective disciplines and who know the facility, the operation, and its inner workings very well. They can "hit the ground running" and start collating data in short order. However, if the audit team members each see, record, and author their findings in their own individualistic way, a disjointed report may follow that lacks consistency of style and quality. Also, a facility employee may have the knowledge and experience to conduct an audit, but lack either the zeal or the probing style to uncover programmatic deficiencies.

There is also a certain amount of peer pressure involved when an internal audit is conducted. It is difficult to interview a co-worker in the morning and find several deficient items with potential regulatory impacts, and then to carpool with that same individual at the end of the day. The combination of peer and social pressures can negatively impact the essential and fundamental purpose of an audit: to evaluate programmatic adequacy and effectiveness and to reduce risk. Additionally, there are competing production pressures. An audit may be important, but customers' orders may repeatedly intervene and push the audit to the back shelf until it loses momentum. After all, without customers there would be no need for an audit, since there would be no business. The other consideration is that the audit team's backlog of routine job-related work may begin to mount up, and the audit is eventually fast-tracked so that life can return to normal.

External Audits

While internal audits may save money, an external audit team has the distinct advantage of being dedicated to the project with minimal interfering job pressures. The firm is hired for this one project and is responsible for delivering a product that is consistent in quality and finished in a timely manner, which helps to ensure that the impetus to excel remains a fresh priority. A report from an unbiased outside firm many times has a certain aura and the advantage of being perceived as presenting the true picture minus any local biases. External firms should dedicate to the project an arsenal of technical personnel who are highly competent. A signed contract ensures the successful completion of the evaluation and report within a specified time frame. These outside experts have no vested interest in reporting only positive findings. In fact, the competitive nature of the business helps to assure a quality product from an in-depth standpoint and the opportunity for repeat business.

Provided that the most important phase of the audit process is management commitment and follow-through, it is essential that senior officials conduct an accurate, open, and in-depth assessment of the true state of affairs. There is nothing more embarrassing to a facility than to undergo an audit that awards high praise, only to have the audit followed up by a regulatory inspection carrying costly fines and negative publicity. A less-than-accurate picture of a facility's compliance status wastes both time and money. Thus, if an audit is to be done, then it must be done correctly.

Preparing for an Audit

A planned audit should begin with a preaudit questionnaire (see Appendix A) in advance of the site visit. The purpose of the questionnaire is to provide the auditor(s) with necessary background information that will help focus on the areas deserving the greatest scrutiny. Properly completed preaudit questionnaires also help lower costs by decreasing the time that the auditor(s) must spend on-site gathering background information. For this reason alone it is absolutely necessary to appoint an on-site coordinator who acts as the key liaison between the facility and the audit team. All audit information should pass through the hands of this coordinator to ensure that the audit flows in a seamless and organized manner. As the need arises, obviously this point-person can delegate fact-finding responsibilities as necessary. However, this person must have overall, in-depth knowledge of the audit issues and time frame so that management can be apprised on short notice.

The preaudit questionnaire serves another important internal function. The act of data collection, collating policies, and completing the forms is an educational process for both the on-site coordinator and the various individuals gathering the documents.(4) Even before the audit team arrives on site, the coordinator will usually have a good estimate of the facility's overall compliance standing from examining the general pattern of responses to items on the questionnaire. Many issues are very evident: either the facility has a written hazard communication plan, or it does not; either there is a written policy governing the disposal of any hazardous chemical wastes generated, or there is not; either there is a written asbestos management plan in place, or there is not; etc. Therefore, even at the early stage of preaudit questionnaire completion, the opportunity exists to give senior management a glimpse of what may be ultimately seen in the final report. There is also a concurrent opportunity to effect some significant remedies that can reduce health, safety, and litigation risks in a timely fashion in advance of receiving the final report.

The Audit Team

The audit team should include qualified individuals experienced in the investigation process, and most importantly in the areas they will be examining. Individuals should have credentials that demonstrate a depth and breadth of knowledge in the specific topics under review. A certified industrial hygienist (CIH) is an individual who has met stringent educational and experience standards and who has successfully completed comprehensive examinations administered by the American Board of Industrial Hygiene. However, a CIH with 10 years' experience conducting asbestos investigations is not necessarily the best choice for a comprehensive trenching, confined space, and lockout/tagout audit. The CIH designation must be viewed in line with the individual's work history and experience. The audit team may also be composed of professional engineers (PE) in the mechanical, civil, and chemical disciplines; certified safety professionals (CSP); meteorologists; archeologists; hydrogeologists; and environmental planners.

The size of the audit team should match the size of the facility, the number of issues under study, and the number of facility individuals who can contribute to the process. One auditor assigned to a 3,000-employee facility to conduct a comprehensive OSHA and EPA audit is an obvious mismatch of resources to scope. Six auditors conducting an examination of OSHA issues at a 250-employee plant does not represent the optimal use of resources either. If the audit team is too small for the facility and the issues involved, then the process is likely to be lengthy and lose momentum. Conversely, if the team is too large, it will not operate efficiently due to an inadequate number of facility employees needed to conduct tours and supply needed information.

Audit Elements

Management Aspects

Perhaps one of the most critical elements of an audit is the assessment of the administration's attitude and level of comprehension of the multitude of regulations requiring compliance. Does senior management have a real-world sense of the number of regulatory mandates in existence, a feel for the level of staffing needed to meet these regulatory requirements, an impression of the current state of affairs for the facility, and a realistic level of commitment to reduce liabilities and risks by maintaining compliance? If upper management's primary concern is limited to product-making and profit, with little sense or even concern that major fines could seriously jeopardize the operation, then the audit results could prove to be extremely shocking. A mentality that accepts new or modified process changes and the associated development and implementation costs, but rejects local exhaust ventilation, medical surveillance, and air sampling costs does not truly appreciate the potential impact of workers' compensation claims and $25,000 per violation, per day environmental fines. If the attitude is based on a "fix it after we get caught" or "it's not worth wasting time or money on now" philosophy, then either a proactive audit would not even have been considered, or it would have been a superficial, rubber-stamp type of assessment made just for appearance.

Proactive management elements include:

- Internal means for evaluating existing policies and procedures for compliance with current regulatory requirements;
- Being aware of worker health and safety affairs by requiring an honest and in-depth status report with a "if it's broke, let's fix it" emphasis during staff meetings;
- Recruiting health and safety personnel and recognizing them as professionals and as part of the facility team; and
- Developing and fostering a safety-oriented culture where every suggestion is considered worthy of consideration, and there is a meaningful awards program that encourages active participation in the movement.

Program Effectiveness

Even when management's commitment to health and safety is strong, the effectiveness of the health and safety program depends on the caliber of the program's personnel and the internal policies and procedures in place to prevent, recognize, evaluate, and control workplace hazards. Programmatic elements that encourage timely and systematic means of surveillance, sampling, remediation, record keeping, and accountability will help ensure the continued quality of the program. If marginally qualified individuals are chosen to staff a health and safety office in order to contain personnel costs, then a marginally effective compliance strategy may emerge due to either lack of personal commitment or competence.

Technical Issues

If the audit is undertaken with the goal of positively impacting worker health and safety in a very direct fashion, then a multitude of technical areas will need to be examined (see Appendix B for a sample audit protocol). Potential hazards exist in the biological, chemical, ergonomic, and physical agent arenas, and audit emphasis will naturally concentrate on those with the highest prevalence.[4]

From a biohazard standpoint, the evaluation should focus on the use of biological organisms that are infectious and pathological to humans. Also of growing significance is the provision of first aid or other work requirements that may result in an exposure to bloodborne pathogens. Effective exposure control procedures incorporating PPE and training are the key elements that prevent workplace illness.

Chemicals are used extensively in a vast number of businesses, and thus the potential for workers to experience chemically induced illnesses or injuries or to file workers' compensation claims is very real. Auditing the effectiveness of a chemical hazard communication program includes assessing the adequacy of training. When the training has been effectively delivered, then recognition of the letters MSDS, knowledge of the location of the list of hazardous materials present and their corresponding material safety data sheets, maintenance of container labels, and familiarity with physical and health hazard terms will be demonstrable.

Ergonomic issues can have a profound effect on worker productivity and manifest themselves in the number of cases recorded on the OSHA 200 Log and Summary of Occupational Injuries and Illnesses, which in turn can have an escalating effect on workers' compensation insurance premiums. Evaluating the adequacy of prevention programs, intervention measures, and back-to-work approaches can lead to decreased employee turnover and the maintenance of a healthy work force.

Physical agents include noise, heat, cold, vibration, lasers, and ionizing/nonionizing radiation. A working hearing conservation program can help safeguard employee health, maintain compliance with OSHA's noise standard, and decrease workers' compensation claims. Noise-induced hearing loss claims due to nonoccupational exposures can be contested more effectively if an audit has revealed a well-managed program with annual training, annual audiograms, and the enforced wearing of approved hearing protection. Vibration can be a nuisance and affect concentration and productivity, or it can cause a disabling condition affecting manual dexterity that may impact employment status. Excess exposures to heat, cold, lasers, and radiation can result in serious personal injuries and specific and general duty clause citations. Regarding ionizing radiation, two very different federal agencies may be involved depending on the source. OSHA has a standard on X-ray radiation, while NRC has jurisdiction over licensed quantities of by-product material. If an NRC license is required for possession and use of a specific sealed or unsealed radioactive material, serious noncompliance can result in a loss of the license. If this radioactive material is used in an important facility operation, dire financial consequences may result. Claims of occupationally induced leukemia and other cancers can be challenged more successfully in the presence of an audit finding that demonstrates that only trained and qualified individuals have access to the controlled usage areas, and monitoring records indicate that contamination is controlled and worker exposures are below detectable, or at least regulatory, limits.

The need for and wearing of PPE can be an important issue from a prevention and injury claim standpoint. If respirators are required because product substitution or engineering controls are not feasible or adequate, then training that stresses the proper cartridge selection, wearing procedures, maintenance, and medical clearance can be evaluated to determine if the practices and the level of protection are consistent with airborne concentrations apt to be present in the work environment. OSHA's PPE standard places a hazard assessment and equipment training requirement on the employer. Are any nonapproved items of PPE brought to work from home, and is a mechanism in place to enforce the wearing of required PPE? If wearing PPE is a job requirement, is the equipment provided to an employee free of charge? Worst case scenarios depicting the improper use of nonapproved PPE, insufficient training, and little enforcement can, aside from the compensation claims, form the basis of costly willful violations. Fines can be magnified depending on the number of employees, the time frame involved, the seriousness of the situation, the facility's compliance history, and the perceived extent of good-faith efforts on the part of the facility to maintain compliance.

Exposure monitoring results form the foundation of important occupational hygiene decision-making involving recommendations for the initiation of medical surveillance programs and installation of engineering controls. Recommendations involving the significant allocation of funds for years to come may use air sampling results to document need. Worker exposure control strategies depend on monitoring results also. The importance of utilizing only qualified individuals who decide where, when, how, and for how long to sample is of utmost importance. Coupled with the need for qualified people is the need to use the proper type of equipment that is calibrated before and after use, and to maintain the operational characteristics of the instruments and pumps. Are samples sent to AIHA-accredited laboratories for analyses to ensure that rigorous quality controls are in place? Accreditation is an important benchmark indicating the increased level of confidence one can have in the laboratory's reported results.

Training programs addressing the various regulatory standards and internal policies and procedures must be delivered in an effective manner by knowledgeable individuals. Have training records been retained that list the worker's name, social security number, department, date, topic, and trainer's name, at a minimum? Was a test administered to demonstrate comprehension and was refresher training provided annually? Spot checks during the intervals between training sessions can indicate the level of retention and thus the need to modify previous methods.

Audit Logistics

An audit normally commences with an opening meeting during which the audit team can discuss any items on the preaudit questionnaire that need further clarification. This is also the time when senior management can reaffirm the facility's commitment to excellence through the systematic evaluation of internal policies and procedures in light of existing health, safety, environmental, and transportation regulations and the firm's own internal operating standards. The meeting discussions should also provide a cursory overview of the intended scope, time frame, progress report/meeting frequency, payment schedule, format of draft and final report, and method of delivery.

A preliminary tour of the facility should be led by the firm's audit coordinator.[4] The tour will enable the auditor(s) to ask any additional questions in light of the preaudit questionnaire and the opening conference. At the conclusion of the tour, a date should be confirmed when a specified auditor(s) will return to conduct an in-depth review. An advance schedule is necessary to allow the coordinator to contact those instrumental facility persons who are best qualified to answer specific questions pertaining to a specific operation. It is vitally important for the coordinator to be able to make the necessary facility contacts sufficiently in advance to confirm meeting dates and times with the audit team. Nothing is more detrimental to the success of an audit than to discover that a key facility person will be away on a scheduled vacation during the site visit. Advance scheduling is important; however, even this approach can be short-circuited by summer vacation requests. On the other hand, the summer months may be the optimal time for some operations that have heavy wintertime demands. Advance scheduling is also an opportune time to garner the active involvement of these key personnel. Optimally, the difficult task of securing employee involvement can be greatly facilitated by a memo from senior management early in the process. Increased levels of morale and a willingness to form a partnership with the administration can be greatly enhanced if the affected employees know the added work will be worth the effort, culminating in meaningful improvements.

At the conclusion of the site visit, a closing meeting is scheduled to discuss the audit findings briefly and to answer any unresolved questions. A date should be determined for the delivery of the draft and final reports. The closing conference should be structured to enable upper management to hear a concise summary of critical issues that require immediate action.[(4)] When the major items are presented in an abbreviated and prioritized format, senior officials can make important resource allocation decisions well in advance of receiving a written report. When the written report does arrive, it should contain no surprises and only serve to recapitulate what was itemized in the closing conference.

Audit Report

The product of the audit is normally a detailed report that addresses the scope, process, checklists, findings, conclusions, and recommendations from the audit. It is often helpful to include a one-page executive summary at the beginning of the report to apprise upper management of the overall status of the facility and any significant recommendations.[(4)] It is common for the audit report to be delivered in draft form to ensure that the information is accurate and that the format and style are consistent with past agreements. Once the draft is sent back to the audit team for any necessary changes, the final report delivery date can be set. It is common for the original contract to specify the number of draft and final reports to be issued to facilitate the review.

Report format and content can vary to suit the needs of the individual facility.[(4)] One approach may be to structure the report to list only the deficiencies. These items may be grouped by severity to facilitate prioritizing corrective actions. This report style has the advantage of being concise and to the point; however, it does not acknowledge the proactive steps that are being taken. Morale can be affected if the only newsworthy material is negative in nature. A different report style can list both the elements that were found to be in compliance and the deficient areas. This format can alert management to all of the problems, while at the same time providing information on the positive aspects that are being handled well by the employees. Therefore, areas in need of attention are denoted at the same time that recognition is given for items that do reduce risks. This type of report can provide senior management with a good overall perspective of facility status, e.g., 63 items that are in good standing versus 11 that need attention. Still another report type may state only a recommended course of action based on the observations made during the facility visit. Recommendations can be generic or specific and serve to point a facility in a certain direction (e.g., replace verbal procedures and agreements with written documentation) or assign responsibility (e.g., for material safety data sheet collection and storage) to one person. Finally, an opinion report can provide the relative degree to which compliance is met by subject area, and include phrases indicating that the facility is "wholly," "substantially," or "minimally" in compliance.[(4)] Whatever format is chosen, it should be stated in the contract, or at least agreed to before the preaudit questionnaire is delivered to the facility for completion. It is difficult for a product to be acceptable to a consumer if the producer and consumer are working from assumptions rather than from explicit contract language.

Postaudit Actions

Following the completion of the site visit, the closing conference, and the acceptance of the final report with copies for senior management, a plan of action to address any deficient areas must be established. If required in the contract, the audit report may be delivered on a user-acceptable database that will facilitate the tracking of each deficiency, the remedial action required, the responsible person, and the completion date. A database can keep the original spirit of the audit alive by permitting periodic status reports to be forwarded to management. This database can serve the extremely valuable checklist function to help ensure that nothing "falls between the cracks." Statistics detailing percent of serious deficiencies corrected and remaining items sorted by priority and deadline provide a tangible measuring stick for facility administrators.

Legal Concerns

There may be legal concerns regarding the confidentiality of audit findings. Questions may arise as to the inclusion of trade secret information or material that may address sensitive issues. If these are important matters, then legal counsel should be sought to determine the distribution and access to this report.[4] There is a continuing discussion as to the facility's vulnerability should the audit report fall into the hands of a regulatory agency. Discussions focus on the fact that an audit report forms a paper trail that can conclusively document that the facility had prior knowledge of violations. Therefore, if a compliance inspection finds these same violations noted in the audit report, they may be viewed as willful violations, which carry a much higher monetary penalty and possibly increased liabilities. These concerns should be addressed by legal counsel, which should be consulted to determine whether the audit findings may be sheltered from public or regulatory scrutiny under an attorney/client privilege or other legal mechanisms. Some states and the EPA have adopted policies that provide a certain amount of protection if an audit is conducted for the purpose of finding and correcting any compliance deficiencies. However, a self-disclosure of audit findings, along with an aggressive compliance strategy and timetable, sent to the appropriate regulatory agency prior to any inspection activity at that facility may be required to enjoy these privileges.

Additional Sources

Huey, M.A.: *Fundamentals of Industrial Hygiene*, 4th Ed. Itasca, IL: National Safety Council, 1996.

References

1. **Rice, P.B.:** The safety professional. In *Fundamentals of Industrial Hygiene*, 4th Ed. Itasca, IL: National Safety Council, 1996. pp. 675–699.
2. **DiNardi, S.R.:** *Calculation Methods for Industrial Hygiene*. New York: Van Nostrand Reinhold, 1995. pp. 124–164.
3. **Huey, M.A.:** The industrial hygiene program. In *Fundamentals of Industrial Hygiene*, 4th Ed. Itasca, IL: National Safety Council, 1996. pp. 749–758.
4. **Leibowitz, A.J. (ed.):** *Industrial Hygiene Auditing: A Manual for Practice*. Fairfax, VA: American Industrial Hygiene Association, 1995.

Appendix A. Pre-Audit Questionnaire

(From *Industrial Hygiene Auditing—A Manual for Practice*, edited by Alan J. Leibowitz, CIH, CSP, AIHA Press, 1994.)

Please fill out this questionnaire as completely as possible. The background information about your facility's operations will be used to tailor the scope of the audit and select suitable audit team members.

1.0 GENERAL INFORMATION

1.1 Identifying Information:

- Unit/Site Name: ____________________
- This questionnaire completed by: NAME: ____________________
- Date: _____/_____/_____ PHONE: ____________________

1.2 Location:

- Mailing Address (Administrative Office): ____________________
- Shipping Address: ____________________
- City: ____________________
- County: ____________________
- State/Province: ____________________
- Zip Code/Country Mail Code: ____________________
- Telephone Number: ____________________
- Telefax Number: ____________________

1.3 Facility Personnel:

Phone #

- General Manager: ____________________
- Site/Facility Manager: ____________________
- Operations or Production Manager: ____________________
- E/S Coordinator: ____________________
- Total number of employees at facility: ____________________
- How many workshifts are operated? ____________________

1.4 Facility Size:

- Facility floor space in square feet (include multiple floors in calculations): ____________________

____

- If unit/facility includes more than one building (at different address), please list all buildings and itemize their floor space in square feet.

____

____

- Site size in acres: ____________________

1.5 Facility Activities:

- Type of activities/processes at site (check all that apply and give approximate number of employees engaged in each).

YES	NO		NUMBER OF EMPLOYEES
____	____	Abrasive blasting	________
____	____	Acid/alkali cleaning	________
____	____	Adhesive bonding	________
____	____	Administrative/office work	________
____	____	Anodizing	________
____	____	Assembly operations	________
____	____	Casting of metal parts	________
____	____	Coating/painting	________
____	____	Degreasing/solvent cleaning	________
____	____	Drilling/machining	________
____	____	Glass/ceramic production	________
____	____	Grinding/polishing/buffing	________
____	____	Heat treating	________
____	____	Ionizing radiation (e.g., X-rays)	________
____	____	Lasers	________
____	____	Metal machining	________
____	____	Metal forging/stamping/forming	________
____	____	Packaging	________
____	____	Plastic formulation/extrusion/blow molding	________
____	____	Plating	________
____	____	Printing	________
____	____	RF/Microwave heating	________
____	____	Shipping/Receiving	________
____	____	Soldering/brazing	________
____	____	Wastewater treatment	________
____	____	Welding	________
____	____	Woodworking	________
____	____	Other activities of concern (specify)	________

- Is facility owned or leased:

Owned by: ____________________ Leased from: ____________________

____________________ ____________________

____________________ ____________________

- Standard Industrial Classification (SIC) Codes under which the facility operates:

____

____

____

____

____

____

____

____

2.0 EMPLOYEE SAFETY/INDUSTRIAL HYGIENE

2.1 Exposures/Controls:

	YES	NO	N/A
• Do you have any potential worker exposure to physical agents such as noise, radiation, heat, etc.?	___	___	___
• Do you have any potential worker exposure to chemical agents?	___	___	___
• Do you have any automatic monitoring or automatic alarm systems to detect hazardous materials?	___	___	___
• Do you routinely or regularly measure air contaminant levels for any particular substances?	___	___	___
• Have employees been informed of their exposure levels?	___	___	___
• Are exposure records kept in a way that lets you estimate exposures for job categories?	___	___	___
• Are all exposures within established limits for chemical and physical agents (e.g., ACGIH TLVs, OSHA PELs)?	___	___	___
• Are any areas designated as limited access "regulated" areas?	___	___	___
• Are any areas posted with hazard warning signs such as "high noise area," "respirators required," or "carcinogen area"?	___	___	___
• Are any ventilation systems used to control hazardous materials exposures?	___	___	___
• Do ventilation systems have:			
– Routine maintenance programs?	___	___	___
– Regular air flow rate tests?	___	___	___
– Regular filter changes?	___	___	___
• Were ventilation systems designed or approved by a qualified engineer or industrial hygienist?	___	___	___
• Are any other engineering controls used for control of hazards (e.g., noise enclosures, controlled-atmosphere control rooms)?	___	___	___
• Are any administrative measures used to limit employee exposures to hazards (e.g., job rotation, temporary job reassignments)?	___	___	___
• Does your facility have an Analytical or Quality Control Laboratory?	___	___	___
• Are any types of personal protective equipment (e.g., gloves, glasses respirators) available?	___	___	___
• Is there a written respirator program at the facility?	___	___	___
• Are there training programs related to hazardous materials?	___	___	___
• Is there a medical surveillance program including pre-employment exams and periodic follow-up exams?	___	___	___
• Is there a hearing conservation program that includes everyone exposed to noise in excess of 85 dBA?	___	___	___

2.2 Safety:	YES	NO	N/A
• Has the facility been reviewed within the past 2 years by OSHA or other internal or external reviewers?	___	___	___
• Does the facility have a safety manual or safety guidelines?	___	___	___
• Is there an up-to-date facility emergency response plan?	___	___	___
• Are records maintained on work-related injuries and illnesses?			
– OSHA record keeping?	___	___	___
– Corporate/division reporting?	___	___	___
– Incident reporting?	___	___	___
– Motor vehicle accident reporting?	___	___	___
– Dispensary log?	___	___	___
• Does your facility operate a motor fleet?	___	___	___
• Is there a written program for motor fleet safety?	___	___	___

2.3 Emergency Response:			
• Does the facility have an Emergency Response Team (ERT) or Fire Brigade (FB)?			
– Is there formal ERT or FB training?	___	___	___
– Are there routine evacuation/fire drills?	___	___	___
• Are there formal programs in place to provide for emergency care of the injured?	___	___	___
• Are there any off-site facilities that fall under your direct control? Please specify.	___	___	___

____

____

• Do you have programs in place for regular inspection of:			
– Fire detection systems?	___	___	___
– Sprinkler control valves?	___	___	___
– Water flow test on sprinkler system?	___	___	___
– Fire pumps?	___	___	___
– Fixed fire extinguishing systems?	___	___	___
– Fire extinguishers?	___	___	___
– Alarm systems?	___	___	___

3.0 E/S PROGRAM DOCUMENTATION

3.1 Employee Safety/Industrial Hygiene:	YES	NO	N/A
• Does the facility have any of its own specific policies, procedures, standards, or guidelines pertaining to:			
– Evaluating work exposures (i.e., chemical, noise, radiation)?	___	___	___
– Calibrating, testing and maintaining industrial hygiene sampling equipment?	___	___	___
– Quality assurance for analysis of industrial hygiene samples?	___	___	___
– Informing employees of industrial hygiene monitoring results?	___	___	___
– Review of Material Safety Data Sheets or other hazard communication information?	___	___	___

3.1 Employee Safety/Industrial Hygiene (continued):

	YES	NO	N/A
– Safety training program?	___	___	___
– Hazardous work permits (i.e., hot work, confined space entry, electrical lockout)?	___	___	___
– Use and maintenance of personal protective equipment other than respirators?	___	___	___
– Contractor on-site safety?	___	___	___
– Testing and maintenance of fire protection equipment?	___	___	___
– Injury, illness, and accident reporting investigation?	___	___	___
– Motor fleet safety?	___	___	___
– Any additional written health and safety procedures that are not required by law?	___	___	___

3.2 Documentation Practices:

- Describe the facility's documentation practices regarding the following:

Governmental EHS inspections ______________________________

____

____

Ongoing correspondence with governmental agencies and personnel concerning regulatory interpretations, permit renewals, other compliance-related guidance sought ______________________________

____

____

____

- Is the information received orally from governmental agencies confirmed with them via written correspondence? ______________________________

____

____

4.0 E/S PROGRAM MANAGEMENT

4.1 Environment & Safety Coordinator:

- Is there a designated ES Coordinator(s)? ______________________________
- What is his or her name and title (other than ES Coordinator)? ______________________________

____

- To whom does he or she report (name and title)? ______________________________

____

- Is the ES Coordinator(s)' position full time? ______________________________

If not, what percentage of time is actually spent on ES-related activities: ______________________________

____

4.1 Environment & Safety Coordinator (continued):

- Does the Coordinator(s)' job description include current ES responsibilities?

- What type of training has the ES Coordinator(s) completed to execute his or her responsibilities?

- Have the ES Coordinator(s)' responsibilities been clearly defined and communicated by management to site personnel?

- Is the ES Coordinator(s)' performance evaluation influenced by the execution of ES responsibilities?

- How does the Coordinator(s) keep up with regulatory developments and company ES policies? How is this information shared with other facility personnel who undertake ES program activities?

4.2 Supervisors, Managers, and Site Personnel:

- Have the responsibilities for ES within each department been clearly defined and communicated by top management to supervisors, managers, and site personnel?

- Has training been conducted for department supervisors and managers for their role and responsibilities for ES?

4.3 Performance Reviews:

- Describe the facility's overall system for reviewing the performance of its ES programs (e.g., facility audits or inspections, ES self-evaluation, review of records and reports) and for identifying departures from established standards and policies (governmental and company).

- Have specific performance goals and measurements for ES been established for all staff and site personnel?

4.4 Project Review:

- Who completes safety project reviews when new products or operational changes are anticipated? When is their review undertaken (e.g., several months prior to project initiation, the week before)?

- What criteria are used to determine when this review should be performed?

4.5 **Communication and Awareness:**

- How are facility and company management kept informed of ongoing ES activities? What mechanisms are used?

Who is notified?

Is all ES program activity communicated to management? If not, what is communicated?

How often do these communications take place?

Who is responsible for communicating this information?

- How are facility and company management made aware of ES problems or incidents (e.g., reportable releases, increase in employee accidents/injuries)? What mechanisms are used?

Who is notified?

What criteria are used for communicating this information? (Are all ES problems communicated to management? If not, what is communicated?)

How often do these communications take place?

Who is responsible for communicating this information?

4.6 **Risk Management:**

- To obtain an understanding of the facility's risk management system, specifically, what are the facility's systems and mechanics for the identification of hazards, determining risk acceptability, developing and implementing risk control systems, and providing for periodic risk review?

5.0 E/S PROGRAM ADMINISTRATION

What person or persons at the facility are responsible for development, implementation, and administration of programs for compliance with applicable governmental and company requirements for each of the following functional areas on a day-to-day basis?

NOTE: Include training and experience level of these staff members for each area of the Environment & Safety (ES) Program. Indicate those ES areas that do not apply to this facility.

		Training Degree Courses, etc.	Experience (# of Years)
5.1	**Equipment Safety:**		
	Overall responsibility		
	______________________	________	________
	Identification and evaluation of machine guarding needs		
	______________________	________	________
	Maintenance of equipment guards		
	______________________	________	________
	Hoist maintenance (i.e., testing, labeling)		
	______________________	________	________
5.2	**Administrative Control Programs** (i.e., hazardous work permits, confined entry procedures):		
	Overall responsibility		
	______________________	________	________
	Development of procedures		
	______________________	________	________
	Issuance of work permits, lock-out tags		
	______________________	________	________
5.3	**Injury, Illness, and Accident Reporting:**		
	Overall responsibility		
	______________________	________	________
	Development of investigation guidelines		
	______________________	________	________
	Determining and recording reportable injuries		
	______________________	________	________
	Coordinating medical services on-site and off-site		
	______________________	________	________

		Training Degree Courses, etc.	Experience (# of Years)
5.4	**Employee Safety Training and Awareness:**		
	Overall responsibility		
	______________________________	__________	__________
	Job safety and training programs		
	______________________________	__________	__________
	First aid training		
	______________________________	__________	__________
	CPR training		
	______________________________	__________	__________
	Supervisor, safety training		
	______________________________	__________	__________
5.5	**Fire Prevention and Life Safety:**		
	Overall responsibility		
	______________________________	__________	__________
	Testing and maintaining fire prevention equipment		
	______________________________	__________	__________
	Managing flammable storage areas		
	______________________________	__________	__________
	Developing and overseeing emergency plans		
	______________________________	__________	__________
5.6	**Hazard Communications:**		
	Overall responsibility		
	______________________________	__________	__________
	Training		
	______________________________	__________	__________
	Review of MSDS		
	______________________________	__________	__________
	Distribution and maintenance of MSDS		
	______________________________	__________	__________

5.7 **Industrial Hygiene:**	Training Degree Courses, etc.	Experience (# of Years)
Overall responsibility		
______________________	________	________
Identification of physical and chemical exposures in the workplace		
______________________	________	________
Measurement of ventilation performance		
______________________	________	________
Calibration of industrial hygiene monitoring devices		
______________________	________	________

Appendix B. Industrial Hygiene Audit Protocol

(From *Industrial Hygiene Auditing—A Manual for Practice,* edited by Alan J. Leibowitz, CIH, CSP, AIHA Press, 1994.)

Facility Name: ______________________________

Date(s) of Audit: ______________________________

Team Members Performing Work: ______________________________

Period Under Review: ______________________________

This protocol serves as a step-by-step guide to collecting information about activities in the area of industrial hygiene at facilities. It might require additions, revisions, or other modifications to meet the needs of facility-specific audit objectives, the facility setting, or other special circumstances.

This protocol is organized to follow a five-step audit process:

1.0 Initiate On-Site Activities

2.0 Examine Details of Environment and Safety Program Activities and Controls

2.1 Injury, Illness, and Accident Reporting
2.2 Hazardous Materials in the Workplace
2.3 Hazardous Physical Agents in the Workplace
2.4 New or Modified Materials or Processes
2.5 Exposure Assessment
2.6 Employee Monitoring
2.7 Record Keeping and Reporting
2.8 Control Systems
2.9 Hearing Protection Programs
2.10 Personal Protective Equipment Programs
2.11 Respirator Programs
2.12 Hazard Communication Programs
2.13 Laboratory Chemical Hygiene Programs

3.0 Assess Apparent Strengths and Weaknesses

4.0 Test and Verify

4.1 Injury, Illness, and Accident Reporting
4.2 Hazardous Materials in the Workplace
4.3 Hazardous Physical Agents in the Workplace
4.4 New or Modified Materials or Processes
4.5 Exposure Assessment
4.6 Employee Monitoring
4.7 Record Keeping and Reporting
4.8 Control Systems
4.9 Hearing Protection Programs
4.10 Personal Protective Equipment Programs
4.11 Respirator Programs
4.12 Hazard Communication Programs
4.13 Laboratory Chemical Hygiene Programs

5.0 Summarize Audit Findings

*Attachment 1 — Chemicals with Special OSHA Requirements
*Protocol Exhibit List
[References: 29 CFR 1910; 29 CFR 1904]

INDUSTRIAL HYGIENE AUDIT PROTOCOL

	Auditor(s) Comments	WP REF
1.0 INITIATE ON-SITE ACTIVITIES		
2.0 EXAMINE DETAILS OF ENVIRONMENT AND SAFETY ACTIVITIES AND CONTROLS Through interviews with facility personnel and review of relevant documents (some of which might have been identified through pre-audit questionnaires), examine the design and operation of facility industrial hygiene activities and controls. This might include confirming answers in the Pre-Audit Questionnaire. When interviewing facility personnel or reviewing key documents to obtain information listed below, consider the following questions: What tasks are undertaken by the facility to comply with applicable requirements? How are these tasks done? When are they done? Are there written or informal procedures for these tasks? Does the facility use any protective measures to abate hazards and meet regulatory and internal requirements? How are these tasks monitored? Is there a system for reviewing the execution of these tasks?		
2.1 Injury, Illness, and Accident Reporting • Facility procedures to investigate and record employee injuries, illnesses, and accidents. • Facility procedures to determine the OSHA recordability of employee injuries, illnesses, and accidents. * Working paper reference: All audits should include detailed documentation of the background behind each finding. The summary worksheets should include a reference to where the applicable information can be found in those worksheets.		
2.2 Hazardous Materials in the Workplace • Facility methods to identify hazardous materials in the workplace (e.g., area monitoring, knowledge of process and resulting exposure).		
2.3 Hazardous Physical Agents in the Workplace • Facility methods to identify hazardous physical agents in the workplace (e.g., area monitoring, knowledge of equipment characteristics, knowledge of process).		
2.4 New or Modified Materials or Processes • Facility procedures used to examine the use of new or modified materials or processes for health and safety implications.		

INDUSTRIAL HYGIENE AUDIT PROTOCOL

	Auditor(s) Comments	WP REF
2.5 Exposure Assessment • Nature of facility programs to quantify employee exposures to hazardous materials and physical agents in the workplace, including frequency, sampling devices, record keeping provisions, and calibration of equipment.		
2.6 Employee Monitoring • Facility procedures or methods to select employees for hazardous material and physical agent exposure monitoring. • Facility practices to ensure that employees targeted for monitoring actually participate in monitoring activities. • Facility practices for determining the cause(s) of employee overexposures to hazardous materials and/or physical agents, if discovered.		
2.7 Record Keeping and Reporting • Facility procedures to inform employees of workplace monitoring results.		
2.8 Control Systems • Facility methods to control employee exposure to hazardous materials and physical agents in the workplace (e.g., ventilation systems, use of respirators, use of restricted access areas). • Nature and frequency of facility inspection and maintenance programs for engineering control systems. • Facility methods to ensure compliance with compound-specific work practice and administrative control programs.		
2.9 Hearing Protection Programs • Facility methods to select employees targeted for a hearing conservation program. • Facility practices to review adequacy of hearing abatement measures. • Facility methods to select hearing protectors.		
2.10 Personal Protective Equipment Programs • Facility programs and practices regarding the selection, use, and maintenance of personal protective equipment.		

INDUSTRIAL HYGIENE AUDIT PROTOCOL

	Auditor(s) Comments	WP REF

2.11 Respirator Programs

- Facility methods to select employees targeted for participation in a respirator protection program.
- Facility methods to select respirators and fit-test and train employees.
- Facility respirator inspection and maintenance program for respirators.

2.12 Hazard Communication Programs

- The nature and scope of the facility's hazard communication program (e.g., labeling system, training program, material safety data sheets).

2.13 Laboratory Chemical Hygiene Programs

- If the facility operates a laboratory where multiple chemicals or chemical procedures are used, the nature and scope of the facility's laboratory chemical hygiene program.

3.0 ASSESS APPARENT STRENGTHS AND WEAKNESSES

4.0 TEST AND VERIFY

Some of the information reviewed under this step might have been obtained during Step 2.0

4.1 Injury, Illness, and Accident Reporting Procedures (29 CFR 1904)

- Test the effectiveness of the facility's injury, illness, and accident reporting system by performing the following:
  - — Review the facility's dispensary log, accident or injury reports and, if applicable, workers' compensation reports. Then prepare a listing of employees who have experienced injuries or illnesses requiring treatment in excess of normal first aid and/or who might reasonably be expected to have experienced an accident resulting in lost time.

4.2 Hazardous Materials in the Workplace (29 CFR 1910.1001-.1048)

(NOTE: Some state OSHA programs have established exposure standards for materials not specifically regulated by federal OSHA [e.g., MBOCA in California]. Auditors should review applicable state regulations prior to completing this section of the protocol.)

INDUSTRIAL HYGIENE AUDIT PROTOCOL

	Auditor(s) Comments	WP REF
— Compare your list with facility information (such as its chemical inventory or exposure monitoring plan) to confirm that all significant hazardous substances present at the facility have been identified. — Prepare a list of substances present at the facility that are regulated under compound-specific (or "vertical") standards. See Attachment 1 for a list of these compounds. If such substances are present at the facility, see 4.6, 4.7, and 4.8 of this protocol for additional auditing requirements. • Interview supervisors and review available documentation to gather any additional information about cases that might be considered "reportable." • Verify whether those accidents/injuries that meet recordability criteria or can be considered significant have been recorded on an OSHA 200 form or its equivalent, and have been reported to appropriate levels of management. (29 CFR 1904). 4.3 Hazardous Physical Agents in the Workplace • Test the facility's program for identifying hazardous physical agents in the workplace by performing the following: — Develop a list of hazardous physical agents in the workplace (e.g., noise, X-ray units, microwave, laser, infrared and ultraviolet sources, other ionizing radiation sources) by reviewing plant records and visually observing facility equipment. — Compare your list with facility information to confirm that the facility has adequately identified hazardous physical agents in the workplace. • Obtain necessary information to confirm that governmental licensing and inspection requirements have been met, when applicable. 4.4 New or Modified Materials or Processes • Based on your understanding of the facility's procedures for review of new materials uses and new or modified processes, perform the following steps: — Inquire whether any process or material changes have been implemented during the review period.		

INDUSTRIAL HYGIENE AUDIT PROTOCOL

	Auditor(s) Comments	WP REF
— From project files in Engineering or Accounting, and/or from Purchasing records, select a sample of process and materials changes. — Confirm that health, safety, and environmental issues were considered by a qualified individual during the planning and approval of the changes and that work conditions meet appropriate standards. 4.5 Exposure Assessment • Based on your understanding of chemical and physical exposures in the workplace, verify the following: — Exposure assessment strategies and procedures are planned and conducted in a systematic, formal manner. — Exposure assessment programs are developed with a documented rationale and program review provisions. — Scope and frequency of the chemical and physical monitoring program is consistent with potential exposure and requirements (review monitoring records and sample strategy and determine status). (29 CFR 1910.95[c], .96[d], .1001–.1048) — Sampling methods are consistent with OSHA requirements or generally accepted NIOSH and ASTM procedures. Note that reports include references to the following: a. proper sampling device; b. calibration and test equipment; c. sample handling; d. sample documentation or field notes. — Exposure evaluations are performed by persons having specific training in exposure assessment. — Sample analysis is performed by a laboratory accredited by the American Industrial Hygiene Association (AIHA). 4.6 Employee Monitoring • From an employee roster, develop a list of employee job titles that should be included in any hazardous material or physical agent monitoring program used by the facility. Verify that your list is equivalent to the facility's list of employees who are in the industrial hygiene monitoring program, and:		

INDUSTRIAL HYGIENE AUDIT PROTOCOL

	Auditor(s) Comments	WP REF
— Confirm that baseline measurements of employee exposure to air contaminants and physical agents have been made. Verify whether exposures of employees to air contaminants and physical agents are within established limits. (29 CFR 1910.95[d], .96[d], .1000–.1048)		
— Verify whether biological monitoring has been performed, where required. Note that test results/records for employees are complete and that the frequency of monitoring is consistent with OSHA requirements. Note any excessive exposure results. (29 CFR 1910.1001–.1048)		
• Verify, by reviewing pertinent records and documents, whether the cause(s) of employee overexposure(s) to hazardous materials and/or physical agents (if discovered) is determined.		
4.7 Record Keeping and Reporting (29 CFR 1910.20)		
• Test the facility's record keeping program to confirm the following:		
— Employees have been informed of monitoring results in a timely manner.		
(NOTE: 29 CFR 1910.20[e][1] requires that access to exposure records be given to employees within 15 working days of their request; 29 CFR 1910.1450[d][4] requires that results arising from monitoring conducted under a chemical hygiene plan be made available to employees within 15 working days of the facility's receipt of results, and 29 CFR 1910.1025[d][8] requires that employees be given monitoring results within 5 days of the facility's receipt of results. Consult specific standards for other timeliness requirements.)		
— Records of worker exposure are being maintained in accordance with regulatory requirements.		
(NOTE: Coordinate this audit step with the completion of the Health protocol.)		
4.8 Control Systems		
• Test the adequacy of engineering control programs for hazardous materials and physical agent exposures in the workplace by reviewing:		
— Documentation of inspection and maintenance operations to verify that control systems are maintained properly;		

INDUSTRIAL HYGIENE AUDIT PROTOCOL

	Auditor(s) Comments	WP REF
— Exposure monitoring systems (if required); — Prohibited processes or operational practices; — Required processes or operational practices; — Regulated area establishment and management; — Methods of compliance; — Engineering controls; — Personal protective equipment; — Hygiene facilities; — Signage; — Training; — Reporting and documentation; — Medical surveillance (coordinate your examination of these activities with the completion of the Health protocol). (29 CFR 1910.1001–.1048) • In instances when employee overexposures to hazardous materials and/or physical agents have been determined, verify that the facility examined the feasibility of engineering controls. Verify whether the feasibility of engineering controls was determined by a qualified person. Confirm that personal protective equipment is used as an exposure control measure only when engineering or administrative controls have been found to be infeasible. (29 CFR 1910.1000[e]) 4.9 Hearing Protection Programs (29 CFR 1910.95) • Test the facility's hearing protection program by performing the following: — Develop a list of areas on site that might be high noise areas by touring the facility and reviewing available documentation. (This might have been obtained during completion of Sections 4.3 and 4.6.) — Obtain and review in-house noise survey documents to determine whether noise level surveys were conducted, where applicable, and that representative exposure data are available. — Confirm, through inquiry or review of facility records, that employees monitored for noise exposure were informed of their exposure.		

INDUSTRIAL HYGIENE AUDIT PROTOCOL

	Auditor(s) Comments	WP REF
— Verify whether the feasibility of engineering and administrative controls were studied by a qualified person and that hearing protection (e.g., ear plugs) is used only as a temporary measure while engineering/administrative controls are being implemented, or in cases in which engineering/administrative controls were found to be infeasible. (29 CFR 1910.95[b][l])		
• If available noise survey data indicate that employees are exposed to 8-hour time-weighted-average noise levels in excess of 85 dBA, confirm that an effective hearing conservation program has been implemented.		
— Select a representative sample of employees who work in high noise areas and confirm that they are included in the program. (29 CFR 1910.95[c])		
— Verify whether program participants have received baseline and, if applicable, annual audiometric testing. (29 CFR 1910.95[g])		
— Verify that individuals' annual audiometric results are compared with baseline results to determine whether standard threshold shifts have been suffered by employees. (29 CFR 1910.95[g][7])		
— Confirm that employees shown to have suffered standard threshold shifts have received the required follow-up procedures as specified in 29 CFR 1910.95(g)(8).		
— Confirm that program participants have received the required hearing conservation training. (29 CFR 1910.95[k][l])		
— Verify whether records of noise monitoring and audiometric test results are maintained and made available to affected employees, as specified by the OSHA standard. (29 CFR 1910.95[m])		
— Test the adequacy of noise protectors or engineering controls by reviewing pre- and post-audit abatement noise measurements. (29 CFR 1910.95[b][1])		
— Evaluate the adequacy of hearing protectors based on their stated noise attenuation ratings. (29 CFR 1910.95[j])		
4.10 Personal Protective Equipment Programs (29 CFR 1910.132.133)		
• Test the facility's personal protective equipment program by confirming that:		

INDUSTRIAL HYGIENE AUDIT PROTOCOL

	Auditor(s) Comments	WP REF
— Procedures and specifications have been established for the purchase, replacement, maintenance and inspection, use, and storage of personal protective equipment.		
— Eye and face equipment is required when there is a reasonable probability of injury. (29 CFR 1910.133[a])		
— Eye protection is provided where machines or operations present the hazard of flying objects, glare, corrosive or toxic liquids, or radiation. (29 CFR 1910.133[a])		
— Gloves, aprons, and other protective clothing is provided that is appropriate for the exposure.		
— Protective footwear and headgear policy is appropriate for the facility's operations.		
— All protective equipment is inspected, clean, in good repair, and stored properly when not in use.		
— Replacement protective equipment is readily available.		
4.11 Respirator Programs (29 CFR 1910.134)		
• Verify the presence and operation of a respirator program by confirming the following:		
— All respirators are provided by the facility and are not provided by employees.		
— Respirators are not used in lieu of engineering controls required by applicable standards.		
— Written programs exist that include standard operating procedures governing the selection and use of respirators, training, fit testing, and chemical exposures. (29 CFR 1910.134[b])		
— Users have been instructed in the proper use of respirators and their limitations and the training has been documented. (29 CFR 1910.134[b])		
— Respirators are cleaned and disinfected regularly. (29 CFR 1910.134[b])		
— Respirators are stored in a convenient, clean, and sanitary location. (29 CFR 1910.134[b])		
— Respirators for emergency use are inspected thoroughly at least once a month and after each use. (29 CFR 1910.134[b])		

INDUSTRIAL HYGIENE AUDIT PROTOCOL

	Auditor(s) Comments	WP REF
— Respirator wearers have been medically approved to wear respirators and, if applicable, receive periodic medical exams (i.e., annual). (29 CFR 1910.134[b][10])		
— The respirator type is specified by a qualified individual. (29 CFR 1910.134[e])		
— Respirator wearers have been fit-tested and the fit-testing is documented. (29 CFR 1910.134[e])		
— Respirators are inspected before and after each use. (29 CFR 1910.134[f])		
— Self-contained breathing apparatus (SCBA) are inspected at least monthly. (29 CFR 1910.134[f][2][ii])		
— Respirators are used when required, and proper types are available. (29 CFR 1910.134[e])		
— Proper specification and use of respiratory protection has been verified by quantitative measurement of exposures to the substance(s) that the respiratory protection is intended to protect against.		
4.12 Hazard Communication Programs (29 CFR 1910.1200; Facility Practice 40.5)		
• Review the facility's hazard communication program to verify whether all applicable requirements have been addressed:		
— Procedures for determining hazards are in writing. (29 CFR 1910.1200[d][6])		
— A written hazard communication program is in place that includes: (29 CFR 1910.1200[e])		
a. a list of the hazardous chemicals present in the workplace;		
b. methods used to inform employees of the hazards of nonroutine tasks and the hazards associated with chemicals in unlabeled pipes;		
c. methods used to inform contractors of the hazardous chemicals they might be exposed to while performing their work;		
d. description of how the requirements for labeling, MSDSs, and employee training will be met.		
— Each container of hazardous chemicals (including piping) in the workplace is labeled, with the identity of the hazardous chemical(s) and appropriate hazard warnings. (29 CFR 1910.1200[f][4])		

INDUSTRIAL HYGIENE AUDIT PROTOCOL

	Auditor(s) Comments	WP REF
— MSDSs are available for each hazardous chemical produced, imported, or used. (29 CFR 1910.1200[g])		
• Verify, by sampling hiring/transfer and training records, whether employees are informed of potential hazards associated with chemicals in their work areas at the time of their initial assignments and whenever a new hazardous material is introduced into their work areas. (29 CFR 1910.1200[h])		
— Employee training includes: (29 CFR 1910.1200[h][2])		
a. methods used to detect the presence or release of a hazardous chemical in the work area.		
b. measures employees can take to protect themselves from these hazards. Details of the facility's hazard communication program, including labeling and MSDSs.		
4.13 Laboratory Chemical Hygiene Programs (29 CFR 1910.1450)		
NOTE: The OSHA definition of "laboratory use of hazardous chemicals" is: Handling or use of hazardous chemicals in which all of the following conditions are met:		
(i) Chemical manipulations are carried out on a "laboratory scale";		
(ii) Multiple chemical procedures or chemicals are used;		
(iii) The procedures involved are not part of a production process, nor in any way simulate a production process;		
(iv) "Protective laboratory practices and equipment" are available and in common use to minimize the potential for employee exposure to hazardous chemicals.		
• If the facility's operations include the laboratory use of hazardous chemicals, verify that:		
— A written Chemical Hygiene Plan has been developed and implemented;		
— A person has been designated as the facility's Chemical Hygiene Officer and, if appropriate, a Chemical Hygiene Committee has been established;		

INDUSTRIAL HYGIENE AUDIT PROTOCOL

	Auditor(s) Comments	WP REF
— Laboratory personnel have been informed of the presence of the Chemical Hygiene Plan (and its details) and the delegation of chemical safety/hygiene responsibilities, and have been trained in safe practices to be implemented when using and storing laboratory chemicals. (See Appendix A of 29 CFR 1910.1450.)		
• Confirm that the facility's Chemical Hygiene Plan includes the following:		
— Standard operating procedures relevant to safety and health considerations;		
— Criteria that are used to determine and implement employee monitoring for hazardous chemical exposure;		
— Provisions for ensuring that fume hoods and other protective equipment are functioning properly;		
— Identification of operations that require specific authorization by a supervisor prior to implementation;		
— Provisions for additional employee protection for work with particularly hazardous substances (i.e., carcinogens and acutely toxic substances).		
• Verify whether fume hoods are evaluated at the time of their installation and at least quarterly thereafter to ensure that air flow into the hood is uniform and that the hood face air velocity is between 60 and 125 linear feet per minute.		
5.0 SUMMARIZE AUDIT FINDINGS		

Industrial Hygiene Audit Protocol

Attachment 1

Chemicals With Special OSHA Requirements

1910.1001	Asbestos
1910.1002	Coal tar pitch volatiles
1910.1003	4-nitrobiphenyl
1910.1004	alpha-Naphthylamine
1910.1006	Methyl chloromethyl ether
1910.1007	3,3′-dichlorobenzidine (and its salts)
	bis-Chloromethyl ether
1910.1008	beta-Naphthylamine
1910.1009	Benzidine
1910.1010	4-Aminodiphenyl
1910.1011	Ethyleneimine
1910.1012	beta-Propiolactone
1910.1013	2-Acetylaminoflourene
1910.1014	4-Dimethylaminoazobenzene
1910.1015	N-Nitrosodimethylamine
1910.1016	Vinyl chloride
1910.1017	Inorganic arsenic
1910.1018	Lead
1910.1025	Benzene
1910.1028	Coke oven emissions
1910.1029	Cotton dust
1910.1043	1,2-dibromo-3-chloropropane
1910.1044	Acrylonitrile
1910.1045	Ethylene oxide
1910.1047	Formaldehyde
1910.1048	

Industrial Hygiene Audit Protocol
Protocol Exhibit List

Facility: ______________________ Date: ______________

Auditor: ______________________

Exhibit Number	Description

Appendix C. Health, Environment, and Safety Program Management Audit Protocol

(From *Industrial Hygiene Auditing—A Manual for Practice,* edited by Alan J. Leibowitz, CIH, CSP, AIHA Press, 1994.)

Facility Name: ______________________

Date(s) of Audit: ______________________

Team Members Performing Work: ______________________

Period Under Review: ______________________

This protocol serves as a step-by-step guide to collecting information about activities in the area of environment and safety program management at facilities. It might require additions, revisions, or other modifications to meet the needs of facility-specific audit objectives, the facility setting, or other special circumstances.

This protocol is organized to follow a five-step audit process:

1.0 Introduction

2.0 Test and Verify
- 2.1 Program Management
- 2.2 Program Reviews

3.0 Summarize Audit Findings
*Protocol Exhibit List

HEALTH, ENVIRONMENT, AND SAFETY PROGRAM MANAGEMENT AUDIT PROTOCOL

	Auditor(s) Comments	WP REF

1.0 INITIATE ON-SITE ACTIVITIES

The purpose of this protocol is to ensure that sufficient data are gathered by team members to evaluate the facility's Health, Environment, and Safety (HES) program management systems. This protocol is to be administered by the team leader in conjunction with team members. Additional information may be obtained from the facility Environment and Safety Program Coordinator.

2.0 TEST AND VERIFY

2.1 Program Management

- Test the effectiveness of the facility's HES program (Are sufficient resources allocated for the program to achieve and maintain compliance? Are staff trained? Are compliance plans and procedures in place?) by performing the following:

  — Document that unit management understands its HES obligations and has clearly assigned responsibility to carry out these activities to qualified individuals.

  — Document training of site HES Coordinator(s) to verify whether sufficient training has been provided to execute job responsibilities.

- Interview team members and the facility coordinator to determine how the facility ensures compliance with both HES practices and applicable governmental regulations. That is, from whom is program guidance obtained? What are the general guidelines or procedures followed? What factors are taken into consideration to ensure compliance with the following program requirements?

  - Exposure assessment
  - Management of engineering controls
  - Emergency preparedness controls
  - Fire prevention
  - Job safety
  - Hazard assessment and communication
  - Training and awareness
  - Medical surveillance programs
  - Equipment safety (guarding)

HEALTH, ENVIRONMENT, AND SAFETY PROGRAM MANAGEMENT AUDIT PROTOCOL

	Auditor(s) Comments	WP REF
• Document and assess the procedures followed by the facility to evaluate materials and procedures (and changes) to control their inherent hazards. Also, verify whether the coordinator maintains an up-to-date inventory of materials by cross-checking the materials list with the MSDS file to verify whether this information is kept up-to-date.		
• Interview medical and first aid personnel to confirm that the Environmental and Safety Coordinator keeps them informed of all workplace hazards to develop appropriate medical surveillance programs.		
• Verify whether the HES coordinator(s) document(s) all compliance activities, etc., and keeps adequate records (e.g., accident investigations, inspection and follow-up reports, ESSEs, monitoring results, reports of potential hazards, and training records).		
• Verify whether the HES coordinator(s) has current copies of national, regional, state and local HES laws, regulations, codes and standards applicable to facility operations.		
• Verify whether the HES coordinator(s) has a system in place to learn of new or forthcoming regulations and requirements.		
• Verify whether accidents and incidents are investigated and analyzed to identify job-related hazards and trends, and initiate corrective actions.		
• Verify whether an inventory of all environmental discharges and wastes has been prepared.		
• Confirm that the facility has systems and/or procedures in place to ensure that timely corrective action is taken, or appropriate responses are made to audit reports, self-evaluations, program reviews, consultant reports, governmental inspections, and safety committee reports.		
• Verify whether the facility has developed and implemented adequate procedures for handling inspections or investigations of its operations, physical plant, and HES program management.		
• Confirm that the facility has implemented a systematic approach for minimizing pollution and controlling hazards to personnel.		
• Verify whether the HES coordinator(s) routinely reports on the status of the unit's HES programs to unit management.		

HEALTH, ENVIRONMENT, AND SAFETY PROGRAM MANAGEMENT AUDIT PROTOCOL

	Auditor(s) Comments	WP REF
2.2 Program Reviews (NOTE: Each year the facility coordinator is required to conduct an in-depth review of the facility's Environment and Safety program and complete an Environment and Safety Self-Evaluation report.) • Annual Reviews Verify whether the facility conducts annual reviews for each location under its control by performing the following: — Review facility records to confirm that the facility has completed Environment and Safety Self-Evaluation reports or their equivalent for previous years. Note any problems recorded in these reports and confirm that they have been corrected in subsequent reports. • Ongoing Inspections — Examine the facility's periodic Environment and Safety program reviews and inspections to verify whether the following areas were covered in-depth and with the frequency required or indicated by the risk and complexity of the facility's operations: • Condition and use of personal protective equipment. • Status of materials inventories and hazard information file. • Exposure monitoring • Health and safety hazards and their controls • Employee training and hazard notification • Technical development of the HES staff • Purchasing controls for hazardous materials • Fire extinguishing and suppression equipment • Unit emergency organizations • Special hazard controls (such as confined spaces entry procedures and hot work permits • Fire alarm systems • Emergency evacuation plans and drills		

HEALTH, ENVIRONMENT, AND SAFETY PROGRAM MANAGEMENT AUDIT PROTOCOL

	Auditor(s) Comments	WP REF
• The status of corrective action plans committed to as a result of audits, inspections, or reviews conducted internally or by external parties • Contractors' compliance with facility health, safety, and environmental requirements • Compliance with applicable environmental laws and regulations — Verify whether the results of these inspections are reported to the facility's manager and the HES executive coordinating committee. — Verify whether inspection reports include or are followed by a corrective action plan that specifies problems, how they will be addressed, the person(s) responsible for corrective action, and the date by which the corrective action will be completed. **3.0 SUMMARIZE AUDIT FINDINGS** See the Environment and Safety Program Audit Field Guide for a description of activities undertaken to complete this audit step.		

Health, Environment, and Safety Program Management
Audit Protocol
Protocol Exhibit List

Facility: ______________________ Date: ____________

Auditor: ______________________

Exhibit Number	Description

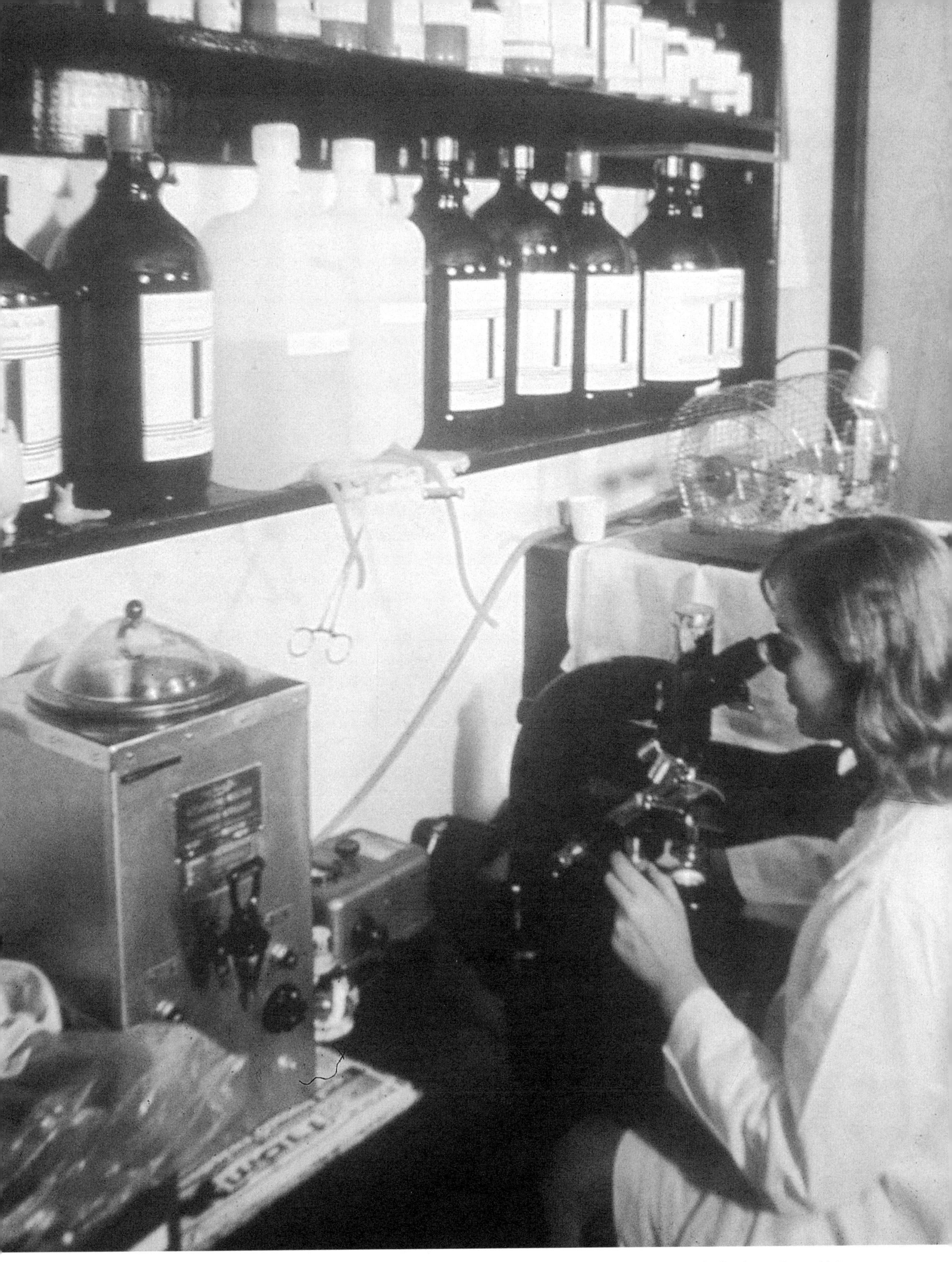

Research laboratories involve a wide variety of exposures to organic chemicals. Several epidemiologic studies have shown higher incidences of a variety of cancers in people with training in organic chemistry, suggesting that exposure to even low levels of multiple organic chemicals may enhance the risk of chronic disease. OSHA has a separate lab standard (29 CFR 1910.1450) that emphasizes good work practices and use of laboratory hoods.

Chapter 39

Outcome Competencies

After completing this chapter, the user should be able to:

1. Define underlined terms used in this chapter.
2. Define quality control.
3. Differentiate between accuracy and precision, and between repeatability, replicability, and reproducibility.
4. Apply quality control concepts developed in this chapter to field sampling activities.
5. Prepare an outline for a laboratory quality control plan.
6. Develop and evaluate the results of an intralaboratory quality control program, including the use of control charts and the evaluation of intra- and inter-analyst data.
7. Develop and evaluate the results of an interlaboratory quality control program.

Key Terms

acceptance testing • accuracy • blank samples • calibration • control chart • corrective action • data validation • document control • duplicate samples • interlaboratory • intralaboratory quality control • performance audit • precision • preventive maintenance • procurement QC • quality • quality assurance • quality control • QC plan • repeatability • replicability • reproducibility • spiked samples • split samples • systems audit • Youden plot

Prerequisite Knowledge

College level chemistry and algebra.

Prior to beginning this chapter, the user should review the following chapters:

Chapter Number	Chapter Topic
6	Principles of Evaluating Worker Exposure
7	Preparation of Known Concentrations of Air Contaminants
8	Principles and Instrumentation for Calibrating Air Sampling Equipment
9	Direct-Reading Instrumental Methods for Gases, Vapors, and Aerosols
10	Sampling of Gases and Vapors
11	Analysis of Gases and Vapors
12	Sampling and Sizing Particles
37	Program Management
38	Surveys and Audits

Key Topics

I. Quality, Quality Control, and Quality Assurance
   A. Quality Control for Sampling Activities
   B. Acceptable Sampling Materials
   C. Calibration
   D. Evaluation of Routine Performance

II. The Laboratory Quality Control Plan
   A. Accuracy and Precision

III. Elements of Intralaboratory Quality Control
   A. The Control Chart

IV. Other Types of Control Charts
   A. Handling Outlier Data
   B. Evaluating Method Accuracy
   C. Evaluating Precision

V. Interlaboratory Testing

39

Quality Control for Sampling and Laboratory Analysis

Reginald C. Jordan, Ph.D., CIH, CQE

Introduction

Goals and Objectives

Many books have been written on some of the specific topics that will be discussed in this chapter. Therefore, this treatment of quality control in sampling and analytical activities will be in the nature of an introduction or overview. However, the material presented here will be augmented by a set of references to assist the user in gaining more insight into this subject area.

Quality, Quality Control, and Quality Assurance

Quality has many definitions. Even within the field quality means different things to different people. Juran[1] defines quality as "freedom from deficiencies," thereby producing a product or service "meeting the needs of customers." So Juran sees a quality product (or service) as one that is defect-free and performs those functions for which it was designed and constructed and produces client satisfaction. "Fitness for use" also describes such a product. Other definitions include the notion of conformance to specifications and/or standards, and client satisfaction at a competitive price. Producing a product that meets client needs is sometimes difficult to do because client needs are not always known—sometimes even by the client. Assistance—sometimes a lot of assistance—may be needed to define the real needs of the client. Once these needs have been determined, they become the product specifications, and, therefore, become the criteria against which the product will be evaluated.

Example

Question: What product does a laboratory produce? Answer: A written analytical report for each sample submitted.

Taylor[2] defines quality control (QC) as "a system of activities whose purpose is to control the quality of a product or service so that it meets the needs of users." So a QC system is that system under which a product is tested to determine whether it meets specifications. This has been referred to as the quality function.

Suppose that some systematic error was being made in the use of the QC system. How would this error be discovered and then corrected? To ensure that the QC system is operating effectively, an additional QC system is implemented to assess the efficacy of the QC system monitoring the product. This additional control system used to monitor the QC system is referred to as a quality assurance (QA) program. So, quality assurance could be defined as QC on QC. The quality hierarchy is pictured in Figure 39.1.

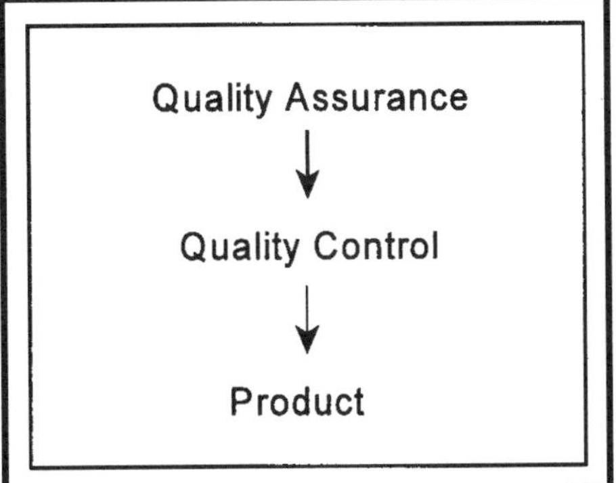

Figure 39.1 — The quality hierarchy.

Example

Question: How do you know whether the QC system is functioning properly? Answer: Through the implementation of a quality assurance program.

QC for Sampling Activities

To be able to draw conclusions about airborne contaminant concentration, the extent to which workers have been exposed or may be exposed in the future, the efficacy of exposure controls, or for other reasons, samples must be collected properly and then analyzed properly. Sample collection will be considered first; QC practices for analytical operations will be discussed later in this chapter.

Of the two processes, sampling and analysis, sampling perhaps generates the largest source of error. This is caused by many factors: environmental factors, such as variability of temperature, relative humidity, barometric pressure, and contaminant concentration; sample collection factors, such as variability in sampler volumetric flow, sampling time, and collection efficiency; human factors, such as personnel selected for monitoring (intentionally or unintentionally) interfering with sample collection, sampling personnel mislabeling samples, contamination of sampling materials, improper shipment of samples to the laboratory, and others. Before an analyst can interpret sampling data meaningfully, a sample must be collected correctly, then analyzed correctly. The variability, a measure of precision expressed as variance, associated with sampling and analysis process can be expressed mathematically as follows:[3]

$$s_T^2 = s_a^2 + s_t^2 + s_s^2 + s_h^2 + s_p^2 + s_c^2 + s_m^2 \quad (1)$$

where:

s_T^2 = total variance
s_a^2 = spatial
s_t^2 = temporal
s_s^2 = sampling
s_h^2 = transport and storage
s_p^2 = sample preparation
s_c^2 = chemical treatment
s_m^2 = measurement

The spatial, temporal, sampling, and transport and storage effects are associated with the sampling process; the sample preparation, chemical treatment, and measurement effects are associated with analysis.

To help minimize those sources of variability in the sampling process that can be minimized, a QC plan for a sampling project needs to be devised, implemented, and complied with. What are some of the elements that should be included in such a quality plan?

A Written Sampling and Analysis Method

A written, validated method must be used for the sampling project to be conducted. Many organizations use (or say they use) a National Institute for Occupational Safety and Health (NIOSH) method or an Occupational Safety and Health Administration reference method for the designated contaminant. These methods are photocopied from some reference and handed out for use in the sampling project. Rarely are these methods followed. Often they are informally modified to suit the needs of the sampling organization; that is, these methods are modified, but nothing is written formally regarding the modifications made to the method. Sometimes parts of these methods are subject to interpretation, and these interpretations are usually left up to field personnel.

What is needed is a clearly written sampling method, or protocol, for use by field personnel. This usually entails taking some published method and reducing it to a specific protocol for the sampling project. This can be an involved process, because every word of the published method must be read and interpreted before personnel go into the field. Further, once the interpreted sampling protocol has been written, at least one training session is needed with all field personnel to review the protocol in detail, so that everyone involved understands what is to be done and how it is to be done.

Acceptable Sampling Materials

An important part of the written sampling and analysis method described immediately above is a protocol for the acceptance testing of as-received supplies and materials. It is not surprising to learn that sampling (and analytical) materials are usually ordered and used without any determination of acceptability for use ("I ordered analytical reagent grade carbon disulfide; the bottle said it was analytical reagent grade CS_2, so it must be OK to use." or "I ordered and received 10 boxes of 25-mm MCE (mixed cellulose ester) filter cassettes. Each box contained a label that stated that the background fiber count was less than 5 fibers per 100 microscope fields, so these cassettes must be acceptable for use."). Perhaps these as-received materials were acceptable for use, but how does one know?

Statistical sampling protocols need to be developed, then implemented for acceptance of as-received materials. These protocols need to be run before as-received materials are placed into use. There is a true story of a sampling project conducted without benefit of acceptance sampling protocol: a technician checked out five boxes of newly procured 25-mm MCE cassettes to use during the course of an asbestos abatement project. Sampling was routine, with the technician analyzing the collected samples at the end of the sampling day. When the technician analyzed the field blanks at the conclusion of the first sampling day, he discovered that the blank fiber counts were in excess of 7 fibers per 100 microscope fields. This indicated that the air samples themselves might be contaminated. The technician thought about the sampling process for the day, and could not point to any specific source of the contamination. But the field blank analysis indicated that the samples were contaminated. He made a mental note to take special care during the next sampling day, so that this situation would not recur. Sampling was conducted very carefully during the second sampling day. When the samples were analyzed that evening, the technician discovered that the field blank counts were again in excess of 7 fibers per 100 fields. The technician became suspicious at this point, having taken special care to avoid contamination. On an impulse, he mounted filter sections from a couple of unused cassettes out of the box, then counted them. To his surprise, the fiber counts were greater than 7 fibers per 100 microscope fields; the as-received cassettes were contaminated. Two days of sample data were questionable because no one tested these materials on receipt to determine whether the order was acceptable. Why should anyone have to discover contaminated sampling materials during the course of a sampling project? An appropriately designed and implemented acceptance testing protocol should prevent this.

Calibration

A protocol for calibration is another important part of the written sampling and analysis method.

Calibration has been defined as "the process of establishing the relationship between the output of a measurement process, or some part thereof, and that of a known input."[4] This protocol should include the following elements.

1. The maximum permissible time interval between multipoint calibrations and calibration checks.
2. The minimum quality of calibration standards to be used. Transfer standards should be 4 to 10 times the accuracy of field equipment.
3. Calibration standard traceability to higher-level standards. All standards used in the calibration process should be traceable to some ultimate standard (e.g., National Institute of Standards and Technology, if such a standard exists). An up-to-date report for each calibration standard should be provided.
4. A detailed procedure for device calibration.
5. A plan for controlling environmental factors.
6. Provisions for adequate documentation of the frequency of calibration and all raw data.

Evaluation of Routine Performance

Method accuracy (defined as closeness to truth) can be evaluated through the use of control samples. Control samples include duplicate samples, split samples, spiked samples, and blank samples.

Duplicate samples are used to evaluate the entire sampling/analysis method. Schlecht et al.[5] refer to duplicate samples as those samples collected "as close as possible to the same physical location, at identical time intervals, with approximately equal volumes sampled" as the actual samples. Schlecht et al. recommend that these samples be collected at a location where the expected concentration of the contaminant is near that of the standard limit. Care must be taken that the duplicate samples are spatially located so that one of the duplicate samples does not affect another. This effect is manifested by the localized airborne concentration of the contaminant and the flow rate with which the contaminant is sampled.

Split samples can be used where duplicate samples could not be collected, and can be obtained using certain types of sampling methods. Bulk sampling methods (e.g., soil, water, paint), in general, lend themselves to the creation of split samples.

Spiked samples are among the most commonly used in industrial hygiene. To create a spiked sample, a known mass of contaminant is placed on the collection medium. A sample volume is devised to make the spike look like an actual sample. One disadvantage of this type of control sample is that there is no matrix effect like one might find when analyzing an actual sample. The situation will determine the suitability of this approach.

There are several types of blank samples that can be used as controls: field blanks, transport blanks, and reagent blanks. A field blank is used to assess the extent to which an actual sample has been contaminated during the collection process. A field blank is treated as though it were an actual sample, except that it is not exposed to the contaminated atmosphere. The field blank should accompany the actual sample(s) through every stage of the sampling process. The mass of contaminant found on the field blank is subtracted from that found on the actual sample(s) before dividing by the air volume sampled in the determination of mass concentration of the contaminant. Each sampling method has a limit on the mass of contaminant permissible on the field blank. A contaminant mass above this limit makes the airborne concentration of contaminant found on actual samples represented by the field blank questionable. A transport blank can be used to assess contamination of the actual sample(s) by the sample container and any preservative used during transport of the sample(s) back to the laboratory as well as storage of the sample in the laboratory prior to analysis. This blank is sometimes distilled, deionized water to which sample preservative is added. A reagent blank, used in a sampling context, is a sample of unexposed reagent. Consider NIOSH Method 3500, formaldehyde by visible light spectrophotometry, in which the atmosphere is sampled for formaldehyde using impingers containing 1% $NaHSO_3$. A sample of unexposed $NaHSO_3$ would constitute a sampling reagent blank. Of course, one blank sample could represent several of these types of blanks.

The Laboratory QC Plan

A written QC plan is fundamental to the operation of any analytical laboratory. It spells out in detail the processes by which data generated by the laboratory will be evaluated, corrected (if necessary), and reported to the customer. It is recommended that the following topics constitute the bare minimum QC plan.

(1) Quality policy: the quality policy is management's written commitment to the production of a product of the highest quality (data, in this case).

> *The responsibility for and commitment to a quality policy belongs to the highest level of management. Quality management is that aspect of the overall management function which determines and implements quality policy. The management of a company should develop and state its corporate quality policy. This policy should be consistent with company policies. Management should take all necessary measures to ensure that its corporate quality policy is understood, implemented, and maintained."*[6]

(2) Document control: There should be a procedure or protocol by which the QC plan is updated and revised, and whose distribution is controlled. While there are several systems in use, one such system includes labeling the top of each page in the QC plan with the section number, revision number, revision date, and the page number (see Figure 39.2).

Section No. 1.4.1
Revision No. 1
Date January 9, 1984
Page 1 of 3

1.4.1 DOCUMENT CONTROL AND REVISIONS

1.4.1.1 ABSTRACT

A quality assurance program should include a system for documenting operating procedures and subsequent revisions. The system used for this Handbook is described and is recommended.

1.4.1.2 DISCUSSION

A quality assurance program should include a system for updating the formal documentation of operating procedures. The suggested system is the one used in this Handbook and described herein. This system uses a standardized indexing format and provides for convenient replacement of pages that may be changed within the technical procedure descriptions.

Figure 39.2 — Document control example.

In addition, a table of revisions may be placed in the front of the QC manual, near the table of contents. The table of revisions should list all revisions by section, revision number, and date.

How are revisions distributed? Each QC manual should be uniquely numbered, and the QC supervisor should maintain a list of those persons to whom QC manuals were distributed, together with the QC manual copy number. Audit of document revisions then becomes more efficient. It is relatively easier to randomly select a sample of persons to whom QC manuals have been distributed, then retrieve those manuals and inspect them to ensure that revisions are current.

(3) Organization: The job duties and responsibilities for each job category within the laboratory should be detailed in this section, together with minimum qualifications, experience levels, and reporting relationship(s).

(4) Training: This section of the QC manual should specify the required level of training for each job category. There are two major types of training: on-the-job and formal (e.g., college courses; continuing education courses; and short, intensive training courses). The minimum levels of training appropriate for each job category should be specified, together with an evaluative instrument for determining what the attendee learned as a result of attendance at such functions. (Parenthetically, this postcourse evaluation of training is done very infrequently. Beyond the qualitative responses of "I didn't learn much. The instructor was boring." or "We should send everyone to this course. I learned so much." there is little in the way of a formalized system for evaluating the effect of training on postcourse job performance.)

(5) Procurement QC: This section should delineate the procedure(s) by which supplies, materials, and capital equipment are procured. Details of how as-received supplies, etc. will be tested to certify the specified quality of those materials should be included. For example, if cases of 25-mm filter cassettes loaded with 0.8 μm pore-size cellulose ester membrane filters are to be purchased to be used for sampling airborne asbestos (NIOSH Method 7400), the procurement document should include a statement that all filter-loaded cassettes come from the same lot of filters as well as a testing mechanism to ensure that the background level of asbestos fibers on the as-received filters does not exceed 5 fibers per 10 microscope fields at some specified level of statistical confidence.

(6) Calibration: The calibration section of the QC manual should contain those elements required to be covered when discussing calibration for a specific sampling or analysis method. This section should contain, for example, frequency of calibration, quality of standards used in calibration, record-keeping protocols, and environmental conditions to be maintained.

(7) Preventive maintenance: The laboratory is most efficient when sampling and analytical instrumentation are maximally operational. To minimize variability and optimize "up time," a system of planned downtime must be developed. Preventive maintenance is performed during this planned downtime; that is, those components most likely to fail are replaced, instruments recalibrated, then the system is brought back online. Preventive maintenance reduces instrument "crashes," or unplanned downtime. Critical to the success of any preventive maintenance system (as well as minimizing unplanned downtime) is the development of a spare-parts inventory. The specification of a spare-parts inventory should be a part of this section.

(8) Sample handling: Protocols should be specified for the selection and use of sampling media in the field, as well as handling of samples, field blanks, and duplicate and replicate samples. Specific procedures should be written for the handling of samples received into the laboratory. These procedures should include, as a minimum, the conditions under which samples are accepted or rejected as received; if accepted, how samples are logged in (how the laboratory sample number is derived and used); how samples are stored prior to analysis; how samples are distributed to analysts; how samples are returned to storage following analysis; how samples are distributed for reanalysis (if such is required); and how long samples are retained before disposal.

(9) Intralaboratory and interlaboratory testing: This is the section in which the interlaboratory and intralaboratory quality control program is found. In the intralaboratory part of the plan, such topics as evaluation of precision and accuracy for within analyst and between analyst data; construction and use of control charts; and use of duplicate, replicate, and/or spiked samples should be found. In the interlaboratory section, such topics as selection of interlaboratory participants, selection of analyte, duration and frequency of interlaboratory testing; and statistical evaluation and reporting of test data should be found.

(10) Data validation: Before reporting test results, data generated should undergo some sort of data validation; that is, calculations for a randomly selected portion of the data should be manually checked (including the use of aliquot factors for samples and subsamples, calibration factors, etc.). This section should contain procedures for performing these types of data validation at specific review levels (e.g., peer, quality assurance, QC, supervisory).

(11) Audits: Every QC program should undergo periodic auditing. The question to be answered, of course, is "Is the QC program effective in producing a quality product?" Audits are of two general types: systems and performance,[7] and can consist of both internal and external audits. A systems audit is essentially a paper audit. If a protocol or procedure calls for specific paperwork to be completed, is that paperwork completed for some randomly selected samples? Are training records current? Have QC manual revisions been incorporated into distributed copies of QC manuals? A performance audit incorporates the quantitative evaluation of data quality through the evaluation of data generated from the analysis of unknown samples. (That is, they are unknown to the analyst. The QC coordinator in the lab provides the samples, inserts them into the sample handling and analytical system, and compares the analytical result with the reference concentration. This process can be blind or double-blind.) The minimum frequency of internal auditing should be specified.

(12) Corrective action: What is the procedure for dealing with nonconformities found during an audit? What is the procedure for dealing with an analytical system suddenly gone out of control? What is the procedure for handling client complaints or inquiries? These procedures are found in the section of the QC plan on corrective action. Essentially, corrective action (CA) should be a closed-loop system; that is, CA should consist of identifying the problem; designation of a person or persons to correct the problem; identifying appropriate corrective actions; instituting the corrective action; evaluating the correction to determine if the CA did, in fact, correct the problem; and finally, placing the previously nonconforming system back on-line.

(13) Quality reports to management: Management should be updated routinely on the state of quality in the organization through the use of short, simple reports. While there are myriad types of reports to issue, this section of the QC manual should specify which types are to be used and how frequently these reports should be issued. Use of graphical summary techniques is effective in conveying information to management without the necessity of writing a long detailed report. It is a certainty that if additional detailed information is needed by managers, it will be requested.

Accuracy and Precision

A laboratory is ethically obligated to produce a report of analysis that is not only accurate and precise, but also is produced at a time when the laboratory is in a state of statistical control.

Accuracy

For a data set of repetitive measurements, measures of central tendency (mean, median, mode) are measures of accuracy. Bias is the difference between the data set mean and the true mean.

Example

Question: Consider two sets of repeated analyses of the same 10 ppm reference standard. Set A includes 10, 11, and 12 ppm; Set B contains 7, 13, 16 ppm. Which is the more accurate data set? Answer: The mean of Set A is 11 ppm. The bias of Set A is 1 ppm (11 ppm 10 ppm). The mean of Set B is 12 ppm. The bias of Set B is 2 ppm (12 ppm 10 ppm). Set A data are more accurate than Set B data.

Precision

The production of accurate data is not enough. Those data must also be precise. Measures of variability (range, variance, standard deviation, coefficient of variation) are to precision what measures of central tendency are to accuracy. For repeated analysis of a sample, replicability (within analyst) and repeatability (between analyst) are measures of intralaboratory precision, while reproducibility (between laboratory) is a measure of interlaboratory precision. Replicability refers to the reanalysis of the same sample by the same analyst using the same instrumentation usually on the same day as the sample was originally analyzed. Repeatability generally is equivalent to replicability with the exception that one of the variables (analyst, instrument, analysis date) is different. For example, reanalysis of the same sample by a different analyst using the same instrumentation on the same analysis date is a measure of repeatability. Reproducibility involves analysis of subsamples of the same sample by different laboratories using their own instrumentation usually, but not always, on different dates. Both measures of precision are important, because analysts within the same laboratory over time become as one analyst; that is, analysts within a laboratory come to think and interpret similarly. When this happens, and it often does happen in statistically well-behaved laboratories, intralaboratory data can become quite internally precise. So, the best measure of method precision would be that from interlaboratory analysis. This is why reproducibility data are so important.

Examples

Question: Referring to Sets A and B in the previous example, which data set is more precise? Answer: The four variability measures are listed in Table 39.1. Notice that for each measure, the Set A data are more precise (less variable) than the Set B data.

Question: Why is it necessary for a measurement to be both accurate and precise? Answer: Consider Figure 39.3. Clearly, the best data are those that are (1) close to the truth (accurate) and (2) closely grouped (precise).

Table 39.1 — Variability Measures for Example Sets A and B

Precision Measure[A]	*Set A*	*Set B*
Range (ppm)	2	9
Variance (ppm^2)	1	21
Std. Dev. (ppm)	1	4.6
CV (unitless)	0.1	0.51

[A]Range = maximum value – minimum value

$$\text{Variance} = s^2 = \frac{\Sigma (x - \bar{x})^2}{n - 1}$$

$$\text{Std. Dev.} = s = \sqrt{\frac{\Sigma (x - \bar{x})^2}{n - 1}}$$

$$CV = \frac{\text{standard deviation}}{\text{mean}}$$

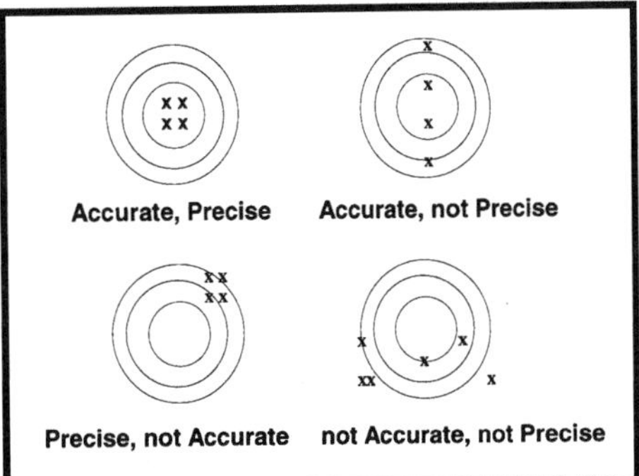

Figure 39.3 — Precision and accuracy.

Elements of Intralaboratory Quality Control

The Control Chart

Walter A. Shewhart[8] is credited with having developed the technique most commonly used to determine the state of process statistical quality control—the control chart. Recall the properties of the normal distribution: it is a symmetrical distribution in which the mean, median, and the mode all have the same value.

Figure 39.4 shows that approximately 68% of the area under the normal distribution curve falls between the mean ±1 standard deviation (SD); approximately 95% of the area under the curve falls between the mean ±2 SD; and approximately 99.73% of the area under the curve falls between the mean ±3 SD. Further recall the properties of the Central Limit theorem:[9] for random samples of size n drawn from a population with mean μ and standard deviation σ, as n increases: (1) the mean of the sampling distribution of means approaches μ, the population mean; (2) the SD of the sampling distribution of means is

$$\sigma_{\bar{x}} = \frac{\sigma}{\sqrt{n}} \tag{2}$$

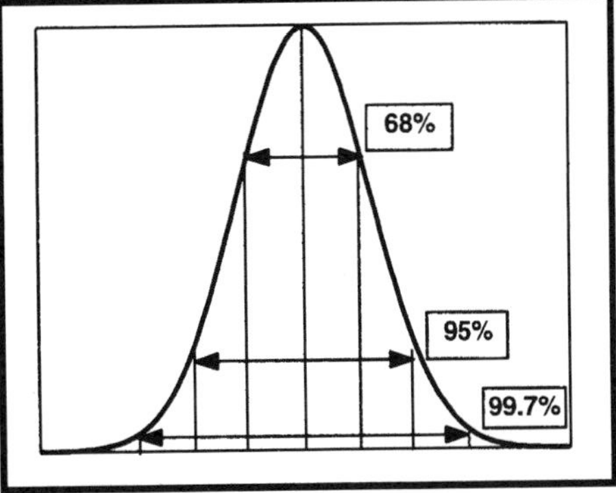

Figure 39.4 — The normal distribution.

the standard error of the mean; and (3) the shape of the distribution of sampling means will approach the normal, as n increases in size. This last property is very interesting; no matter what the distribution of the underlying population of x is, as the sample size n of the xs increases, the distribution of $\bar{x}$s approaches the normal.

Referring once more to Figure 39.4, if the lines that segment the distribution curve by standard deviation are extended, then the curve and the extended lines rotated by 90°, a control chart is formed (see Figure 39.5).

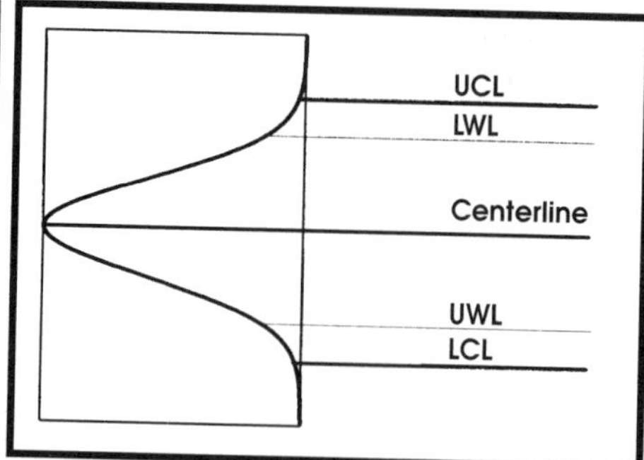

Figure 39.5 — Control chart.

The lines at $\bar{x} \pm 3\sigma_{\bar{x}}$ are called the upper control limit (UCL) and the lower control limit (LCL) ($\bar{x} + 3\sigma_{\bar{x}}$ is the UCL; $\bar{x} - 3\sigma_{\bar{x}}$ is the LCL). The lines at $\bar{x} \pm 2\sigma_{\bar{x}}$ are the upper and lower warning limits. The line at $\bar{x}$ is called the centerline. It is important to note here that QC data (like spike recoveries, check samples, etc.) are plotted on the control chart, not sample data.

In a data-producing system there are two general types of variability: that coming from assignable (or determinate) causes—systematic error—and that coming from unassignable (indeterminate) causes—random error. A process is in a state of statistical control when all causes of systematic error have been minimized or eliminated; leaving only random error. Control chart data inform the user of the state of process quality with respect to systematic error over a previous and up to a specific point in time. For this reason, control charts should be updated contemporaneously with the acquisition of sample data; control charts should be maintained up-to-date because real-time decisions have to be made using them.

Actually, two control charts are needed to evaluate the state of control of a particular analytical process. Recall that a process can be accurate and precise, accurate but not precise, etc. One type of control chart is needed to deal with accuracy and one type is needed to deal with precision. Since accuracy is related to central tendency, a common type of control chart for accuracy plots $\bar{x}$ (a control chart for means). Precision is a measure of variability, and it is commonly monitored by the use of an R chart (a control chart for range). The combination of charts is referred to as an $\bar{x}$-R chart (an xbar and r chart). Range has been used historically in statistical quality control because it is easy to calculate and it can be related to the standard deviation.

The limits for these charts are determined from data acquired at a time when the analytical process was in a known state of statistical control. Optimally, one would like to have as many as 50 data points before determining control chart limits, but in the analytical laboratory setting these data may require a relatively long time to acquire. Trial limits can be established on a smaller data set, with recalculation of the limits on the acquisition of further data.

Refer to Table 39.2, which represents QC data for the replicate analysis of Pb in aqueous solution, nominal 5 ppm concentration. This Pb standard was analyzed four times per day over 5 days. Because this Pb standard was analyzed four times per day by the same analyst using the same instrument, this daily group of data can be referred to as a rational subgroup.

Once the grand mean (that is, the mean of the daily means) and the average range are determined, the control and warning limits can be determined. Factors have been computed for this application and can be found in Table 39.3.

Using the grand mean and average range as determined in Table 39.2 together with factor data in Table 39.3, trial control and warning limits are determined as follows (remember, limits must be recalculated periodically as data are acquired. Since there are only a few days of data, these limits need to be recalculated frequently until the limits stabilize — somewhere around 30 days or so of data).

$$\text{Control Limits} = \bar{\bar{x}} \pm A_2(\bar{R}) \tag{3}$$

Table 39.2 — Data Set 1 — Pb Analysis

	Day 1	*Day 2*	*Day 3*	*Day 4*	*Day 5*	*Grand Mean* $(\bar{\bar{X}})$	*Avg. Range* $(\bar{R})$
	4.94	5.26	5.07	5.03	5.02		
	5.04	5.00	4.86	5.03	5.02		
	5.05	4.99	4.84	4.94	5.03		
	4.98	5.01	4.87	4.88	5.06		
Mean	5.00	5.07	4.91	4.97	5.03	5.00	
Range	0.11	0.27	0.23	0.15	0.04		0.16

Table 39.3 — Control Chart Factors

Subgroup Size, n	A_2	A_3	A_5	d_2	D_3	D_4	D_5	D_6	c_4
2	1.880	2.659	2.224	1.128	0	3.267	0	3.865	0.798
3	1.023	1.954	1.265	1.693	0	2.574	0	2.745	0.886
4	0.729	1.628	0.829	2.059	0	2.282	0	2.375	0.921
5	0.577	1.427	0.712	2.326	0	2.114	0	2.179	0.940
6	0.483	1.287	0.562	2.534	0	2.004	0	2.055	0.952
7	0.419	1.182	0.520	2.704	0.076	1.924	0.078	1.967	0.959
8	0.373	1.099	0.441	2.847	0.136	1.864	0.139	1.901	0.965
9	0.337	1.032	0.419	2.970	0.184	1.816	0.187	1.850	0.969
10	0.308	0.975	0.369	3.078	0.223	1.777	0.227	1.809	0.973

(The standard deviation can be related to the average range by

$$s \approx \frac{\bar{R}}{d_2}$$

Since

$$s_{\bar{x}} = \frac{s}{\sqrt{n}}, \text{ then } 3s_{\bar{x}} = 3\frac{\bar{R}}{d_2\sqrt{n}};$$

therefore,

$$A_2 = \frac{3}{d_2\sqrt{n}} \quad (4)$$

(for a subgroup size of 4, $d_2 = 2.059$, $\sqrt{n} = 2$).

Therefore,

$$A_2 = \frac{3}{2.059 \times \sqrt{4}} = 0.729$$

Look up A_2 in Table 39.3 for a subgroup size of 4. A_2 is 0.729.)

For a subgroup size of 4, $A_2 = 0.729$, therefore

$\bar{x} \pm A_2(\bar{R}) = 5.00 \pm 0.729\ (0.16) =$ 5.00 ± 0.12 ppm

UCL = 5.00 + 0.12 = 5.12 ppm; LCL = 5.00 – 0.12 = 4.88 ppm; centerline = 5.00 ppm.

The factors for the warning limits will not be found in the tables, but it is easy to determine them. Recall the formula for A_2. Substitute 2 for the 3 in the numerator (the warning limits are at 2 SD from the mean). So, the warning limit factor for a subgroup size of 4 is 0.486. Upper warning limit = 5.00 + 0.486(0.16) = 5.08 ppm; lower warning limit = 5.00 – 0.486(0.16) = 4.92 ppm.

The R chart is constructed in much the same manner. Taylor[10] gives formulas for computing the warning and control limits for the range chart as shown in Table 39.4. UCL = (2.282) (0.16 ppm) = 0.365 ppm; upper warning limit = (1.855) (0.16 ppm) = 0.300 ppm.

The Xbar-R chart, then, looks like Figure 39.6.

Now, if future data are plotted on these charts contemporaneously as those data are acquired, the current state of statistical quality can be determined for that analyte. It is highly recommended that QC data be plotted as soon as possible after acquisition.

Example

Question: How do we know if the analytical system is in control or out of control? Answer: There are many rules.[11] Three of the most important are as follows:

(1) 1 point falling outside $\bar{x} \pm 3\sigma$ denotes an out-of-control situation. Assuming a normal distribution of the data, statistically, if 99.73% of all data fall within $\bar{x} \pm 3\sigma$, then 0.27% do not. The probability that the system is in control given that 1 point falls outside $\bar{x} \pm 3\sigma$ is only 27 chances in 10,000, which is too low a probability to confidently state that the system is in control.

(2) 2 consecutive points falling outside $\bar{x} \pm 2\sigma$ denotes an out-of-control situation. The probability that two consecutive points will fall outside $\bar{x} \pm 2\sigma$ given that the system is in control is 25 chances in 10,000: about the same probability as 1 point outside $\bar{x} \pm 3\sigma$.

(3) 8 points consecutively on the same side of the centerline is an out-of-control situation. There are two reasons for this: (1) 8 consecutive points on the same side of the centerline is evidence of something systematic going on in the system, and since we are controlling for systematic error, this constitutes an out-of-control situation; (2) the probability that 8 consecutive points fall on the same side of the centerline,

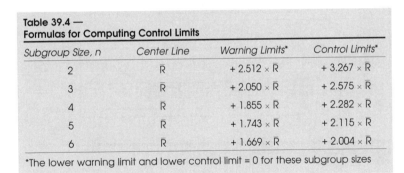

Table 39.4 — Formulas for Computing Control Limits

Subgroup Size, n	*Center Line*	*Warning Limits**	*Control Limits**
2	$\bar{R}$	+ 2.512 × $\bar{R}$	+ 3.267 × $\bar{R}$
3	$\bar{R}$	+ 2.050 × $\bar{R}$	+ 2.575 × $\bar{R}$
4	$\bar{R}$	+ 1.855 × $\bar{R}$	+ 2.282 × $\bar{R}$
5	$\bar{R}$	+ 1.743 × $\bar{R}$	+ 2.115 × $\bar{R}$
6	$\bar{R}$	+ 1.669 × $\bar{R}$	+ 2.004 × $\bar{R}$

*The lower warning limit and lower control limit = 0 for these subgroup sizes

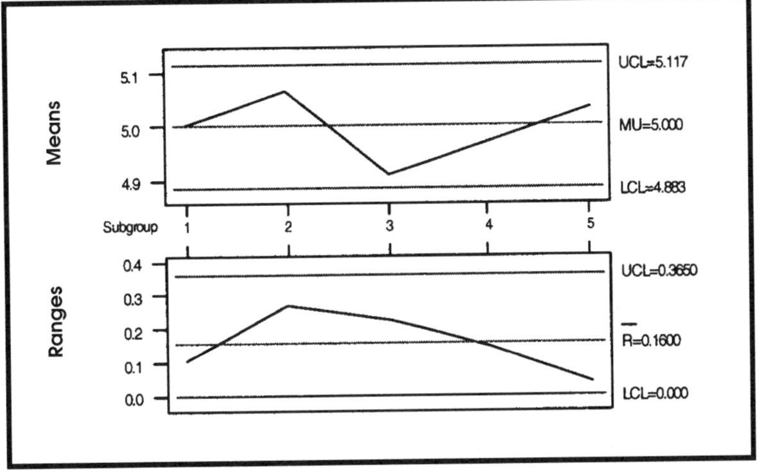

Figure 39.6 — Control chart for Pb analysis, Days 1–5.

given that the system is in control, is equal to 0.0039, or about 40 chances in 10,000. This is clearly a very low probability.

Adding additional data to the control chart is accomplished as follows. Since the limits on the Pb charts created earlier in the chapter are based on a subgroup size of four, it is necessary to plot averages and ranges using the same subgroup size. Examine Data Set 2 representing QC data for Pb analysis for Days 6–10. Combining the data in Table 39.5 with existing Day 1–5 data (and recomputing the limits) yields: grand mean = $\bar{\bar{x}}$ = 4.99 ppm, average range = $\bar{R}$ = 0.18 ppm. The analytical system appears to be in control; that is, it is performing accurately and precisely (see Figure 39.7).

Table 39.5 — Data Set 2, Pb Analysis

	Day 6	Day 7	Day 8	Day 9	Day 10
	4.88	5.01	4.94	5.07	4.87
	5.03	4.87	5.03	5.20	4.86
	5.02	5.02	4.84	5.05	5.05
	4.99	4.86	5.06	4.90	5.01
Mean	4.98	4.94	4.97	5.06	4.95
Range	0.15	0.16	0.22	0.30	0.19

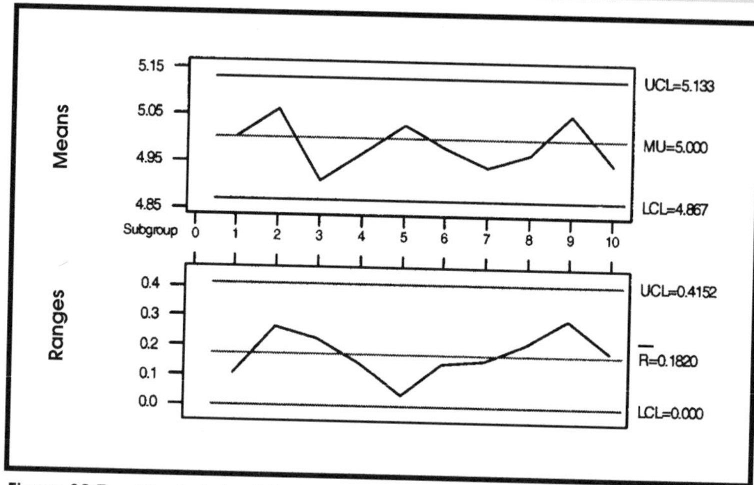

Figure 39.7 — Control chart for Pb analysis, Days 1–10.

If a plotted point indicates an out-of-control situation, the analytical process needs to be stopped, corrective action performed, then the process needs to be restarted (after demonstrating that the corrective action solved the problem).

Much QC work is computerized now, and as a result, many analysts have switched from Xbar-R charts to Xbar-s charts, where s represents the SD. While calculating R and plotting R charts manually is more efficient than calculating SD and s charts manually, the s chart is generally preferable. The s chart does suffer from some of the shortcomings of the R chart. One of those shortcomings is that the control limits for s disregard the fact that the distribution of s is asymmetric even when the distribution of individual values is normal (this is also true of the range). The control limits that are used are as follows.

$$\bar{s} \pm 3\hat{\sigma}_s \tag{5}$$

where $\bar{s}$ = average of subgroup standard deviations, and $\hat{\sigma}_s$ = estimate of the standard deviation of s.

This leads to (see Table 39.3):

$$UCL = B_4\bar{s} \tag{6}$$
$$LCL = B_3\bar{s} \tag{7}$$

Using the Pb data from Days 1–10, the Xbar-s chart looks like Figure 39.8.

Suppose that there is no basis for rational subgrouping of data. One such case would involve the analysis of a single QC sample with each batch of samples. If there is no basis for rational subgrouping (here, the subgroup size is 1), individual data must be plotted on an X chart (an individual's chart). If the subgroup size is 1, then no range can be determined, and therefore, no range chart can be plotted. Hence, no precision information will be available — the only signal that the process has gone out of control will be through a measure of accuracy. In addition, an out-of-control signal may be given by the system because of a quirk in the distribution of data. Remember that due to the Central Limit theorem, the distribution of means of sample size n will tend to be normal (as n increases) regardless of the underlying distribution of the individual data. However, now the individual data must be used. This

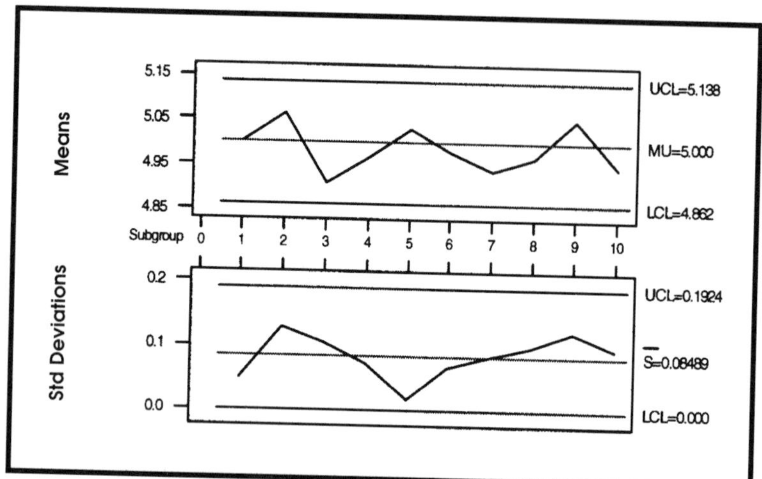

Figure 39.8 — Xbars control chart for Pb data, Days 1–10.

Table 39.6 — I-mR Data

Sample Number	Concentration, ppm	Moving Range
1	29.2	—
2	28.4	0.8
3	29.2	0.8
4	32.9	3.7
5	27.9	5.0
6	26.4	1.5
7	31.8	5.4
8	39.4	7.6
9	28.6	10.8
10	28.0	0.6
11	31.2	3.2
12	37.6	6.4
13	26.9	10.7
14	30.7	3.8
15	31.9	1.2
16	28.9	3.0
17	27.8	1.1

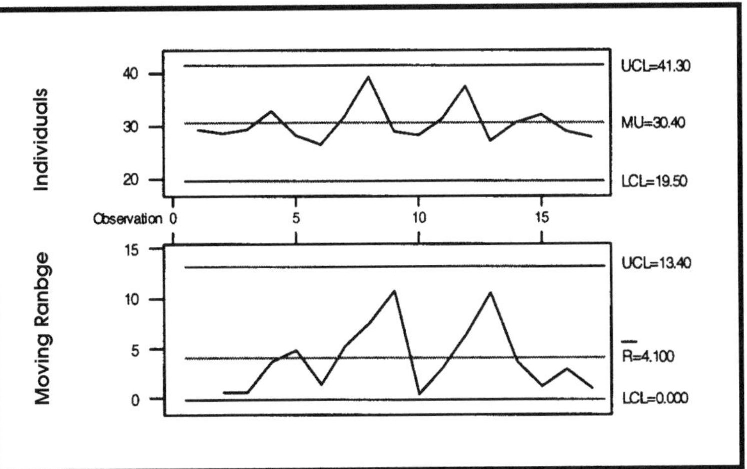

Figure 39.9 — I-mR control chart for Table 39.5 data.

situation can cause unnecessary system downtime for corrective action that results not from a true out-of-control condition, but from a data distribution malady.

One way to deal with individual data is to develop an I-mR chart (that is, an individual's-moving range chart). Examine the data in Table 39.6.[12]

When interpreting charts like that found in Figure 39.9, it is important to note that the point plotted represents two (in this case) individual points which, because of the averaging process, are no longer independent. A plotted point on a moving range chart that is out of control can result from one point, the other point, or both points being out of control. In addition, two consecutive points (because an artificial subgroup size of two was used) outside control limits, not one, designate an out-of-control situation. Runs above or below the centerline do not have the same significance in a moving range chart as they do in an xbar-R chart.[13] Nelson[14] recommends that moving ranges not be plotted because it is very difficult to interpret this control chart when the data are correlated.

Other Types of Control Charts

There are many additional types of control charts, including those used for attribute (accept/reject, go/no go) data. Any reference text discussing statistical quality control[1,13,15–23] can supply all the information needed on these other types. But a laboratory performing metals analysis by atomic absorption spectrophotometry or organic vapor analysis by gas chromatography might be swamped by the handling of all the QC data generated—imagine a control chart for each metal or each organic vapor analyzed. For QC to be successful as a tool for improvement, neither analysts nor QC coordinators can spend all of their time plotting data on countless numbers of charts. The paperwork then becomes the focus, not the process.

Suppose all metals analysis data or all organic vapor analysis data could be plotted on one control chart. This type of chart is called a "short-run" variables control chart. Laboratories often receive, for example, five samples for Pb analysis one day, three samples for Cd analysis the next day, three samples for Be analysis the next day, etc. In short, laboratories receive a few samples for analysis of different metals on different days. The laboratory does not necessarily have to maintain a control chart for Pb, one for Cd, and one for Be. All of these QC data may be plotted on one chart. If the control limits are different for each metal (the most likely case, based on the fact that the analytical instrument may have a different sensitivity for each analyte), then the raw QC data cannot be plotted together. The trick then is to plot coded data. Coded data can be produced by transformation. Recall that

$$LCL_{\bar{x}} < \bar{X} < UCL_{\bar{x}}$$

Then, from the definitions of the control limits,

$$\bar{\bar{X}} - A_2\bar{R} < \bar{X} < \bar{\bar{X}} + A_2\bar{R}$$

Letting $\bar{\bar{X}} = X_{AH}$ (the historical average) and $\bar{R} = R_{AH}$ (the historical range), then subtracting X_{AH} and dividing by R_{AH} yields

$$-3 < \frac{\overline{X} - \overline{X}_{AH}}{\overline{R}_{AH}} < +3 \qquad (8)$$

So if the plot point is $\frac{\overline{X} - \overline{X}_{AH}}{\overline{R}_{AH}}$

on an $\overline{x}$ chart, the control limits will be ±3. The centerline is, of course, zero. This process can be repeated for R. Assuming a normal distribution:

$$D_4\overline{R}_{AH} < \overline{R} < D_3\overline{R}_{AH}$$

Dividing through by $\overline{R}_{AH}$,

$$D_4 < \frac{\overline{R}}{\overline{R}_{AH}} < D_3 \qquad (9)$$

The plot point is $\frac{\overline{R}}{\overline{R}_{AH}}$

on an R chart; the control limits are D_3 and D_4 (for subgroup size < 7, $D_3 = 0$). Again, $\overline{R}_{AH}$ is the historical R.

To illustrate this short-run chart, assume that historical data for some metals' QC standards are as listed in Table 39.7.

Five samples are received 1 day for Pb and Cd analysis. On the next day, six samples are received for Ni analysis. Two QC standards are analyzed with each batch of samples (the concentration for each standard is listed in Table 39.7). The QC data are as shown in Table 39.8.

Table 39.7 — Short-Run Control Chart Data (1)

Metal	$\overline{X}_{AH}$ (ppm)	$\overline{R}_{AH}$ (ppm)	Reference Standard Concentration (ppm)
Pb	5	0.5	5
Cd	10	0.7	10
Ni	2.5	0.2	2.5

Table 39.8 — Short-Run Control Chart Data (2)

Metal	Analytical Result (ppm), QC 1	Analytical Result (ppm), QC2	Average Result (ppm), $\overline{X}$	$\overline{R}$ (ppm)
Pb	5.1	5.4	5.25	0.3
Cd	10.2	10.6	10.4	0.4
Ni	2.5	2.4	2.45	0.1

Table 39.9 — Short-Run Control Chart Plotting Data (3)

Analyte	Plot Point $\overline{X}$ (ppm)	Plot Point R (ppm)	Control Limts, Mean ± A_2	Control Limits, Range 0, D_4
Pb	0.5	0.6	1.880	3.267
Cd	0.57	0.57	1.880	3.267
Ni	−0.5	0.5	1.880	3.267

To determine the plotting points, use the Pb data.

For the mean,

$$\frac{\overline{X} - \overline{X}_{AH}}{\overline{R}_{AH}} = \frac{5.25\text{ ppm} - 5.00\text{ ppm}}{0.5\text{ ppm}} = 0.5$$

For the range,

$$\frac{\overline{R}}{\overline{R}_{AH}} = \frac{0.3\text{ ppm}}{0.5\text{ ppm}} = 0.6$$

The data for Pb, Cd, and Ni are reduced in Table 39.9.

These plot points are plotted on a typical control chart. The advantage of using a short-run control chart is, again, the ability to plot transformed data from various operations on one chart.

Using Spiked Samples

A spiked sample is a sample to which has been added a known amount of analyte. The analysis of these spiked samples can be used to determine the bias and precision of a test method, determine the accuracy of a measurement process in the laboratory, and/or to detect changes in the analytical process.

To use spiked samples effectively, one must know the ranges of concentration of interest and the relationship between recovery and concentration (if recovery is independent of concentration, fewer spiked samples may be necessary).

$$\%\text{ Recovery} = \frac{M_{cs} - M_{co}}{M_{added}} \times 100\% \qquad (10)$$

where:

M_{cs} = measured concentration of the spiked sample
M_{co} = measured concentration of the analyte in the original sample
M_{added} = concentration of the spike

Be careful. When background concentrations are comparable to the level of the spike, estimates of method bias or precision based on recovery data can make a method look worse than it really is. Statistical uncertainty in recovery data needs to be considered before decisions are made regarding the method.

Spike data can be plotted on a control chart, with limits determined based on subgroup size, not on the notion (commonly held) that the data are acceptable if the percentage recovery is, say, 90–110%. This does not permit process improvement, which is the goal of quality control.

Handling Outlier Data

In referring to outliers, the following quotation expresses concisely the nature of, and concerns about, these data.

> *In almost every true series of observations, some are found, which differ so much from the others as to indicate some abnormal source of error not contemplated in the theoretical discussions, and the introduction of which into the investigations can only serve . . . to perplex and mislead the inquirer.*[24]

It is interesting that this quotation was made almost 150 years ago, and, seemingly, outlier data was a concern even then. Now, it is recognized that outliers do not necessarily "perplex" or "mislead," nor are they necessarily "bad" or "erroneous." In some instances researchers may not reject them; they may even welcome them as unexpectedly useful.

An outlier is defined as a data point that "appears to be markedly different from other members of the sample in which it occurs."[25] Notice the use of the word "appears." One should be very careful in deciding that a piece of data may be an outlier. Why? For one reason, the data in question could be an extreme value in the distribution (it happens, you know). It could be also that the data results from some gross deviation from the analytical method used, or a mathematical blunder. So before assuming that some questionable data is outlying, one may wish to investigate the process and the calculations first. If there is no apparent deviation from the method, and if the calculations were completed correctly, it may be that the data in question are outlying. Outlier data are neither discarded nor deleted. These data are flagged in some appropriate manner, then not used in calculations.

The Dixon Ratio test is one that can be used to detect outlier data. It is easy to use and the calculations are simple. The Dixon Ratio test assumes an underlying normal distribution of data. To perform the test (1) rank the data in ascending order, X_1 (lowest) to X_n (highest); (2) select the significance level for rejection; (3) calculate the Dixon Ratio (Tables 39.10 and 39.11 show Dixon Ratio information and calculations); (4) look up the critical value for the appropriate r-value in the Dixon Ratio table (Table 39.12); and (5) if the $r_{calculated} > r_{table}$ for the confidence level selected in Step 2 above, conclude that the suspect point is an outlier; otherwise retain the data point.

Laboratories have an ethical and legal obligation to produce data that are accurate and precise and at a time when the laboratory is in a known state of statistical control. How is accuracy and precision evaluated? The following are some techniques that can be used to accomplish this evaluation.

Evaluating Method Accuracy

Does a particular analytical method used by the laboratory produce accurate (that is, unbiased) data?

Table 39.10 — Dixon Ratio Information

Number of Data Points	*Ratio to be Calculated*
3 to 7	r_{10}
8 to 10	r_{11}
11 to 13	r_{21}
14 to 25	r_{22}

Table 39.11 — Dixon Ratio Calculations

r	*If X_n is Suspect*	*If X_1 is Suspect*
r_{10}	$(X_n - X_{n-1})/(X_n - X_1)$	$(X_2 - X_1)/(X_n - X_1)$
r_{11}	$(X_n - X_{n-1})/(X_n - X_2)$	$(X_2 - X_1)/(X_{n-1} - X_1)$
r_{21}	$(X_n - X_{n-2})/(X_n - X_2)$	$(X_3 - X_1)/(X_{n-1} - X_1)$
r_{22}	$(X_n - X_{n-2})/(X_n - X_3)$	$(X_3 - X_1) / (X_{n-2} - X_1)$

Table 39.12 — Values for Use in the Dixon Ratio Test

Statistic	*Number of Observations*	*Risk of False Rejection* 0.5%	1%	5%	10%
r_{10}	3	0.994	0.988	0.941	0.886
	4	0.926	0.889	0.765	0.679
	5	0.821	0.780	0.642	0.557
	6	0.740	0.698	0.560	0.482
	7	0.680	0.637	0.507	0.434
r_{11}	8	0.725	0.683	0.554	0.479
	9	0.677	0.635	0.512	0.441
	10	0.639	0.597	0.477	0.409
r_{21}	11	0.713	0.679	0.576	0.517
	12	0.675	0.642	0.546	0.490
	13	0.649	0.615	0.521	0.467
r_{22}	14	0.674	0.641	0.546	0.492
	15	0.647	0.616	0.525	0.472
	16	0.624	0.595	0.507	0.454
	17	0.605	0.577	0.490	0.438
	18	0.589	0.561	0.475	0.424
	19	0.575	0.547	0.462	0.412
	20	0.562	0.535	0.450	0.401

Source: Reference 28

Assume that a laboratory has modified a method for the analysis of toluene by gas chromatography by using a new type of column. Does this modified method produce accurate (unbiased) data? To determine this, use the following steps.

Step 1. Make the following decisions.

- What is α? (α is the Type I error; in this case, the risk of concluding the method is biased when it is not)
- What is β? (β is the Type II error; in this case, the risk of concluding the method is accurate when it is not)
- What is the MDB? (MDB is the minimum detectable bias; if the method produces data equal to or greater than the MDB, it should be detected with (1– β) probability)
- What is σ_{TEST}? (σ_{TEST} is the method variability for the standard used for testing)
- What is the standard concentration to be used to determine method accuracy?

Step 2. Determine how many tests must be performed.

$$\text{MDB} = \frac{z \cdot \sigma_{TEST}}{\sqrt{n}}$$

where $z = (z_{\alpha/2} + z_{\beta})$. (Note: z is a z-score, found in z tables in any statistics text.)

$$\text{Rearranging, } n = \frac{z^2 \cdot \sigma^2_{TEST}}{\text{MDB}^2} \qquad (11)$$

If n is not an integer, round it up.

Step 3. Perform the calculated number of tests.

Step 4. Determine $\bar{x}$, the average of the n tests.

Step 5. Determine the method precision for n tests, $\sigma_{\bar{x}} = \sigma_{TEST}/\sqrt{n}$

Step 6. Compare $\bar{x}$ with the standard concentration. If $\bar{x}$ is within $z_{\alpha/2}\sigma_{\bar{x}}$ units of the standard concentration, conclude that the method is accurate. If not, then conclude the method is biased. Bias = $\bar{x}$ standard concentration.

For example, consider the following: α = 0.05, β = 0.10, MDB = 2 ppb, σ_{TEST} = 4 ppb, standard concentration = 10 ppb. If σ_{TEST} is unknown, and the population of test values is normally distributed, use the t-distribution. Determine the (1-α)% confidence interval on the mean: $\bar{X} \pm t_{(\alpha/2,n-1)} s/\sqrt{n}$. If the interval contains the known value and is not "too wide," assume the test is accurate.

$$n = \frac{z^2 \cdot \sigma^2_{test}}{\text{MDB}^2} = \frac{(1.96 + 1.282)^2 \cdot 4^2}{2^2} = 42.04$$

So 43 tests of the 10 ppb standard should be made. Assume that 43 analyses were performed on the 10 ppb standard, and that the mean concentration reported for those 43 tests was 9.62 ppb.

$$\sigma_{\bar{x}} = \sigma_{TEST}/\sqrt{n} = 4 \text{ ppb}/\sqrt{43} = 0.61 \text{ ppb}$$

$$\bar{x} \pm z_{\alpha/2} \cdot \sigma_{\bar{x}} = 9.62 \pm (1.96)(0.61) = 9.62 \pm 1.20 = 8.42 \text{ to } 10.82 \text{ ppb}$$

Since the range 8.42 ppb to 10.82 ppb contains the standard concentration (10 ppb), the method is not biased.

Evaluating Precision

Analyst-to-Analyst Variability

Suppose there are several analysts in the laboratory who analyze charcoal tube samples for organic vapor concentration. Do these analysts produce the same results (statistically) for analysis of identical samples? To determine this for the simplest case of one homogeneous sample use the following steps (if using more than one homogeneous sample, please consult your local statistician).

(1) Let k be the number of analysts. Let each analyst test this one, homogeneous sample m times. Note: for this test, $k \times (m - 1) \geq 50$.

(2) Determine $\bar{X}$ and SD for each analyst.

(3) Pool the variances.

(4) Determine $s_p = \sqrt{\frac{\sum s^2}{n}}$

(5) Determine $s_{\bar{x}}^2$ for the analyst averages.

(6) Calculate

$$F_{calc} = \frac{s_{\bar{x}}^2}{s_p^2/m} \qquad (12)$$

This is an F distribution with $v_1 = k - 1$; $v_2 = k \times (m-1)$

(7) If $F_{calc} > F^{1-\alpha}_{v_1,v_2}$

$$\text{then } \sigma \text{ analyst} = \sqrt{s_{\bar{x}}^2 - \frac{s_p^2}{m}} \qquad (13)$$

(8) If $F_{calc} \ngtr F^{1-\alpha}_{v_1,v_2}$

then analyst-to-analyst variability is not significant as compared to method variability.

As an example, consider the following. Assume one homogeneous sample, k = 5 analysts. Since the total number of tests performed must exceed 50 and there are 5

Table 39.13 — Mean and Standard Deviation for Analyzing 1 Homogenous Sample 12 Times

Analyst	1	2	3	4	5	
$\bar{x}$	0.52	0.47	0.49	0.45	0.53	
s^2	2.5	3.6	1.8	2.3	2.7	x 10^{-4}

analysts, $(m - 1) \geq 10$. Say that each analyst analyzed the one homogeneous sample in a blind study 12 times. Calculations for the mean and standard deviation for each analyst are shown in Table 39.13. Pooled s^2 = .00025 + .00036 + .00018 + .00023 + .00027 = 0.00129. Then

$$s_p = \sqrt{\frac{s^2}{n}} = \sqrt{\frac{0.00129}{5}} = 0.016$$

Determine the standard deviation of the analyst averages. Then square that standard deviation.

$$s_{\bar{x}}^2 = 0.0335^2 = 0.00112$$

Determine

$$F_{calc} = \frac{s_{\bar{x}}^2}{s_p^2/m} = \frac{0.00112}{0.00129/12} = 10.42$$

Determining the number of degrees of freedom: $v_1 = 5 - 1 = 4 \quad v_2 = 5 \cdot (12 - 1) = 55$

Looking up the appropriate F-critical value in the F-tables (in any statistical text)

$$F_{4,55}^{0.95} = 2.55$$

Is $F_{calc} > F_{crit}$? Yes, 10.42 > 2.55. Therefore, there is significant analyst-to-analyst variability. So,

$$\sigma_{analyst} = \sqrt{s_{\bar{x}}^2 - \frac{s_p^2}{m}} = \sqrt{0.00112 - \frac{0.00129}{12}}$$

$$= 0.03$$

Interlaboratory Testing

How do laboratories analyzing the same samples for the same analytes compare with one another? A client has a choice of a few hundred laboratories to which to send a sample. If the same sample could be sent to all of them, they probably would not all report the same results. If not, then how much variability is there between laboratories? How much is acceptable?

To help answer these questions, an interlaboratory test is performed. The American Industrial Hygiene Association conducts several quarterly proficiency testing programs in which many laboratories participate and are evaluated. The Proficiency Analytical Testing Program serves to evaluate laboratories analyzing workplace samples and uses reference samples of several metals, silica, several organic compounds, asbestos, and manmade fibers. The Environmental Lead Proficiency Analytical Testing Program uses lead-spiked paint, soil, dust, and air samples to evaluate those laboratories performing environmental lead analyses. The Asbestos Analyst Testing Program serves those analysts performing asbestos fiber counting in the field. The Bulk Asbestos Proficiency Analytical Testing Program serves those laboratories performing asbestos identification analyses not related to the Asbestos Hazard Emergency Response Act of 1986. The Environmental Microbiology Proficiency Analytical Testing Program is for those laboratories performing analysis of bacterial and fungal samples in support of indoor air quality programs. Laboratories analyze these samples, report the results, and await the statistical analysis of the data, which will determine laboratory proficiency ("yes" or "no"). As a part of some NIOSH methodology,[27] laboratories are required to associate themselves in small groups, generate and analyze test samples, and statistically evaluate the data to determine some facet of interlaboratory variability.

One method that can be used to assess laboratory-to-laboratory variability is what has come to be known as the Youden test.[28] In this test, participating laboratories are sent several of the same samples for analysis. Each laboratory reports the data to the coordinating laboratory for statistical analysis. Each laboratory is ranked for each sample. If seven labs participate in a round-robin, the laboratory with the lowest reported concentration for a given sample is given a "1;" the highest is given a "7." Ranking is performed over each sample. When ranking is complete, scores are totaled and compared with Youden's table. This table produces a score range; lab scores lower than the lower limit of this range are biased low; labs with scores higher than the upper limit are biased high. Examine the data set in Table 39.14 for three labs and seven samples.

The Youden table[29] shown in Table 39.15 gives the approximate 5% two-tailed limits for ranking scores.

The approximate 5% limits for three laboratories analyzing seven samples is 8–20. Laboratories scoring between 8 and 20 are

Table 39.14 — Example Data Set Showing Youden Scores for Three Labs and Seven Samples

	Lab 1	Lab 2	Lab 3
Raw data			
Sample 1	11.6	15.3	21.1
Sample 2	11.0	14.8	20.8
Sample 3	21.0	10.8	15.0
Sample 4	15.0	10.8	20.6
Sample 5	14.9	11.2	20.7
Sample 6	21.3	24.9	29.4
Sample 7	18.8	20.2	25.5
Ranked data			
Sample 1	1	2	3
Sample 2	1	2	3
Sample 3	1	2	3
Sample 4	2	1	3
Sample 5	2	1	3
Sample 6	1	2	3
Sample 7	1	2	3
Total	9	12	21

not significantly different. A laboratory scoring lower than 8 is biased low; and one scoring greater than 20 is biased high. In the example above, Labs 1 and 2 are not significantly different; Lab 3 is biased high.

Large variations in results reported from different laboratories analyzing the same samples might be explained by random errors in the measurements or by systematic errors in the different laboratories. Practically speaking, it is highly probable that systematic errors made by the participating laboratories are the cause for these wide variations. This can be evaluated using another technique for analyzing interlaboratory variability—the "two-sample" or Youden plot.[30] In this graphical technique each participating laboratory is sent two similar samples (A and B) and asked to perform one analysis on each sample. When the results are returned, the results obtained by a laboratory for Sample A are plotted with respect to the results obtained by the same laboratory for Sample B. Median lines are drawn, and outliers are identified and discarded from the data set. If there were no bias in the results, the plotted points should be randomly distributed around the intersection of the two median lines drawn. It is rare that this is observed. Generally what is observed is that the plotted points fall around a line drawn at 45° from the intersection of the two median lines, indicating that each participating laboratory has its own technique (an internal consistency) for analyzing the samples. (Laboratories that report one sample high are more likely to report the second sample high. The converse is true, also.) Through the use of two samples similar in concentration, an estimate of precision can be made. This is done by constructing a perpendicular line from the plotted point to the 45° line.

Table 39.16 shows the data for nine laboratories that analyzed two samples, A and B. The averages for A and B were calculated as was the difference (A – B) and the average difference. In the last column, the average difference was subtracted from (A – B). The absolute value of this difference was then averaged. The data are plotted as shown in Figure 39.10.[31]

Table 39.15 — Youden Table, 5% Two-Tailed Limits on Ranked Scores

Number of Participating Laboratories	Number of Samples Analyzed							
	3	4	5	6	7	8	9	10
3		4	5	7	8	10	12	13
		12	15	17	20	22	24	27
4		4	6	8	10	12	14	16
		16	19	22	25	28	31	24
5		5	7	9	11	13	16	18
		19	23	27	31	35	38	42
6	3	5	7	10	12	15	18	21
	18	23	28	32	37	41	45	49
7	3	5	8	11	14	17	20	23
	21	27	32	37	42	47	52	57

Table 39.16 — Example Data Set for Nine Laboratories Analyzing Two Samples

Lab No.	A	B	A – B	((A – B)) – 0.24)
1	35.1	33.0	2.1	1.86
2	23.0	23.2	– 0.2	0.44
3	23.8	22.3	1.5	1.26
4	25.6	24.1	1.5	1.26
5	23.7	23.6	0.1	0.14
6	21.0	23.1	– 2.1	2.34
7	23.0	21.0	2.0	1.76
8	26.5	25.6	0.9	0.66
9	21.4	25.0	– 3.6	3.84
Avg	24.8	24.5	0.24	1.51

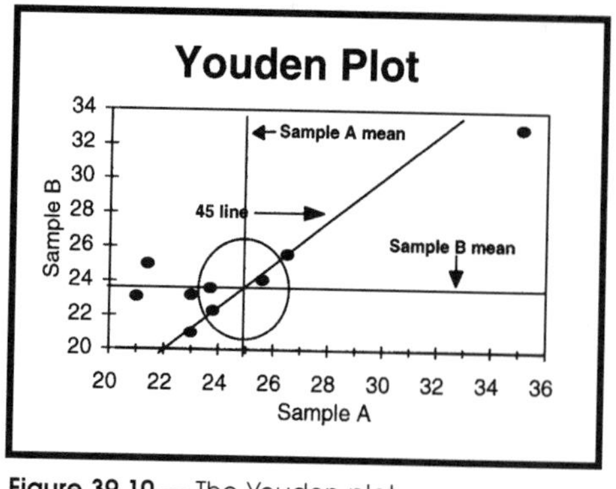

Figure 39.10 — The Youden plot.

The average of the absolute values of (A – B) – 0.24, when multiplied by 0.886, gives an estimate of the standard deviation: 1.51 × 0.886 = 1.34. The estimate of the standard deviation for precision allows the construction of a circle centered on the intersection of the median lines within which any given percentage of points can be expected to fall given that the participating laboratories can eliminate all bias or constant errors. For 95% of points within the circle, the multiple of the standard deviation is 2.448. So, 2.448 × 1.34 = 3.28, which is the radius of the circle. Once the circle is constructed, points falling outside the circle represent laboratories with substantial systematic errors incorporated into their analytical technique.

Summary

In this chapter, it has been demonstrated that a quality product (the concentration of a contaminant resulting from the analysis of a sample collected in the field) is obtained only when a properly collected sample is properly analyzed. Both sampling and analysis must be "in control." For sample collection, the design, implementation, and documented adherence to a legitimate sampling protocol is required. This implies designing a protocol that, when followed, permits the collection of the kind of sample for which the protocol is intended, then following that protocol to the letter. To assure that the samples collected are not altered between the time of collection through analysis at the laboratory, blanks must be employed. Because the laboratory is staffed by humans, and because humans are fallible, duplicate, replicate, and/or spiked samples must be added to the actual samples in such a fashion that the laboratory is not able to discern which are the samples. A technique was discussed that permits the determination of the number of samples to be collected, given a confidence level and permissible bias.

The laboratory must perform analyses and report results only when it can be stated that the results obtained are accurate, precise, and at a time when the laboratory is in a state of known statistical control. The content of a laboratory quality control plan was discussed. Techniques were discussed for the construction and use of Shewhart control charts and short-run control charts, determination of accuracy and precision within-analyst, between-analyst, and between-method, calibration, handling outlier data, and interlaboratory testing.

Acknowledgments

Several people have provided valuable assistance in preparing this chapter. Colleen Becker, Patrick Dunn, Fred Grunder, and Tracy Yerian reviewed the draft and provided thoughtful and considered comments, which substantially improved this work. Denell Deavers provided publishing assistance with courtesy and understanding. Thank you all. And finally to Dr. Salvatore DiNardi: I know you'll understand when I say, "Thanks for your help."

References

1. **Juran, J.M. (ed.):** *Juran's Quality Control Handbook*, 4th ed. New York: McGraw-Hill, 1988. pp. 2.2–2.4.
2. **Taylor, J.K.:** *Quality Assurance for Chemical Measurements*. Chelsea, MI: Lewis Publishers, 1987. p. 2.
3. **Black, S.C.:** Defining control sites and blank sample needs. In *Principles of Environmental Sampling*. Washington, DC: American Chemical Society, 1988. p. 110.
4. **U.S. Environmental Protection Agency (EPA):** *Quality Assurance for Air Pollution Monitoring Systems*, vol. I (EPA-600/9-76-005). Washington, DC: EPA, 1984. Sec. 1.4.12, rev. 1, p. 1.
5. **Schlecht, P.C., J.V. Crable, and W.D. Kelley:** Industrial hygiene. In *Quality Assurance Practices for Health Laboratories*. Washington, DC: American Public Health Association, 1978. p. 801.
6. **ASQC Chemical and Process Industries Divison:** *Quality Management and Quality Systems Elements — Guidelines* (ANSI/ASQC Q94-1987). Milwaukee, WI: American Society for Quality Control, 1987, Sections 4.1 and 4.2. p. 2.
7. **U.S. Environmental Protection Agency (EPA):** *Quality Assurance Handbook for Air Pollution Measurement Systems*, vol. I (EPA-600/9-76-005), Washington, DC: EPA, 1984. Section 1.4.16.
8. **Shewhart, W.A.:** *Economic Control of the Quality of Manufactured Product*. Princeton, NJ: Van Nostrand Reinhold, 1931.
9. **Blaisdell, E.A.:** *Statistics in Practice*. Philadelphia, PA: Saunders College Publishing, 1993. pp. 306–310.
10. **Taylor, J.K.:** *Statistical Techniques for Data Analysis*. Chelsea, MI: Lewis Publishers, 1990. p. 153.

11. See, for example, *Statistical Quality Control Handbook*. Charlotte, NC: Western Electric Company, Inc., 1956. pp. 23–32.
12. **U.S. Environmental Protection Agency (EPA):** Quality Assurance Handbook for Air Pollution Measurement Systems, vol. I (EPA-600/9-76-005). Washington, DC: EPA, 1984. Sec. H, rev. 1, p. 26.
13. **Grant, E.L. and R.S. Leavenworth:** *Statistical Quality Control*, 6th ed. New York: McGraw-Hill, 1988. pp. 318–323.
14. **Nelson, L.S.:** Control charts for individual measurements. *J. Qual. Technol. 14*:172–173 (1982).
15. **Burr, I.W.:** *Elementary Statistical Quality Control*. New York: Marcel Dekker, 1979.
16. **Duncan, A.J.:** *Quality Control and Industrial Statistics*, 3rd ed. Homewood, IL: Richard D. Irwin, 1965.
17. **Ishikawa, K.:** *Guide to Quality Control*, 2nd rev. ed. Tokyo: UNIP-Kraus International Publishers, 1986.
18. **Kume, H.:** *Statistical Methods for Quality Improvement*. Tokyo: AOTS, 1987.
19. **Montgomery, D.C.:** *Introduction to Statistical Quality Control*. New York: John Wiley & Sons, 1996.
20. **Ott, E.R.:** *Process Quality Control*. New York: McGraw-Hill, 1990.
21. **Ryan, T.P.:** *Statistical Methods for Quality Improvement*. New York: John Wiley & Sons, 1989.
22. **Wadsworth, H.M.:** *Modern Methods of Quality Control and Improvement*. New York: John Wiley & Sons, 1986.
23. **Wheeler, D.J.:** *Understanding Statistical Process Control*. Knoxville, TN: SPC Press, 1992.
24. **Peirce, B.:** Criterion for the rejection of doubtful observations. *Astr. J.* 2:161–163 (1852).
25. **American Society for Testing and Materials (ASTM):** *Standard Practice for Dealing with Outlying Observations* (E 178-80). West Conshohocken, PA: ASTM, 1980.
26. **American Society for Testing and Materials (ASTM):** *Standard Practice for Dealing with Outlying Observations* (E 178-80). West Conshohocken, PA: ASTM, 1980. p. 299.
27. See, for example, "Asbestos and other Fibers by Phase Contrast Microscopy" (NIOSH Method 7400), in *NIOSH Manual of Analytical Methods*. Cincinnati, OH: National Institute for Occupational Safety and Health, 1994.
28. **Youden, W.J.:** Measurement agreement comparisons. In *Proceedings of the 1962 Standards Laboratory Conference* (NBS misc. pub. 248). Washington, DC: Government Printing Office, 1962.
29. **Taylor, J.K.:** *Statistical Techniques for Data Analysis*. Chelsea, MI: Lewis Publishers, 1990. p. 170.
30. **Youden, W.J.:** Collection of papers on statistical treatment of data. *J. Qual. Technol. 4*:1–67 (1982).
31. **Linch, A.L.:** Quality control for sampling and laboratory analysis. In *The Industrial Environment—Its Evaluation and Control*. Washington, DC: U.S. Government Printing Office, 1973, p. 293.

This worker is cleaning up after a pulp spill in a paper plant by washing pulp into the sewer. He is equipped with a data-logging electrochemical sensor (belt mount) and a passive diffusion sampling card (collar) to measure chlorine dioxide levels.

Chapter 40

Outcome Competencies

After completing this chapter, the user should be able to:

1. Define underlined terms used in this chapter.
2. Summarize the role of hazard communication in a workplace safety and health program.
3. Describe the three major components of a hazard communication program.
4. Prepare a simple written hazard communication program.

Key Terms

chemical • employee • exposure • hazard warnings • health hazard • identity • physical hazard • responsible party

Prerequisite Knowledge

General occupational hygiene background.

Key Topics

I. Introduction
   A. Role of Hazard Communication in the Workplace
   B. History of the Development of Legal Requirements for Hazard Communication

II. The Federal Hazard Communication Standard
   A. Overview of the Approach
   B. Detailed Description of the Requirements

III. International Hazard Communication

40

Hazard Communication

Jennifer C. Silk

Introduction

Role of Hazard Communication in the Workplace

Occupational hygienists trying to protect exposed workers from hazardous chemicals before the 1980s had to spend a considerable amount of time researching the hazards of the chemicals in the workplaces under their control. While some chemical manufacturers voluntarily transmitted information about their products through labels and material safety data sheets (MSDSs), it was more often the case that the occupational hygienist or other health and safety professional had to conduct a scavenger hunt to obtain the most basic information about the products of concern, particularly if they were proprietary mixtures.

This changed with the advent of the worker right-to-know movement in the 1980s. Worker representatives successfully lobbied state and federal government authorities to ensure that workers exposed to chemicals are apprised of their potential hazards and appropriate precautionary measures. Provisions that implemented this important right for workers had the additional benefit of informing occupational hygienists and other health and safety professionals about hazardous chemicals in the workplaces they are responsible for, thus easing their tasks and improving overall protection.

Today there is increasing evidence that the most beneficial approach to worker safety and health in a workplace is a systematically developed comprehensive safety and health management program. Effective hazard communication is a cornerstone of that comprehensive approach. Without adequate information about the chemicals in use, it is not possible for an occupational hygienist or other professional to design or implement an appropriate protective program for exposed employees.

Providing information to employees as well as to health and safety professionals serves to empower employees to be active participants in an employer's safety and health program. For example, workers who understand why a respirator must be worn when working with a particular chemical are more likely to wear it when needed and to ensure that it is worn properly.

Together, the actions of employers and employees who have the necessary information about the chemicals in their workplaces will reduce the potential for chemical source illnesses and injuries—thus accomplishing the underlying purpose of the federal hazard communication standard (HCS).

History of the Development of Legal Requirements for Hazard Communication

It has been suggested that the first attempts at communicating hazards to users of chemicals can be found in the hieroglyphics in Egyptian tombs.[1] Some of these markings have been interpreted as information about various herbal preparations or medicinal materials. They provided precautions for safe use as well as other information that users of these preparations and materials would find helpful.

Development and transmittal of information about chemical products continued to evolve throughout the centuries that followed. By the 19th century, chemists often provided users with notes regarding chemical properties and safety considerations. Early in the 20th century these became more prevalent and standardized. Labels have been accepted business practice in the chemical industry for many years, including the development of voluntary industry consensus standards on the subject some 40 years ago. MSDSs have also been in use for some time—the Manufacturing Chemists' Association (predecessor to the Chemical Manufacturers Association) made them available as early as 1949.(2)

These activities to provide information about hazardous chemicals were voluntary on the part of manufacturers until the late 1960s. At that time the Bureau of Labor Standards adopted requirements for MSDSs in the maritime industries. These standards were adopted by the Occupational Safety and Health Administration (OSHA) in the early 1970s. OSHA developed a two-page format (the OSHA Form 20) that was used for many years to provide MSDS information in the maritime industries.

Coverage of other industries was a long-term project for OSHA. The agency adopted the HCS in 1983. This was the culmination of nearly 10 years of rulemaking activity. The implementing legislation for OSHA, the Occupational Safety and Health (OSH) Act of 1970,(3) included provisions that addressed labeling of chemicals in the rulemaking authority for the agency. Under Section 6(b)(7) OSHA was required to include in any substance-specific standard addressing toxic substances provisions to prescribe "the use of labels or other appropriate forms of warning as are necessary to insure that employees are apprised of all hazards to which they are exposed, relevant symptoms and appropriate emergency treatment, and proper conditions and precautions of safe use."

OSHA took this substance-specific approach to labeling, including provisions in health standards when promulgated as anticipated under the provisions of the act. It soon became clear, however, that this time-consuming and laborious process was not adequately apprising employees. OSHA's rulemaking process is slow and deliberative, often taking years to complete each individual standard for a chemical substance. On the other hand, the number of chemicals in the workplace that pose potential hazards to exposed employees is large. OSHA has estimated that there are about 650,000 hazardous chemical products in American workplaces. Ultimately, OSHA decided that the lack of information about hazardous chemicals in the workplace was a significant risk to workers and that a way to address them generically needed to be developed.

Thus, in 1974 the agency formed a standards advisory committee to make recommendations to OSHA on how to proceed in developing a standard that addressed labeling and provision of information. This committee, formed under the requirements of the OSH Act, was comprised of members of the public, labor representatives, and management representatives. They completed their report in 1975 and suggested that OSHA needed a standard with requirements for classifying chemicals as to their hazards, labels, MSDSs, and training.(4)

Also in 1975, the National Institute for Occupational Safety and Health (NIOSH) provided a criteria document to OSHA that included similar recommendations. They advised OSHA that a standard including hazard classification, labels, MSDSs, and training was needed to provide employers and employees with information about chemicals.(5)

While these recommendations appear straightforward and were based on the practices of progressive employers, there were many complicated issues involved in implementing them in a mandatory standard. These included which hazards to cover; how to define them; whether the requirements should be performance- or specification-oriented; what chemicals to cover; what employers to cover; and how to address the confidentiality of trade secrets. OSHA had to develop regulatory provisions for these issues to propose a standard. An advance notice of proposed rulemaking was published in 1977 to elicit public comments on these issues. After several years of considering various options, a proposed standard titled Hazards Identification was published in January 1981 by the outgoing Carter administration.

This 1981 proposal diverged from the recommendations OSHA had previously received, since it addressed only hazard classification and labeling. There were no requirements for either MSDSs or training. The chemical industry objected to many of the proposed requirements, and the new Reagan administration withdrew the proposal in February 1981 for further consideration of regulatory alternatives.

In the meantime, employee representatives grew tired of waiting for a federal standard to address worker right-to-know and began lobbying state governments for standards. This was a successful endeavor, and soon it appeared there would be a number of varying requirements for shipping chemicals around the United States. The state standards covered different chemicals, different employers, and had divergent requirements for disclosing information. The specter of 50 different state standards elicited considerable support for a harmonized federal approach.

Thus, OSHA introduced a new proposed standard in March 1982 and completed rulemaking by issuing a final standard in November 1983.[6] It was comprehensive in its coverage of chemicals and hazards but was limited to the manufacturing sector of industry. This scope was immediately challenged in court by worker representatives, and in 1987 OSHA was ordered to expand the scope to cover all employers. OSHA published a new final rule in August 1987 to comply with this order.[7] As a result of various legal and administrative challenges, and the agency's desire to clarify some of the provisions, a third final rule with relatively minor modifications was published in February 1994.[8] The preambles to each of these final rules provide a more detailed history of the proceedings, as well as a summary and explanation of the requirements. These should be consulted for additional information.

The HCS is codified in the *Code of Federal Regulations* (CFR) in several places. The general industry standard can be found at 29 CFR 1910.1200. The construction, maritime, and agriculture industries are also covered in provisions that are identical to the general industry standard, but are found at 29 CFR 1915.1200, 1917.28, 1918.90, and 1926.59. While this chapter describes the provisions of the HCS, it is not intended to be a substitute for consulting the actual regulatory requirements when designing a program for purposes of compliance. In addition to being in CFR, OSHA maintains a home page where the regulations can be accessed, as well as a database of interpretations that can be searched and the compliance instructions given to OSHA compliance safety and health officers to guide them when enforcing the standard. OSHA's home page can be found at http://www.osha.gov.

In addition to the worker right-to-know provisions of the HCS, the U.S. Environmental Protection Agency (EPA) implements community right-to-know provisions under requirements of the Superfund Amendments and Reauthorization Act (SARA) of 1986. Congress mandated under SARA that chemicals required to have MSDSs under OSHA provisions be subject to the requirements of SARA for reporting to communities.

The Federal Hazard Communication Standard

Overview of the Approach

The HCS is unique among OSHA regulations in a number of respects. It covers more workers than any other single health standard—about 35 million of them exposed to hazardous chemicals in over 3.5 million establishments. It includes requirements for evaluating the hazards of all chemicals, preparing written hazard communication programs, labeling containers, providing MSDSs, and training employees. Before examining the specifics of these requirements, it may be helpful to understand the overall design of the standard and the unique characteristics of OSHA's approach to hazard communication.

The HCS covers all industries and all sizes of facilities. The information OSHA collected during its rulemaking process indicated that chemical exposures occur in all types of workplaces and all sizes of facilities. Since the HCS is an information-transmittal standard rather than a standard that establishes specific control measures for a chemical, it appeared that its provisions were also feasible in all types of workplaces and that exposed workers have an equal need for information regardless of the type of work they are performing or the size of their workplace. Unlike some other standards that have different requirements in various industries (e.g., asbestos in construction), OSHA determined that industry-specific differences under the HCS could be reasonably accommodated through flexible implementation and did not require differing provisions.

The HCS is a performance-based standard, establishing goals for compliance but providing minimal specifications for how employers are to reach those goals. This requires employers to use professional judgment to comply and OSHA's enforcement staff to use the same type of judgment to enforce the rule. To regulate all types of workplaces and all sizes of facilities, OSHA determined that it should maintain provisions that are performance-oriented to allow employers the flexibility to implement them in a manner suitable to that particular workplace. There are certain specification aspects in the requirements that help to ensure a consistent approach, however, such as the hazard determination provisions. But each employer is given significant latitude in determining the best way to establish a hazard communication program in his or her workplace. For example, the manner in which training is to be delivered and the amount of time to be spent on training are not specified. Employers may choose from a wide variety of options including toolbox talks, videotapes, and interactive computer programs. The choices made will depend on the needs of the particular work force, the availability of equipment, and the number of hazardous chemicals being addressed.

The HCS is a generic standard, covering 650,000 hazardous chemical products. Hazards are defined, not listed by product, so the scope increases as new products that meet the definition of hazard are developed. During the HCS rulemaking OSHA considered promulgating a list of chemicals to define the scope of coverage for the standard. However, the agency found that (1) there is no list that captures all of the chemicals of concern; and (2) requiring employers to consult a list is an additional burden. Furthermore, a list is fixed in time and does not allow the standard to remain current with actual workplace conditions as products change and new products develop. The criteria-driven approach incorporated in the HCS allows it to remain up-to-date without changing the rule itself.

The HCS includes a downstream flow of information (i.e., chemical manufacturers are required to prepare and provide information about their products to employers using them). In designing the standard, OSHA determined that one of the key problems for employers using chemicals was obtaining information about them. If their suppliers did not provide it voluntarily, it was difficult for employers using the chemicals in their workplaces to ascertain what was in the product and obtain information from some other source. Thus, it appeared clear to the agency that the standard had to take the unique approach of requiring the producers or suppliers to provide the data to their customers. The assessed burdens of the standard showed that this was a much more cost-effective approach overall than to have numerous customers attempting to obtain or develop information about the same product.

It addresses controversial and unique issues, such as trade secret protection and preemption of state laws in states that do not have their own plans. While preemption provisions in the OSH Act address this issue for all standards, it has never been such a key concern in a rulemaking as it was when the HCS was initially promulgated. For suppliers of chemicals, uniformity in requirements for labels and MSDSs was the primary reason for supporting a federal hazard communication standard that preempted requirements in existing state standards, which posed a burden on interstate commerce.

In the area of trade secrets, another key concern of suppliers involved divulging confidential business information to people outside of their firms. This is also an issue that is covered in the OSH Act, but the provisions in the HCS itself were the result of extensive comment in the rulemaking and much negotiation and discussion. The key issue was to balance the needs of workers to be protected with the desire of suppliers to protect their legitimate trade secret concerns.

The HCS depends on people modifying their behavior when they receive information. Employers must use the information to provide better employee protection. Employees must use the information to participate in the protective programs. Together, these actions will result in a decrease of chemically related illnesses and injuries. While the HCS requires the transmittal of information, its sphere of influence on safety and health in the workplace is much broader than the simple receipt of that information. The successful functioning of the standard requires people to act on the information they receive to improve conditions in the workplace — an active process of using the information, not a passive paper trail.

To help ensure that this active process is achieved, the agency examined the communication aspects of existing workplace programs to determine what worked in this regard. As a result, the information transmitted under the standard comes in three forms: labels, MSDSs, and training. The label is a simple snapshot of the hazards—a quick and abbreviated information source that reminds workers and other users that there are hazards and more information is available. The MSDSs provide the comprehensive information available about a chemical and associated precautionary measures. These are reference documents, and in addition to being accessible to workers, they provide the information that health and safety professionals need to design protective programs for exposed workers. And last, employee training is required. It is in this setting that the information can be explained and related to the specific workplace situation. Training helps make the labels and MSDSs effective. These three aspects of the standard are thus interdependent, and effective information transmittal requires all three to work.

The scope of the HCS determined the scope of EPA's community right-to-know requirements under SARA. This greatly expanded the target audience for the information — in particular, the already extensive audience for MSDS information — to emergency responders and local planning authorities.

The HCS was thus a departure for OSHA and the regulated community in many respects. Its implications and impact on worker safety and health are broad, and the standard was written to ensure that the requirements remain current and active as time passes. This appears to have been successful. The HCS is still frequently addressed and challenged in many different fora, particularly in Congress and small business lobbies.[9] However, it also appears to continue to have a significant impact in the workplace and on expectations regarding the information that can and should be made available to chemical users.

Detailed Description of the Requirements

Paragraph (a): Purpose

The stated purpose of the HCS is to ensure that the hazards of all chemicals produced or imported are evaluated and information concerning their hazards is transmitted to employers and employees. The standard further states that this is to be accomplished by means of comprehensive hazard communication programs, including labels, MSDSs, and training. In addition, OSHA also included a paragraph specifically addressing the preemptive authority and intent in this area.

The underlying purpose of the standard is to reduce the incidence of chemical source illnesses and injuries. While providing the right-to-know is an important goal, the use of the information for protection makes the transmittal effective. Working with chemicals without knowing what the hazards are puts the workers at significant risk of developing an adverse health effect. Since these chemicals may have the potential to cause severe acute effects as well as long-term, cumulative damage, it is important to ensure that everyone handling them has information about their hazards and the appropriate precautionary measures to implement.

Paragraph (b): Scope and Application

The scope and application paragraph of any OSHA health standard indicates who and what are covered by the standard. In the case of hazard communication, it describes not only what is covered but provides exemptions for items that are either not covered or are covered in a limited fashion.

First, the standard requires all chemical manufacturers or importers to assess the hazards of chemicals they produce or import. Second, all employers are required to provide exposed employees with information about hazardous chemicals in their workplaces. And third, distributors are required to transmit information to their downstream customers.

The HCS applies to "any chemical which is known to be present in the workplace in such a manner that employees may be exposed under normal conditions of use or in a foreseeable emergency." This is a particularly important part of the scope and application of the standard. It establishes that all chemicals present in the workplace are potentially included in the scope and thus require evaluation (such as byproducts or intermediates of a process). It also indicates that any exposure triggers coverage of the standard; it is not in any way related to exposure above a permissible exposure limit. This includes potential exposure as well as actual exposure, and foreseeable emergencies must be addressed (such as ruptured pipes or containers). On the other hand, if there is no potential for exposure (i.e., the chemical is inextricably bound or in a physical state where exposure is not possible) the chemical is not covered. This provision is thus key to determining which chemicals in a workplace fall under the scope of the HCS.

The HCS includes a series of exemptions or situations of limited coverage. These are based on considerations of special handling circumstances, limited exposure or risk, or coverage by other federal standards.

Laboratory operations and operations where employees handle only chemicals in sealed containers (such as warehousing) have limited coverage. In these work operations, labels on incoming containers must not be removed or defaced, employees must be trained, and MSDSs must be provided on employee request. Written hazard communication programs are not required and employers do not have to ensure there is an MSDS available for every hazardous chemical in the workplace.

A number of federal standards require labeling of various chemical products. Where such labeling requirements are already in place, OSHA is preempted from requiring additional labeling even if the existing labeling does not have all of the information OSHA would require under the HCS. The products currently subject to labeling under other agencies include pesticides, some toxic substances, food, food additives, drugs, cosmetics, medical or veterinary devices, distilled spirits, consumer products or hazardous substances for consumer use, and agricultural or vegetable seeds treated with pesticides. The HCS should be consulted for the specific terms of these labeling exemptions.

There are also a number of situations in which OSHA has determined that application of the HCS is not warranted, either because they are already regulated elsewhere by another federal agency or their hazards occur outside the workplace. These include chemicals that are already regulated as hazardous wastes; tobacco or tobacco products; wood or wood products in terms of flammability or combustibility; articles (where chemicals are bound and are not available for exposure); food or alcoholic beverages in a retail establishment or consumed by workers in the workplace; drugs in solid, final form, packaged for sale to consumers or intended for personal use by employees; cosmetics packaged for sale to consumers or intended for personal consumption by workers; consumer products when used as a consumer would use them with similar exposures; nuisance particulates that pose no physical or health hazards to employees; ionizing and nonionizing radiation; and biological hazards.

It is important to refer to the actual language in the HCS when determining if any of these exemptions applies to specific products or the products in use in specific workplaces. When in doubt, the best rule of thumb is to assume the product is covered if it is hazardous. The exemptions are intended to avoid duplicative coverage or coverage of products where the risk is small and addressed through other means. However, this is an area where OSHA has received many questions, and there are extensive interpretations regarding the extent of coverage in a number of these areas.

Paragraph (c): Definitions

This paragraph of the HCS includes definitions of a number of key terms in the standard. OSHA defines terms used in the standard for a particular purpose and it is important to proper understanding and interpretation to consult these terms when implementing the HCS in the workplace.

The terms include a number of definitions that help define the scope of coverage: chemical manufacturer, chemical, employee, employer, and workplace are examples of terms used that are defined for purposes of this standard. Employee, for example, is defined as "a worker who may be exposed to hazardous chemicals under normal operating conditions or in foreseeable emergencies. Workers such as office workers or bank tellers who encounter hazardous chemicals only in non-routine, isolated instances are not covered."[10]

Thus, this definition includes an important interpretation of the scope of the standard as applied to certain types of employees. Similarly, the term "exposure" means that

> *an employee is subjected in the course of employment to a chemical that is a physical or health hazard, and includes potential (e.g., accidental or possible) exposure. "Subjected" in terms of health hazards includes any route of entry (e.g., inhalation, ingestion, skin contact or absorption).*[10]

This is key to determining which employees and which chemicals are covered by the standard.

Scope questions regarding chemicals covered are addressed in a number of definitions addressing chemical properties, such as health hazard, physical hazard, explosive, oxidizer, etc. These are to be used by chemical manufacturers and importers when evaluating the hazards of their products.

Paragraph (d): Hazard Determination

Under the requirements of the HCS, chemical manufacturers and importers are responsible for evaluating the hazards of the chemicals they produce or import. The producers of the chemicals are in the best position to know what is in the product, the characteristics of the product, the hazards associated with it, and what precautionary measures are appropriate to deal with these hazards. Thus, the standard requires them to generate such information, put it on labels and MSDSs, and provide it automatically to downstream users of the product. Employers who use chemicals are permitted under the HCS to rely on the hazard evaluations performed by their suppliers.

While hazard determination is an area that requires extensive professional judgment in the identification and evaluation of the scientific literature, OSHA decided that certain parameters needed to be established to ensure consistency in approaches. These are described in the hazard determination paragraph, as well as in two mandatory appendices to the standard, Appendix A (Health Hazard Definitions) and Appendix B (Hazard Determination). In addition, OSHA included a nonmandatory appendix describing available sources for information to perform a hazard determination (Appendix C, Information Sources).

The determination of physical hazard potential under the standard is relatively straightforward. Definitions of physical hazards such as flammability tend to be based on objective, measurable criteria (e.g., the flashpoint). Chemical manufacturers and importers generally test for such effects when establishing the characteristics of their products.

The more difficult hazard determinations involve health hazard potential. Chemicals are not generally tested for the full range of health effects, and testing is particularly deficient in the area of chronic effects. Furthermore, there are often disagreements about the interpretation of available data and its applicability to human exposures. This paucity of data and difficulty of interpretation is further complicated by the fact that few employees are exposed to chemical substances; most are exposed to mixtures of those substances. The large majority of hazardous chemical products in a workplace are mixtures unique to a single manufacturer. Few such mixtures have been tested, so a system must be devised to project the hazards of a mixture based on the hazards of its components.

OSHA believes that the threshold data requirements for transmittal of information vs. establishment of specific control measures should be relatively low to ensure that downstream employers and employees get as much information as possible on which to base decisions about protective measures. Thus, OSHA requires chemical manufacturers and importers to consider the existence of one good study to be a sufficient level of evidence for purposes of communicating hazard information. The study must be conducted according to scientific principles and have statistically significant results that indicate the potential for adverse effects. Both human and animal evidence must be evaluated to determine whether the chemical meets the standard's definition of a health hazard.

OSHA anticipated that despite this guidance, there would be differences of opinion regarding the coverage of the standard. Thus, the agency established a "floor list" of chemicals that are to be considered hazardous under the rule in all situations. This list includes all chemicals for which OSHA has adopted a permissible exposure limit (PEL), as well as all those for which the American Conference of Governmental Industrial Hygienists (ACGIH) has adopted a threshold limit value (TLV®). In addition, OSHA was aware that the definition of carcinogenicity would be likely to generate the most controversy under the hazard determination provisions of the rule. To establish a consistent approach, the agency indicated that any chemical found by the National Toxicology Program (NTP) or the International Agency for Research on Cancer (IARC) to be a carcinogen or potential carcinogen was to be considered as such for hazard communication purposes. In addition, however, any chemical for which there is one good study indicating potential carcinogenicity is to be considered carcinogenic for purposes of the standard.

In the area of mixtures, OSHA specified that where a chemical has been tested as a whole to determine its hazards, that information shall be used for purposes of hazard communication. Where such testing has not been done, the rule basically requires the chemical manufacturer or importer to consider the mixture to have the same health hazards as its components. If a health hazard is present in concentrations of 1% or greater, the mixture is presumed to have the same hazard. The exception is for carcinogens, which render the mixture hazardous when present in concentrations of one-tenth of a percent or greater. The rule also has a backup provision for situations where these cutoffs are too high to protect employees. If the component can still exceed the PEL or TLV in those concentrations, or still present a health risk to employees, the mixture is covered when the component is present in the smaller concentrations as well.

The chemical manufacturer or importer is required to document the hazard determination procedures used. While it is not necessary to document each individual hazard determination for a chemical, many producers do so to ensure they can duplicate the decision-making process if questions arise.

Paragraph (e): Written Hazard Communication Program

Employers with hazardous chemicals in their workplaces are required to develop, implement, and maintain a written hazard communication program. The purpose of the written program is to coordinate the hazard communication activities in the workplace and ensure that they are addressed in a comprehensive and consistent manner. It is intended to be a blueprint for action, indicating what's covered, how, and who is responsible for the various components. The program does not need to be lengthy or complicated to accomplish this intent, but anyone reading it should be able to determine how the program works in that facility.

The provisions requiring a written hazard communication program have been cited more than any other requirement of the standard. This often leads people to the conclusion that the cited employers have implemented the other requirements of the standard, and have simply found the written program requirements to be too burdensome, difficult, or unnecessary and have chosen not to comply with them. This conclusion is not accurate, however, since for many years OSHA told its compliance officers to cite the lack of a written program in cases where an employer had done nothing to comply with hazard communication. It is believed to be very unusual for an employer to have complied with all of the other provisions without having a written program.

The written program is required to describe how the employer plans to meet the requirements for labels, MSDSs, and training. It must also include a list of the hazardous chemicals in the workplace. The list may be compiled by work area or for the workplace as a whole. The names used for the list may be common or chemical names, as long as the identity used also appears on the labels for the chemical product and the MSDS. The identity can thus be used to link these three sources of information together.

The written hazard communication program must also include the methods the employers will use to inform employees of the hazards of unusual tasks. During the rulemaking, OSHA received comments that employees are sometimes asked to do tasks that have not been addressed in the training they receive on normal workplace activities. Cleaning out reactor vessels occasionally might be an example of such a task. In this type of situation, the employer must ensure that there is a plan to inform employees of the hazards of this special task and associated protective measures.

Another issue discussed during the rulemaking involves the hazards associated with chemicals contained in unlabeled pipes in workplaces. At one point, OSHA considered requiring that labels be placed on pipes at some regular interval. Manufacturing employers presented information indicating that this approach would be very burdensome, and in many cases employees are stationed in control rooms and only enter the plant areas for specific maintenance or quality control checks. In these situations they can be apprised of the hazards in some other way and achieve the same purpose. OSHA has allowed this to be done, as long as the method is addressed in the written program.

Another issue raised during rulemaking discussions involves protection on multiemployer work sites. An employer may have developed and implemented an adequate program to protect employees from the hazards of the chemicals used by the employer, but if the work takes place on a site where other employers operate, this may not be enough. While construction sites are usually considered the primary example of this situation, nearly every workplace is multiemployer at one time or another. Frequently, employers have contractor employees on site to perform certain tasks like maintenance work. Other repair or service personnel may also be required to work on site, sometimes for long periods of time. Where this occurs, both employers have a responsibility to ensure information is exchanged or made available so the employees are protected from all hazardous chemicals to which they are exposed, not just those generated by their own employer. On a construction site, for example, this may mean that each subcontractor provides a copy of the MSDSs for the chemicals they may bring on site and leaves them in the site trailer. Or perhaps their MSDSs will be accessible from a laptop computer on site. The written program must indicate how this issue will be addressed, but does not specify how the employer must accomplish this. The requirements are flexible so the particular situation in the workplace may be taken into consideration.

The written program is to be made available on request to employees, their designated representatives, OSHA, and NIOSH. Where employees travel between workplaces during a shift, an employer may satisfy the obligation to make the written program available by keeping it at the primary workplace location.

Preparation and implementation of a written hazard communication program should not be viewed as a paperwork exercise. If done right, it ensures that the hazard communication program pieces are properly integrated and produce an effective program. It also serves as a checklist to ensure all the parts have been addressed. The list of chemicals is essentially an inventory of what MSDSs are needed to be in compliance. The written program can thus be an important tool to gauge completeness and compliance and can be part of an employer's assessment of the effectiveness of the approach implemented in a given workplace.

Paragraph (f): Labels and Other Forms of Warning

The HCS transmits hazard information to employees and employers through three communication mechanisms: labels on containers, MSDSs, and training. At one time, the agency considered simply requiring labeling as the only form of information transmittal. However, there were several drawbacks to this approach. A label can convey only a limited amount of information. Besides space considerations, some labels are on containers that are moving and thus only convey the message for a brief period. There is also evidence to indicate that the more information there is on a label, the less likely it is that people will read and act on it. Thus, OSHA determined that while labels have an important role to play in the overall scheme, they do not function well in most situations as the only source of information. Under the HCS, labels are an abbreviated source of information about hazards. This information is reinforced in training, and more information is available through MSDSs.

The HCS requires chemical manufacturers, importers, and distributors to ensure that shipped containers of hazardous chemicals are labeled with the identity of the material, appropriate hazard warnings, and the name and address of the chemical manufacturer, importer, or other responsible party.

As previously described, the identity can be either a chemical or common name, as long as it also appears on the list of hazardous chemicals and the MSDS. The MSDS also contains a list of hazardous ingredients and thus is the primary source for the specific chemical components of a mixture.

Hazard warnings convey the physical or health hazard of the chemical. For example, "potential carcinogen" or "causes lung damage" would be considered appropriate hazard warning statements. Where available, health hazard warnings must convey the target organ effect. It is not considered sufficient to indicate that a chemical is harmful if inhaled, for example. The effect when inhaled is the appropriate hazard warning.

Labels on shipped containers often include other information than that required by OSHA, such as first aid information or precautionary measures. The agency was aware of that when the HCS was adopted and decided to focus on those items that were less frequently present but were nevertheless necessary to worker protection. Other information may be included. There is a voluntary industry consensus standard that provides guidance for label preparation, including statements that can be used to convey information.[11]

Unfortunately, many labels in the workplace today are not designed to communicate information effectively to exposed workers or the employer. They are written to satisfy legal requirements and become cluttered with details that often are not assimilated by the user. Simple labels with direct, easily understood warnings are the most likely to have the desired effect (i.e., workers will modify their behavior to follow safe work practices for the chemical).

Labels or other forms of warning are also required on containers in the workplace. Those containers that are used simply as received with the label from the supplier do not require additional labels. Other containers do require preparation of labels. As with the shipped containers, an identity must be on the label. The workplace labels must also have an appropriate hazard warning. The HCS is a bit more flexible in the warnings permitted on these internal workplace containers. Many different types of in-plant labeling systems have been developed that use numerical rating systems, colors, symbols, and other unique ways of conveying hazards. While they may not provide specific information in some respects, in the context of the overall hazard communication program they can be effective. Employers can use these types of systems as long as the specific information is available immediately in some form in the workplace.

The HCS also allows alternative forms of warning to suffice in lieu of an actual label for stationary process containers. These may include signs, placards, batch tickets, and similar means of conveying the information in writing, as long as the information is readily accessible and clearly identifies the containers to which it applies.

The standard also contains provisions requiring labels to be prominently displayed and legible. With regard to language requirements, the information must be in English, but may also be in other languages if appropriate for the employee audience.

Paragraph (g): MSDSs

While the label serves as a quick reminder of the hazards and the need for appropriate precautionary measures, the MSDS is a detailed reference source that includes all of the pertinent information on a chemical, its characteristics, and ways to handle it safely. These documents may be referred to by exposed employees when they need additional information. But they also serve as reference documents for the employer and for a host of health and safety professionals who provide services to the employee and the employer. Emergency responders, physicians, occupational health nurses, safety engineers, and occupational hygienists all use this document to obtain information to perform their work.

Chemical manufacturers and importers bear the primary responsibility for the development and dissemination of MSDSs. Employers are required to have one for each hazardous chemical in the workplace. The standard is designed to have a downstream flow of information from the producers of the chemical, through the distributors, to the ultimate user.

The HCS does not specify the format in which information is to be disseminated. At the time of the OSHA rulemaking on the HCS, many chemical manufacturers testified that they already prepared and distributed MSDSs for their products and had developed formats they found suitable and appropriate for that purpose. Thus, they supported performance-oriented requirements in the OSHA standard—specification of what information is required, but no particular format. This was the approach OSHA adopted in the final rule.

An MSDS is required to include the following information:

- Identity information, including chemical and common names of hazardous ingredients;
- Physical and chemical characteristics (such as vapor pressure);
- Physical hazards;
- Health hazards, including signs and symptoms of exposure, and medical conditions that may be aggravated by exposure;
- Primary route(s) of entry;
- Exposure limits;
- Carcinogenicity;
- Precautions for safe handling and use, including hygienic practices, protective measures during repair and maintenance, and spill and leak procedures;
- Control measures, such as appropriate engineering controls, work practices, or personal protective equipment;
- Emergency and first aid procedures;
- The date of preparation of the MSDS, or the last change to it; and
- The name, address, and telephone number of a responsible party who can provide additional information or emergency procedures.

MSDSs are to be readily accessible to employees when they are in their work areas during the work shift. This means they are able to consult the MSDS for necessary information at any time during the workday. Many employers are managing their MSDSs electronically and allow access to them through terminals in the work areas or similar electronic means. Electronic access is permitted as long as there is no barrier to obtaining the information.

Many of the users of MSDSs have argued for a standard format to facilitate accessing the information. This has been a particular concern of the emergency responder community, but it is relevant to employees and other users as well. Under the performance-oriented system, the information may appear in a different order and on different pages depending on the manufacturer producing the MSDS. This makes it difficult to find the particular data of concern and may add to the time necessary to obtain information in an emergency.

In response to these types of comments from users, the Chemical Manufacturers Association sponsored development of an American National Standards Institute (ANSI) voluntary industry consensus standard on preparation of MSDSs.[12] Designated as ANSI Z400.1–1993, the MSDS standard outlines a preferred 16-section order of information, as well as providing guidance to preparers on how to design the form to communicate effectively and how to fill in the various sections. Many chemical manufacturers are now using this approach. It is also used in Europe and is permitted to be used in Canada, so it appears to be accepted as an international standard. Standardization of the format makes electronic storage and transmission of MSDS information easier and facilitates management of MSDS collections, as well as improving the utility of the sheets themselves.

Paragraph (h): Employee Information and Training

The final communication component of the HCS is employee information and training. Without training, communication of the information via labels and MSDSs is not likely to be effective. Training ensures that employees understand the information presented to them in written form, have an opportunity to clarify it, and know where they can obtain additional information if necessary. The training requirements in the HCS are performance-oriented. The employer is free to choose the method of delivery (e.g., lecture, interactive computer, or videotape). It is unlikely, however, that a purchased program would meet all of the training requirements. Employers who use such programs must supplement them with site-specific information. No training records are required to document individual employee participation in training.

Designing and implementing an effective training program will be addressed in Chapter 30. Many occupational hygienists are tasked with conducting worker training but have not been trained themselves to do so. This is a serious deficiency, since many people who are technically competent do not have the skills necessary to adequately transmit the information to others.

The HCS requirements are divided into two parts. The first deals with the simple presentation of information, rather than training. Under these requirements, the employer must inform employees of the requirements of the HCS, the location of hazardous chemicals in their work areas, and the location and availability of the written hazard communication program and MSDSs. Such information transmittal is passive.

Training should be a more active process designed to ensure that the employee knows and understands the information being transmitted. Under the HCS, employees are to be trained on how they can detect the presence of hazardous chemicals in their workplace, the physical and health hazards of the chemicals, the measures they can take to protect themselves, and the details of the employer's hazard communication program.

It is important to note that the intent of this training is not to have each employee memorize and be able to repeat all of the information presented. There are written materials available for reference for that purpose. Rather, the training heightens awareness about the existence of hazards in the workplace, the need to handle them appropriately, and how to obtain and use the information available about the hazards, protective measures, and emergency procedures. Employees will be faced with different kinds of labels and forms of MSDSs; the training helps to coordinate these sources of information in a form usable to the worker.

Paragraph (i): Trade Secrets

The HCS requires disclosure of the identity of all hazardous chemicals on the MSDS. However, in some limited circumstances the specific chemical identity of such a chemical may be withheld if it is a bona fide trade secret.

At the time the HCS was promulgated, when MSDSs were being prepared and disseminated voluntarily, trade secret claims for identity information were common. A major issue in the rulemaking process was the question of what constitutes a legitimate trade secret and under what circumstances it can be withheld to protect its secrecy. A special rulemaking just to address this issue followed later.

As a result the HCS includes an appendix that provides a detailed explanation of the types of characteristics that qualify a hazardous chemical identity as a trade secret. An employer must be able to show, for example, that research or extensive developmental processes were required to discover the product or its uses; that extraordinary means have been undertaken to keep the identity secret; and that it would not be a simple analytical process for someone wishing to reverse engineer the product and determine the secret. The specific chemical identity may only be maintained as a trade secret under the HCS if it meets the tests specified in common law for establishing a trade secret.

Nevertheless, even when a chemical identity is a trade secret, protection of exposed workers must be the ultimate concern. Therefore, the HCS requires disclosure of the identity when there is an occupational health need for the information, such as to conduct or assess sampling of the workplace atmosphere to determine employee exposure levels.

In these circumstances the holder of the secret may require the requestor to sign a confidentiality agreement. The HCS specifies the legal constraints for the agreement and a process of adjudication should there be a difference of opinion on the need or the ability to maintain confidentiality.

A considerable amount of the rulemaking process was devoted to this issue, but there have been few complaints about violations of the approach during implementation of the standard. The trade secret provisions have rarely been cited, and as a general rule trade secret claims are rarely seen on MSDSs.

Paragraph (j): Record Keeping

There are no long-term record keeping provisions in the HCS. The written program must be kept current; outdated or changed programs need not be kept. MSDSs must be present in the workplace for those chemicals that are also present. One common misconception about the HCS is that it requires MSDSs to be maintained for 30 years. The Access to Employee Exposure and Medical Records regulation,[13] requires MSDSs to be kept for 30 years if there is no other record of exposure. Where the employer has actual exposure monitoring or generates a record with lists of chemicals, locations, etc., the MSDSs need not be maintained.

Appendices

The provisions of the HCS are accompanied by five appendices that provide additional information. As discussed above, Appendices A and B are related to defining health hazards and making a hazard determination. These are mandatory. Appendix C lists possible references to be consulted when preparing a hazard determination. It is advisory. Appendix D defines a trade secret. And Appendix E walks the reader through the requirements of the standard in lay (nonlegal) language. It is intended to assist the small employer who uses rather than produces chemicals.

Other Issues

This discussion is an abbreviated explanation of the requirements of the HCS. To properly implement the standard in a workplace, the requirements of the rule itself should be reviewed. In addition, there are other sources that provide more detailed guidance for compliance.[14] In response to questions and criticisms related to implementation of the HCS, OSHA asked its National Advisory Committee on Occupational Safety and Health (NACOSH) to consider the issues and provide recommendations for them to the agency. NACOSH submitted a report to OSHA in September 1996.[9] As the recommendations are implemented by OSHA, there will be additional outreach and educational activities to help the public comply. The previously cited OSHA home page has the full text of the NACOSH report and will also have other information about HCS materials developed in response to the NACOSH recommendations as they become available.

International Hazard Communication

Although the United States was the first country to adopt a comprehensive approach to hazard communication (labels, MSDSs, and training), a number of other countries have also adopted such provisions. Canada adopted a very similar approach (the Workplace Hazardous Materials Information System) shortly after the HCS. It, too, incorporates a comprehensive approach of labels, MSDSs, and training. The labels and MSDSs are governed under federal law, while the training is done at the provincial level and may vary in requirements.

The European Union adopted a labeling law for substances in 1967, which has been adapted a number of times since then to broaden its scope. This has been supplemented by a directive requiring labeling of preparations or mixtures and another requiring MSDSs to be prepared. Training may also be required in various member states.

In the area of transport, there are internationally harmonized recommendations prepared under the United Nations Committee of Experts on the Transport of Dangerous Goods. These recommendations address the classification of acute and physical hazards and requirements for labeling and placarding.

While all of these requirements are similar in content and intent, they are dissimilar enough to cause barriers to international trade and confuse chemical users dealing with imported chemicals. The definitions used to classify hazards are different, as are the actual labeling requirements and the treatment of mixtures or preparations. In recognition of these differences and the problems they cause, there is an international activity under way to harmonize requirements by the year 2000. It is being conducted in several international organizations, with the participation of countries and organizations with existing systems as well as representatives of users of chemicals (workers, consumers, environmentalists).

Summary

A properly implemented hazard communication program can form the basis for a comprehensive safety and health program in the workplace. It provides the information needed to design appropriate protective measures and to give workers what they need to take steps to protect themselves.

Each employer with hazardous chemicals in the workplace must have a written hazard communication program. Containers of such chemicals must be labeled, MSDSs must be available for each hazardous chemical in the workplace, and employees must be trained about the hazards and how to obtain and use the hazard information.

References

1. **Kaplan, S.A.:** "Development of Material Safety Data Sheets." Paper presented at American Chemical Society meeting, April 1986.
2. "Chemical Safety Data Sheets Available Through Manufacturing Chemists' Association." *Ind. Hyg. Quart.*, March 1949, p. 22.
3. "Occupational Safety and Health Act," Pub. Law 91–596, Section 2193. 91st Congress, Dec. 29, 1970; as amended, Pub. Law 101–552, Section 3101, Nov. 5, 1990.
4. **Standards Advisory Committee on Hazardous Materials Labeling:** "Report to the Assistant Secretary for Occupational Safety and Health, U.S. Department of Labor." June 6, 1975.
5. **National Institute for Occupational Safety and Health (NIOSH):** *A Recommended Standard—An Identification System for Occupationally Hazardous Materials.* (NIOSH Pub. No. 75–126). Cincinnati: NIOSH, 1975.
6. "Final Rule: Hazard Communications," *Federal Register 48*:53280 (November 1983).
7. "Final Rule: Hazard Communications," *Federal Register 52*:31852 (August 1987).
8. "Final Rule: Hazard Communications," *Federal Register 59*:6126 (February 1994).
9. **National Advisory Committee on Occupational Safety and Health:** "Report to OSHA on Hazard Communication." September 1996.
10. "Hazard Communication Standard," *Code of Federal Regulations* Title 29, Section 1910.1200.
11. **American National Standards Institute (ANSI):** *American National Standard for Hazardous Industrial Chemicals—Precautionary Labeling* (ANSI Z129.1–1994). New York: ANSI, 1994.
12. **American National Standards Institute (ANSI):** *American National Standard for Hazardous Industrial Chemicals—Material Safety Data Sheets—Preparation* (ANSI Z400.1–1993). New York: ANSI, 1993.
13. "Access to Employee Exposure and Medical Records," *Code of Federal Regulations* Title 29, Section 1910.20.
14. **Silk, J.C., and M.B. Kent (eds.):** *Hazard Communication Compliance Manual.* Washington, DC: The Bureau of National Affairs, Inc., 1995.

Outcome Competencies

After completing this chapter, the user should be able to:

1. Define underlined terms used in this chapter.
2. Distinguish between risk and hazard, and between safe and acceptable risk.
3. Explain the factors that influence risk perception and risk acceptance.
4. Describe an effective worker risk communication program.
5. Recognize the importance of worker involvement in effective risk communication program.
6. Quote the essential components of hazard and outrage.
7. Define the concept of acceptable risk.

Key Terms

acceptable risk • hazard • perceived risk • risk • risk assessment • risk communication • risk management • safe

Prerequisite Knowledge

Prior to beginning this chapter, the user should review the following chapters:

Chapter Number	Chapter Topic
1	History and Philosophy of Industrial Hygiene
3	Legal Aspects of the Occupational Environment
6	Principles of Evaluating Worker Exposure
15	Comprehensive Exposure Assessment
17	Risk Assessment in the Workplace
29	Pychology and Occupational Health
30	Worker Education and Training
37	Program Management
40	Hazard Communication
50	Anticipating and Evaluating Trends Influencing Occupational Hygiene

Key Topics

I. Historical View of Risk Communication in the Workplace

II. Terminology

III. Acceptable Risk and Occupational Exposure Limits
  A. Who Decides What Constitutes Acceptable Risk in Occupational Health?

IV. Worker Involvement Leads to Behavioral Change

V. Approaches to Risk Communication
  A. The Importance of Audience Involvement

VI. Sandman's Hazard + Outrage

VII. Risk Comparison

VIII. Implementing the Workplace Risk Communication Program

41

Risk Communication in the Workplace

Jeffrey S. Lee, Ph.D., CIH

Dean R. Lillquist, Ph.D., CIH

Frank J. Sullivan

Introduction

Communication is the single most important skill that an occupational hygienist must have. A good occupational hygienist must be able to communicate with workers, management, the general public, and the media. Such communication goes beyond simply explaining technical matters. The occupational hygienist is typically the unique risk expert with major responsibilities for risk communication.

There is a considerable body of literature available in the area of communication, and a review with a focus on risk communication has been conducted.(1) A smaller body of literature exists for risk communication,(2) but very little research has been reported on risk communication in the workplace. This chapter will focus on the responsibilities of the occupational hygienist for risk communication and, primarily, the interaction between the occupational hygienist and the worker. The chapter will conclude with a list of recommended steps for designing and implementing a workplace risk communication program.

The National Research Council has defined risk communication as "an *interactive* process of exchange of information and opinion among individuals, groups, and institutions. It involves multiple messages about the nature of risk, that express concerns, opinions, or reactions to risk messages or to legal and institutional arrangements for risk management"(3) (emphasis added). Risk communication is a component of risk management, which is the process of selecting among various ways to control risk and implementing the selected controls.

Hazards are an inherent part of life. Every day people are bombarded with information about the hazards they are exposed to from the air, water, food, and basic everyday living. These communications often fail to place the risk from the hazard into a perspective that is understandable and relative. Numerous studies have indicated that individuals perceive certain risks as more likely to cause harm than actual risk and underestimate others.(4) The situation is magnified in the workplace because of the number of hazards and the complexity of associated toxicological information that must be made available to the worker under the regulations.

The primary mission of an occupational hygienist should be to eliminate or reduce risk. Tempering this mission is the need to consider economic and technical feasibility. The occupational hygienist also considers recommended and required occupational exposure limits that may entail some level of risk deemed acceptable by the promulgators.

In many cases the cooperation of the worker is important to optimize the reduction of risk. To achieve maximum cooperation, workers must be significantly enough convinced and motivated by the message that they take positive actions to protect themselves. Ideally, they become interested and concerned enough to be actively involved in suggesting ways to reduce exposure. An effective risk communication program provides the motivational basis for a risk reduction effort involving worker cooperation.

This chapter will not cover the detailed communication requirements of Occupational Safety and Health Administration (OSHA) and

Environmental Protection Agency (EPA) regulations. Although any hazard communication program must contain the elements required by these standards (see Chapter 40), it should also be designed, implemented, and evaluated taking into consideration the factors that influence risk perception and acceptance.

Historical View of Risk Communication in the Workplace

Although unconscionable today, historically occupational hygienists were cautioned against communicating with employees.

Frank Patty, in his second edition of *Patty's Industrial Hygiene and Toxicology*, which was in use up until the mid-1970s, cautioned against talking with employees.

> *It is unpardonable conduct on the part of an [occupational hygienist] to ask an employee a suggestive question such as: "Do you feel all right? Do you get sick often? Does breathing this atmosphere cause you any discomfort or irritation?" The psychological effects of such questioning, which is only one step short of suggesting to a workman that he is ill, are obvious and undesirable. To some workmen the mere fact that tests for airborne toxic materials are being made may indicate that dangerous conditions exist, and careless remarks of the investigator may grow to dire proportions and cause needless alarm among the workers.*
>
> *Although management often finds it advantageous to inform employees in advance of the purpose of an industrial hygiene survey, discussion with employees should be avoided by the industrial hygienist and left to the foreman or other representatives of management. ... To the direct question "What are you doing?" or "What does that instrument do?" a disarming reply should be given. The industrial hygienist might say that he is making routine tests, determining ventilation requirements, studying the efficiency of the exhaust system, measuring solvent loss, or he may make some other similar statement of fact. Any remark about measuring the toxic vapors here would be an ill-considered, and possibly alarming reply. It is not that the workman should be deceived but rather that he should not be alarmed about something that he probably would not fully understand.*[5]

Risk analysis emerged in the mid-1970s as an attempt to provide objective answers about risks. A number of studies were conducted to try to define quantitative criteria to use as a basis for judging the social acceptability of risks. The process of risk benefit analysis was then developed, which further added the quantification of benefits as well as risks, often without consideration of who was taking the risks and who was deriving the benefits.[6] The next development was risk perception, which evaluated how people actually made judgments about risks.

Today, occupational hygienists recognize that effective communication with workers is essential. Workers have a fundamental right to know if hazards are present in their workplace and to understand if risks to their health are present. They should not only know what the occupational hygienist is doing and why, but should also be an important part of the process. Workers, or their representatives, should be active participants in identifying potentially hazardous conditions and helping to ensure that they are being evaluated correctly. Workers must not simply be provided with technical and toxicological information. They must comprehend the degree of risk, be able to prioritize that risk, and be sufficiently motivated to take steps to mitigate that risk. Conversely, it is important for the occupational hygienist to understand and appreciate the factors that influence a worker's understanding and acceptance of risk and that may cause behavioral change. Risk communication in the workplace must be an interactive process to be successful.

Terminology

A hazard is a situation that is capable of causing harm. The bigger the hazard, the greater the capacity to cause harm. A risk is a measure of probability of harm. The greater the risk, the more likely the hazard will cause harm. A completely isolated hazard does not constitute a risk. For example, encapsulated insulation containing asbestos is a hazard, but it is not a risk unless some action makes the asbestos airborne and an individual is exposed. Due to its serious health effects, including cancer, asbestos is a high hazard relative to say, acetone, which has much less serious effects. An individual exposed to high levels of asbestos is at higher risk than an individual who is exposed to low levels of asbestos. Both individuals are exposed to a high hazard, asbestos.

Risk may refer to individual risk or to population risk. For example, the general population may be at relatively low risk of dying of an asbestos-related disease compared with dying from a disease not related to asbestos. The individual risk within the group of dying from an asbestos-related disease would vary considerably depending on exposure.

Perceived risk is the risk that the individual believes exists. In many cases, perceived risk does not equal actual risk, and the major task of the risk communicator is to address and clarify those issues that cause the misconception and reduce what may be termed perception error. When told that asbestos is present in their children's school, parents may perceive there is a major and immediate risk to their children's health when, in fact, the actual risk is likely to be very low.

What then is meant by safe? A dictionary definition of safe is "free from risk." In occupational health, when a situation is safe it is meant that it is free from an unacceptable amount of risk rather than free from all risk. Something is thus safe if its actual risk is judged to be acceptable. Or conversely, acceptable risk is that risk considered safe. What is considered safe by one individual may be considered unsafe by another.

In the above asbestos example, parents may believe the building has an unacceptable risk before the risk communication begins, and as a result of risk communication may conclude that the risk is acceptable. On the other hand, the risk, even when correctly perceived, may never be low enough to be acceptable to them. In such a case the risk communication process has not failed but better risk management is suggested.

Acceptable Risk and Occupational Exposure Limits

With the passage of the Occupational Safety and Health Act (OSH Act) in the United States in 1970, OSHA, through rule making, has essentially been charged with setting acceptable risk levels for physical and chemical agents. The OSHAct charges the secretary of labor with promulgating a standard which most adequately assures, to the extent feasible, on the basis of the best available evidence, that no employee will suffer material impairment of health or functional capacity even if such employee has regular exposure to the hazard dealt with by such standard for the period of his working life.[7]

Feasibility, as used here, has been interpreted to include both economic and technical feasibility. Thus, OSHA permissible exposure limits (PELs) commonly acknowledge a level of risk associated with the standard.

Similarly, the American Conference of Governmental Industrial Hygienists (ACGIH) recommends threshold limit values (TLVs®) for chemical and physical agents, designed to protect "nearly all workers" from adverse health effects, acknowledging that some workers may experience discomfort, may aggravate a pre-existing condition, or may develop an occupational illness at the TLV.[8] Thus, TLVs also acknowledge some level of risk acceptability, although arguably lower than OSHA PELs, since feasibility is not considered.

The U.S. Supreme Court, in the benzene decision (*AFL-CIO v. American Petroleum Institute*), commented that before promulgating any safety and health standard, OSHA must

> *make a finding that the workplaces [where exposures are in excess of the proposed standard] are not safe; however, 'safe' is not the equivalent of 'risk free.' There are many activities that we engage in every day, such as driving a car or even breathing city air, that entail some risk of accident or material health impairment. Nevertheless, few people would consider these activities 'unsafe.' Similarly, a workplace can hardly be considered 'unsafe' unless it threatens the workers with a significant risk of harm.*[9]

In Britain there are two sorts of standards: occupational exposure standards (OESs) and maximum exposure levels (MELs). OESs are established at a level where it is believed that it is possible to establish a no-significant-injury level of exposure. A MEL is set if no such level is known, or if an OES would be too low to be reasonably practical. Reasonably practical is a term defined in British law as a level at which any further measures to reduce risk would be out of proportion to benefit. The employer must control exposures down to a MEL and must control exposures down to an OES as far as reasonably practical. Thus, though not exactly the same, a TLV, which does not consider feasibility, would be comparable to an OES; a PEL, which does consider feasibility, would be comparable to a MEL.[10]

Who Decides What Constitutes Acceptable Risk in Occupational Health?

In the United States OSHA establishes PELs considering actual risk and feasibility.

The rule- making process allows for input from all interested parties, but representatives of labor and industry do not establish the standard.

In Britain a scientific committee, consisting of experts nominated by employers, labor, and independents, establishes an OES considering actual risk and then passes it on to a tripartite (government, labor, and industry) committee, which may endorse the OES or establish a MEL. A MEL must be approved by the secretary of state, and Parliament must be given the opportunity to object, "taking into account the broader issues of cost and benefit and the social acceptability of the residual level of risk."[10]

At the plant level in the United States a company may use PELs, TLVs (where lower), or some combination for their acceptable company standards considering economic and technical feasibility. Company standards need not be established for all chemicals, just those being used. Further, some decision logic might prevail such as having company standards be the lower of the PEL and the TLV except where exposures can be feasibly reduced further as determined by the health and safety committee.

If workers are involved in this process, the risk then becomes more acceptable to them. The degree to which risk is accepted by the individual depends not only on the severity and probability of harm but also on a variety of individual social and psychological mitigating factors. Similarly, these factors are important in trying to motivate individuals to change behavior to reduce their risk. The individual worker may consider the fact that OSHA has determined that their exposure is safe, but most likely such a realization is not sufficient to result in individual risk acceptance or in behavior modification.

Workers do not necessarily expect to work in a risk-free environment. If they are demanding that all hazards be eliminated or contaminants be reduced to undetectable levels far below standards, there are probably other issues that should be identified and addressed. If workers are involved in the company standard-setting process, the result should be a reduction in perception error and a rise in the level of acceptable individual risk.

Risk communication programs often fail to consider what consequences, in terms of benefits or losses, the risk decision represents for the individual. For example, the risks of driving a car, smoking a cigarette, or being knowingly exposed to a chemical in a workplace are all faced because the individual determines that the benefits gained (or maintained) outweigh the perceived risk. These benefits may be tangible (e.g., money) or subjective (e.g., pleasure or ego satisfaction). Commonly, multiple benefits are involved. The benefits of a job, which may also have an increased risk, may overshadow the concern for increased risk.

Worker Involvement Leads to Behavioral Change

A successful risk communication program is likely to lead to successful risk reduction behavior, but such an outcome is generally considered beyond the risk communication process. Risk communication should be considered successful even if it only raises the understanding of the workers to risks to which they are exposed.

Risk reduction in the workplace can almost always be improved through behavioral change of the employee. Employee actions, in almost all situations, can significantly reduce exposures. Often these actions, such as washing hands well when an ingestion hazard exists, are relatively easy to implement following effective risk communication. The motivation for making that behavioral change begins with an understanding of the risk, which in turn is increased from the hazard communication process.

If behavioral change is a desired outcome, the occupational hygienist and the workers should continue the dialogue. Unfortunately, in many instances, the risk communication effort is designed simply to disseminate information with the hope that behavioral change will occur, resulting in risk reduction.

Approaches to Risk Communication

Lundren divided risk communication into three functional lines: crisis communication, consensus communication, and care communication.[11] She focused on how messages are sent and received, how conflicts are managed, and how decisions are made.

Crisis communication is risk communication in the face of extreme or sudden danger. Nuclear meltdown, rising flood waters, and epidemic disease are examples of crises in which information needs to be transferred to the audience quickly. Ideally, such communication should be planned in advance through contingency planning. In the workplace the communication imparted when an alarm goes off is an example of a crisis communication. Crisis communication should move workers to immediate appropriate action. People should evacuate immediately; not wait, not reach a group consensus.

Consensus communication is risk communication to inform and encourage groups to work together to reach a decision about how the risk will be managed. A workplace example would be a safety and health committee working to determine the best way to identify and reduce the number of hazards and degree of risk inherent in a production unit.

Care communication primarily involves communication about health and safety risk and involves two subsets: 1) health care communication pertaining to general health such as information about diet and exercise; and 2) industrial risk communication pertaining to risks incurred on the job.

An occupational hygienist will use all these approaches in successfully communicating risks. Different scenarios will require different elements of these strategies.

The Importance of Audience Involvement

In the 1980s the National Research Council funded an extensive study in effective risk communication.[11] The study concluded that effective risk communication occurred when scientific organizations considered the concerns of nonscientific groups and their opinions about how to manage risks. The lesson is that any form of successful risk communication must incorporate this exchange of information, although how the information is exchanged will vary for each type of risk communication (crisis, care, or consensus). Thus, risk communication in the workplace should involve the workers in identifying the risks and the factors that impact on the perception and acceptance of the risk, and in the managing of the risk.

One approach would be to interview workers to determine how they perceive risk. Risk communication then can be designed to address the gaps in information or misinformation to reduce risk perception error. To really communicate with your audience, you must understand how they think. Risk communications that do not address the key audience concerns or beliefs will fail.[11]

Risk communication should be viewed as a long-term process in which science and societal values are constantly changing. A cycle develops between the organization and the workers. The organization issues information and the workers process it and then issue their own information. The organization then processes this information and responds, thus repeating the cycle.[11] Continuous feedback and interpretation are necessary for the communication to remain effective. Workers must be involved, and the process must be a dialog, not a monologue from management.

Rowan[11] views communication as three challenges.

1. Knowledge challenge: The audience needs to be able to understand the technical information surrounding the risk assessment.
2. Process challenge: The audience needs to feel involved in the risk management process.
3. Communication skills challenge: The audience and those who are communicating the risk need to be able to communicate effectively.

Those who are communicating the risk must meet each of these challenges for risk communication to succeed.

Sandman's Hazard + Outrage

One authority in public risk communication is Peter Sandman. He states that risk should be viewed as the combination of the hazard plus "outrage."[12] Sandman defines hazard to be the expert's risk assessment, which is based in science. He defines outrage as the nonexpert's risk assessment, which includes many psychological factors of risk perception (which earlier researchers also discussed in the literature[13]). If both assessments are in agreement, then there is normally little controversy. Sandman points out that the public pays too little attention to the hazard and the experts pay too little attention to the outrage. If experts ignore the variables that constitute outrage — or worse, label them as irrational and then discount them — distrust and hostility results and ultimately stand in the way of a successful resolution of the problem.

Companies and experts can get into battles if they stubbornly assert their own "rightness," rather than recognizing and dealing with outrage. An example is the case of a manufacturing plant where workers allegedly smelled unusual odors and began passing out. At first only one worker was affected, but within a few weeks several workers at a time began passing out after only a short time in the workplace. Extensive occupational hygiene investigations and occupational medicine examinations were conducted, and all turned out negative. The investigators concluded that the workers were suffering from psychosomatic illness, or "assembly line hysteria." When the workers were presented with the negative results, the investigators made the mistake of stating that since nothing unusual could be found, there was no problem. The workers argued that they could smell an odor; the investigators argued that they could not, and further that their sophisticated air sampling could detect nothing unusual. The investigators immediately lost the workers' trust. The problem got worse, and the plant eventually shut down.

In retrospect, it can be seen that the investigators should have assumed that a hazard existed; they should have involved the workers more in designing and conducting the study, taken positive steps to improve the ventilation, and sought ways to install alarms to alert workers to environmental changes. By assuming that no hazard existed, the investigators indicated a lack of caring. Once people feel outrage due to lack of caring, they do not hear what is being explained.

Workers also often pay too little attention to hazard, and the occupational hygienist also often pays too little attention to outrage. It is important to understand and identify the factors that are making up the outrage portion of the equation in the workplace hazard communication program. It may be necessary to conduct air monitoring even when one is convinced that levels are not significant. Workers may not voice their concerns as openly as the public for fear of reprisal. Instead the hostility is suppressed, further reducing the overall program effectiveness.

In his videotapes and books, Sandman describes the recognized components of public outrage, how risk data should be presented to the public, and how barriers to implementing a public risk communication program within a company can be overcome.(12)

Sandman concludes that when the hazard is high, risk communicators should try to nurture the outrage, and when hazard is low, risk communicators should try to reduce the outrage. This approach is not always best in the workplace, as will be discussed below. Sandman describes 20 outrage components. In the authors' opinion, 17 of these components are important for risk communication in the workplace, although the reader is cautioned that additional research is needed to validate this assumption. These 17 have been prioritized, and the authors have added a discussion pertaining to the workplace.

Trustworthy Sources vs. Untrustworthy Sources. Some companies, authority figures, and organizations are automatically deemed untrustworthy because of their cause, purpose, or position. Others, such as industries known for problems with pollution, are automatically deemed untrustworthy by the general public. The public commonly makes the assumption that an untrustworthy company might be dangerous as well.

The company's image to the outside world may also carry over to the workers within the company. Workers may trust individuals, such as the company occupational hygienist, but their nature is to mistrust "them," the faceless organization. If employees are concerned about "bad air" and management issues a report stating that "breathing this air won't make you sick," there is a tendency not to trust management's message.

In this situation the occupational hygienist is placed in a unique position. Handled correctly, the occupational hygienist can use the situation to build trust with both the workers and upper management. It would be counterproductive to build mistrust, thereby increasing outrage, even if the risk was high.

An admission by the occupational hygienist that "breathing this air entails a very low risk," is generally true and almost always more believable. The occupational hygienist should listen, take samples of environments that cause concern, and report back findings to all involved parties. The workers feel better knowing that their concerns have been heard, and they can see that their concerns have been evaluated. Outrage is diminished.

One of the greatest strengths that an occupational hygienist should have is the trust of both workers and management. Building such trust often takes a long time. They need to demand less trust by making their actions public, collaborative, and accountable. They can earn trust by making themselves accountable. For example, an occupational hygienist should be able to prove that he or she has increased ventilation or reduced exposures to below mutually approved levels.

Immediate vs. Delayed Effects. The general public is more concerned about risks that "lie in wait" rather than immediate risks.

In some cases where the perceived chronic risk is high, the worker may worry about the possible long-term effects, and his outrage may increase. Commonly, however, chronic risks are often discounted. It is not uncommon for the worker to argue: "I've worked here 20 years and never gotten sick; therefore, I don't believe that this represents a risk." Chronic effects are often not associated with workplace exposures because of the latency. Workers may not work at the same place where the initiating exposure occurred. Their physicians, faced with deciding on major treatment, are not concerned with the etiology. Workers may also exhibit some self-denial of the risk. In contrast, if workers are immediately affected they want the condition corrected immediately. If the chronic risk is high and the outrage is low, the occupational hygienist should work to increase the outrage.

No Risk to Future Generations vs. Substantial Risk to Future Generations. The public worries about effects of environmental contamination on future generations.

In the workplace, reproductive toxins are a potential major cause of outrage. The man or woman considering having children, and the pregnant employee in the workplace, are very concerned about any exposure that may be responsible for reproductive effects. Rather than singling out individuals for additional protection, which leads to discrimination issues, the goal should be to try to make the workplace safe for the most susceptible individual. In cases where this is infeasible, the worker must be involved in arriving at an acceptable solution.

Controlled by the Individual vs. Controlled by Others. Most people feel safer driving than being a passenger. Why? Because as a driver they are in control.

In general, individuals like being in control. It is interesting that approximately half the states have chosen to run their own occupational safety and health programs rather than give control to federal OSHA, even though it is much more expensive to the state taxpayers. Similarly, companies would rather be in charge of their occupational safety and health programs, even when more expensive, than to give control to OSHA. Being at the mercy of someone else provokes the most outrage when the someone else is faceless and seemingly unaccountable.

In the workplace, workers usually can play a role in controlling the risks. Occupational hygienists may need to develop methods that help workers feel that they have some control over their individual risk. Techniques may range from opportunities to submit anonymous complaints to formal and empowered safety and health committees.

Responsive Process vs. Unresponsive Process. Unresponsive processes generate outrage.

Occupational hygienists should be immediately responsive and completely open with employees about exposures and any past mistakes. Any attempt at hiding data creates immediate mistrust. Information should be released promptly and completely. Occupational hygienists should deal with workers' concerns compassionately, courteously, and responsively and not with arrogance or defensiveness. When communicating with workers, use of excessive technical jargon should be avoided. It might be useful to provide a nontechnical summary while still providing, or offering to make easily available, the full reports.

Knowable vs. Unknowable. Risk acceptance diminishes when experts disagree. When experts appear uncertain regarding the risk, it also increases anxiety. People really do not care about the most likely consequence of exposure at a certain level; they care more about the worst case scenario or the upper end of the confidence limit. A hazard that is visible or otherwise detectable diminishes fear. Thus, outrage may be reduced if risk can be made more knowable.

In the workplace, direct-reading instruments with alarms may be mounted in appropriate places to give a quantitative indication of exposure levels, thus reducing anxiety. Smoke tubes can be used by employees to ensure that ventilation systems are working. The goal should be to make the workers as knowledgeable as possible.

Not Dreaded vs. Dreaded. Some diseases are more feared than others (e.g., cancer vs. heart disease). It is difficult, generally, to do much to reduce dread, and the best approach is to acknowledge it.

The occupational hygienist may want to increase outrage by dwelling on the serious consequences of high risk exposures.

Familiar vs. Exotic. Familiar risks and familiar surroundings diminish outrage. Workers are more threatened by exposures in high technology industries, such as in clean rooms where the worker does not understand the technology, and the work surroundings are depersonalized. Odors in such situations are much more alarming than the same odors in general industry. Any effort to increase familiarity with the operation and the hazards will probably reduce outrage. Nicknaming chemicals and associating odors or colors with chemicals, so that they are easily remembered or identified, may be useful. In such cases the occupational hygienist should make sure that the worker can go from the nickname to the actual chemical information on the material safety data sheet.

Voluntary vs. Involuntary. People are less outraged by risk when they have a choice of whether to be at risk. There may be considerable disagreement among experts as to the degree of voluntariness of various risk-inherent activities. It could be argued that the public is involuntarily exposed to air and water pollution from their general environment; however, individuals are free to move to a cleaner area.

Similarly, although a worker may voluntarily choose to work in a particular job or industry, to quit may represent an unacceptable hardship.

An individual may voluntarily engage in driving excessively fast and yet be outraged at being exposed to a chemical in the ambient air or in drinking water with a much lower risk of harm.

A worker normally does not volunteer to be exposed to a hazard unless there is some benefit. There may be situations in which workers receive higher pay because they work in hazardous occupations (e.g., coal mining); however, in general, increasing compensation should not be used as an excuse for not controlling exposure. Although compensation for workers who volunteer for high hazard jobs may not be appropriate, the occupational hygienist should clearly describe to the worker the hazards and risks associated with each job, so that workers can consider whether the risk is acceptable to them.

Morally Irrelevant vs. Morally Relevant. To many, pollution is not just harmful, it is wrong. Its moral relevance makes the language of cost-risk and benefit-risk tradeoffs seem callous. In Sandman's words, it is "like a police chief discussing how many child molestations are acceptable given budget restraints."

Like the police chief, the occupational hygienist cannot say that some number of occupational illnesses or accidents are acceptable. Rather, he or she should say that the goal is to get the illness or accident rate to zero. In other words, even if the hygienist does not expect to reach this goal, he or she should want to.

Affects Average Populations vs. Affects Vulnerable Populations. Society tends to worry more about the elderly and children. The public in general worries comparatively little about workers.

For the occupational hygienist in the workplace, worker outrage can occur when a well-liked fellow employee who is hypersusceptible is singled out for unfair treatment. Again, the goal should be to make the workplace safe for all workers.

Fair vs. Unfair. Even if a situation entails more benefits than risks, people who bear most of the risks often reap little of the benefits. This unfairness may provoke outrage. Sandman argues that one should reduce risks when possible, then compensate where risk cannot be reduced or eliminated. He believes that one should ask the risk-takers what is the most appropriate compensation.

In the workplace, instilling a system that compensates individuals based on degree of risk (e.g., high hazard pay) is fraught with difficulties, as discussed above, and should be used only in extreme conditions. Reducing exposure in the workplace is generally a tradeoff between increasing economic and technical feasibility and decreasing risk, and will vary from situation to situation. Any exposure, even one far below standards and that would be extremely difficult to control further, may constitute some amount of risk and may be unacceptable to an individual. Others might find the risk acceptable.

It is not necessary to maintain a work situation that entails a risk that is acceptable to all workers. Outrage will be appropriately reduced if the compensation is fair considering the risk as just one of many pros and cons associated with the job. One good approach is to instill a system that rewards work groups that are more highly exposed for reducing or eliminating exposure.

Victims Statistical vs. Victims Identifiable. People become more concerned when they personally know a victim. Using case studies with real people — ideally, known personally to the audience — will increase the audience's attention and appropriately make them more outraged with risks that are being ignored. This may especially drive home the message that "the same thing could happen to you."

Collective Outrage is Greater Than the Sum of the Individuals' Outrage. Outrage feeds outrage. The more people affected, the greater the outrage.

Safety and health problems in the workplace should be addressed as promptly as possible. The outraged worker interacts with fellow workers, and outrage is collectively magnified. If the hazard is high, individuals should be removed from the work area until the risk can be evaluated and controlled. Even if the hazard is low, consideration should be given to removing individuals who feel they are being affected.

Little Media Attention vs. Substantial Media Attention. The media responds to outrage, but they only report or enhance it. They do not create it. The media should not be ignored, because it may generate considerable public outrage. Try to develop a trust with particular media representatives. Properly presented media coverage can be very helpful in reducing outrage.

Workers are also a part of the public and are influenced by how others perceive the risk. A complete open-door policy should be considered for trusted media, even inviting them into the facility and inviting them to interview employees. An attempt at openness is generally the best policy. Refusing to provide information or to grant access will imply that there is something to hide. When wrong, the company should accept responsibility, make apologies, and promise to increase prevention efforts.

Preventable vs. Not Preventable (Only Reducible). Sandman argues that risks that can be entirely prevented cause more outrage than risks that can only be reduced.

If the toxic substance causing a health effect could be completely eliminated from the workplace, even at a (reasonable) increased cost, rather than simply reduced to levels permitted by regulations, the choice should be to eliminate it.

Substantial Benefits vs. Few Benefits (Foolish Risk). People accept that there often must be risk to achieve societal benefits. In the workplace some risk can be rationalized as part of the job, with the paycheck as the benefit. In some cases the risk is even glamorized in jobs such as steel erection, flying jet fighter planes, and extinguishing oil rig fires. In general, high compensation is an expected benefit from high risk situations where high skill is also a requirement. One should emphasize improving the skill to minimize the risk. An accident means that the job was not conducted satisfactorily.

Sandman makes seven conclusions about hazard and outrage.[11]

1. The public responds more to outrage than to hazard.
2. Activists and the media amplify outrage, but they do not create it.
3. Outraged people pay little attention to hazard data.
4. Outrage is not just a distraction from hazard. Both are legitimate and important.
5. When hazard is high, risk communicators should try to nurture more outrage.
6. When hazard is low, risk communicators should try to reduce the outrage.
7. Companies and agencies usually can do little to reduce outrage until they change their own organizations.

Risk Misconception.[12] Sandman states: In most risk controversies, people do misperceive the hazard and they also feel outraged. The question is which is cause and which is effect. If people are outraged because they misperceive the hazard, the solution is to explain the hazard better. If, however, people misperceive the hazard because they are outraged, then the solution is to find a way to reduce the outrage.

Imagine a roomful of workers listening to an occupational hygienist talking about all the chemicals in the plant and their toxicities. The major complaint the workers have is that the plant is too hot and the smells are unpleasant. The occupational hygienist spends a good deal of time trying to convince the workers that the greatest risks are the potential overexposures to chemicals that are used frequently but are not odoriferous. Overexposure to these chemicals over several years may lead to serious disease. The heat and unpleasant odors are of low relative risk, and further, are much more expensive to control. Imagine that the occupational hygienist is successful, and the workers are convinced that the chemicals are of greater long-term risk. Now the plant manager speaks up. He says that he has only so much money and cannot control everything, and that everyone should vote on what should be controlled first.

How do you think the workers would vote? They would probably vote to first control the heat and annoying odors. As long as the heat and odors generate outrage, and the chemical exposures generate little, individuals are unlikely to re-arrange their priorities. In such situations the occupational hygienist cannot ignore the outrage factors. At a minimum, a long-term plan must include the reasonable abatement of outrage factors and a convincing argument must be made for giving priority to reducing the exposures to the chemical agents down to safe and acceptable levels.

Risk Comparison

Scientists and policymakers are often frustrated by their inability to place a risk into an understandable perspective. Tempers may flare at a company-wide meeting when management says that it will ignore a hazard because it has a risk of less than one in a million of causing cancer and is below existing OSHA standards. And yet the workers voluntarily accept much larger risks when they smoke or drive their cars home. Risks that are seen as serious by regulators (e.g., geological radon in homes) can be met with relative indifference.

The way risk is described is also very important. Workers could be very worried to hear that their occupational exposure would double their risk of cancer. However, the amount of worry might decrease if they learned the risk increase is only from one in a million to two in a million.[14]

It should be noted that quantitatively comparing risks can lead to even greater outrage (e.g., comparing hazardous waste incineration with the risk of smoking[15]) and should be used only with caution considering the nature of the comparison.[16]

Public health professionals may respond to outrage by dismissing the reasons as irrational and concluding that the individuals are unable to understand the scientific aspects of risk. They may rationalize that the individuals are in no position to make decisions about how to deal with the risks that confront them. But when public health professionals make decisions that affect individuals without involving them, they often elicit even angrier responses.[2]

Occupational hygienists must recognize that workers are quite capable of understanding the scientific aspects of risk assessment. The confusion and lack of understanding often is due to the fact that hazards are compared without any attention to exposure. Exposure is absolutely necessary to translate hazard into a risk. Thus, determination of level of risk is at the heart of the occupational hygienist's job.

Too often, occupational hygienists think that the failure of an individual to take proper preventive actions is because he or she does not recognize that there is a risk. The seasoned occupational hygienist can generally recall several examples when the science issues were simply not relevant to the worker. The bottom line is that risk comparison may be useful when communicating to workers, but it should not be used to dismiss the problem.

Implementing the Workplace Risk Communication Program

In summary, the following steps are recommended for designing and implementing a workplace risk communication program.

Obtain Upper Management Support and Commitment. Management at all levels must be committed to preserving the safety and health of employees. Anticipate the concerns that upper management may have about implementing a risk communication program and address the concerns directly (e.g., resources required, legal implications, fear of unnecessarily alarming workers, language difficulties, etc.). Show how the program fits into the company's goals and strategic plan. Define the benefits as much as possible (e.g., reduction in turnover, lowering of health care costs, avoidance of compliance actions, enhancement of company image, etc.).

Develop a Written Risk Communication Program. Make sure that the program is compatible with the OSHA Hazard Communication Rule and any EPA standards that may be applicable. Make sure that it is a compatible part of the overall occupational safety and health plan for the organization. Treat it as a dynamic document that is continually updated and improved. Make sure that it is understandable by all workers.

The Occupational Hygienist Must Be Accountable. Set reasonable and achievable goals and objectives for the risk communication program and track progress. Share these goals, objectives, and progress with the workers. Integrate the hazard communication plan into the occupational safety and health plan and ultimately into the company's strategic plan. It has been said that "if you don't know where you're going, you won't know when you get there." Determine whether the goal is to reduce outrage when the risk is low or to increase outrage when the risk is high.

Establish a Worker-Management Committee. Empower the committee to develop risk reduction policies. Keep the committee small (e.g., five to seven members). Consider each member's contribution equally without regard to status or position within the company. Remember, involvement equals commitment. The committee should have a reasonable discretionary budget to address risk reduction and should be encouraged to prepare proposals to upper management for more expensive controls.

Identify and Prioritize the Potential Risks. Ideally, this should be done by consensus. Begin with the philosophy that it is unacceptable to have any exposures above PELs or TLVs, whichever are the lowest. Acknowledge that there is a lot of uncertainty with available data. Agree that PELs and TLVs do not always protect all workers, and that for any work situation, health effects may occur even though standards are not exceeded. Establish a policy that compliance with government or recommended standards is not good enough, and that the long-term goal is to eliminate all toxic exposures, thereby eliminating all risk, or to reduce exposures and risk as much as possible. Do not ignore or demean the hyper-susceptible worker. If workers claim that something in the workplace is making them ill, presume that they are correct. Make physical changes, such as improving the ventilation system, even when a risk cannot be quantified. Accept outrage components as legitimate contributions to establishing priorities. Involve workers from the very beginning in identifying hazards of concern and establishing priorities for evaluation and control. Not doing so will needlessly expose the occupational hygienist to criticism and mistrust.

Recognize That Effective Risk Communication Must Be Supported by Effective Risk Management. The plan should ensure change. Change does not have to be immediate; workers understand priorities and can accept long-term goals as long as progress is being made. Risk communication is only the first step in reducing or eliminating risk.

Identify Ways That Workers Can Have Some Control That In Turn Result in Lower Exposures. Good hygiene practices, effective use of personal protective equipment, good housekeeping, preventive maintenance, and so forth, are all examples of ways the employee can lower his or her risk.

Identify Risk Communication Modalities. This might include labeling, material safety data sheets, written materials, videos, training courses, open discussion meetings, etc. The program may use a variety of modalities to keep the interest level up. Be sure to allow the maximum possible opportunity for involvement. Consider informal and formal problem-solving exercises.

Work on Instilling Pride in a Safe and Healthy Workplace. Workers should feel responsible for safety and health not only for their own particular job but also for the area in which they work. Encourage workers to understand the technical aspects of the processes. Consider having an open house for workers' families. Involving family members might go a long way toward motivating a worker to work in a safe manner and to be interested in doing his or her part to maintain safe and healthy work sites.

Make Sure Workers Comprehend the Hazards to Which They are Exposed. Check the understanding of the audience. Administer confidential questionnaires. Questionnaires are useful for assessing levels of knowledge as well as opinions and attitudes. Reward the highest scoring individual. Follow up with individual inquiry to make sure workers understand the risk.

Develop an Employee Complaint System. Make sure that there are multiple ways for employees to express their concerns, and that there is at least one way that ensures employee confidentiality. Never belittle or demean concerns. Recognize that some workers are often intimidated by their supervisors and others are simply reluctant to come forward for fear of being labeled as troublemakers or do not understand complex issues. Address worker complaints as quickly and as openly as possible. Mistrust and outrage spread rapidly through a work force.

Consider an Outside Expert to Verify Data and Address Workers' Concerns. Have the workers involved in the selection. Tell the expert to answer all questions openly and honestly. Have the expert make a presentation following his or her study, and allow workers to ask any questions they want.

Conduct Worker Participation Risk Communication Sessions. Allow maximum opportunity for workers to ask questions. Answer all questions as honestly as possible. Take time to understand the audience and make sure that they understand what is being presented. Consider what motivates individuals to risk-reduction behavior.

Measure Effectiveness. Risk communication efforts should undergo some sort of evaluation whenever possible. How can program effectiveness be measured? Have the objectives been met? Do the workers understand relative risk? How is this determined? Do they agree with the program and the means of prioritizing efforts? Do individuals change their behavior as a result of the risk communication?

Summary

Workers have a fundamental right to know whether significant hazards are present in their workplaces that may pose risks to their health. The occupational hygienist must ensure that information communicated about risk to workers is correct, timely, and pertinent; but, equally important, that it is understood and correctly perceived by the worker. To develop and implement an effective risk communication plan, communicate more effectively with workers, and motivate them to take personal actions that reduce risk, the occupational hygienist must appreciate that an individual's concern about a hazard is not always proportional to its relative likelihood. Risk communication in the workplace should be an interactive exchange of information that results in an informed partnership between the worker and the occupational hygienist with the goal of reducing and controlling risk as much as possible.

Additional Sources

Gwin, L.: *Speak No Evil.* New York: Praeger Publishers, 1990.

Hammond, P. and R. Coppock (eds.): *Valuing Health Risks, Costs, and Benefits for Environmental Decision Making, Report of a Conference.* Washington, DC: National Academy Press, 1990.

Hance, B.J., C. Chess, and P.M. Sandman: *Industry Risk Communication Manual, Improving Dialogue with Communities.* Boca Raton, FL: Lewis Publishers, 1990.

Kasperson, R.E. and P.J. Stallens (eds.): *Communicating Risks to the Public.* Boston: Kluwer Academic Publishers, 1991.

Kasperson, R.E. and P.J. Stallens (eds.): *Communicating Risks to the Public, International Perspectives.* Boston: Kluwer Academic Publishers, 1990.

Otway, H.: Experts, risk communication, and democracy. *Risk Anal. 7*:125–129 (1987).

References

1. **Covello, V.T., P.M. Sandman, and P. Slovic:** Risk communication: A review of the literature. *Risk Abs. 3(4)*:171–182 (1986).
2. **Covello, V.T., D.B. McCallum, and M.T. Pavlova:** *Effective Risk Communication, the Role and Responsibility of Government and Non-government Organizations.* New York: Plenum Press, 1989.
3. **National Research Council:** *Improving Risk Communication.* Washington, DC: National Academy Press, 1989.
4. **Mitchell, J.V.:** Perception of risk and credibility at toxic sites. *Risk Anal. 12*:19–26 (1992).
5. **Patty, F.A.:** The industrial hygiene survey and personnel. In *Industrial Hygiene and Toxicology*, vol. 1, 2nd ed. New York: John Wiley & Sons, 1948. p. 45.
6. "Occupational Safety and Health Act." Public Law No. 91-596. Sec. 6(b)(5). 1970.
7. **Baker, F.:** Risk communication about environmental hazards. *J. Pub. Health Pol. 11*:341–357 (1990).
8. **American Conference of Governmental Industrial Hygienists (ACGIH):** *Threshold Limit Values for Chemical Substances and Physical Agents.* Cincinnati, OH: ACGIH, 1996. p. 3.
9. *Industrial Union Department, AFL-CIO v. the American Petroleum Institute et al.*, 488 U.S. 607, 6 Ct. 2844, L. Ed. 2d, 1980.
10. **Ogden, T.L. and M.D. Topping:** Occupational exposure limits for airborne chemicals in Britain. *Appl. Occup. Environ. Hyg. 12*:302–305 (1997).
11. **Lundren, R.:** *Risk Communication, a Handbook for Communicating Environmental, Safety and Health Risks.* Columbus, OH: Battelle Press, 1994. pp. 2–4.
12. **Sandman, P.M.:** *Risk = Hazard + Outrage: A Formula for Effective Risk Communication.* Fairfax, VA: American Industrial Hygiene Association, 1991. [Videotape]
13. **National Research Council:** *Improving Risk Communication.* Washington, DC: National Academy Press, 1989. p. 35.
14. **Daggett, C.J.:** The role of risk communication in environmental gridlock. In *Effective Risk Communication, The Role and Responsibility of Government and Non-government Organizations*, V.T. Covello, D.B. McCallum, and M.T. Pavlova (eds.). New York: Plenum Press, 1989.
15. **Freudenburg, W.R. and J.A. Rursch:** The risks of "putting the numbers in context," a cautionary tale. *Risk Anal. 14*:949–958 (1994).
16. **Gutteling, J.M. and O. Wiegman:** *Exploring Risk Communication.* Boston: Kluwer Academic Publishers, 1996.

Confined spaces contain many hazards that can pose a threat to entry personnel.

Outcome Competencies

After completing this chapter, the user should be able to:

1. Define underlined terms used in this chapter.
2. Recognize confined spaces.
3. Distinguish between types of confined spaces.
4. Recognize confined space hazards.
5. Evaluate confined space hazards.
6. Assess appropriate methods for elimination and control of confined space hazards.
7. Identify critical general operating procedures.
8. Design confined spaces to eliminate the need for entry and/or hazards.

Key Terms

attendant • confined spaces • emergency • entry • entry personnel • entry supervisors • escape • horizontal entry • lower explosive limit • lower flammable limit • oxygen deficient atmospheres • oxygen enriched atmosphere • permit-required confined spaces • rescue • upper flammable limit • vertical entry

Prerequisite Knowledge

College chemistry and physics.

Prior to beginning this chapter, the user should review the following chapters:

Chapter Number	Chapter Topic
1	History and Philosophy of Industrial Hygiene
2	Occupational Exposure Limits
4	Environmental and Occupational Toxicology
6	Principles of Evaluating Worker Exposure
7	Preparation of Known Concentrations of Air Contaminants
8	Principles and Instrumentation for Calibrating Air Sampling Equipment
9	Direct-Reading Instrumental Methods for Gases, Vapors, and Aerosols
10	Sampling of Gases and Vapors
12	Sampling and Sizing Particles
15	Comprehensive Exposure Assessment
16	Modeling Inhalation Exposure
31	General Methods for the Control of Airborne Hazards
32	An Introduction to the Design of Local Exhaust Ventilation Systems
34	Prevention and Mitigation of Accidental Chemical Releases
36	Respiratory Protection
37	Program Management
40	Hazard Communication
41	Risk Communication in the Workplace

Key Topics

I. Confined Space Challenges
II. Identifying Spaces
III. Confined Space Hazards
IV. Programs and Procedures
V. General Rules for Confined Space Operations
VI. Atmospheric Monitoring
VII. Monitoring Devices
VIII. Ventilation
IX. Isolation
X. Communication
XI. After Entry
XII. Emergency Procedures
XIII. Confined Space Design and Reducing Risk

Chapter 42

42

Confined Spaces

R. Craig Schroll, CSP, CET, CHCM

Michael K. Harris, Ph.D.

Introduction

Work in confined spaces frequently amplifies hazards associated with similar work in the open and often entails hazards that are not generally found elsewhere. Failure to anticipate, recognize, evaluate, and control these hazards has resulted in an almost incredible number of preventable deaths and injuries resulting from inappropriate confined space entries. Some perspective on the magnitude of the problem maybe gained from a frequently referenced review of 28,450 accident reports from a three-year period from 1974 to 1977. This work, sponsored by the National Institute of Occupational Safety and Health (NIOSH), reported 276 confined space entry incidents that resulted in 193 deaths and 234 injuries. These data almost certainly underreport the true number of incidents, particularly injuries. Underreporting of industrial accidents was no less a problem 20 years ago than it is today.

Review of accident data by NIOSH and the Occupational Safety and Health Administration (OSHA) indicate that the primary hazards associated with confined space entries can be grouped into five categories: 1) asphyxiation; 2) elevated concentration of toxic contaminants; 3) increased risk of fire or explosion; 4) entrapment and/or engulfment; and 5) other mechanical hazards such as crushing or electrocution.

In order to alleviate or control these risks OSHA has promulgated and enforces several federal regulations that dictate confined space entry procedures. Although these confined space regulations do not drive the task-specific practice of occupational hygiene in the United States to the same extent as the asbestos regulations, their impact may be compared with hazard communication and hazardous waste operations.

There is a temptation to believe that if an employer is in compliance with an appropriate OSHA standard, then everything necessary is being done to protect the health and safety of the affected workers. While this may or may not be so, it is certainly imprudent practice to make such an assumption. OSHA cannot, and does not attempt to, address every conceivable circumstance in writing and promulgating standards. It is simply not feasible to do so. The volume of material required for a truly comprehensive standard would make for a large and awkward document.

Furthermore, the natural evolution of the workplace predicated by changes in technology and culture tends to make any detailed standard out of date even as it is written. As a consequence, the OSHA confined space standards are "performance standards." This means they instruct the employer as to what must be done but not how to do it. The details of compliance and proof thereof are left to the employer.

This regulatory approach is both a boon and a bane to the occupational hygienist addressing confined space issues. It is a boon because the employer is allowed to conduct confined space entries in a manner that best suits the conditions at hand. It is a bane because many employers feel they are provided with inadequate guidance in terms of how to best conduct the entry and remain in compliance with the regulations. However, one may find some degree of solace in that work conducted in compliance with the OSHA regulations will likely meet most of the minimum requirements for healthful and safe working conditions in confined spaces. Prudent practice would be best served by developing and applying procedures that combine the best guidance available from industry practice and OSHA standards, whether or not one has a regulatory responsibility for compliance with a particular standard.

Recognizing the need for guidance for occupational hygienists practicing in the United States, this chapter attempts to address OSHA compliance along with other occupational hygiene issues concerning work in confined spaces. Because the OSHA general industry standard is the most comprehensive confined space entry standard promulgated to date and affects the greatest number of workplaces, this chapter focuses to a great extent on occupational hygiene issues addressed in the OSHA general permit-required confined spaces standard, 29 CFR 1910.146.

Bear in mind that this chapter is not a guideline for guaranteed OSHA compliance. Read the standards carefully; while not all inclusive, they offer valuable information. The reader should take note that there is more than one OSHA confined space entry standard and that more than one standard may apply in the same workplace. The reader should also review standards, understanding that since the publishing of this text, standards could have been added, changed, or deleted.

Confined Space Challenges

Lack of Warnings

One reason that confined space entry is so potentially dangerous is the lack of sensory warning of potential hazards. The most lethal hazard in a confined space is usually a problem with the atmosphere. Many of these atmospheric conditions cannot be detected with the senses. Just looking into the opening to a confined space often reveals no apparent hazard. This too frequently leads to the assumption that there is no hazard.

Previous Experience

Previous experience with confined space entries may also work against those who must use confined spaces. For example, a worker may have had previous experiences when problems that could have happened did not. This tends to make people think that these problems cannot happen in the future. This dilemma is best expressed by the concept of lucky versus smart. When poor safety choices are made and no accident occurs, it is the result of luck; the better approach to safety is to make smart choices.

Previous experience may also work against improvements to work practices developed based on a new understanding of hazards and their control. Always seek continuous improvement in safety. What was done 10 years ago may have been the best way to handle a certain situation then. Over time, methods are developed that allow improved safety practices to be implemented. For example, in the mines many years ago canaries were used to monitor the atmosphere. This should not prevent modern occupational hygienists from using the many technological advances in atmospheric monitoring today.

Identifying Spaces

If one were to walk through an industrial facility or construction work site, how would one recognize a confined space? What are the characteristics? Where are the places where one might find hazards not associated with work in the open? These questions may be answered by the definition taken from the OSHA general industry standard. Note that these are widely applicable but for compliance purposes are not necessarily identical to those in other standards.

A confined space meets *all* of the following criteria:

1. It is large enough and so configured that an employee can bodily enter the space and perform her/his assigned work.
2. It has limited or restricted means for entry or exit.
3. It is not designed for continuous human occupancy.

Examples include but are by no means limited to: storage tanks, process vessels, silos, boilers/fire boxes, open surface tanks, storm drains, pits beneath equipment, storage drums, manholes, cooking vessels, railroad tank cars, and furnaces.

In addition to the criteria above, the construction industry excavation standard identifies any excavation more than 4 feet deep as a work site requiring atmosphere testing. Due to a number of similarities between excavation entries and confined space entries, it is not uncommon practice to include entries into any subgrade work site in a facility's confined space entry program for the sake of administrative convenience.

Confined Space Hazards

Confined spaces may contain many hazards that can pose a threat to entry personnel. This section introduces the major hazards that may be encountered during confined space entry operations. (Hazard elimination and control methods and techniques will be covered later in this chapter.) It is critical that each confined space be evaluated individually to determine that all hazards specific to that space and the work being performed are identified.

Some hazards are relatively obvious, while others may be difficult to identify. General safety hazards that are present in any work environment also need to be considered. Some hazards that pose a threat to entrants originate in the space, others do not. Evaluation must include an overall assessment of all potential hazards associated with the space and the development of techniques and methods to deal with the hazards safely. For ease of discussion the following section divides hazards into categories and individual groupings.

Ideally, hazards should be eliminated. If this is not possible, they should be controlled to a level of acceptable risk.

Atmospheric Hazards

Atmospheric hazards within a confined space can be divided into three major categories: oxygen deficiency or enrichment, flammable gases and/or vapors, and toxic materials. Any one of these could pose a life-threatening situation within the space. They can be present due to conditions within the space or to work being performed there. Atmosphere-related problems create some of the most common safety issues for confined space entry. Studies analyzing confined space accidents also have shown atmospheric hazards to be the most common reason for fatalities. Atmospheric hazards demand detailed evaluation and assessment prior to entering the space and must be continuously monitored during entry operations.

Oxygen Deficiency or Enrichment

Oxygen concentrations within the space should be between 19.5% and 23.5% to be considered safe for entry. When the oxygen concentration within the space falls below 19.5%, the space is considered oxygen deficient. Oxygen deficient atmospheres within the space pose a direct threat to entry personnel and anyone near the opening. According to National Institute for Occupational Safety and Health (NIOSH) statistics, oxygen deficient atmospheres are the most common cause of fatalities in confined spaces. The chart in Figure 42.1 indicates the various effects on an individual in a reduced-oxygen atmosphere.

Oxygen Respiratory Effects

Oxygen %	Effect
20.9	Normal
19.5	Safety limit
17	First sign of hypoxia. Some increase in breathing volume and accelerated heart rate
16	Impaired judgment and breathing
14	Increased breathing volume and heart rate. Impaired attention, thinking and coordination.
12.5	IDLH
10	Very faulty judgement and muscular coordination. Intermittent respiration.
6	Spasmatic breathing, convulsive movements, death in minutes.

Figure 42.1 — Effects to individuals in a reduced-oxygen atmosphere.

Several things can cause oxygen deficient atmospheres. The oxygen may be consumed by reactions in the space. For example, rust is an oxidation process that consumes oxygen. Biological actions such as the breakdown of organic materials through bacterial action may also consume oxygen. Stagnation within the space caused by poor natural airflow can increase the potential for these processes to create a life-threatening atmosphere within the space. Introduction of an inert gas such as nitrogen may also create an oxygen deficiency. The use of an internal combustion engine within the space (for example, a portable engine-driven pump) may consume the available oxygen within the space.

When the oxygen concentration within the space exceeds 23.5%, it is referred to as an oxygen enriched atmosphere. This imbalance can greatly increase the potential for fires or explosions. Chemically, a fire is an oxidation reaction. When enriched levels of oxygen are present, these reactions can be more prone to occur, will be greatly accelerated, and will occur with greater intensity. For example, if the oxygen line on a cutting torch leaks into the space, the resulting oxygen enriched atmosphere will allow any combustible material in the space to be easily ignited and burn with greater than normal intensity.

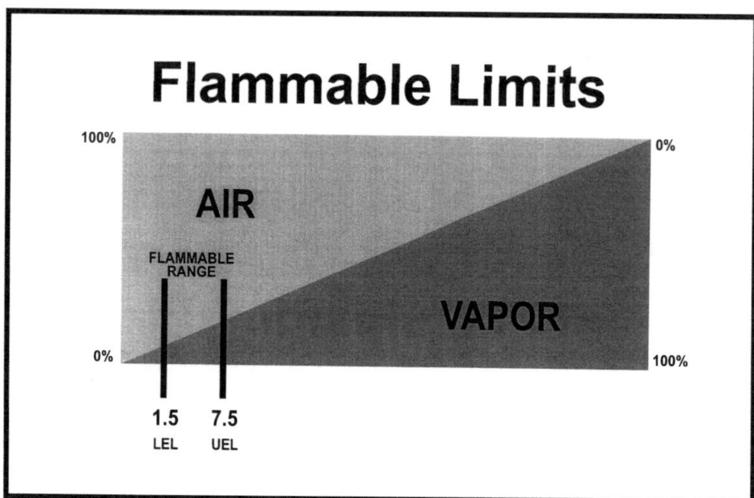

Figure 42.2 — Flammable limits for gasoline vapors.

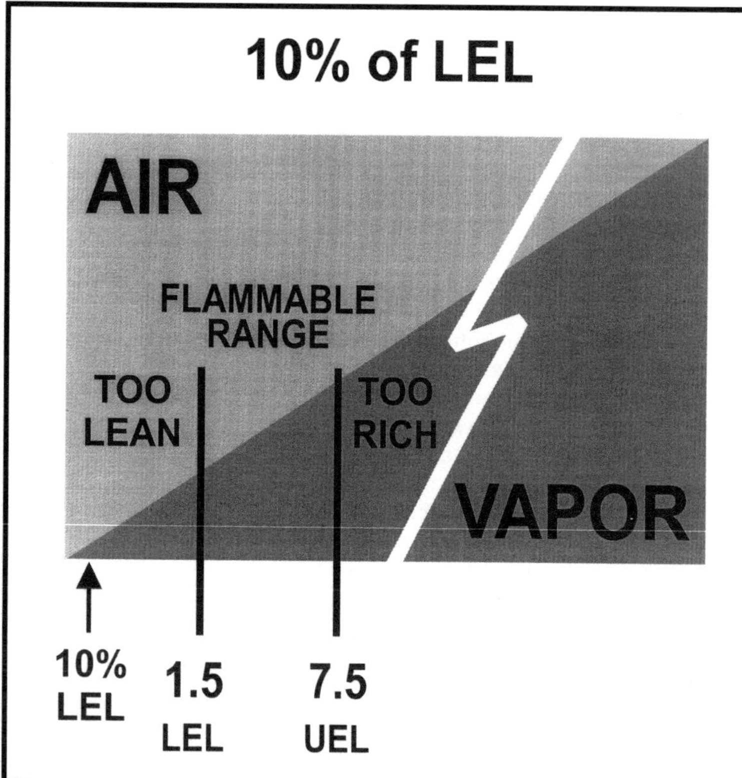

Figure 42.3 — Target range for controlling flammable atmospheres.

Flammable Gases and/or Vapors

Fires and explosions are also leading causes of fatalities in confined space operations. Flammable gases or vapors present two primary hazards. First, these vapors can lead to fires or explosions either within or just outside the confined space. They can also pose inhalation risks. For example, solvent vapors present in a high enough concentration can have a narcotic effect that may lead to drowsiness, reduced capacity to exercise judgment, and other safety problems. Exposure effects may occur at concentrations well below the level where flammability problems could occur.

All flammable vapors and gases have a lower and an upper flammable limit, which together set the boundaries of the flammable or explosive range. The generally accepted safe level for flammable gases or vapors is 10% of the lower explosive limit (LEL). This concept is illustrated in Figure 42.2 using the data for gasoline vapors as an example. These vapors are ignitable only within the flammable range. If an ignition source were introduced into a concentration of vapors below the lower flammable limit, the mixture of gasoline vapors and air would not burn. Below the lower flammable limit the mixture is too lean; not enough fuel vapors are present. The mixture would not ignite above the upper flammable limit because it is too rich, having too many fuel vapors compared with the oxygen in the air. The target for controlling potentially flammable atmospheres is to maintain an environment below 10% of the lower flammable limit (Figure 42.3). This allows a 90% safety margin before the bottom of the flammable range is reached.

Toxic Materials

Toxic gases or vapors may become present within the confined space through three primary methods: product or residue of product in the space, activities or materials taken into the space by the entry crew, and natural decay processes within the space.

Toxic materials pose one of the more difficult challenges in assessing the atmosphere within a space. To monitor adequately for toxic gases or vapors, each must be specifically identified. When the toxic material is present because of the previous contents of the space or operations that are going to be conducted within the space by the entry crew, identifying it is relatively easy. However, toxic gases and vapors generated by decomposition of materials within the space may be more difficult to identify. In open access spaces such as sewers this identification process may become extremely difficult because of the presence of unexpected materials. There is no generally accepted, single, safe value for the accepted level of toxic gas or vapor. The acceptable level of atmospheric contaminants must be assessed based on the knowledge of the specific chemical. A commonly used guide to establish acceptable levels of exposure to toxic materials is the

American Conference of Governmental Industrial Hygienists' guide to threshold limit values (TLVs®). This guide is updated annually and reflects the most current thinking on acceptable exposure levels. The Z Table in U.S. Occupational Safety and Health Administration Standard 1910.1000 establishes the regulatory permissible exposure limits (PELs) within the United States for a wide variety of materials.

Toxic materials may pose not only an inhalation hazard but also a skin contact hazard. Skin contact issues are covered elsewhere in this book (see Chapter 14, "Dermal Exposure and Occupational Dermatosis"). The inhalation hazards can generally be controlled through ventilation or by the wearing of proper respiratory protection.

Detailed information on the evaluation of toxic exposures is included in Chapters 4 and 15 of this book.

Content Hazards

Some hazards within a confined space are due to current contents or remaining hazards from the previous contents of the space. The primary types of content-related hazards include content fill or removal, shifting contents, fluid levels, hazardous materials, and dust.

Content Fill or Removal. Inadvertent introduction of materials into the confined space during an entry operation poses a serious threat to entry personnel. This threat exists even if the material itself is not inherently hazardous. Introduction of nonhazardous contents to the space during an entry poses primarily physical safety hazards. Injuries may occur from contact with materials or engulfment by materials. For example, water poses a potential drowning hazard and grain products or sand may create a suffocation hazard.

The removal of contents from the space may also pose significant risks to entry personnel. This is particularly true with dry flowing materials such as grains, sand, and wood chips. If the entry crew is working on the surface of materials that are removed from the space accidentally during entry operations, personnel may be drawn down into the material and engulfed.

Shifting Contents. Solid granular materials within a confined space may be subject to shifting during an entry that could lead to the entrapment of entry personnel. Walking on the surface of granular materials may cause the entrant to sink into the material. Efforts to pull free may cause the entrant to be drawn down until he or she can no longer get free. Some granular materials are also subject to what is known as bridging. Bridging provides what appears to be a solid surface at the top of the material that is actually not supported from beneath. Entry personnel placing weight on bridged materials will break through and fall to the true supportive surface of the materials, which may be much further below.

Material may also cling to the sides of a confined space. The material may be dislodged by activity in the space and fall on personnel.

Fluid Levels. Fluid levels within a confined space pose two primary challenges to entry personnel. First, the liquid itself may be hazardous. Second, even nonhazardous fluids such as water distort a worker's ability to detect surface changes in the bottom of the space.

Rapidly changing fluid levels may also present a hazard. For example, entries into open storm sewer lines expose personnel to the risk of rapid flooding of the confined space.

Hazardous Materials. Hazardous contents and their residues may present many dangers to entry personnel. Each chemical that may be in the space must be assessed individually for the hazards it may pose. A material safety data sheet should be available in the facility to provide initial information. Other references (for example, the *NIOSH Pocket Guide to Chemical Hazards*) should be available to provide more information on high-risk chemicals. It is good practice to have a copy of the material safety data sheet available at the confined space. Some of the more common categories of hazard include asphyxiants, corrosives, toxics, irritants, and oxidizers.

Dust. Dust poses two primary hazards. First, it may represent a range of respiratory hazards, from nuisance to serious. Respiratory hazards created by dusts are covered elsewhere in this book. Second, dusts may be combustible or explosive. Common materials that may present this risk include coal, corn starch, powdered sugar, and many polymer resins. Accurate field monitoring of dusts at the time of a confined space entry for combustibility hazards is not possible with current technology. If the work environment being assessed contains combustible dusts, a thorough evaluation of the specific dust must be conducted. The *Fire Protection Handbook* and the *National Fire Code®* set contains detailed information on the characteristics of combustible dusts and their appropriate handling.

Other dust hazards may include reduced visibility from ambient conditions or from dust getting into the eyes of entrants, and slippery surface conditions created by dust collecting on the interior surfaces within a confined space.

Dust may result from the contents of the space, the space itself, or operations occurring in the space.

Potential Energy

Hazards in this category involve equipment or devices within the confined space that might operate or release energy that could cause injury to personnel. The most common causes of these hazards include electrical equipment and circuits, and hydraulic, pneumatic, and mechanical equipment and systems.

All potential energy sources that could pose a risk to entry personnel must be identified and eliminated or controlled prior to entering the space.

Electrical. Electrical equipment hazards involve both equipment operation and the electric current present. Switches and circuit controls should be easily identified and well designed for the placement of lockout devices.

With few exceptions, a switch will isolate a circuit. However, it may be difficult to lockout a switch unless it has been specifically designed for lock out. All switches and controls may not be in the immediate area of the device being isolated. Electrical systems generally do not have the capability to store energy, though a few exceptions need to be considered. Systems with capacitors in the circuit must be isolated between the work area and the capacitor. A capacitor may store electricity for a long period, and anything creating a short circuit across the capacitor will cause it to discharge. Systems involving batteries, particularly battery backup power, may be more difficult to identify and isolate. Emergency generators present similar challenges. Also consider any automatic controls such as restart units, level sensors, and pressure sensors.

Hydraulic and Pneumatic. Hydraulic and pneumatic systems also pose the hazard of inadvertent equipment operation and residual pressure on the system. These systems may be more difficult to secure because it is frequently not as easy to identify the appropriate shut-off devices. Many of these systems may be of a loop design that allows flow toward the device from at least two directions. This requires isolation to be established at multiple points for a single system. These devices may also be equipped with internal backup systems, low-pressure sensors, and automatic startup systems that must be identified and secured.

Control of valves, pumps, and compressors is not enough. These systems may hold pressures for an extended period of time and must be bled off.

Mechanical. Mechanical equipment and systems offer the risk of physical injury due to operation of the system or shifting within the system. Mechanical stored energy includes springs that are compressed or stretched and heavy objects held at height.

Mechanical potential energy is the most difficult to identify in many cases. Careful evaluation is required to ensure that this hazard is properly assessed. Usually there are no clearly identified points to secure this energy, which adds to the difficulty.

Environment in the Confined Space

Some ordinary safety hazards may be more difficult to handle when they occur within a confined space. Examples include slippery surfaces, limited lighting, extreme temperatures in the space, and extreme surface temperatures on items within the space.

Limited lighting causes problems in two areas. First, it is difficult to work safely when the crew cannot see properly. Second, apprehension and claustrophobia may increase in poor lighting conditions.

Temperature extremes add stress, thereby increasing fatigue and diminishing concentration on safety issues. Heat stress can become a major problem during work in high heat environments. The three primary types of conditions that personnel will develop from heat exposure are heat cramps, heat exhaustion, and heat stroke. Heats cramps and heat exhaustion, while not pleasant, are not life threatening. Heat stroke can cause death and is a medical emergency.

Cold conditions may also complicate safety within the confined space. Loss of dexterity in the entrant's fingers, for example, makes completing any manual task more difficult.

For additional details on this concern see chapters 24 and 25.

Surface temperature extremes create a contact hazard. This may be a particular problem with metal tanks. Consider items such as steam or cryogenic lines within the space.

Furnaces and ovens are good examples of common confined spaces where surface temperatures will be high. Sufficient time must be allowed for interior surfaces to cool prior to entering the space. Freezers and refrigeration units should be allowed to warm prior to work to reduce the potential for cold contact injuries. If sufficient time to allow for temperature changes is not available, personnel will need proper protective equipment to work around these surface temperature extremes.

Configuration of the Space

Many issues relative to the configuration of the confined space may increase the risk to personnel. Some of the more common challenges include interior shape and slope, interior features, low overhead clearance, drop offs, complex layout, structural integrity, and stability.

The interior shape and slope of the confined space can include many configurations. One of the more common hazardous arrangements is the funnel bottom design common to hoppers.

Low overhead clearance spaces create hazards by forcing personnel to work in an unnatural position. Working in a crouching, kneeling, or stooping position increases fatigue and physical stress and has a negative impact on safety. Crawling through a space also exposes more parts of the body to surface and contact hazards.

Drop offs such as sump pits are a potential fall hazard within confined spaces. The drop into the opening of the space during vertical entry operations also poses a risk of personnel falling into the space. This drop also presents the opportunity for tools, equipment, and other objects to be dropped onto personnel already in the space.

Complex layouts increase the difficulty of access and communications, and may also make the use of external retrieval devices impractical. An access area below a large piece of equipment such as a rolling mill presents these challenges.

Structural integrity is an issue primarily in two types of entries: 1) repair of major damage; and 2) demolition. In either of these situations components of the space itself may collapse, creating a major hazard.

Stability can be a problem with any space that is capable of being moved. Vehicles such as tank trucks and rail cars are obvious examples. Spaces such as tumblers and similar devices where part of the space may be moved even though the entire space is fixed also must be considered.

Nature of the Work

The nature of the work to be performed within the space may increase the hazards of entry operations. Two main issues need to be considered: 1) hazards of the work to be performed; and 2) equipment and supplies taken into the space.

When the work to be performed within the confined space involves hazardous operations, such as cutting, welding, chipping, and abrasive blasting, the hazards may be magnified by the work location. Cutting and welding equipment, electrical tools and equipment, solvents, epoxies, paint, and other such equipment and supplies taken into the space can also significantly increase the hazards.

External Hazards

During entry operations the areas around the space must be assessed for the hazards they may present to personnel during an entry. Traffic hazards would ordinarily be thought of as a concern for crews making entry to manholes on streets and roads. A less obvious hazard is traffic in the form of forklifts and other industrial vehicles in a manufacturing setting. Weather conditions may contribute to heat or cold stress, slip hazards, and visibility problems.

The physical hazard of the operation of nearby machinery and equipment is an issue that must be addressed. Also, the potential for these items to present sources of ignition should the atmosphere within the space be flammable cannot be overlooked. NIOSH statistics reveal that in over half of the fatalities involving fires and explosions, the source of ignition was outside the confined space. Machinery and equipment may also leak contaminants that could enter the space. For example, a propane-powered lift truck with a slight leak at a fitting could introduce propane to the space.

Processes in close proximity to a confined space entry can pose physical contact hazards or provide sources of ignition for flammable vapors or gases. Depending on the type of process involved, hazardous or toxic airborne materials may be released from the process into the confined space work area.

Terrain around the space can add complexity to the entry operation and pose hazards to personnel working in the area. The ideal entry is a large horizontal opening slightly above a wide-open ground or floor area. Many confined spaces offer more difficult challenges. For example, entry may require that the operation be based on a narrow catwalk, well above ground level, which may only be reached by ladder.

Miscellaneous Hazards

Miscellaneous hazards cover many things that do not fit neatly into the categories already discussed. The list could be almost endless, but a few major issues include noise, animals, insects, disease organisms, and radioactivity.

Noise hazards may require the use of hearing protection that may complicate communications between the entrants and attendant. Extreme noise conditions may present hazards from vibration.

Animals such as rodents, pigeons, snakes, skunks, and many others pose hazards from bites, scratches, and in the case of skunks, getting sprayed. Animal waste and dead animals may create atmospheric problems in the space. Another factor to consider is fright. An entrant who is afraid of something encountered in the confined space may be injured or injure others trying to escape.

Insects may present a nuisance or an actual hazard. They can be quite bothersome even when not dangerous, but many insects do bite or sting, and some are poisonous. Particular care is required if a nest of insects is present.

Disease organisms may be present in human waste and other biological materials. Proper protection of entry personnel is essential in these situations.

Radioactive materials pose a special hazard that must be evaluated by competent personnel on a case-by-case basis. If radioactivity is suspected in the confined space a complete evaluation is required (see Chapter 22 for a detailed discussion of these hazards).

Summary of Hazards

This discussion introduced some of the more common hazards involved in confined space operations. It is essential to evaluate the work environment prior to confined space entry in significant detail to determine the specific hazards present. This hazard assessment needs to be comprehensive. Recognition is the critical first step in handling hazards.

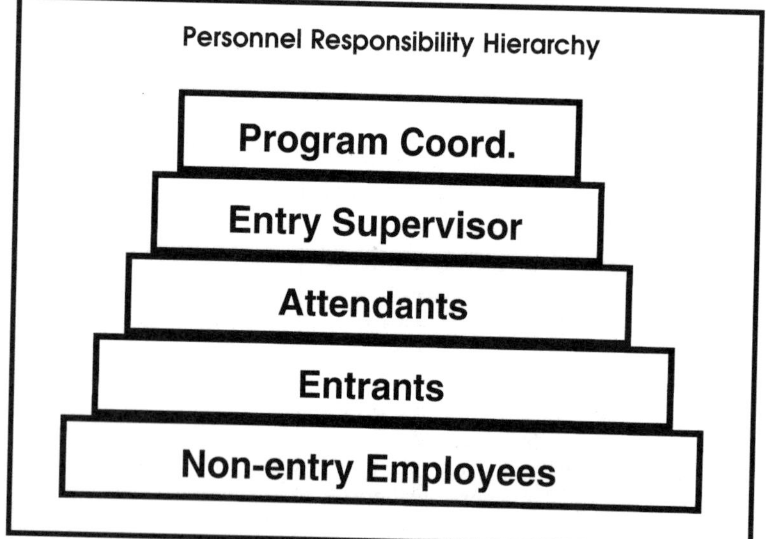

Figure 42.4 — Responsibility levels related to confined space entries.

Programs and Procedures

Confined space operations are covered by numerous regulations and standards throughout the world. Check with local regulatory authorities for the most current information on these regulations and standards when developing programs and procedures.

Personnel Involved in Confined Space Safety Operations

Confined space operations require persons to fulfill specific roles for different levels of involvement. The pyramid shown in Figure 42.4 illustrates progressively greater responsibility related to confined space entries.

Nonentry personnel are involved only to the point of being knowledgeable in recognition of confined spaces and the fact that they are not permitted to enter them.

Entrants are the individuals who go into confined spaces to perform work. Entrants need to be familiar with the hazards they may encounter during confined space entry operations and with the safe use of all of their equipment. They must maintain communications with the attendant and with other entrants. They must also be responsive to the attendant and conditions in and around the space and exit immediately when necessary.

Attendants provide a safety standby outside the confined space anytime there are entrants inside. Attendants must be aware of conditions in the space that may have a behavior effect on the entrants and maintain an awareness of the entrants' conditions. They must also maintain a constantly updated record of the number and specific identity of the entrants within the space. Observation of the interior operation within the space and external areas in close

Training

Entrant/attendant training should cover the following at a minimum.

- Importance of training
- Fatality statistics and case histories
- Goals of training
- Confined space background
  - Types of spaces
  - Lack of sensory warnings
  - Unforgiving nature of confined space situations
- Policy and procedures
- Definitions
  - Confined space
  - Permit-required confined space
  - Nonpermit required confined space
- Recognition of confined spaces
- Duties and responsibilities of entrants
- Duties and responsibilities of attendants
- Duties and responsibilities of entry supervisors
- Confined space hazards
  - Types of hazards
  - Hazard awareness
  - Hazard assessment
- General rules for confined space operations
- Confined space checklist
- Atmospheric monitoring
  - General principles
  - Types of equipment and operation
  - Procedures
  - Methods and techniques
- Ventilation, purging, and inerting
  - Definitions
  - Importance
  - General principles
  - Types of equipment
  - Methods and techniques
  - Hazard control
- Isolation
  - Importance
  - Methods and techniques
- Hazard elimination and control
  - Importance
  - Difference between elimination and control
  - Methods and techniques
- Personal protective equipment
  - Importance
  - Standard items
  - Respiratory protection
  - Special items
  - Proper use
  - Inspection, care and maintenance
- Confined space safety equipment
  - Importance
  - Types
  - Proper use
  - Inspection, care and maintenance
- Lighting
  - Importance
  - Types of equipment
  - Methods and techniques
- Communications
  - Importance
  - Types of equipment
  - Methods and techniques
- Permit system and completing permits
  - Obtaining
  - Completing
  - Posting
  - Canceling
  - Records
  - Review
- Emergency procedures
  - Conditions
  - Notifications
  - Escape
  - Retrieval
  - Rescue
- Returning spaces to service
  - Order and sequence
  - Securing openings
  - Removing isolation
  - Removing lockout/tagout
  - Double check of systems status
- Summary and conclusion

proximity to the space is an important part of the attendant's role. The attendant should prevent entry into the space by unauthorized personnel. Should an emergency occur, the attendant notifies the entrants to escape from the space, operates external retrieval devices, and calls for help when necessary.

The attendant should never enter the space to attempt rescue unless he or she is a qualified member of a rescue crew and has been replaced by another attendant. The attendant role is an important job, and conflicting duties must not be permitted.

During training, entrant and attendant qualifications are typically covered in a combined course so these personnel can trade positions with one another (see sidebar above for a list of minimum training requirements).

Entry supervisors finalize permits, provide a double-check on all safety issues, and authorize entry into the space.

The program coordinator helps ensure consistent safety performance and handles periodic review of completed permits and audits of the entire confined space safety program.

Major operating organizational units must also be involved. Any group that makes confined space entries or has responsibility for confined spaces should be included. This will usually include at a minimum the maintenance, production departments, and outside contractors.

Confined Space Inventory

Prior to the establishment of a confined space program, a thorough evaluation of the facility should be conducted to identify all potential confined spaces. This

CONFINED SPACE SURVEY FORM

Date of survey

Confined Space #

Permit Required Yes No

Location of Space

Description of Space

Possible atmospheric hazards

Possible content hazards

Configuration of space

Unusual hazards

Can be bodily entered?	Yes	No	Hazardous atmosphere?	Yes	No
Limited or restricted entry?	Yes	No	Potential for engulfment?	Yes	No
Not designed for continuous human occupancy?	Yes	No	Internal configuration hazard?	Yes	No
			Other serious safety hazard?	Yes	No

Reasons for entering space & typical activities

Who usually enters space

Frequency of entry

Number of entry points

Alternate Procedures?

Reclassification?

External connections to space

Survey completed by print & sign

Figure 42.5 — Example of a confined space survey form.

CONFINED SPACE ENTRY PERMIT

Permit expires	Date & Time	Confined Space #	Duration of entry	Date & time finished

Location | Description

Purpose of entry

Entrants/Attendants

Entry Supervisor	Work Crew Sup.	Area Sup.	Atmos. Mon.

Hot Work Permit # | Type of Hot Work

Safety Approval | Industrial Hygiene Approval

Alarm Device	Nearest Phone	In-Plant Rescue	Off-site Rescue

Training Qual. | Pre-Entry Briefing | **Return permit to safety department when work is completed.**

ATMOSPHERIC MONITORING

Prior to Entry				Retest			
Oxygen 19.5 - 23.5%	Flammable 10% LEL or less			Oxygen 19.5 - 23.5%	Flammable 10% LEL or less		
Toxic				Toxic			
Chemical Name	MSDS	PEL	Detected	Chemical Name	MSDS	PEL	Detected
Instrument	Calibration			Instrument	Calibration		

ISOLATION of SPACE

Electrical Lockout ☐ Tagout ☐ | **Mechanical** Block Linkage ☐ Disconnect ☐ | **Other**

Piping Lockout ☐ Tagout ☐ Blank ☐ Block & Bleed ☐ | **Hydraulic** Lockout ☐ Tagout ☐ Disc. Lines ☐ Lock Pump & Bleed ☐ | **Pneumatic** Lockout ☐ Tagout ☐ Disc. Lines ☐ Lock Comp. & Bleed ☐

HAZARDS in SPACE

Previous contents | Other

Contents Flammable Irritant Corrosive Toxic Dust Asbestos Solid Liquid Gas | **Configuration** Slippery surfaces Vertical drop Sharp surfaces Temp. - high or low Low Overhead Slope of interior | **Nature of Work** Welding Cutting Grinding Chipping Scraping Spray cleaning

EQUIPMENT REQUIRED

Respiratory protection SCBA ☐ ABA ☐ Sup. Air ☐ Pow. Air ☐ Cart. resp. Full ☐ Half ☐ Cartridge Organic vapor ☐ Acid gas ☐ Organic vapor/Acid gas ☐ Ammonia ☐ HEPA ☐ Dust/Mist ☐

PPE Coveralls ☐ Hardhat ☐ Safety glasses ☐ Safety shoes ☐ Leather gloves ☐ Ear plugs/muffs ☐ Goggles ☐ Welding hood ☐ Welding jacket ☐ Splash suit ☐ Chemical gloves ☐ Chemical boots ☐ Faceshield ☐

Lighting Flashlights ☐ Handlights ☐ Lightsticks ☐ Cord lights ☐ Cords ☐ Portable lights ☐ Generator ☐

All equipment should be explosion-proof & equipped with GFCI

Ventilation Ventilator ☐ 10' sections of duct ☐ 20' sections of duct ☐ Saddlevent ☐ CFM Required ☐

All equipment should be explosion-proof & equipped with GFCI

Entry Equip. Body harness ☐ Retrieval device ☐ Tripod ☐ Anchor point ☐ Access ladder ☐ Personal alert device ☐ Emergency signal ☐ Communications ☐

Rescue Equip. Body harness ☐ Retrieval device ☐ Tripod ☐ Anchor point ☐ Access ladder ☐ Personal alert device ☐ Emergency signal ☐ Communications ☐ SCBA ☐ ABA ☐ Escape mask ☐ Rescue harness ☐ Wristlets ☐

Other

Entry Supervisor (signature)

Figure 42.6 — Example of a confined space entry permit.

inventory should list identifying information for each potential confined space and should be maintained and regularly updated. It is used as a basis for the following steps.

Confined Space Survey

Prior to any entries into a confined space a comprehensive evaluation of the space should be completed to identify hazards and specific areas of concern. The form shown in Figure 42.5 is an example of a tool to be used in this process.

The survey should classify the space based on hazards as either permit-required or nonpermit. This survey will also provide the basis for the development of space-specific operational procedures.

Confined Space Policy

Policies and procedures are essential elements in effective confined space safety programs. A confined space policy should contain at least the following elements: purpose, scope, definitions, confined space program, permit and permit system, operating procedures (general and for specific confined spaces), contractors, emergencies, training, duties, and equipment.

Confined Space Permit

All entries into permit-required confined spaces need a written permit (see Figure 42.6 for an example). Preparation of the space and the permit should generally be completed by the entry crew and double-checked by the entry supervisor. The entry permit should function as a checklist to help ensure that all appropriate safety measure have been taken prior to entry.

Contractor Operations in Confined Spaces

Working with outside contractors often presents special challenges for a confined space safety program. Contractors are a valuable resource for most organizations, and there are many advantages to their use. There can also be significant disadvantages.

The difficulty begins with selecting contractors. There is a large disparity between the best and the worst of outside contractor organizations when it comes to safety performance. Many contractors will do specialized confined space work more safely than could in-house personnel. For example, some of the organizations that specialize in tank cleaning operations have made the investment in the necessary equipment, training, and programs so that they are able to do a better, safer job. The fact that they specialize in this type of operation also means they have had the opportunity to develop a great deal of experience and expertise.

Unfortunately, there are many examples of the opposite end of the spectrum as well, such as the contractor who has no idea how to conduct confined space operations and yet has received a contract to clean a tank.

Contractor involvement in confined space work may be required when specific task expertise or specialized equipment are required. For example, if a vacuum truck is required to complete a pit clean-out operation, it may make the most sense to contract out the entire project. Contractors are also an excellent option if in-house personnel seldom enter confined spaces. By contracting out infrequent entries, many aspects of dealing with confined space issues in-house may be avoided. In this case, leaving this type of work to a contractor with more expertise makes sense.

Contractor Qualification and Selection

The qualification portion of this process should establish the competence of the contractor for confined space work prior to allowing bidding on work involving confined spaces. Several key elements should be obtained from the contractor and reviewed. This process should be a one-time evaluation with periodic reviews for updated material. Although it is a significant time commitment, it can save enormous time and trouble later. The best place to begin is often with organizations that already do contract operations at the facility. Two caveats: never simply collect information and fail to do a thorough review and evaluation; and do not wait to conduct this process until a confined space project is on the horizon.

Obtain the following from the contractor:

- Copy of overall safety program and confined space procedures and permits;
- List of trained entrants, attendants, and entry supervisors, and confined space training records;
- List of other organizations where the contractor performed confined space work (with contact information so these references can be checked);
- Reports on any significant accidents within the last three years and workers' compensation experience modification rate;
- Copies of permits from a few previous confined space entries;

- Hot work procedures and permit if the job involves hot work;
- Emergency procedures the contractor will use;
- Confined space safety equipment list (specific information on make, model, and quantity);
- Calibration records for atmospheric monitoring equipment, inspection and maintenance records for other equipment that requires inspection; and
- A statement from the contractor that all of the above is true and correct.

This may seem excessive, but the majority of difficulties that develop concerning the safety of contractor operations could be completely avoided with the use of an effective selection system. The goal is to qualify a small list of reliable contractors that may be used as a bid list for confined space work.

Once this process is completed, there are job-specific items that must be handled. Inform the contractor that the work will or may involve confined space entry. Provide copies of hazard evaluation of the confined spaces, confined space safety policies and procedures, previous experience with the particular confined space involved, material safety data sheets for any chemicals in the area of the job, and lockout procedures.

Project Planning

At the beginning of the project a brief meeting should be conducted with the contractor and all in-house personnel who may have roles in the safety of the work.

Effective planning can minimize the opportunity for unexpected occurrences during the project. This planning should include clear communications about the scope of the project and the project schedule. The schedule should be designed to minimize the number of entries needed and their duration. Part of the planning process should ensure that all items such as tools, supplies, and equipment required during the confined space entry are available prior to starting the work. This may not be possible on an emergency repair but should be done whenever possible.

During Work

If the project will involve joint operations with in-house personnel and the contractor, these operations must be coordinated. How this will be handled should have been established in the agreement mentioned previously.

Inspect the job site and discuss safety issues with contractor personnel on the first day of the project. Repeat visits should be part of the job monitoring process. Contractor performance and hazards of the work will determine the timing. If the contractor is performing as agreed, fewer visits will be necessary. If safety performance seems less than expected, more frequent visits may be needed to ensure that safety issues are properly addressed.

Plan to visit the job site during any operations that involve significantly increased risk. For example, if part of the project will involve cutting operations in the confined space, a visit during the beginning of that phase would be prudent. During job site visits, talk with the contractor supervisor and employees to determine that the project is progressing as planned from a safety perspective. Observe operations to ensure that the agreed on safety procedures are in fact being used. Review the permit during a visit and compare it to actual operations. Question the contractor about any discrepancies.

After Project Completion

Meet with the contractor to review safety issues that were involved with the job. This meeting should be viewed as primarily a learning experience. Any difficulties encountered during the work should be documented. Any necessary revisions to procedures, training, or other aspects of the confined space program should be completed.

General Rules for Confined Space Operations

The following general rules are the distilled wisdom of confined space operations, incorporating the key concepts that provide for a safe operation. Many individuals and organizations have researched confined space fatalities and their causes. Most notably, NIOSH has done comprehensive research in this area. A review of fatal confined space incidents reveals that each violated at least one of these general rules. When these simple, common sense rules are followed it is extremely unlikely that the confined space entry will result in injury or death. A brief introduction of each of the rules provides the basis for detailed discussion throughout the remainder of the chapter.

Monitor the Atmosphere. Atmospheric monitoring is the first and most critical of the general rules. NIOSH statistics indicate that most fatalities are the result of atmospheric problems. The only reliable method for accurate detection of atmospheric problems is effective monitoring with appropriate instruments. For example, some hazards have characteristic odors and others do not. Some materials may even deaden the sense of smell after short exposure. Even when the presence of a hazard can be detected the extent of the hazard cannot.

Proper Personal Protective Equipment. Proper personal protective equipment (PPE) is essential when the hazards present cannot be eliminated or controlled through other means. PPE must be selected to meet the challenge of the hazard. Personnel must be trained and competent in the proper use of PPE, and supervision must insist on proper use.

Attendant. An outside attendant monitors the safety of the entry operation and is in the best position to prevent an emergency or provide help if one occurs.

Ventilate the Space. Ventilation and related activities are the best options for correcting an atmospheric problem.

Isolate the Space. Isolation of the space should eliminate the opportunity for hazards to be introduced during the entry operation.

Lighting. As discussed earlier, lighting is important for two primary reasons: adequate vision and fear of the unseen.

Eliminate or Control Hazards. All hazards that have been identified during the hazard assessment must be eliminated or controlled prior to entering the space.

Rescue Equipment. Any equipment required for rescue must be available to those designated to use it. External retrieval equipment that may be used by the attendant must be in place. More advanced rescue equipment for entry-type rescues must be available to the designated rescue crew.

Plan for Emergencies. It must be assumed that emergencies will occur. Any variance from acceptable entry conditions as established by the entry permit constitutes an emergency. Even a situation such as a battery failure on atmospheric monitoring equipment should be considered an emergency. Efforts to prevent emergencies need to be constant, but there is a good chance that eventually an emergency will have to be handled. If the entry crew is prepared, it may be handled without a problem. If preparations are not adequate, the emergency may turn into a fatality.

Communications. Effective communications are critical to a safe operation. Any means that ensure contact between the entrants, attendant, and entry supervisor are considered effective.

Atmospheric Monitoring

Atmospheric monitoring is one of the most important parts of safe entry operations for confined spaces. The most lethal hazards historically are problems with the atmosphere. Atmospheric hazards can, and frequently do, kill.

Atmospheric monitoring must be completed prior to entering the confined space. In situations where this is not possible, special procedures must be used. These procedures will usually include the use of respirator protection.

Atmospheric monitoring should be continuous during the entry. When there are people in the space the atmosphere should be monitored. Monitoring must cover all areas of the space where entrants may be exposed to the hazards of the atmosphere. Continuous monitoring should then be established near the breathing zone of the entrants. Another option is to monitor between the entrants and the most likely source of potential atmospheric problems.

Vapor density varies considerably for contaminants that may commonly be found in a confined space. Vapor density is a comparative measure, with air equal to 1. A vapor with a density greater than 1 is heavier than air and would tend to sink in air. A vapor with a vapor density less than 1 is lighter than air and would tend to rise in air. This characteristic indicates where within the space the contaminant is most likely to be found initially. The stratification based on vapor density is most significant while the vapor or gas is being introduced to the space. Given sufficient time, gases will diffuse throughout the space.

Atmospheric Monitoring Process

The flowchart in Figure 42.7 illustrates the basic process of atmospheric monitoring. In confined space entry operations there will always be a need for atmospheric monitoring, so the first point on the flowchart is automatically "yes." Identify the specific potential atmospheric problems likely to be encountered within the space. Oxygen and flammability issues should always be checked. The oxygen sensor may be used in the vast majority of operating environments without specific identification of other issues that may be present. The flammable sensor is typically a broad range sensor that will detect flammability problems without specifically identifying the flammable involved.

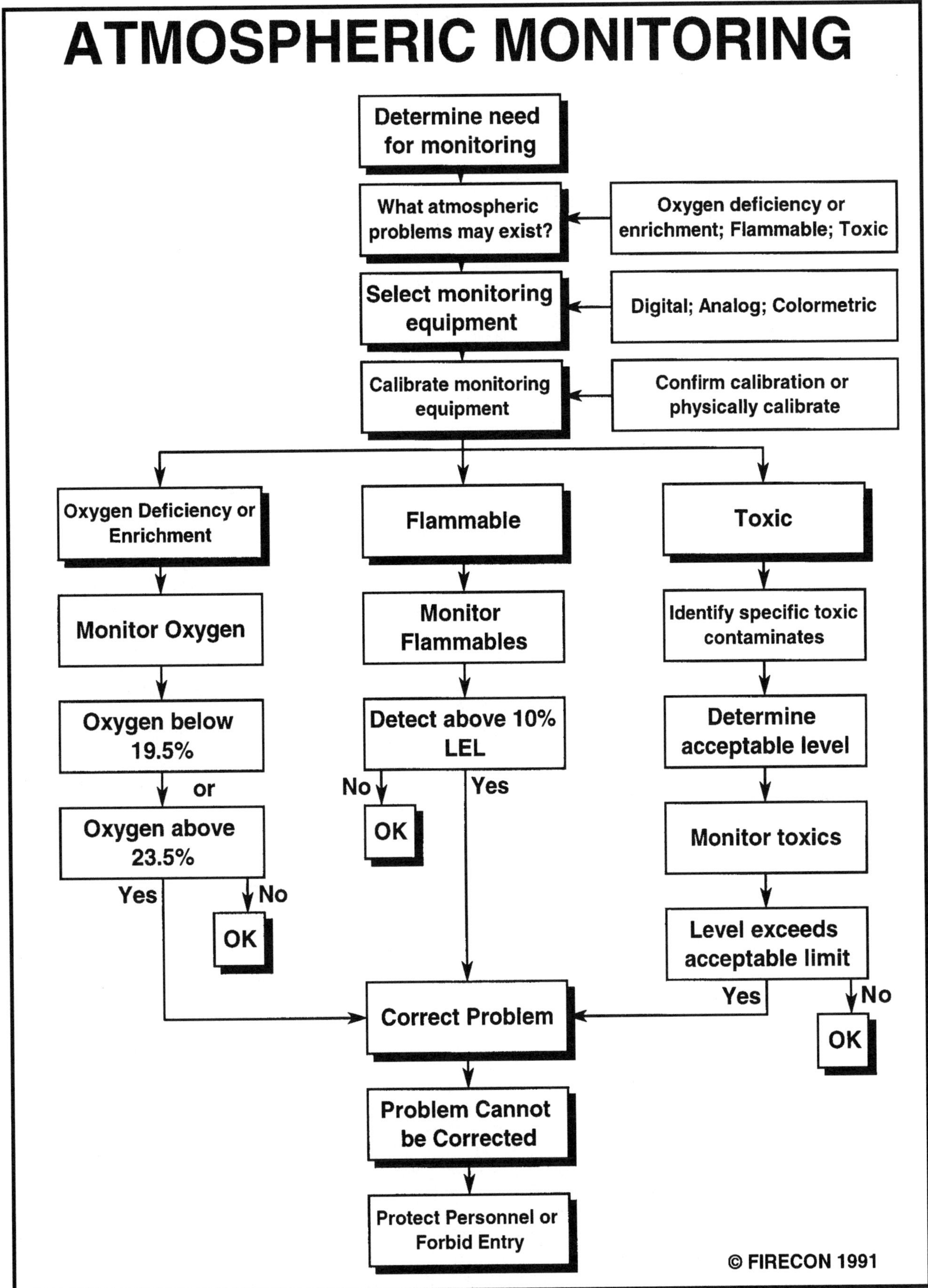
ATMOSPHERIC MONITORING
Determine need for monitoring
What atmospheric problems may exist?
Oxygen deficiency or enrichment; Flammable; Toxic
Select monitoring equipment
Digital; Analog; Colormetric
Calibrate monitoring equipment
Confirm calibration or physically calibrate
Oxygen Deficiency or Enrichment
Flammable
Toxic
Monitor Oxygen
Monitor Flammables
Identify specific toxic contaminates
Oxygen below 19.5%
or
Oxygen above 23.5%
Yes
No
OK
Detect above 10% LEL
No
Yes
OK
Determine acceptable level
Monitor toxics
Level exceeds acceptable limit
Yes
No
OK
Correct Problem
Problem Cannot be Corrected
Protect Personnel or Forbid Entry
© FIRECON 1991

Figure 42.7 — Basic process of atmospheric monitoring.

Toxic sensors are most often chemical specific. This requires that potential toxic contaminants be specifically identified prior to selecting the appropriate monitoring equipment. Each toxic has a specific acceptable level.

After the hazards have been identified, the specific monitoring equipment may be selected based on those hazards.

The most effective approach is to calibrate all the monitoring equipment prior to beginning work on the confined space. Manufacturer recommendations vary on frequency of calibration (monthly is most typical). The ideal situation is to ensure that instruments are calibrated each day of use at the beginning of the shift. At a minimum, confirm that all devices have been properly calibrated within the last month.

The flowchart now divides into oxygen, flammable, and toxic. For oxygen the acceptable range is 19.5% to 23.5%. Normal air will contain 20.9% oxygen. Any variance from normal must be investigated. Oxygen must be the focus of attention first because an oxygen deficiency will cause errors in flammable readings. If the oxygen is within the acceptable range, attention may be focused on flammable gases and vapors. The acceptable range is below 10% of the LEL. For toxic contaminants a range up to the PEL is considered acceptable according to U.S. regulations. The regulations allow exceeding the PEL for materials that are not capable of short-term harm or incapacitation, but using the PEL is the safer choice. The TLV is a more current reference for acceptable limits of exposure.

If problems were discovered during atmospheric monitoring, correct the problem when possible. If the problem cannot be corrected, forbid entry or protect personnel with respiratory protection equipment.

Monitoring Devices

Three major categories of monitoring devices are typically used in confined space work: digital, analog, and colorimetric tubes.

Digital instruments are usually the best choice. They are the easiest to use and to interpret of all the devices. They also offer the widest variety of accessory equipment. They are commonly available in one- to five-sensor models from a variety of manufacturers.

Analog instruments provide a needle and dial readout. These devices are commonly available in one- and two-gas units. There are many excellent devices in this category, but they tend to be more difficult to use and interpret. For example, when testing an atmosphere that exceeds the LEL some types of combustible gas detectors will move quickly to 100% LEL and then drop back to zero.

Because electronic instruments are not available for all toxic contaminants, colorimetric tubes must be used. Generally, tubes are a last resort. These devices are the most difficult to use and interpret of the three types discussed. The major advantage of colorimetric tubes is the wide range of specific chemicals that may be monitored. The major disadvantage is the difficulty in use and margin for error in readings. Chapter 10 contains a more detailed discussion of colorimetric tubes.

Manufacturer's Recommendations and Instructions

Read the manual to develop the knowledge and skill required to effectively use specific instruments. All instruments have particular strengths and limitations. The inability to use and interpret these instruments effectively may lead directly to a fatal accident.

Instrument Capabilities

Gases that can be monitored with electronic devices typically include oxygen, flammables as a percentage of LEL, carbon monoxide, hydrogen sulfide, nitrogen dioxide, sulfur dioxide, ammonia, chlorine, methane, nitric oxide, and hydrogen cyanide.

Flammable Gas Sensors

Flammable gases and vapors are most commonly detected with catalytic diffusion sensors. In this device fine platinum wire is wound into two coils, one active and one reference. The coils are wired into a circuit called a Wheatstone bridge (see Figure 42.8), where voltage is applied to both coils, heating them to approximately 550°C. When a flammable gas or vapor comes into contact with the sensor, the active coil burns the material causing an increase in temperature. Oxygen concentration within the confined space must be within safe limits for flammable readings to be accurate. If the oxygen is deficient, the flammable reading will be lower than the actual conditions within the space. The reference coil is isolated from the flammable gas or vapor with a coating, so it maintains a constant temperature. The response

Table 42.1 — Flammable Detection Conversion Factor Table

Sampled Gas	Calibration Gas Pentane	Methane	Hexane
Acetone	0.9	1.7	0.7
Acetylene	0.7	1.3	0.6
Benzene	1.0	1.9	0.8
Butane	0.9	1.7	0.8
Ethane	0.7	1.3	0.6
Ethanol	0.8	1.5	0.6
Ethylene	0.7	1.3	0.6
Hexane	1.2	2.3	1.0
Hydrogen	0.5	1.0	0.4
Isopropanol	1.0	1.9	0.9
Methane	0.5	1.0	0.4
Methanol	0.6	1.1	0.5
Pentane	1.0	1.9	0.9
Propane	0.8	1.6	0.7
Styrene	1.1	2.2	1.0
Toluene	1.1	2.1	0.9
Xylene	1.3	2.5	1.1

Source: Courtesy of Industrial Scientific Corp. and based on the performance of their sensor.

of one coil and not the other causes a change in resistance in the circuit, which is converted into a readout.

The sensor is made intrinsically safe by using a flame arresting cover over the chamber where the coils are located. Figure 42.9 is a cross-section illustration of an actual sensor.

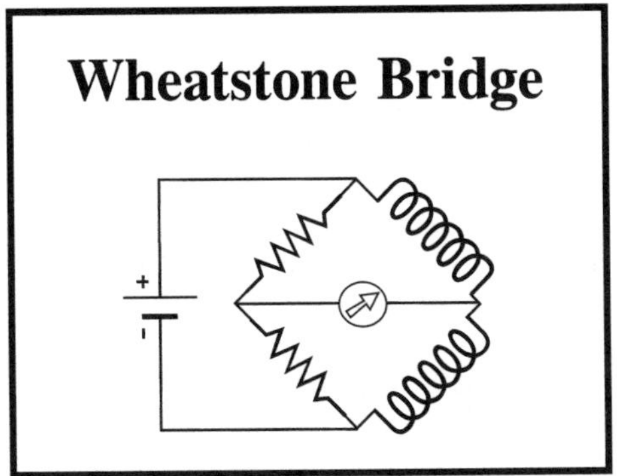

Figure 42.8 — Circuit found in catalytic diffusion sensors.

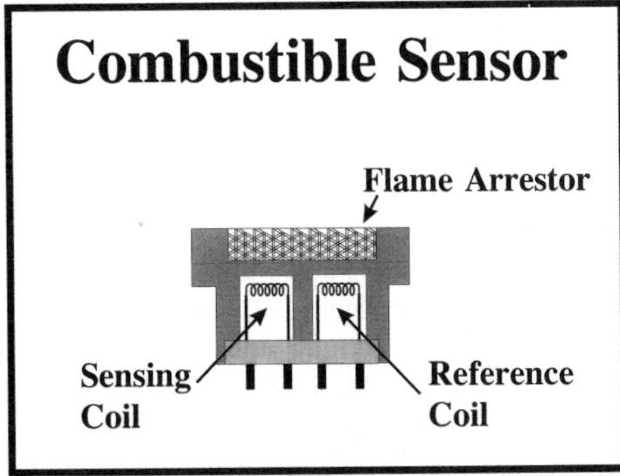

Figure 42.9 — Cross section of a combustible sensor (copyright by FIRECON).

Flammable Detection Conversion Factors

The sensitivity of a flammable sensor depends on the calibration gas used. When the flammable gases or vapors being sampled are significantly different from the calibration gas a conversion may be used to improve the accuracy of the reading.

The multiplier in Table 42.1 should be used for improved accuracy of readings. An instrument calibrated to pentane being used to read acetylene would use a 0.7 conversion factor. For example, a reading of 20% LEL should be converted as follows: 20% LEL reading times 0.7 conversion factor = 14% actual percentage of LEL.

Accuracy of conversion is ±25% of the calculated reading. In the above example the actual percentage LEL of acetylene is between 10.5% and 17.5% LEL. When using the recommended safe cutoff of 10% LEL it is typically not necessary to calculate this conversion. Sufficient safety margin is built in to avoid needing to make conversions. The conversions do allow additional accuracy when conducting specific problem investigations.

Infrared sensors (see Figure 42.10) may be used to monitor flammable gases or vapors when there is an oxygen deficient atmosphere in the space. These sensors do not require oxygen to operate properly.

Flammable sensors may be tuned to a specific gas to provide readings in percentage vapor instead of percentage LEL. This is common in the natural gas and mining industries where instruments are often set up to monitor methane.

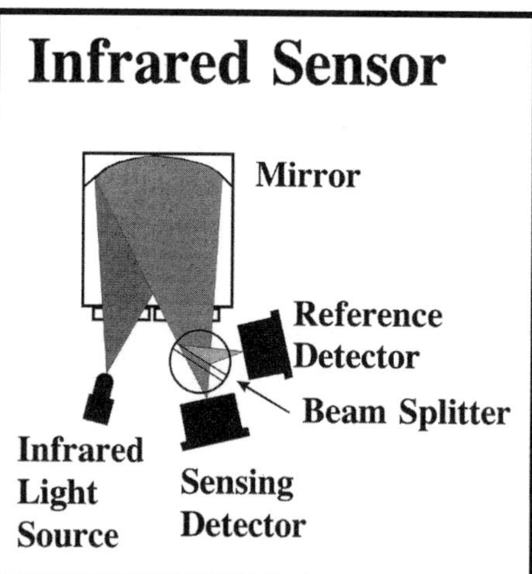

Figure 42.10 — Diagram of infrared sensor (copyright by FIRECON).

Toxic Sensors

Most safety instruments use electrochemical sensors for the measurement of toxic gases. A chemical reaction that releases electrons occurs within the electrolyte when the atmospheric contaminant enters the chamber. The instrument determines the concentration of gas based on the electrons released and provides a reading in parts per million.

Cross-Interfering Gases. Cross-interfering gases cause a sensor to provide false readings. The sensor intended to measure a specific gas such as nitrogen dioxide may also be sensitive to another gas. For example, the nitrogen dioxide sensor when exposed to 100 ppm of chlorine will read 90 ppm. Figure 42.11 is a sensor performance chart.

General Operational Considerations

Accuracy of instruments is typically evaluated based on a comparison of the instrument reading with the actual test atmosphere. Readings should be accurate to within 10%.

Response time is the sensor's ability to react in a specified time to the material it is designed to measure. Instruments should have response times of 90% of reading within 30 seconds.

Measuring ranges are the upper and lower limits of detection for the sensor. For example, an LEL sensor should be capable of 0%–100% LEL readings in 1% increments.

Instruments have temperature and relative humidity operating limits. These may vary significantly among manufacturers. Generally, operating temperature range will be approximately 15°C to 40°C (5°F to 104°F) for continuous exposure, with a somewhat larger range for intermittent exposures. Relative humidity acceptable range is typically 15%–90%.

Instruments should be turned on, allowed to warm up, and zeroed in fresh air prior to approaching the confined space.

Power Supply

Power supply issues have a major impact on the effective use of instruments in the field during confined space operations. Rechargeable batteries power the majority of these units. Service life varies from a low of a few hours to a high of approximately 10 hours. Most units have field changeable batteries. A few manufacturers produce models that may be powered with regular dry cell batteries, usually AA or C size or lithium power cells.

Chargers are available in a variety of configurations. Common chargers usually include 115 volt AC and 12 volt DC. Many are capable of rapid and trickle charging.

Remote Sampling

Remote sampling equipment includes pumps, hand aspirating bulbs, tubing, and probes. These accessories are commonly available in a wide variety of specific configurations.

Maximum draw distance and draw rates vary among manufacturers. Maximum tubing allowed is typically 100 feet or less. Draw rates average 1 ft/sec. Filtration devices are also available for avoiding the intake of dust and fluids.

A dilution tube is a specialty accessory for sampling flammable atmospheres in areas of reduced oxygen concentration with a catalytic bead sensor. The dilution tube allows air, containing oxygen, into the sample stream in measured amounts allowing accurate flammable readings to be obtained using a conversion factor.

Cross-Interfering Gases

↓Sensor	Applied Gas										
	CO	H_2S	SO_2	NO	NO_2	Cl_2	H_2	HCN	HCL	C_2H_4	C_2H_2
CO	100	<10	<10	<30	<15	<10	<60	<15	<3	~75	~150
SO_2	0	0	100	0	~120	<5	0	<50	0	0	~140
NO	0	~35	~5	100	<40	0	0	0	<15	0	<1
H_2S	<10	100	<20	0	~20	~20	<5	0	0	0	<1
Cl_2	0	~20	0	0	120	100	0	0	0	0	<1
NO_2	0	~20	~5	0	100	90	0	<1	0	0	<1
NH_3	0	~130	~70	~20	0	~50	0	~30	~5	0	~5
HCN	<5	~350	~160	0	~120	~55	0	100	~35	~50	<1

Chart courtesy of Industrial Scientific Corp.

Figure 42.11 — Sensor performance chart.

Alarms

Field instruments are usually equipped with audible and visual indications of alarm and fault conditions. Common fault indications are for low battery, battery failure, sensor-related faults, and low flow on sample pumps. Alarm options usually include various remote alarm devices, including audible, visual, and vibrating alarms. The visual and vibrating alarms are particularly useful in high noise areas.

Calibration

Calibration is essential to maintaining the accuracy of atmospheric monitoring equipment. The label on a calibration gas cylinder will indicate the specific contents and the expected reading during calibration. All calibration gas expires, so the label should indicate a manufacture or expiration date.

Calibration should be done at least as often as the manufacturer recommends. This varies, but once per month is the most common. More frequent calibration may be required based on use experience. When the instrument is behaving in a manner that cannot be explained, it should be calibrated to confirm that it is functioning properly or to assist with determining the specific problem.

Calibration records should be maintained on each instrument. At a minimum they should include date, name of person doing the calibration, specific identification of the instrument (serial number or user-assigned unique number), results of calibration, and service performed or needed.

Techniques

Horizontal Entry Situations

Atmospheric monitoring should begin outside the hatch prior to opening the confined space. This allows assessment of any hazards in the area that may have nothing to do with the confined space but may have an impact on safety inside it. This technique will also identify a hazardous atmosphere that may be seeping through the confined space opening.

After determining that it is safe to remove the cover to the opening, initial monitoring inside the space is done with a probe, as illustrated in Figure 42.12. Use of a rigid probe allows sampling farther into the space. Extend the probe up and down as far as reach allows, assessing as much of the space as possible from outside.

Monitoring throughout the space will need to be completed as well. Based on knowledge of the space and results of initial atmospheric monitoring, assess the potential for a hazardous atmosphere within the space. Entry for the remainder of the initial test may need to be completed by an individual using respiratory protection.

Vertical Entry Situations

For vertical entries also monitor outside the space prior to opening the cover. If the cover has an access hole the probe may be used as illustrated in Figure 42.13 for initial monitoring inside the space. When there is no hole the cover must be removed, and monitoring will begin at the opening.

Figure 42.14 illustrates the use of flexible tubing for remote sampling. When this technique is used the tubing must be lowered slowly into the space to allow sampling at all levels. If the sample tube is lowered more rapidly than the draw rate of the sample pump, areas may not be effectively measured.

Low-level obstructions may interfere with monitoring. In this situation if the tubing cannot be placed on the other side of the obstruction from outside the space the monitoring will need to be completed after entry. Divisions of this type within a confined space are relatively common; one excellent example is the baffles within a tank truck.

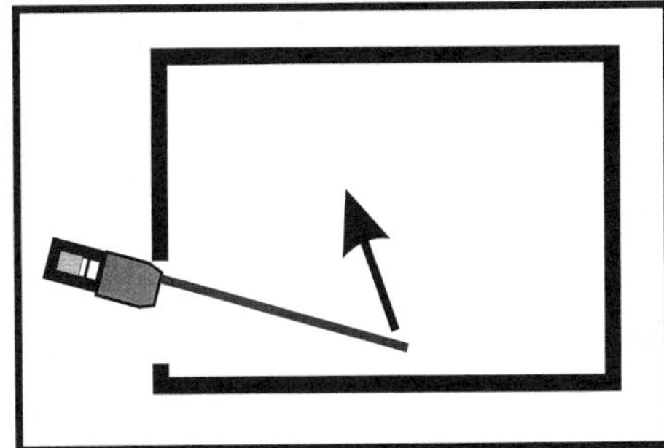

Figure 42.12 — Horizontal entry confined space monitoring with probe.

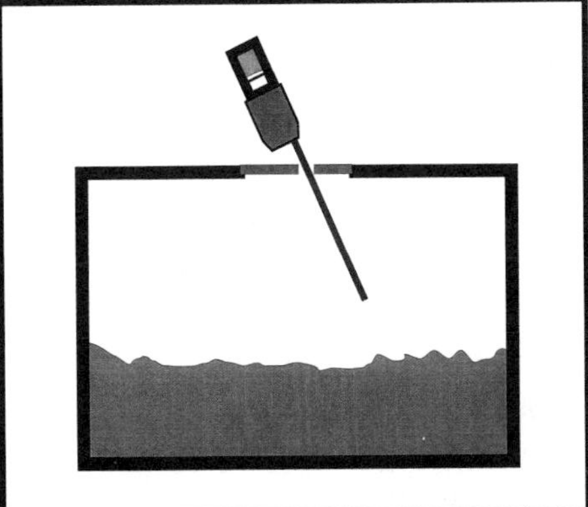

Figure 42.13 — Vertical entry confined space monitoring with probe through opening in hatch.

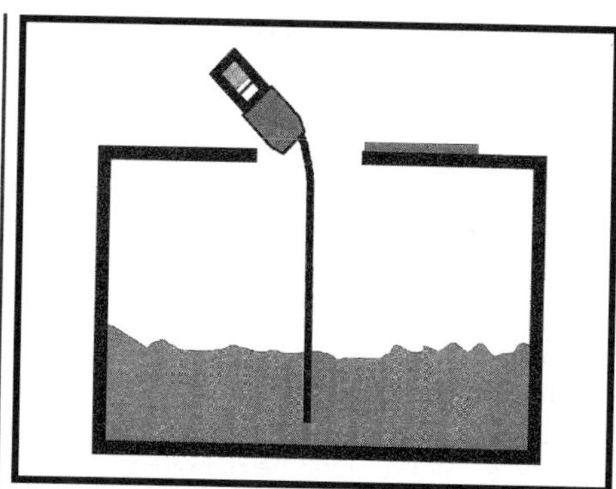

Figure 42.14 — Vertical entry confined space monitoring with tubing.

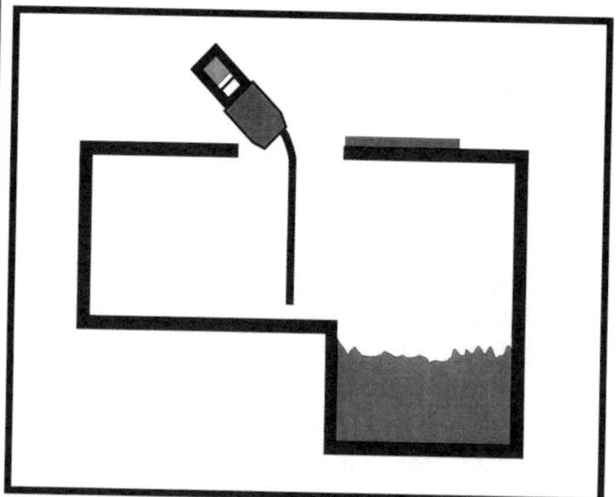

Figure 42.15 — Vertical entry confined space monitoring with tubing impact of different elevations within space.

Different levels within a confined space (see Figure 42.15) can create a similar problem. This situation may be present in a confined space with a depressed drain area or a sump pit.

Ventilation

Ventilation and related activities are used to correct and prevent atmospheric problems that may exist prior to entry or that could occur within the confined space during entry operations.

Prior to ventilating the space it should be cleaned to the extent possible from the outside. Flushing and pressure washing are two possible techniques. Cleaning activities may also include the use of neutralizing solutions. For example, an acid tank will be flushed with a neutralizing solution to eliminate the acid residue hazard prior to ventilating.

Methods of Handling Atmospheric Problems

There are three main types of activities that are intended to handle atmospheric problems within a confined space: ventilation, purging, and inerting.

Ventilation

Ventilation uses air movement and may be accomplished using either forced air or exhaust. Forced air ventilation is the dilution and displacement of a contaminated atmosphere by introducing fresh air. This technique is generally the most effective for confined space work. Exhaust ventilation generally will be used in confined space operations only if a point source of contamination exists. For example, if welding fumes are the problem, local exhaust ventilation set up to extract the fumes where they are created may be effective. Sufficient make-up air intake is critical if exhaust ventilation will be used.

One specialized type of forced air ventilation is the use of compressed air. In working environments where moisture of any type could create problems, ventilation may be accomplished using dry air from a cylinder. For example, some electrical transformers may be equipped with fittings for this type of ventilation.

Purging

Purging involves the introduction of something other than fresh air to the space as an initial displacement agent. Purging will be followed by ventilation to make the space safe for entry.

Inerting

Inerting involves the introduction of a nonreactive gas, typically nitrogen, argon, or carbon dioxide. Inerting trades one hazard for another. It is most often used in spaces where the risk of fire, explosion, or other chemical reaction is great. The inert gas controls these problems but creates an immediately dangerous to life or health oxygen-deficient atmosphere. Entrants must wear air-supplied respiratory protection.

Air-Moving Devices

Fans are available in two primary types, axial and centrifugal. An electric motor or gasoline engine usually powers fans. Other power options are available, such as compressed air, water, or steam. Another common air-moving device is the compressed air horn, which works on the venturi principle. In the chemical and maritime industries hatch ventilators that are designed to fit the opening are commonly used. If the opportunity exists for flammable vapors or gases to be present during ventilation, the equipment used should be explosion-proof.

Air Flow Rate Reductions

No Duct (Rated capacity)	2,000 cfm
15' straight	1,700 cfm
15' 1 90° bend	1,400 cfm
15' 2 90° bends	1,250 cfm
25' straight	1,650 cfm
25' 1 90° bend	1,250 cfm

©FIRECON

Figure 42.16 — Airflow rate reductions based on eight-inch flexible duct and varying bends.

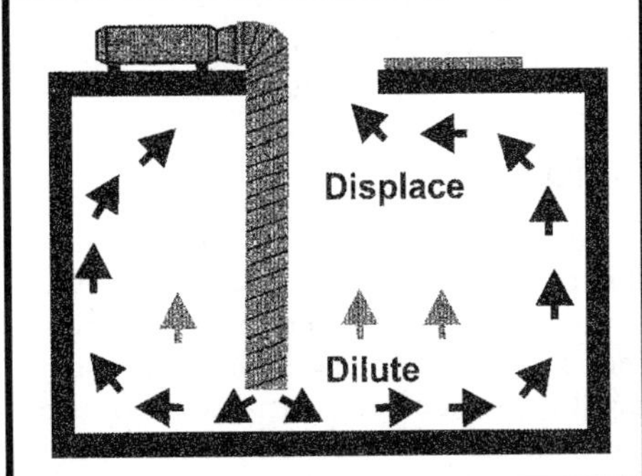

Figure 42.17 — Contaminant dilution and dispursement process.

Procedures

Ventilation must be used whenever a contaminated atmosphere is discovered. It is also good practice to ventilate the space as a precautionary measure even if no problems were encountered with the atmosphere. General space ventilation may also improve the comfort of entrants, an advantage particularly in hot environments or when heated air is used in cold environments.

While many variables must be considered in establishing the amount of ventilation needed, 20 air changes per hour is a good general target. For example, a confined space with a volume of 3000 ft^3 would require a ventilator of at least 1000 ft^3/min capacity to achieve 20 air changes per hour. This level of ventilation may not always be achievable in field applications. The minimum requirement for effective ventilation is that the atmosphere can be maintained at a safe level.

Bonding and grounding should be used while ventilating, as the moving air may generate a static electric potential. This static potential could create an ignition source for flammable gases or vapors.

Techniques

The duct for general space ventilation should extend deep into the confined space. The end of the duct should be approximately 2 feet from the bottom of the space and directed straight down. To avoid recirculation of contaminated air, the inlet of the ventilator should be at least 6 to 8 feet from the center of the opening into the space.

Most manufacturers recommend that no more than 25 feet of duct be used to maintain adequate airflow. The number of bends in the duct should also be minimized. Each bend reduces airflow by increasing turbulence in the airstream. The chart in Figure 42.16 illustrates the approximate reduction in airflow from rated capacity based on a length of 8-inch flexible duct and number of bends.

Airflow introduced to the space will begin to displace and dilute the contaminated atmosphere (see Figure 42.17). However, even with the ideal setup of portable ventilation equipment, 100% efficiency will not be achieved. A likely area for limited airflow is the upper corners of the space.

As ventilation proceeds some of the hazardous atmosphere will be diluted to nonhazardous concentrations, and some will be displaced to the outside of the space. Depending on the level of contaminants at the start of ventilation, some of the displaced atmosphere may exceed safe levels. This escaping atmosphere may be drawn back into the space in an effect called churning.

Obstructions within the space may have a significant effect on the airflow. This can be an especially difficult challenge when the obstruction is at the floor of the space. This area may retain a hazardous atmosphere. The end of the ventilation duct will need to be directed into this area to clear the hazardous atmosphere.

After ventilation operations have been running for at least 15 minutes, the atmosphere in the space may be retested. Fifteen minutes is typically the minimum ventilation time required when using appropriate capacity ventilation equipment. Depending on the size of the space and the ventilation equipment used, this time may need to be significantly extended. During the retest the ventilator should be shut off and the air allowed to stabilize for a few minutes.

For horizontal openings the duct may be extended into the space along the floor. This will begin a circular air movement that will dilute and displace the hazardous atmosphere.

When two openings are available, one may be used to introduce fresh air and the other used for exhaust. The duct does not need to extend as far into the space with this technique. This arrangement is typically referred to as push-pull ventilation.

Isolation

Isolation is the process of securing all interconnections to the space, thus preventing external items from being introduced to the space. Isolation also includes preventing the activation of internal devices such as agitators.

Pipe Connections

Piping connections that may carry gases, liquids, or granular solids must be isolated to ensure that nothing is allowed into the space while entrants are inside. The four most common ways to handle pipe connections are lockout, disconnect, blank, and double block and bleed.

Locks used for lockout should be assigned to the individual for the duration of the job. Many special types of locks are available for this use. Lockout of a single valve should never be trusted as the only means of isolation and should be used in combination with other methods.

Disconnecting piping and offsetting the pipe ends is another option for control of these connections. Many piping arrangements will not allow appropriate offsetting of piping because of the size and configuration of piping.

Blanking is one of the most common methods used for isolation. A blank is a solid plate designed to withstand the maximum possible pressure in the pipe. This blank must cover the entire internal diameter of the pipe and effectively prevent flow past the blank. It is inserted at a flange, and the flange bolts are tightened to hold it in place.

Double block and bleed arrangements must be installed in the piping. These devices are particularly useful for areas that may require frequent isolation of relatively low-hazard materials. The device includes two valves in the main line coming into the confined space, with a bleed pipe and valve located between. During normal operation the bleed valve is closed and possibly locked, and the two main line valves are open. When isolating the space, the two main valves are both closed and locked and the bleed valve is opened and locked. If material leaks past the first valve, it will come out at the bleed-off pipe. This avoids the application of pressure to the second valve and should also alert personnel to the failure of the first valve.

Lockout

Lockout provides the most positive control of items such as electrical connections associated with the confined space. Preparing to shut down the device is the first step. There may be a specific sequence that must be used to accomplish this safely. Equipment-specific procedures should be developed. Next, all controls are placed in the off position. Isolation and lockout devices are applied to the system. Stored energy is released to achieve a zero energy state. Finally, controls are operated to attempt to run the device as a way of confirming that the isolation has been completed effectively. After this test, controls are returned to the off position.

Lighting

Adequate lighting must be provided for the exterior and interior areas of the confined space. Lighting is essential for safe performance and reduces psychological stress for entrants. A backup lighting option should always be available for immediate use inside the confined space.

There are several types of lighting equipment that may be used in confined space work. The most common categories include conventional electrical powered, low voltage, battery operated, and chemical.

Conventional 110-volt lighting provides a good option when an outlet is readily available. This type of equipment is usually the best choice when there are no hazards such as wet operating environments that would prevent it. Ground fault circuit interrupters must always be used.

Low-voltage systems, usually either 12 or 24 volt, are a good selection when shock hazards are a concern. This would include use in metal tanks and wet areas.

Battery-operated flashlights and hand lights are an option for primary lighting if neither of the previous options is practical. Battery-operated devices are also an excellent option for backup lighting.

Chemical light sticks should never be used as the primary lighting but make an excellent backup light source.

Placement

Placement of lighting should not obstruct access to the confined space or create hazards such as cords that can be tripped over.

Ideally, the interior lighting should be placed close to the work area within the space. Trying to light the interior with lighting placed at the opening can create hazards and is generally not as effective.

Another disadvantage of this placement is that entrants cannot see back to the opening well because they are looking directly into the light.

Equipment

If hazards warrant, all lighting equipment must be intrinsically safe. This means that the lighting equipment itself will not contribute to the ignition of a flammable gas, vapor, or dust. Rated equipment is tested to ensure compliance with standards. In the United States the most common testing organizations are Underwriters Laboratories and Factory Mutual. The U.S. Mine Safety and Health Administration standards are some of the most aggressive, and approval under these guidelines is required for items used in mines.

Communication

Communication is a critical part of a safe confined space operation. Communication must be maintained effectively. Several key communication routes must be maintained: entrant to attendant, attendant to entrant, attendant to entry supervisor, and attendant to emergency help.

Communication Methods

Any effective means may be used for communication. Voice communication alone is easiest if it is effective. Where distance or noise make this impossible, other methods must be used. In situations where the entrant and attendant can maintain visual contact, signals may be sufficient for communications. Other options include radios or hard-wired communications systems. Rope pull signals are generally not considered effective. There is too much opportunity for normal movement to be misinterpreted as signals, and if the entrant becomes incapacitated he or she may be unable to pull on the rope to signal a problem.

To call for emergency help, a radio, telephone, or cellular phone is the best option.

Communication Equipment

If flammable hazards exist, all communication equipment must be intrinsically safe.

It is essential that personnel be familiar with the operation of any communication equipment that may be used. Equipment should be tested prior to entry and immediately after entering the space. Failure of communication equipment is an emergency and should be handled accordingly by evacuating the space.

After Entry

Return Space to Service

Properly terminating an entry is a critical part of safe entry operation. All personnel, equipment, and supplies taken into the space must be accounted for when the entry is completed. Secure all openings. Make certain that gaskets and seals are properly placed and in serviceable condition.

Remove all isolation. It will be helpful to use the permit to remind the entry crew of the isolation used. Pipe isolation should be undone first. After all piping is reconnected, the lockout may be removed.

All aspects of the space should be returned to normal operating condition prior to releasing the space to the operating personnel. Failure to return a space to proper operating condition can result in problems during start-up of the process.

Equipment Used During Entry

Clean and Inspect. All equipment used during the entry should be thoroughly cleaned and inspected. Any items damaged during the operation should be removed from service.

Store. All equipment that is in acceptable condition should be returned to proper storage for future use.

Expendable Supplies. Items such as batteries, disposable coveralls, and respirator cartridges are supplies consumed during the entry. These expendable supplies should be replaced.

Reporting. The canceled permit must be turned in after the entry. Also, report on any problems that may have occurred.

Emergency Procedures

What is an Emergency?

An emergency exists when permit conditions are not maintained. It does not require that someone be injured or overcome. For example, if lighting or communications fail, that is an emergency. Another example is an alarm on an atmospheric monitor.

Plan for Emergencies

Planning for emergencies is one of the best defenses against them. An adequate plan requires that a thorough look be taken at the issues involved.

At a minimum the plan should include a briefing of the entire crew on how each individual is expected to respond if an emergency occurs. This briefing must emphasize that the attendant is not to enter the space. The attendant should never enter the space to attempt rescue. The attendant's role is to summon emergency help and use retrieval devices if available. Confined space accident statistics clearly indicate that rescue attempts often lead to additional fatalities.

Emergency Response Options

Escape

This is the most basic emergency response. Escape involves the entrant or the attendant noticing a problem, sounding the alarm, and calling for an evacuation. The entrants leave the space without assistance. During properly planned and conducted operations this should be the only option needed.

Retrieval

The next level of response is also the highest level that may be accomplished by the entry crew. In this case the entrant is unable to escape on his or her own. If a retrieval device is being used, the attendant will call for help first, then retrieve the entrant with the device.

Retrieval devices for horizontal spaces may be just a harness on the entrant and a line attached outside the space. On vertical entries a mechanical-advantage device is required. Two exceptions to this requirement are allowed under Occupational Safety and Health Administration regulations. If the use of the device increases overall risk it is not required. The device may also be eliminated if it will not contribute to a rescue.

If external retrieval will be used the entrant must wear a full body harness. A harness with a single ring in the center of the back and a ring at the top of each shoulder is the most effective. Under certain circumstances wristlets may also be acceptable, but these should be viewed as a last resort. Wristlets used in combination with a full body harness may be the best option in small opening spaces. The wristlets would allow the entrant's arms to be brought out first, allowing retrieval through a smaller opening.

Mechanical-advantage devices are manually operated and use ratios to increase the lifting capability. Only hand-powered devices should be used. A powered device increases the risk of injury to the entrant.

Mechanical-advantage devices are available in cable or rope varieties. Cable devices are usually mounted on a tripod or other fixed point such as davit arms. They are easy to operate and relatively low maintenance. Where flexibility in application is not required, a cable system is a good choice. On the other hand, rope systems offer a great deal more flexibility in application, but they require somewhat more maintenance and may be more difficult to operate.

Self-retracting life-lines may be used to provide fall protection when entry can be made with a ladder or other similar device. Two types are commonly available. One has no retrieval capability and should not be used for confined space work; the other is equipped with a retrieval capability and may be used in a confined space.

In spaces where the entrant must be lowered into the space, a cable winch may be used. Tripods and accessories are often used during confined space entry operations.

Davit arm systems also provide an excellent array of options for retrieval devices. A basic portable davit arm system may be used. A fixed davit arm installation and portable davit arm are an attractive option for frequently entered spaces.

Manhole and tank collars may also be used as davit arm mounts. Adjustable tank collar systems are excellent for tank entry and are available in inside or outside mountings.

Many other special application devices are available; for example, column extraction systems are designed for use on horizontal entries, forklift davit arm systems can be useful in hard to reach locations, and vehicle hitch mount davit arms may be used for field operations.

Rescue

Rescue involves a rescue crew actually entering the space to remove the victim. This can be a high-risk activity if not conducted properly. It must be left to trained and equipped rescue crews. Discussion of this level of rescue is beyond the scope of this book.

Confined Space Design and Reducing Risk

The best way to eliminate the potential for confined space injuries and fatalities is to eliminate the need to enter confined spaces. While this may not be a practical short-term solution, things need to move in this direction. The two primary approaches are either to eliminate the confined space, or eliminate the need to enter the space. The first option really comes down to designing the "confined" out of the space.

Redesigning the work area to eliminate the characteristics that make it a confined space is the ideal approach. This will not work for all, and arguably not even most confined spaces. However, it is possible for some, and the effort should be made to find and use these opportunities. For example, an area containing a tank in a below-grade vault was redesigned to eliminate the features that make it a confined space. Fixed stairs replaced ladder access. Installed ventilation to provide 20 air-changes per hour was built into the vault and fixed lighting was added. Access and potential atmospheric hazards were the only initial issues. The redesign eliminated both of these problems.

The second option, eliminating the need to enter the space, is sometimes easier to apply. Changes in the design of the space may be made that allow personnel to perform functions such as inspection, cleaning, and maintenance from outside the space. Examples include:

- access ports placed to allow scraping and cleaning to be accomplished from outside the space;
- pullout sensor panels that allow sensors to be removed from the space for maintenance or replacement; and
- pull-up pumps that can be raised from outside the space for maintenance or repair.

When complete elimination is not practical, or even possible, the risk may be minimized by reducing the number of entries, their duration, and the number of people needed in the space.

Grouping inspection and preventive maintenance activities within the space into one work project can reduce the number of entries. This is particularly easy to do with a computerized maintenance scheduling system. Start by identifying all the tasks that require entry and then schedule them to come up on the maintenance calendar at the same time.

Reducing the duration of entries is primarily a function of effective planning of the job. Make sure that all of the required items such as tools, parts, and supplies are available and ready before the work crew enters the space.

Reducing the number of people in the space can be a double-edged sword. Fewer people in the space means fewer people at risk, which is good. But reducing the number by too many can overtax the people doing the work. Strike an effective balance between having enough people to do the work safely and eliminating extra people in the space. In most cases keeping the number in the space at any one time as low as possible and rotating crews regularly is the best method. This keeps the number of people small so that if something goes wrong fewer people will need assistance. It also avoids overtaxing a few individuals because crew rotation allows individuals to refresh themselves between work periods. This technique can be particularly useful when heat stress is a concern. Difficult access spaces present the most challenges for this operational method. When access to the work area within the space is difficult, and getting into position to do the work requires a great deal of time and effort, crew rotation will not be an effective technique.

Design improvements can also be made to reduce the opportunity for problems within the space. An example of this approach is using larger and better-positioned openings. Safety features such as installed ventilation and improved isolation designs can also be built in.

Small vertical openings are usually the most challenging, followed closely by small horizontal openings. Vertical entries through a larger opening are usually more difficult than horizontal entries through a larger opening. Modified openings offer an opportunity to reduce the chance of problems during entry by improving the ability of personnel to enter and leave the space easily. However, changing the design of openings is not always possible with current technology. For example, pressure vessel requirements limit the practical size of some openings. Needs of the process may work against larger or better positioned openings in some process vessels. There is, however, much opportunity for improvement within the limits of performance requirements, and safety and health professionals should take advantage of every opportunity.

An excellent example of looking for opportunities to reduce risk by changing a design occurred in a water company. The company had numerous below-grade vaults in its distribution system. Each was equipped with an instrument to monitor chlorine levels. An employee checked each instrument weekly. Initially, this required entry into each below-grade vault. This job was done by one individual (no attendant). Radio contact was used to keep track of the employee's location and to monitor the time spent in the vault, but had a problem occurred there would have been a significant delay before help could be sent. This situation created a potential for many things to go wrong. The piping and instrumentation system was changed to relocate the instruments to grade level above the vaults. This simple design change completely eliminated the need for routine entries into these vaults.

These suggestions cannot be applied to all situations. There is no magic solution. Progress can be made, however, if opportunities are constantly sought. Strive to eliminate the confined spaces by design, eliminate the need to enter those areas that must remain confined spaces, and reduce the risk of those entries that must be made.

Summary

This chapter was designed to allow the reader to improve his or her ability to recognize and identify confined spaces. Detailed information is provided on hazards typically associated with confined space operations. The development of policies and procedures is discussed. Basic general operating procedures are identified and explained. Critical operating procedures such as atmospheric monitoring are covered. Using engineering and administrative controls to eliminate and/or reduce the risk of confined space entry is introduced.

Additional Sources

American Conference of Governmental Industrial Hygienists (ACGIH): *1996 Threshold Limit Values for Chemical Substances and Physical Agents Biological Exposure Indices.* Cincinnati, OH: ACGIH, 1996.

American National Standards Institute (ANSI): *American National Standard Z117.1-1995.* New York: ANSI, 1995.

Code of Federal Regulations Title 29, Part 1910.146."Permit-Required Confined Spaces," 1993.

Harris, M., L. Booher, and S. Carter: *Field Guidelines for Temporary Ventilation of Confined Spaces With an Emphasis on Hotwork.* Fairfax, VA: AIHA Press, 1996.

National Fire Protection Association (NFPA): *Recommended Practice for Classification of Class I Hazardous (Classified) Locations for Electrical Installations in Chemical Process Areas* (NFPA 497A). Boston, MA: NFPA, 1992.

National Fire Protection Association (NFPA): *Recommended Practice for Classification of Class II Hazardous (Classified) Locations for Electrical Installations in Chemical Process Areas* (NFPA 497B). Boston, MA: NFPA, 1991.

National Fire Protection Association (NFPA): *Manual for Classification of Gases, Vapors, and Dusts for Electrical Equipment in Hazardous (Classified) Locations* (NFPA 497M). Boston, MA: NFPA, 1991.

National Fire Protection Association (NFPA): *National Fire Code®*. Boston, MA: NFPA, 1996.

National Fire Protection Association (NFPA): *Fire Protection Handbook*, 18th ed. Boston, MA: NFPA, 1996.

National Institute for Occupational Safety and Health (NIOSH): *Worker Deaths in Confined Spaces.* Cincinnati, OH: NIOSH, 1994.

National Institute for Occupational Safety and Health (NIOSH): "NIOSH Alert: Request for Assistance in Preventing Occupational Fatalities in Confined Spaces." Cincinnati, OH: NIOSH, 1986.

National Institute for Occupational Safety and Health (NIOSH): "NIOSH Alert: Request for Assistance in Preventing Entrapment and Suffocation Carried by the Unstable Surfaces of Stored Grain and Other Materials." Cincinnati, OH: NIOSH, 1988.

National Institute for Occupational Safety and Health (NIOSH): *NIOSH Pocket Guide to Chemical Hazards.* Cincinnati, OH: NIOSH, 1994.

Rose, V.E. and T.W. Krug: Confined Space Entry: An AIHA Protocol Guide. Fairfax, VA: AIHA Press, 1995.

Rekus, J.: *Complete Confined Spaces Handbook.* Ann Arbor, MI: National Safety Council, Lewis Publishers, 1994.

Schroll, R.C.: *Confined Space Safety Manual.* East Earl, PA: FIRECON, 1994.

Industrial hygienists must find creative and innovative ways to use new computer technology to enhance the health and safety of the work force.

Outcome Competencies

After completing this chapter, the user should be able to:

1. Define underlined terms used in this chapter.
2. Explain the components of a personal computer.
3. Explain the value of word processor, database management, spreadsheet, and graphics software.
4. Summarize the strengths and weakness of relational databases.
5. Describe the issues involving computer-based data as they apply to occupational hygiene.
6. Describe the role of computers in training programs.
7. Describe the process for gaining access to the Internet, listservs, the World Wide Web, and newsgroups.

Key Terms

browsers • electronic mail • flat-file database • hardware • interactive multimedia • Internet • Internet service provider • local area network • mainframe computer • multimedia • operating system • personal computers • relational databases • search engines • software • training • viruses • wide area network • workstations • World Wide Web

Prerequisite Knowledge

Prior to beginning this chapter, the user should review the following chapters:

Chapter Number	Chapter Topic
3	Legal Aspects of the Occupational Environment
6	Principles of Evaluating Worker Exposure
30	Worker Education and Training
37	Program Management
38	Surveys and Audits
44	Report Writing
50	Anticipating and Evaluating Trends Infuencing Occupational Hygiene

Key Topics

I. Advantages of Computers
II. Disadvantages of Computers
III. Computers and How They Work
IV. Distribution of Software Uses in Safety and Health
V. Choosing a Computer System
VI. Operating Systems
VII. Networks
VIII. Database Management
IX. Computers and Data Security
  A. Backing Up the Data
X. Training and Multimedia
XI. Document Preparation
XII. Internet and the World Wide Web
  A. Accessing the World Wide Web
  B. Finding Information on the Internet
  C. FTP Use on the Internet
  D. Electronic Mail

43

Use of Computers in Occupational Hygiene

James H. Stewart, Ph.D., CIH

Introduction

In 1976 Apple™ Computer introduced the first personal computers. Personal computers have grown in quantity and functionality to the point that they are now a necessary part of business, government, and education. Computers are involved in virtually all aspects of modern life, such as working, studying, research, banking, and recreation. The challenge to occupational hygienists is to find creative and innovative ways to use this new computer technology to enhance the health and safety of the work force. Computer technology in the context of this chapter involves both hardware and software (see next section for definitions of hardware and software).

The first challenge occupational hygienists face is keeping up with changes in available technology. Each month there are announcements of new and exciting computer tools that may have application in solving one or more problems faced by occupational hygienists. The second challenge is investigating available technologies to find the appropriate computer tool for a particular challenge. Finding the right combination of hardware, software, costs, and benefits, if done correctly, can produce an elegant low-cost solution to real-world occupational hygiene problems. This chapter first reviews the advantages computers bring to occupational hygiene and how computers work and then proceeds to identify areas where computers are currently adding value to occupational hygiene and the workers it supports.

In this chapter the following will be discussed: the terminology of computers and how they work; operating systems and networks; strategies for selecting a computer system; the challenges of confidential data; the Internet and how to use it; and finally, a variety of ways that computers add value in occupational hygiene. At the conclusion of this chapter the reader will have a familiarity with computers and a knowledge of what makes them useful tools in occupational hygiene.

Advantages of Computers

Personal computers are used to collect and analyze data, prepare reports, develop graphical displays of data, and store vast amounts of information. Distinct advantages of computers in occupational hygiene include the following.[1]

Elimination of Duplication. Instead of each department maintaining records, one set of records can be maintained centrally. This requires that one central database be established.

Improved Communications. Networks, including the Internet, have made communications within an organization, between organizations, and with professional societies and government agencies fast and reliable.

Standardization of Data. Computers can be used to require data to be input in a specific form or range. By filtering the data at the time of data entry, quality, and therefore usefulness of the collected data, are improved. Accuracy of data is also enhanced because invalid entries (e.g., invalid department code, wrong date, etc.) can be reduced or eliminated.

Improved Analytical Capabilities. Computers can compile and analyze data

The challenge to occupational hygienists is to find creative and innovative ways to use this new computer technology to enhance the health and safety of the work force.

from many different sources simultaneously. The analysis can be tailored to meet changing needs, and the results can be obtained quickly.

Cost Savings. Although cost savings may be hard to document, they usually take the form of increased productivity, improved analysis, selection of more effective controls, and as a result, lowered exposures, more timely reporting, better tracking of results, and improvements in retrieval of compliance data.

Incorporating computers into the occupational hygiene function can improve the quality of service provided through improved data collection and analysis and more timely reporting.

Disadvantages of Computers

As there are clear advantages to using computers there are also clear disadvantages:

- Cost of purchase of original system and software;
- Cost associated with initial design and setup;
- Cost associated with start-up training;
- Regular training and assessment of the state of the art are essential because of rapidly changing technology;
- Cost of replacing computer systems (and software) because of rapid obsolescence;
- Security of data; and

Table 43.1 — Description of Selected Types of Hardware and Their Functions

Hardware Name	Description of Function
CPU (central processing unit)	Performs calculations, controls overall operation of the computer. May be thought of as the brain of the computer. Processing speeds of 300–500 MHZ will be available by the end of 1997, with much greater speeds in the future. The acronym MHZ means that the CPU can perform 300 million computer operations per second.
Hard disk drive	Physical device that provides long-term storage by writing the information on dish-shaped magnetic media. Writing is accomplished by changing the alignment of the magnetic field of the surface of the hard disk, producing either a zero or a one. (Computers only understand zeros and ones.) The specific series of zeros and ones represents the file contents being stored. When a file is retrieved from the hard disk the series of zeros and ones is read and the file reconstructed. Hard disks can store very large amounts of data (1–2 gigabyte drives are considered average in 1997).
Floppy disk drive	Physical device that reads a thin removable disk encased in plastic used for storing limited amounts of data. Disks currently store 2.88, 1.44, or 1.2 megabytes (MB). The term "floppy" refers to a disk's flexibility.
CD-ROM	CD-ROMs (Compact Disks-Read Only Memory) are high-capacity storage devices for any type of data (e.g., music, databases, text files, graphics). They can be written to only once (hence the term "read only memory"). One CD ROM can hold over 650 MB of data, the equivalent of 451 high-density (3.5-inch) floppy disks. The data are stored by using a laser to mark the surface of a disk. A pit in the surface is read as a zero, a peak is read as a one. To make a CD ROM, the copy is pressed like a record. They can be read many times. CD ROMs are especially useful for displaying training or reference materials where large files are needed (movies, graphics, or even the entire *Federal Register*).
RAM (random access memory)	Memory chips store data for use by the CPU. It is temporary storage and access is very fast (50–60 ns). Processing by the CPU is performed using RAM. Other types of RAM are available, and future advances in memory architecture should provide significant gains in system performance.
Motherboard	A printed circuit board that provides the connections between the hardware devices (CPU, hard disk, floppy disk, video card, etc.).
Bus (PCI, VLB, EISA, ISA)	Peripheral computer interconnect (PCI), VESA local bus (VLB), extended industry standard architecture (EISA), industry standard architecture (ISA) are different electronic methods of transferring information to and from the CPU and various hardware devices. The transfer of information is said to occur on a bus. The faster the bus can operate, the faster information can be transferred. The type of hardware must be matched to the type of bus (slot); for example, a PCI video card must be installed in a PCI slot.
Modem	Modems are used to connect the personal computer to other computers using telephone lines. In the future, coaxial cable television lines may be used. Modems convert digital information from the computer into an analog signal that is transmitted through the telephone lines. The receiving modem then converts the analog signal into digital data the computer can use.

Table 43.1 (continued) — Description of Selected Types of Hardware and Their Functions

Hardware Name	Description of Function
Video card	The video card processes information for presentation on a monitor (VDT).
Tape cassette	Magnetic tape that stores data in a continuous stream. Generally used to maintain a backup copy of information or to transfer large amounts of information.
Removable hard drive	Hybrid of a fixed disk and floppy disk. Generally not flexible, with capacities that are orders of magnitude greater than a floppy disk.
PCMCIA	Personal Computer Memory Card International Association is an industry standard for connecting memory cards and peripheral devices to a portable computer.
SCSI	Small computer systems interface (SCSI) is a type of interface for computers that allows chaining of up to seven devices per SCSI channel. The devices can be hard drives, scanners, or any SCSI device.
Bytes and bits	A group of eight bits comprise one byte. One byte is approximately equal to one character (number, letter, etc.) A kilobyte is 1024 bytes not 1000 bytes. A megabyte is 1,048,576 bytes. RAM and hard disk space are both commonly measured in MB. One billion bytes is called a gigabyte, 1024 megabytes.
Sound card	A type of circuit board that enhances the sound output of the computer. Without a sound card the computer can only make very basic sounds such as beeps and buzzing sounds. Some basic voice simulations can be made, but the sound card makes the voices near original quality. Sound cards can play music, voices, and computer generated sounds.
Viruses (computer)	A program that duplicates itself and spreads from one computer system to another, usually by attaching itself to files transferred between computers. Viruses are often triggered by a date or event, such as writing to the disk. Some viruses delete files or damage hard drives, while others merely print messages on the screen. Antivirus programs are sold commercially and should be used on all computers.

- Dependence on the computer to complete a task.

These are not reasons to avoid computers; however, these problems must be managed.

Computers and How They Work

Hardware is the electronic and physical apparatus necessary to run computer programs (software). The basic personal computer consists of a central processing unit (CPU), random access memory (RAM), a hard disk drive, a floppy disk drive, CD-ROM (compact disk-read only memory), a video card, and a monitor (often called a VDT [video display terminal] or VDU [video display unit]). See Table 43.1 for definitions of terms. If additional functionality is needed in a computer, very often it can be added simply by buying some hardware and attaching it to the basic computer system by inserting it into one of the slots. To improve the overall performance of the standard system, hardware can be exchanged for higher performance parts, such as more RAM, a higher speed CPU, or a faster video card or hard drive. Many possibilities exist to improve performance. Hardware and software suppliers will be happy to suggest performance improvement techniques for your specific computer.

Distribution of Software Uses in Safety and Health

Based on a recent survey,[2] the most common function of software purchased by health and safety professionals dealt with compliance issues. The distribution of software is shown in Table 43.2. Forty-one percent of the products were designated for compliance and regulation access. These data indicate that regulations are stimulating the production of software solutions. The software in turn helps occupational hygienists and others comply with the regulations. The regulatory emphasis on record retention and retrieval has probably made the computer the tool of choice in compliance management.

This view of the importance of software directed at compliance was further supported by Sigler and Lurie.[3] They suggest that databases will expand and interact with other databases; that new data

Based on a recent survey,[2] the most common function of software purchased by health and safety professionals dealt with compliance issues.

Table 43.2 — Estimated Distribution of Software Purchased by Health and Safety Professionals in 1995

Software Type	% Distribution
Compliance	23.3
Regulation access	17.5
Accident tracking	12.5
MSDS	10.8
Employee health	9.5
Accident protection	5.5
Engineering analysis	4.3
Injury/illness analysis	2.2
Personal protective equipment	1.8
Other	7.1

Source: Reference 2.

sources and analyses will be conducted, introducing a "new era in data collection." This new era in environmental health and safety will include data sharing and multi-site remote sensing of exposures/emissions. Specific uses of software include program audits, data analysis, data retention, hazard tracking, database access, Internet access, employee training, professional training, and program management.

Choosing a Computer System

There are five major steps to the successful development and implementation of a computer system:(1)

1. Understand the needs of users of the computer system. In this step a detailed analysis of the uses of the system is conducted. Identify any economic and physical constraints, problems the computer will solve, the data the computer will manage, potential users, and how it will interface with other systems and equipment.
2. Identify and evaluate software and hardware that meet the needs identified in Step 1. Compare off-the-shelf comprehensive packages and custom-built software. Peer contacts, industrial associations, software vendors, and so forth should be contacted to determine the available solutions.
3. Purchase or build the system. In some cases this will mean buying and customizing an off-the-shelf system; otherwise, the system may have to be constructed.
4. Install the hardware and software, train the users, and use the system. Installing the hardware and software may take outside technical support. Training users must be accomplished in a timely manner.
5. Evaluate how the system is performing and how it is meeting the needs identified in Step 1.

After Step 5 is accomplished, a feedback loop to Step 1 should be made. As new needs and weaknesses in the system are identified, they are fed back in at Step 1. The process of continuous improvement proceeds.

The most popular operating systems are those that run on Intel™ or compatible processors, accounting for approximately 94% of environmental health and safety computer users.(2)

Operating Systems

For software to work on a computer, commands must be interpreted into a form that the computer can understand and act on. In fact, before the computer system can perform any requested functions, the operating system must be loaded. The operating system is a piece of software that forms the interface between the user's program and the computer hardware. When a program issues a command to the computer to copy a file, for instance, the operating system first receives the command and then tells the CPU, disk drive, RAM, video display adapter, and so forth what to do and in what sequence. Operating systems tend to be specific for each type of CPU, although there are crossovers. The most popular operating systems are those that run on Intel™ or compatible processors, accounting for approximately 94% of environmental health and safety computer users.(2) Apple holds an estimated 4% of the market, and the remainder is composed of a variety of mainframe operating systems.(2)

A mainframe computer is a very large, powerful computer that can support many simultaneous users. They are the largest computers made today, followed by minicomputers, and then by personal computers. Due to their size and processing capacity, mainframe computers generate large amounts of heat. For this reason they are stored in large temperature-controlled rooms. Mainframe computers have extensive data storage and processing capabilities and are used primarily by large companies, governments, and universities. The operating systems of mainframes are generally different from those of personal computers, so most software designed for personal computers will not run on mainframes. However, mainframes and personal computers can communicate through telephone lines, networks, or the Internet using standard protocols. Once connected, data can be transferred, mail can be sent, training schedules posted, and so forth.

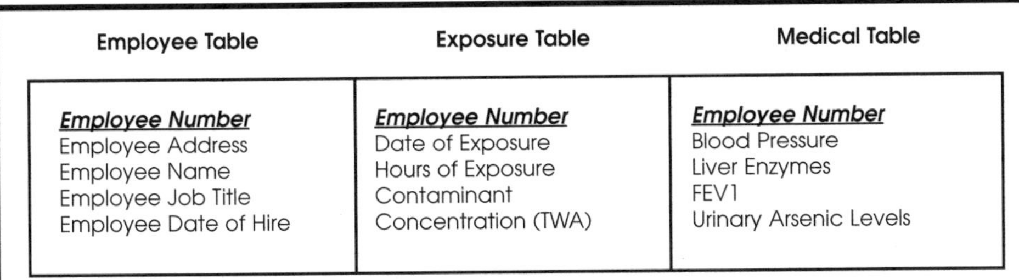

Employee Table
Exposure Table
Medical Table
Employee Number
Employee Address
Employee Name
Employee Job Title
Employee Date of Hire
Employee Number
Date of Exposure
Hours of Exposure
Contaminant
Concentration (TWA)
Employee Number
Blood Pressure
Liver Enzymes
FEV1
Urinary Arsenic Levels

Figure 43.1 — Example of relational database connecting exposure, medical, and employee records using employee number. (Adapted from **Chawes, D.:** Use Relational Databases to Remove Common Errors. *The Synergist*, April 15, 1997.)

<u>**_Employee Number_**</u>
Employee Address
Employee Name
Employee Job Title
Employee Date of Hire
Date of Exposure
Hours of Exposure
Contaminant
Concentration (TWA)

Figure 43.2 — Example of a flat file database for storing exposure information for an employee.

Networks

A network is a group of computers (two or more) connected by cables for the purpose of sharing files, hardware, or software. When a computer is networked, it has access to other computers' hard drives, printers, and so forth as if they were directly connected.

There are two types of networks, the <u>local area network</u> (LAN) and the <u>wide area network</u> (WAN). A local area network usually serves a limited area such as a building, department, or office. The LAN has a central computer called a server that runs the network software and controls access to the files and peripherals, such as printers. The computers linked to the server are called <u>workstations</u>. Each workstation must have a network interface card installed in one of its expansion slots before that computer can have access to the network. The server can be a stand-alone computer whose only function is to run the network, or it can be a computer that is used by someone but is also running the network.

Two or more LANs linked together into a larger network form a WAN. A WAN can be used to connect different divisions within a company, different plants in different cities, and so forth. The links between LANs to form a WAN can be made with modems, satellites, or telephone lines.

To run a network of either type, specific software needs to be loaded and run. This networking software is complex and usually requires a system administrator to supervise the software and hardware forming the network. The system administrator also issues passwords and controls network security. This job function is particularly important when confidential data are placed on a networked computer system (see Computers and Data Security).

For the occupational hygienist, networks provide access to company databases in different locations, electronic mail, even video conferencing.

Database Management

Databases are collections of information. Usually the data in the databases are related in some manner. In occupational hygiene the data might consist of exposure level, method used to take the sample, time of sample, length of sample, type of sample, and so forth. Each piece of data associated with that sample is called a field. All of the fields together comprise a record. If one were to construct a table, the fields would be columns and the rows would be the records. <u>Relational databases</u> relate different types of data, such as sample information, health outcomes, and workers' compensation costs. The key advantage of a relational database over a <u>flat-file database</u> is that data are not duplicated.

For example (in Figure 43.1), to set up a relational database to link worker chemical exposures with health outcomes, one could use a company-issued employee number to relate the data in the chemical exposure file to the medical history files. The result would be a database where analyses could be conducted on data from all three databases. The data from the three different databases are related to the employee through the key field, the employee number. If available, training records, biomonitoring results, and safety performance ratings could also be grouped together in a relational database. The result can be an extremely powerful tool for retrieval and analysis of many types of data. Contrast this to the flat file structure, where for every exposure measurement all the data in the employee database would be stored for each record. It is clear that the database structure in Figure 43.2 will be repetitious, needlessly storing employee data over and over again for each exposure measurement made.

To increase efficiency and improve the quality of the data, a database must be designed with the ultimate use in mind. A list of necessary data must be developed with the end users (the people who will input, retrieve, and analyze the data). The necessary data will include:

1. Data explicitly requested by the end users, such as percentage exposures above the threshold limit value (TLV®), one-half TLV, and percentage reduction since last evaluated;
2. Data needed for reports, such as employee name, badge number, health outcome, work location, sound pressure levels, and direct-reading instrument outputs;
3. Data needed for calculations, such as air sample volume, time sampled, air temperature, pressure, and gravimetric weight;
4. Data to enable relations to other databases, such as badge number, social security number, and plant identifier; and
5. Data to sort and select records, such as work zone, plant name, age, and race.

Constructing a relational database can be a very complex task. Time spent in the beginning designing the system, data, and reports and identifying end user needs will save time and resources in the long term.

Reports are an important part of occupational hygiene. Databases of exposure measurements, job titles, controls, and so forth make generating reports more practical and efficient. When a source of data is very large, such as the Hazardous Substances Databank (HSDB) published by the National Library of Medicine or the National Toxicology Program's database of chemicals, it would simply not be practical to review the data visually and extract all pertinent data for a report. These databases contain information on thousands of chemical agents (4500+ for the HSDB alone), their toxic effects, flammability, physical properties, toxicological study test results, etc. To select the information needed for a specific report would be difficult, if not impossible, to perform in a timely manner without a computer. A computer can be used to extract the specific information needed and then transfer the information for inclusion in the report.

A number of information sources are now available on CD-ROM that previously were available on paper only. For example, the Occupational Safety and Health Administration (OSHA) standards, letters of interpretation, U.S. Environmental Protection Agency (EPA) standards, HSDB, Medline, Toxline, and so forth can all be accessed directly from a CD-ROM or by a modem with a personal computer.

Day-to-day database management activities such as material safety data sheet (MSDS) management, emission data management, medical records management, OSHA record keeping, exposure analysis and measurement, and training records management have been made more efficient through the use of the computer. Commercial products are available to handle all the listed functions. This saves paper and improves compliance reporting and data analysis.

In exposure analysis the computer can be used to analyze trends in the data, select subsets of data for further analysis, identify high- and low-risk groups, etc. Questions like "How many samples taken in the past year at the carbon arc gouging operations in Alabama and Tennessee have indicated exposure to copper fume above the TLV?" are easy to answer using a computer with database management software. This question, and questions like it, are very hard to answer without a computer.

Medical records storage, collection, and retrieval are activities that are easily performed by a computer. Commercial software packages are available to collect and store exposure, medical, and job records. The packages can be customized to a particular site. Analyses from simple OSHA incidence rate determinations to complex epidemiological studies can be performed. The extent of data collection can include laboratory tests, medical exams, office visits, costs, exposure levels, job descriptions, and even the notes from the examination. As always, when dealing in confidential data extreme care must be taken to protect the privacy of individuals (discussed in the section on Backing-Up Computers and Data Security, later in this chapter).

Workers' compensation is a major expense in industry today. Case management is usually a duty of the safety practitioner, nurse, or an insurance department employee. Occupational hygienists are most often brought into cases involving toxic chemicals, ergonomics, radiation exposure, and so forth. These types of cases are often complex, and access to data is necessary for the occupational hygienist to make a recommendation to the case manager. Computers provide a way of collecting data on the particular case or similar cases, performing hazard analyses, conducting literature reviews, and distributing the occupational hygienist's report. After the initial case work is completed, the findings may be used to modify working conditions of similar workers. Identifying these similar workers may involve the use of one or more computers to sort through the various jobs, job titles, job tasks, exposures, and chemicals used.

Computers and Data Security

Information collected and used by occupational hygienists is valuable and often confidential. Securing both the computer and the data it contains, therefore, is of vital importance. Personal computer systems may cost several thousand dollars and are usually of a size that can be carried away. Some personal computers are meant to be portable. As part of the system selection and design, the security of the computer must be addressed. The computer can be bolted down, chained, or locked in a secure area. The data the system contains is even more valuable and more difficult to replace. The data often use employee identifiers and are almost always confidential. Knowing who has access to confidential information and how the information is protected are essential to the system design. These controls must be implemented before the data are placed on the system. There are many ways to limit access to confidential information, such as 1) use of tightly controlled and regularly changed passwords for access, with passwords for employees who leave the company invalidated; 2) not storing confidential data on a networked computer; 3) using specific procedures for confidential data and training employees in such use; and 4) keeping and reviewing logs of access to confidential data.

The issue of confidentiality is fundamental to the continued use of the data. Employees, managers, and other interested stakeholders need to understand how the data will be used and the actions taken to prevent unauthorized access. Informed consent becomes an issue if the data are taken as part of a study. Legal advice should be sought concerning the use and protection of confidential data.

Viruses present a rare but serious threat to the computer and the data it contains. Viruses move between computers by attaching themselves to files. These files could be on a disk provided by a co-worker, downloaded from the Internet, or obtained over the company network. Once the virus gets in the system, it can damage files, format the hard drive, print messages on the screen, etc. The list of effects is long and expanding as each new virus is produced and released by a programmer. Antivirus programs are available commercially. To evaluate which program will work best, the following questions should be considered:

- What types of viruses are detected?
- How many different viruses will it detect?
- Are regular updates to the virus database provided to catch new viruses as they are discovered? On what schedule?
- Can the antivirus program inoculate computers to prevent infection by known viruses?
- Can the infection be removed once discovered?
- Will the antivirus program interfere with other software currently being used?

Even if appropriate precautions are taken a virus may still be found. It is also possible a virus may be found whose disinfection may cause loss of data. Therefore, viruses are just one more reason to back up the data. If a virus is found, the drive can be reformatted and the data restored from the backup (see the following section on Backing Up the Data). Viruses present a real threat but one that can be managed with the appropriate tools.

Backing Up the Data

A more basic concern of data security is backing up the data. As the term implies, backing up the data means that a copy is made of the data and placed in a safe place. A safe place must be a remote, secure, dry, low humidity location with constant room temperature. It is important to keep a copy of critical data off-site in case of hard drive crash, virus infection, fire, flood, or vandalism.

There are software and hardware tools available to automate the making of a back-up copy of the data. Many commercial software programs are designed for this purpose. Hardware solutions include tape, writable CD-ROM, removable hard disks, magneto-optical disks, and external hard disks. In the near future the Internet may provide the means to automatically back up data files to a remote location. When backing up remotely (i.e., over the Internet) steps must be taken to ensure the security and maintenance of confidentiality of the data.

Training and Multimedia

Multimedia is software that incorporates sound, video, text, and sometimes animation. Often the multimedia applications are interactive, that is, it allows or requires the user to press keys or "click" the mouse. This is especially true for training applications. The keys the user presses control the actions of the program. One area where multimedia software is expanding rapidly is employee training. Employees need to be trained in a wide variety of tasks associated with the occupational hygiene aspects of their jobs, including chemical right-to-know; confined space entry; respiratory protection; personal protective equipment; safe work practices; laboratory safety; ergonomics; hearing conservation; or almost any occupational hygiene topic.

Training can be made more effective if the student's attention span is increased and if the student is an active participant in the training.[4] Interactive multimedia can accomplish both of these goals.

Imagine the following worker training scenario. A worker starts a new job on Tuesday. The job she will perform involves the use of hydrochloric acid. Classroom training on the use of acids is available on Mondays only. She needs to be trained in the procedures for working safely with this material, the emergency procedures, first aid, etc. She is sent to the computer training center and immediately begins training; no need to wait for a scheduled training class. She proceeds at her own pace and is finished with training in an hour. To enhance the training she reviewed a movie of the operation with actual sounds integrated into the multimedia training software; correct and incorrect procedures were presented; emergency alarms were demonstrated; and an animation was used to illustrate exposure to hydrochloric acid. After each lesson a question and answer session was conducted with immediate feedback for correct and incorrect answers. The scores for all the lessons were stored in a database. When the worker demonstrated an understanding of the material, a certificate was printed and a copy sent to the supervisor electronically.

The worker in the example received chemical training performed by a computer; therefore, any similar computer or terminal could be used as a training site; no human instructor was required. Records of training times, scores, and so forth were maintained automatically, relieving a significant regulatory burden for the occupational hygienist or training manager.

The learning characteristics demonstrated in this example, such as immediate feedback, audio and visual cues, direct relationship to work, and required interactions, can improve learning and retention of the student.[4–7] Studies have shown that students trained by reading retain only 20% of what they read. If an instructor is added with video the retention rate rises to 40%. Multimedia training includes visual, oral, and task-oriented training (e.g, the actual procedures for handling hydrochloric acid in the example above) and can increase retention rates by 40%.[7,8]

Training time has been shown to be reduced by 40 to 80% when computer-based training was compared with traditional classroom training.[8,9] The net effect of increased retention and less time off the line represent improvements in productivity, or in other words, more output for the same or less input.

Multimedia computer-based training is a developing trend in health and safety. The inherent efficiencies of the computer-based training delivery system will ensure its continued growth. The business and educational arguments are compelling.

- Training can be performed at the employee's discretion;
- The computer is not subjective in its evaluation;
- The computer allows review as many times as it takes to understand the material (in a classroom there is no play-back of the instructor's statements);
- Records are kept with minimal effort;
- Training content can be modified easily to include site-specific information and procedures;
- Production workers need not be removed en masse from the line (they can train one at a time when the pace is slower);
- Management and follow-up of employees who were scheduled for training but who did not attend is minimized because training is available in many locations and at all times;
- New hires can be trained immediately and do not need to wait for a classroom training session with an instructor.

The net effect is that the worker retains more knowledge and is back on the production line more quickly.

There are several restrictions to computer-based training. For example, the example above presumes that a computer is available for training at the site, which may not always be the case. An added restriction is the inability of the student to ask verbal questions. A designated person knowledgeable in the subject matter should be made available to the student to provide the opportunity for verbal questions and answers (which may be a requirement under some regulations).

When a training computer is not available, any computer can still assist immensely in training. For example, a presentation can be prepared on the computer using graphics software that can show animations, colors, pictures, and so forth. The presentation can then be shown on an overhead projector or viewed directly on the screen by a group of employees. In this role the computer is a training support tool.

Document Preparation

Reports are part of the services provided by occupational hygienists. When a typewriter is used to prepare a report and an error is made, the entire page may need to be retyped. When using a computer and word-processing software, the error can be corrected and the document reprinted immediately. Not only can errors be corrected, but suggestions on improved grammar and alternative word selection can also be obtained as the report is being prepared.

The best word processors have spell checkers, grammar checkers, and a thesaurus. They also allow pasting of graphic images, tables, and equations into documents. Import and export capability allows occupational hygienists using different types of software to share files or import graphic images, tables, equations, or text.

It is generally accepted that well-written documents have more impact and allow the reader to understand the issues more quickly and thoroughly. The reader of the material may be a worker, government regulator, plant manager, human resource manager, union steward, or lawyer, all significant customers of the occupational hygienist. The more effective occupational hygienists are in conveying information and analyses, the faster changes can be made in the workplace to improve worker health and safety.

Commercially available products now include safety and occupational hygiene templates for respirator or hearing conservation program management, chemical hazard training, and many others that allow the occupational hygienist to fill in the blanks in the template and produce an almost complete written program. The document can then be further modified and customized to meet the needs of the site. The significant advantage is the time saved in preparing the draft document.

Internet and the World Wide Web

The Internet is a series of interconnected networks of computers. Funded by the Department of Defense Advanced Research Projects Agency in the 1960s, the project sought to create a network of computers that could continue to operate in the event of nuclear war. Further development occurred in the 1970s and 1980s. During this 15–20 year period universities and government agencies signed up in large numbers. In the 1980s the National Science Foundation (NSF) took over administration of the Internet through its contractors. Commercialization began in the early 1990s. In 1997 it was estimated that 13.6 million computers around the world were connected to the Internet.[10] This number is expected to escalate to 122 million by the end of the year 2000. These millions of people from around the world are scattered across an estimated 15,000 networks. The Internet is currently one of the occupational hygienist's greatest information and communication assets. In the future it will be an integral and inextricable part of occupational hygiene.

A word of caution: anyone can place information on the Internet. It is up to the user to assess the quality of the data, the source, when it was last updated, and so forth before using it.

Accessing the World Wide Web

To access the Internet a connection must be made from a computer to the Internet itself. For an individual user it would be very expensive to install and pay the monthly fee for a high throughput line necessary for connection to the Internet. There are companies that provide this connection for a fee. A company providing and managing such connections is called an Internet service provider (ISP). These companies exist because dedicated high-capacity telephone lines are very expensive; managing computer accesses and data traffic is complex; system redundancy is expensive; and security is needed to limit unauthorized access to individual computers (provided by ISPs through the use of software barriers called firewalls).

An individual user must first contact an ISP and obtain an Internet account to access the Internet. Names of ISPs are available in local newspapers, computer magazines, and the local bookstore. After signing an agreement with the ISP, a name and password will be issued specifically to that individual. Software for accessing the Internet can be obtained from the ISP or through software vendors. Directions for installing the software are specific to that software and must be followed exactly.

Large commercial computer networks provide a convenient method of accessing the Internet, connecting to databases, using electronic mail, or joining forums with users who have similar interests. Once the Internet software and modem are installed, the user is ready to begin "surfing" the Internet by dialing the telephone number provided by the ISP and using the name and password assigned.

For users new to the Internet, the *Internet User's Guide for Safety and Health Professionals*(11) provides a good introduction. It also includes 79 pages of site listings of potential interest to safety and health professionals. Getting an Internet account, logging on to an Internet server, finding sites, etc., are all covered in sufficient detail to help the novice computer user become a novice Internet surfer.

Finding Information on the Internet

Finding information on the Internet is sometimes difficult but is greatly assisted by search engines. Because there is a rapid rate of change on the Internet, any listing of search engines will probably be incomplete as soon as it's published. An easy way to obtain a current listing of search engines is through your Internet service provider or by using a combined search capability such as that supplied by various commercial software programs. Search engines are found by using a mouse to click on "Search." The following is a sample of several popular search engines and their current addresses on the World Wide Web (WWW):

- Alta Vista (www.altavista.digital.com)
- Yahoo (www.yahoo.com)
- World Wide Web Worm (www.cs.colorado.edu/home/mcbryan/wwww.html)
- Web Crawler (Webcrawler.cs.washington.edu/webcrawler/webcrawler.html)
- Lycos Search Engine (www.lycos.com)
- Infoseek (www.infoseek.com)

These search engines have databases of key words that correspond to locations on the Internet where information is located. Most searching is accomplished on the WWW using graphical web browsers. Previously, searches were accomplished using nongraphical browsers. The focus of this discussion will be on graphical browsers with background information provided on nongraphical browsers.

Search engines maintain a database of key words used by sites on the WWW. When a search is made, the address of each site with some information related to the key word is returned, sometimes with a short description. For example, in preparation for this manuscript the author used Digital Equipment Corp.'s Alta Vista search engine to find web sites that related to a variety of topics. A web site is a specific address on the WWW. A web page is a graphical hypertext document that provides information about the site, further access to the site, or links to other sites on the Internet. A web page will be displayed if a user connects to that web site. When a search of the WWW is conducted using one of the graphical search engines, the search engine reports back how many hits it had, that is, how many web sites were found that used a particular key word. An example of the results of three searches conducted in March 1997 are listed in Table 43.3. In this example the search engine searched 476,000 Internet servers (computers) and 31 million pages. On an average weekday in March 1997, this particular search engine was accessed over 29 million times.

When viewing web pages a graphical browser is used. A common language of web browsers is HTML (hypertext markup language). Hypertext means that words can be linked to other words or actions. On a typical web page you may find different colors, animation, tables, frames, or data. When the cursor moves over an area that contains a hyperlink (a link to another page, a database, etc.) the cursor will change, usually, from an arrow to the shape of a hand. This change indicates an active area on the screen. When the mouse is clicked, the link is activated and an action initiated, for example, a database is accessed, a definition displayed, a movie run, whatever the web page owner programmed to happen. It is not necessary for the one accessing the web page to be knowledgeable about the programming intricacies.

Table 43.3 — Number of Hits for Selected Words Using an Internet Search Engine, March 1997

Search Words	Hits
Industrial hygiene	100,000
Safety	200,000
Occupational medicine	200,000

The protocol or standard method of navigating the WWW is called "http" (hypertext transfer protocol). All graphical web browsers can use this protocol.

Analogous to the address of a building, each Internet site has an address called a universal resources locator, or URL. A URL is the address of a computer on the Internet. An example is the URL for Harvard University, which is http://www.harvard.edu.

The first part of the address denotes the protocol that is being used at that site. The use of "http://" shows that the site is using hypertext transfer protocol. Other sites may be accessible using file transfer protocol (FTP) or gophers (discussed later in this chapter). Different letters are used to begin these address, such as "ftp://" for FTP sites. In the Harvard example the second part of the address is "www" denoting the World Wide Web and is followed by the name of the site. The name may have several words or letters connected with periods. All parts of the name are essential and must be used to connect to the site. The last part of the URL is a two- or three-letter extension indicating what type of site it is: edu denotes an educational organization, com denotes a commercial site, and org denotes organizations. Other extensions are possible, for example, fr for sites in France and jp for sites in Japan. Whatever the extension is, it must be included to connect to the site. A textual translation of Harvard's URL would be "use hypertext transfer protocol on the world wide web to access the Harvard University web page."

One of the challenges of the Internet is finding just the sites you need. Browsing through 40,000 sites is not feasible. More and more tools are becoming available to help select information by key word, title etc. Below is a sampling of some of the readily available information on the WWW that can be accessed by using the search engines. In addition, in Tables 43.4 and 43.5 a list of useful occupational hygiene web sites is provided.

- MSDSs — hundreds of thousands available;
- Toxicological profiles — available from EPA, Agency for Toxic Substances and Disease Registry (ATSDR), and the National Toxicology Program;
- Literature searches — National Library of Medicine, Medline, Toxline, Chemline, HSDB, etc.;
- Exposure information — National Cancer Institute, Surveillance Exposure and Epidemiology Reports Data;
- OSHA compliance reports, injury and illness rates by industry, regulatory agenda, OSHA Computerized Information Service;
- Training courses — ATSDR physician training for lead and radon; notices for upcoming courses and conferences from offering organizations such as the American Industrial Hygiene Association and the American Conference of Governmental Industrial Hygienists;
- Chat sessions — real-time conferencing by typing messages from the keyboard;
- News groups — groups of Internet users with similar interests such as ergonomics, MSDSs, and chemical safety;
- Complete federal regulations;
- U.S. government agencies — sites provide publications, updates, and perspectives;
- Australian, Finnish, and Swedish Occupational Health and Safety Departments (many other countries also have a presence on the Internet);
- World Health Organization;
- Many universities; and
- Shareware and freeware programs and updates to existing programs; government software from OSHA and EPA.

To illustrate how pervasive (and useful) the WWW is in occupational hygiene, a small sample of web sites has been assembled in Table 43.4, which is a listing of organizations with web pages of interest to occupational hygienists. There are many thousands of sites that pertain to occupational hygiene, so finding needed information is a challenge. This list may be helpful in providing beginning points in the search. All of the listed sites have links to other sites on the Internet.

Table 43.5 illustrates how WWW sites can be organized around a common interest, e.g., ergonomics. As in Table 43.4, all of the listed sites have links to other related sites.

Once a user is connected to the site, information can be downloaded over the Internet. If an address (URL) has changed, simply use one of the search engines to search on the topic of the site in the table.

FTP Use on the Internet

When an FTP site is accessed, a directory list using the operating system is displayed showing subdirectories, files, file

Table 43.4 — Sampling of Organizations with WWW Sites Useful to Occupational Hygienists

Organization	*INTERNET ADDRESS and Description of Content*
Occupational Safety and Health Administration	WWW.OSHA.GOV Source of compliance and rates of injury/illness information, regulatory agenda, regulations
Environmental Protection Agency	WWW.EPA.GOV Environmental regulations, research activities, toxic substances control, chemical-specific information
Agency for Toxic Substances and Disease Registry	WWW.ATSDR1.ATSDR.CDC.GOV:8080/ATSDRHOME.HTML Health effects information concerning chemicals, chemicals released from hazardous waste disposal sites, health study design issues, physician case studies
National Institutes of Health	WWW.NIH.GOV Ongoing research, information on cancer and causation
National Toxicology Program	WWW.NTP-SERVER.NIEHS.NIH.GOV Extensive information on chemicals, reactivity, LD50, flammability, long- and short-term effects
Worksafe Australia	WWW.ALLETTE.COM.AU Chemical search capability; review of programs in Australia
National Institute for Occupational Safety and Health	WWW.CDC.GOV/NIOSH Research studies, health hazard evaluations, extensive links to occupational safety and health resources on the Internet
World Health Organization	WWW.WHO.CH Information on environmental and occupational health with a global perspective; currently, largely nonoccupational health issues
International Agency for Research on Cancer	WWW.IARC.FR Cancer classifications, ongoing research, publications, and ordering information
American Conference of Governmental Industrial Hygienists	WWW.ACGIH.ORG Source of information on TLVs, BEIs (biological exposure indices), chemicals under study, and proposed revisions to TLVs
National Library of Medicine	WWW.NLM.NIH.GOV Over 40 databases containing more than 20 million references to a wide range of health studies. Hazardous substances data bank, Toxline, and Medline are particularly noteworthy. Databases on genetic toxicology, reproductive toxicology, toxic release inventory, AIDS, and cancer are also available. Search software can be downloaded from this site.
American Industrial Hygiene Association	WWW.AIHA.ORG. Information on practice of industrial hygiene. Publications, reference books, and current legislative issues.
American Chemical Society	WWW.ACS.ORG Chemical database searching, chemistry resources, publications
Centers for Disease Control and Prevention	WWW.CDC.GOV Information on environmental chemical risks, biosafety, emergency planning, and programs at the National Center for Environmental Health

Note: These addresses are all accessible with popular web browsers. WWW addresses change regularly. These sites were selected for their usefulness and the minimal chance of an address change.

Table 43.5 — Sample of Internet Sites Based on Particular Health Issues

Health Issue	*INTERNET ADDRESS and Description*
Electromagnetic field health effects	HTTP://INFOVENTURES.COM Extensive current research and findings relating to health and electromagnetic fields
Ergonomics	WWW.HFES.VT.EDU/HFES Human Factors and Ergonomics Society homepage; links to other ergonomics sites
MSDSs	WWW.MSC.CORNELL.EDU A site with links to University of Virginia and University of Utah MSDS master files; extensive access to MSDSs.

Note: Sites can be accessed with popular web browsers.

size, and date of creation. FTP sites contain many useful documents, shareware, and freeware programs.

ARCHIE and VERONICA

To find files at FTP sites, tools such as ARCHIE and VERONICA can be used to search for a specific file or FTP directory. These two search tools are powerful, free, and easy to use. The disadvantage is that you have to know specifically the files you are looking for. As the shift toward searching on the WWW continues, use of these search tools will decline.

Gophers on the Internet

Gopher sites are actually text-sensitive search engines for FTP sites.[12] Gophers are easy to use and provide access to MSDSs and other documents as well as various freeware and shareware programs. Many organizations and universities have numerous gopher sites. However, as more people use the graphical browsers on the WWW these sites will probably wither. (EPA is reducing its gopher offerings in deference to the WWW.)

Electronic Mail (E-Mail)

Electronic mail (e-mail) is a method of sending messages across a network or through telephone lines using a modem. The author of the message addresses the mail to the recipient using an e-mail address, such as joeih@hsph.harvard.edu. This address gives the recipient's name and the name of the network server being used for the recipient's e-mail. The network server name is analogous to the street, city, and state address on a post office-delivered letter. An e-mail address and server name will be provided by the ISP when the Internet access account is established (see Accessing the World Wide Web). E-mail can be sent to one person or as many recipients as the sender desires. There is no paper involved and delivery can be instantaneous. If a report, picture, slide, etc. needs to be sent along with the message, it can be attached to the e-mail.

Two useful forms of group e-mail are newsgroups and listservs. There are approximately 10,000 newsgroups today with new ones being added at the rate of one a day.[13] Newsgroups resemble an open forum for discussion, submission of comments, and getting questions answered. Subscribers can check in to the newsgroup periodically to download the "news" from that particular newsgroup. It is possible to receive large amounts of news from the site during a periodic check.

Finding newsgroups can be difficult. The ISP that provides access to the Internet will have a news server that will store the newsgroup files. When you read the newsgroup files they are actually on the server not the local personal computer. Specific newsgroups can be located by going to the news server and scrolling through the list until the appropriate newsgroup is found.

Within the newsgroup specific issues are raised and discussed electronically. All activity, comments, or messages concerning the specific issue in that newsgroup comprise a "thread." Therefore, downloading a thread on parathion safety would contain all e-mail comments concerning parathion safety. This can be helpful when there are a large number of comments or the comments are complex.

Listservs are analogous to an automatic e-mail system, with e-mail being sent to the entire list of subscribers. To find out which listservs are available on safety, one would send an e-mail message to listserv@listserv.net with the following syntax: *list global safety*. A list of appropriate listservs would be returned via e-mail. To subscribe to one of the listservs, use e-mail with the command *Subscribe listname* as the e-mail message. No other text should be in the e-mail. Listservs are similar to newsgroups except that they use e-mail as the communication protocol. As an example, Duke University has an occupational and environmental medicine listserv that is popular with occupational hygienists. Subscribing to that listserv will generate approximately 20–30 e-mail messages per day (1997).

Summary

After completing this chapter, readers should be able to understand basic terms used to describe computers, networks, computer systems, and the Internet. In addition, they should understand the issues in selecting an occupational hygiene computer system, the challenges of handling confidential data, and the process for gaining access and searching for health and safety information on the Internet.

References

1. **Whyte, A.A.:** Computerizing an industrial hygiene program. In *Fundamentals of Industrial Hygiene*, Plog, J.N. and P.J. Quinland (eds.).

Itasca, IL: National Safety Council, 1996.

2. **Miller, R.K., L.A. Little, and K.L. Moore:** *Software for Industrial Health and Safety.* Norcross, GA: Future Technology Surveys Inc., 1996.
3. **Sigler, D. and J. Lurie:** "Next Generation Environmental, Health and Safety Information Systems" (Session S17-ENT). Presented at the Environmental Conference on Information Technology, Washington, DC, October 1995.
4. **Franchi, J.:** CBT or IVD?—That's the question. *Techtrends 37*:27–30 (1992).
5. **Fletcher, J.D.:** The effectiveness and cost of interactive videodisc instruction. *Mach. Mediat. Learn. 3*:361–385 (1989).
6. **Kulik, J.A.:** Meta-analytic studies of findings on computer-based instruction. In *Technology Assessment in Education and Training,* E.L. Baker and H.F. O'Neil, Jr. (eds.). Hillsdale, NJ: Lawrence Erlbaum Associates, 1994.
7. **Software Publisher's Association:** *Report on the Effectiveness of Technology in Schools, 1990-1994.* San Jose, CA: Software Publisher's Association, 1995.
8. **Oz, E. and D.W. Lodie:** Multimedia for better learning. *J. Syst. Manage.* (1993).
9. **Alexander, M.:** Multimedia focus turns to training. *Computerworld 26*:8 (1992).
10. **Meeker, M. and C. DePuy:** *The Internet Report.* New York: Harper Business, 1997.
11. **Blotzer, M.:** *Internet User's Guide for Safety and Health Professionals.* Schenectady, NY: Genium Publishing, 1996.
12. **Spina, F.:** Safety professionals on the Internet. *Prof. Safety 41(9)*:18–22 (1996).
13. **Lejune, U.A.:** *Netscape & HTML Explorer.* Scottsdale, AZ: Coriolis Group Books, 1995.

The ability to write effective reports helps occupational hygienists achieve their goals of preventing and controlling workplace hazards.

Outcome Competencies

After completing this chapter, the user should be able to:

1. Define underlined terms used in this chapter.
2. Apply the basic elements of effective report writing.
3. Apply the basic elements of effective graphic presentation.
4. Develop strategies for presenting technical information to divergent audiences and levels of expertise.
5. Implement accepted standards of usage and style to achieve clarity and consistency in all written documents.

Key Terms

primary reader • informational report • interpretive report • rewriting • stand-alone graphics

Prerequisite Knowledge

Basic occupational hygiene skills.

Key Topics

I. Defining the Purpose
II. Defining the Primary Reader
III. Developing an Outline
IV. Writing Techniques
V. Presenting Graphs, Tables, and Diagrams

44

Report Writing

Susan M. McDonald, CIH

Introduction

The ability to write effective reports helps occupational hygienists achieve their goals of preventing and controlling workplace hazards. Conversely, the failure to communicate accurately about occupational hygiene matters may have severe consequences: a worker's health or a company's livelihood could be at stake. This chapter describes a report writing process that will assist occupational hygienists in communicating their knowledge and expertise effectively.

Occupational hygienists' reports serve a variety of purposes and readers. They may be used to document worker exposures to hazardous materials, to justify capital expenditures for engineering controls, to support litigation, or to provide a record for regulatory compliance. Potential readers come from divergent backgrounds, including workers, managers, union representatives, other occupational hygienists, nurses, physicians, engineers, government officials, lawyers, and business executives. In addition, reports may be circulated well beyond the original reader and may be used not only to address current issues but also to provide a historical record.

Given this wide range of audiences and purposes, it is critical that occupational hygiene reports convey pertinent information clearly and accurately and at the same time strike a balance between brevity and detail. Today's professionals are inundated with written material, and reports that take too long to get to the point will go unread. This chapter emphasizes presenting the most important information at the beginning of the report, providing supporting details in a logical, organized manner, and writing in a clear, direct style.

Occupational hygiene field work frequently involves gathering large amounts of information. Preparing a well-organized, readable report from such research may seem daunting, unless one approaches writing as a series of steps. To produce effective reports, the writer should 1) define the purpose; 2) define the primary reader; 3) prepare an outline; 4) draft the report; and 5) revise the draft.

Defining the Purpose

Much of the work involved in report writing should take place well before the actual writing begins. Advance preparation can save time by eliminating the need for laborious revisions or additional field work. This book describes methods for collecting occupational hygiene data in a technically valid manner. Preparing for the report writing should begin even as the occupational hygienist develops a strategy for data collection. Scrupulous note taking in the field will serve the occupational hygienist well when writing the report. Nothing should be left to memory. All important or potentially important observations should be recorded at the time they are originally noted.

In preparation for writing, the occupational hygienist should consider the purpose of the report. Is the report simply an update of an ongoing project? Will it include an interpretation of the findings? Will the report be used to provide justification for changing a practice or policy?

Defining the report's purpose assists the writer in determining the scope and format of the report.

Reports usually are intended either to inform or to persuade. The informational report simply presents the facts and generally requires less time and effort than an interpretive report. Examples of informational reports include status reports on a continuing project and survey reports that list findings without developing conclusions or recommendations.

In the interpretive report, the writer not only presents findings but also critically evaluates the results, draws conclusions, and offers recommendations. The purpose of such reports is not merely to inform but also to persuade the reader to accept the writer's viewpoint, to take action, or to make a change in policy or practice. Examples of interpretive reports include technological feasibility studies and comprehensive work site evaluations. The interpretive report is distinguished by its inclusion of expert opinion to guide the reader in making decisions.

It is essential that the writer understand at the outset what type of report is expected. If the primary audience does not want the writer to draw conclusions, the writer should simply present the facts rather than interpret them. For instance, a physician may seek information on a worker's exposure assessment for a medical evaluation. In this case the occupational hygienist should simply present the findings of the occupational hygiene assessment and not provide an opinion on the work-relatedness of a worker's health status. If, on the other hand, the physician seeks the occupational hygienist's expertise in interpreting the exposure assessment findings, the report should include a detailed critical evaluation, including an informed analysis correlating the worker's exposures with potential or known health effects. In this case the occupational hygienist should apply his experience and knowledge in guiding the physician.

Defining the Primary Reader

Knowing the expectations and background of the primary reader is also essential to effective and focused report writing. Trying to write for a generalized or nonspecific audience is likely to be a futile endeavor. Without a specific audience in mind, the occupational hygienist will not know how to present technical information, what terminology to use, or what level of detail to provide.

The primary reader is the person empowered to take action based on the report. Identifying the report's purpose and determining the primary reader usually go hand in hand, since it is the primary reader who explicitly or implicitly establishes the purpose of the report.

Identifying the audience assists the writer in determining the report's tone and language. For example, if the primary reader is also an occupational hygienist, the writer may freely use occupational hygiene terms and notations. However, if the primary reader does not have an occupational hygiene background, the writer must use less technical terms and provide explanations of concepts that are integral to occupational hygiene work. For instance, a reader unfamiliar with the concept of time-weighted average exposures may benefit from an explanation of the calculations involved in determining these results. Instead of simply presenting exposure results, the writer should also assist the reader's understanding by including a glossary or other method of explaining the occupational hygiene concepts used in the report. The challenge is to provide sufficient technical detail without confusing or losing the reader.

Consider the case of the occupational hygienist who must convince an upper-level manager to spend thousands of dollars to install a local exhaust ventilation system to reduce workers' exposures to an airborne hazardous substance. The occupational hygienist must select the most pertinent information and write a report that is brief, convincing, and to the point, perhaps describing the hidden costs of not installing the system. Detailed technical analysis may be included as background, but the first page should summarize the occupational hygienist's position. The writer should proceed under the assumption that the primary reader, in this case a business executive seeking the bottom line only, may not read past the first page of the report. On the other hand, a report on the same topic written for another occupational hygienist or a ventilation engineer would likely include much more technical detail.

The writer needs to be alert to the expectations and background of the reader, to consider what the reader already knows and needs to know. Is the reader knowledgeable about the report's subject? Does the reader expect a detailed analysis or a one-page summary? What action is the reader likely to take as a result of the report?

In addition to considering the primary reader, the writer should determine whether the report may serve as a historical record, or whether it may be used as an exhibit or reference in litigation. It is good practice to be alert to the possibility that any written report may end up as a historical document or legal record. Ensuring the technical validity of the data, documenting all findings, and making clear distinctions between facts and professional opinions are the best methods for developing a report that will withstand legal scrutiny. For guidance on preparing reports as legal documents, the occupational hygienist should consult an attorney or the company's legal department.

Developing an Outline

Identifying the purpose and primary reader assists the writer in determining the scope, format, and language of the report. The next step in the report-writing process is developing an outline, which provides an organization for presenting information in a logical manner. A well-organized report guides the reader effortlessly through even the most complex information. A poorly organized report may never be read before it is filed away indefinitely.

The occupational hygienist who has collected a large amount of data may feel overwhelmed by the prospect of presenting such information in a logical and readable manner. Using an outline allows the writer to approach the report one section at a time, thus helping the writer get started.

Depending on the purpose of the report, the level of detail in the outline may be minimal or extensive. The occupational hygienist may wish to develop a series of outlines to be used for different types of reports that are required on a regular basis. However, in all cases, the basic approach to organization is the same: put the most important information first and provide supporting details in a logical manner. A standard outline that may be applied to all types of reports has the following sections:

- Summary — The summary provides a synopsis of the most important information about the current project. It tells the primary reader what was done and why, provides a synopsis of the results, and conveys what comes, or should come next.
- Background — The background brings the reader up to date, explaining what led to the current project. Depending on the type of report, this section may be subdivided to include an introduction, a further description of the project's purpose, a description of the research methodology, and other information not directly related to the findings or conclusions.
- Findings — This section summarizes the results of the current project. Depending on the report's scope and purpose, this section may contain a detailed description of findings or may simply list the results. This section is distinguished by the factual nature of the information presented, as opposed to the interpretation of findings in the Conclusions section.
- Conclusions — This section correlates the findings presented in the previous section with proposed actions. Depending on the report's purpose, this section may be brief or extensive. For example, a brief informational report may not include detailed interpretations but may propose follow-up work. An interpretive report draws conclusions from the findings and makes recommendations. The amount of detail and degree of analysis in this section depends on the report's scope and the primary reader's expectations. This section presents the occupational hygienist's conclusions and recommendations, which are opinions and should be presented in a way that distinguishes them from facts.

A suggested outline is shown in Figure 44.1 for a generic occupational hygiene survey report, intended as an interpretive report and written for another occupational hygienist. The purpose of the report is to summarize the occupational hygienist's comprehensive work site evaluation, presenting the exposure assessment findings and developing recommendations to improve health and safety conditions at the site. This outline should be seen not as a rigid structure but as an approach to organizing information. Each report, as each occupational hygiene project, is unique, and the report outline should allow for flexibility. The goal is to present the most important information in a way that achieves the desired results. An example of an executive summary is shown in Figure 44.2.

Writing Techniques

Once the report's purpose, audience, and outline have been established, the writer is ready to prepare a first draft. In this draft the writer should not be overly concerned about exact wording or organization. The point of the first draft is to get down all the pertinent information, and then reorganize as necessary to create a logical flow of ideas. (The outline should assist in this regard.)

Report Outline for an Occupational Hygiene Survey

Executive Summary

This section is a one-page summary of the purpose, main activities, findings (e.g., exposure monitoring results), and conclusions. It includes any critical information for the primary reader, such as conclusions correlating workers' health to exposures, and is frequently written after the rest of the report has been prepared.

Background

Introduction. This section includes a more detailed description of the purpose of the survey, a description of the work site and workers, and a summary of previous surveys and findings at the site. It may include subsections covering such information as the health effects associated with identified exposures. It may also be useful to include figures such as process flow diagrams or illustrations depicting the location of workers and engineering controls.

Survey Methods. This section summarizes the industrial hygienist's sampling strategy (e.g., number and types of workers monitored) as well as sampling and calibration methods. It may need to be subdivided depending on the extent of the monitoring or other survey methods. Methods for other exposure assessment (such as sound level measurements) as well as methods for control measure assessment are included here.

Analytical Methods. Methods for laboratory analysis of collected samples are described here (including information on limits of detection), as well as any other pertinent analytical methods.

Findings

Exposure Monitoring Results. Such results may be presented in both text and tabular formats. Information such as analytical limits of detection should be repeated here if necessary to assist the reader in understanding the data presented.

Control Measure Assessments. Measurement results of engineering controls such as local exhaust ventilation are included here. Tables and diagrams may be useful in illustrating the location of ventilation systems and processes with respect to airflow measurements.

Results of Statistical Analysis. Statistical analyses of the sampling results and other findings are presented in this section, along with any associated graphs or tables.

Conclusions

Discussion of Findings. This section contains the occupational hygienist's interpretation of the findings. For instance, this section may present a discussion of the identified correlations between exposures and controls, or between exposures and potential health effects. This section may compare the findings with applicable regulations and professional consensus standards.

Recommendations. This section proposes actions based on the interpretation of the findings. For example, recommendations for additional control measures, medical evaluation of workers, and further exposure monitoring are included here.

Figure 44.1 — Report outline for an occupational hygiene survey.

It is good practice to take a break between writing the draft and completing the final report. Leaving the draft alone for a day or two allows the writer to bring a fresh perspective to the project. Many writers also find it helpful to have colleagues review their reports before preparing the final versions. Such peer readers provide an objective view of the material and may have suggestions for making the report clearer or better organized.

Every good writer knows that the secret to good writing is rewriting. One of the paradoxes of writing is that it takes longer to write a succinct report than it does to write a wordy one. Revising helps to ensure that the final report is free of awkward language or unnecessary words that distract from the content. In addition, thorough proofreading helps to eliminate technical and typographical errors.

This section presents some advice on writing techniques but is not intended as a guide to grammar or usage. Every occupational hygienist's library should contain reference manuals on grammar and usage as well as scientific communication methods (some examples are given in the bibliography following this chapter). The principles of effective communication outlined below apply to all kinds of writing, whether a brief memo or a large technical document:

- Use plain English. Resist the urge to use flowery language or technical jargon. One of the most common writing pitfalls is to use overly technical language in an effort to impress the reader. Rather than impress, the writer may confuse or lose the reader altogether.
- Never use a long word when a short one will do. However, in some cases a long word expresses an entity much more eloquently and precisely than a series of short words. Eloquence should prevail. This advice comes from *The Elements of Style*, by William Strunk, Jr. and E.B. White, a thin volume with indispensable advice about writing.
- Use the active voice whenever possible. For example, rather than write, "It was stated by the safety officer," write, "The safety officer said."
- Eliminate unnecessary words. For example, instead of "call your attention to the fact that," say "remind you"; instead of "emergency situation," say "emergency"; instead of "to provide protection," say "to protect."
- Vary sentence structure and length to avoid sounding choppy or monotonous.
- Be alert to the appearance of the printed page. Reading long stretches of text without paragraph breaks is tedious. Short paragraphs help to make even a complex technical report easier to read.
- Be consistent in the use of abbreviations and notations standard to the occupational hygiene profession. The first use of a term should always be written out, followed by the appropriate abbreviation in parentheses, for example, "0.1 parts per million (ppm)."

• All statements of opinion or fact not directly attributable to the author should be appropriately referenced. Opinions should be presented as such, distinct from facts. In general, conclusions and recommendations are opinions and not facts.

• Avoid generalizations or vague statements. Be as specific as possible and use examples when necessary to illustrate an idea or principle.

• Make the report perfect. There should be no misspellings, no typographical errors, and no errors in calculations. The occupational hygiene report in particular must be technically accurate to be accepted as a legal or historical record.

The widespread use of the personal computer, facsimile machine, and electronic mail has wrought major changes in how professionals, including occupational hygienists, communicate. This technology has made it easier to generate and transmit copious amounts of information via electronic networks. The word processor in particular has accelerated the editing process by allowing a writer to quickly insert, delete, and move text. Yet, effective communication still takes time and effort, and writing improvement comes only with practice. The challenge is using technology to communicate essential information rather than to generate information simply because it is easy to do so.

Sample of an Executive Summary

A preliminary occupational hygiene evaluation of the XYZ Microscopy Unit has determined that the laboratory hood is not working properly and the technicians have been advised to stop all work with glutaraldehyde until the hood is fixed. The unit manager requested that the health and safety officer conduct an investigation after all three technicians recently reported experiencing skin rashes and respiratory problems, effects that may be associated with glutaraldehyde exposure. The evaluation involved observation of work practices and hood face velocity measurements. Airflow across the hood's opening was erratic and barely measurable. In addition to the hood repair, we recommend that the technicians review work practices for safe handling of glutaraldehyde. If the technicians' health problems persist or worsen, they should receive prompt medical evaluations. After the hood is repaired, we will conduct monitoring to assess the technicians' exposure to glutaraldehyde.

Figure 44.2 — Sample of an executive summary.

Presenting Graphs, Tables, and Diagrams

Occupational hygiene information is often presented most clearly in a table, graph, or diagram. A well-chosen illustration can greatly enhance the effectiveness of a report. For instance, tables provide a summary of complex data in a form that is more succinct and illustrative than text. Graphs can illustrate relationships between variables much more efficiently than text. Schematic diagrams can provide a visual guide to process flow or other procedures. However, as with the written portion of the report, the writer must be alert to the expectations and needs of the primary reader when selecting illustrations.

For example, the reader should not have to flip back and forth through the report's pages to view the referenced table or graph. In general, figures should be numbered sequentially and should follow immediately the text in which they are first referenced, rather than being placed in an appendix at the back of the report. The exception is an unusually large table that is not mentioned frequently or only supports the text indirectly. In such a case the information should be placed in an appendix or made available to the reader on request.

All visual displays should be self-explanatory. A good rule of thumb is to consider whether the graph or table would be understandable if it were to become separated from the rest of the report. This practice is particularly important for reports that may ultimately serve as historical or legal documents. Tables and graphs may be copied and circulated separately, another reason to ensure that the figure is self-explanatory. The following guidelines are presented to assist the report writer in preparing stand-alone graphics:

• A complete title must be included on every graph, table, or diagram.

• Each axis on a graph must be labeled clearly and completely.

• A legend containing information pertinent to understanding the data should be provided on the same page as the graph. For example, laboratory conditions such as temperature, pressure, and relative humidity should be included in a legend accompanying a calibration curve.

• Data presented in a table should be appropriately labeled, including units with standard abbreviations.

• Tables that summarize air sampling results should list the analytical method employed and the method's limit of detection. Other information, such as air volume sampled as well as each sample location, should be included wherever necessary to allow the reader to interpret the data presented.

In addition to writing reports, occupational hygienists frequently are expected to make oral presentations of their work. Some guidelines for slide presentations are given in Figure 44.3.

Guidelines for Slide Presentations

- Do not use slides that require an apology for their quality.
- Do not use slides that require you to say, "You probably can't read this from where you're sitting, so I'll read it to you."
- Do not use slides that require you to say, "Ignore the other five columns on this slide and only look at Column C." If the five columns are not necessary to the presentation, do not include them on the slide.
- Use 2 x 2-inch square (35-mm) color slides; they are effective, easy to make, and inexpensive. Color film is also convenient for making slides from black and white copy.
- Use a dark, colored background; i.e., white on black, yellow on black, white on red.
- Plan your slides for a good visual pace in your presentation. Do not leave a slide on the screen after discussing its subject.
- Limit each slide to one main idea.
- Use a slide series for progressive disclosure to clarify your discussion.
- Limit each slide to 15 to 20 words or 25 to 30 elements; include no more than you will discuss.
- Leave space (at least the height of a capital letter) between lines.
- Include titles to supplement rather than duplicate slide data.
- Use several simple slides rather than one complicated slide, especially if you plan to discuss a subject at length.
- Use duplicates if you need to refer to the same slide at several different times in your talk. It is impractical to search for and reshow a slide.
- Mark all slides in the lower left corner when the slide reads correctly on hand viewing. Add sequence numbers.

Using Text in Slides

- Bullets are dramatic and effective; however, avoid using more than six on any one slide.
- Include sufficient margins around text and images on a slide. This allows for differences in slide mounts and duplications.
- Keep space between lines consistent throughout your presentation; avoid crowding words and images on the same slide.
- Make text large enough so that a person at the back of the room can read it easily.
- Do not mix type styles on a single slide.
- Maintain consistency of type throughout your presentation.

From *AIHCE Speaker's Handbook*, American Industrial Hygiene Association, Fairfax, VA, 1996.

Figure 44.3 — Guidelines for slide presentations.

Summary

The method for preparing occupational hygiene reports presented in this chapter may be summarized as follows:

- The occupational hygienist should begin to prepare for the final report at the beginning of the project.
- The writer should identify the report's purpose and primary reader and should work from an outline.
- The writer should begin with a draft and revise it to produce the final report.
- The report should present the most important information first.
- The report should be neat, succinct, readable, and error-free.
- The occupational hygienist should be ever alert to the possibility that any written report may become a legal or historical record.

Additional Sources

Blicq, R.S.: *Guidelines for Report Writers.* Englewood Cliffs, NJ: Prentice-Hall, Inc., 1982.

Roman, K., and J. Raphaelson: *Writing that Works.* New York: HarperPerennial, 1992.

Rubens, P. (ed.): *Science and Technical Writing.* New York: Henry Holt, 1992.

Sabin, W.A.: *The Gregg Reference Manual,* 5th ed. New York: Gregg Division/McGraw Hill, 1977.

Strunk, W., Jr., and E.B. White: *The Elements of Style.* New York: Macmillan, 1959.

By the early 1900s, workers' compensation laws were being passed, providing medical and wage benefits to the injured worker without placing "blame."

Chapter 45

Outcome Competencies

After completing this chapter, the user should be able to:

1. Define underlined terms used in this chapter.
2. Distinguish "occupational safety" from "occupational health."
3. Identify important historical events in the evolution of occupational safety.
4. Review common methods for measuring safety performance.
5. Review safety program elements.
6. Identify main OSHA standards that overlap safety and health.
7. Review qualifications for safety professionals.

Key Terms

common exposure base • continuous improvement • employee involvement • hazard prevention and control • incident rate • instantaneous • management commitment • recordable injury • safety and health training • Standard Industrial Classification (SIC) • Voluntary Protection Program (VPP) • work site analysis

Prerequisite Knowledge

Prior to beginning this chapter, the user should review the following chapters:

Chapter Number	Chapter Topic
1	History and Philosophy of Industrial Hygiene
3	Legal Aspects of the Occupational Environment
26	Biomechanics
28	Ergonomics
33	Evaluating Ventilation Systems
42	Confined Spaces
48	Laboratory Health and Safety

Key Topics

I. Key Historical Events (in the Evolution of Occupational Safety)

II. Measuring Safety Performance

III. Safety Program Elements

IV. OSHA Standards That Overlap Safety and Health

V. Qualifications to Become a Safety Professional

45

Occupational Safety

Daniel S. Markiewicz, CIH, CSP, CHMM

Introduction

This chapter provides an overview of occupational safety. To better understand occupational safety we must first distinguish the term from occupational health. The distinction between occupational safety and occupational health is linked to the definition of occupational injury and occupational illness. The terms "safety" and "injury" have been primarily the work domain of safety professionals, whereas the terms "health" and "illness" have been primarily the work domain of occupational hygienists.

In 1986, the Bureau of Labor Statistics (BLS) published *Recordkeeping Guidelines for Occupational Injuries and Illnesses,* which distinguishes an injury from an illness by one word: instantaneous.(1) Instantaneous may be defined as the point in time it takes to snap one's fingers and refers to a single event. Occupational safety, therefore, is focused primarily on preventing unwanted and unplanned events (i.e., incidents that are instantaneous in nature and that might lead to an employee suffering an injury).

Although on the surface this distinction is clear, there still may be confusion. An example would be a nurse who contracts hepatitis — a classical "illness" — from a needle stick. Since the hepatitis developed from a single instantaneous event, by BLS definition it is an injury. Following this concept, a back problem caused by a single lift would be an injury. Carpal tunnel syndrome caused by the repetitive use of a poorly designed tool would be an illness. Chemical exposures, depending on circumstances, may result in either an injury or illness. These examples help illustrate why there is a blurring of the distinction between a safety professional and an occupational hygienist. There are many areas where the professions meet and overlap; as such, for both professions to be successful they must be complementary.

Occupational safety is a broad topic that encompasses both conditions and acts (e.g., behavior that might lead to injuries). Conditions may be viewed as everything except acts. Major categories for safety conditions include mechanical, thermal, electrical, and chemical energies. In 1931, Heinrich found that 88% of occupational injuries were caused by acts and 10% of occupational injuries were caused by conditions.(2) DeReamer, with a more current view, has shown that conditions and acts are equally responsible for causing occupational injuries.(3) Whether conditions or acts cause more injuries will continue to be debated and both topics will need to be addressed fully for injury prevention efforts to be effective. This overview on occupational safety is not intended to make the reader proficient at preventing or reducing occupational injuries. This chapter's main focus is to highlight key safety concepts and identify areas in which safety professionals and occupational hygienists must work together to help eliminate and minimize worker injury and illness.

Key Historical Events

There are many milestones in safety's long history. Unfortunately, we will be able to briefly cover only a few of the many important historical events that have created and developed the field of occupational safety we know today. Chapter 1 in this

Occupational safety is a major professional discipline related to occupational hygiene. The editor chose not to try to summarize this sibling profession in extensive detail, but rather to provide a brief overview of the field. An appendix to this chapter highlights published occupational safety resources as an example.

This review is provided here to be sure that occupational safety is discussed and its relationship to occupational hygiene highlighted.

Changing Perspectives in Occupational Safety

In his article "Landmarks in the History of Safety," Michael Guarnieri labels the period from 1915 to 1930 the "Investigational Era."[1] This was a time when safety became a research focus and a large body of safety information was published by governmental entities, trade associations, insurance companies, and organizations such as the National Safety Council. Insurance industry research, for one, had two common themes: First, that safety efforts were cost-effective; second, that prevention programs should focus on the psychological factors because most accidents were caused by "workers being unsafe."

According to Guarnieri, the psychological approach to accident prevention was the focus of safety for more than five decades, but this approach began to change in the 1930s when experts could not validate the underlying belief in the existence of "accident-prone personalities." The work of J.J. Gibson and William Haddon, Jr., as cited by Guarnieri,[1] had a profound effect on modern safety. In 1963, Haddon pointed out that injuries can be prevented by controlling energy.[2] This concept formed the basis of the 1966 safety standards for the automotive industry.

Haddon's work reaffirmed earlier work by Gibson.[3] Haddon and Gibson both considered the term "accident" a deterrent to injury-prevention efforts. They theorized that "accident" carries a connotation that the event is random and cannot be prevented. Haddon considered the fundamental problem of injury prevention to be the abnormal energy exchange. The first group of injuries would be those caused by interference with normal whole body or local energy. An example would be a lethal exposure to a toxic gas. The second group of injuries is caused by imparting energy on the body that exceeds local or whole-body injury thresholds. Gibson in his work also related the exchange of mechanical, chemical, thermal, and electrical energy to injuries.

The current trend in occupational safety is a renewal of behavior-based injury prevention programs that focus on "at-risk behaviors." These efforts (along with regulatory compliance programs, voluntary protection programs, and hazard prevention programs) are collectively serving to address workplace injuries.

— *Donald Robinson*

1. **Guarnieri, M.:** Landmarks in the History of Safety. *J. Safety Research 23:* 151-158 (1992).
2. **Haddon, W., Jr.:** A Note Concerning Accident Theory and Research with Special Reference to Motor Vehicle Accidents. *Annals N.Y. Acad. Sci. 107*:635-646 (1935). (As cited in Reference #1.)
3. **Gibson, J.J.:** The Contribution of Experimental Psychology to the Formulation of the Problem of Safety: A Brief for Basic Research. In *Behavioral Approaches to Accident Research*, N.Y. Association for the Aid to Crippled Children, pp. 77-89. (As cited in Reference #1.)

book discusses the history of occupational hygiene and to a limited extent occupational safety.

Since injury and death have occurred since the beginning of mankind, it is important to consider when safety regulations first became a part of society. Grimaldi and Simonds describe the Code of Hammurabi (circa 2100 B.C.), developed by the ruler of Babylon, as one of the earliest known body of laws that addressed the injured or killed worker.[4] The Code of Hammurabi only contained specifications for indemnification and punishment for causing injury or death to a worker. The code did not address prevention, but since it was based on the principle of an "eye for an eye" the code likely acted as a strong motivator for those who controlled workers to engage in injury prevention efforts (i.e., safety activities).

Specific legislation, however, aimed at preventing injury to workers — actually children who worked — is believed to have arisen in England in the late 18th century. Children who worked were made part of an "apprentice system" that provided instruction on how to do their jobs, including how to avoid injuries while they worked. In 1784, an outbreak of fever in the cotton mills in the Manchester area of England drew special attention to the hazardous and long hours being performed by children. Following this event a number of laws were passed to further protect children at work. The most notable act that followed this event, however, was the Health and Morals Act of 1802, which provided regulations governing sanitary and safety conditions and hours worked by children. Taking a lead from English law, the state of Massachusetts in 1877 developed laws concerning employment and safety of children in factories.

Ferguson provides a good review of safety's history in the United States, beginning with the first two decades of the 20th century.[5] In Ferguson's opinion, there is always a desire for people to expect a new century to surpass the last. This is why the early period of the 20th century is labeled the "Progressive Era"; it is a period in which the "Safety First" movement was allowed to grow, according to Ferguson. The Safety First movement recognized the waste of unnecessary injuries to employees and sought means to eliminate these injuries.

Events that capture the public's attention, such as the outbreak of fever in the cotton mills in England in 1784, are a precursor for better controls to a problem. The major event that prompted a cry for better worker safety early in the 20th century was the Triangle Shirtwaist Factory fire in 1911. The fire resulted in the death of 146 workers who could not escape the fire because management at the factory had blocked or locked exits. Although the event influenced the creation and improvement of many codes, standards, and regulations to prevent a similar occurrence eight decades later during a fire at a chicken processing plant in Hamlet, N.C., resulted in the deaths of numerous employees because management had also blocked or locked exits. These events separated by time serve to demonstrate that attention to safety, even for obvious and well-known hazards, must be a continuous process.

Since the creation of the Code of Hammurabi, many employers have been required to indemnify injured workers. Early in the 20th century, workers in the United States were covered by a very restrictive indemnification process. This process would not allow an injured worker to prevail against his or her employer if: 1) the employee "contributed" to his or her injury; 2) if the employee "assumed" the risk associated with the work; or 3) if the injury was caused through the "negligence" of a fellow employee. These restrictions prevented most injured workers from obtaining compensation from their employer.

Following the Triangle Shirtwaist Factory fire, there were changes in laws that made it easier for injured employees to obtain compensation from their employers. In 1911, New Jersey, Wisconsin, and the state of Washington passed workers' compensation laws. Today, every state has a workers' compensation law. Although the application of workers' compensation laws varies from state to state, these laws generally are in the form of an insurance policy in which employers pay a premium for coverage. Because it is financially beneficial for both the company that issues the policy and the company that purchases the policy to have no or few compensable injuries, workers' compensation laws have done much to foster and encourage the implementation of injury prevention activities.

Arguably the greatest event to have shaped the field and profession of occupational safety has been the implementation of the Occupational Safety and Health (OSH) Act of 1970. Before OSHA was created as part of the OSH Act, laws protecting workers were only found on a state-by-state basis or were limited to narrow populations of workers. The OSH Act provided a federal mandate, backed by substantial investigative powers and penalty incentives, for employers to provide and maintain a safe and healthy working environment for nearly every working man and women in the nation. The Act made mandatory national consensus standards for worker safety and health that were in effect prior to 1973. These voluntary standards were developed by organizations such as the American Conference of Governmental Industrial Hygienists (ACGIH), American National Standards Institute (ANSI), and American Petroleum Institute (API). The National Institute for Occupational Safety and Health (NIOSH) was also created by the OSH Act. NIOSH provided for the first time a concerted effort on a national scale to study and find solutions to hazards that might cause injuries to employees.

Few would argue that OSHA has done a substantial job making employers more attentive to safety, but some people nevertheless have questioned whether OSHA has had a significant impact in reducing employee injury and illness. Peterson, after evaluating national occupational injury and illness statistics and subjecting the data to statistical control charts, found there was: 1) no significant change (improvement) from 1981 to 1991 in total (injury and illness) cases; 2) no significant change (improvement) from 1981 to 1991 in lost-time cases; and 3) a significant deterioration in lost workday statistics from 1983 through 1992.[6] Peterson argues that regulatory compliance and preventing occupational injuries have little to do with each other. Roughton and Grabiak agree that OSHA has weaknesses, but they indicate that beginning with changes in 1995 it can build itself into an organization that will place the highest premium on real results rather than reactive activities and processes.[7]

Incidence and Injury

The emphasis in the workplace on product quality has had a profound impact on workplace safety. This has in large measure been attributed to W. Edwards Deming and his 14 points for the transformation of management.[1] In the late 1980s many safety professionals saw in Deming's writing the clear analogy between safety and quality and the ability to use his quality principles to improve workplace safety.

In his book *Employee-Driven Systems for Safe Behavior*, Thomas Krause provides a foundation for applying Deming's work to the science of injury prevention.[2] He makes the analogy between quality defects and injury incidents and describes the application of quality tools to the safety arena. According to Krause, the need to identify at-risk behavior is fundamental to a behavior-based injury prevention program.

W.W. Heinrich[3] and Russell DeReamer[4] are among those who have purported that several incidents can occur before there is an injury. DeReamer points out that "for every accident resulting in an injury there are many similar accidents that cause no injuries whatever" and "for every major injury accident there are numerous similar accidents that result in only minor injuries." Assuming this to be true, if the at-risk behavior can be addressed, we can expect a significant reduction in injury incidents.

— *Donald Robinson*

1. **Aguayo, R.:** *Dr. Deming.* New York: Simon and Schuster, 1990. pp. 124-177.
2. **Krause, T.:** *Employee-Driven Systems for Safe Behavior.* New York: Van Nostrand Reinhold, 1995. pp. 3-42.
3. **Heinrich, W.W.:** *Industrial Accident Prevention.* New York: McGraw-Hill, 1931.
4. **DeReamer, R.:** *Modern Health and Safety Technology.* New York: John Wiley & Sons, 1980. p. 64.

Measuring Safety Performance

Safety performance can be measured in many ways. The Chemical Manufacturers Association (CMA) provides several excellent safety performance measures that can be used by their member companies to demonstrate continuous improvement in safety areas.[8] Demonstrating continuous improvement in safety areas is a goal identified in the CMA's Responsible Care® initiative.

The most basic safety performance measures used by employers are those that were mandated by OSHA shortly after promulgation of the OSH Act. These performance measures are explained and demonstrated in the BLS *Recordkeeping Guidelines for Occupational Injuries and Illnesses*.[1] These guidelines are subject to change, and the latest record keeping requirements by OSHA should always be referenced.

The heart of the record keeping guidelines is a strict adherence to definitions as to what is a recordable injury and what is not. It has already been pointed out in this chapter how the guidelines distinguish between an injury and an illness. Determining if an injury is work-related and whether it is recordable has been a point of considerable confusion among employers who are required to keep injury and illness records.

The "incident rate" is the numerical performance measure required by OSHA. The incident rate is the number of injuries, illnesses, or lost workdays related to a common exposure base of 100 full-time workers. The common exposure base provides for accurate industry comparisons, trend analysis over time, or comparisons among firms regardless of size. The incident rate is calculated as follows:

$$N / EH \times 200{,}000 \qquad (1)$$

where:

N = number of injuries and/or illnesses or lost workdays;
EH = total hours worked by all employees during a calendar year; and
200,000 = base for 100 full-time equivalent workers (working 40 hours/week, 50 weeks/year).

In addition to OSHA requirements for employers to record occupational injuries and illnesses, employers are also required to make periodic reports of deaths, injuries, and illnesses. This periodic reporting becomes part of the Annual Survey of Occupational Injuries and Illnesses. The annual survey produces national occupational injury and illness statistics (averages) at the 4-digit Standard Industrial Classification (SIC) level in most manufacturing industries and at the 2-digit SIC level in most nonmanufacturing industries. Statistics are also produced at the 3-digit level for some high-risk nonmanufacturing industries such as construction.

Occupational injury statistics can be found at both the OSHA and BLS homepages on the Internet (see Appendix A at the end of this chapter). The total cases for nonfatal occupational injury incident rates per 100 full-time workers in the manufacturing industry in 1995 was 9.9. Since this number is an average, some employers in the manufacturing industry will have a lower rate and some will have a higher rate. The incident rate may also be viewed as a percentage of employees that will suffer the degree of injury for which the rate was calculated. As an example, an incidence rate of 9.9 means that nearly 10% of all employees in the manufacturing industry experienced a recordable injury in 1995.

Safety Program Elements

Drivers to prevent injuries to employees include cost savings and cost avoidance, compliance with regulations, meeting laudable goals, and as the president of Jones and Laughlin Steel said in 1948 when asked to justify safe working conditions, "it is the right thing to do."[9] This chapter has already provided hints that simple activity-based safety, such as just complying with regulations, may not be effective in preventing and substantially reducing worker injuries. To prevent and substantially reduce injuries to employees a process, system, or program approach must be taken. These approaches to safety stress the comprehensive nature of the challenge and set forth a pathway, and often a step-by-step pathway, to prevent and reduce injuries to workers. Brown, in describing a systems approach to safety, acknowledges that "more than a knowledge of the principles is necessary to bring about a safe environment. A "procedure" must be followed whereby principles on paper become a reality."[10]

Although there have been many recommendations for ways to prevent and reduce injuries to employees, there is no one clear-cut winner. If there were, everyone would be following it and injuries to employees would be a minor issue, which is simply not the case when we see that nearly 10% of manufacturing employees experienced a recordable injury in 1995.

OSHA through its enforcement and other activities has the most experience in seeing which workplace safety programs have been the most effective in achieving low injury incident rates. In 1982, OSHA developed its Voluntary Protection Program (VPP). The VPP was developed to recognize and promote outstanding safety programs. One of the initial reasons for developing VPP included helping OSHA conserve resources by exempting outstanding safety programs from routine inspections. OSHA could then target its limited resources and attention to work sites with poor safety performances. The early VPP participants were showing lost workday case rates that ranged from 1/5 to 1/3 the rates experienced by average work sites.

Using knowledge gained from VPP participants and other experiences, OSHA in 1989 issued its voluntary Safety and Health Program Management Guidelines.[11] These guidelines consist of program elements which represent a distillation of OSHA's view on what are the most effective practices to protect employees from injuries. These guidelines cover four main elements: 1) management commitment and employee involvement; 2) work site analysis; 3) hazard prevention and control; and 4) safety and health training. Each of these elements is briefly described below.

Management Commitment and Employee Involvement. This element identifies that management commitment provides the motivating force and the resources for organizing and controlling activities within an organization. In effective safety programs, management regards worker safety as a fundamental value no less important than any other organizational purpose. Management commitment can be demonstrated by the following activities:

- Clearly state a work site policy on safety and health so that all personnel will understand the priority of safety and health protection in relation to other organizational values;
- Establish and communicate a clear goal for the safety program and objectives for meeting that goal;
- Provide for and encourage employee involvement in the structure and operation of the safety program;
- Assign and communicate responsibility for all aspects of the safety program;
- Provide adequate authority and resources to responsible parties so that assigned responsibilities can be met;
- Hold people accountable for meeting responsibilities; and
- Review program operations at least annually to evaluate successes and identify deficiencies that should be corrected.

Work Site Analysis. This element involves a variety of work site examinations to identify not only existing hazards but also conditions and operations in which changes might occur to create hazards. Effective management actively analyzes the work and the work site to anticipate and prevent harmful occurrences. To ensure that all hazards are identified, management and employees must:

- Conduct comprehensive baseline work site surveys for safety and periodic comprehensive update safety surveys;
- Analyze planned and new facilities, processes, and equipment for safety hazards;
- Perform routine job hazard analyses, including investigation of accidents;
- Use a reliable system for employees, without fear of reprisal, to notify management about conditions that seem hazardous; and
- Analyze injury and illness trends over time so that patterns with common causes can be identified and prevented.

Hazard Prevention and Control. Where feasible, hazards are prevented by effective design of the job site or job. When it is not feasible to eliminate hazards, they are controlled to prevent unsafe exposure. Elimination or control is accomplished in a timely manner once a hazard or potential hazards is recognized. Hazard prevention and control applies the following measures:

- Engineering techniques are used when feasible and appropriate;
- Procedures for safe work are established, understood, and followed by all affected parties; and
- Provisions are made for personal protective equipment and administrative controls.

Safety and Health Training. Safety and health training addresses the safety and health responsibilities of all personnel concerned with the work site. Training is often most effective when incorporated into other training about performance requirements and job practices. Training is conducted to:

- Ensure that all employees understand the hazards to which they might be exposed and how to prevent harm to themselves and others from exposure to these hazards;

- Permit supervisors to carry out their jobs effectively; and
- Ensure that managers understand their safety and health responsibilities.

From a comprehensive viewpoint, the safety program elements above are not required by OSHA as of mid-1997. OSHA, however, is considering rulemaking that would, at the least, encourage employers to develop and follow basic safety program elements. OSHA requirements for individual aspects of the safety program elements above are covered in the next section of this chapter.

OSHA Standards That Overlap Safety and Health

Some OSHA standards clearly fall into the safety domain. Examples include machine guarding and lockout and tagout requirements. Others (such as chemical-specific standards that include occupational exposure to asbestos, benzene, cadmium, etc.) more clearly align with occupational health and occupational hygiene concerns. There are several standards, however, that overlap safety and health concerns in more or less equal proportions. It is these standards more than any others that should be understood and appreciated by safety professionals and occupational hygienists alike. These standards bring the two professions close together and help demonstrate why both professions must work together if each is to be successful. For purposes of simplicity the standards, with just a few exceptions, will be cited by their general industry (29 CFR 1910) notations. The brief explanation for each standard that follows emphasizes the safety and health connections within that standard. The standards cited are not an inclusive list. Only the most important and significant standards are covered.

Employee Emergency Plans and Fire Prevention Plans (1910.38). Emergency action plans are required to ensure employee safety in the event of a fire or other emergency. The plan must be prepared in writing and reviewed with affected employees. Employers must apprise employees of the fire hazards of the materials and processes to which they are exposed. Fire is predominately a safety issue since burns are, with few exceptions, readily viewed as an injury. Fire also creates noxious and poisonous decomposition products that cross over into health concerns. Because of these reasons the properties of flammable and combustible materials, how they burn, how they are extinguished, and what byproducts burning material may create is a key overlapping safety and health concern. Further, other emergencies may include spills of hazardous materials that may be explosive and/or corrosive indicating safety concerns and toxic indicating a health concern. As such, both safety professionals and occupational hygienists must work together to create effective emergency action plans.

Ventilation (1910.94). Ventilation is used to keep particulates, gases, vapors, and other forms of hazardous materials at safe levels. "Safe" includes the concept of not allowing the concentration of materials to reach flammable or combustible levels (safety issue) as well as to keep materials below toxic levels (health issue). Ventilation also improves housekeeping because it can keep dust or other potentially hazardous debris off the floor.

Occupational Noise Exposure (1910.95). Although most hearing loss is the result of long-term exposure to excessive amount of sound, and thus a health concern, loud noise should also be controlled for safety reasons. Loud or distracting sounds may hinder an employee's ability to hear warning signals or may interfere with an employee being able to clearly understand and communicate safety precautions and instructions.

Subpart H — Hazardous Materials (1910.101–1910.120). For reasons identified above — specifically the fire and corrosive and toxic nature of hazardous materials — this whole category is a strong cross-over for safety and health concerns. Two standards are particularly important, however: process safety management of highly hazardous chemicals (1910.119) and hazardous waste operations and emergency response (1910.120). The standard on process safety management of highly hazardous chemicals incorporates all of the safety program elements previously discussed. Although the standard only addresses employers that have more than a "threshold quantity" of the most hazardous chemicals, this standard sets the tone for proper management of chemicals regardless of the amount or significance of the hazardous properties involved with a chemical. The standard on hazardous waste operations and emergency response clearly is a joint safety and health effort. Both safety and health concerns are stressed as being equally important.

Permit-Required Confined Spaces (1910.146). This is another example of needing to be aware of atmospheric conditions that might be combustible or toxic.

Additional safety concerns include how to extract workers from a confined area using hoists and other equipment.

Subpart L — Fire Protection (1910.155–1910.165). The safety and health connection associated with fire has already been explained. In the standard on portable fire extinguishers (1910.157) safety professionals must define when fighting a fire has gone beyond the "incipient" stage. In other words, when does the smoke from a fire become toxic? Beyond the incipient stage of a fire, employees would need to be better trained and know when and how to wear personal protective equipment. This may then require the creation of fire brigades (1910.156) with safety training. Also, since conductive extinguishing materials such as water may be used to put out a fire, basic electrical safety hazards and precautions must be considered.

Subpart Z — Toxic and Hazardous Substances (1910.1000–1910.1500). Although most of the standards in this category fall more toward occupational hygiene, there are some notable areas in which safety plays a key role. The introduction to this chapter provided the example that hepatitis contracted from a needle stick would be viewed as an injury. Given this example, the entire bloodborne pathogens (1910.1030) standard should be treated as mostly a safety issue. The far-reaching standard on hazard communication (1910.1200) fits both a safety and health pattern. It is fitting that the main vehicle to communicate hazards is identified as a "material *safety* data sheet." Last, the occupational exposure to hazardous chemicals in laboratories (1910.1450) standard also incorporates many safety (e.g., injury prevention principles).

Many other hazards within workplaces are hard to separate as being either a safety or health hazard. Is an ill-fitting glove worn by an employee mostly a safety or health hazard? From a safety point of view the ill-fitting glove may prevent a proper grip and result in an employee dropping a heavy weight onto his or her foot, causing an injury. The ill-fitting glove may also require an employee to squeeze harder to hold parts. Over time carpal tunnel, an illness, might result.

OSHA's standard on lockout and tagout, which addresses the unexpected start-up or operation of equipment, may seem to be entirely a safety issue. But the standard actually is about controlling energy (even energy from stored fluids) and pressurized and hazardous materials. The inadvertent release of these energies could also cause health consequences. Thus, we have another overlap. You likely can imagine many other examples. If all that separates an injury from an illness — and thus safety and health — is the point in time needed for a short snap of one's fingers, then safety professionals and occupational hygienists have much in common.

Qualifications to Become a Safety Professional

According to McLean, since safety is a multidisciplinary profession it draws its workers from many different areas, including education, engineering, psychology, medicine, and biophysics.(12) Most people have knowledge of safety activities being managed and performed at a workplace by the human resource manager, maintenance engineer, or security guard. It is still not uncommon today to find that some injured employee who is unable to return to his or her normal job has taken on safety responsibilities and has become the workplace "safety professional." The term safety professional, then, is a generic term for other titles such as safety engineer, safety manager, safety representative, safety contact, or some other semi-descriptive term that addresses the performance of safety activities. The American Society of Safety Engineers (ASSE) has defined safety professional as "an individual who, by virtue of his specialized knowledge, skill and educational accomplishments, has achieved professional status in the safety field."

Which backgrounds are best suited to safety will be open to debate for a long time. An engineering background may be best suited if prevention and control of physical hazards (i.e., conditions) is a primary objective. If it is true that 80% or more of injuries are the result of unsafe acts, perhaps a person with primarily an education or psychology background would be better suited for the job. Manuele(13) believes that before the practice of safety can be considered a profession, it must have a sound theoretical and practical base for the practice — but safety professionals have not yet agreed on these fundamentals. Since safety professionals are not all cut from the same cloth, qualifications for who is a safety professional and who can practice and manage safety activities will be a situation-by-situation decision made by employers.

Employers looking to hire someone with demonstrated competence in the field of safety can check to see if the safety professional is certified. The Board of Certified Safety Professionals (BCSP) has been issuing certifications in the safety field since 1968. The most demanding of the certifications is the certified safety professional (CSP). To obtain a CSP, a person has to provide evidence of safety work experience and education and pass examinations.

DeClue describes a number of changes made by the BCSP to upgrade the quality of the certification.[14] The first change, in mid-1996, was for the BCSP to become accredited by the National Commission for Certifying Agencies (NCCA). This is important because there are a number of groups and organizations that issue certifications in safety. Some of these are of dubious quality. The NCCA will not accredit a certification that lacks solid merits. Most changes the BCSP took to improve the quality of its certifications took effect in January 1997. The remainder of changes will be effective in January 1998.

A significant quality improvement change is that candidates submitting applications must have an associate degree in safety and health, or a bachelor's degree in any field that has minimum qualification for the academic requirement. The degrees must be from U.S. academic institutions accredited by a regional accrediting body recognized by the Commission on Recognition of Post Secondary Accreditation. The importance of this latter improvement is that there are a number of organizations that issue degrees in safety, and some of these organizations are of dubious quality. A recognized accrediting body for degrees will not accredit an organization that issues degrees of dubious quality. All of these changes should help improve the quality of safety professionals.

References

1. **Bureau of Labor Statistics:** *Recordkeeping Guidelines for Occupational Injuries and Illnesses.* Washington, D.C.: U.S. Department of Labor, Bureau of Labor Statistics, 1986. p. 37.
2. **Heinrich, W.W.:** *Industrial Accident Prevention.* New York: McGraw-Hill, Inc., 1931.
3. **DeReamer, R.:** *Modern Health and Safety Technology.* New York: John Wiley & Sons, Inc., 1980. p. 64.
4. **Grimaldi, J.V., and R.H. Simonds:** *Safety Management.* Homewood, Ill.: Irwin, 1989. pp. 27-28.
5. **Ferguson, D.S.:** Snapshots of Safety's History: What Will They Think of Next? *Professional Safety 41(12):*22-26 (1996).
6. **Petersen, D.:** The Occupational Safety and Health Act of 1970: 25 Years Later. *Professional Safety 41(12):*27-28 (1996).
7. **Roughton, J.E., and L.J. Grabiak:** Reinventing OSHA: Is It Possible? *Professional Safety 41(12):*29-33 (1996).
8. **Chemical Manufacturers Association:** *Resource Guide for Employee Health and Safety Code: Program Performances Measures.* Washington, D.C.: Chemical Manufacturers Association, 1995.
9. **Andrews, E.W.:** Basics of Safety. In *Accident Prevention Manual for Industrial Operations,* 4th Ed. Chicago: National Safety Council. 1959. pp. 6-7.
10. **Brown, D.B.:** *Systems Analysis & Design for Safety.* Englewood Cliffs, N.J.: Prentice-Hall, Inc., 1976. pp. 42-43.
11. **U.S. Department of Labor. Occupational Safety and Health Administration.** *Safety and Health Program Management Guidelines; Issuance of Voluntary Guidelines.* Issued at 54 FR 3904, January 26, 1989.
12. **McLean, W.T.:** The Safety Professional: In *Fundamentals of Industrial Hygiene,* 3rd Ed. Chicago: National Safety Council, 1988. pp. 585-589.
13. **Manuele, F.A.:** Principles for the Practice of Safety. *Professional Safety 42(7):*27-31 (1997).
14. **DeClue, M.C.:** "The Future is Now," BCSP Declares. *Professional Safety 41(12):*21 (1996).

Appendix A

The occupational safety professional should have access to a basic library of safety textbooks. The following books are recommended:

Ashfal, C. Ray: *Industrial Safety and Health Management.* Englewood Cliffs, N.J.: Prentice-Hall, 1990.

Brauer, R.: *Safety and Health for Engineers.* New York: Van Nostrand Reinhold, 1994.

Brown, D.B.: *Systems Analysis & Design for Safety.* Englewood Cliffs, N.J.: Prentice-Hall, 1976.

Chemical Manufacturers Association: *Resource Guide for Employee Health and Safety Code: Program Performances Measures.* Washington, D.C.: Chemical Manufacturers Association, 1995.

DeReamer, R.: *Modern Safety and Health Technology.* New York: John Wiley & Sons, 1980.

Goetsch, D.: *Occupational Safety and Health.* Englewood Cliffs, N.J.: Prentice-Hall, 1996.

Grimaldi, J.V., and R.H. Simonds: *Safety Management.* Homewood, Ill.: Irwin, 1989.

Hammer, W.: *Occupational Safety Management and Engineering.* Englewood Cliffs, N.J.: Prentice-Hall, 1989.

Krause, T.: *Employee-Driven Systems for Safe Behavior.* New York: Van Nostrand Reinhold, 1995.

Manuele, F.: *On the Practice of Safety.* New York: Van Nostrand Reinhold, 1993.

Marshall, G.: *Safety Engineering.* Monterey, Calif.: Brooks/Cole Engineering Division, 1982.

National Safety Council: *Accident Prevention Manual for Industrial Operations,* 11th Ed.

Slote, L.: *Handbook of Occupational Safety and Health.* New York: John Wiley & Sons, 1987. [2-volume set]. Itasca, Ill.: National Safety Council.

The Internet is also a useful source of occupational safety information. There are at least 129 web and gopher sites, 61 mailing lists, and 13 news and discussion groups in this subject area, and these numbers are growing constantly. The instructor's guide that accompanies this book will include specific information on Internet resources.

— Donald Robinson

Chapter 46

Outcome Competencies

After completing this chapter, the user should be able to:

1. Define underlined terms used in this chapter.
2. List the objectives of an occupational health program.
3. Describe medical surveillance programs.
4. Explain the role of the occupational hygienist in a medical surveillance program.
5. Quote the limitations to medical treatment in the occupational setting.
6. List criteria for selecting health care providers.

Key Terms

Americans with Disabilities Act (ADA) • detectable preclinical phase • health history • medical treatment • medical surveillance • occupational history • periodic medical examination • preplacement examination • primary prevention • program management • pulmonary function tests • recommended accommodations • secondary prevention measures • symptom questionnaires • unsuitable workers • work restrictions

Prerequisite Knowledge

Prior to beginning this chapter, the user should review the following chapters.

Chapter Number	Chapter Topic
1	History and Philosophy of Industrial Hygiene
2	Occupational Exposure Limits
3	Legal Aspects of the Occupational Environment
4	Environmental and Occupational Toxicology
5	Epidemiological Surveillance
6	Principles of Evaluating Worker Exposure
13	Biological Monitoring
15	Comprehensive Exposure Assessment
16	Modeling Inhalation Exposure
17	Risk Assessment in the Workplace
31	General Methods for the Control of Airborne Hazards
32	An Introduction to the Design of Local Exhaust Ventilation Systems
37	Program Management
38	Surveys and Audits
40	Hazard Communication
49	Emergency Planning in the Workplace
50	Anticipating and Evaluating Trends Influencing Occupational Hygiene

Key Topics

I. Objectives of an Occupational Health Program

II. Medical Surveillance

III. Integrating the Occupational Hygienist into the Medical Surveillance Program
   - A. Evaluate Workplace Exposures
   - B. Identify Individuals at Risk
   - C. Conduct Medical Examinations for Early Health Effects
   - D. Report Medical Findings to Employees
   - E. Determine Employee Medical Restrictions
   - F. Accommodate Medical Restrictions
   - G. Analyze Potential Relationships Between Medical Results and Occupational Exposure
   - H. Implement and Maintain Effective Workplace Controls

IV. Medical Treatment
   - A. Preparing for Emergency Treatment
   - B. Managing Medical Cases

V. Program Management

VI. Selecting Health Care Professionals

46

Collaborating with the Occupational Physician

Thomas D. Polton

Introduction

Occupational hygiene is most commonly associated with efforts to control exposures to the chemical, physical, and biological causes of occupational disease. Indeed, these so-called primary prevention programs, which range from substituting water-based materials for more toxic chemicals to constructing sound-attenuating barriers around noisy equipment to establishing administrative controls for handling infectious agents, form the cornerstones of any occupational health program.

But primary prevention alone is not sufficient to eliminate workplace illnesses. Instead, a comprehensive occupational hygiene program must also include secondary prevention measures. These are programs designed to identify individuals who are affected by contaminants in the workplace, to prevent further exposure and to provide medical treatment if necessary.

In many cases, responsibility for secondary prevention is put solely in the hands of occupational physicians and other health professionals. But occupational hygienists can also make important contributions to these programs.

A comprehensive occupational health program consists of three efforts:

- Medical surveillance — to identify employees exhibiting signs or symptoms of the early stages of an illness;
- Medical treatment — to provide prompt medical care for all work-related injuries and illnesses; and
- Program management — to integrate the occupational health program with the activities of the occupational hygienist.

The remainder of this chapter will provide a more detailed explanation of these key elements of an occupational medicine program, discuss the complementary roles of the physician and the occupational hygienist, and provide guidance on selecting a qualified occupational health professional.

Objectives of an Occupational Health Program

The first step in establishing an occupational health program is to define the objectives for the medical program. The program's objectives will depend greatly on the industry, its potential hazards, the work force, the resources within the community, and the level of support of the employer. At a minimum, the program must address all regulatory requirements imposed by federal, state, and local authorities. For example, governmental standards require medical surveillance and medical services when employees are exposed to particular contaminants,[1] use certain types of protective equipment,[2] or perform specific tasks.[3]

There are many situations for which employers may decide to provide medical surveillance or establish medical treatment protocols, even though there are not any governmental requirements. For instance, workers may experience heat stress from performing physically demanding work while wearing protective equipment, face the potential for life-threatening emergencies because of accidental chemical releases, or endure continual shoulder pain caused by repetitive tasks. In each of these situations, the governmental standards do not contain specific provisions to address these risks to employee health, yet medical surveillance and, on occasion, medical treatment are indicated. For this reason, the occupational physician and the occupational hygienist are likely to include provisions in the employer's occupational health program to address these situations as well.

Before hiring a health care staff, the employer or the consultant designing the program should assess the availability and quality of the emergency medical care services and medical specialists within the community. For hazards requiring immediate attention, for example, the company medical plan may specify having a physician or nurse on site or training first-aid response teams. This option might not be necessary if medical care is located nearby.

After the medical surveillance and medical treatment programs have been determined, the occupational hygienist and health care provider should develop a written program defining the responsibilities of all the participants and the procedures for maintaining important medical records (see Table 46.1). The program should be reassessed periodically. The next sections discuss in detail the activities and responsibilities of both parties for medical surveillance, medical treatment, and program management in a comprehensive occupational health program.

Medical Surveillance

Medical surveillance is designed to detect an illness or organ dysfunction at an early, treatable phase.[4] Medical examinations are typically offered before employees begin an assignment (preplacement), at a predetermined frequency during that assignment (periodic), and at the conclusion of their employment or the end of the expected latency period of the materials to which workers have been exposed (termination).

The primary goal of the preplacement physical examination is to reveal any medical condition that might put the worker at an increased risk to himself or to others because of work exposures or activities.[5] It should not be used to exclude "unsuitable workers" from work opportunities. Instead, it is the employer's responsibility to develop suitable workplace accommodations whenever feasible. The results of the preplacement examination also provide a baseline for comparing future tests in order to assess the impact of subsequent exposures on the development of illnesses. Some employers also include tests for nonoccupational diseases such as hypertension or breast cancer. Others use the preplacement exam to direct workers to health promotion activities, such as smoking cessation programs.

The preplacement examination also offers an opportunity to educate workers

Table 46.1 — Responsibilities of Health Care Professional and Occupational Hygienist in a Comprehensive Occupational Health Program

Program Element	Health Care Professional	Occupational Hygienist
Medical surveillance	• Perform medical screening to identify early symptoms of occupational disease at a reversible or treatable phase. • Analyze the medical data to identify and communicate medical trends for individuals and groups of workers. • Determine medical restrictions. • Advise on appropriate accommodations for workers to aid their prompt return to work. • Develop and implement wellness and health promotion activities for nonoccupational disorders.	• Characterize employee exposures to hazardous agents. • Identify employees to be offered medical evaluations because exposures exceed designated trigger levels. • Develop and implement appropriate accommodations for workers to aid their prompt return to work.
Medical treatment	• Treat work-related illnesses and injuries. • Coordinate emergency and nonemergency treatment with physicians and hospitals in the community. • Direct the recovery and rehabilitation of employees who have prolonged or complex illnesses or injuries.	• Integrate medical preparedness into emergency response plan. • Develop and implement appropriate accommodations for workers to aid their prompt return to work.
Program Management	• Maintain accessible and confidential individual employee medical and exposure records. • Periodically reassess the medical program and modify it according to revised needs.	

about the potential hazards of the new job and about the health and safety measures being used to avoid illnesses and injuries. Another goal of the exam is to establish a rapport between the worker and the health care professional so that if a potential work-related problem does arise, the employee will feel comfortable discussing it with the health professional.

The periodic medical examination begins with questions concerning any changes in the job, work processes, or exposures that might increase the risk of occupational illness. Medical examinations and testing procedures are designed to detect changes in the health of the employee since the last examination that might indicate a need for a change in the work process or in job placement.[6] The examination may include biologic monitoring tests and a variety of other surveillance techniques.

Employees should have an updated medical examination at the end of their employment or at the end of the disease latency period associated with past exposures. This is a decision that must come from both the occupational physician and the occupational hygienist.

Medical surveillance includes a second important element. This is the analysis of medical data to identify changes in the health status of individual workers, and to identify any patterns or trends among groups of workers.[7] When performed properly, this analysis benefits both the individual worker and the groups of workers who share common exposures and risks.

Individuals benefit when early detection is followed by appropriate intervention (e.g., reducing workplace exposures or providing medical treatment to the employee). Groups of workers benefit when analyses of their aggregate data uncover a link between exposures and illnesses and provide insight into how this link might be broken.

In many cases, analysis of the medical surveillance data leads to the implementation of new primary prevention methods. For example, the employer may re-evaluate the effectiveness of workplace controls or replace hazardous materials with less toxic substances, thereby preventing occupational illness.

Integrating the Occupational Hygienist into the Medical Surveillance Program

An effective medical surveillance program requires communication between the occupational hygienist and the occupational physician. Before the worker receives a medical evaluation, the occupational hygienist should alert the physician to any hazards that the worker has encountered. Following the medical evaluation, the occupational hygienist should be notified of any medical findings that are the result of workplace exposure. The occupational physician should work with the occupational hygienist and the safety professional to accommodate any new work restrictions. The occupational hygienist plays a key role in altering the job to accommodate the worker's restrictions. The occupational hygienist and the occupational physician, with the possible assistance of an epidemiologist, must review the medical data for potential trends or patterns of disease. The relationship between the occupational hygienist and occupational physician is composed of eight steps described in the following sections (see Figure 46.1).

Evaluate Workplace Exposures

A well-designed medical surveillance program begins with a thorough characterization of the hazards encountered by workers. For each work area, process, or

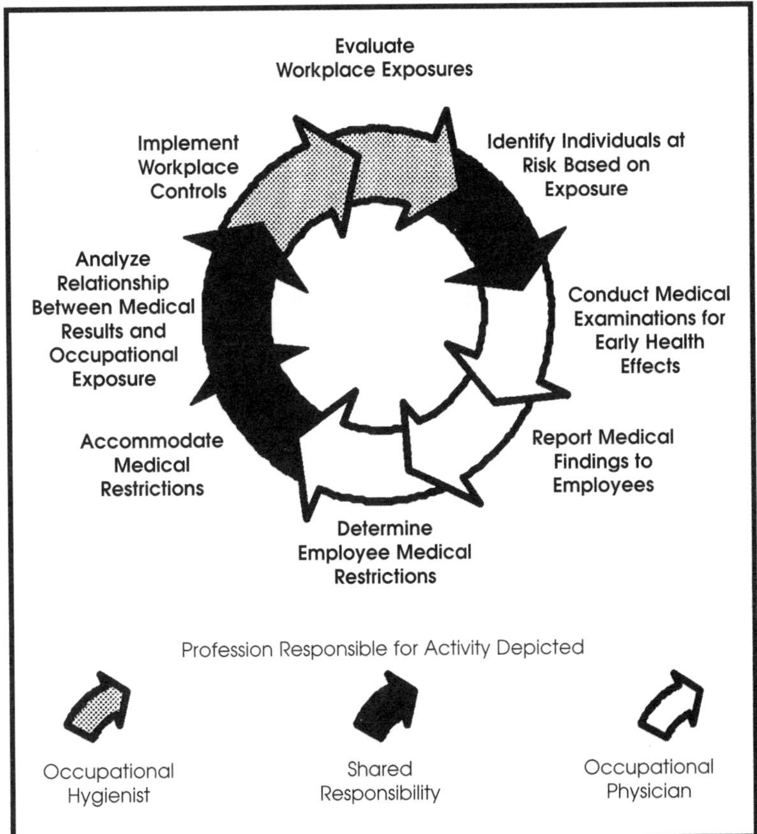

Figure 46.1 — Medical surveillance: The interaction between the occupational hygienist and the occupational physician.

occupation the occupational hygienist completes a workplace exposure assessment. This assessment identifies potential chemical, physical, and biological hazards in the workplace along with details on:

- Likely routes of exposure (inhalation, ingestion, and skin absorption or contact);
- Duration (length of the exposure during a given work shift);
- Frequency (number of times per day, month, or year this exposure occurs); and
- Intensity (measured concentration of contaminant to which the worker is exposed).

The occupational hygienist and the occupational physician review the anticipated health effects, the likely target organs, and the expected disease latency periods associated with these exposures. The occupational physician then uses the information to develop a physical examination program containing all relevant questions that can be added to the basic occupational and health history. The surveillance examination will include all appropriate laboratory tests in a program of biologic monitoring. The designated latency period influences the schedule for further medical evaluations.

The workplace exposure assessment should also contain information on government requirements for medical surveillance. Occupational physicians should be familiar with these regulatory requirements.

It is also important to report information about the workplace controls (engineering, administrative, or personal protective equipment) used to limit worker exposures. This information is not only critical in analyzing any group trends but, as is the case with respirators, it initiates required examinations.(2)

Identify Individuals at Risk

The occupational hygienist and the occupational physician use the findings of the workplace exposure assessments to decide which employees require medical surveillance. Often, the action level or trigger level for initiating an employee medical surveillance program is based on the airborne concentration of contaminants measured in the work areas. The most common trigger levels are as follows:

- *For acutely toxic materials:* a single episode exceeding the ceiling or the 15-minute short-term exposure limit (STEL); and
- *For chronic hazards:* employee exposure to one-half the full shift exposure limit (permissible exposure limit [PEL] or ACGIH threshold limit value [TLV®]) for the contaminant for 30 days or more. (Note that some specific regulations may differ from these general guidelines.)

Excessive exposure through routes other than inhalation may also jeopardize the employee's health. Workers may ingest toxic materials on their hands while eating or smoking in the workplace; they may also absorb solvents directly through the skin. For these situations, the occupational physician may use the results of biological monitoring (analysis of blood, urine, or expired air samples for elevated concentrations of workplace contaminants or their metabolites) as a trigger level for initiating more frequent medical examinations and for the occupational hygienists to intercede with primary prevention measures (see Chapter 13, "Biological Monitoring").

Reports of health complaints or employee symptoms consistent with workplace exposures may be another reason for initiating a medical surveillance program.(8) For instance, employees working with coatings or adhesives containing diisocyanates may suddenly complain about wheezing or dry cough. The medical surveillance program also should include workers exposed to hazardous materials from accidental process releases, chemical spills, or other workplace incidents.

Conduct Medical Examinations for Early Health Effects

The cornerstone of the medical examination is the occupational health history. The initial occupational history for a new employee should consist of the following: a work profile; a listing of symptoms, illnesses, or injuries related to past jobs; questions relevant to the requirements or potential hazards of the new job; and a listing of significant community and home exposures.(9) The medical history during the periodic examination focuses on changes in health status, illnesses, and possible work-related symptoms. Once the history is completed, it is followed by the medical examination and by appropriate medical and laboratory tests.

The selection of medical tests should also take into account both the occupational hygienist's workplace exposure assessments and the availability of reliable measures. Despite the fact the worker's occupational history indicates that he or she has had significant exposure to workplace contaminants, reliable medical tests are not always available. Biological monitoring is advancing rapidly, but it is still able to monitor only a limited number of occupational exposures.

Laboratory analysis of blood or urine, clinical procedures (audiometry, chest X-rays, or pulmonary function tests), symptom questionnaires, and the physical examination are not always definitive. They do, however, represent the best available technology for exercising secondary preventive measures.

The time it takes for symptoms to emerge is crucial in determining whether surveillance tests are feasible. For instance, surveillance testing is inappropriate for acute toxic effects like cyanide poisoning since the onset of symptoms is so rapid. On the other hand, conditions involving a longer onset period may be well-suited to a surveillance program that periodically evaluates a worker's health. The schedule (periodicity) of medical examinations should be based on the expected timing of health effects in relation to exposure. In this regard, both the induction period (the period between first exposure and first manifestation) and the duration of the "detectable preclinical phase" of a health effect are important considerations. Conditions with long induction periods often have long detectable preclinical phases during which testing will be beneficial. Testing, however, is inappropriate before sufficient time has lapsed for the health effect to be detected by the test being used.

The health professional designing the medical surveillance program should develop a plan for how medical data will be interpreted and acted on. Here, again, the occupational physician and the occupational hygienist should work together. This plan should include criteria used to determine when action will be taken in response to medical test results and what that action will be. Workers should be given a written summary of the action plan, in easily understandable terms, before they are asked to provide informed consent for participation in medical surveillance.

The following responses are appropriate when the action threshold for a medical test is exceeded: employee notification; test confirmation either through retesting with the same method or using a more accurate method; medical evaluation and treatment if indicated; and evaluation and control of the occupational exposure. No plan can anticipate every situation that will arise during a surveillance program, and interpretation of individual test results often requires the judgment of the occupational physician.

Report Medical Findings to Employees

Notifying workers in writing of their medical test results is essential, and especially important when abnormalities are detected. The worker should be told the potential significance of the test result; the risk, if any, from continued exposure to the work environment; the recommended changes in work practices or personal habits; and the necessary medical treatment. Nonoccupational medical findings (such as elevated blood pressure) should be referred to the employee's personal physician for treatment.

Determine Employee Medical Restrictions

The occupational physician should inform the employer of an employee's work restrictions and any recommended accommodations. It is important that the medical findings include a specific description of the employee's functional capabilities so that the individual does not perform tasks that jeopardize his or her health. Imprecise recommendations may lead an employer to reassign a worker, taking him or her off a job he is fully capable of performing. For instance, when a physician broadly declares that an employee is "unfit to wear a respirator," the worker will most often be reassigned from any job involving the use of a respirator. Yet, the medical examination may actually have indicated that the employee can work at a job requiring a respirator, provided it is worn occasionally for short periods or only for an emergency escape. This is another example of the requirement for the occupational physician and the occupational hygienist to work together.

The employee must be confident that the medical information will be kept strictly confidential, will be used only for the stated purposes, and will not adversely affect salary or other benefits. Employers may require access to personal information when considering requests for job accommodation or reviewing claims for workers' compensation benefits. The physician should avoid providing the employer with specific details or diagnoses without the worker's permission. The worker's health information also may have to be disclosed to other employees in certain situations.[10] Supervisors and managers may be informed of any necessary work restrictions, and, when appropriate, first aid and safety personnel may be notified if the worker's medical condition might require emergency treatment.

Personnel involved in the occupational health program need to recognize the special confidential status of HIV positivity and drug and alcohol treatment information.[10] They should be aware that a general consent for disclosure of medical records will often be insufficient in these situations and that specific written consent for release of this information must be obtained. This information should only be disclosed in compliance with federal and state law. It is important for workers to feel that their disclosures will be treated in a dignified and confidential manner to assure their agreement with this disclosure of sensitive information.

Accommodate Medical Restrictions

To get the worker back to his or her job, it is often necessary for the occupational hygienist to design accommodations to compensate for the worker's medical restrictions. This may mean implementing workplace controls; developing administrative recommendations to limit the duration, frequency, or intensity of workplace exposures; or providing personal protective equipment to eliminate contact with or exposure to a hazardous agent. If an employer rejects a disabled worker's request to return to work, the burden of proof is on the employer to mount a well-documented defense that proves at least one of the following:[11]

- The worker cannot perform the essential functions of the job, even with reasonable accommodation;
- The employer cannot accommodate the worker by redistributing the work among other employees or by restructuring the job;
- The employer cannot afford the cost of reasonable accommodation;
- Accommodation would substantially change the nature of the business; and/or
- The worker would be a "direct threat" to his or her health and safety or to other workers.

Throughout this process, the employer must verify compliance with the Americans with Disabilities Act (ADA) of 1990 and state laws. Medical removal from a specific job should be considered as an interim measure when all other feasible actions are considered inadequate to protect the affected worker. In most instances, a physician will be required to determine whether an individual should be removed from the job while workplace evaluation and modification are being undertaken. In some circumstances, with frequent clinical monitoring, the individual may safely remain at his or her job during this process. Whenever the occupational physician recommends that a worker should be removed from a specific job, the worker should be informed of the basis for such a recommendation (i.e., the risk and severity of an adverse outcome if the worker remains in the exposed job). Wages and benefits should be protected when individuals are removed from a job because of a medical examination.

Analyze Potential Relationships Between Medical Results and Occupational Exposure

The physician and the occupational hygienist should work together, sometimes with the aid of an epidemiologist, to examine patterns or trends from the occupational hygiene and the medical examination data. To gain the most insight, both individual and group trends should be examined.

In addition to looking at individuals, the team also should examine test results for groups of workers. Examining groups often uncovers links between illness and workplace exposures. The occupational hygienist's workplace exposure assessments are the starting point for this process. By categorizing workers according to their exposures, it may be possible to uncover a relationship between low-level exposure levels and early health effects.

Such analyses cannot be based on a small number of workers. The statistical power of larger groups makes it possible to uncover the impact of exposure to toxic agents at an early stage, where changes in results are subtle but intervention can have its greatest impact.

The medical surveillance program may also reveal that a small group of employees has similar complaints about their health or that they have been diagnosed with similar illnesses. This is known as a disease cluster.[12] It is often the case that these workers also shared something else in common (such as the same occupational or nonoccupational exposure) or that they shared the same work area. Examples of clusters include a group of individuals with cancer, all of whom work in the same location.

The number of cases might be too small to be statistically significant, but the situations require a detailed investigation to determine potential causes. The occupational hygienist may be called on to work with the occupational physician to investigate these situations. Accurate information surrounding the onset of symptoms, the frequency and range of symptoms, symptom intensity and consistency, and pre-existing conditions may help in identifying the source of noncancer clusters. For cancer clusters it is important to gather information on commonality of the organ and cell type and to identify the latency periods.

Implement and Maintain Effective Workplace Controls

All these efforts to identify trends will be meaningless, unless the occupational hygienist acts on his or her findings. To prevent occupational disease, the occupational hygienist must implement workplace controls that will prevent similar problems from occurring in the future. In some cases, workplace evaluation may verify the effectiveness of some controls while discovering deficiencies in others. For instance, the medical surveillance program may confirm that inhalation exposures are adequately controlled for solvent exposures. Reports of reddening of the skin, however, may indicate that a different type of glove is needed to better protect workers' hands and to prevent skin absorption.

To make sure that any improvements are sustained, the occupational hygienist should periodically verify the effectiveness of the workplace controls. Exhaust ventilation systems and other engineering controls must be inspected to measure the adequacy of the face velocities or to maintain the exhaust fans. For administrative controls to work, workers must be supervised to ensure that they follow recommended work practices or adhere to limitations on the maximum duration of an exposure. The occupational hygienist should also test the reliability of any personal protective equipment. This may mean establishing respirator cleaning procedures and schedules for replacing respirator cartridges at the end of their service life and performing fit-testing studies for respirator users.

Occupational illnesses are prevented by controlling exposures. As a corollary to this maxim, we can only be assured that exposures are being controlled when controls are verified.

Medical Treatment

Even with the best prevention programs, occupational injuries and illnesses will occur. To minimize the consequences and disruption associated with these events, it is important to establish protocols for handling emergencies and for providing nonemergency medical care to employees. These plans often involve the use of community medical services (hospitals and physicians) to supplement the employer's own resources. Off-site health care professionals should be taken on plant tours to make them familiar with the employer's operations. Program management includes the necessary preparation for both emergency and nonemergency situations.

Preparing for Emergency Treatment

Timely medical treatment is crucial in emergency situations, such as an accidental chemical release involving toxic chemicals, an employee overcome by heat stress, or an explosion with burn injuries. An emergency treatment plan must address situations that involve not only employees but also contractors, visitors, and community residents who may be affected by an untoward event.[13] The plan should be integrated with the overall site emergency response program designating the roles and responsibilities for both on-site and off-site personnel. Also, it should contain procedures, equipment lists, emergency contacts, and signs and symptoms of exposures. A designated emergency/first aid station should be selected and equipped with supplies that can be used to stabilize patients and treat injuries that require only first aid. Site personnel assisting emergencies should be trained to handle these situations.

The occupational hygienist may be assigned responsibility for briefing off-site medical personnel on the hazardous chemicals used at the operation and for keeping the chemical emergency information files up-to-date. Advanced planning is required for emergency transportation to nearby medical facilities and for having health care professionals available around the clock.

Managing Medical Cases

Medical management programs are important for chronic conditions, especially cumulative trauma disorders and other work disorders due to ergonomic hazards. Like the medical surveillance program, the medical management program is not meant to replace the hazard prevention and control programs but is a necessary complement to them. An occupational physician or occupational health nurse with training in the prevention and treatment of occupational disorders should supervise the program. The medical management program should address the following issues: early recognition and reporting of injuries and illnesses; careful maintenance and interpretation of medical records to identify trends; and medical referral to specialists to ensure that workers receive appropriate treatment; and return-to-work and restricted-duty policies to aid the rehabilitation of employees.(14)

Program Management

Maintenance of complete medical records is essential to comprehensive medical care, and occupational health programs are no exception. Medical records should include not only the results of medical examinations but also personal exposure and biological monitoring results provided to the physician, the purpose of the medical tests given, the interpretation of test results, and the plan for responding to the results. Notifications provided to employers of examination results indicating a need to evaluate the work environment should be maintained in the medical record, as should copies of any notification to the employee, either as an individual, or as a member of a group.

Records containing the results of the medical examinations must be maintained according to law. These records might be needed to compare with future test results for an individual or to further evaluate a suspected toxic effect using aggregate data. According to OSHA, employers must preserve worker medical records for the duration of employment plus 30 years.(15) The records must be made available, on request, to workers, their representatives, and authorized regulatory officials.

Under the OSH Act of 1970, employers must record all work-related injuries that require medical treatment (other than first aid) or involve loss of consciousness, restriction of work or motion, or transfer to another job, as well as diagnosed occupational illnesses.(16) In these cases, the company must file a Supplementary Record of Occupational Injuries and Illnesses, OSHA 101, identifying the details of the incident. The employer must also maintain and post in the workplace an OSHA 200 log of occupational injuries and illnesses.

The records of health and safety incidents form the core information that should be reviewed to assess the effectiveness of the occupational health program. To promote continuous program improvement, the occupational hygienist and health care professional should, at least annually:(17)

- Ascertain that each injury or illness was promptly investigated to determine the cause and make necessary changes in health and safety conditions or procedures;
- Evaluate the efficacy of specific medical testing in the context of potential employee exposures;
- Add or delete medical tests as suggested by current occupational hygiene workplace exposure assessments;
- Review emergency treatment procedures and update lists of emergency contacts and hazardous chemical information files (material safety data sheets [MSDSs]); and
- Evaluate the effectiveness of the medical case management programs and performance of referral physicians.

Selecting Health Care Professionals

A variety of health care professionals may be qualified to assist the occupational physician in providing health care in occupational settings. For instance, occupational health care nurses assess and treat illnesses and injuries. Their primary role in surveillance activities is to assist the occupational physician. Nurse practitioners may perform examinations. Other health care professionals (including physician assistants, licensed practical nurses, and emergency medical technicians) are used to augment occupational health programs. When they identify health problems, they may provide emergency care under the direction or supervision of a physician or an occupational health nurse.

Before hiring or contracting a health care provider for services either on or off site, the capabilities of the individual(s) should be reviewed by the occupational physician. It also is imperative to consider the legal scope of practice unique to each state.(18) The scope of practice refers to the credentials, responsibilities, and limitations of health care practitioners. The health care professional's state licensing or certification board can verify his or her scope of practice for the occupational setting. Physicians receive a standardized medical school education. That still requires further training in occupational and environmental medicine. This additional occupational health training typically includes epidemiology, toxicology, occupational hygiene, recognition and management of occupational illnesses and injuries, research, and general management of a comprehensive occupational health program.

Physicians specialized in occupational medicine are certified by the American Board of Preventive Medicine. The training of other health care professionals is more variable and can be helped by the expertise of the occupational physician in their selection.

Summary

The primary objective of the occupational health program is to integrate occupational hygiene and occupational health, thus leading to the important union of primary and secondary preventive measures. The specific contents of the program vary depending on regulatory requirements, hazards of the workplace and its operations, the resources within the community, and the support of the employer.

To maximize the effectiveness of the occupational health program, the occupational hygienist and the health care professional should work as a team. In a well-organized medical surveillance program, the occupational hygienist characterizes the hazardous exposures. The physician then uses this information to design effective medical evaluations for individuals with significant exposure to chemical, physical, or biological agents. Through preplacement, periodic, and termination examinations, the health care professional uncovers deficient workplace controls and employees with medical restrictions requiring accommodation. Together, the occupational hygienist and occupational physician develop medical treatment programs to minimize the consequences of emergency and nonemergency situations and to ensure that employees receive proper medical care so they can return to work as quickly as possible.

The occupational hygienist and health care professional should periodically review the medical and incident records to identify opportunities to improve the occupational health program. Hazard surveillance is the preferred prevention method over medical surveillance. Eliminating occupational illness relies on effective occupational hygiene programs to recognize, evaluate, and control occupational hazards.

Acknowledgment

The author would like to thank Joseph LaDou, MD, of the University of California–San Francisco for his help in reviewing this chapter.

Additional Sources

"The Americans with Disabilities Act of 1990," 42 USC 12113.

Conway, H., J. Simmons, and T. Talbert: The Purposes of Occupational Medical Surveillance in U.S. Industry and Related Health Findings. *J. Occup. Med. 35(7)*:670-686 (1993).

Ellis, J.: Medical Monitoring: Can We Do Better? *Occup. Health, Safety 65(5)*:19-25 (1996).

LaDou, J. (ed.): *Occupational and Environmental Medicine,* 2nd Ed. Stamford, CT: Appleton & Lange, 1997.

Murphy, L., and W.E. Halperin: Medical Screening and Biological Monitoring: A Guide to the Literature for Physicians. *J. Occup. Environ. Med. 37(2)*:170-184 (1995).

Newman, L.S.: Occupational Illness: Essential Elements of the Occupational History. *N. Eng. J. Med. 333(17)*:1128-1132 (1995).

Polton, T.: Critical Roles for Industrial Hygienists in Medical Surveillance and the Prevention of Occupational Disease. *The Synergist. 6(1)*:24-25 (1995).

Silverstein, M.: Analysis of Medical Screening and Surveillance in 21 Occupational Safety and Health Administration: Support for a Generic Medical Surveillance Standard. *Am. J. Ind. Med. 26*:283-295 (1994).

U.S. Department of Labor Statistics: *A Brief Guide to Recordkeeping Requirements for Occupational Injuries and Illnesses* (OMB No 1220–0029). 1986.

References

1. Title 29 *Code of Federal Regulations* Part 1910.1001, 1003, 1017, 1018, 1025, 1027, 1028, 1029, 1043, 1044, 1045, 1047, 1048, 1050, 1051, 1052.
2. "Respiratory Protection," Title 29 *Code of Federal Regulations* Part 1910.134.
3. "Hazardous Waste Operations and Emergency Response," Title 29 *Code of Federal Regulations* Part 1910.120.

4. **Baker, E.L.:** Role of Medical Screening in the Prevention of Occupational Disease. *J. Occup. Med. 32(9)*:788 (1990).
5. **Goldman, R.H.:** General Occupational Health History and Examination. *J. Occup. Environ. Med. 28(10)*:967 (1986).
6. **Goldman, R.H.:** General Occupational Health History and Examination. *J. Occup. Environ. Med. 28(10)*:971 (1986).
7. **Baker, E.L.:** Role of Medical Screening in the Prevention of Occupational Disease. *J. Occup. Med. 32(9)*:787 (1990).
8. **Silverstein, M.:** Medical Screening, Surveillance, and the Prevention of Occupational Disease. *J. Occup. Med 32*:1035 (1990).
9. **National Institute for Occupational Safety Health (NIOSH):** *Occupational Safety and Health Guidance Manual for Hazardous Waste Site Activities* (DHHS/NIOSH Pub. No. 85–115). Washington, D.C.: U.S. Government Printing Office, 1985. pp. 5-7.
10. **American College of Occupational and Environmental Medicine, Committee on Ethical Practice in Occupational Medicine:** ACOEM Position on the Confidentiality of Medical Information in the Workplace. *J. Occup. Environ. Med. 37(5)*:595 (1995).
11. **Equal Employment Opportunity Commission:** *A Technical Assistance Manual on the Employment Provisions (Title I) of the ADA* [EEOC–M–1A]. 1992.
12. **Sandler, H.M.:** Do You Have a Cluster? *Occup. Hazards 58(7)*:49-50 (1996).
13. **National Institute for Occupational Safety Health (NIOSH):** *Occupational Safety and Health Guidance Manual for Hazardous Waste Site Activities* (DHHS/NIOSH Pub. No. 85–115). Washington, D.C.: U.S. Government Printing Office, 1985. pp. 5-8.
14. **Occupational Safety and Health Administration:** *Ergonomics Program Management Guidelines for Meatpacking Plants.* 1990. p. 5.
15. "Access to Employee Exposure and Medical Records Standard," Title 29 *Code of Federal Regulations* Part 1910.20.
16. "Annual Summary," Title 29 *Code of Federal Regulations* Part 1904.5
17. **National Institute for Occupational Safety Health (NIOSH):** *Occupational Safety and Health Guidance Manual for Hazardous Waste Site Activities* (DHHS/NIOSH Pub. No. 85–115). Washington, D.C.: U.S. Government Printing Office, 1985. pp. 5-10.
18. **Occupational Safety and Health Administration:** *The Occupational Health Professional's Services and Qualifications: Questions and Answers.* Washington, D.C.: U.S. Government Printing Office, 1995. p. 3.

Workers removing chemical containers from a waste site.

Chapter 47

Outcome Competencies

After completing this chapter, the user should be able to:

1. Define underlined terms used in this chapter.
2. Discuss the difference between regulated and nonregulated hazardous wastes.
3. Summarize major federal legislative acts and amendments pertaining to hazardous wastes.
4. Classify hazardous wastes based on characteristics.
5. Draw a schematic diagram of the major phases of hazardous waste management.
6. Classify and summarize major methods for storage, treatment, and disposal of hazardous wastes.
7. Identify contributing factors and potential health effects associated with exposure to hazardous wastes.
8. Identify the role and contrast the phases of occupational health in relation to hazardous waste management.

Prerequisite Knowledge

Prior to beginning this chapter, the user should review the following chapters:

Chapter Number	Chapter Topic
4	Environmental and Occupational Toxicology
5	Epidemiological Surveillance
6	Principles of Evaluating Worker Exposure
9	Direct-Reading Instrumental Methods for Gases, Vapors, and Aerosols
10	Sampling of Gases and Vapors
12	Sampling and Sizing Particles
13	Biological Monitoring
14	Dermal Exposure and Occupational Dermotoses
18	Biohazards in the Work Environment
22	Ionizing Radiation
24	Applied Physiology of Thermoregulation and Exposure Control
25	Thermal Standards and Measurement Techniques
34	Prevention and Mitigation of Accidental Chemical Releases
35	Personal Protective Clothing
36	Respiratory Protection
38	Surveys and Audits
49	Emergency Planning in the Workplace

Key Terms

management • mismanagement • chemical wastes • radiological wastes • biological wastes • generation • storage • transportation • treatment • disposal • impact • occupational health

Key Topics

I. Management vs. Mismanagement

II. Major Applicable Federal Legislative Acts
   A. Legislative Responses to Mismanagement of Hazardous Wastes
   B. Major Federal Enactments and Amendments

III. Definitions of Hazardous Wastes
   A. Regulatory Definition
   B. Generic Definition

IV. Classification and Sources of Hazardous Wastes
   A. Classes of Wastes
   B. Chemical Wastes
   C. Radiological Wastes
   D. Biological Wastes
   E. Sources and Estimating Amounts of Hazardous Waste

V. Major Phases and Program Elements of Hazardous Waste Management
   A. Overview
   B. Generation
   C. Short-Term Storage
   D. Transportation
   E. Treatment, Long-Term Storage, and Disposal

VI. Methods for Treatment of Hazardous Wastes
   A. Concepts
   B. Thermal Treatment
   C. Physical Treatment
   D. Biological Treatment
   E. Chemical Treatment

VII. Methods for Long-Term Storage and Disposal of Hazardous Wastes
   A. Concepts
   B. Disposal Into Air
   C. Disposal Into Water
   D. Disposal and Long-Term Storage on Soil

VIII. Adverse Human Health Impact Due to Hazardous Waste Exposure
   A. Contributing Factors
   B. Impact to Human Health
   C. Risk Factors

IX. The Role of Occupational Health in Hazardous Waste Management
   A. Overview
   B. Recognition
   C. Evaluation
   D. Control

47

Hazardous Waste Management

Michael S. Bisesi, Ph.D., CIH

Christine A. Bisesi, CIH, CHMM

Introduction

For as long as there have been humans, waste materials have been generated and mismanaged. Early problems were mainly associated with human and animal biological wastes. Management of these wastes was not an immediate problem, however, since population densities were low and land area for disposal was vast. Disposal of waste materials in soil and water were likely the earliest practices of waste management.(1)

The primitive management practices were perhaps examples of the adages "out of sight, out of mind" and "the solution to pollution is dilution." In retrospect, however, it is clear that the disposal practices reflected mismanagement rather than management, although this was not realized at the time. As population densities increased, communicable infectious diseases were often correlated with the discharge of human feces on soils used as fields for growing edible crops and into waterways used as sources of drinking and bathing water.(2) The environment eventually demonstrated early signs of intolerance for excessive indiscriminate disposal of wastes.

The generation of chemical wastes in the United States became more prevalent following the initiation of industrialization during the nineteenth century. As industrialization rapidly increased and spread, manufacturing industries were so engrossed in output of product that little attention was focused on the generation of chemical wastes.(3) During the same period, radiological wastes were generated in increasing amounts from military, medical, and research sources.

Although indications of adverse environmental and public health impact were present earlier, it was not until the mid- to late-20th century that it became apparent that the environment had a threshold for the quantity of wastes that could be tolerated. After years of indiscriminate disposal of wastes into the environment, the quality of the air, water, and soil qualitatively and quantitatively exhibited signs of deterioration. In turn, the adverse impacts on the environment and public health were recognized and considered more seriously.

Management vs. Mismanagement

Management of hazardous wastes, regardless of sources, implies controlled and environmentally sound handling during the phases of generation, storage, processing for recovery or reuse, transporting, treating, and discharging into the air and water or discarding onto the soil. The concept of mismanagement, in turn, infers uncontrolled and environmentally unsound or indiscriminate handling of hazardous wastes, whether intentionally or unintentionally.

Several common denominators are essential for ensuring proper hazardous waste management. These include 1) defining and identifying hazardous wastes; 2) listing substances that meet the defined characteristics; 3) anticipating quantities and maintaining inventories; 4) determining appropriate methods for handling; 5) establishing plans for emergency response; and 6) accepting responsibility and liability.

Table 47.1 — Major U.S. Federal Legislative Acts Applicable to Hazardous Waste Management

Act and Citation	*Summary of Major Components*
Solid Waste Disposal Act (1965 as amended1980; 42 USCA 3251et seq.; 6901 et seq.)	Regulates wastes legally classified as solid wastes; foundation for regulation that defined hazardous wastes; amended as the Resource Conservation and Recovery Act of 1976
Resource Conservation and Recovery Act (1976; 42 USCA 6901et seq.)	Regulates wastes legally classified as solid and hazardous wastes; established criteria and lists for classification; permit and manifest to track wastes from initial source to ultimate disposal; identifies toxic flammable, corrosive, and reactive wastes as hazardous waste and pathogenic (medical) wastes as solid waste
Hazardous and Solid Wastes Amendments (1984; 42 USCA 6901 et seq.)	Amendment to RCRA to include corrective action for releases of hazardous substances from RCRA permitted facilities; established schedule to minimize use of land as disposal sites and address underground storage tanks
Medical Waste Tracking Act (1988; 42 USCA 6992 et seq.)	Amendment to RCRA to address medical wastes, including pathogenic or infectious wastes
Uranium Mill Tailings Radiation Control Act (1978 as amended 1988; 42 USCA 7901 et seq.)	Regulates disposal of mill tailings from uranium mining and processing; addresses health and environmental impact, technology
Low-Level Radioactive Waste Policy Act (1980 as amended 1985; 42 USCA 2021b et seq.)	Regulates disposal of low-level radioactive wastes; established standards of performance
Nuclear Waste Policy Act (1982 as amended 1987; 42 USCA 10101 et seq.)	Regulates the disposal/storage of high-level radioactive wastes; established standards of performance and engineering design for geologic repositories used for storage
Surface Mining Control and Reclamation Act (1977; 30 USCA 1201 et seq.)	Regulates surface coal mining operations and associated wastes; established management practices to minimize environmental impact
Comprehensive Environmental Response, Compensation and Liability Act (Superfund; 1980; 42 USCA 9601 et seq.)	Regulates activities involving uncontrolled releases of hazardous substances into the environment; established provisions for response investigation and remediation of waste sites
Superfund Amendments and Reauthorization Act (1986; 42 USCA 9601 et seq.)	Amendment to CERCLA (Superfund); established provisions for remediation activities, schedules, and protecting hazardous waste workers; provisions for inventory and storage of hazardous substances; Title III addresses Community Right-to-Know
Clean Water Act (1971 as amended 1977; 33 USCA 1251 et seq.)	Originally the Federal Water Pollution Control Act of 1948 as amended 1961, 1972, and 1977 to the Clean Water Act; Regulates point and nonpoint discharges into surface waters; established discharge standards and a permit system for generators discharging directly into surface waters (National Pollutant Discharge Elimination System); technology and water quality-based effluent limitations; new source performance standards
Safe Drinking Water Act (1974 as amended 1986; 40 USCA 300f et seq.)	Regulates contaminant levels in public water systems used as potable systems; established standards to prevent contamination of groundwater due to disposal activities such as deep-well injection

Table 47.1 (continued) —

Act and Citation	Summary of Major Components
Marine Protection, Research and Sanctuaries Act (1972; 33 USCA 1401 et seq.)	Regulates the discharge of wastes into the ocean
Oil Pollution Act (OPA) (1990; USCA 2701)	Regulates oil pollution prevention and control
Clean Air Act (1955 as amended 1966, 1970, 1977 and 1990; 42 USCA 7401 et seq.)	Regulates air emissions from new and existing stationary and mobile sources; in 1990 established primary standards to protect health and secondary standards to protect welfare of the environment
Federal Insecticide, Fungicide and Rodenticide Act (1947 as amended 1988; 7 USCA 136 et seq.)	Regulates registration, sale, use, and disposal of pesticides
Toxic Substances Control Act (1976; 15 USCA 2601 et seq.)	Regulates manufacture and registration of chemical products; requires data regarding intended use, potential toxicity to humans, and potential toxicity to environment
Hazardous Materials Transportation Act (1975 as amended 1976, 1994; 49 USCA 1801 et seq.)	Regulates the transport of hazardous materials; established provisions for packaging, labeling, placarding, manifesting, and shipping
Occupational Safety and Health Act (1970; 29 USCA 651 et seq.)	Regulates worker health and safety for general, construction, and maritime industries; established several standards including provisions for workers involved with hazardous waste and emergency activities

Only relatively recently, through enacted and amended legislative acts, have these factors and others been considered part of hazardous waste management.

Major Applicable Federal Legislative Acts

Legislative Responses to Mismanagement of Hazardous Wastes

In response to increased awareness of hazardous waste mismanagement, the regulatory framework was overhauled substantially during the 1970s. Environmental legislation was designated a priority and, accordingly, the regulatory framework was strengthened and new federal agencies such as the U.S. Environmental Protection Agency (EPA) and the Occupational Safety and Health Administration (OSHA) were established. Previously, environmental controls were loosely regulated mainly at state and local levels. At present, over a dozen federal acts[(4)] exist that address some aspect of hazardous wastes (Table 47.1). The statutes require state and local governments to be the primary implementers of the regulatory standards, though federal implementation and authority is permitted where and when deemed necessary.

The focus on environmental issues and legislation has resulted in the development of a prosperous and legitimate waste management industry, a need for increased public awareness, and an increased demand for environmental and occupational health professionals.[(5)] Although major changes have been observed, legislative efforts must continue to focus on policies and practices for decreasing the generation and disposal of hazardous wastes.[(6)] To do so effectively, however, may require additional regulatory and economic incentives for waste reduction.[(7)] For example, regulatory programs offering less complicated multimedia operating permits could ease the burden to generators implementing changes to reduce waste emissions and discharges.[(8)] In addition, waste reduction involving recovery and reuse could result in cost reductions for manufacturing and other facilities.[(9)]

Major Federal Enactments and Amendments

Prior to the changes in the 1970s, hazardous waste impact was connected predominantly with protecting water resources (i.e., Federal Water Pollution Control Act of 1948, as amended to Clean Water Act of 1972) and the air (i.e., Clean Air Act of 1955, as amended 1977). Water protection regulations were expanded with the Marine Protection, Resource and Sanctuaries Act of 1972 and the Safe Drinking Water Act of 1974.

Regulatory controls placed on discharging wastes into the air and water resulted in 1) generators directing wastes to land disposal sites instead; 2) new waste streams as a result of pretreatment requirements for existing waste streams; and 3) construction of municipal treatment facilities, such as wastewater treatment facilities that treated sewage and generated biological sludges.

The Solid Waste Disposal Act of 1965 addressed land disposal, but the regulation was weak in respect to industrial wastes that contributed significantly to the generation of hazardous wastes. The act was later amended via the Resource Conservation and Recovery Act (RCRA) of 1976, which in turn, was amended via the Hazardous and Solid Waste Act of 1984.

The addition of RCRA to the existing regulatory framework strengthened the controls on hazardous wastes discharged to air, water, and soil. RCRA is augmented by additional regulations such as the Surface Mining Control and Reclamation Act of 1977 to address mining run-off; the Uranium Mill Tailings Act of 1978; the Low-Level Radioactive Waste Policy Act of 1980 and the Nuclear Waste Disposal Act of 1982 to address radioactive wastes; and the Medical Waste Tracking Act of 1988 (an amendment to RCRA) to address infectious wastes. In addition, to address the production and use of raw materials that may contribute to the generation of hazardous wastes, the Toxic Substance Control Act of 1976 was enacted and the Federal Insecticide, Fungicide and Rodenticide Act of 1948 was amended.

The Hazardous Material Transportation Act of 1975 was enacted to address issues of transportation of hazardous materials, and the Comprehensive Environmental Response, Compensation and Liability Act (CERCLA, "Superfund") of 1980 remediation of past and future environmental releases of hazardous materials was passed. CERCLA was later amended via the Superfund Amendments and Reauthorization Act (SARA) of 1986.

It was evident that more attention needed to be focused on protecting workers engaged in hazardous waste operations. Although the Occupational Safety and Health Act of 1970 addressed worker health and safety for general, maritime, and construction industries, provisions specific to hazardous waste operations were not mandated until Superfund was amended. Thus, the OSHA standard known as Hazardous Waste Operations and Emergency Response (HAZWOPER) was promulgated in 1987.

Definitions of Hazardous Wastes

Regulatory Definition

Prior to the enactment of RCRA, hazardous wastes in the form of air, water, and soil pollutants were regulated mainly by the Clean Air, Clean Water, and Solid Waste Disposal Acts, respectively. RCRA was the first act, however, in which a definition for "hazardous waste" was provided. The term was defined by Congress as follows and appears in Subpart C of RCRA:

> *solid waste, or a combination of solid wastes, which because of its quantity, concentration or physical, chemical or infectious characteristics may: (i) cause or significantly contribute to an increase in mortality or an increase in serious irreversible, or incapacitating reversible, illness; or (ii) pose a substantial present or potential hazard to human health or the environment when improperly treated, stored, transported, disposed or otherwise managed.*[10]

EPA was subsequently mandated to establish the regulations by which hazardous wastes would be identified. The regulatory definition, although more specific, appears somewhat narrower in scope than that of Congress. For example, the original definition included "physical, chemical and infectious" wastes, whereas the EPA definition focuses only on "chemical" wastes:

> *solid wastes, including solid, semi-solid, liquid and gaseous materials, which (i) contain or are materials listed in the regulations; (ii) exhibit defined characteristics of toxicity, ignitability, corrosivity or reactivity; or, (iii) is not excluded from regulation as a hazardous waste.*[24]

Due to the specificity and exclusions, the RCRA hazardous waste regulations can fail to address waste streams holistically that may be hazardous whether classified as chemical, radiological, or biological. Furthermore, one could state that from the regulatory perspective not all wastes that are potentially or actually hazardous are specifically classified as such. Thus, there are "hazardous wastes" and there are "hazardous wastes as per RCRA."

It should not be inferred at this point that many hazardous waste streams are ignored from a regulatory perspective because the RCRA regulations are too narrow. Instead, it should be understood that there are numerous applicable federal acts that address wastes, but only one, RCRA, specifically defines and designates them as hazardous wastes. The intentions of the other acts, nonetheless, are to regulate hazardous wastes or, indeed, wastes that are hazardous.

Generic Definition

The generic definition of hazardous wastes is intended to encompass all of the regulated wastes and more. From a more generic perspective, hazardous wastes must be defined in more literal and holistic contexts. This is accomplished best by considering the two terms, "hazardous" and "wastes."

The term "hazardous" can be defined in this context as the potential or real tendency — due to toxicity, flammability, corrosiveness, reactivity, radioactivity, or infectivity — to directly or indirectly contaminate and damage the environment and/or cause injury or illness among the inhabitants. The term "wastes" can refer to unused and used solid, semisolid, liquid, or gaseous materials, including byproducts, residuals, and fugitive emissions, subject to both intentional and unintentional discharge without regard for recovery or reuse. Waste materials meant for recovery or reuse would be more appropriately labeled "used." Nonetheless, for the purpose of this discourse, wastes will include materials intentionally and unintentionally discarded, discharged, recovered, or reused.

Thus, the definition of hazardous wastes can imply a different meaning depending on regulatory (e.g., RCRA) or generic usage. The issue of considering wastes hazardous or nonhazardous involves more than simple semantics. Indeed, the issue influences the ability to compile an accurate qualitative and quantitative assessment of the inventory of total hazardous waste generation in the United States and elsewhere.

Table 47.2 — Generic Classification of Hazardous Wastes

Class of Hazardous Waste	*Hazardous Characteristics*
Chemical	toxicity flammability corrosivity reactivity
Radiological	radioactivity
Biological	infectivity

Classification and Sources of Hazardous Wastes

Classes of Wastes

Hazardous wastes are commonly divided into three classes: chemical, physical, and biological. Wastes that emit ionizing radiation are considered physical wastes. Academically, this can be justified based on an understanding of physics. From a practical perspective, however, classifying radioactive wastes as physical wastes may be confusing or vague. Indeed, because sources of ionizing radiation emitted from wastes that are radioactive are actually chemical in origin, then such wastes might more accurately be classified as chemical waste. This classification, although justified, may still evoke confusion. As a result, wastes that emit radiation will be simply classified as radiological wastes. Thus, in the strictest sense and from a practical perspective, the three classes of wastes will be designated as chemical, radiological, and biological (see Table 47.2).

Chemical Wastes

Chemical wastes are either organic (hydrocarbons or substituted hydrocarbons) or inorganic (metallic or nonmetallic elements). The hydrocarbons are composed solely of carbon and hydrogen; substituted hydrocarbons also include functional groups composed of elements such as chlorine, nitrogen, phosphorous, sulfur, or oxygen. Inorganic waste materials are typically in the form of salts, hydrides, and oxides. The hazards associated with organic compounds and the fate of the compounds in the environment depend mainly on their chemical compositions and associated physical properties.

Chemicals of greatest concern include volatile organics, polyaromatic hydrocarbons, organochlorines, heavy metals, and several nonmetallic radicals, such as cyanides, sulfides, nitrates, and phosphates.(26) Chemical wastes are hazardous if they exhibit a single characteristic or combination of characteristics described as toxic, flammable, corrosive, and reactive.

Although distinctions are made among the subclasses of chemical wastes, it should be noted that a waste consisting of even a single chemical component may exhibit a combination of characteristics. Indeed, because any chemical is toxic at a specific dose or concentration, then all flammable, corrosive, and reactive wastes, as well as radiological and some biological wastes, are also toxic. The reverse, however, is not necessarily true. That is, all toxic wastes do not exhibit other hazardous characteristics.

Toxic wastes are those discarded or used materials that may induce biochemical and physiological changes in human systems following either systemic contact via absorption into blood and tissues, or local contact. The changes may be ultimately manifested as adverse effects such as morphological and functional abnormalities, illnesses, and premature deaths among those exposed or their offspring. As suggested above, toxicity is inherent in all compounds. The toxicity of a given agent, however, may be directly attributable to an original parent compound or indirectly attributable to an active metabolite formed via biotransformation in a human system. In addition, it should also be recognized that secondary toxicants can be generated by flammable, corrosive, and reactive wastes in the form of toxic byproducts released from reactions, fires, and explosions.

Flammable wastes include materials that serve as fuels (reducing agents) that can ignite and sustain a chain reaction when combined in a suitable ratio with oxygen (oxidizing agent) in the presence of an ignition source (heat, spark). Flammable wastes are characterized as having a low flash point (i.e., <60°C), which is directly related to vapor pressure. In general, organic compounds vaporize at relatively lower temperatures and, accordingly, are much more sensitive to heat than inorganic compounds. Flammable wastes pose an obvious hazard from potential burns to human tissue. Indirectly during combustion, however, flammable wastes can contribute to the formation of toxic atmospheres due to generation of byproducts such as strong irritants and chemical asphyxiants, as well as by consumption of molecular oxygen during combustion.

Corrosive wastes include those materials that can induce severe irritation and destruction of human tissue on contact due to accelerated dehydration reactions. Corrosive wastes also include materials that can dissolve metal in a relatively short period of time. Typical examples of corrosives are organic and inorganic acids and bases. The strengths of acids and bases and the extremes of pH (i.e., <pH 2 and >pH 12.5, respectively) are directly correlated to the degree of corrosiveness.

Reactive wastes consist of chemically unstable materials typically characterized as either strong oxidizing or reducing agents. Chemical instability results in increased sensitivity to violent reactions, which may result in extremely rapid generation of heat and gases. In turn, this may culminate in ignition, explosion, or emission of toxic byproducts. Some unstable wastes can react with air or water. Other wastes react if the pH of the medium is altered. For example, wastes containing either cyanides or sulfides can react if the pH is increased above 2 or decreased below 12, resulting in generation of toxic hydrogen cyanide or hydrogen sulfide gases.

Radiological Wastes

Radiological wastes consist of radioactive components that emit ionizing radiation. The radioactive components are chemical elements or compounds that are electrochemically unstable due to an imbalance of protons (p^+) or neutrons (n^o) in the atomic nuclei. The elements and compounds undergo natural reactions to achieve stability by emanation of atomic energy in the form of particulates and electromagnetic photons. The particles and photons impart energy in excess of 30 electron volts to the matter with which they interact, resulting in ionization.

Ionizing particulate radiation consists of alpha particles (2 n^o + 2 p^+ in the form of a charged helium nucleus, ${}^4_2He^{+2}$); beta particles (negatron as e^- and positron as e^+); and neutron particles (n^o). Ionizing electromagnetic radiation consists of X-rays and gamma rays. The particulate and electromagnetic forms of ionizing radiation can interact by direct or indirect ionization of macromolecular or cellular components of the human body. Ionizing radiation may induce adverse biochemical and physiological changes manifested as abnormalities, illnesses, or premature deaths among those exposed or their offspring.

Radiological wastes are often designated as high-level or low-level based on the intensity and type of ionizing radiation emitted. Spent nuclear fuels are examples of high-level radioactive wastes; radioisotopes, such as those used in scientific research and medical diagnostics, are examples of low-level radioactive wastes.

Biological Wastes

Biological wastes are commonly referred to as pathogenic or infective wastes. There are at least 193 identified pathogenic biologic agents that may be encountered in the environment.[13] They consist of agents that, if introduced into the human body, may disrupt biochemical and physiological function through infectivity or toxicity. The disruption can result in illness and death if the immune system cannot destroy the biological agents. Infectivity is related to the virulence and the population density of organisms present at a given target site. Toxicity can be induced by biological agents that synthesize and release a chemical toxin. Examples of pathogenic agents include bacteria, rickettsia, fungi, protozoans, helminths, nematodes, viruses, plants, and insects. Unlike radiological and chemical substances, all biological agents except viruses are examples of biotic or living organisms. Viruses are abiotic or nonliving agents composed of biochemicals (i.e., proteins and nucleic acids) that may insert into and disrupt human cells.

Sources and Estimating Amounts of Hazardous Waste

The traditional concept of hazardous waste generation often considered only chemical wastes from industrial manufacturing sources. A more holistic view, however, includes the classes of radiological and biological wastes, as well as additional sources, such as clinical, research, agricultural, business, and household settings.

A variety of statistics are available estimating the volume or mass of hazardous wastes. Many are based on the RCRA definition of hazardous wastes, and so do not include several waste streams such as radioactive and infective wastes. Several factors that influence the ability to accurately estimate the total volume and mass of hazardous wastes generated domestically and internationally[14] include 1) variation in the scope of the estimates; 2) variation in the definitions of hazardous waste; 3) variation in measures for estimating the amount of waste generated or the capacity for storage, treatment, or disposal; and 4) sampling and response problems. Nonetheless, calculations based on one report,[15] assuming a population of 250 million, estimated that approximately 5.67 × 109 metric tons of gaseous, liquid, and solid wastes are generated annually in the United States.

Major Phases and Program Elements of Hazardous Waste Management

Overview

The major phases of hazardous waste management consist of generation; processing for recovery and reuse; transportation; treatment; storage; and disposal (as shown schematically in Figure 47.1). Inherent to the management scheme are auxiliary program elements that include organized occupational hygiene activities and preparedness for emergency response to environmental releases.

Generation

The source or generator of waste materials produces waste streams at a particular facility or site. Generators must qualitatively identify and quantitatively inventory their waste streams[16] to determine whether the wastes are 1) classified as chemical, radiological, or biological; 2) characterized as toxic, corrosive, flammable, reactive, radioactive, or infective; 3) generically classified as hazardous or nonhazardous; and 4) legislatively classified as hazardous or nonhazardous. In addition, it is essential that generators account for the mechanism of management, including 1) on-site discharges to air, water, and land; 2) treatment, reuse, or disposal on site; and 3) transportation for treatment, reuse, or disposal off site. It should be noted that regardless of legislative mandates, waste classifications,

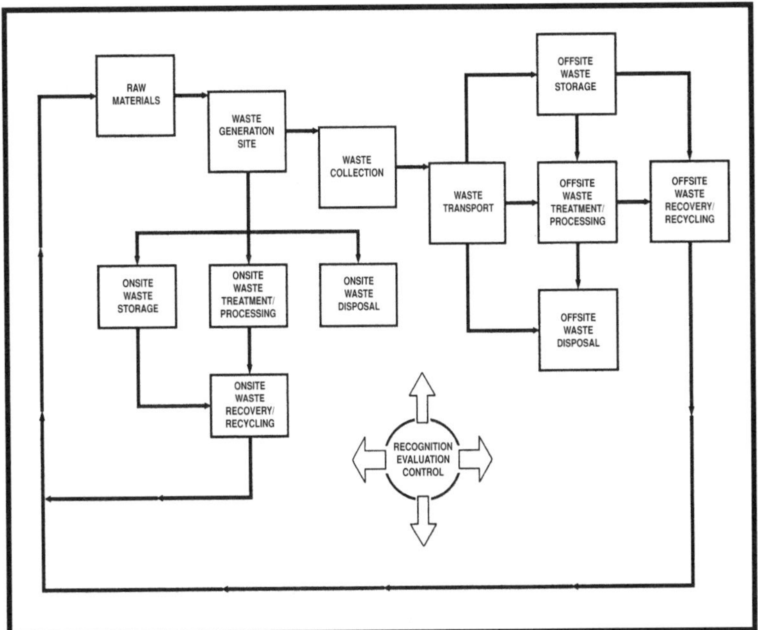

Figure 47.1 — Schematic diagram of the hazardous waste management process.

and quantities of waste, the generator has an ethical responsibility to ensure that potential and real hazardous wastes are managed effectively.

Depending on the nature of the facility and the wastes generated, the wastes may be managed on site. From an economic perspective, on-site recovery, treatment, and disposal is favored because it leaves less waste to transport to off-site facilities.[17] Increased expense of managing generated wastes and the related risk of liability for wastes produced has shifted the emphasis toward reduction of waste production.[18] Process modification and use of less hazardous chemicals[19,20] are two approaches to achieve waste reduction, but several other strategies have been suggested,[21] including on-site recycling of household, office, and industrial wastes.

Short-Term Storage

"Short-term" implies temporary storage by generators during the interim between generation and processing on site or transporting off site to another facility. Storage practices and strategies must focus on 1) the nature of the wastes; 2) use of appropriate containers; 3) isolation of incompatible materials; 4) labeling of containers; 5) containment of material and releases; 6) duration of storage; and 7) overall storage safety.[22,23] These factors must be considered by generators on site as well as the off-site facilities that store wastes prior to processing, treating, disposing, or long-term storing. Numerous containers are used for stationary and mobile storage (see Table 47.3).[24] Selection of storage containers involves consideration of a number of factors to ensure that applicable regulations are met.[25]

Transportation

An alternative to on-site processing is to have pretreated and treated wastes transported off site. Wastes may be transported off site on land using vehicles (see Table 47.4), though this may increase the risk of accidental release of hazardous materials into the environment. Wastes are typically contained in high-volume bulk storage tanks or low-volume storage drums during vehicular transport. The storage containers must be labeled and transport vehicles placarded during transit.

Some waste streams and byproducts of processed waste streams are transported off site via emission into the atmosphere and discharge into surface waters and sewer systems. These are not always considered transport processes. The quality and quantity of wastes generated directly into the air and water must comply with the applicable air and water pollution control regulations.

Changes in treatment technology are constantly seen in response to environmental issues, regulatory controls, and legal liabilities.[26] Related to these changes is the development of a major waste transportation, off-site treatment, and storage industry and frequent public concern regarding the siting of treatment, storage, and disposal (TSD) facilities.[27,28]

Treatment, Long-Term Storage, and Disposal

Treatment alters a waste stream or contaminated waste site to reduce, eliminate, or immobilize hazardous constituents. Numerous treatment technologies are presently available and in development.[20,29] Treatment typically precedes long-term storage and disposal to stabilize the waste streams and reduce the volume and mass, if feasible. Some waste streams, however, may not be amenable to treatment and are stored or disposed as they are.

The purpose of long-term storage and disposal is to segregate waste streams from the environmental surroundings. Long-term storage implies quasipermanent containment until the hazardous characteristic dissipates or technology is developed to eliminate the hazard. Disposal, however, implies disregard for return and, thus, is considered to be permanent storage or release.

Methods for Treatment of Hazardous Wastes

Concepts

Waste reduction is a priority and can be achieved, in part, through source segregation, process modification, end-product substitution, and waste recovery and recycling.[12,30] Source segregation involves separating the hazardous from the relatively nonhazardous components to reduce the volume of a waste. Process modifications can reduce wastes by changing operational parameters, using less hazardous or lower amounts of raw materials, or use of newer, more efficient equipment. End-product substitution focuses on production of commodities that will result in decreased hazardous waste generation during manufacture and a less hazardous product. The concept of waste recovery and recycling reflects the idea that one generator's waste

Table 47.3 —
Summary of Major Container Types for Storing Hazardous Wastes

Container Type	Description	Chemicals Stored
Nonbulk Containers	may consist of single or multiple packaging	
Bags	flexible packaging typically used for solid materials containing up to 100 lb.	cement, fertilizers, and pesticides
Bottles	glass, plastic, metal, or ceramic jugs or jars typically shipped in ounces up to 20 gallons	antifreeze, laboratory reagents, corrosive liquids
Boxes	rigid outside packaging for nonbulk packages for aerosol containers, bottles, and cans	
Multicell packaging	form-fitting box with one or more bottles	specialty chemicals (e.g., hydrochloric and sulfuric acid or solvents)
Carboys	glass or plastic bottles encased in outer packaging of polystyrene boxes, wooden crates, or plywood drums	sulfuric acid, hydrochloric acid, ammonium hydroxide, or potable water
Drums	metal, plastic, fiberboard, or other suitable material; capacity up to 55 gallons. Drums have removable "open" or nonremovable "closed" heads. Open heads attached to drum by ring or lugs. Closed/tight head contains two openings called "bungs" with 2-inch or 3/4-inch diameters; may have liners or linings	solid materials commonly found in open drums and liquid materials in those with closed/tight heads; containers carry a variety of contents depending on chemical characteristics
Jerricans	Metal or plastic packaging with rectangular or polygonal cross-section	antifreeze and other specialty products
Wooden barrels	commonly called "kegs;" made of natural wood and steel; iron or hardwood hoops	distilled spirits
Bulk Containers	Cargo tanks (i.e., tank wagons, tank trucks, tank trailers, or tankers) are bulk transport vehicles	transport liquid, dry, and gaseous chemical products
Nonpressure tanks	usually uninsulated, bottom-unloaded, oval-shaped cargo tanks with cable-operated valve assembly, emergency shut-off system, and safety relief valve (<4 psig); often compartmentalized to carry multiple contents	designed to carry flammable or combustible liquids (e.g., diesel, gasoline)
Low-pressure tanks	usually bottom-unloaded, round-shaped cargo tanks with hydraulic-operated valve assembly, emergency shut off system, fusible links, and safety relief valve (> 25 psig); often compartmentalized to carry multiple contents	designed to carry poisons, certain volatile flammable/combustible liquids, and mild corrosives
Medium-pressure tanks	usually top rear-unloaded, round-shaped cargo tanks with rear external ring stiffeners that carry a smaller capacity; design pressure may range from 35–100 psig	designed to carry highly corrosive products (e.g., sulfuric acid, sodium hydroxide)
High-pressure tanks	usually cylindrical with elliptical ends; usually bottom-unloaded; pressure > 500 psig	designed to carry compressed gases (anhydrous ammonia or LPG) and hazardous liquids
Intermodal containers "Iso-tanks"	bulk containers generally used in international transport; consist of a single noncompartmentalized vessel held by frame for lifting with cranes; most designed as pressure vessels	designed to carry a variety of hazardous chemicals

Table 47.4 — Major Vehicles for Land Transport of Hazardous Wastes

General service tank cars	nonpressure (tank pressures <100 psig) equipped with rupture disk or relief valves mounted in the car body	designed to carry a variety of hazardous materials
Pressure tank cars	designed to handle tank pressures >100 psig and <600 psig; have fittings mounted on top of car with protective housing	designed to carry hazardous materials with high vapor pressures (e.g., ethylene oxide, LPG, anhydrous ammonia, chlorine)
Cryogenic tank cars	designed to handle cryogenic materials	
High-pressure tube cars	tank cars designed to carry high pressure materials	

may be another's raw material. For instance, recovered wastes can be used in manufacturing processes and as supplemental fuels. Emphasis is placed on minimizing processing and, accordingly, determining which wastes can be used in their original state.

In general, waste streams must be subjected to some type of treatment prior to reuse or disposal. Those not recoverable or reusable must be destroyed. In either situation, various treatment methods are applied.

The treatment of hazardous wastes focuses on three major concepts: 1) separation of a hazardous component from a waste stream to reduce the quantity of hazardous waste; 2) elimination of the hazardous component via destruction and/or detoxification of molecules to yield less hazardous compounds; and 3) segregation of an entire hazardous waste stream or contaminated area from noncontaminated areas by immobilization and containment.[18,31] Many treatment processes result in the generation of residuals and byproducts in the form of airborne particulates/aerosols, gases, and vapors; chemical and biological sludges; saturated sorbents; and ashes. These warrant additional precautions[32] as well as treatment and disposal.[18] Treatment methods are classified as thermal, physical, biological, or chemical processes.

Thermal Treatment

Thermal treatment processes involve destruction of hazardous organic components through elevated temperatures with concomitant reduction in volume (see Table 47.5). Common thermal methods include incineration and pyrolysis.[33,34] Some innovative technologies involving a combination of elevated temperature and pressure also have emerged, including volume reduction via supercritical water oxidation.[35]

Thermal processes such as incineration typically generate airborne emissions and residual solids. The composition of these residues depends on the composition of the waste input. Typical residues include carbon dioxide, water, sulfur dioxide, nitrogen oxides, hydrochloric acid, particulates, and ash. Hazardous residues are collected and subjected to additional treatment or disposal.

Physical Treatment

Physical treatment involves nonthermal processes that separate a hazardous component from a waste stream or segregate a hazardous waste stream from a noncontaminated area (see Table 47.6). Thus, theoretically, a hazardous waste stream is converted into separated and more concentrated hazardous and nonhazardous components. The concentrated hazardous component has less volume or mass than the original combined waste stream and is subjected to either additional treatment, long-term storage, or permanent disposal.

Biological Treatment

Biological treatment processes use living organisms to accelerate the decomposition of organic wastes (see Table 47.7). The processes can be conducted under aerobic or anaerobic conditions depending on the type of organisms used as the biotic catalysts.[36] The methods use natural[31] and bioengineered species[37] of microbes as sources of enzymes to catalyze transformational reactions; accordingly, the process could be referred to as biochemical treatment. Although use of aerobic microorganisms is more common, the broad range of applications for anaerobic organisms is

Table 47.5 —
Major Methods for Thermal Treatment of Hazardous Wastes

Process	*Treatment Methods and Summary of Concepts*
Incineration	Hazardous organic components are oxidized via high-oxygen combustion at temperatures ranging from 425 to 1650°C. Efficiency depends on duration of incineration (time), mixing and oxygenation (turbulence), and thermal conditions (temperature).
	Rotary kiln incinerators: Hazardous solid, liquid, and gaseous wastes (particularly solids and semisolids such as contaminated soils and sludges, respectively) are placed in a rotating cylinder that increases exposure of surface area and, in turn, oxygenation and heat transfer.
	Multiple hearth incinerators: Wastes enter at the top of the unit, which consists of a series of alternating shelves (or hearths) and a vertical rotating cylinder. Wastes descend from hearth to hearth as combustion occurs, permitting longer residence time in the incinerator.
	Fluidized bed incinerators: Wastes are injected into the unit and collide with hot, inert granules suspended via forced aeration from a perforated section at the incinerator's base. Thermal interaction is enhanced via conduction of heat from granules to waste. Circulating bed combustion is an alternative for which increased efficiency is claimed via a higher operation velocity and use of fewer and finer granules.
	Liquid injection incinerators: Liquid wastes are atomized via a nozzle as they enter the incinerator, resulting in enhanced treatment efficiency via increased exposure of surface area to elevated temperatures and oxygen.
Pyrolysis	A form of incineration in which combustion occurs in an atmosphere deficient in molecular oxygen. The unit consists of a pyrolyzing chamber operated at 537 to 926°C and a secondary fume incinerator operated at 982 to 1648°C. The advantage of pyrolysis over conventional incineration is that inorganic components are not volatilized, so fewer corrosive air contaminants are generated.
	Boilers and furnaces: Wastes such as flammable petroleum-based solvents are burned in industrial-scale boilers and furnaces as a supplemental fuel.
Molten salt reactors	Liquid, solid, or gaseous wastes are mixed with a molten bath of sodium carbonate or calcium carbonate salt heated to 815 to 980°C. Wastes are degraded via combustion, and corrosive residual acids (e.g., HCl) are neutralized. Residuals include ash plus numerous salts.
Plasma arc reactors	Wastes are exposed to an electrically conductive ionized gas (plasma) heated to temperatures above 28,000°C. Hazardous components are atomized, ionized, and eventually destroyed. An alternative is the microwave plasma reactor in which microwave energy is used as the source of electrons to produce a plasma.
Wet oxidation	Nonhalogenated organic components are destroyed via combustion under highly oxidized conditions in a liquid state. Combustion reactions occur at temperatures ranging from 175 to 349°C and pressures of 68 to 136 atm. Applicable for waste streams too dilute for efficient incineration and too toxic for biological treatment.
Supercritical water oxidation	Organic components are oxidized and degraded when a waste stream is mixed with highly oxygenated water maintained at temperatures exceeding 374°C and pressures exceeding 218 atm.
Vitrification	An in situ process in which soil contaminated with inorganic (including radioactive) components is exposed to intense electrical current. The soil initially melts and subsequently cools into a glass-like (vitrified) solid, immobilizing and thus segregating the surrounding area. Vitrified soil is either left in place or excavated for off-site disposal or storage. The process is also used for other applications such as combining a waste stream with borosilicate glass . The mixture is vitrified and the solid product stored or disposed.
Pasteurization	Wastes containing pathogenic organisms are heated to temperatures in excess 70°C for a given time period to destroy pathogens.

receiving increased attention.[38] In situ bioremediation of media such as contaminated soil and groundwater is rapidly increasing.[39] Studies also suggest that combining mixed communities of biological organisms, dynamically fluctuated temperatures, and geochemical catalysts accelerates biodegradation.[40,41]

Chemical Treatment

Chemical treatment degrades or separates hazardous components through the interaction of a chemical reagent and a waste stream (see Table 47.8). The principle is similar to biochemical treatment in that an exogenous substance is combined with a waste stream to enhance separation via adsorption or degradation via catalysis. However, biological/biochemical treatment involves biotic agents (i.e., living organisms) while chemical treatment involves use of abiotic reagents (i.e., synthetic chemicals). Mixing chemicals with wastes separates hazardous components by coagulation, flocculation, and demulsification of solids and organics.[42] Solvents can be used

Table 47.6 — Major Methods for Physical Treatment of Hazardous Wastes

Process	*Treatment Methods and Summary of Concepts*
Flocculation and sedimentation/filtration	Colloidal solids within a waste stream interact and aggregate (flocculate). Flocculated solids are separated from the waste stream via sedimentation or filtration.
Dissolved air flotation	Insoluble components are separated from a liquid waste stream by aeration under high pressure and subsequent exposure to conditions at atmospheric pressure. Bubbles containing the insoluble components rise to the surface, where they are removed by skimming.
Carbon adsorption	Liquid or gaseous waste streams are passed though beds of porous activated carbon (charcoal) granules. Various organic and inorganic components adsorb to the carbon and are separated from the waste stream.
Resin adsorption/ ion exchange	Liquid or gaseous waste streams are passed though beds of ionized, porous, inert, resin granules. Organic and inorganic ionic components exchange with innocuous ions on the resins and are separated from the waste stream.
Liquid absorption	Gaseous waste streams are scrubbed as they pass through or interact with liquid reagents that absorb specific gaseous components.
Stripping	Volatile components are separated from solid or liquid waste either the passage of air or steam or aeration/atomization, respectively.
Distillation/condensation	Volatile components are separated from a waste stream via evaporation of the material and subsequent condensation and fractional collection of distillate.
Evaporation	Nonvolatile components are separated from a liquid waste stream by evaporation of the liquid phase via boiling to leave a solid nonvolatile residue.
Reverse osmosis	Liquid waste streams are passed at high pressure through a semipermeable membrane, thus separating hazardous inorganic and organic compounds from the waste stream by filtration.
Electrolysis	Organic and inorganic ions (e.g., metals) are separated from a waste stream by the addition of an electric current that influences attraction of negatively and positively charged ions to oppositely charged electrodes.
Electrodialysis	Hazardous ions are separated from a waste stream via electrically enhanced differential diffusion across a semipermeable membrane.
Photolysis	Organic components in a waste stream are exposed to ultraviolet (UV) light to catalyze the destruction of chemical bonds via lysis. **Beta/gamma/UV irradiation:** Wastes containing pathogenic organisms are irradiated with ionizing or nonionizing radiation to destroy the pathogens.

Table 47.7 — Major Methods for Biological Treatment of Hazardous Wastes

Process	*Treatment Methods and Summary of Concepts*
Activated sludge	An aerobic process in which wastewater is pumped through an aeration tank where a suspended community of microorganisms consume and degrade the organic components.
Trickling filter	An aerobic process in which wastewater is sprayed over a fixed medium inhabited by a community of microorganisms that consume and degrade the organic components.
Rotating biological contact	An aerobic process in which wastewater is pumped through a chamber containing a series of continuously rotating disks inhabited by a community of microorganisms that consume and degrade the organic components.
Stabilization pond	A combined aerobic and anaerobic process in which wastewater is pumped into an anthropogenic reservoir where suspended and settled microorganisms consume and degrade the organic components.
Composting	A combined aerobic and anaerobic process suitable for semisolid waste streams such as sludges. Sludges are typically mixed with wood chips and arranged in windrows or piles, where microorganisms consume and degrade the organic components.
Anaerobic digestion	An anaerobic process suitable for semisolid waste streams such as sludges. The waste stream is maintain under anaerobic conditions where a community of anaerobic microorganisms consume and degrade organic components.
Aerobic digestion	An aerobic process suitable for semisolid waste streams such as sludges. The process is similar to the activated sludge process but waste streams are subjected to extended retention time.
In situ bioremediation	Populations of microorganisms, either natural and/or bioengineered species, are introduced to a contaminated soil or groundwater site to consume and degrade hazardous organic components.

Table 47.8 — Major Methods for Chemical Treatment of Hazardous Wastes

Process	*Treatment Methods and Summary of Concepts*
Neutralization	The pH of an acidic or alkaline waste is adjusted to control corrosivity or maintain metallic components in a soluble state via the addition of an acidic or basic reagent.
Precipitation	Addition of a reagent to a waste stream to separate a soluble inorganic component by its conversion to an insoluble precipitate.
Hydrolysis	Addition of enzymatic reagents into a waste stream to catalyze the destruction of hazardous organic components via insertion of water molecules and subsequent lysis of chemical bonds.
Chemical oxidation	Addition of oxidizing reagents (e.g., air, hydrogen peroxide, ozone, perchloric acid, potassium permanganate) to enhance destruction of hazardous organic components via oxidation reactions. Reactions are accelerated via catalysts (e.g., iron, cobalt, UV light).
Chemical reduction	Addition of reducing reagents (e.g., alkalies) into a waste stream to enhance destruction of certain hazardous organic and inorganic components via reduction reactions.
Solvent extraction	Addition of a solvent to a waste stream to separate/remove a hazardous solid or liquid component.
Chemical solidification/ fixation/calcining/ grouting	Addition of reagents (e.g., Portland cement, silicates, fly ash, clay, calcium oxide, asphalt, polyethylene) to a waste stream to induce solidification and immobilize hazardous organic and inorganic components.

to extract hazardous components from waste streams such as process sludges.(43) Chemical catalysts such as iron have been added to accelerate degradation of toxic substances.(44) Hydrogenation technology is emerging as an approach to chemically detoxify organics in liquid waste streams.(45)

Methods for Long-Term Storage and Disposal of Hazardous Wastes

Concepts

Hazardous waste treatment processes typically reduce the quantity of waste and render it less hazardous. Regardless of treatment processes, however, residuals and by-products (e.g., airborne emissions, incinerator ash, chemical and biological sludges, saturated sorbents) are generated. In addition, many hazardous wastes are not amenable to recovery, reuse, and treatment and therefore must be managed in their original states. The options at this phase of the hazardous waste management scheme consist of disposal or long-term storage of nontreated wastes, residuals, and byproducts.

Long-term storage is a quasipermanent method of confining wastes in special structures, geologic repositories, or surface impoundments until the hazards dissipate or treatment and disposal are developed. Long-term storage is a common option for handling radiological wastes.(46) Disposal typically involves a more permanent discharging and discarding of wastes without concern for additional or subsequent handling. Table 47.9 shows the major methods of storage and disposal.

Disposal Into Air

Wastes are discharged into the air via both mobile and stationary sources. Discharges can occur via direct emission from stacks and indirect emissions from volatile compounds.(47) Discharge of

Table 47.9 — Major Methods for Long-Term Storage and Disposal of Hazardous Wastes

Media Classification	*Long-Term Storage and Disposal Method and Summary of Concept*
Air	**Combustion:** Thermal processes (e.g., incineration and pyrolysis, combustion engines, boilers and furnaces, manufacturing processes) that vent airborne emissions into the atmosphere for dispersal and disposal. Waste streams may be altered by natural physical and chemical processes.
	Evaporation: Intentional or unintentional volatilization of hazardous liquids and sublimation of hazardous solids into the atmosphere.
Water	**Ocean dumping:** Transportation of wastes for dispersal and disposal on the ocean floor.
	Surface water discharge: Waste streams are discharged directly into surface waters or indirectly via sewer systems for dilution and disposal; original waste stream may be altered by natural biological, physical, and chemical treatment.
Soil	**Landfilling:** The placement of wastes, exclusive of liquid wastes, into an engineered land burial site. Landfills consist of impervious sides and bottoms to minimize horizontal and vertical leaching into the substrata. Leachate collection and monitoring wells are installed as an extra precaution. Wastes are buried and periodically covered to form discrete compartments or cells. The sites are eventually capped with a final impervious layer on closure. In many respects, landfilling is simply another form of long-term storage.
	Deep-well injection: Liquid wastes are pumped into underground geologic cavities (i.e., vacant aquifers) within the earth for long-term storage.
	Repository storage: Applies predominantly to high-level radioactive wastes. The wastes are placed in engineered steel-lined concrete tanks designed to shield emanating ionizing radiation. An alternative is storage in natural geologic formations within the earth's surface and subsurface.
	Surface impoundment: Discharge and discard of wastes into anthropogenic reservoirs for storage.
	Land application: Liquid and semiliquid wastes are applied via spraying on or injecting into the soil surface in anticipation of degradation and immobilization due to combined biological and physicochemical interaction. The process could be viewed as a combined treatment and disposal process.

hazardous emissions into the air can be considered a combination of disposal and treatment because the wastes are simultaneously discarded and diluted to some extent. The practice of incinerating wastes at sea is another example of this concept. Nonetheless, since wastes disposed of in air can also be transported to other destinations, concerns regarding widespread environmental contamination are justified.

Disposal Into Water

Wastes are discharged into the water via both point (specific) and nonpoint (nonspecific) sources. As with air dispersal, the waste stream is both discharged and diluted. The practice, including dumping wastes into the oceans, is highly debated due to the obvious risks of adverse environmental and public health impact following widespread transport of contaminants. As population growth increases in coastal regions and land disposal sites become more scarce, there is a possibility of a resurgence of ocean dumping.[48]

Disposal and Long-Term Storage on Soil

Although long-term storage and disposal on land for hazardous wastes is discouraged and even banned to some extent (as per the Hazardous and Solid Waste Amendments of 1984), major disposal practices have included use of landfills, surface impoundments, and deep-well injection. Storage of noncontainerized, bulk liquid, regulated hazardous wastes in landfills and disposal on land are prohibited. However, researchers have suggested efficient means of simultaneously treating and disposing of some types of hazardous wastes on land,[42,47] and innovative treatment technologies are being researched and developed as alternatives to landfills.[48]

Adverse Human Health Impact Due to Hazardous Waste Exposure

Contributing Factors

Hazardous wastes, in most cases, exhibit a degree of heterogeneity relative to their constituents. The variations in composition and hazardous properties differ from one operation to another, and to a greater extent, from one generator to another. The waste materials may be in the form of a solid, liquid, gas, or combination. The associated hazards may be radiological, chemical, biological, or a mixture. In addition, exposure to the associated hazards may be via direct or indirect pathways.

The risks associated with hazardous waste exposure may be related to 1) whether the wastes contain pure substances or compounds; 2) whether they contain mixtures of substances that individually might not be as hazardous, but when combined present an increased risk; 3) whether they have properties that in small volumes and masses are innocuous but are hazardous in larger volumes and masses; 4) whether they are in solid, semisolid, liquid, or gaseous form; and 5) the manner in which the wastes are generated, treated, and disposed.[50]

The hazards encountered may be due to direct exposure to a given waste stream or indirect exposure via contact with byproducts or residuals. Indirect exposures also include contact with contaminated food, air, or water along with impact associated with fires and explosions. In addition, the risk of toxicity depends on the mode of exposure and may be exacerbated by simultaneous exposures to multiple contaminants. Thus, to categorize, 1) a waste may be hazardous as generated; 2) may subsequently combine with other contaminants and react to produce magnified and secondary hazards; 3) may pose a greater hazard depending on mode of exposure and subsequent synergistic, potentiation, and additive effects; and 4) may pose a greater hazard by reason of location or method of handling or processing. As a result, stringent precautionary measures and controls should be implemented and maintained during all phases of management. This involves major controls of waste streams from the source of generation to the point of ultimate treatment, long-term storage, or disposal.

Impact on Human Health

Given an understanding of the generic classes of hazardous wastes, it is apparent that flammable, corrosive, and reactive waste materials may cause obvious, immediate effects (e.g., burns and explosions) due to acute or short-term exposures. Toxic wastes, however, may cause both immediate effects (e.g., asphyxiation and irritation) due to acute exposure and insidious, delayed effects (e.g., physiological dysfunction, mutagenesis, and carcinogenesis) due to chronic or long-term exposures.

There is public concern over both kinds of effects. It appears, however, that the public is more intimidated by the threat of delayed adverse effects resulting from exposure to toxic, radiological, and biologic wastes. This is apparently due to the mystery surrounding the potential for delayed effects, as opposed to the more obvious threat of immediate adverse effects one might associate with flammables and reactives. Because of this perception, wastes that should be classified generically as hazardous are often incorrectly qualified from a public perspective as being only toxic.

However, the risk of injury and illness from flammables and reactives is very real. For example, there have been reported fires and explosions at hazardous waste sites due to flammables and reactives.[1,51]

Studies also have investigated increased cancer mortality near hazardous waste sites.[52] Other studies report that exposure to low-level concentrations of airborne contaminants generated from hazardous waste sites may increase the prevalence of illness complaints form the surrounding neighborhood.[53] Another study suggested that residents exposed to higher levels of waste-site chemicals exhibited more adverse neurologic effects than residents with lower exposure.[54] The illnesses experienced also include psychological effects as shown by demonstration of mental distress related to concern over perceived exposure to hazardous waste sites.[55] Equally important for consideration are studies suggesting an influence of reporting biases and possible hypochondriasis among individuals located in the vicinity of hazardous waste sites.[56,57] Unfortunately, epidemiologic studies suggest that there is no model to establish simple cause and effect relationships for these psychosocial issues.[58] Regardless, the threat of exposure has resulted in voluntary and involuntary emigration of residents from their neighborhoods.[59]

Risk Factors

The risk associated with hazardous wastes focuses on four major factors: 1) the hazards inherent to the wastes; 2) the release of wastes from facilities, sites, and vehicles; 3) direct and indirect exposure of humans; and 4) the variability of response and susceptibility to illness, if any, among those humans exposed. The first three factors apply to hazardous wastes in general, regardless of classification. The fourth factor, however, applies only to hazardous wastes subclassified as radioactive, toxic, or infective. There is minimal or no variation in human susceptibility and vulnerability to the primary injuries from events involving flammable, corrosive, or reactive wastes.

Studies conducted on human volunteers continue to demonstrate variation among human responses and susceptibility to toxic, radioactive, and infective waste exposures.[60] Basic toxicological and pathological concepts recognize several contributing factors, including age, weight, gender, health, mode of exposure, and entry path of toxicants. In addition, the problem is complicated by the potential presence of mixtures of components at a given waste site.[61]

Despite the perceived and real risks of occurrence, the number of published health effects related to exposures to hazardous wastes are low.[62] This may be attributable to inaccurate and incomplete assessment.[63] Although some studies concede that there is no demonstration of serious health effects, these studies are influenced by two major factors: 1) many were based on a low number of studies at actual sites; and 2) most focus only on cancer, birth defects, and other clinically defined illness, so subtle health effects may go undetected. Improved mechanisms such as biomarkers to identify changes associated with toxic exposures hold promise for earlier and more accurate assessment.[64]

Thus, qualitatively there are inherent risks to health in relation to hazardous wastes. Quantitatively, however, it is often difficult to assess these risks, especially with regards to toxic, radioactive, and infective exposures and impact. Risk assessment for these types of substances involves several factors: 1) estimating dosage and exposure; 2) considering each entry route to the body to obtain total dose; and 3) accounting for variations of concentrations generated at facilities and sites.[65] In addition, risk assessment must consider variables related to the collective contribution to exposure from multiple sources, including contaminated soil, water, and air. This task, however, is often based on estimation and speculation.

The Role of Occupational Health in Hazardous Waste Management

Overview

Federal regulations such as RCRA influenced the evolution of new industries in the form of controlled hazardous waste treatment, storage, and disposal (TSD) facilities. The enactment of CERCLA, as amended by SARA, influenced the expansion of other related phases of hazardous waste management, such as emergency response to environmental releases and the investigation, characterization, remediation, and restoration of abandoned and mismanaged hazardous waste sites. This increase in hazardous waste facilities and mitigation activities resulted in greater risk of exposure due to the concomitant increased need for human interaction with the facilities and sites.

Environmentally sound operations and accurate assessment at hazardous waste facilities and sites involves a coordinated effort by several disciplines, including environmental scientists, occupational hygienists, toxicologists, health physicists, hydro-geologists, engineers, and safety specialists. All scientific and technical disciplines involve some component of recognition, evaluation, and control of hazards to the environment, human health, and human safety.

Occupational hygiene or occupational health is a discipline of environmental health. The profession specializes in the art and science of recognition, evaluation, and control of chemical, physical, and biological stressors or factors arising in or from the workplace that may cause discomfort or adverse health effects to workers or members of the community.[66] The role occupational health plays in the area of hazardous waste management has been clearly defined.[32, 67–69] When occupational hygiene is applied to hazardous waste management activities, the workplace includes hazardous waste TSD facilities; spill sites; and abandoned or indiscriminately managed disposal sites. Workers include any human personnel, including investigators, involved at the facilities or sites; and members of the community refers to the public.

Recognition

The recognition phase includes anticipation and identification of potential and actual hazards. This phase commences prior to entering a facility or site and continues throughout investigations, evaluations, and general operations as additional data are compiled. Investigation and characterization involve gathering qualitative and quantitative data regarding a given TSD facility or abandoned site. These data include combinations of information, such as 1) the identification, hazardous characteristics, and quantity of wastes spilled, contained, stored, treated, and disposed; 2) the types of unit processes and standard operating procedures; 3) the number of operating personnel actually and potentially exposed; 4) topographic, geologic, and hydrologic features; 5) distance relative to off-site waterways (surface and ground), neighborhoods, or residences; and 6) any historical records.

Knowledge of several waste-handling processes can provide insight into potential exposures. It must be kept in mind, however, that hazardous waste operations present an environment that is somewhat more diversified than a traditional industrial setting, mainly due to the uncertainty and heterogeneity of the wastes and secondary by-products. At abandoned sites it is sometimes a challenge even to locate the wastes[70] since they may have been disposed in bulk quantities onto the soil or disposed of in containers distributed above ground, underground, in surface water, and in structures. Wastes also may have dispersed via evaporation, run-off, and leaching. Accordingly, exposures in the hazardous waste industry are often more difficult to recognize, and, in turn, to evaluate and control.

Assessment of waste facilities and sites can be relatively complex.[71] As much information as feasible is gathered during the recognition phase and is reviewed and evaluated to design the strategy necessary for preliminary on-premise survey and evaluation. In addition, the information is useful in developing preliminary health and safety plans.

Evaluation

Initially, a preliminary survey that includes visual and instrumental monitoring of a facility or site is conducted. Visual monitoring refers to the observation of processes and/or conditions and subjectively and objectively identifying potential and actual hazards. Instrumental monitoring involves the use of monitoring equipment to record qualitative and quantitative environmental surveillance data.

The evaluation of active TSD facilities essentially follows standard occupational hygiene protocol. Preliminary screening is necessary at some spill and abandoned disposal sites because of the increased uncertainty regarding the wastes and conditions present. Nonetheless, the instruments used to evaluate exposures at hazardous waste facilities and sites are the same as those used for general occupational hygiene and ambient environmental surveillance.[72–76]

Preliminary instrumental monitoring focuses on atmospheric screening for combustible gases and vapors, organic and inorganic gases and vapors, ionizing radiation, and oxygen.[77] Due to the uncertainty of the hazards that may be encountered by personnel during initial entry to a site, automated robotic assessment has been suggested[78] and continues to be studied.[79,80] Indeed, perhaps the safest way to evaluate contaminated hazardous waste sites is via remote sensing technologies that permit evaluation from off-site perimeter locations.[81]

The results of the preliminary screening data, in combination with other information, are used to 1) modify health and safety plans; 2) determine appropriate types and levels of controls; 3) designate support, decontamination, and exclusion zones; and 4) plan subsequent monitoring strategy for comprehensive environmental surveillance. Comprehensive environmental surveillance involves monitoring for air-, water-, and soil-borne contaminants, plus characterization of waste streams present. In addition, meteorological data such as temperature, relative humidity, barometric pressure, and wind speed and direction are collected. A summary of common monitoring instrumentation is summarized in Tables 47.10 and 47.11.

Characterization of a site can be influenced by randomness of distribution of wastes in the area.[82] Sampling objectives must be clearly established to collect representative and interpretable samples.[83] Indeed, established protocols, advanced instrumentation, data processing, and quality assurance are essential for compilation of accurate monitoring data.[84]

Inhalation is considered the major mode of entry for contaminants at hazardous waste sites and facilities. As a result, personal external exposures to airborne contaminants are determined by using instantaneous or real-time plus integrated monitoring instruments and methods. Dermal absorption is a secondary mode of exposure, but is difficult to evaluate accurately. A method for estimating dermal absorption via dermal patches impregnated with activated carbon, however, has been suggested.[85] Thermal stress is another major factor that should be evaluated during hazardous waste activities to determine the impact of temperature extremes on workers.[86]

Inhalation exposures due to airborne contaminants should be measured inside and outside of the source of generation within the site. There is a potential for contaminants to migrate off site due to volatilization and disturbance of soil. As a result, monitoring is also conducted at and outside of the site perimeter to determine migratory patterns. The concentrations measured, however, may vary depending on the nature of the operations.

Environmental surveillance data, which can provide some data regarding external exposures, are augmented by medical surveillance data.[87] Medical surveillance can possibly provide data regarding the impact of external exposures or the contribution to internal exposures. As suggested previously, hazardous waste operations and activities present a relatively unique occupational setting and source of potential adverse impact to the public, and a medical surveillance program must be designed accordingly.[88,89] The combination of environmental and medical surveillance data increases the probability of minimizing exposures and establishing a correlation between cause and effect.

The comprehensive set of data is used for several purposes: 1) to determine the extent of contamination and peripheral (horizontal and vertical) migration into the environment; 2) to determine potential and actual exposures to workers engaged in hazardous waste operations; 3) to determine potential and actual exposures to the public; and 4) to design strategy for control and remediation.

Control

Measures for controlling exposures to personnel involve implementation of administrative, engineering, and personal protective equipment. The implementation of administrative controls, such as development of a health and safety plan and standard operating procedures, should be a primary focus. Indeed, even if engineering and personal protective equipment controls are used, their full protective benefit is often compromised by poor work practices.

Engineering controls are most applicable to active TSD facilities and have been shown to reduce airborne levels of air contaminants at waste facilities.[90] Measures would include controlled process design, automation, and ventilation. Engineering controls are not always feasible, however, especially at uncontrolled spill or disposal sites. As a result, personal protective equipment must be used in conjunction with or as an alternative to engineering controls.

Personal protective equipment provides a relatively impermeable but less than 100% efficient barrier between an individual and the contaminated surroundings. A variety of equipment is available to protect the respiratory, dermal, and ocular systems. Procedures have been

Table 47.10 —
Major Instrumentatation for Evaluation of Atmospheric and Meteorologic Parameters at Hazardous Waste Facilities and Sites

Monitored Parameter	*Monitoring Instrument*	*Collection Medium*	*Type of Monitoring*	*Analytical Instrument*
Combustible Gases/Vapors	Combustible Gas Meter		Instantaneous Area	Direct-reading
Oxygen	Oxygen Meter		Instantaneous Area	Direct-reading
Organic Gases/Vapors	Analyzer w/ Flame Ionization Detector (FID)		Instantaneous Area	Direct-reading
	Analyzer w/ Photoionization Detector (PID)		Instantaneous Area	Direct-reading
	Analyzer w/ Infrared Radiation Detector (IR)		Instantaneous Area	Direct-reading
	Manual Piston/ Bellows Pump	Detector Tubes	Instantaneous Area	Direct-reading
	Low/High-Flow Pump Tenax/Chro-mosorb Solid Adsorbents	Charcoal/ Silica Gel Adsorbent Tubes	Integrated Personal/Area	Gas Chromatograph GC Mass Spectrometer/ High-Pressure Liquid Chromatograph
	High-Flow Pump	Impinger/ Bubbler w/ Liquid Absorbents	Integrated Personal/Area	UV/Vis Spectropho-tometer
Organic Aerosols	High-Flow Pump	Fiberglass/ Teflon Filters	Integrated Personal/Area	Gas Chromatograph/ High-Pressure Liquid Chromatograph
Inorganic Gases/Vapors (e.g., Acids)	Manual Piston/ Bellows Pump	Detector Tubes	Instantaneous Area	Direct-reading
	Low/High-Flow Pump	Silica Gel Adsorbent Tubes	Integrated Personal/Area	Ion Chromatograph
	High-Flow Pump	Impinger/ Bubbler w/ Liquid Absorbent	Integrated Personal/Area	UV/Vis Spectro-photometer
Inorganic Aerosols (e.g., Metals; Asbestos	Fibrous Aerosol Monitor (Asbestos/ Total Fibers)		Instantaneous Area	Direct-reading
	X-ray Fluorescence Monitor (Metals)		Instantaneous	Direct-reading
	High-Flow Pump	Cellulose Ester Fiber Filter	Integrated Personal/Area	Atomic Absorption Spectrometer/Induc-tively Coupled Plasma Emission (Metals) . . . Electron (TEM) or Light (PLM/PCM) Microscope (Asbestos)

Table 47.10 (continued) —
Major Instrumentatation for Evaluation of Atmospheric and Meteorologic Parameters at Hazardous Waste Facilities and Sites

Monitored Parameter	*Monitoring Instrument*	*Collection Medium*	*Type of Monitoring*	*Analytical Instrument*
Biological Aerosols	High-Flow Pump	Cellulose Ester Fiber Filter/Impinger w/ Water/Cascade Impactor	Integrated Personal/Area	Incubator/ Light Microscope
Total/ Respirable Aerosols	Suspended Particle Analyzer		Instantaneous Area	Direct-reading
	High-Flow Pump	Polyvinyl Chloride	Integrated Personal/Area	Gravimetric Electrobalance Filter
Ionizing Radiation	Geiger-Mueller Counter		Instantaneous Area	Direct-reading
	Dosimeter Personal		Integrated	Direct-reading
	Film Badges	Film + Filters	Integrated Personal	Densitometer
Noise	Sound Level Meter		Instantaneous	Direct-reading Area
	Audio Dosimeter		Integrated Personal	Direct-reading
Thermal Stress	WBGT Meter		Instantaneous Area	Direct-reading
	Physiological Indice Dosimeters		Instantaneous Area	Direct-reading
Temperature	Thermometer		Instantaneous	Direct-reading
Atmospheric Pressure	Barometer		Instantaneous	Direct-reading
Wind Speed and Direction	Anemometer		Instantaneous	Direct-reading
Relative Humidity	Psychrometer		Instantaneous	Direct-reading

established[75,82] that recommend four levels of personal protection depending on specific environmental surveillance data or perceived conditions (Table 47.12). In the absence of certainty, full-body encapsulation is sometimes warranted. This practice sometimes creates a secondary health hazard, however, due to the increased potential for heat stress.[91]

References

1. **Wilson, D.:** *Handbook of Solid Waste Management.* New York: Van Nostrand Reinhold, 1977.
2. **Jewell, W.J. and B.L. Seabrook:** *A History of Land Application as a Treatment Alternative,* (U.S. EPA publication no. 430/9-79-012) Cincinnati, OH: U.S. EPA Document Distribution Center, 1979.
3. **Harris, C., W.L. Want, and M.A. Ward:** *Hazardous Waste: Confronting the Challenge.* New York: Quorum Books, 1987.
4. *United States Code Annotated.* St. Paul, MN: West Publishing Co., 1995. [Index]
5. **Stern, B.J.:** Education and training of environmental health personnel in health risk assessment and mange-

Table 47.11 —
Major Instrumentation for Evaluation of Water, Soil, and Waste Parameters at Hazardous Waste Facilities and Sites

Monitored Parameter	*Type of Sample*	*Analytical Instrument*
Organic chemical compounds	grab/composite	GC/GC mass spectrometer/ high-pressure liquid chromatograph/ Fourier transform infrared spectrometer
Total organic halide	grab/composite	microcoulometric titrator/neutron bombardment gamma-ray detector
Total organic carbon	grab/composite	carbonaceous analyzer
Total and amenable cyanide	grab/composite	UV spectrophotometer
Sulfide	grab/composite	titration apparatus
Total oil and grease	grab/composite	gravimetric electrobalance
Inorganic chemical compounds (e.g., metals)	grab/composite	atomic absorption spectrometer/ Inductively coupled plasma atomic emission spectrometer
Biological agents (e.g., total coliform)	grab/composite	membrane filter apparatus + incubator + microscope
Solubity (hexane: water partition)	grab/composite	separatory funnel
pH	grab/composite	pH meter
Specific conductance	grab/composite	conductivity meter
Flammability (i.e., ignitability)	grab/composite	Pensky-Martens closed-cup tester/ Setaflash closed-cup tester/Bureau of Explosives closed-drum tester
Corrosivity	grab/composite	resin-flask apparatus + SAE 1020 steel
Reactivity	grab/composite	cyanide release apparatus + UV spectrophotometer
Radioactivity (i.e., gross alpha and beta)	grab/composite	gas-flow proportional counter/ scintillation counter

Table 47.12 —
Levels of Personal Protective Equipment for Hazardous Waste Operations and Activities

Level of Protection	*General Criteria for Use*	*Major Personal Protective Equipment*
A	Highest level of respiratory, dermal, and ocular protection is warranted	• supplied-air respirator • fully encapsulating chemical-resistant suit + boots + gloves
B	Highest level of respiratory protection is warranted, but a lower level of dermal protection	• supplied-air respirator • one- or two-piece chemical resistant suit + boots + gloves
C	Criteria for wearing air-purifying respirators are met; non-immediately dangerous to life or health (IDLH) atmospheres; concentration of oxygen >19.5%	• full- or half-facepiece air-purifying respirator (note: ocular protection is required if half-facepiece is worn) • one- or two-piece chemical resistant suit + boots + gloves
D	No risk of respiratory or dermal dermal exposure; non-IDLH atmospheres; concentration of oxygen >19.5%	• ocular protection • coveralls + boots

ment of hazardous waste. *J. Environ. Health 50*:352–354 (1988).
6. **Ember, L.:** Pollution prevention: study says chemical industry lags. *Chem. Engin. News 73*:6–7 (1995).
7. **Hahn, R.W.:** An evaluation of options for reducing hazardous waste. *Harvard Environ. Law Rev. 12*:201–230 (1988).
8. **Johnson, J.:** New Jersey pilot program eases permit path for companies that cut toxics. *Environ. Sci. Technol. 30*:72A–73A (1996).
9. **Staines, J.:** Tips for managing and reducing waste. *Adhesives Age 38*:44–46 (1995).
10. *Resource Conservation and Recovery Act, U.S. Code Annotated*, Title 42, Pt. 6901 et seq. (1995).
11. "Hazardous Waste Management System," *Code of Federal Regulations*, Title 40, Part 260 (1995).
12. **Office of Technology Assessment:** *Technologies and Management Strategies for Hazardous Waste Control.* Washington, D.C.: U.S. Government Printing Office, 1983.
13. **Dutkkiewicz, J., L. Jablonski, and S.A. Olenchock:** Occupational biohazards: a review. *Am. J. Ind. Med.14*:605–623 (1988).
14. **House of Representatives:** "The condition of information on hazardous waste," by E. Chelimsky. 2nd session, September 24, 1986. Washington, D.C.: U.S. Government Printing Office, 1987. [Congressional subcommittee testimony]
15. **Conservation Foundation:** *America's Waste: Managing for Risk Reduction (State of the Environment: A View Toward the Nineties)*, Conservation Foundation: Washington, D.C., 1987.
16. **Lindgren, G.F.:** *Managing Industrial Hazardous Waste: A Practical Handbook.* Chelsea, MI: Lewis Publishers, 1989.
17. **Royston, M.G.:** *Pollution Prevention Pays.* New York: Pergamon Press, 1979.
18. **Cheremisinoff, P.N.:** Hazardous waste treatment and recovery systems. *Poll. Engin. 20*:52–61 (1988).
19. **Burch, W.M.:** Process modification and new chemicals. *Chem. Engin. Prog. 82*:5–8 (1986).
20. **Gibbs, W.W.:** Ounce of prevention, cleaner chemicals pay, but industry is slow to invest. *Sci. Am. 271*:103–105 (1994).
21. **Freeman, H.M.:** Industrial pollution prevention: a critical review. *J. Air Waste Mgmt Assoc. 42*:618–656 (1992).
22. **Tchobanoglous, G., H. Theisen, and R. Eliassen:** *Solid Wastes.* New York: McGraw Hill, 1977.
23. **Van Valkenburgh, G.:** Storing hazardous wastes safely. *Chem. Engin. 98*:203–204 (1991).
24. **Chemical Manufacturer Association (CMA):** *Guidance for Containers.* Arlington, VA: CMA, 1995.
25. **Bouley, J.:** Drums and containers. *Poll. Engin. 25*:37 (1993).
26. **Muschett, F.D. and M.E. Enowitz:** The changing pollution control industry. *Poll. Engin. 18*:44–47 (1986).
27. **Holton, G.A., J.S. Arendt, and J.J. Rooney:** Addressing public fears related to siting of hazardous waste facilities. *J. Air Poll. Control Assoc. 37*:1202–1206 (1987).
28. **Weidman, P.M. and S. Femers:** Public participation in waste management decision making: analysis and management of conflicts. *J. Haz. Mat. 33*:355–368 (1993).
29. **Belateche, D.H.:** Choose appropriate wastewater treatment technologies. *Chem. Eng. Progress 91*:32–51(1995).
30. **Office of Technology Assessment:** *Superfund Strategies.* Washington, D.C.: U.S. Government Printing Office, 1985.
31. **Hartenstein, R.:** Buffering acid precipitation, reducing soil erosion, and reclaiming toxic soil in the advent of global human carrying capacity. *Int. J. Environ. Studies 30*:287–300 (1986).
32. **Bisesi, M.S.:** Industrial hygiene paramount for worker protection in waste management industry: part I. *Haz. Mat. Waste Mgmt 2*:24–30 (1984).
33. **Cheremisinoff, P.N.:** Thermal treatment technolgies for hazardous wastes. *Poll. Engin. 20*:50–55 (1988).
34. **Dempsy, C.R. and E.T. Oppelt:** Incineration of hazardous waste: a critical review. *J. Air Waste Mgmt Assoc. 43*:25–73 (1993).
35. **Manji, J.F.:** Pressure cooking could solve some hazardous waste problems in industry. *Automation 38*:26 (1991).
36. **Atlas, R.M.:** Bioremediation. *Chem. Engin. News 73*:32–42 (1995).
37. **Garg, S. and D.P. Garg:** Genetic engineering and pollution control. *Chem. Eng. Progress 86*:46–51 (1990).
38. **Stroo, H.F.:** Biotechnology and hazardous waste treatment. *J. Environ. Qual. 21*:167–175 (1992).
39. **Brubaker, G.R.:** In situ bioremediation. *Civil Engin. 65*:38–41 (1995).

40. **Bisesi, M.S. and R. Hartenstein:** Soil and limnologic acidification: a paradox. *Int. J. Environ. Studies 30*:287–291 (1987).
41. **Bisesi, M.S.:** Vermial and microbial management of biological sludges under conditions of dynamic temperature and seasonal changes. *Bio. Wastes 32*:99–109 (1990).
42. **Collett, T.L.:** Enhance hazardous waste treatment using specialty chemicals. *Chem. Engin. Progress 87*:70–74 (1991).
43. **Trowbridge, T.D. and T.C. Holcombe:** Refinery sludge treatment/hazardous waste minimization via dehydration and solvent extraction. *J. Air Waste Mgmt Assoc. 45*:782–788 (1995).
44. **Stinson, S.C.:** Waste recycling plant uses molton metal baths. *Chem. Engin. News 71*:9 (1993).
45. **Gioia, F.:** Detoxification of organic waste liquids by catalytic hydrogenation. *J. Haz. Mat. 26*:243–260 (1991).
46. **Kittel, J.:** Nuclear waste management: issues and progress. *J. Environ. Sci. 27*:34–41 (1984).
47. **United States Environmental Protection Agency:** *The Toxics Release Inventory: Executive Summary.* (U.S. EPA 560/40-89-006) Cincinnati, OH: U.S. EPA Document Distribution Center (1989).
48. **Goldberg, E.D.:** The oceans as waste space: the argument. *Oceanus 24*:12–18 (1981).
49. **Krukowski, J.:** Alternatives to landfilling hazardous waste. *Poll. Engin. 26*:54–56 (1994).
50. **Hagerty, D., J. Heer, and J. Pavoni, J.:** *Solid Waste Management.* New York: Van Nostrand Reinhold, 1973.
51. **Costello, R.J. and M.V. King:** Protecting workers who clean-up hazardous waste sites. *Am. Ind. Hyg. Assoc. J. 43*:12–17 (1982).
52. **Griffith, J., R.C. Duncan, W.B. Riggan and A.C. Pellom:** Cancer mortality in U.S counties with hazardous waste sites and ground water pollution. *Arch. Environ. Health 44*:69–74 (1989).
53. **Ozonoff, D., M.E. Colten, A. Cupples, T. Heeren, et al.:** Health problems reported by residents of a neighborhood contaminated by a hazardous waste facility. *Am. J. Med. 11*:581–597 (1987).
54. **Dayal, H., S. Gupta, N. Treiff, D. Maierson, et al.:** Symptom clusters in a community with chronic exposure to chemicals in two Superfund sites. *Arch. Environ. Health 50*:108–111 (1995).
55. **Edwards, F.L., and A.H. Ringleb:** Exposure to hazardous substances and the mental distress tort: trends, applications, and a proposed reform. Columbia *J. Environ. Law 11*:119–139 (1986).
56. **Roht, L.H., S.W. Vernon, F.W. Weir, S.M. Pier, et al.:** Community exposure to hazardous waste disposal sites: assessing reporting bias. *Am. J. Epidemiol. 122*:418–433 (1985).
57. **Kaye, W.E.:** Recall bias in disease status associated with perceived exposure to hazardous substances. *Ann. Epidemiol 4*:393–397 (1994).
58. **Elliot, S.J., S.H. Taylor, S. Walter, D. Steib, et al.:** Modeling pyschosocial effects of exposure to solid waste facilities. *Soc. Sci. Med. 37*:791–804 (year).
59. **Jacobson, J.L.:** *Abandoning Homelands, Worldwatch Institute Report: State of the World 1989.* New York: W.W. Norton, 1989.
60. **Hattis, D., L. Erdreich, and M. Ballew:** Human variability in susceptibility to toxic chemicals—a preliminary analysis of pharacokinetic data from normal volunteers. *Risk Anal. 7*:415–426 (1987).
61. **Johnson, B.L. and C.T. DeRosa:** Chemical mixtures released from hazardous waste sites: implications for health risk assessment. *Toxicology 105(2–3)*:145–156 (1995).
62. **Levine, R., and D.D. Chitwood:** Public health investigations of hazardous organic chemical waste disposal in the United States. *Environ. Health Perspect. 62*:415–422 (1985).
63. **Smith, M.T., C.S. Lea, and P.A. Buffler:** Human populations changes caused by hazardous waste. *Central Euro. J. Pub. Health 3*:77–79 (1995).
64. **Indulski, J.A. and W. Lutz:** Biomarkers used for the assessment of health hazards in populations living in the vicinity of communal and industrial waste dump sites. *Int. J. Ind. Med. Environ. Health 8*:11–16 (1995).
65. **Corn, M. and P.N. Breysse:** Human exposure estimates for hazardous waste site risk assessment. In *Risk Quantitation and Regulatory Policy* (19 Banbury Report), D.G. Hoel, R.A. Merrill, and F.P. Perea, eds. New York: Cold Spring Harbor Laboratory, 1985.

66. **Radcliffe, J.G., C. W. Frederick, K. Nelson, and E. Wheeler:** Industrial hygiene: definition, scope, function, and organization. *Am. Ind. Hyg. Assoc. J. 20*:428–430 (1959).
67. **Bisesi, M.S.:** Industrial hygiene: paramount for worker protection in waste management industry: part II. *Haz. Mat. Waste Mgmt 2*:22–28 (1984).
68. **Martin, W.F. and S.P. Levine:** *Protecting Personnel at Hazardous Waste Sites.* Boston: Butterworth-Heineman, 1994.
69. **National Institute for Occupational Safety and Health (NIOSH), Occupational Safety and Health Administration, United States Coast Guard, and United States Environmental Protection Agency:** *Occupational Safety and Health Guidance Manual for Hazardous Waste Site Activities.* (NIOSH pub. 85-115) Washington, D.C.: U.S. Government Printing Office, 1985.
70. **Weitzman, D. and L. Cohen:** Industrial hygiene program for hazardous waste site investigations. *Am. Ind. Hyg. Assoc. J. 42*:654 (1981).
71. **Sara, M.A.:** *Standard Handbook for Solid and Hazardous Waste Facility Assessments.* Boca Raton, FL: Lewis Publishers, 1994.
72. **Environmental Protection Agency (EPA):** *Test Methods for Evaluating Solid Waste.* (U.S. EPA SW-846) Washington, D.C.: U.S. Government Printing Office, 1993.
73. **Bisesi, M.S. and J.P. Kohn:** *Industrial Hygiene Evaluation Methods.* Boca Raton, FL: Lewis Publishers/CRC Press, 1995.
74. **Bishop, E.:** *Air monitoring at hazardous waste sites. In Protecting Personnel at Hazardous Waste Sites.* Boston: Butterworth-Heineman, 1994.
75. **Koren, H. and M.S. Bisesi:** *Handbook of Environmental Health and Safety,* vol. II. Boca Raton, FL: Lewis Publishers/CRC Press, 1995.
76. **Simmons, M.S. (ed.):** *Hazardous Waste Measurements.* Chelsea, MI: Lewis Publishers, 1991.
77. **Environmental Protection Agency (EPA):** *Standard Operating and Safety Guides.* Cincinnati, OH: EPA Document Distribution Center, 1988.
78. **Bisesi, M.S.:** "Remote Control Robotic Assessment (RCRA) of Hazardous Waste Sites." Paper presented at the American Industrial Hygiene Conference. Detroit, MI, May 1983.
79. **Osburn, J.F., W.L. Whittaker, and S. Coppersmith:** "Prospects For Robotics in Hazardous Waste Management." Proceedings of the Second International Conference on New Frontiers for Hazardous Waste Management. Pittsburgh, 1987.
80. **Rancatore, R.J. and M.L. Philips:** *Demonstration of Autonomous Air Monitoring Through Robotics.* (EPA 600-S2-89-055) 1990.
81. **Leis, W.M. and F. Bopp III:** Hazardous waste dumps: dangerous as battlefields. *I & CS 66*:43–46 (1993).
82. **Parkhurst, D.F.:** Optimal sampling geometry for hazardous waste sites. *Environ. Sci. Tech. 18*:521–523 (1984).
83. **Sedman, R.M.:** Why did you take that sample? *J. Air Waste Mgmt Assoc.42*:1420–1423 (1992).
84. **Almich, B.P., W.L. Budde, and R.W. Shobe:** Waste monitoring. *Environ. Sci. Tech. 20*:16–21 (1986).
85. **Cohen, B.S. and W. Popendorf:** A method for monitoring dermal exposure to volatile chemicals. *Am. Ind. Hyg. Assoc. J. 50*:216–223 (1989).
86. **Goldman, R.F.:** Heat stress in industrial protective encapsulating garments. In *Protecting Personnel at Hazardous Waste Sites,* W.F. Martin and S.P. Levine, eds. Boston: Butterworth-Heineman 1994.
87. **Schaub, E. and M.S. Bisesi:** Medical and environmental surveillance. *N.J. Med. 91(10)*:715–718 (1994).
88. **Melius, J.M.:** Medical surveillance for hazardous waste workers. *J. Occup. Med. 28*:8 (1986).
89. **American Industrial Hygiene Association/American Conference of Governmental Industrial Hygienists Joint Hazardous Waste Committee:** Proposed criteria for the selection of appropriate medical resources to perform medical surveillance for employees engaged in hazardous waste operations. *Am. Ind. Hyg. Assoc. J. 50*:A870–A872 (1989).
90. **Pederson, B.A.:** Evaluation of chemical exposures in the hazardous waste industry. *J. Air Waste Mgmt Assoc.45*:89–94 (1995).
91. **Paull, J.M. and F.S. Rosenthal:** Heat strain and heat stress for workers wearing protective suits at a hazardous waste site. *Am. Ind. Hyg. Assoc. J. 48*:458–463 (1987).

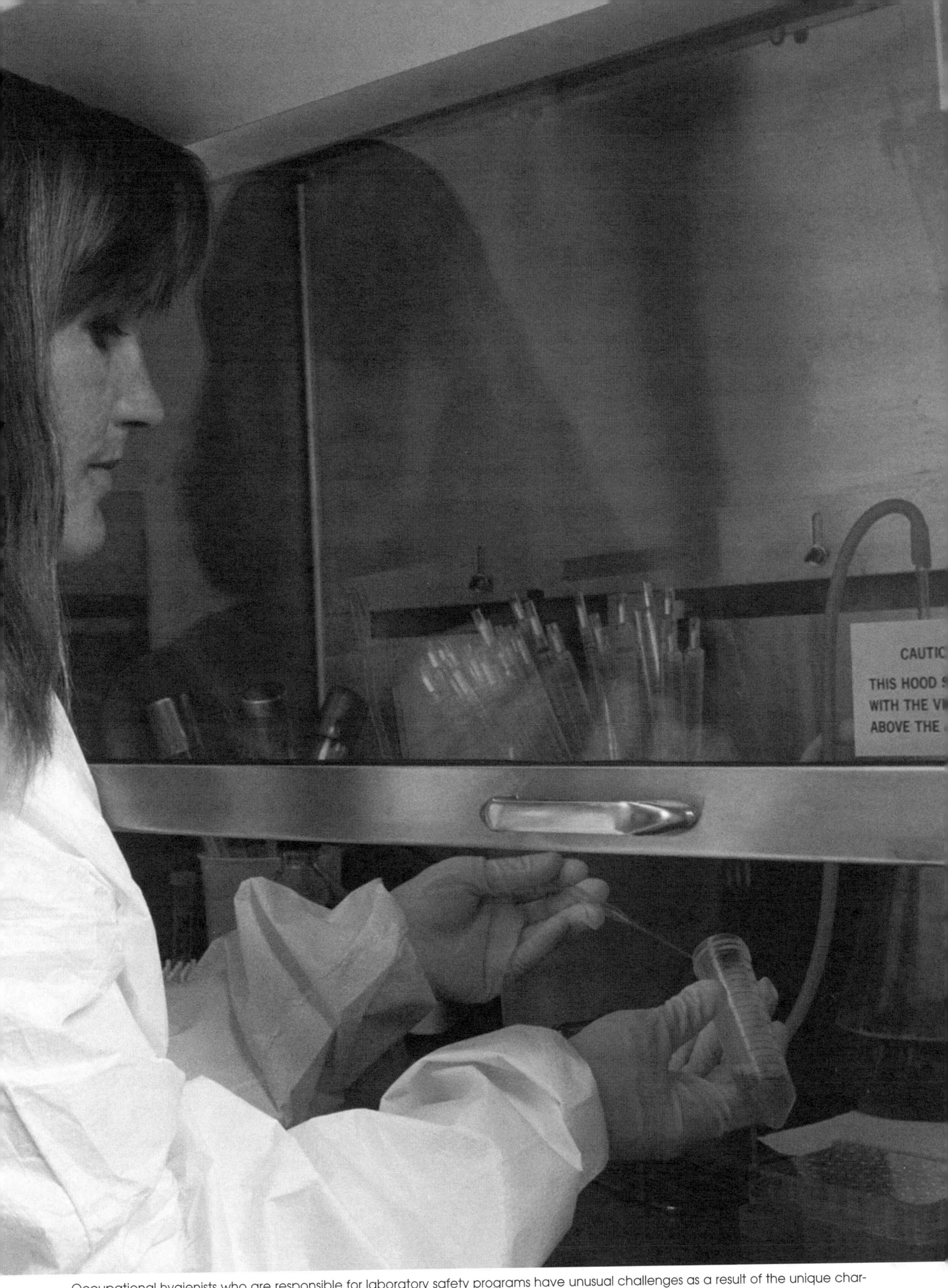

Occupational hygienists who are responsible for laboratory safety programs have unusual challenges as a result of the unique characteristics of the laboratory workplace.

Chapter 48

Outcome Competencies

After completing this chapter, the user should be able to:

1. Define underlined terms used in this chapter.
2. Describe important differences that distinguish laboratories from other workplaces that use hazardous materials.
3. Describe the process used to design and implement a laboratory health and safety program.
4. Explain established risk assessment methods to plan experiments.
5. Assess the hazards assocated with various classes of laboratory chemicals and various laboratory techniques.
6. Recognize unsafe acts and situations in laboratories.
7. Use established risk assessment methods to plan experiments, including selection of appropriate controls.
8. Describe training objectives that comply with regulations.
9. Implement the cradle-to-grave concept to purchase, store, and dispose of chemicals.
10. Describe key elements of OSHA's laboratory standard.

Key Topics

I. Hazardous Materials
  A. Flammable Chemicals
  B. Corrosive Chemicals
  C. Oxidizing Chemicals
  D. Reactive Chemicals
  E. Toxic Chemicals
  F. Hazardous Waste
  G. Radioactive Materials

II. Physical Hazards

III. Hazardous Processes and Equipment
  A. Modified Pressure Techniques
  B. Modified Temperature Techniques
  C. Energy Hazards
  D. Separation Techniques
  E. Anticipation/Recognition
  F. Evaluation/Assessment

IV. Survey/Sampling Strategies

V. Controls
  A. Engineering Controls
  B. PPE
  C. Administrative Controls

VI. Training

VII. Regulations and Guidelines
  A. OSHA Regulations
  B. Related Regulations

Key Terms

chemical hygiene officer • chemical hygiene plan • material safety data sheets (MSDSs) • National Fire Protection Association • performance-oriented standard

Prerequisite Knowledge

Prior to beginning this chapter, the user should review the following chapters:

Chapter Number	Chapter Topic
2	Occupational Exposure Limits
3	Legal Aspects of the Occupational Environment
4	Environmental and Occupational Toxicology
6	Principles of Evaluating Worker Exposure
7	Preparation of Known Concentrations of Air Contaminants
9	Direct-Reading Instrumental Methods for Gases, Vapors, and Aerosols
15	Comprehensive Exposure Assessment
16	Modeling Inhalation Exposure
17	Risk Assessment in the Workplace
20	Noise, Vibration, and Ultrasound
21	Nonionizing Radiation
22	Ionizing Radiation
30	Worker Education and Training
31	General Methods for the Control of Airborne Hazards
32	An Introduction to the Design of Local Exhaust Ventilation Systems
34	Prevention and Mitigation of Accidental Chemical Releases
35	Personal Protective Clothing
36	Respiratory Protection
37	Program Management
38	Surveys and Audits
40	Hazard Communication
45	Occupational Safety
47	Hazardous Waste Management
49	Emergency Planning in the Workplace

48

Laboratory Health and Safety

Stefan Wawzyniecki, Jr., CIH

Fay M. Thompson, Ph.D., CIH

Introduction

The chemical laboratory is a work environment that differs in significant ways from many other work sites. It is important to recognize these differences when evaluating health and safety conditions in the laboratory.

- Laboratories are likely to contain a very large variety but very small quantities of chemicals. Hundreds of chemical compounds, representing many different hazard classes, may be present in a laboratory; however, typical quantities of individual chemicals range from a few milligrams to a kilogram, or perhaps up to five kilograms of organic solvents. Laboratory equipment is small, in keeping with the quantity of chemicals being used.
- The processes carried out in laboratories change frequently, perhaps several times a day. Particularly in research laboratories, processes change rapidly as experimental results lead to decisions to try different materials or techniques. For example, a researcher in a synthetic organic chemistry laboratory may, in one day, use distillation, crystallization, and several separation techniques, either concurrently or sequentially.
- The physical hazards associated with chemical laboratory work vary from insignificant to highly hazardous, including high pressure and high vacuum techniques, very high and very low temperatures, and highly reactive or explosive compounds.
- Individual laboratory workers are often intimately involved in designing the processes they carry out, giving them an unusual level of control over the safety practices associated with their work. Laboratory workers routinely make their own decisions about perrsonal protective equipment (PPE) and engineering control needs associated with individual processes.
- Many standard hazard information sources — in particular, material safety data sheets (MSDSs) — report hazards and recommend controls that are based on operations with quantities of material that are orders of magnitude larger than quantities used in the laboratory, making direct application of the information to the laboratory difficult.

Occupational hygienists responsible for laboratory safety programs have several unusual challenges as a result of these unique laboratory workplace characteristics. The foremost challenge is determining what activities are actually being carried out at various times in a particular laboratory. Direct contact with laboratory workers is essential to this information collection process.

Additionally, evaluating the overall risk associated with highly variable use of chemicals of widely ranging hazards is extremely problematic. Actual exposures are difficult to measure because of limited periods and quantities of use. Another complicating factor is that hazard and toxicity information is not always available for the chemicals used in laboratories. Many specialty chemicals used in laboratories do not have established safe exposure levels. Other parameters for evaluation need to be found.

The following sections of this chapter describe the types of chemicals, processes, and controls that are most likely to be found in chemical laboratories. An understanding of this background information will assist the occupational hygienist in recognizing and evaluating the safety of a laboratory workplace. Appropriate controls can then be specified and implemented.

Hazardous Materials

The chemicals commonly found in laboratories exhibit a number of hazardous characteristics that need to be understood by the occupational hygienist; these include flammability, corrosivity, oxidizing power, reactivity, and toxicity.

Flammable Chemicals

Flammable chemicals occur in both liquid and gaseous states. Flammable liquids are typically present in a laboratory to be used as solvents for chemical reactions or as cleaning agents; therefore, the quantity of flammable liquid may be in the 1-quart to 5-gallon range, enough to present a significant fire hazard. (See section on Controls for recommended storage limits.) The most common solvents are listed in Table 48.1.

Both the flash point and boiling point of a flammable liquid should be reviewed in determining the hazard. This information should be available on the MSDS. The National Fire Protection Association (NFPA) has established a useful classification system for flammable liquids. A Class I flammable liquid is defined as having a flash point below 100°F (37.8°C) and having a vapor pressure not exceeding 40 psi. Subcategories include Class IA, those liquids having flash points below 73°F (22.8°C), and having a boiling point below 100°F; Class IB, those having flash points below 73°F and having a boiling point at or above 100°F; and Class IC, those having flash points at or above 73°F and below 100°F. Class II includes those combustible liquids having flashpoints at or above 100°F and below 140°F. Class IIIA includes those having flashpoints at or above 140°F and below 200°F, and Class IIIB includes those having flashpoints at or above 200°F. Since fire hazard decreases with increasing flashpoint and boiling point, Class IA contains the most problematic chemicals.

Flammable gases are typically present in a laboratory for use as fuels or possibly as reactants. Hydrogen, propane, and acetylene are the most common fuel gases; the flammability of some reactant gases (carbon monoxide, hydrogen sulfide, ammonia, arsine) may be overlooked because of emphasis on their toxicity.

Table 48.1 — Flammable Liquid Solvents

Solvent Class	*Examples*
Aliphatic hydrocarbons	hexane, cyclohexane, pentane, petroleum ether (an alkane, not an ether)
Aromatic hydrocarbons	toluene, xylenes (benzene, once very common, is rarely used as a solvent)
Alcohols	ethanol, methanol, isopropyl alcohol, ethylene glycol
Ethers	diethyl ether, diisopropyl ether, tetrahydrofuran
Ketones	acetone, methylisobutyl ketone, methyl-n-butyl ketone
Esters	ethyl acetate, butyl acetate

Corrosive Chemicals

Corrosive chemicals include a variety of acids and bases and are used in laboratories as reactants, titrants, and occasionally as cleaning chemicals. Highly concentrated acids and bases can readily cause serious damage to skin and eyes, whereas dilute solutions of the same materials may be quite innocuous. Hydrochloric, sulfuric, nitric, and acetic acids are commonly found in laboratories, as are sodium and potassium hydroxide. In concentrated form, both acids and bases have significant potential for generating large amounts of heat when mixed with water or other reagents. Caution must be exercised to avoid splashing or spraying of corrosive materials.

It is important to note that some acids and bases are also strong oxidizers, such as nitric acid and sodium oxide. Also, acetic acid, which is a weak organic acid, is actually a flammable liquid and should be stored as such.

Oxidizing Chemicals

Oxidizing chemicals are capable of providing oxygen (or chlorine or fluorine) in a reaction. Oxidizers must be used under carefully controlled conditions to avoid a fire or other extreme reaction. A solution of potassium dichromate in sulfuric acid has been used as a very strong cleaning solution for glassware; difficulty in disposing of chromium waste has reduced this practice, but old solutions can still be found in some laboratories. Perchloric acid is a strong oxidizer used to digest organic material prior to metals analysis. Peroxides, permanganates, periodates, and perborates also are used as oxidizers in laboratory work, providing a wide range of oxidizing power.

Reactive Chemicals

Reactive chemicals are those materials that are inherently unstable (e.g., explosives) or that react with air or moisture to produce toxic or unstable products, heat, or rapidly expanding gases. These materials need to be handled very carefully, with full cognizance of the conditions that must be avoided. The reader is encouraged to review Bretherick's *Handbook of Reactive Chemical Hazards*[1] for additional information on this subject. The following chemicals are some of the many reactive materials that may be used in the laboratory.

Concentrated (>10%) Solutions of Azides. These can be dangerously explosive; initiation of the explosion can be from heat, friction, or impact. Examples are: acetyl azide, benzoyl azide, and benzenesulfonyl azide.

Acyl Halides. As a group, acyl halides tend to react violently with proton-donating solvents such as water, as well as aprotic solvents (dimethylformamide, dimethyl sulfoxide). The group includes acetyl chloride, benzoyl chloride, and methylchloroformate.

Alkali Metals. This group includes lithium, sodium, and potassium, their alloys, oxides, and hydrides; their reactivity with water has entertained many a high school chemistry class, yet the inherent danger must be recognized. This group is also incompatible with many other chemicals. Reference to the MSDSs is encouraged.

Alkyl Metals. Members of this group such as butyl lithium are highly exothermically reactive to atmospheric oxygen at ambient temperatures, water vapor (even 70% relative humidity conditions in air will cause ignition), and carbon dioxide.

Aqua Regia. A 1:4 mixture by volume of nitric and hydrochloric acids, aqua regia is a powerful oxidant, used for cleaning glassware. It decomposes with evolution of gases, and, if stored in tightly capped bottles, will explode.

Peroxidizable Compounds. These include diisopropyl ether (the most notorious) and diethyl ether, but also include less recognizable solvents such as 1,3-dioxane, and tetrahydrofuran. These ethers by themselves are not particularly reactive, but they react with air to form peroxides that are sensitive to shock and heat.

Metal Fulminates. This group is explosive, with fulminating silver being the most violent member. It is primarily silver nitride; it will explode in solution if stirred.

Metal Azides. These are sensitive explosives on their own, yet, in contact with acids, produce hydrogen azide, a highly toxic and explosive low boiling liquid.

Perchlorates. In their heavy metal form as well as the organic salts, perchlorates are extremely sensitive explosives. Perchloric acid digestions should never be permitted in conventional chemical fume hoods, where the formation of perchlorates could occur. Hoods designed with a wash-down option are required.

Nitric Acid. This chemical deserves special mention, as it is both a corrosive and an oxidizer, and reacts explosively with certain aromatic hydrocarbons, acetone, DMSO, and thiols.

Phosphorous (White). This chemical autoignites at ambient temperatures in air, with resulting irritating vapors to the eyes and respiratory tract.

Picric Acid. Also known as 2,4,6,-trinitrophenol, picric acid is a close relative to TNT (trinitrotoluene). It must be stored wet, as it is explosive when dry. The metal picrates (lead, mercury, copper, and zinc) are sensitive to heat, friction, and impact.

Toxic Chemicals

To quote from the Occupational Safety and Health Administration (OSHA) laboratory standard[2] (discussed later in this chapter):

> *The term "health hazard" includes chemicals which are carcinogens, toxic or highly toxic agents, reproductive toxins, irritants, corrosives, sensitizers, hepatotoxins, nephrotoxins, neurotoxins, agents which act on the hematopoietic systems, and agents which damage the lungs, skin, eye, or mucous membranes.*

All of these toxic characteristics can be found in chemicals used in the laboratory. MSDSs are of some benefit in evaluating the toxicity of particular chemicals, but many of the chemicals used in laboratories have not been even minimally studied. Even when the toxicity of a particular chemical is well understood, the potential exposure to that chemical must be reviewed to make an estimate of the hazard associated with its use. This is also a major challenge, because exposures are highly dependent on how and where the chemical is used and how long the procedure takes. It is often very difficult to sample for airborne chemicals because of the very short period of use.

The National Academy of Sciences/ National Research Council has addressed this issue in detail in its 1995 report, *Prudent Practices in the Laboratory: Handling and Disposal of Chemicals.*[3] Chapter 3 of this publication describes a risk assessment process for laboratory chemical use. Briefly, the process includes the following steps.

1. Identify chemicals to be used and how they will be used.
2. Consult sources of information.
3. Evaluate type of toxicity. If no information is available for a compound, it may be necessary to look for chemical similarities with other compounds of known toxicity.
4. Consider possible routes of exposure.
5. Evaluate quantitative information on toxicity. Classify as highly toxic, moderately toxic, slightly toxic, etc.
6. Select appropriate procedures to minimize exposure.
7. Prepare for contingencies.

Because potential exposure to chemicals in the laboratory is so variable, Point 6 above becomes the critical safety factor; chemical procedures should always be designed to keep exposure as low as reasonably possible.

Chapter 3 of the above reference is also useful for evaluating laboratory chemicals for many other potential hazards.

Hazardous Waste

Chemical laboratories produce hazardous waste. In general, the hazards associated with this waste will be similar to the hazards of the chemicals that make up the waste, with a few notable exceptions. Chemical reactions produce intended products, plus a variety of byproducts. The products are likely to be anticipated and reasonably well understood, whereas there may be very little characterization of the byproducts. Whether product or byproduct, the new materials may be either more or less toxic or hazardous than the original starting materials. Procedures for handling hazardous waste should accommodate the fact that toxicity and hazard may be more significant than anticipated from knowledge of the original chemical components.

Mixing waste from different sources in the same container may also yield unexpected results. Caution is necessary to ensure that incompatible chemicals are not mixed inadvertently.

A third problem to be aware of, both for waste and reagent chemicals, is unlabeled containers. A reasonably common cause of laboratory accidents is opening and emptying an old, unlabeled bottle of waste or unstable chemical. Unlabeled containers should always be handled with extreme caution. Every effort should be made to avoid having unlabeled containers at all. (See Chapter 47, Hazardous Waste Managemen.)

Radioactive Materials

Radioactive materials used in laboratories will, in most cases, fall under the purview of U.S. Nuclear Regulatory Commission rules, or their counterparts in "agreement states," which serve to control the radiation hazard through detailed regulatory requirements. One should realize, however, that laboratory workers who concentrate on controlling radiation hazards sometimes tend to forget that their radioactive isotopes are also chemicals and may also have associated chemical hazards. For example, radiolabeled PCBs (polychlorinated biphenyls) or dioxin may require more control for their chemical toxicity than for their radioactivity.

Physical Hazards

In some instances a laboratory worker may be subject to noise, via the following examples. Certain preparation techniques (grinding, blending, sonicating) may emit annoying frequencies. An instrumentation room may have a constant hum or pitch that may be irritating to a worker. Ventilation systems, especially older fume hoods, may be the source of a constant noise. An assessment should be made of the source, and/or monitoring of the employee using a noise dosimeter may be appropriate (see Chapter 20).

Physical constraints on the lab worker include such activity as lifting and repetitive motion. Many solvents are purchased in 5-gallon containers, and dispensing from a full drum, with back bent and knees locked, can lead to injury. Twisting motions with heavy containers can lead to back strain. Many analytical labs, with high throughput of samples, may require analysts to inject samples constantly, or pipette, requiring repetitive hand motion.

Cuts become a special concern if chemicals are involved, due to the direct contact with the bloodstream. Typical is the example of a mercury thermometer breaking while being handled. Forcing glassware assemblies is another source of cuts and punctures.

Slips and falls due to spills are a physical hazard, yet can be compounded if a chemical is either the spilled entity, or if the worker is carrying an open container of a chemical solvent or solution. The reflex to reach out and grab something may result in laboratory equipment or chemicals being knocked from benches.

Eye strain may occur through constant monitoring of instruments, or through the repeated action of taking readings from a burette. Poor or harsh lighting impacts these activities further.

Hazardous Processes and Equipment

Accidents in laboratories involving equipment may be a result of misuse, or may occur even with proper use, if the equipment has not been well-maintained. As part of a laboratory's internal inspection checklist, one should review the schedule of maintenance procedures required for pieces of equipment. This could be as simple as checking the oil level in a pump or inspecting electrical cords for fraying. After electrical hazards, probably the next most common hazards are those involving working with high pressure (compressed gases), vacuum systems, and extreme temperatures. Ionizing and nonionizing radiation hazards also exist in laboratories, potentially associated with X-ray equipment, lasers, nuclear magnetic resonance instruments, electron microscopes, etc. These hazards are covered in Chapters 21 and 22. Other physical hazards may be from the use of tools (cutting, drilling, mixing), and preventive measures would be similar to those in industry. The reader is encouraged to refer to those chapters addressing such hazards within this text. Instances of long hair being caught in rotating paddle mixers have been documented, and any laboratory using water-cooled equipment always faces the potential of a flood, with the resultant slippery conditions. The remainder of this section will focus on modified pressure techniques, modified temperature techniques, energy hazards, and potential problems associated with separation techniques.

Modified Pressure Techniques

Some processes, such as distillation of a high boiling liquid, are more effective when carried out at reduced pressure, because the reduced pressure also decreases the boiling point. Greatly reduced pressure (high vacuum) systems are also necessary for processes requiring very low concentrations of chemicals (such as vapor deposition). Reduced pressure equipment must be constructed to withstand a large pressure differential between the inside and outside of the system and must be shielded to avoid injury if implosion should occur.

High pressure reaction systems (e.g., hydrogenation cylinders) must also be constructed to hold the pressure that will be developed. In this case barriers against an explosion should be in place.

Modified Temperature Techniques

It is frequently important to control temperature when working with chemicals. Heat may be added to initiate a reaction or increase a reaction rate. Cooling may be used to condense a reaction product so that it can be collected or to keep the reaction from getting out of hand. Heating devices include electric heating mantles, infrared lamps, oil and sand baths, hot air guns, and even warm water. Cooling systems include cold water, ice baths, dry ice-solvent traps, cryogenic liquids, and mechanical refrigeration systems. The bigger the deviation from ambient temperature, the more potential there is for an extreme reaction to occur; therefore, close monitoring of such processes is important. Very rapid temperature changes are also more likely to cause problems than will slower warming or cooling processes. Of special note with cryogenic liquids (e.g., liquid nitrogen) is the fact that they may condense oxygen from the air and cause an explosion if they contact combustible materials.

Energy Hazards

This section will not attempt to detail specific hazards of each powered piece of equipment used in the laboratory. The main thing to recognize is that electric shock can be a major hazard. As little as 10 milliamperes (mA) can result in personal damage, and a current of 80 mA can be fatal. High voltage lines may also be encountered, especially with instruments like atomic absorption spectrophotometers and inductively coupled plasma units.

Electrical safety is governed by the National Electrical Code, sections of which OSHA has adopted in *Code of Federal Regulations* (CFR) Title 29, Part 1910, Subpart S. Although geared to industrial applications, OSHA recognized that any piece of equipment that, through malfunction, disrepair, or poor maintenance, may present a shock hazard or initiate a fire, should be guarded through prudent measures. These measures may include signs, lockout/tagout procedures, and special enclosures.

Basic electrical safety begins with proper insulation, wiring to code by a licensed electrician, grounding, and adequate mechanical devices such as panels and breakers. Extension cords do not meet these requirements and should not be used in laboratories. A particularly important device for laboratory settings is the ground fault interrupter (GFI). A GFI protects an individual by means of creating an imbalance in the circuit if an individual's body and the live wires make contact. This imbalance is detected by the GFI and causes it to break the circuit. Response time by a GFI is measured in milliseconds, which limits the duration of the flow of current. One of the factors for the potential of electrical shock is the duration of the flow of current, and therefore, by minimizing this factor, the hazard is minimized. Other factors include the voltage, the actual current, and the amount of resistivity encountered. Sweaty hands in contact with live wires decreases resistivity, thereby increasing the likelihood of shock.

OSHA's lockout/tagout standard[(4)] is based on common sense—do not work on electrical systems unless they are shut off and de-energized. Many laboratory instruments warn users of the presence of high voltage beneath the instrument's covers. (One commonly misunderstood instrument is the gel electrophoresis unit.) Only trained or qualified staff should attempt repairs. If maintenance requires that the power remain on, the prudent thing to do is ensure that no circuit can be completed through an individual's body. Appropriate tools should be used, but the mere presence of tools for electrical repair does not give anyone the go-ahead to attempt the work.

Instruments that use or generate various types of hazardous radiation include lamps and lasers, X-ray generators, and microwave ovens and furnaces. Along with specific hazards associated with the type of radiation involved (e.g., eye injury from lasers), electrical hazards are always a possibility. Nuclear magnetic resonance instruments, which generate large static magnetic fields, have also become common analytical and research tools.

Separation Techniques

Distillation, extraction, and chromatography techniques are all employed to purify a reaction product or to increase the purity of a solvent. Traditionally these techniques have involved large quantities of solvents, which are often flammable. Although many refinements have taken place in the past 25 years, greatly reducing the scale on which these techniques are carried out, they still account for a large portion of flammable chemical use in the laboratory.

Anticipation/Recognition

The ability to anticipate or recognize hazard potential can be learned, gained through experience, or may be innate in the health and safety officer as well as a worker. Just as a construction worker might anticipate slippery conditions along a muddy path at an outdoor site, he or she might also recognize the potentially hazardous condition of setting up a scaffold in mud. A laboratory worker must also be able to anticipate and recognize dangers in the work setting.

A formal hazard recognition program would incorporate the elements shown in flowchart format in Figure 48.1. The preliminary survey would reveal the obvious hazards. The next two steps on the figure are interconnected—a collection of MSDSs does not necessarily list what chemicals are being stored, nor does an inventory necessarily reveal what MSDSs are on hand. It is important to note that, as part of a laboratory's chemical hygiene plan, a collection of MSDSs could make up one appendix (taking into account space limitations), and a second appendix could be the inventory list. Developing a health and safety plan basically means setting up standard operating procedures—in what manner are the chemicals going to be used, and in what form or concentration? Next, a sampling strategy must be formed, based on number of laboratory workers using the chemicals, their work locations, and whether other factors have an effect on the possible route of exposure. Finally, a baseline survey can be performed,

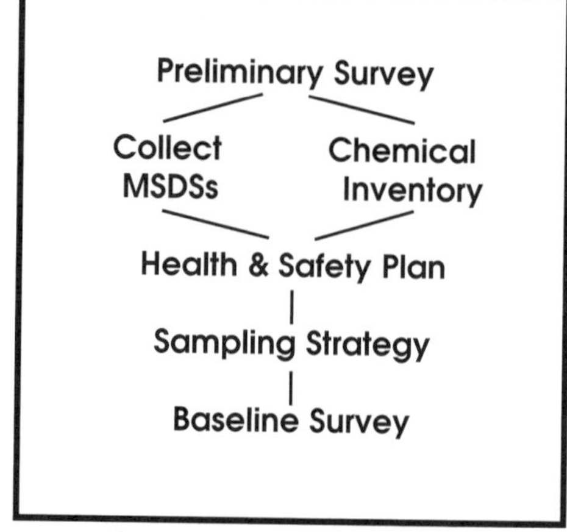

Figure 48.1 — Elements of a formal hazard recognition program.

and based on the outcome, controls incorporated as necessary. For related material, see Chapter 6, Principles of Evaluating Worker Exposure , and Chapter 38, Surveys and Audits.

Evaluation/Assessment

In laboratory operations, evaluation of hazards can be defined as the decision-making procedure that results in an opinion of the degree of hazard found in the laboratory after the recognition process has been completed. On occasion this assessment becomes a judgment call. For instance, if a laboratory's inventory shows only one severe health hazard (poison, or highly toxic by definition), in a single container, is it necessary to provide the "secure" storage area recommended? Or, knowing the premium placed on shelf storage space in labs, would it be allowable to simply provide secondary containment for the one bottle and store it with other chemicals that are known not to be incompatible with the poison? Factors such as laboratory accessibility, number or workers, and unmeasurable variables such as personalities become factors in the evaluation equation.

Determining the existence of a hazard is based on observation and measurement of contaminants arising from the laboratory procedures, in addition to assessment of how well any controls are working. Basic occupational hygiene practice teaches one to compare the results of monitoring with the established guidelines published.

A structured approach to evaluating laboratory hazards follows:

- What are the end-products of any experiments or analyses?
- What are the starting materials?
- What chemicals or external variables (heat, light, irradiation) may be added?
- What is the duration of the procedure?
- Are written experimental procedures in place?
- Are written spill procedures or general cleanup procedures in place?
- Are engineering controls adequate?
- Are safety appliances (showers, eyewash stations) available?
- How are chemicals stored?
- How is the laboratory laid out (aisle space)?

These questions could be incorporated as part of an internal audit overseen by the laboratory supervisor. Laboratory employees, by knowing the audit parameters, can play an active role by requesting that assessment results be addressed.

Survey/Sampling Strategies

Ideally, the occupational hygienist or health and safety officer would attempt to work with the laboratory workers to identify which procedure or experiment causes discomfort or releases contaminants. The difficulty lies in the fact that single room laboratories may be large enough to house many simultaneous experiments, each with its own unique hazard. Alternatively, there may be an exposure that cannot be easily addressed by typical engineering controls, which for laboratories usually is a fume hood. For example, viewing slides under a microscope, which may also require addition of minute volumes of a volatile solvent, may not be possible under a fume hood because of the vibration of the blower. The exposure may not be extreme, it may not be for a duration approaching 8 hours or even the short-term exposure limit, but it may affect the worker. The strategy may be to try to sample the worst case and recommend numerous breaks if the results warrant such action. Another example involves weighing out toxic chemicals that, either because of their potential to become airborne as dust particles or because of their volatility, may pose an inhalation hazard. Again, a fume hood may not solve the problem because of the mechanical vibration, interfering with the sensitive balance. The weighing procedure may occur only once a day; it may last only five minutes. The strategy may be not to sample at all, but to relocate the weighing operation to a dedicated room, under negative pressure, with the room exhaust located such that the worker is not subject to turbulence that could direct the contaminants toward him or her.

Surveys should take into account peak exposures; in an environmental or occupational hygiene analytical laboratory, sample throughput may be maximized in late morning or early afternoon. Sampling should reflect these peaks as well as the daily averages. The reader is encouraged to review Chapter 15 for a more detailed discussion of exposure assessment strategies.

Controls

The primary control for limiting exposure to chemicals in laboratories is the fume hood. This is but one of the methods under the category of engineering controls. Typically these are mechanical devices that reduce or eliminate the hazards by isolation, ventilation, or enclosure, or substitution of a material, piece of equipment, or process. Engineering controls, including adequate provision for chemical storage and basic facility design, are the first line of defense for managing exposures. A second important control parameter is PPE. Certain basic PPE should be used by all laboratory workers; however, the need for more extensive equipment, such as respirators, should be avoided through use of engineering controls whenever possible. A third control mechanism, administrative control, also plays an important role in laboratory safety.

Engineering Controls

Fume Hoods/Laboratory Ventilation. The laboratory fume hood is the critical control technology for laboratories. When properly designed and used, it will control exposure to toxic, flammable, or odorous chemicals.

Chemical fume hoods are of four basic types. Each meets the purpose of a hood, namely, 1) to provide a ventilated enclosure, containing, capturing, and exhausting hazardous chemicals; and 2) to provide a ventilated workplace, to protect personnel from overexposure to those chemicals. (See Chapter 32 for an expanded discussion of local exhaust systems; see also the American Conference of Governmental Industrial Hygienists' *Industrial Ventilation: A Manual of Practice.*[5])

The four types and their brief descriptions follow.

1. Conventional fume hood—the simplest in design and, therefore, the least expensive to purchase. This fume hood moves a constant high volume of air. Operating costs are high, and the face velocities are nonuniform and often excessive. The principal mechanism is that, with the sash fully open, there is a minimum face velocity and, as the sash is lowered, the face velocity increases. The increased velocity can create turbulence inside the hood.
2. Bypass fume hood—also simple in operating principle, with face velocities more maintainable, yet still variable across the hood. Above the sash is the bypass area, which allows air into the hood as the sash is lowered. There is still appreciable wasted energy, because of the relatively large volume of air being exhausted at all times.
3. Auxiliary/make-up air fume hood—having an auxiliary air supply duct positioned above the working area in front of this hood reduces the amount of conditioned room air that is exhausted, resulting in lower operating costs. A disadvantage occurs in regions of temperature extremes if the make-up air is unconditioned, creating worker discomfort, and possible unacceptable conditions inside the hood.
4. Variable air volume hood—controls drive up the initial costs of these hoods, yet the operating costs may be significantly lowered. These hoods are designed to maintain a constant average face velocity as the sash open area changes. The hood only exhausts enough air to meet actual needs.

Fume hood performance can be affected adversely by factors that cause air to be drawn out of the hood into the laboratory or that create turbulence within the hood. Thus, improperly located air diffusers, doors and windows, and traffic patterns will reduce hood effectiveness, as will storage of chemicals and equipment within the hood that block the direction of airflow.

The basic ventilation design premise for the laboratory itself is that the room should be negatively pressurized with respect to the adjacent corridors. This minimizes the potential that contaminants generated within the laboratory space will affect neighboring spaces. Balancing such a ventilation system requires gathering data from all the fume hoods in the building, the room exhausts, windows that may be operable, and make-up (supply) air. In general, laboratory air should not be recirculated.

Some general recommendations that involve laboratory ventilation include:

1. Comfort ventilation, 20 ft^3/min per person in offices (American Society of Heating, Refrigerating and Air-Conditioning Engineers [ASHRAE]);[6]
2. Laboratory ventilation, 6 air changes per hour (ac/hour) (ASHRAE),[6] 4–12 ac/hour (OSHA);[7]
3. Chemical storage rooms, 6 ac/hour (OSHA);[8] and
4. Fume hood face velocity, 60–100 linear feet per minute (OSHA).[7]

Chemical Storage Areas

Because of the large number of individual chemicals typically found in a laboratory, it is tempting to arrange the containers alphabetically on shelves. This will inevitably bring incompatible chemicals into close proximity to each other, which must not be allowed. A primary consideration for chemical storage is providing for segregation of chemicals by hazard class. NFPA defines segregation to mean storage

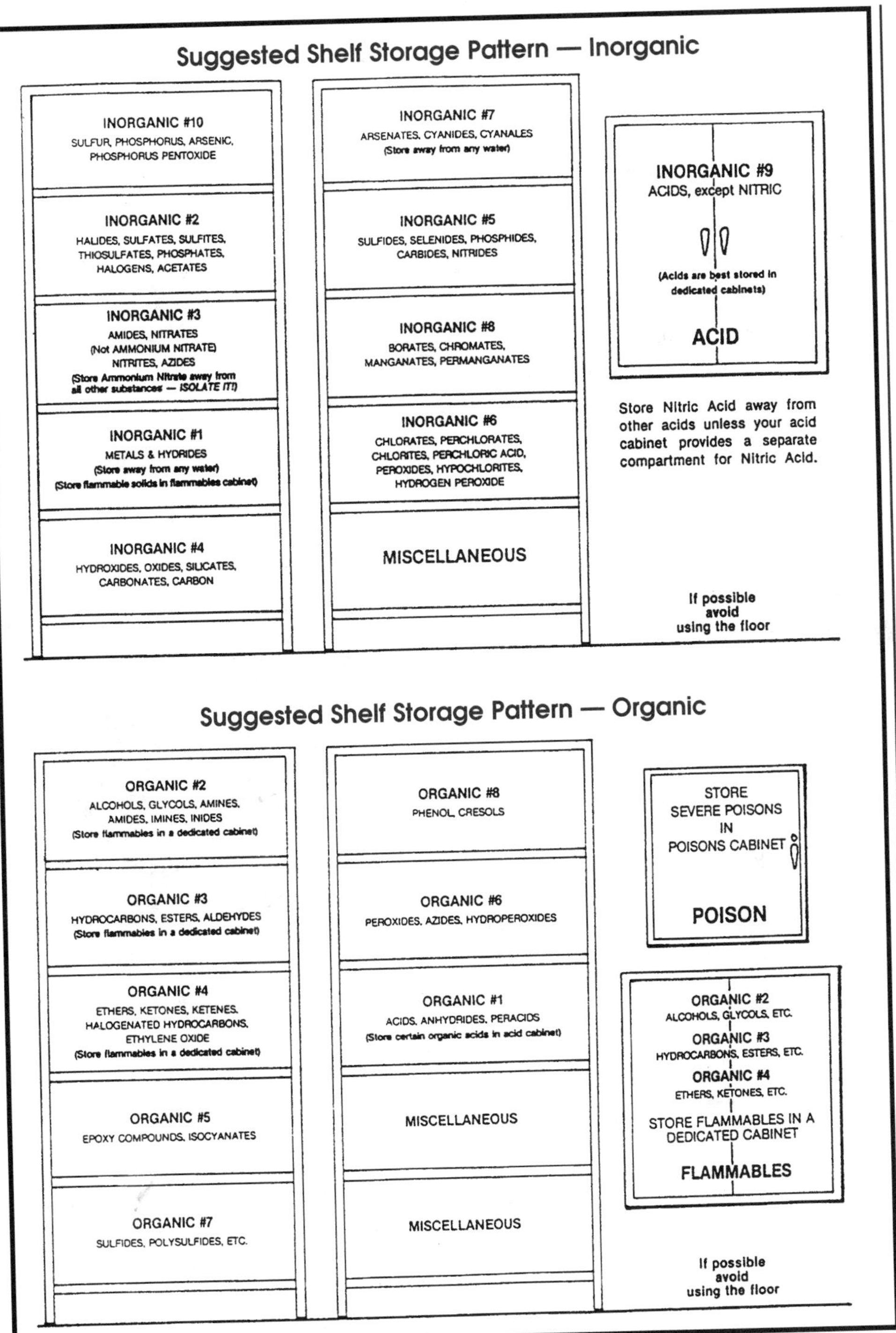

Figure 48.2 — Suggested shelf storage patterns for inorganic and organic chemicals.

source NIOSH Manual of Safety & Health Haz in School Science Lab

within the same room, but physically separated by space. Spacing includes secondary containment (e.g., trays), sills on shelves, or curbs (berms) on floor storage.

In addition to providing for segregation, many other factors must be taken into consideration when designing or evaluating storage space. The construction of the room itself (i.e., its fire rating), existence of a fire suppression system, activities in nonlaboratory spaces (hospital, school, etc.), the type and severity of hazards associated with the chemicals to be stored, the size and number of containers to be stored, provision for secondary containment, and availability of specialized storage cabinets all play a role in determining what constitutes safe storage (see Tables 48.2 and 48.3).

A number of different schemes exist for storing chemicals within a laboratory. For example, the National Institute for

Table 48.2 —
Maximum Quantities of Flammable and Combustible Liquids in Sprinklered Laboratory Units Excluding Flammable Liquid Inside Liquid Storage Areas

		Excluding Quantities in Storage Cabinets or Safety Cans		*Including Quantities in Storage Cabinets or Safety Cans*	
Laboratory Unit Fire Hazard Class	*Flammable Combustible Liquid Class*[A]	*Maximum Quantity per 9.3 m² (100 ft²) of Laboratory Unit*	*Maximum Quantity per Laboratory Unit*	*Maximum Quantity per 9.3 m² (100 ft²) of Laboratory Unit*	*Maximum Quantity per Laboratory Unit*
A	I	38 L (10 gal)	2270 L (600 gal)	76 L (20 gal)	4540 L (1200 gal)
	I, II, and IIIA	76 L (20 gal)	3028 L (800 gal)	150 L (40 gal)	6060 L (1600 gal)
B	I	20 L (5 gal)	1136 L (300 gal)	38L (10 gal)	2270 L (600 gal)
	I, II, and IIIA	38 L (10 gal)	1515 L (400 gal)	76 L (20 gal)	3028 L (800 gal)
C	I	7.5 L (2 gal)	570 L (150 gal)	15 L (4 gal)	1136L (300 gal)
	I, II, and IIIA	15 L (4 gal)	757 L (200 gal)	30 L (8 gal)	1515 L (400 gal)
D	I	4 L (1.1 gal)	284 L (75 gal)	7.5 L (2 gal)	570 L (150 gal)
	I, II, and IIIA	4 L (1.1 gal)	284 L (75 gal)	7.5 L (2 gal)	570 L (150 gal)

[A]Includes Class I flammable liquids and liquified flammable gases.

Table 48.3 —
Maximum Quantities of Flammable and Combustible Liquids in Nonsprinklered Laboratory Units Excluding Flammable Liquid Inside Liquid Storage Areas

		Excluding Quantities in Storage Cabinets or Safety Cans		*Including Quantities in Storage Cabinets or Safety Cans*	
Laboratory Unit Fire Hazard Class	*Flammable Combustible Liquid Class*[A]	*Maximum Quantity per 9.3 m² (100 ft²) of Laboratory Unit*	*Maximum Quantity per Laboratory Unit*	*Maximum Quantity per 9.3 m² (100 ft²) of Laboratory Unit*	*Maximum Quantity per Laboratory Unit*
A	I	38 L (10 gal)	1136 L (300 gal)	76 L (20 gal)	2270 L (600 gal)
	I, II, and IIIA	76 L (20 gal)	1515 L (400 gal)	150 L (40 gal)	3028 L (800 gal)
B	I	20 L (5 gal)	570 L (150 gal)	38L (10 gal)	1136 L (300 gal)
	I, II, and IIIA	38 L (10 gal)	757 L (200 gal)	76 L (20 gal)	1515 L (400 gal)
C	I	7.5 L (2 gal)	284 L (75 gal)	15 L (4 gal)	570 L (150 gal)
	I, II, and IIIA	15 L (4 gal)	380 L (100 gal)	30 L (8 gal)	760 L (200 gal)
D	I	4 L (1.1 gal)	140 L (37 gal)	7.5 L (2 gal)	284 L (75 gal)
	I, II, and IIIA	4 L (1.1 gal)	140 L (37 gal)	7.5 L (2 gal)	284 L (75 gal)

[A]Includes Class I flammable liquids and liquified flammable gases.

ChemAlert* storage codes

A color-coded bar on the label of every Fisher chemical provides an instant guide to storage. The storage code color is also denoted by its initial, and spelled out for additional clarification. The five storage colors and their descriptions are as follows:

RED (R): Flammable. Store in area segregated for flammable reagents.

BLUE (B): Health hazard. Toxic if inhaled, ingested or absorbed through skin. Store in secure area.

YELLOW (Y): Reactive and oxidizing reagents. May react violently with air, water or other substances. Store away from flammable and combustible materials.

WHITE (W): Corrosive. May harm skin, eyes, mucous membranes. Store away from red-, yellow-, and blue-coded reagents above.

GRAY (G): Presents no more than moderate hazard in any of categories above. For general chemical storage.

STORAGE CODE STOP **EXCEPTION:** Denoted by the word "STOP." Reagent incompatible with other reagents of the same color bar. Store separately.

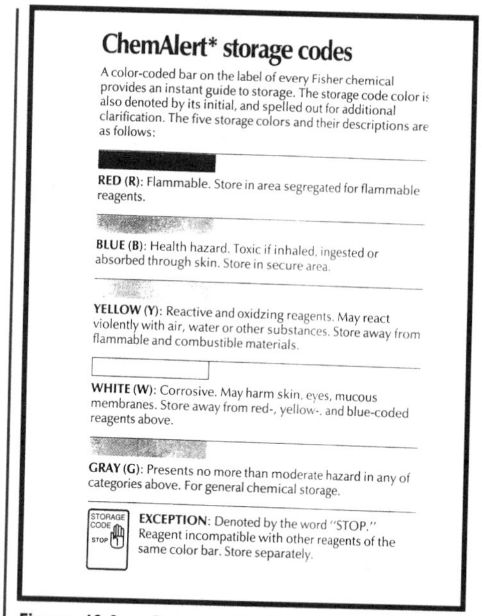

Figure 48.3 — Color-code storage system.

Occupational Safety and Health (NIOSH) *Manual of Safety and Health Hazards in the School Science Laboratory* includes a "Suggested Shelf Storage Pattern" (see Figure 48.2). Chemical distributors have also devised color-coding schemes for their labels, where the color code corresponds to a particular hazard class (see Figure 48.3). The NFPA 704 fire diamond(9) is used by some laboratories as an additional labeling method, both for individual bottles and for posting at the entrances to laboratories, indicating the hazard potential to be found within. Fire department personnel and emergency responders may find this useful; however, they are more likely to assume the worst in the event of an emergency.

Local building and fire codes will determine limitations for various types of chemical storage; laboratory managers and designers must remain in close contact with their local authorities to assure that laboratory facilities meet the applicable regulations. Local codes are generally based on one of three existing regional code publications. Also, NFPA offers guidance on laboratory design and construction through a wide variety of laboratory-related standards, including NFPA 30, Flammable and Combustible Liquids Code; NFPA 45, Fire Protection Standard for Laboratories Using Chemicals (including reference to over 20 other standards); NFPA 49, Hazardous Chemicals Data; NFPA 43A, Code for Storage of Liquid and Solid Oxidizing Materials; NFPA 43C, Code for Storage of Gaseous Oxidizing Materials; and NFPA 56C, Safety Standard for Laboratories in Hospitals.

Facility Design

Safety features should be designed into the laboratory. Safety showers, eyewashes, fire extinguishers or extinguishing systems, general and fume hood ventilation, sinks and ample space for chemical storage, and waste handling are all part of a safe laboratory. Evacuation routes from the laboratory are also important. Laboratories should have two exits that are easily accessed through uncluttered aisles. OSHA, the American National Standards Institute, NIOSH, NFPA, and building/fire codes all have specifications for location and construction of these various safety features. Figure 48.4 demonstrates several basic laboratory designs.

Some laboratories will require additional safety features based on the operations intended to be carried out in the space. Local exhaust for certain instruments (e.g., gas chromatographs or atomic absorption units), wash-down hoods for perchloric acid digestions, explosion-proof refrigerators or freezers, walk-in fume hoods, glove boxes, and constant-temperature rooms all may be required at various times. In rare instances, even a blow-out, pressure-relief panel may be a necessary feature of an isolated laboratory facility.

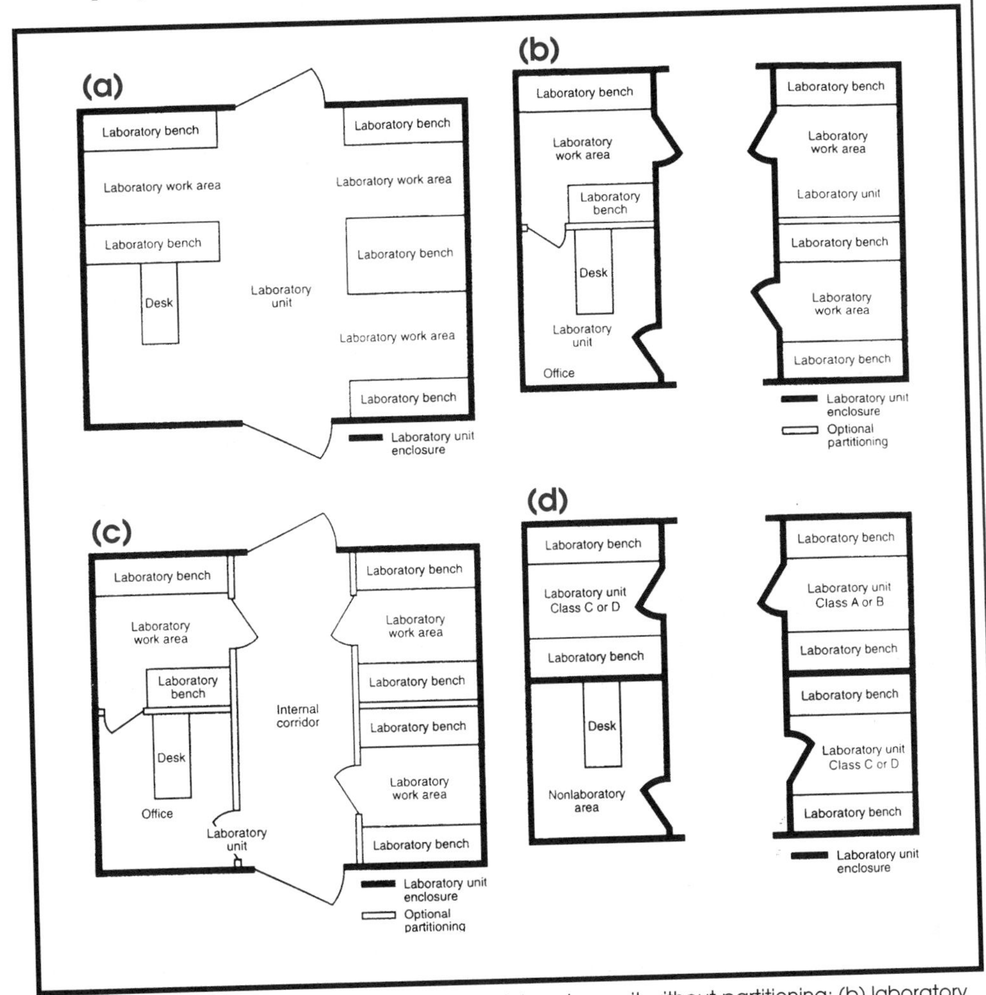

Figure 48.4 — Separation of laboratory areas: (a) laboratory unit without partitioning; (b) laboratory units separated by a required exit corridor; (c) laboratory unit with partitioning; (d) separation of laboratory units and nonlaboratory areas (plans not to scale).

PPE

Because some hazards cannot be completely avoided through engineering, a minimum level of PPE is essential for all work with chemicals. Eye protection is the most important piece of safety equipment. Basic protective clothing includes lab coats or aprons, gloves and closed-toe shoes (no bare feet or open-toed sandals). Because of the variety of chemicals used in the laboratory setting, several types of gloves with different permeation characteristics need to be available.

Face shields and heavy leather coats and gloves may be appropriate for work where high energy chemicals or processes increase the potential for serious injury. In some cases, respiratory protection may be appropriate, such as during change-out of toxic gas cylinders in a central storage vault, but in general it should be possible to use engineering controls to avoid the need for a respirator.

Administrative Controls

Certain types of administrative controls can add considerable assurance that the laboratory will provide a safe working environment. These include establishing basic safety rules, instituting an experiment planning process, establishing restricted areas for certain higher hazard activities, and maintaining an inventory of laboratory chemicals (in some jurisdictions, chemical inventories are required).

Safety Rules

Basic safety rules are useful for all laboratories. These include good housekeeping practices; routine safety inspections; special arrangements for working alone; no practical jokes or horseplay; no unaccompanied visitors; and no eating, drinking, or smoking in the laboratory.

Planning Laboratory Experiments

Experiment planning includes establishing goals for the experiment, evaluating hazards and assessing risks, and determining safe working procedures for the chemicals and equipment involved. Provision for acquisition and storage of chemicals and final disposal of waste should be included in the review. For high hazard work, the institution may choose to require specific approval before work can begin. More advanced programs have developed an experimental process review procedure, especially for work with particularly hazardous substances.

Establishing Restricted Areas

Restricted areas may be useful or required for work with select carcinogens (defined in the OSHA laboratory standard) or with chemicals exhibiting a high degree of acute toxicity. The restricted area may be an entire laboratory, a portion of a lab, a fume hood, or a glove box. Special signs should be used to notify laboratory occupants of the hazards involved.

Chemical Inventories and Inspections

An inventory of laboratory chemicals has several advantages and is required in some states and jurisdictions. In general, an up-to-date inventory allows the chemist or analyst to determine quickly what is on hand. Sharing of resources is another reason for keeping an inventory. Many chemicals remain on shelves unused, yet may be needed in a neighboring laboratory. A recurring inventory also shows which chemicals are not being used, and can facilitate decisions to recycle or dispose of unnecessary material, in some cases significantly reducing potential hazard. With bar coding, computerization, and networking all readily available, keeping an inventory up-to-date is no longer the onerous task it once was.

The inventory process, if performed regularly, works hand in hand with safe storage procedures. By checking dates of receipt, dates of opening, together with condition of container, and legibility of labels, one can demonstrate control over the chemicals. Short shelf-life chemicals can be identified before they deteriorate or become hazardous. The amount of dust on a container raises the dimension of usefulness of the chemical, and whether it should be stored under active or archival storage. The latter must also be inventoried regularly, as chemicals should not be stored indefinitely. Deterioration of specific compounds cannot always be predicted; however, the characteristics of possible degradation reactions of groups of compounds can be recognized. For example, chemicals that are shipped packed under nitrogen are susceptible to oxidation when opened, and dehydration affects those compounds with bonded water molecules. Polymerization may occur over time, as in the case of styrene, and it is only through periodic checks that the determination of the viability of a compound can be made. Visual inspection of stored chemicals should be based on the following criteria.

- Change in color/opaqueness of contents;

- Caking of anhydrous materials;
- Solid residue in a liquid or liquid in a solid;
- Bulging containers (overpressurization);
- Questionable container integrity (rusting, corrosion, etc.);
- Obvious error in labeling; and
- Layering in liquids.

Reporting requirements under the Community Right-To-Know Act[10] are a driving force for inventorying chemicals and hazardous materials, because some of the reportable quantities and threshold planning quantities are for chemicals and amounts that can be found in some laboratories. Local authorities, such as fire departments, working through their local emergency planning committees, may require inventories for their files in the event of a chemical release or other accident involving the laboratory.

Training

One of the most important preventive measures that can be taken to ensure safety in the laboratory is a formal training and information program. Just as laboratory employees need to be trained properly on how to operate an instrument, they must also be familiar with the proper means of handling hazardous materials. No one would feel comfortable running an instrument without first becoming familiar with how to optimize its parameters, yet many laboratory workers approach chemicals without first looking at the MSDS or reviewing the safety information for the material. Ultimate responsibility for offering and implementing training programs lies with the management of the laboratory, but ultimate responsibility for ensuring a safe laboratory environment on a daily basis rests with the laboratory supervisor and worker. These individuals must put laboratory safety into practice.

A comprehensive training program begins with orientation. In some states, supervisors are responsible under law for ensuring the safety of their employees. High throughput laboratories emphasizing the number of analyses produced daily are courting danger if their orientation focuses primarily on analytical techniques and not on safe operating procedures. A new employee's observation that safety is secondary in a laboratory will establish the wrong mind-set at the very beginning.

The initial training should cover these basics.

- The organization's safety policies and procedures;
- Emergency procedures, first aid, and accident reporting;
- Location of the MSDSs;
- Identity of the chemical hygiene officer and how to contact him/her; and
- Location of the chemical hygiene plan.

After the orientation, the employee's responsibilities are outlined, and specific training can be given pertaining to the hazards inherent in the instrumental techniques used, as well as the chemical reactions that may be faced. An ongoing safety training program can be incorporated into part of routine weekly staff meetings, or monthly/quarterly corporate-level meetings. In a university setting, small research groups usually hold seminars highlighting research progress, and safety should become a part of these talks. Departmental safety meetings should be called by the health and safety officer. In either case, topics that should be addressed include what new experiments may be started and what new chemicals may be ordered, followed by distribution and discussion of corresponding MSDSs.

These are ideal opportunities for the laboratory workers to provide input, ask questions, and resolve concerns regarding hazards in the laboratory. If accidents have occurred, answers should be found to the inevitable inquiries about what happened, and how it happened, and most important, how it can be prevented from ever happening again. Training should occur every time a new procedure is introduced. Training on the proper use of safety showers and eyewashes can be combined with discussion of routine maintenance procedures (e.g., flushing the lines).

Training is required under OSHA's hazard communication standard, the laboratory standard, and hazardous waste operations and emergency response regulations. Commercially available programs can be purchased to address the requirement, videos can supplement (but not replace) in-house training, or professionals can be brought in to train large groups. Whatever combination of training techniques is used, some method of evaluation should be part of the program. Safety quizzes may not be popular, but they have the added benefit of providing documentation, in addition to sign-in sheets prior to safety seminars. A measure of the level of comprehension is especially necessary due to the multicultural, talented work force of today.

Regulations and Guidelines

OSHA Regulations

As is the case in many occupations, one of the driving forces that brings attention to health and safety in a laboratory is the regulations. In 1986 OSHA suggested that when chemists work with established health and safety programs, they do not experience any additional occupational risks when compared with the general public; but when working without such a program, they are subject to a higher incidence of occupational diseases, such as cancer. Soon after, OSHA promulgated the laboratory standard, officially known as "Occupational Exposure to Hazardous Chemicals in Laboratories."[(2)] Unlike many other OSHA regulations, this is a performance-oriented standard and is predicated on the expectation that laboratory workers, through education and experience, have at least some level of knowledge about the hazards of chemicals. Laboratory work is also recognized as a very individual activity, affording each worker a greater level of control over safety than is likely to be found in the industrial setting.

OSHA recognizes the unique nature of laboratory work in its definition of "laboratory use of hazardous chemicals" as an occupation in which:

- Containers used for reactions, transfers, and other handling of substances are designed to be easily and safety manipulated by one person;
- Multiple chemical procedures or chemicals are used;
- The procedures involved are not part of a production process; and
- Protective laboratory practices and equipment are available and in common use to minimize the potential for employee exposure to hazardous chemicals.

With the exception of the permissible exposure limits (PELs), the laboratory standard supersedes most of the other standards in Subpart Z of 29 CFR 1910. Thus, the list of substance-specific standards (1910.1001–1048) is applicable to laboratories only for the associated PELs and action levels. The provisions on monitoring and medical surveillance for those specific substances do not apply. The laboratory standard covers monitoring and medical surveillance in general terms for all hazardous chemicals, where the PEL or action level is routinely exceeded. The exception would be a laboratory specializing in anatomy, for example, where substantial amounts of formaldehyde may lead to potential exposure. In this case, the formaldehyde standard (29 CFR 1910.1048) would be applied. Laboratories with only incidental use of formaldehyde would abide by the lab standard.

A chemical hygiene plan (CHP) must be developed to guide laboratory work. It must be a written document that sets forth procedures, equipment, PPE, and work practices that will assure a safe work environment. The CHP must be accessible to all laboratory employees and must be implemented by the employer.

Following are the key elements required in the CHP, followed by further discussion from a practitioner's approach.

Standard Operating Procedures (SOP). These are mandatory for all laboratory activities involving hazardous chemicals. In practice, such procedures can be generalized (e.g., safety glasses must be worn in the lab) and should state the obvious. If safety goggles are not specified for wear when acid is being used, and an accidental splatter injures a laboratory worker, the lack of an SOP is evidence of noncompliance with a performance-based standard.

Designation of a Chemical Hygiene Officer (CHO). This individual is charged with implementing the chemical hygiene plan. The practitioner's approach is to assign names to various responsibilities, including ultimate responsibility being given to upper management, thereby keeping CEOs and presidents involved and committed to a health and safety program. One CHO may suffice for an organization, or many can be named for individual laboratories of diverse research. In addition, a chemical hygiene committee may be formed, which may be already in place as a safety committee.

Employee Training and Provisions for Accessing Safety Information. In practice, a health and safety office is the key for implementation of this element. A well-established health and safety office can be staffed by experts to train laboratory workers in all aspects of the chemical hygiene plan. These professionals can be the consultants on staff who can answer inquiries, evaluate exposures, stay abreast of regulations, interact with regulators, conduct laboratory inspections and act as intermediaries between laboratories and management.

Provisions for Medical Attention and Consultation for Laboratory Employees. In real life it is imperative that health care providers, whether they are on-site doctors or nurses or outside medical facilities, be made aware of laboratory hazards, and to provide them with MSDSs as necessary. Special care is sometimes required, as in the case of hydrofluoric acid burns. Interaction between the health care providers, the health and safety office, the CHO, and laboratory supervisors can be helpful in diagnosing and treating occupational diseases.

Methods of Fume Hood Operation Evaluation and Measures to Ensure the Presence of Safety Equipment. The health and safety office(r) is instrumental in this element as well. Monitoring fume hood performance can be part of a comprehensive laboratory inspection program. Such a program can be placed into a database and can be used for compliance purposes. Internal inspections (within each laboratory) are encouraged and should be on-going. The responsibility can be shared among laboratory workers.

Control Measure Criteria to Minimize/Eliminate Hazardous Chemical Exposure. In practice, this is going to require cooperation among the workers, the supervisors, and the CHO. Training plays a big part here, because to set criteria, knowledge of PELs must be considered when evaluating experimental procedures, and thus addressing whether the potential for exposure exists and whether controls are warranted.

Provisions for Prior Approval Before Attempting Unfamiliar Laboratory Operations, or Other Circumstances Requiring a Supervisor's Approval. The approach best suited to address this element is for the laboratory supervisor or CHO to know the limits of each laboratory worker's expertise in laboratory operations and to oversee those operations, which, because of inherent hazards, require prior notification or approval. Employees must realize that such approval is necessary, and that they are subject to reprimands if it is not acquired.

Provisions for Extra Protective Measures for Employees Working with Particularly Hazardous Chemicals, or "Select Carcinogens." Application of this element to an active laboratory requires everyone in the lab to identify and recognize those reproductive toxins, substances of acute toxicity, and cancer-causing chemicals that are being considered for purchase or are already in the laboratory stockroom.

Because laboratories that use chemicals are likely to generate chemical waste (hazardous waste), it is possible that OSHA's Hazardous Waste Operations and Emergency Response (HAZWOPER) regulation will be applicable to some employees, especially if an emergency response team has been designated for the facility.

Related Regulations

There are many other regulations covering laboratory operations; they will not be discussed in this chapter, but any student or practicing occupational hygienist should become at least somewhat familiar with them. It is frequently the occupational hygienist who is given responsibility for these regulations. These include the Clean Air Act; the Clean Water Act; the Toxic Substances Control Act; the Emergency Planning and Community Right-To-Know Act; radioactive material use regulations (Nuclear Regulatory Commission Title 10); and transportation regulations (Department of Transportation Title 49). Also, the occupational hygienist should be familiar with additional OSHA regulations such as 1910.20, Access to Employee Exposure and Medical Records; 1910.133, Eye and Face Protection; and 1910.134, Respiratory Protection.

Summary

The chemical laboratory is a work environment that differs in significant ways from many other work sites. Laboratories are likely to contain a very large variety but very small quantities of chemicals; laboratory equipment is small in scale, in keeping with the quantity of chemicals being used; processes carried out in laboratories change frequently, perhaps several times a day; physical hazards associated with chemical laboratory work vary from insignificant to highly hazardous; and laboratory workers routinely make their own decisions about PPE and engineering control needs. All of these differences are important in evaluating health and safety conditions in the laboratory.

Chemicals routinely found in laboratories exhibit a number of hazardous characteristics that need to be understood by the occupational hygienist; the most common of these characteristics are flammability, corrosiveness, oxidizing power, reactivity, and toxicity. Also of concern are the unique hazards associated with the chemical waste produced in every laboratory, as well as radioactive materials, which may be present in some locations.

Notice of physical hazards, such as heavy lifting, cuts, slips and falls, eye strain, and noise, should also be taken when evaluating the laboratory workplace.

Hazardous processes and equipment may also be a significant source of injury in a laboratory. Processes may involve unusually high or low temperatures and pressures, deviating significantly from ambient conditions. Electrical hazards may also be associated with equipment, and many devices have moving parts that need to be guarded for safe use.

The primary control for limiting exposure to chemicals in laboratories is the fume hood. Other appropriate engineering controls include adequate provision for chemical storage and basic facility design. A second important control parameter is PPE. Certain basic PPE should be used by all laboratory workers; however, the need for more extensive equipment, such as respirators, should be avoided through use of engineering controls whenever possible. Administrative controls will also help to ensure the safety of the laboratory workplace; these include basic safety rules, an experiment planning process, restricted areas for certain higher hazard activities, inventories of laboratory chemicals, and routine inspections.

One of the most important preventive measures that can be taken to ensure safety in the laboratory is a formal training and information program. At a minimum, training should include the organization's safety policies and procedures, location of hazard information, identification of the organization's CHO, and information about the CHP.

The most relevant OSHA regulation is the laboratory standard, officially known as "Occupational Exposure to Hazardous Chemicals in Laboratories." Unlike many other OSHA regulations, the lab standard is performance-oriented and is predicated on the expectation that laboratory workers, through education and experience, have at least some level of knowledge about the hazards of chemicals. Under the standard, a CHP must be developed to guide laboratory work.

Key elements of the CHP include standard operating procedures for all laboratory activities involving hazardous chemicals; designation of a CHO; employee training and provisions for accessing safety information; provisions for medical attention and consultation for laboratory employees; methods to be used to evaluate fume hood operation and measures to ensure that other equipment designed for safety are in place; criteria for the use of control measures to minimize or eliminate exposure to hazardous chemicals; provisions for prior approval before attempting unfamiliar laboratory operations or other circumstances requiring a supervisor's approval; and provisions for extra protective measures for employees working with particularly hazardous chemicals or select carcinogens.

References

1. **Bretherick, L.:** *Handbook of Reactive Chemical Hazards*, 4th Ed. London: Butterworth, 1990.
2. "Occupational Exposure to Hazardous Chemicals in Laboratories," *Code of Federal Regulations* Title 29, Part 1910.1450. 1990.
3. **National Research Council:** *Prudent Practices in the Laboratory: Handling and Disposal of Chemicals*. Washington, DC: National Academy Press, 1995.
4. "The Control of Hazardous Energy (Lockout/Tagout)," *Code of Federal Regulations* Title 29, Part 1910.147. 1989.
5. **American Conference of Governmental Industrial Hygienists (ACGIH):** Laboratory ventilation. In *Industrial Ventilation: A Manual of Practice*, 22nd Ed. Cincinnati, OH: ACGIH, 1995.
6. **American Society of Heating, Refrigeration and Air Conditioning Engineers (ASHRAE):** Laboratories. In *Applications Handbook*. Atlanta, GA: ASHRAE, 1978.
7. "Occupational Exposure to Hazardous Chemicals in Laboratories," *Code of Federal Regulations* Title 29, Part 1910.1450. 1990. Appendix A.
8. "Flammable and Combustible Liquids," *Code of Federal Regulations* Title 29, Part 1910.106(d).
9. **National Fire Protection Association (NFPA):** *Fire Protection Guide to Hazardous Materials*, Quincy, MA: NFPA, 1991.
10. "Hazardous Chemical Reporting: Community Right-to-Know," *Code of Federal Regulations* Title 40, Part 370. 1987.

An explosion at an ethylene plant in Texas in 1988 where 23 workers were killed and more than 100 were injured. Emergency planning can lessen the impact of workplace and community hazards.

Outcome Competencies

After completing this chapter, the user should be able to:

1. Define underlined terms used in this chapter.
2. List the benefits of developing an emergency action plan.
3. Describe a cross-functional team and its role in emergency planning.
4. Describe the role of the occupational hygienist in emergency response.
5. Describe the process of hazard risk assessment.
6. List the essential elements of an emergency action plan.
7. Design a basic emergency action plan for a familiar facility.
8. Identify and describe community services chartered for emergency response.
9. Describe the benefits of emergency response drills and mock emergencies.
10. Justify emergency readiness for a specific scenario, from minimal to ultimate action plan.

Key Terms

cocoon • codes • consensus standards • credible scenarios • cross-functional teams • emergency planning • emergency response • emergency action plan • fail-safe • HazMat • HAZWOPER • line diagrams • medical surveillance program • nonmandatory guidelines • risk • risk assessment • root cause investigations • standards

Prerequisite Knowledge

Prior to beginning this chapter, the user should review the following chapters:

Chapter Number	Chapter Topic
3	Legal Aspects of the Occupational Environment
8	Principles and Instrumentation for Calibrating Air Sampling Equipment
9	Direct-Reading Instrumental Methods for Gases, Vapors, and Aerosols
10	Sampling of Gases and Vapors
15	Comprehensive Exposure Assessment
17	Risk Assessment in the Workplace
34	Prevention and Mitigation of Accidental Chemical Releases
35	Personal Protective Clothing
36	Respiratory Protection
41	Risk Communication in the Workplace
45	Occupational Safety
46	Collaborating with the Occupational Physician

Key Topics

I. Requirements for Development of EAPs

II. National Fire Protection Association

III. Hazard and Risk Assessment for Emergency Response Planning
   A. Business or Facility Type
   B. Emergency Types
   C. EAP Development
   D. EAP Drills and Preparedness

IV. Occupational Hygienist Roles in Emergency Response
   A. Roles in EAP Development
   B. Roles During Emergency Response

V. Audits of EAPs
   A. General Audit
   B. Occupational Hygiene Aspects of the EAP Audit

49

Emergency Planning in the Workplace

Fred G. Freiberger, CIH, CSP, PE, CmfgE

Susan D. Rogers Ripple, CIH

Introduction

Emergencies can happen in any organization. The extent of anticipation and recognition of risks, planning, and practice before an emergency occurs can determine how serious the impacts will be. This chapter provides a general overview of emergency planning essentials and describes the role of the occupational hygienist in developing a reasonable action plan during emergency response to keep the personal, community, and business impacts of emergencies to a predictable, manageable level. Preparedness is the most important aspect of assuring that an unplanned event or emergency has minimal impact on the enterprise, its workers, the environment, and the surrounding community. Occupational hygienists are an integral part of emergency response, using their knowledge and experience to aid cross-functional teams in the planning of appropriate responses required for unplanned events. The emergency action plan (EAP) is the basis for identifying and assessing the risks associated with potential emergencies, planning for appropriate response and recovery, and training those who would be involved or affected should the emergency occur. Facilities that depend solely on the local community fire, police, and emergency medical services to handle an emergency without including them in emergency response planning may later find an increase in injuries, deaths, and economic losses while time is lost struggling to determine the proper course of action. An EAP provides knowledge to everyone affected so that they may react and respond quickly and safely.

Another important benefit of emergency response planning is the prevention of unplanned events as the potential risks are recognized. As credible scenarios are identified, preventive measures to lessen their impact or prevent their occurrence are of significant value. Many preventive maintenance procedures are put in place as a result of emergency response plans.

In the United States, emergency response programs are driven by several Occupational Health and Safety Administration (OSHA) regulations and by the U.S. Environmental Protection Agency (EPA), as well as by state and local regulations. Generally, if the EAP is written to protect facility employees, company assets, community neighbors, and the environment, the regulatory requirements will likely be met.

It is essential for company management who want to keep emergency impacts minimal to allocate resources to ensure that adequate safety measures and emergency response planning are in place prior to an emergency event. This quality and extent of preparedness for unplanned events is predicated on the risk acceptance level of top management. The EAP is best accomplished by a cross-functional group comprised of personnel designated by the facility's management. The team should include representatives from maintenance, safety, fire, medical, and occupational hygiene groups, who should then interact with and train community emergency response teams. Depending on the nature of the potential catastrophes, representatives from environmental, security, and public affairs could be included in the planning. The occupational hygienist can provide many skills to this team related to evaluation of health risks, developing credible scenarios, personal protective equipment (PPE) selection and planning, scene management, training, drills, and audits of the EAP, as well as providing assistance during actual emergencies.

There are a number of excellent resources available (listed under Additional Sources section of this chapter) that treat the subject of emergency response planning with more detail and explanation, while providing templates for the EAP based on applicability.

Requirements for Development of EAPs

Numerous OSHA and EPA standards require an EAP. Table 49.1 is a limited list of federal standards that may be applicable to facility emergency response planning. All employers with more than 10 employees are required to develop a written EAP covering those designated actions that employers and employees must take to ensure employee safety from fire and other emergencies that could occur at the facility. The written plan must be kept at the workplace and made available for employee review. Employers with 10 or fewer employees may communicate their plans to employees orally and do not need to maintain a written EAP.[1] Several of the OSHA chemical-specific standards in *Code of Federal Regulations* (CFR) Title 29, Part 1910, Subpart Z have reporting requirements for chemical spills or vapor releases, whether or not exposures occur.[2] In addition to federal regulations, state regulations and local requirements add to the complexity of developing an effective EAP.

National Fire Protection Association

The National Fire Protection Association (NFPA) produces consensus standards as national fire codes. There are 275 codes and standards, which are used globally in almost every building, process, service, design, and installation.[3] OSHA standards for fire protection (29 CFR 1910.156 through .165) reference many of these national consensus standards as non-mandatory guidelines that would be considered acceptable in complying with requirements of Title 29, Part 1910, Subpart L.[4] Table 49.2 lists some of the NFPA standards that may be used in emergency planning and response.

Hazard and Risk Assessment for Emergency Response Planning

Business or Facility Type

The nature and location of the business or agency will govern, to a significant extent, whether the emergency will affect customers or the community beyond the physical bounds of the facility itself. Facilities that involve hazardous activities and agents such as chemicals, radiation sources, infectious agents, high energy equipment, or temperature critical processes have a greater probability of unplanned incidents than does a commercial office. Virtually every building, process, service, or installation has the potential for an emergency. The first planning team effort should focus on identification of potential site-specific emergencies. A generic or laundry list pulled from a safety manual may omit critical issues at that particular

Table 49.1 — Limited List of Federal Regulations Requiring Emergency Planning or Reporting

Employee Emergency Action Plans	29 CFR 1910.38
Employee Alarm Systems	29 CFR 1910.165
Hazardous Waste Operations and Emergency Response (HAZWOPER)	29 CFR 1910.120
Process Safety Management of Highly Hazardous Substances	29 CFR 1910.119
Toxic and Hazardous Substances	29 CFR 1910, Subpart Z
Risk Management Programs for Chemical Accident Release Prevention	40 CFR Part 68
Spill Prevention Control and Countermeasures Plan	40 CFR Part 112
Contingency Plan and Emergency Procedures	40 CFR 264.50–264.56, Subpart D
Response Plans for Onshore Oil Pipelines	49 CFR Part 194
Oil Pollution Act of 1990 (OPA)	
Superfund Amendments and Reauthorization Act (SARA) Title III Emergency Planning and Community Right-to-Know Act of 1986	

Table 49.2 — Limited List of NFPA Standards for Use in Emergency Planning or Response

Recommended Practice for Responding to Hazardous Materials Incidents	NFPA 471
Recommended Practice for Disaster Management	NFPA 1600
Standard for Professional Competence of Responders to Hazardous Materials Incidents	NFPA 472
Standard on Industrial Fire Brigades	NFPA 600
Standard on Fire Department Incident Management System	NFPA 1561

facility because of the unique nature of the business, its location, outside resources, or staffing level.

The business activities, materials involved in production, degree of fire protection, and degree of emergency preparedness training should be considered when assessing hazards. The emergency response planning team will rely on the occupational hygienist to recognize the workplace hazards present in that facility and to anticipate the risks to human health and safety should an unplanned event occur. Emergency response planning should include any potential health and safety concerns for neighboring residences and businesses as well as the possibility for transportation incidents and customer emergencies.

Emergency Types

A variety of hazards that could cause emergency situations may be present for any type of business. The risk management (or containment) capability of even the best-prepared emergency plan can be taxed, depending on whether the emergency condition originates from within or outside the facility. Table 49.3 lists four categories of emergency and includes several examples for each classification, but is intended for use as a starting point in the planning process. Site-specific items should be considered in addition to these examples.

Natural disasters, including severe weather, can halt operations without warning, creating as much risk of an emergency as can an explosion. Geographic location governs to what extent natural disasters should be included in the emergency plan. Risk assessment guidance may be obtained from federal, state, county, and city emergency management offices. Hurricanes may be endemic to coastal areas while tornadoes, floods, and winter storms might be more prevalent in other locations. Warnings for these potential catastrophes are the responsibility of the National Weather Service. Some facilities must consider earthquakes, avalanches, and volcanic eruptions in their planning. The Federal Emergency Management Agency can assist the EAP team in assessing the risk of these disasters for a facility in the United States.

Natural disasters have the associated risk of utility outages and interruption of essential services that may cause unsafe conditions to develop, thereby necessitating fail-safe or redundant engineering design for loss prevention. Timely evacuations and process shutdowns can prevent injury and health risks to employees when adequate natural disaster warnings are available. Some facilities may be linked to services such as natural gas, electrical utilities, or pipeline feedstock from off-site suppliers and may be affected by their equipment or process failures. Regardless of the cause of the service interruption (e.g., traffic accident involving a critical utility pole, severe weather, provider equipment failure, etc.), the potential detrimental effect could result in a chain of uncontrollable emergency events such as fires, explosions, or chemical releases. The timing required to replace or repair the service interruption can determine the impact on the facility operations, and appropriate preplanning should account for the risks associated with these outages.

Fires, explosions, radiation exposures, and chemical spills or releases are obvious hazards with high risk potential in some businesses. Although some facilities may plan to evacuate for some of these scenarios, others may determine that the appropriate action would be to "cocoon" in their building with air intake equipment to the building turned off. The occupational hygienist can provide vital insight into the health risks associated with these scenarios to achieve the safest emergency response actions for everyone.

Emergency planning is necessary to direct emergency responders in remote locations regarding viable response actions. Emergency responders might have the potential to respond to high visibility or controversial community complaints about excessive noise, environmental spills, or

Table 49.3 — Categories of Emergency for Emergency Response Planning

Natural Disasters	*Human Error*	*Process Error*	*Equipment Failure*
severe weather	vehicle impact	process overheat	utility outages
floods	confined space entry	process overflow	valve leaks
earthquakes	aircraft crashes	vapor release	chemical spills
avalanches	packaging failures	explosions	fires
volcanic eruptions	public/civil disturbances	community noise	radiation accidents

Note: In reviewing the examples, realize that any of these events may cause personal injuries and illness.

odor complaints. Transportation and shipping accidents can occur anywhere between the manufacturer and its destination. Rail cars, tank trucks, cargo planes, marine vessels, and mail services that handle products may be the unintentional cause or target of an accidental spill or release involving human health risks or environmental damage. Where there is a community health concern, businesses usually rely on their occupational hygienist, medical, toxicology, and public relations personnel to interface with the media and the public to explain the associated health risks of the emergency. Anticipation of these situations may preclude their occurrence, but certainly will make community explanations more credible if appropriate emergency responses are predetermined.

Other types of emergencies may involve personnel health concerns from accidental injury, illness, or life-threatening situations. Biological hazards for responders and nearby workers may exist from contact with body fluids (e.g., blood, vomit, saliva, etc.) from injured victims or sick personnel. Although the safety and health of injured victims is the first priority in emergency response, consideration for the health of emergency responders is also important. Recognition of this potential, anticipation of the correct protective clothing, and training are the responsibilities of the occupational hygienist and medical personnel. The EAP team should consider the provisions required in OSHA's bloodborne pathogens standard, which outlines specific procedures for emergency medical technicians and other medical providers to protect themselves from occupational exposure to blood or other potentially infectious materials (e.g., AIDS, tuberculosis, hepatitis).[5]

A discussion of potential emergencies would not be complete without mention of emergencies caused intentionally by individuals. Emergency response planning for risks associated with terrorism, workplace violence, sabotage, bomb threats, strikes, and civil unrest must be considered. No one expects these scenarios to happen in the workplace, but adequate thought and planning could reduce the risk of injury and economic loss, or better yet, prevent their occurrence.

In listing the types and nature of potential emergencies for the organization, an assessment can be made of impacts affecting people or those events with purely economic consequences. Management must determine the level of risk acceptance for strictly property-impact events. Cost and speed of recovery for the lost or damaged resources will be a factor. Customer consequences also must be considered by management in setting risk acceptance. Damage limited to finished deliverables may be more acceptable to management than impacts on production equipment or creation of long lead-times of the raw material inventory. When planning for an emergency response, business interruption is also an essential emergency planning issue.

EAP Development

The development and practice of the EAP should involve all operating areas of the business or commercial entity, including environmental, health, safety, and occupational hygiene representatives. To be effective the EAP needs to be well understood and known by all on the property. This includes contractors and other on-site visitors.

The hazard and risk assessment should provide the foundation for the development of the EAP. Emergency response planning requires site response teamwork with outside services (e.g., HazMat teams, fire departments, hospitals, police). The appropriate EAP team members, including community responders, should be determined based on the types of potential risks for the facility.

Once the risks are prioritized based on probability of occurrence and impact on workers, community, and environment, credible scenarios should be developed. For example, hurricanes can have devastating impacts on a facility, but if the business location is not in a coastal area, priority for emergency response planning for hurricanes would be less than that for fires or evacuations. Where there is equal probability of risks occurring, then the planning priority should be higher for those with the most impact on the health of workers or community residents, and secondarily for economic losses.

An effective EAP depends on proper coordination between the emergency responders and the assignment of responsibilities. The EAP should document all aspects of an emergency response for each of the various scenarios, including assignment of responsibilities, coordination with outside agencies, procedures to alert employees, escape routes and procedures for accounting for all evacuees, emergency equipment available, and location and training of response personnel.[6] In assigning responsibilities in the plan, keep in mind that variable scheduling, vacations, illness days, and holidays create special demands on response effectiveness. Each risk should be evaluated to determine whether the facility and outside responders have the ability to respond adequately.

In looking at the risk assessment list, realize that any of the events may result in chemical exposures or personal injuries and illness. The location of the effected site will govern to what extent the facility establishes internal first aid capability. In an urban location with ample local medical and ambulance services, a telephone call to 911 may be the only emergency action needed. In a rural setting, well-trained emergency medical technicians and first aid supplies become more critical within the facility. This resource is a valuable consideration in any emergency plan, since outside response may be unreliable depending on the community impact of the emergency. Consider in the development of an EAP the types of medical emergencies that are most probable, may be possible, or are unlikely to occur. A combination of events, including inclement weather and a train derailment, should be well thought out as part of the EAP. Could a nearby fire, traffic accident, or blocked intersection prevent emergency responders from reaching the facility? Could such an event block the only access driveway from the property? In a situation known to the authors, the only community hospital for three towns was located on the fence line of a major chemical company comprised of 87 different manufacturing plants at that location. After assessing the risks imposed by the plant on the hospital staff, patients, and their ability to respond in the event of a major emergency, the company donated land for the relocation of the hospital away from the risks associated with being located near the plant.

Site security during emergencies is an additional concern that should be part of the EAP. This starts with directing evacuees, responders, curious spectators, and the news media to a safe location, well clear of emergency control activities.

The plan should also name an assigned spokesperson to answer questions from community leaders, the public, and the news media. Other personnel should defer all outside questions to the designated individual or spokesperson. Some firms make this a function of the legal staff, public relations, or senior management.

Communication capabilities needed during emergency response for key responders are crucial. Once evacuated from a building, telephones are of no value. Cellular phones, commercial two-way radios, or portable megaphones will permit directions to be communicated to the many people involved. One drawback of this technology is that community members can monitor and misinterpret what is said during emergencies when using these devices. In addition to communicating during an emergency, consideration for summoning emergency responders should be made in case facility telephones cease working. Pagers worn by on-call emergency responders make response time much quicker than in the past, when the city siren was the summoning device.

It may be prudent for a company to help public responders by providing facility tours and a copy of the emergency plan. Familiarity with the site, its operations, and its materials may help ensure a proper response should it become necessary. Such cooperation with the emergency response community improves reliable, competent emergency event handling while building good will.

Interview-based templates for EAP development are available from a number of publishers in a variety of formats (hard copy versus electronic) but are too numerous to list in the Additional Sources section. These templates may not include all the potential scenarios for a facility, but they provide a strong basis for the most common emergency risks. The written EAP can be divided into sections that address emergency plans within the various categories of emergencies, as well as sections for community emergencies, transportation, and customer-related emergencies. The EAP should also include maps of the facility and its exit routes, line diagrams of utilities and process lines, along with locations of shutoff valves and a list of hazardous agents inside the facility. The written EAP must be accessible to all employees at all times, and should be stored for easy retrieval should an emergency arise. Its storage near the likely source of an emergency would be imprudent, since the emergency response team might need to use it for guidance during an unplanned event.

EAP Drills and Preparedness

Once the EAP is complete, the effectiveness of the plan should be tested through drills and mock emergencies regularly (e.g., annually or quarterly), and where possible should include outside responders. Practice through drills provides two important facets to the preparedness for an emergency. First, a drill or mock emergency provides everyone with a training opportunity to know exactly what their response should be for a given situation and to practice their assigned responsibilities. It is much easier to remember what is expected and less likely to cause panic if one has gone through the drill before the emergency occurs. The second point of value is the opportunity to assess the emergency drill for improvements to the EAP. Visitors to the facility may not know what to do, or a key responsibility or need may have been overlooked. Perhaps the training of community responders was inadequate, equipment was unavailable or inaccessible, or perhaps an important aspect of the emergency response was not performed because someone was on vacation or designated as injured in the drill. In the latter situation, primary and two levels of trained alternate responders may be needed. Whatever deficiencies are identified, take the opportunity to find a solution so that they will not be repeated during the real emergency. Once the drill or mock emergency has been evaluated, the EAP should be reviewed and updated with all of the corrective actions and recommendations documented. Also, if there was anything learned that might help prevent future unplanned events, there might be an opportunity to place a preventive action plan together or develop facility safe operating procedures such as sign-in books, or on-call responders.

Occupational Hygienist Roles in Emergency Response

Roles in EAP Development

The occupational hygienist is expected to aid in the development of the EAP by identifying health hazards that might arise in an emergency based on possible scenarios. Familiarity with material safety data sheets (MSDSs) and the likely routes of chemical exposure in each emergency situation will help in consultation with the facility for determining whether to evacuate the premises, cocoon, or remain to mitigate the emergency. MSDSs provide a wealth of information regarding the potential health effects of exposures, the type of PPE to be worn by responders, physical and chemical properties of the substance, and the regulations and requirements that might apply to the facility during an emergency. Another subject available on some MSDSs is proper decontamination and disposal procedures for PPE.

Occupational hygienists may be asked to train emergency responders on aspects of the EAP, such as HAZWOPER, health hazards and risks, odor thresholds, immediately dangerous to life and health guidelines, and proper PPE for response, among other topics. Through consultation, the hygienist should identify emergency responders in the facility who are required to be in a medical surveillance program, directing the content, frequency, and adequacy of examinations and physicals for compliance purposes. The hygienist should identify personnel who must wear respiratory protection during emergency response so that medical certification, respirator training, and respirator fit-testing are conducted according to 29 CFR 1910.134 requirements.

Hygienists also provide expertise in estimating health effects of exposures to workers, community members, or responders with the aid of AIHA Emergency Response Planning Guidelines, emergency exposure guideline levels, or acute exposure guideline levels, given that an exposure concentration could be anticipated. These are guidelines developed by various organizations for use in emergency situations involving a single exposure to substances that might cause adverse health effects.[7] The hygienist should also be familiar with any computerized equipment (e.g., Daisy Mae or MIDAS) used by the facility to estimate or measure chemical release concentrations at various locations within or outside the facility boundaries, because they may assist management or incident command in determining community alert and response.

The hygienist is also responsible for anticipating the proper PPE for re-entry into a mitigated emergency either by direct measurement of remaining concentrations or by the use of dilution ventilation calculations. The prepared occupational hygienist will ascertain and obtain the proper chemical or radiation monitoring equipment and be familiar with its use and limitations prior to emergencies. If the equipment is not available, documenting chemical-specific dilution calculations ahead of time would ensure that the needed data is available if an emergency occurs.

Planned communication procedures used during emergencies should include the hygienist so that constant consultation is available as the emergency unfolds.

Roles During Emergency Response

The occupational hygienist is part of the emergency scene management team once an emergency situation occurs and is consulted by management or incident command throughout. Based on the roles and responsibilities assigned in the EAP, the hygienist's role during emergency response may include performing initial health risk assessments at the scene using direct measurement equipment or vapor concentration calculations. Health risks to community members may be provided to the incident commander or management based on predicted movement of a vapor release. Consultations on proper PPE for unplanned events might be needed, as well as reminders of the need for a buddy system when entering an emergency situation where the concentrations of vapor or smoke are not known. The hygienist might be assigned the duty of actually donning appropriate PPE and entering the emergency scene to evaluate the identity of a chemical in a release situation or to determine whether decontamination and clean up can be performed. Frequently this means using direct reading equipment or media to measure concentrations of chemical vapors, explosive atmospheres, or low oxygen situations. Other response situations may require that the hygienist suit up properly and evaluate fence-line concentrations to verify computerized models of vapor releases for community alerts. Medical providers will need to consult with the hygienist during triage of those injured in the emergency to determine what exposures, if any, might have occurred so that proper treatment can be given.

Once the emergency is over, the hygienist may need to identify workers, responders, or community members affected by the emergency for medical evaluation to comply with OSHA regulations. If a chemical release or spill has occurred, there are may be reporting requirements to federal and state organizations such as OSHA and EPA, whether or not exposures have occurred. This is often the responsibility of the occupational hygienist.

Root cause investigations and assessments of the emergency response from the hygienist's perspective are required as follow-up. There is always educational value in revisiting the emergency to determine what went right, went wrong, and should be anticipated for the next situation. A review and update of the EAP should include any educational experiences from the actual response. Applying these lessons to other emergency scenarios is appropriate and may help identify ways to prevent unplanned events in the future.

Audits of EAPs

General Audit

Audits of the EAP should be made regularly (e.g., annually or more frequent) to re-assess whether it is current. The audit should be made on all aspects of the plan and should include the planning process, the risk assessment, assignment of responder responsibilities, training, drills, reviews, and updates. An audit of community responder facilities may be appropriate to ensure that their availability, proximity, and plans coincide with the facility's expectations as outlined in the EAP. First aid supplies and emergency equipment needed by the responders should be quickly accessible, ready to use, and adequate to meet expectations.[8] There are EAP audit checklists available in the listed resources at the end of this chapter. Regardless of the audit checklist used, the most important aspect of the process is the follow-up with corrective actions and recommendations. These are opportunities to review the EAP with a third party who can provide another perspective on the plan. Documentation of the follow-up completed in response to deficiencies is important to show that the EAP is part of the continuous improvement process.

Occupational Hygiene Aspects of the EAP Audit

The occupational hygiene section of the EAP audit should evaluate the health risk assessment plans, training of workers and responders on health risks, medical surveillance, and PPE elements of emergency planning. Since training is vital to the success of the actual emergency response, it is the most import aspect of the occupational hygiene audit of an EAP. Auditors should determine whether the exact training needs of all applicable regulations have been met (e.g., HAZWOPER training requires 8 to 40 hours). Although training documentation and sign-in sheets may prove that training has been done, the audit should use interviews with responders and facility personnel to evaluate whether the training program was effective.

Another aspect of the occupational hygiene audit should focus on the PPE for the emergency responders to assure that it is appropriate for the hazards that might be encountered. Equipment used for emergency response must be periodically inspected and documented to ensure that it is still available in the designated location, effective, clean, and ready to use. The storage locations must be assessed to determine whether they are accessible during emergency situations or whether they are too close to an identified risk area.[8]

Medical surveillance of emergency responders should be up-to-date and health exams documented. The content of the physicals should be assessed during the audit to determine the quality of the medical examinations. The names of employees for whom medical limitations (e.g., respirators or weight limits) have been determined during medical surveillance should be noted to prevent responder emergencies occurring from unrealistic expectations of their abilities. Remember that medical records are confidential.

Summary

Emergency response preparedness is most effectively achieved through extensive emergency risk assessment, development of the EAP, and the subsequent drills and audits. Emergency response teams are limited in their action by the severity of the unplanned event, emergency preparedness, and adequacy of training. The role of the occupational hygienist in this process is to provide insight in the area of health risk assessment. Particular areas of expertise are health regulations and requirements, PPE, air monitoring, training, and auditing of the EAP. Occupational hygienists may be called on in community emergencies to anticipate the potential exposures and impacts of the emergency on the community. Knowing limitations, as well as knowing what, who, where, when, and how to proceed during emergency response, should significantly lower the impacts of unplanned events on the health of employees and the community as well as reduce financial losses.

Additional Sources

American Society for Testing and Materials (ASTM): *Safe Handling of Hazardous Materials Accidents*, 2nd Ed. Philadelphia: ASTM, 1990.

L.P. Andrews: *Emergency Responder Training Manual for the Hazardous Materials Technician*. New York: Van Nostrand Reinhold, 1992.

W.A. Burgess: *Recognition of Health Hazards in Industry: A Review of Materials and Processes*, 2nd Ed. New York: John Wiley & Sons, 1995.

T.A. Burke (ed.): *Regulating Risk: The Science and Politics of Risk*. Washington, DC: International Life Sciences Institute, 1993.

N.P. Cheremisinoff: *Handbook of Emergency Response and Toxic Chemical Releases: a Guide to Compliance*. Morganville, NJ: SciTech Publishers, 1995.

T.S. Ferry: *Modern Accident Investigation and Analysis*, 2nd Ed. New York: John Wiley & Sons, 1988.

J.W. Hosty: *A Practical Guide to Chemical Spill Response*. New York: Van Nostrand Reinhold, 1992.

R.B. Kelley: *Industrial Emergency Preparedness*. New York: Van Nostrand Reinhold, 1989.

J.D. Kipp and M.E. Loflin: *Emergency Incident Risk Management: A Safety & Health Perspective*. New York: Van Nostrand Reinhold, 1996.

J.P. Kohn, M.A. Friend, and C.A. Winterberger: *Fundamentals of Occupational Safety and Health*. Rockville, MD: Government Institutes, 1996.

P.A. Michaud: *Accident Prevention and OSHA Compliance*. Boca Raton, FL: Lewis Pubishers, 1995.

National Institute for Occupational Safety and Health (NIOSH): *NIOSH Pocket Guide to Chemical Hazards*. Cincinnati, OH: NIOSH, 1994.

National Safety Council (NSC): *Accident Investigation*. Itasca, IL: NSC, 1995.

National Safety Council (NSC): *Accident Prevention Manual for Business and Industry: Engineering and Technology*, 10 Ed., Vol 2. Itasca, IL: NSC, 1996.

National Safety Council (NSC): *Guide for Identifying Causal Factors and Corrective Actions*. Itasca, IL: NSC, 1996.

National Safety Council (NSC): *Study Guide: Accident Prevention manual for Business and Industry*, 10th Ed. NSC, 1996.

T.D. Schneid: *Fire Law: The Liabilities and Rights of the Fire Service*. New York: Van Nostrand Reinhold, 1995.

W.H. Stringfield: *Emergency Planning and Management: Ensuring Your Company's Survival in the Event of a Disaster*. Rockville, MD: Government Institutes, Inc., 1996.

L. Theodore: *Accident and Emergency Management*. New York: John Wiley & Sons, 1989.

U.S. Environmental Protection Agency (EPA): *EPA Training Manual: Emergency Response to Hazardous*

Material Incidents (Manual 165.15). Washington, DC: EPA, 1995.

J. W. Vincoli: *Basic Guide to Accident Investigation and Loss Control*. New York: Van Nostrand Reinhold, 1994.

References

1. "Employee Emergency Plans and Fire Prevention Plans," *Code of Federal Regulations* Title 29, Part 1938.
2. "Toxic and Hazardous Substances," *Code of Federal Regulations* Title 29, Part 1910 Subpart Z.
3. **Stringfield, W.H.:** *Emergency Planning and Management: Ensuring Your Company's Survival in the Event of a Disaster*. Rockville, MD: Government Institutes, Inc., 1996. pp. 15–16.
4. "Fire Protection," *Code of Federal Regulations* Title 29, Part 1910, Subpart L, Appendix B.
5. "Bloodborne Pathogens," *Code of Federal Regulations* Title 29, Part 1910, Subpart Z, 1910.1030.
6. **Leibowitz, A.J.:** *Industrial Hygiene Auditing, A Manual for Practice*. Fairfax, VA: American Industrial Hygiene Association, 1995. p. 40.
7. **American Industrial Hygiene Association (AIHA):** *Emergency Response Planning Guidelines and Workplace Environmental Exposure Level Guides*. Fairfax, VA: AIHA, 1997. p. 4.
8. **Leibowitz, A.J.:** *Industrial Hygiene Auditing — A Manual for Practice*. Fairfax, VA: American Industrial Hygiene Association, 1995. p. 41.

Outcome Competencies

After completing this chapter, the user should be able to:

1. Define underlined terms used in this chapter.
2. Identify societal forces that are driving the future.
3. Explain the impacts of societial forces on occupational and environmental health and safety professionals.
4. Construct scenarios to evaluate potential trend impacts and responses.
5. Apply learning organization concepts into occupational hygiene trends.

Key Terms

business environment • economics • global economy • learning organization • scenario building • systems thinking • teams

Prerequisite Knowledge

Prior to beginning this chapter, the user should review the following chapters:

Chapter Number	Chapter Topic
1	History and Philosophy of Industrial Hygiene
2	Occupational Exposure Limits
3	Legal Aspects of the Occupational Environment
37	Program Management
38	Surveys and Audits
45	Occupational Safety
46	Collaborating with the Occupational Physician

Key Topics

I. Understanding the Forces Shaping the Future
   A. Force I: The End of Communism
   B. Force II: Brainpower Industries
   C. Force III: Demography, The Aging World
   D. Force IV: The Global Economy
   E. Force V: The Multipolar World
   F. Thinking About the Future

II. Thinking About the Future—Building Scenarios
   A. Forecasting
   B. What is a Scenario?

50

Anticipating and Evaluating Trends Influencing Occupational Hygiene

Lawrence R. Birkner, CIH, CSP

Ruth K. McIntyre-Birkner

Introduction

Anticipation, recognition, evaluation, and control are the driving forces of occupational and environmental hygiene. Yet a process to forecast developments in this multidisciplinary field has not been well established. Anticipation has been the most overlooked of these fundamental forces. Lack of a consistent forecasting process has resulted in missed opportunities in asbestos control and late entry into driving regulatory and legislative developments in areas such as ergonomics and lead management. Without the ability to use forecasting tools, occupational and environmental health and safety (OEH&S) professionals are losing opportunities to take advantage of the changing technological, social, and regulatory trends.

The concepts, trends, and techniques discussed in this chapter will give OEH&S professionals a formal framework with which to round out the fundamental activities of the profession. The associated disciplines discussed in this chapter — business planning, organizational behavior, economics, quantitative analysis — are included to integrate the profession into core business processes. In this way, OEH&S professionals can help balance concerns for health, safety, and the environment with economic activity and productivity.

With sources like the Internet, the availability of information is growing at an exponential rate. Information is knowledge, and knowledge is power. The Internet allows communication to transcend hierarchies, structures, and organizations. A recent estimate of knowledge growth suggests that in 1995 knowledge was doubling every six months.[(1)] Technology, computers, and communication systems are driving this explosive growth. The social, economic, and political impact of such rapid knowledge growth and its communication to all corners of the earth are enormous. Some historians[(2)] trace this information explosion to the globalization of the world economy, the collapse of the Soviet Union, and the downsizing of American and European industry. Whatever the cause of these dramatic changes, the more the events, their origins, and their impacts are understood, the better able the occupational and environmental health profession can become to prepare and respond to the future.

The future cannot be credibly predicted; however, events, trends, and historic factors can be identified that will set the stage for what might happen. In the very short term—1 to 5 years—the profession can plan and create strategies for the future it wants to create. The more able OEH&S professionals are to analyze issues and forces that surround their work, and to see how they are part of a system, the greater leverage they will have on their own future. The more flexible the plans, the better able they will be to respond to contingencies and keep those plans on track.

As the next millennium approaches, information can be expected to grow at an even faster pace. Science, technology, economics, social concerns, and politics will influence how and where occupational health professionals work in a global marketplace. Not only must they maintain their respective competencies in health and safety, but they must also expand and integrate learning organization concepts,[(3)] systems thinking, and business techniques into their analytical tool chests.

Peter Senge's approach to the learning organization incorporates five fundamental disciplines into business culture: systems thinking, personal mastery, mental models, shared vision, and team learning. Two of these disciplines are personal (personal mastery and mental models), two require joint efforts (shared vision and team learning), and "The Fifth Discipline" — systems thinking — is the structure within which occupational health professionals operate.

1. *Systems thinking* facilitates seeing, understanding, and solving problems in complex systems by viewing the entire system, not focusing on isolated parts.
2. Achieving *personal mastery* means clarifying what is important and striving to see reality more clearly.
3. *Mental models* are the lenses used to see the world. They are woven into each individual's personal life history and sense of who they are.
4. Shared vision creates a commonality that projects a sense of strategic intent.
5. Team learning involves aligning a team to create the desired outcomes through dialogue — nonjudgmental listening and exploration of complex issues.

Those studying this chapter should identify an issue or event from their work or educational experience that might have had a better outcome if more leveraging information or a better set of analytical tools had been available. Re-create the issue or event and analyze it with tools learned here and see how an event can be influenced for a better outcome. By using this technique, a process to assist in the analysis of future plans and strategies can be internalized.

Understanding the Forces Shaping the Future

Even though change has been at the center of human existence, it has accelerated over the past two decades. As scientific knowledge grows, as technology increases the speed of artificial brainpower, as communication speeds increase, and as transportation becomes quicker and more efficient, change will be lockstep. Socially, the change in human demographics, the rapid economic growth in developing countries, and the entry of the Second World countries (the former Soviet Union countries and their satellites) into the marketplace are unstoppable forces that will drive the global economy.

Technologic, scientific, social, economic, and political forces are changing the relationships between workers and the workplace, changing the way people work, and creating new workplaces and environmental risks. In this section, some of the forces of change will be examined. It is important to recognize that the forces of change that will drive the future of occupational hygiene are exogenous to the profession. They are powerful forces, not easily influenced; therefore, the greater the understanding of them, the more effective the forecasts and responses to issues and trends will be. In his book, *The Future of Capitalism*,[4] economist Lester Thurow uses the earth's tectonic plates as an analogy to describe the powerful economic and social forces driving future events. He identifies five "plates" that are pushing against one another to reshape the economic, political, and social landscape of the globe. These forces will inevitably influence the practice of occupational and environmental hygiene.

Force I: The End of Communism

Capitalism has no competition, and every country that does not have a free market system is moving toward one. The repercussions have been dramatic. Just as the Roman Empire lost its way and began its decline after it had no more territory to conquer, the United States, while not yet lost, has lost communism as its primary motivation for economic, political, and military leadership. The Cold War generated a direction and a motive force for education, for quality research and development, the construction of an industrial infrastructure second to none, and the interstate highway system. These government investments had a powerful impact on the civilian economy — jobs, technological developments, advances in science and medicine. Occupational hygiene was born and thrived in this era. Many advances, such as aerosol technology and respiratory protection, grew from the military's technological demands.[5]

With the end of communism and its threat, government investments slowed and even stopped in many areas. The United States, Western Europe, and Japan are now faced with reduced government expenditures along with increased global competition. Businesses, workers, and massive natural resources have entered the free market. For example, American scientists who earn middle-class salaries in the marketplace will compete with well-educated Soviet scientists who will work for near poverty wages, by U.S. standards.

In western and Japanese economies wages will continue to fall until they equalize in the global marketplace. Industrial efficiencies that permit global competition signify that "mean and lean" are here to stay. New technologies, advances in science and medicine, new materials, productivity pressures, social pressures, and the movement of production to lower wage economies will challenge the creativity and flexibility of the OEH&S professional.

Force II: Brainpower Industries

In the days of the industrial revolution, competitive advantage was defined by the location of an industry and the natural resources that were readily available to it. The steel industry located itself near sources of iron ore and coal and close to transportation centers. Textile mills were frequently located near sources of cotton, wool, or water power. Countries with large financial resources built capital-intensive industries. Today, in the global economy, raw materials are easy to get and technology is easily copied. The only competitive advantage left is knowledge or "brainpower."

Education is the fundamental resource. Cities, states, and countries that excel in cultivating brainpower achieve competitive advantage. For example, in a world where biotechnology is playing a central economic role, developing doctorate holders in biochemistry, biophysics, microbiology, and so forth will provide a country with competitive advantage. Similarly, educating OEH&S professionals to understand and apply their skills to managing new and developing technologies will enhance their competitiveness in the global marketplace.

The rise of brainpower industries has fundamentally altered the relationship between workers and management. While capital (equipment) can be owned, humans cannot be. In a world where the best human brainpower will be independent of company decision makers, and where it can be bought and sold in a virtual marketplace, the ability to invest in and maintain people will drive competitive advantage. OEH&S professionals will have to understand technology-driven exposures, rapid shifts in technology, and be able to work with expansive computerized networks linked by resources like the Internet.

Force III: Demography, The Aging World

The United Nation's World Population Prospect projects that the world's population will grow from 5.7 billion people in 1996 to over 8 billion by 2025.[(6)] This trend will have an enormous impact on the OEH&S profession. People will be working longer, resulting in an older work force with different capabilities and limitations. Workplaces in the United States will be even more diverse, with a high percentage of immigrants whose English skills are marginal and work skills limited.

To address financial shortfalls, major changes will take place in the health care industry. Workers' compensation will be responsive to the changing nature of the work force and the associated injuries and diseases. Entitlements to the elderly will have to change if the United States hopes to avoid bankruptcy or the decay of its educational, research, and development and physical infrastructure. OEH&S professionals can play a key role in designing new systems that mitigate exposures in rapidly changing work, home, and public environments.

Force IV: The Global Economy

Technology, transportation, communication, and the growth of democracy will erase market boundaries in the world economy. Brainpower will be purchased in cyberspace (i.e., via the Internet), a world that, when one is connected, seems and almost feels real, solid, and whole. Products will be designed a continent away from where they are manufactured. Capitalism will work at its best by finding ways and places to build products for a low price and sell them where the largest profit can be made.

As the global economy continues to develop, national regulations will yield to international standards, further solidifying the global market. Over time, International Organization for Standardization (ISO) standards will help ensure that one country does not gain an unfair advantage by building products in an unsafe or environmentally damaging manner. Product safety will be driven by the marketplace. Clearly, the role of the OEH&S professional will be enormous, especially for those practitioners who speak multiple languages, understand cultural differences, and can apply their knowledge to rapidly changing technologies and standards.

Force V: The Multipolar World

The United States is no longer the central force driving the world's economy. The end of capitalism and the postwar growth of the European and Asian Pacific economies will eventually spread economic power throughout the globe. A key issue will be global wage competition that will drive manufacturing to countries where the wage base is lower than in the United States. Even segments of brainpower industries will leave the United States to go to areas where brainpower can be purchased at lower cost. These industries will return to the United States, but only when wage rates are competitive with the rest of the world. Only in those industries and technologies where the United States maintains a global monopoly will U.S. workers be able to demand rates above those set in the world marketplace.

These macroeconomic factors will have a great impact on the OEH&S profession. Practitioners will be working with cutting edge technologies that can earn a premium in the marketplace. Examples include genetic engineering, semiconductor manufacturing, high-end computers, communications, and aerospace. Competitive pressures are likely to support continued growth in demand for occupational and environmental services.

Thinking About the Future

As OEH&S professionals come to understand these driving forces, they can begin to think about the future. They will need better tools for operating in the fast paced, dynamic environment of the 21st century. Linear technical problem solving, which has been effective for much of this profession's history, will not have the power needed to solve the complex problems of the next millennium. Competitive forces will drive the use of teams and learning organizations.

To be effective, the OEH&S professional must command a deep understanding of teams and learning organizations in the business environment, because multidisciplinary teams will be the primary tool for addressing OEH&S problems. New tools to help forecast and manage the future will require the use of teams and team learning. In the next section is a discussion of one such tool — scenario building.

Thinking About the Future—Building Scenarios

In the past, strategic planning was the tool to envision the future. Strategic plans were based on business, economic, and environmental forecasts. As long as the environment was reasonably stable (i.e., the two decades after World War II), forecasting was a fairly linear process. Forecasts were based on the belief that tomorrow's world would be much the same as today's. However, when business planning professionals use linear assumptions and processes in today's world, the results are less than stellar.

The future no longer looks stable, and planning processes must hit moving targets. Humanity lives on the edge of chaos. Uncertainty must be accepted and incorporated into thinking. In the 1970s Pierre Wack and his colleagues at Royal Dutch/Shell began to detect turbulent times and set out to construct a new way to envision the future through scenario building.(7) This process helped Shell predict and deal with the oil price shocks of the 1970s and 1980s.

Scenario building forces both planners and managers to think outside of the traditional comfort zone. This process is not designed for the quiet inner sanctum of one's office. It is a group process involving many organizational levels. Scenarios help planners and managers structure uncertainty when 1) they are based on a sound analysis of reality; 2) they change the decision maker's assumptions about how the world works; and 3) they compel planners and managers to reorganize their mental model of reality. The process is also the subject of a book by Peter Schwartz, *The Art of the Long View*.(8) As Pierre Wack contends, "The art of scenarios is not mechanistic, but organic; whatever we had learned after one step advanced us to the next."

A deep understanding of issues is critical to the success of scenario building and competitive strategy analysis (presented later in this chapter). One cannot hope to build a viable vision or strategy of the future without detailed knowledge of the issues, the politics, the economics, and the societal factors that surround the decisions to be made. Ideally, each member of the planning team has different experiences and areas of competence, providing a multidimensional view of the issues. In addition to an understanding of the business environment, participants must have an idea of where they want to go; that is, they must have a concept of what questions must be answered for the organization. For example, they might need to ask questions similar to the following:

- Should we build a new plant with tighter benzene controls? Will the benzene standard be reduced still further?

- How do we support economic productivity and activity while balancing concerns for human and environmental health?
- How do we weave learning organization strategies into our core competencies?
- Should we structure our safety and health management systems using the concepts of the ISO 9000 and 14000 standards? Will this standard, or other similar standards, eventually drive health and safety?
- Will home offices eventually come under OSHA jurisdiction? Can we allow employees to work at home?
- Will U.S. environmental health and safety regulations continue to strengthen, making certain businesses noncompetitive? Should we build a new plant in the United States or in the Far East?
- What new regulations are on the horizon, and what actions should be taken to influence their development? Will these regulations create a competitive advantage for the United States or for another country, and what action should we take to address the impacts of these regulations?

Questions about the future are endless. A major question that must be faced as one attempts to build scenarios that will help anticipate the future and ultimately to protect the health and safety of the work force is, "What questions should we ask?" The following section will address formulating the right questions and developing a responsive scenario to envision the future.

Forecasting

Forecasts generated by experienced planners are reasonably accurate; however, it is the complacency that comes with being correct that makes forecasting dangerous. Traditionally, forecasts have been built on the assumption that tomorrow will be like today. Over relatively short periods of time, these assumptions work well. Forecasts fail when events occur that were not part of the assumptions — a war, an oil embargo, a major chemical plant explosion, a dramatic research finding that connects an exposure to a disease, a new technology, or an unexpected and dramatic economic shift. Therein lies the forecasting paradox. If a forecast is relied on to plan a strategy, and the forecasting process does not allow for uncertainty, the value of the process diminishes.

The solution lies in accepting uncertainty and embracing it in the forecasting process. It can no longer be assumed that tomorrow will be anything like yesterday. Accepting and integrating uncertainty is the only effective way to envision the future.

What is a Scenario?

Scenarios help structure uncertainties. To be successful, they must be based on a clear understanding of reality and change the decision makers' assumptions about how the world works, while creating a compelling reason to reorganize mental models of reality. It is a process that involves more than formulating a good scenario. It requires an active willingness to face uncertainty, to understand the forces that drive it, and a desire to change how business is conducted based on that understanding.

One of the underlying problems faced by the occupational hygiene and safety community has been professional myopia. This refers to an inability to see beyond the technology or systems currently in use and with which everyone is comfortable, and not seeing the value of building effective working relationships with other staff organizations and line structures. If scenarios are to work, these relationships are critical. Often, OEH&S professionals' views of reality are focused on their technical skills and generally do not contain integrated business views. Additionally, the strategic, organizational, and behavioral changes that are to be made require involvement and concurrence of the line organization. OEH&S professionals must look at business systems and their functions within them.

Step One: Identify a Critical Issue or Decision to be Made

Start by asking a question: What will the decision makers in an organization be thinking about in the near future, and what role does OEH&S play in that decision? Identify the decisions or issues that will have a significant and long-term impact on the business (for example, the impact of a new technology on business and what management decisions need to be made if the technology materializes).

How does one recognize the correct issues or decision? It is important to understand management's issues and what future decisions they will make. If it's an issue that keeps management awake at night, it's the right issue to explore. Some examples might be:

- Should we build the new plant that incorporates a new technology?

- Should a new R&D effort be launched that will explore the impact of the new technology?
- Should we try to maintain control of the technology?
- Should we redesign a product line because of pending regulations?
- Should we use a new environmental law to our competitive advantage? How?
- Can raw material prices force a shift to an alternative, requiring plant redesign?
- Should we preemptively respond to new negative product health data?
- Should we move operations to a country with less stringent safety and health regulations?

Step Two: Identify the Key Forces in the Local Environment

List the key factors that will impact the success or failure of that decision. Consider customers, suppliers, competitors, buyers, substitutes, and new entrants into the market. What will the decision makers want to know as they draw their conclusions and make decisions on options? What constitutes success or failure? What factors will drive the outcomes? Some examples of what decision makers want to know might include the following:

- How long will it take the Occupational Safety and Health Administration to promulgate a new regulation?
- What are the range of compliance options and their costs?
- How must we change our products or processes to address new regulations?
- How will our global competitiveness be influenced by regulatory changes?
- Will we need to retain more occupational health and safety services?
- Can the regulatory changes be used to achieve a competitive advantage?
- Are today's OEH&S professionals equipped to assist in the management of the new regulatory environment?

Some examples of success or failure measures are:

- Market share change — growth or shrinkage in the percentage of a market controlled by a manufacturer or service provider;
- Safety and health budget size — growth or shrinkage of the budget used to protect employees;
- Degree of outsourcing of health and safety service — shift of health and safety services to consultants;
- OEH&S impact on new processes or products — the ability of the OEH&S function to influence workplace controls or product design; and
- Cost of new regulatory compliance requirements — ability to control compliance costs.

Other outcome factor drivers might be:

- Changes in the regulatory environment — increase or decrease in regulatory burden;
- Customers switching to imported products not affected by U.S. regulations — products with a competitive advantage over U.S. products, in part, because they are unburdened by EH&S regulatory compliance costs;
- Cost of product or process redesign to meet new safety requirements—the cost burden of regulatory compliance or market driven requirements;
- Withdrawal from market — exiting from the market because EH&S costs create an inability to compete; and
- Response to the global competition — changes in products or services made to compete in the global marketplace.

Step Three: Driving Forces and Trends in the Macroenvironment

In this step, for each of the issues developed in Step Two, identify and list the trends and forces that will drive those questions or issues in one direction or another. These should include social, demographic, economic, political, environmental, and science and technology forces and trends. Some forces are generally understood, and the outcome is predetermined — i.e., population growth or seasonal weather changes. Some issues are driven by a matter of choice, and some decisions or trends are inevitable. Some trends have a high degree of uncertainty associated with them—such as events in the Middle East, global politics, or the weather a month from now. The objective is to determine the endpoints of the trends and what forces will drive the trends toward the endpoints. This is the most labor intensive component of the analysis, requiring much research and analysis. Some examples of trends include the following.

- Economic growth and growth of market;
- Technologic developments;
- Shift from national to global standards;
- Shift from manufacturing to information technology;
- Changes in society's willingness to accept risk;

- Shift from labor to capital intensive manufacturing processes;
- Major shifts in health and safety regulations (e.g., from specification to performance standards);
- Changes in the education of health and safety professionals;
- Shift in consumer demand for safe products;
- Shifts in global supply of safe products; and
- Demographic shifts.

Step Four: Rank Key Factors and Driving Trends

Using two criteria, the key factors and driving trends need to be ranked by 1) the degree of importance for the success of the issue or decision developed in Step One; and 2) the degree of uncertainty surrounding the factors and trends (see Table 50.1). The purpose is to identify the top two or three factors or trends that are critical to the decision or outcome and that have the greatest uncertainty associated with them.

Scenario building cannot be driven by predetermined elements like population growth in an area, because these elements will be the same for all scenarios, and no useful information is established.

This example focuses analytical attention on shifts — in consumer demand or taste, shifts in global supply of a product, and shifts in energy costs.

Step Five: Selecting the Scenario Framework

In Step Five, the variables for scenario development are identified; that is the critical axes along which scenarios will be constructed. One normally operates in a plane and uses two factors or trends; however, it is possible to operate with three axes, or in hyperspace — with more than three dimensions. Operating in more than one plane makes data interpretation complex and difficult to explain to management.

Successful scenarios are tools that facilitate learning about the issues that underpin the decision-making process. Thus, the objective is to formulate a few scenarios that address the most critical factors and trends that concern decision makers. Complex scenarios, which have too many dimensions, compound decision-making difficulty. Every uncertainty cannot be addressed, so selecting the most critical trends and factors, or scenario drivers, is critical. The ranking process in Step Four is vital.

Once a group of scenario drivers is selected, the next decision is how to juxtapose them against one another. If, for example, four scenario drivers are selected and the decision is made to conduct a two-dimensional analysis, there are six ways to juxtapose the scenario drivers, generating 24 scenario elements. If a three-dimensional analysis is constructed, there will be three ways to conduct the analysis and 32 scenario elements.* Clearly, evaluating

Table 50.1 — Ranking the Importance and Uncertainty of Key Scenario Driving Forces

Trend or Factor	Importance[A]	Uncertainty[A]	Trend Factor/ Rank[A,B,C]
Economic and market growth	10	2	20/6
Technological developments	9	4	36/4
Shift from national to global standards	5	6	30/5
Changes in society's willingness to accept risk	6	10	60/3
Shift from labor to capital intensive manufacturing processes	2	3	6/7
Shifts in health and safety regulations	7	9	63/2
Changes in the education of health and safety professionals	1	5	5/8
Shift in consumer demand for safe products	8	8	64/1
Shift in global supply of safe products	9	7	63/2
Demographic shifts	3	1	3/9

[A]Importance or rank on a scale of 1 to 10, with 1 being of low importance and 10 being of high importance
[B]Trend factor = importance × uncertainty
[C]Trends or factors with equal importance × uncertainty scores receive the same rank.

*Determined by mathematical combination relationship nCr = n!/((r!)(nr)!), where n = number of critical factors or trends and r = number of factors analyzed at one time. C is the mathematical notation for combination.

these elements can be quite complex. Simple is better—creating a powerful reason for selecting the most critical factors or trends for conducting the analysis.

Once the scenarios are formulated, they must be fleshed out and their implications established. Steps Seven and Eight address these areas.

If management, for example, is making a critical decision about moving the location of a manufacturing operation into another country or keeping it in the United States, health and safety regulations may be a key factor driving that decision. They may want to anticipate changes over the next 5 years in the public's willingness to understand and accept environmental and workplace risks and the potential for a shift in the type of regulations promulgated in the United States. Additionally, management will look at various business scenarios in parallel with the health and safety regulatory analysis.

Armed with this information, decisions that drive the company's future can be made with reasonable confidence.

Figures 50.1 and 50.2 show how four scenarios are constructed from two critical factors or trends. Health and safety professionals may wish to construct similar scenarios to integrate health and safety into the management decision process. Figure 50.1 demonstrates what conditions might be needed to achieve the indicated scenario. Figure 50.2 postulates what conditions would be like under each scenario.

The more that is understood about the external environment and its impact on operations, the more effective one can be in developing a useful view of the future. Using this type of analysis aids in understanding how to integrate health and safety into the business process. Scenario matrices can also be developed in prose — in whatever format the author feels comfortable.

Political drive to shift from specification to performance regulations	Willingness to accept risk: Low	Willingness to accept risk: High
High	**D** • Competitive pressure on U.S. industry • New research suggests weakening links between workplace and environmental exposures and disease — lifestyle is driver • Jobs continue to move overseas • Labor union power wanes • Voluntary compliance programs obtain high degree of compliance • Industry has significantly reduced industrial accidents and chemical releases • Public more concerned about societal costs than workplace and environmental risks	**C** • Global economy becomes highly competitive, U.S. has trouble competing • Legislation promulgated that requires EPA and OSHA work to enhance risk knowledge of public • Industry gains public understanding of risk and cost issues • Schools teach students about risk • Public concerned about regulatory costs • Science demonstrates weak link between workplace/environmental exposures and occupational disease • Public starts to understand risk concepts
Low	**A** • Failing U.S. economy • U.S. not competitive • OSHA and EPA very powerful • Labor unions growing rapidly • Research links workplace exposures to new set of diseases • Health care costs out of control • Large regulatory budgets • Public sees regulations the only way to control industry • Industry has high profile accidents and chemical releases	**B** • Mixed economic performance • Regulatory costs, not understanding of risk drives regulatory process • OSHA and EPA use command and control standard setting and enforcement • Government officials more concerned about their power than global competitiveness and economic growth • Labor unions influential but not growing Public believes there are too many industrial accidents and chemical releases • Public is developing an understanding of risk

Figure 50.1— What must happen to drive a scenario.

Step Six: Completing and Summarizing

The next step is to fill in any gaps left as the scenarios were constructed. To do this, one must determine how the key factors and trends identified in Steps Two and Three will impact the scenarios developed in Step Five. This may be written in an analysis matrix. If any of the trends or factors do not differentially impact any of the scenarios (null hypothesis = status quo), they should be removed from the analysis to simplify the results. For the sake of completeness and to prevent someone else from repeating the analysis, discarded factors should be noted and explained.

Figure 50.3 takes two factors and trends and ties them to the scenario.

Summary of Scenarios

Once all trends and factors are evaluated and included in the scenario, this information is formulated into an executive summary with supporting details for management's review. Critical to this summary is the analyst's understanding of the operating environment, since once all the data is gathered, judgment, experience, and competitive forces play central roles in determining which scenario best reflects the future.

Step Seven: Implications

In this step, independent and dependent variables are selected and analyzed, and inferences are made based on different decision matrices. Vulnerabilities or sensitivities are revealed; management may decide to ask a different question or deal with a different focal issue.

Assuming the scenarios were set up realistically, if the decision about a question or focal issue looks good in a number of scenarios, then it is likely to be a reasonable decision. If the decision only looks good in

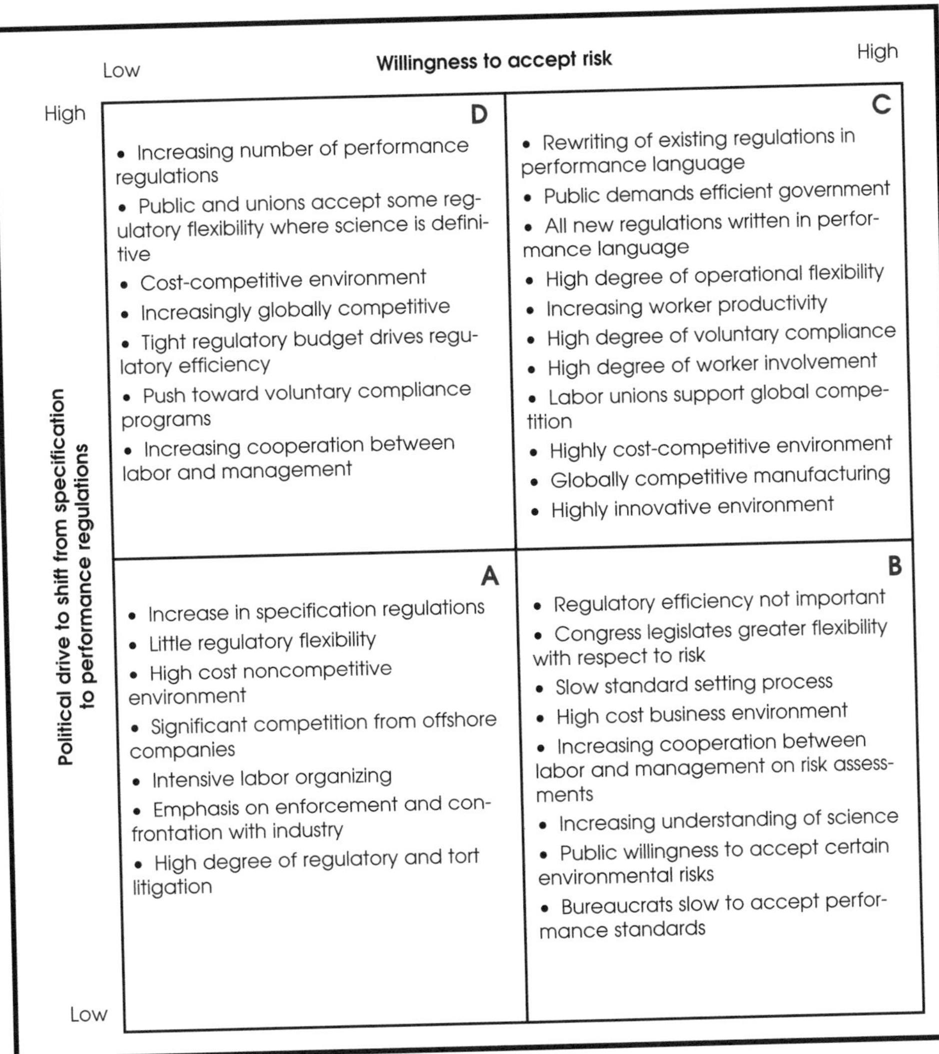

Figure 50.2 — What can be expected under each scenario.

one scenario, then it may have to be rethought or revised, because it may be risky to choose only one view of the future. A decision that looks good in many views has more stability and can withstand a higher degree of uncertainty without betting the company. Figure 50.4 provides an example of how a decision is tested against the developed scenarios.

Implications—Moving to Another Country

As an example, consider the following management decisions (questions): Should manufacturing move to another country because of the difficult health and safety regulatory environment in the United States? If safety and health regulations are central to the decision, how does this question fare in each scenario?

Company sensitivities, ethical issues, and many other business scenarios will become part of the final decision. Decisions are not generally made on the basis of safety and health issues alone, but the analysis is key to fully integrating health and safety into the business decision-making process.

Considering business conditions, driven by global competition in an environment of increased understanding and acceptance of risk, and acceptance of performance oriented standards, the analysis suggests that moving to another country is the wrong decision (Scenarios B, C, and D in Figure 50.4). Only in Scenario A are the prospects for business so poor and costs so high that a move is justified. The forces of global change, discussed earlier in this chapter, are so powerful that the probability of stagnation is unlikely.

Step Eight: Indicators

Once a decision is made and scenarios accepted, it is important to see which scenario most closely tracks events as they

Political drive to shift from specification to performance regulations	Low — **Willingness to accept risk**	— High
High	**D** **Technological Developments** • New monitoring methods are used to automate workplace and environmental exposure monitoring tasks giving employer wider latitude in developing monitoring programs. Regulations now permit the use of this technology, but are reluctant to let tighter technologic controls drive increased risk acceptance. **Demographics** • Younger workers want performance, but older workers like specifications.	**C** **Technological Developments** • New monitoring methods are used to better understand risk and to more precisely set acceptable levels, giving the public greater confidence in the numbers. Industry use of technology adds to competitiveness in performance and risk accepting environment. **Demographics** • Younger workers understand that they can play a larger role in helping define and operate the workplace.
Low	**A** **Technological Developments** • New monitoring technology make exposures easier to measure at lower and lower levels driving acceptable regulatory risk levels lower. Regulatory compliance is more costly. **Demographics** • Labor does not want to change specification regulations to performance regulations for fear that management will not control workplace risks. Labor could lose influence in scenario B, C, D.	**B** **Technological Developments** • While the public is willing to rely on technology to help in risk management, regulators are skeptical because they may play a lesser role in driving standards where risk acceptance is greater. **Demographics** • Younger workers are willing to accept risk while older workers are afraid of the consequences.

Figure 50.3 — Impact of Step Two and Step Three factors on trends.

unfold, so midcourse corrections can be made. From the example, even if the decision is to stay in the United States, but regulatory growth accelerates, the decision on where to build the next plant may be different.

Thus, to build a view of the future, it is important to design metrics or indicators that can track progress and be used by management for future decisions and to adjust existing and future scenarios. Such indicators include:

- Control of Congress;
- Number of pages of new regulations in the *Federal Register*;
- Size of regulatory agency budgets,
- Public opinion polls on risk; and
- Economic indicators of competitiveness.

In the "move" example, different assumptions may be made and issues and trends viewed through different mind models — (deeply ingrained assumptions or pictures that influence how people understand the world and how they take action). Based on their mind-sets, people will come up with different conclusions. The purpose of the example and this chapter is to explain a process that can easily be used to begin to understand and react to the changing environment. The greatest challenge OEH&S professionals face in constructing their futures is adopting a model where ideas can be tested. The Pierre Wack/Peter Schwartz process seems to fit. Data and case studies are abundant, and OEH&S professionals have the skills necessary to do the analyses.

To be most effective the scenarios must be integrated within the planning process of a specific company or organization. A team approach that uses the concepts of team learning and systems thinking will help ensure success. In addition to ensuring success, this approach will enhance the relationships between the OEH&S department and the line function.

Political drive to shift from specification to performance regulations ↓ / Willingness to accept risk →	Low	High
High	**D** **Wrong Decision** • Events are moving in the right direction, creating a more globally competitive environment in the U.S. With this trend and a well-trained, highly productive, quality oriented U.S. work force, a move would most likely damage profits for a significant period. • Emphasis must be placed on building an understanding of risk issues with the work force and public.	**C** **Wrong Decision** • U.S. has become a business-friendly environment where global competition is encouraged. Regulations are performance oriented, providing a high degree of compliance flexibility. • There is a high worker involvement that enhances health and safety performance. • Moving would be costly and an experienced work force lost. • Labor unions have begun to adapt to globally competitive environment.
Low	**A** **Correct Decision** • The U.S. in noncompetitive driven by bureaucratic, inflexible regulations. The public is unwilling to accept any environmental or workplace risks. • Worker involvement is contentious and operating costs are high. • Moving would help the company regain competitiveness. The move is seen as unfortunate, but management feels their hands have been tied by inflexible regulators and a risk-averse public.	**B** **Wrong Decision** • The public has begun to understand environmental risk and is demanding that regulatory changes be made. • Government bureaucrats, fearing loss of power, have been slow to change. • Congress is starting to apply pressure fearing loss of competitiveness. • The future looks bright for domestic companies and moving overseas is becoming less attractive.

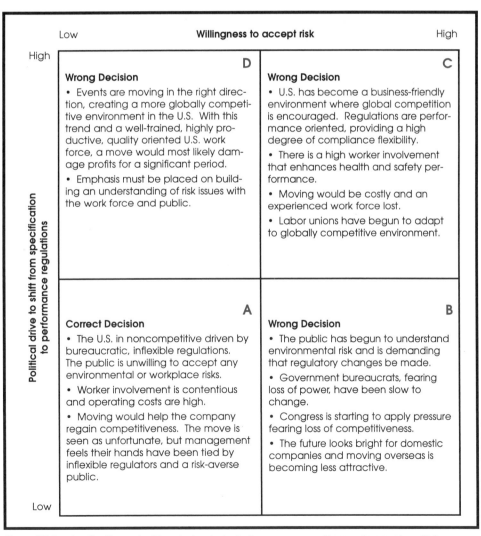

Figure 50.4 — Implications: decisions being tested—the company will move to country with lower health and safety standards.

Conclusion

Analysis of complex issues like scenario building cannot be resolved by a single person. It takes a team effort, team learning, shared vision, a systems approach—all components and practices of a learning organization. A learning organization is continually expanding its capacity to create its future.

The complex, dynamic, and globally competitive nature of business is driving the need for learning organizations. Excelling in a dynamic business environment requires more understanding, knowledge, preparation, and agreement than one person's expertise and experience provides. Multidisciplinary teams are the primary tool for addressing OEH&S problems.

Today, humanity lives on the edge of chaos. Five forces that will shape the future include 1) the free market system; 2) the rise of brainpower industries; 3) the aging population; 4) the global economy; and 5) the multipolar world. These powerful forces are not easily influenced. Therefore, the better they are understood, the more effective forecasts and responses to issues and trends can be. These macroeconomic factors will have a great impact on the OEH&S profession.

While one can only prognosticate about the future, through the use of scenarios and probability modeling, changes can be anticipated, recognized, evaluated, and responded to. Scenario building forces users to think outside of their traditional comfort zones. Scenarios help planners and managers structure uncertainty when 1) the scenarios are based on a sound analysis of reality; 2) they change the decision maker's assumptions about how the world works; and 3) they compel planners and managers to reorganize their mental model of reality.

The first critical step to building scenarios is to ask the question, "What will the decision makers in this particular organization be thinking about in the near future, and what role should OEH&S professionals play in that decision?" The question leads to identification of key forces in the local environment, forecasting, driving forces, and trends in the macroenvironment. This process leads to a ranking of key factors and driving trends, establishing scenarios, and testing assumptions based on probabilities. Once scenarios are accepted, to build a view of the future it is important to design metrics or indicators to track progress so midcourse corrections can be made.

By understanding the macroeconomic forces that shape the future and by studying probabilities through scenario building, OEH&S professionals can provide the leverage needed to balance economic growth and activity with reasonable protection of human health and the environment.

Additional Sources

Fortune Magazine, monthly publication, Tampa, Florida.

The Futurist, bimonthly publication of the World Future Society, Bethesda, Maryland.

Harvard Business Review, Harvard Business School, Boston, Massachusetts.

"The next 25 years." *Industry Week 244*:15, 1995.

Scientific American, edition devoted to science in the year 2020 (273:3, 1995).

References

1. **Verespej, M.A.:** A workforce revolution? *Ind. Week 244*:15 (1995).
2. **McRae, H.:** *The World in 2020.* Boston, MA: Harvard Business School Press, 1994, p. 21.
3. **Senge, P.M.:** *The Fifth Discipline: The Art & Practice of the Learning Organization,* New York: Currency/Doubleday, 1990.
4. **Thurow, L.:** *The Future of Capitalism.* New York: William Morrow and Co., 1996.
5. **American Industrial Hygiene Association (AIHA):** *Respiratory Protection.* Akron, OH: AIHA, 1985.
6. **United Nations (UN):** *World Population Prospect: 1994 Revision* (ST,ESA,SER.A/145). New York: UN Department for Economic and Social Information and Policy Analysis, 1995.
7. **Wack, P.:** *Scenarios: Unchartered Waters Ahead, Strategy—Seeking and Securing Competitive Advantage.* Boston, MA: Havard Business Review Books, 1991.
8. **Schwartz, P.:** *The Art of the Long View.* New York: Currency/Doubleday, 1991.

Periodic Table of Elements

	IA	IIA	IIIB	IVB	VB	VIB	VIIB	VIIIB	VIIIB	VIIIB	IB	IIB	IIIA	IVA	VA	VIA	VIIA	0
1	1 **H** 1.00794																	2 **He** 4.00260
2	3 **Li** 6.941	4 **Be** 9.01218											5 **B** 10.811	6 **C** 12.011	7 **N** 14.00674	8 **O** 15.9994	9 **F** 18.99840	10 **Ne** 20.1797
3	11 **Na** 22.98977	12 **Mg** 24.305											13 **Al** 26.98154	14 **Si** 28.0855	15 **P** 30.97376	16 **S** 32.066	17 **Cl** 35.4527	18 **Ar** 39.948
4	19 **K** 39.0983	20 **Ca** 40.078	21 **Sc** 44.95591	22 **Ti** 47.88	23 **V** 50.9415	24 **Cr** 51.9961	25 **Mn** 54.938	26 **Fe** 55.847	27 **Co** 58.93320	28 **Ni** 58.6934	29 **Cu** 63.546	30 **Zn** 65.39	31 **Ga** 69.723	32 **Ge** 72.61	33 **As** 74.92159	34 **Se** 78.96	35 **Br** 79.904	36 **Kr** 83.80
5	37 **Rb** 85.4678	38 **Sr** 87.62	39 **Y** 88.90585	40 **Zr** 91.224	41 **Nb** 92.90638	42 **Mo** 95.94	43 **Tc** 98.9072	44 **Ru** 101.07	45 **Rh** 102.90550	46 **Pd** 106.42	47 **Ag** 107.8682	48 **Cd** 112.411	49 **In** 114.82	50 **Sn** 118.710	51 **Sb** 121.76	52 **Te** 121.757	53 **I** 126.09447	54 **Xe** 131.29
6	55 **Cs** 132.9054	56 **Ba** 137.327	57 ***La** 138.9055	72 **Hf** 178.49	73 **Ta** 180.9479	74 **W** 183.85	75 **Re** 186.207	76 **Os** 190.2	77 **Ir** 192.22	78 **Pt** 195.08	79 **Au** 196.96654	80 **Hg** 200.59	81 **Tl** 204.3833	82 **Pb** 207.2	83 **Bi** 208.98037	84 **Po** 208.9824	85 **At** 209.9871	86 **Rn** 222.0176
7	87 **Fr** 223.0197	88 **Ra** 226.0254	89 **!Ac** 227.0278	104 **Rf** 261.11	105 **Db** 262.114	106 **Sg** 263.118	107 **Bh** 262.12	108 **Hs** (265)	109 **Mt** (266)	110 **Uun** (269)	111 **Uuu** (272)	112 **Uub** (275)						

*	58 **Ce** 140.115	59 **Pr** 140.90765	60 **Nd** 144.24	61 **Pm** 144.9127	62 **Sm** 150.36	63 **Eu** 151.965	64 **Gd** 157.25	65 **Tb** 158.92534	66 **Dy** 162.50	67 **Ho** 164.93032	68 **Er** 167.26	69 **Tm** 168.93421	70 **Yb** 173.04	71 **Lu** 174.967
!	90 **Th** 232.0381	91 **Pa** 231.0359	92 **U** 238.0289	93 **Np** 237.0482	94 **Pu** 244.0642	95 **Am** 243.0614	96 **Cm** 247.07003	97 **Bk** 247.0703	98 **Cf** 251.0796	99 **Es** 252.083	100 **Fm** 257.0591	101 **Md** 258.10	102 **No** 259.1009	103 **Lr** 262.11

List of Organizations

Acoustical Society of America (ASA)
500 Sunnyside Blvd.
Woodbury, NY 11797
Phone: 516-576-2360
Fax: 516-576-2377

Air and Waste Management Association (A&WMA)
1 Gateway Center, 3rd Floor
Pittsburgh, PA 15222
Phone: 412-232-3444
Fax: 412-232-3450

American Association for the Advancement of Science (AAAS)
1200 New York Ave. NW
Washington, DC 20005
Phone: 202-326-6400
Fax: 202-371-9526

American Association of Occupational Health Nurses (AAOHN)
50 Lenox Pointe
Atlanta, GA 30324-3176
Phone: 404-262-1162
Fax: 404-262-1165

American Board of Industrial Hygiene (ABIH)
6015 W St. Joseph, Suite 102
Lansing, MI 48917-3980
Phone: 517-321-2638

American Chemical Society (ACS)
1155 16th St. NW
Washington, DC 20036
Phone: 202-872-4600
Fax: 202-872-4615

American Conference of Governmental Industrial Hygienists (ACGIH)
Kemper Woods Center
1330 Kemper Meadow Dr., Suite 600
Cincinnati, OH 45240-1634
Phone: 513-742-2020
Fax: 513-742-3355

American National Standards Institute (ANSI)
11 W 42nd St., 13th Floor
New York, NY 10036-8002
Phone: 212-642-4900
Fax: 212-398-0023

American Public Health Association (APHA)
1015 15th St. NW, Suite 300
Washington, DC 20005
Phone: 202-789-5600
Fax: 202-789-5661

American Society of Heating, Refrigerating and Air-Conditioning Engineers (ASHRAE)
1791 Tullie Circle NE
Atlanta, GA 30329-2305
Phone: 404-636-8400
Fax: 404-321-5478

American Society for Testing and Materials (ASTM)
100 Barr Harbor Dr.
West Conshohocken, PA 19428-2959
Phone: 610-832-9500
Fax: 610-832-9635

American Society of Safety Engineers (ASSE)
1800 E Oakton St.
Des Plaines, IL 60018-2187
Phone: 847-699-2929
Fax: 847-296-3769

Board of Certified Safety Professionals (BCSP)
208 Burwash Ave.
Savoy, IL 61874
Phone: 217-359-9263
Fax: 217-359-0055

Centers for Disease Control and Prevention (CDC)
Department of Health and Human Services
Public Health Service
1600 Clifton Rd. NE
Atlanta, GA 30333
Phone: 404-639-3311

Chemical Manufacturers Association (CMA)
1300 Wilson Blvd.
Arlington, VA 22209
Phone: 703-741-5000
Fax: 703-741-6000

Council for Accreditation in Occupational Hearing Conservation (CAOHC)
611 E Wells St.
Milwaukee, WI 53202
Phone: 414-276-5338
Fax: 414-276-3349

Department of Energy (DOE)
Office of Energy Research
ER 60
19901 Germantown Road
Germantown, MD 20874
Phone: 301-903-4944
Fax: 301-903-8583

Department of Health and Human Services (DHHS)
Health Resources and Services Administration (HRSA)
5600 Fishers Lane
Rockville, MD 20857
Phone: 301-443-2216
Fax: 301-443-1246

Environmental Information Association (EIA)
4915 Auburn Ave., Suite 303
Bethesda, MD 20814
Phone: 301-961-4999
Fax: 301-961-3094

Environmental Protection Agency (EPA)
EPA Publications and Information Center
P.O. Box 42419
Cincinnati, OH 45242-0419
Emergency Planning and Community Right-to-Know Hotline: 800-535-0202
SARA and CERCLA Hotline (Nonemergency): 800-424-9346

Health Physics Society (HPS)
1313 Dolley Madison Blvd., Suite 402
McLean, VA 22101-3926
Phone: 703-790-1745
Fax: 703-790-2672

Human Factors and Ergonomics Society (HFES)
P.O. Box 1369
Santa Monica, CA 90406-1369
Phone: 310-394-1811
Fax: 310-394-2410

Illuminating Engineering Society of North America (IESNA)
120 Wall St., 17th Floor
New York, NY 10005-4001
Phone: 212-248-5000
Fax: 212-248-5017

Industrial Safety Equipment Association (ISEA)
1901 N Moore St., Suite 808
Arlington, VA 22209
Phone: 703-525-1695
Fax: 703-528-2148

Institute of Electrical and Electronics Engineers, Inc. (IEEE)
345 E 47th St.
New York, NY 10017-2394
Phone: 212-705-7900
Fax: 212-752-4929

International Occupational Hygiene Association (IOHA)
Suite 2, Georgian House
Great Northern Rd.
Derby
DE1 1LT
United Kingdom
Phone: 011-44-1332-298101
Fax: 011-44-1332-298099

International Organization for Standardization (ISO)
1, rue de Varembe
Case postale 56
CH-1211 Geneve 20
Switzerland
Phone: 011-41-22-749-01-11
Fax: 011-41-22-733-34-30

Library of Congress
Science and Technology Division
Washington, DC 20540
Phone: 202-707-5000

Mine Safety and Health Administration (MSHA)
Department of Labor
4015 Wilson Blvd.
Arlington, VA 22203
Phone: 703-235-1452
Fax: 703-235-4323

National Center for Toxicological Research (NCTR)
3900 NCTR Rd.
Jefferson, AR 72079
Phone: 870-543-7000
Fax: 870-543-7635

National Council on Radiation Protection and Measurements (NCRP)
7910 Woodmont Ave., Suite 800
Bethesda, MD 20814-3095
Phone: 301-657-2652
Fax: 301-907-8768

National Institute for Occupational Safety and Health (NIOSH)
200 Independence Ave. SW
Washington, DC 20201
Phone: 202-401-0721
Tech Info: 800-356-4674

National Institute of Standards and Technology (NIST)
Rte I-270 and Quince Orchard Rd.
Gaithersburg, MD 20899
Phone: 301-975-3058
Fax: 301 926-1630

National Safety Council (NSC)
1121 Spring Lake Dr.
Itasca, IL 60143-3201
Phone: 630-285-1121
Fax: 630-285-1315

National Technical Information Service (NTIS)
Department of Commerce
5285 Port Royal Rd.
Springfield, VA 22161
Phone: 703-487-4650
Fax: 703-321-8547

U.S. Nuclear Regulatory Commission (NRC)
One White Flint North
11555 Rockville Pike
Rockville, MD 20852-2738
Phone: 301-415-7000
Fax: 301-415-7020

Occupational Safety and Health Administration (OSHA)
U.S. Department of Labor
200 Constitution Ave. NW
Washington, DC 20210
Phone: 202-219-8151
Emergency: 800-321-OSHA
Fax: 202-219-5986

Office of Hazardous Materials Safety (OHMS)
Department of Transportation
400 Seventh St. SW
Washington, DC 20590
Phone: 202-366-0656
Fax: 202 366-5713

Society of Manufacturing Engineers (SME)
P.O. Box 930
One SME Dr.
Dearborn, MI 48121-0930
Phone: 313-271-1500
Fax: 313-271-2861

Society of Risk Analysis (SRA)
1313 Dolley Madison Blvd., Suite 402
McLean, VA 22101
Phone: 703-790-1745
Fax: 703-790-2672

Society of Toxicology (SOT)
1767 Business Center Dr., Suite 302
Reston, VA 22190
Phone: 703-438-3115
Fax: 703-438-3113

Underwriters Laboratories, Inc. (UL)
333 Pfingsten Road
Northbrook, IL 60062-2096
Phone: 847-272-8800
Fax: 847-272-8129

About the Authors

Mohamed Y. Beshir, Ph.D., PE, is president and CEO of Modern Textile Industries (Montex) in Cairo, Egypt. He is an industrial engineer and consultant with a strong background in both academia and industry. As a faculty member at Ain-Shams University in Egypt, he served as assistant, then associate, professor. He was also a visiting professor at Texas Tech University for several years. He began his association with Modern Textile Industries as director and co-owner before assuming his present position. Beshir has contributed numerous articles to professional journals and is a member of AIHA, the Institute of Industrial Engineers, the Ergonomics Society, and the Egyptian Engineers Syndicate.

Lawrence R. Birkner, CIH, CSP, is a senior consultant at Atlantic Richfield Company (ARCO) in Los Angeles. Birkner's work focuses on integrating health, safety, and environmental issues into business applications to enhance business performance — primarily for the petrochemical industry. He was awarded an MA in occupational health and safety from New York University and an MBA from Pepperdine University. He has managed numerous corporate-wide programs covering industrial hygiene and safety; environmental health and safety auditing; major incident investigation; and safety and health support for environmental remediation operations. Birkner has taught numerous other occupational hygiene programs, written texts on respiratory protection, and contributes regularly to *Occupational Hazards Magazine.* Birkner is a member of the California OSHA Advisory Board and AIHA, a past president of the American Academy of Industrial Hygiene (AAIH), former member of the AIHA Board of Directors, and a past president of the Southern California AIHA Local Section.

Christine A. Bisesi, CIH, CHMM, is president and owner of Enviro-Health, Inc., an environmental and occupational health and safety consulting firm in Holland, Ohio. Bisesi's work has focused on managing hazardous waste site characterization and remediation projects, performing comprehensive industrial hygiene evaluations, and conducting management and employee education and training. Prior to establishing Enviro-Health, Inc., in 1990, Bisesi worked with OSHA as a health compliance officer and for an engineering firm as environmental remediation consultant. Bisesi received her MS in occupational health and safety from Indiana State University.

Michael S. Bisesi, Ph.D., CIH, is chair and associate professor of the Department of Occupational Health in the School of Allied Health at the Medical College of Ohio in Toledo, Ohio. Bisesi has been jointly appointed to the Department of Pharmacology in the School of Medicine. His laboratory, field research, and consulting interests are environmental toxicology and occupational exposure assessment. He teaches courses in toxicology, monitoring and analytical methods, hazardous materials and wastes, and environmental health. Bisesi has BS and MS degrees in environmental science from Rutgers University and a Ph.D. in environmental science from the SUNY College of Environmental Science and Forestry in association with Syracuse University. He has published several scientific articles and chapters, and authored several industrial hygiene and environmental health and safety texts.

Phillip A. Bishop, Ed.D., is professor of human performance studies and director of the Human Performance Laboratory at the University of Alabama-Tuscaloosa. Bishop's interests include work physiology — specifically, encapsulating protective clothing. Bishop's research covers microclimate cooling, impact of fabrics on heat strain, fluid replacement, fatigue, heat strain measurement and estimation, predictions of heat tolerance, safety and productivity, and worker monitoring. Bishop has consulted on projects for clients such as the U.S. Air Force, Kimberly Clark Corp., and NASA's Exercise Countermeasures Project at Johnson Space Center. He has been awarded multiple contracts and grants and has published numerous papers and abstracts.

Arno S. Bommer is a supervisory consultant at Collaboration in Science and Technology Inc. (CSTI). Bommer's work focuses on the measurement of industrial noise and the design of noise control treatments. He is board-certified by the Institute of Noise Control Engineering (INCE) and currently serves on its board of directors. Bommer's involvement with the Acoustical Society of America (ASA) includes currently serving on its Technical Committee on Noise and as a past president of the Houston chapter. Bommer has also served as president of the Houston MIT alumni club. Bommer's experience includes 15 years of consulting experience since obtaining his BS in architectural design from the Massachusetts Institute of Technology. He has published several articles in publications on acoustics and presented papers at ASA and INCE national meetings.

Patrick N. Breysse, Ph.D., CIH, is an associate professor and director of the Industrial Hygiene Program in the Division of Environmental Health Engineering of the Department of Environmental Health Sciences at the Johns Hopkins University School of Hygiene and Public Health. He is on the editorial review board for the *Applied*

Occupational and Educational Hygiene journal and is a past member of the AIHA Ethics Committee. He has served as a consultant to numerous governmental agencies at the local, state, and national levels, as well as for private organizations and industrial firms.

Robert D. Bruce, PE, is a principal consultant at Collaboration in Science and Technology Inc. (CSTI). Bruce's 30 years of consulting experience have covered a broad range of issues including acoustics, engineering, architecture, oil refinery, chemical plants, power generating facilities, aluminum processing, metal stamping, offshore platforms, textile mills, and food processing. He has served as a consultant to many federal agencies including NIOSH, OSHA, EPA, the U.S. Air Force, U.S. Navy, U.S. Army, and U.S. Coast Guard. Bruce is board-certified by the Institute of Noise Control Engineering (INCE), where he served as a past president. Bruce is a member of AIHA and a Fellow of the Acoustical Society of America (ASA). His publications include several chapters of textbooks, technical papers, and reports.

Harriet A. Burge, Ph.D., is associate professor of environmental microbiology and director of both the Environmental Microbiology Laboratory and the Aerobiology Program at Harvard School of Public Health (HSPH). A member of the research faculty of the University of Michigan Medical School for 25 years, she now holds the title associate research scientist, emeritus, from that institution. Burge has extensive experience in bioaerosol monitoring in both indoor and outdoor environments. She has served on three National Academy of Sciences committees related to indoor air quality concerns. She is principal investigator of several funded research projects, including exposure assessment related to the development of asthma in children; the prevalence of outdoor pollens and fungal spores (including the development of predictive models); bioaerosols, health, and productivity in office workers; and toxigenic fungi in the indoor environment. Burge edited a recently published book on bioaerosols and has published more than 30 peer-reviewed papers on bioaerosols.

D. Jeff Burton, PE, CIH, CSP, is president of IVE, Inc., a training and consulting firm in Salt Lake City, Utah, and adjunct associate professor of mechanical engineering at the University of Utah. A former staff engineer at NIOSH, since 1974 Burton has participated in hundreds of design and evaluation projects related to industrial ventilation, HVAC, and indoor air quality. He is author of more than 50 publications, including five books. Burton also writes columns on ventilation topics for the publications *Applied Occupational and Environmental Hygiene* and *Occupational Health and Safety.* He is active in the field of training, having created both continuing education and home study courses for AIHA and ASHRAE. He is a past member of the national board of directors of AIHA and holds memberships in several other professional societies.

Craig E. Colton, CIH, is a senior technical service specialist for 3M in the Occupational Health and Environmental Safety Division. Colton's work is focused on conducting field tests to determine the performance of respiratory protection equipment. Previously, Colton taught at the OSHA Training Institute, where he served as course chair for the respiratory protection course. Colton has also taught continuing education courses for the University of North Carolina and University of California-Berkeley. Colton has served on the AIHA Respiratory Protection Committee and is a member of the ANSI Z88.12 subcommittee, "Respiratory Protection for Infectious Aerosols." He is a co-editor of AIHA's *Respiratory Protection: A Manual and Guideline* and has published several articles in research journals and trade magazines.

Ronald G. Conrad, CIH, passed away in October 1997, after making extensive contributions to the industrial hygiene field. He was vice president of Environmental Technologies, Inc. (formerly Southwest Environmental) where he was responsible for business and financial operations of the company and technical management for a variety of client industrial hygiene, health physics, and environmental projects. Conrad entered the field of industrial hygiene in 1959 in the U.S. Air Force as a sanitary and industrial hygiene engineer. He completed a 21-year military career, serving as an industrial hygienist, health physicist, human factors engineer, public health officer, and environmental engineer. After retiring from the military, he took a position as manager of the NIOSH Educational Resource Center at the University of North Carolina. In 1984, he left the ERC to join Environmental Technologies. He continued his association with the ERC in continuing education and has designed, developed, and taught more than 38 different industrial hygiene continuing education courses.

Joseph Damiano, CIH, CSP, is senior consultant–environment, health and safety for the Aluminum Company of America (Alcoa) in Pittsburgh, Pennsylvania. Damiano provides leadership and support to Alcoa's businesses and operations in the areas of occupational exposure assessment, industrial hygiene standards of practice, EHS performance metrics, preacquisition assessments, start-up of EHS programs in new ventures, and management of various occupational health issues. Damiano received an MS in industrial hygiene from the University of Cincinnati's Kettering Laboratory, and earned his BS in environmental science at Rutgers University. He has served as chair of the AIHA Exposure Assessment Strategies Committee and the AIHA Management Committee, and is a co-author/co-editor of the second edition of the AIHA book *A Strategy for Assessing and Managing Occupational Exposures.*

Deborah F. Dietrich, CIH, is sales manager and corporate industrial hygienist for SKC Inc. Headquartered in Eighty Four, Pennsylvania, SKC Inc. is involved in the research, development, and manufacture of air sampling equipment. Dietrich manages SKC sales, customer service, and technical support services in North America, South America, the Pacific Rim, and Scandinavia. She also writes technical articles on air sampling for SKC publications and has lectured in more than 30 countries.

Salvatore R. DiNardi, Ph.D., CIH, is the chair of the Environmental Health Science Department and director of the Industrial Hygiene Program at the University of Massachusetts, Amherst. DiNardi, who started his academic career in public health at the university in 1970, teaches Principles of Industrial Ventilation, Principles of Occupational Health, Quantitative Methods in Environmental Health, Industrial Hygiene Laboratory, and Aerosol Science. In addition, he works as a consultant for numerous corporations, including AT&T, Eastman Kodak, Digital Equipment Corporation, DuPont, IBM, and American Cyanamid. Dinardi also consults on continuing education for several universities, including the University of North Carolina, George Washington University, and the University of California-Irvine, and has published two books and more than 40 articles in research journals.

M. Cathy Fehrenbacher, CIH, is chief of the Exposure Assessment Branch in the Office of Pollution Prevention and Toxics at the U.S. Environmental Protection Agency in Washington, D.C. Fehrenbacher's career at the EPA has included the development of regulatory and voluntary programs and policies for effectively and efficiently managing potential risks due to human exposure to chemicals. She also serves on the editorial board of the *AIHA Journal*, the board of directors of the AIHA Potomac Local Section, and is the current chair of the AIHA Exposure Assessment Strategies Committee. She has written or co-authored several publications and chapters on the use of modeling approaches for predicting inhalation and dermal exposure, respiratory protection, and has given numerous presentations and lectures on EPA's programs, methods, and policies for assessing and managing chemical risks.

Fred G. Freiberger, CIH, CSP, PE, CmfgE, is director of Client Operations for Law Engineering and Environmental Services, Inc. Formerly with Liberty Mutual Insurance Company, United Technologies, and IBM, Freiberger's past responsibilities have included risk management, national account emergency contingency recovery planning, aerospace life support systems safety, ergonomics, process safety, product liability management, system safety engineering, and organizational development. He serves as a consultant to business, industry, and government in a wide variety of technical, management, and regulatory matters, and his clients have included the Federal Aviation Administration (FAA), OSHA, U.S. Department of Energy, U.S. Department of Defense, National Aerospace Administration, Product Safety Commission, and NIH. An active guest speaker, conference leader, and university lecturer, Freiberger is a regular contributor to several trade publications.

E. Scott Geller, Ph.D., is a professor of psychology at Virginia Polytechnic Institute and State University and is senior partner of Safety Performance Solutions. He specializes in the application of behavioral science toward solving safety, health, and environmental problems in corporate and community settings. Most of his more than 250 research articles and more than 50 books or book chapters address the development and evaluation of behavior change interventions to improve quality of life. He has recently written two books on the psychology of health and safety, *The Psychology of Safety: How to Improve Behaviors and Attitudes on the Job* and *Practical Behavior-Based Safety: Step-By-Step Methods to Improve Your Workplace*. He is past editor of the *Journal of Applied Behavior Analysis*, current associate editor of *Environment and Behavior*, and is consulting editor for *Behavior and Social Issues* and the *Journal of Organizational Behavior Management.*

Dennis K. George, Ph.D., CIH, is coordinator of the Environmental Science Program and assistant professor of environmental science/industrial at Western Kentucky University. George also teaches courses on industrial ventilation and other topics for the University of North Carolina/Duke University NIOSH Educational Resource Center. George received his MSPH and Ph.D. from the University of North Carolina, after serving with the U.S. Navy as an industrial hygienist. George has published articles in major journals documenting his research in both industrial ventilation and occupational exposure assessment.

Joseph Hamill, Ph.D., is a professor and chair of the Department of Exercise Science and is director of the Biomechanics Laboratory at the University of Massachusetts-Amherst. Hamill teaches biomechanics classes such as Mechanical Analysis of Human Motion, Tissue Mechanics, and Kinesiology. Hamill is a fellow of the American College of Sports Medicine and the Research Consortium, and a member of the American and International Societies of Biomechanics. He has published more than 75 research papers in professional journals and has presented papers at both national and international conferences.

Elizabeth C. Hardin will complete her doctoral degree at the University of Massachusetts-Amherst in December 1997. She has published several research papers and presented papers at both national and international conferences.

Michael K. Harris, Ph.D., is a principal with Hamlin & Harris, Inc., a consulting firm in Baton Rouge, Louisiana, and adjunct professor at Louisiana State University. He has been providing environmental and industrial hygiene services since 1978. Much of his practice has focused on the construction, petrochemical, maritime, and health care industries. Confined space assignments have included designing, implementing, and testing of ventilation systems for confined space work as well as planning and executing industrial hygiene monitoring programs for confined space entries. He is co-author, with Lindsay Booher and Stephanie Carter, of *Field Guidelines for Temporary Ventilation of Confined Spaces, With an Emphasis on Hotwork*, published by the AIHA Press.

R. Timothy Hitchcock, CIH, is a certified industrial hygienist for the IBM Corporation. He holds an MSPH from University of North Carolina and is a member of the

AIHA Nonionizing Radiation Committee, the Executive Committee of the ANSI Z136 Accredited Standard Committee on Laser Safety, the Nonionizing Radiation Committee of the North Carolina Radiation Protection Commission, and the editorial boards of the *Journal of Laser Applications* and *The Synergist.* Hitchcock has authored a number of publications and is active as an instructor in the area of nonionizing radiation protection.

Marion E. Hoyer, Ph.D., is a research fellow in the Environmental and Industrial Health Department of the School of Public Health at the University of Michigan. Hoyer obtained her BS degree in microbiology at the University of Michigan and subsequently worked as a research associate in the Division of Allergy at the University of Michigan Medical Center until returning for her Ph.D. in air quality. Hoyer has taught a course in Principles of Community Air Pollution and has published papers on the characterization of fungal allergens, the mutagenic properties of atmospheric particles, and the long-range transport of toxic air pollutants including heavy metals such as mercury.

Michael A. Jayjock, Ph.D., CIH, is a senior research fellow in the Rohm and Haas toxicology laboratory in Spring House, Pennsylvania. He is also manager of risk assessment for the laboratory, considering himself a practitioner of human health risk assessment with specific expertise in exposure assessment. His primary research interest includes development of better estimating methods and most cost-efficient exposure models. He also advocates use of the scientific method in a more open and conscious construction, codification, and application of the "expert judgment" used by occupational hygienists in their evaluation of exposure and risk. He has published a number of papers and has been active on various committees of AIHA, EPA, the Chemical Manufacturers Association, Canadian National Research Council, and National Academy of Science.

Barbara Johnson, Ph.D. Biographical information is unavailable.

David L. Johnson, Ph.D., CIH, PE, is an associate professor of Occupational and Environmental Health at the University of Oklahoma Health Sciences Center in Oklahoma City. He served as an industrial hygienist, researcher, and teacher with the U.S. Army until his retirement from active duty in 1991. His recent publications have appeared in the *AIHA Journal,* the *Journal of Environmental Health, Aerosols in Medicine, Drug Delivery, Inhalation Toxicology, Particulate Science and Technology,* and the *Journal of the American Association of Occupational Health Nurses.* His research interests include inhalation exposures to hazardous materials in indoor and outdoor environments and aerosol applications in inhalation therapy. He is certified in comprehensive practice by the American Board of Industrial Hygiene and is a Registered Professional Engineer.

Reginald C. Jordan, Ph.D., CIH, CQE, is an assistant professor of industrial hygiene at the Center for Occupational Safety and Health Training, Division of Natural Sciences, Mathematics, and Allied Health at St. Augustine's College in Raleigh, North Carolina. He is a member of AIHA, ACGIH, American Society for Quality Control, ASTM, International Environmetrics Society, Air and Waste Management Association (AWMA), and the American Statistical Association. He has served on the editorial review boards of the journals of the AWMA and AIHA, and has served on the AIHA Laboratory Accreditation and Sampling and Analysis committees. He is consultant to several private corporations, colleges and universities, industrial firms, and governmental agencies.

Dennis R. Klonne, Ph.D., DABT, is a senior toxicologist at Rhône-Poulenc in Research Triangle Park, North Carolina. Klonne received both his MS and Ph.D. degrees from the Kettering Laboratory, Department of Environmental Health, University of Cincinnati Medical College. He began his career in 1980, working in toxicology laboratories and the Bushy Run Research Center of Union Carbide; he joined Rhône-Poulenc in 1989 to work in the corporate toxicology group. He has served on the AIHA WEEL Committee since 1987, holding the positions of chair, vice-chair, and secretary, and also serves on Rhône-Poulenc's Permissible Exposure Limits Committee. He is active in many professional and industry organizations and has published approximately 20 articles in peer-reviewed journals.

Stephan A. Konz, Ph.D., retired from Kansas State University in 1996 after teaching ergonomics for 36 years. The fourth edition of his book *Work Design: Industrial Ergonomics* appeared in 1995, while his work *Facility Design: Manufacturing Engineering* had its second edition in 1994. Both have also been published in Spanish. He also has had approximately 220 articles and chapters published. He continues to be active in professional activities at both the national and international level.

Frank J. Labato, CIH, CSP, is the director of the Department of Environmental Health and Safety at the University of Connecticut. Labato was previously employed at the University of Massachusetts-Amherst as an environmental hygienist and at a uranium enrichment facility in Ohio as an industrial hygienist. In addition, he teaches continuing education industrial hygiene courses for the University of North Carolina. He is a member of AIHA, the American Society of Safety Engineers, and the Campus Safety, Health and Environmental Management Association. Labato is a diplomate of the American Academy of Industrial Hygiene.

Jeffrey S. Lee, MPH, Ph.D., CIH, is an associate professor and director of the Industrial Hygiene Program at the University of Utah and a senior partner in the consulting firm IHI Environmental. Lee has previously worked for both NIOSH and OSHA. He has served as chair of ACGIH, founding president of IOHA, and on the board of directors of AIHA. He has authored more than 30 scientific articles and edited two textbooks in the field of occupational health. Lee teaches courses in Advanced Industrial Hygiene, Research Seminar in Industrial Hygiene, International Issues in Occupational Health, and Agricultural Safety and Health.

Susan B. Lee, CIH, is manager of Environmental Health and Safety at Biogen, Inc., in Cambridge, Massachusetts, where she is responsible for occupational health and safety as well as the environmental protection and radioactive materials compliance programs. Beginning her career as a chemist in the State of Delaware Public Health laboratories, she then spent 10 years as an OSHA industrial hygienist, and has been safety officer for the Childrens Hospital, Boston, and director of occupational and environmental health for the New Hampshire Hospital Association's Workers' Compensation Trust. Lee has served many professional associations, including AIHA, ACGIH, Massachusetts Safety Council Board of Directors, Harvard Medical School Subcommittee on Hospital Safety, and was officer and past chair of the AHIA Biosafety Committee. An active author and presenter, she co-wrote the "Biotechnology" chapter in the fourth edition of the *ILO Encyclopaedia of Occupational Health and Safety.* She is ABIH-certified in the comprehensive practice of industrial hygiene.

Peter S.J. Lees, Ph.D., CIH, is associate professor of environmental health sciences and deputy director of the NIOSH Educational Resource Center at the Johns Hopkins University School of Hygiene and Public Health. He is a long-term member of the AIHA Occupational Epidemiology Committee and has served on other committees for ASTM and the National Cancer Institute (NCI). He has been a consultant to a number of federal, state, and local government agencies, private nonprofit organizations, and industrial firms.

Alan J. Leibowitz, CIH, CSP, is director of Environmental Health and Safety for ITT Defense and Electronics, headquartered in McLean, Virginia. As manager for environmental industrial hygiene and safety, he is responsible for worldwide operations, including the oversight of 22 field professionals. He holds an MS in environmental science from Drexel University and has had extensive post-graduate training at the Columbia University School of Public Health. Leibowitz has held several leadership positions, serving as chair of AIHA's Management Committee, the Electronic Industries Association's Environmental Issues Council, and the Aerospace Industries Association's Environmental Committee. Leibowitz recently edited the AIHA book *Industrial Hygiene Auditing — A Manual for Practice.*

Dean R. Lillquist, Ph.D., CIH, is an assistant professor in the Industrial Hygiene Program at the Rocky Mountain Center for Occupational and Environmen-tal Health, where he instructs, researches, and consults in various aspects of risk analysis, including risk communication. Lillquist obtained his Ph.D. in Environmental Health from Colorado State University in Industrial Hygiene/ Industrial Toxicology and his MSPH in Environmental Public Health from the University of Minnesota. He is ABIH-certified in the comprehensive practice of industrial hygiene.

Jeremiah R. Lynch, CIH, CSP, PE, is a part-time consultant to several agencies and associations, following retirement from a long career as an industrial hygienist. From 1976 to 1996, he worked for Exxon Chemical Company as both a senior environmental health scientist and a manager of industrial hygiene. Earlier, he worked in roles of increasing responsibility for the U.S. Air Force (1954–1962) and NIOSH (1962–1976), ultimately directing the Division of Physical Sciences and Engineering. He is author of approximately 30 publications on air sampling, industrial ventilation, exposure analysis, and other industrial hygiene subjects, and is co-author of the *Handbook of Health Hazard Control in the Chemical Process Industry.* A trustee of the AIH Foundation, Lynch is a past president of AIHA and past chair and honorary member of ACGIH.

S. Zack Mansdorf, Ph.D., CIH, CSP, QEP, is vice president and managing director for Liberty International Risk Services in Boston. He is responsible for the technical and business management of an international consulting practice in occupational safety, industrial hygiene, and environmental management. He was an associate professor of environmental engineering (adjunct) at Cleveland State University from 1983 through 1997, where he taught environmental engineering courses in the graduate school. Mansdorf has published or presented more than 100 papers, is the author or co-author of seven books and manuals, and is the editor and principal author of *The Complete Manual of Industrial Safety.* He served for more than a decade as an editorial board member and senior editor for the *AIHA Journal* and is a past chair of the AIHA Ionizing Radiation Committee. An AIHA past president, Mansdorf is also an AIHA Fellow and a Professional Member of the American Society of Safety Engineers.

Daniel S. Markiewicz, CIH, CSP, CHMM, is a senior industrial hygienist for Aeroquip-Vickers, Inc., Maumee, Ohio. Markiewicz has a BS in environmental health from Ferris State College and an MS in occupational health from the Medical College of Ohio. He is a certified industrial hygienist, certified safety professional, and a certified hazardous materials manager.

Margaret E. McCarthy, Ph.D., is the chair of the Physics Department at Springfield Technical Community College in Springfield, Massachusetts, where she has taught physics since 1974, specializing in medical physics. She is adjunct faculty in environmental health at the University of Massachusetts, Amherst, teaching graduate courses in biological effects of ionizing radiation, and adjunct faculty at Tufts Medical School, teaching residents in diagnostic radiology physics at Baystate Medical Center. She is also a consultant with F.X. Massé Associates, Inc. McCarthy previously worked in the Departments of Medical Physics at medical centers in Pittsburgh and Boston. She is active in the Health Physics Society and a plenary member of AAPM, SNM, and SEGH.

Susan M. McDonald, CIH, is the staff industrial hygienist at the University of Massachusetts-Amherst and consults with clients in government, industry, and education. She is also employed as a senior industrial hygienist at Eastern Research Group, Inc., in Arlington, Virginia, providing technical support to OSHA. Before entering the field of industrial hygiene, she worked as a writing instructor, a newspaper reporter, and an editor at Merriam-Webster Dictionaries.

James D. McGlothlin, Ph.D., CPE, is a director-grade occupational safety and health specialist with the Engineering Control Technology Branch (ECTB) of the Division of Physical Sciences and Engineering of NIOSH. McGlothlin started with NIOSH in 1978 and has been with ECTB since 1985. McGlothlin plans, designs, conducts, and reports on engineering control technology projects that focus on ergonomic solutions to decrease injuries in the occupational environment. He has published more than 90 scientific papers and technical reports, and serves as a course director in ergonomics at Northwestern University School of Engineering, Evanston, Illinois, and at the University of Cincinnati School of Medicine. McGlothlin received his Ph.D. in industrial health, specializing in ergonomics, from the University of Michigan.

Ruth K. McIntyre-Birkner is founder of McIntyre Birkner & Associates, Inc., a consulting firm specializing in adult continuing education. Her programs range from comprehensive reviews in industrial hygiene for those preparing to sit for the Core and Comprehensive Practice Examinations, to teaching supervisors how to conduct effective OSHA compliance training programs. Prior to forming McIntyre Birkner & Associates, Inc., she developed and managed adult education departments for the University of Minnesota Medical School (Continuing Medical Education), St. Paul-Ramsey Hospital, Ramsey Clinic, and the Midwest Center for Occupational Health and Safety (a NIOSH-sponsored Educational Resource Center). She is active in both the Orange County and Southern California Sections of the AIHA and is a contributing editor to *Occupational Hazards Magazine*. She is currently an MBA candidate at Pepperdine University.

Gordon C. Miller, CIH, CHCM, CHMM, REM, QEP, is director of Health, Safety & Environment for Rexam. He is involved in managing the safety, health, and environmental affairs programs for Rexam's Custom Division, providing corporate support for the company's facilities around the world. Prior to joining Rexam, Miller practiced safety, health, and environmental affairs in the industry, university, consulting, and insurance settings. He earned an MSPH in environmental sciences and engineering from the University of North Carolina. Miller is a certified industrial hygienist and qualified environmental professional, as well as holding several other certifications in fields related to safety, health, and environmental affairs.

Charles T. Moritz is a senior consultant at Collaboration in Science and Technology Inc. (CSTI). He holds a master's degree in acoustics from the Pennsylvania State University. He is a board-certified member of the Institute of Noise Control Engineering (INCE) and a member of both the Acoustical Society of America (ASA) and the Audio Engineering Society. He regularly presents papers at ASA and INCE meetings and teaches courses in sound measurement, noise control, and music.

John R. Mulhausen, Ph.D., CIH, is a senior industrial hygiene specialist in the Medical Department of 3M Company. He has experience advising manufacturing and research facilities throughout the world on developing effective industrial hygiene programs. Mulhausen is also an adjunct assistant professor in the Division of Environmental and Occupational Health at the University of Minnesota. He has presented numerous courses and lectures on industrial hygiene, program management, and exposure assessment. He is a past chair of the AIHA Exposure Assessment Strategies Committee and is one of the editors and authors of the committee's monograph "A Strategy for Assessing and Managing Occupational Exposures." He received his Ph.D. in Environmental Health from the University of Minnesota and is certified in the comprehensive practice of industrial hygiene by the ABIH.

Kenneth A. Mundt, Ph.D., is an associate professor and co-director of the Occupational Epidemiology Unit in the Department of Biostatistics and Epidemiology, University of Massachusetts-Amherst. He oversees several industry-based epidemiological surveillance programs and teaches graduate courses in epidemiological research methods, as well as occupational epidemiology. He is active in several national organizations such as the Society for Epidemiologic Research and ACGIH; he also serves on the Policy Committee of the American College of Epidemiology. Mundt's research interests currently include human health effects of occupational exposures to vinyl chloride monomer, chromium and other industrial chemicals; occupational and environmental risk factors for brain and hematopoietic cancers; and the development of methods to be used in active epidemiological surveillance systems.

William E. Murray is the assistant chief of the Health-Related Energy Research Branch at NIOSH. He holds an MS from the University of Rochester and has worked for the U.S. Public Health Service for more than 28 years. Prior to assuming his present position, he served as chief of NIOSH's Radiation Section from 1976 to 1991. Murray represents NIOSH on several national and international committees on health effects and has published extensively in scientific journals and technical reports in the area of nonionizing radiation protection.

Deborah Imel Nelson, Ph.D., CIH, is an associate professor in the School of Civil Engineering and Environmental Science at the University of Oklahoma, where she has taught environmental science, risk assessment, and industrial hygiene since 1989. She holds an MS in environmental science from OU, and a Ph.D. in environmental health from the OU Health Sciences Center. Her previous work includes four years as an OSHA compliance officer and 10 years as an industrial hygiene consultant. Nelson chaired the Risk Assessment Task Group, which produced the AIHA white paper/position paper in 1994. In 1995, she and Neil Hawkins co-founded the Risk Assessment Committee of AIHA, which she served as chair in 1996–1997.

Thomas J. Nelson, CIH, is president of NIHS Inc., an occupational hygiene consulting firm. He began his career as a chemist in the Automotive Products Department of the DuPont Company, where he became the occupational health coordinator for the department and finally a corporate industrial hygienist. During his time at DuPont he became involved in research on workplace performance of respirators and fit testing. He was the chair of the ANSI Z88.2–1992 standard "Practices for Respiratory Protection," and is a past chair of the AIHA Respiratory Protection Committee. He has written several articles on respirator performance and exposure assessment.

Harris Pastides, MPH, Ph.D., professor of Epidemiology and chair of the Department of Biostatistics and Epidemiology at the University of Massachusetts School of Public Health and Health Sciences. He teaches courses at the university and in other venues in epidemiological methods, occupational surveillance, and related topics. His research interests lie in occupational and environmental epidemiology; he has performed research studies on occupational exposures as they relate to the risk of cancer, adverse reproductive outcomes, and musculoskeletal injury. He consults with many companies and labor groups on health issues requiring epidemiological insight. Pastides is a frequent advisor to the World Health Organization and has served on committees of the National Academy of Sciences, the Society for Epidemiologic Research, and the Partnership in Maritime Medicine.

Robert M. Patterson, SM, Sc.D., CIH, is professor of Environmental Health Engineering and director of Graduate Studies in the College of Engineering at Temple University. He earned degrees (SM and Sc.D.) from Harvard University and worked for the Electric Power Research Institute prior to joining Temple's faculty. Patterson is chair of the Physical Agents TLV Committee of ACGIH and a former chair of the AIHA Nonionizing Radiation Committee. He is author or co-author of more than 75 technical publications.

J. Thomas Pierce, Ph.D., is head of the Field Service Division of the Center for Environmental and Occupational Health, University of Kansas Medical Center. He is responsible for maintaining an industrial hygiene service that supports the Medical Center's clinical interests, generating laboratory service for occupational and pulmonary medicine, and providing consultations to Medical Center staff, physicians, and other health care providers throughout the Midwest. Pierce received a doctorate in environmental health from the University of Oklahoma. Currently, his areas of specialization include dermal absorption, environmental analytical chemistry, biological monitoring, and medical surveillance. He is ABIH-certified, as well as being certified by the American Board of Toxicology. He holds a number of professional and honorary memberships, including Phi Kappa Phi, Sigma Xi, and the American Academy of Clinical Toxicology.

Tomas D. Polton is the assistant director for Corporate Health and Safety at Pfizer Inc. Polton has a master's degree in industrial hygiene and safety from the Department of Environmental Health Sciences at the Harvard School of Public Health. Polton has been an active member of the AIHA Occupational Medicine Committee, twice serving as its chair. At the 1995 AIHCE, Polton arranged and presented at a roundtable discussion titled "Medical Surveillance, Screening, and the Prevention of Occupational Disease," which included distinguished representatives from government, academia, industry, labor, and professional associations.

William Popendorf, Ph.D., is a professor of industrial hygiene at Utah State University in the Department of Biology. Originally a mechanical engineer by training, he earned an MPH and Ph.D. in environmental health at the University of California-Berkeley. While a new faculty member at Berkeley, his research focused on the acute health hazards of organophosphate pesticide residues to fruit harvesters. In 1983 he joined the Institute of Agricultural Medicine at the University of Iowa, broadening his research to include both dermal and airborne exposures to inorganic and organic dusts in the foundry, metal working, and agricultural industries.

Shane Stephen Que Hee, Ph.D., M.Sc., L.T.C.L., L.Mus.A., A.Mus. A., F.A.I.C., is a professor in the Department of Environmental Health Sciences of the UCLA School of Public Health and a faculty member of the UCLA Center for Occupational and Environmental Health in Los Angeles. Que Hee teaches courses in biological monitoring, identification and analysis of gases and vapors, identification and analysis of hazardous waste, instrumental methods of analysis, and environmental chemistry, plus laboratories in identification and analysis of gases and vapors, and instrumental methods in environmental health sciences, and field evaluations in environmental hygiene. He has published more than 100 peer-reviewed articles, written two books (including the first textbook on biological monitoring), and has contributed to six U.S. EPA criteria documents. Que Hee currently serves on the NIH study section for epidemiology and disease control and reviews toxicological profiles for ATSDR.

W. Scott Railton, Esq., JD, is the management co-chair of the Occupational Safety and Health Law Committee of the American Bar Association. The author of an OSHA compliance manual, he also writes a legal column concerning OSHA matters for *The Synergist* and has published numerous articles, including a chapter in the American Bar Association's treatise on OSH Law. He helped form the OSHA division in the office of the Solicitor, U.S. Department of Labor, in 1971, and was a chief counsel and acting general counsel of the Occupational Safety and Health Review Commission until 1977. He has practiced OSHA law continuously since then with the law firm of Reed Smith Shaw & McClay, representing employers and industry associations.

Jerry D. Ramsey, Ph.D., PE, is a professor of industrial engineering at Texas Tech University. He served as associate vice president for operations in the Office of the Executive Vice President and Provost for the university from 1977 to 1995. In addition to appointments at OSHA and the U.S.

Consumer Product Safety Commission, he has served as consultant for NIOSH, the U.S. Bureau of International Labor Affairs, and for other industrial, service, consulting, and legal firms. Ramsey serves on the U.S. Technical Advisory Group to International Standards Organization Technical Committee 159, Environmental Ergonomics, and also on several society committees for American Society of Safety Engineers. He is a contributing author to 10 books and more than 90 professional publications.

Susan D. Rogers Ripple, CIH, is manager of the Exposure Criteria Program in the Global Industrial Hygiene Expertise Center for the Dow Chemical Company in Midland, Michigan. Ripple is also the industrial hygienist for more than 1250 researchers in the Midland site research and development function. She acts as the technical focal point for the Industrial Hygiene Emergency Response Team and serves on the AIHA Emergency Response Planning (ERP) Committee. In her risk assessment role, Ripple coordinates the setting of Dow's internal occupational exposure limits and serves on the AIHA Workplace Environmental Exposure Levels (WEEL) Committee. She has also been the industrial hygienist for numerous manufacturing and research facilities at Dow's Texas Operations facility, providing industrial hygiene product stewardship support to Dow's urethane and epoxy resin customers.

Donald Robinson, Ph.D., PE, CSP, is director of Environmental Health and Safety and a faculty member at the University of Massachusetts-Amherst campus, where he teaches industrial safety engineering for both the Mechanical and Industrial Engineering Departments and the Environmental Health Science Department. Before coming to the university, Robinson held various engineering and manufacturing positions at the General Electric Company and was manager of occupational safety for GE in Lynn, Massachusetts. He is also a safety management and engineering consultant for industry, universities, and the legal community. He has published several articles on a broad range of environmental health and safety topics.

R. James Rockwell, Jr., is founder of Rockwell Laser Industries (originally known as Rockwell Associates, Inc.), a consulting firm. He has been active in the field of lasers since 1963, beginning with a 14-year tenure as directing physicist at the University of Cincinnati Medical Center laser research lab. His interests include laser medical system design and safety bioeffects criteria. Rockwell has authored several textbooks and other publications, as well as writing computer software for laser hazard analysis and launching the Lasernet Internet page. He has served since 1968 as chair of the Control Measures Subcommittee, Z136 Committee of the American National Standards Institute. Rockwell has been the Laser Institute of America's president and safety chair and was honored by the LIA in 1984 with the President's Award for more than two decades of service. He was also a keynote speaker of LIA's 1997 International Laser Safety Conference.

Vernon E. Rose, Dr PH, CIH, CSP, PE, is a professor of public health and director of the Deep South Educational Resource Center in the School of Public Health at the University of Alabama-Birmingham. On joining UAB in 1978, he established a graduate educational program in industrial hygiene. Previously, he was an industrial hygienist in NIOSH and for the U.S. Air Force. He also provides consultation to corporations and serves as an expert witness. Rose is a past director and president of the AIHA, has served as a director and chair of ABIH, and as secretary-treasurer of ACGIH. In 1990 he was honored as the ACGIH Meritorious Achievement Awardee for his outstanding long-term contributions to the progress of industrial hygiene. He is the author or co-author of more than 40 articles and book chapters.

Bernard E. Saltzman, Ph.D., CIH, PE, is emeritus professor of environmental health in the Department of Environmental Health, College of Medicine, University of Cincinnati. He was director of the department's training programs in industrial hygiene and air pollution and of its analytical section for many years. He published more than 120 publications in industrial hygiene and air pollution chemistry. Among his many honors are AIHA's Borden Award and honorary membership; the Association of Official Analytical Chemist's Harvey W. Wiley Award; and the Cincinnati Section of the American Chemical Society's Annual Cincinnati Chemist Award. He previously retired from the U.S. Public Health Service as a sanitary engineer director after working in their early industrial hygiene and air pollution programs. He has served as a consultant for NIOSH, OSHA, EPA, National Academy of Sciences–National Research Council, and the World Health Organization.

Margaret C. Samways, LCST, is president of Margaret C. Samways and Associates, in Pittsburgh, Pennsylvania. With 25 years of experience in designing and developing occupational health and safety training, she is presently developing the Model Training Program for Hazard Communication for OSHA and providing instructional methodology assistance for other government and industrial clients. She is active on and past chair of the Communications and Training Methods Committee of AIHA. She served for eight years as an adjunct faculty member of the University of Pittsburgh Graduate School of Education and currently lectures in the Graduate School of Public Health. She has published extensively on the topic of training and has authored chapters on training and communications in *The Handbook of Occupational Safety and Health, Material Safety Data Sheets: The Writers' Desk Reference,* and the *Hazard Communication Compliance Manual.*

A. David Scarchilli, PE, D.E.E., is a vice president at OccuHealth, Inc. (OHI). Scarchilli has more than 20 years of environmental engineering experience in both the private and public sectors. His private sector experience includes assignments at General Electric and Colgate Palmolive/Kendall; his public sector experience includes a stint with the New York State Department of Environmental Conservation, Division of Air Resources. For the past 10 years, he has focused his career in environmental engineering consulting. At OHI, Scarchilli's practice specializes in assisting industry bring their business practices into compliance with environmental regulations. His fields of expertise include environmental auditing and air quality management. He is a registered professional engineer and a diplomate in the American Academy of Environmental Engineers with a specialty certification in air pollution control.

Michelle M. Schaper, Ph.D., is an associate professor of Environmental and Occupational Health at the University of Pittsburgh's Graduate School of Public Health. Her area of specialization is inhalation toxicology, with an emphasis on the development of animal models to evaluate the respiratory effects of airborne chemicals. Schaper teaches toxicology at the university, as well as lecturing at other colleges in the Pittsburgh area. She is also a co-instructor in the professional development course "Introduction to Occupational Toxicology," which is offered each year at the American Industrial Hygiene Conference & Exposition. Schaper has been an author on approximately 45 abstracts (presented at technical meetings) and 40 peer-reviewed papers. She is currently the vice chair of the AIHA Toxicology Committee. She is also a member of the Society of Toxicology and ACGIH.

R. Craig Schroll, CSP, CET, CHCM, is founder and president of FIRECON, a consulting firm that helps clients prevent, plan for, and control emergencies. He has a wide range of experience in improving the effectiveness of clients' loss prevention and control results, having been actively involved in loss prevention and control since 1972. He has worked with groups from many industries and public safety agencies. Craig is a member of numerous safety and related professional organizations and his articles have been published in several trade and professional journals. Schroll is also the author of *The Industrial Fire Protection Handbook.*

Jennifer C. Silk is director of the Office of Technical Programs and Coordination Activities at OSHA, where she was the leader of the team responsible for development of the Hazard Communication Standard (HCS). She has been involved in international harmonization of hazard communication requirements for some years, and chairs the Interorganization Programme for the Sound Management of Chemicals (IOMC) Coordinating Group for the Harmonization of Chemical Classification Systems. In this capacity, she is helping to manage the process of international harmonization currently taking place in various international organizations. She is also chair of the board of directors of the Society for Chemical Hazard Communication and past chair of AIHA's Product Health and Safety and Communication and Training Methods Committees. Silk has lectured and written frequently on hazard communication issues.

Robert D. Soule, Ed.D., CIH, CSP, PE, is a full professor of occupational health and chair of the Safety Sciences Department at Indiana University of Pennsylvania (IUP) in Indiana, Pennsylvania. He teaches courses in industrial hygiene, fire protection and management. His experience includes academia, private industry (environmental health engineer and senior industrial hygienist with the Dow Chemical Company), consultation (vice president for industrial hygiene services for Clayton Environmental Consultants), and government (Pennsylvania/OSHA Consultation Program). Soule works as a consultant for a variety of clients, has published more than 30 articles in varied journals, and has contributed to several current reference books on occupational safety and health.

James H. Stewart, Ph.D., CIH, is president and CEO of Datachem Software, Inc., a company he co-founded with his wife Catherine Stewart. Datachem Software specializes in producing computer-based training programs for environmental health and safety professionals. Stewart is also an instructor at the Harvard University School of Public Health Industrial Hygiene Program, where he teaches graduate courses in industrial hygiene and management and directs several advanced professional review courses. During his more than 20 years in the profession, he has served as a compliance officer with OSHA and in management positions for several multinational companies. He is the author of numerous professional articles in the area of industrial hygiene, risk assessment of carcinogens, mathematical modeling, and management of industrial hygiene programs. During the past five years, Stewart has served on two National Academy of Sciences committees to provide industrial hygiene expertise. Stewart is certified in comprehensive practice of industrial hygiene by ABIH.

Frank J. Sullivan is a graduate student completing final course work at the Rocky Mountain Center for Occupational and Environmental Health. Current interests include risk management and perception in regard to environmental issues. His past professional experience includes six years of consulting as director of Environmental, Health and Safety Service, and technical reviewer for refining, U.S. Department of Defense, mining, milling, and fabrication industries. Sullivan received a NIOSH fellowship from the Rocky Mountain Center, majoring in industrial hygiene.

David L. Swift, Ph.D., recently passed away after a productive career in industrial hygiene. From 1987 to 1993, Swift was the director of the Industrial Hygiene and Safety Program at the NIOSH Education Research Center at the Johns Hopkins University School of Hygiene and Public Health. He was also a professor in the Department of Environmental Health Sciences, Division of Environmental Health Engineering in the Johns Hopkins University School of Hygiene and Public Health, teaching courses on airborne particles, environmental health, and other aerosol related subjects. He was the primary or secondary author of approximately 45 papers or articles published in both national and international publications. He also wrote, co-authored, or edited more than 27 book chapters in the area of industrial hygiene. His work influenced the development of U.S. EPA policies on air pollution regulations, and his professional service included chairing AIHA committees.

Fay Morgen Thompson, Ph.D., CIH, is director of the Department of Environmental Health and Safety and Assistant Professor of Environmental and Occupational Health at the University of Minnesota. Thompson received her Ph.D. in organic chemistry in 1970 and has worked in the environmental health field at the university since that time. Her major responsibility is managing a staff of more than 60 people who provide environmental protection, occupational health and safety, and radiation safety services to the university's 68,000 students and 18,000 faculty and staff statewide. The university's 2500 laboratories occupy a significant portion of the department's effort. Thompson also teaches a graduate level course in hazardous waste management.

Lori A. Todd, Ph.D., CIH, is an associate professor in the Department of Environmental Sciences and Engineering at the University of North Carolina. Todd is investigating new methods for measuring air pollutants and evaluating human exposures. She has developed a new method for visualizing the concentration and movement of chemicals in air that combines computed tomography with optical remote sensing. In 1990, Todd won the Michigan Industrial Hygiene Association Award for her publication on this research. In 1994, President Clinton awarded Todd the Presidential Faculty Fellow Award for this research and for her teaching. In 1995, Todd was a finalist in *Discover Magazine*'s awards for technological innovation. She is an active member of several organizations, including AIHA, AWMA, and ACGIH and is a peer reviewer for many journals, as well as a consultant to several corporations.

Peter F. Waldron, MPH, IHIT, is a senior research assistant in the field of occupational and environmental hygiene at the University of Iowa. Waldron earned an MS in environmental science from the University of Massachusetts in 1984 and worked as both a laboratory manager and for the asbestos abatement industry. During this period he began studying industrial hygiene and was awarded an MPH in 1991. He has worked in both a research capacity as well as an industrial hygiene consultant for the University of Iowa since 1991. Waldron also serves as a member of the university's Confined Space Rescue and Entry Team and as a science/resource adviser to the Johnson County Hazardous Materials Response Team.

Stefan Wawzyniecki, Jr., CIH, is manager of chemical health and safety at the University of Connecticut and has been working in the field of environmental health and safety since 1974. As the chemical hygiene officer at the University of Connecticut, he has participated in offering a HAZWOPER training course. Wawzyniecki is a member of the American Chemical Society, as well as their Division of Chemical Health and Safety, and has presented papers at the society's national meetings. He has consulted for major corporations and educational institutions and was one of the participants in the federal OSHA hearings on the proposed standard on indoor air quality.

Lutz Weber, Ph.D., DABT, is a research associate professor in the Department of Pathology and Laboratory Medicine at the University of Kansas Medical Center. He is also affiliated with GSF–National Center for Environment and Health, a government research agency of the Federal Republic of Germany. Weber's research interest focuses on 2,3,7,8-tetrachlorodibenzo-p-dioxin and related environmental pollutants (specifically, their dermal penetration and developmental toxicity). Between 1974 and 1985 he was a staff scientist as GSF–National Center for Environment and Health in the field of developmental toxicity. Since 1987 he has been affiliated with the University of Kansas Medical Center. He is a member of the Society of Toxicology and is certified by the American Board of Toxicology.

Reviewers

A

Dennis Anderson, Ph.D., CIH, Dow Corning Corporation

B

James D. Banach, Quest Technologies

Edward T. Bartosh, Jr., CIH, Health Risk Associates

William E. Bechtold, Ph.D., Inhalation Toxicology Research Institute

Elliott H. Berger, Aearo Company

Jeffrey S. Birkner, CIH, Moldex-Metric

Roger Briggs, MPH, CIH, DOE

Lisa M. Brosseau, Ph.D., CIH, University of Minnesota

D. Jeff Burton, CIH, CSP, PE, IVE Inc.

C

Mark S. Cameron, IHIT, CA-EPA

Terry F. Carraway, Jr., CIH, CSP, STI, Inc.

Finis L. Cavender, Ph.D., CIH, DABT, Information Ventures, Inc.

Kenneth F. Cherry, CIH, PE

David J. Cochran, University of Nebraska

Michael Collegan, NIOSH

William C. Collier, CIH, Metcalf & Eddy

J. Lindsay Cook, CIH, CSP, Chemical Industry Institute of Technology

Michael Copeland, Ph.D. (candidate),United States Department of the Navy

John P. Curran, CIH, North Carolina Department of Environment, Health and Natural Resources

David M. Cyr, CIH, KRMS-NATLSCO

D

Adolf R. Dassler, Ph.D., MSC, United States Department of the Navy (Retired)

Gary P. DeVany, DeVany Industrial Consultants

Mary C. DeVany, CSP, CHMM, DeVany Industrial Consultants

Deborah F. Dietrich, CIH, SKC Inc.

David Dunham, Ph.D., CSP, PE, CIH, Everest Consulting Associates

Patrick Dunn, CIH, American Analytical Lab

Michael R. Dupuis, CIH, Kimberly-Clark

Joy Erdman, CIH, CSP, United States Department of the Navy

E

William M. Ewing, CIH, Compass Environmental, Inc.

F

Alice C. Farrar, CIH, Clayton Environmental Consultants

Matthew Finucane, CIH, University of Pennsylvania

Lee T. Freeland, CIH

G

Sean Gallagher, NIOSH

Dennis George, Ph.D., CIH, Western Kentucky University

Richard C. Gerlach, Ph.D., CIH, CSP, Halliburton NUS Corporation

Lawrence M. Gibbs, CIH, Stanford University

Sheree L. Gibson, PE, CPE

Manuel R. Gomez, DrPH, CIH

Deborah Rice Gordon, Eastman Kodak

Ross H. Grayson, University of California

Bernard J. Greenspan, Ph.D., Dura Pharmaceuticals

Daniel Groseclose, CIH, CSP, OEHS

Fred I. Grunder, CIH, American Industrial Hygiene Association

H

Rodney G. Handy, Ph.D., Western Kentucky University

Lee D. Hager, James Anderson & Associates, Inc.

Jonathan T. Haney, CIH, New York Departement of Labor

Doan Hanson, Ph.D., MPH, CIH, DOE

Michael K. Harris, Ph.D., IHIT, Hamlin & Harris

Ezekiel Haslam, Jr., CIH, Duke Power Company

Neil C. Hawkins, Ph.D., CIH, Dow Chemical Company

Barbara A. Hebert, CIH, Kansas City Power & Light

Stephen W. Hemperly, CIH, CSP, IBM Almaden Research Center

Robert F. Herrick, SCD, CIH, NIOSH

Paul Hewett, Ph.D., CIH, NIOSH

Kimberly S. Higby, CIH, General Electric

Clifford C. Houk, Ph.D., CIH, Ohio University

Joseph Hurrell, Ph.D., NIOSH

George M. Hutcheson, CIH, John Deere Waterloo Works

I

Jere Ingram, CIH, Clorox Services Company

J

Judy L. Jarrell, Ed.D., University of Cincinnati

David L. Johnson, Ph.D., CIH, University of Oklahoma

Norris D. Johnson, CIH, Westinghouse

LCDR Warren W. Jederberg, MSC, United States Department of the Navy

Ross E. Jones, DABT, Texas Instruments

Reginald C. Jordan, Ph.D., CIH, ETC

K

Chris Keil, Ph.D., Bowling Green State University

L

Joseph LaDou, Ph.D., University of California - San Francisco

Alan J. Leibowitz, CIH, CSP, ITT Defense and Electronics

Lester Levin, CIH, Temple University

Steven P. Levine, Ph.D., CIH, University of Michigan

Virginia C. Lew, CIH, MPH, Lawrence Livermore National Lab

CDR William E. Luttrell, Ph.D., Navy Environmental Health Center

Jeremiah R. Lynch, CIH, CSP, PE

M

Daniel Markiewicz, CIH, Trinova

Mark McGowan, CIH, CSP, Malcolm Pirnie

Members of the AIHA Management Committee

Robert C. Moats, CIH, CSP, PE

Lee E. Monteith, CIH, University of Washington

Warren R. Myers, Ph.D., CIH, West Virginia University

N

Barbara Fox Nellis, Johnson & Johnson

Shirley A. Ness, CIH, CSP, Amoco

O

George R. Osborne, CIH, CSP, Lucent Technologies

P

Eileen Pearlman, CIH, Exxon Biomedical Sciences

Lillian Piantanida, CIH, Piantanida & Walker

John C. Pierdomenico, CIH, E I DuPont De Nemours

William J. Popendorf, Ph.D., CIH, Utah State University

Robert Pristas, CIH, Exxon Biomedical Sciences, Inc.

Vern Putz-Anderson, Ph.D., NIOSH

R

Tena Rallis M.D, University of Utah

Patricia H. Reinke, University of Minnesota

Frank M. Renshaw, Ph.D., CIH, Rohm & Haas

Cindy Lynn Richard, CIH, National Medical Advisory Service

David T. Ridyard, CIH, CSP, CPE, Applied Ergonomics Technologies

James Dylan Romo, CIH, Arco Chemical

Michael E. Russell, CIH, Rust Environmental and Infrastructure

Timothy J. Ryan, CIH, University of Houston

S

Edie Scala-Hampson, CIH, STS Consultants, Ltd.

C. Nelson Schlatter, Ansell Edmont Industrial Inc.

Barry J. Schlegel, CIH, Environmental and Occupational Health Sciences Institute

Henry J. C. Schwartz, MD, United States Department of the Navy

Mary Ann Simmons, Navy Environmental Health Center

Ray Sinclair, NIOSH

Jaswant Singh, Ph.D., CIH, Clayton Environmental Consultants

Robert D. Soule, Ph.D., CIH, CSP, PE, Indiana University of Pennsylvania

Valerie I. Steinberg, Ph.D., University of Massachussettes

David A. Sterling, Ph.D., CIH, ROH, Saint Louis University

Steven W. Strandberg, JD, CIH

Jeffrey O. Stull, International Personal Protection Inc.

Ram S. Suga, CIH, CHMM, Bechtel Environmental Inc.

Frank G. Szopinski, CIH, CSP, WMX Technologies

T

Douglas Taylor, Ph.D., United States Department of the Navy

W. David Taylor, CIH, CSP, R J Reynolds

Carla Treadwell, CIH, United States Department of the Navy

V

Susan Marie Viet, CIH

W

Marjorie Edmonds Wallace, NIOSH

Thurman B. Wenzl, SCD, CIH, NIOSH

Y

Tracy D. Yerian, Sound Analytical Service

Z

John N. Zey, CIH, Central Missouri State University

Neil J. Zimmerman, Ph.D., PE, CIH, Purdue University

Michael H. Ziskin, CHCM, CHMM, Field Safety Corporation

Glossary

Terms included in this glossary are taken principally from the text in this book, and from the following AIHA publications: *Direct-Reading Colorimetric Indicator Tubes Manual,* 2nd Edition (edited by Janet B. Perper and Barbara J. Dawson; published 1993); *The Emergency Response Planning Guidelines and Workplace Environmental Exposure Level Guides Handbook* (updated and published annually); *Extremely Low Frequency (ELF) Electric and Magnetic Fields* (by R. Timothy Hitchcock, Sheri McMahan, and Gordon C. Miller; published 1995); and *Particle Sampling Using Cascade Impactors* (AIHA Aerosol Technology Committee; edited by John Y. Young; published 1995).

The following books were also used in compiling this glossary: *Calculation Methods for Industrial Hygiene* (by Salvatore R. DiNardi; published 1995 by Van Nostrand Reinhold); *Illustrated Dictionary of Environmental Health and Occupational Safety* (by Herman Koren; published 1996 by CRC Press-Lewis Publishers); *McGraw-Hill Dictionary of Scientific and Technical Terms,* 4th Edition (Sybil P. Parker, editor in chief; published 1989 by McGraw-Hill, Inc.); *Terminology of Heating, Ventilation, Air Conditioning, and Refrigeration* (published 1986 by ASHRAE); *Webster's Medical Desk Dictionary* (published 1986 by Merriam-Webster Inc.); and *The Work Environment, Volume Three — Indoor Health Hazards* (edited by Doan J. Hansen; published 1994 by CRC Press-Lewis Publishers).

A

A/V: Standard abbreviation for antivibration.

AAIH: See *American Academy of Industrial Hygiene.*

AAOM: See *American Academy of Occupational Medicine.*

ABC: See *activity-based costing.*

abduction: Movement of an extremity or other body part away from the axis of the body.

ABIH: See *American Board of Industrial Hygiene.*

absolute gain: A ratio of the actual transmitted power density in the main beam to the power density transmitted from an isotropic radiator.

absolute pressure: It is *never* less than zero. The absolute pressure in a system can approach zero, but for the pressure to be negative it would first have to pass through zero which is like absolute temperature on a thermodynamic scale. It is possible to approach but not reach absolute zero pressure.

absolute temperature: Temperature as measured above absolute zero. (Also known as "thermodynamic temperature.")

absolute zero: The temperature at which all molecular motion stops. Either -273.15°C (0 K) in the SI system of units, or -459.67°F (0°R) in the English system. (See also *Kelvin temperature, Rankine temperature.*)

absorbed dose: 1. The mass or moles of exposing compound that actually enters the bloodstream through any external routes of exposure; the absolute bioavailability. 2. Amount of a substance penetrating the exchange boundaries of an organism after contact. Calculated from intake and absorption efficiency and expressed as mg/kg/day.

absorbing medium: A collection medium that allows for the penetration of airborne chemicals into the material where the chemicals will either physically dissolve or chemically react with the collection medium.

absorption: To take in a substance across the exchange boundaries of an organism (skin, lungs, or gastrointestinal tract) and ultimately into body fluids and tissues. (See also *uptake.*)

acceleration: Any gradual speeding up of a process. The time rate of change of velocity.

acceleration due to gravity: The rate of increase in velocity of a body falling freely in a vacuum; value varies with latitude and elevation. The International Standard at sea level and 45 latitude is 9.806 65 meters per second squared (m/sec^2) or 31.174 feet per second squared (ft/sec^2).

accelerometer: An instrument that measures acceleration or gravitational force capable of imparting acceleration.

acceptable air quality: Air in which there are no known contaminants at harmful levels as determined by occupational hygienists, and air with which 80%–90% of the people do not express dissatisfaction based on comfort criteria: temperature, relative humidity, nonhazardous odors, and air movement (draftiness/stuffiness).

acceptable risk: Risk level deemed acceptable by an individual, organization, or society as a whole.

acceptance sampling: The procedures by which decisions to accept or reject a sampled lot or population are made based on the results of a sample inspection. In air pollution work, acceptance sampling could be used when checking a sample of filters for certain measurable characteristics such as pH, tensile strength, or collection efficiency to determine acceptance or rejection of a shipment of filters, or when checking the chemical content of a sample of vials of standard solutions from a lot of vials to be used in an interlaboratory test.

acceptance testing: A systematic procedure to test as-received materials before use to determine whether they are contaminated.

acclimatization: Adaptation of a species or population to a changed environment over several generations. (Also known as "acclimation.")

accuracy: 1. The degree of agreement of a measurement, X, with an accepted reference or true value, T, usually expressed as the difference between the two values, X – T, or the difference as a percentage of the reference or true value, 100(X – T)/T, and sometimes expressed as a ratio, X/T. 2. Measure of the correctness of data, as given by differences between the measured value and the true or specified value. Ideal accuracy is zero difference between measured and true value. 3. Conformity of an indicated value to an accepted standard value, or true value. Quantitatively, expressed as an error or an uncertainty. The property is the joint effect of method, observation, apparatus, and environment. Accuracy is impaired by mistakes, by systematic bias (e.g., abnormal ambient temperature), or by random errors (imprecision). 4. The degree of freedom from error (i.e., the degree of conformity to truth or to a rule). Accuracy is contrasted with precision (e.g., four-place numbers are less precise than six-place numbers; nevertheless, a properly computed four-place number might be more accurate than an improperly computed six-place number). (See also *precision, repeatability*.)

acfm: See *actual cubic feet per minute*.

ACGIH: See *American Conference of Governmental Industrial Hygienists*.

acid: A compound that reacts with an alkali to form a salt and water. It turns litmus paper red and has pH values of 0 to 6.

ACIL: See *American Council of Independent Laboratories*.

acne: A pleomorphic, inflammatory skin disease involving sebaceous follicles of the face, back, and chest and characterized by blackheads, whiteheads, papules, pustules, and nodules.

acoustic trauma: The temporary or permanent hearing loss due to a sudden intense acoustic event such as an explosion.

acoustical absorption: Material added to a workspace environment to reduce noise above 300 Hz.

acrid: Sharp, bitter, stinging, or irritating to the taste or smell.

ACS: See *American Chemical Society*.

action level: 1. In general, the level of a pollutant at which specified actions or counter measures are to be taken. 2. A term used by OSHA in several chemical standards. A level of exposure at which the employer must initiate some actions such as medical monitoring and training. The action level is generally set at 50% of the PEL. 3. This is the concentration or level of an agent at which it is deemed that some specific action should be taken. The action can range from more closely monitoring the exposure atmosphere to making engineering adjustments. In general practice the action level is usually set at one-half of the ACGIH TLV®.

action potential (AP): A transient change in electric potential at the surface of a nerve or muscle cell occurring at the moment of excitation.

activated carbon: Activated carbon is commonly used in gas adsorption. (Also known as "activated charcoal.")

active sampling: The collection of airborne contaminants by means of a forced movement of air by a sampling pump through an appropriate collection device.

activity-based costing: A method of identifying and allocating costs based on measurable activities. There are two elements of ABC: activity analysis and cost object analysis.

actual cubic feet per minute (acfm): Actual cubic feet per minute of gas flowing at existing temperatures and pressure. (See also *scfm*.)

acuity: Pertains to the sensitivity of a bodily organ to perform its function.

acute: Severe, often dangerous effect used to denote an exposure to high concentrations of a contaminant for short duration.

acute effect: An adverse effect (usually) arising from a short exposure (minutes to hours) to a chemical.

acute exposure: Large dose/short time.

acute intake: Intake averaged over a period of less than two weeks.

acute mountain sickness (AMS): Refers to high altitude pulmonary edema and/or high altitude cerebral edema. Symptoms (severe breathlessness and/or chest pain) can rapidly become life-threatening if not treated by an immediate descent to a lower altitude. AMS is subdivided into benign and malignant. (See also *high altitude pulmonary edema, high altitude cerebral edema*.)

acute toxicity: The adverse effects resulting from a single dose or single exposure to a substance. Ordinarily refers to effects occurring within a short time following administration. Terminology and units used for different descriptive categories of toxicity vary. Examples of toxicity classifications as defined by different organizations are:

- LC_{50} inhalation (ppm): extremely toxic = <10; highly toxic = 10–100; moderately toxic = 100–1000; slightly toxic = 1000–10,000; practically nontoxic = 10,000–100,000; relatively harmless = >100,000. [**Hodge, H.C., and J.H. Sterner:** Tabulation of Toxicity Classes. *Am. Ind. Hyg. Quarterly 10*:93 (1949).]
- LC_{50} inhalation (ppm): highly toxic = <200; toxic = 200–2000. [29 CFR 1910.1200, Appendix A — Health Hazard Definitions.]
- LC_{50} inhalation (ppm): extremely toxic = <10; highly toxic = 10–100; moderately toxic = 100–1000; slightly toxic = 1000–5000; practically nontoxic = >5000. [Adapted from E.I. du Pont de Nemours & Co., Haskell Laboratory.]

acute toxicity study: Toxicity study durations of less than five days exposure.

adduct: The product of a reaction between an endogenous macromolecule and an exposing chemical or its metabolite.

adenoma: A benign tumor with glandular structure or of a glandular organ.

administered dose: Mass of a substance given to an organism and in contact with an exchange boundary, expressed as mg/kg/day.

administrative controls: The use of management involvement, training of employees, rotation of employees, air sampling, biological sampling, and medical surveillance to protect individuals.

administrative solution: A managerial rather than an engineering solution to reduce work-related stress.

adsorbent: A material that causes molecules of gases, liquids, or solids to adhere to its internal surfaces without physical or chemical changes to the adsorbent material. Solid materials, such as silica gel and activated alumina, have this property.

adsorbing medium: A collection medium that traps airborne chemicals onto the surface of the material.

adsorption: 1. Surface adherence of a material, which extracts one or more substances present in an atmosphere or mixture of gases and liquids, unaccompanied by physical or chemical change. Condensation of gases, liquids, or dissolved substances on the surfaces of solids. 2. The attachment of molecules or atoms to the surface of another substance; a process whereby one or more components of an interfacial layer between two bulk phases are either enriched or depleted.

aerobic: In presence of air.

aerodynamic diameter: Diameter of a unit-density sphere having the same gravitational settling velocity (terminal velocity) as the particle in question. (See also *cut-off particle diameter, mass median aerodynamic diameter*.)

aerodynamic equivalent diameter: The diameter of a unit density sphere that would exhibit the same settling velocity as the particle in question.

aerosol: Solid or liquid particles of microscopic size dispersed in a gaseous medium, solid or liquid, suspended in air (e.g., dust, fumes, fog, and smoke). The diameter of the particles may vary from micrometers (µm) down to less than 0.01 micrometers, and are fine enough to remain so dispersed for a period of time. (See also *fume*.)

aerosol photometer: The most popular of direct-reading aerosol monitors. They operate by illuminating an aerosol as it passes through a chamber (sensing volume) and by measuring the light scattered by all the particles at a given scattering angle relative to the incident beam. As the number of particles increases, the light reaching the detector increases. The detector can be a solid-state photodiode or a photomultiplier tube.

agent: A chemical, radiological, mineralogical, or biological entity that may cause deleterious effects in an organism after the organism is exposed to it.

Agricola: In *De Re Metallica* (1556), this German scholar described every facet of mining, smelting, and refining, noting prevalent diseases and accidents. He suggested means of prevention including mine ventilation and protective masks. In 1912 his work was translated into English by Herbert and Lou Henry Hoover.

AHU: See *air-handling unit*.

AIChE: See *American Institute of Chemical Engineers*.

AIHA: See *American Industrial Hygiene Association*.

AIME: See *American Institute of Mining, Metallurgical, and Petroleum Engineers*.

air: The atmosphere: the mixture of invisible, odorless, tasteless gases, such as nitrogen and oxygen, that surrounds the earth.

air change: 1. New, cleansed, or recirculated air introduced to a space. 2. A method of expressing the amount of air movement into or out of a building or room, in terms of the number of building volumes, or room volumes, exchanged in unit time.

air cleaner: 1. A device used to remove airborne impurities from air. (See also *air filter*.) 2. A device for removing a chemical hazard from an airstream before discharge to the ambient air. 3. A device to separate contaminants from an airstream. Examples include filters, scrubbers, electrostatic precipitators, cyclones, and afterburners.

air conduction: The process by which sound is transmitted through air from one point to another.

air contaminant: A substance (solid, liquid, or gaseous) not found in the normal composition of the atmosphere.

air ducts: A system of ducts to carry conditioned air to and from rooms.

air exfiltration: Air leaking outward. (See also *air infiltration*.)

air filter: A mechanical device that removes contaminants from an airstream.

air-handling unit (AHU): A device to move and condition (heat, cool, filter, and humidify) air in a central location.

air infiltration: Uncontrolled inward air leakage through cracks and interstices in any building element, and around windows and doors of a building, caused by the pressure effects of wind or the effect of differences in the indoor and outdoor air density.

air monitoring: The sampling for and measuring of contaminants in the air.

air quality: ASHRAE defines acceptable indoor air quality as "air in which there are no known contaminants at harmful concentrations and with which a substantial majority (usually 80%) of the people exposed do not express dissatisfaction." The problem with this definition is that in nonindustrial environments measurable contaminants are rarely present in levels known to be harmful, even when complaints of discomfort and adverse health effects are considerably in excess of the "acceptable" 20%. Until health risks have been established for chronic low-level exposures to both known and currently unrecognized pollutants, one must rely on the second part of this definition for guidance. Indoor air must meet standards that provide for the health and comfort of the majority of occupants.

air sampling: The collection and analysis of samples of air to measure the amounts of various pollutants or other substances in the air or the air's radioactivity.

air temperature: See *dry bulb temperature*.

air velocity: 1. The rate of motion of air in a given direction, measured as distance per unit time. 2. The axial velocity of the air entering or leaving a given effective face area. 3. Units for air velocity are meters per second (m/sec), feet per min (ft/min), or miles per hour (mph).

airborne particles: Impurities as solid or liquid particulate matter from natural or manmade sources. (Also known as "airborne particulates.")

airborne particulate matter: The ACGIH TLV committee has divided this general category into three classes based on the likely deposition within the respiratory tract. Although past practice was to provide TLVs® in terms of total particulate mass, the recent approach is to take into account the aerodynamic diameter of the particle and its site of action. Inhalable particulate mass (IPM) TLVs are designated for compounds that are toxic if deposited at any site within the respiratory tract. The typical size for these particles can range from sub-micrometer size to approximately 100 micrometers (µm). Thoracic particulate mass (TPM) TLVs are designated for compounds that are toxic if deposited either within the airways of the lung or the gas-exchange region.

The typical size for these particles can range from approximately 5–15 µm. Respirable particulate mass (RPM) TLVs are designated for those compounds that are toxic if deposited within the gas-exchange region of the lung. The typical size for these particles is approximately 5 µm or less. It should also be noted that the term "nuisance dust" is no longer used since all dusts have biological effects at some dose. The term "particulates," not otherwise classified, is now being used in place of nuisance dusts; however, the time-weighted average (TWA) of 10 mg/m^3 for IPM is still used, while a value of 3 mg/m^3 for RPM is now recommended.

air-line respirator: An atmosphere-supplying respirator in which the respirable gas is not designed to be carried by the wearer (listed by NIOSH as Type C and CE supplied air respirators).

air-purifying respirator: A respirator in which ambient air is passed through an air- purifying element that removes the contaminant(s). Air is passed through the air-purifying element by means of the breathing action of the respirator wearer or by a blower.

aliphatic: Usually applied to petroleum products derived from a paraffin base and having a straight or branched chain, or saturated or unsaturated molecular structure.

aliquot: Of, pertaining to, or designating an exact divisor or factor of a quantity, especially of an integer; contained exactly or an exact number of times.

alkali: A compound that has the ability to neutralize an acid and form a salt. Turns litmus paper blue and has pH values of 8 to 14. (Also known as "base, caustic.")

allergen: Any antigen (such as pollen, a drug, or food) that induces an allergic state in humans or animals.

allergic reaction: 1. Reaction of the body to chemical and/or biological agents, characterized by bronchoconstriction, nasal congestion, tearing, sneezing, wheezing, coughing, itching rash, and eruptions. 2. Immune response following exposure to a foreign agent or substance in an individual who is hypersensitive to that substance as a result of prior exposures. Examples are some types of skin rashes and asthma.

allergy: A response of a hypersensitive person to chemical and physical stimuli.

alpha particle: A nucleus consisting of two protons and two neutrons at nuclear distances. (Also known as "doubly charged helium ion.")

alveoli: Tiny air sacs of the lungs at the end of a bronchiole, through which gas exchange takes place by which the blood takes in oxygen and gives up its carbon dioxide in the process of respiration.

ambient: Surrounding, encircling, or pertaining to the environment.

ambient air: The air outdoors.

ambient air conditions: Characteristics of the environment (for example, temperature, relative humidity, pressure motion).

ambient monitoring: The monitoring of the workplace environment.

ambient noise: Ambient noise is the total noise within a given environment, being usually a composite of sounds from many sources near and far.

ambient temperature: 1. The temperature of the medium surrounding an object. In a domestic or commercial system having an air-cooled condenser, it is the temperature of the air entering this condenser. 2. The temperature of the medium such as air, water, or earth, into which the heat of equipment is dissipated: (a) For self-ventilated equipment, the ambient temperature is the average temperature of the air in the immediate neighborhood of the equipment; (b) For air or gas-cooled equipment with forced ventilation or secondary water-cooling, the ambient temperature is taken as that of the in-going air or cooling gas; (c) For self-ventilated enclosed (including oil-immersed) equipment considered as a complete unit, the ambient temperature is the average temperature of the air outside the enclosure in the immediate neighborhood of the equipment.

ambient total pressure: The atmospheric pressure at a given location that varies with local weather conditions and altitude; the pressure that would be read by a barometer at that location.

American Academy of Industrial Hygiene (AAIH): A professional association of certified industrial hygienists. Continued listing as a member in good standing of AAIH requires maintenance of certification and active practice of the profession.

American Academy of Occupational Medicine (AAOM): A professional association of physicians working in occupational medicine full time to promote maintenance and improvement of health in the workplace.

American Board of Industrial Hygiene (ABIH): Founded in 1960 by AIHA and ACGIH, ABIH offers a voluntary certification program for industrial hygienists. ABIH considers education, experience, and performance on a two-part examination in granting the titles Industrial Hygienist in Training (IHIT) and Certified Industrial Hygienist (CIH).

American Chemical Society (ACS): A scientific, educational, and professional society of chemists and chemical engineers. Headquarters are in Washington, D.C.

American Conference of Governmental Industrial Hygienists (ACGIH): Founded in 1938, this professional organization of government and university industrial hygienists had a membership of 5400 in 1996. Many of the ACGIH technical committee publications (threshold limit values [TLVs®] and *Industrial Ventilation: A Manual of Recommended Practice*) are recognized worldwide as authoritative sources.

American Council of Independent Laboratories (ACIL): Headquarters are in Washington, D.C.

American Industrial Hygiene Association (AIHA): Founded in 1939, this professional organization of industrial hygienists from the private and public sectors had a membership of more than 13,000 in 1996. AIHA is recognized for its technical committee publications, its proactive role in governmental affairs, and for promoting the profession of industrial hygiene.

American Institute of Chemical Engineers (AIChE): A member of the Accreditation Board for Engineering and Technology, the American National Standards Institute (ANSI), and related organizations. Headquarters are in New York, N.Y.

American Institute of Mining, Metallurgical, and Petroleum Engineers (AIME): This professional organization, founded in 1871 as the American

Institute of Mining Engineers, is dedicated to the exploration, extraction, and production of the Earth's minerals, materials, and energy resources. AIME comprises five separately incorporated units with a combined membership of more than 90,000. Headquarters are in New York, N.Y.

American National Standards Institute (ANSI): A voluntary membership organization that develops consensus standards. Headquarters are in New York, N.Y.

American Public Health Association (APHA): Represents health professionals in more than 40 disciplines in the development of health standards and policies. Headquarters are in Washington, D.C.

American Society for Testing and Materials (ASTM): A nonprofit organization that develops standard testing methods through consensus of volunteers (manufacturers, users, etc.) Headquarters are in West Conshohocken, Pa.

American Society of Heating, Refrigerating and Air-Conditioning Engineers (ASHRAE): A professional society committed to the establishment of standards in heating, refrigeration, and air-conditioning. Headquarters are in Atlanta, Ga.

American Society of Mechanical Engineers (ASME): A technical society of approximately 123,000 members committed to developing safety codes, equipment standards, and educational guidance. Headquarters are in New York, N.Y.

American Society of Safety Engineers (ASSE): A society of safety professionals of approximately 31,000 members committed to improving the workplace through promotion of standards and education. Headquarters are in Des Plaines, Ill.

Ames Test: A test for mutagenicity conducted in the bacterium *Salmonella typhimurium.* (Also known as the "Ames assay.")

amplitude: 1. Angular distance north or south of the prime vertical; the arc of the horizon, or the angle at the zenith between the prime vertical and a vertical circle, measured north or south from the prime vertical to the vertical circle. 2. The maximum absolute value attained by the disturbance of a wave or by any quantity that varies periodically.

ampoule: A small sealed glass vial filled with liquid.

ampoule detector tube: A detector tube consisting of one or more filling layers and a reagent ampoule. The ampoule contains part of the reagent system, which for reasons of stability cannot be placed in a single tube. During use, the ampoule is broken and the contents liberated.

AMS: Acute mountain sickness. See *high altitude pulmonary edema, high altitude cerebral edema.*

AMU: See *atomic mass unit.*

anaerobic: In absence of air.

analogue: Compound of the same structural type.

analytical blank: A blank used as a baseline for the analytical portion of a method. For example, a blank consisting of a sample from a batch of absorbing solution used for normal samples but processed through the analytical system only, and used to adjust or correct routine analytical results. (Also known as "reagent blank.")

analytical limit of discrimination: A concentration above which one can, with relative certainty, ascribe the net result from any analysis to the atmospheric particulate and below which there is uncertainty in the result. One approach to determining a statistical limit is to use a one-sided tolerance limit for the analytical discrimination limit (that is, a level [limit] below which a specified percentage [e.g., 99%] of blank filters analyses fall with a prescribed confidence [e.g., 95%]). [**Note:** "Limits for Qualitative Selection and Quantitative Determination," by L.A. Currie, *Analytical Chemistry 40(3)*:586-593, 1968, contains a detailed discussion of limits of detection.]

analytical methods: Detailed laboratory procedures that specify how to measure the amount of chemicals collected on the sampling media.

anemia: A condition in which the blood is deficient in red blood cells, hemoglobin, or total volume.

anemometer: A device that senses and measures air speed at a point.

anesthetic effect: The loss of the ability to perceive pain and other sensory stimulation.

angstrom (Å): A unit of length equal to 10^{-10} meters or 0.1 nanometers used primarily to express wavelengths of optical spectra. (See also *nanometer.*)

annoyance: A sound level of an intermittent broad-band noise is about 10 dBA above the background sound levels. Tonal sounds and other "attention-getting" sounds might cause annoyance at lower levels.

anoxia: The lack of oxygen (or a significant reduction in oxygen).

ANSI: See *American National Standards Institute* (formerly titled USA Standards Institute; American Standards Association).

antagonists: Skeletal muscles that act to brake or decelerate a limb.

anthropometry: The science of measurement of the body's mass, size, shape, and inertial properties.

anthropometry: The science of measurement of the body's mass, size, shape, and inertial properties.

antibody: The protein that a living organism is stimulated to make from its B lymphocytes when a foreign antigen is present.

anticipation: One of four primary responsibilities of the industrial hygienist. The anticipation of what health hazard problems may occur before a plant, process, or product is introduced. Anticipation depends on and extends the ability to recognize, coupled with a broad and current awareness of developments in the organization and its business, in scientific developments and new technologies, in regulatory areas bearing on the organization's activities, and in other activities that have an impact on the health of workers. (See also *recognition, evaluation,* and *control.*)

antigen: A large macromolecule that triggers an immune response.

antiparticle: A particle with the same mass and spin as the particle itself but an opposite charge and magnetic field.

AP: See *action potential.*

APF: See *assigned protection factor.*

APHA: See *American Public Health Association.*

aphake: Absence of the ocular lens.

apnea: The temporary cessation of breathing.

apparent temperature: An index of heat discomfort during the summer months. It includes the amplifying effect of increasing humidity on the discomfort level. (Also known as "heat index.")

appearance: A description of a substance at room temperature and normal atmospheric conditions.

applied dose: Amount of a substance given to an organism, especially through dermal contact.

aquatic toxicity: The adverse effects to marine life that result from being exposed to a toxic substance.

aqueous: Relating to or resembling water.

area: The cross-sectional area (e.g., duct, window, door, or any space) through which air moves. Units used are square feet (ft^2) or square meters (m^2).

area free: The total minimum opening area in an air inlet or outlet through which air can pass.

area sampling: The collection of airborne chemicals at a fixed position in the work area.

arithmetic mean: The most commonly used measure of central tendency, commonly called the "average." Mathematically, it is the sum of all the values of a set divided by the number of values in the set. (Also known as "average.")

aromatic: Applied to a group of hydrocarbons and their derivatives characterized by a molecular ring structure.

arthrogram: X-ray image of a joint after the injection of a dye or contrast medium.

arthroscope: Lighted surgical tube used to examine the interior of a joint.

ASA: Acoustical Society of America

asbestosis: Pneumoconiosis caused by breathing asbestos dust.

ASHRAE: See *American Society of Heating, Refrigerating and Air-Conditioning Engineers.*

ASME: See *American Society of Mechanical Engineers.*

aspect ratio: A ratio of length to width, greater than 3:1.

asphyxia: Suffocation from lack of oxygen. Chemical asphyxia is produced by a substance, such as carbon monoxide, that combines with hemoglobin to reduce the blood's capacity to transport oxygen. Simple asphyxia is the result of exposure to a substance, such as carbon dioxide, that displaces oxygen.

asphyxiant: 1. A vapor or gas that can cause unconsciousness or death by suffocation (lack of oxygen). Most simple asphyxiants are harmful to the body only when they become so concentrated that they reduce oxygen in the air (normally about 21%) to dangerous levels (18% or lower). Asphyxiation is one of the principal potential hazards of working in confined and enclosed spaces. 2. A chemical that displaces oxygen in the air, potentially resulting in insufficient oxygen to sustain life, especially in poorly ventilated areas. A chemical asphyxiant chemically interferes with the body's ability to take up and transport oxygen; a physical asphyxiant displaces oxygen in the environment.

assay: The quantitative or qualitative evaluation of a hazardous substance; the results of such an evaluation.

ASSE: See *American Society of Safety Engineers.*

assigned protection factor (APF): The expected workplace level of respiratory protection that would be provided by a properly functioning respirator or a class of respirators to properly fitted and trained users.

asthma: A condition marked by recurrent attacks of labored breathing and wheezing resulting from spasms of the upper airways of the lung.

ASTM: See *American Society for Testing and Materials.*

asymptomatic: The lack of identifiable signs or symptoms.

ataxia: A loss of balance with an unsteady gait; a failure of muscular coordination, total or partial.

atm: See *atmosphere.*

atmosphere (atm): A unit of pressure equal to 760 mmHg (mercury) at sea level.

atmosphere-supplying respirator: A class of respirators that supply a respirable atmosphere, independent of the workplace atmosphere.

atmospheric pressure: The pressure exerted in all directions by the atmosphere. At sea level, mean atmospheric pressure is 29.92 in. Hg, 14.7 psi, 407 in. wg, or 760 mmHg.

atom: The smallest unit of an element that still maintains the physical and chemical properties of the element.

atomic mass: The number of protons plus the number of neutrons in the nucleus (the sum of the nucleons).

atomic mass unit (AMU): One-twelfth the mass of the 12-carbon nucleus.

atomic number: The number of protons in the nucleus of a nuclide. (Also known as the "Z number.")

atomic weight: 1. The (weighted) average mass of naturally occurring isotopes of an element. 2. The relative mass of the atom on the basis of $^{12}C \equiv 12$.

atrophy: Wasting of muscles or other tissues in the body caused by a decrease in the number of cells or shrinkage of the cells. Some cases of carpal tunnel syndrome can lead to muscle atrophy.

attendant: An individual stationed outside one or more permit spaces who monitors the authorized entrants and performs all attendant's duties assigned in the employer's permit space program. (See *OSHA 1910.146.*)

attenuation (sound): The reduction, expressed in decibels, of the sound intensity at a designated first location as compared with sound intensity at a second location, which is acoustically farther from the source.

attitude: Manner, disposition, feeling, or position toward a person or thing.

audible range: The frequency range over which normal ears hear — approximately 20 Hz through 20,000 Hz. Above the range of 20,000 Hz, the term "ultrasonic" is used. Below 20 Hz, the term "subsonic" is used.

audiogram: A record of hearing loss or hearing level measured at several different frequencies — usually 500–6000 Hz. The audiogram may be presented graphically or numerically.

audiologist: A person trained in the specialized problems of hearing and deafness.

audiometer: A signal generator or instrument that can be operated manually or automatically for measuring objectively the sensitivity of hearing in decibels referred to audiometric zero. Pure tone audiometers are standard instruments for occupational use.

audiometric testing program: Test records that provide the only data that can be used to determine whether the program is preventing noise-induced permanent threshold shifts (NIPTS). It is an integral part of the hearing conservation program. (See also *noise-induced permanent threshold shift.*)

audit: A systematic check to determine the quality of operation of some function or activity. Audits may be of two basic types: 1) performance audits in which quantitative data are independently obtained for comparison with routinely obtained data in an air pollution measurement system; or 2) system audits are of a qualitative nature and consist of an on-site review of a laboratory's quality assurance (QA) system and physical facilities for air pollution sampling, calibration, and measurement.

auditory: Pertaining to, or involving, the organs of hearing or the sense of hearing.

authoritative occupational exposure limit: An occupational exposure limit set and recommended by credible organizations, such as ACGIH or AIHA. (See also *occupational exposure limit.*)

authority: The power to judge, act, or command.

authorized entrant: An employee who is authorized by the employer to enter a permit space. (See *OSHA 1910.146.*)

autoignition temperature: The lowest temperature at which a flammable gas- or vapor- air mixture will ignite from its own heat source or a contacted heated surface without necessity of spark or flame. Vapors and gases will spontaneously ignite at a lower temperature in oxygen than in air, and their autoignition temperature may be influenced by the presence of catalytic substances.

autoimmunity: An immune state in which antibodies are formed against the person's own body tissues.

automation: See *mechanization.*

autonomic nervous system: A subdivision of the central nervous system (CNS) that transmits signals to the smooth and cardiac muscles associated with the viscera or organ systems.

autoxidation: Slow reaction with air.

availability: The fraction or percentage of time that an item performs satisfactorily (in the reliability sense) relative to the total time the item is required to perform, taking into account its reliability and its maintainability, or the percentage of "up time" of an item or piece of equipment, as contrasted with its percentage of inoperative or "down time."

averaging time: The period over which a function is measured, yielding a time- weighted average (e.g., average concentration of an air pollutant).

aversion response: Blink reflex of the eye.

Avogadro's number: The number of molecules (6.02 × 1023 molecules/mole) contained in one gram molecular weight or one gram molecular volume. (e.g., 28.001 grams of CO = 6.02 × 1023 molecules of CO).

A-weighted response: The simulation of the sensitivity of the human ear at moderate sound levels.

axillary: Of, pertaining to, or near axilla or armpit. Placed or growing in the axis of a branch or a leaf.

B

backpressure: See *pressure drop.*

backup layer: The secondary layer of sorbent material in a sorbent tube that adsorbs chemicals that are not effectively trapped onto the primary collection layer.

bacteria: Small, relatively simple organisms found in soil, water, and the alimentary tract of animals and man. Some cause diseases in man.

baffle: A surface, usually in the form of a plate or wall, used to deflect fluids.

balanced system: A system in which the static pressure of each branch entering a junction is equal or balanced at the junction.

band pressure level: The sound pressure level for the sound contained within a specified frequency band. The reference pressure must be specified.

bandwidths (BW): 1. The difference between the frequency limits of a band containing the useful frequency components of a signal. 2. The range between the low and high cutoff frequencies of an acoustic filter. For measurement of sound the bandwidths generally used are octaves, 1/3 and 1/10 octaves.

bar: 1. A unit of pressure equal to 10^5 pascals or 10^5 newtons per square meter or 10^6 dynes per square centimeter (cm^2). 2. A unit of pressure equal to 10^6 dynes/cm^2 (part of the SI system of units).

barometer: A device for measuring atmospheric pressure using a working fluid, usually mercury. Fabricated from a long glass tube, closed at one end, evacuated, filled with mercury and inverted in a cistern of mercury. The height of the column of mercury is a measure of atmospheric pressure.

barometric effect: Variations in barometric pressure caused by altitude or weather changes.

barometric hazard: From an occupational hygiene perspective, barometric hazards can be categorized as: 1) hypobaric (low pressure) hazards; 2) hyperbaric (high pressure) hazards; and 3) hazards from changes in pressure, predominantly — but not exclusively — decreases in pressure.

barometric pressure: See *atmospheric pressure.*

barotrauma: Injury to air-containing structures (such as the middle ears, sinuses, lungs, and the gastrointestinal tract) due to unequal pressure differences across their walls.

basal cell carcinoma: A locally invasive, rarely metastatic nevoid tumor of the epidermis.

base: A substance that 1) liberates hydroxide (OH) ions when dissolved in water; 2) receives hydrogen ions from a strong acid to form a weaker acid; and 3) neutralizes an acid. Bases react with acids to form salts and water. Bases have a pH greater than 7 and turn litmus paper blue.

baseline survey: The initial evaluation of a health and safety parameter that will be used as the basis for all future comparisons.

basic characterization: The first step of the exposure assessment process. The basic information needed to characterize the workplace, work force, and environmental agents is collected and organized. Information is gathered that will be used to understand the tasks that are being performed, the materials being used, the processes being run, and the controls in place so that a picture of exposure conditions can be made.

batch method: A static generation method for preparing a known mixture of gas or vapor for verification testing of detector tubes. The method uses a volatile liquid to produce a known concentration in a container of known volume. A pure gas, or vapor of known volume and concentration, could be used also.

batch mixture: A fixed quantity of mixture prepared in an appropriate container.

BCSP: See *Board of Certified Safety Professionals.*

Beer-Lambert law: See *Bouguer-Lambert-Beer law.*

behavior: The aggregate of observable responses of an organism to internal or external stimuli.

BEI®: See *biological exposure index.*

bellows pump: A sampling pump that draws a fixed volume of air using an air chamber with flexible sides (i.e., a bellows).

belonging: Someone or something that belongs.

benchmark (BM): A relatively permanent natural or artificial object bearing a marked point with an elevation above or below an adopted datum (e.g., "sea level" is known).

benign: Not malignant or recurrent; often used to describe tumors that might grow in size but do not spread throughout the body.

benign acute mountain sickness (benign AMS): A constellation of symptoms (highlighted by frontal headaches) that can range from discomforting to incapacitating and is precipitated by a rapid ascent but will generally resolve spontaneously within 3 to 5 days.

beta: High speed particle with characteristics of an electron emitted from a nucleus in beta decay.

beta-glucan: Beta-glucans (or β-glucans) form the major portion of most fungal cell walls and may be chemically bound to chitin (and hence insoluble) or may form a soluble matrix in which the chitin fibrils are embedded. Most fungi that are common in indoor environments contain beta-glucans.

bias: A systematic (consistent) error in test results. Bias can exist between test results and the true value (absolute bias or lack of accuracy), or between results from different sources (relative bias). For example, if different laboratories analyze a homogeneous and stable blind sample, the relative biases among the laboratories would be measured by the differences among the results from the different laboratories. If the true value of the blind sample were known, however, the absolute bias or lack of accuracy from the true value would be known for each laboratory. (See also *systematic error.*)

bilateral: Condition that affects both sides of the body. Bilateral carpal tunnel syndrome occurs in both hands.

billion: In the United States, 10^9. In the United Kingdom and Germany, 10^{12}.

bioaccumulation: The accumulation or concentration of material in the body over a period of time.

bioassay: A test to determine the potency of a substance at producing some adverse health effect on a biological system.

biochemical epidemiology: The correlation of chemical markers measured in bodily media with epidemiological variates.

biohazardous waste: Byproducts containing blood, body fluids, or recognizable body parts that present a substantial or potential hazard to human health or the environment when managed improperly.

biological agents: Any of the viruses, microorganisms, and toxic substances derived from living organisms and used as offensive weapons to produce death or disease in humans, animals, and growing plants.

biological effects monitoring: The measurement and assessment of early biological effects — of which the relationship to health impairment has not yet been established — to evaluate exposure and/or health risk compared to appropriate references. It is a measurement of effects, often nonspecific, to evaluate exposure and risk, not health status.

biological exposure index (BEI®): ACGIH's procedures for estimating the amount of a material contained in the human body by measuring it (or its metabolic products) in tissue, body fluids, or exhaled air.

biological extrapolation: Assumption that results of toxicological studies on animals are applicable to humans.

biological half-life: The time required for a living tissue, organ, or organism to eliminate one-half of a substance that has been introduced into it.

biological monitoring: 1. The measurement of chemical markers in body media that are indicative of external exposure to chemical and physical agents. 2. An assessment of overall exposure to chemicals and other materials that are present in the workplace through measurement of the appropriate determinant(s) in biological specimens collected from the worker at a specified time. The determinant can be the chemical itself or its metabolite(s) or a characteristic reversible biochemical change induced by the chemical.

biological safety cabinet: Containment equipment that prevents the release and transmission of biological agents.

biological time constant: The time required for a portion of an absorbed chemical to undergo metabolic changes in the body.

biological wastes: Pathogenic or infective wastes that, if introduced into the human body, might disrupt biochemical and physiological function through infectivity or toxicity.

biologically effective dose: The absolute mass or moles of exposing compound that actually exposes a target organ internally after absorption.

biomarker: The determinant to be measured in a biological system.

biomechanics: A discipline that deals with the mechanical aspects of body motion. This term has been defined by the American Society of Biomechanics as "the application of the principles of mechanics to the study of biological systems." Biomechanics thus uses the knowledge base of anatomy, physiology, and mechanical engineering. The studies of muscle activity and the forces on and within the body are points of interest to the biomechanist.

biopsy: The removal and examination of tissue, cells, or fluids from a living body for examination.

biosafety: The art and science of maintaining a broken chain of infection.

biosafety level (BSL): The rating of biohazard potential described in four degrees of severity: 1) BSL1 agents are low risk and not known to cause disease in healthy adult humans; 2) BSL2 agents are associated with agents known to cause human disease that can be moderately serious, and for which preventive or therapeutic interventions are often available; 3) BSL3 agents are indigenous or exotic with potential for infection following aerosol transmission. Agents are associated with serious or lethal human disease for which preventive or therapeutic interventions may be available; 4) BSL4 organisms are dangerous/exotic agents that pose a high risk of life-threatening disease, and for which preventive or therapeutic interventions are not usually available.

biotechnology: Techniques that use living organisms or parts of organisms to produce a variety of products (from medicines to industrial enzymes) to improve plants or animals or to develop microorganisms to remove toxics from bodies of water, or act as pesticides.

biotransformation: Biotransformation is the process through which toxicants are chemically converted, generally reducing their lipophilicity and increasing their hydrophilicity. (See also *metabolism*.)

blank sample: A sample of a carrying agent (gas, liquid, or solid) that is normally used to selectively capture a material of interest, and that is subjected to the usual analytical or measurement process to establish a zero baseline or background value, which is used to adjust or correct routine analytical results. (Also known as "blank.")

blast gate damper: See *damper, blast gate*.

blood: The red fluid contained in arteries and veins.

blood-brain barrier: A barrier postulated to exist between brain capillaries and brain tissue to explain the relative inability of many substances to leave the blood and cross the capillary walls into the brain tissues.

bloodborne pathogen(s): Pathogenic organism(s) present in human blood, or other potentially infectious body fluids, that can cause disease in humans. These pathogens include, but are not limited to, the hepatitis B virus (HBV) and the human immunodeficiency virus (HIV).

Board of Certified Safety Professionals (BCSP): A professional board establishing minimum academic and experience attainments needed to qualify as a safety professional. The BCSP issues certificates to qualified individuals.

body burden: The total amount of a chemical retained in the body.

boiling point (BP): 1. The temperature at which a liquid changes to a vapor. 2. The temperature at which the vapor pressure of a liquid equals the atmospheric pressure. The temperature at which the vapor pressure of a liquid equals the absolute external pressure at the liquid-vapor interface.

bone conduction: Transmission of sound vibrations to the internal ear via the bones of the skull.

boredom: In the workplace, too little information, which might lead to fatigue.

Botsball: A thermometer that consists of a metal probe with its heat sensor in the center of a 6 cm diameter hollow, copper sphere. The sphere is painted black and covered with a double layer of black cloth that can be saturated with water. The Botsball has a water reservoir of approximately 7 cubic centimeters.

Bouguer-Lambert-Beer law: The intensity of a beam of monochromatic radiation in an absorbing medium decreases exponentially with penetration distance.

Boyle's law: At a constant temperature the volume of a given quantity of any gas varies inversely as the pressure to which the gas is subjected.

BP: See *boiling point*.

Bragg-Gray principle: The amount of ionization produced in a small gas-filled cavity surrounded by a solid absorbing material that is proportional to the energy absorbed by the solid.

brainstorming: A procedure used to find a solution for a problem by collecting all ideas, without regard for feasibility that occurs from a group of people meeting together.

brake horsepower: Brake power expressed in horsepower.

brake power: The actual power delivered by or to a shaft (from the use of a brake to measure power).

branch: 1. In ducts, piping, or conduit another section of the same size or smaller at an angle with the main. Also, the section of pipe from a main to a register or radiator. 2. (computer) A set of instructions that are executed between two successive decision instructions.

branch line: An air supply line connecting the controller and controlled device.

breakthrough: Significant sample loss that occurs when chemicals are not effectively trapped by the collection media.

breakthrough time: The time elapsed from the initial contact of the chemical on the outside surface until detection on the inside surface.

breakthrough volume: That volume of an atmosphere containing two times the permissible exposure limit (PEL) for a specific contaminant that can be sampled at the recommended flow rate before the efficiency of the sampler degrades to 95%.

breast milk: The white viscous fluid extruded from female breasts.

breathing zone (BZ): The volume surrounding a worker's nose and mouth from which he or she draws breathing air over the course of a work period. This zone can be pictured by inscribing a sphere with a radius of about 10 inches centered at the worker's nose.

breathing zone sampling: Air samples that are collected in the breathing zone. NIOSH sometimes uses this term to describe the sample-collection technique in which a second individual collects a sample in an employee's breathing zone.

BRI: See *building-related illness*.

British thermal unit (Btu): The amount of heat energy needed to raise the temperature of 1 pound of water 1°F from 59°F to 60°F. It is defined by the British Standards Institution as 1055.06 joules.

broad-band: A band with a wide range of frequencies.

bronchial tubes: Branches or subdivisions of the trachea (windpipe) that carry air into and out of the lungs. (Also known as "bronchioles.")

bronchoconstriction: Constriction of the bronchial air passages.

browser: A software application used to locate and display Web pages. Three of the most popular browsers are Netscape Navigator®, Microsoft Internet Explorer®, and Spyglass Mosaic®. All of these are graphical browsers, which means that they can display graphics as well as text. Also, most modern browsers can present multimedia information, including sound and video.

BSL: See *biosafety level.*

Btu: See *British thermal unit.*

bubble flowmeter: A device used to calibrate pumps for airflow rate. Air sampled by the pump is measured by the displacement of a soap bubble in a burette. The volume displacement per unit time (i.e., flow rate) can be determined by measuring the time required for the soap bubble to pass between two scale markings that enclose a known volume.

buffer: Any substance in a liquid that tends to resist the change in pH when acid or alkali is added.

building envelope: The outer shell, or the elements of a building, that encloses conditioned spaces, through which thermal energy may be transferred to, or from, outdoors.

building-related illness (BRI): Infectious, allergic, or toxin-induced disease with objective clinical findings related to building occupancy. (Also known as "building-related disease.")

building wake: The zone around a building's envelope.

burn: The visible destruction (or permanent change) in skin, eyes, tissue, etc., at the site of contact.

BW: See *bandwidths.*

BZ: See *breathing zone.*

C

c: See *ceiling.*

CA: See *corrective action.*

calibrate: To check, adjust, or systematically standardize the graduations of a quantitative measuring instrument.

calibration: Establishment of a relationship between various calibration standards and the measurements of them obtained by a measurement system, or portions thereof. The levels of the calibration standards should bracket the range of levels for which actual measurements are to be made.

calibration standard: A standard used to quantitate the relationship between the output of a sensor and a property to be measured. Calibration standards should be traceable to a standard reference material (SRM), certified reference material (CRM), or a primary standard.

calorie: Heat required to raise the temperature of 1 gram of water 1°C, specifically from 4°C to 5°C. Mean calorie = 1/100 part of the heat required to raise 1 gram of water from 0°C to 100°C. Great calorie or kilocalorie = 1000 calories.

calorimeter: 1. A device for measuring heat quantities, such as machine capacity, combustion heat, specific heat, vital heat, heat leakage, etc. 2. A device for measuring quality (or moisture content) of steam or other vapor. 3. Equipment for measuring emitted or absorbed heat quantities.

cancellation: A method for reducing radiation exposure levels. The general principle involves having wires in proximity carry current, and produce fields, that are opposite in phase. The field from one wire then effectively cancels the field from the other at locations removed from the two.

cancer: A cellular tumor that is usually associated with spread throughout the body and can be fatal.

cancer slope factor: See *slope factor.*

canister/cartridge: A container with a filter, sorbent, or catalyst — or combination of these items — which removes specific contaminants from the air passed through the container.

canopy hood: A one- or two-sided overhead hood that receives rising hot air or gas. (See also *receiving hood.*)

capture hood: A ventilation system hood that captures contaminants that are released from a process. A slot-plenum hood is an example of a capture hood.

capture velocity (V_c): 1. Air velocity at any point in front of a capture hood opening necessary to overcome opposing air currents and to capture contaminated air at that point and cause it to flow into the hood and prevent the contamination from reaching the workers' breathing zone. 2. The velocity of air flowing past a contaminant source that is necessary to overcome the opposing air currents and capture or entrain the contaminant at that point and move it into the hood. (Also known as "control velocity.")

carbon dioxide toxicity: Carbon dioxide becomes toxic when it suppresses respiration. The combination of the accumulation of exhaled carbon dioxide at increased pressure (either in the breathing system's dead space or due to a malfunction) can rapidly cause toxic effects.

carcinogen: A substance or agent capable of causing or producing cancer in mammals, including humans. A chemical is considered to be a carcinogen if: a) it has been evaluated by IARC and found to be a carcinogen or potential carcinogen; b) it is listed as a carcinogen or potential carcinogen in the Annual Report on Carcinogens published by the National Toxicology Program (NTP); or c) it is regulated by OSHA as a carcinogen.

carcinogen classification systems: Several different carcinogen classifications have been developed by organizations such as EPA, IARC, and ACGIH. The system used by ACGIH can be summarized as follows: (A1) Confirmed Human Carcinogen — To have this designation a chemical must have human data to support its classification. Examples are nickel subsulfide, bis(chloromethyl) ether, and chromium VI compounds; (A2) Suspected Human Carcinogen — This designation requires relevant animal data in the face of conflicting or insufficient human data. Examples are diazomethane, chloromethyl methyl ether, and carbon tetrachloride; (A3) Animal Carcinogen — To have this designation a chemical must have caused cancer in animal studies by nonrelevant routes of exposure or mechanisms (e.g., kidney tumors produced in male rats from many hydrocarbons, paradichlorobenzene, d-limonene, etc.; hormone-mediated thyroid tumors from compounds such as the ethylene bisdithiocarbamates) or at excessive doses, etc. Furthermore, there are human data available that are in contradiction to the results in animals. Examples are nitrobenzene, crotonaldehyde, and gasoline; (A4) Not Classifiable as a Human Carcinogen — This classification is given to chemicals for which there is inadequate data to say whether it is a potential human carcinogen. It is typically applied to chemicals for which an issue of carcinogenicity has been raised but for which there are insufficient data to answer the question. Examples are pentachloronitrobenzene, phthalic anhydride, and acetone; (A5) Not Suspected as a Human Carcinogen — This classification is given to chemicals that have strong supporting data in humans to show that they are not carcinogens. Data from animals can also be used to support this classification. To date, the only example of a chemical receiving this designation is trichloroethylene.

carcinogenesis: The development of malignant tumors or neoplasms composed of abnormal cells exhibiting uncontrolled growth, invasiveness, and metastasis.

carcinogenic: A substance or material capable of producing cancer.

carcinoma: A malignant tumor of epithelial cell origin (e.g., skin, lung, breast), tending to infiltrate the surrounding tissue and give rise to metastases (tumors at distant sites).

cardiovascular system: See *CVS*.

carpal tunnel: Bony, narrow passage at the intersection of the hand and wrist through which the median nerve and many tendons pass.

carpal tunnel syndrome (CTS): Compression of the median nerve caused by the swelling of tendons in the carpal tunnel. CTS is characterized by numbness, pain, and tingling in the fingers; and clumsiness and loss of grip strength in the hand.

CAS: See *Chemical Abstracts Service*.

cascade impactor: A sampling device that uses a series of impaction stages with decreasing particle cut size so that particles can be separated into relatively narrow intervals of aerodynamic diameter. It is used for measuring aerodynamic size distribution of an aerosol sample.

case control study: A study that starts with the identification of individuals with a disease or adverse health effect of interest and a suitable control group without the disease. The relationship of an attribute to the adverse health effect is examined by comparing how frequently (and at what levels) the attribute is present in the diseased and nondiseased groups.

catalyst: A substance that changes the speed of a chemical reaction but undergoes no permanent change itself.

catchball: The continuing two-way communication at all levels of the policy deployment process, to evaluate lower-level plans and the continuing utility of the original policies.

cavitation: The formation of cavities on a surface of a solid by liquid moving over it with velocity high enough to induce erosion of the surface when the cavity collapses.

cc: Cubic centimeter, a volume of measurement in the metric system that is equal in capacity to one milliliter (mL). One quart is about 946 cc.

CEGL: See *continuous exposure guidance level*.

ceiling (c): 1. The maximum allowable human exposure limit for an airborne substance, not to be exceeded even momentarily. Used in OSHA PELs, ACGIH TLVs®, and NIOSH RELs. 2. The concentration that should not be exceeded during any part of the working exposure. In conventional occupational hygiene practice, if instantaneous monitoring is not feasible, then the ceiling can be assessed by sampling over a 15-minute period, except for chemicals that may cause immediate irritation, even with exposures of extremely short duration.

ceiling limit: Control of exposure to fast-acting substances by value placing a limit on their concentration. Such substances are marked with a "C" in the ACGIH threshold limit value (TLV®) table. The concentration of these substances cannot at any time in the work cycle (except for a 15-minute period) exceed the TLV. Also known as "ceiling value." (See also *threshold limit value*.)

ceiling value: See *ceiling limit*.

Celsius temperature: Temperature scale used with the SI system of units in which the freezing point of water is 0°C, the triple point is 0.01°C, and the boiling point is approximately 100°C. (Formerly referred to as the "centigrade scale.")

center of gravity: A fixed point in a material body through which the resultant force of gravitational attraction acts.

centigrade temperature: See *Celsius temperature*.

central nervous system (CNS): The brain and spinal cord.

central-fan system: A mechanical indirect system of heating, ventilating, or air conditioning in which the air is treated or handled by equipment located outside the rooms served, usually at a control location, and is conveyed to and from the rooms by a fan and distributing ducts.

Certified Industrial Hygienist (CIH): A professional industrial hygienists who by education, experience, and demonstration of knowledge has satisfied the requirements of ABIH and has been designated as a CIH in either the comprehensive practice or chemical aspects of the profession. The CIH designation builds on the IHIT requirements in recognizing long experience and proven professional ability.

certified reference material (CRM): Standards prepared by gas vendors in quantities of at least 10 cylinders for which 1) the average concentration is within 1% of an available standard reference material (SRM); and 2) 2 cylinders are selected at random and audited by the U.S. EPA.

CET: See *corrected effective temperature*.

CFR: See *Code of Federal Regulations*.

chain of custody: A procedure for preserving the integrity of a sample or of data (e.g., a written record listing the location of the sample/data at all times).

channeling: Uneven flow of sampled air through a detector tube because of improper packing.

Charles-Gay-Lussac law: Gases increase in volume for each 1°C rise in temperature. This increase is equal to approximately 1/273.15 of the volume of the gas at 0°C.

chemical: A single molecule or a mixture of molecules.

Chemical Abstracts Service (CAS): An organization under the American Chemical Society. CAS abstracts and indexes chemical literature from throughout the world in "Chemical Abstracts." "CAS Numbers" are used to identify specific chemicals or mixtures.

Chemical Abstracts Service (CAS) number: A concise, unique means of material identification. Each chemical may have more than one synonym but only one CAS number.

chemical agent: Dust, gas vapor, or fume that acts on or reacts with the human physiologic system.

chemical asphyxiant: See *asphyxiant*.

chemical asphyxiation: Toxic reaction wherein chemicals reaching the bloodstream react in such a way as to deprive the body of oxygen.

chemical family: A group of single elements or compounds with a common general name. Example: acetone, methyl ethyl ketone (MEK), and methyl isobutyl ketone (MIBK) are of the "ketone" family; acrolein, furfural, and acetaldehyde are of the "aldehyde" family.

chemical hazard: Exposure to any chemical which, in acute concentrations, has a toxic effect.

chemical name: The name given to a chemical in the nomenclature system developed by the International Union of Pure and Applied Chemistry (IUPAC) or the CAS. The scientific designation of a chemical or a name that will clearly identify the chemical for hazard evaluation process.

chemical pneumonitis: Inflammation of the lungs resulting from inhalation of chemical vapors and characterized by an outpouring of fluid in the lungs.

chemical waste: Chemical waste is either organic (hydrocarbons or substituted hydrocarbons) or inorganic (metallic or nonmetallic elements). The hydrocarbons are composed solely of carbon and hydrogen; substituted hydrocarbons also include functional groups composed of elements such as chlorine, nitrogen, phosphorous, sulfur, or oxygen. Inorganic waste materials are typically in the form of salts, hydrides, and oxides. The hazards associated with organic compounds and the fate of the compounds in the environment depend mainly on their chemical compositions and associated physical properties.

chemicals of potential concern (CoPC): Chemicals that are potentially site-related and whose data are of sufficient quality for use in the quantitative risk assessment.

cholestasis: An acute reaction within the liver where the production and/or secretion of bile is impaired. It is caused by exposure to environmental and occupational agents.

chromatograph: An instrument that can separate and analyze mixtures of chemical substances.

chromosomal aberrations: A change in the normal number, size, or shape of chromosomes.

chromosomes: One of several structures in the nucleus of eukaryotic cells. They contain the genes (or hereditary material) in the form of threads of DNA.

chronic: Applies to persistent, prolonged, repeated exposures and their effects.

chronic daily intake: Intake averaged over a long period of time (seven years to lifetime).

chronic effect: Disease symptom or process of long duration, usually frequent in occurrence, and almost always debilitating.

chronic exposure: Small dose/long time.

chronic mountain sickness: A rare response to prolonged stays at elevation. Its symptoms include those of benign AMS (perhaps also of HACE but not HAPE), but occur only after several years of exposure. This condition is hypothesized to be the cascading result of very high increases in hematocrit. Because the time of response of chronic mountain sickness is so delayed relative to industrial personnel transfers, it is considered herein to be outside the occupational hygienist's realm. (Also known as "Monge's disease.")

chronic reference dose: Applicable for periods of seven years to lifetime.

chronic symptom: Symptom that persists for a long period.

chronic toxicity: Adverse health effects that can occur from prolonged, repeated exposure to relatively low levels of a substance; might have a chronic effect from an acute exposure.

chronic toxicity study: Refers to toxicity study durations of greater than 6 months.

CIH: See *Certified Industrial Hygienist.*

cilia: Tiny hair-like "whips" in the bronchi and other respiratory passages that normally aid in the removal of dust trapped on these moist surfaces.

circadian effects: The routine alterations in physiology related to time of day.

cirrhosis: A condition in which the liver has become hardened and its physiological functions are highly impaired.

class: Collection of related chemical groups or topics.

Class 1: Classification of a laser product's output power and potential hazard during normal use under the Federal Laser Product Performance Standard (See *21 CFR Subchapter J, Part 1040.10*). Class 1 lasers cannot emit laser radiation at known hazard levels.

Class 2: Classification of a laser product's output power and potential hazard during normal use under the Federal Laser Product Performance Standard (See *21 CFR Subchapter J, Part 1040.10*). Class 2 lasers are low-powered devices that emit visible radiation above Class 1 levels but do not exceed 1 mW.

Class 2a: Classification of a laser product's output power and potential hazard during normal use under the Federal Laser Product Performance Standard (See *21 CFR Subchapter J, Part 1040.10*). Class 2a is a special designation based on a 1000-second (16.7- minute) exposure and applies only to lasers that are not intended for viewing such as supermarket laser scanners. The emission from a Class 2a laser is defined such that the emission does not exceed the Class 1 limit for an emission duration of 1000 seconds.

Class 3a: Classification of a laser product's output power and potential hazard during normal use under the Federal Laser Product Performance Standard (See *21 CFR Subchapter J, Part 1040.10*). Class 3a lasers have intermediate power levels (CW = 1 to 5 mW). Class 3a emissions may be "low" or "high" irradiance.

Class 3b: Classification of a laser product's output power and potential hazard during normal use under the Federal Laser Product Performance Standard (See *21 CFR Subchapter J, Part 1040.10*). Class 3b includes moderately powerful lasers (e.g., CW = 5 to 500 mW). In general, Class 3b lasers are not a fire hazard, nor are they generally capable of producing a hazardous diffuse reflection except for conditions of staring at distances close to the diffuser.

Class 4: Classification of a laser product's output power and potential hazard during normal use under the Federal Laser Product Performance Standard (See *21 CFR Subchapter J, Part 1040.10*). Class 4 lasers are high-powered devices (e.g., CW >500 mW). These are hazardous to view under any condition (direct or scattered radiation) and are potential fire and skin hazards.

clean room: An enclosed space environmentally controlled, within specified limits, of airborne particles, temperature, relative humidity, air pressure, and air motion. (Also known as "white room.")

clean space: An open area environmentally controlled as in a clean room.

clean workstation: An open or enclosed work area environmentally controlled as a clean room.

cleanup level: Remedial target for concentration of contaminant in air, water, or soil.

clinical chemistry: Analytical testing of the blood and other body fluids to determine the functioning of various organs and systems.

cluster: A small group of employees have similar complaints about their health or they may have been diagnosed with similar illnesses, generally sharing something in common (such as the same job, exposure, or area of employment).

CMM: Cutaneous malignant melanoma. See *nonmelanoma skin cancer*.

CNS: See *central nervous system*.

CNS depression: A reversible state of stupor or unconsciousness.

CNS effects: Signs and symptoms include drowsiness, dizziness, loss of coherence, and other signs or symptoms of toxic effects on the central nervous system (CNS).

cochlea: The auditory part of the inner ear that contains the nerve cells that when damaged results in hearing impairment.

cocoon: Remaining in a facility during a hazardous situation with air intake equipment to the building turned off.

Code of Federal Regulations (CFR): A collection of the regulations that have been promulgated under U.S. Law.

Code of Professional Ethics for the Practice of Industrial Hygiene: This code provides standards of ethical conduct to be followed by industrial hygienists as they strive toward the goals of protecting employees' health, improving the work environment, and advancing the quality of the profession.

coefficient of variation (CV): A measure of precision calculated as the standard deviation of a set of values divided by the average. It is usually multiplied by 100 to be expressed as a percentage.

cognition: The conscious faculty or process of knowing, of becoming, or of being aware of thoughts or perceptions, including understanding and reasoning.

cohort study: The method of epidemiologic study in which subsets of a defined population can be identified who are, have been, or in the future might be exposed or not exposed — or exposed in different degrees — to a factor or factors hypothesized to influence the probability of occurrence of a given disease or other outcome.

cold trap: A sampling vessel that has been immersed in a cooling system, such as dry ice or liquid nitrogen, to extract the contaminants from the airstream for subsequent analysis.

collaborative tests: The evaluation of a new analytical method under actual working conditions through the participation of a number of typical or representative laboratories in analyzing portions of carefully prepared homogeneous samples. (Also known as "collaborative studies.")

colonization: The state in which infection and establishment of an organism within a host has occurred without resulting in subclinical or clinical disease.

color density tube: A detector tube that uses color intensity (or density) to determine the amount of compound present.

colorimetric (colorimetry): A measuring method that uses a change in color as an indication of the concentration of a compound.

colorimetric detector tube: See *detector tube*.

colorimetric indicator tube: See *detector tube*.

coma: A state of unconsciousness from which the person cannot be aroused by physical stimulation.

combustible: Capable of catching fire or burning; usually a liquid with a flash point at or above 37.8°C (100°F) but below 93.3°C (200°F).

combustible gas indicator: A general survey instrument capable of measuring a wide range of air contaminants but cannot distinguish among them. They are usually used as area samplers to measure concentrations that are immediately dangerous to life and health, and concentrations that are within occupational exposure limits, in the ppm range. (See also *flame ionization detector, photoionization detector [PID]*.)

commissioning: The acceptance process in which an HVAC system's performance is determined, identified, verified, and documented to assure proper operation in accordance with codes, standards, and design intentions.

commitment: The act of pledging or devoting oneself to an issue or course of action.

common exposure base: The unit of 100 full-time workers used to calculate the incident rate. (See also *incident rate*.)

communications plan: The communications plan defines the individuals who must be informed during the evaluation. It is intended to comply with all of the facility security and notification procedures. It includes the team roster that identifies the team members by name and the tasks each is to perform.

comparability: A measure of the confidence with which one data set can be compared to another.

completeness: The amount of valid data obtained from a measurement system compared with the amount that was expected to be obtained under correct normal operations, usually expressed as a percentage.

compliance: Compliance with health and safety regulations.

compliance strategy: One of two general exposure assessment strategies. It usually uses worst-case monitoring with a focus on exposures during the time of the survey. An attempt is made to identify the maximum-exposed worker(s) in a group. One or a few measurements are then taken and simply compared with the occupational exposure limit (OEL). If the exposures of the maximum-exposed workers are sufficiently below the OEL, then the situation is acceptable. This strategy provides little insight into the day-to-day variation in exposures levels and is not amenable to the development of exposure histories that accurately reflect exposures and health risk. (See also *comprehensive strategy*.)

compound: A substance composed of two or more elements joined according to the laws of chemical combination. Each compound has its own characteristic properties different from those of its constituent elements.

compound hood: A hood that has two or more points of significant energy (i.e., static pressure) loss. Examples include slot hoods and multiple opening, lateral draft hoods.

comprehensive exposure assessment: The systematic review of the processes, practices, materials, and division of labor present in a workplace that is used to define and judge all exposures for all workers on all days.

comprehensive strategy: One of two general exposure assessment strategies. It is directed at characterizing and assessing exposure profiles that cover all workers, workdays, and environmental agents. These exposure profiles are used to picture exposures on unmeasured days and for unmeasured workers in the similarly exposed group. In addition to ensuring compliance with OELs, this strategy provides an understanding of the day-to-day distribution of exposures. (See also *compliance strategy*.)

comprehensive survey: A multifaceted examination of all recognized health and safety hazards in a defined work environment. It encompasses a wall-to-wall evaluation of the health and safety factors that may place an employee at increased risk of developing an adverse effect from working in a specific setting.

compressibility: The relative variation of volume with pressure; depends on the process to which a gas is subjected.

computed tomography: A technique in which measurements to create both spatially and temporally resolved concentration distribution maps are generated when a network of intersecting open-path FTIR spectrometers are used in a room.

concentration: 1. The amount of a given substance in a stated unit of measure. Common methods of stating concentration are percent by weight or by volume; weight per unit volume; normality; etc. 2. The quantity of a chemical per unit volume (see *ppm*, *mg/m³*); 10,000 ppm = 1%. In air, the relationship between ppm and mg/m^3 is as follows: ppm × MW = 24.45 × mg/m^3 where ppm = the volume ratio of a chemical in air expressed in parts per million; MW = the chemical's molecular weight; 24.45 = the number of liters occupied by 1 mole of any gas at STP (i.e., 298 K or 25°C [77°F]) and 760 mmHg or 1 atm; mg/m^3 = the chemical's concentration in air expressed in milligrams of chemical per cubic meter of air.

concentration–time (C–T): Two factors on which dosage is based. (See also *dose*.)

concentric: A muscle contraction where the muscle shortens while developing tension.

conceptus: The whole product of conception at any stage of development, from fertilization of the ovum to birth.

condensation: Act or process of reducing from one form to another denser form such as steam to water.

conduction: 1. Transmission of energy by a medium that does not involve movement of the medium itself. 2. Conductive heat transfer when there is direct contact between a hotter and a colder substance.

conductive hearing loss: Any condition that interferes with the transmission of sound to the cochlea.

confidence coefficient: The chance or probability, usually expressed as a percentage, that a confidence interval has of including the population value. The confidence coefficients usually associated with confidence intervals are 90%, 95%, and 99%. For a given sample size, the width of the confidence interval increases as the confidence coefficient increases.

confidence interval: A value interval that has a designated probability (the confidence coefficient) of including some defined parameter of the population.

confidence level: The probability that a stated confidence interval will include a population parameter.

confidence limits: The upper and lower boundaries of a confidence interval.

configuration control: A system for recording the original equipment configuration, physical arrangement, and subsequent changes thereto.

confined space: 1. A space that 1) is large enough and so configured that an employee can bodily enter and perform assigned work; 2) has limited or restricted means for entry or exit (for example, tanks, vessels, silos, storage bins, hoppers, vaults, and pits are spaces that may have limited means of entry); and 3) is not designed for continuous employee occupancy. (See *OSHA 1910.146*.) 2. An enclosure that contains an oxygen deficiency, where the oxygen concentration is less than 19.5%. Examples are underground utility vaults, storage tanks, and large diameter pipes.

congener: Compound with related but not identical structure.

conjugate: The product of reaction of an exposing single chemical or a single metabolite with the endogenous biochemical pathways of the body.

conjunctiva: The delicate membrane that lines the eyelids and covers the exposed surface of the eyeballs.

conjunctivitis: Inflammation of the membrane (conjunctiva) that lines the eyelids and covers the front of the eyeball.

consensus standard: Any generally accepted standard of environmental quality established through input from a number of experts and professional groups knowledgeable about matters pertaining to the subject of the standard.

Conservation of Energy law: Energy can neither be created nor destroyed, and therefore the total amount of energy in the universe is constant.

conservation of mass: In all ordinary chemical changes, the total mass of the reactants is always equal to the total mass of the products.

constant flow: A feature available on air sampling pumps whereby the flow rate will automatically compensate for flow restrictions, thereby ensuring that the flow rate is held constant throughout the sampling period.

constant volume pump: A sampling pump designed to draw a fixed volume of air with each full pump stroke. Examples are the simple squeeze bulb, the bellows pump, and the piston pump.

constrained-layer damping: The use of a laminated construction consisting of one or more sheet metal layers, each separated by and bonded to a viscoelastic layer.

contact dermatitis: An acute or chronic inflammation of the skin resulting from irritation by or sensitizing to some substance coming in contact with the skin.
contact rate: Amount of medium (e.g., water, soil) contacted per unit of time or event.
continuity equation: An equation obeyed by any conserved, indestructible quantity such as mass, electric charge, thermal energy, electrical energy, or quantum-mechanical probability, which is essentially a statement that the rate of increase of the quantity in any region equals the total current flowing into the region.
continuous exposure guidance level (CEGL): Established by the Committee on Toxicology of the National Research Council (NRC) for the U.S. Department of Defense. CEGLs are intended for normal, long-lasting military operations. CEGLs are ceiling concentrations designed to avoid adverse health effects, either immediate or delayed, for exposure periods up to 90 days. (See also *EEGL, SPEGL.*)
continuous flow respirator: An atmosphere-supplying respirator that provides a continuous flow of respirable gas to the respiratory inlet covering.
continuous improvement: See *kaizen.*
continuous operation: Industrial operation in which the final product is produced at or near a continuous rate.
continuous wave (CW): A laser that operates continuously.
contraction: Shortening of the fibers of muscle tissue stimulating the muscle to its active state.
control: One of four primary responsibilities of the occupational hygienist. It is the culmination of the effort in addressing the primary objective of the occupational hygienist: providing a healthful work environment. Current occupational hygiene practice recognizes a hierarchy of controls; in priority order, these are engineering controls, work practices, administrative controls, and as a last resort use of personal protective equipment. (See also *anticipation, recognition,* and *evaluation.*)
control chart: A graphical chart with statistical control limits and plotted values (usually in chronological order) of some measured parameter for a series of samples. Use of the charts provide a visual display of the pattern of the data, enabling the early detection of time trends and shifts in level. For maximum usefulness in control, such charts should be plotted in a timely manner (that is, as soon as the data are available). (Also known as "Shewhart control chart," "statistical control chart.")
control measures: The overall strategy for controlling the environment as well as the specific components that make up that strategy. These include local exhaust and general ventilation, process isolation or enclosure, shielding from heat, ionizing radiation, ultraviolet light, or any other forms of radiant energy, protective clothing, and respiratory protective devices, and other controls.
control velocity: See *capture velocity.*
convection: 1. The transfer of heat by the flow of some liquid or gas. 2. Motion resulting in a fluid from the difference in density and the action of gravity; heat loss or gain by the body to the surrounding atmosphere.
conversion: 1. The process of changing information from one form of representation to another, such as from the language of one type of machine to that of another, or from magnetic tape to the printed page. 2. The process of changing from one data-processing method to another, or from one type of equipment to another. 3. To change from use of one fuel to another.
convulsions: An abnormal, and often violent, involuntary contraction or series of contractions of the muscles.
cooling probe: A probe used in high temperature applications to cool the sample before entry into the detector tube. It sometimes is called a "hot air probe."
CoPC: See *chemicals of potential concern.*
cornea: The transparent structure that forms the exposed surface of the eyeball and allows light to enter the eye.
corneal opacity: A density, spot, or opaque shadow in the cornea.
corrected effective temperature (CET): Amendment to the ET scale that includes allowances for radiation. This scale uses the globe temperature instead of the dry bulb temperature.
corrective action (CA): 1. The act of varying the manipulated process variable by the controlling means in order to modify overall process operating conditions. 2. CA should consist of identifying the problem; designation of a person or persons to correct the problem; identifying appropriate corrective actions; instituting the corrective action; evaluating the correction to determine if the CA did, in fact, correct the problem; and, finally, placing the previously nonconforming system back on-line.
corrosive: 1. A chemical that causes visible destruction of, or irreversible alterations in, living tissue by chemical action at the site of contact. 2. A chemical that causes necrosis of biological tissues.
cosine law: The law that the energy emitted by a radiating surface in any direction is proportional to the cosine of the angle which that direction makes with the normal.
coughing: To expel air from the lungs suddenly with a harsh noise, often involuntarily.
counter-control: When individuality or perceived personal control is made scarce with top-down control, some people will exert contrary behavior in an attempt to assert their freedom. (Also known as "psychological reactance.")
countercurrent heat exchange: Blood flow that helps control heat balance. It occurs when arteries, which carry oxygen-rich blood away from the heart, lie in very close proximity to the veins that bring blood back to the heart.
CPS: cycles per second. See *hertz.*
crepitus: Crackling or grating noise or sensation sometimes made by a joint.
criteria: Standard on which a judgment or decision may be based.
criteria document: Publication of NIOSH-related research on which standards can be based. Documents contain essential parts of a standard, including environmental limits, sampling requirements, labeling, monitoring requirements, medical examinations, compliance methods, protective equipment, record keeping requirements, and other recommendations to OSHA for establishment of a standard.
Criteria for Fatigue Decreased Proficiency (FDP): Boundaries that represent the ability of a person to

work at tasks under vibration exposure without the vibration interfering with the worker's ability to perform.

critical temperature: Saturation temperature corresponding to the critical state of the substance at which the properties of the liquid and vapor are identical; the temperature above which a gas cannot be liquefied by pressure alone.

CRM: See *Certified Reference Material.*

cross-sectional study: A study that examines the relationship between diseases (or other health-related characteristics) and other variables of interest as they exist in a defined population at one particular time.

cross-sensitivity: The tendency of detector tubes to respond to more than one compound.

C–T: See *concentration–time.*

CTD: Cumulative trauma disorder. See *repetitive strain injury.*

CTS: See *carpal tunnel syndrome.*

cubic meter: See m^3.

cumulative dose: Total dose resulting from repeated exposures.

cumulative trauma disorder (CTD): See *repetitive strain injury.*

current density: The basic exposure limit at the lowest radio frequencies (less than 100 kHz).

customer: Anyone who may be affected by the information developed though sampling or other analysis of the work environment. Customers may be internal or external (e.g., regulatory agencies, employees, purchasers of services/products, and company management).

cutaneous: Pertaining to the skin.

cutaneous malignant melanoma (CMM): See *non-melanoma skin cancer.*

cutoff particle diameter: Diameter of a particle that has 50% probability of being removed by the device or stage and 50% probability of passed through; also called "50% cutpoint," "d_{50}," or the "effective cutoff diameter." (See also *aerodynamic diameter*, *mass median aerodynamic diameter.*)

50% cutpoint size (d_{50}): The parameter used to characterize impactor performance. It is the particle size captured by the impactor with 50% efficiency.

CV: See *coefficient of variation.*

CVS: Cardiovascular system; the heart and blood vessels.

CW: See *continuous wave.*

C-weighted response: The simulation of the sensitivity of the human ear at high sound levels.

cyanosis: Blue appearance of the skin, especially on the face and extremities, indicating a lack of sufficient oxygen in the arterial blood.

cycle time: The time required to carry out a cycle; used principally for time and motion studies.

cycles per second (CPS): See *hertz.*

D

d: See *density correction factor.*

d_{50}: See *50% cutpoint size.*

Dalton's law: Also known as Dalton's Law of Partial Pressure. According to this law, at constant temperature the total pressure exerted by a mixture of gases in a definite volume is equal to the sum of the individual pressures that each gas would exert if occupying the same total volume above. Each constituent of a mixture of gases behaves thermodynamically as if it alone occupied the space. The sum of the individual pressures of the constituents equals the total pressure of the mixture.

damage risk criterion: The suggested baseline of noise tolerance. A damage risk criterion may include in its statement a specification of such factors as time of exposure; noise level and frequency; amount of hearing loss that is considered significant; percentage of the population to be protected; and method of measuring the noise.

damaging wrist motion: 1. A bent wrist involving a force. 2. 1000 damaging wrist motions per hour is an upper ergonomic limit. Also, all pinch grips requiring more than eight pounds of pressure are considered dangerous.

damper, blast gate: A sliding damper used in an air-handling system.

data validation: A systematic effort to review data to identify any outliers or errors and thereby cause deletion or flagging of suspect values to ensure the validity of the data for the user. This "screening" process may be done by manual and/or computer methods, and may use any consistent technique such as pollutant concentration limits or parameter relationships to screen out impossible or unlikely values.

dB: See *decibel.*

DC: See *duty cycle.*

de minimis risk: A risk that is so low as to be negligible (i.e., one case of disease per million persons exposed).

deQuervain's disease: A repetitive strain injury (RSI) characterized by inflammation of the tendons and their sheaths that often causes pain at the base of the thumb and the inside of the wrist. (Also known as "deQuervain's syndrome.")

decibel (dB): A dimensionless unit used to express a logarithmic ratio between a measured quantity and a preference quantity. It is commonly used to describe the levels of acoustic intensity, acoustic power sound pressure levels, and hearing threshold when a reference quantity is specified.

decision tree: A flow chart designed to assist detector tube users in determining sampling strategies and interpreting results.

decomposition: Breakdown of a material or substance (by heat, chemical reaction, electrolysis, decay, or other process) into parts, elements, or simpler compounds.

decompression sickness: A condition marked by the presence of nitrogen bubbles in the blood and other body tissues resulting from a sudden fall in atmospheric pressure. (Also known as "decompression illness," "evolved gas dysbarism," "compressed air sickness," "caisson worker's syndrome," "aeroembolism," and "air embolism.")

decontamination: Removal of harmful substances such as noxious chemicals, harmful bacteria or other organisms, or radioactive material from exposed individuals, rooms and furnishings in buildings, or the exterior environment.

defatting: The removal of fat or oils; defatting the skin results in dryness, flakiness, and a whitish appearance.

default value: Standard numbers used in exposure assessment when more specific data is not available (e.g., body weight, media intake).

degradation: A deleterious change in one or more physical properties of a protective material caused by contact with a chemical.

demand respirator: An atmosphere-supplying respirator that admits respirable gas to the facepiece only when a negative pressure is created inside the facepiece by inhalation.

demyelination: Destruction of myelin; loss of myelin from nerve sheaths or nerve tracts.

density: 1. The mass (weight) per unit volume of a substance. For example, lead is much more dense than aluminum. The density of a material (solid, liquid, gas, or vapor) is given by the relationship between the mass of the material and the volume the mass occupies. 2. The ratio of mass to volume of a material, usually expressed in grams (g) per cubic centimeter (cc). At 4°C (39.2°F), 1 cc of water weighs 1 g.

density correction factor (d): A factor to correct or convert air density at any temperature and pressure to equivalent conditions at ACGIH ventilation-defined standard conditions (STP) and vice versa. For example: Actual air density = (0.075 lb/ft^3) × d, where d = 530/460 + T * BP/29.92 in. Hg. [**Note:** T = degrees Fahrenheit; BP = pressure in in. Hg.]

deoxyribonucleic acid (DNA): The molecules in chromosomes that contain genetic information in most organisms.

depressant: A substance that diminishes bodily functions, activity, or instinctive desire.

derivitization: The process of trapping an airborne chemical onto a sorbent material or filter that has been pretreated with a chemical reagent, thereby causing a chemical reaction that produces a stable compound for analysis.

dermal: Pertaining to the skin.

dermal absorption: The transfer of contaminant across the skin and subsequent incorporation into the body.

dermatitis: Inflammation of the skin from any cause. There are two general types of skin reaction: primary irritation dermatitis and sensitization dermatitis.

desiccant: Material that absorbs moisture.

desiccate: Dry intensively.

desorption: Removal of a substance from the surface at which it is absorbed.

desorption efficiency: A measure of how much of a specific analyte can be recovered from a sorbent; typically expressed as a percent of analyte spiked onto the sorbent.

detector tube: A hermetically sealed glass tube containing an inert solid or granular material such as silica gel, alumina, pumice, or ground glass. The most widely used direct-reading devices. (Also known as "colorimetric detector tube," "colorimetric indicator tube," or "length-of-stain tube.")

detector tube system: A measuring device that consists of a pump and a detector tube. Accessories such as stroke counters, hoses, and probes also might be included. The exception to this is a passive dosimeter, which does not use a pump.

determinant: The substrate or indicator to be measured in a biological system.

detonable: Capable of detonation.

developmental reference dose: Likely to be without appreciable risk of developmental effects, applicable to a single exposure event.

developmental toxicity: A harmful effect on the embryo or fetus; embryotoxicity, fetotoxicity, or teratogenicity.

dew point temperature (T_{dp}): 1. The temperature and pressure at which a gas begins to condense to a liquid. 2. The temperature at which air becomes saturated when cooled without addition of moisture or change of pressure; any further cooling causes condensation.

diffuse reflection: A reflection of light, sound, or radio waves from a surface in all directions according to the cosine law.

diffusion: Aerosol particles in a gaseous medium are bombarded by collisions with individual gas molecules that are in Brownian motion. This causes the particles to undergo random displacements known as diffusion. The particle parameter that describes this process is the particle diffusivity (or diffusion coefficient), D_B.

diffusion system: A flow-metering system based on diffusion through a defined space.

diffusive sampling: Passive samplers that rely on the movement of contaminant molecules across a concentration gradient that for steady-state conditions can be defined by Fick's first law of diffusion.

dilution ventilation: A form of exposure control that relies on the dilution of airborne contaminants into workplace air.

dimensional analysis: Dimensional analysis is a technique to manipulate units as numbers.

dimensions of risk: Attributes of risk that affect perception of that risk (e.g., familiarity or voluntariness of exposure).

direct-reading instruments: A tool available to occupational hygienists for detecting and quantifying gases, vapors, and aerosols. (Also known as "real-time monitors.")

disaster: A calamitous event, especially one occurring suddenly and causing great damage or hardship.

disease cluster: An apparent increase in specific outcomes (disease) among individuals linked in time or space (cluster) or by exposure characteristics.

disinfection: The killing of infectious agents (except bacterial spores) below the level necessary to cause infection. Sanitizers are used on inanimate surfaces; antiseptics are used on skin.

dispersion model: A mathematical model to calculate the dispersion or spread of a pollutant from a source.

displacement: 1. The linear distance from the initial to the final position of an object moved from one place to another, regardless of length of path followed. 2. The distance of an oscillating particle from its equilibrium position.

disposal: Final placement or destruction of toxic, radioactive, or other wastes; surplus or banned pesticides or other chemicals; polluted soils; and drums containing hazardous materials from removal actions or accidental releases. Disposal may be accomplished through use of approved secure landfills, surface impoundments, land farming, deep-well injection, ocean dumping, or incineration.

dissipative muffler: A device that absorbs sound energy as the gas passes through it; a duct lined with sound-absorbing material is the most common type.

distress: Acute physical or mental suffering.
distribution: Once a chemical has entered the body, it might need to be transported to other sites to induce toxic effects.
divergence: Laser beam spread.
DMF: Abbreviation for dimethylformamide.
DMSO: Abbreviation of dimethyl sulfoxide.
DNA: See *deoxyribonucleic acid.*
document control: A systematic procedure for indexing the original document (i.e., Revision No. 0) and subsequent revisions (i.e., Revision No. 1, 2, 3...) by number and date of revision.
dominant-lethal study: A mutagenicity study designed to detect an increase in the incidence of embryo lethality related to exposure of a parent to a test agent.
dosage: Mass of substance per mass of body weight (e.g., mg/kg, where mg represents the amount of substance administered and kg is the body weight of the test animal).
dose: A term used interchangeably with dosage to express the amount of energy or substance absorbed in a unit volume of an organ or individual. Dose rate is the dose delivered per unit of time. Concentration of a contaminant multiplied by the duration of human exposure ($D = C \times T$).
dose adjustment: Modification of doses used in animal experimentation to equivalent levels for human beings. The usual method is to calculate the ratio of body weights raised to some power, which is roughly equivalent to the ratio of surface areas; a simple ratio of body weights has also been used.
dose rate: The dose per unit time.
dose response: In general, the relationship between dose and biological change in organisms; the relationship between administered dose or exposure and the biological change in organisms.
dose-effect study: Laboratory experiment in which animals are given varying doses of known or potentially harmful substances over varying periods of time, and the physical effects are measured in order to set exposure limits for these substances in the occupational environment.
dose-response assessment: See *toxicity assessment.*
dose-response curve: 1. Graphic representation relating biologic response to concentration of contaminant and time of exposure. By multiplying these factors, dose is determined. 2. A mathematical relationship between the dose administered or received and the incidence of adverse health effects in the exposed population; toxicity values are derived from this relationship.
dose-response relationship: With increasing dose, greater biological effects (i.e., responses) will be elicited; that is to say a dose-response relationship can be demonstrated.
dosimeter: 1. Instrument used to detect and measure an accumulated dose of radiation. It is usually a pencil-sized chamber with a built-in self-reading meter, used for personnel monitoring. 2. An instrument that measures the accumulated energy to which one might be exposed (i.e., noise, radiation, etc.). (Also known as "dose meter.")
dosimetry: Accurate measurement of doses.
draft: A current of air, when referring to: 1. The pressure difference that causes a current of air or gases to flow through a flue, chimney, heater, or space. 2. A localized effect caused by one or more factors of high air velocity, low ambient temperature, or direction of airflow, whereby more heat is withdrawn from a person's skin than is normally dissipated.
draft coefficient: A coefficient expressing the resistance encountered by a body when moving in a fluid.
Draize Test: Animal testing to assess the potential irritation or corrosion of a material to skin or eyes.
drug: 1. Any substance used internally or externally as a medicine for the treatment, cure, or prevention of disease. 2. A narcotic preparation.
dry bulb temperature (T_{db}): Temperature of air as determined by a standard thermometer. Temperature units are expressed in degrees Celsius (°C), Kelvin (K) (K = Celsius + 273), or degrees Fahrenheit (°F) (F = 9/5 Celsius + 32).
duct: 1. A passageway for conveying gases. 2. A passageway made of sheet metal or other suitable material, not necessarily leak-tight, used for conveying air or other gas at low pressures. 3. The component of the local exhaust ventilation system that carries contaminants from the hood, through the ventilation system, out of the workroom, through an air-cleaning device, and into the ambient environment.
duct distribution: Distribution of air into a room or a building by ductwork.
duct sizing, equal-friction method: A method of calculating duct size so that frictional resistance per unit length is constant.
duct sizing, static-regain method: A method of calculating duct size so that the regain in static pressure between two draw-off points equals the frictional resistance between the points.
duct sizing, velocity-reduction method: A method of calculating duct size so that selected velocities occur in specific duct lengths.
duct system: A series of ducts, elbows, and connectors to convey air or other gases from one location to another.
duct transition section: A section of duct, breeching, or stack used to connect these elements with structures of different cross-sectional dimensions.
duct velocity: Air velocity through the duct cross section. When solid material is present in the air stream, the duct velocity must be equal to the minimum design duct velocity or transport velocity.
duplicate sample: A sample collected in the same location and manner as an actual sample and used to evaluate the entire sampling/analysis method.
dust: 1. Solid particles that are capable of temporary suspension in air or other gases. Usually produced from larger masses through the application of physical forces (for example, handling, crushing, grinding, rapid impact). Typical dusts are rock, ore, metal, coal, wood, and grain. Size ranges are usually between 0.1 µm and 30.0 µm. Particles may be up to 300–400 µm, but those above 20–30 µm usually do not remain airborne. 2. Air suspension (aerosol) of solid particles, usually particle size less than 100 µm. 3. Fine, solid particles; small dust particles may be respirable (less than 10 µm).
duty cycle (DC): The ratio of on-time RF emissions to the total time of operation (on- time plus off-time.

duty cycling (electric): The process of turning off electrical equipment for predetermined periods during operating hours to reduce consumption and demand.

duty factor: See *duty cycle.*

dying back: The condition in which prolonged or repeated exposure to toxics can cause the neural axon to continue to degenerate from the distal to the proximal end.

dynamic blank: A blank that is prepared, handled, and analyzed in the same manner as normal carrying agents except that it is not exposed to the material to be selectively captured. For example, an absorbing solution that would be placed in bubbler tube, stoppered, transported to a monitoring site, left at the site for the normal period of sampling, returned to the laboratory, and analyzed. (Also known as "field blank.")

dynamic calibration: Calibration of a measurement system by use of calibration material having characteristics similar to the unknown material to be measured. For example, the use of a gas containing carbon dioxide of known concentrations in an air mixture could be used to calibrate a carbon dioxide sensor system.

dynamic load: Muscle (local) fatigue is divided into static and dynamic. When muscles are loaded but do not move, it is a static load; when loaded muscles move, it is dynamic.

Dynamic System: A calibration system in which the standard gas solution is generated continually and is flowing.

dyne: The centimeter-gram-second unit of force, equal to the force required to impart an acceleration of 1 centimeter per second (cm/sec) to a mass of 1 gram (g). An arcane term.

dysbaric osteonecrosis: Dysbaric osteonecrosis causes detectable lesions most commonly on the body's long bones. Although its etiology is unknown, this chronic disease might be related to the evolution of gas bubbles that may or may not be diagnosed as decompression sickness.

dysbarism: A condition of the body resulting from the existence of a pressure differential between the total ambient pressure and the total pressure of dissolved and free gases within the body tissues, fluids, and cavities.

dyspnea: Labored or difficult breathing; a symptom.

E

E: See *electric-field strength.*

EAP: See *emergency action plan.*

ear: The entire hearing apparatus, consisting of three parts: the external ear, the middle ear, and the inner ear.

ear protectors: Plugs or muffs designed to keep noise from the ear to preserve hearing acuity.

ear wax: The waxy discharge in the outer ear canal.

earmuff: A type of hearing protector worn outside the ear.

eaters: Organic compounds that may be made by interaction between an alcohol and an acid, and by other means, and includes solvents and natural fats.

eccentric: A muscle contraction where the muscle lengthens while developing tension.

ECD: See *electron capture detector.*

EC_X: See *effective concentration "X."*

eczema: A disease of the skin characterized by inflammation, itching, and formation of scales.

edema: An excessive accumulation of fluid in the cells, tissue spaces, or body cavities due to a disturbance in the fluid exchange mechanism.

E_{eff}: See *effective irradiance.*

EEGL: See *emergency exposure guidance level.*

effective concentration "X" (EC_X): The concentration of a material that has caused a biological effect to X percent of the test animals.

effective temperature (ET): The sensation of warmth or cold felt by the human body.

EHS: See *extremely hazardous substance.*

EL: See *explosive limits.*

electric field: 1. One of the fundamental fields in nature, causing a charged body to be attracted to or repelled by other charged bodies. Associated with an electromagnetic wave or a changing magnetic field. 2. Specifically, the electric force per unit test charge.

electric-field strength (E): The force on a stationary positive charge per unit charge at a point in an electric field. (Also known as "electric vector," "electric field intensity.")

electrochemical detector: An electrochemical detector responds to compounds (such as phenols, aromatic amines, ketones, aldehydes, and mercaptans) that can be readily oxidized or reduced. Electrode systems use working and reference electrodes to quantify analytes over a range of six orders of magnitude.

electrochemical sensors: Sensors used by a variety of instruments dedicated to monitoring specific single gas and vapor contaminants. Electrochemical sensors are available for up to 50 different individual gases, including oxygen, carbon monoxide, nitric oxide, nitrogen dioxide, hydrogen sulfide, hydrogen cyanide, and sulfur dioxide.

electrogoniometer: A device measures joint angles. An electrogoniometer is a potentiometer that is placed at the joint center and two extensions that are attached to the limbs that intersect at the joint. The electrogoniometer is generally interfaced to a computer via an analog-to-digital converter and may be sampled at a very high rate. These devices can be designed to measure rotations about one, two, or three axes at a joint.

electrolyte: A chemical compound that when molten or dissolved in certain solvents, usually water, will conduct an electric current.

electromagnetic radiation: The propagation, or transfer, of energy through space and matter by time-varying electric and magnetic fields.

electromagnetic spectrum: The total range of wavelengths or frequencies of electromagnetic radiation, extending from the longest radio waves to the shortest known cosmic rays. Any location on the spectrum may be characterized by wavelength, frequency, and photon energy.

electromagnetic susceptibility: A problem experienced with sampling equipment due to electromagnetic fields in the environment that might result in errors or malfunctions in operation.

electromyogram (EMG): The detected electrical signal of a muscle contraction.

electromyograph: An instrument that converts the electrical activity of muscles into visual records or sound. Electromyography is used to diagnose carpal tunnel syndrome and other repetitive strain injuries.

electromyography: The study of the electrical signal associated with a muscle contraction.

electron: A subatomic particle that carries a negative charge.

electron capture detector (ECD): Extremely sensitive gas chromatography detector that is a modification of the argon ionization detector, with conditions adjusted to favor the formation of negative ions.

electron equilibrium: A point or area in a radiation detector in which the number of electrons entering equals the number of electrons leaving the local point volume.

electronic mail (e-mail): The transmission of messages over communications networks. The messages can be notes entered from the keyboard or electronic files stored on disk. Most mainframes, minicomputers, and computer networks have an e-mail system. Some electronic-mail systems are confined to a single computer system or network, but others have gateways to other computer systems, enabling users to send electronic mail anywhere in the world. Companies that are fully computerized make extensive use of e-mail because it is fast, flexible, and reliable.

element: Solid, liquid, or gaseous matter that cannot be further decomposed into simpler substances by chemical means.

ELF: See *extremely low frequency.*

elimination: Internal clearance of a marker from an internal organ.

Ellenbog, Ulrich: An Austrian physician who in 1473 described the symptoms of poisoning from lead and mercury and suggested preventive measures.

elutriator: A device used to separate respirable and nonrespirable particulates such as the cyclone or horizontal types.

e-mail: See *electronic mail.*

embryo: The early or developing stage of any organism. In animals, the period of development beginning when the long axis is established and continuing until all major structures are represented.

embryogenesis: During early human development, several cellular phases occur during a process known as embryogenesis. The phases include cell proliferation, cell differentiation, cell migration, and organogenesis.

embryotoxin: A material harmful to the developing embryo.

emergency: 1. A sudden, urgent, usually unforeseen occurrence or occasion requiring immediate action. 2. Any occurrence (including any failure of hazard control or monitoring equipment) or event internal or external to the permit space that could endanger entrants. (See *OSHA 1910.146.*)

emergency action plan (EAP): The basis for identifying and assessing the risks associated with potential emergencies, planning for appropriate response and recovery, and training those who would be involved or affected should the emergency occur. Facilities that depend solely on the local community fire, police, and emergency medical services to handle an emergency without including them in emergency response planning might later find an increase in injuries, deaths, and economic losses while time is lost struggling to determine the proper course of action. An EAP provides knowledge to everyone affected so that they may react and respond quickly and safely.

emergency exposure guidance level (EEGL): Established by the Committee on Toxicology of the National Research Council (NRC) for the U.S. Department of Defense. EEGLs provide guidelines for military personnel operating under emergency conditions in which circumstances are peculiar to military operations. The EEGL is a ceiling guidance level for single emergency exposures usually lasting 1 hour to 24 hours — an occurrence expected to be infrequent in the lifetime of an individual. (See also *CEGL, SPEGL.*)

emergency planning: Planning for an unintended event.

emergency response: Responding to an unintended event.

Emergency Response Planning Guideline 1 (ERPG–1): The maximum airborne concentration below which nearly all individuals could be exposed for up to 1 hour without experiencing more than mild, transient adverse health effects or without perceiving a clearly defined objectionable odor.

Emergency Response Planning Guideline 2 (ERPG–2): The maximum airborne concentration below which nearly all individuals could be exposed for up to 1 hour without experiencing or developing irreversible or other serious health effects or symptoms that could impair an individual's ability to take protective action.

Emergency Response Planning Guideline 3 (ERPG–3): The maximum airborne concentration below which nearly all individuals could be exposed for up to 1 hour without experiencing or developing life-threatening health effects.

Emergency Response Planning Guidelines (ERPGs): Values intended to provide estimates of concentration ranges above which one could reasonably anticipate observing adverse health effects (see *ERPG–1; ERPG–2; ERPG–3*). The term also refers to the documentation that summarizes the basis for those values. The documentation is contained in a series of guides produced by the Emergency Response Planning Committee of AIHA.

EMF: Acronym for "electric magnetic fields" or "electromagnetic fields."

EMG: See *electromyogram.*

emission: 1. Material released into the air either by a primary source or a secondary source, as a result of a photochemical reaction or chain of reactions. 2. Any radiation of energy by means of electromagnetic waves, as from a radio transmitter. 3. A discharge of fluid from a living body.

emphysema: A lung disease resulting from the enlargements of the alveoli accompanied by destruction of normal tissue.

employee: A worker who might be exposed to hazardous chemicals under normal operating conditions or in foreseeable emergencies. Workers such as office workers or bank tellers who encounter hazardous chemicals only in unusual isolated instances are not covered.

encephalopathy: 1. Any disease of the brain. 2. Acute toxic encephalopathy is characterized by headaches, irritability, poor coordination, seizures, coma, and death. Causative agents include carbon monoxide, organic solvents such as carbon disulfide, and metals (including lead and manganese). 3. Chronic toxic encephalopathy is characterized by a gradual loss of memory and psychomotor control, dementia, and motor disorder. Associated toxicants include arsenic, lead, manganese, and mercury.

enclosing hood: A hood that either completely or partially encloses the contaminant emission source.

end-exhaled breath: The exhaled breath forced from the lungs after natural exhalation. (Also known as "alveolar exhaled breath.")

endogenous: Intrinsic; found naturally in the living system under study.

endotherm: Absorption of heat.

endotoxin: 1. A toxin that is produced within a microorganism and can be isolated only after the cell is disintegrated. 2. A lipopolysaccharide that forms the outer cell wall of gram-negative bacteria.

energy: 1. The capacity to do work. 2. Having several forms that may be transformed from one to another, such as thermal (heat), mechanical (work), electrical, and chemical.

engineering controls: Process change, substitution, isolation, ventilation, source modification.

engineering solution: The use of machines to reduce work-related stress.

entry: The action by which a person passes through an opening into a permit-required confined space. Entry includes ensuing work activities in that space and is considered to have occurred as soon as any part of the entrant's body breaks the plane of an opening into the space. (See *OSHA 1910.146*.)

entry loss: Loss in pressure caused by a fluid stream flowing into a pipe, duct hood, or vessel.

entry personnel: Any individuals directly involved in a confined space entry. This may include the entrant, attendant, and entry supervisor. The designated attendant does not actually enter the space. (Also known as "entry crew.")

entry supervisor: The person (such as the employer, foreman, or crew chief) responsible for determining whether acceptable entry conditions are present at a permit space where entry is planned, for authorizing entry and overseeing entry operations, and for terminating entry as required by this section. (See *OSHA 1910.146*.)

environmental conditions: Natural or controlled conditions of air and radiation prevailing around a person, an object, a substance, etc.

environmental monitoring: Program in which samples of air contaminants or energy measurements are taken and which establishes the level of worker exposure to such agents.

environmental quality: Any standard specifying lower limits for contaminants, chemical or physical agents, and/or resulting stresses to the human body in order to maintain a particular, healthful, and safe environment in which to work.

enzyme: An agent that catalyzes a biological reaction that is not itself consumed in the reaction.

epicondylitis: An repetitive strain injury (RSI) characterized by swelling and pain in the tendons and muscles around the elbow joint. (Also known as "tennis elbow.")

epidemiological surveillance: An ongoing systematic analysis and interpretation of the distribution and trends of illness, injury, and/or mortality in a defined population, relative to one or more indicators of workplace hazards or risks. (Also known as "rate-based surveillance.")

epidemiology: The science that deals with the incidence distribution and control of disease in a population.

equation of state: An equation that relates the pressure (P), volume (V), and thermodynamic temperature (T), and an amount of substance (n). The simplest form is the Ideal Gas law: $PV = nRT$.

equilibrium: Condition in which a particle, or all the constituent particles of a body, are at rest or in unaccelerated motion in an inertial reference frame. (Also known as "static equilibrium.")

equivalent chill temperature: The expression of windchill reflecting the cooling power of wind on exposed flesh.

equivalent length: The resistance of an appurtenance in a conduit through which the fluid flows; expressed as the number of feet of straight conduit of the same diameter that would have the same resistance.

ergonomics: The application of human biological sciences with the engineering sciences to achieve optimum mutual adjustment of people and their work, the benefits measured in terms of human efficiency and well-being. (Also known as "human factors engineering.")

ERPGs: See *Emergency Response Planning Guidelines*.

error: The difference between an observed or measured value and the best obtainable estimate of its true value.

erythema: Redness of the skin.

escape: The act of entrants exiting the space by their movement during an emergency condition.

eschar: A dry scab that forms as a result of a burn or other corrosive action.

estimated risk: Prediction of risk level.

ET: See *effective temperature*.

etiology: The study or knowledge of the causes of disease.

evaluation: One of four primary responsibilities of the occupational hygienist. The examination and judgment of the amount, degree, significance, worth, or condition of something. Evaluation perhaps uses more "art" in its implementation and than any of the other occupational hygiene responsibilities. (See also *anticipation*, *recognition*, and *control*.)

evaporation rate: 1. The ratio of the time required to evaporate a measured volume of a liquid to the time required to evaporate the same volume of a reference liquid under ideal test conditions. The higher the ratio, the slower the evaporation rate. 2. The rate at which a material is converted from the liquid or solid state to the vapor state; may be expressed relative to the evaporation rate of a known material, usually n-Butyl acetate (with an evaporation rate of 1.0 by definition). Faster evaporation rates are >1, and slower evaporation rates are <1.

excretion: Appearance of a marker outside the body.

excursion: Deviation from a definite path. A movement above or below a norm.
excursion factor: Maximum extent to which an ACGIH threshold limit value (TLV®) can be exceeded.
exfoliation: Peeling or flaking of skin.
exhaust air: Air discharged from any conditioned space.
exotherm: Liberation of (reaction) heat.
expiration date: The date beyond which the manufacturer will no longer assure the stability and reliability of the tube.
explosive: A chemical that causes a sudden, almost instantaneous release of pressure, gas, and heat when subjected to sudden shock, pressure, or high temperature.
explosive limits: See *flammable limits.*
exposure: 1. As it pertains to air contaminants, it is the state of being exposed to a concentration of a contaminant. 2. Subjection of an employee in the course of employment to a chemical that is a physical or health hazard, and includes potential (e.g., accidental or possible) exposure. "Subjected" in terms of health hazards includes any route of entry (e.g., inhalation, ingestion, skin contact, or absorption). 3. Contact of an organism with a chemical or physical agent, quantified as the amount of chemical available at the exchange boundaries of the organism and available for absorption. Usually calculated as the mean exposure and some measure of maximum exposure. 4. The amount of an environmental agent that has reached the individual (external dose) or has been absorbed into the individual (internal dose or absorbed dose).
exposure assessment: Determination or estimation (qualitative or quantitative) of the magnitude, frequency, duration, and route of exposure.
exposure event: An incident of contact with a chemical or physical agent. It can be defined by time or incident.
exposure limit: A limit established to prevent an adverse health effect.
exposure limit value: General term designating any standard or measurement restricting human exposure to harmful or toxic agents.
exposure modeling: A mathematical model used to compute a worker's exposure.
exposure pathway: Path a chemical or physical agent takes from source to exposed organism. Consists of 1) source or release; 2) transport medium (possible); 3) exposure point; and 4) exposure route.
exposure point: Location of potential contact between an organism and a chemical or physical agent.
exposure profile: Graphic presentation of data on exposure of workers to contaminants in industry.
exposure rating: An estimate of exposure level relative to the occupational exposure limit (OEL).
exposure route: The way an organism comes into contact with a chemical or physical agent (e.g., ingestion, inhalation, dermal contact, etc.).
exposure surveillance: The systematic and ongoing characterization of chemical or physical agents in the occupational setting, often to determine how many employees have been exposed and to what extent over time. This approach may be especially useful in those instances where there is an absence of health outcomes known to be associated with potential hazardous exposures in the workplace. (Also known as "hazard surveillance.")
extensor: Muscle that straightens (extends) a limb.
exterior hood: A hood that is located adjacent to a contaminant source but does not enclose it.
external quality control: The activities that are performed occasionally, usually initiated and performed by persons outside of normal routine operations (such as on-site system surveys, independent performance audits, interlaboratory comparisons) to assess the capability and performance of a measurement process.
extrapolation: The process of estimating unknown values from known values.
extremely hazardous substance (EHS): As defined by the U.S. Environmental Protection Agency in the Superfund Amendment and Reauthorization Act (SARA) Title III.
extremely low frequency (ELF): An order of magnitude band designation usually applied to the part of the electromagnetic spectrum between 30 Hz and 300 Hz.
extrinsic safety: Plants that require external features to control hazards. Those features should be arranged to minimize the undesirable consequences of a release.

F

F: See *frequency.*
face velocity: See *hood face velocity.*
factor of safety: Ratio of a normal working condition to the ultimate conditions, such as, in strength of materials, ratio of working stress to ultimate strength.
Fahrenheit temperature: The temperature scale in which at standard atmospheric pressure the boiling point of water is 212°F and the freezing point is 32°F. (See also *Rankine temperature.*)
fail-safe: Redundant engineering designed to prevent loss in case of a disaster.
FAM: See *fibrous aerosol monitor.*
fan: 1. A device for moving air by two or more blades or vanes attached to a rotating shaft. Provides the energy to move air through the system. 2. The local exhaust ventilation system component that provides the energy required by a specific design to move air through the system.
far field: In noise measurement, this refers to the distance from the noise source where the sound pressure level decreases 6 dB for each doubling of distance. (Also known as "free field," "Inverse Square law.")
fate: Destiny of a chemical or biological pollutant after release into the environment. It involves temporal and spatial considerations of transport, transfer, storage, and transformation. (Also known as "environmental fate.")
Fatigue Decreased Proficiency: See *Criteria for Fatigue Decreased Proficiency.*
fault tree analysis: A graphical method of performing a failure modes and effects analysis.
FDP: See *Criteria for Fatigue Decreased Proficiency.*
feces: Solid/liquid waste excreted from the anus.
feedback: The reaction of some results of a process serving to alter or reinforce the character of that process.
fempto: A prefix meaning 10^{-15}.
fetotoxin: A material harmful to the fetus.
fetus: The unborn offspring of an animal in the postembryonic period, after all major structures are represented.

FFT spectrum analyzer: It divides the audible frequency spectrum into even smaller bands. This type of analysis can be used to identify tones that can be traced to specific pieces of equipment.

fiber: An elongated particle having an aspect ratio (i.e., a ratio of length to width of greater than 3:1). A fiber may be naturally occurring (such as plant fibers and asbestiform silicate minerals) or synthetic (such as vitreous or graphite fibers).

fibrillation: Rapid uncoordinated contractions of the heart that are ineffective in pumping blood.

fibrosis: The development in an organ of excess fibrous connective tissue usually as a reparative or reactive process.

fibrous: Made up of fibers or fiber-like tissue.

fibrous aerosol monitor (FAM): Modified light-scattering monitor that is a direct- reading device designed to measure airborne concentrations of fibrous materials (such as asbestos and fiber glass) with a length-to-diameter aspect ration greater than three. Results are reported as a fiber count rather than mass concentration.

FID: See *flame ionization detector.*

field data sheets: A marked-up floor plan or system schematic to show where the field data is to be collected, the methodology for collecting the data, and the individual responsible for collecting the data. The use of field data sheets ensures that the field data is collected in a concise manner, that all of the data required to meet the project objectives are recorded, and that future data are collected in a similar manner for trend analysis and correlation purposes.

filter: 1. A device to remove solid material from a fluid. 2. A capacitor and/or inductor placed in a series/parallel combination across a DC line to remove the effects of the AC signal or to decrease the ripple voltage in a DC power supply. 3. A component used in respirators to remove solid or liquid aerosols from the inspired air.

filter bank: Interchangeable frame or cylinder containing a filtering material. Most HVAC system filters remove solid material (particles) from the air stream. (Also known as "filter cell," "filter cartridge," "filter unit," "filter element.")

fire point: The lowest temperature at which a volatile, combustible material can evolve enough vapors to support combustion.

fit check: A test conducted by the wearer to determine whether the respirator is properly seated to the face.

fit factor: A quantitative measure of the fit of a particular respirator to a particular individual.

fit-test: The use of a challenge agent to evaluate the fit of a respirator on an individual.

fixators: Skeletal muscles that generally stabilize and control the bony structures that constitute the joint.

flame ionization detector (FID): A device in which the measured change in conductivity of a standard flame (usually hydrogen) due to the insertion of another gas or vapor is used to detect the gas or vapor.

flame photometric detector (FPD): The flame photometric detector is used to measure phosphorus- and sulfur-containing compounds such as organophosphate pesticides and mercaptans. The FPD measures phosphorus- or sulfur-containing compounds by burning the column effluent in a hydrogen-air flame with an excess of hydrogen. Sulfur and phosphorus compounds emit light above the flame; a filter optimized to pass light at 393 nm is used to detect sulfur compounds; a filter optimized to pass light at 535 nm is used to detect phosphorus compounds. A photomultiplier tube is then used to quantify the amount of light passing through the selective filter.

flammable: The capability of a substance to be set on fire or support combustion easily. Flammable is not synonymous with "inflammable."

flammable limits: 1. The range of flammable vapor or gas-air mixture between the upper and lower flammable limits. 2. The minimum and maximum concentrations in air of a flammable gas or vapor at which ignition can occur. Concentrations below the lower flammable limit (LFL), also referred to as the lower explosive limit (LEL), are too lean to burn; concentrations above the upper flammable limit (UFL), also referred to as the upper explosive limit (UEL), are too rich to burn.

flammable liquid: Any liquid having a flash point below 37.8°C (100°F).

flange: A surface at and parallel to the hood face that provides a barrier to unwanted airflow from behind the hood.

flash blindness: A temporary effect in which visual sensitivity and function is decreased severely in a very short period. (Also known as "blinding glare.")

flash point: 1. Minimum temperature to which a product must be heated for its vapors to ignite momentarily in a flame when operating under standardized conditions. (See also *fire point*). 2. In a vacuum cooling chamber, that pressure corresponding to the vapor pressure at the product temperature and below which water vaporizing commences. 3. The temperature at which the vapor above a volatile liquid forms a combustible mixture with air.

flat response: A response that looks at the entire audible frequency spectrum without applying any weighting.

flat-file database: A relatively simple database system in which each database is contained in a single table. In contrast, relational database systems can use multiple tables to store information, and each table can have a different record format. Relational systems are more suitable for large applications, but flat databases are adequate for many small applications.

flatus: Sudden excretion of internal gas/vapor from the mouth or the anus.

flexion: Act of bending, especially of a joint.

flexor: Muscle that bends (flexes) a limb.

flow chart: A graphical representation of the progress of a system for the definition, analysis, or solution of a data-processing or manufacturing problem in which the symbols are used to represent operations, data or material flow, and equipment, and lines and arrows represent interrelationships among the components. (Also known as a "control diagram," "flow diagram," "flow sheet.")

flow rate standards: Measurements based on discrete standards, such as the platinum kilogram (known as "K-20") for mass and the platinum meter bar for length. For flow rate measurements there is no off-the-shelf identity standard such as a gallons per minute or a liters per second. To supply a fundamental basis for any flow rate measurement the identity standard must be one that is derived.

flow-dilution system: A system that continuously mixes accurately metered flows of a test component with a diluent (e.g., clean air).

fluids: A state of matter that flows under pressure (i.e., gas and liquid states).

fluorescence detector: The fluorescence detector measures the emission of light produced by fluorescing eluents and is extremely sensitive to highly conjugated aromatic compounds such as PAHs. Some analytical methods use derivitization reagents to fluoresce the analyte. In these methods a light source raises the fluorescent analyte to an unstable higher energy level, which quickly decays in two or more steps, emitting light at longer wavelengths. The basic detector components are a lamp; a flow cell with windows at a 90° angle for the column effluent; filters or diffraction gratings to select the excitation and emission wavelengths; and a photomultiplier tube or other light-measuring device.

flux: The absorption rate per unit area.

fog: General term applied to visible aerosols in which the dispersed phase is liquid: formation by condensation is implied.

foot candle: A unit of illumination. The illumination at a point on a surface that is one foot from, and perpendicular to, a uniform point source of one candle.

force: An interaction of two objects that produces a change in the state of motion of an object. A force may cause an object to move, accelerate or decelerate it, change its direction, or stop it from moving.

force couple: The pair of forces arranged to produce pure rotation (angular motion).

force platform: A force platform measures the ground reaction force (GRF) applied to the body by the surface of the platform.

force ratio: See *mechanical advantage*.

formable earplug: A type of hearing protector that is formed by the user prior to being inserted in the ear canals.

formative evaluation: Gathering information on adequacy and using this information as a basis for further development. (Also known as "process evaluation.")

formula: The scientific expression of the chemical composition of a material.

Fourier transform infrared spectrometry: A spectroscopic technique in which all pertinent wavelengths simultaneously irridate the sample for a short period and the absorption spectrum is found by mathematical manipulation of the Fourier transform so obtained.

FPD: See *flame photometric detector*.

free-body diagram: A conceptual drawing of the forces and moments acting on the system.

free-layer damping: A layer of nonhardening viscoelastic material (usually in the form of tapes, sheets, mastics, or sprays) is adhered to the surface. (Also known as "extensional damping.")

freezing point: For a particular pressure, the temperature at which a given substance will solidify or freeze upon removal of heat. The freezing point of water is 32°F (0°C).

frequency (F): The time rate of repetition of a periodic phenomena. The frequency is the reciprocal of the period and is sometimes called pitch.

frequency of sound: Rate of oscillation or vibration; units are 1 cycle per second (cps) or 1 hertz (Hz).

fresh air: Air taken from the outside ambient environment. The fresh air must satisfy the EPA Ambient Air Quality Standards. (Also known as "outdoor air," "outside air.")

fresh-air make up: The volume of outside air introduced into an occupied space to replace the air removed from a space. (Also known as "replacement air.")

friction factor: 1. A coefficient used to calculate friction forces due to fluid flow. 2. Quotient of the tangential force exerted by a fluid on a surface (per unit area) by half the product of the density and the square of the velocity.

friction loss: Pressure loss due to friction between a flowing fluid and its contact surface.

frictional resistance: The resistance of fluid flow due to friction between the fluid and the contact surface it flows past.

fritted glass bubblers: In fritted glass bubblers, air passes through formed porous glass plates and enters the liquid in the form of small bubbles.

frostbite: Injury to skin and subcutaneous tissues, and in severe cases also to deeper tissues, from exposure to extreme cold.

full work cycle: Amount of time required to complete a task or process; may be less or more than an 8-hour shift.

full-shift sampling: The collection of survey information over the course of an 8-hour day to determine the actual 8-hour time-weighted average (TWA) exposure concentration. It is the preferred method to use to accurately characterize an operation for the gathering of baseline and routine monitoring data and to document regulatory compliance.

fume: 1. Minute solid particles generated by condensation from the gaseous state, generally after volatilization (evaporation) from melted substances, such as welding, and often accompanied by a chemical reaction, such as oxidation. Examples are iron oxide from welding; lead oxide from soldering; and copper oxide from smelting. Size ranges are usually between 0.001 μm and 1.0 μm. 2. Very small, airborne particles commonly formed by condensing vapors from burning or melting materials. (See also *aerosol, dust*.)

functional analysis: A mathematical analysis that examines each aspect of the measurement system (sampling and analysis) in order to quantitate the effect of sources of error. A functional analysis is usually performed prior to a ruggedness test to determine those variables that should be studied experimentally.

fundamental unit: Mass, length, and time.

G

g: See *gram*.

g/kg: Grams per kilogram body weight; an expression of dosage.

gage: 1. An instrument for measuring pressure, flow, or level. 2. A scale of measurement for sheet metal thickness, wire and drill diameters, etc. (Also known as "gauge.")

gamma ray: A photon emitted from a nucleus.

gas: Any material in the gaseous state at 25°C and 760 mmHg. Normally, a formless fluid, it expands to fill the space or enclosure. Gases can be changed to the liquid or solid state only by the combined effect of increased pressure and decreased temperature. Examples are welding gases, internal combustion engines exhaust gases, and waste gases from refining or sewage (such as hydrogen sulfide, waste anesthesia gases, hydrogen, and ammonia). Size ranges are usually less than 0.0005 µm. Within acceptable limits of accuracy, satisfies the perfect gas law.

gas chromatograph (GC): Highly sophisticated instrument that identifies the molecular composition and concentrations of various chemicals in water and soil samples. (Also known as "mass spectrometer.")

gas chromatography: A separation technique involving passage of a gaseous moving phase through a column containing a fixed adsorbent phase; it is used principally as a quantitative analytical technique for volatile compounds.

gas constant (R): The coefficient (R) in the Ideal Gas law equation given by PV = nRT; the constant factor in the equation of state for ideal gases. In the SI system of units, R = 0.08205 liters * atmosphere/mole (K); in the I-P system, 21.85 cubic feet * inches mercury/pounds mole (°R).

gas narcosis: Gas narcosis is caused by nitrogen in normal air during dives of more than 120 feet (35 meters). Helium, substituted for nitrogen in "mixed gas diving," can cause an effect called "high pressure nervous syndrome" beyond 500 fsw.

gas solubility: The extent that a gas dissolves in a liquid to produce a homogenous system.

gas toxicities: Gas toxicities caused by oxygen and carbon dioxide. The damage of lack of oxygen to the lung and brain (central nervous system [CNS]) will vary with time of exposure and depth. While a carbon dioxide partial pressure of 15–40 mmHg will stimulate the central respiratory sensor, concentrations >80 mmHg suppress respiration.

gaseous exchange: In the alveoli, the absorption of oxygen and concomitant removal of waste gases.

gastrointestinal tract: The mouth, esophagus, stomach, intestines, and related organs.

gauge pressure: The difference between two absolute pressures, one of which is usually atmospheric pressure.

gauss: The cgs (centimeter-gram-second) unit of magnetic flux density.

gavage: Introduction of a test agent through a tube passed into the stomach.

GC: See *gas chromatograph.*

general duty clause: Section 5(a)(1) of the Administrative Procedures Act (see *29 CFR 1905*) imposes the following general duty: "Each employer—(1) shall furnish to each of his employees employment and a place of employment which are free from recognized hazards that are causing or are likely to cause death or serious physical harm to his employees."

general motion: A combination of both translation and rotation. (See also *rotation, translation.*)

general reliability: The capability of an item or system to perform a required function under stated conditions for a stated period of time.

general ventilation: Ventilation systems designed primarily for temperature, humidity, and odor control; health hazard protection is secondary. (Also known as "comfort ventilation.")

generation: A group of organisms having a common parent or parents and comprising a single level in line of descent.

genetic engineering: A process of inserting new genetic information into existing cells in order to modify any organism for the purpose of changing one of its characteristics.

genetic mutation: A mutation that involves expression of the aberration in the offspring of the exposed individual due to alteration of germ or sex cells (male spermatozoa and/or female ova).

genotoxic chemical: An electrophilic (i.e., electron-deficient) compound that has an affinity for genetic information, specifically the electron-dense (i.e., nucleophilic) DNA.

genotoxin: A material harmful to the genetic material.

geometric mean (σ_g): Mathematically, the geometric mean (σ_g) can be expressed in two equivalent ways or in words, the nth root of the product of all values in a set of n values, or the antilogarithm of the arithmetic mean of the logarithms of all the values of a set of n values.

geometric standard deviation (GSD): A measure of dispersion in a lognormal distribution. The value will always be equal to or greater than one. Typically, lognormal distribution is used to describe particle size distribution of an aerosol sample. (See also *lognormal distribution.*)

geometry: The physical orientation of a detector to the radiation source.

globe temperature (T_g): The measure of radiant heat.

globe thermometer: It exchanges heat with the environment by radiation, convection, and conduction. It stabilizes when the heat exchange by radiation is equivalent to that by convection and conduction: normally 15–20 min is required for a globe that is 15 cm in diameter.

GLR: See *graphic level recorder.*

good samaritan doctrine: A legal theory in which persons who bring a personal injury action against a defendant who normally does not owe a duty of care to the plaintiff show that defendant owed a duty to the plaintiff, that they negligently failed to discharge that duty, and that the plaintiff's injury was caused by such negligence.

grab sample: A sample taken within a short time period, generally to determine the contaminants at a specific time or during a specific event.

grab sampling: The direct collection of an air-contaminant mixture into a device such as a sampling bag, syringe, or evacuated flask over a few seconds or minutes.

gram (g): A unit of mass; defined as the mass of 1 cc of water at 4°C (39.2°F).

gram mole: The amount of substance represented by one gram molecular weight or one gram molecular volume (mole).

gram molecular weight: The sum of the individual atomic weights of all the atoms in a molecule (express mass in units of grams, g).

graphic level recorder (GLR): An instrument for providing a written record of the sound levels of particular events as a function of time.

gravimetric: Of or pertaining to measurement by weight.

gravimetric analysis: A process in which a known volume of aerosol-laden air is drawn through a filter of known initial weight, then reweighing the filter to determine the mass captured.

GSD: See *geometric standard deviation.*

H

H: See *magnetic-field strength.*

HSI: See *heat stress index.*

Haber's law: The Haber relationship expresses the constancy of the product of exposure concentration and exposure duration (Ct = K, where C represents exposure concentration, t is time, and K is constant). The Haber relationship does not hold over more than small differences in exposure time.

HACE: See *high altitude cerebral edema.*

hair: The flexible shaft of distinct coloring that protrudes from the skin surface.

half-time (pseudo first order): $t_{0.5} = 0.693/k$ where k is the pseudo first-order process rate constant in units of time.

halo formation: A visible ring or line of excess particles that are deposited on the substrate or around the nozzle. The presence of the halo formation indicates incorrect sample flow through the nozzle and is very undesirable.

halocarbon: Partially or fully halogenated hydrocarbon.

Hamilton, Dr. Alice: An American physician and social reformer. She entered the field of industrial medicine in 1910 and not only presented substantial evidence of the relationship between toxins and ill health, but also provided solutions to the problems she encountered. In 1919 she became the first woman faculty member at Harvard University and in 1943 wrote her autobiography *Exploring the Dangerous Trades.* She has been called the "Mother of American occupational medicine."

hand-arm vibration syndrome (HAVS): A condition contracted after prolonged exposure to hand-arm vibration with symptoms that include intermittent tingling and/or numbness of the fingers and finger blanching (turning white). With additional vibration exposure, the symptoms of HAVS become more severe and include increasing stiffness of the finger joints, loss of manipulative skills, and loss of blood circulation, which can lead to gangrene and tissue necrosis.

HAPE: See *high altitude pulmonary edema.*

hapten: A simple substance that reacts like an antigen *in vitro* by combining with antibody; may function as an allergen when linked to proteinaceous substances of the tissue.

hardware: Hardware is the electronic and physical apparatus necessary to run computer programs (i.e., software).

HAVS: See *hand-arm vibration syndrome.*

Hawthorne Effect: A tendency for employees to do the job in a nonroutine manner while being observed.

HAZAN: See *hazard analysis.*

hazard: Source of risk.

hazard analysis (HAZAN): A generic term for a variety of quantitative hazard and risk analysis methods. A logical sequence for the systematic examination of a facility is to use hazard identification and ranking techniques first. If there are possible scenarios that could lead to unacceptable consequences, then qualitative techniques (HAZOP) can be applied next. When there is a qualitative hazard, HAZAN methods can be used to estimate the quantitative probability of adverse events. How much analysis is worthwhile is a function of the consequence of the adverse event and the difficulty in preventing it.

hazard and operability (HAZOP) survey: One of the most common and widely accepted methods of systematic qualitative hazard analysis. It is used for both new and existing facilities and can be applied to a whole plant, a production unit, or a piece of equipment. It uses as its database the usual sort of plant and process information and relies on the judgment of engineering and safety experts in the areas with which they are most familiar. The end result, therefore, is reliable in terms of engineering and operational expectations, but it is not quantitative and might not consider the consequences of complex sequences of human errors.

hazard distance: The linear distance from the antenna at which the field intensity is reduced to the exposure limit.

hazard evaluation: Evaluation based on data concerning concentration of a contaminant and duration of exposure.

hazard identification: Determining whether a chemical can cause adverse health effects in humans and what those effects might be.

hazard index: Sum of more than one hazard quotient for multiple substances and/or multiple exposure pathways. Calculated separately for chronic, subchronic, and shorter-duration exposures.

hazard prevention and control: Hazard prevention and control applies the following measures: 1) engineering techniques are used when feasible and appropriate; 2) procedures for safe work are established, understood, and followed by all affected parties; and 3) provisions are made for personal protective equipment and administrative controls.

hazard quotient: Ratio of a single substance exposure level over a specified period to a reference dose (RfD) for that substance derived from a similar exposure period.

hazard ratio: A number obtained by dividing the airborne concentration of a contaminant by its exposure limit.

hazard warning: Any words, pictures, symbols, or combinations thereof appearing on a label or other appropriate form of warning that convey the specific physical or health hazard(s) — including target organ effects — of the chemical(s) in the container(s). (See the definitions for *physical hazard* and *health hazard* to determine the hazards that must be covered.)

HazMat: Universally recognized abbreviation for hazardous materials. HAZMAT is also used.

HAZOP: See *hazard and operability survey.*

HAZWOPER: Acronym for hazardous waste operations and emergency response.

head: 1. Historically, a unit of pressure. 2. In fluid statics and dynamics, a vertical linear measure.

Health Advisories (1-day or 10-day): Concentrations of contaminants in drinking water at which adverse, noncarcinogenic health effects would not be expected to occur to a child exposed for one or10 days. [**Note:** Issued by the U.S. EPA Office of Drinking Water.]

health care professional: Occupational physicians are qualified to design, manage, supervise, and deliver health care in occupational settings. Some services may be provided by other practitioners such as occupational health nurses, physician assistants, licensed practical nurses, or emergency medical technicians.

Health Effects Assessment Summary Table (HEAST): Tabular presentation of toxicity information and values for certain chemicals that have been evaluated by different U.S. EPA programs. As data in HEAST might not have received the level of validation required by IRIS, some chemicals not listed in IRIS may appear in HEAST.

health hazard: A chemical for which there is statistically significant evidence based on at least one study conducted in accordance with established scientific principles that acute or chronic health effects might occur in exposed employees. The term "health hazard" includes chemicals that are carcinogens, toxic or highly toxic agents, reproductive toxins, irritants, corrosives, sensitizers, hepatotoxins, nephrotoxins, neurotoxins, agents that act on the hematopoietic system, and agents that damage the lungs, skin, eyes, or mucous membranes. Appendix A of the OSHA hazard communication standard provides further definitions and explanations of the scope of health hazards covered by this section, and Appendix B describes the criteria to be used to determine whether a chemical is to be considered hazardous for purposes of this standard.

health inspections: An inspection conducted by those who are classified as occupational hygienists.

health surveillance: The measurement of chemical markers in body media that are indicative of adverse and nonadverse health effects.

healthy worker effect: A phenomenon observed initially in studies of occupational diseases; workers usually exhibit lower overall death rates than the general population because severely ill and disabled are ordinarily excluded from employment.

hearing conservation: The program for preventing or minimizing noise-induced deafness through audiometric testing, measurement of noise, engineering control, and ear protection.

hearing level: A measurement of hearing acuity. The deviation in decibels of an individual's threshold from the zero reference of the audiometer.

hearing loss: The deviation of hearing acuity from normal.

hearing protection device (HPD): Equipment worn to reduce the sound exposure of persons either before engineering or administrative noise controls can be administered or when these controls are not yet feasible. (See also *earmuff, formable earplug, preformable earplug, semi- insert.*)

HEAST: See *Health Effects Assessment Summary Table.*

heat: Energy transferred by a thermal process.

heat balance: A statement that shows the changes in a system from heat and work input to output losses.

heat capacity: The amount of heat necessary to raise the temperature of a given mass 1 degree; numerically, the mass multiplied by the specific heat.

heat strain: The body's response to heat stress.

heat stress: 1. The external heat load placed on the body due to the characteristics of the environment. 2. The burden, or load of heat, that must be dissipated if the body is to remain in thermal equilibrium.

heat stress index (HSI): A composite measure used for the quantitative assessment of heat stress.

hedonic tone: A category judgment of the relative pleasantness or unpleasantness of an odor. Perception of hedonic tone is influenced by subjective experience, frequency of occurrence, odor character, odor intensity, and duration. (See also *odor threshold.*)

helium oxygen saturation diving: A decompression schedule specifying a set rate of feet per hour. For example, an ascent from saturation diving at 340 ft would require 120 hours (or five days). On the other hand, saturation diving allows more working time per day, greatly reducing the total time for long jobs in addition to avoiding the hazards of multiple compressions and decompressions.

helmet: A hood that offers head protection against impact and penetration.

hematopoietic: Pertaining to or affecting the formation of blood cells.

hematuria: The presence of blood or blood cells in the urine.

hemoglobin: The red coloring matter of the blood which carries the oxygen.

Henry's law: The equilibrium concentration of a gas dissolved into a liquid will equal the product of the partial pressure of the gas times its solubility in the liquid.

HEPA filter: High efficiency particulate air filter: a filter capable of removing very small particles from the airstream. A HEPA filter is capable of trapping particulate material in the size of 0.3 microns (or greater) from the air with a minimum efficiency of 99.97%.

hepatic: Pertaining to the liver.

hepatitis: The inflammation of the liver; commonly of viral origin but also occurring in association with syphilis, typhoid fever, malaria, toxemias, and parasitic infestations.

hepatotoxicant: A material harmful to the liver.

hertz (Hz): Unit of frequency equal to one cycle per second. (See also *frequency of sound.*)

hidromeiosis: The small reduction in sweat production once the skin surface is wetted with sweat.

high altitude cerebral edema (HACE): A sickness with symptoms that include many benign AMS symptoms but are differentiated by disturbed consciousness (irrationality, disorientation, and even hallucinations), abnormal reflex and muscle control (ataxia, bladder dysfunction, and even convulsions), and/or perhaps most characteristically papilloedema (swelling of the optic disc). (See also *acute mountain sickness.*)

high altitude pulmonary edema (HAPE): The edema in HAPE is characterized by the release of large quantities of a high protein fluid into the lung. Differential symptoms, which are often denied by the patient, include severe breathlessness and chest pain, with or without the above symptoms of benign AMS. Symptoms of patients with HAPE will rapidly progress to a dry cough, production of a foamy pink sputum, audible bubbling and gurgling sounds while breathing, and cyanosis of the lips and extremities. (See also *acute mountain sickness.*)

high performance liquid chromatography (HPLC): Laboratory method used to separate organic molecules using a liquid phase.

homeotherm: An endotherm that maintains a constant body temperature as do most mammals and birds.

homolog: 1. One of a series of compounds, each of which is formed from the one before it by the addition of a constant element; any chemical structurally similar to another chemical. 2. Compound of the same (organic) series. (Also known as "homologue.")

hood: 1. A device that encloses, captures, or receives emitted contaminants. 2. A respiratory inlet covering that completely covers the head and neck and may cover portions of the shoulders.

hood centerline: A line from the center or the hood face extending perpendicularly outward.

hood face velocity (V_f): Air velocity at the hood face opening of an enclosing hood (e.g., a laboratory hood).

horizontal entry: A confined space entry that requires the entrant to enter through an opening in the side of the confined space.

horsepower: The unit of power in the I-P system indicating work done at the rate of 550 foot- pounds per second, or 745.7 watts. (See also *brake horsepower.*)

House of Quality: A tool that correlates customer needs with process design requirements.

HPD: See *hearing protection device.*

HPLC: See *high performance liquid chromatography.*

human factors: Plants that are designed so that human operators can affect or control hazards.

human health risk assessment: The evaluation by occupational hygienists of the potential risk of exposure to the health of workers.

humidity: Water vapor within a given space.

Hunter or Lewis reflex: A physiological aspect of cold tolerance in which the finger tips, palms, toes, sole of foot, ear lobe, and parts of the face react to the cold exposure by occasional vasodilation that periodically rewarms peripheral tissues without the loss of excess heat that would occur if higher temperatures were maintained constantly in these areas. (See also *vasoconstrict, vasodilation.*)

HVAC system: HVAC (heating, ventilation, and air-conditioning) is the distribution system that heats, ventilates, cools, humidifies, dehumidifies, and cleanses air in a building or building zone, principally for the comfort, health, and safety of the occupants.

hydrocarbons: The basic building blocks of all organic chemicals which are composed solely of carbon and hydrogen.

hygrometer: An instrument to measure humidity in the atmosphere.

hygroscopic: Readily absorbing or retaining moisture.

hygroscopicity: The tendency to absorb water vapor.

hyperbaric: Pertaining to an anesthetic solution with a specific gravity greater than that of the cerebrospinal fluid.

hyperemia: An increased blood flow or congestion of blood anywhere in the body.

hypergolic: Ignites on contact.

hyperplasia: An abnormal increase in the number of normal cells composing a tissue or organ.

hypersensitivity: A state of heightened responsiveness in which the body reacts to a foreign agent or substance more strongly than normal; generally results from prior exposures to the agent or substance.

hypersensitivity diseases: Diseases that result from specific immune system responses to environmental challenges. There are two general categories: 1) the IgE-mediated diseases (asthma, allergic rhinitis, or hayfever); 2) hypersensitivity pneumonitis, which is mediated by IgG and the cellular immune system. All hypersensitivity diseases require an initial series of sensitizing exposures during which the immune system becomes activated. Symptoms occur on subsequent exposures in response to stimulation of the previously activated immune response. Most cases of hypersensitivity disease are caused by proteins or glycoproteins, although some highly reactive chemicals can bind to larger molecules to cause hypersensitivity pneumonitis.

hypersusceptibility: Greater than normal sensitivity to certain substances.

hypobaric: Pertaining to an anesthetic solution with a specific gravity lower than that of the cerebrospinal fluid.

hyponatremia: Subnormal or reduced blood sodium levels.

hypothermia: Condition of reduced body temperature in homeotherms.

hypotonic: 1. Pertaining to subnormal muscle strength or tension. 2. Referring to a solution with a lower osmotic pressure than physiological saline.

hypoxia: Occurs when there is an insufficient amount of oxygen delivered to the tissues.

Hz: See *hertz.*

I

IARC: See *International Agency for Research on Cancer.*

I_c: See *contact current.*

IC: See *ion chromatography.*

ice point: 1. Equilibrium temperature of ice and water (usually at standard atmospheric pressure). 2. Temperature at which water freezes under normal atmospheric pressure, 14.696 psig, 32°F (101.325 kPA, 0°C).

ideal gas: A gas with internal energy and enthalpy that depends solely on temperature and that is defined by the perfect gas equation, Pv = nRT. (Also known as "perfect gas.")

Ideal Gas law: An equation of state, a relationship between the pressure, volume, and thermodynamic temperature of a gas (PV = nRT). (Also known as the "Perfect Gas law.")

identity: Any chemical or common name indicated on the material safety data sheet (MSDS) for a chemical. The identity used permits cross-references to be made among the required list of hazardous chemicals, the label, and the MSDS.

IDLH: See *immediately dangerous to life and health.*

IDP: See *integrated product development.*

IHIT: See *industrial hygienist in training.*

I_i: See *induced current.*

illuminance: The luminous flux crossing a surface of a given area.

im/clo ratio: The ratio of clo to the impermeability-to-water-vapor of clothing. The higher the clo, the warmer the clothing. The higher the im/clo ratio, the greater the problems in evaporating sweat.

immediately dangerous to life and health (IDLH): Any atmosphere that poses an immediate hazard to life or poses immediate irreversible debilitating effects on health.

immune response: Chemical/cellular response of the body to an antigen.

immunosuppression: Suppression of an immune response by the use of drugs or radiation. Toxic interactions also can cause suppression of the immune response. Immunosuppression decreases an individual's resistance and increases vulnerability to infection and proliferation of neoplastic or other mutated cells.

impact: A forceful collision between two bodies that is sufficient to cause an appreciable change in the momentum of the system on which it acts. (Also known as "impulsive force.")

impaction: The state of impacting.

impaction plate: A supporting surface on which a substrate is placed for collection of certain cutoff size particles. (See also *impactor stage, substrate, substrate coating.*)

impactor stage: One in a series of intercepting devices in the cascade impactor to collect certain cutoff size particulates. The stage includes the impaction plate a support surface where a substrate can be placed, the nozzle for the subsequent stage, and an O-ring for sealing gaps around the edge of the stage. (See also *impaction plate, substrate, substrate coating.*)

impeller: The rotating part of a device (fan, blower, compressor, or pump).

impervious: Incapable of being passed through or penetrated.

impinge: To impact, hit, strike, collide, or push against.

impingement: Method of measuring air contaminants in which particulates are collected by their collision against some other material; also refers to the way in which particulate matter collects inside the respiratory tract.

impingers: Small glass bottles normally filled with a specific liquid that will absorb airborne chemicals when air containing the contaminant is bubbled through it.

incentive/reward programs: A program designed to achieve a desired outcome in worker behavior by rewarding an individual or group.

Inch-Pound System of Units: The I-P system is the de facto engineering standard set of units used in the United States.

inch-pound units: See *Inch-Pound System of Units.*

incident rate: The numerical performance measure required by OSHA. It is the number of injuries, illnesses, or lost workdays related to a common exposure base of 100 full-time workers. It is calculated as N / EH ˇ 200,000 where N = number of injuries and/or illnesses or lost workdays; EH = total hours worked by all employees during a calendar year; and 200,000 = the base for 100 full-time equivalent workers (working 40 hours/week, 50 weeks/year).

indicating layer: The colorimetric reactive portion of the detector tube.

induced current (I_i): A current produced in a conductor by a time-varying magnetic field.

industrial hygiene survey: Systematic analysis of a workplace to detect and evaluate health hazards and recommend methods for their control.

industrial hygienist: Professional hygienist primarily concerned with the control of environmental health hazards that arise out of or during the course of employment. (See also *Certified Industrial Hygienist, Qualified Industrial Hygienist.*)

Industrial Hygienist in Training (IHIT): A degree indicating partial fulfillment of CIH certification. The IHIT designation recognizes special education in and knowledge of the basic principles of industrial hygiene.

industrial ventilation (IV): The equipment or operation associated with the supply or exhaust of air, by natural or mechanical means, to control occupational hazards in the industrial setting.

inert chemical: Not having active properties.

inert dust: Dust that does not chemically react with other substances.

inert gas: A gas that neither experiences nor causes chemical reaction, nor undergoes a change of state in a system or process (e.g., nitrogen or helium mixed with a volatile refrigerant).

infection: 1. Invasion of the body by a pathogenic organism with or without disease manifestation.
2. Pathological condition resulting from invasion of a pathogen.

inflammation: A form of tissue reaction to injury that is often marked by pain, heat, redness, and swelling.

informational report: A simple presentation of the facts that generally requires less time and effort than an interpretive report. Examples of informational reports include status reports on a continuing project and survey reports that list findings without developing conclusions or recommendations.

infrared: The region of the electromagnetic spectrum including wavelengths from 0.78 microns to about 300 microns.

infrared gas analyzer: A direct-reading instrument that is versatile, can quantify hundreds of chemicals, and is capable of being used for continuous monitoring, short-term sampling, and bag sampling. It is often used in indoor air investigations to measure the buildup of carbon dioxide. (Also known as "infrared gas monitor.")

infrared radiation (IR): Wavelengths of the electromagnetic spectrum that are longer than those of visible light and shorter than radio waves; infrared wavelengths measure 10^{-4} cm to 10^{-1} cm.

ingestion: Introduction of substances into the digestive system.

inhalable fraction: The fraction of total workplace aerosol actually entering the respiratory tract.

inhalation: The breathing in of a substance in the air (e.g., gas, vapor, particulate, dust, fume, mist).

inhibitor: A chemical added to another substance to prevent an unwanted chemical change.

initiation: 1. Triggering off explosion or decomposition.
2. The formation of the DNA- carcinogen adduct is the first step of carcinogenesis.

innervation ratio: The number of muscle fibers innervated by a single motor neuron, which varies from 1:1900 (1 neuron per 1900 muscle fibers) as in the gastrocnemius to 1:15 (1 neuron per 15 muscle fibers) as in the extraocular muscles. The lower the innervation ratio, the finer the control of the muscle force.

inorganic: Term used to designate compounds that generally do not contain carbon.

inspired air: Air drawn in during the breathing process.

inspiring: A management method in which leaders share authority and acknowledge contributions of others, enabling others to feel and act like leaders. This facilitates as sense of ownership for the team members and helps the group achieve peak performance.

instantaneous: Instantaneous may be defined as long as it takes to snap one's fingers and refers to a single event. Occupational safety, therefore, is focused primarily on preventing unwanted and unplanned events (i.e., accidents that are instantaneous in nature that might lead to an employee suffering an injury).

instantaneous sampling: Sampling done at one particular time either by a direct-reading instrument or by trapping a definite volume of air for analysis.

instructional objectives: A road map for the development of training content and what training format should be used. They are also a means of measurement that allows for the evaluation of whether performance does, or does not, reach the desired goal.

instructional systems design: The systematic approach to instructional technology that emphasizes the importance of a training needs assessment, the specification of instructional objectives, precisely controlled learning experiences to achieve these objectives, and criteria for performance; and evaluative information based on performance measures.

instructional technology: The theory and practice of design, development, utilization, management, and evaluation of processes and resources for learning.

intake: Measure of exposure expressed as mass of substance in contact with the exchange boundary per unit body weight per unit time (i.e., mg/kg/day).

integrated product development (IPD): A systematic approach to the multifunctional, concurrent design of products and their related processes. It includes manufacturing and support of the products through their life cycle.

Integrated Risk Information System: U.S. EPA database containing verified reference doses (RfDs), slope factors, and current health and regulatory information. It is the EPA's preferred source of toxicity information for Superfund. Known by the acronym IRIS.

integrated sampling: The passage of a known volume of air through an absorbing or adsorbing medium to remove the desired contaminants from the air during a specified period.

interactive multimedia: Interactive multimedia is software that allows or requires the user to press keys or "click" the mouse to control the actions of the software. (See also *multimedia.*)

interception: The state of intercepting.

interference: The term applied when a contaminant, other than the target gas or vapor, reacts with a reagent in the detector tube to produce erroneous results.

interlaboratory quality control (QC): A systematic procedure for selecting interlaboratory participants, analyte, duration, and frequency of interlaboratory testing, and evaluating statistics and reporting of test data to ensure the quality of test results.

internal body temperature: The temperature of the body's internal organs.

internal occupational exposure limit: An occupational exposure limit (OEL) formally set by an organization for its private use. (See also *occupational exposure limit.*)

internal quality control: The routine activities and checks (such as periodic calibrations, duplicate analyses, and use of spiked samples) included in normal internal procedures to control the accuracy and precision of a measurement process. (See also *quality control.*)

International Agency for Research on Cancer: Headquarters are in Geneva, Switzerland. Known by the acronym IARC.

International System of Units: Le Systeme International d'Unites (the SI system) divides units into three categories: base units, supplementary units, and derived units. The base units consist of seven well-defined and dimensionally independent quantities. These include length, mass, time, thermodynamic temperature, electric current, amount of a substance, and luminous intensity.

Internet: The Internet is a series of interconnected networks of computers. Unlike online services, which are centrally controlled, the Internet is decentralized by design. Each Internet computer, called a host, is independent. Its operators can choose which Internet services to provide to its local users and which local services to make available to the global Internet community.

Internet service provider (ISP): A company that provides and manages connections from users to the Internet.

interpreting: A management skill in which a leader sifts through all available data to choose that which is most important to the organization's future.

interpretive report: A report that presents findings and also critically evaluates the results, draws conclusions, and offers recommendations. The purpose of such reports is not merely to inform but also to persuade the reader to accept the writer's viewpoint, to take action, or to make a change in policy or practice. Examples of interpretive reports include technological feasibility studies and comprehensive work site evaluations. The interpretive report is distinguished by its inclusion of expert opinion to guide the reader in making decisions.

interstitial: Situated between the cellular components of an organ or structure.

intrabeam viewing: The viewing condition in which the eye is exposed to all or part of a laser beam.

intralaboratory quality control (QC): A systematic procedure for evaluating the precision and accuracy for within analyst and between analyst data, constructing and using control charts, and using duplicate, replicate, and/or spiked samples to ensure the quality of test results.

intrinsic safety: Plants that are designed so that departures from normal tend to be self- correcting or at most lead to minor events rather than major disasters. Plants designed to be forgiving and self-correcting are inherently safer than plants where equipment has been added to control hazards or where operators are expected to control them.

intrinsically safe: A feature available on air sampling equipment that ensures that pumps are not an explosive hazard in specific environments.

inverse dynamics: Since all of the forces and moments that cause the moment are calculated by evaluating the resulting motion itself, the technique generally used for this calculation is known as an inverse dynamics approach.

Inverse Square law: The propagation of energy through space is inversely proportional to the square of the distance it must travel. An object 3 feet away from an energy source receives 1/9 the energy as an object 1 foot away.

***in vitro*:** Literally means "in glass"; experimental work done on cell cultures.

***in vivo*:** Literally means "in life"; an experiment that was conducted in the living organism.

ion chromatography (IC): A chromatographic technique that separates mixtures of ions.

ionization potential: The energy per unit charge needed to remove an electron from a given kind of atom or molecule to an infinite distance; usually expressed in volts. Also know as "ion potential."

I-P units (inch-pound units): Units using inches, pounds, and other designations; as opposed to SI units in the metric system. Examples are foot, Btu, horsepower, and gallon.

IR: See *infrared radiation.*

IREQ: See *required clothing insulation.*

IRIS: See *Integrated Risk Information System.*

irradiance: The amount of radiant power per unit area that flows across or on to a surface. (Also known as "radiant flux density.")

irreversible injury: An injury or effect that is not reversible once the exposure has been terminated.

irritant: A chemical, which is not corrosive, that causes a reversible inflammatory effect on living tissue by chemical action at the site of contact.

irritation: An inflammatory response or reaction of tissues resulting from contact with a material.

isobar: Two or more nuclides with the same atomic mass but different atomic numbers.

isokinetic sampling: A sampling condition in which air flowing into an inlet has the same velocity and direction as ambient airflow. (See also *subisokinetic sampling; superisokinetic sampling, stack sampling.*)

isolation: Separating employees from hazardous operations, processes, equipment, or environments (e.g., use of control rooms, physically separating employees and equipment, and barriers placed between employees and hazardous operations.)

isomer: 1. A metastable state of a nucleus that exists longer than 10^6 sec and decays by gamma emission to a nuclide with the same atomic number and same atomic mass. 2. A molecule that has the same number and kind of atoms as another molecule, but has a different arrangement of the atoms.

isometric: A muscle contraction in which the muscle remains at the same length while developing tension.

isothermal: A process at constant temperature.

isotope: 1. Two or more nuclides with the same atomic number but different atomic mass. 2. Atoms of the same element that differ in atomic weight.

ISP: See *Internet service provider.*

iterative risk assessment: A process in which increasingly complex and data-rich risk assessments are conducted.

IV: See *industrial ventilation.*

J

J: See *joule.*

jaundice: A yellow discoloration of the skin, mucous membranes, and eyes most often due to abnormalities of the liver or rapid destruction of blood cells.

jet: See *nozzle, jet.*

job rotation: In job-rotation, people shift jobs periodically during the day.

joule (J): In the SI system of units, the unit of energy equal to the work done when a current of 1 ampere is passed through a resistance of 1 ohm for 1 second; a unit of energy equal to the work done when the point of application of a force of 1 newton is displaced 1 meter in the direction of the force. (See also *newton.*)

K

kaizen: The philosophy of total quality requires a paradigm shift, away from a reluctance to tamper with processes that appear to be working to one stating that all processes are imperfect and an organization must strive for continuous improvement. This philosophy ensures that the organization will never be satisfied with less than optimal performance in any of its processes. Kaizen indicates that every process can and should be continually evaluated and improved, in terms of time required, resources used, resultant quality, and other aspects relevant to the process.

kcal: See *kilocalorie.*

Kelvin scale (absolute): The fundamental temperature scale, also called the absolute or thermodynamic scale, in which the temperature measure is based on the average kinetic energy per molecule of a perfect gas. The zero of the Kelvin scale is -273°C.

Kelvin temperature: The SI system of units absolute temperature scale (K), on which the triple point of water is 273.16 K and the boiling point is approximately 373.15 K (1 K = 1°C). The Kelvin is the fraction 1/273.16 of the temperature of the thermodynamic triple point of water.

kg: Kilogram, or 1000 grams.

kilocalorie (kcal): Amount of heat required to raise the temperature of 1000 grams 1°C.

kilogram: See *kg.*

kilogram mole: The amount of substance represented by 1 kilogram molecular weight or 1 kilogram molecular volume (mol).

kilogram molecular weight: The sum of the individual atomic weights of all the atoms in a molecule (express mass in units of kilograms [kg]).

kinematics: A branch of mechanics that concerns the description of a body's motion in space without an explanation of the cause of the observed motion.

kinesiology: The study of movement.

kinetic energy: Energy due to motion.

kinetics: A branch of mechanics that concerns the underlying causes of movement rather than the result of the movement.

K_m: See *partition coefficient.*

L

L: See *liter.*

laboratory quality control (QC) plan: A written plan that details the processes by which data generated by the laboratory will be evaluated, corrected (if necessary), and reported to the customer.

lacrimation: The excessive secretion and discharge of tears.

lagging: Asbestos and magnesia plaster that is used as a thermal insulation on process equipment and piping.

Lambertian surface: An ideal, perfectly diffusing surface for which the intensity of reflected radiation is independent of direction.

laminar flow: Gas flow with a smooth nonturbulent pattern of streamlines, with no streamline looping back on itself; usually occurs at very low Reynolds numbers. (See also *turbulent flow.*)

LAN: See *local area network.*

laser: Light amplification through stimulated emission of radiation resulting when the rate at which energy is added to a monochromatic beam is greater than that at which it is extracted and light amplification results.

laser radiation: Optical radiation that propagates in the form of a beam and has various special properties, including low divergence, monochromaticity, and coherence.

Laser Safety Officer (LSO): The LSO, as described in *American National Standard for the Safe Use of Lasers* (ANSI Z136.1–1993), is one who administers the overall laser safety program. ANSI Z136.1 defines the LSO as one with "the authority to monitor and enforce the control of laser hazards and effect the knowledgeable evaluation and control of laser hazards."

latency: A delay between the onset of exposure to a hazardous agent and the onset of illness attributable to that agent.

latency period: The time that elapses between exposure and the first manifestation of damage. (Also known as the "latent period.")

LC: See *lethal concentration.*

LC_{50}: The airborne concentration of a given substance that when inhaled over a period of time will kill 50% of the animals under test.

LC_{Lo}: See *lowest lethal concentration.*

LC_X: See *lethal concentration "X."*

LD: See *lethal dose.*

LD_{50}: The oral dose required to produce death in 50% of the exposed species, usually within the first 30 days.

LD_{Lo}: See *lowest lethal dose.*

LD_X: See *lethal dose "X."*

leak test: A procedure to determine whether the pump or the detector tube is leaking.

Legionella: A genus of bacteria, some species of which have caused a type of pneumonia called "Legionnaires' Disease."

LEL: Lower explosive limit. See *lower flammable limit.*

length of stain: The length of color change in a detector tube. The length of stain is proportional to the concentration of the contaminant in the sampled atmosphere.

length-tension relationship: This relationship describes the maximal force that a muscle, muscle fiber, or sarcomere can exert and its length.

lesion: A structural or functional alteration due to injury or disease.

lethal concentration "X" (LC_X): The concentration that was lethal to X percent of test animals. It may be expressed, for example, as LC_{50}, LC_{10}, etc.; these would represent the concentrations producing deaths in 50%, 10%, etc., of the exposed animals, respectively.

lethal concentration (LC): LC_{50} indicates atmospheric concentration of a substance at which half of a group of test animals die after a specified exposure time. LC_0 indicates atmospheric concentration at which no deaths occur.

lethal dose "X" (LD_X): The dose that was lethal to X percent of test animals. The amount of a chemical, per unit of body weight, that will cause death in X percent of test animals. Most commonly used as LD_{50}, the dose producing deaths in 50% of the dosed animals.

lethal dose (LD): LD_{50} indicates a dose that kills half of a group of test animals. LD_0 indicates a dose at which no deaths occur.

leukopenia: A decrease below normal in the amount of leukocytes in the blood.

LEV: See *local exhaust ventilation.*

level: Logarithm of the ratio of one quantity to a reference quantity of the same kind. The base of the logarithm, the reference quantity, and the kind of level must be specified.

levers: The musculoskeletal system is a system of levers in which the bones are rigid bodies and the muscles are the force actuators. Levers consist of a resisting load, a fulcrum (axis of rotation), and a motive force (applied by muscle).

LFL: See *lower flammable limit.*

L_i: See *sound intensity level.*

lifetime cancer risk estimate: The result of the exposure and toxicity assessments of a carcinogen. Represents the upper bound of the probability of an individual developing cancer as a result of lifetime exposure to the chemical.

ligament: 1. Cord-like tissue that connects bone to bone. 2. A flexible, dense white fibrous connective tissue joining, and sometimes encapsulating, the articular surfaces of bones.

limit of detection (LOD): A stated limiting value designating the lowest concentration that can be detected and that is specific to the analytical procedure used. (See also *minimum detectable level.*)

limit of quantification (LOQ): A stated limiting value designating the lowest concentration that can be quantified with confidence and that is specific to the analytical procedure used.

liquid: A state of matter in which the substance is a formless fluid that flows in accord with the law of gravity.

liter (L): An SI unit of volume limited to capacity in dry measure and fluid measure (both liquids and gases). No prefixes except milliliter (mL) or micrometer (μm) should be used. (Also known as "litre," "cubic decimeter.")

LOAEL: See *lowest observable adverse effect level.*

local area network: A local area network (LAN) usually serves a limited area such as a building, department, or office. The LAN has a central computer called a server that runs the network software and controls access to the files and peripherals, such as printers.

local exhaust ventilation (LEV): An industrial ventilation system that captures and removes emitted contaminants before dilution into the workplace ambient air can occur.

LOD: See *limit of detection.*

logic chart: A flow chart designed to assist detector tube users in interpreting the results of a series of measurements.

lognormal distribution: 1. Particle size distribution characterized by a bell-shaped or Gaussian distribution shape when plotted on a logarithmic size scale. (See also *geometric standard deviation*.) 2. The distribution of a random variable that has the property that the logarithms of its values are normally distributed.

loose-fitting facepiece: A respiratory inlet covering that is designed to form a partial seal with the face, does not cover the neck and shoulders, and may or may not offer head protection against impact and penetration.

LOQ: See *limit of quantification*.

loss: Usually refers to the conversion of static pressure to heat in components of the ventilation system (i.e., "the hood entry loss.")

low-dose extrapolation models: Mathematical models used to extend a dose-response curve beyond known data points representing higher dose levels. A number of models, including the single-hit, multi-hit, multi-stage, and linearized multi-staged models, seem to fit the known data points equally well but result in widely varying estimates of response at lower dose levels.

lower boundary of working range: Refers to the contaminant concentration that may be quantitated at a specific air volume when the mass of contaminant is equal to the LOQ.

lower explosive limit (LEL): See *flammable limits, lower flammable limit*.

lower flammable limit (LFL): The minimum concentration, as a percentage, of flammable gas or vapor mixed with air that can be ignited. Also referred to as the lower explosive limit (LEL). See *flammable limits, upper flammable limit*.

lowest lethal concentration (LC_{Lo}): The lowest concentration of a substance in air that has been reported to have caused death in test animals.

lowest lethal dose (LD_{Lo}): The lowest dose of a substance reported to have caused death in test animals.

lowest toxic concentration (TC_{Lo}): The lowest concentration of a substance shown to have produced an adverse health effect in humans or test animals.

lowest toxic dose (TD_{Lo}): The lowest dose reported to cause toxic effects in humans or test animals.

lowest observable adverse effect level (LOAEL): Lowest exposure level at which there are statistically or biologically significant increases in frequency or severity of adverse effects between the exposed population and its appropriate control group.

L_p: See *sound pressure level*.

LSO: See *Laser Safety Officer*.

lumen: The light flux on 1 square foot (ft^2) of an area, every part of which is 1 foot from a point source having a luminous intensity of one candle.

luminance: 1. Brightness as perceived by the eye (just visible wavelengths). 2. The ratio of the luminous intensity in a given direction of an infinitesimal element of a surface containing the point under consideration, to the orthogonally projected area of the element on a plane perpendicular to the given direction.

M

m^3: Cubic meter; 1 cubic meter of air is equal to 1000 liters of air.

MA: See *mechanical advantage*.

macromolecules: High molecular weight biochemicals such as proteins, phospholipids, glycosides, nucleic acids, and their mixed analogs like glycolipids, lipoproteins, and chromatin (nuclear protein/DNA complex).

magnetic field: One of the elementary fields in nature; it is found in the vicinity of a magnetic body or current-carrying medium and, along with electric field, in a light wave.

magnetic flux density: A vector quantity that is used as a quantitative measure of magnetic field. The force on a charged particle moving in the field is equal to the particle's charge times the cross product of the particle's velocity with the magnetic induction. (Also known as "magnetic displacement," "magnetic induction," "magnetic vector.")

magnetic-field strength (H): An auxiliary vector field, used in describing magnetic phenomena, whose curl, in the case of static charges and currents, equals the free current density vector, independent of the magnetic permeability of the material. (Also known as "magnetic field intensity," "magnetic force.")

main: 1. Pipe or duct for distributing or collecting flowing fluid from various branches. 2. The regulated compressed air piped to pneumatic controls.

mainframe computer: A mainframe computer is a very large powerful computer with extensive data storage and processing capabilities that can support many simultaneous users. They are the largest computers made today, followed by minicomputers, and then by personal computers.

maintainability: The probability that an item that has failed (in the reliability sense) can be restored (i.e., repaired or replaced) within a stated period of time.

MAK: Maximum allowable concentration.

malformation: Defective or abnormal formation.

malignant: Tending to become progressively worse and to lead to death; often used to describe tumors that grow in size and also spread throughout the body.

management commitment: Management commitment can be demonstrated by the following activities: 1) clearly state a work site policy on safety and health so that all personnel will understand the priority of safety and health protection in relation to other organizational values; 2) establish and communicate a clear goal for the safety program and objectives for meeting that goal; 3) provide for and encourage employee involvement in the structure and operation of the safety program; 4) assign and communicate responsibility for all aspects of the safety program; 5) provide adequate authority and resources to responsible parties so that assigned responsibilities can be met; 6) hold people accountable for meeting responsibilities; and 7) review program operations at least annually to evaluate successes and identify deficiencies that should be corrected.

management of hazardous waste: The implied control and environmentally sound handling of hazardous waste during the phases of generation, storage, processing for recovery or reuse, transporting, treating, and discharging into the air and water or discarding onto the soil.

manifold: A holder for more than one detector tube. When attached to the sampling pump, it permits simultaneous sampling.

manometer: An instrument for measuring pressure; essentially a U-tube partially filled with a liquid (usually water, mercury, or a light oil), so constructed that the amount of displacement of the liquid indicates the pressure being exerted on the instrument.

marker: The determinant to be measured in human body media.

maser: Microwave amplification by stimulated emission of radiation. When used in the term "optical maser," it is often interpreted as molecular amplification by stimulated emission of radiation.

mass: A quantitative measure of a body's resistance to being accelerated. Equal to the inverse of the ratio of the body's acceleration to the acceleration of a standard mass under otherwise identical conditions.

mass loading: The situation in which an accelerometer that is too heavy will weigh down the surface and give inaccurate results. To avoid mass loading, the general rule is that the accelerometer's mass should be no more than one-tenth (1/10) of the effective mass of the surface to which it is mounted. (See also *accelerometer*.)

mass median aerodynamic diameter (MMAD): Aerodynamic diameter of the particles that falls in the middle of the distribution of mass, the size of the particles for which half of the total mass is contributed by smaller particles and half by larger particles. (See also *aerodynamic diameter, cutoff particle diameter*.)

mass spectrometer: See *gas chromatograph*.

material safety data sheet (MSDS): A document containing information on hazardous ingredients, their properties, and precautions for use for a specific chemical substance.

maximally exposed individual (MEI): The single individual with the highest exposure in a given population.

McCready, Benjamin W.: American physician who wrote *On the Influence of Trades Professions and Occupations in the United States in the Production of Disease* (1837). His monograph is generally recognized as the first work on occupational medicine in the United States

MCS: See *multiple chemical sensitivity*.

mean: A statistical description of the "average." Mathematically, it is the sum of the results divided by the number of results. (See also *arithmetic mean, geometric mean*.)

mean free path: The average distance traveled between collisions by the molecules in a gas or vapor.

mean radiant temperature: The mean radiant temperature is the temperature of an imaginary black enclosure, of uniform wall temperature, that provides the same radiant heat loss or gain as the environment measured. It can be approximated from readings of globe temperature, dry bulb temperature, and air velocity.

measures of central tendency: Measures of the tendency of values in a set of data to be centered at some location. Measures of central tendency are, for example, the median, the mode, the arithmetic mean, and the geometric mean.

measures of dispersion or variability: Measures of the differences, scatter, or variability of values of a set of numbers. Commonly used measures of the dispersion or variability are the range, the standard deviation, the variance, and the coefficient of variation (or relative standard deviation).

mechanical advantage: The ratio of the force produced by a machine such as a lever or pulley to the force applied to it. (Also known as "force ratio.")

mechanization: 1. The replacement of human or animal labor by machines. 2. To produce or reproduce by machine.

MED: See *minimum erythemal dose*.

median: The middle value of a set of data when the set of data are ranked in increasing or decreasing order. If there are an even number of values in the set, the median is the arithmetic average of the two middle values.

median nerve: The nerve that runs down the center of the front of the arm and through the carpal tunnel. Swollen tendons pressing on the median nerve cause the symptoms of carpal tunnel syndrome.

Medical Literature Analysis and Retrieval System (Medlars®): The computerized system of databases and data banks offered by the National Library of Medicine. It is comprised of two computer subsystems: ELHILL® and TOXNET®.

medical monitoring: The measurement of chemical markers in body media known to be indicative of adverse health effects.

medical removal: Transfer of employees from jobs entailing exposures until they are sufficiently recovered to return to the work area without risk of impairment to health. Medical removal should never be used to avoid correction of excessive exposures.

medical screening: The performance of tests or procedures aimed at the early identification of subclinical or clinical disease. Usually, medical screening consists of the application of tests or procedures that provide an indication of the presence or absence of disease but are not as sensitive or specific as diagnostic tests.

medical surveillance: 1. A system for the identification and management of individual cases of illness, as well as for the early recognition of latent disease (screening) and for the provision of health promotion activities. 2. The measurement of chemical markers in body media that may be indicative of external exposure to chemical and physical agents and/or of potentially adverse effects. 3. The systematic collection, analysis, and evaluation of health data in the workplace to identify cases, patterns, or trends suggesting an adverse effect on workers' health. Medical surveillance can be used to evaluate the effectiveness of control activities.

medical testing: The medical testing of workers to detect organ dysfunction or disease before an individual would normally seek medical care and while intervention is still beneficial. Tests may indicate the presence of a disease, or an early sign of illness, and the need for additional testing.

medical treatment: The medical care given to work-related injuries or illnesses.

MEI: See *maximally exposed individual*.

melting point (MP): For a given pressure, the temperature at which the solid and liquid phases of the substance are in equilibrium.

menses: The discharge from the vagina during menstruation.

metabolic heat: Heat generated by the body's physical and chemical processes.

metabolism: Energy resulting from physical and chemical changes that are constantly occurring in the body. Term used for heat stress evaluation. (See also *heat stress.*)

metabolite: The stable reduction/oxidation (redox) product of catabolism of an exposing chemical.

metallic oxide semiconductor (MOS) sensor: A solid-state sensor used to detect ppm and combustible concentrations of gases. It can be used to detect a variety of compounds including nitro, amine, alcohol, and halogenated hydrocarbons, as well as a limited number of inorganic gases.

metastable: An excited state of a nucleus, indicated with a letter "m" in the mass number, which will decay by an isomeric transition with the emission of a gamma.

meter: A basic unit of length or distance in the SI system of units. (Also known as "metre.")

methemoglobinemia: The presence of hemoglobin in the oxidized state in the blood.

methods study: An analysis of the methods in use, of the means and potentials for their improvement, and of reducing costs.

mg: See *milligram.*

mg/kg: Milligrams per kilogram body weight; an expression of dosage.

mg/m^3: Air sampling measurement in milligrams (of contaminant) per cubic meter (of air).

microbar: A unit of pressure, commonly used in acoustics, which equals 1 micropascal (μPa). A reference point for the decibel is 20 micronewtons per meters squared (mN/m^2).

microbe: A microorganism, especially a bacterium of a pathogenic nature.

microbiology: The study of microorganisms, including algae, bacteria, fungi, viruses, and protozoa.

microclimate: The conditions such as temperature, humidity, and motion of air within an enclosure or outdoor limited area.

microenvironment: A well-defined place (such as a home, office, or car) in which a chemical or biological agent is present in a uniform manner.

micrometer: A unit of length, one millionth of a meter (one thousandth of a millimeter). (Also known as "micrometre.")

micron: 1. Greek mu (μ); often used interchangeably for micrometer, but micrometer is the preferred usage. 2. A unit of length equal to 10^{-4} centimeters, approximately 1/26,000 of an inch.

micronucleus: A small nucleus; in eukaryotic organisms, micronuclei are produced when chromosomes are broken and the fragments are left in the cytoplasm following cell division.

microsecond: One millionth of a second.

microwave: Any electromagnetic radiation having a wavelength in the approximate range of from 1 millimeter to 1 meter; the region of the electromagnetic spectrum between infrared and short wave-radio lengths.

microwave radiation: A subset of RF radiation that occupies the spectral region between 300 GHz and 300 MHz.

midstream urine: A urine sample taken with the first couple of mL discarded to eliminate potential microorganisms or semen.

MIG: Metal inert gas; a type of welding.

milestones: Interim goals in the process of eliminating a problem that helps evaluate the success of the solution.

milliamp: One-thousandth of an amp.

milligram (mg): One-thousandth of a gram.

milliliter (mL): One-thousandth of a liter.

millimeter (mm): One-thousandth of a meter, or one-tenth of a centimeter.

Mine Safety and Health Administration (MSHA): Established by the Mine Safety and Health Act of 1977, this agency is part of the U.S. Department of Labor. MSHA has the responsibility to set standards for and conduct inspections of working conditions in mines and mining-related industries.

minimum detectable level: The limit of detection for an analytical method is the minimum concentration of the constituent or species of interest that can be observed by the instrument and distinguished from instrument noise with a specified degree of probability. For example, one approach used is to make repeated measurements of the extractant liquid (trace metal analyses) and calculate the standard deviation of the results and hence the desired statistical tolerance limit for instrumental noise (e.g., an upper 99% limit at 95% confidence). (See also *limit of detection.*)

minimum duct transport velocity: According to the ACGIH *Industrial Ventilation Manual*, "When solid material is present in the airstream, the duct velocity must be equal to or greater than the minimum air velocity required to move the particles in the air stream."

minimum erythemal dose (MED): The lowest dose of ultraviolet (UV) light to cause an erythemal response (skin reddening). From 250–304 nanometers (nm) ranged from 14–47 millijoules per cubic centimeters (mJ/cm^2) and increased dramatically above 313 nm.

miscible: Susceptible to being mixed; soluble in all proportions.

mismanagement of hazardous waste: The uncontrolled and environmentally unsound or indiscriminate handling of hazardous wastes, whether intentionally or unintentionally.

mist: A dispersion of suspended liquid particles, many large enough to be individually visible to the unaided eye. Generated by condensation from the gaseous to the liquid state or by breaking up a liquid into a dispersed state, such as splashing, foaming, or atomization. Mist forms when a finely divided liquid is suspended in the atmosphere. Examples are the oil mist produced during cutting and grinding operations, acid mists from electroplating, acid or alkali mist from picking operations, and paint spray mist from spraying operations. Size ranges are between 0.01 μm and 10.0 μm.

mixed exhaled breath: The breath that is naturally exhaled without forcing.

mixing box: A compartment in which two air supplies mix before being discharged. (Also known as "blending box" or "mixing unit.")

mixing factor: A dimensionless quantity used to adjust the volume of air moving in a space for poor air distribution.

mixture: Any combination of two or more chemicals if the combination is not, in whole or part, the result of a chemical reaction.

mL: See *milliliter.*

mm: See *millimeter.*

MMAD: See *mass median aerodynamic diameter.*

mmHg: Abbreviation for millimeter(s) of Mercury

mobilizing: Enlisting available resources to reach an identified goal and ultimately achieve the organization's vision.

mode: The value or values occurring most frequently in a sample of data.

modifying factor: Used in converting NOAELs/LOAELs to reference doses (RfDs). Range from >0 to 10, reflects professional judgment of uncertainties not addressed by uncertainty factors.

moist air: ASHRAE defines moist air as "a binary (or two component) mixture of dry air and water vapor. The amount of water vapor varies from zero (dry air) to a maximum that depends on temperature and pressure."

molar gas volume: The volume (usually liters) occupied by 1 mole (usually gram mole) of gas.

mole: Mass of a substance represented by 1 molecular weight or 1 molecular volume. If the mass is in pounds, the unit is a pound mole; in grams, the unit is a gram mole, in kilograms the unit is a kg mole or mol.

molecular volume: The volume occupied by 1 molecular weight of a gas (either kilogram molecular weight or gram molecular weight).

molecular weight: Weight (mass) of a molecule based on the sum of the atomic weights of the atoms that make up the molecule.

molecule: Generally the smallest particle of an element or a compound capable of retaining the physical properties and chemical identity of the substance in mass.

moment arm: The distance from the fulcrum to each force. (See also *levers.*)

moment of force: See *torque.*

moment of inertia: The measure of resistance to rotational change.

monitor: Periodic or continuous determination of the amount of contamination present in an occupied region; used as a safety measure for purposes of health protection.

monitoring instruments: A broad range of scientific equipment used for the purposes of collecting and/or measuring chemical levels.

monodisperse: Composed of particles with a single size or a small range of sizes. (See also *polydisperse.*)

monodisperse aerosol: An aerosol with a uniform size distribution having a geometric standard deviation of less than 1.1.

Monte Carlo: A repeated random sampling from the distribution of values for each of the parameters in a generic exposure or dose equation to derive an estimate of the distribution of exposures or doses in the population.

Monte Carlo analysis: A method that obtains a probabilistic approximation to the solution of a problem by using statistical sampling techniques.

MOS: See *metallic oxide semiconductor.*

motor neuron: An efferent nerve cell. For a muscle to reach its active state it must receive a signal from the nervous system via a nerve referred to as a motor neuron, at which time the muscle exerts tension on its skeletal attachments.

motor unit: A single motor neuron and the muscle fibers that it innervates. A muscle may have many motor units, ranging from a few hundred to up to a thousand per muscle.

mottling: Colored spots or blotches.

MP: See *melting point.*

mppcf: Million particles per cubic foot (of air).

MS: Mass spectrometer. See *gas chromatograph.*

MSDS: See *material safety data sheet.*

MSHA: See *Mine Safety and Health Administration.*

mucous membrane: The mucous-secreting membranes lining the hollow organs of the body (i.e., eyes, nose, mouth, etc.).

multi-layer detector tube: A detector tube construction containing several filling layers. In addition to the indicating layer the tube contains one or more prelayers, which act as a filter to remove interfering substances or for chemical conversion of the gas or vapor being measured. Such tubes may be used to determine qualitatively the classes of compounds present in the atmosphere sampled.

multimedia: Multimedia is software that incorporates sound, video, text, and sometimes animation.

multiple chemical sensitivity (MCS): A condition resulting from exposure to toxic chemicals that affect the immune system, leading to multiple sensitivities to other chemicals and/or foods. Symptoms of MCS, which may be similar to those of sick building syndrome (SBS), are often attributed to exposure to trace amounts of chemicals (especially those with perceptible odor) in indoor air.

multiple particle optical monitor: A real-time dust monitor used to measure aerosol concentrations.

mutagen: A substance or agent capable of altering the genetic material in a living cell.

mycotoxins: Secondary products of fungal metabolism. The chemical structures of mycotoxins are quite diverse, ranging from that of moniliformin ($C_4H_2O_3$) to complex polypeptides with molecular weights higher than 2000.

N

N: See *newton.*

nail: The horny covering at the upper tip of fingers and toes.

nanometer: Unit of measurement for radiation wavelengths. One nanometer (nm) equals 10^{-6} millimeters or 10 angstrom units. (See also *angstrom.*)

narcosis: Stupor or unconsciousness produced by chemical substances.

NAS: See *National Academy of Sciences.*

nasopharyngeal region: The space behind the posterior nasal orifices, above a horizontal plane through the lower margin of the palate.

National Academy of Sciences (NAS): A private, honorary organization of scholars in scientific and engineering research serving as advisory agency to the federl government.

National Cancer Institute (NCI): One of the National Institutes of Health designed to expand existing scientific knowledge on cancer causes and prevention as well as on the diagnosis, treatment, and rehabilitation of cancer patients.

National Fire Protection Association (NFPA): An international organization that promotes fire protection and prevention. NFPA establishes safeguards (standards, etc.) against loss of life and property by fire. Headquarters are in Quincy, Mass.

National Institute for Occupational Safety and Health (NIOSH): Established by the Occupational Safety and Health Act of 1970, NIOSH is part of the Centers for Disease Control and Prevention within the U.S. Department of Health and Human Services. NIOSH, based in Cincinnati, Ohio, traces its origins to 1914 when the U.S. Public Health Service organized a division of Industrial Hygiene and Sanitation. NIOSH's responsibilities include research and recommending occupational health and safety standards.

National Institute of Standards and Technology (NIST): An agency of the U.S. Department of Commerce's Technology Administration. Originally established by Congress in 1901 as the National Bureau of Standards; its name was changed in 1988. NIST's primary mission is to promote U.S. economic growth by working with industry to develop and apply technology, measurement, and standards. Headquarters are in Gaithersburg, Md.

National Research Council (NRC): The NRC develops and publishes emergency exposure limits such as CEGLs, EEGLs, and SPEGLs. Headquarters are in Washington, D.C.

National Safety Council (NSC): The NSC is a nonprofit, international public service organization dedicated to improving the safety, health, and well-being of populations throughout the world. Total membership exceeds 18,500. Headquarters are in Itasca, Ill.

National Toxicology Program (NTP): Overseen by the U.S. Department of Health and Human Services.

natural wet bulb: One of two types of wet bulb measurement that relies on ambient air motion. (See also *psychrometric, wet-bulb globe temperature.*)

natural wet bulb temperature (T_{nwb}): The temperature measured when the wetted wick covering the sensor is exposed only to naturally occurring air movements.

NCEL: New chemical exposure limit.

NCI: See *National Cancer Institute.*

near field: In noise measurement, this refers to a field in the immediate vicinity of the noise source where the sound pressure level does not follow the Inverse Square law.

necrosis: 1. Death of a cell or group of cells as a result of injury, disease, or other pathological state. 2. Tissue death.

negative-pressure device: See *negative-pressure respirator.*

negative-pressure respirator: A respirator in which the air pressure inside the respiratory inlet covering is negative during inhalation with respect to the ambient air pressure.

neoplasm: See *tumor.*

nephrotoxicant: A substance harmful to the kidney.

net force: The sum of all concurrent forces.

neural: Relating to the nervous system.

neuropathy: Functional disturbance or pathology of the nervous system; may be central (affecting the brain or spinal cord) or peripheral (affecting nerves outside the brain and spinal cord).

neurotoxicant: A substance harmful to nerves or the brain.

neurotransmitter: A susbtance (such as norepinephrine or acetylcholine) that transmits nerve impulses across a synapse.

neutral (handshake) position: Keeping the wrist straight to avoid joint deviation.

neutrino: Massless particle traveling at the speed of light; created in isobaric decay.

new effective temperature: The new effective temperature is similar to the ET scale, but it uses as its basis an environment at 50% relative humidity instead of 100% relative humidity.

newton (N): In the meter-kilogram-second system, the unit of force required to accelerate a mass of 1 kilogram 1 meter per second; equal to 100,000 dynes. (See also *dyne.*)

NFPA: See *National Fire Protection Association.*

NHZ: See *nominal hazard zone.*

NIC: See *notice of intended change.*

NIOSH: See *National Institute for Occupational Safety and Health.*

NIPTS: See *noise-induced permanent threshold shift.*

NIR: See *nonionizing radiation.*

NIST: See *National Institute of Standards and Technology.*

nitrogen narcosis: Narcosis caused by gaseous nitrogen at high pressure in the blood. Produced in divers breathing air at depths of 100 ft (30 m)or more.

nitrogen-phosphorus detector: The nitrogen-phosphorus detector is highly sensitive and selective for nitrogen and phosphorous compounds, including amines and organophosphates. The detector is similar in principle to the FID, except that ionization occurs on the surface of an alkali metal salt, such as cesium bromide, rhobidium silicate, or potassium chloride. (Also known as a "thermionic" or "alkali flame detector.")

NITROX: Enriched oxygen mixtures used to increase total dive time.

NITTS: See *noise-induced temporary threshold shift.*

nm: See *nanometer.*

NMSC: See *nonmelanoma skin cancer.*

no observable adverse effect level (NOAEL): Exposure level at which there are no statistically or biologically significant increases in frequency or severity, or any adverse effects between the exposed population and its appropriate control group.

no observable effect level (NOEL): That quantity of a chemical to which laboratory animals are chronically exposed (expressed in parts per million [ppm] in their diets or mg/kg of body weight) that produces no effect when compared with control animals.

NOAEL: See *no observable adverse effect level.*

NOEL: See *no observable effect level.*

noise: Unwanted sound; unwanted because it can cause annoyance, interfere with speech or communication, and/or cause hearing impairment.

noise enclosure: Equipment that reduces the noise of a sound source by completely surrounding the source with a barrier material.

noise level: For airborne sound, unless specified to the contrary, noise level is the weighted sound pressure level called sound level; the weighting must be indicated.

noise reduction: See *attenuation.*

Noise Reduction Rating (NRR): A single-number rating of the hearing protection. The higher the NRR, the higher the attenuation for a specific ideal situation (laboratory-fit of HPD).

noise-induced hearing loss: Hearing loss due to excessive exposure to noise.

noise-induced permanent threshold shift (NIPTS): A permanent loss in hearing sensitivity due to the destruction of sensory cells in the inner ear. This damage can be caused by long-term exposure to noise or by acoustic trauma.

noise-induced temporary threshold shift (NITTS): A temporary loss in hearing sensitivity. This loss can be a result of the acoustic reflex, short-term exposure to noise, or simply neural fatigue in the inner ear. With NITTS, hearing sensitivity will return to the pre-exposed level in a matter of hours or days (without continued excessive exposure).

nominal hazard zone: The NHZ is the zone around the laser where the beam intensity exceeds the exposure limit.

nonbeam hazard: A hazardous agent, other than the beam, generated by the use of lasers (e.g., electricity, airborne contaminants, plasma radiation, fires, and explosions).

noncarcinogen: A chemical that exerts adverse health effects other than cancer.

noncompliance: Noncompliance with health and safety regulations.

nonflammable: Not easily ignited, or if ignited, not burning rapidly.

nongenotoxic chemicals: Toxic substances that do not interact directly with DNA.

nonionizing radiation (NIR): Photons with energies less than 12.4 eV are considered to have insufficient energy to ionize matter, and are nonionizing in nature. The nonionizing spectral region includes the ultraviolet (UV), visible, infrared (IR), radio-frequency (RF), and extremely low frequency (ELF) spectral regions.

nonmandatory guidelines: OSHA standards for fire protection (29 CFR 1910.156 through .165) reference many of the NFPA's national consensus standards as nonmandatory guidelines that would be considered acceptable in complying with requirements of Title 29, Part 1910, Subpart L.

nonmelanoma skin cancer : NMSC is a type of skin cancer; and basal cell carcinoma. Four variables have been implicated in NMSC: 1) lifetime sun exposure; 2) the intensity and duration of the UV-B component in sunlight; 3) genetic predisposition; and 4) other factors unrelated to sunlight, such as exposure to ionizing radiation and polycyclic aromatic hydrocarbons. (Also known as "cutaneous malignant melanoma.")

nonstochastic effect: The effect of radiation exposure on a population in which there is an assumed threshold dose that increases in effect with an increase in severity with an increase in dose. (See also *stochastic effect*).

nonthreshold: Characterizes a dose-response curve that passes through the origin of the graph indicating that any exposure will increase the probability of cancer occurrence. Used by the U.S. EPA for carcinogens. (See also *threshold.*)

nontraditional workplace: A nontraditional workplace is usually the typical office, schools, shopping malls and other public places. They are not normally considered workplaces where exposure occurs. These are locations that often suffer poor indoor air quality. Domestic residential environments may also be considered a nontraditional workplace.

normal distribution: An important symmetric continuous probability distribution characterized completely by two parameters: the mean and the standard deviation.

normal temperature and pressure (NTP): 298.15 K and 1 atmosphere, 760 mmHg.

notice of intended change (NIC): This term is unique to ACGIH. Chemicals appearing on the NIC list for at least one year serve as notice that a chemical has a TLV® proposed for the first time or that a current TLV is being changed. This procedure allows ample time for those with data or comments to come forth.

noxious: Harmful to health.

nozzle, jet: The opening through which aerosol particles travel during sampling. Nozzles or jets restrict or confine the flow of air carrying the aerosol; nozzle sizes affect the airstream velocities (and the collection efficiency of the cascade impactor).

NRC: See *National Research Council.*

NRR: See *Noise Reduction Rating.*

NSC: See *National Safety Council.*

NTP: See *National Toxicology Program.*

NTP: See *normal temperature and pressure.*

nuclear binding force: That amount of energy required to separate the nucleus into its individual nucleons beyond the nuclear distance.

nuclear force: Short-range fundamental force in nature that is attractive and holds nucleons together at nuclear range.

nucleon: Particle found in the nucleus; either a proton or a neutron.

nuclide: General term referring to any nucleus plus the orbital electrons (number of protons plus number of neutrons) in each energy state.

nuisance dust: Generally innocuous dust, not recognized as the direct cause of a serious pathological condition.

numerical extrapolation: Extension of dose-response relationship to dose levels below those observed in toxicological or epidemiological studies.

O

o.d.: Outside diameter.

occluded: Closed, shut, or blocked.

occupational cancer: Cancer caused by exposure to chemical or physical agents in the work environment.

occupational disease: Disease associated with a work environment, usually caused by a specific agent.

occupational exposure limit (OEL): A health-based workplace standard to protect workers from adverse exposure (e.g., PELs, TLVs®, RELs, WEELs, etc.).

occupational health: A discipline of environmental health. The profession specializes in the art and science of recognition, evaluation, and control of chemical, physical, and biological stressors or factors arising in or from the workplace that may cause discomfort or adverse health effects to workers or members of the community.

occupational health and safety (OHS) plan: A plan that at minimum should include specific instructions on personal protective equipment (PPE); lockout/tagout (LO/TO); confined space entry (CSE); security; spills and contingency plans; emergency procedures; phone numbers of the facility emergency coordinator, police, fire, ambulance and nearest hospital; locations of first aid kits; eye washes and showers; and the facility contractor OHS policy.

occupational hygiene audit: A process used to evaluate periodically the existence and the effectiveness of biological, chemical, ergonomic, and physical health and safety program elements present in a workplace.

occupational hygiene survey: An activity carried out by a qualified individual to measure and evaluate various biological, chemical, ergonomic, and physical parameters in the workplace or environment to determine their standings in relation to established health and safety standards.

Occupational Safety and Health Act (OSH Act) of 1970: In 1970, the U.S. Congress passed the OSH Act "to assure safe and healthful working conditions for working men and women; by authorizing enforcement of the standards developed under the Act; by assisting and encouraging the states in their efforts to assure safe and healthful working conditions; by providing for research, information, education, and training in the field of occupational safety and health; and for other purposes." (Public Law 91–596) ,

Occupational Safety and Health Administration (OSHA): Established by the OSH Act of 1970, OSHA is located within the U.S. Department of Labor. The agency's responsibilities include promulgating occupational safety and health standards and inspecting workplaces to ensure compliance with these standards.

Occupational Safety and Health standards: Occupational Safety and Health standards are defined at § 3(8) of the OSH Act to mean standards that require the adoption of "practices, means, methods, operations, or processes, reasonably necessary or appropriate to provide safe or healthful employment and places of employment."

occupied space: Any space inside a building occupied at various times.

octave bands: 1. A frequency range in which the ratio of upper to lower frequency is 2:1. 2. A measurement of the broad range of frequencies humans can hear. Frequencies are normally divided into nine octave bands. An octave is defined as a range of frequencies extending from one frequency to exactly double that frequency. Each octave band is named for the center frequency (geometric mean) of the band.

OD: See *optical density*.

odor character: Used to describe a substance's odor (e.g., "fishy," "rancid," "bananas").

odor threshold: In general, the lowest concentration of gas or a material's vapor that can be detected by odor. The detection threshold is the lowest concentration of odorant that will elicit a sensory response in the olfactory receptors of a specified percentage of a given population. The recognition threshold is the minimum concentration that is recognized as having a characteristic odor quality by a specified percentage of a given population. (See also *odor character, hedonic tone.*)

OEL: See *occupational exposure limit*.

off-specification: Low quality. (Also known as "off-spec.")

ohm (Ω): The unit of electrical resistance in the rationalized meter-kilogram-second system of units, equal to the resistance through which a current of 1 amphere will flow when there is a potential difference of 1 volt across it.

Ohm's law: The current, I, through an electrical circuit is related directly to the difference in potential, V, and inversely related to the resistance, R, or I (amperes) = V/R (volts/ohms).

OHS: Occupational health and safety. See *occupational health and safety plan*.

olfactory: Relating to the sense of smell.

olfactory fatigue: The numbing of the olfactory nerve endings.

OPC: See *optical particle counter*.

operating plan: A plan that generally describes where the organization would like to be in one to two years.

operating system: The operating system is a piece of software that forms the interface between the user's program and the computer hardware.

optical density (OD): The quantity used to specify the ability of protective eyewear to attenuate optical radiation, where OD = $\log_{10}$ (ML/EL).

optical particle counter (OPC): A single-particle, direct-reading instrument that uses monochromatic light such as a laser or light-emitting diode or a broad band light source such as a tungsten filament lamp to illuminate aerosols.

optical radiation: A broad category of radiation that includes: ultraviolet (UV), visible, and infrared (IR).

optimism: A disposition or tendency to look on the more favorable side of happenings or possibilities.

optimum risk: A risk level that balances the cost of the risk with the cost of risk mitigation.

organic: Term used to designate chemicals that contain carbon. To date nearly 1 million organic compounds have been synthesized or isolated. Many occur in nature; others are produced by chemical synthesis.

organic peroxide: An organic compound that contains the bivalent -O-O structure and may be considered a structural derivative of hydrogen peroxide where one or both of the hydrogen atoms has been replaced by an organic radical.

orifice: A small opening that controls flow rate of gases or liquids. Orifices are used in some hand-held detector tube pumps.

orifice plate: A plate with a relatively sharp-edged opening or orifice used to measure fluid flow rates based on pressure difference between the two sides of the plate.

origin: The point on a graph that represents zero on both the vertical and horizontal axes or lines.

O-ring: A ring made of rubber or latex to seal gaps around the impaction stages, and to facilitate proper seating of the stage in the cascade impactor.

oscillate: To move back and forth in a steady uninterrupted rhythm; to vary between alternate extremes, usually within a definable period.

OSH Act (of 1970): See *Occupational Safety and Health Act of 1970*.

OSHA: See *Occupational Safety and Health Administration.*

OSHA Compliance: Sampling being done for the purpose of ensuring that airborne chemical levels in the workplace are within allowable limits specified by the Occupational Safety and Health Administration.

OSHRC: Occupational Safety and Health Review Commission.

otologist: A physician who specializes in diseases of the ear.

outer shell: The skin and muscles of the body.

outlier: An extreme value that questionably belongs to the group of values with which it is associated. If the chance probability of its being a valid member of the group is very small, the questionable value is thereby "detected" and may be eliminated from the group based on further investigation of the data.

overload: In the workplace, too much information, which might lead to fatigue.

oxidant: Oxidizing agent (Also known as "electron sink").

oxidation: In a literal sense, oxidation is a reaction in which a substance combines with oxygen provided by an oxidizer or oxidizing agent.

oxidizer: A chemical other than a blasting agent or explosive that initiates or promotes combustion in other materials, causing fire either by itself or through the release of oxygen or other gases.

oxidizing agent: A chemical that produces oxygen readily and gains electrons during the reaction; may start or assist the combustion of other materials.

oxygen deficiency: An atmosphere having less than the percentage of oxygen found in normal air. Normally, air contains about 21% oxygen at sea level. When the oxygen concentration in air is reduced to approximately 16%, many individuals become dizzy, experience a buzzing in the ears, and have a rapid heartbeat. OSHA indicates 19.5% as the lower limit of oxygen acceptable in industry.

oxygen toxicity: Harmful effects of breathing oxygen at pressures greater than atmospheric.

oxygen-deficient atmosphere: An atmosphere containing less than 19.5% oxygen by volume. (See *OSHA 1910.146.*)

oxygen-enriched atmosphere: An atmosphere containing more than 23.5% oxygen by volume. (See *OSHA 1910.146*)

ozone: A colorless gas with a characteristic odor that is produced in ambient air during the photochemical oxidation of combustion products such as the nitrogen oxides and hydrocarbons. It can also result from the operation of electrical motors, photocopy machines, and electrostatic air cleaners in occupational environments.

P

Pa: See *pascal.*

PAH(s): See *polycyclic aromatic hydrocarbons.*

parameter: Limit of consideration in a study or investigation.

pareto analysis: An analysis tool based on the ratio that often 80% of the problems in a process are the result of only 20% of the potential causes. Pareto analysis is used to isolate and identify areas of significant concern from a group of many potential concerns.

partial pressure: Pressure of a gas in a mixture equal to the pressure that it would exert if it occupied the same volume alone at the same temperature. (See also *Dalton's law.*)

particle: A small discrete object, often having a density approaching the intrinsic density of the bulk material. It may be chemically homogenous or contain a variety of chemical species. It may consist of solid or liquid material or both. (See also *aerosol.*)

particle bounce: Rebound of particles that fail to adhere after impacting on a collecting surface.

particle diffusivity (D_B): Aerosol particles in a gaseous medium are bombarded by collisions with individual gas molecules that are in Brownian motion. This causes the particles to undergo random displacements known as diffusion. The particle parameter that describes this process is the particle diffusivity, D_B. (Also known as "diffusion coefficient.")

particle size distribution: A relationship expressing the quantity of a particle property associated with particles in a given size range.

particulate: Particle of solid or liquid matter.

partition coefficient (K_m): 1. In the equilibrium distribution of a solute between two liquid phases, the constant ratio of the solute's concentration in the upper phase to its concentration in the lower phase. 2. The ratio at equilibrium of penetrant concentration in the SC to that in the vehicle.

parts per billion by volume (ppb): Sometimes written as ppbv or ppb v/v; an expression of concentration as a volume ratio, usually parts of contaminant per billion parts of air. It may also be used to express dietary exposure concentration as a weight ratio.

parts per million by volume (ppm): Sometimes written as ppmv or ppm v/v; an expression of concentration as a volume ratio, usually parts of contaminant per million parts of air. It may also be used to express dietary exposure concentration as a weight ratio.

part-time work: In part-time work, the job is split among several people, each of whom works part of the shift.

PAS: See *photoacoustic spectroscopy.*

pascal (pa): 1. A unit of pressure equal to the pressure resulting from a force of 1 newton acting uniformly over an area of 1 square meter (m^2). 2. A unit of pressure equal to kg/m sec^2 (part of the internationally adopted SI system of units).

passive dosimeter: A sample collection device based on the mass transport of the air contaminant to the sorbent by gaseous diffusion. It can incorporate direct-reading colorimetry to determine the concentration of a chemical in the air.

passive sampling: The collection of airborne gases and vapors at a rate controlled by a physical process such as diffusion through a static air layer or permeation through a membrane without the active movement of air through an air sampler.

pathway: The course a chemical or pollutant takes from the source to the organism exposed.

PCB: Polychlorinated biphenyl.

PEL: See *permissible exposure limit.*

PEL–C: See *permissible exposure limit–concentration.*

PEL–STEL: See *permissible exposure limit–short-term exposure limit.*

PEL–TWA: See *permissible exposure limit–time-weighted average.*

penetration: The flow of a chemical through zippers, weak seams, pinholes, cuts, or imperfections in the protective clothing on a nonmolecular level.

perceived risk: The risk that the individual believes exists.

percent volatile: Percent volatile by volume is the percentage of a liquid or solid (by volume) that will evaporate at an ambient temperature of 70°F (unless some other temperature is specified). Examples are butane, gasoline, and paint thinner (mineral spirits) are 100% volatile; their individual evaporation rates vary, but in time each will evaporate completely.

performance audit: A quantitative analysis or check with a material or device with known properties or characteristics. The audit is performed by a person different from the routine operator/analyst using audit standards and audit equipment different from the calibration equipment. Such audits are conducted periodically to check the accuracy of a project measurement system. Some performance audits might require the identification of specific elements or compounds, in lieu of, or in addition to, a quantitative analyses. For some performance audits it may be impractical or unnecessary to have a person different than the routine operator/analyst; in these cases the routine operator/analyst must not know the concentration or value of the audit standards until the audit is completed. The other conditions of the audit must still be met (that is, the audit standards must be different from the calibration standards, and the audit device must be different from the calibration device).

performance measures: The observable demonstration that the individual has learned a desired instructional objective.

performance standards: 1. Regulatory requirements limiting the concentrations of designated organic compounds, particulate matter, and hydrogen chloride in emissions from incinerators. 2. Operating standards established by the U.S. EPA for various permitted pollution control systems, asbestos inspections, and various program operations and maintenance requirements. 3. Performance standards state the object to be obtained or the hazard to be abated.

perfusion: The pumping of a fluid through a tissue or organ by way of an artery.

periodic motion: Any motion that is repeated at regular intervals. (See also *oscillate.*)

periodic vibration: Vibration is considered periodic if the motion of a particle repeats itself considerably over time.

periodicity: The frequency at which tests are given.

peripheral nervous system (PNS): The autonomic nervous system, the cranial nerves, and the spinal nerves including their associated sensory receptors.

peripheral neuropathy: Any disease affecting the peripheral nervous system.

permeability: A factor, characteristic of a material, that is proportional to the magnetic induction produced in a material divided by the magnetic-field strength; it is a tensor when these quantities are not parallel.

permeation: The movement of a chemical through a protective clothing barrier that has no visible holes.

permeation method: A method for preparing a known mixture of a low concentration gas for verification testing of detector tubes. Gas permeates the walls of a gas permeable vessel (permeation tube) into a mixing solution, where it is combined with the diluent gas (usually purified air). The test gas concentration is calculated from the permeation rate, the flow rate of the diluent gas, and the thickness of the walls of the permeation tube.

permeation rate: The rate of movement (mass flux) of the chemical through the barrier. The permeation rate is normally reported in mass per unit area per unit time (e.g., μg/cm/min) after equilibrium is reached and may be normalized for thickness.

permeation tube: A plastic tube in which is sealed a liquefied gas or volatile liquid that slowly permeates through the walls at a constant rate.

permissible concentration: Official term that replaces "threshold limit value." (See also *threshold limit value.*)

permissible dose: Amount of radiation that may be received by an individual within a specified period with expectation of no significantly harmful result.

permissible exposure limit (PEL): Established by OSHA (see *29 CFR 1910.1000, Subpart Z*). The permissible concentration in air of a substance to which nearly all workers may be repeatedly exposed 8 hours a day, 40 hours a week, for 30 years without adverse effects. (See also *PEL–C, PEL–STEL, PEL–TWA.*)

permissible exposure limit–concentration (PEL–C): An acceptable ceiling concentration. An employee's exposure to any material in 29 CFR 1910.1000, Table Z1 — the name of which is preceded by a C — shall at no time exceed the ceiling value given for that material in the table.

permissible exposure limit–short-term exposure limit (PEL–STEL): Short-term exposure limit is the employee's 15-min TWA exposure that shall not be exceeded at any time during a workday unless another time limit is specified in a parenthetical notation below the limit.

permissible exposure limit–time-weighted average (PEL–TWA): An 8-hour time-weighted average. An employee's exposure to any material listed in 29 CFR 1910.1000, Table Z2, in any 8-hour work shift of a 40-hour workweek shall not exceed the 8-hour time-weighted average limit given for that material in the table.

permissible heat exposure threshold limit values: A limit designed to provide a work temperature and wet bulb globe temperature (WBGT) combination so that 95% of the workers would not have a deep body temperature exceeding 38°C. Work load categories are light work <200 kcal/hour (233 watts), moderate work 200–350 kcal/hour (233–407 watts), and heavy work 350–500 kcal/hour (407–581 watts).

permit-required confined space: A confined space that has one or more of the following characteristics: 1) contains or has the potential to contain a hazardous atmosphere; 2) contains a material that has the potential for engulfing an entrant; 3) has an internal configuration such that an entrant could be trapped or asphyxiated by inwardly converging walls or by a floor that slopes downward and tapers to a smaller cross section; or 4) contains any other recognized serious safety or health hazard. Also known as a "permit space." (See *OSHA 1910.146.*)

permittivity: A fundamental quantity that describes the interaction of matter with the electric field. Permittivity of tissue depends on the water content of the tissue and the frequency of the electric field.

personal computer: The basic personal computer consists of a central processing unit (CPU), random access memory (RAM), a hard disk drive, a floppy disk drive, CD-ROM (computer disk, read-only memory), a video card, and a monitor (often called a VDT [video display terminal] or VDU [video display unit]).

personal mastery: An employee's commitment to one's own learning, as demonstrated by a desire to focus one's energies by continually aligning one's personal growth activities with the needs of the organization.

personal protective equipment (PPE): Equipment (e.g., gloves, eye protection, respirators) designed to protect individuals from biohazards.

personal sampler: Air-sampling instrument developed in the United States for estimating exposure of individual workers to air contaminants.

personal sampling: The collection of airborne chemicals in the worker's breathing zone done by having the worker wear the sampling equipment throughout the workday.

personnel enclosure: When there are multiple noisy sound sources in a room and a low number of operators, it is sometimes useful to enclose the employees instead of the equipment.

perturbation: Any effect that makes a small modification in a physical system.

PF&ID: See *process flow and instrumentation diagram.*

pH: 1. The symbol relating the hydrogen ion (H+) concentration to that of a given standard solution. A pH of 7 is neutral. Numbers increasing from 7 to 14 indicate greater alkalinity. Numbers decreasing from 7 to 0 indicate greater acidity. 2. Means used to express the degree of acidity or alkalinity of a solution with neutrality indicated as 7.

phased sampling scheme: The conduction of a second sequential set of sampling is based on results of a first set.

phosphenes: The sensation of flashes of light within the eye caused by EMF.

photoacoustic spectroscopy (PAS): A spectroscopic technique for investigating solid and semisolid materials, in which the sample is placed in a closed chamber filled with a gas such as air and illuminated with monochromatic of any desired wavelength, with intensity modulated at some suitable acoustic frequency; absorption of radiation results in a periodic heat flow from the sample, which generates sound that is detected by a sensitive microphone attached to the chamber.

photoconjunctivitis: See *photokeratitis.*

photoionization detector: The PID is a portable, general survey instrument used for detecting leaks, surveying plants to identify problem areas, evaluating source emissions, monitoring ventilation efficiency, evaluating work practices, and determining the need for personal protective equipment for hazardous waste site workers.

photokeratitis: An injury to the eye that results from acute, high-intensity exposure to UV-B and UV-C. Commonly referred to as "arc eye" or "welder's flash" by workers, this injury results from exposure of the unprotected eye to a welding arc or other artificial sources rich in UV-B and UV-C. Sunlight exposure produces these sequelae only in environments where highly reflective materials are present, such as snow ("snow blindness") or sand. (Also known as "photoconjunctivitis.")

photon energy: Photon energy describes the energy possessed by electromagnetic energy when characterized as discrete bundles, as described by quantum theory. The unit of photon energy is the Joule (J) or the electron volt (eV).

photosensitivity: Reaction to light that causes the skin to become sensitive. Photosensitivity includes two types of reactions: phototoxicity and photoallergy. Phototoxicity is more common and affects all individuals if the UV dose or the dose of the photosensitizer is high enough. Photoallergy is an acquired altered reactivity in the exposed skin resulting from an immunologic response. An agent, such as many medications, some sunscreen agents, plants (e.g., figs, parsley, limes, parsnips, and pinkrot celery), and industrial photosensitizers (including coal tar, pitch, anthracene, naphthalene, phenanthrene, thiophene, and many phenolic agents) can instigate photosensitivity.

physical asphyxiant: See *asphyxiant.*

physical hazard: A chemical for which there is scientifically valid evidence that it is a combustible liquid, a compressed gas, explosive, flammable, an organic peroxide, an oxidizer, pyrophoric, unstable (reactive), or water-reactive.

physical work capacity (PWC): The maximum amount of oxygen that an individual can consume per minute.

physiological heat exposure limit: An exposure limit based on a physiological response.

physiology: The science and study of the functions or actions of living organisms.

PID: See *photoionization detector.*

piezoelectric mass sensor: A sensor based on the principle that when crystalline materials are mechanically stressed by compression or tension they produce a voltage proportional to the stress. When these crystals are subjected to an electric current, they oscillate, and the natural vibrational frequency depends on the thickness and density of the crystal.

pilot tube: A small bore tube inserted into a flowing stream with its orifice facing the stream to measure total pressure. The term is often used for a double tube instrument from which the flow velocity can be calculated with one orifice facing the flowing stream to register total pressure and the other perpendicular to the stream to register static pressure.

pion: The exchange particle in nucleus.

piston pump: A hand-held sampling pump that draws a fixed volume of air. It operates by pulling and locking a piston into position while the sample is drawn into the detector tube. The piston is released after the sampling time has elapsed.

pitch: See *frequency.*

Planck's constant: Constant of proportionality relating the quantum of energy that can be possessed by radiation to the frequency of that radiation; value is approximately 6.625×10^{-27} erg/sec. (See also *quantum.*)

plane wave: 1. Wave in which the wavefront is a plane surface; a wave whose equiphase surfaces form a family of parallel planes. 2. To understand a plane wave, consider an ideal point-source antenna that emits radiation in an isotropic pattern. The radiation pattern would be spherical (uniform in all directions), so near this antenna a receiver detects curvature in the approaching field. However, if removed sufficiently far from the source, some distance into the far field, a receiver would sample only a very small area of an immense curved wavefront. In the local region of space occupied by the receiver, it would detect a flat, or planar front; hence, the name, plane wave.

plasma: The liquid that does not contain the cellular components of blood on sitting or mild centrifugation of a blood sample.

plenum: A large air compartment or chamber connected to the slot that functions to distribute the static pressure (i.e., suction) evenly across the slot area.

plenum chamber: 1. An air compartment at a pressure slightly above atmospheric. 2. In an air distribution system, that part of the casing, or an air chamber attached to the furnace, from which the air duct system or direct discharge heat outlet delivers heated or cooled air.

plenum velocity: Air velocity in the plenum. Provide a uniform air distribution with slot types hoods, the maximum plenum velocity should be one-half the slot velocity or less.

Pliny the Elder: A Roman scholar who in 50 A.D. identified the use of animal bladders to prevent the inhalation of lead dust and fume.

pneumoconiosis: A chronic disease of the lungs resulting from the inhalation of various kinds of dusts. The pneumoconioses that include siderosis (iron oxide), silicosis (free silica), asbestosis (asbestos), etc., generally require a period of years for development.

PNS: See *peripheral nervous system*.

policy deployment: The process by which a company develops policies, including improvement targets, and deploys them throughout the organization. This deployment is performed in a manner that permits the operating organizations to establish supporting goals and targets, along with a method to measure performance.

pollutants (air): 1. Generally, any substance introduced into the environment that adversely affects the usefulness of a resource. 2. Any substance in air that could, in high enough concentration, harm man, other animals, vegetation, or material. Pollutants may include almost any natural or artificial composition of airborne matter capable of being airborne. They may be in the form of solid particles, liquid droplets, gases, or in combination thereof. Generally, they fall into two main groups: 1) those emitted directly from identifiable sources; and 2) those produced in the air by interaction between two or more primary pollutants, or by reaction with normal atmospheric constituents, with or without photoactivation. Exclusive of pollen, fog, and dust, which are of natural origin, about 100 contaminants have been identified and fall into the following categories: solids, sulfur compounds, volatile organic chemicals, nitrogen compounds, oxygen compounds, halogen compounds, radioactive compounds, and odors.

polycyclic aromatic hydrocarbons (PAHs): Organic compounds, usually formed from incomplete combustion, that might pose a risk of cancer.

polydisperse: Composed of particles with a range of sizes.

polymerization: A chemical reaction in which one or more small molecules combine to form larger molecules. A hazardous polymerization is a reaction that takes place at a rate that releases large amounts of energy. If hazardous polymerization can occur with a given material, it will be cited in the material safety data sheet (MSDS).

poor warning properties: A substance with odor, taste, or irritation effects that are not detectable or not persistent at concentrations at or below the exposure limit.

popliteal height: The distance from the floor to the joint in back of the knee.

portal of entry: See *route of entry*.

positive beta ray: The high speed particle with characteristic of a positron emitted from a nucleus in positive beta decay.

positive-pressure device: See *positive-pressure respirator*.

positive-pressure respirator: A respirator in which the pressure inside the respiratory inlet covering is normally positive with respect to ambient air pressure.

positron: The positive electron.

Pott, Sir Percival: English physician who in 1775 described the relationship between scrotal cancer and work as a chimney sweep as related to the lack of hygiene measures.

power density: The amount of power per unit area in a radiated microwave or other electromagnetic field, usually expressed in units of watts per square centimeter.

powered air-purifying respirator (PAPR): An air-purifying respirator that uses a blower to force the ambient atmosphere through air-purifying elements to the inlet covering.

ppb: See *parts per billion by volume*.

PPE: See *personal protective equipment*.

PPE controls: The use of PPE to protect individuals.

ppm: See *parts per million by volume*.

precision: The degree of agreement of repeated measurements of the same type using a specified sampling device; usually expressed as the coefficient of variation or relative standard deviation of replicate measurements.

pre-classifier: A device that removes particles ahead of a particle sensor or sampler, often similar to the particle removal occurring ahead of the respiratory region of interest. (Also known as "pre-collector," "pre-cutter," or "pre-separator.")

pre-collector: See *pre-classifier*.

pre-cutter: See *pre-classifier*.

prelayer: A layer in a detector tube preceding the indicating layer. Prelayers usually are nonindicating and are used to remove humidity, control interferences, or provide actual reaction components.

premolded earplug: A type of hearing protector that is preformed and simply inserted in the ear.

presbycusis: The hearing loss normally occurring due to age because of the degeneration of the nerve cells due to the ordinary wear and tear of the aging process.

pre-separator: See *pre-classifier*.

pressure: 1. Thermodynamically, the normal force exerted by a homogeneous liquid or gas, per unit of area, on the wall of the container. 2. Force exerted per unit area.

pressure drop: 1. Loss in pressure (as from one end of a refrigerant line to the other) from friction, static, heat, etc. 2. The differential pressure across some element of a system, such as a valve or orifice. (Also known as "backpressure.")

pressure-demand respirator: A positive-pressure atmosphere-supplying respirator that admits respirable gas to the facepiece when the positive pressure is reduced inside the facepiece by inhalation.

preventive maintenance: A systematic procedure by which laboratory components that are most likely to fail are replaced and instruments recalibrated to ensure that the laboratory operates at maximum efficiency.

primary barriers: Protection of the worker and environment in the immediate area of potential exposure. Primary barriers include biosafety cabinets; sealed centrifuge rotors; glove boxes; high efficiency particulate aerosol (HEPA)-filtered animal enclosures, etc.; and PPE (especially gloves, eye protection, and sometimes respirators) are important primary barriers.

primary prevention: Actions taken to prevent initiation of the disease process by removing or reducing risk factors.

primary reader: The person empowered to take action based on the report.

primary standard: 1. A measurement device that is directly traceable to the National Institute of Standards and Technology (NIST). Examples of traceable primary standards for volume are soap bubble flow meters and spirometers. 2. A material having a known property that is stable, that can be accurately measured or derived from established physical or chemical constants, and that is readily reproducible. 3. Devices such as flowmeters that base their calibration on direct and measurable linear dimensions such as the length and diameter of a cylinder.

prime movers: Skeletal muscles that are responsible for the primary action at the joint.

process: All activities that produce an output for a customer.

process change: Changing a process to make it less hazardous (e.g., paint dipping in place of paint spraying).

process controls: Emission control devices installed on a process or piece of equipment.

process hood: A device to capture, enclose or receive hazards from a process. Hoods enclose a process and capture or receive contaminants.

procurement quality control (QC): A systematic procedure by which supplies, materials, and capital equipment are procured and tested to certify the specified quality of those materials.

proficiency testing: Special series of planned tests to determine the ability of field technicians or laboratory analysts who normally perform routine analyses. The results may be used for comparison against established criteria, or for relative comparisons among the data from a group of technicians or analysts.

progeny of 222-radon: The first four decay products — 218-Po, 214-Pb, 214-Bi, and 210-Po.

program audit: A survey that looks at the management system in place to ensure that health hazards remain under continuing control.

program management: The keeping of records and program assessment efforts to evaluate the effectiveness of the comprehensive occupational health program.

promotion: The mutated precancerous cell can divides and proliferates in the second step of carcinogenesis.

promulgated standard: Standard that has been enacted or otherwise made into law or regulation by authority.

pronation: 1. Turning the palm downward or toward the back. 2. Eversion of the foot.

propagation: Spread or transmission of decomposition, flame, or explosion.

prospective epidemiology: The use of ongoing epidemiological studies to determine if there is an association between past exposures and the current health conditions.

protocol: In science, the rules and outline of an experiment.

prudent avoidance: An approach in which one chooses a low-cost, easily accomplished method to reduce exposure but makes no concerted effort in this regard. An example is the choice between walking in the vicinity of a high-strength field source or taking an equally effective, alternative route farther away from the source.

psi: Pounds per square inch.

psychology: The science that deals with the functions of the mind and the behavior of an organism in relation to its environment.

psychosomatic: A physical disorder that is caused by or notably influenced by the emotional state of the patient.

psychrometer: An instrument consisting of wet and dry bulb thermometers for measuring relative humidity.

psychrometric: One of two types of wet bulb measurement that relies on artificial ventilation. (See also *natural wet bulb, wet-bulb globe temperature.*)

psychrometric chart: A graphical representation of the properties of moist air, usually including wet and dry bulb temperatures, specific and relative humidities, enthalpy, and density.

psychrometric wet bulb temperature (T_{wb}): Temperature measured by a thermometer on which the sensor is covered by a wetted cotton wick that is exposed to forced movement of the air. Accuracy of wet bulb temperature measurements requires using a clean wick, distilled water, and proper shielding to prevent radiant heat gain. (Also known as "wet bulb temperature.")

PTS: Permanent threshold shift. (See also *temporary threshold shift.*)

pulmonary: Relating to or associated with the lungs.

pulmonary irritation: Irritation of the lungs.

pulmonary region: The region containing the respiratory bronchioles, alveolar ducts, and alveolar sacs across which gas exchange occurs.

pulmonary system: A subsystem of the cardiovascular system (CVS) in which blood from the right ventricle is pumped to the lungs, where carbon dioxide is removed and oxygen is added before the blood is returned to the left side of the heart.

pulvation: The act of particles being emitted or induced to become airborne. [**Note:** This term was coined by Hemeon.]

pump: A mechanical device used in air monitoring to draw the sample gas through a collection device.
pure tone: The sound energy that is characterized by its singleness of frequency.
PWC: See *physical work capacity*.
pyrogenic: Changes in body temperature due to fever.
pyrolyzer: A device that thermally decomposes certain gases and vapors, releasing constituents that then can react with the indicating layer in the detector tube.
pyrophoric: A chemical that will ignite spontaneously in air (and occasionally friction) at a temperature of 13°F or below.

Q

Q: See *volume flow rate*.
QA: See *quality assurance*.
QC: See *quality control*.
QCM: See *quartz crystal microbalance*.
QFD: See *quality function deployment*.
Qualified Industrial Hygienist: A person qualified to practice industrial hygiene by virtue of educational qualifications and examination by the American Board of Industrial Hygiene (ABIH).
qualitative: Pertaining to kind or type (name).
qualitative fit-test: A pass/fail fit-test that relies on the subject's sensory response to detect the challenge agent.
quality: The totality of features and characteristics of a product or service that bear on its capability to satisfy a given purpose. For air pollution measurement systems, the product is air pollution measurement data, and the characteristics of major importance are accuracy, precision, completeness, and representativeness. For air monitoring systems, "completeness," or the amount of valid measurements obtained relative to the amount expected to have been obtained, is a very important measure of quality. The relative importance of accuracy, precision, and completeness depends on the particular purpose of the user.
quality assurance: A system for integrating the quality planning, quality assessment, and quality improvement efforts of various groups in an organization to enable operations to meet user requirements at an economical level. In air pollution measurement systems, quality assurance (QA) is concerned with all of the activities that have an important effect on the quality of the air pollution measurements as well as the establishment of methods and techniques to measure the quality of the air pollution measurements. The more authoritative usage differentiates between "quality assurance" and "quality control" — quality control being "the system of activities to provide a quality product" and quality assurance being "the system of activities to provide assurance that the quality control system is performing adequately."
quality assurance program plan: An orderly assembly of management policies, objectives, principles, and general procedures by which an agency or laboratory outlines how it intends to produce data of acceptable quality.
quality assurance project plan: An orderly assembly of detailed and specific procedures by which an agency or laboratory delineates how it produces quality data for a specific project or measurement method. A given agency or laboratory would have only one quality assurance program plan, but would have a quality assurance project plan for each of its projects (group of projects using the same measurement methods; for example, a laboratory service group might develop a plan by analytical instrument since the service is provided to a number of projects).
quality audit: A systematic examination of the acts and decisions with respect to quality in order to independently verify or evaluate compliance to the operational requirements of the quality program or the specification or contract requirements of the product or service, or to evaluate the adequacy of a quality program.
quality control (QC): The system of activities designed and implemented to provide a quality product.
quality control reference sample: A material used to assess the performance of a measurement or portions thereof. It is intended primarily for routine intralaboratory use in maintaining control of accuracy and would be prepared from or traceable to a calibration standard. (Also known as "working standard.")
quality function deployment (QFD): A systematic means of ensuring that the demands of the customer are accurately translated into action within the supplier organization.
quantitative: Pertaining to the amount (mass).
quantitative fit-test: A fit-test that uses an instrument to measure the challenge agent inside and outside the respirator.
quantum: Invisible unit of energy equal for radiation of frequency to the product h , where h is Planck's constant. (See also *Planck's constant*.)
quartz: Crystalline silicone dioxide. The main constituent in sandstone and some igneous rocks. (Also known as "free silica.")
quenching: The repressing of optical radiation in a detector caused by heat or ultraviolet and not by radiation.
queue time: On a flowchart, the amount of time for each step and time between steps.

R

R: See *gas constant*.
radiance: Radiometric brightness (all optical wavelengths).
radiant exposure: The photobiological dose of radiant energy.
radiant temperature: The temperature resulting from the body-absorbing radiant energy.
radiation: The transmission of energy in the form of electric and magnetic fields (i.e., electromagnetic waves). Also known as "radiant energy."
radio frequency (RF): A frequency at which coherent electromagnetic radiation of energy is useful for communication purposes. Roughly the range from 10 kilohertz (kHz) to 100 gigahertz (gHz).
radioactivity: The spontaneous change in nucleonic configuration and/or energy content of a nucleus.
radiological wastes: Wastes that consist of radioactive components that emit ionizing radiation.
radiometer: 1. A receiver for detecting microwave thermal radiation and similar weak wide-band signals that resemble noise and are obscured by receiver noise. 2. An instrument for measuring radiant energy.

radionuclide: An unstable nucleus.

radiowave: An electromagnetic wave produced by reversal of current in a conductor at a frequency in the range from about 10 kilohertz (kHz) to about 300,000 megahertz (mHz).

radius of gyration: The square root of the ratio of the moment of inertia of a body about a given axis to its mass.

radon: A noble gas that emits alpha particles with a half-life of 3.8 days. It is a decay product of radium 226, which is a decay product of the uranium 238 series. Radon equilibrates rapidly with its decay products so that without a replenishing source, significant concentrations cannot be maintained.

radon daughters: Radon daughters, two of which emit alpha particles, have half-lives of less than 30 min.

Ramazzini, Bernardino: Italian physician and educator and author of *De Mobis Artificum Diatriba (Disease of Workers)* visited shops and workplaces to study the working conditions and health problems of various trades. He suggested that physicians ask of their patients, "Of what trade are you?" He is recognized as the "father of occupational medicine."

random error: Variations of repeated measurements that are random in nature and individually not predictable. The causes of random error are presumed to be indeterminate or nonassignable. The distribution of random errors is generally presumed to be normal (Gaussian).

random noise: Random noise is an oscillation in which instantaneous magnitude is not specified for any given instant of time.

random sample: A sample obtained in such a manner that all items or members of the lot, or population, have an equal chance of being selected in the sample. In air pollution monitoring the population is usually defined in terms of a group of time periods for which measurements are desired. For 24-hour samplers, the population is usually considered as all of the 365 (or 366) 24-hour calendar day periods in a year. For continuous monitors, the population is often considered as all of the hourly average values obtained (or that could have been obtained) during a particular period, usually a calendar year. For either 24-hour or continuous monitors, a single air pollution result from a site could be a sample of the conceptually infinite population of values that might have been obtained at the given site for all possible combinations of equipment, materials, personnel, and conditions that could have existed at that site and time.

random vibration: A varying force acting on a mechanical system which may be considered to be the sum of a large number of irregularly timed small shocks; induced typically by aerodynamic turbulence, airborne noise from rocket jets, and transportation over road surfaces.

range: The difference between the maximum and minimum values of a set of values. When the number of values is small (i.e., 8 or less), the range is a relatively sensitive (efficient) measure of variability. As the number of values increases above 8, the efficiency of the range (as an estimator of the variability) decreases rapidly. The range or difference between two paired values is of particular importance in air pollution measurements, since in many situations duplicate analyses or measurements are performed as a part of the quality assurance program.

Rankine temperature: An absolute temperature scale conventionally defined by the temperature of the triple point of water equal to 491.68°R, with 180 divisions between the melting point of ice and the boiling point of water under standard atmospheric pressure (1°R = 1°F).

Raynaud's Phenomenon of Occupational Origin: See *hand-arm vibration syndrome.*

Raynaud's syndrome: A type of repetitive strain injury (RSI) considered to be caused by persistent heavy vibration (for example, from a jackhammer or bicycle handlebars) or exposure to extreme cold, characterized by pain, tingling, numbness, and pale skin on the fingers or toes. (Also known as "constitutional cold fingers," "dead finger.")

RC: Reduced comfort. See *reduced comfort resonance.*

reaction: A chemical transformation or change. The interaction of two or more substances to form new substances.

reactivity: Chemical reaction with the release of energy. Undesirable effects (such as pressure buildup, temperature increase, formation of noxious, toxic, or corrosive byproducts) may occur because of the reactivity of a substance to heating, burning, direct contact with other materials, or other conditions in use of in storage.

reagent grade: A chemical reagent that meets standards for purity set by the American Chemical Society. These reagents are produced for laboratory use.

real-time monitors: See *direct-reading instruments.*

reasonable maximum exposure (RME): Used in conservative exposure assessment calculations. Based not on worst-case scenario, but on 90% or 95% upper confidence limits on input parameters.

receiving hood: 1. A ventilation system hood that receives contaminants that are thrown or directed into the hood. A canopy hood is an example of a receiving hood. 2. A hood designed and positioned to take advantage of any initial velocity or motion imparted to the contaminant by the generating process.

reciprocity: Mutual exchange.

recognition: One of four primary responsibilities of the industrial hygienist. The line separating "anticipation" and "recognition" is not always a clear one. Some have distinguished them on the basis of whether the situation being examined actually exists. If it is still in a conceptual phase, the process being applied is considered to be "anticipation." Then it is assumed that, in the recognition phase, the facility exists. This is a somewhat arbitrary distinction; anticipation of hazards can and does occur with existing facilities and recognition of hazards can take place when the facility is in a planning stage. (See also *anticipation, evaluation,* and *control.*)

recognition process: A process used to recognize workplace hazards.

recombinant DNA: The new DNA that is formed by combining pieces of DNA from different organisms or cells.

recommended alert limits: Heat exposure limits for unacclimatized workers.

recommended exposure limit (REL): An occupational exposure limit recommended by NIOSH as being protective of worker health and safety over a working lifetime. The REL is used in combination with engineering and work practice controls, exposure and medical monitoring, labeling, posting, worker training, and personal protective equipment. It is frequently expressed as a TWA exposure for up to 10 hours/day during a 40-hour work week; may also be expressed as 1) a short-term exposure limit (STEL) that should never be exceeded and is to be determined in a specified sampling time (usually 15 minutes); or 2) a ceiling (C) limit that should never be exceeded even instantaneously unless specified over a given time period.

recommended standard: A standard that has been recommended to authority for promulgation or enactment into law.

recordable injury: A work-related injury reported by an employer under the OSH Act of 1970.

redox compound: Compound with reducing and oxidizing features.

reduced comfort (RC) resonance: The boundaries concerned with preservation of comfort during vibration exposure.

reducing agent: In a reduction reaction (which always occurs simultaneously with an oxidation reaction) the reducing agent is the chemical or substance that 1) combines with oxygen; or 2) loses electrons to the reaction. (See also *oxidation*.)

reductant: Reducing agent (electron source).

re-entrainment: Resuspension of particles after they have been settled onto a collecting surface.

reference dose (RfD): U.S. EPA toxicity value for evaluating noncarcinogenic effects resulting from exposures at Superfund sites. An estimate (with uncertainty spanning an order of magnitude or greater) of daily exposure level for humans, including sensitive subpopulations, that is likely to be without an appreciable risk or deleterious effects during a lifetime.

reflective listening: A management tool by which one summarizes and repeats the position of the other party.

regulatory occupational exposure limit: An occupational exposure limit set and recommended by government agencies. (See also *occupational exposure limit*.)

regulatory standards: Regulatory standards are issued by a governmental body that is authorized by law to issue, and usually enforce compliance with, such standards; there is no voluntary or discretionary option for the user to determine whether or not to comply.

REL: See *recommended exposure limit*.

relational database: A type of database that stores data in the form of a table. Relational databases are powerful because they require few assumptions about how data is related or how it will be extracted from the database. As a result, the same database can be viewed in many different ways. Another feature of relational systems is that a single database can be spread across several tables. This differs from flat-file databases, in which each database is self-contained in a single table. Nearly all full-scale database systems for personal computers use a relational database. Small database systems, however, use other designs that provide less flexibility in posing queries.

relative error: An error expressed as a percentage of the true value or accepted reference value. All statements of precision or accuracy should indicate clearly whether they are expressed in absolute or relative sense. (This gets complicated when the absolute value is itself a percentage as is the case for many chemical analyses.)

relative humidity (RH): The ratio of the quantity of water vapor present in the air to the quantity that would saturate it at any specific temperature.

relative permittivity: Permittivity values that have been normalized to a fundamental constant, the permittivity of space, 8.85×10^{-12} farads/meter. Sometimes called the "dielectric constant." (See also *permittivity*.)

relative standard deviation (RSD): See *coefficient of variation*.

renal: Relating to or associated with the kidney.

repeatability: The precision, usually expressed as a standard deviation, measuring the variability among results of measurements at different times of the same sample at the same laboratory. The unit of time should be specified, since within a day repeatability would be expected to be smaller than between day repeatability.

repetitive strain injury (RSI): A general term for a group of soft-tissue injuries caused by overuse of the hand, wrist, and arm experienced by people who regularly work on computers, play certain musical instruments, or work in construction, mining, clerical, and food preparation jobs. RSIs result when repetitive motions damage the nerves of the hands, arms, shoulders, or neck, or affect the joints, muscles, ligaments, tendons, or tendon sheaths of these areas. Some common types of RSI are carpal tunnel syndrome, thoracic outlet syndrome, tendinitis, tenosynovitis, and deQuervain's disease. Symptoms range from tingling, numbness, and pain in the neck, shoulders, arms, or hands to reduced grip strength and clumsiness. (Also known as "repetitive motion injury," "cumulative trauma disorder," or "overuse injury.")

replacement air: Air supplied to a space to replace exhausted air. (Also known as "compensating air" or "make-up air.")

replicability: The precision, usually expressed as a standard deviation, measuring the variability among replicates.

replicates: Repeated but independent determinations of the same sample, by the same analyst, at essentially the same time and same conditions. Care should be exercised in considering replicates of a portion of an analysis and replicates of a complete analysis. For example, duplicate titrations of the same digestion are not valid replicate analyses, although they may be valid replicate titrations. Replicates may be performed to any degree (e.g., duplicates, triplicates).

representative sample: A sample taken to represent a lot or population as accurately and precisely as possible. A representative sample may be either a completely random sample or a stratified sample depending on the objective of the sampling and the conceptual population for a given situation.

reproducibility: The precision, usually expressed as a standard deviation, measuring the variability among results of measurements of the same sample at different laboratories.

reproductive toxicity: A harmful effect on the adult reproductive system.

required clothing insulation (IREQ): An index that assumes a minimal cold tolerance level for skin temperature of 30°C and skin wetness of 0.06 in a stationary standing man. With these assumptions the amount of insulation required to obtain thermal balance can be calculated.

required sweat rate: The international standard (ISO 7933) for the amount of sweat required to maintain thermal balance.

rescue: The activities associated with a trained rescue crew entering a confined space to retrieve an injured or incapacitated entrant.

resorption: In mammalian species that give birth to litters, the intrauterine tissue remaining at the site of death of an embryo or fetus.

respirable dust: Term used to indicate particulate matter that can be inhaled. Generally considered to be 5 µm or less in aerodynamic diameter.

respirable fraction: Particles that penetrate the pulmonary region.

respiratory inlet covering: That portion of a respirator that connects the wearer's respiratory tract to an air-purifying device or respirable gas source, or both. It may be a facepiece, helmet, hood, suit, or mouthpiece/nose clamp.

respiratory system: The breathing system (including mouth, nose, larynx, trachea, lungs, etc.) and associated nerves and blood supply.

responsible party: Someone who can provide additional information on the hazardous chemical and appropriate emergency procedures, if necessary.

retention time: In gas chromatography, the time at which the center, or maximum, of a symmetrical peak occurs on a gas chromatogram.

retinal hazard region: The spectral region where the retina is at greatest risk.

return air: Air extracted from a space and returned to that space, usually after passing through filters and the air-handling unit.

reverberant field: An area where sound levels will remain constant regardless of additional distance from the source. The sound level in the reverberant field depends on the sound power level of the sound source and the amount of sound absorption in the room.

reversible behavior disruption: A sensitive measure of RF exposure. In general, behavioral changes are thermal effects attributed to significant increases in body temperature due to absorbed RF energy.

rework: The portion of the output that must undergo additional processes to correct any unacceptable characteristics introduced during initial production.

rewriting: The process of revising a report to ensure that the final version is free of awkward language or unnecessary words that distract from the content.

Reynolds number: A dimensionless number proportional to pipe or duct diameter, velocity, and density of fluid, and inversely proportional to the viscosity of the fluid. A Reynolds number above 2500 indicates turbulent flow; less than 2500, it indicates streamlined flow.

RfD: See *reference dose.*

RH: See *relative humidity.*

risk: Probability and magnitude of harm. For exposures to chemicals, risk is a function of both exposure and toxicity.

risk agent: The chemical or physical agent that is the source of the risk.

risk analysis: A quantitative analysis used to determine risk.

risk assessment: The process of determining, either quantitatively or qualitatively, the probability and magnitude of an undesired event and estimating the cost to human society or the environment in terms of morbidity, mortality, or economic impact.

risk characterization: The last step in the risk assessment process characterizes the potential for adverse health effects and evaluates the uncertainty involved.

risk communication: The exchange of information about health or environmental risks among risk assessors and managers, the general public, news media, interest groups, etc.

risk estimate: Different expressions of risk may have different implications: 1) individual lifetime risk — the risk of an individual developing the adverse health effect sometime during the remaining lifespan; 2) population or societal risk – the integration of the individual lifetime risk over the exposed population; 3) relative risk – the probability of developing a specific adverse health effect given exposure to a risk agent compared with the same probability given no exposure to the agent; 4) standardized mortality ratio (SMR) – death rate due to a specific cause in an exposed population compared with the death rate in the general population; used often in occupational epidemiology studies; 5) loss of life expectancy — individual lifetime risk multiplied by 36 years equals the average remaining lifetime.

risk factor: Characteristic (e.g., race, sex, age, obesity) or variable (e.g., smoking, occupational exposure level) associated with increased probability of a toxic effect.

risk management: Control of risks to acceptable levels through the application of various remediation techniques.

risk perception: The level of risk an individual, organization, or society associates with a risk agent.

RME: See *reasonable maximum exposure.*

rms: See *root-mean-square.*

robust design: A technique for making the utility of the final product insensitive to variations in the manufacturing process. In EHS terms this might mean that no matter how different the input chemicals are the final product will have no increased exposure potential from one batch to the next.

root cause: The activity that if corrected or eliminated will solve the identified problem.

root-mean-square (rms): The square root of the arithmetic mean of the squares of a set of values.

rotameter: A flowmeter, consisting of a precision-bored, tapered, transparent tube with a solid float inside.

rotation: Angular motion.

roughness factor: The ratio of size of projections from the surface of a pipe or duct, to the diameter of the pipe or duct.

route of entry: Means by which a chemical enters the body (e.g., ingestion, inhalation, dermal absorption, or injection).

routinely collected data: Various types of exposure and health data that are collected routinely in the occupational setting for use in surveillance programs.

RPE: Rating of Perceived Exertion.

RSD: Relative standard deviation. See *coefficient of variation.*

RSI: See *repetitive strain injury.*

ruggedness testing: A special series of tests performed to determine the sensitivity (hopefully, to confirm the insensitivity) of a measurement system to variations of certain factors suspected of affecting the measurement system.

S

SA: See *specific absorption.*

safe: In occupational health, when a situation is safe it is meant that it is free from an unacceptable amount of risk rather than free from all risk. Something is thus safe if its actual risk is judged to be acceptable.

Safety Equipment Institute (SEI): An organization that conducts a voluntary third-party certification program of safety equipment, including detector tubes and pumps.

safety and health training: Safety and health training addresses the safety and health responsibilities of all personnel concerned with the work site. Training is often most effective when incorporated into other training about performance requirements and job practices. Training is conducted to 1) ensure that all employees understand the hazards to which they might be exposed and how to prevent harm to themselves and others from exposure to these hazards; 2) permit supervisors to carry out their jobs effectively; and 3) ensure that managers understand their safety and health responsibilities.

safety inspections: An inspection conducted by those who are classified as safety specialists.

saliva: Watery fluid in the mouth.

sample: A subset or group of objects or things selected from a larger set, called the "lot" or "population." The objects or things may be physical such as specimens for testing or they may be data values representing physical samples. Unless otherwise specified, all samples are assumed to be randomly selected. Usually, information obtained from the samples is used to provide some indication or inference about the larger set. Samples rather than the population are examined usually for reasons of economy — the entire population under consideration is usually too large or too inaccessible to evaluate. In cases in which destructive testing is performed, sampling is a must — otherwise the entire population would be consumed. In many situations, the population is conceptually infinite and therefore impossible to check or measure.

sample breakthrough: Most commercially available sampling tubes consist of two sections of sorbent separated by glass wool or polyurethane. In charcoal tubes the second or backup section is usually one-third of the total weight of the charcoal. These two sections are desorbed and analyzed separately in the laboratory. As a guideline, if 25% or less of the amount of contaminant collected on the front section is found on the backup section, significant loss of the compound (breakthrough) has probably not occurred. If greater than 25% is detected, breakthrough is evident and results should be reported as "breakthrough, possible sample loss."

sampler capacity: A predetermined conservative estimate of the total mass of contaminant that can be collected on the sampling medium without loss or overloading; typically two-thirds of the mass of contaminant on the sorbent at the breakthrough volume.

sample volume: The amount of air pulled through the collection device by a pump during sampling.

sampling media: Devices used to collect airborne chemicals for subsequent analysis, such as sorbent tubes, bags, and filters.

sampling strategy: A sampling plan that includes, but is not limited to, an assessment of the type and number of samples to be collected, the methods to be used, the accuracy of the methods, and the overall objectives for sampling.

SAR: See *specific absorption rate.*

sarcoma: A malignant tumor that develops from connective tissue cells.

SC: See *stratum corneum.*

SCBA: See *self-contained breathing apparatus.*

SCE(s): See *sister-chromatid exchange(s).*

scenario building: A method that forces both planners and managers to think outside the traditional comfort zone. It is a group process involving many organizational levels.

scfm: Air volume flow rate at standard conditions.

scope of practice: The legal specifications for credentials, responsibilities, and limitations of health care practitioners generally developed by state authorities.

SE: See *shielding effectiveness.*

search engine: A program that searches documents for specified keywords and returns a list of the documents where the keywords were found. Although search engine is really a general class of programs, the term is often used to specifically describe systems that enable users to search for documents on the World Wide Web. Typically, a search engine works by sending out a spider to fetch as many documents as possible. Another program, called an indexer, then reads these documents and creates an index based on the words contained in each document. Each search engine uses a proprietary algorithm to create its indices such that, ideally, only meaningful results are returned for each query.

sebum: The waxy excretion on the skin surface.

secondary barriers: Protection of the external environment, including nonlaboratory work areas and the outside community. Secondary containment is achieved by a combination of facility design (differential pressurization of work areas, HEPA-filtered exhaust ventilation, sterilization of effluent liquids, etc.) and work practices and procedures.

secondary prevention: Early intervention in a disease process to reverse or limit its progression.

secondary standard: Devices such as flowmeters that trace their calibration to primary standards and maintain their accuracy with reasonable care and handling in operation.

sedimentation: The movement of an aerosol particle through a gaseous medium under the influence of gravity.

SEG: See *similar exposure group.*

self-contained breathing apparatus (SCBA): An atmosphere-supplying respirator in which the respirable gas source is designed to be carried by the wearer.

self-efficacy: The self-confidence that a person can successfully accomplish what he or she sets out to do. (Also known as "self-effectiveness.")

self-esteem: Respect for or a favorable impression of oneself.

semen: The viscous creamy fluid obtained from the penis on ejaculation.

semi-insert: A type of hearing protector that is a cross between earplugs and earmuffs; earplug-like devices attached to the ends of a headband that are pressed into the ear canal.

sensation: A mental condition or physical feeling resulting from stimulation of a sense organ or internal bodily changes.

sensitive volume: The part of a radiation detector that actually detects radiation.

sensitivity: The proportion of individuals with the disease that a test measure identifies correctly.

sensitivity analysis: In uncertainty analysis, comparison of risk estimates based on the means and upper bounds of the probability distributions of the input variables.

sensitization: The phase (usually 10 days) in which a person becomes allergic to an antigen.

sensitizer: A foreign agent or substance that is capable of causing an immune response in an individual. In most cases, initial exposure results in a normal response, but repeated exposures lead to progressively strong and abnormal responses.

sensorineural hearing loss: The type of hearing loss that affects the inner ear.

serum: The liquid that does not contain the cellular components of blood on coagulation.

server: A computer or device on a network that manages network resources. For example, a file server is a computer and storage device dedicated to storing files. Any user on the network can store files on the server. A print server is a computer that manages one or more printers, and a network server is a computer that manages network traffic. A database server is a computer system that processes database queries.

shaping: Using the interpreted data, leaders shape a vision for the future that reflects the organization's values and goals. (See also *interpreting.*)

shared vision: A level beyond simply signing onto the personal vision of a charismatic leader. It is the ability to translate such a vision throughout the organization in a manner that fosters a genuine commitment.

shelf life: The period in which detector tubes remain stable when stored in accordance with a manufacturer's specifications.

shielding effectiveness (SE): Shielding effectiveness is used to describe the capability of a material as a shield. SE is a function of losses resulting from absorption, reflection, and internal reflection.

shock: 1. Clinical manifestations of circulatory insufficiency, including hypertension, weak pulse, tachycardia, pallor, and diminished urinary output. 2. A pulse or transient motion or force lasting thousandths to tenths of a second that is capable of exciting mechanical resonances.

short-term exposure limit (STEL): 1. Maximum concentration for continuous 15-minute period. Allowed four times a day, with at least 60 minutes between exposures. 2. Used in reference to the OSHA PEL–STEL and ACGIH's TLV–STEL. The STEL represents a time-weighted average (TWA) exposure that should not be exceeded for any 15-minute period. 3. STELs are recommended when exposures of even short duration to high concentrations of a chemical are known to produce acute toxicity. It is the concentration to which workers can be exposed continuously for a short period of time without suffering from 1) irritation; 2) chronic or irreversible tissue damage; or 3) narcosis of sufficient degree to increase the likelihood of accidental injury, impaired self-rescue, or reduced work efficiency. A STEL is defined as a 15-minute TWA exposure that should not be exceeded at any time during a workday, even if the overall 8-hour TWA is within limits, and it should not occur more than four times per day. There should be at least 60 minutes between successive exposures in this range. If warranted, an averaging period other than 15 minutes can also be used.

short-term public emergency guidance level (SPEGL): Established by the Committee on Toxicology of the National Research Council (NRC) for the U.S. Department of Defense. The SPEGL is a suitable concentration for unpredicted, single, short-term, emergency exposure of the general public. In contrast to the EEGL, the SPEGL takes into account the wide range of susceptibility of the general public. (See also *EEGL, CEGL.*)

SI metric units: 1. Le Systeme International d'Unites; the international metric system. 2. An internationally accepted system of metric units based on mass in kilograms, length in meters, and time in seconds.

SIC: See *Standard Industrial Classification Code.*

sick building syndrome: The term "sick building syndrome" has come into common use for those problems where excessive comfort and health-related symptoms are present that are clearly related to building occupancy, but that are not associated with objective clinical signs.

significance: Importance; consequence. Part of the evaluation process. (See also *evaluation.*)

significant figures: Numerical figures containing as many digits (other than location zeroes) as are contained in the least exact factor used in their determination. [**Note:** Do not confuse significant figures with the 10 to 16 decimal places that appear on the display of scientific and engineering calculators.]

SIL: See *speed interference level.*

silica gel: A regenerative absorbent consisting of the amorphous silica. Used in dehydrating and in drying and as a catalyst carrier.

similar exposure group (SEG): A group of workers having the same general exposure profile for the agent(s) being studied because of the similarity and frequency of the tasks they perform, the materials and processes with which they work, and the similarity of the way they perform the tasks.

simple asphyxiation: Coating or blockage of passageways in the lungs so that oxygen cannot reach the alveoli or be absorbed into the bloodstream.

sister-chromatid exchange(s) SCE(s): An exchange of genetic material between the chromatid pairs of a chromosome during cell division.

skeletal variant: A minor alteration of the skeletal structure that will not adversely affect function or length of life of the organism and that occurs normally in a small proportion of animals.

skin: When used with an exposure guideline such as an OSHA PEL or ACGIH TLV®, it indicates that the substance may be absorbed in toxic amounts through the skin or mucous membranes.

SKIN designation: A designation that emphasizes the role of the skin in contributing to overall exposures to chemicals and does not denote toxic effects on the skin.

skin notation: Denotes the possibility that dermal absorption may be a significant contribution to the overall body burden of the chemical (that is, the airborne OEL might not be adequate to protect the worker because the compound also readily penetrates the skin. Other toxicity endpoints on skin such as irritation, dermatitis, and sensitization are not sufficient to warrant the skin notation. In practice, the skin notation is given to compounds with a dermal LD_{50} less than 1000 mg/kg, or if there are other data indicating that repeated dermal exposure results in systemic toxicity.

SLM: See *sound level meter*.

slope factor: Plausible upper-bound estimate of the probability of a response per unit intake of a chemical over a lifetime. Used to estimate an upper-bound probability of an individual developing cancer as a result of a lifetime of exposure to a particular level of a potential carcinogen.

slot: See *slot hood*.

slot hood: A hood that has an opening (i.e., a slot) that has a width to length ratio of 0.2 or less; used to provide uniform air distribution and adequate capture velocity over a finite length of contaminant generation.

slot velocity: Air velocity through the openings into a slot-plenum hood. Primarily used to obtain a uniform air velocity across the edge of a tank. The correct selection of the slot velocity will produce a uniform capture velocity.

slurry: Pourable mixture of solid and liquid.

smog: Smoke and fog: extensive atmospheric contamination by aerosols arising from a combination of natural and anthropogenic sources.

smoke: 1. Carbon or soot particles less than 1.0 µm in size. These small, gas-phase particles created by incomplete combustion, consist predominantly of carbon and other combustible materials. Smoke generally contains droplets and dry particles. Size ranges are usually between 0.01 µm and 1.0 µm. 2. A mixture of dry and liquid particles generated by incomplete combustion of an organic material, combined with and suspended in the gases from combustion.

soap bubble burette: A tube with a defined volume in which a soap film (bubbles) are injected to measure flow rate. It is a flow calibration device for sampling pumps. Modern technology has computerized the measurement of a flow rate with electronic timing sensors in the bubble burette/meter. (Also called a "bubble meter.")

social psychology: The study of the manner in which the attitudes, personality, and motivations of the individual influence, and are influenced by, the structure, dynamics, and behavior of the social group with which the individual interacts.

software: Computer instructions or data. Anything that can be stored electronically is software. The storage devices and display devices are hardware. Software is often divided into two categories: 1) systems software — includes the operating system and all the utilities that enable the computer to function; 2) applications software — includes programs that do real work for users. For example, word processors, spreadsheets, and database management systems fall under the category of applications software.

solubility in water: Susceptible of being dissolved; may be expressed as the percentage of a material (by weight) that will dissolve in a solvent. The following terms are used to express solubility: negligible = <0.1%; slight = 0.1%–1.0%; moderate = 1.0%–10.0%; appreciable = 10.0%–99.0%; complete = soluble in all proportions.

solution: Mixture in which the components lose their identities and are uniformly dispersed. All solutions are composed of a solvent (water or other fluid) and the substance dissolved called the "solute." Air is a solution of oxygen and nitrogen. A true solution is homogeneous as salt in water.

solvent: A substance that dissolves other substances, most commonly water but often an organic compound.

solvent extraction: The process of extracting adsorbed chemicals from sorbent material through the use of solvents.

somatic mutation: A mutation that is characterized by manifestation of the damage in the exposed individual due to alteration of somatic or body cells.

somatic nervous system: A subdivision of the central nervous system (CNS) that transmits signals to the skeletal muscles for movement.

SOP: See *standard operating procedure*.

sorbent tube: A small glass tube normally filled with two layers of a solid sorbent material that will adsorb specific chemicals for subsequent elution and laboratory analysis.

sound absorption coefficient: The ratio of sound energy absorbed to that arriving at a surface or medium. (Also known as "acoustic absorption coefficient" or "acoustic absorptivity.")

sound analyzer: A device for measuring the band pressure or pressure-spectrum level of a sound as a function of frequency.

sound intensity analyzer: Analyzers used to identify specific noise sources and to determine compliance with purchase specifications limiting the sound power level.

sound intensity level (L_i): A vector quantity having both magnitude and direction. Although the L_i does not correspond to the loudness of a sound, it is useful as an analytical measurement and is necessary to determine accurately the sound power level.

sound level meter (SLM): An instrument used to measure noise and sound levels in a specified manner;

the meter may be calibrated in decibels or volume units and includes a microphone, an amplifier, an output meter, and frequency-weighting networks.

sound power level: The sound pressure level or sound intensity level data can be used to calculate the sound power level of a sound source. Sound power is analogous to the electrical power rating of a light bulb and is also measured in watts.

sound pressure level (L_p): The level, in decibels, of a sound is 20 times the logarithm to the base 10 of the ratio of the measured pressure of this sound to the reference pressure. The reference pressure is 0.0002 dynes/cm^2.

sound shadow: Noise that slips through sound-blocking barriers.

source modification: Changing a hazard source to make it less hazardous (e.g., wetting dust particles or lowering the temperature of liquids to reduce off-gassing and vaporization).

SP: See *static pressure.*

span gas: The concentration of a calibration gas mixture needed to span a desired concentration range of a chemical specific detector.

span vapor: The concentration of a calibration vapor mixture needed to span a desired concentration range of a chemical specific detector.

spatial averaging: A method of estimating average exposure of the whole body by collecting densitometric data across the vertical dimension of a simple linear model of the body.

specific absorption (SA): A quantity that represents RF exposure within tissues. The SA is the time integral of the sound absorption rate (SAR) and, as such, is the RF dose. It is the energy absorbed per unit mass of tissue, with units of joules per kilogram (J/kg). (See also *specific absorption rate.*)

specific absorption rate (SAR): A quantity that represents RF exposure within tissues. The SAR is the rate at which energy is absorbed per unit mass, or dose rate. It is the fundamental quantity of the exposure criteria, and the dosimetric quantity of choice in studies of biological effects. It is generally expressed in units of watts per kilogram (W/kg), representing the power deposited in a unit mass. The unit, W/kg, is also used for metabolic rate, where the resting metabolic rate of an adult human being is about 1 W/kg. The SAR depends on the electric-field strength in tissues (E_i), the electrical conductivity of tissues, and the density of tissues.

specific activity: The activity per unit mass of a radionuclide.

specific gravity: 1. The ratio of the mass of a unit volume of a substance to the mass of the same volume of a standard substance at a standard temperature. Water at 39.2°F (4°C) is the standard usually referred to for liquids; for gases, dry air (at the same temperature and pressure as the gas) is often taken as the standard substance. 2. The weight of a material compared with the weight of an equal volume of water is an expression of the density (or heaviness) of a material. Insoluble materials with specific gravity of less than 1.0 will float in (or on) water. In soluble materials with a specific gravity greater than 1.0 will sink in water. Most, but not all, flammable liquids have specific gravity less than 1.0 and, if not soluble, will float on water — an important consideration for fire suppression.

specific reliability: The probability that an item will perform a required function under stated conditions for a stated period of time.

specification standards: Description of the specific means of hazard abatement. (See also *performance standards.*)

specificity: The proportion of people without the disease that a test measure identifies correctly.

spectrophotometer: An instrument used for comparing the relative intensities of the corresponding colors produced by chemical reactions.

spectrophotometry: A procedure to measure photometrically the wavelength range of radiant energy absorbed by a sample under analysis; can be visible light, ultraviolet light, or X-rays.

spectrum (noise): The distribution in frequency of the magnitudes (and sometimes phases) of the components of the wave. Spectrum also is used to signify a continuous range of frequencies, usually wide in extent, within which waves have some specified common characteristics.

specular reflection: Reflections may be specular or diffuse, or a combination of the two. Specular reflection, which may be regarded as a type of direct viewing, occurs when the beam is incident on a mirror-like surface. This depends on the wavelength and the dimension of the surface irregularities of the reflector (i.e., it is a wavelength-dependent phenomenon). Specular reflections occur when the size of the surface irregularities of the reflecting surface are smaller than the incident wavelength.

speed interference level (SIL or PSIL): The speech interference level of a noise is the arithmetic average, in decibels, of the sound pressure levels of the noise in the three octave bands of frequency 600–1200, 1200–2400, 2400–4800 Hz for SIL and the preferred center frequencies of 500, 1000, and 2000 Hz for PSIL.

SPEGL: See *short-term public emergency guidance level.*

spiked sample: A normal sample of material (gas, solid, or liquid) to which is added a known amount of some substance of interest. The extent of the spiking is unknown to those analyzing the sample. Spiked samples are used to check on the performance of a routine analysis or the recovery efficiency of a method.

spontaneously combustible: A material that ignites as a result of retained heat from processing, or that will oxidize to generate heat and ignite, or that absorbs moisture to generate heat and ignite.

spot cooling: Cooling the air of a limited portion of an enclosed space.

sprain: Injury in which some fibers of a ligament are overstretched or torn but not completely severed. When the ligament is torn in two, the injury is called a severe sprain.

sputum: Watery fluid with solids excreted from the throat and upper lungs on expectoration.

squamous: Scaly or plate-like.

squamous cell carcinoma: A carcinoma of epithelial origin.

squeeze bulb pump: A sampling pump that draws a fixed volume of air using a squeeze bulb.

SRM: See *standard reference material.*

SRS: See *standard reference sample.*

stack: 1. A structure that contains a flue, or flues, for the discharge of gases. 2. The vertical train of a system of soil, waste, or vent piping extending through one or more stories. 3. A device used to discharge air into the ambient environment and away from the building wake.

stack sampling: A collection of aerosol samples in exhaust air ducts such as stacks or chimneys. (See also *isokinetic sampling, subisokinetic sampling, superisokinetic sampling*.

stand-alone graphics: Visual displays in a report that are self-explanatory.

standard: Any rule, principle, or measure established by authority.

standard air: In ventilation, dry air at 70°F, 29.92 in. Hg (21°C, 760 mmHg) [ACGIH]; air at 68°F, 29.92 in. Hg, 50% relative humidity (20°C, 760 mmHg) [ASHRAE]. [**Note:** Chapter 31 in this book uses the ACGIH STP in all calculations.]

standard air decompression: A schedule applicable to either scuba or surface air-supplied divers who completely decompress either in the water or in a diving bell or chamber before reaching surface pressures. The maximum recommended air dive is 190 ft for 40 minutes with emergency dives to 300 ft for 180 minutes.

standard ambient temperature: Datum condition for the rating of equipment (usually 20°C).

standard atmosphere: Standard atmospheric pressure is the barometric pressure at standard conditions, 25°C and sea level.

standard conditions: 1. A set of physical, chemical, or other variables of a substance or system that defines an accepted reference state or forms a basis for comparison. 2. Operating conditions of a refrigeration system corresponding to some standard (e.g., evaporating temperature, condensing temperature, subcooling temperature, superheat).

standard deviation: 1. The positive square root of the variance of a distribution; the parameter measuring the spread of values about the mean. 2. The positive square root of the expected value of the square of the difference between a random variable and its mean.

standard error: The standard deviation of the distribution of a sample statistic.

Standard Industrial Classification (SIC) Code: A classification system used to identify places of employment by major types of activity.

standard operating procedure (SOP): A written document that details an operation, analysis, or action whose mechanisms are thoroughly prescribed and that is commonly accepted as the method for performing certain routine or repetitive tasks.

standard reference material (SRM): A material produced in quantity, of which certain properties have been certified by the National Institute of Standards and Technology (NIST) or other agencies to the extent possible to satisfy its intended use. The material should be in a matrix similar to actual samples to be measured by a measurement system or be used directly in preparing such a matrix. Intended uses include 1) standardization of solutions; 2) calibration of equipment; and 3) auditing the accuracy and precision of measurement systems.

standard reference sample (SRS): A carefully prepared material produced from or compared against an SRM (or other equally well-characterized material) such that there is little loss of accuracy. The sample should have a matrix similar to actual samples used in the measurement system. These samples are intended for use primarily as reference standards 1) to determine the precision and accuracy of measurement systems; 2) to evaluate calibration standards; and 3) to evaluate quality control reference samples. They may be used "as-is" or as a component of a calibration or quality control measurement system. Examples: An NIST-certified sulfur dioxide permeation device is an SRM. When used in conjunction with an air dilution device, the resulting gas becomes an SRS. An NIST-certified nitric oxide gas is an SRM. When diluted with air, the resulting gas is an SRS.

standard temperature and pressure (STP): 1. Measured volumes of gases are generally recalculated to 0°C and 760 mm pressure. 2. Standard temperature and pressure; 298 K (25°C) and 760 mmHg (1 atm).

standardization: A physical or mathematical adjustment or correction of a measurement system to make the measurements conform to predetermined values. The adjustments or corrections are usually based on a single-point calibration level.

standards in naturally occurring matrix: Standards relating to the pollutant measurement portions of air pollution measurement systems may be categorized according to matrix, purity, or use. Standards in a naturally occurring matrix include Standard Reference Materials and Standard Reference Samples.

standing wave: A wave in which the ratio of an instantaneous value at one point to that at any other point does not vary with time. (Also known as "stationary wave.")

static: Without motion or change.

static calibration: The artificial generation of the response curve of an instrument or method by use of appropriate mechanical, optical, electrical, or chemical means. Often a static calibration checks only a portion of a measurement system. For example, a solution containing a known amount of sulfite compound would simulate an absorbing solution through which has been bubbled a gas containing a known amount of sulfur dioxide. Use of the solution would check out the analytical portion of the pararosaniline method but would not check out the sampling and flow control parts of the bubbler system.

static load: Muscle (local) fatigue is divided into static and dynamic. When muscles are loaded but do not move, it is a static load; when loaded muscles move, it is dynamic.

static pressure (SP): 1. The potential pressure exerted in all directions by a fluid at rest. For a fluid in motion, it is measured in a direction normal (at right angles) to the direction of flow; thus, it shows the tendency to burst or collapse the pipe. When added to velocity pressure, it gives total pressure. 2. The pressure developed in a duct by a fan; SP exerts influence in all directions; the force in inches of water measured perpendicular to flow at the wall of the duct; the difference in pressure between atmospheric pressure and the absolute pressure inside a duct, cleaner, or other equipment. 3. The fan-generated differential pressure between the inside of the local exhaust ventilation (LEV) system and the outside ambient pressure; represents the energy that is continually delivered to the system by the fan.

static pressure (SP) loss: The amount of kinetic energy to overcome an obstruction or fitting in a ventilation system.

statistical control chart: See *control chart.*

statistical control chart limits: The limits on control charts that have been derived by statistical analysis and are used as criteria for action or for judging whether a set of data does or does not indicate lack of control.

steatosis: An acute reaction within the liver where lipid droplets may accumulate and produce what is known as fatty liver.

STEL: See *short-term exposure limit.*

sterilization: The removal or destruction of all microorganisms, including pathogenic and other bacteria, vegetative forms, and spores.

stochastic effect: 1. Biological effect from radiation exposure on a population based on random statistics in which there is assumed no threshold with an increase in the probability of occurrence with an increase in dose. 2. Effect for which the probability of occurrence depends on the absorbed dose. Heredity effects and cancer induced by radiation are considered to be stochastic effects. This means that, even for an individual, there is no threshold of dose below which the effect will not appear, but the chance of experiencing the effect increases with increasing dose. (See also *nonstochastic effect.*)

Stokes diameter: In measuring aerosol sedimentation behavior, the shape, size, and density of particles is often unknown. It is convenient to discuss particle size in terms of the diameter of a spherical particle of the same density that would exhibit the same behavior as the particle in question, or the Stokes diameter (d_{Stk}).

storage: Temporary holding of waste pending treatment or disposal, as in containers, tanks, waste piles, and surface impoundments.

STP: See *standard temperature and pressure.*

strain: 1. Change in length of an object in some direction per unit undistorted length in some direction, not necessarily the same; the nine possible strains from a second-rank tensor. 2. Injury in which some of the fibers of a muscle are overstretched or torn but not completely severed. When the muscle is torn in two, the injury is called a severe strain. Also, a general term used to describe a minor tear in a ligament or tendon.

strategic plan: A plan that generally describes where the organization would like to be in three to five years.

stratified sample: A sample consisting of various portions that have been obtained from identified subparts or subcategories (strata) of the total lot, or population. Within each category or strata, the samples are taken randomly. The objective of taking stratified samples is to obtain a more representative sample than that which might otherwise be obtained by a completely random sampling. The idea of identifying the subcategories or strata is based on knowledge or suspicion of (or protection against) differences existing among the strata for the characteristics of concern. The identification of the strata is based on knowledge of the structure of the population, which is known or suspected to have different relationships with the characteristic of the population under study. Opinion polls or surveys use stratified sampling to ensure proportional representation of the various strata (e.g., geographic location, age group sex, etc.). Stratified sampling is used in air monitoring to ensure representation of different geographical areas, different days of the week, and so forth. (Also known as "stratified random sample.")

stratum corneum (SC): The outer layer of flattened keratinized cells of the epidermis.

strength-of-evidence: A method of evaluating the evidence of carcinogenicity of a chemical in which both positive and negative evidence are considered. (See also *weight-of-evidence.*)

stresses: Environmental stimuli that represent potential adverse impacts on the health of workers. Stresses may include chemical and physical agents as well as ergonomic and psychological stress.

stressors: Stimuli that disrupts the homeostasis of an organism.

subacute toxicity study: Toxicity study durations of five days to 14 days.

subchronic intake: Intake averaged over a period of two weeks to seven years.

subchronic reference dose: Applicable for time periods of two weeks to seven years.

subchronic toxicity study: Toxicity study durations of 15 days to six months.

subcutaneous: Beneath the skin.

subharmonics: A sinusoidal quantity having a frequency that is an integral submultiple of the frequency of some other sinusoidal quantity to which it is referred; a third subharmonic would be one-third the fundamental or reference frequency.

subisokinetic sampling: A sampling condition in which the air flowing into an inlet has a lower velocity than the ambient airflow. Subsequently, the sample collected tends to be biased with larger particles. (See also *isokinetic sampling, stack sampling, superisokinetic sampling.*)

substitution: Substituting a less hazardous material, equipment, or process for a more hazardous one (e.g., use of soap and water in place of solvents, use of automated equipment in place of manually operated equipment).

substrate: A collection element that is placed onto the impaction stage for collection of aerosol samples. It can be glass plate, aluminum foil, plastic film, or membrane filter. (See also *impaction plate, impactor stage, substrate coating.*)

substrate coating: Materials that are applied onto certain substrates to capture or stabilize particles that are impacted during sampling. (See also *isokinetic sampling, stack sampling, superisokinetic sampling.*)

SUMMA® canister: A pre-evacuated stainless steel canister that acts as an air collection vessel; the interior surface is electrochemically passivated using the SUMMA process to prevent reaction of the sample with the canister.

summative (product) evaluation: Gathering information on adequacy and using this information to make decisions about utilization.

superisokinetic sampling: A sampling condition in which the air flowing into an inlet has a higher velocity than the ambient airflow. Subsequently, the sample collected tends to be biased with smaller particles. (See also *isokinetic sampling, stack sampling, subisokinetic sampling.*)

supination: 1. Turning the palm upward. 2. Inversion of the foot.

supplier: The provider of that output for a customer.

supply air: Air supplied through ducts to an occupied space.

surface supplied helium-oxygen decompression: A schedule that can involve stops breathing either the bottom supplied mixture, a 40% oxygen mixture, or pure oxygen. The maximum dive on this schedule is 380 ft for 120 minutes.

survey: An episodic investigation of a particular situation and, usually, for a specific purpose.

sweat: The watery fluid excreted on the skin surface during high physical activity or in hot, humid environments.

symptom: Any condition accompanying or resulting from an exposure, a disease, or a disorder.

synergism: Cooperative action of substances whose total effects is greater than the sum of their separate effects.

synergists: Skeletal muscles that generally stabilize and control the bony structures that constitute the joint.

system: 1. An organized collection of parts united by regular interaction. 2. A heating or refrigerating scheme or machine, usually confined to those parts in contact with a heating or refrigerating medium. 3. An arbitrarily chosen group of materials and devices set apart for analytical study.

system audit: A systematic on-site qualitative review of facilities, equipment, training, procedures, record keeping, data validation, data management, and reporting aspects of a total (QA) system 1) to arrive at a measure of capability of the measurement system to generate data of the required quality; and/or 2) to determine the extent of compliance of an operational QA system to the approved QA Project Plan.

systematic error: The condition of a consistent deviation of the results of a measurement process from the reference or known level. The cause for the deviation, or bias, may be known or unknown, but is considered "assignable." By assignable it is meant that if the cause is unknown, it should be possible to determine the cause. (See also *bias.*)

systemic: Spread throughout the body, affecting all body systems and organs, not localized in one spot or area.

systemic effects: Adverse effects other than at the site of contact.

systemic system: A subsystem of the cardiovascular system (CVS) in which blood from the left ventricle is pumped to the body arteries. Oxygen and nutrients are removed and carbon dioxide and metabolic waste products are added before the blood is returned in the veins to the right side of the heart. Nutrients are added from the intestines. Fat-soluble wastes are biotransformed in the liver into water-soluble wastes and put back into the blood. The kidney then eliminates water-soluble wastes through urine. Some wastes are eliminated in the intestines (feces).

systems thinking: The ability to see all processes as a system of interrelated events. It allows for the identification and correction of unacceptable patterns.

T

Taguchi experiments: A method for evaluating several different elements of a process at the same time, as opposed to classic experimental design, which focuses on time- and resource-consuming technical analysis of one factor at a time.

tapered-element oscillating microbalance (TEOM): An instrument that determines aerosol mass using resonance oscillation.

tare: A deduction of weight, made in allowance for the weight of a container or medium; for example, the initial weight of a filter.

target: Either the nucleus of the nuclide to be bombarded by a subatomic particle or photon or an anode of an X-ray producing machine.

target concentration: A preliminary estimate of the airborne concentration of the contaminant of interest relative to the purpose of testing.

target organs: 1. The organ of the body most affected by exposure to a particular substance. 2. The body organs that are affected by exposure to a hazardous chemical, physical, or biological agent.

target velocity: The target velocity is the average of the range of transport velocity and is used at the beginning of a transport design problem. Once a duct diameter is selected, the precise actual velocity is then calculated.

TC_{Lo}: See *lowest toxic concentration.*

T_{db}: See *dry bulb temperature.*

TD_{Lo}: See *lowest toxic dose.*

T_{dp}: See *dew point temperature.*

team learning: The situation in which a team develops the ability to think as a group.

teams: Groups associated in a joint action.

technical audits: The examination of present status of health hazards.

temperature: That property of a body that determines the flow of heat. Heat will flow from a warm body to a cold body.

temporal effect: An effect over time.

temporary threshold shift (TTS): The hearing loss suffered as the result of noise exposure, all of which is recovered during an arbitrary period of time. (The loss that is not recovered is called permanent threshold shift [PTS].)

tendinitis: Inflammation of a tendon and some of the adjacent muscle tissue.

tendon: Fibrous tissue, similar to a ligament, that attaches muscle to bone.

tenosynovitis: A type of repetitive strain injury (RSI) characterized by swelling of the outside sheath of a tendon. (Also known as "synovitis.")

TEOM: See *tapered-element oscillating microbalance.*

teratogen: A substance that causes birth defects in the offspring.

teratogenesis: The process by which toxic compounds called teratogens can induce perinatal aberrations in an exposed embryo or fetus, yet because of relatively low concentrations may pose no significant hazard of toxicity to the mother.

T_g: See *globe temperature.*

Thackrah, Charles T.: English physician who published *The Effects of the Principal Arts, Trades, and Professions and of Civic States and Habits of Living on Health and Longevity* in 1831. His views on disease and prevention helped stimulate factory and health legislation in England. The British Factories Acts of 1864, 1878, and 1901 were the forerunners of modern occupational safety and health legislation.

thenar eminence: The ball of flesh at the base of the thumb.

thermal balance: The heat exchange between the human body and the environment.

thermal conductivity: The heat flow across a surface per unit area per unit time, divided by the negative of the rate of change of temperature with distance in a direction perpendicular to the surface.

thermal conductivity detector: The thermal conductivity detector is the most universal gas chromatographic detector because it can measure most gases and vapors. It has low sensitivity compared with the other detectors, however, and is used primarily for analysis of low molecular weight gases such as carbon monoxide, carbon dioxide, nitrogen, and oxygen. This detector measures the differences in thermal conductivity between the column effluent and a reference gas, made of uncontaminated carrier gas. The most common carrier gas for this detector is helium because it is inert and has a very low molecular weight. The column effluent and the reference gas pass through separate detector chambers that contain identical electrically heated filaments. Energy is transferred from the filament when analyte molecules strike the heated filament and rebound with increased energy. Heat loss will therefore be directly proportional to the number of collisions per unit time. Differences in thermal conductivity between gases is proportional to the rate of diffusion to and from the filament. Since diffusion is inversely proportional to molecular weight, lighter molecules will have higher thermal conductivities. Compounds with molecular weights greater than the reference gas will conduct more heat away from the filament than the pure, low molecular weight reference gas, thereby reducing the electrical resistance to the filament. The difference in resistance between the two filaments is amplified and recorded.

thermal desorption: The process of extracting adsorbed chemicals from sorbent material through the use of heat.

thermal drift: Drift caused by an internal heating of equipment during normal operation or by changes in external ambient temperature.

thermal effects on safety behavior: The relationship between unsafe work behavior and ambient temperatures, described as a U-shaped curve, with the minimum unsafe behavior rate occurring in the preferred temperature zone of 17°C (63°F) to 23°C (73°F) wet bulb globe temperature (WBGT), and with the unsafe behavior rate increasing when ambient temperatures increase or decrease from this range.

thermal stress: A combination of air temperature, radiant heat exchange, air movement, and the partial pressure of water vapor that makes the environment stressful.

thermodynamic properties: Basic qualities used to define the condition of a substance (e.g., temperature, pressure, volume, enthalpy, entropy).

thermodynamics: The science of the relation of heat to other forms of energy.

thermoluminescent dosimetry (TLD): The method of quantifying radiation dose by use of material that stores the radiation energy in electron traps. The material is later subjected to heat that releases the energy in the form of optical photons.

thermometer: An instrument that measures temperature.

THF: Abbreviation for Tetrahydrofuran.

thoracic fraction: Particles that enter the tracheobronchial region generally smaller than 10 µm.

thoracic outlet syndrome: A type of repetitive strain injury (RSI) that occurs when both the arteries and nerves going from the neck to the arms are compressed. Signs include coldness and weakness in the forearm, hand, and fingers, and numbness and pain in the entire arm.

threshold: Characterizes a dose-response curve that does not pass through the origin of the graph, indicating there is an exposure level that will not result in an increase in the probability of occurrence of an adverse health effect. Used by the U.S. EPA for noncarcinogens and developmental toxins. (See also *nonthreshold.*)

threshold limit value (TLV®): Used by the American Conference of Governmental Industrial Hygienists (ACGIH) to designate degree of exposure to contaminants and expressed as parts of vapor or gas per million parts of air by volume at 25°C and 760 mmHg pressure, or as approximate milligrams of particulate per cubic meter of air (mg/m^3). (See also *permissible concentration.*) An exposure level under which most people can work consistently for 8 hours a day, day after day, with no harmful effects. TLVs® are listed as either an 8-hour TWA or a 15-minute STEL. ,

threshold limit value–time-weighted average (TLV–TWA): The time-weighted average concentration for a normal 8-hour workday and a 40-hour workweek to which nearly all workers may be exposed repeatedly, day after day, without adverse effects.

tidal movement: Volume of air inspired of expired during each respiratory cycle.

tiered risk assessment: See *iterative risk assessment.*

tight-fitting facepiece: A respiratory inlet covering that is designed to form a complete seal with the face. A half facepiece (includes quarter-masks, disposable masks, and masks with elastomeric facepieces) covers the nose and mouth; a full facepiece covers the nose, mouth, and eyes.

time-response relationship: When a particular toxic effect is closely monitored as a function of time, it is possible to develop a time-response relationship. This curve, like the dose-response curve, contains valuable information that can reveal the onset of response, the time of maximum response, the duration of the response (i.e., sustained or transient), and possible recovery from the exposure.

time-weighted average (TWA): 1. Average exposure for an individual over a given working period, as determined by sampling at given times during the period. 2. The most frequently used exposure guideline term; the average concentration over a workday (8 hours for OSHA PELs and ACGIH TLVs®, up to 10 hours in a 40-hour workweek for NIOSH RELs). 3. This is the fundamental concept of most occupational exposure limits (OELs). It is usually presented as the average concentration over an 8-hour workday for a 40-hour workweek; however, this implies that concentrations will be both above and below the average value. The ACGIH TLV committee has recommended excursion limits to prevent concentrations from severely exceeding the average value. The proposed excursion limits are that exposures should typically not exceed the TWA by more than threefold and for a period not exceeding 30 minutes during the workday. Even if the TWA is not exceeded for the work shift, in no case should the excursion be more than fivefold the TWA value.

time-weighted average (TWA) concentration: Refers to concentrations of contaminants that have been weighted for the time duration of sample. A sufficient number of samples are needed to permit a time-weighted average concentration throughout a complete cycle of operations or throughout the work shift.

time-weighted exposure: Average over a given working period of a person's exposure, as determined by sampling at given times during the period.

tinnitus: A ringing, roaring, or hissing sound in one or both ears.

tissue equivalent: A material with an effective Z equal to that of tissue.

TLD: See *thermoluminescent dosimetry.*

TLD chip: The dosimeter made from thermoluminescent material, either CaF^2 or LiF.

TLV®: See *threshold limit value.*

TLV–C: A ceiling exposure limit that should not be exceeded for even an instant.

TLV–STEL: A short-term exposure limit; a 15-minute time-weighted average exposure that should not be exceeded.

TLV–TWA: An 8-hour time-weighted average (TWA) exposure limit.

TNT: Abbreviation for Trinitrotoluene.

T_{nwb}: See *natural wet bulb temperature.*

tolerance: An adaptive state characterized by diminished responses to the same dose of a chemical.

tolerance limits: A particular type of confidence limit used frequently in quality control work in which the limits apply to a percentage of the individual values of the population.

torque: The product of a force and the perpendicular distance from the line of action of the force to the axis of rotation. The unit of torque in the SI system is the newton-meter (Nm).

total absorption: The amount of sound absorption in an enclosure is quantified as its total absorption.

totally encapsulated chemical protective (TECP) suit: A full body garment that completely encloses the wearer by itself or in combination with a respirator, gloves, and boots. The suit is constructed of protective clothing material that isolates the body from direct contact with potentially hazardous chemicals.

total pressure (TP): The sum of the static pressure and the velocity pressure.

total quality: A management methodology that emphasizes the improvement of the processes by which businesses operate and products are produced.

toxic agent: Substance potentially or actually poisonous to the human body.

toxic reaction: Alteration of a biologic system or organ due to the action of toxic agents.

Toxic Substances List: Annual compilation of known toxic substances prepared by NIOSH and containing about 25,000 names representing some 11,000 different chemical substances.

toxicity: A relative property of a chemical agent; refers to a harmful effect on some biologic mechanism and the condition under which this effect occurs.

toxicity assessment: Both qualitative and quantitative data are developed in a toxicity assessment: 1) a description of the types of health effects that might be expected to occur in humans; and 2) some estimate of toxicity, such as the dose required to cause these health effects. Conceptually, this estimate of toxicity is based on a dose-response curve. The outcome of the toxicity assessment may be a slope factor or an exposure limit. Slope factors are based on the slope of the dose-response curve and can be used to develop a probabilistic risk estimate.

toxicity value: Numerical expression of a substance's dose-response relationship. The most common values used in Superfund risk assessments are reference doses (RfDs) for noncarcinogenic effects and slope factors for carcinogenic effects.

toxicologic effect: Harmful or poisonous effect of a chemical agent.

toxicologist: Specialist in the science that deals with poisons and their effects.

toxicology: Scientific study of poisons, their actions, their detection, and the treatment of conditions produced by them.

toxicoses: Toxicoses include cancer, asphyxiation, skin rash, and mucous membrane irritation. Most of the common pollutants are toxins that exert their effects in a dose-response way, for the most part without regard to host susceptibility. Many are inflammatory agents that act directly on the contacted cells. Some are absorbed and exert effects on organ systems distant from the site of exposure. Some cross the blood-brain barrier to cause central nervous system (CNS) effects.

TP: See *total pressure.*

traceability: A documented chain of comparisons connecting a working standard (in as few steps as practical) to a national (or international) standard such as a standard maintained by the National Institute of Standards and Technology (NIST).

tracheitis: Inflammation of the trachea.

tracheobronchial region: The region consisting of the trachea and conducting airways (bronchi and bronchioles).

traditional workplace: The traditional workplace as viewed from an occupational hygiene perspective, includes foundries, machine shops, chemical plants, manufacturing areas, assembly lines, and oil refineries.

training: The education or instruction of a person. Multimedia software is an effective training tool.

training needs assessment: The determination of an instructional need, triggered either by external or internal factors, leads to an analysis of the training task or training needs assessment. It includes identifying needs, determining to what extent the problem can be classified as instructional in nature, identifying constraints, resources and learner characteristics, and determining goals and priorities.

transducer: A device such as a microphone capable of being actuated by waves from one or more transmission system or media and of supplying related waves to one or more other transmission systems or media.

transient: Something lasting only a very short period of time.

translation: Straight line motion.

transmission loss: The ratio, expressed in decibels, of the sound energy incident on a structure to the sound energy that is transmitted. The term is applied both to building structures (walls, floors; etc.) and to air passages (muffler, ducts, etc.).

transport: The movement of chemicals within one environmental compartment or from one compartment to another.

transport velocity: The transport velocity or the minimum design transport velocity required to move particles in an air stream.

transportation: The removal of wastes from a plant to an off-site location.

treated filter: A filter that has been coated with a layer of chemical reagent to improve collection of specific chemicals for subsequent laboratory analysis.

treatment: 1. Any method, technique, or process designed to remove solids and/or pollutants from solid waste, wastestreams, effluents, and air emissions.
2. Methods used to change the biological character or composition of any regulated medical waste so as to substantially reduce or eliminate its potential for causing disease.

trigger-level: An exposure condition that, when exceeded, requires the affected employees to be offered medical surveillance.

Trilinear chart of the nuclides: The organization of nuclides by atomic number and number of neutrons to yield isotopes, isobars, and isotones.

true risk: The actual value of the risk level.

TTS: See *temporary threshold shift*.

tumor: Abnormal mass of tissue that might be or might not be malignant.

turbulent flow: Chaotic flow with streamlines looping back on themselves; less "well-behaved" and predictable than laminar flow. (See also *laminar flow*.)

TWA: See *time-weighted average*.

8-hour TWA: Average concentration to which an employee is actually exposed over an 8-hour day. See *time-weighted average*.

T_{wb}: See *psychrometric wet bulb temperature*.

twin detector tube: A detector tube construction consisting of a combination of a pretube and an indicating tube, joined by an intermediate sleeve and a piece of shrunk-on tubing.

U

UEL: See *upper flammable limit*.

UFL: See *upper flammable limit*.

UL: Underwriters' Laboratories, Inc.

ultrasonics: The technology of sound at frequencies above the audio range.

ultrasound: Sound with a frequency above about 20,000 Hz, the upper limit of human hearing.

ultraviolet (UV): Those wavelengths of the electromagnetic spectrum which are shorter than those of visible radiation and larger than X-rays, 10^{-5} cm to 10^{-6} cm wavelength. (Also known as "black light.")

ultraviolet (UV) absorbance detector: The ultraviolet (UV) absorbance detector measures the UV light absorbance of the column effluent. It is especially sensitive to aromatic hydrocarbons. The basic components of the UV detector are a UV lamp, a flow-cell with a UV-transparent window for the column effluent, and a photodiode or other light-measuring device. Some UV detectors can be operated at only one wavelength, typically 254 nm, while others can also operate at additional wavelengths using fluorescent waveplates. A variable UV detector can be tuned to any wavelength within its operating range.

ultraviolet radiation: Wavelengths of the electromagnetic spectrum that are shorter than those of visible light and longer than X-rays; wavelengths measure 10^{-5} cm to 10^{-6} cm.

uncertainty: Deviation in predicted values from the actual values. May result from lack of data or variability in the data.

unit: 1. A factory-made encased assembly of the functional elements indicated by its name, such as air-conditioning unit, room-cooling unit, humidifying unit, etc. 2. A portion or subassembly of a computer that accomplishes some operation or function.

Universal Precautions: Guidelines established by the Centers for Disease Control and Prevention (CDC) instructing health care professionals to take blood and body fluid precautions for *all* patients regardless of their bloodborne infection status. These guidelines, known officially as "Universal Blood and Body Fluid Precautions," are part of a 1987 CDC report titled "Recommendations for Prevention of HIV Transmission in Health-Care Settings."

unstable: Tending toward decomposition or other unwanted chemical change during normal handling or storage.

unstable reactive: A chemical that in the pure state, or as produced or transported, will vigorously polymerize, decompose, condense, or become self-reactive under conditions of shocks, pressure, or temperature.

upper explosive limit (UEL): See *upper flammable limit*.

upper flammable limit (UFL): The percentage by volume of a flammable gas or vapor that is the maximum level ignitable. Also referred to as "upper explosive limit."

upper measurement limit: The useful limit (mg of analyte per sample) of the analytical instrument.

upper respiratory tract: The mouth, nose, sinuses, and pharynx.

uptake: Mass potentially breathed in/out after a specific time period.

urine: The watery nonviscous excretion voided by the penis in males and by the urethra in females.

urticarial reaction: 1. Hives or nettle rash. 2. A skin condition characterized by the appearance of intensely itching wheals or welts with elevated, usually white centers and a surrounding area of erythema.

UV: See *ultraviolet*.

UV-A: The UV divisions are named regions. UV-A is the blacklight region.

UV-B: The UV divisions are named regions. UV-B is the erythema region. UV-B and UV-C regions are often called actinic UV, because they are capable of causing chemical reactions.

UV-C: The UV divisions are named regions. UV-C is the germicidal region. UV-B and UV-C regions are often called actinic UV, because they are capable of causing chemical reactions.

V

vacuum: State in which the gas pressure is lower than atmospheric pressure.

vacuum UV: The 100–180 nm region is known as the vacuum UV region because these wavelengths are readily absorbed in air.

validated sampling and analysis method: A method that has met critical accuracy requirements when tested throughout the working range.

validity: See *accuracy*.

values: Ideals, customs, or institutions that arouse an emotional response in a given group or individual.

vapor: Gaseous phase of a substance ordinarily liquid or solid at 25°C and 760 mmHg. Evaporation is the process by which a liquid changes to a vapor state, and mixes with the surrounding atmosphere. Solvents with low boiling points volatilize readily. Examples of substances that emit vapors are trichloroethylene, methylene chloride, and mercury. Size ranges are usually less than 0.005 µm. It is near equilibrium with its liquid phase and for occupational hygiene calculations it follows the gas laws.

vapor density: 1. The ratio of the mass of a vapor or gas to the mass of an equal volume of air at the same temperature; an expression of the density of the vapor or gas. 2. The weight of a vapor or gas compared to the weight of an equal volume of air is an expression of the density of the vapor or gas. Materials lighter than air have vapor densities less than 1.0 (e.g., acetylene, methane, hydrogen). Materials heavier than air (e.g., propane, hydrogen sulfide ethane, butane, chlorine, sulfur dioxide) have vapor densities greater than 1.0. All vapors and gases will mix with air, but the lighter materials will tend to rise and dissipate (unless confined). Heavier vapors and gases are likely to concentrate in low places – along or under floors, in sumps, sewers, and manholes, in trenches and ditches – where they may create fire or health hazards.

vapor pressure (VP): 1. The pressure exerted by a vapor. If a vapor is confined and accumulates over its liquid or solid at a constant temperature, it reaches a maximum pressure called the saturated vapor pressure. [**Note:** For a specific liquid or solid with a constant volume of vapor above it, the vapor pressure depends only on the temperature.] 2. The pressure exerted by a saturated vapor above its own liquid in a closed container. When quality control tests are performed on products, the test temperature is usually 100F, and the vapor pressure is expressed as pounds per square inch (psig or psia), but vapor pressures reported as MSDSs are in millimeters of mercury (mmHg) at 68°F, unless stated otherwise. Three facts are important to remember: a) vapor pressure of a substance at 100°F will always be higher than the vapor pressure of the substance at 68°F; b) vapor pressures reported as MSDSs in mmHg are usually very low pressures; 760 mmHg is equivalent to 14.7 pounds per square inch; c) the lower BP of a substance, the higher its vapor pressure. 3. A measure of the tendency of a liquid to form a gas (usually a function of temperature); the pressure exerted when a solid or liquid is at equilibrium with its own vapor; normally reported in mm of mercury (mmHg). The higher the vapor pressure, the more volatile the chemical.

variability: Source of uncertainty in risk assessment due to the fact that many parameters are best described not as point values but as probability distributions.

variance: 1. The square of the standard deviation. 2. Mathematically, the sample variance is the sum of squares of the differences between the individual values of a set and the arithmetic average of the set, divided by one less than the number of values.

vasoactive: Affecting blood vessel size. It can influence heat loss and blood supply and thereby contribute to hypothermia or frostbite.

vasoconstrict: Blood vessel constriction.

vasodilation: Blood vessel dilation.

V_c: See *capture velocity*.

velocity: 1. The time rate of distance moved. 2. A vector quantity that denotes the simultaneous time rate and direction of linear motion. The designer must fully understand the different velocities of importance in exhaust ventilation design. 3. The time rate of movement of air in meters per second (feet per minute).

velocity pressure (VP): The pressure created by moving air.

ventilation (control): The process of supplying or removing air by natural or mechanical means to or from any space. Such air may or may not have been conditioned.

verification: The establishment of the reliability of a measurement method by comparison with one or more reference methods.

vertical entry: A confined space entry that requires the entrant to descend into the confined space from above or below.

vertical standards: An industry-specific standard that regulates an industry in addition to OSHA general industry standards. For example the construction industry and the longshoring industry are subject to both general and vertical standards.

V_f: See *hood face velocity*.

vibration isolator: It decouples structures, such as a human hand from a pneumatic tool or a driver from a vehicle, by using "soft" connections. Vibration energy is absorbed by the isolator instead of being transmitted to the user.

vibration transmissibility ratio: The vibration transmissibility ratio is the ratio of the vibration output (vibration at the seat) to the vibration input (vibration at the floor of the vehicle). A transmissibility ratio of 1 means there is no change in the level of vibration between the input and output; a ratio >1 indicates an amplification of the original vibration; a ratio <1 indicates a reduction or attenuation of the original vibration.

vibration-induced damage: Bodily damage caused by excessive exposure to vibration.

vibration-induced white finger (VWF): See *hand-arm vibration syndrome.*

videography: A tool used to conduct a kinematic analysis that uses the digitization of key anatomical landmarks into Cartesian coordinates that are recorded on a storage device.

virus: A software program that attaches itself to another program in computer memory or on a disk, and spreads from one program to another. Viruses may damage data, cause the computer to crash, display messages, or lie dormant.

viscosity: The tendency of a fluid to resist internal flow without regard to its density.

visible radiation: Electromagnetic radiation with wavelengths capable of causing the sensation of vision, ranging approximately from 4000 angstroms (extreme violet) to 7700 angstroms (extreme red). (Also known as "light radiation.")

vision statement: A statement that identifies where an organization wants to be in the future. In broad terms it establishes the end point for an improvement process and allows the development of a plan to move from an unacceptable present to a beneficial future.

VOC: See *volatile organic compound.*

vol: See *volume.*

volatile organic compound (VOC): Any organic compound that participates in atmospheric photochemical reactions.

volatility: 1. The tendency or ability of a liquid to vaporize. Such liquids as alcohol and gasoline, because of their well-known tendency to evaporate rapidly, are called volatile liquids. 2. A measure of how quickly a substance forms a vapor at ordinary temperatures.

volatilize: Readily convertible to a vapor or gaseous state.

voltage: The electrical potential energy difference per unit charge between two points.

volume (vol): A measure of the size of a body or definite region in three-dimensional space. It is equal to the least upper bound of the sum of the volumes of nonoverlapping cubes that can be fitted inside the body or region, where the volume of a cube is the cube of the length of one of its sides.

volume flow rate (Q): The quantity of air flowing in cubic feet per minute (ft^3/min) or cubic meters per second (m^3/sec).

voluntary guidelines: Guidelines are published by consensus organizations and are published for general voluntary use; the decision to use the guidelines is totally a discretionary one considered by the user.

Voluntary Protection Program (VPP): Developed in 1982 by OSHA to recognize and promote outstanding safety programs.

vortex coolers: An open-loop system in which both a hot and cold stream of air from a compressed air source is generated.

VP: See *vapor pressure.*

VP: See *velocity pressure.*

VPP: See *Voluntary Protection Program.*

VWF: Vibration-induced white finger. See *hand-arm vibration syndrome.*

W

WAN: See *wide area network.*

water intoxication: See *hyponatremia.*

water vapor pressure: The pressure exerted by water vapor at a specific temperature.

water-reactive: A chemical that reacts with water to release a gas that is either flammable or presents a health hazard.

watts per square: Energy measurement for radiation stated in terms of.

wavelength (Λ): 1. The distance between the ends of one complete cycle of a wave. 2. Wavelength is the descriptor used for UV, visible, and IR radiation. 3. The distance in the line of advance of a wave from any point to a like point on the next wave. It is usually measured in angstroms, microns or nanometers.

WBGT: See *wet bulb globe temperature.*

WEEL(s)®: See *workplace environmental exposure level(s).*

WEEL guides: See *workplace environmental exposure level guides.*

weighting network (sound): Electrical networks (A, B, C) that are associated with sound level meters. The C network provides a flat response over the frequency range 20–10,000 Hz of interest while the B and A networks selectively discriminate against low (below I KC) frequencies.

weight-of-evidence: U.S. EPA classification system for characterizing the evidence that an agent is a human carcinogen or developmental toxin. It relies mainly on positive evidence. (See also *strength-of-evidence*).

wet bulb globe temperature (WBGT): 1. The combination of the effect of the four main thermal components affecting heat stress: air temperature, humidity, air velocity, and radiation, as measured by the dry bulb (T_{db}), natural wet bulb (T_{nwb}), and globe (T_g) temperatures. 2. Temperature calculated as the sum of 0.7 natural wet bulb + 0.2 black globe + 0.1 dry bulb. (See also *natural wet bulb, psychrometric.*)

wet bulb temperature: See *psychrometric wet bulb temperature.*

wet globe temperature (WGT): A temperature reading that combines air temperature, humidity, air velocity, and radiation into a single reading.

wet kata thermometer: A thermometer where the bulb covered by a wetted wick and heated in water and then allowed to cool in the environment being measured. The time required for the liquid in the wet kata to cool between the two marks can be measured with a stop watch, and this time can be used with a calibration sheet to determine the cooling power of the air.

WGT: See *wet globe temperature.*

Wheatstone bridge: A four-arm bridge circuit, all arms of which are predominantly resistive; used to measure the electrical resistance of an unknown resistor by comparing it with a known standard resistance.

white room: See *clean room.*

whole-body vibration: The exposure of the entire body to workplace vibrations. Whole-body vibration can cause both physiological and psychological effects ranging from fatigue and irritation to motion sickness (kinetosis) and to tissue damage.

wide area network (WAN): Two or more LANs linked together into a larger network form a wide area network (WAN). A WAN can be used to connect different divisions within a company, different plants in different cities, and so forth. The links between LANs to form a WAN can be made with modems, satellites, or telephone lines.

wind-chill index: The cooling effect of any combination of temperature and wind, expressed as the loss of body heat in kilogram calories per hour per square meter of skin surface; it is only an approximation because of individual body variations in shape, size, and metabolic rate.

window: The entry point on a radiation detector for radiation into sensitive volume.

WMSDs: See *Work-Related Musculoskeletal Disorders.*

work: Exertion or effort directed to produce or accomplish something.

work plan: A plan that ensures that all required data are efficiently collected and properly documented. It should include field data sheets and a communications plan.

work site analysis: Examination of the work site to identify not only existing hazards but also conditions and operations in which changes might occur to create hazards. Effective management actively analyzes the work and the work site to anticipate and prevent harmful occurrences. To ensure that all hazards are identified, management and employees must: 1) Conduct comprehensive baseline work site surveys for safety and periodic comprehensive update safety surveys; 2) Analyze planned and new facilities, processes, and equipment for safety hazards; 3) Perform routine job hazard analyses, including investigation of accidents; 4) Use a reliable system for employees, without fear of reprisal, to notify management about conditions that seem hazardous; and 5) Analyze injury and illness trends over time so that patterns with common causes can be identified and prevented.

working fluids: Pressure is indicated by the height of a working fluid in the barometer or manometer. The most commonly used working fluid for a barometer is mercury. Other fluids can be used but are not as convenient.

working occupational exposure limit: An informal occupational exposure limit set by a occupational hygienist based on whatever information may be available to differentiate acceptable from unacceptable exposures. Working OELs are sometimes stated in ranges (e.g., 0.1 to 1.0 mg/m^3) or incorporate large safety factors to account for uncertainty. (See also *occupational exposure limit.*)

working range: The range of contaminant concentration (mg/m^3) that may be accurately quantified at specified air volumes (liters) by a specific method.

workplace environmental exposure level (WEEL®) guides: Exposure guidelines developed by AIHA intended to protect the health and safety of workers exposed to hazardous substances or conditions.

workplace exposure assessment: The exposure characterization of the worker's potential exposure to hazardous chemical, physical or biological agents. The information included should include the agents, their likely routes of exposure, and intensity, duration, and frequency of the exposure.

Work-Related Musculoskeletal Disorders (WMSDs): The specific term "work-related musculoskeletal disorders" refers to: 1) musculoskeletal disorders to which the work environment and the performance of work contribute significantly; or 2) musculoskeletal disorders that are made worse or longer lasting by work conditions. These workplace risk factors, along with personal characteristics (e.g., physical limitations or existing health problems) and societal factors, are thought to contribute to the development of WMSDs.

workstation: The computers linked to a server.

World Wide Web (WWW): A system of Internet servers that support specially formatted documents. The documents are formatted in a language called HTML (HyperText Markup Language) that supports links to other documents, as well as graphics, audio, and video files. This means you can jump from one document to another simply by clicking on hot spots. Not all Internet servers are part of the Web.

worst-case-scenario: A method of conducting an exposure assessment in which the most conservative value of each input parameter is selected. (See also *reasonable maximum exposure.*)

WWW: See *World Wide Web.*

XYZ

Youden plot: Large variations in results reported from different laboratories analyzing the same samples might be explained by random errors in the measurements or by systematic errors in the different laboratories. Practically speaking, it is highly probable that systematic errors made by the participating laboratories are the cause for these wide variations. This can be evaluated using another technique for analyzing interlaboratory variability—the "two-sample" or Youden plot. In this graphical technique each participating laboratory is sent two similar samples (A and B) and asked to perform one analysis on each sample. When the results are returned, the results obtained by a laboratory for Sample A are plotted with respect to the results obtained by the same laboratory for Sample B. Median lines are drawn, and outliers are identified and discarded from the data set. If there were no bias in the results, the plotted points should be randomly distributed around the intersection of the two median lines drawn. It is rare that this is observed. Generally what is observed is that the plotted points fall around a line drawn at 45 from the intersection of the two median lines, indicating that each participating laboratory has its own technique (an internal consistency) for analyzing the samples. (Laboratories who report one sample high are more likely to report the second sample high. The converse is true, also.) Through the use of two samples similar in concentration, an estimate of precision can be made. This is done by constructing a perpendicular line from the plotted point to the 45 line.

zero gas: A gas containing less than 1 ppm sulfur dioxide.

zoonoses: See *zoonotic infection*.

zoonotic infection: Disease transmissible from animals to man.

zygomycosis: Any infection caused by fungi of the class *zygomycetes*.

zygote: A cell produced by the joining of two gametes that are either sex or germ cells.

Index

A

D

F

G

H

J

K

L

N

O

P

S

T

U

V

W

X

Y

Z

Design by James Myers

Composed at AIHA in QuarkXpress® with electronic output to film.

Typefaces
Body Text: 9/10, Palatino
Heads: 12/13, Avant Garde

Printed by Jarboe Printing Company
Cover: 80# White Litho coated one side and laminated; PMS# 3165, PMS# 8283 metallic, and black inks.
Text: 50# White Opaque, PMS #3165 and black inks.
Case bound and sewn with reinforced end sheets.

The inside pages are of 20% post-consumer waste. The entire book has been printed with alcohol-free soybean inks.